Y0-BYB-343

INDIRECT FOOD ADDITIVES *and* POLYMERS

Migration and Toxicology

INDIRECT FOOD ADDITIVES *and* POLYMERS

Migration and Toxicology

VICTOR O. SHEFTEL, M.D., PH.D., D.SC.

Library of Congress Cataloging-in-Publication Data

Sheftel, Victor O.
Indirect food additives and polymers : migration and toxicology / by Victor O. Sheftel.
p. cm.
Includes bibliographical references and index.
ISBN 1-56670-499-5 (alk. paper)
1. Plastics—Toxicology. 2. Food contamination. 3. Food—Packaging.
4. Polymers—Toxicology. I. Title.
RA1242.P66 S543 2000
615.9′54—dc21
99-089582
CIP

Direct all inquiries to CRC Press LLC, 2000 N.W. Corporate Blvd., Boca Raton, Florida 33431.

Lewis Publishers is an imprint of CRC Press LLC

International Standard Book Number 1-56670-499-5
Library of Congress Card Number 99-089582
Printed in the United States of America 1 2 3 4 5 6 7 8 9 0
Printed on acid-free paper

PREFACE

Substances which in the U.S. have generally been called ***Indirect Food Additives*** are in fact ***Food Contact Substances***. Predominantly, they are ***polymeric materials*** but also rubber, cellulose, metal, glass, paper and paperboard, etc.

Plastics or ***Polymeric Materials*** (PM) appear to be an important source of chemical contamination of food and the environment, along with industrial wastes and pesticides.

Because of their unique characteristics, plastics are becoming the most important packaging material for food products. The packaging industry is the largest user of plastics; more than 90% of flexible packaging is made of plastics. For the last several decades, the safety assessment of plastics intended for use in contact with foodstuffs or drinking water has continued to present a serious challenge for industry and regulatory agencies.

Toxicology of Plastics studies potential hazards of the materials and their ingredients for human health and develops recommendations for manufacture and safe use of such materials.

Since it is a comparatively new and insufficiently investigated branch of applied toxicology, the experimental data obtained in this field hitherto have not been collected and generalized. One cannot find a large body of migration data in the current literature. Incomplete and fragmentary information on the subject can be found in the following sources: *Practical Toxicology of Plastics*, CRC Press, 1968; *Les Matieres Plastiques dans l'Industrie Alimentaire*, Paris, 1972, by R. Lefaux; *Industrial Hazards of Plastics and Synthetic Elastomers*, J. Jarvisalo, P. Pkaffli, and H. Vainio, Eds., Alan R. Liss, Inc. (1984). Unfortunately all these books are out-of-date.

This handbook is an attempt to provide comprehensive information on the toxic effects of plastics and their ingredients that enter the body mainly by the oral route and thus it may serve as a sort of encyclopedia for specialists and practitioners in this field. Basic toxicological and other scientific data necessary to identify, characterize, measure, and predict hazards of plastic-like materials use have been assembled from the scientific literature and from regulatory national and international documents.

The contents of this handbook factually overstep the limits of toxicology of plastics because they comprise information concerning many of the widespread food and water contaminants: heavy metals, solvents, monomers, plasticizers, etc. Because toxic properties of PM depend on toxic properties of the substances released by them, this book will be of use when assessing toxic properties not only of the existing plastics but also of future materials containing ingredients that have already undergone toxicological evaluation.

The handbook includes the thoroughly reviewed American and European toxicology literature as well as the screened Russian toxicology data, often unknown in the West but reasonably fit to interpretation. Since the toxic properties of PM are determined by the toxic properties of the substances released by them, toxic hazard assessment of ingredients migrating into food and water is the essential part either of new material development or of the regulatory decision-making process.

It should be borne in mind that assessment of toxic potential is an extremely complicated task even for those who have many years of experience in toxicology testing. Such assessment requires examining and scrutinizing of complex data that describe the toxicology of a substance, selecting the appropriate valid information from often conflicting or incomplete data, and arriving at a conclusion on the relevance of the information to human health risk.

For these and other reasons, it has not been possible to assess exactly a relative validity either of data obtained before GLP implementation or of Russian toxicology data. On the other hand, a toxicology profile will be incomplete today if only findings of the last decade are taken into account. Since Russian toxicology developed separately from that of the West, a lot of data collected in Russia are unique, and cannot be ignored, even if they do not completely conform to GLP requirements. In every case, references should enable easy identification of the place and year of each research cited.

The following remarks concerning presentation of the data in this handbook are to be made:

1. For each chemical, the following data are provided when available:

- Substance Prime Name
- Molecular or Structural Formula
- Molecular Mass
- Synonyms
- CAS number
- RTECS number
- Properties (sometimes also Composition)
- Applications and Exposure
- Migration Data
- Acute Toxicity
- Repeated Exposure
- Short-term Toxicity
- Long-term Toxicity
- Immunotoxicity or Allergenic Effect
- Reproductive Toxicity (Embryotoxicity, Teratogenicity, and Gonadotoxicity)
- Mutagenicity (*In vivo* cytogenicity and *In vitro* genotoxicity)
- Carcinogenicity (including IARC, U.S. EPA, NTP and other cancer classifications)
- Chemobiokinetics
- Standards, Guidelines, Regulations, Recommendations
- References

2. The oral route of administration is intended throughout the book unless otherwise specified. In the absence of oral toxicity data, information from inhalation or dermal toxicity studies, as well as from administration via *i/p, i/v*, or other routes, is presented.

3. Descriptions of the common toxic effects are subdivided into acute, repeated, short-term, and long-term (chronic) toxicity: **acute toxicity** refers to the result of a single oral exposure; **repeated exposure** refers to a length of exposure of about 2 weeks to 2 months; **short-term toxicity** refers to a period of treatment not less than 3 to 4 months; **long-term toxicity** refers to a length of exposure not less than 6 months.

4. A quantitative assessment of functional accumulation is given. **Coefficient of accumulation** (on the lethal level) has been determined by one of the following three methods: the method of Yu. S. Kagan and V. V. Stankevich (designated as "by Kagan") stipulates administration of the agent to experimental animals at equal daily doses of 1/5 to 1/50 LD_{50} for 2 to 4 months; the method of R. K. Lim et al. (designated as "by Lim") stipulates administration of the substance at gradually increasing doses, beginning with 1/10 LD_{50} for no more than 4 weeks; the method of S. N. Cherkinsky et al. (designated as "by Cherkinsky") stipulates administration of 1/5 LD_{50} for 20 days.

5. Data on certain long-term or delayed effects, in particular carcinogenicity, may not present appropriate information about safe levels. These data are usually obtained at high dose levels, and recommendations for carcinogenicity evaluation have been computed from hypothetical mathematical models that cannot be verified experimentally.

6. There are three usually acknowledged **carcinogenicity classifications**: IARC, U.S. EPA, and NTP. All substances tested could be subdivided according to **IARC weight of evidence** for carcinogenicity:

1 - *Human carcinogens*
2A - *Probable carcinogens*
2B - *Possible carcinogens*
3 - *Not classified*
4 - *Probably not carcinogenic to humans.*

According to the **U.S. EPA weight of evidence** for carcinogenicity the classification is as follows:

A - *Human carcinogens*

B1 and **B2** - *Probable human carcinogens*

C - *Possible human carcinogens*

D - *Not classified*

E - *No evidence of carcinogenicity in humans.*

NTP categorization according to the weight of experimental evidence is presented by the following groups:

CE - *Clear evidence of carcinogenic activity*

SE - *Some evidence of carcinogenic activity*

EE - *Equivocal evidence of carcinogenic activity*

NE - *No evidence of carcinogenic activity*

IS - *Inadequate study of carcinogenic activity.*

Earlier NTP designations are as follows:

P - *Positive*

E -*Equivocal*

N - *Negative.*

Designations of this categorization are displayed in the following order: *Male Rats - Female Rats - Male Mice - Female Mice*. Absence of the data is designated as XX. Rodent carcinogens with significantly elevated tumor rate at some dose(s) below MTD are marked with * (according to J. K. Haseman and A. Lockhart, 1993). It is now clear that tumor induction can arise in a variety of ways including not only a DNA-reactive genotoxic mode of action, but also non-DNA-reactive non-genotoxic cytotoxic and non-genotoxic-mitogenic modes of action. Initial risk assessment approaches that recognized this distinction identified a chemical carcinogen as either genotoxic or non-genotoxic, with no middle ground (B. E. Butterworth and M. S. Bogdanffy, 1999).

7. Within chapters, ingredients are placed in the English alphabetical order of their prime names, ignoring special characters such as Greek letters or numerals. Abbreviation of a substance described in an entry consists predominantly of its first letter. If abbreviation consists of two or more letters, it is specified. Toxicity data are transformed into a special, newly developed format to facilitate the use of available toxicology information to evaluate potential migration levels of plastic ingredients into food or drinking water.

8. WHO, EU, U.S., and some other available national standards, guidelines, and recommendations, taken from the following sources are presented: *WHO Guidelines for Drinking-Water Quality*, 1996; *Council Directive of 15 July 1980*, *Official Journal of European Communities*, 30, 8, August 1980; *Drinking Water Regulations and Health Advisories* by the U.S. EPA, 1998; *Commission Directive 90/128/EEC* of 23 February 1990 relating to plastics materials and articles intended to come in contact with foodstuffs, *Official Journal of European Communities*, 33 (L75), 19, 1990; *Commission Directive 96/11/EC* of March 5, 1996 (amending Directive 90/128/EEC), *Official Journal of European Communities*, L61, 12.03.1996; *Code of Federal Regulations, Food and Drugs*, 21 CFR Part 175-179, 1998; *Sanitary Standards of Maximum Allowable Concentrations of Harmful Chemicals in Water Bodies* set by Ministry of Health, Russia, Appendix 2, Sanitary Rules and Standards, No 4630-88, 1988 (revised in 1994 and 1995). References to the U.S. FDA are cited according to CFR.

9. Definitions of abbreviations:

ADI - Acceptable Daily Intake (an estimate of the amount of a substance in food that can be ingested daily over a lifetime by humans without appreciable risk)

BW - Body Weight

CFR - U.S. Code of Federal Regulations

DNA - Deoxyribonucleic acid

DWEL - Drinking Water Equivalent Level (a lifetime exposure concentration in drinking water protective of adverse, non-cancer health effects)

ET_{50} - Median Time from ingestion up to death of animals after LD_{50} administration
EU - European Union
FAO - Food and Agricultural Organization of the United Nations
GLP - Good Laboratory Practice
GRAS - Generally Recognized As Safe
IARC - International Agency for Research on Cancer
IPCS - International Program of Chemical Safety
JECFA - Joint FAO/WHO Expert Committee on Food Additives
K_{acc} - Coefficient of Accumulation (a quantitative assessment of functional accumulation)
LOAEL - Lowest-Observed-Adverse-Effect-Level (the lowest level of exposure at which an adverse effect was observed)
LOEL - Lowest-Observed-Effect-Level (the lowest level of exposure at which any biological effect was observed)
MAC - Maximum Allowable Concentration in water bodies (maximum permissible level of a contaminant protective of adverse, non-cancer health effects)
MCL - Maximum Contaminant Level in drinking water
MCLG - Maximum Contaminant Level Goal in drinking water
MPC - Maximum Permissible Concentration in food (maximum permissible level of a contaminant in food, *Russia*)
MTD - Maximum Tolerable Dose, see **ADI**
n/m - not monitored
NOAEL - No-Observed-Adverse-Effect-Level (the greatest concentration or amount of chemical found by experiment or observation that causes no detectable adverse alteration of morphology, functional capacity, growth, development, or lifespan of the target organism)
NOEL - No-Observed-Effect-Level (the greatest concentration or amount of chemical found by experiment or observation that causes no detectable change in morphology, functional capacity, growth, development, or lifespan of the target organism)
NTP - U.S. National Toxicology Program
organolept. - organoleptic criterion
PML - Permissible Migration Level from polymeric materials to food or water *(Russia)*
PTWI - Provisional Tolerable Weekly Intake [a provisional estimate of the amount of a substance (contaminant) in food or drinking water that can be ingested weekly over a lifetime without appreciable health risk]
QM - Maximum Permitted Quantity of the residual substance in the material or article
RfD - Reference Dose (an estimate of a daily exposure to the human population that is likely to be without appreciable risk of deleterious effects over lifetime)
RNA - Ribonucleic acid
RTECS - Registry of Toxic Effects of Chemical Substances
SML - Specific Migration Limit in food or food simulant
TDI - Tolerable Daily Intake [an estimate of the amount of a substance (contaminant) in food or drinking water that can be ingested daily over a lifetime without appreciable health risk]
UF - Uncertainty factor (a factor applied to the NOEL to derive ADI)
WHO - World Health Organization

ALT - alanine aminotransferase
AST - aspartate aminotransferase
BW - body weight
CA - chromosome aberrations
CNS - central nervous system
DLM - dominant lethal mutations
DNA - deoxyribonucleic acid

ECG - electrocardiogram
EEG - electroencephalogram
GC - gas chromatography
GI - gastro-intestinal
Hb - hemoglobin
HPLC - high-performance liquid chromatography
i/g - intragastric administration
i/m - intramuscular injection
i/p - intraperitoneal injection
i/v - intravenous injection
LDH - lactate dehydrogenase
MetHb - methemoglobin
MS - mass spectrometry
NS - nervous system
ppm - parts per million
ppb - parts per billion
RNA - ribonucleic acid
s/c - subcutaneous injection
SCE - sister chromatid exchanges
STI - summation threshold index

10. The following reference numbers relate to the most repeatedly used literature sources:

- 01 - Alexander, H. C., McCarty, W. A., and Bartlett, E. A., Aqueous odor and taste threshold values of industrial chemicals, *J. Am. Water Work Assoc.*, 74, 595, 1982.
- 02 - Amoore, J. E., and Hautala, E., Odor as an aid to chemical safety: Odor threshold compared with threshold limit values and volatilities for 214 industrial chemicals in air and water dilution, *J. Appl. Toxicol.*, 3, 272, 1983.
- 03 - Patty, F. A., Ed., *Industrial Hygiene and Toxicology*, 3rd ed., John Wiley & Sons, New York, 1982.
- 04 - Lefaux, R., *Chimie et Toxicologie des Matieres Plastiques*, Paris, 1964.
- 05 - Lefaux, R., *Les Matieres Plastiques dans l'Industrie Alimentaire*, Paris, 1972.
- 06 - *Registry of Toxic Effects of Chemical Substances*, 1985-86 ed., User's Guide, D. V. Sweet, Ed., U. S. Dept. Health and Human Services, Washington, D.C., 1987.
- 07 - Sheftel, V. O., *Toxic Properties of Monomers and Additives*, E. Inglis, and S. Dunstall, Eds., Rapra Technology Ltd., UK, 1990.
- 08 - Broitman, A. Ya., *Basic Problems of the Toxicology of Synthetic Antioxidants Intended for Stabilizing Plastics,* Author's abstract of thesis, Leningrad, 1972, 48 (in Russian).
- 09 - Gold, L. S., Manley, N. B., Slone, T. H., Garfinkel, G. B., Rohrbach, L., and Ames, B. N., The fifth plot of the Carcinogenic Potency Database: results of animal bioassays published in the general literature through 1988 and by the National Toxicology Program through 1989, *Environ. Health Perspect.*, 100, 65, 1993.
- 010 - *Identification and Treatment of Tastes and Odors in Drinking Water*, Mallevialle, J. and Suffet, I. H., Eds., AWWA Research Foundation, Lyonnaise des Eaux-Dumez, 1987, 292.
- 011 - *Industrial Hygiene and Toxicology*; Patty, F. A., Ed., 2nd ed., Volume II, Toxicology, Interscience Publishers, New York, 1963.
- 012 - Sax, N. I., *Dangerous Properties of Industrial Materials*, 6 th ed., Van Nostrand Reinold, New York, 1984.
- 013 - Mortelmans, K., Haworth, S., Lawlor, T., Speck, W., Tainer, B., and Zeiger, E., *Salmonella* mutagenicity tests: II. Results from the testing of 270 chemicals, *Environ. Mutagen.*, 8 (Suppl. 7), 1, 1986.

- 014 - American Conference of Governmental Industrial Hygienists, *Documentation of the Threshold Limit Values and Biological Exposure Indices*, 5th ed., Cincinnati, OH, 1986.
- 015 - Zeiger, E., Anderson, B., Haworth, S., Lawlor, T., Mortelmans, K., and Speck, W., *Salmonella* mutagenicity tests: III. Results from the testing of 255 chemicals, *Environ. Mutagen*, 9 (Suppl. 9), 1, 1987.
- 016 - Zeiger, E., Haseman, J. K., Shelby, M. D., Margolin, B. H., and Tennant, R. W., Evaluation of four *in vitro* genetic toxicity tests for predicting rodent carcinogenicity: Confirmation of earlier results with 41 additional chemicals, *Environ. Mutagen.*, 16 (Suppl. 18), 1, 1990.
- 017 - *Hazardous Substances in Industry*, Handbook for chemists, engineers, and physicians, Lazarev, N. V., Levina, E. N., and Gadaskina, I. D., Eds., 7th ed., vol. 1, 2 and 3, Khimya, Leningrad, 1976-1977.
- 018 - Verschuren, K., *Handbook of Environmental Data of Organic Chemicals*, 2nd ed., Van Nostrand Reinhold Co., New York, NY, 1983.
- 019 - Gosselin, R. E., Smith, R. P., Hodge, H. C., and J. E. Braddock, *Clinical Toxicology of Commercial Products*, 5th ed., Williams & Wilkins, Baltimore, 1984.
- 020 - *The Merck Index, An Encyclopedia of Chemicals, Drugs, and Biologicals*, 11th ed., Merck & Co, Inc., U.S., 1989.
- 021 - Parke, D. V., *The Biochemistry of Foreign Compounds*, Pergamon Press, Oxford, 1968.
- 022 - Gosselin, R. E., Hodge, H. C., Smith, R. P., and Gleason, M. N., *Clinical Toxicology of Commercial Products*, 4th ed., Williams & Wilkins, Baltimore, 1976.
- 023 - Kel'man, G. Ya., *Toxic Properties of Chemical Additives for Polymeric Materials*, Meditsina, Moscow, 1974, 53 (in Russian).
- 024 - Carpenter, C. P., Weil, C. S., and Smyth, H. F., Range-finding toxicity data: List VIII, *Toxicol. Appl. Pharmacol.*, 8, 31 , 1974.
- 025 - Haworth, S., Lowler, T., Mortelmans, K., and Speck, W., *Salmonella* mutagenicity test results for 250 chemicals, *Environ. Mutagen.*, 5 (Suppl. 1), 3, 1983.
- 026 - Oppenheimer, B. S., Oppenheimer, E. T., Danishefsky, J., Stiut, A. P., and Eirich, F. R., Further studies of polymers as carcinogenic agents in animals, *Cancer. Res.*, 15, 333, 1955.
- 027 - Takahashi, A., Problems of hygiene maintenance for food coming into contact with rubber and plastics products, *Intern. Polymer Sci. Technol.*, 3, 93, 1976.
- 028 - *Handbook on the Hygiene of the Use of Polymers,* K. I. Stankevich, Ed., Zdorov'ya, Kiyv, 1984, 192 (in Russian).
- 029 - Marhold, J., *Introduction into Industrial Toxicology. Organic Compounds*, Avicenum, Prague, Czechoslovakia, 1986, 1013 (in Czech).

In order to reduce an enormous bulk of references, some of them are given only in the text (mainly in the *Acute Toxicity* and *Chemobiokinetics* sections) in a shorter form, that is the name of the author(s) and year of publications. References to IARC Monographs in the text indicate volume and page numbers.

This handbook is intended for specialists in industry as well as for health care providers, legislators, regulators, scientists, and practitioners of occupational and environmental medicine, various national and international agencies and organizations, national and local governmental authorities and consumer associations, and movements such as Greenpeace, etc.

Application of the appropriate database should present complete and necessary information for many of those concerned with research, application, and legislation relevant to toxic hazards of packaging materials and food-contact coatings and articles.

The Author has made every effort to ensure that the information presented in this book is accurate and up-to-date. Nevertheless, despite reasonable screening and evaluation of presented data, inclusion herein does not imply endorsement of the cited literature. No claims of assurance or liability for the inaccuracy of information presented are assumed, either by the Author or the Publisher. Final evaluation of the references included is the responsibility of the readers.

I owe a special debt of thanks to my wife. Her understanding, cooperation, and support through the preparation of this book are sincerely appreciated.

Victor O. Sheftel, MD, PhD, DSc
Jerusalem, 1999

ABOUT THE AUTHOR

Dr. Victor O. Sheftel is the Chief Environmental Toxicologist of the Ministry of Health, State of Israel, Jerusalem.

Dr. Sheftel received his M.D. degree at the Kiyv Medical Institute, Ukraine. By 1966 he made a doctorate (PhD in hygiene) and after this he was a senior researcher and a team manager at the All-Union Research Institute of Hygiene and Toxicology of Pesticides, Polymers and Plastic Materials (Kiyv, USSR). There he engaged in development of the modern toxicology of plastics and was a leading scientist in this field in Russia.

By 1978 he earned a post-doctoral degree (Doctor of Sciences, equivalent to professor) from the Sysin Research Institute of Environmental Hygiene (Moscow) in the Academy of Medical Sciences of the former USSR.

Dr. Sheftel has published 14 monographs and over 50 scientific papers (in the U.S., Great Britain, Russia, and China), mainly related to the toxicological evaluation of food- and water-contact materials and contaminants, the methodology of regulatory process in this realm, and other actual problems of general and applied toxicology. He has recommended a differential approach to the choice of criteria of reliability in toxicological studies (1992).

In 1990 Dr. Sheftel emigrated to Israel. At present, he deals mainly with the problems of drinking water quality and the development of a unique database of toxic effects of food- and water-contact materials.

Table of Contents

INTRODUCTION

HARMFUL SUBSTANCES IN PLASTICS

As it stands today, it is impossible not only during production but also in everyday life to prevent the population from coming into contact with plastics. In any case, approximately 70 to 80% of food is packaged in various polymeric materials (PM).

Unfortunately, PM appear to be a potential source of the release of chemicals into the environment; they may have a variety of effects on human health as a consequence of water, air, or skin contamination. The principal hazardous factor associated with the use of PM remains the possible contamination of food.

The absence of acute poisonings with fatal outcome does not prove the safety of synthetic packaging materials. Nevertheless, it must be remembered that we do not completely realize the real contribution of PM to the actual contamination of food.

It is true that PM ingredients do not act like pesticides (or a variety of other highly bioactive substances), and one can hardly expect immediate and pronounced clinical manifestations of their toxic action. The occurrence of acute toxicity due to PM used in contact with food and drinking water is most unlikely, since only trace quantities of toxic substances are likely to migrate. However, it would be a great underestimation to consider PM ingredients (indirect food additives) as presenting no real public health threat. It is well known that chronic effects may be observed as the result of repeated ingestion of a number of small doses, each in itself insufficient to cause an immediate acute reaction but in the long term having a cumulative toxic effect. Thus, PM and other widely used chemicals have introduced a problem of protracted action of low concentrations of chemicals upon human health.

PM as well as many other materials are likely to be a depot of organic (sometimes also inorganic) compounds which, during the lifespan, are discharged into the environment, polluting various contact media such as food, water, air, skin surface, etc. (whenever food or drinks contact a solid surface, the resulting food contaminants could be called migrants). PM ingredients have a potential to migrate from the packaging or wrapping materials into the food in measurable amounts and thereby become *indirect food additives* (though these substances are never deliberately added to foods!). These migrants are appropriately regulated in Parts 175 through 179 of Volume 21 of the Code of Federal Regulations where appropriate regulations covering *indirect food additives* occupy more than 265 pages. US CFR lays out general safety requirements for all indirect food additives covering the safe use of food-contact PM. Under the *Food, Drug, and Cosmetic Act* (FD&CA), an industry must show that a new PM having indirect contact with food, such as packaging material or can coating material, is safe for the intended use.

The Food and Drug Administration (FDA) assesses the initial safety of packaging materials for food contact. If substances are shown to migrate from the packaging into the food, toxicological studies are conducted in order to establish safety standards. Appropriate toxicological information along with the results of animal toxicity tests is submitted to the *Food and Drug Administration* for review as a part of a *Food Additive Petition.*

Thus, testing migration of constituents from PM and articles and toxicological evaluation of these extractable contaminants is essential in selecting materials for use in contact with food by verifying that even if migration occurs, there would be no known toxic hazard to the consumer. EU food legislation as well as FDA requirements underscores that verification of compliance of migration into foodstuffs with the migration limits must be carried out under the most extreme conditions of time and temperature foreseeable in actual use. PM and articles which fail to meet the requisite standards shall not be used in the course of storage, preparation, packing, sale or serving of food for human consumption.

It is known that food contact applications are numerous and include the use of plastics, cellulose, paper, aluminum foil, glass, rubber, printing inks, and coatings. PM, in particular, are widely used in contact with foodstuffs, namely, in food processing equipment, food utensils, and as food packaging.

PM are manufactured by polymerization or polycondensation of one or more monomers and/or other starting substances. As basic polymers, the following compounds are most widely used (21 CFR):

*Vinyl resinous substances (Polyvinyl acetate, Polyvinyl alcohol, Polyvinyl butyral, Polyvinyl chloride, Polyvinyl formal, Polyvinylidene chloride, Polyvinyl pyrrolidone, Polyvinyl stearate, a number of Polyvinyl chloride copolymers);

*Styrene polymers (Polystyrene, α-Methylstyrene polymer, Styrene copolymers with acrylonitrile and α-Methylstyrene);

*Polyethylene and its copolymers;

*Polypropylene and its copolymers;

*Acrylics and their copolymers;

*Elastomers (Butadiene-acrylonitrile copolymer, Butadiene-acrylonitrile-styrene copolymer, Butadiene-styrene copolymer, Butyl rubber, Chlorinated rubber, 2-Chloro-1,3-butadiene/ neoprene, Natural rubber, Polyisobutylene, Rubber hydrochloride, Styrene-isobutylene copolymer).

In the manufacture of PM, numerous *additives* are used depending on the type of produced polymer. These *additives* include *plasticizers, antioxidants, catalysts, suspension* and *emulsifying agents, stabilizers* and *polymerization inhibitors, pigments, fillers, etc*. These additives are bound either chemically or physically into the polymer and may be present in their original or an altered form. In addition, the polymerization process may leave trace quantities of residual monomer or low-molecular-mass polymer in the PM. It is therefore necessary to specify the purity of the polymer to be used in the preparation of PM intended for food and/or drinking water contact use. Subsequently, PM contain a spectrum of polymers of different molecular mass, side effects and residues of all the auxiliary chemicals.

The migration potency of PM depends predominantly on the presence of unpolymerized monomers, residues of additives, reaction and transformation products of starting ingredients, and destruction products (about 3,000 PM components are listed by EU Commission as conceivable migrants). With time, the structure of PM could be changed as a result of destruction and aging processes, and of leaching or evaporation of PM ingredients or of their interaction products.

All additives are liable to break down during processing; some, such as antioxidants, are intended to do so to fulfill their function. A number of PM release not only additives into the environment but also monomers that could be present in PM as residues or have appeared as a result of destructive processes. As a matter of fact, PM appears to be a complicated and mobile system that is more or less stable, depending on its age, manufacture technology, and conditions of actual use.

Potential migrants encompass a large group of substances with differing molecular mass and physical properties. In some cases it is impossible to assess accurate amounts of ingredients migrating from PM into contact media. Migration levels can be considerably affected by destruction processes aging of plastics, and by the presence of unbound low-molecular-mass compounds. The extent to which migration occurs will depend upon such factors as the contact area, the rate of transfer, the type of PM, the temperature, and the contact time. The migration of substances from PM into food is also related to the type of food packaged in PM. Alcoholic beverages and edible fats and oils will extract substances more readily than dry food such as cereals.

The high-molecular-mass polymer itself does not pose a toxic hazard, being inert and essentially insoluble in food. *Monomers* are very reactive and biologically aggressive. Some of them have been shown to cause allergic effects, to damage the liver and reproductive functions, and to induce carcinogenicity.

Plasticizers are used to assist processing and impart flexibility to plastics. They intersperse around the polymer molecules and prevent them from bonding to each other so tightly that they form a rigid substance. Plasticizers may lower the melting point of the plastic, and many have relatively low melting points themselves. Plasticizers can be present in food packaging materials in significant amounts and have the potential to migrate into food. The migration of plasticizers can be aggravated by heat and by the presence of a food into which the plasticizing chemical will dissolve (for example, oil, acid or alcohol).

Migration increases with length of contact time and temperature, and levels of migration are highest when there is direct contact between PM and foods with high fat content at the surface.

In the food packaging and food processing industries the plasticizers most often used are *di(2-ethylhexyl) adipate* (DEHA), polymeric species, *epoxidized soybean oil* (ESBO) and *acetyl tributyl citrate* (ATBC) in packaging films and *di(-2ethylhexyl) phthalate* (DEHP), *diisodecyl phthalate*, and *diisooctyl phthalate* in closure seals for containers.

Those currently used in polyvinyl chloride (PVC) include *phthalates, phosphates, aliphatic dibasic acid esters,* and *polyesters*. The plasticizers most widely used in PVC for food contact applications are DEHA, DEHP, and polymeric species. Different plastics may contain different levels of phthalate plasticizers. ATBC is commonly used in vinylidene chloride copolymers; ESBO is also used in polyvinyl chloride and vinylidene chloride copolymer films.

The main function of *stabilizers* is to prevent destruction of PM when it is heated or exposed to ultraviolet radiation. *Antioxidants* are introduced to avoid undesirable oxidation. Stabilizers and antioxidants are not bound to polymeric macromolecules and could be easily leached into contact liquid media. Thermal stabilizers contribute to food contamination with their own residues.

Catalysts and *hardeners* are usually present in the finished product, they are used to aid in the polymer formation reaction. Catalysts of polycondensation (e.g., alkali and acids) seem to be less aggressive in comparison with catalysts used in polymerization.

A number of PM ingredients listed for use in U.S. and EU Regulations have been shown to migrate into the foodstuffs and simulant media under ordinary conditions at hazardous concentrations. When evaluating safety of a food-contact article it is often impossible to perform worst-case calculations (i.e., based on 100% migration) to demonstrate that the regulatory limits will not be exceeded under the intended conditions of use. In such cases, migration testing is required to establish compliance with specific migration limits. Modern chromatography techniques including capillary gas chromatography and high-performance liquid chromatography (HPLC) together with other highly selective detectors (FID, Electron capture, Nitrogen-phosphorous, MS for GC, UV, electrochemical, fluorescent, Mass Spectrometry for HPLC) ensure separation, identification, and precise determination of the majority of toxic substances migrating from plastics to contact media at the levels required for safety evaluation.

Advances in analytical chemistry have made it possible in many cases to decode the complex set of chemicals released from PM. Analytical chemistry has gradually become a method of routine monitoring the safety properties of plastics and can estimate or measure directly PM contamination.

Thus far, the United States and the European Union have developed procedures to assure the public that the food packaging materials in use are as safe as they can be. The majority of these Regulations is a listing of substances that can be used in food-contact applications.

In order to prevent or eliminate the risk of health hazard to a population exposed to PM, U.S. CFR lays out general safety requirements for all indirect food additives covering the safe use of food-contact materials. It regulates the use of these materials and articles in contact with food in accordance with the prescribed conditions. The FDA regulations are not conventional regulations prescribing a course of action, and do not govern manufacturing process, catalysts, or reaction control agents.

EU Legislation on PM and articles intended to come in contact with foodstuffs presents a strict list system according to which all substances used in the manufacture of a food packaging material will eventually have to appear and is published by the European Union authorities. EU legislation prescribes an overall migration limit in food and food simulant media and, in some cases, specific migration limits and maximum permitted quantities of the residual substance in PM and articles. This Legislation includes provisions applicable when checking migration limits, and a positive list of monomers and other starting substances that may be used in the manufacture of materials and articles intended to come into contact with foodstuffs.

Neither the CFR nor the EU legislation distinctly explains why they are publishing records of those PM ingredients for which there are no established specific migration limits. The positive lists (lists of approved PM ingredients) produce a certain unfavorable effect, creating an illusion that the PM composition of listed ingredients is safe. This illusion, shared by many, is erroneous. Sometimes the legislator must admit reluctantly that "such a list **would offer no tangible benefit** (bolded by author - V. S.) in terms of safeguarding human health" (Commission Directive 90/128/EEC). Then what benefit can it offer? Positive lists neither contribute to the problem nor help with the solution. Regulators in many countries including the U.S., Canada, Denmark, and Switzerland believe that positive lists appear to be the wrong approach and that we would be better off using a case-by-case system when a clearance is granted only on the basis of the specific conditions and is applicable only to the stated conditions of use (L. Borodinsky, 1997).

Some approaches based on an artificial hypothesis aimed at adopting a level of toxicological insignificance, have been suggested in order to avoid toxicity testing of all migrants that might be present (the overall migration test and the so-called threshold of regulation). Unfortunately, so far there is little scope for these approaches since each such recommendation needs to have a firm experimental base or thorough scanning of the existing literature on migration and toxicity. Nevertheless, in the U.S., a threshold of regulation set at 0.5 ppb has been adopted by the FDA for chemicals used in food packaging materials.

Successful regulation of food-contact materials seems to be possible with the help of the newest analytical methods and achievements of modern experimental and regulatory toxicology. All relevant information should be integrated into the risk assessment process on a case-by-case basis. A correct strategy in toxicology of plastics in many cases comprises precise chemical analysis of potential contamination of food, water, or simulant media under specified conditions. Obtained in this way, analytical results must be compared with available toxicology data and safety standards.

In the Author's opinion (and those who have labored long in the area will notice this), this handbook provides an opportunity to make an advancement in the regulatory toxicology of plastics, specifically to implement the modern toxicology approach instead of the application of an out-of-date limit of total extractives and positive lists.

CHAPTER 1. MONOMERS

ACETALDEHYDE

Molecular Formula. C_2H_40
M = 44.05
CAS No 75-07-0
RTECS No AB1925000
Abbreviations. AA (acetaldehyde), PET (polyethylene terephthalate).

Synonyms. Acetic aldehyde; Ethanal; Ethylaldehyde.

Properties. Colorless liquid with a pungent odor of rotten apples. Readily miscible with water and alcohol. Odor perception threshold is reported to be 0.034 mg/l,[02] taste perception threshold is 0.21 mg/l.

Applications. May be used in the production of synthetic rubbers, alkyd resins, and epoxy compounds intended for use in contact with food.

Exposure. AA occurs in common dietary components such as vegetables, fruits, and beverages, and in tobacco smoke. It is a food additive.

Migration Data. PET contains detectable amounts of AA, which is able to migrate from the polymer into liquid media. By the help of a static headspace GC method, AA was found in carbonate mineral water and lemonade. AA concentration ranged between 11 and 7447 μg/l, while the contents of AA in the PET packages ranged from 1.1 to 3.8 μg/g.[1] The amount of AA migrating from PET at 40°C reached a constant level after 4 days, which was about 10% of the residual value of AA. On increasing the temperature by 20°C, this level was raised up to 50%. The migration was found not to be sufficiently high to influence the taste of this soft drink. A negative effect of the taste may be recognized with mineral waters and soda when they are exposed to 40°C or more over longer periods of time.[2]

Acute Toxicity. LD_{50} is 1.93 g/kg BW for rats,[3] and 1.2 g/kg BW for mice. The treated animals displayed adynamia and labored respiration followed by convulsions. Death occurs within 3 to 10 min. after administration.

Repeated Exposure revealed very low accumulation of AA because of its rapid decomposition in the body (7 to 8 mg/min in rabbits).[4] Rats received doses of 25, 125, and 675 mg/kg BW via their drinking water over a period of 4 weeks. Food and liquid intake was decreased in the top dose group. Hyperkeratosis of the forestomach in the top dosed rats was the only adverse effect observed. The NOAEL of 125 mg/kg BW was identified in this study.[5]

Long-term Toxicity. Rats received 0.05% AA in drinking water (approximately, 40 mg/kg BW) for 6 months. There was an increase in collagen synthesis in the liver.[6] Rats were dosed by gavage with 10 and 100 mg/kg BW for 6 months. The treatment affected CNS functions and increased arterial pressure.

Reproductive Toxicity. AA can cross the placenta.[7]

Embryotoxicity. Rats were given AA on days 10 to 12 of gestation. Embryotoxic effects comprised a great number of fetal resorptions, edema, microcephaly, hemorrhaging, retardation of fetal growth, and other lesions including skeletal abnormalities. The treatment resulted in reduced placenta weights and umbilical cord length.[8]

Inhalation exposure to 5.0 mg AA/m^3 produced embryotoxic effect in rats. Morphology changes in the placenta are also reported.[9] However, according to other data, administration of AA on days 9 through 12 of gestation caused no embryonic effects or growth retardation.[10]

Teratogenicity. Skeletal abnormalities were observed in the rat following *in utero* exposure to AA.[11]

Gonadotoxicity. AA did not induce abnormal sperm, nor did it influence the frequency of meiotic micronuclei.[12]

Mutagenicity.

Observations in man. No data are reported on the genetic and related effects of AA in humans.

In vitro genotoxicity. AA is capable of inducing gene mutations at the *hprt* locus in human cells[13] and produced DNA single- and double-strand breaks in human lymphocytes.[14] AA is a well-known clastogen and CA and SCE inducer in cultured human and hamster cells.[15,16] It induced micronucleated polychromatic erythrocytes dose-dependently in male CD-1 mouse bone marrow cells.[17] AA was shown to be negative in the *Salmonella* mutagenicity assay; it caused DNA cross-links and CA in human cells *in vitro*.

In vivo cytogenetics. AA increased the incidence of SCE in bone marrow cells of mice and hamsters and induced CA in rat embryos exposed *in vivo*.[17]

Carcinogenicity. Inhalation exposure increased the incidence of GI tract tumors and carcinomas of the nasal cavity in rats and of the larynx in hamsters.[18]

Carcinogenicity classifications. An IARC Working Group concluded that there is sufficient evidence for the carcinogenicity of AA in *experimental animals* and there were no adequate data available to evaluate the carcinogenicity of AA in *humans*.

The Annual Report of Carcinogens issued by the U.S. Department of Health and Human Services (1998) defines AA to be a substance which may reasonably be anticipated to be carcinogen, i.e., a substance for which there is a limited evidence of carcinogenicity in *humans* or sufficient evidence of carcinogenicity in *experimental animals*.

IARC: 2B;

U.S. EPA: B2;

EU: 3.

Chemobiokinetics. AA is absorbed and metabolized to *acetic acid* by NAD-dependent aldehyde dehydrogenase.[19] According to Casier and Polet (1959), AA is likely to play a role in the acetylation of coenzyme AA and in subsequent synthesis of *cholesterol* and *fatty acids*. Urinary excretion appeared to be nonexistent.

Regulations.

U.S. FDA (1998) regulates AA as a direct food additive. It is considered to be *GRAS* for its intended use (1) as a synthetic flavoring substance and adjuvant. AA is also approved for use (2) as a component in the manufacture of phenolic resins in molded articles intended for repeated use in contact with non-acid food (*pH* above 5) in accordance with the conditions prescribed in 21 CFR part 177.2410.

EU (1992). AA is available in the *List of authorized monomers and other starting substances which shall be used for the manufacture of plastic materials and articles intended to come into contact with foodstuffs (Section* A).

Great Britain (1998). AA is authorized without time limit for use in the production of polymeric materials and articles in contact with food or drink or intended for such contact.

Recommendations. Joint FAO/WHO Expert Committee on Food Additives (1999). ADI: No safety concern.

Standards. *Russia* (1995). MAC and PML: 0.2 mg/l (organolept., odor).

References:

1. Linssen, G., Reitsma, H., and Cozynsen, G., Static headspace gas chromatography of acetaldehyde in aqueous foods and polythene terephthalate, *Z. Lebensm. Untersuch. Forsch.*, 201, 253, 1995.
2. Eberhartinger, S., Steiner, I., Washuttl, J., and Kroyer, G., The migration of acetaldehyde from polyethylene terephthalate bottles for fresh beverages containing carbonic acid, *Z. Lebensm. Unters. Forsch.*, 191, 286, 1990.
3. Smyth, H. F., Carpenter, C. P., and Weil, C. S., Range-finding toxicity data: list IV, *AMA Arch. Ind. Health Occup. Med.*, 4, 119, 1951.
4. Tsai, L. M., On the problem of acetaldehyde metabolism in the body, *Gig. Truda Prof. Zabol.*, 12, 33, 1962 (in Russian).
5. Til, H. P., Woutersen, R. A., and Feron, V. J., Evaluation of the oral toxicity of acetaldehyde and formaldehyde in a 4-week drinking water study in rats, *Food Chem. Toxicol.*, 26, 447, 1988.

6. Bankowski, E., Pawlicka, E., and Sobolewski, K., Liver collagen of rats submitted to chronic intoxication with acetaldehyde, *Molec. Cell. Biochem.*, 121, 37, 1993.
7. Blakley, P. M. and Scott, W. J., Determination of the proximate teratogen of the mouse fetal alcohol syndrome, *Toxicol. Appl. Pharmacol.*, 72, 364, 1984.
8. Sreenathan, R. N., Padmanabhan, R., and Singh, S., Teratogenic effects of acetaldehyde in the rat, *Drug Alcohol. Depend.*, 9, 339, 1982.
9. Taskayev, I. I., in *Structural Principles and Regulation of Compensation Adaptation Reactions*, Omsk Medical Institute, Omsk, 1986, 72 (in Russian).
10. Ali, F. and Persaud, T. V. N., Mechanism of fetal alcohol effects: role of acetaldehyde, *Exp. Pathol.*, 33, 17, 1988.
11. Fadel, R. A. and Persaud, T. V., Skeletal development in the rat following *in utero* exposure to ethanol and acetaldehyde, *Teratology*, 41, 553, 1990.
12. Lahdetie, J., Effects of vinyl acetate and acetaldehyde on sperm morphology and meiotic micronuclei in mice, *Mutat. Res.*, 202, 171, 1988.
13. Sai-Mei He and Bdanbert, Acetaldehyde-induced mutations at the *hprt* locus in human lymphocytes *in vitro, Environ. Mutagen.*, 16, 57, 1990.
14. Singh, N. P. and Khan, A., Acetaldehyde: genotoxicity and cytotoxicity in human lymphocytes, *Mutat. Res.*, 337, 9, 1995.
15. Bohlke, J. U., Singh, S., and Goedde, H. W., Cytogenetic effects of acetaldehyde in lymphocytes of Germans and Japanese: SCE, clastogenic activity, and cell cycle delay, *Human Genet.*, 63, 285, 1983.
16. Wangenheim, J. and Bolcsfoldi, G., Mouse lymphoma L5178Y thymidine kinase locus assay of 50 compounds, *Mutagenesis*, 3, 193, 1988.
17. Morita, T., Asano, N., Awogi, T., et al., Evaluation of the rodent micronucleus assay in the screening of IARC carcinogens (Groups 1, 2A and 2B). The summary report of the 6 collaborative studies by CSGMT/JEMS.MMS, *Mutat. Res.*, 389, 3, 1997.
18. Woutersen, R. A., Appelman, L. M., van Garderen-Hoetmer, A., and Feron, V. J., Inhalation toxicity of acetaldehyde in rats. III. Carcinogenicity study, *Toxicology*, 41, 213, 1986.
19. Brien, J. F. and Loomis, C. W., Pharmacology of acetaldehyde, *Canad. J. Physiol. Pharmacol.*, 61, 1, 1983.

ACETONITRILE

Molecular Formula. C_2H_3N
M = 41.05
CAS No 75-05-8
RTECS No AL7700000
Abbreviation. AN.

Synonyms. Acetic acid nitrile; Cyanomethane; Ethanenitrile; Ethylnitrile; Methanecarbonitrile; Methylcyanide.

Properties. Colorless liquid with an ether-like odor. In water solution AN is hydrolyzed to form acetamide and acetic acid. May be reduced to ethyl amine. Readily miscible with water and alcohol. Odor perception threshold is 2.4 mg/l.[1] According to other data, odor perception threshold is 0.75 mg/l at 60°C. However, odor perception threshold of 300 mg/l is also reported.[02]

Applications. A monomer and a high-polarity organic solvent used in the production of food-contact materials. Used also in consumer articles such as cosmetics, and in various chemical industries and laboratories.

Acute Toxicity. LD_{50} is 3.8 to 3.9 g/kg BW in rats, 0.2 to 0.33 g/kg BW in mice, and 0.14 to 0.35 g/kg BW in guinea pigs.[1,2] According to Hashimoto,[3] LD_{50} is 170 to 520 mg/kg BW in mice, which are shown to be one of the most susceptible animals to AN. Rats exhibited different age sensitivity to acute poisoning: in young animals (BW up to 50 g), LD_{50} was 200 mg/kg BW, while in adult rats (80 to 100 g), it was 3900 mg/kg BW, and in old animals (300 to 400 g), it was 4400 mg/kg BW.[4] Poisoning was accompanied by adynamia or agitation, coordination disorder, convulsions, dyspnea, depression of reflexes, hypothermia, etc. Lung emphysema was found to develop in 3 hours after administration of AN. Death occurs as a result of respiratory arrest.

Repeated Exposure failed to reveal cumulative properties. Guinea pigs were dosed by gavage with 1/5 and 1/20 LD_{50} for 75 days. The treatment resulted in decreased CO_2 production, reticulocytosis, leukocytosis, and increased contents of ascorbic acid in the liver, kidneys, and adrenal glands.[1] Gross pathology examination revealed dystrophic changes in the viscera and reduced spleen relative weights.

Long-term Toxicity. In a 6-month study, guinea pigs were exposed to the oral doses of 0.7 mg/kg and 3.5 mg/kg BW. The treatment with the higher dose caused reduced catalase activity, increased relative weights of the adrenals, and elevated ascorbic acid levels in the liver and spleen. Histology examination revealed moderate dystrophic changes in the visceral organs.[1] U.S. EPA based an ADI of 18 μg/kg BW on the rat NOEL of 18.3 μg/kg BW derived from the subchronic inhalation study of Pozzani et al.[5,6]

Reproductive Toxicity.

Embryotoxicity. Sprague-Dawley rats were administered the doses of 125 to 175 mg/kg BW on days 6 to 19 of gestation. Maternal toxicity and embryotoxicity but not teratogenicity effects were observed at the high dose level.[3] According to Smith et al., AN at maternally toxic dosages had no adverse postnatal effects.[7]

Sprague-Dawley rats were exposured to AN by inhalation during days 6 to 20 of gestation at concentrations of 900 to 1800 ppm. Embryolethality was observed only at the highest concentration.[8]

Teratogenicity. Inhalation by pregnant animals may produce malformations in the offspring (axial skeletal disorders) at maternal toxic levels.[3]

Significant teratogenic effects (exencephalia, medullary hernias, fusion of the ribs) and increased embryolethality were found in golden hamsters after ingestion of 100 to 400 mg AN/kg BW on day 8 of gestation. The treatment produced a reduction in BW of fetuses. According to Willhite,[9] these malformations were likely to occur because of the release of CN^- during AN metabolism. However, no teratogenic effect had been observed in the study.[10]

Rats were given 125, 190 or 275 mg AN/kg BW by gavage for 14 days. The LOEL of 275 mg/kg BW was established. Munro et al. (1996) suggested the calculated NOEL of 190 mg/kg BW for this study.

Mutagenicity.

In vitro genotoxicity. AN was negative in *Salmonella typhimurium* strains *TA98, TA100, TA1535* and *TA1537* with and without *S9* fraction. Weak positive response was observed in SCE without metabolic activation. There was a small increase in CA in the the presence, but not in the absence of metabolic activation. AN was positive in micronucleus test.[3,12]

Carcinogenicity classification.

NTP: EE - NE - NE - NE (inhalation).

Chemobiokinetics. AN is readily absorbed after ingestion. Cytochrome *P-450* IIE1 is a probable catalyst in oxidation of AN to *cyanide* by microsomes.[13] Freeman and Hayes believe the metabolism of AN occurs by cytochrome *P-450*-dependent pathway and not by a nucleophilic substitution reaction with glutathione.[14] The toxic effects are attributable to the metabolic release of *cyanide*, but the symptoms of poisoning may be delayed due to slow hepatic metabolism.[3]

Standards. ***Russia*** (1995). MAC and PML: 0.7 mg/l (organolept., odor).

References:

1. Rubinsky, N. D., *Acetonitrile and Succinonitrile, Their Potential Chemical Hazards, and Conditions of Draining to Reservoirs,* Author's abstract of thesis, Khar'kov, 1969, 16 (in Russian).
2. Trubnikov, N. A., (*Toxicological Characteristics of Acetonitrile*, Author's abstract of thesis, Yaroslavl', 1966, 14 (in Russian).
3. Hashimoto, K., *Toxicology of acetonitrile*, Abstract, *Sangyo-Igaku,* 33, 463, 1991 (in Japanese).
4. Kimura, E. T., Ebert, D. M., and Dodge, P. W., Acute toxicity and limits of solvent residue for sixteen organic solvents, *Toxicol. Appl. Pharmacol.,* 19, 699. 1971.
5. *Health and Environmental Effects Profile for Acetonitrile*, Prep. by the Office of Health & Environ. Assess., Environ. Criteria & Assess. Office, Cincinnati, OH, for the U.S. EPA Office of Solid Waste & Emerg. Response, Washington, D. C., 1985.

6. Pozzani, U. C., Carpenter, C. P., Palm, P. E., Weil, C. S., and Noir, J. H., III, An investigation of the mammalian toxicity of acetonitrile, *J. Occup. Med.*, 1, 634, 1959.
7. Smith, M. K., Zenick, H. H., and George, E. L., Reproductive toxicology of disinfection by-products, *Environ. Health Perspect.*, 69, 177, 1986.
8. Saillenfait, A. M., Bonnet, P., Guenier, J. P., and de Ceaurriz, J., Relative developmental toxicities of inhaled aliphatic mononitriles in rats, *Fundam. Appl. Toxicol.*, 20, 365, 1993.
9. Willhite, C. C., Developmental toxicology of acetonitrile in the Syrian golden hamster, *Teratology*, 27, 313, 1983.
10. Johannsen, F. R., Levinskas, G. J., Berteau, P. E., et al., Evaluation of the teratogenic potential of three aliphatic nitriles in the rat, *Fundam. Appl. Toxicol.*, 7, 33, 1986.
11. Munro, I. C., Ford, R. A., Kennepohl, E., and Sprenger, J. G., Correlation of structural class with No-Observed-Effect Levels: A proposal for establishing a threshold of concern, *Food Chem. Toxicol.*, 34, 829, 1996.
12. *Toxicology and Carcinogenesis Studies of Acetonitrile in F344/N Rats and B6C3F$_1$ Mice (Inhalation Studies),* NTP Technical Report Series No 447, Research Triangle Park, NC, April 1996.
13. Feierman, D. E. and Cederbaum, A. I., Role of cytochrome P-450 IIE1 and catalase in the oxidation of acetonitrile to cyanide, *Chem. Res. Toxicol.*, 2, 359, 1989.
14. Freeman, J. J. and Hayes, E. P., The metabolism of acetonitrile to cyanide by isolated rat hepatocytes, *Fundam. Appl. Toxicol.*, 8, 263, 1987.

ACROLEIN

Molecular Formula. C_3H_40
M = 56.07
CAS No 107-02-8
RTECS No AS1050000

Synonyms and **Trade Name.** 2-Propenal; Acraldehyde; Acrylaldehyde; Allyl aldehyde; Aqualin; 2-Propen-1-on.

Properties. Colorless, volatile, transparent liquid with a pungent odor. A. is fairly soluble in water (200 g/l at 25°C)[02] and in organic solvents (ethanol, diethyl ether, etc.). Odor perception threshold is reported to be 0.2 mg/l[1] or 0.11 mg/l.[02]

Applications. A. is an intermediate in the production of polymers and copolymers of acrylic acid and acrylonitrile, and cycloaliphatic epoxy resins intended for use in contact with food.

Exposure. The product of degradation of synthetic polymers. A. could be detected in cigarette smoke and in volatile components of some foods.

Acute Toxicity. LD_{50} values in Wistar rats, mice and rabbits are 39 to 56, 28 and 7.0 mg/kg BW, respectively.[1] However, according to other data, LD_{95} for Charles River rats is 11.2 mg/kg BW, and 5 out of 10 rats died following administration of a single oral dose of 10 mg/kg by gavage.[2,3]

Repeated Exposure. Mice were dosed with 1.5 mg/kg BW for a month. The treatment caused a decrease in the food consumption and morphological changes in the liver and kidneys. The NOEL appeared to be 0.17 mg/kg BW.[4]

Long-term Toxicity.

Observations in man. Chronic human exposure is unlikely due to severe irritating properties of A.[5]

Animal studies. In a 6-month study, disorders of the kidney (protein in the urine) and liver (shortening of the prothrombin time) functions were found in rats.[5] Histology examination revealed pneumonia and reduced relative liver weights. In a 1-year study, beagle dogs received 0.1, 0.5, and 1.5 mg A./kg BW (as 0.1% aqueous solution in gelatin capsules). After 4 weeks the highest dose was increased to 2.0 mg A./kg BW. The major test effect was frequent vomiting after dosing. This was considered to be an adaptive effect. Serum albumin, calcium and total protein values were found to be depressed only in high-dosed animals.[6]

Sprague-Dawley rats were exposed to 0.05, 0.5, and 2.5 mg/kg BW in water (10 ml/kg BW) for 102 weeks. The only effects noted were a consistent depression of creatinine phosphokinase levels, which was difficult to explain, and consistent increases in early cumulative mortalities. There was no significantly increased incidence of microscopic lesions in treated rats, whether neoplastic or non-neoplastic.[7]

Reproductive Toxicity. A. is not found to be a selective reproductive toxicant in the rat. It produced toxic effects down to a dose level of 3.0 mg/kg BW.[8]

Embryotoxicity. A. is a metabolite of cyclophosphamide which is a known embryotoxic agent. A. produces the embryotoxic effect in rats, inhibiting fetal growth[8] but only at maternal toxicity level. It was also found to cause embryotoxic effects in whole embryo culture.

The data on the toxic effect of A. on chick embryos are contradictory.[9-12]

Teratogenicity. There is no solid evidence that A. produces fetal malformations. Schmidt et al.[13] found no teratogenic activity in Sprague-Dawley rats. A. was not found to be a developmental toxicant or teratogen in New Zealand white rabbits at doses not toxic to the dams (up to 2.0 mg/kg BW). The bigger doses (4.0 and 6.0 mg/kg BW) produced high incidences of maternal mortality, spontaneous abortions, resorptions, and gastric ulcerations.[14]

Allergenic Effect. Contact sensitizing potential is not shown to be found in maximized guinea-pig assays.[15]

Mutagenicity. Assays on the mutagenic potential revealed conflicting results. A. is unlikely a direct-acting mutagen.[3]

Observations in man. There are no data available on the genetic and related effects.

In vivo cytogenetics. In the dominant lethal assay, mice were exposed *i/p* with 1.5 to 2.2 mg/kg BW. The treatment produced no increase in the incidence of early death or pre-implantation losses.

A. was found to be a substance of relatively low DNA reactivity in a *Dr. melanogaster* assay.[16]

In vitro genotoxicity. A. was negative in *Salmonella* mutagenicity assay (NTP-92). It was positive in mutagenicity assay when measured in *V79* cells as resistance to 6-thioguanidine.[17]

Carcinogenicity. In a 104-week study, F344 rats were exposed to A. in their drinking water at a concentration of 625 ppm. No decrease in survival or increase in tumor incidence was reported. 5 out of 25 female rats developed adrenal cortical adenomas; 2 out of 20 rats had neoplastic nodules in the adrenal cortex. The authors did not consider the study to be a definitive carcinogenicity bioassay.[1]

Parent et al. have revised the tissues from this study and found no proof of A. carcinogenicity.[18] Sprague-Dawley rats received 10 mg A./l in their drinking water for 102 weeks (equivalent to 0.05 to 2.5 mg/kg BW). The only effect was consistent depression of creatinine phosphokinase levels. No microscopic lesions in the treated rats, whether neoplastic or non-neoplastic were noted.[19] In other chronic studies, A. was given by gavage to Sprague-Dawley rats and CD-1 mice. The treatment did not produce carcinogenic response. In addition, the authors failed to observe any significant systemic effect other than increased mortality and retardation of BW gain.[19]

Carcinogenicity classifications. An IARC Working Group concluded that there is inadequate evidence for the carcinogenicity of A. in *experimental animals* and in *humans*.

IARC: 3;

U.S. EPA: C.

Chemobiokinetics. After ingestion, A. is found to be readily absorbed in the GI tract of experimental animals. Its transport throughout the body is very low.

Being a highly reactive compound, A. reacts with the substances present in the tissues at the site of the contact. A. can combine with glutathione. It inhibits metabolism of xenobiotics. A. is likely to be converted into *acrylic acid.* According to Linhart et al., the major and minor urinary metabolites after inhalation and *i/p* administration in Wistar (Charles River, Germany) rats were 3-*hydroxypropylmercapturic acid* and 2-*carboxyethylmercapturic acid*, respectively.[20]

Regulations. ***U.S. FDA*** (1998) approved the use of A. (1) as an ingredient of resinous and polymeric coatings for polyolefin films to be safely used as a food-contact surface and subject to certain limitations in accordance with the conditions prescribed in 21 CFR part 175.320; (2) as a slimicide in the manufacture of paper and paperboard products in contact with food as prescribed in 21 CFR part 176.300. A. may be used (3) as an etherifying agent in the manufacture of food additives or of "modified food starch" in an amount not exceeding that reasonably required to reach the intended effect (GMP) or not exceeding 4.0% when added alone, or 0.6% when added with vinyl acetate.

Standards. ***Russia*** (1995). PML: 0.02 mg/l.

References:

1. Lijinsky, W. and Reuber, M. D., Chronic carcinogenesis studies of acrolein and related compounds, *Toxicol. Ind. Health,* 3, 337, 1987.
2. Sprince, H., Parker, C. M., and Smith, G. G., Comparison of protection by *L*-ascorbic acid, *L*-cysteine, and adrenergic-blocking agents acetaldehyde, acrolein, and formaldehyde toxicity: implications in smoking, *Agents Actions*, 9, 407, 1979.
3. Draminski, W., Eder, E., and Henschler, D., A new pathway of acrolein metabolism in rats, *Arch. Toxicol.*, 52, 243, 1983.
4. Goeva, O. E., Maximum allowable concentration of acrolein in water reservoirs, in *Proc. Leningrad Sanitary Hygiene Medical Institute*, Leningrad, Issue No 81, 1965, 58 (in (in Russian).
5. Beauchamp, R. O., Andjelkowich, D. A., Kleigerman, A. D., et al., A critical review of literature on acrolein toxicity, *CRC Crit. Rev. Toxicol.,* 14, 309, 1985.
6. Parent, R. A., Caravello, H. E., Balmer, M. F., Shellenberger, T. E., and Long, J. E., One-year toxicity of orally administered acrolein to the beagle dogs, *J. Appl. Toxicol.,* 12, 311, 1992.
7. Parent, R. A., Caravello, H. E., and Long, J. E., Two-year toxicity and carcinogenicity study of acrolein in rats, *J. Appl. Toxicol.*, 12, 131, 1992.
8. Parent, R. A., Caravello, H. E., and Hoberman, A. M., Reproductive study of acrolein in two generations of rats, *Fundam. Appl. Toxicol.*, 19, 228, 1992.
9. Mirkes, P. E., Fantel, A. G., Greenaway, R. N., et al., Teratogenicity of cyclophosphamide metabolites: phosphoramide mustard, acrolein, and 4-ketocyclophosphamide in rat embryos culture *in vitro*, *Toxicol. Appl. Pharmacol.,* 58, 322, 1981.
10. Slued, V. L. and Hales, B. F., The embryolethality and teratogenicity of acrolein in cultured rat embryos, *Teratology*, 34, 155, 1986.
11. Chibber, G. and Gilani, S. H., Acrolein and embryogenesis: an experimental study, *Environ. Res.*, 39, 44, 1986.
12. Kankaanpaa, E. E., Hemminki, K., and Vainio, H., Embryotoxicity of acrolein, acrylonitrile and acrylic acid in developing chick embryos, *Toxicol. Lett.*, 4, 93, 1979.
13. Schmidt, B. P. and Schon, H., The post-implantation rodent embryo culture system: a potential pre-screen in teratology, *Experientia*, 37, 675, 1981.
14. Parent, R. A., Caravello, H. E., Christian, M. S., and Hoberman, A. M., Developmental toxicity of acrolein in New Zealand white rabbits, *Fundam. Appl. Toxicol.*, 20, 248, 1993.
15. Susten, A. S. and Breitenstein, M. J., Failure of acrolein to produce sensitization in the guinea pig maximization test, *Contact Dermat.*, 2, 299, 1990.
16. Vogel, E. W. and Nivard, N. J., Performance of 181 chemicals in a *Drosophila* assay predominantly monitoring interchromosomal mitotic recombination, *Mutagenesis*, 8, 57, 1993.
17. Smith, R. A., Cohen, S. M., and Lawson, T. A., Acrolein mutagenicity in the V79 assay, *Carcinogenesis*, 11, 497, 1990.
18. Parent, R. A., Caravello, H. E., and Long, J. E., Two-year toxicity and carcinogenicity study of acrolein in rats, *J. Appl. Toxicol.*, 12, 131, 1992.
19. Parent, R. A., Caravello, H. E., and Long, J. E., Eighteen month oncogenicity study of acrolein in mice, *J. Am. Coll. Toxicol.*, 10, 647, 1992.
20. Linhart, I., Frantik, L., Vodickova, L., Vosmanska, M., Smejkal, J., and Mitera, J., Biotransformation of acrolein in rat: Excretion of mercapturic acids after inhalation and intraperitoneal injection, *Toxicol. Appl. Pharmacol.*, 136, 155, 1996.

ACRYLAMIDE

Molecular Formula. C_3H_5NO

M = 71.08

CAS No 79-06-1
RTECS No AS3325000
Abbreviations. AA (acrylamide); PAA (polyacrylamide).

Synonyms. Acrylic acid, amide; Ethylenecarboxamide; 2-Propenamide.

Properties. Colorless, odorless crystals. Water solubility is 204 g/100 ml, soluble in alcohol. When hydrolyzed, AA forms acrylic acid. Concentrations of 0.01 to 0.5 mg/l give water neither color or opalescence nor any unrequired odor. A concentration of 0.001 to 0.01 mg/l gives water no foreign taste.[1]

Applications. Used in the manufacture of articles intended for use in contact with food.

AA is predominantly used as a monomer in the production of *polyacrylamide* (PAA), a number of copolymers and synthetic rubber. Polymers of AA are used in the production of lacquers, paints, adhesives, and dentistry compositions. PAA are also used as grouting agents in the construction of drinking water reservoirs and wells.

Exposure. Residual AA monomer occurs in PAA coagulants used in the treatment of drinking water. Generally, the maximum authorized dose of polymer is 1.0 mg/l corresponding to a maximum theoretical concentration of 0.0005 mg/l monomer in water. Practical concentration may be lower by a factor of 2 to 3. This applies to the anionic and non-anionic PAA, but residual levels from cationic PAA may be higher. Additional human exposure might result from food owing to the use of PAA in food processing.

Acute Toxicity. LD_{50} is reported to be 120 to 180 mg/kg BW in rats, 100 to 170 mg/kg BW in mice, 150 to 170 mg/kg BW in guinea pigs, and 150 to 280 mg/kg BW in rabbits.[1-3,05] In 12 hours after administration of 50 to 200 mg/kg BW, poisoned animals displayed impaired hind limb functioning, convulsions, and diffuse damage to different sections of the NS. Gross pathology examination revealed circulatory disturbances in the parenchymatous organs and brain matter as well as signs of parenchymatous dystrophy of the liver and kidneys. The treatment caused marked lysis of tigroidal matter in the cerebellum neurons.[2]

The acute effect threshold for the change in STI value appeared to be 40 mg/kg BW. Cats seem to be more sensitive to AA than rats and monkeys.

Repeated Exposure.

Observations in man. Five individuals were exposed through ingestion and external use of well water contaminated with AA. Signs of intoxication included confusion, disorientation, memory disturbances, hallucinations, and truncal ataxia. All exposed persons recovered completely within 4 months.[4]

Animal studies. AA exhibited pronounced cumulative properties: administration of 1/5 and 1/20 LD_{50} resulted in K_{acc} of 2.2 and 2.4, respectively (by Kagan). According to Lim method, K_{acc} appeared to be 1.07 in mice, and 1.56 in rats.[1]

AA produces prominent effects in the neuromuscular domain. It is shown to reduce the amount of large-diameter fibers in the peripheral nerves and to cause their degeneration. Doses of 50 to 100 mg/kg BW impair functioning of the hind legs, but the front legs remain unimpaired even with a 400 mg/kg BW dose.[3] Repeated administration of doses up to 50 mg/kg BW by any route produced neuropathy of the peripheral nerves in the majority of laboratory animals including non-antropoidal monkeys. Changes observed seem to be reversible but secondary effects include atrophy of the skeletal muscles. Administration of 300 mg AA/kg BW produced muscular weakness in the limbs and the loss of reflexes. A dose of 100 mg AA/kg BW in the feed caused retardation of BW gain.

Addition of 10 to 50 mg AA/kg BW to the diet of young rats had no clear toxic effect (IARC 19-93). The doses of 1.0, 4.0, or 60 mg/kg BW were administered *i/p* to Long-Evans rats for 13 weeks. Treatment with A. caused time- and dose-related changes in muscle tone and equilibrium, and produced axonal degeneration in peripheral nerves and within long tracts of the spinal cord.[5]

Short-term Toxicity. Young rats were given a dose of 31 mg/kg BW for 3 months. The treatment caused retardation of BW gain and anemia. Gross pathology examination revealed edema of the mucosa and submucosa of the stomach and small intestine, moderate epithelial sloughing in the large intestine, dystrophic changes in the brain, parenchymatous dystrophy in the liver and kidneys, and increased relative weights of the liver.[6,7] Rats exposed to AA solution at dose levels of 5.0 to 20 mg/kg BW for 3 months became weak and dragged their hind legs. The treatment caused retardation of BW gain. Histology examination revealed the loss of axones, and degeneration and demyelination of the peripheral nerves.[8] Administration of 10 mg/kg BW to male rats 3 times a week for 13 weeks resulted in retardation of BW gain observed seven weeks after the onset of treatment.[9]

In one study, macaques received an overall dose of 320 to 450 mg/kg BW until toxic effect developed.[8] Animals displayed reduced BW gain preceded by neurological disturbances, including loss of equilibrium, tremors, reduction in activity, and coordination disorder. The vibration sensitivity threshold was increased, but there was no change in sensitivity to electrostimulation. A dose of 0.2 mg/kg BW appeared to be ineffective in a 3-month experiment in rats.[19]

Long-term Toxicity. Rats were fed 100 to 400 mg AA/kg BW for 6 to 48 weeks. The treatment disrupted nervous conduction in the hind legs and caused degenerative changes in the peripheral NS. Recovery began a long time after administration had ceased (IARC 19-99). Baboons received 10 to 20 mg AA/kg BW with their drinking water for 29 to 192 days. The treated animals displayed weakness and ataxia of their limbs, weakness of the eye muscles, and later, development of tetraplegia. Histology examination revealed an extensive effect on the peripheral nerves.[11] The NOAEL in rat chronic toxicity studies was reported to be 0.2 mg/kg[8] or 0.3 mg/kg BW.[05]

Immunotoxicity. Novikov and Ostapova found no sensitizing effect during acute or chronic adminstration of AA.[7,12]

With repeated administration of 12 mg AA/kg BW there were changes in the immune reactivity. A dose of 0.1 mg/kg BW caused an increase in the amount of plaque-forming cells, but a dose of 0.01 mg/kg BW had no such effect.

Reproductive Toxicity.

Gonadotoxicity. AA affects germ cells and impairs reproductive function. The dose of 500 mg/kg BW led to a reduction in the testicle weights. A marked degeneration of the seminiferous tubules was observed.[13,14] The NOEL of 1.5 mg/kg BW was identified based on the absence of DLM in male sperm. Rumyantzev et al.[2] did not observe a gonadotoxic effect at dose levels up to 0.05 mg A./kg BW.

Embryotoxicity. No adverse effect of AA on the offspring development was found in the study where Porton rats were given AA during gestation with their feed at doses of 0.2 to 0.4 g AA/kg BW with the simultaneous administration of an *i/v* dose of 0.1 g/kg BW on days 9, 14, or 21 of gestation.[5]

Maternal toxicity at the highest dose level was demonstrated in mice gavaged with 3.0 to 45 mg/kg BW on days 6 to 17 of gestation, and in rats given 2.5 to 15 mg/kg BW on days 6 to 20 of gestation. BW was decreased and hind limb splaying occurred in mice only.[15] In another study, the NOEL for embryotoxic effect was found to be 2.0 mg/kg BW.[1]

Swiss CD-1 mice were provided drinking water containing 3.0, 10, or 30 ppm AA during and after a 14-week cohabitation. The last litter was reared and dosed after weaning until mating at 2.5 months of age, with same level of compound given to the parents. In the F_o generation, the treatment caused an 11% decrease in pup number without measurable neurotoxicity. Female fertility was not affected. Larger changes were observed in the fertility-related endpoints in the F_1 mice than in the F_o generation. There was no concomitant change in organ weights or sperm parameters.[16]

Teratogenicity. No fetal abnormalities were shown after oral administration of high doses.[17,18]

Mutagenicity. Despite the fact that AA is structurally similar to the mutagens and carcinogens acrylonitrile and vinyl chloride, and is capable to react with nucleophilic sites or compounds, published data on its mutagenicity are contradictory.

In vivo cytogenetics. In the dominant lethal assay, mice were exposed *i/p* to 1.5 to 2.2 mg/kg BW. The treatment produced no increase in the incidence of early death or pre-implantation losses.[19] AA is reported to produce mutagenic effect in the germ cells of mice which were given it at a dose level of 500 mg/kg in the diet for 3 weeks.[13]

AA induced gene mutations in mammalian cells and CA *in vitro* and *in vivo.*[5] It appeared to be a clastogenic agent inducing CA, DLM, SCE, and unscheduled DNA synthesis. There was no increase of micronuclei in CD-1 mice.[20]

Shelby et al. found AA to produce DLM and translocations in early spermatozoa and late spermatides of the mouse, but not at earlier stages.[21-23] DLM studies revealed fertility effects that could be explained by a male-mediated increase in postimplantation loss. Dominant lethality occurred without structural effects on the reproductive system and without detectable neural histopathology.[16] According to Hurt et al.,[24] AA caused unscheduled DNA synthesis in pachytene spermatocytes of rats. AA produced a DNA repair response in the *in vivo* spermatocyte but not in either the *in vitro* or *in vivo* hepatocyte DNA repair assay.[23]

Batiste-Allentorn et al.[25] reported that AA produced a significant increase in *Dr. melanogaster* wing spot somatic mutation and recombination assay. However, according to Vogel and Nivard, AA was found to be a substance of relatively low DNA reactivity in a *Dr. melanogaster* assay.[26]

In vitro genotoxicity. According to Knaap et al.,[27] AA exhibits no mutagenic activity in different test systems, including *Salmonella* mutagenicity assay and test on gene mutations in murine lymphoma cells. Glycidamide, a mutagenic metabolite of AA, induced DNA repair in the *in vitro* hepatocyte DNA repair assay and produced a strong unscheduled DNA synthesis response.[23]

Carcinogenicity. There was an increased incidence of lung adenomas in mice exposed to average daily doses of 2.7 to 10.7 mg/kg BW.[28]

In a 2-year study, rats were administered 0.5 to 2.0 mg/kg BW via drinking water. Increased rates of scrotal mesotheliomas, mammary gland tumors, thyroid adenomas, uterine adenocarcinomas, clitoral gland adenomas, and oral papillomas were found.[29] AA is considered to be a genotoxic carcinogen (WHO, 1991).

Carcinogenicity classifications. An IARC Working Group concluded that there is sufficient evidence for the carcinogenicity of AA in *experimental animals* and there is inadequate evidence for the carcinogenicity of AA in *humans*.

The Annual Report of Carcinogens issued by the U.S. Department of Health and Human Services (1998) defines AA to be a substance which may reasonably be anticipated to be a carcinogen, i.e., a substance for which there is a limited evidence of carcinogenicity in *humans* or sufficient evidence of carcinogenicity in *experimental animals*.

IARC: 2A;

U.S. EPA: B2;

EU: 2.

Chemobiokinetics. Following ingestion, AA is readily absorbed from the GI tract and widely distributed in the body fluids. Accumulation in the liver and kidneys as well as in the male reproductive system has been demonstrated. AA is likely to cross the placenta. Less than 1.0% of the administered dose accumulates in the NS. AA is found mainly in the muscles and skin. Male rats were given 50 mg ^{14}C- AA/kg BW. Rapid absorption from the GI tract was reported. After 28 hours the highest activity was found in gastric contents, followed by stomach, lung, bone marrow, and skin. After 144 hours the availability of the highest total activity was lung > pancreas >adrenal >esophagus. Kidney is the major route of elimination. 10% of activity was excreted in feces.[30]

AA is electrophilic and, subsequently, undergoes reactions with nucleophiles. In rats, biotransformation of AA occurs through glutathione conjugation and decarboxylation. The major metabolite is *N-acetyl-S-(3-amino-3-oxypropyl)cysteine*, which accounted for 48% of the oral dose.[31] Despite much research it is still unclear whether the parent compound or a metabolite is responsible for the observed toxic effects.[32]

AA is mainly excreted as metabolites in the urine and bile.

Regulations.

U.S. FDA (1998) regulates AA as an indirect food additive, a component of food-contact surfaces for single and repeated use. PAA has been approved for various applications. MPC in food contact PAA solutions: 0.02% (1990); MPC for PAA food additive: 0.05 to 0.2%.

U.S. FDA has established regulations for use of AA: It can be used (1) in washing or to assist in the peeling of fruits and vegetables using lye if the concentration does not exceed 10 mg/l in the wash water and if no more than 0.2% AA is present; (2) in adhesives as a component (monomer) of articles intended for use in packaging, transporting, or holding food in accordance with the conditions prescribed in 21 CFR part 175.105; as (3) a component (monomer) of the uncoated or coated food-contact surface of paper and paperboard intended for use in producing, manufacturing, packaging, processing, preparing, treating, packing, transporting, or holding dry food in accordance with the conditions prescribed in 21 CFR part 176.180; (4) as a monomer in the manufacture of semirigid and rigid acrylic and modified acrylic plastics in the manufacture of articles intended for use in contact with food in accordance with the conditions prescribed in 21 CFR part 177.1010. (5) AA-sodium acrylate resins can be used as boiler water additives in the preparation of articles that will be in contact with food, if the water contains not more than 0.05% by weight of AA; (6) PAA can be used as a film former in the imprinting of soft-shell gelatin capsules if no more than 0.2% of the monomer is present; (7) homopolymers and copolymers of AA may be safely used as food packaging adhesives, providing the amount used does not exceed that "reasonably required to accomplish the intended effect"; (8) AA-acrylic acid resins may be safely used as components in the production of paper or paperboard used for packaging food, providing the resin contains less than 0.2% residual monomer and that the resin does not exceed 2.0% by weight of the paper or paperboard.

EU (1990). AA is available in the *List of authorized monomers and other starting substances which may continue to be used for the manufacture of plastic materials and articles intended to come into contact with foodstuffs pending a decision on inclusion in Section A* (*Section B*).

Great Britain (1998). AA is authorized without time limit for use in the production of polymeric materials and articles in contact with food or drink or intended for such contact. The specific migration of this substance shall be not detectable (when measured by a method with a detection limit of 0.01 mg/kg).

In ***Germany***, PAA flocculents must not contain more than 0.05% of the monomer.[11] The addition of PAA is limited to 1.0 mg/l in treated water. The concentration of the monomer will therefore not exceed 0.0005 mg/l in finished water. The level of PAA used in foodstuff packaging is limited to 0.3% and the level of residual AA monomer in polyacrylamide to 0.2%.

Recommendations.

WHO (1996). Guideline value for drinking water: 0.0005 mg/l. Generally, AA cannot be detected at concentrations of 0.001 mg/l or less, but can be controlled by product specification.[33]

U.S. EPA (1999): Health Advisory for a longer-term exposure is 0.04 mg/l.

The UK Committee on Carcinogenicity (1993) recommended that the level of AA monomer in PAA used for drinking water treatment should be reduced to the lowest practicable level.[34]

Standards.

EU (1990). SML: Not detectable (Detection limit is 0.01 mg/kg).

U.S. EPA (1999). MCL: TT; MCLG: zero.

Russia (1995). MAC and PML: 0.01 mg/l.

References:

1. Kozeyeva, E. E., *Production and Experimental Investigations on Hygienic Assessment of Acrylamide*, Author's abstract of thesis, Moscow, 1980, 23 (in Russian).
2. Rumyantsev, G. I., Novikov, S. M., Kozeeva, E. E., et al., Experimental study of biological effects of acrylamide, *Gig. Sanit.*, 9, 38, 1980 (in Russian).
3. Tilson, H. and Cabe, P., The effects of acrylamide given acutely or in repeated doses on fore- and hindlimb function of rats, *Toxicol. Appl. Pharmacol.*, 47, 253, 1979.
4. Igusu, H., Goto, I., Kawamura, Y., et al., Acrylamide encephaloneuropathy due to well water pollution, *J. Neurol. Neurosurg. Psychiatr.*, 38, 581, 1975.
5. Zenick, H., Hope, E., and Smith, M. K., Reproductive toxicity associated with acrylamide treatment in male and female rats, *J. Toxicol. Environ. Health*, 17, 457, 1986.
6. Loshchenkova, I. F., in *Physiology of Autonomic Nervous System*, Kuibyshev, Issue No 1, 1979, 321 (in Russian).
7. Ostapova, I. F., *Pharmaceutical Kinetics, Toxicology, and Pharmaceutical Dynamics of Acrylamide, Methacrylamide, and Diacetone Acrylamide*, Author's abstract of thesis, Yaroslavl', 1981, 333 (in Russian).
8. Burek, J. D., Albee, R. R., Beyer, J. E., et al., Subchronic toxicity of acrylamide administered to rats in the drinking water followed by up to 144 days of recovery, *J. Environ. Pathol. Toxicol.*, 4, 157, 1980.
9. Moser, V. C., Antony, D. C., Sette, W. F., and MacPhail, R. C., Comparison of subchronic neuro-toxicity of 2-hydroxyethyl acrylate and acrylamide in rats, *Fundam. Appl. Toxicol.*, 18, 343, 1992.
10. Maurissen, J. P., Weiss, B., Davis, H. T., et al., Somatosensory thresholds in monkeys exposed to acrylamide, *Toxicol. Appl. Pharmacol.*, 71, 266, 1983.

11. Hopkins, A., The effect of acrylamide on the peripheral nervous system of the baboon, *J. Neurol. Neurosurg. Psychiatr.*, 33, 805, 1970.
12. Novikov, S. M., *Problems of Occupational Safety and Industrial Toxicology in Production and Use of Acrylamide*, Author's abstract of thesis, Moscow, 1974, 23 (in Russian).
13. Shiraishi, Y., Chromosome aberrations induced by monomeric acrylamide in bone marrow and germ cells of mice, *Mutat. Res.*, 57, 313, 1978.
14. McCollister, D. D., Oyen, F., and Rowe, V. K., Toxicology of acrylamide, *Toxicol. Appl. Pharmacol.*, 6, 172, 1964.
15. Field, E. A., Price, C. J., Sleet, R. B., et al., Developmental toxicity evaluation of acrylamide in rats and mice, *Fundam. Appl. Toxicol.*, 14, 502, 1990.
16. Chapin, R. E., Fail, P. A., George, J. D., Grizzle, T. B., Heindel, J. J., Harry, G. J., Collins, B. J., and Teague, J., The reproductive and neural toxicities of acrylamide and three analogues in Swiss mice, evaluated using the continuous breeding protocol, *Fundam. Appl. Toxicol.*, 27, 9, 1995.
17. Edwards, P. M., The insensitivity of the developing rat fetus to the toxic effects of acrylamide, *Chem.-Biol. Interact.*, 12, 13, 1976.
18. Kankaanpaa, J., Elovaara, E., Hemminki, K., et al., Embryotoxicity of acrolein, acrylonitrile, and acrylamide in developing chick embryos, *Toxicol. Lett.*, 4, 93, 1979.
19. Epstein, S. S., Arnold, E., Andrea, J., Bass, W., and Bishop, Y., Detection of chemical mutagens by the dominant lethal assay in the mouse, *Toxicol. Appl. Pharmacol.*, 23, 288, 1972.
20. Dearfield, K. L., Abernathy, C. O., Ottley, M. S., et al., Acrylamide: its metabolism, developmental and reproductive effects, genotoxicity, and carcinogenicity, *Mutat. Res.*, 195, 45, 1988.
21. Shelby, M. D., Cain, K. T., Hughes, L. A., et al., Dominant lethal effects of acrylamide in male mice, *Mutat. Res.*, 173, 35, 1986.
22. Shelby, M. D., Cain, K. T., Cornett, C. V., et al., Acrylamide: induction of heritable translocations in male mice, *Environ. Mutagen.*, 9, 363, 1987.
23. Butterworth, B. E., Eldridge, S. R., Sprankle, C. S., Working, P. K., Bentley, K. S., and Hurt, M. E., Tissue-specific genotoxic effects of acrylamide and acrylonitrile, *Environ. Molec. Mutagen.*, 20, 148, 1992.
24. Hurt, M. E., Bentloy, K., and Working, P. K., Effects of acrylamide and acrylonitrile on unscheduled DNA synthesis in rat spermatocytes, *Environ. Mutagen.*, 9, 49, 1987.
25. Batiste-Allentorn, M., Xamena, N., Creus, A., and Marcos, A., Genotoxic evaluation of ten carcinogens in the *Drosophila melanogaster* with spot test, *Experientia*, 51, 73, 1995.
26. Vogel, E. W. and Nivard, N. J., Performance of 181 chemicals in a *Drosophila* assay predominantly monitoring interchromosomal mitotic recombination, *Mutagenesis*, 8, 57,1993.
27. Knaap, A., et al., in *The 14th Annual Conf. Europ. Soc. Environ. Mutagens.*, Abstracts, Moscow, 1984, 247 (in Russian).
28. Bull, R. J., Robinson, M., Laurie, R. D., et al., Carcinogenic effects of acrylamide in SENCAR and A/J mice, *Cancer Res.*, 44, 107, 1987.
29. Johnson, K. A., Gorzinski, S. J., Bodner, K. M., et al., Chronic toxicity and oncogenicity study on acrylamide incorporated in the drinking water of Fischer 344 rats, *Toxicol. Appl. Pharmacol.*, 85, 154, 1986.
30. Kady, A. M., Kardos, S., Johnson, L. R., Friedman, M., and Abdel-Rahman, M. S., Pharmacokinetics of acrylamide after oral exposure in male rats, in *Toxicologist*, Abstracts of SOT 1996 Annual Meeting, Abstract No 204, Issue of *Fundam. Appl. Toxicol.*, 30, Part 2, March 1996.
31. Miller, M. J., Carter, D. E., and Sipes, I. G., Pharmacokinetics of acrylamide in Fischer 344 rats, *Toxicol. Appl. Pharmacol.*, 63, 36, 1982.
32. Calleman, C. J., Bergmark, E., and Costa, L. G., Acrylamide is metabolized to glycidamide in the rat: evidence from hemoglobin adduct formation, *Chem. Res. Toxicol.*, 3, 406, 1990.
33. *Revision of the WHO Guidelines for Drinking Water Quality*, Report on the Consolidation Meeting on Organics and Pesticides, Medmenham, UK, January 30-31, 1992, 11.

34. *1992 Annual Report of the Committee on Toxicity, Mutagenicity, Carcinogenicity of Chemicals in Food, Consumer Products and the Environment,* Dept. of Health, HMSO, London, 1993, 72.

ACRYLIC ACID

Molecular Formula. $C_3H_4O_2$
M = 72.07
CAS No 79-10-7
RTECS No AS4375000
Abbreviation. AA.

Synonyms. Acroleic acid; Ethylenecarboxylic acid; Propenoic acid; Vinylformic acid.

Properties. Colorless, fuming, corrosive liquid with an acrid odor. Readily soluble in alcohol and water, insoluble in ether. Odor perception threshold is 0.57 or 0.094 mg/l.[010] Taste (practical threshold of 50 mg/l) and odor disappeared entirely after 24 hours.[1]

Applications and **Exposure.** Used in the manufacture of acrylic and modified acrylic polymeric materials and articles intended for use in contact with food or drink. Used in the production of various acrylic esters, raw materials for a wide variety of detergents, water treatment chemicals, dispersants, cross-linked hydroscopic polymers, paints and coatings, and adhesives, and thus human exposure may be expected to occur.

Acute Toxicity. LD_{50} is found to be 830 and 250 mg/kg BW for mice and rabbits, respectively. The LD_{50} values of crystallized AA. in rats have been reported to range from 193 to 350 mg/kg BW[1,2,03] with estimated LD_{50} to lie within the range of 2.1 to 3.2 g/kg BW. Administration of 0.5 to 1.0 mmol AA/kg BW can cause hyperglycemia development in 4 hours after poisoning (Vodicka et al., 1985). Poisoned animals displayed CNS inhibition and convulsions. Death within a few days. Histopathology lesions consisted of liver congestion and spleen enlargement.[3]

Repeated Exposure revealed cumulative properties.

Short-term Toxicity. High mortality was noted in Wistar rats dosed by gavage with 150 or 375 mg AA/kg BW for 3 months. Gross pathology examination revealed a dose-dependent pronounced irritation in the forestomach and glandular stomach with ulcerations and necroses.[4]

In a 3-month drinking water study, AA was administered to F344 rats in their drinking water at dose levels of 0.75, 0.25, or 0.083 g/kg BW. At the highest dose level, BW gain as well as food and water consumption were reduced. There were neither significant treatment-related histopathological changes nor statistically significant changes in reproductive indices. The NOEL of 83 mg/kg BW was established in F344 rats.[5]

Long-term Toxicity. In a 12-month study, Wistar rats were given drinking water containing 120 to 5000 ppm AA (providing doses 9 to 331 mg/kg BW). Feed and water consumption was reduced in the high-dosed groups. There was no indication on systemic toxicity and/or any carcinogenic potential.[4]

Reproductive Toxicity.

Embryotoxicity and ***teratogenicity.*** Female rats were exposed to dose levels of 0.0023 to 0.0075 ml/kg BW by *i/p* injections on days 5, 10, and 15 of gestations. The treatment caused resorptions, gross and skeletal malformations, fetal death, and decreased fetal size.[6] In the above-cited study, no adverse treatment-related effects on reproductive function were noted. The NOAEL of 460 mg/kg BW for reproductive effects was proposed in this study.

In a two-generation reproductive study, Wistar rats received 500, 2500, or 5000 ppm AA in drinking water (53, 240 or 460 mg/kg BW, respectively) throughout the premating, gestational, and lactation periods. A decrease in water consumption was noted in rats exposed to 240 and 460 mg/kg BW. There was a reduction in BW level in the high-dose F_1 parental generation and in the F_2 pups. Histological lesions were found in the forestomach and glandular stomach in exposed rats. The NOAEL of 240 mg/kg BW was established in this study for histological changes in the stomach.[6]

In a two-generation reproduction toxicity study, Wistar rats received AA in drinking water at least 70 days prior to mating, through mating, gestation, lactation, and to weaning. Dose-related signs of developmental toxicity were observed in F_1 and F_2 pups at 2500 and 5000 ppm AA in the form of retarded growth and some delay in the eye/auditory canal opening in F_2 pups. There was no evidence of adverse effect on pup morphology. The NOAEL for fertility and reproductive performance of the parents was 5000 ppm; the NOAEL for developmental toxicity appeared to be 500 ppm.[7]

Mutagenicity.

In vivo cytogenetics. The rapid clearance of AA in animals and the weight of evidence of genetic toxicity tests in both somatic and germ cells *in vivo* indicate that AA lacks mutagenic potential.[8]

In vitro genotoxicity. However, AA caused a concentration-dependent increase in mutant frequency, producing gross CA in mouse lymphoma cells.[9] Negative results were found in the *Salmonella* mutagenicity assay (NTP-90). AA did not induce micronuclei, unscheduled DNA synthesis, and morphological transformations in cultured Syrian hamster embryo fibroblast cells.[10,11]

Carcinogenicity. Wistar rats were given AA in their drinking water at concentrations of 120, 400 or 1200 ppm (8, 27, or 78 mg/kg BW) over 26 (males) or 28 (females) months. The study did not reveal any toxic changes or indications of a carcinogenic potential.[4]

Carcinogenicity classification. An IARC Working Group concluded that there are no adequate data avaiable to evaluate the carcinogenicity of AA in *experimental animals* and in *humans*.

IARC: 3.

Chemobiokinetics. AA metabolism involves *beta-* and *omega*-oxidation in the body; it takes part in β-alanine synthesis. AA alters the processes of phosphorilation. The principal route of detoxication of AA in mammals comprises rapid incorporation into a mitochondrial pathway for *propionic acid* catabolism that results in the release of CO_2 and possible bioincorporation as *acetate*. Administered dose is excreted as CO_2 (60% to 80%) within 2 to 8 hours of oral dosing in rats.[7]

Regulations.

U.S. FDA (1998) listed AA as an ingredient (1) in the manufacture of resinous and polymeric coatings for polyolefin films to be safely used in the food-contact surface in accordance with the conditions prescribed in 21 CFR part 175.300; (2) in homopolymers and copolymers in the production of paper and paperboard intended for contact with dry food of the type identified in 21 CFR part 176.170 (c); (3) (monomer) in the manufacture of semirigid and rigid acrylic and modified acrylic plastics in the manufacture of articles intended for use in contact with food in accordance with the conditions prescribed in 21 CFR part 177.1010; (4) as polymers which may be safely admixed, alone or in mixture with other permitted polymers, as modifiers in semirigid and rigid vinyl chloride plastic food-contact articles prepared from vinyl chloride homopolymers and/or from vinyl chloride copolymers in accordance with the conditions specified in 21 CFR; and (5) in polyethylenephthalate polymers that may be safely used as articles, or components of plastics intended for use in contact with food in accordance with the conditions prescribed in 21 CFR part 177.1630.

EU (1992). AA is available in the *List of authorized monomers and other starting substances which shall be used for the manufacture of plastic materials and articles intended to come into contact with foodstuffs (Section A)*.

Great Britain (1998). AA is authorized without time limit for use in the production of polymeric materials and articles in contact with food or drink or intended for such contact.

Standards. *Russia* (1995). MAC and PML: 0.5 mg/l.

References:

1. Carpenter, C. P., Weil, C. S., and Smyth, H. F., Jr., Range-finding toxicity data: list VIII, *Toxicol. Appl. Pharmacol.*, 28, 313, 1974.
2. Miller, M. L., Acrylic acid polymers, in *Encyclopedia of Polymer Science and Technology, Plastics, Resins, Rubbers, Fibers*, N. M. Bikales, Ed., Interscience, New York, 1, 197, 1964.
3. Klimkina, N. V., Boldina, Z. N., and Sergeyev, A. N., Experimental substantiation of the maximum allowable content for acrylic acid in water bodies, in *Industrial Pollution of Water Bodies*, S. N. Cherkinsky, Ed., Meditsina, Moscow, Issue No 9, 1969, 171 (in Russian).
4. Helwig, J., Deckardt, K., and Freisberg, K. O., Subchronic and chronic studies of the effects of oral administration of acrylic acid to rats, *Food Chem. Toxicol.*, 31, 1, 1993.
5. DePass, L. R., Weil, C. S., and Frank, F. R., Acrylic acid CH_2=CH-COOH, Subchronic and Reproductive Toxicity studies on acrylic acid in the drinking water of the rat, *Drug Chem. Toxicol.*,6,1, 1983.

6. Singh, A. R., Lawrence, W. H., and Autian, J., Embryonic-fetal toxicity and teratogenic effects of a group of methacrylic esters in rats, *J. Dent. Res.*, 51, 1632, 1972.
7. Hellwig, J., Gembardt, C., and Murphy, S. R., Acrylic acid: Two-generation reproduction toxicity study in Wistar rats with continuous administration in the drinking water, *Food Chem. Toxicol.*, 35, 859, 1997.
8. Finch, L. and Frederick, C. B., Rate and route of oxidation of acrylic acid to carbon dioxide in rat liver, *Fundam. Appl. Toxicol.*, 19, 498, 1992.
9. Moore, M. M., Amtower, A., Doerr, C. L., Brock, K. H., and Dearfield, K. L., Genotoxicity of acrylic acid, methyl acrylate, ethyl acrylate, methyl methacrylate, and ethyl methacrylate in L5178Y mouse lympohoma cells, *Environ. Molec. Mutagen.*, 11, 49, 1988.
10. Wiegand, H. J., Schiffmann, D., and Henschler, D., Non-genotoxicity of acrylic acid and *n*-butyl acrylate in a mammalian cell system (SHE cells), *Arch. Toxicol.*, 63, 250, 1989.
11. McCartny, K. I., Thomas, W. C., Aardema, M. J., et al., Genetic toxicology of acrylic acid, *Food Chem. Toxicol.*, 30, 505, 1992.

ACRYLIC ACID, BUTYL ESTER

Molecular Formula. $C_7H_{12}0_2$
M = 128.17
CAS No 141-32-2
RTECS No UD3150000
Abbreviation. AABE.

Synonyms. *n*-Butyl acrylate; Butyl-2-propenoate; 2-Propenoic acid, butyl ester.

Properties. Colorless liquid with an unpleasant odor. Water solubility is 1.4 g/l at 20°C, soluble in ethanol, diethyl ether, and acetone. Odor perception threshold is reported to be 0.17 mg/l[1] or 0.0078 mg/l.[02] Taste perception threshold is 0.005 mg/l.[1]

Applications. Used in the manufacture of a variety of acrylic and modified acrylic polymeric materials and articles, lacquers, and dyes intended for use in contact with food or drink.

Acute Toxicity. LD_{50} is 4.8 g/kg BW in rats, and 5.4 g/kg BW in mice.[2-4] A single administration of 520 mg/kg BW dose in oil solution to rats produced no edema of the gastric mucosa (unlike methyl- or ethylacrylates) but did cause edema when the same dose was administered in an aqueous solution.[5]

Repeated Exposure revealed no pronounced cumulative properties.[4]

Long-term Toxicity. Rabbits were administered a daily dose of AABE for 1 to 16 months. The treatment caused changes in the blood and bone marrow, changes in the ratio of blood serum protein fractions, and dystrophic changes in the liver.[6]

Allergenic Effects.

Animal studies. AABE was shown to be a sensitizing agent in the guinea pig maximization test.[7]

Reproductive Toxicity.

No ***fetotoxic*** or ***teratogenic effects*** were observed in the inhalation study (700 and 1310 mg/m^3) in Sprague-Dawley rats. An increase in postimplantation embryolethality was noted.[8]

Mutagenicity.

In vitro genotoxicity. AABE was negative in *Salmonella* mutagenicity assay; it did not produce CA in bone marrow cells or Chinese hamster ovary cells.[9]

Carcinogenicity. AABE appeared to be negative in dermal carcinogenicity study in mice.[10]

Carcinogenicity classification. An IARC Working Group concluded that there is inadequate evidence for the carcinogenicity of AABE in *experimental animals* and there were no adequate data available to evaluate the carcinogenicity of AABE in *humans*.

IARC: 3.

Chemobiokinetics.

In *in vitro studies* AABE was found to be rapidly eliminated from rat blood and hepatic homogenate due to hydrolysis under the action of unspecific enzymes. The disappearance of AABE from the blood may also occur because of its combination with erythrocytes.

Animal studies. After *i/p* administration to rats, 6.0% AABE were passed with the urine over the course of 24 hours in the form of *mercapturic acid*.[11]

According to Sapota, following oral administration to rats, a great part of administered 2,3-^{14}C-AABE underwent rapid metabolism and excretion with expired air (more than 70% of the dose) and urine (15 to 22%). Accumulation of radioactivity occurs predominantly in the liver and kidneys.[12]

According to Linhart et al., among metabolites excreted in urine, several physiological carboxylic acids were determined. There was a significant increase in *3-hydroxypropanoic acid*, and slight increase in *lactic* and *acetic acid*. Two mercapturic acids, namely, *N-acetyl-S-(2-carboxyethyl) cysteine* and the corresponding *N-acetyl-S- [(2-alkoxycarbonyl)ethyl]cysteine*, were also found.[13]

Regulations.

U.S. FDA (1998) permits the use of AABE (1) as a monomer or co-monomer in the manufacture of acrylate polymers or copolymers, in accordance with the conditions prescribed in 21 CFR part 175.210; (2) in the manufacture of semirigid and modified rigid acrylic and vinyl chloride plastics for food-contact articles in accordance with the conditions prescribed in 21 CFR part 177.1010; (3) as a component of adhesives for food-contact surface, in accordance with the conditions prescribed in 21 CFR part 175.105; (4) as an ingredient of resinous and polymeric food-contact coatings for polyolefin films for food-contact surface of articles intended for use in producing, manufacturing, packing, processing, preparing, treating, packaging, transporting, or holding food of the type identified in 21 CFR part 176.320; (5) as a component (monomer) of the uncoated or coated food-contact surface of paper and paperboard intended for use in producing, manufacturing, packaging, processing, preparing, treating, packing, transporting, or holding dry food in accordance with the conditions prescribed in 21 CFR part 176.180; (6) in the manufacture of cross-linked polyester resins for repeated use in contact with food, subject to the provisions prescribed in 21 CFR part 177.2420; (7) as polymers which may be safely admixed, alone or in mixture with other permitted polymers, as modifiers in semirigid and rigid vinyl chloride plastic food contact articles prepared from vinyl chloride homopolymers and/or from vinyl chloride copolymers in accordance with the conditions specified in the CFR; and (8) in the manufacture of polyethylene phthalate polymers to be safely used as articles or components of articles intended for use in contact with food in accordance with the conditions prescribed in 21 CFR part 177.1630.

EU (1990). AABE is available in the *List of authorized monomers and other starting substances which shall be used for the manufacture of plastic materials and articles intended to come into contact with foodstuffs (Section A).*

Great Britain (1998). AABE is authorized without time limit for use in the production of polymeric materials and articles in contact with food or drink or intended for such contact.

Standards. *Russia* (1995), MAC and PML: 0.01 mg/l (organolept., taste).

References:

1. Klimkina, N. V. et al., in *Reports on the Methods for Removing Harmful Substances from Gaseous Emissions and Industrial Effluents*, Abstracts, Dserzhinsk, 1967, 37 (in Russian).
2. Chernyshova, V. V. et al., in *Chemical Industry,* Series: *Toxicology and Sanitary Chemistry of Plastics*, Issue No 2, 1979, 22 (in Russian).
3. Carpenter, C. P., Weil, C. S., and Smyth, H. F., Range-finding toxicity data: list VIII, *Toxicol. Appl. Pharmacol.*, 28, 313, 1974.
4. *Toxicology and Sanitary Chemistry of Plastics*, Abstracts, NIITEKHIM, Moscow, Issue No 2, 1979, 22 (in Russian).
5. Ghanayem, B. J., Maronpot, R. R., and Mattehews, H. B., Ethyl acrylate-induced gastric toxicity. II. Structure-toxicity relationship and mechanisms, *Toxicol. Appl. Pharmacol.*, 80, 323, 1985.
6. Ivanov, V. A. et al., in *Problems of Toxicology and Hygiene of Synthetic Rubber Production*, Voronezh, 1968, 67 (in Russian).
7. van der Walle, H. B. and Bensink, T., Cross reaction pattern of 26 acrylic monomers on guinea pig skin, *Contact Dermat.*, 8, 376, 1982.
8. Merkle, J. and Klimish, H.-J., *n*-Butyl acrylate: prenatal inhalation toxicity in the rat, *Fundam. Appl. Toxicol.*, 3, 443, 1983.

9. Engelhardt, G. and Klimish, H.-J., *n*-Butyl acrylate cytogenic investigation in the bone marrow of Chinese hamsters and rats after 4-day inhalation, *Fundam. Appl. Toxicol.,* 3, 640, 1983.
10. de Pass, L. W., Fowler, E. N., Meckley, D. R., et al., Dermal oncogenicity bioassays of acrylic acid, ethyl acrylate, and butyl acrylate, *J. Toxicol. Environ. Health,* 14, 115, 1986.
11. Miller, R. R., Young, J. T., Kociba, R. J., et al., Chronic toxicity and oncogenicity bioassay of inhaled ethyl acrylate in Fischer 344 rats and B6C3F_1 mice, *Drug Chem. Toxicol.*, 8, 1, 1985.
12. Sapota, A., The dynamics of distribution and excretion of butyl-(2,3-^{14}C)-acrylate in male Wistar albino rats, *Pol. J. Occup. Med.*, 4, 55, 1991.
13. Linhart, I., Vosmanska, M., and Smejkal, J., Biotransformation of acrylates. Excretion of mercapturic acids and changes in urinary carboxylic acid profile in rat dosed with ethyl and 1-butyl acrylate, *Xenobiotica*, 24, 1043, 1994.

ACRYLIC ACID, DECYL ESTER

Molecular Formula. $C_{13}H_{24}O_2$
M = 212.37
CAS No 2156-96-9
RTECS No AS7400000
Abbreviation. AADE.

Synonyms. *n*-Decyl acrylate; 2-Propenoic acid, decyl ester.

Properties. Very slightly soluble in water.

Applications. Used in the manufacture of a variety of acrylic and modified acrylic polymeric materials and articles, lacquers, and dyes intended for use in contact with food or drink.

Acute Toxicity. LD_{50} is 6.46 g/kg BW in rats.

Regulations. ***U.S. FDA*** (1998) approved the use of AADE in adhesives as a component of articles intended for use in packaging, transporting, or holding food in accordance with the conditions prescribed in 21 CFR part 175.105.

Reference:

Smyth, H. F., Carpenter, C. P., Weil, C. S., Pozzani, U. C., and Striegel, J. A., Range-finding toxicity data, list VI, *Am. Ind. Hyg. Assoc. J.*, 25, 95, March-April 1962.

ACRYLIC ACID, ETHYL ETHER

Molecular Formula. $C_5H_8O_2$
M = 100.12
CAS No 140-88-5
RTECS No AT0700000
Abbreviation. AAE.

Synonyms. Ethyl acrylate; Ethyl propenoate; 2-Propeonic acid, ethyl ester.

Properties. Colorless liquid with an acrid odor. Solubility in water is 2.0% at 20°C, soluble in ethanol. Odor perception threshold is reported to be 0.2 to 6.7 mg/l.[02,07,010]

Applications. Used in the manufacture of acrylic and modified acrylic plastics and in surface coatings, as a co-monomer in the production of several polymers, in the preparation of paints, textiles, paper coatings intended for use in contact with food or drink. AAE has been found as a residual monomer in polyethylacrylate and in aqueous polymer latexes.

Acute Toxicity. LD_{50} is reported to be 1000 to 2800 mg/kg BW in rats,[1,2] 1800 mg/kg BW in mice,[3] and 280 to 420 mg/kg BW in rabbits.[4] Poisoning affected the CNS functions and vessel permeability. In rabbits, dyspnea, cyanosis and convulsive movements were observed.[03] Gross pathology examination revealed irritation of the intestinal walls as well as lesions of the liver and kidneys.

Administration of 200 to 400 mg/kg BW to rats led to ataxia as well as hypodynamia and bradycardia indicating a fall in blood catecholamine content.[5] Male F344 rats received a single oral administration of up to 4.0% AAE in their feed. Poisoning increased the weights of the forestomach, reflecting edema and inflammation, but that of glandular stomach did not change. No DNA damage was observed with these doses.[6]

Repeated Exposure revealed moderate cumulative properties in mice. K_{acc} is 2.76 (by Lim). Two or four consecutive oral daily doses of 200 mg AAE/kg BW caused mucosal edema associated with vesicle formation, mucosal hyperplasia, submucosal edema, and inflammation in forestomach of male F344 rats.[1] Administration of 31.5 mg/kg BW to rabbits for 35 days caused no toxic effect.[4]

In a 14-day study, rats and mice were given 100 to 800 mg AAE/kg BW. Rats developed abdominal adhesions in response to administration of 600 mg and 800 mg AAE/kg BW in vegetable oil. Stomach lesions were found at 400 mg/kg BW dose level; gastritis was noted at 400 to 600 mg/kg BW in mice. Doses up to 200 mg/kg BW did not produce such effect.[7, 8]

Short-term Toxicity. F344 rats were given EA in their drinking water over a period of 3 months. A reduction in the consumption of drinking water and feed was noted at a dose of 83 mg/kg BW. No definite toxicity or any clear target organ was determined in this study (DePass et al., 1980 and 1983).

Long-term Toxicity. No treatment-related lesions were found in female, but not male Wistar rats received 2000 mg AAE/l in their drinking water for two years. Retardation of BW gain was the only sign noted. No effect was seen in dogs that were given 1.0 g/kg diet.[9] However, details on survival and results of pathology examination in this study seem to be insufficient for evaluation (IARC 19-61).

Rats and mice exposed to 100 to 200 mg/kg BW for two years exhibited inflammation and epithelial hyperplasia of the stomach walls.[6]

Allergenic Effect. 7.9 % of patients exposed to AAE developed allergic reactions.[10]

Reproductive Toxicity.

Embryotoxicity. Rats were injected *i/p* with 1/10 to 1/3 LD_{50} on day 5.0, 10, and 15 of pregnancy. The treatment increased embryo mortality and reduced fetal weights. A dose-effect relationship was noted.[11] Exposure to 25 to 400 mg/kg BW on days 7 through 16 of pregnancy led to an increase in maternal BW. Although 100 to 400 mg/kg BW doses significantly increased embryo mortality, the litter size was only slightly affected. Skeletal ossification was retarded.[12]

Teratogenic effect was noted on inhalation exposure.[13]

Mutagenicity.

In vitro genotoxicity. AAE was negative in *Salmonella* mutagenicity assays (Ashby and Tennant, 1991). It caused concentration-dependent increase in mutant frequency, producing gross CA in mouse lymphoma cells,[14] and CA and SCE in Chinese hamster ovary cells (NTP-85). Doses of 0.0075 and 0.015 mg/ml caused dose-dependent increases of micronuclei and CA in the cultured Chinese hamster lung cells.[15]

In vivo cytogenetics. AAE was negative in the micronucleus assay in BDF_1 mouse bone marrow cells examined 24 hours after *i/p* and oral treatment with up to 1000 mg/kg BW which killed all the animals.[16] A reduced polychromatic erythrocyte to total erythrocyte ratio was observed. Przybojewska et al. (1984), however, reported positive results in *Balb/c* mice but Ashby et al. (1989) could not confirm them.

AAE was not mutagenic in *Dr. melanogaster*.

Carcinogenicity. F344 rats and $B6C3F_1$ mice were gavaged with 100 and 200 mg AAE/kg BW bolus doses in vegetable oil for 103 weeks. Squamous cell papillomas and forestomach carcinomas were revealed in mice and rats of both sexes. A dose-dependent relationship was established. There was no apparent effect on other tissues.[7] The lesions developed only at the dosing site, without systemic toxicity. Evidence for carcinogenicity was greater in males than in females.

Recent studies have not confirmed these results. Doses of 110 to 160 mg/kg BW given in drinking water did not evoke oncogenic response.[17] In order to identify early forestomach lesions, F344 male rats received 100 mg AAE/kg BW for two weeks. Histology examination revealed the increased cell proliferation that together with the induced foreign body reactions may contribute to the previously demonstrated carcinogenic effect of AAE on the rat forestomach.[8]

The gavage administration of 200 mg/kg BW to male F344 rats for 6 and 12 months caused a sustained increase in forestomach epithelial hyperplasia. No neoplasms developed after 6-month treatment. They were observed in rats dosed for 12 months. No gross lesions in the liver of rats were noted.[18]

Carcinogenicity classifications. An IARC Working Group concluded that there is sufficient evidence for the carcinogenicity of AAE in *experimental animals* and there were no adequate data available to evaluate the carcinogenicity of AAE in *humans*.

The Annual Report of Carcinogens issued by the U.S. Department of Health and Human Services (1998) defines AAE to be a substance which may reasonably be anticipated to be carcinogen, i.e., a substance for which there is a limited evidence of carcinogenicity in *humans* or sufficient evidence of carcinogenicity in *experimental animals*.

IARC: 2B;

U.S. EPA: B2;

NTP: P* - P* - P - P (gavage).

Chemobiokinetics. AAE is readily absorbed and rapidly hydrolyzed into *acrylic acid* and *ethanol* in the blood and liver, which do not circulate in the body.[7] Although toxic at high concentrations, AAE is rapidly metabolized and detoxified at low concentrations. It may bind with non-protein *SH*-groups in the erythrocytes.[19] According to other data,[17] ^{14}C-AAE could react with both glutathione and protein.

A single gavage dose of 200 mg/kg BW has been shown to cause severe glutathione depletion in the forestomach with less depletion observed in the glandular stomach and duodenum.[20] AAE is metabolized by cellular carboxylesterases and by conjugation with glutathione. *Mercapturic acid* has been shown to be a minor urinary metabolite.[21]

According to Linhart et al., among metabolites excreted in urine, several physiological *carboxilic acids* were determined. There was a significant increase in *3-hydroxypropanoic acid*, and slight increase in *lactic* and *acetic acid.* Two mercapturic acids, namely, *N-acetyl-S-(2-carboxyethyl) cysteine* and the corresponding *N-acetyl-S-[(2-alkoxycarbonyl)ethyl]cysteine*, were also found.[22]

Regulations.

The Council of Europe (1981) included AAE in a list of artificial flavoring substances that may be added to foodstuffs without hazard to public health at a level of 1.0 mg/kg in food and 0.2 mg/l in beverages.

EU (1990). AAE is available in the *List of authorized monomers and other starting substances which shall be used for the manufacture of plastic materials and articles intended to come into contact with foodstuffs (Section A).*

U.S. FDA (1999) considered AAE to be a *GRAS* synthetic flavoring substance or food adjuvant. FDA regulates the use of AAE (1) in adhesives as a component of articles (monomer) intended for use in packaging, transporting, or holding food in accordance with the conditions prescribed in 21 CFR part 175.105; as (2) a component (monomer) of the uncoated or coated food-contact surface of paper and paperboard intended for use in producing, manufacturing, packaging, processing, preparing, treating, packing, transporting, or holding dry food in accordance with the conditions prescribed in 21 CFR part 176.180; (3) as a monomer in the manufacture of semirigid and rigid acrylic and modified acrylic plastics in the manufacture of articles intended for use in contact with food in accordance with the conditions prescribed in 21 CFR part 177.1010; (4) as a monomer in the manufacture of polyethylenephthalate polymers used as, or components of plastics (films, articles, or fabric) intended for use in contact with food in accordance with the conditions prescribed in 21 CFR part 177.1630; and (5) in the manufacture of resinous and polymeric coatings for polyolefin films for food-contact surface of articles intended for use in producing, manufacturing, packing, processing, preparing, treating, packaging, transporting, or holding food in accordance with the conditions prescribed in 21 CFR part 175.320.

AAE *polymers or copolymers* are permitted for use (1) as components of adhesives for food-contact surface in accordance with the conditions prescribed in 21 CFR part 175.105; (2) in resinous and polymeric coatings in a food-contact surface of articles intended for producing, manufacturing, packing, transporting, or holding food in accordance with the conditions prescribed in 21 CFR part 175.300; (3) in paper and paperboard in contact with dry, aqueous, or fatty foods in accordance with the conditions prescribed in 21 CFR part 176.170; (4) in semirigid and rigid acrylic and modified acrylic plastics in accordance with the conditions prescribed in 21 CFR part 177.1010; and (5) in cross-linked polyester resins for repeated use in articles or components of articles coming in contact with food in accordance with the conditions prescribed in 21 CFR part 177.2420.

Great Britain (1998). AAE is authorized without time limit for use in the production of polymeric materials and articles in contact with food or drink or intended for such contact.

Standards. ***Russia*** (1995). MAC and PML: 0.005 mg/l (organolept., odor).

References:

1. Klimkina, N. V., Smirnova, R. D., Savelova, V. A., et al., Sanitary-toxicological characteristics of plastics components administered in the body in water solutions, in *Hygiene and Toxicology of High- Molecular-Mass Compounds and of the Chemical Raw Material Used for Their Synthesis*, Proc. 4th All-Union Conf., S. L. Danishevsky, Ed., Khimiya, Leningrad, 1969, 224 (in Russian).
2. Pozzani, U. C., Weil, C. S., and Carpenter, C. P., Subacute vapor toxicity and range-finding data for ethyl acrylate, *J. Ind. Hyg. Toxicol.*, 31, 311, 1949.
3. Tonji, H. and Hashimoto, K., Structure-toxicity relationship of acrylates and methacrylates, *Toxicol. Lett.*, 11, 125, 1982.
4. Treon, J. F., Sigmon, H., Wright, H., et al., The toxicity of methyl and ethyl acrylate, *J. Ind. Hyg. Toxicol.*, 31, 317, 1949.
5. Sobezak, Z. and Baransky, B., *Bromatol.Chem. Toksykol.*, 12, 405, 1979 (in Polish).
6. Morimoto, K., Tsuji, K., Osawa R., and Takahashi, A., DNA damage test in forestomach squamous epithelium of F344 rat following oral administration of ethyl acrylate, Abstract, *Eisei Shikenjo Hokoku*, 108, 125, 1990 (in Japanese).
7. *Carcinogenesis Studies of Ethyl Acrylate in F344 Rats and B6C3F$_1$ Mice (Gavage Studies)*, NTP Technical Report Series No 259, Research Triangle Park, NC, NIH Publ. 87-2515, 1983.
8. Ghanayem, B. J., Maronpot, R. R., and Mattehews, H. B., Ethyl acrylate-induced gastric toxicity. II. Structure-toxicity relationship and mechanisms, *Toxicol. Appl. Pharmacol.*, 80, 323, 1985.
9. Borzelleca, J. F., Larson, P. C., and Hennigar, G. R., Studies on the chronic oral toxicity of monomeric ethyl acrylate and methyl methacrylate, *Toxicol. Appl. Pharmacol.*, 6, 29, 1964.
10. Kanerva, L., Allergy caused by acrylate compounds, *Hum. Exp., Toxicol.*, No 1, 996, 1997.
11. Singh, A. R., Lawrence, W. H., and Autian, J., Embryonic-fetal toxicity and teratogenic effects of a group of methacrylate esters in rats, *J. Dent. Res.*, 51, 1632, 1972.
12. Pietrowicz, D., Owecka, A., and Baranski, B., Disturbances in rat embryonic development due to ethyl acrylate, *Zwierzeta. Lab.*, 17, 67, 1980 (in Polish).
13. John, J. A., Deacon, M. M., Murray, J. S., et al., Evaluation of inhaled allyl chloride and ethyl acrylate for embryotoxic and teratogenic potential in animals, *Toxicologist*, 1, 147, 1981.
14. Moore, M. M., Amtower, A., Doerr, C. L., Brock, K. H., and Dearfield, K. L., Genotoxicity of acrylic acid, methyl acrylate, ethyl acrylate, methyl methacrylate, and ethyl methacrylate in L5178Y mouse lymphoma cells, *Environ. Molec. Mutagen.*, 11, 49, 1988.
15. Ishidate, M. *The Databook of Chromosomal Aberration Test In Vitro on 587 Chemical Substances Us ing a Chinese Hamster Fibroblast Cell Line*, The Realize Inc., Tokyo, 197.
16. Morita, T., Asano, N., Awogi, T., Sasaki, Yu. F., et al. Evaluation of the rodent micronucleus assay in the screening of IARC carcinogens (Groups 1, 2A and 2B). The summary report of the 6th collaborative study by CSGMT/JEMS-MMS, *Mutat. Res.*, 389, 3, 1997.
17. Frederick, C. B., Potter, D. W., Chang-Mateu, I. M., et al., A biologically-based pharmacokinetic model for oral dosing of ethyl acrylate, *Toxicologist*, 9, 237, 1989.
18. Ghanayem, B.I, Sanchez, I. M., Maronpot, R. R., Elwell, M. R., and Matthews, H. B., Relationship between the time of sustained ethyl acrylate forestomach hyperplasia and carcinogenicity, *Environ. Health Perspect.*, 101 (Suppl. 5), 277, 1993.
19. Potter, D. W. and Tran, Thu-Ba, Rates of ethylacrylate binding to glutathione and protein, *Toxicol. Lett.*, 62, 275, 1992.
20. Stott, W. T. and McKenna, M. J., Hydrolysis of several glycol ether acetates and acrylates by nasal mucosal carboxylesterase *in vitro*, *Fundam. Appl. Toxicol.*, 5, 399, 1985.
21. De Bethizy, J. D., Udinsky, J. R., Scribner, H. E., et al., The disposition and metabolism of acrylic acid and ethyl acrylate in male Sprague-Dawley rats, *Fundam. Appl. Toxicol.*, 8, 549, 1987.
22. Linhart, I., Vosmanska, M., and Smejkal, J., Biotransformation of acrylates. Excretion of mercapturic acids and changes in urinary carboxylic acid profile in rat dosed with ethyl and 1-butyl acrylate, *Xenobiotica*, 24, 1043, 1994.

ACRYLIC ACID, 2-ETHYLHEXYL ESTER

Molecular Formula. $C_{11}H_{20}O_2$
M = 184.31
CAS No 103-11-7
RTECS No AT0855000
Abbreviation. AAEHE.

Synonyms. 2-Ethylhexyl acrylate; 2-Ethylhexanol acrylate; 2-Ethylhexyl 2-propeonate; Octyl acrylate; 2-Propenoic acid, 2-ethylhexyl ester.

Properties. Clear, colorless liquid with a sharp pungent odor. Solubility in water is 50 mg/l at 20°C. Threshold concentrations for taste and odor appear to be 0.03 and 0.02 mg/l, respectively. There was no film or foam formation in water solution.[1]

Applications. AAEHE is used as a co-monomer with vinyl acetate for polymer emulsion paints, as a co-monomer with vinyl chloride for latex paints, as a monomer for plastics, protective coatings, paper treatment, and in the production of water-based paints intended for use in contact with food or drink.

Acute Toxicity. LD_{50} is 6.7 to 10.5 g/kg BW in rats, and 4.4 to 4.5 g/kg BW in mice. Ingestion of large doses resulted in CNS inhibition and respiratory disorder.[2,3]

Repeated Exposure. Rats received 1/10, 1/50 and 250 LD_{50} in sunflower oil solution for 30 days. The treatment with 670 mg/kg BW affected CNS, and true cholinesterase and catalase activity. The dose of 27 mg/kg BW was considered the LOAEL in this study.[1]

Long-term Toxicity. Rats were given 0.3, 3.0, and 15 mg/kg BW for 26 weeks. Changes were observed in the liver and blood, in true cholinesterase activity. The NOAEL in this study was found to be 0.3 mg/kg BW.[1]

Carcinogenicity. Skin applications caused tumor in site of treatment.[4]

Carcinogenicity classification. An IARC Working Group concluded that there is limited evidence for the carcinogenicity of AAEHE in *experimental animals* and there is inadequate evidence for carcinogenicity of AAEHE in *humans.*

IARC: 3.

Regulations.

U.S. FDA (1998) approved the use of AAEHE in resinous and polymeric coatings used as the food-contact surfaces of articles intended for use in producing, manufacturing, packing, processing, preparing, treating, packaging, transporting, or holding food in accordance with the conditions prescribed in 21 CFR part 175.300.

Great Britain (1998). Acrylic acid, *n*-octyl ether is authorized without time limit for use in the production of polymeric materials and articles in contact with food or drink or intended for such contact.

Standards. ***Russia*** (1995). Recommended MAC: 0.02 (organolept., odor).

References:

1. Piskarev, Yu. G., Tikhomirov, Yu. P., and Chumakova, T. F., Experimental substantiation of the 2-ethylhexyl acrylate maximum allowable concentration in surface waters, *Gig. Sanit.*, 3, 16, 1993 (in Russian).
2. Lomonova, G. V., Data on substantiation of MAC for 2-ethylhexyl acrylate in the workplace air, *Gig. Truda Prof. Zabol.*, 9, 52, 1982 (in Russian).
3. Chesnokova, N. P., Bersudsky, S. S., and Bryll, G. Ye., in *Hygienic Aspects of Environmental Health Protection*, Saratov, 1979, 61 (in Russian).
4. De Pass, L. R., Maronpot, R. R., and Weil, C. S., Dermal oncogenicity bioassays of monofunctional and multifunctional acrylates and acrylate-based oligomers, *J. Toxicol. Environ. Health*, 16, 55, 1985.

ACRYLIC ACID, ISOBUTYL ESTER

Molecular Formula. $C_7H_{12}O_2$
M = 128.19
CAS No 106-63-8
RTECS No AT2100000
Abbreviation. AAIB.

Synonyms and **Trade Name.** Isobutyl acrylate; Isobutyl propenoate; *Z*-Methylpropyl acrylate; 2-Propenoic acid, 2-methylpropyl ester.

Properties. Clear liquid with a sharp, fragrant odor. Soluble in alcohol. Solubility in water: 1.8 g/l.

Applications. Monomer and comonomer in the production of acrylic surface coatings and synthetic resin irtended for use in contact with food or drink.

Acute Toxicity. LD_{50} is 7.07 ml/kg BW in rats,[1] and 6.1 g/kg BW in mice.[2]

Mutagenicity. AAIB was negative in the four *Salmonella typhimurium* strains *TA98, TA100, TA1535* and *TA1537* in the the presence and absence of rat or hamster liver *S9* fraction.[016]

Regulations. ***Great Britain*** (1998). AAIB is authorized without time limit for use in the production of polymeric materials and articles in contact with food or drink or intended for such contact.

References:

1. Carpenter, C. P., Weil, C. S., and Smyth, H. F., Range-finding toxicity data: list VIII, *Toxicol. Appl. Pharmacol.*, 28, 313, 1974.
2. Tanii, H. and Hashimoto, K., Structure-toxicity relationship of acrylates and methacrylates, *Toxicol. Lett.*, 11, 125, 1982.

ACRYLIC ACID, METHYL ESTER

Molecular Formula. $C_4H_6O_2$

M = 86.09

CAS No 96-33-3

RTECS No AT2800000

Abbreviation. AAME.

Synonyms and **Trade Names.** Curithane 103; Methoxycarbonyl ethylene; Methyl acrylate; Methyl propenoate; 2-Propeonic acid, methyl ester.

Properties. Colorless, transparent liquid with an acrid odor and sharp, sweet, fruity odor. Water solubility is 5.2% at 20°C, soluble in alcohol. Odor perception threshold is reported to be 0.01 mg/l[1] or 0.0021 mg/l.[02]

Applications. Co-monomer in the production of all types of vinyl monomers. Used for "internal plasticization" of rigid plastics, in preparation of surface coatings, paper coatings and adhesives.

Acute Toxicity. LD_{50} was established at the level of 230 to 545 mg/kg BW for rats, 826 mg/kg BW for mice, and 200 mg/kg BW for rabbits. Administration of 280 mg/kg BW to rabbits resulted in the lethal effect preceded by dyspnea, cyanosis, convulsions, and drop of temperature.[1,2,03]

Long-term Toxicity. Manifestations of the toxic action included effects on the CNS, cardiovascular system, GI tract, and urinary ducts.[3] In a 6-month study, the treatment affected liver functions, acid-alkali balance of the blood and urine, and conditioned reflex activity of rats.[3]

Mutagenicity.

In vitro genotoxicity. AAME did not show mutagenic activity in *Salmonella typhimurium* and in *E. coli*;[4] however, it caused concentration-dependent increase in mutant frequency, producing gross CA in mouse lymphoma cells,[5] and in the culture of Chinese hamster lung cells.[6]

Carcinogenicity. No tumor was found in a 2-year inhalation study (concentrations of 53 to 475 mg/m^3) in rats.[7]

Carcinogenicity classification. An IARC Working Group concluded that there is inadequate evidence for the carcinogenicity of AAME in *experimental animals* and there were no adequate data available to evaluate the carcinogenicity of AAME in *humans*.

IARC: 3.

Chemobiokinetics. 2 hours after oral administration of 34 mg ^{14}C-tagged AAME/kg BW to guinea pigs, it was discovered in the liver, bladder, and brain of the animals. AAM hydrolyzed to *acrylic acid* in rat liver, kidneys, and lung homogenates. Connecting to *SH*-groups seems to be important for AAME detoxication metabolism in guinea pigs. AAME is removed with the urine in the form of *thioether*.[8]

According to Sapota,[9] following oral administration of methyl-2,3-^{14}C-acrylate to male Wistar rats, the monomer has been readily absorbed and distributed over all tissues.[9] The main excretion routes for CO_2 are with the expired air and with urine. In the urine, the following metabolites were determined: *N-acetyl-(S-methylcarboxyethyl) cystein* and *N-acetyl-S-(2-carboxyethyl) cystein.*[9,10]

Regulations.

EU (1990). AAME is available in the *List of authorized monomers and other starting substances which shall be used for the manufacture of plastic materials and articles intended to come into contact with foodstuffs (Section A)*.

U.S.FDA (1998) listed AAME, its polymers and copolymers for use (1) as a monomer in adhesives as a component of articles intended for use in packaging, transporting, or holding food in accordance with the conditions prescribed in 21 CFR part 175.105; (2) in the manufacture of resinous and polymeric coatings in a food-contact surface of articles intended for use in producing, manufacturing, packing, transporting, or holding food as prescribed in 21 CFR part 175.300; (3) as a solvent in polyester resins for food-contact surfaces; (4) in paper and paperboard in contact with dry, aqueous or fatty foods as prescribed in 21 CFR part 176.170 and 176.180; (5) in semirigid and rigid acrylic and modified acrylic plastics in accordance with the conditions prescribed in 21 CFR part 177.1010; (6) in the manufacture of cross-linked polyester resins for repeated use as articles or component of articles coming in the contact with food, subject to the provisions prescribed in 21 CFR part 177.2420; and (7) as a monomer in the manufacture of polyethylene-phthalate polymers to be safely used as articles, or components of articles intended for use in contact with food in accordance with the conditions prescribed in 21 CFR part 177.1630.

Great Britain (1998). AAME is authorized without time limit for use in the production of polymeric materials and articles in contact with food or drink or intended for such contact.

Standards. ***Russia*** (1995). MAC and PML: 0.02 mg/l (organolept., odor).

References:

1. Paulet, G. and Vidal, M., On the toxicity of some acrylic and methacrylic esters of acrylamides and polyacrylamides, *Arch. Mal. Prof. Med. Trav. Sec. Soc.*, 36, 58, 1975.
2. Tanii, H. and Hashimoto, K., Structure-toxicity relationship of acrylates and methacrylates, *Toxicol. Lett.*, 11, 125, 1982.
3. Klimkina, N. V. et al., in *Methods of Removing Harmful Substances from Gas Emission and Industrial Effluents,* Sci. Conf., Preprints of Papers, Dzerzhinsk, 1967, 37 (in Russian).
4. Waeggemaekers, T. H. J. M. and Bensink, M. P. M., Non-mutagenicity of 27 aliphatic acrylate esters in the *Salmonella*-microsome test, *Mutat. Res.*, 137, 95, 1984.
5. Moore, M. M., Amtower, A., Doerr, C. L., Brock, K. H., and Dearfield, K. L., Genotoxicity of acrylic acid, methyl acrylate, ethyl acrylate, methyl methacrylate, and ethyl methacrylate in L5178Y mouse lymphoma cells, *Environ. Molec. Mutagen.*, 11, 49, 1988.
6. Ishidate, M. et al., in *Data Book of Chromosomal Aberration Tests in Vitro on 587 Chemical Substances Using Chinese Hamster Fibroma Cell Line*, The Realize Inc., Tokyo, 1983, 334.
7. Klimish, H.-J. and Reininghaus, W., Carcinogenicity of acrylates: Long-term inhalation studies on methyl acrylate and *n*-butyl acrylate in rats, Abstract No 211, *Toxicologist*, 4, 53, 1984.
8. Seutter, E. and Rijnties, N., Whole-body autoradiography after systemic and topical administration of methyl acrylate in the guinea-pig, *Arch. Dermatol. Res.*, 270, 273, 1981.
9. Sapota A., The disposition and metabolism of methyl acrylate in male Wistar albino rats, *Pol. J. Occup. Med. Environ. Health*, 6, 185, 1993.
10. Miller, R. R., Ayres, J. A., and Ramphy, L. W., McKenna M. J., Metabolism of acrylate ester in rat tissue homogenates, *Fundam. Appl. Toxicol.*, 1, 410, 1981.

ACRYLIC ACID, MONOESTER with ETHYLENE GLYCOL

Molecular Formula. $C_5H_8O_3$

M = 116.13

CAS No 818-61-1

RTECS No AT1750000

Abbreviation. AAEG.

Synonyms and **Trade Name.** 2-(Acryloyloxy)ethanol; Bisomer 2-HEA; Ethylene glycol, acrylate; 2-Hydroxyethyl acrylate; 2-Propenoic acid, 2-hydroxyethyl ester.

Applications. A monomer and a cross-linking agent in the production of resins and adhesives intended for use in contact with food or drink.

Acute Toxicity. LD_{50} is 600 mg/kg BW in mice.[1]

Short-term Toxicity. The doses of 3.0, 20 or 60 mg AAEG/kg BW were administered *i/p* to Long-Evans rats for 13 weeks. AAEG exposure caused changes in muscular function, but no neuropathological lesions were detected. Authors stated that the neurotoxic potential of AAEG appeared to be minimal.[2]

In a 100-day feeding study, rats received approximately 15, 50, or 150 mg AAEG/kg BW. The treatment caused only retardation of BW gain and some changes in organ weights (McCollister et al., 1967).[2] F344 rats were exposed to 2.5 or 50 mg AAEG/kg BW by *i/p* injections for 13 weeks. Sciatic nerve damage including "axonal swelling, ovoids, tomaculum formation, and degeneration and corrugation of myelin" was observed. Damage in peripheral nerves was found in all rats of the high-dose group (Osborne et al., 1980).[2]

Allergic reactions in patients with AAEG exposure were 12.1%.[3]

Mutagenicity. AAEG produced positive mutagenic response in L5178Y mouse lymphoma cells.[4]

Regulations.

U.S. FDA (1998) approved the use of AAEG in adhesives as a component of articles intended for use in packaging, transporting, or holding food in accordance with the conditions prescribed in 21 CFR part 175.105.

EU (1990). AAEG is available in the *List of authorized monomers and other starting substances which shall be used for the manufacture of plastic materials and articles intended to come into contact with foodstuffs (Section A)*.

Great Britain (1998). AAEG is authorized without time limit for use in the production of polymeric materials and articles in contact with food or drink or intended for such contact.

References:

1. Tanii, H. and Hashimoto, K., Structure-toxicity relationship of acrylates and methacrylates, *Toxicol. Lett.*, 11, 125, 1982.
2. Moser, V. C., Antony, D. C., Sette, W. F., and MacPhail, R. C., Comparison of subchronic neurotoxicity of 2-hydroxyethyl acrylate and acrylamide in rats, *Fundam. Appl. Toxicol.*, 18, 343, 1992.
3. Kanerva, L., Allergy caused by acrylate compounds, *Hum. Exp., Toxicol.*, 1, 996, 1997.
4. Dearfield, K. L., Millis, C. S., Harrington-Brock, K., Doerr, C. L., and Moore, M. M., Analysis of the genotoxicity of nine acrylate/methacrylate compounds in L5178Y mouse lymphoma cells, *Mutagenesis*, 4, 381, 1989.

ACRYLIC ACID, 5-NORBORNEN-2-YLMETHYL ESTER

Molecular Formula. $C_{11}H_{14}O_2$
M = 178.25
CAS No 95-39-6
RTECS No AT3850000

Synonym and Trade Name. Bicyclo(2.2.1)-hept-5-ene-2-methylol acrylate; Cyclol acrylate; 2-Norbornene-2-methanol acrylate; 5-Norbornene-2-methylolacrylate.

Properties. Colorless liquid. Soluble in organic solvents. Monomer used for cross-linking of vinyl acetate emulsions polymer systems.

Applications. A difunctional monomer particularly suited for cross-linking of vinyl acetate emulsions polymer systems.[07]

Acute Toxicity. LD_{50} is 1.41 ml/kg BW in rats.

Reference:

Carpenter, C. P., Weil, C. S., and Smyth, H. F., Range-finding toxicity data: list VIII, *Toxicol. Appl. Pharmacol.*, 28, 313, 1974.

ACRYLONITRILE

Molecular Formula. C_3H_3N
M = 53.06
CAS No 107-13-11
RTECS No AT5250000
Abbreviation. AN.

Synonyms. Cyanoethylene; 2-Propenenitrile; Vinylcyanide.

Properties. Clear, colorless liquid with a slight odor. Forms acrylamide when added to water. Water solubility is 73 g/l at 25°C.[02] The threshold of the change of water organoleptic properties is 0.01 to 0.05 mg/l (does not affect the color of water).[1] Odor perception threshold is 9.1 mg/l.[02]

Applications. AN is used in the production of acrylic and modified acrylic resins and rubbers intended for use in contact with food or drink.

Exposure. Estimated production is in the range of 30 million to 1.5 billion pounds per year. Main human exposure occurs in the chemical industry.

Migration. Free AN can migrate into food from plastics when they are used as packaging materials in the food industry (migration levels up to 0.15 mg/kg).

Migration of AN into aqueous, hexane, alcohol (20 and 40%) and acid extracts from butadiene-nitrile rubber amounted to the concentration of 1.4 mg/l.[2]

According to other data, migration of AN from the ABC-polymers (residual AN level 12 mg/kg) was found to be 0.109 mg/l into 8.0% ethanol solution (exposure 24 hours, 49°C) and 0.306 mg/l into water (exposure 10 days, 49°C). From the materials containing 49 mg residual AN/kg, migration was 0.65 and 2.6 mg/l, respectively. Study of the migration of AN into water from seven ABS-polymers showed that a linear relationship exists between the concentration of AN in the polymer and the amount of AN migrating, for a given set of exposure conditions.[3]

Acute Toxicity. The LD_{50} values are reported to be 78 to 150 mg/kg BW in rats, 20 to 35 mg/kg BW in mice, 50 to 90 mg/kg BW in guinea pigs, and 93 to 100 mg/kg BW in rabbits.[1,05] Toxic symptoms included agitation immediately after exposure followed by depression. Death was within 5 hours from respiratory arrest. Administration of single oral doses of 20 to 80 mg/kg BW to male Sprague-Dawley rats caused dose-dependent acute neurotoxic effects cholinemetic in nature. Signs of intoxication included salivation, lacrimation, chromodacryorrhea, polyuria, miosis, vasodilatation in face, ears and extremities, increased gastric secretion, and diarrhea. This phase of intoxication is followed by the second phase developing hours after poisoning and including depression, convulsions, and respiratory failure.[4]

A single administration of 7.5 to 15 mg AN/kg BW produced CNS inhibition, a reduction in blood serum content of *SH*-groups and in body temperature.[5] Gross pathology examination revealed integumental hyperemia of the visceral organs, especially of the spleen, and brain degenerative changes. Hemorrhages in the GI tract of animals were caused by immediate irritating effect of AN.

Repeated Exposure failed to reveal cumulative properties. Exposure to high doses of AN caused unspecific changes which are characteristic of the general adaptation syndrome: neutrophilic leukocytosis with relative lymphopenia. A rapid increase in the content of hormones, enzymes, and biologically active substances was observed.

Rat received AN via drinking water. Consumption of AN at a concentration of 0.5 mg/l for 21 days caused an increase in the glutathione content of the liver. Concentrations of 0.01 to 0.2% AN in water or equivalent *i/g* doses given for 7 to 60 days reduced the concentration of plasma corticosterone and caused cortical atrophy in the adrenal glands. Gross pathology examination revealed mucosal hyperplasia in the stomach and duodenum.[6] Administration of AN threshold doses produced hematological changes such as hypochromic anemia, granulocytopenia, and lymphocytosis. A dose-dependence was observed.[7]

Short-term Toxicity. Rats received 0.1% aqueous solution of AN (1.0 mg/l) for 13 weeks. The treated animals displayed emaciation (adult rats) and reduced BW gain (young animals).[05]

Long-term Toxicity. In a 6-month study, pneumonia, chronic gastritis, and gastric polyps were observed in rabbits and rats.[8] Negligible if any effect was noted in rats exposed to 0.5 mg/l in drinking water for two years.[05]

Immunotoxicity. The sensitizing effect of AN administration was found. A strong contact sensitizing potential is found in maximized guinea-pig assays.[9]

Reproductive Toxicity.

Embryotoxic and ***teratogenic effects*** were found in rats gavaged with 65 and 25 mg AN/kg BW, respectively, during gestation. 10 mg AN/kg BW dose level was found to be ineffective.[10] Mehrota *et al.* reported no postnatal ill effects in rats given 5.0 mg AN/kg BW.[10] Pregnant Charles-Wistar rats received 5.0 mg AN/kg BW from days 5 to 21 of gestation. Functional development of pups was not affected. Nevertheless, the treatment induced changes in biogenic amines, which could give rise to future neurological effects.[12] Sprague-Dawley rats were exposured to AN by inhalation during days 6 to 20 of gestation at concentrations of 12 to 100 ppm. Fetotoxicity was observed after exposure to 25 ppm in the the presence of overt signs of maternal toxicity.[13] AN was found to produce teratogenic effects in hamsters, causing encephalocele and exencephaly.[14] It did not produce malformations in mice.[15]

Gonadotoxicity. Mice received oral doses of 10 mg/kg BW for 60 days. The treatment caused a significant decrease in the activity of testicular sorbitol dehydrohenase and acid phosphatase, and an increase in that of lactate dehydrogenase and β-glucuronidase. Histological investigation revealed degeneration of the seminiferous tubules and a decrease in the sperm counts of the epididymal spermatozoa.[16]

Mutagenicity.

Observations in man. AN enhanced the frequency of CA in lymphocytes of the exposed workers.

In vitro genotoxicity. Genotoxicity assays of AN have indicated no particular carcinogenic mechanism. Positive reverse mutagenesis in *Salmonella typhimurium HisG46* base substitution tester strains by AN is attributable to AN metabolite *cyanoethylene oxide*. Cyanoethylene oxide, a mutagenic metabolite of AN, did not induce DNA repair in *in vitro* hepatocyte DNA repair assay.[17]

Other *in vitro* genotoxicity test assays of AN have yielded mixed results, without consistent effect of metabolic activation.[18] Some positive genotoxicity data for AN appear to result from artifacts or from non-DNA-reactive mechanisms.[19] AN induced SCE, mutation, and unscheduled DNA synthesis but not CA in human cells *in vitro* (IARC 19-73). AN produced cell transformation in several test systems and inhibited intercellular communication in Chinese hamster *V79* cells.[20]

In vivo cytogenetics. It was found to be a substance of relatively low DNA reactivity in a *Dr. melanogaster* assay.[21] AN did not induce DLM in male mice and rat germ cells[17] or micronuclei in mice, or CA in rat bone-marrow cells. It bound covalently to rat liver DNA *in vivo* and induced unscheduled DNA synthesis in rat liver but not brain (IARC 19-73). AN does not induce DNA repair in either the *in vitro* or *in vivo* hepatocyte DNA repair assays.

Zhurkov[22] believes there are no data indicating mutagenic activity of AN for mammals. Nevertheless, according to Ahmed et al., AN can act as multipotent genotoxic agent by alkylating DNA in testicular tissue and may affect the male reproductive function in rats. Covalent binding of radioactivity to testicular tissue DNA was studied after a single oral dose of 46.5 mg 2,3-^{14}C-AN/kg BW. A significant decrease in DNA synthesis was observed at 0.5 hours after treatment.[23]

Carcinogenicity. AN can produce a carcinogenic effect through alkylation of DNA in the extrahepatic target tissues, stomach and brain.[24] The comparative genotoxicity of *vinyl chloride* and AN indicates that despite other similarities, they cause rodent tumors by different mechanisms.[18]

Rats were exposed to 5.0 mg AN/kg BW[3] times for a year. After 131 weeks from the start of the experiment, there was a certain increase in the rate of fibroadenomas and carcinomas of the mammary gland in the test animals as compared to the controls.[25] *I/g* administration caused tumors of the brain, squamous cell carcinoma of the stomach and Zymbal gland, cancer of the tongue, small intestine, and mammary gland.[26]

Sprague-Dawley rats received 20, 100 or 500 ppm AN in their drinking water for 2 years. Drinking water with the concentration of 500 ppm AN caused accelerated mortality and a decrease in water consumption. An increase in incidence of Zymbal's gland carcinomas and forestomach papillomatous proliferations was observed in this study.[27]

Carcinogenicity classifications. An IARC Working Group concluded that there is sufficient evidence for the carcinogenicity of AN in *experimental animals* and there is limited evidence for the carcinogenicity of AN in *humans*.

The Annual Report of Carcinogens issued by the U.S. Department of Health and Human Services (1998) defines AN to be a substance which may reasonably be anticipated to be carcinogen, i.e., a substance for which there is a limited evidence of carcinogenicity in *humans* or sufficient evidence of carcinogenicity in *experimental animals*.

IARC: 2B;

U.S. EPA: B1;

EU: 2.

Chemobiokinetics. There is little evidence of AN accumulation in animal tissues following prolonged exposure. The radioisotope method showed the accumulation of AN in the erythrocytes and the liver.[28,29] However, according to Burka, following gavage administration of 0.87 mmol 2-^{14}C-AN/kg BW to male F344 rats, the monomer has been well absorbed and distributed to all major tissues.[30]

Distribution in the body is studied by whole-body autoradiography after AN being administered orally. Uptake of radioactivity was seen in the blood, liver, kidney, lung, and adrenal cortex.[31] Metabolism of organic cyanides in the body culminates in their transformation into *thiocyanates*, and their concentration in the urine of animals increases after poisoning.[32] *S-(2-cyanoethyl)mercapturic acid* is likely to be the main metabolite of AN in rats. AN combines with erythrocytes to form an integral molecule. *Thiocyanate* (SCN^-) and *hydroxyethyl mercapturic acid* are also found to be the major metabolites of AN *in vivo*. The character of AN excretion depends on the route of administration and species. Ninety percent AN is removed with the urine, and only traces of AN are found in feces. Excretion via expired air is negligible.[31-33]

Regulations.

EU (1990). A. is available in the *List of authorized monomers and other starting substances which shall be used for the manufacture of plastic materials and articles intended to come into contact with foodstuffs* (Section A).

U.S. FDA (1998) has banned the use of AN copolymers for beverage containers and proposed to limit its migration from other food-contact materials:

- 0.003 mg/inch ^{2} in the case of single-use articles having a volume to surface ratio of >10 ml/inch^2 of food contact surface when extracted to equilibrium at 120°F, and in the case of repeated-use articles when extracted at a time equivalent to initial batch usage;

- 0.3 ppm in the case of single-use articles having a volume to surface ratio of <10 mg/inch^2 calculated on the basis of the volume of the container when extracted at equilibrium at 120°F.

AN has been approved for use (1) in adhesives as a component (monomer) of articles intended for use in packaging, transporting, or holding food, in accordance with the provisions prescribed in 21 CFR part 175.105; (2) in the manufacture of resinous and polymeric coatings for polyolefin films for food-contact surface of articles intended for use in producing, manufacturing, packing, processing, preparing, treating, packaging, transporting, or holding food, in accordance with the provisions prescribed in 21 CFR part 175.300; as (3) a component (monomer) of the uncoated or coated food-contact surface of paper and paperboard intended for use in producing, manufacturing, packaging, processing, preparing, treating, packing, transporting, or holding dry food of the type identified in 21 CFR part 176.170 (c); (4) as a monomer in the manufacture of semirigid and rigid acrylic and modified acrylic plastics in the manufacture of articles intended for use in contact with food, subject to the provisions of 21 CFR part 177.1010; (5) as a monomer in the manufacture of polyethylenephthalate polymers used as, or components of plastics (films, articles, or fabric) intended for use in contact with food, in accordance with the provisions prescribed in 21 CFR part 177.1630; and (6) in polymers which may be safely admixed, alone or in mixture with other permitted polymers, as modifiers in semirigid and rigid vinyl chloride plastic food-contact articles prepared from vinyl chloride homopolymers and/or from vinyl chloride copolymers in accordance with the conditions specified in 21CFR.

U.S. FDA is reviewing all AN Regulations to align them with current carcinogenicity data. The review includes an advisory opinion and a subsequent proposal for use of a new type of container that limits migration of AN.

Great Britain (1998). AN is authorized without time limit for use in the production of polymeric materials and articles in contact with food or drink or intended for such contact. The specific migration of this substance shall be not detectable (when measured by a method with a DL of 0.02 mg/kg, analytical tolerance included).

Standards.

EU (1990). SML: not detectable (detection limit is 0.02 mg/kg).

U.S. EPA (1999). MCLG: zero.

Russia (1995). MAC: 2.0 mg/l, PML: 0.02 mg/l.

References:

1. Zabezhinskaya, N. A and Brook, Ye. S., in *Protection of Water Reservoirs against Pollution by Industrial Liquid Effluents*, S. N. Cherkinsky, Ed., Medgiz, Moscow, Issue No 4, 1960, 147 (in Russian).
2. Medvedev, V. I., Hygienic properties of vulcanizates made on the base of butadiene-nitrile rubber with the help of sulfurorganic vulcanization accelerators, in *Hygiene and Toxicology of High-Molecular-Mass Compounds and of the Chemical Raw Material Used for Their Synthesis*, Proc. 6th All-Union Conf., B. Yu. Kalinin, Ed., Leningrad, Khimiya, 1979, 81 (in Russian).
3. Lickly, T. D., Markham, D. A., and Rainey, M. L., The migration of acrylonitrile from acrylonitrile/ butadiene/ styrene polymers into food-simulating liquids, *Food Chem. Toxicol.*, 29, 25, 1991.
4. Chanayem, B. I., Farooqui, M. Y., Elshabrawy, O., Mumtaz, M. M., and Ahmed, A. E., Assessment of the acute acrylonitrile-induced neurotoxicity in rats, *Neurotoxicol. Teratol.,* 13, 499, 1991.
5. Stasenkova, K. P., Bondarev, G. I., and Murav'iova, S. I., Evaluation of toxicity of acrylonitrile, *Kauchuk i Rezina,* 3, 29, 1978 (in Russian).
6. Szabo, S., Bailey, K. A., Boor, P. J., et al., Acrylonitrile and tissue glutathione - differential effect of acute and chronic interaction, *Biochem. Biophys. Res. Commun.*, 79, 32, 1977.
7. Kuznetzov, P. P. and Shustov. V. Ya., Influence of acrylonitrile on blood indices, *Gig. Truda Prof. Zabol.*, 7, 41, 1986 (in Russian).
8. Grushko, Ya. M., *Harmful Organic Compounds in the Liquid Industrial Effluents*, Khimiya, Leningrad, 1976, 23 (in Russian).
9. Bakker, J. G., Jongen, S. M., Van Neer, F. C., and Neis, J. M., Occupational contact dermatitis due to acrylonitrile, *Contact Dermat.*, 24, 50, 1991.
10. Murray, F. J., Schwetz, B. A., Nitsche, K. D., et al., Teratogenicity of acrylonitrile given to rats by gavage or by inhalation, *Food Cosmet. Toxicol.,* 16, 547, 1978.
11. Mehrota, J., Khannal, V. K., Hussain, R., et al., Biochemical and developmental effects in rats following in utero exposure to acrylonitrile: a preliminary report, *Ind. Health*, 26, 251, 1988.
12. Szabo, S., Reynolds, E. S., and Unger, S. H., Structure-activity relations between alkyl nucleophilic chemicals causing duodenal ulcer and adrenocortical necrosis, *J. Pharmacol. Exp. Ther.,* 223, 68, 1982.
13. Saillenfait, A. M., Bonnet, P., Guenier, J. P., and de Ceaurriz, J., Relative developmental toxicities of inhaled aliphatic mononitriles in rats, *Fundam. Appl. Toxicol.*, 20, 365, 1993.
14. Willhite, C. C., Ferm, V. H., and Smith, R. F., Teratogenic effects of aliphatic nitriles, *Teratology*, 23, 317, 1981.
15. Scheufler, H, Experimental testing of chemical agents for embryotoxicity, teratogenicity and mutagenicity-ontogenic reactions of the laboratory mouse to these injections and their evaluation - a critical analysis method, *Biol. Rundsch.*, 14, 227, 1976.
16. Tandon, R., Saxena, D. K., Chandra, S. V., Seth, P. K., and Srivastava S. P., Testicular effects of acrylonitrile in mice, *Toxicol. Lett.*, 42, 55, 1988.
17. Butterworth, B. E., Eldridge, S. R., Sprankle, C. S., Working, P. K., Bentley, K. S., and Hurtt, M. E., Tissue-specific genotoxic effects of acrylamide and acrylonitrile, *Environ. Molec. Mutagen.*, 20, 148, 1992.
18. Lambotte-Vandepaer, M. and Duverger-van-Bogaert, M., Genotoxic properties of acrylonitrile, *Mutat. Res.,* 134, 49, 1984.
19. Whysner, J., Ross, P. M., Conaway, C. C., Verna, L. K., and Williams, G. M., Evaluation of possible genotoxic mechanisms for acrylonitrile tumorigenicity, *Regul. Toxicol. Pharmacol.,* 27, 217, 1998.

20. Leonard, A., Garny, V., Poncelet, F., and Mercier, M., Mutagenicity of acrylonitrile in mouse, *Toxicol. Lett.*, 7, 329, 1981.
21. Vogel, E. W. and Nivard, N. J., Performance of 181 chemicals in a *Drosophila* assay predominantly monitoring interchromosomal mitotic recombination, *Mutagenesis*, 8, 57, 1993.
22. Zhurkov, V. S., Shram, R. L., and Dugan, A. M., Analysis of mutagenic activity of acrylonitrile, *Gig. Sanit.*, 1, 71, 1983 (in Russian).
23. Ahmed, A. E., Abdel-Rahman, S. Z., and Nour al Deen, A. M., Acrylonitrile interaction with testicular DNA in rats, *J. Biochem. Toxicol.*, 7, 5, 1992.
24. Farooqui, M. Y. G. and Ahmed, A. E., Molecular interaction of acrylonitrile and potassium cyanide with rat blood, *Chem. Biol. Interact.*, 38, 145, 1982.
25. Maltoni, C., Ciliberti, A., and Carretti, D., Experimental contributions in identifying brain potential carcinogens in the petrochemical industry, *Ann. N. Y. Acad. Sci.*, 381, 216, 1982.
26. Bigner, D. D., Bigner, S. H., Burger, P. C., et al., Primary brain tumors in Fisher 344 rats chronically exposed to acrylonitrile in their drinking water, *Food Chem. Toxicol.*, 24, 129, 1986.
27. Gallagher, G. T. et al., *J. Am. Coll. Toxicol.*, 7, 603, 1988.
28. Sokal, J. A. and Klyszejko-Stefanovicz, L., Nicotin amideadenine dinucleotides in acute poisoning with some toxic agents, *Lodz. Tow. Nauk. Prac. Wydz.*, 3, 104, 1972 (in Polish).
29. Sapota, A., The disposition of ^{14}C-acrylonitrile in rats, *Xenobiotica*, 12, 259, 1982.
30. Burka, L. T., Sanchez, I. M., Ahmed, A. E., and Ghanayem, B. I., Comparative metabolism and disposition of acrylonitrile and methacrylonitrile in rats, *Arch. Toxicol.*, 68, 611, 1994.
31. Sondberg, E. Ch. and Salnina, P., Distribution of 1-^{14}*C*-acrylonitrile in rat and monkey, *Toxicol. Lett.*, 6, 187, 1980.
32. Ahmed, A. E. and Petel, K., Acrylonitrile: *in vivo* metabolism in rats and mice, *Drug Metab. Dispos.*, 9, 219, 1981.
33. Kopecky, J., Zachardova, D., Gut, I., et al., Metabolism of acrylonitrile in rats *in vivo*, *Prac. Lek.*, 31, 203, 1979 (in Czech).

ADIPIC ACID

Molecular Formula. $C_6H_{10}O_4$
M = 146.15
CAS No 124-04-9
RTECS No AU8400000
Abbreviation. AA.

Synonyms. Adipinic acid; 1,4-Butanedicarboxylic acid; 1,6-Hexanedioic acid.

Properties. White to yellowish odorless odor. Water solubility is 1.4% at 15°C, readily soluble in alcohol. AA has no effect on water color, odor, or foaming. Organoleptic perception threshold is 200 mg/l.[1]

Applications. Monomer in the production of polyamides intended for use in the food industry, a curing agent for epoxy resins, etc.

Acute Toxicity. In mice, LD_{50} is 4.2 g/kg BW.[2] According to Kropotkina et al. (1981), rats tolerate this dose. AA is a polytropic toxin affecting primarily the parenchymatous organs, NS and enzyme systems and disrupting exchange processes.

Repeated Exposure failed to reveal cumulative properties. Rats tolerated administration of 10 g/kg BW. The treatment with 42 to 420 mg/kg BW resulted in retardation of BW gain, a reduction in the STI value and in the total blood serum protein, and changes in the enzyme activity. Gross pathology examination revealed changes in the liver, kidneys, and GI tract.[3]

Long-term Toxicity. In a 6-month study, rabbits were dosed by gavage with 5.0 mg *sodium adipate*/kg BW. The increased catalase activity and reduced cholinesterase activity were noted in the blood serum.[4] AA was found to produce an irritating effect on the kidneys.

Reproductive Toxicity. Rats were given oral doses of 2.0 to 250 mg AA/kg BW for 13 days. Munro et al. (1996) suggested the calculated NOEL of 250 mg/kg BW for this teratogenicity study.[5]

Mutagenicity.

In vitro genotoxicity. AA did not produce gene mutation in *E.coli* assay with metabolic activation (GENE-TOX, *U.S. EPA*).

Chemobiokinetics. Following ingestion, AA and its polyesters are poorly absorbed. AA metabolism involves its oxidation. AA metabolites are discovered in the urine: *urea, glutamic, lactic, β-ketoadipic, and citric acids,* etc.[6]

Regulations.

U.S. FDA (1998) approved the use of AA (1) in the manufacture of cross-linked polyester resins which may be safely used as articles or components of articles intended for repeated use in contact with food, in accordance with the provisions prescribed in 21 CFR part 177.2420; (2) in the manufacture of polyurethane resins safely used as food-contact surfaces of articles intended for use in contact with dry food, in accordance with the provisions prescribed in 21 CFR part 177.1680; and (3) in the manufacture of resinous and polymeric coatings of food-contact surfaces of articles intended for use in producing, manufacturing, packing, transporting, or holding food, in accordance with the provisions prescribed in 21 CFR part 175.300. Affirmed as *GRAS.*

EU (1990). AA is available in the *List of authorized monomers and other starting substances which shall be used for the manufacture of plastic materials and articles intended to come into contact with foodstuffs (Section A).* Adipic anhydride is available in the *List of authorized monomers and other starting substances which may continue to be used for the manufacture of plastic materials and articles intended to come into contact with foodstuffs pending a decision on inclusion in Section A* (*Section B*).

Great Britain (1998). AA is authorized without time limit for use in the production of polymeric materials and articles in contact with food or drink or intended for such contact.

Recommendations.

EU (1995). AA is a food additive generally permitted for use in foodstuffs (Maximum level: 2.0 g/kg).

Joint FAO/WHO Expert Committee on Food Additives. ADI: 5.0 mg/kg BW.

Standards. *Russia* (1995). MAC: 2.0 mg/l.

References:

1. Andreyev, I. A., The method for rapid elaboration of hygienic norms of the content of chemical compounds (as exemplified by adipinic and sebacic acids) in water bodies, *Gig. Sanit.*, 7, 10, 1985 (in Russian).
2. Kropotkina, M. A., Galitskaya, V. A., Garkavenko, O. S., et al., Toxicology of dicarbonic acids, in *Hygiene and Toxicology of High-Molecular-Mass Compounds and of the Chemical Raw Material Used for Their Synthesis*, Proc. 6th All-Union Conf., B. Yu. Kalinin, Ed., Leningrad, 1979, 224 (in Russian).
3. Novikov, Yu. V., Andreyev, I. A., Ivanov, Yu. V., et al., Hygienic standardization of sebacic and adipic acids in water bodies, *Gig. Sanit.*, 9, 72, 1983 (in Russian).
4. Savelova, V. A. and Brook, Ye. S., in *Protection of Water Reservoirs against Pollution by Industrial Liquid Effluents*, S. N. Cherkinsky, Ed., Meditsina, Moscow, Issue No 6, 1964, 118 (in Russian).
5. Munro, I. C., Ford, R. A., Kennepohl, E., and Sprenger, J. G., Correlation of structural class with No-Observed-Effect Levels: A proposal for establishing a threshold of concern, *Food Chem. Toxicol.*, 34, 829, 1996.
6. Rusoff, J. J., Baldwin, R. R., Domingues, F. J., et al., Intermediary metabolism of adipic acid, *Toxicol. Appl. Pharmacol.*, 2, 316, 1960.

ADIPIC ACID, compound with 1,6-HEXANEDIAMINE

Structural Formula. $NH(CH_2)_6NHCO(CH_2)_4CO$

M = 262.39

CAS No 15511-81-6

3323-53-3

RTECS No AV1940000

Synonyms and **Trade Names.** Hexamethylenediamine adipate; Hexamethylenediammonium adipate; Hexanedioic acid, compound with 1,6-hexanediamine (1:1); Nylon-6,6 salt; Salt AG.

Properties. Colorless, oily crystals with a faint specific odor. Solubility in water is up to 50% (room temperature). Concentrations of 5.0 and 1.0 g/l give water a foreign odor and taste, respectively.[1]

Applications. A monomer in the production of polyamide 6,6 (Anid).

Acute Toxicity. 5.0 g/kg BW dose is reported to be lethal to both rats and rabbits. According to other data, LD_{50} is 6.7 g/kg BW for rats, and 1.14 to 3.6 g/kg BW for mice. Single administration of 0.5 or 2.0 g/kg BW doses to rats caused a transient decrease in the body temperature, labored breathing, and leukopenia.[1,2]

Repeated Exposure failed to reveal cumulative properties in the study when a 1.0 g/kg BW dose was administered for 56 days.[2]

Short-term Toxicity revealed moderate cumulative properties. Ten administrations of 700 mg/kg BW to mice led to a pronounced toxic effect. Three mice out of 20 died. Rats were dosed with 300 and 500 mg/kg BW for 40 days. None of the animals died but they displayed agitation, aggressiveness, retardation of BW gain, decrease in cholinesterase activity and in the content of serum *SH*-groups. Gross pathology examination revealed dystrophic changes in the liver and kidneys.[1]

Long-term Toxicity. Administration of 5.0 mg/kg BW to rabbits for 7 months resulted in decreased activity of blood cholinesterase and increased content of protein and *SH*-groups in the blood. Doses of 0.05 and 0.5 mg/kg BW appeared to be ineffective. There were no morphological changes in the visceral organs.[1]

Regulations. ***EU*** (1990). A. is available in the *List of authorized monomers and other starting substances which may continue to be used for the manufacture of plastic materials and articles intended to come into contact with foodstuffs pending a decision on inclusion in Section A* (*Section B*).

Standards. ***Russia*** (1995). PML: 1.0 mg/l.

References:

1. Brook, Ye. S. and Klimkina, N. V., in *Protection of Water Reservoirs against Pollution by Industrial Liquid Effluents*, S. N. Cherkinsky, Ed., Meditsina, Moscow, Issue No 7, 1965, 56 (in Russian).
2. Babayev, D. A., Toxicity of hexamethylenediamine adipate, *Voprosy Pitania*, 1, 54, 1981 (in Russian).

4-ALLYLOXY-3,5-DICHLOROBENZOIC ACID, ALLYL ESTER

Molecular Formula. $C_{10}H_{12}O_5$
M = 212.2
CAS No 91954-47-1
RTECS No DG0995600

Synonyms. Allyl-α,α'-(allyloxycarbonyloxy)acrylate; 2-(Allyloxycarbonyloxy)propenoic acid, allyl ester; Benzoic acid 4-(allyloxy)-3,5-dichloroallyl ester.

Properties. Colorless, transparent liquid with a slight acrylate odor. Poorly soluble in water.

Applications. Used as a co-monomer in the production of polymers intended for use in contact with food or drink.

Acute Toxicity. LD_{50} is reported to be 390 mg/kg BW in rats, and 309 mg/kg BW in mice.[1] Poisoning is accompanied by labored breathing, adynamia, and changes in the GI tract. Death within 3 days.

Repeated Exposure failed to reveal pronounced cumulative effect. K_{acc} is 5.3 (by Lim). Inhalation exposure affected the parenchymatous organs.

Allergenic Effect. A. seems to be a sensitizer on repeated skin applications.

Reproductive Toxicity. Inhalation exposure affected the reproductive function: gonadotoxic effect was evident in rats.[2]

Mutagenicity.

In vivo cytogenetics. Inhalation exposure produced a mutagenic effect in the somatic mammalian cells.[3,4]

References:

1. Oskerko, E. F. and Klimova, A. I., Toxic effect of allyl-α,α'-allyl(oxycarbonyloxy) acrylate, in *Hygiene and Toxicology of High-Molecular-Mass Compounds and of the Chemical Raw Material Used for Their Synthesis*, Proc. 6th All-Union Conf., B. Yu. Kalinin, Ed., Leningrad, 1979, 279 (in Russian).

2. Kasatkin, A. N., in *Morphological Methods for Investigations in Hygiene and Toxicology*, Moscow, 1983, 48 (in Russian).
3. Oskerko, E. F. and Kasatkin, A. N., Some mechanism of reproduction function impairment in rats on exposure to allyl-α,α'- allyloxycarbonyloxyacrylate, *Gig. Truda Prof. Zabol.*, 12, 44, 1987 (in Russian).
4. Gavrilyuk, Yu. I., Study of mutagenic load in human populations, *Gig. Sanit.*, 12, 36, 1987 (in Russian).

11-AMINOUNDECANOIC ACID

Molecular Formula. $C_{11}H_{23}NO_2$
M = 201.35
CAS No 2432-99-7
RTECS No YQ2293000
Abbreviation. AA.

Synonym. 11-Aminoundecylic acid

Properties. White, crystalline substance.

Applications. Used in undecane (or Nylon-11) production.

Acute Toxicity. Male rats tolerate a dose of 21.5 g/kg BW for 14 days. One female out of 5 died from a dose of 14.7 g/kg BW, and all females died following exposure to 21.0 g/kg BW.[1]

Long-term Toxicity. Mice and rats were administered 7.5 and 15 g/kg BW for 2 years. The treated animals displayed dose-dependent changes consisting of transitional epithelium hyperplasia in the renal pelvis and bladder in rats of both sexes, changes in the cortical and medullary layers of the kidneys in female rats, and mineralization of the kidneys and vacuolization of the hepatocytes in mice.[1]

Mutagenicity.

In vivo cytogenetics. It appeared to be negative in the sex-linked recessive mutation test in *Dr. melanogaster.*[2]

In vitro genotoxicity. AA did not induce CA in bone marrow cells of mice (NTP-94).

Carcinogenicity. B6C3F$_1$ mice were exposed to dietary levels of 7.5 and 15 g/kg BW for 2 years. Only males developed malignant lymphomas. In male rats given the same doses, transitional-cell bladder carcinomas were found. A dose-dependent hyperplasia of the bladder and renal pelvis cells was observed in rats. Nodular neoplasms were found in the liver of males.[1,3]

Carcinogenicity classifications. An IARC Working Group concluded that there is limited evidence for the carcinogenicity of AA in *experimental animals* and there were no adequate data available to evaluate the carcinogenicity of AA in *humans*.

IARC: 3;

NTP: P* - N - E - P (feed).

Regulations. ***EU*** (1990). AA is available in the *List of authorized monomers and other starting substances which shall be used for the manufacture of plastic materials and articles intended to come into contact with foodstuffs (Section A)*. SML: Not Detectable (DL = 0.01 mg/kg).

U.S. FDA (1998) permits the use of Nylon 11 resins in articles intended for use in contact with food. They are also approved for use as components of side seam cements intended for a single use in contact with food (only of the identified types), subject to the conditions prescribed in 21 CFR part 175.300.

Great Britain (1998). AA is authorized without time limit for use in the production of polymeric materials and articles in contact with food or drink or intended for such contact. The specific migration of this substance shall not exceed 5.0 mg/kg.

References:

1. *Carcinogenesis Bioassay of 11-Aminoundecanoic Acid*, NTP Technical Report Series No 216, Research Triangle Park, NC, 1982.
2. Yoon, J. S., Mason, J. M., Valencia, R., et al., Chemical mutagenesis testing in *Drosophila*. IV. Results of 45 coded compounds tested for the NTP, *Environ. Mutagen*, 7, 349, 1985.

3. Dunnick, J. K., Huff, J. E., Haseman, J. K., et al., Lesions of the urinary tract produced in Fischer 344 rats and B6C3F$_1$ mice after chronic administration of 11-aminoundecanoic acid, *Fundam. Appl. Toxicol.*, 3, 614, 1983.

ANILINE

Molecular Formula. C_6N_7N
M = 93.12
CAS No 62-53-3
RTECS No BW6650000

Synonyms and **Trade Name.** Aminobenzene; Aminophen; Anyvim; Benzenamine; Phenylamine.

Properties. Colorless, oily liquid, darkening rapidly in air and light, with an aromatic odor. Water solubility is 34 g/l, or 37 g/l,[02] readily soluble in alcohol and fats. Odor threshold concentration is reported to be 50 mg/l,[1] or 70 mg/l,[2] or 2.0 mg/l.[3]

A concentration of 5.0 mg/l gives water a yellow color.[4]

Applications. Used in the manufacture of plastics, vulcanizates, and dyes.

Migration of A. into water and simulant media from vulcanizate based on synthetic polyisoprene rubber (SKI-3) was investigated (1 hour to 5 days; 20 and 100°C; 0.1 to 2.0 cm^{-1}). A. (as a product of diphenyl guanidine transformation) was determined at the level up to 0.2 mg/l).[028]

Acute Toxicity. LD_{50} is found to be 300 to 750 mg/kg BW in rats, and 1075 mg/kg BW in mice. According to other data, LD_{50} is 460 mg/kg for mice and 250 mg/kg BW for cats.[5]

There was no correlation between the toxic effect severity and methemoglobin formation. Even lethal doses did not produce *metHb* in mice and rabbits, but it was found in the blood of guinea pigs, cats, and dogs. Poisoning was followed by body tremor. In 20 to 30 min animals became irresponsive to the external stimuli. In 2 to 3 hours, the ears, tails, and legs turned grayish-blue. Persistent anemia developed, followed by reduction in blood viscosity and disturbance in protein exchange. Gross pathology examination revealed a brown color of all organs and tissues.

Repeated Exposure failed to reveal cumulative properties on administration of 1/5 LD_{50} for 5 weeks. Rats were orally exposed to 50 mg/kg BW for 14 days. The treatment resulted in elevated relative weights of the visceral organs, dystrophic and necrobiotic changes in hepatocytes and in the spleen.[6]

Short-term Toxicity. Male Sprague-Dawley rats were given A. at a concentration of 600 ppm in their drinking water for 3 months. The main toxic effect was the formation of *metHb.* Toxicity to the hemopoietic system comprised a 65% increase in leukocyte count in 30 days, whereas no changes were recorded at later time points. Erythrocyte count appeared to be significantly decreased. Histopathological examination revealed striking changes in the spleen, including marked red pulp expansion and light brown pigment of heme origin. Testes size was slightly decreased in 60 days. There was no evidence of neoplasia.[7]

Reproductive Toxicity.

Pregnant rats were gavaged with 100 mg/kg BW on days 7 to 20 of gestation. The treatment caused no ***teratogenic effect***.[8] Rat dams were given 260 mg/kg BW on days 12 to 14 or 15 to 17 of gestation. Fetal ventricular septal defects were shown to develop.[9]

Allergenic Effect. A strong contact sensitizing potential was found in maximized guinea-pig assays.[10]

Mutagenicity.

Observations in man. No data are available on the genetic and related effects in humans.

***In vitro genotoxicity*.** Negative results were reported for SCE assay in human cells *in vitro*. Syrian hamster embryo cells and virus-infected Fischer rat embryo cells were not transformed, but BALB/c 3T3 cells were. A. induced SCE and CA but not DNA strandbreaks or unscheduled DNA synthesis in mammalian cells *in vitro* (IARC 4-27; IARC 27-39).[11]

A. was negative in the *Salmonella typhimurium* strain *TA98* and *TA100* both in the absence and the presence of *S9*.[12]

***In vivo cytogenetics*.** A. induced SCE but not micronuclei in the bone-marrow cells of mice treated *in vivo*; DNA strand breakage was induced in the liver and kidney of rats *in vivo*.

Carcinogenicity.

Observations in man. Increased risk of bladder cancer was strongly associated ($p < 0.001$) with increased occupational exposure to aniline. Simultaneously, *o*-toluidine, a well-known carcinogen, was detected in the air in this study.[13]

Animal studies. An oral study in $B6C3F_1$ mice given food containing 6,000 or 12,000 ppm A. *hydrochloride* for 103 weeks did not show any increase in tumor incidence in males and females. In rats given dietary concentrations of 3,000 or 6,000 ppm A. *hydrochloride*, statistically significant dose-related trends in the incidence of fibrosarcomas, sarcomas and hemangiosarcomas of the spleen and peritoneal cavity were observed.[14]

In a 2-year study, CD-F rats received 200, 600, or 2000 ppm A. *hydrochloride* in their diet. The exposure revealed an increase in the incidence of primary splenic sarcomas in males in 2000 ppm-treated animals. Structural changes observed included stromal hyperplasia and fibrosis of the splenic red pulp, which may represent a precursor lesion of sarcoma.[15]

Wistar rats were administered 0.03, 0.06, or 0.12% A. alone or in combination with the comutagen norharman in the drinking water for 80 weeks. The incidence of forestomach papillomas was low and not dose-related, no cocarcinogenic effect on the urinary bladder or other organs examined was found.[16]

Rats were administered 22 mg A. *hydrochloride*/day for their lifetime. 50% animals died at day 450 and 100% died at day 750. Histological examination revealed no tumors in the bladder, liver, spleen or kidney.[17]

Carcinogenicity classifications. An IARC Working Group concluded that there is limited evidence for the carcinogenicity of A. in *experimental animals* and there were inadequate data available to evaluate the carcinogenicity of A. in humans.

IARC: 3;

U.S. EPA: B2;

EU: 3;

NTP: P* - P - N - N (*A. hydrochloride*, feed).

Chemobiokinetics. Formation of *metHb* could result from the oxidation of *Hb* by A. metabolites such as *phenylhydroxyaniline, 2-aminophenol* and *4-aminophenol* of which phenylhydroxyaniline has been demonstrated to be the principal mediator of A.-induced methemoglobinemia in rats.[18] A. is metabolized via hydroxylation of the aromatic ring. Metabolites (predominantly, *aminophenols, phenylsulfamic acid*, and *acetanilide*) are excreted with the urine in the form of conjugates. See also *Parke*, 1973.[19]

Erythrocytes of rats preferentially bind aniline. Administration of 100 mg A./kg BW led to A. accumulation in the spleen due to scavenging of damaged erythrocytes. The deposition of debris from erythrocytes results in hemosiderosis which in turn induces a fibrotic response in the spleen that may be involved in sarcoma production.[20]

Standards. ***Russia*** (1995). MAC and PML: 0.1 mg/l.

References:

1. Kuper, A. I. and Ozerova, V. F., in *Industrial Pollution of Water Reservoirs*, S. N. Cherkinsky, Ed., Meditsina, Moscow, Issue No 8, 1967, 156 (in Russian).
2. Zoetman, B., *Organoleptic Assessment of Water Quality*, Stroyizdat, Moscow, 1984, 160 (in Russian, translation from English).
3. Grushko, Ya. M., *Harmful Organic Compounds in the Liquid Industrial Effluents*, Khimiya, Leningrad, 1976, 35 (in Russian).
4. Obukhov, V. P., On the problem of norm-setting for aniline in water bodies, *Gig. Sanit.*, 9, 16, 1954 (in Russian).
5. Jacobson, K. H., Acute oral toxicity of mono- and di-alkyl-ring-substituted derivatives of aniline, *Toxicol. Appl. Pharmacol.*, 22, 153, 1972.
6. Agranovsky, M. Z., Experimental study of lipamide as a means of prevention of aniline methemoglobinemia, *Gig. Sanit.*, 10, 96, 1973 (in Russian).
7. Khan, M. F., Kaphalia, B. S., Boor, P. J., et al., Subchronic toxicity of aniline hydrochloride in rats, *Arch. Environ. Contam. Toxicol.*, 24, 368, 1993.

8. Price, C. G., Tyl, R. W., Marks, T. A., et al., Teratologic and postnatal evaluation of aniline hydrochloride in the Fischer 344 rat, *Toxicol. Appl. Pharmacol.*, 77, 465, 1985.
9. Nakatsu, T., Kanamori, H., Matsumoto, K., and Kusanagi, T., Cardiovascular malformations in rat fetuses induced by administration of aniline hydrochloride, Abstract, *Congen. Anomal.*, 33, 278, 1993.
10. Basketter, D. A. and Scholes, E. W., Comparison of the local lymph node assay with the guinea pig maximization test for the detection of a range of contact allergens, *Food Chem. Toxicol.*, 30, 65, 1992.
11. Pis'ko, G. T. et al., in *Proc. 4th Conf. Geneticists & Breeders*, Naukova Dumka, Kiyv, 1981, 143 (in Russian).
12. Assmann, N., Emmrich, M., Kampf, G., and Kaiser, M., Genotoxic activity of important nitro-bezenes and nitroanilines in the Ames test and their structure-activity relationship, *Mutat. Res.*, 12, 395, 1997.
13. Ward, E., Carpenter, A., Markowitz, S., et al., Excess number of bladder cancer in workers exposed to o-toluidine and aniline, *J. Natl. Cancer Inst.*, 83, 501, 1991.
14. *Bioassay for Aniline Hydrochloride for Possible Carcinogenicity*, Natl. Cancer Institute Technical Report Series No 130, DHEW Publ. No 78-1385, Bethesda, MD, 1991.
15. *104-week Chronic Toxicity Study in Rats: Aniline Hydrochloride*, Final Report, Chem. Ind. Institute of Toxicology, 1982.
16. Hagiwara, A. M., Arai, M., Hirose, M., Nakanowatari, J., Tsuda, H., and Ito, N., Carcinogenic effects of norharman in rats treated with aniline, *Toxicol. Lett.*, 6, 71, 1980.
17. Druckrey, H., Beitrage zur Pharmakologie cancerogener Substanzen, Versuche mit Anilin, *Arch. Exp. Pathol. Pharmakol.*, 210, 137, 1950.
18. Harrison, J. H. and Jollow, D. J., Role of aniline metabolites in aniline-produced hemolitic anemia, *J. Pharmacol. Exp. Ther.*, 238, 1045, 1986.
19. Parke, D. B., *Biochemistry of Foreign Compounds*, Meditsina, Moscow, 1973, 269 (in Russian, translation from English).
20. Bus, J. S. and Sun, J., Accumulation of covalent binding of radioactivity in rat spleen after ^{14}C-aniline *HCl* administration, *Pharmacologist*, 21, 221, 1979.

AZELAIC ACID

Molecular Formula. $C_9H_{16}O_4$
M = 188.22
CAS No 123-99-9
RTECS No CM1980000
Abbreviation. AA.

Synonyms and **Trade Names.** Anchoic acid; Emerox 1110; Heptanedicarboxilic acid; Lepargylic acid; Nonanedioic acid.

Properties. Monoclinic prismatic needles. Freely soluble in alcohol and boiling water.[020]

Applications. A monomer in the manufacture of plastics intended for use in contact with food or drink.

Acute Toxicity. LD_{50} was found to be more than 5.0 g/kg BW in rats.

Chemobiokinetics. Six male volunteers were treated topically and one week later orally with 1.0 g AA as aqueous microcristalline suspension. About 60% of the dose had been excreted unchanged with the urine.

Regulations.

EU (1990). AA and azelaic anhydride are available in the *List of authorized monomers and other starting substances which shall be used for the manufacture of plastic materials and articles intended to come into contact with foodstuffs* (*Section A*).

U.S. FDA (1998) approved the use of AA (1) in adhesives as a component of articles intended for use in packaging, transporting, or holding food in accordance with the conditions prescribed in 21 CFR part 175.105; and (2) as a component of polyethylene phthalate polymers (films, articles, or fabrics) intended for use in contact with food in accordance with the conditions prescribed in 21 CFR part 177.1630.

Great Britain (1998). AA is authorized without time limit for use in the production of polymeric materials and articles in contact with food or drink or intended for such contact.

Reference:

Tauber, U., Weiss, C., and Matthes, H., Percutaneous absorption of azelaic acid in humans, *Exp. Dermatol.*, 1, 176, 1992.

1,2,4-BENZENETRICARBOXYLIC ACID, TRI-2-PROPENYL ESTER

Molecular Formula. $C_{18}H_{18}O_6$
M = 330.36
CAS No 2694-54-4
RTECS No DC2075000

Synonyms and **Trade Name.** 1,2,4-Benzenetricarboxylic acid, triallyl ester; 1,2,4-Triallyl trimellitate; Triam 705; Trimellitic acid triallyl ester.

Properties. Viscous, light-yellow liquid. Readily soluble in organic solvents.

Applications. Used in the manufacture of unsaturated polyester resins.

Acute Toxicity. LD_{50} for rats is found to be 2.4 g/kg BW.

Repeated Exposure failed to reveal cumulative properties. Rats were dosed by gavage with 1/5 LD_{50} for a month. The treatment reduced the *Hb* level and erythrocyte count in the peripheral blood. There were no changes in liver, kidney and NS functions.

Reference:

Volodchenko, V. A. and Sadokha, E. P., Toxic effects of triallyl trimellitate, *Gig. Truda Prof. Zabol.*, 11, 58, 1985 (in Russian).

BIS(2,3-EPOXYCYCLOPENTYL) ETHER

Molecular Formula. $C_{10}H_{14}O_3$
M = 182.24
CAS No 2386-90-5
RTECS No RH9100000

Synonym. 2,2'-Oxybis(6-oxabicyclo[3.1.0]hexane).

Properties. Homogenous liquid at 43°C and more. A mixture of liquids and solids at lower temperatures. Water solubility is < 0.3% at 30°C.

Applications. Was developed as a high-performance component and modifier of epoxy resins.

Acute Toxicity. In rats, LD_{50} is 2.14 ml/kg BW (IARC 47-234).

Mutagenicity.

In vitro genotoxicity. B. appeared to be mutagenic to *Salmonella* in the presence and absence of an exogenous metabolic system. It increased the frequency of SCE in human lymphocytes *in vitro.*

In vivo cytogenetics. B. induced micronuclei in mice.[1]

Carcinogenicity. B. produced a certain number of skin tumors in animals after skin applications.[2]

References:

1. Xie, D. and Dong, S., Mutagenicity of bis(2,3-epoxy cyclopentyl)ether, *Acta Acad. Med. Sci.*, 6, 210, 1984.
2. Holland, J. M., Gosslee, D. G., and Williams, N. J., Epidermal carcinogenicity of bis(2,3-epoxycyclopentyl) ether, 2,2-bis(*p*-glycidyloxyphenyl)propane and *m*-phenylenediamine in male and female C3H and C57BL/6 mice, *Cancer Res.*, 39, 1718, 1979.

BIS(METHYLCYCLO)PENTADIENE

Molecular Formula. $C_{12}H_{14}$
M = 160.28
CAS No 26472-00-4
RTECS No PC1075000

Synonyms. Methyl-1,3-cyclopentadiene, dimer; 3a,4,7,7a-Tetrahydrodimethyl-4,7-methanoindene.

Properties. Liquid with an unpleasant specific odor. Soluble in ethyl alcohol and other organic solvents.

Applications. Used in the synthesis of rubbers and resins intended for use in contact with food or drink.

Acute Toxicity. LD_{50} is reported to be 10 g/kg BW for rats and 7.7 g/kg BW for mice. Species sensitivity was not found. Poisoning was accompanied by ataxia and convulsive twitching of the muscles.

Repeated Exposure revealed moderate cumulative properties.

Reference:

Ivanov, N. G., Mel'nikova, L. V., Kliachkina, A. M, et al., Toxicity and hazard study and characterization of biological effect of methylcyclopentadiene dimer on the body, *Gig. Truda Prof. Zabol.*, 5, 45, 1981 (in Russian).

BISPHENOL A

Molecular Formula. $C_{15}H_{16}O_2$
M = 228.29
CAS No 80-05-7
RTECS No SL6300000
Abbreviation. BPA.

Synonyms. Bis(*p*-hydroxyphenyl)propane; Diane; 4,4'-Dihydroxydiphenylolpropane; 4,4'-Dimethylmethylenediphenol; Diphenylolpropane; 4,4'-Isopropylidenediphenol.

Properties. Grayish or colorless, crystalline powder. Water solubility is 0.04%. Readily soluble in alcohol, insoluble in oils. Odor perception threshold is 50 mg/l, taste perception threshold is 0.25 mg/l (astringent).[1] BPA forms chlorophenol odors (threshold concentration 0.01 mg/l).

Applications. Monomer in the production of numerous chemical products including polycarbonates and epoxy, phenolic, ethoxylene, ion-exchange resins, corrosion-resistant unsaturated polyester-styrene resins, reinforced pipes, food packaging materials and vulcanizates intended for use in contact with food or drink. A thermal stabilizer of polyvinyl chloride.

Migration of BPA to food from molded discs prepared from a composite of BPA-derived polycarbonate resins was determined using food-simulating solvents and time and temperature conditions recommended by FDA. The study demonstrated that no detectable BPA was found in the extracts obtained under FDA's most severe default testing condition.

The potential dietary exposure to BPA from use of polycarbonate resins was determined to be less than 0.25 ppb.[2] Migration into water from epoxy coatings (exposure 7 days, 37°C) was 0.004 mg/l.[3] Migration of BPA from baby feeding bottles made of polycarbonate was not detectable in infant feed using a very sensitive method of liquid chromatography with fluorescence detection with a 0.03 mg/kg detection limit.[4]

Acute Toxicity. Oral LD_{50} values were reported to range from 4.24 to 12.0 g/kg BW in rats, and from 2.4 to 12 g/kg BW in mice. In rabbits and guinea pigs, LD_{50} is 4.0 g/kg BW.[1] Alcohol enhances toxicity of BPA. Poisoning is characterized by a transient agitation followed by CNS inhibition and labored respiration, coordination disorder, and convulsions.[5]

Repeated Exposure failed to reveal cumulative properties. Two rats out of 10 died after oral exposure to 1.0 g/kg BW for 10 days. Histology examination revealed fatty dystrophy of the liver, parenchymatous dystrophy of the renal tubular epithelium, and irritation of the spleen pulp.[5]

Rats and rabbits were given a dose of 0.5 g BPA/kg BW for 2 months. The treatment led to a reduction in BW gain, and to an increase in the acid resistance of erythrocytes and in the content of oxidized glutathione. The level of free phenols in the urine was increased. Rabbits displayed erythropenia and elevated liver and spleen weights.[1]

Short-term Toxicity. Male and female $B6C3F_1$ mice received BPA at dose levels of 0.2 to 4.0% in the diet for 13 weeks. It was concluded that the maximum tolerated dose (MTD) is 0.2% in diet, because the dose level of 0.5% proved to exert significant hematological toxicity.[6]

Long-term Toxicity. A 6-month study in rats and rabbits revealed signs of anemia. The treatment affected the NS, activity of *SH*-groups of tissue proteins, and detoxication system of the body.[1]

Reproductive Toxicity.

Embryotoxicity. CD-1 mice received 0.25 to 1.0% BPA in the diet for 18 weeks. The treatment caused a decrease in mean number of litters per pair and in mean number of live pups per litter in 0.5% and 1.0% treated groups.[7] Sprague-Dawley rats and CD-1 mice were dosed by gastric intubation with 160 to 640 mg/kg and 500 to 1250 mg/kg BW, respectively, on days 6 to 15 of gestation. Higher doses produced fetal toxicity in mice, but not in rats and did not alter fetal morphological development in either species.[8] Rats were treated *s/c* with 300 mg BPA/kg BW from postnatal day 1 to 5. All male and female rats showed normal reproductive function and no histopathologic abnormalities of reproductive organs. These results indicated that treatment with BPA at a fairly high dose was ineffective if given postnatally to male and female rats.[9]

Teratogenic effect is reported in rats, but not in mice. Administration of BPA to pregnant animals resulted in hydrocephaly and impaired ossification in the offspring.[10]

Gonadotoxicity. BPA has an estrogenic effect. The xenoestrogen BPA has been shown to mimic estrogen both *in vivo* and *in vitro*. Treatment of F344 rats with approximately 0.3 mg BPA/kg BW for 3 days resulted in hypertrophy, hyperplasia, and mucus secretion in the uterus and hyperplasia and cornification of the vaginal epithelium. Continuous exposure to mg levels of BPA is sufficient for exerting estrogenic actions.[11] Han-Wistar albino rats were exposed to drinking water that contained 0.01, 0.1, 1.0, or 10 ppm BPA, 7 days per week, for a total of 10 weeks. No treatment-related effects on growth or reproductive endpoints were observed in adult females exposed to any concentration of BPA. Similarly, no treament-related effects were observed on the growth, survival, or reproductive parameters (including testes, prostate and preputial gland weights, sperm count, daily sperm production, or testes histopathology) of male offspring from dams exposed to BPA during gestation and lactation.[12] These results do not confirm the previous low-dose observations reported as preliminary results by Sharpe et al. [13]

Mutagenicity.

In vitro genotoxicity. BPA is reported to be negative in *Salmonella* mutagenicity bioassay and mouse lymphoma assay; it did not cause CA in cultured mammalian cells.[016,025]

For assessment of *in vitro* carcinogenicity and related activity of BPA, the abilities of this compound to induce cellular transformation and genetic effects were examined simultaneously using the Syrian hamster embryo (SHE) cell model. BPA showed marked cell-transforming and genotoxic activities in cultured mammalian cells and potential carcinogenic activity.[14]

Carcinogenicity. In a long-term carcinogenicity bioassay rats were given 1000 and 2000 ppm in their diet, mice were given 1000 or 5000 ppm (males) and 5000 or 10000 ppm (females). BPA did not appear to be carcinogenic, however, increased incidence of leukemia in male rats and lymphoma in male mice was associated with the test chemical.[15]

Carcinogenicity classification.

NTP: E - N - N - N (feed).

Chemobiokinetics. BPA metabolism occurs through partial conversion into *phenols*, increasing their urinary content in a free and bound form. BPA is passed from the body unaltered in the urine and feces in the form of *glucuronides*.[16]

Administration of a single or multiple dose of 200 mg BPA/kg BW to CD-1 male rats produced two major and several minor adducts in the liver DNA. BPA is oxidized to *BPA-o-quinone* in the the presence of activation system. It is capable of binding covalently to DNA. One of the DNA-binding metabolites may be *BPA-o-quinone*.[17]

Regulations.

EU (1990). BPA is available in the *List of authorized monomers and other starting substances which shall be used for the manufacture of plastic materials and articles intended to come into contact with food-stuffs (Section A)*.

U.S. FDA (1998) approved the use of BPA in the manufacture of 4,4'-isopropylidenediphenol-epichlorohydrin thermosetting epoxy resin (finished articles containing the resins shall be thoroughly cleansed prior to their use in contact with food) in accordance with the conditions specified in the 21 CFR part 177.1440.

Great Britain (1998). BPA is authorized without time limit for use in the production of polymeric materials and articles in contact with food or drink or intended for such contact. The specific migration of this substance shall not exceed 3.0 mg/kg.

Standards.
EU (1990). SML: 3.0 mg/kg.
Russia (1995). MAC and PML: 0.01 mg/l (organolept., taste).

References:

1. Fedyanina, V. N., *Study of the Effect of Epichlorohydrin and Diphenylolpropane upon the Body*, Author's abstract of thesis, Novosibirsk, 1970, 22 (in Russian).
2. Howe, S. R. and Borodinsky, L., Potential exposure to bisphenol A from food-contact use of polycarbonate resins, *Food Addit. Contam.*, 15, 370, 1988.
3. Krat, A. V., Kesel'man, I. M., and Sheftel', V. O., Sanitary-chemical evaluation of polymeric articles in water-supply constructions, *Gig. Sanit.*, 10, 18, 1986 (in Russian).
4. Mountfort, K. A, Kelly, J., Jickells, S. M., and Castle, L., Investigations into the potential degradation of polycarbonate baby bottles during sterilization with consequent release of bisphenol A, *Food. Addit. Contam.*, 14, 737, 1997.
5. Stasenkova, K. P., Shumskaya, N. I., Greenberg, A. Ye., et al. Comparative evaluation of toxicity of bisphenol A and its derivatives, in Hygiene and Toxicology of High-Molecular-Mass Compounds *and of the Chemical Raw Material Used in Their Synthesis*, Proc. 4th All-Union Conf., S. L. Danishevsky, Ed., Khimiya, Leningrad, 1969, 180 (in Russian).
6. Furukawa, F., Nishikawa, A., Mitsui, M., Sato, M., Suzuki, J., Imazawa, T., and Takahashi, M., A 13-week subchronic study of bisphenol A in $B6C3F_1$ mice, Abstract, *Eisei Shikenjo Hokoku*, 112, 89, 1994 (in Japanese).
7. Reel, J. R., George, J. D., Lawton, A. D., et al., *Bisphenol A: Reproduction and Fertility Assessment in CD-1 Mice when Administered in the Feed*, Final study report, NTP/NIENS contract No ES-2-504, NTIS Accession N PB86103207, 1985.
8. Morrissey, R. E., George, J. D., Price, C. J., et al., The developmental toxicity of bisphenol A in rats and mice, *Fundam. Appl. Toxicol.*, 8, 571, 1987.
9. Nagao, T., Saito, Y., Usumi, K ., Kuwagata, M., and Imai, K., Reproductive function in rats exposed neonatally to bisphenol A and estradiol benzoate, *Reprod. Toxicol.*, 13, 303, 1999.
10. Bond, G. P. et al., Reproductive effects of bisphenol A, *Toxicol. Appl. Pharmacol.*, 23, 1980.
11. Steinmetz, R., Mitchner, N. A., Grant, A., Allen, D. L., Bigsby, R. M., and Ben-Jonathan, N., The xenoestrogen bisphenol A induces growth, differentiation, and c-fosgene expression in the female reproductive tract, *Endocrinology*, 139, 2741, 1998.
12. Cagen, S. Z., Waechter J. M., Jr, Dimond, S. S., Breslin, W. J., Butala, J. H., Jekat, F. W., Joiner, R. L., Shiotsuka, R. N., Veenstra, G. E., and Harris, L. R., Normal reproductive organ development in wistar rats exposed to bisphenol A in the drinking water, *Regul. Toxicol. Pharmacol.*, 30, 130, 1999.
13. Sharpe, R. M., Majdic, G., Fisher, J., Parte, P., Millar, M., and Saunders, P. T. K., Effects on testicular development and function, *10th Int. Congress Endocrinol*,, S23, 1996.
14. Tsutsui, T., Tamura, Y., Yagi, E., Hasegawa, K., Takahashi, M., Maizumi, N., Yamaguchi, F., and Barrett, J. C., Bisphenol A induces cellular transformation, aneuploidy and DNA adduct formation in cultured Syrian hamster embryo cells, *Int. J. Cancer*, 75, 290, 1998
15. *Carcinogenesis Bioassay of Bisphenol A in F344 Rats and $B6C3F_1$ Mice (Feed Study)*, NTP Technical Report Series No 215, Litton Bionetics, Inc., 1982.
16. Knaak, J. and Sullivan, L., Metabolism of bisphenol A in the rat, *Toxicol. Appl. Pharmacol.*, 8, 175, 1966.
17. Atkinson, A. and Roy, D., *In vivo* DNA adduct formation by bisphenol A, *Environ. Molec. Mutagen.*, 26, 60, 1995.

BISPHENOL A, DIGLYCIDYL ETHER

Molecular Formula. $C_{21}H_{24}0_4$
M = 340.45
CAS No 1675-54-3
RTECS No TX3800000

Abbreviation. BADGE.

Synonyms and **Trade Names.** Araldite; 2,2-Bis(4-hydroxyphenyl)propane bis(2,3-epoxypropyl) ether; EPI-Rez; EPON; Epotuff; ERL Bakelite; 2,2'-[(1-Methylethylidene)bis(4,1-phenyleneoxymethylene)] bis-(oxirane); Oligomer 340.

Composition. A mixture of monomer, dimer, trimer, and tetramer.

Properties. A medium viscosity, unmodified, liquid epoxy resin. Poorly soluble in water. Soluble in acetone and toluene.

Applications. The most common active component of epoxy resins, which are used in protective coatings, in reinforced plastic laminates in composites and adhesives intended for use in contact with food or drink. BADGE is used in the manufacture of lacquers for coating inside of food and beverage cans.

Migration Data. When used in acrylic-epoxy adhesive, BADGE was found to migrate into meat-and-vegetable-filled pastry at the level of 1.26 $\mu g/cm^2$, and into food-simulating liquid (Miglyol 812) at the level of 5.67 $\mu g/cm^2$.[1] The overall and specific migration of BADGE from epoxy system was investigated in 3 water-based food simulants. Hydrolysis of BADGE was observed in all of these simulants, giving more polar products.[2] Migration of BADGE into two samples of pizzas was 0.1 to 0.7 mg/kg when pizzas were cooked in their packaging according to on-pack instructions. This pizza was found to be packaged with susceptors containing BADGE at between 700 and 800 mg/kg (1.8 to 2.0 mg/dm^2).[3] Migration of oligomers of BADGE was investigated in canned foods, such as sea foods in oil, meat products, and soups. A major component released from bisphenol-A resins was identified by the LC analysis as the cyclo-(bisphenol-A monoglycidyl ether) dimer and was commonly present in foods at concentrations of around 1.0 mg/kg.[4]

BADGE was found at levels exceeding 1.0 mg/kg in seven of the 15 canned anchovy samples and five of the 22 sardine samples. Analysis for BADGE was conducted by HPLC with fluorescence detection with confirmation of BADGE identity by GC/MS analysis using selected ion monitoring. The results clearly showed that BADGE concentrations in the oil were about 20 times higher than in the drained fish. BADGE levels in the samples of canned sardines and anchovies determined one year later were found to be below 1.0 mg/kg. In the other retail canned foods, BADGE was not detectable (DL = 0.02 mg/kg) or detected at concentrations well below the tempcrary SML of 1 mg/kg.[5]

Acute Toxicity. LD_{50} is 19.6 ml/kg BW for rats.[6]

Repeated exposure. Addition of 5.0% BADGE in the diet caused death of rats in 2 weeks.[7]

Long-term Toxicity. Rats received 0.2% BADGE in their diet for 26 weeks. The treatment caused no signs of intoxication other than increase in kidney/body weight ratio. Addition of 1.0% BADGE in the diet produced retardation of BW gain and no other obvious effects.[7]

Allergenic Effects.

Observations in man. BADGE has been shown to cause allergenic contact dermatitis in humans. The workers exposed to epoxy resins in different factories all gave positive reactions in a patch test.[8]

Animal studies. Skin sensitization in experimental animals was established. BADGE sensitized all animals in a guinea pigs skin maximization test and was classified as an extreme allergen.[9]

Reproductive Toxicity.

Embryotoxicity. BADGE induced prenatal toxicity in New Zealand white rabbits following dermal exposure at up to 300 mg/kg BW. Doses of 85 or 125 mg/kg BW were injected *i/p* to rats on days 0 through 14 of gestation. At the higher dose very few rats became pregnant. At the lower dose there was a significant increase in dilated cerebral ventricles and growth retardation.[10] Rats and mice were gavaged with up to 1250 and 640 mg/kg BW, respectively, during organogenesis. No effects were found in the rat. In the mouse, maternal BW was reduced and resorptions were increased.[11]

Teratogenic effect was not observed.[12]

Mutagenicity.

Observations in man. CA were found in the peripheral lymphocytes of workers exposed to BADGE.

In vitro genotoxicity. BADGE was found to be negative in *Salmonella* mutagenicity assay.[13]

Carcinogenicity. BADGE was tested only by skin applications in mice. It caused an increased incidence of epidermal tumors.[14]

Carcinogenicity classification. An IARC Working Group concluded that there is limited evidence for the carcinogenicity of BADGE in *experimental animals* and there were no data available to evaluate the carcinogenicity of BADGE in *humans*.

IARC: 3

Chemobiokinetics. BADGE is rapidly metabolized in mice via the epoxide groups to form the corresponding *bisdiol*. Metabolites including conjugates are excreted in feces and urine.[15]

Regulations.

U.S. FDA (1998) permits the use of BADGE epoxy resins as components of coatings that may come into contact with food in accordance with the conditions prescribed in 21 CFR part 177.1650.

Great Britain (1998). BADGE is authorized without time limit for use in the production of polymeric materials and articles in contact with food or drink or intended for such contact. The quantity of BADGE in the finished plastic material or article shall not exceed 1.0 mg/kg or the specific migration of this substance shall not be detectable (when measured by a method with DL of 0.02 mg/kg, analytical tolerance included).

Recommendations. *EC Scientific Committee for Food* (1996) temporarily increased the SML applying to BADGE to 1.0 mg/kg.

References:

1. Begley, T. H., Biles, J. E., and Hollifield, H. C., Migration of an epoxy adhesive compound into a food-simulating liquid and food from microwave susceptor packaging, *J. Agric. Food Chem.*, 39, 1944, 1991.
2. Simal-Gandara, J., Lopez-Mahia, P., Paseiro-Losada, P., and Simal-Lozano, J., Overall migration and specific migration of bisphenol A diglycidyl ether monomer and *m*-xylylenediamine hardener from an optimized epoxy-amine formulation into water-based food simulants, *Food Addit. Contam.*, 10, 555, 1993.
3. Sharman, M., Honeybone, C. A., Jickells, S. M., and Castle, L., Detection of residues of the epoxy adhesive component bisphenol A diglycidyl ether (BADGE) in microwave susceptors and its migration into food, *Food Addit. Contam.*, 12, 779, 1995.
4. Biedermann, M. and Grob, K., Food contamination from epoxy resins and organosols used as can coatings: analysis by gradient NPLC, *Food Addit. Contam.*, 15, 609, 1998.
5. Summerfield, W., Goodson, A., and Cooper, I., Survey of bisphenol A diglycidyl ether (BADGE) in canned foods, *Food Addit. Contam.*, 15, 818, 1998.
6. Weil, C. S., Condra, N., Haun, C., and Striegel, J. A., Experimental carcinogenicity of representative epoxides, *Am. Ind. Hyg. Ass. J.*, 24, 305, 1963.
7. Hine, C. H., Kodama, J. K., Anderson, H. H., Simonson, D. W., and Wellington, J. S., The toxicity of epoxy resins, *Arch. Ind. Health*, 17, 129, 1958.
8. Fregert, S. and Thorgeirsson, A., Patch testing with low molecular oligomers of epoxy resins in humans, *Contact Dermat.*, 3, 301, 1977.
9. Zakova, N., Froehlich, E., and Hess, R., Evaluation of skin carcinogenicity of technical 2,2-bis (*p*-glycidyloxyphenyl) propane in CF1 mice, *Food Chem. Toxicol.*, 23, 1081, 1985.
10. Hardin, B. D., Bond, G. P., Sikov, M. R., Andrew, F. D., Beliles, R. P., and Niemeier, R. W., Testing of selected workplace chemicals for teratogenic potential, *Scand. J. Work Environ. Health*, 7, 66, 1981.
11. Morrissey, R. E., George, J. D., Price, C. J., Tyl, R. W., Marr, M.C., and Kimmel, C. A., The developmental toxicity of bisphenol A in rats and mice, *Fundam. Appl. Toxicol.*, 8, 571, 1987.
12. Breslin, W. J., Kirk, H. D., and Johnson, K. A., Teratogenic evaluation of diglycidyl ether of bisphenol A (DGEBPA) in New Zealand white rabbits following dermal exposure, *Fundam. Appl. Toxicol.*, 10, 736, 1988.
13. Wade, M. J., Moyer, J. W., and Hine, C. H., Mutagenic action of a series of epoxides, *Mutat. Res.*, 66, 367, 1979.
14. Holland, J. M., Gosslee, D. G., and Williams, H. J., Epidermal carcinogenicity of bis(2,3-epoxycyclopentyl) ether, 2,2-bis(*p*-glycidyloxyphenyl)propane and *m*-phenylenediamine in male and female C3H and C57BL/6 mice, *Cancer Res.*, 39, 1718, 1979.

15. Climie, I. J. G., Hutson, D. H., and Stydin, G., Metabolism of the epoxy resin component 2,2-bis (*p*-glycidyloxyphenyl)propane, the diglycidyl ether of bisphenol A in the mouse, *Xenobiotica*, 11, 401, 1981.

BROMOETHYLENE

Molecular Formula. C_2H_3Br
M = 106.96
CAS No 593-60-2
RTECS No KU8400000
Abbreviation. BE.

Synonyms. Bromethene; Vinyl bromide.

Properties. Gas under normal atmospheric conditions, a colorless liquid under pressure. BE has a characteristic odor. Poorly soluble in water, soluble in ethanol, acetone, and chloroform.

Applications. An intermediate in the manufacture of polymers and copolymers.

Acute Toxicity. LD_{50} is approximately 500 mg/kg BW (50% solution in corn oil) in male rats.[1]

Mutagenicity.

In vitro genotoxicity. BE was found to be positive in *Salmonella* mutagenicity assay (NTP-92; Lijinsky and Andrews, 1980).

In vivo cytogenetics. BE did not induce micronucleated polychromatic erythrocytes in male CD-1 mice after *i/p* injection or oral gavage of up to 2000 mg/kg BW.[2] It was positive in *Dr. melanogaster*.[3,4]

Carcinogenicity. Does not produce tumors in mice in response to skin application.[5]

Carcinogenicity classifications. An IARC Working Group concluded that there is sufficient evidence for the carcinogenicity of BE in *experimental animals* and there were no adequate data available to evaluate the carcinogenicity of BE in *humans*.

IARC: 2A;
U.S. EPA: B2.

Chemobiokinetics. BE is metabolized since bromide ion is found in the blood of exposed animals. Metabolism probably proceeds through epoxidation, with subsequent conjugation to macromolecules and other biological compounds.[03]

References:

1. Leong, B. K. J. and Torkelson, T. R., Effects of repeated inhalation of vinyl bromide in laboratory animals with recommendations for industrial handling, *Am. Ind. Hyg. Assoc.*, 31, 1, 1970.
2. Morita, T., Asano, N., Awogi, T., Sasaki, Yu. F., et al. Evaluation of the rodent micronucleus assay in the screening of IARC carcinogens (Groups 1, 2A, and 2B). The summary report of the 6th collaborative study by CSGMT/JEMS-MMS, *Mutat. Res.*, 389, 3, 1997.
3. Bartsch, H., Malaveille, C., Barbin, A., et al., Alkylating and mutagenic metabolites of halogenated olefins produced by human and animal tissues, Abstract No 67, *Proc. Am. Ass. Cancer Res.*, 17, 17, 1976.
4. Vogel, E. W. and Nivard, N. J., Performance of 181 chemicals in a *Drosophila* assay predominantly monitoring interchromosomal mitotic recombination, *Mutagenesis*, 8, 57, 1993.
5. van Duuren, B. L., Chemical structure, reactivity, and carcinogenicity of halohydrocarbons, *Environ. Health Perspect*., 21, 17, 1977.

3-BROMOPROPENE

Molecular Formula. C_3H_5Br
M = 120.99
CAS No 106-95-6
RTECS No UC7090000
Abbreviation. BP.

Synonym. Allyl bromide; 1-Bromo-2-propene; 3-Bromopropylene; Bromallylene.

Properties. Colorless liquid with a pungent specific odor. Poorly soluble in water. Miscible with alcohol and ether.

Application. Used in the manufacture of plastics and dyestuff.

Acute Toxicity. LD_{50} is reported to be 270 mg/kg BW for rats, 190 mg/kg BW for mice, and 30 mg/kg BW for guinea pigs. Poisoning is accompanied by a transient motor agitation followed by depression and loss of balance. Death at the narcosis level within 24 hours.[1]

Mutagenicity.

In vitro genotoxicity. BP was positive in *Salmonella* mutagenicity assay.[2]

Chemobiokinetics. *S-Allyl-L-cystein* is found in the urine of rats treated with BP.[3] Other metabolites identified in rats dosed *s/c*, were *allylmercapturic acid, S-allylcystein S-oxide,* and, probably, 3-*hydroxy-propylmercapturic acid.*[4]

References:

1. Shugaev, B. B. and Buckhalovsky, A. A., Toxicity of allyl bromide, allyl iodide, resorcin dinitro-diphenyl ether, methyl resorcin dibenzoate, *Gig. Sanit.*, 6, 89, 1985 (in Russian).
2. Schiffmann, D. et al., Induction of unscheduled DNA synthesis in HeLa cells by allylic compounds, *Cancer Lett.*, 20, 263, 1983.
3. Kaye, C. M. et al., Metabolic formation of mercapturic acids from allyl halides, *Xenobiotica,* 2, 129, 1972.
4. *Foreign Compounds Metabolism in Mammals*, The Chemical Society, London, vol 3, 1975, 433.

1,3-BUTADIENE

Molecular Formula. C_4H_6
M = 54.09
CAS No 106-99-0
RTECS No EI9275000

Synonyms and **Trade Names.** Biethylene; Bivinyl; Divinyl; Erythrene; Pyrrolylene; Vinyl ethylene.

Properties. A colorless gas with a garlic and horseradish odor. Solubility in water is 1.3 g/l at 5°C or 735 mg/lg at 20°C, soluble in alcohol. Odor perception threshold is reported to be 0.45 mg/l[010] or 0.0014 mg/l.[02]

Applications. A chemical intermediate and a polymer component in the manufacture of synthethic rubbers and plastics intended for use in contact with food or drink. Used as comonomer for ABS resins and styrene-butadiene latexes, neoprene elastomers, and other polymers or copolymers. Monomer in the prodution of a wide range of polymers and copolymers: styrene-butadiene, polybutadiene, butadiene-acrylonitrile, acrylonitrile-butadiene-styrene.

Exposure. Due to its volatile nature, uptake of 1,3-B. occurs almost exclusively by inhalation and absorption through the respiratory system.

Acute Toxicity. LD_{50} is 5.48 g/kg BW in rats, and 3.2 g/kg BW in mice.[1,2]

Short-term Toxicity. Rats were administered the dose of 100 mg/kg BW for 3 months. Manifestations of toxic action included adynamia, decreased BW gain, and depressed erythropoiesis and activity of acetylcholinesterase and cholinesterase. Inhalation of 1250 ppm over a period of 6 to 24 weeks led to changes in the spleen and thymus.[2] 1,3-B. was found to affect the CNS.

Immunotoxicity. Inhalation exposure to 1,3-B. resulted in the transient changes in the immune system function.[2]

Long-term Toxicity. In chronic inhalation carcinogenicity study in $B6C3F_1$ mice and Sprague-Dawley rats, mice appeared to be considerably more sensitive than rats.[3]

Reproductive Toxicity. Inhalation exposure to 1,3-B. caused no reduction in fertility of rats, guinea pigs, and rabbits.1,3-B. might produce tumors and atrophy of the gonads of mice.[4] Effect on reproduction was observed in Swiss mice but not in Sprague-Dawley rats.[5,6]

Mutagenicity. 1,3-B. appeared to be both a somatic and germ cell mutagen in mammals, possibly including humans.[7,8] It is metabolized in the body into two major genotoxic metabolites *3,4-epoxybutene* and *1,2:3,4-diepoxybutane*.

Observations in man. No genotoxic effects were found in cytogenetic studies in peripheral blood lymphocytes of exposed workers, although one report described elevated *hprt* variant levels.[9]

In vivo cytogenetic effects of 1,3-B. were observed in bone marrow cells of B6C3F_1 mice (but not of rats) exposed to it by inhalation. The exposure induced significant increase in SCE, micronuclei, and CA.[10-12] 1,3-B. did not induce unscheduled DNA synthesis in rats or mice hepatocytes following *in vivo* exposure.[13] It caused sperm head abnormalities in mice, but not in rats, produced micronuclei and SCE.[8,12]

It was positive in DLM study in rodents suggesting 1,3-B. heritable risk.

1,3-B. did not induce sex-linked recessive mutations in *Dr. melanogaster.*[12]

In vitro genotoxicity. 1,3-B. was positive in *Salmonella typhimurium* mutagenicity assay with strain *TA1535*; mutagenic activity was not observed in *Salmonella typhimurium* strains *TA100, TA97* and *TA98.*[10] 1,3-B. is assumed to be mutagenic for human embryo cells.[4] It was negative in SCE studies in human whole blood lymphocytes cultures treated in the the presence of rat, mouse, or human liver *S9* metabolic activation.[12,13]

Carcinogenicity.

Observations in man. Workers displayed increased risk for leukemia (IARC 54-273). The study of a 1,3-B.-exposed cohort consisting of 364 workers from Union Carbide butadiene production units in West Virginia revealed an increased risk of death from lymphosarcoma and reticulosarcoma (Ward et al., 1995).

Animal studies. Inhalation exposure induced a dose-related increase in the incidence of tumor development in hematopoietic system, heart, lung, forestomach, liver, Harderian gland, preputial gland, brain, female gonads, and kidney.[1,12,14] As a result of inhalation exposure of mice to 625 and 1250 ppm 1,3-B., an increased incidence of myocardial hemoangiosarcomas, malignant lymphomas, tumors of the lungs and stomach was reported.[15]

Mice are about 1000 times more sensitive than rats. The greater sensitivity of mice, compared to rats, to the carcinogenicity of 1,3-B. is linked to higher rates of its metabolism to butadiene diepoxide by mice than rats. An increased rate of thyroid, pancreas and mammary gland tumors was found in rats.[16]

Carcinogenicity of 1,3-B. is likely to be the result of its biotransformation to epoxide (see ***Chemobiokinetics***).

Carcinogenicity classifications. An IARC Working Group concluded that there are inadequate data available to evaluate the carcinogenicity of 1,3-B. in *humans* and there is sufficient evidence for the carcinogenicity of 1,3-B. in *experimental animals*.

The Annual Report of Carcinogens issued by the U.S. Department of Health and Human Services (1998) defines 1,3-B. to be a substance which may reasonably be anticipated to be carcinogen, i.e., a substance for which there is a limited evidence of carcinogenicity in *humans* and sufficient evidence of carcinogenicity in *experimental animals*.

IARC: 2A;

U.S. EPA: B2;

ARC: 1

EU: 2;

NTP: XX - XX -CE* - CE* (inhalation).

Chemobiokinetics. In rats, mice and monkeys, 1,3-B. undergoes transformation into epoxides. It is metabolized by liver microsomes to *butadiene oxide*, with subsequent conversion into *3-butane-1,2-diol* by microsome epoxyhydrolases. Four other metabolites were found as a result of incubation of butadiene oxide with NADP and microsomes.[17]

According to Maniglier-Poulet et al., 1,3-B. is metabolized to *butadiene monooxide* by myeloperoxidase and by mouse and human bone marrow cells. Such a metabolism is stimulated by hydrogen peroxide suggesting a peroxide-mediated process. Cytochrome *P-450* is not involved in butadiene monooxide formation.[18] Human and rat liver tissues produce much less of 1,3-B. monoepoxide than do those of B6C3F_1 mice. The human liver tissue rapidly detoxifies the metabolite, while the mouse liver tissue very slowly degrades 1,3-B. monoepoxide to non-toxic products.[19]

Regulations.

U.S. FDA (1998) regulates 1,3-B. as an indirect food additive. Polymers or copolymers of *1,3-Butadiene* are allowed to be used (1) in adhesives as components of articles coming in contact with food in accordance with the conditions prescribed in 21 CFR part 175.105; (2) in pressure-sensitive adhesives in accordance with the conditions prescribed in 21 CFR part 175.125; (3) in resinous or polymeric coatings and rubber articles intended for repeated use in accordance with the conditions prescribed in 21 CFR part 175.300; (4) in the manufacture of cellophane for packaging food in accordance with the conditions prescribed in 21 CFR part 177.1200; (5) in the manufacture of rubber articles intended for repeated use in producing, manufacturing, packaging, processing, preparing, treating, packing, transporting, or holding food in accordance with the conditions prescribed in 21 CFR part 177.2600; (6) as a coating or component of coating, limited to a level not to exceed 1.0% by weight of paper and paperboard; (7) in semirigid and rigid acrylic polymers in repeated use articles in accordance with the conditions prescribed in 21 CFR part 177.1010; (8) in acrylontrile-styrene-butadiene copolymers used in closures with sealing gaskets that may be safely used on containers in accordance with the conditions prescribed in 21 CFR part; and (9) in textiles and textile fibers intended for use in contact with food in accordance with the conditions prescribed in 21 CFR part 177.2800.

Great Britain (1998). 1,3-B. is authorized without time limit for use in the production of polymeric materials and articles in contact with food or drink or intended for such contact. The quantity of 1,3-B. in the finished polymeric material or article shall not exceed 1 mg/kg or the specific migration of this substance shall not be detectable (when measured by a method with DL of 0.02 mg/kg, analytical tolerance included).

Standards.

EU (1990). Maximum Permitted Quantity in materials and articles: 1.0 mg/kg; SML: not detectable (detection limit is 0.02 mg/kg).

Russia (1995). MAC: 0.05 mg/l (organolept., odor).

References:

1. *Toxicology and Carcinogenesis Studies of 1,3-Butadiene in B6C3F$_1$ Mice (Inhalation Studies)*, NTP Technical Report Series No 288, Research Triangle Park, NC, 1984.
2. Thurmond, L. M., Lauer, L. D., House, R. V., et al., Effect of short-term inhalation exposure to 1,3-butadiene on murine immune functions, *Toxicol. Appl. Pharmacol.*, 86, 170, 1986.
3. Himmelstein, M. W., Acouavella, J. F., Recio, L., Medinsky, M. A., and Bond, J. A., Toxicology and epidemiology of 1,3-butadiene, *Crit. Rev. Toxicol.*, 27, 1, 1977.
4. Rosenthal, S. L., The reproductive effects assessment group's report on the mutagenicity of 1,3-butadiene and its reactive metabolites, *Environ. Mutagen.*, 7, 933, 1985.
5. Hackett, P. L., Sikov, M. R., Mast, T. J., et al., *Inhalation Developmental Toxicicology Studies of 1,3-Butadiene in The Rat*, Final Report No NIH-401-ES-40131, Pacific Northwest Laboratory, Richland, WA, 1987.
6. Morrissey, R. E., Schwetz, B. A., Hackett, P. L., et al., Overview of reproductive and developmental toxiсity studies of 1,3-butadiene in rodents, *Environ. Health Perspect.*, 86, 79, 1990.
7. Jacobson-Kram, D. and Rosenthal, S. L., Molecular and genetic toxicology of 1,3-butadiene, *Mutat. Res.*, 339,121, 1995.
8. de Meester, C., Genotoxic properties of 1,3-butadiene, *Mutat. Res.*, 195, 273, 1988.
9. Adler, I. D., Cochrane, J., Osterman-Golkar, S., Skopek, T. R., Sorsa, M., and Vogel, E., 1,3-Butadiene working group report, *Mutat. Res.*, 330, 101, 1995.
10. Recio, L., Pluta, L., Bond, J., and Sisk, S., *Environ. Molec. Mutagen.*, 23 (Suppl. 23), 56, 1994.
11. Shelby, M. D., Results of NTP-sponsored mouse cytogenetic studies on 1,3-butadiene, isoprene, and chloroprene, *Environ. Health Perspect.*, 86, 71, 1990.
12. *Toxicology and Carcinogenesis Studies of 1,3-Butadiene in B6C3F$_1$ Mice (Inhalation Studies)*, NTP Technical Report Series No 434, NTIS Accession No PB94-101631, Research Triangle Park, NC, May 1993.
13. Arce, G. T., Vincent, D. R., Cunningham, M. J., Choy, W. N., and Sarrif, A. M., *In vitro* and *in vivo* genotoxicity of 1,3-butadiene and metabolites, *Environ. Health Perspect.*, 86, 75, 1990.
14. Melnick, R. L., Huff, J., Chou, B. J., et al., Carcinogenicity of 1,3-butadiene in C57Bl/6 x C3HF$_1$ mice at low exposure concentrations, *Cancer Res.*, 50, 6592, 1990.
15. Huff, J. E., Melnik, R. L., Holleveld, H. R., et al., Multiple organ carcinogenicity of 1,3-butadiene in B6C3F$_1$ mice after 60 weeks of inhalation exposure, *Science*, 227, 548, 1985.

16. Owen, P. E., Pullinger, D. H., Glaister, J. R., et al., 1,3-Butadiene: Two-year inhalation toxicity/carcinogenicity study in the rat (Abstract No P34), in *The 26th Congr. Eur. Soc. Toxicol.*, H. Hanhijarvi, Ed., 16-19 June, 1985, University of Kuopio, Kuopio, 1985, 69.
17. Malvoisin, E. and Roberfroid, M., Hepatic microsomal metabolism of 1,3-butadiene, *Xenobiotica*, 12, 137, 1982.
18. Maniglier-Poulet, C., Cheng, X., Ruth, J. A., and Ross, D., Metabolism of 1,3-butadiene to butadiene monooxide in mouse and human bone marrow cells, *Chem.-Biol. Interact.*, 97, 119, 1995.
19. Csanady, G. A., Guengerich, F. P., and Bond, J. A., Comparison of the biotransformation of 1,3-butadiene and its metabolite, butadiene monoepoxide, by hepatic and pulmonary issues from humans, rats and mice, *Carcinogenesis*, 13, 143, 1992.

BUTADIENE DIEPOXIDE

Molecular Formula. $C_4H_6O_2$
M = 86.10
CAS No 1464-53-5
RTECS No EJ8225000

Synonym and **Trade Names.** Bioxirane; Butadiene dioxide; Butane epoxide; Diepoxide; Dioxybutadiene; Erythritol anhydride.

Applications. Used in the manufacture of materials intended for use in contact with food or drink.

Acute Toxicity. LD_{50} is 78 mg/kg BW in rats.[1]

Mutagenicity.

In vitro genotoxicity. BDE did not induced unscheduled DNA synthesis in rat and mouse hepatocytes *in vitro*.[2] It was mutagenic in *Salmonella typhimurium TA100* strain.[3]

Carcinogenicity. The greater sensitivity of mice, compared to rats, to the carcinogenicity of 1,3-butadiene is linked to higher rates of its metabolism to BDE by mice than rats.

Carcinogenicity classifications. An IARC Working Group concluded that there is sufficient evidence of carcinogenicity of BDE in *experimental animals* and there are no data available to evaluate the carcinogenicity BDE in *humans*.

IARC: 2B
U.S. EPA: B2

Chemobiokinetics. BDE is a metabolite of 1,3-butadiene.

References:

1. Smyth, H. F., Carpenter, C. P., Weil, C. S., et al., Range-finding toxicity data, List V, *Arch. Ind. Hyg. Occup. Med.*, 10, 61, 1954.
2. Arce, G. T., Vincent, D. R., Cunningham, M. J., Choy, W. N., and Sarrif, A. M., *In vitro* and *in vivo* genotoxicity of 1,3-butadiene and metabolites, *Environ. Health Perspect.*, 86, 75, 1990.
3. Wade, M. J., Moyer, J. W., and Hine, C. H., Mutagenic action of a series of epoxides, *Mutat. Res.*, 66, 367, 1979.

1,2-BUTANEDIOL

Molecular Formula. $C_4H_{10}O_2$
M = 90.14
CAS No 584-03-2
RTECS No EK0380000
Abbreviation. 1,2-BD.

Synonyms. 1,2-Butylene glycol; 1,2-Dihydroxybutane.

Properties. Clear viscous liquid.

Applications. Used in the production of polyester resins.

Acute Toxicity. LD_{50} is found to be about 16 g/kg BW in rats. Ingestion of large doses caused narcosis and irritation of the GI tract. Gross pathology examination revealed marked congestion in the kidneys.[011]

Repeated Exposure. Young female rats tolerated a basic diet containing up to 30% of calories being replaced by 1,2-BD. 40% replacement caused death of animals in 11 to 29 days.[011]

1,3-BUTANEDIOL

Molecular Formula. $C_4H_{10}O_2$
M = 90.14
CAS No 107-88-0
RTECS No EK0440000

Synonyms. Butane-1,3-diol; 1,3-Butylene glycol; β-Butylene glycol; 1,3-Dihydroxybutane; Methyl-3-methyleneglycol.

Properties. Colorless, viscous, hygroscopic liquid. Soluble in water, acetone, and ethanol.

Applications. Used in the production of polyester plasticizers; a humictant for cellophane intended for use in contact with food or drink.

Acute Toxicity. The LD_{50} values are reported to be 18.6 to 22.8 g/kg BW for rats, 13 g/kg BW for mice, and 11 g/kg BW for guinea pigs.[1]

Repeated Exposure. Exposure to high dietary levels (about 200 mg/kg BW) produced ketosis in man and rats.[1] No signs of nephrotoxicity or CNS depression were observed on 1,3-B. administration.[2]

Long-term Toxicity. In a 2-year study, Sprague-Dawley rats received 1.0, 3.0, or 10% in feed, dogs were given 0.5, 1.0, or 3.0% in feed. There were no gross or microscopic lesions found in the treated animals attributive to chemical exposure.[5]

In a 2-year feeding study in dogs in which they were exposed to the levels of up to 30 g/kg BW, no adverse effects were reported.[1]

Reproductive Toxicity.

In a 5-generation study, Wistar rats were fed a diet containing 5.0, 10, or 24.% 1,3-B. ***Embryolethal effect*** was noted: no pups were obtained in the high-dose level group of the 5th series of litters. Slight ***teratogenic action*** was observed.[3]

Mutagenicity.

In vivo cytogenetics. Mutagenic potential was not demonstrated in the DLM assay.[3]

Carcinogenicity. No tumors were found in the described studies.[3]

Chemobiokinetics. Following ingestion, 1,3-B. is destroyed in the body of rabbits. Small amounts of *dicarboxilic acid* are discovered in the urine.[03] 1,3-B. binds to alcohol dehydrohenase more efficiently than ethylene glycol and is orally less toxic than ethylene glycol and ethanol.[2]

Studies in perfused livers from meal-fed and starved rats showed that most of the metabolism of R-1,3-B. is accounted for by conversion to the physiological *ketone bodies R-3-hydroxybutyrate* and *acetoacetate*. Only 29 to 38% of S-1,3-B. uptake is accounted for by conversion to the *physiological ketone bodies S-3-hydroxybutyrate, lipids* and CO_2.[4]

Regulations.

EU (1990). 1,3-B. is available in the *List of authorized monomers and other starting substances which shall be used for the manufacture of plastic materials and articles intended to come into contact with foodstuffs (Section A).*

U.S. FDA (1998) regulates the use of 1,3-B. as a secondary direct food additive permitted in food for human consumption. It may also be used (1) as a component of adhesives for food-contact surface in accordance with the conditions prescribed in 21 CFR part 175.105; (2) in the manufacture of cellophane for packaging food in accordance with the conditions prescribed in 21 CFR part 177.1200; (3) in the manufacture of closures with sealing gaskets for food containers in accordance with the conditions prescribed in 21 CFR part 177.1210; (4) in the manufacture of cross-linked polyester resins for repeated use in contact with food in accordance with the conditions prescribed in 21 CFR part 177.2420; (5) in the manufacture of polyurethane resins for contact with dry food in accordance with the conditions prescribed in 21 CFR part 177.1680; (6) in the manufacture of resinous and polymeric coatings for polyolefin films to be safely used as a food-contact surface in accordance with the conditions prescribed 21 CFR part 175.320; (7) as a stabilizer for polymers in accordance with the conditions prescribed in 21 CFR part 178.2010; (8) in the manufacture of rubber articles intended for repeated use in producing, manufacturing, packaging, processing, preparing, treating, packing, transporting, or holding food in accordance with the conditions prescribed in 21 CFR part 177.2600; and (9) as a defoaming agent that may be safely used in the manufacture of paper and paperboard intended for use in producing, manufacturing, packing, transporting, or holding food in accordance with the conditions prescribed in 21 CFR part 176.170. 1,3-B. may be used in food in accordance with the conditions prescribed in 21 CFR part 173.220.

Great Britain (1998). 1,3-B. is authorized without time limit for use in the production of polymeric materials and articles in contact with food or drink or intended for such contact.

Recommendations. Joint FAO/WHO Expert Committee on Food Additives (1980). ADI: 4.0 mg/kg BW.

References:

1. The 23rd Report of the Joint FAO/WHO Expert Committee on Food Additives, Technical Report Series No 648, WHO, Geneva, 1980, 15.
2. Cox, S. K., Ferslew, K. E., and Boelen, L. J., The toxicokinetics of 1,3-butylene glycol versus ethanol in the treatment of ethylene glycol poisoning, *Vet. Hum. Toxicol.*, 34, 36, 1992.
3. Hess, F. G., Cox, G. E., Bailey, D. E., Parent, R. A., and Becci, P. J., Reproduction and teratology study of 1,3-butanediol in rats, *J. Appl. Toxicol.*, 1, 202, 1981.
4. Desrochers, S., David, F., Garneau, M., Jette, M. N., and Brunengraber, H., Metabolism of *R*- and *S*-1,3-butanediol in perfused livers from meal-fed and starved rats, *Biochem. J.*, 285, 647, 1992.
5. Scala, R. A. and Paynter, O. E., Chronic oral toxicity of 1,3-butanediol, *Toxicol. Appl. Pharmacol.*, 10, 160, 1967.

1,4-BUTANEDIOL

Molecular Formula. $C_4H_{10}O_2$

M = 90.2

CAS No 110-63-4

RTECS No EK0525000

Synonyms. 1,4-Butylene glycol; 1,4-Dihydroxybutane; Methyl-3-methylene glycol; Tetramethylene glycol.

Properties. Glycerine-like, sweetish liquid with a faint odor. Readily miscible with water and 95% ethanol. Odor perception threshold is 15.0 mg/l.[1]

Applications. Used as a monomer in the production of polyurethanes; an intermediate product of the synthesis of the blood substitute *poly-N-vinyl-2-pyrrolidone*.

Acute Toxicity. LD_{50} is reported to be 2060 mg/kg BW for mice, 1525 mg/kg BW for rats, 1200 mg/kg BW for guinea pigs, and 2530 mg/kg BW for rabbits.[1] All rats and mice died within 1 to 2 days after exposure to the doses of 500 and 1000 mg/kg BW, respectively.[2] Poisoning is accompanied by a transient agitation with subsequent depression and respiratory disorders. Gross pathology examination revealed hemorrhages and congestion of the visceral organs, dystrophic changes in the renal tubular epithelium.

Repeated Exposure failed to reveal evidence of accumulation in rats and guinea pigs. Wistar rats were dosed with 5.0, 50, or 500 mg 1,4-B./kg BW by oral gavage for 28 consecutive days. The treatment caused an overall low degree of systemic toxicity. At the highest dose level, there was a slight increase in activity of sorbitol dehydrogenase and ALT in males. Altered hematology analysis, mild to moderate inflammation of the liver, characterized by proliferation of bile ducts and periportal infiltrations with fibroblasts and mononuclear cells were observed in treated animals. Histological changes were seen only at the highest dose.[3]

Long-term Toxicity. Mice were dosed by gavage with 30.3 mg/kg BW for 6 months. The treatment caused a reduction in the blood serum cholinesterase activity and dysbalance in serum proteins. Decrease in the blood *SH*-group content was also noted. There was a reduction in the activity and content of cholinesterase in the organs, and increased activity of transaminase of the blood serum.[1]

Reproductive Toxicity. Doses of 1.0 to 600 mg/kg BW were administered by gavage in water to timed pregnant Swiss albino mice on gestation days 6 through 15. There was no maternal mortality, but CNS intoxication, and lowered body and liver weights were observed after dosing in the 300 and 600 mg/kg BW groups.[4]

Embryotoxicity. Significant reduction in live fetal weight occurred at 300 and 600 mg/kg BW.

Teratogenicity. No external or visceral chemical-dependent malformations, but an increasing trend in skeletal malformations (missing or branched ribs and fused thoracic vertebrae), were observed primarily in the 600 mg/kg BW group.[4]

Allergenic Effect. No allergic contact dermatitis was developed in guinea pigs on skin applications of 1,4-B.[5]

Carcinogenicity. 1,4-B. is unlikely to be carcinogenic in animals (NTP).

Chemobiokinetics. In the NTP studies, 1,4-B. was found to be metabolized rapidly to *β-hydroxybutyric acid*, which is also the end metabolite of *β-butyrolactone*. According to Testa and Jenner, 1,4-B. is metabolized by hepatic NAD-dependent alcohol dehydrogenase to *aldehyde* or *ketone*.[6] Because of rapid and extensive conversion of both compounds to *β-hydroxybutyric acid*, the toxicological profile of 1,4-B. reflects that of *β-butyrolactone* with lack of toxic or carcinogenic potential being demonstrated. *β-Hydroxybutyric* acid readily crossed the blood-brain barrier and produced neuropharmacological effect.[7]

Regulations.

EU (1990). 1,4-B. is available in the *List of monomers and other starting substances which may continue to be used for the manufacture of plastic materials and articles intended to come into contact with foodstuffs pending a decision on including in Section A.*

U.S. FDA (1998) regulates the use of 1,4-B. (1) in adhesives as a component of articles intended for use in packaging, transporting, or holding food in accordance with the conditions prescribed in 21 CFR part 175.105; (2) as an ingredient of polyurethane resins for contact with dry food of the type identified in 21 CFR part 176.170 (c); (3) as a defoaming agent that may be safely used in the manufacture of paper and paperboard intended for use in producing, manufacturing, packing, transporting, or holding food, subject to the provisions prescribed in 21 CFR part 176.200; and (4) in cross-linked polyester resins to be safely used as articles or components of articles intended for repeated use in contact with food, subject to the provisions prescribed in 21 CFR part 177.2420.

Great Britain (1998). 1,4-B. is authorized up to the end of 2001 for use in the production of polymeric materials and articles in contact with food or drink or intended for such contact.

Standards. *Russia* (1995). MAC and PML: 5.0 mg/l.

References:

1. Knyshova, S. P., Biological effect and hygienic standardization of 1,4-butynediol and 1,4-butanediol, *Gig. Sanit.*, 1, 37, 1968 (in Russian).
2. Stasenkova, K. P., in *Toxicology of New Industrial Chemical Substances*, A. A. Letavet, Ed., Meditsina, Moscow, Issue No 7, 1965, 5 (in Russian).
3. Jedrychowski, R. A, Gorny, R., Stetkiewicz, J., and Stetkiewicz, I., Subacute oral toxicity of 1,4-butanediol in rats, *Pol. J. Occup. Med.*, 3, 421, 1990.
4. Price, C. J., Marr, M. C., Myers, C. B., Heindel, J. J., and Schwetz, B. A., Developmental toxicity evaluation of 1,4-butanediol (BUTE) in Swiss mice, *Teratology*, 47, 433, 1993.
5. Jedrychowski, R. A., Stetkiewicz, J., and Stetkiewicz, I., Acute toxicity of 1,4-butanediol in laboratory animals, *Pol. J. Occup. Med.*, 3, 415, 1990.
6. Testa, B. and Jenner, P., *Drug Metabolism: Chemical and Biochemical Aspects*, Marcel Dekker, Inc., New York, 1976, 310.
7. *Metabolism, Disposition, and Toxicity of 1,4-Butanediol*, NTP Summary Report Series No 54, Research Triangle Park, NC, NIH Publ. 96-3932, 1996.

N-tert-BUTYLACRYLAMIDE

Molecular Formula. $C_7H_{13}NO$

M = 127.21

CAS No 107-58-4

RTECS No AS3460000

Abbreviation. BAA.

Synonym. *N*-(1,1-Dimethylethyl)-2-propenamide.

Applications. A monomer in the synthesis of polymers intended for use in contact with food or drink.

Acute Toxicity. LD_{50} is 940 mg/kg BW in mice.[1]

Chemobiokinetics. BAA is metabolized by microsomal enzymes with a NADFH-generating system and by hepatic glutathione *S*-transferases as well.[2]

Regulations. ***U.S. FDA*** (1998) regulates the use of BAA as a monomer in adhesives as a component of articles intended for use in packaging, transporting, or holding food in accordance with the conditions prescribed in 21 CFR part 175.105.

References:

1. Hashimoto, K., Sakamoto, Y., and Tanii, H., Neurotoxicity of acrylamide and related compounds and their effects on male gonads in mice, *Arch. Toxicol.*, 47, 179, 1981.
2. Tanii, H. and Hashimoto, K., Studies on *in vitro* metabolism of acrylamide and related compounds, *Arch. Toxicol.*, 48, 157, 1981.

BUTYLENE

Molecular Formula. C_4H_8
M = 56.11
CAS No 25167-67-3
RTECS No EM2893000

Synonyms. 1-Butene; α-Butylene.

Properties. Colorless gas. Water solubility is 425 mg/l. Rapidly volatilizes from the open surface of water. Odor perception threshold is 0.2 mg/l. Has no effect on the color and transparency of water.

Applications. Monomer for polyolefins.

Acute Toxicity. Rats tolerate administration of 0.5 ml of 350 mg B./l solution without signs of toxicity.

Repeated Exposure. Accumulation of B. is impossible because of its rapid removal from the body. Mice were dosed by gavage with 3.75 mg/kg BW for 4 months. The treatment caused no abnormalities in behavior, BW gain, or oxygen consumption. Gross pathology examination revealed no changes in the visceral organs and their relative weights.

Long-term Toxicity. Rats were fed a dose of 0.05 mg/kg BW for 6 months. In the course of this experiment, no changes were observed in general condition or BW gain of the treated animals. No changes were found in the phagocytic activity of leukocytes, the activity of cholinesterase in the blood serum, or in conditioned reflex activity.

Chemobiokinetics. Direct chemical interaction between B. and biological media in the body is unlikely to occur. B. is rapidly removed unchanged via the lungs.

Regulations.

EU (1990). B. is available in the *List of authorized monomers and other starting substances which may continue to be used for the manufacture of plastic materials and articles intended to come into contact with foodstuffs pending a decision on inclusion in Section A* (*Section B*).

U.S. FDA (1998) regulates B. as a component of adhesives to be safely used in food-contact surfaces as prescribed in 21 CFR part 175.105.

Great Britain (1998). *But-1-ene* is authorized without time limit for use in the production of polymeric materials and articles in contact with food or drink or intended for such contact.

Standards. ***Russia*** (1995). MAC and PML: 0.2 mg/l (organolept., odor).

Reference:

Amirkhanova, G. F., Latypova, Z. V., and Tupeyeva, R. B., in *Protection of Water Reservoirs against Pollution by Industrial Liquid Effluents*, S. N. Cherkinsky, Ed., Meditsina, Moscow, Issue No 7, 1965, 28 (in Russian).

BUTYRIC ACID, VINYL ESTER

Molecular Formula. $C_6H_{10}O_2$
M = 114.16
CAS No 123-20-6

RTECS No ET700000

Abbreviation. BAVE.

Synonyms. Butanoic acid, ethenyl ester; Vinyl butanoate; Vinyl butyrate.

Applications. Used in the production of synthetic materials intended for use in contact with food or drink

Acute Toxicity. LD_{50} is 8.53 g/kg BW in rats (Union Carbide Data Sheet, 1970).

Regulations. ***U.S. FDA*** (1998) approved the use of BAVE as a monomer (1) in adhesives as a component of articles (a monomer) intended for use in packaging, transporting, or holding food, in accordance with the provisions prescribed in 21 CFR part 175.105; and as (2) a component (a monomer) of the uncoated or coated food-contact surface of paper and paperboard intended for use in producing, manufacturing, packaging, processing, preparing, treating, packing, transporting, or holding dry food of the type identified in 21 CFR part 176.170 (c).

ε-CAPROLACTAM

Molecular Formula. $C_6H_{11}NO$

M = 113.16

CAS No 105-60-2

RTECS No CM3675000

Synonyms. Aminocaproic lactame; 2-Azacycloheptanone; Cyclohexanoneisooxyme; Hexahydro-2*H*-azepin-2-one; 6-Hexanelactam; 2-Perhydroazepinone.

Properties. White, hygroscopic, crystalline solid. Water solubility is 525 g/100 ml at 25°C, highly soluble in ethanol and chloroform. Organoleptic perception threshold is reported to be 360 mg/l.[1]

Applications. Monomer in the production of polycaprolactam commonly known as *Nylon 6, Perlon, capron*, etc.

Exposure. The annual U.S. production of C. is over 500,000 tons.

Migration Data. C. readily migrates to water and food simulants due to its high water solubility: Concentrations detected up to 5.0 to 10 mg/l.[2] C. and its cyclic oligomers were extracted with boiling water from *nylon 6* films used in boil-in-bag food packaging. This resulted in the loss of up to 1.0 to 5.0% of the original nylon film weight.[3] Migration of C. from the protecting plastic envelope through the PVC barrier into the intravenous solution was found at concentrations of 1.2 to 15 mg C./l.[4]

Migration of low-molecular-mass compounds of caprolactam (LMC) from textile materials with different percentages of polyamide silk (PAS) was studied. Summed up contents of LMC of caprolactam was established in the first extracts of 5.0 to 7.0 mg/l. Monomer contents in the first extracts reached 1.15 mg/l, decreasing in the following ones to complete disappearance. At the same time dimers, trimers, etc., oligomer content of caprolactam progressively increased. The the presence of triethanolamine was established in the extracts.[5] Migration of C. from Nylon "microwave and roasting bags" was investigated. C. was released at amount of 5.0 to 35.5 mg/bag at cooking temperature.[6]

Acute Toxicity. LD_{50} values are reported to be 0.93 g/kg BW in mice, and 1.0 g/kg BW in rabbits.[7] LD_{50} is 2.1 g/kg BW in male, and 2.5 g/kg BW in female B6C3F$_1$ mice, and 1.6 and 1.2 g/kg BW, respectively, in male and female F344 rats.[8] According to Polushkin (1974), LD_{50} appeared to be 0.58 g/kg BW in rats.[9] Poisoning is accompanied by severe convulsions and increased diuresis. Sensitivity to C. decreases in the following order: mice, rabbits, rats, guinea pigs.[10]

Repeated Exposure to daily doses of 50 to 100 mg C./kg BW caused retardation in BW gain, changes in hematology, a relative increase in serum γ-globulin. Kidney and liver damage was also noted.[014]

Short-term Toxicity. A target organ or tissue for C. toxic action was not found. Administration of 1.0 g/kg BW to rats resulted in a decrease of their body temperature to 34 to 35°C and a 10 to 20% BW loss. The treatment caused 50% mortality in rats within 4 to 11 days. Rats tolerate administration of 10 mg/kg BW for 30 days without evident toxic effect.[7] Ingestion of C. in the form of 1.0% paste for 8 weeks caused no changes in behavior, hematology analysis, or skeletal tissues.[11]

Rats were given 50 or 250 mg C./kg BW in their diet for 90 days. The LOEL of 250 mg/kg BW, and the NOEL of 50 mg/kg BW for hematological effects were established in this study (Powers et al., 1984).

Long-term Toxicity. Rats were given drinking water containing 0.5% C. for 12 months. There were no changes in their general condition, hematological indices, or skeletal tissues but a slight decrease in BW gain.[11] In a 6-month study, the dose of 500 mg/kg BW caused a decrease of BW gain, anemia, and reticulocytosis in rabbits. Histology examination revealed changes in the gastric mucosa, brain congestion, perivascular and pericellular edema.[1]

In a reproduction study, F344 rats were fed diets containing 1000, 5,000, or 10,000 ppm C. for three generations. Retardation of BW gain and a decrease in consumption of food were observed at two higher concentrations. A slight increase in the severity of spontaneous nephropathy was seen in males in the high-dose group of the first parental concentration. The NOAEL of 1000 ppm C. (50 mg C./kg BW) was established in this study.[12]

Allergenic Effects. Caused dermal irritation and sensitization. Guinea pigs developed allergy after inhalation of C. In rats, the same effect was noted after oral treatment.[13]

Reproductive Toxicity. Controversial data are reported.

Observations in man. C. is shown to produce an increased rate of various complications during pregnancy in women exposed occupationally.[14]

Embryotoxicity.

Animal studies. C. was added to the diet of rats at doses of up to 10 g/kg BW. No adverse effect in the parent animals was noted. Decreased BW, food consumption, and nephropathy were the only signs of C. toxicity.[15] There was no effect on reproduction in rats treated with 0.5% aqueous solution of C. for a year.[11] However, in other studies, C. has been shown to modify reproductive function and produce a fetotoxic effect.[16]

Mice were treated by oral intubation with 6.5 to 6.7 mg/kg BW dose of ^{14}C-caprolactam and displayed rapid transfer of the radioactivity across the placenta with near complete elimination from the fetal and maternal compartments in 24 hours.[17]

In a 3-generation reproductive study, F344 rats received 1.0, 5.0, or 10 ppm C. in the diet. There was no embryotoxic or teratogenic effect in newborns.[18]

Teratogenic effect was not found either in rats or in rabbits.[19]

Gonadotoxicity. There was no significant increase in the frequency of abnormal sperm in mice given C. orally.[20]

Mutagenicity.

In vitro genotoxicity. C. showed contradictory results in the cultured human lymphocytes[7,21] and no activity in a large number of *in vitro* tests. Significant increase in the incidence of CA in cultured Chinese hamster lung cells was observed only with the dose of 12 mg/ml. C. appeared to be negative in mouse lymphoma and *Salmonella* mutagenicity assays, and in micronuclear test. It did not cause SCE.[11,016]

In vivo cytogenetics. In the mouse spot test, pregnant animals were dosed with 0.5 or 0.7 g C./kg BW. The treatment increased the frequency of gene mutations that produced color spots in the adult pelts of animals exposed *in utero*.[22,016] C. induced sex-linked recessive lethals in *Dr. melanogaster*.

Carcinogenicity. No carcinogenic effect was observed in mice and rats after dietary exposure. $B6C3F_1$ mice were fed a ration with 7.5 and 15 g C./kg BW doses for 103 weeks. F344 rats were exposed to 3.75 and 7.5 g/kg BW for 103 weeks. Slight reduction in BW gain in animals of both sexes was noted. No treatment-related tumors were observed.[8]

Carcinogenicity classifications. An IARC Working Group concluded that there is evidence suggesting lack of carcinogenicity of C. in *experimental animals* and there were no adequate data available to evaluate the carcinogenicity of C. in *humans*.

IARC: 4;

NTP: N - N - N - N (feed).

Chemobiokinetics. After oral administration, ^{14}C-caprolactam was readily absorbed from the stomach and distributed throughout the body including the fetuses.[17] Efficient elimination by the kidneys and liver was observed. When a dose of 300 mg/kg BW was administered to rats for 2 months C. was found in the liver (0.046 to 0.077 mg/100 g), kidneys (0.066 to 0.143 mg/100 g), adipose tissue (0.113 to 0.253 mg/100 g), and blood (0.07 mg/100 g).[23]

Rabbits metabolize C. almost entirely. High oral doses of C. may serve as a non-specific nitrogen source that can modify patterns of hepatic amino-acid metabolism. After *i/p* administration to rats, C. is removed partially unchanged, partially in the form of *ε-aminocaproic acid.*[24]

When 0.18 mg/kg of ^{14}C-tagged C. is administered to rats, 80% are removed with the urine during 24 hours, 3.5% with the feces, 1.5% with exhaled air. C. and two of its metabolites were found in the urine.

Regulations.

EU (1990). C. is available in the *List of authorized monomers and other starting substances which shall be used for the manufacture of plastic materials and articles intended to come into contact with foodstuffs (Section A).*

U.S. FDA (1998) permits the use of C. (ethylene-ethyl acrylate) graft polymers as a component of side-seam cements intended for use in contact with food. *Nylon 6* may be used for processing, handling, and packaging food.

Great Britain (1998). C. is authorized without time limit for use in the production of polymeric materials and articles in contact with food or drink or intended for such contact. The specific migration of C. alone or together with C. sodium salt shall not exceed a total of 15 mg/kg.

Standards.

EU (1990). SML: 15 mg/kg.

Russia (1995). PML in food: 0.5 mg/l, PML in drinking water: 0.3 mg/l.

References:

1. Savelova, V. A., in *Protection of Water Reservoirs against Pollution by Industrial Liquid Effluents*, S. N. Cherkinsky, Ed., Medgiz, Moscow, Issue No 4, 1960, 156 (in Russian).
2. Sheftel', V. O. and Katayeva, S. Ye., *Migration of Harmful Substances from Polymeric Materials*, Khimiya, Moscow, 1978, 168 (in Russian).
3. Barkby, C. T. and Lawson, G., Analysis of migrants from nylon 6 packaging films into boiling water, *Food Addit. Contam.*, 10, 541, 1993.
4. Iordanova, I. and Lolova, D., Migration of chemical substances out of polyamide textile materials, *Probl. Khig.*, 6, 119, 1981 (in Bulgarian).
5. Soto-Valdez, H., Gramshaw, J. W., and Vandenburg, H. J., Determination of potential migrants present in Nylon 'microwave and roasting bags' and migration into olive oil, *Food Addit. Contam.*, 14, 309, 1997.
6. Ulsaker, G. A. and Teien, G., Identification of caprolactam as a potential contaminant in parenteral solutions stored in overwrapped PVC bags, *J. Pharmacol. Biomed. Anal.*, 10, 77, 1992.
7. Lomonova, G. V., Toxicity of caprolactam, *Gig. Truda Prof. Zabol.*, 10, 54, 1966 (in Russian).
8. *Carcinogenesis Bioassay of Carpolactam in F344 Rats and B6C3F$_1$ Mice (Feed Study),* NTP Technical Report Series No 214, Research Triangle Park, NC, 1982.
9. Polushkin, B. V., Toxic properties of caprolactam monomer, *Pharmacologia i Toxicologia*, 27, 234, 1974 (in Russian).
10. Savelova, V. A. et al., Substantiation of the maximum allowable concentration for caprolactam in water, *Gig. Sanit.*, 1, 80, 1962 (in Russian).
11. Bornmann, G. und Loeser, A., Monomer-Polymer: Studie uber Kunststoffe aus ε-Caprolactam oder Bisphenol, *Arzneimittel-Forsch.*, 9, 9, 1959.
12. Serota, C. G., Hoberman, A. M., and Gad, S. C., A three-generation reproduction study with caprolactam in rats, in *Proc. Symp. Ind. Approach Chem. Risk Assess.*, Caprolactam Relat. Compd. Case Study, Ind. Health Found., Pittsburgh, PA, 1984, 191.
13. Baida, N. A. and Khomak, S. A., Allergenic effect of caprolactam in mammals, *Vrachebnoye Delo*, 2, 104, 1988 (in Russian).
14. Martynova, A. P., Lotis, V. N., Khadzhiyeva, E. D., et al., Occupational hygiene of women engaged in the production of capron fiber, *Gig. Truda Prof. Zabol.*, 11, 9, 1972 (in Russian).
15. Serota, D. A., Hoberman, A. M., Friedman, M. A., et al., Three-generation reproduction study with caprolactam on rats, *J. Appl. Toxicol.*, 8, 285, 1988.

16. Khadzhieva, E. D., Effects of caprolactam on female reproductive functions, in *Problem of Hygienic Standardization and Examination of the Delayed Effects*, 1972, 68 (in Russian).
17. Waddell, W. J., Marlowe, C., and Friedman, M. A., The distribution of [^{14}C]caprolactam in male, female and pregnant mice, *Food Chem. Toxicol.*, 22, 293, 1984.
18. Serota, D. G., Hoberman, A. M., Friedman, M. A., and Gad, S. C., Three-generation reproductive study with caprolactam in rats, *J. Appl. Toxicol.*, 84, 285, 1988.
19. Gad, S. C., Robinson, K., Serota, D. G., et al., Developmental toxicity in the rat and rabbits, *J. Appl. Toxicol.*, 7, 317, 1987.
20. Salamone, M. F., Abnormal sperm assay tests on benzoin and caprolactam, *Mutat. Res.*, 224, 385, 1989.
21. Krassov, S. V., Ivanov, V. P., Zhurkov, V. S., et al., Mutagenic risk of caprolactam, *Gig. Sanit.*, 7-8, 64, 1992 (in Russian).
22. Neuhauser-Klaus, A. and Lehmacher, W., The mutagenic effect of caprolactam in the spot test with (TxHT) F_1 mouse embryos, *Mutat. Res.*, 224, 369, 1989.
23. Sheftel', V. O., Batuyeva, L. N., and Sova, R. Ye., Comparative investigation of toxic action under constant and decreasing dose regimen, *Gig. Sanit.*, 8, 97, 1978 (in Russian).
24. Goldblatt, M. W., Farguharson, M. E., Bennett, G., et al., ε-Caprolactam, *Brit. J. Ind. Med.*, 11, 1, 1954.

CARBAMIC ACID, ETHYL ESTER

Molecular Formula. CH_4N_2O
M = 89.11
CAS No 51-79-6
RTECS No FA8400000
Abbreviation. CAE.

Synonyms and **Trade Names.** Ethyl carbamate; Ethy lurethane; Leucethane; Pracarbamine; Urethane.

Properties. Colorless, almost odorless, columnar crystals or white granular powder. Solutions are neutral. Water solubility is 2.0 g/l, soluble in ethanol, chloroform, and olive oil.

Applications. Used as a monomer and chemical intermediate in preparation of amino resins intended for use in contact with food.

Exposure. Humans are exposed to CAE in food and alcoholic beverages. Currently, CAE is mainly found as a natural trace constituent in some alcoholic beverages and in fermented food items,[1] in tobacco leaves and in tobacco smoke. Daily intake by adults is ~20 mg/kg BW. According to the U.S. EPA, CAE is not currently being used for commercial purposes due to its potent toxicity.[016] Concentrations of CAE measured in various foods and beverages are less than 5.0 µg/l in wines and beers, and 0.9 µg/kg in white bread.[2]

Acute Toxicity. LD_{50} is reported to be 2.5 g/kg BW in mice.[3]

Repeated Exposure. 1.0 ml of 1.0% solution injected in male rats for 30 days and in females for 45 days caused degranulation of liver and lung microsomes by approximately 15% and 30%, respectively. Cholesterol levels declined by more than 50%.[4]

Short-term Toxicity. F344/N rats and B6C3F_1 mice received 110 to 10,000 ppm CAE in their drinking water or in 5.0% ethanol for 13 weeks. Concentrations of 1,100 ppm CAE or greater induced lymphoid and bone-marrow cell depletion and hepatocellular lesions and increased severity of nephropathy and cardiomyopathy in rats. The treatment with these concentrations caused lung inflammation, alveolar and bronchial epithelial hyperplasia, alveolar/bronchiolar adenoma, nephropathy, lymphoid and bone marrow cell depletion, seminiferous tubule degeneration, and ovarian atrophy and follicular degeneration in mice. Administration of CAE in 5.0% ethanol did not enhance the frequency of micronucleated erythrocytes induced in the peripheral blood of mice.[5]

Immunotoxicity. B6C3F_1 mice received daily *i/p* injections of CAE (total dose of 4.0 mg CAE/g) over a 14-day period. The treatment depressed splenic lymphoproliferative response to Con A (a *T*-lymphocyte mitogen) and reduced number of primary antibody plaque-forming cells following sheep erythrocyte or *E. coli* lipopolysaccharide challenge.[6]

Reproductive Toxicity.

Gonadotoxicity. Administration of CAE in 5.0% ethanol over a period of 13 weeks exacerbated ovarian atrophy in mice. In male mice, epididymal spermatozoal motility was found to be significantly lower.[5]

Teratogenicity. CAE administered *i/p* to hamsters on the 8th day of pregnancy produced defects in the palate, brain, eye, extremities, and omphalocele at doses of 5 to 17 mM/g BW.[7] Teratogenic effect in mice was found to be strain dependent.[8] CAE-induced malformations were observed also in rats and hamsters.[9,10] Suppressed humoral immune functions were noted in the offspring of rats treated during organogenesis with 1.0 mg/kg and 2.0 mg/kg BW.[11]

Mutagenicity. CAE undergoes metabolic conversion to vinyl carbamate and other electrophilic derivatives that may contribute to its genotoxicity.[12] CAE is reported to show genotoxic potential *in vitro* and *in vivo.*

In vitro genotoxicity. In the absence of cytochrome *P450* enzymes, CAE produces no genotoxic effect in *Salmonella typhimurium, Saccharomyces cerevisiae,* and in human lymphoblastoid TK6 cells. In the presence of an activating system, CAE was found to be mutagenic in *Salmonella typhimurium* strain *TA100* but not in strains *TA98* and *TA102.* No significant mutagenicity has been noted in human lymhoblastoid TK6 cells.[13]

In vivo cytogenetics. Maternal and fetal SCE responses are observed in Swiss Webster mice.[14] Treatment of parent mice before mating gives rise to birth defects in the offspring.[15] Such effect in offspring was not observed by Russell et al.[16] in mice treated with up to 1750 mg CAE/kg BW.

CAE induced sex-linked recessive lethal mutations and reciprocal translocations in *Dr. melanogaster.*[17]

Carcinogenicity. CAE was discovered as a carcinogen in 1943 and was shown to induce tumors in various organs in rodents.

An increase in the lung adenoma response in strain A/J mice was observed after *i/p* and oral administration of CAE. [18] Quantitative analysis of the total tumor incidences after chronic exposure of rats and mice to 0.1 to 12.5 mg CAE/kg BW in the drinking water showed a dose-related increase. The main target organs were the mammary gland and the lungs (mice only). A "virtually safe dose" (10^{-6}) is 20 to 80 ng CAE/kg BW.[19] Doses of 100 to 2000 mg/kg BW have been shown to induce tumors in rats, mice, and hamsters after oral administration. A tumor incidence of 40 to 100% is typical after administration of 100 to 1000 mg/kg BW in the drinking water, depending on the duration of treatment (IARC 7-111).

Mice were treated twice a week with the dose of 2.0 mg/mouse (100 mg/kg BW) by gastric intubation either in water or in 0.2 ml 40% ethanol. The the presence of ethanol as a solvent enhanced pulmonary adenoma development after 4 or 6 months.[20] However, according to data presented by Kristiansen et al. and Altman et al., ethanol did not affect neoplasm development in mice.[21,22] CAE is also an effective transplacental carcinogen. In ICR/Jcl and strain A mice, lung and liver tumors were induced in the offspring following acute or multiple exposure of dams to CAE.[23]

Carcinogenicity classifications. An IARC Working Group concluded that there is sufficient evidence for the carcinogenicity of CAE in *experimental animals* and there were no case reports of epidemiological studies to evaluate the carcinogenicity of CAE in *humans.* CAE has been recognized as a rodent and non-human primate carcinogen, while IARC has determined that alcoholic beverages are human carcinogens. Perhaps the tumorigenicity of CAE might be inhibited by the presence of ethanol.

The Annual Report of Carcinogens issued by the U.S. Department of Health and Human Services (1998) defines CAE to be a substance which may reasonably be anticipated to be carcinogen, i.e., a substance for which there is a limited evidence of carcinogenicity in *humans* or sufficient evidence of carcinogenicity in *experimental animals.*

IARC: 2B;

EU: 2.

Chemobiokinetics. After oral administration to rats and mice, CAE is completely absorbed from the GI tract and rapidly distributed throughout the body. In animals, CAE is degraded to CO_2, H_2 and NH_3 with intermediate formation of *ethanol.* A quantitatively minor pathway involves a two-step oxidation of the ethyl group to *vinyl carbamate* and *epoxyethyl carbamate*; the postulated electrophilic moiety that reacts with

DNA is excreted in the urine and feces as volatile organics. Mice eliminated CAE as CO_2 more rapidly than rats.[24]

Regulations.

U.S. FDA has prohibited the use of CAE in drugs and food.[016]

References:

1. Zimmerli, B. and Schlatter, J., Ethyl carbamate: analytical methodology, occurrence, formation, biological activity and risk assessment, *Mutat. Res.*, 259, 325, 1991.
2. Ough, C. S., Ethylcarbamate in fermented beverages and foods. I. Naturally occurring ethylcarbamate, *J. Agric. Food Chem.*, 24, 323, 1976.
3. Osswald, H., Zur Frage der Mitosehemmung Mehrwertiger Carbaminsaureester am Ehlich-Carcinom, *Arzneimittel-Forsch.*, 9, 595, 1959.
4. Dani, R. et al., Effects of toxic doses of urethane on rat liver and lung microsomes, *Toxicol. Lett.*, 15, 61, 1983.
5. *Toxicity Studies of Urethane in Drinking Water and Urethane in 5% Ethanol Administered to F344/N Rats and B6C3F$_1$ Mice*, NTP Technical Report Series No 52, NIH Publ. 96-3937 (Po C. Chan), Research Triangle Park, NC, March 1996.
6. Luster, M. I., Dean, J. H., Boorman, G. A., Dieter, M. P., and Hayes, H. T., Immune functions in methyl and ethyl carbamate treated mice, *Clin. Exp. Immunol.*, 50, 223, 1982.
7. Di Paolo, J. A. and Elis, J., The comparison of teratogenic and carcinogenic effects of some carbamate compounds, *Cancer Res.*, 27, 1696, 1967.
8. Nito, S. et al., Effect of H-2 histocompatibility gene complex on urethane and ethylnitrosourea (ENU)-induced malformation in mice, *Congen. Anom.*, 27, 61, 1987.
9. Hall, E. K., Developmental anomalies in the eye of the rat after various experimental procedures, *Anat. Rec.*, 116, 383, 1953.
10. Ferm, V. H., Severe developmental malformations, *Arch. Pathol.*, 81, 174, 1966.
11. Luebke, R. W., Riddle, M. M., Rogers, R. R., et al., Immune function in adult C57BL6J mice following exposure to urethane pre- or postnatally, *J. Immunopharmacol.*, 8, 243, 1986.
12. Gupta, R., and Dani, H. M., *In vitro* formation of organ-specific ultimate carcinogens of 4-dimethylaminoazobenzene and urethane by microsomes, *Toxicol. Lett.*, 45, 49, 1989.
13. Hubner, P., Groux, P. M., Weibel, B., Sengstag, C., Horlbeck, J., Leong-Morgenthaler, P. M., and Luthy, J., Genotoxicity of ethyl carbamate (urethane) in *Salmonella*, yeast and human lymphoblastoid cells, *Mutat. Res.*, 390, 11, 1997.
14. Neeper-Bradley, T. L. and Conner, M. K., Intralitter variation in fetal sister chromatid exchange responses to the transplacental carcinogen ethyl carbamate, *Environ. Mutagen.*, 14, 90, 1989.
15. Nomura, T., X-ray- and chemically induced germ-line mutation causing phenotypic anomalies in mice, *Mutat. Res.*, 198, 309, 1988.
16. Russell, L. B., Hunsicker, P. R., Oakberg, E. F., Cummings, C. C., and Schmoyer, R. L., Tests for urethane induction of germ-cell mutations and germ-cell killing in the mouse, *Mutat. Res.*, 188, 335, 1987.
17. Foureman, P., Mason, J. M., Valencia, R. and Zimmering, S., Chemical mutagenesis testing in *Drosophila*. IX. Results of 50 coded compounds tested for the National Toxicology Program, *Environ. Molec. Mutagen.*, 23, 51, 1994.
18. Stoner, G. D., Greisiger, E. A., Schut, H. A. J., et al., A comparison of the lung adenoma response in strain A/J mice after intraperitoneal and oral administration of carcinogens, *Toxicol. Appl. Pharmacol.*, 72, 313, 1984.
19. Shlatter, J. and Lutz, W. K., Carcinogenic potential of ethyl carbamate (urethane): risk assessment at human dietary exposure levels, *Food Chem. Toxicol.*, 28, 205, 1990.
20. Griciute, L., Influence of ethyl alcohol on experimental carcinogenesis, in *Proc. 5th Conf. Oncol. Eston., Latv., Litov. SSR*, October 26-28, 1980, Tallinn, 1981, 170 (in Russian).
21. Kristiansen, E., Clemmensen, S., and Meyer, O., Chronic ethanol intake and reduction of lung tumors from urethane in strain A mice, *Food Chem. Toxicol.*, 28, 35, 1990.

22. Altman, H.-J., Dusemund, B., Goll, M., and Grunov, W., Effect of ethanol on the induction of lung tumors by ethyl carbamate in mice, *Toxicology*, 68, 195, 1991.
23. Nomura, T., An analysis of the changing response to the developing mouse embryo in relation to mortality, malformation, and neoplasm, *Cancer Res.*, 34, 2217, 1974.
24. Nomeir, A. A., Iounnou, Y. M., Sanders, J. M., and Matthews, H. B., Comparative metabolism and disposition of ethyl carbamate (urethane) in male Fischer 344 rats and male B6C3F$_1$ mice, *Toxicol. Appl. Pharmacol.*, 97, 203, 1989.

CARBONIC ACID, CYCLIC ETHYLENE ESTER

Molecular Formula. $C_3H_4O_3$
M = 88.07
CAS No 96-49-1
RTECS No FF9550000

Synonyms. 1,3-Dioxolane-2-one; Ethylene carbonate; Ethylene glycol carbonate; Glycol carbonate.

Properties. Yellowish-white, crystalline substance.

Applications. Used in the production of polyester resins (*Lavsan* type) and foam polymeric materials intended for use in contact with food.

Acute Toxicity. LD_{min} for mice is reported to be 8.0 g/kg BW. LD_{50} is 10 g/kg BW.

Repeated Exposure failed to reveal cumulative properties.

Reference:

Krasnokutskaya, L. M., in *Hygiene and Toxicology*. Proc. Sci. Conf. Junior Hygienists, L. I. Medved', Ed., Zdorov'ya, Kiyv, 1967, 118 (in Russian).

CARBONIC ACID, CYCLIC VINYLENE ESTER

Molecular Formula. $C_3H_2O_3$
M = 86.05
CAS No 872-36-6
RTECS No FG3325000
Abbreviation. CAVE.

Synonyms. 1,3-Dioxol-2-one; Vinylene carbonate.

Properties. Yellow-white, crystalline substance. Soluble in water and alcohol.

Applications. A monomer in the manufacture of materials and articles intended for use in contact with food or drink.

Acute Toxicity. LD_{50} is 1.86 g/kg BW in rats, and 0.5 g/kg BW in mice. Animals given the doses of 0.2 to 0.8 g/kg BW displayed labored breathing with asphyxia and CNS damage (loss of coordination, paresis of the legs, and clonic-tonic convulsions), and the absence of nutritive and defensive reflexes. Histology examination revealed congestion in the visceral organs, fatty dystrophy of the liver cells, and swelling of the endothelium of the renal ducts. There were also mild hyperplasia of the interfollicular cells and deposits of dark brown pigment in the spleen.[1]

Repeated Exposure revealed pronounced cumulative properties and marked toxic action of CAVE. Five administrations of 1/5 LD_{50} produced 100% rat mortality. Animals that received 1/10 LD_{50} died later. Gross pathology examination revealed increased relative weights of the visceral organs while histology changes were the same as with acute poisoning.[1]

Carcinogenicity. Oral administration to Charles River CD rats caused no carcinogenic effect.[2]

References:

1. Krasnokutskaya, L. M., Data on toxicity of vinylene carbonate, in *Hygienic Aspects of the Use of Polymeric Materials and Articles Made of Them*, L. I. Medved', Ed., All-Union Research Institute of Hygiene and Toxicology of Pesticides, Polymers and Plastic Materials, Kiyv, 1969, 386 (in Russian).
2. Weisburger, E. K., Ulland, B. M., Nam, J., Gart, J. J., and Weisburger, J. H., Carcinogenicity tests of certain environmental and industrial chemicals, *J. Natl. Cancer Inst.*, 67, 75, 1981.

CHLORENDIC ACID

Molecular Formula. $C_9H_4C_{16}0_4$
M = 388.83
CAS No 115-28-6
RTECS No RB9000000
Abbreviation. CA.

Synonyms and **Trade Names.** Hexachloroendomethylene tetrahydrophthalic acid; 1,4,5,6,7,7-Hexachloro-5-norbornene-2,3-dicarboxilic acid; HET acid.

Properties. White crystalline solid. Water solubility is 0.3% at 25°C, readily soluble in alcohol.

Applications. A fire-retardant monomer in production of coatings, unsaturated polyester resins, and epoxy resins intended for use in contact with food or drink.

Acute Toxicity. LD_{50} is 1.17 g/kg BW in rats.[012]

Repeated Exposure. $B6C3F_1$ mice and F344/N rats were fed 3,100 to 50,000 ppm CA in their diet for 14 days. Deaths occurred only in rats and in male mice given the highest dose. Gross pathology examination revealed no changes in the *viscera*.[1]

Short-term Toxicity. $B6C3F_1$ mice and F344/N rats were given the diet containing 1250 to 20,000 ppm CA (mice) and 620 to 10,000 ppm CA (rats) for 13 weeks. All animals survived. The treatment caused decreased mean BW at the highest doses, and increased incidences of liver lesions (rats: centrilobular cytomegaly, mitotic alterations, bile duct hyperplasia; mice: centrilobular cytomegaly, mitotic alterations, coagulative necrosis). The NOELs of 250 mg/kg BW in mice and 62.5 mg/kg BW in rats were established in this study.[1]

Long-term Toxicity. $B6C3F_1$ mice and F344/N rats were fed CA over a period of 103 weeks. The doses tested were 27 or 56 mg/kg BW in male rats, 39 or 66 mg/kg BW in female rats. Male mice received in feed 89 or 185 mg/kg BW, females received 100 or 207 mg/kg BW. The treatment caused retardation of BW gain in high dose rats, and increased incidences of non-neoplastic lesions of the liver: cystic degeneration in male rats, granulomatous inflammation, pigmentation, and bile duct hyperplasia in female rats. Lesions in the liver included coagulative necrosis in male mice, and mitotic alteration in high dose female mice.[1]

Reproductive Toxicity. Charles River CD rats were exposed orally by gavage at dose levels of up to 400 mg C. anhydride/kg BW on gestation days 6 to 15. Maternal toxicity but not teratogenic effects were observed in this study (Goldenthal et al., 1978).[2]

Mutagenicity.

In vitro genotoxicity. CA was negative in *Salmonella* mutagenicity assay, and was mutagenic in mouse lymphoma cell forward assay.[1] CA induced mutations in mammalian cells (IARC 48-50).

Carcinogenicity. In the above described study,[1] neoplastic nodules of the liver and acinar cell adenomas of the pancreas were found in male rats and high dose female rats. Increased incidences of neoplastic nodules and of hepatocellular carcinomas of the liver were seen in high dose female rats. There were increased incidences of hepatocellular adenomas and of hepatocellular carcinomas in male mice.[1]

Carcinogenicity classifications. An IARC Working Group concluded that there is sufficient evidence for the carcinogenicity of CA in *experimental animals* and there were no data available to evaluate the carcinogenicity of CA in *humans*.

The Annual Report of Carcinogens issued by the U.S. Department of Health and Human Services (1998) defines CA to be a substance which may reasonably be anticipated to be carcinogen, i.e., a substance for which there is a limited evidence of carcinogenicity in *humans* or sufficient evidence of carcinogenicity in *experimental animals*.

IARC: 2B;

NTP: CE* - CE - CE* - NE (feed).

Chemobiokinetics. After oral administration of radiolabelled CA to rats, it was rapidly distributed throughout the body and rapidly metabolized. Liver is the major site of deposition of CA metabolites. A smaller amount was found in blood, muscle, skin, and kidneys. Eliminated through bile and feces. In urine, there was less than 6.0% of total dose. Within 1 day, more than 75% of dose was excreted with feces.[3]

Regulations.

EU (1990). CA is available in the *List of monomers and other starting substances which may continue to be used for the manufacture of plastic materials and articles intended to come into contact with foodstuffs pending a decision on inclusion in Section A.*

Great Britain (1998). CA is authorized without time limit for use in the production of polymeric materials and articles in contact with food or drink or intended for such contact.

The specific migration of CA shall be not detectable (when measured by a method with a DL of 0.01 mg/kg).

References:

1. *Toxicology and Carcinogenesis Studies of Chlorendic Acid in F344/N rats and B6C3F$_1$ mice (Feed Studies)*, NTP Technical Report Series No 304, Research Triangle Park, NC, April 1987.
2. *Chlorendic Acid and Anhydride,* Environmental Health Criteria Series No 185, WHO/IPCS, Geneva, 1966, 56.
3. Decad, G. M. and Fields, M. T., Disposition and excretion of chlorendic acid in Fischer 344 rats, *J. Toxicol. Environ. Health*, 9, 911, 1982.

3-CHLORISOPRENE

Molecular Formula. C_5H_7Cl

M = 102.56

Synonym. 2-Methyl-3-chloro-1,3-butadiene.

Properties. Colorless liquid.

Applications. A monomer in the production of synthetic rubbers.

Acute Toxicity. LD_{50} is 4.85 g/kg BW in rats, and 2.8 g/kg BW in mice. Poisoned animals experienced excitation with subsequent prolonged CNS inhibition and narcosis. Gross pathology examination revealed congestion in the viscera, hemorrhages and edema in the lungs. An increase in the relative weights of the visceral organs was reported.

Repeated Exposure revealed little evidence of accumulation. Rats were given 1/20 and 1/50 LD_{50} for a month. The treatment affected liver antitoxic function. The maximum effect occurs with the smallest dose.

Reference:

Arutyunyan, D. G. and Gizhlaryan, M. S., Toxicity of chlorisoprene, *J. Exp. Clinic. Med.,* Acad. Sci. Armenia, Yerevan, 20, 397, 1980 (in Russian).

CHLOROPRENE

Molecular Formula. C_4H_5Cl

M = 88.54

CAS No 126-99-8

RTECS No EI9625000

Synonym. 2-Chloro-1,3-butadiene.

Properties. Colorless, inflammable, volatile, rapidly polymerizing liquid with a strong odor. Solubility in water is 0.5 mg/l, readily soluble in alcohol. Odor perception threshold is 0.1 mg/l at 20°C and 0.05 mg/l at 60°C.[1] C. is very unstable in water. When 0.1 mg/l is introduced, within as little as 15 to 16 seconds its specific odor is undetectable due to its volatility.

Applications. A monomer in the polychloroprene synthetic rubber manufacturing. Widely used in the production of elastomers intended for use in contact with food or drink; on copolymerization with acrylonitrile and metacrylonitrile.

Acute Toxicity. LD_{50} is found to be 250 mg/kg BW in rats, and 260 mg/kg BW in mice. Doses of 400 and 500 mg/kg BW caused total mortality in ıats and mice. Poisoning was accompanied by the CNS inhibition.[2] Gross pathology examination revealed congestion, hemorrhages, and dystrophic changes in the CNS and visceral organs.

Repeated Exposure produced little evidence of accumulation.

Long-term Toxicity. In a 6-month study, rats were gavaged with 15 mg/kg BW dose. Gross pathology examination revealed congestion in all the visceral organs and cloudy swelling in the renal tubular epithelium.[1]

Male rats were exposed to C. via their drinking water for 6 months.[3] The treatment caused an increase in the activity of β-galactosidase in the liver, blood serum, and seminal secretions, and of inosindiphosphatases in the gonads.

Reproductive Toxicity.

Gonadotoxicity. Long-term exposure to C. affected the estrus cycle and ovary structure of rats.[4] The activity of maleate dehydrogenase was reduced in the liver and gonads, sperm mobility time was decreased.[3]

Embryotoxicity and ***teratogenicity.*** A significant embryolethal effect was observed in rats. Fetal malformations comprised hydrocephalies and hemorrhages into the thoracic and abdominal cavities.[5]

Mutagenicity.

Observations in man. CA was found in the lymphocytes of workers exposed to C. (IARC Suppl. 6-164).

In vitro genotoxicity. In regard to C. genotoxicity, controversial data are reported. Freshly distilled C. appeared to be not mutagenic in *Salmonella TA 100* assay. However, mutagenic effect increased with increasing age of the C. distillates, probably because of the appearance of C. byproducts (cyclic C. dimers).[6] The mutagenicity of C. in Ames test appeared to be due to the the presence of decomposition products. Mutagenicity to Salmonella in the the presence and absence of metabolic activation increased with increasing age of this monomer.[6]

In vivo cytogenetics. Cytogenicity effects of C. were investigated in bone marrow cells of $B6C3F_1$ mice exposed to it by inhalation. Inhalation of C. gave negative results for cytogenetic endpoints assessed in bone marrow cells (SCE, micronuclei and CA).[7] C. did not cause CA, SCE, or micronucleus induction (NTP-92). Mutagenic activity was demonstrated in DLM assay in rats, CA were noted in the bone-marrow cells of mice treated *in vivo* (IARC Suppl. 7-160).

Carcinogenicity. The chemical structure of C. is similar to those of apparently carcinogenic vinyl chloride and 1,3-butadiene.

Observations in man. Exposure to C. (from polychloroprene latex and glue) was associated with increases in the risk of dying from cancer of the liver and leukemia. This was seen over a 14-year follow-up period in a group of almost 5200 shoe manufacturing workers (Bulbulyan et al., 1998).

Animal studies. C. was given orally to pregnant rats, and their offspring were treated by gavage for the life span. No difference in tumor incidence was found in this study.[8] C. has been predicted to be a rodent carcinogen by the computer automated structure evaluation (CASE) and multiple computer automated structure evaluation (MULTICASE) system.[9]

Carcinogenicity classifications. An IARC Working Group concluded that there is inadequate evidence for the carcinogenicity of C. in *experimental animals* and in *humans*.

IARC: 3.

ARC: 2

Chemobiokinetics. Mechanism of C. toxic action is linked to the formation of peroxides, which reinforce the lipid peroxidation process.[2] Under C. intoxication, the quantity of lipid peroxides in the tissues (particularly in the liver) is increased. Evidently, C. is effectively rendered harmless by conjugation with *SH*-glutathione, which does not react directly with C. but with its *epoxy metabolites*, formed with the help of microsomal enzymes.[10] After a single administration of 20 mg C./kg and 80 mg C./kg BW to rats, the additional excretion of *thioesters* in the urine was increased.

Regulations. ***U.S. FDA*** (1998) approved the use of C. in adhesives as a component of articles intended for use in packaging, transporting, or holding food in accordance with the provisions prescribed in 21 CFR part 175.105.

Standards. ***Russia*** (1995). MAC and PML: 0.01 mg/l.

References:

1. Khachatryan, M. K. and Asmangulyan, T. A., in *Protection of Water Reservoirs Against Pollution by Industrial Liquid Effluents*, S. N. Cherkinsky, Ed., Medgiz, Moscow, Issue No 4, 1960, 169 (in Russian).

2. Semerdzhyan, L. V. and Mkhitaryan, V. G., Toxic effects of chloroprene, *J. Exp. Clin. Med.*, Acad. Sci. Armenia, Yerevan, 16, 3, 1976 (in Russian).
3. Votyakov, A. V., in *Proc. Sanit.-Hyg. Res. Inst.*, Georgian SSR, 14, 65, 1978 (in Russian).
4. Melik-Alaverdyan, N. O., Studies on chloroprene toxicity, *Bull. Exp. Biol.* (Yerevan), Issue No 6, 1965, 107 (in Russian).
5. Sal'nikova, L. S. and Fomenko, V. N., Comparative characteristics of chloroprene embryotoxicity depending on exposure regimen at different routes of administration in the body, *Gig. Truda Prof. Zabol.*, 7, 30, 1975 (in Russian).
6. Westphal, G. A., Blaszkewicz, M., Leutbecher, M., Muller, A., Hallier, E., and Bolt, H. M., Bacterial mutagenicity of 2-chloro-1,3-butadiene (chloroprene) caused by decomposition products, *Arch. Toxicol.*, 68, 79, 1994.
7. Shelby, M. D., Results of NTP-sponsored mouse cytogenetic studies on 1,3-butadiene, isoprene, and chloroprene, *Environ. Health Perspect.*, 86, 71, 1990.
8. Ponomarkov, V. and Tomatis, L., Long-term testing of vinylidene chloride and chloroprene for carcinogenicity in rats, *Toxicology*, 37, 136, 1980.
9. Zhang, Y. P., Sussman, N., Macina, O. T., Rosenkranz, H. S., Klopman, G., Prediction of the carcinogenicity of a second group of organic chemicals undergoing carcinogenicity testing, *Environ. Health Perspect.*, 104S, 1045, 1996.
10. Summer, K. H. and Greim, H., Detoxification of chloroprene (2-chloro-1,2-butadiene) with gluthatione in rats, *Biochem. Biophys. Res. Commun.*, 96, 566, 1980.

CRESOLS

Molecular Formula. C_7H_8O
M = 108.14
CAS No 1319-77-3

Synonyms.

2-Methylphenol	3-Methylphenol	4-Methylphenol
2-Cresol	3-Cresol	4-Cresol
2-Hydroxy toluene	3-Hydroxy toluene	4-Hydroxy toluene
o-Cresylic acid	*m*-Cresylic acid	*p*-Cresylic acid
CAS No 95-48-7	CAS No 108-39-4	CAS No 106-44-5
RTECS No GO6300000	RTECS No GO6125000	RTECS No GO6475000

Properties. *o*-Cresol occurs as crystals (or liquid) darkening in air; *m*-cresol is a colorless liquid with a phenol odor; *p*-cresols are crystals with a phenol odor. Water solubility (g/100 ml) is 3.1 (40°C), 2.35 (20°C) and 2.4 (40°C), respectively. Dicresol is a mixture of 65% *m*-C. and 35% *p*-C. Tricresol is a technical grade product containing a mixture of the isomers. Cresols are soluble in alcohol. Odor perception threshold is 0.05 mg *o*-C./l, 0.68 mg[010] or 0.001 mg *m*-C./l, and 0.002 mg *p*-C./l.[1] According to other data,[02] odor perception threshold of *m*-C. is 0.037 mg/l.

Applications. Mainly used in the production of cresol-aldehyde and phenolic resins intended for use in contact with food or drink. Tricresol is also used in the production of heat-resistant lacquers and molding powders. Dicresol is used in the synthesis of phosphate plasticizers (tricresyl phosphate, cresyldiphenyl phosphate), phenol-formaldehyde resins, and stabilizers. *p*-C. is used in large quantities in the production of ionol and antioxidant 2246, which are used to stabilize rubbers, plastics, vulcanizates, and foodstuffs.

Exposure. Concentrations of cresols in the urine of workers employed in the distillation of the phenolic fraction of tar and those of non-exposed male workers were found to be for *p*-C.: 58.6 and 25.7 mg/l, respectively; for *o*-C.: 76.9 and 68.1 mg/l, respectively.[2]

Acute Toxicity.

Observations in man. Symptoms of acute toxicity following ingestion of 1.0 to 60 ml of cresol include involuntary muscle movements followed by paresis, GI tract disturbances, renal toxicity, initial CNS stimulation followed by depression, etc.[03,019]

Animal studies. The LD_{50} values for *o*-, *m*-, and *p*-C. administered as 10% oil solution are 344, 828, and 344 mg/kg BW (mice) and 1470, 2010, and 1460 mg/kg BW (rats). The LD_{50} of dicresol (10% oil solution) is 1625 mg/kg BW for rats,[3] that of *o*-C. being 940 mg/kg BW for rabbits, 100 to 500 mg/kg BW for minks, and 300 to 500 mg/kg BW for polecats.[07]

Symptoms of poisoning are similar to those produced by phenol, i.e., hematuria, agitation, and convulsions. The irritating effect of cresols can be increased by an alkaline medium. Gross pathology examination of mice exposed to *p*-C. revealed dystrophic changes in the myocardium, liver and kidney cells, as well as parenchymatous dystrophy of the renal tubular epithelium.[4]

Repeated Exposure. Exposure to cresols has been associated with hemolysis, methemoglobinemia, and acute Heinz-body anemia.[5] F344 rats and $B6C3F_1$ mice were given *o*-, *m*-, and *p*-C. or *m/p*-C. (60:40) at concentrations of 300 to 30,000 ppm in the diet for 28 days. Some mice given *o*-C. at a concentration of 30,000 ppm or *m*- C. and *p*-C. at concentrations of 10,000 or 30,000 ppm, respectively, died before the end of the studies.[6]

Concentration of 3,000 ppm increased relative liver and kidney weights without microscopic changes. Atrophic changes in the nasal and forestomach epithelium were presumably considered to be a specific effect of *p*- and *m/p*-C. and a direct result of the irritant effects.[6]

Short-term Toxicity. In a 90-day study, Sprague-Dawley rats were gavaged with 50, 150, or 450 mg *m*-C./kg BW. The treatment with the highest dose caused retardation of BW gain in males; effects on CNS included significant increase in incidence of salivation, tremors, and urination. The NOAEL of 50 mg/kg BW for systemic toxicity in rats was established in this study.[7-9]

In a 90-day study, Sprague-Dawley rats were gavaged with 50, 175 or 600 mg *p*-C./kg BW. The treatment with the highest dose caused combined mortality and significant reduction in BW and food consumption; CNS effects included lethargy, ataxia, coma, dyspnea, tremor, and convulsions within 15 to 30 minutes after dosing. The NOAEL of 50 mg/kg BW for systemic toxicity in rats was established in this study.[7-9]

In another 90-day study, Sprague-Dawley rats were gavaged with doses of 50 to 600 mg *o*-C./kg BW. The treatment with the highest dose damaged CNS functions, namely, increased salivation, urination, tremors, lacrimation, palpebral closure, and rapid respiration. Changes were noted in neurobehavioral reactions. The NOAEL of 50 mg/kg BW for systemic toxicity in rats was established in this study.[7-9]

In NTP 13-week feeding studies in rats and mice, the isomers exhibited generally similar patterns of toxicities, with the *o*-C. being somewhat less toxic than *m*- or *p*-C., *o*-C., or *m/p*-C. (60:40) were added to the diet at concentrations of 30,000 ppm to rats, 20,000 ppm (*o*-cresol), or 10,000 ppm (*m/p*-cresol) to mice. There were no serious changes in hematology, clinical chemistry, and urinalysis. A deficient hepatocellular function and forestomach hyperplasia were found in mice given diets containing higher concentration of *o*-cresol.[6]

Rats receiving *o*-cresol in their drinking water (0.3 g/l) were sacrificed after 5 to 20 weeks of administration. No adverse effects were observed but homogenates of the cerebrum had increased RNA content, decreased glutathione, and lower azoreductase activity.[10]

Long-term Toxicity. Rats exposed to 1.0 mg *p*-cresol/kg BW for 7 months developed decreased BW gain, reduced oxygen consumption and uropepsin level in the urine, and a change in conditioned reflex activity.[1] Hamsters received the diet containing 1.5% *p*-C. for 20 weeks. The treatment caused an increase in the incidence of mild-to-moderate forestomach hyperplasia.[11]

Reproductive Toxicity. Minks were fed 5.0, 25, or 105 mg *o*-C./kg BW (males) and 10, 40, or 190 mg *o*-C./kg BW (females) in their diet for 2 months before mating and through weaning. The highest doses were associated with paternal toxicity.[12]

Gonadotoxicity.

Observations in man. Women exposed occupationally to varnishes containing tricresol are reported to have gynecological problems.[13]

Animal studies. There were no adverse effects on male reproductive function, but the oestrus cycle was lengthened in rats and mice receiving higher concentrations of *o*-cresol, and in rats receiving *m/p*-cresol in a 13-week study.[6] An increase in testis weight was noted in ferrets dosed with 2520 and 4536 mg *o*-C./kg diet.[14]

Embryotoxicity. *p*-C. caused maternal toxicity at 410 mg/kg BW, but failed to elicit effects on post-implantation loss or litter weight at any dose tested.[15] In continuous breeding protocol study, CD-1 Swiss mice received 0.25, 1.0, or 1.5% and *m*- plus *p*-C. or 0.05, 0.2, and 0.5% of *o*-C. in their feed for 14 weeks. 1.5% mixture significantly reduced litter size and adjusted pup weight, 1.0 and 1.5% dose level affected pre- and post-weaning growth and survival. *o*-C. at doses up to 0.5% (550 mg *o*-C./kg BW) did not affect reproductive parameters in either generation.[16]

Mutagenicity.

In vitro genotoxicity. Cresol isomers are not mutagens in bacteria. They were not found to exhibit mutagenic potential in *Salmonella* either in the the presence or absence of mammalian liver homogenates.[17]

In vivo cytogenetics. None of the isomers increased SCE in mouse bone marrow, lung, or liver cells *in vivo.*[15,18] *p*- and *m*-cresol produced CA in Wistar rats given 0.001 to 1.0 LD_{50}.[18] *p*-Cresol induced unscheduled DNA synthesis in human lung fibroblast cells in the the presence of hepatic homogenates. It was negative in cell transformation assay with the mouse fibroblast cell line C3H1OT/2 (Crowley and Margard, 1978).

Carcinogenicity. Simultaneous oral administration of 1.0 mg benzo[a]pyrene and 1.0 mg *o*-C. in CC57Br mice increased the incidence of tumors, the multiplicity of tumors, and the degree of malignancy; latency of tumor incidence was shortened.[19]

Carcinogenicity classification.

U.S. EPA: C.

Chemobiokinetics. Following oral administration cresols are readily absorbed from the GI tract of experimental animals. *p*-Cresol is removed with the urine of rabbits in the form of glucuronide (60%) and sulfate (15%) conjugates. About 10% are oxidized to *p-hydroxybenzoic acid*, while traces of *p*-cresol are hydroxylated to 4-*methylpyrocatechole* (3,4-*dihydroxy toluene*).[021]

In addition to urinary excretion, cresols undergo enterohepatic circulation.[03] *p*-Cresol is a normal constituent of the human urine with levels of excretion ranging from 16 to 74 mg/day.[03,5]

Regulations.

U.S. FDA (1998) has established allowable levels of p-cresol in food products. Cresols may be safely used (1) in the manufacture of resinous and polymeric coatings for the food-contact surface of articles intended for use in producing, manufacturing, packing, processing, preparing, treating, packaging, transporting, or holding food in accordance with the provisions prescribed in 21 CFR part 175.300; and (2) in the manufacture of phenolic resins used as food-contact surface of molded articles intended for repeated use in contact with nonacid food (*pH* above 5) in accordance with the provisions prescribed in 21 CFR part 177.2410.

Great Britain (1998). *o*-, *m*-, and *p*-Cresols are authorized without time limit for use in the production of polymeric materials and articles in contact with food or drink or intended for such contact.

EU (1990). Cresols are available in the *List of authorized monomers and other starting substances which shall be used for the manufacture of plastic materials and articles intended to come into contact with foodstuffs (Section A)*.

Standards. ***Russia*** (1995). MAC (*m*- and *p*-cresols): 0.004 mg/l.

References:

1. Budeev, N. I. et al., in Problems of Water Supply in Sanitary Protection of Reservoirs, *Proc. Republ. Sci. Conf., Perm'*, 1969, 114 (in Russian).
2. Bieniek, G., Concentrations of phenol, *o*-cresol, and 2,5-xylenol in the urine of the workers employed in the distillation of the phenolic fraction of tar, *Occup. Environ. Med.*, 51, 354, 1994.
3. Uzhdavini, E. R., Astafyeva, N. K., Mamayeva, A. A., and Bakhtizina, G. Z., Inhalation toxicity of *o*-cresol, in *Proc. Research Institute Hygiene Occup. Diseases*, Ufa, Issue No 7, 1972, 115 (in Russian).
4. Pavlenko, M. N., and Kuznetsova, V. A., in *Problems of Hygiene*, Coll. Sci. Works Research Sanitary Hygiene Institute, Novosibirsk, Issue No 15, 1965, 71 (in Russian).
5. Cote, M. A., Lyonnais, J., and Leblond, P. F., Acute Heinz-body anemia due to severe cresol poisoning: Successful treatment with erythrocytophoresis, *Can. Med. Assoc. J.*, 130, 1319, 1984.
6. *Toxicity Studies of Cresols in F344/N Rats and B6C3F$_1$ Mice (Feed Studies)*, NTP Technical Report No 9, Research Triangle Park, NC, U.S. Dept. Health and Human Services, 1981, 128.

7. *o, m, p-Cresol. 90-day Oral Subchronic Toxicity Studies in Rats*, Office of Solid Waste, U.S. EPA Washington, D. C., 1986.
8. *o-, m-, p-Cresol. 90-day Oral Subchronic Neurotoxicity Studies in Rats*, U.S. EPA Office of Solid Waste, Washington, D. C., 1987.
9. Henck, J. W., Traxler, D. J., Dietz, D. D., et al., Neurotoxic potential of *ortho-, meta-*, and *para*-cresol, *Toxicologist*, 7, 246, 1987.
10. Savolainen, H., Toxic effects of oral *o*-cresol intake on the rat brain, *Res. Comm. Chem. Pathol. Pharmacol.*, 25, 357, 1979.
11. Hirose, M., Inoue, T., Asamoto, M., Tagawa, Y., and Ito, N., Comparison of the effects of 13 phenolic compounds in induction of proliferative lesions of the forestomach and increase in the labeling indices of the glandular stomach and urinary bladder epithelium of Syrian golden hamsters, *Carcinogenesis*, 7, 1285, 1986.
12. Hornshaw, T. C., Aulerich, R. J., and Ringer, R. K., Toxicity of *o*-cresol to mink and European ferrets, *Environ. Toxicol. Chem.*, 5, 713, 1986.
13. Syrovadko, O. N. and Malysheva, Z. V., Working conditions and their effect on some specific functions of women engaged in the manufacture of enamel-insulated wires, *Gig. Truda Prof. Zabol.*, 4, 25, 1977 (in Russian).
14. Cheng, M. and Kligerman, A. D., Evaluation of genotoxicity of cresols using sister-chromatid exchange (SCE), *Mutat. Res.*, 137, 51, 1984.
15. Kavlock, R. J., Structure-activity relationship in the developmental toxicity of substituted phenols: *In vivo* effects, *Teratology*, 41, 43, 1990.
16. Izard, P. A., Fail, P. A., George, J. D., Grizzle, T. B., and Heindel, J. J., Reproductive toxicity of cresol isomers administered in feed to mouse breeding pairs, Abstract, *Toxicologist*, 12, 198, 1992.
17. Florin, I., Rutberg, M., Curvall, M., and Enzell, C. R., Screening of tobacco smoke constituents for mutagenicity using Ames' test, *Toxicology*, 18, 219, 1980.
18. Ekshtat, B. Ya. and Isakova, G. K., in Papers Sci. Conf. Siberian Hygienists and Physicians, Novosibirsk, 1969, 101 (in Russian).
19. Yanysheva, N. Ya., Balenko, N. V., Chernichenko, I. A., and Babiy, V. F., Peculiarities of carcinogenesis under simultaneous oral administration of benzo[a]pyrene and *o*-cresol in mice, *Environ. Health Perspect.*, 101, (Suppl. 3), 341, 1993.

CROTONIC ACID

Molecular Formula. $C_4H_60_2$
M = 86.09
CAS No 3724-65-0
RTECS No GO2800000
Abbreviation. CA.

Synonyms. 2-Butenoic acid; 3-Methylacrylic acid; Solid crotonic acid.

Properties. Colorless solid or needle-like crystals. Readily soluble in water, solubility in ethanol is 52.5% at 25°C.

Applications. Used in the manufacture of copolymers with vinyl acetate, in lacquers and paper sizing, in soft agents for synthetic rubber intended for use in contact with food or drink..

Acute Toxicity. LD_{50} is 1.0 g/kg BW in rats (Smyth and Carpenter, 1944).

Chemobiokinetics. CA is metabolized to β-*hydroxybutyryl-coenzyme A* by crotonase that is available in the liver and other tissues.[011]

Regulations.

EU (1990). CA is available in the *List of authorized monomers and other starting substances which may continue to be used for the manufacture of plastic materials and articles intended to come into contact with foodstuffs pending a decision on inclusion in Section A* (*Section B*).

U.S. FDA (1998) approved the use of CA (monomer) in (1) adhesives as a component of articles intended for use in packaging, transporting, or holding food in accordance with the conditions prescribed in

21 CFR part 175.105; and as (2) a component (monomer) of the uncoated or coated food-contact surface of paper and paperboard intended for use in producing, manufacturing, packaging, processing, preparing, treating, packing, transporting, or holding dry food in accordance with the provisions prescribed in 21 CFR part 176.170.

Great Britain (1998). CA is authorized up to the end of 2001 for use in the production of polymeric materials and articles in contact with food or drink or intended for such contact.

CYCLOHEXANOL

Molecular Formula. $C_6H_{12}O$
M = 100.16
CAS No 108-93-0
RTECS No GV7875000
Abbreviation. CH.

Synonyms and **Trade Names.** Anol; Cyclohexyl alcohol; Hexahydrophenol; Hexaline; Hydroxycyclohexane.

Properties. Colorless, slightly oily liquid with an odor of camphor and amyl alcohol. Solubility in water is 3.6% (20°C), soluble in alcohol. Odor perception threshold is 3.5 or 2.8 mg/l.[02]

Applications. Used in the production of caprolactam and polyamide fibers intended for use in contact with food or drink.

Acute Toxicity. For rats, LD_{100} appeared to be 1.5 g/kg BW. LD_{50} for mice is 1.24 g/kg BW. Poisoning is accompanied by ataxia and muscular weakness. Animals become irresponsive to painful stimuli. Death is preceded by deep narcosis.[1]

Repeated Exposure failed to reveal cumulative properties. An increase of the relative liver weights and stimulation of microsomal enzyme and diphenyl-4-hydroxylase activity were found in animals given 455 to 1500 mg CH/kg BW for 7 days (Lake et al., 1982). Rats were gavaged with a 400 mg/kg BW dose. The treatment had no effect on BW gain, general condition, and behavior. There were changes in the glycogen-forming function of the liver, oxidation-reduction processes in the tissues and in hematology indices.[1]

Long-term Toxicity. In a 6-month study, rabbits were exposed to 2.0 mg/kg and 20 mg/kg BW doses. The treatment resulted in decreased BW gain, impaired liver function, and reduced catalase activity. Gross pathology examination revealed changes in the viscera.[1]

Reproductive Toxicity.

Gonadotoxicity. Testicular atrophy, loss of type A spermatogonia, spermatocytes, and spermatozoa were reported in rats and gerbils given 15 mg/kg BW for 21 to 37 days. Shrinkage were noted in seminiferous tubule and Leidig cells. A decrease in RNA, protein, sialic acid, and glycogen of testes, epididymides, and seminal vesicles was observed. The treatment caused an increase in testicular cholesterol and alkaline phosphatase.[2]

Mutagenicity.

In vitro genotoxicity. Equivocal results are reported in *Salmonella* mutagenicity assay.[025]

Chemobiokinetics. CH is metabolized by hepatic NAD-dependent alcohol dehydrogenase.[3] When fed to rabbits, CH is mostly excreted as *glucuronide conjugate* (60% of the dose administered), a small amount is oxidized further into *trans-cyclohexane-1,2-diol* (6%).[021]

Regulations. ***U.S. FDA*** (1998) listed CH for use (1) in adhesives as a component of articles intended for use in packaging, transporting, or holding food in accordance with the provisions prescribed in 21 CFR part 175.105; (2) as a component of the uncoated or coated food-contact surface of paper and paperboard that may be safely used for producing, manufacturing, packing, transporting, or holding dry food in accordance with the provisions prescribed in 21 CFR part 176.170; and (3) as a component of a defoaming agent intended for articles which may be used in producing, manufacturing, packing, processing, preparing, treating, packaging, transporting, or holding dry food in accordance with the provisions prescribed in 21 CFR part 176.200.

Standards. ***Russia*** (1995). MAC and PML: 0.5 mg/l.

References:

1. Savelova, V. A. and Brook, Ye. S., in *Protection of Water Reservoirs Against Pollution by Industrial Liquid Effluents*, S. N. Cherkinsky, Ed., Medgiz, Moscow, Issue No 5, 1962, 78 (in Russian).

2. Tyagi, A., Joshi, B. C., Kumar, S., and Dixit, U. P., Antispermatogenic activity of cyclohexanol in gerbil (*Meriones hurrianae Jerdon*) and house rat (*Rattus ruffercens*), *Indian J. Exp. Biol.*, 17, 1305, 1979.
3. Testa, B. and Jenner, P., *Drug Metabolism, Chemical and Biochemical Aspects*, Marcel Dekker, Inc, New York, 1976, 310.

CYCLOHEXANONE

Molecular Formula. $C_6H_{10}O$
M = 98.15
CAS No 108-94-1
RTECS No GW1050000

Synonyms. Cyclohexyl ketone; Ketohexamethylene; Pimelic ketone.

Properties. Colorless, oily liquid with a sweet, sharp odor. Miscible with water (50 g/l at 30°C) and most organic solvents: ethanol, diethyl ether, chloroform. Odor perception threshold is reported to be 0.12 mg/l,[010] 1.0 to 2.0 mg/l[1] or 8.3 mg/l.[02]

Applications. C. is primarily used as an intermediate in the production of nylon, and as an additive and an excellent solvent for a variety of products intended for use in contact with food or drink: rubbers, polyurethane lacquers, natural and synthetic resins and gums, as well as some biomedical polymers, etc; C. is also used in the production of adipic acid.

Acute Toxicity. The LD_{50} values are 2.07 g/kg BW for male, and 2.1 g/kg BW for female mice, and 1.8 g/kg BW for female rats.[2] Rabbits seem to be less sensitive. Death occurs with signs of narcosis and without an excitation stage.[3] Autopsy revealed edema with hemorrhages in the lung parenchyma, peritoneal and intestinal congestion in mice, suggesting an irritant effect.

Repeated Exposure. C. induced diverse reactions by virtue of a general vascular or tissue reaction and CNS inhibition.[2] Mice were exposed to 280 mg/kg BW for 25 days. The treatment had no effect on their general condition and BW gain.

Short-term Toxicity. Mice were given 0.4 to 47.0 g C./l of their drinking water for 13 weeks. 1/3 to 2/3 of the animals in the highest dose group died during experiment. Other animals exhibited retardation of BW gain; C. concentration of 47 g/l produced focal liver necrosis and thymus hyperplasia.[3]

Mice and rats was administered C. in water for 95 to 175 days. The treatment caused no effect on survival, but decreased BW at the highest concentration of 7,000 ppm in rats. 14,000 ppm or higher levels decreased BW in mice. The highest dose decreased survival (50,000 ppm C.). The NOEl could be calculated at the level of 665 mg/kg BW.[4]

Long-term Toxicity. F344 rats and $B6C3F_1$ mice received C. as a solution in drinking water at concentrations of 3,300 or 6,500 ppm (rats), and 6,500; 13,000; or 25,000 ppm (mice). Retardation of BW gain and increased mortality were observed at higher doses. The NOAEL of 3,300 ppm C. was established in this study (462 mg/kg BW).[5]

Allergenic Effect. C. did not induce skin allergy in the guinea pig maximization test.[6]

Reproductive Toxicity.

Embryotoxicity. 1.0%-dietary administration to mice for several generations was reported to affect the viability and growth of the first-generation males and females. There were no such effects in the 2nd generation.[7] CD-1 mice were exposed by oral intubation to 800 mg/kg BW on days 8 to 12 of gestation.

Reproductive function was not affected[8]

Mutagenicity.

In vitro genotoxicity. C. was found to be negative in *Salmonella* mutagenicity assay.[025]

It exhibited ***cytogenetic effect*** in human cultured lymphocytes at concentrations of 0.0005 to 0.01 mg/100 ml of the culture medium.[9]

Carcinogenicity. $B6C3F_1$ mice received 6.5 or 13 g C./l in their drinking water for 104 weeks. A slight increase in the incidence of tumors that occur commonly in this strain was found only in animals given the lower dose. In F344 rats given 3.3 or 6.5 g C./l in their drinking water for 104 weeks a slight increase in the incidence of adrenal cortical adenomas occurred in males treated with the lower dose.[5]

Carcinogenicity classification. An IARC Working Group concluded that there is inadequate evidence for the carcinogenicity of C. in *experimental animals* and there were no adequate data available to evaluate the carcinogenicity of C. in *humans*.

IARC: 3.

Chemobiokinetics. Data on ability of C. to accumulate in the body appear to be contradictory. Repeated *i/v* administration (overall dose of 284 mg/kg BW) provided no evidence of enzyme induction.[10] C. is likely to be metabolized to cyclohexanol, which is conjugated with glucuronic acid and excreted mainly in the urine (almost 60%); very little C. or free cyclohexanol is found in the urine.[11,021]

Standards. *Russia* (1995). MAC and PML: 0.2 mg/l.

References:

1. Vertebnaya, P. I. and Mozhayev, Ye. A., in *Protection of Water Reservoirs against Pollution by Industrial Liquid Effluents*, S. N. Cherkinsky, Ed., Medgiz, Moscow, Issue No 4, 1960, 76 (in Russian).
2. Gupta, P. K., Lawrence, W. H., Turner, J. E., et al., Toxicological aspects of cyclohexanone, *Toxicol. Appl. Pharmacol.*, 49, 525, 1979.
3. Novgorodova, L. G., Savelova, V. A., and Sergeev, A. N., in *Industrial Pollution of Water Reservoirs*, S. N. Cherkinsky, Ed., Meditsina, Moscow, Issue No 8, 1967, 111 (in Russian).
4. *Summary and Experimental Design of Subchronic Studies of Cyclohaxanone*, Natl. Cancer Institute, Final Report, 1979.
5. Lijnsky, W. and Kowatch, M., A chronic toxicity study of cyclohexanone in rats and mice (NCI study), *J. Natl. Cancer Inst.*, 77, 941, 1986.
6. Bruze, M., Bomon, A., Bergqvist-Karlson, A., et al., Contact allergy to cyclohexanone resin in humans and guinea pigs, *Contact Dermat.*, 18, 46, 1988.
7. Gondry, E., Studies on the toxicity of cyclohexylamine, cyclohexanone and cyclohexanol, metabolites of cyclomate, *J. Exp. Toxicol.*, 5, 227, 1972.
8. Chernoff, N. and Kavlock, R. J., A teratology test system which utilizes postnatal growth and viability in the mouse, in *Short-term Bioassay in the Analysis of Complex Environmental Mixtures*, M. Wasters, S. Sandhy, J. Lewtas, et al., Eds., Plenum, New York, 1983, 417.
9. Dyshlovoi, V. D., Boiko, N. L., Shemetun, A. M., et al., Cytogenetic effect of cyclohexanone, *Gig. Sanit.*, 5, 76, 1981 (in Russian).
10. Martis, L., Tolhurst, T., Koeferl, M. T., et al., Disposition kinetics of cyclohexanone in beagle dogs, *Toxicol. Appl. Pharmacol.*, 55, 545, 1980.
11. Greener, Y., Martis, L., and Indacochea-Redmond, N., Assessment of the toxicity of cyclohexanone administered intravenously to Wistar and Gunn rats, *J. Toxicol. Environ. Health.*, 10, 385, 1982.

DICHLORO-1,3-BUTADIENE

Molecular Formula. $C_4H_4Cl_2$

M = 122.99

CAS No 28577-62-0

RTECS No EI9799000

Synonym. Dichlorobutadiene.

Properties. Volatile, readily mobile liquid. Organoleptic threshold concentration is 2.0 mg/l.[1]

Applications. Used in the production of chloroprene rubber and latex.

Acute Toxicity. LD_{50} is reported to be 280 to330 mg/kg BW in rats, and 150 mg/kg BW in mice. Poisoning is accompanied by symptoms of narcosis. Gross pathology examination revealed congestion in the visceral organs, necrotic foci in the liver, and local damage to the gastric mucosa. There were no morphological findings in animals that survived the toxic effect.[1,2]

Repeated Exposure revealed moderate cumulative properties. K_{acc} is 6 (by Lim).

Reproductive Toxicity. Inhalation exposure to about 2.0 mg/m^3 is likely to be the threshold concentration for effects on reproduction in rats.[2]

Standards. *Russia* (1995). MAC and PML: 0.03 mg/l.

References:

1. Gizhlaryan, M. S., *Hygiene Regulations for Chlorine-substituted Butenes and Butadienes and Relationship Between Their Structure and Toxicity*, Author's abstract of thesis, Kiyv, 1985, 34 (in Russian).
2. Grigoryan, A. O., in *Conf. Junior Scientists*, Proc. Research Institute Occup. Med., Acad. Sci. USSR, Moscow, 1965, 79 (in Russian).

1,3-DICHLORO-2-BUTENE

Molecular Formula. $C_4H_6C_{l2}$
M = 125.01
CAS No 926-57-8
RTECS No EM4760000
Abbreviation. DCB.

Properties. Colorless, volatile liquid with a strong odor. Poorly soluble in water. Odor perception threshold is 0.05 mg/l. Heating increases intensity of odor.[1]

Applications. Used in the synthesis of rubber.

Acute Toxicity. In mice, LD_{50} was found to be about 0.5 g/kg BW;[2] in rats, it was 3.1 mmol/kg BW. Poisoning was accompanied by changes in the CNS, myocardium, liver, and kidney. Histology examination revealed signs of fatty and parenchymatous dystrophy in the viscera.[3]

Long-term Toxicity. In a 6-month study, rabbits were dosed by gavage with 1.0 mg/kg and 10 mg/kg BW doses. The treatment resulted in increased glucose and pyruvic acid blood levels. Histology examination revealed changes in the liver and kidneys.[1,4]

Mutagenicity. C57B mice received a dose of 100 mg DDB, consisting of 1,3-dichlorobutane, 1,3- and 3,3-dichlorobutene, per kg BW. The exposure caused cytogenic effect on bone marrow cells of the mice.[2] The significant increase in frequency of CA by 5th day of the treatment was observed in mice, given *i/g* 10 mg DCB/kg BW over a period of 20 days.[5]

Standards. ***Russia*** (1995). MAC and PML: 0.05 mg/l (organolept., odor).

References:

1. Khachatryan, M. K., in *Protection of Water Reservoirs against Pollution by Industrial Liquid Effluents*, S. N. Cherkinsky, Ed., Medgiz, Moscow, Issue No 5, 1962, 44 (in Russian).
2. Ekshtat, B. Y. et al., Cytogenetic effect of DDB compounds, *Gig.* Sanit., 12, 22, 1971 (in Russian).
3. Petrosyan, F. K. and Gizhlaryan, M. S., Local and skin-resorptive effect of chemicals applied in chloroprene rubber manufacture, in experiment, *Gig. Truda Prof. Zabol.*, 4, 51, 1984 (in Russian).
4. Frolov, I. N., in *Proc. Republ. Conf. Hygienic Studies in 1966-1967*, F. F. Erisman Research Sanitary Hygiene Institute, Moscow-Stavropol', 1969, 219 (in Russian).
5. Matveeva, V. G. Kerkis Yu. Ya., and Ekshtat, B. Ya., Mutagenic effect in mammalian bone marrow cells following DDB pesticide prolonged administration, *Gig. Sanit.*, 1, 94, 1973 (in Russian).

1,1-DICHLOROETHYLENE

Molecular Formula. $C_2H_2Cl_2$
M = 96.9
CAS No 75-35-4
RTECS No KV9275000
Abbreviation. 1,1-DCE.

Synonyms. 1,1-Dichloroethene; Vinylidene chloride (VDC).

Properties. Clear liquid with a sweet odor. Solubility in water is 2.5 g/l at 25°C. 1,1-DCE is readily soluble in alcohol, miscible with the majority of organic solvents. Odor perception threshold is 1.5 mg/l.[02]

Applications. A co-monomer in the production of a number of polymers used to manufacture food containers and packaging (*Saran* film). Copolymerizes with other vinyl monomers such as acrylonitrile, alkylacrylates, methacrylates, vinyl acetate, and vinyl chloride. A solvent.

Exposure. In the U.S., a daily drinking water exposure has been estimated the maximum as high as 0.001 mg/l.

Migration from food packaging materials is expected to be very low due to limited release and rapid degradation.

Acute Toxicity. In adult rats, LD_{50} was 200 to 1800 mg/kg BW,[1] in young and fasted rats, it was about 50 mg/kg BW.[2] In mice and dogs, these values were 200 and 5750 mg/kg BW, respectively.[3]

Short-term Toxicity. Liver and kidney damage was generally observed in animals after a high dose exposure. Male Swiss OF-1 mice were exposed orally to DCE. The treatment caused damage to about 50% of the proximal tubules.[4] Other changes noted are CNS depression and sensitization of the heart. Vacuolization of hepatocytes was observed in rats receiving 5.0 to 40 mg/kg BW with their drinking water for 3 months.[5]

Long-term Toxicity. Sprague-Dawley rats were given 9.0, 14, and 30 mg/kg BW (females), or 7.0, 10, and 20 mg/kg BW (males) with their drinking water for 2 years. The most sensitive endpoint was liver damage. No consistent treatment-related biochemical changes were observed in any parameter measured. The only abnormal histopathology reported was mid-zonal fatty accumulation in the livers of both sexes receiving the highest dose. No liver degeneration was observed. A NOAEL of 100 ppm (9.0 mg/kg BW) in females was identified, based upon a trend toward increased fatty deposition in the liver. Changes and hypertrophy of the liver cells in females at all doses and male at the highest dose were reported.[6,7]

After giving doses of 0.5 to 20 mg/kg BW by gavage in corn oil for 1 year no adverse effects were observed on post-mortem examination of Sprague-Dawley rats.[8] The dose of 5.0 mg/kg BW given in corn oil (but not 1.0 mg/kg BW) for two years caused renal inflammation in F344/N rats. Liver necrosis was found at the dose level of 10 mg/kg BW in male mice and of 2.0 mg/kg BW in female mice.[9]

Reproductive Toxicity.

Embryotoxicity. Concentrations of 50 to 200 ppm in drinking water did not affect reproductive capacity in a three-generation study using Sprague-Dawley rats.[10] Rats received up to 500 ppm in the diet for 2 years. The treatment caused no increase in embryolethality and did not affect fetal weight. Approximately 60 to 70% 1,1-DCE was actually consumed.[11]

Teratogenic effects were not shown in rats following ingestion of 1,1-DCE in their drinking water at the concentration of 200 mg/l on days 6 through 15 of gestation.[12]

Mutagenicity.

Observations in man. There are no data available on the genetic and related effects of 1,1-DCE in humans (IARC 39-211).

In vivo cytogenetics. Adverse effects were not shown in the DLM assay in mice or rats,[4,13,14] and using *V79* Chinese hamster ovary cells.[15]

In vitro genotoxicity. However, according to other data, 1,1-DCE induced CA and SCE in Chinese hamster ovary cells in the the presence of *S9* mix prepared from rat liver. It was negative in micronucleus test using mouse bone marrow, fetal liver and blood.[16] 1,1-DCE did not induced CA in the bone marrow cells of rats treated *in vivo*. It induced unscheduled DNA synthesis in mice but not in rat hepatocytes. In the short-term tests, positive responses for DNA alkylation, repair, and synthesis were observed. 1,1-DCE showed mutagenic effects in *Salmonella* and in *E. coli*.[12,17]

Carcinogenicity. The compound is structurally similar to *vinyl chloride*. Oral administration to mice (2.0 to 10 mg/kg BW) and rats (50 mg/kg BW) every week caused no statistically significant tumor growth, though tumors that were not seen in the controls appeared at a variety of sites in the treated rats.[8] Malignant neoplasms were found in the kidneys, lungs, and mammary glands of mice inhaling 1,1-DCE.[7]

Carcinogenicity classifications. An IARC Working Group concluded that there is limited evidence for the carcinogenicity of 1,1-DCE in *experimental animals* and there is inadequate evidence for the carcinogenicity of 1,1-DCE in *humans*.

IARC: 3;

U.S. EPA: C;

NTP: N - E - N - E (gavage).

Chemobiokinetics. 1,1-DCE is completely absorbed after ingestion and excreted within 72 hours. The highest concentrations were discovered in the liver and kidneys. 1,1-DCE is transformed into

1,1-dichloroethylene oxide and *chloroacetyl chloride*, which are probably responsible for the carcinogenic effect in mice.[3] The *monochloroacetic acid* was also found to be a metabolite, and it can be conjugated with glutathione.[3]

In male mice, but not in rats and female mice, 1,1-DCE caused renal tumor after inhalation. According to Speerschneider and Dekant, renal bioactivation of 1,1-DCE occurs due to male-specific expression of cytochrome *P-450 2El*.[18] According to Ban et al., DCE undergoes biotransformation by *NADPH*-cytochrome *P-450* to several reactive species, which conjugate with glutathione. Further modification by *beta*-lyase and cysteine conjugate *S*-oxidase led to formation of reactive metabolite(s).[19] 1,1-DCE cytotoxic effects are associated with cytochrome *P-450*-dependent formation of metabolite(s) that had bound covalently to tissue macromolecules.[20] A greater proportion is removed unaltered through the lungs.[4]

Regulations.

EU (1990). VDC is available in the *List of authorized monomers and other starting substances which shall be used for the manufacture of plastic materials and articles intended to come into contact with foodstuffs (Section A)*.

U.S. FDA (1998) permits the use of VDC copolymers as components of articles intended for use in contact with food, including (1) adhesives in accordance with the conditions prescribed in 21 CFR part 175.105. 1,1-DCE may be used (2) in the production of resinous and polymeric coatings for polyolefin films in accordance with the conditions prescribed in 21 CFR part 175.320; as (3) a monomer in the production of vinylidene chloride copolymer coatings that applied on nylon or polycarbonate film intended for use as food-contact surfaces; as (4) a component (a monomer) of the uncoated or coated food-contact surface of paper and paperboard intended for use in producing, manufacturing, packaging, processing, preparing, treating, packing, transporting, or holding dry food (providing the finished copolymers contain at least 50 wt% polymer units derived from vinylidene chloride); (5) as polymers which may be safely admixed, alone or in mixture with other permitted polymers, as modifiers in semirigid and rigid vinyl chloride plastic food contact articles prepared from vinyl chloride homopolymers and/or from vinyl chloride copolymers in accordance with the conditions specified in the CFR; (6) in the manufacture of cellophane to be safely used for packaging food in accordance with the conditions prescribed in 21 CFR part 177.1200; (7) as a monomer in the manufacture of semirigid and rigid acrylic and modified acrylic plastics in the manufacture of articles intended for use in contact with food in accordance with the conditions prescribed in 21 CFR part 177.1010; (8) as a monomer in the manufacture of polyethylenephthalate polymers used as, or components of plastics (films, articles, or fabric) intended for use in contact with food in accordance with the conditions prescribed in 21 CFR part 177.1630; (9) packaging materials for use during the irradiation of prepackaged food (providing the film contains not less than 70% by weight of VDC and has a viscosity of 0.5-1.5 cP); and as (10) polymer modifiers in semirigid and rigid vinyl chloride plastics.

VDC-methyl acrylate copolymers may be used in contact with food providing that (1) less than 15% by weight of the polymer units are derived from methyl acrylate; (2) the average molecular mass of the copolymer is not less than 50,000; and (3) the residual VDC will not exceed 10 mg/kg.

Recommendations. ***WHO*** (1996). Guideline value for drinking water: 0.02 mg/l.

Standards.

EU (1990). Proposed limits in food: 0.05 mg/kg, Maximum Permitted Quantity in materials and articles: 5.0 mg/kg

U.S. EPA (1999). MCL and MCLG: 0.007 mg/l.

Russia (1995). PML: 0.0006 mg/l.

References:

1. Ponomarkov, V. and Tomatis, L., Long-term testing of vinyliden chloride and chloroprene for carcinogenicity in rats, *Oncology*, 37, 136, 1980.
2. Andersen, M. E. and Jenkins, L. R., Oral toxicity of 1,1-dichloroethylene in the rat: Effects of sex, age and fasting, *Environ. Health Perspect.*, 21, 157, 1977.
3. Jones, B. D. and Hathway, D. E., Differences in metabolism of vinylidene chloride between mice and rats, *Brit. J. Cancer*, 37, 411, 1978.

4. Simmon, V. F., Kauhanen, K., and Tardiff, R. G., Mutagenic activity of chemicals identified in drinking water, *Dev. Toxicol. Environ. Sci.*, 2, 249, 1977.
5. Norris, J. M., Toxicologic and pharmacokinetic studies on inhaled and ingested vinylidene chloride in laboratory animals, in *Proc. Techn. Assoc. Paper. Synth. Conf.*, Atlanta, GA, TAPPI, 1977.
6. Rampy, L. W., Quast, J. F., Humiston, C. G., et al., Interim results of two-year toxicological studies in rats of vinylidene chloride incorporated in the drinking water or administered by repeated inhalation. *Environ. Health Perspect.*, 21, 33, 1977.
7. Quast, J. F., Humiston, C. G., Wade, C. E., et al., A chronic toxicity and oncogenicity study in rats and subchronic toxicity study in dogs on ingested vinylidene chloride, *Fundam. Appl. Toxicol.*, 3, 55, 1983.
8. Maltoni, C., Cotti, G., and Chieco, P., Chronic toxicity and carcinogenicity bioassays of vinylidene chloride, *Acta Oncol.*, 5, 91, 1984.
9. *Carcinogenesis Bioassay of Vinylidene Chloride in F344 Rats and B6C3F$_1$ Mice (Gavage Study)*, NTP Technical Report Series No 124, U.S. PHS, NTP-80-2, NIH Publ. No 82-1784, 1982.
10. Nitschke, K. D., Smith, F. A., Quast, J. F., et al., A three-generation rat reproductive toxicity study of vinylidene chloride in the drinking water, *Fundam. Appl. Toxicol.*, 3, 75, 1983.
11. Alumot, E., Nachtomi, E., Mandal, E., and Holstein, P., Tolerance and acceptable daily intake of chlorinated fumigant in the rat diet, *Food Cosmet. Toxicol.*, 14, 105, 1976.
12. Murray, F. J., Nitschke, K. D., Rampy, L. W., and Schwetz, B. A., Embryotoxicity and fetotoxicity of inhaled or ingested vinylidene chloride in rats and rabbits, *Toxicol. Appl. Pharmacol.*, 49, 189, 1979.
13. Short, R. D., Minor, J. L., Winston, J. M., and Lee, C. C., A dominant lethal study in male rats after repeated exposure to vinyl chloride or vinylidene chloride, *J. Toxicol. Environ. Health*, 3, 965, 1977.
14. Anderson, D. et al., Dominant lethal studies with the halogenated vinyl chloride and vinylidene chloride in male CD-1 mice, *Environ. Health Perspect.*, 21, 71, 1977.
15. Drevon, C. and Kuroki, T., Mutagenicity of vinyl chloride, vinylidene chloride and chloroprene in V79 Chinese hamster cells, *Mutat. Res.*, 67, 173, 1979.
16. Sawada, M., Sofuni, T., and Ishidate, M., Cytogenetic studies on 1,1-dichloroethylene and its two isomers in mammalian cells *in vitro* and *in vivo*, *Mutat. Res.*, 187, 157, 1987.
17. Bartsch, H., Malaveille, C., Barbin, A., et al., Tissue-mediated mutagenicity of vinylidene chloride and 2-chlorobutadiene in Salmonella typhimurium, *Nature*, 155, 641, 1975.
18. Speerschneider, P. and Dekant, W., Renal tumorogenicity of 1,1-dichloroethene in mice: the role of male-specific expression of cytochrome P-450 2El in the renal bioactivation of 1,1-dichloroethene, *Toxicol. Appl. Pharmacol.*, 130, 48, 1995.
19. Ban, M., Hettich, D., Huguet, N., and Cavelier, L., Nephrotoxicity mechanism of 1,1-dichloroethylene in mice, *Toxicol Lett.*, 78, 87, 1995.
20. Lee, R. P. and Forkert, P. G., *In vitro* biotransformation of 1,1-dichloroethylene by hepatic cytochrome P-450 2El in mice, *J. Pharmacol. Exp. Ther.*, 270, 371, 1994.

DICYCLOPENTADIENE

Molecular Formula. $C_{10}H_{12}$
M = 132.21
CAS No 77-73-6
RTECS No PC1050000
Abbreviation. DCPD.

Synonyms. Biscyclopentadiene; Cyclopentadiene; 1,3-Cyclopentadiene; 3a,4,7,7a-Tetrahydro-4,7- methanoindene.

Properties. Colorless, crystalline powder with a pungent, nauseous odor. Water solubility is 26 g/l. Miscible with alcohol and ether. Odor perception threshold in water is 0.006 mg/l[1] or 0.011 mg/l.[010]

Applications. Used as a monomer in the production of synthetic rubber intended for use in contact with food or drink.

Acute Toxicity. In rats, LD_{50} of pure DCPD is 670 mg/kg BW[1] or 480 mg/kg BW;[2] LD_{50} of the technical-grade product is 1000 mg/kg BW.[3] Poisoning resulted in general congestion, hyperemia, and focal hemorrhage in different organ tissues. Pathological changes were observed in the kidney, intestine, stomach, bladder, and particularly, the lung.[03] Death within 1 to 2 days preceded by clonic convulsions.

Repeated Exposure failed to reveal cumulative properties. Rats tolerated administration of 1/10 to 1/50 LD_{50} for 30 days. DCPD has a polytropic toxic action (see ***Norborndiene***). The LOAEL was estimated to be 5.0 mg/kg BW.[2]

In 28-day study, F344 rats received 8.0, 40, or 200 mg/kg BW. At the highest dose tested, there were changes in BW gain and organ weights, as well as histologically revealed hypertrophy of the adrenal cortex, and foamy cytoplasm in hepatocytes. The NOEL of 8.0 mg/kg BW was identified in this study.[4]

Short-term Toxicity. There were no signs of intoxication in rats given 0.4 and 4.0 mg DCPD/kg BW, or in mice given 0.15 and 1.5 mg DCPD/kg BW for 90 days.[4]

Reproductive Toxicity. No toxic ***effect on embryos*** was observed. A dose of 0.1 mg/kg BW had no ***gonadotoxic effect***.[2]

Mutagenicity.

In vitro genotoxicity. DCPD was found to be negative in *Salmonella typhimurium* assay in the presence or absence of exogenous metabolic activation.[015]

Chemobiokinetics. Following oral administration, labeled ^{14}C-DCPD was found in the bladder, bile, and fatty tissue. Unchanged DCPD is exhaled or excreted via the kidneys. Metabolites were also detected in the urine (Ross and Dacre, 1977).

Regulations.

EU (1990). DCPD is available in the *List of monomers and other starting substances which may continue to be used for the manufacture of plastic materials and articles intended to come into contact with foodstuffs pending a decision on inclusion in Section A* (*Section B*).

Great Britain (1998). DCPD is authorized up to the end of 2001 for use in the production of polymeric materials and articles in contact with food or drink or intended for such contact.

Standards. ***Russia*** (1995). MAC: 0.015 mg/l (organolept., odor), PML: 0.002 mg/l.[4]

References:

1. Taradin, Ya. I., Sanitary-toxicological characteristics of dicyclopentadiene, in *Toxicology and Hygiene of Petroleum Chemistry Products*, Yaroslavl, 1972, 197 (in Russian).
2. Zholdakova, Z. I., Sil'vestrov, A. E., and Mikhailovsky, N. I., Substantiation of maximum allowable concentration for norbornene, norbornadiene and dicyclopentadiene in water bodies, *Gig. Sanit.*, 2, 77, 1986 (in Russian).
3. Krat, A. V., in *Hygiene of Application and Toxicology of Pesticides and Polymeric Materials*, L. I. Medved', Ed., Kiev, 13, 98, 1983 (in Russian).
4. Satoh, M., Okamiya, H., Furukawa, F., et al., Twenty-eight day repeated dose toxicity test of dicyclopentadiene in F344 rats, Abstract, *Eisei Shikenjo Hokoku,* 108, 71, 1990 (in Japanese).

N-(1,1-DIMETHYL-3-OXOBUTYL) ACRYLAMIDE

Molecular Formula. $C_9H_{15}NO_2$

M = 169.25

CAS No 2873-97-4

RTECS No AS3475000

Synonyms. Diacetone acrylamide; *N*-(1,1-Dimethyl-3-oxobutyl)-2-propenamide; *N*-[2-(2-Methyl-4-oxopentyl)]acrylamide.

Properties. White crystalline solid. Readily soluble in water and in organic solvents.

Applications. Used in the manufacture of coatings, laminates, sealers, adhesives, etc. intended for use in contact with food or drink. Cross-linking agent for polyesters.

Acute Toxicity. LD_{50} is 1.77 g/kg BW in rats. Poisoning caused CNS inhibition, convulsions, changes in respiratory system.[1] LD_{50} is 1.3 g/kg BW in mice.[2]

Repeated Exposure produced no neurotoxic effect.[2]

Regulations. ***U.S. FDA*** (1998) approved the use of D. in adhesives as a component of articles intended for use in packaging, transporting, or holding food in accordance with the conditions prescribed in 21 CFR part 175.105.

References:

1. Acute Toxicity Data, *J. Am. Coll. Toxicol.*, Part B, 1, 113, 1990.
2. Hashimoto, K., Sakamoto, J., and Tanii, H., Neurotoxicity of acrylamide and related compounds and their effects on male gonads in mice, *Arch. Toxicol.*, 47, 179, 1981.

DIMETHYL TEREPHTHALATE

Molecular Formula. $C_{10}H_{10}O_4$
M = 194.19
CAS No 120-61-6
RTECS No WZ1225000
Abbreviation. DMTP.

Synonyms. 1,4-Benzenedicarboxylic acid, dimethyl ester; Dimethyl 1,4-benzenedicarboxylate; Dimethyl *p*-phthalate; Methyl-4-carbomethoxybenzoate; Terephthalic acid, dimethyl ester.

Properties. White, crystalline powder. Poorly soluble in cold water. Solubility in hot water is 3.32 g/l. Soluble in alcohol. Odor perception thresholds are 13.4 mg/l (20°C) and 1.8 mg/l (60°C).[1]

Applications. A monomer in the production of polyesters, mainly of polyethylene terephthalate (PET) intended for use in contact with food or drink.

Migration of total levels of PET oligomers was detected by using an analytical approach that involves hydrolysis of oligomers to terephthalic acid, methylation, and detection as DMTP. Total levels of migration of PET oligomers were 0.02 to 2.73 mg/kg depending on the food and temperature attained during cooking.[2]

Acute Toxicity. LD_{50} exceeded 5.0 g/kg BW in guinea pigs.[04] Male rats tolerate a 6.6 g/kg BW dose, and mice did not die from a dose of 10 g/kg BW. Poisoning is accompanied with tremor and increased motor activity (followed by depression in mice). Gross pathology examination revealed no histopathological lesions.[1]

Repeated Exposure to high levels (5.0%) in the diet of rats for 28 days revealed BW loss, marked reduction in food consumption and high mortality (for unclear reasons). Ten administrations of 2.0 to 4.0 g/kg BW doses did not alter cholinesterase activity in the blood serum and produced no histopathological lesions in the visceral organs of rats. There were no changes in blood analysis.[04]

Short-term Toxicity. Oral exposure of rats to 1.0% DMTP emulsion in the feed for 96 days, and to 50 and 200 mg/kg BW for 115 days resulted in decreased BW gain.[1] A 500 mg/kg dose administered for 35 to 39 days caused no toxic effect.[3] 28-day-old male rats were fed 3.0% DMTP in their diet for 2 weeks. All the animals developed bladder stones. DMTP concentrations less than 1.5% produced no such effect. Histology examination revealed epithelial hyperplasia.[4]

Long-term Toxicity. 0.5 and 7.5 mg/kg BW doses caused no histopathological changes in the visceral organs of rats, and did not affect their conditioned reflex activity.[1] F344 rats and $B6C3F_1$ mice received 2.5 or 5.0 ppm DMTP in their feed for 103 weeks. The treatment caused no effect on the mean BW, and induced no signs of toxic action in the rats. Survival was not affected in rats and mice.[5]

Allergenic Effect.

Animal studies. DMTP produced a sensitizing effect. 0.075 mg/kg BW and higher doses administered *i/g* for 30 days caused the development of allergic and autoallergic changes manifested by specifically increased degranulation of peripheral blood basophils. There was an increase in the number of autoimmune hemolysis plaques and in the formation of antibodies against both hapten and tissue antigen.[6]

Reproductive Toxicity.

Inhalation study in pregnant rats revealed no ***embryotoxic*** or ***teratogenic effect***.[7]

Mutagenicity.

In vivo cytogenetics. DMTP was positive in *Dr. melanogaster* and in micronuclear test in mice following a single *i/p* administration of 65 and 195 mg DMTP/kg BW.[8]

In vitro genotoxicity. DMTP was not found to produce CA and SCE in Chinese hamster ovary cells (NTP-85). DMTP was not genotoxic in *Salmonella* mutagenicity assay, DNA single-strand break assays, in primary rat hepatocytes, unscheduled DNA synthesis in *HeLa* cells, CA and in micronucleus assays in human peripheral blood lymphocytes.[9]

Carcinogenicity. DMTP showed no carcinogenic potential in the rat and mouse feeding study at doses of 5.0 g/kg BW.[5,10]

Carcinogenicity classification.

NTP: N - N - E - N (feed).

Chemobiokinetics. Following single and repeated ingestion by rabbits, ^{14}C-DMTP is readily absorbed and excreted from the body. There was no evidence of tissue accumulation.[11]

Regulations.

EU (1990). DMTP is available in the *List of authorized monomers and other starting substances which shall be used for the manufacture of plastic materials and articles intended to come into contact with foodstuffs (Section A).*

Great Britain (1998). DMTP is authorized without time limit for use in the production of polymeric materials and articles in contact with food or drink or intended for such contact.

Standards. *Russia* (1995). MAC: 1.5 mg/l (organolept., odor).

References:

1. Krassavage, W. I., Yanno, F. J., and Terhaar, C. J., Dimethyl terephthalate: Acute toxicity, subacute feeding and inhalation studies in male rats, *Am. Ind. Hyg. Assoc. J.*, 34, 455, 1973.
2. Castle, L., Mayo, A., Crews, C., et al., Migration of poly(ethylene terephthalate) (PET) olygomers from PET plastics into foods during microwave and conventional cooking and into bottled beverages, *J. Food Prot.*, 52, 337, 1989.
3. *Problems of Environmental Hygiene*, D. N. Kalyuzhny, Ed., Kiyv, Issue No 6, 1966, 94 (in Russian).
4. Chin, T. Y., Tyl, R. W., Papp, J. A., et al., Chemical urolithiasis. I. Characteristics of bladder stone induction by terephthalic acid and dimethylterephthalate in weanling Fischer 344 rats, *Toxicol. Appl. Pharmacol.*, 58, 307, 1981.
5. *Bioassay of Dimethyl Terephthalate for Possible Carcinogenicity,* NTP Technical Report Series No 121, Research Triangle Park, NC,1979.
6. Vinogradov, G. I., Vinarskaya, E. I., and Antomonov, M. Yu., Hygienic assessment of dimethyl terephthalate allergenic activity at inhalation and oral intake, *Gig. Sanit.*, 5, 7, 1986 (in Russian).
7. Krotov, Ya. A. and Chebotar', N. A., Studies of embryotoxic and teratogenic effects of some industrial chemicals produced in dimethyl terephthalate manufacture, *Gig. Truda Prof. Zabol.*, 6, 40, 1972 (in Russian).
8. Goncharova, R. I., in *The 14th Annual Conf. European Society Environ. Mutagens*, Abstracts, Moscow, 1984, 199 (in Russian).
9. Monarca, S., Pool-Zobel, B. L., Rizzi, R., Klein, P., Schmezer, P., Piatti, E., Pasquini, R., De Fusco, R., and Biscardi, D., *In vitro* genotoxicity of dimethyl terephthalate, *Mutat. Res.*, 262, 85, 1991.
10. Haseman, J. K. and Clark, A. M., Carcinogenicity results for 114 laboratory animal studies used to assess the predictivity of four *in vitro* genetic activity assays for rodent carcinogenicity, *Environ. Mutagen.*, 16 (Suppl. 18), 15, 1990.
11. Moffitt, N., Clary, J. J., Lewis, T. R., Blanck, M. D., and Perone, V. B., Absorption, distribution and excretion of terephthalic acid and dimethylterephthalate, *Am. Ind. Hyg. Assoc. J.*, 36, 633, 1975.

DIVINYL BENZENE

Molecular Formula. $C_{10}H_{10}$

M = 130.18

CAS No 1321-74-0

RTECS No CZ9370000
Abbreviation. DVB.

Synonyms. Diethenyl benzene; Vinyl benzene; Vinyl styrene.

Properties. Solubility in water 50 mg/l.

Applications. Polymerization monomer for special synthetic rubbers, styrene-vinylstyrene ion-exchange resins, and polyesters intended for use in contact with food or drink. Monomer and cross-linking agent for acrylic polymers.

Acute Toxicity. DVB appears to resemble that of styrene. 10 rats were given orally 2.5 ml DVB at 1:1 ratio in olive oil. Half of animals died.[03] DVB exhibited narcotic action.

Carcinogenicity. There is concern about the potential carcinogenicity of DVB because it can be oxidized by cytochrome P450 to *styrene-7,8-epoxide*.

Chemobiokinetics. DVB is structurally similar to styrene and is likely to be biotransformed by the same metabolic pathways.

Regulations.

U.S. FDA (1998) approved the use of DVB (monomer) (1) in adhesives as a component of articles intended for use in packaging, transporting, or holding food in accordance with the conditions prescribed in 21 CFR part 175.105; (2) as a component (monomer) of the uncoated or coated food-contact surface of paper and paperboard intended for use in producing, manufacturing, packaging, processing, preparing, treating, packing, transporting, or holding dry food in accordance with the conditions prescribed in 21 CFR part 176.180; (3) as a monomer in the manufacture of semirigid and rigid acrylic and modified acrylic plastics in the manufacture of articles intended for use in contact with food in accordance with the conditions prescribed in 21 CFR part 177.1010; and (4) in surface lubricants employed in the manufacture of metallic articles that contact food in accordance with the conditions prescribed in 21 CFR part 177.1010.

EU (1990). DVB is available in the *List of authorized monomers and other starting substances which may continue to be used for the manufacture of plastic materials and articles intended to come into contact with foodstuffs pending a decision on inclusion in Section A* (*Section B*).

Great Britain (1998). DVB is authorized up to the end of 2001 for use in the production of polymeric materials and articles in contact with food or drink or intended for such contact.

Reference:

Morgan, D. L., Mahler, J. F., Wilson, R. E., Moorman, M. P., Price, H. C., and O'Connor, R. W., Toxicity of divinylbenzene-55 for B6C3F$_1$ mice in two-week inhalation study, *Fundam. Appl. Toxicol.*, 39, 89, 1997.

EPICHLOROHYDRIN

Molecular Formula. C_3H_5ClO
M = 92.52
CAS No 106-89-8
RTECS No TX4900000
Abbreviation. ECH.

Synonyms. 1-Chloro-2,3-epoxypropane; 3-Chloro-1,2-epoxypropane; (Chloromethyl)oxyrane; α-Chloropropylene oxide; Glycerol epichlorohydrin; Glycidyl chloride.

Properties. Colorless, mobile liquid with an irritant, chloroform-like odor. Solubility in water is 66 g/l at 20°C. ECH is hydrolyzed in the aqueous media, causing difficulties for drinking water monitoring. Odor perception threshold is reported to be 1.0 mg/l[1] or 3.0 mg/l.[02] Irritation threshold is 0.1 mg/l.

Applications. ECH is widely used in the manufacture of food-contact unmodified epoxy resins, elastomers, water treatment resins, ion exchange resins, plasticizers, stabilizers, solvents.

Migration into food and drinking water from plastics is possible but is expected to be low. Migration into water from epoxy coatings (exposure 7 days, 37°C) was determined at the level of 0.05 mg/l.[2]

Acute Toxicity.

Observations in man. Ingestion of ECH may cause death in man due to respiratory insufficiency.

Animal studies. Oral doses of 325 and 500 mg/kg BW caused renal damage, vacuolization and fatty degeneration in the liver of the test animals. Necrotic foci were found in the GI tract.[3] LD_{50} values appeared to be 140 to 260 mg/kg BW in rats and mice, 345 mg/kg BW in rabbits, and 280 mg/kg BW in guinea pigs.[1,4]

According to other data,[3] ECH exhibits lower acute toxicity: rats and mice tolerate 250 mg/kg BW without visible symptoms but all of them die from the dose of 325 mg/kg BW. Poisoned animals displayed adynamia, labored respiration, marked cutaneous hyperemia, ataxia, tremor, and abdominal swelling. Histopathological changes in the lungs, liver, kidneys, suprarenals, and thyroid were found.

Repeated Exposure failed to reveal signs of accumulation. Administration of 1/10 LD_{50} for 30 days did not cause mortality. However, according to other data,[5] a nephrotoxic effect with a tendency to accumulation, as well as a change in enzyme activity in the kidneys was found to develop. It exerted a narcotic action and irritated the skin and mucous membranes. Rabbits were given a dose of 80 mg/kg BW for 2 months. The treatment caused retardation of BW gain, reduction in leukocyte count and glutathione content in the blood, hypercholesterinemia, and an increase in the relative weights of the kidneys and suprarenals.[1]

Damage to the kidney was demonstrated after administration of 40 and 80 mg/kg BW.[6,7] Sprague-Dawley rats received ECH by gavage in distilled water for 10 consecutive days at dose levels of 3.0, 7.0, 19, or 46 mg/kg BW. The treatment did not affect mortality, toxicity was evident at the higher doses: losses in BW gain and organ weights, reductions in food and water consumption, hematology and histology analyses. Significant dose-related changes in mucosa, hyperplasia, and hyperkeratosis were noted in the fore-stomach as a primary target organ. The LOAEL of 3.0 mg/kg BW was established in this study.[8]

Short-term Toxicity. In the above study, Sprague-Dawley rats received ECH by gavage in distilled water for 90 days at dose levels of 1.0, 5.0, and 25 mg/kg BW. The NOAEL of 1.0 mg/kg BW was established in this study.[8]

Long-term Toxicity. Biological activity of ECH is likely to be determined by the presence of chlorine in its molecule. ECH affects the CNS, liver and kidney functions, oxidation processes, neuroendocrine control and the blood system. In a two-year study, weanling Wistar rats were given ECH in drinking water by gavage (doses of 2.0 mg/kg and 10 mg/kg BW). Gradual increase in mortality in males with clinical symptoms including dyspnea and weight loss, decrease in leukocyte count at dose levels of 1.43 and 7.14 mg/kg BW was observed. The LOAEL of 1.4 mg/kg BW for forestomach hyperplasia was identified.[9]

Reproductive Toxicity.

Gonadotoxicity.

Observations in man. No significant alteration of sperm motility were observed in workers occupied in ECH production for more than 5 years and exposed by inhalation to less than 1.0 ppm of ECH.[10]

Animal studies. 5 daily doses of 50 mg/kg BW or one single dose of 100 mg/kg BW caused permanent sterility in rats.[11] Antifertility effects in male rats after single oral or *i/p* dose of 50 mg/kg BW were reported.[12]

Administration of 10 mg/kg BW for 3 months or 15 mg/kg BW for 12 days also reduced the fertilizing ability of male rats, but a dose of 2.0 mg/kg BW had no effect.[7,13] The sperm of rats which received 25 or 50 mg ECH/kg BW showed an increased percentage of abnormal sperm heads at the higher dose and a reduced number of sperm heads at the lower dose, while no changes were observed in the weights and microscopic picture of the testes.[14]

Embryotoxicity and ***teratogenicity.*** Administration of ECH to Sprague-Dawley rats and CD-1 mice on days 6 to 15 of pregnancy produced no embryotoxic and teratogenic effect even at doses causing 10% maternal mortality (up to 160 mg/kg BW); BW was reduced and liver weight increased in pregnant mice.[15]

Mutagenicity. ECH has been shown to be genotoxic *in vitro* and *in vivo*.

Observations in man. An increase in incidences of CA has been observed in workers exposed to this compound by inhalation, although the studies are difficult to interpret (IARC, Suppl. 6-286).

In vitro genotoxicity. ECH causes gene mutations in all cell systems, as well as CA in eukaryotic cells, and should be considered as potentially dangerous to man. ECH induced CA, SCE, and unscheduled DNA synthesis in human cells *in vitro*. Weakly positive results were obtained in a cell transformation assay in C3H 10T1/2 cells. It induced CA, SCE, mutation and DNA strand breaks in rodent cells *in vitro* (IARC 11-131).

Animal studies. ECH induced SCE in bone marrow cells but not micronuclei or DLM in mice treated *in vivo*; equivocal findings were found for CA. ECH has been shown to be a substance of relatively low DNA reactivity in *Dr. melanogaster* assay.[16]

Carcinogenicity. Positive results were obtained in several carcinogenicity bioassays with rats exposed via multiple routes of administration. Exposure by ingestion (by gavage or via drinking water) caused forestomach tumors, and exposure via inhalation resulted in tumors of the nasal cavity. Tumors were shown to be present only at the site of administration where ECH was highly irritant. After administration of 80 mg/kg BW for 3 months, papillomata and squamous cell carcinoma were found in 2 out of 5 rats. Administration of 20 to 100 mg/kg BW in distilled water to Wistar rats for 2 years caused stomach tumors to develop.[7]

18 male Wistar rats were given 18, 39, and 89 mg/kg BW with their drinking water for 81 weeks: forestomach tumors at two higher doses and prestomach hyperplasia at all three doses were found. Mortality in animals increased up to 33 to 45%.[17] No carcinogenic effect was found using a rapid method of organotypical culture of pulmonary and renal embryonic tissue.[18]

Carcinogenicity classification. An IARC Working Group concluded that there is sufficient evidence for the carcinogenicity of ECH in *experimental animals* and there were inadequate data available to evaluate the carcinogenicity of ECH in *humans*.

The Annual Report of Carcinogens issued by the U.S. Department of Health and Human Services (1998) defines ECH to be a substance which may reasonably be anticipated to be carcinogen, i.e., a substance for which there is a limited evidence of carcinogenicity in *humans* or sufficient evidence of carcinogenicity in *experimental animals.*

IARC: 2A.

Chemobiokinetics. ECH is a bifunctional alkylating agent. In the gastric juice, it is converted to *chlorohydrin.* The rate of conversion declines with ECH concentration. ECH is rapidly and extensively absorbed following oral, inhalation or dermal exposure. It is rapidly removed from the blood and is, therefore, not likely to be accumulated during chronic exposure. ECH may bind to the cellular nucleophiles such as glutathioneutathione. The major metabolites in the urine were identified as *N-acetyl-S-(3-chloro-2-hydroxypropyl)-L-cysteine* formed by conjugation with glutathione, and α-chlorohydrin.[19]

Oxalic acid that is probably a metabolite of ECH is responsible for its renal toxicity. Up to 40% of the dose is removed with the urine in 72 hours and less than 4.0% in the feces. Up to 20% of the labeled material is excreted as CO_2 in the expired air in 4 hours, indicating the rapid conversion of ECH in the body.[20]

Regulations.

EU (1990). ECH is available in the *List of authorized monomers and other starting substances which shall be used for the manufacture of plastic materials and articles intended to come into contact with foodstuffs (Section A).*

Great Britain (1998). ECH is authorized without time limit for use in the production of polymeric materials and articles in contact with food or drink or intended for such contact. A quantity of this substance in the finished plastic material or article shall not exceed 1.0 mg/kg.

In ***Japan,*** JWWA standardized epoxide resin coating condition of temperature, humidity, times, etc., according to the resin components so as to be less than 0.001 mg/l of resin ingredients released from the coating into water (JWWA K135, 1989).

Recommendations.

WHO (1996). Provisional Guideline value for drinking water: 0.0004 mg/l.[21]

U.S. EPA (1999). Health Advisory for longer-term exposure: 0.07 mg/l.

Standards.

Russia (1995). MAC and PML: 0.01 mg/l.

U.S. EPA (1999). MCL: TT (treatment technique), MCLG zero; the maximum residual content in floculating agent shall not exceed 0.01% at the maximum usage rate of 20 ppm of polymer (level of ECH would not exceed 0.0022 mg/l).

References:

1. Fedyanina, V. N., A Study of Epichlorohydrin and Diphenylolpropane Effect on the Body, Author's abstract of thesis, Novosibirsk, 1970, 22 (in Russian).

2. Krat, A. V., Kesel'man, I. M., and Sheftel', V. O., Sanitary-chemical evaluation of polymeric articles used in water-supply constructions, *Gig. Sanit.*, 10, 18, 1986 (in Russian).
3. Kremneva, S. N. and Tolgskaya, M. S., in *Toxicology of New Industrial Chemical Substances*, A. A. Letavet and I. V. Sanotsky, Eds., Medgiz, Moscow, Issue No 2, 1961, 28 (in Russian).
4. John, J. A., Quast, J. F., Murray, F. J., et al., Inhalation toxicity of epichlorohydrin: effects on fertility in rats and rabbits, *Toxicol. Appl. Pharmacol.*, 68, 415, 1983.
5. Pallade, S., Dorobantu, M., Bernsten, I., et al., De quelques modifications de l'active enzymatique dans l'intoxication par l'epichlorhydrine, *Arch. Mal. Prof. Med. Trav. Secur. Soc.*, 31, 365, 1970.
6. Lawrence, W. H., Malik, M., and Autian, J., Toxicity profile of epichlorohydrin, *J. Pharmacol. Sci.*, 61, 1712, 1972.
7. van Esch, G. J. and Wester, P. W., *Epichlorohydrin*, 1982, 25.
8. Daniel, F. B., Robinson, M., Olson, G. R., and Page, N. P., Toxicity studies of epichlorohydrin in Sprague-Dawley rats, *Drug Chem. Toxicol.*, 19, 41, 1996.
9. Wester, P. M., van der Heijden, C. A., Bisschop, A., et al., Carcinogenicity study with epichlorohydrin by gavage in rats, *Toxicology*, 36, 325, 1985.
10. Milby, T. H., Whorton, M. D., Stubbs, H. A., Ross, C. E., Joyner, R. E., and Lipshutz, L. I., Testicular function among epichlorohydrin workers, *Brit. J. Ind. Med.*, 38, 372, 1981.
11. Cooper, E. R. A., Jones, A. R., and Jackson, H., Effects of α-chlorohydrin and related compounds on the reproductive organs and fertility of the male rat, *J. Reprod. Fertil.*, 38, 379, 1974.
12. Jones, A. R., Davies, P., Edwards, K., et al., Antifertility effects and metabolism of α-epichlorohydrin in rat, *Nature* (London), 224, 83, 1969.
13. Hahn, J. D., Post-testicular antifertility effects of epichlorhydrin and 2,3-epoxypropanol, *Nature* (London) 226, 87, 1970.
14. Cassidy, E. R., Jones, A. R., and Jackson, H., Evaluation of testicular sperm head counting technique using rats exposed to dimetoxyethyl phthalate, glycerol α-monochlorohydrin, epychlorohydrin, formaldehyde, or methylmethanesulfonate, *Arch. Toxicol.*, 53, 71, 1983.
15. Marks, T., Gerling, F. S., and Staples, R. E., Teratogenic evaluation of epichlorohydrin in the mouse and rat and glycidol in the mouse, *J. Toxicol. Environ. Health*, 9, 87, 1982.
16. Konishi, T., Kawabata, A., and Denda, A., Forestomach tumors induced by orally administered epichlorohydrin in male Wistar rats, *Gann*, 71, 922, 1980.
17. Bokaneva, S. A., *Epichlorohydrin, Its Toxico-medical Characteristics and Significance in the Hygiene Regulations for New Epoxy Resins*, Authors' abstract of thesis, Moscow, 1980, 18 (in Russian).
18. Gingell, R., Beatty, P. W., Mitschke, H. R., et al., Evidence that epychlorohydrin is not a toxic metabolite of 1,2-dibromo-3-chloropropane, *Xenobiotica*, 17, 229, 1987.
19. Weigel, W. W., Plotnick, H. B., and Conner, W. L., Tissue distribution and excretion of ^{14}C-epichlorohydrin in male and female rats, *Res. Commun. Chem. Pathol. Pharmacol.*, 20, 275, 1978.
20. *Revision of the WHO Guidelines for Drinking Water Quality*, Report on the Consolidation Meeting on Organics and Pesticides, Medmenham, UK, January 30-31, 1992, 11.

ETHYLENE

Molecular Formula. C_2H_4
M = 28.05
CAS No 74-85-1
RTECS No KU5340000

Synonyms and **Trade Name.** Acetene; Elayl; Ethene.

Properties. Colorless gas with a faint odor of ether. Solubility in water is 20 mg/l (20°C) and 250 mg/l (0°C). Rapidly volatilizes from the open surface of water. Odor perception threshold is reported to be 0.039 mg/l,[02] 0.5 mg/l[1] or even 260 mg/l.[010] Does not affect the color or clarity of water.

Applications. A monomer in the production of polyolefins and copolymers intended for use in contact with food or drink.

Acute Toxicity. Mice tolerate administration of 0.5 ml of a solution with a concentration of 150 mg E./l without changes in their behavior.

Repeated Exposure. Accumulation is impossible because of E. rapid excretion from the body.

Short-term Toxicity. Mice were dosed by gavage with 3.75 mg/kg BW for 4 months. The treated animals displayed no changes in behavior or in BW gain and oxygen consumption. Gross pathology examination revealed no changes in the relative weights or in the histological structure of the visceral organs.[1]

Long-term Toxicity. Rats were given 0.05 mg/kg BW for 6 months. The treatment produced no abnormalities in behavior, BW gain, leukocyte phagocytic activity, or in cholinesterase and conditioned reflex activity.[1]

Mutagenicity.

In vivo cytogenicity. F344 rats and B6C3F_1 mice were exposed by inhalation to up to 3000 ppm E. for 6 hours a day, 5 days a week, over a four-week period. No genotoxic effect was found using the bone marrow micronucleus test.[2]

Carcinogenicity. Experiments proved E. to be converted in certain species, notably mice and rats, into the carcinogenic and mutagenic ethylene oxide.[3] Carcinogenic effect of E. of endogenous origin is suggested.[4] Whether such an effect is possible with oral administration of E. is not clear. No toxic or carcinogenic effects were found after inhalation of 300 to 3000 ppm.[5]

Carcinogenicity classification. An IARC Working Group concluded that there was sufficient evidence for the carcinogenicity of E. in *experimental animals* and limited evidence for the carcinogenicity of E. in *humans*.

IARC: 3.

Chemobiokinetics. It is unlikely that there is a direct chemical interaction between E. and biological media. E. is not broken down in the body. It seems to be rapidly excreted via the lungs.

Regulations.

EU (1990). E. is available in the *List of authorized monomers and other starting substances which shall be used for the manufacture of plastic materials and articles intended to come into contact with foodstuffs (Section A).*

U.S. FDA (1998) permits use of E.-containing polymers in products in contact with food: (1) in adhesives as a component of articles (monomer) intended for use in packaging, transporting, or holding food in accordance with the conditions prescribed in 21 CFR part 175.105; and as (2) a component (monomer) of the uncoated or coated food-contact surface of paper and paperboard intended for use in producing, manufacturing, packaging, processing, preparing, treating, packing, transporting, or holding dry food in accordance with the conditions prescribed in 21 CFR part 176.180.

Great Britain (1998). E. is authorized without time limit for use in the production of polymeric materials and articles in contact with food or drink or intended for such contact.

Standards. ***Russia*** (1995). MAC and PML: 0.5 mg/l (organolept., odor).

References:

1. Amirkhanova, G. F., Latypova, Z. V., and Tupeyeva, R. B., in *Protection of Water Reservoirs Against Pollution by Industrial Liquid Effluents*, S. N. Cherkinsky, Ed., Meditsina, Moscow, Issue No 7, 1965, 28 (in Russian).
2. Vergnes, J. S. and Pritts, L. M., Ethylene: Genotoxic potential using the bone marrow micronucleus test following four weeks of vapor exposure, *Environ. Molec. Mutagen.*, 23 (Suppl. 23), 69, 1994.
3. Filser, J. G. and Bold, H. M., Exhalation of ethylene oxide by rats on exposure to ethylene, *Mutat. Res.*, 120, 57, 1983.
4. Kokonov, M. T., Ethylene - an endogenous substance in tumor-carrier, in *Problems of Medical Chemistry*, 2, 158, 1960 (in Russian).
5. Rostron, C., Ethylene metabolism and carcinogenicity, *Food Chem. Toxicol.*, 24, 70, 1987.

ETHYLENEIMINE

Molecular Formula. C_2H_5N

M = 43.07

CAS No 151-56-4
RTECS No KX5075000
Abbreviation. EI.

Synonyms. Aziridine; Azacyclopropane; Dihydroazirene; Dimethyleneimine; Vinylamine.

Properties. Colorless liquid with an amine odor. Readily soluble in ether and acetone, soluble in water and ethyl alcohol and other organic solvents. Odor perception threshold is 170 mg/l.[02]

Applications. EI has industrial and laboratory applications. It is a monomer in the production of polyethyleneimine and copolymers. An ingredient of different coatings and adhesives used in contact with food or drink, ion exchange resins, polymer stabilizers, and surfactants. It is also used in cosmetics. On polymerization, EI forms a thermoplastic water-soluble mass used for clarification of water (polyethyleneimine).

Acute Toxicity. LD_{50} is reported to be 15 mg/kg BW in rats, and to range from 4.2 to 8.4 mg/kg BW in mice, guinea pigs, and dogs. Poisoning is accompanied by vomiting and atony. Death occurs from respiratory arrest. Administration of 10 mg/kg BW increases the urea nitrogen content in the blood.[1, 05]

Repeated Exposure usually results in general weakness and renal damage. The NS can remain unaffected.[2]

Reproductive Toxicity.

Gonadotoxic effects are likely to be specific for EI. At doses which cause no toxic effect, it produces and maintains sterility in males and females and interrupts pregnancy. EI is shown to interfere with the process of RNA synthesis in the testes.[3]

Embryotoxicity and ***teratogenicity.*** Inhalation exposure to EI affected reproduction. The LOAEL for embryotoxicity and teratogenicity effects is reported to be 1.0 mg/kg BW in rats.[4]

Mutagenicity. EI seems to be the most powerful chemical mutagen. It is exclusively genotoxic in all test systems investigated: in microorganisms, insects, and cultured cells of mammalian bone marrow by DLM method, and in human cell culture.[5,6] The mechanism by which EI exerts a mutagenic effect involves its interference with the metabolism of a pyrimidine precursor of DNA.

Carcinogenic Effect is reported as a result of inhalation or skin application.[7] OSHA considered EI to be an occupational carcinogen.

Chemobiokinetics. ^{14}C-labeled EI administered *i/p* can be detected in the urine, both in a free form and as an unidentified metabolite. Accumulation occurs in the liver, kidneys and spleen. EI content in the fatty tissues appeared to be negligible.[05]

Regulations.

EU (1990). EI is available in the *List of authorized monomers and other starting substances which shall be used for the manufacture of plastic materials and articles intended to come into contact with foodstuffs (Section A).*

Great Britain (1998). EI is authorized without time limit for use in the production of polymeric materials and articles in contact with food or drink or intended for such contact. The specific migration of this substance shall be not detectable (when measured by a method with a DL of 0.01 mg/kg).

Standards. ***EU*** (1990). SML: not detectable (detection limit is 0.01 mg/kg).

References:

1. Carpenter, C. P., Smith, H. F., and Shaffer, C. B., Acute toxicity of ethyleneimine to small animals, *J. Ind. Hyg. Toxicol.*, 30, 2, 1948.
2. Gelembitsky, P. A. et al., *Chemistry of Ethyleneimine*, Nauka, Moscow, 191, 1966 (in Russian).
3. Golubovich, Ye. Ya. and Orlyanskaya R. L., Data on the mechanism of ethyleneimine gonadotoxic action, in *Toxicology of New Industrial Chemical Substances*, N. F. Ismerov and I. V. Sanotsky, Eds., Meditsina, Moscow, Issue No 14, 1975, 16 (in Russian).
4. Bespamyatnova, A. V., Zaugol'nikov, S. D., and Sukhov, Yu. Z., Embryotoxic and teratogenic effects of ethyleneimine, *Pharmacologia i Toxicologia*, 33, 347, 1970 (in Russian).
5. Bochkov, N. P., Sram, R. J., Kuleshov, N. P., et al., System for the evaluation of the risk from chemical mutagens for man: basic principles and practical recommendations, *Mutat. Res.*,38, 191, 1976.
6. Glatt, H. et al., Arene imines, a new class of exceptionally potent mutagens in bacterial and mammalian cells, *Cancer Res.*, 48, 249, 1984.

7. Zayeva, G. N, Fyodorova, V. I., Timofievskaya, L. A., et al., Blastomogenic effect of ethyleneimine, in *Basic Problems of Delayed Consequences of Exposure to Industrial Poisons*, Proc. Research Institute Occup. Diseases, A. K. Plyasunov and G. M. Pashkova, Eds., Moscow, 1976, 140 (in Russian).

ETHYLENE OXIDE

Molecular Formula. C_2H_4O
M = 44.1
CAS No 75-21-8
RTECS No KX2450000
Abbreviation. EtO.

Synonyms and **Trade Names.** Dihydrooxirene; Dimethylene oxide; 1,2-Epoxyethane; Ethene oxide; Oxacyclopropane; Oxane; Oxirane.

Properties. Colorless, incombustible gas. EtO has a high solubility in water (270 g/l at 25°C) but will evaporate to a great extent. Dissolves in water and alcohol. Odor perception threshold is 140 mg/l [02] or 260 mg/l.[010]

Applications. EtO is used in the production of polyethylene oxide, polyethylene terephthalate, non-ionogenic surfactants, di- and triethylene glycols. EtO is used for the sterilization of medical equipment, and in the production of polyester fibers. A major source of exposure is its use as a sterilant.

Acute Toxicity. LD_{50} is reported to be 330 mg/kg BW for male rats, 280 mg/kg BW for female rats, and 365 mg/kg BW for mice. When 200 mg/kg BW was administered in olive oil, all five rats died.[1,2] According to other data, LD_{50} of the aqueous solution is 270 mg/kg BW in guinea pigs.[2] Oral administration produced ataxia, prostration, labored breathing, convulsions in rats and mice.[3]

Repeated Exposure. Wistar rats received by gavage 22 doses of 3.0 to 30 mg EtO/kg BW and 15 doses of 100 mg EtO/kg BW in olive oil during 21 days. No effects have been reported on mortality rate, growth, hematology, blood-urea nitrogen, organ weights, gross and histopathology findings. With the 100 mg/kg BW dose there was a marked BW loss, gastric irritation and a slight liver damage.[1]

Reproductive Toxicity.

Gonadotoxicity.

Observations in man. There is an increased risk of spontaneous abortion among spouses of workers exposed to low levels of EtO.[4]

Animal studies. EtO caused gonadotoxic effects in males: reduced sperm number and sperm motility and an increased time to traverse a linear path. Decreased sperm count and motility were noted in male monkeys exposed to 0.9 or 180 mg EtO/m^3 for 24 months.[5]

Embryotoxicity. EtO produced depression of fetal weight gain, as well as fetal death, and fetal malformations.

Teratogenicity. The levels needed to cause fetal effects approach the doses needed to produce maternal toxicity.[6]

Mutagenicity. EtO appeared to be an effective mutagen in a variety of organisms, namely, microorganisms, insects, mammalian cells, inducing both cell mutations and carcinogenicity. It easily crosses the blood-testis barrier and attacks the germ cells, causing mutations especially in the late spermatid and early spermatozoal stages.[7,8]

In vitro genotoxicity. EtO induced point mutations in *Salmonella typhimurium* strains and SCE in *S. cerevisiae RS112.*[9]

Observations in man. EtO is a potential human mutagen for both somatic and germ cells. Significant increase in *Hb* alkylation, in the rate of CA and SCE in peripheral lymphocytes and, in a single study, micronuclei in erythrocytes have been found in workers exposed occupationally to EtO. No oral data are available.

In vivo cytogenetics. Following chronic exposure, SCE and CA were observed in blood lymphocytes of monkeys.[5] CA were noted in the bone marrow cells of rats exposed by inhalation.[10] The dose of 150 mg/kg BW was positive in DLM assay.[11]

F344 rats and B6C3F_1 mice were exposed by inhalation to up to 3000 ppm EtO for 6 hours/5 days a week over a four-week period. Genotoxic potential was studied using the bone marrow micronucleus test. EtO caused substantial increase in test indices up to 0.79% micronucleus erythrocytes in rats, and 0.72% in mice.[12]

Carcinogenicity.

Observations in man. Epidemiological studies revealed an association between EtO inhalation and excessive risk of carcinogenicity, but both studies have limitations. There are no epidemiology reports on effects of consuming EtO-treated foods.

Animal studies. Sprague-Dawley rats were placed on regimens of 7.5 and 30 mg EtO/kg in salad oil by gavage twice a week over 110 weeks. Observation covered the lifespan of these animals. An increased rate of tumor development was observed only in the forestomach. Metastases in 10 rats and 2 fibrosarcomas were found with the dose of 30 mg/kg.[13]

In another study, 50 rats were fed EtO-fumigated feed for 2 years. EtO residues in the feed were 50 to 1400 mg/kg.[14] After pathology examination, there was no evidence of excess tumors. Both these studies have flaws, and neither is definitive, they appear to indicate a lack of carcinogenic potential of EtO by the dietary route of exposure.

Carcinogenicity classifications. An IARC Working Group concluded that there is limited evidence for the carcinogenicity of EtO in *experimental animals* and inadequate evidence for the carcinogenicity of EtO in *humans*.

The Annual Report of Carcinogens issued by the U.S. Department of Health and Human Services (1998) defines EtO to be a substance which may reasonably be anticipated to be carcinogen, i.e., a substance for which there is a limited evidence of carcinogenicity in *humans* or sufficient evidence of carcinogenicity in *experimental animals*.

IARC: 1;
U.S. EPA: B1;
EU: 2;
NTP: XX - XX - CE* - CE* (inhalation).

Chemobiokinetics. After absorption equivalent doses of EtO are distributed throughout the body. The degree of alkylation of proteins and DNA varies slightly between the different organs and blood. In man and rodents, the half-life of EtO has been estimated to be 9 to 10 min. EtO is metabolized including hydrolysis to *1,2-ethandiol* and *conjugation with glutathione*, and is excreted primarily via the urine.[15]

Regulations.

EU (1990). EtO is available in the *List of authorized monomers and other starting substances which shall be used for the manufacture of plastic materials and articles intended to come into contact with foodstuffs (Section A).*

U.S. FDA (1998) regulates EtO as a direct and indirect food additive under FD&CA and has proposed maximum residue limits for the compound in drug products and medical devices. FDA sets a 50 ppm tolerance level for EtO in food; the tolerance limitations for EtO are set only for residues that remain after treatment. FDA is re-evaluating its established regulations governing EtO residues with regard to recent toxicity data and information concerning the formation of 1,4-dioxane.

FDA approved the use of EtO for the following purposes: (1) as a component of the uncoated or coated food-contact surface of paper and paperboard intended for use in producing, manufacturing, packaging, processing, preparing, treating, packing, transporting, or holding dry food in accordance with the conditions prescribed in 21 CFR part 176.180; as (2) a defoaming agent in the manufacture of paper and paperboard intended for use in packaging, transporting, or holding food in accordance with the conditions prescribed in 21 CFR part 176.200; (3) as an etherifying agent in the production of modified industrial starch, provided the level of reacted EtO in the finished product does not exceed 3.0%; (4) as a fumigant in sizing used as a component of paper and paperboard in contact with dry foods; and (5) as a fumigant for species and other processed natural seasoning materials, except mixtures to which salt has been added, in accordance with the prescribed conditions.

Standards.

EU (1990). Maximum Permitted Quantity in the material or article: 1.0 mg/kg. EU legislation prohibits the presence of EtO in cosmetics.

Great Britain (1998). EtO is authorized without time limit for use in the production of polymeric materials and articles in contact with food or drink or intended for such contact. The quantity of EtO in the finished plastic material or article shall not exceed 1.0 mg/kg.

Canada (1980). Residues in food must be below 0.1 ppm.

References:

1. Hollingsworth, R. L., Rowe, V. K., Oyen, F., et al., Toxicity of ethylene oxide determined in experimental animals, *Arch. Ind. Health,* 13, 217, 1956.
2. Smyth, H. F., Seaton, J., and Fisher, L., Single dose toxicity of some glycols and derivatives, *J. Ind. Hyg. Toxicol.,* 23, 259, 1941.
3. Woodard, G. and Woodart, M. Toxicity of residues from ethylene oxide gas sterilization, in *Proc. HIA Technical Symposium*, Health Industry Association, Washington, D. C., 1971.
4. Lindbohm, M.-L., Hemminki, K., Bonhomme, M. G., Anttila, A., Rantala, K., Heikkila, P., and Rosenbeearg, M. J., Effect of paternal occupational exposure on spontaneous abortions, *Am. J. Ind. Med.*, 81, 1029, 1991.
5. Lynch, D. W., Lewis, T. R., Moormaqn, W. J., Sabharwal, P. S., and Burg, J. A., Toxic and mutagenic effects of ethylene oxide and propylene oxide on spermatogenic functions in cynomolgus monkeys, *Toxicologist*, 3, 60, 1983.
6. LaBorde, J. B. and Kimmel, C. A., The teratogenicity of ethylene oxide administered intravenously to mice, *Toxicol. Appl. Pharmacol.*, 56, 16, 1980.
7. Lee, I. P. and Dixon, R. L., Factors influencing reproduction and toxic effects on male gonads, *Environ. Health Perspect.,* 24, 117, 1978.
8. Dellarco, V. L., Generoso, W. M., Sega, G. A., Fowle, J. R., III, and Jacobson-Kram, D., Review of the mutagenicity of ethylene oxide, *Environ. Molec. Mutagen.*, 16, 85, 1990.
9. Agurell, E., Cederberg, H., Ehrenberg, L., Lindahl-Kiessling, K., Rannung, U., and Tornqvist, M., Genotoxic effects of ethylene oxide and propylene oxide: a comparative study, *Mutat. Res.*, 250, 229, 1991.
10. Strekalova, Ye. M. and Golubovich, Ye. Ya., Mutagenic action of ethylene oxide on germinal and somatic cells of male rats, in *Toxicology of New Industrial Chemical Substances*, N. F. Ismerov and I. V. Sanotsky, Eds., Meditsina, Moscow, Issue No 14, 1975, 11 (in Russian).
11. Generoso, W. M., Cain, K. T., Crishna, M., et al., Heritable translocation and dominant-lethal mutation induction with ethylene oxide in mice, *Mutat. Res.*, 73, 133, 1980.
12. Vergnes, J. S. and Pritts, L. M., Ethylene: Genotoxic potential using the bone marrow micronucleus test following four weeks of vapor exposure, *Environ. Molec. Mutagen.*, 23 (Suppl. 23), 69, 1994.
13. Dunkelberg, H., Carcinogenicity of ethylene oxide and 1,2-propylene oxide upon intragastric administration to rats, *Brit. J. Cancer.,* 46, 924, 1982.
14. Bar, F. und Griepentrog, F., Langzeitfutterungs-versuch an Ratten mit Athylenoxidbegastem futter, *Bundesgesundheitblatt*, 11, 105, 1969.
15. *Ethylene oxide*, Environmental Health Criteria No 55, WHO, Geneva, 1985, 80.

5-ETHYL-2-PICOLINE

Molecular Formula. $C_8H_{11}N$
M = 121.19
CAS No 104-90-5
RTECS No TJ6825000
Abbreviation. EP.

Synonyms. Aldehydine; 2-Methyl-5-ethylpyridine.

Properties. Liquid with a pungent odor. Water solubility is 1.2%, soluble in alcohol. Odor perception threshold is 0.5 to 1.0 mg/l.[1,2]

Applications. Used in the production of synthetic rubbers and ion exchange resins.

Acute Toxicity. In rats, LD_{50} is 368 mg/kg BW; in mice, it is 282 mg/kg BW.[1] Other values are also reported: 1460 and 1680 mg/kg BW, respectively.[03]

Repeated Exposure. Rats received 1.0 mg EP/kg BW for 30 days. The treatment lowered the level of total lipids and increased that of phospholipids in the blood serum, liver, and lung.[3]

Short-term Toxicity. The exposure produced absolute neutropenia and relative lymphocytosis. Rats were gavaged with 0.1 or 1.0 mg EP/kg BW for 4 months. The treatment affected protein metabolism in ganglion cells of rat cerebellar cortex, caused reversible chromatolysis and a decrease in glutamate dehydrogenase activity.[4]

Long-term Toxicity. Rats were treated orally with 0.1 or 1.0 mg EP/kg BW for up to 6 months. The treatment affected neurons of the sensory and motor sections of spinal cord of rats.[4] The same dosing in rats caused a decrease in peroxidase and alkaline phosphatase activity, as well as in glycogen in neutrophils (Baranets et al., 1979).

Reproductive Toxicity. The treatment does not affect gestation but causes an increase in embryolethality.

References:

1. Zyabbarova, S. A., in *Proc. Leningrad Sanitary Hygiene Medical Institute*, Leningrad, Issue No 68, 1961, 138 (in Russian).
2. Rosen, A. A. et al., *Water Poll. Control Fedn.*, 35, 777, 1963.
3. Konvai, V. D. and Berezovskaya, M. P., in *Sci. Proc. Omsk State Medical Institute*, Omsk, Issue No 83, 1972, 112 (in Russian).
4. Lyutikova T. M., in *Sci. Proc. Omsk State Medical Institute*, Omsk, Issue No 126, 1977, 75 (in Russian).

5-ETHENYL-2-METHYLPYRIDINE

Molecular Formula. C_8H_9N
M = 119.17
CAS No 140-76-1
RTECS No UT2975000
Abbreviation. EMP.

Synonyms. 2-Methyl-5-vinylpyridine; 5-Vinyl-2-picoline.

Properties. Colorless to faintly opalescent liquid. Water solubility is 0.98%. Odor perception threshold is 0.5 mg/l (practical threshold 1.0 mg/l). Concentration of 10.0 mg/l does not alter the transparency or color of water. Average odor threshold concentration is 0.04 mg/l.[1]

Applications. M. is used in the manufacture of rubber and ion exchange resins.

Acute Toxicity. In rats, LD_{50} is reported to be 0.51 g/kg BW (Sanotsky et al., 1982) or 1.5 g/kg BW; in mice it is 1.25 g/kg BW.[2]

Repeated Exposure. Rats were given doses of 300 mg EMP/kg BW. The treatment inhibited overall function of the hypothalamic-hypophyseal neurosecretory system and disturbed the secretion of neurohormones into the blood stream.[3]

Short-term Toxicity. Rats were gavaged with 0.1 or 1.0 mg EMP/kg BW for 4 months. The treatment affected protein metabolism in ganglion cells of rat cerebellar cortex, caused reversible chromatolysis and a decrease in glutamate dehydrogenase activity.[4]

Long-term Toxicity. The studies revealed changes in structural elements of the blood-brain histohematic barrier, endothelium and epithelium of the vascular plexus, and *Kupffer* cells, in the contents and distribution of nucleotides, proteins, and enzymes.[5]

In a 6-month study, CNS activity was depressed in rabbits exposed to 3.0 to 5.0 mg/l in their drinking water. Concentration of 0.2 mg/l was reported to produce no adverse effects.[6]

Reproductive Toxicity.

Embryotoxicity. Repeated exposure did not affect gestation. Nevertheless, the offspring die on days 3 to 4 after birth.[6]

Pregnant rats received the doses of 157 mg/kg BW. The treatment damaged liver structure in the offspring of the treated rats, indicating placental permeability to this pyridine derivative.[7]

Chemobiokinetics. Absorbed from the GI tract.

Standards. ***Russia***. Recommended PML: 0.05 mg/l.

References:

1. Baker, K. A., Threshold odors of organic chemicals, *J. Am. Water Work Assoc.*, 55, 913, 1963.
2. Dukhovnaya, A. I., *Experimental Assessment and Substantiation of Approaches to Hygienic Standards of Certain Alkyl Derivatives of Pyridine*, Author's abstract of thesis, Moscow, 1972, 21 (in Russian).
3. Zybina, L. S., in All-Union Institute Sci.-Technical Information, Deposit. Doc. Issue, VINITI 2667-78, 64, 5, 1979 (in Russian).
4. Lyutikova T. M., in *Sci. Proc. Omsk State Medical Inst.,* 126, 76, 1977 (in Russian).
5. Lyutikova, T. M. and Taskayev, I. I., in All-Union Institute Sci.-Technical Information, Deposit. Doc. Issue of All-Union Institute Sci.-Technical Information No 149-79, 95, 8, 1979 (in Russian).
6. Zyabbarova, S. A., in *Problems of Commune Hygiene*, Proc. Leningrad Sanitary Hygiene Medical Institute, Leningrad, Issue No 68, 1961, 138 (in Russian).
7. Nikiforova, A. A. and Taskaev, I. I., in *React. Plast. Epit. Connect. Tissue Norm. Exp. Pathol.*, Proc. Sci. Histol. Conf., 1974, 196 (in Russian).

2-ETHYL-2-(HYDROXYMETHYL)-1,3-PROPANEDIOL

Molecular Formula. $C_6H_{14}O_3$

M = 134.20

CAS No 77-99-6

RTECS No TY6470000

Abbreviation. EHP.

Synonyms and **Trade Names.** Ethriol; Ethyltrimethylolmethane; Hexaglycerine; 1,1,1-Tri(hydroxymethyl)propane; Trimethylolpropane; Tris(hydroxymethyl)propane.

Properties. White powder or flakes. Freely soluble in water and alcohol.

Applications. Used in the production of alkyd resins, polyurethane foams and coatings intended for use in contact with food or drink.

Acute Toxicity. LD_{50} is 14.1 g/kg BW in rats, 13.7 g/kg BW in mice. Poisoning is accompanied by CNS depression, dyspnea, and respiratory arrest.

Regulations.

EU (1990). EHP are available in the *List of authorized monomers and other starting substances which shall be used for the manufacture of plastic materials and articles intended to come into contact with foodstuffs (Section A)*.

U.S. FDA (1998) approved the use of EHP (1) in the manufacture of polyurethane resins which may be used as food-contact surface of articles intended for use in contact with bulk quantities of dry food of the type identified in 21 CFR part 176.170 (c); (2) in resinous and polymeric coatings used as the food-contact surfaces of articles intended for use in producing, manufacturing, packing, processing, preparing, treating, packaging, transporting, or holding food for use as a pigment dispersant at levels not to exceed 0.45% by weight of the pigment and in accordance with the conditions prescribed in 21 CFR part 175.300. and (3) in the manufacture of cross-linked polyester resins which may be used as articles or components of articles intended for repeated use in contact with food, subject to the provisions prescribed in 21 CFR part 177.2420.

Great Britain (1998). EHP is authorized without time limit for use in the production of polymeric materials and articles in contact with food or drink or intended for such contact. The specific migration of this substance shall not exceed 6.0 mg/kg.

References:

1. Stankevich, V. V., Data on substantiation of the maximum permissible concentration of trimethylolpropane (Extrol) in the air of industrial sites, *Gig Sanit.*, 5, 107, 1967 (in Russian).

2. Krupitskaya, I. D., Pikuleva, M. M., and Ranov, A. I., Experience in studying the efficacy of measures for preventing the pollution of the Miass River with phenols, *Gig. Sanit.*, 5, 88, 1967 (in Russian).

FORMALDEHYDE

Molecular Formula. CH_2O
M = 30.3
CAS No 50-00-0
RTECS No LP8925000
Abbreviation. FA.

Synonyms and **Trade Names.** Methylaldehyde; Methanal; Formaline; Formic aldehyde; Formol; Methylene oxide; Oxymethylene.

Properties. Colorless gas with a pungent, suffocating odor. Freely miscible with water (550 g/l at 25°C). 35 to 40% solution is called *formaline*, which occurs often with 10 to 15% methanol added to prevent polymerization. FA is extremely active chemically. Odor perception threshold is reported to be 0.6 mg/l,[02] taste perception threshold is 50 mg/l.[1] FA as a monomer may exist only in the air. In the water solutions, FA occurs in a form of *methylene glycol.*

Applications. 80% of the FA production are used for plastic and resin manufacture: urea-formaldehyde resins, phenolic resins, pentaerythritol and polyacetal resins, and melamine resins intended for use in contact with food.

Migration from polyacetal plastic fittings into drinking water is reported. Release of FA from dishes and bowls made of melamine-formaldehyde resin was studied. Concentrations, found in 4.0% acetic acid at the temperature of 60 to 95°C after 30 min. extraction, were 0.5 to 3.0 ppm FA. The release of FA was affected predominantly by temperature.[2]

Migration of FA into food-simulating solvents from cups made of *melamine resins* were studied under various conditions. Migration of FA from the cups being used at a cafeteria was undetectable when the cups were kept at 60°C for 30 min. with 4.0% acetic acid.[3]

FA was identified in extracts of paper and board materials of Polish and foreign production intended for food contact. Levels of migration never exceeded 1.0 mg/dm^2, with the exception of the recycled paper from which it was 1.77 mg/dm^2.[4]

Exposure. FA is present as a natural component in fresh and preserved fish, seafood, honey, roasted foods and many fruits and vegetables: tomatoes (5.7 to 7.7 ppm), apples (17.3 to 22.3 ppm), spinach (3.3 to 7.3 ppm), and carrots (6.7 to 10 ppm). FA can also be found in a number of consumer products including some cosmetics and air fresheners. In addition, there is an exposure to FA deriving from food containers or food-packaging materials and to FA used as preservative in the food industry.[5] Concentrations of FA ranging from 3.0 to 23 mg/kg have been reported in a variety of foodstuffs.

FA is present in tobacco smoke and it is a widespread contaminant of the air of dwellings and public buildings and of the atmosphere.

Acute Toxicity. Ingestion of 100 to 200 ml of *formaline* is likely to cause a fatal poisoning. LD_{50} (37% solution) appears to be 523 mg/kg BW.[6] Since FA is rapidly metabolized into formic acid, a severe acidosis may develop producing a local corrosive action on the upper GI tract. Thus, ulceration, necrosis, perforation, hemorrhage occur, frequently resulting in death some days later.[7]

LD_{50} is 400 to 800 mg/kg BW in mice, and rats and 260 mg/kg BW in guinea pigs.[8] Poisoning of animals is accompanied by immediate marked excitation with a subsequent narcotic effect. Death occurs in the first 2 to 3 hours. The survivors remain apathetic and drowsy for 2 weeks, with no appetite.[9]

Repeated Exposure failed to reveal cumulative properties. Habituation developed. Administration of 1/5 LD_{50} to rats resulted in 20% animal mortality on day 14 to 19 of administration. The survivors suffered from exhaustion. Gross pathology examination revealed parenchymatous dystrophy of the liver and kidneys and circulatory disturbances.[9] Rats were given the doses of 5.0 to 125 mg/kg BW via their drinking water for 4 weeks. Food and liquid intake was decreased in the top dose group. Yellow discolorations of the fur, decreased protein and albumin levels in the blood plasma, thickening of the limiting ridge and hyperkeratosis in the forestomach, and focal gastritis in the glandular stomach were also noted at this dose level. No treatment-related effects were observed in animals given 5.0 and 25 mg FA/kg BW.[10]

Short-term Toxicity.

Observations in man. The NOAEL in humans who ingested FA daily over a period of 13 weeks appears to be 200 mg.[11]

Animal studies. Rats and dogs received oral doses of FA over a period of 3 months. The only adverse effect was depression of BW gain when doses of 100 mg/kg BW and higher were administered.[12]

Long-term Toxicity.

Observations in man. Epidemiological investigations carried out hitherto have not made it possible to obtain reliable data about carcinogenic, mutagenic, teratogenic, embryotoxic, and neurotoxic danger of FA.

Animal studies. Aldehydes due to their instability in water are highly unlikely to have a chronic effect. Histopathology changes in the GI tract were observed in a study where adverse effects were exhibited only with the 300 mg/kg BW dose.[13] Wistar rats were given FA in their drinking water over a period of 2 years at doses of 1.2 to 82 mg/kg BW (males) and 1.8 to 109 mg/kg BW (females). Adverse effects were demonstrated only in the highest dose groups. The principal target organs appeared to be the stomach (thickening of the mucosal wall) and kidney (renal papillary necrosis in both sexes).[14] A concentration of 0.5 mg/l produced no effect on conditioned reflex activity in rats.[1] NOELs claimed by the authors are reported in the following table.

Table. Formaldehyde NOELs in short-term and long-term toxicity studies.[7]

Species	*Length of exposure*	*Medium tested*	*NOEL, mg/kg BW*	*References*
Rat	90 days	Drinking water	100	6
Dog	90 days	Diet	75	2
Rat	24 months	Drinking water	15 - 21	9
Rat	24 months	Drinking water	10	8

TDI/ADI appears to be 0.15 mg/kg BW. Such ADI is in agreement with a 20-mg figure calculated on the basis of the NOAEL equal to 200 mg/kg BW demonstrated in humans.[11] Considering that FA daily intake calculated by Owen et al.[15] (11 mg approximately) is in this range and that the liver may convert 22 mg FA to CO_2 in 1 min, it can be inferred that occasional ingestion of food containing free FA or its precursor *hexamethylene tetramine*, in the order of few ppm, would not cause any harmful effect. It is likely that at low concentrations exogenous FA is handled in the body in a way that is not significantly different from that of endogenous FA.

Reproductive Toxicity.

Gonadotoxicity. A significant increase in sperm abnormalities was found in rats after a single oral administration of 200 mg/kg BW.[1] No effect on sperm was shown in male mice given FA by gavage at a dose of 100 mg/kg BW for 5 days.[16] Rats received 5 *i/p* injections. The treatment caused an increase in induction of sperm head abnormalities at 0.125 to 0.5 mg/kg BW. Increase in DLM was seen in female rats sired by males exposed to FA. Reduction of fertile matings was observed in females mated 1 to 7 days after treatment of males with FA.[17]

Teratogenicity.

Observations in man. There was no increase in spontaneous abortion or birth defects observed in women exposed to FA before and during pregnancy when compared to those employees not exposed.[18]

Animal studies. A group of about 30 pregnant CD-1 mice was given oral doses of 74 to 185 mg/kg BW on days 6 to 15 of gestation. No teratogenic effects were shown.[19]

Embryotoxicity. The dose of 540 mg/kg BW was administered to pregnant mice on days 8 to 12 of gestation. No embryotoxic effects were demonstrated.[20] Pregnant beagle dogs were fed FA at dose levels of 3.1 to 9.4 mg/kg BW from day 4 to day 56 after mating. There were no effects on reproductive performance or on health of the offspring.[21]

A noticeable decrease in weight was found in the offspring of Sprague-Dawley female rats given FA in drinking water at a concentration of 2500 mg/l since day 15 of pregnancy.[22] Male rats treated with FA at the level of 0.1 mg/l in drinking water exhibited no changes in reproductive function.[23]

Immunotoxicity. Inhalation exposure and oral administration of FA are found to produce allergenic effect.[24] The threshold sensitizing concentration of 0.5 mg/m^3 is not accompanied by any change in the number and functional activity of lymphocytes, which allows it to be assessed as non-allergenic.[18]

Observations in man. Human allergic reactions including photodermatosis, urticaria, rhinitis, asthma, eosinophilia, and anaphylactic shock in hemodialyzed patients were considered to be rare systemic hypersensitivity reactions (risk 1:1000).[26]

Animal studies. A strong contact sensitizing potential is shown in maximized guinea-pig assays.[25]

Mutagenicity of FA was reported by Rapoport as early as 1946.[27]

Observations in man. FA caused an increase in the frequency of CA and SCE in peripheral lymphocytes and in cultured human lymphocytes,[28-30] but negative results also have been reported.[31] FA induced DNA-protein cross-links, unscheduled DNA synthesis, and mutation in the human cells *in vitro.*[32]

In vitro genotoxicity. FA binds readily to proteins, RNA, and single-stranded DNA including DNA-protein cross-links and breaks in single-stranded DNA.[33] FA affects prokaryotic and eukaryotic cells *in vitro*. It would appear that FA reacts readily with macromolecules in cells primarily at the point of exposure and does not reach other sites in the body at sufficient concentration to produce detectable effects.

The level of CA and SCE in a culture of Syrian hamster ovary cells is increased significantly but FA does not induce micronuclei and CA in bone marrow cells (dose of 25 mg/kg BW) or CA in spleen cells of mice.[34] It also induced transformation of mouse C3H 10TI/2 cells and DNA strand breaks and DNA-protein cross-links in rodent cells *in vitro*.

FA increases DNA synthesis and the number of micronuclei and nuclear abnormalities in the rat epithelial cells.[35]

In vivo cytogenetics. As Basler et al. indicate, no mutagenic effect is found in mammals *in vivo.*[36] It has been shown to be genotoxic in *Dr. melanogaster* (IARC 29-345).

Carcinogenicity. The mechanism of FA carcinogenicity is not completely understood, although its reactivity with macromolecules is well known. It is thought that the mutagenic and carcinogenic effects of FA appear both as a result of direct damage to the DNA molecule and as a consequence of the inhibition of its repair.

FA has been carcinogenic in rats and mice by inhalation at concentrations which caused irritation of the nasal epithelium. Positive results were demonstrated in rodents following inhalation exposure: after a 24-month inhalation exposure (5.6 ppm up to 14.3 ppm), rats developed an unusual incidence of nasal mucosa tumors, although in mice the same concentration of FA induced a non-significant increase of this pathology.[37,38]

Oral and dermal studies have given little indication of carcinogenic potential of FA. In spite of FA being cytotoxic to the mucosa of the glandular and non-glandular part of the stomach, it is unlikely that FA induces a significant increase of gastric tumors or tumors at other sites at the dose of 100 mg/kg BW according to the findings of several long-term carcinogenicity studies.[13,39]

Only one study reports the occurrence of gastrointestinal tumors in the long-term bioassay: in a 2-year study, Sprague-Dawley rats received FA in drinking water at concentrations of 10 to 1500 mg/l. A dose-dependent increase in the incidence of leukemia was noted.[22] However, since the doses administered were higher than 100 mg/kg BW, the results cannot be considered to be of physiological significance. It was concluded [6] that the glandular stomach mucosa in rats is more resistant to FA than forestomach mucosa. This is particularly relevant because the target tissue in the rat has no human counterpart.

Humans seem to be less sensitive to the carcinogenic effect of inhaled FA than rodents.[40] The weight of evidence indicates that FA is not carcinogenic by the oral route.

Carcinogenicity classifications. An IARC Working Group concluded that there is sufficient evidence for the carcinogenicity of FA (gas) in *experimental animals* and there is limited evidence to evaluate the carcinogenicity of FA in *humans*.

The Annual Report of Carcinogens issued by the U.S. Department of Health and Human Services (1998) defines FA (gas) to be a substance which may reasonably be anticipated to be carcinogen, i.e., a substance for which there is a limited evidence of carcinogenicity in *humans* or sufficient evidence of carcinogenicity in *experimental animals.*

IARC: 2A (inhalation);

U.S. EPA: B1;

EU: 3.

Chemobiokinetics. FA is a normal cell metabolite in mammalian systems. Under normal conditions, the level of FA is very low in animal and human tissues because of its rapid oxidation to CO_2 (70 to 80%). FA is recycled in single-carbon biosynthetic pathway and hence a small proportion of FA may be retained in the body. Under normal conditions, FA absorbed by the GI tract mucosa would remain as the *aldehyde* for only a very short period of time, either binding to macromolecules at the site of absorption or being metabolized by oxidative enzymes to *formic acid* and subsequently *carbon dioxide* and *water.* In rats and mice given a single oral dose, about 65% of radiolabel was excreted in the urine and feces, and 26% was eliminated in the expired air.[41]

Regulations.

EU (1992). FA is available in the *List of authorized monomers and other starting substances which shall be used for the manufacture of plastic materials and articles intended to come into contact with foodstuffs (Section A).*

U.S. FDA (1998) regulates FA as an indirect food additive under FD&CA. It is listed in the CFR for use (1) in adhesives as a component of articles intended for use in packaging, transporting, or holding food in accordance with the conditions prescribed in 21 CFR part 175.105; (2) as a component of paper and paperboard for contact with dry, aqueous, and fatty food (only as preservative for coating formulations) in accordance with the conditions prescribed in 21 CFR part 176.170; (3) as a constituent of acrylate ester copolymer coating which may be used as a food-contact surface of articles intended for packaging and holding food, including heating of prepared food, subject to the provisions prescribed in 21 CFR part 175.210; (4) a component of a defoaming agent intended for articles which may be used in producing, manufacturing, packing, processing, preparing, treating, packaging, transporting, or holding food; (5) as a defoaming agent in the manufacture of paper and paperboard intended for use in packaging, transporting, or holding food in accordance with the conditions prescribed in 21 CFR part 176.200; (6) in the production of phenolic resins for the surface of molded articles intended for repeated use in contact with non-acid food (*pH* above 5.0) in accordance with the conditions prescribed in 21 CFR part 177.2410; (7) as an adjuvant substance in the manufacture of foamed plastics intended for use in contact with food, subject to the provisions prescribed in 21 CFR 178.3010; and (8) as a substance employed in the production of or added to textiles and textile fibers intended for use in contact with food in accordance with the conditions prescribed in 21 CFR part 177.2800. It is approved for the use (9) as a preservative in various aspects of food production. Most of these relate to its the presence in a number of food packaging products, but it is also approved for inclusion (10) in defoaming agents containing dimethyl polysiloxane, which are used in food processing; the level of formaldehyde is limited to a maximum of 1.0% of the dimethyl polysiloxane content.

FDA also approved the use of FA (11) as an additive in the manufacture of animal feeds based on animal fats and oilseed meals in order to improve the handling characteristics of the feed. The dried feed may contain a maximum of 1.0% FA.

Great Britain (1998). FA is authorized without time limit for use in the production of polymeric materials and articles in contact with food or drink or intended for such contact. The specific migration of FA shall not exceed 15 mg/kg.

FA is permitted by *Italian* law in the production of two Italian cheeses, Grana Podano and Provolone (the level of FA in the final product not to exceed 0.5 mg/kg). The Italian law was passed by the *EU.*[22] FA is permitted for use in a limited number of cosmetics in the European Union, with the following authorized concentrations in the finished cosmetic product: 5.0% in nail hardeners; 0.2% as a preservative, and 0.1% in mouth hygiene products (EU, 1976).

Recommendations. ***WHO*** (1996). Guideline value for drinking water: 0.9 mg/l.
Standards. ***Russia*** (1995). MAC and PML: 0.05 mg/l; PML in food: 5.0 mg/l.
References:

1. Nazarenko, I. V., in *Protection of Water Reservoirs against Pollution by Industrial Liquid Effluents*, S. N. Cherkinsky, Ed., Medgiz, Moscow, Issue No 4, 1960, 76 (in Russian).
2. Sugita, T., Ishiwata, H., and Yoshihira, K., Release of formaldehyde and melamine from tableware made of melamine-formaldehyde resin, *Food Addit. Contam.*, 7, 21, 1990.
3. Ishiwata, H., Inoue, T., and Tanimura, A., Migration of melamine and formaldehyde from tableware made of melamine resin, *Food Addit. Contam.*, 3, 63, 1986.
4. Lewandowska, I., Biernat, U., Jurkiewicz, M., and Stelmach, A., Formaldehyde determination in packing paper for food products by the colorimetric method with acetylacetone, *Rocz. Panstw. Zakl. Hig.*, 45, 221, 1994 (in Polish).
5. Cassidy, S. L., Dix, K. M., and Jenkins, T., Evaluation of a testicular sperm head counting technique using rats exposed to dimethoxyethyl phthalate, glycerol-monochlorohydrin, epichlorohydrin, formaldehyde, or methyl methanesulfonate, *Arch. Toxicol.*, 53, 71, 1983.
6. Bartone, N. F., Grieco, R. V., and Herr, B. S., Corrosive gastritis due to ingestion of formaldehyde without esophageal impairment, *JAMMA*, 203, 50, 1968.
7. Restani, P. and Corrado, L. G., Oral toxicity of formaldehyde and its derivatives, *Crit. Rev. Toxicol.*, 21, 315, 1991.
8. Smyth, H. F., Jr., Seaton, J., and Fisher, L., The single dose toxicity of some glycols and derivatives, *J. Ind. Hyg. Toxicol.*, 23, 259, 1941.
9. Pomerantzeva, N. S., in *Coll. Sci. Proc. Research Hygiene Institute*, Kuibyshev, Issue 6, 1971, 103 (in Russian).
10. Til, H. P., Woutersen, R. A., and Feron, V. J., Evaluation of the oral toxicity of acetaldehyde and formaldehyde in a 4-week drinking water study in rats, *Food Chem. Toxicol*, 26, 447, 1988.
11. Zurlo, N., In: *Occup. Health and Safety*, McGraw-Hill, San Francisco, 1, 574, 1971.
12. Johannsen, F. R., Levinskas, G. V., and Tegeris, A. S., Effects of formaldehyde in the rat and dog following oral exposure, *Toxicol. Lett.*, 30, 1, 1986.
13. Tobe, M., Naito, K., Caldwell, W. M., et al., Chronic toxicity study of formaldehyde administered orally to rats, *Toxicology*, 56, 79, 1988.
14. Til, H. P., Woutersen, R. A., Feron, V. J., et al., Two-year drinking-water study of formaldehyde in rats, *Food Chem. Toxicol.*, 27, 77, 1989.
15. Owen, B. A., Dudney, C. S., Tan, E. L., et al., Formaldehyde in drinking water: comparative hazard evaluation and an approach to regulation, *Regul. Toxicol. Pharmacol.*, 11, 220, 1990.
16. Ward, J. B. Jr., Hokanson, J. A., Smith, E. R., et al., Sperm count morphology and fluorescent body frequency in autopsy service workers exposed to formaldehyde, *Mutat. Res.*, 130, 417, 1984.
17. Odeigah, P.G.C., Sperm head abnormalities and dominant lethal effects of formaldehyde in albino rats, *Mutat. Res.*, 389, 141, 1997.
18. Duyeva, L. A., Problems of protection of human health and environment against harmful chemical factors, in *The First All-Union Toxicol. Conf.*, Abstracts, Rostov-na-Donu, 1986, 227 (in Russian).
19. Marks, T. A., Worthy, W. C., and Staples, R. E., Influence of formaldehyde and Sonacide (potentiated acid gluturaldehyde) on embryo and fetal development of mice, *Teratology*, 22, 51, 1980.
20. Seidenberg, J. M., Anderson, D. G., and Becker, R. A., Validation of an *in vivo* developmental toxicity screen in the mouse, *Teratogen. Carcinogen. Mutagen.*, 6, 361, 1987.
21. Hurni, H. and Ohder, H., Reproduction study with formaldehyde and hexamethylenetetramine in beagle dogs, *Food Cosmet. Toxicol.*, 11, 459, 1977.
22. Soffritti, M., Maltoni, C., Maffei, F., et al., Formaldehyde: an experimental multipotential carcinogen, *Toxicol. Ind. Health.*, 5, 699, 1989.
23. *EEC Council Directive of 30 January 1978*, Commission of EC, Amend. for the 13th Directive 64/54/EEC.

24. Vinogradov, G. I., *Chemical Allergens in the Environment and Their Effects on Human Health*, Rev. Inform., Moscow, 1985, 58 (in Russian).
25. Bakker, J. G., Jongen, S. M., Van Neer, F. C., and Neis, J. M., Occupational contact dermatitis due to acrylonitrile, *Contact Dermat.*, 24, 50, 1991.
26. Garnier, R., Rousselin, X., and Rosenberg, N., Toxicite de l'aldehyde formique. Une revue bibliographique, *Cah. Note Doc., Secur. Hyg. Trav.*, 134, 63, 1989.
27. Rapoport, I. A., Carbonyl compounds and the chemical mechanism of mutation, *Reports. Acad. Sci.* USSR, 54, 65, 1946 (in Russian).
28. Miretskaya, L. M. and Shvartsman, P. Y., Studies of chromosome aberrations in human lymphocytes under the influence of formaldehyde, I. Formaldehyde treatment of lymphocytes *in vitro, Tsitologia (Cytology)*, 24, 1056, 1982 (in Russian).
29. Bauchinger, M. and Schmid, E., Cytogenetic effects in lymphocytes of formaldehyde workers of a paper factory, *Mutat. Res.*, 158, 195, 1985.
30. Kreiger, R. A. and Garry, V. F., Formaldehyde-induced cytotoxicity and sister-chromatid exchanges in human lymphocyte cultures, *Mutat. Res.*,120, 51, 1983.
31. Thompson, E. J., Shackleton, S., and Harrington, J. M., Chromosome aberrations and sister-chromatid exchange frequencies in pathology staff occupationally exposed to formaldehyde, *Mutat. Res.*, 141, 89, 1984.
32. Ma, T. M. and Harris, M. M., Review of genotoxicity of formaldehyde, *Mutat. Res.*, 196, 37, 1988.
33. Hemminki, K. et al., Spontaneous abortions in hospital staff engaged in sterilizing instruments with chemical agents, *Brit. Med. J.*, 285, 1461, 1982.
34. Migliore, L. et al., Micronuclei and nuclear anomalies induced in the gastrointestinal epithelium of rats treated with formaldehyde, *Mutagenesis*, 4, 327, 1989.
35. Natarajan, A. T., Darroudi, F., Bussmann, C. J. M., and van Kestern-van Leeuwen, A. C., Evaluation of the mutagenicity of formaldehyde in mammalian cytogenic assays *in vivo* and *in vitro*, *Mutat. Res.*, 122, 355, 1983.
36. Basler, A., Hude, W. V. D., and Scheutwinkel-Reiche, M., Formaldehyde-induced SCE *in vitro* and the influence of the exogenous metabolizing systems *S9* mix and primary rat hepatocytes, *Arch. Toxicol.*, 58, 10, 1985.
37. Svenberg, J. A., Kerns, W. D., Mitchell, R. I., et al., Induction of squamous cell carcinomas of the rat nasal cavity by inhalation exposure to formaldehyde vapor, *Cancer Res.*, 40, 3398, 1980.
38. Kerns, W. D., Pavkov, K. L., Donofrio, D. J., et al. Carcinogenicity of formaldehyde in rats and mice after long-term inhalation exposure, *Cancer Res.*, 43, 4382, 1983.
39. Takahashi, M., Hasegawa, R., Fyrukawa, F., et al., Effects of ethanol, potassium metabisulfite, formaldehyde and hydrogen peroxide on gastric carcinogenesis in rats after initiation with *N*-methyl-*N'*-nitro-*N*-nitrosoguanidine, *Japan. J. Cancer Research* (Gann), 77, 118, 1986.
40. Squire, R. A. and Cameron, L. I., An analysis of potential carcinogenic risk from formaldehyde, *Regul. Toxicol. Pharmacol.*, 4, 107, 1984.
41. Galli, C. L., Ragusa, C., Resmini, P., et al., Toxicological evaluation in rats and mice of the ingestion of the cheese made from milk with added formaldehyde, *Food Chem. Toxicol.*, 21, 313, 1983.

FUMARIC ACID, BIS(2-ETHYLHEXYL) ESTER

Molecular Formula. $C_{20}H_{36}O_4$
M = 340.56
CAS No 141-02-6
RTECS No LT0525000

Synonyms and **Trade Name.** Bis(2-ethylhexyl) fumarate; 2-Butenedioic acid, bis(2-ethylhexyl) ester; Di(2- ethylhexyl)fumarate; Dioctyl fumarate; (2-Ethylhexyl) fumarate; Fumaric acid, bis(2-ethylhexyl) ester; RC Comonomer DOF.

Applications. A monomer for use in the production of polymeric materials and articles in contact with food or drink.

Acute Toxicity. LD_{50} is 29.2 g/kg BW in rats.

Regulations. ***U.S. FDA*** (1998) approved the use of F. in adhesives as a component of articles intended for use in packaging, transporting, or holding food in accordance with the conditions prescribed in 21 CFR part 175.105. as a component of the uncoated or coated food-contact surface of paper and paperboard intended for use in producing, manufacturing, packaging, processing, preparing, treating, packing, transporting, or holding dry food of the type identified in 21 CFR part 176.180.

Reference:

Smyth, H. F., Carpenter, C. P., Weil, C. S., et al., Range-finding toxicity data, List V, *Arch. Ind. Hyg. Occup. Med.,* 10, 61, 1954.

FUMARIC ACID, DIBUTYL ESTER

Molecular Formula. $C_{12}H_{20}O_4$
M = 228.32
CAS No 105-75-9
RTECS No LT1225000
Abbreviation. FADBE.

Synonyms and **Trade Names.** 2-Butenedioic acid, dibutyl ester; Dibutyl fumarate; RC Comonomer DBF; Stafex DBF.

Applications. A monomer for use in the production of polymeric materials and articles in contact with food or drink.

Acute Toxicity. LD_{50} is 8.53 g/kg BW in rats.

Regulations. ***U.S. FDA*** (1998) approved the use of FADBE in adhesives as a component of articles intended for use in packaging, transporting, or holding food in accordance with the conditions prescribed in 21 CFR part 175.105.

Reference:

Smyth, H. F., Carpenter, C. P., and Weil, C. S., Range-finding toxicity data: list IV, *Arch. Ind. Hyg. Occup. Med.*, 4, 119, 1951.

2-FURALDEHYDE

Molecular Formula. $C_5H_4O_2$
M = 96.09
CAS No 98-01-1
RTECS No LT7000000

Synonyms and **Trade Names.** 2-Formylfuran; Fural; 2-Furancarbox aldehyde; Furfural; Furfuraldehyde; Furfurole; Pyromucic aldehyde.

Properties. Colorless, clear, oily liquid with a characteristic odor of bitter almond, rapidly turning brown in air. Solubility in water is 83 g/l (20°C), readily soluble in alcohol. Odor perception threshold is 1.86 mg/l[1] or 3.5 mg/l.[02]

Applications. Used in the manufacture of plastics and rubbers, in the production of furfuryl alcohol, tetrahydrofuran and furan-phenolic corrosion-resistant polymers; a solvent in refining nitrocellulose, cellulose acetate, etc.

Exposure. F. has been defined in 150 foods, including fruits, vegetables, beverages, bred and bread products.

Acute Toxicity. The LD_{50} values are reported to range from 65 to 127 mg/kg BW in rats, 400 to 425 mg/kg BW in mice, 541 mg/kg BW in guinea pigs, 800 mg/kg BW in rabbits, and 950 mg/kg in dogs. Poisoning is accompanied by CNS depression, changes in behavior and GI tract functioning. Death occurs in the first 5 to 6 hours. Gross pathology examination showed circulatory disturbances, liver and spleen enlargement, and hyperemia of the gastric mucosa.[1]

Repeated Exposure failed to reveal cumulative properties. K_{acc} appeared to be about 20 (by Cherkinsky). Administration of 2.5 and 25 mg/kg BW to rats for 5 weeks caused enhanced BW gain. Hepatic cirrhosis was observed following single or repeated oral administration to male Wistar rats.[2]

In a 16-day study, rats received 15 to 240 mg F./kg BW, and mice were given 25 to 400 mg F./kg BW. The treatment did not induce retardation of BW gain or compound-related histological lesions.[3]

Short-term toxicity. F. was administered at doses of 11 to 180 mg/kg to rats, and at doses of 75 to 1200 mg/kg to mice over a period of 13 weeks. High mortality at high doses, increase in liver weight, and inflammation of the liver (in mice) were observed.[3]

Long-term Toxicity. Rats were dosed by gavage for 6 months. Administration of 75 mg/kg BW caused reduction in the blood chlorides contents and *Hb* levels to the end of the experiment.[1]

Mutagenicity.

In vitro genotoxicity. F. was positive in *Salmonella typhimurium* strain *TA100*. Produced a doubling of the SCE frequency, and a 3-point monotonic increase with at least the highest dose at the $p < 0.1$ significance level.[4]

Mutagenic effect does not depend on the presence of an activation system or of benzo[a]pyrene.[5] Cultured Chinese hamster ovary cells were exposed for 3 hours to F. F. induced a relatively high frequency of chromatid breaks and chromatid exchanges in the absence of a liver microsomal activation preparation. The clastogenic activities of F. was increased when *S9* mixture was added.[9]

Carcinogenicity. In a 2-year study, F344 rats were given 30 and 60 mg/kg BW doses. The treatment caused dysplasia with fibrosis, a precursor lesion to cholangiocarcinomas in males. In mice, it increased the incidence of hepatocellular adenomas and forestomach papillomas in females.[3]

Carcinogenicity classification. An IARC Working Group concluded that there is limited evidence for the carcinogenicity of F. in *experimental animals* and there is inadequate evidence to evaluate the carcinogenicity of F. in *humans.*

The Annual Report of Carcinogens issued by the U.S. Department of Health and Human Services (1998) defines F. to be a substance which may reasonably be anticipated to be carcinogen, i.e., a substance for which there is a inadequate evidence of carcinogenicity in *humans* or limited evidence of carcinogenicity in *experimental animals.*

IARC: 3;

NTP: SE - NE - CE - CE (gavage).

Chemobiokinetics. F. is rapidly converted to *furoic acid, furanacrylic acid,* and *furoylglycine*, most likely by non-mixed-function oxidase-dependent pathways.[6] *Furoylglycine* and *furanacryloylglycine* were identified as the major urinary metabolites in rats and mice. An additional minor polar metabolite was excreted by male rats and mice. F. is eliminated in the urine mainly as a *furoic acid.*

F. is distributed primarily to the liver and kidney, brain being a tissue with the lowest F. concentration. The initial step in F. metabolism involves the oxidation to *furoic acid*, which is excreted unchanged and decarboxylated to form $^{14}C\text{-}CO_2$, conjugated with *glycine*, or condensed with *acetic acid.* The urine appeared to be the major route of elimination (up to 90%), followed by feces (2.0 to 4.0%) and exhaled CO_2 (7.0%).[7,8]

Regulations.

U.S. FDA (1998) regulates F. as a component of adhesives intended for use in packaging, transporting, or holding food in accordance with the conditions prescribed in 21 CFR part 175.105. It is considered to be *GRAS* if used as a flavor ingredient.

EU (1990). F. is available in the *List of monomers and other starting substances which may continue to be used for the manufacture of plastic materials and articles intended to come into contact with foodstuffs pending a decision on inclusion in Section A* (*Section B*).

Standards. ***Russia*** (1995). MAC and PML: 1.0 mg/l (organolept., odor).

References:

1. Kuznetsov, P. I., in *Proc. Omsk State Med. Inst.*, Omsk, Issue No 69, 1967, 38; *Gig. Sanit.*, 5, 7, 1966 (in Russian).
2. Shimizu, A. and Kanisawa, M., Experimental studies on hepatic cirrhosis and hepatocarcinogenesis. I. Production of hepatic cirrhosis by furfural administration, *Acta Pathol. Japonica*, 36, 1027, 1986.
3. *Toxicology and Carcinogenesis Studies of Furfural in Fischer 344 Rats and B6C3F$_1$ Mice*, NTP Technical Report Series No 382, Research Triangle Park, NC, 1989.

4. Tucker, J. D., Auletta, A., Cimino, M. C., et al., Sister-chromatid exchange: second report of the Gene-Tox program, *Mutat. Res.*, 297, 101, 1993.
5. Zdzienicka, M., Tudek, B., Zielenska, M., et al., Mutagenic activity of furfural in *Salmonella typhimurium* TA 100, *Mutat. Res.*, 58, 205, 1978.
6. Irwin, R. D., Enke, S. B., and Prejean, J. D., Urinary metabolites of furfural and furfuryl alcohol in Fischer 344/N rats, *Toxicologist*, 5, 240 (Abstract 960), 1985.
7. Parkash, M. K. and Caldwell, J., Metabolism and excretion of [^{14}C]furfural in the rat and mouse, *Food Chem. Toxicol.*, 32, 887, 1994.
8. Nomeir, A. A., Silveira, D. M., McComish, M. F., and Chadwick, M., Comparative metabolism and disposition of furfural and furfuryl alcohol in rats, *Drug Metab. Dispos.*, 20, 198, 1992.
9. Stich, H. F., Rosin, M. P., Wu, C. H., and Powrie, W. D., Clastogenicity of furans found in food, *Cancer Lett.*, 13, 89, 1981.

FURAN

Molecular Formula. C_2H_4O
M = 68.07
CAS No 110-00-9
RTECS No LT8524000

Synonyms and **Trade Names.** Divinylene oxide; 1,4-Epoxy-1,3-butadiene; Furfuran; Oxacyclopentadiene; Tetrole.

Properties. Colorless liquid with a specific odor. Solubility in water is 2.3% (20°C), readily soluble in alcohol. Odor perception threshold is 50 to 75 mg/l. F.'s sweetish taste in water is detected at higher concentrations.[1]

Applications. Used in the production of furan resins for corrosion-resistant polymeric materials. A solvent.

Acute Toxicity. After administration of 2.0 g/kg BW, death occurs in half an hour. A 0.5 g/kg BW dose caused death of rats in 3 to 8 days. Poisoned animals displayed excitation with subsequent CNS inhibition. Death occurs from respiratory arrest. Gross pathology examination failed to reveal evident damage but congestion was noted in the vessels of the visceral organs and in the bone marrow. Visceral organ degeneration develops as a result of F. doses which do not lead to rapid death.[1]

Repeated Exposure revealed cumulative properties. Administration of 100 mg/kg BW causes death in rats in 3 to 10 days with signs of emaciation and jaundice. Liver changes are of the parenchymatous hepatitis type. Necrosis in the proximal renal tubules and reduction in *N*-acetylglucosoaminidase in the urine were found to develop in mice (Wiley et al.).

F344 rats and B6C3F$_1$ mice were given F. in corn oil by gavage for 16 days. Mottled and enlarged livers were found in rats which received 20 to 40 and more mg F./kg BW. There were no chemical-related lesions in mice by the end of the study.[2]

Oral gavage for 3 weeks at bioassay doses of 4.0, 8.0, and 15 mg F./kg BW elicited hepatotoxicity in a dose-related manner through a toxic metabolite, inducing cytolethality through apoptosis.[3]

Short-term Toxicity. Rats were given 50 mg/kg BW for 3 months. The treatment caused atrophic liver cirrhosis with typical pseudopathic lobules and a reduction in the vascular flow. In a 4-month study, dystrophic changes in the liver were noted. A dose of 0.1 mg/kg BW was reported to be ineffective for peripheral blood formula, prothrombine time, serum cholinesterase activity, and conditioned reflex activity in rats.[1]

In a 13-week gavage study, dose-dependent toxic lesions of the liver were observed in rats and mice that received 4.0 to 60 mg F./kg BW. Kidney lesions, thymic, testicular or ovarian atrophy were found in rats exposed to 30 to 60 mg F./kg BW.[2]

Allergenic Effect. A slight response was observed in experiments on guinea pigs.[4]

Mutagenicity.

In vitro genotoxicity. F. was shown to be weakly mutagenic in the *Salmonella* assay[2] and negative in mouse lymphoma cells (NTP-86); it produced CA and SCE in Chinese hamster ovary cells (NTP-88). Cultured Chinese hamster ovary cells were exposed to F. for 3 hours. F. induced a relatively high frequency of chromatid breaks and chromatid exchanges in the the presence of a liver microsomal activation preparation.[5]

In vivo genotoxicity. F. did not induce unscheduled DNA synthesis in hepatocyte isolated after single gavage treatment of male F344 rats with 5.0, 30, or 100 mg F./kg BW, or male B6C3F$_1$ mice treated with doses of 10 to 200 mg F./kg BW.[6]

Carcinogenicity. F. is a potent rodent hepatocarcinogen that probably acts through a non-genotoxic mechanism involving hepatotoxicity and regenerative hepatocyte proliferation. F. induces cytolethality through apoptosis, which has been suggested to play a key role in carcinogenesis in B6C3F$_1$ mice.[3]

F. is found to be more hepatotoxic than furfural or furfuryl alcohol, causing extensive distortion of gross liver structure and substantial microscopic pathology. In a 90-day study, male F344 rats were given daily doses of 30 mg F./kg BW. The treated animals developed marked cholangiofibrosis and cholangiohepatitis and, when subsequently maintained without further treatment for an additional 6 to 18 months, the cholangiofibrosis progressed to yield a 100% incidence of cholangiocarcinomas.[7]

Administration of 2.0, 4.0, and 8.0 mg/kg BW for 2 years by gavage caused 86 to 100% incidence of cholangiosarcomas at all doses in male rats with occasional metastases. F. produced hepatocellular adenomas and carcinomas in mice, given 8 mg/kg or 15 mg/kg BW by gavage in corn oil for 104 weeks.[2]

Carcinogenicity classifications. An IARC Working Group concluded that there is sufficient evidence for the carcinogenicity of F. in *experimental animals* and there is inadequate evidence to evaluate the carcinogenicity of F. in *humans*.

The Annual Report of Carcinogens issued by the U.S. Department of Health and Human Services (1998) defines F. to be a substance which may reasonably be anticipated to be carcinogen, i.e., a substance for which there is a inadequate evidence of carcinogenicity in *humans* or sufficient evidence of carcinogenicity in *experi mental animals*.

IARC: 2B;

NTP: CE* - CE* - CE* - CE* (gavage).

Chemobiokinetics. Oral administration of F. decreased hepatic *P-450* content and the activity of *P-450*-dependent enzymes. Covalent binding of F. to microsomal protein indicates that *P-450 llEl* preferentially catalyzes the metabolic activation of F. CO_2 is found to be the major metabolite. Ten metabolites detected in the urine indicate opening of the F. ring, followed by extensive carbon oxidation.[8,9]

Standards. ***Russia*** (1995). MAC and PML: 0.2 mg/l.

References:

1. Rubets, V. I., Substantiation of MAC for furan in water bodies, in *Protection of Water Reservoirs against Pollution by Industrial Liquid Effluents*, S. N. Cherkinsky, Ed., Medgiz, Moscow, Issue No 7, 1965, 219 (in Russian).
2. *Toxicology and Carcinogenesis of Furan in Fischer 344 Rats and B6C3F$_1$ Mice, (Gavage Studies)*, NTP Technical Report Series No 402, Research Triangle Park, NC, January 1993.
3. Fransson-Steen, R., Goldsworthy, T. L., Kedderis, G. L., and Maronpot, R. R., Furan-induced liver cell proliferation and apoptosis in female B6C3F$_1$ mice, *Toxicology*, 118, 195, 1997
4. Chernousov, A. D., On allergenic properties of furan compounds, *Gig. Sanit.*, 6, 28, 1974 (in Russian).
5. Stich, H. F., Rosin, M. P., Wu, C. H., and Powrie, W. D., Clastogenicity of furans found in food, *Cancer Lett.*, 13, 89, 1981.
6. Wilson, D. M., Goldsworthy, T. L., Popp, J., and Butterworth, B. E., Evaluation of genotoxicity, pathological lesions and cell proliferation in livers of rats and mice treated with furan, *Environ. Molec. Mutagen.*, 19, 209, 1992.
7. Maronpot, R. R., Giles, H. D., Dykes, D. J., and Irwin, R. D., Furan-induced cholangiocarcinomas in Fischer 344 rats, *Toxicol. Pathol.*, 19, 561, 1991.
8. Burka, L. T., Washborn, K. D., and Irwin, R. D., Disposition of ^{14}C-furan in the male Fischer 344 rats, *J. Toxicol. Environ. Health*, 34, 245, 1991.
9. Kedderis, G. L., Carfagna, M. A., Held, S. D., Batra, R., Murphy, J. E., and Gagas, M. L., Kinetic analysis of furan biotransformation by F-344 rats *in vivo* and *in vitro, Toxicol. Appl. Pharmacol.*, 123, 274, 1993.

HEXACHLORO-1,3-BUTADIENE

Molecular Formula. C_4Cl_6
M = 260.79
CAS No 87-68-3
RTECS No EJO700000
Abbreviation. HCBD.

Synonyms. 1,1,2,3,4,4-Hexachloro-1,3-butadiene; Perchlorobutadiene.

Properties. Clear, colorless, oily liquid. Solubility in water is 3.2 mg/l at 20°C, readily soluble in organic solvents. Odor perception threshold is 0.007 to 0.03 mg/l, taste perception threshold is 0.048 mg/l.[1]

Applications. HCBD is used in the production of synthetic rubbers for contact with food or drink and as a solvent for elastomers.

Exposure. The estimated daily intake by man seems to be adequate to avoid adverse non-carcinogenic effects on the kidneys.

Acute Toxicity.

Observations in man. Ingestion of high doses led to irritation or corrosion of the mouth, throat, and GI tract tissues.

Animal studies. LD_{50} is 200 to 400 mg/kg BW in female rats, 500 to 670 mg/kg BW in male rats,[1] and 50 to 200 mg/kg BW in mice.[2] Poisoning is accompanied by nephrotoxic changes: increased relative kidney weights and blood urea nitrogen as well as renal tubular cell necrosis.[3] Death may occur in several days or even months after exposure.

Repeated Exposure. B6C3F_1 mice were given HCBD at the dose levels of 30 to 3000 mg/kg BW in the diet for 15 days. All mice in the two highest dose groups died by day 7. Toxic responses (primarily in the highest dose groups) included abnormal clinical signs (lethargy, hunched posture, rough hair coats, light sensitivity, and/or coordination disturbances), gross and histopathological changes in the liver, kidneys, and testes.[4]

Wistar rats were exposed for 14 days to 3.0 to 27 mg HCBD/kg BW.[5] The only lesion observed was degeneration of the renal tubular epithelial cells. Wistar rats were fed hexachloro-1,3-butadiene for 4 weeks.[6] The treatment damaged epithelial cells of the proximal tubules. It caused slight growth depression in males at a minimum nephrotoxic effect level, slightly increased kidney weights in males and/or caused light histopathological changes in the kidneys at a minimum nephrotoxic effect level.

Short-term Toxicity. The kidney was found to be a target organ in short-term studies. Rats given HCBD by gavage for 13 weeks exhibited increased liver and kidney weights, some renal function disorders, and degeneration and hyperplasia of the proximal tubular epithelial cells.[4,7] Japanese quails were fed diets containing 0.3 to 30.0 mg HCBD/kg BW for 90 days. No adverse effects were noted.[1]

In a 13-week study, the NOEL was identified to be 1.0 and 2.5 mg/kg BW for female and male rats, respectively,[5] and 4.9 mg/kg BW for B6C3F_1 mice.[4]

Long-term Toxicity. In a 2-year feeding study, the NOAEL of 0.2 mg/kg BW was identified for renal toxicity in rats.[4,8] TDI was calculated to be 0.0002 mg/kg BW.

Reproductive Toxicity.

Gonadotoxicity. Decrease in sperm motility was observed in mice dosed with 0.1 to 16.8 mg/kg BW for 13 weeks.[4]

Embryotoxicity. Rats were fed the doses of 0.2 to 20 mg/kg BW for 90 days prior to mating, 125 days during mating, and subsequently throughout gestation (22 days) and lactation. There were no treatment-related effects on pregnancy or neonatal survival. No toxic effects were observed in neonates at given doses of 0.2 to 2.0 mg/kg BW, at higher doses histological kidney changes were reported.[1]

Teratogenic effects were found at a dose level of 8.1 mg/kg BW given to pregnant rats throughout gestation.[5,7] Rats were exposed to 10 mg HCBD/kg BW by *i/p* injection on days 1 to 15 of gestation. The treatment caused soft tissue anomalies and external hydrocephaly in pups from mothers treated with HCBD.[9]

Mutagenicity.

In vitro genotoxicity. Contradictory results have been obtained in bacterial assays for point mutations; HCBD is reported to be negative in *Salmonella* mutagenicity assay.[025] However, several metabolites gave positive results. The percentage of unscheduled DNA synthesis caused by HCBD was increased in Syrian hamster embryo fibroblasts both with and without metabolic activation.[10] HCBD is considered to be a genotoxic compound.

Carcinogenicity. Administration in the diet for 2 years caused renal tubular adenomas and adenocarcinomas in rats at doses causing renal injury (20 mg/kg BW).[10]

Carcinogenicity classifications. An IARC Working Group concluded that there is limited evidence for the carcinogenicity of HCBD in *experimental animals* and there were no adequate data available to evaluate the carcinogenicity of HCBD *in humans*.

IARC: 3;

U.S. EPA: C.

Chemobiokinetics. Following a single oral administration and absorption, HCBD is distributed mainly in the adipose tissue and liver. In rats and mice, HCBD is easily absorbed and metabolized via conjugation with glutathione followed by biliary excretion of *S-(1,2,3,4,4-pentachlor-1,3-butadienyl) glutathione*.[11] This conjugate can be further metabolized to nephrotoxic derivatives. The principal route of excretion is in the bile.[12] According to Payan et al., these glutathione conjugates undergo biliary recycling, being excreted in the urine as the corresponding mercapturates.[13]

Recommendations.

WHO (1996). Guideline value for drinking water: 0.0006 mg/l.

U.S. EPA (1999). Health Advisory for a longer-term exposure is 0.4 mg/l.

Standards.

U.S. EPA (1999). MCLG: *zero*.

Russia (1995). MAC in drinking water: 0.01 mg/l (organolept., odor), MAC in food: 0.01 mg/kg.

References:

1. Murzakayev, F. G., Toxicity data for hexachlorobutadiene and its intermediates, *Pharmacologia i Toxi cologia*, 26, 750, 1963 (in Russian).
2. Badayeva, L. N., Structural and metabolic index of the postnatal neurotoxicity of organochloride pesti cides, in *Reports of Ukrainian Acad. Sci.*, Kiyv, 1983, 55 (in Ukrainian).
3. Hook, J. B., Ishmael, J., and Lock, E. A., Nephrotoxicity of hexachloro-1,3-butadiene in the rat: the effect of monooxygenase inducers, *Toxicol. Appl. Pharmacol.*, 65, 373, 1984.
4. *The Toxicity Studies of Hexacloro-1,3-butadiene in B6C3F$_1$ Mice (Feed Studies)*, NTP Technical Report, NIH Publ. No 91-3120, 1991, 22.
5. Harleman, J. H. and Seinen, W., Short-term toxicity and reproduction studies in rats with hexachloro-(1,3)-butadiene, *Toxicol. Appl. Pharmacol.*, 47, 1, 1979.
6. Jonker, D., Woutersen, R. A., van Bladeren, P. J., Til, H. P., and Feron, V. J., Subacute (4-week) oral toxicity of a combination of four nephrotoxins in rats: comparison with the toxicity of the individual compounds, *Food Chem. Toxicol.*, 31, 125, 1993.
7. Gehring, P. J. and MacDougall, D., *Review of the Toxicity of Hexachlorobenzene and Hexachlorobutadiene*, Dow Chemical Co., Midland, MI, 1971.
8. Kociba, R. J., Keyes, D. G., Jersy, G. C., et al., Results of a two-year chronic toxicity study with hexachlorobutadiene in rats, *J. Am. Ind. Hyg. Assoc.*, 38, 589, 1977.
9. Harris, S. J., Bond, G. P., and Niemeir, R. W., The effects of 2-nitropropane, naphthalene, and hexachlorobutadiene on fetal rat development, *Toxicol. Appl. Pharmacol.*, 48, A69, 1979.
10. Schiffman, D., Reichert, D., and Henschler, D., Induction of morphological transformation and unscheduled DNA synthesis in Chinese hamster embryo fibroblasts by hexachlorobutadiene and its putative metabolite pentachlorobutenoic acid, *Cancer Lett.*, 23, 297, 1984.
11. Dekant, W., Schrenk, D., Vamvakas, S., et al., Metabolism of hexachloro-1,3-butadiene in mice: *in vivo* and *in vitro* evidence for activation by glutathione conjugation, *Xenobiotica*, 18, 803, 1988.
12. Nash, J. A., King, L. D., Lock, E. A., et al., The metabolism and disposition of hexachloro-1,3-butadiene in the rat and its relevance to nephrotoxicity, *Toxicol. Appl. Pharmacol.*, 73, 124, 1984.
13. Payan, J. P., Fabry, J. P., Beydon, D., et al., Biliary excretion of hexachlorobutadiene and its relevance to tissue uptake and renal excretion in male rats, *J. Appl. Toxicol.*, 11, 437, 1991.

HEXAFLUOROPROPENE

Molecular Formula. C_3F_6
M = 150.03
CAS No 116-15-4
RTECS No UDO350000
Abbreviation. HFP.

Synonyms. Hexafluoropropylene; Perfluoropropylene.

Properties. Colorless gas, poorly soluble in water.

Applications. HFP is used in the synthesis of polymers intended for contact with food.

Acute Toxicity. Only data on inhalation studies are available. Depression and clonic spasms are seen as poisoning symptoms in rodents.[1]

Repeated Exposure. Only data on inhalation studies are available. Severe kidney damage was revealed but no clinical signs or hemato- and microscopic tissue damage were observed.[1]

Mutagenicity. Showed negative results in *Salmonella* mutagenicity assay either with or without metabolic activation.

Chemobiokinetics. Following ingestion HFP may be absorbed as it was noted in animal models and in man. The greatest residual amount was found in the kidney, bone, and lungs. Approximately 90% of the degradation product is not volatile. Essentially no HFP was found in the urine. The product of metabolism included *piruvic acid, ammonia*, and unidentified reactive metabolite(s).[2]

Regulations.

U.S. FDA (1998). The material is cleared for food related use as a perfluorocarbon resin in packaging materials.

Great Britain (1998). HFP is authorized without time limit for use in the production of polymeric materials and articles in contact with food or drink or intended for such contact. The specific migration of this substance shall be not detectable (when measured by a method with a DL of 0.01 mg/kg).

References:

1. Kennedy, G. L., Toxicology of fluorine-containing monomers, *CRC Crit. Rev. Toxicol.*, 21, 149, 1990.
2. Green, T. and Odum, J., Structure/activity studies of the nephrotoxic and mutagenic action of cys-tein conjugates of chloro-and fluoroalkenes, *Chem.-Biol. Interact.*, 54, 15, 1985.

1,6-HEXANEDIAMINE

Molecular Formula. $C_6H_{18}N_2$
M = 116.21
CAS No 124-09-4
RTECS No MO1180000
Abbreviations. HMDA or HDA.

Synonyms. 1,6-Diaminohexane; Hexamethylenediamine.

1,6-HEXANEDIAMINE DIHYDROCHLORIDE

Molecular Formula. $C_6H_{16}N_2.2ClH$
M = 189.16
CAS No 6055-52-3
RTECS No MO1312000
Abbreviation. HDDC.

Synonyms. Hexamethylenediamine dichloride; 1,6-Diamino-*N*-hexane dichloride.

Note. The exact form of compound is not specified in the majority of literature reports on HDA (HMDA) studies (see *Properties*).

Properties. Colorless leaflets (or lustrous crystals) with a pungent ammonia odor. HDA (HMDA) absorbs water and CO_2 from the atmosphere. Readily soluble in water and alcohol. In its solid form, HDDC crystallizes as needles from water or ethanol. The technical product is supplied as a 70% aqueous solution. Aqueous solutions of HDA are highly basic. 1,6-HDA dihydrochloride (HDDC) is formed by the neutralization of HDA with hydrochloric acid. Odor perception threshold is 500 mg/l. Aqueous solutions are colorless and transparent.[1]

Applications. Used in the production of nylon-type polyamide resins (especially *Nylon 66*), in the synthesis of polyurethane coatings, wet-strength resins, and polyamide adhesives intended for contact with food. Curing agent for epoxy resins.

Exposure. HDA has not been found in U.S. or European drinking water supplies (NAS, 1977).

Acute Toxicity. LD_{50} is 750 to 980 mg/kg BW in rats and 1110 mg/kg BW in rabbits.[2,3,03] Clinical signs of intoxication include weakness, malaise, salivation, diarrhea, tremors, and BW loss. Five guinea pigs given 0.02 g HDA orally each day died within 20 to 70 days.[4]

According to other data, LD_{50} is 582 mg/kg BW in rats, 665 mg/kg BW in mice, and 620 mg/kg BW in rabbits.[1] Lung emphysema is reported to develop rapidly, and 2 to 5 min. after HDA is administered a yellow fluid is discharged from the mouth and nose of the animal. Death occurs in 10 min. with clonic and tonic convulsions. Dissection reveals serous emphysema in the lungs and congestion in the liver, kidneys, and spleen; the gastric mucosa seems to be edematous but without necrosis and hemorrhaging.

Repeated Exposure. Mice and rats were given the dose of 100 mg/kg BW. Toxic effects developed after 2 to 3 administrations. The majority of the animals died after 6 to 7 administrations, which indicates pronounced cumulative properties. With doses of 10 and 30 mg/kg BW, toxic effects developed after 2 to 3 weeks. An increase in BW gain ceased in the fourth week, and in the sixth to eighth week, there was a 10 to 35% BW loss, changes in the peripheral blood formula, an increase in the activity of cholinesterase, a change in the ratio of protein fractions in the blood serum, and a reduction in the content of *SH*-groups in the blood.[1] In a 2-week study in F344/N rats (0.75 to 6.7 mg/ml) and B6C3F_1 mice (0.2 to 3.0 mg/ml), no clinical signs of toxicity and no gross or microscopic pathologic changes were found.[5]

Short-term Toxicity. Sprague-Dawley rats given daily doses of up to 500 mg/kg in their feed for 13 weeks experienced no changes in BW gain and clinical chemistry parameters examined.[3]

In a 13-week inhalation study, the NOAEL for respiratory damage was 5.0 mg/m^3 for rats and mice.[5]

Immunotoxicity. In a 12-month study, rats were given HMDA in their drinking water at concentrations of 0.1 to 10.0 mg/l. The inhibition of antibody production and reduction in the volume of the splenic lymphoid tissue were observed in approximately 40% of animals.[6] Doses of 0.1 and 1.0 mg HDA/kg BW, but not 10 mg HDA/kg BW, given in drinking water for 1 to 2 years caused an increase in the mitotic index of the lymphoid tissues.[7]

Reproductive Toxicity.

Embryotoxicity. No adverse effect on reproduction of rats was noted following inhalation exposure.[5] There was no treatment-related mortality in Sprague-Dawley rats given the diet containing average daily doses of 50, 150, and 500 mg/kg BW over two generations. BW and litter size were slightly reduced in the high dose group. There was no adverse effect on survival during lactation; doses up to 150 mg/kg BW did not affect reproduction or fertility in rats.[8]

According to David and Heck, administration of 200 mg/kg BW dose to pregnant F344 rats did not affected litter size.[9]

Teratogenic effect was not observed in mice and rats.[3,10]

Gonadotoxicity. Inhibition of ornithine decarboxylase is reported in rats given HNDA.[10,11]

Mutagenicity.

In vitro cytogenetics. HDA was not mutagenic in four strains of *Salmonella;* it did not increase SCE or CA in cultured Chinese hamster ovary cells.

Animal studies. Negative results were obtained in micronuclei test in mice *in vivo*.[5]

Chemobiokinetics.

Observations in man. In six human volunteers, orally given HDA was completely excreted with the urine within 15 hours. The primary urinary metabolites found were *6-aminohexanoic acid* and *N-acetyl-1,6-HDA*. Unchanged HMDA and 6-aminohexanoic acid were rapidly excreted.[12]

Regulations.

U.S. FDA (1998) listed HMDA for use as a component in the manufacture of resinous and polymeric coatings that may be safely used as a food-contact surface of articles intended for use in producing, manufacturing, packing, transporting, or holding food in accordance with the conditions prescribed in 21 CFR part 175.300.

EU (1999). HMDA is available in the *List of authorized monomers and other starting substances which shall be used for the manufacture of plastic materials and articles intended to come into contact with foodstuffs (Section* A).

Great Britain (1998). HMDA is authorized without time limit for use in the production of polymeric materials and articles in contact with food or drink or intended for such contact. The specific migration of this substance shall not exceed 2.4 mg/kg.

Standards.

EU (1990). SML: 2.4 mg/kg.

Russia (1995). MAC and PML: 0.01 mg/l.

References:

1. Brook, Ye. S., Klimkina N. V., and Panov, P. V., in *Protection of Water Reservoirs against Pollution by Industrial Liquid Effluents*, S. N. Cherkinsky, Ed., Meditsina, Moscow, Issue No 7, 1965, 69 (in Russian).
2. Vernot, E. H., MacEwen, J. D., Haus, C. C., et al., Acute toxicity and skin corrosion data for some organic and inorganic compounds and aqueous solutions, *Toxicol. Appl. Pharmacol.*, 42, 417, 1977.
3. Johannsen, F. R. and Levinskas, G. J., Toxicological profile of orally administered 1,6-hexanediamine in the rat, *J. Appl. Toxicol.*, 7, 259, 1987.
4. Ceresa, C. and De Blasiis, M., Experimental research on intoxication with hexamethylenediamine, *Med. Lavoro*, 4, 78, 1950.
5. *Toxicity Studies of 1,6-Hexanediamine Dihydrochloride Administered by Drinking Water and Inhalation to F344/N Rats and B6C3F$_1$ Mice*, Draft NTP Technical Report, U.S. Dept. Health & Human Service, NIH, 1992, 74.
6. Shubik, V. M., Nevstruyeva, M. A., Kal'nitsky, S. A., et al., A comparative study of changes in immunological reactivity during prolonged exposure to radioactive and chemical substance in drinking water, *J. Hygiene, Epidemiol., Microbiol., Immunol.*, 22, 408, 1978.
7. Ponomareva, T. V., and Merkushev, G. N., Effects of some nonradioactive and radioactive chemical compounds on the structure of the spleen, *Arch. Anat. Histol. Embryol.*, 74, 47, 1978 (in Russian).
8. Short, R. D., Johannsen, F. R., and Schardein, J. L., A two generation reproduction study in rats receiving diets containing hexamethylenediamine, *Fundam. Appl. Toxicol.*, 16, 490, 1991.
9. David, R. M. and Heck, H. D., Localization of 1,6-[^{14}C]diamino hexane (HMDA) in the prostate and the effects of HMDA on early gestation in Fischer 344 rats, *Toxicol. Lett.*, 17, 49, 1983.
10. Manen, G. A., Hood, R. D., and Farina, J., Ornithine decarboxylase inhibition and fetal growth retardation in mice., *Teratology*, 28, 237, 1983.
11. Guha, S. K. and Janne, J., Inhibition of ornitine decarboxylase *in vivo* in rat ovary, *Biochem. Biophys. Res. Commun.*, 75, 136, 1977.
12. Brorson, T., Soapring, G., Sandstrom, J. F., and Stenberg, J. L., Biological monitoring of isocyanates and related amines. I. Determination of 1,6-hexamethylenediamine in hydrolyzed human urine after oral administration of HDA, *Int. Arch. Occup. Environ. Health*, 62, 79, 1990.

HEXAMETHYLENE DIISOCYANATE

Molecular Formula. $C_8H_{12}N_2O_2$

M = 248.31

CAS No 822-06-0

RTECS No MO1740000

Abbreviation. HMDI.

Synonyms and **Trade Names.** Desmodur G; 1,6-Diisocyanato hexane; Hexane diisocyanate; Isocyanic acid, hexamethylene ester; Tolnate HD.

Properties. Pale yellow, slightly volatile liquid with a pungent unpleasant odor (threshold concentration 0.001 mg/l). Reacts with water and alcohol.

Applications. A cross-linking agent or hardener in the production of polyurethane materials intended for use in contact with food.

Acute Toxicity. LD_{50} is found to be 738 to 960 mg/kg BW in rats, 350 to 1080 mg/kg BW in mice, 1100 mg/kg BW in cats (Kimmerle et al.,1982). Ingestion caused irritation of the mouth, pharynx and GI tract, accomplished by headache, nausea and vomiting.

Long-term toxicity. HMDI inhibits acetylcholinesterase activity. Signs of nephrotoxicity were observed in mice with chronic feeding at dose levels of 108 mg/kg BW (NTP-86).

Mutagenicity. Was negative in *Salmonella* mutagenicity assay with and without exogenous metabolic activation system (Anderson, 1980).

Chemobiokinetics. In the the presence of water, diisocyanates rapidly disassociate to form a *primary amine, urea,* and *chlorine dioxide*. This led to denaturation of proteins, loss of enzyme function, and/or formation of *haptens* and immunological reactivity (Tornling et al., 1990).

Regulations.

EU (1990). HMDI is available in the *List of authorized monomers and other starting substances which shall be used for the manufacture of plastic materials and articles intended to come into contact with foodstuffs (Section A)*.

U.S. FDA (1998) listed HMDI for use in the manufacture of polyurethane resins that may be safely used as the food-contact surface of articles intended for use in contact with bulk quantities of dry food in accordance with the conditions prescribed in 21 CFR part 177.1680.

Great Britain (1998). HMDI is authorized without time limit for use in the production of polymeric materials and articles in contact with food or drink or intended for such contact. The quantity in the finished plastic material or article of any substance within, or any combination of substances within shall not exceed 1 mg/kg (expressed as isocyanate moiety).

Standards. ***EU*** (1990). MPQ in the finished material or article: 1.0 mg/kg (expressed as isocyanate moiety).

References:

1. von Burg, R., Hexamethylene diisocyanate, *J. Appl. Toxicol.*, 13, 435, 1993.
2. Sei, A., *Monomers for Polycondensation*, Mir, Moscow, 1976, 358 (Russian translation from English).

N-(HYDROXYMETHYL) ACRYLAMIDE

Molecular Formula. $C_4H_7NO_2$
M = 101.11
CAS No 924-42-5
90456-67-0
RTECS No AS3600000
Abbreviation. HMA.

Synonyms and **Trade Names.** *N*-(Hydroxymethyl)-2-propenamine; *N*-Methylolacrylamide; Monomethylolacrylamide; *N*-Methanolacrylamide; Uramine T 80.

Properties. White, odorless, crystalline substance. Readily soluble in water. Concentrated solutions are colorless and transparent and do not cloud when left to stand. Taste perception threshold is 560 mg/l.[1]

Applications. Monomer in the production of plastics, organic glass, lacquers, and enamels intended for use in contact with food or drink. A cross-linking agent used in adhesives, resins, latex films, and sizing agents.

Acute Toxicity. The LD_{50} values are reported to be 312 to 474 mg/kg BW in rats, 400 to 420 mg/kg BW in mice, and 328 mg/kg BW in rabbits. Poisoning results in NS affection.[1,2]

Repeated Exposure revealed no cumulative properties in mice given 1/5 and 1/10 LD_{50} for a month.

Male Porton rats received 1800 ppm HMA via the diet (27 mg/animal/day) for one week and thereafter 900 ppm (18.6 mg HMA) for 4 weeks. Four *i/p* doses of 50 mg HMA/kg BW were also given. The treatment caused slight ataxia. The disability became moderate at seven weeks.[3]

F344 rats and B6C3F_1 mice received 200 to 400 mg/kg BW in water by gavage for 16 days. All rats that received 400 mg/kg BW died within 4 days. Administration of the dose of 200 mg/kg BW led to retardation of BW gain, caused ataxia, muscle tremors and hyperirritability in rats. Histology investigation revealed hyperplasia of the bronchiolar and tracheal epithelium, displasia of the nasal and tracheal epithelium, centrilobular hepatocellular necrosis, lymphoid depletion of the spleen, etc. Analogous changes were observed in mice.[4]

Short-term Toxicity. In the above described studies, rats were given doses of 12.5 to 200 mg/kg BW for 13 weeks. All rats that received 100 and 200 mg/kg BW died before the end of the studies. The animals exhibited hind limb ataxia, which progressed to hind limb paralysis. Retardation of BW gain was noted. Histology investigation revealed axon filament and myelin sheath degeneration of the brain stem, spinal cord, and/or peripheral nerves in rats given 25 mg/kg BW and higher doses. Hepatocellular necrosis and thymic lymphocytic necrosis were found in mice given a dose of 200 mg/kg BW.[4]

Long-term Toxicity. In a 6-month rat study, the CNS and liver functions were found to be the target systems.[2] In a 2-year study, survival and BW gain in F344 rats receiving doses of 6.0 or 12 mg/kg BW, and in B6C3F_1 mice receiving up to 50 mg/kg BW, were minimally affected. Neither biologically important clinical signs of intoxication nor non-neoplastic lesions were noted in treated animals.[5]

Reproductive Toxicity.

Embryotoxicity. Swiss CD-1 mice were provided with drinking water containing 60, 180, or 360 ppm HMA during and after a 14-week cohabitation. The last litter was reared and dosed after weaning until mating at 2.5 months of age with same level of compound given to the parents. In the F_0 generation, the treatment caused the largest decrease (26%) in pup number during cohabitation together with a small effect on grip strength. No change was noted in F_0 female reproduction. Slightly greater toxic effect was observed in the F_1 generation, concomitant with increased consumption of HMA.[6]

Gonadotoxicity. Repeated oral doses of HMA caused testicular atrophy in mice. Histological investigation revealed degeneration of the epithelial cells of the seminiferous tubules with interstitial cells being normal. The treatment also induced neuropathy.[7]

DLM studies revealed the fertility effects that could be explained by a male-mediated increase in postimplantation loss. Dominant lethality occurred without structural effects on the reproductive system and without detectable neural histopathology.[6]

Mutagenicity.

In vitro genotoxicity. Did not induce gene mutations in *Salmonella typhimurium*. HMA induced both SCE and CA in Chinese hamster ovary cells with and without metabolic activation.[4,8]

Carcinogenicity. In NTP study, F344 rats and B6C3F_1 mice received doses of 6.0 or 12 mg/kg BW and 25 or 50 mg/kg BW, respectively, administered in water by gavage for 103 weeks. The treatment produced no neoplastic lesions in rats. In male mice, there was increased incidence of neoplasms of the Harderian gland, liver, and lung. In addition, compound-related ovarian atrophy was observed in female mice.[4]

Carcinogenicity classification.

IARC: 3;

NTP: NE - NE - CE - CE (gavage).

Chemobiokinetics. Following *i/v* injection to rats at doses of 140 mg/kg BW, HMA was distributed over all the body fluids within a few minutes. HMA is hydrolyzed in the body with *formaldehyde* formation (0.26 mg formaldehyde after complete hydrolysis of 1.0 mg HMA). According to Edwards, breakdown of HMA to *acrylamide* was negligible *in vitro* and *in vivo*. HMA metabolism occurs in the hepatic enzyme system of the mouse by glutathione *S*-transferase. HMA caused a rapid decrease in liver glutathione *in vivo* and *glutathione conjugates* were excreted in rats in the bile.[9,10]

Regulations.

U.S. FDA (1998) regulates HMA for use (1) in adhesives as a component of articles (a monomer) intended for use in packaging, transporting, or holding food in accordance with conditions prescribed in 21 CFR part 175.105; and (2) as a monomer in the production of semirigid and rigid acrylic and modified acrylic plastics intended for use in contact with food in accordance with the conditions prescribed in 21 CFR part 177.1010.

EU (1990). HMA is available in the *List of monomers and other starting substances, which may continue to be used pending a decision on inclusion in Section A.*

Great Britain (1998). HMA is authorized without time limit for use in the production of polymeric materials and articles in contact with food or drink or intended for such contact. The specific migration of this substance shall be not detectable (when measured by a method with a DL of 0.01 mg/kg).

Standards. *Russia* (1995). MAC and PML: 0.1 mg/l.

References:

1. Barnes, J. M., Observations of the effects on rats of compounds related to acrylamide, *Brit. J. Ind. Med.*, 27, 147, 1970.
2. Stryzhak, E. K., *Experimental Data on Hygienic Standards for Methacrylic Acid and Amides in Water Bodies*, Author's abstract of thesis, Moscow, 1967, 24 (in Russian).
3. Edwards, P. M., Neurotoxicity of acrylamide and its analogs and effects of these analogs and other agents on acrylamide neuropathy, *Brit. J. Ind. Med.*, 32, 31, 1975.
4. *Toxicology and Carcinogenesis Studies of N-Methylolacrylamide in F344/N Rats and B6C3F₁ Mice*, NTP Technical Report Series No 352, Research Triangle Park, NC, September 1989.
5. Bucher, J. R., Huff, J., Haseman, J. K., Eustis, S. L., Peters, A., and Toft, J. D., Neurotoxicity and carcinogenicity of *N*-methylolacrylamide in F344 rats and B6C3F$_1$ mice, *J. Toxicol. Environ. Health*, 31, 161, 1990.
6. Chapin, R. E., Fail, P. A., George, J. D., Grizzle, T. B., Heindel, J. J., Harry, G. J., Collins, B. J., and Teague, J., The reproductive and neural toxicities of acrylamide and three analogues in Swiss mice, evaluated using the continuous breeding protocol, *Fundam. Appl. Toxicol.*, 27, 9, 1995.
7. Hashimoto, K., Sakamoto, J., and Tanii, H., Neurotoxicity of acrylamide and related compounds and their effects on male gonads in mice, *Arch. Toxicol.*, 47, 179, 1981.
8. Hashimoto, K. and Tanii, H., Mutagenicity of acrylamide and its analogues in *Salmonella typhimurium*, *Mutat. Res.*, 158, 129, 1985.
9. Edwards, P. M., Distribution and metabolism of acrylamide and its neurotoxic analogs in rats, *Biochem. Pharmacol.*, 24, 1277, 1975.
10. Tanii, H. and Hashimoto, K., Studies on *in vitro* metabolism of acrylamide and related compounds, *Arch. Toxicol.*, 48, 157, 1981.

N-(HYDROXYMETHYL)-2-METHYL-2-PROPENAMIDE

Molecular Formula. $C_5H_9N0_2$
M = 115.15
CAS No 923-02-4
RTECS No UC6380000
Abbreviation. HMMPA.

Synonyms. *N*-(Hydroxymethyl)-2-methylacrylamide; Methylolmethacrylamide.

Applications. Monomer in the production of plastics, organic glass, lacquers, and enamels intended for use in contact with food or drink.

Acute Toxicity. LD_{50} is 0.3 g/kg BW in rats, and 0.4 g/kg BW in mice.

Regulations.

EU (1992). HMMPA is available in the *List of authorized monomers and other starting substances which may continue to be used for the manufacture of plastic materials and articles intended to come into contact with foodstuffs pending a decision on inclusion in Section A* (*Section B*).

Great Britain (1998). HMMPA is authorized up to the end of 2001 for use in the production of polymeric materials and articles in contact with food or drink or intended for such contact.

Reference:

Information from the Soviet Toxicological Center, *Gig. Truda Prof. Zabol.*, 9, 50, 1988 (in Russian).

INDENE

Molecular Formula. C_9H_8
M = 116.17
CAS No 95-13-6

RTECS No NK8225000

Synonym and **Trade Names.** *1H*-Indene; Inden; Indonaphthene.

Properties. Liquid or yellow needles. Insoluble in water. Soluble in all proportions in alcohol.

Applications. Monomer in the synthesis of coumarone-indene resins.

Acute Toxicity. 2.5 ml of 50% mixture in olive oil was absolutely lethal to rats. Liver, lungs, and GI tract were affected.[014] Inhalation exposure to I. vapor at concentrations of 800 to 900 ppm for 6 to 7 hours caused severe liver necrosis and kidney damage.[014]

Allergenic Effect. Sensitizer, can cause allergic dermatitis.[04]

Chemobiokinetics. Can be oxidized to *trans*-1,2-*dihydroxyindane* by liver microsomes of rats and rabbits (Leibman and Ortiz, 1968).

Regulations. ***EU*** (1990). I. is available in the *List of monomers and other starting substances, which may continue to be used pending a decision on inclusion in Section A*.

ISOBUTYLENE

Molecular Formula. C_4H_8

M = 56.11

CAS No 115-11-7

RTECS No UD0890000

Synonyms. Isobutene; 2-Methylpropene.

Properties. Colorless gas. Water solubility is 375 mg/l. Volatilizes rapidly from the open surface of water. Odor perception threshold is 0.5 mg/l. I. has no effect on the color or transparency of water.[1]

Applications. I. is used in the manufacture of rubber, plastic polymers, adhesives intended for contact with food, and in the production of antioxidants.

Acute Toxicity. Mice tolerate administration of 8.4 mg/kg BW dose (given as a 0.5 ml of I. solution at a concentration of 335 mg/l) without any manifestations of the toxic action.[1]

Short-term Toxicity. Mice were dosed by gavage with 3.75 mg/kg BW for 4 months. The treatment caused no changes in the behavior, BW gain or oxygen consumption. Gross pathology examination revealed no changes in the relative weights of the visceral organs or their histological structure. Accumulation is impossible because of its rapid removal from the body.[1]

Mutagenicity.

In vitro cytogenetics. 5.0 to 100% I. (gas) is found to be negative in *Salmonella* mutagenicity assay with or without metabolic activation.[2] It exhibited no mutagenic effect in mouse lymphoma cells. I. showed no mutagenic effect in micronucleus test using human lymphocytes, while its metabolite, epoxide 2-methyl-1,2-epoxypropane, induced a statistically significant dose-dependent increase in the number of micronuclei.[3]

Chemobiokinetics. I. is rapidly excreted unchanged through the lungs. According to other data, it is metabolized to epoxide *2-methyl-1,2-epoxypropane*, or is converted by hepatic monooxygenase(s) to the epoxide, *2,2-dimethyloxirane*.[3,4] Two urinary metabolites, *isobutenediol* and 2-*hydroxyisobutyric acid*, were identified in F344 rats following inhalation exposure to I.[5] The epoxidation is initiated by epoxide hydrolase and glutathione *S*-transferase and is cytochrome *P-450*-dependent.[6]

Regulations.

U.S. FDA (1998) listed I. for use (1) as a component (a monomer) of adhesives to be safely used in food-contact surfaces in accordance with the conditions prescribed in 21 CFR part 175.105; and (2) as a component (a monomer) of the uncoated or coated food-contact surface of paper and paperboard that may be safely used in producing, manufacturing, packing, transporting, or holding dry food in accordance with the conditions prescribed in 21 CFR part 176.180.

Great Britain (1998). I. is authorized without time limit for use in the production of polymeric materials and articles in contact with food or drink or intended for such contact.

Standards.

Russia (1995). MAC and PML: 0.5 mg/l (organolept., odor).

References:

1. Amirkhanova, G. F., Latypova, Z. V., and Tupeyeva, R. B., Hygienic substantiation of MAC for isobutylene in water bodies, *Protection of Water Reservoirs against Pollution by Industrial Liquid Effluents*, S. N. Cherkinsky, Ed., Meditsina, Moscow, Issue No 7, 1965, 28 (in Russian).
2. Staab, R. J. and Sarginson, N. J., Lack of genetic toxicity of isobutylene gas, *Mutat. Res.*, 130, 259, 1984.
3. Jorritsma, U., Cornet, M., van Hummelen, P., Bolt, H. M., and Vercruysse, A., Comparative mutagenicity of 2-methylpropene (isobutene), its epoxide 2-methyl-1,2-epoxypropane and propylene oxide in the *in vitro* micronucleus test using human lymphocytes, *Mutagenesis*, 10, 101, 1995.
4. Bolt, H. M., Interspecies differences in metabolism and kinetics of 1,3-butadiene, isobutene and styrene, *IARC Sci. Publ.*, 127, 37, 1993.
5. Henderson, R. F., Sabourin, P. J., Bechtold, W. E., Steinberg, B., and Chang, I. Y., Disposition of inhaled isobutene in F344/N rats, *Toxicol. Appl. Pharmacol.*, 123, 50, 1993.
6. Cornet, M., Sonck, W., Callaerts, A., Csanady, G., Vercruysse, A., Laib, R. J., and Rogiers, V., *In vitro* biotransformation of 2-methylpropene (isobutene): epoxide formation in mice liver, *Arch. Toxicol.*, 65, 263, 1991.

ISOCYANIC ACID, 3,3'-DIMETHOXY-4,4'-BIPHENYL ESTER

Molecular Formula. $C_{16}H_{12}N_2O_4$
M = 296.30
CAS No 91-93-0
RTECS No NQ8800000

Synonyms. Dianizidine diisocyanate; 4,4'-Diisocyanato-3,3'-dimethoxy-1,1'-biphenyl; 3,3'-Dimetoxybenzidine-4,4'-diisocyanate; 3,3'-Dimetoxy-4,4'-biphenylene diisocyanate.

Properties. Gray to brown powder soluble in esters.

Applications. I. can be used in isocyanate-based adhesive systems and as a component of the polyurethane elastomers intended for contact with food.

Mutagenicity.

In vitro genotoxicity. I. was mutagenic to *Salmonella typhimurium* strain *TA98* in the the presence but not in the absence of an exogenous metabolic system; it was not found to be mutagenic to strains *TA1535, TA1537,* or *TA100.*[1]

Carcinogenicity. There were no treatment-related tumors in B6C3F$_1$ mice gavaged initially with 1.5 and 3.0 g I./kg BW for 22 weeks, and subsequently given dietary concentrations of 22 or 44 g I./kg over a period of 56 weeks. A statistically significant increase in the combined incidence of leukemia and malignant lymphomas was observed in F344 rats together with a treatment-related increase in the incidence of tumors of the skin and Zymbal gland.[2]

Carcinogenicity classifications. An IARC Working Group concluded that there is limited evidence for the carcinogenicity of I. in *experimental animals* and there were no adequate data available to evaluate the carcinogenicity of I. in *humans*.

IARC: 3;

NTP: P - P - N - N (feed).

Regulations. ***U.S. FDA*** (1998) approved the use of I. in the manufacture of polyurethane resins which may be used as food-contact surface of articles intended for use in contact with bulk quantities of dry food of an appropriate type in accordance with the conditions prescribed in 21 CFR part 177.1680.

References:

1. Haworth, S., Lawlor, T., Mortelmans, K., et al., *Salmonella* mutagenicity test results for 250 chemicals, *Environ. Mutagen.*, Suppl. 1, 3, 1983.
2. *Bioassay of 3,3'-Dimethoxybenzidine 4,4'-diisocyanate for Possible Carcinogenicity*, Natl. Cancer Institute Technical Report Series No 128, Bethesda, MD, 1979.

ISOPRENE

Molecular Formula. C_5H_8

M = 68.12
CAS No 78-79-5
RTECS No NT4037000

Synonyms. Isopentadiene; β-Methylbivinyl; β-Methylbutadiene; 3-Methyl-1,3-butadiene.

Properties. Colorless, mobile, volatile, combustive liquid or gas with a pungent unpleasant odor. Natural rubber is an I. polymer that can be depolymerized by heating to yield I. monomer. Water solubility is 1200 mg/l at 20°C, odor perception threshold is 0.005 mg/l; the concentration of 1000 mg/l does not alter the color or taste of water.[1]

Applications. Used in the manufacture of synthetic rubber (polyisoprene and butyl rubber). Copolymer for synthetic elastomers.

Exposure. I. is the major endogenous hydrocarbon exhaled in human breath. Concentration of I. in human blood has been reported to be between 15 and 70 ηmol/l.[2]

Acute Toxicity. Doses of 4.5 to 5.0 g/kg BW appeared to be lethal for rats. LD_{50} for rats is 2.04 g/kg (females) and 1.4 g/kg BW (males). LD_{50} for mice is 1.4 g/kg (females) and 1.7 g/kg BW (males).[3] Toxic effects include NS depression, slow movements, and labored breathing.

Repeated Exposure. 10 daily administrations of 25 mg/kg BW produced no effect on the general condition, BW gain, or behavior of mice and rats. Decreased catalase activity was noted.

With a 2.5 mg/kg BW dose, the glycogen-forming function of the liver was disturbed, catalase activity was reduced, and there were changes in conditioned reflex activity. Histology examination revealed no specific changes in the visceral organs.[1]

Long-term Toxicity. Following 26-week inhalation exposure to up to 7000 ppm I., non-neoplastic lesions in mice included spinal cord degeneration, testicular atrophy, degeneration of the olfactory epithelium, and epithelial hyperplasia of the forestomach.[4]

Reproductive Toxicity. I. did not produce ***teratogenic effect*** after inhalation exposure in rats and mice.[5]

Mutagenicity.

In vitro genotoxicity. I. did not induce CA and SCE in Chinese hamster ovary cells (NTP-85), and showed no mutagenic activity in *Salmonella* mutagenicity assay.

In vivo cytogenetics. According to later data, I. appeared to be genotoxic in mammalian assays.[7] I. induced an increase in frequency of SCE and micronuclei when its cytogenetic effects were investigated in bone marrow cells of $B6C3F_1$ mice exposed to I. by inhalation.[6]

Carcinogenicity. In mice, inhalation exposure resulted in increased incidences of benign and malignant tumors of the lung, liver, and forestomach and of Harderian gland adenomas.[4]

Carcinogenicity classifications. An IARC Working Group concluded that there is sufficient evidence for the carcinogenicity of I. in *experimental animals* and there were inadequate data available to evaluate the carcinogenicity of I. in *humans*.

IARC: 2B.

NTP: CE - SE – XX - XX

Chemobiokinetics. Distribution in the brain and parenchymatous organs was similar but the level in the fatty tissues was substantially higher than that in the brain, liver, kidney, or spleen.[8] When incubated in the the presence of microsomes of the murine liver, the main products of biotransformation of I. are *3,4-epoxy-3-methyl-1-butene* and *3,4-epoxy-2-methyl-1-butene*. In the the presence of microsomes, the latter underwent further epoxidation with the formation of mutagenic *isoprene oxide*.[9] Rats eliminate 6 times more I. metabolites in the urine than mice at low inhalation exposure concentrations (NTP-91).

Regulations.

U.S. FDA (1998) listed I. for use as a component (a monomer) in the uncoated or coated food-contact surface of paper and paperboard that may be safely used in producing, manufacturing, packing, transporting, or holding dry food in accordance with the provisions prescribed in 21 CFR part 176.180.

EU (1990). I. is available in the *List of monomers and other starting substances which may continue to be used for the manufacture of plastic materials and articles intended to come into contact with foodstuffs pending a decision on inclusion in Section A* (*Section B*).

Great Britain (1998). I. is authorized up to the end of 2001 for use in the production of polymeric materials and articles in contact with food or drink or intended for such contact.

Standards. *Russia* (1995). MAC and PML: 0.005 mg/l (organolept., odor).

References:

1. Klimkina, N. V., Hygienic norm-setting of harmful substances from the production of synthetic isoprene rubber in water bodies, *Gig. Sanit.*, 6, 8, 1959 (in Russian).
2. Cailleux, A., Cogny, M., and Allain, P., Blood isoprene concentrations in humans and in some animal species, *Biochem. Med. Metab. Biol.*, 47, 157, 1992.
3. Repina, E. F., in *Occupational and Environmental Hygiene and Health Protection in Workers of the Oil Industry*, Moscow, 105, 1987 (in Russian).
4. Melnick, R. L., Sills, R. C., Roycroft, J. H., Chou, B. J., Ragan, H. A., and Miller, R. A., Isoprene, an endogenous hydrocarbon and industrial chemical, induces multiple organ neoplasia in rodent after 26 weeks of inhalation exposure, *Cancer Res.*, 54, 5333, 1994.
5. Mast, T. J., Rommereim, R. I., Weigel, R. J., et al., Inhalation developmental toxicity of isoprene in mice and rats, *Toxicologist*, 10, 42, 1990.
6. Shelby, M. D., Results of NTP-sponsored mouse cytogenetic studies on 1,3-butadiene, isoprene, and chloroprene, *Environ. Health Perspect.*, 86, 71, 1990.
7. Tice, R. R. et al., Chloroprene and isoprene: cytogenetic studies in mice, *Mutagenesis*, 3, 141, 1988.
8. Shugaev, B. B., Concentrations of hydrocarbons in tissues as a measure of toxicity, *Arch. Ind. Health.*, 18, 878, 1969.
9. Gervasi, P. G. and Longo, V., Metabolism and mutagenicity of isoprene, *Environ. Health Perspect.*, 86, 85, 1990.

MALEIC ACID

Molecular Formula. $C_4H_4O_4$
M = 116.08
CAS No 110-16-7
RTECS No OM9625000
Abbreviation. MAc.

Synonyms. *cis*-Butenedioic acid; *cis*-1,2-Ethylene dicarboxylic acid; Toxilic acid.

MALEIC ANHYDRIDE

Molecular Formula. $C_4H_2O_3$
M = 98.06
CAS No 108-31-6
RTECS No ON3675000
Abbreviation. MAn.

Synonyms. *cis*-Butanedioic anhydride; 2,5-Furandione; Maleic acid, anhydride; Toxilic anhydride.

Properties. MAn occurs as a white, crystalline powder with a pungent odor. Readily soluble in water, forming MAc (1.0 mg of anhydride is equivalent to 1.18 mg of acid). Odor and taste perception threshold appeared to be 1.0 mg/l.[1]

Applications. Used as a monomer in the production of polyester plastics intended for contact with food.

Acute Toxicity. LD_{50} is 708 mg/kg BW in rats, and 2400 mg/kg BW[2] or 465 mg/kg BW in mice.[014] LD_{50} in Sprague-Dawley rats is reported to be 900 mg/kg BW (administration in the water) or 1050 mg/kg BW (gavage in corn oil). Clinical manifestations of poisoning in rats included decreasing food consumption and activity, increasing weakness, collapse and death. Gross pathology examination revealed hemorrhagic areas in the lung and liver and acute GI tract inflammation. There were no other lesions in the visceral organs.[3] However, according to Everett et al., a single oral dosage of MA produced nephrotoxic effect including acute tubular necrosis in male and female beagle dogs.[4]

Long-term Toxicity. F344 rats were fed 10, 32, or 100 mg MAn/kg BW in their diet for 2 years. Slight retardation of BW gain was noted in animals treated with two higher doses. The NOAEL of 10 mg/kg BW was established in this study.[5]

Rabbits were dosed by gavage with 2.5 mg/kg BW for 6 months. The treatment affected liver glycogen-forming function and leukocyte phagocytic activity. Histology examination revealed signs of parenchymatous dystrophy in the liver, kidneys, spleen, and GI tract mucosa.[1]

Allergenic Effect.

Observations in man. Human allergic reactions including asthma and alveolitis were considered to be rare systemic hypersensitivity reactions (risk 1:1000).[6,7]

Animal studies. Moderate allergenic effect has been observed in the tests on guinea pigs.[8,9]

Reproductive Toxicity.

Embryotoxicity. Short et al.[10] observed no treatment-related fetal effects in rats given 140 mg MAn/kg BW on days 6 to 15 of gestation. In a multi-generation study, CD rats were gavaged with 20, 55, or 150 mg/kg BW doses administered in corn oil until sacrifice. The highest dose caused an increase in mortality and renal damage in F_0 and F_1 parent rats. The lowest adverse effects were observed in this study with the 20 mg/kg BW dose.[11]

Mutagenicity.

In vitro genotoxicity. MAc was negative in *Salmonella* mutagenicity assay when reacted with chlorine in aqueous solution. In 1:1 methanol/water solution, MAc appeared to be substantially mutagenic.[12] MAn was negative in *Salmonella* mutagenicity assay.[025]

Regulations.

EU (1990). MAc and MAn are available in the *List of authorized monomers and other starting substances which shall be used for the manufacture of plastic materials and articles intended to come into contact with foodstuffs (Section A).*

U.S. FDA (1998) approved the use of MAn (1) in adhesives as a component of articles (a monomer) intended for use in packaging, transporting, or holding food in accordance with the conditions prescribed in 21 CFR part 175.105; (2) as a component of paper and paperboard for contact with aqueous and fatty food (polymer with ethyl acrylate and vinyl acetate, hydrolyzed) and dry food (a deposit control additive to the sheet-forming operation at a level not to exceed 0.075% by weight of the dry paper and paperboard); (3) in the manufacture of resinous and polymeric coatings for the food-contact surface of articles intended for use in producing, manufacturing, packing, processing, preparing, treating, packaging, transporting, or holding food in accordance with the conditions prescribed in 21 CFR part 175.300; (4) (a monomer) in the manufacture of closures with sealing gaskets for food containers intended for use in producing, manufacturing, packaging, processing, preparing, treating, packing, transporting, or holding dry food in accordance with the conditions prescribed in 21 CFR part 177.1210; and (5) in the manufacture of the polyurethane resins to be safely used in the food-contact surface of articles intended for use in contact with bulk quantities of dry food of an appropriate type in accordance with the conditions prescribed in 21 CFR part 177.1680.

MAc is listed for use (6) in the manufacture of resinous and polymeric coatings to be safely used as a food-contact surface of articles intended for use in producing, manufacturing, packing, transporting, or holding food in accordance with the conditions prescribed in 21 CFR part 175.300; (7) as an ingredient in the manufacture of cellophane for food packaging (1.0%) in accordance with the conditions prescribed in 21 CFR part 177.1200; (8) as a component of adhesives for food-contact surface in accordance with the conditions prescribed in 21 CFR part 175.105; (9) as a component of the uncoated or coated food-contact surface of paper and paperboard intended for use in producing, manufacturing, packaging, processing, preparing, treating, packing, transporting, or holding aqueous and fatty foods in accordance with the conditions prescribed in 21 CFR part 176.21; and (10) in the manufacture of cross-linked polyester resins for repeated use as articles or components of articles coming in contact with food in accordance with the conditions prescribed in 21 CFR part 177.2420.

Great Britain (1998). MAc and MAn are authorized without time limit for use in the production of polymeric materials and articles in contact with food or drink or intended for such contact. The specific migration of MAc and MAn shall not exceed 30 mg/kg.

Standards.

EU (1990). SML: 30 mg/kg (expressed as maleic acid).

Russia (1995). MAC and PML: 1.0 mg MAc/l (organolept., odor).

References:

1. Lisovskaya, E. V., Dyatlovitskaya, F. G., Tomashevskaya, L. V., et al., Experimental data on substantiation of MAC for maleic anhydride in water bodies, *Gig. Sanit.*, 4, 11, 1963 (in Russian).
2. Kirk-Othmer's *Encyclopedia of Chemical Technology*, 3rd ed., M. Grayson and D. Eckroth, Eds., John Wiley & Sons, Inc., New York, 1978, 14, 786.
3. Randall, D. J., Healy, C. E., Acute Toxicity Data, Part B, *J. Am. Coll. Toxicol.*, 1, 75, 1990.
4. Everett, R. M., Descotes, G., Rollin, M., Greener, Y., Bradford, J. C., Benziger, D. P., and Ward, S. J., Nephrotoxicity of pravadoline maleate (WIN 48098-6) in dogs: evidence of maleic acid-induced acute tubular necrosis, *Fundam. Appl. Toxicol.*, 21, 59, 1993.
5. *Chronic Dietary Administration of Maleic Anhydride*, U.S. EPA, vol 1, Narrative, Microfiche No OTS 1283-0277, 1983.
6. Topping, M. D., Venables, K. M., Luczynska, C. M., Howe, W., and Newman Taylor, A. J., Specificity of the human IgE response to inhaled acid anhydrides, *J. Allergy Clin. Immunol.*, 77, 834, 1986.
7. Venables, K. M., Low-molecular-mass chemicals, hypersensitivity, and direct toxicity: the acid anhydrides, *Brit. J. Ind.. Med.*, 46, 222, 1989.
8. Kryzhanovskaya, M. V., Vinogradov, G. I., and Vrodzinskaya, E. V., Comparative assessment of combined and isolated activity of chemical allergens, *Gig. Sanit.*, 10, 6, 1968 (in Russian).
9. Stevens, M. A., Use of the albino guinea-pig to detect the skin-sensitizing ability of chemicals, *Brit. J. Ind. Med.*, 24, 189, 1967.
10. Short, R. D., Johannsen, F. R., Levinskas, G. J., Rodwell, D. E., and Schardein, J. L., Teratology and multigeneration reproduction studies with maleic anhydride in rats, *Fundam. Appl. Toxicol.*, 7, 359, 1986.
11. *Three-generation reproduction study in rats (modified to a 2-generation study),* U.S. EPA,. Microfiche No OTS 0206655, Doc. ID 878214777, 1982. Available in *IRIS*.
12. Nazar, M. A., Rapson, W. H., Brook, M. A., May, S., and Tarhanen, J., Mutagenic reaction products of aqueous chlorination of catechol, *Mutat. Res.*, 89, 45, 1981.

MELAMINE

Molecular Formula. $C_3H_6N_6$
M = 126.13
CAS No 108-78-1
RTECS No OS0700000

Synonyms. Cyanuramide; Cyanurotriamine; Isomelamine; 2,4,6-Triamino-1,3,5-triazine.

Properties. Colorless-to-white, monoclinic crystals or prisms. Water solubility is 9.5% at 20°C. Readily soluble in hot water and alcohol. In aqueous solutions M. is transformed into *cyanuric acid*. Decomposes at 600°C to form *cyanamide*. A concentration of 3.5 g/l does not alter the taste, odor, color, or transparency of water.[1]

Applications. M. is used in the manufacture of melamine-formaldehyde amino resins, cross-linked polymeric materials. It is also used in laminates, surface coating resins, plastic molding compounds, lacquers, and adhesives as a monomer, stabilizer, etc. Melamine-based amino resins are used in housewares (dinnerware, table coverings, etc.).

Migration Data. Release of M. from dishes and bowls made of melamine-formaldehyde resin was studied. Concentrations found in 4.0% acetic acid at the temperature of 60 to 95°C after a 30 min. extraction period, were 0.2 to 1.19 ppm M. The release of M. was affected predominantly by temperature.[2] Migration of M. into food-simulating solvents from cups made of melamine resins were studied under various conditions. Low migration of M. was observed in any unused cup kept at 60°C for 30 min., 26°C for 1 hour, or cooled at 20°C for several days. The highest migration level of M. into 4.0% acetic acid was about 40 ppm when the migration test was repeated 7 times at 95°C for 30 min. Migration of M. from the cups being used at a cafeteria was 0.4 to 0.5 ppm.[3]

Acute Toxicity. LD_{50} is 3.2 and 3.8 g/kg BW in male and female F344 rats, respectively. In $B6C3F_1$ mice given M. in corn oil by gavage, LD_{50} was found to be 3.3 g/kg BW (males) and 7.0 g/kg BW (females).[4] Toxic doses produced a diuretic effect, and crystals of *dimelamine-monophosphate* appear in the urine. Diuretic effect is reported in rats and dogs. According to other data, both mice and rats tolerate 10 g/kg BW, and this dose as well as a three-fold administration of 5.0 g/kg BW to rats for 6 hours caused no changes in their behavior.[1]

Repeated Exposure revealed pronounced cumulative properties.[5] K_{acc} is 1.34 (by Kagan). F344 rats and $B6C3F_1$ mice were fed diets containing 5.0 to 30 g/kg for 14 days. The treatment produced a hard crystalline solid in the urinary bladder in the majority of male rats receiving dietary levels of 10 g/kg or more, and in all treated male mice. In females, this effect was found at dietary levels 20 g/kg or more; it was noted in rats and in 40% of mice given dietary levels of 30 g/kg.[4]

Short-term Toxicity. In 13-week feeding study, rats and mice were given diets containing 6.0 to 18.0 g/kg over a period of 13 weeks.[4] Stones and ulceration were found in the urinary bladders of animals receiving diets containing 12 to 15 g/kg.

Long-term Toxicity. Rats and mice were administered 2.0 g/kg BW doses for 6 months. The treatment resulted in reduced phagocytic index, increased cholinesterase activity and blood urea level. Histology examination revealed mild dystrophic changes in the liver, kidneys, and myocardium, and a slight hyperplastic reaction in the spleen of some animals. The NOAEL of 0.1 g/kg BW was identified in this study.[1]

Rats received M. at the dietary level of 1000 ppm for a period of 2 years; dogs were given it at the level of 30,000 ppm for 1 year. Animals demonstrated no signs of toxicity. In two to three month after beginning the treatment, dogs showed M. crystalluria.[03]

Allergenic Effect. Skin applications produced a sensitizing effect.

Reproductive Toxicity.

Embryotoxicity and ***Teratogenicity.*** No toxic effect or gross malformations were observed in fetuses of pregnant rats treated *i/p* with 70 mg/kg BW on gestation day 4, 5 or 7, 8 or 11, 12.[6]

Mutagenicity.

Observations in man. With regard to its mutagenic effect, M. is believed to present no risk in production and application.

In vivo cytogenicity. M. was not mutagenic to *Dr. melanogaster*.

In vitro genotoxicity. M. was negative in *Salmonella* mutagenicity assay. Negative results were observed also in Chinese hamster ovary cell assay, in a culture of rat hepatocytes, or in micronuclear test.[7,8]

Carcinogenicity. Male F344/N rats given the diet containing M., developed transitional-cell tumors of the urinary bladder. Rats were administered doses of 2.25 and 4.5 g/kg BW (males), 4.5 and 9.0 g/kg BW (females) for 103 weeks. Survival was significantly reduced in the high dose males. There was a dose-related increased incidence of bladder stones in male rats.[4] There seems to be a relationship between bladder stone formation in 7 out of 8 rats and bladder tumor development in this study.[9]

Carcinogenicity classifications. An IARC Working Group concluded that there is inadequate evidence for the carcinogenicity of M. in *experimental animals* and there were no adequate data available to evaluate the carcinogenicity of M. in *humans*.

IARC: 3;

NTP: P - N - N - N (feed).

Chemobiokinetics. M. can be accumulated for a short time in the kidneys and bladder. 24 hours after administration of ^{14}C-tagged M., it is not found in the tissues.[7] In rats and humans, M. is a metabolite of the antineoplastic agent *hexamethylmelamine*. M. can be decomposed to *urea* in the body. However, M. is not metabolized in male F344 rats.[10]

No metabolites were found in the urine and blood plasma. In 6 hours after oral administration of 250 mg/kg BW, 50% of the dose were found in the urine of rats.[11] 93% of the administered dose were removed via the urine.[10] M. increased diuresis and the removal of fine crystals of low-toxicity *dimelamine monophosphate* with the urine.[05]

Regulations.

EU (1990). M. is available in the *List of authorized monomers and other starting substances which shall be used for the manufacture of plastic materials and articles intended to come into contact with foodstuffs (Section A).*

U.S. FDA (1998) regulates the use of M. (1) as a component of adhesives for food-contact surface in accordance with the conditions prescribed in 21 CFR part 175.105. FDA approved the use of melamine polymers (2) as ingredients of resinous or polymeric coatings for polyolefin films as the basic polymer or modified with methanol surface in accordance with the conditions prescribed in 21 CFR part 175.320; (3) in epoxy resins as the basic polymer in accordance with conditions prescribed in 21 CFR part 175.320; (4) in paper or paperboard for contact with dry food in accordance with the conditions prescribed in 21 CFR part 176.180; (5) in cellophane as the basic polymer in accordance with the conditions prescribed in 21 CFR part 177.1200; or (6) in the modified form as a resin to anchor coatings to the substrate.

Melamine-formaldehyde resins may be used as surfaces in contact with food in molded articles, providing the yield of chloroform-soluble extractives does not exceed 80 $\mu g/cm^2$ of food contact surface under specific solvent and temperature parameters.

Great Britain (1998). M. is authorized without time limit for use in the production of polymeric materials and articles in contact with food or drink or intended for such contact. The specific migration of this substance shall not exceed 30 mg/kg.

Standards.

EU (1990). SML: 30 mg/kg.

Russia (1995). PML in the food 1.0 mg/l; PML in drinking water *n/m*.

References:

1. Gabrylevskaya, L. N. et al., in *Environmental Factors and Their Significance for Health of Population*, D. N. Kaluzhny, Ed., Zdorov'ya, Kiyv, Issue No 2, 1970, 115 (in Russian).
2. Sugita, T., Ishiwata, H., and Yoshihira, K., Release of formaldehyde and melamine from tableware made of melamine-formaldehyde resin, *Food Addit. Contam.*, 7, 21, 1990.
3. Ishiwata, H., Inoue, T., and Tanimura, A., Migration of melamine and formaldehyde from tableware made of melamine resin, *Food Addit. Contam.*, 3, 63, 1986.
4. *Carcinogenesis Bioassay of Melamine in F344/N Rats and B6C3F$_1$ Mice (Feed Study)*, NTP Technical Report Series No 245, Research Triangle Park, NC, 1983.
5. Babayan, E. A. and Alexandryan, A. V., *Pharmacologia i Toxicologia*, 14, 122, 1973 (in Russian).
6. Thiersch, J. B., Effect of 2,4,6-triamino-'*S*'-triasine, 2,4,6-tris(ethyleneimino)-*S*-triasine and *N,N',N''*-triethylene phosphoramide on rat litter in utero, *Proc. Soc. Exp. Biol. Med.*, 94, 36, 1957.
7. Mast, R. W., Jeffcoat, A. R., Sadler, B. M., et al., Metabolism, disposition, and excretion of ^{14}C-melamine in male Fischer 344 rats, *Food Chem. Toxicol.*, 21, 807, 1983.
8. Mirsalis, J., Tysen, C. K., and Butterworth, B. E., Detection of genotoxic carcinogens in the *in vivo* and *in vitro* hepatocyte DNA repair assay, *Environ. Mutagen.*, 4, 553, 1982.
9. Melnick, R. L., Boorman, G. A., Haseman, J. K., et al., Urolithiasis and bladder carcinogenicity of melamine in rodents, *Toxicol. Appl. Pharmacol.*, 72, 292, 1984.
10. Mast, R. W., Jeffcoat, A. R., Sadler, B. M., et al., Metabolism and excretion of ^{14}C-melamine in Fischer 344 rats, 3rd Int. Congress on Toxicology, Abstracts, San Diego, *Toxicol. Lett.*, 18 (Suppl. 1), 68, 1983.
11. Lipschitz, W. L. and Stockey, E., Mode of action of three new diuretics: melamine, adenine and formoguanidine, *J. Pharmacol. Exp. Ther.*, 83, 235, 1945.

METHACRYLIC ACID

Molecular Formula. $C_4H_6O_2$

M = 86.09

CAS No 79-41-4

RTECS No OZ2975000

Abbreviation. MA.

Synonyms. α-Methacrylic acid; 2-Methylpropenoic acid.

Properties. Readily mobile, colorless liquid with an acetic acid odor. Soluble in water (readily soluble in hot water) and in alcohol. Odor perception threshold of MA stabilized by hydroquinone (0.1%) is 106 mg/l. Concentrations up to 10 mg MA/l do not give water any foreign taste.[1]

Applications. Used as a monomer in the production of acrylic plastics intended for contact with food and drink.

Acute Toxicity. LD_{50} is reported to be 1.06 g/kg BW in rats, 1.3 g/kg BW in mice, and 1.2 g/kg BW in rabbits. A local irritating and resorptive effect is observed. Poisoning is accompanied by an increase in general debility, decreased BW gain, and reduction in diuresis.[2] Gross pathology examination indicated dystrophic changes in the parenchymatous organs and inflammation and edema along the GI tract.

Repeated Exposure failed to reveal pronounced cumulative properties. K_{acc} is 6 (by Cherkinsky). Exposure to 270 mg/kg BW dose produced toxic effect on the liver cells and decreased the antitoxic function of the liver in mice. Habituation to MA was not observed.[1] Rats received 50 to 1000 mg MA/kg BW by gavage for 5 days. The treatment reduced food intake and caused BW loss to the end of the study. Doses of 5.0 or 10 mg/kg BW given for 10 treatments produced no above described effects. There were no changes in hematology, serum clinical chemistry, or gross pathology findings. Histopathological examination revealed alveolar hemorrhage and lipid granuloma in the lungs, and moderate to severe granularity of liver cytoplasm.[03]

Long-term Toxicity. The treatment caused a considerable reduction in the acidity level and variation in the chloride content of the blood in rabbits, diffuse damage to the liver, and a number of pathomorphological changes in the visceral organs, primarily in the liver and kidneys.[1]

Reproductive Toxicity. MA caused malformations and growth retardation in *in vitro* rats embryocultures at concentrations ranging from 1.2 to 2.1 mM. At these concentrations, MA produced concentration-dependent decreases in growth parameters, i.e., crown-rump length, number of somites, and embryo protein content. Malformed embryos were characterized primarily by abnormal neurulation. Less frequent abnormalities included hypoplasia of the prosencephalon, edema, malpositioned heart, abnormal flexion, and dilated otic vesicles. An increase in methacrylic acid-induced cell death was reported.[3]

Mutagenicity.

In vitro genotoxicity. MA was negative in *Salmonella* mutagenicity assay.[025]

Regulations.

U.S. FDA (1998) listed MA (1) in adhesives as a component of articles a monomer intended for use in packaging, transporting, or holding food in accordance with the conditions prescribed in 21 CFR part 175.105; (2) as an ingredient in the manufacture of semirigid of semirigid and rigid acrylic and modified acrylic plastics that may be safely used as articles intended for use in contact with food in accordance with the conditions prescribed in 21 CFR part 177.210; (3) as a monomer in the manufacture of polyethylene phthalate polymers that may be safely used as articles or components of plastics intended for use in contact with food in accordance with the conditions prescribed in 21 CFR part 177.1630; and (4) in the manufacture of cross-linked polyester resins which may be used as articles or components of articles intended for repeated use in contact with food in accordance with the conditions prescribed in 21 CFR part 177.2420.

Great Britain (1998). MA is authorized without time limit for use in the production of polymeric materials and articles in contact with food or drink or intended for such contact.

Standards. ***Russia*** (1995). MAC and PML: 1.0 mg/l.

References:

1. Klimkina, N. V., Ekhina, R. S., Sergeev, A. N., et al., Data on hygiene substantiation of MAC for methacrylic acid in water bodies, *Gig. Sanit.*, 8, 13, 1973 (in Russian).
2. *Toxicology of High-Molecular-Mass Compounds*, Proc. Sci. Conf., S. L. Danishevsky, Ed., Leningrad, 1961, 46 (in Russian).
3. Rogers, G. J., Greenaway, J. C., Mirkes, P. E., et al., Methacrylic acid as a teratogen in rat embryo culture, *Teratology*, 33, 113, 1986.

METHACRYLIC ACID, BUTYL ESTER

Molecular Formula. $C_8H_{14}O_2$

M = 142.2

CAS No 97-88-1

RTECS No OZ3675000

Abbreviation. MAB.

Synonyms. Butyl methacrylate; Butyl-2-methylpropenoate.

Properties. Colorless, transparent liquid with a characteristic unpleasant odor. Water solubility is 10.2% (20°C). Odor perception threshold is 0.022 mg/l.[1]

Applications. Monomer in the production of acrylic and methacrylic polymers and copolymers intended for use in contact with food.

Acute Toxicity. LD_{50} is reported to be 16 to 25 g/kg BW for rats, 12.9 to 15.8 g/kg BW for mice, and 25 g/kg BW for rabbits.[1,2] Poisoning was accompanied by slower movements, depression, reddening of the nose, ears, and paws, dyspnea, irritation of the GI tract. Death is preceded by loss of coordination and reflexes, side position, and labored respiration.

Repeated Exposure revealed mild cumulative properties of MAB during administration of 1/10 and 1/5 LD_{50} to mice for 30 days.[1]

Short-term Toxicity. Rats tolerated administration of 0.9 g/kg BW for 4 months. The animals developed de crease of BW gain from the 7th week of the experiment and increased liver weights.[2]

Long-term Toxicity. Rats and rabbits were dosed by gavage with up to 5.0 mg MAB/kg BW for 8 to 9 months.[1] The treatment affected predominantly the liver, oxidizing enzymes, and erythrocytes. Changes in the CNS function were also noted.

Reproductive Toxicity. Similar to other acrylates MAB produced embryotoxic and teratogenic effects in rats.[3]

Embryotoxicity. MAB was administered to pregnant rats on days 5, 10, and 15 of gestation at doses of up to 1/3 *i/p* LD_{50} (0.78 ml/kg BW). The treatment increased the incidence of hemangiomas and resorptions. Fetal weight was reduced.[4]

Allergenic Effect.

Animal studies. N-MAB was found to be a strong sensitizer when tested by skin application in olive oil or ethanol.[5] Nevertheless, according to Parker and Turk, methacrylates tested in five different sensitization protocols in guinea pigs appeared to be ineffective.[6]

Mutagenicity.

In vitro genotoxicity. MAB was negative when tested in four *Salmonella typhimurium* strains.[015]

Chemobiokinetics. MAB is metabolized by saponization into *butyl alcohol* and *methacrylic acid.* Methacrylic acid may form *acetyl coenzyme A derivative*, which then enters normal lipid metabolism.[03]

Regulations.

U.S. FDA (1998) approved the use of MAB as a monomer (1) in adhesives as a component (monomer) of articles intended for use in packaging, transporting, or holding food in accordance with the conditions prescribed in 21 CFR part 175.105; (2) in the manufacture of resinous and polymeric coatings in a food-contact surface, subject to conditions prescribed in 21 CFR part 175.300; (3) as a monomer in the manufacture of semirigid and rigid acrylic and modified acrylic plastics in the manufacture of articles intended for use in contact with food in accordance with conditions prescribed in 21 CFR part 177.1010; (4) in cross-linked polyester resins to be safely used as articles or components of articles intended for repeated use in contact with food, subject to the provisions prescribed in 21 CFR part 177.2420; as (5) a component (monomer) of the uncoated or coated food-contact surface of paper and paperboard intended for use in producing, manufacturing, packaging, processing, preparing, treating, packing, transporting, or holding dry food 176.180 as prescribed in 21 CFR part; (6) as polymers which may be safely admixed, alone or in mixture with other permitted polymers, as modifiers in semirigid and rigid vinyl chloride plastic food contact articles prepared from vinyl chloride homopolymers and/or from vinyl chloride copolymers in accordance with the conditions specified in the CFR; (7) (monomer) in the manufacture of polyethylene phthalate polymers used as, or components of plastics (films, articles, or fabric) intended for use in contact with food; and (8) as a solvent in the production of polyester resins.

Great Britain (1998). MAB is authorized without time limit for use in the production of polymeric materials and articles in contact with food or drink or intended for such contact.

EU (1990). MAB is available in the *List of authorized monomers and other starting substances which shall be used for the manufacture of plastic materials and articles intended to come into contact with foodstuffs (Section A).*

Standards. ***Russia*** (1995). MAC and PML: 0.02 mg/l (organolept., odor).

References:

1. Klimkina, N. V., Ekhina, R. S., and Sergeev, A. N., Experimental data on the MAC of methyl and butyl ether of methacrylic acid, *Gig. Sanit.,* 4, 6, 1976 (in Russian).
2. Shepelskaya, N. R., Hygienic comparative study of some methyl methacrylate-based materials, *Gig. Sanit.*, 1, 93, 1976. (in Russian).
3. Singh, A. R., Lawrence, W. H., and Autian, J., Embryonic-fetal toxicity and teratogenic effects of a group of methacrylic esters in rats, *J. Dent. Res.*, 51, 1632, 1972.
4. Shepard, T. H., *Catalog of Teratogenic Agents*, 5th ed., The Johns Hopkins Univ. Press, Baltimore, MD, 1986, 369.
5. Chung, C. W. and Giles, A. L., Sensitization potential of methyl, ethyl and *n*-butyl methacrylates and mutual cross-sensitivity in guinea pigs, *J. Invest. Dermatol.*, 68, 187, 1977.
6. Parker, D. and Turk, J. L., Contact sensitivity to acrylate compounds in guinea pigs, *Contact Dermat.*, 9, 55, 1983.

METHACRYLIC ACID, DIESTER with TRIETHYLENE GLYCOL

M = 286.36
CAS No 109-16-0
102770-39-8
RTECS No OZ4100000
Abbreviation. MADTEG.

Synonym and **Trade Names.** Polyester TGM 3; 2-Methyl-2-propenoic acid, 1,2-ethanediylbis(oxy-2,1-ethanediyl) ester; Tedma; Triethylene glycol, dimetacrylic ester.

Properties. Clear liquid of low viscosity of a yellow or green to dark-green or dark-brown color and with a styrene-like odor. Contains up to 3.5% toluene and 0.71% hydroquinone. Readily soluble in alcohol.

Gives water an aromatic odor and a specific astringent taste. Taste perception threshold is lower than odor perception threshold. The practical taste threshold of purified MADTEG is reported to be 22 mmol/l; that of the technical grade product is 23.2 mmol/l. Chlorination does not enhance the odor. The lower limits of reliability of the taste perception threshold are 7.74 mmol/l (purified substance) and 8.49 mmol/l (technical grade product).[1]

Applications. MADTEG is used for copolymerization with methyl methacrylate and other monomers to obtain unfusible and insoluble heat-resistant polymers, and also in the manufacture of unsaturated polyesters intended for contact with food. A plasticizer for polar rubbers: acrylonitrile-butadienes and chloroprenes. Used in the production of glass-reinforced plastics.

Acute Toxicity. LD_{50} of pure or technical grade MADTEG is 10.84 g/kg BW in rats, and 10.75 g/kg BW in mice. Changes in behavior were observed. Administration produced general inhibition, apathy and urinary and fecal incontinence. Death occurred in 1 to 2 days from respiratory failure.[1]

Repeated Exposure. Rats were given 1/10, 1/50, or 1/250 LD_{50} for 2 months. The treatment revealed low ability to cumulate toxic effects. K_{acc} = 10. The highest dose tested (1/10 LD_{50}) produced disturbances of oxidation-reduction processes and of liver function.

Long-term Toxicity. Administration of 1/250 and 1/1250 LD_{50} caused anemia and leukopenia, as well as a decline in serum cholinesterase and peroxidase activity in rats.[1]

Allergenic Effect was not observed on *i/p* injection of up to 8.0 mg/kg BW to guinea pigs. MADTEG did not induce contact sensitivity in several different sensitization protocols in guinea pigs.[2]

Chemobiokinetics. In the body, MADTEG is readily converted by saponification to *alcohol* and *methacrylic acid.* The acid may form an *acetyl-coenzyme A derivatives*, which then enter the normal lipid metabolism.[03]

Regulations. ***U.S. FDA*** (1998) approved the use of MADTEG in the manufacture of resinous and polymeric coatings for the food-contact surface of articles intended for use in producing, manufacturing, packing, processing, preparing, treating, packaging, transporting, or holding food (for use only as polymerization cross-linking agent in side seam cements for containers intended for use in contact with food only of the identified types), subject to the conditions prescribed in 21 CFR part 175.300.

Standards. ***Russia*** Recommended MAC: 0.2 mg/l.

References:

1. Manenko, A. K., Kravets-Bekker, A. K., Sakhnovskaya, N. N., et al., Hygienic substantiation of maximum allowable concentration for triethylene glycol dimethacrylate in water bodies, *Gig. Sanit.*, 4, 17, 1982 (in Russian).
2. Parker, D. and Turk, J. L., Contact sensitivity to acrylate compounds in guinea pigs, *Contact Dermat.*, 9, 55, 1983.

METHACRYLIC ACID, ETHYL ESTER

Molecular Formula. $C_6H_{10}O_2$
M = 114.16
CAS No 97-63-2
RTECS No OZ4550000
Abbreviation. MAE.

Synonyms and **Trade Name.** Ethyl 2-methylacrylate; Ethyl 2-methyl-2-propeonate; 2-Methyl-2-propenoic acid, ethyl ester; Rhoplex AC-33.

Applications. Monomer in the production of acrylic and methacrylic polymers and copolymers intended for use in contact with food.

Acute Toxicity. LD_{50} is 7.8 g/kg BW in mice,[1] and 14.8 g/kg BW in rats. Poisoning is followed by muscle weakness, coma, respiratory depression.[2]

Reproductive Toxicity.

Embryotoxicity. MAE was administered to pregnant rats on days 5, 10, and 15 of gestation at doses of up to 1/3 *i/p* LD_{50} (0.4 ml/kg BW). The treatment increased the incidence of hemangiomas and resorptions. Fetal weight was reduced.[3]

Allergenic Effect.

Animal studies. MAE appeared to be a strong sensitizer when tested by skin application in olive oil or ethanol. It showed allergic reactions in 7.0% patients.[4] Nevertheless, according to Parker and Turk, methacrylates tested in five different sensitization protocols in guinea pigs were found to be ineffective.[5]

Mutagenicity.

In vitro genotoxicity. MAE exhibits mutagenic action in lymphocytes.[6]

Regulations.

EU (1990). MAE is available in the *List of authorized monomers and other starting substances which shall be used for the manufacture of plastic materials and articles intended to come into contact with foodstuffs (Section A).*

U.S. FDA (1998) approved the use of MAE(1) in adhesives as a component of articles (monomer) intended for use in packaging, transporting, or holding food in accordance with the conditions prescribed in 21 CFR part 175.105; as (2) a component (monomer) of the uncoated or coated food-contact surface of paper and paperboard intended for use in producing, manufacturing, packaging, processing, preparing, treating, packing, transporting, or holding dry food in accordance with the conditions prescribed in 21 CFR part 176.170; (3) as a monomer in the manufacture of semirigid and rigid acrylic and modified acrylic plastics in the manufacture of articles intended for use in contact with food in accordance with the conditions prescribed in 21 CFR part 177.1010; and (4) as a monomer in the manufacture of polyethylenephthalate polymers used as, or components of plastics (films, articles, or fabric) intended for use in contact with food in accordance with the conditions prescribed in 21 CFR part 177.1630.

Great Britain (1998). MAE is authorized without time limit for use in the production of polymeric materials and articles in contact with food or drink or intended for such contact.

References:

1. Tanji, H. and Hashimoto, K., Structure-activity relationship of acrylates and methylacrylates, *Toxicol. Lett.*, 11, 125, 1982.
2. Deichman, W., Toxicity of methyl, ethyl and *N*-butyl methacrylate, *J. Ind. Hyg. Toxicol.*, 23, 343, 1941.

3. Shepard, T. H., *Catalog of Teratogenic Agents*, 5th ed., The Johns Hopkins Univ. Press, Baltimore, MD, 1986, 369.
4. Kanerva, L., Allergy caused by acrylate compounds, *Hum. Exp. Toxicol.*, 1, 996, 1997
5. Parker, D. and Turk, J. L., Contact sensitivity to acrylate compounds in guinea pigs, *Contact Dermat.*, 9, 55, 1983.
6. Moore, M. M, Amtower, A., Doerr, C. L., Brock, K. H., and Dearfield, K. L., Genotoxicity of acrylic acid, methyl acrylate, ethyl acrylate, methyl methacrylate, and ethyl methacrylate in L5178Y mouse lymphoma cells, *Environ. Molec. Mutagen.*, 11, 49, 1988.

METHACRYLIC ACID, 2-ETHYLHEXYL ESTER

Molecular Formula. $C_{12}H_{22}O_2$
M = 198.34
CAS No 688-84-6
RTECS No OZ4630000
Abbreviation. MAEH.

Synonyms. 2-Ethyl-1-hexyl methacrylate; 2-Ethylhexyl methacrylate.

Properties. Colorless liquid with an acrid odor. Soluble in alcohol. Water solubility amounted to 5600 ppm at 20°C.

Applications. Used as a comonomer in the production of acrylic polymers for surface coating resins intended for use in contact with food and acrylic emulsion polymers for polishes.

Acute Toxicity. LD_{50} *i/p* is 2.6 g/kg BW in mice.[1]

Chemobiokinetics. MAEH undergoes metabolism by saponification into the alcohol and methacrylic acid.

Regulations. ***U.S. FDA*** (1998) approved the use of MAEH as (1) basic component of acrylic and modified acrylic plastics, semirigid and rigid, which may be safely used as articles intended for single and repeated use in food-contact surfaces in accordance with the conditions prescribed by 21 CFR part 177.1010; and (2) in adhesives as a component of articles intended for use in packaging, transporting, or holding food in accordance with the conditions prescribed in 21 CFR part 175.105.

Reference:

Mir, G. N., Lawrence, W. H., and Autian, J., Toxicological and pharmacological actions of methacrylate monomers. I. Effects on isolated, perfused rabbit heart, *J. Pharmacol. Sci.*, 62, 778, 1973.

METHACRYLIC ACID, 2-HYDROXYETHYL ESTER

Molecular Formula. $C_6H_{10}O_3$
M = 130.14
CAS No 868-77-9
RTECS No OZ4725000
Abbreviation. MEHE.

Synonyms and **Trade Name.** Ethylene glycol methacrylate; Glycol methacrylate; 2-Hydroxyethylmethacrylate; 2-(Methacryloyloxy)ethanol; 2-Methyl-2-propenoic acid, 2-hydroxyethyl ester; Monomer MG-1.

Properties. Clear mobile liquid. Miscible with water, soluble in organic solvents.

Applications. A cross-linking monomer in the production of acrylic resins and enamels intended for use in contact with food or drink; a basic material for contact lenses. Used also in medicine, as a dental filler, etc.

Acute Toxicity. LD_{50} is 5.05 g/kg BW in rats, 3.28 g/kg BW in mice, 4.68 g/kg BW in guinea pigs. Administration of large doses led to coma development.[1]

Long-term Toxicity. Rats received 0.5, 2.5 or 12.5 mg MEHE/kg BW over a period of 8 months. Treatment with the highest dose caused retardation of BW gain, changes in CNS and biochemical parameters. There were treatment-related findings in the liver, spleen, and myocardium either at necropsy or at histological examination. The NOAEL of 0.5 mg/kg BW was established in this study.[2]

Reproductive Toxicity.

Embryotoxicity. Rats were exposed orally to MEHE for 35 weeks prior to mating. The treatment caused an increase in postimplantation mortality and fetal death (in the group of treated females), pre-implantation mortality, fetotoxicity, fetal death (in the group of treated males). The NOAEL for embryotoxic action appeared to be 0.5 mg/kg BW.[2]

No ***terata*** was found in animals of all doses tested.[2]

Allergenic Effect.

Observations in man. MEHE produces contact sensitization in workers in printing industry.[019] Allergic reactions in patients with MEHE exposure accounted to 11%.[3]

Animal studies. Methacrylates did not sensitize guinea pigs.[4]

Regulations.

U.S. FDA (1998) approved the use of MEHE (1) as a monomer in the manufacture of semirigid and rigid acrylic and modified acrylic plastics in the manufacture of articles intended for use in contact with food in accordance with the conditions prescribed in 21 CFR part 177.1010; and (2) in adhesives as a component of articles intended for use in packaging, transporting, or holding food in accordance with the conditions prescribed in 21 CFR part 175.105.

EU (1990). MENE is available in the *List of authorized monomers and other starting substances which may continue to be used for the manufacture of plastic materials and articles intended to come into contact with foodstuffs pending a decision on inclusion in Section A (Section B).*

Great Britain (1998). MEHE is authorized without time limit for use in the production of polymeric materials and articles in contact with food or drink or intended for such contact.

References:

1. Datsenko, I. I., Doloshitsky, S. L., Dychok, L. A., et al., Hygienic regulation of ethylene glycol monomethacrylic ether in reservoir water, *Gig. Sanit.*, 9, 75, 1989 (in Russian).
2. Vyshemirskaya, L. D., Doloshitsky, S. L., and Romanov, R. V., Study of thresholds and late effects of chronic exposure to ethylene glycol monomethacrylate ether, *Gig. Sanit.*, 11, 81, 1987 (in Russian).
3. Kanerva, L., Allergy caused by acrylate compounds, *Hum. Exp., Toxicol.*, 1, 996, 1997.
4. Parker, D. and Turk, J. L., Contact sensitivity to acrylate compounds in guinea pigs, *Contact Dermat*, 9, 55, 1983.

METHACRYLIC ACID, 2-HYDROXYPROPYL ESTER

Molecular Formula. $C_7H_{12}O_3$

M = 144.19

CAS No 923-26-2

RTECS No OZ4750000

Abbreviation. MAHP.

Synonyms. β-Hydroxypropyl methacrylate; 2-Hydroxypropyl-2-methyl-2-propenoate; 2-Methyl-1,2-propanediol, monomethacrylate; 2-Methyl-2-propenoic acid, 2-hydroxypropyl ester; Methacrylic acid, monoester with 1,2-propanediol.

Properties. Clear liquid with a slight acrylic odor. Poorly soluble in water, soluble in common organic solvents.

Applications. Monomer for acrylic resins, comonomer in cross-linked thermoset-acrylic coating resins and in the production of elastomers intended for use in contact with food or drink. Dental filling material.

Acute Toxicity. LD_{50} is 6.0 ml/kg BW (Schwach and Hofer, 1978) or about 8.0 g/kg BW in mice.[1]

Allergenic Effect. MAHP showed allergic reactions in 11.6 % patients.[2]

Chemobiokinetics. Small quantities of methacrylates may readily be metabolized by saponification into the alcohol and methacrylic acid.

Regulations.

U.S. FDA (1998) approved the use of MAHP in the manufacture of semirigid and rigid acrylic and modified acrylic plastics used as articles intended for use in contact with food in accordance with the conditions prescribed in CFR part 177.1010; and (2) as optional substances that may be used including polymers derived by copolymerization with one or more minor monomers including hydroxypropyl methacrylate. HPMA is limited to use only in plastic articles intended for repeated use in contact with food.

Great Britain (1998). MAHP was authorized up to 1998 for use in the production of polymeric materials and articles in contact with food or drink or intended for such contact.

References:

1. Tanii, H. and Hashimoto, K., Structure-toxicity relationship of acrylates and methacrylates, *Toxicol. Lett.*, 11, 125, 1982.
2. Kanerva, L., Allergy caused by acrylate compounds, *Hum. Exp. Toxicol.*, 1, 996, 1997.

METHACRYLIC ACID, ISOBUTYL ESTER

Molecular Formula. $C_8H_{14}O_2$
M = 142.22
CAS No 97-86-9
RTECS No OZ4900000
Abbreviation..

Synonyms. Isobutyl methacrylate; 2-Methylpropylmethacrylate.

Properties. Liquid. Water solubility is about 0.6 g/l at 25°C. Soluble in alcohol (>10%).

Applications. Monomer in the production of acrylic resins intended for use in contact with food or drink.

Acute Toxicity. LD_{50} is 12 g/kg BW in mice.[1]

Reproductive Toxicity.

Embryotoxic and ***teratogenic effects.*** Female rats were exposed to MAIB at dose levels of 0.14 to 0.47 ml/kg BW by *i/p* injections on days 5, 10, and 15 of gestations. The treatment caused resorptions, gross and skeletal malformations, fetal death, and decreased fetal size.[2]

Mutagenicity.

In vitro genotoxicity. MAIB was ineffective when tested over a wide range of doses in four *Salmonella typhimurium* strains in the the presence and absence of liver *S9* fraction.[015]

Regulations.

EU (1992). MAIB is available in the *List of authorized monomers and other starting substances which shall be used for the manufacture of plastic materials and articles intended to come into contact with foodstuffs (Section A*).

Great Britain (1998). MAIB is authorized without time limit for use in the production of polymeric materials and articles in contact with food or drink or intended for such contact.

References:

1. Tanii, H. and Hashimoto, K., Structure-activity relationship of acrylates and methacrylates, *Toxicol. Lett.,* 11, 125, 1982.
2. Singh, A. R., Lowrence W. H., and Autian, J., Embryonic-fetal toxicity and teratogenic effects of a group of methacrylate esters in rats, *J. Dental Res.*, 51, 1632, 1972.

METHACRYLIC ACID, ISOPROPYL ESTER

Molecular Formula. $C_7H_{12}O_2$
M = 128.19
CAS No 4655-34-9
RTECS No OZ5020000
Abbreviation. MAIP.

Synonyms. Isopropyl methacrylate; 1-Methylethyl-2-methyl-2-propeonate.

Applications. Monomer in the production of acrylic resins intended for use in contact with food or drink.

Acute Toxicity. LD_{50} is 196 mg/kg BW in mice.

Regulations.

EU (1992). MAIP is available in the *List of authorized monomers and other starting substances which shall be used for the manufacture of plastic materials and articles intended to come into contact with foodstuffs (Section A*).

Great Britain (1998). MAIP is authorized without time limit for use in the production of polymeric materials and articles in contact with food or drink or intended for such contact.

Reference:

Tanii, H. and Hashimoto, K., Structure-activity relationship of acrylates and methacrylates, *Toxicol. Lett.*, 11, 125, 1982.

METHACRYLIC ACID, METHYL ESTER

Molecular Formula. $C_4H_8O_2$
M = 100.12
CAS No 80-62-6
RTECS No OZ5075000
Abbreviation. MAM.

Synonyms. Methyl methacrylate; Methyl-2-methyl-2-propenoate.

Properties. Colorless liquid. Water solubility is 1.9% at 20°C. Miscible with alcohol at any ratio. Forms methacrylic acid and methyl alcohol when hydrolyzed. Odor perception threshold is 0.45 mg/l[1] or 0.025 to 0.05 mg/l.[02,010]

Applications. Used for latex coatings and in the production of acrylic polymeric materials, including copolymers with acrylonitrile, α-methyl styrene, and butadiene intended for use in contact with food or drink.

Acute Toxicity. LD_{50} was found to be 8.7 g/kg BW for rats and rabbits, 3.6 to 5.2 g/kg BW for mice, 6.3 g/kg BW for guinea pigs, and 5.0 g/kg BW for dogs. The minimum lethal dose is 6.6 g/kg BW for rabbits.[1,2]

Repeated Exposure. Rats received 1/5 and 1/10 LD_{50} for 30 days. At the end of the experiment, LD_{50} was administered to animals. Slight signs of accumulation were manifested.[1]

Long-term Toxicity. Rats were dosed by gavage with a total dose of 8.125 g MAM/kg BW (administered twice a week for 8 months). The treatment caused a reduction in the contents of glycoproteins and albumins and an increase in the activity of leucinoaminopeptidase and β-glucuronidase in the blood serum, and reversible toxic effects on the liver. The kidneys were much less affected (Motoc et al., 1971).

In a 8 to 9-month rat study, there was liver damage; oxidizing enzymes and erythrocytes were affected. The functional condition of the brain cortex was impaired.[1]

Reproductive Toxicity.

Embryotoxicity and ***Teratogenicity.*** Female rats were exposed to MAM at dose levels of 0.13 to 0.44 ml/kg BW by *i/p* injections on days 5, 10, and 15 of gestations. The treatment caused resorptions, gross and skeletal malformations, fetal death, and decreased fetal size.[3] There was no teratogenic effect when polymethylmetacrylate was used for implantation in humans.

Gonadotoxic effect of MAM was found in mice and rats in response to inhalation exposure above MAC.[4] Administration of 1/20 LD_{50} to rats for 2 months did not affect the estrus cycle.[5]

Allergenic Effect. MAM appeared to be a strong sensitizer, when tested by skin application in olive oil or ethanol. Nevertheless, according to Parker and Turk, methacrylates tested in five different sensitization protocols in guinea pigs were found to be ineffective.[6]

Mutagenicity.

Observations in man. No mutagenic effect was established in humans with polymethylmethacrylate implants.

In vitro genotoxicity. MAM appeared to be positive in *Salmonella typhimurium* mutagenicity bioassay. It caused a concentration-dependent increase in mutant frequency, producing gross CA in mouse lymphoma cells.[7]

Carcinogenicity. In a 2-year study, Wistar rats were given drinking water with concentrations of 6.0 to 2000 mg MAM/l. The treatment did not cause tumor development.[8] IARC considers that the carcinogenic properties of MAM are not proved.

Carcinogenicity classifications. An IARC Working Group concluded that there is sufficient evidence suggesting lack of carcinogenicity of MAM in *experimental animals* and there is inadequate data available to evaluate the carcinogenicity of MAM in *humans*.

IARC: 3;

NTP: NE - NE - NE - NE (inhalation).

Chemobiokinetics. Accumulation in the blood, brain, and lungs was observed following acute inhalation exposure. MAM seems to affect the capillary network of the lung tissue (Raje et al., 1985). It is evidently completely oxidized in the body or hydrolyzed to form *alcohol* that undergoes further oxidation to CO_2 and H_2O. The main part of the dose is removed with the exhaled air in the form of CO_2, irrespective of the route of MAM administration. A smaller part of MAM escapes in the form of *methylmelonate, succinate*, and possibly, *β-hydroxy- isobutyrate*.[9] According to the data available, MAM metabolites were not found in the urine.[10] MAM's relatively weak toxicity for rats is due to the fact that following oral administration the monomer is rapidly converted to *pyruvic acid*.

Regulations.

U.S. FDA (1998) regulates the use of MAM (1) as a monomer in adhesives as a component of articles intended for use in packaging, transporting, or holding food in accordance with the conditions prescribed in 21 CFR part 175.105; (2) as an ingredient in the manufacture of resinous and polymeric coatings of food-contact surface, subject to the conditions prescribed in 21 CFR part 175.300; (3) in acrylic and modified acrylic plastics for single and repeated use in contact with food in accordance with the conditions prescribed in 21 CFR part 177.1010; (4) in cross-linked polyester resins for repeated use as articles or components of articles coming in contact with food, subject to the provisions prescribed in 21 CFR part 177.2420; (5) as a component (a monomer) of the uncoated or coated food-contact surface of paper and paperboard intended for use in producing, manufacturing, packaging, processing, preparing, treating, packing, transporting, or holding dry food of the type identified in 21 CFR part 176.170 (c); (6) as a monomer in the manufacture of semirigid and rigid acrylic and modified acrylic plastics in the manufacture of articles intended for use in contact with food in accordance with the conditions prescribed in 21 CFR part 177.1010; (7) as a component (a monomer) of paper and paperboard used in contact with food in accordance with the conditions prescribed in 21 CFR part 176.180; the MAM contents in styrene-MAM copolymers used as components of paper and paperboard in contact with fatty foods were limited to 0.5%; and (8) as a monomer in the manufacture of polyethylene phthalate polymers used as, or components of plastics (films, articles, or fabric) intended for use in contact with food in accordance with the conditions prescribed in 21 CFR part 177.1630.

EU (1990). MAM is available in the *List of authorized monomers and other starting substances which shall be used for the manufacture of plastic materials and articles intended to come into contact with food-stuffs (Section A)*.

Great Britain (1998). MAM is authorized without time limit for use in the production of polymeric materials and articles in contact with food or drink or intended for such contact.

Standards. ***Russia*** (1995). MAC and PML in drinking water: 0.01 mg/l; PML in food: 0.25 mg/l.

References:

1. Klimkina, N. V., Ekhina, R. S., and Sergeev, A. N., Experimental data on MAC of methyl and butyl ether of methacrylic acid, *Gig. Sanit.*, 4, 6, 1976 (in Russian).
2. Deichmann, W., Toxicity of methyl, ethyl and *n*-butyl methacrylate, *J. Ind. Hyg. Toxicol.*, 23, 343, 1941.
3. Singh, A. R., Lawrence, W. H., and Autian, J., Embryonic-fetal toxicity and teratogenic effects of a group of methacrylate esters in rats, *J. Dent. Res.*, 51, 1632, 1972.
4. Solovyeva, M. S., Smirnova, Ye. S., and Blagodatin V. M., Effect of methyl methacrylate on the specific female functions and sex functions of experimental animals, in *Basic Problems of Delayed Consequences of Exposure to Industrial Poisons*, Proc. Research Institute Occup. Diseases, A. K. Plyasunov and G. M. Pashkova, Eds., Moscow, 1976, 149 (in Russian).
5. Sheftel', V. O., *Hygienic Aspects of the Use of Polymeric Materials in Water Supply*, Author's abstract of thesis, All-Union Research Institute of Hygiene and Toxicology of Pesticides, Polymers and Plastic Materials, Kiyv, 1977, 158 (in Russian).
6. Parker, D. and Turk, J. L., Contact sensitivity to acrylate compounds in guinea pigs, *Contact Dermat.*, 9, 55, 1983.
7. Moore, M. M., Amtower, A., Doerr, C. L., Brock, K. H., and Dearfield, K. L., Genotoxicity of acrylic acid, methyl acrylate, ethyl acrylate, methyl methacrylate, and ethyl methacrylate in L5178Y mouse lymphoma cells, *Environ. Molec. Mutagen.*, 11, 49, 1988.

8. Borzelleca, J. F., Larson, P. S., Henniger, G. R., et al., Studies on the chronic oral toxicity of monomeric ethylacrylate and methyl methacrylate, *Toxicol. Appl. Pharmacol.*, 6, 29, 1964.
9. Bratt, H. and Hathway, D. E., Fate of methyl methacrylate in rats, *Brit. J. Cancer*, 36, 114, 1977.
10. Pantucek, M., Methylmethacrylate metabolism, *Food Cosmet. Toxicol.*, 8, 105, 1970.

METHACRYLIC ACID, PROPYL ESTER

Molecular Formula. $C_7H_{12}O_2$
M = 127.17
CAS No 2210-28-8
RTECS No OZ5250000
Abbreviation. MAP.

Synonyms. 2-Methyl-2-propeonic acid, propyl ester; *N*-Propyl methacrylate; Propyl 2-methyl-2-propenoate.

Properties. Insoluble in water, soluble in alcohol (>10%).

Applications. Monomer in the production of acrylic resins intended for use in contact with food or drink.

Acute Toxicity. LD_{50} *i/p* is 1.12 ml/kg BW in mice.

Chemobiokinetics. MAP is metabolized into the *N-propyl alcohol* and *methacrylic acid.*

Regulations.

EU (1990). MAP is available in the *List of authorized monomers and other starting substances which shall be used for the manufacture of plastic materials and articles intended to come into contact with foodstuffs (Section A*).

Great Britain (1998). MAP is authorized without time limit for use in the production of polymeric materials and articles in contact with food or drink or intended for such contact.

Reference:

Mir, G. N., Lawrence, W. H. and Autian, J., Toxicological and pharmacological actions of methacrylate monomers. II. Effects on isolated guinea pig ileum, *J. Pharmacol. Sci.*, 62, 1258, 1973.

METHYL-1,3-CYCLOPENTADIENE, DIMER

Molecular Formula. $C_{12}H_{14}$
M = 158.26
Abbreviation. MCD.

Methyl-1,3-cyclopentadiene

Molecular Formula. C_6H_8
M = 80.14
CAS No 26519-91-5
RTECS No GY1500000
Abbreviation. MC.

Properties. MCD occurs as a liquid with unpleasant specific odor. Soluble in organic solvents.

Applications. MCD is used in the production of synthetic rubber and resins intended for use in contact with food.

Acute Toxicity. LD_{50} is 7.7 g/kg BW in mice, 10 g/kg BW in rats for MCD, and 2.7 g/kg BW in mice for MC. Species sensitivity is not pronounced. Manifestations of toxic action included motive disorder and convulsive twitching of the muscles.

Repeated Exposure to MCD revealed moderate cumulative properties. K_{acc} is 4.1 (by Lim).

Reference:

Ivanov, N. G., Mel'nikova, L. V., Kliachkina, A. M., and Germanova, A. L., Toxicity, hazard and nature of the biological action of methylcyclopentadiene dimer, *Gig. Truda Prof. Zabol.*, 5, 45, 1981 (in Russian).

2-METHYL-1,3-DIOXOLANE

Molecular Formula. $C_4H_8O_2$

M = 88.10
CAS No 497-26-7
RTECS No JI3509000

Synonyms. Acetaldehyde ethylacetal; Ethylene ethylidene ether; Glycol ethylidene ether.

Properties. Colorless, transparent liquid with a specific odor. Readily soluble in water. Odor perception threshold is 1.13 mg/l, taste perception threshold is slightly higher. M. does not alter the transparency of water or its color and does not froth.

Applications. Used in the manufacture of plastics.

Acute Toxicity. LD_{50} is 2.9 to 4.7 g/kg BW in rats, and 3.5 g/kg BW in mice. Poisoned animals displayed muscle weakness, and behavioral changes. Histology examination revealed congestion in the vessels, parenchymatous dystrophy of the renal tubular epithelium, liver, and brain.

Repeated Exposure failed to reveal evident cumulative properties. Negligible morphological changes in the liver, kidneys, and NS were shown to develop.

Long-term Toxicity. Rats were orally exposed to 50 mg/kg BW over a period of 5 months. The treatment resulted in retardation of BW gain, an increase in STI value and protein amount in the urine. These indices were normal in 1 month after the end of the treatment. The LOAEL of 5.0 mg/kg BW was established in this study for a tendency of protein level in the urine to be higher.

Standards. ***Russia*** (1995). MAC and PML: 1.0 mg/l (organolept, odor).

Reference:

Buzina, A. Z. and Rudi, F. A., Substantiation of maximum allowable concentrations for vinyl ethers of glycols in water bodies, *Gig. Sanit.*, 3, 12, 1977 (in Russian).

N,N'-METHYLENEBISACRYLAMIDE

Molecular Formula. $C_7H_{10}N_2O_2$
M = 154.19
CAS No 110-26-9
RTECS No AS3678000
Abbreviation. MBA.

Synonyms. *N,N'*-Methylenebis(2-propenamide); *N,N'*-Methylidenebisacrylamide.

Applications. MBA is used in the production of synthetic materials for contact with food.

Acute Toxicity. LD_{50} is 390 mg/kg BW in rats, and 380 mg/kg BW in mice. Poisoning caused tremor and changes in respiratory system.[1]

Repeated Exposure. MBA produced no neurotoxic effect.[2]

Reproductive Toxicity.

Embryotoxicity. Swiss CD-1 mice were provided with drinking water containing 10, 30, or 60 ppm MBA during and after a 14-week cohabitation. The last litter was reared and dosed after weaning until mating at 2.5 months of age with the same level of compound given to the parents. The treatment reduced the number of live pups and their adjusted weight, with no neurotoxicity and no change in F_o female reproduction. The greater adverse effect was noted in the F_1 generation.[3]

Gonadotoxicity. Male mice were dosed with 200 mg/kg BW for 16 days prior to mating. The treatment affected gonads of treated animals.[2] Male mice were given 200 mg/kg BW one day prior to mating. Gonadotoxic effect was observed.[4]

DLM studies revealed the fertility effects that could be explained by a male-mediated increase in postimplantation loss. Dominant lethality occurred without structural effects on the reproductive system and without detectable neural histopathology.[3]

Mutagenicity. MBA induced DLM and heritable translocations in maturing sperm. There was an increased incidence of developmental anomalies including fetal death and phenotypic defects.[5]

Regulations. ***U.S. FDA*** (1998) approved the use of MBA in adhesives as a component of articles intended for use in packaging, transporting, or holding food in accordance with the conditions prescribed in 21 CFR part 175.105.

References:

1. Acute Toxicity Data, *J. Am. Coll. Toxicol.*, Part B, 1, 111, 1990.
2. Hashimoto, K., Sakamoto, J., and Tanii, H., Neurotoxicity of acrylamide and related compounds and their effects on male gonads in mice, *Arch. Toxicol.*, 47, 179, 1981.
3. Chapin, R. E., Fail, P. A., George, J. D., Grizzle, T. B., Heindel, J. J., Harry, G. J., Collins, B. J., and Teague, J., The reproductive and neural toxicities of acrylamide and three analogues in Swiss mice, evaluated using the continuous breeding protocol, *Fundam. Appl. Toxicol.*, 27, 9, 1995.
4. Sakamoto, J. and Hashimoto, K., Effects of *N,N'*-methylene-bis-acrylamide (MBA) on mouse germ cells, sperm count and morphology, and testicular pathology, *Arch. Toxicol.*, 62, 54, 1988.
5. Rutledge, J. C., Cain, K. T., Kyle, J., Cornett, C. V., Cacheiro, N. L. A., Witt, K., Shelby, M. D., and Generoso, W. M., Increased incidence of developmental anomalies among descendants of carriers of methylenebisacrylamide-induced balanced reciprocal translocations, *Mutat. Res.*, 229, 161, 1990.

2-METHYL-5-HEXEN-3-YN-2-OL

Molecular Formula. $C_7H_{10}O$
M = 110.16
CAS No 690-94-8
RTECS No MP9880500

Synonyms. 1,1-Dimethyl-4-penten-2-yn-1-ol; Dimethyl(vinyl)ethynylcarbinol; Nazarov carbinol; (Vinylethynyl)dimethylcarbinol.

Properties. Colorless liquid with a characteristic pungent odor. Almost insoluble in water, soluble in alcohol.

Applications. Used to produce carbinol resins (copolymers of M. with different vinyl compounds) and in the manufacture of carbinol lacquers.

Acute Toxicity. The LD_{50} values are reported to be 600 mg/kg BW in rats and guinea pigs, 590 mg/kg BW in mice, and 800 mg/kg BW in rabbits. Gross pathology examination revealed congestion in the *viscera*, enlargement of the liver and spleen, swelling of the stomach and intestines.

Repeated Exposure revealed no cumulative properties.

Allergenic Effect. No sensitizing effect after M. application is reported.

Reference:

Data on Hygiene Assessment of Pesticides and Polymers, F. F. Erisman Research Sanitary Hygiene Institute, Moscow, 1977, 100 (in Russian).

2-METHYL-2-PROPENAMIDE

Molecular Formula. C_4H_7NO
M = 85.11
CAS No 79-39-0
RTECS No UC6475000
Abbreviation. MPA.

Synonyms. Methacrylic acid, amide; 2-Methylacrylamide; α-Methyl acrylic amide; Methacrylamide; Methacrylic amide.

Properties. Colorless, crystalline substance. Readily soluble in water. Forms methacrylic acid when hydrolyzed. Taste perception threshold is 1.0 g/l.[1]

Applications. Used as a monomer in the production of a number of copolymers and polymers intended for use in contact with food or drink.

Acute Toxicity. LD_{50} is 1220 to 1540 mg/kg BW in rats, and 480 to 570 mg/kg BW in mice. Poisoning is accompanied by NS impairment (ataxia, spasms resembling multiple sclerosis). Death occurs in 1 to 2 days. Acute effect threshold is reported to be 200 mg/kg BW.[1,2]

Long-term Toxicity. In a 6-month study in rabbits and rats, CNS and liver functions were mainly affected.[1]

Repeated Exposure revealed no cumulative effects in experiments on mice. K_{acc} is 4.7 (by Cherkinsky). Dogs tolerate 100 mg/kg BW for several months, but 200 mg/kg BW produced a severe effect on the NS.[05]

Reproductive Toxicity.

Embryotoxicity. Swiss CD-1 mice were provided drinking water containing 24, 80, or 240 ppm MPA during and after a 14-week cohabitation. The last litter was reared and dosed after weaning until mating at 2.5 months of age with the same level of compound given to the parents. In both generations, neurotoxicity or reproductive toxicity effects were not noted.[3]

Gonadotoxicity. DLM studies revealed the fertility effects that could be explained by a male-mediated increase in postimplantation loss. Dominant lethality occurred without structural effects on the reproductive system and without detectable neural histopathology.[3]

Regulations.

EU (1990). MPA is available in the *List of monomers and other starting substances which may continue to be used for the manufacture of plastic materials and articles intended to come into contact with foodstuffs pending a decision on inclusion in Section A* (*Section B*).

U.S. FDA (1998) approved the use of MPA as a monomer in the manufacture of semirigid and rigid acrylic and modified acrylic plastics in the manufacture of articles intended for use in contact with food in accordance with the conditions prescribed in 21 CFR part 177.210.

Great Britain (1998). MPA is authorized until the end of 2001 for use in the production of polymeric materials and articles in contact with food or drink or intended for such contact.

Standards. *Russia* (1995). MAC and PML: 0.1 mg/l.

References:

1. Strizhak, E. K., *Experimental Data on Hygiene Standards for Methacrylic Acid Amides in Water Bodes*, Author's abstract of thesis, Moscow, 1967, 15 (in Russian).
2. Porokhova, L. A., Data on substantiation of MAC for methacrylamide in the industrial zone air, *Gig. Sanit.*, 10, 74, 1980 (in Russian).
3. Chapin, R. E., Fail, P. A., George, J. D., Grizzle, T. B., Heindel, J. J., Harry, G. J., Collins, B. J., and Teague, J., The reproductive and neural toxicities of acrylamide and three analogues in Swiss mice, evaluated using the continuous breeding protocol, *Fundam. Appl. Toxicol.*, 27, 9, 1995.

2-METHYL-2-PROPENENITRILE

Molecular Formula. C_4H_5N
M = 67.09
CAS No 126-98-7
RTECS No UD1400000
Abbreviation. MPN.

Synonyms. 2-Cyanopropene; Isocrotononinitrile; Isopropene cyanide; Isopropenylnitrile; Methacrylic acid, nitrile; Methacrylonitrile.

Properties. Colorless, transparent, volatile liquid with a specific odor. Solubility in water is 3.5 to 4.5%. Readily miscible with alcohol. Water solutions are colorless and transparent, and when left to stay for a long time do not cloud or form any residue. Solutions exhibit unpleasant specific odor, which becomes more pungent with increasing temperature. Odor perception threshold is 1.05 mg/l (practical threshold 3.17 mg/l). Taste perception threshold is significantly higher.[1]

Applications. MPN is used in a variety of organic processes related to the manufacture of polymers. It is used in the production of polymerization plastics, synthetic rubbers, elastomers, homo- and copolymers, and coatings intended for use in contact with food.

Exposure. MPN is found in cigarette smoke, and it is known to liberate cyanide *in vivo*.

Acute Toxicity. The minimum lethal dose in mice appeared to be 15 mg/kg BW. Administration of 115 mg/kg BW in oil resulted in the death of rats within 24 hours after treatment.[2] The LD values vary in rats from 240 mg/kg BW[3,4] to 25 to 50 mg/kg BW;[05] in mice, LD_{50} is reported to be 17 mg/kg BW.[5] However, according to other data, LD_{50} was found to be 184 mg/kg BW in mice, 167 mg/kg BW in rats, and 216 mg/kg BW in guinea pigs.[1] Poisoning is accompanied by NS inhibition, tremor, cyanosis, and convulsions. An increase in the content of thiocyanates in the blood serum and urine of rats is found. Death occurs within

1 to 3 days from respiratory arrest. Gross pathohistology examination failed to reveal pronounced changes. Toxicity of MPN is likely to be much higher than that of orally and *i/p* administered acrylonitrile.

Repeated Exposure failed to reveal cumulative properties when MPN was administered at doses of 1/5 and 1/10 LD_{50}. Habituation developed.[1] K_{acc} is identified to be 5.6 (by Lim).[6] Rats received MPN in sunflower oil during pregnancy. Within 1 hour following ingestion of 50 mg/kg or 100 mg/kg BW (1/2 LD_{50}), rats displayed dose-related mild to severe conditions including ataxia, trembling, convulsions, salivation and irregular breathing. The rats recovered from these signs at various times depending on the dose administered.[7]

Long-term Toxicity. Manifestations of the toxic action included changes in cholinesterase and serum oxidase activity, an increase in the serous index of the urine, and disorder of the hemocoagulative function of the liver in rats. Gross pathology examination revealed mild dystrophic changes in the visceral organs.[1]

Reproductive Toxicity.

Embryotoxicity. MPN caused fetotoxicity in Sprague-Dawley rats following inhalation exposure at a concentration of 100 ppm during days 6 to 20 of gestation.[8] Rats were administered 50 mg MPN/kg BW during the first week of pregnancy (group 1), 50 mg MPN/kg BW (group 2) and 100 mg MPAN/kg BW (group 3) during the second week of pregnancy. BW gain was affected in all the treated animals, pregnancy was disrupted in groups 1 and 3 (100%) and in group 2 (84%). Edema of uterine tubes was noted.[7]

Mutagenicity. MPN was negative in *Salmonella* mutagenicity assay, in *Dr. melanogaster*, and in the mutation test in mice lymphoma cells.[9]

Chemobiokinetics. MPN is a potent neurotoxin, it depletes glutathione both *in vivo* and *in vitro*.[10] F344 rats were exposed to MPN by gavage with drinking water at the doses of 1.15, 11.5 or 115 mg/kg BW. MPN was rapidly absorbed from the GI tract and distributed to all major tissues. Labeled MPN is found to be primarily exhaled in the expired air.[2] Tissue distribution and characteristics of MPN metabolism to *cyanide* are reported.[11]

The resemblance in features of acute poisoning to those observed following treatment with inorganic cyanides is an indirect confirmation of detachment of *cyanide* groups of MPN in the body. Two pathways of MPN metabolism are indicated: hydrolysis, and then thiocyanate formation; conjugation of other metabolic products with glucuronic acid and their removal in the form of glucuronides.[6]

The majority of MPN-derived radioactivity is eliminated in the exhaled air as CO_2 or in the urine. When exposed to 1.15 to 11.5 mg/kg BW, 60 to 70% MPN were exhaled as ^{14}C-CO_2. Two other components were identified to be unchanged MPN and *acetone*.[2,5] The major urinary metabolite of MPN results from conjugation of the epoxide with glutathione.[12] A small quantity of cyanide ions is passed with the urine within 18 hours in the form of *thiocyanate ions*. Urine metabolites detected were *N-acetyl-S-(2-cyanopropyl)-L-cystein*, *N-acetyl-S-(2-hydroxy-propyl)-L-cystein*, and *desoxyuridine isomer*.[5] Apparent saturation of metabolism was seen at 115 mg/kg BW, the highest dose studied (NTP-91).

Regulations.

EU (1990). MPN is available in the *List of authorized monomers and other starting substances which shall be used for the manufacture of plastic materials and articles intended to come into contact with foodstuffs (Section A)*.

U.S. FDA (1998) listed MPN (1) as an ingredient (a monomer) in the manufacture of semirigid and rigid acrylic and modified acrylic plastics to be safely used as articles intended for use in contact with food in accordance with the conditions prescribed in 21 CFR part 177.1010; (2) in the manufacture of resinous and polymeric coatings for food-contact surface in accordance with the conditions prescribed in 21 CFR part 175.300; and (3) as a monomer in the manufacture of polyethylenephthalate polymers used as, or components of plastics (films, articles, or fabric) intended for use in contact with food in accordance with the conditions prescribed in 21 CFR part 177.1630.

Great Britain (1998). MPN is authorized without time limit for use in the production of polymeric materials and articles in contact with food or drink or intended for such contact. The specific migration of this substance shall be not detectable (when measured by a method with a DL of 0.02 mg/kg (analytical tolerances included).

Standards. ***EU*** (1990). SML: not detectable (detection limit 0.02 mg/kg, analytical tolerance included).

References:

1. Loskutov, N. F. and Piten'ko, N. N., Characteristics of the toxic action and safe levels of crotononitrile and isocrotononitrile in water bodies, *Gig. Sanit.*, 4, 10, 1972 (in Russian).
2. Ghanayem, B. I., Sanchez, I. M., and Burka, L. T., Effects of dose, strain, and dosing vehicle on methacrylonitrile disposition in rats and identification of a novel-exhaled metabolite, *Drug Metab. Dispos.*, 20, 643, 1992.
3. McOmie, W. A., Comparative toxicity of methacrylonitrile and acrylonitrile, *J. Ind. Hyg. Toxicol.*, 31, 113, 1949.
4. Pozzani, U. C., Kinkead, E. R., King, J. M., et al., The mammalian toxicity of methacrylonitrile, *Am. Ind. Hyg. Assoc. J.*, 29, 202, 1968.
5. Ghanayem, B. I., Sanchez, I. M., and Burka, L. T., Investigation of methacrylonitrile metabolism and the metabolic basis for the differences in its toxicity in rats and mice, *J. Pharmacol. Exp. Ther.*, 269, 581, 1994.
6. Kurzaliev, S. A., Peculiarities of the toxic effect of methacrylonitrile, *Gig. Truda Prof. Zabol.*, 5, 35, 1985 (in Russian).
7. Farooqui, M. Y. N. and Villarreal, M. I., Maternal toxicity of methacrylonitrile in Sprague-Dawley rats, *Bull. Environ. Contam. Toxicol.*, 48, 696, 1992.
8. Saillenfait, A. M., Bonnet, P., Guenier, J. P., and de Ceaurriz, J., Relative developmental toxicities of inhaled aliphatic mononitriles in rats, *Fundam. Appl. Toxicol.*, 20, 365, 1993.
9. Knaap, A. et al., in *The 14th Annual Conf. Eur. Soc. Environ. Mutagen.*, Preprints of Papers, Moscow, 1984, 247 (in Russian).
10. Farooqui, M. Y. N., Cavazos, R., Villarreal, M. I., et al., Toxicity and tissue distribution of methacrylonitrile in rats, *Ecotoxic. Environ. Safety*, 20, 185, 1990.
11. Day, W. W., Cavazos, R., and Farooqui, M. Y. N., Interaction of methacrylonitrile with glutathione, *Res. Commun. Chem. Pathol. Pharmacol.*, 62, 267, 1988.
12. Burka, L. T., Sanchez, I. M., Ahmed, A. E., and Ghanayem, B. I., Comparative metabolism and disposition of acrylonitrile and methacrylonitrile in rats, *Arch. Toxicol.*, 68, 611, 1994.

α-METHYL STYRENE

Molecular Formula. C_9H_{10}
M = 118.18
CAS No 98-83-9
RTECS No WL5075300
Abbreviation. α-MS.

Synonyms. Isopropenylbenzene; 1-Methyl-1-phenylethylene; 2-Phenylpropene.

Properties. Colorless or slightly yellow liquid with a pungent specific odor. Water solubility is 560 mg/l at 25°C, soluble in alcohol. Odor perception threshold is reported to be 0.11 mg/l[1] or 0.043 to 0.052 mg/l,[02,010] taste perception threshold is 0.08 mg/l. Practical thresholds are 0.2 mg/l and 0.14 mg/l, respectively.[1]

Applications. Monomer in the production of plastics and rubbers intended for use in contact with food.

Migration Data. A study of extracts from vulcanizates based on α-MS rubber (38-100°C; 1, 24, 120 hours; 0.5, 0.1, and 0.01 cm^{-1}) was carried out with the use of distilled water and milk simulating media, lactic acid prodicts, and wine. Release of α-MS reached 0.03 mg/l.[2]

Acute Toxicity. LD_{50} in rats is 4.9 g/kg BW.[3] According to other data, LD_{50} is 10 g/kg BW in rats, and 5.0 g/kg BW in mice.[1] A dose of 3.0 g/kg BW seems to be LD_{30} in mice and produced considerable morphological damage. Agitation, ataxia, tremor, and convulsions preceded death. Gross pathology examination revealed severe congestion in the visceral organs, especially in the lungs. Lethal doses caused dystrophic changes in the parenchymatous organs. Toxic encephalopathy developed.[4]

Repeated Exposure. In a 1-month study, 1/10 LD_{50} in mice as well as a 0.5 g/kg BW dose in rats caused retardation of BW gain. Decrease in the leukocyte count and increased level of protein and coproporphyrins in the urine were also found in rats. Histology examination revealed mild congestion and circulation disorders in the parenchymatous organs and brain, as well as bronchitis.[5]

Long-term Toxicity. No changes in the blood morphology and activity of blood cholinesterase were observed in rats given up to 0.5 mg/kg BW for 6 months. The highest dose produced reticulocytosis.[3]

Reproductive Toxicity. No teratogenicity or maternal toxicity was observed in rats after *i/p* administration of 250 mg/kg BW on days 1 to 15 of gestation. The treatment caused some fetal toxicity.[6]

Allergenic Effect. Sensitization was found to develop in guinea pigs on subcutaneous and cutaneous applications.[7]

Chemobiokinetics. Oral administration results in disorder of calcium exchange. One of the metabolism products of α-MS in the body is *atrolactic (α-hydroxy-α-phenylpropionic) acid* in the urine (normally absent). *α-Methyl-α-phenyl ethylene glycol* is also formed and is eliminated with the urine in the form of glucuronide. α-MS is oxidized with participation of microsomal enzymes of the liver to products of higher toxicity than α-MS itself. When up to 10 mg/kg BW is administered, α-MS is completely metabolized.

Regulations.

U.S. FDA (1998) regulates the use of α-MS (1) as a component of adhesives to be safely used in food-contact surface in accordance with the conditions prescribed in 21 CFR part 175.105; (2) as a component (a monomer) of the uncoated or coated food-contact surface of paper and paperboard intended for use in producing, manufacturing, packaging, processing, preparing, treating, packing, transporting, or holding dry food of the type identified in 21 CFR part 176.170 (c); (3) in cross-linked polyester resins for repeated use as articles or component of articles coming in contact with food, subject to the provisions prescribed in 21 CFR part 177.2420; (4) as a monomer in the manufacture of semirigid and rigid acrylic and modified acrylic plastics in the manufacture of articles intended for use in contact with food in accordance with the conditions prescribed in 21 CFR part 177.1010; (5) as polymers which may be safely admixed, alone or in mixture with other permitted polymers, as modifiers in semirigid and rigid vinyl chloride plastic food contact articles prepared from vinyl chloride homopolymers and/or from vinyl chloride copolymers in accordance with the conditions specified in the 21 CFR; and (6) in the manufacture of resinous and polymeric coatings for articles intended for use in producing, manufacturing, packing, transporting, or holding food), subject to the conditions prescribed in 21 CFR part 175.300.

EU (1990). α-MS is available in the *List of authorized monomers and other starting substances which may continue to be used for the manufacture of plastic materials and articles intended to come into contact with foodstuffs pending a decision on inclusion in Section A* (*Section B*).

Great Britain (1998). α-MS is authorized up to the end of 2001 for use in the production of polymeric materials and articles in contact with food or drink or intended for such contact.

Standards. *Russia* (1995). MAC and PML: 0.1 mg/l (organolept., taste).

Reference:

1. Wolf, M. A., Rowe, V. K., McCollister, D. D., et al., Toxicological studies of certain alkylated benzenes and benzene, *Am. Med. Ass. Arch. Ind. Health.*, 14, 387, 1956.
2. Sokol'nikov, Ye. A., *Hygiene Assessment of Vulcanizates Based on SKMS 30 ARKM 15 Synthetic Rubber and Used in Contact with Foodstuffs*, Author's abstract of thesis, Kiyv, 1987, 23 (in Russian).
3. Oglesnev, G. A., Experimental substantiation of MAC for α-methyl styrene in water bodies, *Gig. Sanit.*, 4, 24, 1964 (in Russian).
4. Veselova, V. V. and Ogleznev, G. A., in *Proc. Ryazan' Medical Institute*, Ryazan', Issue No 80, 1963, 95 (in Russian).
5. Aizvert, L. G., *Data on Toxicity of α-Methyl Styrene*, Author's abstract of thesis, Alma-Ata, 1979, 26 (in Russian).
6. Hardin, B. D., Bond, G. P., Sikov, M. R., et al., Testing of selected workplace chemicals for teratogenic potential, *Scand. J. Work Environ. Health*, 7, 66, 1981.
7. Akhmetov, V. M. and Maksimov, G. G., in *Proc. Ryazan' Medical Institute*, Ryazan', Issue No 80, 1963, 56 (in Russian).

MONOMER FA

Molecular Formulas. $C_8H_8O_2$

M = 136.15

CAS No 28438-99-5
RTECS No QB7695000

Synonyms and **Trade Names.** FA monomer; Furfural-acetone adduct; Furfural-acetone monomer; Furfurylidene.

Composition. Mixture of furfurylidene and difurfurylidene acetone, which also contains deeper condensation products of an unknown composition.

Properties. A thick, dark brown liquid with the odor of furfural.

Applications. Used in the production of furan resins.

Acute Toxicity. LD_{50} is 1070 mg/kg BW in rats, and 980 mg/kg BW in mice, and 285 mg/kg BW in rabbits. Poisoning is accompanied by adynamia and ataxia, and plethora of the visceral organs. Gross pathology examination revealed excretion of serosanguinous liquid from the nasopharynx, congestion in the visceral organs, and focal hemorrhages in the lungs.[1] According to other data, 50% rats died after administration of 370 mg/kg BW, and LD_{100} appeared to be 600 mg/kg BW.[2]

Repeated Exposure failed to reveal marked cumulative properties in rats. Rats tolerated administration of 20 to 100 mg/kg BW. 50 and 100 mg/kg BW produced changes in the hematology analyses, cholinesterase activity, and blood sugar content. A 20 mg/kg BW dose was found to be ineffective.[1]

Allergenic Effect is shown to be moderate in guinea pigs.[3]

References:

1. Tuzhilina, L. A., Experimental data on toxicity of monomer FA, *Gig. Truda Prof. Zabol.*, 5, 53, 1974 (in Russian).
2. Kondrashkin, G. I., in *Problems of Theoretical Medicine*, Cheboksary, 1972, 83 (in Russian).
3. Chernousov, A. D., On allergenic properties of furan compounds, *Gig. Sanit.*, 6, 28, 1974 (in Russian).

See ***Furan.***

1,3-PENTADIENE

Molecular Formula. C_5H_8
M = 68.13
CAS No 504-60-9
RTECS No RZ2464000
Abbreviation. 1,3-PD.

Synonym and **Trade Name.** 1-Methylbutadiene; Piperylene.

Properties. Colorless liquid, miscible with alcohol. Water solubility exceeded 341 mg/l at room temperature.

Applications. Comonomer in the production of polymers intended for use in contact with food.

Reproductive Toxicity. Rats received up to 1.0 g 1,2-PD/kg BW from two weeks before mating through the fourth day postpartum. The treatment did not result in adverse reproductive effects.[1]

Mutagenicity.

In vitro genotoxicity. 1,3-PD was tested using variations of the *Salmonella*/microsome assay. It was found to be mutagenic and extremely toxic to the tested bacteria with an amount as low as 2 μg/plate causing cellular death.[2]

Regulations. ***EU*** (1990). 1,3-PD is available in the *List of monomers and other starting substances, which may continue to be used pending a decision on inclusion in Section A* (*Section B*).

References:

1. Lington, A. W. and Beyer, B. K., Evaluation of a combined reproductive/developmental toxicity screening study in rats on 1,3-pentadiene, Abstract, *Toxicologist*, 13, 76, 1993.
2. Liewen, M. B. and Marth, E. H., Evaluation of 1,3-pentadiene for mutagenicity by the *Salmonella*/mammalian microsome assay, *Mutat. Res.*, 157, 49, 1985.

PENTAERYTHRITOL

Molecular Formula. $C_5H_{12}O_4$
M = 136.15

CAS No 115-77-5
RTECS No RZ2490000
Abbreviation. PER.

Synonym. Pentaerythrite 2,2'-di(hydroxymethyl)-1,3-propanediol.

Properties. Colorless crystals. Water solubility is 7.1% at 20°C and 19.3% at 55°C. Poorly soluble in ether and acetone. Taste perception threshold is 3.0 g/l.

Applications. A monomer. Used in the production of polymeric materials intended for use in contact with food.

Acute Toxicity. LD_{50} is 19.5 g/kg BW for rats, 11.3 g/kg BW for guinea pigs, 18.5 to 25.5 g/kg BW for mice and rabbits.[1,021] Poisoning is accompanied by a narcotic effect. Death occurs in 2 to 4 hours after administration. Gross pathology examination revealed hemorrhaging zones in the gastric mucosa and intestines, dystrophic changes in the liver.[1] However, according to other data, even 40 g/kg BW caused no changes but had laxative effect in the rat.[05]

Repeated Exposure. Rats of both sexes were dosed with 1.0 g PER/kg BW for 28 days. Morphology and biochemistry investigation revealed no changes in the blood composition and blood serum. There were no histology findings as well. The dose of 1.0 g/kg BW was considered to be ineffective.[2]

Short-term Toxicity. PER is known to affect the motor system of the intestines and to regulate lipid exchange. Mice and rats were given 1/5 to 1/10 LD_{50} for 20, 30, and 90 days. The exposure increased the blood histamine level and lengthened the time of barbiturate narcosis.[1]

Long-term Toxicity.

Observations in man. PER is used in France as a medicine, a daily dose of 2.5 to 15 g being harmless during prolonged use.

Animal studies. Rats and rabbits were exposed to doses of up to 5.0 mg PER/kg BW for 7 months. The treatment revealed a number of changes in the functional condition of the brain cortex cells, in mediator and nuclear exchange, in immune reactivity, etc.[1]

Chemobiokinetics. PER is readily absorbed in the intestines and excreted unchanged by the kidneys, which distinguishes it from other polyols.[05] PER is partly oxidized in the body, partly released in the urine in the form of *aldehydes* and *ketones*. In human volunteers, 85% ingested PER was eliminated unchanged in the urine.[014]

Regulations.

EU (1990). PER is available in the *List of authorized monomers and other starting substances which shall be used for the manufacture of plastic materials and articles intended to come into contact with foodstuffs (Section A).*

U.S. FDA (1998) approved the use of PER (1) in the manufacture of resinous and polymeric coatings for the food-contact surface of articles intended for use in producing, manufacturing, packing, processing, preparing, treating, packaging, transporting, or holding food in accordance with the conditions prescribed in 21 CFR part 175.300; (2) in the manufacture of resinous and polymeric coatings for polyolefin films for food-contact surface of articles intended for use in producing, manufacturing, packing, processing, preparing, treating, packaging, transporting, or holding food in accordance with the conditions prescribed in 21 CFR part 175.320; as (3) a defoaming agent in the manufacture of paper and paperboard intended for use in packaging, transporting, or holding food in accordance with the conditions prescribed in 21 CFR part 176.200; and (4) in the manufacture of cross-linked polyester resins that may be safely used as articles or components of articles intended for repeated use in contact with food in accordance with the conditions prescribed in 21 CFR part 176.200. PER and its stearate ester may be used (4) as stabilizers in rigid polyvinyl chloride and/or in rigid vinyl chloride copolymers provided that the total amount of PER and/or PER stearate (calculated as free PER) does not exceed 0.4% by weight of such polymers.

Great Britain (1998). PER is authorized without time limit for use in the production of polymeric materials and articles in contact with food or drink or intended for such contact.

Standards. ***Russia*** (1995). MAC and PML: 0.1 mg/l.

Reference:

1. Plitman, S. I., Experimental validation of the MAC for pentaerythritol and xylitol in water bodies, *Gig. Sanit.*, 2, 25, 1971 (in Russian).

2. Hayashi, S., Toyoda, K., Furuta, K., Uneyama, C., Kawanishi, T., and Imaida, K., Study of pentaerythritol toxicity at repeated 28-day administration to F344 rats, *Bull. Natl. Inst. Hyg. Sci.*, 110, 32, 1992 (in Japanese).

PHENOL

Molecular Formula. C_6H_6O
M = 94.12
CAS No 108-95-2
RTECS No SJ3325000

Synonyms and **Trade Names.** Benzenol; Carbolic acid; Hydroxybenzene; Oxybenzene; Phenyl alcohol; Phenyl hydroxide.

Properties. White, crystalline solid that liquefies after absorption of water from air. Turns red in the air, particularly in the light. Has an acrid odor and a sharp burning taste. Water solubility is 82 g/l at 15°C; with water forms a hydrate, $C_6H_5OH.H_2O$. Soluble in acetone and ethanol. Technically P. (carbolic acid) is a reddish-brown, sometimes black, liquid. Odor perception threshold for P. is 5.9 mg/l[010] or 7.9 mg/l,[02] for chlorophenols it is 0.001 mg/l. The taste and odor of P. are detected more rapidly at low water temperatures when phenol oxidation and volatility are decreased. In Japan and in the Netherlands, odor threshold values are regulated at the levels of 0.012 and 0.010 ppm, respectively.[1]

Applications. A basic feedstock for production of phenolic resins (phenolaldehyde plastics), bisphenol A and caprolactam. Disinfectant and antiseptic.

Exposure. P. derivatives may, by *in vivo* conversion, form a source of endogenous human P. exposure. P. is available in smoked meat and fish products and in outdoor air. Hazardous contamination is unlikely to occur because of the conspicuous P. smell and taste. Chlorophenol formation in chlorinated drinking water greatly enhances smell and taste. Concentration of P. in the urine of the workers employed in the distillation of the phenolic fraction of tar and that of non-exposed male workers was found to be 87.3 and 11.7 mg/l, respectively.[2]

Acute Toxicity. LD_{50} is 340 to 650 mg/kg BW for rats, 300 to 430 mg/kg BW for mice, and 100 mg/kg in cats. Species sensitivity in laboratory animals is negligible. Cats are the most sensitive, guinea pigs the most resistant. Poisoning is accompanied by respiratory function impairment. A local irritating effect of P. was noted.[3,4,018] Gross pathology examination revealed lesions in the parenchymatous organs and also in the large arteries.

Repeated Exposure failed to reveal cumulative properties of P. K_{acc} is identified to be about 20 (by Cherkinsky). Tremors were observed only after the first administration of a 120 mg/kg BW dose, and this dose appeared to be lethal to all rats within 11 days.[5]

There were no changes in the indices studied following oral administration of 0.5 mg P./kg BW to rats for 3 weeks.[6] Male CD-1 mice were exposed to 4.7 to 95.2 mg P./l in drinking water for 4 weeks. The treatment did not produce any overt clinical signs of toxicity. Peripheral red blood counts and hematocrit decreased.[7]

Rats were dosed by gavage with 8.0 to 80 mg P./kg BW for 1.5 months. The treatment did not affect blood morphology, blood ALT, AST, cholinesterase and peroxidase activity.[8]

Short-term Toxicity. Mice and rats received 0.1 to 10.0 g P./l in their drinking water for 13 weeks. Only the highest concentration (calculated to be equal to 2.0 g P./kg BW for mice and 1.0 g P./kg BW for rats) resulted in decreased BW.[6]

Long-term Toxicity. Adequate studies are not available. Limit could be based on organoleptic data. F344 rats and $B6C3F_1$ mice received 2.5 or 5.0 g/l in their drinking water for 103 weeks. A dose-related reduction in BW and water consumption only was observed.[9] Rats given 1600 ppm P. in their drinking water exhibited no decrease in BW.[10]

In a multigenerational reproductive study, rats were given P. in the water at concentrations up to 5.0 g/l (400 mg/kg BW). No retardation of BW gain was observed in treated animals.[11]

Reproductive Toxicity. In a feeding study, P. levels in placenta and embryo were equivalent to those in maternal serum.[12]

Embryotoxicity. Sprague-Dawley rats were exposed to 30 to 120 mg P./kg BW by oral intubation on days 6 to 15 of gestation. There was no maternal toxicity or teratogenicity but fetal growth was retarded at the 120 mg/kg BW level. CD-1 mice were dosed with 70 to 280 mg/kg BW by oral intubation on days 6 to 15 of gestation. Maternal and fetal toxicity but not teratogenicity were observed. Dose-dependence was noted.[13]

Immunotoxicity. Male CD-1 mice were exposed to 4.7 to 95.2 mg P./l in drinking water for 4 weeks. The treatment with the highest dose suppressed the stimulation of cultured splenic lymphocytes by lipopolysaccharide, pokeweed mitogen, and phytohemagglutinin, and the response in mixed lymphocyte cultures. Antibody production response to the *T* cell-dependent antigen was suppressed, and levels of neurotransmitters were lowered in several brain regions.[7] Skin sensitization was not observed in guinea pigs.[14]

Mutagenicity.

Observations in man. CA was observed in the peripheral lymphocytes of 50 workers exposed to formaldehyde, styrene, and P.[15]

In vitro genotoxicity. P. induced SCE and mutations in cultured human cells. Positive results were noted in mouse lymphoma cells test.[9] P. showed weak capability to induce micronuclei in bone marrow cells of pregnant mice and, transplacentally, in fetal liver cells.[16] It did not produced DNA damage in cultured animal cells (IARC 47-275). P. was negative in *Salmonella* mutagenicity as say.[025]

Carcinogenicity. F344 rats and B6C3F_1 mice received P. at concentrations of 2.5 or 5.0 g/l in their drinking water for 103 weeks. There was no treatment-related increase in the incidence of tumors. A dose-related reduction in BW gain and in water consumption only was observed. In male rats, there was an increase in leukemia at the lower dose but not at the higher dose.[9] These data are considered to be inadequate for evaluation of carcinogenicity in experimental animals (IARC 47-275).

Carcinogenicity classifications. An IARC Working Group concluded that there is inadequate evidence for the carcinogenicity of P. in *experimental animals* and in *humans*.

IARC: 3;

U.S. EPA: D;

NTP: N - N - N - N (water).

Chemobiokinetics. P. metabolism is likely to involve binding with proteins, oxidation of the aromatic ring, and its breakdown to CO_2 (average amount of CO_2 is 10%, of *hydroquinone* 10% and of *pyrocatechine* 1.0%). Unchanged phenol and diphenols bind with sulfur and glucuronic acids.[17] According to Kenyon et al., sulfatation seems to be the dominant pathway at all dose levels, but decreased as a percent of the excreted dose with a concomitant increase in glucuronidation as the dose level increased. The major urinary metabolites of P. were *P. sulfate, P. glucuronide*, and *hydroquinone glucuronide*.[18] Elimination of P. and its oxidation products occurs rapidly via the urine, predominantly in the bound state. The main metabolites found in the urine are *p-cresol*, *phenol*, traces of *resorcinol*, *hydroquinone*, etc. It was suggested that by the oral route there was a potential for a threshold of biological activity as P. is rapidly conjugated and detoxified through the glutathione pathway.[19]

Regulations.

U.S. FDA (1998) regulates P. for use (1) in adhesives as a component of articles intended for use in packaging, transporting, or holding food in accordance with the conditions prescribed in 21 CFR part 175.105; (2) in the manufacture of resinous and polymeric coatings for the food-contact surface of articles intended for use in producing, manufacturing, packing, processing, preparing, treating, packaging, transporting, or holding food in accordance with the conditions prescribed in 21 CFR part 175.300; and (3) in the manufacture of phenolic resins for use as a food-contact surface of molded articles intended for repeated use in contact with nonacid food (*pH* above 5) in accordance with the conditions prescribed in 21 CFR part 177.2410.

EU (1990). P. is available in the *List of authorized monomers and other starting substances which shall be used for the manufacture of plastic materials and articles intended to come into contact with foodstuffs (Section A).*

Great Britain (1998). P. is authorized without time limit for use in the production of polymeric materials and articles in contact with food or drink or intended for such contact.

Recommendations. ***U.S. EPA*** (1999). Health Advisory for a longer-term exposure is 20 mg/l.

Standards.

EEC (1980). 0.5 mg/l.

Russia (1995). MAC and PML in drinking water: 0.001 mg/l, PML in food: 0.05 mg/l. In the absence of water chlorination, the recommended PML: 1.0 mg/l.

References:

1. Hoshika, Y., Imamura, T., Muto, G., Van Gemert, L. J., Don, J. A., and Walpot, J. I., International comparison of odor threshold values of several odorants in Japan and in the Netherlands, *Environ. Res.*, 61, 78, 1993.
2. Bieniek, G., Concentrations of phenol, *o*-cresol, and 2,5-xylenol in the urine of the workers employed in the distillation of the phenolic fraction of tar, *Occup. Environ. Med.*, 51, 354, 1994.
3. Kostovetsky, Ya. I. and Zholdakova, Z. I., On hygienic norm-setting of phenol in water bodies, *Gig. Sanit.*, 7, 7, 1971 (in Russian).
4. Nagorny, P. A., *Industrial Toxicology of Phenol Formaldehyde Resins, Phenol and Formaldehyde, and Industrial Hygiene in Areas of Exposure*, Author's abstract of thesis, Kiyv, 1981, 32 (in Russian).
5. Schlicht, M. P., Moser, V. C., Sumrel, B. M., Berman, E., and MacPhail, R. C., Systemic and neurotoxic effects of acute and repeated phenol administration, Abstract No 1047, *Toxicologist*, 12, 274, 1992.
6. Kretov, I. A. et al., in *Hygienic Aspects of Environmental Protection*, Moscow, Issue No 2, 1974, 88 (in Russian).
7. Hsieh, G. C., Sharma, R. P., Parker, R. D., and Coulombe, R. A., Immunological and neurobiochemical alterations induced by repeated oral exposure of phenol in mice, *Eur. J. Pharmacol.*, 228, 107, 1992.
8. Korolev, A. A., Aibinder, A. A., Bogdanov, M. V., et al., Hygiene and toxicity profile of phenol destruction products formed during ozone treatment of water, *Gig. Sanit.*, 8, 6, 1973 (in Russian).
9. *Bioassay of Phenol for Possible Carcinogenicity,* NTP Technical Report Series No 203, Research Triangle Park, NC, U.S. Dept. Health Human Service, 1980.
10. Heller, V. G. and Pursell, L., *J. Pharmacol. Exp. Ther.*, 63, 99, 1938 (cit. in Deichmann and Oesper, 1940. See #11).
11. Deichmann W., and Oesper, P., Ingestion of phenol - effects on the albino rat, *Ind. Med.*, 9, 296, 1940.
12. Gray, J. A. and Kavlock, R. J., A pharmacokinetic analysis of phenol in the pregnant rat: deposition in the embryo and maternal tissues, Abstract, *Teratology*, 41, 561, 1990.
13. Price, C. J., Ledoux, T. A., Reel, J. R, et al., Teratologic evaluation of phenol in rats and mice, *Teratology*, 33, 92C, 1986.
14. Rao, K. S., Betso, J. E., and Olson, K. J., A collection of guinea pig sensitization test results grouped by chemical class, *Drug Chem. Toxicol.*, 4, 331, 1981.
15. Mierauskiene, J. R. and Lekevicius, R. K., Cytogenetic studies of workers occupationally exposed to phenol, styrene and formaldehyde, Abstract No 60, *Mutat. Res.*, 147, 308, 1985.
16. Cirrani, R., Barale, R., Marrazzini, A., and Loprieno, N., Benzene and the genotoxicity of its metabolites. I. Transplacental activity in mouse fetuses and in their dams, *Mutat. Res.*, 208, 61, 1988.
17. Gadaskina, I. D. and Filov, V. A., in *Conversion and Determination of Industrial Organic Poisons in the Body*, Meditsina, Leningrad, 1972, 193 (in Russian).
18. Kenyon, E. M., Seeley, M. E., Janszen, D., and Medinsky, M. A., Dose-, route-, and sex-dependent urinary excretion of phenol metabolites in B6C3F$_1$ mice, *J. Toxicol. Environ. Health.*, 44, 219, 1995.
19. *1994 Annual Report of the Committees on Toxicity, Mutagenicity, Carcinogenicity of the Chemicals in Food, Consumer Products and the Environment*, DoH, HMSO, London, 1995, 61.

PROPYLENE

Molecular Formula. C_3H_6

M = 42.08

CAS No 115-07-1
RTECS No UC6740000

Synonyms. 1-Propene; Methylethylene.

Properties. A colorless gas with aromatic odor. Water solubility is 350 mg/l at 25°C. Volatilizes rapidly from the free surfaces. Odor perception threshold is 22.5 mg/l,[010] 0.5, or 0.028 mg/l.[02] P. does not alter the color or transparency of water.

Applications. A monomer. Used as a starting material in the manufacture of polypropylene plastics intended for use in contact with food and other basic chemicals, such as acrylonitrile, isopropyl alcohol, and propylene oxide.

Acute Toxicity. Rats tolerated administration of 0.5 ml of a solution with concentration of 340 mg P./l without behavioral changes.[1]

Repeated Exposure. Accumulation is impossible because of rapid removal from the body.[1]

Short-term Toxicity. Mice were administered a dose of 3.75 mg/kg BW for 4 months. The treatment caused no changes in the behavior of animals, BW gain, or oxygen consumption. Gross pathology examination failed to reveal alterations in the relative weights and histology of the visceral organs.

Long-term Toxicity. In a 6-month study, there were no changes in behavior, BW gain, phagocytic activity of the leukocytes, and cholinesterase activity in the blood serum after oral administration of 0.05 mg/kg BW.[1]

Mutagenicity.

In vitro genotoxicity. Negative results were noted in *Salmonella* mutagenicity assay (NTP-90).

Carcinogenicity. Inhalation exposure to 8.6 mg/m^3 and 17.2 mg/m^3 produced no carcinogenic effect in mice and rats.[2]

Carcinogenicity classifications. An IARC Working Group concluded that there is inadequate evidence for the carcinogenicity of P. in *experimental animals* and there are inadequate data available to evaluate the carcinogenicity of P. in *humans.*

IARC: 3;

NTP: NE - NE - NE - NE (inhalation).

Chemobiokinetics. A direct chemical interaction between P. and biological media of the body is unlikely. Elimination occurs rapidly in an unaltered form through the lungs.

Regulations.

U.S. FDA (1998) approved the use of P. in adhesives as a component (monomer) of articles intended for use in packaging, transporting, or holding food in accordance with conditions prescribed in 21 CFR part 175.105.

EU (1990). P. is available in the *List of authorized monomers and other starting substances which shall be used for the manufacture of plastic materials and articles intended to come into contact with foodstuffs (Section A).*

Great Britain (1998). P. is authorized without time limit for use in the production of polymeric materials and articles in contact with food or drink or intended for such contact.

Standards. *Russia* (1995). MAC and PML: 0.5 mg/l (organolept., odor).

References:

1. Amirkhanova, G. F., Latypova, Z. V., and Tupeyeva, R. B., in *Protection of Water Reservoirs against Pollution by Industrial Liquid Effluents*, S. N. Cherkinsky, Ed., Meditsina, Moscow, Issue No 7, 1965, 28 (in Russian).
2. Quest, J. A., Tomaszewski, J. E., Haseman, J. K., et al., Two-year inhalation toxicity study of propylene in F344 rats and $B6C3F_1$ mice, *Toxicol. Appl. Pharmacol.*, 76, 288, 1984.

PROPYLENE OXIDE

Molecular Formula. C_3H_6O
M = 58.08
CAS No 75-56-9
RTECS No TZ2975000

Abbreviation. PO.

Synonyms. 1,2-Epoxypropane; Methyloxirane; Propenoxide.

Properties. Colorless, volatile liquid with a sweet, ether odor. Readily soluble in water (40.5% at 20°C) and alcohol but is likely to evaporate to a great extent. Odor perception threshold is 9.9 mg/l,[010] 11.3 mg/l,[1] or 31 mg/l;[02] taste perception threshold is 0.43 mg/l.

Applications. Initial product for the synthesis of polyoxypropylene, propylene glycol and its esters and ethers, polyesters and polyurethane resins and foams, etc. A solvent for resins. Food additive. Fumigant for medical plastics and foodstuffs.

Acute Toxicity. No adverse effects have been reported due to the ingestion of PO and its reaction products with food.[1] The oral LD_{50} was reported to be 380 to 1140 mg/kg BW in rats, 440 to 630 mg/kg BW in mice, 660 to 690 mg/kg BW in guinea pigs, and 1245 mg/kg BW in rabbits. Poisoned animals displayed agitation followed by depression, increased salivation, and diarrhea. Death is preceded by short-term clonic-tonic convulsions. There were certain changes in hematology and biochemistry analyses, an increase in the content of chlorides in the blood. Increased level of free histamine in the blood is considered to be the main mechanism of toxic effect development. Gross pathology examination revealed damage to the stomach mucosa and liver in rats. The liver cells exhibited edema and signs of fatty dystrophy.[2-4]

Repeated Exposure revealed slight cumulative properties. Administration of 0.2 g/kg BW as a 10% PO solution in oil for 24 days produced no toxic effect in rats.[1] However, according to other data,[4] oral doses of 25 and 100 mg/kg BW given to rats for 1.5 months affected the peripheral blood and CNS functions, the liver, and kidneys.

Long-term Toxicity. In a 6-month rat study, effect on of the hematological indices was noted. The treatment produced phase changes in the chloride content of the blood and protein fractions of the blood serum, and other lesions.[4]

Allergic contact dermatitis was diagnosed in 3 cases of exposure to a PO solution.[5]

Reproductive Toxicity. Embryotoxic and teratogenic effects were not observed when pregnant rabbits were exposed to PO through inhalation.

Embryotoxicity. Pregnant rats given 260 mg PO/kg BW during 2 weeks of gestation exhibited higher embryotoxicity and lower offspring BW compared to the controls. Doses of 1/5 and 1/20 LD_{50} given to rats for 6 months also had a toxic effect on embryos.[4]

Teratogenicity. Teratogenic properties of PO were not established, although there was a reduction in size and BW, retardation of the ossification process, and distortion of the ribs in the offspring.[6] Hardin et al.[7] did not note teratogenic effects in rats and rabbits.

Gonadotoxicity. There were no sperm abnormalities in monkeys following inhalation exposure.[8] Reduced sperm motility, damage to spermatocytes, and reduced fertility were found in rats treated with oral LD_{50} (520 mg/kg BW).[4]

In female rats, a single oral dose of 260 mg/kg BW caused disturbance of the oestral cycle without affecting the generation function.[4] Male Wistar rats were exposed to PO by *i/p* injections three days a week for 6 weeks. Administration of 186 mg PO/kg BW decreased epididymal weight, sperm count, and tail of the epididymis. Significant increase in abnormal sperm was noted. There were no changes in the serum testosterone concentration.[9]

Mutagenicity. PO is structurally related to *ethylene oxide.*

In vitro genotoxicity. PO was shown to produce mutagenic effect in assays employing microorganisms and insects. According to Rapoport,[10] it induced a significant increase of revertants in *Salmonella typhimurium* strains *TA1535* and *TA100*.[11] According to Agurell et al.[12], it induced point mutations in *Salmonella typhimurium* strains and SCE in *S. cerevisiae RS112*. PO caused CA in mammalian cells *in vitro*, in particular, chromatid gaps and breaks.[13,14]

In vivo cytogenetics. An increase in micronuclei was observed following *i/p* administration of 300 mg/kg BW. The dose of 150 mg/kg BW or less did not produce such effect.[8] No CA and SCE were shown in monkeys exposed via inhalation.[15]

Negative results were observed in DLM assay in male mice following administration of 50 and 250 mg/kg BW for 2 weeks prior to mating. Slight mutagenic activity in *in vivo* experiments was attributed to rapid detoxication of PO in the liver of mammals.[13]

PO was mutagenic to *Dr. melanogaster.*[10]

Carcinogenicity. Rats were gavaged with 15 and 60 mg/kg BW doses in salad oil for 112 weeks. There was an increased incidence of squamous cell carcinoma of the forestomach; malignant tumors were found mainly at the site of entry into the body.[16] Female Wistar rats exposed by inhalation to 300 ppm for 28 months exhibited an increase in the rate of both benign and malignant mammary tumors.[17]

Carcinogenicity classifications. An IARC Working Group concluded that there is sufficient evidence for the carcinogenicity of PO in *experimental animals* and there were no adequate data available to evaluate the carcinogenicity of PO in *humans.*

The Annual Report of Carcinogens issued by the U.S. Department of Health and Human Services (1998) defines PO to be a substance which may reasonably be anticipated to be carcinogen, i.e., a substance for which there is a limited evidence of carcinogenicity in *humans* or sufficient evidence of carcinogenicity in *experimental animals.*

IARC: 2B;
U.S. EPA: B2;
EU: 2;
NTP: SE - SE - CE - CE (inhalation).

Chemobiokinetics. Following absorption PO is found in the blood. It is metabolized to *formaldehyde,* the content of which in the blood after administration of 1/5 and 1/20 LD_{50} was 0.137 to 0.24 mg (0.067 mg in the controls).[4] PO metabolism is described as transformation into *S-(2-hydroxy-1-propyl)glutathione* under the action of glutathione-epoxide-transferase with subsequent transformation into *1,2-propanediol* under the action of epoxidehydrolase. Diol can subsequently be oxidized into *lactic* and *pyruvic acids.*[2]

Regulations.

U.S. FDA (1998) regulates the use of PO as a direct and indirect food additive. It may be used in products which come in contact with food; the use of PO is approved for the following purposes: (1) as a etherifying agent in the production of modified food starch (at the levels of not more than 25% PO, by itself or in combination with various sources of active chlorine and/or oxygen); (2) as a defoaming agent in the manufacture of paper and paperboard intended for use in packaging, transporting, or holding food in accordance with the conditions prescribed in 21 CFR part 176.200; (3) as slimicides used in the manufacture of paper and paperboard in accordance with the conditions prescribed in 21 CFR part 176.300; (4) as a reactant in the production of lubricants with incidental food contact in accordance with the conditions prescribed in 21 CFR part 178.3570; and (5) as a package fumigant for certain fruit products and as a fumigant for bulk quantities of several food products, provided residues of PO or propylene glycol do not exceed specified limits.

EU (1990). PO is available in the *List of authorized monomers and other starting substances which shall be used for the manufacture of plastic materials and articles intended to come into contact with foodstuffs (Section A).*

Great Britain (1998). PO is authorized without time limit for use in the production of polymeric materials and articles in contact with food or drink or intended for such contact. The quantity of PO in the finished polymeric material or article shall not exceed 1 mg/kg).

Standards.

EU (1990). MPQ in the material or article: 1.0 mg/kg.

Russia (1995). MAC: 0.01 mg/l.

References:

1. Rowe, V. K. Holligworth, R. L., Oyen, F., McCollister, D. D., and Spencer, H. C., Toxicity of propylene oxide determined in experimental animals, *Arch. Ind. Health,* 13, 228, 1956.
2. Grushko, Ya. M., *Harmful Organic Compounds in the Liquid Industrial Effluents,* Khimiya, Leningrad, 1976, 162 (in Russian).

3. Smyth, H. F., Seaton, J., and Fisher, L., The single dose toxicity of some glycols and derivatives, *J. Ind. Hyg Toxicol.*, 23, 259, 1941.
4. Antonova, V. I., Zommer, E. E., Kuznetzova, A. D., et al., Toxicology of propylene oxide and regulations for surface water, *Gig. Sanit.*, 7, 76, 1981 (in Russian).
5. Jensen, O., Contact allergy to propylene oxide and isopropyl alcohol in a skin disinfectant sevab, *Contact Dermat.*, 7, 148, 1981.
6. Meylan, W., Papa, L., De Rosa, C. T., et al., Chemical of current interest - propylene oxide: health and environmental effect profile, *Toxicol. Ind. Health*, 2, 219, 1986.
7. Hardin, B. D., Niemeier, R. W., Sikov, M. R., et al., Reproductive toxicological assessment of the epoxides ethylene oxide, propylene oxide, butylene oxide, and styrene oxide, *Scand. J. Work Environ. Health*, 9 (2 Spec. No), 94, 1983.
8. Hardin, B. D., Schuler, R. L., McCormic, P. M., et al., Evaluation of propylene oxide for mutagenic activity in 3 *in vivo* test system, *Mutat. Res.*, 117, 337, 1983.
9. Omura, M., Tanaka, A., Mori, K., Hirata, M., Zao, M., and Inoue, N., Dose-dependent testicular toxicity of propylene oxide in rats induced by repeated intraperitoneal injections, Abstract, *Fukuoka Igaku Zasshi*, 85, 204, 1994 (in Japanese).
10. Rapoport, I. A. Alkylation of gene molecule, *Reports Acad. Sci. USSR*, Issue No 59, 1948, 1183 (in Russian).
11. Castelain, P., Criado, B., Cornet, M., Laib, R., Rogiers, V., Kirsch-Volders, M., Comparative mutagenicity of structurally related aliphatic epoxides in a modified *Salmonella*/microsome assay, *Mutagenesis*, 8, 387, 1993.
12. Agurell, E., Cederberg, H., Ehrenberg, L., Lindahl-Kiessling, K., Rannung, U., and Tornqvist, M., Genotoxic effects of ethylene oxide and propylene oxide: a comparative study, *Mutat. Res.*, 250, 229, 1991.
13. Bootman, J., Lodge, D. C., and Whalley, H. E. Mutagenic activity of propylene oxide in bacterial and mammalian systems, *Mutat. Res.*, 67, 101, 1979.
14. Dean, B. J. and Hodson-Walker, G., An *in vitro* chromosome assay using cultured rat liver cells, *Mutat. Res.*, 64, 329, 1979.
15. Lynch, D. W., Lewis, T. R., Moorman, W. J., et al., Sister chromatid exchanges and chromosome aberrations in lymphocytes from monkey exposed to ethylene oxide and propylene oxide by inhalation, *Toxicol. Appl. Pharmacol.*, 76, 85, 1984.
16. Dunkelberg, H., Carcinogenicity of ethylene oxide and 1,2-propylene oxide upon intragastric administration to rats, *Brit. J. Cancer*, 46, 924, 1982.
17. Kuper, C. F., Reuzel, P. G. J., Feron, V. J., et al., Chronic inhalation toxicity and carcinogenicity study of propylene oxide in Wistar rats, *Food Chem. Toxicol.*, 26, 159, 1988.

SEBACIC ACID

Molecular Formula. $C_{10}H_{18}O_4$
M = 202.25
CAS No 111-20-6
RTECS No VS087500
Abbreviation. SA.

Synonym. Decanedioic acid.

Properties. White or cream, crystalline powder. Water solubility is 1.0 g/l at 17°C and 20 g/l at 100°C. Readily soluble in alcohol. Has no effect on the taste or odor of water; no frothing effect was noted. According to Andreyev,[1] organoleptic perception threshold is 250 mg/l.

Applications. Used in the production of cold-resistant plasticizers, polyester and polyamide resins, and polyurethanes intended for use in contact with food; a stabilizer for alkyd resins, antiscorching agent in the production of food-grade resins.

Migration from SA-containing rubber samples into distillated water (40°C, for 24 hours) has been studied by Prokof'yeva.[2] Concentrations of 0.025 to 0.03 mg SA/l are reported.

Acute Toxicity. LD_{50} is 3.4 g/kg BW in rats, and 6.0 g SA/kg BW in mice. However, according to Krapotkina et al.,[3] administration of 11 mg/kg BW appeared to be not lethal to mice, rats, or rabbits, but in 2 days the animals experienced significant BW loss.

Short-term Toxicity. Oral administration of 110 and 1000 mg/kg BW to rats resulted in the reduction in BW gain and in STI value, a change in the enzymatic activity in the blood serum and liver homogenate. A reduction in protein content of the blood serum and pathological changes in the GI tract, liver, and kidneys were also observed. The NOEL of 11 mg/kg BW was identified in this study.[1]

Long-term Toxicity. In a 7-month study, the treatment caused similar changes in rats.[4]

Reproductive Toxicity.

The NOEL for ***gonadotoxic effect*** is likely to be 110 mg/kg BW.[1]

Teratogenicity. No malformations were observed in offspring of rats and rabbits fed 0.5 or 1.0 g SA/kg BW, respectively.[5]

Chemobiokinetics.

Observations in man. 30 to 46% of the dose is removed with the urine.

Animal studies. Disodium sebacate was administered to rats at the dose of 25 μCi of ^{14}C-labeled compound. About 35% of administered radioactivity were delivered in the urine as unchanged *sebacate*; 25% were identified as ^{14}C-CO_2 in expired air. Sebacate half-life is the longest in adipose tissues and in the liver, being sites of likely biotransformation.[6]

Regulations.

EU (1990). Sebacic anhydride is available in the *List of monomers and other starting substances which may continue to be used for the manufacture of plastic materials and articles intended to come into contact with foodstuffs pending a decision on inclusion in Section A.*

U.S. FDA (1998) regulates the use of SA (1) in adhesives as components of articles intended for use in packaging, transporting, or holding food in accordance with the conditions prescribed in 21 CFR part 175.105; (2) in the manufacture of resinous and polymeric coatings for articles and surfaces coming in contact with food in accordance with the conditions prescribed in 21 CFR part 175.300; (3) in the manufacture of resinous and polymeric coatings for polyolefin films for the food-contact surface of articles intended for use in producing, manufacturing, packing, processing, preparing, treating, packaging, transporting, or holding food in accordance with the conditions prescribed in 21 CFR part 175.320; (4) in the manufacture of cross-linked polyester resins which may be safely used as articles or components of articles intended for repeated use in contact with food in accordance with the conditions prescribed in 21 CFR part 177.2420; and (5) as a component of polyethylene phthalate polymers (films, articles, or fabrics) intended for use in contact with food in accordance with the conditions prescribed in 21 CFR part 177.1630.

Great Britain (1998). SA is authorized without time limit for use in the production of polymeric materials and articles in contact with food or drink or intended for such contact.

Standards. ***Russia*** (1995). MAC and PML: 1.5 mg/l.

References:

1. Andreyev, I. A., The method for rapid elaboration of hygienic norms of the contents of chemical compounds (as exemplified by adipic and sebacic acids) in water bodies, *Gig. Sanit.*, 7, 10, 1985 (in Russian).
2. Prokof'yeva, L. G., Detection methods and migration levels of carbonic acids - rubber vulcanization retardants, in *Hygiene of Use, Toxicology of Pesticides and Polymeric Materials*, Coll. Works All-Union Research Institute of Hygiene and Toxicology of Pesticides, Polymers and Plastic Materials, A. V. Pavlov, Ed., Kiyv, Issue No 14, 1984, 114 (in Russian).
3. Krapotkina, M. A., Garkavenko, O. S., and Abramova, A. M., Clinical and experimental estimation of toxic effects and derivation of MAC for adipic acid in the work place air, *Gig. Truda Prof. Zabol.*, 5, 48, 1980 (in Russian).
4. Novikov, Yu. V., Andreyev, I. A., Ivanov, Yu. V., et al., Hygienic standardization of sebacic and adipic acids in water bodies, *Gig. Sanit.*, 9, 72, 1983 (in Russian).
5. Greco, A. V., Mingrone, G., Mastromattei, E. A., Finotte, E., and Castogneto, M., Toxicity of disodium sebacate, *Drugs Exp. Clin. Res.*, 16, 531, 1990.

6. Tataranni, P. A., Mingrone, G., De Gaetano, A., Raguso, C., and Greco, A. V., Tracer study of metabolism and tissue distribution of sebacic acid in rats, *Ann. Nutr. Metab.*, 36, 296, 1992.

SEBACIC ACID, compound with 1,6-HEXANEDIAMINE (1:1)

Structural Formula. $NH(CH_2)_6NHCO(CH_2)_8CO$
M = 318.52
CAS No 6422-99-7
RTECS No VS1450000

Synonyms and **Trade Names.** Decanedioic acid, compound with 1,6-hexanediamine; Hexamethylenediamine sebacate; Hexamethylenediammonium sebacate; Nylon-6,10 salt; Salt SG; Sebacic acid, compound with 1,6-hexanediamine.

Properties. Light, odorless, sweetish, amorphous powder with a faintest rosy tinge. Poorly soluble in water. The threshold organoleptic concentration is 1.0 g/l.[1]

Applications. Used as a monomer in the production of polyamide 6,8 (Nylon 6,10).

Acute Toxicity. LD_{50} is found to be 1.88 g/kg BW for mice, and 11 g/kg BW for rats. Treatment-related effects included convulsions, mucosal hyperemia, and nasal bleeding in some animals 15 minutes after administration. Death occurred within 3 days after exposure, survivors recovered in 7 to 9 days.[1]

Repeated Exposure revealed no cumulative properties. Rats were dosed with 1.3 g/kg BW dose for 45 days. The treatment resulted in reduced BW gain, changes of hematological indices and of gastric function.[2]

Regulations. ***EU*** (1990). S. is available in the *List of authorized monomers and other starting substances which may continue to be used for the manufacture of plastic materials and articles intended to come into contact with foodstuffs pending a decision on inclusion in Section A* (*Section B*).

References:

1. Zhakov, Yu. A., in *Protection of Water Reservoirs against Pollution by Industrial Liquid Effluents*, S. N. Cherkinsky, Ed., Medgiz, Moscow, Issue No 4, 1960, 230 (in Russian).
2. Babayev, D. A. *Scientific Principles of Hygienic Study of Polymers Intended for Use in Food Industry and Supply Engineering*, Author's abstract of thesis, Kiyv, 1980, 22 (in Russian).

STYRENE

Molecular Formula. C_8H_8
M = 104.15
CAS No 100-42-5
RTECS No WL3675000

Synonyms and **Trade Names.** Cinnamene; Cinnamol; Ethylenyl benzene; Phenylethylene; Styrol; Styrolene; Vinylbenzene.

Properties. Colorless, viscous liquid with a characteristic penetrating unpleasant odor. On exposure to air or light undergoes polymerization and oxidation. Water solubility is 125 mg/l at 20°C and 320 mg/l at 25°C, soluble in alcohol. Odor perception threshold is 0.05 to 0.12 mg/l, or 0.011 mg/l;[02,010] taste perception threshold is 0.06 mg/l.[1]

In Japan and in the Netherlands, odor threshold values are regulated at the levels of 0.033 and 0.016 ppm, respectively.[2]

Applications. S. is used as a monomer in the production of various plastics, resins and vulcanizates intended for use in contact with food: styrene-butadiene rubber, acrylonitrile-butadiene-styrene (ABS) polymer, and styrene-acrylonitrile copolymer (SAN) resins. It is also used as a cross-linking agent and a solvent for unsaturated polyester resins.

Migration Data. S. has been found in food packaged in polystyrene containers (in yogurt up to 0.02 mg/l),[3] and in water (0.05 to 0.1 mg/l).[4,5] It was shown to migrate from *ABS*-polymers into water: 0.05 mg S./l for 24 hours at 37°C.[5]

Migration of S. from thermoset polyester cookware into foods and fatty food simulant olive oil was investigated under high temperature test condition. S. in foods cooked in thermoset polyester articles was in the range <0.5 to 5.0 μg/kg and 5.0 to 30 μg/kg where the polyester contained 9.0 and 380 mg/kg residual monomer, respectively. Testing for 2 hours at 175°C into olive oil resulted in significantly higher migration of S. than seen for other foods, although there was a marked decrease in migration on repeat use.[6] Studies of S. migration from thermoformed polystyrene foam articles were completed using food oil as the simulant medium. Migration was proportional to the square root of time of exposure.[7]

Migration of S. in corn oil from high impact polystyrene at 40°C was found to be 170 and 390 ppb after incubation for 3 and 10 days, respectively; in oil in water emulsion it was 170 and 250 to 340 ppb; and in fatty foods it was 150 ppb and 200 to 350 ppb.[8] After incubation for 15 days, migration of S. from high impact polystyrene was determined at the levels of 0.73 mg/dm^2 (in water), 12 mg/dm^2 (in 50% ethanol), 6.9 mg/dm^2 (in olive oil), and 3.4 mg/dm^2 (in 3.2% fat milk).[9] Release of S. from polystyrene food contact wares into *n*-heptane was not observed at the detection limit of 0.1 ppm.[10]

The level of S. migration from polystyrene cups collected from retail markets in Belgium, Germany, and the Netherlands was investigated in different food systems including water, milk (0.5, 1.55, and 3.6% fat), cold beverages (apple juice, orange juice, carbonated water, cola, beer, and chocolate drink), hot beverages (tea, coffee, chocolate, and soup) (0.5, 1.0, 2.0, and 3.6% fat), take out foods (yogurt, jelly, pudding, and ice-cream), as well as aqueous food simulants (3.0% acetic acid, 15, 50 and 100% ethanol) and olive oil. S. migration was found to be strongly dependent upon the fat content and storage temperature. Drinking water gave migration values considerably lower than all of the fatty foods. Ethanol at 15% showed a migration level equivalent to milk or soup containing 3.6% fat. Maximum observed migration for cold or hot beverages and take out foods was 0.025% of the total S. in the cup. Food simulants were responsible for higher migration (0.37% in 100% ethanol).[11]

Exposure. Concentrations of S. reported in drinking water are 0.01 to 0.05 mg/l. Concentration of S. in the blood in general population was found to be 217 ηg/l (Brugnone et al., 1994).

Acute Toxicity. In rats, LD_{50} varies from 5.0 g/kg to 8.0 g/kg BW.[12] Poisoned animals displayed NS impairment, convulsions, loss of reflexes, cyanosis, and body temperature drop. On administration of the lethal oral doses, rats became comatose and died. Gross pathology examination revealed hepatic dystrophic changes and, incidentally, renal changes,[13] but also diffuse damage to the upper sections of the CNS. The maximum tolerated doses appear to be 0.5 mg/kg BW for mice, and 2.0 g/kg BW for rats. These doses caused no appreciable morphological changes.[1]

Repeated Exposure. Male rats were given 1.0 ml/kg BW by oral intubation for 15 days. There was an increase in the serotonin and noradrenaline contents in the brain but no changes in the amount of dophamine were marked. The monoaminooxidase activity was suppressed but that of ACE was unaltered.[14]

Doses of 100 mg/kg and 200 mg/kg BW administered to rats over the same period of time caused no disruption of behavioral reactions. An increase was observed in the serotonin level in the hippocampus and hypothalamus as a result of exposure to the 200 mg/kg BW dose.[14]

Short-term Toxicity. Doses up to 500 mg/kg BW caused irritation of the esophagus and stomach and hyperkeratosis of the forestomach. No hematology changes were observed in the short-term oral studies in rats.[13] The dose of 400 mg/kg BW given orally for 100 days produced elevated levels of hepatic AST and ALT. In addition, significantly decreased activity of hepatic acid phosphatase and of other enzymes was found. Histopathological examination revealed liver focal necrosis, which was supported by the biochemical analysis described earlier.[15]

Long-term Toxicity. In a 6-month oral toxicity study, the NOEL of 133 mg/kg BW was suggested.[12] In the more recently reported experiment, beagle dogs were given S. in a peanut oil suspension by gavage for up to 561 days.[16] The NOAEL of 200 mg/kg BW was identified in this study. Parameters investigated in another oral rat study were clinical signs, mortality, growth, food and water intake, hemograms, clinical chemistry, urinalysis, gross necropsy, and histopathology. The NOEL was found to be 125 mg/l of drinking water that corresponds to 7.7 mg/kg BW for males and 12 mg/kg BW for females.[17]

Reproductive Toxicity.

Embryotoxicity. Negative results were reported in rats following oral dosing with 90 to 150 mg/kg BW.[12] No adverse effects were observed in Sprague-Dawley rats in the 3-generation reproductive study: there were no treatment-related changes in rats exposed for 2 years to 125 or 250 ppm in drinking water.[18]

Teratogenicity. The oral teratogenicity study in rats did not reveal maternal toxicity, teratogenic or embryotoxic effect at dose levels up to and including 300 mg/kg BW.[19] However, according to Chernoff et al.,[20] treatment-related anomalies in litters (enlarged renal pelvis) were found in Sprague Dawley rats exposed to S. by oral gavage on days 6 to 15 of gestation. Teratogenic effect for chicken embryos, rats and rabbits, as well as fetotoxicity in mice and hamsters following inhalation was reported in some studies.[19,21]

Inhalation of S. at low concentrations affects the gonads, embryogeny, and the offspring of mammals. S. penetrates through the placenta into the milk of feeding females.[22]

Gonadotoxicity.

Observations in man. The motility of human spermatozoa was reported to be inhibited *in vitro* by S.-based polymers.[23] Reduced libido and potency were noted in 58 out of 143 workers (40%), and a decreased sperm motility was found in 23% of the workers employed in a reinforced plastic plant.[24] No significant increased risk for spontaneous abortions among wives of workers assumed to be highly exposed to S.[25,26]

Animal studies. Data available on the impact of S. on male reproduction are contradictory, but animal fertility is not likely to be affected at doses up to 250 ppm in the diet.[27] Rats were given 100 mg/kg and 200 mg/kg BW for 60 days. The higher dose produced a significant decrease in epididymal spermatozoa count. S. has been shown to alter the activities of enzymes associated with specific cell type of testis.[28] Reduced epididymal sperm counts were observed in rats orally dosed with 400 mg/kg BW for 2 months.[20]

Immunotoxicity. Mice were orally exposed to mixture of S. and dioctyl phthalate in ground nut oil at the doses of 0.02, 0.03, or 0.05 LD_{50} for 4 weeks. Exposure to low doses of mixture may modulate some of the immune functions as compared to exposure to either chemical alone.[30]

Mutagenicity. S. requires metabolic activation to produce genotoxic action. Mutagenic effect is due to the formation of *styrene oxide* (*q.v.*) in the body. S. is transformed into styrene oxide in the blood of humans under the action of oxyhemoglobin of erythrocytes.

In vitro genotoxicity. S. is mutagenic in a variety of test systems, results being sometimes rather equivocal. It induces gene mutations in prokaryotic and eukaryotic microorganisms, and in mammalian cells *in vitro.* Reports on chromosomal abnormalities are contradictory. When positive results were observed, mainly high doses were used.[31] Scott and Preston came to the conclusion that S. and S oxide can induce CA and SCE *in vitro*, but the chromosome-damaging ability of S. is only manifested if test conditions favor its metabolic activation over inactivation.[32]

In vivo cytogenetics. According to Simula and Priestly,[33] S. produced weak genotoxic responses in the bone marrow micronucleus, sperm morphology, and SCE assays in Swiss mice and Porton rats.

S. induced gene mutations in *Dr. melanogaster.*

Observations in man. The increases in cytogenetic effects in some studies on S. workers are probably attributable to the the presence of other chromosome-damaging agents in the workplace and/or to inadequate investigations.

Carcinogenicity studies in mice and rats with various routes of administration did not provide the evidence of S. being a carcinogen. The evidence of genotoxicity in short-term animal tests and in humans occupationally exposed to S. along with the data on the metabolite styrene-7,8-oxide seems to be supportive of carcinogenicity of S. However, S. showed low DNA and protein binding activities *in vivo.*[34]

In the oral carcinogenicity study with $B6C3F_1$ mice, a significantly increased incidence of lung tumors (adenoma + carcinomas) was seen in males at the highest dose level (300 mg/kg BW administered in corn oil). However, in this study the control group was rather small.[35]

In a study with F344 rats,[22] the doses of 500 to 2000 mg/kg BW given in corn oil did not cause any significantly increased tumor incidence. Sprague-Dawley rats received doses of 50 and 250 mg/kg BW in olive oil over a period of 52 weeks. The study was terminated after 140 weeks. No significantly increased tumor incidences were observed.[36]

A 2-year oral toxicity/carcinogenicity study in conjunction with a 3-generation reproduction study was carried out.[17] Rats received 0.125 and 250 mg S./l in their drinking water. No increase in tumor incidence was observed. BW loss was noted in a group of females exposed to 250 mg S./l. No other treatment-related effects were seen. Long-term effects of oral administration were reported. S. was given in olive oil to pregnant female O20 mice (1350 mg/kg BW), C57BL mice (300 mg/kg BW), and BD IV rats (1350 mg/kg

BW) on day 17 of gestation. Their offspring were treated weekly throughout their lifespan. The incidence of tumors occurring at sites other than the lung was higher in the untreated mice than in the S.-treated animals. Very high doses, which caused earlier mortality, were used in this study.[37]

Carcinogenicity classifications. An IARC Working Group concluded that there is limited evidence for the carcinogenicity of S. in *experimental animals* and there were inadequate data available to evaluate the carcinogenicity of S. in *humans*.

IARC: 2B;

U.S. EPA: C;

NTP: N - N - E - N (gavage).

Chemobiokinetics. S. is readily absorbed and distributed in the body tissues, predominantly in fats. Following inhalation exposure, its content in paranephric fatty cells is 10 times higher than in any other tissue.[38] S. is shown to enter the liver through the portal vein. It is believed that S. metabolism occurs under the action of liver monooxidase with participation of cytochrome *P-450*. The monomer is transformed into *styrene oxide*, which is covalently combined to the macromolecules of hepatocytes and then metabolized to several metabolites, the major ones being *mandelic* and *phenylglyoxylic acids*. Both of these metabolites have been detected in the urine of exposed rodents and humans. The nature of the S. metabolites varies according to the species of mammals.[39] Unchanged S. in rats is excreted with the exhaled air.

Regulations.

U.S. FDA (1998) approved the use of S. (1) as a component of paper and paperboard for contact with dry food (for S. and methyl methacrylate copolymers in contact with fatty foods, S. content in the copolymer is limited to 0.5%) in accordance with the conditions prescribed in 21 CFR part 176.180; (2) as a monomer in the manufacture of semirigid and rigid acrylic and modified acrylic plastics intended for use in food-contact articles in accordance with the conditions prescribed in 21 CFR part 177.1010; (3) in adhesives as a component of articles (a monomer) intended for use in packaging, transporting, or holding food in accordance with the conditions prescribed in 21 CFR part 175.105; (4) in the manufacture of resinous and polymeric coatings that may be safely used as a food-contact surface of articles intended for use in producing, manufacturing, packing, transporting, or holding food in accordance with the conditions prescribed in 21 CFR part 175.300; (5) in polyethylene phthalate polymers to be safely used as articles or components of plastics intended for use in contact with food in accordance with the conditions prescribed in 21 CFR part 177.1630; (6) in the manufacture of rubber articles intended for repeated use in producing, manufacturing, packaging, processing, preparing, treating, packing, transporting, or holding food in accordance with the conditions prescribed in 21 CFR part 177.2600; (7) in the manufacture of cross-linked polyester resins used as articles or components of articles intended for repeated contact with food in accordance with the conditions prescribed in 21 CFR part 177.2420; and (8) as polymers which may be safely admixed, alone or in mixture with other permitted polymers, as modifiers in semirigid and rigid vinyl chloride plastic food contact articles prepared from vinyl chloride homopolymers and/or from vinyl chloride copolymers in accordance with the conditions specified in the 21 CFR.

Great Britain (1998). S. is authorized without time limit for use in the production of polymeric materals and articles in contact with food or drink or intended for such contact.

Recommendations.

WHO (1996): Guideline value for drinking water is 0.02 mg/l. The levels of 0.004 to 2.6 mg/l are likely to give rise to consumer complaints of foreign odor and taste.

U.S. EPA (1999). Health Advisory for a longer-term exposure is 7.0 mg/l.

Standards.

U.S. EPA (1999). MCL and MCLG: 0.1 mg/l.

Russia (1995). MAC and PML: 0.1 mg/l organolept., odor).

References:

1. Taradin, Ya. I., in *Protection of Water Reservoirs against Pollution by Industrial Liquid Effluents*, S. N. Cherkinsky, Ed., Medgiz, Moscow, Issue No 3, 1959, 137 (in Russian)

2. Hoshika, Y., Imamura, T., Muto, G., Van Gemert, L. J., Don, J. A., and Walpot, J. I., International comparison of odor threshold values of several odorants in Japan and in the Netherlands, *Environ. Res.*, 61, 78, 1993.
3. Wharton, F. D. and Levinskas, G. I., *Chemistry and Industry*, 11, 470, June 1976.
4. Katayeva, S. Ye., Regulation of the application of polymeric materials in water supply systems used for economic and drinking purposes, *Gig. Sanit.*, 10, 8, 1988 (in Russian).
5. Petrova, L. I., *Investigation of Possible Use of Polystyrene Plastics of Different Composition in Contact with Foodstuffs*, Author's abstract of thesis, Leningrad, 1979, 20 (in Russian).
6. Jickells, S. M., Gancedo, P., Nerin C., Castle, L., and Gilbert, J., Migration of styrene monomer from thermoset polyester cookware into foods during high temperature applications, *Food Addit. Contam.*, 10, 567, 1993.
7. Lickly, T. D., Lehr, K. M., and Welsh, G. C., Migration of styrene from polystyrene foam food-contact articles, *Food Chem. Toxicol.*, 33, 475, 1995.
8. Linssen, J. P. H. and Raityma, J. C. E., Comparison of migration of styrene monomer from high impact polystyrene in oil, in water emulsion and fatty foods, *J. Polymer Engin.*, 15, 133, 1995-1996.
9. Baner, A. L., Franz, R., and Viringer, O., Alternative fatty food simulants for polymer migration testing: experimental confirmation, *J. Polymer Engin.*, 15, 161, 1995 -1996.
10. Ito, S., Hosogai, T., Sakurai, H., Tada, Y., Sugita, T., Ishiwata, H., Takada, M., Determination of volatile substances and leachable components in polystyrene food contact wares, Abstract, *Eisei Shikenjo Hokoku,* 110, 85, 1992 (in Japanese).
11. Tawfik, M. S. and Huyghebaert, A., Polystyrene cups and containers: styrene migration, *Food. Addit. Contam.,* 15, 592, 1998.
12. Wolf, M. A., Rowe, V. K., McCollister, D. D., et al., Toxicological studies of certain alkylated benzenes and benzene, *Am. Med. Ass. Arch. Ind. Health*, 14, 387, 1956.
13. van Apeldoorn, M. E., van der Heijden, C. A., Heijena-Markus, E., et al., *Styrene Criteria Documents,* Air Effects Project No. 668310, National Institute of Public Health and Environmental Protection, The Netherlands, 1985.
14. Hussain, R., Srivastava, S. P., Mushtaq, M., et al., Effect of styrene on levels of serotonin, noradrenaline, dopamine and activity of acetyl cholinesterase and monoamine oxidase in rat brain, *Toxicol. Lett.*, 7, 47, 1980.
15. Srivastava, S. P., Das, M., Mushtaq, M., et al., Metabolism and genotoxicity of styrene, *Adv. Exp. Med. Biol.*, 136A, 1982.
16. Quast, J. F., Humiston, C. G., Kalnins, R. V., et al., *Results of Toxicity Study of Monomeric Styrene Administered to Beagle Dogs by Oral Intubation for 19 Months*, Final report. Dow Chemical Co., Midland, MI, 1979.
17. *Toxicology Study on Styrene Incorporated in Drinking Water of Rats for Two Years in Conjunction with a Three-Generation Reproduction Study, Styrene*, Revised Final Report Weeks 1-105, Litton Bionetics to Chemical Manufacturers Association, vol 1, 1980.
18. Beliles, R. P., Butala, J. H., Stack, C. R., Makris, S., Chronic toxicity and three-generation reproduction study of styrene monomer in the drinking water of rats, *Fundam. Appl. Toxicol.*, 5, 855, 1985.
19. Hemminki, K., Paasivirta, J., Kurnirinne, T., et al., Alkylation products of DNA bases by simple epoxides, *Chem.-Biol. Interact.*, 30, 259, 1980.
20. Chernoff, N., Setzer, R. W., Miller, D. B., Rosen, M. B., and Rogers, J. M., Effects of chemically induced maternal toxicity on prenatal development in the rat, *Teratology*, 42, 651, 1990.
21. Murray, F. J., John, J. A., Balmer, M. F., et al., Teratologic evaluation of styrene given to rats and rabbits by inhalation or by gavage, *Toxicology*, 11, 335, 1978.
22. Ragul'e, N., Problems concerning the embryotropic effects of styrene, *Gig. Sanit.*, 11, 85, 1974 (in Russian).
23. Singh, H., Jabbal, M. S., Ray, A. R., and Vasudevan, P., Effect of anionic polymeric hydrogels on spermatozoa motility, *Biomaterials*, 5, 307, 1984.

24. Neshkov, N. S. and Noskov, A. M., Effects of toxic components of the fiber glass-reinforced plastics on the higher nervous activity and sexual function of males, *Gig. Truda Prof. Zabol.*, 12, 92, 1976 (in Russian).
25. Lindbohm, M.-L., Hemminki, K., Bonhomme, M. G., Anttila, A., Rantala, K., Heikkila, P., and Rosenberg, M. J., Effect of paternal occupational exposure on spontaneous abortions, *Am. J. Ind. Med.*, 81, 1029, 1991.
26. Taskinen, H., Anttila, A., Lindbohm, M.-L., and Sallmen, M., Spontaneous abortions and congenital malformations among the wives of men occupationally exposed to organic solvents, *Scand. J. Work Environ. Health*, 15, 345, 1989.
27. Brown, N. A., Reproductive and developmental toxicity of styrene, *Reprod. Toxicol.*, 5, 3, 1991.
28. Srivastava, S., Seth, P. K., and Srivastava, S. P., Effect of styrene on testicular enzymes of growing rat, *Indian J. Exp. Biol.*, 30, 399, 1992.
29. Srivastava, S., Seth, P. K., and Srivastava, S. P., Effect of styrene administration on rat testis, *Arch. Toxicol.*, 63, 43, 1989.
30. Dogra, R. K., Khanna, S., Srivastava, S. N., Shukla, L. J., Chandra, K., Saxena, G., and Shanker, R., Immunomodulation due to coexposure to styrene and dioctyl phthalate in mice, *Immunopharmacol. Immunotoxicol.*, 15, 491, 1993.
31. *Styrene*, IPCS, Environmental Health Criteria No 26, WHO, Geneva, 1983.
32. Scott, D. and Preston, R. J., A re-evaluation of the cytogenetic effects of styrene, *Mutat. Res.*, 318, 175, 1994.
33. Simula, A. P. and Priestly, B. G., Species differences in the genotoxicity of cyclophosphamide and styrene in three *in vivo* assays, *Mutat. Res.*, 271, 49, 1992.
34. Phillips, D. H. and Farmer, P. B., Evidence for DNA protein binding by styrene and styrene oxide, *CRC Crit. Rev. Toxicol.*, Suppl. 24, 35, 1994.
35. *Bioassay of Styrene for Possible Carcinogenicity*, Natl. Cancer Institute Technical Report Series No 185, NIH Publ. No 79-1741, 1979.
36. Maltoni, C., *Study of the Biological Effects (Carcinogenicity Bioassay) of Styrene*, Report of Bologna Tumour Center Department of Experimental Oncology, 1978.
37. Ponomarkov, V. and Tomatis, L., Effects of long-term oral administration of styrene to mice and rats, *Scand. J. Work Environ. Health*, 4 (Suppl. 2), 127, 1978.
38. Withey, J. R., The toxicology of styrene monomer and its pharmacokinetics and distribution in the rat, *Scand. J. Work Environ. Health*, 4 (Suppl. 2), 31, 1978.
39. Bond, J. A., Review of the toxicology of styrene, *CRC Crit. Rev. Toxicol.*, 19, 227, 1989.

TEREPHTHALIC ACID

Molecular Formula. $C_8H_6O_4$
M = 166.14
CAS No 100-21-0
RTECS No WZ0440000
Abbreviation. TA.

Synonyms and **Trade Name.** 1,4-Benzenedicarboxylic acid; Dicarboxybenzene; 4-Formylbenzoic acid; *p*-Phthalic acid; Tepfthol.

Properties. White powder. Poorly soluble in water. Odor perception threshold is about 100 mg/l, taste per ception threshold is 150 mg/l.[1]

Applications. Used in the production of polyethylene terephthalate.

Migration Data. Levels of migration of polyethylene terephthalate oligomers from polyethylene terephthalate plastics are 0.02 to 2.73 mg/kg depending on the food and temperature attained during cooking.[2]

Acute Toxicity. In mice, LD_{50} is found to be more than 5.0 g/kg BW;[3] in rats, LD_{50} is more than 6.4 g/kg BW.[018] In 20 min after administration of 0.5 and 5.0 g/kg BW to mice there was irritation of the upper respiratory tract and motor stimulation. Lethal dose is not less than 10 g/kg BW. Death occurs as a result of the blood and lymph circulation disorder.[4] Histology examination revealed congestion in the visceral organs and parenchymatous dystrophy in the renal tubular epithelium.

Repeated Exposure. Oral exposure to 2.0 to 4.0 g/kg BW for 10 days did not affect serum cholinesterase activity. Gross pathology examination failed to reveal changes in the visceral organs of rats. Dietary intake of TA may result in bladder stone formation.

According to Chin et al.,[5] for stones to form under the action of TA a critical saturated concentration of TA and calcium in the urine is necessary, stone formation evidently being preceded by the development of hyperplasia of the bladder epithelium.

Long-term Toxicity. In a 6-month study, the doses of 0.125 mg/kg and 15 mg/kg BW produced no changes in conditioned reflex activity as well as in the histological structure of the visceral organs of rats.[1]

Reproductive Toxicity.

Embryotoxicity. ^{14}C-TA has been transported to the fetus following its administration to pregnant rats.[6] Neonatal rats received 5.0% TA in the diet of their dams and did not develop calculi.

TA produced no ***teratogenic effect*** in rats.[7]

Chemobiokinetics. TA is rapidly absorbed and excreted from the body. Accumulation in the tissues seems to be negligible. TA is found in the kidneys of humans and rats. It induced calculi and transitional cell hyperplasia in urinary bladders in rats.[8] Analysis of urine by HPLC method revealed no evidence of ^{14}C-TA metabolism.[6]

Up to 30% of the *i/v* dose administered to chickens is removed unchanged via the urine.[9]

Regulations.

U.S. FDA (1998) approved the use of TA (1) in the manufacture of resinous and polymeric coatings for the food-contact surface of articles intended for use in producing, manufacturing, packing, processing, preparing, treating, packaging, transporting, or holding food, in accordance with the conditions prescribed in 21 CFR part 175.300; (2) in the manufacture of resinous and polymeric coatings for polyolefin films for food-contact surface of articles intended for use in producing, manufacturing, packing, processing, preparing, treating, packaging, transporting, or holding food, in accordance with the conditions prescribed in 21 CFR part 175.320; (3) in the manufacture of cross-linked polyester resins which may be used as articles or components of articles intended for repeated use in contact with food, in accordance with the conditions prescribed in 21 CFR part 177.2420; and (4) as a component of polyethylene phthalate polymers (films, articles, or fabrics) intended for use in contact with food in accordance with the conditions prescribed in 21 CFR part 177.1630.

EU (1990). TA is available in the *List of authorized monomers and other starting substances which shall be used for the manufacture of plastic materials and articles intended to come into contact with food-stuffs (Section A).*

Great Britain (1998). TA is authorized without time limit for use in the production of polymeric materials and articles in contact with food or drink or intended for such contact. The specific migration of this substance shall not exceed 7.5 mg/kg.

Standards.

EU (1990). SML: 7.5 mg/kg.

Russia. PML: 5.0 mg/l.

References:

1. Prusakov, V. M., in *Problems of Environmental Hygiene*, D. N. Kaluzhny, Ed., Kiyv, Issue No 6, 1966, 94 (in Russian).
2. Castle, L., Mayo, A., Crews, C., et al., Migration of poly(ethylene terephthalate) (PET) oligomers from PET plastics into foods during microwave and conventional cooking and into bottled beverages, *J. Food Prot.*, 52, 337, 1989.
3. Hoshi, A. R., Yanai, R., and Kuretani, K., Toxicity of terephthalic acid, *Chem. Pharmacol. Bull.*, 16, 1655, 1968.
4. Sanina, Yu. P., in *Toxicology of New Industrial Chemical Substances*, N. F. Ismerov and I. V. Sanotsky, Eds., Meditsina, Moscow, Issue No 7, 1965, 91 (in Russian).

5. Chin, T. I., Tyl, R. W., Papp, J. A., et al., Chemical urolithiasis. I. Characteristics of bladder stone induction by terephthalic acid and dimethylterephthalate in weanling Fischer 344 rats, *Toxicol. Appl. Pharmacol.*, 58, 307, 1981.
6. Wolkowski-Tyl, R. and Chin, T. Y., Effect of selected therapeutic agents on urothiasis induced by terephthalic acid in the male weanling Fischer 344 rat, *Fundam. Appl Toxicol.*, 3, 552, 1983.
7. Ryan, B. M., Hatoum, N. S., and Jernigan, J. D., A segment II inhalation teratology study of terephthalic acid in rats, *Toxicologist*, 10, 40, 1990.
8. Heck, H. D. and Tyl, R. W., The induction of bladder stones by terephthalic acid, dimethyl terephthalate, and melamine (2,4,6-triamino-*S*-triazine) and its relevance to risk assessment, *Regul. Toxicol. Pharmacol.*, 5, 294, 1985.
9. Tremaine, L. M. and Queblemann, A. J., The renal handling of terephthalic acid, *Toxicol. Appl. Pharmacol.*, 77, 165, 1985.

TETRAFLUOROETHYLENE

Molecular Formula. C_4F_4

M = 100.02

CAS No 116-14-3

RTECS No KX4000000

Abbreviation. TFE.

Synonyms. Perfluoroethene; Perfluoroethylene; Tetrafluoroethene; 1,1,2,2-Tetrafluoroethylene.

Properties. Colorless gas. Poorly soluble in water.

Applications. Used primarily in the synthesis of polytetrafluoroethylene polymers for contact with food.

Acute Toxicity. Only data on inhalation studies are available. Irritation to the respiratory tract and the lungs along with kidney injury were reported.[1]

Short-term Toxicity. Only data on inhalation studies are available. No clinical signs of toxicity were seen other than kidney damage, and to a lesser extent, changes in the lungs, colon, hematopoietic system, and endocrine glands.[2]

Mutagenicity. Based on the structure-mutagenicity relationship, TFE was not supposed to exhibit genetic activity.[3]

Carcinogenicity classifications. An IARC Working Group concluded that there were no adequate data available to evaluate the carcinogenicity of TFE in *experimental animals* and in *humans.*

IARC: 2B;

ARC: 2

Chemobiokinetics. An increase of urinary fluoride level was determined following inhalation of 3,500 ppm for 30 min. It returned to normal when exposure was discontinued.[4] TFE was metabolized to *S-(1,1,2,2- tetrafluoroethyl)-L-cysteine* by rat liver fractions *in vitro*. Compound may undergo conjugation with glutathione followed by degradation of the *S*-conjugate to produce cytotoxicity. Reactions with renal cells and reactive thiols produce cell injury but reaction with DNA is not likely.[2]

Regulations.

U.S. FDA (1998) regulates TFE for use in the production of perfluorocarbon-cured elastomers intended for use as articles or components of articles for repeated use in contact with nonacid food (*pH* above 5.0).

Great Britain (1998). TFE is authorized without time limit for use in the production of polymeric materials and articles in contact with food or drink or intended for such contact. The specific migration of this substance shall not exceed 0.05 mg/kg.

References:

1. Kennedy, G. L., Toxicology of fluorine-containing monomers, *CRC Crit. Rev. Toxicol.*, 21, 149, 1990.
2. Lock, E. A., Studies on the mechanism of nephrotoxicity and prenephrocarcinogenicity of halogenated alkenes, *CRC Crit. Rev. Toxicol.*, 19, 23, 1988.
3. Jones, R. B. and Mackrodt, W. C., Structure-mutagenicity relationships for chlorinated ethylenes: a model based on the stability of the metabolically derived epoxides, *Biochem. Pharmacol.*, 31, 3710, 1982.

4. Ding, R. S. and Kwon, B. K.. The inhalation toxicity of perolysis products of polytetrafluoroethylene heated below 500°C, *Am. Ind. Hyg. Ass. J.*, 29, 19, 1968.

TETRAHYDROFURAN

Molecular Formula. C_4H_8O
M = 72.10
CAS No 109-99-9
RTECS No LU5950000
Abbreviation. THF.

Synonyms and **Trade Names.** Butylene oxide; Cyclotetramethylene oxide; 1,4-Furanidine; Epoxybutane; Oxacyclopentane; Oxolane; Tetramethylene oxide.

Properties. Colorless, mobile liquid with a pungent ether odor. Readily soluble in water. Odor perception thresholds of 3.0 mg/l or 1.0 mg/l[1] were reported. Coloring occurs at a concentration of 100 mg/l.

Applications. THF has wide industrial, research, and consumer applications. It is used to produce polyethers and polytetramethylene oxide and also as an industrial solvent, primarily in the dissolution of plastic resins such as polyvinyl chloride and vinylidene chloride food-contact copolymers.

Acute Toxicity. LD_{50} is 2.8 to 3.0 g/kg BW in rats, 2.5 g/kg BW in mice, and 2.3 g/kg BW in guinea pigs. LD_{50} was lower in 14-day old rats than in the young rats.[1-3] Three to five min. after poisoning animals became motionless and experienced labored breathing, hyperemia, and cyanosis of the skin. Gross pathology examination revealed severe congestion in the visceral organs as well as inflammation, necrosis, and hemorrhage of the GI tract, injury to the kidney tubules, inflammation of the liver.[2] Hepatocellular dysfunction induced by acute poisoning at high doses is reported to be a primary sign of intoxication.[4]

Repeated Exposure revealed evidence of functional accumulation when THF was inhaled. Mice seem to be more sensitive to THF than rats.

Long-term Toxicity. In a 6-month study, a dose of 5.0 mg/kg BW was found to be the NOAEL.[5]

Immunotoxic effects were not observed in tests on guinea pigs due to the impossibility of THF to combine with proteins *in vivo*.[6]

Reproductive Toxicity. Rats were exposed to up to 5,000 ppm on days 6 to 9 of gestation, and mice inhaled THF on days 6 to 17. The NOAEL for maternal toxicity appeared to be 1800 ppm for both rats and mice; for developmental toxicity it appeared to be 1800 ppm for rats, and 600 ppm for mice.[7]

Mutagenicity. THF was negative in *Dr. melanogaster*[8,9] and in *Salmonella*/microsome assay.[10]

Chemobiokinetics. THF is widely distributed in fatty tissues and can persist for days to a week after inhalation exposure in laboratory animals.[11] THF is metabolized by mammalian hepatic enzymes to *oxygen* followed by cleavage to a *straight-chain fatty acid*.[12,13] THF is an inhibitor of a number of cytochrome *P-450*-dependent mixed function oxidase activities. It enhances the toxic action of a number of compounds by stimulation of the more rapid absorption of reactive metabolites.[4]

Regulations.

U.S. FDA (1998) approved THF for use (1) as a component of adhesives intended for use in food-contact articles, subject to the provisions prescribed in 21 CFR part 175.105. THF may be safely used in the fabrication of articles intended for packing, transporting, or storing foods; (2) in the manufacture of resinous and polymeric coatings for polyolefin films for food-contact surface of articles intended for use in producing, manufacturing, packing, processing, preparing, treating, packaging, transporting, or holding food, subject to the provisions prescribed in 21 CFR part 175.320; as (3) a slimicide in the manufacture of paper and paperboard that contact food, subject to the provisions prescribed in 21 CFR part 176.300; and (4) as a solvent in the casting of film from a solution of polymeric resins of vinyl chloride, vinyl acetate, vinylidene chloride, or polyvinylchloride copolymers. The residual amount of THF in the film should not exceed 1.5% by weight. THF may be safely used (5) in the manufacture of cellophane for packaging food (residue limit 0.1%), subject to the provisions prescribed in 21 CFR part 177.1200.

EU (1990). THF is available in the *List of authorized monomers and other starting substances which shall be used for the manufacture of plastic materials and articles intended to come into contact with foodstuffs (Section A).*

Great Britain (1998). THF is authorized without time limit for use in the production of polymeric materials and articles in contact with food or drink or intended for such contact. The specific migration of this substance shall not exceed 0.6 mg/kg.

Standards.

EU (1990). SML in food: 0.6 mg/kg.

Russia (1995). MAC: 0.5 mg/l.

References:

1. Pozdnyakova, A. G., in *Hygiene Problems of Populated Locations*, Proc. Leningrad Sanitary Hygiene Research Institute, Leningrad, Issue No 81, 1965, 91 (in Russian).
2. Sax, N. Y., *Hazardous Chemicals, Information Annual,* Van Nostrand Reinhold Information Service, NY, 1986, 1, 640.
3. Kimura, E. T., Ebert, D. M., and Dodge, P. W., Acute toxicity and limits of solvent residue for sixteen organic solvents, *Toxicol. Appl. Pharmacol.*, 19, 699, 1971.
4. Moody, D. E., The effect of tetrahydrofuran on biological systems: does a hepatotoxic potential exist? *Drug Chem. Toxicol.*, 14, 319, 1991.
5. Teplyakova, E. V. et al., in *Proc. Leningrad Sanitary Hygiene Research Institute*, Leningrad, Issue No 105, 1974, 51 (in Russian).
6. Chernousov, A. D., On allergenic properties of furan compounds, *Gig. Sanit.*, 6, 28, 1974 (in Russian).
7. Mast, T. J., Weigel, R. J., Bruce, R., et al., Evaluation of the potential developmental toxicity in rats and mice following inhalation exposure to tetrahydrofuran, *Fundam. Appl. Toxicol.*, 18, 255, 1992.
8. Valencia, R., Mason, J. M., Woodraff, R. C., et al., Chemical mutagenesis testing in *Drosophila, Environ. Mutagen.*, 7, 325, 1985.
9. Vogel, E. W. and Nivard, N. J., Performance of 181 chemicals in a *Drosophila* assay predominantly monitoring interchromosomal mitotic recombination, *Mutagenesis*, 8, 57, 1993.
10. Maron, D., Katzenellenbogen, J., and Ames, B. N., Compatibility of organic solvents with the *Salmonella*/microsome test, *Mutat. Res.*, 88, 343, 1981.
11. Elovara, E. et al., Burden and biochemical effects of extended tetrahydrofuran vapor inhalation of three concentration levels, *Acta Pharmacol. Toxicol.* (Copenh.), 54, 221, 1984.
12. Fujita, T. and Suzuki, Z., Enzyme studies on the metabolism of the tetrahydrofurfuryl mercaptan moiety of thiamin tetrahydrofurfuryl disulfide, *J. Biochem.*, 74, 733, 1973.
13. Stasenkova, K. P. and Kochetkova, T. A., Comparative toxicity evaluation of a series of furan compounds, in *Toxicology of New Industrial Chemical Substances*, A. A. Letavet, Ed., Medgiz, Moscow, Issue No 5, 1963, 21 (in Russian).

2,4- and 2,6-TOLUENE DIISOCYANATES

Molecular Formula. $C_9H_6N_2O_2$

M = 174.15

CAS No 26471-62-5

RTECS No NQ949000

Abbreviation. TDI.

2,4-Toluene diisocyanate

CAS No 584-84-9

RTECS No CZ6300000

Synonyms and **Trade Names.** 2,4-Diisocyanato-1-methylbenzene; Desmodur T100; Diisocyantomethylbenzene; Diisocyanatotoluene; Hylene-T; Isocyanic acid, methyl-*m*-phenylene ester; Methyl-*m*-phenylene diisocyanate; Methylphenylene isocyanate; Mondur-TD; Nacconate-100; Rubinate TDI.

2,6-Toluene diisocyanate

CAS No 91-08-7

RTECS No CZ6310000

Synonyms. 1,3-Diisocyanato-2-methylbenzene; 2,6-Diisocyanato-1-methylbenzene; 2,6-Diisocyanato-toluene; 2-Methyl-*m*-phenylene diisocyanate; 2-Methyl-*m*-phenylene isocyanate; Isocyanic acid, 2-methyl-*m*-phenylene ester; *m*-Tolylene diisocyanate.

Properties. Transparent, colorless or slightly yellow liquids with a sharp, pungent odor. Darken on exposure to light; are rapidly polymerized by basis. In water, 92% TDI hydrolyzes with the formation of diaminotoluenes. Soluble in diethyl ether, acetone and other organic solvents. Miscible with olive oil.

Applications. TDI are used in the production of rigid polyurethane foam, urethane rubbers, and various food-contact polymeric materials. 2.6-TDI can be used as a component of coatings and elastomer systems.

Migration Data. A study of extracts (a pre-polymer based on liquid SKD-PG rubber and *2,4-TDI*-3,5; cocatalyst: triethylenediamine) was carried out (20 and 37°C; 1 to 3 days; 4.1 and 0.05 cm^{-1}). Migration of TDI from unwashed specimens into water amounted to 0.01 to 0.02 mg/l. After specimens had been aired, the release of TDI decreased to trace amounts, and in extracts obtained with a specific surface of 0.05 cm^{-1} this substance was not found. With 0.05 cm^{-1} the migration does not occur. An adverse effect of SKU-7L elastomer on water quality was observed, especially at 85°C. TDI content amounted to 0.005 to 0.01 mg/l).[1] TDI was noted to migrate from polyurethane adhesive *Styk* in contact water at 80°C (0.55 mg/l).[2]

Acute Toxicity. LD_{50} is 5.5 to 7.5 g/kg BW in rats, and 4.7 g/kg BW in mice.[3] Poisoning is accompanied by a short-term agitation with subsequent adynamia and respiratory disorder. Death within 3 to 4 days.[4]

Repeated Exposure revealed pronounced cumulative properties.[3] K_{acc} is 2.87 (by Kagan). Rats were administered the dose of 1.5 g/kg BW for 10 days. Damage to the liver and GI tract was observed.

Short-term Toxicity. Rats were exposed to the oral doses of up to 240 mg TDI/kg BW by gavage for a 13-week period. Mild to moderate bronchopneumonia was reported to develop.[5] Administration of 1/20 LD_{50} (247 mg/kg BW) to rats for 4 months caused dysbalance in protein fractions of the blood serum and deviations in the leukocyte count. Erythrocyte count and *Hb* level in the blood were decreased.[4]

Long-term Toxicity. F344 rats and B6C3F$_1$ mice were dosed with a mixture of TDI in corn oil for 105 or 106 weeks. Female rats and mice were given 60 or 120 mg/kg BW; male rats received 30 or 60 mg/kg BW and male mice received 120 or 240 mg/kg BW. Depression of BW gain was found in rats at all doses tested. Dose-related increase in acute bronchopneumonia was observed. In mice, there was cytomegaly of the kidney tubular epithelium in males. TDI elicited cumulative toxic response. A decrease in survival in rats and lower survival in high dose male mice were noted.

Immunotoxicity. TDI produces sensitization in animals and humans.

Observations in man. Frequent systemic hypersensitivity reactions to diisocyanates, including asthma, pneumonia, interstitial fibrosis, were reported (risk 1:100).

Animal studies. A systemic immune response to TDI conjugated with the dog serum albumin was observed in dogs exposed to 1.0 mg/kg BW. TDI was delivered as an aerosol intratracheally once every second week for 41 weeks.[6] Antibodies to TDI were produced in guinea pigs exposed by inhalation, dermally or by *i/p* injections.[7]

Mutagenicity.

In vivo cytogenetics. TDI induced sex-linked recessive lethal mutations and reciprocal translocations in germ cells of *Dr. melanogaster*.[8]

In vitro genotoxicity. TDI is reported to show equivocal results in *Salmonella* mutagenicity bioassay, it was found to be positive in the mouse lymphoma assay, and caused CA and SCE.[7,9,016]

Carcinogenicity. Rats received the dietary doses of 30 and 60 mg/kg (males), 60 and 120 mg/kg (females) for 2 years. The treatment produced a dose-dependent increase in the rate of subcutaneous fibromas, fibrosarcomas and adenomas of the pancreas in males. In females there were nodular neoplasms in the liver and fibroadenomas in the mammary glands. Mice were given 60 and 120 mg/kg (females), 120 and 240 mg/kg (males) in the diet for 105 weeks. The treatment caused hemangiomas, hemangiosarcomas and hepatic cellular adenomas to develop in the females. No growth of tumors was observed in males.[5,10,11] Inhalation of TDI is not reported to cause tumors.

Carcinogenicity classifications. An IARC Working Group concluded that there is sufficient evidence for the carcinogenicity of TDI in *experimental animals* and there were no data available to evaluate the carcinogenicity of TDI in *humans*.

The Annual Report of Carcinogens issued by the U.S. Department of Health and Human Services (1998) defines TDI to be a substance which may reasonably be anticipated to be carcinogen, i.e., a substance for which there is a limited evidence of carcinogenicity in *humans* or sufficient evidence of carcinogenicity in *experimental animals.*

IARC: 2B;

ARC: 2;

NTP: P - P* - N - P (gavage).

Chemobiokinetics. It is suggested that following exposure to TDI the chemical would hydrolyze at aqueous tissue surfaces to give rise to *toluenediamine* (TDA), a mutagen and rodent carcinogen. Toxicokinetic studies revealed significant levels of TDA following oral dosing with TDI due to the acidic stomach environment.[12] In the above described study,[11] main accumulation of TDI was observed in the stomach, caecum, large intestine, bladder, and urine. Major metabolites of 2,4-TDI were found to be identical with those from the metabolism of the carcinogen *2,4-diaminotoluene* whereas the major metabolite of 2,6-TDI was identified as *2,6-bis(acetylamino) toluene.*

Rats received a single oral dose of 60 mg ^{14}C-labeled 2,4-TDI/kg BW. More than 93% of radioactivity was recovered in the urine (about 8.0%), feces (81%), cage wash, and tissues. 16 to 39% of the quantitated urinary metabolites exist as acid-labile conjugates.[13]

On ingestion in corn oil by rats, 2,6-TDI formed polymers in the GI tract. Excretion occurs predominantly via the feces and urine.[10]

Regulations.

U.S. FDA (1998) has determined that 2,4- and 2,6-TDI isomers may be used (1) as components of adhesives that come in contact with food in accordance with the conditions prescribed in 21 CFR part 175.105; and (2) in the manufacture of polyurethane resins which may be used as a food-contact surface of articles intended for use in contact with bulk quantities of dry food of an appropriate type in accordance with the conditions prescribed in 21 CFR part 177.1680.

EU (1990). 2,4- and 2,6-TDI are available in the *List of authorized monomers and other starting substances which shall be used for the manufacture of plastic materials and articles intended to come into contact with foodstuffs (Section A).*

Great Britain (1998). TDI are authorized without time limit for use in the production of polymeric materials and articles in contact with food or drink or intended for such contact. The quantity in the finished polymeric material or article of any substance within, or any combination of substances within shall not exceed 1.0 mg/kg (expressed as isocyanate moiety).

Standards. ***EU*** (1990). MPQ in finished material or article: 1.0 mg/kg (expressed as isocyanate moiety).

References:

1. Krat, A.V., Kesel'man, I. M., and Sheftel', V. O., Sanitary-chemical evaluation of polymeric articles in water-supply constructions, *Gig. Sanit.*, 10, 18, 1986 (in Russian).
2. Kupyrov, V. N., Kaplina, T. V., Gakal, R. K., Vinarskaya, E. I., and Starchenko, S. N., Hygienic evaluation of films intended for the waterproofing of unit prefabricated swimming pools, *Gig. Sanit.*, 5, 91, 1978 (in Russian).
3. Zapp, J. A., Hazards of isocyanates in polyurethane foam plastic production, *Arch. Ind. Health*, 15, 324, 1957.
4. Shuba, P. A., in *Safe Use and Toxicology of Pesticides and Polymeric Materials*, Proc. All-Union Research Institute of Hygiene and Toxicology of Pesticides, Polymers and Plastic Materials, Kiyv, Issue No 14, 1984, 122 (in Russian).
5. *Carcinogenesis Studies of Commercial Grade 2,4(86%)- and 2,6(14%)-Toluene Diisocyanate in F344/N Rats and B6C3F$_1$ Mice (Gavage Studies)*, NTP Technical Report Series No 251, Research Triangle Park, NC, 1983.
6. Patterson, R., Zeiss, C. R., and Harris, K. E., Immunological and respiratory responses to airway challenges of dogs with toluene diisocyanate, *J. Allergy Clin. Immunol.*, 71, 604, 1983.

7. Karol, M. H., Study of guinea pig and human antibodies to toluene diisocyanate, *Am. Resp. Dis.*, 122, 965, 1980.
8. Foureman, P., Mason, J. M., Valencia, R., and Zimmering, S., Chemical mutagenesis testing in *Drosophila*, X. Results of 70 coded chemicals tested for the National Toxicology Program, *Environ. Molec. Mutagen.*, 23, 208, 1994.
9. Anderson, M. and Styles, J. A., Appendix II, The bacterial mutation test, *Brit. J. Cancer*, 37, 924, 1978.
10. *Toxicology and Carcinogenesis Studies of Commercial Grade 2,4(80%)- and 2,6(20%)-Toluene Diisocyanate in F344/N Rats and B6C3F$_1$ Mice (Gavage Studies)*, NTP Technical Report Series No 251, NIH Publ. No 86-2507, Research Triangle Park, NC, 1986.
11. Dieter, M. P., Boorman, G. A., Jameson, C. W., Matthews, H. B., and Huff, J. E., The carcinogenic activity of commercial grade toluene diisocyanate in rats and mice in relation to the metabolism of the 2,4- and 2,6-TDI isomers, *Toxicol. Ind. Health*, 6, 599, 1990.
12. Doe, J. E. and Hoffman, H. D., Toluene diisocyanate: an assessment of carcinogenic risk following oral and inhalation exposure, *Toxicol. Ind. Health*, 11, 13, 1995.
13. Timmchalk, C., Smith, F. A., and Bartels, M. J., Route-dependent comparative metabolism of ^{14}C-toluene 2,4-diisocyanate and ^{14}C-toluene 2,4-diamine in Fischer 344 rats, *Toxicol. Appl. Pharmacol.*, 124, 181, 1994.

1,1,2-TRICHLORO-1,3-BUTADIENE

Molecular Formula. $C_4H_3Cl_3$
M = 157.42
CAS No 2852-07-5
RTECS No EJ1070000

Properties. Colorless liquid with a specific unpleasant odor. Water solubility is 220 mg/l at 25°C.

Applications. Used as a monomer in the production of polytrichlorobutadiene, in adhesive compositions.

Acute Toxicity. LD_{50} is 680 mg/kg BW in rats, and 1000 mg/kg BW in mice. Poisoning is accompanied by convulsions and labored breathing. A narcotic effect was marked.

Repeated Exposure failed to reveal any evidence of accumulation.

Allergenic Effect was not observed in a guinea pig study.

Reproductive Toxicity. Embryotoxic and gonadotoxic effects were revealed in the inhalation exposure studies.

Reference:

Bal'ian, V. V., Kazarian, A. S., and Gizhlaryan, M. S., Establishment of the embryotoxic and gonadotoxic threshold of 1,1,2-trichlorobutadiene-1,3 and 1,1,2,3-tetrachlorobutadiene-1,3 in rats, *J. Exp. Clin. Med.*, Yerevan, 19, 60, 1979 (in Russian).

UREA

Molecular Formula. CH_4N_2O
M = 60.07
CAS No 57-13-6
RTECS No YR6250000

Synonyms and **Trade Names.** Carbamide; Carbonic acid, diamide; Isourea; Pseudourea.

Properties. Colorless, odorless, crystalline prisms or granules. Readily soluble in water and alcohol. Undergoes hydrolysis when heated in aqueous solutions of acids or alkalis. U. does not render to water a considerable odor and does not alter its color, but does give it an astringent taste.[1] However, according to other data, taste and odor perception threshold is 80 mg/l.[2]

Applications. U. is used in the manufacture of urea-formaldehyde food-contact polymers and plastics, adhesives, synthetic fibers, dyes, etc.

Exposure. U. is a naturally occurring constituent of the body.

Acute Toxicity. Rats tolerate doses of up to 12 g/kg BW. LD_{50} is found to be 14.3 g/kg BW or 16.3 g/kg BW in rats, and 11.0 g/kg BW in mice.[1] In rabbits, LD_{50} was not attained.

Repeated Exposure revealed slight cumulative properties. Rats developed U. blood levels up to 40 to 45 mg% after dietary intake of a 2.0 g/kg BW dose for a month. A dose of 50 mg/kg BW did not alter the balance of U.[1,2]

Long-term Toxicity. High doses are reported to affect bioenergetic and cholinergic processes and cause changes in the CNS and visceral structure. The concentration of 10 mg/l in drinking water was found to be safe.[3]

Reproductive Toxicity.

Observations in man. Effect on fertility in women is reported.

Embryotoxicity and ***Teratogenicity.***

Animal studies. Administration of 2.0 g/kg BW to rats on day 12, and to mice on day 10 of gestation caused no increase in the defect rates or resorptions.[4]

Mutagenicity.

In vitro genotoxicity. U. could produce DNA damage in the culture of human leukocytes. It was found to be negative in *Salmonella* mutagenicity assay, but gave positive/negative results in a CA test carried out on cultured Chinese hamster cells.[5]

Carcinogenicity. Negative results are reported in C57B1/6 mice given 674 to 6750 mg U./kg BW and in F344 rats receiving 225 to 2250 mg U./kg BW in their diet for a year.[6]

Chemobiokinetics. U. is a nitrogenous compound that serves as a major excretory product for protein-derived wastes in humans. 92% of administered ^{15}N-labeled U. was excreted in the urine of young pigs after 48 hours and 1.9% during subsequent 48 hours.[7] U. is an excretory end-product of amino acid metabolism in mammals. The formation of U. occurs in the liver. Average daily urinary excretion of U. in adults is likely to be about 20.6 g.

Regulations.

EU (1990). U. is available in the *List of authorized monomers and other starting substances which shall be used for the manufacture of plastic materials and articles intended to come into contact with foodstuffs (Section A)*.

U.S. FDA (1998) approved the use of U. (1) in the manufacture of resinous and polymeric coatings to be safely used in food-contact surface in accordance with the conditions prescribed in 21 CFR part 175.300; and (2) in the manufacture of cellophane for packaging food in accordance with the conditions prescribed in 21 CFR part 177.1200. Affirmed as *GRAS.*

Great Britain (1998). U. is authorized without time limit for use in the production of polymeric materials and articles in contact with food or drink or intended for such contact.

Recommendations. Joint FAO/WHO Expert Committee on Food Additives (1993). Use at levels of up to 3.0% in chewing gum is not of toxicological concern.

Standards. ***Russia*** (1988). Recommended MAC: 10 mg/l.[2]

References:

1. Mazaev, A. P. and Skachkova, I. V., Substantiation of the conditions of draining for urea-containing industrial sewage into water bodies, *Gig. Sanit.*, 10, 7, 1966 (in Russian).
2. Kotova, N. I., Data on substantiation of MAC of carbamate in the work-place air, *Gig. Truda Prof. Zabol.*, 3, 43, 1986 (in Russian).
3. Mironets, N. V., Savina, R. V., Kucherov, I. S., et al., MAC of urea in reclaimed potable water and its biological action, *Kosmicheskaya Biologia i Aviakosmicheskaya Medicina*, 22, 63, 1988 (in Russian).
4. Teramoto, S., Kaneda, M., Aoyama, H., et al., Correlation between the molecular structure of *N*-alkylureas and *N*-alkylthioureas and their teratogenic properties, *Teratology*, 23, 335, 1981.
5. Ishidate, M. and Odashima, S., Chromosome tests with 134 compounds in Chinese hamster cells *in vitro* - a screening for chemical carcinogens, *Mutat. Res.*, 48, 337, 1977.

6. Fleischman, R. W., Hayden, D. W., Smith, E. R., et al., Carcinogenesis bioassay of acetamide, hexanamide, adipamide, urea and *p*-toluurea in mice and rats, *J. Environ. Pathol. Toxicol.*, 1, 149, 1980.
7. Grimson, R. E., Bowland, J. P., and Milligan, L. P., Use of nitrogen-15 labeled urea utilization by pigs, *Canad. J. Anim. Sci.*, 51, 103, 1971.

VINYL ACETATE

Molecular Formula. $C_4H_6O_2$
M = 86.09
CAS No 108-05-4
RTECS No AK0875000
Abbreviation. VA.

Synonyms. Acetic acid, ethenyl ester; Acetic acid, ethylene ether; Acetic acid, vinyl ester; 1-Acetooxy-ethylene; Vinyl ethanoate.

Properties. Colorless, transparent liquid with a pungent odor. Water solubility is 2.5% at 20°C. Soluble in ethanol, ether, acetone, chloroform, and carbon tetrachloride. Odor threshold concentrations are 0.25 mg/l,[1] 0.12 mg/l,[010] or 0.088 mg/l.[02] Solutions with a concentration of up to 50 mg/l do not alter the *pH*.

Applications. Used for the production of polyvinyl acetate homopolymer emulsions and resins, polyvinyl alcohol, used as a copolymer with ethylene in adhesives, and in paper and paper board coatings intended for use in contact with food.

Acute Toxicity. LD_{50} is 1.6 g/kg BW in mice,[1] and 2.9 g/kg BW in rats.[2]

Repeated Exposure revealed slight cumulative effect when mice were administered 0.3 g/kg BW for 3 weeks.

Short-term Toxicity. Rats were given VA in the drinking water for up to 3 months at initial concentrations of 200, 1,000, and 5,000 ppm (Gale, 1980). Water consumption was reduced, retardation of BW gain was observed in animals given 5,000 ppm dose, but there were no other signs of toxicity.[3]

Long-term Toxicity. VA was given to Crl:CD(SD)BR rats in drinking water from the time of gestation. Target concentrations were 200, 1,000, and 5,000 ppm. F_0 rats were given VA for 10 weeks and then mated. Offspring was treated for up to 104 weeks. There was no evidence of systemic target organ toxicity. The treatment caused a decrease in water consumption (1,000 and 5,000 ppm groups), a decrease in food consumption, and retardation of BW gain (5,000 ppm group). There were no effects on clinical chemical, hematological, or urinalyses parameters. The pathology evaluation revealed no compound-related effects on organ weight, non-neoplastic lesions, or neoplastic lesions. The NOEL was found to be 200 ppm in this study.[3]

Reproductive Toxicity.

Gonadotoxicity. Administration of 500 mg VA/kg BW significantly increased the frequency of sperm abnormalities, while lower doses did not. These doses did not influence the frequency of meiotic micronuclei in mice. VA caused a dose-dependent decrease in sperm production and a reduction of testicular weight at 500 and 125 mg/kg BW.[4]

Embryotoxicity. Maternal toxicity and retardation in embryonic BW gain were observed in rats treated with high oral doses of VA.[5]

Teratogenicity. Exposure to 1,000 ppm on days 6 to 15 of gestation resulted in minor skeletal defects and smaller weights of rat fetuses. No adverse fetal effects were observed in the offspring of rats receiving 5000 ppm VA in their drinking water. Crl:CD(SD)BR rats received 200, 1,000, or 5,000 ppm VA via their drinking water over 2 generations. There were no evidence of maternal or developmental toxicity. The NOAEL of 1000 ppm VA, and even greater than 5,000 ppm VA, was reported by the executors of this study.[6,7]

Mutagenicity.

In vivo cytogenetics. According to Gale (1980), VA was shown to be negative by the oral route of administration, especially when tested at non-lethal levels.[11]

In vitro genotoxicity. VA effectively induces CA in mammalian cells, as well as micronuclei in bone marrow erythrocytes and DNA-DNA cross-links.[4,8,9] It was tested for ability to induce SCE in cultured (72 hours) human lymphocytes with a 48-hour treatment, starting at 24 hours after culture initiations. VA induced a clear dose-dependent increase in the number of SCE. The product of transformation of VA, *acetaldehyde* (*q.v.*), leads to the same cytogenic effect as VA in a culture of lymphocytes. It is likely to be a metabolite resonsible for the SCE induction.[10] VA was negative in *Salmonella* mutagenicity assay.

Carcinogenicity. F344 rats received 1.0 and 2.5 g/kg BW with their drinking water for 2 years. The treatment resulted in increased animal mortality. Tumor types were similar in the control and test animals. Thyroid adenoma and uterine carcinoma were found with the higher dose.[11] However, significant decomposition of VA in the drinking water made the data difficult to interpret. IARC does not consider these results to be completely conclusive. In the recent study, VA was not shown to be oncogenic when administered to rats in their drinking water.[3]

Carcinogenicity classification. An IARC Working Group concluded that there is inadequate evidence for the carcinogenicity of VA in *experimental animals* and there were no adequate data available to evaluate the carcinogenicity of VA in *humans*.

IARC: 2B.

Chemobiokinetics. After oral administration or inhalation, VA was incorporated metabolically into hepatic DNA; specific alkylation products were not detected.[12] VA rapidly breaks down in the body to form *acetic acid* and *vinyl alcohol*. It is assumed also to be metabolized (hydrolyzed) in mammalian cells via conversion to *acetal dehyde* and *acetic acid*. After VA inhalation rats exhaled *acetaldehyde*.

Regulations.

U.S. FDA (1998) approved the use of *VA-monomer* as a component of surfaces in contact with food, including (1) adhesives used as components of articles intended for use in packaging, transporting, or holding food, in accordance with the provisions prescribed in 21 CFR part 175.105; (2) in the manufacture of resinous and polymeric coatings in a food-contact surface of articles intended for use in producing, manufacturing, packing, transporting, or holding food, in accordance with the provisions prescribed in 21 CFR part 175.300; (3) as a component of the uncoated or coated food-contact surface of paper and paperboard intended for use in producing, manufacturing, packaging, processing, preparing, treating, packing, transporting, or holding dry food dry food of the type identified in 21 CFR part 176.170 (c); (4) in the production of cellophane to be safely used for packaging food, in accordance with the provisions prescribed in 21 CFR part 177.1200; and (5) as substances employed in the production of or added to textiles and textile fibers, in accordance with the provisions prescribed in 21 CFR part 177.2800. Subsequently, polyvinylacetate has been approved for use as a component of surfaces in contact with food, including adhesives, resinous and polymeric coatings, components of paper or paperboard in contact with dry, fatty or aqueous foods, constituents of cellophane, and textiles and textile fibers. *VA-Vinyl chloride* copolymer resins are regulated for use as components of resinous and modified acrylic plastics. *Copolymers of ethylene - VA - Vinyl alcohol* are approved for use in articles in contact with food, with limitations on thickness based on polymer composition. *Ethylene - VA copolymers* may also be used in this way. *VA - Crotonic acid copolymer*, the surface of polyolefin film that is in contact with food, may be used as a coating.

Great Britain (1998). VA is authorized without time limit for use in the production of polymeric materials and articles in contact with food or drink or intended for such contact. The specific migration of this substance shall not exceed 12 mg/kg.

Standards.

EU (1990). SML: 12 mg/kg.

Russia (1995). MAC and PML: 0.2 mg/l.

References:

1. Goeva, O. E., in *Proc. Leningrad Sanitary Hygiene Medical Institute*, Leningrad, Issue No 81, 1965, 64 (in Russian).
2. Dernehl, C. U., Clinical experiences with exposures to ethylene amines, *Ind. Med. Surg.*, 20, 541, 1951.
3. Bogdanffy, M. S., Tyler, M. B., Vinegar, M. B., et al., Chronic toxicity and oncogenicity study with vinyl acetate in the rat: *in utero* exposure in drinking water, *Fundam. Appl. Toxicol.*, 23, 206, 1994.
4. Lahdetie, J., Effect of vinyl acetate and acetaldehyde on sperm morphology and meiotic micronuclei in mice, *Mutat. Res.* 202, 171, 1988.

5. Clarry, J. J., Chronic and reproduction toxicological studies on vinyl acetate, Status report, *Ann. N.Y. Acad. Sci.*, 534, 255, 1988.
6. Mebus, C. A., Carpanini, F. M. B., Rickard, R. W., Cascieri, T. C., and Vinegar, M. V., A two-generation reproduction study in rats receiving drinking water containing vinyl acetate, *Fundam. Appl. Toxicol.*, 24, 206, 1995.
7. Hurtt, M. E., Vinegar, M. B., Rickard, R. W., Cascieri, T. C., and Tyler, T. R., Developmental toxicity of oral and inhaled vinyl acetate in the rat, *Fundam. Appl. Toxicol.*, 24, 198, 1994.
8. Norppa, H., Yursi, F., Pfaffi, P., et al., Chromosome damage induced by vinyl acetate through *in vitro* formation of acetaldehyde in human lymphocytes and Chinese hamster ovary cells, *Cancer Res.*, 45, 4816, 1985.
9. Maki-Paakkanen, J. and Norppa, A., Induction of micronuclei by vinyl acetate in mouse bone marrow cells and cultured human lymphocytes, *Mutat. Res.*, 190, 41, 1987.
10. Siri, P., Jarventaus, H., and Norppa, H., Sister chromatid exchanges induced by vinyl esters and respective carboxylic acids in cultured human lymphocytes, *Mutat. Res.*, 279, 75, 1992.
11. Lijinsky, W. and Reuber, M. D., Chronic toxicity studies of vinyl acetate in Fischer rats, *Toxicol. Appl. Pharmacol.*, 68, 43, 1983.
12. Simon, P., Fisher, J. G., and Bolt, H. M., Metabolism and pharmacokinetics of vinyl acetate, *Arch. Toxicol.*, 57, 19, 1985.

VINYL CHLORIDE

Molecular Formula. C_2H_3Cl
M = 62.49
CAS No 75-01-4
RTECS No KU9625000
Abbreviations. VCM (monomer); PVC (polymer).

Synonyms. Chloroethene; Chloroethylene; Monochloroethylene; VC-monomer.

Properties. Colorless gas with a chloroform odor. Solubility in water is 1.1 g/l at 25°C. Odor perception threshold is reported to be 2.0 mg/l or 3.4 mg/l.[021]

Applications. A monomer in the production of PVC and VCM copolymers intended for use in contact with food.

Exposure. Residual VCM levels in food and drinks are now estimated to be well below 0.01 mg/kg. Maximum exposures from food and drinks are estimated to be less than 0.0001 mg/day. According to FDA data,[1] a total maximum daily intake from oil, liquor, and wine bottles, food packaged in PVC, or VCM-VDC copolymer films, and other sources is 0.025 µg. People who have PVC water pipes that have not been treated adequately to remove VCM may ingest 0.06 to 2.8 µg VCM/day from drinking water (ATSDR, 1988). Nevertheless, the average daily intake of VCM from the diet is predicted to be essentially zero.[016]

Migration Data. Residual VCM levels ranging from 0.3 to 913 ppb were determined in polymer samples taken from tubing, blood bags, food packaging films, bottles, and unprocessed resin. VCM was confirmed by GSC-mass spectrometry, using selecting ion recording with full mass scans to identify components eluting near VCM.[2] There are indications of possible migration of VCM from PVC. The maximum amount of VCM is released from rigid PVC into fat-containing products and alcoholic drinks (VCM content in PVC < 30 mg/kg): 0.21 mg/l in dry Martini, 012 mg/l in gin, 0.25 mg/l in Sherry, 0.19 mg/l in whisky.[3] The authors do not consider given concentrations to be health hazards. Migration of VCM into water amounts to 0.01 to 0.2 mg/l (exposure from 1 week to 12 months). With VCM content of the polymer of 10 mg/kg, 0.03 mg/l of the monomer was released into water from PVC bottles within 3 months, but with VCM content of the polymer of 1.0 mg/kg no monomer was released into water.[4,5]

Ando and Sayato point out that with time there is a greater probability of VCM reacting with chlorine dissolved in water and its transformation into chloroacetic anhydride, chloroacetic acid, etc. There is no migration of VCM into water when its content in PVC is 2.0 µg/kg.[6]

There is no evidence that VCM can migrate from PVC into food or water in a quantity that can be a health hazard and stay in food or water for a long enough time to cause undesirable effects. In contrast to workplace conditions where VCM inhaled with the air can accumulate in the blood and form metabolites with a carcinogenic and mutagenic effect, such accumulation is impossible with occasional entry of VCM into human body in trace amounts with water or food.[7]

VCM tetramer migration has been studied as a representative oligomer that has the potential for migration from PVC packaging. Tetramer levels in PVC bottles for retail beverages ranged from 70 to 190 mg/kg. No tetramer migration was observed into the following simulant media: distilled water, 3.0% acetic acid, 15% ethanol and olive oil (detection limit of 5.0 to 10 μg/kg).[8]

Acute Toxicity. In rats, LD_{50} is reported to be 500 mg/kg BW (Dow Chemical Co.).

Repeated Exposure. Rats were orally exposed to 300 mg/kg BW dose for 4 weeks. Increased liver and kidney weights as well as histopathological changes in these organs were found (Lefaux, 1975).

Short-term Toxicity. Rats were administered 100 and 300 mg/kg BW in solution of soybean oil for 13 weeks. Hematological indices and clinical chemistry were affected. Gross pathology examination revealed elevated organ weights but failed to detect morphological liver changes.[9] The dose of 30 mg/kg BW in a 3-month study appeared to be ineffective in Wistar rats (Lefaux, 1975).

Long-term Toxicity. Wistar rats received VCM in their feed and *i/g* as a solution of soyabean oil at doses of 1.7, 14.1, and 300 mg/kg BW for 144 weeks.[9] The treatment increased mortality in the test animals. Two higher doses shortened prothrombin time and increased hemopoietic function of the spleen. All doses caused morphol ogy changes in the liver.

Immunotoxic effect was registered in workers employed in VCM production.[10]

Reproductive Toxicity. The data available to date cannot be considered conclusive, and the mechanism for the possible reproductive related risk posed by paternal exposure to VCM is not certain.[1]

Embryotoxicity.

Observations in man. In contrast to previous publications, a recent study did not reveal significant association between paternal exposure to VCM and miscarriage.[11]

Animal studies. VCM was ineffective in the offspring of mice, rats, and rabbits at maternal toxicity levels. However, other experiments have suggested some signs of embryotoxicity of VCM in rats and mice. A concentration of 4.8 mg/m^3 was found to produce a selective embryotoxic effect manifested by the loss of permeability of the fetal vessels and functional abnormalities in the NS, liver, and peripheral blood, changes in the relative weight of the visceral organs of the rat offspring.[12]

Teratogenicity.

Observations in man. Epidemiology study did not reveal a relationship between parental occupational exposure to VCM and birth defects in the offspring.[13]

Animal studies. Teratology studies, after VCM inhalation, have been carried out in mice, rats, and rabbits. No significant effects on malformations or anomaly rates resulted from the exposure to 130 to 6470 mg VCM/m^3. No studies for VCM administered by the oral route are available.[14,15]

Gonadotoxicity. Chronic inhalation of VCM at concentrations of 4.8 and 35 mg/m^3 had a toxic effect on the gonads.[16]

Mutagenicity. VCM caused a significant degree of DNA binding in short-term studies.

Observations in man. A number of cytogenic studies have demonstrated an increased frequency of CA and SCE in the peripheral blood lymphocytes of workers exposed to VCM at levels of 5.0 to 500 ppm (13 to 1300 mg VCM/m^3).[16,17] Two studies reported negative results for SCE in exposed workers, while in another study a weak positive response was found.

In vivo cytogenetics. VCM appeared to be mutagenic in various test systems (IARC 19-231). It induced CA, SCE and micronuclei in rodents exposed *in vivo* but did not induce mutations in the mouse spot test or DLM in rats or mice. VCM brought about sex-linked recessive lethal mutations, but not aneuploidy, heritable translocations or DLM in *Dr. melanogaster*.[17]

In vitro genotoxicity. VCM induced mutations in Chinese hamster cells and unscheduled DNA synthesis in rat hepatocytes *in vitro* and caused transformation of BALB/c 3T3 cells and virus-infected Syrian hamster cells.

Carcinogenicity. VCM has been shown to have carcinogenic effects in humans and animals. In 1974, a connection was reported between VCM and the instances of a rare tumor (angiosarcoma of the liver) in human beings and animals.[18] It is likely that VCM is not carcinogenic, and that it is a product of its metabolic bioactivation (*chloroethylene oxide*) that primarily exhibits carcinogenic activity. It is still not clear whether or not this product is the only carcinogenic metabolite of VCM. Rhomberg offered new evidence of the occurrence of VCM-induced angiosarcomas outside the liver. Furthermore, he suggested from his analysis that PVC and its monomers are not the only polymeric materials that may contribute to an induction of angiosarcomas in humans [19]

Observations in man. VCM has been associated with tumors of the liver, brain, lung, and hematolymphopoietic system. All studies refer to inhalation exposure, and positive findings come from an industrial population exposed to high concentrations of VCM. Worldwide there have now been several dozen proved cases in people exposed to very high concentrations of VCM over an average period of 20 years before it was recognized as a health hazard. The cause-effect relationship between VCM exposure and angiosarcoma of the liver is commonly accepted. Opinions vary on whether or not it is related to other tumors. In the Russian VCM and PVC industry, the monomer has been found to have a positive effect on the rate of malignant neoplasms (Fedotova, 1984).

Simonato et al.[20] have found a nearly threefold increase in liver cancer among VCM workers which clearly correlated to the time of initial exposure, duration of employment, and estimated quantitative exposure. Wong *et al.* carried out an investigation in cohort consisting of 10,173 men who had worked for at least one year in jobs involving exposure to VCM. The study confirmed significant mortality excesses in angiosarcoma, cancer of the liver and biliary tract, and cancer of the brain and other CNS tissues.[21]

According to IARC (Suppl. 7-373), VCM can be also associated with hepatocellular carcinoma, brain tumors, lung tumors, and malignancies of the lymphatic and hematopoietic tissues. On the other hand, there is insufficient evidence to establish any relationship between exposure to VCM and an increased incidence of cancer of the brain, lung, and lymphatic or hematopoietic tissues.[1]

According to Doll,[22] no positive evidence of a hazard of any type of cancer other than angiosarcoma of the liver has been found except for the possibility of a small hazard of lung cancer when exposure was heavy. Human data on carcinogenic risk following oral exposure to VCM are not available. The carcinogenic risk of VCM administered orally should not be exaggerated.[23]

Animal studies. There is a sufficient evidence for VCM-induced carcinogenicity in animals. When administered by inhalation, it induced angiosarcomas of the liver in rats, mice, and hamsters, Zymbal gland tumors in rats and hamsters, nephroblastomas in rats, and pulmonary and mammary gland tumors in mice.

Sprague-Dawley rats were given doses of 3.35, 16.65, and 50 mg/kg BW as a solution in olive oil 4 to 5 times a week for 52 weeks.[24] The treatment caused angiosarcomas of the liver, nephroblastomas, thymus tumors, etc. A dose dependence was noted. These studies proved VCM to be undoubted carcinogen for animals, causing the development of tumors in different organs and various types of tumors in the same organ. In the opinion of Maltoni, not even a dose of 0.3 mg/kg BW is the threshold.[24]

In studies of Feron et al.,[9,25] development of both hepatoangiosarcomas and angiosarcomas of the lungs and abdominal cavity and mesotheliomas and adenocarcinomas of the mammary glands of rats was observed at doses of 5.0 mg/kg BW and more. Wistar rats were exposed to 0.014, 0.13, and 1.3 mg/kg BW in their diet. A variety of VCM-related liver lesions was found in the high-dose group. There was no evidence that feeding with VCM affected the incidence of tumors in organs other than the liver. The NOAEL for tumor induction in the rat was identi fied to be 0.13 mg/kg BW.

The cancer risk of oral daily intakes of 0.00002 or 0.0001 mg VCM/person/day is estimated to be negligibly small.[26] Such intakes can be considered virtually safe. Assuming that the number of carcinomas in other sites may be equal to that of hepatic angiosarcomas,[14] in the estimates for excess risk it was considered that a 10^{-5} risk occurs as a result of a lifetime exposure to 1.0 mg VCM/person/day.

Carcinogenicity classifications. An IARC Working Group concluded that there is sufficient evidence for the carcinogenicity of VCM in *experimental animals* and in *humans*.

The Annual Report of Carcinogens issued by the U.S. Department of Health and Human Services (1998) defines VCM to be a known carcinogen, i.e., a substance for which the evidence from human studies indicates that there is a casual relationship between exposure to the substance and *human cancer*.

IARC: 1;

U.S. EPA: A;

EU: 1.

Chemobiokinetics. After ingestion of small doses of VCM with water and food, it is rapidly excreted from the body, and therefore metabolites evidently have no time to form. Only persistent the presence of VCM in the industrial environment enables it to be accumulated in the blood and to form carcinogenic metabolites.[23]

The metabolism of VCM is a dose-dependent and saturable process. VCM is metabolized in the microsomal mixed function oxidase system, forming *chloroethylene oxide* which can be transformed spontaneously to *chloroacetaldehyde*. These two metabolites (particularly chloroethylene oxide) are highly reactive and mutagenic. A major route of metabolism of chloroacetaldehyde involves oxidation to *chloroacetic acid*. Chloroethylene oxide, chloroacetaldehyde, and chloroacetic acid can be conjugated to glutathione and/or cysteine and excreted in the urine.[27] It is known that metabolism of chemical substances in small animals is more rapid than in large animals. Therefore, the accumulation of carcinogenic metabolites of VCM is more rapid in small animals, and there is an increased risk of angiosarcomas developing in their liver compared with a human being.[28] Low doses of VCM administered by gavage are metabolized and eliminated primarily in the urine. In contrast, higher doses are mainly excreted unchanged via the lung.

Regulations. PVC can have residues of VCM in the finished plastic film. These levels have been reduced to insignificant amounts since the late 1970s.

The Codex Committee on Food Additives and Contaminants has proposed guideline levels of 1.0 mg/kg for VCM in PVC packages. Materials and articles must not pass on to foodstuffs any VCM detectable by the method possessing a detection limit of 0.01 mg/kg.

U.S. FDA eliminated the use of VCM in drug products and proposed to alert food manufacturers to the need for monitoring packaging materials that may contain it. Later this proposal was withdrawn by FDA's Center for Food Safety and Applied Nutrition. VCM may be used (1) in adhesives as a component of articles (as monomer) intended for use in packaging, transporting, or holding food in accordance with the conditions prescribed in 21 CFR part 175.105; (2) as a component (a monomer) of the uncoated or coated food-contact surface of paper and paperboard intended for use in producing, manufacturing, packaging, processing, preparing, treating, packing, transporting, or holding dry food of the type identified in 21 CFR part 176.170 (c); (3) as a monomer in the manufacture of semirigid and rigid acrylic and modified acrylic plastics in the manufacture of articles intended for use in contact with food in accordance with the conditions prescribed in 21 CFR part 177.1010; (4) as a monomer in the manufacture of polyethylenephthalate polymers used as, or components of plastics (films, articles, or fabric) intended for use in contact with food in accordance with the conditions prescribed in 21 CFR part 177.1630; and (5) as a monomer in the production of PVC pipes in water supply.

EU (1989 and 1992). VCM is available in the *List of authorized monomers and other starting substances which shall be used for the manufacture of plastic materials and articles intended to come into contact with foodstuffs (Section A)*.

Great Britain (1998). VCM is authorized without time limit for use in the production of polymeric materials and articles in contact with food or drink or intended for such contact.

Recommendations. ***WHO*** (1996). Guideline value for drinking water: 0.005 mg/l.

Standards.

U.S. EPA (1999). MCL: 0.002 mg/l; MCLG: *zero*.

Russia (1995). MAC: 0.05 mg/l; PML in food 0.01 mg/l.

References:

1. *The Mutagenicity and Carcinogenicity of Vinyl Chloride: A Historical Review and Assessment*, ECETOC Technical Report No 31, European Chemical Industry Ecology and Toxicology Center, Brussels, 1988.

2. Dennison, J. L., Breder, C. V., McNeal, T., Snyder, R. C., Roach, J. A., and Sphon, J. A., Head-space sampling and gas-solid chromatographic determination and confirmation of greater than or equal to 1 ppb vinyl chloride residues in polyvinyl chloride food packaging, *J. Ass. Off. Ann. Chem.*, 61, 813, 1978.
3. Davies, I. W. and Perry, R., *Environ. Pollut. Manag.*, 5, 22, 1975.
4. Foster, D. A., *Soc. Cosmet. Chem.*, 27, 456, 1976.
5. Daniels, G. A. and Proctor D. E., *Med. Package*, 48, 45, 1975.
6. Ando, M. and Sayato, W., *Water Res.*, 18, 315, 1984.
7. Petersen, J. H., Lillemark, L., and Lund, L., Migration from PVC cling films compared with their field of application, *Food Addit. Contam.*, 14, 345, 1997.
8. Castle, L., Price, D., and Dawkins, J. V., Oligomers in plastics packaging. Part 1: Migration tests for vinyl chloride tetramer, *Food Addit. Contam.*, 13, 307, 1996.
9. Feron, V. J., Speek, A. J., Willems, M. I., et al., Observation on the oral administration and toxicity of vinyl chloride in rats, *Food Cosmet. Toxicol.*, 13, 633, 1975.
10. Peneva, M. and Kis'ova, Kr., Repercussion of chronic vinyl chloride effect on some immunological parameters, *Khigiene i Zdraveopazvane*, 2, 29, 1987 (in Bulgarian).
11. Mur, J. M., Mandereau, L., Deplan, F., et al., Spontaneous abortion and exposure to vinyl chloride, *Lancet*, 339, 127, 1992.
12. Sal'nikova, L. S. and Kitzovskaya, M. A., Effect of vinyl chloride on embryogenesis in rats, *Gig. Truda Prof. Zabol.*, 3, 46, 1980 (in Russian).
13. Theriault, G. et al., Evaluation of the association between birth defects and exposure to ambient vinyl chloride, *Teratology*, 27, 359, 1983.
14. *Vinyl chloride*, Air Quality Guidelines for Europe, WHO, Regional Office for Europe, WHO Regional Publications, European Series No 23, Copenhagen, 1987, 158.
15. *Vinyl chloride*, Joint FAO/WHO Expert Committee on Food Additives, 1986, 197.
16. Fomenko, V. N. et al., in *Hygienic Significance of Low-Intencity Factors in Industry and in Populated Areas*, Moscow, 1983, 60 (in Russian).
17. Muratov, M. M. and Gus'kova, S. I., Problem of vinyl chloride mutagenicity, *Gig. Sanit.*, 7, 111, 1978 (in Russian).
18. Greech, J. and Johnson, M., Angiosarcoma of the liver in the manufacture of polyvinyl chloride, *J. Occup. Med.*, 16, 150, 1974.
19. Rhomberg, W., Exposure to polymeric materials in vascular soft-tissue sarcomas, *Int. Arch. Occup. Environ. Health.*, 71, 343, 1998.
20. Simonato, L., L'Albo, K., Andersen, A., et al., A colaborative study of cancer incidence among vinyl chloride workers, *Scand. J. Environ. Health.*, 17, 159, 1991.
21. Wong, O., Whorton, M. D., Foliart, D. E., and Ragland, D., An industry-wide epidemiological study of vinyl chloride workers, 1942-1982, *Am. Ind. Med.*, 20, 317, 1991.
22. Doll, R., Effects of exposure to vinyl chloride, *Scand. J. Work Environ. Health*, 14, 61, 1988.
23. Sheftel', V. O., On the risk of vinyl chloride migration into water and foodstuffs, *Gig. Sanit.*, 2, 63, 1980 (in Russian).
24. Maltoni, C., Lefemine, C., Gilberti, A., et al., Carcinogenicity bioassays of vinylchloride monomer: a model of risk. Assessment on an experimental basis, *Environ. Health Perspect.*, 41, 3, 1981.
25. Feron, V. J., Hendriksen, C. F. M., Speek, A. J., et al., Lifespan oral toxicity study of vinyl chloride in rats, *Food Cosmet. Toxicol.*, 19, 317, 1981.
26. Til, H. P., Feron, V. J., and Immel, H. R., Lifetime (149 weeks) oral carcinogenicity study of vinyl chloride in rats, *Food Chem. Toxicol.*, 29, 713, 1991.
27. *Criteria Document over Vinylchloride (Vinyl chloride criteria document),* The Hague, Ministerie van Volkshuisvesting Ruimtelijke Ordening en Milienbeheer, 1984 (Publikatiereeks Luckt, No 34) (in Dutch).
28. Dietz, F. K., Ramsey, J. C., and Watanabe, P. G., Relevance of experimental studies to human risk, *Environ. Health Perspect.*, 52, 9, 1983.

4-VINYL-1-CYCLOHEXENE

Molecular Formula. C_8H_{12}
M = 108.18
CAS No 100-40-3
RTECS No GW6650000
Abbreviation. VCHE

Synonyms. 1,3-Butadiene, dimer; 4-Ethenyl-1-cyclohexene.

Properties. Colorless, transparent liquid with a pungent aromatic odor. Water solubility is 95 mg/l at 20°C, soluble in diethyl ether and petroleum ether. Contained in commercial polybutadiene rubber.

Applications. An intermediate in the production of plastics and rubbers intended for use in contact with food; a diluent for epoxy resins; a co-monomer in styrene polymerization. An antioxidant.

Acute Toxicity. LD_{50} was reported to be 2.63 g/kg BW in Carworth-Wistar rats.[1] According to other data, LD_{50} is 7.0 g/kg BW in rats. The maximum tolerated dose is 4.0 g/kg BW.[2] A dose of 14.0 g/kg BW kills all the animals. Poisoning is accompanied by clonic convulsions and digestive disorder. A dose of 5.0 g/kg BW caused erythropenia, reticulocytosis, and thrombocytosis, and an increase in the activity of catalase and cholinesterase in the blood.[2]

Repeated Exposure. $B6C3F_1$ mice were given oral doses of 200 and 400 mg/kg BW. Survival among high-dosed animals was poor.[3] In a 14-day study, doses of 300 to 5,000 mg/kg BW were administered by gavage in corn oil. All rats and most mice died on ingestion of up to 1250 mg/kg BW but there were no gross or histopathology effects.[4]

Short-term Toxicity. In a 13-week study, mice received up to 1.2 g VCHE/kg BW by gavage; rats were given doses of up to 0.8 g/kg BW according to the same schedule. There was hyaline droplet degeneration of the proximal convoluted tubules of the kidney, severity of which appeared to be dose related in the rats. No compound-related gross or histopathology effects were evident in female rats or in male mice.[3,4]

Long-term Toxicity. Prolonged exposure to VCHE caused high mortality in mice and rats.[3]

Reproductive Toxicity.

Gonadotoxicity. VCHE is shown to be ovarian toxicant in mice but not in rats. It is known to be metabolized to *mono-* and *diepoxides*. According to Doerr et al., diepoxide is essential for induction of ovarian atrophy in $B6C3F_1$ mice.[5] A reduction in the number of primary and mature Graafian follicles was seen in the ovaries of $B6C3F_1$ mice given the 1.2 g/kg BW dose over a period of 3 months.[3]

VCHE at the doses up to 500 mg/kg BW reduced the gamete pool in both ovary and testes of $B6C3F_1$ mice but did not affect reproduction in either F_o or F_1 generation.[6]

Mutagenicity.

In vitro genotoxicity. VCHE was not mutagenic to *Salmonella* in the the presence or absence of exogenous metabolic system.[3]

Carcinogenicity. $B6C3F_1$ mice and F344 rats were given 200 and 400 mg VCHE/kg BW in corn oil by gastric intubation for 103 weeks. No clear evidence of carcinogenicity was noted in rats. Treatment-related increase in the incidence of granular-cell and mixed-cell tumors of the ovary were found in mice. High mortality was observed in male mice; increases in the incidence of lung tumors and lymphomas were statistically significant only by life-table analysis.[3] These carcinogenicity studies were considered to be inadequate due to extensive and early mortality (in rats and male mice) as well as lack of conclusive evidence of carcinogenic effect.[7]

Carcinogenicity classifications. An IARC Working Group concluded that there is sufficient evidence for the carcinogenicity of VCHE in *experimental animals* and there were inadequate data available to evaluate the car cinogenicity of VCHE in *humans*.

IARC: 2B;

NTP: IS - IS - IS - CE* (gavage).

Chemobiokinetics. Following oral administration, the highest concentrations are found in adipose tissue.[11] Rat and mice liver and lung microsomes metabolized VCHE to VCHE-1,2-epoxide.[8] Administration of a 0.5 g/kg BW dose to mice caused cytochrome *P-450*, cytochrome *b5, NADPH*-cytochrome-*c*-reductase and aminopyrine-*N*-demethylase and epoxide hydrolase to be induced, and rapidly depleted hepatic glutathione levels, suggesting that glutathione is probably involved in metabolism of VCHE.[9]

Wistar rat and Swiss mouse liver microsomal mixed-function oxidase metabolize VCHE to *4-vinyl-1,2-epoxycyclohexane*, *4-epoxyethylcyclohexene* and traces of *4-epoxyethyl-1,2-epoxy-cyclohexane*, which are further hydrolyzed to the corresponding *diols*.[10] Excretion occurs with the urine and expired air.[11]

References:

1. Smith, H. F., Carpenter, C. P., and Weil, C. S., Range-finding toxicology data, List VII, *Am. Ind. Hyg. Ass. J.*, 30, 470, 1969.
2. Taradin, Ya. I., Pulyakhin, G.T., and Fetisova, L. N., Toxicological characteristics of divinyl oligomers - vinylcyclohexene and cyclododecatriene, in *Hygiene and Toxicology of High- Molecular-Mass Compounds and of the Chemical Raw Material Used for Their Synthesis*, Proc. 4th All-Union Conf., S. L. Danishevsky, Ed., Khimiya, Leningrad, 1969, 235 (in Russian).
3. *Toxicology and Carcinogenicity Studies of 4-Vinylcyclohexene in Fischer 344/N Rats and B6C3F$_1$ Mice (Gavage Studies)*, NTP Technical Report Series No 303, Research Triangle Park, NC, 1985.
4. Collins, J. J. and Manus, A. G., Toxicological evaluation of 4-vinylcyclohexen. I. Prechronic (14 day) and subchronic (13 week) gavage studies in Fischer 344 rats and B6C3F$_1$ mice, *J. Toxicol. Environ. Health*, 21, 493, 1987.
5. Doerr, J. K., Hooser, S. B., Smith, B. J., and Sipes, I. G., Ovarian toxicity of 4-vinylcyclohexene and related olefins in B6C3F$_1$ mice: role of diepoxides, *Chem. Res. Toxicol.*, 8, 963, 1995.
6. Grizzle, T. B., George, J. D., Fail, P. A., et al., Reproductive effects of 4-vinylcyclohexene in Swiss mice assessed by a continuous breeding protocol, *Fundam. Appl. Toxicol.*, 22, 122, 1994.
7. Collins, J. J., Montali, R. J., and Manus, A. G. Toxicology evaluation of 4-vinylcyclohexene. II. Induction of ovarian tumors in female B6C3F$_1$ mice by chronic oral administration of 4-vinylcyclohexene, *J. Toxicol. Environ. Health*, 21, 507, 1987.
8. Keller, D. A., Carpenter, S. C., Cagen, S. Z., Reitman, F. A., *In vitro* metabolism of 4-vinylcyclohexene in rat and mouse liver, lung, and ovary, *Toxicol. Appl, Pharmacol.*, 144, 36, 1997.
9. Giannarini, C., Citty, L., Gervasi, G., et al., Effect of 4-vinylcyclohexene and its main oxirane metabolite on mouse hepatic microsomal enzymes and glutathione levels, *Toxicol. Lett.*, 8, 115, 1981.
10. Watabe, T., Hiratsuka, A., Ozawa, N., et al., A comparative study on the metabolism of *d*-limonene and 4-vinylcyclohex-1-ene by hepatic microsomes, *Xenobiotica*, 11, 333, 1981.
11. Smith, B. J., Carter, D. E., and Sipes, I. G., Comparison of the disposition and *in vitro* metabolism of 4-vinylcyclohexene in the female mouse and rat, *Toxicol. Appl. Pharmacol.*, 105, 364, 1990.

VINYLETHYNYL-*p*-HYDROXYPHENYL DIMETHYLMETHANE

Molecular Formula. $C_{13}H_{14}O$

M = 186.25

Synonym. 2-(1,1-Dimethyl-4-penten-2-inyl)phenol.

Applications. The base for the synthesis of resins used in the manufacture of anticorrosion coatings (lacquers).

Acute Toxicity. In mice, LD_{50} is 1.42 g/kg BW. The day after poisoning, mice experienced BW loss by 11 to 26%. Death within 6 days. Gross pathology examination revealed inflammation of the GI tract.

Repeated Exposure failed to reveal cumulative properties.

Reference:

Galitskaya, V. A., in *Proc. 12th Sci. Pract. Conf. of Junior Hygienists and Sanitarians*, Moscow, 1969, 139 (in Russian).

VINYLISOPROPYL ACETYLENE

Molecular Formula. C_7H_{10}

M = 92.15

Properties. Liquid with a specific odor. Poorly soluble in water.

Applications. A monomer in the production of copolymeric chloroprene rubber.

Acute Toxicity. In rats, LD_{50} is 1.8 g/kg BW. Poisoning is accompanied by a short-term agitation followed by narcosis.

Reference:

Gizhlaryan, M. S. and Khechumov, S. A., Toxicological characteristics of vinylisopropyl acetylene, hexachlorobutane, and pentachlorobutene, *Gig. Sanit.*, 10, 89, 1983 (in Russian).

5-VINYL-2-METHYLPYRIDINE

Molecular Formula. C_8H_9N
M = 119.17

Properties. Colorless liquid. Water solubility is 0.98%. Odor perception threshold (rating 1) is 0.1 mg/l, practical odor threshold (rating 2) 0.5 to 1.0 mg/l. Does not alter transparency or color of water. Average odor threshold concentration is 0.04 mg/l (Baker).

Applications. A monomer. Used in the production of rubber and ion-exchange resins intended for use in contact with food.

Acute Toxicity. LD_{50} is 1.5 g/kg BW in rats, and 1.25 g/kg BW in mice.[1]

Long-term Toxicity. Rabbits received V. in their drinking water at concentrations of 3.0 to 5.0 mg/l. Inhibition of CNS function is reported.[2]

Reproductive Toxicity. Repeated administration produced no visible signs of toxicity in pregnant rats, but their offspring died in the first 3 to 4 days after birth.

References:

1. Dukhovnaya, A. I., *Experimental Assessment and New Approach to Hygiene Standards for Certain Alkyl Derivatives of Pyridine*, Author's abstract of thesis, Moscow, 1972, 18 (in Russian).
2. Zyabbarova, S. A., in *Proc. Leningrad Sanitary Hygiene Medical Institute*, Leningrad, Issue No 68, 1961, 138 (in Russian).

4-VINYLPYRIDINE

Molecular Formula. C_7H_7N
M = 105.14
CAS No 100-43-6
RTECS No UU1045000
Abbreviation. VP.

Synonyms. 2-Ethenylpyridine; γ-Vinylpyridine

Properties. Colorless liquid with a pyridine odor. Soluble in organic solvents. Solubility in water is 25 g/l at 20°C,[012] very soluble in alcohol.

Applications. A monomer. Co-monomer with butadiene and styrene for food-contact latex adhesives.

Acute Toxicity. LD_{50} is found to be 250 mg/kg BW in rats, 160 or 420 mg/kg BW in mice, and 320 mg/kg BW in guinea pigs. Poisoned animals displayed motor excitation followed by depression. Aggressiveness is sometimes observed. The animals experience labored breathing, and die on the second day. According to other data, manifestations of toxic action included weakness, ataxia, vasodilatation, respiratory disorder, and convulsions.[03]

Repeated Exposure failed to reveal cumulative properties K_{acc} is 6.1 (by Lim).

Allergenic Effect. The NOEL appeared to be 0.5 to 1.0 ml on skin applications in an acetone solution.

Reference:

Shugaev, B. B. and Bukhalovsky, A. A., Toxicological characteristics of vinylpyridine, *Gig. Truda Prof. Zabol.*, 6, 56, 1984 (in Russian).

1-VINYL-2-PYRROLIDINONE

Molecular Formula. C_6H_9NO
M = 111.16
CAS No 88-12-0

RTECS No UY6107000

Abbreviation. VP.

Synonyms. 1-Ethenyl-2-pyrrolidinone; *N*-Vinylbutyrolactam; Vinylpyrrolidone; 1-Vinyl tetrahydropyrrol-2-one.

Properties. Colorless, transparent liquid. Soluble in water (gives a colloidal solution), in alcohols, ethers, and esters.

Applications. VP is used to manufacture water-soluble and insoluble forms of polyvinyl pyrrolidone for pharmaceutical and cosmetic industries and to make copolymers with other monomers intended for use in contact with food. VP is used in cast films adherent to glass, metals, plastics.

The only possibility of **exposure** of the general public relates to the emission of VP in polymers.

Acute Toxicity. In rats, LD_{50} is 1.37 to 1.47 g/kg BW. Poisoning is accompanied by signs of narcotic and irritating effects.

Repeated Exposure revealed moderate cumulative properties.

Short-term Toxicity. Rats were exposed by gavage to 40 mg/kg BW for 3 months. The treatment caused liver weight to increase and liver lesions to develop.[1]

Reproductive Toxicity effects are not found after inhalation exposure to a concentration of 5.0 mg/m^3.[2]

Mutagenicity. Several *in vitro* and *in vivo* tests have shown negative results, and no binding to DNA of the liver was found in rats following an *i/p* injection of VP.[3] VP is found to be positive in *Salmonella typhimurium* and *Dr. melanogaster* assays, and in tests for gene mutations in mice lymphoma cells.[4]

Carcinogenicity classification. An IARC Working Group concluded that there are no adequate data available to evaluate the carcinogenicity in *experimental animals* and in *humans.*

IARC: 3.

Chemobiokinetics. Following ingestion VP is mainly distributed in the liver and small intestine. It is partially it is excreted in the urine in an acetate form, but it is mostly (88%) combined with water-soluble acid compounds.[5] Following *i/v* injection, ^{14}C-VP was cleared from the blood with a half-life of about 2 hours. Unchanged VP accounted for less than 0.6% of the dose administered. Main metabolites were not identified.[3]

Regulations.

EU (1990). VP is available in the *List of monomers and other starting substances which may continue to be used for the manufacture of plastic materials and articles intended to come into contact with foodstuffs pending a decision on inclusion in Section A* (*Section B*).

U.S. FDA (1998) regulates VP for use in adhesives as a component of articles (a monomer) intended for use in packaging, transporting, or holding food in accordance with the provisions prescribed in 21 CFR part 175.105.

References:

1. Klimisch, H.-J., Deckardt, K., Gembardt, C., Hilderbrand, B., Kuttler, K. and Roe, F. J. C., Subchronic inhalation and oral toxicity of *N*-vinylpyrrolidone-2. Studies in Rodents, *Food Chem Toxicol.,* 35, 1061, 1997.
2. Kvasov, A. R., in *Proc. Rostov-na-Donu Medical Institute*, Rostov-na-Donu, Issue No 17, 1974, 124 (in Russian).
3. Greim, H., *Gesundheitssghadliche Arbeitsstoffe; Toxicologisch-arbeitsmedizinische Bergrundung von MAK-Werten,* 1-20, Lieferung 1994, pp. N1-14, VCH-Verlagsgesellschaft mbH, Weinheim, 1994.
4. Knaap, A. et al., in *The 14th Ann. Conf. Eur. Soc. Environ. Mut.*, Abstracts, Moscow, 1984, 247 (in Russian).
5. McClanahan, J. S., Lin, Y. C., and Digenis, G. A., Disposition of *N*-vinyl-2-pyrolidone in the rat, *Drug Chem. Toxicol.*, 7, 129, 1984.

VINYLTOLUENES (mixed isomers)

Molecular Formula. C_9H_{10}

M = 118.18

CAS No 25013-15-4
RTECS No WL5075000
Abbreviation. VT.

Synonyms and **Trade Names.** Ethenyl methyl benzene; Methyl ethenyl benzenes; Methyl vinyl benzenes; Tolylethylene.

	CAS No	RTECS No
***m*-Methyl styrene**	100-81-1	WL5075800
***o*-Methyl styrene**	611-15-4	WL5075900
***p*-Methyl styrene**	622-97-9	WL5076000

Properties. Oily liquids with a pungent odor. Technical-grade VT are mixtures of isomers. Water solubility is approximately 100 mg/l at 25°C. Odor perception threshold is 0.42 mg/l.[02] According to other data, organoleptic threshold is 0.005 mg/l.[01]

Applications. Monomers. Used in the production of plastics, synthetic rubbers, film materials intended for use in contact with food. Chemical intermediate for unsaturated polyester resins, and modified alkyd resins. Used in quick-dry coatings.

Acute Toxicity. In young rats, LD_{50} is 4.0 g/kg BW.[1] The same LD_{50} was identified for the mixture of *m*-VT (55 to 70%) and *p*-VT (30 to 45%). Poisoning is accompanied by excitation followed by depression, side position, clonic convulsions, hyperemia of paws and tails. High doses cause narcosis in mice.[2] LD_{50} is 3.16 g/kg BW in mice.[03]

Long-term Toxicity. VT were given by gastric intubation in olive oil to Sprague-Dawley rats (10, 50 or 250 mg/kg BW, 78 weeks) and Swiss mice (50, 250 or 500 mg/kg BW, 107 weeks). There was no effect on BW gain; survival of males was reduced.[3]

Reproductive Toxicity.

Embryotoxicity and ***Teratogenicity*.** Pregnant rats received 250 mg/kg BW. The treatment induced fetotoxicity but caused no malformations in offspring.[4] Embryotoxic effect is observed after inhalation exposure of guinea pigs.

Mutagenicity.

***In vitro genotoxicity*.** VT caused no CA and SCE in human lymphocytes *in vitro*[5] and in Chinese hamster ovary cells (NTP-85). No mutagenic activity was found in the *Salmonella* bioassay, and in the tests for gene mutations in mouse lymphoma cells.[6]

***In vivo cytogenetics*.** VT were negative in *Dr. melanogaster* bioassay.[6]

Carcinogenicity. Sprague-Dawley rats were given 10 to 500 mg *p*-VT/kg BW by gavage in olive oil for 108 weeks. The study was terminated when the survival rate reached 50% in at least one experimental group. No increase in the incidence of tumors was observed.[3]

***Carcinogenicity classification*.** An IARC Working Group concluded that there is evidence suggesting lack of carcinogenicity of VT in *experimental animals* and there were inadequate data available to evaluate the carcinogenicity of in *humans*.

IARC: 3.

Chemobiokinetics. VT are the structural analogues of styrene. They are metabolized via reactive intermediates which bind hepatic non-protein *thiols*; in the rat liver VT caused a dose-dependent decrease in hepatic glutathione with a concomitant excretion of *thioethers* in the urine.[7] However, repeated daily *i/p* administrations were not followed by any appreciable increase in metabolite excretion, i.e., no enzyme induction phenomenon was apparent.[8]

Regulations.

U.S. FDA (1998) approved the use of VT (1) in the manufacture of resinous and polymeric coatings for food-contact surface of articles intended for use in producing, manufacturing, packing, transporting, or holding food, subject to the provisions of 21 CFR part 175.300; and (2) in the production of cross-linked polyester resins for repeated use as articles or components of articles coming in contact with food, subject to the provisions of 21 CFR part 177.2420. (3) *p-Methylstyrene* is a monomer of poly(*p*-methylstyrene) and rubber-modified poly(*p*-methylstyrene) which may be safely used as components of articles intended for use in contact with food, subject to the provisions of 21 CFR part 177.1635.

EU (1990). *o*-VT and *m*-VT is available in the *List of monomers and other starting substances which may continue to be used for the manufacture of plastic materials and articles intended to come into contact with foodstuffs pending a decision on inclusion in Section A* (*Section B*).

References:

1. Wolf, M. A., Rowe, V. K., McCollister, D. D., Hollingsworth, R. L., and Oyen, F., Toxicological studies of certain alkylated benzenes and benzene, *AMA Arch. Ind. Health*,14, 387, 1956.
2. *Hazardous Chemicals. Hydrocarbons, Halogenated hydrocarbons*, Handbook, V. A. Filov, Ed., Khimia, Leningrad, 1990, 207 (in Russian).
3. Conti, B., Maltoni, C., Perino, G., and Ciliberti, A., Long-term carcinogenicity bioassays on styrene administered by inhalation, ingestion and injection and styrene oxide administered by ingestion in Sprague-Dawley rats, and para-methylstyrene administered by ingestion in Sprague-Dawley rats and Swiss mice, *Ann. N. Y. Acad. Sci.*, 534, 203, 1988.
4. Hardin, B. D., Bond, G. P., Sikov, M. R., et al., Testing of selected workplace chemicals for teratogenic potential, *Scand. J. Work Environ. Health*, 7, 66, 1981.
5. Norppa, H., The *in vitro* induction of sister chromatid exchanges and chromosome aberrations in human lymphocytes by styrene derivatives, *Carcinogenesis*, 2, 237, 1981.
6. Knaap, A. et al., in *The 14th Ann. Conf. Eur. Soc. Environ. Mutagen.*, Abstracts of reports, Moscow, 1984, 247 (in Russian).
7. Vainio, H. and Heinonen, T., Metabolism and toxicity of vinyltoluene, The 3rd Int. Congress on Toxicology, San Diego, Abstracts, *Toxicol. Lett.*, 18 (Suppl. 1), 151, 1983.
8. Bergemalm-Rynell, K. and Steen, G., Urinary metabolites of vinyltoluene in the rat, *Toxicol. Appl. Pharmacol.*, 62, 19, 1982.

CHAPTER 2. PLASTICIZERS

ADIPIC ACID, BENZYLOCTYL ESTER

Molecular Formula. $C_{25}H_{32}O_4$
M = 348.47
CAS No 3089-55-2

Synonyms. Benzyloctyl adipate; Hexanedioic acid, octylphenylmethyl ester.

Properties. Viscous liquid. Poorly soluble in water, soluble in oil.

Applications. Used as a plasticizer in the production of polyvinyl chloride, nitro-, ethyl and benzyl cellulose, chlorinated rubber, etc. Renders the materials resistance to low temperatures.

Acute Toxicity. In rats, rabbits and dogs, the LD_{50} is 20 ml/kg BW (24-hour observation) and 10 ml/kg BW (7-day observation).

Long-term Toxicity. In a 1-year study, rats were given 0.5 ml/kg and 1.0 ml/kg BW twice a week. The treatment produced no effect on growth, hematology, or reproductive function of animals.[05]

Standards. ***Russia*** (1995). Recommended PML: *n/m.*

ADIPIC ACID, BIS[2-(2-BUTOXYETHOXY)ETHYL] ESTER

Molecular Formula. $C_{22}H_{42}O_8$
M = 434.64
CAS No 141-17-3
RTECS No AU8420000

Synonyms and **Trade Names.** Adipic acid, bis[2-(2-butoxyethoxy)ethyl]ester; Dibutoxyethoxyethyl adipate; Hexanedioic acid bis[2-(2-butoxyethoxy)ethyl] ester; Wareflex.

Applications. Plasticizer for cellulose nitrate and polyvinyl acetate. Plasticizer and softener for natural rubbers.

Acute Toxicity. In rats and mice, LD_{50} is 6.0 to 7.0 g/kg BW. Poisoned animals exhibited distinct signs of a narcotic effect.

Repeated Exposure revealed development of habituation. K_{acc} is 10 (by Lim). Poisoning is accompanied by excitation, followed by depression 1 to 2 hours after repeated administration.

Regulations. ***U.S. FDA*** (1998) approved the use of A. in the manufacture of rubber articles intended for repeated use in producing, manufacturing, packaging, processing, preparing, treating, packing, transporting, or holding food in accordance with the conditions prescribed by 21 CFR part 177.2600.

Reference:

Timofiyevskaya, L. A. and Kuz'mina, A. N., Mechanism of neuroparalytic effect of phthalic ethers, *Gig. Truda Prof. Zabol.*, 1, 50, 1983 (in Russian).

ADIPIC ACID, BIS(2-BUTOXYETHYL) ESTER

Molecular Formula. $C_{18}H_{34}O_6$
M = 346.52
CAS No 141-18-4
RTECS No AU8450000

Synonyms and **Trade Names.** Adipic acid, dibutoxyethyl ester; Adipol BCA; Butyl cellosolve adipate; Di(2-butoxyethyl) adipate; Hexanedioic acid, bis (2-butoxyethyl) ester; Staflex DBEA.

Properties. Liquid with a characteristic odor.

Applications. Used as a plasticizer for vinyl resins.

Acute Toxicity. LD_{50} *i/p* is 0.6 g/kg BW.[011]

ADIPIC ACID, BUTYLBENZYL ESTER

Molecular Formula. $C_{17}H_{24}O_4$
M = 292.41
CAS No 4121-13-5
RTECS No AU8417000

Synonyms and **Trade Name.** Adimoll BB; Butylbenzyl adipate; Butyl benzyladipinate; Hexanedioic acid, butyl phenylmethyl ester.

Properties. Clear yellowish liquid with a characteristic odor. Poorly soluble in water, readily soluble in alcohol.

Applications. Used as a plasticizer in the production of polyvinyl chloride and other food-contact plastic materials.

Acute Toxicity. In rats, the LD_{50} is 19.4 g/kg BW. Poisoning is accompanied by excitation with subsequent inhibition, lassitude and impairment of motor coordination. Four to five hours before death the animals do not react to external stimuli. Five to seven days after a single exposure, signs of intoxication have completely disappeared in surviving animals. Mild anemia was found.

Reference:

Nikolayeva, G. A., Data on hygienic assessment of polyvinylchloride materials plasticized with ethers of adipinic acid, in *Hygiene Aspects of the Use of Polymeric materials and Articles Made of Them,* Coll. Works, All-Union Research Institute of Hygiene and Toxicology of Pesticides, Polymers and Plastic Materials, L. I. Medved', Ed., Kiyv, 1969, 36 (in Russian).

ADIPIC ACID, BUTYLENEGLYCOL ESTER

Molecular Formula. $C_{10}H_{16}O_4$
M = 200.24
Abbreviation. BGA.

Synonym. Butyleneglycol adipate.

Applications. Used as a plasticizer in the production of polyvinyl chloride and other food-contact plastic materials.

Long-term Toxicity. Rats and dogs were dosed with 1.5 mg/kg BW and 10 mg/kg BW for 2 years. The treatment produced no effect on BW gain, food consumption, or mortality. Gross pathology examination failed to reveal macro- and microchanges in the structure of the visceral organs and tissues.

Reproductive Toxicity. In a 3-generation study, rats were administered 1.0 to 10 mg BGA/kg BW. Reproductive function or development of the offspring were not affected.

Regulations. ***U.S. FDA*** (1998) approved the use of BGA terminated with a 16% by weight mixture of myristic, palmic, and stearic acids, at levels not exceeding 33% by weight of polyvinyl chloride homopolymers used in contact with food (except foods that contain more than 8.0% of alcohol) at temperatures not to exceed room temperature.

Reference:

Fancher, O. E., Kennedy, G. L., Plank, J. B., et al., Toxicity of a butylene glycol adipic acid polyester, *Toxicol. Appl. Pharmacol.,* 26, 58, 1973.

ADIPIC ACID, DIALKYL ESTER

Molecular Formula. $ROOC(CH_2)_4COOR$, where $R = C_7H_{15} - C_9H_{19}$
Abbreviations. DAA, PVC (polyvinyl chloride).

Synonyms. Adipic acid, dialkyl esters, a mixture; Dialkyl adipate.

Composition. A mixture of dialkyl esters of adipic acid.

Properties. Oily liquid with a yellowish color.

Applications. DAA is used for plasticizing PVC and its copolymers, rendering them greater elasticity and low temperature resistance than di(2-ethylhexyl) phthalate. One of the principal plasticizers of PVC.

Acute Toxicity. In rats, LD_{50} (DAA-C_{810}) is about 40 g/kg BW; in mice, it is 8.0 to 12 g/kg BW. Poisoned animals are depressed, with unkempt fur.[1,2]

Repeated Exposure revealed moderate cumulative properties. K_{acc} is 8.33 (by Lim).

Short-term Toxicity. Rats were exposed to 0.125 to 1.0% DAA in their feed for 98 days. Treatment at higher dietary levels produced a reduction in food consumption and blood *Hb* level.[2]

Regulations. ***U.S. FDA*** (1998) approved the use DAA-C_{789} in which the C_{789}-alkyl groups are derived from linear α-olefins by the oxoprocess. It is listed in CFR for exclusive use as a component of resinous and polymeric coatings:

- at levels not to exceed 24% by weight of the permitted PVC homo- and/or copolymers used in contact with non-fatty foods (thickness <0.005 inch);
- at levels not exceeding 24% by weight of the permitted PVC homo-/or copolymers used in contact with fatty foods having a fat and oil content not exceeding a total of 40% by weight (thickness of polymers <0.005 inch);
- at levels not exceeding 35% by weight of the permitted PVC homo-/or copolymers used in contact with nonfatty foods (thickness of polymers <0.002 inch);
- at levels not exceeding 35% by weight of the permitted PVC homo-/or copolymers used in contact with fatty foods having a fat and oil content not exceeding a total of 40% by weight (thickness of polymers <0.002 inch);

U.S. FDA also approved the use of DAA-C_{789} (1) in adhesive as a component of articles intended for use in packaging, transporting, or holding food in accordance with the conditions prescribed by 21 CFR part 175.105; and (2) as components of the uncoated or coated food-contact surface of paper and paperboard intended for use in producing, manufacturing, packing, transporting, or holding dry, aqueous and fatty food in accordance with the conditions prescribed by 21 CFR part 176.170 and 176.180. DAA-C_{810} made from C_{6810} or $C_{8\text{-}10}$ synthetic fatty alcohols are listed in CFR to be used:

- at levels not exceeding 24% by weight of the permitted PVC homo-/or copolymers used in contact with non-fatty foods (thickness <0.005 inch);
- at levels not exceeding 24% by weight of the permitted PVC homo-/or copolymers used in contact with fatty foods having a fat and oil content not exceeding a total of 40% by weight (thickness of polymers <0.005 inch);
- at levels not exceeding 35% by weight of the permitted PVC homo-/or copolymers used in contact with non-fatty foods (thickness of polymers <0.002 inch);
- at levels not exceeding 35% by weight of the permitted PVC homo-/ or copolymers used in contact with fatty foods having a fat and oil content not exceeding a total of 40% by weight (thickness of polymers <0.002 inch).

Standards. ***Russia.*** Recommended PML: *n/m.*

References:

1. Mel'nikova, N. N., Toxicologic characteristics of plasticizer dialkyladipate-810, *Gig. Truda Prof. Zabol.,* 12, 57, 1984 (in Russian).
2. Gaunt, I. F., Grasso, P., Landsdown, A. B., et al., Acute (rat and mouse) and short-term (rat) toxicity studies on dialkyl (C_{789}) adipate, *Food Cosmet. Toxicol.,* 7, 35, 1969.

ADIPIC ACID, DIBUTYLCARBITOL ESTER

Synonyms. Dibutylcarbitol adipate; Diethylene glycol, dibutyl ether, adipate.

Properties. Colorless, oily liquid.

Applications. Used as a plasticizer in the production of food-contact plastics.

Acute Toxicity. LD_{50} is 7.0 g/kg BW in rats, and 6.0 g/kg BW in mice. Poisoning is accompanied by distinct manifestations of narcotic effect.

Repeated Exposure failed to reveal cumulative properties; development of habituation was noted. Poisoning was accompanied by excitation. CNS inhibition occurs in 1 to 2 hours after repeated administration.

Reference:

Timofiyevskaya, L. A. and Kuz'mina, A. N., Mechanism of neuroparalytic effect of phthalic ethers, *Gig. Truda Prof. Zabol.,* 1, 50, 1983 (in Russian).

ADIPIC ACID, DIBUTYL ESTER

Molecular Formula. $C_{14}H_{26}O_4$
M = 258.40
CAS No 105-99-7
RTECS No AV0900000
Abbreviation. DBA.

Synonyms. Adipic acid, dibutyl ester; Butyl adipate; Dibutyl adipate; Hexanedioic acid, dibutyl ester.

Properties. Clear, yellowish liquid with a characteristic odor. Solubility in water is 0.025% (20°C).

Applications. Used as a plasticizer in the production of polyvinyl chloride, ethyl and nitrocellulose, polystyrene and synthetic rubber intended for contact with food or drink.

Exposure. Estimated daily intake is 1.5 mg.[1]

Acute Toxicity. LD_{50} for rats is found to be 12.9 g/kg BW,[05] and for mice it is about 17 g/kg BW. Symptoms of narcotic action are present. Animals became immobile, irresponsive to food and water, with rapid, shallow respiration, and diarrhea. Survivors recovered on days 5 to 7 after a single exposure. Gross pathology examination revealed mild anemization of the visceral organs.[2]

Repeated Exposure. Rats were given 1/10 LD_{50} for 2 months. The treatment caused labored respiration, neutrophilia (with the leukocyte formula shifted to the left) and monocytosis, eosinophilia, and dystrophic changes in the liver.[1]

Short-term Toxicity revealed mild toxic effect in dogs given a DBA isomer, *diisobutyl adipate* (DIBA), for 3 months. The NOAEL was considered to be 2.5% DBA in the diet.[3]

Long-term Toxicity. In a two-year rat study, the NOEL was identified to be about 0.5% DIBA by weight in the diet.[3]

Reproductive Toxicity. DBA produced teratogenic effect in rats.[4]

Mutagenicity.

In vitro genotoxicity. *Diisononyl adipate* is shown to be negative in *Salmonella* mutagenicity assay and in mouse lymphoma cells. It does not cause morphological transformations in the culture of Syrian hamster embryo cells.[5]

Allergenic Effect was not found.

Standards. ***Russia.*** PML in food and water: *n/m.*

References:

1. Dubinina, O. H. et al., in *Industrial Hygiene and Protection of Health of Workers in the Petroleum and Petrochemical Industries*, Moscow, 1982, 181 (in Russian).
2. Nikolayeva, G. A., Data on hygienic assessment of polyvinylchloride materials elasticized with ethers of adipinic acid, in *Hygienic Aspects of Use of Polymeric Materials and Articles Made of Them*, L. I. Medved‘, Ed., All-Union Research Institute of Hygiene and Toxicology of Pesticides, Polymers and Plastic Materials, Kiyv, 1969, 36 (in Russian).
3. Weil, C. S. and McCollister, D. D., Relationship between short- and long-term feeding studies in designing an effective toxicity test, *J. Agric. Food Chem.*, 11, 486, 1963.
4. Singh, A. R., Lawrence, W. H., and Autian, J., Embryonic-fetal toxicity and teratogenic effects of adipic acid esters in rats, *J. Pharmacol. Sci.*, 62, 1596, 1973.
5. McKee, R. H., Lington, A. W. Traul, K. A., et al., An evaluation of the genotoxic potential of diisonony ladipate, *Environ. Mutagen.*, 6, 461, 1984.

ADIPIC ACID, DIDECYL ESTER

Molecular Formula. $C_{26}H_{50}O_4$
M = 426.68
RTECS No AV1030000
Abbreviation. DDA.

Synonym. Didecyl adipate (mixed isomers).

Properties. Colorless or light-amber colored liquid with a weak characteristic odor. Solubility in water is about 0.01% at 20°C. Slightly leached by water and detergents.

Applications. Used as a plasticizer in the production of polyvinyl chloride, nitro- and ethylcellulose, polystyrene and chlorinated rubber intended for contact with food or drink.

Acute Toxicity. LD_{50} appeared to be 20.5 ml/kg BW in mice, and 21 g/kg BW in rats.[04]

Repeated Exposure. No changes were found in rats dosed with 0.5% DDA in the diet for 30 days.

Regulations.

EU (1990). DDA is available in the *List of authorized monomers and other starting substances which may continue to be used for the manufacture of plastic materials and articles intended to come into contact with foodstuffs pending a decision on inclusion in Section A* (*Section B*).

U.S. FDA (1998) approved the use of DDA in the manufacture of rubber articles intended for repeated use in producing, manufacturing, packaging, processing, preparing, treating, packing, transporting, or holding food in accordance with the conditions prescribed by 21 CFR part 177.2600.

ADIPIC ACID, DI(2-ETHYLHEXYL) ESTER

Molecular Formula. $C_{22}H_{42}O_4$

M = 370.58

CAS No 103-23-1

RTECS No AU9700000

Abbreviations. DEHA, DOA (dioctyl phthalate), PVC (polyvinyl chloride), PVDC (polyvinylidene chloride).

Synonyms and **Trade Names.** Bis(2-ethylhexyl) adipate; Bisoflex DOA; Di(2-ethylhexyl) adipate; Dioctyl adipate; Flexol A 26; Hexanedioic acid, bis(2-ethylhexyl) ester; Hexanedioic acid, dioctyl ester; Kodaflex DOA; Monoplex DOA; Octyl adipate; Staflex DOA; Uniflex DOA; Vestinol OA.

Properties. Colorless or light-amber liquid with a characteristic odor. Solubility in water is 0.025% at 20°C, soluble in alcohol.

Applications. Used as a plasticizer in the production of plastic materials. DEHA combines well with polyvinyl chloride and its copolymers as well as with the majority of polar thermoplastics intended for contact with food or drink. Gives the materials a good low-temperature resistance (down to -60°C).

Exposure. Food is the most important source of human exposure. Assessment of the dietary uptake of DEHA in a limited population study has been made in 112 individuals from 5 different geographical locations in the UK. A skewed distribution was determined with a median value of 2.7 mg/day. This is similar to an estimated maximum daily intake of 8.2 mg/day established by UK Ministry of Agriculture, Fisheries and Food.[1]

DEHA has been occasionally found in drinking water at concentrations of 0.001 to 0.005 mg/l.

Migration Data. 60% of the films declared for use in contact with fatty foods showed too high overall migration. In most instances DEHA made up about 80% of the total amount of plastic constituents migrating to isooctane.[2]

Migration of DOA plasticizer from plasticized PVC and PVDC/PVC (*Saran*) films into both olive oil and distilled water during microwave heating was studied. Migration of DOA into olive oil reached equilibrium after heating for 10 min. at full power (604.6 mg DOA/l). Migration into distilled water was 74.1 mg/l after 8 min at full power.[3]

The following DEHA migration levels from PVC films were reported: 1.0 to 20 mg/kg in fresh chicken, 8.0 to 48 mg/kg in cooked chicken, 5.0 to 9.0 mg/kg in fresh beef, and 9.0 to 94 mg/kg in sandwiches. DEHA migration from "low migration" PVC film plasticized with a mixture of DEHA and a polymeric plasticizer into samples of cheese at storage time of 2 hours at 5°C was found to be at the level of 45 mg DEHA/kg cheese. After a 10-day contact period, DEHA amount reached 150 mg/kg. Use of this film may lead to consumer intakes of DEHA close to or above the tolerable daily intake of 0.3 mg/kg BW as defined by the EU Scientific Committee for Food.[5]

PVC films were exposed to the official food simulant media, olive oil, or to isooctane. DEHA was determined by combined capillary GC-MS. A migration exceeding the specific migration limit of 4 mg/cm^2 was found in 42 films (89% of the samples) and these films were deemed to be illegal according to their present declared field of application as given by their labelling.[6]

Acute Toxicity. In rats, LD_{50} is reported to be 9.0 to 45 g/kg BW.[6] Poisoning is accompanied by excitation with subsequent general inhibition, apathy, and motor coordination disorder.[7,8] Hours before death, the animals are comatose. In the survivors, signs of intoxication have completely disappeared in 5 to 7 days. No deaths occurred when rats were given a 2.5 g/kg BW dose.

In mice, the LD_0 values ranged from 15 g/kg BW in males to 24.6 g/kg BW in females.[8] It was reported to be 8.4 g/kg BW in rabbits and 12.2 g/kg BW in guinea pigs; the latter tolerated a 4.5 g/kg BW dose, but all died at 18 g/kg BW. Death occurs after progressive paralysis.[05] Doses higher than 6.0 g/kg BW induced *peroxisome proliferation* (PP) in the liver of rodents.

Repeated Exposure. Repeated administration of high doses to rodents produced hepatomegaly, hepatic *PP*, induction of liver catalase and enzymes involved in the oxidation of fatty acids as well as hypolipidemia.[9]

Exposure to 1.0% DEHA in the feed for 7 weeks resulted in reversible decrease of triglyceride and cholesterol concentration in the blood plasma. Exposure of mice to 5.0% DEHA in the diet for a month caused retardation of BW gain. Doses of 1.0 and 1.8 g/kg BW increase the content of *SH*-groups in the blood serum.[10]

Short-term Toxicity. In a 13-week study, F344 rats and $B6C3F_1$ mice were exposed to dietary concentrations up to 25 g/kg. Treatment with a high dose caused retardation of BW gain in all the animals, while the lower doses produced this effect only in male rats and female mice. No compound-related increased mortality, histopathologi cal changes or reduction in feed consumption was noted in this study.[8]

Long-term Toxicity. Introduction of 2.0% DEHA in the diet increased liver size, induced hypolipidemia and hepatic *PP* in male F344 rats.[11] F344 rats and $B6C3F_1$ mice received DEHA at dietary levels of 12 and 25 g/kg BW. The treatment caused no effect on longevity. Retardation of BW gain was observed in mice and rats in the high dose group. Histology examination revealed no changes in the viscera of rats and mice with the exception of mice livers where tumors were found.[8]

Rats were exposed orally to DEHA for 6 months. The treatment with the highest dose caused retardation of BW gain. Administration of 100 and 200 mg/kg BW led to a rise in the content of free *SH*-group in the blood serum. The NOAEL was identified to be 10 mg/kg BW.[10]

Reproductive Toxicity.

Embryotoxicity. Wistar rats were given DEHA in the diet from 10 weeks pre-mating up to 36 days postpartum. The treatment with 12 g DEHA/kg diet resulted in reduction of total litter weights, BW gain of pups, and mean litter size. Histology examination failed to reveal changes in the pups.[12]

ICR mice were exposed to a single *i/p* dose of 0.5 to 10.0 ml/kg BW that produced antifertility and mutagenic effect, as indicated by reduced percentage of pregnancies and increased number of early fetal deaths. Reduction in the number of implantations was also noted. However, these findings were questioned[13] since the study lacked data on the number of pregnancies per treated male and the number of *corpora lutea* per female.

Teratogenicity study in the rat has demonstrated a dose-dependent increase in minor skeletal defects (slightly poorer ossification) in addition to an increased incidence of ureter abnormalities. The NOEL was identified to be 28 mg/kg BW. ADI (TDI) of 0.3 mg/kg BW was calculated for this NOEL.[13]

Mutagenicity. DEHA was tested for genotoxic activity in a large number of tests. With one exception, these studies have shown that DEHA is not a genotoxic agent.

In vitro genotoxicity. Clastogenic activity was not found in the *in vitro* assay in human leukocytes. DEHA was negative in mouse lymphoma assay and in *Salmonella* test.[016] Genotoxic potential of DEHA was determined in primary cultures of adult rat hepatocytes. No genotoxic effect was observed on addition up to 200 μmol DEHA.[14]

In vivo cytogenetics. However, positive results were shown in a DLM study.[15] Orally administered DEHA does not bind covalently to mouse liver DNA.[16]

Carcinogenicity. A study in mice where high doses of DEHA (up to 25,000 ppm in the diet) were fed for the lifetime showed an increased rate of hepatic tumors though a similar study in rats,[8] which were exposed to lower doses than mice, did not show any increase in the rate of hepatic tumors. The NOEL was not defined because of the very high doses applied (1.8 to 3.8 g/kg BW).

DEHA is a hepatic peroxysomal proliferator in rodents, and carcinogenicity of DEHA may be explained by its activity as peroxysomal proliferator. In contrast to a previous carcinogenicity bioassay,[5] Keith et al. revealed a higher sensitivity of F344 rats than B6C3F_1 mice to hepatic PP caused by DEHA.[17] In spite of assuming that PP is associated with cancer development, the dose-dependency of PP would lead to the conclusion that the threshold exists for DEHA carcinogenicity. Such a threshold can be established at 0.01 mg/kg BW, being a NOEL based on peroxisomal parameter in the rat.[7,18,19]

Carcinogenicity classifications. An IARC Working Group concluded that there is limited evidence for the carcinogenicity of DEHA in *experimental animals* and there were no adequate data available to evaluate the carcinogenicity of DEHA in *humans*.

IARC: 3;

U.S. EPA: C;

NTP: N - N - P - P* (feed).

Chemobiokinetics.

Observation in man. Six male volunteers received 46 mg labeled on the ethyl site-chains DEHA. No parent molecule was found in plasma. Metabolite *2-ethylhexanoic acid* was detected. Minor metabolites were *2-ethyl-5-hydroxyhexanoic acid, 2-ethylhexanedioic acid, 2-ethyl-5-ketohexanoic acid* and *2-ethyl-hexanol.*[21] Healthy men received 45 mg DEHA, tagged with deiterium. Main metabolites in the urine were found to be 2-ethylhexanoic acid and 2-ethyl-5-hydroxyhexanoic acid, which were detected also in the blood plasma. Metabolites were found in the urine during the first day at a quantity of about 13%.[25]

Animal studies. Rapid and almost complete absorption from the GI tract is reported. The highest proportion of a single *i/g* dose was recovered in the stomach and intestine.[20] Studies indicated little if any prolonged retention of DEHA or its metabolites in the blood and tissues after oral administration.[13]

DEHA is rapidly hydrolyzed. *Adipic acid* and *2-ethylhexanol* have been observed as DEHA metabolites. 2-ethylhexanoic acid is reported to be a urinary metabolite of DEHA in humans. It seems to be a useful marker metabolite for assessing DEHA intake.[1]

DEHA inhibits the synthesis of cholesterol from acetone. It has a hypocholesterolic action and affects the profile of the phospholipids synthesized in the liver. Bell reported that DEHA and di(2-ethylhexyl) phthalate had a similar effect.[22] Guest et al. found that B6C3F_1 mice dosed orally with 50 or 500 mg/kg BW excreted 91, 7.0, and 1.0 to 2.0% of the radioactive dose in the urine, feces, and expired air.[23] According to other data, DEHA is excreted in the bile and in the urine.[24]

Regulations.

U.S. FDA (1998) regulates DEHA (1) as a plasticizer in polymeric substances used in manufacture of articles which are used in contact with food, and up to 50% may occur as a component of the following products when used in contact with food in accordance with the conditions prescribed in 21 CFR part 178.3740; (2) in adhesives as a component of articles intended for use in packaging, transporting, or holding food in accordance with the conditions specified in the 21 CFR part 175.105; (3) in the production of cellophane in accordance with the conditions specified in the 21 CFR part 177.1200; (4) in the production of closures with sealing gaskets (up to 2.0%) for food containers in accordance with the conditions specified in the 21 CFR part 177.1210; (5) in water-insoluble hydroethyl cellulose film in accordance with the conditions specified in the 21 CFR part 177.1400; and (6) in the production of rubber articles (up to 30%) intended for repeated use in accordance with the conditions specified in the 21 CFR part 177.2600. DEHA may be used (7) in films for contacting non-fatty foods at all temperatures and for fatty foods stored in a refrigerator. The level of DEHA allowed depends on film thickness.

In ***England,*** DEHA may be used at up to 35% in flexible PVC for all applications.

In ***Germany,*** DEHA may be used at levels of up to 35% in general purpose film. This film may not be used in contact with powdered and fine grain foods, milk and milk derivatives (including cheese), fatty foods and foods containing alcohol or essential oils. DEHA may be used at up to 22% in films for wrapping fresh meat provided migration into the meat does not exceed 60 ppm.

In ***Belgium, the Netherlands, France*** and ***Spain,*** DEHA may be used in all food contact applications provided that under normal conditions of use its concentration in the foodstuff does not exceed 60 ppm.

In ***Italy,*** DEHA may be used in film for wrapping fruit and vegetables, either fresh or dried. DEHA may be used in sealing gaskets and closures for virtually all foodstuff except meat and milk.

Recommendations. ***WHO*** (1996). Guideline value for drinking water: 0.09 mg/l.

Standards.

U.S. EPA (1999). MCL and MCLG: 0.5 mg/l.

Great Britain. (1990). Maximum Intake: 8.2 mg/day.[2] A specific migration limit of DEHA is reported to be 3.0 mg/dm^2.[2,23]

Russia (1995). PML: *n/m*.

References:

1. Loftus, N. J., Woollen, B. H., Steel, G. T., Wilks, M. F., and Castle, L., An assessment of the dietary uptake of di-2-(ethylhexyl) adipate (DEHA) in a limited population study, *Food Chem. Toxicol.*, 32, 1, 1994.
2. Petersen, J. H., Lillemark, L., and Lund, L., Migration from PVC cling films compared with their field of application, *Food Addit. Contam.*, 14, 345, 1997.
3. Badeka, A. B. and Kontominas, M. G., Effect of microwave heating on the migration of dioctyladipate and acetyltributylcitrate plasticizers from food-grade PVC and PVDC/PVC films into olive oil and water, *Z. Lebensmit. Unters. Forsch.*, 202, 313, 1996.
4. *Plasticizers: Continuing Surveillance*. The Thirtieth Report of Steering Group on Food Surveillance, The Working Party on Chemical Contaminants from Food Contact Materials: Sub-Group on Plasti cizers, Food Surveillance, Paper No 30, London, HMSO, 1990, 55.
5. Petersen, J. H., Naamansen, E. T., and Nielsen, P. A., PVC cling film in contact with cheese: health aspects related to global migration and specific migration of DEHA, *Food Addit. Contam.*, 12, 245, 1995.
6. Petersen, J. H. and Breindahl, T., Specific migration of di(2-ethylhexyl)adipate (DEHA) from plasticized PVC film: results from an enforcement campaign, *Food Addit. Contam.*, 15, 600, 1998.
7. Smyth, H. F., Carpenter, C. P., and Weil, C. S., Range-finding toxicity data: list IV, *AMA Arch. Ind. Hyg.*, 4, 119, 1951.
8. *Carcinogenesis Bioassay of Di(2-ethylhexyl) Adipate in F344 Rats and $B6C3F_1$ Mice (Feed Study)*, NTP Technical Report Series No 212, Research Triangle Park, NC, 1982.
9. *A 21-day feeding study of di(2-ethylhexyl)adipate to rats: effect on the liver and liver lipids,* BIBRA, Report No 0542/1/85, 1986.
10. Andreyeva, G. A., in *Safe Application and Toxicology of Pesticides, and Clinical Poisoning*, L. I. Medved', Ed., All-Union Research Institute of Hygiene and Toxicology of Pesticides, Polymers and Plastic Materials, Kiyv, 9, 373, 1971 (in Russian).
11. Reddy, J. K., Reddy, M. K., Usman, M. I., et al., Comparison of hepatic peroxisome proliferative effect and its implication for hepatocarcinogenicity of phthalate esters, di(2-ethylhexyl) phthalate, and di(2-ethylhexyl)adipate with a hypolipidimic drug, *Environ. Health Perspect.*, 65, 317, 1986.
12. *DEHA: Fertility Study in Rats*, Imperial Chemical Industries, Report No CTL/P. 2229, 1989.
13. *DEHA: Teratogenicity Study in the Rat*, Imperial Chemical Industries, ICI Report No CTL/ P. 2119, 1988.
14. Reisenbichler, H. and Eckl, P. M., Genotoxic effects of selected peroxisome proliferators, *Mutat. Res.*, 286, 135, 1993.
15. Singh, A. R., Lawrence, W. H., and Autian, J., Dominant lethal mutation and antifertility effects of di(2-ethylhexyl)adipate and diethyl adipate in mice, *Toxicol. Appl. Pharmacol.*, 32, 566, 1975.
16. von Daniken, A., Lutz, W. K., Jackh, R., et al., Investigation on the potential for binding of di(2-ethylhexyl phthalate (DEHP) and di(2-ethylhexyl)adipate (DEHA) to liver DNA *in vivo*, *Toxicol. Appl. Pharmacol.*, 73, 373, 1984.
17. Keith, Y., Cornu, M. C., Canning, P. M., Foster, J., Lhuguenot, J. C., and Elcombe, C. R., Peroxysomal proliferation due to di(2-ethylhexyl)adipate, 2-ethylhexanol and 2-ethylhexanoic acid, *Arch. Toxicol.*, 66, 321, 1992.
18. Final Report on the Safety Assessment of Dioctyl Adipate and Diisopropyl Adipate, *J. Am. Coll. Toxicol.*, 3, 101, 1984.

19. *Toxicological Effects of Diethylhexyladipate*, Midwest Research Institute, Final Report, MRI Project No 7343-B, CMA Contract No PE-14.0-BIO-MRI, 1982.
20. Takahashi, T., Tanaka, A., and Yamaha, T., Elimination, distribution and metabolism of di(2-ethylhexyl) adipate (DEHA) in rats, *Toxicology*, 22, 223, 1981.
21. Loftus, N. J., Laird, W. J., Steel, G. T., Wilks, M. F., and Woollen, B. H., Metabolism and pharmacokinetics of deuterium-labeled di-(2-ethylhexyl)adipate (DEHA) in humans, *Food Chem. Toxicol.*, 31, 609, 1993.
22. Bell, F. P., Di(2-ethylhexyl)adipate (DEHA): effect on plasma lipids and hepatic cholesterol genesis in the rat, *Bull. Environ. Contam. Toxicol.*, 32, 20, 1984.
23. Guest, D., Pallas, F., Northup, S., et al., Metabolic studies with di(2-ethylxehyl)adipate in the mouse, *Toxicologist*, 5, 237, 1985.
24. Bergman, K. and Albanus, L., Di(2-ethylxehyl) adipate: Absorption, autoradiographic distribution and elimination in mice and rats, *Food Chem. Toxicol.*, 25, 309, 1987.
25. Steel, G. T., Woollen, B. H., Loftus, N. I., et al., Biological monitoring of exposure to plasticizers, Abstr., Brit. Toxicol. Soc. Meet., Edinburg 18-20 Sept., 1991, *Hum. Exp. Toxicol.*, 11, 387, 1992.

ADIPIC ACID, DIHEXYL ESTER

Molecular Formula. $C_{18}H_{34}O_4$
M = 314.0
CAS No 110-33-8
RTECS No AV1150000

Synonyms. Adipic acid, dihexyl ester; Dihexyl adipate; Dihexyl hexanedioate; Hexanedioic acid, dihexyl ester.

Properties. Yellowish, oily liquid.

Applications. Used as a plasticizer in the production of polyvinyl chloride and other polymeric materials intended for contact with food.

Acute Toxicity. In rats, LD_{50} was not attained; in mice it appeared to be about 20.0 g/kg BW. Poisoning gives way to inhibition with subsequent excitation. Complete adynamia occurs a day after treatment.

Repeated Exposure revealed no signs of accumulation. K_{acc} exceeded 10 (by Lim).

Standards. ***Russia*** (1995). Recommended MAC: 0.25 mg/l.

Reference:

Mel'nikova, N. N., Toxicity of plasticizer dihexyl adipinate, *Gig. Sanit.*, 12, 57, 1984 (in Russian).

ADIPIC ACID, DIISOBUTYL ESTER

Molecular Formula. $C_{14}H_{26}O_4$
M = 258.40
CAS No 141-04-8
RTECS No AV1480000
Abbreviation. DIBA.

Synonyms and **Trade Name.** Adipic acid, bis(2-methylpropyl) ester; Diisobutyl adipate; Ftaflex DIBA; Hexanedioic acid, bis(2-methylpropyl) ester; Isobutyladipate.

Properties. Colorless and odorless liquid, insoluble in water. Soluble in alcohols.

Applications. Plasticizer compatible with most natural and synthetic polymers.

Reproductive Toxicity. DIBA was less teratogenic in rats than other phthalates when administered during gestation in reasonably large doses.

Regulations. ***U.S. FDA*** (1998) approved the use of DIBA (1) in adhesives as a component of articles intended for use in packaging, transporting, or holding food in accordance with the conditions prescribed by 21 CFR part 175.105; and (2) in the manufacture of resinous and polymeric coatings for the food-contact surface of articles intended for use in producing, manufacturing, packing, processing, preparing, treating, packaging, transporting, or holding food in accordance with the conditions prescribed by 21 CFR part 175.300.

Reference:

Singh, A. R., Lawrence, W. H., and Autian, J., Embryonic-fetal toxicity and teratogenic effects of adipic acid esters in rats, *J. Pharmacol. Sci.*, 62, 1596, 1973.

ADIPIC ACID, DIISOOCTYL ESTER

Molecular Formula. $C_{22}H_{45}O_4$
M = 370.64
CAS No 1330-86-5
Abbreviation. DIOA.

Synonyms and **Trade Names.** Adipic acid, diisooctyl ester; Adipol 10A; Diisooctyl adipate; Dimethyl heptyl adipate; Hexanedioic acid, diisooctyl ester; Isooctyladipate; PX 208.

Properties. Colorless or pale-amber liquid with a slight aromatic smell.

Applications. Used as a plasticizer in the production of polyvinyl chloride and other food-contact polymeric materials.

Toxicity is found to be negligible. Guinea pigs received up to 14 ml/kg BW without visible signs of toxic effects for several hours after poisoning. Half the animals died during three weeks.[05] Doses of 5.0 ml/kg BW did not cause mortality in guinea pigs; administration of 20 ml/kg BW dose caused death of animals in the ensuing weeks. Death was preceded by progressive enfeeblement and muscular atrophy.[04]

Regulations. ***U.S. FDA*** (1993) approved the use of DIOA (1) as a plasticizer for rubber articles intended for repeated use in contact with food (up to 30% by weight of the rubber product) in accordance with the conditions prescribed by 21 CFR part 177.2600, and (2) as an ingredient for closures with sealing gaskets on containers intended for use in producing, manufacturing, packing, processing, preparing, treating, packaging, transporting or holding food (up to 2.0%) in accordance with the conditions prescribed by 21 CFR part 177.1210.

ADIPIC ACID, DIVINYL ESTER

Molecular Formula. $C_{10}H_{14}O_4$
M = 198.22
CAS No 4074-90-2
RTECS No AV1800000
Abbreviation. DVA.

Synonyms. Divinyl adipate; Hexanedioic acid, diethenyl eater.

Properties. Clear, colorless liquid. Poorly soluble in water, readily soluble in alcohol. Odor and taste perception thresholds are 1.0 and 0.5 mg/l, respectively; practical organoleptic threshold is 2.0 mg/l. Concentrations of up to 20 mg/l affect neither color nor transparency of water.

Applications. Used as a plasticizer to give particular low-temperature resistance to polymeric materials.

Acute Toxicity. LD_{50} is found to be 6.4 g/kg BW in rats and mice, and 4.3 g/kg BW in guinea pigs and rabbits. Poisoning is accompanied by excitation with subsequent depression; the animals became irresponsive to tactile or pain stimuli.

Repeated Exposure revealed marked cumulative properties. K_{acc} is 4.4 (by Cherkinsky). Treatment with 1/20 and 1/10 LD_{50} produced reticulocytosis, an increase with subsequent decline in liver cholinesterase activity, a decline in *SH*-group content, a rise in liver glycogen content, in the relative weights of the suprarenals, and a reduction of their ascorbic acid level. Gross pathology examination revealed inflammatory and dystrophic changes in the viscera.

Long-term Toxicity. In a 7-month rat study, the toxic manifestations consisted of reticulocytosis, reduction in phagocytic leukocyte activity, increase with subsequent decline in the blood cholinesterase activity. Dystrophic and inflammatory changes were found on autopsy.

Regulations.

EU (1990). DVA is available in the *List of authorized monomers and other starting substances which may continue to be used for the manufacture of plastic materials and articles intended to come into contact with foodstuffs pending a decision on inclusion in Section A* (*Section B*).

Great Britain (1998). DVA is authorized up to the end of 2001 for use in the production of polymeric materials and articles in contact with food or drink or intended for such contact.

Standards. *Russia.* Recommended PML: 0.2 mg/l.

Reference:

Mironets, N. V., Comparative toxicological characteristics of the plasticizers vinylmethyl adipate and divinyl adipate and their hygienic standards in water bodies, *Gig. Sanit.*, 10, 88, 1970 (in Russian).

ADIPIC ACID, mixture with GLUTARIC and SUCCINIC ACID, ISOOCTYL ISODECYL ESTER

Molecular Formula. $C_{21}H_{40}O_4$
M = 356.0
RTECS No AV1930000

Synonym. Isooctyl isodecyl nylonate.

Applications. Used as a plasticizer in the manufacture of succinic acid, isooctyl isodecyl ester of plastic materials.

Acute Toxicity. Rats tolerate administration of high doses. LD_{50} was not attained. In mice, it was found to be 8.0 to 10 g/kg BW.

Short-term Toxicity. Rats were exposed to the doses of 8.0 to 400 mg/kg feed for 3.5 months. The treatment caused no decrease in BW gain, changes in food consumption, or hematology analyses. Doses of 100 mg/kg and 400 mg/kg diet caused an increase in the content of urinary ascorbic acid, in the relative kidney weights, and in microsomal enzyme activity of the liver. The weights of the suprarenals and ovaries were elevated in female rats. The NOEL appeared to be 25 mg/kg BW.

Reference:

Gaunt, G. F., Colley, J., Grasso, P., et al., Acute (rat and mouse) and short-term (rat) toxicity studies on isooctylisodecyl nylonate, *Food Cosmet. Toxicol.*, 7, 115, 1969.

ADIPIC ACID, METHYLVINYL ESTER

Molecular Formula. $C_9H_{14}O_4$
M = 186.21
CAS No 2969-87-1
RTECS No AV2000000
Abbreviation. VMA.

Synonyms. Adipic acid, vinylmethyl ester; Hexanedioic acid, methyl ester; Vinylmethyl adipate.

Properties. A clear, colorless liquid. Poorly soluble in water, readily soluble in alcohol. Aromatic odor and taste are detectable at a concentration of 2.0 mg/l. Practical organoleptic threshold is in the range of 5.0 to 5.5 mg/l. A concentration of up to 20 mg VMA/l does not affect the color or transparency of water.[1]

Applications. Used as a plasticizer in the production of polymeric materials; gives plastics a special low-temperature resistance.

Acute Toxicity. In rats, mice, and rabbits, LD_{50} is 6.2 g/kg BW.[1] Lethal doses produce excitation with subsequent CNS inhibition and death. Guinea pigs are insensitive to AAVME: a single administration caused none of them to die.

Repeated Exposure revealed moderate cumulative properties. K_{acc} is 6.4 (by Cherkinsky). A decrease in leukocyte phagocytic activity and in ascorbic acid contents in the suprarenals was found in rats that received 1/20 LD_{50}. Histology examination showed inflammatory and dystrophic changes in the viscera.[1]

Mutagenicity.

In vitro genotoxicity. VMA was negative in mutagenicity assay in *Dr. melanogaster*.[2]

Standards. *Russia* (1995). MAC and PML: 0.2 mg/l.

References:

1. Mironets, N. V., Comparative toxicological characteristics of the plasticizers vinylmethyl adipate and divinyl adipate and their hygienic standards in water bodies, *Gig. Sanit.*, 10, 88, 1970 (in Russian).

2. Sheftel', V. O., Shquar, L. A., and Naumenko, G. M., The use of some genetic methods in hygienic studies, *Vrachebnoye Delo,* 7, 120, 1969 (in Russian).

ALKYLSULFONIC ACID, PHENOLCRESOL ESTER

Trade Name. Mesamol.

Applications. Used as a plasticizer in polyvinyl chloride production. Gives light stability, at a maximum, if it is combined with dinonyl phthalate.

Acute Toxicity. In rats, rabbits, and dogs, LD_{50} is 25 g/kg BW.[05]

Long-term Toxicity. No manifestations of the toxic effect, pathomorphology changes, or impairment of reproductive function were observed in rats given 0.5 and 1.0 ml/kg BW twice a week for a year.[05]

3-AMINO-2-BUTENOIC ACID, THIODI-2,1-ETHANEDIYL ESTER

Molecular Formula. $C_{12}H_{20}N_2O_4S$
M = 288.40
CAS No 13560-49-1
RTECS No EM9103000

Synonyms. 3-Aminocrotonic acid, thiodiethylene ester; 2,2'-Thidiethanolbis (3-aminocrotonate); Thiodiethylene glycol, γ-aminocrotonate.

Properties. Yellowish powder with an ammonia smell. Poorly soluble in water and organic solvents.

Applications. Used as a plasticizer for polyvinyl chloride (2.0%) rendering transparency to it.

Acute Toxicity. LD_{50} exceeds 7.24 g/kg BW for rats. Large doses cause diarrhea but do not affect BW gain. No histology or gross pathology changes were revealed.[05]

Repeated Exposure. Rats tolerate 1.0% A. in their diet without any signs of toxicity. 5.0% level caused an increase in the relative weights of the liver, kidneys, and thyroid.

Long-term Toxicity. Rats tolerate ingestion of 5.0 mg/kg BW for 2 years without signs of toxicity.[05]

Reproductive Toxicity and **Carcinogenicity.** The dose applied in the long-term study did not affect the reproductive function and had no blastomogenic effect.

1-AMINO-2-PROPANOL

Molecular Formula. C_3H_9NO
M = 75.13
CAS No 78-96-6
RTECS No UA5775000
Abbreviation. AP.

Synonyms and **Trade Name.** β-Aminoisopropyl alcohol; 1-Aminopropan-2-ol; 2-Hydroxypropylamine; Isopropanolamine; 1-Methyl-2-aminoethanol; Threamine.

Properties. Liquid with a slight ammonia odor. Readily soluble in water.

Applications. Used as emulsifying agent, as a plasticizer in the production of food-packaging materials; insecticide.

Acute Toxicity. LD_{50} is 4.0 g/kg BW in male rats, and 1.7 g/kg BW in female rats. Poisoning induced severe CNS inhibition, coordination disorder, hypothermia, and diarrhea.[1]

Repeated Exposure revealed weak cumulative properties.

Long-term Toxicity. The NOAEL of 0.017 mg/kg BW was established in rats and guinea pigs.[2]

Allergenic Effect. Weak sensitization has been observed.

Reproductive Toxicity. Rats received the dose of 280 mg/kg BW on a day prior to mating. The treatment produced effects on fertility, and an increase in pre-implantation mortality.[3]

Regulations. ***U.S.FDA*** (1998) approved the use of AP in adhesives as a component of articles intended for use in packaging, transporting, or holding food in accordance with the conditions prescribed by 21 CFR part 175.105.

Standards. ***Russia*** (1995). MAC: 0.44 mg/l.

References:

1. Burkatskaya, Ye. N., Karpenko, V. N., Pokrovskaya T. N., Ivanova Z. V., Medved', I. L., and Verich, G. Ye., Data on norm-setting maximum allowable concentrations of mono-, di-, and triisopropanolamine in the workplace air, *Gig. Truda Prof. Zabol.*, 7, 46, 1986 (in Russian).
2. Toporkov, V. V., Hygienic substantiation of maximum allowable concentrations of mono-, di-, and triisopropanolamine in water bodies, *Gig. Sanit.*, 3, 79, 1980 (in Russian).
3. *Proc. Leningrad Sanitary Hygiene Medical Institute*, Leningrad, 140, 79,1981 (in Russian).

AZELAIC ACID, BIS(2-ETHYLHEXYL) ESTER

Molecular Formula. $C_{25}H_{48}O_4$
M = 412.64
CAS No 103-24-2
RTECS No CM2000000
Abbreviation. DEHA.

Synonyms and **Trade Names.** Bis(2-ethylhexyl) azelate; Di(2-ethylhexyl) azelate; Dioctyl azelate; Nonanedioic acid, bis(2-ethylhexyl) ester; Plastolein 9058; Staflex DOX; Truflex DOX.

Properties. Colorless, odorless liquid. Poorly soluble in water, soluble in alcohol.

Applications. Used as a plasticizer in the production of polyvinyl chloride, polystyrene and other plastics intended for contact with food or drink. Compatible with nitro-, ethyl-, and acetobutyrate cellulose, styrenebutadiene copolymers, polychloroprene, and butadiene-nitrile-acrylic rubber.

Acute Toxicity. LD_{50} is reported to be 8.7 to 9.1 g/kg BW in rats, and 11.5 g/kg BW in mice.[04] Poisoning is accompanied by depression which is followed with adynamia, labored breathing, and spasms of the separate muscle groups. Death is preceded by lethargy.

Repeated Exposure failed to reveal cumulative properties. Habituation develops. K_{acc} is 6 (by Lim). The treatment caused signs of intoxication to develop, including excitation and irritation of the upper respiratory tract mucosa and eyes. Recovery occurs within days.[1]

Short-term Toxicity. There were no symptoms of intoxication in rats dosed with 0.5 to 2.5 g/kg BW over 3 months.[05]

Regulations. ***U.S. FDA*** (1993) approved the use of DEHA as a plasticizer in polymeric substances and as a component of adhesives intended for use in contact with food:

- at levels not exceeding 24% by weight of the permitted polyvinyl chloride homo- and/or copolymers used in contact with nonfatty, nonalcoholic food (thickness of polymers < 0.003 inch);
- at levels not exceeding 24% by weight of the permitted polyvinyl chloride homo- and/or copolymers used in contact with fatty, nonalcoholic food having a fat and oil content not exceeding a total 30% by weight (thickness of polymers < 0.003 inch).

Standards. ***Russia.*** Recommended PML in drinking water: *n/m.*

Reference:

Timofiyevskaya, L. A., Toxicity of di(2-ethylhexyl) azelainate, *Gig. Truda Prof. Zabol.*, 1, 52, 1983 (in Russian).

AZELAIC ACID, DIHEXYL ESTER

Molecular Formula. $C_{21}H_{40}O_4$
M = 356.54
CAS No 109-31-9
RTECS No CM2100000 or CM2000000
Abbreviation. DHA.

Synonyms and **Trade Name.** Azelaic acid, dihexyl ester; Di-*n*-hexyl azelate; Nonanedioic acid, dihexyl ester; Plastolein 9050.

Properties. Liquid.

Applications. Used as a plasticizer to produce vinyl copolymers, and *Saran* film intended for contact with food, giving them low-temperature resistance; as a plasticizer for cellulose and polypropylene food wrap.

Acute Toxicity. LD_{50} is found to vary from 15 to 24.0 g/kg BW in rats, from 15 to 45 g/kg BW in mice, and from 6.0 to 10 g/kg BW in guinea pigs.[04]

Long-term Toxicity. Addition of up to 15% DHA to the diet caused retardation of BW gain and an increase in kidney weights. Dogs were dosed with 0.1 to 3.0 g DHA/kg BW. The treatment produced no changes in the general condition of animals or in the microscopic structure of the visceral organs.[05]

Chemobiokinetics. DHA metabolism comprises its hydrolysis to form *azelaic acid* and *hexyl alcohol.* Azelaic acid is known to be rapidly excreted with the urine in dog, rabbit, and man. Another pathway of DHA decomposition in the body occurs via formation of *adipic acid*, from which *succinic acid* is obtained through β-oxidation. Both acids have no effect on the kidneys.

Regulations. ***U.S. FDA*** (1998) regulates DHA for use only (1) in polymeric substances used in contact with nonfatty food; (2) in the production of polymeric substances used in contact with fatty food and limited to use at levels not exceeding 15% by weight of such polymeric substances; and (3) at levels of 15 to 24% by weight of the permitted vinyl chloride homo- and/or copolymers used in contact with fatty food having a fat and oil content not exceeding 30% by weight (the average thickness of such polymers should not exceed 0.003 inch).

Standards. ***Russia.*** Recommended PML: *n/m.*

Reference:

1. Hodge, H. C., Maynard, E. A., Downs, E. L., et al., Chronic oral toxicity studies of di-*n*-hexyl- azelate in rats and dogs, *Toxicol. Appl. Pharmacol.*, 4, 247, 1962.

BENZOIC ACID, BENZYL ESTER

Molecular Formula. $C_{14}H_{12}O_2$
M = 212.26
CAS No 120-51-4
RTECS No DG4200000
Abbreviation. BABE.

Synonyms and **Trade Names.** Ascarbin; Ascarbiol; Benylate; Benzyl alcohol benzoic ester; Benzyl benzoate.

Properties. Leaflets or clear, colorless, oily liquid with an aromatic odor and burning taste. Insoluble in water, miscible with alcohol.

Applications. Camphor substitute in celluloid and plastic pyroxylin compounds. Plasticizer for cellulose acetate and nitrocellulose intended for food-contact surfaces.

Acute Toxicity. LD_{50} is reported to be 1.7, 1.4, 1.8, and 1.0 g/kg BW in rats, mice, rabbits, and guinea pigs, respectively (Draize et al., 1948). Ingestion of large doses induced progressive coordination disorder, excitation, convulsions, and death.[020]

Reproductive Toxicity.

Teratogenicity. Pregnant rats consumed commercial flour diets supplemented with 0.04 or 1.0% BABE from day 0 of gestation to day 21 postparturition. The treatment caused no effect on fetuses that was proved by absence of external, skeletal, or visceral abnormalities.[1] There were no *terata* in the offspring of mice injected with 10% BABE in castor oil on days 1 to 12 of gestation.[2]

Chemobiokinetics. Following ingestion, BABE undergoes metabolic transformation into *hippuric acid.*

Regulations. ***U.S. FDA*** (1998) approved the use of BABE as a constituent of food-contact adhesives which may be safely used as component of articles intended for use in packaging, transporting, or holding food in accordance with the conditions prescribed by 21 CFR part 175.105.

Standards. ***Russia*** (1995). MAC: 0.4 mg/l.

References:

1. Morita, S. et al., Safety evaluation of chemicals for use in household products. II. Teratological studies on benzyl benzoate and 2-(morpholinothio)benzothiazole in rats, *Ann. Rep. Osaka City Inst Public Health Environ Sci.*, 43, 90, 1981.

2. Eibs, H. G., Spielmann, H., and Hagele, M., Teratogenic effects of cyproterone acetate and medroxyprogesterone treatment during the pre- and postimplantation period of mouse embryos, *Teratology*, 25, 27, 1982.

BENZOIC ACID, BUTYL ESTER

Molecular Formula. $C_{11}H_{14}O_2$
M = 178.25
CAS No 136-60-7
RTECS No DG4925000
Abbreviation. BABE.

Synonyms and **Trade Names.** Anthrapole; Benzoic acid, *n*-butyl ester; *n*-Butyl benzoate.

Properties. Thick, colorless, oily liquid with a mild odor. Insoluble in water, soluble in alcohol, miscible with oils.

Applications. Plasticizer. Camphor substitute in plastics. Solvent for cellulose ethers. Used in food industry.

Acute Toxicity. LD_{50} is 5.14 g/kg BW in rats.

Regulations. ***U.S. FDA*** (1998) approved BABE for use only as a constituent of food-contact adhesives which may be safely used as a component of articles intended for use in packaging, transporting, or holding food in accordance with the conditions prescribed by 21 CFR part 175.105.

Reference:

Smyth, H. F., Carpenter, C. P., Weil, C. S., et al., Range-finding toxicity data, List V, *Arch. Ind. Hyg. Occup. Med.*, 10, 61, 1954.

BENZOIC ACID, DIESTER with DIETHYLENE GLYCOL

Molecular Formula. $C_{18}H_{18}O_5$
M = 314.36
CAS No 120-55-8
RTECS No ID6650000
Abbreviation. DEGDB.

Synonyms and **Trade Names.** Benzoflex 2-45; Benzoyloxyethoxyethyl benzoate; Dibenzoyl diethyleneglycol ester; Diethylene glycol, dibenzoate; 2,2'-Oxybisethanol dibenzoate.

Properties. Liquid. Soluble in water.

Applications. Plasticizer for cellulose acetate butyrate. Used in the production of food-contact cellulose nitrate resins, ethyl cellulose resins, polymethyl methacrylate resins, polyvinyl acetate, butyral and chloride.

Acute Toxicity. LD_{50} is 2.83 g/kg BW in rats.

Regulations. ***U.S. FDA*** (1993) approved the use of DEGDB as (1) a component of adhesives intended for use in contact with food in accordance with the conditions prescribed by 21 CFR part 175.105, and as (2) a component of the uncoated or coated food-contact surface of paper and paperboard intended for use in producing, manufacturing, packaging, processing, preparing, treating, packing, transporting, or holding dry, aqueous, and fatty foods in accordance with the conditions prescribed by 21 CFR part 176.170 and 176.180.

Reference:

Smyth, H. F., Carpenter, C. P., Weil, C. S., Pozzani, U. C., and Striegel, J. A., Range-finding toxicity data: List VI, *Am. Ind. Hyg. Assoc. J.*, 23, 95, 1962.

BENZOMONOBUTYLAMIDE

Molecular Formula. $C_{11}H_{15}NO$
M = 177.27
Abbreviation. BBA.

Synonym. *N*-Butylbenzamide.

Properties. White crystals with a faint odor. Poorly soluble in water, soluble in organic solvents.

Applications. Used as a plasticizer in the production of polyamides and vinyl resins intended for contact with food or drink.

Short-term Toxicity. Wistar rats were exposed to 0.1 and 3.0% BBA in the diet. The treatment caused retardation of BW gain. A dose level of 0.01% BBA in the diet was ineffective.[05]

2,2'-BIPHENYLDICARBOXYLIC ACID, DIBUTYL ESTER

Molecular Formula. $C_{22}H_{26}O_4$

M = 354.45

Abbreviation. DBDP.

Synonym. Dibutyl diphenate.

Properties. Oily liquid. Poorly soluble in water; mixes with oil at all ratios.

Applications. Used as a plasticizer in the production of food-contact plastics.

Acute Toxicity. LD_{50} is 17.0 g/kg BW in rats, and 38.0 g/kg BW in mice. Toxicity of DBDP increases when it is given in oil solutions: in this case, LD_{50} is 5.4 g/kg BW for rats, and 17.5 g/kg BW for mice. Gross pathology examination revealed pulmonary congestion, hemorrhages, and edema, as well as congestion in other visceral organs.[1, 2]

Repeated Exposure. Rats and mice were gavaged with 1/10 LD_{50} of the pure chemical and its oil solution every other day for 1 month. The treatment caused retardation of BW gain and increased liver relative weights.

Short-term Toxicity. Similar changes were noted in rats and mice exposed to 1/20 LD_{50} for 135 days. Histology examination failed to find any lesion in the visceral organs.[2]

Standards. ***Russia.*** Recommended PML: *n/m.*

References:

1. Druzhinina, V. A. and Kochetkova, T. A., in *Current Problems of Environmental Hygiene*, F. F. Erisman Research Sanitary Hygiene Institute, A. P. Shitzkova, Ed., Moscow, 1976, 53 (in Russian).
2. Komarova, Ye. N., Krynskaya, I. L., and Robachevskaya, Ye. G., Data on toxicology of ethers of diphenoic acid, in *Toxicology and Hygiene of High-Molecular-Mass Compounds and of the Chemical Raw Material Used in Their Synthesis*, Proc. 3rd All-Union Conf., S. L. Danishevsky, Ed., Khimiya, Moscow-Leningrad, 1966, 89 (in Russian).

BIS[2-(2-BUTOXYETHOXY)ETHOXY]METHANE

Molecular Formula. $C_{17}H_{36}O_6$

M = 336.53

CAS No 143-29-3

RTECS No PA3400000

Synonyms and **Trade Names.** Bis(butylcarbitol) formal; Cryoflex; 5,8,11,13,16,19-Hexaoxatricosane; Dibutylcarbitol formal.

Properties. Yellowish liquid.

Applications. A plasticizer and softener in the production of polyurethanes, polyacrylates, natural and synthetic rubbers intended for contact with food or drink.

Acute Toxicity. In rats, LD_{50} is 1.75 g/kg BW or 3.05 g/kg BW; in mice, it is 2.7 g/kg BW. Interspecies susceptibility is not evident. CNS depression occurred before the death of animals.

Repeated Exposure failed to reveal cumulative properties. K_{acc} exceeded 10 (by Lim). Habituation develops.

Reference:

Timofiyevskaya, L. A., Toxicity and hazard of some plasticizers, *Gig. Sanit.*, 5, 87, 1981 (in Russian).

CARBONIC ACID, CYCLIC PROPYLENE ESTER

Molecular Formula. $C_4H_6O_3$

M = 102.10

CAS No 108-32-7

RTECS No FF9650000

Abbreviation. PC.

Synonyms and **Trade Names.** Arconate 5000; Carbonic acid, cyclic propylene ester; Cyclic propylene carbonate; 4-Methyl-1,3-dioxolan-2-one; 1-Methylethylene carbonate; 1,2-Propanediol carbonate; Propylene carbonate; Texacar PC.

Properties. Colorless and odorless liquid, soluble in water and alcohol.

Applications. Plasticizer and solvent.

Acute Toxicity. LD_{50} is 20.7 g/kg BW in mice.

Regulations. ***U.S. FDA*** (1998) approved the use of PC in adhesives as a component of articles intended for use in packaging, transporting, or holding food in accordance with the conditions prescribed by 21 CFR part 175.105.

Reference:

Acute Toxicity Data, *J. Am. Coll. Toxicol.*, 6, 23, 1987.

CARBONIC ACID, DIPHENYL ESTER

Molecular Formula. $C_{13}H_{10}O_3$

M = 214.23

CAS No 102-09-0

RTECS No FJ0500000

Abbreviation. DPC.

Synonyms. Diphenyl carbonate; Phenyl carbonate.

Properties. Lustrous needles. Practically insoluble in water, soluble in hot alcohol.[020]

Applications. A plasticizer and solvent in the production of polycarbonate resins and nitrocellulose.

Carcinogenicity. Mice received DPC for 78 weeks. The treatment caused an increase in tumor incidence in the liver and respiratory system (lungs, thorax). Equivocal tumorigenic agent by *RTECS* criteria.

Regulations.

EU (1990). DPC is available in the *List of authorized monomers and other starting substances which may continue to be used for the manufacture of plastic materials and articles intended to come into contact with foodstuffs pending a decision on inclusion in Section A* (*Section B*).

Great Britain (1998). DPC is authorized up to the end of 2001 for use in the production of polymeric materials and articles in contact with food or drink or intended for such contact.

Reference:

Natl. Technical Information Service PB223-159.

CHLORINATED PARAFFINS

CAS No 63449-39-8

RTECS No RV0450000

Abbreviation. CP.

Synonyms and **Trade Names.** Cerechlor; Chlorinated paraffins and hydrocarbon waxes; Chlorocosane; Chloroparaffins; Chlorez; Chlorowax.

Properties. Light-yellow to amber, thick, viscous, odorless liquids. Viscosity and density of CP increase with the chlorine content. CP differ in the amount of chlorine contained in the molecule (from 28 to 70%). There are many grades of CP. CP of short and medium chain length are oily liquid materials at ambient temperature. The long-chain highly chlorinated paraffins are solid waxy materials. At 65 to 70% of chlorine, they are whitish solids. They have an odor when the chlorine content exceeds or is equal to 55%. Insoluble in water and alcohol, soluble in plasticizers, vegetable oil, and fat.

Applications. Used as plasticizers for polyvinyl chloride, rubber, and other food-contact polymeric material. Used also in production of polyethylene sealants and detergents. Flame-retardants in plastics.

Acute Toxicity. Doses that produced no intestinal obstruction failed to cause injury in treated animals.[022] Rats fed 1.0 g for 42 days developed no toxic effect.[04]

Short-term Exposure. Weanling Sprague-Dawley rats were fed diets containing 5.0 to 5000 ppm of medium-chain CP for 13 weeks. The treatment produced an increase in relative liver and kidney weights as well as some biochemical changes at the highest dose level. Histological examination revealed minimal changes in the kidney proximal tubules of males fed the 5000 ppm diet and in the inner medulla tubules of female rats fed the 500 and 5000 ppm diets. 50-ppm dose level was likely to be the LOEL.

Regulations. ***U.S. FDA*** (1998) approved the use of CP (1) as cross-linking agent in polysulfide polymer-polyepoxy resins used as the surface contacting dry food in accordance with the conditions prescribed by 21 CFR part 177.2420; and (2) in the manufacture of resinous and polymeric coatings for the food-contact surface of articles intended for use in producing, manufacturing, packing, processing, preparing, treating, packaging, transporting, or holding food (Types I and II) in accordance with the conditions prescribed by 21 CFR part 175.300.

Reference:

Poon, R., Lecavalier, P., Chan, P., Viau, C., et al., Subchronic toxicity of a medium-chain chlorinated paraffin in the rat, *J. Appl. Toxicol.,* 15, 455, 1995.

CHLORINATED PARAFFINS, 48.5% chlorination

Abbreviation. CP.

Acute Toxicity. In mice, LD_{50} of CP containing 48.5% chlorine and 0.0024% iron is 26 g/kg BW.

Repeated Exposure failed to reveal cumulative effect in mice given 2.6 g/kg BW for a month. Consumption of 1.0 g CP with the diet over 6 weeks caused no signs of intoxication in rats.[05]

Reference:

Abasov, D. M., in *Proc. Azerbaidjan Research Institute Occup. Hygiene*, Baku, Issue No 5, 1970, 180 (in Russian).

CHLORINATED PARAFFINS $C_{10\text{-}13}$, 58% chlorination

Chlorinated paraffins C_{12}***, 60% chlorination***

CAS No 108171-26-2

Abbreviation. CP.

Acute Toxicity. LD_{50} exceeds 4.0 g/kg BW.[1] LD_{50} for CP C_{12}, 59% chlorination, exceeds 21.5 ml/kg BW in rats.[2]

Repeated Exposure. F344 rats were given doses of 900 to 27,300 ppm in the diet or 30 to 3000 mg/kg BW by gavage over a period of 14 days. The liver was found to be the target organ (its weight was increased at dose levels of 100 mg/kg BW and above). Histology examination revealed hepatocellular hypertrophy. The NOEL was considered to be 30 mg/kg BW.[3,4]

In a 16-day study, F344 rats were dosed with 7.5 g/kg BW (CP C_{12}, 60% chlorination). The treatment caused animal mortality. $B6C3F_1$ mice died when they received 3.75 g/kg BW and more. The livers of animals were found to be enlarged.[5] Male rats and mice were administered Chlorowax 500C (short-chain CP with 58% chlorination) and Cereclor 56L (short-chain CP with 56% chlorination) in corn oil at dose levels of 10 to 1000 mg/kg BW for 14 days. The NOEL for hepatic peroxisome proliferation, as determined by the CN^- insensitive palmitoyl co-enzyme A β-oxidation assay, was calculated as 184 and 600 mg/kg BW or 180 and 120 mg/kg BW for rats and mice, respectively. Rats administered the highest dose showed a depressed plasma thyroxine level, with a concomitant increase in the plasma concentrations of thyroid-stimulating hormone. The decreased plasma thyroxine levels appeared to be the result of increased plasma thyroxine glucuronidation.[6]

Short-term Toxicity. In a 90-day study, F344 rats were exposed to the CP doses of 10 to 625 mg/kg BW (in the diet or by gavage). The treatment showed the kidney and parathyroid glands to be the target organs in addition to the liver. No overt signs of toxicity were noted. There was an increase in the liver and kidney weights at doses 100 mg/kg BW and above. Thyroid-parathyroid weights were increased at 625 mg/kg BW. Microscopically, hepatocellular hypertrophy was shown to develop. No treatment-related microscopic changes were found in the tissues. The NOEL by oral route appeared to be 10 mg/kg BW.[3,4]

In a 90-day study, rats tolerated the doses up to 5.0 g/kg BW, mice tolerated up to 2.0 g/kg BW. An increase in the liver weight was noted. Histology examination revealed liver hypertrophy.[5]

Reproductive Toxicity. F344 rats received 100 to 200 mg/kg BW, rabbits were fed 10 to 100 mg/kg BW. The doses of 30 mg/kg BW and 100 mg/kg BW caused maternal toxicity in rats, a 100 mg/kg dose increased post-implantation loss, early and late resorptions, and decreased the number of viable fetuses per dam. No effects on dams or fetuses were noted at the lowest dose. Rabbits were more sensitive, but no teratogenic effects were observed in any dose group. The doses of 28 to 1000 ppm in the diet were administered to young ducks. The NOEL for reproduction was considered to be 166 ppm in the diet.[3,4] However, significance of these studies is questioned by IARC Workgroup.

Mutagenicity.

In vivo cytogenetics. 250 to 2500 mg/kg BW was administered orally (by gavage) to F344 rats for 5 consecutive days. The exposure did not increase the frequency of chromosomal or chromatid aberrations in the bone marrow cells. CP was shown to be negative in DLM assay: being administered at dose levels of 250 to 2000 mg/kg BW, it did not induce any increase in early fetal deaths or decrease in viable embryos during the ten matings of the study.[3]

In vitro genotoxicity. CP C_{10-13}, 50% chlorination,[1] and CP C_{12}, 60% chlorination,[3] were not found to be mutagenic in several strains of *Salmonella* with and without metabolic activation. Chlorowax 500C was found to be negative in *Salmonella* mutagenicity assay in the presence and absence of *S9* metabolic activation system.[015]

Carcinogenicity. The doses of 312 and 625 mg/kg BW (CP C_{12}, 60% chlorination, length of exposure 104 weeks) produced clear evidence of carcinogenicity, namely, increased incidence of hepatocellular neoplasms, adenomas or adenocarcinomas of the kidney tubular cells in male F344 rats, follicular cell adenomas or carcinomas of the thyroid gland in female rats.[3] In B6C3F_1 mice, an increase in the incidence of hepatocellular tumors and of alveolar/bronchiolar carcinomas in males, and of follicular-cell tumors of the thyroid gland in females are reported.[5] As usual, in the NTP studies, extremely large doses to provide adequate information were used. These doses could produce a marked liver and kidney toxicity.[3] Anyway, carcinogenicity of CP may occur through a non-genotoxic mechanism, according to mutagenicity data.

Carcinogenicity classifications. An IARC Working Group concluded that there is sufficient evidence for the carcinogenicity of the CP (C_{12}, 60% chlorination) in *experimental animals* and there were no data available to evaluate its carcinogenicity in *humans*.

The Annual Report of Carcinogens issued by the U.S. Department of Health and Human Services (1998) defines CP (C_{12}, 60% chlorination) to be a substance, which may reasonably be anticipated to be carcinogen, i.e., a substance for which there is limited evidence *of carcinogenicity* in *humans or sufficient evidence of carcinogenicity* in *experimental animals.*

IARC: 2B;

NTP: CE* - CE* - CE* - CE* (C_{12}, 60%, gavage).

Chemobiokinetics. A part of the dose was absorbed, but the main part was excreted with the feces. Radiolabel assay has shown distribution mainly in the liver, fatty tissue, and ovary.[3]

References:

1. Birtley, R. D. N., Conning, D. M., Daniel, J. W., et al., The toxicological effects of chlorinated paraffins in mammals, *Toxicol. Appl. Pharmacol.*, 54, 514, 1980.
2. Howard, P. H., Santodonato, J., and Saxena, J., *Investigation of selected potential environmental contaminants: chlorinated paraffins,* EPA-560/2-75-007; PB 248634, Washington, D.C., 1975.
3. Serrone, D. M., Birtley, R. D. N., Weigand, W., et al., Toxicology of chlorinated paraffins, *Food Chem. Toxicol.*, 25, 553, 1987.
4. *Toxicology and Carcinogenesis Studies of Chlorinated Paraffins (C_{12}, 60% chlorine average content),* NTP Technical Report Series No 308, U.S. Dept. Health & Human Service, Publ. Health Service, NIH Publ., 1986.
5. Bucher, J. R., Alison, R. H., Montgomery, C. A., et al., Comparative toxicity and carcinogenicity of two chlorinated paraffins in Fischer 344/N rats and B6C3F_1 mice, *Fundam. Appl. Toxicol.*, 9, 454, 1987.

6. Wyatt, I., Coutts, C. T., and Elcombe, C. R., The effect of chlorinated paraffins on hepatic enzymes and thyroid hormones, *Toxicology*, 77, 81, 1993.

CHLORINATED PARAFFINS C_{14-17}, 52% chlorination

Abbreviation. CP.

Acute Toxicity. LD_{50} exceeds 4.0 g/kg BW.[1]

Repeated Exposure. In a 14-day study, dietary administration of 150 to 15000 ppm increased liver weights in F344 rats. Histology examination revealed diffuse hepatocellular hypertrophy at 5,000 ppm and 15,000 ppm in the diet. Dietary level of 500 ppm was considered to be the NOEL.[2,3]

Short-term Toxicity. In a 90-day study, increased liver and kidney weights at doses of 100 mg/kg and higher were observed. At 625 mg/kg BW dose level, thyroid and parathyroid weights were increased in male rats and adrenal weights were increased in both males and females. Histology examination revealed hepatocellular hypertrophy in the liver in the high dose group. The NOEL was identified to be 10 mg/kg BW.[2,3]

Reproductive Toxicity. Despite a high level in the ovary, no morphological changes were observed in this organ. Rats and rabbits were given 500 to 5000 mg/kg BW and 10 to 100 mg/kg BW, respectively. There were signs of maternal toxicity in the rats at the high dose level. Rabbits seem to be more sensitive. No teratogenicity effect was noted.[2]

Mice were fed doses of 100 to 6250 ppm in their feed for 28 days before and during mating, and in the case of females, continuously up to postnatal day 21. Pups were given the same diet as their parents from weaning until the pups were 70 days of age. No impairment of reproductive function was noted.[2] However, the significance of these studies is questioned by IARC Workgroup.

Mutagenicity.

In vivo cytogenetics. CP was administered orally to F344 rats by gavage for 5 consecutive days at doses of 500 to 5000 mg/kg BW. The exposure did not increase the frequency of chromosomal or chromatid aberrations in bone marrow cells.

In vitro genotoxicity. CP C_{14-17}, 52% chlorination, was not found to be mutagenic in several strains of *Salmonella* with and without metabolic activation.[1]

Carcinogenicity. No evidence of carcinogenicity was noted in male rats. In females, there was an increased incidence of adrenal gland medullary neoplasms.[3]

The liver seems to be the target organ in this study, but no carcinogenicity in the liver was reported in the NTP Report. An increased incidence of malignant lymphoma was noted in $B6C3F_1$ mice. However, the induction of lymphomas in mice as an index of carcinogenic activity might be questioned. In any case, this provides no clear evidence of carcinogenic potential.[2]

Chemobiokinetics. Absorption from the GI tract is limited: a part of the dose was absorbed, but mostly it was excreted with the feces. Radiolabelled assay has shown distribution mainly in the liver, adipose tissue, and ovary. Fecal excretion includes both unabsorbed material and radiolabelled material excreted in the bile. Tissue concentrations were the highest initially in the liver and kidney and later in the adipose tissue and ovary.[2,3]

References:

1. Birtley, R. D. N., Conning, D. M., Daniel, J. W., et al., See ***Chlorinated paraffins*** C_{10-13}, 58% ***chlorination***.
2. Serrone, D. M., Birtley, R. D. N., Weigand, W., et al., See ***Chlorinated paraffins*** C_{10-13}, 58% ***chlorination***.
3. *Toxicology and Carcinogenesis Studies of Chlorinated Paraffins (C_{23}, 43% chlorine, average content),* NTP Technical Report No 305, U.S. Dept. Health & Human Service, Publ. Health Service, NIH, 1986.

CHLORINATED PARAFFINS C_{18}, 50-53% chlorination

Chemobiokinetics. Sprague-Dawley rats were treated orally with 0.5 g ^{14}C-labeled CP (C_{18}, 50-53% chlorination)/kg BW. Approximately 86% of the orally administered dose of [^{14}C]*polychlorooctadecane* was recovered.

References:

Yang, J. J., Roy, T. A., Neil, W., Krueger, A. J., and Mackerer C. R., Percutaneous and oral absorption of chlorinated paraffins in the rat, *Toxicol. Ind. Health.*, 3, 405, 1987.

CHLORINATED PARAFFINS C_{20-30}, 43% chlorination

Chlorinated paraffins C_{23}, 43% chlorination

CAS No 108171-27-3

Abbreviation. CP.

Acute Toxicity. LD_{50} exceeds 4.0 g/kg BW.[1] Rats tolerated 10 ml/kg BW (CP C_{24}, 40% chlorination).[2]

Repeated Exposure. F344 rats received by gavage 30 to 3000 mg/kg BW for 14 days. No treatment-related effects were found in organ weights or in tissue histology.[3]

Male rats and mice were administered CP 40G (medium-chain CP with 40% chlorination) in corn oil at dose levels of 10 to 1000 mg/kg BW for 14 days. The NOEL for hepatic peroxisome proliferation, as determined by the CN-insensitive palmitoyl co-enzyme A β-oxidation assay, was calculated as 473 mg/kg BW and 252 mg/kg BW for rats and mice, respectively. Rats administered the highest dose showed a depressed plasma thyroxine level, with a concomitant increase in the plasma concentrations of thyroid-stimulating hormone. The decreased plasma thyroxine levels appeared to be the result of increased plasma thyroxine glucuronidation.[4]

Short-term Toxicity. In a 90-day study, F344 rats and $B6C3F_1$ mice were given CP C_{23}, 43% chlorination, by gavage at dose levels of 3750 mg/kg BW and 7500 mg/kg BW, respectively, in corn oil. No manifestations of the toxic action was noted in the treated animals.[5]

In another 90-day study, 100 to 3750 mg CP/kg BW administered by gavage produced no adverse effect on BW gain, water or food consumption or clinical biochemistry indices. There were the treatment-related effects (inflammatory changes and necrosis on histology examination) on the liver in female rats at all doses, but no damage was observed in the livers of males.[3]

Similar hepatic lesions were found in female F344 rats in a 90-day study, and in male rats receiving much larger doses of Chlorowax 40 for 6 to 12 months.[6]

Long-term Toxicity. In a 2-year oral rat study, only lymphocytic infiltration and granulomatous inflammation in the liver mesenteric and pancreatic lymphoid nodes were found. Mice given 5000 mg/kg BW dose displayed no non-neoplastic lesions.[5]

Reproductive Toxicity.

Teratogenicity. Rats and rabbits were given doses of 500 to 5000 mg/kg BW. There were signs of maternal toxicity in rats at high dose level without any fetal malformations. Rabbits exhibited similar sensitivity. No teratogenic response was shown.[3] However, significance of these studies was questioned by the IARC Work Group.

Mutagenicity.

In vivo cytogenetics. CP was administered orally by gavage to F344 rats for 5 consecutive days at doses of 500 to 5000 mg/kg BW. The exposure did not increase the frequency of chromosomal or chromatid aberrations in bone marrow cells.[3]

In vitro genotoxicity. CP C_{20-30}, 42% chlorination, is not found to be mutagenic in several strains of *Salmonella* with and without metabolic activation.[1]

Carcinogenicity. $B6C3F_1$ mice were dosed with 2500 and 5000 mg CP C_{23}, 43% chlorination/kg BW, for 103 weeks. There was an increase in the incidence of malignant lymphomas in males. F344 rats received 1875 mg/kg and 3750 mg/kg (males) and 100 mg/kg, 300 mg/kg, and 900 mg/kg BW (females) for 103 weeks. The treatment caused phaeochromocytomas of the adrenal medulla in females.[5]

Carcinogenicity classification (C_{23}, 43% chlorination).

NTP: NE - EE - CE - EE (gavage).

Chemobiokinetics. A higher level of radioactivity was detected in the ovary than in the blood or adipose tissue during the first 7 days after administration of the labeled material.[3]

References:

1. Birtley, R. D. N., Conning, D. M., Daniel, J. W., et al., See ***Chlorinated paraffins*** C_{10-13}, 58% ***chlorination***.
2. Howard, P. H., Santodonato, J., and Saxena, J., See ***Chlorinated paraffins*** C_{10-13}, 58% ***chlorination***.
3. Serrone, D. M., Birtley, R. D. N., Weigand, W., et al., See ***Chlorinated paraffins*** C_{10-13}, 58% ***chlorination***.
4. Wyatt, I., Coutts, C. T., and Elcombe, C. R., The effect of chlorinated paraffins on hepatic enzymes and thyroid hormones, *Toxicology*, 77, 81, 1993.
5. Bucher, J. R., Alison, R. H., Montgomery, C. A., et al., See ***Chlorinated paraffins*** C_{10-13}, 58% ***chlorination***.
6. Bucher, J. R., Montgomery, C. A., Thompson, R., et al., Hepatic lesion associated with administration of Chlorowax 40 to F344/N rats, *Toxicologist*, 5, 158, 1984.

CHLORINATED PARAFFINS C_{22-26}, 70% chlorination

Abbreviation. CP.

Acute Toxicity. LD_{50} exceeds 4.0 g/kg BW.[1] Rats tolerate 50 g/kg BW (CP C_{24}, 70% chlorina tion).[2]

Repeated Exposure. F344 rats received CP for 14 days by dietary administration of 150 to 15000 ppm. No treatment-related effects were found in organ weights or in tissue histology.[3]

Short-term Toxicity. Dietary administration of 100 to 3750 mg/kg BW for 90 days caused a slight decrease in BW gain at the highest dose level. Increased ALT and AST activity and liver weights in the high-dose group were noted. Hepatocellular hypertrophy and cytoplasmic fat vacuolation was found to develop.[3]

Reproductive Toxicity. Rats received doses of 500 to 5000 mg/kg BW, and rabbits were administered 100 to 1000 mg/kg BW. Neither rats nor rabbits exhibited maternal toxicity or developmental abnormalities.[3] However, the significance of these studies is questioned by the IARC Workgroup.

Mutagenicity.

In vivo cytogenetics. CP was administered orally by gavage to F344 rats for 5 consecutive days at doses of 500 to 5000 mg/kg BW. The exposure did not increase the frequency of chromosomal or chromatid aberrations in the bone marrow cells.

In vitro genotoxicity. CP C_{23}, 43% chlorination,[4] and CP C_{10-20}, 70% chlorination,[5] are not found to be mutagenic in several strains of *Salmonella* with and without metabolic activation.

Chemobiokinetics. A small part of radiolabelled material was absorbed after oral administration. The highest content of radioactivity was noted in the liver. Radioactivity did not appear to concentrate in the ovary unlike other CP. The NOEL of 900 mg/kg BW was identified in this study.[3]

References:

1. Birtley, R. D. N., Conning, D. M., Daniel, J. W., et al., See ***Chlorinated paraffins*** C_{10-13}, 58% ***chlorination***.
2. Howard, P. H., Santodonato, J., and Saxena, J., See ***Chlorinated paraffins*** C_{10-13}, 58% ***chlorination***.
3. Serrone, D. M., Birtley, R. D. N., Weigand, W., et al., See ***Chlorinated paraffins*** C_{10-13}, 58% ***chlorination***.
4. NTP (1986), See ***Chlorinated paraffins*** C_{10-13}, 58% ***chlorination***.
5. Meijer, J., Rundgren M., Astrom, A., et al., Effects of chloroparaffins on some drug-metabolizing enzymes in the rat liver and in the Ames test, *Adv. Exper. Med. Biol.*, 136, 821, 1981.

2-CHLOROETHANOL PHOSPHATE (3:1)

Molecular Formula. $C_6H_{12}C_{13}O_4P$

M = 365.51

CAS No 115-96-8

RTECS No KK2450000

Abbreviation. CEP.

Synonyms. Phosphoric acid, tri(2-chloroethyl) ether; Tri(2-chloroethyl)phosphate.

Properties. Clear, colorless liquid of low volatility with a faint odor. Solubility in water is 0.7%, soluble in nu merous organic solvents.

Applications. Used as a plasticizer in the production of cellulose derivatives. A plasticizer for fire-resistant lacquers (coatings) and plastics, intended for contact with food, based on ethyl cellulose, polyester resins, polyacrylates, and polyurethanes.

Acute Toxicity. LD_{50} is reported to be 0.2 to 0.4 g/kg BW,[05] 0.63 to 0.74 g/kg BW,[1] or 6.8 g/kg BW[2] in rats, 0.74 g/kg BW in mice, and 1.0 g/kg BW in guinea pigs.[1] According to other data, LD_{50} in rats is 1.25 g/kg BW (sex unspecified), or 0.5 g/kg BW in males and 0.43 to 0.8 g/kg BW in females.[3]

CEP produces a weak anticholinesterase action. Poisoning is accompanied by tonic and tetanus-like convulsions but not paralysis. The skin of animals becomes cyanotic and they assume side position. Death within 1 to 2 days. Gross pathology examination revealed celiectasia, visceral congestion, and tuberosity of the spleen.

Repeated Exposure revealed slight cumulative properties. CEP exhibits a polymorphous toxic action with the CNS as the primary target followed by the liver, kidneys, and myocardium. Rats were given oily solution of CEP (dose level of 6.3 and 63 mg/kg BW) for 1.5 months. No deaths occurred. The greater dose resulted in decreased activity of some enzymes (AST, ALT, lactate dehydrogenase, creatinekinase, *X*-hydroxybutyrate dehydrogenase) and creatinine content in the blood serum and other changes. The dose of 6.3 mg/kg BW caused less pronounced reversible changes.[1]

There were no chemical-related deaths, differences in final mean BW, or histopathological lesions in rats receiving 22 to 350 mg CEP/kg BW or in mice receiving 44 to 700 mg CEP/kg administered in corn oil by gavage for 12 doses over 16 days.[4,5] Serum cholinesterase activity in female rats receiving 175 or 350 mg CEP/kg BW was reduced slightly.

Short-term Toxicity. Administration of 350 mg/kg BW by oral gavage over a period of 16 weeks resulted in necrosis of pyramidal neurons in the CA I region of the hippocampus of Fischer 344 rats but not in $B6H3F_1$ mice.[4,5]

Long-term Toxicity. F344 rats received 44 and 88 mg CEP/kg BW by gavage for up to 103 weeks, $B6C3F_1$ mice received 175 and 350 mg CEP/kg BW for 66 weeks. The principal toxic effect occurred in the brain and kidney.[5]

Reproductive Toxicity. ***Embryo-*** and ***gonadotoxic effect*** was not demonstrated with the 6.3 mg/kg BW dose.[1] Wistar rats were gavaged with 50 to 200 mg/kg BW suspended in olive oil on days 7 to 15 of gestation. There were no changes in maternal body weight, food consumption, on general appearance up to 100 mg/kg BW level. A dose of 200 mg/kg BW caused weakness, decreased maternal food consumption; 7 out of 30 dams died. No malformations were registered at any dose, there was normal development of the offspring of all groups.[6]

Mutagenicity.

In vitro genotoxicity. CEP was negative in *Salmonella* mutagenicity assay.[025]

In vivo cytogenetics. CEP is not found to react with DNA *in vivo,*[7] it did not cause an increased number of SCE in Chinese hamster *V79* cell line and DLM in rats after inhalation exposure.[5,8,9]

Carcinogenicity. In the above described study,[4] renal tubular hyperplasia and adenoma were observed. Renal neoplasms were found in 10% of low-dose and in 50% of high-dose male rats. Mice were less sensitive.

Carcinogenicity classifications. An IARC Working Group concluded that there is inadequate evidence for the carcinogenicity of CEP in *experimental animals* and there were no data available to evaluate the carcinogenicity of CEP in *humans.*

IARC: 3;

NTP: CE - CE - EE - EE (gavage).

Chemobiokinetics. *In vitro* metabolism by rat and human liver slices led to formation of *bis(2-chloroethyl)hydrogen phosphate* and *2-chloroethanol* and three unidentified minor metabolites. CEP metabolism was approximately twice as rapid in rat liver slices as in human liver slices (NTP-92).

Regulations. ***U.S. FDA*** (1998) listed CEP for use in adhesives as a component of articles intended for use in packaging, transporting, or holding food in accordance with the conditions prescribed by 21 CFR part 175.105.

Standards. ***Russia.*** Proposed MAC and PML: 1.0 mg/l.

References:

1. Zaitsev, N. A. and Skachkova, I. N., Hygienic regulation of tri(chloroethyl) phosphate in water, *Gig. Sanit.*, 9, 77, 1989 (in Russian).
2. Dvorkin, E. A., in *Proc. 2nd Conf. Junior Scientists,* Leningrad Research Institute Occup. Diseases, Leningrad, 1948, 43 (in Russian).
3. Ulsamer, A. G., Osterberg, R. E., McLaughlin, J., et al., Flame retardant chemicals in textiles, *Clin. Toxicol.,* 17, 101, 1980.
4. Matthews, H. B., Eustis, S. L., and Haseman, J., Toxicity and carcinogenicity of chronic exposure to tris(2-chloroethyl)phosphate, *Fundam. Appl. Toxicol.*, 20, 477, 1993.
5. *Toxicology and Carcinogenicity Studies of Tris(2-chloroethyl) Phosphate in F344/N Rats and B6C3F$_1$ Mice (Gavage Studies),* NTP Technical Report Series No 391, Research Triangle Park, NC, May 1991.
6. Kawashima, K., Tanaka, S., Nakaura, S., et al., Effect of oral administration of tris(2-chlorethyl) phosphate to pregnant rats on prenatal and postnatal developmental, *Bull. Natl. Inst. Hyg. Sci.*, 101, 55, 1983.
7. Lown, J. W., Joshua, A. V., and Melaughlin, L. W., Novel antitumor nitrosoureas and related compounds and their reactions with DNA, *J. Med. Chem.*, 23, 798, 1980.
8. Sala, M., Gu, Z. G., Meons, G., et al., *In vivo* and *in vitro* biological effects of the flame retardant tris(2,3-dibromopropyl)phosphate and tris(2-chlorethyl)phosphate, *Eur. J. Cancer,* 18, 1337, 1982.
9. Shepelskaya, N. R. and Dyshinevich, N. Ye., Experimental study of the gonadotoxic effect of tris(2-chlorethyl) phosphate, *Gig. Sanit.*, 6, 20, 1881 (in Russian).

1-CHLORONAPHTHALENE

Molecular Formula. $C_{10}H_7Cl$

M = 162.62

CAS No 90-13-1

RTECS No QJ2100000

Abbreviation. CN.

Synonym. α-Chloronaphthalene.

Properties. Clear or slightly yellowish liquid without precipitate. Solubility in water is 20 mg/l. Readily soluble in organic solvents. Threshold concentration for odor is 0.01 mg/l. 1-CN gives water a sweetish, astringent taste, the perception threshold for which is higher than for odor. Half-life in aqueous environment is 1 to 2 days.

Applications. CN is used in the production of plasticizers.

Acute Toxicity. LD_{50} is reported to be 2.2 g/kg BW in rats, 1.1 g/kg BW in mice, and 2.0 g/kg BW in guinea pigs. Poisoning is accompanied by apathy and adynamia. Death occurs in 3 days.

Repeated Exposure revealed evident cumulative properties.

Long-term Toxicity. Rats were dosed by gavage. The treatment produced changes in the kidneys similar to those of subacute glomerulonephritis. Cholinesterase activity was reduced.

Allergenic Effect. See ***Naphthalene***.

Chemobiokinetics. CN was shown to be converted by pretreated liver microsomes *in vitro* to active intermediate which formed covalent complexes with ^{35}*S-labeled glutathione* (Brodie *et al.*, 1971).

Standards. ***Russia*** (1995): MAC and PML 0.1 mg/l (organolept., odor).

Reference:

Matorova, N. I., Data on the substantiation of the maximum allowable concentration for naphthalin and naphthalene in water bodies, *Gig. Sanit.*, 11, 78, 1982 (in Russian).

CITRIC ACID, ACETYL TRIBUTYL ESTER

Molecular Formula. $C_{20}H_{34}O_6$
M = 402.54
CAS No 77-90-7
RTECS No TZ8330000
Abbreviation. ATBC.

Synonyms and **Trade Name.** 2-Acetoxy-1,2,3-propanetricarboxylic acid, tributyl ester; Acetylcitric acid, tributyl ester; Acetyl tributyl citrate; Citrolex A.

Properties. Colorless liquid with faint sweet, herbaceous odor. Poorly soluble in water, soluble in alcohol. High concentrations induce mild fruity flavor. Odor perception threshold is 50 mg/l. Taste perception threshold is 25 mg/l.

Applications. ATBC is used as a plasticizer for flexible packaging films, in particular, vinylchloride-vinylidene chloride copolymer films, and cellulose. It has been cleared for use in the production of food-contact surfaces of resinous and polymeric coatings and in paper/paperboard for use in contact with fatty foods.

Exposure. Daily intake is reported to be 1.5 mg.[1]

Migration in the cheese wrapped in vinylidene chloride copolymer films (exposure 5 days, temperature 5°C) was found at the level of 6.1 ppm or 2.0 to 8.0 mg/kg, and into wrapped cake, at the level of 3.2 ppm.[2,3] Migration from plasticized vinylidene chloride-vinyl chloride *copolymer* film in fatty or aqua-type foods was determined at the levels from 0.4 mg/kg after minimal contact during microwave cooking of a soup to 79.8 mg/kg for use of the film during the microwave cooking of peanut-containing cookies.[4] Migration of ATBC plasticizer from plasticized polyvinylidene chloride-polyvinyl chloride (Saran) films into both olive oil and distilled water during microwave heating was studied. The amount of ATBC migrating into olive oil after heating for 10 min was 73.9 mg ATBC/l, into distilled water it was 4.1 mg/l after heating for 8 min.[5]

No differences in migrated amounts between ATBC irradiated with γ-radiation and non-irradiated samples of food-grade polyvinyl chloride and polyvinylidene chloride-polyvinyl chloride were observed. The amount of ATBC that migrated into olive oil at 4 to 5°C after 97 hours of contact was non-detectable (<1.0 mg/l). Concentrations of ATBC at 20°C, after 29 and 94 hours were 3.3 and 5.1 mg/l, respectively.[6]

Acute Toxicity. Rats and mice tolerate doses of up to 20 g ATBC/kg BW without signs of intoxication. Some animals died when given 22 to 24 g/kg BW.[7,8] LD_{50} is reported to be 31.4 g/kg BW in rats.[9]

Repeated Exposure. Rats were dosed by gavage with 0.125 to 2.5 g ATBC/kg BW. The treatment caused retardation of BW gain, a reduction in STI, and an increase in the relative weights of the liver.[8] Rats received 10% ATBC in their diet for 8 weeks. The treatment produced no marked signs of toxicity. A decrease in BW gain (because of diarrhea) was reported.[04]

Short-term Toxicity. Rats and mice were exposed to 0.4 g/kg BW and 1.0 g/kg BW for 4 months. The administration caused no changes in general condition or in BW gain of animals. No effect of ATBC was found on blood coagulation time, content of blood serum calcium, prothrombin time or detoxication- and protein-forming functions of the liver. There were no changes in the relative weights of the visceral organs.[7]

Long-term Toxicity. Rats and mice received 50 and 250 mg ATBC/kg BW as a milk solution with their feed. The 250 mg/kg dose caused an increase in BW gain and a reduction in the STI value in mice. A reduction in blood peroxidase activity of rats, followed by an increase, was observed. There were no deviations in the indices studied at the end of the experiment. A 50 mg/kg BW dose was identified to be the NOEL.[8]

Reproductive Toxicity.

Gonadotoxicity. Exposure to ATBC produced no significant effect on the male gonads. Only some exfoliation of the spermatogenic epithelium was reported.[9]

Embryotoxicity. An increase in the fetal weight and size and also in the weight of the placenta was noted in a long-term study with administration of a 250 mg/kg BW dose. Growth and development of the progeny were unchanged.[9]

Mutagenicity.

In vitro cytogenetics. ATBC was shown to be negative in *Salmonella* mutagenicity assay and in a clastogenicity assay in rat lymphocytes, but a study of gene mutations in mammalian cells was considered inadequate.[10]

Regulations.

U.S. FDA (1998) listed ATBC for use (1) as a plasticizer in food-packaging materials in accordance with the conditions prescribed by 21 CFR part 178.3740; (2) in adhesives as a component of articles intended for use in packaging, transporting, or holding food in accordance with the conditions prescribed by 21 CFR part 175.105; (3) in the manufacture of resinous and polymeric coatings to be safely used as a food-contact surface of articles in accordance with the conditions prescribed by 21 CFR part 175.300; and (4) in the manufacture of resinous and polymeric coatings for polyolefin films for food-contact surface of articles intended for use in producing, manufacturing, packing, processing, preparing, treating, packaging, transporting, or holding food in accordance with the conditions prescribed by 21 CFR part 175.320.

British Food Agency required labeling of food cling wraps.

Standards. *Russia* (1995). MAC and PML: *n/m.*

References:

1. Anonymous, Plasticizers migration in food, *Food Chem. Toxicol.*, 29, 139, 1991.
2. *Plasticizers: Continuing Surveillance,* The Thirtieth Report Steering Group on Food Surveillance, The Working Party on Chemical Contaminants from Food Contact Materials: Sub-Group on Plasticizers, *Food Surveillance*, Paper No 30, London: HMSO, 55, 1990.
3. Castle, L., Mercer, A. J., Startin, J. R., and Gilbert, J., Migration from plasticized films into foods. 3. Migration of phthalate, sebacate, citrate and phosphate esters from films used for retail food packaging, *Food Addit. Contam.*, 5, 9, 1988.
4. Castle, L., Jickells, S. M., Sharman, M., et al., Migration of the plasticizer acetyltributyl citrate from plastic film into foods during microwave cooking and domestic use, *J. Food Prot.*, 51, 916, 1988.
5. Badeka, A. B. and Kontominas, M. G., Effect of microwave heating on the migration of dioctyl adipate and acetyl tributyl citrate plasticizers from food-grade PVC and PVDC/PVC films into olive oil and water, *Z. Lebensmit. Unters. Forsch.*, 202, 313, 1996.
6. Goulas, A. E., Kokkinos, A., und Kontominas, M. G., Effect of gamma-radiation on migration behavior of dioctyl adipate and acetyl tributyl citrate plasticizers from food-grade PVC and PVDC/PVC films into olive oil, *Z. Lebensm. Unters. Forsch.*, 201, 74, 1995.
7. Bidnenko, L. I., in *Development of Technology for the Production of Rigid PVC and Packaging Based on It*, Proc. All-Union Conf., Kiyv, 1973, 39 (in Russian).
8. Larionov, L. N. and Cherkasova, T. E., Hygiene evaluation of acetyl tributyl citrate, *Gig. Sanit.*, 4, 102, 1977 (in Russian).
9. Casarett and Doull's *Toxicology*, J. Doull, C. D. Klassen, and M. D. Amdur, Eds., New York, Macmillan Publ. Co., 1980, 549.
10. *1994 Annual Report of the Committees on Toxicity, Mutagenicity, Carcinogenicity of the Chemicals in Food, Consumer Products and the Environment*, DoH, HMSO, London, 1995, 61.

CITRIC ACID, ACETYL TRIETHYL ESTER

Molecular Formula. $C_{14}H_{22}O_8$
M = 318.36
CAS No 77-89-4
RTECS No GE8225000
Abbreviation. ATEC.

Synonyms and **Trade Name.** Acetyl triethyl citrate; Citroflex A2; Triethyl acetyl citrate.

Applications. ATEC is used predominantly as a plasticizer in food-packaging materials.

Acute Toxicity. LD_{50} is 7.0 g/kg BW in rats.[027]

Regulations. *U.S. FDA* (1998) approved the use of ATEC (1) as a plasticizer in food-packaging materials in accordance with the conditions prescribed by 21 CFR part 178.3740; (2) in adhesives as a component of articles intended for use in packaging, transporting, or holding food in accordance with the conditions prescribed by 21 CFR part 175.105; (3) in the manufacture of resinous and polymeric coatings for the food-contact surface of articles intended for use in producing, manufacturing, packing, processing, preparing, treating, packaging, transporting, or holding food in accordance with the conditions prescribed by 21 CFR part 175.300; and (4) in the manufacture of resinous and polymeric coatings for polyolefin films for food-contact surface of articles intended for use in producing, manufacturing, packing, processing, preparing, treating, packaging, transporting, or holding food in accordance with the conditions prescribed by 21 CFR part 175.320.

CITRIC ACID, TRIETHYL ESTER

Molecular Formula. $C_{12}H_{20}O_7$
M = 276.29
CAS No 77-93-0
RTECS No GE8050000
Abbreviation. TEC.

Synonyms and **Trade Name.** Citroflex 2; Ethyl citrate; 2-Hydroxy-1,2,3-propanetricarboxylic acid, triethyl ester; Triethyl citrate.

Properties. Colorless, odorless, oily liquid with a bitter taste. Readily soluble in water (6.9%). Miscible with alcohol and ethers.

Applications. Used as a plasticizer in the production of polyvinyl acetate and cellulose citrate.

Exposure. Usual concentration in detergents is 0.015%.

Acute Toxicity. LD_{50} is found to be 5.9 to 8.0 g/kg BW in rats, and 4.0 g/kg BW in cats. Poisoning is accompanied with convulsions and respiratory disorders. Changes in behavior and the GI tract were observed.

The toxic action of TEC is likely to occur due to the binding of calcium in the body fluids.[011] Similar data were obtained in relation to *acetyl triethyl citrate*. However, its LD_{50} in rats is half that of TEC. TEC produced complete loss of blood pressure when administered in toxic doses.[1]

Repeated Exposure. Small amounts of TEC in the feed are harmless for rats and cats. Rats were given 0.5, 1.0, and 2.0% TEC in their diet for 6 or 8 weeks. The treatment did not affect BW gain or hematology analysis. Gross and histological examination failed to reveal changes in the viscera.[2]

Long-term Toxicity. According to results of a 2-year feeding study, rats can tolerate TEC at a dose of up to 2.0 g/kg BW. Dogs tolerated up to 0.25 ml/kg BW for six months without any adverse effect.[3]

Allergenic Effect.

Observations in man. TEC showed no sensitization reactions in a maximization test that was carried out on 22 volunteers (Epstein, 1975).

Reproductive Toxicity. The doses of 0.5 to 10 mg/kg BW are not ***teratogenic*** for chick embryos (Verett, 1980).

Mutagenicity.

In vitro genotoxicity. TEC is not mutagenic in bacterial tests.[2]

Chemobiokinetics. TEC is likely to be hydrolyzed *in vivo* to yield *citrate* and *ethyl alcohol* by usual biochemical routes.

Regulations.

U.S. FDA (1998) listed TEC for use (1) in the manufacture of resinous and polymeric coatings for the food-contact surface of articles intended for use in producing, manufacturing, packing, processing, preparing, treating, packaging, transporting, or holding food in accordance with the conditions prescribed by 21 CFR part 175.300; and (2) as a plasticizer in resinous and polymeric coatings for polyolefin films to be safely used as a food-contact surface of articles intended for use in producing, manufacturing, packing, transporting, or holding food in accordance with the conditions prescribed by 21 CFR part 175.320. Affirmed as *GRAS.*

EU (1995). TEC is a food additive generally permitted for use in dried egg white.

Recommendations. Joint FAO/WHO Expert Committee on Food Additives (1989). ADI for man: 10 mg/kg BW.

Standards. ***Russia.*** Recommended PML: *n/m.*

References:

1. Casarett and Doull's *Toxicology*, J. Doull, C. D.Klassen, and M. D. Amdur, 2nd ed., Macmillan Publ. Co., New York, 1980, 620.

2. Finkelstein, M. and Gold, H., Toxicology of the citric acid esters: tributylcitrate, acetyltributylcitrate, triethyl citrate, and acetyltriethylcitrate, *Toxicol. Appl. Pharmacol.*, 1, 283, 1959.
3. *The 23rd Report of Joint FAO/WHO Expert Committee on Food Additives*, Technical Report Series No 648, WHO, Geneva, 18, 1980.

DEHYDROACETIC ACID

Molecular Formula. $C_8H_8O_4$
M = 168.16
CAS No 520-45-6
RTECS No UP8050000
Abbreviation. DHA.

Synonyms and **Trade Names.** 2-Acetyl-5-hydroxy-3-oxo-4-hexenoic acid; 3-Acetyl-6-methylpyrandone-2,4; 3-Acetyl-6-methyldihydropyrandione-2,4; Methylacetopyronone.

Properties. White or cream, odorless and tasteless crystalline powder. Solubility in water is 1.0 g/l at 25°C. Poorly soluble in alcohol and olive oil.

Applications. Plasticizer.

Acute Toxicity. LD_{50} is 1.0 g/kg BW in rats.[020] Large doses caused vomiting. Kidney function was affected. In monkeys, anorexia, vomiting, weakness, stupor, ataxia, and convulsions were found to develop following high dose exposure.[019]

Short-term Toxicity.

Observations in man. Administration of a dose of 0.01 g/kg BW for 150 days caused no observable changes.[019]

Allergenic Effect was not found on skin application.[019]

Chemobiokinetics. Main metabolites are as follows: *triacetic acid lactone*, *a hydroxydihydroacetic acid* and possibly the *salt of triacetic acid lactone 3-carboxylic acid* (Spencer, 1982).

Regulations. ***U.S. FDA*** (1998) approved the use of DHA in adhesives as a component of articles intended for use in packaging, transporting, or holding food in accordance with the conditions prescribed by 21 CFR part 175.105. DHA is a food additive permitted for direct addition to food for human consumption, as long as (1) the quantity added to food does not exceed the amount reasonably required to accomplish its intended physical, nutritive, or other technical effect in food, and (2) when intended for use in or on food it is of appropriate food grade and is prepared and handled as a food ingredient in accordance with the conditions prescribed in 21 CFR part 172.130.

DEHYDROACETIC ACID, SODIUM SALT

Molecular Formula. $C_8H_7O_4.Na$
M = 190.14
CAS No 4418-26-2
RTECS No UP8225000
Abbreviation. DAS

Synonyms and **Trade Names.** 3-Acetyl-6-methyl-2*H*-pyran-2,4(3*H*)-dione, sodium salt; DHA-sodium; Harven; 3-(Hydroxyethylidene)-6-methyl-2*H*-pyran-2,4(3*H*)-dione, sodium salt; Prevan; Sodium dehydroacetate.

Properties. White tasteless powder. Soluble in water (33% at 25°C), olive oil (less than 0.1%), and ethanol (1.0% at 25°C).[020]

Applications. Used as a plasticizer, a preservative in food-packaging adhesives and cosmetic products.

Acute Toxicity. LD_{50} is 500 mg/kg BW in rats, and 1050 mg/kg BW in mice. Poisoning damaged kidney function, led to vomiting, ataxia, and convulsions.[020] Following ingestion of high doses, monkeys exhibited anorexia, vomiting, weakness, stupor, ataxia, and convulsions.[1,019]

Reproductive Toxicity.

Embryotoxicity and ***teratogenicity.*** Rats received 50 mg/kg BW on days 6 to 17 of pregnancy. The treatment caused fetotoxicity and musculoskeletal abnormalities.[2]

Mutagenicity.

In vivo cytogenetics. DAS was found to induce micronuclei after a single *i/p* injection.[1]

In vitro genotoxicity. Did not produce SCE and CA.[3]

Regulations. ***U.S. FDA*** (1998) approved the use of DAS in adhesives as a component of articles intended for use in packaging, transporting, or holding food in accordance with the conditions prescribed by 21 CFR part 175.105.

References:

1. Hayashi, M., Kishi, M., Sofuni, T., and Ishidate, M., Micronucleus tests in mice on 39 food additives and eight miscellaneous chemicals, *Food Chem. Toxicol.*, 26, 487, 1988.
2. Latt, S. A., Aillen, J., Bloom, S. E., Carrano, A., Falke, E., Kram, D., Schneider, E., Schreck, R., Tice, R., Whitfield, B., and Wolff, S., Sister-chromatid exchanges: a report of the GENE-TOX Program, *Mutat. Res.*, 87, 17, 1981.
3. Ishidate, M., Sofuni, T., Yoshikawa, K., Hayashi, M., Nohmi, T., Sawada, M., and Matsuoka, A., Primary mutagenicity screening of food additives currently used in Japan, *Food Chem. Toxicol.*, 22, 623, 1984

DIDECYL GLUTARATE

Molecular Formula. $C_{25}H_{50}O_4$
M = 412.56
CAS No 3634-94-4
RTECS No MA3740000
Abbreviation. DDG.

Synonyms. Glutaric acid, didecyl ester; Pentanedioic acid, didecyl ester.

Properties. Light yellow solid.

Applications. Used as a plasticizer in the production of food-contact film materials and synthetic leathers.

Acute Toxicity. LD_{50} was not attained in rats and mice. LD_{16} appeared to be 60 g DDG/kg BW being identified by administration of divided doses to mice. Poisoning produced decreased BW gain, adynamia, and unkemptness.

Repeated Exposure revealed moderate cumulative properties. K_{acc} is 4.2 (by Lim). The treatment caused depression, adynamia, and reduced skin turgor in rats given the dose of 1/10 LD_{16}.

Reference:

Timofiyevskaya, L. A., Toxicity of dialkyl phthalate on mixture of alcohols C_7-C_8-C_9 (DAP-789), *Gig. Sanit.*, 10, 89, 1982 (in Russian).

DIDECYL PHTHALATE

Molecular Formula. $C_{28}H_{46}O_4$
M = 446.67
CAS No 84-77-5
RTECS No TI0900000
Abbreviation. DP.

Synonym and **Trade Name.** 1,2-Benzenedicarboxilic acid, didecyl ester; Decyl phthalate; Phthalic acid, didecyl ester; Vinicizer 105.

Properties. Colorless, odorless, viscous liquid. Poorly soluble in water.

Applications. Used as a plasticizer in the production of polyvinyl chloride; used in plastisol manufacture.

Acute Toxicity. LD_{50} exceeded 64 ml/kg BW in rats.

Regulations. ***U.S. FDA*** (1998) regulates the use of DP as (1) a plasticizer in rubber articles intended for repeated use in producing, manufacturing, packing, processing, treating, packaging, transporting, or holding food (total not to exceed 30% by weight of the rubber products) in accordance with the conditions prescribed by 21 CFR part 178.3740; and as (2) an adjuvant in the preparation of slimicides in the manufacture of paper and paperboard that may be safely used in contact with food in accordance with the conditions prescribed by 21 CFR part 176.300.

Reference:

Smyth, H. F., Carpenter, C. P., Weil, C. S., Pozzani, U. C., and Striegel, J. A., Range-finding toxicity data, list VI, *Am. Ind. Hyg. Assoc. J.*, 25, 95, March-April 1962.

DIETHYLENE GLYCOL

Molecular Formula. $C_4H_{10}O_3$

M = 106.12

CAS No 111-46-6

RTECS No ID5950000

Abbreviation. DEG.

Synonyms and **Trade Names.** Bis(2-hydroxyethyl)ether; Carbitol; DiEG; Diglycol; Digol; 2,2'-Dihydroxyethyl ester; Ethylene diglycol; Glycol ethyl ether; 2,2'-Hydroxydiethanol; 2,2'-Oxybisethanol; 2,2'-Oxyethanol.

Properties. Thick, colorless liquid with a faint odor. Mixes with water and alcohol at all ratios. Odor perception threshold is 3.28 g/l, taste perception threshold is 2.05 g/l.[1] According to other data, organoleptic properties threshold is 240 mg/l.[01]

Applications. DEG is used as a plasticizer in the production of food-contact materials based on regenerated cellulose; it is a solvent or a component of synthetic coatings, lacquers and cosmetics.

Exposure. In 1985, DEG was reported to be found in certain German and Austrian wines and juices, but no cases of intoxication in man were reported.[2] Levels found in contaminated wine, usually less than 3.0 g/l, would seem unlikely to represent hazard to those consuming part or even all of a bottle on a one-of time basis.[3] A content of 0.5 to 1.0 mg/kg is considered to be permissible in wine (Altman, 1986).

Migration Data. Chocolates, boiled sweets, toffees, cakes, and meat pies were wrapped in regenerated cellulose films that contained various mixtures of glycol softeners. DEG levels in the food samples were below 10 mg/kg.[4]

Acute Toxicity. DEG is likely to be less toxic than ethylene glycol (see above).

Observations in man. Man is more sensitive to DEG than rodents. Poisonings are known as a result of the occasional consumption of products containing DEG. Lethal dose is reported to be about 1.0 ml/kg BW.[5] Poisoning with DEG produces acidosis and nephrotoxicity.[6]

Animal studies. LD_{50} values are reported to be 13.3 to 23.7 g/kg BW in mice, 12.5 to 32 g/kg BW in rats, 2.69 to 4.4 g/kg BW in rabbits, 7.8 to 14.0 g/kg BW in guinea pigs, 3.3 g/kg BW in cats, and 9.0 g/kg BW in dogs.[1,7,06] Poisoning is accompanied by a short period of excitation and aggression with subsequent inhibition, disturbance of motor coordination, refusal of food, and vomiting. An important signs of DEG acute intoxication is renal damage (increased thirst and urination at the early stage followed by severely reduced urine production with heavy protein excretion). Labored breathing and coma were observed prior to death in 1 to 5 days. Gross pathology examination revealed hydropic degeneration, particularly in the renal tubular epithelium and in the centrilobular portion of the liver, punctate hemorrhages in the stomach and intestinal walls.

Repeated Exposure. When a 3.1 g/kg BW dose was given to rats for 20 days, no cumulative effect was found.[1] DEG produced metabolic acidosis, hydropic degeneration of the tubuli, oliguria, anuria, accumulation of urea nitrogen, and death in uremic coma.[2] DEG was found to be a CNS and cardiac depressant of low toxicity; it produced arrhythmia and fibrillation.[7]

Short-term Toxicity. Kidneys are likely to be a target organ in the picture of DEG intoxication. DEG produced hydropic swelling and degeneration of the epithelial cells, and the development of necrosis. Calcium oxalate crystals were formed in the bladder, leading to the onset of hematuria. A 0.5 mg/kg BW dose given to rats for 4 months had no harmful effect on the experimental animals.[8]

Long-term Toxicity. DEG was shown to produce renal damage, formation of bladder calcium oxalate stones, and liver damage in a number of species, including man. In many studies, toxic effect could be influenced by MEG (monoethylene glycol) contamination. Bladder stones were formed at dietary levels above 20 g/kg food. In an unpublished BIBRA study,[3] rats were given 0.4 to 4.0% DEG in the diet for 99 days, or 2.0% in the diet for 225 days. DEG used contained less than 0.01% MEG (the lowest reported MEG contamination). At 4.0% concentration (3.0 g/kg BW in males and 3.7 g/kg BW in females), DEG caused the death of 6 out 15 males with the signs of kidney damage. Dietary levels of 0.4% DEG (about 300 mg/kg BW) produced oxalate crystals in the urine, particularly in females, and mild defect in kidney function in the males, but no histological damage. At 100 mg/kg BW dose, the only treatment-related finding was the presence of a small amount of oxalic acid in the urine. The NOAEL of 100 mg/kg BW, therefore, seems justified (presence of an urinary metabolite was not considered by authors as a toxic finding). Bearing in mind that man might be very much more susceptible than the rat to DEG short-term toxicity, a choice of conservative uncertainty factor (*uncertainty factor* of 200) would lead to a TDI of 0.5 mg/kg BW.

Allergenic Effect is moderate (skin tests).[9]

Reproductive Toxicity.

Embryotoxicity and ***Teratogenicity***. Administration of 11.2 g/kg BW on days 7 to 14 of gestation to CD-1 mice caused 4.0% maternal mortality and reduced pup BW gain on postpartum days 1 to 3.[10] CD-1 mice were dosed orally via drinking water with a total dose of approximately 8.0 ml/kg (9.0 g/kg BW). Decreased fetal weights and some malformations were observed.[11] DEG appeared to be not embryotoxic or teratogenic in mice and rats except at maternally toxic doses.[12,13]

Gonadotoxicity. DEG exhibits a very pronounced gonadotoxic effect, similar to that of other glycols, with resulting reduction of sperm motility time, increased number of immobile forms, and lowered resistance to *NaCl* solution. DEG considerably decreases the spermatogenesis index, the number of cells with generative changes increases, and cytochrome-*c*-oxidase activity in the testicular tissues and alkaline phosphatase activity in the tissues of the epididymis are also increased (0.5 mg/kg BW).[14]

Carcinogenicity. Incidences of malignant bladder tumors in rats and mice due to DEG exposure are reported. Sanina described the development of malignant mammary gland tumors in rats on chronic exposure to DEG.[15] Hiasa et al. found no carcinogenic and promoting effects of DEG on renal tumorogenesis in rats.[16]

Chemobiokinetics. Oral doses of 1.0 and 5.0 ml ^{14}C-DEG/kg BW given to rats were rapidly and almost completely absorbed and distributed from the blood into kidneys, brain, spleen, liver, and muscle fat. The main metabolite is *2-hydroxyethoxyacetate*. 73 to 96% of ^{14}C-DEG is excreted with the urine, 0.7 to 2.2% with the feces.[2] According to Mathew et al., greater than half of the dose administered to rats was excreted unchanged, with 10 to 30% of the dose appearing as a single metabolite.[17]

Regulations.

U.S. FDA (1998) approved the use of DEG (1) as a component of adhesives intended for use in articles coming into the contact with food in accordance with the conditions prescribed by 21 CFR part 175.105; (2) in the manufacture of resinous and polymeric coatings for the food-contact surface of articles intended for use in producing, manufacturing, packing, processing, preparing, treating, packaging, transporting, or holding food in accordance with the conditions prescribed by 21 CFR part 175.300; (3) in the manufacture of resinous and polymeric coatings for polyolefin films for food-contact surface of articles intended for use in producing, manufacturing, packing, processing, preparing, treating, packaging, transporting, or holding food in accordance with the conditions prescribed by 21 CFR part 175.320; (4) in the manufacture of cross-linked polyester resins which may be used as articles or components of articles intended for repeated use in contact with food in accordance with the conditions prescribed by 21 CFR part 177.2420; (5) as a defoaming agent in the manufacture of paper and paperboard intended for use in packaging, transporting, or holding food in accordance with the conditions prescribed by 21 CFR part 176.210; and (6) as a substance employed in the production of or added to textiles and textile fibers intended for use in contact with foods in accordance with the conditions prescribed by 21 CFR part 177.2800.

Great Britain (1998). DEG is authorized without time limit for use in the production of polymeric materials and articles in contact with food or drink or intended for such contact. The specific migration of DEG alone or together with ethylene glycol shall not exceed 30 mg/kg.

Standards.

Russia (1995). MAC and PML: 1.0 mg/l.

EU (1990). SML: 30 mg/kg alone or with ethylene glycol.

References:

1. Plugin, V. P., Ethylene glycol and diethylene glycol as a goal of hygiene standard-setting in the sanitary protection of water bodies, *Gig. Sanit.*, 3, 16, 1968 (in Russian).
2. Heilmair, R., Lenk, W., and Lohr, D., Toxicokinetics of diethyleneglycol (DEG) in the rat, *Arch. Toxicol.*, 67, 655, 1993.
3. Hesser, L., Diethylene glycol toxicity, *Food Chem.Toxicol.*, 24, 261, 1986.
4. Castle, L., Cloke, H. R., Crews, C., and Gilbert, J., The migration of propylene glycol, mono-, di-, and triethylene glycols from regenerated cellulose film into food, *Z. Lebensmit. Untersuch. Forsch.*, 187, 463, 1988 (in German).
5. Casarett and Doull's *Toxicology*, J. Doull, C. D. Klassen, and M. D. Amdur, Eds., 3rd ed., New York, Macmillan Co., 1989, 272.
6. Hebert, J. L. et al., Acute human and experimental poisoning with diethylene glycol, *Sem. Hosp.* Paris, 59, 344, 1983.
7. Shiedeman, F. E. and Procita, L., The pharmacology of the monomethyl ethers of mono-, di- and tripropylene glycol in the dog with observations on the auricular fibrillation produced by these compounds, *J. Pharmacol. Exp. Ther.*, 102, 79, 1951.
8. Sheftel', V. O., Bardik, Yu. V., and Petrusha, V. G., On effect of harmful substances administered to animals in decreasing regimen, *Gig. Sanit.*, 12, 73, 1982 (in Russian).
9. Ivanova, T. P., Sensitizing effect of some chemicals migrating from polymeric materials, in *Hygienic Aspects of the Use of Polymeric Materials*, Proc. 2nd All-Union Meeting on Health and Safety Monitoring of the Use of Polymeric Materials in Construction, Kiyv, 1976, 102 (in Russian).
10. Schuler, R. L., Hardin, B. D., Niemeier, R. W., et al., Results of testing 15 glycol ethers in a short-term *in vivo* reproductive toxicity assay, *Environ. Health Perspect.*, 57, 141, 1984.
11. Williams, J., Reel, J. R., George, J. D., and Lamb, J. C., Reproductive effects of diethylene glycol and diethylene glycol monoethyl ether in Swiss CD-1 mice assessed by continuous breeding protocol, *Fundam. Appl. Toxicol.*, 14, 622, 1990.
12. Neeper-Bradley, T. L., Fisher, L. C., Tarasi, D. J., Fowler, E. H., and Ballantyne, B., Developmental toxicity evaluation of diethylene glycol (DEG) administered by gavage to CD (Sprague-Dawley) rats, *Toxicologist*, 12, 102, 1992.
13. Bates, H. K., Price, C. J., Marr, M. C., et al., Developmental toxicity study of diethylene glycol (DEG) in CD-1 mice, *Toxicologist*, 11, 340, 1991.
14. Byshovets, T. F., Barilyak, I. R., Korkach, V. I., et al., Gonadotoxic effect of glycols, *Gig. Sanit.*, 9, 84, 1987 (in Russian).
15. Sanina, Yu. P., On diethylene glycol toxicity, *Gig. Sanit.*, 2, 36, 1968 (in Russian).
16. Hiasa, Y., Kitahori, Y., Morimoto, J., Konishi, N., and Ohshima, M., Absence of carcinogenic and promoting effects of diethylene glycol on renal tumorogenesis in rats, *J. Toxicol. Pathol.*, 3, 97, 1990.
17. Mathew, J. M., Parker, M. K., and Mathews, H. B., Metabolism and disposition of diethylene glycol in rat and dog, *Drug Metab. Dispos. Biol. Fate Chem.*, 19, 1066, 1991.

DIETHYLENE GLYCOL, DIESTER with BUTYL PHTHALATE

CAS No 7483-25-2
RTECS No ID6730000
Abbreviation. DDGB.

Synonym. Dibutyl(diethylene glycol bisphthalate).

Applications. Used as a plasticizer in the production of moisture-proof coatings on cellulose film intended for contact with food or drink.

Acute Toxicity. In rats, LD_{50} of a commercial sample (80% DDGB, 15% dibutyl phthalate) exceeds 11.2 g/kg BW. In mice, LD_{50} is 10.3 g/kg BW. Poisoning is accompanied by changes in motor activity and in the liver. Animals become comatose. The acute toxicity of DDGB is greater when it is administered as a 50% solution in arachis oil.

Short-term Toxicity. Rats were fed diets containing 0.25 and 2.5% DDGB. The treatment caused retardation in BW gain. Heart and liver enlargement was found in males at dose levels of 1.0% and 2.0% in the feed. The highest dose produced oxaluria in all treated animals. The NOAEL is not reported.

Chemobiokinetics. Oxaluria seems to be a direct result of *in vivo* liberation of *ethylene glycol* (a known producer of oxalate stones in the bladder).

Reference:

Hall, D. E., Austin, P., and Fairweather, F. A., Acute (mouse and rat) and short term (rat) toxicity studies on dibutyl (diethylene glycol bisphthalate), *Food Cosmet. Toxicol.,* 4, 383, 1966.

DIPROPYLENE GLYCOL

Molecular Formula. $C_6H_{14}O_2$
M = 134.20
CAS No 110-98-5
25265-71-8
RTECS No UB8765000
Abbreviation. DPG.

Synonym. Oxybispropanol.

Properties. A colorless, slightly viscous liquid. Miscible with water and alcohol.

Applications. Plasticizer and solvent. Used in the production of polyesters, alkyd resins, and reinforced plastics intended for contact with food. Plasticizer and solvent. Used also as a stabilizer in cosmetics and in fragrances in the U.S.

Exposure. Production volume of DPG reaches nearly 100 million pounds used in the U.S. annually. Usual concentration in detergents is 0.015%.

Acute Toxicity. LD_{50} is 14.8 ml/kg BW in rats.[011]

Short-term Toxicity. Rats received 12% DPG in the diet for 15 weeks. The treatment resulted in depression of running activity. Moderate degenerative changes in kidneys were found.[1] No signs of intoxication were observed in rats receiving 5.0% DPG in their drinking water for 2.5 months. The concentration of 10% DPG in drinking water caused death in some animals. Histology examination revealed hydropic degeneration of kidney tubular epithelium and liver parenchyma.[1]

In NTP 13-week feeding studies, DPG was found to be of low to moderate toxicity. The most adverse effect occurred at dose levels of 2.0 to 8.0% in the drinking water. At the highest dose, mortality, hepatocellular lesions including atypical foci, and adenoma were seen in rats. Findings in mice included only an increase in liver weights.[2]

Long-term Toxicity. Chronic exposure revealed minimal liver damage.[1] According to Shiedeman and Procita, DPG is CNS and cardiac depressant of low toxicity; it produced arrhythmia and fibrillation.[3]

Allergenic Effect.

Observations in man. DPG showed no sensitization reactions in a maximization test that was carried out on 25 volunteers (Epstein, 1974).

Chemobiokinetics. After ingestion of the doses of 5.0 ml DPG/kg BW, it is eliminated from the blood in about 24 hours. It is not utilized by the liver or stored as glycogen.[4]

Regulations.

U.S. FDA (1998) approved DPG for use (1) as a constituent of food-contact adhesives which may safely be used as a component of articles intended for use in packaging, transporting, or holding food in accordance with the conditions prescribed by 21 CFR part 175.105; (2) in the manufacture of resinous and polymeric coatings for polyolefin films for food-contact surface of articles intended for use in producing, manufacturing, packing, processing, preparing, treating, packaging, transporting, or holding food in accordance with the conditions prescribed by 21 CFR part 175.320; (3) as an intermediate in production of cross-linked polyester resins which may be used as articles or components of articles intended for repeated use in contact with food in accordance with the conditions prescribed by 21 CFR part 177.2420, (4) as a component of a defoaming agent intended for articles which may be used in producing, manufacturing, packing, processing, preparing, treating, packaging, transporting, or holding food, subject to the provisions prescribed in 21 CFR part 176.200; and

(5) as a component of the uncoated or coated food-contact surface of paper and paperboard intended for use in producing, manufacturing, packaging, processing, preparing, treating, packing, transporting, or holding aqueous and fatty foods in accordance with the conditions prescribed by 21 CFR part 176.170.

EU (1990). DG is available in the *List of authorized monomers and other starting substances which shall be used for the manufacture of plastic materials and articles intended to come into contact with foodstuffs (Section* A).

Great Britain (1998). DG is authorized without time limit for use in the production of polymeric materials and articles in contact with food or drink or intended for such contact.

References:

1. Kesten, H. D., Mulinos, M. G., and Pomerantz, L., Pathologic effects of certain glycols and related compounds, *Arch. Pathol.*, 27, 447, 1939.
2. *NTP Fiscal Year 1995 Annual Report*, Publ. Health Service, Dept. Health & Human Services, December 1995, 73.
3. Shiedeman, F. E. and Procita, L., The pharmacology of the monomethyl ethers of mono, di- and tripropylene glycol in the dog with observations on the auricular fibrillation produced by these compounds, *J. Pharmacol. Exp. Ther.*, 102, 79, 1951.
4. Browning, E., *Toxicity and Metabolism of Industrial Solvents*, American Elsevier, New York, 1965, 486.

2-[2-(2-ETHOXYETHOXY)ETHOXY]ETHANOL

Molecular Formula. $C_8H_{18}O_4$
M = 178.26
CAS No 112-50-5
RTECS No KK8950000

Synonyms. Ethoxytriethylene glycol; Ethoxytriglycol; Ethyltriglycol; Triethylene glycol, ethyl ether; Triethylene glycol, monoethyl ether; 3,6,9-Trioxaundecan-1-ol.

Applications. Used as a plasticizer in the production of food-contact plastics.

Acute Toxicity. In mice, LD_{50} is 4.2 g/kg BW. Administration of the lethal doses caused immediate CNS inhibition with subsequent adynamia in 1 to 2 hours. The furry coat became moist and labored breathing was noted. Mice given high doses refused feed and died in 1 to 2 days. Main manifestations of poisoning included adynamia, convulsions and respiratory arrest. Gross pathology examination revealed congestion, circulatory disturbances, parenchymatous and fat dystrophy in the visceral organs.

Repeated Exposure revealed slight cumulative properties. K_{acc} is 4.5 (by Lim).

Reference:

Toxicology and Sanitary Chemistry of Polymerization Plastics, Coll. Works, B. Yu. Kalinin, Ed., Leningrad, 1984, 48 (in Russian).

2-ETHYLHEXANOIC ACID, diester with TRIETHYLENE GLYCOL

Molecular Formula. $C_{22}H_{42}O_6$
M = 402.64
CAS No 94-28-0
RTECS No MO7725000

Synonyms and **Trade Name.** Flexol plasticizer 3GO; Triethylene glycol, bis(ethylhexanoate); Triethylene glycol, di(2-ethylhexoate).

Properties. Colorless liquid. Soluble in organic solvents.

Applications. Used as a plasticizer in the production of polyvinyl butyrate film and other food-contact polymeric materials.

Acute Toxicity. LD_{50} is reported to be 12.5 g/kg BW in rats, 16 g/kg BW in mice, and 21 g/kg BW in guinea pigs. Species sensitivity is not evident. Death occurs in the state of lethargy.

Repeated Exposure failed to reveal signs of accumulation. K_{acc} is 6 (by Lim). Habituation develops.

Reference:

Timofiyevskaya, L. A., Toxicity and hazards of some plasticizers, *Gig. Sanit.*, 5, 87, 1981 (in Russian).

2-ETHYL-1-HEXANOL PHOSPHATE

Molecular Formula. $C_{24}H_{51}O_4P$
M = 434.65
CAS No 78-42-2
RTECS No MP0770000
Abbreviation. EHP.

Synonyms and **Trade Name.** Phosphoric acid, tris(2-ethylhexyl) ester; Flexol TOF; Triethylhexyl phosphate; Trioctyl phosphate.

Properties. Thick, clear viscous liquid. Water solubility exceeds 0.1 g/l at 20°C, soluble in alcohol.

Applications. Used as a plasticizer in the production of polyvinyl chloride and synthetic rubber compounds giving it low-temperature resistance. A solvent, an antifoaming and dispersing agent for plastisols.

Acute Toxicity. EHP appears to be a substance of very slight toxicity. LD_{50} exceeded 35 g/kg BW in rats, and about 46 g/kg BW in rabbits.[1] Some animals die only after administration of 46 g/kg dose.[05] However, Akhlustina[1] reported the LD_{50} to be 7.2 g/kg BW in mice; a dose of 10 g/kg BW in rats had no lethal effect. After administration of the lethal dose mice exhibited disturbances in respiratory and GI tract function. Death occurs within 18 to 24 hours. Gross pathology examination revealed visceral congestion, marked perivascular and pericellular brain edema, and parenchymatous dystrophy of the liver and kidneys. EHP causes no neurotoxic (demyelinization) effect.

Repeated Exposure failed to reveal cumulative properties. Administration of 1/10 LD_{50} caused no mortality in rats and mice. There is skin erosion. Changes are noted in the STI, blood cholinesterase activity, leukocyte count, urinary hippuric acid content, and the glycemic curve.[1]

Short-term Toxicity. Exposure to 0.43 g/kg BW (rats) for 30 days revealed no signs of intoxication. A 1.55 g/kg BW dose caused retardation of BW gain.[05]

Mutagenicity.

In vitro genotoxicity. EHP is found to be negative in *Salmonella* mutagenicity bioassay in the presence or absence of metabolic activation.[2]

Carcinogenicity. F344/N rats and $B6C3F_1$ mice received EHP in corn oil for 103 weeks. Rats were gavaged with 2000 and 4000 mg/kg BW (males) or 1000 and 2000 mg/kg BW (females); mice received (500 and 1000). The treatment caused increased incidence of pheochromocytomas of the adrenal glands in male rats, and increased incidence of liver carcinoma in female mice.[2]

Carcinogenicity classification.

NTP: EE - NE - NE - SE (gavage).

Chemobiokinetics. Following inhalation exposure to a labeled aerosol, EHP is detected in the brain, liver, and stomach contents. EHP is metabolized in at least one other compound in rats. After 2 days, the content of EHP excreted in the feces is significantly higher than in the urine.

Regulations. ***U.S. FDA*** (1998) approved the use of EHP (1) in adhesives used as components of articles intended for use in packaging, transporting, or holding food in accordance with the conditions prescribed by 21 CFR part 175.105; and (2) as a defoaming agent that may be safely used in the manufacture of paper and paperboard intended for use in producing, manufacturing, packing, transporting, or holding food in accordance with the condi tions prescribed by 21 CFR part 176.210.

References:

1. Akhlustina, L. V., in *Hygiene and Toxicology of Polymeric Construction Materials*, A. N. Bokov, Ed., Rostov-na-Donu Medical Institute, Rostov-na-Donu, 1973, 353 (in Russian).
2. *Toxicology and Carcinogenesis Studies of Tris(2-ethylhexyl)phosphate in F344/N Rats and $B6C3F_1$ Mice (Gavage Studies)*, NTP Technical Report Series No 274, Research Triangle Park, NC, NIH Publ. No 83-027.

GLYCEROL TRIHEPTANOATE

Molecular Formula. $C_{24}H_{44}O_6$
M = 428.68
CAS No 620-67-7
RTECS No MB2660000
Abbreviation. GT.

Synonym and **Trade Names.** Glycerin, triheptyl ether; Heptanoic acid, 1,2,3-propanetriyl ester; Trienanthoin; Triheptanoin; Triheptanoic glyceride; Triheptylin.

Properties. Odorless, light, almost colorless liquid.

Applications. Used as a plasticizer in polyvinyl chloride production. Gives good low-temperature resistance to plastic materials. Used as a component of plastisols for reducing the viscosity of pastes.

Acute Toxicity. LD_{50} *i/v* is 320 mg/kg BW in mice. Poisoning caused muscle contraction or spasticity and changes in respiratory system. Urine volume increased.

Long-term Toxicity. No retardation of BW gain was observed in Wistar rats that consumed the diet with 0.05, 0.1, and 1.0 g GT/kg BW (Le Breton).

Reproductive Toxicity was not noted.

Carcinogenicity. A 200 mg/kg BW dose administered for 21 months caused no tumor growth.[05]

Reference:

Acta Physiol. Scand., 40, 338, 1957.

HEXACHLOROBENZENE

Molecular Formula. C_6Cl_6
M = 284.81
CAS No 118-74-1
RTECS No DA2997500
Abbreviation. HCB.

Synonyms and **Trade Names.** Anticarie; Bant-cure; Bant-no-more; Pentachlorophenyl chloride; Perchlorobenzene.

Properties. Needles. Solubility in water is 0.007 mg/l at 20°C, sparingly soluble in cold alcohol, soluble in hot alcohol. Odor perception threshold is 0.06 mg/l. Heating does not increase odor intensity.

Applications. Used as a plasticizer and a solvent in the production of high-molecular-mass compounds and as a fungicide.

Acute Toxicity. LD_{50} is reported to be 1.7 to 4.0 g/kg BW in rats and mice; in guinea pigs and rabbits, it is 1.0 g/kg BW.[1]

Repeated Exposure. HCB is highly lipophilic and thus has a propensity to bioaccumulate.[2] K_{acc} is reported to be 1.0 (by Kagan). Degree of accumulation depends on the dose administered and on the length of the period during which HCB is given.[3]

Wistar rats were given 100 mg/kg BW every other day for 6 weeks. The treatment caused porphyria to develop as a result of disturbance of polyporphyrin biosynthesis in the liver.[4]

After five administrations of the same dose of HCB in corn oil to Sprague-Dawley rats, hepatic porphyria in female rats was found to develop after a delay period of 6 weeks, whereas toxicity was minimal in male rats.[5]

A total dose of 1500 mg HCB/kg BW in corn oil was administered to Sprague-Dawley rats as 50 mg/kg BW for 6 weeks or 100 mg/kg BW for 3 weeks. In males, HCB caused porphyria measured as urinary uroporphyrin and hepatic porphyrin levels. This total dose given to females for 3 weeks was not porphyrinogenic. A total dose of 600 mg/kg BW induced porphyria after 6 weeks. Minimally effective cumulative dose inducing porphyria was determined to be 400 mg/kg BW.[6]

Short-term Toxicity.

Observations in man. A misuse of HCB in Turkey in 1955 to 1959 caused about 3000 cases of *porphyria cutanea tarda* due to consumption of bread made from grain treated with HCB with a mortality rate of 10%.[1]

Animal studies. HCB administration results in hepatomegaly, porphyria, interference with hematopoiesis and CNS function. HCB hepatotoxicity is greatest in hamsters and rats; it gives rise to peliosis and necrosis with hemosiderosis, vasicular damage. Manifestations of the toxic action are highly dependent on species, dose, and time of exposure. Pigs were given 50 mg/kg BW in their feed for 90 days. The treatment caused an increase in coproporphyrin excretion, the induction of microsomal liver enzymes, histology changes, increased relative weights of the liver, and death. A dose of 0.05 mg/kg BW is considered to be ineffective.[7]

Gross pathology examination of rats fed more than 2.0 mg/kg BW in their diet for 15 weeks revealed changes in the liver and spleen. A dose of 0.5 mg/kg BW was ineffective.[8] Charles River rats were fed diets providing 0.5, 2.0, 8.0, or 32 mg/kg BW for up to 15 weeks. HCB was dissolved in corn oil. Females appeared to be more susceptible (decreased survival, splenomegaly, and ataxia at higher dose). The "apparent" NOEL of 0.5 mg/kg was established by authors.[9]

Third-litter sows were given HCB at concentrations of 1.0 and 20 ppm in their food throughout gestation and nursing. Accumulation in the fatty tissue was noted at the higher dose. Animals displayed neutrophilia, gastric irritation, fatty replacement of Brunner's gland, and pancreatic periductal fibrosis.[10]

Long-term Toxicity. Rats were dosed with 1.0 mg/day for a year. The treatment caused exhaustion, digestive and metabolic disorders, and death. In dogs, the same dose produced nodular hyperplasia of the lymphoid tissue of the GI tract.[3]

Allergenic Effect. Contact sensitizing potential is not found in maximized guinea-pig assays.[11] HCB is shown to cause immunodepression.

Reproductive Toxicity.

Teratogenic effect is not found as a result of administration of 120 mg/kg BW to pregnant Wistar rats at the time of organogenesis.[10,12] However, Andrews and Courtney observed hydronephrosis in the newborns of CD rats and CD-1 mice treated with 10 mg HCB/kg BW on days 15 to 20 and 6 to 16 of gestation, respectively.[13]

Embryotoxicity. Rats received 60 to 140 mg/kg BW with their feed. The treatment produced a significant dose-related increase in the mortality of the 1st and 2nd generation of the offspring.[14]

HCB transport across the placenta has been described in rats and mice.[8] A dose-related accumulation of HCB in the tissues, placenta, and fetal gall bladder was revealed in the treated pregnant hamsters and guinea pigs.

Gonadotoxicity. HCB has been detected in human ovary and human ovarian follicular fluid. It causes destruction of ovarian primordial germ cells in association with systemic toxicity.[15] Monkeys received oral doses of 0.1, 1.0, and 10.0 mg/kg BW by gelatin capsule for 90 days. The treatment caused dose-dependent HCB accumulation in serum and other tissues without any change in the serum estradiol response to human menopausal gonadotropin, oocyte recovery, oocyte maturation, oocyte fertilization *in vitro*, and early embryo cleavage rate.[15]

Mutagenicity.

In vivo cytogenetics. HCB does not cause DLM in rats exposed by gavage.[16]

In vitro genotoxicity. HCB is not shown to be genotoxic in *Salmonella typhimurium* and *E. coli* and human peripheral blood lymphocytes *in vitro*.[17] According to Lawlor et al., HCB did not revert histidine auxotrophs of *Salmonella typhimurium.*[18] It caused neither increase in incidence of CA in Chinese hamster cell culture nor in bacteria (IARC, Suppl. 6-331).

Carcinogenicity. In a 2-generation rat study, significant linear trends in the rate of parathyroid adenomas, neoplastic liver nodules, and other changes in the liver and kidneys were observed.[19] Carthew and Smith believed that accumulation of iron in the liver would strongly potentiate the development of hepatic tumors.[20]

Later HCB was reported to induce liver cancer in female rats.[5] Female Agus and Wistar rats received 100 ppm HCB in arachis oil in the diet for 90 weeks. The treatment produced 100% incidence of liver tumors.[21] Swiss rats were given dietary levels of 50, 100, or 200 ppm HCB for up to 120 weeks. An increase in dose-related hepatomas was reported in this study.[22]

In a 2-generation feeding study, parental Sprague-Dawley rats were fed 0.32 to 0.40 ppm HCB in the diet for 3 months. Following mating, females were maintained on the diet through pregnancy and lactation.

Pups received 0.32 to 0.40 ppm HCB in the diet for 130 weeks. Significant increases in the incidence of neoplastic liver nodules and adrenal pheochromocytomas were noted in F_1 females in the high dose group. F_1 males exhibited increased incidence of parathyroid tumors.[20]

Short-term exposure (15 weeks) to 300 ppm HCB did not significantly increase the rate of tumor incidence in mice.[22] No tumors were found in ICR mice that received 10 or 50 ppm HCB in the diet for 24 weeks. Hypertrophy of the hepatocentrylobular region was noted.[23] Hamsters received 4.0, 8.0, or 16 mg/kg BW in the diet throughout their lifetime. A significant dose-dependent increase in the incidence of hepatomas and thyroidal hemoangioepitheliomas and adenomas was found in the treated animals.[24]

Carcinogenicity classifications. An IARC Working Group concluded that there is sufficient evidence for the carcinogenicity of HCB in *experimental animals* and there was inadequate evidence for the carcinogenicity of HCB in *humans*.

The Annual Report of Carcinogens issued by the U.S. Department of Health and Human Services (1998) defines HCB to be a substance which may reasonably be anticipated to be carcinogen, i.e., a substance for which there is a limited evidence of carcinogenicity in *humans* or sufficient evidence of carcinogenicity in *experimental animals*.

IARC: 2B;

U.S. EPA: B2;

EU: 2.

Chemobiokinetics. Absorbed HCB is distributed mainly to the liver, kidneys, and brain. Fat contains 500 times more HCB than blood. HCB concentrates in the endocrine tissues in addition to the fat. Its residue levels in dosed rats (50 mg/kg BW) were significantly greater in the periovarian fat compared to the thyroid gland and in the thyroid compared to adrenal and ovary glands. HCB treatment had no effect on circulating levels of oestradiol or on the free *thyroxin index*.[25] Widespread fibrin deposits in the livers of rats chronically exposed to HCB have been confirmed using an antibody to rat fibrin.

HCB is biotransformed to *tetra-* and *pentachlorobenzenes* and *pentachlorophenol*, as well as to *sulfur-containing metabolites*. Biliary excretion of pentachlorothiophenol, a metabolite originating from glutathione conjugation of HCB, was higher in male than female rats.[5] Nine weeks after the start of administration of 0.5 g HCB/kg BW, an equilibrium was established between the HCB intake into the body and its excretion.[4] There was a high accumulation tendency in the rhesus monkeys given a daily oral dose of 0.11 mg ^{14}*C*-HCB for 15 months. The excretion showed a very slow approach to a saturation level.[26] However, HCB is poorly absorbed and excreted predominantly unchanged with the feces.[27] Small amounts of HCB are metabolized by the intestinal microflora. HCB excretion was observed for many months after administration had ceased.[4] It is readily transferred in the milk of lactating dams to their suckling neonates.

Regulations. ***EU*** (1990). Banned for certain uses owing to its effects on health and the environment.[28]

Recommendations.

WHO (1996). Guideline value for drinking water: 0.001 mg/l.

U.S. EPA (1999). Health Advisory for a longer-term exposure is 0.2 mg/l.

Standards.

U.S. EPA (1999): Proposed MCL: 0.001 mg/l, MCLG: zero.

Russia (1995): MAC and PML: 0.001 mg/l. PML in wheat grain: 0.01mg/kg.

References:

1. Peters, H. A., Hexachlorobenzene poisoning in Turkey, *Fed. Proc.*, 35, 2400, 1976.
2. Foster, W.G., McMahon A., Jarrel, J. F., et al., Hexachlorobenzene suppresses circulating progesterone concentrations during the lutea phase in the cynomologus monkey, *J. Appl. Toxicol.*, 12, 13, 1992.
3. Gralla, E. J., Fleischman, R. W., Luthra, Y. K., et al., Toxic effects of hexachlorobenzene after daily administration to beagle dogs for one year, *Toxicol. Appl. Pharmacol.*, 20, 227, 1977.
4. Koss, G., Seubert, S., Seubert, A., et al., Studies on the toxicology of hexachlorobenzene. V. Different phases of porphyria during and after treatment, *Arch. Toxicol.*, 52, 13, 1983.
5. D'Amour, M. and Charbounean, M., Sex-linked differences in hepatic glutathione conjugation of hexachlorobenzene in the rat, *Toxicol. Appl. Pharmacol.*, 112, 229, 1992.

6. Krishnan, K., Brodeur, J., Charbounean, M., et al., Development of an experimental model for the study of hexachlorobenzene-induced hepatic porphyria in the rat, *Fundam. Appl. Toxicol.*, 17, 433, 1990.
7. den Tonkelaar, E. M., Verschuuren, H. G., Bankovska, J., et al., Hexachlorobenzene toxicity in pigs, *Toxicol. Appl. Pharmacol.*, 43, 137, 1978.
8. Courtney, K. D., Andrews, J. E., and Graddy, M. A., Placental transfer and fetal deposition of hexachlorobenzene in the hamster and guinea pig, *Environ. Res.*, 37, 239, 1985.
9. Kuiper-Goodman, T., Grant, D. L., Moodie, C. A., Korsrud, G. O., and Munro, I. C., Subacute toxicity of hexachlorobenzene in the rat, *Toxicol. Appl. Pharmacol.*, 40, 529, 1977.
10. Hansen, L. G., Simon, J., Dorn, S. B., et al., Hexachlorobenzene distribution in tissues of swine, *Toxicol. Appl. Pharmacol.*, 51, 1, 1979.
11. Magnusson, B. and Kligman, A. M., The identification of contact allergens by animal assay. The guinea-pig maximization assay, *J. Invest. Dermatol.*, 52, 268, 1969.
12. Khera, K. S., Teratogenicity and dominant lethal studies of hexachlorobenzene in rats, *Food Cosmet. Toxicol.*, 12, 471, 1986.
13. Andrews, J. E. and Courtney, K. D., Hexachlorobenzene-induced renal maldevelopment in CD-1 mice and CD rats, *Hexachlorobenzene*, Proc. Int. Symp., C. R. Morris and J. R. P. Cabral, Eds., *IARC Sci. Publ.*, 77, 381, 1986.
14. Kitchin, K. T., Linder, R. E., Scotti, T. M., et al., Offspring mortality and maternal lung pathology in female rats fed hexachlorobenzene, *Toxicology*, 23, 33, 1982.
15. Jarrell, J. F., McMahon, A., Villeneuve, D., Franklin, C., Singh, A., Valli, V. E., and Bartlett, S., Hexachlorobenzol toxicity in the monkey promordial germ cell without induced porphyria, *Reprod. Toxicol.*, 7, 41, 1993.
16. Simon, G. S., Tardiff, R. G., and Borzelleca, J. F., Failure of hexachlorobenzene to induce dominant lethal mutations in the rat, *Toxicol. Appl. Pharmacol.*, 47, 415, 1979.
17. Siekel, P., Chalupa, J., Beno, J., et al., A geno-toxicological study of hexaclorobenzene and pentachloranizole, *Teratogen., Carcinogen., Mutagen.*, 11, 55, 1991.
18. Lawlor, T., Haworth, S. R., and Voytek, P., Evaluation of the genetic activity of nine chlorinated phenols, seven chlorinated benzenes, and three chlorinated hexanes, *Environ. Mutagen.*, 1, 143, 1979.
19. Arnold, D. L., Moodie, C. A., Charbonneau, S. M., et al., Long-term toxicity of hexachlorobenzene in the rat and the effect of dietary vitamin A, *Food Chem. Toxicol.*, 23, 779, 1985.
20. Carthew, P. and Smith, A. G., Pathological mechanisms of hepatic tumor formation in rats exposed chronically to dietary hexachlorobenzene, *J. Appl. Toxicol.*, 14, 447, 1994.
21. Smith, A. G. and Cabral, J. R., Liver-cell tumors in rats fed hexachlorobenzene, *Cancer Lett.*, 11, 169, 1980.
22. Cabral, J. R. P., Mollner, T., Raitano, F., and Shubik, P., Carcinogenesis of hexachlorobenzene in mice, *Int. J. Cancer*, 23, 47, 1979.
23. Shirai, T., Miyata, Y., Nakanishi, K., Murasaki, G., and Ito, N., Hepatocarcinogenicity of polychlorinated terphenyl (PCT) in ICR mice and its enhancement by hexachlorobenzene (HCB), *Cancer Lett.*, 4, 271, 1978.
24. Cabral, J. R. P., Shubik, P., Mollner, T., et al., Carcinogenic activity of hexachlorobenzene in hamsters, *Nature*, 269, 510, 1977.
25. Foster, W. G., Pentick, J. A., McMahon, A., and Lecavalier, P. R., Body distribution and endocrine toxicity of hexachlorobenzene (HCB) in the female rat, *J. Appl. Toxicol.*, 13, 79, 1993.
26. Rozman, K., Mueller, W., Coulston, F., et al., Long-term feeding study of hexachlorobenzene in rhesus monkeys, in *The 16th Annual Meeting Soc. Toxicol.*, Abstracts, Toronto, March 27-30, 1977, 173.
27. Parke, D.B., *Biochemistry of Foreign Compounds*, Moscow, Meditsina, 1973, 234 (in Russian, translation from English).
28. List of Chemicals Banned or Severely Restricted to Certain Uses by Community Legislation Owing to Their Effects on Health and the Environment, Annex I, *Off. J. Eur. Comm.*, No L251/19, 1992.

HYDRAZINE

Molecular Formula. H_4N_2
M = 32.05
CAS No 302-01-2
RTECS No MU7175000

Hydrazine sulfate

Molecular Formula. $H_4N_2.H_2O_4S$
M = 130.14
CAS No 10034-93-2
Abbreviations. HS (Hydrazine sulfate), HH. (Hydrazine hydrate).

Synonyms and **Trade Names.** Anhydrous hydrazine; Diamide; Diamine; Hydrazine base.

Properties. At ordinary temperature and pressure, H. is a caustic, fuming, hygroscopic liquid with a faint ammonia-like odor. HH, the principal compound produced, contains 64% H. by weight H. forms HH with water, with a composition of $NH_2NH_2.H_2O$. Both substances are colorless, clear liquids, which fume in the air. Mixes with all common solvents and with water at all ratios. Odor perception threshold for H. is 160 mg/l.[02] A concentration of 250 mg HH/l creates an odor with rating 2 intensity.

Applications. HH. is used as a wetting agent, a softener, a polymerization catalyst, and a foaming agent in the production of plastics.

Acute Toxicity. The LD_{50} values are reported to be 60 to 130 mg/kg BW in rats, 60 to 80 mg/kg BW in mice, 26 to 40 mg/kg BW in guinea pigs, and 55 mg/kg BW in rabbits.[1,2]

Lethal doses produce excitation, CNS inhibition and convulsions. Histology examination revealed hepatic infiltration, dystrophic changes in the kidneys, and hemorrhages and edema in the lungs.

Repeated Exposure revealed pronounced cumulative properties. H. affects the CNS, liver, and kidneys. 5.0 to 20 mg H. (free base)/kg powdered diet was given to mice and rats for 3 to 10 days. No animals died, but they exhibited weakness. Metamitochondria or fatty vacuolization were induced at 10 mg/kg and 20 mg/kg dose levels.[3]

Short-term Toxicity. Golden hamsters were administered H. by gavage (60 and 100 daily doses, equivalent to 0.74 and 0.68 mg H. over 15 and 20 weeks). Toxic effects in animals consisted of liver lesions, reticuloendothelial cell proliferation, cirrhosis, bile-duct proliferation, degenerative fibrous cells in hyalinized tissue.[4]

Long-term Toxicity. Mayore *et al.* indicated disruption of lipid metabolism to be the prime cause of H. toxic damage to the liver.[5] This is manifested morphologically as a diffused fatty dystrophy of hepatocytes. Accumulation of amino acids occurs in the plasma as a result of their reduced content in the liver, whose ability to convert amino acids to glycogen is impaired.

Mice were given 150 daily doses of HS (0.98 to 8.0 mg/kg BW) in drinking water by gavage. Animals were followed up for their lifetime. Brown degeneration of the adrenals was mentioned as a marked non-neoplastic lesion.[6] Gross pathology examination revealed parenchymatous dystrophy of the liver.

Guinea pigs received 0.5 mg/kg BW. The treatment lowered blood *Hb* content and concentrations of *SH*-groups in the total blood and serum. Reticulocytosis and an increase in the urinary urobilin concentration were noted. Liver and thyroid functions were disturbed.[1]

Reproductive Toxicity.

Embryotoxicity. In rats and mice, H. given orally produced adverse effects on fetuses and embryos at doses that were toxic for mothers. Adverse effects included increased resorptions, reduced fetal weight, increased perinatal mortality, increased incidences of litters and fetuses with abnormalities in the 10 mg/kg BW group.[7]

In the offspring of Syrian golden hamsters subtle postnatal changes were observed after oral administration up to 170 mg H./kg BW on day 12 of gestation.[8]

Reproductive functions of rats (fertility of females, number of newborns, resorption of embryos) were not affected by administration of 13 mg H./kg BW for 30 days prior to mating.[9]

Teratogenicity. Skeletal abnormalities were found in chicks after administration of 0.03 to 0.2 mg on the third day of incubation.[10] Concentration of 0.002 mg/l was identified to be ineffective for reproduction and teratogenicity in rats.[11]

Gonadotoxicity. Oral ingestion of H. at a concentration of 0.82% in the diet resulted in marked embryotoxic and gonadotoxic effects in rats. Destruction of gonadal epithelium was observed in male rats.[11]

Mutagenicity.

In vivo cytogenetics. H. did not induce CA, DLM, and micronuclei in mice, but produced CA in rats *in vivo.*[12,13]

LacZ transgenic mice received a single oral dose of up to 400 mg/kg BW. No dose induced any lacZ mutation in lung, liver or bone marrow.[14] However, according to other data, H. induced micronucleated reticulocytes marginally in male ICR mice 48 hours after a single *i/p* treatment. It did not induce DLM or micronuclei in bone marrow cells of mice treated *in vivo.*[15]

H. causes gene mutations in *Dr. melanogaster.*

In vitro genotoxicity. H. is a weak mutagen. It induces unscheduled DNA synthesis in human cells, causes gene mutations in bacteria; in a culture of Chinese hamster cells, it increases the frequency of CA and SCE and causes DNA damage in *in vitro* tests. H. caused no gene mutations in mouse lymphoma cells (Myhr and Caspary, 1988). *H. sulfate* is also known to be an *in vitro* mutagen. It was reported to be mutagenic in *Salmonella typhimurium* strain *TA102* assay in three different laboratories.[15,16]

Carcinogenicity.

Animal studies. Increased mortality was observed in rats given HS by gavage for 68 weeks (14.6 mg/kg BW for male rats and 9.7 mg/kg BW for female rats in drinking water). Pulmonary tumors, adenomas, and adenocarcinomas were found.[17] An increase in hepatocarcinoma incidence in mice was noted in these studies. An increase in the incidence of lung tumor was observed in Swiss mice given HS in drinking water (0.012%) throughout their lifetimes.[18]

In another mouse study, the dose of 0.27 mg/mouse/day was administered to animals for their lifetimes. Lung adenocarcinomas occurred in 30 out of 34 male mice and in 21 out of 27 females.[19]

No tumors were observed in golden hamsters receiving 0.57 mg H./day for up to maximum 110 weeks;[20] similar results are also reported.[4]

Observations in man. A preliminary report of an epidemiological study of men engaged in H. manufacture revealed no unusual excess of cancer.[15]

Carcinogenicity classifications. An IARC Working Group concluded that there is sufficient evidence for the carcinogenicity of H. and HS in *experimental animals* and there is inadequate evidence for the carcinogenicity of H. and HS in *humans.*

The Annual Report of Carcinogens issued by the U.S. Department of Health and Human Services (1998) defines H. and HS to be substances which may reasonably be anticipated to be carcinogen, i.e., substances for which there is a limited evidence of carcinogenicity in *humans* or sufficient evidence of carcinogenicity in *experimental animals.*

IARC: 2B;

U.S. EPA: B2;

EU: 2.

Chemobiokinetics. H. is rapidly absorbed and rapidly distributed to and eliminated from most tissues. It may compete to slow down the formation of *glutamine* and *urea* by combining with *glutamic acid, carbamyl phosphate,* or *amino acid* precursor of the urea cycle, as a result of which *ammonia* is released.[21]

In mice and rats, a part of the absorbed H. is excreted unchanged, and a part as labile conjugates or as *acid-hydrolysable derivatives* via the urine. When H. is metabolized, a significant amount of *nitrogen* is produced, which is excreted via the lungs.[13]

Regulations. ***U.S. FDA*** (1998) regulates H. under FD&CA requiring that steam in contact with food contain no H.

Standards. ***Russia*** (1995). MAC and PML: 0.01 mg/l.

References:

1. Ekshtat, B. Ya., Experimental substantiation of the maximum allowable concentration for hydrazine-hydrate and phenyl hydrazine in water reservoirs, in *Industrial Pollutants of Water Reservoirs*, Meditsina, Moscow, Issue No 9, 1969, 186 (in Russian).
2. Witkin, L. B., Acute toxicity of hydrazine and some of its methylated derivatives, *Arch. Ind. Health*, 13, 51, 1956.
3. Wakabayashi, T., Horiuchi, M., Sakaguchi, M., et al., Induction of megamitochondria in the mouse and rat livers by hydrazine, *Exp. Mol. Pathol*., 39, 139, 1983.
4. Biancifiory, C., Pulmonary and liver tumors from low doses of hydrazine sulfate in BALB/c/Cb/Se mice, *Lav. 1st Anat. Histol. Patol.*, Univ. Studi Perugia, 30, 89, 1970.
5. Mayore, A. Ya. et al., in *Metabolic Aspects of the Effect of Industrial Chemicals on the Body*, Krasnoyarsk, 126, 1982 (in Russian).
6. Biancifiori, C., Hepatomas in CBA/Cb/Se mice and liver lesions in golden hamsters induced by hydrazine sulfate, *J. Natl. Cancer Inst.*, 44, 943, 1970.
7. Keller, W. C., Toxicity assessment of hydrazine fuels, *Aviat. Space Eviron. Med.*, 59, 100, 1988.
8. Schiller, C. M., Walden, R., and Kee, T. E., Effects of hydrazine and its derivatives on the development of intestinal brush border enzymes, *Toxicol. Appl. Pharmacol*., 49, 305, 1979.
9. Savchenkov, M. F. and Samoilova, T. J., Effect of hydrazine nitrate on reproductive function of albino rats, in *Problems of Limitation of Environmental Pollutant Circulation*, Ufa, 1984, 82 (in Russian).
10. Stoll, R., Bodit, F., and Marand, R., Sur l'action de divers antimitotiques apparentes a l'hydroxyuree et a la semicarbazide chez l'embryon de poulet, *CR Soc. Biol.*, 1963-1967, Paris, 1967, 161.
11. Dymin, V. V., Denisov, V. L., Andropova, S. N., et al., Influence of hydrasine administered by different routes on reproductive function of animals, *Gig. Sanit.*, 9, 25, 1984 (in Russian).
12. Speit, G., Wick, C., and Wolf, M., Induction of SCE by hydroxylamine, hydrazine, and isoniazid and their inhibition by cystein., *Hum. Genet*., 54, 155, 1980.
13. *Hydrazine*, Environmental Health Criteria, WHO, Geneva, 1987, 90.
14. Douglas, G. R., Ginderich, J. D., and Soper, L. M., Evidence for *in vivo* non-mutagenicity of the carcinogen hydrazine sulfate in target tissues of *lacZ* transgenic mice, *Carcinogenesis,* 19, 801, 1995.
15. Morita, T., Asano, N., Awogi, T., Sasaki, Yu. F., et al. Evaluation of the rodent micronucleus assay in the screening of IARC carcinogens (Groups 1, 2A and 2B). The summary report of the 6 collaborative studies by CSGMT/JEMS-MMS, *Mutat. Res*., 389, 3, 1997.
16. Muller, W., Engelhart, G., Herbold, B., Jackh, R., and Jung, R., Evaluation of mutagenicity testing with *Salmonella typhimurium TA102* in three different laboratories, *Enviro. Health Perspect.*, 101 (Suppl. 3), 33, 1993.
17. Severi, L. and Biancifiori, C., Hepatic carcinogenesis in CBA/Cb/Se mice and Cb/Se rats by isonicotinic acid hydrazine and hydrazine sulfate, *J. Natl. Cancer Inst*., 41, 331, 1968.
18. Toth, B., Lung tumor induction and inhibition of breast adenocarcinomas by hydrazine sulfate in mice, *J. Natl. Cancer Inst.*, 42, 469, 1969.
19. Menon, M. M. and Bhide, S. V., Perinatal carcinogenicity of isoniazide in Swiss mice, *J. Cancer Res. Clin. Oncol.*, 105, 258, 1983.
20. Toth, B., Tumourogenesis studies with 1,2-dimethylhydrazine dihydrochloride, hydrazine sulfate, and isonicotinic acid in golden hamsters, *Cancer Res.*, 32, 804, 1972.
21. Floyd, W. N., The importance of ammonia in the metabolite effects of hydrazine, *Aviat. Space Environ. Med*., 51, 899, 1980.

2-HYDROXY-1,2,3-PROPANETRICARBOXYLIC ACID, TRIBUTYL ESTER

Molecular Formula. $C_{18}H_{32}O_7$
M = 360.44
CAS No 77-94-1
RTECS No TZ8608000

Abbreviation. TBC.

Synonyms. *n*-Butyl citrate; Citric acid, tributyl ester; Tributyl citrate.

Properties. Colorless or pale-yellow, odorless liquid. Poorly soluble in water. Miscible with most organic solvents.

Applications. Used as a plasticizer in the polyvinyl chloride production. Solvent for nitrocellulose, lacquers intended for contact with food.

Acute Toxicity. Single administration of 10 to 30 ml/kg BW caused no signs of intoxication.[1] According to other data, TBC caused complete loss of blood pressure when administered in toxic doses. TBC was shown to produce local anesthetic action in rabbits.[2]

Repeated Exposure failed to show signs of accumulation. TBC is likely to be of low toxicity due to its insolubility in the body fluids.

Rats received 5.0% TBC in their diet for 8 weeks. The treatment did not affect growth of animals. However, 10% dose level of TBC reduced BW gain and caused diarhhea. Histology examination revealed no changes in the viscera.[1] Compare to ***Acetyl tributyl citrate***.

Regulations. ***U.S. FDA*** (1998) approved the use of TBC as a constituent of food-contact adhesives which may be safely used as a component of articles intended for use in packaging, transporting, or holding food in accordance with the conditions prescribed by 21 CFR part 175.105.

Standards. *Russia.* Recommended PML: *n/m.*

References:

1. Finkelstein, M. and Gold, H., Toxicology of the citric acid esters: tributylcitrate, acetyltributylcitrate, triethylcitrate, and acetyltriethylcitrate, *Toxicol. Appl. Pharmacol.,* 1, 283, 1959.
2. Casarett and Doull's *Toxicology*, J. Doull, C. D. Klassen, and M. D. Amdur, Eds., Macmillan Publ. Co., New York, 1980, 549.

ISOPHTHALIC ACID

Molecular Formula. $C_8H_6O_4$
M = 166.14
CAS No 121-91-5
RTECS No NT2007000
Abbreviation. IPA.

Synonyms. Benzene-1,3-dicarboxylic acid; *m*-Benzenedicarboxylic acid; Isophthalate; *m*-Phenylenedicarboxylic acid; *m*-Phthalic acid.

Properties. White, odorless, finely crystalline powder. Water solubility is 0.013 g/100 g at 25°C, solubility in hot water is 2.2 g/l, soluble in alcohol.

Applications. Used in the production of lacquers and coatings intended for contact with food. An ingredient in the manufacture of amide resins and polyvinyl chloride plasticizers.

Acute Toxicity. LD_{50} is 10.8 g/kg BW in rats, and 9.58 g/kg BW in mice. Primary target organ and tissue are the liver and CNS.[1]

Repeated Exposure revealed pronounced cumulative properties in rats.

Allergenic Effect was found in *in vitro* experiments and skin tests.[2]

Regulations.

U.S. FDA (1998) regulates the use of IPA (1) in adhesives as a component of articles intended for use in packaging, transporting, or holding food in accordance with the conditions prescribed by 21 CFR part 175.105; (2) in the manufacture of resinous and polymeric coatings in a food-contact surface of articles intended for use in producing, manufacturing, packing, transporting, or holding food in accordance with the conditions prescribed by 21 CFR part 175.300; and (3) in the manufacture of cross-linked polyester resins to be safely used as articles or components of articles intended for repeated use in contact with food in accordance with the conditions prescribed by 21 CFR part 177.2420.

EU (1990). IPA is available in the *List of monomers and other starting substances which may continue to be used for the manufacture of plastic materials and articles intended to come into contact with foodstuffs pending a decision on inclusion in Section A* (*Section B*).

Great Britain (1998). IPA is authorized up to the end of 2001 for use in the production of polymeric materials and articles in contact with food or drink or intended for such contact.

Standards. ***Russia*** (1995). MAC and PML: 0.1 mg/l.

References:

1. Zhukova, T. V., Il'chenko, G. Ya., and Beregovykh, T. M., Substantiation of the maximum allowable concentration for isophthalic acid in industrial zone air, in *Hygiene and Toxicology of High- Molecular-Mass Compounds and of the Chemical Raw Material Used for Their Synthesis,* Proc. 6th All-Union Conf., B. Yu. Kalinin, Ed., Leningrad, 1979, 217 (in Russian).
2. Zhukova, T. V. et al., in *Toxicology of New Chemicals and Occupational Hygiene in Their Production and Application*, Rostov-na-Donu, 1974, 113 (in Russian).

MALEIC ACID, BIS(2-ETHYLHEXYL) ESTER

Molecular Formula. $C_{20}H_{36}O_4$
M = 340.56
CAS No 142-16-5
RTECS No ON0160000
Abbreviation. DEHM.

Synonyms and **Trade Names.** 2-Butenedioic acid, bis(2-ethylhexyl) ester; Bis(2-ethylhexyl) maleate; Di- (2-ethylhexyl) maleate; Dioctyl maleate; RC co-monomer DOM.

Properties. Liquid. Insoluble in water.

Applications. Co-monomer and internal plasticizer in the production of vinyl acetate, acrylates and methacrylates.

Acute Toxicity. LD_{50} is 14 g/kg BW in rats.

Chemobiokinetics. Maleates are readily hydrolyzed by the mammalians.[03]

Regulations. ***U.S. FDA*** (1998) approved the use of DEHM (1) in adhesives as a component of articles (monomer) intended for use in packaging, transporting, or holding food in accordance with the conditions prescribed by 21 CFR part 175.105; and (2) as a component (monomer) of the uncoated or coated food-contact surface of paper and paperboard intended for use in producing, manufacturing, packaging, processing, preparing, treating, packing, transporting, or holding dry food in accordance with the conditions prescribed by 21 CFR part 176.180.

Reference:

Smyth, H. F., Carpenter, C. P., and Weil, C. S., Range-finding toxicity data: List III, *J. Ind. Hyg. Toxicol.*, 31, 60, 1949.

MALEIC ACID, DIBUTYL ESTER

Molecular Formula. $C_{12}H_{20}O_4$
M = 228.24
CAS No 105-76-0
RTECS No ON0875000
Abbreviation. DBM.

Synonyms and **Trade Name.** 2-Butanedioic acid, dibutyl ester; Dibutyl maleate; Staflex DBM.

Properties. Viscous liquid. Poorly soluble in water, soluble in alcohol.

Applications. Used as a plasticizer in the production of food-contact plastics.

Acute Toxicity. LD_{50} is 2.7 to 4.4 g/kg BW for rats, and 4.4 to 9.3 g/kg BW for mice.[1-3] Gross pathology examination showed severe distension of the stomach and intestine, with congestion in their serous membrane and mesentery. Histology examination revealed congestion in all the visceral organs with pulmonary hemorrhages in some animals.

Repeated Exposure revealed no marked cumulative properties. Rats and mice were dosed by gavage with 1/5 and 1/10 LD_{50}. The treatment caused exhaustion, increase in the relative weights of the kidneys, as well as death of some animals. Gross pathology examination failed to reveal changes in the viscera.[1]

Short-term Toxicity. Rats and mice were orally exposed to 1/100 and 1/50 LD_{50} (in an oil solution). Administration caused no changes in enzyme and conditioned reflex activity, in BW gain, hematological indices, and urinalysis. Histology examination failed to reveal changes in the viscera.[1]

Regulations.

U.S. FDA (1998) approved the use of DBM (1) in adhesives as a component of articles intended for use in packaging, transporting, or holding food in accordance with the conditions prescribed by 21 CFR part 175.105; and (2) as a component (monomer) of the uncoated or coated food-contact surface of paper and paperboard intended for use in producing, manufacturing, packaging, processing, preparing, treating, packing, transporting, or holding dry, aqueous and fatty foods in accordance with the conditions prescribed by 21 CFR part 176.170 and 176.180.

EU (1990). DBM is available in the *List of monomers and other starting substances which may continue to be used for the manufacture of plastic materials and articles intended to come into contact with foodstuffs pending a decision on inclusion in Section A (Section* B).

Great Britain (1998). DBM is authorized up to the end of 2001 for use in the production of polymeric materials and articles in contact with food or drink or intended for such contact.

Standards. *Russia.* Recommended PML: *n/m.*

References:

1. Mikhailets, I. B. and Robachevskaya, Ye. G., Toxicologic characteristics of some maleic acid ethers, in *Hygiene and Toxicology of High-Molecular-Mass Compounds and of the Chemical Raw Material Used for Their Synthesis*, Proc. 4th All-Union Conf., S. L. Danishevsky, Ed., Khimiya, Leningrad, 1969, 226 (in Russian).
2. Druzhinina, V. A. and Kochetkova, T. A., in *Current Problems of Environmental Hygiene*, A. P. Shitzkova, Ed., F. F. Erisman Research Sanitary Hygiene Institute, Moscow, 1976, 53 (in Russian).
3. Smyth, H. F., Carpenter, C. P., Weil, C. S., et al., Range-finding toxicity data, List V, *Arch. Ind. Hyg. Occup. Med.*, 10, 61, 1954.

MALEIC ACID, DIMETHYL ESTER

Molecular Formula. $C_6H_8O_4$
M = 144.14
CAS No 624-48-6
RTECS No EM6300000
Abbreviation. DMM.

Synonyms. 2-Butenedioic acid, dimethyl ester; Dimethyl maleate.

Properties. A clear, colorless liquid.

Applications. Used as an intermediate in the production of food-contact copolymers and films.

Acute Toxicity. LD_{50} is 1.4 g/kg BW in rats, and 1.34 g/kg BW in mice.[1]

Allergenic Effect. The results of the maximization test in guinea pigs revealed a clear sensitizing potential.[2]

Mutagenicity.

In vitro genotoxicity. Negative results were observed in *Salmonella* mutagenicity assay at concentrations up to 5.0 mg DMM/plate.[2]

In vivo cytogenetics. Mice given 1.0 g DMM/kg BW by gavage exhibited no clastogenic effect.[2]

Regulations. *EU* (1990). DMM is available in the *List of authorized monomers and other starting substances which may continue to be used for the manufacture of plastic materials and articles intended to come into contact with foodstuffs pending a decision on inclusion in Section A* (*Section B*).

References:

1. Smyth, H. F., Carpenter, C. P., and Weil, C. S., Range-finding toxicity data, List VI, *Am. Ind. Hyg. Assoc. J.*, 23, 95, 1962.
2. Heimann, K.G., Jung, R., and Kieczka, H., Maleic acid dimethylester: evaluation of dermal toxicity and genotoxicity, *Food Chem. Toxicol.*, 29, 575, 1991.

MINERAL OILS

CAS No 8012-95-1
RTECS No PY8030000
Abbreviation. MO.

Trade Names. Adepsine oil; Alboline; Food grade mineral oils; Heavy mineral oil; Irgawax 361; Kaydol; Liquid paraffin; Liquid petrolatum; Liquid vaseline; Paraffin oil; White mineral oil.

Composition. Mineral (petroleum) oils are a complex mixture of methanenaphthenic, aromatic, and naphtheno-aromatic compounds. They contain 0.2 to 2.0% of sulphur (elemental or in the form of H_2S). Purified oil contains it in organic forms. The components of liquid paraffins are predominantly of two hydrocarbon types: paraffinic, which are branched-chain alkanes; and naphthenics, which are alkanes containing one or more saturated cyclic structures.

White mineral oil is refined to meet specified requirements (*U.S. Pharmacopeia XX*, 1980). Analogue of Shell 33 oil is characterized by a high degree of removal of aromatic hydrocarbons.

Properties. Clear, colorless, odorless, viscous liquids. Naphthene compressor oil-40 (NCO-40) occurs as a colorless liquid with a specific faint odor. High-purity mineral oil (HPMO) is a light-straw, oily liquid with an indefinite odor of petroleum products. Insoluble in water and alcohol. Miscible with most fixed oils.

Applications. Used in the production of polyethylene and as plasticizers for high-impact polystyrene, polyvinyl chloride, ABC plastics, polyamides, polyurethanes, ethyl cellulose intended for contact with food. Liquid paraffins are used for cosmetics and ointments for pharmaceuticals.

Migration Data. Polystyrene plasticized with compressor oil does not alter the characteristics of water or simulant media in contact with it. Polystyrene and acrylonitrile/butadiene/styrene (ABS) containers (for individual serving portions of milk, cream, butter, margarine, and spreads) were found to contain 1.0 to 4.0% mineral oil. Migration into the foodstuffs appeared to be <5.0 to 15 mg/kg. Jickells et al. studied migration from 105 samples of wine bottle corks (treated with mineral wax or mineral oil) from 11 different countries. Mineral carbon contamination of the wine was <0.2 mg/l.[1]

Jute fibers were treated with about 5.0 to 7.0% of a high boiling mineral oil fraction (batching oil). Contamination of nuts, coffee, cocoa products, and rice transported in such jute bags ranged between about 5.0 and 500 ppm.[3]

Exposure. MO was determined in canned sea foods. Concentrations was mostly around 100 mg/kg, but reached 820 mg/kg. The predominant source of these MO materials was lubricating oil used for can manufacturing.[2]

Acute Toxicity. Administration to mice and rats produced no functional or morphologic damage. No manifestations of the toxic effect were found on administration of 25 g NCO-40 or HPMO/kg BW to rats and mice. No acute effect threshold was established.[4,5] However, according to other data, *i/g* administration of petroleum oil to mice caused a toxic effect.[6] LD_{50} is reported to be 22 g/kg BW.[06]

Repeated Exposure failed to reveal cumulative properties. Mice and rats received 2.0 g/kg BW for 2 months. The treatment produced reversible changes in motor activity.

F344 rats and Sprague-Dawley rats were fed paraffinic white oil at 0.2 or 2.0% of the diet for 30, 61, or 92 days. No adverse effects or unscheduled deaths were reported. Mesenteric lymph nodes were enlarged. F344 rats appeared to be more sensitive compared to Sprague-Dawley rats.[7]

Short-term Toxicity. Liquid paraffins were shown to cause granulomas in the liver and mesenteric lymph nodes in F344 rats at dietary levels up to 2.0%.[8,9] Such findings were not observed in previous subchronic and chronic feeding studies of various mineral hydrocarbons at 2.0 to 10% dietary levels, conducted in Long-Evans, Sprague-Dawley, FDRL, BD I, BD III, and Wistar rats.[10,12]

JECFA reviewed 90-day dietary studies with MPO in rats.[13] Hematology changes, MO deposition in the liver and spleen were reported. Mineral oils tested were N10 (A), N15 (H), P15 (H), N70 (A), N70 (H), P70 (H) and P100 (H) at dose levels of 2.0 to 2000 mg/kg BW.[8] The P100 (H) oil produced no effects but did accumulate in the liver to a small extent at the highest dose level. The P70 (H) oil accumulated in the liver, kidney, and mesenteric lymph nodes. It increased incidence of pigmented macrophages in the lymph nodes when administered at the highest dose level. The typical effects observed included focal histiocytosis, increase in the weight of the liver, lymph nodes, spleen and kidney, granulomas or microgranulomas in the liver, hematological changes typical of a mild, chronic inflammatory reaction. Biochemical changes exhibited mild hepatic damage.

F344 rats were given free access to diets containing 10,000 to 20,000 ppm of oleum-treated white oil (OTWO) or hydrotreated white oil (HTWO) for 90 days. The treatment caused dose-related hematological, clinical chemical, and pathological changes, more marked in female than in male rats. Gross pathology examination revealed mesenteric lymph nodes being enlarged. There was a significant increase in liver, kidney, and spleen weights. Multifocal lipogranulomata were found in mesenteric lymph node and liver. No changes were observed in rats fed OTWO or HTWO at doses of 0.65 and 6.4 mg/kg BW, respectively.[8]

Table. NOELs and ADI for mineral oils tested in 90-day studies in F344 rats.[13]

Substances	*NOEL (mg/kg BW)*	*ADI (mg/kg BW)*
N10(A) oil	2	0 - 0.01 temporary
N15(H) oil	<2	0 - 0.01 temporary
P15(H) oil	2	0 - 0.01 temporary
N70(A) oil	2	0 - 0.01 temporary
N70(H) oil	2	0 - 0.01 temporary
P70(H) oil	200	0 - 0.01 temporary
P100(H) oil	2000	0 - 20

Long-Evans rats and beagle dogs were fed the dietary doses of 300 and 1500 ppm (*w/w*) of highly refined white mineral oils. No treatment-related toxic effects were observed. Gross and histopathological examination did not reveal any microscopic or macroscopic changes. There was no evidence of oil lipid deposition in liver, spleen, or mesenteric lymph nodes.[10]

Fleming et al. recently concluded that the reported lesions seen in F344 rats following highly refined MO ingestion would appear to have no significance for human disease.[14]

Long-term Toxicity. Rats and mice were dosed by gavage with 50 and 200 mg MO/kg BW for 10 months. No harmful effect was found.[5]

In a 6-month study, guinea pigs were administered IS-45 oil at a dose of 0.5 g/kg BW. The treatment caused an increase in the phagocytic index and in the γ-globulin blood serum content.[9]

Seven white oils differing in crude oil type (naphthenic or paraffinic), refined method, and viscosity, and 3 waxes (paraffinic-derived, hydrogenation- or percolation-refined with a range of molecular masses) were fed in the diet of rats at doses of around 2, 18, 180, or 1900 mg/kg BW. No treatment-related biology effect was seen with the higher molecular-sized hydrocarbons (microcrystalline waxes and the higher viscosity oils). Paraffin waxes and low- to mid-viscosity oils produced pathological changes, mainly in the liver and lymph nodes. The effects were inversely related to mol. mass, viscosity, and melting point, but appeared to be independent of oil type processing.[15]

F344 rats received medium-viscosity liquid paraffin at dietary doses of 2.5 or 5.0% for 104 weeks. The treatment was not accompanied by clinical signs of toxicity, with an increase in mortality or hematology findings.[16]

Reproductive Toxicity. No ***gonadotoxic*** action is reported, the reproductive function was not affected.

Mutagenicity.

Observations in man. An increased frequency of CA was observed in the peripheral blood lymphocytes of glass workers exposed to mineral oil mists (Sram, Hola, Kotesovec, 1985).

In vitro genotoxicity. Two insulation oils from highly refined mineral-base oils induced transformation of Syrian hamster embryo cells and enhanced transformation of mouse C3H 10TI/2 cells.Unused new, re-refined and used crankcase oils induced transformation in Syrian hamster embryo cells (IARC 33-87).

Carcinogenicity. The carcinogenicity of MO depends on the source and formulation of the petroleum. Analysis of MO used for medicinal and cosmetic purposes reveals the presence of several carcinogenic polycyclic aromatic hydrocarbons (PAH).

Observations in man. Considerable mortality or morbidity from stomach cancer was seen in workers exposed to MO (IARC, Suppl. 7-252).

Animal studies. The macroscopy and histopathology examination of the Swiss mice treated by skin application for up to 18 months showed a definite tumorigenic skin effect of the aromatic extract and distillate. Gradiski et al. pointed out that the observed effect was associated with the PAH concentration in the samples of white petroleum oil.[6] No statistically significant increase in the incidence of any tumor type was observed in F344 rats given medium-viscosity liquid paraffin at dietary doses of 2.5 or 5.0% for 104 weeks.[16]

Carcinogenicity classifications. An IARC Working Group concluded that there is sufficient evidence for the carcinogenicity of untreated and mildly treated oils in *experimental animals* and in *humans*.

IARC: 1.

An IARC Working Group concluded that there is inadequate evidence for the carcinogenicity of highly refined oils in *experimental animals* and in *humans*.

IARC: 3.

Chemobiokinetics. MO is poorly absorbed from the GI tract. Emulsification probably enhanced intestinal absorption.[17] Regular ingestion may interfere with absorption of fat soluble vitamins. Treatment with 150 ml MO caused marked decrease in carotene and tocopherol blood levels in cows. Administration of MO produces vitamin K deficiency in rats.[18]

Regulations.

U.S. FDA (1998) regulates MO as a direct and indirect food additive. MO may be used (1) in adhesives used as components of articles intended for use in packaging, transporting, or holding food in accordance with the conditions prescribed by 21 CFR part 175.105; (2) as a component of resinous and polymeric coatings of food-contact surface in accordance with the conditions prescribed by 21 CFR part 175.300; (3) in cellophane for food packaging in accordance with the conditions prescribed by 21 CFR part 177.1200; (4) as a component of a defoaming agent that may be safely used as a component of articles and in the manufacture of paper and paperboard intended for use in contact with food in accordance with the conditions prescribed by 21 CFR part 176.210; (5) as an ingredient of paper and paperboard for contact with dry food; (6) in acrylate ester copolymer coating to be safely used as a food-contact surface of articles intended for packaging and holding food, including heating of prepared food, subject to the provisions prescribed in 21 CFR part 175.210; (7) in production of hot-melt strippable food coatings which may be applied to food, subject to the provisions prescribed in 21 CFR part 175.230; (8) as lubricants with incidental food contact intended for used on machinery used for producing, manufacturing, packaging, processing, preparing, treating, packing, transporting, or holding food, subject to the provisions prescribed in 21 CFR part 178.3570; (9) as a substance employed in the production of or added to textiles and textile fibers intended for use in contact with food in accordance with the conditions prescribed by 21 CFR part 177.2800; and (10) as a plasticizer for rubber articles intended for repeated use in contact with food up to 30% by weight of the rubber product, alone or in combination with waxes, petroleum, total not exceeding 45% by weight of the rubber articles that contain at least 20% by weight of the ethylene-propylene copolymer elastomer.

FDA also regulates MO as additives in animal feed. *U.S. FDA* recommended warning labels for drugs containing MO that are taken internally, and classifies over-the-counter drug products containing MO as *GRAS*. White MO may be used (11) in the manufacture of resinous and polymeric coatings for the food-contact surface of articles intended for use in producing, manufacturing, packing, processing, preparing, treating, packaging, transporting, or holding food in accordance with the conditions prescribed by 21 CFR part 175.300; (12) as a component of the uncoated or coated food-contact surface of paper and paperboard intended for use in producing, manufacturing, packaging, processing, preparing, treating, packing, transporting, or holding aqueous and fatty foods in accordance with the conditions prescribed by 21 CFR part 176.170; (13) in the manufacture of cellophane for packaging food in accordance with the conditions prescribed by 21 CFR part 177.1200; and (14) in plasticizers for polymeric substances in the manufacture of articles or components of articles intended for use in producing, manufacturing, packaging, processing, preparing, treating, packing, transporting, or holding food in accordance with the conditions prescribed by 21 CFR part 178.3740.

Japan (1992). Liquid paraffins are permitted as food additives when used to aid dough release from cutters or pans in bread making.

Recommendations. Joint FAO/WHO Expert Committee on Food Additives (1991). ADI for mineral oil (food grade): not specified.

Standards. ***Russia*** (1995). Proposed PML: *n/m.*

References:

1. Jickells, S. M., Nichol, J., and Castle, L., Migration of mineral hydrocarbons into foods. V. Miscellaneous applications of mineral hydrocarbons in food contact materials, *Food Addit. Contam.*, 11, 333, 1994.
2. Grob, K., Huber, M., Boderius, U., and Bronz, M., Mineral oil material in canned foods, *Food Addit. Contam.*, 14, 83, 1997.
3. Grob, K., Lanfranchi, M., Egli, J., and Artho, A., Determination of food contamination by mineral oil from Jute sacks using coupled LC-GC, *J. Assoc. Off. Anal. Chem.*, 74, 506, 1991.
4. *Toxicology and Sanitary Chemistry of Plastics*, Abstracts, NIITEKHIM, Moscow, Issue No 1, 1979, 28 (in Russian).
5. Komarova, Ye. N., Toxicological characteristics of the high-refined mineral oil used as a plasticizer, in *Hygiene and Toxicology of High-Molecular-Mass Compounds and of the Chemical Raw Material Used for Their Synthesis,* Proc. 6th All-Union Conf., B. Yu. Kalinin, Ed., Leningrad, 1979, 263 (in Russian).
6. Gradiski, D., Vinit, J., Zissu, D., et al., The carcinogenic effect of a series of petroleum-derived oils on the skin of mice, *Environ. Res.*, 32, 258, 1983.
7. Firriolo, J. M., Morris, C. F., Trimmer, G. W., Twitty, L. D., Smith, J. H., and Freeman, J. J., Comparative 90-day feeding study with low-viscosity white mineral oil in Fischer-344 and Sprague-Dawley-derived CRL:CD rats, *Toxicol. Pathol.*, 23, 26, 1995.
8. Baldwin, M. K., Berry, P. H., Esdaile, D. J., Linnett, S. L., Martin, J. G., et al., Feeding studies in rats with mineral hydrocarbon food grade white oils, *Toxicol. Pathol.*, 20, 426, 1992.
9. Krasovsky, G. N. and Friedland, S. A., Toxicology characteristics of mineral oils as flotation reagent for iron ores, *Gig. Sanit.*, 7, 17, 1969 (in Russian).
10. Smith, J. H., Bird, M. G., Lewis, S. C., Freeman, J. J., Hogan, G. K., and Scala, R. A., Subchronic feeding study of four white mineral oils in dogs and rats, *Drug Chem. Toxicol.*, 18, 83, 1995.
11. Hulse, M., Klan, M. J., Noreyko, J. M., et al., in *American Petroleum Institute (API) Mineral Oil Review*, API, Washington, D. C., 1992.
12. McKee, R. H., Plutnick, R. T., and Traul, K. A., Assessment of the potential reproductive and subchronic toxicity of EDS coal liquids in Sprague-Dawley rats, *Toxicology*, 46, 267, 1987.
13. *Evaluation of Certain Food Additives and Contaminants,* 44th Report of the Joint FAO/WHO Expert Committee on Food Additives, Geneva, 1995, 18.
14. Fleming, K. A., Zimmerman, H., and Shubik, P., Granulomas in the livers of humans and Fischer rats associated with the ingestion of mineral hydrocarbons: A comparison, *Regul. Toxicol. Pharmacol.*, 27, 75, 1998.
15. Smith, J. H., Mallett, A. K., Priston, R. A., Brantom, P. G., Worrell, N. R., Sexsmith, C., and Simpson, B. J., Ninety-day feeding study in Fischer-344 rats of highly refined petroleum-derived food-grade white oils and waxes, *Toxicol. Pathol.*, 24, 214, 1996.
16. Shoda, T., Toyoda, K., Uneyama C., Takada K., and Takahashi, M., Lack of carcinogenicity of medium-viscosity liquid paraffin given in the diet to F344 rats, *Food Chem. Toxicol.*, 35, 1181, 1997.
17. *The Pharmacological Basis of Therapeutics*, A. G. Gilman, L. S. Goodman & A. Gilman, Eds., 6th ed., Macmillan Publ. Co., Inc., New York, 1980, 1009.
18. Russoff, I. S., *Handbook of Veterinary Drugs*, Springer Publ. Co, New York, 1974, 366.

9-OCTADECANOIC ACID, BUTYL ESTER

Molecular Formula. $C_{22}H_{42}O_2$
M = 338.64
CAS No 142-77-8
RTECS No RG3711000
Abbreviation. BO.

Synonyms and **Trade Names.** Butyl oleate; Oleic acid, butyl ester; Plasthall 503; Uniflex BYO.

Properties. Yellow or light-colored, oleaginous liquid with a mild odor. Insoluble in water, soluble in alcohol, miscible with vegetable and mineral oil.

Applications. A plasticizer for polyvinyl chloride resins, polystyrene, cellulose, *etc.* A softener for natural and synthetic rubber. A solvent and a lubricant. Water-resisting agent. Used in the preparation of food-contact coatings.

Regulations. ***U.S. FDA*** (1998) approved the use of BO in the manufacture of rubber articles intended for repeated use in producing, manufacturing, packaging, processing, preparing, treating, packing, transporting, or holding food in accordance with the conditions prescribed by 21 CFR part 177.2600.

OLEIC ACID, 2,3-EPOXYPROPYL ESTER

Molecular Formula. $C_{21}H_{38}O_3$
M = 338.59
CAS No 5431-33-4
RTECS No RKO700000

Synonyms. 2,3-Epoxy-1-propanol oleate; Glycidol oleate; Glycidyl oleate; Glycidyl octadecenoate; Oleic acid, 2,3-epoxypropyl ester; Oleic acid glycidyl ester.

Applications. A plasticizer.

Acute Toxicity. LD_{50} is 3.52 g/kg BW in rats.[027]

OLEIC ACID, 2-METHOXYETHYL ESTER

Molecular Formula. $C_{21}H_{40}O_3$
M = 340.61
CAS No 111-10-4
RTECS No RK0893000
Abbreviation. MEO.

Synonyms. Methoxyethyl oleate; Methyl cellosolve oleate.

Properties. Oily liquid with a characteristic odor. Insoluble in water (25°C), soluble in alcohol.

Applications. Used as a plasticizer in the production of polyvinyl chloride, polyvinyl butyral, ethyl cellulose, chlorinated, natural and synthetic rubber, intended for contact with food or drink. Gives materials transparency and elasticity.

Acute Toxicity. LD_{50} in young rats (BW of 70 to 75 g) is identified to be 16 g/kg BW.

Long-term Toxicity. Rats received 0.01 to 1.25% MEO in their diet for a year. More than a half of animals died at the dose level of 1.25% MEO in the diet by the end of the 9th month of the experiment; lower levels caused retardation of growth. Gross pathology examination revealed kidney stones. Addition of 0.01% MEO to the diet appeared to be harmless.

Chemobiokinetics. MEO is shown to be hydrolyzed by lipase *in vitro*. It is likely to be hydrolyzed in the body similar to fats.

Reference:

Smith, C. C. (Cincinnati), Toxicity of butylstearate, dibutylsebacate, dibutylphthalate, and methoxyethyloleate, *Arch. Ind. Hyg. Occup. Med.*, 4, 310, 1953.

PALM OILS

CAS No 8002-75-3
RTECS No RJ36960000

Abbreviation. PO.

Synonym. Palm butter.

Applications. Softeners in rubber processing.

Properties. Reddish-yellow to dark dirty red, fatty mass with faint odor of violet.[020] Insoluble in water.

General Toxicity. Harmful only under unusual conditions or overwhelming dosage.[012]

Reproductive Toxicity. Rats were fed red PO at the level of 10% of their diet. No adverse reproductive effects were reported.[1]

Teratogenicity. Rats received daily doses of 1.0 to 3.0 ml PO on days 5 through 15 of pregnancy. Malformations observed included exencephaly, eye defects, and cleft palate. The defects are likely to be attributed to the high carotene contents (32 to 48 mg carotene/100 ml PO), although other studies with carotene have not associated it with an elevation of congenital anomalies.[2]

Allergenic Effect. PO appeared to be a mild allergen.[012]

Regulations. ***U.S. FDA*** (1998) approved the use of PO in the manufacture of resinous and polymeric coatings for the food-contact surface of articles intended for use in producing, manufacturing, packing, processing, preparing, treating, packaging, transporting, or holding food in accordance with the conditions prescribed by 21 CFR part 175.300.

References:

1. Manorama, R., Chinnasamy, N., and Rukmini, C., Multigeneration studies on red palm oil, and on hydrogenated vegetable oil containing mahua oil, *Food Chem. Toxicol.*, 31, 369, 1993.
2. Singh, J. D., Palm oil induced congenital anomalies in rats, *Congen. Anom.*, 20, 139, 1980.

1,2-PENTANEDIOL

Molecular Formula. $C_5H_{12}O_2$

M = 104.17

CAS No 5343-92-0

RTECS No SA0455000

Abbreviation. 1,2-PD.

Synonym. 1,2-Dihydroxypentane.

Properties. Viscous liquid.

Applications. A plasticizer.

Acute Toxicity. LD_{50} in 12.7 g/kg BW in rats, 7.4 g/kg BW in mice, 3.7 g/kg BW in rabbits, 5.2 g/kg BW in guinea pigs. CNS inhibition, followed by excitation, and muscle weakness were observed.

Regulations. ***EU*** (1990). 1,2-PD is available in the *List of authorized monomers and other starting substances which may continue to be used for the manufacture of plastic materials and articles intended to come into contact with foodstuffs pending a decision on inclusion in Section A* (*Section B*).

Reference:

Ivanov, A. V., Hygienic standardization of reagent BB-2 and its ingredients as water body pollutants, *Gig. Sanit.*, 9, 14, 1973 (in Russian).

1,5-PENTANEDIOL

Molecular Formula. $C_5H_{12}O_2$

M = 104.17

CAS No 111-29-5

RTECS No SA0480000

Abbreviation. 1,5-PD.

Synonyms. 1,5-Dihydroxypentane; Pentamethylene glycol; Pentane-1,5-diol; 1,5-Pentylene glycol.

Properties. Viscous, oily liquid with a bitter taste. Miscible with water and alcohol.

Applications. Used as a plasticizer for cellulose products and adhesives.

Acute Toxicity. LD_{50} is 2.0 g/kg BW in rats, 6.3 g/kg BW in mice and rabbits, and 4.6 g/kg BW in guinea pigs. Animals developed CNS inhibition and muscle weakness.[2]

Regulations. ***EU*** (1990). 1,5-PD is available in the *List of authorized monomers and other starting substances which may continue to be used for the manufacture of plastic materials and articles intended to contact with foodstuffs pending a decision on inclusion in Section A* (*Section B*).

References:

1. Frankenfeld, J. W., Moham, R. R., and Squibb, R. L., Preservation of grain with aliphatic 1,3-diols and their esters, *J. Agric. Food. Chem.*, 23, 418, 1975.
2. Ivanov, A. V., Hygienic standardization of reagent BB-2 and its ingredients as water body pollutants, *Gig. Sanit.*, 9, 14, 1973 (in Russian).

PHOSPHORIC ACID, BIS(2-ETHYLHEXYL)PHENYL ESTER

Molecular Formula. $C_{22}H_{37}O_4P$
M = 396.51
CAS No 16368-97-1
RTECS No TB7900000
Abbreviation. DEHPP.

Synonym. Bis(2-ethylhexyl)phenyl phosphate.

Properties. Clear, viscous liquid of low volatility. Poorly soluble in water.

Applications. Used as a plasticizer in the production of food-contact plastic materials.

Acute Toxicity. In mice given DEHPP in oil solution, LD_{50} is 6.5 g/kg BW; for pure substance LD_{50} appeared to be 9.1 g/kg BW. Rats and guinea pigs tolerate the administration of 15 g/kg BW. Poisoning is accompanied by depression followed by excitation, aggressiveness, diarrhea, and BW loss. Some animals experienced convulsions. In the survivors there is reversible alopecia in the area of the back and haunches.

Repeated Exposure showed weak cumulative properties. Exposure to DEHPP by Lim method revealed a mild cholinesterase action, mainly affecting the NS and parenchymatous organs. The treatment caused a reduction in BW, increased CNS excitability, a dysbalance in the blood serum protein fractions, an increased content of residual nitrogen in the blood serum, and the appearance of erythrocytes in the urine. Histology examination of the visceral organs revealed changes in the brain, liver, and kidneys. Increased relative weights of the kidneys and liver were noted.

Regulations. ***U.S. FDA*** (1998) approved the use of DEHPP (1) as a component of adhesives intended for use in contact with food in accordance with the conditions prescribed by 21 CFR part 175.105, and (2) as an ingredient for resinous and polymeric coatings for polyolefin films to be safely used as a food-contact surface of articles in accordance with the conditions prescribed by 21 CFR part 175.320.

Reference:

Kalinina, N. I., Toxicity of organophosphorous plasticizers tributyl phosphate and di(2-ethylhexyl)-phenyl phosphate, *Gig. Truda Prof. Zabol.*, 8, 30, 1971 (in Russian).

PHOSPHORIC ACID, BUTYL CELLOSOLVE ESTER

Molecular Formula. $C_{18}H_{39}O_7P$
M = 286.47
CAS No 78-51-3
RTECS No KJ9800000
Abbreviation. BEP.

Synonyms. 2-Butoxyethanol phosphate; Tri(2-butoxyethyl) phosphate.

Properties. Liquid with a sweetish odor. Solubility in water is 1.1 g/l, soluble in fats.

Applications. Used as a fire-resistant and light-stable plasticizer in the production of vinyl resins, rubber, nitrocellulose and cellulose acetate, and synthetic rubber intended for contact with food or drink.

Exposure. BEP has been detected at *ppb* level in underground water.

Acute Toxicity. LD_{50} is 2.4 to 3.0 g/kg BW[04] or 5.0 g/kg BW in rats, and 3.94 g/kg BW in mice.[1] Poisoning with lethal doses is characterized by an anticholinesterase action. Signs of intoxication include adynamia, ataxia, clonic-tonic spasms, disturbance of the rhythm and rate of respiration, salivation and tremor. According to Laham et al.,[2] a single administration of 1.0 to 3.3 g/kg BW to Sprague-Dawley female rats, and 1.0 to 9.0 g/kg BW to males affects the peripheral NS. There was a significant reduction in caudal nerve conduction velocity. Light and electron microscopic examination revealed degenerative changes in myelinated and unmyelinated fibers.

Repeated Exposure failed to reveal cumulative properties. K_{acc} is 7 (by Lim).[1] A 14-week oral toxicity study was conducted in Wistar rats (diet containing 0.03, 0.3, or 3.0% BEP). The higher concentration group showed decreased BW gain. Serum cholinesterase activity was decreased and serum γ-glutamyl transferase activity increased in both sexes in the 3.0% group. Liver is the target organ for BEP toxicity. The NOEL of BEP in the diet was found to be 0.03% (males: 20 mg/kg BW; females: 22 mg/kg BW).[5]

Short-term Toxicity. Sprague-Dawley rats were dosed with 0.25 or 0.5 ml/kg BW for 18 weeks. The treatment caused a decline in cholinesterase activity in male but not in female rats. Histology examination revealed necrotic changes in the myocardium with inflammatory cell infiltration.[2]

Long-term Toxicity. Male and female Sprague-Dawley rats were administered 0.25 and 0.5 ml BEP/kg BW by gavage for 18 weeks. Histology examination revealed cardiac lesions including myocardial necrosis with inflammatory cell infiltration.[4]

Allergenic Effect is not observed.

Mutagenicity.

In vitro genotoxicity. BEP was negative in *Salmonella* mutagenicity assay.[025]

Chemobiokinetics. Products of hydrolysis (by plasma and tissue enzymes) were excreted in urine. Oxidation was also involved in the metabolism.[3]

Regulations. ***U.S. FDA*** (1998) approved the use of BEP (1) in adhesives as a component of articles intended for use in packaging, transporting, or holding food in accordance with the conditions prescribed by 21 CFR part 175.105; (2) as a defoaming agent in the manufacture of paper and paperboard intended for use in packaging, transporting, or holding food in accordance with the conditions prescribed by 21 CFR part 176.210; and (3) as a constituent of acrylate ester copolymer coating which may safely be used as a food-contact surface of articles intended for packaging and holding food, including heating of prepared food in accordance with the conditions prescribed by 21 CFR part 175.210.

References:

1. Gomonkov, P. I., et al., in *Current Problems of Industrial Hygiene and Occupational Pathology*, Voronezh, 1975, 46 (in Russian).
2. Laham, S., Szabo, J., Long, G., et al., Dose-response toxicity of tributoxyethyl phosphate orally administered to Sprague-Dawley rats, *Am. Ind. Hyg. Assoc. J.*, 46, 442, 1985.
3. *The Pharmacological Basis of Therapeutics*, A. G. Gilman, T. W. Rall, A. S. Nies, and P. Taylor, Eds., 8th ed., Pergamon Press, New York, 1990, 139.
4. Laham, S., Broxup, B. R., and Long, G. W., Subchronic oral toxicity of tributoxyethyl phosphate in the Sprague-Dawley rat, *Arch. Environ. Health*, 40, 12, 1985.
5. Saitoh, M., Umemura, T., Kawasaki, Y., Momma, J., Matsushima, Y., et al., Subchronic toxicity study of tributoxyethyl phosphate in Wistar rats, Abstract, *Eisei Shikenjo Hokoku*, 112, 27, 1994 (in Japanese).

PHOSPHORIC ACID, DIBUTYL PHENYL ETHER

Molecular Formula. $C_{14}H_{23}O_4P$

M = 286.31

CAS No 2528-36-1

RTECS No TB9626600

Abbreviation. DBPP.

Synonym. Dibutyl phenyl phosphate.

Properties. Oily liquid. Solubility in water is 7.0 g/l. Odor perception threshold is 157 mg/l, taste perception threshold is 48 mg/l.

Applications. Used as a plasticizer in the production of food-contact polymeric materials.

Acute Toxicity. LD_{50} for the rat varies from 2.14 g/kg BW to 2.2 g/kg BW; in mice, it is in the range of 0.87 to 1.79 g/kg BW; in rabbits and guinea pigs, it is 1.2 g/kg BW. Poisoned animals display excitation with subsequent symptoms of CNS inhibition.[1,2] According to other data, LD_{50} is 2.62 g/kg BW in rats and 5.0 g/kg BW in rabbits.[3] No evidence of delayed neurotoxicity or demyelinization was observed in adult chicken hens that were given oral doses of 1.34 g/kg BW on days 1 and 21.[014]

Repeated Exposure revealed no evident accumulation (studied by Lim method). Gross pathology examination showed reactive changes in the parenchymatous organs in the form of lymphoid infiltration of the kidneys and liver, as well as confluent hyperplasia of the splenic lymphoid follicles.[2]

Rats were gavaged with 105 and 210 mg/kg BW for 45 days. The treatment produced a reduction in oxygen consumption, a change in leukocyte formula, and a reduction in the content of pyruvic acid and *SH*-groups in the blood serum. There was an increase in the urinary levels of free phenols and sulfates.[1]

Short-term Toxicity. Sprague-Dawley rats were given 5.0, 50, and 250 mg/kg BW in their diet for 3 months.[4] There was a decrease in erythrocyte count, hematocrit, and *Hb* level, increased absolute and/or relative liver weights with concomitant reduction in hepatocyte vacuolization and increased fatty accumulation. The NOAEL appeared to be 5.0 mg/kg BW.

DBPP was administered to rats at dietary concentrations equivalent to 50, 150, and 500 mg/kg BW for 90 days. The treatment decreased BW gain and food consumption at 500 mg/kg BW dose level. Increased liver weight and liver weight to BW ratio and decreased lung weights were observed in the mid and/or high dose groups. A reduction of platelets was reported in all treatment groups; elevations of biochemical parameters (i.e., albumin, protein, and cholesterol) were reported in the 500 mg/kg BW group. Histopathology examination revealed lesions, including multifocal hepatocellular degeneration, bladder transitional cell hyperplasia, and interstitial cell hypertrophy of the ovary in most treatment groups. No effects were seen in rats given DBPP in the diet at a dosage of 5.0 mg/kg BW.[3]

Long-term Toxicity. Rabbits and rats were given 0.16 and 40 mg/kg BW for 6 months. The higher dose caused changes in the oxidation processes. There were hemodynamic disturbances and dystrophic changes in the visceral organs and necrotic foci in the myocardium, liver, and kidneys.[1]

Reproductive Toxicity. In a two-generation study,[3] Sprague-Dawley rats were given 5.0, 50, and 250 mg/kg BW in the diet. BW was lower in high-exposed adult animals but mating and fertility indices were unaltered. Survival among high-exposed pups appeared to be decreased. Histopathology changes consisting of mononuclear cell infiltration and transitional epithelial hyperplasia were noted in the urinary bladder. The NOAEL of 5.0 mg/kg BW was identified for these effects.[1]

Teratogenicity. No teratogenic or fetotoxic effects were observed in the offspring of rats gavaged with DBPP at a dosage of 3.0, 30, or 300 mg/kg BW on days 6 through 15 of gestation. No maternal toxic effects were observed at any treatment level.[3]

Mutagenicity.

In vitro genotoxicity. DBPP was negative in five *Salmonella* strains and one strain of *Saccharomyces yeast* assays, in *in vitro* induction of l5178y TK mouse lymphoma cells, and in a hepatocyte primary culture/DNA repair assay.[3]

In vivo cytogenetics. Groups of male and female F344 rats given *i/p* injections of 40, 200, or 400 mg/kg BW (males) or 60, 300, or 600 mg/kg BW (females) had no evidence of a clastogenic effect in bone marrow cells.[3]

Chemobiokinetics. DBPP is likely to be hydrolyzed in the body to form *phenol metabolites*. Conjugation of the metabolites with *glucuronic* and *sulphuric acids* occurs in the liver and, in this form, they are excreted in the urine.

Standards. ***Russia.*** MAC: 1.5 mg/l; Recommended PML: 3.0 mg/l.

References:

1. Vorobyova, L. V., in *Problems of Environment Protection, Occupational Safety and Professional Diseases*, Leningrad, 1972, 30; Proc. Perm‘ Polytech. Institute, Perm’, Issue No 141, 1973, 109 (in Russian).
2. Bezsmertny, V. E., Mashbitz, F. D., Rotenberg, Yu. S., et al., Toxicity of dibutylphenyl phosphate, *Gig. Truda Prof. Zabol.*, 4, 46, 1971 (in Russian).
3. *Threshold Limit Values and Biological Exposure Indices*, 5th ed., American Conference of Governmental Industrial Hygienists, Doc., 1988, 176.

4. Healy, C. E., Nair, R. S., Lemen, J. K., et al., Subchronic and reproduction studies with dibutylphenyl phosphate in Sprague-Dawley rats, *Fundam. Appl. Toxicol.*, 16, 117, 1991.

PHOSPHORIC ACID, DIPHENYL 2-ETHYLHEXYL ESTER

Molecular Formula. $C_{20}H_{27}O_3P$
M = 362.41
CAS No 1241-94-7
RTECS No TC6125000
Abbreviation. DPEHP.

Synonyms and **Trade Names.** Diphenyl 2-ethylhexyl phosphate; Ethylhexyl diphenylphosphate; 2-Ethyl-1-hexanol ester with diphenyl phosphate; Octicizer; Phosphoric acid, 2-ethylhexyl diphenyl ester; Santicizer 141.

Properties. Viscous liquid with a faint odor. Solubility in water is up to 2.0 mg/l at 25°C.

Applications. Used as a plasticizer in the production of polyvinyl chloride and vinyl polymers. Renders material elasticity at low temperatures, light stability and resistance to combustion. Mixes well with acrylonitrile-butadiene-synthetic rubber, nitro- and ethyl cellulose, cellulose acetobutyrate, polymethyl methacrylate, polystyrene, and polyvinyl butyral.

Migration Data. Migration of plasticizers from nitrocellulose-coated regenerated cellulose film (RCF) purchased from retail and take-away outlets was studied. Levels of the plasticizer found to migrate from RCF were as high as 0.5 to 53 mg diphenyloctyl phosphate/kg in confectionery, meat pies, cake, and sandwiches[1]

Acute Toxicity. LD_{50} was not attained. Rats and rabbits tolerated a single administration of 24 g/kg BW. Higher doses caused diarrhea. The same dose caused only emaciation in rats. The animals recovered. Histology examination failed to reveal changes in the viscera.[2]

Long-term Toxicity. No mortality or decrease in BW gain was noted in rats exposed to 0.125% DPEHP in the diet for 2 years. No signs of intoxication were observed in dogs fed the diet with 1.5% DPEHP for the same period of time. Gross pathology examination failed to reveal changes in the viscera.

Reproductive Toxicity.

Teratogenicity. Charles River rats were dosed with 300 to 3000 mg/kg BW by gavage on days 6 through 15 of gestation. Maternal BW was reduced on administration of higher doses. No signs of teratogenicity were noted.[3]

Chemobiokinetics. DPEHP is shown to be poorly absorbed in the intestine. Its metabolism in the body occurs through hydrolysis to form *phenol metabolites*.

Observations in man. Intake of 5.0 ml/kg BW by volunteers enabled *phenol* (a product of DPEHP hydrolysis) in all forms to be detected in the urine and feces.[1] 96% DPEHP is excreted as phenol derivatives.

Animal studies. Male Wistar rats received a single oral dose of 10 mg ^{14}C-DPEHP which was distributed all over the body tissues. DPEHP is shown to have a relatively high affinity for adipose tissue and liver. Major urinary metabolites included *diphenyl phosphate* and *phenol*. Total liberation with the urine (primarily) and feces occurred in 7 days.[4]

Regulations. ***U.S. FDA*** (1998) approved the use of DPEHP (1) in the manufacture of resinous and polymeric coatings in a food-contact surface in accordance with the conditions prescribed by 21 CFR part 175.300; and (2) in the manufacture of resinous and polymeric coatings for polyolefin films to be safely used as a food-contact surface of articles intended for use in producing, manufacturing, packing, transporting, or holding food in accordance with the conditions prescribed by 21 CFR part 175.320.

Standards. *Russia*. Recommended PML: *n/m*.

References:

1. Castle, L., Mercer, A. J., Startin, J. R., and Gilbert, J., Migration from plasticized films into foods. 3. Migration of phthalate, sebacate, citrate and phosphate esters from films used for retail food packaging, *Food Addit. Contam.*, 5, 9, 1988.
2. Treon, J. F., Cappel, J., and Sigmon, H., Toxicity of 2-ethylhexyl diphenylphosphate, metabolic fate in man and animals, *Arch. Ind. Hyg. Occup. Med.*, 8, 268, 1953.

3. Robinson, E. C., Hammond, B. G., Johnnsen, F. R., et al., Teratogenicity studies of alkylary phosphate ester plasticizers in rats, *Fundam. Appl. Toxicol.*, 7, 138, 1986.
4. Nishimaki-Mogami, T., Minegishi, K., and Sato, M., Isolation and identification of metabolites of 2-ethylhexyl diphenyl phosphate in rats, *Arch. Toxicol.*, 61, 259, 1988.

PHOSPHORIC ACID, DIPHENYL TOLYL ESTER

Molecular Formula. $C_{19}H_{17}O_4P$
M = 240.33
CAS No 26444-49-5
RTECS No TC5520000

Synonyms and **Trade Name.** Cresyl diphenyl phosphate; Diphenyl cresyl phosphate; Diphenyl tolyl phosphate; Methyl phenyl diphenyl phosphate; Phosphoric acid, cresyl ester; Santicizer 140.

Composition. Commercial P. preparation was found to contain triphenyl phosphate (35%), P. (45%), dicresyl phenyl phosphate (18%) and tricresyl phosphate (2.0%).[1]

Properties. Clear, transparent liquid with a very mild odor. Miscible with water, soluble in organic solvents.

Applications. Used as a plasticizer and flame retardant in the production of polyvinyl chloride, polystyrene, polycarbonate, butadiene rubber, cellulose acetate and acetobutyrate, and other food-contact polymeric material.

Acute Toxicity. Administration of 4.0 g/kg BW caused no mortality in rats. In 2 to 3 days after administration the animals became lethargic with profuse diarrhea. Death was preceded with signs of paralysis. Gross pathology examination revealed capillary atonia with edema, and brain hemorrhages. Toxicity depends on the presence of *o*-cresol.[2] According to other data, LD_{50} is 6.4 to 12.8 g/kg BW in rats and mice. These doses do not cause paralyses.[05]

Repeated exposure to small oral doses of 2.5 mg/kg BW caused no clinical and only doubtful histological signs of neuropathy in hens. Neurotoxic esterase levels were depressed during two months.[3]

Short-term Toxicity. P. is a neurotoxic agent. Adult animals are shown to be more sensitive than young. The levels of neurotoxic esterase were depressed to about of 40% of normal in the brain and to 55% of norm in the spinal cord of hens. No clinical signs of neuropathy were found.[3]

Long-term Toxicity. Chronic feeding of P. to hens does not clinically cause observable delayed neurotoxic effects prior to a point when inhibition of neurotoxic esterase of brain and spinal cord reaches 70 to 90%.[4]

Reproductive Toxicity.

Gonadotoxicity and ***Teratogenicity.*** P. (100 or 300 mg/kg BW) was administered orally to male Sprague-Dawley rats for 3 weeks. The higher dose decreased weight of epidydimis, sperm motility and viability. An increase in morphological abnormalities was noted at 100 mg/kg BW.[5]

Mutagenicity.

In vitro genotoxicity. P. was shown to be negative in *Salmonella* mutagenicity assay.[015]

Regulations. ***U.S. FDA*** (1998) approved use of P. as a constituent of food-contact adhesives which may be safely used as a component of articles intended for use in packaging, transporting, or holding food in accordance with the conditions prescribed by 21 CFR part 175.105.

References:

1. Vainiotalo, S., Verkkala, E., Savolainen, H., Nickels, J., and Zitting, A., Acute biological effects of commercial cresyldiphenyl phosphate in rats, *Toxicology*, 44, 31, 1987.
2. Mallette, F. S., Studies on toxicity and skin effects of compounds used in rubber and plastics industries: accelerators, activators, and antioxidants, *Arch. Ind. Hyg. Occup. Med.*, 5, 311, 1952.
3. Lotti, M. and Johnson, M. K., Repeated small doses of neurotoxic organophosphate. Monitoring of neurotoxic esterase in brain and spinal cord, *Arch. Toxicol.*, 45, 263, 1980.
4. Johnson, M. K. and Lotti, M., Delayed neurotoxicity caused by chronic feeding of organophosphates requires a high point of inhibition of neurotoxic esterase, *Toxicol. Lett.*, 5, 99, 1980.

5. Matsuura, I., Hoshino, N., Wako, Y., Tani, E., Satou, T., Aoyama, R., and Ikeda, Y., Sperm parameter studies on three testicular toxicants in rats, Abstract of Papers, 35th Annual Meeting of the Japanese Teratology Society, Tokyo, 1995, *Teratology*, 52, P33, October 1995.

PHOSPHORIC ACID, PHENYLDI(2-ETHYLHEXYL) ESTER

Synonym. Phenyldi(2-ethylhexyl) phosphate

Properties. Clear, viscous liquid of low volatility. Poorly soluble in water.

Applications. Used as a plasticizer in polyvinyl chloride production.

Acute Toxicity. In mice, LD_{50} of oil solution is found to be 6.5 g/kg BW; for pure P., it is 9.1 g/kg BW. Rats and guinea pigs tolerate the dose of 15 g/kg BW. Administration led to CNS inhibition with subsequent excitation, increased aggressiveness, diarrhea, emaciation and, in some animals, convulsions. Survivors exhibited reversible alopecia on the back and haunches.

Repeated Exposure. P. produces mild anticholinesterase action. The treatment affected mainly the NS and parenchymatous organs.

Reference:

Kalinina, N. I., Toxic effects of phenyldi(2-ethylhexyl)phosphate, *Gig. Truda Prof. Zabol.*, 8, 30, 1971 (in Russian).

PHOSPHORIC ACID, TRIBUTYL ESTER

Molecular Formula. $C_{12}H_{27}O_4P$

M = 266.32

CAS No 126-73-8

RTECS No TC7700000

Abbreviation. TBP.

Synonym. Tributyl phosphate.

Properties. Colorless, oily liquid with a sharp odor. Solubility in water is 397 mg/l at 19°C, completely soluble in mineral oil. Miscible with organic solvents. Odor perception threshold is 0.014 mg/l, taste perception threshold is 0.019 mg/l.[1]

Applications. TBP is used as a primary plasticizer in the manufacture of nitrocellulose, plastics, and vinyl resins intended for contact with food or drink; a solvent for cellulose esters, lacquers, and natural gums.

Acute Toxicity. LD_{50} is 1.4 to 3.35 g/kg BW in rats, 0.9 to 1.24 g/kg BW in mice, and 1.8 g/kg BW in chickens.[1-3]

Administration of 1.0 mg TBP/kg BW is accompanied by CNS excitation without convulsions, as well as its transient impairment.[4] Approximately the same dose causes impairment of the renal function in rabbits, with the appearance of protein in the urine. LD_{50} of *triisobutyl phosphate*, which is of similar toxicity, appears to be in the range of 3.2 to 6.0 g/kg BW.[05]

Repeated Exposure revealed moderate cumulative properties. K_{acc} exceeded 3. TBP exerts a neurotoxic effect on the peripheral nerves; it produces a weak anti-cholinesterase action, affecting mainly the NS and parenchymatous organs. The kidney seems also to be a target organ.[2]

Delayed neuropathy was observed following oral administration of 0.42 ml/kg BW dose for 14 days. In this study, no axonal degeneration or other overt signs of toxicity were reported.[6]

Short-term Toxicity. Wistar male rats were fed 0.5 and 1.0% TBP in the diet for 10 weeks.[5] BW gain and food consumption in the treated animals were significantly lower than those in the controls. There were no differences in cholinesterase activity in the serum but in the brain it was significantly increased. The blood coagulation time was found to be prolonged. In a 18-week study, doses of 200 mg/kg BW and 300 mg/kg BW produced dif fuse hyperplasia of the urinary bladder epithelium in Sprague-Dawley rats.[7]

Long-term Toxicity. The treatment of rats and rabbits caused dystrophic changes in the liver, necrosis of separate cells or of groups of cells, and fatty dystrophy. Functional effects are not reported.[1]

Reproductive Toxicity. CD rats received 200, 700, or 3000 ppm TBP in the diet for 10 weeks and then were mated within groups for 3 weeks with continued exposure. There was no evidence of reproductive toxicity, of reproductive organ pathology, or of effects on gestation or lactation at any dose tested. The

NOAEL for reproductive toxicity was at least 3000 ppm and the NOAEL for postnatal toxicity was approximately 200 ppm.[12]

Teratogenicity. TBP is reported to be slightly teratogenic in chickens at high dose levels. A dose of 0.42 mg/kg BW (14 days) causes degenerative changes in the seminiferous tubules to develop.[7]

According to Schroeder et al., TBP produced no teratogenic effect in rats and rabbits.[8]

Mutagenicity.

In vivo cytogenetics. TBP was negative in *Dr. melanogaster.*

In vitro genotoxicity. TBP was shown to be negative in *E. coli* test. Equivocal results were reported in *Salmonella* mutagenicity assay.[9,014]

Carcinogenicity. Male Sprague-Dawley rats received 200, 700, or 3000 ppm TBP in their diet for 10 weeks. Scanning electron microscope examination of the urine showed no increased or abnormal crystalluria, urinary precipitate, or calculi. Ulceration and hemorrhage into bladder lumen and consequent diffuse papilllary and nodu- lar hyperplasia were seen at two highest doses. TBP produces tumors of bladder urothelium in rats at high doses (700 and 3000 ppm), with greater effects in males than in females.[13]

Chemobiokinetics. Following ingestion, TBP is distributed mainly to the blood, GI tract, and liver.[10] More than 50% of an orally administered dose was absorbed within 24 hours.[11]

TBP undergoes oxidation of the butyl moieties. Oxidized methyl groups are removed as *glutathione* conjugates and subsequently excreted as *N-acetyl cysteine derivatives*. Excretion occurs predominantly via the urine.[2]

Regulations. ***U.S. FDA*** (1998) regulates the use of TBP (1) in adhesives as a component of articles intended for use in packaging, transporting, or holding food in accordance with the conditions prescribed by 21 CFR part 175.105; and (2) as a defoaming agent that may be safely used in the manufacture of paper and paperboard intended for use in producing, manufacturing, packing, transporting, or holding food in accordance with the conditions prescribed by 21 CFR part 176.210.

Standards. ***Russia*** (1995). MAC: 0.01 mg/l (organolept., taste).

References:

1. Zyabbarova, S. A. and Teplyakova, E. V., Data on hygienic characteristics of tributyl phosphate, *Gig. Sanit.*, 7, 100, 1968 (in Russian).
2. Kalinina, G. I. and Peresadov, V. P., in *Proc. Volgograd Medical Institute*, Volgograd, Issue No 23, 1970, 153 (in Russian).
3. *Tri-n-butyl Phosphate*, Environmental Health Criteria No 112, WHO, IPCS, Geneva, 1991, 80.
4. Sabine, J. C. and Hayes, F. N., Anticholinesterase activity of tributyl phosphate, *Arch. Ind. Hyg. Occup. Med.*, 6, 174, 1952.
5. Oishi, H., Oishi. S., and Hiraga, K., Toxicity of tri-*n*-butyl phosphate, with special reference to organ weights, serum components and cholinesterase activity, *Toxicol. Lett.*, 6, 81, 1980; Toxicity of several phosphoric acid esters in rats, *Toxicol. Lett.*, 13, 29, 1982.
6. Laham, S. and Long, G., Subacute oral toxicity of tri-*n*-butyl phosphate in the Sprague-Dawley rat, *J. Appl. Toxicol.*, 4, 150, 1984.
7. Laham, S., Long, G., and Broxup, B., Induction of urinary bladder hyperplasia in Sprague-Dawley rats orally administered tri-*n*-butyl phosphate, *Arch. Environ. Health*, 40, 301, 1985.
8. Schroeder, R. E., Gehart, J. M., and Kneiss, J., Developmental toxicity studies of tributyl phosphate (TBP) in the rat and rabbit, *Teratology*, 43, 455, 1991.
9. Gafieva, Z. A. and Chudin, V. A., Evaluation of the mutagenic activity of tributyl phosphate in *Salmonella typhimurium*, *Gig. Sanit.*, 9, 81, 1986 (in Russian).
10. Khalturin, G. V. and Andryushkeyeva, N. I., Toxicokinetics of tributyl phosphate following single and chronic intragastric intake by rats, *Gig. Sanit.*, 2, 87, 1986 (in Russian).
11. Suzuki, T., Sasaki, K., Takeda, M., et al., Metabolism of tributylphosphate in male rats, *J. Agric. Food Chem.*, 32, 603, 1984
12. Tyl, R. W., Gerhart, J. M., Myers, C. B., Marr, M. C., Brine, D. R., Seely, J. C., and Henrich, R. T., Two-generation reproductive toxicity study of dietary tributyl phosphate in CD rats, *Fundam. Appl Toxicol.*, 40, 90, 1997.

13. Arnold, L. L., Christenson, W. R., Cano, M., John, M. K., Wahle, B. S., and Cohen, S. M., Tributyl phosphate effects on urine and bladder epithelium in male Sprague-Dawley rats, *Fundam. Appl. Toxicol.*, 40, 247, 1997.

PHOSPHORIC ACID, TRIETHYL ESTER

Molecular Formula. $C_6H_{15}O_4P$
M = 182.16
CAS No 78-40-0
RTECS No TC7900000
Abbreviation. TEP.

Synonyms. Ethyl phosphate; Triethyl phosphate.

Properties. Colorless liquid, soluble in water and alcohol.

Applications. Used as a plasticizer in the production of food-contact plastics. Blends well with polyvinyl chloride and its copolymers. Gives products high elasticity and low temperature resistance. TEP is used to make artificial leather and film materials.

Acute Toxicity. LD_{50} is reported to be 1.6 g/kg BW in guinea pigs.[1] In rats and mice, LD_{50} is 1.4 g/kg BW. Administration of the lethal doses led to excitation with subsequent CNS inhibition, motor coordination disorder, paresis of the hind legs, and respiratory disturbances. Death occurs in 24 hours. Gross pathology examination revealed visceral congestion, particularly in the liver, and hyperemia of the gastric mucosa.[2]

Repeated Exposure revealed no cumulative properties.[2]

Mutagenicity.

In vitro genotoxicity. TEP is found to be negative in *Salmonella* mutagenicity assay.[015]

Regulations. ***U.S. FDA*** (1998) approved the use of TEP as a constituent of food-contact adhesives which may be safely used as a component of articles intended for use in packaging, transporting, or holding food in accordance with the conditions prescribed by 21 CFR part 175.105.

Standards. ***Russia*** (1995). MAC and PML: 0.3 mg/l.

References:

1. Deichmann, W. B., *Toxicology of Drugs and Chemicals*, Academic Press, Inc., New York, 1969, 605.
2. Pyatlin, V. N., in *Proc. Kuibyshev Research Institute Epidemiol. Hygiene*, Kuibyshev, Issue No 5, 1968, 117 (in Russian).

PHOSPHORIC ACID, TRIPHENYL ESTER

Molecular Formula. $C_{18}H_{15}O_4P$
M = 326.31
CAS No 115-86-6
RTECS No TC8400000
Abbreviation. TPP.

Synonym and **Trade Names.** Celluflex TPP; Disflamoll TP; Phosphoric acid, triphenyl ester; Phosflex TPP; Triphenyl phosphate.

Properties. Crystalline, slightly aromatic solid. Water solubility is about 2.1 mg/l, moderately soluble in ethanol. A marked change is noted in the taste and smell of solutions stored in containers made of etrol, which has been plasticized with TPP.

Applications. Used as a plasticizer in the production of cellulose acetate articles and in lacquers manufacture. Because of its limited compatibility and low effectiveness, it is used in blends with other plasticizers. A flame retardant and a solvent.

Exposure. Average daily intake in food was found to be several ηg/kg BW (U.S. FDA).

Acute Toxicity. LD_{50} is 3.5 to 10.8 g/kg BW in rats, 1.3 to 5.0 g/kg BW in mice, more than 4.0 g/kg BW in guinea pigs, and about 8.0 g/kg BW in rabbits.[1] After administration animals are depressed. The urine and expired air smell of TPP. Death occurs within 4 to 5 days. Gross pathology examination revealed distention of, and inflammatory changes in, the stomach and intestine, a clay-like consistency of the liver and kidneys, and brain hemorrhages.[2,3] Sutton et al. reported 3.0 to 4.0 g/kg BW dose to be harmless in rats and mice.[4]

Repeated Exposure failed to reveal a cumulative effect.[2,3] The early studies[5] on neurotoxicity of TPP have been questioned. In a 35-day feeding study in male Holtzman rats (doses applied 1.0 and 5.0 g/kg BW), retardation of BW gain and increase in liver weights were found only with a 5.0 g/kg BW dose, but no hematological changes were noted. A pronounced reduction in the BW of rabbits on repeated administration of 2.0 g/kg BW was observed.[4] Being ineffective for the CNS of rats and young chickens, TPP caused limb paralysis in cats.[05]

Short-term Toxicity. Oral administration of 380 to 1900 mg/kg BW for 3 months to rats caused no deaths; there was no evidence of abnormal growths; cholinesterase activity was unaltered.[014]

Long-term Toxicity. Rats were exposed in the diet to the levels of 5.0, 10, and 100 mg/animal/day for 6 months. The treatment caused no retardation of BW gain.[05]

Immunotoxicity. Sprague-Dawley rats were exposed to the levels of 2.5 to 10 g TPP/kg diet for 120 days. No significant effect on the humoral response was found. The only effects noted were a decreased rate of growth at high levels of TPP and increases in the levels of α- and β-globulins suggestive of increased hepatic activity.[6]

Reproductive Toxicity. There were no overt maternal toxicity or embryotoxicity in the Sprague-Dawley rats after dietary exposure to 166 to 690 mg/kg BW over a period of 91 days, including mating and gestation period.[7]

Mutagenicity. There were negative results in several *in vitro* studies.[1]

Chemobiokinetics. TPP is likely to be hydrolyzed to form *phenol metabolites*, which are excreted in the urine. TPP is quite possibly excreted unchanged by the same route.

Regulations. ***U.S. FDA*** (1998) approved the use of TPP (1) in adhesives as a component of articles intended for use in packaging, transporting, or holding food in accordance with the conditions prescribed by 21 CFR part 175.105; and (2) in the manufacture of cross-linked polyester resins to be safely used in articles or as a component of articles coming in contact with food in accordance with the conditions prescribed by 21 CFR part 177.2420.

Standards. ***Russia*** (1995). PML in food: 0.5 mg/kg. PML: 1.0 mg/l.

References:

1. *Triphenyl Phosphate*, Environmental Health Criteria No 111, WHO, IPCS, Geneva, 1991, 80.
2. Antonyuk, O. K., Hygienic evaluation of the plasticizer triphenyl phosphate added to polymer composition, *Gig. Sanit.*, 8, 98, 1974 (in Russian).
3. Antonyuk, O. K., in *Problems of Industrial Hygiene, Occupational Pathology and Toxicology in the Production and Use of Organophosphorus Plasticizers*, Meditsina, Moscow, 1983, 96 (in Russian).
4. Sutton, W. L., Terhaar, C. J., Miller, F. A., et al., Studies on the industrial hygiene and toxicology of triphenyl phosphate, *Arch. Environ. Health*, 1, 45, 1960.
5. Smith, M. I., Evolve, E., and Frazier, W. H., Pharmacological action of certain phenol esters with special reference to the etiology of the so-called ginger paralysis., *Public Health Rep.*, 45, 2509, 1930.
6. Hinton, D. M., Jessop, J. J., Arnold, A., Albert, R. H., and Hines, F. A., Evaluation of immunotoxicity in subchronic feeding study of triphenyl phosphate, *Toxicol. Ind. Health*, 3, 71, 1987.
7. Welsh, J. J., Collins, T. F. X., Whitby, K. E., et al., Teratogenic potential of triphenyl phosphate in Sprague-Dawley (Spartan) rats, *Toxicol. Ind. Health*, 3, 357, 1987.

PHOSPHORIC ACID, TRI(2-PROPOXYETHYL) ESTER

Molecular Formula. $C_{15}H_{33}O_7P$

M = 356.38

Synonyms. Propylglycol phosphate; Propyl cellosolve phosphate; Tri(2-propoxyethyl) phosphate.

Properties. Colorless liquid with a faint odor.

Applications. Used as a plasticizer in the production of cellulose acetate.

Acute Toxicity. In Wistar rats, LD_{50} of T. mixed with pure ethyl alcohol (small amounts to reduce viscosity) appeared to be 4.0 to 6.0 g/kg BW. Rats tolerated doses of 2.0 to 2.5 g/kg BW.

Long-term Toxicity. No signs of intoxication or histology changes were noted in rats dosed with up to 0.9 g T./kg diet for 21 months. The treatment caused no growth retardation.[04,05]

Reproductive Toxicity. At the dose level of 50 to 368 mg T./kg BW it did not affect development of the progeny over 3 to 5 generations.

Carcinogenicity. In treated animals, tumor growth was not observed.[05]

Standards. ***Russia.*** PML: *n/m.*

PHOSPHORIC ACID, TRITOLYL ESTER

Molecular Formula. $C_{21}H_{21}O_4P$
M = 368.36
CAS No 78-30-8
RTECS No TD0175000
Abbreviation. o-TCP.

Synonyms and **Trade Names.** Celluflex 179C; Cresyl phosphate; Durad; Flexol; Plasticizer TCP; Kronitex; Lindol; Phosphoric acid, tricresyl ester; Phosflex 179A; *o*-Tricresyl phosphate; Tris(tolyloxy) phosphine oxide; Tritolyl phosphate.

Properties. The technical product is a mixture of isomers which differ in that the CH_3- group occupies the *ortho-*, *meta-*, or *para*-position in the cresyl radicals. Light-yellow, oily liquid. Solubility is 60 mg/l in water, 0.014% in human sweat at 36°C; readily soluble in fats and oils. Renders water a specific odor. The practical threshold corresponds to a concentration of 60 mg technical TCP/l. At this concentration it does not affect the color of water, does not give it a foreign taste, or alter the *pH.*

Applications. Used commercially as a plasticizer in the production of ethyl- and nitrocellulose, polyvinyl chloride, polystyrene and chlorinated rubber; flame retardant, lubricant, and gasoline additive.

Acute Toxicity. *o*-TCP is known as a CNS toxicant, being a cause of many associated poisonings and polyneuritis with paralysis of the extremities.[05]

Observations in man. Accidental human exposure to a single large dose results in GI tract disturbances varying from slight to severe nausea and vomiting, accompanied by abdominal pain and diarrhea.[1] Consumption of food contaminated by *o*-TCP may be followed by GI symptoms, although in some cases polyneuropathy is the first evidence of poisoning. Histopathology findings show axonal degeneration. Severe symptoms may be observed following ingestion of 0.15 g *o*-TCP, while some individuals failed to show any toxic effect after ingesting 1.0 to 2.0 g.

Animal study. The LD_{50} values are reported to be 2.4 g/kg BW (the product containing 37% *o*-isomer) and 7.5 g/kg BW (1.1% *o*-isomer) in mice; 3.7 g/kg BW[2] or 0.1 g/kg BW in rabbits; and 0.5 g/kg BW in dogs.[3] Guinea pigs are less sensitive but chickens are more so. Poisoning is accompanied by immediate general excitation with subsequent short-lasting general inhibition, motor coordination disorder and tremor. In guinea pigs, within days there are urinary incontinence and intestinal dysfunction. Paralysis and a decline in cholinesterase activity is observed. The lowest dose causing paralysis is 10 to 30 mg/kg BW.

A single oral dose of 50 to 500 mg/kg BW induced delayed neuropathy in chickens, whereas doses of 840 mg/kg BW or more could produce spinal cord degeneration in Long-Evans rats. Administration of a single oral dose of 500 mg/kg BW produced no detectable clinical signs in the Japanese quail. Oral doses resulted in only sparse Fink-Heimer silver-impregnated degeneration in the white matter of the cerebellum with no degeneration noted in any other region of the brain, while injection of *o*-TCP caused widespread degeneration in large numbers of brainstem nuclei and tracts and in all cerebellar foliae and deep nuclei.[4]

Repeated Exposure.

Observations in man. Exposure to small cumulative doses results in delayed neurotoxicity that gradually proceeds after a latent period of 3 to 28 days.[1] TCP poisonings have occurred throughout the world,[5] the major outbreaks took place in 1930 and 1931 (the U.S., 50,000 cases), in 1957 (Morocco, about 10,000 cases) and in 1962 (India, more than 400 cases).

Animal study. *o*-TCP is the only isomer out of three isomers of TCP that produces delayed neurotoxicity in a wide range of experimental species (it is classified as a dying-back neuropathy). Degenerative changes occur in the distal axons and extend with time toward the cell bodies.

Wistar rats were given a mixture of isomers (doses of 5.0 g TCP/kg in the diet) for 9 weeks. There was an increase in liver weights. No hematology changes were observed, but in the plasma, total protein, urea, cholesterol, and glutamate-pyruvate transaminase were significantly increased.[6] CD-1 mice were fed the diet containing up to 0.875% *o*-TCP. No clinical signs of intoxication were reported.[7]

Short-term Toxicity. In a 3-month study in Sprague-Dawley rats, no histopathology changes were found after oral administration of 30 to 1000 mg/kg BW doses.[8]

Long-term Toxicity. Though short-term symptoms of *o*-TCP ingestion might involve vomiting, abdominal pain, and diarrhea, characteristically delayed, long-term symptoms are neurologic, frequently leading to paralysis and pyramidal signs.[1] As a result of NTP chronic oral exposure only non-neoplastic lesions were reported. The treatment caused cytoplasmic vacuolization in the adrenal gland and interstitial hyperplasia in the ovary in female rats, ceroid pigmentation of adrenal and ceroid pigmentation, clear cell foci, fatty change in the liver in male mice.[9]

Immunotoxicity. Immune system appears to be a sensitive target for *o*-TCP. Immunodepression can be a consequence of toxic chemical stress associated with organophosphate-induced cholinergic stimulation. A suppression of humoral and cell-mediated immune responses was reported in Wistar rats exposed to 20 to 100 ppm *o*-TCP in the diet for 6 weeks.[10]

Reproductive Toxicity.

Embryotoxicity. Chronic feeding prior to and during pregnancy caused a decrease in litter size and pup viability in rats.[11] Rats received by gavage 350 mg/kg BW on days 6 through 18 of gestation. There were no adverse fetal effects observed even on treatment with near lethal doses.[12]

TCP causes ***gonadotoxic effect***; it may produce morphologic damage to the testes and ovaries, and morphologic changes in the sperm. It interferes with spermatogenic processes and sperm motility directly and not via androgenic mechanism or decreased vitamin *E* availability.[13] F344 rats were given doses of 10 to 100 mg/kg BW for 63 days. Testicular pathological changes were seen at doses above 25 mg/kg BW. The LOEL of 10 to 24 mg/kg BW was identified in this study.[14]

Doses of 200 mg/kg BW and 400 mg/kg BW were administered in corn oil to male Long-Evans rats by gavage. Sperm was injured at a dose level of 400 mg/kg BW and there was a dose-dependent increase in abnormal sperm morphology. CD-1 mice were given 0.5 to 2.0 g/kg in the diet over a period of 98 days. A crossover mating trial revealed impaired fertility in both males and females exposed to 2.0 g/kg in the diet.[7] Doses close to the NOAEL do not affect the function and microstructure of the ovaries.[15]

F344 rats were administered 150 mg/kg BW in corn oil for 1 to 21 days. The treatment caused initial effect on Sertoli cells. Spermatogenesis was affected as is seen from the decrease in sperm density and increase in necrotic spermatids.[14]

Teratogenicity. *o*-TCP is unlikely to be teratogenic in rats. Administration of o-TCP on day 18 and 19 of gestation did not cause abnormalities in fetuses from pregnant Wistar rats.[15]

Mutagenicity.

In vitro cytogenetics. *o*-TCP was negative in *Salmonella* mutagenicity assay (NTP-88).

Carcinogenicity classification.

NTP: NE - NE - NE - NE (feed).

Chemobiokinetics. In cats, TCP was widely distributed throughout the body, the highest concentration found in the sciatic nerve, a target tissue. TCP is absorbed in the small intestine and passes into the fat tissues. Other tissues with high concentrations of TCP and its metabolites were liver, kidney, and gallbladder.

TCP is metabolized via three pathways. The first is hydroxylation of one or more of the methyl groups; the second is dearylation of the *o*-cresyl groups. The third is further oxidation of the hydroxymethyl to *aldehyde* and *carboxylic acid.*[1]

Excretion occurs via the urine and feces (85%), and also in the bile and expired air.[16]

Standards. ***Russia*** (1995). MAC: 0.005 mg/l.

References:

1. *Tricresyl Phopsphate*, Environmental Health Criteria No 110, WHO, IPCS, Geneva, 1990, 122.
2. Johanssen, F. R., Wright, P. L., Gordon, D. E., et al., Evaluation of delayed neurotoxicity and dose-response relationship of phosphate esters in the adult hen, *Toxicol. Appl. Pharmacol.*, 41, 291, 1977.
3. *Hygiene Problems in the Production and Use of Polymeric Materials*, Proc. F. F. Erisman Research Sanitary Hygiene Institute, Moscow, 1969, 195 (in Russian).
4. Varghese, R. G., Bursian, S. J., Tobias, C., and Tanaka, D., Organophosphorus-induced delayed neurotoxicity: a comparative study of the effects of tri-*ortho*-tolyl phosphate and triphenyl phosphite on the central nervous system of the Japanese quail, *Neurotoxicology*, 16, 45, 1995.
5. Inoue, N., Fujishiro, K., Mori, K. et al., Tri-*ortho*-cresyl phosphate poisoning: A review of human cases, Abstract, *Sangyo Ika Daigaku Zasshi*, 10, 433, 1988 (in Japanese).
6. Oishi, H., Oishi, S., and Hiraga, K., Toxicity of several phosphoric acid esters in rats, *Toxicol. Lett.*, 13, 29, 1982.
7. Chapin, R. E., George, J. D., and Lamb, J. C., Reproductive toxicity of tricresyl phosphate in a continuous breeding protocol in Swiss (CD-1) mice, *Fundam. Appl. Toxicol.*, 10, 344, 1988.
8. Saito, C., Kato, T., Taniguchi, H. et al., Subacute toxicity of tricresylphosphate (TCF) in rats, *Pharmacometrics*, 8, 107, 1974.
9. *NTP Fiscal Year Annual Plan*, Technical Report Series No 433, May 1994.
10. Banerjee, B. D., Saha, S., Ghosh, K. K., and Handy, P., Effect of tricresyl phosphate on humoral and cell-mediated immune responses in albino rats, *Bull. Environ. Contam. Toxicol.*, 49, 312, 1992.
11. Carlton, B. D. et al., Examination of the reproductive effects of tricresyl phosphate administered to Long-Evans rats, *Toxicology*, 46, 321, 1987.
12. Tocco, D. R., Randal, J. L., York, R. G., and Smith, M. K., Evaluation of the teratogenic effects of tri-*ortho*-cresyl phosphate in the Long-Evans rats, *Fundam. Appl. Toxicol.*, 8, 291, 1987.
13. Somcuti, S. G., Lapadula, D. M., Chapin, R. E., et al., Reproductive tract lesions resulting from subchronic administration (63 days) of tricresylphosphate in male rats., *Toxicol. Appl. Pharmacol.*, 89, 49, 1987.
14. Somcuti, S. G., Lapadula, D. M., Chapin, R. E., et al., Light and electron microscopic evidence of TOCP-mediated testicular toxicity in Fischer 344 rats, *Toxicol. Appl. Pharmacol.*, 107, 35, 1992.
15. Pashkova, A., in *Coll. Works, Kuibyshev Hygiene Epidemiology Research Institute*, Kuibyshev, Issue No 5, 1968, 44 (in Russian).
16. Kurebayshi, H., Tanaka A., and Yamaha, T., Metabolism and disposition of the flame retardant plasticizer, tri-*p*-cresyl phosphate, in the rat, *Toxicol. Appl. Pharmacol.*, 77, 395, 1985.

PHTHALIC ACID

Molecular Formula. $C_8H_6O_4$
M = 166.14
CAS No 88-99-3
RTECS No TH9625000
Abbreviation. PAc.

Synonym. 1,2-Benzene-*o*-dicarboxylic acid; Benzene-1,2-dicarboxylic acid; *o*-Benzenedicarboxylic acid; *o*-Dicarboxybenzene; *o*-Phthalic acid.

PHTHALIC ANHYDRIDE

Molecular Formula. $C_8H_4O_3$
M = 148.12
CAS No 85-44-9
RTECS No TI3150000
Abbreviation. PAn.

Synonyms and **Trade Name.** 1,2-Benzene-*o*-dicarboxylic anhydride; 1,3-Isobenzofurandione; Phthalic acid, anhydride; Phthalide.

Properties.

Phthalic anhydride. Flaky crystals of white to light-brown color with an odor of naphthalene; readily volatizes. Poorly soluble in cold water, readily soluble in hot water with PAc formation (1.0 mg PAn is equivalent to 1.12 mg of PAc). Soluble in alcohol.

Phthalic acid. White crystals. Solubility in water is 0.57 g/100 ml at 20°C and 18 g/100 ml at 99°C. Odor and taste perception threshold is 56 to 57 mg/l.[1]

Applications. Widely used in the production of plasticizers and pigments, and also as an anti-scorching agent, and as a light- and heat-stabilizer of polyolefins. PAn is a monomer for synthetic resins such as glyptal, the alkyd resins, and the polyester resins intended for contact with food.

Acute Toxicity. LD_{50} of PAn was identified to be 1.1 to 4.0 g/kg BW in rats, and 1.5 or 2.0 g/kg BW in mice. According to other data, LD_{50} is 2.2 g/kg BW in mice. The doses of 2.5 to 5.0 g/kg BW cause death in rats from necrosis of the renal tubules.[2,012] LD_{50} of PAc is reported to be 8.0 g/kg BW.[020]

Exposure to the lethal doses results in adynamia, dyspnea, damp, ungroomed fur. Gross pathology examination revealed distension of the stomach and intestine with ulcerations of their walls, and pulmonary hemorrhages.[2]

Repeated Exposure failed to reveal cumulative properties. A dose of 680 mg/kg BW caused the majority of rats to die within a month.[3] Guinea pigs received a 500 mg/kg BW dose every 2 to 3 days for 42 days. The treatment caused no mortality or retardation of BW gain. The blood parameters were unchanged. Histology examination revealed signs of surface necrosis in the GI tract.[2]

Short-term Toxicity. Cats were given 68 mg PAn/kg BW for 90 days. Administration caused meteorism (tympanites) and diarrhea, and some excitation of the NS without other signs of intoxication.[4] Rabbits were given the dose of 20 mg/kg BW for 3 months. PAn administration produced leukocytosis and increased aldolase activity. On gross pathology examination there was moderate parenchymatous dystrophy of the liver cells with slight perivascular lymphoid infiltration.[4]

Long-term Toxicity. Rats and rabbits were given PAc for 6 months. The treatment resulted in increased bilirubin content (rabbits) and a decreased number of thrombocytes. Gross pathology examination showed dystrophic and reactive changes in the liver, kidneys, stomach, and intestine.[1]

Reproductive Toxicity.

Embryotoxic Effect (congenital defects) was shown in chickens after administration of 3.0 to 20 mg PAn/kg into the yolk sac.[5]

Pregnant rats were given PAc at a dose of 1.25, 2.5, or 5.0% in the diet on day 7 through day 16 of pregnancy. Average daily intakes of PAc were 1020 mg/kg BW for the 1.25% group, 1760 mg/kg BW for the 2.5% group, and 2980 mg/kg BW for the 5.0% group. Maternal toxicity occurred in the 2.5 and 5.0% groups as can be seen by significant decreases in the maternal BW gain and food consumption during the administration period. Neither deaths nor clinical signs of embryotoxicity were noted in any groups. Significant decreases in the weight of male fetuses and number of ossification centers of the caudal vertebrae were found in the 5.0% group (about 3.0g/kg BW).[6]

Teratogenicity. PAc received research attention in teratology because of structural similarities to *thalidomide*. In the above-cited study, morphological examinations of fetuses revealed no evidence of teratogenesis.[6] According to Smith et al. and Kohler et al., testing in pregnant mice and rabbits showed no evidence of teratogenicity.[12,13]

Gonadotoxicity. A gonadotoxic effect of chronic inhalation exposure to PAn was observed at concentrations of 0.2 to 1.0 mg/m^3.[2,7]

Allergenic Effect.

Animal studies. A strong contact sensitizing potential was found in maximized guinea-pig assays.[8] Exposure to 0.5 mg PAn dust/m^3, a concentration below the current threshold limit value of 6.0 mg/m^3, can sensitize animals to produce allergic antibody.[9]

Observations in man. Human allergic reactions including asthma and allergic rhinitis are considered to be rare systemic hypersensitivity reactions (risk 1:1000).[10]

Mutagenicity.

In vitro genotoxicity. PAc was negative in any of the strains of *Salmonella typhimurium* tested, with or without *S9* metabolic activation (Agarwal et al., 1995).

Carcinogenicity. F344 rats received 25,000 and 50,000 ppm PAn for 32 weeks. Because of excessive depression in the amount of BW gained in the treated groups, the doses for the males were then reduced to 12,500 and 25,000 ppm, respectively, and the doses for the females were reduced to 6,250 and 12,500 ppm. The treatment was continued for 72 weeks. No significant increase in tumor incidence occurred in the animals in this study.[11]

Carcinogenicity classification.

NTP: N - N - N - N (PAn, feed).

Regulations.

EU (1990). PAc and PAn are available in the *List of authorized monomers and other starting substances which shall be used for the manufacture of plastic materials and articles intended to come into contact with foodstuffs (Section* A).

U.S. FDA (1998) approved PAn for use (1) as an accelerator for rubber articles intended for repeated use in contact with food at quantities up to 1.5% by weight of the rubber product in accordance with the conditions prescribed by 21 CFR part 177.2600; (2) in the manufacture of resinous and polymeric coatings for the food-contact surface of articles intended for use in producing, manufacturing, packing, processing, preparing, treating, packaging, transporting, or holding food in accordance with the conditions prescribed by 21 CFR part 175.300; (3) in the manufacture of resinous and polymeric coatings for polyolefin films for food-contact surface of articles intended for use in producing, manufacturing, packing, processing, preparing, treating, packaging, transporting, or holding food in accordance with the conditions prescribed by 21 CFR part 175.320; (4) in adhesives as a component of articles intended for use in packaging, transporting, or holding food in accordance with the conditions prescribed by 21 CFR part 175.105; (5) in the manufacture of cross-linked polyester resins which may be used as articles or components of articles intended for repeated use in contact with food in accordance with the conditions prescribed by 21 CFR part 177.2420; and (6) in polyurethane resins used as the food-contact surface of articles intended for use in contact with dry food in accordance with the conditions prescribed by 21 CFR part 177.1680. PAa-modified hydrolyzed soy protein isolate may be used (7) as a component of the uncoated or coated food-contact surface of paper and paperboard intended for use in producing, manufacturing, packaging, processing, preparing, treating, packing, transporting, or holding dry food of the type identified in 21 CFR part 176.170 (c).

Great Britain (1998). PAc and PAn are authorized without time limit for use in the production of polymeric materials and articles in contact with food or drink or intended for such contact. The specific migration of PAc and PAn alone or together with terephthalic acid or terephthalic acid dichloride shall not exceed 7.5 mg/kg.

Standards.

EU (1990). SML: 7.5 mg/kg.

Russia (1995). PML: 0.5 mg/l.

References:

1. Meleshchenko, K. F., *Prevention of the Pollution of Water Bodies by Liquid Effluents of Chemical Plants,* Zdorov'ya, Kiyv, 1971, 70 (in Russian).
2. In *Toxicology of New Industrial Chemical Substances*, A. A. Letavet, Ed., Medgiz, Moscow, 2, 63, 1961 (in Russian).
3. Pludro, G., Karlowski, K., Mankowska, M., et al., Toxicological and chemical studies of some epoxy resins and hardeners, I. Determination of acute and subacute toxicity of phthalic acid anhydride, 4,4'-diaminodiphenylmethane and of the epoxy resin: Epilox EG-34, *Acta Pol. Pharmacol.,* 26, 352, 1969.
4. Zhilova, N. A. and Kasparov, A. A., Phthalic anhydride and *N*-nitrosodiphenylamine (vulcalent A), in *Toxicology of New Chemical Substances Used in Rubber and Tyre Industry,* Z. I. Israel'son, Ed., Medtsina, Moscow, 1968, 157 (in Russian).

5. Verrett, M. J., Mutcher, M. K., Scott, W. F., et al., Teratogenic effect of capstan and related compounds in the developing chicken embryo, *Ann. NY Acad. Sci.*, 160, 334, 1969.
6. Ema, M., Miyawaki, E., Harazono, A., and Kawashima, K., Developmental toxicity evaluation of phthalic acid, one of the metabolites of phthalic acid esters, in rat, *Toxicol. Lett.*, 93, 109, 1997.
7. Protsenko, E. I., Gonadotropic action of phthalic anhydride, *Gig. Sanit.*, 1, 105, 1970 (in Russian).
8. Basketter, D. A. and Scholes, E. W., Comparison of the local lymph node assay with the guinea-pig maximization test for the detection of a range of contact allergens, *Food Chem. Toxicol.*, 30, 65, 1992.
9. Sarlo, K., Clark. E. D., Ferguson, J., Zeiss, C. R., and Hatoum, N., Induction of type I hypersensitivity in guinea pigs after inhalation of phthalic anhydride, *J. Allergy Clin. Immunol.*, 94, 747, 1994.
10. Venables, K. M., Low-molecular-weight chemicals, hypersensitivity, and direct toxicity: the acid anhydrides, *Brit. J. Ind. Med.*, 46, 222, 1989.
11. *Bioassay of Phthalic Anhydride for Possible Carcinogenicity*, Natl. Cancer Institute Technical Report Series No 159, 1979.
12. Smith, R. L., Fabro, S., Shumacher, H. J., and Williams, R. T., Studies on the relationship between chemical structure and embryotoxic activity of thalidomide and related compounds, in *Embryopathic Activity of Drugs*, J. M. Robson, F. M, Sullivan, and R. L. Smith, Eds., Little Brown, 1965, 194.
13. Kohler, F., Meise, W., and Ockenfels, H., Teratological testing of some thalidomide metabolites, *Experientia*, 27, 1149, 1971.

PHTHALIC ACID, BUTYLBENZYL ESTER

Molecular Formula. $C_{19}H_{20}O_4$
M = 312.35
CAS No 85-68-7
RTECS No TH9990000
Abbreviation. BBP.

Synonyms. 1,2-Benzenedicarboxylic acid, butylphenyl methyl ester; Butylbenzyl phthalate; Phthalic acid, benzylbutyl ester.

Properties. Colorless, oily liquid. Solubility in water is 2.9 mg/l.

Applications. Used as a plasticizer in the production of polymeric materials. It gives softness and flexibility to a wide variety of synthetic polymers. Compatible with synthetic polymers: polyvinyl chloride, polyacrylates, polyvinyl acetate and nitrocellulose. Plasticizer in the production of vinyl floor tiles and adhesives, in food containers and wrapping materials. Increases the water- and aging-resistance of materials.

Migration of BBP from aluminum foil-paper laminates in contacted butter and margarine was found to be about 0.05 mg/kg.[1] Migration of the plasticizer from nitrocellulose-coated regenerated cellulose film (RCF) purchased from retails and take-away outlets was studied. Levels of BBP found to migrate from RCF were as high as 0.5 to 53 mg/kg in confectionery, meat pies, cake, and sandwiches.[2]

Acute Toxicity. LD_{50} in rats seems to be more than 10 g/kg BW when BBP is given in the feed, and 2.33 g/kg BW when it is given *i/v*.[3] Other current data are also reported: in $B6C3F_1$ mice given BBP in corn oil, LD_{50} was found to be 4.17 g/kg BW (females) and 6.16 g/kg BW (males).[4]

Repeated Exposure. Administration of high doses caused exhaustion, adynamia or aggression, tonic convulsions with subsequent pareses and paralyses of the extremities. Gross pathology examination revealed irritation of the lower respiratory pathways, dystrophic changes in the liver and brain, and spinal cord lesions with demyelinization of the antero-lateral columns and peripheral nerves.[5]

F344 rats were exposed to BBP for 14 days (0.625 to 5.0% in the diet). BW gain and thymus weights were reduced in the 2.5 and 5.0% dose groups. Enlargement of the liver and kidneys, thymic atrophy, and morphological changes in these organs were observed. Prolonged exposure to BBP can affect hematopoietic system.[4,6]

Short-term Toxicity. In a subchronic study, rats were dosed with 300 to 600 mg/kg BW. The treatment resulted in reduced BW gain, pareses of the rear extremities, and death, all to a greater extent in males. Different breeds of animals of a single species exhibited different toxicity. In dogs ingesting diets with 1.0 to

5.0% BBP for 90 days, there were no alterations in urinary or hematology parameters. No gross or histopathological effects were found.[7]

In a 90-day study, depressed BW gain and testicular degeneration were noted, but no compound-related mortality occurred in rats given the doses of 25 g/kg BW.[4]

Long-term Toxicity. Male F344 rats were given BBP in feed for 26 weeks at doses of 30 to 550 mg BBP/kg BW. No clinical findings were noted. Gross pathology examination revealed focal hyperplasia in the pancreas in male rats, and transitional epithelial hyperplasia in the urinary bladder of female rats.[8]

Reproductive Toxicity.

Embryotoxicity. Rats were dosed with 250, 500, and 1000 mg/kg BW for 14 days before and during mating period (up to 14 days). Females received BBP for 6 additional days after delivery. There were no clinical signs of toxicity. The treatment produced effect on BW gain and spermatogenesis. An embryotoxic effect was observed, but there were no congenital defects in newborns.[9] Pregnant rats were given a single dose of BBP by gastric intubation at a dose of 1000 mg/kg BW on one of days 13 to 15 of pregnancy and at 1500 mg/kg BW on one of days 6 to 16 of pregnancy. An increase in post-implantation embryolethality was found in pregnant rats given on one of days 6 to 16, except for day 7.[10] In the above-cited study, embryotoxic effects were not observed at two lower doses.[8] Pregnant Wistar rats were exposed to BBP (2.0% in the diet) during pregnancy. Decrease in food consumption and BW in pregnant rats was noted.

Administration during pregnancy (on days 0 to 20) caused complete resorption of all the implanted embryos; administration on days 0 to 7 and 7 to 16 produced increased post-implantation loss.[11,12]

Teratogenicity effect was noted after a single dosing of BBP. Deformity of the cervical vertebrae frequently was observed after administration of BBP on day 7 of pregnancy. Cleft palate and fusion of the sternebrae were found exclusively, after administration of BBP on day 151.[10] Exposure on days 7 to 16 of pregnancy revealed striking teratogenicity: cleft palate in 95% of fetuses and fusion of sternebrae were predominantly observed. Administration during the first and the second half of pregnancy produced embryolethality and teratogenicity effects, respectively.[11,12]

Gonadotoxicity. BBP produced a direct toxic action on the testes with secondary effects on other reproductive organs. The weights of testes, epididymis, prostate, and seminal vesicles were reduced in rats exposed to 2.5 and 5.0% BBP in their feed.[7] Histology examination showed atrophy of the testes, prostate, and seminal vesicles, atrophy of the thymus and epididymis in 2.5 and 5.0% dose groups. Plasma testosterone concentration was decreased at a higher dose level. Ability to cause testicular atrophy depends on the route of administration and the species. Mice seem to be less sensitive than rats and guinea pigs for the effect on the gonads. In dogs, testicular lesions were not observed even with the high doses.[3]

Male F344 rats were given 20, 200, or 2200 mg BBP/kg in their feed for 10 weeks. There were no clinical findings. Gross pathology examination revealed degeneration of the seminiferous tubule germinal epithelium observed at the highest dose. The fertility indices were significantly affected at the highest dose.[8]

Ashby et al.[13] did not observe an effect of BBP administration in drinking water at the concentration of 1.0 mg/l on AP rats pup testis weight and testicular sperm count at postnatal day 90. This is in contrast to reduction in these measurements at the same level of exposure reported earlier by Sharpe et al.[14]

Male rats were exposed to BBP during gestation or during the first 21 days of postnatal life. BBP was administered via the drinking water. The treatment generally had no major adverse effect or BW. Exposure to BBP at a concentration of 1.0 mg/1 resulted in a small (5 to13%) but significant reduction in mean testicular size. The treatments had no adverse effect on testicular morphology but caused reductions of 10 to 21% in daily sperm production.[15]

Mutagenicity.

In vitro genotoxicity tests appeared to be uniformly negative. BBP was negative in several strains of *Salmonella typhimurium* with or without *S9* metabolic activation enzymes, in mammalian cell systems with and without *S9* activation, as well as in mouse lymphoma cells, SCE and CA in Chinese hamster ovary cells with and without *S9* activation system.[8]

In vivo cytogenetics. BBP was negative in *Dr. melanogaster*. After administration of BBP either in feed or by injections BBP was positive in mouse bone marrow SCE test. CA were induced in bone marrow cells of male mice injected with 5.0 g BBP/kg BW.[8]

Carcinogenicity. No differences in tumor incidence were reported in the control animals and in B6C3F$_1$ mice given doses of 6.0 or 12 g/kg BW for 103 weeks. F344 rats given the same doses over the same period showed increased incidence of myelomonocytic leukemia in high-dosed females.[4,16]

F344/N rats received 120, 240, or 500 mg/kg BW (males) and 300, 600, or 1200 mg/kg BW (females). Increased incidence of pancreatic acinar cell adenoma and of acinar cell adenoma or carcinoma (combined) were noted in male F344 rats. There was marginally increased incidence of pancreatic acinar cell adenoma and/or transitional epithelial papilloma of the urinary bladder in female F344 rats.[8]

Carcinogenicity classifications. An IARC Working Group concluded that there is inadequate evidence for the carcinogenicity of BBP in *experimental animals* and there were no data available to evaluate the carcinogenicity of BBP in *humans*.

IARC: 3.

U.S. EPA: C;

NTP: IS - P - N - N (feed).

Chemobiokinetics. F344 rats were given the doses of 20 to 2000 mg ^{14}C- BBP/kg BW. Plasticizer was found to be rapidly metabolized with subsequent excretion as *glucuronides* in the urine and feces. The main route of excretion of metabolites is biliary. These metabolites are reabsorbed and ultimately eliminated in the urine.[17]

Regulations. ***U.S. FDA*** (1998) regulates BBP as a component of the following materials used in contact with food products provided that the BBP contains not more than 1.0% by weight of dibenzylphthalate and provided further that the finished food-contact article, when extracted with the solvent or solvents characterizing the type of food and under the conditions of time and temperature characterizing the condition of its intended use will yield net chloroform-soluble extractives not to exceed 0.5 mg/inch: (1) in adhesives as a component of articles intended for use in packaging, transporting, or holding food in accordance with the conditions prescribed by 21 CFR part 175.105; (2) as a component of the uncoated or coated food-contact surface of paper and paperboard intended for use in producing, manufacturing, packaging, processing, preparing, treating, packing, transporting, or holding dry food in accordance with the conditions prescribed by 21 CFR part 176.180; (3) in the manufacture of cross-linked polyester resins which may be used as articles or components of articles intended for repeated use in contact with food in accordance with the conditions prescribed by 21 CFR part 177.2420; (4) in plasticizers for polymeric substances in the manufacture of articles or components of articles intended for use in producing, manufacturing, packaging, processing, preparing, treating, packing, transporting, or holding food in accordance with the conditions prescribed by 21 CFR part 178.3740; and (5) in the manufacture of resinous and polymeric coatings for polyolefin films to be safely used as a food-contact surface of articles intended for use in producing, manufacturing, packing, transporting, or holding food in accordance with the conditions prescribed by 21 CFR part 175.300.

References:

1. Page, B. D. and Lacroix, G. M., The occurrence of phthalate ester and di-2-ethylhexyl adipate plasticizers in Canadian packaging and food sampled in 1985-1989: a survey, *Food Addit. Contam.*, 12, 129, 1995.
2. Castle, L., Mercer, A. J., Startin, J. R., and Gilbert, J., Migration from plasticized films into foods. 3. Migration of phthalate, sebacate, citrate and phosphate esters from films used for retail food packaging, *Food Addit. Contam.*, 5, 9, 1988.
3. Hammond, B. G., Levinskas, G. J., Robinson, E., et al., A review of subchronic toxicity of butylbenzyl phthalate, *Toxicol. Ind. Health,* 3, 79, 1987.
4. *Bioassay of Butylbenzyl Phthalate for Possible Carcinogenicity,* DHHS Publ. No NIN 80-1769, NTP Technical Report Series No 81, U.S. Dept. Health & Human Services, Washington, D.C., 1981.
5. Aldyreva, M. V. and Gafurov, S. A., *Industrial Safety in the Production of Synthetic Leathers,* Medgiz, Moscow, 1980, 138 (in Russian).
6. Agarwal, D. K., Maronpot, R. R., Lamb, J. C., et al., Adverse effects of butylbenzyl phthalate on the reproductive and hematopoietic systems of male rats, *Toxicology*, 35, 189, 1985.

7. Hammond, B. G., Toxicology of butylbenzyl phthalate, Abstract No 414, *Toxicologist*, 1, 14, 1981.
8. *Toxicology and Carcinogenesis Studies of Butyl Benzyl Phthalate in F344/N Rats (Feed Studies)*, NTP Technical Report Series No 458, Research Triangle Park, NC, June, 1995.
9. Piersma, A. H., Veerhoef, A., and Dortant, P. M., Evaluation of the OECD 421 reproductive toxicity screening test protocol using butyl benzyl phthalate (BBP), *Hum. Exp. Toxicol.*, 14, 689, 1995.
10. Ema, M., Miyawaki, E., and Kawashima, K., Developmental effects of plasticizer butyl benzyl phthalate after a single administration in rats, *J. Appl. Toxicol.*,19, 357, 1999.
11. Ema, M., Itami, T., and Kawasaki, H., Effect of period of exposure on the developmental toxicity of butyl benzylphthalate in rats, *J. Appl. Toxicol.*, 12, 57, 1992.
12. Ema, M., Itami, T., and Kawasaki, H., Embryolethality and teratogenicity of butyl benzylphthalate in rats, *J. Appl. Toxicol.*, 12, 179, 1992.
13. Ashby, J., Tinwell, H., Lefevre, P. A., Odum, J., Paton, D., Millward, S. W., Tittensor, S., and Brooks, A. N., Normal sexual development of rats exposed to butyl benzyl phthalate from conception to weaning, *Regul. Toxicol. Pharmacol.*, 26, 102, 1997.
14. Sharpe, R. M., Fisher, J. S., Millar, M. M., Jobling, S., and Sumpter, J. P., Gestational and lactational exposure of rats to xenoestrogens results in reduced testicular size and sperm production, *Environ. Health Perspect.*, 103, 1136, 1995.
15. Sharpe, R. M., Fisher, J. S., Millar, M. M., Jobling, S., and Sumpter, J. P., Gestational and lactational exposure of rats to xenoestrogens results in reduced testicular size and sperm production, *Environ. Health Perspect.*, 103, 1136. 1995.
16. Tarone, R. E., Chu, K. C., and Ward, J. M., Variability in the rates of some common naturally occurring tumors in Fischer 344 rats and (C57BL/6N x C3H/HeN)F1 (B6C3F$_1$) mice, *J. Natl. Cancer Inst.*, 66, 1175, 1981.
17. Eigenberg, D. A., Bozigian, H. P., Carter, D. E., et al., Distribution, excretion and metabolism of butylbenzyl phthalate in the rat, *J. Toxicol. Environ. Health*, 17, 445, 1986.

PHTHALIC ACID, BUTYLDECYL ESTER

Molecular Formula. $C_{22}H_{34}O_4$
M = 362.56
CAS No 89-19-0
RTECS No TI0527000
Abbreviation. BDP.

Synonyms and **Trade Names.** 1,2-Benzenedicarboxylic acid, butyldecyl ester; Butyldecyl phthalate; Decylbutyl phthalate; Plasticizer BDP; PX 114.

Acute Toxicity. LD_{50} is 21 g/kg BW in rats.

Regulations. ***U.S. FDA*** (1998) approved the use of BDP in adhesives as a component of articles intended for use in packaging, transporting, or holding food in accordance with the conditions prescribed by 21 CFR part 175.105.

Reference:

Smyth, H. F., Carpenter, C. P., Weil, C. S., Pozzani, U. C., Striegel, J. A., and Nycum, J. S., Range-finding toxicity data: List VII, *Am. Ind. Hyg. Assoc. J.*, 30, 470, 1969.

PHTHALIC ACID, BUTYL ESTER, ESTER with BUTYL GLYCOLATE

Molecular Formula. $C_{18}H_{24}O_6$
M = 326.37
CAS No 85-70-1
RTECS No TI0535000
Abbreviation. BPBG.

Synonyms. Butylphthalyl butyl glycolate; Phthalic acid, butyl(butoxycarbonylmethyl) ester.

Properties. Colorless liquid without taste and odor. Solubility in water is 0.0012% (30°C).

Applications. Used as a plasticizer in the production of polyvinyl chloride, polyvinyl acetate, and other food-packaging materials to make them heat- and light-resistant.

Acute Toxicity. In rats, LD_{50} is found to be 14.6 g/kg BW.[1] Administration of 2.1 g/kg BW dose to rabbits and 4.7 g/kg BW dose to rats produced no toxic effects.[04]

Long-term Toxicity. Young rats were exposed to 0.02, 0.2, or 2.0% of BPBG in their feed for a year. Only the highest dose level caused retardation of BW gain. No pathohistological changes were observed in the viscera.[04,05]

Mutagenicity.

In vitro genotoxicity. BPBG gave negative results in CA test carried out on cultured Chinese hamster cells.[2]

Carcinogenicity. Sherman rats received BPBG in their diet at the levels of 0.2 g/kg BW, 2.0 g/kg BW, and 20 g/kg BW for 2 years. No tumors were found, but a majority of the rats were sacrificed before the end of observation period.[3]

Chemobiokinetics. BPBG is readily absorbed from the GI tract. Male Sprague-Dawley rats received BPBG in their diet at a dose of 50 mg/kg for 21 to 28 days before oral administration of a single dose of BPBG labeled with ^{14}C in carbonyl group. BPBG was excreted mainly in the urine.[4]

Regulations.

U.S. FDA (1998) regulates the use of BPBG (1) in adhesives as a component of articles intended for use in packaging, transporting, or holding food in accordance with the conditions prescribed by 21 CFR part 175.105; (2) in the manufacture of resinous and polymeric coatings for the food-contact surface of articles intended for use in producing, manufacturing, packing, processing, preparing, treating, packaging, transporting, or holding food in accordance with the conditions prescribed by 21 CFR part 175.300; and (3) in the manufacture of resinous and polymeric coatings for polyolefin films for food-contact surface of articles intended for use in producing, manufacturing, packing, processing, preparing, treating, packaging, transporting, or holding food in accordance with the conditions prescribed by 21 CFR part 175.320.

Permitted in some ***EU*** countries (***Italy*** and ***The Netherlands***) for use in materials coming into contact with food products.

Standards. *Russia.* Recommended PML: *n/m.*

References:

1. Smyth, H. F. and Carpenter, C. P., Further experience with the range finding test in the industrial toxicological laboratories, *J. Ind. Hyg. Toxicol.*, 30, 63, 1948.
2. Ishidate, M. and Odashima, S., Chromosome tests with 134 compounds in Chinese hamster cells *in vitro* - a screening for chemical carcinogens, *Mutat. Res.*, 48, 337, 1977.
3. Wilbourn, J. and Montesano, R., An overview of phthalate ester carcinogenicity testing results: the past, *Environ. Health Perspect.*, 45, 127, 1982.
4. Ikeda, G. J., Sapienza, P. P., Convillion, J. L., Farber, T. M., Smith, C. P., Inskeep, P. B., Marks, E. M., Cerra, F. E., and van Loon, E. J., Distribution and excretion of two phthalate esters in rats, dogs and miniature pigs, *Food Cosmet. Toxicol.*, 16, 409, 1978.

PHTHALIC ACID, DIALKYL(C_{789}) ESTER

Molecular Formula. o-$C_6H_4(COOR)_2$, where *R* is alkyl radical C_7-C_9 or C_6-C_8

CAS No 83968-18-7

RTECS No TI0585000

Abbreviation. DAA-C_{789}.

Synonyms. Dialkyl(C_{789}) phthalate; Phthalic acid and alcohols, dialkyl esters.

Properties. Clear, oily liquid with a specific weak odor. Poorly soluble in water. Soluble in organic solvents. Odor perception threshold is 0.46 g/l, taste perception threshold is 0.3 g/l. Foam-forming ability threshold is 20 mg/l.[1]

Applications. DAA-C_{789} is used for plasticizing polyvinyl chloride and its copolymers, being similar to di(2-ethylhexyl) phthalate in plasticizing properties. It is widely used for making food-contact film materials and artificial leather.

Acute Toxicity. In rats and mice, LD_{50} appeared to be more than 20.0 g/kg BW. In 24 hours after poisoning, the animals experienced depression, labored breathing and adynamia, motor coordination disorder, paresis of the extremities and exhaustion. However, according to other data, a single dose of 20 g/kg BW produced no deaths or clear signs of poisoning in rats and mice.[2,3]

Repeated Exposure revealed moderate cumulative properties K_{acc} is 5 (by Lim).[3] Rats tolerate 3.0 g/kg BW doses for 10 days,[2] but 2.5 g/kg BW administered to Wistar rats for 7 to 21 days increased the liver size and changed its enzyme activity.[4] DAA-C_{789} is found to enhance xenobiotic metabolism in females but to depress it in males. Degenerative changes are noted in the liver cells of males but not of females.

Short-term Toxicity. Rats tolerated administration of 10 to 500 mg/kg BW doses for 4 months.[2] Addition of 0.5 and 1.0% DAA-C_{789} to the diet of rats resulted in increased absolute and relative weights of the liver and kidneys. The relative weights of the brain, spleen, heart, and other organs were unchanged. Histology examination revealed the amount of hemosiderin in the spleen to be increased at the maximum dose.[3]

Addition of 1.0% DAA-C_{789} to the diet for 90 days caused retardation of BW gain and reduction in blood *Hb* level, erythrocyte count, and hematocrit.[5]

Allergenic Effect is not observed on skin application (Braun et al., 1970).

Reproductive Toxicity.

Gonadotoxicity. Wistar rats were exposed s/c to 4.0 mg DAA-C_{789}/kg BW. The exposure caused sperm damage and reduced sperm cell motility. The relative weights of the testes were unchanged.[6] According to other data, following oral administration of DAA-C_{789}, the testes were reduced in size and weight and there was atrophy of the seminiferous tubules.

Regulations. ***U.S. FDA*** (1998) regulates DAA-C_{789} as a component of adhesives to be safely used in food-contact surfaces. It may be used only in polymeric substances at levels not to exceed 24 to 35% by weight of the permitted vinyl chloride homo- and/or copolymers used in contact with non-fatty foods depending on average thickness of such polymers.

Standards. Russia (1995). MAC: 2.0 mg/l.

References:

1. Zaitsev, N. A., Korolev, A. A., Baranov, Yu. B., et al., Hygienic regulation of diethyl phthalate, di-*n*-hexyl phthalate and dialkyl phthalate-810 in water medium, *Gig. Sanit.*, 9, 26, 1990 (in Russian).
2. Statsek, N. K. et al., in *Pharmacology and Toxicology*, Republ. Issue, Kiyv, 10, 151, 1980; *Gig. Sanit.*, 6, 35, 1975 (in Russian).
3. Timofiyevskaya, L. A., Toxicity of dialkylphthalate in mixture of alcohols C_7-C_8-C_9 (DAP_{789}), *Gig. Sanit.*, 10, 89, 1982 (in Russian).
4. Mangham, B. A., Foster, J. R., and Lake, B. G., Comparison of the hepatic testicular effects of orally administered di(2-ethylhexyl) phthalate and dialkyl $C_{7\text{-}9}$ phthalate in the rat, *Toxicol. Appl. Pharmacol.*, 61, 205, 1981.
5. Gaunt, I. F., Colley, J., Grasso, P., et al., Acute (rat and mouse) and short-term (rat) toxicity studies on dialkyl $C_{7\text{-}9}$ phthalate (a mixture of phthalate esters of alcohols having 7 to 9 carbon atoms), *Food Cosmet. Toxicol.*, 6, 609, 1968.
6. Bainova, M. et al., Toxicology effects of dialkyl phthalates, *Reports Bulg. Acad. Sci.*, 35, 121, 1982 (in Bulgarian).

PHTHALIC ACID, DIALKYL($C_{8\text{-}10}$) ESTER

Molecular Formula. o-$C_6H_4(COOR)_2$, where *R* - alkyl radical C_8-C_{10}

RTECS No TI0570000

Abbreviation. DAP.

Synonyms. Dialkyl(C_8-C_{10}) phthalate; Phthalic acid, dialkyl-C_{810} ester.

Properties. Clear, oily liquid with a specific weak odor. Poorly soluble in water. Soluble in organic solvents. DAP does not color water but imparts a slight aromatic odor and a sweetish, astringent taste.

Applications. DAP is used for plasticizing polyvinyl chloride and its copolymers, being similar to di(2-ethylhexyl) phthalate in plasticizing properties.

Acute Toxicity. LD_{50} is reported to be higher than 30 g/kg BW in male and more than 18 g/kg BW in female rats, 20.5 g/kg BW in mice, and 17 g/kg BW in guinea pigs. Poisoning is accompanied by ataxia, muscle weakness, and liver damage.

Repeated Exposure. Male rats were given *i/g* oily solutions of DAP for 30 days. There was a polymorphism of the toxic action with predilection to affect the liver function (hydrocarbon, protein, and lipid metabolism), myocardium, CNS, and kidneys (at the dose of 1.0, but not 0.1 g/kg BW). Decreased activity of several enzymes is reported.

Reproductive Toxicity.

Embryotoxic effect was not found when 1.0 g DAP/kg BW dose was given to rats.

Standards. ***Russia*** (1990). Recommended MAC: 0.3 mg/l, recommended PML: 12.0 mg/l.

Reference:

Zaitsev, N. A., Korolev, A. A., Baranov, Yu. B., et al., Hygienic regulation of diethyl phthalate, di-*n*-hexyl phthalate, and dialkyl phthalate-810 in water medium, *Gig. Sanit.*, 9, 26, 1990 (in Russian).

PHTHALIC ACID, DIALKYL(C_{10}-C_{13}) ESTER

Molecular Formula. $C_6H_4(COOR)_2$

Synonyms. Dialkyl(C_{10}-C_{13}) phthalate; *o*-Phthalic acid, ester mixed with C_{10}-C_{13} alcohols.

Properties. Clear, light-yellow liquid.

Applications. Used as a plasticizer in the production of polyvinyl chloride.

Acute Toxicity. Rats and mice tolerate the maximum dose of 50 g/kg BW (divided administrations). Poisoned animals experienced depression, adynamia, and after a week, decreased BW and skin turgor.

Repeated Exposure failed to reveal cumulative properties: no rats died on administration of 5.0 g/kg BW by Lim method. Signs of toxicity included CNS depression. Animals recovered on day 4 or 5 despite continuing administration.

Reference:

Timofiyevskaya, L. A. and Kuz'mina, A. N., Mechanism of neuroparalytic effect of phthalic ethers, *Gig. Truda Prof. Zabol.*, 1, 50, 1983 (in Russian).

PHTHALIC ACID, DI(2-BUTOXYETHYL) ESTER

Molecular Formula. $C_{20}H_{30}O_6$

M = 366.44

CAS No 117-83-9

RTECS No TI0175000

Abbreviation. DBEPE.

Synonyms. Bis(2-butoxyethyl) phthalate; Butylglycol phthalate; Butyl cellosolve phthalate Di(2-butoxyethyl) phthalate.

Properties. Liquid with a faint, characteristic odor. Solubility in water is 0.03% (25°C). Resistant to UV-rays and hydrolysis.

Applications. Used as a plasticizer for vinyl resins and in the production of plastisols, chlorinated rubber, ethyl cellulose, polystyrene, and polymethyl methacrylate intended for contact with food or drink.

Acute Toxicity. LD_{50} appeared to be 6.0 ml/kg BW for guinea pigs, and 8.38 g/kg BW for rats. Acute poisoning of guinea pigs (3.2 ml/kg BW) is accompanied by ataxia, lassitude, and loss of reflex activity.[05]

Repeated Exposure revealed marked cumulative properties: 50% animals died from a total dose of 4.5 ml/kg BW.[05]

In a 30-day study, retardation of BW gain and loss of appetite were observed in rats given 1.7 g/kg BW in the diet. Toxic effects could be a result of hydrolysis of DBEPE to *butoxyethanol*. The dose of 0.5 g/kg BW produced no evident toxic effects.[04]

Reproductive Toxicity.

Teratogenicity effect is observed following administration of 0.1 to 0.025 ml into the yolk sac of the chick embryo. CNS lesions appeared in the postnatal period.[1,2]

Regulations. *U.S. FDA* (1998) regulates DBEPE as a component of adhesives intended for use in articles coming into contact with food in accordance with the conditions prescribed by 21 CFR part 175.105.

References:

1. Haberman, S., Guess, W. L., Rowan, D. F., et al., Effects of plastics and their additives on human serum proteins, antibodies and developing chick embryos, *Soc. Plastic Eng. J.*, 24, 62, 1968.
2. Bower, R. K., Haberman, S., and Mintin, P. D., Teratogenic effects in chick embryo caused by esters of phthalic acid, *J. Pharmacol. Ther.*, 171, 314, 1970.

PHTHALIC ACID, DIBUTYL ESTER

Molecular Formula. $C_{16}H_{22}O_4$
M = 278.35
CAS No 84-74-2
RTECS No TI0875000
Abbreviation. DBP.

Synonyms and **Trade Names.** 1,2-Benzenedicarboxylic acid, dibutyl ester; *n*-Butyl phthalate; Celluflex DBP; Dibutyl phthalate; Phthalic acid, dibutyl ester; Plasticizer DBP; Staflex DBP.

Properties. Colorless, odorless, oily liquid. Solubility in water is 0.1% (20°C). Aromatic odor and bitter taste perception threshold is 5.0 mg/l; practical threshold is 10 mg/l.

Applications. Used as a plasticizer in the production of polyvinyl chloride, polymethyl metacrylate, polyvinyl acetate, and cellulose esters intended for contact with food or drink. It makes the materials light-stable. A solvent for chlorinated rubbers.

Migration from polyvinyl chloride (DBP contents 0.04%, exposure 8 days) into potato snacks was reported to be 3.5 mg/kg, into coated candy (0.5% DBP, 5 days) 1.23 mg/kg, and in covered chocolate (0.21% DBP, 5 days) 0.9 mg/kg.[1] DBP migration from aluminum foil - paper laminates in contacted butter and margarine was found to be 10.6 µg/kg.[2] Specific migration of DBP from epoxy resin composed of bisphenol A diglycidyl ether, 4,4'-methylene dianiline and additives (plasticizers: DBP, dioctyl phthalate; accelerator: salicylic acid; inorganic fillers) diminished greatly as the curing temperature increased.[3] Migration of the plasticizer from nitrocellulose-coated regenerated cellulose film (RCF) purchased from retails and take-away outlets was studied.

Levels of plasticizer found to migrate from RCF were as high as 0.5 to 53 mg DBP/kg in confectionery, meat pies, cake, and sandwiches.[4] Migration of plasticizers from printing ink (of PP packaging film) into foods has been studied. It was demonstrated that transfer can occur of components from the ink on the outer surface of the film onto the inner food-contact surface. The migration of plasticizer increased with storage time of the wrapped product: for DBP, levels increased from 0.2 to 6.7 mg/kg over the period from 0 to 180 days. Migration of DBP from snack products and biscuits wrapped in printed polypropylene film reached 0.02 to 14.1 mg/kg.[5]

Acute Toxicity. LD_{50} is about 10 g/kg BW in rats, 5.0 to 20 g/kg BW in mice, and 8.0 to 10 g/kg BW in chickens.[6,03] Guinea pigs and rabbits are found to be less sensitive. The acute effect threshold appeared to be 26 mg/kg BW for behavior.[6]

Repeated Exposure revealed moderate cumulative properties. K_{acc} is 7 (by Lim). Following oral exposure to the dose of 3.0 g/kg BW, animals died in 3 to 14 days. Histology examination revealed stomach paresis and intestinal and mesenteric lesions.[7] Oral administration of ^{14}C-labeled DBP failed to demonstrate accumulation of radioactivity in the gonads.[8]

Short-term Toxicity. Wistar rats were gavaged with 120 mg/kg BW for 3 months. This treatment caused lesions in the GI tract mucosa, development of pneumonia and endometritis, increase in liver weight. A dose of 1200 mg/kg BW caused death of 5.0% animals.[9]

In a 13-week study, Fischer rats received 2,500 to 40,000 ppm DBP in their diet. No death occurred. Hepatomegaly was found in males (5,000 ppm or greater) and in females (10,000 ppm or greater), hypocholesterolemia was noted at 20,000 ppm, as well as elevation of alkaline phosphatase. Morphological lesions were discovered in liver and testes. Histological findings included hepatocellular cytoplasmic alterations.[10]

Long-term Toxicity. No signs of intoxication were observed when 0.05% DBP was added to the diet of rats for a year. Addition of 0.125% DBP to the feed caused 15% mortality in rats.[8] Retardation of BW gain, leukocytosis, and NS affection were observed in rats given 20 mg DBP/kg BW.[7]

Immunotoxicity effects are revealed at the dose level of 200 mg/kg BW. Tissue antibodies determine the cytotoxic effect on the fetus and progeny.[11]

Reproductive Toxicity.

Embryotoxicity. DBP is capable of passing through the placenta, it accumulates in the brain and subcutaneous tissues. It is distributed in the fetal blood and in the amniotic fluid.[12] Exposure of Wistar rats to 120 and 600 mg/kg in the diet for 3 months had no effect on the number of litters. Administration of these doses during pregnancy resulted in decreased litter size and increased neonatal mortality.[9,13] Administration of 250 mg/kg BW dose throughout the gestation period produced no selective fetotoxicity.[14]

However, Zinchenko[11] reported an embryotoxic effect when rats were given 20 and 200 mg/kg BW throughout the pregnancy. CD-1 mice of both sexes were dosed with 0.03 to 1.0% DBP in their diet for 7 days prior to and during a 98-day cohabitation period. A reduction in the number of litters per pair, and of live pups per litter, and in the proportion of pups born alive was found at the 1.0% dose but not at a lower dose level. A crossover mating trial demonstrated that female mice but not males were affected by DBP.[15] Though high doses of DBP are reported to be embryotoxic, the NOEL in mice is more than 2000 times the estimated level of human intake via the food chain.[15]

DBP was administered in the diet of F344 rats during gestation and lactation, and to the pups postweaning for 4 additional weeks at concentrations of 1,250 to 20,000 ppm. At higher dietary levels, a decrease in BW gains and hepatomegaly was observed. No gross lesions were revealed at necropsy. No signs of degeneration of the germinal epithelium were noted in testes of rats.[10]

Teratogenicity. Shiota et al. have found the development of neural cord abnormalities in the fetuses of mice given 400 mg/kg BW and above during pregnancy.[15] DBP was found to produce greater adverse reproductive developmental effects on the second than on the first generation when it was studied in a continuous breeding protocol in Sprague-Dawley rats at levels of 0.1, 0.5, or 1.0% DBP in the diet.[16]

Neural tube defects, namely, exencephaly and spina bifida, were found in this study. The NOEL for embryotoxic effect appeared to be 70 mg/kg BW or 1.0% DBP in the feed; the NOEL for teratogenic effect appeared to be 0.2% in the feed.

Gonadotoxicity. DPB is known to be a testicular toxicant.[17] Wistar rats received a single *i/g* administration of 2.4 g DBP/kg BW. The treatment caused testicular atrophy in males including Heinze body formation together with iron liberation from *Hb*. In its turn, this leads to a decrease in iron concentration in the blood and gonads.[19]

Oral administration of 0.5 and 1.0 g/kg BW significantly reduced relative testes weights within 6 and 4 days, respectively. Perinatal exposure to 10,000 ppm DBP in the diet of F344 rats followed by 13-week feeding caused hepatocellular cytoplasmic alterations consistent with glycogen depletion, increased number of peroxisomes degeneration of the germinal epithelium in testes of rats (at the dose level of 10,000 ppm), changes in testicular zinc concentration were observed at 40,000 ppm). Testicular zinc concentration and serum testosterone values were lower at 20,000 and 40,000 ppm DBP in the diet. The treatment also caused marked loss of germinal epithelium.[10]

Monobutyl phthalate, the major urinary metabolite, causes even more marked reduction in the testes weights. The testicular lesion produced by DBP in immature rats is characterized by early sloughing of spermatids and spermatocytes and severe vacuolization of Sertoli cell cytoplasm.[20] Urinary zinc levels following DBP treatment are increased.[21,22]

Mutagenicity.

In vivo cytogenetics. No mutagenic effect is noted in *Dr. melanogaster*.[21]

In vitro genotoxicity. DBP is shown to cause mutations in mouse lymphoma cells (NTP-88). It was positive in *Salmonella* mutagenicity assay.[23]

Carcinogenicity. In peroxisome induction study, DBP is found to be no more carcinogenic than the weak liver carcinogen di(2-ethylhexyl)phthalate.[24]

In the long-term study,[04] oral doses of 100, 300, and 500 mg/kg BW administered for 15 to 21 months revealed no signs of tumor growth over 6 generations in rats. It was noted that the doses tested correspond to 60, 180, and 300 mg/kg BW in man.

Carcinogenicity classification.

U.S. EPA: D.

Chemobiokinetics. DBP metabolism involves hydrolysis to *the monoester* with subsequent oxidation of the remaining alkyl chain. Main metabolites: *phthalic acid, monobutyl phthalate, mono(3-hydroxybutyl), mono(4-hydroxybutyl)phthalate. Phthalic acid* was found to be a minor metabolite (less than 5.0% to the total urinary metabolites).

No accumulation of DBP or its metabolites was found in rats given feed containing 1.0 mg DBP/kg for 3 months.[21] According to Jaeger and Rubin,[16] DBP is unlikely to be fully hydrolyzed in the body. Possibly, it may accumulate, mainly in the blood and lungs.[23] Excretion of ^{14}C-labeled DBP occurs mainly through the urine.

Regulations. ***U.S. FDA*** (1998) approved the use of DBP (1) in adhesives used as components of articles intended for use in packaging, transporting, or holding food in accordance with the conditions prescribed by 21 CFR part 175.105; (2) in the manufacture of cross-linked polyester resins to be safely used as articles or components of articles intended for repeated use in contact with food in accordance with the conditions prescribed by 21 CFR part 177.2420; (3) as a plasticizer for rubber articles intended for repeated use in contact with food (content up to 30% by weight of the rubber product) in accordance with the conditions prescribed by 21 CFR part 177.2600; (4) as an ingredient of cellophane for food packaging alone or in combination with other phthalates where total phthalates do not exceed 5.0% in accordance with the conditions prescribed by 21 CFR part 177.1200; (5) as a component of the uncoated or coated food-contact surface of paper and paperboard intended for use in producing, manufacturing, packing, transporting, or holding aqueous and fatty food in accordance with the conditions prescribed by 21 CFR part 176.170; (6) in the production of resinous and polymeric coatings in a food-contact surface of articles intended for use in producing, manufacturing, packing, transporting, or holding food only for containers having a capacity of 1,000 gallons or more when such containers are intended for repeated use in contact with alcoholic beverages containing up to 8.0% alcohol in accordance with the conditions prescribed by 21 CFR part 175.300; and (7) in slimicide in the manufacture of paper and paperboard that may be safely used in contact with food in accordance with the conditions prescribed by 21 CFR part 176.300.

Standards. ***Russia*** (1995). PML in food: 0.2 mg/l, PML in drinking water: 0.1 mg/l.

References:

1. *Plasticizers: Continuing Surveillance.* The Thirtieth Report Steering Group on Food Surveillance, The Working Party on Chemical Contaminants from Food Contact Materials: Sub-Group on Plasticizers, Food Surveillance, Paper No 30, London: HMSO, 1990, 55.
2. Page, B. D. and Lacroix, G. M., Studies into the transfer and migration of phthalate esters from aluminium foil-paper laminates to butter and margarine, *Food Addit. Contam.*, 9, 197, 1992.
3. Lambert, C., Larroque, M., Lebrun, J. C., and Gerard, J. F., Food-contact epoxy resin, co-variation between migration and degree of cross-linking, *Food Addit. Contam.*, 14, 199, 1997.
4. Castle, L., Mercer, A. J., Startin, J. R., and Gilbert, J., Migration from plasticized films into foods. 3. Migration of phthalate, sebacate, citrate and phosphate esters from films used for retail food packaging, *Food Addit. Contam.*, 5, 9, 1988.
5. Catsle, L., Mayo, A., and Gilbert, J., Migration of plasticizers from printing ink into foods, *Food Addit. Contam.*, 6, 437, 1989.
6. Balynina, E. S. and Timofiyevskaya, L. A., On the problem of application of behavioral reactions in toxicity studies, *Gig. Sanit.*, 7, 54, 1978 (in Russian).
7. Komarova, E. N., Toxic properties of some additives for plastics, *Plast. Massy*, 12, 30, 1976 (in Russian).
8. Gangolli, S. D., Testicular effects of phthalate esters, *Environ. Health Perspect.*, 45, 77, 1982.
9. Nikanorov, M., Mazur, H., and Piekacz, H., Effect of orally administered plasticizers and polyvinyl chloride stabilizers in the rat, *Toxicol. Appl. Pharmacol.*, 25, 253, 1973.

10. *Toxicity Studies of Dibutyl Phthalate Administered in Feed to F344/N Rats and B6C3F$_1$ Mice* (D. S. Marsman), NTP Technical Report Series No 30, NIH Publ. 95-3353, Research Triangle Park, NC, April 1995, 108.
11. Zinchenko, T. M., Studies of autoallergenic effect of dibutylphthalate and di(2-ethylhexyl) phthalate, *Gig. Sanit.*, 2, 80, 1986 (in Russian).
12. Klimova, T. D., in *Effect of Occupational Factors on the Specific Functions of Female Body*, Sverdlovsk, 1978, 101 (in Russian).
13. Peters, J. W. and Cook, R. M., Effect of phthalate esters on reproduction in rats, *Environ. Health Perspect.*, Exp. Issue No 3, 1973, 91.
14. Timofiyevskaya, L. A., Nikitenko, T. K., Govorchenko, V. I., et al., Correlation between general and specific effects of biological activity of phthalic esters, in *Basic Problems of Delayed Effects of Exposure to Industrial Poisons*, Proc. Research Institute Occup. Diseases., A. K. Plyasunov and G. M. Pashkova, Eds., Moscow, 1976, 39 (in Russian).
15. Shiota, K., Chou, M. I., and Nishimura, H., Embryotoxic effects of di(2-ethylhexyl) phthalate (DEHP) and di-*n*-butylphthalate (DBP) in mice, *Environ. Res.*, 54, 342, 1980.
16. Jaeger, R. J. and Rubin, R. J., Plasticizers from plastic devices: Extraction, metabolism, and accumulation by biological systems, *Science*, 170, 460, 1970.
17. Wine, R. N., Li, L. H., Barnes, L. H., Gulati, D. K., and Chapin, R. E., Reproductive toxicity of di-*n*-butylphthalate in a continuous breeding protocol in Sprague-Dawley rats, *Environ. Health Perspect.*, 105, 102, 1997.
18. Lamb, J. C., Chapin, R. E., Teague, J., et al., Reproductive effects of four phthalic acid esters in the mouse, *Toxicol. Appl. Pharmacol.*, 88, 255, 1987.
19. Fukuoka, M., Kobayashi, T., and Haykawa, T., Mechanism of testicular atrophy induced by di-*n*-butyl phthalate in rats. VI. A possible origin of testicular iron depletion, *Biol. Pharmacol. Bull.*, 17, 1609, 1994.
20. Cater, B. R., Cook, M. W., Gangolli, S. D., et al., Studies on dibutyl phthalate-induced testicular atrophy in the rat: Effect on zinc metabolism, *Toxicol. Appl. Pharmacol.*, 41, 609, 1977.
21. Sheftel', V. O., Zinchenko, T. M., and Katayeva, S. Ye., Sanitary toxicology of phthalates as water pollutants, *Gig. Sanit.*, 8, 64, 1981 (in Russian).
22. Pashkova, G. A., Comparative risk assessment of gonadotoxic effects of phthalic and organophosphorous plasticizers, in *Basic Problems of Delayed Effects of Exposure to Industrial Poisons*, Proc. Research Institute Occup. Diseases., A. K. Plyasunov and G. M. Pashkova, Eds., Moscow, 1976, 43 (in Russian).
23. Seed, J. L., Mutagenic activity of phthalic esters in bacterial liquid suspension assay, *Environ. Health Perspect.*, 45, 111, 1982.
24. Barber, E. D., Astill, B. D., Moran, E. J., Schneider, B. F., Gray, T. J., Lake, B. G., and Evans, J. G., Peroxisome induction studies on seven phthalate esters, *Toxicol. Ind. Health*, 3, 7, 1987.

PHTHALIC ACID, DECYLOCTYL ESTER

Molecular Formula. $C_{26}H_{42}O_4$
M = 418.68
CAS No 119-07-3
RTECS No TI0550000
Abbreviation. DOP.

Synonyms and **Trade Names.** 1,2-Benzenedicarboxylic acid, decyloctyl ester; Dinopol 235; Decyloctyl phthalate; Octyldecyl phthalate; Phthalic acid, decyloctyl ester; Polycizer 532 or 562; Staflex 500.

Properties. Clear liquid with a mild characteristic odor. Phthalate esters are soluble to various extents in many common organic solvents and oils, but have a relatively low solubility in water.

Applications. Used mainly in the manufacture of vinyl resins for contact with food and drink.

Acute Toxicity. LD_{50} is 45.2 ml/kg BW in rats.

Chemobiokinetics. Readily absorbed from the intestinal tract and through the skin. Distributed quite rapidly in various organs and tissues, predominantly in the liver, kidney, and bile.

Regulations. ***U.S. FDA*** (1998) approved the use of DOP (1) in adhesives as a component of articles intended for use in packaging, transporting, or holding food in accordance with the conditions prescribed by 21 CFR part 175.105; (2) in resinous and polymeric coatings used as the food-contact surfaces of articles intended for use in producing, manufacturing, packing, processing, preparing, treating, packaging, transporting, or holding food in accordance with the conditions prescribed by 21 CFR part 175.300; (3) in the manufacture of rubber articles intended for repeated use in producing, manufacturing, packaging, processing, preparing, treating, packing, transporting, or holding food in accordance with the conditions prescribed by 21 CFR part 177.2600; and (4) in resinous and polymeric coatings used as the food-contact surfaces of articles intended for use in producing, manufacturing, packing, processing, preparing, treating, packaging, transporting, or holding food in accordance with the conditions prescribed by 21 CFR part 175.300.

Reference:

1. Smyth, H. F., Carpenter, C. P., Weil, C. S., Pozzani, U. C., Striegel, J. A., and Nyaun, J. S., Range-finding toxicity data: List VII, *Am. Ind. Hyg. Assoc. J.*, 30, 470, 1969.

PHTHALIC ACID, DICYCLOHEXYL ESTER

Molecular Formula. $C_{20}H_{26}O_4$
M = 330.43
CAS No 84-61-7
RTECS No TI0889000
Abbreviation. DCHP.

Synonym. Dicyclohexyl phthalate.

Properties. White, crystalline powder with a characteristic faint odor. Poorly soluble in water, soluble in alcohol and fats.

Applications. Used as a plasticizer in the production of nitro- and benzyl cellulose; combines with polyvinyl chloride, polyvinyl butyral, polystyrene, and acrylic plastics intended for contact with food or drink.

Migration from polyvinyl chloride (DCHP content 0.33%, exposure 5 days) into potato snacks was reported to be 6.2 mg/kg, into coated candy (0.50% dibutyl phthalate, exposure 5 days) 0.1 mg/kg, and in covered chocolate (0.21% dibutyl phthalate, exposure 5 days) 0.38 mg/kg.[1]

Migration of DCHP from nitrocellulose-coated regenerated cellulose film (RCF) purchased from retails and take-away outlets was studied. Levels of the plasticizer found to migrate from RCF were as high as 0.5 to 53 mg DCHP/kg in confectionery, meat pies, cakes, and sandwiches.[2] Migration of plasticizers from printing ink (of PP packaging film) into foods has been studied. It was demonstrated that transfer of components from the ink on the outer surface of the film can occur onto the inner food-contact surface. For DCHP this transfer amounted to 6.0% of the total amount of plasticizer available in the printing ink system. The migration of the plasticizer increased with storage time of the wrapped product.[3]

Acute Toxicity. LD_{50} is not attained with a single administration. After seven administrations, LD_{50} is likely to be about 30 ml/kg BW.[4] LD_{33} is less than 60 g/kg BW. Poisoning is accompanied by adynamia, general inhibition and lack of response to external stimuli.[5]

Repeated Exposure failed to reveal cumulative properties. K_{acc} exceeded 10 (by Lim). Habituation develops. Rats given 0.5 to 2.5 g/kg BW for 7 days exhibited liver enlargement, increase in 7-ethoxy-cumarin-*o*-diethylase activity and in the content of *P-450* cytochrome in the hepatic microsomes in young animals.[5]

Long-term Toxicity. Rats were exposed to the oral doses of 0.5 and 1.0 mg DCHP/kg BW in a 20% solution in vegetable oil for 1 year. There were no changes in biochemistry and hematology analyses. Gross pathology examination failed to reveal changes in the viscera or the progeny.

Reproductive Toxicity, Carcinogenicity. In a long-term study the doses of 5.0, 10, and 100 mg DCHP/kg BW given in the feed for 18 months produced no effect on reproduction or any carcinogenic activity in rats.[04,05]

Chemobiokinetics. DCHP seems to be a weak, medicinal type inducer of metabolism in the rat liver.[5]

Regulations. ***U.S. FDA*** (1998) approved the use of DCHP (1) as an ingredient in the manufacture of cellophane for food packaging, alone or in combination with other phthalates where total phthalates do not exceed 5.0% in accordance with the conditions prescribed by 21 CFR part 177.1200; (2) as a component of the uncoated or coated food-contact surface of paper and paperboard intended for use in producing, manufacturing, packaging, processing, preparing, treating, packing, transporting, or holding aqueous and fatty foods in accordance with the conditions prescribed by 21 CFR part 176.210; (3) as a component of adhesives for articles intended for use in contact with food, for use only alone or in combination with other phthalates in accordance with the conditions prescribed by 21 CFR part 175.105; (4) in plasticizers for polymeric substances in the manufacture of articles or components of articles intended for use in producing, manufacturing, packaging, processing, preparing, treating, packing, transporting, or holding food in accordance with the conditions prescribed by 21 CFR part 178.3740; and (5) in plastic film or sheet prepared from polyvinylacetate, polyvinyl chloride, and/or vinyl chloride copolymers at temperatures not to exceed room temperature, containing no more than 10% by weight of the total phthalates, calculated as phthalic acid.

Standards. ***Russia.*** Recommended PML in drinking water: *n/m.*

References:

1. *Plasticizers: Continuing Surveillance.* The Thirtieth Report Steering Group on Food Surveillance, The Working Party on Chemical Contaminants from Food Contact Materials: Sub-Group on Plasticizers, Food Surveillance, Paper No 30, London: HMSO, 1990, 55.
2. Castle, L., Mercer, A. J., Startin, J. R., and Gilbert, J., Migration from plasticized films into foods. 3. Migration of phthalate, sebacate, citrate and phosphate esters from films used for retail food packaging, *Food Addit. Contam.*, 5, 9, 1988.
3. Catsle, L., Mayo, A., and Gilbert, J., Migration of plasticizers from printing ink into foods, *Food Addit. Contam.*, 6, 437, 1989.
4. Timofiyevskaya, L. A., Toxicity of dialkyl phthalate on mixture of alcohols C_7-C_8-C_9 (DAP-789), *Gig. Sanit.*, 10, 89, 1982 (in Russian).
5. Lake, B. G., Foster, J. R., Collins, M. A., et al., Studies on the effects of orally administered dicyclohexyl phthalate in the rat, *Acta Pharmacol. Toxicol.*, 51, 217, 1982.

PHTHALIC ACID, DIDODECYL ESTER

Molecular Formula. $C_{32}H_{54}O_4$
M = 502.75
CAS No 2432-90-8
RTECS No TI0930000
Abbreviation. DDP.

Synonyms. Didodecyl phthalate; Dilauryl phthalate.

Properties. Colorless, odorless, oily liquid. Poorly soluble in water, soluble in ethanol.

Applications. Used mainly as a polyvinyl chloride plasticizer, rendering high elasticity of materials. Exhibits low volatility.

Acute Toxicity. LD_{50} was not attained even after administration of 15 g/kg BW to mice and rats. The calculated LD_{50} is 29.8 g/kg BW for rats, and above 7.9 g/kg BW for mice.[1] According to other data, LD_{50} exceeded 50 g/kg BW in rats, and 50 g/kg BW in mice.[2]

Repeated Exposure failed to reveal cumulative properties when 2.0 and 5.0 g/kg BW were administered to rats for 2 months.[1]

Regulations. ***U.S. FDA*** (1998) regulates the use of DDP as an adjuvant in the preparation of slimicides in the manufacture of paper and paperboard that may be safely used in contact with food.

Standards. ***Russia.*** PML in food: 2.0 mg/l.

References:

1. Antonyuk, O. K., On the toxicity of didodecyl phthalate, *Gig. Truda Prof. Zabol.*, 11, 51, 1973 (in Russian).

2. Timofievskaya, L. A., Balynina, E. S., and Ivanova, N. I., Patterns of the toxic action and an accelerated setting of standards for a number of *o*-phthalic acid esters, *Gig. Truda Prof. Zabol.*, 7, 52, 1988 (in Russian).

PHTHALIC ACID, DIETHYL ESTER

Molecular Formula. $C_{12}H_{14}O_4$
M = 222.24
CAS No 84-66-2
RTECS No TI1050000
Abbreviation. DEP.

Synonyms. 1,2-Benzenedicarboxylic acid, diethyl ester; Diethyl phthalate; Ethyl phthalate; Phthalic acid, diethyl ester.

Properties. Colorless, oily liquid, odorless or with a faint odor. Solubility in water is 0.15% at 20°C, soluble in alcohol and fats. Does not color water but imparts a slight aromatic odor and a sweetish, astringent taste to it. Odor perception threshold is 1.62 g/l; taste perception threshold is 7.0 g/l; foam-forming ability threshold is 80 mg/l.[1]

Applications. DEP is used in the production of cellulose esters (cellulose acetate films) intended for contact with food. Blends well (up to 70%) with cellulose nitrate as well as with polyacrylates and polymethyl methacrylates. It is used extensively as a denaturant for cosmetic alcohol and in hairspray preparations.

Migration Data. Migration of 2.0 to 4.0 mg DEP/kg from cellulose acetate film purchased from retails and take-away outlets into quiches was determined.[2]

Acute Toxicity. The LD_{50} is reported to vary from 9.5 or 10.3 g/kg BW (males) to 10.8 g/kg BW (females) in rats, and 6.2 to 9.5 g/kg BW in mice; it was 3.0 g/kg BW in guinea pigs and 1.0 g/kg BW in rabbits.[1,3,05] The poisoning is accompanied by depression, apathy, adynamia, side position, respiratory distress, and convulsions. Death occurs within 2 to 3 days. Gross pathology examination revealed congestion in the abdominal cavity organs and cerebral meninges.

Repeated Exposure revealed marked cumulative properties: rats died when given 280 to 1400 mg/kg BW. Male rats were exposed *i/g* to oily solutions of 100 and 1000 mg/kg BW over a period of 30 days. The treatment showed polymorphism of the action with predilection to affect the liver function (hydrocarbon, protein, and lipid metabolism), myocardium, CNS, and kidneys (1000 mg/kg BW dose). There was also a decreased activity of several enzymes. A pronounced dose-dependence in the toxic action of DEP was found.[1]

Short-term Toxicity. Rats were given the diets containing 0.2, 1.0, or 5.0% DEP for 16 weeks. Food consumption was decreased at the highest dose level. Hematology analyses, water intake, serum enzyme levels, urinary cell-excretion rate, and histology examination showed no significant changes.[4]

Reproductive Toxicity.

Embryotoxicity. High embryolethality is reported as a result of the exposure to high doses given one or three times during pregnancy.[5] However, embryotoxicity was not observed following oral administration of 100 mg/kg BW.[1]

CD-1 mice of both sexes were dosed for 7 days prior to and during a 98-day cohabitation period. These animals were given the diets with 0.25 to 2.5% DEP. There was no apparent effect on reproductive function despite significant effects on BW gain and liver weights.[6,7]

Teratogenic effect was found to develop in the chick embryos.[5] Such effect was not found in the offspring of CD rats given 0.25 to 2.5% DEP in the diet on gestation days 6 through 15.[7] Timed-mated rats received 0.25, 2.5, and 5.0% DEP in their feed from gestation day 6 to 15 (corresponds to approximately 0.2 to 4.0 g DEP/kg BW). Ingestion of the feed with 5.0% DEP reduced BW gain during treatment and during gestation. The NOAEL for maternal toxicity was 0.25% DEP in the feed. DEP treatment caused no effect on any parameter of embryo/fetal development, except an increased incidence of supernumerary ribs in the 5.0% DEP group.

This results do not support the conclusion of other investigators that DEP is a potent developmental toxicant.[8]

Gonadotoxicity. Rats given 7.2 mmol/kg BW dose for 4 days exhibit gonadotoxic effect. The treatment did not alter the zinc content in the testis and urine.[9]

Mutagenicity.

In vitro genotoxicity. DEP appeared to be positive in *Salmonella typhimurium TA 100* mutagenicity assay,[10] it has been shown to be toxic to cultured cells. However, according to other data, DEP was negative in *Salmonella typhimurium* strains *TA98, TA100, TA1535,* and *TA1537* with and without *S9* fraction. It induced SCE in Chinese hamster ovary cells in the presence, but not in the absence of metabolic activation.[11]

Carcinogenicity. In the NTP study, DEP was administered to rats and mice by topical applications. The treatment caused fatty degeneration to develop in the liver. Cases of mammary gland fibroadenoma were noted in rats. In mice, liver adenomas or carcinomas were observed.[11]

Carcinogenicity classifications.

U.S. EPA: D;

NTP: SE - NE - EE - EE.

Chemobiokinetics. DEP metabolism involves hydrolysis to the *monoester* with subsequent oxidation of the remaining alkyl chain. *Phthalic acid* was found to be a minor metabolite (less than 5.0%) among the total urinary metabolites. The principal metabolite is *monoethyl phthalate*.[3] A small part of ^{14}C-labeled DEP was passed with the urine in 24 hours after application. After 72 hours activity was detected in the lungs, heart, gonads, spleen, and brain but not in the fatty tissue.[12]

Regulations. ***U.S. FDA*** (1998) regulates the use of DEP (1) in adhesives as a component of articles intended for use in packaging, transporting, or holding food in accordance with the conditions prescribed by 21 CFR part 175.105; (2) as an ingredient in the manufacture of acrylic and modified acrylic plastics in articles intended for use in contact with food in accordance with the conditions prescribed by 21 CFR part 177.1010; (3) in the manufacture of resinous and polymeric coatings for the food-contact surface of articles intended for use in producing, manufacturing, packing, processing, preparing, treating, packaging, transporting, or holding food in accordance with the conditions prescribed by 21 CFR part 175.300; and (4) in the manufacture of resinous and polymeric coatings for polyolefin films to be safely used as a food-contact surface of articles intended for use in producing, manufacturing, packing, transporting, or holding food in accordance with the conditions prescribed by 21 CFR part 175.320.

Standards. ***Russia*** (1995). Recommended MAC: 3.0 mg/l.

References:

1. Zaitsev, N. A., Korolev, A. A., Baranov, Yu. B., et al., Hygienic regulation of diethyl phthalate, di-*n*-hexyl phthalate and dialkyl phthalate-810 in water bodies, *Gig. Sanit.*, 9, 26, 1990 (in Russian).
2. Castle, L., Mercer, A. J., Startin, J. R., and Gilbert, J., Migration from plasticized films into foods. 3. Migration of phthalate, sebacate, citrate and phosphate esters from films used for retail food packaging, *Food Addit. Contam.*, 5, 9, 1988.
3. Antonyuk, O. K., in *Hygiene and Toxicology*, Proc. Sci. Conf., Kiyv, 1967, 90 (in Russian).
4. Brown, D., Butterworth, K. R., Gaunt, I. F., et al., Short-term oral toxicity study of diethylphthalate in the rats, *Food Chem. Toxicol.*, 16, 415, 1978.
5. Peters, I. W. and Cook, R. M., Effect of phthalate esters on reproduction in rats, *Environ. Health Perspect.*, Exp. Issue No 3, 1973, 91.
6. Lamb, J. C., Chapin, R. E., Teague, J., et al., Reproductive effects of 4 phthalic acid esters in the mouse, *Toxicol. Appl. Pharmacol.*, 88, 255, 1987.
7. *Developmental Toxicity Evaluation of Diethyl Phthalate Administered to CD Rats on Gestation Days 6 through 15,* NTP Final Report Series No 336, Research Triangle Park, NC, NTP-88-336, NIEHS, 1988.
8. Field, E. A., Price, C. J., Sleet, R. B., George, J. D., Marr, M. C., Myers, C. B., Schwetz, B. A., and Morrissey, R. E., Developmental toxicity evaluation of diethyl and dimethyl phthalate in rats, *Teratology*, 48, 33, 1993.
9. Foster, P. M. D., Lake, B. G., Cook, M. W., et al., Structure-activity requirements for the induction of testicular atrophy by butyl phthalates in immature rats: effects on testicular zinc content, *Adv. Exp. Med. Biol.*, 136A, 445, 1982.

10. Seed, J. L., Mutagenic activity of phthalic esters in bacterial liquid suspension assay, *Environ. Health Perspect.*, 45, 111, 1982.
11. *Toxicology and Carcinogenesis Studies of Diethylphthalate in F344/N Rats and B6C3F₁ Mice (Dermal Studies) with Dermal Initiations (Promotion Study of Diethylphathale and Dimethylphthalate in Male Swiss Mice),* NTP Technical Report Series No 427, Research Triangle Park, NC, May 1995.
12. Autian, J., Toxicity and health threats of phthalate esters: review of the literature, *Environ. Health Perspect.*, 4, 3 1973.

PHTHALIC ACID, DI(2-ETHYLHEXYL) ESTER

Molecular Formula. $C_{24}H_{38}O_4$
M = 106.12
CAS No 117-81-7
RTECS No TI0350000
Abbreviations. DEHP, MEHP [mono(2-ethylhexyl) phthalate], PVC (polyvinyl chloride), PP (peroxisome proliferation).

Synonyms and **Trade Names.** 1,2-Benzenedicarboxylic acid, bis(2-ethylhexyl) ester; Bis(2-ethylhexyl) phthalate; Compound 889; Di(2-ethylhexyl) orthophthalate; Di(2-ethylhexyl) phthalate; Ergoplast FDO; 2-Ethylhexyl phthalate; Vestinol AH; Vinicizer 80; Witcizer 312.

PHTHALIC ACID, DIOCTYL ESTER

Molecular Formula. $C_{24}H_{38}O_4$
M = 106.12
CAS No 117-84-0
RTECS No TI1125000
Abbreviation. DOP.

Synonym and **Trade Names.** Bisoflex 81 or DOP; Dioctyl phthalate; Flexol DOP; Kodaflex DOP; Palatinol AH; Sicol 150; Staflex DOP; Truflex DOP.

Properties. Clear, colorless, oily liquid with a faint odor. DEHP forms colloidal solutions. Because of this phenomenon the "true solubility in water" is believed to be 0.025 to 0.05 mg/l. Detection threshold for odor and bitter-salt taste is 2.5 mg/l.

Applications. Used as a principal plasticizer in the production of PVC and its copolymers. It gives the materials high elasticity and low-temperature resistance (-45°C), which are combined with resistance to ultraviolet radiation. It is also used as a plasticizer for cellulose nitrate, polystyrene, and other polymers intended for contact with food or drink. Because of its solubility in fats, DEHP is used in the production of packaging materials for vegetables, fruit, juices, and milk.

Exposure. DEHP and total phthalate ester plasticizer levels were determined in milk, cream, butter, and cheese samples from a variety of sources from three European countries (UK, Norway, and Spain). In the samples of Norwegian milk, total phthalates were found between 0.12 and 0.28 mg/kg. On processing, the DEHP was concentrated in the cream at levels up to 1.93 mg/kg whereas low-fat milk contained from < 0.01 to 0.07 mg/kg. Retail dairy products from Spain were contaminated with < 0.01 to 0.55 mg/kg DEHP. UK pooled milk samples contained <0.01 to 0.09 mg DEHP/kg and 0.06 to 0.32 mg total phthalate/kg. UK cream samples contained 0.2 to 2.7 mg DEHP/kg and 1.8 to 19.0 mg total phthalate/kg. The authors indicated that the level found in these products was too high to have resulted solely from milk and must therefore have arisen in other ways.[1]

U.S. ATSDR (1988) quoted annually a U.S. FDA survey of various foods which showed that DEHP levels in most foods were less than 1.0 mg/kg (margarine, cheese, meat, cereal, eggs, milk, white bread, etc.). The likely human intake in Europe seems to be approximately 20 to 32 mg/person/annually.

Migration levels in food and water are 0.05 to 2.5 mg/l. In man, after transfusion of blood kept in a DEHP-plasticized PVC container, the plasticizer content in the spleen, liver, lungs, and omentum was 2.5 to 27 mg/l.[2] DEHP was found to be migrating from PVC blood tubing into human blood during hemodialysis in 16 patients with end-stage renal failure.[3]

Migration of epoxidized soybean oil from plasticized PVC was found to be 700 mg/dm^2 in ethanol (3-day contact period, 40°C), 250 mg/dm^2 in ethanol-water (1:1, 7 days, 40°C), 625 mg/dm^2 in isooctane (2 days, 20°C), and 650 mg/dm^2 in sunflower oil (10 days, 40°C).[4,5]

Migration of DEHP from plasticized PVC tubing used in commercial milking equipment reached 30 to 50 μg/kg. Retail whole milks from the UK contained 35 μg/kg DEHP.[6] Migration of DEHP from Canadian food-contacting films in different store-wrapped and ready-to-eat foods was observed at levels as high as 310 μg/g (cheese). DEHP was found in butter and margarine as migrants from the aluminum foil-paper laminates; and DEHP was determined in pies at 1.8 μg/g as a migrant from pie carton windows.[7] Migration of DEHP from a flexible PVC container into intravenous cyclosporine solutions was studied. By 48 hours, nearly 33 mg of DEHP had leached into the solution.[8]

Migration of phthalates from PVC into water has been the subject of many investigations.[9-11] No differences in migrated amounts between food-grade PVC irradiated with γ-radiation and non-irradiated samples of food-grade PVC were observed. After 47 hours of contact, DOP migration was noted at the level of 302.8 mg/l.[12] At 20°C, traces of plasticizer DOP migrate into water from PVC water pipes containing 8 parts of plasticizer, and at 37°C level of migration amounted to 0.1 to 0.16 mg/l.[11]

The introduction of fillers (chalk, barium sulfate, etc.) into plastisols reduces migration of plasticizers, but increased content of epoxided soybean oil in PVC (> 3.0%) increases migration of DOP.

Acute Toxicity.

Observations in man. Two volunteers were dosed with 10 or 5.0 g DEHP, respectively. 10 g caused mild gastric disturbances while 5.0 g did not.[13]

Animal studies. LD_{50} is reported to be 30 to 33 g/kg BW in rats, 6.4 to 17.7 g/kg BW in mice, and 34 g/kg BW in rabbits. However, no mortality in mice was reported even with the dose of 60 g/kg BW.[07] The acute action threshold for behavioral effect is 47 mg/kg BW and the no effect dose is reported to be 19 mg/kg BW.[14] Single oral administration of 2.0 g/kg BW to dogs apparently caused no toxicity.

Repeated Exposure revealed cumulative properties. K_{acc} is 2.3 to 2.7 (by Lim). Liver and testes appeared to be the main target organs. DEHP can cause functional hepatic damage, as reflected by morphology changes, alterations in the activity of energy-linked enzymes as well as changes in metabolism of lipids and carbohydrates. The most striking effect is proliferation of hepatic peroxisomes (PP).[14]

Observations in man. A month after *i/v* administration of approximately 150 mg DEHP per week, dialysis patients were reported to have no liver changes that might have been resulted from their treatment. Peroxisomes were reported to be significantly higher.[15]

Animal studies. Administration of the dose of 2.0 g/kg BW for 21 days increases hepatic alcohol dehydrogenase, microsomal protein, and cytochrome P-450 concentrations in rats.[16] The NOEL of 25 mg/kg BW was identified for PP based on changes in peroxisomal-related enzyme activities and/or ultramicroscopic changes in a 14-day gavage study in Sprague-Dawley rats (effect level 100 mg/kg BW).[17,18] The NOEL of 2.5 mg/kg BW was observed in a similar 7-day study; the NOEL of 10 mg/kg BW was defined in a 3-week study.[19] In addition to the effects on the liver associated changes in the kidney and thyroid of Wistar rats were reported.[20]

Short-term Toxicity. Doses equal or more than 50 mg/kg BW caused a significantly dose-related increase in liver weights, a decrease in serum triglyceride and cholesterol levels and dose-related microscopic changes in the liver, i.e., periportal accumulation of fat and mild centrilobular loss of glycogen. A significant increase in hepatic peroxisomal enzymes activities and in the number of peroxisomes in the liver was found.[17,19]

Rats fed 20 to 60 mg DEHP/kg BW for 104 days exhibited growth retardation and increased liver and kidney weights. No histopathology findings were noted in the visceral organs.[21]

Female Sprague-Dawley rats were given 5.0, 50, 500, or 5000 ppm DEHP or DOP in the diet for 13 weeks. Hepatomegaly, mild to moderate seminiferous tubule atrophy and Sertoli cell vacuolation were observed only in males and rats of both sexes administered highest DEHP doses. At the highest dose, DOP caused a significant increase in liver ethoxyresorufin-O-deethylase activity while DEHP did not. Both substances caused mild histological changes in the thyroid and the liver. The NOAEL was judged to be 3.7 mg DEHP/kg BW (50 ppm in the diet) and 36.8 mg DOP/kg BW (500 ppm in the diet).[22]

In the Japanese study, DEHP was administered orally at 100, 500, and 2500 mg/kg BW to marmosets for 13 weeks. The treatment did not cause hepatic PP, testicular atrophy, or pancreatic acinar cell hyperplasia.[23]

Long-term Toxicity.

Observations in man. Humans are less sensitive for chemically induced PP compared to rodents. Provisional ADI appeared to be 0.025 mg/kg BW.[23]

Animal studies. PP in the liver seems to be the most sensitive effect and the rat appears to be the most sensitive species. There appeared to be a threshold for PP and peroxisome-associated enzyme induction. The NOEL for the induction of PP in the rat is 2.5 mg/kg BW.[24]

In a 2-year feeding study, F344 rats and $B6C3F_1$ mice were fed diets containing 6.0 to 12 g/kg BW (rats) and 3.0 to 6.0 g/kg BW (mice) for 103 consecutive weeks. Hepatic and testicular effects were observed. Scientific interpretation of these long-term studies seems to be problematic because of use of MTD (maximum tolerated dose).[22]

In a 1-year feeding study in guinea pigs, there were no signs of hepatic changes. In a 2-year experiment, rats received the diet containing 0.4 and 0.13% DEHP. Liver morphology was altered as evidenced by excessive pigmentation, congestion, and some fatty degeneration.[25] No hepatic histology alterations were reported with 0.02% DEHP administered in the diet.[26] Hepatic effect of DEHP remains to be resolved.

Reproductive Toxicity. Reproduction effect is dose-dependent and is influenced by the duration of phthalate administration.

Gonadotoxicity. DEHP can cause testicular damage. These effects, consisting of atrophy, tubular degeneration, inhibition or cessation of spermatogenesis, were seen in mice, rats, guinea pigs, and ferrets. The testicular injury was accompanied by decreased zinc content in the gonads and increased urinary excretion of zinc.[27,28]

At the dose level of 5.0 g/kg BW, testosterone level in blood was decreased. The dose of 20 g/kg BW reduced content of triglycerides and cholesterol in the sperm. Doses of 0.2 and 2.0 g/kg BW caused testes atrophy and decreased zinc contents in them.[29-31]

Ovarian toxicity of DEHP was identified in the female rat. Adult, regularly cycling Sprague-Dawley rats were dosed daily with 2.0 g DEHP/kg BW in corn oil by gavage for 1 to 12 days. DEHP exposure resulted in prolonged estrous cycles. Microscopic evaluation of the ovaries revealed that 7 of 10 DEHP-exposed rats had not ovulated by vaginal estrus, whereas all control rats had ovulated by vaginal estrus. The authors concluded that exposure to DEHP resulted in hypoestrogenic anovulatory cycles and polycystic ovaries in adult female rats.[32]

Rats were *in utero* exposed to 1.0 g DEHP/kg BW. The treatment significantly decreased activities of testicular sorbitol dehydrogenase and acid phosphatase and increased γ-glutamyl transpeptidase, lactate dehydrogenase and β-glucuronidase activities at early ages. A decrease in the sperm count of the epididymal spermatozoa was also noted in the sexually matured animals of DEHP exposed group. The authors concluded that *in utero* exposure to DEHP may affect the normal development of testes.[33]

Administration of 0.15 to 2.0 g/kg BW to rats for 2 weeks caused reduction of spermatozoa count in epidydimis. Decreased weights of the testes were observed at the maximum dose level. Increased activity of β-glutamyltranspeptidase, LDH (at all doses), β-glucuronidase (at maximum dose), decreased activity of acid phosphatase and sorbitoldehydrogenase (1.2 g/kg BW) were found in the testicular tissue.[29] Complete suppression of fertility in mice was seen at 430 mg/kg BW; it was significantly reduced at 140 mg/kg BW. The NOEL of 15 mg/kg BW was identified in this study.[34]

Embryotoxicity. There is evidence that DEHP and its metabolite, MEHP, are embryolethal to rodents. Oral administration produced a significant reduction in placental weight.

DEHP was administered in the diet to pregnant animals at levels of 0.5, 1.0, 1.5, or 2.0% (Fischer 344 rats) and at 0.025, 0.05, 0.1, or 0.15% (CD-1 mice). No treatment-related embryotoxic effects including developmental abnormalities were observed in mice at 0.025% or in rats at 0.5% DEHP dietary levels.[35]

At oral doses equal to and higher than 200 mg/kg BW, decreased fetal weights and an increased number of resorptions were observed in rats. Maternal toxicity was observed with doses up to 340 mg/kg and 500 mg/kg BW.[16]

A single administration on day 7 of gestation caused a reduction in BW of living mouse fetuses. However, no significant changes in the number of live fetuses and no gross or skeletal abnormalities are reported.[36]

Female Long-Evans rats were exposed to different concentrations of DEHP suspended in drinking water from day 1 of pregnancy to day 21 after delivery. Perinatal exposure to DEHP produced no statistically significant changes in the BW gain of offspring. Conversely, it produced a significant decrease in kidney and testes relative weight (organ/BW) with a significant increase in relative liver weight. Signs of histological damage in kidneys, liver, and particularly testes, were observed.[37]

Embryos were cultured for 44 hours beginning on gestational day 10 in rat serum to which DEHP was added. The plasticizer decreased growth and development at all tested concentrations higher than 0.5%. These results suggest that DEHP itself is able to alter normal embryonic growth and development; however the high embryotoxic concentrations are unlikely to be attained *in vivo*.[38]

Teratogenicity. There is evidence that DEHP and its metabolite, MEHP, are teratogenic to rodents. Rats were less susceptible to DEHP-related adverse effects on fetal development than mice.

In mice given oral doses of 1.0 g DEHP/kg BW on days 7, 8, and 9 of gestation, the rate of re - sorptions and malformed fetuses was increased significantly. Fetal weights were also significantly lower. Anterior neural tube defects (anencephaly and exencephaly) were the malformations most commonly produced.

DEHP appeared to be highly embryotoxic and teratogenic in mice when given orally but not i/p. The difference in metabolism, disposition, or excretion by the route of administration may be responsible for the difference in DEHP teratogenicity. Although MEHP is a principal metabolite of DEHP and is several times more toxic than DEHP to adult mice, it seems that MEHP and its metabolites are not teratogenic in ICR mice.[39]

In the above cited study,[35] increased resorptions and decreased number of live fetuses/litter were observed in rats and mice at the higher doses. In mice, the number and percentage of fetuses malformed per litter (open eye, exophthalmia, exencephaly, short, constricted, or no tail, major vessel malformations, andfused or branched ribs) were elevated at 0.05, 0.1, and 0.15% dietary levels. The authors concluded that DEHP was not teratogenic at any dose tested in Fischer 344 rats when administered in the feed throughout gestation but did produce maternal and other embryotoxic effects at 1.0, 1.5, and 2.0% dietary levels. In CD-1 mice, the treatment with DEHP resulted in an increased incidence of malformations at doses which produced maternal and other embryotoxic effects and at a dose (0.05%) which did not produce significant maternal toxicity.[35]

A dose of 10 ml DEHP/kg BW was given to pregnant mice on days 9 or 10 of gestation. The rates of live fetuses were more than 90%. Gross and skeletal abnormalities in the live fetuses occurred with 2.5 or 7.5 ml DEHP/kg BW given orally on days 7 or 8 of gestation, respectively. Similar toxic effects were observed with the administration of MEHP. The gross malformations included exencephaly, open eyelid and club foot. Skeletal abnormalities occurred in the skull, cervical and/or thoracic bones.[40]

In carefully designed experimental protocol DEHP lacks birth defect actions. Teratogenicity occurs principally in mice at doses that are exceedingly high. Fetal mortality, fetal resorption, decreased fetal weight, neural tube defects and skeletal disorders *viz.* exencephaly, spina bifida, open eyelid, exophthalmia, major vessel malformations, club-foot, and delayed ossification were seen in several DEHP teratogenicity studies. The NOAEL for these effects in mice was found to be 0.025% DEHP in the diet (equal to 35 mg/kg BW).[36] The LOEL in mice appeared to be 0.05 ml/kg BW. However, according to other data, teratogenic effects were not observed in Fischer 344 rats even at dose levels of 0.5 to 2.0% DEHP in the diet (equal to 250 to 1000 mg/kg BW).[36,41,42]

Immunotoxicity. Male rats were *i/p* exposed to 0.002 to 0.012 LD_{50} DOP. The treatment affected B-lymphocyte population by suppression of the synthesis of cellular DNA, thus modulating the immune system.[43] Mice were orally exposed to a mixture of styrene and DOP in ground nut oil at the doses of 0.02,

0.03, or 0.05 LD_{50} for 4 weeks. Exposure to low doses of mixture may modulate some of the immune functions as compared to exposure to either chemical alone.[44]

Mutagenicity. The total weight-of-evidence clearly indicates that DEHP is not genotoxic.[45] In the majority of mutagenicity studies *in vitro* and *in vivo* DEHP showed negative results, *viz.* no induction of gene-mutations in bacterial system, eukariotic systems or mammalian systems *in vitro*, no induction of CA in somatic or germ cells *in vitro* and no induction of CA in somatic or germ cells *in vivo*.

In vivo cytogenicity. DEHP or MEHP were administered to pregnant Syrian golden hamsters on day 11 of gestation; this was followed by the cultivation of embryonic cells for 15 to 20 days. Both DEHP and MEHP induced 8AG/6TG-resistant mutation, CA and morphological transformation in the embryonic cells of the Syrian golden hamster.[46]

DEHP caused dominant lethal mutations in mice after systemic but not oral administration.

In vitro genotoxicity. DEHP was tested for mutagenicity in *Salmonella typhimurium* strains TA98, TA100, TA1535, and TA1537 without metabolic activation and in the presence of rat and hamster liver *S*9 metabolic activation systems. No mutagenic activity was seen.[47] However, according Anderson et al., DEHP appeared to be a weak direct acting mutagen in *Salmonella typhimurium* strain *TA100* in the presence of rat liver microsomes (*S*9 mix).[48]

MEHP was found to cause CA in CHO cells but showed no effect in SCE and hypoxanthine guanine phosphoribosyl assay.[48]

MEHP but not DEHP produced a dose-dependent DNA damaging effect to *B. subtilis* in *Rec*-assay. MEHP produced not only the mutation in *E. coli* but also SCE in Chinese hamster *V*79 cells. It also induced 8AG/6TG-resistant gene mutations and CA in the *V*79 cells.[46]

No chemically induced DNA damage or repair was observed *in vivo* or *in vitro* in rat or human hepatocytes under any DEHP concentrations applied. However, an increase in the percentage of cells in *S*-phase in the animals given DEHP was observed. DEHP does not exhibit direct genotoxic activity in the animals even with a treatment regimen which eventually produced tumors in a long term bioassay.[49]

No evidence for a covalent interaction of DEHP with DNA, no induction of single-strand breaks in DNA or unscheduled DNA repair by DEHP were reported. However, DEHP induced aneuploidy in eukariotic cells *in vitro* and cell transformations in mammalian cells *in vitro*.[50-53]

Carcinogenicity. DEHP was tested in mice and rats using higher doses by oral administration; it significantly increased the incidence of benign and malignant liver-cell tumors in animals of both species, and a dose-response relationship was observed.

An oral 2-year NTP study in mice revealed increased incidence of hepatocellular carcinoma in males and females at both dose levels (3.0 and 6.0 g/kg in the diet). Rats administered 6.0 or 12.0 g/kg in the diet for 2 years showed also an increased incidence of hepatocellular carcinomas and neoplastic nodules in the liver.[22] Effects of DEHP in the liver, including the induction of neoplasms, were reversible. The relationship of mononuclear cell leukemia to DEHP is questionable due to unusually low incidence in the male controls.

The NTP results were confirmed and extended by Cattley et al. (1987). In this study rats were dosed with 0.1, 0.3, or 1.2% DEHP in conventional diets. The treatment caused a 30% increase in liver tumors at the high dose level of 1.2% (600 mg/kg BW) compared with absence of tumor incidence in the controls.[54] A clear increase in liver tumor incidence in rats fed diets containing 2.0% DEHP (1000 mg/kg BW/day) was observed by Rao et al.[55,56] Most of the tumorigenic effects seen in these two carcinogenicity studies were noted in the high-dose animals only.

In the most recent study, DEHP was administered to $B6C3F_1$ mice at doses of 19.2, 98.5, 292.2, and 1266.1 mg/kg BW in the diet of males and at doses of 23.8, 116.8, 354.2, and 1458.2 mg/kg BW in the diet of females over a period of 104 weeks.[57] The NOEL and LOEL values for systemic effects were 19.2 and 116.8 mg/kg diet, respectively, with no evidence of carcinogenicity.

In the rat study (David et al., 1996), F344 rats received dietary concentrations of 0.01, 0.05, 0.25, and 1.25% (5.8, 28.9, 146.6, and 789 mg/kg BW, respectively, for males, and 7.3, 36.1, 181.7, and 938.5 mg/kg BW, respectively, for females) over a period of up to 104 weeks. The NOEL and LOEL for systemic effects were dietary concentrations of 0.05 and 0.25%, respectively, in males and females. No increase in serum levels of liver enzymes was found at any concentration.[58] The NOEL of 20 mg/kg BW from the David et al. (1997) mouse study is about fivefold lower than the NOEL for cancer (both adenomas and carcinomas) observed in the same study (116.8 mg/kg BW for females and 98.6 mg/kg BW for males).[57]

Though the precise mechanism of rodent hepatocarcinogenesis of DEHP is not yet completely understood, there is a clear evidence that DEHP exerts its tumorigenic response through increased cell proliferation and hepatocellular hyperplasia. The most recent studies clearly indicate a nonlinear dose-response curve below which tumors are not induced.[59]

Two different protocols were used to evaluate the initiating potential of DEHP in the liver using histochemically defined foci as the endpoint. In both experiments the results were negative. Exposure to DEHP at the level of 1.2% in the diet for 2 years resulted in elevation of hepatic peroxisomal enzymes while DNA replication, an indication of cell proliferation, was not affected in hepatocytes. The number of foci was not elevated in the DEHP group compared to the controls, even though a low incidence of rats with liver tumors occurred in the treated group. The authors concluded that these data, as well as other published results, suggest that DEHP and other PP chemicals have unique effects on the development of hepatic neoplasms. The absence of altered foci after chronic administration or in initiation-promotion protocols distinguishes DEHP from both classic liver carcinogens and promoters.[60]

The data from the latest rodent bioassays with DEHP show the $B6C3F_1$ mouse to be the most sensitive species with NOELs for both PP and increased liver weight of 19.2 mg/kg BW for males and 23.8 mg/kg BW for females (average of about 20 mg/kg BW).

Rats and mice appeared to be uniquely responsive to the morphological, biochemical, and chronic carcinogenic effects of PP, while guinea pigs, dogs, non-human primates, and humans are essentially nonresponsive or refractory; Syrian hamsters exhibit intermediate responsiveness. The differences in the metabolic fate of DEHP in rodents compared to primates are reported by many authors. No PP was observed in marmosets after feeding with 2.0 g DEHP/kg BW for 14 days whereas in a parallel study in rats, marked PP was observed. *In vivo* rat studies did not show tumor initiating or promotion activity or sequential carcinogenic activity of DEHP in the liver.[34]

These differences are explained, in part, by marked interspecies variations in the expression of peroxisome proliferator-activated receptor. Since humans are quite refractory to PP, it is concluded that the hepatocarcinogenic response of rodents to DEHP is not relevant to human cancer risk at any anticipated exposure level.[59]

An increased incidence of liver tumors in mice and rats in chronic bioassays is hypothesized to be caused by the prolonged hepatocellular PP and the enhanced production of the peroxisomal metabolic by-product, *hydrogen peroxide*. Non-human primates and humans appear to be far less sensitive to PP than mice and rats.[61]

Carcinogenicity classifications. An IARC Working Group concluded that there is sufficient evidence for the carcinogenicity of DEHP in *experimental animals* and there were no adequate data available to evaluate the carcinogenicity of DEHP in *humans*.

The Annual Report of Carcinogens issued by the U.S. Department of Health and Human Services (1998) defines DEHP to be a substance which may reasonably be anticipated to be carcinogen, i.e., a sub- stance for which there is a limited evidence of carcinogenicity in *humans* or sufficient evidence of carcinogenicity in *experimental animals*.

According to Doull et al. (1999), the current U.S. EPA classification of DEHP as a B2 (probable human carcinogen) is based on outdated information. The most appropriate and conservative point of reference for assessing margins of exposure should be 20 mg/kg BW, which is the mouse NOEL for PP and increased liver weight. Exposure of the general human population to DEHP is approximately 30 μg/kg BW, the major source being from residues in food. Doull et al. (1999) believe that new toxicology data and a considerable amount of new mechanistic evidence allow reconsideration of the cancer classification of DEHP under U.S. EPA's proposed new cancer risk assessment guidelines. DEHP should be classified as unlikely to be a human carcinogen under any known conditions of human exposure.[59]

IARC: 2B;

U.S. EPA: B2;

NTP: P - P* - P* - P* (feed).

Chemobiokinetics. In rats, DEHP is well absorbed from the GI tract after oral administration. Metabolism involves hydrolysis to the *monoester* with subsequent oxidation of the remaining alkyl chain. *Phthalic acid* was found to be a minor metabolite (less than 5.0%) to the total urinary metabolites. Dirven et al. used the method for biological monitoring of exposure to DEHP, which included enzymatic hydrolysis, ether extraction, and derivatization with triethyloxonium tetrafluoroborate. Four metabolites are determined in urine samples, namely, *mono(2-ethylhexyl) phthalate, mono(5-carboxy-2-ethylpentyl) phthalate, mono(2-ethyl-5-oxohexyl) phthalate*, and *mono(2-ethyl-5-hydroxyhexyl) phthalate*. Hydrolysis to *mono(2-ethylhexyl) phthalate* (MEHP) with release of *2-ethylhexanol* largely occurs prior to intestinal absorption.[62] The highest levels of the dose were found in the liver and adipose tissue. No or little accumulation in rats was observed. Estimated half-life for DEHP and its metabolites in rats is 3 to 5 days for fat and 2 days for other tissues. A single oral dose is excreted in rats unchanged in feces within 4 days, whereas the excretion rate for continuous feeding is not known. In mice and rats, urinary metabolites consist primarily of terminal oxidation products.[62]

Regulations.

U.S. FDA (1998) approved the use of DEHP (1) as a component of adhesives intended for use in contact with food in accordance with the conditions prescribed by 21 CFR part 175.105; (2) as a plasticizer in polymeric substances in accordance with the conditions prescribed in 21 CFR part 178.3740; (3) in the manufacture of rubber articles intended for repeated use in producing, manufacturing, packing, processing, treating, packaging, transporting, or holding food (total not to exceed 30% by weight of the rubber products) in accordance with the conditions prescribed by 21 CFR part 177.2600; (4) in the manufacture of semirigid and rigid acrylic and modified acrylic plastics in the manufacture of articles intended for use in contact with food in accordance with the conditions prescribed by 21 CFR part 177.1010; (5) in the manufacture of resinous and polymeric coatings in a food-contact surface of articles intended for use in producing, manufacturing, packing, transporting, or holding food in accordance with the conditions prescribed by 21 CFR part 175.300; (6) in surface lubricants employed in the manufacture of metallic articles that contact food in accordance with the conditions prescribed by 21 CFR part 178.357; (7) as a cellophane ingredient for packaging food, alone or in combination with other phthalates where total phthalates do not exceed 5.0% in accordance with the conditions prescribed by 21 CFR part 177.1200; and (8) as a defoaming agent that may be safely used in the manufacture of paper and paperboard intended for use in producing, manufacturing, packing, transporting, or holding food in accordance with the conditions prescribed by 21 CFR part 176.210.

In ***Belgium, The Netherlands,*** and ***Spain,*** DEHP may be used in all applications provided the level migrating into the foodstuff is less than 40 ppm. In ***France*** it may not be used to contact alcoholic beverages and fatty foods. In ***Italy*** the possible applications are limited in exactly the same way as for DEHA.

In ***Germany*** DEHP may be used at up to 35% in general purpose film with the same foodstuff restrictions as for DEHA. It may also be used in conveyor belts intended to contact food. In this application the level may be up to 12% for fatty foods including milk products and up to 40% for other foodstuffs. In ***Great Britain*** it may be used at up to 40% in items for contacting non-fatty foods.

A limit of 3.0% DEHP content for all materials used to store foodstuffs is approved in the ***U.S.***[48]

Recommendations.

WHO (1996). Guideline value for drinking water: 0.008 mg/l.

EU Scientific Committee for Food (1991). TDI: 0.025 mg/kg BW.

Standards.

U.S. EPA (1999). MCL: 0.006 mg/l, MCLG: zero.

Russia (1995). PML: 0.1 mg/l.

References:

1. Sharman, M., Read, W. A., Castle, L., and Gilbert, J., Levels of di(2-ethylhexyl)phthalate and total phthalate esters in milk, cream, butter and cheese, *Food Addit. Contam.*, 11, 375, 1994.
2. *Di(2-ethylhexyl) phthalate*, Environmental Health Criteria, WHO, IPCS, Geneva, 1992, 142.
3. Ono, K., Tatsukawa, R., and Wakimoto, T., Migration of plasticizer from hemodialysis blood tubing. Preliminary report, *JAMA*, 234, 948, 1975.

4. Castle, L., Mayo, A., and Gilbert, J., Migration of epoxidized soya bean oil into foods from retail packaging materials and from plasticized PVC film used in the home, *Food Addit. Contam.*, 7, 29, 1990.
5. Feigenbaum, A. E., Hamdani, M., Ducruet, V. J., and Riquet A.-M., Classification of interactions: volatile simulants, global and specific migration, *J. Polymer Engin.*, 15, 47, 1995-1996.
6. Castle, L., Gilbert, J., and Eklund, T., Migration of plasticizer from poly(vinyl chloride) milk tubing, *Food Addit. Contam.*, 7, 591, 1990.
7. Page, B. D. and Lacroix, G. M., The occurrence of phthalate ester and di-2-ethylhexyl adipate plasticizers in Canadian packaging and food sampled in 1985-1989: a survey, *Food Addit. Contam.*, 12, 129, 1995.
8. Venkataramanan, R., Burckart, G. J., Ptachinski, R. J., Blaha, R. et al., Leaching of diethyhexyl phthalate from polyvinyl chloride bags into intravenous cyclosporine solution, *Am. J. Hospit. Pharmacol.*, 43, 2800, 1986.
9. Sheftel', V. O. and Grinberg, I. M., Migration of harmful chemical substances from polyvinyl chloride materials used in water supply, *Gig. Sanit.*, 8, 78, 1979 (in Russian).
10. Sheftel', V. O. and Katayeva, S. E., *Migration of Harmful Substances from Polymeric Materials*, Moscow, Khimiya, 1978, 146 (in Russian).
11. Zinchenko, T. M., *Hygienic Evaluation of Phtalate Plasticizers of PVC*, Author's abstract of thesis, Kyiv, 1988, 20 (in Russian).
12. Goulas, A. E., Kokkinos, A., and Kontominas, M. G., Effect of gamma-radiation on migration behavior of dioctyl adipate and acetyl tributyl citrate plasticizers from food-grade PVC and PVDC/PVC films into olive oil, *Z. Lebensm. Unters. Forsch.*, 201, 74, 1995.
13. IARC monographs, *Some Industrial Chemicals and Dyestuffs*, Lion, vol 29, 269, 1982.
14. Schaffer, C. B., Carpenter, C. P., and Smyth, H. F., Acute and subacute toxicity of di(2-ethylhexyl) phthalate with note upon its metabolism, *J. Ind. Hyg. Toxicol.*, 27, 130, 1945.
15. Thomas, R. D., Ed., *Drinking Water and Health*, Board on Toxicology and Environmental Health Hazards, Commission on Life Sciences, National Research Council, National Acad. Press., Washington, D.C., 6, 338, 1986.
16. Thomas, J. A. and Thomas, J. M., Biological effects of di(2-ethylhexyl) phthalate and other phthalic acid esters, *CRC Crit. Rev. Toxicol.*, 13, 283, 1984.
17. Lake, B. G., Pels Rijcken, W. R., Gray, T. J. B., et al., Comparative studies on the hepatic effects of di- and mono-*n*-octyl phthalates, di(2-ethylhexyl) phthalate and clofibrate in the rat, *Acta Pharmacol. Toxicol.*, 54, 167, 1986.
18. Morton, S. J., *The Hepatic Effects of Dietary Di(2-ethylhexyl) Phthalate*, PhD Thesis, The John Hopkins University, Ann Arbor, Michigan, 1979, 135.
19. Barber, E. D., Astill, B. D., Moran, E. J., et al., Peroxisome induction studies on seven phthalate esters, *J. Toxicol. Environ. Health*, 3, 7, 1987.
20. Hinton, R. H., Mitchell, F. E., Mann, A., et al., Effects of phthalic acid esters on the liver and thyroid, *Environ. Health Perspect.*, 70, 195, 1986.
21. Poon, R., Lecavalier, P., Mueller, R., Valli, V. E., Procter, B. G., and Chu, I., Subchronic oral toxicity of di- *n*-octyl phthalate and di(2-ethylhexyl)phthalate in the rat, *Food Chem. Toxicol.*, 35, 225, 1997.
22. NTP Technical Report Series No 217 *Carcinogenesis Bioassay of Di(2-ethylhexyl)phthalate in F344/N Rats and B6C3F$_1$ Mice,* NIH Publ. No. 82-1773, U.S. Dept of Health & Human Services, National Institute of Environmental Health Sciences, Research Triangle Park, NC, 1982.
23. Kurata, Y., Kidachi, F., Yokoyama, M., Toyota, N., Tsuchitani, M., and Katoh, M., Subchronic toxicity of di(2-ethylhexyl)phthalate in common marmosets: lack of hepatic peroxisome proliferation, testicular atrophy, or pancreatic acinar cell hyperplasia, *Toxicol. Sci.*, 42, 49, 1998.
24. Seth, P. K., Hepatic effects of phthalate esters. *Environ. Health Perspect.*, 45, 27, 1982.
25. Carpenter, D., Weil, C. S., and Smyth, H. F., Chronic oral toxicity of di(2-ethylhexyl) phthalate for rats, guinea pigs and dogs, *AMA Arch. Ind. Hyg. Occup. Med.*, 8, 219, 1953.

26. Nikanorov, M., Mazur, H., and Piekacz, H., Effect of orally administered plasticizers and polyvinyl chloride stabilizers in the rat, *Toxicol. Appl. Pharmacol.*, 25, 253, 1973.
27. Gangolli, S. D., Testicular effects of phthalate esters, *Environ. Health Perspect.*, 45, 77, 1982.
28. Gray, T. J. and Gangolli, S. D., Aspects of testicular toxicity of phthalic esters, *Environ. Health Perspect.*, 65, 229, 1986.
29. Parmar, D., P., Srivastava, S. P., Seth, P. K., et al., Effect of di(2-ethylhexyl) phthalate on spermatogenesis in adult rats, *Toxicology*, 42, 47, 1986.
30. Agarwal, D. K., Eustis, S., Lamb, J. C., et al., Influence of dietary zinc on di(2-ethylhexyl) phthalate testicular atrophy and zinc depletion in adult rats, *Environ. Health Perspect.*, 84, 12, 1986.
31. Saxena, D. K., Srivastava, S. P., Chandra, S. V., et al., Testicular effects of di(2-ethylhexyl) phthalate: histochemical and histopathological alterations, *Ind. Health*, 23, 191, 1985.
32. Davis, B. J., Maronpot, R. R., and Heindel, J. J., Di-(2-ethylhexyl) phthalate suppresses estradiol and ovulation in cycling rats, *Toxicol. Appl. Pharmacol.*, 128, 216, 1994.
33. Tandon, R., Seth, P. K., and Srivastava, S. P., Effect of *in utero* exposure to di(2-ethylhexyl)phthalate on rat testes, *Indian. J. Exp. Biol.*, 29, 1044, 1991.
34. Food Additives Series No 24, *Toxicological Evaluation of Certain Food Additives and Contaminants,* The 33rd Meeting of Joint FAO/WHO Expert Committee on Food Additives, WHO, Geneva, 1989, 222.
35. Tyl, R. W., Price, C. J., Marr, M. C., and Kimmel, C. A., Developmental toxicity evaluation on dietary di(2-ethylhexyl) phthalate in Fischer 344 rats and CD-1 mice, *Fundam. Appl. Toxicol.*, 10, 395, 1988.
36. Tomita, I., Nakamura, Y., Yagi, Y., et al., Teratogenicity/fetotoxicity in mice, *Environ. Health Perspect.*, 45, 71, 1982.
37. Arcadi, F. A., Costa, C., Imperatore, C., Marchese, A., Rapisarda, A., Salemi, M., Trimarchi, G. R., and Costa, G., Oral toxicity of bis(2-ethylhexyl) phthalate during pregnancy and suckling in the Long-Evans rat, *Food Chem. Toxicol.*, 36, 963, 1998.
38. Hansen, D. K. and Grafton, T. F., Evaluation of di(2-ethylhexyl)phthalate-induced embryo-toxicity in rodent whole-embryo culture, *J. Toxicol. Environ. Health., 43, 361, 1994.*
39. Shiota, K. and Mima, S., Assessment of the teratogenicity of di(2-ethylhexyl)phthalate and mono(2-ethylhexyl)phthalate in mice, *Arch. Toxicol.*, 56, 263, 1985.
40. Yagi, Y., Nakamura, Y., Tomita, I., Tsuchikawa, K., Shimoi, N., Teratogenic potential of di- and mono-(2-ethylhexyl)phthalate in mice, *J. Environ. Pathol. Toxicol.*, 4, 533, 1980.
41. Onda, H. et al., Effect of phthalate ester on reproductive performance in rats, *Japan. J. Hyg.*, 31, 507, 1976.
42. Nakamura, Y., Yagi, Y., Tomita, I., et al., Teratogenicity of di(2-ethylhexyl) phthalate in mice, *Toxicol. Lett.*, 4, 113, 1979.
43. Dogra, R. K. S., Khanna, S., Nagale, S. L., et al., Effect of dioctylphthalate on immune system of rat, *Indian. J. Exp. Biol.*, 23, 315, 1985.
44. Dogra, R. K., Khanna, S., Srivastava, S. N., Shukla, L. J., Chandra, K., Saxena, G., and Shanker, R., Immunomodulation due to coexposure to styrene and dioctyl phthalate in mice, *Immunopharmacol. Immunotoxicol.*, 15, 491, 1993.
45. Budroe, J. D. and Williams, G. M., Genotoxicity studies of peroxisome proliferators, in *Peroxisome: Biology and Importance in Toxicology and Medicine,* G. G. Gibson and B. G. Lake, Eds., Taylor and Francis, London, 1993, 525.
46. Tomita, I., Nakamura, Y., Aoki, N., and Inui, N., Mutagenic/carcinogenic potential of DEHP and MEHP, *Environ. Health Perspect.*, 45, 119, 1982.
47. Zeiger, E., Haworth, S., Mortelmans, K., and Speck, W., Mutagenicity testing of di(2-ethylhexyl)phthalate and related chemicals in *Salmonella*, *Environ. Mutagen.*,7, 213, 1985.

48. Anderson, D., Yu, T. W., and Hincal, F., Effect of some phthalate esters in human cells in the comet assay, *Teratog. Carcinog. Mutagen.*, 19, 275, 1999.
49. Butterworth, B. E., Bermudez, E., Smith-Oliver, T., Earle., L, Cattley, R., Martin, J., Popp, J. A., Strom, S., Jirtle, R., and Michalopoulos, G., Lack of genotoxic activity of di(2-ethylhexyl)phthalate (DEHP) in rat and human hepatocytes, *Carcinogenesis,* 5, 1329, 1984.
50. Turnbull, D. and Rodricks, J. V., Assessment of possible carcinogenic risk to humans resulting from exposure to di(2-ethylhexyl)phthalate (DEHP), *J. Am. Coll. Toxicol.*, 4, 11, 1985.
51. Putman, D. L., Moore, W. A., Schechtman, L. M., et al., Cytogenic evaluation of di(2-ethylhexyl) phthalate and its major metabolites in Fischer 344 rats, *Environ. Mutagen.*, 5, 227, 1983.
52. Tomita, I., Nakamura, Y., Aoki, N., et al., Mutagenic/carcinogenic potential of DEHP and MEHP, *Environ. Health Perspect.*, 45, 119, 1982.
53. Zeiger, E., Howarth, S., Speck, W., et al., Phthalate esters testing in the NTP's environmental mutagenesis test development program, *Environ, Health Perspect.*, 45, 99, 1982.
54. Cattley, R. C., Conway, J. G., and Popp, J. A., Association of persistent peroxisome proliferation and oxidative injury with hepatocarcinogenicity in female F344 rats fed di(2-ethylhexyl) phthalate for two years, *Cancer Lett.*, 38, 15, 1987.
55. Rao, M. S., Usuda, N., Subbarao, V., and Reddy, J. K., Absence of gamma-glutamyl transpeptidase activity in neoplastic lesions induced in the liver of male F344 rats by di(2-ethylhexyl) phthalate, a peroxisome proliferator, *Carcinogenesis,* 8, 1347, 1987.
56. Rao, M. S., and Reddy, J. K., An overview of peroxisome proliferator-induced hepatocarcinogenesis, *Environ. Health Perspect.*, 93, 205, 1991.
57. David, R. M., Moore, M. R., Cifone, M. A., Finney, D. C., and Guest, D., Correlation of peroxisome proliferation and oncogenicity of di(2-ethylhexyl) phthalate in mice, *Fundam. Appl. Toxicol.*, 36 (Suppl.), 173, 1997.
58. David, R. M., Moore, M. R., Cifone, M. A., Finney, D. C., and Guest, D., Further investigations of the oncogenicity of di(2-ethylhexyl) phthalate in rats, *Fundam. Appl. Toxicol.*, 30 (Suppl.), 204, 1996.
59. Doull, J., Cattley, R., Elcombe, C., Lake, B. G., Swenberg, J., Wilkinson, Ch., Williams, G., and van Gemert, M., A cancer risk assessment of di(2-ethylhexyl)phthalate: application of the new U.S. EPA risk assessment guidelines, *Regul. Toxicol. Pharmacol.*, 29, 327, 1999.
60. Popp, J. A., Garvey, L. K., and Cattley, R. C., *In vivo* studies on the mechanism of di(2-ethylhexyl) phthalate carcinogenesis, *Toxicol. Ind. Health,* 3, 151, 1987.
61. Stott, W. T., Chemically induced proliferation of peroxisomes: Implications for risk assessment, *Regul. Toxicol. Pharmacol.*, 8, 125, 1988.
62. Dirven, H. A., van den Broek, P. H., and Jongeneelen, F. J., Determination of four metabolites of the plasticizer di(2-ethylhexyl)phthalate in human urine samples, *Int. Arch. Occup. Environ. Health,* 64, 555, 1993.

PHTHALIC ACID, DIHEPTYL ESTER

Molecular Formula. $C_{22}H_{34}O_4$
M = 362.53
CAS No 3648-21-3
RTECS No TI1090000
Abbreviation. DHP.

Synonym. Diheptyl phthalate.

Properties. Colorless, almost odorless liquid. Solubility in water is 0.1 mg/l.

Applications. Used as a plasticizer in the production of food-contact vinyl polymers.

Acute Toxicity. LD_{50} is 29.6 g/kg BW in rats.[1]

Repeated Exposure. F344 rats received DHP at dose levels 0.2, 1.0, or 5.0 g/kg BW for 28 days. The treatment caused no mortality. Retardation of BW gain was seen at the dose of 5.0 g/kg BW, blood biochemistry revealed significant increases in albumin and A/G ratio in males at 0.2 g/kg BW and more, and in albumin and total protein in females treated with 1.0 g/kg BW or more. Increases in liver and kidney weights were noted in both sexes treated with 1.0 g/kg BW or more. Histological examination revealed swelling and necrosis of hepatocytes in males treated with two higher doses. Males of the 5.0 g/kg BW group showed atrophy of the seminiferous tubules accompanied with loss of spermatogenesis.[1]

Reproductive Toxicity.

Gonadotoxicity. Rats were exposed orally to 7.2 mmol/kg BW dose for 4 days. The treatment caused testicular atrophy, an increase in the excretion of zinc in the urine, and a reduction in the zinc content of the testes.[2] In the above described study, a decrease in testes weight, edema and necrotization of hepatocytes were reported in animals given 1.0 and 5.0 g DHP/kg BW. Atrophy of seminal ducts and cessation of spermatogenesis were reported.[1]

Teratogenic effect in mice was reported by Nakashima et al.[3] However, Plasterer et al. observed no congenital defects and no increase in lethality in the offspring of mice given 3.5 g DMP/kg BW on days 7 to 14 of gestation.[4]

Chemobiokinetics. DHP is rapidly metabolized to the *heptyl phthalate* which may undergo further oxidation of the alcohol side chain. Probably, the half hydrolysis of diester substrates is due to the anionic charge of the free carbonyl group inhibiting formation of the enzyme-substrate complex.

References:

1. Matsushima, Y., Onodera, H., Mitsumori, K., Maekawa, A., Kurokawa, Y., and Takahashi, M., Study of diheptyl phthalate toxicity after repeated administration for 28 days, *Bull. Natl. Inst. Hyg. Sci.*, 110, 26, 1992 (in Japanese).
2. Foster, P. M. D., Lake, B. G., Cook, M. W., et al., Structure-activity requirements for the induction of testicular atrophy by butyl phthalates in immature rats: effects on testicular zinc content, *Adv. Exper. Med. Biol.*, 136A, 445, 1982.
3. Nakashima, K., Kishi, K., Nishikiori, M., et al., Teratogenicity of di-*n*-heptyl phthalate, *Teratology*, 16, 117, 1977.
4. Plasterer, M. R., Bradshaw, W. S., Booth, G. M., et al., Developmental toxicity of nine selected compounds following prenatal exposure in the mouse: Naphthalene, *p*-nitrophenyl, sodium selenite, dimethyl phthalate, ethylene thiourea, and four glycol ether derivatives, *J. Toxicol. Environ. Health*, 15, 25, 1985.

PHTHALIC ACID, DIHEXYL ESTER

Molecular Formula. $C_{20}H_{30}O_4$
M = 334.0
CAS No 84-75-3
RTECS No TI1100000
Abbreviation. DHP.

Synonym. Dihexyl phthalate.

Properties. Does not color water but imparts a slight aromatic odor and a sweetish, astringent taste to it. Odor and taste perception thresholds are 2.8 and 1.0 mg/l, respectively. Foam-forming ability threshold is 20 mg/l.[1]

Applications. Used as a plasticizer in the production of food-contact vinyl polymeric materials.

Acute Toxicity. In rats, LD_{50} is 20 g/kg BW or even more than 30 g/kg BW. In mice and guinea pigs, it is 22.5 and 12 g/kg BW, respectively.[1] Oral administration to rats revealed accumulation in the tissue of large fat droplets of DHP. There was necrosis and an increased activity of glucose-6-phosphatase in the central lobular part of the liver.[2]

Repeated Exposure. Administration of 2.0% DHP in the diet of rats for 21 days caused hepatomegaly, decreased liver glycogen content and the foci of necrosis and fatty accumulations. Endoplasmic reticulum proliferation, increased content of triiodine tironine, and decreased content of tiroxine in the blood serum are reported in this study.[3] Rats received 20 g DHP/kg diet for 3 weeks. The treatment caused slight liver enlargement and changes in enzyme activities. No lesions were found in the kidneys and pancreas.[2]

Male rats were *i/g* exposed to oily solutions (100 and 1000 mg/kg BW doses) for 30 days. Polymorphism of the toxic action with predilection to affect the liver function (hydrocarbon, protein and lipid metabolism), myocardium, CNS and kidneys was observed with the 1000 mg/kg BW dose. Decreased activity of several enzymes was found.[1]

Reproductive Toxicity.

Embryotoxicity. CD-1 mice of both sexes were given diets with 0.3 to 1.2% DHP for 7 days prior to and during a 98-day cohabitation period. A dose-related adverse effect was found on the number of litters per pair, and of live pups per litter, and on proportion of pups born alive. A crossover mating study demonstrated that both sexes were affected.[4]

According to other data, at a dose level of 1.0 g/kg BW embryotoxic effect was not observed.[1]

Gonadotoxicity. DHP was found to produce testicular injury. The testicular lesion produced by DHP in immature rats is characterized by early sloughing of spermatides and spermatocytes and severe vacuolization of *Sertoli cell* cytoplasm.[5]

Chemobiokinetics. DEP is metabolized to *ethyl phthalate* by nonspecific esterases in the intestinal mucosa and other tissues.

Regulations. ***U.S. FDA*** (1998) listed DHP for use (1) as plasticizers for polymeric substances in the manufacture of articles or components of articles intended for use in producing, manufacturing, packaging, processing, preparing, treating, packing, transporting, or holding food in accordance with the conditions prescribed in 21 CFR part 178.3740; and (2) in adhesives as a component of articles intended for use in packaging, transporting, or holding food in accordance with the conditions prescribed by 21 CFR part 175.105.

Standards. ***Russia*** (1994). Recommended PML: 0.5 mg/l (organolept., taste).

References:

1. Zaitsev, N. A., Korolev, A. A., Baranov, Yu. B., et al., Hygienic regulation of diethyl phthalate, di-*n*-hexylphthalate and dialkyl phthalate-810 in water bodies, *Gig. Sanit.*, 9, 26, 1990 (in Russian).
2. Mann, A. H., Price, S. C., Mitchell, E. F., et al., Comparison of short-term effects of di(2-ethyl hexyl)phthalate, di(*n*-hexyl)phthalate, and di(*n*-octyl)phthalate in rats, *Toxicol. Appl. Pharmacol.*, 77, 116, 1985.
3. Hinton, R. H., Mitchell, F. E., Mann, A., et al., Effects of phthalic acid esters on the liver and thyroid, *Environ. Health Perspect.*, 70, 195, 1986.
4. Lamb, J. C., Chapin, R. E., Teague, J., et al., Reproductive effects of four phthalic acid esters in the mouse, *Toxicol. Appl. Pharmacol.*, 88, 255, 1987.
5. Cater, B. R., Cook, M. W., Gangolli, S. D., et al., Studies on dibutyl phthalate induced testicular atrophy in the rat: Effect on zinc metabolism, *Toxicol. Appl. Pharmacol.*, 41, 609, 1977.

PHTHALIC ACID, DIISOBUTYL ESTER

Molecular Formula. $C_{16}H_{22}O_4$

M = 278.35

CAS No 84-69-5

RTECS No TI1225000

Abbreviation. DIBP.

Synonym. Diisobutyl phthalate.

Properties. Colorless, almost odorless liquid.

Applications. Used as a plasticizer in the production of polyvinyl chloride and other polymers intended for contact with food.

Acute Toxicity. LD_{50} is 15 to 25 g/kg BW in rats, 12.8 g/kg BW in mice, and 10.0 g/kg BW in guinea pigs.[03]

Short-term Toxicity. Rats received 5.0% DIBP in their diet for several months. The treatment caused no animal mortality.

Long-term Toxicity. Dogs received 2.0 g DIBP/kg BW in their diet for several months. The treatment caused no mortality and produced no toxic effect.[03]

Reproductive Toxicity.

DIBP produced ***teratogenic effect*** in rats.[1]

Gonadotoxicity. It induced severe atrophy of the testes. Ingestion by rats caused high testosterone and low zinc concentrations in the testes.[2]

Mutagenicity.

In vitro genotoxicity. DIBP was negative in *Salmonella* mutagenicity assay.[3]

Chemobiokinetics. Ingested DIBP is readily absorbed. The vehicle can play an important role in the absorption, distribution, and elimination of the plasticizer (U.S. EPA Document, 1980). In Sprague-Dawley rats, approximately equal parts of administered DIBP were eliminated with urine and feces. In beagle dogs, a major part of ^{14}C-DIBP was liberated in feces; in miniature pigs, DIBP is excreted predominantly in the urine.[4]

Regulations. ***U.S. FDA*** (1998) regulates the use of DIBP (1) in adhesives used as components of articles intended for use in packaging, transporting, or holding food in accordance with the conditions prescribed by 21 CFR part 175.105; (2) in the manufacture of cellophane to be safely used for packaging food, alone or in combination with other phthalates where total phthalates do not exceed 5.0% in accordance with the conditions prescribed by 21 CFR part 177.1200.

References:

1. Singh, A. R., Lawrence, W. H., and Autian, L., Teratogenicity of phthalate esters in rats, *J. Pharmacol. Sci.*, 61, 51, 1972.
2. Oishi, S. and Hiraga, K., Effect of phthalic acid esters on mouse testes, *Toxicol. Lett.*, 5, 413, 1980.
3. Seed, J. L., Mutagenic activity of phthalic esters in bacterial liquid suspension assay, *Environ. Health Perspect.*, 45, 111, 1982.
4. Ikeda, G. J., Sapienza, P. P., Convillion, J. L., et al., Distribution and excretion of two phthalate esters in rats, dogs, and miniature pigs, *Food Cosmet. Toxicol.*, 16, 409, 1978.

PHTHALIC ACID, DIISODECYL ESTER

Molecular Formula. $C_{28}H_{46}O_4$

M = 446.74

CAS No 26761-40-0

RTECS No TI1270000

Abbreviation. DIDP.

Synonyms and **Trade Names.** 1,2-Benzenedicarboxylic acid, diisodecyl ester; Bis(isodecyl) phthalate; Phthalic acid, bis(8-methylnonyl) ester; Palatinol Z; Vestinol DZ.

Properties. Soluble in alcohols.

Applications. Plasticizer for polyvinyl chloride for calendered film, sheet, coated fabrics, in the manufacture of vinyl swimming pools, plasticizer for cellulose ester plastics.

Acute Toxicity. LD_{50} is 64 g/kg BW in rats.[1] LD_0 is 22.5 ml/kg BW in rabbits.[2]

Reproductive Toxicity.

Embryotoxicity. In a two-generation reproduction study, Sprague-Dawley rats received DIDP at concentrations of 0.2, 0.4, or 0.8% DIDP in their diet. There were no treatment-related differences in clinical observations, survival, reproductive organ weights or histopathology, male mating, male and female fertility, female fecundity or female gestational indices in either generation of the rats. Some evidence of reduced BW gain in the high-dose parental females was seen during the gestation and postpartum periods. DIDP did not produce evidence of reproductive or fertility effects under the conditions of this study.[3]

Rats were gavaged with 40, 200, or 1000 mg/kg BW from gestation days 6 to 15. DIDP showed fetal effects of borderline significance at 1000 mg/kg BW.[4]

Teratogenicity. DIDP was administered by gavage to mated Sprague-Dawley rats at doses of 100, 5000, and 1000 mg/kg BW on gestation days 6 through 15. Only mild maternal and developmental effects were observed at the maternally toxic exposure level of 1.0 g/kg BW. Maternal BW gain and food consumption were significantly reduced during the exposure period. No treatment-related morphological changes were noted for exception of an increased frequency of seventh cervical and rudimentary lumbar ribs. Both maternal and developmental NOAELs were therefore established at 0.5 g/kg BW. Authors concluded that DIDP is not teratogenic or a selective developmental toxicant.[5]

Allergic contact dermatitis from DIDP has been reported.[6]

Mutagenicity.

In vitro genotoxicity. DIDP was shown to be negative in *Salmonella typhimurium* mutagenicity assay with and without *S9* metabolic activation system.[7]

Carcinogenicity. In a peroxisome induction study, DIDP is found to be no more carcinogenic than the weak liver carcinogen di(2-ethylhexyl) phthalate.[8]

Regulations. *U.S. FDA* (1998) approved the use of DIDP (1) in adhesives as a component (monomer) of articles intended for use in packaging, transporting, or holding food in accordance with the conditions prescribed in 21 CFR part 175.105; (2) in the manufacture of closures with sealing gaskets for food containers intended for use in producing, manufacturing, packaging, processing, preparing, treating, packing, transporting, or holding dry food in accordance with the conditions prescribed by 21 CFR part 177.1210; and (3) in the manufacture of rubber articles intended for repeated use in producing, manufacturing, packaging, processing, preparing, treating, packing, transporting, or holding food in accordance with the conditions prescribed by 21 CFR part 177.2600.

References:

1. Shibko, S. I., Blumenthal, H., Toxicology of phthalic acid esters used in food-packaging material, *Environ. Health Perspect.*, 3, 131, 1973.
2. Krauskopf, L. G., Studies on the toxicity of phthalates via ingestion, *Environ. Health Perspect.*, 3, 61, 1973.
3. Nikiforov, A. I., et al., *Toxicologist,* 36, 359, 1997.
4. Hellwig, J., Freudenberger, H., and Jackh, R. Differential prenatal toxicity of branched phthalate esters in rats, *Food Chem. Toxicol.,* 35, 501, 1997
5. Waterman, S. J., Ambroso, J. L., Keller, L. H., Trimmer, G. W., Nikiforov, A. I., and Harris, S. B. Developmental toxicity of diisodecyl and diisononyl phthalates in rats, *Reprod. Toxicol.,* 13, 131, 1999.
6. Hills, R. J. and Ive, F. A., Allergic contact dermatitis from diisodecyl phthalate in a polyvinyl chloride identity band, *Contact Dermat.*, 29, 94, 1993.
7. Seed, J. L., Mutagenic activity of phthalic esters in bacterial liquid suspension assay, *Environ. Health Perspect.,* 45, 111, 1982.
8. Barber, E. D., Astill, B. D., Moran, E. J., Schneider, B. F., Gray, T. J., Lake, B. G., Evans, J. G., Peroxisome induction studies on seven phthalate esters, *Toxicol. Ind. Health*, 3, 7, 1987.

PHTHALIC ACID, DIISONONYL ESTER

Molecular Formula. $C_{26}H_{42}O_4$

M = 418.68

CAS No 28553-12-0

105009-97-0

41375-91-1

RTECS No CZ3850000

Abbreviation. DINP.

Composition. DINP is an isomeric mixture of diesters of *o*-phthalic acid and primary aliphatic branched chain alcohols, predominantly in the C_9 range.

Synonyms and **Trade Names.** 1,2-Benzenedicarboxylic acid, diisononyl ester; Diisononyl phthalate; Palatinol N; Sansocizer DINP; Vestinol NN.

Properties. Solubility in water is 0.2 mg/l.

Applications. DINP is widely used a plasticizers to impart softness and flexibility to normally rigid polyvinyl chloride products. It is used in the manufacture of food-contact adhesives, plastisoles, and nitrocellulose.

Exposure. During the past 2 years, concern has been voiced by public interest groups and regulatory agencies in Europe, Canada, and the United States regarding the potential adverse health effects of DINP migrating from children's toys during mouthing activities.[1]

Migration. Estimates of DINP migration from soft polyvinyl chloride materials have been obtained from a variety of *in vitro* methods (simulated saliva and controlled agitation) as well as *in vivo* methods (controlled chewing) that more closely resemble child chewing and mouthing activities. Recent estimates by the Consumer Product Safety Commission (CPSC) suggest that maximum exposures occur in infants 3 to 12 months of age. These exposure values are 17,500 to 70,000 and 1100 to 4200 times, respectively, lower than the chronic rodent NOAEL for DINP and 175 to 700 and 11 to 42 times lower than the corresponding ADI of 1.0 to 4.0 mg/kg BW.[1]

Exposure. Daily intake by all routes is unlikely to be more than 1.0 mg/kg/day.

Acute Toxicity. DINP exhibited low order of acute toxicity in rats and dogs.[2,3]

Short-term Toxicity. Rats given DINP at high dietary level doses of more than 1000 mg/kg BW exhibited significant increases in liver weight and also increases in the number of peroxisomes and the quantity of peroxisomal enzymes in the liver without developing histopathological changes.[2,4]

Long-term Toxicity. F344 rats were fed DINP at dietary levels of 0.03, 0.3, or 0.6 wt% for two years. Treatment with two higher doses caused slight decreases in food consumption and BW gain, slight increase in mortality, and dose-related increase in relative organ weights of the liver and kidney. The NOAEL appeared to be 0.03 wt% or approximately 17 mg/kg BW in this study. There was no peroxisome induction in the livers of treated rats.[5]

F344 rats were fed diets containing 500, 1500, 6000, and 12,000 ppm DINP (equivalent to 29 to 733 mg/kg BW in males, and 36 to 885 mg/kg BW in females). The treatment with higher doses caused a dose-related decrease in BW and an increase in relative liver and kidney weights, liver hypertrophy, mineralization of renal papilla (males) and renal tubule pigment. The reversibility of these effects were noted in animals fed DINP diet (high dose) for 78 weeks. The NOAELs of 358 and 442 mg/kg BW, based on neoplastic response observed at the high dose, were established for males and females, respectively.[1,6] Results consistent with these from the rat studies were obtained in 2-year study in $B6C3F_1$ mice.[1,7]

Because DINP exerts its effects on rodent liver through a known threshold-based mechanism of little, if any, relevance to humans, a highly conservative risk assessment can be conducted using a NOAEL uncertainty factor approach. Chronic rodent NOELs range from about 100 to 400 mg/kg BW. They were based on end points such as increased liver weight and changes in liver pathology that are early indicators of peroxisome proliferation but should not be considered adverse. Application of a 100-fold UF yields ADI ranging from 1.0 to 4 mg/kg BW.[1]

Reproductive Toxicity.

Embryotoxicity. DINP has been added to the diet of parental male and female Sprague-Dawley rats at concentrations of 0.2, 0.4, or 0.8%. The treatment caused no changes in survival, reproductive organ weights or histopathology, male mating, male and female fertility, female fecundity, or female gestational indices in either generation. There were no significant differences in survival, clinical observation for offspring in F_1 or F_2 generations. DINP did not produce evidence of reproductive or fertility effects under the conditions of this study.[8]

Rats were gavaged with 0.04, 0.2, or 1.0 g/kg BW from gestation days 6 to 15. DINP showed fetal effects of borderline significance at 1.0 g/kg BW.[9]

Teratogenicity. DINP were administered by gavage to mated Sprague-Dawley rats at doses of 0.1, 0.5, and 1.0 g/kg BW on gestation days 6 through 15. Only mild maternal and developmental effects were observed at the maternally toxic exposure level of 1.0 g/kg BW. Maternal BW gain and food consumption were significantly reduced during the exposure period. No treatment-related morphological changes were noted except for an increased frequency of seventh cervical and rudimentary lumbar ribs. Both maternal and developmental NOAELs were therefore established at 0.5 g/kg BW. Authors concluded that DINP is not teratogenic or a selective developmental toxicant.[10]

Gonadotoxicity. DINP did not affect sperm in the testis or epididymis after chronic exposure even at the dose of 8000 ppm in the diet.[1,6]

Mutagenicity. DINP appeared to be non-genotoxic in a battery of bacterial and mammalian cell assyas.[11-13]

In vitro genotoxicity. DINP was negative in *Salmonella* mutagenicity assay with and without *S9* metabolic activation system.[015]

Carcinogenicity. Prolonged high-level exposure of rodents to DINP leads to an increased incidence of liver tumors (adenomas and renal cell carcinomas in male rats). This neoplasm is believed to proceed via α-2u-globulin mechanism, one that is specific for male rats and regarded as not relevant to humans. The chronic cancer and noncancer effects of DINP on rodent liver are consistent with its known action as a peroxisome proliferator.[1,6]

In a peroxisome induction study, DINP is found to be no more carcinogenic than the weak liver carcinogen di(2-ethylhexyl) phthalate.[14] There were no treatment-related nonneoplastic or neoplastic lesions found in the above-cited study.[5] However, signs of mononuclear cell leukemia were seen in the two higher doses (these findings are unlikely to be relevant to man).

Chemobiokinetics. DINP is readily hydrolyzed to *isononyl phthalate* and correspondent alcohol. These compounds can be absorbed to yield a number of oxidized metabolites.[15,16]

Regulations. ***U.S. FDA*** (1998) approved the use of DINP in plasticizers for polymeric substances in the manufacture of articles or components of articles intended for use in producing, manufacturing, packaging, processing, preparing, treating, packing, transporting, or holding food in accordance with the conditions prescribed in 21 CFR part 178.3740 only at levels not exceeding 43% by weight of the permitted polyvinyl chloride homo-/or copolymers used in contact with food only of the specified type at temperature not exceeding room temperature (thickness of polymers exceeded 0.005 inch).

Recommendations.

US Consumer Product Safety Commission (December 1998) decided against recommending a ban of soft plastic toys but asked toy manufacturers voluntarily to remove phthalates from teethers and rattles.[17]

EU Directive (1999) proposed to ban all phthalates except DINP in soft polyvinyl chloride toys. DINP can only be used in toys intended to be mouthed by children under 36 months of age if its rate of migration does not exceed 1.2 mg/10 cm^2 area in 3 hours.[18] Nevertheless, a review and risk assessment published by Wilkinson and Lamb (1999) showed with a high degree of confidence, that the toxicology of DINP and its rate of migration from soft polyvinyl chloride toys during child mouthing activities does not present a significant risk of adverse health effects in children. The scientific evidence supports the continued use of DINP as a plasticizer in children's products.[1]

Canada (1998). PTDI: 292 μg/kg BW.

References:

1. Wilkinson, C. F. and Lamb, J. C., IV, The potential health effects of phthalate esters in children's toys: A review and risk assessment, *Regul. Toxicol. Pharmacol.*, 30, 140, 1999.
2. Bird, M. G., Lington, A. W., and Cockrell, B. A., Subchronic and chronic oral studies of diisononyl phthalate (DINP) in F-344 rats: Effects on hepatic peroxisome induction, *Toxicologist*, 7, 56, 1986.
3. Lington, A. W., Gray, T. J. B., Evans, J., Lake, B., and Moran, B., Short-term feeding studies assessing the testicular effects of nine plasticizers in the F344 rat, *Acta Pharmacol. Toxicol.*, 73 (Suppl. 11), 132, 1993.
4. Shellenberger, T. E., Kowalski, J. J., Unwin, S., Grandjean, C., Carter, J., and Hodgson, J. R., Comparative 28-day oral toxicity of selected phthalate esters, *Toxi cologist*, 3, 157, 1983.
5. Lington, A. W., Bird, M. G., Plutnik, R. T., Stubblefield, W. A., and Scala, R. A., Chronic toxicity and carcinogenicity evaluation of diisononyl phthalate in rats, *Fundam. Appl. Toxicol.*, 36, 79, 1997.
6. Butala, J. H., Moore, M. R., Cifone, M. A., Bankston, J. R., and Astill, B., Oncogenicity study of di(isononyl)phthalate in rats, *Toxicologist*, 30, A1031, 202, 1996.
7. Butala, J. H., Moore, M. R., Cifone, M. A., Bankston, J. R., and Astill, B., Oncogenicity study of di(isononyl)phthalate in mice, *Toxicologist*, 36, A879, 173, 1997.

8. Nikiforov, A. I., Keller, L. H., and Harris, S. B., Lack of transgenerational reproductive effects following treatment with diisononyl, phthalate (DINP), in *Toxicologist*, SOT 1996 Annual Meeting, Abstract No 608, Issue of *Fundam. Appl. Toxicol.*, 30, Part 2, March 1996.
9. Hellwig, J., Freudenberger, H., and Jackh, R. Differential prenatal toxicity of branched phthalate esters in rats, *Food Chem. Toxicol.*, 35, 501, 1997
10. Waterman, S. J., Ambroso, J. L., Keller, L. H., Trimmer, G. W., Nikiforov, A. I., and Harris, S. B. Developmental toxicity of di-isodecyl and di-isononyl phthalates in rats, *Reprod. Toxicol.*, 13, 131, 1999
11. *In vitro chromosomal aberration assay in Chinese hamster ovary cells* (unpublished data), Exxon Biomedical Sciences, Inc. (EBSI), 1966, *cit. in* Lington et al., 1997.
12. Omori, Y., Recent progress in safety evaluation studies on plasticizers and plastics and their controlled use in Japan, *Environ. Health Perspect.*, 17, 203, 1976.
13. Zeiger, E., Haworth, S., Mortelmans, K., and Speck, W., Mutagenicity testing of di(2-ethylhexyl)phthalate and related chemicals in *Salmonella, Environ. Mutagen.*, 7, 213, 1985.
14. Barber, E. D., Astill, B. D., Moran, E. J., Schneider, B. F., Gray, T. J., Lake, B. G., and Evans, J. G., Peroxisome induction studies on seven phthalate esters, *Toxicol. Ind. Health*, 3, 7, 1987.
15. El-Hawari, M., Murrill, E., Stoltz, M., Pallas, F., Lington, A. W., and Baldwin, J. K., Disposition and metabolism of diisononyl phthalate (DINP) in Fischer 344 rats: Single dosing studies, *Toxicologist*, 5, 237, 1985.
16. Lington, A, W., Baldwin, J. K., Murril, E., Stoltz, M., Pallas, F., and El-Hawari, M., Disposition and metabolism of diisononyl phthalate (DINP) in Fischer 344 rats, Multiple dosing studies, *Toxicologist*, 5, 238, 1985.
17. Babich, M. A., *The Risk of Chronic Toxicity Associated with Exposure to Diisononyl Phthalate (DINP) in Children Products*, Directorate for Epidemiology and Health Sciences, US Consumer Product Safety Commission, Washington, D.C., 1998.
18. *Draft Proposal to Amend Directive 76/769/EEC*, European Parliament and the Council of the European Union, 1998.

PHTHALIC ACID, DIISOOCTYL ESTER

Molecular Formula. $C_{24}H_{38}O_4$
M = 390.64
CAS No 27554-26-3
RTECS No TI1300000
Abbreviation. DIOP.

Synonym. Diisooctyl phthalate.

Properties. Oily, colorless liquid with a characteristic weak odor. Solubility in water is 0.1% (25°C).

Applications. Used as a plasticizer in the production of polyvinyl chloride and other polymeric materials intended for contact with food.

Acute Toxicity. LD_{50} is reported to be 22 g/kg BW in rats[1] and 2.8 g/kg BW in mice.[2]

Reproductive Toxicity and **Carcinogenicity** effects were not found during 15 to 21-month oral exposure of Wistar rats. The doses of DIOP tested in this study are 100 to 500 mg/kg in the diet.[05]

Chemobiokinetics. DIOP was added to the diet of Sprague-Dawley rats, beagle dogs, and miniature pigs at a dose of 50 mg/kg BW for 21 to 28 days. Before oral administration, a single dose was given, labeled with ^{14}C. Half a dose was excreted in feces, half in the urine. In the dogs, excretion occurred predominantly in the feces, in pigs, mainly via the urine.[3]

Note. The isomer *di(1-methylheptyl) phthalate*, known incorrectly by the name *dicapryl phthalate*, is also of low toxicity; daily exposure to 2.5 g/kg BW caused no signs of intoxication in rats.

Regulations. ***U.S. FDA*** (1998) approved the use of DIOP (1) in adhesives as a component of articles intended for use in packaging, transporting, or holding food in accordance with the conditions prescribed by 21 CFR part 175.105; and (2) as an ingredient in the manufacture of resinous and polymeric coatings of food-contact surface of articles intended for use in producing, manufacturing, packing, transporting, or holding food in accordance with the conditions prescribed by 21 CFR part 175.300.

References:

1. Krauskopf, L. G., Studies on the toxicity of phthalates via ingestion, *Environ. Health. Perspect.*, 3, 61, 1973.
2. Antonyuk, O. K., Toxicity of phthalic esters. Review of the literature, *Gig. Truda Prof. Zabol.*, 1, 32, 1975 (in Russian).
3. Ikeda, G. J., Sapienza, P. P., Convillion, J. L., et al., Distribution and excretion of two phthalate esters in rats, dogs, and miniature pigs, *Food Cosmet. Toxicol.*, 16, 409, 1978.

PHTHALIC ACID, DI(2-METHOXYETHYL) ESTER

Molecular Formula. $C_{14}H_{18}O_6$
M = 282.28
CAS No 117-82-8
RTECS No TI1400000
Abbreviation. DMEP.

Synonyms. 1,2-Benzenedicarboxylic acid, bis(2-methoxyethyl) ester; Bi(2-metoxyethyl) phthalate; Di(2-methoxyethyl) phthalate; Phthalic acid, bis(2-methoxyethyl) ester; Phthalic acid, methylglycol ester; 2-Methoxyethyl phthalate; Methyl cellosolve phthalate.

Properties. Colorless, oily liquid with a faint odor. Water solubility is 0.8% (20°C), insoluble in organic oils.

Applications. Used as a plasticizer in the production of nitro- and acetyl cellulose, polyvinyl acetate, polyvinyl chloride, and polyvinylidene chloride intended for contact with food or drink, giving polymeric materials good light resistance.

Acute Toxicity. LD_{50} is reported to be 2.75 g/kg BW (mixed with a small amount of ethyl alcohol) or 4.4 g/kg BW in rats, 3.2 to 6.4 g/kg BW in mice, and 1.6 to 3.2 g/kg BW in guinea pigs.[03] Administration caused no visible signs of poisoning. Animals that did not die in the first 36 hours recovered. Histology examination failed to reveal changes in the viscera (IARC 19-392).

Long-term Toxicity. Rats were given 0.3 to 0.9 g/kg BW in their diet for 21 months. Treatment with the highest dose caused decrease of BW gain. There were no visible signs of poisoning and no deviations in the relative weights of the visceral organs.

Reproductive Toxicity.

DMEP is known to be the most ***teratogenic*** and ***embryotoxic*** of all the phthalates. The rat embryos do not hydrolyze DMEP to *monoester*. However, after a single *i/v* administration to pregnant Wistar-Porton rats on day 14 of gestation, ester and diester are found in the fetal tissues. DMEP and *2-methoxyethanol*, derived from DMEP, seem to be teratogens.[1]

DMEP was administered to Wistar rats on day 12 of gestation. The treatment caused hydronephrosis, heart defects, and short limb and tail. This suggests DMEP to be a proximate teratogen.[2] Embryotoxic and teratogenic effects of DMEP are reported by Singh et al.[2] DMEP passes rapidly through the placenta. It causes decreased zinc content in the fetal tissues.

Chemobiokinetics. DMEP is likely to be hydrolyzed to *2-methoxyethanol*, which in turn is metabolized to *methoxyacetic acid*.[3]

References:

1. Campbell, J., Holt, D., and Webb, M., Dimethoxyethyl phthalate: Teratogenicity of the diester and its metabolites in the pregnant rats, *J. Appl. Toxicol.*, 4, 35, 1984.
2. Singh, A. R., Lawrence, W. H., and Autian, J., Teratogenicity of phthalate esters in rats, *J. Pharmacol. Sci.*, 61, 51, 1972.
3. Ritter, E. J., Scott, W. J., Randall, J. L., et al., Teratogenicity of dimethoxyethyl phthalate and its metabolites methoxyethane and methoxyacetic acid in the rat, *Teratology*, 32, 25, 1985.

PHTHALIC ACID, DIMETHYL ESTER

Molecular Formula. $C_{10}H_{10}O_4$
M = 194.19

CAS No 131-11-3
RTECS No TI1575000
Abbreviation. DMP.

Synonyms and **Trade Names.** Avolin; 1,2-Benzenedicarboxylic acid, dimethyl ester; Dimethyl 1,2-benzenedicarboxylate; Dimethylbenzene-*o*-dicarboxylate; Dimethyl *o*-phthalate; Methyl phthalate; Mipax; Palatinol M; Phthalic acid, dimethyl ester; Repeftal; Solvanom; Solvarone; Unimoll DM.

Properties. Clear, colorless, oily liquid without or with a faint odor. Solubility in water is 0.45% (20°C). Soluble in alcohol and fats. Taste threshold concentration is 3.5 mg/l.[1]

Applications. DMP is used as a plasticizer for cellulose esters, particularly for cellulose acetate and, less frequently, for cellulose nitrate. It is used in mixtures with other plasticizers, for example, with triphenyl phosphate, triacetine, diethyl- and dibutyl phthalates. DMP is used as a plasticizer for polyvinyl chloride, polyvinyl acetate, rubber, polyacrylates,and polymethyl methacrylates intended for contact with food or drink.

Acute Toxicity. LD_{50} is 5.5 to 7.2 g/kg BW in mice, 6.9 to 8.7 g/kg BW in rats, 5.2 g/kg BW in rabbits, and 2.4 to 2.9 g/kg BW in guinea pigs.[1-4,03] The toxicity manifestations included depression, apathy and adynamia. Three to four days after exposure the animals develop labored respiration and convulsions and die within 2 to 3 hours. Gross pathology examination revealed congestion in the visceral organs and cerebral meninges.

Repeated Exposure revealed marked cumulative properties. Administration of 1/5 and 1/20 LD_{50} has shown K_{acc} to be 2.9 (by Lim).[5] The treatment with 1/10 LD_{50} and 1/50 LD_{50} resulted in decreased BW gain, functional disturbances in the liver, kidneys and NS, changes in hematology analysis, in the morphology and relative weights of the visceral organs.[6]

In a 30-day study, increased ascorbic acid content in the suprarenals was marked on day 5 but not on day 30 of the exposure. Animals display reduction in liver and brain cholinesterase activity, behavioral changes, and erythrocytosis. Doses below 1/50 LD_{50} do not produce histological changes. The NOEL was identified to be 0.17 mg/kg BW.[1]

Long-term Toxicity. Young rats received 8.0% DMP in their feed. The treatment caused chronic renal impairment. A dose of 4.0% caused retardation of BW gain while a 2.0% dose appeared to be harmless.[05]

Reproductive Toxicity.

Gonadotoxicity. DMP is likely to produce no testicular injury.[7] A 4-day administration of 7.2 mmol/kg BW to rats had no gonadotoxic effect. *I/g* administration at the LD_{50} level did not alter spermatozoa motility time.[8]

Embryotoxicity. Implantation process was affected after DMP administration on days 3, 6, and 9 of gestation. The treatment caused a reduction in the fetal weights and sizes; bleeding during parturition led to maternal death in several cases.[9,10] Significant embryolethality and death of neonates were observed on dermal application of 1.25 g DMP/kg BW in pregnant rats.[2]

Timed-mated rats received 0.25, 2.5, and 5.0% DMP in their feed from gestation days 6 to15 (that corresponds to approximately 0.2 to 4.0 g DMP/kg BW). The treatment with 5.0% DMP increased relative maternal liver weight and reduced BW gain during treatment. The NOAEL for maternal toxicity was 1.0% DMP in the feed. DMP treatment caused no effect on any parameter of embryo/fetal development. This results do not support the conclusion of other investigators that DMP is a potent developmental toxicant.[11]

Teratogenicity. No teratogenic effects were reported in mice and rats.[12] According to other data, DMP is shown to produce an increased incidence of skeletal abnormalities in neonatal rats.[13]

Allergenic Effect was not observed.[1] Contact sensitizing potential of dimethyl isophthalate is not found in maximized guinea-pig assays.[14]

Mutagenicity.

In vitro genotoxicity. DMP was found to be positive in *Salmonella typhimurium* strain *TA 100* mutagenicity assay.[15] According to other data, DMP was negative in *Salmonella typhimurium* strains *TA98, TA100, TA1535,* and *TA1537* with and without *S9* fraction. It induced SCE in Chinese hamster ovary cells in the presence, but not in the absence of metabolic activation.[16]

In vivo cytogenetics. Skin application of 20 to 40% DMP solution or oral exposure to pure DMP (200 to 2000 mg/kg BW) for 1 to 1.5 months produced neither cytogenic effect nor increased DLM and changes in fertility of animals.[8]

There was no increase in the frequency of DLM in mice given 1250 mg DMP/kg BW orally for 5 weeks even 10 weeks after experiment onset.[17]

Carcinogenicity. No carcinogenic effect was observed on dermal application.[16] Nevertheless, Kozumbo et al. considered DMP to represent a great mutagenic and carcinogenic hazard.[15]

Carcinogenicity classification.

U.S. EPA: D.

Chemobiokinetics. Metabolism involves hydrolysis to the monoester with subsequent oxidation of the remaining alkyl chain. *Monomethyl phthalate, phthalic acid, benzoic acid,* and *formaldehyde* are found in the blood and urine of rats after skin application of a dose of 1.5 mg/kg BW.[18] *Phthalic acid* was found to be a minor metabolite (14%) among the total urinary metabolites.

Regulations.

U.S. FDA (1998) regulates the use of DMP (1) in adhesives as a component of articles intended for use in packaging, transporting, or holding food in accordance with the conditions prescribed by 21 CFR part 175.105; (2) as a solvent in polyester resins intended for use in articles coming in contact with food; (3) as an ingredient of acrylic and modified acrylic plastics for single and repeated use in contact with food in accordance with the conditions prescribed in 21 CFR part 177.1010; and (4) in the manufacture of cross-linked polyester resins which may be used as articles or components of articles intended for repeated use in contact with food in accordance with the conditions prescribed by 21 CFR part 177.2420.

Great Britain (1998). *Dimethyl isophthalate* is authorized without time limit for use in the production of polymeric materials and articles in contact with food or drink or intended for such contact. The specific migration of *dimethyl isophthalate* shall not exceed 0.05 mg/kg.

Standards. *Russia* (1995). MAC: 0.3 mg/l; PML: 0.5 mg/l. MAC for dimethyl isophthalate: 0.1 mg/l.

References:

1. Shatinskaya, I. G., *Comparative Description of Methods of Studying Accumulation in Resolving Problems of the Safe Regulation of Harmful Chemicals*, Author's abstract of thesis, Kiyv, 1986, 24 (in Rus sian).
2. Gleyberman, S. E. et al., in *Coll. Sci. Papers of Research Institute of Vaccines and Sera*, Moscow, 24, 303, 1975 (in Russian).
3. Antonyuk, O. K., in *Hygiene and Toxicology*, Proc. Conf. Junior Scientists, Kiyv, 1967, 90 (in Russian).
4. Woodward, K. N., Smith, A. M., Mariscotti, S. P., and Tomlinson, N. J., *Review of the Toxicity of the Esters of o-Phthalic Acid (Phthalate Esters),* Toxicity Review, 14, Health and Safety Executive, HMSO, London, 1988, 103.
5. Timofiyevskaya, L. A., Nikitenko, T. K., Govorchenko, V. I., et al., Correlation between general and specific effects of biological activity of phthalic esters, in *Basic Problems of Delayed Effects of Exposure to Industrial Poisons,* Proc. Research Inst. Occup. Diseases., A. K. Plyasunov and G. M. Pashkova, Eds., Moscow, 1976, 39 (in Russian).
6. Stankevich, V. V. and Zarembo, O. K., in *Current Problems of Applications of Polyvinylchloride Materials*, All-Union Instit. Med. Information, 1978, 4 (in Russian).
7. Foster, P. M. D., Lake, B. G., Cook, M. W., et al., Structure - activity requirements for the induction of testicular atrophy by butyl phthalates in immature rats: Effect on testicular zinc content, *Adv. Exp. Med. Biol.*, 136, 445, 1981.
8. Pashkova, G. A., Comparative risk assessment of gonadotoxic effects of phthalic and organophosphorous plasticizers, in *Basic Problems of Delayed Effects of Exposure to Industrial Poisons*, Proc. Research Institute Occup. Diseases., A. K. Plyasunov and G. M. Pashkova, Eds., Moscow, 1976, 43 (in Russian).
9. Peters, L. W. and Cook, R. M., Effect of phthalate esters on reproduction in rats, *Environ. Health Perspect.*, Exp. Issue No 3, 91, 1973.

10. Bower, R. K., Haberman, S., and Minton, P. D., Teratogenic effects in the chick embryo caused by esters of phthalic acid, *J. Pharmacol. Exp. Ther.*, 171, 312, 1970.
11. Field, E. A., Price, C. J., Sleet, R. B., George, J. D., Marr, M. C., Myers, C. B., Schwetz, B. A., and Morrissey, R. E., Developmental toxicity evaluation of diethyl and dimethyl phthalate in rats, *Teratology*, 48, 33, 1993.
12. Plasterer, M. R., Bradshaw, W. S., Booth, G. M., et al., Developmental toxicity of nine selected compounds following prenatal exposure in the mouse. Naphthalene, *p*-nitrophenol, sodium selenite, dimethyl phthalate, ethylene thiourea, and four glycol ether derivatives, *J. Toxicol. Environ. Health*, 15, 25, 1985.
13. Singh, A. R., Lawrence, W. H., and Autian, J., Teratogenicity of phthalate esters in rats, *J. Pharmacol. Sci.*, 61, 51, 1972.
14. Basketter, D. A. and Scholes, E. W., Comparison of the local lymph node assay with the guinea-pig maximization test for the detection of a range of contact allergens, *Food Chem. Toxicol.*, 30, 65, 1992.
15. Kozumbo, W. J., Kroll, R., and Rubin, R. J., Assessment of the mutagenicity of phthalate esters, *Environ. Health Perspect.*, 45, 103, 1982.
16. *Toxicology and Carcinogenesis Studies of Diethylphthalate in F344/N Rats and B6C3F$_1$ Mice (Dermal Studies) with Dermal Mixtations (Promotion study of Diethylphathale and Dimethylphthalate in Male Swiss Mice),* NTP Technical Report Series No 427, Research Triangle Park, NC, May 1995.
17. Yurchenko, V. V. and Gleyberman, S. E., *Med. Parasitol. & Parasitol. Diseases*, 1, 58, 1980 (in Russian).
18. Surina, T. Ya., Gleiberman, S. E., and Nikolayev, G. M., Metabolism of dimethyl phthalate on skin applications, *Med. Parazitol. & Parazitol. Diseases*, 4, 67, 1984 (in Russian).

PHTHALIC ACID, DINONYL ESTER

Molecular Formula. $C_{26}H_{42}O_4$
M = 418.6
CAS No 84-76-4
RTECS No TI1800000
Abbreviation. DNP.

Synonyms and **Trade Names.** 1,2-Benzenedicarboxylic acid, dinonyl ester; Bisoflex 91 and DNP; Dinonyl 1,2-benzenedicarboxylate; Di-*n*-nonyl phthalate; Phthalic acid, dinonyl ester Unimoll DN.

Properties. Colorless, almost odorless, oily liquid. Poorly soluble in water, soluble in alcohol.

Applications. Used as a plasticizer for polyvinyl chloride and other plastics intended for contact with food, and in the manufacture of plastizols and pastes.

Migration Data. Diethyl ether extracts of food-contact polyvinyl chloride were analyzed by GC-MS methods: di(2-ethylhexyl) phthalate, dinonyl phthalate and other phthalates were present in relatively large quantities (10 times higher than the internal standards). However, 68% of the extracts contained no peaks higher than the internal standards.[1]

Acute Toxicity. LD_{50} was 18 to 21.5 g/kg BW in rats, mice, and guinea pigs.[2,3]

Repeated Exposure failed to reveal cumulative properties. K_{acc} is 10 (by Lim).[3]

There were no toxic manifestations in chickens on five administrations of a 1.0 mg/kg BW dose.[2]

Short-term Toxicity. Cats were given 2.0 mg/kg BW oral doses for 3 months. The treatment did not cause retardation of BW gain, changes in the hematology analysis, or in liver and kidney functions.[4]

Chemobiokinetics. DNP was readily absorbed, distributed, metabolized, and excreted by all three routes of exposure (as all phthalates).

Regulations. ***U.S. FDA*** (1998) approved *diisononyl phthalate* for use as a plasticizer in polymeric substances in the manufacture of food-contact articles only at levels not exceeding 43% by weight of the permitted polyvinyl chloride homo-/or copolymers used in contact with food only of the specified type at temperature not exceeding room temperature (thickness of polymers exceeded 0.005 inch).

References:

1. van Lierop, J. B., Enforcement of food packaging legislation, *Food. Addit. Contam.*, 14, 555, 1997.
2. Fishbein, I. and Albro, P. W., Chromatographic and biological aspects of the phthalate esters, *J. Chromatogr.*, 70, 365, 1972.
3. Timofiyevskaya, L. A., Nikitenko, T. K., Govorchenko, V. I., et al., Correlation between general and specific effects of biological activity of phthalic esters, in *Basic Problems of Long-term Effects of Exposure to Industrial Poisons*, Proc. Research Inst. Occup. Diseases., A. K. Plyasunov and G. M. Pashkova, Eds., Moscow, 1976, 39 (in Russian).
4. Hoffmann, H. T., Z. *Arbeitsmed.-Arbeitsschutz.*, 11, 240, 1961.

PHTHALIC ACID, DIPHENYL ESTER

Molecular Formula. $C_{20}H_{14}O_4$
M = 320.36
CAS No 84-62-8
RTECS No TI1935000
Abbreviation. DPP.

Synonyms. 1,2-Benzenedicarboxylic acid, diphenyl ester, Diphenyl phthalate; Phenyl phthalate.

Properties. White powder with a yellowish tint. Poorly soluble in water, alcohol and fats.

Applications. Used as a plasticizer in the manufacture of nitro- and ethyl cellulose, polystyrene, phenol and vinyl resins intended for contact with food or drink. Forms strong films with nitro- and ethyl cellulose.

Acute Toxicity. LD_{50} is 8.0 g/kg BW in rats.

Short-term Toxicity. Dogs were given 1.2 and 5.0% DPP in their diet for 90 days. The higher dose level caused decreased BW gain due to loss of appetite in animals (Erikson, 1965).

Chemobiokinetics. Intestinal absorption is insignificant. 90% DPP is passed with the feces, 4.0 to 5.0% DPP is found in the urine as *phthalic acid* and the remainder in the form of *free* or *bound phenols.*

Regulations. ***U.S. FDA*** (1998) approved the use of DPP (1) in adhesives as a component of articles intended for use in packaging, transporting, or holding food in accordance with the conditions prescribed by 21 CFR part 175.105; (2) in the manufacture of cellophane for food packaging in accordance with the conditions prescribed by 21 CFR part 177.1200; (3) for use alone or in combination with other phthalates, in plastic film or sheet prepared from polyvinyl acetate, polyvinyl chloride, and/or vinyl chloride copolymers. Such film or sheet shall be used in contact with food at temperatures not to exceed room temperature and shall contain no more than 10% by weight of the total phthalates, calculated as phthalic acid; and as (4) plasticizers for polymeric substances in the manufacture of articles or components of articles intended for use in producing, manufacturing, packaging, processing, preparing, treating, packing, transporting, or holding food in accordance with the conditions prescribed by 21 CFR part 178.3740.

Reference:

Shibko, S. I. and Blumenthal, H., Toxicology of phthalic acid esters used in food-packaging material, *Environ. Health Perspect.*, 3, 131, 1973.

PHTHALIC ACID, DIUNDECYL ESTER

Molecular Formula. $C_{30}H_{50}O_4$
M = 474.80
CAS No 3648-20-2
RTECS No TI1980000
Abbreviation. DUP.

Synonyms and **Trade Name.** 1,2-Benzenedicarboxylic acid, diundecyl ester; Diundecyl phthalate; Santicizer 711.

Applications. A plasticizer in the manufacture of food-contact polyvinyl chloride plastics.

Properties. Resistant to migration from polymers (Kayser et al., 1982).

Acute Toxicity. LD exceeded 20 g/kg BW in rats.

Repeated Exposure. F344 rats received 0.3, 1.2, and 2.5% DUP in their diet for 21 days. The treatment produced a significant decrease in serum triglycerides and cholesterol for 1.2 and 2.5% dietary males; an increase in fat deposition has been observed in liver of males. The only effect on hepatic lipids in females appeared to be a reduction of the normal distribution of periportal neutral lipid.[1]

Mutagenicity.

In vitro genotoxicity. Negative in four *Salmonella typhymurium* strains *TA98, TA 100, TA 1535*, and *TA 1537* in the presence and absence of liver fraction *S9*.[015]

Carcinogenicity. In peroxisome induction study, DUP was found to be no more carcinogenic than the weak liver carcinogen di(2-ethylhexyl) phthalate.[2]

Chemobiokinetics. Following absorption, DUP has been quickly distributed over the body, and readily excreted.[03]

Regulations. ***U.S. FDA*** (1998) approved the use of DUP as a component of poly(*p*- methylstyrene) and rubber-modified poly(*p*-methylstyrene) intended for use in contact with food, subject to the provisions of 21 CFR part 177.1635.

References:

1. *A 21-day Feeding Study of Diundecyl Phthalate to Rats: Effects on the Liver and Liver Lipids,* The British Ind. Biol. Res. Assoc., 1985. Available on Fiche No OTS0509538, EPA Doc. No 408526207.
2. Barber, E. D., Astill, B. D., Moran, E. J., Schneider, B. F., Gray, T. J., Lake, B. G., and Evans, J. G., Peroxisome induction studies on seven phthalate esters, *Toxicol. Ind. Health*, 3, 7, 1987.

PHTHALIC ACID, ETHYL ESTER, ESTER with ETHYL GLYCOLATE

Molecular Formula. $C_{14}H_{16}O_6$

M = 280.27

CAS No 84-72-0

RTECS No TI2060000

Abbreviation. EPEG.

Synonyms and **Trade Name.** Carbethoxymethyl ethyl phthalate; Diethyl *o*-carboxybenzoyl oxyacetate; 2-Ethoxy-2-oxoethyl-1,2-benzenedicarboxylic acid, ethyl ester; Ethyl carbethoxymethyl phthalate; Ethyl-(ethoxycarbonylmethyl) phthalate; Ethylphthalyl ethylglycolate; Santicizer E-15.

Properties. Almost colorless, oily liquid with a very faint odor. Water solubility is 0.17% (30°C).

Applications. Used as a plasticizer for polyvinyl chloride, cellulose acetate, nitrocellulose, ethylcellulose, polyvinyl acetate, polyvinylidene chloride, acrylonitrile rubbers intended for contact with food or drink. Renders materials heat- and light-resistant.

Acute Toxicity. LD_{50} was not attained.

Repeated Exposure. Administration of 10% EPEG in the diet caused death in rats within 7 to 15 days.[1]

Long-term Toxicity. Rats received 0.05, 0.5, and 5.0% EPEG in their feed for 2 years. Only the highest dose caused mortality. The lower doses exerted no visible harmful effect. Dogs were given up to 250 mg/kg BW for 12 months. No retardation of BW gain or signs of poisoning were reported. The LOEL of 2500 mg/kg BW, and the NOEL of 250 mg/kg BW were established in this study.[2]

Chemobiokinetics. Calcium oxalate crystals were found in the kidneys.

Regulations. ***U.S. FDA*** (1998) approved the use of EPEG (1) in adhesives as a component of articles intended for use in packaging, transporting, or holding food in accordance with the conditions prescribed by 21 CFR part 175.105; (2) in the manufacture of resinous and polymeric coatings in a food-contact surface of articles intended for use in producing, manufacturing, packing, transporting, or holding food in accordance with the conditions prescribed by 21 CFR part 175.300; (3) in the manufacture of resinous and polymeric coatings for polyolefin films for food-contact surface of articles intended for use in producing, manufacturing, packing, processing, preparing, treating, packaging, transporting, or holding food in accordance with the conditions prescribed by 21 CFR part 175.320.

References:

1. Fishbein, L. and Albro, P. W., Chromatographic and biological aspects of phthalic esters, *J. Chromatogr.*, 70, 365, 1972.
2. *Toxicological Evaluation of some Enzymes, Modified Starches and Certain Other Substances*, The 15th Meeting of the Joint FAO/WHO Expert Committee on Food Additives, WHO Food Additives Series No 1, Geneva, 1972.

PHTHALIC ACID, HEPTYLNONYL ESTER

Molecular Formula. $C_{24}H_{38}O_4$

M = 390.54

Abbreviation. HNP.

Synonym and **Trade Name.** Alfanole; Heptylnonyl phthalate.

Properties. Clear, colorless liquid.

Applications. Used as a plasticizer in the polyvinyl chloride production.

Acute Toxicity. Administration of 20 g/kg BW dose of *Alfanole* caused diarrhea in rats and mice.

Short-term Toxicity. Rats exposed to 60 mg/kg BW for 3 months developed no signs of toxicity.

Long-term Toxicity, Reproductive Toxicity, and **Carcinogenicity.** Wistar rats were gavaged with doses of 0.5, 1.0, and 3.0 mg HNP/day for 15 to 21 months. Gross pathology examination failed to reveal treatment-related lesions. In a 5-generation study, reproductive function of animals was unaffected. No carcinogenic effect was reported.[05]

Reference:

Gaunt, J., Colley, J., Grasso, P., et al., Acute (rat and mouse) and short-term (rat) toxicity studies on dialkyl-79 phthalate (a mixture of phthalate esters of alcohols having 7-9 carbon atoms), *Food Cosmet. Toxicol.*, 6, 609, 1968.

PHTHALIC ACID, METHYLCYCLOHEXYL ESTER

Molecular Formula. $C_{15}H_{18}O_4$

M = 262.33

Synonym. Methylcyclohexyl phthalate

Properties. Viscous, light-yellow liquid with a faint odor. Poorly soluble in water, soluble in alcohol.

Applications. Used as a plasticizer (up to 25%) in the production of nitrocellulose, polyvinyl chloride, polyvinyl acetate, polystyrene, polymethyl methacrylate, intended for contact with food or drink. Used to obtain cellophane.

Toxicity is likely to be insignificant.

PLASTIAZAN-1, PLASTIAZAN-25, PLASTIAZAN-30

Abbreviation. P-1, P-25, P-30.

Composition. A mixture of the esters of C_5-C_6 or C_7-C_9 fatty acids and ethylene glycol.

Properties. Oily liquids with a faint odor.

Applications. Used as plasticizers in the production of polyvinyl chloride; substitutes for dibutyl and dioctyl phthalates.

Acute Toxicity. In mice, the LD_{50} values are 25.5 g/kg BW (P-1), 19.2 g/kg BW (P-25), and 15.4 g/kg BW (P-30); in rats these values are 50, 35, and 30 g/kg BW, respectively. Poisoning is accompanied by adynamia with subsequent death within 2 to 5 days. The acute action threshold of P-1 is 1.5 g/kg BW for rabbits and 1.0 g/kg BW for mice.

Repeated Exposure. Mice were dosed by gavage with 1/10 LD_{50} for a month. P-1 and P-30 caused an increase in muscular strength from the second decade of the experiment. The neuro-muscular excitability threshold was increased.

Reference:

Toxicologic Evaluation of Some New Plasticizers, Additives and Oil Coolants, Azerb. State Publisher, Baku, 1979, 21 (in Russian).

PLASTIAZAN-41

Composition. A mixture of the esters of C_7-C_9 fatty acids and 2-phenoxyethanol.

Applications. Used as a plasticizer in the production of polyvinyl chloride and perchlorovinyl.

Acute Toxicity. The LD_{50} is 16.5 g/kg BW in mice, and 13.7 g/kg BW in rats. Poisoning is accompanied by apathy and adynamia. Death is within a week. Histology examination showed no changes in the viscera.

Repeated Exposure failed to reveal cumulative properties.

Reference:

Toxicologic Evaluation of Some New Plasticizers, Additives and Oil Coolants, Azerb. State Publisher, Baku, 1979, 25 (in Russian).

PLASTIAZAN-60

Composition. A mixture of monoesters of ethylene glycol and synthetic C_7-C_9 fatty acids as a result of these acids interacting with ethylene chlorohydrin.

Applications. Used as a plasticizer in the production of polyvinyl chloride; substitute for dibutylsebacate.

Acute Toxicity. No animals died after single administration. Poisoning is accompanied by apathy and adynamia. Histology examination showed no changes in the viscera.

Repeated Exposure. Daily administration of the maximum possible doses resulted in death of half the animals from a total dose of 151 g/kg BW.

Reference:

Toxicologic Evaluation of Some New Plasticizers, Additives and Oil Coolants, Azerb. State Publisher, Baku, 1979, 24 (in Russian).

PLASTICIZER PPA-4

Composition. Adipic acid, propylene glycol and butyl alcohol, derivative (polyester).

Properties. Viscous liquid. Water solubility is 0.25% at 20°C.

Applications. Used as a plasticizer in the production of polyvinyl chloride.

Acute Toxicity. LD_{50} is 11.5 g/kg BW in rats, and 13.5 g/kg BW in mice. Poisoning is accompanied by excitation with subsequent general inhibition. Death in the deep narcotic conditions.

Repeated Exposure revealed moderate cumulative properties. K_{acc} is 2.77 (by Lim). Habituation developed.

Reference:

Timofiyevskaya, L. A. and Mel'nikova, N. N., Toxicity of polyester plasticizers PPA, *Gig. Truda Prof. Zabol.*, 10, 89, 1982 (in Russian).

PLASTICIZER PPA-7

Properties. Polyester. Viscous liquid.

Applications. Used as a plasticizer in the production of polyvinyl chloride.

Acute Toxicity. LD_{50} is 14.0 g/kg BW in rats, and 8.5 g/kg BW in mice. Poisoning is accompanied by excitation with subsequent general inhibition. Death in the deep narcotic conditions.

Repeated Exposure revealed slight cumulative properties. K_{acc} is 4.22 (by Lim). The treatment led to retardation of BW gain, excitation, then decline in mobility. Habituation developed.

Reference:

Timofiyevskaya, L. A. and Mel'nikova, N. N., Toxicity of polyester plasticizers PPA, *Gig. Truda Prof. Zabol.*, 10, 89 1982 (in Russian).

PLASTICIZER PPA-12

Properties. Polyester. Viscous transparent liquid. Poorly soluble in water, readily soluble in alcohol.

Applications. Used as a plasticizer in the production of polyvinyl chloride.

Acute Toxicity. In rats, LD_{50} is 27 g/kg BW. Poisoning is accompanied by excitation. By the end of the first day after administration, there was adynamia, labored breathing, ataxia. Death occurs in lethargy condition.

Repeated Exposure failed to reveal cumulative properties. K_{acc} is 10 (by Lim). Habituation developed.

Reference:

Timofiyevskaya, L. A. and Mel'nikova, N. N., Toxicity of polyester plasticizers PPA, *Gig. Truda Prof. Zabol*., 10, 89, 1982 (in Russian).

1,2-PROPANEDIOL

Molecular Formula. $C_3H_8O_2$
M = 76.09
CAS No 57-55-6
RTECS No TY2000000
Abbreviation. 1,2-PD.

Synonyms. 1,2-Dihydroxypropane; 2-Hydroxypropanol; Isopropylene glycol; Methylethylene glycol; 1,2-Propylene glycol.

Properties. Colorless, odorless, viscous liquid with a sweet taste. Mixes with water and alcohol at all ratios. Organoleptic perception threshold is 340 mg/l.[01]

Applications. Used as a plasticizer and a solvent in the production of high-molecular-mass compounds.

Migration Data. Chocolates, boiled sweets, toffees, cakes and meat pies were wrapped in regenerated cellulose films that contained various mixtures of glycol softeners. It was shown that higher levels of migration occurred for PD than for triethylene glycol and the presence of coating reduced the migration of both softeners.[10]

Acute Toxicity. LD_{50} is 26.4 g/kg BW in rats, and 20.3 g/kg BW in mice.[1] Accidental administration of 6.0 ml 1,2-PD/kg BW to horses caused severe depression, ataxia, malodorous breath and feces.[11]

Repeated Exposure. Rats were given 5.0 and 10% aqueous solutions of 1,2-PD for 5 weeks. Administration resulted in increased liver weight and blood glucose level. The treatment reduced blood urea concertration and the erythrocyte count but did not affect the weights of other visceral organs.[2]

Male Sprague-Dawley rats received undiluted 1,2-PD at a dose of 4.0 mg/kg BW by gastric intubation for 30 days. The treatment produced no significant differences in plasma concentration of total phospholipids, cholesterol, triglycerides and free fatty acids, and in the liver concentration of phospholipids, triglycerides, and gangliosides. Liver total cholesterol was moderately, but significantly, increased.[3]

Short-term Toxicity. PDs are CNS and cardiac depressants of low toxicity; they produce arrhythmia and fibrillation.[9]

Rats received 3.28 g 1,2-PD/kg BW for 3 months. The treatment caused no impairment of their general condition and BW gain or alteration in the hematology analysis and renal function. Histology examination failed to reveal any changes in the visceral organs.[1]

Rats were given 50,000 ppm 1,2-PD in their diet for 105 days. The NOEL of 2500 mg/kg BW was established in this study.[5]

Long-term Toxicity. Dogs were dosed with 5.0 mg/kg BW for 2 years. The treatment increased the rate of erythrocyte hemolysis and lowered *Hb* content (reversible changes). A dose of 2.0 mg/kg BW appeared to be ineffective.[4]

No toxic manifestations were found to develop in rats given 200 to 2100 mg/kg BW for 2 years. The 25 mg/kg BW dose is suggested to be acceptable in man.[5]

Reproductive Toxicity.

Embryotoxicity. Use of 1,2-PD in cryopreservation of early human embryos revealed no adverse effect on embryo survival.[6]

Rats were fed 2.45 or 4.9% 1,2-PD in their diet. Doses of up to 1.77 ml/kg BW caused no significant effect on growth rate. Histologically, very slight liver damage, but no renal pathology was noted.[04]

Rats received 0.2 ml of 10%-solution during the first 10 days of pregnancy. The treatment produced no adverse fetal effects.[7]

Teratogenicity. 1,2-PD was not found to be teratogenic (Shepard, 1986).

Allergenic Effect was not observed.

Mutagenicity.

In vitro genotoxicity. 1,2-PD produced a negative response in the SCE test system.[8] It was negative in *Salmonella* mutagenicity assay.[025]

Carcinogenicity. Negative results are reported in rats.[09]

Chemobiokinetics. 1,2-PD metabolism is likely to occur through partial oxidation in the body to *lactic acid* with subsequent formation of *glucuronic acid.* A significant part of ingested 1,2-PD appeared unchanged in the urine.

Regulations.

EU (1990). 1,2-PD is available in the *List of authorized monomers and other starting substances which shall be used for the manufacture of plastic materials and articles intended to come into contact with foodstuffs (Section* A).

U.S. FDA (1998) regulates the use of PD (1) as a component of adhesives used in food-contact surface in accordance with the conditions prescribed by 21 CFR part 175.105; (2) as a defoaming agent that may be safely used as a component of food-contact articles in accordance with the conditions prescribed by 21 CFR part 176.200; (3) in the manufacture of resinous and polymeric coatings for the food-contact surface of articles intended for use in producing, manufacturing, packing, processing, preparing, treating, packaging, transporting, or holding food in accordance with the conditions prescribed by 21 CFR part 175.300; (4) in the manufacture of resinous and polymeric coatings for polyolefin films to be safely used as a food-contact surface in accordance with the conditions prescribed by 21 CFR part 175.320; (5) in the manufacture of polyurethane resins to be safely used in articles intended for use in contact with bulk quantities of dry food in accordance with the conditions prescribed by 21 CFR part 177.1680; (6) in the manufacture of cross-linked polyester resins for repeated use in articles or as components of articles coming in the contact with food in accordance with the conditions prescribed by 21 CFR part 177.2420; (7) as a plasticizer for rubber articles intended for repeated use (up to 30% by weight of the rubber product) in accordance with the conditions prescribed by 21 CFR part 177.2600; as (8) a defoaming agent in the manufacture of paper and paperboard intended for use in packaging, transporting, or holding food in accordance with the conditions prescribed by 21 CFR part 176.210; and (9) as a substance employed in the production of or added to textiles and textile fibers intended for use in contact with food in accordance with the conditions prescribed by 21 CFR part 177.2800. PG (MW above 1000) and PG alginate are listed (10) as components of paper and paperboard for use in contact with aqueous and fatty foods in accordance with the conditions prescribed by 21 CFR part 176.170.

1,2-PD is used in foods at levels not to exceed current GMP that results in maximum levels of 5.0% for alcdholic beverages, 2.5% for frozen dairy products, and 2.0% for all other food categories.

Great Britain (1998). 1,2-PD is authorized without time limit for use in the production of polymeric materials and articles in contact with food or drink or intended for such contact.

Recommendations. Joint FAO/WHO Expert Committee on Food Additives considered the dose of 20 mg/kg BW as being safe to man.

Standards. *Russia*. PML: *n/m.*

References:

1. Guchok, V. M. and Zborovskaya, E. A., in *Cryobiologia i Cryomedicina*, Kiyv, 8, 46, 1981 (in Russian).
2. Vaille, Ch., Debray, C., Koze, C., et al., Hyperglycemic action of propylene glycol, *Ann. Pharmacol. Franc.*, 29, 577, 1971.
3. Hoenig, V. and Werner, F., Is propylene glycol an inert substance? *Toxicol. Lett.*, 5, 389, 1980.
4. Weil, C. S., Woodside, M. D., Smyth, H. F., et al., Results of feeding propylene glycol in the diet to dogs for two years, *Food Cosmet. Toxicol.*, 9, 479, 1971.
5. Gaunt, I. F., Carpanini, F. M., Grasso, P., et al., Long-term toxicity of propylene glycol in rats, *Food Cosmet. Toxicol.*, 10, 151, 1972.

6. Lassalle, B. et al., Human embryo features that influence the success of cryopreservation with the use of 1,2-propanediol, *Fertil. Steril.*, 44, 645, 1985.
7. El-Shabrawy, O. A. and Arbid, M., Evaluation of some drug solvents for teratological investigations in rats, *Egypt. J. Vet. Sci.*, 24, 143, 1988.
8. Tucker, J. D., Auletta, A., Cimino, M. C., et al., Sister-chromatid exchange: second report of the Gene-Tox Program, *Mutat. Res.*, 297, 101, 1993.
9. Shiedeman, F. E. and Procita, L., The pharmacology of the monomethyl ethers of mono-, di- and tripropylene glycol in the dog with observations on the auricular fibrillation produced by these compounds, *J. Pharmacol. Exp. Ther.*, 102, 79, 1951.
10. Castle, L., Cloke, H. R., Crews, C., and Gilbert, J., The migration of propylene glycol, mono-, di-, and triethylene glycols from regenerated cellulose film into food, *Z. Lebensmit. Untersuch. Forsch.*, 187, 463, 1988.
11. McClanahan, S., Hunter, J., Murphy, M., and Valberg, S., Propylene glycol toxicosis in a mare, *Vet. Hum. Toxicol.*,40, 294, 1998.

1,3-PROPANEDIOL

Molecular Formula. $C_3H_8O_2$
M = 76.09
CAS No 504-63-2
RTECS No TY2010000
Abbreviation. 1,3-PD.

Synonyms. 2-Deoxyglycerol; 1,3-Dihydroxypropane; 2-(Hydroxymethyl)ethanol; Propane-1,3-diol; 1,3-Propanediol; 1,3-Propylene glycol; β-Propylene glycol; Trimethylene glycol.

Properties. Colorless to pale yellow-brown, viscous liquid with a brackish irritating taste. Miscible with water and alcohol.

Acute Toxicity. LD_0 is 10 g/kg BW in rats, 3.0 g/kg BW in cats. Ingestion caused CNS inhibition.[1] Accidental administration of 6.0 ml PG/kg BW to horses caused severe depression, ataxia, and malodorous breath and feces.[2]

Reproductive Toxicity.

Teratogenicity. 1,3-PD was not teratogen in rats and rabbits.[3]

Mutagenicity.

***In vivo cytogenetics*.** Cross-linking of DNA was observed in the liver and testes of rats fed 1,3-PD.[4]

Chemobiokinetics. According to Newman et al. (1940), 1,3-PD is converted to *lactic acid* and hence is burned or converted to *glycogen*. Summerfield and Tappel noted cross-linking of DNA in liver and testes of rats fed *1,3-propylenediol*.[4] 1,3-PD was converted to *malondialdehyde in vivo*, and the latter is the reactive species that caused toxic effect.

Regulations.

EU (1990). 1,3-PD is available in the *List of authorized monomers and other starting substances which may continue to be used for the manufacture of plastic materials and articles intended to come into contact with foodstuffs pending a decision on inclusion in Section A* (*Section B*).

U.S. FDA (1998) affirmed 1,3-PD as *GRAS*.

Great Britain (1998). 1,3-PD is authorized up to the end of 2001 for use in the production of polymeric materials and articles in contact with food Addition of 1.0% DAA-C_{789} to the diet for 90 days caused retardation of BW gain and reduction in blood *Hb* level, erythrocyte count, and hematocrit.[5] or drink or intended for such contact.

References:

1. van Winkle, W., Toxicity and actions of trimethylene glycol, *J. Pharmacol. Exp. Ther.*, 72, 227, 1941.
2. McClanahan, S., Hunter, J., Murphy, M., and Valberg, S., Propylene glycol toxicosis in a mare, *Vet. Hum. Toxicol.*, 40, 294, 1998.

3. Kimmel, C. A., LaBorde, J. B., and Hardin, B. D., Reproductive and developmental toxicology of selected epoxides, in *Toxicology and the Newborn*, Chapter 13, S. Kasem and M. J. Reasor, Eds., El sevier Sci. Publish., Amsterdam, 1984, 270.
4. Summerfield, F. W. and Tappel, A. L., Cross-linking of DNA in liver and testes of rats fed 1,3-propanediol, *Chem.-Biol. Interact.*, 50, 87, 1984.

PROPIONIC ACID, ester with TRIETHYLENE GLYCOL

Molecular Formula. $C_{12}H_{22}O_6$
M = 272.30

Synonym. Triethylene glycol, propionate.

Applications. Used as a plasticizer in the production of food-contact polymeric materials based on ethyl cellulose.

Acute Toxicity. LD_{50} is found to be 14.1 g/kg BW in rats, and 9.5 g/kg BW in mice. Administration led to digestive upset and impairment of the CNS function. Death within 1 to 2 days. Gross pathology examination revealed hemodynamic disturbances.[05]

Short-term Toxicity. There was no mortality or signs of intoxication in rats given 1/20 LD_{50} for 4 months.

Long-term Toxicity. Rats received 1/100 and 1/1000 LD_{50} over a period of 10 months. The treatment caused unbalance in the ratio of blood serum protein fractions and in the albumen content thereof.[05] The dose equal to 1/300 LD_{50} is likely to be ineffective.

RICINOLEIC ACID, METHYL ESTER, ACETATE

Molecular Formula. $C_{21}H_{38}O_4$
M = 354.59
CAS No 140-03-4
RTECS No VJ3410000

Synonyms and **Trade Name.** Flexricin P-4; Methyl 12-acetoxy-9-octadecenoate; Methyl 12-acetoxy oleate; Methyl acetyl ricinooleate.

Applications. A plasticizer.

Acute Toxicity. LD_{50} is 34.9 g/kg BW in mice.[027]

Regulations. ***U.S. FDA*** (1998) approved the use of RAMA (1) in resinous and polymeric coatings used as the food-contact surfaces of articles intended for use in producing, manufacturing, packing, processing, preparing, treating, packaging, transporting, or holding food in accordance with the conditions prescribed in 21 CFR part 175.320; and (2) in adhesives as a component (monomer) of articles intended for use in packaging, transporting, or holding food in accordance with the conditions prescribed in 21 CFR part 175.105.

SEBACIC ACID, BIS(2-ETHYLHEXYL) ESTER

Molecular Formula. $C_{26}H_{50}O_4$
M = 426.69
CAS No 122-62-3
RTECS No VS1000000
Abbreviation. DEHS.

Synonyms and **Trade Names.** Bis(2-ethylhexyl) decanedioate; Bis(2-ethylhexyl) sebacate; Bisoflex; Di(2-ethylhexyl) sebacate; Dioctyl sebacate; DOS; Monoplex DOS; Octyl sebacate; Plexol; PX 438.

Properties. Pale, straw-colored or oily, colorless liquid with mild odor. Soluble in alcohol. Solubility in water is 0.02% at 20°C.[04]

Applications. DEHS seems to be one of the best low-temperature resistant plasticizers to produce polyvinyl chloride and polyvinyl acetate, and other elastomers intended for contact with food or drink; it is compatible with nitro- and ethyl cellulose and with polyvinylidene chloride. Plasticizer for polymethyl methacrylate.

Acute Toxicity. LD_{50} is 26.2 g/kg BW in rats, and 19.6 g/kg BW in mice. Poisoning is accompanied by adynamia and lethargy; next day there are coordination disorder, dyspnea, and diarrhea. Death occurs in 2 to 4 days. Gross pathology examination revealed dystrophic and necrotic changes in the liver, myocardium, spleen and brain.[1]

Repeated Exposure failed to reveal accumulation in rats. Administration of up to 4.0 g/kg BW caused no mortality in animals even when the LD_{100} was exceeded.[2]

Long-term Toxicity. Retardation of growth or signs of intoxication were not found in rats given 500 mg/kg diet for 6 months and 200 mg/kg diet for 16 months. Gross pathology and histology examination revealed no abnormalities in the viscera.[05] Wistar rats received 200 mg DEHS/kg diet for 19 months. There were no findings related to the treatment with DEHS either at necropsy or at histological examination.[04]

Reproductive Toxicity.

Teratogenicity. No developmental abnormalities were reported in Wistar rats after exposure to 200 mg DEHS/kg diet.[04]

Carcinogenicity. Weanling female Sprague-Dawley rats were fed doses of 200 and 500 mg DEHS/kg BW, given 3 times weekly by gavage for 11 consecutive weeks after initiation with a single oral dose. The treatment did not produce a promoting effect in rat liver foci bioassay.[3]

Regulations.

U.S. FDA (1998) regulates the use of DEHS (1) as a component in resinous and polymeric coatings of food-contact surface in accordance with the conditions prescribed by 21 CFR part 175.300; (2) in surface lubricants employed in the manufacture of metallic articles that contact food in accordance with the conditions prescribed by 21 CFR part 178.3400; (3) as a plasticizer for rubber articles intended for repeated use in contact with food (up to 30% by weight of the rubber product) in accordance with the conditions prescribed by 21 CFR part 177.2600; (4) in adhesives used as components of articles intended for use in packaging, transporting, or holding food in accordance with the conditions prescribed by 21 CFR part 175.105; and (5) as an ingredient for closures with sealing gaskets on containers intended for use in producing, manufacturing, packing, processing, preparing, treating, packaging, transporting, or holding food (up to 2.0%) in accordance with the conditions prescribed by 21 CFR part 177.1210.

In ***France*** and ***Italy*** DEHS is a permitted constituent of plastics that will come into contact with foodstuffs.[05]

Standards. ***Russia*** (1995). MAC and PML: 0.1 (organolept., turbidity). PML in food: 4.0 mg/l.

References:

1. Kustov, V. V. et al., in *Toxicity of Synthetic Lubricants*, F. F. Erisman Research Sanitary Hygiene Institute, Moscow, 60, 1977 (in Russian).
2. Radeva, M. and Dinoeva, S., in *Proc. Research Hygiene Institute*, Sofia, 10, 105, 1966 (in Bulgarian).
3. Oesterle, D. and Deml, E. J., Promoting activity of di(2-ethylhexyl)phthalate in rat liver foci bioassay, *Cancer Res. Clin. Oncol.*, 114, 133, 1988.

SEBACIC ACID, DIBUTYL ESTER

Molecular Formula. $C_{18}H_{34}O_4$

M = 314.48

CAS No 109-43-3

RTECS No VS1150000

Abbreviation. DBS.

Synonym. Dibutyl sebacate.

Properties. Pale-yellow, oily liquid with the odor of ether. Solubility in water is 0.005% at 25°C, soluble in alcohol.

Applications. Used as a plasticizer in the production of polyvinyl chloride and its copolymers, acetobutyrate cellulose, polyvinyl butyral, chlorinated rubber, polyvinylidene chloride intended for contact with food or drink. Compatible with nitro- and ethyl cellulose, polystyrene, and synthetic rubber. Renders the materials highly flexible with low temperature resistance.

Migration Data. DBS has been found to pass from polyvinylidene chloride packaging films to the food.[1] Migration of plasticizers from nitrocellulose-coated regenerated cellulose film (RCF) purchased from retails and take-away outlets was studied. Levels of the plasticizer found to migrate from RCF were as high as 76 to 137 mg DBS/kg in processed cheese and cooked meats.[2]

Acute Toxicity. LD_{50} is reported to be 16 to 32 g/kg BW in rats, and 18 to 25.5 g/kg BW in mice.[3,4,04] General condition and BW gain in the survivors did not differ from that in the controls. Histology examination revealed pulmonary hemorrhage, edema and congestion in the myocardium, brain, intestinal wall and stomach.

Repeated Exposure. There were no signs of toxicity in mice dosed with 0.9 to 3.6 g/kg BW, or rats dosed with 0.9 to 34 g/kg BW for 1.5 months.[4]

Short-term Toxicity. Decreased BW gain, reticulocytosis, decline in the erythrocyte count, and blood *Hb* level were observed in rats given 600 mg/kg BW for 13 weeks.[5]

Long-term Toxicity. Exposure of rats to 1/50 LD_{50} for 9 months produced no changes;[3] 0.01, 1.25, and 6.25% DBS in the diet caused no signs of intoxication to develop in two-year studies in rats.[04]

Reproductive Toxicity. Decreased BW gain in the second generation of rats was found with the maximum dose in the diet.[5]

Mutagenicity.

In vivo cytogenicity. Cytogenic effect is not reported in the somatic (brain) cells of mice given doses of 6.0 to 24 g/kg BW.[5]

Chemobiokinetics. DBS is rapidly hydrolyzed by lipase *in vitro*. It is suggested to be hydrolyzed in the body in the same way as fats.[05]

Regulations. ***U.S. FDA*** (1998) regulates the use of DBS (1) in adhesives as a component of articles intended for use in packaging, transporting, or holding food in accordance with the conditions prescribed by 21 CFR part 175.105; (2) as a plasticizer for rubber articles intended for repeated use in contact with food (content up to 30% by weight of the rubber product) in accordance with the conditions prescribed by 21 CFR part 177.2600; (3) in the manufacture of resinous and polymeric coatings for the food-contact surface of articles intended for use in producing, manufacturing, packing, processing, preparing, treating, packaging, transporting, or holding food in accordance with the conditions prescribed by 21 CFR part 175.300; (4) as a component of resinous and polymeric coatings for polyolefin films for food-contact surface in accordance with the conditions prescribed by 21 CFR part 175.320; and (5) as a component of the uncoated or coated food-contact surface of paper and paperboard intended for use in producing, manufacturing, packing, transporting, or holding aqueous and fatty food in accordance with the conditions prescribed by 21 CFR part 176.170.

Standards. ***Russia.*** PML in food: 4.0 mg/l, PML in drinking water: *n/m*.

References:

1. Baba, T., Hosokawa M., Yamada A., Studies on food contaminants from plastic containers and packages. Migration of plasticizers from polyvinylidene chloride packaging films to food, *Ann. Rep. Osaka City Institute Publ. Health Environ. Sci.*, 50, 61, 1988.
2. Castle, L., Mercer, A. J., Startin, J. R., and Gilbert, J., Migration from plasticized films into foods. 3. Migration of phthalate, sebacate, citrate and phosphate esters from films used for retail food packaging, *Food Addit. Contam.*, 5, 9, 1988.
3. Smith, C. C., Toxicity of butylstearate, dibutylsebacate, dibutylphthalate, and methoxyethyloleate, *Arch. Ind. Hyg. Occup. Med.*, 4, 310, 1953.
4. Komarova, E. N., Toxic properties of some additives for plastics, *Plast. Massy*, 12, 30, 1976 (in Russian).
5. Ilyukevich, Z. F. et al., in *Scientific and Technical Progress as Medical Prophylaxis*, Part 1, Moscow, 1971, 60 (in Russian).

STABILOIL-18

Applications. Used as a plasticizer in rubber manufacture.

Acute Toxicity. In rats and mice, LD_{50} was not attained. Administration of 10 g/kg BW dose caused no mortality or evident signs of intoxication.

Long-term Toxicity. Rats were dosed by gavage with 20 mg and 100 mg S./kg BW for 10 months. The treatment caused no retardation of BW gain, impairment of NS, liver and kidney functions. There were no changes in the blood formula.

Carcinogenic Effect is not observed.

Reference:

Toxicology of the Components of Rubber Mixes and of Rubber and Latex Articles, Centr. Research Institute Petrol. Chem., Moscow, 1974, 21 (in Russian).

STEARIC ACID, BUTYL ESTER

Molecular Formula. $C_{22}H_{44}O_2$
M = 340.57
CAS No 123-95-5
RTECS No WI2900000
Abbreviation. BS.

Synonym. Butyl stearate.

Properties. Colorless or pale-yellow liquid, almost odorless at the temperatures above 20°C. Soluble in alcohol, miscible with vegetable oils.

Applications. Used as a plasticizer in the production of polystyrene and other food-contact polymeric materials.

Migration from ABC-polymers into sunflower oil at 20°C over a period of 3 months is reported to be 0.04 mg/l.[1] Migration of BS from general purpose polystyrene into water (60°C; 7 months) amounts to 0.6 mg/l, into 40% alcohol (20°C; 5 months; 4.0 cm^{-1}) to 1.14 mg/l, and into 96% alcohol under the same conditions to 9.26 mg/l.[2]

Acute Toxicity. Young rats (BW of 60 to 75 g) tolerate doses of 32 g/kg BW.[3]

Repeated Exposure failed to reveal cumulative properties. Rats and mice were exposed to 1.0 to 4.0 g/kg BW for 1.5 months. No toxic effect was noted to develop.[4]

Long-term Toxicity. Rats received 250 to 6250 mg/kg in their feed. No changes in growth, hematology, or histology of the visceral organs are reported.[2] Administration of 20 and 400 mg/kg BW produced no toxic effect.[4]

Reproductive Toxicity. Addition to the feed did not affect the reproductive function in rats.[3]

Regulations. ***U.S. FDA*** (1998) listed BS for use (1) in adhesives as a component of articles intended for use in packaging, transporting, or holding food in accordance with the conditions prescribed by 21 CFR part 175.105; (2) in the manufacture of resinous and polymeric coatings for the food-contact surface of articles intended for use in producing, manufacturing, packing, processing, preparing, treating, packaging, transporting, or holding food in accordance with the conditions prescribed by 21 CFR part 175.300; (3) in the manufacture of resinous and polymeric coatings for polyolefin films for food-contact surface of articles intended for use in producing, manufacturing, packing, processing, preparing, treating, packaging, transporting, or holding food in accordance with the conditions prescribed by 21 CFR part 175.320; (4) as a plasticizer for rubber articles intended for repeated use in contact with food (content up to 30% by weight of the rubber product); (5) as a component of a defoaming agent intended for articles which may be used in producing, manufacturing, packing, processing, preparing, treating, packaging, transporting, or holding food in accordance with the conditions prescribed by 21 CFR part 176.200; (6) in cross-linked polyester resins to be safely used as articles or components of articles intended for repeated use in contact with food in accordance with the conditions prescribed by 21 CFR part 177.2420; (7) in the manufacture of rubber articles intended for repeated use in producing, manufacturing, packaging, processing, preparing, treating, packing, transporting, or holding food in accordance with the conditions prescribed by 21 CFR part 177.2600; (8) as a component of the uncoated or coated food-contact surface of paper and paperboard intended for use in producing, manufacturing, packing, transporting, or holding dry, aqueous, and fatty food in accordance with the conditions prescribed by 21 CFR parts 175.170 and 175.180; and (9) as defoaming agent in the processing of food in accordance with the conditions prescribed in 21 CFR part 173.320.

Recommendations. Joint FAO/WHO Expert Committee on Food Additives (1999). ADI: No safety concern.

Standards. ***Russia.*** PML in drinking water: *n/m.*

References:

1. *Toxicology and Sanitary Chemistry of Polymerization Plastics*, Coll. Sci. Proc., B. Yu. Kalinin, Ed., Leningrad, 1984, 64 (in Russian).
2. Petrova, L. I., *Investigation of Possible Use of Polystyrene Plastics of Different Composition in Contact with Foodstuffs*, Author's abstract of thesis, Leningrad, 1979, 20 (in Russian).
3. Smith, C. C., Toxicity of butylstearate, dibutylsebacate, dibutylphthalate, and methoxyethyloleate, *Arch. Ind. Hyg. Occup. Med.*, 4, 310, 1953.
4. Komarova, E. N., Toxic properties of some additives for plastics, *Gig. Sanit.*, 12, 30, 1976 (in Russian).

SUBERIC ACID, DI(2-ETHYLHEXYL) ESTER

Molecular Formula. $C_{24}H_{46}O_4$

M = 398.7

Synonym. Di(2-ethylhexyl) suberate.

Properties. Colorless liquid.

Applications. Used as a plasticizer in the production of food-contact polymeric materials.

Acute Toxicity. LD_{50} is 8.8 g/kg BW in rats, and 12.5 g/kg BW in mice. Poisoning is accompanied by excitation with subsequent depression and respiratory distress. Animals died in the lethargy condition.

Repeated Exposure revealed marked cumulative properties. K_{acc} is 0.6 (by Lim).

Reference:

Timofiyevskaya, L. A., Toxicity of di(2-ethylhexyl)suberate, *Gig. Truda Prof. Zabol.*, 1, 52, 1983 (in Russian).

TEREPHTHALIC ACID, BIS(2-ETHYLHEXYL) ESTER

Molecular Formula. $C_{24}H_{38}O_4$

M = 390.57

CAS No 6422-86-2

RTECS No WZ0883500

Abbreviation. DEHT.

Synonyms and **Trade Name.** Benzene dicarboxylic acid, bis(2-ethylhexyl) ester; Bis(2-ethylhexyl) terephthalate; Kodaflex DOTP; Terephthalic acid, dioctyl ester.

Properties. Colorless liquid.

Applications. Used as a low-volatile plasticizer in the production of polyvinyl chloride and other polymeric materials intended for contact with food or drink.

Acute Toxicity. LD_{50} is reported to exceed 3.2 g/kg BW in rats and mice.[1]

According to other data, LD_{50} is not attained in mice, but some animals died after administration of 20 g/kg BW. Poisoning was accompanied by excitation followed by CNS inhibition.[2]

Short-term Toxicity. In a 90-day study, rats were given doses of 560 to 620 mg/kg BW. Microscopic examination of a wide range of tissues and an examination of peroxisome levels in the liver revealed only minor changes in the erythrocytes and slight increase in liver weights. The NOAELs of 280 and 310 mg/kg BW were established in this study for the males and females, respectively.[3]

Mutagenicity.

In vivo cytogenetics. Rats were fed DEHT in the diet at a dose level of 2.0 g/kg, and their urine was negative in *Salmonella* mutagenicity assay in the presence or absence of metabolic activation.[1]

In vitro genotoxicity. Both DEHT and its metabolite, *mono(2-ethylhexyl) phthalate*, were found to be negative in *Salmonella* test, the Chinese hamster ovary cell/hypoxantine guanine phosphoribosyl transferase mutagenicity assay, and in *in vitro* CA assay using Chinese hamster ovary cells.[1]

Chemobiokinetics. Sprague-Dawley rats were given by gavage a single dose of 100 mg ^{14}C-DEHT/kg BW. Radioactivity was primarily recovered in the feces (predominantly, unchanged DEHT) and urine [principally *mono(2-ethylhexyl)terephthalate* and metabolic products of *2-ethylhexanol*]. Small amounts of DEHT were detected in the liver and fat. Urine metabolites identified were *terephthalic acid, oxidized metabolites of 2-ethylhexanol* and *mono(2-ethylhexyl) terephthalate*, and *glucuronic* and *sulphuric acid conjugates*.[4]

DEHT was hydrolyzed in rat gut homogenate fractions *in vitro* to *2-ethylhexanol* and *terephthalic acid.*

References:

1. Barber, E. D., Genetic toxicology testing of di(2-ethylhexyl) terephthalate, *Environ. Molec. Mutagen.*, 23, 228, 1994.
2. Timofiyevskaya, L. A., Toxicity of di(2-ethylhexyl) terephthalate, *Gig. Sanit.*, 8, 91, 1982 (in Russian).
3. Barber, E. D. and Topping, D. C., Subchronic 90-day toxicology of di(2-ethylhexyl) terephthalate, *Food Chem. Toxicol.*, 33, 971, 1995.
4. Barber, E. D., Fox, J. A., and Giordano, C. J., Hydrolysis, absorption and metabolism of di(2-ethylhexyl) terephthalate in the rat, *Xenobiotica*, 24, 441, 1994.

p-TOLUENE SULFONAMIDE

Molecular Formula. C_7H_9NOS
M = 155.23
CAS No 70-55-3
RTECS No XT5075000
Abbreviation. TSA.

Synonyms and **Trade Names.** Celludol; 4-Methylbenzene sulfonamide; *p*-Methylbenzene sulfonamide; *p*-Toluene sulfonylamide; *p*-Tolyl sulfonamide; *p*-Tosylamide; Tosylamide.

Properties. Odorless, white crystalline substance. Poorly soluble in cold water (0.25% at 20°C), more readily soluble in hot water (50% at 90°C).

Applications. Used as a plasticizer in cellulose acetate and nitrocellulose manufacture, compatible with polyvinyl butyral.

Toxicity appeared to be insignificant.[05]

Mutagenicity.

In vitro genotoxicity. Is not found to be mutagenic in *Salmonella*/microsome test, and in micronucleus test in mice.[1]

In vivo genotoxicity. TSA was not mutagenic in *Dr. melanogaster.*

Chemobiokinetics. Following ingestion by dogs and rats, TSA has been oxidized to *p-sulfamoylbenzoic acid* by methyl group oxidation. In rats, up to 90% TSA are rapidly eliminated unchanged in the urine.[2-4]

Regulations.

EU (1990). TSA is available in the *List of authorized monomers and other starting substances which may continue to be used for the manufacture of plastic materials and articles intended to come into contact with foodstuffs pending a decision on inclusion in Section A* (*Section B*).

U.S. FDA (1998) approved mixture of *o*-TSA and *p*-TSA for use (1) as a constituent of food-contact adhesives which may be safely used as a component of articles intended for use in packaging, transporting, or holding food if used in quantities not exceeding limits of good manufacturing practice in accordance with the conditions prescribed by 21 CFR part 175.108; and (2) in the manufacture of resinous and polymeric coatings for the food-contact surface of articles intended for use in producing, manufacturing, packing, processing, preparing, treating, packaging, transporting, or holding food in accordance with the conditions prescribed by 21 CFR part 175.300.

References:

1. Eckhardt, K. et al., Mutagenicity study of Remsen-Fahleberg saccharin and contaminants, *Toxicol. Lett.*, 7, 51, 1980.

2. Williams, R. T. *Detoxication Mechanisms*, Vol 1, Chapman and Hall Ltd., London, 1959.
3. Minegishi, K. et al., Metabolism of saccharin and the related compounds in rats and guinea pigs, *Chem. Pharmacol. Bull.*, 20, 1351, 1972.
4. Ball, L. M. et al., The fate of saccharin impurities: the excretion and metabolism of (^{14}C)toluene-4-sulfonamide and 4-sulfamoyl (^{14}C)benzoic acid in rats, *Xenobiotica*, 8, 183, 1978.

p-TOLUENESULFONIC ACID, ETHYL ESTER

Molecular Formula. $C_9H_{12}O_3$
M = 200.27
CAS No 80-40-0
RTECS No XT6825000
Abbreviation. TSAE.

Synonyms and **Trade Name.** Ethyl-*p*-methyl benzenesulfonate; Ethyl PTS; Ethyl-*p*-toluene sulfonamide; 4-Methylbenzenesulfonic acid, ethyl ester.

Properties. Monoclinic crystals. Soluble in water insoluble in alcohol.

Applications. Plasticizer for cellulose acetate.

Acute Toxicity. LD_{50} *i/p* is 1.0 g/kg BW in mice.

Mutagenicity.

In vivo cytogenetics. TSAE induced an increase in frequency of mutations in the *i/p* host-mediated assay.[1]

It was not mutagenic in a sex-linked recessive lethal test in *Dr. melanogaster* when fed to the adult flies. This is likely a result of metabolic deactivation, presumably in the gut. Injection, however, did produced significant effect.[2]

Carcinogenicity. BD rats were administered *s/c* 65 mg TSAE/kg BW weekly for 50 weeks; 3/11 rats developed local sarcomas after 600 days.[3]

Regulations. ***U.S. FDA*** (1998) approved the use of TSAE in adhesives as a component of articles intended for use in packaging, transporting, or holding food in accordance with the conditions prescribed by 21 CFR part 175.105.

References:

1. Simmon, V. F., Rozencranz, H. S., Zeiger, E., and Poirier, L. A., Mutagenic activity of chemical carcinogens and related compounds in the intraperitoneal host-mediated assay, *J. Natl. Cancer Inst.*, 62, 911, 1979.
2. Zijlstra, J. A. and Vogel, E. W., Metabolic inactivation of mutagens in *Drosophila melanogaster, Mutat. Res.*, 198, 73, 1988.
3. *Chemical Carcinogens*, C. E. Searle, Ed., ACS Monograph 173, *Am. Chem. Soc.*, Washington, D.C., 1976, 168.

p-TOLYL-1-NAPHTHYLMETHANE

Abbreviation. TNM.

Properties. Crystals. Render an aromatic odor to water. Odor and taste perception threshold appeared to be 0.96 mg/l; practical threshold is 1.92 mg/l.

Applications. Used as a plasticizer in the polyvinyl chloride production.

Acute Toxicity. In rats, LD_{50} is 10 g TNM/kg BW. Poisoning is accompanied by adynamia, respiratory and GI tract disturbances.

Long-term Toxicity. No signs of intoxication were noted in rats given doses of 100 and 200 mg TNM/kg BW for 6 months. There were no changes in hematology parameters or in serum *SH*-group level.

Standards. ***Russia.*** Recommended MAC: 1.0 mg/l (organolept.).

Reference:

Voloshchenko, O. I. and Chekal', V. N., in *Hygiene of Populated Areas*, D. N. Kaluzhny, Ed., Kiyv, Issue No 9, 1970, 1960 (in Russian).

TRIACETIN

Molecular Formula. $C_9H_{14}O_6$
M = 218.20
CAS No 102-76-1
RTECS No AK3675000

Synonyms and **Trade Names.** Fungacetin; Glycerin triacetate; Glycerol triacetate; Glyceryl triacetate; Glyped; 1,2,3-Propanetriol, triacetate; Vanay.

Properties. Colorless liquid with very faint, fruity odor and mild, sweet taste. T. is a water-soluble triglyceride. Solubility in water is about 60 g/l at 25°C. It is also soluble in alcohol.

Applications. Cellulose acetate plasticizer in manufacture of cigarette filters. Plasticizer for cellulose nitrate. Solvent in manufacture of celluloid. T. may have a role as a parenteral nutrient.

Exposure. Used in fragrances in the U.S. Usual concentration in detergents is 0.005%.

Migration Data. Aqueous extracts of film plasticized with T. and triethylene glycol dipropionate are transparent, colorless, and have no sediment. The extracts exhibit a foreign odor and sourish taste, irrespective of temperature or exposure. T. migrates at the level of 1.1 to 1.6 mg/l and triacetin degradation product (acetic acid) migrates in a quantity of 0.5 to 2.0 mg/l.[6]

Acute Toxicity. LD_{50} is 3.0 g/kg BW in rats,[1] and 1.1 g/kg BW in mice. T. caused spastic paralysis and stiffness, it altered sleep time.[2] Ingestion of T. led to general weakness and ataxia. T. may produce slight hemolysis *in vitro.*[011]

According to other data, LD_{50} is 6.4 to 12.8 g/kg BW in rats, and 3.2 to 6.4 g/kg BW in mice.[03]

Repeated Exposure. Growing rats received T. for 30 days in the diet in which it supplied 25.9% of metabolizable energy. The treated rats showed a BW gain greater than that of control animals.[3]

Short-term Toxicity. In 3-month feeding studies, growing rats were found to tolerate up to 20% T. in the diet, but diets containing 60% T. caused marked retardation of BW gain and considerable increase in deaths.[4]

Long-term Toxicity. Laboratory rats tolerated diets consisting of 50% T. Systemic acidosis is likely to be a possible consequence.[019]

Allergenic Effect.

Observations in man. T. showed no sensitization reactions in a maximization test that was carried out on 33 volunteers (Epstein, 1976).

Chemobiokinetics. T. has been administered *i/v* to mongrel dogs. The majority of infused T. underwent intravascular hydrolysis, and the majority of the resulting *acetate* is oxidized.[5] T. was found to be hydrolyzed by human intestinal lipase. If hydrolyzed, systemic acidosis is a possible consequence.[019]

Regulations. ***U.S. FDA*** (1998) approved the use of T. (1) in the manufacture of resinous and polymeric coatings for polyolefin films for food-contact surface of articles intended for use in producing, manufacturing, packing, processing, preparing, treating, packaging, transporting, or holding food in accordance with the conditions prescribed by 21 CFR part 175.320; and (2) as a component of adhesives to be safely used in food-contact surface in accordance with the conditions prescribed by 21 CFR part 175.105. T. is considered to be *GRAS* when used in accordance with GMP in accordance with the conditions prescribed by 21 CFR part 182.1073. Under these conditions, T. may be used in animal drugs, feeds, and related products.

Recommendations. Joint FAO/WHO Expert Committee on Food Additives: ADI is not specified.

References:

1. *Encyclopedia of Occupational Health and Safety,* vol 1 & 2, Int. Labour Office, Geneva, 1983, 973.
2. *Proc., Fedn. of Am. Soc. Exp. Biol.*, 22, 368, 1963.
3. Vermorel, M., Utilization energetique de la triacetine par le rat en croissance, *Annls. Biol. Anim. Biochim. Biophys.*, 8, 453, 1968.
4. Shapira, J., Mandel, A. D., Quattrone, P. D., and Bell NL, Current research on regenerative systems, *Life Sci. Space Res.*, 7, 123, 1968.
5. Bleiberg, B., Beers, T. R., Persson, M., and Miles, J. M., Metabolism of triacetin-derived acetate in dogs, *Am. J. Clin. Nutr.*, 58, 908, 1993.

6. Kas'an, V. N., in Proc. IV All-Union Sci. Techn. Conf. *Synthesis and Applications of Polymeric Materials Used in Food Industry*, Minsk, 1980, 128 (in Russian).

1,2,4-TRICHLOROBENZENE

Molecular Formula. $C_6H_3C_{l3}$
M = 181.45
CAS No 120-82-1
RTECS No DC2100000
Abbreviation. TCB.

Synonym. *unsym*-Trichlorobenzene.

Properties. Colorless liquid or crystals. Solubility in water is ~26 mg/l at 25°C. Odor and taste perception thresholds in water are 0.03 and 0.01 mg/l.[01] Sparingly soluble in alcohol. According to other data, odor perception threshold is 0.064 mg/l.[02] Heating does not noticeably increase odor intensity. Solution coloration does not change at a concentration of 1.0 g/l.[01]

Applications. Used as a plasticizer and a solvent in the manufacture of food-contact plastics.

Acute Toxicity. In rats and mice, LD_{50} is 750 mg/kg BW. Poisoning through the feed in sublethal doses affects the liver without causing porphyria and jaundice.[1]

Repeated Exposure. In rhesus monkey, daily oral doses of 173.6 mg/kg BW were lethal within 20 to 30 days. Doses of up to 25 mg/kg BW were without effect.[2]

Short-term Toxicity. In a 90-day study, in which rats were given oral doses of 10 to 40 mg 1,2,4-TCB/kg BW, hepatic cytochrome P-450, and cytochrome-*C*-reductase were induced.[3] A slight increase in the relative liver weight was found at 40 mg/kg BW but no changes in hemoglobin, hematocrit, or liver histopathology were evident at any dose level. In a 120-day study with rhesus monkeys, daily oral doses of 1 to 25 mg/kg BW did not produce any evidence of toxicity.[4]

At 125 mg/kg BW dose level, temporary BW loss and some induction of cytochrome P-450 were found. In a 13-week feeding study in rat, the NOAEL appeared to be 100 ppm in the diet or approximately 7.7 mg/kg BW.[5] The liver was primarily affected. ADI/TDI is considered to be 0.0077 mg/kg BW.

Long-term Toxicity. Rats and rabbits were given 0.06 mg TCB/l of aqueous solution. No changes in the hematology analyses, in the balance of inorganic to total sulfates in the urine in rabbits, and in conditioned reflex activity in rats were reported in this study.[6]

Rats were dosed by gavage for 5.5 months. The treatment led to a decline in blood *Hb* content, and to eosinophilia and reticulocytosis. Gross pathology examination revealed congestion in all the visceral organs and an ill-defined fatty dystrophy in the liver.[7]

Reproductive Toxicity.

Embryotoxicity and ***Teratogenicity.*** Pregnant rats received up to 600 mg/kg BW by gavage on days 6 through 15. The treatment caused no fetotoxic action or abnormalities in offspring.[8] There was no evidence of teratogenic effects in Sprague-Dawley rats given oral doses of 75 to 300 mg 1,2,4-TCB/kg BW on days 6 to 15 of gestation.[9]

Mutagenicity. 1,2,4-TCB was negative in *Salmonella* mutagenicity assay.[025]

Carcinogenicity. 1,2,4-TCB was not shown to be a carcinogen to animals.

Carcinogenicity classification.

U.S. EPA: D.

Chemobiokinetics. Following oral administration to rats and rhesus monkeys, TCB is distributed primarily to the adipose tissue and may undergo extensive enterohepatic recirculation.[10] Due to rapid excretion, repeated exposure to TCB does not change considerably its level in the body tissues and no accumulation was noted. 20% TCB was found in the GI tract, 6.0% in the fatty tissue, and 1.7% in muscles.[11]

The effective half-life in the fatty tissue is 60 days. In rabbits, the major urinary metabolites are *trichlorophenols, 3,4,5-trichlorocatechol* and *mercapturic acid derivatives.*[12,13] In rats, 1,2,4-TCB is metabolized primarily to *2,3,5-* and *3,4,5-trichlorobenzene mercapturic acids*. Smaller quantities of *trichlorothiophenols* and *trichlorophenols* are also found in the urine.[10] TCB is excreted in the urine and to a lesser extent in the feces.

Recommendations.

WHO (1996): Guideline value for drinking water is 0.02 mg/l for total TCB. However, this value exceeds the lowest reported odor threshold in water. The levels of 0.005 to 0.05 mg/l are likely to give rise to consumer complaints of foreign odor and taste.

U.S. EPA (1999). Health Advisory for longer-term exposure is 0.5 mg/l.

Standards. ***U.S. EPA*** (1999). MCL and MCLG: 0.07 mg/l.

References:

1. Brown, V. K. H., Muir, C., and Thorpe, E., The acute toxicity and skin irritant properties of 1,2,4-trichlorobenzene, *Ann. Occup. Hyg.*, 12, 209, 1969.
2. Smith, C. C., Cragg, S. T., and Wolfe, G. F., Subacute toxicity of 1,2,4-trichlorobenzene in subhuman primates, *Fedn. Proc.*, 37, 248, 1978.
3. Carlson, G. P. and Tardiff, R. G., Effect of chlorinated benzenes on the metabolism of foreign organic compounds, *Toxicol. Appl. Pharmacol.*, 36, 383, 1976.
4. Crass, S. T., Wolfe, G. F., and Smith, C. C., Toxicity of 1,2,4-trichlorobenzene in rhesus monkey: Comparison of two *in vivo* methods for estimating P-450 activity, *Toxicol. Appl. Pharmacol.*, 45, 340, 1978.
5. Cote, M., Chu, I., Villeneuve, D. C., et al., Trichlorobenzenes: results of a thirteen week study in the rat, *Drug Chem. Toxicol.*, 11, 11, 1988.
6. Gurfein, L. N. and Pavlova, Z. K., in *Protection of Water Reservoirs against Pollution by Industrial Liquid Effluents*, S. N. Cherkinsky, Ed., Medgiz, Moscow, Issue No 4, 1960, 117 (in Russian).
7. Meleshchenko, K. F., On the problem of hygienic principles of trichlorobenzene standardization in water bodies, *Gig. Sanit.*, 3, 13, 1960 (in Russian).
8. Black, W. D., Valli, V. E. O., Ruddick, J. A., and Villeneuve, D. C., Assessment of teratogenic potential of 1,2,3-, 1,2,4- and 1,2,5-trichlorobenzenes in rats, *Bull. Environ. Contam. Toxicol.*, 41, 719, 1988.
9. Ruddick, J. A., Black, W. D., Villeneuve, D. C., et al., A teratological evaluation following oral administration of trichloro- and dichlorobenzene isomers to the rat, *Teratology*, 27, 73, 1983.
10. *Toxic Evaluation of Halogenated Aromatic Compounds in the Context of the Contamination of Underground Waters*, WHO, Working Party, Copenhagen, Moscow, 1981, 21 (in Russian).
11. Khalturin, G. V. and Andryushkeyeeva, M. N., Transformations of trichlorobenzene in rats following single and repeated intragastrical administration, *Gig. Sanit.*, 2, 86, 1988 (in Russian).
12. Jondorf, W. R., Parke, D. V., and Williams, R. T., The metabolism of halobenzenes, 1,2,4- and 1,3,5-trichlorobenzenes, *Biochem. J.*, 61, 512, 1955.
13. Kohli, J., Jones, D., and Safe, S., The metabolism of higher chlorinated benzene isomers, *Canad. J. Biochem.*, 54, 203, 1976.

TRIETHYLENE GLYCOL

Molecular Formula. $C_6H_{14}O_4$

M = 150.18

CAS No 112-27-6

RTECS No YE4550000

Abbreviation. TEG.

Synonym and **Trade Name.** Triglycol; Ethylene glycol, di-2-hydroxyethyl ester.

Properties. Glycerin-like, colorless, odorless liquid with a burning taste. Miscible with water and alcohol, very hygroscopic. A concentration up to 1.0 g/l does not affect color, odor, and taste of water. At levels up to 500 mg/l, foam formation does not occur. According to other data, organoleptic perception threshold is 700 mg/l.[01]

Applications. TEG is used in the manufacture of solvents and plasticizers. An antistatic.

Migration Data. Chocolates, boiled sweets, toffees, cakes, and meat pies were wrapped in regenerated cellulose films that contained various mixtures of glycol softeners. It was shown that higher levels of migration occurred for propylene glycol than for TEG and the presence of coating reduced the migration of both softeners.[1]

Acute Toxicity.

Observations in man. LD_{50} is reported to be 5.0 g/kg BW.

Animal studies. LD_{50} is in the range of 15.5 to 17.5 g/kg BW in rats, 18.5 to 20.8 g/kg BW in mice, 7.9 to 14.7 g/kg BW in guinea pigs, and 8.4 to 9.5 g/kg BW in rabbits (Stenger et al., 1968). Toxic manifestations comprise convulsions, exophthalmia, dyspnea, ataxia and hematuria. Gross pathology examination revealed pronounced swelling of the renal tubular epithelium, vacuolar dystrophy, and focal hepatocyte necrobiosis, a marked congestion in the viscera.[2]

Repeated Exposure failed to reveal cumulative properties. Female rats were dosed by gavage with 1.0 g/kg BW for 20 days. The treatment produced changes in the liver, kidney, and CNS function. Main signs of intoxication appeared to be an increase in blood *Hb* level and eosinophil count and in serum ALT activity, a decline in the activity of cytochrome-*c*-oxidase, and a rise in the STI value. Gross pathology examination revealed congestion of the viscera, parenchymatous dystrophy in the liver and kidney.[2] Rats received 3.0% TEG in their drinking water for 30 days without any signs of toxic effect. Concentration of 5.0% was effective.[03] TEG has been found to be CNS and cardiac depressant of low toxicity; it produced arrhythmia and fibrillation.[3]

Long-term Toxicity. Rats received up to 5.0 mg TEG/kg BW. The treatment lowered the activity of transaminase in the liver, led to disruption of amino acids conversion process and to subsequent restriction of their involvement in the synthesis of proteins, glycogen, and fatty acids.[4] Administration of 5.0 mg/kg BW to rats affected kidney and liver functions (a decline in some enzyme activity, reduction of glycogen content).[2] Rats were exposed to the doses of 3.0 to 4.0 g/kg diet for 2 years without any signs of toxic effect.[03]

Reproductive Toxicity.

Gonadotoxicity. TEG is found to disturb energy exchange in the testicular tissue and produce dystrophic changes in the spermatogenic epithelium (see ***Diethylene glycol***). TEG caused a reduction in sperm motility, it increased the number of immobile forms and lowered their resistance. It induced DLM and damaged the germinal cells of male rats, but produced little effect on the female gonads.[4]

Swiss CD-1 mice received 1.5 to 3.0% TEG in their drinking water. The treatment reduced live pup weight, but did not affect mating or fertility of animals.[5]

Embryotoxicity. A dose of 11 g/kg BW reduced pup weight gain and increased the number of dead pups per litter.[6]

Teratogenic effect was not observed in mice, rats, hamsters, and rabbits administered 1.0 to 4.0 ml TEG/kg BW throughout pregnancy.[6] However, according to other data, the LOEL for teratogenic effects was found to be 5.0 mg/kg BW in mice gavaged with TEG for 10 days (Neeper-Bradley et al., 1994). Munro et al. suggested the calculated NOEL of 0.5 mg/kg BW for this study.[7]

Mutagenicity.

In vitro genotoxicity. TEG was shown to be positive in *Salmonella* mutagenicity assay (NTP-91).

Carcinogenicity. Negative results are reported in male rats.[09]

Chemobiokinetics. Intestinal absorption is very low. TEG does not cause oxalate stone formation in the bladder and kidneys. Almost entirely excreted unchanged with the urine and feces (90 to 97%).[8]

Regulations. ***U.S. FDA*** (1998) regulates the use of TEG (1) in the manufacture of resinous and polymeric coatings for the food-contact surface of articles intended for use in producing, manufacturing, packing, processing, preparing, treating, packaging, transporting, or holding food in accordance with the conditions prescribed by 21 CFR part 175.300; (2) in plasticizers for polymeric substances in the manufacture of articles or components of articles intended for use in producing, manufacturing, packaging, processing, preparing, treating, packing, transporting, or holding food in accordance with the conditions prescribed by 21 CFR part 178.3740; (3) in the manufacture of cellophane for food packaging in accordance with the conditions prescribed by 21 CFR part 177.1200; (4) as a component of polyester resins for coatings not exceeding a coating weight of mg/inch^2 and those intended for contact under the specified conditions; and (5) as a component of adhesives (diethyleneglycol content in TEG not to exceed 0.1%) in accordance with the conditions prescribed by 21 CFR part 175.105.

EU (1990). TEG is available in the *List of authorized monomers and other starting substances which shall be used for the manufacture of plastic materials and articles intended to come into contact with foodstuffs (Section A).*

Great Britain (1998). TEG is authorized without time limit for use in the production of polymeric materials and articles in contact with food or drink or intended for such contact.

Standards. *Russia* (1995). MAC: 0.5 mg/l; PML: *n/m.*

References:

1. Castle, L., Cloke, H. R., Crews, C., and Gilbert, J., The migration of propylene glycol, mono-, di-, and triethylene glycols from regenerated cellulose film into food, *Z. Lebensmit. Untersuch. Forsch.*, 187, 463, 1988.
2. Tolstopyatova, G. V., Korcach, V. I., Barilyak, I. R., et al., Hygienic regulations of tri-, tetra-, and pentaethylene glycol in water bodies, *Gig. Sanit.*, 12, 77, 1987 (in Russian).
3. Shiedeman, F. E. and Procita, L., The pharmacology of the monomethyl ethers of mono-, di- and tripropylene glycol in the dog with observations on the auricular fibrillation produced by these compounds, *J. Pharmacol. Exp. Ther.*, 102, 79, 1951.
4. Korkach, V. I. and Spitkovskaya, L. D., The effect of triethylene glycol on biochemical processes in animals, *Gig. Sanit.*, 5, 91, 1986 (in Russian).
5. Bossert, N. L., Reel, J. R., Lawton, A. D., George, J. D., Lamb, J. C., Reproductive toxicity of triethylene glycol and its diacetate and dimethyl ether derivatives in continuous breeding protocol in Swiss CD-1 mice, *Fundam. Appl. Toxicol.*, 18, 602, 1992.
6. Schuler, R. L., Hardin, B. D., Niemeier, R. W., Booth, G., Hazelden, K., Piccirillo, V., and Smith, K., Results of testing fifteen glycol ethers in a short-term *in vivo* reproductive toxicity assay, *Environ. Health Perspect.*, 57, 141, 1984.
7. Munro, I. C., Ford, R. A., Kennepohl, E., and Sprenger, J. G., Correlation of structural class with No-Observed-Effect Levels: A proposal for establishing a threshold of concern, *Food Chem. Toxicol.*, 34, 829, 1996.
8. McKennis, H., Turner, R. A., Turnbull, L. B., et al., The excretion and metabolism of triethylene glycol, *Toxicol. Appl. Pharmacol.*, 4, 441, 1962.

TRIPROPYLENE GLYCOL

Molecular Formula. $C_9H_{20}O_4$

M = 192.29

CAS No 24800-44-0

RTECS No YK6825000

Abbreviation. TPG.

Synonyms. 2-[2-(2-Hydroxypropoxy)propoxy]-1-propanol; [(1-Methyl-1,2-ethanediyl)bis(oxy)]bispropanol.

Properties. Colorless and odorless liquid. Soluble in water and alcohol.

Applications. An intermediate in the manufacture of resins. A plasticizer for hydroxypropyl cellulose. A solvent for gum and resins.

Acute Toxicity. LD_{50} is 3.0 g/kg BW in rats.[011]

Toxic Action. TPG is CNS and cardiac depressant of low toxicity; it produced arrhythmia and fibrillation.

Regulations. *EU* (1990). TPG is available in the *List of authorized monomers and other starting substances which shall be used for the manufacture of plastic materials and articles intended to come into contact with foodstuffs (Section A).*

Reference:

Shiedeman, F. E. and Procita, L., The pharmacology of the monomethyl ethers of mono-, di- and tripropylene glycol in the dog with observations on the auricular fibrillation produced by these compounds, *J. Pharmacol. Exp. Ther.*, 102, 79, 1951.

TRIXYLYL PHOSPHATE

Molecular Formula. $C_{24}H_{27}O_4P$
M = 410.45
CAS No 25155-23-1
RTECS No ZE8320000
Abbreviation. TXP.

Synonyms. Phosphoric acid, trixylyl ester; Phosphate xylenol; Trixylenyl phosphate.

Composition. A mixture of isomers differing in the position of two CH_3- groups in the xylyl radicals.

Properties. Dark, oily liquid. Poorly soluble in water, readily soluble in oil. Perception threshold for the effect on the organoleptic properties of water is 0.05 mg/l.

Applications. Used as a plasticizer in the production of food-contact plastics.

Acute Toxicity. LD_{50} values of *m*-TXP and *p*-TXP are 24 and 25 g/kg BW, respectively.[1]

According to other data, the maximum tolerated dose in mice is 5.0 g/kg BW, and LD_{50} is 12 g/kg BW; rats tolerate 20 g/kg BW. Administration of lethal doses led to weakness and adynamia. The fur of animals becomes disheveled. Histology examination revealed diffuse congestion in the renal tissue.[2]

Administration in the vegetable oil increases toxicity. LD_{50} of di-3,5-xylylphenylphosphate for mice is 7.4 g/kg BW. Administration did not cause paralysis.[3]

Repeated Exposure. TXP has marked cumulative properties and is retained almost entirely in the body. Rats received 1/30 LD_{50} for a month. The treatment led to exhaustion and decline in cholinesterase activity. Gross pathology examination revealed areas of congestion in the brain, gliosal nodules, invagination of the vessel walls, nerve cell changes, and fatty dystrophy of the renal tubular epithelium.[2] No paralyses were observed in chickens in the course of forty daily administrations of 1.0 g/kg BW into the crop of rooster.[3]

Mutagenicity.

In vitro genotoxicity. TXP, mixed isomers, were negative in four *Salmonella typhimurium* strains with and without metabolic activation system.[015]

Standards. ***Russia*** (1995). MAC and PML: 0.05 mg/l (organolept., odor).

References:

1. Zilber, Yu. D., Toxicology of phosphororganic plasticizers, in *Problems of General and Industrial Toxicology,* Leningrad, 1965, 135 (in Russian).
2. Ayzenshtadt, V. S. and Minyukova, G. D., in *Proc. Research Institute Epidemiol. Hyg.,* Kuibyshev, 5, 94, 1968 (in Russian).
3. Dvorkin, E. A., in *Problems of Occupational Hygiene, Occupational Pathology and Toxicology in Phosphorganic Plasticizer Production and Use*, Moscow, 1973, 80 (in Russian).

VASELINE OIL

Abbreviation. VO.

Composition. A mixture of hydrocarbons. Purified fractions obtained after distillation of kerosene. Polycyclic aromatic hydrocarbons, in particular benzo[a]pyrene, are found in technical grade VO. These compounds are not present in VO that had undergone double purification.

Properties. A colorless oily liquid without odor or taste. Poorly soluble in water.

Applications. Used as a plasticizer in the production of food-contact rubber and plastics.

Acute Toxicity. 15 g/kg BW of technical grade VO caused no mortality in rats and mice.[1]

Repeated Exposure revealed slight cumulative properties.

Long-term Toxicity. No signs of intoxication or carcinogenicity effect were noted in rats given 0.1 g/kg BW over a period of 10 months.[2]

Carcinogenicity. No carcinogenic action was observed on subdermal administration to rats in the groin region.

Regulations. ***Russia.*** Medical grade VO is provisionally permitted by the Ministry of Health for use in the production of rubbers intended for contact with food products.

Standards. ***Russia.*** PML: *n/m*.

References:

1. Vysheslavova, M. Ya., Toxicity of vaseline oil, *Voprosy Pitania,* 3, 73, 1976 (in Russian).
2. *Toxicology of the Components of Rubber Mixes and of Rubber and Latex Articles*, Centr. Research Institute Petrol. Chem., Moscow, 1974, 20 (in Russian).

XYLITOL

Molecular Formula. $C_5H_{12}O_5$
M = 152.15
CAS No 87-99-0
RTECS No ZF0800000

Trade Names. Klinit; Pentane pentol; Xylit.

Properties. Colorless hygroscopic crystals. Water solubility is 30% (20°C). Soluble in alcohol. Has a cooling, sweet taste. Renders water a foreign taste at a concentration of 1.0 g/l.[1]

Applications. Used as a plasticizer in the production of cellophane and other food-contact plastics.

Exposure. X. is a sweet five carbon sugar alcohol that has been recommended as a sugar substitute for special dietary uses.

Acute Toxicity.

Observations in man. Human volunteers were served single oral doses of X. either incorporated with water or within test meals. The safety of small amounts of X. in the diet appeared to be evident.[2]

Animal studies. LD_{50} is 17.3 or 22 g/kg BW in rats, 12.5 g/kg BW in mice, and 25 g/kg BW in rabbits.[2,3] Adaptation to dietary xylitol did not alter these values.[4] Lethal doses produce narcotic action. Death was within 6 hours. Gross pathology examination revealed hemorrhagic foci in the mucosa of the stomach and intestine.[1]

Repeated Exposure. Long-Evans male rats were fed a X.-containing diet for 4 weeks. The amount of polyol in the diet was increased from 5.0% to the final 20% level within 3 weeks. The treatment produced retardation of BW gain and a 4-fold increase in the titratable acid excretion. Urinary *pH* was lowered (from 6.6 to 5.6). Increase in daily urine volumes by 49% was noted.[5]

Short-term Toxicity. Mice and rats were dosed by gavage with 1/10 and 1/5 LD_{50} for 20, 30 and 90 days. The treatment a rise in the blood histamine level, and a shortening of the period of barbiturate narcosis.[1] A reversible enlargement of the caecum was reported.[3]

In subchronic feeding studies in Wistar and Sprague-Dawley rats, and CD-1 and NMRI mice, the major observation was the reversible enlargement of the caecum. Higher amounts of X. in the diet may cause transient diarrhea and GI disturbances.[3,4]

Long-term Toxicity. Rats and rabbits received X. by gavage for 7 months. The treatment caused disturbances in the functional state of the cerebral cortex cells, in mediator and nuclear metabolism.[1]

Wistar rats, CD-1 and NMRI mice were given oxalic acid via gastric gavage or with water at the dose of 0.625 g X./kg BW. There was an increase in the absorption and urinary excretion of dietary *oxalate* in X.-treated mice.[4] Animals were found capable of adapting 20% dietary X.[2]

Rats were given 100 mg X./kg BW in their diet for 330 days.[2] Munro et al. suggested this dose to be a NOEL for this study.[6]

Immunotoxicity. In a long-term study, administration of X. did not affect immunoreactivity of rats.

Reproductive Toxicity. In the 3-generation reproduction study, 20% X. in the diet caused no impairment in NMRI mice.[3]

Mutagenicity.

In vitro genotoxicity. Produced a doubling of the SCE frequency, and a 3-point monotonic increase with at least the highest dose tested in this study at the $p < 0.001$ significance level.[7]

Carcinogenicity. An increased incidence of bladder tumors was observed in male mice but not in rats fed diets containing 10 or 20% X. This increase has been associated with epithelial hyperplasia and urinary bladder calculi consisting mainly of calcium oxalate.[3]

Chemobiokinetics. X. metabolism occurs through intensive oxidation within a few hours after administration. Neither X. nor its metabolites are found in the urine. Increased amounts of *methylmalonic* *acid* and *2-oxoglutaric acid* are noted in the urine of polyol-fed rats. The urinary excretion of *citric acid* and *maleic acid* was also increased significantly. The increased levels of urinary organic acids may be explained in terms of impaired mitochondrial oxidation of these acids and impaired conversion of *methymalonic acid* to *succinic acid*.[5]

Wistar rats were given *oxalic acid* via gastric gavage either with water or with 625 mg X./kg BW. The author indicated an increase in the absorption and urinary excretion of dietary oxalate in xylitol-treated mice.[3]

Regulations. ***U.S. FDA*** (1998) approved the use of X. in foods for special dietary uses in accordance with the conditions prescribed in 21 CFR part 172.395.

Recommendations.

EU (1995). X. is a food additive generally permitted for use in liqueurs.

Joint FAO/WHO Expert Committee on Food Additives. ADI is not specified.

Standards. *Russia.* Recommended MAC and PML: 1.0 mg/l.

References:

1. Plitman, S. I., Experimental validation of the Maximum Allowable Concentration of pentaerythritol and xylitol in water bodies, *Gig. Sanit.*, 2, 25, 1971 (in Russian).
2. *Summary of Toxiciological Data of Certain Food Additives,* 21st Meeting of the Joint FAO/WHO Expert Committee on Food Additives, WHO Food Additives Series No 12, 1978.
3. Salminen, S., Bridges, J. M., and Marks, V., Study on the toxicity and biological properties of xylitol in rats, mice and man, in The 3rd Int. Congress on Toxicology, San Diego, Abstracts., *Toxicol. Lett.*, 18 (Suppl. 1), 37, 1983.
4. Salminen S., Bridges, J. W., and Marks, V., Studies on the toxicological and biological properties of xylitol in rats, mice and man, Abstract, *Toxicologist*, 13, 37, 1993.
5. Hamalainen, M. M., Organic aciduria in rats fed high amounts of xylitol or sorbitol, *Toxicol. Appl. Pharmacol.*, 90, 217, 1987.
6. Munro, I. C., Ford, R. A., Kennepohl, E., and Sprenger, J. G., Correlation of structural class with No-Observed-Effect Levels: A proposal for establishing a threshold of concern, *Food Chem. Toxicol.*, 34, 829, 1996.
7. Tucker, J. D., Auletta, A., Cimino, M. C., et al., Sister-chromatid exchange: Second Report of the GENE-TOX Program, *Mutat. Res.*, 297, 101, 1993.

CHAPTER 3. STABILIZERS AND ANTIOXIDANTS

ACETIC ACID, COPPER (2^+) SALT

Molecular Formula. $C_4H_6O_4.Cu$
M = 181.64
CAS No 142-71-2
RTECS No AG3480000
Abbreviation. CA.

Synonyms and **Trade Names.** Acetic acid, cupric salt; Copper (II) acetate; Copper diacetate; Crystals of Venus; Cupric acetate; Neutral verdigris.

Properties. Dark-green, monoclinic, odorless crystals. Readily soluble in cold (72 g/l) and hot (200 g/l) water, and in alcohol.

Applications. A stabilizer for polyurethanes and nylons. Used as human drugs (in tablets), as well as a sanitizer, disinfectant, bactericide. Effective catalyst in the polymerization of styrene, acrylonitrile, and vinyl pyridines. A preservative for cellulose ethers.

Acute Toxicity. LD_{50} is 0.5 g/kg BW in rats, and 0.2 g/kg BW in mice. Poisoning caused CNS inhibition and convulsions.[1] The kidney lesions included hematuria, proteinuria, oliguria, and/or uremia.[2]

Repeated Exposure. Rats received 0.125% CA in their drinking water for 30 days. There were no signs of liver necrosis. Meanwhile, there was impairment of liver function: reduced bile and bromsulfalein secreting capacities (Gaeta et al., 1980).

Long-term Toxicity. Rats received the diet containing 5,000 ppm CA for 16 months. Heavy accumulation occurred in the liver and kidney. Rats received 1250 ppm in the drinking water for 2 years. Accumulation was noted in the liver, kidneys, brain, and bowels.[3]

Reproductive Toxicity.

Embryotoxicity manifestations were observed in chick embryos.[4] Wistar rats received CA in their drinking water before and during pregnancy. Manifestations of embryotoxic action included moderate retardation of growth and differentiation, especially of the neural tube. Numbers of ossification centers in the vertebrae, sternum and forelimb, and hindlimb phalanges were found reduced in fetuses and newborn rats.[5]

Mutagenicity.

In vivo cytogenetics. Cu^{2+} produced a weak but significant increase in frequency of CA and SCE in rats and mice receiving it in toxic doses.[6]

Chemobiokinetics. High doses of CA affect the liver function due to the great capacity for biliary excretion of copper and intracellular sequestration of the metal in a non-toxic form.[2] Histolological examination and a copper analysis revealed copper deposition in the liver and kidney with a subsequent inflammatory reaction in these organs of the pregnant rats exposed to CA in their drinking water.[5]

Regulations. ***U.S. FDA*** (1998) approved the use of CA as an antioxidant and/or stabilizer in polymers used in the manufacture of articles or components of articles intended for use in producing, manufacturing, packing, processing, preparing, treating, packaging, transporting, or holding food, subject to the provisions and definitions set in 21 CFR part 178.2010.

References:

1. *Indian J. Pharmacol.*, 23, 153, 1991.
2. Friberg, L., Nordberg, G. F., Kessler, E., and Vouk V. B., Eds., *Handbook of the Toxicology of Metals*, 2nd ed., vol 2, Elsevier Science Publishers B. V., Amsterdam, 1986, 246.
3. U.S. EPA, *Health Issue Assessment: Copper,* EPA/600/8-87/001, 1987, 33.
4. Korhonen, A., Hemminki, K., and Vainio, H., Embryotoxicity of industrial chemicals on the chicken embryo: dithiocarbamates, *Teratogen., Carcinogen., Mutagen.*, 3, 2, 163, 1983.

5. Haddad, D. S., al-Alousi, L. A., and Kantarjian, A. H., The effect of copper loading on pregnant rats and their offspring, *Funct. Dev. Morphol.*, 1, 17, 1991.
6. Nadeyenko, V. G., Goldina, I. R., Dyachenko, O. Z., and Pestova, L. V., Comparative informative value of chromosomal aberrations and sister chromatid exchanges in the evaluation of environmental metals, *Gig. Sanit.*, 10, 3, 1997 (in Russian).

5-ACETOAMINOTETRAZOLE

Molecular Formula. $C_3H_5N_5O$

M = 127.12

Trade Name. Tetrazole.

Properties. White crystals soluble in water and alcohol when heated.

Applications. Used as a stabilizer in the manufacture of plastic.

Acute Toxicity. In mice, LD_{50} is 5.24 g/kg BW. In 2 to 3 hours after poisoning, the treated animals exhibited CNS inhibition with convulsions. Death occurs in 1 to 3 days.

Repeated Exposure. Rats received 30 administrations of 2.0 mg A./kg BW dose. The treatment did not affect hematology analyses and liver and kidney functions. An increase in STI value and reduction in the relative heart weights were noted.

Reference:

Gnezdilova, A. I. and Volodchenko, V. A., Toxicological evaluation of 5-acetoaminotetrazole, *Gig. Truda Prof. Zabol.*, 12, 58, 1983 (in Russian).

4-ALK(C_7-C_9)OXY-2-HYDROXYBENZOPHENONE

Molecular Formula. $C_6H_5COC_6H_3OHOR$

Composition. A mixture of products where $R = C_7H_{15}$ - C_9H_{19}.

Trade Name. Benzone OA.

Properties. Viscous, light-orange liquid. Almost insoluble in water, soluble in ethanol.

Applications. Used as a light stabilizer of polyolefins, polystyrene, pentaplast, polyvinyl chloride, cellulose acetobutyrate and other polymers.

Acute Toxicity. LD_{50} is reported to be 10 g/kg BW in rats, and 8.7 g/kg BW in mice.

Repeated Exposure failed to reveal cumulative effect. Mice tolerate 0.75 to 3.0 g/kg BW, and rats 1.0 g/kg BW doses without signs of intoxication.

Long-term Toxicity. Treatment with 150 and 200 mg/kg BW (mice) or 200 mg/kg BW (rats) doses revealed no changes in BW gain, hematology analyses, conditioned reflex activity, and structure of the visceral organs.

Allergenic Effect not found.

Standards. ***Russia*** (1995). MAC and PML: *n/m.*

Reference:

Mikhailets, I. B., Toxicity of some stabilizers for plastics, *Plast. Massy*, 12, 41, 1976 (in Russian).

N-ALKYL(C_7-C_9)-*N'*-PHENYL-*p*-PHENYLENEDIAMINE

Molecular Formula. C_6H_5NH - C_6H_4NHR

M = 282.42 to 310.47

Composition. A mixture of products where R is C_7H_{15} to C_9H_{19}.

Properties. Thick, dark-brown liquid. Insoluble in water, readily soluble in alcohol.

Applications. Used as a thermostabilizer of butadiene-styrene and polyisoprene rubber.

Acute Toxicity. LD_{50} is 4.0 g/kg BW in rats, and 3.16 g/kg BW in mice. Poisoning is accompanied by CNS inhibition. Histology examination revealed fatty dystrophy and perivascular infiltration by lymphohistiocytic elements in the liver, moderate fatty dystrophy of the kidneys, and diffuse infiltration by the intersti tial cells of the intermuscular spaces of the myocardium.

Reference:

Vorob'yeva, R. S. and Mezentseva, N. V., in *Toxicology of New Chemical Substances Used in Rubber and Tyre Industry*, Z. I. Israel'son, Ed., Meditsina, Moscow, 1968, 35 (in Russian).

2-AMINO-4-*p*-HYDROXYPHENYLAMINO-1,3,5-TRIAZINE

Molecular Formula. $C_9H_9N_5$

M = 203.21

Properties. White, odorless powder. Poorly soluble in water and alcohol.

Applications. Used as a thermo- and light-stabilizer of polyamides and polyurethanes.

Acute Toxicity. LD_{50} was reported to be 520 mg/kg BW in mice and 1666 mg/kg BW in rats. In rats, manifestations of the toxic effect develop in 2 days and include edema and cyanosis of the tail and ears and ulcers on the paws. Death occurs within 3 to 12 days. In mice, the lethal dose caused weakness, adynamia, and hind leg paresis in 10 to 12 hours after administration. Death occurred in 2 to 4 days. Gross pathology examination revealed distention of the stomach and vascular, intestine and liver congestion.[1]

Repeated Exposure revealed evident cumulative properties. K_{acc} is 1.4 (by Kagan). Rats were exposed to A. for 2 months. In this study, 10 out of 11 rats given 166 mg/kg BW died by the end of administration period while only 7 out of 18 animals treated with 83 mg/kg BW died. Rats tolerated administration of 33 mg/kg BW without visible signs of intoxication. The first and second groups exhibited exhaustion, changed hematology indices, and increased protein content in the urine. The animals in the third group exhibited these changes only to the end of the treatment.[2]

References:

1. Vlasyuk, M. G., in *Hygiene and Toxicology*, Proc. Sci. Conf. Junior Hygienists, L. I. Medved', Ed., Kiyv, 100, 1967 (in Russian).
2. Statsek, N. K. and Vlasyuk, M. G., Data on toxicology of new stabilizers of polymers, derivatives of 4-oxydiphenylamine, arensulfone derivatives of aminophenols, amino-arylamino-triazines and alkylated salycilic acids, in *Hygienic Aspects of the Use of Polymeric Materials and Articles Made of Them*, L. I. Medved', Ed., All-Union Research Institute of Hygiene and Toxicology of Pesticides, Polymers and Plastic Materials, Kiyv, 1969, 314 (in Russian).

2-AMINO-4-*p*-NAPHTHYLAMINO-1,3,5-TRIAZINE

Molecular Formula. $C_{13}H_{11}N_5$

M = 237.27

Properties. White, odorless powder. Poorly soluble in water and alcohol.

Applications. Used as a thermo- and light stabilizer of polyamides and polyurethanes.

Acute Toxicity. LD_{50} was reported to be 1.35 g/kg BW in mice and 5.34 g/kg BW in rats.[1] General weakness and reduced pain response developed in animals in 1 to 5 days after poisoning. Death within 1 to 10 days. Gross pathology examination revealed congestion in the visceral organs and brain.[2]

References:

1. Vlasyuk, M. G., in *Hygiene and Toxicology*, Proc. Sci. Conf. Junior Hygienists, L. I. Medved', Ed., Kiyv, 100, 1967 (in Russian).
2. Statsek, N. K. and Vlasyuk, M. G., Data on toxicology of new stabilizers of polymers, derivatives of 4-oxydiphenylamine, arensulfone derivatives of aminophenols, amino-arylamino-triazines and alkylated salycilic acids, in *Hygienic Aspects of the Use of Polymeric Materials and Articles Made of Them*, L. I. Medved', Ed., All-Union Research Institute of Hygiene and Toxicology of Pesticides, Polymers and Plastic Materials, Kiyv, 1969, 314 (in Russian).

4-(*o*-AMINOPHENYLIMINOMETHYL)-2,6-DI-*tert*-BUTYLPHENOL

Molecular Formula. $C_{21}H_{28}N_2O$

M = 324.5

Synonym. *N*-(3,5-Di-*tert*-butyl-4-hydroxybenzylidene)-*o*-phenylenediamine.

Applications. Used as a stabilizer in the plastic manufacture.

Acute Toxicity. In mice, the LD_{50} is 15.0 g/kg BW.[08] The treatment caused slight alteration of osmotic stability of the erythrocytes and mild methemoglobinemia. Histology examination revealed circulatory disturbances in the visceral organs, necrotic changes in the intestinal mucosa and parenchymatous dystrophy of the renal tubular epithelium.

Short-term Toxicity. Rats received 100 mg/kg BW for 3.5 months. The treatment caused mild methemoglobinemia and stimulation of conditioned reflex activity. Retardation of BW gain was noted. Gross pathology examination revealed superficial necrosis of the villi and desquamation of the intestinal mucosa epithelium.[08]

ANTIOXIDANT VS-1

Properties. Solid, aminophenol-type resin. Almost insoluble in water. Readily soluble in alcohol.

Applications. Used as a stabilizer in the rubber manufacture.

Acute Toxicity. LD_{50} was not attained in rats. A 36 g/kg BW dose caused no mortality. The only signs of poisoning observed were a loss of tonus and refusal of food. Full recovery was noted in 3 to 4 days. Administration of A. as a 20% suspension in a 3.0% starch solution produced no evident toxic effect.

Short-term Toxicity. Rats were given 0.1 and 1.0 g A./kg BW for 4 months. The treatment affected liver function, but no changes in blood enzyme activity were found.

Reference:

Taradin, Ya. I. and Kuchmina, N. Ya., in *Current Problems of Industrial Hygiene and Occupational Pathology*, Voronezh, 74, 1975 (in Russian).

BARIUM compounds

	CAS No	RTECS No
Barium acetate	543-80-6	AF4550000
Molecular Formula. $C_4H_6O_4.Ba$		
M = 255.44		
Synonym. Acetic acid, barium salt; Barium diacetate.		
Barium caprylate	4696-54-2	RHO367200
Molecular Formula. $C_{16}H_{30}O_4.Ba$		
M = 425.82		
Barium chloride	10361-37-2	CQ8750000
Molecular Formula. $BaCl_2$		
M = 208.24		
Barium chloride dihydrate	10326-27-9	CQ8751000
Molecular Formula. $BaCl_2.2H_2O$		
M = 244.28		
Barium peroxide	1304-29-6	CR0175000
Molecular Formula. BaO_2		
M = 169.36		
Synonyms. Barium binoxide; Barium dioxide; Barium superoxide.		
Barium stearate	6865-35-6	WI2840000
Molecular Formula. $C_{36}H_{70}O_4.Ba$		
M = 704.40		
Barium sulfate	7727-43-7	CRO600000
Molecular Formula. $O_4S.Ba$		
M = 233.40		
Barium-Cadmium laurate	15337-60-7	OE9805000
Barium-Cadmium stearate	1191-79-3	WI2830000
Molecular Formula. $C_{72}H_{140}O_8.BaCd$		
M = 1383.86		

Properties.

Barium chloride. Colorless crystals (usually + H_2O). Solubility in water is 36.2 g/10 g at 20°C, insoluble in al cohol. Threshold for effects on the organoleptic properties of water is 4.0 mg/l.[1]

Barium-cadmium laurate contains 13.8 to 14.5% cadmium and 8.0 to 9.2% barium.

Barium stearate. White, amorphous powder. Insoluble in water and ethanol.

Barium peroxide. White or grayish-white, odorless heavy powder. Poorly soluble in cold water (about 15 g/l at 0°C)

Barium-cadmium stearate. Finely dispersed, white or yellow powder. Difficult to wet and poorly soluble in water.

Barium sulfate. White, odorless powder. Solubility in water is 0.22 mg/100 g at 18°C and 0.41 mg/100 g at 100°C.

Barium acetate and *barium nitrate* are soluble in water, but *barium chromate, fluoride, oxalate, phosphate*, and *sulfate* are quite insoluble.

Applications. Barium compounds are used in plastics and rubber, in ceramic glazes and enamels, in glass-making, and in cosmetics. Some barium salts are used as stabilizers of polyvinyl chloride and other plastics. Barium sulfate is used as a filler for rubber. Barium acetate is used as a catalyst and a lubricant. Barium peroxide is used as a bleaching and oxidizing agent and decolorizer.

Exposure. Food and water seem to be the main sources of exposure to barium. The long-term mean dietary intake is about 1.0 mg barium/day. Intake from the air is negligible.

Acute Toxicity.

Observations in man. Soluble barium salts possess high acute toxicity. 0.2 to 0.5 g/kg BW doses ingested by man caused acute poisoning, and 3.0 to 4.0 g/kg BW or even 0.8 to 0.9 g/kg BW doses are fatal. The symptoms of intoxication include general weakness, dyspnea, and impaired cardiac activity. Toxicity of barium salts is likely to be a function of their aqueous solubility; depending on the dose and solubility of the barium compounds, death may occur in a few hours or a few days. Barium causes strong vasoconstriction by its direct stimulation of the smooth muscles, and convulsions and paralysis following stimulation of the CNS.[03]

Animal studies. Main manifestations of the acute toxic action comprise cardiovascular, GI, NS, hematopoietical and skeletal muscle changes.

Barium acetate. LD_{50} *i/v* is 23.3 and 11.3 mg/kg BW in ICR and Swiss-Webster mice, respectively.[2]

Barium chloride. In rats, LD_{50} is reported to be 118 to 150 mg/kg BW[2,06] or 400 mg/kg BW;[3] the last value was recently confirmed: 419 mg/kg BW in male rats, and 408 mg/kg BW in females. Signs of poisoning develop in 20 to 30 minutes after administration and include apathy and adynamia with subsequent smooth muscle spasm which manifests itself as vomiting, diarrhea, and ejaculation. Paresis of the hindlimbs and convulsions are observed in some rats. Death is preceded with complete adynamia. A single administration of 300 mg/kg BW decreased kidney weights.[4]

Barium caprylate. LD_{50} is reported to be 1.0 g/kg BW in rats, 1.1 g/kg BW in mice, and 1.25 to 1.9 g/kg BW in guinea pigs.[5] The hematopoietic system appeared to be the target tissue.

Barium nitrate. LD_{50} *i/v* is 20.1 and 8.5 mg/kg BW in ICR and Swiss-Webster mice, respectively.[6]

Barium peroxide. LD_{50} *s/c* is 50 mg/kg BW in mice.

Barium stearate. LD_{50} is 1.5 to 4.0 g/kg BW in rats, 2.3 to 5.5 g/kg BW in mice, and 1.9 to 3.6 g/kg BW in guinea pigs.[5,7,8] Poisoning is accompanied by immobility, apathy, and refusal of food. Death occurs in 2 to 8 days. Gross pathology examination revealed parenchymatous and fatty dystrophy of the liver and kidneys.

Barium-cadmium stearate. LD_{50} is 1.98 g/kg BW in mice, a dose of 0.6 g/kg BW causes no mortality. In rats, LD_{50} is 3.17 g/kg BW. Gross pathology examination showed lesions in the GI tract and testes and dystrophic changes in the heart, liver, and kidneys.

Barium sulfate. Pure *barium sulfate* is non-toxic since it is hardly absorbed at all, but the technical product often contains poisonous admixtures of *barium carbonate* and *barium chloride*.

Barium-cadmium laurate. LD_{50} is 1.7 g/kg BW in rats and 0.5 g/kg BW in mice.

Repeated Exposure.

Observations in man. The toxicity of barium is determined primarily by its reaction with calcium and potassium salts, as well as by the neurotropic character of the action of barium and its compounds.

Repeated Exposure (*observations in man*) to barium chloride seems to have caused recurrent outbreaks of "Pa-Ping" disease (a transient paralysis resembling familial periodic paralysis) in China.[9]

In a short-term study of a small number of volunteers, there was no consistent indication of adverse cardiovascular effects following exposure up to 10 mg barium/l in water.[10]

Animal studies. Pronounced cumulative properties of the inorganic barium salts are reported.[3] Barium chloride has K_{acc} equal to 1.8. In a 10-day study, survival in rats given 300 mg/kg BW dose was substantially lower.[4]

Rats were given 100 ppm barium chloride in their feed for a month.[11] The treatment caused no cardiomyopathy or increase in the blood pressure. Rats received barium chloride dihydrate in the drinking water at concentrations up to 2,000 ppm (100 mg barium/kg BW) for 15 days. No chemical-related deaths, differences in final mean BW and organ weights, hematology, neurobehavioral parameters, or clinical signs of toxicity were noted. In the same 15-day study in mice, analogous results were obtained.[12]

Barium-cadmium stearate has marked cumulative properties. On administration of 1/5, 1/10, and 1/20 LD_{50}, K_{acc} appeared to be 1.89, 1.54, and 1.1, respectively. There was, however, little evidence of cumulative properties of barium stearate.

Short-term Toxicity. No clinical signs of toxicity (except for a decrease in the relative weights of the adrenals at the highest dose) or microscopic alterations are seen in rats given tap water containing up to 250 mg barium chloride/l for 13 weeks.[13]

No adverse histology and hematology changes, or effects on serum enzymes were found in rats given the doses of 1.7 to 45.7 mg barium chloride/kg BW via drinking water over the same period of time. No effects on the blood pressure were reported in the rat studies after a 20-week exposure to a dose of 15 mg barium chloride/kg BW.[13] There was, however, an increase in the systolic blood pressure of rats exposed to relatively low concentrations of barium chloride in drinking water. The NOAEL of 2000 ppm was identified in a 3-month study in F344/N rats and $B6C3F_1$ mice given barium chloride in their drinking water.[15]

Sprague-Dawley rats received drinking water containing up to 1000 ppm barium for 16 weeks. No histological or cardiovascular effects were noted. BW gain and food consumption was unaltered. Ultrastructural changes in the kidney glomeruli and the presence of myelin figures were observed only at the 1000 mg/l concentration in drinking water.[16]

Long-term Toxicity. Available epidemiology studies have not shown cardiovascular effect in humans, the available animal studies have demonstrated that hypertension is associated with exposure to barium.

Observations in man. In the most sensitive epidemiological study conducted to date, there were no significant differences in blood pressure or the prevalence of cardiovascular disease between a population exposed to drinking water containing 7.3 mg barium/l compared to one ingesting water containing 0.1 mg barium /l. Thus, the NOAEL in this study appeared to be 7.3 mg barium /kg BW.[17] This value is within the range of that derived from the toxicology studies in animals.

Animal studies. Significant increases in the mean systolic blood pressure were observed in rats exposed to drinking water containing barium chloride over a long period of time: 5.1 mg/kg BW for 16 months and 0.5 mg/kg BW for 8 months.[18,19] The NOAEL was identified to be 0.51 mg/kg BW,[18] and ADI/TDI hence appeared to be 0.05 mg/kg BW (uncertainty factor was considered to be 10, since the results of a well-conducted epidemiology study indicate that humans are not more sensitive than rats to barium in drinking water).

More recent studies of the same authors[13] failed to reveal changes in BW gain, appearance, selected weights, or morphology of rats given 1.0 to 100 ppm barium chloride in drinking water for 16 months. Drinking water containing 10 ppm barium chloride given for 8 months caused no cardiomyopathy or increase in the blood pressure. Sprague-Dawley rats received drinking water containing up to 250 ppm barium for 46 weeks (females) and 68 weeks (males). No histological or cardiovascular effects were noted. BW gain and food consumption were unaltered.

No effects on the blood pressure were reported in the rat studies after a 20-week exposure to a dose of 15 mg barium chloride/kg BW.[13]

Reproductive Toxicity.

Some barium salts may cross the placenta barrier and produce adverse reproductive and teratogenic effects.

Gonadotoxic effect was reported to occur at a dose level of 0.5 mg/kg BW and the NOAEL of 0.005 mg/kg BW was identified for total reproductive effects.[20]

Mutagenicity.

In vitro genotoxicity. A dose of 0.5 mg barium chloride/kg BW was found to be genotoxic (increase in the number of CA without affecting the mitotic index).[20] Barium did not increase the frequency of mutations in repair-deficient strains of *Bac.subtilis*[21] and did not induce errors in viral DNA transcription *in vitro*.[22]

Barium chloride dihydrate was negative in *Salmonella* assay, but it was mutagenic in mouse lymphoma cells. It did not induce SCE or CA in cultured Chinese hamster ovary cells with or without *S9* fraction.[12]

Carcinogenicity. Carcinogenic effect was not observed in extremely limited lifetime bioassay in rats and mice exposed to 5.0 mg barium/l in drinking water, based on gross examination only of tumors at autopsy.[23] There was no evidence of carcinogenicity in the 2-year NTP study when rats and mice received up to 2500 ppm barium chloride dihydrate in their drinking water.[12]

Carcinogenicity classifications.

U.S. EPA: D;

NTP: NE - NE - NE - NE (barium chloride dehydrate, water).

Chemobiokinetics. Not only soluble barium salts but insoluble barium compounds may also be absorbed to a significant extent.[24] A degree of absorption of Ba^{2+} from the GI tract depends on the solubility of the compound, species, the contents of the GI tract, diet, and age. Moreover, insoluble barium salts may be partially solubilized in the acid medium of the stomach. Less than 10% of an ingested quantity seems to be absorbed in adults. In spite of barium sulfate poor solubility, it is not insoluble and no data indicate that dissolved barium sulfate is not absorbed from the GI tract.

Barium is rapidly distributed in the blood plasma, principally to the bones, but it may accumulate in the kidney, liver, and myocardium as well. [25] Its metabolism is similar to that of calcium; however, unlike it, barium has no known biological function.

Excretion of barium compounds occurs mainly with the feces and to a lesser extent with the urine, elimination varying according to the route of administration and the solubility of the compound.[26]

Regulations. ***U.S. FDA*** (1998) approved the use of barium compounds (1) in resinous and polymeric coatings for food-contact surface (barium sulfate) in accordance with the conditions prescribed by 21 CFR part 175.300; (2) as a filler for rubber articles intended for repeated use in contact with food (barium sulfate) in accordance with the conditions prescribed by 21 CFR part 177.2600; (3) in adhesives as a component of articles intended for use in packaging, transporting, or holding food (barium acetate, barium peroxide, and barium sulfate) in accordance with the conditions prescribed by 21 CFR part 175.105; (4) as colorant only of the uncoated or coated food-contact surface of paper and paperboard intended for use in producing, manufacturing, packaging, processing, preparing, treating, packing, transporting, or holding aqueous and fatty foods (barium sulfate) in accordance with the conditions prescribed by 21 CFR part 176.170; and (5) as a catalyst in the manufacture of phenolic resins for food-contact surface of molded articles intended for repeated use in contact with nonacid food (*pH* above 5.0) (barium hydroxide) in accordance with the conditions prescribed by 21 CFR part 177.2410; and (6) as a component of the uncoated or coated food-contact surface of paper and paperboard intended for use in producing, manufacturing, packaging, processing, preparing, treating, packing, transporting, or holding dry food barium metaborate in accordance with the conditions prescribed by 21 CFR part 176.180.

Recommendations. ***WHO*** (1996). Guideline value for drinking water: 0.7 mg/l.

Standards.

EU (1982). MAC in drinking water: 0.1 mg/l.

U.S. EPA (1999). MCL and MCLG: 2.0 mg/l; MPC for colors that may be used in food, drugs, and/or cosmetics (1986): 0.5 g/kg.

Russia (1995). MAC and PML in food and drinking water: 0.1 mg/l.

Canada (1994). MAC in drinking water 0.1 mg/l.

References:

1. Khachidze, Sh. G., in *Protection of Water Reservoirs against Pollution by Industrial Liquid Effluents*, S. N. Cherkinsky, Ed., Medgiz, Moscow, Issue No 4, 1960, 54 (in Russian).
2. Grushko, Ya. M., *Harmful Organic Compounds in the Liquid Industrial Effluents*, Khimiya, Leningrad, 1976, 34 (in Russian).

3. Akinfieva, T. A. and Gerasimova, I. L., Comparative toxicity of some barium compounds, *Gig. Truda Prof. Zabol.*, 6, 45, 1984 (in Russian).
4. Borzelleca, J. F., Condie, L. W., Egle, J. L., et al., *J. Am. Coll. Toxicol.*, 7, 675, 1988.
5. Mitin, L. S., Comparative toxicity investigation of some barium compounds, *Gig. Sanit.*, 11, 91, 1974 (in Russian).
6. Syed, I. B. and Hosain, F., Determination of LD_{50} of barium chloride and allied agents, *Toxicol. Appl. Pharmacol.*, 22, 150, 1972.
7. Antonovich, L. A. and Bake, M. Ya., Toxicity of barium stearate, *Pharmacologia i Toxicology*, 49, 117, 1986 (in Russian).
8. Antonovich, L. A. and Sprudzhas, D. P., Synthesis and study of physiologically active substances, in *Proc. Republ. Sci. Conf.*, December 14, 1984, Vil'nus, 1984, 12 (in Russian).
9. Shankle, R. and Keane, J. R., Acute paralysis from barium carbonate, *Arch. Neurol.*, 45, 579, 1988.
10. Wones, R. G., Stadler, B. L., and Frohman, L. A., Lack of effect of drinking water barium on cardiovascular risk factor, *Environ. Health Perspect.*, 85, 113, 1990.
11. Perry, H. M., Kopp, S. J., Perry, E. F., et al., Hypertension and associated cardiovascular abnormalities induced by chronic barium feeding, *J. Toxicol. Environ. Health*, 28, 373, 1989.
12. *Toxicology and Carcinogenesis Studies of Barium Chloride Dehydrate in F344/N Rats and $B6C3F_1$ Mice (Drinking Water Studies)*, NTP Technical Report Series No 432, Research Triangle Park, NC, January 1994.
13. Tardiff, R. G., Robinson, M., and Ulmer, N. S., Subchronic oral toxicity of $BaCl_2$ in rats, *J. Environ. Pathol. Toxicol.*, 4, 267 1980.
14. McCauley, P. T., Douglas, B. H., Laurie, R. D., et al., Investigations into the effects of drinking water barium on rats, in *Advances in Modern Environ. Toxicology*, Princeton Publ. Co., NY, 9, 197, 1985.
15. Dietz, D. D., Elwell, M. R., Davis, W. E., et al., Subchronic toxicity of barium chloride dianhydride administered to rats and mice in the drinking water, *Fundam. Appl. Toxicol.*, 19, 527, 1992.
16. McCauley, P. T., Douglas, B. H., Laurie, R. D., Bull, R. J., Investigation into the effect of drinking water barium on rats, *Environ. Health Perspect.*, 9, 1985.
17. Brenniman, G. R. and Levy, P. S., Epidemiological study in Illinois drinking water supplies, in *Advances in Modern Environ. Toxicology*, Princeton Publishing Co, NY, 9, 1985, 231.
18. Perry, H. M. Jr., Kopp, S. J., Erlanger, M. W., et al., Cardiovascular effects of chronic barium ingestion, in *Trace Substances Environ. Health.*, D. D. Hemphil, Ed., University of Missouri, Columbia, 16, 155, 1983.
19. Perry, H. M. Jr., Perry, E. F., Erlanger, M. W., et al., Barium-induced hypertension, inorganics in drinking water and cardiovascular disease, in *Advances in Modern Environ. Toxicology*, E. Calabrese, Ed., Princeton Publ. Co., NY, Chapter 20, 1985, 221.
20. Krasovsky, G. N. and Sokolovsky, N. G., Genetic effects of heavy metals, *Gig. Sanit.*, 9, 56, 1979 (in Russian).
21. Nishioka, H., Mutagenic activities of metal compounds in bacteria, *Mutat. Res.*, 31, 185, 1975.
22. Loeb, L., Sirover, M., and Agarval, S., Infidelity of DNA synthesis as related to mutagenesis and carcinogenesis, *Adv. Exp. Med. Biol.*, 91, 103, 1978.
23. Schroeder, H. A. and Mitchener, M., Life-term effects on mercury, methyl mercury and nine other trace elements on mice, *J. Nutr.*, 105, 452, 1975.
24. McCauley, P. T. and Washington, I. S., Barium bioavailability as the chloride, sulfate, or carbonate salt in the rat, *Drug Chem. Toxicol.*, 6, 209, 1983.
25. *Drinking Water and Health*, Natl. Acad. Sci., vol 1, Natl. Res. Council, Washington, D. C., 1977.
26. *Barium*, Health and Safety Guide, IPCS, WHO, Geneva, 1991, 28.

BENZOIC ACID, CALCIUM SALT TRIHYDRATE

Molecular Formula. $C_{14}H_{16}O_7.Ca$

M = 336.36

CAS No 2090-05-3

Synonym. Calcium benzoate.

Properties. Orthorhombic crystals or powder. Solubility in water is 2.67 g/100 g at 0°C. Highly soluble in boiling water.

Applications. Used as a stabilizer in polyvinyl chloride production. A food preservative (fats, fruit juices, etc.).

Acute Toxicity. LD_{50} is found to be 4.0 g/kg BW in rats, and 2.3 g/kg BW in mice. Manifestations of the toxic effect are absent. Death within a few days. Survivors do not differ from the controls.

Repeated Exposure failed to reveal evident cumulative properties. The treatment with a dose of 0.4 g/kg BW for 2 months led to mortality of less than 50% of animals. Gross pathology examination revealed parenchymatous dystrophy of the cardiac and hepatic cells with the loss of separate cells.

Short-term Toxicity. In a 4-month study, 80 mg/kg BW dose caused no retardation of BW gain or changes in hematology analyses, in the liver and kidney functions, and in the relative weights of the visceral organs.

Regulations. ***U.S. FDA*** (1998) approved the use of calcium benzoate in the manufacture of antioxidants and/or stabilizers for polymers which may be used in the manufacture of articles or components of articles intended for use in producing, manufacturing, packaging, processing, preparing, treating, packing, transporting, or holding food, subject to the provisions prescribed in 21 CFR part 178.2010.

Recommendations. Joint FAO/WHO Expert Committee on Food Additives. ADI: 5.0 mg/kg BW.

Standards. ***Russia.*** PML: *n/m.*

Reference:

Belova, G. B., Evaluation of calcium dibenzoate toxicity, in *Hygiene and Toxicology of High-Molecular-Mass Compounds and of the Chemical Raw Material Used for Their Synthesis*, Proc. 4th All-Union Conf., S. L. Danishevsky, Ed., Khimiya, Leningrad, 1969, 198 (in Russian).

2-BENZOTHIAZOLETHIOL, SODIUM SALT

Molecular Formula. $C_7H_5NS_2$

M = 189.23

CAS No 2492-26-4

RTECS No DL6825000

Abbreviation. BTTS.

Synonyms and **Trade Name.** 2-Benzothiazolethiol, sodium salt; 2-Benzothiazolethione, sodium salt; Duodex; 2-Mercaptobenzothiazole, sodium salt; Sodium 2-benzothiazolethiol; Sodium mercaptobenzothiozolate.

Properties. Light-amber liquid (50% aqueous solution). Soluble in water.

Applications. Termal stabilizer for methyl methacrylate copolymers, acrylonitrile polymers, polyester fibers, anion exchange resins, polyoxypropylene, etc. Intermediate in the manufacture of natural and synthetic rubber vulcanization accelerators.

Acute Toxicity. Ingestion caused tremors, convulsions, severe depression, and hematuria in male rats. Mortality was observed at the doses above 300 mg/kg BW. Hemorrhage of the stomach occurred in all rats that died.

Mutagenicity.

In vitro genotoxicity. BTTS was negative in *Salmonella* mutagenicity assay (NTP-94).

Carcinogenicity. Mice received daily oral administration of 100 mg/kg BW dose in 0.5% gelatin for 18 months. The treatment failed to cause a significant increase in tumors incidences.[1]

Chemobiokinetics. Following *i/p* treatment of mice with ^{35}S-mercapto-labeled 2-MBT, it underwent conjugation with glutathione and with glucuronic acid. Radiolabelled *inorganic sulfate* was also identified in the urine (Colucci et al., 1965).

Regulations. ***U.S. FDA*** (1998) approved the use of BTTS (1) in adhesives as a component of articles intended for use in packaging, transporting, or holding food in accordance with the conditions prescribed by 21 CFR part 175.105; as (2) a component of a defoaming agent intended for articles which may be used in producing, manufacturing, packing, processing, preparing, treating, packaging, transporting, or holding food, subject to the provisions prescribed in 21 CFR part 176.200; (3) as a slimicide in the manufacture of paper and paperboard that contact food, subject to the provisions prescribed in 21 CFR part 176.300; (4) in the production of animal glue which may be safely used as a component of articles intended for use in producing, manufacturing, packaging, processing, preparing, treating, packing, transporting, or holding food, subject to the provisions prescribed in 21 CFR part 178.3020; (5) as an adjuvant substance in the manufacture of foamed plastics intended for use in contact with food, subject to the provisions prescribed in 21 CFR part 178.3010; and (6) to be employed in the production of textile and textile fibers intended for use in producing, manufacturing, packaging, processing, preparing, treating, packing, transporting, or holding food in accordance with the conditions prescribed by 21 CFR part 177.2800.

Reference:

1. Innes, J. R. M., Ulland, B. M., Valerio, M. G., et al., Bioassay of pesticides and industrial chemicals for tumorigenicity in mice. A preliminary note, *J. Natl. Cancer Inst.*, 42, 1101, 1969.

1H-BENZOTRIAZOLE

Molecular Formula. $C_6H_5N_3$
M = 119.12
CAS No 95-14-7
RTECS No DM1225000

Synonyms and **Trade Name.** 1,2,3-Aminoazaphenylene; Benzene azimide; 1,2,3-Benzotriazole; Cobratec 99; 2,3-Diazaindole.

Properties. White to light tan, crystalline powder. Sparingly soluble in water, soluble in alcohol.

Applications. Used as a stabilizer in the manufacture of plastics.

Acute Toxicity. LD_{50} is 560 mg/kg BW in rats, 615 mg/kg BW in mice,[1] and 500 mg/kg BW in guinea pigs. Manifestations of toxic action included cyanosis and CNS inhibition.

Mutagenicity.

In vitro genotoxicity of B. has been studied in *Salmonella typhimurium* and *E. coli.*[2]

Carcinogenicity. F344 rats were given 6,700 or 12,100 ppm B. by administering in the feed for 78 weeks. Additional observation period was of 26 to 27 weeks. B6C3F_1 mice were dosed with 11,700 or 23,500 ppm B. for 104 weeks. Under conditions of this bioassay, there was no convincing evidence of B. carcinogenicity in B6C3F_1 mice or F344 rats.[3]

Carcinogenicity classification.

NTP: E - E - N - E (feed).

Regulations. ***U.S. FDA*** (1998) approved the use of B. in surface lubricants employed in the manufacture of metallic articles that contact food.

Standards. ***Russia*** (1995). MAC and PML: 0.1 mg/l.

References:

1. National Technical Information Service, OTS0516797 and AD-A067-313.
2. Dunkel, V. C., Zeiger, E., Brusick, D., McCoy, E., McGregor, D., Mortelmans, K., Rosenkranz, H. S., and Simmon, V. F., Reproducibility of microbial mutagenicity assays: II. Testing of carcinogens and noncarcinogens in *Salmonella typhimurium* and *Escherichia coli*, *Environ. Mutagen.*, 7 (Suppl 5), 1, 1985.
3. *Bioassay of 1H-Benzotriazole for Possible carcinogenicity*, NTP Technical Report Series No 88, Research Triangle Park, NC, 1978.

2-(*2H*-BENZOTRIAZOL-2-YL)-*p*-CRESOL

Molecular Formula. $C_{13}H_{11}N_3O$
M = 225.27
CAS No 2440-22-4
RTECS No GO6860000

Synonyms and **Trade Names.** Benazole P; 2-(2*H*-Benzotriazol-2-yl)-4-methyl phenol; 2-(2'-Hydroxy-5'- methylphenyl)benzotriazole; Tinuvin P.

Properties. Pale-yellow, crystalline powder. Poorly soluble in water, soluble in alcohol.

Applications. Used as a thermostabilizer in the production of polyamides, a light stabilizer for polystyrene, polyvinyl chloride, polymethacrylate, polypropylene, and cellulose acetate.

Acute Toxicity. In mice, LD_{50} is 6.5 g/kg BW. Lethal doses resulted in adynamia with subsequent death on day 2. Gross pathology examination revealed pulmonary hemorrhages and fatty liver dystrophy. According to other data, mice and rats tolerated administration of 5.0 and 10 g/kg BW in vegetable oil suspension.[1,2]

Repeated Exposure revealed slight cumulative properties. The treatment with 1/5 LD_{50} resulted in decreased BW gain, lowered threshold of neuro-muscular excitability and impairment of the detoxifying liver function. Gross pathology examination revealed initial signs of fatty dystrophy in the liver, signs of congestion in the kidneys, and local hemorrhages in the spleen pulp.[023]

Long-term Toxicity. In a 6-month study, rats given 120 mg/kg BW exhibited slight erythrocyte hemolysis. Gross pathology examination revealed desquamation of the epithelium of the small intestine.[1]

Mice and rats were given 10 mg/kg and 50 mg/kg BW for 12 months. Gross pathology examination failed to reveal changes in the visceral organs. A 50 mg/kg BW dose appeared to be the NOAEL.[2]

Reproductive Toxicity effects are not observed in rats exposed to a dose of 5.0 mg/kg BW for 2 years over 3 generations.[05]

Allergenic Effect. B. has a contact sensitizing ability.[3]

Carcinogenicity. Tumor growth was noted in a 2-year oral study in rats.[05]

Regulations. ***U.S. FDA*** (1998) regulates B. for use (1) as a component of non-food articles at levels not to exceed 0.25% by weight of the rigid polyvinyl chloride and/or rigid vinyl chloride copolymers; (2) in polystyrene that is limited to use in contact with dry food in compliance with CFR requirements at levels not to exceed 0.25% by weight of polystyrene and/or rubber-modified polystyrene polymers intended to contact non-alcoholic food, provided that the finished basic rubber-modified polystyrene polymers in contact with fatty foods shall contain not less than 90% by weight of the total polymer units derived from styrene monomers; (3) at levels not to exceed 0.5% by weight of the polycarbonate resins, provided that the finished polycarbonate resins contact food only of specified types and under the specified conditions of use; and (4) at levels not to exceed 0.5% by weight of the ethylene-1,4-cyclohexylene dimethylene terephthalate copolymers and of ethylene phthalate polymers that contact food only under the conditions specified in 21 CFR.

Standards. ***Russia*** (1994). PML in food: 2.0 mg/l.

References:

1. Putilina, L. V., Data on toxicity of some aromatic amines and azomethines, *Gig. Truda Prof. Zabol.*, 3, 49, 1966 (in Russian).
2. *Toxicology and Sanitary Chemistry of Plastics*, Abstracts, NIITEKHIM. Moscow, Issue No 1, 1979, 25 (in Russian).
3. Ikarashi, Y., Tsuchiya, T., and Nakamura, A., Contact sensitivity to Tinuvin P in mice, *Contact Dermat.*, 30, 226, 1994.

p-BENZYLIDENAMINOPHENOL

Molecular Formula. $C_{13}H_{11}NO$

M = 197.24

CAS No 588-53-4

RTECS No SJ7650000

Synonyms. 4-(Benzylideneamino)phenol; Benzylidene-4-hydroxyaniline; *N*-Benzylidene-*p*-hydroxyaniline; 4-(Phenylmethyleneamino)phenol.

Properties. Colorless, crystalline substance. Poorly soluble in water, readily soluble in alcohol.

Applications. Used as a stabilizer in the production of plastics.

Acute Toxicity. In mice, LD_{50} is 3.5 g/kg BW in rats.[1] Administration of the lethal doses slightly affected osmotic stability of the erythrocytes but caused methemoglobinemia. Gross pathology examination revealed circulatory disturbances in the visceral organs, necrotic changes in the small intestine mucosa, and parenchymatous dystrophy of the renal tubular epithelium.[029]

Short-term Toxicity. Rats received 100 mg/kg BW for 3.5 months. Manifestations of the toxic effect included retardation of BW gain, slight methemoglobinemia, and stimulation of conditioned reflex activity. Gross pathology examination revealed necrosis of the *villi* and sloughing of the epithelium of the small intestine mucosa.[08]

p-(BENZYLOXY)PHENOL

Molecular Formula. $C_{13}H_{12}0_2$
M = 200.25
CAS No 103-16-2
RTECS No SJ7700000
Abbreviation. BOP.

Synonyms and **Trade Names.** Agerite; Benzyl hydroquinone; Monobenzone; Pigmex.

Properties. White and odorless crystalline powder. Solubility in boiling water is about 10 g/l, soluble in alcohol.

Applications. Antioxidant in the production of polyvinyl chloride, polypropylene, and polypropylene oxide. Used as a rubber antioxidant, stabilizer, polymerization inhibitor, and chemical intermediate.

Acute Toxicity. LD_{50} exceeded 3.2 g/kg BW in rats (sodium salt).[03] LD_{50} *i/p* is 4.5 g BOP/kg BW.[020]

Repeated Exposure. Guinea pigs ingested BOP at the level of 160 mg/kg BW for 2 months. The treatment did not cause toxic changes. The only change noted was a depigmentation of the hair.[03]

Carcinogenicity. Oral administration over a period of 18 month caused lung and liver tumors in mice. Equivocal carcinogenic agent by *RTECS* criteria.

Chemobiokinetics. BOP may readily be metabolized in the same manner as phenol.

Regulations. *U.S. FDA* (1998) approved the use of BOP (1) in adhesives as a component of articles intended for use in packaging, transporting, or holding food in accordance with the conditions prescribed in 21 CFR part 175.105; and (2) in the manufacture of cross-linked polyester resins which may be used as articles or components of articles intended for repeated use in contact with food in accordance with the conditions prescribed in 21 CFR part 177.2420.

Reference:

Natl. Technical Information Service PB223-159.

2,5-BIS(*tert*-BUTOXY)-2,5-DIMETHYLHEXANE

Molecular Formula. $C_{16}H_{34}O_2$
M = 258.50
RTECS No MO0720000

Applications. Used as a stabilizer in the manufacture of polymeric materials.

Acute Toxicity. LD_{50} is 2.4 g/kg BW in mice.[027]

2,2'-BIS(6-*tert*-BUTYL-*p*-CRESYL)METHANE

Molecular Formula. $C_{23}H_{32}0_2$.
M = 340.55
CAS No 119-47-1
RTECS No PA3500000

Synonyms and **Trade Names.** Advastab 405; Agidol 2; Antioxidant 2246; Bisalkofen BP; Bis(3-*tert*-butyl-2-hydroxy-5-methylphenyl) methane; CAO-5; 2,2'-Methylenebis(6-*tert*-butyl-*p*-cresol); 2,2'-Methylenebis(6-*tert*- butyl-4-methylphenol).

Properties. White, crystalline powder. Almost insoluble in water, readily soluble in alcohol.

Applications. Used as a thermo- and light-stabilizer in the production of natural and synthetic rubber, polyolefins, pentaplast, and polystyrene.

Acute Toxicity. Adult rats tolerate administration of 25 g/kg BW, young rats 36 g/kg BW and young and adult mice 30 g/kg BW without manifestations of the toxic effect. An increased thirst was noted.[1]

According to Stasenkova et al.,[2] however, the LD_{50} seems to be 5.0 g/kg BW in rats, and 11 g/kg BW in mice. Marked CNS depression and refusal of food were accompanied by severe weight loss.

Repeated Exposure failed to reveal marked cumulative properties. When technical and pure B. were administered over a period of one month, the animals died from the total doses of 15.5 and 28 g/kg BW, respectively. K_{acc} (by Lim) appeared to be 3.0 and 6.2 for technical grade product and pure chemical, respectively.[2]

Sprague-Dawley male rats received 1.135 mmol % B. in the diet for 1 week. The treatment produced little change in the liver lipids.[3]

Short-term Toxicity. Mice received 250 mg/kg BW for 17 weeks. This treatment caused slight stimulation of conditioned reflex activity. Gross pathology examination revealed mild changes in the gastric and intestinal mucosa.[08]

Long-term Toxicity. Rats were dosed by gavage with 1.0 g/kg BW for 5.0 months. The treatment did not produce clear symptoms of intoxication. Slightly decreased motor activity and BW gain, and transient leukopenia were noted. Gross pathology examination revealed dystrophic changes in the epithelial cells of the liver, myocardium, renal tubular epithelium and bronchi. There was parenchymatous dystrophy in the lungs. Capillary congestion was present in all the organs.[1] In a 10-month experiment,[08] rats received a 50 mg/kg BW dose of pure substance in the diet. Manifestations of the toxic action included reversible functional changes in the NS (STI), liver, and oxygen consumption. A dose of 10 mg/kg BW caused changes in the detoxicant function of the liver. Gross pathology examination failed to reveal significant changes in the visceral organs.[2]

Reproductive Toxicity.

Embryotoxicity. Rats were dosed with 250 mg/kg BW (in the diet) through pregnancy. The treatment produced an increase in post-implantation mortality.[5]

Gonadotoxicity. A decrease in fertility of rats is observed after prolonged administration of doses exceeding 50 mg/kg BW.

Carcinogenicity. B. is shown to cause an inhibiting effect on the induction of tumors by benzo[a]pyrene.[4]

Allergenic Effect is not observed.

Regulations. ***U.S. FDA*** (1998) regulates B. for use (1) at levels not to exceed 0.1% by weight of the olefin polymers used in articles that contact food; (2) at levels not to exceed 1.0% by weight of the polyoxymethylene copolymers; and (3) at levels not to exceed 0.5% by weight of the polyoxymethylene homopolymers in compliance with existing regulation (CFR). M. may be used (4) in adhesives as a component of articles intended for use in packaging, transporting, or holding food in accordance with the conditions prescribed by 21 CFR part 175.105; (5) in the manufacture of rubber articles intended for repeated use in producing, manufacturing, packaging, processing, preparing, treating, packing, transporting, or holding food in accordance with the conditions prescribed by 21 CFR part 177.2600; and (6) in the manufacture of antioxidants and/or stabilizers for polymers which may be used in the manufacture of articles or components of articles intended for use in producing, manufacturing, packaging, processing, preparing, treating, packing, transporting, or holding food in accordance with the conditions prescribed by 21 CFR part 178.2010.

Standards. ***Russia.*** PML in food: 4.0 mg/l.

References:

1. Vlasyuk, M. G., Hygienic evaluation of bisalkofen BP and rubbers stabilized with it, in *Hygiene and Toxicology of High-Molecular-Mass Compounds and the Chemical Raw Material Used in Their Synthesis*, Proc. 4th All-Union Conf., S. L. Danishevsky, Ed., Khimiya, Leningrad, 1969, 57 (in Rus sian).
2. Stasenkova, K. P., Shumskaya, N. I., Sheveleva, T. A., et al., Toxicity of bisalkofen BP, a stabilizer for polymeric materials, *Kauchuk i Rezina*, 1, 24, 1977 (in Russian).
3. Takahashi, O. and Hiraga, K., Effects of four bisphenolic antioxidants on lipid contents of rat liver, *Toxicol. Lett.*, 8, 77, 1981.

4. Braun, D. D., Effect of polyolefin antioxidants on induction of neoplasms by benzo(a)pyrene, *Gig. Sanit.*, 6, 18, 1975 (in Russian).
5. Telford, I. R., Woodruff, C. S., and Linford, R. H., Fetal resorption in the rat as influenced by certain antioxidant, *Am. J. Anat.*, 110, 29, 1962.

BIS-3-(3,5-DI-*tert*-BUTYL-4-HYDROXYPHENYL)PROPIONIC ACID, ESTER with DIETHYLENE GLYCOL

Molecular Formula. $C_{38}H_{58}O_7$

M = 626.87

Trade Name. Fenozan 28.

Properties. White, crystalline powder. Poorly soluble in water.

Applications. Used as a stabilizer in the production of polyolefins and other polymeric materials.

Acute Toxicity. Rats tolerate administration of the 10 g/kg BW dose without visible manifestations of the toxic effect.

Repeated Exposure. Young rats tolerated daily administration of B. (overall dose 90 g/kg BW). The treat ment did not affect BW gain or other indices of body condition.

Chemobiokinetics. Significant amount of B. is unlikely to be absorbed from the GI tract into the blood. Excreted unchanged from the body.

Reference:

Lapik, A. S., Pavlova, M. P., Sorokina, I. V., et al., Toxicology of Fenozan 28, in *Hygiene and Toxicology of High-Molecular-Mass Compounds and of the Chemical Raw Material Used for Their Synthesis*, Proc. 6th All-Union Conf., B. Yu. Kalinin, Ed., Leningrad, 1979, 269 (in Russian).

3,5-BIS(1,1-DIMETHYLETHYL)-4-HYDROXYBENZENEPROPANOIC ACID, 2,2-BIS{3-[3,5-BIS(1,1-DIMETHYLETHYL)-4-HYDROXYPHENYL]-1-OXOPROPOXY)METHYL}-1,3-PROPANEDIYL ESTER

Molecular Formula. $C_{73}H_{108}O_{12}$

M = 1177.67

CAS No 6683-19-8

RTECS No DA8340000

Abbreviation. PE.

Synonyms and **Trade Names.** 3,5-Bis(1,1-dimethylethyl)-4-hydroxybenzenepropanoic acid, 2,2'-bis{3-[3,5-bis(1,1-dimethylethyl)-4-hydroxyphenyl)-1-oxopropoxy]methyl}-1,3-propanedyl ester; 3,5-Di-*tert*- butyl-4-hydroxyhydrocinnamic acid, neopentanetetrayl ester; Fenosan 23; Irganox 1010; Irganox 1040; Pentaerythrite tetra-3-(3,5-di-*tert*-butyl-4-hydroxyphenyl) propionate; Tetralkofen BPE.

Properties. White, crystalline powder. Poorly soluble in water.

Applications. Used as an antioxidant and thermostabilizer for polypropylene, polyethylene, impact resistant polystyrene, poly-4-methylpentene. A stabilizer for natural and synthetic rubber, polyvinyl chloride. A copolymer of acrylonitrile with butadiene and styrene, polyacetals, alkyde resins, polyamides, and polyesters.

Migration from high-density polypropylene into the higher alcohols appeared to be slightly greater into 95% ethanol than into cooking oil, and slightly less migration was observed into 50% ethanol than into cooking oil. Migration was greater from polypropylene than from high-density polyethylene.[1] PE was found to migrate rapidly from ethylene-vinyl acetate copolymer (EVA) into *n*-heptane, 100% ethanol, and corn oil. The rate of migration into these media was greater from EVA than from low-density polyethylene (LDPE). Little migration of PE was to aqueous media, while migration from LDPE into such media was relatively high.[2]

Acute Toxicity. LD_{50} was not attained. Rats tolerate administration of 5.0 g/kg BW, mice did not die following a single intake of 7.5 and 10 g/kg BW. Neither retardation of BW gain was reported[08] nor were there gross pathology findings.[3]

Repeated Exposure failed to reveal cumulative properties.[4] Rats were exposed to the overall dose of 53 g/kg BW during a month. In the group of 10 rats one animal died. No changes in BW gain, blood morphology, relative weights of the visceral organs were noted. K_{acc} exceeded 10.

Fifty administrations of 2.0 and 5.0 g/kg BW to mice and rats over a period of 2 months caused no mortality or reduction in BW gain. Histology examination revealed no changes in the visceral organs.[08] Forty administrations of 0.5 and 2.0 g/kg BW to mice and of 0.5 g/kg BW to rats showed no gross pathology to develop.[3]

Long-term Toxicity. Doses of 100 mg/kg and 200 mg/kg BW given to mice as a suspension in sunflower oil caused decrease of BW gain. Histology examination revealed no lesions in the organs and tissues.[3, 08]

Reproductive Toxicity. The dose of 100 mg/kg BW given over a period of 24 months did not affect reproductive function of rats.[05]

Carcinogenicity. In the above cited study, no tumor development was observed.[05]

Regulations. ***British Standard*** (1992). PE is listed as an antioxidant for polyethylene and polypropylene compositions used in contact with foodstuffs or water intended for human consumption (max. permitted level for the final compound 0.5%).

Standards. ***Russia.*** PML: *n/m.*

References:

1. Lickly, T. D., Bell, C. D., and Lehr, K. M., The migration of Irganox 1010 antioxidant from high-density polyethylene and polypropylene into series of potential fatty-food simulants, *Food Addit. Contam.*, 7, 805, 1990.
2. Schwope, A. D., Till, D. E., Ehbtholt, D. J., Sidman, K. R., et al., Migration of Irganox 1010 from ethylene-vinyl acetate films to foods and food-simulating liquids, *Food Chem. Toxicol.*, 25, 327, 1987.
3. Zimnytskaya, L. P., Kalinin, B. Ye., and Robachevskaya, Ye. G., Toxicity of some light- and thermostabilizers of plastics, in *Hygiene and Toxicology of High-Molecular-Mass Compounds and of the Chemical Raw Material Used for Their Synthesis*, Proc. 4th All-Union Conf., S. L. Danishevsky, Ed., Khimiya, Leningrad, 1969, 193 (in Russian).
4. Mel'nikova, V. V., Toxicity of Fenozane 23, *Gig. Sanit.*, 12, 92, 1987 (in Russian).

3,5-BIS(1,1-DIMETHYLETHYL)-4-HYDROXYBENZENEPROPANOIC ACID, THIODI- 2,1-ETHANEDIYL ESTER

Molecular Formula. $C_{38}H_{58}O_6S$
M = 643.02
CAS No 41484-35-9
RTECS No DA8342500

Synonyms and **Trade Names.** Bis[(3,5-di-*tert*-butyl-4-hydroxyphenyl)ethoxycarbonylethyl] sulfide; Feno- san 30; Irganox 1035; Thiodiethylene glycolbis-3-(3,5-di-*tert*-butyl-4-hydroxyphenyl) propionate; Thioglycolate.

Properties. White crystalline powder. Poorly soluble in water, soluble in alcohol.

Applications. Used as a thermostabilizer in the production of polyolefins and other plastics.

Acute Toxicity. In male and female rats and in female mice, LD_{50} is reported to be in the range of 6.3 to 23.8 g/kg BW. Male mice tolerate Fenozan administration even at higher dose levels. Acute action threshold (for STI) is 1.2 g/kg BW in female mice, 2.5 g/kg BW in male mice, and 0.3 g/kg BW in female rats. STI appeared to be unchanged in male rats.

Repeated Exposure revealed no cumulative properties. Administration of low doses (even 10 LD_{50} in total) did not cause animal mortality. BW gain was unaltered. Gross pathology examination revealed thyroid changes in rats.

Short-term Toxicity. Rats tolerate administration of an overall dose equivalent to ten LD_{50}. Changes in the thyroid structure were noted. Mice exhibited no reduction of BW gain and had no pathomorphological lesions.

Long-term Toxicity. Mice received 50 and 200 mg/kg BW for 10.5 months without toxic manifestations. An increase in the relative weights of the thyroid is reported in rats.

Reproductive Toxicity.

Gonadotoxicity. Reproductive function or the gonads were not found to be affected following skin application of B.

Allergenic Effect was not observed in acute experiments on guinea pigs.

Standards. ***Russia.*** PML in drinking water: 3.0 mg/l; PML in food: *n/m*.

Reference:

Toxicology and Sanitary Chemistry of Plastics, Abstracts, NIITEKHIM, Moscow, Issue No 4, 1979, 26 (in Russian).

N,N′-BIS(1,4-DIMETHYLPENTYL)-*p*-PHENYLENEDIAMINE

Molecular Formula. $C_{18}H_{24}N_2$

M = 268.44

CAS No 793-24-8

76600-84-5

RTECS No ST0900000

Synonym and **Trade Names.** Antioxidant 4020; Diafen FDMB; *N*-(1,3-Dimethylbutyl)-*N′*-phenyl-*p*-phe- nylenediamine; Santoflex 13; Vulkanox 4020.

Properties. Crystalline powder. Poorly soluble in water, soluble in alcohol.

Applications. Used as a stabilizer in the production of synthetic rubber and vulcanizates.

Acute Toxicity. LD_{50} is found to be 2.5 g/kg BW in rats, and 3.2 g/kg BW in mice. High doses produced a narcotic effect. An acute action threshold appeared to be 1.0 g/kg BW.[1]

Repeated Exposure failed to reveal cumulative properties. No animal mortality was noted, but there were retardation of BW gain and impairment of the liver and kidney functions.[1]

Allergenic Effects. B. is shown to cause allergic contact dermatitis.[2]

Reproductive Toxicity.

Teratogenicity. B. does not produce a significant increase in congenital anomalies in rabbits at maternal doses up to 30 mg/kg BW, in spite of maternal toxicity observed at these doses.

References:

1. Stasenkova, K. P. and Kochetkova, T. A., Toxicity of butyric anhydride, in *Toxicology of the Components of Rubber Mixes and of Rubber and Latex Articles*, Central Research Institute Petrol. Chem., Moscow, 1970, 32 (in Russian).
2. Hansson, C., Allergic contact dermatitis from *N*-(1,3-dimethylbutyl)-*N′*-phenyl-*p*-phenylene-diamine and from compounds in polymerized 2,4-trimethyl-1,2-dihydroquinoline, *Contact Dermat.*, 30, 114, 1994.

N,N′-BIS(1-ETHYL-3-METHYLPENTYL)-*p*-PHENYLENEDIAMINE

Molecular Formula. $C_{22}H_{40}N_2$

M = 332.64

CAS No 139-60-6

RTECS No SS8450000

Synonym and **Trade Names.** *N,N′*-Bis(1-ethyl-3-methylpentyl)-*p*-phenylenediamine; *N,N′*-Bis(5-methyl-3-heptyl)-*p*-phenylenediamine; Santoflex 17.

Applications. Used as a stabilizer in the manufacture of plastics and rubber.

Acute Toxicity. LD_{50} is 0.75 mg/kg/BW[1] or 2.4 g/kg BW in rats, and 0.8 g/kg BW in mice.[027]

Reference:

1. Izmerov, N. F., Sanotsky, I. V., and Sidorov, K. K., *Toxicometric Parameters of Industrial Toxic Chemicals under Single Exposure*, USSR/SCST Commission for UNEP, Center Int. Projects, Moscow, 1982 (in Russian).

BIS(ISOOCTYLOXYCARBONYLMETHYLTHIO)DIBUTYLSTANNANE

Molecular Formula. $C_{28}H_{36}O_4SSn$

M = 575.44

CAS No 25168-24-5

60083-70-7

RTECS No WH6719000

Synonyms and **Trade Name.** Dibutyltin-*S,S'*-bis(isooctylmercaptoacetate); Dibutyltin-*S,S'*-bis (iso-octylthioglycolate); BTS-70.

Properties. Oily liquid with a strong, unpleasant odor. Poorly soluble in water. Odor perception threshold is 0.05 mg/l, taste perception threshold is 0.23 mg/l.[1]

Applications. Used as a stabilizer in the production of polyvinyl chloride plastics.

Acute Toxicity. LD_{50} is 570 mg/kg BW in mice, and 500 mg/kg BW in rats.[2] According to other data, LD_{50} in rabbits, male and female rats is found to be 500, 340, and 220 mg/kg BW, respectively.[1]

Administration of high doses caused refusal of food and diarrhea in the animals. Rats become photophobic, adynamic, and highly aggressive. Death occurred on days 1 to 11 after poisoning. Gross pathology examination revealed distension of the intestine, hyperemia with necrotic areas in the intestinal walls. There were also gall bladder engorgement and inflammation of the bile duct walls.

Repeated Exposure revealed evident cumulative properties. K_{acc} is 0.3 (by Cherkinsky). Male rats received 8.5, 17, and 34 mg/kg BW. The treatment affected the content of *SH*-groups in the blood. The highest dose caused changes in STI and hematology analyses. No changes were found in enzyme activity.[1]

Short-term Toxicity. Histology examination revealed parenchymatous dystrophy of the renal tubular epithelium, cellular breakdown in the lung tissue, the loss of parenchyma cells in the liver, and lymphocyte and plasma cell infiltration in rats given 0.5 and 5.0 mg/kg BW for 3 months.[2]

Long-term Toxicity. In a 7-month study, changes were observed only in conditioned and unconditioned reflex activity.

Reproductive Toxicity.

Gonadotoxic effect in rats is not observed.[1] Wistar rats were given 120 mg/kg BW on day 9 and 13 of pregnancy.[2] The dose of 0.01 mg/kg BW appeared to be ineffective for ***embryotoxic*** and ***teratogenic effects***.

Teratogenicity. *Dibutyltins* exhibited a general cytotoxic effect on rabbit articular and growth-plate chondrocytes that could be interpreted as possible effects on skeletal growth and development.[3]

Standards. ***Russia*** (1995). MAC: 0.01 mg/l; PML: 0.001 mg/l.

References:

1. Igumnov, A. S., Data on the Maximum Allowable Concentration for dibutyltin dithioglycolate in water bodies, *Gig. Sanit.*, 2, 13, 1975 (in Russian).
2. Sheftel', V. O., Permissible levels of migration for some organotin polyvinyl chloride stabilizers, in *Hygienic Aspects of the Use of Polymeric Materials*, Proc. 2nd All-Union Meeting on Health and Safety Monitoring of the Use of Polymer Materials in Construction, Kiyv, 1976, 36 (in Russian).
3. Webber, R. J., Dollins, S. C., Harris, M., and Hough, A. J., Effect of alkyltins on rabbit articular and growth-plate chondrocytes in monolayer culture, *J. Toxicol. Environ. Health,* 16, 229, 1985.

BIS(ISOOCTYLOXYCARBONYLMETHYLTHIO)DIOCTYLSTANNANE

Molecular Formula. $C_{36}H_{72}O_4SnS_2$

M = 751.89

CAS No 26401-97-8
67053-64-9

RTECS No WH6723000

Synonym and **Trade Name.** Advastab 17 MO; Di-*n*-octyltin-*S,S'*-bis(isooctylmercaptoacetate); Di-*n*-octyltin bis(isooctylthioglycolate).

Composition. Contains 11 to 12% tin and 25% epoxidized soya oil. A light-yellow, viscous liquid.

Properties. Poorly soluble in water, readily soluble in plasticizers.

Applications. A starting substance for stabilizers used in the manufacture of clear polyvinyl chloride food containers, rigid pipes, and flexible membranes.

Migration from polyvinyl chloride film to food and food-simulating liquids is found to be continuous with no indication of cessation.[1] Migration of organotins from polyvinyl chloride articles into tetra-hydrofuran, xylene, and methylene chloride was investigated. Methylene chloride extracted >97% of the total extractable organotins in two extractions (between <0.8 and 8.8 mg octyltins/g solvent). In industrial application pipe samples, the levels were 0.7 to 3.0 mg octyltins/g polyvinyl chloride.[2]

It was shown that the level of migration is lower when the stabilizer content in the composition is reduced. The maximum release was observed at 60°C after 5 days contact of polyvinyl chloride with water. At 20°C, 0.36 to 0.52 mg/l was released after 24 hours.[3]

Acute Toxicity. LD_{50} was reported to be 1.1 g/kg BW in mice, and 1.5 g/kg BW [4] or 2.1 g/kg BW[5] in rats. Poisoning is characterized by the symptoms of CNS impairment with adynamia and respiratory difficulty. Administration of 4.0 g/kg BW into the stomach of mice resulted in dilatation, ishemia of the stomach walls, and intestinal wall hyperemia.[6]

Repeated Exposure revealed reasonably evident cumulative properties in rats on daily administration of 1/5 LD_{50} for 20 days.[7] K_{acc} is 6.9 (by Cherkinsky). Wistar rats were dosed by gavage with 100 mg/kg and 200 mg/kg BW. All the animals died in 10 days from acute inflammation of the GI tract. The first administration impaired mobility, caused drowsiness, diarrhea, and a reduced BW gain.

Short-term Toxicity. Rats were given a 20 mg/kg BW dose for 3 months. The treatment caused 15% animal mortality.[8]

Long-term Toxicity. Rats were given 20 and 200 mg B./kg BW as a suspension in olive oil for 90 days, and also with the feed (0.02% in the diet) for 7 to 18 months. Treatment-related effects included decreased BW gain, increased liver and kidney weights, enhanced ALT and AST activity, changes in the ratio of the serum protein fractions.[9]

Allergenic Effect was found when 1/3000 LD_{50} had been applied to the skin of guinea pigs.[6]

Reproductive Toxicity.

Embryotoxicity. A reduction in the number of embryos and increased fetal mortality are reported in rats administered B. before mating and during pregnancy.[9] The dose of 21 mg/kg BW caused an embryotoxic effect.[7]

Teratogenicity. Dioctyltins exhibited a general cytotoxic effect on rabbit articular and growth-plate chondrocytes that could be interpreted as possible effects on skeletal growth and development.[10]

Mutagenicity.

In vitro genotoxicity. A mixture of organotins (though principally comprised of B.) exhibited mutagenicity in one of five strains of *Salmonella typhimurium* when tested in the absence of metabolic activation but was negative in its presence.[11]

Chemobiokinetics. B. seems to be slowly absorbed in the GI tract. As all thioalkyltin compounds, it is likely to undergo diamination and hydroxylation in the body and is partially excreted in the feces.

Regulations. ***U.S. FDA*** (1998) permits the use of B. alone or in combination, at levels not to exceed a total of 3.0% by weight in resin used to stabilize plastic bottles and films that are in contact with food.

Standards. ***Russia.*** Recommended PML in food: 0.03 mg/l, PML in drinking water: 0.5 mg/l.

References:

1. Schwope, A. D., Till, D. E., Ehutholt, D. J., et al., Migration of an organotin stabilizer from PVC film to food and food-simulating liquids, *Deutsch. Leben-Rundsch.*, 82, 277, 1986.
2. Forsyth, D. S, Dabeka, R., Sun, W. F., and Dalglish, K., Specification of organotins in poly(vinyl chloride) products, *Food Addit. Contam.*, 10, 531, 1993.
3. Sheftel', V. O., *Hygiene Aspects of the Use of Polymer Materials in the Water Supply*, Thesis diss. DSc, All-Union Research Institute of Hygiene and Toxicology of Pesticides, Polymers and Plastic Materials, Kiyv, 1977, 70 (in Russian).
4. Klimmer, O. R. und Nebel, I. U., Experimentelle Untersuchungen zur Frage der Toxicitat einiger Stabilizatoren in Kunststoffen aus Polyvinylchlorid, *Arzneim.-Forsch.*, 10, 44, 1960.
5. Tverskaya, M. Ya., Ivanova, T. P., Safris, Ye. S., et al., Studies of harmful effects and migration potential of some sulfur-containing dioctyltin derivatives, in *Hygienic Aspects of the Use of Polymeric Materials*, Proc. 2nd All-Union Meeting on Health and Safety Monitoring of the Use of Polymer Materials in Construction, Kiyv, 1976, 159 (in Russian).
6. Pelikan, Z., Cerny, E., and Polster, M., Toxic effects of some di-*n*-octyltin compounds in white mice, *Food Cosmet. Toxicol.*, 8, 655, 1970.

7. Levitskaya, V. N., in *14th Sci. Session on Chemistry & Technology of Organic Compounds Sulph. Min. Oil*, Abstracts of Reports, Riga, 1976, 96 (in Russian).
8. Nikanorov, M., Mazur, H., and Piekacz, H., Effects of orally administered stabilizers in the rat, *Toxicol. Appl. Pharmacol.*, 26, 253, 1973.
9. Mazur, H., Effect of oral administration of dioctyltin bis-isooctyl thioglycolate and dibenzyltin bis-isooctylthioglycolate on rat body, *Roczn. Panstv. Zakl. Hig.*, 22, 39, 1971 (in Polish).
10. Webber, R. J., Dollins, S. C., Harris, M., and Hough, A. J., Effect of alkyltins on rabbit articular and growth-plate chondrocytes in monolayer culture, *J. Toxicol. Environ. Health,* 16, 229, 1985.
11. Zeiger, E., Anderson, B., Haworth, S., et al., *Salmonella* mutagenicity tests: IV. Results from the testing of 300 chemicals, *Environ. Molec. Mutagen.*, 11 (Suppl. 12), 1, 1988.

2,6-BIS(α-METHYLBENZYL)-*p*-CRESOL

Molecular Formula. $C_{23}H_{24}O$
M = 316.47
CAS No 1817-68-1
RTECS No GO6863500

Synonyms and Trade Names. Alkofen MBP; 2,6-Bis(1-phenylethyl)-4-methyl phenol; Ionol 6; 4-Methyl-2,6-bis(1-phenylethyl)phenol.

Properties. Viscous, oil-like liquid. Poorly soluble in water.

Applications. Used as a thermo- and light-stabilizer in the synthesis of vulcanizates based on natural, butadiene, and styrene-butadiene rubber; a thermostabilizer of polyolefins, polyvinylchloride, and cellulose esters.

Acute Toxicity. In mice, LD_{50} is 4.3 g/kg BW. Manifestations of the toxic action include excitation with subsequent CNS inhibition. M. exerts an irritating and necrotizing effect on the tissues.[029]

Long-term Toxicity. Rats were exposed to 215 mg/kg BW for about 5 months. The treatment resulted in CNS excitation and impairment of kidney function. Histology examination revealed necrosis of the gastric and intestinal mucosa and dystrophy of the renal tubular epithelium.[08]

Reproductive Toxicity. Affects fish reproductive processes but does not affect the development of chick embryos.

1,1'-BIS[(2-METHYL-4'-HYDROXY-5-*tert*-BUTYL)PHENYL]PENTANE

Molecular Formula. $C_{27}H_{40}O_2$
M = 396.67

Applications. Used as an antioxidant in the production of polyolefins and synthetic rubber.

Acute Toxicity. LD_{50} is approximately 17 g/kg BW in rats. Administration does not affect the GI tract (Haskel).

Repeated Exposure. Rats received 0.05% B. in their feed for 90 days. The treatment resulted in fatty infiltration of the liver. A dose of 0.005% in the diet had no toxic effect.[05]

2,2'-BIS(*p*-PHENYLAMINOPHENOXY)DIETHYL ETHER

Molecular Formula. $C_{28}H_{28}N_2O_3$
M = 440.53

Trade Name. Bisphenam AO.

Properties. Finely dispersed, odorless, gray powder. Insoluble in water and alcohol.

Applications. Used as a stabilizer in the production of polyamides and polyurethanes.

Acute Toxicity. Rats tolerate a single 10 g/kg BW dose without signs of intoxication.

Repeated Exposure caused 50% mortality in rats given 1.0 g/kg BW by day 36 of the exposure. Symptoms of intoxication were pronounced. 0.5 g/kg BW dose produced no visible signs of intoxication.

Short-term Toxicity. No animals given 0.5 g B./kg BW died in a 3-month experiment.

Reference:

Zyabbarova, S. A., in *Protection of Water Reservoirs against Pollution by Industrial Liquid Effluents*, S. N. Cherkinsky, Ed., Medgiz, Moscow, Issue No 4, 1960, 138 (in Russian).

N,N'-BIS(2,2,6,6-TETRAMETHYL-4-PIPERIDYL)ETHYLENEDIAMINE

Molecular Formula. $C_{20}H_{42}N_4$

M = 338.58

Synonym and **Trade Name.** *N,N'*-Bis(2,2',6,6'-tetramethylpyridyl)ethylenediamine; C-245.

Applications. Used as a stabilizer in the production of plastics and rubber.

Acute Toxicity. LD_{50} is reported to be 2850 or 3940 mg/kg BW in rats, and 550 mg/kg BW in mice.[1,2] Poisoning is accompanied by agitation and frequent clonic spasms lasting 2 to 3 min. Death within 24 hours.

Repeated Exposure revealed marked cumulative properties. K_{acc} is 1.6 (by Lim). Rats given doses of 190 and 570 mg/kg BW exhibited NS depression and a change in biochemistry analyses including decreased catalase and blood cholinesterase activity. Gross pathology examination revealed changes in the visceral organs and an increase in the relative weights of the brain, heart, and spleen.[2]

Rats were given 1/20 LD_{50} for a month. The treatment caused erythropenia, a reduction in BW gain and shortening of the latent period of the electrodefensive reflex.[1]

References:

1. Sarycheva, V. N. and Khanbekova, S. K., in *Methodological Aspects of Investigation of the Biological Action of Chemical Compounds*, A. P. Shitzkova, Ed., Proc. F. F. Erisman Research Sanitary Hygiene Institute, Moscow, 1982, 40 (in Russian).
2. Volodchenko, V. A. and Gnezdilova, A. G., Toxicology expertise of the product C-245, *Gig. Sanit.*, 6, 90, 1983 (in Russian).

BIS(TRIFLUOROACETOXY) DIBUTYLSTANNANE

Molecular Formula. $C_{12}H_{18}F_6O_4Sn$

M = 458.9

CAS No 52112-09-1

RTECS No WH6734000

Synonyms. Dibutyltinbis(trifluoroacetate); Stannous dibutylditrifluoroacetate.

Applications. Used as a stabilizer in the production of polyvinyl chloride plastics.

Exposure. See ***Dibutyldichlorostannane***.

Migration of organotins from polyvinyl chloride articles (clear food container, rigid pipe, and flexible membrane) into tetrahydrofuran, xylene, and methylene chloride was investigated. Methylene chloride extracted >97% of the total extractable organotins in two extractions. There were found to be between <0.3 and 4.7 mg butyltins/g solvent. The level of butyltins in potable water pipe samples was between <0.9 and 5.99 mg/g. In industrial application pipe samples these levels were between <13 and 1.5 mg butyltins/g polyvinyl chloride.[1]

Acute Toxicity. In rats, LD_{50} is reported to be 53.6 mg/kg BW.[2]

Repeated Exposure revealed cumulative properties. On administration of 1/5 or 1/25 LD_{50} to mice for 20 days, K_{acc} (by Cherkinsky) appeared to be 3.6 to 5.4 and 1.5 to 1.6, respectively. Anemia and leukocytosis were observed in rats that received 0.25 mg/kg BW for 30 days.[2]

Long-term Exposure. In a 6-month study, significant increase in the cholinesterase activity was noted in rabbits. The catalase activity has fallen sharply by the end of the experiment.[2]

Reproductive Toxicity.

Teratogenicity. Dibutyltins exhibited a general cytotoxic effect on rabbit articular and growth-plate chondrocytes that could be interpreted as possible effects on skeletal growth and development.[3]

Mutagenicity.

In vivo cytogenetics. A mutagenic effect in bone marrow cells is reported in mice on a single administration of LD_{50}.

Chemobiokinetics. Excretion from the body was slow.[2]

References:

1. Forsyth, D. S, Dabeka, R., Sun, W. F., and Dalglish, K., Specification of organotins in poly(vinyl chloride) products, *Food Addit. Contam.*, 10, 531, 1993.
2. Belyaeva, N. N., Bystrova, T. A., Revazova, Iu. A., et al., Comparative evaluation of toxicity and mutagenicity of organotin compounds, *Gig. Sanit.*, 5, 10, 1976 (in Russian).
3. Webber, R. J., Dollins, S. C., Harris, M., and Hough, A. J., Effect of alkyltins on rabbit articular and growth-plate chondrocytes in monolayer culture., *J. Toxicol. Environ. Health,* 16, 229, 1985.

o-tert-BUTYL-*p*-CRESOL

Molecular Formula. $C_{11}H_{16}O$
M = 164.27
CAS No 2409-55-4
RTECS No GO7000000
Abbreviation. BC.

Synonyms. 2-*tert*-Butyl-4-methylphenol; 2-(1,1-Dimethylethyl)-4-methylphenol; 4-Methyl-2-*tert*-butylphenol.

Properties. Colorless, crystalline powder. Poorly soluble in water, soluble in alcohol.

Applications. Used as a thermostabilizer in the production of plastics.

Acute Toxicity. LD_{50} is reported to be 0.8 g/kg BW in mice, 2.5 g/kg BW in rats, and 1.2 g/kg BW in guinea pigs.[1]

Repeated Exposure. CFE female rats were given doses of 150 mg/kg BW by gastric intubation for 7 days. Significant increase in hexobarbitone oxidase and aminopyrine demethylase activity, and a 5.4% increase in liver weight was observed.[2]

Mutagenicity.

In vivo cytogenetics. BC induced unscheduled DNA synthesis in rats exposed orally to BMP for 8 weeks.[3]

Regulations. ***U.S. FDA*** (1998) listed BC as an antioxidant in semirigid and rigid acrylic and modified acrylic plastics to be safely used as articles intended for use in contact with food in accordance with the conditions prescribed in 21 CFR part 178.2010.

References:

1. Shumskaya, N. I. and Naidenko, N. F., in *Toxicology of New Industrial Chemical Substances*, N. F. Ismerov and I. V. Sanotsky, Eds., Moscow, Issue No 12, 1971, 124 (in Russian).
2. Gilbert, D. et al., Effect of substituted phenols on liver weights and liver enzymes in the rat structure-activity relations, *Food Cosmet. Toxicol.*, 7, 603, 1969.
3. Shibata, M.-A., Yamada, M., Hirose, M., Asakawa, E., Tatematsu, M., and Ito, N., Early proliferative responses of forestomach and glandular stomach of rats treated with five different phenolic antioxidants, *Carcinogenesis*, 11, 425, 1990.

1,4-BUTYLENE GLYCOL, β-AMINOCROTONATE

M = 278.3

Properties. Powder. Poorly soluble in water.

Applications. Used as a stabilizer in the production of polyvinyl chloride.

Acute Toxicity. In rats, LD_{50} significantly exceeds the 4.0 g/kg BW dose.[05]

Repeated Exposure. Rats received 4.0% B. in their diet. The treatment resulted in decreased BW gain due to deterioration of the organoleptic properties of the feed.

Reproductive Toxicity effects are not observed.

Standards. ***Russia.*** Recommended PML: *n/m*.

tert-BUTYLHYDROQUINONE

Molecular Formula. $C_{10}H_{14}O_2$
M = 166.24

CAS No 1948-33-0
RTECS No MX4375000.
Abbreviation. TBHQ.

Synonym and **Trade Name.** 2-(1,1-Dimethylethyl)-1,4-benzenediol; Sustane.

Application. Used as an antioxidant in the manufacture of food-contact materials and cosmetic products, and as a polymerization inhibitor for various polyunsaturated polyesters.

Acute Toxicity. LD_{50} is 790 mg/kg BW in guinea pigs.[1]

Short-term Toxicity. F344 rats received up to 780 mg/kg in their diet for 13 weeks following weaning. No biologically significant changes and clinical pathology were observed. Gross pathology examination revealed increased incidence of kidney cysts and inflammation.[2]

B6C3F$_1$ mice received up to 9.0 g TBHQ/kg diet for 13 weeks. The treatment caused no biologically significant differences in organ weights. There were increased incidences in severity of mucosal hyperplasia in the forestomach in the higher dose animals.[3]

Long-term Toxicity. In a 6-month study with rats and 2-year dog study, a dose level of 0.5% TBHQ in the diet caused no significant changes in enzyme activities in the liver.

In a 117-week feeding study, dogs received up to 5.0 g TBHQ/kg feed. The NOAEL of 37.5 mg/kg BW for hematological changes was established in this study.[2] This NOEL was re-evaluated and confirmed on the level of 72 mg/kg BW by Joint FAO/WHO Expert Committee on Food Additives in 1999.

Reproductive Toxicity.

Embryotoxicity. Rats received up to 0.5% TBHQ in the diet during active organogenesis. No fetal lesions were noted.[4]

Teratogenicity. In the above-cited study, there were no visceral or skeletal abnormalities in rats with dose levels up to 0.5% in the diet.[4]

Mutagenicity.

In vitro genotoxicity. TBHQ did not show genotoxic potential in *Salmonella* with and without *S9* mix in the reverse mutagenicity assay. It was positive in CA test *in vitro* using Chinese hamster fibroblast cells only in the presence of *S9 mix*[3] and sometimes significantly increased the frequency of SCE.[3,5]

No increase in the frequency of micronucleated erythrocytes was observed in the bone marrow of male mice treated with TBHQ.[3]

In vivo cytogenetics. DLM test in rats was negative (Krasavage, 1984).[6]

Carcinogenicity. In the above-cited study, there was no increased incidence of neoplasms in rats that received up to 225 to 240 mg/kg BW and in mice given up to 600 to 680 mg/kg in the diet.[3] TBHQ did not induce forestomach tumors in rodents with dose levels of 0.5 and 0.25%.[6]

It has been predicted to be a rodent carcinogen by the computer-automated structure evaluation (CASE) and multiple-computer-automated structure evaluation (MULTICASE) system.[8]

Chemobiokinetics. TBHQ is oxidatively metabolized to *tert-butylquinone* both enzymatically and by autoxidation. TBHQ is a principal metabolite of *2-tert-butyl-4-hydroxyanisole*. Metabolic studies in the rat and dog showed rapid absorption from the GI tract and rapid excretion in the urine, predominantly as the *4-O-sulfate conjugate*, with smaller quantities of *4-O-glucuronide conjugate*, and minor quantities of unchanged TBHQ.[2]

^{14}C-ring-labeled TBHQ was diluted with non-radioactive TBHQ and administered orally to F344 rats. An average of 72.9 and 10.6% of ^{14}C-ring-labeled TBHQ was recovered in the urine and feces, respectively, of male rats, and 77.3 and 8.2% in the urine and feces, respectively, of female rats in 4 days. No significant sex-related differences were found in excretion, tissue distribution or urinary metabolites of TBHQ-derived radiolabel. When administered *i/p,* TBHQ quickly disappeared from tissue *in vivo*. The highest concentrations of radiolabel were found in the liver and kidneys. The urinary metabolites consisted of conjugated TBHQ and unidentified polar substances.[7]

Regulations. 10 countries have accepted the use of TBHQ as an antioxidant.

U.S. FDA (1998) approved the use of TBHQ alone or in combination with BHA and/or BHT if a total antioxidant content of the food does not exceed 0.02% of the oil or fat content.

Recommendations. Joint FAO/WHO Expert Committee on Food Additives (1999). ADI: No safety concern.

References:

1. Food Additives Series No 8, WHO, Geneva, 1975, 26.
2. van Esch, C. J., Toxicology of *tert*-butylhydroquinone (TBQO), *Food Chem. Toxicol.*, 24, 1063, 1986.
3. *Toxicicology and Carcinogenesis Studies of t-Butylhydroquinone in F344/N Rats and B6C3F$_1$ Mice (Feed Studies),* Draft NTP Technical Report Series No 459, Research Triangle Park, NC, June 1995.
4. Matsuoka, A., Matsui, M., Miyata, N., Sofuni, T., and Ishidate, M., Mutagenicity of 3-*tert*-butyl-4- hydroxyanisole (BHA) and its metabolites in short-term tests *in vivo*, *Mutat. Res.*, 241, 125, 1990.
5. Rogers, C. G., Boyes, B. G., Matula, T. I., and Stapley, R., Evaluation of genotoxicity of *tert*-butylhydroquinone in an hepatocyte-mediated assay with V79 Chinese hamster lung cells and in strain D7 of *Saccharomyces cerevisiae*, *Mutat Res.*, 280, 17, 1992.
6. Krasavage, W. J., Evaluation of the teratogenic potential of tertiary butylhydroquinone (TBQH) in the rat, *Teratology*, 16, 31, 1977.
7. Ikeda, G. J., Sapienza, P. P., and Ross, I. A., Distribution and excretion of radiolabelled *tert*-butylhydroquinone in Fischer 344 rats, *Food Chem. Toxicol.*, 36, 907, 1998.
8. Zhang, Y. P., Sussman, N., Macina, O. T., Rosenkranz, H. S., and Klopman, G., Prediction of the carcinogenicity of a second group of organic chemicals undergoing carcinogenicity testing, *Environ. Health Perspect.*, 104S, 1045, 1996.

2(3'-*tert*-BUTYL-2'-HYDROXY-5'-METHYLPHENYL)-5'-CHLOROBENZO-TRIAZOLE

Molecular Formula. $C_{17}H_{18}ClN_3O$

M = 315.80

Trade Name. Tinuvin 326.

Properties. Pale-yellow, crystalline powder. Poorly soluble in water and alcohol.

Applications. Used as a light-stabilizer in the production of polypropylene, polyethylene, polybutylene, etc; stabilizer for unsaturated polyesters and vinyl plastics.

Acute Toxicity. Mice tolerate 15 g/kg BW; rats do not die after administration of 10 g/kg BW.

Repeated Exposure. Rats and mice tolerate exposure to 1.0 to 3.0 g/kg BW without any deviation in their general condition or structure of the viscera.

Long-term Toxicity. Treatment with 150 mg/kg and 300 mg/kg BW (mice) and 200 mg/kg BW (rats) produced no effect on BW gain, hematology analyses, blood enzyme composition, and conditioned reflex activity. Gross pathology examination revealed no changes in the visceral organs. Unlike 3,5-(*di-tert-butyl-2-hydroxy- phenyl)-5-chlorbenzo-triazole* or *Tinuvin* 327 (*q.v.*), B. in which a methyl group replaces one *tert*-butyl group is virtually non-toxic.

Allergenic Effect was not observed.

Standards. ***Russia***. PML in drinking water: *n/m*.

Reference:

Mikhailets, I. B., Toxic properties of some stabilizers for plastics, *Plast. Massy*, 12, 41, 1976 (in Russian).

BUTYLHYDROXYOXOSTANNANE

Molecular Formula. $C_4H_{10}O_2Sn$

M = 208.83

CAS No 2273-43-0

RTECS No WH6770000

Abbreviation. BHOS.

Synonyms. 1-Butanestannonic acid; Butyltin hydroxide oxide; Butylhydroxytin oxide.

Applications. Used for stabilization of clear polyvinyl chloride materials intended for the manufacture of food-contact containers, rigid pipes, and flexible membranes.

Migration of organotins from polyvinyl chloride articles into tetrahydrofuran, xylene, and methylene chloride was investigated. Methylene chloride extracted >97% of the total extractable organotins in two extractions. They were found between <0.3 and 4.7 mg butyltins/g solvent. The level of butyltins in potable water pipe samples was between <0.9 and 5.99 mg/g. In industrial application pipe samples there were between <13 and 1.5 mg butyltins/g polyvinyl chloride.

Acute Toxicity. LD_{50} *i/v* is 180 mg/kg BW in mice.

Regulations. ***U.S. FDA*** (1998) approved the use of BHOS in resinous and polymeric coatings used as the food-contact surfaces of articles intended for use in producing, manufacturing, packing, processing, preparing, treating, packaging, transporting, or holding food in accordance with the conditions prescribed in 21 CFR part 175.300.

Reference:

Forsyth, D. S, Dabeka, R., Sun, W. F., and Dalglish, K., Specification of organotins in poly(vinyl chloride) products, *Food Addit. Contam.*, 10, 531, 1993.

N-BUTYLIDENEANILINE

Molecular Formula. $C_{10}H_{13}N$

M = 147.21

Properties. A thick, dark-brown liquid, poorly soluble in water.

Applications. Used as a thermostabilizer for vulcanizates based on butyl rubber and neoprene.

Acute Toxicity. LD_{50} is 2.2 g/kg BW in mice and 2.6 g/kg BW in rats. Manifestations of the toxic action include depression and immobility. Bloody diarrhea was noted in 2 days. Histology examination revealed tiny hemorrhagic foci in the lungs, small areas of destruction and inflammatory type microfoci in the liver, and signs of engorgement in the kidney medulla.[023]

Repeated Exposure. The treatment with 1/10 LD_{50} produced pronounced functional changes in the liver and a decline in catalase activity and gas exchange indices. Gross pathology examination revealed hepatocyte granularity, tiny hemorrhagic foci in the lungs, bronchitis, and engorgement and congestion in the brain and spleen.[023]

4,4'-BUTYLIDENEBIS(6-*tert*-BUTYL-*m*-CRESOL)

Molecular Formula. $C_{26}H_{38}O_2$

M = 382.64

CAS No 85-60-9

RTECS No GO7050000

Synonyms and **Trade Names.** 1,1-Bis(2-methyl-4-hydroxy-5-*tert*-butylphenyl)butane; 4,4'-Butylidene-bis-(3-methyl-6-*tert*-butylphenol); Santowhite.

Properties. White powder.

Applications. Antioxidant for natural and synthetic rubber and latex.

Acute Toxicity. LD_{50} is 17 g/kg BW in rats.[04]

Repeated Exposure. Rats received B. powder at a level of 1.135 mmol% for 1 week. The treatment induced fatty liver and increased triglyceride, diglyceride, and cholesteryl-ester concentrations. Marked decrease was noted in the levels of plasma triglycerides, total cholesterol, and nonesterified fatty acids.[1]

Regulations. ***U.S. FDA*** (1998) approved the use of B. (1) in the manufacture of rubber articles intended for repeated use in producing, manufacturing, packaging, processing, preparing, treating, packing, transporting, or holding food in accordance with the conditions prescribed in 21 CFR part 177.2600; and (2) in the manufacture of antioxidants and/or stabilizers for polymers which may be used in the manufacture of articles or components of articles intended for use in producing, manufacturing, packaging, processing, preparing, treating, packing, transporting, or holding food in accordance with the conditions prescribed in 21 CFR part 178.2010.

Reference:

Takahashi, O. and Hiraga, K., Effects of four bisphenolic antioxidants on lipid contents of rat liver, *Toxicol. Lett.*, 8, 77, 1981.

2-*tert*-BUTYL-6-METHYL-4-α-METHYLBENZYLPHENOL

Molecular Formula. $C_{19}H_{24}O$

M = 268.43

Applications. Used as a stabilizer in the production of plastics.

Acute Toxicity. In mice, LD_{50} is 1.9 g/kg BW.[08] Gross pathology examination revealed congestion, edema, and sloughing of the esophageal and gastric mucosa, necrosis of the small intestine walls, and parenchymatous dystrophy of the renal tubular epithelium.

Short-term Toxicity. Rats given 75 mg B./kg BW for 4 months exhibited stimulation of the cortical areas of the CNS. Histology examination revealed congestion, cellular infiltration and edema of the gastric mucosa, congestion and necrosis of the small intestine mucosa, and parenchymatous dystrophy of the renal tubular epithelium.[08]

4-*tert*-BUTYL PYROCATECHOL

Molecular Formula. $C_{10}H_{14}O_2$

M = 166.24

CAS No 98-29-3

RTECS No UX1400000

Abbreviation. BPC.

Synonyms and **Trade Name.** 4-*tert*-Butyl-1,2-benzenediol; 4-*tert*-Butylcatechol; 4-(1,1-Dimethylethyl)-1,2-benzenediol; Synox TBC.

Applications. Used as a stabilizer in the manufacture of plastics.

Acute Toxicity. LD_{50} is 2.8 g/kg BW in rats.[1]

Mutagenicity.

In vitro genotoxicity. 4-*tert*-Butyl catechol was active at the lowest doses in the mouse lymphoma L5178Y cell assay for mutagens without supplementary metabolic activation system.[2]

Regulations. ***U.S. FDA*** (1998) approved the use of BPC in adhesives as a component of articles intended for use in packaging, transporting, or holding food in accordance with the conditions prescribed in 21 CFR part 175.105.

References:

1. Smyth, H. F., Carpenter, C. P., Weil, C. S., et al., Range-finding toxicity data, List V, *Arch. Ind. Hyg. Occup. Med.*, 10, 61, 1954.
2. McGregor, D. B., Riach, C. G., Brown, A., Edwards, I., Reynolds, D., West, K., and Willington, S., Reactivity of catecholamines and related substances in the mouse lymphoma L5178Y cell assay for mutagens, *Environ. Molec. Mutagen.*, 11, 523, 1988.

BUTYLTRI(LAUROYLOXY) STANNANE

Molecular Formula. $C_{40}H_{78}O_6Sn$

M = 773.87

CAS No 25151-00-2

RTECS No WI3000000

Synonyms. Butyltin tri(dodecanoate); Butyltin trilaurate.

Exposure. Residues of monobutyltin compounds were not detected in most foods. Concentrations found in rice, sunflower oil, peanut oil, butter, cheese, or chicken meat were less than 3.0 mg/kg wet weight of these foodstuffs.[1]

Applications. Used as a stabilizer in the production of polyvinyl chloride plastics.

Migration of organotins from polyvinyl chloride articles (clear food container, rigid pipe, and flexible membrane) into tetrahydrofuran, xylene, and methylene chloride was investigated. Methylene chloride extracted >97% of the total extractable organotins in two extractions. They were found to be between <0.3 μg and 4.7 mg butyltins/g solvent. The level of butyltins in potable water pipe samples was found to be between <0.9 μg and 5.99 mg/g. In industrial application pipe samples they were detected at the levels of <13 μg to 1.5 mg butyltins/g polyvinyl chloride.[2]

Acute Toxicity. In rats, LD_{50} is 325 mg/kg BW.

Repeated Exposure. Cumulative properties were evident in mice given 65 and 13 mg/kg BW for 20 days, and cumulative effect increased as the dose was reduced. Exposure to 5.0 mg/kg BW for 30 days led to development of signs of anemia and leukocytosis.[3]

Long-term Toxicity. Rabbits received 2.0 mg/kg BW for 6 months. The treatment did not affect hematology analyses and serum protein fractions ratio in animals.[3]

Mutagenicity.

In vivo cytogenetic investigation revealed mutagenic effect in the bone marrow cells of mice following a single administration of LD_{50}.

References:

1. Kannan, K., Tanabe, S., and Tatsukawa, R., Occurrence of butyltin residues in certain foodstuffs, *Bull. Environ. Contam. Toxicol.*, 55, 310, 1995.
2. Forsyth, D. S, Dabeka, R., Sun, W. F., and Dalglish, K., Specification of organotins in poly(vinyl chloride) products, *Food Addit. Contam.*, 10, 531, 1993.
3. Belyaeva, N. N., Bystrova, T. A., Revazova, Yu. A., et al., Comparative evaluation of toxicity and mutagenicity of organotin compounds, *Gig. Sanit.*, 5, 10, 1976 (in Russian).

6-*tert*-BUTYL-2,4-XYLENOL

Molecular Formula. $C_{12}H_{18}O$

M = 178.27

CAS No 1879-09-0

RTECS No ZE6825000

Synonyms and **Trade Name.** Antioxidant A; 2-*tert*-Butyl-4,6-dimethylphenol; 6-*tert*-Butyl-2,4-dimethylphenol; Topanol A.

Properties. Yellow-orange liquid. Soluble in methanol.

Applications. Thermostabilizer in the production of polyethylene and polyamides.

Acute Toxicity. LD_{50} is 530 mg/kg BW in mice. Poisoning caused impairment in the respiratory system; acute pulmonary edema was observed. There was also ulceration or bleeding from small intestine.

Reference:

J. Am. Pharmacol. Assoc., Sci., Ed., 38, 366, 1949.

CADMIUM compounds

	CAS No	RTECS No
Cadmium caprylate	2191-10-8	RH0370000
Molecular Formula. $C_{16}H_{30}O_4.Cd$		
M = 398.82		
Cadmium chloride	10108-64-2	EV0175000
Molecular Formula. $CdCl_2$		
M = 183.30		
Cadmium oxide	1306-19-0	EV1925000
Molecular Formula. CdO		
M = 128.40		
Cadmium stearate	2223-93-0	RG1050000
Molecular Formula. $C_{36}H_{70}O_4.Cd$		
M = 679.35		
Cadmium sulfide	1306-23-6	EV3150000
Molecular Formula. CdS		

M = 144.46

Cadmium sulfate 10124-36-4 EV2700000

Molecular Formula. $O_4S.Cd$

M = 208.46

Properties. Many *cadmium* salts are readily soluble in water. The aqueous solutions are colorless and odorless. The taste perception threshold of a solution is 25 mg/l, and of a suspension of insoluble cadmium salts, 2.0 mg/l.

Cadmium caprylate. White, very light powder. Poorly soluble in water.

Cadmium stearate. White powder. Poorly wettable with water, slightly soluble in ethanol, and soluble in acids up to 1.0 mg/l.

Cadmium sulfide. Light-yellow or orange crystals. Solubility in water is 130 mg/l at 18°C.

Applications. Cadmium salts are used as polyvinyl chloride stabilizers.

Cadmium sulfide and selenide are commonly used as pigments in plastics. Cadmium is used as a stabilizer.

Cadmium stearate is used as a thermostabilizer of polyvinyl chloride and as a lubricant. Cadmium is used in the production of plastics and synthetic products.

Barium-cadmium laurate and *barium-cadmium stearate.*

See ***Barium compounds.***

Exposure. Food is the main source of exposure. The daily intake varies from 0.01 to 0.035 mg for non-smoking people in unpolluted areas.[1] Cigarette smoking has been shown to significantly elevate the amount of cadmium in the body. Smoking one pack of cigarettes results in the inhalation of 2.0 to 4.0 μg of cadmium.[2]

Migration Data. Release of cadmium from plastic tableware into 4.0% acetic acid or 0.1 N *HCl* (60°C or 95°C; 30 min) was examined. No migration of cadmium from samples containing up to 6863 ppm cadmium was observed.[3] Cadmium was detected in teabag tissue made of paper and board materials and in unbleached Kraft paper at the level of 0.3 mg/kg.[4]

Acute Toxicity.

Observations in man. An estimated oral lethal dose for humans is 25 to 150 mg/kg BW; a dose without effect in adults is 3.0 mg/person.[5]

Animal studies. In mice and rats, LD_{50} values vary from 60 to more than 5000 mg/kg BW for different cadmium compounds; in dogs, it is 150 to 600 mg/kg BW. In rats, the LD_{50} of different cadmium compounds is as follows (mg/kg BW): cadmium oxide, 72; cadmium sulfate, 88; cadmium chloride, 94; cadmium nitrate, 100 to 200; cadmium iodide, 166; cadmium sulfide, 2425. Following acute poisoning, gross pathology examination revealed dystrophic changes in the testicular cells of surviving animals.[6]

Acute cadmium poisoning mainly led to proteinuria associated with renal tissue damage, as well as glucosuria and phosphaturia. Impairment of the enzyme system was reported. Major effects result in desquamation of the epithelium and necrosis of the gastric and intestinal mucosa, and dystrophic changes in the liver, heart, kidneys, and testicular tissues.

Cadmium caprylate. LD_{50} is reported to be 950 mg/kg BW in rats and 300 mg/kg BW in mice. Testicular tissue is severely affected. Gross pathology examination revealed damage to the spermatogenic epithelium, necrotic and dystrophic changes in the liver and kidney, as well as necrosis and ulceration of the gastric mucosa. A dose-dependence in degree of lesions was observed.

Cadmium stearate. LD_{50} is reported to be 1225 or 1500 mg/kg BW in rats, and 590 mg/kg BW in mice.[7] Gross pathology examination revealed circulatory disturbances and signs of gastritis and enteritis, and in rats, dystrophic changes in the testes, with loss of, and damage to, the sex cells. A dose of 300 mg/kg BW caused a persistent extension of the prothrombine time which does not return to normal even after some days.[8]

Repeated Exposure. Cadmium compounds are toxic to a wide range of organs and tissues, with the kidney, liver, lung, and testis being clear targets in rodents.

Cadmium chloride. Rabbits were dosed with 30 μg cadmium chloride /kg BW for 15 days. The treatment produced absolute erythrocytosis and appearance of anantocytes (in 64% of samples). A stable eosinophylia and a decrease in neutrofile count were observed.[9]

Cadmium caprylate. Exhibited evident cumulative properties. K_{acc} is 0.8 (by Lim). Rats received 0.1 LD_{50} for 1 to 3 weeks. The treatment caused a reduction in the liver glycogen, an increase in the lactic acid content of the blood, enhanced succinic acid dehydrogenation and inhibition of glutamic acid dehydrogenation, and a reduction in the activity of alkaline phosphatase and other enzymes.[10]

Short-term Toxicity. With respect to cadmium compounds' toxicity, the kidney is likely to be a target organ and renal tubular dysfunction is the critical effect. When a critical concentration of cadmium in the kidney cortex is reached, tubular dysfunction will occur.

Cadmium chloride. In the Rhesus monkey study, the NOEL of 2.0 mg/kg BW administered in the diet was identified with respect to impaired tubular resorption. The renal effect was demonstrated in rats given 10 mg cadmium/l of drinking water. Bone effects were frequently demonstrated at doses of 10 to 30 mg/kg the diet or 10 mg/l in drinking water and above. In addition, effects on the cardiovascular system, liver, hematopoietic and immune system have been reported.[5]

Cadmium stearate. When 1/20 and 1/10 LD_{50} are given, cumulative properties are evident.

Long-term Toxicity.

Observations in man. The kidney seems to be the most sensitive organ in man for chronic exposure, where cadmium compounds affect the resorption function. This is followed by an increase in the urinary excretion of low-molecular-mass proteins, known as tubular proteinuria.[5]

Among people living in contaminated areas in Japan and exposed to cadmium compounds with food and drinking water, many cases of *Itai-itai* disease (osteomalacia with various grades of osteoporosis accompanied by severe renal tubular disease) and low-molecular-mass proteinuria have been reported.[1]

Animal studies. Rabbits were given food containing cadmium chloride for 200 days (the dose of 14.9 mg Cd^{2+}/kg BW). Histology examination revealed the liver tissue to show hyperplasia, cellular infiltration and reduction in glycogen content. Considerable interstitial fibrosis and necrosis in the proximal sections of the tubules were found in the kidneys.[11] Male rats received cadmium chloride in their drinking waterover a period of 6 months. Changes in calcium metabolism and bone structure characteristics of osteomalacia were found to develop.[12]

2.5% absorption of cadmium from food and 5.0% from water has been assumed by U.S. EPA (1985). The NOAEL for chronic cadmium exposure was considered to be 0.005 and 0.01 mg cadmium/kg BW from water and food, respectively (these levels would result in 0.2 mg cadmium/g wet weight human renal cortex, the highest renal level not associated with significant proteinuria). RfD of 0.0005 mg cadmium/kg BW (water) or 0.001 mg cadmium/kg BW (food) was calculated.[13]

Reproductive Toxicity. Cadmium compounds may be accumulated in the placenta; they penetrate through placenta to embryo tissues. They appeared to be toxic to the human placenta.[14]

Gonadotoxicity.

Observations in man. Epidemiological studies failed to find any alteration of semen quality and fertility in the subjects exposed to cadmium compounds.[15,16] Gennart et al. investigated the fertility of smelters exposed to cadmium and found no difference in the rate of live births between exposed and unexposed workers.[17]

Animal studies. Testicular tissues seem to be a primary target for different cadmium salts on single or repeated administration. Histology examination revealed changes in the testes, including necrosis, reduction in the functional activity of the spermatozoa, and depression of the process of spermatogenesis.[6,18]

High doses and chronic administration provoked vascular changes and ischemic necrosis in the testes.[19] A single low dose has been shown to have selective effects on sperm formation, impairing the releases of sperm from the seminiferous epithelium in rats.[20] Rats received cadmium in their diet at high doses ranging from 7 to 14 mg cadmium/kg BW for 90 to 120 days. The treatment produced pathologic alterations in testis with loss of reproductive capacity.[21,22] A decrease in sperm motility and spermatogenesis index was found to develop in rats continuously exposed via the diet to 0.5 to 5 mg cadmium chloride/kg BW.[23]

Cadmium oxide affected sperm motility and was positive in vaginal cytology test (NTP-91). A decreased fertility in animals were noted only on cadmium exposure at near-fatal doses.[24]

The estrous cycle is reported to be affected in females. Histology examination showed changes in the microstructure and hemorrhages in the ovaries. Mild testicular changes were seen in rats after oral administration of 50 mg/kg BW over a period of 15 months.[5]

Electron microscope examination failed to reveal effect on spermatogenous function following treatment with the dose of 0.012 mg/kg BW.[25]

In a multi-generation study in rats, dose levels of up to 100 mg/kg, the diet did not affect reproduction, while in 4-generation studies, 1.0 mg/l in drinking water and 0.125 mg/kg the diet, respectively, affected fertility in mice and rats.[5]

Embryotoxicity is likely to be explained by the effect on the mother and the placenta.[26] Embryotoxic effects were observed only at dose levels which induced maternal toxicity. Cadmium caused an increase in embryolethality. Wistar rats received 40 mg cadmium chloride/kg BW on days 7 to 12 of pregnancy. The treatment reduced the number of live embryos and the fetal weights. An increased number of resorptions was also reported.[27,28]

Teratogenicity. Cadmium salts produce teratogenic effect in golden hamsters, rats, and birds. At the high levels of exposure, cadmium induced a wide spectrum of developmental abnormalities: lesions of nervous tube, skeletal anomalies, hydrocephaly, delay in gonads development, etc.

According to Chernoff, cadmium chloride produced teratogenic effects in rats. Absence of the forelimbs and hydropericardium seems to be the most frequent abnormalities.[29] Mechanism of cadmium teratogenicity is unknown. Cadmium potential for teratogenicity in regard to humans is not yet proved definitively.[28]

Immunotoxicity. Cadmium compounds strengthen or suppress the host resistance to infectious agents or to tumors depending on the species or strain of animal. With acute exposure to low doses of cadmium, humoral immune reactions are suppressed or unaltered. Chronic exposure to high or moderate doses suppresses antibody production. Cadmium exerts its main toxic effect on the macrophages, greatly suppressing a number of their functions.

Rats received 25 mg cadmium acetate/l in their drinking water for 2 weeks and during immunization with a 10% suspension of ram erythrocytes. The treatment did not affect significantly blood antibody reactions against ram erythrocytes as compared to the control. There were no differences from the controls with respect to the inclusion of 3*H*-thymidine in the rat blood cells.[5]

The data about the allergenic effects of cadmium compounds are therefore contradictory.[30]

Mutagenicity.

Observations in man. Studies on the cytogenic effects in human peripheral lymphocytes following *in vivo* exposure to cadmium both orally and by inhalation did not show increased frequencies of CA. These findings contrast markedly with the positive results obtained on workers exposed in zinc smelting plants and on people environmentally intoxicated by Cd^{2+} (IARC 11-39).

In vivo cytogenetics. In DLM test, a single *i/g* administration of 2.0 mg cadmium/kg BW (MTD) caused an increase in overall, pre- and postimplantation mortality of mouse embryos, indicating damage at the late stages of the spermatogenesis cycle.[31]

According to other data, in the dominant lethal assay, mice were exposed *i/p* with 5.4 to 7.0 mg cadmium chloride/kg BW. The treatment produced no increase in the incidence of early death or pre-implantation losses.[32]

Cadmium compounds did not produce DLM effects in mice or rats nor did they increase the frequencies of CA or micronuclei in mice treated *in vivo*. Cadmium compounds induced aneuploidy in hamsters but not in mice. Cadmium oxide did not induce micronuclei in erythrocytes of mice exposed by inhalation.[22,32]

In vitro genotoxicity. *Cadmium* compounds induced DNA single-strand breaks in human and rodent cells. They did not induce SCE in human cells *in vitro*; they induced transformations of cultured rodent cells in several test systems and induced CA but not SCE in human[33] and rodent cells *in vitro* (IARC 11-39).

Cadmium oxide, chloride, and sulfate were shown to be negative in *Salmonella typhimurium* mutagenicity assay.[22,34] Cadmium chloride was positive when tested by *rec-assay*, cadmium nitrate produced CA[35] and was found to be negative when tested by *rec-assay*.[36]

Carcinogenicity.

Observations in man. Ingestion-specific human data are not available. Epidemiology studies in man living in polluted areas and exposed orally to cadmium via the diet have not shown an increased cancer risk.

Animal studies. No evidence for carcinogenicity in mice and rats (dose levels 1.0 to 50 mg/kg diet) has been found from oral studies.[37] However, according to recent studies,[38] oral cadmium exposure may be clearly associated with tumors of the prostate (concentration of 50 ppm in drinking water, exposure of 77 weeks), testes (200 ppm), and hematopoietic system in rats. There is evidence for the carcinogenicity of cadmium compounds in experimental animals by the inhalation route of exposure (at concentrations about 0.0125 mg/m^5 and higher).[29]

Carcinogenicity classifications. An IARC Working Group concluded that there is inadequate evidence for the carcinogenicity of cadmium and certain cadmium compounds *in humans* and there is sufficient evidence for the carcinogenicity of cadmium and the following cadmium compounds in *experimental animals*: cadmium chloride, cadmium oxide, cadmium sulfate and cadmium sulfide.

The Annual Report of Carcinogens issued by the U.S. Department of Health and Human Services (1998) defines cadmium and certain cadmium compounds to be substances which may reasonably be anticipated to be carcinogen, i.e., substances for which there is a limited evidence of carcinogenicity in *humans* or sufficient evidence of carcinogenicity in *experimental animals*.

IARC: 1;
U.S. EPA: B1;
ARC: 1;
EU: 2.

Chemobiokinetics. Absorption of cadmium is greater from water than from food. Absorption via the GI tract is influenced by the solubility of the cadmium compounds ingested, the type of diet, and the nutritional status of the animal or person involved. Absorbed cadmium enters the blood stream and is transported via blood to different organs and tissues. It may bound to a low-molecular-mass plasma protein, i.e., metallothionein.

At the molecular level, cadmium disrupts the utilization of essential metals: calcium, zinc, selenium, chromium, iron. A shortage of these metals, and also of proteins and vitamins, enhances the toxicity of cadmium and increases its absorption in the intestine and retention in different organs in the form of a complex with metalthioneins.[39]

Cadmium accumulates primarily in the kidneys and liver (50 to 85% of the body burden), it has a long biological half-life in man, viz. 10 to 30 years. The critical cadmium concentration in the renal cortex at which a 10% prevalence of low-molecular-mass proteinurea in the general population would occur is about 200 mg/kg (WHO, 1991).

An absorption rate for dietary cadmium seems to be 5.0% and daily excretion rate appears to be 0.005% (*JECFA*).

Regulations. Cadmium release from glazed ceramic tableware has been regulated as follows:

EU (1984), mg/dm²

Flatware < 25 mm internal depth	0.07
Hollow-ware	0.3
Cookware, packaging & storage vessels	0.1

ISO (1981,1986), mg/l:

Flatware	0.17
Small Hollow-ware	0.5
Large Hollow-ware	0.25
Cookware	0.5

UK (1986, 1988):

Flatware < 25 mm internal depth	0.07 mg/dm^2
Hollow-ware	0.3 mg/l
Cookware, packaging & storage vessels	0.1 mg/l

U.S. FDA (1988, 1992) µg/ml (mg/l):

Flatware	(average of 6 units)	0.5
Small Hollow-ware	(any one of 6 units)	0.5
Large Hollow-ware		0.25

Recommendations.

WHO (1989). PTWI was therefore set at the level of 7.0 mg/kg BW (~0.4 to 0.5 mg/person/week).

WHO (1996). Guideline value for drinking water: 0.003 mg/l.

Joint FAO/WHO Expert Committee on Food Additives concluded that if levels of cadmium in the renal cortex are not to exceed 50 mg/kg, a total intake should not exceed 1.0 μg/kg BW/day.

Standards.

U.S. EPA (1999). MCL and MCLG: 0.005 mg/l.

EEC (1980). MAC: 0.005 mg/l.

Canada (1989). MAC: 0.005 mg/l.

Russia (1995). MAC and PML: 0.001 mg/l; MAC in milk and diary produce: 0.03 mg/kg; MAC in meat and poultry: 0.05 mg/kg; MAC in mineral water 0.01 mg/kg; MAC in children's food 0.02 mg/kg.

References:

1. *Cadmium*, Environmental Health Criteria, WHO, IPSC, Geneva, 1992, 280.
2. Hallenbeck, H., Human health effects of exposure to cadmium, *Experentia*, 40, 136, 1984.
3. Hosogai, T., Ito, S., Tada, Y., Sakurai, H., Sugita, T., Ishiwata, H., and Takeda, M., Migration test of lead and cadmium from plastic wares in contact with food, Abstract, *Eisei Shikenjo Hokoku*, 110, 83, 1992 (in Japanese).
4. Castle, L., Offen, C. P., Baxter, M. J., and Gilbert, J., Migration studies from paper and board food packaging materials. 1. Compositional analysis, *Food Addit. Contam*, 35, 14, 1997.
5. Krajnc, E. I., van Gestel, C. A. M., Mulder, H. C. M., et al., *Integrated Criteria Document. Cadmium Effects*, Appendix to Report No 758476004, Natl. Inst. Publ. Health and Environ. Protection, The Netherlands, 1987.
6. Shabalina, L. P. Experimental study on toxicity of barium-cadmium stearate, *Gig. Sanit.*, 2, 98, 1970 (in Russian).
7. Maksimova, N. S., Toxicity of cadmium stearate, *Plast. Massy*, 12, 33, 1976 (in Russian).
8. Zille, L. N. and Shabalina, L. P., in *Hygiene Science - Practice*, Materials Sci. Conf., 1st Moscow Medical Institute, Moscow, 1972, 97 (in Russian).
9. Malyariva, M. A., Rubekin, E. A., and Smirnova, I. M., Impact of cadmium compounds available in the feed on hematological parameters of animals, in *Proc. Sci. Works All-Rus. Research Institute Veterinary Sanitary Hygiene Ecology*, 94, 59, 1994 (in Russian).
10. Larionova, T. I., Zille, L. N., and Vorob'yeva, R. S., The influence of *Ba-Cd* laurate on some fermentative processes in the liver, *Gig. Truda Prof. Zabol.*, 3, 50, 1974 (in Russian).
11. Stowe, H. D., Wilson, M., and Goyer, R. A., Clinical and morphological effects of oral cadmium toxicity in rabbits, *Arch. Pathol.*, 94, 389, 1972.
12. Likutova, I. V., *Specific Effect of Inorganic Cadmium Compounds on Health and Its Hygienic Assessment*, Author's abstract of thesis, 1989, 25 (in Russian).
13. *Drinking Water Criteria Document on Cadmium* (*Final draft*), U.S. EPA, Office of Drinking Water, Washington, D. C., 1985.
14. Boadi, W. Y., Urbach, J., Brandes, J. M., and Yannai, S., *In vitro* exposure to mercury and cadmium alters term human placental membrane fluidity, *Toxicol. Appl. Pharmacol.*, 116, 17, 1992.
15. Kazantzis, G., Flynn, F. V., Spwage, J. S., and Trott, D. G., Renal tubular malfunction and pulmonary emphysema in cadmium pigment workers, *Q. J. Med.*, 32, 165, 1963.
16. Saaranen, M., Kantola, M., Saarikoski, S., and Vanha-Perttula, T., Human seminal plasma cadmium: comparison with fertility and smoking habits, *Andrologia*, 21, 140, 1989.
17. Gennart, J.-P., Buchet, J.-P., Roels, H., Ghyselen, P., Ceulemans, E., and Lauwerys, R., Fertility of male workers exposed to cadmium, lead, or manganese, *Am. J. Epidemiol.*, 135, 1208, 1992.
18. Bomhard, E., Vogel, O., and Loser, E., Chronic effects of single and multiple oral and subcutaneous cadmium administration on the testes of Wistar rats, *Cancer Lett.*, 36, 307, 1987.
19. Gouveia, M. A., The testes in cadmium intoxication: morphological and vascular aspects, *Andrologia*, 20, 225, 1988.
20. Hew, K. W., Ericson, W. A., and Welsh, M. J., A single low cadmium dose failure of spermiation in the rat, *Toxicol. Appl. Pharmacol.*, 121, 15, 1993.

21. Saxena, D. K., Murphy, R. C., Singh, C., and Chandra, S. V., Zinc protects testicular injury induced by concurrent exposure to cadmium and lead in rats, *Res. Commun. Chem. Pathol. Pharmacol.*, 317, 64, 1989.
22. Saygi, S., Deniz, G., Kustal, O., and Vural, N., Chronic effects of cadmium on kidney, liver, testis, and fertility of male rats, *Biol. Trace Elem. Res.*, 31, 209, 1991.
23. Krasovsky, G. N., Varshavskaya, S. P., and Borisov, A. I., Toxic and gonadotropic effects of cadmium and boron relative to standards of the subtsances in drinking water, *Environ. Health Perspect.*, 69, 13, 1976.
24. Kotsonis, F. N. and Klassen, C. D., Toxicity and distribution of cadmium administered to rats at sublethal dose, *Toxicol. Appl. Pharmacol.*, 41, 667, 1977.
25. Polyakova, N. A., in *The Endocrine System of the Body and Toxic Factors of the Environment*, Abstracts, 2nd All-Union Conf., Leningrad, 1980, 162 (in Russian).
26. Baranski, B., Stetkiewicz, I., Trzciuka-Ochocka, M., et al., Teratogenicity, fetal toxicity and tissue concentrations of cadmium administered to female rats during organogenesis, *J. Appl. Toxicol*, 2, 255, 1982.
27. Chernoff, N., The teratogenic effects of cadmium in the rat, *Toxicol. Appl. Pharmacol.*, 22, 313,1972.
28. Webster, W. S., The teratology and developmental toxicity of cadmium, *Issues and Review, Teratology,* New York - London, 5, 255, 1990.
29. Oldiges, H., Hochrainer, P., and Glaser, U., Long-term inhalation study with Wistar rats and four cadmium compounds, *Toxicol. Environ. Chem.*, 19, 217, 1989.
30. Wahlberg, J. E. and Fregert, S., Guinea-pig maximization test, in *Models in Dermatology*, N. H. I. Maibach and N. Lowe, Eds., Karger, Basel, 2, 225, 1985.
31. Volkova, N. A., Karplyuk, I. A., and Yemel'yanova, Ye. V., Studies of mutagenic activity of cadmium by DLM test, *Voprosy Pitania*, 2, 24, 1995.
32. Epstein, S. S., Arnold, E., Andrea, J., Bass, W., and Bishop, Y., Detection of chemical mutagens by the dominant lethal assay in the mouse, *Toxicol. Appl. Pharmacol.*, 23, 288, 1972.
33. Jokhadze, T. A. and Lezhava, T. A., *In vivo* and *in vitro* studies of chromosome structural mutations induced by heavy metals' salts during aging, *Genetica*, 30, 1630, 1994 (in Russian).
34. De Flora, S., Zanacchi, P., Camoirano, A., Bennicelli, C., and Badolati, G. S., Genotoxic activity and potency of 135 compounds in the Ames reversion test and in a bacterial DNA-repair test, *Mutat. Res,* 161, 133, 1984.
35. Oehever, F., Chromosome breaks influenced by chemicals, *Heredity*, Suppl. 6, 95, 1953.
36. Nioshioka, H., Mutagenic activities of metal compounds in bacteria, *Mutat. Res.*, 31, 185, 1975.
37. Loser, E., A two year oral carcinogenicity study with cadmium on rats, *Cancer Lett.,* 9, 191, 1980.
38. Waalkes, M. P. and Rehm, S., Carcinogenicity of oral cadmium in the male Wistar rat: Effect of chronic dietary zinc deficiency, *Fundam. Appl. Pharmacol.*, 19, 512, 1992.
39. Nath, R., Prasad, R., Palinal, V. K., et al., Molecular basis of cadmium toxicity, *Progr. Food Nutr. Sci.,* 8, 109, 1984.

CALCIUM compounds

M = 40.08
CAS No 7440-70-2
RTECS No EV8040000

Properties.

Calcium-zinc stabilizer. White-yellowish paste.

Calcium chloride technical grade occurs as hygroscopic white-gray scales.

Applications.

Calcium-zinc stabilizer. Thermostabilizer for polyvinyl chloride materials intended for packaging of dry foods.

Acute Toxicity. LD_{50} of 50% water solution of calcium chloride technical grade was found to be 1344 mg/kg BW in rats, and 210 mg/kg BW in mice. LD_{50} of 50% water solution of calcium chloride inhibited by phosphates was 1554 mg/kg BW in rats. Administration of lethal doses caused an increase in motor activity followed by CNS inhibition, convulsions. Death occurs within a day after poisoning.[1]

Repeated Exposure.

Calcium chloride technical grade. Rats were given by gavage 1/20 LD_{50} for 30 days. Functional and material accumulation was noted. K_{acc} was 3.45. The treatment affected blood coagulative system, caused cardiotoxic effect and a decrease in general immunity response. Gross pathology examination revealed multiple punctate hemorrhages in stomach and intestinal mucosa.[1]

Reproductive Toxicity.

Embryotoxicity. Pregnant rats received drinking water containing 60 or 200 mg calcium hydrocarbonate, lactate, or glycerophosphate/l. This treatment did not affect length and weight of fetuses.[2] High concentration of calcium salts in the diets of rats and mice significantly reduced fetal weights. No signs of teratogenic activity were noted in these species,[3,4] but abnormal ossification was reported in sheep.[5]

Chemobiokinetics.

Observations in man. Dietary calcium salts have not been significantly absorbed while women were nursing. The extra calcium needed for production of breast milk has been taken from lowered excretion in the urine and resorption from bone.[6]

Regulations. ***U.S. FDA*** (1998) approved the use of calcium naphthenate (1) in the manufacture of cross-linked polyester resins which may be used as articles or components of articles intended for repeated use in contact with food in accordance with the conditions prescribed in 21 CFR part 177.2420; calcium silicate and sulfate may be used as (2) as colorants in the manufacture of articles or components of articles intended for use in producing, manufacturing, packaging, processing, preparing, treating, packing, transporting, or holding food, subject to the conditions prescribed in 21 CFR part 178.3297; calcium glycerophosphate may be used as (3) as a stabilizer in food packaging materials in accordance with the conditions prescribed in 21 CFR part 178.2010; it is affirmed as *GRAS* (4) when used as an emulsifying agent in accordance with GMP. Calcium diacetate, glycerophosphate, oxide, and phosphate are affirmed as *GRAS*.

Recommendations. ***EU*** (1995). Calcium acetate and chloride are food additives generally permitted for use in foodstuffs.

References:

1. Sukhanov, V. V., Petul'ko, S. N., Bolonova, L. N., and Yulish, N. R., Toxicological evaluation of calcium chloride and containing products, *Gig. Truda*, 5, 51, 1990 (in Russian).
2. Kuzanova, I. Z. and Rakhmanin, Yu. A., Experimental study of embryotoxic effect of various calcium compounds administered orally into the body, *Gig. Sanit.*, 8, 25, 1990 (in Russian).
3. Fairney, A. and Weir, A. A., The effect of abnormal maternal plasma calcium levels on the offspring of rats, *J. Endocrinol.*, 48, 337, 1970.
4. Liebgott, B. and Srebrolow, G., Fetal toxicity caused by excessive maternal dietary calcium, *J. Canad. Dent. Assoc.*, 55, 129, 1989.
5. Corbellini, C. N., Krook, L., Nathaniels, P. W., and Kallfetz, F. A., Osteochondrosis in fetuses of ewes overfed calcium, *Calcif. Tissue Int.*, 48, 37, 1991.
6. Specker, B. L., Viera, N. E., O'Brien, K. O., et al., Calcium kinetics in lactating women with low and high calcium intakes, *Am. J. Clin. Nutr.*, 59, 593, 1994.

CHLOROTRIETHYL STANNANE

Molecular Formula. $C_6H_{15}ClSn$

M = 241.35

CAS No 994-31-0

RTECS No WH6840000

Abbreviation. CTES.

Synonyms. Chlorotriethyl tin; Triethyltin chloride.

Properties. Colorless liquid.

Applications. Used as a stabilizer in the production of polyvinyl chloride materials.

Acute Toxicity.

Observations in man. Consumption of the drug containing triethyltin as an impurity resulted in 209 patients who developed neurological symptoms: increased intracranial pressure as a consequence of cerebral edema; 110 patients died.[1]

Animal studies. LD_{50} values have been reported to be 3.45 mg/kg BW in male, and 2.88 mg/kg BW in female rats, 20 mg/kg BW in mice, 15 mg/kg BW in guinea pigs, and 7.0 mg/kg BW in rabbits. Lethal doses affected the NS and caused animals to die on days 2 to 15.[07] Changes in behavior and in conditioned reflex activity were observed in rats following *i/a* administration of a 3.0 mg/kg BW dose.[2]

Repeated Exposure revealed cumulative properties. Rats received CTES via drinking water with a concentration of 20 mg/l for 28 days. The treatment affected muscle tonus and caused hindlimb paralysis. Lower doses led to a reduction in motor activity. This phenomena appeared after two weeks and disappeared after one month.[2,3] Rats fed the diets containing 15 ppm triethyltin (0.75 mg/kg BW) for 2 weeks exhibited retardation of BW gain, decreased relative weights of the thymus and spleen but increased relative adrenal weights (at 50 ppm level). 30% animals died while survivors exhibited brain edema.[4] CTES has a neurotoxic effect.

Reproductive Toxicity.

Teratogenicity. CTES exhibited a general cytotoxic effect on rabbit articular and growth-plate chondrocytes that could be interpreted as possible effects on skeletal growth and development.[5]

Chemobiokinetics. Following absorption in the GI tract, CTES is distributed predominantly to the liver, kidneys, and brain. Edema is noted in the white matter of the brain. The mechanism by which tri-substituted *tin* compounds exert a toxic effect consists in inhibition of oxidative phosphorilation in the mitochondria as a result of the penetration therein of tin and *chlorine* ions, and also in the suppression of mitochondrial ATP-ase activity.

References:

1. Alajouanine, T., Derobert, B. L., and Thieffry, S., Etude clinique d'ensemble de 210 cas d'intoxication par les sels organiques de l'etain, *Revue Neurol.*, 8, 85, 1958.
2. Wenger, G. R. and McMillan, D. E., Effect of triethyltin-SO_4 in mice responding under a multiple schedule of food presentation, *Toxicologist*, 24, 1985.
3. Reiter, L., Kidd, K., Heavner, G., et al., Behavioral toxicity of acute and subacute exposure to triethyltin in the rat, *Neurotoxicology*, 97, 1981.
4. Snoeij, N. J., van Iersel, A. A. J., Penninks, A. H., et al., Toxicity of triorganotin compounds: comparative *in vivo* studies with a series of trialkyltin compounds and triphenyltin chloride in male rats, *Toxicol. Appl. Pharmacol.*, 1, 274, 1985.
5. Webber, R. J., Dollins, S. C., Harris, M., and Hough, A. J., Effect of alkyltins on rabbit articular and growth-plate chondrocytes in monolayer culture, *J. Toxicol. Environ. Health,* 16, 229, 1985.

N-CYANOETHYL-*p*-PHENETIDINE

Molecular Formula. $C_{11}H_{14}N_2O$

M = 190.25

CAS No 72060-06-1

RTECS No SI7025000

Synonym. 3-(*p*-Ethoxyphenylamino)propionitrile.

Properties. White powder. Poorly soluble in water.

Applications. Used as a stabilizer in the production of polyolefins and polyethylene terephthalate.

Acute Toxicity. In rats, LD_{50} is 4.6 g/kg BW. The lethal doses cause hindlimb paresis. Animals assume side position in 15 to 20 min after administration. Death occurs in 24 hours.

Repeated Exposure. Rats were dosed with 460 mg/kg BW for 1 month. The treatment caused anemia and an increase in the relative weights of the liver; the content of *SH*-groups in the blood serum was reduced.

Reference:

Vasilenko, N. M., Volodchenko, V. A., Nakonechny, A. A., et al., Comparative toxicity evaluation of *p*-phenytidine and cyanoethyl-*p*-phenytidine, *Pharmacologia i Toxicologia*, 35, 367, 1972 (in Russian).

[(DIBENZYLSTANNYLENE)DITHIO]DIACETIC ACID, DIISOOCTYL ESTER

Molecular Formula. $C_{34}H_{52}O_4SSn$
M = 707.67
CAS No 28675-83-4
RTECS No AG5975000

Synonyms. Dibenzyltin-*S,S'*-diisooctylthioglycolate; Dibenzyltin-*S,S'*-bis-(isooctylmercaptoacetate).

Applications. Used as a stabilizer in the production of polyvinyl chloride plastics.

Acute Toxicity. In rats, LD_{50} is reported to be 2.0 g/kg[1] or 1.25 g/kg BW.[2]

Repeated Exposure failed to reveal cumulative properties. An increased relative weight of the liver was found at the dose level of 200 mg/kg BW; a rise in the activity of ALT and AST was observed when the dose of 100 mg/kg BW was administered.[3]

Short-term Toxicity. In a 3-month study, Wistar rats received 180 mg D./kg the diet. Poisoning resulted in 15% animal mortality. The relative liver weights were increased.[4]

Long-term Toxicity. In a 1.5-year study, the liver, enzymatic activity, and serum proteins contents were predominantly affected.[3]

Reproductive Toxicity.

Embryotoxicity. An increase in embryolethality and a reduction in the number of embryos were found in rats given D. before mating and during pregnancy.[4] Doses of 18 and 90 mg/kg BW given to rats throughout the pregnancy led to a reduction in litter size and to an increase in neonatal mortality.[5]

References:

1. Klimmer, O. R. and Nebel, I. U., Experimentelle Untersuchungen zur Frage der Toxicitat einiger Stabilizatoren in Kunstoffen aus Polyvinylchlorid, *Arzneimittel.-Forsch.*, 10, 44, 1960.
2. Homrowski, S., Studies on the toxicity of additives applied in domestic production of plastics. I. Acute toxicity of some stabilizers and plasticizers, *Rocz. Panstw. Zakl. Hig.*, 18, 283, 1967 (in Polish).
3. Mazur, H., Effect of oral administration of dioctyltin bisisooctylthioglycolate and dibenzyltin bis-isooctylthioglycolate on rat body, *Roczn. Panstv. Zakl. Hig.*, 22, 39, 1971 (in Polish).
4. Mazur, H., Effect of oral administration of dioctyltin bis-isooctylthioglycolate and dibenzyltin bis-isooctylthioglycolate on rat body. II. Influence on fertility and fetal development, *Roczn. Panstv. Zakl. Hig.*, 5, 509, 1971 (in Polish).
5. Nikonorow, M., Mazur, H., and Piekacz, H., Effect of orally administered stabilizers in the rat, *Toxicol. Appl. Pharmacol.*, 26, 253, 1973.

DIBUTYLBIS(LAUROYLOXY)STANNANE

Molecular Formula. $C_{32}H_{64}O_4Sn$
M = 631.65
CAS No 77-58-7
RTECS No WH7000000

Synonyms and **Trade Names.** Advastab 52; Dibutylbis(lauroyloxy)tin; Dibutylbis[(1-oxododecyl)oxy]-stannate; Dibutyltin dilaurate.

Properties. Soft crystals or yellow liquid. A clear, oily liquid at higher temperatures. Poorly soluble in water but gradually hydrolyzes.

Applications. A stabilizer in the production of polyvinyl chloride and chlorinated rubber; a light-stabilizer for polystyrene; a light- and thermostabilizer of polyamides and foam plastics; a catalyst in the cold vulcanization of organosilicon rubber.

Exposure. See ***Dibutyldichlorostannane.***

Migration of organotins from polyvinyl chloride articles (clear food container, rigid pipe, and flexible membrane) into tetrahydrofuran, xylene, and methylene chloride was investigated. Methylene chloride extracted >97% of the total extractable organotins in two extractions. They were between <0.3 μg and 4.7 mg butyltins/g solvent. The level of butyltins in potable water pipe samples was between <0.9 μg and 5.99

mg/g. In industrial application pipe samples they were between <13 μg and 1.5 mg butyltins/g polyvinyl chloride.[1]

Acute Toxicity. The oral LD_{50} was reported to be 1.7 g/kg BW in mice, and 13 g/kg BW in rats. A single administration of 5.0 g/kg BW caused liver and stomach damage. Evidence of neurotoxicity was reported: rats were given oral doses of 20 to 80 mg/kg BW in 0.2 ml of groundnut oil for 3 consecutive days. Administration caused decrease in spontaneous locomotor activity and learning, and decreased level of the neurotransmitters, noradrenaline, dopamine and serotonin, particularly in the hypothalamus and frontal cortex.[2] D. produced a significant increase in polyamine levels in selected brain areas.[6]

Repeated Exposure revealed marked cumulative properties in mice given 85 to 340 mg/kg BW, and in rats receiving 65 to 130 mg/kg BW. The treatment caused reduced BW gain and a change in the weight coefficients of the visceral organs, CNS excitation, and lesions in the stomach, intestines, and liver.[3]

15 to 40 mg D./kg BW were fed to rabbits over a period of 6 weeks. Indication of liver damage through elevated serum glutamic-oxaloacetic transaminase, glutamic-pyruvic transaminase, and lactate dehydrogenase was found.[4]

Rats received oral doses of 17.5 mg/kg BW for 15 days. 20% of animals died during the experiment and the survivors showed reduced BW gain. Some hepatic enzymes had been inhibited.[5]

Reproductive Toxicity.

Teratogenicity. Dibutyltins exhibited a general cytotoxic effect on rabbit articular and growth-plate chondrocytes that could be interpreted as possible effects on skeletal growth and development.[7]

Pregnant rats were treated orally on day 8 of gestation with 0.08 mmol D./kg BW. DBTL has been shown to cause malformations such as cleft mandible, ankyloglossia, fused ribs, etc. in rat fetuses.[8]

Regulations. ***U.S. FDA*** (1998) approved the use of D. (1) in adhesives as a component of articles intended for use in packaging, transporting, or holding food in accordance with the conditions prescribed by 21 CFR part 175.105; (2) as a curing (cross-linking) catalyst for silicone adhesives used as components of articles intended for use in packaging, transporting, or holding food (the maximum amount of tin catalyst used shall be that required of effect optimum cure but shall not exceed 1.0% siloxane resins solids); (3) as a catalyst in polyurethane resins for the food-contact surface of articles intended for use in contact with dry food alone or in combination, at levels not to exceed a total of 3.0% by weight in the resin used; and (4) in the manufacture of resinous and polymeric coatings in a food-contact surface of articles intended for use in producing, manufacturing, packing, transporting, or holding food in accordance with the conditions prescribed by 21 CFR part 175.300.

Standards. ***Russia*** (1995). MAC and PML: 0.01 mg/kg.

References:

1. Forsyth, D. S, Dabeka, R., Sun, W. F., and Dalglish, K., Specification of organotins in poly(vinyl chloride) products, *Food Addit. Contam.*, 10, 531, 1993.
2. Alam, M. S., Hussain, R., Srivastava, S. P., et al., Influence of dibutyltin dilaurate on rain neurotransmitter systems and behavior in rats, *Arch. Toxicol.*, 61, 373, 1988.
3. Maksimova, N. S., Krynskaya, I. L., and Yevsyukov, V. I., Toxicity of organotin and cadmium stabilizers, *Plast. Massy*, 12, 33, 1976 (in Russian).
4. Tanaka, S., Experimental studies on the effects of dibutyltin dilaurate, dibutyltin dichloride and dimethyltin dichloride to serum enzymes and lipids of rabbits, Abstract, *Tokyo Ika Daigaku Zasshi,* 38, 607, 1980 (in Japanese).
5. Mushtaq, M., Mukhtar, H., Datta, K. K., et al., Toxicological studies of a leackable stabilizer di-*n*-butyltin dilaurate (DBTL): Effects on hepatic drug metabolizing enzyme activities, *Drug Chem. Toxicol.,* 4, 75, 1981.
6. Knalig, M. A., Husain, R., Seth, P. K., and Srivastava, S.P., Effect of dibutyltin dilaurate on regional brain polyamines in rats, *Toxicol. Lett.*, 55, 179, 1991.
7. Webber, R. J., Dollins, S. C., Harris, M., and Hough, A. J., Effect of alkyltins on rabbit articular and growth-plate chondrocytes in monolayer culture,. *J. Toxicol. Environ. Health,* 16, 229, 1985.
8. Noda, T., Morita, S., and Baba, A., Teratogenic effects of various di-*n*-butyltins with different anions and butyl(3-hydroxybutyl)tin dilaurate in rats, *Toxicology*, 85, 149, 1993.

2,6-DI-*tert*-BUTYL-*p*-CRESOL

Molecular Formula. $C_{15}H_{24}O$
M = 220.39
CAS No 128-37-0
RTECS No GO7875000
Abbreviation. BHT.

Synonyms and Trade Names. Advastab; Agidol 401; Alkofen BP; 2,6-Bis(1,1-dimethylethyl)-4-methylphenol; Butylated hydroxytoluene; Butylhydroxytoluene; CAO-3; Ionol.

Properties. White, crystalline solid. Solubility in water is 0.2 mg/l at 20°C, soluble in alcohols, food oils, and fats.

Applications. BHT is an effective, widely used, low-cost synthetic phenolic antioxidant in production of plastics (mainly polyolefins), cosmetics, rubber, elastomers. A thermostabilizer for polyethylene, polypropylene, polyesters, polystyrene, polyurethanes, polyvinyl chloride, and copolymers of vinyl and vinylidene chloride.

Exposure. It cannot be excluded that ADI for BHT is exceeded in all ages and sex groups.

Migration Data. BHT was identified in solvent extracts obtained from paper and board materials intended for food contact. Level was generally below 1.0 mg/kg paper.[1]

A study of extracts from vulcanizates based on synthetic polyisoprene rubber (water and simulant media; 0.1 and 0.5 cm^{-1}; 1 and 24 hours) revealed that BHT migrated in a quantity of 0.1 mg/l.[2]

Acute Toxicity. The LD_{50} values are reported to be 1.7 to 1.97 g/kg BW in Wistar rats, 0.94 to 2.1 g/kg BW in cats, 2.1 to 3.2 g/kg BW in New Zealand rabbits, and 10.7 g/kg BW in guinea pigs.[3]

According to other data, LD_{50} is 2.45 g/kg BW in rats, and 2.0 g/kg BW in mice. With a single administration there is depression, loss of mobility, and hypothermia.[08] According to other data, LD_{50} is more than 5.0 g/kg BW (Takagi *et al.*) or even 10 g/kg BW in rats.[4,5]

Repeated Exposure. In a 4-week study, Wistar rats were fed diets containing 0.2, 1.0, or 5.0% BHT. At 1.0 and 5.0% concentration levels, retardation of BW gain, a decrease in *Hb* and triglyceride level and cholinesterase activity, and an increase in amylase activity were observed in treated rats. There were no changes in urinalysis. Histological examination revealed testicular atrophy and a decrease in spermatogenesis.[5]

BHT significantly decreased peroxisomal enzymes, cyanide-insensitive palmitoyl-CoA oxidizing activity, and catalase activity in the livers of rats which received 0.2, 1.0, or 5.0% BHT in their feed for 4 weeks. In microsomal enzyme assay, *NADPH* cytochrome-*c*-reductase activity was significantly increased, while *P-450*, cytochrome *b5* levels, aminopyrine *N*-demethylase, and benzo[a]pyrene hydroxylase activities were not significantly increased in the livers of rats fed 1.0% or 5.0% BHT for 4 weeks.[6] Administration of 1/5 LD_{50} to rats (in the diet) caused growth retardation, a reduction in blood peroxidase, catalase and cholinesterase activity, and an increase in the relative weights of the liver and in its fat content.[7] The NOEL for (transient) hemorrhagic effects following oral administration is of about 5.0 mg/kg BW in a short-term experiment with rats.[8]

Short-term Toxicity. In a 12-week study, Wistar rats were fed diets containing 0.2, 1.0, or 5.0% BHT. At 1.0% and 5.0% concentration levels, along with the effects observed in 4-week study, vacuolization of parathyroid gland cells was noted in female rats. The LOAEL of 171 mg/kg BW in males and 180 mg/kg BW in females was established in this study.[8] According to other data, administrations of 50 mg/kg BW to rats caused functional and morphological changes.[9]

Rats were fed diets containing 0.12 to 3.0% BHT for 12 weeks. The treatment caused severe suppression of BW gain, atrophy of ovaries, thymus atrophy, and bone marrow hypoplasia in animals which consumed 0.6 and 3.0% BHT in their diet. Hematology analyses confirmed BHT hepatotoxicity.[10]

Long-term Toxicity.

Observations in man. At low concentrations, BHT is permitted in food; it has been a part of the human diet for many years without evidence of adverse effect (*WHO*, 1987).

Animal studies. On prolonged exposure, there is an increase in diuresis, liver weights, and the activity of cytochrome *P-450*-dependent macrosomal liver enzymes. Pulmonary hypertrophy and hyperplasia are noted.[11]

Daily ingestion of fat containing 4.0% BHT (calculated at 4.0 mg/kg BW) by dogs for a year produced no adverse effects.[7] According to other data, addition of 1.0% BHT to the diet resulted in enhanced formation of oxidizing enzymes (BHT-oxidase) and of other enzymes (amino-purinemethylase) which are involved in nitrogen metabolism.[12] The lifespan of animals kept under unfavorable living conditions is in this case extended.

Rats were fed diets containing 0.01, 0.03, or 0.1% BHT for 18 months. The treatment caused retardation of BW gain. Histological examination revealed testicular atrophy and decrease of spermatogenesis only at 0.1% BHT concentration. No neoplastic response was noted. The NOAEL of 0.035% in the diet (12.7 mg/kg BW for males and 15.1 mg/kg BW for females) was established in this study.[10]

Reproductive Toxicity.

Gonadotoxicity. At high dose levels, BHT is shown to induce sperm abnormalities in mice. Rats were fed diets containing 0.12 to 3.0% BHT for 12 weeks. Histological examination revealed testicular atrophy and decrease of spermatogenesis in all treated males.[10]

Embryotoxicity. Mice were given 0.5 to 20% BHT in the diet from weaning for the lifespan. Animals were allowed to mate and produce offspring. The number of live-born progeny was reduced by about 10% in the high dose group but no other signs of toxicity in the parents or offspring were noted.[13]

In a three-generation study, mice received 0.015 to 0.405% BHT in their diet from 5 weeks of age of the F_o generation to the weaning of the F_2 generation. The treatment caused no consistent significant effects on number of litters, litter size, and litter weight at birth for each generation. An increase in BW gain was noted. BHT caused little adverse effect on reproductive and neurobehavioral parameters in mice.[14]

Mice were exposed to 750 mg BHT/kg BW in arachis oil administered by gavage throughout gestation (5 dams), or before and throughout gestation (7 dams). Normal litter was produced.[15]

Wistar SPF rats received 500 mg BHT/kg BW in the diet for 13 weeks before conception and throughout gestation and lactation. No adverse effects in the offspring were observed.[16] In a 90-day reproductive study, rats were given 25, 100, or 500 mg/kg in their diet. The LOEL was found to be 100 mg/kg BW. Munro et al. (1996) suggested the calculated NOEL of 25 mg/kg BW for this study.[17]

In the dominant lethal assay, mice were exposed *i/p*. The treatment produced no increase in the incidence of early death or pre-implantation losses.[18]

Teratogenic effects were not observed in rats and mice.[15]

Mutagenicity. BHT does not cause point mutations, it also lacks clastogenic potential.

In vitro genotoxicity. No evidence of genotoxicity in the hepatocyte primary culture/DNA repair test, the *Salmonella*/microsome test, the adult rat liver epithelial cell/hypoxanthine-guanine phosphoribosyl transferase test was reported. BHT is shown to cause mutation in Chinese hamster *V79* cells.[19-22]

Carcinogenicity. BHT is likely to be a potent enhancer of mutagenicity and carcinogenicity induced by chemicals (IARC 40-123). There is some evidence of carcinogenicity in the liver of rodents.[23,24] However, no increase in neoplasm incidences at any site was noted in F344 rats fed up to 12 g BHT/kg diet for 76 weeks.[25] The mechanism of BHT carcinogenic activity is not well understood. Inhibitory activity in relation to carcinogenicity was shown.[26,27]

Carcinogenicity classifications. An IARC Working Group concluded that there is limited evidence for the carcinogenicity of BHT in *experimental animals* and there were no adequate data available to evaluate the carcinogenicity of BHT in *humans*.

IARC: 3;

NTP: N - N - N - N (feed).

Chemobiokinetics.

Observations in man. Only traces of BHT are found in the blood and there are no metabolites 24 hours after ingestion. In the urine, there are derivatives of BHT: the acid and glucuronide.[11]

Animal studies. After ingestion, BHT is widely distributed to various tissues (primarily to the small intestine, stomach, liver, kidney). Lung showed no greater accumulation than most of the other tissues.[28]

The predominant metabolic pathway involves oxidation of the 4-methyl group.[29] In rabbits, the urinary metabolites consisted of *glucouronides, sulfates,* and *free phenols* excreted in the urine and feces. Unchanged BHT was isolated from the feces but not from the urine.[30]

Regulations. According to *U.S.* concept, BHT is *GRAS* as a result of its relatively low toxicity.

U.S. FDA (1998) approved BHT for use (1) in adhesives as a component of articles intended for use in packaging, transporting, or holding food in accordance with the conditions prescribed in 21 CFR part 175.105; (2) as a defoaming agent (not to exceed 0.1% defoamer) in the manufacture of paper and paperboard intended for use in packaging, transporting, or holding food in accordance with the conditions prescribed in 21 CFR part 176.210; (3) in the manufacture of resinous and polymeric coatings of articles coming in the contact with food in accordance with the conditions prescribed in 21 CFR part 175.300; (4) in preparation of pressure-sensitive adhesives which may be used as the food-contact surface of labels and/or tapes applied to raw fruit and raw vegetables; (5) in the manufacture of rubber articles intended for repeated use in producing, manufacturing, packaging, processing, preparing, treating, packing, transporting, or holding food in accordance with the conditions prescribed in 21 CFR part 177.2600; (6) as an antioxidant in food in accordance with the conditions prescribed in 21 CFR part 172.115; and (7) in the manufacture of closures with sealing gaskets for food containers in accordance with the conditions prescribed in 21 CFR part 177.1210.

Recommendations. Joint FAO/WHO Expert Committee on Food Additives (1995). ADI: 0.3 mg/kg BW.[31]

Standards.

EU (1986). ADI: 0.05 mg/kg.[22]

Russia. PML in food: 2.0 mg/l.

References:

1. Castle, L., Offen, C. P., Baxter, M. J., and Gilbert, J., Migration studies from paper and board food packaging materials. 1. Compositional analysis. *Food Addit. Contam.,* 14, 35, 1997
2. *Hygiene and Toxicology of High-molecular-Mass Compounds and of the Chemical Raw Material Used for Their Synthesis.* Proc. 6th All-Union Conf., B. Yu. Kalinin, Ed., Leningrad, Khimiya, 1979, 64 (in Russian).
3. Deichmann, W. B., Clemmer, J. J., Rakoczy, R., et al., Toxicity of ditertiary butylmethylphenol, *Arch. Ind. Health,* 11, 93, 1955.
4. Takagi, A., Takada, K., Sai, K., Ochiai, T., Matsumoto, K., Sekita, K., Momma, J., Aida, Y., Saitoh, M., Naitoh, K., et al., Acute, subchronic and chronic toxicity studies of synthetic antioxidant, 2,2'-methylenebis(4-methyl-6-*tert*-butylphenol) in rats, *J. Toxicol. Sci.*, 19, 77, 1994.
5. Takagi, A., Momma, J., Aida, Y., Takada, K., Suzuki, S., Naitoh, K., and Tobe, M., Toxicity studies of a synthetic antioxidant, 2,2'-methylenebis(4-ethyl-6-*tert*-butylphenol) in rats. 1. Acute and subchronic toxicity, *J. Toxicol. Sci.*, 17, 135, 1992.
6. Takagi, A., Kawasaki, N., Momma, J., Aida, Y., Ohno, Y., Hasegawa, R., and Kurokawa, Y., Toxicity studies of a synthetic antioxidant, 2,2'-methylenebis(4-ethyl-6-*tert*-butylphenol) in rats. 2. Uncoupling effect on oxidative phosphorylation of liver mitochondria, *J. Toxicol. Sci.,* 18, 49, 1993.
7. Karplyuk, K. A., Toxicity of BHT, in *Voprosy Pitaniya*, 26, 378, 1971 (in Russian).
8. *Report Sci. Committee Food Commission of EC on Antioxidants*, CS/ANT/20-Final, 1987.
9. Gilbert, D. and Goldberg, L., BHT oxidase. A liver-microsomal enzyme induced by the treatment of rats with BHT, *Food Cosmet. Toxicol.,* 5, 481, 1967.
10. Anonymous, *Food Chem. Toxicol.*, 29, 73, 1991.
11. Jori, A., Toxico-kinetic aspects of butylated hydroxytoluene: a review, *Ann. 1st Super. Sanit.*, 19, 271, 1983.
12. Hirose, M., Shibata, M., Hagiwara, A., et al., Chronic toxicity of butylated hydroxytoluene in Wistar rats, *Food Cosmet. Toxicol.*, 19, 147, 1981.
13. Johnson, A. R., A re-examination of the possible teratogenic effects of butylated hydroxytoluene and its effect in the reproductive capacity of the mouse, *Food Cosmet. Toxicol.*, 3, 371, 1965.
14. Tanaka, T., Oishi, S., and Takahashi, O., Three-generation toxicity study of butylated hydroxytoluene administered to mice, *Toxicol. Lett.*, 66, 295, 1993.
15. Clegg, D. J., Absence of teratogenic effect of butylated hydroxyanisole and butylated hydroxytoluene in rats and mice, *Food Cosmet. Toxicol.*, 3, 387, 1965.

16. Meyer, O. and Hansen, E., Behavioral and developmental effects of butylated hydroxytoluene dosed to rats *in utero* and in the lactation period, *Toxicology*, 16, 247, 1980.
17. Munro, I. C., Ford, R. A., Kennepohl, E., and Sprenger, J. G., Correlation of structural class with No-Observed-Effect Levels: A proposal for establishing a threshold of concern, *Food Chem. Toxicol.*, 34, 829, 1996.
18. Epstein, S. S., Arnold, E., Andrea, J., Bass, W., and Bishop, Y., Detection of chemical mutagens by the dominant lethal assay in the mouse, *Toxicol. Appl. Pharmacol.*, 23, 288, 1972.
19. Williams, G. M., McQueen, C. A., and Tong, C., Toxicity studies of BHA and BHT. I. Genetic and cellular effects, *Food Chem. Toxicol.*, 28, 793, 1990.
20. Pashin, Y. V. and Bahitova, L. M., Inhibition of the mutagenicity of benzo[a]pyrene in the V79/HGPRT system by bioantioxidants, *Mutat. Res.*, 137, 57, 1984.
21. Ishidate, M., Yoshikawa, K., and Sofuni, T., A primary screening for mutagenicity of food additives in Ja pan, *Mutat. Toxicol.*, 3, 82, 1980.
22. Bomhard, E. M., Bremmer, J. N., and Herbold, B. A., Review of the mutagenicity/ genotoxicity of butylated hydroxytoluene, *Mutat. Res.*, 277, 187, 1992.
23. Clayson, D. B., Iverso, F., Nera, E., et al., Histopathological and autoradiopathological studies on the forestomach of Fischer-344 rats treated with BHA and related chemicals, *Food Chem. Toxicol.*, 24, 1171, 1986.
24. Chapp, N. K., Tyndall, R. L., Satterfield, L. C., et al., Selective sex-related modification of diethylnitrosamine-induced carcinogenesis in BALB/c mice by concomitant administration of butylated hydroxytoluene, *J. Natl. Cancer Inst.*, 61, 177, 1978.
25. Williams, G. M., Wang, C. K., and Iatropoulos, M. J., Toxicity studies of BHA and BHT. II. Chronic feeding studies, *Food Chem. Toxicol.*, 28, 799, 1990.
26. Hocman, G., Chemoprevention of cancer: phenolic antioxidants (BHT, BHA), *Int. J. Biochem.*, 7, 639, 1988.
27. Wattenberg, L. W., Chemoprevention of cancer, *Cancer Res.*, 45, 1, 1985.
28. Daugherty, J. P., Beach, L., Franks, H., et al., Tissue distribution and excretion of radioactivity in male and female mice after a single administration of [^{14}C]butylated hydroxytoluene, *Res. Commun. Subst. Abuse.*, 1, 99, 1980.
29. Daniel, J. M., Gage, J. C., Jones, D. I., et al., Excretion of butylated hydroxyanizole and butylated hydroxytoluene by man, *Food Cosmet. Toxicol.*, 5, 475, 1967.
30. Akagi, M. and Aoki, I., Studies on food additives. VI. Metabolism of 2,6-di-*tert*-butyl-*p*-cresol in a rabbit. Determination and paper chromatography of a metabolite, *Chem. Pharm. Bull.*, 10, 101, 1962.
31. *Evaluation of Certain Food Additives and Contaminants*, The 44th Report of the Joint FAO/WHO Expert Committee on Food Additives, Geneva, 1995, 5.

DIBUTYLDICHLOROSTANNANE

Molecular Formula. $C_8H_{18}C_{l2}Sn$
M = 303.85
CAS No 683-18-1
RTECS No WH7100000
Abbreviation. DBDCS.

Synonyms. Dibutylstannous dichloride; Dibutyltin chloride; Dibutyltin dichloride.

Properties. Light-brown, crystalline substance. Solubility in water is 80 mg/l. Odor perception threshold is 3.0 mg/l.

Applications. Heat and light stabilizer of polyvinyl chloride. Like other ogranotin compounds, D. prevents degradation of polymer during melting and forming of the resin into its final product. A catalyst in polyurethane foam products. A vulcanizing agent for silicone rubber.

Exposure. Residues of dibutyltin compounds were not detected in most foods. Concentrations found in rice, sunflower oil, peanut oil, butter, cheese, or chicken meat were less than 1.0 mg/kg wet weight of these foodstuffs.[1]

Migration of organotins from polyvinyl chloride articles (clear food container, rigid pipe, and flexible membrane) into tetrahydrofuran, xylene, and methylene chloride was investigated. Methylene chloride extracted >97% of the total extractable organotins in two extractions. There were between <0.3 μg and 4.7 mg butyltins/g solvent. The level of butyltins in potable water pipe samples was between <0.9 μg and 5.99 mg/g. In industrial application pipe samples there were between <13 μg and 1.5 mg butyltins/g polyvinyl chloride.[2]

Acute Toxicity. LD_{50} is found to be 35 mg/kg BW in mice, 112 to 182 mg/kg BW in rats, 125 mg/kg BW in rabbits, and 190 mg/kg BW in guinea pigs.[3] Lethal doses caused motor disturbances in the poisoned animals. Pathology changes were detected in the bile duct of rats after acute oral dose of 20 to 160 mg/kg BW.[4]

DBDCS affects cytochrome P-450 activity in the small intestine, liver, and kidney of rats in 24 hours after a single oral dose administration.[5] Hepatotoxicity that was evaluated by the activity of ornithine carbamyl transferase in serum was noted in mice in 24 hours after oral administration of 60 mmol dibutyltin compound per kg BW.[6]

A single administration of 5.0 to 35 mg/kg BW to rats resulted in reduced thymus weights but produced no effect on spleen weights. Recovery was completed in 9 days posttreatment.[7]

Repeated Exposure revealed pronounced cumulative properties of DBDCS. A 20 mg/kg BW dose caused 100% mortality in rats by day 13. 1/10 LD_{50} resulted in apathy, depression of BW gain, and reduced skeletal muscle tone from days 2 to 5 of the treatment.[3]

A 50 to 150 mg/kg diet (2.5 to 7.5 mg/kg BW) administered for 2 weeks induced marked changes in the lymphoid tissues including dose-related reduction in thymus weights and lymphocyte depletion in the thymus.[8]

Short-term Toxicity. Rats were fed dietary levels of 10 to 80 ppm (0.5 to 4.0 mg/kg BW) for 90 days. Effect included depression of BW gain at the highest dose. Thymus weights were not monitored, but no pathology changes were revealed on section of this organ. Anemia was caused in rats with the dose of 80 mg/kg in diet but not by 40 mg/kg dose. The NOEL appeared to be 2.0 mg/kg BW.[9]

Immunotoxicity. The immunotoxic potential of dialkyltins in species other than the rat is less clear. Decreased weights of thymus and *burza Fabricii* were found in rats and chickens exposed over a period of 10 to 14 days.[10] However, Seinen[6] did not find any evidence of lymphoid toxicity in mice, guinea pigs, or Japanese quail fed DBDCS.

Bressa et al. consider tri-*n*-butyltin chloride to be less immunotoxic than tri-*n*-butyltin oxide.[11]

Reproductive Toxicity.

Embryotoxicity. In a 3-month study, the doses of 20 and 40 mg/kg BW caused an embryolethal effect.[12] Wistar rats were given 10 or 15 mg DBDCS/kg BW on days 7 to 8 of pregnancy. The treatment caused retardation of BW gain in females and an increase in resorptions and post-implant mortality. DBDCS caused decreased BW of embryos.[13]

In vitro exposure of cultured rat embryos to DBDCS interferes with normal development of embryos.[14]

Teratogenicity. No teratogenic effect was reported in above-cited study.[12] However, gastric intubation with 5.0 to 10 mg/kg and 20 mg/kg BW on days 7 to 9 of pregnancy resulted in significant dose-dependent increase in the incidence of fetuses with malformations (anatomy of tail, anal atresia, club foot, deformity of the vertebral column, defects of the ribs and anophtalmia) even at a dose which did not induce apparent toxicity in maternal rats.[15,16]

On gestation day 8, pregnant rats were orally exposed to 80 mmol DBDCS/kg BW. The treatment caused an increase in number of fetuses with external malformations, such as cleft mandible, ankyloglossia, and fused ribs.[17]

In the above-cited study, DBDCS was found to cause visceral malformations and skeletal abnormalities.[13]

Teratogenic potential was observed in rat embryos in culture.[18]

Mutagenicity.

In vitro genotoxicity. Positive results were found in *in vitro* mammalian cell mutation assays and in *Salmonella typhimurium TA 98* and *TA 100.* DBDCS did not require the presence of an exogenous metabolizing system to express genotoxic activity.[19-21] DBDCS showed high SOS-inducing potency in the SOS-chromotest with *E. coli*. It is recognized as a genotoxic chemical by the *rec-assay*.[22]

Chemobiokinetics. At low concentrations, DBDCS blocks α-ketoacid dehydrogenase and, at high concentrations, causes disruption of oxidative phosphorylation.[23] It is excreted unchanged in the bile (60%) and as *monoethyltin* (25%) and *diethyltin* (5.0%) in the urine.[24]

Regulations. ***U.S. FDA*** (1998) approved the use of DBDCS as a catalyst in the manufacture of polyurethane resins for articles intended for use in contact with dry food alone or in combination, at levels not to exceed a total of 3.0% by weight in resin used.

Standards. ***Russia*** (1995). MAC and PML: 0.002 mg/l.

References:

1. Kannan, K., Tanabe, S., and Tatsukawa, R., Occurrence of butyltin residues in certain foodstuffs, *Bull. Environ. Contam. Toxicol.*, 55, 310, 1995.
2. Forsyth, D. S, Dabeka, R., Sun, W. F., and Dalglish, K., Specification of organotins in poly(vinyl chloride) products, *Food Addit. Contam.*, 10, 531, 1993.
3. Mazayev, V. T. and Korolev, A. A., Experimental substantiation of maximum allowable concentration for dichlorodibutyltin in water reservoirs, in *Industrial Pollution of Water Reservoirs*, S. N. Cherkinsky, Ed., Meditsina, Moscow, Issue No 9, 1969, 15 (in Russian).
4. Barness, J. M. and Stoner, H. B., Toxic properties of some dialkyl and trialkyl tin salts, *Brit. J. Ind. Med.*, 15, 15, 1958.
5. Rosenberg, D. W. and Kappas, A., Actions of orally administered organotin compounds on metabolism and cytochrome P-450 content and function in the intestinal epithelium, *Biochem. Pharmacol.*, 38, 1155, 1989.
6. Ueno, S., Susa, N., Furukawa, Y., and Sugiyama, M., Comparison of hepatotoxicity caused by mono-, di- and tributyltin compounds in mice, *Arch. Toxicol.*, 69, 30, 1994.
7. Snoeij, N. J., Penninks, A. H., and Seinen, W., Dibutyltin and tributyltin compounds induce thymus atrophy in rats due to a selective action on thymic lymphoblasts, *Int. J. Immunopharmacol.,* 10, 891, 1988.
8. Seinen, W., Voss, J. G., van Spanje, I., et al., Toxicity of organotin compounds, *Toxicol. Appl. Pharmacol.*, 42, 197, 1977.
9. Gaunt, I. F., Colley, J., Grasso, P., et al., Acute and short-term toxicity studies on di-*n*-butyltin dichloride in rats, *Food Cosmet. Toxicol.*, 6, 599, 1968.
10. Renhof, M., Kretzer, V., Schurmeyer, T., et al., Toxicity of organotin compounds in chicken and rats, *Arch. Toxicol.*, Suppl. 4, 148, 1980.
11. Bressa, G., Hinton, R. H., Price, S. C., Isbir, M., Ahmed, R.S., and Grasso, P., Immunotoxicity of tri-*n*-butyltin oxide (TBTO) and tri-*n*-butyltin chloride (TBTC) in the rat, *J. Appl. Toxicol.,* 11, 397, 1991.
12. Nikonorow, M., Mazur, H., and Piekacz, H., Effect of orally administered stabilizers in the rat, *Toxicol. Appl. Pharmacol.*, 26, 253, 1973.
13. Ema, M., Kurosaka, R., Hiro, A., and Ogawa, Y., Comparative developmental toxicity of butyltin trichloride, dibutyltin dichloride and tributyltin chloride in rats, *J. Appl. Toxicol.*, 5, 297, 1995.
14. Ema, M., Iwase, T., Iwase, Y., Ohyama, N., and Ogawa, Y., Change of embryotoxic susceptibility to di-*n*-butyltin dichloride in cultured rat embryos, *Arch. Toxicol.*, 70, 742, 1996.
15. Ema, M., Itamy, T., Kawasali, H., et al., Toxicology susceptible period for teratogenicity of di-*n*-butyltin dichloride in rats, *J. Appl. Toxicol.*, 73, 81, 1992.
16. Ema, M., Itamy, T., Kawasali, H., et al., Teratogenicity of di-*n*-butyltin dichloride in rats, *Toxicol. Lett.*, 58, 347, 1991.
17. Noda, T., Morita, S., and Baba, A., Teratogenic effects of various di-*n*-butyltins with different anions and butyl(3-hydroxybutyl)tin dilaurate in rats, *Toxicology*, 85, 149, 1993.
18. Iwase, T., Ema, M., Iwase, Y., Inazawa, K., and Ogawa, Y., Study on the sensitivity of rat embryos to teratogenic potential of di-*n*-butyltin dichloride in culture started at different developmental

stages, Abstracts of paper, 35th Annual Meeting of the Japanese Teratology Society, Tokyo, 1995, *Teratology*, 52, A17, October 1995.
19. Li, A. P., Dahl, A. R., and Hill, J. O., *In vitro* cytotoxicity and genotoxicity of dibutyltin dichloride and dibutylgermanium dichloride, *Toxicol. Appl. Pharmacol.*, 64, 482, 1982.
20. Westendorf, J. and Marquardt, H., DNA interaction and mutagenicity of the plastic stabilizer di-*n*-octyl dichloride, *Arzneimittel-Forsch.*, 36, 1263, 1986.
21. Hamasaki, T., Sato, T., Nagase, H., and Kito, H., The mutagenicity of organotin compounds as environmental pollutants, *Mutat. Res.*, 300, 265, 1993.
22. Hamasaki, T., Sato, T., Nagase, H., et al., The genotoxicity of organotin compounds in SOS chromotest and rec-assay, *Mutat. Res.*, 280, 195, 1992.
23. Penninks, A. H., Verschuren, P. M., and Seinen, W., Di-*n*-butyltin dichloride uncouples oxidative phosphorylation in rat liver mitochondria, *Toxicol. Appl. Pharmacol.*, 70, 115, 1983.
24. Bridges, J. W., Davies, D. S., and Williams, R. T., The fate of ethyltin and diethyltin derivates in the rat, *Biochem. J.*, 105, 1261, 1967.

2,6-DI-*tert*-BUTYL-α-(DIMETHYLAMINO)-*p*-CRESOL

Molecular Formula. $C_{17}H_{29}NO$
M = 263.47
CAS No 88-27-7
RTECS No GO7887000

Synonym and **Trade Name.** Agidol 3; 4-[(Dimethylamino)methyl-2,6-bis(1,1-dimethylethyl)phenol.

Properties. Yellowish powder with a pleasant specific odor. Practically insoluble in water, soluble in organic solvents.

Applications. Used as a stabilizer for polymeric materials and white rubber.

Acute Toxicity. LD_{50} is 1.03 g/kg BW^{027} or 1.65 g/kg BW in rats, and 0.4 g/kg BW in mice. Administration of the high doses is accompanied by a narcotic effect (lethargy).

Repeated Exposure revealed evident cumulative properties: K_{acc} is 1.4 (by Kagan).

Reference:

Sekretarev, V. I. et al., in *Current Problems of Toxicology and Hygiene of Pesticides and Polymeric Materials Use*, Abstracts of the All-Union Conf., Kiyv, Sept. 17-19, 1985, 1985, 160 (in Russian).

N,N'-DIBUTYL-*N,N'*-DINITROSO-*p*-PHENYLENEDIAMINE

Molecular Formula. $C_{14}H_{22}N_4O_2$
M = 278.40
CAS No 19433-82-0
RTECS No SS9100000

Synonym and **Trade Name.** *N,N'*-Dibutyl-*N,N'*-dinitroso-1,4-benzenediamine; AH-278.

Properties. Dark-brown powder. Poorly soluble in water, readily soluble in oils.

Applications. Used as a stabilizer in the production of rubber and vulcanizates.

Acute Toxicity. In rats, LD_{50} is found to be 2.6 g/kg BW. Mice and rabbits did not die after oral administration of 5.0 g/kg BW. Administration of the high doses led to narcotic effect development. Subsequently, anemia and methemoglobinemia were observed.

Repeated Exposure revealed marked cumulative properties on administration of 1/5, 1/10, and 1/20 LD_{50}. The NS and liver functions, and blood composition were mainly affected.

Reference:

Stasenkova, K. P., Kochetova, T. A., Sklovskaya, M. A., et al., in *Toxicology of New Industrial Chemical Substances*, A. A. Letavet, Ed., Meditsina, Moscow, Issue No 11, 1969, 95 (in Russian).

2,2-DIBUTYL-1,3,2-DIOXASTANNEPIN-4,7-DIONE

Molecular Formula. $C_{12}H_{20}O_4Sn$
M = 347.01

CAS No 78-04-6
RTECS No JH4735000

Synonym and **Trade Names.** Advastab DBTL; Advastab 340; Dibutylstannylene maleate; Dibutyltin maleate; Irgastab T 150; Stanclere DBTL.

Properties. White amorphous powder.

Applications. Light- and thermostabilizer for polyvinyl chloride and other galogen-vinyl polymers.

Migration of organotins from polyvinyl chloride articles (clear food container, rigid pipe, and flexible membrane) into tetrahydrofuran, xylene, and methylene chloride was investigated. Methylene chloride extracted >97% of the total extractable organotins in two extractions. They were found between <0.3 and 4.7 mg butyltins/g solvent. The level of butyltins in potable water pipe samples was between <0.9 and 5.99 mg/g. In industrial application pipe samples they were between <13 and 1.5 mg butyltins/g polyvinyl chloride.[1]

Acute Toxicity. 470 mg/kg BW dose was found to be lethal in mice.[2]

Reproductive Toxicity.

Teratogenicity. Dibutyltins exhibited a general cytotoxic effect on rabbit articular and growth-plate chondrocytes, that could be interpreted as possible effects on skeletal growth and development.[3]

References:

1. Forsyth, D. S, Dabeka, R., Sun, W. F., and Dalglish, K., Specification of organotins in poly(vinyl chloride) products, *Food Addit. Contam.*, 10, 531, 1993.
2. Schafter, E. W. and Bowles, W. A., Acute oral toxicity and repellency of 933 chemicals to house and deer mice, *Arch. Environ. Contam. Toxicol.*, 14, 111, 1985.
3. Webber, R. J., Dollins, S. C., Harris, M., and Hough, A. J., Effect of alkyltins on rabbit articular and growth-plate chondrocytes in monolayer culture, *J. Toxicol. Environ. Health,* 16, 229, 1985.

2,5-DI-*tert*-BUTYLHYDROQUINONE

Molecular Formula. $C_{14}H_{22}O_2$
M = 222.36
CAS No 88-58-4
RTECS No MX5160000
Abbreviation. DBHQ.

Synonym and **Trade Names.** Dibug; 2,5-Di(*tert*-butyl)-1,4-hydroquinone; Nonflex Alba; Santovar O.

Properties. Light-gray crystalline powder. Poorly soluble in water, soluble in hot alcohol.

Applications. Used as a stabilizer of polyolefins, polyformaldehyde, and synthetic rubber.

Acute Toxicity. In mice, the LD_{50} is reported to be 15 g/kg BW; however, according to other data, it is 4.3 g/kg BW. In rats, it is 0.8 g/kg BW. No manifestations of the toxic effect were found in rats given 5.0 g/kg BW. Drowsiness and adynamia developed after administration of larger doses. Gross pathology examination revealed focal bronchopneumonia, pulmonary vascular stasis, and congestion in the liver.[023]

Repeated Exposure failed to reveal cumulative properties or signs of intoxication in rats. BW gain was decreased and restored to the end of the second month of treatment. The treatment caused inhibition of antitoxic and oxidizing functions of the liver, the balance of protein fractions in the blood serum was changed, and erythrocyte catalase activity reduced. Gross pathology examination revealed congestion in the brain vessels, fatty dystrophy of the liver, parenchymatous dystrophy of the kidneys, and bronchopneumonia. Mice exhibited the same changes with the 2.0 g/kg BW dose administered for 2 months.[023]

Female Wistar rats were treated orally for 5 days with 80 mg DBHQ/kg BW. There was a decrease in BW in the treated rats, adoption of a prone position, salivation, lacrimation, and abnormal gait and/or muscle weakness. Ultrastructurally, neurotoxicity characterized by loss of synaptic vesicles and mitochondria in the motor endplates, and by destruction of the motor terminals was detected in the lumbrical muscles of the treated rats.[1]

Regulations. ***U.S. FDA*** (1998) approved the use of DBHQ (1) in adhesives as a component of articles intended for use in packaging, transporting, or holding food in accordance with the conditions prescribed by 21 CFR part 175.105; as (2) a defoaming agent in the manufacture of paper and paperboard intended for use in packaging, transporting, or holding food in accordance with the conditions prescribed by 21 CFR part 176.210; and (3) in the manufacture of cross-linked polyester resins which may be used as articles or components of articles intended for repeated use in contact with food in accordance with the conditions prescribed by 21 CFR part 177.2420.

Standards. ***Russia.*** PML: *n/m.*

Reference:

Mitsumori, K., Imazawa, T., Onodera, H., Takahashi, M., Kitajima, S., Inoue, T., and Kurokawa, Y., Ultrastructural changes in motor endplates of the lumbrical muscles of rats induced by a microsomal Ca2+ ATPase inhibitor, 2,5-di(*tert*-butyl)-1,4-hydroquinone, *Arch. Toxicol.*, 72, 115, 1998.

3,5-DI-*tert*-BUTYL-4-HYDROXYBENZYL ALCOHOL

Molecular Formula. $C_{15}H_{24}O_2$
M = 236.39
CAS No 88-26-6
RTECS No DO0750000

Synonym and **Trade Names.** 2,6-Di-*tert*-butyl-4-hydroxymethylphenol; Ionox 100; Antioxidant 754.

Applications. Used as a stabilizer in the production of plastics and rubber.

Acute Toxicity. Rats tolerated administration of 7.0 g/kg BW without visible manifestations of the toxic effect.[1]

Long-term Toxicity. Rats received a dose of 175 mg/kg BW in their diet for 2 years. The treatment affected food consumption and BW gain. No changes in hematology analyses and enzyme activity were found. A dose of 1.75 mg/kg BW is considered to be acceptable in man.

Reproductive Toxicity. In the above described study, there were no deviations in the development of progeny over 3 generations or in the reproductive function.[1]

Mutagenicity. Caused DNA inhibition in human lymphocytes at concentration of 25 μmol/l.[2]

Chemobiokinetics. D. is readily absorbed and its metabolism occurs rapidly in rats, rabbits, and dogs. *3,5-di-tert-butyl-4-hydroxybenzoic acid* and conjugates with glucuronides are found in the urine.

References:

1. Dacre, J. C., Toxicologic studies with 2,6-di-*tert*-butyl-4-hydroxymethylphenol in the rat, *Toxicol. Appl. Pharmacol.*, 17, 669, 1970.
2. Daugherty, J. P., Davis, S., and Yielding, K. L., Inhibition by butylated hydroxytoluene of excision repair synthesis and semiconservative DNA synthesis, *Biochem. Biophys. Res. Commun.*, 80, 963, 1978.

3,5-DI-*tert*-BUTYL-4-HYDROXYHYDROCINNAMIC ACID, OCTADECYL ESTER

Molecular Formula. $C_{35}H_{62}O_3$
M = 530.87
CAS No 2082-79-3

Synonyms and **Trade Names.** Alkofen BP 18; 3,5-Bis(1,1-dimethylethyl)-4-hydroxy-benzenepropanoic acid, octadecyl ester; 3,5-Di-*tret*-butyl-4-hydroxyhydrocinnamic acid, octadecyl ester; Irganox 1076.

Properties. Yellowish or white, crystalline, free flowing odorless powder. Poorly soluble in water and in vegetable oil.

Applications. Used as an antioxidant and thermostabilizer in the production of polypropylene; as a stabilizer for polyethylene, polyamides, polystyrene, polyvinyl chloride and polyesters.

Migration from polypropylene films into food simulants (water, 3.0% acetic acid, 95% ethanol, olive oil, and heptane) has been reported. PP films (50, 100, and 200 microns thick) were exposed to simulants at temperature-time conditions simulating migration under long-term storage.[1]

Acute Toxicity. Administration of 5.0 g/kg BW to rats, and 7.5 g/kg and 10 g/kg BW to mice caused retardation of BW gain. LD_{50} in rats exceeded 15 g/kg BW. Gross pathology examination failed to reveal changes in the viscera.[2,05,08]

Repeated Exposure. Forty administrations of 0.5 and 2.0 g/kg BW doses to mice and 0.5 g/kg BW dose to rats led to increased excitability in the treated animals. There were no gross pathology findings.[08]

Short-term Toxicity. Mice received doses of 100 and 200 mg/kg BW in suspension in sunflower oil. The treatment resulted in retardation of BW gain. Histology examination failed to reveal any structural changes.[08]

Reproductive Toxicity and **Carcinogenicity.** Rats given orally 5.0 mg/kg BW for two years did not exhibit any reproductive abnormalities or tumor growth.[05]

Mutagenicity.

In vitro genotoxicity. Photoreaction products of T. were negative in *Salmonella* mutagenicity assay.[3]

Chemobiokinetics. Ingestion of T. produced liver enlargement and induced hepatic microsomal xenobiotic metabolism, including cytochrome P-450, mixed function oxidase enzymes, and UDP-glucuronosyltransferase in rats.[4]

Regulations. ***British Standard*** (1992). T. is listed as an antioxidant for polyethylene and polypropylene compositions used in contact with foodstuffs or water intended for human consumption (max. permitted level for the final compound, 0.5%).

Standards. ***Russia.*** Recommended PML: *n/m.*

References:

1. Garde, J. A., Catala, R., and Gavara, R., Global and specific migration of antioxidants from polypropylene films into food simulants, *J. Food Prot.*, 61, 1000, 1998.
2. Zimnytskaya, L. P., Kalinin, B. Ye., and Robachevskaya, Ye. G., On toxicity of some light- and thermostabilizers of plastics, in *Hygiene and Toxicology of High-Molecular-Mass Compounds and of the Chemical Raw Material Used for Their Synthesis*, Proc. 4th All-Union Conf., S. L. Danishevsky, Ed., Khimiya, Leningrad, 1969, 193 (in Russian).
3. Yoshida, Y., Mutagenicity of photoreaction products of 2,6-di-*tert*-butylcyclohexadienone and photoreaction products of octadecyl-3-(1-*tert*-butylperoxy-3,5-di-*tert*-butyl-4-oxocyclohexa-2,5-dien-1-yl) propionate, Abstract, *Osaka-Furitsu Koshu Eisei Hen*, 12, 95, 1981 (in Japanese).
4. Lake, B. G. et al., The induction of rat hepatic microsomal xenobiotic metabolism by N-octadecyl-(3',5'-di-*tert*-butyl-4'-hydroxyphenyl)propionate, *Food Cosmet. Toxicol.*, 18, 47, 1980.

2,6-DI[3-*tert*-BUTYL-2-HYDROXYMETHYL(PHENYL)]-4-METHYLPHENOL

Molecular Formula. $C_{29}H_{36}O_3$

M = 432.58

Applications. Used as a stabilizer in the manufacture of polymeric materials..

Acute Toxicity. Mice tolerate administration of 10 g/kg BW without visible toxic effect and changes in the visceral organs.

Repeated Exposure. Mice received 6.0 g/kg BW over 18 days. The treatment did not cause mortality or affect the general state and BW gain. No histology changes were found. Exposure to 430 mg/kg BW over 6 months did not result in the development of gross pathology changes in the visceral organs. CNS excitation is found in response to administration of a 1420 mg/kg BW dose during the same period of time. Gross pathology examination revealed congestion and edema of the gastric mucosa, hyperplasia of the spleen pulp, and parenchymatous dystrophy of the renal tubular epithelium.[08]

Standards. Russia. PML: *n/m.*

2-(3,5'-DI-*tert*-BUTYL-2'-HYDROXYPHENYL)-5-CHLORO-*2H*-BENZOTRIAZOLE

Molecular Formula. $C_{20}H_{24}ClN_3O$

M = 357.88

Trade Names. Benazole BH; Tinuvin 327.

Properties. Light-yellow, crystalline powder. Poorly soluble in water and alcohol.

Applications. Used as a light stabilizer in the production of polyolefins and also as a stabilizer for polyacetate and cellulose esters.

Acute Toxicity. Mice given 7.5 g/kg and 10 g/kg BW, and rats administered 5.0 g/kg BW exhibited only decrease in BW gain.

Repeated Exposure. Up to 50% mice died in response to 40 administrations of 0.5 g/kg and 2.0 g/kg BW. The treatment caused a reduction in BW gain, increased excitability, an increase in the relative weights of the liver and kidneys, hypertrophy, and parenchymatous and fatty dystrophy of the liver, discomplexing of liver proteins.

Long-term Toxicity. Mice were fed 100 and 200 mg/kg BW given in suspension in sunflower oil. Retardation of BW gain and an increase in excitability were noted. Mortality reached 45% in animals fed 200 mg/kg BW, and 64% in animals fed 100 mg/kg BW. Gross pathology examination revealed pronounced liver hypertrophy.

Reproductive Toxicity and **Carcinogenicity.** Reproductive function of rats given 5.0 mg/kg BW for 2 years had not been affected over 3 generations. The treatment did not cause tumor growth.[05]

Reference:

Zimnitskaya, L. P., Kalinin, B. Yu., and Robachevskaya, Ye, G., On toxicity of some light- and thermostabilizers of plastics, in *Hygiene and Toxicology of High-Molecular-Mass Compounds and of the Chemical Raw Material Used for Their Synthesis*, Proc. 4th All-Union Conf., S. L. Danishevsky, Ed., Khimiya, Leningrad, 1969, 193 (in Russian).

2,6-DI-*tert*-BUTYL-4-METHOXYMETHYLPHENOL

Molecular Formula. $C_{16}H_{26}O_2$
M = 250.42
CAS No 87-97-8
RTECS No SK8350000
Abbreviation. DBM.

Synonym and **Trade Names.** Agidol 42; Antioxidant 762; 3,5-Di-*tert*-butyl-4-hydroxybenzyl alcohol, methyl ester.

Properties. White, crystalline powder turning yellow when exposed to light. Poorly soluble in water, soluble in alcohol.

Applications. Stabilizer of the synthetic rubber; a thermostabilizer of polyolefins.

Acute Toxicity. In rats, the LD_{50} is 10.6 g DBM/kg BW. The stabilizer exhibited a narcotic effect. Death within 3 days. Gross pathology examination revealed congestion in the visceral organs and brain.

Repeated Exposure. Investigation of cumulative properties showed 4 out of 10 rats die after 4 to 6 administrations in 16 days.

Reference:

Stasenkova, K. P., Shklovskaya, M. L., and Sergeyeva, L. G., Comparative evaluation of toxicity of antioxidants for phenolic resins, in *Toxicology of New Industrial Chemical Substances*, N. F. Ismerov and I. V. Sanotsky, Eds., Meditsina, Moscow, Issue No 13, 1973, 154 (in Russian).

DIBUTYLTHIOXOSTANNANE

Molecular Formula. $C_8H_{18}SSn$
M = 264.92
CAS No 4253-22-9
RTECS No WH7195000

Synonyms and **Trade Names.** Austrostab 206; Dibutyltin mercaptide; Dibutyltin sulfide; Mark 954.

Properties. Yellowish, oily, foul-smelling liquid. Odor perception threshold is 0.12 mg/l; taste perception threshold is higher.[1]

Applications. Used as a stabilizer in the production of polyvinyl chloride and vinyl chloride copolymers.

Exposure. See ***Dibutyldichlorostannane.***

Migration of organotins from polyvinyl chloride articles (clear food container, rigid pipe, and flexible membrane) into tetrahydrofuran, xylene and methylene chloride was investigated. Methylene chloride extracted >97% of the total extractable organotins in two extractions. They were found between <0.3 μg and

4.7 mg butyltins/g solvent. The levels of butyltins in potable water pipe samples were between <0.9 μg and 5.99 mg/g. In industrial application pipe samples they were between <13 μg and 1.5 mg butyltins/g polyvinyl chloride.[2]

Acute Toxicity. LD_{50} is 1.2 g/kg BW in mice, and 1.0 g/kg BW in rats. A single oral administration of 5.0 g/kg BW caused liver and stomach damage.[3] However, a significantly lower LD_{50} is also reported: 24 mg/kg BW in mice, 145 to 180 mg/kg BW in rats, and 150 mg/kg BW in rabbits. In this study, NS depression has been noted.[1]

Repeated Exposure revealed marked cumulative properties in mice administered 60 to 240 mg/kg BW and in rats given 50 to 100 mg/kg BW. A dose of 60 mg/kg BW produced no mortality in mice. The treatment caused retardation of BW gain, CNS excitation, changes in the relative weights of the visceral organs, and lesions in the stomach, intestine, and liver.[3] However, Mazayev and Shlepnina[1] observed significant animal mortality on repeated administration of a few milligrams.

Reproductive Toxicity.

Teratogenicity. Dibutyltins exhibited a general cytotoxic effect on rabbit articular and growth-plate chondrocytes that could be interpreted as possible effects on skeletal growth and development.[4]

Mutagenicity.

In vivo cytogenetics. There was an increased incidence of CA in bone marrow cells and the changed mitotic index of the intestinal mucosa of rats as a result of 1.0 mg/kg BW oral dosing for 6 months. The LOAEL for genotoxic activity was identified to be 0.1 mg/kg BW.[1]

Standards. *Russia* (1995). MAC and PML: 0.02 mg/l.

Not recommended for plastics coming in contact with foodstuffs.

References:

1. Mazayev, V. T. and Shlepnina, T. G., Experimental data on hygienic standardization of dibutyltin sulfide in water bodies, *Gig. Sanit.*, 8, 10, 1973 (in Russian).
2. Forsyth, D. S, Dabeka, R., Sun, W. F., and Dalglish, K., Specification of organotins in poly(vinyl chloride) products, *Food Addit. Contam.*, 10, 531, 1993.
3. Maksimova, N. S., Krynskaya, I. L., and Yevsyukov, V. I., Toxicity of organotin and cadmium stabilizers, *Plast. Massy*, 12, 33, 1976 (in Russian).
4. Webber, R. J., Dollins, S. C., Harris, M., and Hough, A. J., Effect of alkyltins on rabbit articular and growth-plate chondrocytes in monolayer culture, *J. Toxicol. Environ. Health*, 16, 229, 1985.

DICHLORODIETHYL STANNANE

Molecular Formula. $C_4H_{10}C_{12}Sn$

M = 247.73

CAS No 866-55-7

RTECS No NH7200000

Abbreviation. DCDES.

Synonyms. Dichlorodiethyltin; Diethyldichlorostannane; Diethylstannyl dichloride; Diethyltin chloride.

Acute Toxicity. Minimal lethal dose in rats amounted to 160 mg/kg BW. The poisoning affected liver functions.[1]

Repeated Exposure. DCDES decreased cytochrome *P-450* activity but induced hemoxygenase activity in the liver and kidney of rats.[2]

Doses of 2.5 to 7.5 mg/kg BW administered for 2 weeks induced pronounced changes in the lymphoid tissues including dose-related reduction in thymus weights and lymphocyte depletion in the thymus.[3]

Reproductive Toxicity. Administration of 10 mg DCDES/kg BW to adult rats and to neonates induced weight loss but there were no indications of neurotoxicity.[4]

Chemobiokinetics. DCDES is poorly absorbed in the GI tract in rats,[5] it is distributed mainly to the liver, and excreted primarily with the bile.[6]

References:

1. *Brit. J. Ind. Med.*, 15, 15, 1958.

2. Rosenberg, D. W., Drummond, G. S., and Kappas, A., The influence of organometals on heme metabolism, *In vivo* and *in vitro* studies with organotins, *Molec. Pharmacol.*, 21, 150, 1981.
3. Seinen, W., Vos, J. G., van Spanje, I., et al., Toxicity of organotin compounds. II. Comparative *in vivo* and *in vitro* studies with various organotin and organolead compounds in different animal species with special emphasis on lymphocyte cytotoxicity, *Toxicol. Appl. Pharmacol.*, 42, 197, 1977.
4. Bouldin, T. W., Goines, N. D., Bagnell, C. R., et al., Pathogenesis of trimethyltin neuronal toxicity. Ultrastructural and cytochemical observations, *Am. J. Pathol.*, 104, 237, 1981.
5. Mazayev, V. T., Tikhonova, Z. I., and Shlepnina, T. G., Distribution and elimination of tin taken up into the body in the form of organotin compounds, *J. Hyg. Epid. Microbiol. Immunol.*, 20, 392, 1976 (in Russian).
6. Bridges, J. W., Davies, D. S., and Williams, R. T., The fate of ethyltin and diethyltin derivatives in the rat, *Biochem. J.*, 105, 1261, 1967.

3,5-DICHLOROSALICYLIC ACID, PHENOL ESTER

Molecular Formula. $C_{13}H_9Cl_2O_3$
M = 284.02

Synonym. Phenyl-3,5-dichlorosalicylate.

Applications. Used as a stabilizer in the production of plastic materials.

Acute Toxicity. In mice, LD_{50} is 780 mg/kg BW.[08] Gross pathology examination revealed congestion in the visceral organs, edema and detachment of the gastric and esophageal mucosa, parenchymatous dystrophy of the renal tubular epithelium.

Repeated Exposure failed to reveal cumulative properties. Mice and rats tolerate administration of 40 to 160 mg/kg BW for 2 months. The administration did not affect general condition and BW gain. Histology examination of the visceral organs in the treated animals revealed no changes as compared to the controls.

Long-term Toxicity. In a 6-month study, rats received 16 mg/kg and 160 mg/kg BW. Changes in conditioned reflex activity are reported. Gross pathology examination revealed congestion in the viscera, lesions of the GI tract walls, and parenchymatous dystrophy of the renal tubular epithelium.[08]

3,3'-(DICHLOROSTANNYLENE)BISPROPANOIC ACID, DIBUTYL ESTER

Molecular Formula. $C_{14}H_{26}C_{12}O_4Sn$
M = 447.99
CAS No 61470-33-5
RTECS No UA2459110

Synonym. Bis-(β-carbobutoxyethyl)tin dichloride.

Applications. Used as a stabilizer for polyvinyl chloride plastics.

Repeated Exposure. In a 14-day study, rats were given D. in their feed. A dietary level of 450 ppm did not affect BW gain, but at the 1350-ppm level, growth retardation and decrease in the relative liver weights were evident. D. also reduced the relative weights of the thymus and spleen. A diminished amount of liver glycogen was the only treatment-related histopathologic change observed.

Immunotoxicity. D. shows a particularly high degree of lymphotoxicity *in vitro*, but in contrast to dialkyltins the estertins do not induce lymphocytotoxicity when administered *in vivo* and are less toxic than dibutyl or dioctyltins.

Chemobiokinetics. In lymphocyte metabolism studies, D. induced a dose-dependent stimulation of glucose consumption.

Reference:

Penninks, A. H. and Seinen, W., Comparative toxicity of alkyltin and estertin stabilizers, *Food Chem. Toxi col.*, 20, 909, 1982.

3,3'-(DICHLOROSTANNYLENE)DIPROPIONIC ACID, DIMETHYL ESTER

Molecular Formula. $C_8H_{14}Cl_2O_4Sn$
M = 328.36

CAS No 10175-01-6
RTECS No UF1300000

Synonyms. Bis-(β-carbomethoxyethyl)tin dichloride; Dimethyl-3,3'-(dichlorostannylene) dipropionate.

Applications. Used as a stabilizer for polyvinyl chloride plastics.

Repeated Exposure. In a 14-day feeding study in rats, a dietary level of 450 ppm did not affect BW gain, but at the 1350 ppm level, growth retardation and a decrease in the relative weight of the liver were noted. A diminished amount of liver glycogen was the only treatment-related histopathologic change observed.

Immunotoxicity. In contrast to dialkyltins the estertins do not induce lymphocytotoxicity when administered *in vivo* and are less toxic than dibutyl or dioctyltins.

Chemobiokinetics. In lymphocyte metabolism studies, D. induced a dose-dependent stimulation of glucose consumption.

Reference:

Penninks, A. H. and Seinen, W., Comparative toxicity of alkyltin and estertin stabilizers, *Food Chem. Toxicol.*, 20, 909, 1982.

2,4-DIHYDROXYBENZOPHENONE

Molecular Formula. $C_{13}H_{10}O_3$
M = 214.23
CAS No 131-56-6
RTECS No DJ0700000
Abbreviation. DHBP.

Trade Name. Advastab 48.

Properties. White, crystalline powder. Soluble in ethanol, poorly soluble in water.

Applications. Effective stabilizer for lacquer coatings, polyolefins, polyvinyl chloride, etc.

Acute Toxicity. 8.6 g/kg BW appeared to be a lethal dose in rats. The poisoning caused CNS inhibition and a decrease in food intake. GI tract disturbance, hypermotility, and diarrhea were noted.

Short-term Toxicity. Rats received 600 mg/kg BW for 90 days. The treatment caused changes in liver weight, pigmented and nucleated red blood cells, changes in erythrocyte count.[1]

Mutagenicity. DHBP was not found to be mutagenic in *Salmonella typhimurium* strains *TA100, TA98, TA1535,* and *TA 1538.*[2]

References:

1. Homrowski, S., Studies on the toxicity of additives applied in the domestic production of plastics. 3. Acute and subacute toxicity of some benzophenone derivatives, *Rozn. Panstw. Zakl. Hig.*, 19, 179, 1968 (in Polish).
2. Bonin, A. M. et al., UV-absorbing and other sun-protecting substances: Genotoxicity of 2-ethylhexyl-*p*-methoxycinnamate, *Mutat. Res.,* 105, 303, 1982.

2,6-DIISOBORNYL-4-METHYLPHENOL

Molecular Formula. $C_{27}H_{40}O$
M = 380.59

Trade Name. Alkofen DIP.

Applications. Used as a stabilizer in the production of polyethylene and polypropylene.

Acute Toxicity. Mice tolerate a dose of 7.7 g/kg BW. Gross pathology examination revealed no morphology changes in the visceral organs.

Long-term Toxicity. Rats were dosed by gavage with 1.5 g/kg BW for 6.5 months. The treatment caused an impairment of conditioned reflex activity. Gross pathology examination revealed congestion in the visceral organs, edema and desquamation of the gastric and esophageal mucosa, and parenchymatous dystrophy of the renal tubular epithelium.[08]

Reproductive Toxicity. Affects the reproduction of fish but does not affect the development of chick embryos.

Standards. ***Russia.*** Recommended PML: *n/m.*

3,5-DIISOPROPYL SALICYLAMIDE

Molecular Formula. $C_{13}H_{19}NO_2$

M = 221.29

Properties. Gray, finely dispersed, odorless powder. Poorly soluble in water.

Applications. Used as a stabilizer of polyamides and polyurethanes.

Acute Toxicity. Rats and mice tolerate administration of a 10 g/kg BW dose without any manifestations of toxic action.

Repeated Exposure. 0.5 and 1.0 g/kg BW doses produced consistent mortality and marked toxicity effects in rats. The animals tolerated administration of 0.2 g/kg BW for two months without visible signs of toxicity.

Reference:

Statsek, N. K. and Vlasyuk, M. G., Data on toxicology of new stabilizers of polymers, derivatives of 4-oxydiphenylamine, arensulfone derivatives of aminophenols, amino-arylamino-triazines and alkylated salycilic acids, in *Hygienic Aspects of the Use of Polymeric Materials and Articles Made of Them*, L. I. Medved', Ed., All-Union Research Institute of Hygiene and Toxicology of Pesticides, Polymers and Plastic Materials, Kiyv, 1969, 314 (in Russian).

2,6-DIISOPROPYL-4-*p*-TOLUENE SULFAMIDOPHENOL

Molecular Formula. $C_{19}H_{25}NO_3S$

M = 347.49

Properties. Finely dispersed, odorless, gray powder. Poorly soluble in water.

Applications. A stabilizer. Used in the production of polyamides and polyurethanes.

Acute Toxicity. Rats and mice tolerate administration of a 10 g/kg BW dose without manifestations of the toxic action.

Repeated Exposure. The doses of 0.5 and 1.0 g/kg BW caused 25% animal mortality during 5 weeks of treatment and produced visible symptoms of intoxication.

Short-term Toxicity. Rats tolerate administration of a 0.5 g/kg BW dose during 3 months of experiment without any manifestations of toxic action.

Reference:

Statsek, N. K. and Vlasyuk, M. G., Data on toxicology of new stabilizers of polymers, derivatives of 4-oxydiphenylamine, arensulfone derivatives of aminophenols, amino-arylaminotriazines and alkylated salycilic acids, in *Hygienic Aspects of the Use of Polymeric Materials and Articles Made of Them*, L. I. Medved', All-Union Research Institute of Hygiene and Toxicology of Pesticides, Polymers and Plastic Materials, Kiyv, 1969, 314 (in Russian).

4,4'-DIMETHOXY-4-DIPHENYLAMINE

Molecular Formula. $C_{14}H_{15}NO_2$

M = 229.30

CAS No 101-70-2

RTECS No DU9085000

Synonyms and **Trade Names.** Bis(*p*-anisylamine); Bis(*p*-methoxyphenyl)amine; Difenam MO; 4-Methoxy-*N*-(4-methoxyphenyl)benzeneamine; Termoflex A.

Properties. White, crystalline disks. Poorly soluble in water, readily soluble in hot alcohol.

Applications. Light- and thermostabilizer for synthetic rubber. D. is used in combination with other antioxidants to prevent rubber aging and fatigue.

Acute Toxicity. In rats and mice, LD_{50} is 2.5 g/kg BW. Administration of the lethal doses to mice was accompanied by agitation and body tremor. Some animals exhibited hard, labored breathing with subsequent convulsions. Death occurred within days. Gross pathology examination revealed changes in the liver, kidneys, and spleen.[023]

Repeated Exposure. No cumulative properties were evident in rats dosed by gavage with 410 mg/kg BW for 2 months. The treatment affected CNS and liver functions; gas exchange was decreased.[023]

Mutagenicity.

In vitro genotoxicity. D. exhibited mutagenic activity in Chinese hamster lung cells.[1]

Regulations. ***U.S. FDA*** (1998) approved the use of D. in the manufacture of rubber articles intended for repeated use in producing, manufacturing, packaging, processing, preparing, treating, packing, transporting, or holding food in accordance with the conditions prescribed by 21 CFR part 177.2600.

Reference:

Sofuni, T., Matsuoka, A., Sawada, M., Ishidate, M., Zeiger, E., and Shelby, M. D., A comparison of carcinogenesis induction by 25 compounds tested by two Chinese hamster cell (CHL and CHO) systems in culture, *Mutat. Res.*, 241, 175, 1990.

3,3'-DI-α-METHYLBENZYL(4,4'-ISOPROPYLIDENEDIPHENOL) (1) and 3,5,3'-TRI-α-METHYLBENZYL(4,4'-ISOPROPYLIDENEDIPHENOL (2), mixture

Molecular Formula. R = $C_6H_3OHC(CH_3)_2CH_2OHR'R$,
where (I) R = H; R' = R'' = $CH(CH_3)C_6H$
M = 436.63
(II) R = R' = R'' = $-CH(CH_3)C_6H$
M = 540.79

Synonym and **Trade Name.** 3,3'-Di-α-methylbenzyldiphenylol propane (1) and 3,5,3-Tri-α-methylbenzyldiphenylolpropane (2), mixture; AO-21.

Properties. A viscous resin; poorly soluble in water, readily soluble in alcohol.

Applications. Used as a stabilizer in the production of epoxy, phenol, and ion exchange resins; an antioxidant for rubber.

Acute Toxicity. No visible manifestations of the toxic effect are found in rats and mice following administration of the 5.0 g/kg BW dose.

Repeated Exposure failed to reveal cumulative properties. Rats tolerated administration of 1.0 g/kg and 5.0 g/kg BW for 10 days. Histology examination revealed moderate fatty dystrophy of the liver, parenchymatous dystrophy of the renal tubular epithelium, and signs of irritation of the spleen pulp.

Reference:

Stasenkova, K. P., Shumskaya, N. I., Greenberg, A. Ye., et al., Comparative evaluation of toxicity of bisphenol A and its derivatives, in *Hygiene and Toxicology High-Molecular-Mass Compounds and of the Chemical Raw Material Used for Their Synthesis*, Proc. 4th All-Union Conf., S. L. Danishevsky, Ed., Khimiya, Leningrad, 1969, 180 (in Russian).

4-α,α -DIMETHYLBENZYL PYROCATECHOL

Molecular Formula. $C_{15}H_{16}O_2$
M = 228.32

Properties. Crystalline powder. Solubility in water is 0.001%.

Applications. Used as a stabilizer in the production of plastic materials.

Acute Toxicity. In mice, LD_{50} is 15 g/kg BW.[08]

Short-term Toxicity. Mice were dosed by gavage with 8.75 mg/kg BW for 4 months. The treatment caused some growth retardation. Gross pathology examination revealed necrosis in the stomach and intestinal walls, parenchymatous dystrophy of the renal tubular epithelium, and hyperplasia of the spleen pulp.

(1,1-DIMETHYLETHYL)-4-METHOXYPHENOL

Molecular Formula. $C_{11}H_{16}O_2$
M = 180.27
CAS No 25013-16-5
RTECS No G07875000
Abbreviation. BHA.

Synonyms and **Trade Names.** Butylated hydroxyanisole; *tert*-Butyl hydroxianisole; *tert*-Butyl-4-methoxyphenol; Ionol CP; Topanol OC.

Properties. White or slightly yellow waxy solid. Has a faint odor. Produced in the form of small white tablets. A mixture of two isomers (*o*- and *m*-isomers at the ratio 85:15). Insoluble in water even at 50°C. Soluble in ethanol, other alcohols, fats, and oils.

Applications. The primary use for BHA is as an antioxidant and preservative in cosmetics, food packaging, and rubber. It is a synthetic food antioxidant used to prevent oils, fats, and shortening from oxidative deterioration and rancidity.[1] Antioxidant for polyolefins. Fifty countries permitted BHA as a food additive. Synthetic phenolic antioxidant.

Exposure. Estimated average daily dietary intake (in the Netherlands) does not exceed the ADI except in extreme cases in one to six-year-olds.

Acute Toxicity. In rats, LD_{50} is 2.2 g/kg BW; in mice, it is 2.0 g/kg BW.[2] Rodents given BHA diet are less vulnerable to the acute toxicity.[3]

Short-term Toxicity. In a 90-day study, rats were given BHA in their diet. Consumption of the diets containing 2.0% BHA (Wistar Han/BGA rats) or 0.5 to 2.0% BHA (Fischer 344 rats) for a month caused superficial necrosis, ulceration and hyperplasia of the squamous epithelium of the forestomach. The LOEL for these effects was found to be 1.8 g/kg BW. Munro et al. (1996) suggested the calculated NOEL of 115 mg/kg BW for this study.[4]

Long-term Toxicity.

Observations in man. At the low concentrations permitted in food, BHA has been a part of the human diet for many years without evidence of adverse effect (WHO,1987). Oral administration of the ADI (0.5 mg/kg BW) to man and 400 times higher dose to the rat (200 mg/kg) resulted in plasma BHA levels within one order of magnitude. It indicates that the current ADI might not be sufficient to protect man from possible adverse effects.

Animal studies. F344 rats were fed 0.5 and 2.0% BHA in the diet over a period of 104 weeks. Increased incidence of hyperplasia of the forestomach was observed in a dose-dependent manner. There were similar findings in hamsters. The NOEL of 62.5 mg/kg BW was identified for the induction of proliferative changes in the rat fore stomach.[5]

Monkeys received 125 or 500 mg/kg BW in corn oil by gavage for 17 weeks without adverse effects demonstrated. BHA given orally to monkeys at about maximum tolerated dose failed to induce the massive changes noted with rats given 2.0% dietary BHA.[6]

Reproductive Toxicity.

BHA exhibits slight ***embryotoxicity*** but no ***teratogenicity*** (rabbits, pigs, Rhesus monkeys). The dose of 750 mg/kg BW given on days 1 to 20 of gestation or for 70 days before conception and through gestation caused no abnormalities in rats.[7]

ICI-SPF mice were given the dose of 500 mg/kg BW 7 days before conception and until day 18 of gestation. Maternal mortality reached 25% but no signs of embryotoxicity or teratogenicity were found.[8]

Pregnant Danish Landrace swine were fed diets containing 50 to 400 mg/kg from insemination to day 110 of gestation; BW was lower in dams fed 400 mg/kg BW but no other significant signs of maternal toxicity, embryotoxicity, or teratogenicity were reported.[9]

Pregnant New Zealand SPF rabbits were administered doses of 50 to 400 mg/kg BW by gavage on days 7 to 18 of gestation; this produced no treatment-related effects in dams or offspring.[8]

Pregnant Danish Landrace swine were fed diets countering up to 3.7% BHA for 16 weeks: no treatment-related lesions of the forestomach or glandular stomach were reported.[99]

Mutagenicity. There is no indication that BHA is genotoxic.

In vitro genotoxicity. No evidence of genotoxicity in the hepatocyte primary culture/DNA repair test, the *Salmonella*/microsome test, the adult rat liver epithelial cell/ hypoxanthine-guanine phosphoribosyl transferase test, and also in the Chinese hamster ovary cells and SCE test was reported for BHA.[10-12]

Carcinogenicity. F344 rats were fed 100 to 6000 ppm in the diet for 76 weeks. No increase in neoplasms at any site was noted; 12,000 ppm BHA in the diet resulted in a small increase in squamous cell papilloma of the non-glandular squamous portion of the stomach.[13]

Administration of up to 2.0% BHA in the diet for 104 weeks induced benign and malignant tumors of the forestomach in F344 rats.[14,15]

The mechanism of BHA carcinogenic activity is not understood. BHA is known to have inconsistent functions on carcinogenesis including prevention and initiation. Kanazawa and Mizuno assumed that both functions of BHA were introduced by the derivatives formed after the reaction with gastric components such as nitrile in the stomach. Furthermore, since one of the mutagens derived from BHA is easily detoxified in our bodies and another is converted to a desmutagen, BHA appears to be one of the favorable chemicals for humans.[16] Inhibitory activity of BHA in relation to carcinogenicity has been demonstrated.[17-19]

On the other hand, Verhagen et al. hypothesized that BHA gives rise to tumor formation in rodent forestomach by inducing heritable changes in DNA.[20] BHA was also shown to be a potent enhancer of carcinogenicity induced by chemicals (IARC 40-123).

Carcinogenicity classification. An IARC Working Group concluded that there is sufficient evidence for the carcinogenicity of BHA in *experimental animals* and there were no adequate data available to evaluate the carcinogenicity of BHA in *humans*.

The Annual Report of Carcinogens issued by the U.S. Department of Health and Human Services (1998) defines BHA to be a substance which may reasonably be anticipated to be carcinogen, i.e., a substance for which there is a limited evidence of carcinogenicity in *humans* or sufficient evidence of carcinogenicity in *experimental animals*.

IARC: 2B.

Chemobiokinetics. Considerable differences in BHA chemobiokinetics following oral administration of high and low doses were observed in the rat and man, respectively.[21]

BHA is rapidly absorbed and metabolized and accumulates in the fatty tissue but has not been found elsewhere. Main metabolites were 4-*O*-conjugates: the *O*-sulfate and the *O-glucuronide* (astil). BHA can be oxidized by tissue peroxydases to a derivative that can bind to DNA. Excretion occurs in the urine; 95% of the administered dose is excreted in days in the form of glycuronides.[22,23]

BHA was eliminated in the urine predominantly as conjugated BHA together with smaller amount of conjugated *tert-butylhydroquinone*. Free BHA were found only in the urine or feces of rats (about 40%) but not of man.

Regulations.

U.S. FDA (1998) regulates BHA as a direct and indirect food ingredient and considers it to be *GRAS* when total content is not more than 0.2% the total fat or oil content of the food. BHA is listed for use (1) as a defoaming agent in the processing of food in accordance with the conditions prescribed in 21 CFR part 173.320; (2) as an antioxidant in food in accordance with the conditions prescribed in 21 CFR part 172.110; (3) as an antioxidant in the production of rubber articles intended for repeated use in contact with food (up to 1.5 wt% rubber product) in accordance with the conditions prescribed in 21 CFR part 178.2010; (4) in adhesives used as components of articles intended for use in packaging, transporting, or holding food in accordance with the conditions prescribed in 21 CFR part 175.105; (5) as a defoaming agent (not to exceed 0.1% defoamer) in the manufacture of paper and paperboard intended for use in packaging, transporting, or holding food in accordance with the conditions prescribed in 21 CFR part 176.210; (6) in the manufacture of resinous and polymeric coatings for the food-contact surface of articles intended for use in producing, manufacturing, packing, processing, preparing, treating, packaging, transporting, or holding food in accordance with the conditions prescribed in 21 CFR part 175.300; BHA may be used (7) in the manufacture of closures with sealing gaskets for food containers in accordance with the conditions prescribed in 21 CFR part 177.1210; (8) as lubricants with incidental food contact intended for used on machinery used for producing, manufacturing, packaging, processing, preparing, treating, packing, transporting, or holding food in accordance with the conditions prescribed in 21 CFR part 178.3570; and (9) in preparation of pressure-sensitive adhesives which may be used as the food-contact surface of labels and/or tapes applied to raw fruit and raw vegetables.

British Standard (1992). BHA is listed as an antioxidant for polyethylene and polypropylene compositions used in contact with foodstuffs or water intended for human consumption (max. permitted level for the final compound 0.2%).

Recommendations.

Joint FAO/WHO Expert Committee on Food Additives. Temporary ADI: 0.3 mg/kg BW.

EU (1987). ADI: 0 to 0.5 mg/kg.

References:

1. Lehman, A. J., Fitzhugh, O. G., Nelson, A. A., et al., The pharmacological evaluation of antioxidants, *Advan. Food Res.*, 3, 197, 1951.
2. Miranda, C. L., Reed, R. L., Cheek, P. R., et al., Protected effect of hydroanisole against the acute toxicity of monocrotaline in mice, *Toxicol. Appl. Pharmacol.*, 59, 424, 1981.
3. Altmann, H. J., Wester, P. W., Metthiaschk, G., et al., Induction of early lesion in the forestomach of rats by 3-*tert*-butyl-4-hydroxyanisole, *Food Cosmet. Toxicol.*, 23, 723, 1985.
4. Munro, I. C., Ford, R. A., Kennepohl, E., and Sprenger, J. G., Correlation of structural class with No-Observed-Effect Levels: A proposal for establishing a threshold of concern, *Food Chem. Toxicol.,* 34, 829, 1996.
5. Butylated hydroxyanizole, WHO, Joint FAO/WHO Expert Committee on Food Additives, Food Additives Series No 21, 1986, 3.
6. Iverson, F., Truelove, Y., Nera, E., et al., An 85-day study of butylated hydroxyanisole in the cynomologus monkey, *Cancer Lett.,* 26, 43, 1985.
7. Clegg, D. J., Absence of teratogenic effect of butylated hydroxyanisole and butylated hydroxytoluene in rats and mice, *Food Cosmet. Toxicol.*, 3, 387, 1965.
8. Hansen, E. and Meyer, O., A study of teratogenicity of butylated hydroxyanisole in rabbits, *Toxicology*, 10, 195, 1978.
9. Olsen, P., The carcinogenic effect of butylated hydroxyanisole on the stratified epithelium of the stomach in rat versus pig, *Cancer Lett.*, 21, 115, 1983.
10. Williams, G. M., McQueen, C. A., and Tong, C., Toxicity studies of BHA and BHT, I. Genetic and cellular effects, *Food Chem. Toxicol.,* 28, 793, 1990.
11. Rogers, C. G., Nayak, B. N., and Herouks-Metcalf, C., Lack of induction of sister chromatide exchanges and mutations to 6-thioguanine resistance in V79 cells by butylated hydroxyanisole with and without activation by rat or hamster hepatocytes, *Cancer Lett.*, 27, 61, 1985.
12. Abe, S. and Sasaki, M., Chromosomal aberrations and sister chromatide exchanges in Chinese hamster cells exposed to various chemicals, *J. Natl. Cancer Inst.,* 58, 1635, 1977.
13. Williams, G. M., Wang, C. K., and Iatropoulos, M. J., Toxicity studies of BHA and BHT. II. Chronic feeding studies, *Food Chem. Toxicol.,* 28, 799, 1990.
14. Ito, N., Fukushima, S., Hagiwara, A., et al., Carcinogenicity of butylated hydroxyanisole and butylated hydroxytoluene in Fischer 344 rats, *J. Natl. Cancer Inst.*, 70, 343, 1983.
15. Ito, N., Fukushima, S., and Tsuda, H. Carcinogenicity and modification of the carcinogenic response by butylated hydroxyanisole, butylated hydroxytoluene, and other antioxidants, *CRC Crit. Rev. Toxicol.*, 15, 109, 1985.
16. Hocman, G., Chemoprevention of cancer: phenolic antioxidants (BHT, BHA), *Int. J. Biochem.,* 7, 639, 1988.
17. Wattenberg, L. W., Chemoprevention of cancer, *Cancer Res.*, 45, 1, 1985.
18. Ito, N. and Hirose, M., The role of antioxidants in chemical carcinogenesis, *Gann*, 78, 1011, 1987.
19. Rahimtula, A., *In vitro* metabolism of 3-*tert*-butyl-4-hydroxyanisole and its irreversible binding to proteins, *Chem.-Biol. Interact.,* 45, 125, 1983.
20. Verhagen, H., Schilderman, P. A., and Kleinjans, J. C., Butylated hydroxyanisole in perspective, *Chem.-Biol. Interact.,* 80, 109, 1991.
21. Verhagen, H., Thijssen, H. H., ten Hoor, F., and Kleinjans, J. C., Disposition of single oral doses of buty lated hydroxyanysole in man and rat, *Food Chem. Toxicol.*, 27, 151, 1989.
22. Truhaut, R., *L'Alimentation et la Vie*, 50, 57, 1962.
23. Kanazawa, K. and Mizuno, M., Butylated hydroxyanizole produces both mutagenic and desmutagenic derivatives under gastric conditions, *Int. J. Tissue React.*, 14, 211, 1992.

2,5-DI-*tert*-PENTYLHYDROQUINONE

Molecular Formula. $C_{16}H_{26}O_2$
M = 250.42
CAS No 77-74-3
RTECS No MX6300000
Abbreviation. DPH.

Synonyms and **Trade Names.** 2,5-Bis(1,1-dimethylpropyl)-1,4-benzenediol; 2,5-Di-*tert*-amylhydroquinone; Ditag; Santovar A.

Properties. Cream-colored powder. Poorly soluble in water.

Applications. Used as a stabilizer in the production of synthetic rubber and polyolefin fibers; a staining protector in rubber.

Acute Toxicity. LD_{50} is 2.0 g/kg BW in rats and rabbits.[027] In mice, LD_{50} is reported to be 2.2 g/kg BW or, according to other data, 11.0 g/kg BW. Lethal doses cause apathy with subsequent hypodynamia and death in 3 days. Gross pathology examination revealed signs of pneumonia and inflammatory and destructive changes in the liver. Rats tolerated a dose of 5.0 g/kg BW with no adverse effects.[1]

Repeated Exposure revealed slight cumulative properties. Rats were dosed by gavage with 1/5 LD_{50} for 2 months. The treatment inhibited the detoxifying, oxidative and protein forming functions of the liver. Some animals developed adynamia and died. Histology examination revealed fatty dystrophy of the liver cells, acinous hepatitis, and vascular changes in the lungs, spleen, myocardium, and GI tract. In mice, given higher doses (1.0 g/kg BW), similar changes are found.[023]

Regulations. ***U.S. FDA*** (1998) approved the use of DPH (1) in adhesives as a component of articles intended for use in packaging, transporting, or holding food in accordance with the conditions prescribed by 21 CFR part 175.1052; and (2) in the manufacture of rubber articles intended for repeated use in producing, manufacturing, packaging, processing, preparing, treating, packing, transporting, or holding food in accordance with the conditions prescribed by 21 CFR part 177.2600.

Reference:

Volodchenko, V. A. and Gnezdilova, A. I., Toxicology profile of 2,5-di-*tert*-amylhydroquinone, *Gig. Truda Prof. Zabol.*, 12, 57, 1983 (in Russian).

DIPHENYLAMINE

Molecular Formula. $C_{12}H_{11}N$
M = 169.23
CAS No 122-39-4
RTECS No JJ7800000
Abbreviation. DPA.

Synonym and **Trade Name.** Biphenylamine; Fenam; *N*-Phenylbenzeneamine.

Properties. White or yellowish crystalline powder with a weak floral odor. Solubility in water is 30 mg/l. Soluble in alcohol. Odor perception threshold is 0.05 mg/l.

Applications. DPA is used in the production of plastics, paints, dyes, and lacquers. A stabilizer for formaldehyde copolymers, polyolefins, epoxy resins, polyvinyl chloride, etc. A stabilizer for polyoxyethylene.

Acute Toxicity. LD_{50} is reported to be 11.5 g/kg BW in rats, and 2.9 g/kg BW in mice.[017] However, according to Korolev et al.,[1] LD_{50} is 2.0 g/kg BW in rats, and 1.75 g/kg BW in mice. Manifestations of the toxic action included symptoms of CNS impairment and cyanosis development. Up to 13% met*Hb*, up to 3.0% *sulfHb*, Heinz bodies, and lower *Hb* level were found in the blood[017] in 1.5 to 55 hours following administration of 0.5 LD_{50}. LD_{33} is reported to be 0.6 g/kg BW. Gross pathology examination revealed dehydratation and acute nephrosis. No signs of intoxication in survivors were noted.

Repeated Exposure revealed low to moderate cumulative properties. 1/5, 1/25, and 1/125 LD_{50} given to rats for 25 days caused no animal mortality. 1/25 LD_{50} appeared to be the NOEL for change in the erythrocyte count and peroxidase activity. Male mice were given 10 weekly doses of 1.4 g/kg BW by gavage. This treatment resulted in 50% survival for longer than 4 months. Severe kidney damage (all normal structure in the renal cortex had been lost), and changes in the liver morphology are reported.[2]

F344 rats received 98% pure DPA in olive oil by gavage at doses of 111, 333, or 1000 mg/kg BW for 28 days. Treatment with two lower doses caused reduction in absolute and relative organ weights. Administration of 1000 mg/kg BW led to a decrease in BW gain during the treatment period and 2 weeks thereafter. Thymus, testes, and ovaries were reduced in weights; reduction in *Hb* level, elevated leukocyte count, and erythropenia in females were noted. Histological changes were observed. The LOAEL of 111 mg/kg BW for liver weight changes was established in this study.[3]

Renal papillary necrosis and necrosis of the *pars recta* were observed in male Syrian hamsters, male Sprague-Dawley rats, and male Mongolian gerbils orally treated with 600 to 800 mg DPA/kg BW.[4]

Long-term Toxicity. Rats were dosed by gavage with 5.0 mg/kg BW for 6 months. The treatment did not affect conditioned reflex activity, liver excretory function, peroxidase and ceruloplasmin activity, content of *SH*-groups in the blood.[1] The NOAEL of 2.5 mg/kg BW was established in chronic toxicity studies in dogs.[5]

Rats were given doses of 3.1 or 31 mg DPA.hydrochloride/kg BW in their diet for 730 days. The LOEL for renal effects was found to be 31 mg/kg BW; the NOEL appeared to be 3.1 mg/kg BW.

Allergenic Effect was not found at the maximum dose tested, namely, 0.5 g/kg BW.

Reproductive Toxicity.

Gonadotoxic effect was not observed at a dose of 0.5 mg/kg BW. The functional state of the spermatozoa as well as the morphology of seminiferous tubular epithelium were not affected.[1]

Teratogenicity. Rats received 2.5% DPA in their diet during the last six days of pregnancy. The treatment produced cystic dilatation of the renal collecting ducts and degeneration of the proximal tubules in the fetuses.[6]

Mutagenicity.

In vitro genotoxicity. No mutagenic effect was found in Ames test.[7]

Carcinogenicity. Rats received dietary levels of 0.025 to 1.0% (15,000 ppm) DPA for 226 days. The treatment caused renal cysts formation (0.1 to 1.0%), but no tumors occurred. When crystals were encapsulated in collodion, bladder papillomas developed in 125 days.[8]

F344 rats were fed diets containing 0.1 or 0.3% *biphenyl hydrochloride* for 104 to 106 weeks. The treatment caused no tumors to develop. Hemangiosarcoma of the circulatory system was noted in the treated animals.[9]

Carcinogenicity classifications.

U.S. EPA: D;

NTP: N - N - E - P (*2-Biphenylamine hydrochloride*).

Chemobiokinetics. A single oral dose of DPA and *N*-nitorosodiphenylamine was administered to male albino rats. Poisoning caused functional changes of the xenobiotic metabolism: an increase in the total content of cytochrome *P-450* and the activity of *NADPH*-cytochrome *P-450* reductase as well as a marked elevation of the activity of microsomal glutathione *S*-transferase. An increase in the number of *SH*-groups unbound to the protein was also noted.[10]

According to Alexander et al. (1965), DPA has been converted to *4-hydroxydiphenylamine* in rat and man. An antioxidant DPA actually promotes lipid hydroperoxide formation and oxygen consumption while markedly inhibiting generation of thiobarbituric acid reactive substances. DPA abolishes formation of *cyclooxygenase-dependent conversion products of arachidonate*.[11]

Weanling male Wistar rats received drinking water containing 0.2% DPA for 9 months. The treatment caused higher *in vitro* lipoperoxidation, free lysosomal enzyme activities, and cytosolic superoxide dismutase activity in liver. DMA may induce toxicity through some free radical reactions.[12]

Regulations. ***U.S. FDA*** (1998) approved the use of DPA (1) in the manufacture of paper and paperboard prior to the sheet-forming operation; (2) as a component of the uncoated or coated food-contact surface of paper and paperboard intended for use in producing, manufacturing, packaging, processing, preparing, treating, packing, transporting, or holding dry, aqueous and fatty foods in accordance with the conditions prescribed by 21 CFR part 176.170 and 176.180; (3) as an antioxidant for fatty based coatings and adjuvants provided it is used at a level not to exceed 0.005% by weight of the coating solids; and (4) in the manufacture of resinous and polymeric coatings for food-contact surface of articles intended for use in producing, manufacturing, packing, transporting, or holding food in accordance with the conditions prescribed by 21 CFR part 175.300.

Standards. ***Russia*** (1995). MAC and PML: 0.05 mg/l (organolept., odor).

References:

1. Korolev, A. A., Arsen'yev, M. V., and Vitvitskaya, B. R., et al., Experimental data on hygiene substantiation of MAC for diphenylamine and diphenyldiethylurea in water bodies, *Gig. Sanit.*, 5, 21, 1976 (in Russian).
2. Kronevy, T. and Holmberg, B., Acute and subchronic kidney injuries in mice induced by diphenylamine, *Exp. Pathol.*, 17, 77, 1979.
3. Yoshida, J. N., Shimoji, K., Furuta, K., et al., Twenty-eight day repeated dose toxicity testing of diphenylamine in F344 rats, Abstract, *Eisei Shikenjo Hokoku*, 107, 56, 1989 (English translation).
4. Lenz, S. D. and Carlton, W. W., Diphenylamine-induced renal papillary necrosis and necrosis of the pars recta in laboratory rodents, *Vet. Pathol.*, 27, 171, 1990.
5. Thomas, J. O., Ribelin, W. E., Woodward, J. R., and Deeds, F., The chronic toxicity of diphenylamine for dogs, *Toxicol. Appl. Pharmacol.*, 11, 184, 1967.
6. Crocker, J. F. S. and Vernier, R. L., Chemically induced polycystic disease in the newborn, Abstract, *Pediatr. Res.*, 4, 448, 1970.
7. Florin, I. et al., Screening of tobacco smoke constituents for mutagenicity using the Ames' test, *Toxicol ogy*, 15, 219, 1980.
8. Threshold Limit Values, *Am. Conf. Govern. Ind. Hyg., Inc.*, 4th ed., Cincinnati, Ohio, 1980, 157.
9. Abdo, K. M., Murthy, A. S., Haseman, J. K., Dieter, M. P., Hildebrandt, P., and Huff, J. E., Carcinogenesis bioassay in rats and mice fed diets containing 2-biphenylamine hydrochloride, *Fundam. Appl. Toxi col.*, 2, 201, 1982.
10. Semak, I. V. and Pikulev, A. T., The functional status of the xenobiotic biotransformation system in poisoning animals with diphenylamine and *N*-nitorosodiphenylamine, *Biochimia*, 58, 1562, 1993 (in Russian).
11. Sugihra, T., Rao, G., and Hebbel, R. P., Diphenylamine: an unusual antioxidant, *Free Radic. Biol. Med.*, 14, 381, 1993.
12. Darad, R., De, A. K., and Aiyar, A. S., Toxicity of nitrite and dimethylamine in rats, *Toxicol. Lett.*, 17, 125, 1983.

DIPHENYLAMINE and ACETONE, high-temperature condensation product

Trade Name. Akrin MD.

Composition. A mixture of products of the high-temperature condensation of diphenylamine with acetone. Dark brown, viscous liquid. Insoluble in water, partially soluble in alcohol.

Properties. Dark-brown, viscous liquid. Insoluble in water, partially soluble in alcohol.

Applications. Used as a stabilizer in the manufacture of synthetic rubber and latexes, a thermostabilizer for polypropylene, polyvinyl chloride and polyamides.

Acute Toxicity. In mice, LD_{50} is 23 g/kg BW.[08] High doses affect the osmotic stability of erythrocytes and cause methemoglobinemia. Histology examination revealed circulatory disturbances in the visceral organs, necrotic changes in the small intestine mucosa, and parenchymatous dystrophy of the renal tubular epithelium.

Short-term Toxicity. Rats were dosed with 250 mg/kg BW for 3.5 months. Manifestations of the toxic action included decreased BW gain, mild methemoglobinemia, and stimulation of conditioned reflex activity. Gross pathology examination revealed surface necrosis of the *villi* and desquamation of the epithelium of the gastric and intestinal mucosa.[08]

N,N'-DIPHENYLETHYLENEDIAMINE

Molecular Formula. $C_{14}H_{16}N_2$

M = 212.32

CAS No 150-61-8

RTECS No KV4800000

Synonym and **Trade Name.** 1,2-Dianilinoethane; Nonox DED.

Properties. Cream-colored powder. Insoluble in water, readily soluble in ethanol.

Applications. Stabilizer for butadiene-styrene and olefine rubber and latex. Thermostabilizer for ethylene-propylene copolymers.

Acute Toxicity. 0.5 g/kg BW dose appeared to be lethal for rats.

Reference:

Dieke, S. H., Allen, G. S., and Richter, C. P., The acute toxicity of thioureas and related compounds to wild and domestic Norway rats, *J. Pharmacol. Exp. Ther.*, 90, 260, 1947.

N,N'-DIPHENYL-*p*-PHENYLENEDIAMINE

Molecular Formula. $C_{18}H_{16}N_2$
M = 260.36
CAS No 74-31-7
RTECS No ST2275000
Abbreviation. DPPD.

Synonyms and **Trade Names.** Antioxidant DIP; 1,4-Dianilinobenzene; Diafen FF; *N,N'*-Diphenyl-1,4-benzenediamine; Nonox DPPD.

Properties. Dark-gray, crystalline powder. Poorly soluble in water, slightly soluble in alcohol.

Applications. Used as a stabilizer in the production of synthetic rubber, polyamides, and polyethylene; a polymerization inhibitor; a stabilizer for foodstuffs.

Acute Toxicity. LD_{50} is 2.37 g/kg BW in rats, and 18.5 g/kg BW in mice.[08]

Short-term Toxicity. Mice were dosed by gavage with 120 mg/kg BW 3 times a week for 4 months. Gross pathology examination revealed hemodynamic disorders and parenchymatous dystrophy in the kidneys. Sloughing of the mucosal epithelium was noted in the small intestine.[08]

Long-term Toxicity. F344 rats were fed the diet containing 0.5 and 2.0% DPPD for 104 weeks. There was a dose-dependent reduction of BW gain but no lower survival rate. Hematology indices and urinalysis showed no remarkable treatment-related changes.[1]

Reproductive Toxicity.

Embryotoxic action and effect on fertility are reported. Rats were dosed with 0.5 g/kg BW (in the diet) through the pregnancy. The treatment produced an increase in postimplantation mortality.[2]

Mutagenicity.

In vitro genotoxicity. DPPD appeared to be positive in *Salmonella* mutagenicity assay and in the test of gene mutations in hamster lung cells.[3]

Carcinogenicity. In a 2-year feeding study, there was no increase in tumor incidence.

Chemobiokinetics. Calcium deposition in the kidney of males was the only significant histology change relating to treatment.

Regulations. ***U.S. FDA*** (1998) approved the use of DPPD (1) in adhesives used as components of articles intended for use in packaging, transporting, or holding food in accordance with the conditions prescribed by 21 CFR part 175.105; and (2) as a component of the uncoated or coated food-contact surface of paper and paperboard intended for use in producing, manufacturing, packaging, processing, preparing, treating, packing, transporting, or holding aqueous and fatty foods in accordance with the conditions prescribed by 21 CFR part 176.170.

References:

1. Hasegawa, R., Fukushima, S., Hagiwara, A., et al., Long-term feeding study of *N,N'*-diphenyl-*p*-phenylenediamine in Fischer 344 rats, *Toxicology*, 54, 69, 1989.
2. Telford, I. R., Woodruff, C. S., and Linford, R. H., Fetal resorption in the rat as influenced by certain antioxidants, *Am. J. Anat.*, 110, 29, 1962.
3. Sofuni, T., Matsuoka, A., Sawada, M., Ishidate, M., Zeiger, E., and Shelby, M. D., A comparison of carcinogenesis induction by 25 compounds tested by two Chinese hamster cell (CHL and CHO) systems in culture, *Mutat. Res.*, 241, 175, 1990.

DODECYLBENZENESULFONIC ACID, SODIUM SALT

Molecular Formula. $C_{18}H_{30}O_3S.Na$
M = 348.52
CAS No 25155-30-0
RTECS No DB6825000
Abbreviation. DBS.

Synonyms and **Trade Names.** Detergent HD-90; Dodecylbenzene sodium sulfonate; Sodium laurylbenzene sulfonate; Sulframin; Witconate 1238, 1250 or 60B.

Applications. Photostabilizer for polyethylene, polypropylene fibers, etc. An intermediate in paper, rubber, and polymer processing.

Properties. White to light-yellow flakes, granules, or powder. Water solubility is 800 mg/l at 20°C.

Migration Data. Level of migration into contact liquid media including water, acetic acid, alcohol, safflower oil, or heptane depends on the type of plastic. Elution was low in polystyrene and high in polyolefins (polyethylene).[1]

Acute Toxicity. LD_{50} is 438 mg/kg BW in rats. Manifestations of the toxic action included lacrimation, CNS inhibition, hypermotility, and diarrhea. LD_{50} is 1.33 g/kg BW in mice. Poisoning caused paralysis, convulsions, hypermotility in GI tract.[2]

Short-term Toxicity. Rats were fed DBS for 90 days with no effects on growth, food consumption and utilization, survival, blood and urine analysis, organ and body weight. Histology examination failed to reveal any changes.[3]

Reproductive Toxicity.

Gonadotoxicity. Spontaneous bleeding in testis, nasal, and abdominal cavities was observed in 50% of the ICR male mice fed 0.9% *linear alkylbenzenesulfonic acid* after 70 days of the treatment (Masubuchi et al., 1976).

Mutagenicity.

In vivo cytogenetics. Oral exposure of mice to 0.5 LD_{50} did not cause CA in the bone marrow cells (Inoue et al., 1977).

In vitro genotoxicity. Induced cell mutations in cultured hamster lung cells, but produced no effect on frequency of SCE.[4]

Chemobiokinetics. Male rats were fed 1.4 mg ^{14}C-labeled DBS/kg BW for 5 weeks. Elimination occurred predominantly via feces and urine. Low levels of ^{14}C-DBS-derived residues were detected in all tissues. All fecal and renal ^{14}C- DBS-derived activity consisted of highly polar metabolites.[4]

Regulations. ***U.S. FDA*** (1998) approved the use of DBS (1) in resinous and polymeric coatings used as the food-contact surfaces of articles intended for use in producing, manufacturing, packing, processing, preparing, treating, packaging, transporting, or holding food in accordance with the conditions prescribed in 21 CFR part 175.300; (2) as an adjuvant substance in the manufacture of foamed plastics intended for use in contact with food, subject to the provisions prescribed in 21 CFR part 178.3010; and (3) in the manufacture of textiles and textile fibers used as articles or components of articles intended for use in producing, manufacturing, packing, processing, preparing, treating, packaging, transporting, or holding food, subject to the provisions prescribed in 21 CFR part 177.2800.

References:

1. Horacek, J. and Uhde, W. J., Plastics from the aspect of hygiene. A contribution to the estimation of ultraviolet absorbers and light stabilizers from the aspect of hygiene, *J. Hyg., Epidemiol., Microbiol., Immunol.*, 24, 133, 1980.
2. *The Annual Report of Tokyo Metropolitan Res. Lab. Publ. Health*, 24, 397, 1972.
3. Kay, J. H., Kohn, F. E., and Calandra, J. C., Subacute oral toxicity of biodegradable, linear alkylbenzene sulfonate, *Toxicol. Appl. Pharmacol.*, 7, 812, 1965.
4. Lay, J. P., Klein, W., and Korte, F., Elimination and biodistribution studies of [^{14}C]dodecylbenzene sulfonate in rats, following low dosing in the daily diet and a single *i/p* administration, *Toxicol. Lett.*, 17, 187, 1983.

9,10-EPOXYOCTADECANOIC ACID, 2-ETHYLHEXYL ESTER

Molecular Formula. $C_{26}H_{50}O_3$
M = 410.76
CAS No 141-38-8
RTECS No RQ1600000

Synonyms and **Trade Name.** Drapex 2,3; (2-Ethylhexyl)epoxystearate; Octylepoxystearate.

Properties. Liquid of low viscosity, insoluble in water. Odor and taste perception threshold is 1.0 mg/l.

Applications. Used as a light- and thermostabilizer with plasticizing properties in the production of plastics. Gives high low-temperature resistance to polyvinyl chloride.

Acute Toxicity. LD_{50} was not attained when the dose of 14 g/kg BW was administered to rats. In mice, LD_{50} is found to be 11.8 g/kg BW.

Short-term Toxicity. Rats were dosed by gavage with 600 mg/kg BW dose for 3 months. Manifestations of the toxic action were not evident. In 2 months after onset of administration, there was leukocytosis and an increase in the relative liver weights.

Standards. ***Russia.*** PML: 1.0 mg/l (organolept., odor and taste).

Reference:

Sheftel', V. O., Toxic effects of some polyvinyl chloride stabilizers, in *Hygiene and Toxicology of High-Molecular-Mass Compounds and of the Chemical Raw Materials Used for Their Synthesis*, Proc. 6th All-Union Conf., B. Yu. Kalinin, Ed., Leningrad, 1979, 297 (in Russian).

2,3-EPOXYPROPANOL

Molecular Formula. $C_3H_6O_2$
M = 74.08
CAS No 556-52-5
RTECS No UB4375000
Abbreviation. EP.

Synonyms and Trade Names. Allyl alcohol oxide; 2,3-Epoxy-1-propane; Glycide; Glycidol; Glycidyl alcohol; 3-Hydroxy-1,2-epoxypropane; 3-Hydroxypropylene oxide; Oxiranemethanol.

Properties. Colorless, oily liquid. Miscible with water and alcohol at any ratio.

Applications. Used in the manufacture of epoxy resins and as a stabilizer in the production of vinyl polymers.

Acute Toxicity. In rats, LD_{50} is 850 mg/kg BW. Poisoning is accompanied by excitation with subsequent CNS inhibition causing tremor and twitching of the facial muscles. In mice, LD_{50} is 430 mg/kg BW. Death occurs within 3 days.[1]

Repeated Exposure revealed moderate cumulative properties. In a 16-day gavage study, all mice and rats which received 600 mg EP/kg BW died. Lesions in testicular and epididymal tissues were found in rats given 300 mg EP/kg BW. Focal dimyelination in the medulla and thalamus of the brain was observed in female mice that received 300 mg EP/kg BW.[2]

Short-term Toxicity. In a 13-week gavage study, reduced survival and BW gain, and histopathologic lesions in the brain and kidney are reported in rats exposed to 200 or 400 mg/kg BW and in mice given 150 or 300 mg/kg BW.[2]

Reproductive Toxicity.

Gonadotoxicity. Wistar rats receiving 200 mg/kg BW became sterile in 3 weeks.[3]

Sperm count and sperm motility were reduced in male rats and mice that received EP in A 13-week gavage study.[2]

Teratogenicity effect was not found in rats given 200 mg/kg BW dose although the litter of one female was of significantly lower weights.[4] However, limbs malformations and low set ears are reported by Slott and Hales, following injection of 1.0 mg EP into rat amniotic sacs. The effect was not noted at the lower doses tested.[5]

Mutagenicity.

In vivo cytogenetics. EP induced sex-linked recessive lethal mutations and reciprocal translocations in germ cells of *Dr. melanogaster*.[6]

In vitro genotoxicity. EP is shown to be positive in several test systems including *Salmonella* mutagenicity assay.[7] It produced a doubling of the SCE frequency, and a 3-point monotonic increase with at least the highest dose at the p <0.001 significance level.[8]

Carcinogenicity. Syrian golden hamsters were given by gavage 12 mg EP twice a week for 60 weeks. The total dose per animals was 1.45 g or 20 mmol EP. More tumors were seen in the EP-treated hamsters than in controls, but the spleen was the only notable target organ, and the number of animals with spleen hemangiosarcomas was small.[9]

In a 2-year gavage study, the following carcinogenic effects were reported: increased incidence of mesotheliomas of the tunica vaginalis; fibroadenomas of the mammary gland; gliomas of the brain; neoplasms of the forestomach, intestine, skin, Zymbal gland, and thyroid gland in male F344 rats; increased incidence of fibroadenomas and adenocarcinomas of the mammary gland; gliomas of the brain; neoplasms of the oral mucosa, forestomach, clitoral gland, and thyroid gland; leukemia in female F344 rats; increased incidence of neoplasms of the Harderian gland, forestomach, skin, liver, and lung in male $B6C3F_1$ mice, and of the Harderian gland, mammary gland, uterus, subcutaneous tissue, and skin in female $B6C3F_1$ mice.[2]

F344 rats were exposed to 37.5 or 75 mg G./kg BW, and $B6C3F_1$ mice were given 25 mg or 50 mg G./kg BW by oral gavage in water for 2 years. The treatment caused dose-related increase in incidence of neoplasms in numerous tissues in both rats and mice. Survival was markedly reduced. In male rats, mesothelioma was arising in the tunica vaginalis and frequently metastasizing to the peritoneum. In mice, there was an increase in incidence of neoplasms of the Harderian gland, the forestomach (in males), and in mammary gland in females.[11]

Carcinogenicity classification. The Annual Report of Carcinogens issued by the U.S. Department of Health and Human Services (1998) defines EP to be a substance which may reasonably be anticipated to be carcinogen, i.e., a substance for which there is a limited evidence of carcinogenicity in *humans* or sufficient evidence of car cinogenicity in *experimental animals*.

NTP: CE* - CE* - CE - CE* (gavage).

Chemobiokinetics. EP is readily absorbed and distributed after gavage administration. It is converted to *α-chlorohydrin* by the action of *HCl* in the stomach that probably resulted in its reproductive effect. *α-chlorohydrin* has been metabolized and excreted in urine as *α-chloroacetic acid*. EP is likely to be converted to *glycerol* by liver epoxide hydrase.[10] EP-derived radioactivity was excreted primarily in the urine, to a lesser extent in the expired air as CO_2, and in the feces (NTP-92).

References:

1. Mel'nikova, L. V. and Klyachkina, A. M., Study of glycidol toxicity and characterization of its harmful effects, in *Toxicology of New Industrial Chemical Substances*, N. F. Ismerov and I. V. Sanotsky, Eds., Meditsina, Moscow, Issue No 15, 1979, 97 (in Russian).
2. *Toxicology and Carcinogenesis Studies of Glycidol in F344/N Rats and $B6C3F_1$ (Gavage Studies)*, NTP Technical Report Series No 374, Research Triangle Park, NC, 1990.
3. Cooper, E. R., Johns, A. R., and Jackson, H., Effects of α-chlorohydrin and related compounds on the reproductive organs and fertility of the male rat, *J. Reprod. Fertil.*, 38, 379, 1974.
4. Marks, T. A., Gerling, F. S., and Staples, R. E., Teratogenic evaluation of epichlorohydrin in the mouse and rat and glycidol in the mouse, *J. Toxicol. Environ. Health*, 9, 87, 1982.
5. Slott, V. L. and Hales, B. F., Teratogenicity and embryolethality of acrolein and structurally related compounds in rats, *Teratology*, 32, 65, 1985.
6. Foureman, P., Mason, J. M., Valencia, R., and Zimmering, S., Chemical mutagenesis testing in *Drosophila*, X. Results of 70 coded chemicals tested for the National Toxicology Program, *Environ. Molec. Mutagen.*, 23, 208, 1994.
7. Wade, M. J., Mayer, J. W., and Hine, C. H., Mutagenic action of series of epoxides in *Salmonella*, *Mutat. Res.*, 66, 367, 1979.

8. Tucker, J. D., Auletta, A., Cimino, M. C., et al., Sister-chromatid exchange: second report of the Gene-Tox program, *Mutat. Res.*, 297, 101, 1993.
9. Lijinsky, W. and Kovatch, R. M., A study of the carcinogenicity of glycidol in Syrian hamsters, *Toxicol. Ind. Health*, 8, 267, 1992.
10. Petel, J. M., Wood, J. C., and Leibman, K. C., The biotransformation of allyl alcohol and acrolein in cat liver and lung preparation, *Drug Metab. Dispos.*, 8, 305, 1980.
11. Irwin, R. D., Eustis, S. L., Stefanski, S., and Haseman, J. K., Carcinogenicity of glycidol in F344 rats and B6C3F$_1$ mice, *J. Appl. Toxicol.*, 16, 201, 1996.

6-ETHOXY-1,2-DIHYDRO-2,2,4-TRIMETHYLQUINOLINE

Molecular Formula. $C_{14}H_{19}NO$
M = 217.34
CAS No 91-53-2
RTECS No VB8225000
Abbreviation. EQ.

Synonyms and **Trade Names.** Ethoxyquin; Quinol ED; Santoflex AW; Santoquin; 2,2,4-Trimethyl-6-ethoxy-1,2-dihydroquinoline.

Properties. Thick, viscous, dark-brown or yellow liquid. Poorly soluble in water, soluble in ethanol.

Applications. Used as a stabilizer of vulcanizates based on natural and synthetic rubber. An antioxidant.

Acute Toxicity. LD_{50} is found to be 0.8 g/kg BW or 3.15 g/kg BW in rats, and 1.73 g/kg BW[020] or 3.0 g/kg BW in mice. Exposure caused general inhibition and adynamia. Death occurs in 7 days following administration of the lethal doses. Histology examination revealed fatty dystrophy of the liver, parenchymatous dystrophy of the renal tubular epithelium, and focal bronchopneumonia.[023]

Repeated Exposure. Rats were administered 1/5 LD_{50} for 2 months. Retardation of BW gain, signs of anemia and a decrease in gas exchange indices were noted. Histology examination showed bronchopneumonia and early signs of liver fatty dystrophy.[023]

Short-term Toxicity. EQ induced hemorrhage similar to BHT.[1]

Long-term Toxicity. A substance of low chronic toxicity. Rats received 0.2% EQ in the diet. The treatment caused transient depression of growth. Gross pathology examination revealed changes in the liver, kidney, and thyroid gland in many males but none in females.[2]

F344 rats and mice were fed a diet containing 0.5% EQ. Males appeared to be more susceptible to the toxic action of EQ. In males, there was damage to the renal cortex and proteinuria. Treatment of weanlings led to the development of extensive papillary necrosis. There was very little evidence of nephrotoxicity in adult female rats on exposure to EQ at 0.5% EQ in the diet for 26 weeks.[3,4]

Reproductive Toxicity.

Teratogenicity. Rats received 125 to 500 mg EQ/kg BW from days 6 to 15 of gestation. No adverse effects were noted in fetuses.[5] No signs of developmental toxicity at the dose of 950 mg/kg BW were observed in rabbits.[7]

Mutagenicity. EQ caused an elevation of DNA synthesis in the urothelium.[7]

Carcinogenicity. Continuous consumption of the diet containing 0.2% EQ resulted in some tumor development in rats.[2]

Chemobiokinetics. EQ is metabolized through deethylation which produced *6-hydroxy-2,2,4-trimethyl-1,2-dihydroquinoline* and *2,2,4-trimethyl-6-quinoline*. Other reactions, including hydroxylation to four different hydroxylated metabolites and one dihydroxylated metabolite, take place.[8]

Regulations. ***U.S. FDA*** (1998). EQ is a food additive permitted for direct addition to food for human consumption in accordance with the conditions prescribed in 21 CFR part 172.140.

Recommendations. ***FAO/WHO*** (1969). ADI: 0.06 mg/kg BW.

References:

1. Takahashi, O. and Hiraga, K., The relationship between hemorrhage induced by butylated hydroxytoluene and its antioxidant properties or structural characteristics, *Toxicol. Appl. Pharmacol.*, 46, 811, 1978.

2. Garner's *Veterinary Toxicology*, Rev. by E. G. C. Clarke and M. L. Clarke, 3rd ed., Williams and Wilkins, Baltimore, 1967, 286.
3. Manson, M. M., Green, J. A., Wright, B. J., and Caryhew, P., Degree of ethoxyquin-induced nephrotoxicity in rat in dependent on age and sex, *Arch. Toxicol.*, 66, 51, 1992.
4. Hard, G. C. and Neal, G. E., Sequential study of the chronic nephrotoxicity induced by dietary administration of ethoxyquin in Fischer 344 rats, *Fundam. Appl. Toxicol.,* 18, 278, 1992.
5. Khera, K. S., Teratologic assessment of maleic hydrazide and daminozide, and formulations of ethoxyquin, thiabendazole and naled in rats, *J. Environ. Sci. Health.,* 14, 563, 1979.
6. DeSesso, J. M. and Goeringer, G. C., Ethoxyquin and nordihydroguaiaretic acid reduce hydroxyurea developmental toxicity, *Reprod. Toxicol.*, 4, 267, 1990.
7. Shibata, M.-A., Yamada, M., Tanaka, H., Kagawa, M., and Fukushima, S., Changes in urine composition, bladder epithelial morphology, and DNA synthesis in male F344 in response to ingestion of bladder tumor promoters, *Toxicol. Appl. Pharmacol.*, 99, 37, 1989.
8. Skaare, J. U. and Solheim, E., Studies on the metabolism of the antioxidant ethoxyquin, 6-ethoxy-2,2,4-trimethyl-1,2-dihydroquinoline, in the rat, *Xenobiotica*, 9, 649, 1979.

2-(EXO-2-BORNYL)-*p*-CRESOL

Molecular Formula. $C_{17}H_{24}O$
M = 244.41
RTECS No GO6865000

Synonyms and **Trade Name.** Alkofen IP; Isobornyl cresol; 2-Isobornyl-4-methylphenol.

Composition. Contains admixtures of 2-isobornyl-5-methyl-phenols and 4-isobornyl-3-methyl-phenols.

Properties. Light-yellow, viscous liquid. Almost insoluble in water, readily soluble in vegetable oils and alcohol.

Applications. Used as a stabilizer in the production of polyethylene, polypropylene, polyvinyl chloride, and ABS plastics.

Acute Toxicity. In mice, the LD_{50} is 4.0 to 4.8 g/kg BW. Poisoning is accompanied by CNS inhibition, weakness, and subsequent death in 1 to 2 days. Gross pathology examination revealed congestion in the visceral organs, edema and sloughing of the esophageal and gastric mucosa, necrosis of the intestinal walls, and parenchymatous dystrophy of the renal tubular epithelium.[08]

Long-term Toxicity. Rats were exposed orally to 175 mg/kg BW dose 44 times during 4 months. Manifestations of the toxic action included lymphocytosis and impairment of conditioned reflex activity. Gross pathology examination revealed an increase in the relative weights of the liver and kidneys and circulatory disturbances.[08]

Reproductive Toxicity. Does not affect the development of chick embryos.

FATTY ACID, EPOXIDIZED SOYBEAN OIL

CAS No 8013-07-8
68333-67-5
RTECS No LH8010000
Abbreviation. ESO.

Trade Names. Advaplast 39; Drapex 6,8; Epoxom; Flexol EPO; Soya resin.

Composition. ESO is either a polymer with formaldehyde or with the latter and phenol. It consists of triglycerides of oleic, linoleic, linolenic, and saturated fatty acids, and small amounts of free fatty acids, phospholipids, and sterols.

Properties. Clear, viscous, light-yellow liquid. Polysiccative oil with a high fatty acid content. Insoluble in water, soluble in fats.

Applications. Used as a light- and thermostabilizer with plasticizing properties for polyvinyl chloride and chlorinated rubber (in combination with other substances). A plasticizer in polyvinyl chloride gaskets in lids for glass jars used for packaging of ready-cooked baby food, including purees of beef, pork, fish, poultry, berries, and vegetables. ESO is used predominantly in the production of foods and paints. It is used in conjunction with organic acid salts. The stabilizing value of ESO is proportional to the epoxy index. The more highly epoxidized vegetable oils produce a better stabilizing effect but are correspondingly more toxic.

Migration of ESO from plasticized polyvinyl chloride was found to be 200 mg/dm^2 in ethanol (3-day contact period, 40°C), 125 mg/dm^2 in isooctane (2 days, 20°C), and 50 mg/dm^2 in sunflower oil (10 days, 40°C) (data from the graph).[1]

Levels of ESO in fresh retail meat samples wrapped in PVC film ranged from less than 1.0 to 4.0 mg/kg, but were higher (up to 22 mg/kg) in retail cooked meat. Migration into sandwiches and rolls from take-out outlets ranged from less than 1.0 to 27 mg/kg depending on factors such as the type of filling and the length of contact time prior to analysis. When the film was used for microwave cooking in direct contact with food, levels of ESO from 5.0 to 85 mg/kg were observed, whereas when the film was employed only as a splash cover for re-heating foods, ESO levels ranged from 0.1 mg/kg to 16 mg/kg.[2]

The migration of ESO has been determined in baby food using a GS/MS analytical procedure with a detection limit of 1.5 mg/kg. Residues of ESO were found in all dishes except in blueberries. The levels ranged from < 1.5 to 50.8 mg/kg. It was concluded that the presented levels of ESO in the baby food are only due to migration from the lids and not of natural origin.[3]

Exposure. Daily intake is estimated to be 0.4 mg/day.[4]

Acute Toxicity. In rats, LD_{50} is found to be 28.7 or even 40 g/kg BW.

Long-term Toxicity. Rats consumed 1.0% ESO and more in their feed over a period of 1 year. The treatment lowered the nutritional protein utilization coefficient and increased basal metabolism and water consumption. Accumulation of the epoxy acids was noted in the fat deposits. Retardation of BW gain became evident.

Dogs received 5.0% ESO in the diet. Manifestations of the toxic action included an increase in the relative liver and kidney weights. There are no special features in the results of hematology analyses or in visceral organ histology (Larsen).[04] No signs of intoxication were observed in rats given 1.4 g/kg BW twice a week for 16 months and in dogs treated for 12 months (Krause).[04]

In a 2-year rat feeding study, no evidence of systemic toxicity has been found after administration of 100 mg/kg BW. Therefore, TDI appears to be 1.0 mg/kg BW.[5,6]

Allergenic Effect. ESO is likely to be possible sensitisizer because it contains di- and triepoxide residues.[03]

Mutagenicity. Chemicals containing epoxide groups are known to cause mutagenic effect.

In vitro genotoxicity. However, ESO is not found to be mutagenic in the *Salmonella* test.

Carcinogenicity. No evidence of carcinogenicity was reported following administration of a 100 mg/kg BW dose.[5]

Regulations. ***U.S. FDA*** (1998) approved the use of ESO (1) as a component of adhesives intended for use in food-contact surface in accordance with the conditions prescribed in 21 CFR part 175.105; (2) in the manufacture of closures with sealing gaskets for food containers in accordance with the conditions prescribed in 21 CFR part 177.1210; (3) as a plasticizer in polymeric substances used in the manufacture of food-contact articles in accordance with the conditions prescribed in 21 CFR part 178.3740; (4) as a plasticizer in the manufacture of resinous and polymeric coatings (iodine number maximum 14; oxirane oxygen content 6.0% minimum), as the basic polymer in accordance with the conditions prescribed in 21 CFR part 175.300; and (5) in the manufacture polysulfide polymer-polyepoxy resins in articles safely used as the dry food-contact surface in accordance with the specified conditions prescribed in 21 CFR part 177.1650. Soybean oil may be used (6) as a defoaming agent in the manufacture of paper and paperboard intended for use in packaging, transporting, or holding food in accordance with the conditions prescribed in 21 CFR part 176.210.

References:

1. Feigenbaum, A. E., Hamdani, M., Ducruet, V. J., and Riquet A.-M., Classification of interactions: volatile simulants, global and specific migration, *J. Polymer Engin.*, 15, 47, 1995-1996.

2. Castle, L., Mayo, A., and Gilbert, J., Migration of epoxidized soya bean oil into foods from retail packaging materials and from plasticized PVC film used in the home, *Food Addit. Contam.*, 7, 29, 1990.
3. Hammarling, L., Gustavsson, H., Svensson, K., Karlsson, S., and Oskarsson, A., Migration of epoxidized soya bean oil from plasticized PVC gaskets into baby food, *Food Addit. Contam.*, 15, 203, 1998.
4. Kieckebush, W. et al., Fette, Seifen, *Austrichmittel*, Bd. 65, 11, 919, 1963.
5. Anonymous, Plasticizers migration in food, *Food Chem. Toxicol.*, 29, 139, 1991.
6. *1994 The Annual Report of the Committees on Toxicity, Mutagenicity, Carcinogenicity of the Chemi cals in Food, Consumer Products and the Environment*, DoH, HMSO, London, 1995, 61.

GALLIC ACID, BUTYL ESTER

Molecular Formula. $C_{11}H_{14}O_5$
M = 226.25
CAS No 1083-41-6
RTECS No LW7600000

Synonyms. Butyl gallate; Butyl-3,4,5-trihydroxybenzoate.

Properties. Grayish, odorless powder, rather bitter taste. Poorly soluble in water, soluble in alcohol.

Applications. Stabilizer of polyvinyl chloride and linear polyesters.

Acute Toxicity. In mice, LD_{50} is 860 mg/kg BW. Gross pathology examination revealed congestion, lesions of the GI tract mucosa and parenchymatous dystrophy of the renal tubular epithelium.

Repeated Exposure failed to show cumulative properties. No mice died during administration of 45 mg/kg, 90 mg/kg, and 170 mg/kg BW for 2 months. There was retardation of BW gain and some changes found on histology examination of the visceral organs.

Short-term Toxicity. No manifestations of the toxic action or morphology changes were observed in rats given 200 mg/kg BW for 4.5 months.

Reproductive Toxicity. Affects the reproductive processes in fish.[08]

Reference:

Karplyuk, I. A., On the problem of the possible use of dodecyl gallate as an antioxidant of dietary fats, *Gig. Sanit.*, 12, 34, 1962 (in Russian).

GALLIC ACID, OCTYL ESTER

Molecular Formula. $C_{15}H_{22}O_5$
M = 282.37
CAS No 1034-01-1
RTECS No LW8225000
Abbreviation. OG.

Synonyms. Octyl gallate; 3,4,5-Trihydroxybenzoic acid, *n*-octyl ester.

Acute Toxicity.

Observations in man. In humans, after drinking beer containing 20 mg/l OG, severity of erythema and edema in the oral mucosa was greater than after drinking untreated beer.[3]

Animal studies. In rats, LD_{50} is 4.7 g/kg BW.[1] According to other data, it is 2.7 g/kg BW in male and 2.0 to 2.3 g/kg BW in female rats.[2]

Short-term Toxicity. In a 13-week study in rats using dietary levels of 1.0, 2.5, and 5.0 g OG/kg BW feed, the only effect observed was a slight elevation of serum glutamic-oxalacetic transaminase. No histology changes were found. A similar result in dogs was observed at 5.0 g/kg feed but not at 1.0 or 2.5 mg/kg feed.[3]

In another 13-week study in pigs, no effect on growth, hematology analyses, organ weights, or histology was noted after using a dose of 2.0 g/kg feed.[4]

Long-term Toxicity. Rats of two generations received 100 mg OG/kg BW. The treatment caused a slight hypochromic anemia. Temporary ADI of 0 to 0.1 mg/kg BW was established, based on a NOEL of 17.5 mg/kg BW derived in this study.[5]

Reproductive Toxicity.

Embryotoxicity. In 3-generation studies in rats, no adverse effect was observed as a consequence of consumption of 350 to 5000 mg OG/kg feed.[4,6,7]

In the 2-generation test, rats were dosed with 1.0 and 5.0 g/kg feed. Dose-dependent reduction in implantation sites as well as a reduction in the number of corpora lutea in F_2 parent animals was observed. In the third-litter pups of the second generation, the incidence of gross kidney alterations was increased at the 5.0 g/kg dietary level. No histology changes are found.[3] Munro et al. (1996) suggested the calculated NOEL at the level of 89 mg/kg BW for this study.[8]

Allergenic Effect. OG caused experimental sensitization in guinea pigs.[9]

Carcinogenicity. No increase in tumor incidence was noted in rats given 350 to 5000 mg/kg feed over a period of 2 years.[4]

Chemobiokinetics. Octyl and propyl gallates are metabolized similarly. Van Esch detected the unchanged compound as a major urinary component in rats.[4,10]

Recommendations. Joint FAO/WHO Expert Committee on Food Additives (1997): No ADI allocated.

References:

1. Joint FAO/WHO Expert Committee on Food Additives, Food Additives Series No 10, WHO, Geneva, 1986, 45.
2. Brun, R., Contact eczema due to an antioxidant of margarine (gallate) and change of occupation, *Dermatologica*, 140, 390, 1970.
3. van der Heijden, C. A., Janssen, P. J. C. M., and Strik, J. J. T. W., Toxicology of gallates: a review and evaluation, *Food Chem. Toxicol.*, 24, 1067, 1986.
4. Van Esch, G. J., The toxicity of the antioxidants propyl, octyl and dodecyl gallate, *Voeding*, 16, 683, 1955.
5. *Evaluation of Certain Food Additives and Contaminants,* WHO Techn. Report Series No 868, 46th Report of JECFA, Geneva, 1997, 69.
6. Sluis, K. J. H., The higher alkyl gallates as antioxidants, *Food Mf.*, 26, 99, 1951.
7. Hazleton Lab. Inc., Unpublished report, 1970., cit. in *Evaluation of Mercury, Lead, Cadmium and the Food Additives Amaranth, Diethylpyrocarbonate, and Octyl Gallate,* The 16th Meeting of the Joint FAO/WHO Expert Committee on Food Additives, WHO Food Additives Series No 4, Geneva, 1972.
8. Munro, I. C., Ford, R. A., Kennepohl, E., and Sprenger, J. G., Correlation of structural class with No-Observed-Effect Levels: A proposal for establishing a threshold of concern, *Food Chem. Toxicol.,* 34, 829, 1996.
9. Hausen, B. M. and Beyer, W., The sensitizing capacity of the antioxidants propyl, octyl, and dodecyl gallate and some related gallic acid esters, *Contact Derm.*, 26, 253, 1992.
10. Koss, G. and Koransky, W., Enteral absorption and biotransformation of the food additive octyl gallate in the rat, *Food Chem. Toxicol.*, 20, 591, 1982.

GALLIC ACID, PROPYL ESTER

Molecular Formula. $C_{10}H_{12}O_5$
M = 212.22
CAS No 121-79-9
RTECS No LW8400000
Abbreviation. PG.

Synonyms and **Trade Name.** Progallin P; Propyl gallate; 3,4,5-Trihydroxybenzoic acid, *n*-propyl ester.

Properties. Colorless, odorless, white to pale brownish-yellow, crystalline powder having a slightly bitter taste. Solubility is 3.5 g/l in water at 25°C, 1030 g/l in alcohol, and 830 g/l in ether.

Applications. PG is used as an antioxidant for rubber, acetobutyrate cellulose, polyethylene, and polyethers. It is used in cosmetics and in the preparation of packaging materials for food products, and as a food additive.

Acute Toxicity. In mice, rats, hamsters and rabbits, LD_{50} varies from 2.0 to 3.8 g/kg BW.[1] In cats, it is 0.4 g/kg BW.

Repeated Exposure In a 4-week feeding study, increased liver enzyme activities were found at a dose level of 5.0 g/kg feed; no effect was noted at 1.0 g/kg feed.[2]

Short-term Toxicity. Exposure to 117 to 5000 mg/kg feed caused no adverse effects.[1] No changes were observed in mice after 90-day consumption of the diet containing 5.0 or 10 g PG/kg.[3]

In a 13-week study, rats received 490, 1910, or 7455 mg/kg feed. At a high-dose level, changes in hematology analysis and increased activity of hepatic ethoxyresorufin-*O*-diethylase were noted. Pathological changes were observed in the spleen. The NOAEL of 1910 mg PG/kg feed (135 mg/kg BW) was determined in this study.[4]

Long-term Toxicity. Rats were given 0.001 to 2.3% PG in the diet for 730 days. The LOEL was established to be 864 mg/kg BW (Orten et al., 1948).[4] Munro et al. (1996) suggested the calculated NOEL of 86 mg/kg BW for this study.[5]

In mice that received 5.0 and 10 g/kg feed, only a reduced spleen weight at a 10 g/kg dietary level was found.[3]

Allergenic Effect. In guinea pigs, sensitizing properties were observed, especially after intradermal applications. Sensitization did not occur when there was oral pre-exposure.[6]

PG was considered to be a moderate sensitizer.[7]

Reproductive Toxicity.

Embryotoxicity. A three-generation reproduction study revealed no treatment-related effects when 350 to 5000 mg/kg was added to the diet of rats.[8] Rats were dosed with 500 mg/kg the diet through the pregnancy. The treatment produced an increase in postimplantation mortality.[9]

No ***teratogenic effect*** occurred in rats treated with up to 25 g/kg in the diet.[10]

Mutagenicity.

In vitro genotoxicity. PG is found to be negative in *Salmonella* mutagenicity assay.[11] It did not produce mutagenic effect in *in vitro* cytogenic tests in human embryonic lung cells. DNA damage was observed in isolated rat hepatocytes exposed to non-cytotoxic concentrations (0.5 mM PG). Supplementary studies indicated, however, that the effects may have been due to compound-induced apoptosis (programmed cell death), a non-genotoxic phenomenon.[12]

In vivo cytogenetics. PG failed to demonstrate genotoxic effect in DLM test in rats, and in *in vivo* bone marrow metaphase analysis in Sprague-Dawley rats.[13]

Carcinogenicity. F344 rats and $B6C3F_1$ mice were exposed to the dietary levels of 5 to 20 g/kg feed. No dose-related increase in tumor incidences was observed.[14]

Oral administration of PG was found to decrease the incidence of tumors induced in rats by dimethylbenz[a]anthracene.[15]

F344 rats and $B6C3F_1$ mice were maintained on diets containing 0.6 or 1.2% PG for 103 weeks. The treatment produced no effect on survival, but decreased BW gain. Tumors of the preputial gland, islet-cell tumors of the pancreas, and phaeochromocytoma of the adrenal gland occurred at a significantly higher incidence in the low-dose male rats. Nevertheless, under conditions of this study, PG was not considered to be carcinogenic to F344 rats and $B6C3F_1$ mice.[16]

Carcinogenicity classification.

NTP: E - N - E - N (feed).

Chemobiokinetics. Following oral administration, more than 70% PG is absorbed in the intestines. Biokinetics of PG apparently differs from that of octyl and dodecyl gallates which are absorbed and hydrolyzed to a lesser degree than the PG. In rats and rabbits, PG is hydrolyzed to *gallic acid* and undergoes further methylation, yielding *4-o-methylgallic acid* as a biotransformation product.

Excretion occurs in the urine either as the free compound or as the *glucuronide conjugate*. In rats, PG may be eliminated via the urine as unchanged ester.[17,18]

Regulations. *U.S. FDA* (1998) regulates the use of PG (1) in the manufacture of resinous and polymeric coatings in a food-contact surface of articles intended for use in producing, manufacturing, packing, transporting, or holding food in accordance with the conditions prescribed by 21 CFR part

175.300; and (2) in preparation of pressure-sensitive adhesives which may be used as the food-contact surface of labels and/or tapes applied to raw fruit and raw vegetables in accordance with the conditions prescribed by 21 CFR part 175.105. Affirmed as *GRAS.*

In ***the U.S.***, permitted use level of PG (as total antioxidants) is 0.02% for all foods.

Recommendations. Joint FAO/WHO Expert Committee on Food Additives (1997). ADI: 1.4 mg/kg BW.

Standards. ***The Netherlands***. Maximum Permitted Level: 0.01% in fats and oils (total antioxidants).

References:

1. *Joint FAO/WHO Expert Committee on Food Additives*, Food Additives Series No 10, WHO, Geneva, 1995, 45.
2. van der Heijden, C. A., Janssen, P. J. C. M., and Strik, J. J. T. W., Toxicology of gallates: a review and evaluation, *Food Chem. Toxicol.*, 24, 1067, 1986.
3. Dacre, J. C., Long-term toxicity study of *n*-propyl gallate in mice, *Food Cosmet. Toxicol.*, 12, 125, 1974.
4. *Toxicological Evaluation of Certain Food Additives and Contaminants,* The 41st Meeting of the Joint FAO/WHO Expert Committee on Food Additives. WHO Food Additives Series No 32, 1993.
5. Munro, I. C., Ford, R. A., Kennepohl, E., and Sprenger, J. G., Correlation of structural class with No-Observed-Effect Levels: A proposal for establishing a threshold of concern, *Food Chem. Toxicol.,* 34, 829, 1996.
6. *Propyl Gallate,* Antioxidant No 10, *COLIPA Monograph* CSC/386/83, 1983.
7. Hausen, B. M. and Beyer, W., The sensitizing capacity of the antioxidants propyl, octyl, and dodecyl gallate and some related gallic acid esters, *Contact Dermat.*, 26, 253, 1992.
8. Van Esch, G. J., *The Toxicity of the Antioxidants Propyl, Octyl and Dodecyl Gallate*, Voeding, 16, 683, 1955.
9. Telford, I. R., Woodruff, C. S., and Linford, R. H., Fetal resorption in the rat as influenced by certain antioxidants, *Am. J. Anat.*, 110, 29, 1962.
10. Tanaka, S., Kawashima, K., Nakaura, S., et al., Effect of dietary administration of propyl gallate during pregnancy on prenatal and postnatal development of rats, *J. Food Hyg. Soc. Jap.*, 20, 378, 1979.
11. Rosian, M. and Stich, H., Enhancing and inhibiting effect of propyl gallate on cocarcinogen-induced mutagenesis, *J. Environ. Pathol. Toxicol.*, 4, 159, 1980.
12. Nakagawa, Y., Moldeus, P., and Moore, G., Propyl gallate-induced DNA fragmentation in isolated rat hepatocytes, *Arch . Toxicol.,* 72, 33, 1997.
13. *Carcinogenesis Bioassay of Propyl Gallate in F344/N Rats and B6C3F$_1$ Mice (Feeding Study*), NTP Technical Report Series No 240, Research Triangle Park, NC, 1982.
14. Abdo, K. M., Huff, J. U. E., Haseman, J. K., et al., Carcinogenesis bioassay of propyl gallate in F344 rats and B6C3F$_1$ mice, *J. Am. Coll. Toxicol.,* 2, 425, 1983.
15. King, M. M. and McCay, P. B., Modulation of tumor incidence and possible mechanisms of inhibition of mammary carcinogenesis by dietary antioxidants, *Cancer Res*., Suppl. 43, 2485S, 1983.
16. Abdo, K. M., Huff, J. E., Haseman, J. K., and Alden, C. J., No evidence of carcinogenicity of *D*-mannitol and propyl gallate in F344 rats and B6C3F$_1$ mice, *Food Chem. Toxicol.*, 24, 1091, 1986.
17. Dacre, J. C., Metabolic pathways of the phenolic antioxidants, *J. N. Z. Inst. Chem.*, 24, 161, 1960.

GLUCITOL

Molecular Formula. $C_6H_{14}O_6$

M = 182.17

CAS No 50-70-4

RTECS No LZ4290000

Synonym and **Trade Names.** Diakarmon; *D*-Glucitol; *L*-Gulitol; Hexahydric alcohol; Sorbit; Sorbitol; Sorbol.

Properties. Odorless, white granules, powder or crystalline powder having a sweet taste (~60% as sweet as sugar, w/w) with a cooling sensation in the mouth. Hygroscopic. Readily soluble in water, poorly soluble in cold ethyl alcohol.

Applications. Used in the manufacture of propylene glycol, synthetic plasticizers and resins.

Exposure. A stabilizer in the polyvinyl chloride production. G. occurs naturally in many plant materials.

Acute Toxicity. In rats, mice, and rabbits, LD_{50} values are in excess of 15 mg/kg BW.[1] The only treatment-related effects from high doses are soft stools and a laxative action. If adequate time is allowed for adaptive changes in the microflora of the lower bowel, then animals have been shown to tolerate diets containing up to 20% G. without exhibiting overt laxation.

Repeated Exposure. Increase in daily urine volumes was caused by 63% G. in the diet.[2]

Short-term Toxicity. G. is added to food (for example, of diabetics) and causes no side effects since it does not raise the blood glucose level, or raises it to a considerably lesser extent than glucose.

Long-term Toxicity. In a 2-year feeding study of xylitol in rats, a group exposed to 20% G. was included as the control.[1] Induced changes consisted of caecal enlargement, reduced BW gain up to week 78, decreased efficiency of food utilization, increased water intake, and urinary output, etc. Adrenal medullar hyperplasia and decreased absolute thyroid weights were also reported.

JECFA (1982) concluded that a diet containing as much as 20% G. "produced gross dietary imbalance, which may produce metabolic imbalance".

Reproductive Toxicity. G. is shown to be negative in unpublished teratogenicity studies in rats, rabbits, mice, and hamsters.[1] Charles River rats were fed diets with 2.5, 5.0, and 10% G. Treatment was associated with no consistent adverse effect on any index of reproductive performance. No abnormal pups were observed in any generations.

MacKenzie and Hauck[1] concluded that G. administered in the diet to 3 successive generations of rats at levels up to 10% had no adverse effect on growth or reproductive performance.

Mutagenicity.

In vivo cytogenetics. G. gave positive response in DLM assay in mice.[1]

In vitro genotoxicity. G. appeared to be negative in a majority of tests (*Salmonella* mutagenicity assay, the rat bone marrow test).[1]

Chemobiokinetics. The absorption of G. occurs by passive diffusion and is slower than that of glucose or fructose. It is metabolized basically in the liver and only very small quantities are found in the blood or urine.[3] Two metabolic pathways are described: oxidation after conversion to *glucose* or direct oxidation of *fructose*. Herz considered G. converted first to fructose and then to glucose.[4]

Increased amounts of *methylmalonic acid* and *2-oxoglutaric acid* are found in the urine of polyol-fed rats. The urinary excretion of *citric acid* and *maleic acid* is also increased significantly. The increased levels of urinary organic acids in the polyol-fed rats may be explained in terms of impaired mitochondrial oxidation of these acids and impaired conversion of *methymalonic acid* to *succinic acid*.[2]

Regulations.

U.S. FDA (1998) regulates G. for use (1) in the manufacture of resinous and polymeric coatings in a food-contact surface in accordance with the conditions prescribed in 21 CFR part 175.300; (2) in the manufacture of resinous and polymeric coatings for polyolefin films to be safely used as a food-contact surface of articles intended for use in producing, manufacturing, packing, transporting, or holding food in accordance with the conditions prescribed in 21 CFR part 175.320; (3) as a defoaming agent in the manufacture of paper and paperboard intended for use in packaging, transporting, or holding food in accordance with the conditions prescribed in 21 CFR part 176.210; and (4) in the manufacture of cross-linked polyester resins to be safely used as articles or components of articles intended for repeated use in contact with food in accordance with the conditions prescribed in 21 CFR part 177.2420. Considered to be *GRAS* as a substance mi grating to food from paper and paperboard products.

EU (1990). G. is available in the *List of authorized monomers and other starting substances which shall be used for the manufacture of plastic materials and articles intended to come into contact with foodstuffs (Section A)*.

EU (1995). G. is a food additive generally permitted for use in foodstuffs.

Recommendations. Joint FAO/WHO Expert Committee on Food Additives. ADI is not specified.

References:

1. MacKenzie, K. M. and Hauck, W. N., Three-generation reproduction study of rats ingesting up to 10% sorbitol in the diet, and a brief review of the toxicological status of sorbitol, *Food Chem. Toxicol.*, 24, 191, 1986.
2. Hamalainen, M. M., Organic aciduria in rats fed high amounts of xylitol or sorbitol, *Toxicol. Appl. Pharmacol.*, 90, 217, 1987.
3. Adcock, L. H. and Gray, C. H., The metabolism of sorbitol in the human subject, *Biochem. J.*, 65, 554, 1957.
4. Hers, H. G., The conversion of fructose-1-^{14}C and sorbitol-1-^{14}C to liver and muscle glycogen in the rat, *J. Biol. Chem.*, 214, 373, 1955.

N-HEPTYL-*p*-ANIZIDINE

Molecular Formula. $C_{14}H_{23}NO$

M = 221.33

Trade Name. AN-4.

Composition. A product of condensation of 1-heptanol and *p*-anizidine.

Properties. Clear, oily liquid with an unpleasant odor. Insoluble in water, soluble in alcohol.

Applications. Used as an ageing retarder and antiozonant in the synthesis of vulcanizates and rubber.

Acute Toxicity. The LD_{50} is 1.5 g/kg BW in mice, and 2.1 g/kg BW in rats. In two hours after administration of the lethal doses, mice exhibited labored respiration and nasal bleeding. Survivors exhibited drowsiness and adynamia. In rats, poisoning is accompanied by excitation, increased salivation, and vomiting. Histology examination revealed fatty dystrophy of the liver and perivascular inflammatory infiltration. Parenchymatous dystrophy of the tubules and focal agglomeration of lymphohistiocytes were revealed in the kidneys while the lungs had signs of edema and congestion.[023]

Repeated Exposure. Rats were dosed by gavage with 1/10 LD_{50} for 2 months. The treatment caused general depression, adynamia, decreased BW gain, CNS inhibition, anemia, and reduced gas exchange. Gross pathology examination revealed changes in the lungs, liver, kidneys, and spleen.[023]

2,2',4,4',6,6'-HEXABROMODIPHENYLAMINE

Molecular Formula. $C_{12}H_5Br_6N$

M = 652.56

Trade Name. Difenam 06.

Properties. White, crystalline powder with a yellowish tint. Poorly soluble in water and alcohol.

Applications. Used as a stabilizer in the production of plastics.

Acute Toxicity. In rats and mice, LD_{50} was not attained.

Repeated Exposure. Rats were dosed with a 2.0 g/kg BW dose over a period of 1 month. The treatment caused leukocytosis, impairment of liver protein formation, a decline in total protein, and an increase in the relative weights of the visceral organs.

Reference:

Gnezdilova, A. I. and Volodchenko, V. A., Toxicity effects of hexabromodiphenylamine, *Gig. Truda Prof. Zabol.*, 12, 59, 1983 (in Russian).

HEXABUTYLDISTANNOXANE

Molecular Formula. $C_{24}H_{54}OSn_2$

M = 596.16

CAS No 56-35-9

RTECS No JN8750000

Abbreviation. TBTO.

Synonyms and **Trade Names.** Bis(tributyltin) oxide; Hexabutyldistannoxane; Hexabutylditin; Oxybis-(tributyltin); Tributyltin oxide.

Properties. Colorless to yellow liquid with a weak odor. Solubility in hot water is 1000 ppm and <20 ppm at 20°C.

Applications. Used in the production of flame resistant polyester; a curing agent. TBTO in particular is an effective biocide preservative. It has been used as a polyvinyl chloride stabilizer and an antifouling agent in marine paints, representing up to 20% of total ingredients.

Exposure. Food chain accumulation and bioconcentration of alkyltin compounds have been demonstrated in crabs, oysters and salmon exposed to TBTO.

Migration of organotins from polyvinyl chloride articles (clear food container, rigid pipe, and flexible membrane) into tetrahydrofuran, xylene, and methylene chloride was investigated. Methylene chloride extracted >97% of the total extractable organotins in two extractions. They were found between <0.3 μg and 4.7 mg butyltins/g solvent. The level of butyltins in potable water pipe samples was between <0.9 μg and 5.99 mg/l. In industrial application pipe samples, levels were found between <13 μg and 1.5 mg butyltins/g polyvinyl chloride.[1]

Acute Toxicity. The ranges reported for acute oral LD_{50} in rodents are 10 to 234 mg/kg BW. In rats, LD_{50} is 55 mg/kg BW. In mice, LD_{50} is 122 mg/kg BW.[2, 03]

Repeated Exposure revealed material and functional accumulation in mice given TBTO for 20 days. K_{acc} is 3.5 to 5.4 (1/5 LD_{50}) and 1.5 to 1.6 (1/25 LD_{50}).[3] Wistar rats fed dietary levels of 5.0 to 320 mg/kg BW (0.25 to 16 mg/kg BW) exhibited decreased food and water consumption and clinical signs of toxicity. Doses of 4.0 mg/kg and 16 mg/kg BW caused thymus weight reduction. At a dose level of 16 mg/kg BW histopathologic changes in the thymus, mesenteric lymph nodes, and liver were noted. Hematology effects were also found.[4]

Weanling rats received 0.5 to 50 mg/kg diet for at least 28 consecutive days. No treatment-related effects were noted on clinical chemistry and hematology parameters.[5]

Rats were given 5.0 or 25 ppm TBTO in their feed. The TBTO ingestion decreased BW gain and food consumption at 25 ppm in exposed animals. Histological examination of animals treated with TBTO for 7 days revealed atrophy of the thymus with severe lymphocyte depletion in the cortex. After 28 days of exposure, most of the lesions reversed and the thymus became markedly smaller than in control rats. Accumulation predominantly occurs in the liver and kidneys.[6]

Short-term Toxicity. In a 13-week feeding study in rats, the NOEL was established to be 0.2 mg/kg BW.[7]

Long-term Toxicity. Doses of 0.5, 5.0, and 50 mg/kg diet administered to Wistar rats for 106 weeks produced increased mortality, reduction of BW, hematology changes: anemia, lymphocytopenia, thrombocytosis. Decreased kidney function and increased plasma enzyme activity were also noted.[8]

At 50 mg/kg dose, the ovaries, adrenals, spleen (females), heart (males), pituitary, liver and kidneys were increased in weight, but the thyroid weight was decreased in females. After 2 years, only the thyroid changes were still present. The NOAEL appeared to be 0.5 mg/kg BW.[8]

In a 6-month study, rabbits received 0.04 mg TBTO/kg BW. The treatment caused no statistically reliable changes in hematology analyses, serum proteins, and protein fraction contents, or in CNS condition of animals.[3]

Male monkeys, Macaca fascicularis, received an oral dose of 0.16 mg/kg BW for 22 weeks. There were no signs of toxic action or changes in BW gain. Hematology analyses showed a decrease in leukocyte count, no changes in clinical chemistry, or total tin concentration in blood were noted.[9]

TDI of 5.0 or 0.25 mg/kg BW has been calculated for reduction in lymphoid organ weights (thymus) or lymphoid function.[10]

Reproductive Toxicity.

Embryotoxicity. Pregnant mice received 5.0, 20, or 40 mg/kg BW on days 6 to 15 of gestation. The treatment with 40 mg TBTO/kg BW caused a significant reduction of maternal BW gain and embryotoxic effect. Gross pathology examination revealed a dose-dependent decrease in spleen weight and an increase in placental weight.[11]

In a later study of these authors carried out under the same conditions, low TBTO embryotoxicity was reported.[12] They concluded that TBTO maternal toxicity (an effective dose of 5.0 mg/kg BW) is much stronger than its embryo-fetotoxic potential. Non-specific alterations of hematology parameters are reported in this study.[13]

An oral dose of 16 mg/kg BW administered to pregnant Long-Evans rats on days 6 to 20 of gestation resulted in retarded fetal growth.[14] A reduction in the numbers of live births and effect on postnatal growth and behavior were observed at doses of 10 mg/kg BW and above. Maternal toxicity was also evident at these dose levels.

No evidence of ***teratogenicity*** was found in a rabbit study at doses of up to 2.5 mg/kg BW.[15]

In NMRI mice given 35 mg TBTO/kg BW on days 6 to 15 of gestation, an increase in embryolethality was revealed. Developmental abnormalities were increased in a dose-dependent fashion at a 11.7 mg/kg BW dose level and above. However, there was also the evidence of maternal toxicity at this and higher dose levels.[16]

Immunotoxicity. TBTO appeared to be immunotoxic at a dose level that did not affect other organs and systems. The thymus is a target organ of TBTO immunotoxicity. Immunotoxicity of TBTO in rodent species is predominantly expressed by *T*-lymphocyte deficiency resulting from a depletion of cortical thymocytes. Male rats were administered a single oral dose of 30 or 60 mg/kg BW. The poisoning did not cause severe toxicity or overt clinical signs. Reduction in relative thymus weights and increased DNA fragmentation were noted in the isolated thymocytes. Dibutyltin, a major metabolic dealkylation product of tributyltin, failed to significantly stimulate apoptosis when added to the isolated thymocytes *in vitro*.[17]

When rats were exposed to 20 or 80 ppm in their feed for 6 weeks, TBTO was found to cause changes in the blasto-transformation processes of the thymus and spleen cells. There was a decline in the contents of *T*-lymphocytes and a rise in the contents of *B*-lymphocytes in the spleen. The number of viable cells in the thymus and spleen was reduced.[18]

Rats received 0.5 or 5.0 ppm TBTO in the diet for 2 years. This treatment produced the immunotoxic effect.[19]

There were no treatment-related effects on plaque-forming cell and delayed type hypersensitivity response, whereas thymic atrophy and impaired clearance of *Listeria monocytogenes* were noted in weanling rats given dietary concentration of 50 mg TBTO/kg diet for 28 consecutive days. The NOEL of 0.5 mg/kg BW for this effect was established under the conditions of this study.[5,23]

Mutagenicity.

In vivo cytogenetics. Administration of a single LD_{50} caused mutagenic effect in mouse bone marrow cells. Mice given single oral doses of 31 to 125 mg/kg BW failed to show an increased incidence of micronuclei in bone marrow polychromatic erythrocytes.[7]

In vitro genotoxicity. TBTO is recognized as a genotoxic chemical by the *rec-assay*;[20] it produced mutations in Chinese hamster ovary cells[19] but did not induce reverse mutations in *Salmonella typhimurium*.[16] However, according to Hamasaki et al.,[21] TBTO appeared to be mutagen in *Salmonella TA 100.*

Carcinogenicity. In a 106-week feeding study in rats, the incidence of benign tumors of the pituitary was significantly elevated at 0.5 and 50 mg TBTO/kg in the diet (2.5 mg/kg BW). At a dietary level of 50 mg TBTO/kg, an increase in the incidence of adrenal and parathyroid adenomas was noted. This increase was not dose-related.[8]

Rats were given 50 ppm TBTO in the diet for 2 years. The treatment increased the incidence of tumors of endocrine origin.[19]

Chemobiokinetics. When administered with drinking water, TBTO is distributed primarily to the kidneys with only low levels in the blood.[22] Tributyltin compounds may be converted to *dibutyltin derivatives* and metabolites. Given by *i/p* injection to mice, the major part of the dose was rapidly eliminated in the feces.

Regulations.

U.S. FDA (1998) regulates the use of TBTO in adhesives as a component of articles intended for use in packaging, transporting, or holding food (for use as a preservative only) in accordance with the conditions prescribed in 21 CFR part 175.105.

The use of triorganotin compounds in antifouling paints has been banned in the ***UK*** and ***France***, and ***EU*** ban has been proposed.

Standards. ***Russia*** (1995). MAC and PML: 0.0002 mg/l.
Recommendations. ADI for oral exposure: 0.0003 mg/kg BW.[24]
References:

1. Forsyth, D. S, Dabeka, R., Sun, W. F., and Dalglish, K., Specification of organotins in poly(vinyl chloride) products, *Food Addit. Contam.*, 10, 531, 1993.
2. Oakley, S. D. and Fawell, J. K., Toxicity of selected organotin compounds to mammalian species, *Water Research Centre*, Report ER 1307-M, 1986.
3. Belyaeva, N. N., Bystrova, T. A., Revazova, Iu. A., et al., Comparative evaluation of toxicity and mutagenicity of organotin compounds, *Gig. Sanit.*, 5, 10, 1976 (in Russian).
4. Krajnc, E. I., Wester, P. W., Loeber, J. G., et al., Toxicity of bis(tri-*n*-butyltin)oxide in the rat. I. Short-term effects on general parameters and on the endocrine and lymphoid systems, *Toxicol. Appl. Pharmacol.*, 75, 363, 1984.
5. Verdier, F., Viraf, M., Schweinfurth, H., and Descotes, J., Immunotoxicity of bis(tri-*n*-butyltin)oxide in rat, *J. Toxicol. Environ. Health.*, 32, 307, 1991.
6. Bressa, G., Hinton, R. H., Price, S. C., Isbir, M., Ahmed, R. S., and Grasso, P., Immunotoxicity of tri-*n*-butyltin oxide (TBTO) and tri-*n*-butyltin chloride (TBTC) in the rat, *J. Appl. Toxicol.*, 11, 397, 1991.
7. Schweinfurth, H. A., Toxicology of tributyltin compounds, in 'Tin and Its Uses', *Quarterly J. Int. Tin Res. Inst.*, 143, 9, 1985.
8. Wester, P. W., Krajnc, E. I., van Leeuwen F. X. R., et al., Chronic toxicity and carcinogenicity of bis-(tri-*n*-butyltin)oxide (TBTO) in the rat, *Food Chem. Toxicol.*, 28, 179, 1990.
9. Karrer, D., Baroncelli, S., Ciaralli, L., and Turillazzi, P. G., Effect of subchronic bis(tri-*n*-butyltin)oxide (TBTO) oral administration on hematological parameters in monkeys: a preliminary report, *Food Chem. Toxicol.*, 30, 715, 1992.
10. Penninks, A. H., The evaluation of data-derived safety factors for bis(tri-*n*-butyltin)oxide, *Food Addit. Contam.*, 10, 351, 1993.
11. Baroncelli, S., Karrer, D., and Turillazzi, P. G., Embryotoxic evaluation of bis(tri-*n*-butyltin) oxide (TBTO) in mice, *Toxicol. Lett.*, 50, 257, 1990.
12. Baroncelli, S., Karrer, D., and Turillazzi, P. G., Oral bis(tri-*n*-butyltin)oxide in pregnant mice. I. Postnatal influence on maternal behavior or postnatal mortality, *J. Toxicol. Environ. Health*, 46, 355, 1995.
13. Karrer, D., Baroncelli, S., and Turillazzi, P. G., Oral bis(tri-*n*-butyltin)oxide in pregnant mice. II. Alteration in hematological parameters, *J. Toxicol. Environ. Health*, 46, 364, 1995.
14. Crofton, K. M., Dean, K. F., Boncec, V. M., et al., Prenatal or postnatal exposure to bis(tri-*n*-butyltin)oxide in the rat: postnatal evaluation of teratology and behavior, *Toxicol. Appl. Pharmacol.*, 97, 113, 1989.
15. Schweinfurth, H. A. and Gunzel, P., The tributyltins: mammalian toxicity and risk evaluation for humans, in *Ocean 87, The Ocean "An International Workplace"*, 1987, 1421.
16. Davis, A., Barale, R., Brun, G., et al., Evaluation of the genetic and embryotoxic effects of bis(tri-*n*-butyltin)oxide, a broad spectrum pesticide in multiple *in vivo* and *in vitro* short-term tests, *Mutat. Res.*, 188, 65, 1987.
17. Raffray, M. and Cohen, G. M., Thymocyte apoptosis as a mechanism for tributyltin-induced thymic at rophy *in vivo. Arch. Toxicol.*, 67, 231, 1933.
18. Vos, J. G., de Klerk, A., Krajnc, E. I., Kruizinga, W., van Omen, B., and Rozing, J., Toxicity of bis(tri-*n*-butyltin)oxide in rat. II. Suppression of thymus-dependent immune responses and of parameters of nonspecific resistance after short-term exposure, *Toxicol. Appl. Pharmacol.*, 75, 387, 1984.
19. Boyer, I. J., Toxicity of dibutyltin, tributyltin and other organotin compounds to humans and to experimental animals, *Toxicology*, 55, 253, 1989.
20. Hamasaki, T., Sato, T., Nagase, H., et al., The genotoxicity of organotin compounds in SOS chromotest and rec-assay, *Mutat. Res.*, 280, 195, 1990.

21. Hamasaki, T., Sato, T., Nagase, H., and Kito, H., The mutagenicity of organotin compounds as enviromental pollutants, *Mutat. Res.*, 300, 265, 1993.
22. Evans, W. N., Cardarelli, N. F., and Smith, D. J., Accumulation and excretion of [^{14}C]bis(tri-*n*-butyltin)oxide in mice, *J. Toxicol. Environ. Health*, 5, 871, 1979.
23. Vos, J. G., De Klerk, A., Krajnc, E. I., Van Loveren, H., and Rozing, J., Immunotoxicity of bis(tri-*n*-butyltin)oxide in the rat: effects on thymus-dependent immunity and on nonspecific resistance following long-term exposure in young versus aged rats, *Toxicol. Appl. Pharmacol.*, 105, 144, 1990.
24. *Tributyltin Oxide*, Concise International Chemical Assessment Document No 14, IOMS, WHO, Geneva, 1999, 29.

HYDROXYBIS(STEARATO)ALUMINUM

Molecular Formula. $C_{36}H_{71}AlO_5$
M = 611.05
CAS No 300-92-5
RTECS No BD0962000
Abbreviation. HA.

Synonyms. Aluminum distearate; Hydroxybis(octadecanoato-*O*)-aluminum; Aluminum hydroxydistearate.

Properties. White powder. Insoluble in water and alcohol.

Applications. Lubricant in plastics industry. Stabilizer for food-packaging materials. Pigment. Used also in cosmetics.

Toxicity. See ***Aluminum compounds.***

Regulations. ***U.S. FDA*** (1998) approved the use of HA. (1) in adhesives as a component of articles intended for use in packaging, transporting, or holding food in accordance with the conditions prescribed by 21 CFR part 175.105; (2) as a stabilizer in food packaging materials in accordance with the conditions prescribed by 21 CFR part 181.29; and (3) as a colorant in the manufacture of articles or components of articles intended for use in producing, manufacturing, packing, processing, preparing, treating, packaging, transporting, or holding food, subject to the provisions and definitions set in 21 CFR part 178.3297.

p-HYDROXYDIPHENYLAMINE

Molecular Formula. $C_{12}H_{11}NO$
M = 185.24
CAS No 122-37-2
RTECS No SJ6950000

Synonyms and **Trade Name.** *p*-Anilinophenol; Diphenam O; *N*-Phenyl-*p*-aminophenol.

Properties. White crystalline powder with a characteristic aromatic odor. Insoluble in water, soluble in ethanol.

Applications. Used as an antioxidant and stabilizer in the production of nitrocellulose and synthetic rubber.

Acute Toxicity. LD_{50} is reported to be 1.22 or 3.3 g/kg BW in rats, and 2.2 to 2.3 g/kg BW in mice. Within 3 minutes after poisoning narcosis is found to develop with retention of corneal reflexes, paresis of the hindlimbs, respiratory distress, and cyanosis of the mucous membranes. The *metHb* level is noted to be up to 53 to 56%. Death occurs in 3 days.

Repeated Exposure failed to reveal cumulative properties. Administration of 0.5 LD_{50} produced anemia and an increase (up to 12%) in *metHb* level and in the content of creatinine and residual nitrogen in the blood.

Immunotoxicity. A pronounced sensitizing effect was found in rabbits.

Reference:

Volodchenko, V. A., Toxicity of *p*-hydroxydiphenylamine, *Gig. Truda Prof. Zabol.*, 3, 53, 1983 (in Russian).

4-(2-HYDROXYETHOXY)DIPHENYLAMINE

Molecular Formula. $C_{14}H_{15}NO_2$

M = 229.3

Synonym. 2-(*p*-Phenylaminophenoxy)ethanol.

Properties. Finely dispersed, odorless, gray powder. Insoluble in water.

Applications. Used as a stabilizer in the production of polyamides and polyurethanes.

Acute Toxicity. LD_{50} is reported to be 3.0 g/kg BW in rats, and 1.75 g/kg BW in mice. Administration of high doses led to a sharp decline in muscle tonus, to paresis, ataxia of the hind legs, and lethargy. Death was preceded by adynamia and shallow respiration.

Repeated Exposure revealed evident cumulative properties. K_{acc} is 2.4 (by Kagan). The treatment with 0.4 g/kg BW resulted in almost total mortality. The lower doses caused anemia and leukocytosis and changes in the hippuric acid synthesis. Other manifestations of the toxic effect included changes in the duration of hexenal sleep, prothrombin time, residual blood nitrogen, diuresis, urinary protein, and the relative weights of the visceral organs.

Reference:

Zabezhinskaya, N. A., Brook, Ye. S., et al., in *Protection of Water Reservoirs against Pollution by Industrial Liquid Effluents*, S. N. Cherkinsky, Ed., Medgiz, Moscow, Issue No 4, 1960, 147 (in Russian).

2-HYDROXY-4-METHOXYBENZOPHENONE

Molecular Formula. $C_{14}H_{12}O_3$

M = 228.26

CAS No 131-57-7

RTECS No DJ1575000

Abbreviation. HMBP.

Synonyms and **Trade Name.** Advastab 45; Benzophenone-3; 2-(2-Hydroxy-4-methoxyphenyl)phenyl-methanone; 4-Methoxy-2-hydroxybenzophenone; Oxybenzone.

Properties. Light-yellow or white, crystalline powder. Almost insoluble in water, soluble in alcohol.

Applications. HMBP is one of the UV-absorbing agents used in industry and medicine. Used in the production of plastic surface coatings, polymers, and in cosmetics. It is used in the manufacture of such products as lipsticks, hair sprays, hair dyes, shampoos, and detergent bars, and sunscreen lotions. Light stabilizer for polystyrene, pentaplast, polyvinyl chloride, cellulose acetobutyrate, polyamide, and polyolefin fibers, and paints and varnishes based on perchlorovinyl resins.

Acute Toxicity. LD_{50} is reported to be 7.4 g/kg BW or even more than 12.8 g/kg BW in rats.[1]

Repeated Exposure. In a 14-day study, F344 rats were given feed containing 3,125 to 50,000 ppm HMBP. The treatment caused an increase in the liver and kidney weights. Renal lesions (consisting of dilated tubules and regeneration of tubular epithelial cells) were noted in high-dosed rats. Marked hepatocyte cytoplasmic vacuolization was observed in the liver. The NOAEL of 6,250 ppm HMBP. in the diet is identified in rats for microscopic kidney lesions. Histology examination revealed kidney damage in mice given 50,000 ppm HMBP in their diet for 14 days.[2]

Short-term Toxicity. Rats were fed 0.5 to 1.0% HMBP in the diet for 90 day. Animals displayed retardation of BW gain, nephrotoxicity, reduction in the relative weights of the hypophysis, thymus, heart, and adrenals.[3]

In a 13-week study, F344 rats and $B6C3F_1$ mice received 3125 to 50,000 ppm in their feed. Rats displayed retardation of BW gain, increased liver and kidney weights, and kidney lesions (papillary degeneration or necrosis and inflammation). Rats were fed 0.02 to 1.0% HMBP over a period of 90 days. Treatment with the two highest doses depressed BW gain in the animals. Females given 0.5 or 1.0% HMBP for 6 weeks exhibited lowered *Hb* level and leukocytosis, with an increase in lymphocytes and a decrease in neutrophils. After 12 weeks of treatment, anemia and lymphocytosis, with a reduction in granulocytes were noted, 1.0% dose level caused degenerative nephrosis in the kidneys. Administration to males affects testes, epididymis, sperm duct.[1]

Reproductive Toxicity.

Gonadotoxicity. In the NTP-92 study, a markedly lower epididymal sperm density and an increase in the length of the estrous cycle at the end of the experiment were observed in rats given 50,000 ppm HMBP in the diet. The NOAEL was not reached but it has to be more than 23 mg HMBP/kg BW.[2]

In a 13-week study, dermally applied HMBP caused no gonadotoxic effect in male B6C3F_1 mice at dosages as high as 400 mg/kg.[4]

Allergenic Effect. Capable to cause hypersensitivity contact reactions.[5]

Mutagenicity.

In vitro genotoxicity. HMBP was negative in five *Salmonella typhimurium* strains.[6] In NTP studies, HMBP was found to be weakly mutagenic in *Salmonella* (with metabolic activation). Equivocal results were found in SCE in Chinese hamster ovary cells. In a 13-week mice study, SCE and CA were noted in Chinese hamster ovary cells in presence of a metabolic activation system. No increase in the frequency of micronucleated erythrocytes in the blood of mice was reported.[2]

In vivo cytogenetics. HMBP was shown to be not genotoxic *in vivo*. Sprague-Dawley rats received by gavage single oral doses up to 5.0 g/kg BW or a dose of 5.0 g/kg BW for 5 consecutive days. The treatment caused no increase in CA.[7]

Chemobiokinetics. HMBP is absorbed from the GI tract. Following administration of ^{14}C-HMBP, two major components found in the bile are *glucuronides* and *demethylated* HMBP; the third metabolite was probably a *sulfate ester* of *hydroxylated* HMBP. Excretion occurs primarily through the kidney.

After oral administration of ^{14}C-HMBP, nine radioactive components are noted in the urine.[8] Male Sprague-Dawley rats were exposed orally to the dose of 100 mg/kg BW. HMBP absorption occurred rapidly. Distribution of HMBP was observed mainly in the liver, kidney, and testis. Conjugation with glucuronic acid was the major systemic elimination route. Urine appeared to be the main route of excretion, followed by feces.[9]

Standards. *Russia.* PML in food: 2.0 mg/l.

References:

1. Lewerenz, H.-J., Lewerenz, G., and Plass, R., Acute and subchronic toxicity of UV-absorber MOB in rats, *Food Cosmet. Toxicol.*, 10, 41, 1972.
2. *Toxicity Studies of 2-Hydroxy-4-methoxybenzophenone Administered Topically and in Dosed Feed to F344/N Rats and B6C3F_1 Mice*, NTP Technical Report, U.S. Dept. Health & Human Service, Washington, D. C., October 1992, 52.
3. Cosmetic Ingredient Review Expert Panel, Final report on the safety assessment of the benzophenones-1, -3, -4, -5, -9, and -11, *J. Am. Coll. Toxicol.*, 2, 35, 1983.
4. Daston, G. P., Gettings, S. D., Karlton, B. D., et al., Assessment of the reproductive toxic potential of dermally applied 2-hydroxy-4-methoxybenzophenone to male B6C3F_1 mice, *Fundam. Appl. Toxicol.*, 20, 120, 1993.
5. Holzle, E. and Plewig G., Photoallergenic contact dermatitis by benzophenone containing sunscreen preparations, Abstract, *Hautarzt*, 33, 391, 1982.
6. Bonin, A. M. et al., UV-absorbing and other sun-protecting substances: genotoxicity of 2-ethylhexyl *p*-methoxycinnamate, *Mutat. Res.*, 105, 303, 1982.
7. Robinson, S. H., Odio, M. R., Thompson, E. D., Aardema, M. J., and Kraus, A. L., Assessment of the *in vivo* genotoxicity of 2-hydroxy-4-methoxybenzophenone, *Environ. Molec. Mutagen.*, 23, 312, 1994.
8. El Dareer, S. M., Kalin, J. R., Tillery, K. F., et al., Disposition of 2-hydroxy-4-methoxybenzophenone in rats, dosed orally, intravenously, or topically, *J. Toxicol. Environ. Health*, 19, 491, 1986.
9. Kadry, A. M., Okereke, C. S., Abdel-Rahman, M. S., Friedman, M. A., and Davis, R. A., Pharmacokinetics of benzophenone-3 after oral exposure in male rats, *J. Appl. Toxicol.*, 15, 97, 1995.

2-HYDROXY-4-(OCTYLOXY)BENZOPHENONE

Molecular Formula. $C_{21}H_{26}O_3$

M = 326.47

CAS No 843-05-6
RTECS No DJ1595000

Synonym and **Trade Names.** Advastab 46; Benzon OO; Biosorb; Octabenzone.

Applications. A stabilizer for plastics.

Acute Toxicity. LD_{50} appeared to be more than 10 g/kg BW in rats, and 13 g/kg BW in mice.[1,027]

Short-term Toxicity. Rats tolerated dietary levels of 18,000 ppm for 90 days except for changes in males' kidney weight. A dietary level of 6,000 ppm or less caused no toxic effect on dogs.[1] Kidney, urethra, and bladder were predominantly damaged.

Chemobiokinetics. Most of the ingested H. was recovered unchanged with the feces. The small portions was excreted as a glucouronide in the urine. The inertness to absorption may be due to the presence of the hydroxyl and octyl groups in the aromatic ring.[1]

Reference:

Patwel, Y M. and Levinskas, G. J., Toxicity and metabolism of 2-hydroxy-4-*n*-octooxybenzophenone (CYASORB UV 531 Light Absorber), *Toxicol. Appl. Pharmacol.*, 12, 315, 1968.

2-HYDROXY-4-PROPOXYPHENYLTHENOYL KETONE

Molecular Formula. $C_{15}H_{14}O_4$
M = 258.29

Applications. Used as a stabilizer in plastics manufacture.

Acute Toxicity. LD_{50} is found to be 10 g/kg BW in rats, and 10.2 g/kg BW in mice.

Repeated Exposure revealed no cumulative properties. Administration of 0.5 and 2.0 g/kg BW to mice was not accompanied by any signs of intoxication.

Long-term Toxicity. Mice and rats were dosed by gavage with 100 and 200 mg/kg BW. The treatment produced no changes in the BW gain or hematology analyses. Gross pathology examination failed to reveal changes in the viscera.

Allergenic Effect is not found.

Reference:

Mikhailets, I. B., Toxic properties of some stabilizers for plastics, *Plast. Massy,* 12, 41, 1976 (in Russian).

HYDROXYTRIPHENYL STANNANE

Molecular Formula. $C_{18}H_{16}OSn$
M = 367.03
CAS No 76-87-9
RTECS No WH8575000
Abbreviation. HTPS.

Synonyms and **Trade Name.** Fentin hydroxide; Triphenylstannanol; Triphenyltin hydroxide; Vanacide KS.

Properties. Solubility in water: 1.0 mg/l (at *pH* 7), solubility in ethanol: 10 g/l.

Applications. HTPS is used to prevent aging in plastics. Exhibits antimicrobial and antifungal properties.

Acute Toxicity. In Sprague-Dawley rats, LD_{50} varies from 171 mg/kg BW (males) to 268 mg/kg BW (females).[1] Other LD_{50} values reported were 540 mg/kg BW for rats, 710 mg/kg BW for mice, 680 mg/kg BW for guinea pigs, and 500 to 1000 mg/kg BW for rabbits.[2,3]

Short-term Toxicity. Rats tolerate administration of 1.0 mg/kg BW (in the feed) for 3 months without signs of toxicity.

Long-term Toxicity. In a 2-year rat study, the NOEL was identified to be about 0.1 mg/kg BW.[2] Decrease of the blood *Hb* content and leukocyte count was noted at the dose level of 50 mg/kg diet.

Reproductive Toxicity. Conflicting results are reported.

Gonadotoxicity. Administration of 200 mg/kg BW for 276 days caused overt decrease in fertilization capacity in males but their progeny was found to be normal.[4]

Administration of HTPS chloride to rats in the feed, at a dose level of 20 mg/kg BW for 4 weeks, caused impairment of spermatogenesis, necrotic changes in the cell components, and obliteration of the seminiferous tubules. Mitosis was interrupted. Spermatogenesis was restored 70 days after the end of administration.[5]

Rats were given the dose of 5.0 mg/kg BW for 90 days. There was a reduction of testis size, delayed maturation of the germinal epithelium, but a three-generation study failed to show any adverse effect on reprodution.[2]

Teratogenicity. Sprague-Dawley rats were exposed to HTPS by oral gavage on days 6 to 15 of gestation. The treatment produced developmental toxicity.[6] Hamsters received HTPS by gavage. Developmental anomalies such as hydronephrosis, hydrocephalus, and delayed ossification were observed in the pups at 5.08 mg/kg BW and higher (Carlton and Howard, 1982).[7]

Allergenic Effect was not observed following skin application to guinea pigs.

Mutagenicity. HTPS was not found to be positive in most *in vitro* and *in vivo* genotoxicity tests.[8]

Carcinogenicity. F344 rats and B6C3F$_1$ mice received doses of 37.5 or 75 ppm in the feed for 78 weeks. After the treatment, there was an additional observation period of 26 weeks. Significant mortality was observed only in male mice treated. No increase in the incidence of tumors was noted under conditions of this study. Probably, the doses used in this study appear to be less than the MTD.[9]

Carcinogenicity classification.

NTP: N - N - N - N.

Chemobiokinetics. In a 3-month study rats were administered 11 mg HTPS/kg BW. To the end of this period the main part of the administered dose was found in the kidneys, brain, heart, and lungs.[10] Administered orally to rats, HTPS is eliminated predominantly via feces, with smaller amounts in the urine. Metabolites include *di-* and *monophenyl tin* as well as a significant portion of non-extractable bound residues (the *sulfate conjugates of hydroquinone, catechol*, and *phenol*).

After oral administration, about 80% the dose is excreted in 10 days. Half-life in the brain is 5 days.[11]

References:

1. Marks, M. J., Winek, C. L., and Shanor, S. P., Toxicity of triphenyltin hydroxide (Vanacide KS), *Toxicol. Appl. Pharmacol.*, 14, 627, 1969.
2. FAO/WHO, Evaluations of some pesticide residues in food, Rome, 1971, 327.
3. Winek, C. L., Marks, M. J., Shanor, S. P., et al., Acute and subacute toxicology and safety evaluation of triphenyltin hydroxide (Vanacide KS), *Clin. Toxicol.*, 13, 281, 1978.
4. Gaines, T. B. and Kimbrough, R. D., Toxicity of triphenyltin hydroxide to rats, *Toxicol. Appl. Pharmacol.*, 12, 397, 1968.
5. Snow, R. L. and Hays, R. L., Phasic distribution of seminiferous tubules in rats treated with triphenyltin compounds, *Bull. Environ. Contam. Toxicol.*, 31, 658, 1983.
6. Chernoff, N., Setzer, R. W., Miller, D. B., Rosen, M. B., and Rogers, J. M., Effects of chemically induced maternal toxicity on prenatal development in the rat, *Teratology*, 42, 651, 1990.
7. *Triphenyltin compounds*, Concise International Chemical Assessment Document No 13, IOMS, WHO, Geneva, 1999, 40.
8. *Pesticide residues in food - 1991*, Evaluations 1991, Part II - Toxicology, WHO, Geneva, 1992, 173.
9. *Bioassay of Triphenyltin Hydroxide for Possible Carcinogenicity*, NTP Technical Report Series No 139, Research Triangle Park, NC, 1978.
10. 10. Cremer, J. E., The metabolism *in vitro* of tissue slices from rats given triethyltin compounds, *Biochem. J.*, 67, 87, 1957.
11. Heath, D. F., The retention of triphenyltin and dieldrin and its relevance to the toxic effects of multiple dosing, in *Radioisotopic Detection of Pesticide Residues*, Vienna, *Proc. Panel*, 1965, 18, 1967.

N-ISOPROPYL-*N'*-PHENYL-*p*-PHENYLENEDIAMINE

Molecular Formula. $C_{15}H_{18}N_2$

M = 226.35

CAS No 101-72-4
RTECS No ST265000

Synonym and Trade Names. Antioxidant 4010 NA; Diafen FP; 2-Isopropylaminodiphenylamine; Santoflex IP.

Properties. White, crystalline powder that turns pink on storage. Soluble in ethanol, almost insoluble in water.

Applications. Used as a stabilizer in the production of synthetic rubber, antioxidant and anti-ozone agent for vulcanizates, thermostabilizer for polyethylene, polystyrene and polyamides.

Acute Toxicity. LD_{50} is reported to be in the range of 0.8 to 1.1 g/kg BW for rats, and of 3.0 to 3.9 g/kg BW for mice.[1,2,08] Administration of the high doses led to lethargy. The treatment caused methemoglobinemia and slight changes in the osmotic resistance of erythrocytes. Histology examination revealed circulatory disturbances in the visceral organs, necrotic changes in the small intestine mucosa and in the renal tubular epithelium.

Repeated Exposure failed to reveal cumulative properties. Rats received 180 mg/kg BW (as a suspension in 1.0 ml of 3.0% starch solution) for 35 days. The treatment caused no mortality in animals or pronounced changes in their general condition. BW gain seems to be slightly increased.[1] In rats, manifestations of the toxic effect were characterized by leukocytosis and CNS inhibition, an increase in urinary protein amount, a reduction in the percentage of bromosulfalein excretion. The relative weights of the liver, kidneys, spleen, and testes were increased.

Short-term Toxicity. Rats were given 200 mg/kg BW for 4 months. The treatment caused retardation of BW gain, slight methemoglobinemia, and an increased conditioned reflex activity. Gross pathology findings included surface necrosis of the *villi* and desquamation of the intestinal mucosa epithelium.[08]

Rabbits were dosed with 20 mg/kg BW over a period of 4 months. Poisoning was accompanied by liver and kidney lesions.[2]

Allergenic Effect. I. showed positive patch test reactions in patients.[3]

Carcinogenicity. I. is related to the aromatic aminocompounds which are likely to have a potential to form carcinogenic substances in the body. While excreted through the kidneys, they might give rise to bladder malignant tumors.

Weak carcinogenic activity is reported in rats that received I. in the diet for 2 years.[4]

References:

1. Mel'nikova, L. V., Comparative toxicity of isopropylaminodiphenylamine and products of its manufacture, in *Toxicology of New Industrial Chemical Substances*, A. A. Letavet, Ed., Meditsina, Moscow, Issue No 8, 1966, 126 (in Russian).
2. Zhilova, N. A. and Kasparov, A. A., *N*-Phenyl-*N*-isopropylparaphenylene amine, in *Toxicology of New Chemical Substances Used in Rubber and Tyre Industry*, Z. I. Israel'son, Ed., Meditsina, Moscow, 175, 1968 (in Russian).
3. Kaniwa, M. A., Isama, K., Nakamura, A., Kantoh, H., Itoh, M., Ichikawa, M., Hayakawa, R., Identification of causative chemicals of allergic contact dermatitis using a combination of patch testing in patients and chemical analysis. Application to cases from industrial rubber products, *Contact Dermat.*, 30, 20, 1994.
4. Pliss, G. V., *An Experimental Study of the Carcinogenic Effect of Aminocompounds*, Author's abstract of thesis, 1966, 32 (in Russian).

LAST A

Composition. A complex stabilizer based on barium amylphenolate, alkylphenol, and triphenylphosphate.

Properties. Colorless or creme-colored, highly mobile liquid. Poorly soluble in water, soluble in ethanol.

Applications. Used as a secondary thermostabilizer of polyvinyl chloride together with calcium-cadmium-, barium-cadmium-zinc- and epoxy-stabilizers.

Acute Toxicity. LD_{50} is reported to be 5.9 g/kg BW in rats, and 8.7 g/kg BW in mice.[1] According to other data, LD_{50} is 11 g/kg BW in rats, and 5.7 g/kg BW in mice. High dose administration led to convulsions. Gross pathology examination revealed lesions of the GI tract.[1,2]

Repeated Exposure revealed evident cumulative properties according to the data published by Rumyantsev et al. (1981) but in contrast to other data.[2]

Short-term Toxicity. No reduction in BW gain or change in STI value was observed in rats exposed to the oral dose of 550 mg/kg BW for 3 months. After 2 months of treatment, leukocytosis and an increase in the relative renal weights were noted. No mortality was observed.

Allergenic Effect was observed (Rumyantsev et al., 1981).

References:

1. Krynskaya, I. L., Yevsyukov, V. I., and Maksimova, N. S., Toxicity of some stabilizers for PVC-compositions, *Plast. Massy*, 12, 45, 1976 (in Russian).
2. Sheftel', V. O., Toxic effects of some polyvinyl chloride stabilizers, in *Hygiene and Toxicology of High-Molecular-Mass Compounds and of the Chemical Raw Materials Used for Their Synthesis*, Proc. 6th All-Union Conf., B. Yu. Kalinin, Ed., Leningrad, 1979, 297 (in Russian).

LAST B-94

Composition. A complex stabilizer containing co-precipitated calcium and zinc stearates, epoxidized soya oil (see above), glycerin, topanol, and trihexylcyanurate.

Properties. White, viscous paste.

Applications. Used as a thermostabilizer in the production of rigid and plasticized polyvinyl chloride materials.

Acute Toxicity. Rats and mice tolerate administration of 10 g/kg BW.

Repeated Exposure failed to reveal cumulative properties. Rats were gavaged with the dose of 2.0 g/kg BW. An increase in the relative liver weights was found.

Long-term Toxicity. Mice were given 200 mg/kg BW for 10 months. The treatment produced a temporary increase in motor activity and in the relative lung weights in mice.

Allergenic Effect was not observed.

Reference:

Krynskaya, I. L., Yevsyukov, V. I., and Maksimova, N. S., Toxicity of some stabilizers for PVC-compositions, *Plast. Massy*, 12, 45, 1976 (in Russian).

LAST DP-4

Composition. Mechanical mixture of co-precipitated barium and cadmium laurates (1:2), zinc laurate, glycerin, epoxidized soya oil (see above), bisphenol A (see above), and an ethylene glycol ester based on phthalic and methacrylic acids.

Properties. White, hygroscopic powder with the odor of vegetable oil.

Applications. Used as a thermo- and light-stabilizer in the production of vinyl plastics, and of plasticized polyvinyl chloride and its copolymers.

Acute Toxicity. LD_{50} is found to be 3.5 g/kg BW in rats, and 2.1 g/kg BW in mice. A mixture of *Last DP-4* and *Last A* (1:5) has LD_{50} of 2.0 g/kg BW in mice, and of 2.5 g/kg BW in rats. Gross pathology examination revealed damages to the GI tract.

Repeated Exposure revealed marked cumulative properties. K_{acc} is 1.75. Some male mice died when given a mixture of *Last DP-4* and *Last A* (1:5) at a dose of 400 mg/kg BW. The treatment caused an impairment of liver and kidney functions. Doses of 100 and 200 mg/kg BW produced an increase in the relative liver weights only.

Long-term Toxicity. Some of the male rats died when exposed to 10 and 50 mg *Last DP-4*/kg BW for 10 months. Toxicity of the stabilizer may be enhanced by the co-precipitated *Ba-Cd laurate* contained in the stabilizer.

Allergenic Effect was not noted.

Reference:

Krynskaya, I. L., Yevsyukov, V. I., and Maksimova, N. S., Toxicity of some stabilizers for PVC-compositions, *Plast. Massy*, 12, 45, 1976 (in Russian).

LEAD compounds

	CAS No	RTECS No
Lead dioxide	1309-60-0	OG0700000
Molecular Formula. $O_2.Pb$		
M = 239.19		
Synonym. Lead brown.		
Basic lead carbonate	598-63-0	OF9275000
Molecular Formula. $CO_3.Pb$		
M = 267.20		
Lead stearate	7428-48-0	WM3000000
Molecular Formula. $C_{18}H_{36}O_2.Pb$		
M = 1734.87		
Basic lead stearate	1072-35-1	
Tribasic lead sulfate	7446-14-2	OG4375000
Molecular Formula. $O_4S.Pb$		
M = 303.25		
Basic (dibasic) lead phosphite	1344-40-7	OG2880000
Molecular Formula. $HO_5P.Pb_2.1/2H_2O$		
M = 742.56		
Lead chloride	7758-95-4	OF9450000
Molecular Formula. $PbCl_2$		
M = 278.09		
Lead nitrate	10099-74-8	OG2100000
Molecular Formula. $N_2O_6.Pb$		
M = 331.21		

Properties and **Applications.** Lead chloride, lead nitrate, and lead acetate are soluble in water; lead oxide and lead dioxide are poorly soluble in water. Lead concentration of 2.0 to 4.0 mg/l causes a suspension to form in water. Concentrations of 100 to 300 mg lead nitrate, lead acetate, and lead chloride/l give water an acidic taste.[1] Lead compounds were the commonest stabilizers of polyvinyl chloride.

Lead dioxide. Brown, amorphous or fine crystalline powder. Poorly soluble in water and ethanol. Used as a vulcanizing agent for acrylonitrile-butadiene and polysulfide rubber and for butyl rubber.

Basic lead carbonate. White, amorphous powder. Poorly soluble in water. Used as a thermostabilizer of polyvinyl chloride and vinyl chloride copolymers. To reduce toxicity, it is used in the form of paste. A pigment. Use is declining of late.

Lead stearate. White or yellowish powder. Limited solubility in ethanol, poorly soluble in water. Used as a thermostabilizer and a lubricant for polyvinyl chloride and vinyl chloride copolymers. Used in combination with basic lead salts.

Basic lead stearate. White or lightly tinted, finely dispersed powder. Poorly soluble in water and alcohol. Used as a thermostabilizer of polyvinyl chloride.

Tribasic lead sulfate. White, fine crystalline powder. Poorly soluble in water, fats and oils. Used as a thermo- and light-stabilizer of polyvinyl chloride in combination with organotin stabilizers, etc.

Basic (dibasic) lead phosphite. White, crystalline powder. Poorly soluble in water. Used as a thermo- and light-stabilizer for opaque polyvinyl chloride.

Basic lead phthalate. Fine white powder, not melting at 300°C. Poorly soluble in water. Used as a thermo- and light-stabilizer for polyvinyl chloride, with plasticizing properties.

Exposure. Main intake is via food and water. Mean daily intake of 0.003 to 0.004 mg/kg BW was not associated with an increase of lead levels in blood or in the body burden of lead, while an intake of 0.005 mg/kg BW or more resulted in lead retention.[2] In recent years, most countries have attempted to reduce human lead exposure.

Migration Data. Release of lead from plastic tableware into 4.0% acetic acid or 0.1 N *HCl* (60°C or 95°C; 30 min) was examined. No migration from samples containing up to 391 ppm lead was observed.[3]

Lead was present in 9/10 samples of paper and board materials intended for food contact, ranging from 0.3 to 5.9 mg/kg.[4]

When over 2.5% lead stabilizer is introduced into polyvinyl chloride, the amount of lead washed out is proportional to its content in the composition, and it is washed out only from the surface layer of pipes. The bulk of lead is washed out during the first 2 to 3 days, later the process slows down. Intense flow of water and its temperature results in a more intense washing out of lead stabilizers. Content of CO_2 content dissolved in water has a considerable effect on washing out lead salts from polyvinyl chloride pipes. Even small amounts of CO_2 can promote stability of lead on account of formation of almost insoluble basic lead carbonate. On the other hand, $Pb(HCO_3)_2$ which readily passes into solution is formed with higher CO_2 concentrations.

Acute Toxicity. In guinea pigs, LD_{min} for different salts is (mg/kg BW): lead carbonate, 1.0; lead chloride, 1.5-2.0; lead nitrate, 2.0; lead oxide, 2.0; lead phthalate, 8.0. No animals died after administration of even 20 g lead stearate/kg BW.[5]

Repeated Exposure. Rats were given 20 doses of 2.0 mg lead/kg BW and below. Signs of lead poisoning were not found to develop.[6]

Short-term Toxicity.

Observations in man. Such characteristic manifestations as dullness, restlessness, irritability, poor attention span, headaches, etc. were demonstrated in adults at lead blood levels of 0.1 mg to 0.12 mg/100 ml and in children at 0.08 mg to 0.1 mg/100 ml.

Lead is toxic to both the CNS and peripheral NS inducing sub-encephalopathic neurological and behavioral effects and electrophysiological evidence of effects on the NS function in children with both levels well below 0.03 mg/100 ml. Lead toxic effects in the kidney are reported to be the most insidious.

Animal studies. A dose of 0.5 mg lead acetate/kg BW (calculated as lead) caused a reduction in the δ-aminolevulinic acid dehydratase activity after 2 to 4 administrations to the newborn monkeys. In 2 to 6 weeks the concentration of protoporphyrins in the erythrocytes increased from 0.03 to 0.05 mg/100 ml to 0.2 to 0.3 mg/100 ml. Changes in the EEG and in motor responses were observed only after 3 to 3.5 months. Female Wistar rats were given 1.0 or 0.1% lead acetate in drinking water for 2 or 4 months, respectively. Exposure to 0.1% lead acetate induced a blood level of 376 μg/l without any signs of nephrotoxicity.[8]

Long-term Toxicity.

Observations in man. Lead is a cumulative general toxicant. Fetuses, infants, children up to six years of age and pregnant women are the most susceptible to its adverse health effects. Main manifestations of poisoning include tiredness, sleepiness, irritability, headaches, GI tract symptoms in adults at blood levels of 0.05 to 0.08 mg/100 ml. After 1 to 2 years of exposure, muscle weakness, GI disturbances, and symptoms of peripheral neuropathy were observed at lead blood levels of 0.04 to 0.06 mg/100 ml.[9] Chronic renal nephropathy was not observed at blood levels of 0.04 mg/100 ml.[10]

A significant association between lead blood level of 0.007 to 0.034 mg/100 ml and high diastolic blood pressure was found in people aged 21 to 55.[11]

Lead interferes with the activity of several major enzymes involved in the biosynthesis of hem.[9] Lead-induced anemia is the result of two separate processes: inhibition of hem synthesis and acceleration of erythrocyte destruction.[12]

Animal studies. Lead can severely affect the CNS and produce significant behavioral and cognitive deficits as demonstrated in the monkey study (postnatal exposure to lead for 29 weeks in an amount resulting in blood levels ranging from 0.0109 to 0.033 mg/100 ml).[13] Apparently, lead acts on globin synthesis in animals in the same way as in man and may inhibit cytochrome *P-450* formation.

Chronic lead nephrotoxicity is irreversible and is typically accompanied by interstitial fibrosis, both hyperplasia and atrophy of the tubules, glomerulonephritis, and ultimately, renal failure.[14]

Reproductive Toxicity. Reproductive effects associated with lead poisoning have been reported for a long time.[15] Effects of low-level lead exposure on growth and development are reported by Schwartz.[16]

Gonadotoxicity.

Observations in man. Lead produces gonadal dysfunction in men, including depressed sperm counts. This effect is associated with blood lead levels of 0.04 to 0.05 mg/100 ml.[17]

Lerda compared 38 workers exposed to lead in a battery factory to 30 unexposed subjects. A significant decrease in sperm count and proportions of motile sperms, as well as an increase in abnormal sperm were found to develop in the lead workers.[18] Decreased sperm counts and motility and an increased percentage of abnormal sperms were found to be correlated with mean lead blood levels in ten workers exposed to lead for 10 years in the printing industry.[19]

Animal studies. In rats, blood levels ranging from 0.040 to 0.060 mg lead/100 ml affect sperm counts and cause testicular atrophy in males;[20,21] they impair the estrous cycles in females.[22,23]

Lead salts caused morphologic sperm abnormalities in mice but not in rabbits. Rats received 0.6% lead acetate in their drinking water *ad libitum*. In males, the treatment produced a significant decrease in male sex organ weights. Serum testosterone levels were significantly suppressed. In females exposed prepubertally, delayed vaginal opening and disrupted estrous cycling were noted. Suppression of circulating estradiol occurred in the rats exposed beginning *in utero*. Prepubertal growth in both sexes was suppressed by 25% in *in utero* group.[24]

Male CF-1 mice received two concentrations of lead (0.25 and 0.5%) via drinking water for 6 weeks. The low lead dose significantly reduced the number of sperm within the epididymis, while the high dose reduced both the sperm count and percentage of motile sperm and increased the percentage of abnormal sperm within the epididymis. There was no significant effect on testis weight; however, the high-dose treatment significantly decreased the epididymis and seminal vesicle weights as well as overall BW gain.[25]

Embryotoxicity.

Observations in man. Lead salts pass through the placenta and may cause fetal death. Placental transfer of lead occurs in humans as early as the 12th week of gestation and continues throughout development.

Animal studies. Mice that received lead carbonate on days 7 to 8 of pregnancy exhibited NS defects developing in the offspring.

Rats and mice were dosed by gavage with doses of 7.14 and 71.4 mg lead acetate/kg BW on days 5 to 15 of pregnancy. The treatment did not affect the course of pregnancy or exert any embryotoxic effect.[26]

Teratogenic effects of lead acetate are reported in rats, hamsters and primates, but not in sheep, mice, and rabbits.[27]

Immunotoxicity. Lead poisoning produced qualitative and quantitative changes in the protein spectrum of the blood. Lead produced an immunodepressant effect in experimental animals, revealing itself through inhibition of antibody formation[28] and macrophage function, a reduction in thymus weights, and by suppression of cell-mediated reactions.[29,30]

Mutagenicity.

Observations in man.

In vivo cytogenicity. Studies of CA in people exposed to lead have shown conflicting results: positive reports have been published concerning workers in lead-battery industries and lead smelters, but other studies of workers under comparable conditions and of children exposed to high levels in the environment have given negative results.[31]

Lead salts did not induce CA in human lymphocytes *in vitro*.

In vivo cytogenetics. A few studies in rodents treated with lead salts *in vivo* have shown small but significant increases in the frequency of CA and micronuclei in bone marrow cells. Most studies showed no increase. SCE and unscheduled DNA synthesis were not induced in animals treated with lead salts. Lead acetate showed contradictory results (Ishidate et al., 1988): no micronucleus induction was observed after double *i/p* treatment of male ddY mice although a reduced polychromatic erythrocyte to total erythrocyte ratio was observed. Lead acetate induced micronuclei in rats and mice *in vivo* (Mavournin et al., 1990).

Lead oxide gave negative response in male ddY mouse bone marrow after *i/p* treatment with up to 2000 mg/kg BW.[31]

Lead sulfide and lead nitrite were mutagenic when added to Chinese hamster ovary *V79* cells (possible indirect mechanism).[17, 32]

In vitro genotoxicity. Lead salts were negative in induction of CA in human lymphocytes *in vitro*. Gene mutations in cultured mammalian cells are observed only at toxic concentrations. Lead acetate and sulfide were negative in *Salmonella* mutagenicity assay (NTP-94) and when tested by *rec-assay*.[33]

Carcinogenicity. Although renal tumors have been induced in experimental animals exposed to high concentrations of lead compounds in the diet, there is the evidence from the studies in humans that adverse neurotoxic effects other than cancer may occur at very low concentrations of lead and a guideline value derived on this basis would also be protective for carcinogenic effects.[2]

Observations in man. Epidemiology studies showed either negative results or only very small excess mortalities from cancer. Whether or not lead has a role in renal adenocarcinoma in humans is not clear.[34] A statistically significant excess of cancers of the digestive system was found in a study of battery plant workers in the UK, spanning 1925 to 1976. A significant excess of stomach and respiratory cancers was seen in a study of U.S. battery plant workers.[31]

Animal studies. Evidence of carcinogenicity in animals is inconclusive because of the limited number of studies, the small cohort size, and lack of sufficient consideration for potential conforming variables.[35]

Renal tumors were found in rats, mice, hamsters exposed orally to lead acetate, subacetate, or phosphate; 5 to 2000 ppm in the diet (0.3 to 105 mg/kg BW) were fed to rats for 2 years. Renal carcinogenicity (adenocarcinomas) occurs on a background of proximal tubular cell hyperplasia, cytomegaly, cellular dysplasia. Renal tumors developed in 5 out of 50 male rats at 500 ppm, 10 out of 20 at 1000 ppm, and 16 out of 20 at 2000 ppm, but only at 2000 ppm (7 out of 20) in female rats. Histopathology findings of this study were not reported.[31,32] Tumors have not been shown to occur below the Maximum Tolerated Dose of 200 ppm lead in drinking water.[36] This is the maximum dose not associated with any morphologic or functional evidence of renal toxicity in rats fed a diet containing adequate levels of calcium.[37]

Carcinogenicity classifications. An IARC Working Group concluded that there is sufficient evidence for the carcinogenicity of inorganic lead compounds in *experimental animals* and there were inadequate data available to evaluate the carcinogenicity of inorganic lead compounds in *humans*.

The Annual Report of Carcinogens issued by the U.S. Department of Health and Human Services (1998) defines lead acetate and phosphate to be substances which may reasonably be anticipated to be carcinogen, i.e., substances for which there is a limited evidence of carcinogenicity in *humans* or sufficient evidence of carcinogenicity in *experimental animals*.

Lead inorganic:
IARC: 2B;
U.S. EPA: B2.
Lead acetate:
IARC: 2B;
U.S. EPA: B2
EU: 3.
Lead chromate:
IARC: 1;
EU: 3.
Lead phosphate:
IARC: 2B;
U.S. EPA: B2;

Chemobiokinetics. Pb^{2+} absorption from the GI tract occurs slowly (up to 10%) and depends predominantly on the solubility of its compounds. Young children absorb 4 to 5 times as much lead as adults and the biological half-life is considerably longer in children than in adults. In children, up to 50% of lead accumulates in the bones (half-life of 20 years), liver, and kidneys. Intake of more than 1.0 mg/day led to the rise in lead blood concentration that followed a logarithmic course. Inhibition of the activity of δ-aminolevulinic acid dehydratase (one of the major enzymes involved in the biosynthesis of hem) in children has been observed at blood lead levels as low as 0.005 mg/100 ml, though adverse effects are not associated with its inhibition at this level.

Lead also interferes with calcium metabolism and significant decreases in calculated levels of 1,2,5-dihydroxycholecalciferol have been observed in children at blood lead levels in the range from 0.012 mg to 0.12 mg/100 ml, with no evidence of a threshold.[2]

Lead interacts with renal membranes and enzymes and disrupts energy production, calcium metabolism, glucose homeostasis, ion-transport processes, and renin-angiotensin system.[14] Inorganic lead that is absorbed but not retained is excreted unchanged via the kidneys or through the biliary tract.[13]

Regulations. Lead release from glazed ceramic tableware has been regulated as follows:

EU (1984):	mg/dm^2
Flatware < 25 mm internal depth	0.8
Hollowware	4.0
Cookware, packaging & storage vessels	1.5
ISO (1981,1986):	mg/l:
Flatware	1.7
Small hollowware	5.0
Large hollowware	2.5
Cookware	5.0
UK (1986, 1988):	
Flatware < 25 mm internal depth	0.8 mg/dm^2
Hollowware	4.0 mg/l
Cookware, packaging and storage vessels	1.5 mg/l
U.S. FDA (1998):	μg/ml (mg/l):
Flatware (average of 6 units)	3.0
Small hollowware (any one of 6 units)	2.0
Cups and Mugs (any one of 6 units)	0.5
Large hollowware (any one of 6 units)	1.0
Pitchers (any one of 6 units)	0.5

Recommendations.

Joint FAO/WHO Expert Committee on Food Additives (1986): PTWI: 0.0025 mg/l for infants and for chil dren (that is, ~3.5 mg/kg/day, ½ the adult value).

WHO (1992). Guideline value for drinking water: 0.01 mg/l (on the basis that lead is a cumulative poison and that there should be no accumulation of body burden of lead; since infants are considered to be the most sensitive subgroup of the population, this guideline will also be protective for other age groups).[2]

The International Organization for Standardization adopted limits of release for lead from ceramic foodware, varying from 1.7 mg/dm^2 for flatware to 2.5 to 5.0 mg/l for hollowware.

Standards.

U.S. EPA (1999). MCL (at tap): 0.0015 mg/l, MCLG: zero.

EEC (1980). MAC: 0.05 mg/l.

Canada (1989). MAC: 0.01 mg/l,

Russia. State Standard for Drinking Water: 0.01 mg/l; MAC in milk and diary products: 0.1 mg/kg; MAC in meat and poultry: 0.5 mg/kg; MAC in mineral water 0.1 mg/kg; MAC in children's food 0.05 mg/kg.

References:

1. Zaitseva, A. F., in *Protection of Water Reservoirs against Pollution by Industrial Liquid Effluents*, S. N. Cherkinsky, Ed., Medgiz, Moscow, Issue No 2, 1954, 44 (in Russian).
2. Ryu, J. E., Ziegler, E., Nelson. S., et al., Dietary intake of lead and blood lead concentration in early infancy, *Am. J. Disease Child.*, 137, 886, 1983.
3. Hosogai, T., Ito, S., Tada, Y., Sakurai, H., Sugita, T., Ishiwata, H., and Takeda, M., Migration test of lead and cadmium from plastic wares in contact with food, Abstract, *Eisei Shikenjo Hokoku*, 110, 83, 1992 (in Japanese).
4. Castle, L., Offen, C. P., Baxter, M. J., and Gilbert, J., Migration studies from paper and board food packaging materials. 1. Compositional analysis. *Food Addit. Contam.*, 14, 35, 1997.
5. Grushko, Ya. M., *Hazardous Organic Compounds in the Liquid Industrial Effluents*, Khimiya, Leningrad, 1976, 104 (in Russian).

6. Golubovich, Ye. Ya., Avkhimenko, M. M., and Chirkova, Ye. M., Biochemical and morphological changes in the gonads of male rats treated with small doses of lead, in *Toxicology of New Industrial Chemical Substances*, Meditsina, Moscow, Issue No 10, 1968, 64 (in Russian).
7. Willes, R. F., Lok, E., Truelove, J. F., et al., Retention and tissue distribution of $^{210}Pb(NO_3)_2$ administered orally to infant and adult monkeys, *J. Toxicol. Environ. Health*, 3, 395, 1977.
8. Viskocil, A., Semecky, V., Fiala, Z., Cizkova, M., and Vian, C., Renal alteration of female rats following subchronic lead exposure, *J. Appl. Toxicol.*, 15, 257, 1995.
9. U.S. EPA, *Air Quality Criteria for Lead*, Report EPA-600/8-83/028F, 1986.
10. Lilis, R., Fischbein, A., Diamond, S., et al., Lead effect among secondary lead smelter workers with blood lead below 0.08 mg/100 ml, *Arch. Environ. Health.*, 32, 256, 1977.
11. Pirkle, J. L., Schwartz, J., Landis, J. R., et al., The relationship between blood lead level and blood pressure and its cardiovascular risk implications, *Am. J. Epidemiol.*, 121, 246, 1985.
12. Moore, M. R., Hematological effects of lead, *Science of the Total Environment*, 71, 419, 1988.
13. Rice, D. C., Primate research: relevance to human learning and development, *Developm. Pharmacol. Ther.*, 10, 314, 1987.
14. Nolan, C. V. and Shaikh, Z. A., Lead nephrotoxicity and associated disorders: biochemical mechanisms, *Toxicology*, 73, 127, 1992.
15. Bell, J. U. and Thomas, J. A., Effects of lead on mammalian reproduction, in *Lead Toxicity*, R. L. Singhal and J. A. Thomas, Eds., Urban & Schwartzberg, Baltimore/Zurich, 1980, 167.
16. Schwartz, J., Low level health effects of lead: Growth, development and neurological disturbances, in *Human Lead Exposure*, H. L. Needleman, Ed., CRC Press, Boca Raton, FL, 1992, 233.
17. Assenato, G., Paci, C., Baser, M. F., et al., Sperm count suppression without endocrine dysfunction in lead-exposed men, *Arch. Environ. Health*, 4, 387, 1986.
18. Lerda, D., Study of sperm characteristics in persons occupationally exposed to lead, *Am. J. Ind. Med.*, 22, 567, 1992.
19. Chowdhury, R., Male reproductive impairment by lead exposure, *Proc. Int. Congr. Occup. Health*, Nice, September 26 to October 1, 1993.
20. Sokol, R. Z., Madding, C. E., and Swerdloff, R. S., Lead toxicity and hypothalamic-pituitary-testicular axis, *Biol. Reprod.*, 33, 722, 1985.
21. Sokol, R. Z., Reversibility of the toxic effect of lead on the male reproductive axis, *Reprod. Toxicol.*, 3, 175, 1989.
22. Hilderbrandt, D. C., Der, R., Griffin, W. T., and Fahim, M. S., Effect of lead acetate on reproduction, *Am. J. Obstetrics Gynecol.*, 115, 1058, 1973.
23. Chowdhury, A. R., Dewaan, A., and Ghandi, D. N., Toxic effect of lead on the testes of rats, *Biomed. Biochim. Acta.*, 43, 95, 1984.
24. Ronis, M. J., Badger, T. M., Shema, S. I., Robersome, P. K., and Sheikh, F., Reproductive toxicity and growth effects in rats exposed to lead at different periods during development, *Toxicol. Appl. Pharmacol.*, 136, 361, 1996.
25. Wadi, S. A. and Ahmad, G., Effects of lead on the male reproductive system in mice, *J. Toxicol. Environ. Health*, 56, 513, 1999.
26. Kennedy, G. L., Arnold, D. W., and Calandra, J. C., Teratogenic evaluation of lead compounds in mice and rats, *Food Cosmet. Toxicol.*, 13, 629, 1975.
27. Schardein, J. L., *Chemically Induced Birth Defects*, 2nd ed., rev. and expanded, Marcel Decker, Inc., New York, 1993, 843.
28. Tovmasyan, V. S., Effect of chronic lead poisoning on antibody-forming function, *Yerevan J. Exper. Clin. Med.*, 17, 46, 1977.
29. Blakley, B. R. and Archer, D. L., The effect of lead acetate on the immune response in mice, *Toxicol. Appl. Pharmacol.*, 61, 18, 1981.
30. Zelikoff, I. T., Li, J. H., Hartwig, A., et al., Genetic toxicology of lead compounds, *Carcinogenesis*, 9, 1727, 1988.

31. Morita, T., Asano, N., Awogi, T., Sasaki, Yu. F., et al. Evaluation of the rodent micronucleus assay in the screening of IARC carcinogens (Groups 1, 2A and 2B). The summary report of the 6th collaborative study by CSGMT/JEMS-MMS, *Mutat. Res.*, 389, 3, 1997.
32. Oehever, F., Chromosome breaks influenced by chemicals, *Heredity*, Suppl. 6, 95, 1953.
33. Nioshioka, H., Mutagenic activities of metal compounds in bacteria, *Mutat. Res.*, 31, 185, 1975.
34. Goyer, R.A., Update of lead exposure and effects, Symposium overview, *Fundam. Appl. Toxicol.*, 18, 1, 1992.
35. *Revision of the WHO Guidelines for Drinking-Water Quality*, Report of the 2nd Review Group Meeting on Inorganics, Brussels, Belgium, 14-18 October, 1991, 6.
36. *Evaluation of the Potential Carcinogenicity of Lead and Lead Compounds*, U.S. EPA/600/8-89/045A, Office of Health & Environ. Assessment, Washington, D. C., 1989.
37. Goyer, R. A., Leonard, D. L., Moore, J. F., et al., Lead dosage and the role of the intranuclear inclusion body, *Arch. Environ. Health*, 20, 705, 1970.

MALEIC ACID, DIOCTYLSTANNANE DI(MONOISOBUTYL) ESTER

Molecular Formula. $C_{32}H_{56}O_8Sn$
M = 687.57
CAS No 15571-59-2

Properties. Oily, yellowish liquid with an unpleasant odor. Poorly soluble in water. Odor perception threshold is 0.1 mg/l, taste perception threshold is 0.4 mg/l.

Applications. A starting substance for stabilizers used in the manufacture of clear polyvinyl chloride food containers, rigid pipes, and flexible membranes.

Migration of organotins from polyvinyl chloride articles into tetrahydrofuran, xylene, and methylene chloride was investigated. Methylene chloride extracted >97% of the total extractable organotins in two extractions (between <0.8 μg and 8.8 mg octyltins/g solvent). In industrial application pipe samples, the levels found were 0.7 to 3.0 mg octyltins/g polyvinyl chloride.[1]

Acute Toxicity. LD_{50} is 2.7 g/kg BW in rats, and 2.25 g/kg BW in rabbits and guinea pigs.[2]

Repeated Exposure revealed evident cumulative properties. After 8 to 10 administrations of 1/5 and 1/20 LD_{50}, clinical signs of intoxication became evident in rats. The treatment resulted in convulsions immediately followed by death; 1/40 to 1/10 LD_{50} produced an effect on the blood content of *SH*-groups and a decline in the STI value. There were no changes in the enzyme systems of animals.[2]

Long-term Toxicity. In a 6-month study in rats, manifestations of the toxic action included deviations in the hematology analyses and a reduction in the *SH*-groups level in the blood as well as pronounced physiological changes: adynamia, impairment in conditioned reflex activity, a decrease in muscle tonus and in reactions to painful and tactile stimuli.[2]

Reproductive Toxicity.

Teratogenicity. Doctyltins exhibited a general cytotoxic effect on rabbit articular and growth-plate chondrocytes, that could be interpreted as possible effects on skeletal growth and development.[3]

Standards. ***Russia*** (1995). MAC and PML: 0.02 mg/l.

References:

1. Forsyth, D. S, Dabeka, R., Sun, W. F., and Dalglish, K., Specification of organotins in poly(vinyl chloride) products, *Food Addit. Contam.*, 10, 531, 1993.
2. Golovanov, O. V., Substantiation of maximum allowable concentration for PVC octyl stabilizers in water bodies, *Gig. Sanit.*, 5, 36, 1975 (in Russian).
3. Webber, R. J., Dollins, S. C., Harris, M., and Hough, A. J., Effect of alkyltins on rabbit articular and growth-plate chondrocytes in monolayer culture, *J. Toxicol. Environ. Health*, 16, 229, 1985.

D-MANNITOL

Molecular Formula. $C_6H_{14}O_6$
M = 182.20
CAS No 69-65-8

RTECS No OP2060000

Synonyms and **Trade Names.** Cordycepic acid; Hexahydric alcohol; Manna sugar; Mannite; Osmosal.

Properties. White, odorless, crystalline powder with a sweet taste. A stereo-isomer of sorbitol.

Applications. Used as a stabilizer in the production of polyvinyl chloride plastics, artificial resins and plasticizers.

Exposure. M. occurs naturally in algae, fungi, bacteria and variety of higher plants.[1]

Acute Toxicity.

Observations in man. A laxative effect could be initiated with the single oral dose greater than 20 g.[2]

Animal studies. LD_{50} is reported to be 17.3 g/kg BW in rats, and 22.0 g/kg BW in mice.[1] High doses administered to mice are followed by CNS inhibition, degeneration of the GI tract mucosa, and diarrhea.

Repeated Exposure. M. is used as an osmotic diuretic. The daily dose in man is 50 mg/kg BW (unrestricted).

Short-term Toxicity. Rhesus monkeys were given 3.0 g M. in their diet for 3 months (IARC 19-237). No signs of intoxication were noted.

Long-term Toxicity. F344 rats and $B6C3F_1$ mice of each sex were fed a diet containing 2.2 and 5.0% M. for 103 weeks. The treatment produced no effect on survival or BW gain of rats and mice. The incidence of dilatation of the gastric fundic gland was higher (46%) in treated female rats than in the controls (12%). There was a mild nephrosis that was considered to be related to the treatment.[3]

Reproductive Toxicity.

Embryotoxicity. *I/v* injection of hypertonic M. solution to pregnant rabbits caused reduction in fetal weights with occasional hemorrhages to the limbs.[4]

Mutagenicity.

In vitro genotoxicity. M. showed negative results in *Salmonella* mutagenicity assay, rat bone marrow studies, and cultured human cells.[5]

Carcinogenicity. F344 rats and $B6C3F_1$ mice were maintained on diets containing 2.5 or 5.0% M. for 103 weeks. There was no increase in tumor incidence. Under conditions of this study, M. was not considered to be carcinogenic to F344 rats and $B6C3F_1$ mice.[3,7]

Carcinogenicity classification.

NTP: N - N - N - N (feed).

Chemobiokinetics. Following oral administration, M. is rapidly metabolized. In humans, a part of absorbed dose was excreted intact in the urine and the rest of the absorbed material was oxidized to *carbon dioxide* and *water*.[6]

Regulations.

EU (1995). M. is a food additive generally permitted for use in foodstuffs.

U.S. FDA (1998) listed M. as a food additive and regulates it for use (1) in the manufacture of resinous and polymeric coatings for food-contact surface of articles intended for use in producing, manufacturing, packing, transporting, or holding food in accordance with the conditions prescribed by 21 CFR part 175.300; (2) in the manufacture of resinous and polymeric coatings for polyolefin films for food-contact surface of articles intended for use in producing, manufacturing, packing, processing, preparing, treating, packaging, transporting, or holding food in accordance with the conditions prescribed by 21 CFR part 175.320; and (3) in cross-linked polyester resins used as articles or components of articles coming in the contact with food in accordance with the conditions prescribed by 21 CFR part 177.2420.

Recommendations. Joint FAO/WHO Expert Committee on Food Additives. ADI is not speci fied.

Standards. ***Russia.*** PML: *n/m.*

References:

1. Joint FAO/WHO Expert Committee on Food Additives, *Toxicological Evaluation of Some Antimicrobials, Antioxidants, Emulsifiers, Stabilizers, Flour-treatment Agents, Acids and Bases*, FAO Nutr. Mutag. Report, Series No 40A, BC, 1967, 160.
2. Ellis, F. W. and Krantz, J. C., Sugar alcohols. XXII.Metabolism and toxicity studies with mannitol and sorbitol in man and animals, *J. Biol. Chem.*, 141, 147, 1941.

3. Abdo, K. M., Haseman, J. K., and Boorman, G., Absence of carcinogenic response in F344 rats and B6C3F$_1$ mice given *D*-mannitol in the diet for two years, *Food Chem. Toxicol.*, 21, 259, 1983.
4. Petter, C., Lesions of the extremities provoked in the fetus of the rat by intravenous injection of hypertonic mannitol into the mother, *C. R. Soc. Biol.*, 161, 1010, 1967.
5. *Carcinogenesis Bioassay of D-Mannitol in F344 Rats and B6C3F$_1$ Mice*, NTP Technical Report Series No 52, Dept. Health & Human Services, NIH Publ. No 82-1792, Washington, D.C., 1981.
6. Nasrallah, S. M. and Iber, F. L., Mannitol absorption and metabolism in man, *Am. J. Med. Sci.*, 258, 80, 1969.
7. Abdo, K. M., Huff, J. E., Haseman, J. K., and Alden, C. J., No evidence of carcinogenicity of D-mannitol and propyl gallate in F344 rats and B6C3F$_1$ mice, *Food Chem.Toxicol.*, 24, 1091, 1986.

METHACRYLIC ACID, 2-(DIMETHYLAMINO) ETHYL ESTER

Molecular Formula. $C_8H_{15}NO_2$
M = 157.24
CAS No 2867-47-2
RTECS No OZ4200000

Synonyms and **Trade Name.** Ageflex FM-1; β-Dimethylaminoethyl methacrylate; 2-(Dimethylamino) ethanol methacrylate.

Properties. Liquid soluble in water.

Applications. A stabilizer; used also in the manufacture of acrylic polymers, coatings and adhesives, ion exchange resins, and emulsifying agents.

Acute Toxicity. LD_{50} is 1.75 g/kg BW in rats.[1] Ingestion of high doses led to suppression of brain electrical activity and thus produced clonic-tonic convulsions.[2]

Chemobiokinetics. M. undergoes metabolism by saponification into the *alcohol* and *methacrylic acid.*

Regulations.

U.S. FDA (1998) approved the use of M. as a basic component of acrylic and modified acrylic plastics, semirigid and rigid, which may be safely used as articles intended for single and repeated use in food-contact surfaces in accordance with the conditions prescribed by 21 CFR part 177.1010.

Great Britain (1998). M. is authorized up to the end of 2001 for use in the production of polymeric materials and articles in contact with food or drink or intended for such contact.

References:

1. Izmerov, N. F., Sanotsky, I. V., and Sidorov, K. K., *Toxicometric Parameters of Industrial Toxic Chemicals under Single Exposure*, USSR/SCST Commission for UNEP, Center Int. Projects, Moscow, 1982, 55 (in Russian).
2. Mitina, L. V. et al., in *Pharmacol. Toxicol. New Products Chem. Industry*, Proc. 3rd Republ. Conf., 1975, 179 (in Russian).

o-(α-METHYLBENZYL)METHYLPHENOL

Molecular Formula. $C_{15}H_{16}O$
M = 212.31

Synonym. 2-(α-Methylbenzyl)-*p*-cresol.

Properties. Yellowish liquid. Soluble in alcohol, forms stable emulsions with water.

Applications. Used as a stabilizer in the production of plastics, latexes, and synthetic rubber.

Acute Toxicity. In mice, LD_{50} is 2.0 g/kg BW. Histology examination revealed congestion, edema, and sloughing of the esophageal and gastric mucosa, necrosis of the small intestine walls, and parenchymatous dystrophy of the renal tubular epithelium.[08]

α-METHYL BENZYLPHENOLS, a mixture

Molecular Formula. $C_6H_2OHR_1R_2R_3$, where
1)$R_1 = R_2 = H$; $R_3 = -CH(CH_3)C_6H_5$
M = 198.26

2)$R_1 = R_3 = H$; $R_2 = -CH(CH_3)C_6H_5$
M = 198.26
3)$R_1 = H$; $R_2 = R_3 = -CH(CH_3)C_6H_5$
M = 302.42
4)$R_1 = R_2 = R_3 = -CH(CH_3)C_6H_5$
M = 406.27

Trade Names. Agidol 20; Alkofen MB.

Composition. A product of alkylation of phenol by styrene.

Properties. Viscous liquid with a color ranging from light-yellow to dark-brown and with unpleasant odor. Poorly soluble in water, soluble in alcohol and in aqueous solutions of acids and alkalis. Forms stable emulsions with water.

Applications. Used as a stabilizer in the production of synthetic rubber and polyolefins; an antioxidant for vulcanizates.

Acute Toxicity. In rats, LD_{50} is 3.5 g/kg BW. Administration of high doses is characterized by a narcotic effect. Gross pathology examination revealed inflammatory changes in the GI tract, fatty dystrophy of the renal tubular epithelium, and a significant increase of polynuclear cell number in the spleen.[1]

Repeated Exposure failed to reveal cumulative properties.

Long-term Toxicity. In a 6-month study, mild retardation of BW gain and CNS stimulation were found in rats given the 175 mg/kg BW dose.

Reproductive Toxicity. Skin applications to rats caused growth retardation and a decline in birth rate in succeeding generations not exposed to the product. The treatment produced an embryotoxic effect in guinea pigs (blind, unviable offspring).

Allergenic Effect was not observed in experiments on guinea pigs.

Reference:

Broitman, A. Ya., Putilina, L. V., and Robachevskaya, Ye. G., Toxicology of stabilizers for plastics, in *Toxicology and Hygiene of High-Molecular-Mass Compounds and of the Chemical Raw Material Used for Their Synthesis*, Proc. 2nd All-Union Conf., A. A. Letavet and S. L. Danishevsky, Eds., Leningrad, 1964, 72 (in Russian).

α-METHYLBENZYL PHENYLPHOSPHITES, mixture

Trade Names. Antioxidant 6; Phosphite P-24.

Properties. Light-yellow, viscous liquid. Poorly soluble in water, soluble in alcohol.

Applications. Used as a thermostabilizer in the production of polyethylene, polypropylene, and ethylene-propylene copolymers, as well as a stabilizer of butadiene-styrene and butadiene-nitrile synthetic rubber.

Acute Toxicity. In mice, LD_{50} is 20 g/kg BW (laboratory sample) and 5.6 g/kg BW (industrial sample).[08] Manifestations of the toxic action included excitation and clonic convulsions. Recovery occurred in 2 to 3 hours after administration. Gross pathology examination revealed a stretched stomach and flabby walls of the small intestine.

Repeated Exposure. Mice were given 1/5 and 1/20 LD_{50} (industrial sample) for 3 weeks. The treatment caused no animal mortality.

Short-term Toxicity. Mice were treated with 400 mg/kg BW (laboratory sample) for 4 months. BW gain and conditioned reflex activity were affected. Gross pathology examination failed to reveal changes in the viscera.[08]

Long-term Toxicity. In a 6-month study, a 55 mg/kg BW dose (industrial sample) appeared to be ineffective in mice. Gross pathology examination failed to reveal changes in the visceral organs.[08]

Reproductive Toxicity. Affects the reproductive processes in fish.

4-α-METHYLBENZYL PYROCATECHOL

Molecular Formula. $C_{14}H_{14}O_2$
M = 214.26

Properties. Crystalline powder. Solubility in water is 0.001%.

Applications. Used as a stabilizer in the manufacture of plastics.

Acute Toxicity. Histology examination of animals that died after a single administration of the lethal dose showed congestion, lesions in the mucosa of the GI tract walls, and parenchymatous dystrophy of the renal tubular epithelium. No morphology changes were found in the survivors.

Long-term Toxicity. Rats were dosed by gavage with 62.5 mg/kg BW for 4 months. The treatment caused changes in the conditioned reflex activity. Gross pathology examination showed lesions in the GI tract mucosa as well as parenchymatous dystrophy of the renal tubular epithelium.[08]

METHYLDIMETHOXYPHENYL SILANE

Molecular Formula. $C_9H_{14}O_2Si$

M = 154.23

Synonym. Methylphenyl dimethoxysilane.

Properties. Clear, colorless liquid with a faint odor. Hydrolyzes in presence of moisture to form methyl alcohol.

Applications. Used as a stabilizer in the manufacture of plastics.

Acute Toxicity. LD_{50} was reported to range from 680 mg/kg to 892 mg/kg BW in mice. Lethal doses led to general inhibition and impairment of motor coordination. In the survivors, signs of intoxication disappear in 2 weeks.

Reference:

Perel', S. S., Experimental studies of toxicity of methylphenyl dimethoxysilane, *Gig. Truda Prof. Zabol.*, 2, 50, 1978 (in Russian).

4-METHYLDI(1-METHYLHEPTADECYL)PHENOL

Molecular Formula. $C_{43}H_{80}O$

M = 613.11

Applications. Used as a stabilizer in the manufacture of polyolefins.

Acute Toxicity. LD_{50} is found to be 6.4 ml/kg BW in rats, and 3.2 ml/kg BW in mice.

Repeated Exposure. 1.0% solution of the product added to the feed of puppies appeared to be harmless (Terhaer, 1967).

4,4'-METHYLENEBIS(2,6-BIS(1,1-DIMETHYLETHYL)PHENOL

Molecular Formula. $C_{29}H_{44}O_2$

M = 424.66

CAS No 118-82-1

RTECS No SL9650000

Abbreviation. MB-1.

Synonyms and Trade Names. Antioxidant 702; Bis(3,5-di-*tert*-butyl-4-hydroxyphenyl methane); Ionox 220, 4,4'-Methylenebis(2,6-di-*tert*-butylphenol).

Properties. Light-yellow, crystalline powder. Poorly soluble in water, of limited solubility in alcohol.

Applications. MB-1 is used as a stabilizer in the production of polyolefins; as an antioxidant for vulcanizates; it is used also in the production of epoxy-, phenol-, and ion exchange resins.

Acute Toxicity. Rats and mice tolerate administration of 5.0 g/kg BW.[1]

Repeated Exposure. Rats and mice tolerate administration of 1.0 and 5.0 g/kg BW for 10 days. Histology examination revealed moderate fatty dystrophy of the liver cells, renal tubular epithelium, and signs of irritation of the spleen pulp.[1]

Administration to male Sprague-Dawley rats at the level of 1.135 mmol% for one week induced fatty liver and increased triglyceride, diglyceride and cholesteryl-ester concentration 7.1-, 5.8-, and 6.1-fold, respectively.[2]

Long-term Toxicity. In a 2-year dog study, the treatment caused ulceration or bleeding from the large intestine. Changes in the liver included hepatitis, cirrhosis, post-necrotic scarring. Mortality was observed among treated animals.[3]

Regulations. ***U.S. FDA*** (1998) approved the use of MB-1 (1) in the manufacture of antioxidants and/or stabilizers for polymers which may be used in the manufacture of articles or components of articles intended for use in producing, manufacturing, packaging, processing, preparing, treating, packing, transporting, or holding food in accordance with the conditions prescribed by 21 CFR part 178.2010; and (2) in adhesives used as components of articles intended for use in packaging, transporting, or holding food in accordance with the conditions prescribed by 21 CFR part 175.105 at levels not to exceed:

- 0.25% by weight of the petroleum hydrocarbon resins;
- 0.25% by weight of the terpene resins;
- 0.5% by weight of the polyethylene provided that the polyethylene and product contact foods only of the types identified in CFR;
- 0.5% by weight of polybutadiene used in rubber articles provided that the rubber end-product contacts food only of the types identified in CFR;
- 0.5% by weight of the polybutadiene used in rubber articles in compliance with correspondent regulations (CFR).

References:

1. Stasenkova, K. P., Shumskaya, N. I., Greenberg, A. Ye., et al., Comparative evaluating toxicity of bisphenol A and its derivatives, in *Hygiene and Toxicology of High-Molecular-Mass Compounds and of the Chemical Raw Material Used for Their Synthesis*, Proc. 4th All-Union Conf., S. L. Danishevsky, Ed., *Khimiya*, Leningrad, 1969, 180 (in Russian).
2. Takahashi, O. and Hiraga, K., Effect of 4 bisphenolic antioxidants on lipid contents of rat liver, *Toxicol. Lett.*, 8, 77, 1981.
3. Natl. Technical Information Service OTS0536073.

2,2'-METHYLENEBIS(6-*tert*-BUTYL-4-ETHYLPHENOL)

Molecular Formula. $C_{25}H_{36}O_2$
M = 368.61
CAS No 88-24-4
RTECS No SL9800000

Synonyms and **Trade Names.** Agidol 7; Antioxidant 425; Bis(3-*tert*-butyl-5-ethyl-2-hydroxyphenyl) methane; 2,2'-Methylenebis[6-(1,1-dimethylethyl)]-4-ethyl-phenol; 2,2-Methylenebis(4-ethyl-6-*tert*-butyl) phenol.

Properties. White, crystalline powder with a weak characteristic odor. Insoluble in water, soluble in ethanol and other organic solvents.

Applications. Stabilizer and antioxidant for synthetic rubber (siloxane, urethane, chloroprene, butadiene, and butyl). Thermostabilizer for polyethylene.

Acute Toxicity. LD_{50} exceeded 10 g/kg BW in male and female Wistar rats.[1]

Repeated Exposure. Sprague-Dawley rats were administered 1.135 mmol% M. for 1 week. The treatment induced increased triglyceride, diglyceride, and cholesteryl-ester concentrations in the liver. Change in liver lipids was also noted.[2]

Short-term Toxicity. Wistar rats were fed diets containing 0.2, 1.0, or 5.0 % M. and examined at 4 and 12 weeks. The treatment depressed BW gain, triglyceride level, and cholinesterase activity and increased amylase activity. Decreased *Hb* levels were noted at the highest doses. Histopathology examination revealed testicular atrophy and decreased spermatogenesis in male rats fed 1.0 or 5.0% M. Vacuolization of parathyroid gland cells was observed in female rats fed 1.0 and 5.0% M. for 12 weeks. The LOAEL of 171 mg/kg in male rats and 180 mg/kg BW in female rats was established in this study.[1]

Long-term Toxicity. Wistar rats were fed diets containing 0.03, 0.1, or 0.3% M. for 18 months. Survival of the treated animals was unaffected. The treatment caused a decrease in BW gain. Addition of 0.3% M. in the feed caused slight anemia and an increase in the level of blood urea nitrogen. Kidney damage in males and effects on the parathyroid of females were observed. Histological examination of the visceral organs revealed changes in parathyroid gland and kidney. The NOAEL of 12 mg/kg BW for the males was derived in this study. Females were affected even at the lowest tested dose of 15 mg/kg BW.[3]

Reproductive Toxicity.

Embryotoxicity. Rats were dosed with 0.25 g/kg BW through the pregnancy. The treatment produced an increase in post-implantation mortality.[4]

Gonadotoxicity. M. produced testicular damage.[5]

Carcinogenicity. No neoplastic findings were noted in the above-cited study.[3]

Regulations. ***U.S. FDA*** (1998) approved the use of M. (1) in adhesives as a component of articles intended for use in packaging, transporting, or holding food in accordance with the conditions prescribed by 21 CFR part 175.105; (2) in the manufacture of rubber articles intended for repeated use in producing, manufacturing, packaging, processing, preparing, treating, packing, transporting, or holding food in accordance with the conditions prescribed by 21 CFR part 177.2600; and (3) in the manufacture of antioxidants and/or stabilizers for polymers which may be used in the manufacture of articles or components of articles intended for use in producing, manufacturing, packaging, processing, preparing, treating, packing, transporting, or holding food in accordance with the conditions prescribed by 21 CFR part 178.2010.

References:

1. Takagi, A., Momma, J., Aida, Y., Takada, K., Suzuki, S., Naitoh, K., Tobe, M., Hasewaga, R., and Kurokawa, Y., Toxicity studies of a synthetic antioxidant, 2,2'-methylenebis (4-ethyl-6-*tert*-butylphenol) in rats. 1. Acute and subchronic toxicity, *J. Toxicol. Sci.*, 17, 135, 1992.
2. Takahashi, O. and Hiraga, K., Effects of four bisphenolic antioxidants on lipid contents of rat liver, *Toxicol. Lett.*, 8, 77, 1981.
3. Takada, K., Sai, K., Momma, J., Aida, Y., Saruki, S., Naitoh, K., Tobe, M., Hasegawa, R., and Kurokawa, Y., Chronic oral toxicity of synthetic antioxidants 2,2'-methylenebis (4-ethyl-6-*tert*-butylphenol) in rats, *J. Appl. Toxicol.*, 16, 15, 1996.
4. Telford, I. R., Woodruff, C. S., and Linford, R. H., Fetal resorption in the rat as influenced by certain antioxidants, *Am. J. Anat.*, 110, 29, 1962.
5. Takagi, A., Kawasaki, N., Momma, J., Aida, Y., Ohno, Y., Hasegawa, R., and Kurokawa, Y., Toxicity studies of a synthetic antioxidant, 2,2'-methylenebis(4-ethyl-6-*tert*-butylphenol) in rats. 2. Uncoupling effect on oxidative phosphorylation of liver mitochondria, *J. Toxicol. Sci.*, 18, 49, 1993.

2,2'-METHYLENEBIS(6-*tert*-BUTYL-4-METHYLPHENYL)-α-NAPHTHYL PHOSPHITE

Molecular Formula. $C_{33}H_{37}O_3P$

M = 512.63

Trade Name. Stafor-10.

Properties. White powder. Soluble in alcohol.

Applications. Used as a *templene* stabilizer.

Acute Toxicity. Rats and mice tolerated administration of 5.0 g/kg BW without signs of the toxic action.

Repeated Exposure of mice to 250 to 1000 mg/kg BW and rats to 500 mg/kg BW did not produce any functional or morphologic changes.

Long-term Toxicity effects were not observed in mice and rats fed 25 mg/kg and 100 mg/kg BW for 5.5 months.

Standards. ***Russia.*** PML: *n/m* when the content of M. in the polymeric material is not more than 0.3%.

Reference:

Toxicology and Sanitary Chemistry of Plastics, Abstracts, NIITEKHIM, Moscow, Issue No 1, 1979, 18 (in Russian).

2,2'-METHYLENEBIS-[4-METHYL-6-(1-METHYLCYCLOHEXYL)] PHENOL

Molecular Formula. $C_{29}H_{40}O_2$

M = 420.64

CAS No 77-62-3

Synonyms and **Trade Names.** Bisalkofen MCP; Bis-2-hydroxy-5-methyl-3-(1-methylcyclohexyl) phenyl methane; 6,6'-Bis(1-methyl cyclohexyl)-2,2'-methylenedi-*p*-cresol; Nonox WSP.

Properties. White, crystalline powder. Soluble in alcohol, poorly soluble in water.

Applications. Used as a thermostabilizer in the synthesis of polyethylene, polyesters, and impact-resistant polystyrene.

Acute Toxicity. Rats and mice tolerate administration of 10 g/kg BW. A dose of 3.2 g/kg BW had no toxic effect. The treatment produced no morphology changes in the viscera.[05, 08, 023]

Repeated Exposure failed to reveal cumulative properties. Rats tolerated administration of 0.5 to 2.0 g/kg BW over a period of 2 months. No morphological changes were found.[023]

Long-term Toxicity. Rats received 5.0% M. in their diet for 2 years. The treatment resulted in liver damage, hepatocyte degeneration, and proliferation of the bile duct cells of rats. A 0.5% level of M. in the diet had no toxic effect.[05]

No signs of intoxication were observed in dogs fed 0.01 g/kg BW for 2 years but a 5.0 g/kg BW dose caused severe liver damage (Hodge et al.). Excitation of the cortical areas of the CNS and increased BW gain are reported in rats given 2.0 g/kg BW for 5 months. No morphologic changes were found in the visceral organs.[08]

Regulations.

U.S. FDA (1998) approved the use of M. (1) in the manufacture of closures with sealing gaskets that may be safely used on containers intended for use in producing, manufacturing, packing, processing, treating, packaging, transporting, or holding food (1.0%) in accordance with the conditions prescribed by 21 CFR part 177.1210; and (2) as an antioxidant and/or stabilizer in polymers used in the manufacture of food-contact articles (for use at levels not to exceed 0.2% by weight of the polyethylene) in accordance with the conditions prescribed by 21 CFR part 178.2010.

British Standard (1992): M. is listed as an antioxidant for polyethylene and polypropylene compositions used in contact with foodstuffs or water intended for human consumption (max. permitted level for the final compound 0.2%).

Standards. *Russia*. Recommended PML: *n/m*.

2,2'-METHYLENEBIS(6-NONYL-*p*-CRESOL)

Molecular Formula. $C_{33}H_{52}O_2$
M = 480.85
CAS No 7786-17-6
RTECS No GP2150000

Synonym and **Trade Name.** 2,2'-Methylenebis(4-methyl-6-nonylphenol); Nauga white.

Properties. Viscous, amber-colored liquid. Insoluble in water.

Applications. Thermostabilizer for polyolefins, latexes, and synthetic rubber.

Acute Toxicity. Substance of low if any toxicity. LD_{50} is found to be more than 30 g/kg BW in rats.

Regulations. *U.S. FDA* (1998) approved the use of M. in the manufacture of rubber articles intended for repeated use in producing, manufacturing, packaging, processing, preparing, treating, packing, transporting, or holding food in accordance with the conditions prescribed by 21 CFR part 177.2600.

4,4'-[(METHYLIMINO)BIS(METHYLENE)]BIS[2,6-BIS(1,1-DIMETHYL-ETHYL)]PHENOL

Molecular Formula. $C_{31}H_{49}NO_2$
M = 467.71

Synonym and **Trade Name.** *N,N'*-Bis(3,5-di-*tert*-butyl-4-hydroxybenzyl)methylamine; FA-15.

Properties. White, crystalline powder with a lemon tint. Insoluble in water.

Applications. Used as a stabilizer of plastics, butadiene-acrylonitrile synthetic rubber, and *cis*-1,4-isoprene rubber; a thermostabilizer of rubber.

Repeated Exposure revealed no evident cumulative properties. Rats received 1.0 g M./kg BW for 4 weeks. The treatment caused a considerable animal mortality, while a 0.5 g/kg BW dose caused a decrease in the blood erythrocyte count, the serum level of *SH*-groups, and vitamin C in the liver, an increase in relative liver weights, and a positive thymol test.

Reference:

Volodchenko, V. A. and Sadokha, Ye. R., Toxicity profiles of the new stabilizers for polymeric materials, *Gig. Truda Prof. Zabol.*, 7, 58, 1976 (in Russian).

4,4',4"-(1-METHYL-1-PROPANYL-3-YLIDENE)TRIS[2-(1,1-DIMETHYL ETHYL)-5-METHYLPHENOL]

Molecular Formula. $C_{37}H_{52}O_3$
M = 544.89
CAS No 1843-03-4
RTECS No SM1157000

Synonyms and **Trade Names.** 4,4',4"-(1-Methyl-1-propanyl-3-ylidene)tris(6-*tert*-butyl-*m*-cresol; Topanol CA; 6,6',6"-Tri-*tert*-butyl-4,4',4"-(1-methylpropan-1-yl-3-ylidene)tri-*m*-cresol; Trisalkofen BMB; 1,1,3-Tris-(5- *tert*-butyl-4-hydroxy-2-methylphenol)butane.

Properties. White, crystalline powder. Solubility in water is 0.03% at 25°C, soluble in alcohol.

Applications. Used as an antioxidant in the production of polypropylen,; as a stabilizer for polyethylene and polyvinyl chloride, and for synthetic rubber.

Acute Toxicity. LD_{50} is 14.0 g/kg BW in rats, and 16.1 g/kg BW in mice. Poisoning was accompanied by CNS inhibition. Gross pathology examination revealed gastric paresis.[08]

Repeated Exposure revealed marked cumulative properties; 90% mice died after 3 administrations of 6.0 g/kg BW. The dose of 3.0 g/kg BW produced 50% mortality after 4 administrations.[08] However, according to other data, administration of 2.0 g/kg BW to female rats for 2 weeks failed to reveal manifestations of toxicity.[05]

Long-term Toxicity. General condition, hematology analyses, and conditioned reflex activity were not affected in mice given the doses of 230 or 690 mg/kg BW three times a week for 6 months. Capacity to work was lowered and relative weights of the heart and spleen were increased. Gross pathology examination failed to reveal changes in the viscera.[08]

Reproductive Toxicity. Affects reproduction in fish.[08]

Chemobiokinetics. After ingestion, M. is poorly absorbed in the GI tract of experimental animals. It is removed predominantly in the feces (98% the dose).

Standards.

***Russia*.** Recommended PML: *n/m*.

British Standard (1992). M. is listed as an antioxidant for polyethylene and polypropylene compositions used in contact with foodstuffs or water intended for human consumption [max. permitted level for the final compound 0.25% (non-fatty foods) or 0.1% (fatty foods)].

p-(2-NAPHTHYLAMINO)PHENOL

Molecular Formula. $C_{16}H_{14}NO$
M = 235.29
CAS No 93-45-8
RTECS No SM1750000

Synonym and **Trade Names.** *p*-Hydroxyneozone; *N*-(*p*-Hydroxyphenyl)-2-naphthylamine.

Properties. Light-gray, fine crystalline powder. Poorly soluble in water, soluble in alcohol.

Applications. Used as a stabilizer in the synthesis of rubber, vulcanizates, and polypropylene.

Acute Toxicity. LD_{50} is found to be 5.2 g/kg BW in rats, and 2.2 g/kg BW in mice.[1] Manifestations of the toxic action included marked degenerative changes and congestion in the liver, lungs, and spleen.[023]

Repeated Exposure. Rats received 1/5 LD_{50} for 2 months. This treatment caused mild functional disturbances, retardation of BW gain, an increase of arterial pressure and neuro-muscular excitability. Histology examination revealed changes in the liver and kidneys.[1,023]

Reference:

Shumskaya, N. I. and Stasenkova, K. P., *Toxicology of the Components of Rubber Mixes and of Rubber and Latex Articles*, Central Research Institute Petrol. Chem., Moscow, 1970, 26 (in Russian).

OCTADECANOIC ACID, ZINC SALT

Molecular Formula. $C_{36}H_{70}O_4.Zn$
M = 632.33
CAS No 557-05-1
RTECS No ZH5200000
Abbreviation. ZS.

Synonyms. Dibasic zinc stearate; Zinc distearate; Zinc stearate.

Properties. White, amorphous powder with a slight characteristic odor. Repels water (poorly soluble). Insoluble in ethanol.

Applications. ZS is used in the rubber and plastic industry. Waterproofing agent, heat and light stabilizer, lubricant, filler; antifoamer; antisticking agent for elastomers; acid acceptors for vulcanization.

Acute Toxicity. Rats tolerate administration of 6.0 g/kg BW as 50% emulsion in sunflower oil. This dose caused only a reduction in the catalase index. Rats and mice tolerate a 5.0 g/kg BW dose without any sign of intoxication.[1-3,012]

Repeated Exposure. Female rats received 250 mg ZS/kg BW for 7 weeks. The total dose (10.5 g/kg BW) seems to be more than 12 times greater than that consumed over the same period with the diet. No manifestations of the toxic action were observed.[4]

Long-term Toxicity. Rats and mice tolerate administration of 4.0 and 100 mg/kg BW without manifestations of the toxic effect. Gross pathology examination failed to reveal lesions in the viscera.[3]

Regulations. ***U.S. FDA*** (1998) approved the use of ZS (1) as a component of the uncoated or coated food-contact surface of paper and paperboard intended for use in producing, manufacturing, packaging, processing, preparing, treating, packing, transporting, or holding dry food in accordance with the conditions prescribed in 21 CFR part 176.180; (2) as antioxidants and/or stabilizers in polymers used in the manufacture of food-contact articles in accordance with the conditions prescribed in 21 CFR part 178.2010; (3) in articles made of urea-formaldehyde resins or as components of these articles intended for use in contact with food in accordance with the conditions specified in 21 CFR part 177.1900; and (4) in the manufacture of phenolic resins for food-contact surfaces of molded articles intended for repeated use in contact with non-acid food (*pH* above 5.0) in accordance with the conditions prescribed in 21 CFR part 177.2410. Affirmed as *GRAS.*

Standards. See ***ZINC compounds***.

References:

1. Vorob'yova, R. S., Spiridonova, V. S., and Shabalina, L. P., Hygienic assessment of PVC stabilizers containing heavy metals, *Gig. Sanit.*, 4, 18, 1981 (in Russian).
2. Spiridonova, V. S. and Shabalina, L. P., Experimental study on zinc nitrate and zinc carbonate toxicity, *Gig. Truda Prof. Zabol.*, 3, 50, 1986 (in Russian).
3. Komarova, E. N., Toxic properties of some additives for plastics, *Plast. Massy*, 12, 30, 1976 (in Russian).
4. Korostelev, V., V. and Taradin, Ya. I., Toxicology evaluation of the technical grade zinc stearate, in *Hygiene and Toxicology of High-Molecular-Mass Compounds and of the Chemical Raw Material Used for Their Synthesis*, Proc. 4th All-Union Conf., S. L. Danishevsky, Ed., Khimiya, Leningrad, 1969, 196 (in Russian).

(OCTYLOXYCARBONYLETHYLTHIO)DIOCTYL STANNANE

Molecular Formula. $C_{22}H_{42}O_4S_2Sn$
M = 553.36

Synonyms. Dioctyltin-*S,S'*-ethylenebismercaptoacetate; Dioctyltin-*S,S'*-ethylenebis thioglycolate.

Properties. A viscous, yellowish liquid, poorly soluble in water. Taste perception threshold is 0.1 mg/l; odor perception threshold is 0.4 mg/l.

Applications. A starting substance for stabilizers used in the manufacture of clear polyvinyl chloride food containers, rigid pipes, and flexible membranes.

Migration of organotins from polyvinyl chloride articles into tetrahydrofuran, xylene, and methylene chloride was investigated. Methylene chloride extracted >97% of the total extractable organotins in two extractions (between <0.8 μg and 8.8 mg octyltins/g solvent). In industrial application pipe samples, there were 0.7 to 3.0 mg octyltins/g polyvinyl chloride.[1]

Acute Toxicity. LD_{50} is found to be 640 mg/kg BW in male and 840 mg/kg BW in female rats, 340 mg/kg BW in rabbits, and 450 mg/kg BW in guinea pigs.

Repeated Exposure revealed pronounced cumulative properties. After 5 to 7 administrations of 1/10 and 1/20 LD_{50}, clinical signs of intoxication became evident in rats. Administration of 1/10 LD_{50} caused 80% animal mortality. The treatment led to a reduction in the erythrocyte count and STI.[2]

Long-term Toxicity. Rats were dosed by gavage with 0.1 mg/kg BW for 6 months. The treatment caused adynamia, decrease in the muscle tonus and in a response to painful and tactile stimuli as well as changes in hematology analyses, a decline in the *SH*-group content and in the STI to the end of the experiment.[2]

Reproductive Toxicity.

Teratogenicity. Doctyltins exhibited a general cytotoxic effect on rabbit articular and growth-plate chondrocytes that could be interpreted as possible effects on skeletal growth and development.[3]

Standards. Russia (1995). MAC and PML: 0.002 mg/l.

References:

1. Forsyth, D. S, Dabeka, R., Sun, W. F., and Dalglish, K., Specification of organotins in poly(vinyl chloride) products, *Food Addit. Contam.*, 10, 531, 1993.
2. Golovanov, O. V., Substantiation of maximum allowable concentration for PVC octyl stabilizers in water bodies, *Gig. Sanit.*, 5, 36, 1975 (in Russian).
3. Webber, R. J., Dollins, S. C., Harris, M., and Hough, A. J., Effect of alkyltins on rabbit articular and growth-plate chondrocytes in monolayer culture, *J. Toxicol. Environ. Health,* 16, 229, 1985.

5-OXIDE-2,2-DIOCTYL-1,3,2-OXATHIASTANNOLANE

Molecular Formula. $C_{16}H_{34}SnS$
M = 377.25
CAS No 15535-79-2
RTECS No RP4440000

Synonyms and **Trade Name.** Dioctyltin sulfide; Dioctyltin thioglycolate; Stanklear 176.

Composition. A sulphur-containing compound of dioctyltin.

Properties. Yellowish, oily liquid with an unpleasant odor (perception threshold 0.5 mg/l at 60°C).[1]

Applications. A starting substance for stabilizers used in the manufacture of clear polyvinyl chloride food containers, rigid pipes, and flexible membranes.

Migration of organotins from polyvinyl chloride articles into tetrahydrofuran, xylene, and methylene chloride was investigated. Methylene chloride extracted >97% of the total extractable organotins in two extractions (between <0.8 μg and 8.8 mg octyltins/g solvent). In industrial application pipe samples, there were 0.7 to 3.0 mg octyltins/g polyvinyl chloride.[2]

Acute Toxicity. LD_{50} is 1.75 g/kg BW for rats, and 2.25 g/kg BW for mice.[1]

Repeated Exposure revealed evident cumulative properties. K_{acc} is 1.45 (by Cherkinsky).[3] On administration of 1/20 and 1/5 LD_{50}, all rats died in 3 weeks.

Reproductive Toxicity.

Teratogenicity. Doctyltins exhibited a general cytotoxic effect on rabbit articular and growth-plate chondrocytes that could be interpreted as possible effects on skeletal growth and development.[4]

Standards. *Russia* (1976). PML: 0.05 mg/l.

References:

1. Sheftel', V. O., Permissible levels of migration for some organotin polyvinylchloride stabilizers, in *Hygienic Aspects of the Use of Polymeric Materials,* Proc. 2nd All-Union Meeting on Health and Safety Monitoring of the Use of Polymer Materials in Construction, Kiyv, 1976, 36 (in Russian).
2. Forsyth, D. S, Dabeka, R., Sun, W. F., and Dalglish, K., Specification of organotins in poly(vinyl chloride) products, *Food Addit. Contam.*, 10, 531, 1993.

3. Sheftel', V. O., Sahm, Z. S., Martson', L. V., and Kalinichenko, L. T., On permissible migration levels of PVC stabilizers, *Gig. Sanit.*, 3, 109, 1977 (in Russian).
4. Webber, R. J., Dollins, S. C., Harris, M., and Hough, A. J., Effect of alkyltins on rabbit articular and growth-plate chondrocytes in monolayer culture., *J. Toxicol. Environ. Health,* 16, 229, 1985.

OXIME 2-BUTANONE

M = 87.13
CAS No 96-29-7
RTECS No EL9275000
Abbreviation. OB.

Synonym. Methyl ethyl ketoxime.

Applications. MEKO is extensively used as an antioxidant in the production of paints, resins, and adhesives.

Exposure. Production and importation in the U.S. appears to be more than 5 million pounds per year.

Acute Toxicity. LD_{50} is 2.3 to 3.7 g/kg BW in rats.[1]

Repeated Exposure. In a 90-day oral gavage exposure study, OB was found to induce hemolytic anemia with compensatory hemopoiesis in CD rats.[1]

Reproductive Toxicity of OB has been studied in two-generation studies. F_o generation of Sprague-Dawley rats was administered OB in water, by gavage, at the doses of 10, 100, or 200 mg/kg BW for 10 weeks with vaginal cytology evaluation of F_o females during the last 3 weeks of the pre-breed period. Dosing has been carried out during mating, gestation, and lactation. F_1 weanlings were dosed for 11 weeks. Adult toxicity was observed in both generations at all doses. Parental deaths occurred at 200 mg/kg BW; signs of embryotoxic action were noted at 100 and 200 mg/kg BW. At 10 mg/kg BW, only adult liver spleen histological effects were present. The NOAEL for reproductive and postnatal toxicity was at least 200 mg/kg BW in this study.[2]

Mutagenicity. OB has been shown to be positive in the mouse lymphoma gene mutation test in the absence, but not in the presence of liver enzymes; it was negative in several *in vitro* and *in vivo* mutagenicity assays.[3,4]

References:

1. *Workplace Environ. Exposure Level Guide - Methyl Ethyl Ketoxime*, Am. Ind. Hyg. Assoc., 1990.
2. Tyl, R. W., Gerhart, J. M., Myers, C. B., Marr, M. C., Brine, D. R., Gilliam, A. F., Seely, J. C., Derelanco, M. J., and Rinehart, W. E., Reproductive toxicity evaluation of methylethyl ketoxime by gavage in CD rats, *Fundam. Appl. Toxicol.,* 31, 149, 1996.
3. Rogers-Beck, A. M., Lawlor, T. E., Cameron, T. P., and Dunkel, V. C., Genotoxicity of 6 oxime compounds in the Salmonella/mammalian-microsome assay and mouse lymphoma TK-assay, *Mutat. Res.*, 204, 149, 1988.
4. *Methylethylketoxime,* U.S. EPA, Final Test Rule, *Federal Register*, 54, 37799, Sept. 13, 1989.

PHENOL, STYRENE and HEXAMETHYLENETETRAMINE, product of high temperature polycondensation

Trade Name. BFSA.

Properties. Poorly soluble in water, soluble in organic solvents.

Applications. A stabilizer for polymeric materials used for packaging foodstuffs, with antioxidant and thermostabilizing properties.

Acute Toxicity. In rats, LD_{50} is reported to be 2.2 g/kg BW for males and 2.75 g/kg BW for females. Lethal doses caused adynamia, loss of motor coordination and coat sheen. Animals die in 2 to 6 days. Gross pathology examination revealed acute venous congestion of the viscera and parenchymatous dystrophy of hepatocytes.

Reference:

Radeva, M., Penkov, A., and Donchev, N., Acute oral toxicity of BFSA-1, supplement to plastics intended for packing of foodstuffs, *Khigiene i Zdraveopazvane*, 27, 262, 1984 (in Bulgarian).

PHENOTHIAZINE

Molecular Formula. $C_{12}H_9NS$

M = 199.28

CAS No 92-84-2

RTECS No SN5075000

Synonyms and **Trade Names.** Agrazine; Dibenzo-1,4-thiazine; Nemazine; Phenosan; Thiodiphenylamine.

Applications. Antioxidant for epichlorohydrin elastomers and polyethylene oxide. Polymerization inhibitor for chloroprene.

Properties. Grayish-green to greenish-yellow powder, granules or flakes with slight odor. Insoluble in water, slightly soluble in alcohol.

Acute Toxicity.

Observations in man. Probable oral lethal dose is in the range of 0.5 to 5.0 g/kg BW.[019]

Animal studies. LD_{50} is 4.0 g/kg BW in rabbits, 0.5 g/kg BW in cattle,[029] 5.0 g/kg BW in rats.[1] Average or large oral doses may cause abdominal cramps and tachycardia. Poisoning caused GI and skin irritation, and kidney damage.[2]

Allergic contact dermatitis was found to develop without light.[2]

Mutagenicity.

In vitro genotoxicity. The highest ineffective dose tested in five *Salmonella typhimurium* strains in the presence and absence of liver *S9* was 10 g/plate.[013]

Chemobiokinetics. Because of P. low solubility, the rate of absorption from the GI tract is dependent on particle size;[3] 30 to 50% of the dose administered orally is excreted unchanged. Some part of P. is converted into soluble derivatives secreted in the urine, part of these derivatives appear in bile and in milk of lactating animals.[4] P. has been metabolized into *3-hydroxyphenothiazine* (Fishman et al., 1965) and to *P. sulfate*. According to Mitchell, rat, mouse, and gerbil excreted P. in conjugated form with *leukophenothiazone sulfate* as major metabolite.[5]

Regulations. ***U.S. FDA*** (1998) approved the use of P. (1) in adhesives as a component of articles intended for use in packaging, transporting, or holding food in accordance with the conditions prescribed by 21 CFR part 175.105; (2) in the manufacture of resinous and polymeric coatings for polyolefin films for the food-contact surface of articles intended for use in producing, manufacturing, packing, processing, preparing, treating, packaging, transporting, or holding food in accordance with the conditions prescribed by 21 CFR part 175.320; (3) for use only as a polymerization-control agent; and (4) in the manufacture of rubber articles intended for repeated use in producing, manufacturing, packaging, processing, preparing, treating, packing, transporting, or holding food in accordance with the conditions prescribed by 21 CFR part 177.2600.

References:

1. *Documentation of the Threshold Limit Values and Biological Exposure Indices,* 6th ed., Vol 1, 2 and 3, American Conference of Governmental Industrial Hygienists, Inc., 1991, 1209.
2. *Documentation of the Threshold Limit Values*, American Conference of Governmental Industrial Hygienists, 4th ed., Cincinnati, Ohio, 1980, 329.
3. *Encyclopedia of Occupational Health and Safety*, vol 1 and 2, Int. Labour Office, Geneva, Switzerland, 1983, 1676.
4. Clarke, M. L., Harvey, D. G., and Humphreys, D. J., *Veterinary Toxicology*, 2nd ed., Bailliere Tindall, London, 1981, 198.
5. Mitchell, S. C., Comparative metabolism of phenothiazine in the rat (*Rattus Norvegicus*), mouse (*Mus Musculus*), hamster (*Mesocricetus Auratus*), and gerbil (*Gerbillus Gerbillus*), *Comp. Biochem. Physiol.*, 67C, 199, 1980.

N-PHENYL-*N'*-CYCLOHEXYL-*p*-PHENYLENEDIAMINE

Molecular Formula. $C_{18}H_{22}N_2$

M = 266.39

CAS No 101-87-1
RTECS No ST3500000

Synonyms and **Trade Names.** *N*-Cyclohexyl-*N'*-phenyl-*p*-phenylenediamine; Diafen FC; Flexzone 6H; Ozonone 6H; *N*-Phenyl-*N'*-cyclohexylphenylenediamine-1,4; Santoflex CP Vulkacit 4010.

Properties. Light-gray powder with a pleasant odor, darkened on storage. Solubility in water is 0.0015% (12°C), soluble in alcohol.

Applications. Used as a stabilizer in the production of synthetic rubber, polypropylene, and polyacetaldehyde.

Acute Toxicity. LD_{50} exceeded 2.0 g/kg BW in rats.[027] In mice, LD_{50} is 3.9 g/kg BW.[08] Manifestations of the toxic effect following a single administration included methemoglobinemia and slight changes in the osmotic resistance of erythrocytes. Histology examination revealed circulatory disturbances in the visceral organs, necrotic changes in the intestinal mucosa, and parenchymatous dystrophy of the renal tubular epithelium. In some animals, nephritis culminating in the nephro sclerosis was observed.

Short-term Toxicity. Rats were gavaged with 60 mg/kg BW for 4.5 months. The treatment resulted in changes in BW gain, methemoglobinemia, stimulation of conditioned reflex activity. Gross pathology examination revealed superficial necrosis of the villi and desquamation of the epithelium of the small intestine mucosa.[08]

Reproductive Toxicity. Affects reproduction in fish.

Regulations. ***U.S. FDA*** (1998) approved the use of P. in the manufacture of rubber articles intended for repeated use in producing, manufacturing, packaging, processing, preparing, treating, packing, transporting, or holding food in accordance with the conditions prescribed by 21 CFR part 177.2600.

2-PHENYL-1,3-INDANDIONE

Molecular Formula. $C_{14}H_{12}N$
M = 194.26
CAS No 83-12-5
RTECS No NK6125000

Synonyms and **Trade Names.** Indion; 2-Phenyl-1,3-diketohydrindene; 2-Phenyl-1*H*-Indene-1,3(2*H*)-dione; Phenylen; Phenylindole.

Properties. Almost white, crystalline powder. Very poorly soluble in water (1.0 mg/l at 20°C), solubility in vegetable oils is 3.0%, in alcohol 5.6 mg/l (15°C).

Applications. Used as a stabilizer in the production of plastic materials.

Acute Toxicity.

Rats tolerate administration of 2.5 and 0.5 g/kg BW without any signs of intoxication. LD_{50} is found to be 6.0 to 10 g/kg BW.[05] Catand rabbits tolerate doses of 3.0 and 5.0 g/kg, respectively. In mice, LD_{50} is 0.5 g/kg BW (P. was given in 60% aqueous suspension) (Cicolella et al., 1975). Administration of the lethal doses led to constricted pupils, diarrhea, tousled fur, peripheral vasodilatation, and sedation. Leukocytosis and reduced activity of a number of blood enzymes, and decreased levels of serum proteins were observed. Gross pathology examination revealed abdominal cavity effusion, liver hypertrophy, spleen enlargement, and testicular atrophy. Dystrophic changes in the liver and ovaries, renal damage, necrosis in the epididymis, and erythroblastosis in the spleen were shown to develop.

Repeated Exposure.

Observations in man. Ingestion of P. at dose level of 2.5 mg/kg BW over a period of 17 days caused signs of intermittent nephritis. Urine volume decreased.[1]

Animal studies. Repeated exposure failed to reveal cumulative properties. Rats were dosed by gavage with 500 mg/kg BW over a period of 24 days. The treatment resulted in reduced BW gain. No signs of the toxic action were observed in rats fed the diet with 5.0% P. for 2 months. Administration of 50 mg/kg BW to these animals caused leukocytosis, and after 2 months they developed hypo- or normochronic anemia and relative neutrophilia. Gross pathology examination revealed signs of chronic hepatitis, generalized nephropathy, glomerulonephritis, and other changes (Cicolella et al., 1975).

Long-term Toxicity. 1.0% P. added to the standard dry feed of the young rats (50 g of weight) for 2 years produced no toxic effect. Histology examination failed to reveal changes in the endocrine glands, skeleton, and brain. Wistar rats consumed 1.0% P. in their diet for a year without evident manifestations of toxicity.[05] Rats were given 48-hour extracts of P.-stabilized polyvinyl chloride for a year and exhibited no signs of intoxication and no deviations in the standard body parameters over three subsequent generations.[2,3]

Chemobiokinetics. P. is not completely absorbed following oral administration to experimental animals; 65 to 95% ingested dose were excreted in the feces.

References:

1. *Brit. Med. J.*, 1, 1655, 1963
2. Bornmann, G. and Loeser, A., Toxicity study of 2-phenylindole as a stabilizer of PVC-plastics, *Arch. Toxikol.*, 21, 1, 1965.
3. Nothdurft, H. und Mohr, H. J., Zur Vertaglichkeit von 2-Phenylindol aufgrand langfristiger, *Arch. Toxikol.*, 20, 220, 1964.

N-PHENYL-1-NAPHTHYLAMINE

Molecular Formula. $C_{16}H_{13}N$
M = 219.29
CAS No 90-30-2
RTECS No QM4500000
Abbreviation. PNA-1.

Synonym and **Trade Names.** Antioxidant PAN; Neozone A; Nonox AN; *N*-Phenyl-α-naphthylamine.

Properties. Yellowish, crystalline powder. Poorly soluble in water, soluble in alcohol.

Applications. Used as a stabilizer for various synthetic rubber. A thermostabilizer for vulcanizates and polyethylene.

Acute Toxicity. In mice, LD_{50} is 1.8 g/kg BW. A single high dose of PNA-1 caused methemoglobinemia and mild changes in the osmotic stability of the erythrocytes. Histology examination revealed circulatory disturbances, necrotic changes in the intestinal mucosa, and parenchymatous dystrophy of the renal tubular epithelium.[08]

Short-term Toxicity. Rats were gavaged with 50 mg PNA-1/kg BW for 3.5 months. The treatment led to retardation of BW gain, mild methemoglobinemia, and stimulation of conditioned reflex activity. Superficial necrosis and desquamation of the epithelium of the small intestine mucosa were found on autopsy.[08]

Reproductive Toxicity. Affects the reproduction of fish but not the development of chick embryos.[08]

Carcinogenicity. Admixture of β-naphthylamine may render PNA-1 carcinogenicity. ICR mice were dosed with repeated *i/v* injections of both technical and pure PNA-1 in dimethylsulfoxide. The treatment resulted in a high percentage of malignant tumors (as well as in the case of NPA-2). Similar results are found in the studies with TA-1 mice.[1] NPA-1 is reported to cause tumors in the lungs, kidney, ureter, and bladder. Is considered to be carcinogenic by *RTECS* criteria.

Chemobiokinetics. The similar carcinogenic potential of PNA-1 and PNA-2 suggests other routes of metabolic activation besides dephenylation for both chemicals.[1]

Regulations. ***U.S. FDA*** (1998) approved the use of PNA-1 as an antioxidant in rubber articles intended for repeated use in producing, manufacturing, packing, processing, treating, packaging, transporting, or holding food (total not to exceed 5.0% by weight of the rubber product) in accordance with the conditions prescribed by 21 CFR part 177.2600.

Reference:

Wang, H., Wang, D., and Dzeng, R., Carcinogenicity of *N*-phenyl-1-naphthylamine and *N*-phenyl-2-naphthylamine in mice, *Cancer Res.*, 44, 3098, 1984.

N-PHENYL-2-NAPHTHYLAMINE

Molecular Formula. $C_{16}H_{13}N$
M = 219.29
CAS No 135-88-6

RTECS No QM4550000
Abbreviation. PNA-2.

Synonyms and **Trade Names.** 2-Anilinonaphthalene; 2-Naphthyl phenylamine; Neozone D; Nonox D; *N*-Phenyl-β-naphthylamine.

Properties. Light-gray powder. Poorly soluble in water and in cold alcohol, soluble in hot ethanol.

Applications. Formerly used as a stabilizer in the production of the synthetic rubber, as a thermostabilizer for polyethylene, ethylene-vinyl acetate coplymer, and polyisobutylene. An anti-aging agent for vulcanizates.

Migration Data. Investigation of aqueous, hexane, alcohol (20 and 40%), and acid extracts revealed migration of PNA-2 (0.5 mg/l) from vulcanizates based on butadiene-nitrile rubber.[8]

Acute Toxicity. LD_{50} is 3.73 g/kg BW in rats, and 1.45 g/kg BW in mice.[023] Manifestations of the toxic action in mice included reduced muscle tonus, drowsiness, apathy, and CNS depression; in rats, there was general apathy and a slight reduction in motor activity. Guinea pigs tolerate 1.0 g PNA-2/kg BW.[2] The same dose caused all rabbits to die.

Repeated Exposure. Guinea pigs tolerated six administrations (at weekly intervals) of 1.0 g/kg BW without retardation of BW gain. Poisoning was accompanied by a slight rise of temperature. Gross pathology examination revealed moderate congestion in the visceral organs, parenchymatous dystrophy of the renal tubular epithelium, and pneumonia.[1]

All rabbits died given 1.0 g PNA-2/kg BW for 5 to 15 days.[2] This dose killed 50% treated rats. Functional and morphology changes were noted. According to other data, there were only the signs of intoxication in rats given 1.75 g/kg BW for a month. None of the animals died.

In a 14-day study, F344 rats given diets containing 50,000 ppm PNA-2 died before the end of the studies. Feeding diets containing 12,500 ppm PNA-2 considerably affected BW gain of the exposed animals; no compound-related toxicity in mice given feed containing up to 20,000 ppm was reported.[7]

Short-term Toxicity. Rats were dosed by gavage with 100 and 1000 mg/kg BW for 4 months. The treatment caused anemia and an increase in the urinary albumen content.[3] In a 13-week study, a dietary level of 40,000 ppm PNA-2 increased mortality rate. Nephropathy occurred in rats as indicated by renal tubular epithelial degeneration and hyperplasia. Liver enlargement and lower BW gain were observed in mice and rats.[7]

Long-term Toxicity. A transient liver and kidney dysfunction was found in rats given 100 mg/kg BW over a period of 12 months. A 20 mg/kg BW dose appeared to be ineffective.[2]

In 2-year study, rats and mice received dietary concentrations of 2,500 ppm and 5,000 ppm PNA-2. The kidney was the principal target for toxic effects predominantly in high dose female rats.[7]

Reproductive Toxicity. SNK mice were dosed with 9.0 mg PNA-2/animal during the entire period of gestation and postnatally. The treatment caused malignant tumors in the offspring.

Embryotoxic action was not found.[4]

Allergenic Effect was noted.

Mutagenicity.

In vitro genotoxicity. Naphthylamines are shown to be genetically active compounds that exhibit post-metabolic activity. However, PNA-2 did not cause CA in cultured cells and is reported to be negative in *Salmonella* strains *TA98, TA100, TA1535,* or *TA1537* mutagenicity assay in the presence or absence of metabolic activation and in Chinese hamster ovary cells in the absence of metabolic activation.[7]

In vivo cytogenetics. PNA-2 produced sperm abnormalities in mice. Some other positive findings are revealed to justify their ability to induce genotoxic effects in mammals. Naphthylamines appeared to be positive in *Dr. melanogaster* assay.

Carcinogenicity. In the above-cited study, no compound-related renal neoplasms were observed in rats and male mice. There was equivocal evidence of carcinogenicity for female mice.[7]

Rats received 20 mg PNA-2/kg BW in the antenatal, and then in postnatal periods of ontogenesis. An increase in the incidence of malignant and benign neoplasms was observed.[5] At present there is no sufficient proof of carcinogenicity of PNA-2 and no grounds for banning it. Admixture of β-*naphthylamine* may render PNA-2 carcinogenic but a product with the minimum admixture should be used. According to Russian recommendations, materials may be regarded as safe from the point of view of carcinogenicity when the technical product contains 0.005% or less of β-*naphthylamine*.[6]

See ***N-Phenyl-1-naphthylamine.***

Carcinogenicity classifications. An IARC Working Group concluded that there is limited evidence for the carcinogenicity of PN-2 in *experimental animals* and there were inadequate data available to evaluate the carcinogenicity of PN-2 in *humans*.

IARC: 3;

NTP: NE - NE - NE - EE.

Chemobiokinetics. In man, PNA-2 metabolism may result in formation of the known urinary bladder carcinogen β-*naphtylamine*.

In dogs, elimination occurs mainly in the feces.

See ***N-Phenyl-1-naphthylamine.***

Regulations. ***U.S. FDA*** (1998) approved the use of PNA-2 (1) in the manufacture of resinous and polymeric coatings for polyolefin films to be safely used as a food-contact surface in accordance with the conditions prescribed by 21 CFR part 175.320; (2) as a component of the uncoated or coated food-contact surface of paper and paperboard intended for use in contact with aqueous and fatty food (for use only as an antioxidant in dry rosin size and limited to use at a level not to exceed 0.4% by weight of the dry rosin size) in accordance with the conditions prescribed by 21 CFR part 176.170; (3) in the manufacture of rubber articles intended for repeated use in producing, manufacturing, packaging, processing, preparing, treating, packing, transporting, or holding food in accordance with the conditions prescribed by 21 CFR part 177.2600; and (4) as an antioxidant in rubber articles intended for repeated use in producing, manufacturing, packing, processing, treating, packaging, transporting, or holding food (total not to exceed 5.0% by weight of the rubber product) in accordance with the conditions prescribed by 21 CFR part 177.2600.

Standards. ***Russia*** (1986). PML in food and water: 0.2 mg/l.

References:

1. Verkhovsky, G. Ya., *Toxicity of Some Chemicals Used in the Production of Vulcanizates*, Author's abstract of thesis, Moscow, 1965, 14 (in Russian).
2. Shumskaya, N. I., Tolgskaya, M. S., and Ivanov, V. N., On toxicity of Neozone D for prolonged exposure upon the body, in *Hygiene and Toxicology of High-Molecular-Mass Compounds and Chemical Raw Materials Used in Their Synthesis*, Proc 4th All-Union Conf., S. L. Danishevsky, Ed., Khimia, Moscow, 1969, 195 (in Russian).
3. Taradin, Ya. I. and Kuchmina, N. Ya., in *Current Problems of Industrial Hygiene and Occupational Pathology*, Voronezh, 1975, 74 (in Russian).
4. Sal'nikova, L. C., Vorontsov, R. S., Pavlenko, G. I., et al., Mutagenic, embryotropic and blastomogenic effect of Neozon D (phenyl-β-naphthylamine), *Gig. Truda Prof. Zabol.*, 9, 57, 1979 (in Russian).
5. Sal'nikova, L. S., Long-term effects of Neozone D, in *Hygiene and Toxicology of High-Molecular-Mass Compounds and of the Chemical Raw Materials Used for Their Synthesis*, Theses VI All-Union Conf., B. Yu. Kalinin, Ed., Leningrad, 1979, 285 (in Russian).
6. Pliss, G. B. and Zabezhinsky, M. A., Carcinogenic hazard of Neozone D (*N*-phenyl-β-naphthylamine), *Problemy Oncologii*, 29, 91, 1983 (in Russian).
7. *Toxicology and Carcinogenesis Studies of N-Phenyl-2-naphthylamine in F344 Rats and B6C3F$_1$ Mice (Feed Studies)*, NTP Technical Report Series No 333, Research Triangle Park. NC, January 1988.
8. Medvedev, V. I., Hygienic properties of vulcanizates obtained on the base of butadiene-nitrile rubber with help of organosulphuric vulcanization accelerators, in *Hygiene and Toxicology of High-Molecular-Mass Compounds and of Chemical Raw Material Used for Their Synthesis*, Proc. 6th All-Union Conf., B. Yu. Kalinin, Ed., Leningrad, Khimiya, 1979, 81 (in Russian).

o-PHOSPHORIC ACID, DIALLYL-α-NAPHTHYL ESTER

Properties. Colorless liquid. Poorly soluble in water. Readily soluble in vegetable oil and ethanol.

Applications. Used as an antioxidant.

Acute Toxicity. In rats, LD_{50} is 350 mg/kg BW. Manifestations of the toxic effect included apathy, impaired motor coordination, reduced pain sensitivity. Death occurs within 1 to 5 days.

Repeated Exposure revealed cumulative properties.

Reference:

Buzina, A. Z. et al., in *Problems of Industrial Hygiene and Occupational Diseases in Chemical Industry Workers*, Alma-Ata, Issue No 32, 1978, 66 (in Russian).

PHOSPHORIC ACID, TRIMETHYL ESTER

Molecular Formula. $C_3H_9O_4P$
M = 140.08
CAS No 512-56-1
RTECS No TC8225000
Abbreviation. TMP.

Synonyms. Methyl phosphate; Trimethoxyphosphine oxide; Trimethyl phosphate.

Properties. Colorless liquid.

Applications. A stabilizer and antioxidant.

Acute Toxicity. LD_{50} is 840 mg/kg BW in rats, 1470 mg/kg BW in mice, 1050 mg/kg BW in rabbits.[012] According to other data, LD_{50} is 1275 mg/kg BW in rabbits and 1700 mg/kg BW in guinea pigs.[1]

Short-term Toxicity. Beagle dogs were treated orally with 1.0 to 2.0 ml TMP for 1 to 4 months. Organophosphate neuropathy was associated by functional and morphological changes including severe distally accentuated degeneration of long spinal tracts and peripheral nerve fibers.[2]

Long-term Toxicity. Wistar received 1to100 mg TMP/kg BW through their drinking water up to 30 months. The most important toxic effect was neurotoxicity, consisting of degeneration and loss of nerve fibers in the peripheral nerves and the spinal cord, associated with myopathic changes, and occurring at 50 to 100 mg/kg BW. The NOAEL of 1.0 mg/kg BW for suppression of BW gain in males was established in this study.[3]

Reproductive Toxicity.

Gonadotoxicity. TMP induced sterility in mice, rats, and rabbits. It affected primarily epididymal spermatozoa, probably by action of sperm motility.[4,5]

Rats were given 400 mg/kg BW for 5 weeks. The major changes observed included aggregate of multinucleated giant cells and maturation arrest of spermatid level, which appear immediately after administration of TMP.[5]

Mutagenicity.

In vivo cytogenetics. Administration of 0.5 or 1.0 g/kg BW caused mutagenic effects in mice.[6]

Carcinogenicity. In the above-cited study, TMP was not found to be carcinogenic to Wistar rats.[3]

References:

1. Browning, E., *Toxicity and Metabolism of Industrial Solvents*, R. Snyder, Ed., 2nd ed., vol 2, Nitrogen and Phosphorus Solvents, Elsevier, Amsterdam-New York-Oxford, 1990.
2. Schaeppi, U., Krinke G., and Kobel, W., Prolonged exposure to trimethylphosphate induces sensory motor neuropathy in the dog, *Neurobehav. Toxicol. Teratol.*, 6, 39, 1984.
3. Bomhard, E. M., Krinke, G. J., Rossberg, W. M., and Skripsky, I., Trimethylphosphate: a 30-month chronic toxicity/carcinogenicity study in Wistar rats with administration in drinking water, *Fundam. Appl. Toxicol.*, 40, 75, 1997.
4. Harbison, R. D., Dwivedi,C., and Evans, M. A., A proposed mechanism for trimethylphosphate-induced sterility, *Toxicol. Appl. Pharmacol.*, 35, 481, 1976.
5. Cho, N. H. and Park, C., Effects of dimethylphosphonate (DMMP) and trimethylphosphate (TMP) on spermatogenesis of rat testis, *Yonsei Med. J.*, 35, 198, 1994.
6. Epstein, S. S., Bass, W., Arnold, E., and Bishop, Y., Mutagenicity of trimethyl phosphate in mice, *Science*, 168, 584, 1970.

PHOSPHORUS ACID, BIPHENYL-2(2-ETHYLHEXYL) ESTER

Molecular Formula. $C_{20}H_{27}O_3P$
M = 346.41

Synonyms. Biphenyl(2-ethylhexyl) phosphite; 2-Ethylhexyldiphenyl phosphite.

Applications. Used as a stabilizer in the manufacture of polyolefins, polyvinyl chloride, polystyrenes and other polymeric materials.

Acute Toxicity. LD_{50} is 8.2 g/kg BW in rats, and 2.1 g/kg BW in mice. Manifestations of the toxic action include apathy and tremor and paresis of the extremities. Gross pathology and histology examination revealed congestion in the visceral organs and damage to the myelin sheaths of the peripheral neurons.

Repeated Exposure. Rats received 85 mg P./kg BW for 45 days. The treatment caused reduction of BW gain, CNS dysfunction, development of anemia, decreased cholinesterase activity, and impairment of liver excretory function.

Reference:

Bol'shakov, A. N. and Baranov, V. I., Materials for substantiating MAC of 2-ethylhexyldiphenyl phosphate in the workplace air, *Gig. Truda Prof. Zabol.*, 1, 41, 1979 (in Russian).

PHOSPHOROUS ACID, 2-ETHYLHEXYL DIPHENYL ESTER

Molecular Formula. $C_{20}H_{27}O_3P$
M = 346.44
CAS No 15647-08-2
RTECS No TG8800000

Synonym and **Trade Name.** Diphenyl(2-ethylhexyl)phosphite; Forstab K-201.

Properties. Colorless or slightly colored liquid with a specific odor. Soluble in ethanol.

Applications. Stabilizer for polyolefins, polyvinyl chloride, polystyrene, polypropylene, and other polymers.

Acute Toxicity. LD_{50} is 8.2 g/kg BW in rats, and 2.1 g/kg BW in mice. Poisoning is accompanied with weakness, tremor, paresis of extremities. Gross pathology examination revealed congestion in the visceral organs. Myelin membranes of peripheral nerves were affected.

Repeated Exposure. Rats received 85 mg/kg BW for 45 days. The treatment affected CNS, excretorial function of the liver, caused anemia and retardation of BW gain.

Reference:

Bol'shakov, A. M. and Baranov, V. I., Data for establishing the MPEL of 2-ethylhexyl diphenyl-phosphite in the air of a work area, *Gig. Truda. Prof. Zabol.*, 1, 41, 1979 (in Russian).

PHOSPHOROUS ACID, TRIS(2-ETHYLHEXYL) ESTER

Molecular Formula. $C_{24}H_{51}O_3P$
M = 387.67
CAS No 301-13-3
RTECS No TH2090000

Synonyms. Trioctylphosphite; Tris(2-ethylhexyl) phosphite.

Properties. Odorless, light-yellow, oily liquid. Poorly soluble in water, readily soluble in alcohol.

Applications. Used as a plasticizer and stabilizer in the production of polyvinyl polymers, polyesters, polystyrenes, and polyolefins.

Acute Toxicity. LD_{50} is reported to be 6.4 g/kg BW for male and 9.3 g/kg BW for female rats, 7.0 to 7.6 g/kg BW for mice, and 8.5 g/kg BW for rabbits. There was no great difference in sensitivity depending of species, sex, or age. The main symptoms of poisoning were affection of the CNS and peripheral NS dysfunction: adynamia, apathy, tremor, and hindlimb paralysis. Death occurs within days. Histology examination revealed diffuse hemorrhagic and atelectatic foci in the lungs, and congestion in the liver, kidneys, and spleen.[1,2]

Repeated Exposure revealed moderate cumulative properties. K_{acc} (by Lim) appeared to be 4.2 (mice) and 5.1 (rats).[2] Male rats received the dose of 250 mg/kg BW. The treatment caused a reduction in the peroxide resistance of the erythrocytes, increased STI and activity of blood cholinesterase, sorbitol dehydrogenase, lactate dehydrogenase, malate-, dehydro- and creatine phosphokinase. Histology examination showed lesions in the NS and visceral organs.[1]

Allergenic Effect is observed following skin applications to guinea pigs.[2]

References:

1. Baranov, V. I. and Bol'shakov, A. M., On substantiation of the MAC for 3(2-ethylhexyl) phosphite in workplace air, *Gig. Truda Prof. Zabol.*, 12, 51, 1981 (in Russian).
2. Golubev, A. A., Andreyeva, N. B., Dvorkin, E. A., et al., Toxicological characteristics of trioctylphosphite and triphenylphosphite, *Gig. Truda Prof. Zabol.*, 10, 38, 1973 (in Russian).

PHOSPHOROUS ACID, TRINONYL ESTER

Molecular Formula. $C_{27}H_{57}O_3P$

M = 460.81

Synonym. Trinonyl phosphite.

Applications. Used as an antioxidant in the production of isoprene, polyester, and polyamide polymers.

Acute Toxicity. In mice, LD_{50} is 14 g/kg BW. A long alkyl chain has determined low toxicity (Phillips and Marks, 1961).

PHOSPHOROUS ACID, TRI-*p*-NONYLPHENYL ESTER

Molecular Formula. $C_{45}H_{69}O_3P$

M = 689.02

CAS No 58968-53-9

RTECS No TQ8020000

Synonyms and **Trade Names.** Naugard P; Phosphite NF; Polygard; Tri(mixed mono- and dinonylphenyl)phosphite; Tri(*p*-nonylphenyl)phosphite.

Properties. Light-yellow, syrup-like viscous liquid. Hydrolyzes in water, soluble in alcohol.

Applications. Stabilizer for synthetic rubber, impact-resistant polystyrene, acrylonitrile-butadiene-styrene copolymers, polyurethane elastomers, polycarbonate and polyvinyl chloride.

Acute Toxicity. In mice, LD_{50} is reported to be 28 g/kg BW (English product), 5.0 g/kg BW (Germany), 4.2 g/kg BW (Japan), and 9.3 g/kg BW (Russia). A short period of stimulation is followed by inhibition, narcosis, and death. Gross pathology examination revealed plethora, edema, and sloughing of the esophageal and gastric mucous membrane. Necrosis of the wall of the small intestine was also observed.[1] According to other data, LD_{50} is 19.5 g/kg BW in rats.[027]

Short-term Toxicity. Mice were given 100 mg/kg and 500 mg/kg BW dose by gavage for 4 months. The treatment with the higher dose produced a slight change in BW gain and stimulation of the cortical areas of CNS. Gross pathology examination revealed necrosis of the mucosa of the small intestine.[08] The lower dose tested caused changes in the relative weight of the thyroid and spleen and dystrophic changes in liver tissue.

Standards. ***Russia*** (1994). PML in water 3.0 ng/l.

Reference:

1. *Toxicology and Sanitary Chemistry of Plastics*, Sci. and Techn. Abstracts, NIITEKHIM, Issue No 1, 1979, 35 (in Russian).

PHOSPHOROUS ACID, TRIPHENYL ESTER

Molecular Formula. $C_{18}H_{15}O_3P$

M = 310.29

CAS No 101-02-0

RTECS No TH1575000

Abbreviation. TPP.

Synonym and **Trade Names.** Irgastab TPP; Phosphite F; Triphenyl phosphite.

Properties. Colorless crystals or clear, oily liquid with a light odor of phenol. Poorly soluble in water, soluble in organic solvents.

Applications. Used as an antioxidant and stabilizer for synthetic rubber; a secondary stabilizer for polyolefins, polyorganosiloxanes, and epoxy resins; a thermo- and light-stabilizer of cellulose ethers.

Acute Toxicity. LD_{50} is reported to be 1.5 g/kg BW in rats, and 1.3 g/kg BW in mice.[1,2] When TPP is given in an oil suspension, LD_{50} is 1.92 g/kg and 1.08 g/kg BW, respectively (Talakin et al., 1987). According to Krasovsky et al., LD_{50} is 444 mg/kg BW in rats.[8]

Lethal doses led to an immediate excitation with subsequent CNS inhibition, body tremor, equilibrium disturbances, and convulsions. Death occurs within 2 to 3 days. Histology changes included circulatory disturbances and peripheral nerve degeneration (granular breakdown of myelin). White Leghorn hens received a single *s/c* injection of 1.0 g TPP/kg BW. Spinal cord and brainstem were predominantly affected. In addition, exposure to TPP also damages order centers responsible for processing and integrating sensomotor, visual, and auditory information.[6] According to Fioroni et al., administration of even 60 mg TPP/kg BW (*i/v*) caused neuropathy in hens.[7]

Repeated Exposure failed to reveal cumulative properties. TPP was shown to be a potent neurotoxicant; it produced a characteristic delayed neurotoxicity.[3]

Rats exposed to repeated *s/c* injections of TPP displayed dysfunctional changes. Two *s/c* administrations of 1.2 g/kg BW caused tail rigidity and hindlimb paralysis. Gross pathology examination showed lesions in the spinal cord.[4] The same total dose administered *s/c* during two weeks caused inhibition of neurotoxic esterase in rats.[5]

Allergenic Effect is reported on skin application.

Chemobiokinetics. Cats were administered *i/p* 0.3 ml ^{32}P-labeled TPP/kg BW. The treatment caused considerable hydrolysis, and only a small concentration of phosphorus acid in the nervous tissue may be related to degenerative changes.[04]

Administration of TPP affects mitochondrial metabolism in skeletal muscle, especially red muscle of chickens.[2]

Standards. ***EU*** (1990). SML: 2.4 mg/kg.

References:

1. Bol'shakov, A. M. and Baranov, V. I., On hygiene norm-setting of the MAC for triphenylphosphite at workplace, *Gig. Tryda Prof. Zabol.*, 9, 55, 1977 (in Russian).
2. Golubev, A. A., Andreeva, N. B., Dvorkin, E. A., et al., Toxicity characteristics of trioctyl phosphite and triphenyl phosphite, *Gig. Truda Prof. Zabol.*, 10, 38, 1973 (in Russian).
3. Kouno, N., Katou, K., Yamauchi, T., et al., Delayed neurotoxocity of triphenylphosphite in hens: pharmacokinetic and biochemical studies, *Toxicol. Appl. Pharmacol.*, 100, 440, 1989.
4. Veronesi, B., Padilla, S. S., and Newland, D., Biochemical and neurological assessment of triphenylphosphite in rats, *Toxicol. Appl. Pharmacol.*, 83, 203, 1986.
5. Padilla, S. S., Grizzle, S. B., and Lyerby, D., Triphenyl phosphite: *in vivo* and *in vitro* inhibition of rat neurotoxic esterase, *Toxicol. Appl. Pharmacol.*, 87, 249, 1987.
6. Tanaka, D., Bursian, S. J., and Lehning, E. J., Neuropathological effects of triphenyl phosphite on the central nervous system of the hen (*Gallus domesticus*), *Fundam. Appl. Toxicol.*, 18, 72, 1992.
7. Fioroni, F., Moretto, A., and Lotti, M., Triphenylphosphite neuropathy in hens, *Arch. Toxicol.*, 69, 705, 1995.
8. Krasovsky, G. N., Shtabsky, B. M., Gzjegotsky, M. R., and Zholdakova, Z. I., Hygienic significance of percutaneous absorption of chemical water pollutants, *Gig Sanit.*, 12, 13, 1981 (in Russian).

4,4'-[1,4-PIPERAZINEDIYLBIS(METHYLENE)]BIS[2,6-BIS(1,1-DIMETHYL ETHYL)]PHENOL

Molecular Formula. $C_{34}H_{50}N_2O$

M = 518.86

Synonym and **Trade Name.** *N,N'*-Bis(3,5-di-*tert*-butyl-4-hydroxybenzyl)piperidine; Phenol-85.

Properties. White powder. Contains 99% of basic substance. Poorly soluble in water.

Applications. Uncolored stabilizer of impact-resistant polystyrene and polystyrene plastics, of chloroprene, natural and fluorinated rubber.

Acute Toxicity. In rats, LD_{50} is 10 g/kg BW.

Reference:

Volodchenko, V. A. and Gnezdilova, A. I., Toxcicology expertise of Phenol-85, *Gig. Truda Prof. Zabol.,* 12, 57, 1983 (in Russian).

PLASTAB K-107

RTECS N TP0350000

Composition. K-107 contains barium, cadmium, and zinc salts of synthetic fatty acids (SFA) of C_5 - C_6 and C_7 - C_9 fractions (30%), of a complex ester based on synthetic fatty acids of C_7 - C_9 fractions and alcohols of C_7 - C_9 fractions (30%), of organic phosphites (22.5%), and of isodecyl alcohol (17.5%).

Properties. P. is a viscous liquid, dark yellow in color and partly soluble in water.

Applications. Used as a light- and thermostabilizer in polyvinyl chloride production.

Acute Toxicity. In rats, LD_{50} is 5.33 g/kg BW.

Repeated Exposure failed to reveal cumulative properties. Rats were given 30 doses of 1/10 LD_{50} of K-107 by gavage. The treatment caused reversible anemia, sulfohemoglobinemia, reticulocytosis, leukocytosis, a reduction in the *SH*-group content in the blood, decrease of blood cholinesterase activity, and other changes. Reduction of BW gain and increased relative weights of the liver, spleen, heart, and suprarenals are reported.

Reproductive Toxicity.

Gonadotoxic effect included a reduction in sperm count and motility time, increase in the number of dead forms, obliteration of the seminal vesicles in some animals with lime deposited at sites where cell elements had died.

Chemobiokinetics. On single and repeated administration, there was accumulation of cadmium but not of zinc in the kidneys and testes.

Reference:

Zvezdai, V. I., Alferova, L. I., Hypko, S. E., et al., Toxicological characteristics of Plastab K-107, *Gig. Sanit.,* 6, 85, 1987 (in Russian).

PLASTAB K-445

CAS No 111566-35-9

RTECS No TP0500000

Composition. K-445 contains *Ba-Cd-Zn* salt of synthetic fatty acids of C_7 - C_9 fractions (~30%), Forstab K-201 (~55%), ethylcellosolve (~15%).

Properties. A straw-colored liquid. Reacts with water and hydrolyzes to form phosphites and phenol. Soluble in alcohol.

Applications. Used as a stabilizer in the production of plastic materials; a complex stabilizer for polyvinyl chloride plastisols.

Acute Toxicity. LD_{50} is 2.2 g/kg BW in rats, and 1.6 g/kg BW in mice.

Repeated Exposure revealed cumulative properties. Rats received 1/10 LD_{50} for a month. The treatment affected liver and kidney functions. Reduction in STI and BW gain, and an increase in the relative weights of the visceral organs were reported. Histology examination revealed signs of moderate parenchymatous dystrophy in the liver and kidneys.

Reproductive Toxicity.

Gonadotoxic effect was not observed.

Reference:

Zvezdai, V. I. et al., Toxicity of Plastab K-445, *Gig. Truda Prof. Zabol.,* 6, 57, 1984 (in Russian).

POLY(1,2-DIHYDRO-2,2,4-TRIMETHYLQUINOLINE)

Molecular Formula. $[\sim C_{12}H_{15}N\sim]_n$

CAS No 26780-96-1

RTECS No TQ2625000

Synonym and **Trade Names.** Acetonanil; Algerite MA; Antioxidant HS; Homopolymer 1,2-dihydro-2,2,4-trimethylquinoline; Trimethyldihydroquinoline polymer.

Applications. An antioxidant and an antidegradation agent for rubber.

Acute Toxicity. LD_{50} is 2.0 g/kg BW in rabbits.[027]

PYROCATECHOL PHOSPHOROUS ACID, 2,6-DI-*tert*-BUTYL-4-METHYL PHENYL ESTER

Molecular Formula. $C_{21}H_{27}O_3P$

M = 358.40

Synonym. 2,6-Di-*tert*-butyl-4-methylphenyl pyrocatechol phosphite.

Applications. Used as a stabilizer in the manufacture of polymeric materials.

Acute Toxicity. Mice tolerated administration of 10 g/kg BW dose. Gross pathology examination revealed cellular infiltration of the submucosal and muscular layers of the stomach wall and necrosis in the small intestine.

Repeated Exposure. Administration of 7.0 g/kg BW oral dose for a month caused 100% animal mortality.

Long-term Toxicity. Mice received a dose of 850 mg/kg BW for 3.5 months. The treatment produced changes in BW gain and stimulation of the CNS cortical areas. Gross pathology examination revealed necrotic changes in the GI tract mucosa and signs of parenchymatous dystrophy in the renal tubular epithelium.[08]

PYROCATECHOL PHOSPHOROUS ACID, ISOPROPYL ESTER

Molecular Formula. $C_9H_{11}O_3P$

M = 198.15

Synonym. Isopropylpyrocatechol phosphite.

Applications. Used as a stabilizer in the production of plastics.

Acute Toxicity. In mice, LD_{50} is 2.2 g/kg BW.[08] Gross pathology examination revealed circulatory disturbances in the visceral organs, particularly in the lungs, necrotic changes in the gastric and small intestine mucosa, and parenchymatous dystrophy of the renal tubular epithelium.

Short-term Toxicity. Mice were given 200 mg P./kg BW for 3.5 months. The treatment caused mild changes in BW gain and CNS activity. Gross pathology examination revealed necrotic changes in the esophageal and intestinal mucosa, parenchymatous dystrophy of the renal tubular epithelium, and fatty dystrophy of the liver cells.[08]

PYROCATECHOL PHOSPHOROUS ACID, α-NAPHTHYL ESTER

Molecular Formula. $C_{16}H_{11}O_3P$

M = 282.24

Synonym. α- Naphthylpyrocatechol phosphite.

Properties. A solid.

Applications. Used as a stabilizer in the production of polyamides, polyolefins, and polyethylene terephthalate.

Acute Toxicity. In mice, LD_{50} is 1.4 g/kg BW.[08] Histology examination revealed circulatory disturbances in the visceral organs, necrotic changes in the gastric and intestinal mucosa, and parenchymatous dystrophy in the renal tubular epithelium.

Long-term Toxicity. Mice received 12 mg/kg BW for 3.5 months. The treatment caused threshold changes in BW gain and conditioned reflex activity. Gross pathology examination revealed necrotic lesions in the esophageal and intestinal mucosa, parenchymatous dystrophy of the renal tubular epithelium, and fatty dystrophy of the liver cells.[08]

PYROCATECHOL PHOSPHOROUS ACID, β-NAPHTHYL ESTER

Molecular Formula. $C_{16}H_{11}O_3P$

M = 282.24

Synonym. β-Naphthylpyrocatechol phosphite.

Applications. Used as a stabilizer of plastics.

Acute Toxicity. LD_{50} is 1.3 g/kg BW in rats, and 0.7 g/kg BW in mice.[08] Gross pathology examination revealed edema and cell infiltration of the submucosal and muscle layers of the stomach wall and necrosis of the intestinal mucosa.

Repeated Exposure. No cumulative effect was observed in rats given doses of 170 mg/kg and 520 mg/kg BW for a month. Rats were dosed by gavage with 130 to 260 mg/kg BW for 2 months; 50 to 80% of mice died. Animals tolerated a dose of 65 mg/kg BW.

Long-term Toxicity. Mice were given 13 and 65 mg/kg BW for 6 months. The LOAEL was identified to be 13 mg/kg BW for BW gain and effect on the NS.[08]

Reproductive Toxicity. Does not affect the development of chick embryos.

PYROCATECHOL PHOSPHOROUS ACID, 2,4,6-TRI-*tert*-BUTYLPHENOLIC ESTER

Molecular Formula. $C_{24}H_{33}O_3P$

M = 400.47

Synonyms and **Trade Names.** Alkofen B phosphite; 2,4,6-Tri-*tert*-butylphenylpyrocatechol phosphite.

Applications. Used as a stabilizer in the production of polymeric materials.

Acute Toxicity. LD_{50} is reported to be 3.0 g/kg BW in rats, and 1.5 g/kg BW in mice.[08] Gross pathology examination revealed edema and cellular infiltration of the submucosal and muscle layers of the stomach wall. Necrotic areas are found in the small intestine mucosa.

Repeated Exposure. Partial dose-dependent mortality was observed in rats given a 100 mg/kg BW dose over a period of 2 months. Mice tolerate administration of 75 to 300 mg/kg BW. Lethal doses caused copious hemorrhages in the limbs, tips of the ears, tail, scrotum, and mesentery but did not alter blood cholinesterase activity.[08]

Reproductive Toxicity. Affects the reproductive processes in fish.

8-QUINOLINOL

Molecular Formula. C_9H_7NO

M = 145.15

CAS No 148-24-3

RTECS No VC4200000

Synonyms and **Trade Names.** 1-Azanaphthalene-8-ol; Bioquin; Fennosan H 30; 8-Hydroxyquinoline; Hydroxybenzopyridine; Hydroxyphenopyridine; 8-Quinol; Phenopyridine; Quinophenol.

Properties. White crystals or white crystalline powder with a phenolic odor. Freely soluble in alcohol and water (1 part in 1500 parts water).

Applications. A stabilizer in nylon production. Corrosion inhibitor.

Acute Toxicity. LD_{50} is 1.2 g/kg BW in rats. Mice given large doses exhibited confusion, respiratory difficulty, hind leg paralysis, and violent convulsions with death within 2 hours. Smaller fatal doses caused anorexia, malaise, general indifference to sound and light, and eventually death.[019]

Mutagenicity.

In vivo cytogenetics. It did not produce effects indicative of genotoxic activity in rats given by gavage a single or two successive doses (equal to 1/2 LD_{50}). The following end points were investigated: the frequency of both micronucleated polychromatic erythrocytes in the bone marrow and micronucleated hepatocytes (after partial hepatectomy); the *in vivo - in vitro* induction of DNA fragmentation, as measured by the alkaline elution technique, and unscheduled DNA synthesis, as measured by autoradiography, in hepatocyte primary cultures.[1]

Q. was tested for its genotoxicity in CD1 male mice by using a bone marrow micronucleus assay. CD1 male mice were *i/p* treated in single injections with three dose levels (25, 50, and 100 mg/kg BW) with corn oil as solvent vehicle. Q. induced a statistically significant increase in the number of micronucleated normochromatic erythrocytes at all three doses following 24-hour treatment.[2]

In vitro genotoxicity. Q. was shown to be devoid of unscheduled DNA synthesis activity (Ashby et al., 1989). It has been known as a genotoxin in several *in vitro* systems. In the above-cited study, it was ineffective in the two latter end points after *in vitro* exposure of hepatocytes to log-spaced subtoxic concentrations.[1]

Q. was reported to be mutagenic to *Salmonella typhimurium* in the presence of activation systems.[3]

Carcinogenicity. Male and female F344/N rats and B6C3F$_1$ mice received 1500 to 3000 ppm 8-Q. in the feed for 103 weeks. The treatment produced no evidence of carcinogenicity. Survival was similar to that of controls, with slight decreases in appetite and BW with the high dose level.[4]

Carcinogenicity classifications. An IARC Working Group concluded that there are inadequate data available to evaluate the carcinogenicity of Q. in *experimental animals* and there are no data available to evaluate its carcinogenicity in *humans*.

IARC: 3.

NTP: NE - NE - NE - NE

Chemobiokinetics. In rats, Q. was metabolized to *glucuronide* and *sulfate conjugates*. More glucuronide than sulfate was excreted in urine and only glucuronide was excreted in bile.[5]

References:

1. Allavena, A., Martelli, A., Robbiano, L., and Brambilla, G., Evaluation in a battery of *in vivo* assays of four *in vitro* genotoxins proved to be noncarcinogens in rodents, *Teratogen. Carcinogen. Mutagen.*, 12, 31, 1992.
2. Hamoud, M.A., Ong, T., Petersen, M., and Nath, J., Effects of quinoline and 8-hydroxyquinoline on mouse bone marrow erythrocytes as measured by the micronucleus assay, *Teratogen. Carcinogen. Mutagen.*, 9, 111, 1989.
3. Hollstein, M., Talcott , R., and Wei, E., Quinoline: conversion to a mutagen by human and rodent liver, *J. Nat. Cancer Inst.*, 60, 405, 1978.
4. *Toxicology and Carcinogenesis Studies of 8-Hydroxyquinoline in F344/N Rats and B6C3F$_1$ Mice (Feed Studies)*, NTP Technical Report Series No 276, NIH Pub. No 85-25321985, April 1985.
5. Kiwada, H., Hayashi, M., Fuwa, T., Awazu, S., and Hanano, M., The pharmacokinetic study on the fate of 8-hydroxyquinoline in rat, *Chem. Parmacol. Bull.*, 25, 1566, 1977.

SALICYLIC ACID, 4-*tert*-BUTYLPHENYL ESTER

Molecular Formula. $C_{17}H_{18}O_3$

M = 270.33

CAS No 87-18-3

RTECS No VO2020000

Synonyms and **Trade Names.** 4-*tert*-Butylphenyl salicylate; 2-Hydroxybenzoic acid, 4-(1,1-dimethylethyl) phenyl ester; Salicylic acid, 4-*tert*-butylphenyl ester; Salol B; TBS inhibitor.

Properties. Light crystalline powder with a slight odor resembling that of salol. Solubility in water is up to 0.1% at 20°C.

Applications. Used as a light-stabilizer and plasticizer in the production of polyvinyl chloride, polyesters, polystyrene, polyolefins, cellulose esters, and polyurethanes. Used to make films for packaging food products.

Acute Toxicity. In rats, LD_{50} exceeds 1.2 g/kg BW; in mice, it is 2.9 g/kg BW.[029] Histology examination revealed congestion, edema and sloughing of the GI tract mucosa, and parenchymal dystrophy of the renal tubular epithelium.[05, 08]

Long-term Toxicity. Approximately 1.0 g/kg diet caused some retardation of BW gain in rats. Dogs were not affected with 0.5 g/kg administered in their feed.[05]

Rats received 145 mg/kg BW over a period of 5 months. Functional manifestations of the toxic effect consisted of stimulation of the CNS cortical areas. Gross pathology examination revealed congestion in the visceral organs, cellular infiltration and edema of the gastric mucosa, necrosis of the small intestine mucosa, parenchymal dystrophy of the renal tubular epithelium, and hyperplasia of the spleen pulp.[08]

Regulations. ***U.S. FDA*** (1998) approved the use of BPS as a plasticizer in food-packaging materials in accordance with the conditions prescribed by 21 CFR part 178.3740.

SALICYLIC ACID, PHENYL ESTER

Molecular Formula. $C_{13}H_{10}O_3$
M = 214.23
CAS No 118-55-8
RTECS No VO6125000
Abbreviation. PS.

Synonyms and **Trade Name.** 2-Hydroxybenzoic acid, phenyl ester; Phenyl salicylate; Salol.

Properties. White small crystals or crystalline powder with a pleasant aromatic odor. Solubility in water is 0.0015% at 12°C, soluble in alcohol.

Applications. Used as a light-stabilizer of polyesters, cellulose ethers, polyvinyl chloride, polyvinylidene chloride, polystyrenes, polyolefins, and polyurethanes. An antiseptic for the GI and urinary tracts.

Migration of PS from orthodontic appliances made of cold-cured resins was reported.[1]

Acute Toxicity. In rats and rabbits, LD_{50} is reported to be 3.0 g/kg BW.[2]

Repeated Exposure. The maximum curing daily dose in man is 6.0 g/kg BW.

Reproductive Toxicity. PS was reported to be ***teratogenic*** in rats.[2]

Chemobiokinetics. In the intestine, PS is broken down by lipase or simply by the alkaline environment into *phenol* and *salicylic acid.*

Standards. ***EU*** (1990). SML: 2.4 mg/kg.

References:

1. Lygre, H., Klepp, K. N., Solheim, E., and Gjerdet, N. R., Leaching of additives and degradable products from cold-cured orthodontic resins, *Acta Odontol. Scand.*, 52, 150, 1994.
2. Nagahama, M., Akiyama, N., and Miki, T., Experimental production of malformations with salicylates (acetyl salicylate and phenyl salicylate) in rats, *Congenital Anom.*, 5, 35, 1965 (cit. in Schardein, J. L., *Chemically Induced Birth Defects*, Marcel Dekker, New York, 1993, 142).

SODIUM BISULFITE (1:1)

Molecular Formula. $HNaO_3S$
M = 104.07
CAS No 7631-90-5
RTECS No VZ2000000
Abbreviation. SBS.

Synonym. Sodium acid sulfite.

Properties. White crystal or crystalline powder with a slight odor of sulfur dioxide. Soluble in 3.5 parts of cold water, in about 70 parts of alcohol.[020]

Applications. A stabilizer and antioxidant. Used in coagulation of rubber latex.

Acute Toxicity. LD_{50} is 2.0 g/kg BW.[014]

Reproductive Toxicity.

Gonadotoxicity. Single or repeated *i/p* injections at doses up to 1.0 g/kg BW did not produce cytotoxic effect on spermatogonia in mice.[1]

Mutagenicity. SBS induced SCE in Chinese hamster cells.[2]

Carcinogenicity classification. An IARC Working Group concluded that there is limited evidence for the carcinogenicity of bisulfites in *experimental animals* and there was inadequate evidence to evaluate the carcinogenicity of SBS in *humans.*

IARC: Group 3.

Regulations. ***U.S. FDA*** (1998) approved the use of SBS (1) in adhesives as a component (monomer) of articles intended for use in packaging, transporting, or holding food in accordance with conditions prescribed in 21 CFR part 175.105, and (2) in the manufacture of cellophane for packaging food in accordance with the conditions prescribed in CFR part 177.1200. Considered as *GRAS.*

References:

1. Bhattacharjee, D., Shetty, T. K., and Sunderman, K., Effects on the spermatogonia of mice following treatment with sodium bisulfite, *J. Environ. Pathol. Toxicol.*, 3, 189, 1980.

2. McRae, W. D. and Stich, H. F., Induction of sister chromatid exchanges in Chinese hamster cells by the reducing agents bisulfite and ascorbic acid, *Toxicology* 13, 167, 1979.

STABILIZER *CaZn*-5

Composition. A complex composition of calcium, magnesium and zinc salts, carbonic acids, epoxy stabilizers and other co-stabilizers.

Properties. A dark, rather cloudy liquid. Soluble in higher alcohols and acetone. Poorly soluble in water.

Applications. Used as a thermostabilizer for articles made of plasticized polyvinyl chloride.

Acute Toxicity. In mice, LD_{50} is 2.7 g/kg BW. Administration of the lethal doses produced immediate depression in the treated animals.

Repeated Exposure. Rats exposed to 1.5 g/kg BW for a month exhibited retardation of BW gain, anemia, leukocytosis, and an increase in blood clotting time.

Reference:

Volodchenko, V. A. and Gnezdilova, A. I., Toxicological evaluation of *CaZn* stabilizers, *Gig. Truda Prof. Zabol.*, 12, 57, 1983 (in Russian).

STABILIZER *CaZn*-113

RTECS No WH3539550

Composition. *See* ***CaZn-5.***

Trade Name. CAZ-113

Properties. Flaky product with a yellow-white or yellowish-brown color.

Applications. Used as a thermo- and light-stabilizer in the production of physiologically harmless molding materials made of polyvinyl chloride elastomers.

Acute Toxicity. LD_{50} is 11.0 g/kg BW in rats, and 7.0 g/kg BW in mice. Poisoning is accompanied by general depression.

Repeated Exposure. Rats received a dose of 1.5 g/kg BW for a month. The treatment produced erythropenia, impairment of kidney function, an increase in residual nitrogen blood level, and decrease in the creatinine concen tration in the urine. STI and BW gain were reduced.

Reference:

Volodchenko, V. A. and Gnezdilova, A. I., Toxicological evaluation of *CaZn* stabilizers, *Gig. Truda Prof. Zabol.*, 12, 57, 1983 (in Russian).

STABILIZER *CaZn*-120

Composition. *CaZn*-120 contains calcium and zinc stearates, epoxy oil, and other co-stabilizers.

Properties. Tenacious, yellowish-white paste, which is dispersed in ordinary solvents and softening agents.

Applications. Used as a thermo- and light-stabilizer in the production of physiologically harmless molding materials made of polyvinyl chloride elastomers, for articles to be used in contact with food, and in the production of medical devices.

Acute Toxicity. LD_{50} is not attained.

Repeated Exposure. Rats received a dose of 1.5 g/kg BW for a month. Manifestations of the toxic action included erythropenia, changes in kidney functions (decrease in the urinary creatinine concentration and increase in blood residual nitrogen) and decreased STI were noted.

Reference:

Volodchenko, V. A. and Gnezdilova, A. I., Toxicological evaluation of *CaZn* stabilizers, *Gig. Truda Prof. Zabol.*, 12, 57, 1983 (in Russian).

STABILIZERS *CKC-K*

CAS No 111566-56-4

RTECS No VX5497500

Composition. Complex stabilizers, containing salts of barium, cadmium and zinc: (1) CKC-K-11 [barium, cadmium and zinc laurates]; (2) CKC-K-17 [barium, cadmium and zinc laurates and stearates]; (3) CKC-K-22 [barium, cadmium and zinc laurates, stearates and caprilates, bisphenol A, forstab]; (4) CKC-K-14 [lead laurate and stearate, 22 to 41%, dioctyl phthalate, 41 to 56%]; (5) CKC-K-22 FM.

Trade Names. SKS-K 22FM; Stabilizer SKS-K 22FM.

Properties. Pastes.

Applications. Mixed complex polyvinyl chloride stabilizers.

Acute Toxicity. LD_{50} is 4.2 g/kg BW (1, 2) and 10 g/kg BW (3, 4). LD_{50} for (5) is 5.1 g/kg BW in rats, and 2.0 g/kg BW in mice. CKC-K-22 FM caused a 38% decrease of blood acetylcholinesterase activity. Toxicity of the pastes is likely to be attributed to the fatty acid salts.[1]

Repeated Exposure revealed marked cumulative properties. K_{acc} is 1.2 (by Lim). Administrations of 1/10 LD_{50} of CKC-K-22 FM for a month led to retardation of BW gain and relative weights of some visceral organs. Hematology analysis, liver, kidney, and NS functions were affected. Histology examination revealed parenchymatous dystrophy of the liver and renal tubular epithelium and accumulation of protein elements in the lumen of the straight renal tubules, and myocardial dystrophy.[2]

Allergenic Effect. (1) shows a weak sensitizing effect, (3) does not.[1]

Reproductive Toxicity.

Gonadotoxicity. CKC-K-22 FM is found to produce changes in the functional state of spermatozoa, acute increase in the number of dead forms, reduction in the quantity of spermatozoa in the epididymis, and a decrease in their acid resistance. Morphological changes are also noted. The spermatogenesis index and testicular relative weights are reduced and occlusion of the seminal vesicles and deposition of lime therein are noted. The same lesions were observed following administration of CKC-K-101 and CKC-K-103.[3]

References:

1. Vorob'yeva, R. S., Spiridonova, V. S., and Shabalina, L. P., Hygienic assessment of PVC stabilizers containing heavy metals, *Gig. Sanit.*, 4, 18, 1981 (in Russian).
2. Zvezdai, V. I. et al., Toxicity of CKC-K stabilizers, *Gig. Truda Prof. Zabol.*, 6, 57, 1984 (in Russian).
3. Zvezdai, V. I. et al., in *Endocrine System and Harmful Environmental Factors*, Proc. 2nd All-Union Conf., Leningrad, 1983, 171 (in Russian).

STABILIZER *CKC-K-20P*

Composition. CKC-K-20P contains potassium and zinc salts of synthetic fatty acids fractions C_{17}-C_{20} (calcium : zinc = 1:1), 30%; dioctyl phthalate, 50%; pentrol, 20%.

Properties. Homogenous, red, paste-like mass. Poorly soluble in water, partly soluble in alcohol.

Applications. Used as a thermostabilizer for polyvinyl chloride compositions when they are processed into film materials and artificial leathers.

Acute Toxicity. LD_{50} is not attained. Almost all animals tolerated administration of 10 g/kg BW.

Repeated Exposure. Rats were dosed by gavage with 2.0 g/kg BW for a month. Administration affected predominantly hematology and urinalyses: reduction of total and oxidized *Hb*, as well as reticulocytosis, erythropenia, shortening of blood clotting time, increase in the relative density of the urine and its urea content.

Reproductive Toxicity.

Gonadotoxic effects included functional and morphological changes in the testes of rats and an increase in their calcium content.

Reference.

Zvezdai, V. I., Alferova, L. I., Zovsky, V. N., et al., Toxicity of thermostabilizers CKC K-20 and CKC-20P for PVC, *Gig. Sanit.*, 10, 89, 1987 (in Russian).

STABILIZER *SO-3*

Properties. White, crystalline, odorless powder of low water solubility.

Applications. Used as a stabilizer of polymeric materials including those intended for contact with food. Makes it possible to replace Santonox-type antioxidants.

Acute Toxicity. Rats and mice tolerate administration of up to 15 g/kg BW without any manifestations of toxicity and retardation of BW gain.

Short-term Toxicity. There were no treatment-related changes in the functional and morphology parameters of rats that received 500 mg/kg BW for 3 months.

Long-term Toxicity. Rats tolerate ingestion of 100 mg/kg BW in oil suspension for 8 months without any detectable signs of intoxication.

Reproductive Toxicity.

Embryotoxicity. In a 4-month study, a 100 mg/kg BW dose did not affect the development of fetus or progeny in rats.

Chemobiokinetics. Poorly absorbed following oral administration to experimental animals and is not found in the blood and urine. It remains in the GI tract for 1 to 3 days and is excreted unchanged in the feces.

Standards. ***Russia.*** PML in food: *n/m*.

Reference:

Toxicology and Sanitary Chemistry of Polymerization Plastics, Coll. Works, B. Yu. Kalinin, Ed., Leningrad, 1984, 55 (in Russian).

STEARIC ACID, ALUMINUM SALT

Molecular Formula. $C_{54}H_{105}O_6.Al$
M = 877.57
CAS No 637-12-7
RTECS No WI2820000
Abbreviation. AS.

Synonyms. Aluminum stearate; Aluminum tristearate; Octadecanoic acid, aluminum salt.

Properties. White powder. Practically insoluble in water. When freshly made, soluble in alcohol.[020]

Applications. A colorant and a stabilizer. Used in the food, cosmetic, and pharmaceutical industries.

Toxicity. See ***Aluminum compounds.***

Regulations. ***U.S. FDA*** (1998) approved the use of AS (1) in adhesives as a component of articles intended for use in packaging, transporting, or holding food in accordance with the conditions prescribed by 21 CFR part 175.105; and (2) as a stabilizer in food packaging materials in accordance with the conditions prescribed by 21 CFR part 181.29.

STEARIC ACID, CALCIUM SALT

Molecular Formula. $C_{36}H_{70}O_4.Ca$
M = 607.14
CAS No 1592-23-0
RTECS No WI3000000

Synonyms. Calcium distearate; Calcium stearate; Octadecanoic acid, calcium salt.

Composition. Calcium stearate is a mixture of calcium salts of different fatty acids consisting mainly of calcium stearate and calcium palmitate with a minor proportion of other fatty acids.

Properties. White, fatty, crystalline powder. Unstable in presence of lipolythic microorganisms, for which it is a nutrient medium. Poorly soluble in water, up to 4.0 mg/100 g at 15°C. Insoluble in alcohol, slightly soluble in hot vegetable and mineral oil.

Applications. Used as a thermostabilizer of polyvinyl chloride, can be used as an internal lubricant in the plastic manufacture.

Acute Toxicity. Rats and mice tolerate administration of the 5.0 g/kg BW dose without signs of intoxication.

Repeated Exposure. A 250 mg/kg BW dose increases BW gain in rats. Calcium compounds are likely to suppress oxidation processes in the body.

Long-term Toxicity. Mice were dosed by gavage with 20 and 100 mg calcium stearate/kg BW for 9 months. The treatment resulted in CNS inhibition due to the known physiologic features of calcium. Gross pathology examination failed to find morphological changes in the visceral organs.

Chemobiokinetics.

Observations in man. Calcium makes up about 1.6% of human total BW. A relatively small amount of calcium lost from the body is excreted in the urine. The major part of calcium eliminated from the body is excreted in the feces.

Regulations. ***U.S. FDA*** (1998) approved the use of calcium stearate (1) as a lubricant in phenolic resins used as the food-contact surface of molded articles intended for repeated use in contact with non-acid food (*pH* above 5.0); (2) in the manufacture of antioxidants and/or stabilizers for polymers which may be used in the manufacture of articles or components of articles intended for use in producing, manufacturing, packaging, processing, preparing, treating, packing, transporting, or holding food in accordance with the conditions prescribed in 21 CFR part 178.2010; and (3) as a plasticizer in the manufacture of rubber articles intended for repeated use in producing, manufacturing, packing, processing, treating, packaging, transporting, or holding food (total not to exceed 30% by weight of the rubber products) in accordance with the conditions prescribed in 21 CFR part 178.3740. Calcium stearate is affirmed as GRAS.

Standards. ***Russia.*** PML: *n/m.*

Reference:

Komarova, E. N., Toxic properties of some additives for plastics, *Plast. Massy*, 12, 30, 1976 (in Russian).

TETRAETHYLSTANNANE

Molecular Formula. $(C_2H_5)_4Sn$
M = 234.194
CAS No 597-64-8
RTECS No WH8625000
Abbreviation. TES.

Synonym. Tetraethyltin.

Properties. Colorless, oily liquid with a specific, sharp odor. Solubility in water is 10 to 12 mg/l at 20°C. Odor perception threshold is 0.2 mg/l; practical threshold is 0.5 mg/l. The metallic taste perception threshold is 1.5 mg/l.[1]

Applications. Used as a stabilizer in the polyvinyl chloride material manufacture.

Acute Toxicity. LD_{50} is reported to be 9.0 to 15 mg/kg BW in rats, 40 mg/kg BW in mice and guinea pigs, and 7.0 mg/kg BW in rabbits.[1,2]

Administration of the lethal doses caused BW loss, adynamia, convulsions, muscle weakness with subsequent death in 5 to 7 days in a comatose condition.

Repeated Exposure revealed marked cumulative properties. Rats given 1/10 LD_{50} died in 2 to 3 weeks. Manifestations of the toxic action included adynamia, muscle weakness, refusal of food, and exhaustion.[1,2]

Long-term Toxicity. Anemia and reticulocytosis are observed in rats. Other toxic effects include a change in the activity of some enzymes and in *SH*-group content in the blood serum.

Teratogenicity and **Carcinogenicity.** There are no experimental data to confirm these properties of tin compounds.[3]

Chemobiokinetics. TES metabolism in isolated hepatocytes includes formation of ethane and ethylene, the latter being the principal metabolite (95%). No toxic effects on hepatocytes are noted.[4]

Standards. ***Russia*** (1994). MAC and PML: 0.0002 mg/l.

References:

1. Skachkova, I. N., Hygienic substantiation of Maximum Allowable Concentration for tetraethyltin in water bodies, *Gig. Sanit.*, 4, 11, 1967 (in Russian).
2. Mazayev, V. G., Korolev, A. A., and Skachkova, I. N., Experimental study of toxic effect of tetraethyltin and dichlorodibutyltin in the animals, *J. Hyg. Epidemiol. Microbiol. Immunol.* (Prague), 15, 104, 1971 (in Russian).
3. *Handbook of the Toxicology of Metals*, L. Friberg, G. F. Nordberg, E. Kessler, and V. B. Vouk, Eds., 2nd ed., vol 2, Elsevier Sci. Publ., Amsterdam, 1986, 588.

4. Wiebkin, P., Prough, R. A., and Bridges, J. W., The metabolism and toxicity of some organotin compounds in isolated rat hepatocytes, *Toxicol. Appl. Pharmacol.*, 62, 409, 1982.

N-(2,2',6,6'-TETRAMETHYL-4-PIPERIDINYL)-2-[(2,2',6,6'-TETRAMETHYL-4-PIPERIDINYL)AMINO]ACETAMIDE

Molecular Formula. $C_{20}H_{40}N_4O$
M = 352.55
CAS No 93547-60-5
RTECS No AC8770600

Synonym and **Trade Name.** Diacetam-537; 2,2',6,6'-Tetramethylpiperidylamide, aminoacetic acid.

Properties. White, crystalline powder. Partially soluble in water.

Applications. Used as a stabilizer in the production of plastic materials. An effective light-stabilizer.

Acute Toxicity. LD_{50} is 5.09 g/kg BW in rats, and 0.89 g/kg BW in mice. Lethal doses caused the development of clonic-tonic convulsions and general inhibition. Death within 1 to 3 days.

Repeated Exposure revealed moderate cumulative properties. K_{acc} is 4.2 (by Lim). Anemia developed in rats exposed by gavage to the 250 mg/kg BW dose. Kidneys, liver, and CNS functions were not affected.

Reference:

Gnezdilova, A. I. and Volodchenko, V. A., Toxcity of diacetam-5, *Gig. Truda Prof. Zabol.*, 12, 57, 1983 (in Russian).

N-(2,2',6,6'-TETRAMETHYL-4-PIPERIDINYL)-3-(2,2',6,6'-TETRAMETHYL-4-PIPERIDINYLAMINO)PROPANAMIDE

Molecular Formula. $C_{21}H_{38}N_4O$
M = 362.63
CAS No 76505-58-3
RTECS No TX1498400

Synonym and **Trade Name.** Diacetam-5; 2,2',6,6'-Tetramethylpiperidylaminopropionic acid, 2,2',6,6'-tetramethylpiperidylamide.

Properties. White, crystalline powder. Partially soluble in water.

Applications. Used as a light-stabilizer in the production of polymeric materials.

Acute Toxicity. LD_{50} is found to be 4.18 g/kg BW in rats, and 0.77 g/kg BW in mice. Manifestations of the toxic action include convulsions and general inhibition. Death occurs in the first three days.

Repeated Exposure revealed moderate cumulative properties. K_{acc} is 4.28 (by Lim). Anemia, leukocytosis, and increased STI were observed at the end of a one-month treatment in rats given the dose of 210 mg/kg BW.

Reference:

Gnezdilova, A. I. and Volodchenko, V. A., Toxcity of diacetam-5, *Gig. Truda Prof. Zabol.*, 12, 57, 1983 (in Russian).

2,2'-THIOBIS(6-*tert*-BUTYL-4-CHLOROPHENOL)

Molecular Formula. $C_{20}H_{24}C_{12}O_2S$
M = 399.40
CAS No 6012-78-8
RTECS No SM9950000

Synonym and **Trade Name.** Bis(3-*tert*-butyl-2-hydroxy-5-chlorophenyl)sulfide; Product AN-9.

Properties. Odorless, white powder. Poorly soluble in water, readily soluble in alcohol.

Applications. Used as an anti-ozone and anti-scorching agent in the production of vulcanizates and rubbers.

Acute Toxicity. LD_{50} is reported to be 6.2 g/kg BW in rats, and 3.5 g/kg BW in mice. High doses led to adynamia and diarrhea development within an hour after administration. Rats die in 2 days, mice in 5 days. Gross pathology examination revealed congestion in the liver, vacuolization of hepatocytes, early signs of fatty degeneration and lymphoid infiltration. Congestion in the kidneys and lobular pneumonia in the lungs were found to develop.

Repeated Exposure failed to reveal cumulative properties. Rats were dosed by gavage with 1/5 LD_{50} for 2 months. The treatment resulted in lethargy and adynamia, retardation of BW gain, increase in the neural excitability, decreased blood erythrocyte count, and a change in the liver antitoxic function. 50% mortality was noted after 34 days of experiment.

Reference:

Vorob'yeva, R. S., General data about methods of investigation of toxic properties of rubber ingredients, in *Toxicology of New Chemical Substances Used in Rubber and Tire Industry*, Z. I. Israel'son, Ed., Meditsina, Moscow, 1968, 23 (in Russian).

4,4'-THIOBIS(6-*tert*-BUTYL-*o*-CRESOL

Molecular Formula. $C_{22}H_{30}O_2S$
M = 358.54
CAS No 96-66-2
RTECS No GP3200000
Abbreviation. TBBC.

Synonyms and **Trade Names.** Antioxidant 736; Bis(3-*tert*-butyl-2-hydroxymethylphenyl) sulfide; Tioalkofen BM; 4,4'-Thiobis [2-(1,1'-dimethylethyl)-6- methylphenol].

Properties. Colorless or white crystalline powder. Poorly soluble in water, soluble in alcohol.

Applications. Used as a thermostabilizer in the synthesis of vulcanizates based on natural and butyl rubbers, polyolefins, pentaplast, and polyvinyl chloride.

Acute Toxicity. In rats, LD_{50} is found to be 6.34 g/kg BW.[027] Gross pathology examination failed to reveal changes in the visceral organs.[08]

Short-term Toxicity. Rats were dosed by gavage with 3.25 g/kg BW for 4 months. Administration of TBBC caused no morphology changes. The only treatment-related effect observed was the change in conditioned reflex activity.[08]

Reproductive Toxicity. TBBC affects the reproductive processes in fish but not the development of chick embryos.[08]

Carcinogenicity. TBBC is likely to produce an inhibiting effect upon tumor induction by benzo[a]pyrene.[1]

Standards. ***Russia.*** PML in food: 4.0 mg/l.

Reference:

1. Braun, D. D., Effect of polyolefins antioxidants on induction of neoplasms by benzo[a]pyrene, *Gig. Sanit.*, 6, 18, 1975 (in Russian).

4,4'-THIOBIS(6-*tert*-BUTYL-*m*-CRESOL)

Molecular Formula. $C_{22}H_{30}O_2S$
M = 358.54
CAS No 96-69-5
RTECS No GP3150000
Abbreviation. TBBC.

Synonyms and **Trade Names.** Bis(3-*tert*-butyl-4-hydroxy-6-methylphenyl)sulfide; Santonox R; 4,4'-Thiobis(6-*tert*-butyl-3-methylphenol); 1,1-Thiobis(2-methyl-4-hydroxy-5-*tert*-butylbenzene); Tioalkofen BM 4.

Properties. Santonox (technical product) is a light-gray, crystalline powder with a slightly aromatic odor. Purified product is fine white crystals. Solubility in water is 0.08%, soluble in alcohol.

Applications. An antioxidant for synthetic and natural rubber and latexes, a stabilizer for high- and low-pressure polyethylene food-packaging films. Used as a light-stabilizer of polystyrene, polyvinyl chloride, polypropylene and cellulose acetate; a thermostabilizer of polyamides, etc.

Acute Toxicity. In mice, LD_{50} is 4.88 g/kg BW. Lethal doses caused adynamia. Death occurs in 2 days. Rats appeared to be less sensitive; administration of up to 5.0 g/kg BW caused no evident signs of intoxication. Gastroenteritis (severe diarrhea) was found to develop in rats before death[04] Gross pathology examination revealed coarctation and fatty dystrophy of the liver.[023]

Repeated Exposure. F344 rats received doses of 95 to 365 mg TBBC/kg BW (males) or 82 to 270 mg TBBC/kg BW (females) for 15 days. The treatment with higher doses caused diarrhea, renal papillary and tubular necrosis.[1]

B6C3F$_1$ mice were gavaged with TBBC in corn oil at doses of 10, 100, or 200 mg/kg BW for 14 consecutive days. There were no signs of overt toxicity or marked effects on serum chemistry. An increase in the liver and kidney weights was noted. Histological examination of the liver revealed mild focal hydropic degeneration, mild hepatitis, and a slight increase in the number of Kupffer cells (at the high dose treatment only). No microscopic changes were observed in the spleen, and no other organs were affected.[2]

Rats received 0.05 and 0.25% TBBC in the diet for 30 days. The higher dose caused retardation of BW gain and enlarged livers.[05]

Rats were dosed by gavage with 1/30 LD_{50} for 2 months. The treatment caused an increase in alkaline phosphatase activity in the blood and inhibition of phagocytic activity.

Rats were given 1/5 LD_{50}; this killed a half of animals by day 41 of the treatment. Manifestations of the toxic action included retardation of BW gain, lowering of the threshold of neuro-muscular excitability, and impairment of the detoxifying function of the liver.[023] Histology examination revealed early signs of fatty dystrophy in the liver, signs of congestion in the lungs, and hemorrhagic foci in the spleen pulp.

Short-term Toxicity. F344 rats received doses of 15 to 315 mg TBBC/kg (males) or 15 to 325 mg TBBC/kg BW (females) for 13 weeks. All rats survived to the end of the study. Histopathological findings included hypertrophy of Kuppfer cells, bile duct hyperplasia, and necrosis of hepatocytes. No clinical findings were noted in a 13-week study in mice.[1]

In a 3-month rat study, administration of 0.05 and 0.005% TBBC in the diet caused a decrease in BW gain (with the higher dose). Gross pathology examination failed to reveal changes in the viscera.[05]

Long-term Toxicity. Rats received 0.05 and 1.25% TBBC in their diet; mice were given 0.025, 0.05, and 1.0% TBBC. The treatment caused only non-neoplastic lesions including hypertrophy of Kupffer cells and cytoplasmic vacuolization (in male rats), nephropathy, combined fibroadenoma, adenoma, or carcinoma in mammary gland of female rats. Fatty changes, clear cell foci, and combined adenoma or carcinoma were seen in the male mice liver.[1]

Allergic contact dermatitis developed on the hands and/or face of patients after exposure to latex gloves. The patients had a positive patch test reaction to TBBC.[3]

Mutagenicity.

In vitro genotoxicity. TBBC was negative in several strains of *Salmonella typhimurium* with or without *S9* metabolic activation enzymes. SCE were induced, but no increases in CA were noted in cultured Chinese ovary cells with and without *S9* activation system.[1]

Carcinogenicity. In 2-year feed studies, there was no evidence of carcinogenic activity in F344/N rats administered 500, 1000, or 2500 ppm or in B6C3F$_1$ mice administered 250, 500, or 1000 ppm TBBC/kg BW.[1]

Carcinogenicity classification.

NTP: NE - NE - NE - NE (feed).

Chemobiokinetics. ^{14}C-labeled TBBC is shown to be incompletely absorbed in male rats. Rate of absorption is proportional to the administered dose. Distribution occurs mainly in the liver, blood, muscles, and adipose tissues. The major metabolites appeared to be *glucuronide conjugates* of the parent compound. It is rapidly cleared from all tissues except adipose one.[4]

Regulations.

U.S. FDA (1998) approved the use of TBBC in the manufacture of antioxidants and/or stabilizers for polymers which may be used in the manufacture of articles or components of articles intended for use in producing, manufacturing, packaging, processing, preparing, treating, packing, transporting, or holding food in accordance with the conditions prescribed in 21 CFR part 178.2010.

British Standard (1992). TBBC is listed as an antioxidant for polyethylene and polypropylene compositions used in contact with foodstuffs or water intended for human consumption (max. permitted level for the final compound 0.25%).

References:

1. *Toxicology and Carcinogenesis Studies of 4,4'-Thiobis(6-t-butyl-m-cresol) in F344/N Rats and B6C3F$_1$ Mice (Feed Studies),* NTP Technical Report Series No 435, Research Triangle Park, NC, December, 1994.
2. Munson, A. E., White, K. L., Barnes, D. W, Musgrove, D. L., Lysy, H. H., and Hoslappe, M. P., An immunotoxicological evaluation of 4,4'-thiobis-(6-*tert*-butyl-*m*-cresol) in female B6C3F$_1$ mice. 1. Body and organ weights, hematology, serum chemistries, bone marrow cellularity, and hepatic microsomal parameters, *Fundam. Appl. Toxicol.*, 10, 691, 1988.
3. Rich, P., Belozer, M. L., Norris, P., and Storrs, F. J., Allergic contact dermatitis to two antioxidants in latex gloves: 4,4'-thiobis(6-*tert*-butyl-*m*-cresol) (Lowinox 44S36) and butylhydroxyanisole, *J. Am. Acad. Dermatol.*, 24, 37, 1991.
4. Birnbaum, L. S., Eastin, W. C., Johnson, L., et al., Disposition of 4,4'-thiobis(6-*tert*-butyl- *m*-cresol) in rats, *Drug Metab. Dispos.,* 11, 537, 1983.

2,2'-THIOBIS(6-ISOBORNYL-4-METHYLPHENOL)

Molecular Formula. $C_{34}H_{46}O_2S$

M = 514.77

Properties. White, crystalline powder. Poorly soluble in water, soluble in alcohol.

Applications. Used as a stabilizer in the production of polymeric materials.

Acute Toxicity. Rats and mice tolerate 10 g/kg BW and 15 g/kg BW, respectively.

Repeated Exposure failed to reveal cumulative properties. Manifestations of the toxic action were not found on administration of 1.0 to 3.0 g/kg BW.

Long-term Toxicity. In a 7-month study, rats were given 200 mg T./kg BW, and mice 150 and 300 mg T./kg BW. The treatment did not affect BW gain, hematology analyses, or conditioned reflex activity. Histology examination failed to reveal any difference from the controls.

Allergenic Effect was not observed.

Standards. *Russia.* Recommended PML: *n/m.*

Reference:

Mikhailets, I. B., Toxic properties of some stabilizers for plastics, *Plast. Massy,* 12, 41, 1975 (in Russian).

4,4'-THIOBIS(2-METHYL-6-CYCLOHEXYL)PHENOL

Molecular Formula. $C_{26}H_{34}O_2S$

M = 378.6

Applications. Used as a stabilizer in the production of polymeric materials.

Acute Toxicity. In mice, LD_{50} is 4.8 g/kg BW. Gross pathology examination revealed congestion in the viscera.[08]

2,2'-THIOBIS(4-METHYL-6-α-METHYLBENZYLPHENOL)

Molecular Formula. $C_{30}H_{30}O_2S$

M = 454.63

Trade Names. Thioalkofen C; Thioalkofen MBP.

Properties. White, crystalline powder. Poorly soluble in water, soluble in ethanol.

Applications. Used as a thermo- and light-stabilizer in the production of polyvinyl chloride and polyolefins.

Acute Toxicity. Mice tolerate a 15 g/kg BW dose. Gross pathology examination showed no lesions.

Repeated Exposure failed to reveal cumulative properties.

Long-term Toxicity. In a 6-month study, rats were dosed with 0.43 and 1.42 g/kg BW. The treatment caused only stimulation of the CNS cortical areas. There were no pathologic changes in the viscera.[08]

Reproductive Toxicity. Does not affect the reproduction in fish.

THIOCARBANILIDE

Molecular Formula. $C_{13}H_{14}N_2S$

M = 228.33

CAS No 102-08-9

RTECS No FE1225000

Abbreviation. TCA.

Synonyms and **Trade Names.** *sym*-Diphenylthiourea; Stabilizer C; Sulfocarbanilide; Vulkacit CA.

Properties. Crystalline leaflets. On decomposition may form phenyl mustard oil. Poorly soluble in water, freely soluble in alcohol and ether.

Applications. Used primarily as a thermostabilizer in the production of polyvinyl chloride and as a vulcanization accelerator for rubber.

Acute Toxicity. LD_{50} is 1500 mg/kg BW in rats and rabbits,[020] and 720 mg/kg BW in cats.

Long-term Toxicity. Wistar rats received 0.1% TCA in the diet over a period of one year.[05] The treatment caused no toxic effect. Histology examination revealed no changes in the visceral organs.

Allergenic Effect. TCA was positive in a new sensitive mouse lymph node assay.[2]

Reproductive Toxicity.

Embryotoxicity. Sprague-Dawley rats were administered TC by gavage from days 6 to 20 of gestation. Fetotoxic effect was found to develop at 100 mg/kg BW and above. Administration of the 200 mg/kg BW dose caused embryolethal action.[3]

Teratogenicity. No increase in the incidence of malformations was reported in the above described study. Embryotoxic and teratogenic effect was observed following injection of TCA into chick eggs.[4]

Carcinogenic properties are not observed.

Chemobiokinetics. In rabbits, absorbed TCA is not desulfurized. It is metabolized to 1-(p-hydroxyphenyl)-3-phenylthiourea (Williams, 1961).

Regulations. ***U.S. FDA*** (1998) approved the use of TCA (1) in adhesives as a component of articles intended for use in packaging, transporting, or holding food in accordance with the conditions prescribed by 21 CFR part 175.105; (2) in the manufacture of rubber articles intended for repeated use in producing, manufacturing, packaging, processing, preparing, treating, packing, transporting, or holding food in accordance with the conditions prescribed by 21 CFR part 177.2600; and (3) in the manufacture of antioxidants and/or stabilizers for polymers which may be used in the manufacture of articles or components of articles intended for use in producing, manufacturing, packaging, processing, preparing, treating, packing, transporting, or holding food in accordance with the conditions prescribed by 21 CFR part 178.2010.

References:

1. Grant, W. M., *Toxicology of the Eye*, 2nd ed., Charles C. Thomas, Springfield, IL, 1974, 419.
2. Ikarashi, Y., Ohno, K., Momma, J., Tsuchiya, T., and Nakamura, A., Assessment of contact sensitivity of four thiourea rubber accelerators: comparison of two mouse lymph node assays with the guinea pig maximization test, *Food Chem. Toxicol.*, 32, 1067, 1994.
3. Saillenfait, A. M., Sabate, J. P., Langonne, I., and de Ceaurriz, J., Difference in the developmental toxicity of ethylenethiourea and three *N,N'*-substituted thiourea derivatives in rats, *Fundam. Appl. Toxicol.*, 17, 399, 1991.
4. Korhonen, A., Hemminki, K., and Vainio, H., Embryotoxicity of industrial chemicals on the chicken embryo: thiourea derivatives, *Acta Pharmacol. Toxicol.* (Copenh.), 51, 38, 1982.

THIODIACETIC ACID, DI-[(3,5-DI-*tert*-BUTYL-4-HYDROXYPHENYL) PROPYL] ESTER

Molecular Formula. $C_{32}H_{58}O_6S$

M = 642.9

Trade Name. Tioalkofen P.

Properties. Highly dispersed white, powder. On heating, soluble in alcohols, poorly soluble in water.

Applications. Used as a stabilizer in the plastic manufacture.

Acute Toxicity. LD_{50} is 1.98 g/kg BW in rats, and 1.38 g/kg BW in mice. Lethal doses led to CNS excitation including restlessness, convulsions, and subsequent adynamia. Death within 2 to 4 days.

Reference:

Kirichek, L. T. and Derkach, Z. M., Acute toxicity of new light-stabilizers, *Gig. Sanit.*, 6, 87, 1985 (in Russian).

4,4'-THIODIMETHYLENEBIS(2,6-DI-*tert*-BUTYLPHENOL)

Molecular Formula. $C_{30}H_{46}O_2S$
M = 470.75

Synonym. TB-3.

Properties. White, crystalline powder. Poorly soluble in water and alcohol.

Applications. Used as a stabilizer in the production of polymeric materials.

Acute Toxicity. Rats and mice tolerate administration of 10 g/kg BW.[08] Gross pathology examination revealed no changes in the viscera.

Repeated Exposure failed to reveal marked cumulative properties. One-month exposure to 1.5 g/kg BW caused 30% mortality in mice. Administration of a 10 g/kg BW dose for a month appeared to be non-lethal for rats. Gross pathology and histology examination revealed no lesions in the visceral organs.

Short-term Toxicity. Mice were dosed by gavage with 75 and 150 mg/kg BW over a period of 4 months. The treatment was shown to cause only excitation of the CNS cortical areas in animals. Gross pathology examination revealed no changes in the visceral organs.[08]

Standards. ***Russia.*** Recommended PML: *n/m*.

4,4'-THIODIPHENOL

Molecular Formula. $C_{12}H_{10}O_2S$
M = 218.28
CAS No 2664-63-3
RTECS No SN0800000

Synonym. 4,4'-Dihydroxydiphenyl sulfide.

Properties. Light, fine crystalline powder. Poorly soluble in water, soluble in alcohol.

Applications. Used as a stabilizer in the production of polymeric materials.

Acute Toxicity. LD_{50} is reported to be 6.4 g/kg BW in rats, and 5.5 g/kg BW in mice. Poisoning is accompanied by lethargic condition. Death occurs in 3 days. Gross pathology examination revealed congestion in the visceral organs.

Repeated Exposure. All the animals died within 16 days during treatment by Lim method.

Reference:

Stasenkova, K. P, Shklovskaya, M. L., and Sergeyeva, L. G., Comparative evaluation of toxicity of antioxidants for phenolic resins, in *Toxicology of New Industrial Chemical Substances*, N. F. Ismerov and I. V. Sanotsky, Eds., Meditsina, Moscow, Issue No 13, 1973, 154 (in Russian).

THIODIPROPIONIC ACID, DIALKYL ESTER

Molecular Formula. $C_{26}H_{50}O_4S$ - $C_{28}H_{58}O_4S$
M = 458.74 to 490.82

Synonym. Dialkyl-3,3'-thiodipropionate.

Properties. White, crystalline powder or clear, colorless, oily liquid. Poorly soluble in water, soluble in alcohol.

Applications. Antioxidant with plasticizing properties. A synergistic additive for various thermo- and light stabilizers of polypropylene and polyvinyl chloride. Stabilizer of packaging film for food products.

Acute Toxicity. Rats and mice tolerate the dose of 20 g/kg BW.

Repeated Exposure failed to reveal cumulative properties. Rats were dosed with 200 mg/kg BW and mice with 150 to 300 mg/kg BW. The treatment produced no changes in the CNS, hematology analyses, or blood enzyme activity. Histology examination failed to reveal changes in the visceral organs.

Allergenic Effect. Does not cause sensitization when applied to the skin of guinea pig.

Standards. ***Russia***. PML: *n/m.*

Reference:

Mikhailets, I. B., Toxic properties of some stabilizers for plastics, *Plast. Massy*, 12, 41, 1976 (in Russian).

3,3'-THIODIPROPIONIC ACID, DIDODECYL ESTER

Molecular Formula. $C_{30}H_{58}O_4S$
M = 514.94
CAS No 123-28-4
RTECS No UF8000000

Synonym and **Trade Names.** Advastab 800; Didodecyl-3,3'-thiodipropionate; Dilauryl-3,3'-thiodipropionate; 3,3'-Thiodipropionic acid, didodecyl ester.

Properties. White, crystalline powder. Poorly soluble in water.

Applications. Used as a sinergistic additive to thermo- and light-stabilizer in the production of polyolefins; an antioxidant with plasticizing properties.

Acute Toxicity. Maximum tolerated dose in mice and rats is reported to be 2.0 g/kg BW[05] or even 15 g/kg BW.[08] Gross pathology examination failed to reveal changes in the visceral organs.

Repeated Exposure. In a 100-day study, dogs were given a 3.0% mixture of thiodipropionic acid and T. (1:9) in their diet; the mixture was first heated to 190°C. Gross pathology examination failed to reveal changes in the visceral organs. Mice received 20 administrations of 625 mg/kg BW. The treatment caused no mortality or morphological changes in the visceral organs.[08]

Long-term Toxicity. In a 2-year study, rats received 1.0 and 3.0% T. in their diet. This treatment caused retardation of growth. 0.5 to 1.3% T. added to the diet for 2 years had no toxic effect on growth, survival rate, or structure of the viscera in rats.[05]

In a 6-month study, rats were dosed by gavage with 430 and 1420 mg T./kg BW. The treatment caused excitation of the CNS cortical areas only. There were no gross pathology changes in the visceral organs.[08]

Reproductive Toxicity. Affects fish reproduction processes and the development of chick embryos.[08]

Chemobiokinetics. Lauryl thiodipropionate is readily absorbed from the GI tract in mammals and removed from the body in the urine in the form of 3,3'-thiodipropionic acid. According to Lefaux, T. is absorbed with great difficulties, and eliminated predominantly with feces, partially with urine.[05]

Regulations. ***U.S. FDA*** (1998) approved the use of T. in the manufacture of resinous and polymeric coatings for the food-contact surface of articles intended for use in producing, manufacturing, packing, transporting, or holding food in accordance with the conditions prescribed by 21 CFR part 175.300. Affirmed as *GRAS.*

Recommendations. Joint FAO/WHO Expert Committee on Food Additives. ADI: 3.0 mg/kg BW.

3,3'-THIODIPROPIONIC ACID, DIOCTODECYL ESTER

Molecular Formula. $C_{42}H_{84}O_4S$
M = 683.30
CAS No 693-36-7
RTECS No UF8010000

Synonyms and **Trade Name.** Advastab 802; Distearyl thiodipropionate.

Properties. White flakes insoluble in water.

Applications. An antioxidant for polyethylene, polypropylene, and rubbers. A plasticizer.

Acute Toxicity. Mice and rats tolerated ingestion of 2.0 g/kg BW.[05]

Long-term Toxicity. Signs of toxicity were observed in rats given 3.0% T. in their feed for 2 years. Administration of 3.0% T. in the diet caused retardation of BW gain.[05]

Chemobiokinetics. T. is poorly absorbed from the GI tract of dogs, rabbits, and rats. Excreted with the urine and feces.[05]

Regulations. ***U.S. FDA*** (1998) approved the use of T. in the manufacture of resinous and polymeric coatings for the food-contact surface of articles intended for use in producing, manufacturing, packing, processing, preparing, treating, packaging, transporting, or holding food in accordance with the conditions prescribed by 21 CFR part 175.300.

THIODIVALERIC ACID, DIALKYL ESTER

Molecular Formula. $[RCOO(CH_2)_4]_2S$, where R = alkyl radical

Synonym. Dialkyl-3,3'-thiodivaleriate.

Applications. Used as a stabilizer in the manufacture of food-contact plastics.

Toxicity and **Allergenic Effect.** See ***Thiodipropionic acid, dialkyl ester***.

Standards.

Russia. PML: *n/m*.

British Standard (1992). Didodecyl 3,3'-thiodipropionate, ditetradecyl 3,3'-thiodipropionate, dioctadecyl 3,3'-thiodipropionate are listed as antioxidants for polyethylene and polypropylene compositions used in contact with foodstuffs or water intended for human consumption (max. permitted level for the final compound is 1.0%).

Reference:

Mikhailets, I. B., Toxic properties of some stabilizers for plastics, *Plast. Massy*, 12, 41, 1976 (in Russian).

THIOHYDROXYETHYLENEDIOCTYL STANNANE

M = 506.23

Properties. Odorless, white, crystalline powder. Poorly soluble in water.

Applications. Used as a stabilizer in the production of polyvinyl chloride materials.

Acute Toxicity. LD_{50} is found to be 2.23 g/kg BW in rats, 1.8 g/kg BW in mice, and 2.02 g/kg BW in guinea pigs. Lesions are observed following a single administration of 1/5 LD_{50}.

Repeated Exposure revealed evident cumulative properties. K_{acc} is 0.7 (by Kagan). Animals given 1/20 LD_{50} and some of those given 1/100 LD_{50} died within a month.[1]

Reproductive Toxicity.

Embryotoxic effect was not observed.[2]

Standards. *Russia*. PML in food: 0.05 mg/l.

References:

1. Levitskaya, V. N., in *Proc. 4th Conf. Junior Scientists*, Abstracts, Donetsk, 1983, 250 (in Russian).
2. Levitskaya, V. N. and Badayeva, L. N., Effect of some organotin stabilizers on the ultrastructure of hepatocytes in rat study, *Gig. Sanit.*, 6, 74, 1984 (in Russian).

o-TOLUIDINE HYDROCHLORIDE

Molecular Formula. $C_7H_9N_2.ClH$

M = 143.62

CAS No 636-21-5

RTECS No XU7350000

Abbreviation. THC.

Synonyms. 1-Amino-2-methylbenzene hydrochloride; *o*-Aminotoluene hydrochloride; 1-Methyl-2-aminobenzene hydrochloride; 2-Methylaniline hydrochloride; 2-Methylbenzenamine hydrochloride.

Properties. THC occurs as a white, crystalline powder. Very soluble in water and ethanol.

Applications. Used in the manufacture of various dyes, and as an antioxidant in the production of rubber.

Exposure. In 1983, U.S. production value of THC was 11 to 21 million pounds. Main routes of potential human exposure are inhalation and dermal contact.

Acute Toxicity. LD_{50} is 2950 mg/kg BW in Osborne-Mendel rats, and 900 mg/kg BW in Sprague-Dawley rats.[1,2]

Short-term Toxicity. F344 rats and $B6C3F_1$ mice received THC in their feed at concentrations ranging from 1,000 to 50,000 ppm for 7 weeks. The treatment caused retardation of BW gain. Mortality in rats was noted at the highest dose. Histological examination revealed small amount of renal and splenic pigmentation in rats in the 12,500 ppm groups and in mice in 50,000 ppm groups.[3]

Long-term Toxicity. Male F344 rats were administered THC in their feed at doses of 5,000 ppm. The treatment caused mesothelial hyperplasia and mesothelioma after 13 or 26 weeks of dietary exposure. Liver effects included minimal hemosiderin accumulation in Kupffer cells that decreased during the recovery period. Small placental glutathione *S*-transferase-positive foci of cellular alteration were noted. There was an increase in the incidence of hematopoiesis, hemosiderin accumulation, and capsular fibrosis in the spleen. An increase in spleen weights and hyperplasia of the transitional epithelium in the urinary bladder were also reported.[3]

Mutagenicity. THC was genotoxic in a variety of *in vivo* and *in vitro* assays, namely, in bacterial mutation tests, yeast aneuploidy tests, and *Drosophila* somatic recombination assays.[4]

In vivo cytogenetics. THC induced micronuclei in bone marrow cells of male F344 rats but not of male $B6C3F_1$ administered 400 mg THC/kg BW by *i/p* injections.[5] It induced SCE but not CA or micronuclei in bone marrow cells of $B6C3F_1$ administered THC by *i/p* injections (150 and 600 mg THC/kg BW).[6]

Carcinogenicity. Male F344 rats were fed 0.028 mol THC/kg diet for 72 weeks. The treatment caused tumors of the bladder and liver, peritoneal tumors and fibroma of the skin and spleen.

o-THC induced more mammary tumors (13/30) than *o-nitrosotoluene,* one of its metabolites (3/30). The results indicate that *N*-oxidation is important in the induction of bladder and liver tumors.[7]

F344 rats received 3,000 and 6,000 ppm THC, $B6C3F_1$ mice received 1,000 or 3,000 ppm in their feed for about two years. In rats, there was an increase in the incidence of benign and malignant tumors (sarcomas) of the spleen, mesotheliomas of the abdominal cavity or scrotum in males, and transitional-cell carcinomas of the urinary bladder in females. The treatment caused an increased incidence of fibromas of the subcutaneous tissue in males and fibroadenomas or adenomas of the mammary gland in females. In mice, an increased incidence of hemangiosarcomas in males and hepatocellular carcinomas or adenomas in females was also noted.[3]

Carcinogenicity classifications. An IARC Working Group concluded that there is sufficient evidence for the carcinogenicity of THC in *experimental animals* and there is inadequate evidence for the carcinogenicity of THC in *humans.*

IARC: 2B;

U.S. EPA: B2.

Chemobiokinetics. Male Sprague-Dawley rats received a single oral dose of 500 mg ^{14}C-THC (methyl carbon labeled)/kg BW. Metabolism was primarily on the aromatic ring. More than a half of the administered dose was eliminated as phenol conjugates and about 40% of the administered dose was excreted as parent compound.[8]

Elimination occurred predominantly in the urine.

References:

1. Lindstrom, H. V., Bowie, W. C., Wallace, W. C., Nelson, A. A., and Fitzhugh, O. G., The toxicity and metabolism of mesidine and pseudocumidine in rats, *J. Pharmacol. Exp. Ther.,* 167, 223, 1969.
2. Jacobson, K. H., Acute oral toxicity of mono- and di-alkyl ring-substituted derivatives of aniline, *Toxicol. Appl. Pharmacol.*, 22, 153, 1972.
3. *Bioassay of o-Toluidine Hydrochloride for Possible Carcinogenicity*, NCI Technical Report Series No 153, Research Triangle Park, NC, NIH Publ. 79-1709, 1979.
4. Danford, N., The genetic toxicology of *ortho*-toluidine, *Mutat. Res.*, 258, 207, 1991.
5. *Comparative Toxicity and Carcinogenicity Studies of o-Nitrotoluene and o-Toluidine Hydrochloride Administered in Feed to male F344/N Rats,* NTP Technical Report Series No 44, Rsearch Triangle Park, NC, NIH Publ. 96-3936 (M. R. Elwell), March 1996, 74.

6. McFee, A. F., Jauhar, P. P., Lowe, K. W., MacGregor, J. T., and Wehr, C. M., Assays of three carcinogen/non-carcinogen chemicals pairs for *in vitro* induction of chromosome aberrations, sister chromatid exchanges and micronuclei, *Environ. Molec. Mutagen*, 14, 207, 1989.
7. Hecht, S. S., El-Bayoumy, K., Rivenson, A., and Fiala, E., Comparative carcinogenicity of *o*-toluidine hydrochloride and *o*-nitrosotoluene in F-344 rats, *Cancer Lett.*, 16, 103, 1982.
8. Cheever, K. L., Richards, D. E., and Plotnick, H. B., Metabolism of *ortho-*, *meta-*, and *para*-toluidine in the adult male rat, *Toxicol. Appl. Pharmacol.*, 56, 361, 1980.

TRIBUTYL(METHACRYLOYLOXY) STANNANE

Molecular Formula. $C_{16}H_{32}O_2Sn$
M = 375.18
CAS No 2135-70-6
RTECS No WH8692000
Abbreviation. TBMS.

Synonym. Tin tributylmetacrylate.

Properties. Colorless or light-yellow, oily liquid with a distinct unpleasant odor. Taste perception threshold is 0.43 mg/l, odor perception threshold is 0.064 mg/l, practical threshold is 1.154 mg/l.

Applications. TBMS is used as a stabilizer in the production of polyvinyl chloride materials.

Exposure. Residues of tributyltin compounds were not detected in most of foods. Concentrations found in rice, sunflower oil, peanut oil, butter, cheese, or chicken meat were less than 1.0 mg/kg wet weight of these foodstuffs.[1]

Acute Toxicity. LD_{50} is reported to be 210 mg/kg BW in rats, 150 mg/kg BW in guinea pigs, and 150 mg/kg BW in rabbits. Manifestations of the toxic action included lethargy, loss of appetite, photophobia, adynamia and, in many animals, paresis of the hindlimbs.[2] Hepatotoxicity that was evaluated by the activity of ornithine carbamyl transferase in serum was noted in mice in 24 hours after oral administration of 180 mmol tributyltin compound per kg BW.[3]

Repeated Exposure revealed cumulative properties. Rats were given 5.0, 10, and 20 mg/kg BW for three weeks. Lethality was observed from the second day of the treatment. Muscular weakness, paresis of the hindlimbs, abdominal swelling, lacrimation, and salivation were observed. All the doses tested affected hematology analyses; a 20 mg/kg BW dose affected unconditioned reflex activity (STI).[2]

Long-term Toxicity. In a 6-month study, rats received TBMS in an oil solution. The treatment caused a decrease in the excretory function of the liver and an increase in the vitamin C content of the suprarenals. Higher doses caused depression of BW gain.[2]

Standards. ***Russia*** (1991). MAC and PML: 0.0002 mg/l.

References:

1. Kannan, K., Tanabe, S., and Tatsukawa, R., Occurrence of butyltin residues in certain foodstuffs, *Bull. Environ. Contam.Toxicol.*, 55, 310, 1995.
2. Tsai, V. N., Maximum allowable concentration of tributyltin methacrylate in water bodies, *Gig. Sanit.*, 4, 42, 1975 (in Russian).
3. Ueno, S., Susa, N., Furukawa, Y., and Sugiyama, M., Comparison of hepatotoxicity caused by mono-, di- and tributyltin compounds in mice, *Arch. Toxicol.*, 69, 30, 1994.

N,N',N''-TRIBUTYLTHIOUREA

Molecular Formula. $C_{13}H_{28}N_2S$
M = 244.44
Abbreviation. TBTU.

Trade Name. Santowhite TBTU.

Properties. Brown, oily liquid. Poorly soluble in water, readily soluble in alcohol.

Applications. Used as an antiozone agent in the production of vulcanizates and rubber.

Acute Toxicity. LD_{50} is 3.0 g/kg BW in rats, and 4.3 g/kg BW in mice. Administration caused a narcotic effect. Death within 1 to 3 days.

Repeated Exposure failed to reveal cumulative properties. Administration of TBTU by Lim method caused no mortality in rats. The treatment slows down hippuric acid synthesis and shows a tendency to decrease BW gain.

Reference:

Stasenkova, K. P, Sergeyeva, L. G., Ivanov, V. N., and Fomenko, V. N., Assessment of toxic and embryotoxic action of *N*-butyl-2-dibutylthiourea, in *Toxicology of New Industrial Chemical Substances*, A. A. Letavet and I. V. Sanotsky, Eds., Meditsina, Moscow, Issue No 13, 1973, 82 (in Russian).

3,4,5-TRIHYDROXYBENZOIC ACID, DODECYL ESTER

Molecular Formula. $C_{19}H_{30}O_5$
M = 338.49
CAS No 1166-52-5
RTECS No DH9100000

Synonyms. Dodecyl gallate; Gallic acid, dodecyl ester; Gallic acid, lauryl ester.

Applications. Used as a stabilizer.

Acute Toxicity. In rats, LD_{50} is 6.6 g/kg BW.[1]

Repeated Exposure. Feeding a diet containing 2.0 g/kg BW for 70 days produced no effect on BW gain in rats.[1,2] However, there was no survival in rats after feeding 25 or 50 g/kg BW for 10 days.[2,3]

Short-term Toxicity. In a 13-week study, there was no adverse effect in pigs consuming a dietary level of 2.0 g/kg BW.[3,4]

In a 150-day study, rats were administered 50 mg T./kg BW by gavage. A temporary ADI of 0 to 0.05 mg/kg BW was established, based on a NOEL of 10 mg/kg BW derived in this study.[5]

Long-term Toxicity. No signs of toxicity were found in rats fed 350 to 2000 mg/kg BW in the feed.[4] Male Wistar rats were exposed by gavage to 10, 50, and 250 mg/kg BW for 150 days. Both higher doses produced changes in serum lipids and enzymes, and reduction in the spleen weight. Gross pathology examination revealed changes in the liver, kidney, and spleen. The NOAEL appeared to be 10 mg/kg BW.[6]

Reproductive Toxicity. Rats were dosed via their feed with 350 to 5000 mg/kg BW over three generations. Growth retardation and loss of litters were found, probably due to underfeeding at 5.0 g/kg BW dose level; slight hypochromic anemia was observed at 2000 mg/kg BW.[4]

Allergenic Effect. T. caused experimental sensitization in guinea pigs.[7]

Carcinogenicity. In the above described study, no increase in tumor incidence was reported.[4]

Recommendations. Joint FAO/WHO Expert Committee on Food Additives. (1997): No ADI allocated.

References:

1. Tollenaar, F. D., Prevention of rancidity in edible oils and fats with special reference to the use of antioxidants, *Proc. Pacific. Sci. Congr.*, 5, 92, 1957.
2. Joint FAO/WHO Expert Committee on Food Additives, Food Additives Series No 10, WHO, Geneva, 1979, 45.
3. van der Heijden, C. A., Janssen, P. J. C. M., and Strik, J. J. T. W., Toxicology of gallates: a review and evaluation, *Food Chem. Toxicol.*, 24, 1067, 1986.
4. van Esch, G. J., The toxicity of the antioxidants propyl, octyl and dodecyl gallate, *Voeding*, 16, 683, 1955.
5. Mikhailova, Z. N., et al., Toxicological studies of the long term effects of the antioxidant dodecyl gallate on albino rats, *Voprosy Pitaniya*, 2, 49, 1992 (in Russian).
6. Hausen, B. M. and Beyer, W., The sensitizing capacity of the antioxidants propyl, octyl, and dodecyl gallate and some related gallic acid esters, *Contact Dermat.*, 26, 253, 1992.
7. *Evaluation of Certain Food Additives and Contaminants*, Techn. Report Series No 868, 46th Report of Joint FAO/WHO Expert Committee on Food Additives, WHO, Geneva, 1997, 69.

2',4',5'-TRIHYDROXYBUTYROPHENONE

Molecular Formula. $C_{10}H_{12}O_4$
M = 196.22

CAS No 1421-63-2
RTECS No EU5425000
Abbreviation. THBP.

Properties. Yellow-tan crystals, slightly soluble in water, soluble in alcohol.

Applications. Antioxidant for polyolefins and paraffin waxes, food additive. Antioxidant for vegetable oils and animal fats.

Acute Toxicity. LD_{50} *i/p* is 200 mg/kg BW in mice.[1]

Mutagenicity.

In vitro genotoxicity. THBP was found to be positive in *Salmonella* mutagenicity bioassay without metabolic activation system (*S9* fraction).[2]

Regulations. ***U.S. FDA*** (1993) approved the use of THBP (1) in adhesives as a component of articles intended for use in packaging, transporting, or holding food in accordance with the conditions prescribed in 21 CFR part 175.105; and (2) as an antioxidant in the manufacture of food-packaging materials in accordance with the conditions prescribed in 21 CFR part 178.2010. 0.005% is maximum addition to food.

References:

1. Natl. Technical Information Service AD277-689.
2. The 17th Annual Meeting of the Environmental Mutagen Society, Baltimore, Maryland, April 9-13, 1986, Abstracts, *Environ. Mutagen.*, 8 (Suppl 7), 1, 1986.

1,3,5-TRIMETHYL-2,4,6-TRIS(3',5'-*tert*-BUTYLHYDROXYBENZYL) BENZENE

Molecular Formula. $C_{54}H_{78}O_3$
M = 775.32
CAS No 1709-70-2
99346-90-4
RTECS No DC3750000

Synonym and **Trade Name.** Agidol 40; Antioxidant 330; AO-40; Ionox 330; 2,4,6-Tri-(3',5'-di-*tert*-butyl-4'-hydroxybenzyl)mesitylene.

Properties. White, crystalline powder. Poorly soluble in water. Solubility in oil is 4.0%.

Applications. Used as an antioxidant in the production of polyolefins and polyamides.

Acute Toxicity. Administration of 10 g/kg BW does not cause death in rats. Poisoning is characterized by a weak narcotic effect and adynamia in 1 to 3 hours after administration. Acute effect threshold appeared to be 1.0 g/kg BW for STI and body temperature.[1] LD_{50} is found to be 1.5 g/kg BW in rats.[027]

Repeated Exposure. Administration of 1.0 g/kg BW by Lim method did not cause functional changes or death. There was BW gain decrease and reduction in diuresis and hippuric acid contents in the urine after exposure to sodium benzoate.[1]

Short-term Toxicity. Rats received 1.0% T. in their diet for 90 days. The treatment caused no hematology and histology changes in rats.[2]

Long-term Toxicity. A dose of 100 mg/kg BW caused functional changes in the CNS, liver, and kidneys of rats. Pathological changes were noted in the stomach. A dose of 20 mg/kg BW is considered to be the NOAEL.[1]

Reproductive Toxicity. A dose of 5.0 mg/kg BW was subthreshold in terms of effect upon the body of pregnant animals, fetus, and progeny; in terms of ***gonadotoxic effect***, a 100 mg/kg BW dose was ineffective.[1]

Allergenic Effect was not found.[1]

Chemobiokinetics. T. is not absorbed in the body. After administration, ^{14}C-labeled T. is detected unchanged in the feces. Urine, expired air, bones, and other tissues show absence of radioactivity (Wright et al., 1964).

Regulations. ***U.S. FDA*** (1998) approved T. for use (1) at levels not to exceed 0.5% by weight of the polymers except nylon resins identified in 21 CFR; (2) at levels not to exceed 1.0% by weight of nylon resins identified in 21 CFR; as (3) a stabilizer in polymers used in the manufacture of articles intended for use in producing, manufacturing, packing, processing, preparing, treating, packaging, transporting, or holding food, subject to the provisions prescribed in CFR part 178.2010.

Standards. ***Russia.*** Recommended PML in milk: 1.25 mg/l.[1]

References:

1. Stasenkova, K. P. and Samoilova, L. M., Data on substantiation of permissible migration level for stabilizer AO-40 (Agidol) from rubbers and plastics, *Gig. Sanit.*, 9, 31, 1978 (in Russian).
2. Stevenson, D. E., Chambers, P. L., and Hunter, C. G., Toxicity studies with 2,4,6-tri-(3',5'-di-*tert*-butyl- 4'-hydroxybenzyl)mesitylene in the rat, *Food Cosmet. Toxicol.*, 3, 281, 1965.

2,4,6-TRIS(1,1-DIMETHYLETHYL)PHENOL

Molecular Formula. $C_{18}H_{30}O$
M = 262.43
CAS No 732-26-3
RTECS No SN3570000
Abbreviation. TDMEP.

Synonym and **Trade Names.** Agidol 1; Alkofen B; Antioxidant P-21; P-23; 2,4,6-Tri-*tert*-butylphenol.

Properties. White, crystalline powder with a yellowish tint. Solubility in water is 0.00005%. Soluble in ethanol.

Applications. Used as a stabilizer in the production of polyolefins, impact resistant polystyrene and synthetic rubber, and also of acetyl cellulose and acetobutyrate cellulose etrols.

Acute Toxicity. In mice, the LD_{50} is 1.6 g/kg BW.[1] Histology examination revealed congestion, edema and desquamation of the esophageal and gastric mucosa, necrosis of the small intestine wall, and parenchymatous dystrophy of the renal tubular epithelium.

Short-term Toxicity. In a 4-month rat study, a dose of 88 mg/kg BW caused slight CNS excitation. Histology examination revealed parenchymatous dystrophy of the renal tubular epithelium.[1]

Long-term Toxicity. Wistar rats were fed diets containing 30 to 1000 ppm TDMEP for up to 24 months. The treatment with doses of 100 to 1000 ppm TDMEP in the diet caused slight microcytic anemia, changes in some biochemical parameters of liver function, and focal necrosis of hepatocytes. Elevations of serum phospholipids and cholesterol levels were noted.[2]

Reproductive Toxicity.

Embryotoxic effect was not observed in the experiment with chick embryos. Uncertain evidence of ***teratogenicity*** is reported.

Mutagenicity. TDMEP is structurally close to butylated hydroxytoluene that exhibits a mutagenic effect in the form of pathologic changes in the sperm. It exerts a modulating effect as related to mutagens and teratogens and antimutagen action as related to benzo[a]pyrene.[3]

Carcinogenicity. In the above described study no neoplastic responses are reported.[2]

Chemobiokinetics. TDMEP is shown to be rapidly absorbed in the GI tract of Sprague-Dawley fasting rats. It appears immediately in the blood where its maximum concentration is attained in 15 to 60 min. TDMEP or its metabolites are not found in the urine of rats; one metabolite is found in the feces and bile.[4]

Standards. ***Russia.*** PML in food: 2.0 mg/l.

References:

1. Broitman, A. Ya., Comparative toxicity of some stabilizers applied to raise thermo- and photoresistance of polymeric materials, *Gig. Truda. Prof. Zabol.*, 2, 20, 1962 (in Russian).
2. Matsumoto, K., Ochiai, T., Sekita, K., Uchida, O., Furuya, T. and Kurokawa, Y., Chronic toxicity of 2,4,6-tri-*tert*-butylphenol in rats, *J. Toxicol. Sci.*, 16, 167, 1991.
3. Malkinson, A. M., Review: Putative mutagens and carcinogens in foods. III. Butylated hydroxytoluene (BHT), *Environ. Mutagen.*, 5, 353, 1983.
4. Takahasi, O. and Hiraga, K., Metabolic studies in the rat with 2,4,6-tri-*tret*-butylphenol: a haemorrhagic antioxidant structurally related to BHT, *Xenobiotika*, 13, 319, 1983.

1,1,1-TRIS(HYDROXYMETHYL)ETHANE

Molecular Formula. $C_5H_{12}O_3$
M = 120.2

CAS No 77-85-0

Abbreviation. THME.

Synonyms and **Trade Names.** 2-(Hydroxymethyl)-2-methyl-1,3-propanediol; Methyltrimethylolmethane; Pentaglycerine; Pentaglycerol; Trimethylolethane.

Properties. Colorless crystals. Readily soluble in water and alcohol.

Applications. Used in the manufacture of alkyd surface coatings and polyester resins (polyurethane foams). Heat stabilizer for polyvinyl chloride resins.

Regulations. ***U.S. FDA*** (1998) approved the use of THME in the manufacture of cross-linked polyester resins which may be used as articles or components of articles intended for repeated use in contact with food in accordance with the conditions prescribed in 21 CFR part 177.2420.

ZINC compounds

	CAS No	RTECS No
Zinc acetate hydrate	5970-45-6	ZG8750000
Molecular Formula. $C_4H_6O_4.Zn._2H_2O$		
M = 219.51		
Zinc caprylate	557-09-5	RH0790000
Molecular Formula. $C_{16}H_{30}O_4.Zn$		
M = 351.84		
Synonym. Octanoic acid, zinc salt.		
Zinc chloride	7646-85-7	ZH1400000
Molecular Formula. $Cl_2.Zn$		
M = 136.27		
Trade Name. Butter of zinc.		
Zinc oxide	1314-13-2	ZH4810000
Molecular Formula. ZnO		
M = 81.38		
Trade Name. Flowers of zinc; Philosopher's wool; Zinc white.		
Zinc sulfate	7733-02-0	ZH5260000
Molecular Formula. $O_4S.Zn$		
M = 161.43		

Synonyms and **Trade Names.** White vitriol; Zinc vitriol; White cooperas; CI Pigment White 7; Zinc blende; Zinc monosulfide.

Properties.

Zinc *acetate dihydrate.* Crystals with faint acetous odor and astringent taste. Readily soluble in water and alcohol.

Zinc caprylate. Crystalline substance. Sparingly soluble in boiling water.

Zinc nitrate. Colorless crystalline solid.

Zinc oxide. White or cream, fine and soft powder with faint odor. Hexagonal crystals. Poorly soluble in water and ethanol.

Zinc sulfate monohydrate. Powder or granules. Soluble in water, poorly soluble in alcohol.

Zinc sulfide. Colorless hexagonal or cubic crystals. Solubility in water is 7.0 mg/l at 18°C.

Zinc salts, such as zinc chloride, zinc nitrate, zinc sulfate are soluble in water. Levels in drinking water above 3.0 mg/l give an undesirable astringent taste and may result in discoloration. The threshold perception concentration for the effect on the organoleptic properties of water is 5.0 mg/l. Cohen et al. showed that 5.0% of the population distinguish water not containing zinc from water containing it (as zinc sulfate) at a concentration of 4.3 mg/l.[1] At a concentration more than 5.0 mg/l, water becomes opalescent, and an oily film may form on boiling.

Zinc hydrosulfite. White amorphous solid with the slight odor of sulphur dioxide. Readily soluble in water (400 g/l at 20°C).

Applications. Zinc compounds are widely used as stabilizers.

Zinc oxide is used in the rubber manufacture, and as a pigment.

Zinc nitrate is used as amino resin catalyst.

Zinc sulfate monohydrate is used for bleaching paper and in the manufacture of other zinc salts.

Zinc sulfide is a pigment in the production of plastics and paper.

Zinc hydrosulfite is used as a bleaching agent.

Exposure. The diet is normally the principal source of zinc exposure. Its content in a typical mixed diet of North American adults has been reported to vary between 10 and 15 mg zinc/day.

Migration of zinc into aqueous and citric acid extracts of nitrile rubber SKN-26 [0.5 cm^{-1}; 100, 37, and 20°C (infusion); 1 and 24 hours] under the most rigorous simulant conditions amounted to 0.3 to 0.4 mg/l.[2] Investigation of extracts (distilled water and 3.0% lactic acid solution; 0.5 cm^{-1}; 38 to 40°C, then 28°C; 1 hour) revealed migration of zinc ions (0.53 to 5.2 mg/l) from vulcanizates based on nitrile rubber SKN-26MP. Investigation of aqueous extracts of nitrile rubber SKN-26M (0.5 cm^{-1}; 100°C, 4 hours) revealed migration of zinc ions up to 2.5 mg/l.[3]

Acute Toxicity.

Observations in man. A concentration in water of 0.7 to 2.3 g zinc/l can cause vomiting in humans.[03]

Animal studies. Animals exhibited general signs of acute metal poisoning. Manifestations of toxic action included pulmonary distress, fever, chills, and gastroenteritis. Generally, a protective mechanism occurs after consumption of more than 500 mg zinc, corresponding to 2.0 g zinc in the form of sulfate.[4]

Zinc chloride. LD_{50} is identified to be 350 to 480 mg/kg BW in rats, and 250 mg/kg BW in mice. The acute action threshold (for anemia, adynamia, and labored breathing) appeared to be 0.1 g/kg BW.

Zinc sulfate. LD_{50} is found to be 2.95 g/kg BW in rats.

Zinc acetate hydrate. LD_{50} is 2.5 g/kg BW in rats.

Zinc caprylate. LD_{50} is 2.37 g/kg BW in mice, in rats it was not attained.

Zinc nitrate. LD_{50} is 0.34 g/kg BW in mice, 1.56 g/kg BW in male rats, and 1.4 g/kg BW in female rats.

Zinc oxide. In mice, LD_{50} is 7.95 g/kg BW. Acute effect threshold is 1.0 g/kg BW. Toxic action manifestations included increased blood *Hb* concentration, changed motor activity, and reduced ceruloplasmin activity in the blood plasma. According to other data, administration of zinc oxide caused intense gastroenteritis, since severe irritation and even corrosion of the mucosa of the stomach follow the formation of zinc chloride in the stomach by its reaction with the hydrochloric acid of the gastric juice.[5-8,012]

Repeated Exposure. Accumulation of the zinc salts is unlikely to occur due to delayed absorption and rapid excretion from the body. Adult male rabbits were exposed to lead (0.2%), zinc (0.5%), or to both lead and zinc (0.2% and 0.5%) which were given as acetates in their drinking water for 2 or 4 weeks. Zinc did not prevent peripheral neuropathy or metabolic alterations caused by lead.[9] Retardation of BW gain was noted at the dose levels of 250 mg and 1000 mg zinc/kg BW.

Short-term Toxicity. Mice and rats were dosed by gavage with zinc sulfate at dose levels of 0.3, 3.0, and 30 g/kg in their feed for 13 weeks. The maximum dose caused decreased BW gain, reduction in food and water consumption (mice), erythrocyte count, enzyme activity, cholesterol and glucose content. Gross pathology examination revealed morphology changes in the GI tract, spleen and kidneys, and decreased relative weights of the visceral organs. The NOAEL appeared to be 458 mg/kg BW in male mice, 479 mg zinc/kg BW in female mice, and 240 BW mg/kg in rats.[10]

In a 4-month study in rats, the LOEL is reported to be 3.5 mg zinc oxide/kg BW.[11] Zinc content is found to be the highest in the blood, bones, liver, and kidneys indicating a dose-related effect.[12]

Long-term Toxicity.

Observations in man. The major consequence of chronic ingestion of zinc for medical purposes is copper deficiency and anemia as a result of the intestinal interaction of zinc and copper.[13] Long-term administration of 150 mg daily produced the same result,[14] but some opposite results were reported after equal doses given over a period of 6 weeks to healthy volunteers.[15]

Animal studies. Zinc toxicosis has been observed in various mammalian species, including ferrets, sheep, cattle, horses, and dogs.[16] Signs of toxicosis among 95 calves began to appear when calves were fed approximately 1.2 to 2.0 g zinc/day and exposed to a cumulative zinc intake of 42 to 70 g/calf.[17]

In a 6-month study, male rats were dosed by gavage with 5.0 mg zinc chloride/kg BW. In this study carried out in Russia, the treatment is reported to cause changes in the activity of liposomal enzymes in the liver, gonads, and to a lesser extent, in the kidneys.[18]

Rats were given 25 mg zinc chloride/kg BW for 9 months. Manifestations of the toxic action included retardation of BW gain at the beginning of the study, a persistent lowering of the STI and of motor activity, and an increase in the blood erythrocyte count. This dose also caused a significant and persistent rise in arterial pressure and changes in ECG. Histology examination revealed foci of vacuolar, large fatty dystrophy in the liver. The NOAEL of 1.0 mg/kg BW was established in this study.[19]

Rats were dosed with 5.0 mg zinc oxide/kg BW for 6 months. Histology examination revealed hyaline cylinders in the kidneys and moderate hyperplasia in the white pulp of the spleen. A dose of 5.0 mg/kg BW was considered to be the LOAEL.[17]

Concentrations up to 0.25% zinc sulfide in diet do not produce signs of toxicity in rats. At levels above this homeostatic mechanism breaks down.[20]

Decreases in serum copper concentrations were noted in rats exposed to daily use of 50 mg zinc/day supplements over a period of 6 to 10 weeks. Adverse effects of supplemental zinc (50 to 200 mg/day) were seen on copper bioavailability. Based on these data, 50 mg zinc/kg/day supplemental zinc is considered a minimal LOAEL for effects on human health. RfD of 0.3 mg zinc/kg BW can be derived (except infants and children). Support of zinc RfD is provided by data that indicate possible adverse effects of zinc supplementation (50 to 300 mg/day) on serum high-density lipoprotein concentrations (a risk factor for cardiovascular disease).[21]

Immunotoxicity. Zinc is essential for the function of the immune system.[22]

An intake of 150 mg zinc by healthy males twice a day in food and drink for 1.5 months caused a reduction in lymphocyte and bacterial phagocytic activity.[23] A high zinc content in water and food depressed the immune system, whereas *i/v* administration of 1.0 mg/kg BW stimulated the immune response.[24] Zinc slows down histamine release and alleviates allergic effect.

Reproductive Toxicity.

Gonadotoxicity. High dietary levels of zinc sulfate decreased the conception rate in rats, but no embryotoxic effect was observed.[25]

Rats received 5.0 mg zinc chloride/kg BW in their drinking water for six months. The treatment caused a 40% reduction in the activity of the lactate dehydrogenase in the seminal fluid and a 43% reduction of pyruvic acid activity. The gonadotoxic effect of zinc could be explained by its selective accumulation in the sperm.[26]

In a 4-month study, functional and morphological changes in spermatogenesis were observed in rats given 2.5 mg zinc or zinc oxide/kg BW.[27]

Embryotoxicity. Addition of 0.4% zinc to the diet of pregnant animals caused a decrease in the fetal weights and reduced the activity of cytochrome oxidase in the liver. There were no deviations in the levels of zinc, copper, iron, calcium, and magnesium in the fetal tissues. Before mating and during gestation zinc led to a decrease in the number of implantation sites in rabbits.[28]

Embryolethal effect was noted in rats given drinking water containing 10 mg zinc chloride/l during pregnancy.[29]

Mutagenicity.

Observations in man. Subtoxic doses of zinc salts are reported to cause severe CA in human lymphocyte culture and may be the cause of the mutagenic effect found among workers in the zinc industry experiencing calcium deficiency in their food intake.[30] However, there are no direct indications that zinc is a mutagen in man or in mammals.

In vivo cytogenetics. Mutagenic effect was noted in rats given 5.0 mg zinc/kg BW (6 months) and 100 mg zinc/kg BW (1 month).[15] Zinc chloride does not increase the percentage of DLM.

In vitro genotoxicity. Zinc sulfate was negative in *Salmonella* mutagenicity assay and in micronucleus test.[31]

Carcinogenicity.

Observations in man. No increased mortality was shown in any type of cancer among zinc refinery workers.

Animal studies. Zinc compounds are unlikely to be carcinogenic, no matter what route of administration, except by intratesticular injection, was applied.[32]

However, according to other data, pulmonary reticulosarcomata, seminomata and testicular tumors were observed 18 to 24 months after administration of zinc metal and zinc chloride to rats *i/p, i/a, s/c* and into the testis.[33] Tumor localization was not related to the route of administration. Forelimb fibromata are reported to occur in 3 out of 5 rats given zinc chloride with water at a dose of 5.0 mg/kg BW for 16 to 18 months.[34]

Carcinogenicity classifications. An IARC Working Group concluded that there are no data available to evaluate the carcinogenicity of zinc compounds in *experimental animals* and in *humans*.

IARC: D (zinc chromate);

U.S. EPA: D (zinc chloride).

Chemobiokinetics. Man absorbs about 20 to 30% of the dietary zinc. Absorption occurs throughout the small intestine; jejunum has the highest rate of absorption. Zinc is found in breast milk and is readily absorbed by neonate.[36,37]

According to Engstrom et al., zinc is readily absorbed in the GI tract of newborn and sexually mature rats. In the GI tract, zinc compounds are shown to form carbonates, which are poorly soluble and slowly absorbed. Increased zinc intake with food does not considerably raise zinc content in the blood because of its relatively rapid excretion in the feces.[38]

Zinc is distributed mainly in the liver, kidneys, bones, retina, prostate, and muscle.[35] It is shown to be accumulated in the mitochondria of the liver and affect mitochondrial function.

If free gastric acidity is high, the ingestion of zinc sulfide may result in their decomposition to hydrogen sulfide in stomach, with subsequent systemic toxicity.[019]

In rats given a daily dose of 1.4 mg zinc chloride, content of zinc in the liver increased in the first 17 days but more prolonged administration resulted in normalization of its concentration. A mechanism by which zinc absorption is reduced is associated with the exhaustion of metallothionein synthesis in the intestinal mucosa.[39] Erythrocytes contain most of zinc in blood; 10 to 20% is found in plasma.[40]

The excretion takes place via GI tract (1 to 2 mg/day) and via urine (about 0.5 mg/day) and by sweating (about 0.5 mg/day).[36] The bulk of zinc taken orally is excreted in the feces (90%) and the remainder in the urine (10%). On average, 3.0 to 4.0 mg of zinc are excreted in 24 hours.

Regulations. ***U.S. FDA*** (1998) approved the use of zinc compounds (1) in the production of resinous and polymeric coatings in accordance with the conditions prescribed in 21 CFR part 175.300; and (2) as the components of adhesives to be safely used in a food-contact surface in accordance with the conditions prescribed in 21 CFR part 175.105.

Zinc oxide is considered as *GRAS* and may be used (3) as a component of waterproof coatings at a level not to exceed 1.0% by weight of the dry product; and (4) as a colorant of the uncoated or coated food-contact surface of paper and paperboard intended for use in producing, manufacturing, packaging, processing, preparing, treating, packing, transporting, or holding aqueous and fatty foods in accordance with the conditions prescribed in 21 CFR part 176.170.

Zinc hydroxide may be used (5) as a defoaming agent in the manufacture of paper and paperboard intended for use in packaging, transporting, or holding food in accordance with the conditions prescribed in 21 CFR part 176.210.

Zinc sulfide may be used (6) as filler for preparation of rubber articles intended for repeated use in contact with food; as a component of food-contact adhesives in accordance with the conditions prescribed in 21 CFR part 177.2600; (7) as lubricants with incidental food contact intended for used on machinery used for producing, manufacturing, packaging, processing, preparing, treating, packing, transporting, or holding food.

Zinc hydrosulfide is regulated (8) as a substance employed in the production of or added to textiles and textile fibers intended for use in contact with food in accordance with the conditions prescribed in 21 CFR part 177.2800.

Zinc salicylate is listed for use (9) in rigid polyvinyl chloride and/or in rigid vinyl chloride copolymers provided that total salicylates (calculated as the acid) do not exceed 0.3% by weight of such polymers; (10) as antioxidants and/or stabilizers in polymers used in the manufacture of food-contact articles in accordance with the conditions prescribed in 21 CFR part 178.2010; and (11) in urea-formaldehyde resins in molded articles in accordance with the conditions prescribed in 21 CFR part 177.1900.

Zinc chromate and *zinc oxide* are listed as (12) colorants in the manufacture of rubber articles intended for repeated use; total use is not to exceed 10% by weight of the rubber article in accordance with the conditions prescribed in 21CFR part 178.3297.

Zinc-calcium stearate may be used (13) in the manufacture of closures with sealing gaskets for food containers intended for use in producing, manufacturing, packaging, processing, preparing, treating, packing, transporting, or holding dry food in accordance with the conditions prescribed in 21 CFR part 177.1210.

Zinc octoate may be used (14) as a component of the uncoated or coated food-contact surface of paper and paperboard intended for use in producing, manufacturing, packaging, processing, preparing, treating, packing, transporting, or holding aqueous and fatty foods in accordance with the conditions prescribed in 21 CFR part 176.170.

Zinc chloride, sulfate and *oxide* are affirmed as *GRAS*.

Recommendations.

WHO (1996). Guideline value in drinking water 3.0 mg/l (aesthetic criterion). Taking into account recent studies on humans, this concentration would not be likely to present a hazard to health but could give rise to consumer complaints of taste and appearance of the water.

Joint FAO/WHO Expert Committee on Food Additives (1982) proposed a provisional maximum TDI of 0.3 mg/kg to 1.0 mg /kg BW.

Standards.

U.S. EPA (1999). SMCL: 5.0 mg/l.

Canada (1988). MAC: 5.0 mg/l (organolept.).

Russia. State Standard for drinking water: 5.0 mg/l; MAC in milk and diary produce: 15 mg/kg; MAC in meat and poultry: 100 mg/kg; MAC in mineral water 5.0 mg/kg; MAC in children's food 5.0 mg/kg.

References:

1. Cohen, J. M. et al., Taste threshold concentrations of metals in drinking water, *J. Am. Water Work Assoc.,* 52, 660, 1960.
2. Znamensky, N. N. and Peskova, A. V., Study of migration of 3,5 benzopyrene from paraffin packaging in diary products, in *Hygiene Aspects of the Use of Polymer Materials and of Articles Made of Them*, L. I. Medved', Ed., All-Union Research Institute of Hygiene and Toxicology of Pesticides, Polymers and Plastic Materials, Kiyv, 1969, 62 (in Russian).
3. Khoroshilova, N. V. and Ol'pinskaya, A. Z., Hygienic evaluation of resins, intended for manufacture of milking apparatus, *Gig. Sanit*, 10, 95, 1988 (in Russian).
4. Prasad, A. S., Deficiency of zinc in man and its toxicity, in *Trace Elements in Health and Disease*, Vol 1, *Zinc and Copper*, A. S. Prasad, Ed., Acad. Press, New York, 1976.
5. Vorob'yova, R. S., Spiridonova, V. S., and Shabalina, L. P., Hygienic assessment of PVC stabilizers containing heavy metals, *Gig. Sanit.,* 4, 18, 1981 (in Russian).
6. Spiridonova, V. S. and Shabalina, L. P., Experimental study on zinc nitrate and zinc carbonate toxicity, *Gig. Truda Prof. Zabol*., 3, 50, 1986 (in Russian).
7. Komarova, E. N., Toxic properties of some additives for plastics, *Plast. Massy*, 12, 30, 1976 (in Russian).
8. *Poisoning Toxicology, Symptoms, Treatments*, J. M. Arena and R. H. Drew, Eds., 5th ed., Charles C. Thomas Publ., Springfield, IL, 1986, 350.
9. Hietanen, E., Aitio, A., Koivusaari, U., et al., Tissue concentrations and interaction of zinc with lead toxicity in rabbits, *Toxicology*, 25, 113, 1982.
10. Keizo, M., Masahiro, H., Kunitoshi, M., et al., *J. Pest. Sci*., 8, 327, 1981.
11. Uvarenko, A. P., in *Proc. Khar'kov Medical Institute*, Khar'kov, Issue No 114, 1974, 156 (in Russian).
12. Llobet, J. M., Domingo, J. L., Colomina, M. F., et al., Subchronic oral toxicity of zinc in rats, *Bull. Environ. Contam. Toxicol.,* 41, 36, 1988.

13. Solomons, N. W., Zinc and Cooper, in *Modern Nutrition in Health and Disease*, M. E. Shils and V. R. Young, Eds., Lea & Febinger, Philadelphia, 1988.
14. Festa, M. D., Anderson, H.-L., Dowdy, R. P., et al., Effect of zinc intake on copper excretion and retention in man, *Am. J. Clin. Nutr.*, 41, 285, 1985.
15. Samman, S. and Roberts, D. C. K., The effect of zinc supplements on plasma zinc and copper levels and the reported healthy symptoms in healthy volunteers, *Med. J. Aust.*, 146, 246, 1987.
16. Torrance, A. G. and Fulton, R.., Jr., Zinc-induced hemolytic anemia in a dog, *J. Am. Vet. Med. Assoc.*, 191, 443, 1987.
17. Graham, T. W., Goodger, W. J., Christiansen, V., et al., Economic losses from an episode of zinc toxicosis on a California veal calf operation using a zinc sulfate-supplemented milk replacer, *J. Am. Vet. Med. Assoc.*, 190, 668, 1987.
18. Merkur'yeva, R. V., Koras, G. N., Koganova, Z. I., et al., Biochemical studies on enzyme activity of lysosomes in tissues and biological fluids in the course of oral zinc treatment, *Gig. Sanit.*, 10, 17, 1979 (in Russian).
19. Shumskaya, N. I., Mel'nikova, V. V., Zhilenko, V. N., et al., Hygienic assessment of zinc ions in rubber extracts contacting with food products, *Gig. Sanit.*, 4, 89, 1986 (in Russian).
20. Casarett and Doull's *Toxicology*, Doull, J., Klassen, C. D., and Amdur, M. D., Eds., 2nd ed., McMillan Publishing Co., New York, 1980.
21. Donohue, J. M., Cantilli, R., and Steele, C. L., The reference dose for zinc, in *Toxicologist*, SOT 1996 Annual Meeting, Abstract No 941, Issue of *Fundam. Appl. Toxicol.*, 30, Part 2, March 1996.
22. Renoux, G., Renoux, M., and Guillaumin, J. M., Immunopharmacology and immunotoxicity of zinc diethyldithiocarbamate, *Int. J. Immunopharmacol.*, 10, 489, 1988.
23. Chandra, R. K., Excessive intake of zinc impairs immune response, *J. Am. Med. Assoc.*, 252, 1443, 1984.
24. Fishchenko, L. Ya., in *Coll. Problems of Immunology*, Zdorov'ya, Kiyv, Issue No 4, 1969, 66 (in Russian).
25. Pal, N. and Pal, B., Zinc feeding and conception in the rat, *Int. J. Vitam. Nutr. Res.*, 57, 437, 1987.
26. Leonov, V. A. and Dubina, T. L., in *Zinc in the Body of Man and Animals*, Nauka i Tekhnika, Minsk, 1971, 120 (in Russian).
27. Arckhangel'skaya, L. N. et al., in *Proc. All-Union Toxicol. Conf.*, Abstracts, November 25-27, 1980, Moscow, 1980, 119 (in Russian).
28. Zipper, J., Medel, M., and Prager, R., Suppression of fertility by intrauterine copper and zinc in rabbits. A new approach to intrauterine contraception, *Am. J. Obstet. Gynec.*, 105, 529, 1969.
29. Dinerman, A. A., *Role of Environmental Contaminants in Impaired Embryo Development*, Meditsina, Moscow, 1980, 192, (in Russian).
30. De Knudt, G. and Deminatti, M., Chromosome studies in human lymphocytes after *in vitro* exposure to metal salts, *Toxicology*, 10, 67, 1978.
31. Gocke, E., King, M.-T., Eckhardt, K., and Wild, D., Mutagenicity of cosmetic ingredients licensed by the EEC, *Mutat Res.*, 90, 91, 1981.
32. Sunderman, F. W., Carcinogenicity and anticarcinogenicity of metal compounds, in *Environ. Carcinogenesis*, Elsevier/North-Holland, Amsterdam, 1979, 165.
33. Dvizhkov, P. P., Blastomogenic properties of industrial metals and its compounds, *Arch. Pathol.*, 29, 3, 1967 (in Russian).
34. Krasovsky, G. N. et al., in *Metals. Hygienic Aspects of the Assessment and Improvement of the Environment*, Coll. Works, Moscow, 1983, 66 (in Russian).
35. Prasad, A. S., Clinical and biochemical manifestations of zinc deficiency in human subjects, *J. Am. Coll. Nutr.*, 4, 65, 1985.
36. Lee, H. H., Prasad, A. S., Brewer, C. J., et al., Zinc absorption in the human small intestine, *Am. J. Physiol.*, 256, G87, 1989.
37. Casey, C. E., Neville, M. C., Hambridge, K. M., Studies in human lactation: secretion of zinc, copper and manganese in human milk, *Am. J. Clin. Nutr.*, 49, 773, 1989.

38. Friedland, S. A., in *Protection of Water Reservoirs against Pollution by Industrial Liquid Effluents*, S. N. Cherkinsky, Ed., Medgiz, Moscow, 4, 277, 1960 (in Russian).
39. Dubina, T. L. and Kirichenko, L. I., in *The Biological Role of Trace Elements and Their Use in Agriculture and Medicine*, Coll. Works, Ivano-Frankovsk, Issue No 1, 1978, 112 (in Russian).
40. Sted, L. and Cousins, R. J., Kinetics of zinc absorption by luminally and vascularly perfused rat intestine, *Am. J. Physiol.*, 248, 646, 1985.
41. Donohue, J. M., Cantilli, R., and Steele, C. L., The reference dose for zinc, in *Toxicologist*, SOT 1996 Annual Meeting, Abstract No 941, Issue of *Fundam. Appl. Toxicol.*, 30, Part 2, March 1996.

CHAPTER 4. CATALYSTS, INITIATORS, HARDENERS, CURING and CROSS-LINKING AGENTS

ACETIC ACID, COBALT SALT

Molecular Formula. $C_4H_6O_4.Co$
M = 177.03
CAS No 71-48-7
RTECS No AG3150000
Abbreviation. CA.

Synonyms. Bis(acetato)cobalt; Cobalt acetate; Cobalt diacetate; Cobaltous acetate.

Applications. Oxidation catalyst.

Properties. Light-pink crystals. CA tetrahydrate has a vinegar-like odor. Readily soluble in water.

Acute Toxicity. Following administration to rats by gastric intubation, LD_{50} was found to be 503 mg/kg BW.[1] Acute effects included sedation, diarrhea, and decreased body temperature. Gross pathology examination revealed hemorrhages and dystrophic changes in the liver, kidney, and heart.

Reproductive Toxicity.

CA was not reported to be *teratogenic* in hamsters and rats, respectively.[2]

Mutagenicity. CA enhanced viral transformation frequency, and the number of transformed foci among treated cells.[3]

Carcinogenicity classification. An IARC Working Group concluded that there is evidence suggesting lack of carcinogenicity of cobalt compounds in *experimental animals* and there were inadequate data available to evaluate the carcinogenicity of cobalt compounds in *humans*.

IARC: 2B.

Chemobiokinetics. Absorption from the GI tract in rats was found to vary between 11 and 34% depending on administered dose value.[4] The highest concentration was found in the liver, a low concentration was noted in the kidney, pancreas, and spleen.[5,6]

Cobalt specifically reacts with *SH*-groups of *dihydrolipoic acid*, thus preventing oxidative decarboxylation of *pyruvate* to *acetylcoenzyme A*.[7]

Regulations. ***U.S. FDA*** (1998) listed cobalt compounds as (1) a component of the uncoated or coated food-contact surface of paper and paperboard intended for use in producing, manufacturing, packaging, processing, preparing, treating, packing, transporting, or holding aqueous and fatty foods in accordance with the conditions prescribed in 21 CFR part 176.170; and as (2) a component of adhesives for food-contact surface intended for use in packaging, transporting, or holding food in accordance with the conditions prescribed in 21 CFR part 175.105.

References:

1. Speijers, G. J. A., Krajnc, E. I., Berkvens, J. M., et al., Acute oral toxicity of inorganic cobalt compounds in rats, *Food Chem. Toxicol.*, 20, 311, 1982.
2. Leonard, A. and Lauwerys, R., Mutagenicity, carcinogenicity and teratogenicity of cobalt metal and cobalt compounds, *Mutat. Res.*, 239, 17, 1990.
3. Casto, B. C., Meyers, J., and DiPaolo, A., Enhancement of viral transformation for evaluation of the car cinogenic or mutagenic potential of inorganic metal salts, *Cancer Res.*, 39, 193, 1979.
4. Taylor, D. M., The absorption of cobalt from the gastrointestinal tract of the rat, *Physiol. Med. Biol.*, 6, 445, 1962.
5. Stenberg, T., The distribution in mice of radioactive cobalt administered by two different methods, *Acta Odontol. Scand.*, 41, 143, 1983.

6. Clyne, N., Lins, L.-E., Pehrsson, S. K., et al., Distribution of cobalt in myocardium, skeletal muscle and serum in exposed and unexposed rats, *Trace Elem. Med.*, 5, 52, 1988.
7. Alexander, C. S., Cobalt-beer cardiomyopathy, *Am. J. Med.*, 53, 395, 1972.

ACETIC ACID, ZINC SALT

Molecular Formula. C_4H_6OZn
M = 183.47
CAS No 557-34-6
RTECS No AK1500000

Synonyms and **Trade Name.** Dicarbomethoxyzinc; Zinc acetate; Zinc diacetate.

Zinc acetate, dihydrate.
CAS No 5970-45-6
RTECS No ZG8750000

Synonym. Zinc acetate hydrate.

Applications. Cross-linking agent for polymers.

Properties.

Zinc acetate. White granules with faint vinegar odor and astringent taste. 1.0 g zinc acetate dissolves in 2.3 ml water or in 1.6 ml boiling water; 1.0 g zinc acetate dissolves in 30 ml alcohol or in 1.0 ml boiling alcohol.[020]

Zinc acetate hydrate. Crystals with faint acetous odor and astringent taste. Readily soluble in water and alcohol. Levels in drinking water above 3.0 mg/l give an undesirable astringent taste and may result in discoloration. The threshold perception concentration for the effect on the organoleptic properties of water is 5.0 mg/l.

Acute Toxicity.

Zinc acetate. Large doses produced violent vomiting, purgation, and abdominal pain and collapse. Wistar rats received 1.0 and 3.0 g/kg BW. The treatment decreased hematocrit, caused an increase in erythrocyte count, and reduction in the percentage of neutrophils and monocytes.[1]

Zinc acetate hydrate. LD_{50} is 2.5 g/kg BW in rats.[2]

Repeated Exposure. Rats were fed sublethal doses of zinc acetate for 30 days. The treatment caused no depletion of glycogen in the liver. Zinc appeared to be a better stimulator of glycogen storage than was copper.[3]

Short-term Toxicity. Female rats received 160 to 640 mg zinc acetate/kg BW in their drinking water for 3 months. The highest dose caused death of some animals. Some other manifestations of toxicity were also evident.[4] Rats received orally 10 to 15 mg zinc acetate daily for 4 months. The treatment produced no signs of toxicity.[03]

Mutagenicity.

In vitro genotoxicity. Concentrations of 0.07 to 0.21 mg zinc acetate/ml are shown to increase the number of CA in a culture of human leukocytes.[5] However, there are no direct indications that zinc is a mutagen in man or in mammals.

Chemobiokinetics. Absorption by the GI tract is variable in animals and poor in humans. Accumulation occurred in the liver and pancreas. Some regulation of intake and output of *zinc* probably takes place in the intestines. In rats and mice, metallothionein, a low-molecular-mass cytoplasmic metalloprotein, takes considerable part in this process. Excreted predominantly with feces. Urinary excretion is negligible.[6,7,03]

Regulations. ***U.S. FDA*** (1998) considered zinc acetate to be *GRAS*. It may be used in adhesives as a component of articles intended for use in packaging, transporting, or holding food in accordance with the conditions prescribed in 21 CFR part 175.105.

References:

1. Mengo, M. S., Lopez, C., Frasquet, I., Ocon, C. D., and de Armino, M. V., Changes in various hematological parameters following treatment with zinc acetate, *Sangre. Barc.*, 35, 227, 1990 (in Spanish).

2. Komarova, E. N., Toxic properties of some additives for plastics, *Plast. Massy*, 12, 30, 1976 (in Russian).
3. Rana, S. V. S., Prakashi, R., Kumar, A., and Sharma, C. B., A study of glycogen in the liver of metal-fed rats, *Toxicol. Lett.*, 29, 1, 1985.
4. Llobet, J. M., Domingo, J. L., Colomina, M. F., et al., Subchronic oral toxicity of zinc in rats, *Bull. Envi ron. Contam. Toxicol.*, 41, 36, 1988.
5. Voroshilin, S. I., Plotko, E. G., Fink, T. V., et al., Cytogenic effects of inorganic compounds of tungsten, zinc, cadmium and cobalt on human and animal somatic cells, *Cytologia i Genetika*, 12, 241, 1978 (in Russian).
6. Browning, E., *Toxicity of Industrial Metals*, 2nd ed., Appleton Century Crofts, New York, 1969, 349.
7. *Handbook of the Toxicology of Metals*, 2nd ed., L. Frieberg, G. F. Nordberg, E. Kessler, and V. B. Vouk, Eds., vol 1 and 2, Elsevier Sci., Publ. B. V., Amsterdam, vol 2, 1986, 669.

See ***Zinc compounds***.

ACETYL PEROXIDE

Molecular Formula. $C_4H_6O_4$
M = 118.09
CAS No 110-22-5
RTECS No AP8500000
Abbreviation. AP.

Synonym. Diacetyl peroxide.

Properties. Colorless crystals. Slightly soluble in water.

Applications. Polymerization initiator and catalyst.

Carcinogenicity. On oral exposure, mice exhibited lymphomas development including Hodgkin's disease.[1] Equivocal tumorigenic agent by RTECS criteria.

Regulations. ***U.S. FDA*** (1998) approved AP for use as a constituent of food-contact adhesives which may be safely used as component of articles intended for use in packaging, transporting, or holding food in accordance with the conditions prescribed in 21 CFR part 175.105.

Reference:

Radiation Research, Supplement, New York, 1-7, 1959-67.

ACRYLIC ACID, 2,3-EPOXYPROPYL ESTER

Molecular Formula. $C_6H_8O_3$
M = 128.14
CAS No 106-90-1
RTECS No AS9275000
Abbreviation. GA.

Synonyms and **Trade Names.** 2,3-Epoxypropyl acrylate; Glycidyl acrylate; Glycidyl propenate; 2,3-Epoxy-1-propanol acrylate.

Applications. Cross-linking agent. An epoxy resin diluent. Monomer for polymers used in adhesives and coatings intended for use in contact with food or drink.

Properties. Insoluble in water.

Acute Toxicity. LD_{50} is 210 mg/kg BW in rats.[1]

Mutagenicity.

In vitro genotoxicity. GA was found to be positive in *Salmonella* mutagenicity assay.[2] Induced SCE in Chinese hamster lung cells.[3]

Regulations. ***EU*** (1990). GA is available in the *List of authorized monomers and other starting substances which may continue to be used for the manufacture of plastic materials and articles intended to come into contact with foodstuffs pending a decision on inclusion in Section A* (*Section B*).

References:

1. Smyth, H. F., Carpenter, C. P., Weil, C. S., Pozzani, U. C., and Striegel, J. A., Range-finding toxicity data: List VI, *Am. Ind. Hyg. Assoc. J.*, 23, 95, 1962.
2. Canter, D. A., Zeiger, E., Haworth, S., Lawlor, T., Mortelmans, K., and Speck, W., Comparative mutagenicity of aliphatic epoxides in *Salmonella*, *Mutat. Res.*, 172, 105, 1986.
3. Hude, W., Carstensen, S., and Obe, G., Structure-activity relationship of epoxides: induction of sister-chromatide exchanges in Chinese hamster *V79* cells, *Mutat. Res.*, 249, 55, 1991.

ALUMINUM compounds

	CAS No	RTECS No
Al	7429-90-5	BD0330000
Aluminum chloride (hexahydrate)	7446-70-0	BD0525000
Molecular Formula. $AlCl_3.6H_2O$		
M = 133.33		
Aluminum oxide	1344-28-1 12522-88-2 53809-96-4 74871-10-6	BD1200000
Molecular Formula. Al_2O_3		
M = 101.96		
Trade Name. Alumina.		
Aluminum hydroxide	1302-29-0 12252-70-9 51330-22-4 21645-51-2	BD0940000
Molecular Formula. AlH_3O_3		
M = 78.01		

Synonyms and **Trade Names.** Aluminum hydrate; Aluminum trihydrate; Alkagel; Antacid; Fluagel; Vanogel.

Aluminum trialkyl

Molecular Formula. AlR_3

Synonyms. Triethylaluminum; Triisobutylaluminum.

Properties.

Aluminum trialkyl. Colorless liquids at room temperature. Alkylaluminum halides are colorless, volatile liquids. Bond cleaved by water and alcohol.

Aluminum oxide is in the form of colorless crystals. Solubility in water is 0.098 mg% (18-37°C).

Aluminum hydroxide. White monoclinic crystals or white powder, balls or granules, insoluble in water and alcohol.

Aluminum chloride. White when a pure substance, ordinarily gray or yellow to greenish. Combines with water with explosive violence and liberation of much heat. Freely soluble in many organic solvents.[020] Gives water a bitter astringent taste, with perception threshold at a concentration of 0.1 mg/l; odor perception threshold is 0.5 mg/l (calculating on the aluminum ion). Odor perception threshold for aluminum nitrate appeared to be the concentration of 0.1 mg/l. The turbidity index threshold is 0.5 mg/l.

Applications.

Aluminum chloride and *nitrate, trialkyl aluminum*, etc. are used as catalysts in the production of polymeric materials (Ziegler-Natta polymerization catalysts, etc.).

Aluminum hydroxide is used as an emulsifier, ion-exchanger, detergent, in waterproofing fabrics, paper and lubricating compositions, as a desiccant powder, used in packaging materials, active filler in paper, plastics, rubber and cosmetics. It is a fire-retardant for polyethylene, polyvinyl chloride, polyester and latex composition.

Aluminum oxide is used as a filler and catalyst.

Aluminum chloride is used as an acid catalyst in the manufacture of rubber, and as a lubricant.

Exposure. Exposure to aluminum compounds comprises their use as food containers and packaging constituents, pharmaceutical preparations, food additives, and coagulants in drinking water treatment.

The major sources of dietary aluminum include several foods with aluminum additives (grain products, processed cheese, and salt) and several that are naturally high in aluminum (tea, herbs, and spices). Normally, the addition of aluminum to the diet from cooking utensils is minimal and has no toxicological significance. Storage of Coca-Cola in internally lacquered aluminum cans resulted in aluminum levels below 0.25 mg/l. Antacids increase daily intake by 0.2 to 3 g, i.e., ~ 100 times the normal intake from foodstuffs.[1]

Aluminum is found in drinking water at concentrations up to 2.5 mg/l. Concentrations and types of aluminum complexes may vary with the *pH* of water. Aluminum is present in treated drinking water in the form of low- molecular-mass reactive species.

In Switzerland, where most pans are currently made of stainless steel or teflon-coated aluminum, the average contribution for the use of aluminum utensils to the daily aluminum intake of 2 to 5 mg from the diet is estimated to be less than 0.1 mg.[2] Average dietary intake of aluminum compounds is 9.0 to 36 mg/day. It is safe to use alumi num ware for the preparation of food.

Daily intakes of aluminum, as reported prior to 1980, were 18 to 36 mg/day. More recent data, which are probably more accurate, indicate intakes of 9.0 mg/day for teenage and adult females and 12 to 14 mg/day for teenage and adult males.[3]

Migration Data. Aluminum that may migrate from aluminum utensils is probably not a major or consistent source of this element.

Acute Toxicity.

Observations in man. Aluminum may accumulate in tissue of children receiving oral aluminum.[4,5]

Animal studies. Sensitivity to different aluminum compounds in acute experiments in rats, guinea pigs, and rabbits are shown to be almost equal. Korolev and Krasovsky reported LD_{50} values for certain aluminum salts.

Table. LD_{50} values (g/kg BW) for aluminum salts.[6]

	Aluminum chloride		***Aluminum nitrate***		***Aluminum sulfate***	
Species	in terms of a salt	in terms of *Al*	in terms of a salt	in terms of *Al*	in terms of a salt	in terms of *Al*
Rats	0.32	0.18	0.28	0.21	0.41	0.32
Mice	0.39	0.20	0.37	0.19	0.82	0.41
Guinea pigs					0.49	0.40

There are other data reported: LD_{50} of $Al(NO_3)_3.9H_2O$ is found to be 3.6 g/kg BW in rats, and 3.98 g/kg BW in mice; that of $AlCl_3.6H_2O$ are 3.3 g/kg BW and 1.99 g/kg BW, respectively; LD_{50} of $Al(SO_4)_3.18H_2O$ is identified to be more than 9.0 g/kg BW in rats and mice. Death occurred in 1 to 4 days.[7]

Lethal doses of aluminum salts exhibit a broad spectrum of toxic action, affecting the CNS, blood, liver, kidneys, and gonads. Aluminum compounds are shown to cause porphyria in rats (but not in man). Alkaline phosphatase levels in rats decreased within 3 hours following oral administration of 17 mg aluminum chloride/kg BW.[8]

Repeated Exposure. Aluminum salts are not found to have evident cumulative properties. Exposure to the doses of 200 to 600 mg/kg BW for 28 days shortens the time rats can remain on a revolving column and increases their activity in the open space.[9]

Rats fed a diet containing 2,665 mg aluminum sulfate/kg BW (133 mg/kg BW) exhibited a negative phosphorus balance.[10] Male Sprague-Dawley rats were fed diets containing basic sodium aluminum phosphates and aluminum hydroxide (mean doses of 67 to 302 mg/kg BW) for 28 days. The treatment caused no changes in BW gain, hematology, and clinical chemistry. Accumulation of aluminum in the bone was not found.[11]

The U.S. National Academy of Sciences[12] developed a suggested NOAEL for 1-day and 7-day exposure to 35 mg/l and 5.0 mg/l, respectively. These values were based on an 18-day study in rats receiving an oral dose of 1.0 g/kg BW. After 18 days, decreased liver glycogen and *coenzyme A* levels were reported.[13] Histological changes in the liver and spleen, and elevated aluminum levels were noted in the heart and spleen of rats that received drinking water containing aluminum nitrate (calculated doses of 54 mg/kg BW and 108 mg/kg BW). The NOAEL of 27 mg/kg BW was identified in this study.[14]

In a 1-month study, rats were dosed with aluminum via drinking water. The treatment produced changes in serum alkaline phosphatase activity, contents of adenine nucleotides in the blood, and the functional state of the spermatozoa. The LOAEL was established to be 17 mg/kg BW for rats and mice, and 9.0 mg/kg BW for rabbits.[15]

Short-term Toxicity. Young and adult rats were fed a sucrose diet with addition of aluminum hydroxide. A decrease in serum triglyceride was observed in rats fed with 2000 ppm aluminum hydroxide in the diet for 67 days.[16]

Long-term Toxicity.

Observations in man. For many years aluminum was not considered to be toxic in humans. However, recent reports suggest that exposure to aluminum may pose a health hazard. Aluminum has been associated with dialysis osteomalacia and dialysis encephalopathy in persons with chronic renal failure who have undergone long-term hemodialysis. Its role in other neurological disorders remains controversial (Alzheimer's disease, amyotropic lateral sclerosis, and Parkinsonism dementia in Guam).

Animal studies. Available data are controversial. In a 7-month rat study, signs of general toxicity were reported only at a dose level of 5.0 mg/kg BW, but not at lower doses. An impairment of conditioned reflex activity, changes in behavior and in serum alkaline phosphatase activity were observed in rats given the dose of 2.5 mg/kg BW.[8,15]

Rats exposed to aluminum salt at a concentration of 5.0 mg/l in their drinking water had no deviations in clinical and biochemical indices.[17] The treatment did not affect their lifespan.

Rats received 5.0 and 20 mg aluminum/kg BW in their drinking water over a period more than 6 months. Aluminum concentration in blood plasma was 2- and 28-fold more than in control animals, respectively, 0.2 and 2.9 mg aluminum/g tissue. Aluminum concentrations increased in bones, liver, and kidneys. The dose of 20 mg/kg BW caused pathological changes in the kidneys and brain. Neurofibrill degeneration like in Alzheimer's disease was observed.[18]

Allergenic Effects. Addition of aluminum to the feed led to the development of allergic asthma in laboratory animals. Aluminum stearate is not reported to exhibit such effect.[10] Contact sensitizing potential is not found in maximized guinea-pig assays with aluminum chloride.[19]

Reproductive Toxicity. Aluminum compounds cross the placenta and accumulate in fetal tissues.[20] Aluminum may be a developmental toxin depending on the route of administration and/or the solubility of a compound.

Embryotoxicity. Aluminum exposure has been associated with neurotoxicity in mouse dams and their offspring. Golub et al.[22] reported weight loss and neurotoxicity on days 12 to 15 postpartum in dams fed 0.5 or 1.0 g aluminum lactate/kg (75 or 150 mg/kg BW). Offspring of rats given oral doses of 155 or 192 mg aluminum/kg BW from day 8 of gestation through parturition experienced a delay in weight and neuromotor development.[23]

Pregnant Swiss mice were given by gavage 30 mg aluminum hydrate/kg on gestation days 6 to 15. No embryotoxic or fetotoxic effects, no malformations or variations related to the treatment were observed.[24] When aluminum (0.5 to 1.0 mg/g) was added to the diets of pregnant rats from day 6 through 19 of gestation, there was an increase only in the resorption rate in those animals receiving 1.0 mg/kg BW.[22] Pregnant Swiss mice were given by gavage up to 216 mg aluminum hydroxide/kg BW on gestation days 6 to 15. No embryotoxic or teratogenic effects were observed.[25]

Teratogenic effects in rats were caused by aluminum fluoride, but not by aluminum lactate and aluminum trichloride. Oral administration of maternally toxic doses of aluminum nitrate caused dose-related abnormalities in the skeletal tissues of fetal rats.[26,27]

Gonadotoxicity.

Observations in man. The health of workers consuming daily (in food and beer) 300 to 400 mg aluminum (calculated as elementary aluminum) was monitored for 3 years. No signs of harmful effect in their health or (in women) in relation to sexual function were noted.[21]

Animal studies. In a 6-month study, rats were exposed to 2.5 mg aluminum chloride/kg BW dose. A gonadotoxic effect was found in these animals. Meanwhile, consumption of Al^{3+} in drinking water at concentrations of 0.5 to 2.0 g/l for 90 days did not affect the reproductive function of male rats.[28,29]

Mutagenicity.

In vitro genotoxicity. Aluminum compounds are shown to produce CA[30] and alter cell division in animals and humans. Aluminum chloride was negative when tested by *rec*-assay.[31]

In vivo cytogenetics. The dose of 0.025 mg/kg BW was reported to be ineffective in the DLM test.[24]

Carcinogenicity. A concentration of 5.0 mg/l in drinking water presents no risk of tumor development in rats.[17]

Chemobiokinetics. Metabolism of aluminum in humans is not well understood; it was once considered to be poorly absorbed (less than 1.0%).

Observations in man. Ingestion by volunteers of up to 125 mg/day did not reveal its accumulation in the body. Nevertheless, the new data suggest that 12 to 31% of the oral intake may be absorbed.[32]

Animal studies. Young (21 days old), adult (8 months), and old (16 months) rats were exposed to 50 and 100 mg aluminum/kg BW administered as aluminum nitrate in drinking water for a period of 6 months. Brain concentrations were found to be higher in young rats. Urinary aluminum levels of old rats tended to increase.[33] Male rats were exposed to 10, 50, or 500 mg labile aluminum/l in acidic drinking water (pH =3) for 9 weeks. It was found that labile aluminum in drinking water is complexed by feed constituents in the stomach of the rat *in vivo*, thus causing a non-detectable absorption of aluminum at 10 mg aluminum/l. An increased absorption of aluminum at 50 and 500 mg aluminum/l was associated with a saturation of the aluminum-binding capacity of feed components in the lumen of the stomach, causing the appearance of labile aluminum.

Glynn *et al.* showed that the presence of labile aluminum in drinking water does not necessarily result in a high aluminum absorption when the water is ingested, since the bioavailability of labile aluminum is dependent both on the amount and composition of aluminum-binding components present in the GI tract at the time of ingestion of the water. It is thus not possible to predict the body burden of aluminum in humans just by measuring the aluminum concentrations in drinking water.[34]

Absorption of aluminum occurs predominantly in the stomach and duodenum, in an acidic environment. The normal distribution of aluminum is the greatest in the lungs, followed by bone and, to a lesser extent, soft tissues and brain.[35] It was also found in the liver, muscles, and testes.[36] Probably, aluminum salts could exhibit their toxic effect due to some diminution in phosphorus absorption and a speeding-up of phosphorylation reactions. Aluminum is specifically absorbed by hepatocyte nuclei. The DNA of the nuclei have a specific affinity for aluminum. Most of the absorbed aluminum appears to be rapidly excreted with the urea (<0.02 mg/day).

Regulations. ***U.S. FDA*** (1998) approved the use of aluminum salts (1) in resinous and polymeric coatings for the food-contact surface of articles intended for use in producing, manufacturing, packing, transporting, or holding food (aluminum butyrate) in accordance with the conditions prescribed in 21 CFR part 175.300; (2) as components of adhesives applied in contact with foodstuffs in accordance with the conditions prescribed in 21 CFR part 175.105; (3) in cellophane to be safely used for packaging food (aluminum hydroxide and silicate) in accordance with the conditions prescribed in 21 CFR part 177.1200; (4) as a defoaming agent in acrylate ester copolymer coating to be safely used as a food-contact surface of articles intended for packaging and holding food, including heating of prepared food (aluminum hydroxide and stearate) subject to the provisions prescribed in 21 CFR part 176.200; (5) in the manufacture of rubber articles intended for repeated use in producing, manufacturing, packaging, processing, preparing, treating, packing, transporting, or holding food (aluminum hydroxide and silicate) in accordance with the conditions prescribed in 21 CFR part 1177.2600; (6) as colorants in the manufacture of articles intended for use in contact with food (aluminum, aluminum hydrate, aluminum mono-, di-, and tristearate, aluminum silicate (mica and China clay), subject to the provisions and definitions set in CFR part 178.3297; (7) as a component of acrylate ester copolymer coating which may be used as a food-contact surface of articles intended for packaging and holding food, including heating of prepared food, subject to the provisions prescribed in CFR part 175.210 (aluminum stearate); (8) as a component of the uncoated or coated food-contact surface of paper and paperboard intended for use in producing, manufacturing, packing, transporting, or holding aqueous and fatty food (aluminum acetate) in accordance with the conditions prescribed in 21 CFR part 176.170; and (9) as defoaming agent in the processing of food in accordance with the conditions prescribed in 21 CFR part 173.320 (aluminum stearate). Aluminum sulfate, ammonium sulfate, potassium sulfate, sodium sulfate, hydroxide, and oleate are considered to be *GRAS.*

Standards.

U.S. EPA (1999). Secondary MCL: 0.05 to 0.2 mg/l.

EEC (1980). MAC: 0.2 mg/l.

Russia. State Drinking Water Standard: 0.5 mg/l.

Recommendations.

WHO (1996): Guideline value for drinking water is 0.2 mg/l (this level is likely to give rise to consumer complaints of depositions, discoloration).

Joint FAO/WHO Expert Committee on Food Additives (1988) established PTWI of 7.0 mg/kg BW based on studies of aluminum phosphate (acidic). The chemical form of aluminum in drinking water is different.

References:

1. Friberg, L., Nordberg, G. F., Kessler, E., Vouk, and V. B., in *Handbook of the Toxicology of Metals*, 2nd ed., vol 2, Elsevier Science Publishers B. V., Amsterdam, 1986, 377.
2. Muller, J. P., Steinegger, A., and Schlatter, C., Contribution of aluminum from packaging materials and cooking utensils to the daily aluminum intake, *Z. Lebensm. Unters. Forsch.,* 197, 332, 1993.
3. Pennington, J. A., Aluminum contents of foods and diets, *Food Addit. Contam.,* 5, 161, 1988.
4. Grisword, W. R., Reznik, V., Mendoza, S. A., Trauner, D., and Alfrey, A. C., Accumulation of aluminum in nondialyzed uremic child receiving aluminum hydroxide, *Pediatrics*, 71, 56, 1983.
5. Hurley, J. K., Bowel obstruction occurring in a child during treatment with aluminum hydroxide gel, *J. of Pediatrics*, 92, 592, 1978.
6. Korolev, A. A. and Krasovsky, G. N., Hygienic substantiation of MAC for a new reagent aluminum oxychloride in drinking water, *Gig. Sanit.*, 4, 12, 1978 (in Russian).
7. Llobet, J. M., Domingo, J. L., Gomez, M., et al., Acute toxicity studies of aluminum compounds: antidotal efficacy of several chelating agents, *Toxicol. Appl. Pharmacol.*, 60, 280, 1987.
8. Krasovsky, G. N., Vasyukovich, L. Y., and Charyev, O., Experimental study of the biological effects of lead and aluminum following oral administration, *Environ. Health Perspect.*, 30, 47, 1979.
9. Antonovich, L. A. and Sprudzha, D. R., Synthesis and study of pharmacologically active substances, Abstract, *Republ. Sci. Conf.,* December 14, 1984, Vil'nus, 1984, 12 (in Russian).
10. Ondreicka, R., Ginter, E., and Kortus, J., Chronic toxicity of aluminum in rats and mice and its effects on phosphorus metabolism, *Brit. J. Ind. Med.*, 23, 305, 1966.
11. Hicks, J. S., Haskett, D. S., and Sprague, G. L., Toxicity and aluminum concentration in bone following dietary administration of two sodium aluminum phosphate formulations in rats, *Food Chem. Toxicol.*, 25, 533, 1987.
12. *Drinking Water and Health*, U.S. Natl. Acad. Sci., Natl. Acad. Press., Washington, D.C., vol 4, 1982, 155.
13. Kortus, J. The carbohydrate metabolism accompanying intoxication by aluminum salts in rat, *Experien tia*, 23, 912, 1967.
14. Gomez, M. et al., Short-term oral toxicity study of aluminum in rats, *Arch. Pharmacol. Toxicol.*, 12, 144, 1986.
15. Krasovsky, G. N. et al., in *Materials of the 2nd Joint Soviet-American Symposium on Environ. Health Protection*, Moscow, 1977, 34 (in Russian).
16. Sugawara, C., Sugawara, N., Kiyosawa, H., and Migake, H., Decrease in serum triglyceride in normal rat fed with 2000 ppm alum diet for 67 days. II. Feeding young and adult rats a sucrose diet

with addition of aluminum hydroxide and aluminum potassium sulfate, *Fundam. Appl. Toxicol.*, 10, 616, 1988.
17. Elliott, H. L., Dryborgh, F., Fell, G. S., et al., Aluminum toxicity during regular hemodialysis, *Brit. Med. J.*, 1, 1101, 1978.
18. Somova, L. I., Missankov, A., and Khan, M. S., Chronic aluminum intoxication in rats: Dose-dependent morphological changes, Abstract, *Meth. Find. Exp. Clin. Pharmacol.*, 19, 599, 1997.
19. Magnusson, B. and Kligman, A. M., The identification of contact allergens by animal assay. The guinea-pig maximization assay, *J. Invest. Dermatol.*, 52, 268, 1969.
20. Hall, G. S. and Carr, M. J., *Al, Ba, Si* and *St* in amniotic fluid by emission spectrometry, *Clin. Chem.*, 29, 1318, 1983.
21. Evenstein, Z. M., Allowable contents of aluminum in the food intake of adults, *Gig. Sanit.*, 2, 109, 1971 (in Russian).
22. Golub, M., Gershwin, M., Donald, J., et al., Maternal and developmental toxicity of chronic aluminum exposure in mice, *Fundam. Appl. Toxicol.*, 8, 346, 1987.
23. Bernuzzi, V., Desor, D., and Lehr, P. R., Effects of prenatal aluminum exposure on neuromotor maturation in the rat, *Neurobehavior. Toxicol. Teratol.*, 8, 115, 1986.
24. Colomina, M. T., Gomez, M., Domingo, J. L., Corbella, J., Lack of maternal and developmental toxicity in mice given high doses of aluminum hydroxide and ascorbic acid during gestation, *Pharmacol. Toxi col.*, 1994, 74, 236.
25. Domingo, J. L., Gomez, M., Bosque, M. A., and Corbella, J., Lack of teratogenicity of aluminum hydroxide in mice, *Life Sci.*, 45, 243, 1989.
26. Schardein, J. L., *Chemically Induced Birth Defects*, 2nd Ed., rev. and expanded, Marcel Decker, Inc., 1993, 843.
27. Paternain, J. L. et al., Embryotoxic and teratogenic effects of aluminum nitrate in rats upon oral administration, *Teratology*, 38, 253, 1988.
28. Dickson, R. L. et al., *Proc. 2nd Soviet-American Symposium on Environ. Health Protection*, Moscow, 1977, 41 (in Russian).
29. Dickson, R. L., Sherins, R. L., and Lee, I. P., Assessment of environmental factors affecting male fertility, *Environ. Health Perspect.*, 30, 53, 1979.
30. Oehever, F., Chromosome breaks influenced by chemicals, *Heredity*, Suppl. 6, 95, 1953.
31. Nioshioka, H., Mutagenic activities of metal compounds in bacteria, *Mutat. Res.*, 31, 185, 1975.
32. Lione, A., Aluminum toxicology and the aluminum containing medications, *J. Pharmacol. Ther.*, 29, 255, 1985.
33. Gomez, M., Sanchez, D. J., Llobet, J. M., Corbella, J., and Domingo, J. L., The effect of age on aluminum retention in rats, *Toxicology,* 116, 1, 1997.
34. Glynn, A. W., Sparen, A., Danielsson, L. G., Sundstrom, B., and Jorhem, L., Concentrtion-dependent absorption of aluminum in rats exposed to labile aluminum in drinking water, *J. Toxicol. Environ. Health,* 56, 501, 1999.
35. Ganrot, P. O., Metabolism and possible health effects of aluminum, *Environ. Health Perspect.*, 65, 363, 1986.
36. Talakina, N. V., Novikov, Yu, V., Plitman, S. I., et al., Setting norms of the aluminum in drinking water of various hardness, *Gig. Sanit.*, 12, 75, 1988 (in Russian).

AMINE CURING AGENT UP-604/1

Properties. Powder of a light-gray to dark-brown color. Poorly soluble in alcohol.

Applications. Used for rapid curing of epoxy resins by cyandiamide and guanidine.

Acute Toxicity. Rats tolerate administration of the maximum doses. The acute effect threshold was established to be 5.0 mg/kg BW for decreasing the spleen relative weights. Gross pathology examination revealed fatty dystrophy of the liver cells and parenchymatous dystrophy of the myocardial muscle cells.

Allergenic Effect is observed in response to skin application in guinea pigs.

Reference:

Talakin, Yu. N. et al., Toxicological characteristics of new amine hardeners UP-604/1, *Gig. Truda Prof. Zabol.*, 4, 57, 1986 (in Russian).

AMINE CURING AGENT UP-605/1R

Molecular Formula. $C_6H_7BF_3N$
M = 160.95
CAS No 660-53-7
RTECS No ED7360000

Synonyms. Aniline, compound. with boron fluoride (1:1); (Benzenamine)trifluoroboron.

Composition. 50% solution of a boron trifluoride-aniline complex in diethylene glycol. Contains 21.1% BF_3 and 28.9% aniline.

Properties. Soluble in water and alcohol.

Applications. Used for curing epoxy resins at 40 to 60°C.

Acute Toxicity. In rats, LD_{50} is 1200 mg/kg BW (348 mg/kg BW in terms of aniline). Lethal doses led to convulsions with subsequent death within 1 to 5 days. Gross pathology examination revealed dystrophic changes in the parenchymatous organs of animals given a single oral dose of 250 to 500 mg/kg BW.

Repeated Exposure revealed cumulative properties. K_{acc} is 1.8 (by Kagan). The treatment caused circulatory disturbance. Gross pathology examination findings included dystrophic changes in the liver, renal tubular epithelium, myocardium, nerve cells, and brain.

Allergenic Effect was evidenced by histology findings of lymphoid/eosinophil infiltration of the mucosa and submucosa of the esophagus, stomach and duodenum, hyperplasia of the peribronchial lymphoid tissue, allergic vasculitis in the lungs, and focal pulmonary emphysema.

Reference:

Talakin, Yu. N., Krivobok, G. K., Nekrasova, I. R., et al., Toxicologic characteristics of new amine hardeners UP-605/1R and UP-605/3R, *Gig. Truda Prof. Zabol.*, 12, 54, 1983 (in Russian).

AMINE CURING AGENT UP-605/3R

Molecular Formula. $C_7H_9BF_3N$
M = 174.98
CAS No 696-99-1
RTECS No ED7365000

Synonyms. Benzenemethanamine, compound with trifluoroborane (1:1); Benzylamine, compound with boron fluoride (1:1).

Composition. 50% solution of BF_3-benzylamine complex in diethylene glycol. Contains 19.4% BF_3 and 30.6% benzylamine.

Properties. Soluble in water and alcohol.

Applications. Used as a curing agent for epoxy resins at 120 to 140°C.

Acute Toxicity. In rats, LD_{50} is 1052 mg/kg BW (326 mg/kg BW in terms of benzylamine). Poisoning is accompanied by deviations in the renal function. Histology examination: see ***UP-605/1R.***

Repeated Exposure revealed pronounced cumulative properties. K_{acc} is 1.3 (by Kagan). See also ***UP-605/1R***. Gross pathology examination showed foci of myocardial micronecrosis with deposition of calcium salts, and tiny calcerous blocks in the renal tissue.

Allergenic Effect. See ***UP-605/1R.***

Reference:

Talakin, Yu. N., Krivobok, G. K., Nekrasova, I. R., et al., Toxicologic characteristics of new amine hardeners UP-605/1R and UP-605/3R, *Gig. Truda Prof. Zabol.*, 12, 54, 1983 (in Russian).

AMINE CURING AGENT UP-606/2

Molecular Formula. $C_{15}H_{27}N_3O$
M = 265.45

CAS No 90-72-2
RTECS No SN3500000

Synonym and **Trade Name.** Alkofen MA; Anchor K 54; Araldite DY 061 or DY 064, Hardener HY 960; 2,4,6-Tris(dimethylaminomethyl)phenol; α, α', α''-Tris(dimethylamino) mesitol.

Properties. Colorless, oily liquid with the odor of ammonia. Soluble in alcohol and cold water, almost insoluble in hot water.

Applications. A low-temperature curing agent used in the synthesis of epoxy resins. May be used for curing polyurethanes, polyamides, and polysulfides.

Acute Toxicity. In rats, LD_{50} is 1.2 g/kg BW. Poisoning is accompanied with immediate (in 30 to 40 minutes after administration of the lethal doses) respiratory impairment. In 24 hours there is hind limb paralysis and death.

Repeated Exposure failed to reveal cumulative properties. Rats tolerated administration of 1/10 LD_{50} for two months without impairment of the liver or kidney functions.

Standards. *Russia.* PML in drinking water: 4.0 mg/l.

Reference:

Volodchenko, V. A. and Sadokha, E. R., Toxicity of amine curing agent UP-606/2, *Pharmacologia i Toxicologia*, 37, 130, 1974 (in Russian).

AMINOESTER CURING AGENT DTB-2

CAS No 66555-02-0
RTECS No JU9496200

Composition. A product of the reaction of diethylenetetramine with butyl methacrylate.

Properties. An oily, dark-brown liquid with a specific odor. Poorly soluble in water, readily soluble in alcohol.

Applications. A curing agent in the production of epoxy resins.

Acute Toxicity. LD_{50} is 3.8 g/kg BW for mice, and 4.6 g/kg BW for rats. Manifestations of the toxic action include agitation with subsequent convulsions and skeletal muscular paralysis. Gross pathology examination revealed visceral congestion and hyperemia of the intestinal mucosa.[1]

Repeated Exposure revealed slight cumulative properties. K_{acc} is 5.0 (by Kagan).

Long-term Toxicity. Rats were dosed by gavage with 46 mg/kg BW for 10 months. The exposure caused changes in hematological analysis and conditioned reflex activity. A dose of 4.6 mg/kg BW produced alterations in the activity of a number of enzymes.[2]

Allergenic Effect. Mild sensitizing effect is observed in guinea pigs following administration of 1/5 LD_{50}.[3]

Standards. Russia. PML: 0.1 mg/l.

References:

1. Sokolovsky, N. V., Some Hygienic Aspects of the Use of Anticorrosion Coatings in Construction, in *Hygiene and Toxicology of High-Molecular-Mass Compounds and of the Chemical Raw Material Used for Their Synthesis*, Proc. 6th All-Union Conf., B. Yu. Kalinin, Ed., Leningrad, 1979, 131 (in Russian).
2. Sokolovsky, N. V. and Larionov, V. G., Hygienic norm-setting for aminoether curing agent DTB-2 migrating from the anticorrosive coatings as referred to food hygiene, *Gig. Sanit.*, 3, 21, 1982 (in Russian).
3. Sokolovsky, N. V. and Larionov, V. G., Sensitizing effects of some sulfonates used in production of the concrete pipes, in *Hygienic Aspects of the Use of Polymeric Materials,* Proc. 3rd All-Union Meeting on New Methods of Hygienic Monitoring of the Use of Polymers, December 2-4, 1981, K. I. Stankevich, Ed., Kiyv, 1981, 336 (in Russian).

N-[(2-AMINOETHOXY)METHYL]BENZENAMINE

Molecular Formula. $C_8H_{14}N_2O$
M = 154.24
CAS No 22683-30-3

Synonyms and Trade Name. *N*-[(2-Aminoethoxy)methyl]aniline; *N*-[(2-Aminoethoxy) methyl] benzenamine; Product 40-A.

Properties. Poorly soluble in water and oil.

Applications. Used as a curing agent in the synthesis of epoxy resins.

Acute Toxicity. In rats, LD_{50} is 625 mg/kg BW. Administration of 1.0 g/kg BW caused 100% mortality in animals. Lethal doses caused adynamia, apathy, and respiratory disorder. Gross pathology examination revealed necrosis of the GI tract mucosa with multiple hemorrhages, mild pulmonary congestion, and parenchymatous dystrophy of the liver and kidneys.

Reference:

Shumskaya, N. I., Volkova, A. P., et al., in *Toxicology of New Industrial Chemical Substances*, A. A. Letavet and I. V. Sanotsky, Eds., Meditsina, Moscow, Issue No 9, 1967, 132 (in Russian).

N-AMINOETHYLPIPERAZINE

Molecular Formula. $C_6H_{15}N_3$
M = 129.24
CAS No 140-31-8
RTECS No TK8050000

Synonyms and Trade Name. (2-Aminoethyl)piperazine; *N*-(2-Aminoethyl)piperazine; *N*-(β-Aminoethyl)piperazine; USAF DO-46.

Properties. Liquid. Soluble in water.

Applications. Used in the production of polyurethane foams. A cross-linking agent.

Acute Toxicity. In rats, LD_{50} is 3.05 g/kg BW.

Allergenic Effect. A reliable increase in the level of platelet formation in the blood was produced by admini stration of 1/1000 LD_{50}.

Reproductive Toxicity.

Gonadotoxicity. A decline in sperm motility time was noted in rats given a dose of 600 mg/kg BW for 28 days.

Embryotoxicity. Male rats were dosed with 0.1 and 0.01 LD_{50} for 20 days before their mating with untreated females. The treatment caused a two-fold (but unreliable) rise in pre- and post-implantation fetal mortality. Ingestion of 1/1000 LD_{50} by pregnant rats increased engorgement of the mesenteric blood vessels which became cyanotic. Accumulation of amniotic fluid in the tubal cavity and sites of placenta detachment was found. Hematomata in the inter-scapular region were noted in some embryos.

Standards. ***Russia***. Recommended PML: 0.6 mg/l.

Reference:

Mazaev, V. T., Troenkina, L. B., and Gladun, V. I., The effect of aliphatic amines on reproduction of ani mals, *Gig. Sanit.*, 10, 66, 1986 (in Russian).

AMINOIMIDAZOLINE CURING AGENTS UP-0636/1

CAS No 91450-36-1
RTECS No YQ8578600

Composition. Diethylenetriamine and methyl methacrylate (MMA), which are the main impurities in the finished products, are used in the manufacture of this agent. UP-0636/1 is based on methyl esters of soya oil and MMA.

Properties. Dark, clear liquids soluble in alcohol and water.

Acute Toxicity. LD_{50} are found to be 2.5 g/kg BW. Manifestations of the toxic action in survived animals included exudative inflammation of the GI tract. Gross pathology examination revealed dystrophic changes and circulatory disturbances in the viscera and brain.

Allergenic Effect was found with the help of allergenic test battery including radio-allergenic- sorbent test, eosinophil count, etc. in 40% guinea pigs exposed to 10% solution.

Reference:

Talakin, Yu. N., Chernykh, L. V., Nekrasova, I. A., Ivanova, L. A., and Kuznetsova, E. Ya., Toxicological characteristics of aminoimidazole hardeners of epoxy resins, *Gig. Truda Prof. Zabol.*, 6, 49, 1984 (in Russian).

AMINOIMIDAZOLINE CURING AGENTS UP-0636/2

CAS No 91933-08-3
RTECS No YQ8578700

Composition. Diethylenetriamine and methyl methacrylate (MMA), which are the main impurities in the finished products, are used in the manufacture of this agent. UP-0636/2 is based on synthetic fatty acids of the C_7-C_9 fractions and MMA.

Properties. Dark, clear liquids soluble in alcohol and water.

Acute Toxicity. LD_{50} are found to be 0.5 g/kg BW. Manifestations of the toxic action in survived animals included exudative inflammation of the GI tract. Gross pathology examination revealed dystrophic changes and circulatory disturbances in the viscera and brain.

Allergenic Effect was found with the help of allergenic test battery including radio-allergenic-sorbent test, eosinophil count, etc. in 100% guinea pigs exposed to 1.0% solution.

Reference:

Talakin, Yu. N., Chernykh, L. V., Nekrasova, I. A., Ivanova, L. A., and Kuznetsova, E. Ya., Toxicological characteristics of aminoimidazole hardeners of epoxy resins, *Gig. Truda Prof. Zabol.*, 6, 49, 1984 (in Russian).

AMINOSHALE CURING AGENT ASF-10

Composition. A product of condensation of phenols (alkylresorcinols) contained in a shale modifier, formal dehyde, and polyethylenepolyamine.

Applications. Used as a curing agent and modifier in the production of epoxy resins.

Acute Toxicity. The LD_{50} values are reported to be 5.76 g/kg BW for rats, and 6.75 g/kg BW for mice. Lethal doses cause tremor, loss of appetite, and adynamia.

Repeated Exposure failed to reveal cumulative properties in rats dosed with 1/5 LD_{50}.

Reference:

Petrusha, V. G. and Kesel'man, I. M., Toxicological evaluation of the amino-shale curing agent used in the production of epoxy-coating, in *Hygiene and Toxicology of High-Molecular-Mass Compounds and of the Chemical Raw Material Used for Their Synthesis*, Proc. 6th All-Union Conf., B. Yu. Kalinin, Ed., Leningrad, 1979, 153 (in Russian).

ANTIMONY compounds

Antimony oxide

Molecular Formula. O_3Sb_2
M = 291.50
CAS No 1309-64-4
RTECS No CC5650000

Antimony potassium tartrate

Molecular Formula. $C_8H_{10}K_2O_{15}Sb_2$
CAS No 28300-74-5
RTECS No CC6825000
Abbreviation. APT.

Synonyms. Tartare emetic; Tartrated antimony; Potassium antimonyltar trate.

Properties. Sb_2O_3 is in the form of colorless crystals. Solubility of Sb_2O_3 in water is 1.6 mg/100 g (15°C). Taste perception threshold for Sb^{3+} and Sb^{5+} compounds is 0.6 mg/l.

Applications. Antimony compounds are used as flame retardants for plastics and elastomers, as regeneration activators for vulcanizates, as pigments.

Antimony trioxide is used as a polycondensation catalyst in Lavsan (polyethylene terephthalate) production and a filler for increasing the fire-resistance of vulcanizates.

Antimony sulfate could be applied as a vulcanization accelerator and a dye for vulcanizates. *APT* is used worldwide as an anti-schistosomal drug; it was efficacious in humans only if administered *i/v* at a near-lethal dose of 36 mg/kg BW.[1]

Exposure. Dissolved antimony in natural waters is found mostly as Sb^{3+} or Sb^{5+} hydroxides. Environmental exposure occurs via water, air, food, dermal contact and urban dust, with drinking water as the most widespread route.[2] Estimated dietary intake for adults is about 0.018 mg/day.[3] Concentrations found in drinking water were less than 0.005 mg/l.[4]

Migration Data. Antimony was found to migrate from food-contact polyethylene terephthalate in olive oil at the level of 4.0 µg/kg. The level which migrated from polymer to simulant media was less than a proposed limit of migration.[5]

Acute Toxicity.

Observations in man. Doses of 0.1 to 1.0 g of antimony compounds are lethal to man.[6] Acute intoxication is characterized by abdominal and muscular pains, diarrhea, and hemoglobinuria. Poisoning of 250 children has been described who have drunk lemonade contaminated with antimony and suffered from vomiting and diarrhea. Sb^3 is likely to be more toxic than Sb^{5+}.

Animal studies. Exposure to antimony compounds affects the CNS with nausea and vomiting development. Gross pathology examination revealed mucosal irritation and inflammation, particularly in the distal region of the large intestine; 115 to 120 mg/kg BW doses of APT (calculated as Sb^{3+}) are lethal to rabbits; 450 to 600 mg BW doses cause vomiting in cats.

In mice, LD_{50} for antimony trioxide (calculated as Sb^{3+}) is found to be 209 mg/kg BW (according to Smyth and Carpenter, it is more than 20 g/kg BW), that of antimony oxide is 978 mg/kg BW.[7]

In dogs, 33 mg antimony tartrate/kg BW causes emesis. Cats appeared to be more sensitive to the emetic effect.[8]

Repeated Exposure. BW gain was decreased in rats on daily administration of 20 mg APT/kg BW. In a 1-month study, the toxic effect was evident in rabbits given antimony oxide at a dose of 150 mg/kg BW.[9]

Four rabbits were dosed with antimony tartrate at a dose level of 5.6 mg antimony/kg BW for 7 to 22 days. The treatment slightly increased the level of non-protein nitrogen in the blood and urine. Gross and histology examination revealed hemorrhagic lesions in the kidney cortex, stomach, and small intestine, liver atrophy with fat accumulation and congestion. This dose is suggested as a LOAEL.[10]

F344 rats and $B6C3F_1$ mice received APT in drinking water containing 16 to 168 mg/l (rats) and 59 to 407 mg/l (mice) for 2 weeks. No mortality or histopathological lesions were observed. There were treatment-related lesions in the liver and forestomach of most mice given 407 mg/kg BW.[1]

Short-term Toxicity. In a 91-day study, rats were exposed to two antimony-containing pigments with oral daily doses of 36 mg/kg BW and 22 mg/kg BW, respectively. Adverse effects were not observed.[10]

No manifestations of the toxic action were noted in rats given the doses of 0.1 to 4.0 mg antimony compounds/kg BW with food over a period of 3.5 months.[8] Sprague-Dawley rats received a soluble APT in drinking water at concentrations of 0.5, 5.0, 50, or 500 ppm for 13 weeks. In the highest dose group, mild adaptive histological changes were observed in the thyroid, liver and pituitary gland of both sexes, and in the spleen of male rats and thymus of female rats. The NOAEL of 0.5 ppm antimony in drinking water (0.06 mg antimony/kg BW) was established on the basis of the histological and biochemical changes observed in rats.[2]

Long-term Toxicity. CD-1 mice received 5.0 ppm APT in their drinking water for 18 months. Life spans of mice were significantly reduced in both males and females.[11]

In a life-time study in which rats were exposed to APT via drinking water at a dose of 0.43 mg/kg BW, effects observed were decreased longevity and altered blood levels of glucose and cholesterol (LOAEL). ADI/TDI appeared to be 0.43 mg/kg BW.[12] However, Lynch et al. (1999) indicate that this report on which are based the U.S. EPA reference dose value and a number of state, national, and international drinking water criteria for antimony, has severe inadequacies in study conduct making it uninterpretable and inappropriate for characterization of APT toxicity.[24]

According to Elinder and Frieberg, even consumption of 100 mg Sb^{3+}/kg or Sb^{5+} oxides/kg BW by dogs and cats for several months produced no toxic effect in the animals.[13]

Reproductive Toxicity.

Embryotoxicity. Inhalation exposure to 0.27 mg antimony oxide/m^3 caused a pronounced embryotoxic effect in rats. Concentration of 0.082 mg/m^3 produced maternal toxicity.[14]

The treatment with tartar emetic reduced litter viability and the overall lifespan in rats by 15% compared to the controls.[6]

Teratogenicity. No developmental effects were observed in ewe offspring whose mothers were given APT at a dose of 2.0 mg/kg BW for 45 days or throughout gestation.[15]

Five *i/m* injections of Sb^{5+} to Wistar rats during gestation caused no abnormalities in fetuses.[16]

Belyayeva found no teratogenic effect of antimony trioxide in rats.[17]

Gonadotoxicity. Antimony compounds are shown to impair the generative function and disturb the sexual cycle of females, although their general condition remains satisfactory. Ability to conceive and fertility are significantly lowered. In males, signs of impaired spermatogenesis are noted and fertilizing ability is reduced.[2]

Inhalation exposure to antimony trioxide over 2 months resulted in sterility and fewer offspring in rats.[17]

Mutagenicity.

In vitro genotoxicity. Antimony chloride and potassium and sodium antimony tartrates induce CA in cultured human leukocytes.[18]

Antimony chloride was negative when tested by *rec*-assay.[19]

Carcinogenicity. An increase in the incidence of benign and malignant tumors was not observed after lifetime exposure to an average daily dose of 0.43 mg APT/kg BW via drinking water in Charles River CD mice.[12]

Carcinogenicity classifications. An IARC Working Group concluded that there is sufficient evidence for the carcinogenicity of antimony trioxide in experimental animals and there is inadequate evidence for the carcinogenicity of antimony trioxide in humans.

Antimony

U.S. EPA: D.

Antimony trioxide (inhalation):

IARC: 2B;

EU: 2.

Antimony trisulfide

IARC: 3.

Chemobiokinetics. Following ingestion, antimony compounds are absorbed slowly from the intestine of experimental animals. The highest tissue levels (approximately 50%) were detected in the intestine and bones.[20] However, according to Edel et al., antimony compounds are found to be accumulated in the thyroid, liver, kidneys, spleen, and in the membrane structure of erythrocytes.[21]

Antimony compounds do not bind to blood proteins. Their transfer may occur from maternal to fetal blood.[22]

Antimony compounds are excreted in the urine with only low levels in the feces. APT is poorly absorbed and relatively non-toxic when given orally.[1,23]

Regulations. ***U.S. FDA*** (1998) approved the use of antimony oxide in adhesives as a component of articles intended for use in packaging, transporting, or holding food in accordance with the conditions specified in the 21 CFR part 175.105.

Recommendations. ***WHO*** (1996). Provisional Guideline value for drinking water: 0.005 mg/l.

Standards.

U.S. EPA (1999). MCL and MCLG: 0.006 mg/l.

EEC (1980). MAC: 0.01 mg/l.

Czechoslovakia. Permissible levels: 0.3 mg/kg in foods, 0.05 mg/l in drinks.

Russia (1995). MAC: 0.05 mg/l in water and milk, 0.3 mg/kg in meat and bread products, 0.5 mg/kg in fish.

References:

1. *Toxicity Studies of Antimony Potassium Tartrate in F344/N Rats And B6C3F$_1$ Mice (Drinking Water and Intraperitoneal Injection Studies)*, NTP Technical Report Series No 11, Research Triangle Park, NC, NTIS PB93-149714, March 1992.
2. Poon, R., Chu, I., Lecavalier, P., Valli, V. E., Foster, W., Gupta, S., and Thomas, B., Effects of antimony on rats following 90-day exposure via drinking water, *Food Chem. Toxicol.*, 36, 21, 1998.
3. *Toxicological Profile for Antimony and Its Compounds*, Draft U.S. Dept. Health & Human Service, ATSDR, Atlanta, GA, 1990.
4. *Revision of WHO Guidelines for Drinking Water Quality*, Report of the First Review Group Meeting on Inorganics, Bilthoven, March 18-22, 1991, 2; Medmenham, United Kingdom, January 27-29, 1992.
5. Fordham, P. J., Gramshaw, J. W., Crews, H. M., and Castle, L., Element residues in food contact plastics and their migration into food simulants, measured by inductively coupled plasma-mass spectrometry, *Food Addit. Contam.*, 12, 651, 1995.
6. Grushko, Ya. M., *Poisonous Metals*, Meditsina, Moscow, 1972, 130 (in Russian).
7. Gudzovsky, G. A., *Antimony and Its Compounds*, Sci. Review, Int. Project Center GKNT, Moscow, 1984, 28 (in Russian).
8. Flury, F., Zur Toxicologie des Antimons, *Arch. Exp. Pathol. Pharmacol.*, 126, 87, 1927.
9. Arzamastsev, E. V., in *Results Sci. Studies,* A. N. Sysin Research General Comm. Hygiene Institute, Acad. Med. Sci. USSR, Moscow, 20; *Gig. Sanit.*, 12, 16, 1964 (in Russian).
10. Bomhard, E., Loser, E., Dornemann. A., et al., Subchronic oral toxicity and analytical studies on nickel rutile yellow and chrome rutile yellow with rats, *Toxicol. Lett.,* 14, 189, 1984.
11. Kanisawa, M. and Schroeder, H. A., Life term studies on the effect of trace elements on spontaneous tu mor in mice and rats, *Cancer Res.*, 29, 892, 1969.
12. Schroeder, H. A., Mitchener, M., and Nason, A. P., Zirconium, niobium, antimony, vanadium, and lead in rats: life-term studies, *J. Nutr.*, 100, 59, 1970.
13. Elinder, C. and Frieberg, L., in *Handbook on Toxicology of Metals*, Amsterdam, 1980, 283.
14. Grin', N. V., Govorunova, N. N., Bessmertny, A. N., et al., Experimental study of embryotoxic effect of antimony oxide, *Gig. Sanit.*, 10, 85, 1987 (in Russian).
15. James, L. F., Lazar, V. A., and Binns, W., Effect of sublethal doses of certain minerals on pregnant ewes and fetal development, *Am. J. Vet. Res.*, 27, 132, 1966.
16. Casals, J. B., Pharmacokinetic and toxicological studies of antimony dextran glycoside (RL-712), *Brit. J. Pharmacol.*, 46, 281, 1972.
17. Belyayeva, A. P., The effect of antimony on reproductive function, *Gig. Truda Prof. Zabol.*, 11, 32, 1967 (in Russian).
18. Paton, G. R. and Allison, A. C., Chromosome damage in human cell cultures induced by metal salts, *Mutat. Res.,* 16, 332, 1972.
19. Nioshioka, H., Mutagenic activities of metal compounds in bacteria, *Mutat. Res.*, 31, 185, 1975.
20. Gerber, G. B., Maes, J., and Eykens, B., Transfer of antimony and arsenic to the developing organism, *Arch. Toxicol.*, 49, 159, 1982.
21. Edel, J. et al., *Heavy Metals Environ.*, Proc. Int. Conf., Heidelberg, September 1983, 180.
22. Leffler, P. and Nordstroem, S., *Metals in Maternal and Fetal Blood*, Governmental reports, Announcements, and Indexes, Issue 9, 1983 (in Swedish).
23. Pribyl, E., On the nitrogen metabolism in experimental subacute arsenic and antimony poisoning, *J. Biol. Chem.*, 74,775, 1927.
24. Lynch, B. S., Capen, C. C., Nestmann, E. R., Veenstra, G., and Deyo, J. A., Review of sub-chronic/chronic toxicity of antimony potassium tartrate, *Regul. Toxicol. Pharmacol.,* 30, 9, 1999.

2,2'-AZOBIS(2-METHYLPROPIONITRILE)

Molecular Formula. $C_8H_{12}N_4$

M = 164.24

CAS No 78-67-1
RTECS No UG0800000

Synonyms and **Trade Names.** 2,2'-Azobis(isobutyronitrile); Azodiisobutyronitrile; 2,2'-Dicyano-2,2'-azopropane; Porophor N.

Properties. Fine, white powder with a mild odor. Solubility in ethanol is 20.4 g/l at 20°C.[020] Miscible with water.

Applications. Used as a porofor and a blowing agent for elastomers in the manufacture of plastics. Catalyst for vinyl resins and curing agent in the manufacture of unsaturated polyester resins.

Migration Data. Concentration of tetramethyl succinonitrile (TMSN), the main decomposition product of A., in polyvinyl chloride products used for food packaging ranged from below the detection limit, 0.05 mg/kg, up to 523 mg/kg. Release from pieces of PVC bottles (40°C, 120 days) achieved 1.0 to 5.0 μg TMSN/kg in the olive oil.[1]

Acute Toxicity. LD_0 is 670 mg/kg BW in rats,[2] and 0.7 g/kg BW in mice. Oral administration caused excitation followed by tremor, labored breezing, and death from asphyxia (Rusin).[017]

Chemobiokinetics. In the body, A. forms *HCN* which is found in the blood, liver, and brain of treated animals (Rusin).[017]

Regulations. ***U.S. FDA*** (1998) approved A. for use as (1) a constituent of food-contact adhesives which may be safely used as component of articles intended for use in packaging, transporting, or holding food in accordance with the conditions prescribed in 21 CFR part 175.105; (2) in the manufacture of cross-linked polyester resins which may be used as articles or components of articles intended for repeated use in contact with food in accordance with the conditions prescribed in 21 CFR part 177.2420; and as (3) a component of the uncoated or coated food-contact surface of paper and paperboard intended for use in producing, manufacturing, packaging, processing, preparing, treating, packing, transporting, or holding aqueous and fatty foods in accordance with the conditions prescribed in 21 CFR part 176.170.

References:

1. Ishiwata, H., Inoue, T., and Yoshihira, K., Tetramethyl succinonitrile in polyvinyl chloride products for food and its release into food-simulating solvents, *Z. Lebensm. Unters. Forsch.*, 185, 39, 1987.
2. Deichmann, W. B., *Toxicology of Drugs and Chemicals*, Acad. Press, Inc., New York, 1969, 117.

1,2-BENZENEDICARBOXYLIC ACID, DI-2-PROPENYL ESTER

Molecular Formula. $C_{14}H_{14}O_4$
M = 246.28
CAS No 131-17-9
RTECS No CZ4200000
Abbreviation. DAP.

Synonyms. Allyl phthalate; Diallyl phthalate; Phthalic acid, diallyl ester.

Properties. Colorless oily liquid with a mild odor. Soluble in mineral oil and organic solvents. Solubility in water is 182 mg/l at 25°C.

Applications. Used as a cross-linking agent in the production of unsaturated resins, plasticizer and carrier for adding catalysts and pigments to polyesters, a stabilizer and a monomer in the processing of thermosetting plastics.

Acute Toxicity. In rats and mice, LD_{50} is reported to be 0.7 to 1.7 g/kg BW. Poisoning caused CNS inhibition.[1]

Repeated Exposure. DAP is the most toxic phthalate ester manufactured. It is more hepatotoxic to rats than to mice.[2,3] Hepatic necrosis was found to develop in F344 rats but not in B6C3F$_1$ mice dosed orally with up to 400 mg DAP/kg BW for 14 days. Changes were observed in the respiratory system or GI tract. An increase in mortality was noted in the treated animals.[1]

Short-term Toxicity. In a 13-week study, doses of 200 and 400 mg DAP/kg BW caused death, or a reduction in BW gains. The liver was found to be a target organ of DAP toxicity. Gross pathology examination revealed signs of hepatitis and cirrhosis manifested as a periportal hepatocellular necrosis and fibrosis in rats.[1]

Long-term Toxicity. Doses of 150 and 300 mg DAP/kg BW administered in corn oil by gavage caused the development of chronic inflammation and hyperplasia of the forestomach in mice.[1]

Mutagenicity.

In vitro genotoxicity. DAP is reported to be negative in *Salmonella* mutagenicity assay. It caused CA and SCE in cultured cells and was shown to be positive in the mouse lymphoma assay.[4]

Carcinogenicity. In NTP studies, oral exposure (by gavage) showed equivocal evidence of DAP carcinogenicity in mice and no or equivocal evidence in rats.[1]

Carcinogenicity classification.

NTP: NE - EE - XX -XX (gavage);

XX - XX - E - E (gavage).

Chemobiokinetics. Toxicity probably results from allyl alcohol cleaved from DAP.[2] F344 rats and B6C3F$_1$ mice were given 1.0, 10, or 100 mg ^{14}*C*-DAP/kg BW. In rats, 25 to 30% of DAP radioactivity was excreted as CO_2, and 50 to 70% appeared in the urine within 24 hours. In mice, 6 to 12% of DAP was eliminated as CO_2 and 80 to 90% was excreted in the urine within 24 hours. Main metabolites found in the urine of mice and rats appeared to be *monoallyl phthalate, allyl alcohol, 3-hydroxy-propylmercapturic acid*, and an unidentified polar metabolite.[3]

Regulations.

EU (1990). DAP is available in the *List of monomers and other starting substances which may continue to be used for the manufacture of plastic materials and articles intended to come into contact with foodstuffs pending a decision on inclusion in Section A* (*Section B*).

U.S. FDA (1998) approved DAP for use (1) in adhesives as a component (monomer) of articles intended for use in packaging, transporting, or holding food in accordance with the conditions prescribed in 21 CFR part 175.105, and as (2) a component (monomer) of the uncoated or coated food-contact surface of paper and paperboard intended for use in producing, manufacturing, packaging, processing, preparing, treating, packing, transporting, or holding dry food in accordance with the conditions prescribed in 21 CFR part 176.180.

Great Britain (1998). DAP is authorized without time limit for use in the production of polymeric materials and articles in contact with food or drink or intended for such contact.

The specific migration of this substance shall be not detectable (when measured by a method with a DL of 0.01 mg/kg).

Standards.

EU (1990). SML: Not detectable (DL = 0.01 mg/kg).

Russia (1995). MPL: 0.002 mg/l (organolept., odor).

References:

1. *Toxicology and Carcinogenesis Studies of Diallylphthalate in F344/N Rats (Gavage Study)*, NTP Technical Report Series No 284, Research Triangle Park, NC, August 1985.
2. Autian, J., Toxicity and health threats of phthalate esters, Review of literature., *Environ. Health Perspect.*, 4, 3, 1973.
3. Eigenberg, D. A., Carter, D. E., Schram, K. H., and Sipes, I. G., Examination of the differential hepatotoxicity of diallyl phthalate in rats and mice, *Toxicol. Appl. Pharmacol.*, 1, 12, 1986.
4. Seed, J. L., Mutagenic activity of phthalic esters in bacterial liquid suspension assay, *Environ. Health Perspect.*, 45, 111, 1982.

1,2,4,5-BENZENETETRACARBOXYLIC, 1,2:4,5 DIANHYDRIDE

Molecular Formula. $C_{10}H_2O_6$

M = 218.10

CAS No 89-32-7

RTECS No DB9300000

Synonyms. *1H,3H*-Benzo(1,2-c:4,5-c')difuran-1,3,5,7-tetrone; Pyromellitic acid, dianhydride; Pyromellitic anhydride.

Properties. Triclinic plates (*Pyromellitic acid*) or powder. Solubility in water is 15 g/l, freely soluble in alcohol.

Applications. Used as a curing agent in the production of epoxy and polyester resins. Used in the manufacture of aromatic polyamides (synthetic fibers manufacture) and plasticizers.

Acute Toxicity. LD_{50} is reported to be 2.25 to 3.5 g/kg BW in rats, 2.3 to 2.65 g/kg BW in mice, and 1.6 g/kg BW in guinea pigs.[1,2] Poisoning is accompanied by CNS depression including behavioral changes (ataxia, somnolence, adynamia, muscle weakness, etc.). Death within the first three days.

Repeated Exposure revealed little evidence of cumulative properties.[2,3]

However, Kondratyuk et al. believe B. to have pronounced cumulative potential. Rats were given 1/7 LD_{50} for 1.5 months.[1] Manifestations of the toxic action included CNS inhibition, erythropenia, and changes in the kidney relative weights. Disturbances in hydrocarbon- phosphate and nitrite-protein metabolism as well as alteration of the renal function are reported in rats gavaged with 8.8 to 220 mg/kg BW for 1 month.

Long-term Toxicity. Female rats exposed to 4.4 and 8.8 mg/kg BW for 6 months developed changes in the activities of a number of enzymes (cholinesterase, AST and ALT, cytochromeoxidase, etc.). Gross pathology and histology changes were evident.[1]

Allergenic Effect. B. exhibits mild sensitizing properties on skin application of 0.2 mg/kg BW to guinea pigs.

Reproductive Toxicity.

Embryotoxicity. No such effect was observed in rats given oral doses of 44 and 220 mg/kg BW.

Gonadotoxicity. In rats exposed to the oral dose of 44 mg/kg BW, a weak effect on gonad morphology was observed. The LOAEL for this effect appeared to be the 8.8 mg/kg BW dose.

Mutagenicity.

Observations in man. Workers exposed to B. demonstrated reliably increased occurrence of aberrant metaphases.[3]

Regulations. ***U.S. FDA*** (1993) approved the use of B. in ethylene terephthalate copolymers at a level not to exceed 0.5% by weight of the finished copolymer which may be used under the conditions of use described in 21 CFR part 176.170.

Standards. ***Russia***. Recommended PML: 0.4 mg/l.

References:

1. Kondratyuk, V. A., Gun'ko, L. M., Pereima, V. Ya., et al., Hygienic substantiation of the maximum allowable concentration for pyromellitic dianhydride in water bodies, *Gig. Sanit.*, 11, 79, 1986 (in Russian).
2. Karimullina, N. K., Filippova, Z. Kh., and Lisnyansky, Ye. Z., Toxicity of pyromellitic dianhydride, in *Hygiene and Toxicology of High-Molecular-Mass Compounds and of the Chemical Raw Material Used for Their Synthesis*, Proc. 4th All-Union Conf., S. L. Danishevsky, Ed., Khimiya, Leningrad, 1969, 230 (in Russian).
3. Victorova, T. V., Khustnutdinova, E. K., Lobanov, V. V., and Rafikov, Kh. S., An analysis of the chromosome aberrations in the peripheral blood lymphocytes of workers in the manufacture of pyromellitic dianhydride, *Med. Truda Prom. Ecol.*, 9, 24, 1994 (in Russian).

BENZOIC ACID

Molecular Formula. $C_7H_6O_2$

M = 122.13

CAS No 65-85-0

RTECS No DG0875000

Abbreviation. BA.

Synonyms and **Trade Name.** Benzenecarboxylic acid; Benzeneformic acid; Benzenemethanoic acid; Phenylcarboxylic acid; Retarder AB.

Properties. Monoclinic tablets, plates, or leaflets with faint pleasant odor. Soluble in water up to 0.21% at 17.5°C.

Applications. Curing agent for epoxy resins. Vulcanization retarder. Used as an ultraviolet absorber in production of plastics, plasticizers, and alkyd resins.

Acute Toxicity. In rats, LD_{50} is 3.3 g/kg BW. Manifestations of toxic action included narcosis, tremor and convulsions.

Repeated Exposure.

Observations in man. Administration of 0.5 to1.0 g/day (approximately 14 mg/kg BW) for 44 consecutive days or for 3 months produced no evident signs of toxicity in humans. However, according to Gerlach,[1] irritation, discomfort, weakness, and malaise were observed in individuals given oral bolus doses of less than or equal to 1.75 g/day (approximately, 25 mg/kg BW) over 20-day period.

Animal studies. F344 rats received 1.81, 2.09, or 2.4% sodium benzoate, and B6C3F_1 mice received 2.08, 2.50, or 3.00% sodium benzoate in their diet for 10 days. In male rats of the 2.4% group, relative liver and kidney weight, serum levels of albumin, total protein, and γ-glutamyl transpeptidase were significantly increased; enlarged hepatocytes with glassy cytoplasm were seen. In male mice of the 3.0% group, absolute liver weights and serum cholesterol and phoshpolipid levels were increased, and enlargement, vacuolation and necrosis of hepatocytes were evident.[2]

Short-term Toxicity. Rats received 0.5% or 1.0% BA in the diet (Kieckebush and Lung, 1960). Munro *et al.* (1996) suggested the calculated NOEL of 887 mg/kg BW for this study.[3]

Long-term Toxicity. Administration of 80 mg/kg BW for 18 months caused no adverse effects on BW gain, survival, or gross or microscopic pathology in rats and mice.[4] Mice were fed 40 mg BA/kg BW for 1.5 year. The treatment caused a decrease in resistance to stress and reduction in food and water intake.[5] Rats received 1.5% BA in their diet (750 mg/kg BW). The treatment decreased food intake and BW gain. The dose of 500 mg/kg BW was considered ineffective (Marquardt, 1960). ADI is considered to be 4.0 mg/kg BW (U.S. EPA, 1991).

BA was inactive in a human estrogen receptor yeast estrogenicity assay. It was not estrogenic in a range of test protocols and dose levels.[6]

Reproductive Toxicity.

Teratogenicity. Sodium benzoate is shown to cause malformations in rat fetuses.[7]

Mutagenicity.

In vitro genotoxicity. Inconclusive data are noted in *Salmonella* mutagenicity assay.[8]

Allergenic Effects were observed in a worker exposed to BA during his work. Oral exposure induced anaphylactic shock and caused milder reactions when eating food contaminated with BA (Pevny et al., 1981). Positive reactions to BA were found in almost 50% patients with asthma.[9]

Chemobiokinetics. BA is readily and completely absorbed from the GI tract. It undergoes conjugation with glycine in the liver to form *hippuric acid* which is rapidly excreted in the urine.[5]

Regulations.

U.S. FDA (1998) approved BA for use as (1) a constituent of food-contact adhesives which may be safely used as component of articles intended for use in packaging, transporting, or holding food in accordance with the conditions prescribed in 21 CFR part 175.105, and (2) in the manufacture of resinous and polymeric coatings as the food-contact surface of articles intended for use in producing, manufacturing, packing, processing, preparing, treating, packaging, transporting, or holding food in accordance with the conditions prescribed in 21 CFR part 175.300. Affirmed as *GRAS.*

EU (1990). BA is available in the *List of authorized monomers and other starting substances which shall be used for the manufacture of plastic materials and articles intended to come into contact with foodstuffs (Section* A).

Great Britain (1998). BA is authorized without time limit for use in the production of polymeric materials and articles in contact with food or drink or intended for such contact.

Recommendations. Joint FAO/WHO Expert Committee on Food Additives (1997). ADI: 5.0 mg/kg BW.[10]

Standards. ***Russia*** (1995). MAC: 0.6 mg/l.

References:

1. Gerlach, V., in *Physiological Activity of Benzoic Acid and Sodium Benzoate*, 1909.
2. Fujitani, T., Short-term effect of sodium benzoate in F344 rats and B6C3F_1 mice, *Toxicol. Lett.,* 69, 171, 1993.

3. Munro, I. C., Ford, R. A., Kennepohl, E., and Sprenger, J. G., Correlation of structural class with No-Observed-Effect Levels: A proposal for establishing a threshold of concern, *Food Chem. Toxicol.*, 34, 829, 1996.
4. Ignat'yev, A. D., Experimental data contributing to a hygienic characterization of the combined effect produced by some food presentations, *Voprosy Pitania*, 3, 61, 1965 (in Russian).
5. Shtenberg, A. J. and Ignat'ev, A. D., Toxicological evaluation of some combination of food preservatives, *Food Cosmet. Toxicol.*, 8, 369, 1970
6. Ashby, J., Lefevre, P. A., Odum, J. et al., Failure to confirm estrogenic activity for benzoic acid and clofibrate: Implications for list of endocrine-disrupting agents, *Regul. Toxicol. Pharmacol.*, 26, 96, 1997.
7. Minor, J. L. and Becker, B. A., A comparison of teratogenic properties of sodium salicylate, sodium benzoate, and phenol, *Toxicol. Appl. Pharmacol.*, 19, 373, 1971.
8. Rossman, T. G., Molina, M., Meyer, L., Boone, P., Klein, C. B., Wang, Y., Li, F., Lin, W. C., and Kinney, P. L., Performance of 133 compounds in the lambda prophase induction endpoint of the microscreen assay and comparison with *S. typhimurium* mutagenicity and rodent carcinogenicity, *Mutat. Res.*, 260, 349, 1991.
9. Martindale *The Extra Pharmacopeia*, J. E. F. Reynolds and A. B. Prasad, Eds., 28th ed., The Pharmaceutical Press, London, 1982, 1283.
10. *Evaluation of Certain Food Additives and Contaminants,* WHO Techn. Report Series No 868, 46th Report of Joint FAO/WHO Expert Committee on Food Additives, Geneva, 1997, 69.

BENZOIC ACID, DIMETHYLRESORCINOLIC ESTER

Molecular Formula. $C_{21}H_{17}O_4$
M = 333.38

Properties. White powder with a pinkish tint and no odor or taste. Poorly soluble in water, soluble in acetone and ethanol.

Applications. Used as a modifier in the manufacture of plastics and vulcanizates.

Acute Toxicity. Both rats and mice tolerate administration of 20 g/kg BW dose without any sign of intoxication. In mice, the acute action threshold for reducing spontaneous mobility appeared to be 2.5 g/kg BW.

Repeated Exposure. Rats were dosed by gavage with 2.0 g/kg BW for a month. Liver or kidney functions were not altered. Gross pathology examination revealed parenchymatous dystrophy in the hepatocytes and in the renal tubular epithelium.

Reference:

Shugaev, B. B. and Bukhalovsky, A. A., Toxicity of allyl bromide, allyl iodide, resorcin dinitrodiphenyl ether, methylresorcin dibenzoate, *Gig. Sanit.*, 6, 89, 1985 (in Russian).

BENZOPHENONE

Molecular Formula. $C_{13}H_{10}O$
M = 182.23
CAS No 119-61-9
RTECS No DI9950000

Synonyms and **Trade Names.** Benzoylbenzene; Diphenyl ketone; Diphenylmethanone; α-Oxydiphenyl methane; Phenyl ketone.

Properties. White prisms with geranium-like odor. Insoluble in water. 1.0 g. B. dissolves in 7.5 ml alcohol.

Applications. Ultraviolet curing agent, a polymerization inhibitor for styrene. A fragrance and flavor enhancer.

Migration of B. from the printing ink of a paperboard sleeve during microwave heating of a pre-cooked meal has been described.[1] B. was identified in solvent extracts obtained from paper and board materials intended for food contact. Level was generally below 1.0 mg/kg paper.[2]

B.-based ultraviolet absorbents (up to 0.5% of a composition) for commodities of polystyrene showed a very slight tendency to migrate into aqueous, acid, or dilute alcoholic foods. Very high migration levels were noted for sunflower oil or 50% ethanol. Migration of 2-hydroxy-4-octoxybenzophenone into fat-free foods was observed at the toxicologically insignificant level; nevertheless, these ultraviolet sorbents should not be used for the ultraviolet stabilization of plastics designed for packaging fat-containing foods.[3]

Acute Toxicity. LD_{50} exceeded 10 g/kg BW in rats,[3] and 2.9 g/kg BW in mice.[4]

Repeated Exposure. Rats received 100 or 500 mg B./kg in their diet for 4 weeks. As a result of the treatment, pigmented or nucleated erythrocytes, and changes in serum composition were observed. Histopathology investigation revealed hepatocellular enlargement with an associated clumping of cytoplasmatic basophilic clusters around the central vein.[5]

Short-term Toxicity. Rats received B. in their diet at dietary levels of 20 mg/kg for 90 days and 100 or 500 mg/kg for 28 days. Absolute and relative liver and kidney weights were increased at higher dose levels. In these animals, histology investigation revealed hepatocellular enlargement. The NOEL was found to be 20 mg/kg in a 90-day study.[5]

In a NTP 13-week oral study, exposure to B. caused hepatocellular hypertrophy in rats and mice, and evidence of cholestatic liver injury and renal damage in rats. Marked induction of hepatic CYP 450 IIB was observed in rats and mice.[6]

Allergenic Effect. B. is capable to cause hypersensitivity contact reactions.[7]

Carcinogenicity. No carcinogenic effects have been initiated by oral administration to female Swiss mice.[7]

Regulations. ***U.S. FDA*** (1998) approved the use of B. as a synthetic flavoring substance or adjuvant in food in accordance with conditions indicated in 21 CFR part 172.515.

References:

1. Gilbert, J., Castle, L., Jickells, S. M., and Sharman, M., Current research on food contact materials undertaken by the UK Ministry of Agriculture, Fisheries and Food, *Food Addit. Contam.*, 11, 231, 1994.
2. Castle, L., Offen, C. P., Baxter, M. J., and Gilbert, J., Migration studies from paper and board food packaging materials. 1. Compositional analysis. *Food Addit. Contam.*, 14, 35, 1997
3. Uhde, W. J. and Woggon, H., New results of migration behavior of benzophenone-based UVG absorbents from polyolefi ns in foods, *Nahrung*, 2, 185, 1976.
4. Caprino, L., Togna, G., and Mazzei, M., Toxicological studies of photosensitizer agents and photodegradable polyolefins, *Eur. J. Toxicol. Environ. Hyg.*, 9, 99 1976
5. Burdock, G. A., Pence, D. H., and Ford, R. A., Safety evaluation of benzophenone, *Food Cosmet. Toxicol.*, 29, 741, 1971.
6. *NTP Fiscal Year 1995, Annual Report*, Publ. Health Service, Dept. Health & Human Services, December 1995, 72.
7. Holzle, E. und Plewig G., Photoallergic contact dermatitis by benzophenone containing sunscreen prepa rations, Abstract, *Hautarzt*, 33, 391, 1982.

BENZOYL PEROXIDE

Molecular Formula. $C_{14}H_{10}O_4$
M = 242.23
CAS No 94-36-0
RTECS No DM8575000
Abbreviation. BP.

Synonym and **Trade Name.** Acetoxyl; Dibenzoyl peroxide; Perbenyl.

Properties. White or yellowish, fine hygroscopic powder. Very poorly soluble in water, solubility in alcohol is 12 g/l. On hydrolysis forms *benzoic acid* and *benzoyl hydroxyperoxide*.

Applications. Used as a vulcanizing agent in the production of vulcanizate mixes based on natural and synthetic rubber and polyester resins. Also used in the production of polyesters, glass-reinforced plastics, and polystyrene foam. Polymerization catalyst for vinyl, acrylic, and styrene plastics and for unsaturated polyesters. It is also used as a decolorant of flour and milk in cheese manufacture.

Exposure. Phthalates and BP containing hardeners may be a potential source of benzene exposure to consumers as well as for the industry workers.[1]

Acute Toxicity. LD_{50} is 6.4 g/kg BW in rats, and 2.1 g/kg BW in mice. Lethal doses caused immediate motor coordination disorder with subsequent CNS inhibition and respiratory distress. Death occurs in 24 hours.[2,3] According to other data, LD_{50} exceeded 950 mg/kg BW for rats (IARC 36-272).

Repeated Exposure failed to reveal signs of accumulation.[2]

Short-term Toxicity. Rats and mice were given 1/50 and 1/10 LD_{50} for 4 months. The treatment with the higher dose of BP caused adynamia and significant retardation of BW gain.[3] In a 3-month study in dogs, there were no manifestations of toxicity when 160 mg DBP/kg was adminis tered in the diet.[4]

Long-term Toxicity. Rats and mice were fed for 120 and 80 weeks, respectively, up to 1000 times greater quantities of BP than that consumed by humans in the diet. No signs of intoxication were found to develop in the animals. Exposure to 280 and 2800 mg/kg feed led to some retardation of BW gain in rats. It is thought that a dose up to 40 mg BP/kg diet is harmless to man.[4]

Allergenic Effect is likely to be observed as skin reactions in man after contact with BP.[5]

Reproductive Toxicity.

Gonadotoxicity. Prolonged administration of 2800 mg BP/kg feed to rats caused testicular atrophy, which may be associated with the destruction of tocopherol in the feed. There were no such changes in mice.[4]

Teratogenic effect as well as late death is observed in chick embryos.[6]

Mutagenicity.

In vitro genotoxicity. BP is neither mutagenic in *Salmonella* nor induces CA in cultured Chinese hamster lung cells.[7] It does not seem to be directly genotoxic.

In vivo cytogenetics. BP was negative in DLM assay in Swiss mice (*i/p* injection of 54 or 62 mg/kg BW).[8]

Carcinogenicity. Addition of milk containing 28 to 2800 mg BP/kg to the feed of rats for 120 weeks and to the feed of mice for 80 weeks produced no carcinogenic effect.[4] BP was shown to promote carcinogenic activity in the mouse skin (Slaga et al., 1981).

Rudra and Singh consider it to be a tumor promoter, mediating its effect via genetic and epigenetic mechanisms.[9] However, Kraus et al. concluded that there is no evidence to support an association between BP and the development of skin cancer in humans.[10]

Carcinogenicity classification. An IARC Working Group concluded that there is inadequate evidence for the carcinogenicity of BP in *experimental animals* and in *humans*.

IARC: 3.

Chemobiokinetics. *Benzoic acid* appeared to be a major metabolite. Renal clearance of the benzoic acid metabolite is sufficiently rapid in the Rhesus monkey to preclude its hepatic conjugation with glycine.[11]

Regulations.

U.S. FDA (1998) approved BP for use as a direct and indirect food additive (1) in adhesives used as components of articles intended for use in packaging, transporting, or holding food in accordance with the conditions prescribed in 21 CFR part 175.105; (2) as a preservative of the uncoated or coated food-contact surface of paper and paperboard intended for use in producing, manufacturing, packing, transporting, or holding aqueous and fatty food in accordance with the conditions prescribed in 21 CFR part 176.170; (3) as a catalyst in the formulation of polyester resins, subject to the provisions prescribed in 21 CFR part 177.2420; and (4) as an accelerator in the production of rubber articles intended for repeated use in accordance with the conditions prescribed in 21 CFR part 177.2600. BP may be also used (6) as a bleaching agent for flour and for milk used in the preparation of certain cheeses and lecithin; (7) as an ingredient in the preparation of hydroxylated lecithin. In 1982, it was proposed that BP be affirmed as *GRAS* as a direct human food ingredient when used as a bleaching agent following conditions of use of current GMP.

In the ***UK***, BP may be added to all types of flour or bread (except whole meal) at up to 50 mg/kg flour.

Standards. ***Russia***. PML in drinking water: *n/m*.

References:

1. Rastogi, S. C., Formation of benzene by hardeners containing benzoyl peroxide and phthalates, *Bull. Environ. Contam. Toxicol.*, 53, 747, 1994.
2. *Occupational Exposure to Benzoyl Peroxide*, NIOSH, Criteria for a Recommended Standard, Publ. No 77-166, Washington, D. C., 1977.
3. Antonyuk, O. K., Experimental data on toxicity of benzoyl peroxide and triphenyl phosphate, in *Hygienic Aspects of the Use of Polymeric Materials and Articles Made of Them*, L. I. Medved', Ed., All-Union Research Institute of Hygiene and Toxicology of Pesticides, Polymers and Plastic Materials, Kiyv, 1969, 311 (in Russian).
4. Sharratt, M., Frazer, A. C., and Forbes, O. C., Study of the biological effects of benzoyl peroxide, *Food Cosmet. Toxicol.*, 2, 527, 1964.
5. Poole, R. L., Griffith, J. F., and MacMillan, F. S. K., Experimental contact sensitization with benzoyl per oxide, *Arch. Dermatol.*, 102, 635, 1970.
6. Korhonen, A., Hemminki, K., and Vainio, H., Embryotoxic effects of eight organic peroxides and hydrogen peroxide on three-day chicken embryos, *Environ. Res.*, 33, 54, 1984.
7. Ishidate, M., Sofuni, T., and Yoshikawa, K., A primary screening for mutagenicity of food additives in Japan, *Mutagen. Toxicol.*, 3, 80, 1980 (in Japanese).
8. Epstein, S. S., Arnold, S., Andrea, J., Bass, W., and Bishop, Y., Detection of chemical mutagens by the dominant lethal assay in the mouse, *Toxicol. Appl. Pharmacol.*, 23, 288, 1972.
9. Rudra, N. and Singh, N., Insight into mechanism of action of benzoyl peroxide as a tumor promoter, *Indian J. Physiol. Pharmacol.*, 41, 109, 1997.
10. Kraus, A. L, Munro, I. C., Orr, J. C., Binder, R. L., LeBoeuf, R. A., and Williams, G. M., Benzoyl peroxide: An integrated human safety assessment for carcinogenicity, *Regul. Toxicol. Pharmacol.*, 21, 87, 1995.
11. Nacht, S. et al., Benzoyl peroxide: Percutaneous penetration and metabolic deposition, *J. Am. Acad. Dermatol.*, 4, 31, 1981.

2-BENZYL-2-THIOPSEUDOUREA, MONOHYDROCHLORIDE

Molecular Formula. $C_8H_{10}N_2S.ClH$
M = 202.72
CAS No 538-28-3
RTECS No UM0738000

Synonyms. Benzyl isothiouronium chloride; Benzylthiouronium chloride; 2-Thio-2-benzylpseudourea hydrochloride.

Properties. White powder with a specific odor. Readily soluble in water and organic solvents.

Applications. Used as a regulator of the emulsion polymerization of chloroprene (instead of tertiary dodecylmercaptan).

Acute Toxicity. In rats, LD_{50} is 27 mg/kg BW. Death occurs within a day after poisoning. High doses led to retardation of BW gain in the survivors in the first 2 to 6 days. According to other data, LD_{50} is 150 mg/kg BW in rats (Gizhlaryan and Khechumov, 1983).

2,5-BIS(*tert*-BUTOXY)-2,5-DIMETHYLHEXANE

Molecular Formula. $C_{16}H_{34}O_2$
M = 258.50
RTECS No MO0720000

Applications. Cross-linking agent.

Acute Toxicity. LD_{50} *i/p* is 2.4 g/kg BW in mice.[027]

BIS(2,4-DICHLOROBENZOYL)PEROXIDE

Molecular Formula. $C_{14}H_6C_{l4}O_4$
M = 379.99

CAS No 133-14-2
RTECS No SD7880000

Synonym and **Trade Name.** 2,4-Dichlorobenzoyl peroxide; Silopren CL-40.

Properties. White gygroscopic crystalline powder. Insoluble in water.

Applications. Initiator of the polymerization of polyvinyl chloride, acrylates, styrene. Hardener for unsaturated polyether resins. Structurizing agent for siloxane crude rubber. Used in the vulcanization of rubber intended for medical use.

Acute Toxicity. LD_{50} *i/p* is 225 mg/kg BW in mice.[027]

1,4-BUTANEDIAMINE

Molecular Formula. $C_4H_{12}N_2$
M = 88.18
CAS No 110-60-1
RTECS No EJ6800000
Abbreviation. BDA.

Synonyms. Butylenediamine; 1,4-Diaminobutane; Putrescine; Tetramethylenediamine.

Acute Toxicity. The dose of 1.6 g/kg BW is lethal for mice.[1,019]

Reproductive Toxicity. Mice received *i/p* injections of 314 mg BDA/kg BW on 12th day of pregnancy. The treatment produced fetotoxicity and developmental abnormalities in muscular skeletal sys tem.[2]

Mutagenicity. BDA caused unscheduled DNA synthesis in mouse liver cells, and DNA inhibition in mouse ascitis tumor cells.[3]

Chemobiokinetics. Administration to healthy rats induced rapid and significant changes in carbohydrate and lipid metabolism. Target organs of BDA toxicity are the liver, muscle, and probably adipose tissue. A lower dose of 100 g/kg BW produced effects similar to those induced by glucagon and adrenaline.[4]

Regulations.

EU (1990). BDA is available in the *List of monomers and other starting substances which may continue to be used pending a decision on inclusion in Section A.*

Great Britain (1998). BDA is authorized without time limit for use in the production of polymeric materials and articles in contact with food or drink or intended for such contact.

References:

1. Schafter, E. W., Bowles, W. A., Acute oral toxicity and repellency of 933 chemicals to house and deer mice, *Arch. Environ. Contam. Toxicol.*, 14, 111, 1985
2. Manen, C. A., Hood, R. D., and Farina, I., Ornothine decarboxylase inhibition and fetal growth retarda tion in mice, *Teratology*, 28, 237, 1983.
3. Abstract, *Acta Medicia Okayama*, 33, 149, 1979.
4. Shelepov, V. P., Chekulayev, V. A., and Pasha-zade, G. R., Effect of putrescine on carbohydrate and lipid metabolism in rats, *Biomed. Sci.*, 1, 591, 1990.

2-BUTANONE PEROXIDE

Molecular Formula. $C_8H_{16}O_4$
M = 176.21
CAS No 1338-23-4
RTECS No EL9450000
Abbreviation. BP.

Synonyms and **Trade Names.** Butanox LPT; Di(1-methylethylketone)peroxide; Di(1-methylpropylidene)diperoxide; Lupersol; Methyl ethyl ketone peroxide; Trigonox M 50.

Properties. Colorless, oily liquid with a pleasant odor.

Applications. Used in the manufacture of acrylic resins, as a hardening agent for fiberglass-reinforced plastics, as a polymerization initiator and curing agent for unsaturated polyester resins. It is commercially available as a 40 to 60% solution in dimethyl phthalate (DMP).

Acute Toxicity.

Observations in man. Intake of 50 ml of Norox catalyst (60% BP, cyclohexanone and dimethyl phthalate; see above) caused fatal poisoning.[1]

Animal studies. LD_{50} is 470 to 484 mg/kg BW in rats, and 260 mg/kg BW in mice.[1,2] Manifestations of the toxic action included GI tract bleeding, abdominal burns, necrosis, perforation of the stomach, stricture of the esophagus, severe metabolic acidosis, rapid hepatic failure, respiratory insufficiency.[3,4]

However, according to other data, LD_{50} for rats is found to be 6.86 ml/kg BW.[020]

Repeated Exposure. Rats received oral doses of BP 3 times a week for 7 weeks. Administration of 1/5 LD_{50} caused all five rats to die. Gross pathology examination revealed mild liver damage with glycogen depletion.[5]

Skin lesions were noted following dermal exposure. Topical administration of BP + DMP resulted in necrotic inflammatory and regenerative skin lesions limited to the application site. BP is highly irritating and corrosive to the skin and mucous membranes.[6]

Reproductive Toxicity. BP administered into the air chamber was toxic to 3-day chicken embryos: lethality and malformations were noted in embryos.[7]

Mutagenicity.

In vitro genotoxicity. BP was not shown to be positive in *Salmonella* mutagenicity assay (BP and DMP 45:55 w/w). It induced SCE and CA in Chinese hamster ovary cells.

In vivo cytogenetics. No increase in the frequency of micronucleated erythrocytes was observed in the peripheral blood samples obtained from mice to the end of the 13-week study.[7]

Carcinogenicity. BP induced malignant lymphomas in C57Bl mice (treatment route and regimen of admini stration were not specified).[8]

The dose of 10 mg BP/mouse showed a weak tumor-promoting activity when applied topically to the skin of mice that had been UV-irradiated.[9]

Chemobiokinetics. BP could be detoxified by glutathione peroxidase, thus preventing oxidative deterioration of cells.[10]

Regulations. ***U.S. FDA*** (1998) has regulated BP as an indirect food additive. It is permitted (1) as a catalyst in the production of resins to be used at levels not exceeding 2.0% of the finished resin; (2) as an optional adjuvant substance employed to facilitate the production of the cross-linked polyester resins intended for repeated use in contact with food; and (3) in cross-linked polyester resins to be safely used as articles or components of articles intended for repeated use in contact with food in accordance with the conditions prescribed in 21 CFR part 177.2420.

References:

1. Ovcharov, V. G. and Loit, A. O., On the problem of toxicity of some organic peroxides, in *Toxicology and Hygiene of High-Molecular-Mass Compounds and of the Chemical Raw Material Used for Their Synthesis*, Proc. 2nd All-Union Conf., A. A. Letavet and S. L. Danishevsky, Eds., Leningrad, 1964, 74 (in Russian).
2. Floyd, E. P. and Stockinger, H. E., Toxicity studies of certain organic peroxides and hydroperoxides, *Am. Ind. Hyg. Assoc. J.*, 19, 205, 1958.
3. Mittleman, R. E., Romig, L. A., and Gressman, E., Suicide by ingestion of methyl ethyl ketone peroxide, *J. Forensic Sci.*, 31, 312, 1986.
4. Korhunen, P. J., Ojanpera, I., Lalu, K., et al., Peripheral zonal hepatic necrosis caused by accidental ingestion of methyl ethyl ketone peroxide, *Hum. Exp. Toxicol.*, 9, 197, 1985.
5. Sittig, M., *Handbook of Toxic and Hazardous Chemicals and Carcinogens*, Noyes Publ., Park Ridge, NJ, 1985.
6. *Toxicity Studies of Methyl Ethyl Ketone Peroxide in Dimethyl Phthalate (45:55) Administered Topically to Fischer-344/N Rats and B6C3F$_1$ Mice*, NTP Technical Report, U.S. Dept. Health and Human Services, Publ. Health Service, NIH, 1992, 58.
7. Korhonen, A., Hemminki, K., and Vainio, H., Embryotoxic effects of eight organic peroxides and hydrogen peroxide on three-day chicken embryos, *Environ. Res.*, 33, 54, 1984.

8. Kotin, P. and Falk, H. L., Organic peroxides, hydrogen peroxide, epoxides and neoplasma, *Radiat. Res.*, Suppl. 3, 193, 1963.
9. Logani, M. K., Sambuco, C. P., Forbes, P. D., et al., Skin-tumor promoting activity of a potent lipid peroxidizing agent, *Food Chem. Toxicol.*, 22, 879, 1984.
10. Condell, R. A. and Tappel, A. L., Evidence for suitability of glutathione peroxide as a protective enzyme: studies of oxidative damage, renaturation, and proteolysis, *Arch. Biochem. Biophys.*, 223, 407, 1983.

tert-BUTYL HYDROPEROXIDE

Molecular Formula. $C_4H_{10}O_2$
M = 90.14
CAS No 75-91-2
RTECS No EQ4900000
Abbreviation. TBHP.

Synonym and **Trade Name.** 1,1-Dimethylethyl hydroperoxide; Perbutyl H.

Properties. Water-white liquid. Soluble in water and ethanol.

Applications. Used in the manufacture of adhesives, plastics, rubber, and elastomers. Curing agent for thermoset polyesters. Initiator in the PVC production.

Acute Toxicity. LD_{50} is 406 mg/kg BW in rats.[1] LD_{50} of 70% TBHP is found to be 560 mg/kg BW in rats. The dose of 0.6 ml/kg BW caused 1/10 rats to die. Poisoning induced depression and lacrimation. Administration of 0.8 to 1.6 ml/kg BW caused loss of righting, hypotermia, and hematuria.[2]

LD_{50} *i/p* is 246 mg/kg BW in mice.[3] Manifestations of toxic effect included severe depression, coordination disorder, and death due to respiratory arrest.[012]

Mutagenicity.

In vivo cytogenetics. In the dominant lethal assay, mice were exposed *i/p* to 15 to 75 mg/kg BW. The treatment produced no increase in the incidence of early death or pre-implantation losses.[4]

TBHP was positive when administered at a dose of 2000 ppm in the feed of *Dr.melanogaster*.[5]

In vitro genotoxicity. In a concentration range of 0.1 to 1.0 mmol, TBHP induced chromosomal cultural aberrations, consisting mainly of chromosomal gaps and breaks.[6] It gives rise to DNA damage profile in *Salmonella typhimurium* cells.[7]

Chemobiokinetics. TBHP can be metabolized to much less genotoxic material.

Regulations. ***U.S. FDA*** (1998) approved the use of TBHP (1) in adhesives as a component of articles intended for use in packaging, transporting, or holding food in accordance with the conditions prescribed in 21 CFR part 175.105; and (2) as a component of the uncoated or coated food-contact surface of paper and paperboard intended for use in producing, manufacturing, packaging, processing, preparing, treating, packing, transporting, or holding aqueous and fatty foods in accordance with the conditions prescribed in 21 CFR part 176.170.

References:

1. Floyd, E. P. and Stockinger, H. E., Toxicity studies of certain organic peroxides and hydroperoxides, *Am. Ind. Hyg. Assoc. J.*, 19, 205, 1958.
2. Patty's *Industrial Hygiene and Toxicology*, 4 ed., vol 2, Part A, G. D. Clayton and F. E. Clayton, Eds., John Wiley & sons, Inc., New York, 1993, 580.
3. Izmerov, N. F., Sanotsky, I. V., and Sidorov, K. K., *Toxicometric Parameters of Industrial Toxic Chemicals under Single Exposure*, USSR/SCST Commission for UNEP, Center Int. Projects, Moscow, 1982, 30 (in Russian).
4. Woodruff, R. C., Mason, J. M., Valencia, R., and Zimmerling, S., Chemical mutagenesis testing in *Drosophila*. V. Results of 53 coded compounds tested for National Toxicology Program, *Environ. Mutagen.*, 7, 677, 1985.
5. Ochi, T., Effect of iron chelators and glutathione depletion on the induction and repair of chromosomal aberrations by *tert*-butyl hydroperoxide in cultured Chinese hamster cells, *Mutat. Res.*, 213, 243, 1989.

6. Epstein, S. S., Arnold, E., Andrea, J., Bass, W., and Bishop, Y., Detection of chemical mutagens by the dominant lethal assay in the mouse, *Toxicol. Appl. Pharmacol.*, 23, 288, 1972.
7. Epe, B., Hegler, J., and Wild, D., Identification of ultimate DNA damaging oxygen species, *Environ. Health Perspect.*, 88, 111, 1990.

tert-BUTYL PEROXIDE

Molecular Formula. $C_8H_{18}0_2$
M = 146.26
CAS No 110-05-4
RTECS No ER2450000
Abbreviation. tert-BP.

Synonyms. Bis(*tert*-butyl) peroxide; Bis(1,1'-dimethylethyl) peroxide.

Properties. Clear, water-white or yellow liquid. Solubility in water is about 0.01%, soluble in organic solvents.

Applications. Important initiator for high-temperature and high-pressure polymerization of ethylene and halogenated ethylenes. Curing and vulcanization agent. Catalyst for polystyrene, acrylonitrile polymers, silicone rubbers, polyethylene.

Acute Toxicity. LD_{50} is 1.9 g/kg BW in mice. Clinical signs of intoxication included tremor and coordination disorder, convulsions, respiratory arrest, and decrease in the body temperature (Ovcharov).[017]

Mutagenicity.

In vitro genotoxicity. *tert*-BP is reported to be negative in *Salmonella* mutagenicity assay with and without metabolic activation.[1]

Regulations. ***U.S. FDA*** (1998) regulates the use of *tert*-BP (1) in the manufacture of rubber articles intended for repeated use in producing, manufacturing, packaging, processing, preparing, treating, packing, transporting, or holding food in accordance with the conditions prescribed in 21 CFR part 177.2600; and (2) as a component of the uncoated or coated food-contact surface of paper and paperboard intended for use in producing, manufacturing, packaging, processing, preparing, treating, packing, transporting, or holding aqueous and fatty foods in accordance with the conditions prescribed in 21 CFR part 176.170.

Reference:

1. Yamaguchi, T. and Yamashita, Y., Mutagenicity of hydroperoxides of fatty acids and some hydrocarbons, *J. Agric. Biol. Chem.*, 44, 1675, 1980.

1-*tert*-BUTYLPEROXY-1-METHACRYLOYLOXYETHANE

Molecular Formula. $C_{10}H_{18}O_4$
M = 202.28

Properties. A liquid with an unpleasant odor. Poorly soluble in water, readily soluble in alcohol.

Applications. Used as a polymerization initiator.

Acute Toxicity. The LD_{50} values are 9.0 g/kg BW for mice, and 8.2 g/kg BW for rats. Gross pathology examination revealed injury to the GI tract walls and signs of corrosive gastritis and acute enteritis. Glomerulonephritis was found to develop. Gross pathology examination did not reveal changes in the viscera after administration of 2.5 to 5 g/kg BW doses.

Reference:

Krynskaya, I. L., Yevsyukov, V. I., and Sukhareva, L. V., Toxicological characteristics of peroxides used in plastics production, in *Environmental Protection in Plastic Industry and Safety of the Use of Plastics*, T. N. Zelenkova and B. Yu. Kalinin, Eds., Plastpolymer, Leningrad, 1978, 133 (in Russian).

p-CHLOROMETHYLBENZOIC ACID, PEROXIDE

Properties. White, odorless powder. Poorly soluble in water.

Applications. Used in polymerization and high-temperature vulcanization.

Acute Toxicity. The maximum doses given caused some animal mortality due to mechanical obstruction of the GI tract.

Allergenic Effect was not observed.

Reference:

Matreshin, A. V. and Klimov, V. S., in *Medical Evaluation of the Environment in the Middle Volga Region*, Moscow, 1982, 115 (in Russian).

CHROMIUM compounds

	CAS No	RTECS No
Chromium chloride	10049-05-5	GB5250000
Molecular Formula. $CrCl_3$		
M = 122.90		
Chromium hydroxide	1308-38-9	GB6475000
Molecular Formula. $H_3O_3.Cr$		
M = 152.0		
Chromium nitrate	13548-38-4	GB6280000
Molecular Formula. CrN_3O_9		
M = 238.03		
Synonyms. Chromic nitrate; Chromium trinitrate.		
Chromium oxide	1333-82-0	GB6650000
Molecular Formula. Cr_2O_3		
M = 100.0		
Chromium sodium oxide	7775-11-3	GB2955000
Molecular Formula. $CrO_4.2Na$		
M = 161.98		
Cromium sulfate	14489-25-9	WS6985000
Molecular Formula. CrO_4S		
M = 724.42		

Synonyms and **Trade Name.** Chromosulphuric acid; Sulfuric acid, chromium salt; UN 2240.

Properties. In general, Cr^{6+} salts are more soluble than those of Cr^{3+}, making Cr^{6+} relatively mobile. Cr^{3+} is shown to oxidize to Cr^{6+} in drinking water systems in the presence of chlorine at concentrations similar to those used to disinfect drinking water.

Chromium chloride, nitrate, and *sulfate* are readily soluble in water; the chromates and dichromates are also soluble. The threshold concentration of Cr^{3+} for tint is 1.0 mg/l, and for taste, 4.0 mg/l.

Chromium hydroxide at a concentration of 0.5 mg/l does not noticeably increase the turbidity of water; 1.0 mg/l reduces transparency to 15 cm in terms of Snellin scale.

Chromium oxide. Green powder or fine hexagonal crystals. Practically insoluble in water and alcohol.

Chromium trioxide. Red or brown prismatic, odorless crystals, flakes or ground powder. Solubility in water exceeded 60 mg/100 g at 100°C. Soluble in alcohol.

Chromium sodium oxide. Yellow, odorless, rhombic bipyramidal crystals. Aqueous solution is alkaline. Soluble in water (83 g/l at 30°C), slightly soluble in alcohol

Applications. Chromium compounds are used in the manufacture of catalysts and stabilizers, and in the production of ceramic glazes, colored glass and paints. Chromium oxide is used in dyeing polymers, colorant for latex paints. Catalyst in the production of polyethylene.

Exposure. Concentrations in drinking water are usually less than 0.005 mg/l of total chromium, although concentrations of 0.06 to 0.12 mg/l also were reported. In general, food appears to be the major source of intake,[1] but in some cases drinking water may provide about 70% of the total chromium intake.

Migration Data. Chromium was present in paper intended for food contact at concentrations of 1.1 to 7.8 mg/kg.[2]

Acute Toxicity.

Observations in man. Ingestion of 1.0 to 5.0 g of "chromate" (not further specified) results in severe acute effects such as GI tract impairment, hemorrhagic diathesis and convulsions. Death may occur with a clinical picture of cardiovascular shock.[3]

Animal studies. In rats, LD_{50} varies from 50 to 250 mg/kg BW for Cr^{6+}, and from 185 to 615 mg/kg BW for Cr^{3+}, based on tests with (di)chromates and chromic compounds, respectively.[3]

In mice, LD_{50} of aluminum-chromium-potassium catalyst (type A-30) has been reported to be about 450 mg/kg BW (calculated as chromium oxide); in rats, it is 792 mg/kg BW. In the case of a ferrochrome catalyst, the LD_{50} is 695 mg/kg BW; in the case of a zinc-chromium catalyst (36% chromium oxide, 63% zinc oxide), LD_{50} is 2060 mg/kg BW in rats and 1110 mg/kg BW in mice.[4]

Mortality in guinea pigs has been reported to occur at a dose level of 600 mg chromium sulfate/kg BW. LD_{50} of chromium nitrate is about 3.0 g/kg BW in rats and mice.

Repeated Exposure. According to Turebayev et al.,[5] chromium compounds may accumulate in the reticuloendothelial system and affect hemopoiesis. A decline in diuresis with water consumption remaining the same is noted in Wistar rats given drinking water with Cr^{6+} concentration of 700 mg/l for 4 weeks. The treatment did not alter concentration of glucose, ketones, nitrites, and bilirubin in the urine. Proteinuria developed and motor activity was lowered. No changes in diuresis and motor activity were observed with a concentration of 70 mg/l.[6]

Mice were given $K_2Cr_2O_3$ in their drinking water at a concentration of 10 mg/l (calculated as chromium) for 30 days. The exposure resulted in the altered phagocyte activity and reduced bactericidal capacity of blood; a 1.0 mg/l concentration exerted no harmful effect.[7]

Short-term Toxicity. Rats were given non-hydrated chromic oxide pigment in their feed over a period of 90 days. The only effect observed was a dose-related decrease (no statistics applied) in the liver and spleen weights at a dose equivalent to 480 mg/kg and 1,210 mg Cr^{3+}/kg BW.[8]

Long-term Toxicity. Cr^{6+} does not produce health effects in animals except of fairly high doses. The lack of oral toxicity at lower doses is likely to be due to the reductive conditions of the stomach converting ingested Cr^{6+} to Cr^{3+} prior to systemic absorption.[9]

Rabbits were dosed by gavage with 50 mg Cr^{3+}/kg BW. The treatment did not affect general condition, BW gain, and hematological indices.[10] A brief rise in blood sugar level was noted during the 3rd month. Gross pathology examination failed to reveal considerable lesions to the visceral organs. 1.9 to 5.5 mg/kg BW doses of mono- and dichromates given for 29 to 685 days caused no signs of intoxication in dogs, cats, and rabbits (IARC 23-300). A cardiotoxic and gerontological effect of chromium compounds is reported.[11]

The latter is characterized by the change in the metabolic reactions of glucose conjugates and by morphological and functional changes in older animals. In a 1-year study, Sprague-Dawley rats were given Cr^{6+} and Cr^{3+} compounds at the concentration of 25 mg/l (2.5 mg/kg BW) in their drinking water. The only effect observed was some accumulation of chromium in different tissues.[12] From this study, the NOEL of 2.5 mg/kg BW and TDI of 0.005 mg/kg BW were identified (considering an uncertainty factor of 500, because of limitation of the study). But the results of a lifetime study[8] suggest that the NOEL for Cr^{3+} may be considerably higher.

Harlan SD rats were fed a stock diet to which was added 5.0, 25, 50, or 100 mg chromium/kg diet (as chloride or picolinate) for 24 weeks (highest levels are about 30 mg/kg BW). The treatment produced no changes in BW gain or hematology analyses. Chromium concentrations were increased in the liver and kidney. The study demonstrated the lack of toxicity of Cr^{3+} at levels that are much more than 10,000 times the upper limit of the estimated safe and adequate daily dietary intake for humans.[13]

Immunotoxicity. Chromium has been associated with systemic autoimmune disease.[14] Such reactions as asthma and contact dermatitis are reported to be induced by certain chromium compounds.

Reproductive Toxicity. Young mice were exposed to chromium chloride (Cr^{3+}) and potassium dichromate (Cr^{6+}) in drinking water for 12 weeks. The ingestion caused adverse effect on fertility and reproduction of male and female mice (Cr^{3+} and Cr^{6+}).[15]

Embryotoxicity. A single administration of Cr^{6+} to female rats at critical periods of embryonic development had a pronounced embryotoxic (but not teratogenic) effect.[16] However, survival and growth of NMRI mice were not affected in a 3-generation study where animals were given drinking water containing 135 mg Cr^{6+}/l.[17]

Dinerman believes that Cr^{6+} at the MAC level has no effect on embryonic development.[13] A dose of 0.008 mg/kg BW leads to reduction in the number of live neonates and to an increase in the overall pre- and postimplantation mortality. The LOEL is reported to be 0.25 mg Cr^{3+}/kg and 0.025 mg Cr^{6+}/kg BW for embryotoxic effect and 0.25 mg Cr^{6+} and Cr^{3+}/kg BW for gonadotoxicity.[19]

Teratogenicity. Administration of chromium compounds to pregnant rats on day 3 of gestation causes spontaneous changes in the neuroglia and neural tube defects in all embryos. A teratogenic effect with Cr^{3+} is obtained only at high doses, it being unclear whether this is the result of an effect on the fetus or on the body of the mother. Embryotoxic and teratogenic effect was observed in pure-bred and mongrel hamsters exposed to 8.0 mg/kg BW dose on day 8 of gestation.[20]

Addition of 5.0% insoluble, non-hydrated chromic oxide pigment to their feed did not result in embryo- and fetotoxicity or teratogenicity.[7] Teratogenic effects caused by chromium chloride, chromium trioxide, and chromium potassium dichromate were observed in rats.[6,21]

Gonadotoxicity. *I/p* administration of Cr^{6+} but not Cr^{3+} for five consecutive days to rats induced testicular atrophy and a reduction in epididymal sperm number after 60 days.[22] Conception rate in rats appears to be reduced following inhalation or food intake of Cr^{6+} or Cr^{3+} compounds for 2 months.[23]

Mutagenicity. Cr^{6+} compounds are positive in a wide range of *in vitro* and *in vivo* genotoxicity tests, whereas Cr^{3+} compounds are not. The explanation is that Cr^{6+} readily penetrates cell membranes in contrast to Cr^{3+}. The threshold for DLM effect is identified at the 0.25 mg/kg dose level for Cr^{6+} and Cr^{3+}.[17]

Cr^{6+}.

Observations in man. People occupationally exposed to Cr^{6+} compounds had elevated incidences of CA and SCE in their peripheral blood lymphocytes; reports on SCE induction were conflicting (IARC 23-205). In human cells *in vitro* these compounds caused CA, SCE and DNA damage.

In vivo cytogenetics. Cr^{6+} induced DLM, CA, and micronuclei in rodents treated *in vivo*. In cultured rodent cells, it induced transformation, CA, SCE, mutation, and DNA damage (IARC 2-100, IARC 23-205).

In vitro genotoxicity. SC induced DNA single-strand breaks in Chinese hamster *V79* cells in a concentration-dependent manner.[24] Chromium sodium oxide (2.5 to 7.5 μM) induced mutations at the HGPRT locus, but only within a very limited concentration range.[25] Several helants reduce or eliminate the mutagenicity of Cr_2O_3.[26]

Cr^{3+}.

Observations in man. No data are available on the genetic and related effects in humans. Conflicting results were obtained for the induction of CA in human lymphocytes *in vitro*, and neither SCE nor unscheduled DNA synthesis were induced in human cells.

In vivo cytogenetics. Mutagenic potential of Cr^{3+} is 100 times lower than that of Cr^{6+}. Cr^{3+} did not induce micronuclei in bone-marrow cells of mice treated *in vivo*. Conflicting results were obtained concerning the induction of CA, mutation and SCE in cultured rodent cells. Insoluble crystalline chromium oxide induced SCE and mutation in cultured Chinese hamster cells which were shown to contain particles of the test material (IARC 2-100; IARC 23-205).

Carcinogenicity. A reduction of Cr^{6+} to its lower oxidation states and related free-radical reactions play an important role in carcinogenesis.[27]

Observations in man. In epidemiology studies, an association was demonstrated between occupational exposure to Cr^{6+} compounds and mortality due to lung cancer, especially in the chromate-producing industry. Data on lung cancer risk in other chromium-associated occupational settings and for cancer at other sites than the lungs seem to be insufficient.[3] There is also some evidence of an increased hazard of GI tract cancer.

Animal studies. Cr^{6+} was reported to be carcinogenic via inhalation. The limited data available do not provide evidence for carcinogenicity via the oral route. In a rat study with the oral route of exposure to 2.0 to 5.0% insoluble, non-hydrated chromium oxide in the feed, no increase in tumor incidence was observed.[8] There is a limited evidence for the carcinogenicity of cobalt-chromium alloy in rats (IARC, 1980).

Carcinogenicity classifications. An IARC Working Group concluded that there is inadequate evidence for the carcinogenicity of chromium and trivalent chromium compounds in humans and there is limited evidence for the carcinogenicity of hexavalent chromium compounds in *humans*. An IARC Working Group concluded that there is inadequate evidence for the carcinogenicity of chromium and most trivalent chromium compounds in *experimental animals* and there is sufficient evidence for the following hexavalent chromium compounds in experimental animals: calcium chromate, chromium trioxide, chromium trioxide, zinc chromate, and others.

The Annual Report of Carcinogens issued by the U.S. Department of Health and Human Services (1998) defines chromium and certain chromium compounds to be known carcinogens, i.e., the substances for which the evidence from human studies indicates that there is a casual relationship between exposure to the substance and *human cancer.*

***Chromium total*:**

IARC: 3;

U.S. EPA: D.

***Cr^{6+}*:**

IARC: 1;

U.S. EPA: A;

U.S. ARC: 1;

EU: 2.

***Cr^{3+}*:**

IARC: 3.

Chemobiokinetics. Absorption after oral exposure is relatively low and depends on the speciation. Cr^{6+} is more readily absorbed from the GI tract than Cr^{3+} and is capable of penetrating the cellular membranes.

Adult male volunteers ingested a liter of deionized water containing concentrations ranging from 0.1 to 10 mg Cr^{6+}/l. The treatment caused a dose-related increase in urinary chromium excretion. Red blood cell and plasma chromium concentrations became elevated in certain individuals at the highest doses. Authors suggested that the ingested Cr^{6+} was reduced to Cr^{3+} before entering the bloodstream, since the chromium concentration in the erythrocytes dropped rapidly postexposure. Safe drinking water concentrations of Cr^{6+} are based in part on presumed capability of human gastric juices to rapidly reduce Cr^{6+} to non-toxic Cr^{3+} prior to systemic absorption. The Cr^{6+} drinking water standard of 0.1 mg Cr^{6+}/l is below the reductive capacity of the stomach.[9]

After entering the blood system, Cr^{6+} is likely to be selectively concentrated in the liver (the principal sites of chromium accumulation) and erythrocytes, where it undergoes metabolic inactivation. Cr^{6+} caused a sustained increase in red blood cell chromium levels, a specific marker of systemic uptake of Cr^{6+}. [9]

Other sites of chromium accumulation seem to be the kidneys and bones. According to other data, absorbed chromium is accumulated predominantly in the lung, which may contain 2 to 3 times the concentrations of other tissues.[28]

Cr^{3+} is a more stable oxidation state, and under physiological conditions it may form complexes with ligands such as nucleic acids, proteins and organic acids. Biological membranes are thought to be impermeable to Cr^{3+}, although phagocytosis of particulate Cr^{3+} can occur. The saliva and gastric juice in the upper alimentary tract of mammals, including humans, have a varied capability to reduce Cr^{6+} with the gastric juice having a notably high capacity. Thus, the normal body physiology provides detoxification for Cr^{6+}. Chromium is mainly excreted in the urine with some excretion through the bile and feces.[28]

Elevated urine chromium concentrations can be used as a general marker of exposure to chromium.

Regulations. ***U.S. FDA*** regulates the use of chromium as an indirect food additive and the use of chromium oxide in drugs and cosmetics. It may be used as a colorant in olefin polymers in the manufacture of articles intended for use in contact with food according to the provisions prescribed in 21 CFR part 178.3297.

Chromium (Cr^{3+}) complex of N-ethyl-N'-heptadecylfluorooctane sulfonyl glycine may be used (1) as a component of paper for packaging dry food in accordance to the provisions prescribed in 21 CFR part 178.3297.

Chromium caseinate, chromium potassium sulfate, chromium sulfate, chromium sodium oxide, and *chromium nitrate* may be used (2) as components of adhesive for food-contact surface intended for use in packaging, transporting, or holding food in accordance with the conditions prescribed in 21 CFR part 175.105.

Chromium potassium sulfate may be used as an adjuvant substance in the manufacture of foamed plastics intended for use in contact with food, subject to the provisions prescribed in 21 CFR part 178.3010.

Recommendations.

WHO (1996) sets a guideline value for chromium total (for analytical problems). Provisional Guideline value of 0.05 mg/l for chromium total was recommended as well as re-evaluation of chromium, when additional infor mation is available.

U.S. EPA (1999): Health Advisory for longer-term exposure is 0.02 mg chromium total/l.

Standards.

EU (1989). MAC for chromium total: 0.05 mg/l.

U.S. EPA (1999). MCL and MCLG for chromium total: 0.1 mg/l. The National Research Council in the National Academy of Sciences (NAS, 1989) established safe and adequate daily dietary intake for Cr^{3+} to range from 0.05 to 0.2 mg/day. The lower limit is based on the absence of deficiency symptoms in individuals consuming an average of 0.05 mg chromium/day.[29]

Russia (1995). MAC: 0.5 mg Cr^{3+}/l and 0.05 mg Cr^{6+}/l.

Canada (1989). MAC for chromium total: 0.05 mg/l.

References:

1. *Revision of the WHO Guidelines for Drinking-Water Quality*, Report of the First Review Group Meet ing on Inorganics, Bilthoven, the Netherlands, March 18-22, 1991, 5-6.
2. Castle, L., Offen, C. P., Baxter, M. J., and Gilbert, J., Migration studies from paper and board food packaging materials. 1. Compositional analysis. *Food Addit. Contam.*, 14, 35, 1997.
3. Borneff, I., Engelhardt, K., Griem, W., et al., Carcinogenic substances in water and soil, *Arch. Hyg.*, 152, 45, 1968.
4. Shustova, M. N. and Samoylovich, L. N., Toxicological characteristics of alumochromopotassium catalyst of A-30 type, *Gig. Truda Prof. Zabol.*, 10, 52, 1971 (in Russian).
5. Turebayev, K. et al., in *Industrial Hygiene, Occupational Pathology and Toxicology in the Chemical Industry and Non-Ferrous Metallurgy in Kazakhstan*, Alma-Ata, 1984, 215 (in Rus sian).
6. Diaz-Mayans, J., Laborda, R., Nunez, A., et al., Hexavalent chromium effects on motor activity and some metabolic aspects of Wistar albino rats, *Comp. Biochem. Physiol.*, 83, 191, 1986.
7. Karimova, R. I., Talayeva, U. G., and Vasyukovich L. Y., Laboratory experiments on the joint effects of metals and enterobacteria, *Gig. Sanit.*, 12, 58, 1982 (in Russian).
8. Janus, J. A. and Krajnc, E. I., Appendix to Report No 758701001, Integrated Criteria Document, *Chromium: Effects*, Natl. Institute Public Health. Environ. Protec., Bilthoven, the Netherlands, 1989.
9. Finley, B. L., Kerger, B. D., Katona, M. W., Gargas, M. L., Corbett, G. C., and Paustenbach, D. J., Human ingestion of chromium (VI) in drinking water. Pharmacokinetics following repeated exposure, *Toxicol. Appl. Pharmacol.*, 142, 151, 1997.
10. Ginzburg, F. I. and Faidysh, E. V., in *Protection of Water Reservoirs against Pollution by Industrial Liquid Effluents*, S. N. Cherkinsky, Ed., Medgiz, Moscow, Issue No 2, 1954, 101 (in Russian).
11. Krasovsky, G. N. et al., in *Current Biochemical Methods in Environmental Hygiene*, G. I. Sidorenko and R. V. Mercur'yeva, Eds., Moscow, 1982, 68 (in Russian).
12. MacKenzie, R. D., Byerrum, R. U., Decker, C. F., et al., Chronic toxicity studies. II. Hexavalent and trivalent chromium administered in drinking water to rats, *Am. Med. Assoc. Arch. Ind. Health*, 18, 232, 1958.
13. Anderson, R. A., Bryden, N. A., and Polansky, M. M. Lack of toxicity of chromium chloride and chromium picolinate, in *Toxicologist*, Abstracts of SOT 1996 Annual Meeting, Abstract No 1532, Is sue of *Fundam. Appl. Toxicol.*, 30, Part 2, March 1996.
14. Bagazzi, P. E., Autoimmunity caused by xenobiotics, *Toxicology*, 119, 1, 1991.
15. Elbetieha, A. and Al-Hamood, M. H., Long-term exposure of male and female mice to trivalent and hexavalent chromium compounds: effect on fertility, *Toxicology*, 116, 39, 1997.
16. Rozhdestvenskaya, N. A., in *Hygiene Aspect of Environment Protection*, Moscow, Issue No 6, 1978, 154 (in Russian).
17. Ivancovic, S. and Preussmann, R., Absence of toxic and carcinogenic effects after administration of high doses of chromic oxide pigment in subacute and long-term feeding experiments in rats, *Food Cosmet. Toxicol.*, 13, 347, 1975.

18. Dinerman, A. A., *The Role of Environmental Pollutants in Disturbance of Embryonic Development*, Meditsina, Moscow, 1980, 192 (in Russian).
19. Smirnov, M. E., Late effects of water-containing hexa- and trivalent chromium on the body, *Gig. Sanit.*, 7, 33, 1985
20. Schardein, J. L., *Chemically Induced Birth Defects*, 2nd ed., Rev. and expanded, Marcel Decker, Inc., New York, 1993, 843.
21. Gale, T. F., The embryotoxic response to maternal chromium trioxide exposure in different strains of hamsters, *Environ. Res.*, 29, 196, 1982.
22. Ernst, E., Testicular toxicity following short-term exposure to tri- and hexavalent chromium: an experi mental study in the rat, *Toxicol. Lett.*, 51, 269, 1990.
23. Ozernaya, Z. A., in *Industrial Hygiene and Human Health Protection*, Minsk, 1976, 120 (in Russian).
24. Sugiyama, M., Costa, M., Nakagawa, N., Hidaka, T., and Ogura, R., Stimulation of polyadenosine diphosphoribose synthesis by DNA lesions induced by sodium chromate in Chinese hamster V79 cells, *Cancer Res.*, 48, 1100, 1988.
25. Sugiyama, M., Lin, X., and Costa, M., Protective effect of vitamin E against chromosomal aberrations and mutation induced by sodium chromate in Chinese hamster V79 cells, *Mutat. Res.*, 260, 19, 1991.
26. Gentile, J. M., Hyde, K., and Schubert, J., Chromium genotoxicity as influenced by complexation and rate effects, *Toxicol. Lett.*, 7, 439, 1981.
27. Shi, X., Chiu, A., Chen, C. T., Halliwell, B., Castranova, V., and Vallyathan, V., Reduction of chromium(VI) and its relationship to carcinogenesis, *J. Toxicol. Environ. Health. B Crit. Rev.*, 2, 87, 1999.
28. *Handbook on the Toxicity of Inorganic Compounds*, H. G. Seiler, H. Siegel, and A. Siegel, Eds., Marcel Dekker, Inc., New York, 1988, 243.
29. *Federal Register*, vol 56, No 20, January 30, 1991, 3537.

CHROMIUM PHOSPHATE adhesives

Chromium phosphate
CAS No 7789-04-0
RTECS No GB6870000

Table. Composition and acute toxicity of compounds on the basis of monosubstituted chromium phosphates.

	Content, %						*Rats*		*Mice*	
Substances				*Cu*	*Al*		***LD_{50}, g/kg, calculated***			
	Cr^3	Cr^6	*Ni*	-	-	P_2O_5	as mol	as *Cr*	as mol	as *Cr*
Chromium phosphate $Cr(H_2PO_4)_3 H_2O$	9.0	-	-	2.8	-	33.1	1.57	0.14	4.3	0.39
Chromium copper phosphate	7.7	-	-	-	-	36.3	1.37	0.105	0.66	0.051
Chromium nickel phosphate	4.0	-	7.1	-	-	34.0	0.87	0.035	0.31	0.012

Table (continuation).

	Content, %						***Rats***		***Mice***	
Substances				*Cu*	*Al*		***LD_{50}, g/kg, calculated***			
Chromium nickel phosphate	4.0	-	7.1	-	-	34.0	0.87	0.035	0.31	0.012
Chromium calcium nickel phosphate	0.68	-	1.3	-	3.2	9.8	0.87	0.005	0.44	0.003
Calcium aluminum phosphate	1.9	3.8	-			30.4	0.60	0.034	0.45	0.026

Acute Toxicity. Gross pathology examination of the treated animals revealed acute gastritis with heavy infilt- ration of the submucosal layer by leukocytes, hyperemia, and hemorrhages. Other histological findings included dystrophic changes of the liver and kidney cells and increased hemosiderin deposition in the spleen (see *Table*).

Repeated Exposure revealed functional accumulation only in the case of chromium-nickel phosphates. K_{acc} is 3.1 (by Lim). All other chemicals show little accumulation. Presence of *chromium* in the molecule seems to be essential to determine the toxic effect of the compounds.

Reference:

Sigova, N. V. and Blokhin, V. A., Toxicity characteristics of chromium phosphate adhesives, *Gig. Truda Prof. Zabol.*, 6, 48, 1984 (in Russian).

COBALT compounds

Co

CAS No 7440-48-4

RTECS No GF8750000

Applications. Cobalt compounds are used primarily as catalysts and pigments.

Exposure. Cobalt compounds occur in vegetables via uptake from soil, and vegetables account for the major part of human dietary intake of cobalt. It is an important component of vitamin B_{12}.

Acute Toxicity. Following administration to rats by gastric intubation, LD_{50} (mg/kg BW) for different anhydrous cobalt compounds appeared to be cobalt oxide, 202; cobalt phosphate, 387; cobalt chloride, 418; cobalt sulfate, 424; cobalt nitrate, 434; cobalt acetate, 503.[1] Acute effects included sedation, diarrhea and decreased body temperature. Gross pathology examination revealed hemorrhages and dystrophic changes in the liver, kidney, and heart. According to Christensen and Luginbyh,[2] LD_{50} of cobalt chloride for rats is about 180 mg/kg BW. According to Smyth et al.,[3] LD_{50} of cobalt oxide is 1.7 g/kg BW in rats.

Repeated Exposure. Dietary components can greatly modify the toxicity of ingested cobalt. Oral administration of 2.5 to 10 mg cobalt/kg BW caused polycythemia in rats.[4,5]

Rabbits were dosed by gavage with 3.0 mg cobalt nitrate/kg BW for 2 months. Th blood system was predominantly affected. The treatment caused an increase in erythrocyte count and *Hb* level, reticulocytosis, and hyperplasia of the red bone marrow.[6]

Short-term Toxicity. Weanling rats died because of administration of daily doses of 1.0 and 0.5 mg cobalt for 35 months when fed a milk diet. Rats tolerated 1.0 mg cobalt/day in their water for 14 weeks.[03]

Long-term Toxicity.

Observations in man. Actual cases of acute or chronic poisoning are not common. Excessive ingestion is characterized by congestive heart failure and polycythemia and anemia.[7] Syndrome of cardiomyopathy was described in subjects consuming large volumes of beer containing added cobalt.[8,9]

Animal studies. Cobalt intoxication caused labored respiration, tremors and convulsions. Toxic nephritis was found to develop.[019]

Rats received 5.0 mg cobalt chloride/kg BW (calculated as cobalt) for 6 months. An impairment of hemopoi esis (true polycythemia) with subsequent thymal hyperplasia was noted.[10]

Allergenic Effect. Cobalt may provoke allergic dermatitis.[11]

Observations in man, including occupational asthma sometimes with rhinitis, urticaria, and pneumonitis, are considered to be occasional systemic hypersensitivity reactions (risk 1:100).[12,13]

Animal studies. A strong contact sensitizing potential is shown in maximized guinea pig as says.[14]

Reproductive Toxicity.

Gonadotoxicity. A dose of 265 mg cobalt chloride/kg BW given in the diet for 98 days, providing an initial dose of 20 mg/kg BW, induced degenerative and necrotic changes in the seminiferous tubules of rats.[15] Cobalt chloride reduced fertility in male mice.

Embryotoxicity. Rats received 12 to 48 mg cobalt chloride/kg BW by gavage from day 14 of gestation through day 21 of lactation. The treatment caused a decrease in the number of litters, growth and survival of the offspring.[16] Significant increase in the early embryonic losses was found in mice.[17]

Concentrations of 0.05 to 5.0 mg cobalt chloride/l in drinking water given to rats before or during pregnancy caused embryotoxicity and death.[18]

Teratogenicity. No developmental toxicity was observed in the offspring of rats given daily doses of 25 to 100 mg cobalt chloride/kg BW by gavage on days 6 to 15 of gestation.[19]

Cobalt acetate and cobalt chloride are not reported to be teratogenic in hamsters and rats, respectively.[20]

Mutagenicity. Genetic toxicology of cobalt and cobalt compounds has been reviewed.[20,21] In procaryotic assays, cobalt salts were generally non-mutagenic. They had a weak or no genetic effect in bacteria.

In vitro genotoxicity. In mammalian cells, cobalt compounds caused DNA strand breaks, SCE and aneuploidy, but not CA. Cobalt acetate enhanced viral transformation frequency, and the number of transformed foci among treated cells.[22] Cobalt sulfide induced morphological transformation in Syrian hamster embryo cells.

In vivo cytogenetics. Cobalt chloride induced CA in laboratory bred mice *in vivo* after single oral administrations of 1/10 to 1/40 of the lethal dose. The effect appeared to be dose dependent. Clastogenic effect of cobalt chloride is reported.[23]

Carcinogenicity. Workers occupationally exposed to cobalt compounds exhibited significantly enhanced risk for lung cancer. Interpretation of the available evidence for the carcinogenicity of cobalt compounds in animals is difficult (IARC, 1991). When administered to rats via *i/p* injections, cobalt oxide (76.7% cobalt) has a weak carcinogenic effect. Cobalt chloride appeared to induce carcinogenic effect in mice after oral administration of subtoxic doses.[24]

Carcinogenicity classification. An IARC Working Group concluded that there is evidence suggesting lack of carcinogenicity of cobalt compounds in *experimental animals* and there were inadequate data available to evaluate the carcinogenicity of cobalt compounds in *humans*.

IARC: 2B.

Chemobiokinetics. Absorption from the GI tract in rats was found to vary between 11 and 34% depending on administered dose value.[25] The highest concentration was found in the liver, a low concentration was noted in the kidney, pancreas, and spleen.[26,27]

Cobalt chloride was given to male F344 rats orally at 33.3 mg cobalt/kg BW. The study revealed the small extent of absorption following oral dosing. Heme oxygenase studies following oral administration did not result in an increase in activity over controls. Ayala-Fierro *et al.* concluded that the extent of cobalt absorption across the GI tract is incomplete, and that the concentration administered and the route of exposure may determine its systemic toxicity.[28]

Cobalt specifically reacts with *SH*-groups of dihydrolipoic acid, thus preventing oxidative decarboxylation of pyruvate to acetylcoenzyme A.[29] After a low dose of 2.8 mg/kg BW the tissue levels of cobalt compounds were not different from those in the control animals. The half-life of cobalt compounds in plasma was approximately 24 hours. Approximately 20% cobalt naphthenate administered orally was eliminated in the urine.[9]

Regulations. ***U.S. FDA*** (1998) listed cobalt compounds (1) as ingredients in resinous and polymeric coatings for food-contact surface in accordance with the conditions prescribed in 21 CFR part 175.300.

Cobalt naphthenate and *cobalt acetate* may be used as (2) a component of the uncoated or coated food-contact surface of paper and paperboard intended for use in producing, manufacturing, packaging, processing, preparing, treating, packing, transporting, or holding aqueous and fatty foods in accordance with the conditions prescribed in 21 CFR part 176.170; *Cobalt naphthenate* may be used as (3) an accelerator in the manufacture of cross-linked polyester resins for use as articles or components of articles intended for repeated use in contact with food in accordance with the conditions prescribed in 21 CFR part 177.2420; *Cobalt acetate* may be used as (4) a component of adhesives for food-contact surface in accordance with the conditions prescribed in 21 CFR part 175.105; and (5) as an ingredient of paper and paperboard intended for use in contact with dry food in accordance with the conditions prescribed in 21 CFR part 176.180.

Cobalt aluminate may be safely used as a colorant in the manufacture of articles intended for use in contact with food only (6) in resinous and polymeric coatings in accordance with the conditions prescribed in 21 CFR part 175.300 and (7) in melamine-formaldehyde resins in accordance with the conditions prescribed in 21 CFR part 177.1460; (8) in ethylene-vinylacetate copolymers in accordance with the conditions prescribed in 21 CFR part 177.1950; and (9) in food-contact surfaces of urea-formaldehyde resins in molded articles in accordance with the conditions prescribed in 21 CFR part 177.1900.

Standards. ***Russia*** (1995). MAC and PML: 0.1 mg cobalt/l.

References:

1. Speijers, G. J. A., Krajnc, E. I., Berkvens, J. M., et al., Acute oral toxicity of inorganic cobalt compounds in rats, *Food Chem. Toxicol.*, 20, 311, 1982.
2. Christensen, H. E. and Luginbyh, T. T., *Toxic Substances*, List 218, Public Health Service, NIOSH, Rockville, MD, 1974.
3. Smyth, H. F., Carpenter, C. P., Weil, C. S., Pozzani, U. C., Striegel, J. A., and Nyaun, J. S., Range-finding toxicity data: List VII, *Am. Ind. Hyg. Assoc. J.*, 30, 470, 1969.
4. Orten, J. M. and Bucciero, M. C., The effect of cystein, histidine, and methionine on the production of polycythemia by cobalt, *J. Biol. Chem.*, 176, 961, 1948.
5. Oskarsson, A., Reid, M. C., and Sunderman, F. W., Effects of cobalt chloride, nickel chloride, and nickel subsulfide upon erythropoiesis in rats, *Ann. Clin. Lab. Sci.*, 11, 165, 1981.
6. Yastrebov, A. P., in *Problems Physiol. Exp. Ther.*, 9, 34, 1965 (in Russian).
7. Calabrese, E. J., Canada, A. T., and Sacco, C., Trace elements and public health, *Ann. Rev. Public Health*, 6, 131, 1985.
8. Taylor, A. and Marks, V., Cobalt: A review, *J. Hum. Nutr.*, 32, 165, 1976.
9. *NTP Fiscal Year Annual Plan*, 1991, 100.
10. Gus'kova, V. N., Gurfein, A. N., and Pavlova, Z. K., in *Protection of Water Reservoirs against Pollution by Industrial Liquid Effluents,* S. N. Cherkinsky, Ed., Medgiz, Moscow, Issue No 3, 1959, 182 (in Russian).
11. Camarasa, J. M. G., Cobalt contact dermatitis, *Acta Dermat. Venerol.*, 47, 287, 1967.
12. Cugell, D. W., Morgen, W. K. C., Perkins, D. G., and Rubin, A., The respiratory effects of cobalt, *Arch. Int. Med.*, 150, 177, 1990.
13. Shirakawa, T., Kusaka, Y., Fujimara, N., Goto, S., Kato, Heki, S., and Morimoto, K., Occupational asthma from cobalt sensitivity in workers exposed to hard metal dust, *Chest* 95, 29, 1989.
14. Basketter, D. A. and Scholes, E. W., Comparison of the local lymph node assay with the guinea-pig maximization test for the detection of a range of contact allergens, *Food Chem. Toxicol.*, 30, 65, 1992.
15. Corrier, D. E., Mollenhauer, H. H., Clark, D. E., et al., Testicular degeneration and necrosis induced by dietary cobalt, *Vet. Pathol.*, 22, 610, 1985.
16. Domingo, J. L., Patternain, J. L., Llobet, J. M., et al., Effects of cobalt on postnatal development and late gestation in rats upon oral administration, *Rev. Esp. Fisiol.*, 41, 293, 1988 (in Sparish).
17. Pedigo, N. G., George, W. J., and Andersen, M. B., The effect of acute and chronic exposure to cobalt on male reproduction in mice, *Reprod. Toxicol.*, 2, 45, 1988.

18. Nadeenko, V. G., Lenchenko, V. G., Saichenko, S. P., Arkhipenko, T. A., and Radovskaya, T. L., Embryotoxic action of cobalt administered *per os*, *Gig. Sanit.*, 2, 6, 1980 (in Russian).
19. Paternain, J. L., Domingo, J. L., and Corbella, J., Developmental toxicity of cobalt in the rat, *J. Toxicol. Environ. Health*, 24, 193, 1988.
20. Leonard, A. and Lauwerys, R., Mutagenicity, carcinogenicity and teratogenicity of cobalt metal and cobalt compounds, *Mutat. Res.*, 239, 17, 1990.
21. Beyersmann, D. and Hartwig, A., The genetic toxicology of cobalt, *Toxicol. Appl. Pharmacol.*, 115, 137, 1992.
22. Casto, B. C., Meyers, J., and DiPaolo, A., Enhancement of viral transformation for evaluation of the carcinogenic or mutagenic potential of inorganic metal salts, *Cancer Res.*, 39, 193, 1979.
23. Palit, S., Sharma, A., and Taluker, G., Chromosomal aberrations induced by cobaltous chloride in mice *in vivo*, *Biol. Trace. Elem. Res.*, 29, 139, 1991.
24. Steinhoff, D. and Mohr, U., On the question of carcinogenic action of Co-containing compounds, *Exp. Pathol.*, 41, 169, 1991.
25. Taylor, D. M., The absorption of cobalt from the gastrointestinal tract of the rat, *Physiol. Med. Biol.*, 6, 445, 1962.
26. Stenberg, T., The distribution in mice of radioactive cobalt administered by two different methods, *Acta Odontol. Scand.*, 41, 143, 1983.
27. Clyne, N., Lins, L.-E., Pehrsson, S. K., et al., Distribution of cobalt in myocardium, skeletal muscle and serum in exposed and unexposed rats, *Trace Elem. Med.*, 5, 52, 1988.
28. Ayala-Fierro, F., Firriolo, J. M., and Carter, D. E., Disposition, toxicity, and intestinal absorption of cobaltous chloride in male Fischer 344 rats, *J. Toxicol. Environ. Health*, 56, 571, 1999.
29. Alexander, C. S., Cobalt-beer cardiomyopathy, *Am. J. Med.*, 53, 395, 1972.

CYANOGUANIDINE

Molecular Formula. $C_2H_4N_4$
M = 84.08
CAS No 461-58-5
RTECS No ME9950000
Abbreviation. CG.

Synonym. Dicyandiamide.

Properties. Colorless, crystalline substance. Solubility in water is 2.26% (13°C). Gives no color to water and does not affect its transparency. Taste perception threshold is 10 mg/l.

Applications. Used as a curing agent for epoxy resins and in the production of plastics and melamine.

Acute Toxicity. Mice and rats tolerate administration of 15 g/kg BW. Lethal doses caused respiratory center stimulation and vasodilatation in the skin. Hematology analysis was unchanged. Cases of fatal poisonings are not reported. However, according to Hold (IARC 19-206), LD_{50} is 4.0 g/kg BW in mice, and 3.0 g/kg BW in rabbits.[1]

Repeated Exposure revealed no cumulative properties.

Short-term Toxicity. Male and female F344 rats were fed CRF-1 powder diet containing 1.25, 2.5, 5.0, and 10 CG for 13 weeks. Inhibition of BW gain was more marked in the 10% group. Histopathological examination revealed intranuclear eosinophilic inclusion bodies in the proximal tubular epithelium of the kidney in both sexes of the 10% group. This level is considered unequivocally toxic.[2]

Long-term Toxicity. Rats were dosed with 1.0 and 2.0 g/kg BW for 6 months. The treatment produced some changes in cholinesterase activity, reduction in phagocytic activity, and a rise in the blood urea level in rats. Gross pathology examination revealed mild dystrophy of the liver and kidney. The NOAEL of 50 mg/kg BW was identified in this study. F344 rats were fed CRF-1 pulverized diets containing 2.5 or 5.0% CG for up to 2 years. Retardation of BW gain was noted in treated animals. No changes were observed during histological examination.[3]

Allergenic Effect. CG seems to be a contact sensitizer.

Carcinogenicity. No carcinogenic potential of CG was found during this study.[3]

Chemobiokinetics. CG does not exhibit properties of cyanide compounds and does not form CN^- ion. It is broken down in the body, forming *cyanamide* and *urea*.

Regulations. ***U.S. FDA*** (1998) approved the use of CG (1) in resinous and polymeric coatings for food-contact surface in accordance with the conditions prescribed in 21 CFR part 175.300; (2) as a modifier for aminoresins and as a fluidizing agent in starch and protein coatings for paper and paperboard intended for use in contact with dry food; (3) in adhesives used as components of articles intended for use in packaging, transporting, or holding food in accordance with the conditions prescribed in 21 CFR part 175.105; (4) in the manufacture of rubber articles intended for repeated use in producing, manufacturing, packaging, processing, preparing, treating, packing, transporting, or holding food in accordance with the conditions prescribed in 21 CFR part 177.2600; (5) as an antioxidant in the manufacture of articles or components of articles intended for use in contact with food, subject to the provision to use CG only at levels not to exceed 1.0% by weight of polyoxymethylene copolymer; (6) as a stabilizer (total amount of stabilizers not to exceed 2.0%) in polyoxymethylene copolymer which may be used as an article or a component of articles intended for food-contact use; (7) in the manufacture of antioxidants and/or stabilizers for polymers which may be used in the manufacture of articles or components of articles intended for use in producing, manufacturing, packaging, processing, preparing, treating, packing, transporting, or holding food in accordance with the conditions prescribed in 21 CFR part 178.2010; as (8) a component of the uncoated or coated food-contact surface of paper and paperboard intended for use in producing, manufacturing, packaging, processing, preparing, treating, packing, transporting, or holding aqueous and fatty foods in accordance with the conditions prescribed in 21 CFR part 176.170; and (9) as a retarder of rubber articles intended for repeated use in food-contact articles (total not to exceed 10% by weight of the rubber product).

Standards. ***Russia*** (1995). MAC and MPL: 10 mg/l (organolept., taste).

References:

1. Gabrilevskaya, L. N., Laskina, V. P., and Faidysh, Ye. V., Experimental substantiation of the maximum allowable concentration for dicyandiamide in water reservoirs, in *Industrial Pollution of Water Reservoirs*, S. N. Cherkinsky, Ed., Meditsina, Moscow, Issue No 9, 1969, 96 (in Russian).
2. Matsushima, Y., Onodera, H., Ogasawara, H., Kitaura, K., Mitsumori, K., Maekawa, A., and Takahashi, M., Subchronic oral toxicity study of cyanoguanidine in F344 rats, Abstract, *Eisei Shikenjo Hokoku,* 109, 61, 1991 (in Japanese).
3. Yasuhara, K., Shimo, T., Mitsumori, K., et al., Lack of carcinogenicity of cyanoguanidine in F344 rats, *Food Chem Toxicol.*, 35, 475, 1997.

CYANURIC ACID

Molecular Formula. $C_3H_3N_3O_3$
M = 129.09
CAS No 108-80-5
RTECS No XZ1800000
Abbreviation. CA.

Synonyms. Isocyanic acid; Pseudocyanuric acid; *sym*-Triazine-2,4,6-triol; Tricyanic acid; Trihydroxy-cyanidine; 2,4,6-Trihydroxy-*S*-triasine.

MONOSODIUM CYANURATE

Molecular Formula. $C_3H_3N_3O_3.Na$
M = 152.08
CAS No 2624-17-1
RTECS No XZ1912000
Abbreviation. MC.

Properties. CA occurs as a colorless, odorless crystalline powder with a slightly bitter taste. Solubility in water is 2.7 to 5.0 g/l, poorly soluble in hot alcohol. A concentration of 6.0 mg/l does not change organoleptic properties of water.

Applications. CA derivatives are used as cross-linking agents in polymerization. CA is used in swim- ming pools to lower the rate of photochemical reduction of chlorine, hypochlorous acid, and hypochlorite ion.[019]

Acute Toxicity. LD_{50} of CA is reported to be 3.37 g/kg BW in rats, and 7.7 g/kg BW in mice.[1] However, according to other data, rats and rabbits tolerate administration of a 10 g/kg BW dose without toxic manifestations.[2,03] High doses of CA are found to cause adynamia, motor coordination disorder, and hematuria in the treated animals. Death occurs in 1 to 3 days.

Repeated Exposure revealed cumulative properties. K_{acc} is 1.85 (by Kagan). No evidence of bioaccumulation was noted in rats and dogs following repeated oral administration of 5.0 mg MC/kg BW for 15 days.

Short-term Toxicity. Three dogs were fed 8.0% MC in their diet for 2 years. Two dogs died during the study. Gross pathology and histology examination revealed kidney fibrosis, focal dilatation and epithelial proliferation of Bellini's ducts.[3]

Bladder calculi with accompanying hyperplasia were observed in a few animals from the group exposed orally to MC at a dose of 5375 mg/l for 90 days. Hematological indices, urinalysis or histopathology were not altered in animals exposed to 896 and 1792 mg/kg BW (Monsanto Material Safety Data Sheet, 1992).

Long-term Toxicity. In a 6-month study, administration of 0.8% MC in the diet induced no evident adverse effects in three dogs.[3]

Oral administration of 30 mg CA/kg BW to guinea pigs and rats for 6 months caused dystrophic changes in the kidneys; the dose of 3.0 mg CA/kg BW produced no effect in animals.[3]

In a 2-year study, MC was administered to mice and rats in their drinking water. No considerable signs of toxicity (except for an increased incidence of calculi in the bladders of high-dosed male rats) or treatment-related oncogenic effects were observed. The NOAELs were considered to be the concentrations of 2.4 g/l in rats and 5.38 g/l in mice (Monsanto Material Safety Data Sheet, 1992).

Reproductive Toxicity.

Teratogenicity. No increase in skeletal or external defects was observed in the offspring of rats given 500 mg CA/kg BW with or without calcium hypochlorite on days 6 to 15 of gestation.[3] No teratogenic effects were noted in the offspring of rats (doses up to 5.0 g/kg BW) or rabbits (doses up to 0.5 g/kg BW) exposed by gavage to MC on days 6 to 15 (rats) or on days 6 to 18 (rabbits) of gestation.[1]

Embryotoxicity. No adverse treatment-related effects on reproductive performance were observed following MC administration to male and female rats in their drinking water at concentrations of 0.4 to 5.38 g/l throughout 3 consecutive generations.[4] An increased incidence of calculi in the urinary bladder of high-dosed F_2 parent males was noted. The histology examination of the urinary bladder revealed microscopic changes attributed to chronic irritation from the calculi (Monsanto Material Safety Data Sheet, 1992).

Mutagenicity.

In vitro genotoxicity. No mutagenic activity of MC was observed in *Salmonella* test, and in *in vitro* mouse lymphoma cell point mutation assay. It did not induced SCE in Chinese hamster ovary cells *in vitro*.[5] CA was also negative in *Salmonella* mutagenicity assay.[025]

In vivo cytogenetics. No induction of rat bone marrow cell clastogenesis was observed *in vivo*.[5]

Carcinogenicity. Rats were fed 150 to 300 mg CA/kg BW. Cysticerian sarcomas were observed between 19 to 25 month. Fibroadenoma of mammary gland was found in 2 females. Mice received oral doses of 280 to 310 mg CA/kg BW. The treatment caused myeloid leukosis in 2/14 mice that have survived for 23 months.[3]

Chemobiokinetics.

Observations in man. CA was shown to be rapidly and quantitatively eliminated unchanged in excreta following oral ingestion by human volunteers.

Animal studies. In rats and dogs, MC was completely absorbed after oral administration. It is rapidly excreted unchanged with the urine following a single oral dose of 5.0 mg/kg BW. The radiolabelled CA is rapidly eliminated unchanged in the urine of rats.

Standards. ***Russia*** (1995). MAC and PML: 25 mg/l (organolept., taste).

References:

1. Babayan, E. A. and Alexandryan, A. V., Toxicity of cyanuric acid, *Pharmacologia i Toxicologia*, 49, 122, 1986 (in Russian).
2. Bidnenko, L. I., Problem of toxicity of cyanuric acid and triallyl isocyanurate, in *Hygienic Aspects of the Use of Polymeric Materials and Articles Made of Them*, L. I. Medved', Ed., All-Union

Research Institute of Hygiene and Toxicology of Pesticides, Polymers and Plastic Materials, Kiyv, 1969, 299 (in Russian).
3. Canelli, E., Chemical, bacteriological and toxicological properties of cyanuric acid and chlorinated isocyanurates as applied to swimming pool disinfection, *Am. J. Publ. Health*, 64, 155, 1974.
4. Wheeler, A. G. et al., Three generation reproduction study in rats administered cyanurate, *Toxicologist*, 5, 189, 1985.
5. Hammond, B. G. et al., Absence of mutagenic activity for monosodium cyanurate, *Fundam. Appl. Toxicol.*, 5, 655, 1985.

DIACETOXYDIBUTYL STANNANE

Molecular Formula..$C_{12}H_{24}O_4.Sn$
M = 351.05
CAS No 1067-33-0
RTECS No WH6880000
Abbreviation. DADBS.

Synonyms. Bis(acetyloxy)dibutylstannate; Diacetoxydibutyltin; Dibutyltin diacetate.

Properties. Colorless or yellow clear liquid with a slight acetic acid odor. Insoluble in cold and hot water, soluble in alcohol.

Applications. Stabilizer for polymers, in particular for polyvinyl chloride especially if colorlessness and transparency are required. Catalyst for silicone and urethane foams.

Acute Toxicity. In mice, LD_{50} is reported to be 110 mg/kg BW[1] or 46 mg/kg BW; in rats, LD_{50} is 32 mg/kg BW.[012]

Repeated Exposure. Rats received 0.25 LD_{50} for 10 days. Light and electron microscopy investigation revealed liver lesions. Swelling of granular and agranular endoplasmic reticulum followed soon after the mitochondrial changes were observed. Progressive congestion of bile canaliculi occurred, due in part to cellular and mcrovilli swelling.[1]

Reproductive Toxicity.

Embryotoxicity. DADBS caused maternal toxicity only at the highest dose level when administered orally to rats throughout pregnancy.[2] There was an increase in the incidence of dead or resorbed fetuses and total fetal resorption as well as an increase in external malformations.[3]

Teratogenicity. Dibutyltins exhibited a general cytotoxic effect on rabbit articular and growth-plate chondrocytes that could be interpreted as possible effects on skeletal growth and devel opment.[4]

Following administration of 1.7 to 15 mg/kg BW to Wistar rats on days 7 to 17 of gestation, dose-dependent thymic atrophy in the pregnant animals was reported. The treatment caused an increase in incidence of fetuses with external malformations, such as cleft mandible, cleft lower lip, ankyloglossia (tongue-tie), and schistoglossia (cleft tongue) in a dose-dependent manner.[3]

Immunotoxicity. Dibutyltin compounds are known to be potent thymolytic and immunotoxic agents in rats.

Mutagenicity.

In vivo cytogenetics. DADBS was negative in *Dr. melanogaster*.

In vitro genotoxicity. DADBS was found to be negative in *Salmonella typhimurium* and a large battery of mutagenicity assays but produced base-pair substitution in one of the bacterial strains tested.[5-7]

Positive results were shown in mouse lymphoma cells and in Chinese hamster ovary cells (NTP-85).

Carcinogenicity. F344 rats were exposed to 66.5 or 133 ppm in their diet, while $B6C3F_1$ mice received 76 or 152 ppm in the feed for 78 weeks. Increased incidence of hepatocellular adenomas in female mice and both hepatocellular adenomas and carcinomas in male mice were found but these were not statistically significant.[6]

Carcinogenicity classification.

NTP: N - IS - N - N (feed).

Chemobiokinetics. Following injection, the highest concentrations were observed in liver and kidney of mice and rats.[3]

DADBS undergoes monooxygenase or nonenzymatic cleavage to *butyltin derivatives*. Several metabolites were identified in the liver and feces of mice treated with DBTA by gavage,[8] *n-butyltin trichloride* being one of the main metabolites.[3]

In mice, the major part of the dose is excreted in the feces unchanged; small quantities of the dose are eliminated in the expired air as *carbon dioxide* and *butene*.[2]

Regulations. ***U.S. FDA*** (1998) approved the use of DADBS in polyurethane resins for the food-contact surface of articles intended for use in contact with dry food alone or in combination, at levels not to exceed a total of 3.0% by weight in the resin used, subject to the provisions described in 21 CFR part 177.1680.

References:

1. Casarett and Doull's *Toxicology*, J. Doull, C. D. Klassen, and M. D. Amdur, 2nd ed., Macmillan Publ. Co., New York, 1980, 550.
2. Noda, T., Morita, S., Shimizu, M., et al., Safety evaluation of chemicals for use in household products. (VII) Teratology studies of di-*n*-butyltin diacetate in rats, *Annu. Rep. Osaka. Inst. Publ. Health Environ. Sci.*, 50, 66, 1988.
3. Noda, T., Yamano, T., Summizu, M., Saitoh, M., Nakamura, T., Yamada, A., and Morita, S., Comparative teratogenicity of di-*n*-butyltin diacetate and *n*-butyltin trichloride in rats, *Arch. Environ. Contam. Toxicol.*, 23, 216, 1992.
4. Webber, R. J., Dollins, S. C., Harris, M., and Hough, A. J., Effect of alkyltins on rabbit articular and growth-plate chondrocytes in monolayer culture, *J. Toxicol. Environ. Health,* 16, 229, 1985.
5. Tennant, R. W., Stasiewicz, S., and Spalding, J. W., Comparison of multiple parameters of rodent carcinogenicity and *in vitro* genetic toxicity, *Environ. Mutagen.*, 8, 205, 1986.
6. *Bioassay of Dibutyltin Diacetate for Possible Carcinogenicity*, NTP Technical Report Series No 183, Natl. Cancer Institute, DHEW Publ. 79-1739, Washington, D. C., 1979.
7. Woodruff, R. C., Mason, J. M., Valencia, R., and Zimmerling, S., Chemical mutagenesis testing in *Drosophila.* V. Results of 53 coded compounds tested for National Toxicology Program, *Environ. Mutagen.*, 7, 677, 1985.
8. Boyer, I. J., Toxicity of dubutyltin, tributyltin and other organotin compounds to humans and to experi mental animals, *Toxicology*, 55, 253, 1989.

1,4-DIAZABICYCLO(2,2,2)OCTANE

Molecular Formula. $C_6H_{12}N_2$
M = 112.17
CAS No 280-57-9
RTECS No HM0354200
Abbreviation. DACO.

Synonym. Triethylenediamine.

Properties. Yellowish, hygroscopic, crystalline powder with a specific, penetrating odor. Readily soluble in water. Taste and odor perception threshold appears to be 60 mg/l.[1]

Applications. Used as a catalyst and a foaming agent in the production of polyurethane foams.

Migration from the polyurethane coating into water (3-day exposure at a temperature of 37°C) was defined at the level of 0.5 mg/l.[2]

Acute Toxicity. LD_{50} is reported to be 3.3 g/kg BW in rats, 2.25 g/kg BW in guinea pigs, and 1.1 g/kg BW in rabbits. No differences are evident in the sex or species sensitivity. Poisoning is accompanied mainly by NS impairment with tension in the muscles and extremities and clonic convulsions with subsequent death in the first 3 to 5 hours after poisoning. The acute action threshold for the effect on STI appeared to be 1.55 g DACO/kg BW. Gross pathology examination revealed distention of the stomach and small intestine, edema of the brain matter, and myocardial flaccidity.[1]

Repeated Exposure revealed slight cumulative properties. Administration of 1/5 and 1/10 LD_{50} produced only mild signs of intoxication on day 3.

Long-term Toxicity. Rats were dosed by gavage with a 30 mg/kg BW oral dose for 6 months. The treatment affected the liver functions.[1]

Reproductive Toxicity. No selective ***embryotoxic effect*** is reported.[3]

Gonadotoxic effect is found only when high doses, causing maternal toxicity, are administered.[4]

Mutagenicity.

In vitro genotoxicity. DACO was found to be mutagenic in the *Salmonella typhimurium* mutation assay. It was considered active in the induction of SCE in Chinese hamster ovary cells. It produced significant increases in the amount of unscheduled DNA synthesis activity and was considered positive in inducing primary DNA damage in this assay. In a micronucleus study with Swiss-Webster mice, no clastogenic activity was observed with DACO.[5]

Standards. ***Russia*** (1995). MAC and PML: 6.0 mg/l.

References:

1. Troenkina, L. B., Hygienic substantiation of the MAC for triethylenediamine and monooxyethyl-piperasine in water bodies, *Gig. Sanit.,* 5, 67, 1980 (in Russian).
2. Krat, A. V., Kesel'man, I. M., and Sheftel', V. O., Sanitary-chemical evaluation of polymeric articles used in water-supply constructions, *Gig. Sanit.*, 10, 18, 1986 (in Russian).
3. Martson', L. V. and Sheftel', V. O., Teratogenic effects of polymeric materials and methods of their investigation, in *Hygienic Aspects of the Use of Polymeric Materials*, Proc. 2nd All-Union Meeting on Health and Safety Monitoring of the Use of Polymeric Materials in Construction, Kiyv, 1976, 104 (in Russian).
4. Mazayev, V. T., Troenkina, L. B., and Gladun, V. I., The effect of aliphatic amines on reproduction in animals, *Gig. Sanit.*, 10, 66, 1986 (in Russian).
5. Leung, H. W., Evaluation of the genotoxic potential of alkyleneamines, *Mutat. Res.,* 320, 31, 1994.

DI-*tert*-BUTYLPEROXYSUCCINATE

Molecular Formula. $C_{12}H_{22}O_6$

M = 262.34

Synonym. Diperoxysuccinic acid, di-*tert*-butyl ester.

Properties. White powder. Poorly soluble in water, readily soluble in alcohol.

Applications. Used as a polymerization initiator, in the form of 50% solution in dibutyl phthalate.

Acute Toxicity. No rats died when a single oral dose of 4.0 g/kg BW was administered. In mice, LD_{50} is 5.3 g/kg BW for females and 7.7 g/kg BW for males, but it appeared to be much higher if D. is given to mice as a 50% solution in dibutyl phthalate. In such a case, LD_{50} was 16.5 g/kg BW in females, while males tolerated 10 g/kg BW. There were no manifestations of the toxic action or retardation of BW gain in the survivors compared with the controls during a 3-week follow-up. Gross pathology examination revealed no changes in the viscera.[1]

Mutagenic Effect is found on exposure to disuccinyl peroxide (succinic acid peroxide).[2]

References:

1. Krynskaya, I. L., Yevsyukov, V. I., and Sukhareva, L. V., Toxicological characteristics of peroxides used in plastics production, in *Environmental Protection in Plastic Industry and Safety of the Use of Plastics*, T. N. Zelenkova and B. Yu. Kalinin, Eds., Plastpolymer, Leningrad, 1978, 133 (in Russian).
2. Luzzarti, D. and Chevallier, M. R., Comparison of the lethal and mutagenic action of an organic peroxide and radiation on *E. coli*, *Ann. Inst. Pasteur*, 93, 366, 1957.

2-(DIETHYLAMINO)ETHANOL

Molecular Formula. $C_6H_{15}NO$

M = 117.22

CAS No 100-37-8

RTECS No KK5075000

Abbreviation. DEAE.

Synonyms. *N,N'*-Diethylethanolamine; Diethyl-β-hydroxyethylamine; 2-Hydroxytriethylamine.

Properties. Viscous, hygroscopic liquid. Water soluble; miscible with alcohol. Takes up CO_2 from air.

Applications. DEEA is used as a resin-curing agent and a chemical intermediate.

Acute Toxicity. In rats, LD_{50} of non-neutralized DEAE is 1.3 g/kg BW, that of neutralized DEAE is 5.6 g/kg BW.[012]

Repeated inhalation **exposure** to 10 ppm DEAE produced no toxic effects. Concentration of 301 ppm caused significant mortality.[1]

Short-term Toxicity. In a 14-week inhalation rat study, there were no significant changes in hematology analyses or neurobehavioral parameters.[1]

Long-term Toxicity. In a 6-month dietary study, neutralized DEAE was given to rats in their drinking water. The treatment produced a decrease of BW gain and increased kidney relative weights. Neither renal lesions nor other treatment-related effects were found.[2]

Mutagenicity.

In vitro genotoxicity. Is shown to be negative in *Salmonella* mutagenicity assay.[015]

Carcinogenicity. Oncogenic properties of aliphatic amines are still questioned. They do not appear to cause carcinogenic action without nitrosation.

Chemobiokinetics. Easily absorbed from GI tract. Up to 60% DEAE accumulated in the liver, CNS, and spinal cord. Eliminated within first 24 hours predominantly via kidneys.[3]

References:

1. Hinz, J. P., Thomas, J. A., and Ben-Dyke, R., Evaluation of the inhalation toxicity of diethylethanolamine (DEEA) in rats, *Fundam. Appl. Toxicol.*, 18, 418, 1992.
2. Kornish, H. H., Oral and inhalation toxicity of 2-diethylaminoethanol, *Am. Ind. Hyg. Assoc. J.*, 26, 479, 1965.
3. Schulte, K. E., Dreymann, E., and Mollmann, H., Resorption, Verteilung in den Organen und Metabolizirung von Diathylaminoathanol nach Applikation at Ratten, *Arzneimittel-Forsch.*, 22, 1381, 1972.

DIETHYLENETRIAMINE

Molecular Formula.. $C_4H_{13}N_3$
M = 103.17
CAS No 111-40-0
RTECS No IE1225000
Abbreviation. DETA.

Synonyms. Aminoethylethanediamine; 2,2'-Diaminodiethylamine.

Properties. Colorless, oily liquid with a pungent specific odor. Readily miscible with water. Odor perception threshold is 0.7 mg/l at 20°C and 0.2 mg/l at 60°C. Taste perception threshold is 1.3 mg/l.[1]

Applications. Used in the synthesis of epoxy and ion exchange resins.

Acute Toxicity. For rats, mice, and rabbits LD_{50} is in the range of 970 to 1200 mg/kg BW. However, according to other data,[07] DETA is less toxic on a single administration, and LD_{50} is 2.33 g/kg BW. Guinea pigs seem to be the most sensitive animals: LD_{50} is 600 mg/kg BW. Poisoned animals displayed behavioral changes and convulsions.

Repeated Exposure failed to reveal cumulative properties of DETA. Guinea pigs tolerated administration of 120 mg/kg BW for a month.

Short-term Toxicity. F344 rats were fed a diet containing 1000, 7500, or 15000 ppm DETA hydrochloride for 90 consecutive days. The treatment caused a decrease in BW, but no treatment-related clinical signs, or gross pathology of histological findings were noted at any dose level. The NOEL of 1000 ppm (70 mg/kg BW) was established in this study.[2]

Long-term Toxicity. In a 6-month study, guinea pigs were dosed by gavage with 0.6 mg DETA/kg BW. The treatment did not cause retardation of BW gain; it did not affect the contents of *metHb* in the blood and ascorbic acid in the liver, or the activity of liver enzyme systems. Administration of 10 mg/kg BW to rabbits over a 6-month period caused a reduction in prothrombin blood level and an increase in the activity of glutaminate-oxalate and glutaminate-pyruvate transaminase. The dose of 1.0 mg/kg BW was identified to be the NOEL in this study.[1]

Mutagenicity. DETA could produce a mutagenic action due to contamination with ethyleneimine and other highly active mutagens. It was evaluated for potential genotoxic activity using a battery of *in vitro* and *in vivo* assays.

In vitro cytogenetics. It was not mutagenic in *Salmonella typhimurium* mutation assay, and inactive in the Chinese hamster ovary gene mutation assay. It also showed no effect in SCE assay with or without metabolic activation and upon unscheduled DNA synthesis with hepatocytes.[3-5]

Chemobiokinetics. Excretion occurs mainly with feces and urine.

Regulations.

U.S. FDA (1998) approved the use of DETA (1) as a modifier for aminoresins for use in paper and paperboard articles coming in the contact with aqueous, fatty, and dry food; (2) in the manufacture of resinous and polymeric coatings intended for use in food-contact surface of articles in accordance with the conditions prescribed in 21 CFR part 175.300; (3) as a defoaming agent that may be safely used in the manufacture of paper and paperboard intended for use in producing, manufacturing, packing, transporting, or holding food in accordance with the conditions prescribed in 21 CFR part 176.210, and (4) as a component of the uncoated or coated food-contact surface of paper and paperboard intended for use in producing, manufacturing, packaging, processing, preparing, treating, packing, transporting, or holding dry food in accordance with the conditions prescribed in 21 CFR part 176.180 .

EU (1990). DETA is available in the *List of monomers and other starting substances which may continue to be used for the manufacture of plastic materials and articles intended to come into contact with foodstuffs pending a decision on inclusion in Section* A.

Great Britain (1998). DETA is authorized without time limit for use in the production of polymeric materials and articles in contact with food or drink or intended for such contact. The specific migration of this substance shall not exceed 5.0 mg/kg.

Standards. *Russia* (1995). MAC and PML: 0.2 mg/l (organolept., odor).

References:

1. Trubko, E. I. and Teplyakova, E. V., Studies on diethylenetriamine hygienic norm-setting in water bodies, *Gig. Sanit.*, 7, 103, 1972 (in Russian).
2. Leung, H. W. and Miller, J. P., Effects of diethylenetriamine dihydrochloride following 13 weeks of dietary dosing in Fischer 344 rats, *Food Chem. Toxicol.*, 35, 481, 1997.
3. Hulla, J. E., Rogers, S. J., and Warren, J. R., Mutagenicity of a series of polyamines, Environ. Mutagen., 3, 332, 1981.
4. Leung, H. W., Evaluation of the genotoxic potential of alkyleneamines, *Mutat. Res.*, 320, 31, 1994.
5. Hedenstedt, A., Rannug, U., Ramel, C., et al., Mutagenicity and metabolism studies on 12 thiuram and dithiocarbamate compounds used as accelerators in the Swedish rubber industry, *Mutat. Res.*, 68, 313, 1979.

DIMETHYLAMINE

Molecular Formula. C_2H_7N

M = 45.09

CAS No 124-40-3

RTECS No IP8750000

Abbreviation. DMA.

Synonym. *N*-Methylmethanamine.

Properties. A gas, easily converted into a colorless liquid on cooling or under pressure, with the odor of ammonia or rotting fish. DMA is soluble in water (550 g/l at 25°C) and in ethanol. The aqueous solutions exhibit an alkaline reaction (unstable methyl ammonium hydroxide is formed). With acids, DMA forms salts which are non-volatile, odorless, water-soluble solids. The odor perception threshold is 0.67 mg/l; the practical threshold is 1.0 mg/l[1] or according to other data, 0.29 mg/l.[02] The taste perception threshold is much higher.

Applications. DMA is predominantly used as a curing agent or emulsifier. It is also found to be a degradation product in rubber materials which are made using dialkyldithiocarbamine acids derivatives as vulcanization accelerators.

Reported migration levels of 0.1 to 0.2 mg/l usually depend of rubber and contact media compositions.[1]

Acute Toxicity. LD_{50} is 698 mg/kg BW in rats, 316 mg/kg BW in mice, and 240 mg/kg BW in guinea pigs and rabbits. When DMA is neutralized by *HCl*, LD_{50} for rats, guinea pigs, and rabbits appeared to be 1.0, 1.07, and 1.6 g/kg BW, respectively.

The poisoning is accompanied by excitation with subsequent adynamia and motor coordination disorder. Gross pathology examination revealed extensive hemorrhages in the stomach and intestinal wall.[1] Amines are known to affect the activity of the mixed oxidase, function and metabolism of biogenic amines, this culminating in the impaired neurohumoral control.

Repeated Exposure failed to reveal cumulative properties. Guinea pigs and rabbits were dosed by gavage with 1/10 LD_{50} for a month. The treatment affected blood *Hb* level, cholinesterase activity, serum urea, and ascorbic acid contents in the visceral organs. There was an increase in the urinary coproporphyrin and in the relative liver weights.[2]

Long-term Toxicity. Weanling male Wistar rats received drinking water containing 0.2% DMA for 9 months. The treatment caused higher *in vitro* lipoperoxidation, free lysosomal enzyme activities, and cytosole superoxide dismutase activity in liver. Authors concluded that induced toxicity developed through some free radical reactions.[3] Temporary rise of urea level in the blood serum, an increase in the leukocyte count, and a decline in amount of ascorbic acid in the suprarenals were noted in guinea pigs.[2]

Reproductive Toxicity.

Embryotoxicity. No maternal or fetal effects were found in mice given *i/p* 2.5 or 5.0 mM DMA/kg BW on days 1 through 17 of gestation.[4]

Addition of DMA to culture of mouse embryos damaged their growth and viability.[5]

Mutagenicity.

In vivo cytogenetics. Genotoxic effect has not been observed in mice.[6]

In vitro genotoxicity. DMA gave negative results in CA test carried out on cultured Chinese hamster cells.[7]

Carcinogenicity. Following ingestion of sodium nitrite and DMA by mice, a marked dose-dependent inhibition of liver nuclear DNA synthesis was noted. This study suggests the possibility of *in vivo* biosynthesis of carcinogenic *nitrosamines*.[03]

Following inhalation exposure to DMA and nitrogen peroxide, their reaction products are reported to exhibit carcinogenic potential.[8]

Chemobiokinetics. A small amount of DMA undergoes oxidative transformation. In mammals, the secondary amines, to which DMA refers, are more resistant to deamination, compared with the primary. They may be deaminated or excreted unchanged. D. is a potential substrate for *in vivo* nitrosation to form *N-nitrosodimethylamine*, a potent carcinogen. DMA is predominantly eliminated via the urine, and to a lesser extent it is excreted with the bile or secreted into the intestine, where it may be reabsorbed.[9]

When ^{14}C-DMA was given *i/p*, recovery of radioactivity in the urine was essentially complete. Thus, exhalation, fecal excretion, and accumulation in tissues of DMA and its metabolites were negligible.[10] According to Zhang et al., in rats and mice given 20 mmol ^{14}C-DMA/kg BW, radioactivity was rapidly excreted, predominantly with the urine. Unmetabolized DMA came to 9.0% of radioactivity, the only metabolite determined was *methylamine* (3.4 to 4.5% of the total radioactivity excreted).[11]

Standards. ***Russia.*** MAC and PML: 0.1 mg/l.

References:

1. Duchovnaya, I. S., Grushevskaya, N. Yu., and Kazarinova, N. F., Methods of detection of secondary aliphatic amines in sanitary-chemical investigations of rubber, in *Hygiene of Use, Toxicology of Pesticides and Polymeric Materials*, Coll. Works, All-Union Research Institute of Hygiene and Toxicology of Pesticides, Polymers and Plastic Materials, A. V. Pavlov, Ed., Kiyv, Issue No 14, 1984, 112 (in Russian).

2. Dzhanashvili, G. D., Hygienic substantiation of the maximum contents of dimethylamine in water bodies, *Gig. Sanit.*, 6, 12, 1967 (in Russian).
3. Darad, R., De, A. K., and Aiyar, A. S., Toxicity of nitrite and dimethylamine in rats, *Toxicol. Lett.*, 17, 125, 1983.
4. Guest, I. and Varma, D. R., Developmental toxicity of methylamines in mice, *J. Toxicol. Environ. Health*, 32, 319, 1991.
5. Varma, D. R., Guest, I., Smith, S., and Mulay, S., Dissociation between maternal and fetal toxicity of methyl isocyanates in mice and rats, *J. Toxicol. Environ. Health*, 30, 1, 1990.
6. Friedman, M. A., Miller, G., Sengupta, M., et al., Inhibition of mouse liver protein and nuclear RNA synthesis following combined oral treatment with sodium nitrile and dimethylamine or methylbenzylamine, *Experientia*, 28, 21, 1972.
7. Ishidate, M. and Odashima, S., Chromosome tests with 134 compounds in Chinese hamster cell *in vitro* - a screening for chemical carcinogens, *Mutat. Res.*, 48, 337, 1977.
8. Benemansky, V. V., Prusakov, V. M., and Dushutin, K. K., Biological activity of reaction products of dimethylamine and hydrogen dioxide during short-term exposures in animals, *Gig. Sanit.*, 12, 15, 1979 (in Russian).
9. Ishiwata, H., Iwata, R., and Tanimura, A., Intestinal distribution, absorption and secretion of dimethylamine and its biliary and urinary excretion in rats, *Food Chem. Toxicol.*, 22, 649, 1984.
10. Smith, J. L., Wishnok, J. S., and Deen, W. M., Metabolism and excretion of methylamines in rats, *Toxi col. Appl. Pharmacol.*, 125, 296, 1994.
11. Zhang, A. O., Mitchell, S. C., and Smith, R. L., Fate of dimethylamine in the rat and mouse, *Xenobiotica*, 24, 1215, 1994.

N,N'-DIMETHYLANILINE

Molecular Formula. $C_8H_{11}N$
M = 121.20
CAS No 121-69-7
RTECS No BX4725000
Abbreviation. DMA.

Synonyms. Dimethylaminobenzene; *N,N'*-Dimethylphenylamine.

Properties. Pale-yellow oily liquid with a strong tar odor. Freely soluble in alcohol. Solubility in water is 1454 ppm at 25°C.

Applications. Catalytic hardener and solvent. DMA is used as a polymerizing agent and polymerization accelerator in the manufacture of bone cements and prosthetic devices.

Acute Toxicity. LD_{50} is 1.35 g/kg BW in male Carworth-Wistar rats (Smyth et al., 1962). A single dose of 50 mg DMA/kg BW caused *metHb* formation.[011]

Repeated Exposure. F344/N rats and $B6C3F_1$ mice were gavaged in corn oil with doses of 94 to 1,500 mg DMA/kg BW for 2 weeks. Death of rats and mice occurred at 750 or 1,500 mg DMA/kg. Lower doses caused decrease in BW gain. Clinical signs of intoxication included cyanosis (in rats) and lethargy and tremors (in rats and mice). Dose-related splenomegaly was found in nearly all doses.[1]

Short-term Toxicity. F344/N rats and $B6C3F_1$ mice were gavaged in corn oil with doses of 31.25 to 500 mg DMA/kg BW for 13 weeks. No chemical-related deaths occurred in any groups of animals. Dose-related decreases in BW gain were noted in the higher doses. Manifestations of the toxic action included lethargy in rats and mice and cyanosis in rats. In addition, there were tremors in rats and mice, and dose-related splenomegaly at all doses. Compound-related extramedullary hematopoiesis and hemosiderosis were observed in the kidney or testes of rats and liver and spleen of the rats and mice. An LOAEL of 31.25 mg/kg BW was suggested in this study.[1,2]

Long-term Toxicity. F344/N rats received 3.0 or 0.3 mg DMA/kg BW, $B6C3F_1$ mice were given 15 or 30 mg DMA/kg BW in corn oil by gavage for 2 years. The treatment caused fatty metamorphosis and fibrosis in the spleen of high dose male rats, and spleen hemosiderosis and hematopoiesis in rats.[1]

Weanling male Wistar rats received drinking water containing 0.2% DMA for 9 months. The treatment caused higher *in vitro* lipoperoxidation, free lysosomal enzyme activities, and cytosole superoxide dismutase activity in liver. Authors concluded that induced toxicity developed through some free radical reactions.[3]

Reproductive Toxicity. Pregnant CD-1 mice received 365 mg/kg BW in corn oil by gavage on days 6 through 13 of gestation. The treatment caused no toxic effect on the dams or in their offspring.[4]

Mutagenicity.

In vitro genotoxicity. DMA was found to be negative in *Salmonella* mutagenicity assay. It produced a positive response in the mouse lymphoma assay; it was found to be weakly positive in inducing DNA damage; it proved clearly positive in inducing numerical chromosome alterations. DMA induced SCE in Chinese hamster ovary cells.[1,5]

Carcinogenicity. 2-year NTP study in F344/N rats and B6C3F$_1$ mice revealed increased incidence of sarcomas of the spleen or osteosarcomas (combined) in male rats. There was no or equivocal evidence of carcinogenicity in the other tested groups of animals.[1]

Carcinogenicity classification.

NTP: SE - NE - NE - NE (gavage).

Chemobiokinetics. DMA is demethylated in rat liver. Urinary metabolites in dogs are *4-aminophenol, 4-(methylamino)phenol, 4-(dimethylamino)phenol, 2-aminophenol,* and *N-methylaniline*.[6,7]

Regulations. ***U.S. FDA*** (1998) approved the use of DMA in the manufacture of cross-linked polyester resins which may be used as articles or components of articles intended for repeated use in contact with food in accordance with the conditions prescribed in 21 CFR part 177.2420.

References:

1. *Toxicology and Carcinogenesis Studies of N,N'-Dimethylaniline in F344/N Rats and B6C3F$_1$ Mice (Gavage Studies)*, NTP Technical Report Series No 360, Research Triangle Park, NC, October 1989.
2. Abdo, K. M., Jokinen, M. P., and Hiles, R., Subchronic (13 weeks) toxicity studies of *N,N'*-dimethylaniline administered to Fischer 344 rats and B6C3F$_1$ mice, *J. Toxicol. Environ. Health*, 29, 77, 1990.
3. Darad, R., De, A. K., and Aiyar, A. S., Toxicity of nitrite and dimethylamine in rats, *Toxicol. Lett.*, 17, 125, 1983.
4. Hardin, B. D., Schuler, R. L., Burg, J. R., Booth, G. M., Hazelden, K. P., MacKenzie, K. M., Piccirillo, V. J., and Smith, K. N., Evaluation of 60 chemicals in a preliminary developmental toxicity test, *Teratogen. Carcinogen. Mutagen.*, 7, 29, 1987.
5. Taningher, M., Pasquini, R., and Bonatti, S., Genotoxicity analysis of *N,N'*-dimethylaniline and *N,N'*-dimethyl-*p*-toluidine, *Environ. Molec. Mutagen.*, 21, 349, 1993.
6. Kiese, M. and Renner, G., Urinary metabolites of *N,N'*-dimethylaniline produced by dogs, *Naunyn-Schmiedeberg's Arch. Pharmacol.*, 283, 143, 1974.
7. Birner, G. and Neumann, H.-G. (1988) Biomonitoring of aromatic amines. II. Hemoglobin binding of some monocyclic aromatic amines, *Arch. Toxicol.*, 62, 110, 1988.

α,α'-DIMETHYLBENZYL HYDROPEROXIDE

Molecular Formula. $C_9H_{12}0_2$

M = 152.19

CAS No 80-15-9

RTECS No MX2450000

Abbreviation. DMBHP.

Synonyms and **Trade Names.** Bis-α,α'-(dimethylbenzyl) peroxide; Cumyl hydroperoxide; Dicumyl peroxide; Diisopropylbenzene peroxide; α,α'-Dimethylbenzyl hydroperoxide; Isopropylbenzene hydroperoxide; Percum.

Properties. A colorless, oily liquid with an odor reminiscent of ozone, or light-yellow crystalline substance. Poorly soluble in water, readily soluble in alcohol and vegetable oils. Odor perception threshold is 3.0 mg/l.[1]

Applications. Used in the production of vinyl chloride and acrylates as a polycondensation and polymerization initiator; an initiator in the synthesis of unsaturated polyester resins. Polymerization catalyst in the production of polystyrenes. Used in vulcanization and plasticization of natural and synthetic rubber and for curing epoxy resins. A cross-linked agent for silicone rubber, polyethylene foam and polystyrene foam. Acetophenone is found to be a degradation product of DMBHP (Novitskaya, 1984).

Acute Toxicity. LD_{50} is reported to be 800 to 1270 mg/kg BW in rats, and 342 to 350 mg/kg BW in mice.[2] Single lethal doses affect the CNS, liver, and kidney. Poisoned animals displayed general inhibition, drowsiness, and dyspnea; death occurs from respiratory failure. Gross pathology examination revealed parenchymatous dystrophy of the liver, kidney, and myocardium as well as hyperemia of the gastric and intestinal mucosa, congestion in the visceral organs, and liver enlargement.[3]

Repeated Exposure revealed little evidence of cumulative properties.[1]

Short-term Toxicity. Rats given a 100 mg/kg BW dose for 3 months displayed no changes in the blood formula or in catalase and cholinesterase activity.[1] In the course of treatment animals became lethargic and drowsy.

Long-term Toxicity. In a 7-month study, rabbits were given doses of up to 25 mg DMBHP/kg BW. The treatment affected different liver functions. Vitamin C blood level was decreased while carotine level increased. Cholinesterase activity tended to be reduced.[1]

Reproductive Toxicity.

Embryotoxic and ***teratogenic effects*** are observed in chick embryos.[4]

Mutagenicity.

In vitro genotoxicity. DMBHP is found to be negative in the *Salmonella* mutagenicity bioassay (NTP-92).

In vivo cytogenetics. No DLM in the male sex cells is noted in rats given 0.25 mg/kg BW for 2 months.[5] Genotoxic effect is noted in *Dr. melanogaster.*[6]

Carcinogenicity. Incomplete data are reported concerning carcinogenic effect following *s/c* injections (administered amount unspecified).[7]

Regulations. ***U.S. FDA*** (1998) approved the use of DMBHP (1) in adhesives as a component of articles intended for use in packaging, transporting, or holding food in accordance with the conditions prescribed in 21 CFR part 175.105; (2) in the manufacture of cross-linked polyester resins to be safely used as articles or components of articles intended for repeated use in contact with food in accordance with the conditions prescribed in 21 CFR part 177.2420; and (3) as an accelerator of rubber articles intended for repeated use in producing, manufacturing, packing, processing, treating, packaging, transporting, or holding food (total not to exceed 1.5% by weight of the rubber product) in accordance with the conditions prescribed in 21 CFR part 177.2600.

Standards. ***Russia***. MAC and PML: 0.5 mg/l.

References:

1. Smirnova, R. D. and Kosmina, L. F., Experimental data on substantiation of MAC for isopropylbenzene hydroperoxide in water bodies, *Gig. Sanit.*, 12, 17, 1971 (in Russian).
2. Ovcharov, V. G. and Loit, A. O., Problem of toxicity of some organic peroxides, in *Toxicology and Hygiene of High-Molecular-Mass Compounds and of the Chemical Raw Material Used for Their Synthesis*, Proc. 2nd All-Union Conf., A. A. Letavet and S. L. Danishevsky, Eds., Leningrad, 1964, 74 (in Russian).
3. Orlova, F., in *Proc. Kuibyshev Research Hygiene Epidemiology Institute*, Kuibyshev, Issue No 5, 1968, 107 (in Russian).
4. Korhonen, A., Hemminki, K., and Vainio, H., Embryotoxic effects of eight organic peroxides and hydrogen peroxide on three-day chicken embryos, *Environ. Res.*, 33, 54, 1984.
5. Martson', L. V. and Sheftel', V. O., Investigation of the genetic effects of cumyl hydroperoxide by the dominant lethal mutation test, in *Hygienic Aspects of the Use of Polymeric Materials*, Proc. 3rd All-Union Meeting on New Methods of Hygienic Monitoring of the Use of Polymers, K. I. Stankevich, Ed., Kiyv, December 2-4, 1981, 328 (in Russian).

6. Sheftel', V. O., Shquar, L. A., and Naumenko, G. M., Application of some genetic methods in hygiene studies, *Vrachebnoy Delo*, 7, 128, 1969 (in Russian).
7. Anonymous, Peroxides, genes and cancer, *Food Chem. Toxicol.*, 23, 957, 1985.

N,N'-DIMETHYLCYCLOHEXYLAMINE

Molecular Formula. $C_8H_{17}N$
M = 127.26
CAS No 98-94-2
RTECS No GX1198000

Synonyms and **Trade Name.** Cyclohexanamine; *N*-Cyclohexyldimethylamine; (Dimethylamino)cyclohexane; Polycat 8.

Properties. Colorless, highly volatile liquid with an unpleasant odor. Poorly soluble in water, readily soluble in oil and alcohol.

Applications. Used as a catalyst in the production of polyurethane foams.

Acute Toxicity. LD_{50} is reported to be 450 mg/kg BW in rats, 320 mg/kg BW in mice,[1] 720 mg/kg BW in guinea pigs, and 620 mg/kg BW in rabbits (Schmidt et al., 1983).

Poisoning is accompanied by clonic-tonic convulsions with subsequent death in the first two hours after administration. D. exhibits no after-effect and does not produce methemoglobinemia in contrast to the structurally similar cyclohexylamine.

Reproductive Toxicity. No gonadotoxic effect was observed after inhalation exposure to D.

Mutagenicity.

***In vivo cytogenetics*.** An increase in the number of CA in bone marrow cells was observed in rats as a result of D. inhalation at a concentration of 92 mg/m^3.

Reference:

1. Smirnova, E. S., Kasatkin, A. N., Obryadina, G. I., et al., Toxic properties of *N,N'*-dimethylcyclohexylamine, *Gig. Truda Prof. Zabol.*, 5, 54, 1984 (in Russian).

α,α'-DIMETHYL-4-CYCLOHEXYLBENZYL, HYDROPEROXIDE

Molecular Formula. $C_{15}H_{22}O_2$
M = 234.33

Synonym. *p*-Isopropyl cyclohexylbenzene peroxide.

Properties. Crystals. The technical product is a yellowish-brown liquid with an unpleasant odor. Solubility in water is 20 mg/l, readily soluble in alcohols and ethers.

Applications. Used as a polymerization initiator.

Acute Toxicity. In mice, LD_{50} is 2.6 g/kg BW. The dose of 3.5 g/kg BW causes 100% mortality in animals. Poisoning is accompanied by adynamia.

Reference:

Taradin, Ya. I, Kuchmina, N. Ya., Fetisova, L. N., et al., Toxicological characteristics of isopropylcyclohexylbenzene and isopropyl cyclohexylbenzene hydroperoxide, in *Hygiene and Toxicology of High- Molecular-Mass Compounds and of the Chemical Raw Material Used for Their Synthesis*, Proc. 4th All-Union Conf., S. L. Danishevsky, Ed., Khimiya, Leningrad, 1969, 169 (in Russian).

ELAMINE-65

Composition. Product of the alkylation of hexamethylenediamine by isoprene.

Properties. A clear liquid.

Applications. Used as an elasticizing curing agent for epoxy resins.

Acute Toxicity. In mice, LD_{50} is found to be 1.6 g/kg BW.

Repeated Exposure failed to reveal cumulative properties. Rats were exposed to 160 mg/kg BW oral dose for a month. The treatment caused irritation of the GI and urinary tracts. Hematology changes and CNS inhibition were found to develop.

Reference:

Volodchenko, V. A., Vasilenko, N. M., and Sadokha, Ye. R., Data on toxicology of elamine-65 and 3,3'-dimethyl-4,4'-diaminodiphenylmethane, in *Hygiene and Toxicology of High-Molecular-Mass Compounds and of the Chemical Raw Material Used for Their Synthesis*, Proc. 4th All-Union Conf., S. L. Danishevsky, Ed., Khimiya, Leningrad, 1969, 219 (in Russian).

(EPOXYETHYL)BENZENE

Molecular Formula. C_8H_9O
M = 120.2
CAS No 96-09-3
RTECS No CZ9625000
Abbreviation. StO.

Synonyms. Epoxystyrene; Phenylethylene; Phenyloxirane; Styrene oxide; Styrene-7,8-oxide.

Properties. Colorless to straw-colored liquid with a pleasant odor. Water solubility is 0.28%, miscible with alcohol. Odor threshold concentration is 0.063 mg/l.[010]

Applications. Used in the production of epoxy plastics as a catalyst, a cross-linking agent, and a reactive diluent for epoxy resins, and also for manufacture of coatings.

Migration Data. StO could in principle be present in polystyrene food packs as a contaminant formed by the oxidation of styrene monomer. Concentrations of StO were measured in base resins and samples of polystyrene articles intended for food contact. StO was not detected in the resins (limit of detection 0.5 mg/kg) but was found in 11 out of 16 packaging samples at up to 2.9 mg/kg. Calculated migration levels expected in packaged foods were from 0.002 to 0.15 µg SO/kg foods. Hydrolysis of the epoxide group gave rise to the *diol* as the principal product. Ring opening in aqueous ethanol simulant media gave the diol and also the glycol monoethyl ether. Instability of StO led to formation of hydrolysis products that are less toxic than parent epoxide.[1]

Acute Toxicity. The LD_{50} values are reported to be 2.0 to 4.3 g/kg BW in rats, 2.8 g/kg BW in rabbits, and 2.0 g/kg BW in guinea pigs.[2,3]

Reproductive Toxicity.

Embryotoxicity. Inhalation of StO by rats, rabbits, and hamsters had a certain toxic effect on embryos, increasing the frequency of ossification defects of the sternebrae and occipital bones, but caused no teratogenic effect.[4,5]

Mutagenicity.

In vitro genotoxicity. In mammalian cell cultures, including Chinese hamster cell lines, mouse cell lines, Wistar rat hepatocytes and human lymphocytes, StO was genotoxic, producing mutations, CA, micronuclei, and anaphase bridges.[6-8] In human lymphocytes (whole blood cultures), StO was clustogenic and produced CA, micronuclei, and SCE. StO (~5.0 mg/plate) induced mutations (spot test) in *Salmonella typhimurium*.[9,10]

In vivo cytogenetics. StO had a low level of DNA binding activity in experimental animals.[11] In rodents exposed to StO, CA were seen in mice and rats but not in Chinese hamsters.[6-8]

StO initiated CA in the bone marrow cells of CD-1 mice given in doses of 50 to 1,000 mg/kg BW.[12]

Carcinogenicity. Administration of 275 or 550 mg StO/kg BW (rats) and 375 and 700 mg StO/kg BW (mice) in corn oil by gavage 3 times a week for 2 years caused a high incidence of squamous cell carcinomas or papillomas of the forestomach in both rats and mice treated. No neoplasm were observed in control animals. StO was found to be carcinogenic to rats and mice in this study.[13]

Male and female Sprague-Dawley rats were administered 50 and 250 mg/kg BW by gavage for 52 weeks.[14-16] There was a high incidence of forestomach epithelial tumors, including papillomas, *in situ* carcinomas, and invasive carcinomas. A dose response was demonstrated for these tumors. No tumors were reported in the control animals. StO was administered by oral gavage in corn oil to male CD rats at two dose levels: 1.65 or 240 mg/kg BW. The treatment did not result in detectable production of DNA adducts in an *in vivo* situation.

Cantoreggi and Lutz[17] suggested that StO tumorigenic potency depends on strong tumor promotion by high-dose cytotoxicity followed by regenerative hyperplasia.

Carcinogenicity classifications. An IARC Working Group concluded that there is sufficient evidence for the carcinogenicity of StO in *experimental animals* and there are inadequate data available to evaluate the carcinogenicity of StO in *humans*.

IARC: 2A;

EU: 2.

Chemobiokinetics. StO is the main metabolite of *styrene* (q.v.), formed under the action of cytochrome *P-450*-dependent monoxygen of the liver. StO metabolism may occur via the formation of *conjugates with glutathione*. Thioether metabolites of StO, which are precursors of the *mercapturic acids*, failed to show mutagenic activity.[13] StO is also found to be transformed into *phenylethylene glycol* by microsomal epoxyhydrases of the liver, kidneys and other organs.[16]

Metabolites are removed by the kidneys.

Regulations. ***U.S. FDA*** (1998) approved the use of StO as (1) a cross-linked agent for epoxy resins in coatings for containers with a volume of 1000 gallons (3785 liters) or more intended for repeated use in contact with alcoholic beverages containing up to 8.0% ethanol by volume; (2) a component of paper and paperboard for contact with dry food in accordance with the conditions prescribed in 21 CFR part 176.180; (3) in cross-linked polyester resins for repeated use as articles or components of articles coming in contact with food in accordance with the conditions prescribed in 21 CFR part 177.2420; (4) in the manufacture of resinous and polymeric coatings to be safely used in food-contact surface in accordance with the conditions prescribed in 21 CFR part 175.300.

References:

1. Philo, M. R., Fordham, P. J., Damant, A. P., and Castle, L., Measurement of styrene oxide in polystyrenes, estimation of migration to foods, and reaction kinetics and products in food simulants, *Food Chem. Toxicol.*, 35, 821, 1977.
2. Smyth, H. F., Carpenter, C. P., Weil, C. S., et al., Range-finding toxicity data: List VII, *Am. Ind. Hyg. Assoc. J.*, 30, 470, 1969.
3. Weil, C. S., Condra, N., Haun, C., et al., Experimental carcinogenicity and acute toxicity of representative epoxides, *Am. Ind. Hyg. Assoc. J.*, 24, 305, 1963.
4. Sikov, M. R. et al., Reproductive toxicology of inhaled styrene oxide in rats and rabbits, *J. Appl. Toxicol.*, 6, 155, 1986.
5. Kimmel, C. A. et al., Reproductive and developmental toxicology of selected epoxides, in *Toxicology and the Newborn*, Ch. 13, S. Kacem and M. J. Reasor, Eds., Elsevier Sci. Publ., Amsterdam, 1984, 270.
6. Norppa, H. and Vainio, H., Genetic toxicity of styrene and some of its derivatives, *Scand. J. Work Environ. Health*, 9, 108, 1983.
7. Vainio, H., Norppa, H., Hemminki, K., et al., Metabolism and genotoxicity of styrene, in *Biological Reactive Intermediates-LL: Chemical Mechanisms and Biological Effects*, R. Snyder, D. V. Parke, J. Kocisis, et al., Eds., Plenum Press, New York, Part A, 1982, 207.
8. Vainio, H., Norppa, H., and Belevedere, G., Metabolism and mutagenicity of styrene oxide, *Progr. Clin. Biol. Res.*, 141, 215, 1984.
9. Milvy, P. and Garro, A., Mutagenic activity of styrene oxide, a presumed styrene metabolite, *Mutat. Res.*, 40, 15, 1976
10. de Meester, C., Poncelet, F., Roberfroid, M., et al., Mutagenic activity of styrene and styrene oxide. A preliminary study, *Arch. Int. Physiol. Biochim.*, 85, 398, 1977.
11. Phillips, D. H. and Farmer, P. B., Evidence for DNA protein binding by styrene and styrene oxide, *CRC Crit. Rev. Toxicol.*, Suppl. 24, 35, 1994.
12. Loprieno, N., Presciuttini, S., Sbrana, I., Stretti, G., Zaccaro, L., Abbondolo, A., Bonatti, S., Fioro, R., and Mazzaccaro, A., Mutagenicity of industrial compounds. VII. Styrene and styrene oxide: II. Point mutations, chromosome aberrations and DNA repair induction analyses, *Scand. J. Work Environ. Health*, 4 (Suppl. 2), 169, 1978.
13. Lijnsky, W., Rat and mouse forestomach tumors induced by chronic oral administration of styrene oxide, *J. Natl. Cancer Inst.*, 77, 471, 1986.

14. Maltoni, C., Failla, G., and Kassapidis, G., First experimental demonstration of the carcinogenic effects of styrene oxide, *Med. Lavoro*, 5, 358, 1979.
15. Maltoni, C., Early results of the experimental assessment of the carcinogenic effects of one epoxy solvent: styrene oxide, in *Occup. Health Hazards of Solvents*, vol 2., A. Englund, K. Ringen, and M. A. Mehlman, Eds., Princeton Sci. Publ., Princeton, NJ, 1982, 97.
16. Conti, B., Maltoni, C., Perino, G., and Ciliberti, A., Long-term carcinogenicity bioassays on styrene administered by inhalation, ingestion and injection and styrene oxide administered by ingestion in Sprague-Dawley rats, and *para*-methylstyrene administered by ingestion in Sprague-Dawley rats and Swiss mice, *Ann. NY Acad. Sci.*, 534, 203, 1988.
17. Cantoreggi, S. and Lutz, W. K., Investigation of the covalent binding of styrene-7,8-oxide to DNA in rat and mouse, *Carcinogenesis*, 13, 193, 1992.
18. Pogano, D. A., Yagen, B., Hernandez O., et al., Mutagenicity of (R) and (S) styrene-7,8 -oxide and the intermediary mercapturic acid metabolites formed from styrene-7,8-oxide, *Environ. Mutagen.*, 5, 575, 1982.
19. Oesch, F., Jerina, D. M., Daly, J. W., et al., Induction, activation, and inhibition of epoxide hydrolase, *Chem.-Biol. Interact.*, 6, 189, 1973.

1,2-ETHANEDIAMINE

Molecular Formula. $C_2H_8N_2$
M = 60.10
CAS No 107-15-3
RTECS No KH8575000
Abbreviation. EDA.

Synonyms. β-Aminoethylamine; 1,2-Diaminoethane; Dimethylenediamine; Ethylenediamine.

Properties. Colorless, oily, yellow viscous hygroscopic liquid with a penetrating odor of ammonia. Mixes with water and alcohol at all ratios. Threshold perception concentration for odor change is 0.21 mg/l, practical odor perception threshold is 0.85 to 1.0 mg/l.[1, 010] Taste perception threshold is 0.8 mg/l (practical threshold is 2.5 mg/l). According to other data,[01] organoleptic perception threshold is 12 mg/l.

Applications. EDA is used in the synthesis of polyamides, in papermaking, and as a curing agent and hardener for epoxy resins. A solvent for resins.

Acute Toxicity. LD_{50} was reported to be 0.7 to 1.4 g/kg BW in rats, and 0.45 g/kg BW in mice and guinea pigs.[2,3,05] According to other data, the median lethal doses for rats and mice are 1.2 and 1.0 g/kg BW, respectively.[1] LD_{50} of EDA dihydrochloride was found to be 3.25 g/kg BW in rats, and 1.62 to 1.77 g/kg BW in guinea pigs.[3]

Lethal doses led to excitation followed by CNS inhibition and, often, convulsions. Asthmatic syndrome was shown to develop. Manifestations of the toxic action included a reduction in the activity of mono- and diamino- oxidase in the liver and catalase and peroxidase activity. Alterations in serotonin metabolism are noted.

Repeated Exposure revealed moderate cumulative properties. K_{acc} is 5.4 (by Lim). Rats were gavaged with 1/10 LD_{50}. The treatment resulted in disorder of liver and kidney functions, and in an increase of their relative weights. The contents of γ-globulin in the blood were increased.[1]

Rats and mice received EDA in the diet at dose levels up to 2.7 g/kg for 7 days. The treatment depressed BW gain and some visceral organs weight at the highest dose level.[3]

Short-term Toxicity. In a 3-month study, rats were given 0.05 to 1.0 g EDA dihydrochloride/kg BW. The highest dose caused retardation of BW gain in F344 rats. No increase in mortality was observed. Gross and histopathology examination revealed dose-related increase in hepatocellular pleomorphism and mild hepatocellular degeneration.[3]

Long-term Exposure. F344 rats were dosed with 20, 100, or 350 mg EDA.2*HCl*/kg diet (equivalent to 9.0, 45, or 158 mg EDA/kg BW) for 2 years. The liver was found to be a target tissue, with changes in the size and shape of hepatocytes and their nuclei. The NOAEL of 9.0 mg EDA/kg BW was established in this study.[4]

Reproductive Toxicity.

Embryotoxicity. Pregnant CD-1 mice were given EDA at dose of 400 mg/kg BW in water by gavage on days 6 to 13 of gestation and were allowed to deliver. The treatment caused reductions in the birth weight and BW gain in the offspring of treated dams. No maternal toxicity was noted.[5]

Ingestion of up to 178 mg *EDA* dihydrochloride/kg BW on 6 to 19 gestation days produced neither maternal toxicity nor embryotoxicity effects in New Zealand white rabbits.[6]

No evidence of ***teratogenicity*** was found in rats treated with up to 1.0 g EDA dihydrochloride/kg BW on days 6 through 15 of gestation.[7]

Allergenic Effect. A strong contact sensitizing potential was found in maximized guinea-pig assays.[8] Systemic hypersensitivity reactions in humans, including asthma, rhinitis, fever, anaphylactic shock when used as an additive in aminophylline solution or in rubber production, were considered to be occasional (risk 1:100).

Mutagenicity.

EDA was evaluated for potential genotoxic activity using a battery of *in vitro* and *in vivo* assays.

In vitro genotoxicity. Equivocal data are reported in *Salmonella* mutation assay. According to Haworth et al., EDA appeared to be weakly mutagenic in four *Salmonella typhimurium* strains, both in the presence and absence of *S9* homogenate fraction.[9]

EDA was inactive in the Chinese hamster ovary gene mutation assay and in SCE assay with or without metabolic activation. It exhibited no positive effects upon unscheduled DNA synthesis with Sprague-Dawley rat hepatocytes.[10-12]

In vivo cytogenetics. F344 rats were given 0.05 to 0.5 g EDA/kg BW for 23 days. Genotoxic effect was not observed in DLM assay.[9] EDA did not induce micronuclei in bone marrow cells in Swiss-Webster mice (*in vivo*).[11]

It was negative in *Dr. melanogaster* recessive lethal mutation assay.[13]

Carcinogenicity. In the above-cited study, no neoplastic effects were observed in Fischher 344 rats.[4]

Chemobiokinetics. After ingestion and absorption in the GI tract of animals, EDA is metabolized by *N*-acetylation which may also occur in man. The principal metabolites in the rat urine are *N-acetyl* and *N,N'-di-acetyl derivatives*. A small amount of *hippuric acid* is also found. The major urinary and fecal metabolite after oral administration in rats is *N-acethylethylenediamine*, which may subsequently undergo further metabolism to EDA, *aminoacetaldehyde, ethanolamine*, and eventually *carbon dioxide*. Tagged EDA is passed via the urine (47%) and expired air (18%).[14,15] The basic metabolite *N-acetylethylenediamine* provided 50% urinary radioactivity.

Wistar rats were treated with ^{14}C-EDA.2*HCl* at dose levels of 5.0, 50, and 500 mg/kg BW. EDA was mostly found in the thyroid, bone marrow, liver, and kidneys. Urinary excretion was shown to be the main route of elimination (42%); 5.0 to 32% radioactivity were excreted via feces, 9.0% via expired air in the form of CO_2.[16]

Regulations.

U.S. FDA (1998) regulates the use of EDA (1) in resinous and polymeric coatings in a food-contact surface of articles intended for use in producing, manufacturing, packing, transporting, or holding food in accordance with the conditions prescribed in 21 CFR part 175.300; as (2) a component of the uncoated or coated food-contact surface of paper and paperboard that may be safely used in producing, manufacturing, packing, transporting, or holding dry food in accordance with the conditions prescribed in 21 CFR part 176.180; as (3) a constituent of food-contact adhesives which may be safely used as a component of articles intended for use in packaging, transporting, or holding food in accordance with the conditions prescribed in 21 CFR part 175.105; (4) in the production of animal glue which may be safely used as a component of articles intended for use in producing, manufacturing, packaging, processing, preparing, treating, packing, transporting, or holding food in accordance with the conditions prescribed in 21 CFR part 178.3120; and (5) as an adjuvant in the preparation of slimicides in the manufacture of paper and paperboard that may be safely used in contact with food in accordance with the conditions prescribed in 21 CFR part 176.300.

Great Britain (1998). EDA is authorized without time limit for use in the production of polymeric materials and articles in contact with food or drink or intended for such contact. The specific migration of this substance shall not exceed 12 mg/kg.

Standards.

EU (1990). SML: 12.0 mg/kg.

Russia (1995). MAC and PML: 0.2 mg/l (organolept., odor).

References:

1. Dubinina, O. N. and Fukalova, L. A., in *Proc. F. F. Erisman Research. Sanitary Hygiene Institute*, A. P. Shitskova, Ed., Moscow, Issue No 10, 1979, 141 (in Russian).
2. Tiunov, L. S. et al., in *Current Problems of Clinical and Experimental Toxicology*, Gorky, 1971, 116 (in Russian).
3. Yang, R. H., Garman, R. R., Maronpot, R. R., McKelvey, J. A., Well, C. S., and Woodside, M. D., Acute and subchronic toxicity of ethylenediamine in laboratory animals, *Fundam. Appl. Toxicol.*, 3, 512, 1983.
4. Yang, R. H., Garman, R. R., Maronpot, R. R., Mirro, E., and Woodside, M. D., Chronic toxicity/carcinogenicity study of ethylenediamine in Fischher 344 rats, *Toxicologist*, 4, 53, 1984.
5. Hardin, B. D., Schuler, R. L., Burg, J. R., Booth, G. M., Hazelden, K. P., MacKenzie, K. M., Piccirillo, V. J., and Smith, K. N., Evaluation of 60 chemicals in a preliminary developmental toxicity test, *Teratogen. Carcinogen. Mutagen.*, 7, 29, 1987.
6. Price, C. J., George, J. D., Marr, M. C., et al., Developmental toxicity evaluation of ethylenediamine in New Zealand white rabbits, Program, Abstracts, Teratol. Soc. 33rd Annual Meeting, Tuscon, Ariz., June 28- July 1, 1993, *Teratology*, 47, 432, 1993.
7. DePass, L. R., Yang, R. S. H., and Woodside, M. D., Evaluation of the teratogenicity of ethylenediamine dihydrochloride in Fischer 344 rats by conventional and pair-feeding studies, *Fundam. Appl. Toxicol.*, 9, 687, 1987.
8. Eriksen, K., Sensitization capacity of ethylenediamine in the guinea pig and induction of unresponsiveness, *Contact Dermat.*, 5, 293, 1979.
9. Haworth, S., Lawlor, T., Mortelmans, M., Speck, W., and Zeiger, E., *Salmonella* mutagenicity test results for 250 chemicals, *Environ. Mutagen.*, 10 (Suppl. 5), 94, 1993.
10. Hedenstedt, A., Genetic health risk in the rubber industry: Mutagenicity studies on rubber chemicals and process vapors, in *Prevention of Occupational Cancer*, Int. Symposium, Int. Labour Office, Geneva, 1982, 476.
11. Leung Hon-Wing, Evaluation of the genotoxic potential of alkyleneamines, *Mutat. Res.*, 320, 31, 1994.
12. Slesinsky, R. S., Guzzie, P. J., Hengler, W. G., Watanabe, P. G., Woodside, M. D., and Yang, R. S. H., Assessment of genotoxic potential of ethylenediamine: *In vitro* and *in vivo* studies, *Mutat. Res.*, 124, 299, 1983
13. Zimmerling S., Mason, J. M., Valencia, R., and Woodruff, R. C., Chemical mutagenesis testing in Drosophila. 2. Results of 20 coded compounds tested for the National Toxicology Program, *Environ. Mutagen.*, 7, 87, 1985.1985
14. Coldwell, J. and Cotgreave, I. A., *Drug Determinant Therapeutic and Forensic Contexts*, New-York - London, 1984, 47.
15. Yang, R. S. H. and Tallant, M. J., Metabolism and pharmacokinetics of ethylenediamine in the rat following oral, endotracheal or intravenous administration, *Fundam. Appl. Toxicol.*, 2, 252, 1982.
16. Raymond, S. H., Yang, R. H., and Tallant, M. J., Metabolism and pharmacokinetics of ethylenediamine in the rat following oral, endotracheal or intravenous administration, *Fundam. Appl. Toxicol.*, 2, 252, 1982.

(ETHYLENEDINITRILO)TETRAACETIC ACID

Molecular Formula. $C_{10}H_{16}N_2O_8$

M = 292.28

CAS No 60-00-4
RTECS No AH4025000
Abbreviation. EDTA.

Synonyms. Edetic acid; Ethylenediamine, tetraacetic acid.

(ETHYLENEDINITRILO)TETRAACETIC ACID, DISODIUM SALT

Molecular Formula. $C_{10}H_{14}N_2O_8.Na$
M = 336.24
CAS No 139-33-3
RTECS No AH4375000
Abbreviation. EDTAS.

Synonyms and **Trade Name.** Disodium salt of EDTA; Ethylenediamine tetraacetic acid, disodium salt; Trilone B.

Properties. EDTAS is a white crystalline powder with a rather salty taste. Solubility in water is 10% at 20°C. It does not decompose in water and forms complexes with metals.[05] Water solubility is 0.5 g EDTA/l (25°C).

Applications. EDTA and its salts have a number of applications in medicine and pharmacology. EDTA is used to remove calcium from the human body, and serves as an anticoagulant and as a detoxicant after poisoning by heavy metals. It is used as a food additive to bind metal ions. EDTA and its salts are used as components in the production of food-contact paper and paperboard. EDTAS is used in synthetic rubber manufacture.

Exposure. EDTA is found in the environment as metal complexes. The levels recorded in natural water are usually less than 0.1 mg/l.

Acute Toxicity. LD_{50} of EDTAS is reported to be 2.0 g/kg BW,[05] but according to other data, it was not attained. After a single administration of 2.5 g/kg BW, rabbits became apathetic, refused feed, and suffered from severe diarrhea. Poisoning is accompanied by acceleration of the erythrocyte sedimentation rate, and reduction in serum calcium and serum cholinesterase activity. Death occurred within 5 days.[05]

Repeated Exposure revealed no accumulation of EDTA. Clinical experience with EDTA for treating metal poisonings has provided evidence of its safety in man. Rats were given a saturated solution of EDTAS (a total dose of 5.8 g/kg BW). The treatment caused 50% animal mortality in the first 2 or 3 days. Administration of 9.0 g/kg BW total dose resulted in 100% mortality in rats.[05] Oral administration of doses of 5, 10, and 15 mg EDTAS/kg BW on 5 consecutive days did not induce any obvious signs of toxicity.[1]

Short-term Toxicity. Zinc deficiency may develop as a consequence of zinc complexed by EDTA.[2] This was reaffirmed in Dutch workers (Janssen et al., 1990).

Long-term Toxicity. In a 7-month study, rabbits were given the oral dose of 2.1 mg EDTAS/kg BW. The treatment did not alter hematology analyses, cholinesterase activity, or serum calcium level in animals.[05]

Reproductive Toxicity.

Embryotoxicity. Rats received EDTA in the diet at about 950 mg/kg BW and by gastric intubation at 1250 mg/kg BW. The treatment revealed signs of maternal and embryotoxicity.[3] The treatment of male mice with 10 mg EDTAS/kg BW for 5 consecutive days induced no increase in the incidence of post-implantation embryonic deaths, except for a marginal but statistically insignificant increase during weeks 2 and 3 of mat ing.[1]

Teratogenicity. Equivocal data are reported in regard to developmental effects in animals orally exposed to EDTA or its salts. A variety of anomalies was observed in above-cited study (Janssen et al., 1990).

Teratogenic effect was found in rats given 2.0 to 3.0% EDTA in the diet after day 6 of gestation.[1] The treatment produced cleft palate, brain and eye defects and skeletal abnormalities in fetuses. However, Schardein et al. did not observe adverse fetal changes in rats gavaged with up to 1.0 g EDTA or its sodium and calcium salts per kg BW on days 7 to 14 of gestation.[4]

Gonadotoxicity. Administration of EDTAS at doses of 5.0, 10, and 15 mg/kg BW for 5 consecutive days did not produce any evident effect on either the testicular or epididymal weights and histology. No appreciable alterations were observed in the caudal sperm counts and there was no treatment-related increase in the incidence of sperm-head abnormalities.[1]

Mutagenicity.

In vitro genotoxicity. EDTA led to morphological changes of chromatin and chromosome structure in plant and animal cells. It showed weak activity in the induction of gene mutations. EDTA affected inhibition of DNA synthesis in primary cultures of mammalian cells and inhibited unscheduled DNA synthesis. Although EDTA produced a whole set of genetic effects, it seems to be a harmless compound to man as far as genotoxicity is concerned.[5] In the bone marrow micronucleus assay acute doses of 5 to 20 mg EDTAS/kg BW induced a dose-dependent increase in the incidence of micronucleated polychromatic erythrocytes at a 24-hour sampling.[1]

Carcinogenicity. EDTA showed no carcinogenic potential when administered even at high doses. F344 rats and $B6C3F_1$ mice received 3,750 and 7,500 ppm trisodium EDTA in their feed for 103 weeks. No compound-related signs of clinical toxicity were noted. Although a variety of tumors occurred among test and control animals of both species, no tumors were related to treatment. Since survival was satisfactory and showed no consistent variation among test and control groups, the absence of treatment-related tumors could not be attributed to early mortality.[6]

Carcinogenicity classification.

NTP: N - N - N - N. (Trisodium EDTA)

Chemobiokinetics. EDTA is poorly absorbed from the gut; it seems to be metabolically inert. Feeding studies in rats and dogs gave no substantial evidence of EDTA interference with mineral metabolism in either species. Nevertheless, EDTA is known to be capable of chelating metals both in liquid media and in animal body. It has been suggested that EDTA may enter kidney cells and, by interfering with zinc metabolism, exacerbate the toxicity of cadmium.[7]

Regulations. ***U.S. FDA*** (1998) listed EDTA and its sodium and/or calcium salt (1) in food in accordance with the conditions prescribed in 21 CFR parts 172.120 and 172.135; (2) as a component of the uncoated or coated food-contact surface of paper and paperboard intended for use in producing, manufacturing, packaging, processing, preparing, treating, packing, transporting, or holding aqueous and fatty foods in accordance with the conditions prescribed in 21 CFR part 176.170; (3) as a substance employed in the production of or added to textiles and textile fibers intended for use in contact with food in accordance with the conditions prescribed in 21 CFR part 177.2800; (4) in the manufacture of cellophane to be safely used for packaging food in accordance with the conditions prescribed in 21 CFR part 177.1200; and (5) in the manufacture of resinous and polymeric coatings for the food-contact surface of articles intended for use in producing, manufacturing, packing, processing, preparing, treating, packaging, transporting, or holding food (EDTAS) in accordance with the conditions prescribed in 21 CFR part 175.300.

Recommendations.

WHO (1996). Provisional guideline value of EDTA for drinking water: 0.2 mg/l (in view of the possibility of complexation with zinc the proposed guideline value was derived assuming that a child weighing 10 kg drinks 1.0 liter of water).[8]

Joint FAO/WHO Expert Committee on Food Additives (1974) proposed ADI of 0 to 2.5 mg $CaNa_2$EDTA/kg BW (0 to 1.9 mg/kg BW as the free acid) as a food additive, but not in food of children under 2 years.

Russia. Recommended PML (EDTAS): 3.5 mg/l.

Standards. ***Russia*** (1995). MAC and PML (EDTAS): 4.0 mg/l.

References:

1. Muralidhara, K., Assessment of *in vivo* mutagenic potency of ethylenediaminetetraacetic acid in albino mice, *Food Chem. Toxicol.*, 12, 845, 1991.
2. Swenerton, H. and Hurley, L. S., Teratogenic effects of chelating agent and their prevention by zinc, *Sci ence*, 173, 62, 1971.
3. Kimmel, C. A., Effect of route of administration on the toxicity and teratogenicity of EDTA in the rat, *Toxicol. Appl. Pharmacol.*, 40, 299, 1977.
4. Schardein, J. L., Sakowski, J., Petrere, J., et al., Teratogenesis studies with EDTA and its salts in rats, *Toxi col. Appl. Pharmacol.*, 61, 423, 1981.

5. Heindorff, K., Aurich, O., Michaelis, A., and Rieger, R., Genetic toxicology of ethylenediamine tetraacetic acid (EDTA), *Mutat. Res.*, 115, 149, 1983.
6. *Bioassay of Trisodium Ethylenediaminetetraacetate Trihydrate (EDTA) for Possible Carcinogenicity,* NTP Technical Report Series No 11, Research Triangle Park, NC, 1977.
7. Dieter, H. H., Implication of heavy metal toxicity related to EDTA exposure, *Toxicol. Environ. Chem.*, 27, 91, 1990.
8. *Revision of the WHO Guidelines for Drinking Water Quality,* Report on the Consolidation Meeting on Organics and Pesticides, Medmenham, UK, January 30-31, 1992, 11.

1,1',1'',1'''-(ETHYLENEDINITRILO)TETRAKIS-2-PROPANOL

Molecular Formula. $C_{14}H_{32}N_2O_4$
M = 292.48
CAS No 102-60-3
RTECS No UB5604000

Properties. Viscous, water-white liquid. Miscible with water, soluble in alcohol.

Synonym and **Trade Name.** 1,1',1'',1''' -Tetrakis(2-hydroxypropyl)ethylenediamine; Quadrol.

Applications. A cross-linking agent and catalyst, a plasticizer; used in epoxy resin curing.

Acute Toxicity. LD_0 is 0.5 g/kg BW in rats.[1] LD_{50} was found to be about 4.0 g/kg BW in mammals.[2]

Regulations. ***U.S. FDA*** (1998) approved the use of E. (1) as a curing agent in the manufacture of the polyurethane resins used as the food-contact surface of articles intended for use in contact with bulk quantities of dry food of the type identified in 21 CFR part 176.170 (c); and (2) in polyvinyl alcohol films intended for use in contact with food in accordance with the conditions prescribed in 21 CFR part 177.1670.

References:

1. Diss. Abstracts Int., B, *The Sci. Engin.*, 52, 762, 1991.
2. *Pharm. Chemicals Handbook*, 1991, 258.

FATTY ACIDS (synthetic), DIACYL(C_6-C_9)PEROXIDES, mixture

Properties. Colorless liquid with a specific odor.

Applications. Used as an efficient initiator in the production of low-density polyethylene.

Acute Toxicity. LD_{50} is 28 g/kg BW in male and 15.2 g/kg BW in female rats; in mice, it is found to be 10.1 and 9.0 g/kg BW, respectively. Histology examination revealed vascular engorgement and slight cerebral edema. The acute action threshold was not attained.

Repeated Exposure. Rats and mice tolerate administration of 0.5 and 2.5 g/kg BW for 2 months without visible changes in their general condition or retardation of BW gain. Gross pathology examination failed to reveal changes in the viscera.

Reference:

Toxicology and Sanitary Chemistry of Plastics, Abstracts, NIITEKHIM, Moscow, Issue No 1, 1979, 20 (in Russian).

1-HEXADECANONE

Molecular Formula. $C_{16}H_{34}0$
M = 242.50
CAS No 36653-82-4
RTECS No MM0225000
Abbreviation. HD.

Synonyms and **Trade Names.** Alcohol C-16; Cetal; Cetyl alcohol; Cetylol; Ethol; 1-Hexadecyl alcohol; Palmityl alcohol.

Properties. Solid or leaf-like colorless crystals or liquid with a faint odor and mild taste. Insoluble in water, slightly soluble in alcohol.

Applications. Used as a monomer lubricant in suspension polymerization, in the production of plasticizers and detergents.

Exposure. Used in fragrances in the U.S. Usual concentration in perfume is 0.5%.

Acute Toxicity. LD_{50} is 3.2 to 6.44 g/kg BW in mice, and 6.4 to 12.8 g/kg BW in rats.[03]

Allergenic Effect.

Observations in man. HD showed no sensitization reactions in a maximization test that was carried out in 26 volunteers (Epstein, 1976).

Chemobiokinetics. HD is oxidized in rats to the corresponding *fatty acid, palmitic acid.*[1] Following ingestion at a dose level of 2.0 g/kg BW, HD is partly absorbed and metabolized, about 20% of the dose being recovered unchanged in the feces.[2]

Regulations.

U.S. FDA (1998) approved use of HD (1) in adhesives as a component of articles intended for use in packaging, transporting, or holding food in accordance with the conditions prescribed in 21 CFR part 175.105; (2) in the manufacture of resinous and polymeric coatings for the food-contact surface of articles intended for use in producing, manufacturing, packing, processing, preparing, treating, packaging, transporting, or holding food in accordance with the conditions prescribed in 21 CFR part 175.300; as (3) a component of a defoaming agent intended for articles which may be used in producing, manufacturing, packing, processing, preparing, treating, packaging, transporting, or holding food in accordance with the conditions prescribed in 21 CFR part 176.200, and (4) in the manufacture of cellophane for packaging food in accordance with the conditions prescribed in 21 CFR part 177.1200.

EU (1990). HD is available in the *List of authorized monomers and other starting substances which shall be used for the manufacture of plastic materials and articles intended to come into contact with foodstuffs (Section* A).

Great Britain (1998). HD is authorized without time limit for use in the production of polymeric materials and articles in contact with food or drink or intended for such contact.

References:

1. Williams, R. T., *Detoxication Mechanisms. The Metabolism and Detoxification of Drugs, Toxic Substances and Other Organic Compounds*, 2nd ed., Chapman & Hall Ltd., London, 1959, 61.
2. McIsaac, W. M. and Williams, R. T., The metabolism of spermaceti, *W. Afr. J. Biol. Chem.*, 2, 42, 1958.

LAUROYL PEROXIDE

Molecular Formula. $C_{24}H_{46}O_4$
M = 398.70
CAS No 105-74-8
RTECS No OF2625000
Abbreviation. LP.

Synonyms. Bis(1-oxododecyl)peroxide; Didodecanoyl peroxide; Dodecanolyl peroxide; Dilauroyl peroxide.

Properties. Fatty, white, coarse powder with a faint odor. Poorly soluble in water, readily soluble in alcohol and vegetable oils.

Applications. Used as a polymerization catalyst in the production of polystyrenes, in vulcanization and plasticization of natural and synthetic rubber, for curing synthetic resins, in the production of polyesters, and as an initiator in the suspension polymerization of vinyl chloride, ethylene, and acrylates.

Acute Toxicity. LD_{50} was not attained in rats and mice. A 10 g/kg BW dose caused 30% animal mortality. High doses produce CNS inhibition, drowsiness, and dyspnea. Death occurs with increasing asphyxia. Gross pathology examination revealed visceral congestion and liver enlargement. Parenchymatous dystrophy of the liver, kidneys and myocardium, and disturbed hemopoiesis were also noted.

Repeated Exposure. Cumulative properties are not pronounced. A dose of 3.0 g/kg BW administered to rats for 10 days caused 50% animal mortality. Leukocytosis with subsequent leukopenia was found to develop. Death occurs in 2.5 to 3 months after the treatment with signs of anemia. Gross pathology examination showed the visceral organs being pallid and the liver, kidneys, and spleen being enlarged. In a 1-month study, the NOAEL appeared to be 1.0 g/kg BW.[1]

Reproductive Toxicity. LP caused malformation in chick embryos over a range of concentrations from 0.25 to 0.50 mmol/egg.[2]

Mutagenicity.

In vitro genotoxicity. LP is not shown to be mutagenic to *Salmonella typhimurium* strains *TA98* or *TA100* in the presence of an exogenous metabolic system.[3]

Carcinogenicity. In the mouse and rat studies which were considered inadequate for evaluation of complete carcinogenicity (IARC 36-319), LP did not produce the evidence of carcinogenic effect following subcutaneous administration or skin application.

Carcinogenicity classification. An IARC Working Group concluded that there is inadequate evidence for the carcinogenicity of LP in *experimental animals* and there are no data available for the carcinogenicity of LP in *humans*.

IARC: 3.

Regulations. ***U.S. FDA*** (1998) regulates LP for use (1) in adhesives as a component of articles intended for use in packaging, transporting, or holding food in accordance with the conditions prescribed in 21 CFR part 175.105; (2) as a polymerization catalyst in components of the uncoated or coated food-contact surface of paper and paperboard intended for use in producing, manufacturing, packaging, processing, preparing, treating, packing, transporting, or holding aqueous and fatty food of the type identified in 21 CFR part 176.170 (c), and (3) as a catalyst, at a level of 1.5% by weight max, in cross-linked polyester resins to be safely used for repeated contact with food, subject to the provisions prescribed in 21 CFR part 177.2420.

References:

1. Orlova, F. I., in *Proc. Research Epidemiol. Hygiene Institute*, Kuibyshev, Issue No 5, 1968, 107 (in Russian).
2. Korhonen, A., Hemminki, K., and Vainio, H., Embryotoxic effects of eight organic peroxides and hydrogen peroxide on three-day chicken embryos, *Environ. Res.*, 33, 54, 1984.
3. Yamaguchi, T. and Yamashita, Y., Mutagenicity of hydroperoxides of fatty acids and some hydrocarbons, *Agric. Biol. Chem.*, 44, 1675, 1980.

METHACRYLIC ACID, DIESTER with TRIETHYLENE GLYCOL

Molecular Formula. $CH_2 = C(CH_3)COOR_3COC(CH_3) = CH_2$, where R = ~$CH_2$-$CH_2$-O~

M = 281 to 286

CAS No 109-16-0

Abbreviation. TGM-3.

Synonyms and **Trade Names.** 1,2-Bis[2-(methacryloyloxy)ethoxy]ethane; Ethylenebis (oxyethylene) methacrylate; Polyester TGM-3; Triethylene dimethacrylate; Triethylene glycol, dimethacrylate.

Properties. Colorless, transparent, non-volatile liquids of different viscosity or solid, colorless, crystalline products. TGM-3 is the product of condensation of triethylene glycol and methacrylic acid.

Applications. Cross-linking agent for thermoset and UV-curable acrylics and vinyls.

Acute Toxicity. For mice, LD_{50} is reported to be 2.3 g/kg BW. Gross pathology examination revealed plethora and, rarely, necrosis of gastric and intestinal mucosa. Some animals developed hemorrhagic pneumonia.

Allergenic Effect.

Observations in man. Allergenic effect was noted in patients occupationally sensitized to dental resin products containing TGM-3.[2]

Animal studies. TGM-3 was found to induce contact sensitivity, using 5 different sensitization protocols.[3]

References:

1. Dikshtein, E. A. et al., *Gig. Truda Prof. Zabol.*, 1, 63, 1981 (in Russian)
2. Kanerva, L., Estlander, T., and Jolanki, R., Allergic contact dermatitis from dental composite resins due to aromatic epoxy acrylates and aliphatic acrylates, *Contact Dermat.*, 20, 201, 1989.
3. Parker, D. and Turk, J. L., Contact sensitivity to acrylate compounds in guinea pigs, *Contact Dermat.*, 9, 551, 1983.

METHACRYLIC ACID, OXYDIETHYLENE ESTER

Molecular Formula. $C_{12}H_{18}O_5$
M = 242.27
CAS No 23-58-1
Abbreviation. DGM-2.

Synonyms and **Trade Names.** Diethylene glycol bis(methacrylate); Diethylene glycol, dimethacrylate; 2-Methyl-2-propenoic acid, oxydi-2,1-ethanediyl ester.

Properties. Colorless liquid with an ester-like odor.

Applications. Cross-linking monomer in the production of synthetic coatings.

Allergenic Effect. In workers of printing industry, DGM-2 is found to produce contact sensitization.

Chemobiokinetics. DGM-2 is metabolized by saponification into *alcohol* and *methacrylic acid.*[03]

Regulations. ***U.S. FDA*** (1998) approved the use of DGM-2 as a monomer in the manufacture of semirigid and rigid acrylic and modified acrylic plastics in the manufacture of articles intended for use in contact with food in accordance with the conditions prescribed in 21 CFR part 177.1010.

Reference:

Malten, K. E. and Bende, J. M., 2-Hydroxyethyl methacrylate and di- and tetraethylene glycol dimethacrylate: contact sensitizers in a photoprepolymer printing plate procedure, *Contact Dermat.*, 5, 214, 1979.

4,4'-METHYLENEBIS-2-CHLOROBENZENAMINE

Molecular Formula. $C_{13}H_{12}C_{l2}N_2$
M = 267.17
CAS No 101-14-4
RTECS No CY1050000
Abbreviation. MOCA.

Synonyms and **Trade Name.** Di(4-amino-3-chlorophenyl)methane; 3,3'-Dichloro-4,4'-methylene-dianiline; 4,4'-Diamino-3,3'-dichlorodiphenyl methane; Methylenebis-*o*-chloroaniline; Millionate M; *p,p'*-Methylenebis (α-chloroaniline); 4,4'-Methylenebis (2-chloroaniline).

Properties. Light-gray or cream colored, crystalline powder. Solubility in water is 13.6 mg/l, soluble in alcohol.

Applications. Used as a curing agent in the manufacture of plastics and for vulcanization of isocyanate-containing polyurethane rubber, polyurethane foams, and glass-reinforced plastics.

Acute Toxicity. The LD_{50} values are reported to be 2.1 g/kg BW in rats and 0.88 g/kg BW in mice. In mice, poisoning is accompanied by adynamia, apathy, loss of appetite. Death occurs in 3 days. Manifestations of the toxic action in rats are not very pronounced and death occurs in 2 days.[023] The LOEL for methemoglobinemia formation appears to be 83 mg/kg BW. Gross pathology examination revealed distention of the stomach and intestine, traces of blood in the urinary bladder, and pleural effusion in the thorax. Histology examination detected fine-drop adiposis of the liver, tiny foci of inflammatory infiltration, and circulatory disturbances in the visceral organs.

Repeated Exposure failed to reveal evident cumulative properties. Administration of 1/5 LD_{50} caused 50% mortality only on day 52. Manifestations of the toxic action comprised decreased BW gain, impairment of liver functions and gas exchange intensity. Gross pathology examination revealed morphological changes in the liver, myocardium, lungs, and spleen.[023]

Mutagenicity. MOCA form adducts with DNA both *in vitro* and *in vivo.*

Observations in man. An increased frequency of SCE was seen in a small number of exposed workers.[1]

In vivo cytogenetics. Biologically insignificant micronucleated polychromatic erythrocyte increase was observed in CD-1 mice after a single *i/p* treatment with 160, 320, or 640 mg/kg BW (80% of the LD_{50}). Double treatment at 640 mg/kg BW gave negative responses. In a mutagen sensitive mouse strain, MS/Ae, a significant induction of micronucleated reticulocytes was observed. MOCA induced DNA damage in animals treated *in vivo.*[1]

In vitro genotoxicity. induced DNA damage in cultured mammalian and human cells. Gene mutations were induced in bacteria and cultured mammalian cells.[2] MOCA is comprehensively genotoxic in *in vitro* tests and in bacteria.[3,4]

Carcinogenicity. MOCA produces a variety of tumors in the liver, mammary glands, and urinary bladder.[4,5] According to Russfield et al., it induced lung and liver tumors in rats and mice.[6] Oral administration of MOCA to female beagle dogs produced transitional cell carcinoma of the urinary bladder.[1]

Carcinogenicity classification. An IARC Working Group concluded that there is sufficient evidence for the carcinogenicity of MOCA in *experimental animals* and there were inadequate data available to evaluate the carcinogenicity of MOCA in *humans*.

The Annual Report of Carcinogens issued by the U.S. Department of Health and Human Services (1998) defines MOCA to be a substance which may reasonably be anticipated to be carcinogen, i.e., a substance for which there is limited evidence of carcinogenicity in *humans* or sufficient evidence of carcinogenicity in *experimental animals*.

IARC: 2A.

Chemobiokinetics. The doses of 28.1 mM ^{14}C-MOCA/kg BW were administered for 28 consecutive days. Distribution appears to be the highest in the liver, kidney, lung, spleen, testes, urinary bladder.[7]

According to Tobes et al., administration of 0.49 mg ^{14}C-MOCA/kg BW led to accumulation of radioactivity in the small intestine, liver, adipose tissue, etc.[8]

The major route of excretion seems to be via the feces. Approximately 1/3 of the administered radioactivity was excreted in the urine. Only 1.0 to 2.0% was identified as MOCA. At least 9 other metabolites were separated by reversed phase HPLC.[9]

Regulations. ***U.S. FDA*** (1999) prohibits the use of MOCA in human food.

References:

1. Morita, T., Asano, N., Awogi, T., Sasaki, Yu. F., et al., Evaluation of the rodent micronucleus assay in the screening of IARC carcinogens (Groups 1, 2A and 2B). The summary report of the 6th collaborative study by CSGMT/JEMS-MMS, *Mutat. Res.*, 389, 3, 1997.
2. Myhr, B. C. and Caspary, W. J., Evaluation of the L5178Y mouse lymphoma cell mutagenesis assay: intralaboratory results for 63 coded chemicals tested at Litton Bionetics, Inc., *Environ. Molec. Mutagen.*, 12 (Suppl. 13), 103, 1988.
3. Mori, H., Yoshimi, N., Sugie, S., et al., Genotoxicity of epoxy resin hardeners in the hepatocyte primary culture/DNA repair test, *Mutat. Res.*, 204, 683, 1988.
4. McQueen, C. A. and Williams, G. M., Review of genotoxicity and carcinogenicity of 4,4'-methylenedianiline and 4,4'-methylenebis-2-chloroaniline, *Mutat. Res.*, 239, 133, 1990.
5. Stula, E. F., Barnes, J. R., Sherman, H., et al., Urinary bladder tumors in dogs from 4,4'-methylene bis(2-chloroaniline) (MOCA), *J. Environ. Pathol. Toxicol.*, 1, 31, 1977.
6. Russfield, A. B., Homburger, F., Boger, E., et al., The carcinogenicity effect of 4,4'-methylene bis-2-chloroaniline in mice and rats, *Toxicol. Appl. Pharmacol.*, 31, 47, 1975.
7. Chever, K. L., DeBord, D. G., and Swearengin, T. E., 4,4'-methylene bis(2-chloroaniline) (MOCA): The effect of multiple oral administration route, and phenobarbital induction on macromolecular adduct formation in rat, *Fundam. Appl. Toxicol.*, 16, 71, 1991.
8. Tobes, M. C., Brown, L. E., Chin, B., et al., Kinetics of tissue distribution and elimination of 4,4'-methylene bis(2-chloroaniline) in rats, *Toxicol. Lett.*, 17, 69, 1983.
9. Farmer, P. B., Rickard, J., and Robertson, S., The metabolism and distribution of 4,4'-methylene bis(2-chloroaniline) (MBOCA) in rats, *J. Appl. Toxicol.*, 1, 317, 1981.

4,4'-METHYLENEBIS(CYCLOHEXYLAMINE)

Molecular Formula. $C_{13}H_{26}N_2$

M = 210.37

CAS No 1761-71-3

RTECS No GX1530000

Synonyms. 1,4-Bis(*p*-aminocyclohexyl)methane; 4,4'-Diaminodicyclohexylmethane; 4,4'-Methylene-dicyclohexanamine.

Properties. White or brown, wax-like solid. Poorly soluble in water.

Applications. Used as an amine-curing agent in the production of epoxy resins. Used in the synthesis of polyamide resins and in rubber production.

Acute Toxicity. LD_{50} is 270 mg/kg BW in rats, and 197 mg/kg BW in mice. Administration of lethal doses is accompanied by NS depression, motor coordination disorder, tremor of the head and extremities, and clonic-tonic convulsions.[1] According to Kennedy, LD_{50} is found to be between 670 and 1000 mg/kg BW in rats.[2]

Short-term Toxicity. Rats tolerate a 3-month administration of 1/10 and 1/5 LD_{50} without any sign of intoxication.[1]

Long-term Toxicity. Dogs were given oral doses of 50 mg/kg BW for 18 months. The treatment caused kidney and liver damage, as well as a local irritation of GI tract.[2]

Allergenic Effect. M. appeared to be a weak dermal sensitizer in guinea pigs.[2]

Carcinogenicity. Rats received doses of 25 to 100 mg/kg BW (8 times/10 days, 17 times/4 weeks, or 20 to 22 times/4 weeks). The treatment at the high doses caused a decrease in BW gain and loss of muscular strength of limbs. Though scleroderma-like change was not observed in the skin histologically, the authors believe that M. is one of the causative agents that induced toxic symptoms like those of collagen disease.[3]

References:

1. Statsek, N. K., Toxicological characteristics of 4,4'-diaminodicyclohexylmethane and 4,4'-diamino-diphenylmethane, in *Toxicology and Hygiene of High-Molecular-Mass Compounds and of the Chemical Raw Material Used for Their Synthesis*, Proc. 2nd All-Union Conf., A. A. Letavet and S. L. Danishevsky, Eds., Leningrad, 1964, 60 (in Russian).
2. Kennedy, G. L., Toxicity of 1,4-bis(aminocyclohexyl) methane, *J. Appl. Toxicol.*, 11, 367, 1991.
3. Ohshima, S., Shibata, T., Sasaki, N., et al., Subacute toxicity of an amine-curing agent for epoxy resin, Abstract, *Sango Igaku*, 26, 197, 1984 (in Japanese).

4,4'-METHYLENEDIANILINE

Molecular Formula. $C_{13}H_{12}N_2$
M = 198.206
CAS No 101-77-9
RTECS No BY5425000
Abbreviation. MDA.

Synonyms. Bis(aminophenyl)methane; Dianilinemethane; 4,4'-Diaminodiphenylmethane; 4,4'-Diphenylmethanediamine; 4,4'-Methylenebisbenzenamine.

4,4'-METHYLENEDIANILINE DIHYDROCHLORIDE

Molecular Formula. $C_{13}H_{14}N_2.2ClH$
M = 271.21
CAS No 13552-44-8
RTECS No BY5426000
Abbreviation. MDAC.

Synonym. 4,4'-Methylenebisbenzenamine, dihydrochloride.

Properties. MDA is a crystalline solid or pearly leaflets; colorless to pale-yellow thin flakes or lumps with a faint amine-like odor (IARC 39-348). Water solubility is 1.0 g/l at 25°C, highly soluble in ethanol, acetone, and diethyl ether.

Applications. MDA is used in the production of synthetic rubber, dyes, and polyamide resins. A curing agent in the synthesis of epoxy resins. A vulcanization accelerator for chloroprene rubber, a vulcanization activator. An antioxidant for vulcanizates based on natural and synthetic rubber. A monomer in the production of polydiisocyanates.

Exposure. MDA has been available commercially since the 1920s. It is used mainly as an intermediate in 4,4'-methylenediphenyl diisocyanate production and as a curing agent for epoxy resins. Exposure occurs during the production of 4,4'-methylenedianiline and the use of 4,4'-methylenediphenyl diisocyanate resins.

Acute Toxicity.

Observations in man. Collective poisoning with bread made of flour contaminated with MDA has been reported. MDA caused hepatobiliary damage in exposed humans (Shoental).

Animal studies. LD_{50} is 665 mg MDA/kg BW in rats, and 264 mg MDA/kg BW in mice.[1] Lethal doses produced CNS inhibition, motor coordination disorder, tremor of the head and extremities, and clonic-tonic convulsions. Rats got a single oral administration of 25 to 225 mg MDA/kg BW. The treatment caused dose-dependent elevation in serum ALT and β-glutamyl transferase activity, and in serum bilirubin concentration. Decrease in bile flow and elevation of liver weight were also noted. Histologically, multifocal, necrotizing hepatitis with neutrophil infiltration was found.[2]

Sprague-Dawley rats were administered MDA by biliary cannulas. MDA rapidly diminished bile flow and altered the secretion of biliary constitutients. Thus, it was found to be highly injurious to biliary epithelial cells.[3]

Repeated Exposure revealed moderate cumulative properties. Administration of 1/10 and 1/5 LD_{50} produced some rat mortality. Short-term excitation and aggressiveness are followed by adynamia.[4] Hepatotoxic effects were described in rats gavaged with 8 to 600 mg MDA/kg BW for 10 days (necrotizing cholangitis, periportal necrosis, and glycogen loss at the 200 mg/kg BW dose level and higher).[5]

A dose-related reduction in BW gain was found in male F344 rats exposed to 1.6 and 3.2 g MDAC/l in their drinking water for 14 days.[6]

There were increased weights of the adrenal gland, uterus, and thyroid gland in ovariectomized female Sprague-Dawley rats given 14 daily doses of 150 mg MDAC/kg BW by gavage.[7]

Short-term Toxicity. Atrophy of the liver parenchyma and increased relative spleen weights associated with hyperplasia of the lymphatic system were observed in Wistar rats given 83 mg/kg BW for 12 weeks.[8]

Cirrhosis occurred in all 21 male rats exposed to doses of 38 mg/kg BW by gavage for 17 weeks.[9] Bile-duct hyperplasia was observed in rats following oral administration (in the diet or in drinking water).[6,10,11]

Long-term Toxicity. In a 40-week study, bile-duct proliferation was found in male Wistar rats fed a diet containing 1.0 g MDA/kg feed. The hepatic parenchyma was replaced by proliferating bile ducts, and eventually portal cirrhosis developed. These alterations reversed when treatment was discontinued.[12]

Mineralization of the kidney was revealed in rats and mice exposed to MDA in their drinking water for 2 years.[6]

Dogs were given 50 mg MDA/kg BW for up to 18 months. The treatment caused kidney and liver damage, along with local irritation of the GI tract.[13]

Allergenic Effect. MDA is a weak dermal sensitizer in guinea pigs.[13]

Mutagenicity.

In vitro genotoxicity. MDA was shown to be mutagenic toward *Salmonella typhimurium* strains *TA98* and *TA100* in the presence of *S9* mix.[14] It induced DNA damage in Chinese hamster V79 cells in the presence of an exogenous metabolic system and in the liver of rats. This chemical induced gene mutations in mouse lymphoma cells (Zeiger et al., 1990, 1992).

In vivo cytogenetics. MDA produced an increase in the incidence of SCE in the bone marrow of mice treated *in vivo.*[15] It caused DNA damage in rat liver *in vivo.* A statistically significant and dose-dependent induction of micronucleated reticulocytes, as well as a marginal response, was observed in male CD-1 mice after a single *i/p* treatment at up to 80% of the LD_{50} and a negative response after double treatment.[16]

Carcinogenicity.

MDA and MDAC were tested for carcinogenicity by oral administration in mice, rats and dogs. In rats, treatment-related increases in the incidences of thyroid follicular-cell carcinomas and hepatic nodules were observed in males, and thyroid follicular-cell adenomas occurred in females. In a study in rats in which 4,4'-methylenedianiline was administered orally in conjunction with a known carcinogen, the incidence of thyroid tumors was greater than that produced by the carcinogen alone.

$B6C3F_1$ mice were given 150 or 300 mg MDAC/l in the drinking water for 103 weeks. Treatment-related increase in the incidence of thyroid follicular-cell adenomas and hepatocellular neoplasms was observed. Similar effect was found in F344 rats under the same conditions. The incidence of thyroid follicular-cell carcinomas in the high-dose group of animals was significantly increased as compared to the controls.[1,17]

Carcinogenicity classifications. An IARC Working Group concluded that there is sufficient evidence for the carcinogenicity of MDA and its dihydrochloride in *experimental animals* and there were no adequate data avail able to evaluate the carcinogenicity of MDA in *humans*.

The Annual Report of Carcinogens issued by the U.S. Department of Health and Human Services (1998) defines MDA to be a substance which may reasonably be anticipated to be carcinogen, i.e., a substance for which there is limited evidence of carcinogenicity in *humans* or sufficient evidence of carcinogenicity in *experimental animals*.

IARC: 2B

NTP: P* - P - P* - P* (MDA dihydrochloride, water).

Chemobiokinetics. MDA was metabolized *in vivo* and its *N-acetyl* and *N,N'-diacetyl* metabolites were found in the urine.[14]

Observations in man. 4,4'-MDA and *N-acetyl-MDA* were found in the urine of workers exposed to 4,4'-MDA.[18]

Regulations. ***U.S. FDA*** (1998) has approved the use of 4,4'-MDA (1) as a catalyst and a cross-linking agent for epoxy resins intended for use in contact with food; (2) as a catalyst for the polyurethane resins intended for use in contact with food; and (3) in the manufacture of resinous and polymeric coatings in a food-contact surface for use only in coatings for containers with a capacity of 1000 gallons (37.85 liter) or more, when the containers are intended for repeated use in contact with alcoholic beverages containing up to 8.0% alcohol by volume in accordance with the conditions prescribed in 21 CFR part 177.1680; (4) in the manufacture of 4,4'-isopropyl- idenediphenol-epichlorohydrin thermosetting epoxy resins (finished articles containing the resins shall be thoroughly cleansed prior to their use in contact with food) in accordance with the conditions prescribed in 21 CFR part 177.2280; (5) in articles made of mineral-reinforced nylon resins or as components of these articles intended for repeated use in contact with nonacidic food (*pH* above 5.0) and at use temperature not exceeding 212°F in accordance with the conditions specified in the 21 CFR part 177.2355; and (6) as an antioxidant for rubber articles intended for repeated use in contact with food (up to 5.0% by weight of the rubber product) in accordance with the conditions prescribed in 21 CFR part 177.2600.

References:

1. Statsek, N. K., Toxicological characteristics of 4,4'-diaminodicyclohexylmethane and 4,4'-diaminodiphenylmethane, in *Toxicology and Hygiene of High-Molecular-Mass Compounds and of the Chemical Raw Material Used for Their Synthesis*, Proc. 2nd All-Union Conf., A. A. Letavet and S. L. Darishevsky, Eds., Leningrad, 1964, 60 (in Russian).
2. Bailie, M. B., Mullaney, T. P., and Roth, R. A., Characterization of acute 4,4'-methylene dianiline hepa totoxicity in the rat, *Environ. Health Perspect.*, 101, 130, 1993.
3. Kanz, M. F., Kaphalia, L., Kaphalia, B. S., Romagnoli, E., and Ansari, G. A., Methylenedianiline: acute toxicity and effects on biliary function, *Toxicol. Appl. Pharmacol.*, 117, 88, 1992.
4. *Protection of Water Reservoirs against Pollution by Industrial Liquid Effluents*, S. N. Cherkinsky, Ed., Medgiz, Moscow, Issue No 4, 1960, 60 (in Russian).
5. Gohike, R. and Schmidt, P., 4,4'-Diaminodiphenylmethane: Histological, enzyme histological and autoradiographic investigations in acute and subacute experiments in rats with and without additional heat stress, *Int. Arch. Arbeitsmed.*, 32, 217, 1974.
6. *Carcinogenesis Studies of 4,4'-Methylenedianiline Dihydrochloride in Fischer-344/N Rats and B6C3F$_1$ Mice (Drinking Water Studies),* NTP Technical Report Series No 248, Research Triangle Park, NC, 1983.
7. Tullner, W. W., Endocrine effects of methylenedianiline in the rat, rabbit and dog, *Endocrinology*, 66, 470, 1960.
8. Pludro, G., Karlowski, K., Mankowska, M., et al., Toxicological and chemical studies of some epoxy resins and hardeners. I. Determination of acute and subacute toxicity of phthalic acid anhydride, 4,4'-diaminodiphenylmethane and of the epoxy resin: Epilox EG-34, *Acta Pol. Pharmacol.*, 26, 352, 1969.

9. Munn, A., Occupational bladder tumors and carcinogens: Recent developments in Britain, in *Bladder Cancer*, Symposium, W. Deichmann and K. Lampe, Eds., Aesculapius, Birmingham, AL, 1967, 187-193.
10. Miyamoto, J., Okuno, Y., Kadota, T., et al., Experimental hepatic lesions and drug metabolizing enzymes in rats, *J. Pest. Sci.*, 2, 257, 1977.
11. Gohike, R., 4,4'-Diaminodiphenylmethane in a chronic experiment, *Z. Gesumte Hyg.*, 24, 159, 1978.
12. Fukushima, S., Shibata, M., Hibino, T., et al., Intrahepatic bile duct proliferation induced by 4,4'-diaminodiphenylmethane in rats, *Toxicol. Appl. Pharmacol.*, 48, 145, 1978.
13. Kennedy, G. L., Toxicity of 1,4-bis(aminocyclohexyl)methane, *J. Appl. Toxicol.*, 11, 367, 1991.
14. Tanaka, K., Ino, T., Sawahata, T., Marui, S., Igaki, H., and Yashima, H., Mutagenicity of *N*-acetyl and *N,N'*-diacetyl derivatives of 3 aromatic amines used as epoxy-resin hardeners, *Mutat. Res.*, 143, 11, 1985.
15. McQueen, C. A. and Williams, G. M., Review of genotoxicity and carcinogenicity of 4,4'-methylenedianiline and 4,4'-methylenebis-2-chloroaniline, *Mutat. Res.*, 239, 133, 1990
16. Morita, T., Asano, N., Awogi, T., Sasaki, Yu. F., et al. Evaluation of the rodent micronucleus assay in the screening of IARC carcinogens (Groups 1, 2A and 2B). The summary report of the 6 collaborative studies by CSGMT/JEMS-MMS, *Mutat. Res.*, 389, 3, 1997.
17. Weisburger, E. K., Murthy, A. S. K., Lilja, H. S., et al., Neoplastic response of Fischer-344 rats and B6C3F$_1$ mice to the polymer and dyestuff intermediates, 4,4'-methylenebis (*N,N'*-dimethyl) benzenamine, 4,4'-oxydianiline, and 4,4'-methylenedianiline, *J. Natl. Cancer Inst.*, 72, 1457, 1984.
18. Schutze, D., Sepai, O., Lewalter, J., Miksche, L., Henschler, D., and Sabbioni, G., Biomonitoring of workers exposed to 4,4'-methylenedianiline or 4,4'-methylenediphenyl diisocyanate, *Carcinogenesis*, 16, 573, 1995.

2-METHYL-2-PROPENOIC ACID, 2-PROPENYL ESTER

Molecular Formula. $C_7H_{10}O_2$
M = 126.17
CAS No 96-05-9
RTECS No UD3483000

Synonyms and **Trade Name.** Allyl methacrylate; Ageflex AMA; Methacrylic acid, allyl ester.

Applications. Cross-linking agent; used in dental plastics, lenses, adhesives, fiber reinforced plastics, acrylic ester rubbers, etc.

Acute Toxicity. LD_{50} is 430 mg M./kg BW in rats.[1] However, according to other data, LD_{50} is 70 mg/kg BW in rats and 57 mg/kg BW in mice.[2]

Repeated Exposure. In dermal toxicity study, no overt signs of toxicity or abnormal behavior were seen among the New Zealand white rabbits treated at doses 25, 50, or 100 mg/kg BW for 4 weeks.[3]

Allergenic Effect. M. did not produce contact sensitivity reactions in guinea pigs.[4]

Chemobiokinetics. M. undergoes metabolism by saponification into *allyl alcohol* and *methacrylic acid.*

Regulations.

U.S. FDA (1998) approved the use of M. as a basic component of acrylic and modified acrylic plastics, semirigid and rigid, which may be safely used as articles intended for single and repeated use in food-contact surfaces in accordance with the conditions prescribed by 21 CFR part 177.1010.

EU (1990). M. is available in the *List of authorized monomers and other starting substances which may continue to be used for the manufacture of plastic materials and articles intended to come into contact with foodstuffs pending a decision on inclusion in Section A* (*Section B*).

Great Britain (1998). M. is authorized up to the end of 2001 for use in the production of polymeric materials and articles in contact with food or drink or intended for such contact.

References:

1. Smyth, H. F., Carpenter, C. P., Weil, C. S., Pozzani, U. C., Striegel, J. A., and Nycum, J. S., Range-finding toxicity data: List VII, *Am. Ind. Hyg. Assoc. J.*, 30, 470, 1969.

2. Smirnova, E. S, Oskerko, E. F., Kasatkin, A. N., and Beloborodova, I. Yu., Characteristics of the biological effect and toxicometry of allyl ether of methacrylic acid, *Gig. Truda Prof. Zabol.*, 8, 59, 1990 (in Russian).
3. Siddiqui, W. H. and Hobbs, E. G., Subchronic dermal toxicity study of allyl methacrylate in rabbits, *Drug Chem. Toxicol.*, 5, 165, 1982.
4. Parker, D. and Turk, J. L., Contact sensitivity to acrylate compounds in guinea pigs, *Contact Dermat.*, 1, 55, 1983.

NAPHTHENIC ACID, COBALT SALT

Molecular Formula. $[RCOO]_2Co$, where *R* - alkyl C_8 -C_9
CAS No 61789-51-3
RTECS No QK8925000
Abbreviation. CN.

Synonym and **Trade Name.** Cobalt naphthenate; Naftolite.

Applications. CN is used to enhance the adhesion of sulfur-vulcanized rubber to steel and other metals, and as a catalyst.

Acute Toxicity. LD_{50} is 3.9 g/kg BW in rats.[1]

Mutagenicity.

In vitro genotoxicity. CN was negative in *Salmonella* mutagenicity assay (NTP-94).

Carcinogenicity classification. An IARC Working Group concluded that there is evidence suggesting lack of carcinogenicity of cobalt compounds in *experimental animals* and there were inadequate data available to evaluate the carcinogenicity of *cobalt* compounds in *humans.*

IARC: 2B.

Chemobiokinetics. Absorption from the GI tract in rats was found to vary between 11 and 34% depending on administered dose value.[2]

The highest concentration was found in the liver; a low concentration was noted in the kidney, pancreas, and spleen.[3,4] Cobalt specifically reacts with *SH*-groups of dihydrolipoic acid, thus preventing oxidative decarboxylation of pyruvate to acetylcoenzyme A.[5]

Approximately 20% cobalt naphthenate administered orally was eliminated in the urine.[6]

Regulations. ***U.S. FDA*** (1998) listed CN as an accelerator in the manufacture of cross-linked polyester resins for use as articles or components of articles intended for repeated use in contact with food in accordance with the conditions prescribed in 21 CFR part 177.2420.

References:

1. *AMA Arch. Ind. Health*, 12, 477, 1955.
2. Taylor, D. M., The absorption of cobalt from the gastrointestinal tract of the rat, *Physiol. Med. Biol.*, 6, 445, 1962.
3. Stenberg, T., The distribution in mice of radioactive cobalt administered by two different methods, *Acta Odontol. Scand.*, 41, 143, 1983.
4. Clyne, N., Lins, L.-E., Pehrsson, S. K., et al., Distribution of cobalt in myocardium, skeletal muscle and serum in exposed and unexposed rats, *Trace Elem. Med.*, 5, 52, 1988.
5. Alexander, C. S., Cobalt-beer cardiomyopathy, *Am. J. Med.*, 53, 395, 1972.
6. *NTP Fiscal Year Annual Plan*, 1991, 100.

4,4'-OXYDIANILINE

Molecular Formula. $C_{12}H_{12}N_2O$
M = 200.26
CAS No 101-80-4
RTECS No BY7900000
Abbreviation. ODA.

Synonyms. 4-Aminophenyl ether; Bis(*p*-aminophenyl) ether; 4,4'-Diaminobiphenyloxide; 4,4'-Diaminodiphenyl ether; 4,4'-Oxybisbenzenamine; Oxydi-*p*-phenylenediamine.

Properties. Colorless, odorless powder. Poorly soluble in water, readily soluble in alcohol and acetone.

Applications. Used in the production of plasticizers and straight polyamide resins, provides high-temperature resistance. Epoxy resin hardener. Antioxidant.

Acute Toxicity. LD_{50} ranges from 570 to 725 mg/kg BW in rats; it is 685 mg/kg BW in mice and 650 mg/kg BW in guinea pigs.[1-3] Produces a narcotic effect, causes formation of *metHb*. The animals die within 5 days.

Repeated Exposure revealed pronounced cumulative properties.[2] Rats and mice were exposed to 3.0 g/kg dose in their diet. 0 to 20% rats and 40 to 80% mice were still alive on day 14 of the treatment.[4]

Administration of 72.5 mg/kg BW for 15 days caused decreased blood *Hb* level, and increased spleen and adrenal weights in rats.[1] Liver and kidney injuries were noted in mice and rats.

Short-term Toxicity. In a 90-day study, F344 rats and $B6C3F_1$ mice received dietary concentrations up to 2.0 g/kg. The dose of 0.6 g/kg caused decrease in BW gain. Animals fed dietary concentrations of 0.6 g/kg (rats) and 1.0 g/kg (mice) and above displayed diffuse parenchymatous dystrophy, pituitary hyperplasia, seminiferous tubular degeneration, and atrophy of the prostate and seminal vesicles.[5]

Long-term Toxicity. In a 2-year study, F344 rats were given doses of 0.2 to 0.5 g ODA/kg feed, while $B6C3F_1$ mice received 0.15 to 0.8 g/kg. Decrease in BW gain, thyroid follicular-cell and pituitary hyperplasia were observed.[4]

In a 6-month rat study, a toxic effect on the cardiovascular system was found; dose-dependence was noted.[6] Rats and mice were given orally 5.0 and 25 mg ODA per animal, respectively. The treatment caused renal lesions of the nephrosis type with a nephritic component.[7]

Reproductive Toxicity.

Gonadotoxicity. Chronic exposure of male rats to *Diamin P* (4,4'-diaminodiphenyl ether of resorcin) by inhalation caused higher percentage of dead spermatozoa and spermatogonium diseases.[8]

Mutagenicity.

In vitro genotoxicity. ODA was shown to be mutagenic towards *Salmonella typhimurium* strains *TA98* and *TA100,* both in the absence and presence of *S9* mix.[9]

Carcinogenicity. Treatment with 1/4 LD_{50} given for 14 days inhibited the growth of spontaneous mammary tumors and transplanted tumors in mice.[10] In the above-cited studies, no evidence of carcinogenic activity was observed.[7]

Mice received 0.15 to 0.8 g/kg, rats were given 0.2 to 0.5 g/kg for 103 weeks. The treatment produced liver-cell tumors and malignant follicular-cell tumors of the thyroid.[4]

Administration to rats (a total dose given 14.4 g/kg BW) caused cirrhosis and necrosis in the liver. Incidence of death from malignant hepatomas was increased.[11]

Carcinogenicity classifications. An IARC Working Group concluded that there is sufficient evidence for the carcinogenicity of ODA in *experimental animals* and there were no adequate data to evaluate the carcinogenicity of ODA in *humans*.

The Annual Report of Carcinogens issued by the U.S. Department of Health and Human Services (1998) defines ODA to be a substance which may reasonably be anticipated to be carcinogen, i.e., a substance for which there is a limited evidence of carcinogenicity in *humans* or sufficient evidence of carcinogenicity in *experimental animals.*

IARC: 2B;

NTP: P* - P* - P* - P* (feed).

Chemobiokinetics. ODA was metabolized *in vivo* and its *N-acetyl* and *N,N'-diacetyl* metabolites were found in the urine.[9]

Standards. ***Russia*** (1995). MAC: 0.03 mg/l.

References:

1. Lapik, A. S., Makarenko, A. A., and Zimina, L. N., Toxicological characteristics of 4,4'-oxydianiline, *Gig. Sanit.,* 10, 110, 1968 (in Russian).
2. Ivanov, N. G. and Mel'nikova, L. V., Comparative investigation of toxicity and mode of action of 4,4'-diaminodiphenyl sulfone and 4,4'-diaminodiphenyl oxide, in *Toxicology of New Industrial*

Chemical Substances, N. F. Ismerov and I. V. Sanotsky, Eds., Meditsina, Moscow, Issue No 14, 1975, 118 (in Russian).
3. Ismerov, N. F., Sanotsky, I. V., and Sidorov, K. K., *Toxicometric Parameters of Industrial Toxic Chemicals under Single Exposure*, Center of Int. Projects, GKNT, Moscow, 1982, 43 (in Russian).
4. *Bioassay of 4,4'- Oxydianiline for Possible Carcinogenicity*, NCI Technical Report Series No 205, DHEW Publ. No NIH 80- 1761, U.S. Dept. Health, Education & Welfare, Washington, D.C., 1980.
5. Hayden, D. W., Wage, G. G., and Handler, A. H., The goitrogenic effect of 4,4'-oxydianiline in rats and mice, *Vet. Pathol.*, 15, 649, 1978.
6. Kondratyuk, V. A., Gnat'yuk, M. S., and Volkov, K. S., On the cardiotoxic effect of 4,4'-diaminodiphenyl ether, *Gig. Sanit.*, 6, 31, 1986 (in Russian).
7. Dzhioev, F. K., Carcinogenic activity of 4,4'-diaminodiphenyl ester, *Voprosy Onkologii*, 21, 69, 1975 (in Russian).
8. Vasilenko, N. M., Khipko, S. E., Paranich, L. I., and Omel'chenko, O. A., Study of the reproductive function in albino rats after inhalation of 4,4'-diaminodiphenyl ether of resorcin, *Gig. Truda Prof. Zabol.*, 10, 26, 1991 (in Russian).
9. Tanaka, K., Ino, T., Sawahata, T., Marui, S., Igaki, H., and Yashima, H., Mutagenicity of *N*-acetyl and *N,N'*-diacetyl derivatives of 3 aromatic amines used as epoxy-resin hardeners, *Mutat Res.*, 143, 11, 1985.
10. Boyland, E., Experiments on the chemotherapy of cancer. VI. The effect of aromatic bases, *Biochem. J.*, 40, 55, 1946.
11. Steinholf, D., Kancerogene Wirkung von 4,4'-Diamino diphenylather bei Ratten, *Naturwissenschaften*, 64, 344, 1977.

PEROXYACETIC ACID

Molecular Formula. $C_2H_4O_3$
M = 76.05
CAS No 79-21-0
RTECS No SD8750000
Abbreviation. PAA.

Synonyms. Acetic peroxide; Acetyl hydroperoxide; Ethaneperoxoic acid; Peracetic acid.

Properties. Colorless liquid with an acrid odor. Very soluble in water and ethanol.

Applications. Polymerization catalyst and bleaching agent for textile and paper; used in the production of caprolactam and epoxy resin, and in food processing. PAA is used as a bactericide and a fungi cide.

Acute Toxicity. LD_{50} is 1.54 g/kg BW in rats,[1] 210 mg/kg BW in mice.[2]

Carcinogenicity. The skin of female ICR Swiss mice was painted with PAA. Evidence of tumor promoting activity has been noted. The complete carcinogenic activity of PAA in the absence of 7,12-dimethyl[a]antracene might be questioned. Equivocal tumorigenic agent by RTECS criteria.[3]

Regulations. ***U.S. FDA*** (1998) approved the use of PAA as a component of sanitizing solutions used on food-processing equipment and utensils, and on other food-contact articles as specified in 21 CFR part 178.1010.

References:

1. *Toxic and Hazardous Industrial Chemicals Safety Manual*, Int. Techn. Information Institute, Tokyo, Japan, 1988, 400.
2. Lyarsky, P. P., Gleiberman, S. E., Pankratova, G. P., Yaroslavskaya, L. A., and Yurchenko, V. V., Toxicological and hygiene characteristics of decontaminating preparations produced on the basis of hydrogen peroxide and its derivatives, *Gig. Sanit.*, 6, 28, 1983 (in Russian).
3. Bock, F. G., Myers, H. K., and Fox, H. W., Carcinogenic activity of peroxy compounds, *J. Natl. Cancer Inst.*, 55, 1359, 1975.

PEROXYBENZOIC ACID, *tert*-BUTYL ESTER

Molecular Formula. $C_{11}H_{14}O_3$
M = 194.23
CAS No 614-45-9
RTECS No SD9450000
Abbreviation. tert-BPB.

Synonyms and **Trade Names.** Benzenecarboperoxoic acid, 1,1-dimethylester; *tert*-Butylperbenzoate; Trigonox C.

Properties. Colorless, oily liquid with a pungent odor. Poorly soluble in water, soluble in oils and alcohol.

Applications. *tert*-BPB is used almost exclusively as a free-radical initiator in the polymer industry, and also as a vulcanization agent or catalyst in polymerization of unsaturated resins such as polystyrene and polyvinyl chloride; a curing agent for polyester resins. A cross-linking agent for polystyrene foam.

Acute Toxicity. In mice, LD_{50} is 914 mg/kg BW. Lethal doses cause adynamia and respiratory disturbances with subsequent death. Gross pathology examination revealed congestion in the visceral organs. Hemorrhages and necrotic areas were seen in the gastric and intestinal mucosa. The liver had a yellowish hue.[1]

Repeated Exposure. *tert*-BPB exhibits little if any toxicity in rodents. F344 rats were exposed to doses of 70 to 1112 mg/kg BW for 14 days (administered by gavage in corn oil). The highest doses caused no signs of toxicity but only increased stomach weights, forestomach epithelial hyperplasia, ulceration, and acute inflammation.[2]

Short-term Toxicity. In a 13-week rat and mice study, doses up to 500 mg/kg BW were administered by gavage. Depression of BW gain in the highest dose group, dose-dependent increase in the rate of hyperplasia of the forestomach mucosa epithelium and of stomach weights were noted. The NOAEL of 30 mg/kg BW was identified for forestomach lesions.[2]

Reproductive Toxicity. Inhalation exposure at the toxic level produced no gonadotoxic effect. Caused malformations in chick embryos.[3]

Mutagenicity.

In vitro genotoxicity. *tert*-BPB is found to be mutagenic in *Salmonella* mutagenicity assay with and without metabolic activation. It induced SCE and CA in Chinese hamster ovary cells *in vitro*.

In vivo cytogenetics. *tert*-BPB did not induce formation of micronuclei in the peripheral blood of mice in a 13-week study.[2]

Chemobiokinetics. *tert*-BPB degrades rapidly in the body, initial degradation products being *benzoic acid* and *tert-butanol.*

Regulations. ***U.S. FDA*** (1998) regulates use of *tert*-BPB in the manufacture of cross-linked polyester resins which may be used as articles or components of articles intended for repeated use in contact with food in accordance with the conditions prescribed in 21 CFR part 177.2420.

References:

1. Sanotsky, I. V., Ivanov, N. G., Germanova, A. L., et al., Evaluation of the toxic and gonadotoxic action of *tert*-butylperbenzoate, in *Toxicology of New Industrial Chemical Substances*, A. A. Letavet and I. V. Sanotsky, Eds., Moscow, Issue No 10, 1968, 55 (in Russian).
2. *Toxicity Studies of tert-Butylperbenzoate Administered by Gavage to F344/N Rats and B6C3F$_1$ Mice*, NTP Technical Report Series No 15, Research Triangle Park, NC, NIH Publ. No 92-3134, July 1992.
3. Korhonen, A., Hemminki, K., and Vainio, H., Embryotoxic effects of eight organic peroxides and hydrogen peroxide on three-day chicken embryos, *Environ. Res.*, 33, 54, 1984.

PEROXYDISULFURIC ACID, DIPOTASSIUM SALT

Molecular Formula. $H_2O_8S_2.2K$
M = 272.34
CAS No 7727-21-1
RTECS No SE0400000

Abbreviation. PAPS.

Synonyms. Dipotassium persulfate; Potassium peroxydisulfate; Potassium persulfate.

Properties. White or colorless, odorless triclinic crystals. Readily soluble in water, insoluble in alcohol.

Applications. An initiator in emulsion polymerization.

Acute Toxicity. LD_{50} is 800 mg/kg BW in rats.[014]

Regulations. ***U.S. FDA*** (1998) approved the use of PAPS in rubber articles intended for repeated use in producing, manufacturing, packing, processing, preparing, treating, packaging, transporting, or holding food, subject to the provisions and definitions set in 21 CFR part 177.2600.

m-PHENYLENEBISMETHYLAMINE

Molecular Formula. $C_8H_{12}N_2$
M = 136.22
CAS No 1477-55-0
RTECS No PF8970000

p-PHENYLENEBISMETHYLAMINE

CAS No 539-48-0
RTECS No PF9000000
Abbreviation. PBMA.

Synonyms. 1,3- and 1,4-Benzenedimethanamine; 1,3- and 1,4-Bis(aminomethyl)benzene; *m*- and *p*-Phenylenebis(methylamine); ω,ω'-Diamino-*m*-(*p*)-xylene; *m*- and *p*-Xylenediamine.

Properties. *m*-PBMA is a liquid of low volatility. It is soluble in water and mixes with alcohol. *p*-PBMA is a white, crystalline substance.

Applications. Used as curing agents in the production of epoxy resins.

Acute Toxicity. LD_{50} is 1.3 to 2.0 g/kg BW in rats and 0.47 to 1.4 g/kg BW in mice. Lethal doses affect the NS. Pronounced excitation is followed by apathy, unsteady gait, and respiratory disturbances. Gross pathology examination revealed visceral congestion.[1,05] LD_{50} of *m*-PBMA is 0.93 g/kg BW in rats.[014]

Repeated Exposure revealed marked cumulative properties (Grigorova et al., 1971).

Reproductive Toxicity.

Acute ***embryotoxic effect*** was found to develop as a result of administration of 1/5 LD_{50} to rats. 1/100 LD_{50} produced no ***teratogenic effect***.[05]

Allergenic Effect was observed on ingestion and skin application of *m*-PBMA.

Observations in man. *m*-PBMA. was found to be a very potent sensitizer. Patch tests revealed allergic cross reactions to benzylamine (in all four patients observed) and to ethylenediamine (in two patients).[2]

Regulations.

EU (1990). *m*-PBMA. is available in the *List of authorized monomers and other starting substances which shall be used for the manufacture of plastic materials and articles intended to come into contact with foodstuffs (Section A*). SML into food or food simulants is 0.05 mg/kg. *p*-PBMA is available in the *List of authorized monomers and other starting substances which may continue to be used for the manufacture of plastic materials and articles intended to come into contact with foodstuffs pending a decision on inclusion in Section A* (*Section B*).

U.S. FDA (1998) approved the use of *m*-PBMA and *p*-PBMA in resinous and polymeric coatings at a level not to exceed 3.0% and 0.6%, respectively, by weight of the resin when such coatings are intended for repeated use in contact with foods only of the types identified in 21 CFR part 175.300.

Great Britain (1998). *m*-PBMA is authorized without time limit for use in the production of polymeric materials and articles in contact with food or drink or intended for such contact. The specific migration of this substance shall not exceed 0.05 mg/kg.

References:

1. Ivanov, V. I., Loit, A. O., and Matyukhin, N. Ya., Toxicology of *m*- and *p*-xylenediamine - the raw materials in the production of synthetic fibers, in *Hygiene and Toxicology of High-Molecular-Mass Compounds and of the Chemical Raw Material Used in Their Synthesis*, Proc. 4th All-Union Conf., S. L. Danishevsky, Ed., Khimia, Leningrad, 1969, 171 (in Russian).

2. Richter, G. and Kadner, H., Allergic contact eczema caused by *m*-xylenediamine in the polyurethane silk production, *Dermat. Beruf. Umwelt.*, 38, 117, 1990.

m-PHENYLENEDIAMINE

Molecular Formula. $C_6H_8N_2$
M = 108.14
CAS No 108-45-2
RTECS No SS7700000
Abbreviation. *m*-PDA.

Synonyms. *m*-Aminoaniline; *m*-Benzenediamine; *m*-Diaminobenzene.

Properties. Colorless or faintly tinted rhombic crystals, which acquire a violet color in the light and have an unpleasant odor. Readily soluble in water and alcohol.

Migration from the epoxy coatings (*ED-16* resin) into water (24-hour exposure at 20°C) was determined at the level of 0.013 mg/l; migration level from the epoxy coatings (*ED-20* resin) into 96% ethanol (10-day exposure at 50°C) was found to be 0.15 mg/l.[028]

Applications. Used as a curing agent for epoxy compositions. Antioxidant in a variety of applications.

Acute Toxicity. LD_{50} is in the range of 280 to 350 mg/kg BW in rats, 65 to 72 mg/kg BW in mice, 437 mg/kg BW in rabbits, and 450 mg/kg BW in guinea pigs. There are certain differences in species sensitivity. Lethal doses caused immediate respiratory disturbances and altered the NS function. Death occurred in 3 to 48 hours after administration. *MetHb* formation was not observed. Gross pathology examination revealed lung and brain edema and granularity of the liver tissue.[1,2] A threshold dose for acute effect on coproporphyrin concentration in the urine is 10 mg/kg BW.

Repeated Exposure revealed slight cumulative properties.[1] Possibility of functional accumulation is reported despite there being no fatal consequences.[2] K_{acc} (by Kagan) appeared to be 3.3 (administered dose 0.5 mg/kg BW), or 1.0 (administered dose was 0.1 mg/kg BW).

Short-term Toxicity. Rats received 2.0, 6.0, or 10 mg *m*-PDA/kg BW in aqueous solution for 3 months. The treatment caused no retardation of BW gain, no changes in food consumption, ophthalmology, hematology, or blood and urine biochemistry analyses. A significant increase in relative and absolute liver weights along with degenerative liver lesions was noted in animals given 18 mg/kg BW dose. In females receiving this dose, a significant increase in relative kidney weights was noted. The NOAEL of 6.0 mg/kg BW was established in this study.[3]

Long-term Toxicity. Rats were dosed by gavage with 10 mg/kg BW for 7 months. Methemoglobinemia was not noted in animals. The NOAEL in rats appeared to be 0.02 mg/kg BW for the effect on hematology parameters and for coproporphyrin contents in the urine.[2] $B6C3F_1$ mice were exposed to 0.02 or 0.04% *m*-PDA in their drinking water for 78 weeks. There was no increase in mortality; BW was found to be significantly lower in the groups of high-dosed females and males.[4]

Allergenic Effect was observed in a chronic toxicity study in rats and guinea pigs.[2]

Reproductive Toxicity.

Teratogenicity. Sprague-Dawley rats received *m*-PDA by gavage on days 6 through 15 of gestation at doses of 45, 90, and 180 mg/kg BW. The treatment caused significant reduction in mean maternal BW gain at a dose level of 180 mg/kg BW. No teratogenic effect was noted.[6,7]

Mutagenicity. The mutagenicity of *m*-PDA was remarkably enhanced by oxidation; major mutagenic oxidation product was *2,7-diaminophenazine* (Watanabe et al., 1990).

In vivo cytogenicity. There was no early pregnancy loss observed in dominant lethal assay in rats given up to 20 mg/kg BW by injection.[5]

In vitro genotoxicity. *m*-PDA appeared to be positive in *Salmonella* mutagenicity assay,[8] and in Chinese hamster ovary and lung cells with and without *S9* activation.[9]

Carcinogenicity. A single case of sarcoma was reported in the group of 5 rats given s/c injections of 1/10 LD_{50} every other day for 11 months.[10]

Male Charles River CD rats and ICR mice were fed *m*-PDA *hydrochloride* at doses of 1.0 to 4.0 g/kg of their diets. No increase in tumor incidence was noted in this study.[11]

The maximum tolerated dose, tested in the study severely inhibited BW gain but did not produce carcinogenic effect. *m*-PDA in this study showed no carcinogenic potential in B6C3F_1 mice when administered in drinking water.[3]

Carcinogenicity classification. An IARC Working Group concluded that there is inadequate evidence of carcinogenicity of *m*-PDA in *experimental animals* and there are no data available to evaluate the carcinogenicity *m*-PDA in *humans*.

IARC: 3.

Chemobiokinetics. *m*-PDA is excreted rapidly, unchanged, with little absorption (IARC 16-119).

Regulations.

EU (1990). *m*-PDA is available in the *List of authorized monomers and other starting substances which shall be used for the manufacture of plastic materials and articles intended to come into contact with foodstuffs (Section A)*.

U.S. FDA (1998) approved *m*-PDA for use in the manufacture of 4,4'-isopropylidenediphenol epichlorohydrin thermosetting epoxy resins (finished articles containing the resins shall be thoroughly cleansed prior to their use in contact with food) in accordance with the provisions prescribed in 21 CFR part 177.1440.

Great Britain (1998). *m*-PDA is authorized without time limit for use in the production of polymeric materials and articles in contact with food or drink or intended for such contact. The quantity of *m*-PDA in the finished polymeric material or article shall not exceed 1.0 mg/kg.

Standards.

EU (1990). MPQ in finished material or article: 1.0 mg/kg.

Russia (1995). MAC and PML in drinking water: 0.1 mg/l; PML in food: 0.005 mg/l.

References:

1. Sardarova, L. G., in *Proc. Conf. of Junior Scientists*, Res. Inst. Hyg. Occup. Med., Ufa, 1968, 133 (in Russian).
2. Myannik, L. V., *Toxicological and Hygienic Characteristics of m-Phenylenediamine and of Epoxy Coatings Cured with It Used in the Food Industry*, Author's abstract of thesis, Kiyv, 1981, 17 (in Russian).
3. Hofer, H., Hruby, R., Hruby, E., et al., Ninety-day toxicity study with *m*-phenylene diamine in rats, *Oestrr. Foeschungszent*, Seibersdorf (Berlin), 4155, 1, 1982.
4. Amo, H., Matsuama, M., Amano, H., et al., Carcinogenicity and toxicity study of *m*-phenylenediamine administered in the drinking water to (C57BL/6xC3H/He)F_1 mice, *Food Chem. Toxicol.*, 26, 893, 1988.
5. Burnett, C., Loehr, R., and Corbett, J., Dominant lethal mutagenicity study on hair dyes, *J. Toxicol. Environ. Health*, 2, 657, 1977.
6. Picciano, J. C., Morris, W. E., Kwan, S., and Wolf, B. A., Evaluation of the teratogenic and mutagenic potential of the oxidative dyes, 4-chlororesorcinol, *m*-phenylenediamine, and pyrogallol, *J. Am. Coll. Toxicol.*, 1, 325, 1983.
7. Picciano, J. C., Morris, W. E., and Wolf, B. A., The absence of teratogenic effects of several oxidative dyes in Sprague-Dawley rats, *J. Am. Coll. Toxicol.*, 1, 125, 1982.
8. Ames, B. N., Kammen, H. O., and Yamasaki, E., Hair dyes are mutagenic: Identification of a variety of mutagenic ingredients, *Proc. Natl. Acad. Sci. USA*, 72, 2423, 1975.
9. Sofuni, T., Matsuoka, A., Sawada, M., Ishidate, M., Zeiger, E., and Shelby, M. D., A comparison of chromosome aberration induction by 25 compounds tested by two Chinese hamster cell (CHL and CHO) systems in culture, *Mutat. Res.*, 241, 175, 1990.
10. Weisburger, E. K., Russfield, A. B., Homburger, F., et al., Testing of twenty one environmental aromatic amines or derivatives for long-term toxicity or carcinogenicity, *J. Environ. Pathol. Toxicol.*, 2, 325, 1978.
11. Saruta, N., Yamaguchi, S., and Matsuoka, T., Sarcoma produced by subdermal administration of *m*-phenylenediamine and *m*-phenylenediamine hydrochloride, Abstract, *Kyushu J. Med. Sci.*, 13, 175, 1962 (in Japanese).

p-PHENYLENEDIAMINE

Molecular Formula. $C_6H_8N_2$
M = 108.14
CAS No 106-50-3
RTECS No SS8050000
Abbreviation. *p*-PDA

Synonyms and **Trade Names.** *p*-Aminoaniline; *p*-Benzenediamine; *p*-Diaminobenzene; Santoflex LC; Ursol D.

Properties. Almost colorless, thin leaflets rapidly darken in air. Solubility in water is 3.24 g/100 ml, soluble in alcohol. Water color-change threshold is 0.1 mg/l. A concentration of 1.0 g/l is odorless, and that of 0.1 g/l is colorless. The color of *p*-PDA solutions increases in the course of storage.[1]

Applications. *p*-PDA is used predominantly as a curing agent in the production of epoxy materials and coatings, a vulcanizing agent and accelerator of rubber, an antioxidant.

Acute Toxicity. LD_{50} is reported to be 133 to 180 mg/kg BW in rats, and 145 mg/kg BW in guinea pigs. Lethal doses led to liver damage and *metHb* formation.[1,2] However, according to Lloyd et al.,[3] LD_{50} is found to be 98 mg/kg BW in rats.

Repeated Exposure. F344 rats received up 0.2% *p*-PDA in the diet. The treatment appeared to be ineffective; 0.4% 1,4-PDA concentration in the diet caused pronounced toxic effect.[4]

Short-term Toxicity. Mice were fed dietary levels of up 0.4% *p*-PDA for 12 weeks. The animals exhibited decreased BW gain and increased relative weights of the liver and kidneys. Some of the animals died. Doses of 0.2% and less produced no remarkable toxicity.[4]

Long-term Toxicity. The effect upon glycogen-forming function of the liver is reported in rabbits.[1,2] In rats exposed to 0.1% *p*-PDA in the feed, increased splenic and body relative weights were noted (Katsumi et al.).

Reproductive Toxicity.

Embryotoxicity. There was no early pregnancy loss observed in dominant lethal assay in rats given by injection up to 20 mg/kg BW.[5]

Teratogenicity. No malformations were observed in a reproductive toxicity study when *p*-PDA was given by gavage to Sprague-Dawley rats.[6]

Allergenic Effect. *p*-PDA has a pronounced sensitizing potential. Sensitization seems to be caused not by *p*-PDA itself but by its partial oxidation product *p-quinonediamine*. Some researchers reported a sensitizing effect on dermal application (Kurlyandsky et al., 1987).

Mutagenicity.

In vivo cytogenetics. Inadequate results were obtained in the *in vivo* micronuclei assay in bone marrow cells of rats exposed orally to *p*-PDA. Five CD-mice were injected *i/p* with *p*-PDA. Negative response was noted in *in vivo* mouse bone marrow micronucleus assay by checking frequency of micronucleated polychromatic erythrocytes.[7]

In vitro genotoxicity. Following oral exposure, *azo* dyes are metabolized to *aromatic amines* by intestine microflora or liver azoreductases. Aromatic amines are further metabolized to genotoxic compounds by mammalian microsomal enzymes. Many of these aromatic amines are presented mainly by *p*-PDA and *benzidine moieties* and are mutagenic in *Salmonella.*[10] Treatment of *p*-PDA solution with hydrogen peroxide revealed a strong mutagenic effect in *Salmonella typhimurium* under the conditions of metabolic activation.[8]

p-PDA was positive in Chinese hamster ovary and lung cells with and without *S9* mix.[9]

Carcinogenicity. *p*-PDA caused no carcinogenic response in F344 rats or $B6C3F_1$ mice given 625 and 1250 ppm in their feed. In a 80-week feeding study in F344 rats, a concentration of 0.1% *p*-PDA was found to be a NOEL for carcinogenic effect.[4]

Carcinogenicity classification. An IARC Working Group concluded that there is inadequate evidence for the carcinogenicity of *p*-PDA in *experimental animals* and there are no data to evaluate the carcinogenicity of *p*-PDA in *humans*.

IARC: 3.

NTP: N - N - N - N (feed).

Chemobiokinetics. Following ingestion, *p*-PDA is nearly completely absorbed in F344 rats and B6C3F$_1$ mice. The highest radioactivity was present in the muscles, skin, and liver.[11] Metabolism occurs through oxidation to form *quinone diamine*. In addition, *p*-PDA is partially acetylated to a *diacetyl derivative; p*-PDA or its metabolites do not bind covalently to hepatic DNA.

Rats and mice rapidly cleared radioactivity from their tissues. In 24 hours only 10 to 15% radioactivity was still present in the animal body. *p*-PDA is excreted predominantly through the urine (70 to 90%) and also in the feces (10 to 20%). *p*-PDA metabolites are 95% radioactivity in the urine.

Standards. ***Russia*** (1995). MAC and PML: 0.1 mg/l.

References:

1. Zhakov, Yu. A., Experimental study on hygiene substantiation of maximum allowable concentration of ursol D in water bodies, *Gig. Sanit.*, 5, 14, 1958 (in Russian).
2. Zhakov, Yu. A., in *Protection of Water Reservoirs against Pollution by Industrial Liquid Effluents*, S. N. Cherkinsky, Ed., Medgiz, Moscow, Issue No 4, 1960, 230 (in Russian).
3. Lloyd, G. K., Ligett, M. P., Kynoch, S. R., et al., Assessment of the acute toxicity and potential irritancy of hair dye constituents, *Food Cosmet. Toxicol.*, 15, 607, 1977.
4. Imaida, K., Ishihara, Y., Nishio, O., et al., Carcinogenicity and toxicity tests on *p*-phenylenediamine in Fischer-344 rats, *Toxicol. Lett.*, 16, 259, 1983.
5. Barnett, C., Loehr, R., and Corbett, J., Dominant lethal mutagenicity study on hair dyes, *J. Toxicol. Environ. Health,* 2, 657, 1977.
6. Re, T. A. Loehr, R. F., Rodwell, D. E., D'Aleo, C. J., and Burnett, C. M., The absence of teratogenic hazard potential of *p*-phenylenediamine in Sprague-Dawley rats, *Fundam. Appl. Toxicol.*, 1, 421, 1981.
7. Soler-Niedziela, L., Shi, X., Nath, J., and Ong, T., Studies on three structurally related phenylenediamines with the mouse micronucleus assay system, *Mutat. Res.*, 259, 43, 1991.
8. Nishi, K. and Nishioka, H., Light induces mutagenicity of hair-dye *p*-phenylenediamine, *Mutat. Res.*, 104, 347, 1982.
9. Sofuni, T., Matsuoka, A., Sawada, M., Ishidate, M., Zeiger, E., and Shelby, M. D., A comparison of chromosome aberration induction by 25 compounds tested by two Chinese hamster cell (CHL and CHO) systems in culture, *Mutat. Res.* 241,175, 1990.
10. Chung, K. T., and Cerniglia, C. E., Mutagenicity of azo dyes: structure-activity relationship, *Mutat. Res.,* 277, 201, 1992.
11. Ioannou, Y. M. and Matthews, H. B., *p*-Phenylenediamine: comparative disposition in male and female rats and mice, *J. Toxicol. Environ. Health*, 16, 299, 1985.

2,2'-(PHENYLIMINO)DIETHANOL

Molecular Formula. $C_{10}H_{15}NO_2$
M = 181.26
CAS No 120-07-0
RTECS No KM2100000
Abbreviation. PIDE.

Synonyms and **Trade Names.** Agidol AF-2; Aminophenol curing agent AF-2; *N,N'*-Bis(2-hydroxyethyl)aniline; Diethanolaminobenzene; Diethanolaniline; *N,N'*-Dioxyethylaniline; Phenyldethanolamine.

Composition. Product of interaction of phenol, formaldehyde, and ethylenediamine.

Properties. Yellow caking crystalline solid. Poorly soluble in water, freely soluble in organic sol vents.

Applications. Used in the production of monomers. A hardener.

Acute Toxicity. LD_{50} is found to be 3.7 to 4.1 g/kg BW or, according to other data, 0.98 g/kg BW in rats, and 1.2 to 2.3 g/kg BW in mice. Immediately after poisoning, animals experienced adynamia, motor coordination disorder, labored respiration, and tonic convulsions. Death occurs within 2 days.[1,2]

In 24 hours after administration of 2.0 g PIDE/kg BW, blood *metHb* level was found to be up to 10%; *sulfHb* level was 0.7%; STI was decreased. Acute action threshold (for STI) appeared to be 80 mg/kg.[2]

Repeated Exposure revealed moderate cumulative properties. K_{acc} is 9.1 (by Lim). The CNS, liver, blood system were predominantly affected. The treatment caused a decrease in STI and in bromsulfalein retention test, and an increase in the relative liver and spleen weights. *MetHb* and *sulfHb* formation was noted. Histology examination revealed hemosiderosis and atrophy of the spleen white pulp, congestion, and parenchymatous dystrophy in the liver.[2]

Reproductive Toxicity. Administration of 1/20 LD_{50} caused embryotoxic and mild gonadotoxic effect.[1]

Allergenic Effect is not noted on skin applications in guinea pigs.[2]

References:

1. Dregval', G. F., Kharchenko, T. F., Kirpichev, V. P., et al., Sanitary-chemical and toxicological studies of the aminophenol hardener agidol AF-2, *Gig. Sanit.*, 11, 23, 1990 (in Russian).
2. Bukhalovsky, A. A. and Bitkina, A. V., Toxicity of phenyldiethanolamine, *Gig. Truda Prof. Zabol.*, 6, 90, 1991 (in Russian).

PIPERIDINE

Molecular Formula. $C_5H_{11}N$
M = 86.14
CAS No 110-89-4
RTECS No TM3500000

Synonyms. Hexahydroperidine; Pentamethylenimine.

Properties. Colorless liquid with a penetrating ammonia-like odor. Readily miscible with water and alcohols.

Applications. Used as a curing agent in the synthesis of epoxy resins. P. derivatives are used as vulcanization accelerators for rubber.

Acute Toxicity.

Observations in man. General weakness, muscular paralysis, labored breathing, and asphyxia are noted in individuals who ingested 30 to 60 mg P./kg BW.[05]

Animal studies. In rats, LD_{50} is 50 mg/kg BW.[1] Water solutions of P. are rather less toxic. LD_{50} of 8.0% aqueous solution is 371 mg/kg BW.[03]

A dose of about 250 mg/kg BW decreased the neuromuscular excitability threshold as well as body temperature and increased the arterial pressure. Gross pathology examination revealed acute vascular disturbances, necrosis of the gastric mucosa, and pneumonia as well as signs of parenchymatous dystrophy of the liver and renal tubular epithelium.

According to other data, LD_{50} is 400 mg/kg BW in rats, 30 mg/kg BW in mice, 145 mg/kg BW in rabbits.[2]

Repeated Exposure revealed pronounced cumulative properties in the course of treatment with 1/10 LD_{50}. K_{acc} appeared to be 0.6 (by Kagan).

Short-term Toxicity. Rats were given 0.08%, 0.16 to 0.62% P. in their diet for 84 days. The treatment produced organ weight changes. A LOEL of 138 mg/kg BW, and a NOEL of 69 mg/kg BW were established in this study.[3]

Reproductive Toxicity.

Embryotoxic effect appeared to be non-specific.[03]

Mutagenicity.

In vitro genotoxicity. P. was found to be negative in direct bacterial, host-mediated, and microsomal mutagenesis.[2]

Carcinogenicity. Negative results are reported in rats.[09]

Chemobiokinetics. P. is readily absorbed in the GI tract. It has been isolated and identified in animal and human urine.[04]

Regulations. ***U.S. FDA*** (1998). P. is a food additive permitted for direct addition to food for human consumption as a synthetic flavoring substance and adjuvant in accordance with the conditions prescribed in 21 CFR part 172.515.

Standards. ***Russia***. PML: 0.07 mg/l.

References:

1. Bazarova, L. A. and Osipenko, N. I., Comparative evaluation of toxicity of piperidine and hexamethyleneimine in acute and subacute studies, in *Toxicology of New Industrial Chemical Substances*, A. A. Letavet and I. V. Sanotsky, Eds., Meditsina, Moscow, Issue No 9, 1967, 91 (in Russian).
2. *Toxicity and Metabolism of Industrial Solvents*, R. Snyder, Ed., 2nd ed., vol 2, Nitrogen and Phosphorus Solvents, Elsevier, Amsterdam-New York-Oxford, 1990, 254.
3. Munro, I. C., Ford, R. A., Kennepohl, E., and Sprenger, J. G., Correlation of structural class with No-Observed-Effect Levels: A proposal for establishing a threshold of concern, *Food Chem. Toxicol.*, 34, 829, 1996.

POLYETHYLENEPOLYAMINES

Structural Formula. $H[\sim NHCH_2CH_2\sim]_nNH_2$

M = 146 (average)

RTECS No TQ7840000

Abbreviation. PEPA.

Trade Name. Vulkacit TR.

Composition. Mixtures of variable composition containing ethylenediamine (n = 1), triethylenetetramine (n = 3) up to 90%, diethylenetriamine (n = 2) and other more complex amines.

Properties. Oily liquids of yellow to dark-brown color with a penetrating odor. Readily soluble in water (see ***Ethylenediamine***). Odor perception threshold is 5.6 mg/l, taste perception threshold is 10 mg/l.[1]

Applications. Used as curing agents in the synthesis of epoxy and ion exchange resins and rubber; a vulcanization accelerator for rubber.

Migration of PEPA from polyurethane coating was reported.[2]

Acute Toxicity. LD_{50} is reported to be 2.0 to 3.0 g/kg BW in rats, 1.8 to 2.0 g/kg BW in mice, and 0.82 mg/kg BW in guinea pigs.[1,3] Gross pathology examination revealed necrotic lesions in the stomach, as well as liver affection, circulatory disturbances, and changes in the biogenic amine level. Administration of a 4.0 g/kg BW dose to mice caused increased excitability with subsequent adynamia, labored breathing, disappearance of tendon reflexes, and convulsions. Gross pathology examination revealed visceral congestion and necrotic foci in the GI tract mucosa.[4]

Repeated Exposure failed to reveal cumulative properties. K_{acc} is 11.8 (by Cherkinsky).[3]

Administration of 1/10 to 1/5 LD_{50} to rats and mice for 45 days caused increased oxygen consumption, reticulocytosis, a decline in cholinesterase activity, and other changes.[1]

Short-term Toxicity. Rats tolerate administration of 1/100 to 1/5 LD_{50} for 3 months. BW gain, hematology analyses and STI are unchanged.[3]

Long-term Toxicity. No manifestations of the toxic action were found in rats and rabbits exposed to 0.5 mg/kg BW via drinking water for 6 months.[3,5] Higher doses caused changes visible on gross pathology examination: increased liver and kidney weights, congestion, hemorrhages, and granular dystrophy in the parenchymatous organs.[1]

Reproductive Toxicity.

Gonadotoxicity. In a 6-month study, sloughing of spermatogenic epithelium as well as testicular dystrophic changes were revealed in rats exposed to the 0.5 mg/kg BW dose.[6] The treatment disturbed the estrous cycle and caused decreased sperm motility.[7]

Embryotoxicity. Reduced fetal weights and increased mortality were observed.[7] However, according to other data,[6] administration of 1/100 and 1/5 LD_{50} to pregnant female rats had no embryotoxic effect.

Standards. ***Russia*** (1995). MAC and PML: 0.005 mg/l.

References:

1. Antonova, V. I., Salmina, Z. A., and Vinokurova, T. V., Toxicity of polyethylenepolyamine and substantiation of its maximum allowable concentration in water bodies, *Gig. Sanit.*, 2, 32, 1977 (in Russian).

2. Kupyrov, V. N., Kaplina, T. V., Gakal, R. K., Vinarskaya, E. I., and Starchenko, S. N., Hygienic evaluation of films intended for the waterproofing of unit prefabricated swimming pools, *Gig. Sanit.*, 5, 91, 1978 (in Russian).
3. Sheftel', V. O. and Tsendrovskaya, V. A., Application of mathematical design of experiment to hygienic and chemical studies of epoxy coatings, *Gig. Sanit.*, 10, 66, 1974 (in Russian).
4. Shumskaya, N. I., in *Toxicology of New Industrial Chemical Substances*, A. A. Letavet and I. V. Sanotsky, Eds., Medgiz, Moscow, Issue No 5, 1963, 35 (in Russian).
5. Ivanova, E. V. and Suleimanov, S. M., in *Urgent Problems of Occupational Hygiene and Pathology*, Voronezh, 1975, 51 (in Russian).
6. Sheftel', V. O., *Hygienic Aspects of the Use of Polymeric Materials in Water Supply*, Author's abstract of thesis, All-Union Research Institute of Hygiene and Toxicology of Pesticides, Polymers and Plastic Materials, Kiyv, 1977, 156 (in Russian).
7. Antonova, V. I. and Salmina, Z. A., Reproductive function on *i/g* administration of polyethylenepoliamine, *Gig. Sanit.*, 8, 78, 1977 (in Russian).

POLYIMIDAZOLINE CURING AGENT

Composition. A condensation product of acid ethyl ethers, soya oil, and triethylenetriamine containing polyamidazole groups.

Applications. Used as a curing agent in the manufacture of epoxy resins.

Acute Toxicity. In rats, LD_{50} is 2.3 g/kg BW. Manifestations of the toxic action include increased neuromuscular excitability, labored breathing, weakening of tendon reflexes, hyperemia of the skin and conjunctiva, and profuse salivation and lacrimation followed by hypodynamia, a reduction in muscle tone of the extremities, diarrhea, and death. Necrotic foci are observed on gross pathology examination.

Reference:

Stavreva, M., in *Topical Questions Concerning the Safe Use of Pesticides in Different Climate and Geographic Zones*, Aistan, Yerevan, 1976, 173 (in Russian).

POLYOXYPROPYLENEDIAMINE

Molecular Formula. $[\sim C_3H_6O\sim]_nC_6H_{16}N_2O$
CAS No 9046-10-0
RTECS No TR3702500

Synonyms. α,ω-Diaminopoly propylene glycol; Poly(oxy-α-(2-aminomethylethyl)-ω-(2-aminomethyl-ethoxy).

Properties. Colorless, slightly opalescent, non-volatile liquid. Readily soluble in water and alcohol.

Applications. Used as a curing agent in the production of epoxy resins and compounds, foams, glass fiber lubricants, etc.

Acute Toxicity. LD_{50} is 242 mg/kg BW in rats. Poisoning induced convulsions or effect on seizure threshold. Ulceration or bleeding from stomach an hemorrhage were noted.

Reference:

Natl. Technical Information Service OTS0570544.

POLYOXYPROPYLENEDIAMINE *DA 200*

RTECS No MD0912200
M = 200
Abbreviation. DA 200.

Synonym. Polypropylene glycol, bis(2-aminopropyl) ether.

Properties. Colorless, slightly opalescent, non-volatile liquid. Readily soluble in water and alcohol.

Applications. DA 200 is used as a curing agent in the production of epoxy resins and compounds, foams, glass fiber lubricants, etc.

Acute Toxicity. LD_{50} is 627 mg/kg BW in rats. Poisoning caused excitement and dyspnea.

Reference:

Yavorovsky, A. P., Bogorad, V. S., and Yefremova, L. D., Comparative toxicological-hygienic assessment of polyoxypropylenamines and substantiation of their MAC in the work zone air, *Gig. Truda Prof. Zabol.*, 11, 45, 1989 (in Russian).

POLYOXYPROPYLENEDIAMINE *DA 500*

Molecular Formula. $C_{10}H_{24}N_2O_3[\sim C_3H_6O\sim]_{2(n-1)}$

M = 480 to 500

RTECS No MD0912300

Abbreviation. DA 500.

Synonym. Polypropylene glycol, bis(2-aminopropyl) ether.

Properties. Colorless, slightly opalescent, non-volatile liquid. Readily soluble in water and alcohol. Threshold concentration for foam formation is 2.5 mg/l. Odor perception threshold is 16.4 mg/l, the practical threshold being 30 mg/l. Chlorination does not provoke foreign odors or tastes.[1]

Applications. DA 500 is used as a curing agent in the production of epoxy resins and compounds, foams, glass fiber lubricants, etc.

Acute Toxicity. LD_{50} is reported to be 300 to 570 mg/kg BW in rats, 140 to 230 mg/kg BW in mice, 94 mg/kg BW in guinea pigs, and 172.5 mg/kg BW in rabbits.[1,2]

Administration of the lethal dose is accompanied by rapid disappearance of the investigatory reflex, coordination disorder, CNS depression, and clonic-tonic convulsions with subsequent death. Serous nasal dis charges were noted in some animals. ET_{50} appeared to be 34 hours.

Repeated Exposure revealed pronounced cumulative properties. K_{acc} is 0.84 (by Kagan).[1]

Male rats were dosed with 1/10 LD_{50} for 1.5 months. The treatment affected the red and white cell germinal centers of the bone marrow (decline in *Hb* level, erythropenia, leukocytosis). Hepatotropic effect and renal impairment were observed.[1,2]

Long-term Toxicity. In a 6-month study in male rats, symptoms of hypochromic anemia and leukocytosis were revealed as a result of an increased number of eosinophils and monocytes, a decline in whole blood cholinesterase activity, and a fall in the serum urea concentration.[1]

Allergenic Effect. DA-500 has a weak sensitizing potential.[2] The NOAEL for such effects appeared to be 0.015 mg/kg BW.[1]

Reproductive Toxicity.

Gonadotoxicity. A dose of 1.5 mg/kg BW produced an increased number of abnormal spermatozoa in rats. The NOAEL for gonadotoxic effect appeared to be 0.15 mg/kg BW.[1]

Embryotoxicity. Administration of the 1.5 mg/kg BW dose to pregnant rats resulted in an increase of post-implantation and total embryolethality. The 0.15 mg/kg BW dose caused only post-implantation mortality. Both doses increased the embryonal heart relative weights.[1]

Mutagenicity.

In vivo cytogenetics. Mutagenic effect was not found in rats exposed to the NOAEL established for general toxicity action (0.015 mg/kg BW).[1]

Standards. ***Russia.*** Recommended MAC: 0.3 mg/l.

References:

1. Ivanova, I. F., Experimental derivation of MAC for polyoxypropylenes in water, *Gig. Sanit.*, 12, 73, 1986 (in Russian).
2. Yavorovsky, A. P. and Bogorod, V. S., Toxicity of polyoxypropylenediamine DA-500 and polyoxypropylenetriamine TA 1100, *Gig. Sanit.*, 8, 94, 1986 (in Russian).

POLYOXYPROPYLENEDIAMINE *DA 1000*

M = 1000

RTECS No MD0912320

Abbreviation. DA 1000.

Synonym and **Trade Name.** Polypropylene glycol, bis(2-aminopropyl) ether.

Properties. Colorless, slightly opalescent, non-volatile liquid. Readily soluble in water and alcohol.

Applications. DA 1000 is used as a curing agent in the production of epoxy resins and compounds, foams, glass fiber lubricants, etc.

Acute Toxicity. LD_{50} is reported to be 68.9 mg/kg BW in rats, 37.4 mg/kg BW in mice, 86.5 mg/kg BW in rabbits, and 86.5 mg/kg BW in guinea pigs. Poisoning caused excitement and dyspnea.

Reference:

See ***Polyoxypropylenediamine DA 200.***

POLYOXYPROPYLENETRIAMINE

Molecular Formula. $C_{12}H_{29}N_3O_3(C_3H_6O)_{3(n-1)}$

M = 1050 to 1100

Synonym. α-Hydro-ω-(2-aminomethylethoxy)poly[oxy(methyl-1,2-ethanediyl)] ether with 2-ethyl-2-(hydroxymethyl)-1,3-propanediol (3:1).

Properties. A non-combustible, colorless, slightly opalescent, practically non-volatile liquid. Readily soluble in water and alcohol. Threshold concentration for foam formation is 0.18 mg/l. Odor perception threshold is 14.9 mg/l; practical threshold is 30.7 mg/l.[1] Chlorination of aqueous solutions does not provoke foreign odors or tastes.

Applications. Used as a curing agent in the production of epoxy resins and compounds, foams, glass fiber lubricants, etc.

Acute Toxicity. LD_{50} is 90 to 147 mg/kg BW in rats, 17 to 48 mg/kg BW in mice, 55 mg/kg BW in guinea pigs, and 52 mg/kg BW in rabbits.[1, 2]

See ***Polyoxypropylenediamine.***

Repeated Exposure revealed evident cumulative properties. K_{acc} is 0.21 (by Kagan).[1]

See ***Polyoxypropylenediamine.***

Long-term Toxicity.

See ***Polyoxypropylenediamine.***

Allergenic Effect. The NOEL for this effect is reported to be 0.0015 mg/kg BW.

Reproductive Toxicity.

Gonadotoxic effect was not observed in male rats.

Embryotoxicity. According to Yavorovsky and Bogorod, oral administration of doses up to 1.5 mg/kg BW did not cause embryolethality. The relative weights of the heart and liver were reduced in embryos when doses of up to 1.5 mg/kg BW were applied.

Mutagenicity. See ***Polyoxypropylenediamine.***

Standards. ***Russia***. Recommended MAC: 0.03 mg/l.

References:

1. Ivanova, I. F., Experimental substantiation of maximum allowable concentration for polyoxypropyleneamines in water, *Gig. Sanit.*, 12, 73, 1986 (in Russian).
2. Yavorovsky, A. P. and Bogorod, V. S., Toxicity of polyoxypropylenediamine DA 500 and polyoxypropylenetriamine TA 1100, *Gig. Sanit.*, 8, 94, 1986 (in Russian).

POLYOXYPROPYLENETRIAMINE *TA 750*

M = 750

CAS No 39423-51-3

RTECS No MD0912375

Abbreviation. TA 750.

Synonym. Polypropylene glycol, tris(2-aminopropyl) ether.

Acute Toxicity. LD_{50} is 189 mg/kg BW in rats. Poisoning caused excitement, dyspnea, and changes in respiratory system.

Reference:

See ***Polyoxypropylenediamine DA 200.***

POLYOXYPROPYLENETRIAMINE *TA 1100*

M = 1100
RTECS No MD0912380
Abbreviation. TA 1100.

Synonym. Polypropylene glycol, tris(2-aminopropyl) ether.

Acute Toxicity. LD_{50} is 31.5 mg/kg BW in rats, 16.9 mg/kg BW in mice, 22.5 mg/kg BW in rabbits and guinea pigs. Poisoning caused excitement, dyspnea, and changes in respiratory system.[1]

Long-term Toxicity. In 6-month rat study, changes in leukocyte count were reported.[2]

References:

1. See ***Polyoxypropylenediamine DA 200.***
2. Ivanova, I. F., Experimental establishment of the maximum permissible concentrations of polyoxypropyleneamines in water, *Gig. Sanit.*, 12, 73, 1986 (in Russian).

POLYOXYPROPYLENETRIAMINE *TA 1500*

M = 1500
RTECS No MD0912400

Synonym. Polypropylene glycol, tris(2-aminopropyl) ether.

Acute Toxicity. LD_{50} is 161 mg/kg BW in rats, 20.8 mg/kg BW in mice, 124 mg/kg BW in rabbits, and 186 mg/kg BW in guinea pigs. Poisoning caused excitement, dyspnea, and changes in respiratory system.

Reference:

See ***Polyoxypropylenediamine DA 200.***

1,3-PROPANEDIAMINE

Molecular Formula. $C_3H_{10}N_2$
M = 74.13
CAS No 109-76-2
RTECS No TX6825000
Abbreviation. 1,3-PDA.

Synonyms. 1,3-Diaminopropane; 1,3-Propylenediamine; Trimethylene diamine.

Properties. Extremely hygroscopic, strongly alkaline liquid. Very soluble in water.

Acute Toxicity. LD_{50} is 350 µl/kg BW in rats.[1]

Reproductive Toxicity. Mice received *i/p* 264 mg/kg BW on day 12 of pregnancy. Fetotoxicity and developmental abnormalities in muscular skeletal system were noted.[2]

References:

1. Smyth, H. F., Carppenter, C. P., Weil, C. S., Pozzani, U. C., and Striegel, J. A., Range-finding toxicity data, list VI, *Am. Ind. Hyg. Assoc. J.*, 25, 95, March-April 1962.
2. Manen, C. A., Hood, R. D., and Farina, I., Ornothine decarboxylase inhibition and fetal growth retarda tion in mice, *Teratology*, 28, 237, 1983.

4,4'-SULFONYLDIANILINE

Molecular Formula. $C_{12}H_{12}O_2N_2S$
M = 248.31
CAS No 80-08-0
RTECS No BY8925000
Abbreviation. SDA.

Synonyms and **Trade Name.** Bis(4-aminephenyl)sulfone; Dapsone; 4,4'-Diaminodiphenyl sul fone.

Properties. A powdery substance. Water solubility is 15 mg/l at 20°C, soluble in alcohol and acetone. Odor and taste threshold concentrations exceed 15 mg/l.[1]

Applications. Used as a curing agent for epoxy resins and as an antioxidant.

Acute Toxicity. LD_{50} is 1000 to 1400 mg/kg BW (LD_0 being 18 to 32 mg/kg BW) in rats, 375 mg/kg in mice, and 357 mg/kg BW in cats. Poisoning is accompanied by narcosis.[2] Death occurs within 5 days.

Repeated Exposure revealed pronounced cumulative properties. K_{acc} is 1.4 (by Kagan).

Long-term Toxicity. Rats were dosed by gavage with 5.0 mg/kg BW for 6 months. The treatment produced a steady change in the functional condition of the liver and an increase in lactate dehydrogenase activity. Methemoglobinemia and changes in conditioned reflex activity were observed.[1]

Mutagenicity. SDA was negative in the DNA repair test with rat hepatocytes and in *Salmonella typhimurium* strains *TA98* and *TA100* in the presence of *S9* mix.[3,4]

Carcinogenicity. BDIV rats received oral doses of 100 mg SDA/kg BW for 104 weeks. The study revealed spleen sarcomas in males, and an increase in morbidity from *C*-cell carcinomas in both sexes. Tumors appeared after lifetime treatment with maximum tolerated doses.[5] According to Mori et al., there is no sound proof of SDA carcinogenicity.[3]

Carcinogenicity classifications. An IARC Working Group concluded that there is limited evidence for the carcinogenicity of SDA in *experimental animals* and there were inadequate data available to evaluate the carcinogenicity of SDA in *humans*.

IARC: 3.

NTP: P* - N - N - N (feed).

Chemobiokinetics. SDA is slowly but almost completely absorbed from the GI tract. It is rapidly distributed all over the body but predominantly in the liver and kidneys. It is retained in circulation for a long time. SDA and its major metabolite undergo diacetylations in humans but not in dogs. Main urinary metabolites that were found in the urine are *mono-N-glucuronide* and *mono-N-sulfamate* as well as other unidentified metabolites. *Monohydroxylamine* is also reported to be a major urinary metabolite.[6,7] According to Tanaka et al., *N-acetyl* and *N,N'-diacetyl* metabolites were found in the urine.[4]

In the presence of rat liver microsomes, SDA was the most potent former of *metHb* in human erythrocytes.[8]

See ***4,4'-Methylenedianiline***

Standards. ***Russia*** (1995). MAC and PML: 1.0 mg/l.

References:

1. Tsyganovskaya, L. Kh. et al., Substantiation of MAC for 4,4'-diaminodiphenyl sulfone in water bodies, *Gig. Sanit.*, 9, 23, 1978 (in Russian).
2. Ivanov, N. G. and Mel'nikova, L. V., Comparative investigation of toxicity and mode of action of 4,4'-diaminodiphenyl sulfone and 4,4'-diaminodiphenyl oxide, in *Toxicology of New Industrial Chemical Substances*, N. F. Ismerov and I. V. Sanotsky, Eds., Meditsina, Moscow, Issue No 14, 1975, 118 (in Russian).
3. Mori, H., Yoshimi, N., Sugie, S. et al., Genotoxicity of epoxy resin hardeners in the hepatocyte primary/DNA repair test, *Mutat. Res.*, 204, 683, 1988.
4. Tanaka, K., Ino, T., Sawahata, T., Marui, S., Igaki, H., and Yashima, H., Mutagenicity of *N*-acetyl and *N,N'*-diacetyl derivatives of 3 aromatic amines used as epoxy-resin hardeners, *Mutat Res.*, 143, 11, 1985.
5. Grticiute, L. and Tomatis, L., Carcinogenicity of dapsone of mice and rats, *Int. J. Cancer,* 25, 123, 1980.
6. Goodman and Gilman's *The Pharmacological Basis of Therapeutics*, A. G. Gilman, L. S. Goodman, and A. Gilman, Eds., 6th ed., MacMillan Publishing Co., Inc., New York, 1980, 1215.
7. *Foreign Compound Metabolism in Mammals*, vol 3, The Chemical Society, London, 1975, 297.
8. Mahmud, R., Tingle, M. D., Maggs, J. L, Cronin, M. T. D., Dearden, J. C., and Park, B. K., Structural basis for the hemotoxicity of dapsone: the importance of the sulfonyl group, *Toxicology,* 117, 1, 1997.

SULPHURIC ACID, AMMONIUM IRON SALT

Molecular Formula. $O_8S_2H_8N_2.Fe$

M = 284.07

CAS No 10045-89-3

RTECS No WS5890000

Abbreviation. AFS.

Synonyms. Ammonium ferrous sulfate; Ammonium iron sulfate.

Applications. A polymerization catalyst.

Acute Toxicity.

Observations in man. As little as 1 to 2 g iron may cause death, but 2 to 10 g are usually ingested in fatal cases.[1]

Animal studies. LD_{50} is at the level of 0.5 to 5.0 g/kg BW in rats.[2]

Chemobiokinetics. Free circulating iron damages systemic blood vessels.[3] Iron was shown to cross the placenta and concentrate in the fetus. In cases of overload iron is excreted predominantly in the urine and feces. A part of it is excreted via the bile.[4]

In cells Fe^{2+} is converted to Fe^{3+} in ferritin, the latter not being absorbed until the cell is physiologically "depleted". In the blood stream Fe^{2+} could be quickly oxidized by dissolved oxygen to Fe^{3+}, which complexes with specific *iron*-transport β1-globulin.[03]

Regulations. ***U.S. FDA*** (1998) approved the use of AFS in the production of cellophane intended for packaging food in accordance with the conditions prescribed in 21 CFR part 177.1200. Optional substances used in the base sheet and coating may include AFS.

Standards. ***U.S. EPA*** (1999). MCL: 0.3 mg/l */Fe/*

References:

1. Goodman and Gilman's *The Pharmacological Basis of Therapeutics,* 8th ed., Gilman, A. G., Rall, T. W., Nies, A. S., and Taylor, P., Eds., Pergamon Press, New York, 1990, 1291.
2. U.S. Coast Guard, Department of Transportation, CHRIS - *Hazardous Chemical Data*, vol 2, U.S. Government Printing Office, Washington, D.C., 1984-5.
3. Ellenhorn, M. J. and Barceloux, D.G., *Medical Toxicology - Diagnosis and Treatment of Human Poisoning,* Elsevier Science Publishing Co., Inc., New York, 1988, 1024.
4. Casarett and Doull's *Toxicology*, 2nd ed., Macmillan Publishing Co., New York, 1980, 446.

TEREPHTHALOYL CHLORIDE

Molecular Formula. $C_8H_4Cl_2O_2$

M = 203.02

CAS No 100-20-9

RTECS No WZ1797000

Abbreviation. TPC.

Synonyms. 1,4-Benzenedicarbonyl chloride; *p*-Phenylene dicarbonyl dichloride; *p*-Phthaloyl chloride; Terephthalic acid, chloride; Terephthalic acid, dichloroanhydride; Terephthalic dichloride.

Properties. White crystals or colorless needles with a sharp odor. Solubility in water is 8.0 mg/l at 20°C. Readily dissolves in most organic solvents. Odor perception threshold in water is 0.026 mg/l. Does not affect taste, color, or transparency of water.

Applications. A cross-linked agent in the synthesis of polyurethanes, polyacrylates, and polysulsulfide rubbers. Used also in the dye manufacture, and in the production of synthetic fibers, resins, films.

Acute Toxicity. LD_{50} is 2.14 g/kg BW in mice, 5.73 g/kg BW in rats, and 0.95 g/kg BW in rabbits. Poisoning produced somnolence and respiration disorder. Animals treated with the high doses died in 3 to 5 days (mice), 2 to 4 days (rats) or 7 to 10 days (rabbits).[1]

Repeated Exposure. Rats received daily doses equal to 1/10, 1/50, 1/250 LD_{50} for 45 days. TPC was found to have pronounced cumulative properties. Signs of intoxication included depressed CNS and respiratory functions.[1]

Long-term Toxicity. Rats and rabbits received 1/250 and 1/1250 LD_{50} by gastric intubation as a suspension in sunflower oil for 6 and 8 months, respectively. In rats, the treatment caused retardation of BW gain, a decrease in glucose and *Hb* level, erythrocyte and leukocyte count and increased γ-globulin level in the blood, lowered activity of aldolase in the blood serum. Histological investigation revealed structural changes and hemodynamic disturbances in the brain, morphological changes in the liver, kidney, myocardium. Treatment with a lower dose produced less evident changes in rats and rabbits. A decrease in absolute and relative weights of liver, lungs, spleen, and heart was observed only with this dose. NOAELs of 2.29 and 0.38 mg/kg BW were identified for rats and rabbits, respectively.[1]

Reproductive Toxicity.

Gonadotoxicity. Male and female rats received a total dose of about 4.0 mg/kg BW during 26 weeks prior to mating. In males, spermatogenesis was affected. In females, there were maternal effects, namely, changes or disturbances in estrus cycle.[2]

Mutagenicity.

In vivo cytogenetics. Oral administration of TPC to rats induced CA in the bone marrow cells. NOEL for the mutagenic effect appeared to be 22.9 mg/kg BW in rats.[1]

Regulations. ***EU*** (1990). TPC is available in the *List of monomers and other starting substances which may continue to be used pending a decision on inclusion in Section A*.

References:

1. Devyatka, D. G., Stepanyuk, G. I., Makats, V. G., Pushkar', M. S., Bogachuk, G. P., Korolik, A. G., and Devyatka, O. D., Experimental substantiation of maximum allowable concentration for terephthalyl chloride in water bodies used for domestic purposes, *Gig. Sanit.*, 12, 35, 1984 (in Russian).
2. Deichmann, W. B., *Toxicology of Drugs and Chemicals*, Academic Press, Inc., New York, 1969, 475.

N,N,N',N'-TETRAMETHYL-1,2-ETHYLENEDIAMINE

Molecular Formula. $C_6H_{16}N_2$
M = 116.24
CAS No 110-18-9
RTECS No KV7175000
Abbreviation. TMEDA.

Synonyms. 1,2-Bis(dimethylamino)ethane; *N,N,N',N'*-Tetramethylethanediamine.

Properties. Colorless liquid with slight ammoniac odor. Soluble in water and organic solvents.

Applications. Used in preparation of epoxy curing agents, and in polyurethane formation.

Reproductive Toxicity. TMEDA has not been found to affect the fertility of eggs produced by female quail after their mates were given a single oral dose of about 50% of the estimated LD_{50} value.

Mutagenicity. Was found to be negative in *Salmonella* mutagenicity assay.[015]

2-THIOUREA

Molecular Formula. CH_4N_2S
M = 73.13
CAS No 62-56-6
RTECS No YU2800000
Abbreviation. TU.

Synonyms. β-Naphthylthiourea; β-Thiopseudourea; Isothiourea; Thiocarbamide.

Properties. White powder or crystals with bitter taste. Solubility in water is 10 g/110 ml at 25°C, soluble in alcohol.

Applications. Catalyst and intermediate for manufacture of rubber accelerators and flame-retardant resins.

Acute Toxicity. LD_{50} is 125 mg/kg BW in rats.[1]

Reproductive Toxicity.

Embryotoxicity. Rats were given 2.0 g/kg BW on day 12, mice received 0.1 g/kg BW on day 10 of gestation. The treatment caused no increase in the defect rate but induced a slight increase in resorption levels. TU caused an increase in post-implantation mortality (i.e., dead and/or resorped implants).[2]

Teratogenicity. TU was found to be teratogenic in rats exposed to a 0.2% solution in their drinking water.[3]

Mutagenicity.

In vitro genotoxicity. TU was not found mutagenic in *Salmonella typhimurium* strains *TA1535, TA1536, TA1537, TA1538, TA98,* and *TA100*.[4] It showed no DNA-modifying activity with normal and DNA polymerase-deficient *E. coli* strains.[1]

Carcinogenicity. When administered in the drinking water, TU caused liver, thyroid and Zymbal gland tumors in rats (IARC 7-95). 21 female C3H mice were given the diet containing 0.25 to 0375% TU for 63 weeks. No thyroid tumors were found to develop.[5]

Carcinogenicity classifications. An IARC Working Group concluded that there is sufficient evidence for the carcinogenicity of TU in *experimental animals* and there are no adequate data to evaluate the carcinogenicity of TU in *humans*.

IARC: 2B;

U.S. EPA: 2A.

References:

1. Rosenkranz, H. S. and Poirier, L. A., Evaluation of the mutagenicity and DNA-modifying activity of carcinogens and noncarcinogens in microbial systems, *J. Natl. Cancer Inst.*, 62, 873, 1979.
2. Teramoto, S., Kaneda, M., Aoyama, H., and Shirasu, Y., Correlation between the molecular structure of *n*-alkylureas and *n*-alkylthioureas and their teratogenic properties. *Teratology,* 23, 335, 1981.
3. Kern, M. et al., Teratogenic effect of 2'-thiourea in the rat, *Acta Morphol. Acad. Sci. Hung.,* 28, 259, 1980.
4. Simmon, V. F., *In vitro* mutagenicity assays of chemical carcinogens and related compounds with *Salmonella typhimurium*, *J. Natl. Cancer Inst.*, 62, 893, 1979.
5. Dalton, A. J., Morris, H. P., and Dubnik, C. S., Morphologic changes in the organs of female C3H mice after long-term ingestion of thiourea and thiouracil, *J. Natl. Cancer Inst.*, 9, 201, 1948.

TIN inorganic compounds

Sn

CAS No 7440-31-5

RTECS No YXP7320000

Properties. The most important inorganic compounds of tin are oxides, chlorides, fluorides, and halogenated sodium stannates and stannites. Some tin salts are water soluble.

Applications. Tin is a component of a number of catalysts: organotin compounds are applied as polyvinyl chloride stabilizers. Tin is used principally in the production of coatings in the food industry.

Migration Data. Food, and particularly canned food, therefore represents the major route of human exposure to tin compounds. Higher concentrations are found in canned food as a result of dissolution of the tin coating or tin plate. Acidity of the food, the presence of oxidants, time and temperature of storage, and the presence of air in the headspace can influence this process. Tin concentrations in foodstuffs in unlacquered cans frequently exceed 0.1 mg/g (WHO, 1989).

Exposure. Although tin is a natural component of food products, its physiological significance is unknown. An intake of 1 to 30 mg tin/day consumed in the food is 1000 times greater than the amount consumable in water.

Acute Toxicity.

Observations in man. The main adverse effect with excessive levels of tin compounds in food, such as canned fruit, has been acute gastric irritation. Tin compounds act as an irritant for the GI tract mucosa. Vomiting, diarrhea, fatigue, and headache were seen in a lot of cases following the consumption of canned products (tin compounds concentrations as low as 150 mg/kg in one incidence involving canned beverages and 250 mg/kg BW in other canned foods).[1]

Animal studies. In the mouse and rat, LD_{50} values are 250 and 700 mg/kg BW for tin chloride, 2.14 and 2.2 g/kg BW for tin sulfate, and 2.7 and 4.35 g/kg BW for sodium stannate hydrate, respectively. Both rats and mice tolerate administration of 10 g tin oxide/kg BW (II, IV). The rabbit is less sensitive: LD_{50} is 10 g/kg BW.[2] High doses seem to affect the CNS, producing ataxia.

The species-related differences illustrated by the LD_{50} values are apparent even at low levels of exposure. The cat appeared to be more sensitive to oral administration of tin compounds than either the dog or the rat, only in cats were vomiting and diarrhea observed after oral administration of tin-containing fruit beverages.[1]

Poisoning by inorganic salts of tin compounds is characterized by a brief period of excitation, which gives way to general inhibition. Intoxication is manifested through transient digestive upset and apathy.

Poisoning symptoms disappear in 2 to 3 days, depending on the dose. At autopsy, distention of the stomach was found.[3]

Repeated Exposure.

Observations in man. In nine male volunteers consuming packaged military rations (tin compounds contents ranging from 13 to 204 mg/kg BW) for a successive 24-day period and in other cases with human volunteers who ate canned food with tin compound contents varying from 250 to 700 mg/kg BW over a period of 6 to 30 days, no toxic effects were noted.[2] Toxic signs following the consumption of tin-containing food could be seen only after administration of tin concentrations of about 1400 ppm and higher.[4]

Animal studies. Hb concentration in the blood of rats was decreased significantly after feeding a diet containing 150 mg tin compounds/kg.[5]

Short-term Toxicity. In the 4-week and/or 13-week feeding studies, rats were given various tin salts or tin oxide at dose levels of 50 to 10000 mg/kg BW. Doses of 3.0 g/kg BW and more caused anemia, changes in tissue enzyme activities, and extensive damage to the liver and kidney. These findings were especially observed with the more soluble tin salts like chloride, *o*-phosphate, and sulfate.[1,6]

Biochemical effects attributable to tin intoxication have been observed even after oral administration of 1.0 and 3.0 mg/kg BW.[5] These doses reflect 10 and 30 ppm tin in the diet. The relative weight of the femur, calcium concentration, lactic dehydrogenase and alkaline phosphatase activity in the serum appeared to be significantly decreased in rats given the highest dose of tin chloride. The NOEL was considered to be less than 0.6 mg/kg BW in this study.

Long-term Toxicity.

Observations in man. There is no evidence of adverse effects in man associated with chronic exposure to tin compounds.

Animal studies. Toxic effects can be caused by the ingestion of rather high doses of tin. Rats were exposed to 200 to 800 mg tin chloride/kg food for 115 weeks. No histopathology changes were observed. The NOAEL of 20 mg/kg BW was identified in this study.[1]

No effects on survival or retardation of BW gain were observed in mice that received 1.0 and 5.0 g tin chlorostannate/l or 5.0 g tin oleate/l over a period of 1 year.[1]

Immunotoxicity. *I/p* administration of 167 mole tin chloride/kg BW for 3 days caused a decline in the immune response to sheep erythrocytes. Liver weight was increased and production of antibody-forming cells slowed down.[7]

Reproductive Toxicity. Tin has not been shown to be *teratogenic* or fetotoxic in mice, rats, and hamsters.

Gonadotoxicity. Testicular degeneration was shown in rats administered 10 mg tin chloride/kg BW in the feed for 13 weeks.[6] This tin salt at doses 200 to 800 mg/kg feed did not affect reproductive performance of rats, but transient anemia was observed in the offspring prior to weaning.[1]

Embryotoxicity. Low transplacental transfer of tin compounds was observed after feeding of different tin salts (concentration of 500 mg/kg diet) to pregnant rats, although no effects were seen in the fetuses.[1]

Mutagenicity.

In vitro genotoxicity. Tin chloride produced a negative response in the SCE test system.[8]

Carcinogenicity bioassays did not show an increase in tumor incidence in mice and rats except in one occasion.[1,9]

Chemobiokinetics. Tin inorganic compounds are poorly absorbed from the GI tract. Only a few percents of sodium stannate hydrate, for example, are absorbed. In the presence of the citric acid, often found in fruit juice, absorption of tin is increased.[10] About 50% the dose was shown to be absorbed by man, if 0.11 mg of tin is ingested with the diet.[11] Tin compounds do not accumulate in the tissues. Administration in a 1.2 mg tin/l solution for 36 days was not accompanied by accumulation in the body. The highest tissue concentration was found in the bones (principal site of distribution).

Metabolism of tin compounds is preceded by alkylation and dealkylation processes. Methylation of tin is ini tiated by sedimentation microorganisms, and it occurs along with abiotic methylation.

Tin compounds are rapidly excreted from the body primarily in the urine and feces (up to 99% of administered dose).

Regulations. ***Russia***. The tin content in canned food is regulated as 100 to 200 mg/kg for condensed milk, meat, fish, and vegetable products.

Recommendations.

WHO (1982, 1989) concluded that due to the low toxicity of inorganic tin a tentative Guideline value could be derived three orders of magnitude higher than the normal tin concentration in drinking water. Therefore, the presence of tin in drinking water does not present a hazard to human health. For that reason, the establishment of a numerical Guideline value was not deemed necessary (1991).

Joint FAO/WHO Expert Committee on Food Additives. ADI: 2.0 mg/kg BW. PWTI: 14 mg/kg BW.

Standards. ***Czechoslovakia***. 10 mg/kg is permitted in milk.

References:

1. *Toxicological Evaluation of Certain Food Additives and Contaminants*, Food Additives Series No 17, Joint FAO/WHO Expert Committee on Food Additives, 1982, 297.
2. *Tin and Organotin Compounds* - a Preliminary Review, Environmental Health Criteria No 15, WHO, IPCS, Geneva, 1980.
3. Bessmertny, A. N. and Grin', N. V., Acute toxicity studies on inorganic tin compounds for the purpose of hygienic norm-setting, *Gig. Sanit.*, 6, 82, 1986 (in Russian).
4. Benoy, C. J., Hooper, P. A, and Schneider, R., The toxicity of tin in canned fruit juice and solid foods, *Food Cosmet. Toxicol.*, 9, 645, 1971.
5. Yamaguchi, M., Saito, R., and Okada, S., Dose-effect of inorganic tin on biochemical indices in rats, *Toxi cology*, 16, 267, 1980.
6. de Groot, A. P., Feron, V. J., and Til, H. P., Short-term toxicity studies on some salts and oxides of tin in rats, *Food Cosmet. Toxicol.*, 11, 19, 1973.
7. Hayashi, O., Chiba, M., Kikuchi, M., et al., The effect of stannous chloride on the humoral immune response in mice, *Toxicol. Lett.*, 21, 279, 1984.
8. Tucker, J. D., Auletta, A., Cimino, M. C., et al., Sister-chromatid exchange: second report of the Gene-Tox Program, *Mutat. Res.*, 297, 101, 1993.
9. *Carcinogenesis Bioassay of Stannous Chloride*, NTP Technical Report, DHHS Publ., NIH No 81-1787, Bethesda, MD, 1982.
10. Kojima, S., Saito, K., and Kiyozumi, M., Studies on poisonous metals: IV. Absorption of stannic chloride from rat alimentary tract and effect of various food components on its absorption, Abstract, *Yakugaku Zasshi*, 98, 495, 1978 (in Japanese).
11. Johnson, M. A. and Greger, J. L., Effects of dietary tin on calcium metabolism of adult males, *Am. J. Clin. Nutr.*, 35, 655, 1982.

TIN CHLORIDE (1:2)

Molecular Formula. Cl_2Sn

M = 189.59

CAS No 7772-99-8

RTECS No XP8700000

Abbreviation. TC.

Synonyms. Stannochlor; Stannous chloride; Tin dichloride; Tin protochloride.

Properties. White, orthorhombic, odorless crystalline mass or flakes, or colorless to brown solid of fatty appearance. Readily soluble in ethanol (544 g/l at 23°C) and in water (900 g/l at 20°C).

Applications. An intermediate and a catalyst in the production of dyes and plastics.

Acute Toxicity.

Observations in man. Acute poisoning occurred following ingestion of fruit juices containing concentration of tin above 250 mg tin/l.

Animal studies. LD_0 is 60 mg/kg BW in guinea pigs, and 500 mg/kg BW in dogs. LD_{50} is 700 mg/kg BW in rats, 250 mg/kg BW in mice, and 10 g/kg BW in rabbits.[1,2] However, according to other data, LD_{50} is 1.2 g/kg BW in mice.[3]

Repeated Exposure. Rats were given six oral doses of 10 mg/kg BW at 12-hour intervals. The treatment caused significant decrease in alkaline phosphatase and hepatic phosphorylase activities, serum calcium concentration, and femoral calcium contents. Dose of 3.0 mg/kg BW (x6) produced significant decrease of calcium contents in the epiphysis of the femur. Bone appeared to be a critical organ of TC toxicity.[4,5]

Short-term Toxicity. Weanling male rats received oral doses of 0.3, 1.0, and 2.0 mg tin/kg BW at 12-hour intervals for 90 days. The highest dose level caused significant decrease of the relative weights of the femur, calcium concentration, lactic dehydrohenase and alkaline phosphatase activities in the serum, succinate dehydrohenase activity in the liver, and calcium contents and acid phosphatase activity in the femoral diaphysis and epiphysis. Authors considered the NOEL for inorganic tin on biochemical indices in this study to be lower than 0.6 mg tin/kg BW.[6]

Long-term Toxicity. In six-month study, rats have been fed ad libitum with a diet containing 0.4 or 0.8 g tin/100 g of dry food. γ-Radioactivity measurements showed that tin compounds do not practically clear the digestive barrier. At the end of the treatment, there was significant diminution of hematocrit, *Hb* and serous iron, particularly at the highest dose tested. Histological examination revealed a total irritation of GI tract.[7]

Reproductive Toxicity. Chronic tin poison caused testicular degeneration, severe pancreatic atrophy, acute bronchopneumonia, and enteritis in rats.[3]

Mutagenicity.

In vitro genotoxicity. TC was found to be negative in L5178Y mouse lymphoma cells, and positive in Chinese hamster ovary cells.[8,9] TC appeared to be capable of inducing and/or producing lesions in DNA in *B. coli.*[10]

Chemobiokinetics. Rats were administered 7 doses of 2.0 mg tin/kg BW, given *s/c* every other day. About 60% of ^{113}Sn was retained in the body. Tin is distributed predominantly in the skin and hair. A 3-fold increase of the contents of zinc was noted in the liver while decreases were found in the spleen, heart, brain, lungs, and particularly in the muscles. A significant decrease was found in copper contents in the blood and brain.[11]

Regulations. *U.S. FDA* (1998) approved the use of TC (1) in the manufacture of resinous and polymeric coatings for the food-contact surface of articles intended for use in producing, manufacturing, packing, processing, preparing, treating, packaging, transporting, or holding food in accordance with the provisions prescribed in 21 CFR part 175.300; and (2) in the manufacture of rubber articles intended for repeated use in producing, manufacturing, packaging, processing, preparing, treating, packing, transporting, or holding food in accordance with the provisions prescribed in 21 CFR part 177.2600.

TC is affirmed as *GRAS* when used in accordance with GLP.

References:

1. FAO Nutrition Meeting Reports, Series No 50A, 1972, 101.
2. Abdernalden's *Handbuch der Biologischen Arbeitsmethoden*, vol 4, 1420, 1935.
3. Venugopal, B. and Luckey, T. D., *Metal Toxicity in Mammals*, vol 2, Plenum Press, New York, 1978, 184.
4. Yamaguchi, M., Sugii, K., and Okada, S., Inorganic tin in the diet affects the femur in rats, *Toxicol. Lett.*, 9, 207, 1981.
5. Yamaguchi, M., Kitade, M., and Okada, S., The oral administration of stannous chloride in rats, *Toxicol. Lett.*, 5, 275, 1980.
6. Yamaguchi, M., Saito, R., and Okada, S., Dose-effect of inorganic tin on biochemical indices in rats, *Toxi cology*, 16, 267, 1980.
7. Fritsch, P., de Saint Blanquat, G., and Derache, R., Nutritional and toxicological impacts of administered inorganic tin for six months in rats, *Toxicol. Eur. Res.*, 1, 253, 1978.
8. *Annual NTP Plan*, 1987 Fiscal Year, NTP-87-001, 1986, 2.
9. Rossman, T. G., Molina, M., Meyer, L., Boone, P., Klein, C. B., Wang, Z., Li, F., Lin, W. C. and Kinney, P. L., Performance of 133 compounds in the lambda prophage induction endpoint of the Microscreen assay and a comparison with *S. typhimurium* mutagenicity and rodent carcinogenicity assays, *Mutat. Res.*, 260, 349, 1991.

10. Bernardo-Filho, M., da Cunha, M. C., de Valsa, J. O., de Araujo, A. C., da Silva, C. P. and de Fonseca, A. S., Evaluation of potential genotoxicity of stannous chloride: inactivation, filamentation and lysogenic induction of *Escherichia Coli, Food Chem. Toxicol.*, 32, 477, 1994.
11. Chmielnicka, J., Szymanska, J. A., and Sniec, J., Distribution of tin in the rat and disturbances in the metabolism of zinc and cooper due to repeated exposure to *$SnCl_2$, Arch. Toxicol.*, 47, 263, 1981.

TITANIC ACID, TETRABUTYL ESTER

Molecular Formula. $C_{16}H_{36}O_4.Ti$
M = 340.42
CAS No 5593-70-4
RTECS No XR1585000
Abbreviation. TBT.

Synonyms and **Trade Name.** Butyl alcohol, titanium (4^+) salt; Butyl orthotitanate; Butyl titanate; Tetrabutyl titanate; Titanium tetrabutoxide; Titanium tetrabutylate; Titanium tetrakis(butoxide); Tyzor TBT.

Applications. Cross-linking agent. Condensation catalyst. Used to improve adhesion of paints, rubber, and plastics to metal surfaces.

Properties. Colorless to light-yellow liquid with a weak, alcohol-like odor. Soluble in most organic solvents.

Acute Toxicity. LD_{50} is 3.1 g/kg BW in rats.[029]

Chemobiokinetics. Soluble titanium compounds are readily absorbed from the GI tract.[1]

The highest concentration of titanium compounds are found in the blood, brain and parenchymatous organs.[1]

Most ingested titanium is eliminated unabsorbed. In man, titanium is probably excreted with the urine at a rate of about 10 µg/l. However, high urinary losses of 0.41 and 0.46 mg/day (30-day mean), respectively, were reported.

The mechanism of excretion and the possible amount of *titanium* excreted by the intestinal route are unknown.[1]

Regulations. *U.S. FDA* (1998) approved the use of TBT (1) in the manufacture of resinous and polymeric coatings for the food-contact surface of articles intended for use in producing, manufacturing, packing, processing, preparing, treating, packaging, transporting, or holding food in accordance with the conditions prescribed in 21 CFR part 175.300; and (2) in the manufacture of resinous and polymeric coatings for the food-contact surface of articles intended for use in producing, manufacturing, packing, processing, preparing, treating, packaging, transporting, or holding food in accordance with the conditions prescribed in 21 CFR part 175.320.

Reference:

1. *Handbook of the Toxicology of Metals*, L. Friberg, G. F. Nordberg, E. Kessler, and V. B. Vouk, Eds., 2nd ed., vol 2, Elsevier Sci., Publ. B. V., Amsterdam, 1986, 594.

TITANIUM CHLORIDE

Molecular Formula. $Cl_4.Ti$
M = 189.70
CAS No 7550-45-0
RTECS No XR1925000

Synonym. Titanium tetrachloride.

Properties. Colorless, mobile liquid with a penetrating acid odor, which fumes in the air. Gives no odor to water, but a faintly acidic, slightly astringent taste (rating 2) is determined at 4.5 mg/l. It is hydrolyzes, whereupon Titanium hydroxide precipitates out, turning water milky. Color change threshold is 12.5 mg/l.[1]

Applications. A component of Ziegler-Natta polymerization catalysts.

Acute Toxicity. The LD_{50} values are 150 mg/kg BW for mice, and 472 mg/kg BW for rats. Rabbits and guinea pigs are more sensitive: LD_{50} is 100 mg/kg BW.[1]

Repeated Exposure. Mice tolerate administration of 15 mg/kg BW for a month. The STI is found to be lowered in animals given 50 mg/kg BW for 10 days. Rats received 5.0 to 20 g/kg BW over a period of 2 months. The treatment caused no changes in catalase activity, hematology analysis, or oxygen consumption. The higher dose caused an increase in blood cholinesterase activity.[1]

Long-term Toxicity. In a 6-month study, guinea pigs were given 20 and 5.0 mg/kg BW, rabbits received 5.0 and 1.0 mg/kg BW. The treatment with the higher dose caused retardation of BW gain and decreased blood cholinesterase activity in guinea pigs. Gross pathology examination revealed lymphohistiocyte infiltrations and an increased number of *Kupffer cells* in the liver parenchyma. The foci of necrosis were observed in the intestinal mucosa. In rabbits, a dose of 1.0 mg/kg BW did not cause evident changes either in the functional state or in the histological structure of the visceral organs.[1]

Mice were given 5.0 mg/kg BW of soluble titanium salts in drinking water throughout their life-span. The treatment did not affect the latter and, furthermore, BW of animals exposed to titanium was found to be greater than that of the controls (Schroeder et al., 1964).

Reproductive Toxicity. Titanium is shown to pass through the placenta into the fetal body.

Embryotoxicity. Rats and mice were given 5.0 mg soluble titanium salts/l in drinking water. The treatment caused a reduction in the number of newborns in the third generation. The male/female ratio in the progeny was reduced.[2]

There are no data on the teratogenic effect.

Mutagenicity. Titanium did not show mutagenic activity[3] or cytotoxicity (Takeda et al., 1989).

Chemobiokinetics. Titanium is found in the brain of healthy people and in human embryos. Following ingestion, only about 3.0% of administered dose were absorbed in the GI tract. Distribution occurs predominantly in the lungs, and then in the kidneys and liver. Bones are the principle site of storage. Following inhalation exposure, titanium is immediately detected in large amounts in the blood of rats.[4] It is excreted with the urine and feces.

Standards. ***Russia.*** PML (Ti^{4+}): 4.0 mg/l (organolept., taste).

References:

1. Selyankina, K. P. and Nekrasova, E. B., in *Industrial Pollution of Water Reservoirs*, S. N. Cherkinsky, Ed., Meditsina, Moscow, Issue No 8, 1967, 233 (in Russian).
2. Berlin, M. and Nordman, O., Titanium, in *Handbook on Toxicology of Metals*, Amsterdam, 1980, 627.
3. Hise, A. W. et al., *Trace Metals in Health and Disease*, Raven Press, New York, 1979, 55.
4. Gurfein, L. N., Pavlova, Z. K., et al., in *Toxicology of New Industrial Chemical Substances*, Medgiz, Moscow, Issue No 4, 1962, 128 (in Russian).

TOLUENE-2,4-DIAMINE

Molecular Formula. $C_7H_{10}N_2$
M = 122.17
CAS No 95-80-7
RTECS No XS9625000
Abbreviation. 2,4-TDA.

TOLUENE-2,6-DIAMINE

CAS No 823-40-5
RTECS No XS9750000
Abbreviation. 2,6-TDA.

Synonyms. 2,4- and 2,6-Diaminotoluenes; Methylphenylenediamine.

Properties. Colorless crystals. Freely soluble in hot water, alcohol, and ether.

Applications. Used as curing agents in the epoxy resin production. Rubber antioxidants, intermediates in the manufacture of toluene diisocyanates.

Migration Data. 2,6-TDA was found in urine of patients implanted with polyurethane-covered breast implants at a concentration of about 1.0 μg/l.[1]

According to Sepai et al., following implantation of polyurethane-covered breast prostheses, levels of 2,4-TDA and 2,6-TDA rose to above 4.0 and 1.5 μg/l plasma, respectively. Elevated levels were found up to 2 years post-operation.[2]

Acute Toxicity. In rats, LD_{50} of 2,4- and 2,6-TDA (mixture) is 270 to 300 mg/kg BW.[3,4] Poisoning is accompanied by CNS inhibition and methemoglobinemia.

Repeated Exposure. Male F344 rats were dosed by gavage with 70 mg 2,4-TDA/kg BW for 5 days. Activities of microsomal P-450-dependent emzymes were depressed, while there was a pronounced increase in that of epoxide hydrolase.[5]

Mice received 25 to 100 mg 2,4-TDA/kg BW by gavage for 14 days. The treated animals showed 42% increase in liver weight and a slight decrease in spleen weight. Histological examination revealed the liver to be the major target organ. The dose of 100 mg 2,4-TDA/kg BW induced moderate centrilobular necrosis. No changes were observed in the spleen, lungs, thymus, kidney, or mesenteric lymph nodes.[6]

Long-term Toxicity. Renal toxicity was found to develop following oral administration of 2,4-TDA for 2 years. Oral ingestion of 50 and 100 mg/kg BW accelerated the development of chronic renal disease in F344 rats. The treatment also resulted in decreased survival.[7]

Immunotoxicity. 2,4-TDA was found to be hepatotoxic. In the above described study, it perturbs differentiation and maturation of leukocytes. Natural killer cell activity was depressed, phagocytosis by splenic macrophages was inhibited. There was a decrease of host resistance to the bacteria.[6]

Reproductive Toxicity.

Gonadotoxicity. 2,4-TDA appeared to be a potent reproductive toxicant in the male rats when given in the diet at the dose of 15 mg/kg BW for 10 weeks. Its toxic action included an effect on spermatogenesis (66% reduction), decreased weight of seminal vesicles and epididymis, as well as a diminished level of circulating testosterone, and elevation of serum-luteinizing hormone. The effect persisted after an additional 11 weeks on a normal diet and in addition, profound testicular atrophy was noted.[8]

Embryotoxicity and ***teratogenicity.*** In mouse and rat studies, TDA altered testicular DNA synthesis and produced *Sertoli* cell damage that could be a reason of subsequent reproductive effects.[9]

2,6-TDA was found to be embryotoxic in rats and rabbits, it caused malformations in rats. The NOAEL for these effects is identified to be 10 mg 2,6-TDA/kg BW in rats and 30 mg 2,6-TDA/kg BW in rabbits.[10]

Mutagenicity.

In vitro genotoxicity. 2,6-TDA gave positive response in *Salmonella* mutagenicity assay only in the presence of metabolic activation systems.[11] 2,4- and 2,6-TDA were weakly positive in the micronucleous test; however, with 2,4-TDA this weak effect was only detectable at very toxic doses.[12]

In vivo cytogenetics. A single *i/p* administration of 250 mg 2,4-TDA/kg BW induced approximately 6500 times more DNA adducts than 2,6-TDA injection.[13]

2,6-TDA caused a significant increase over controls in the amount of DNA damage and repair displayed by hepatocyte cultures obtained from rats given two 0.5 LD_{50} separated by a 24-hour interval.[14]

According to other data, 2,4- and 2,6-TDA were negative in mammalian assays.[15]

Both isomers are shown to be positive in *Dr. melanogaster* assay.

Carcinogenicity. 2,4-TDA produced tumors in rodents: administration in the diet at a dose more than 79 mg/kg BW caused subcutaneous and mammary gland tumors in rats and hepatocellular and vascular tumors in mice.[16]

2,6-TDA was not carcinogenic in rats and mice of both sexes.[17,18] There is a positive correlation between increased cell proliferation and hepatocarcinogenesis induced by the isomers.[19]

Carcinogenicity classifications. An IARC Working Group concluded that there is sufficient evidence for the carcinogenicity of 2,4-TDA in *experimental animals* and there were no adequate data available to evaluate the carcinogenicity of 2,4-TDA in *humans.*

The Annual Report of Carcinogens issued by the U.S. Department of Health and Human Services (1998) defines 2,4-TDA to be a substance which may reasonably be anticipated to be carcinogen, i.e., a substance for which there is limited evidence of carcinogenicity in *humans* or sufficient evidence of carcinogenicity in *experimental animals.*

IARC: 2B;

U.S. EPA: B2;

NTP: N - N - N - N (2,4-TDA dihydrochloride, feed).

Chemobiokinetics. Following ingestion, 2,4- and 2,6-TDA are readily absorbed, metabolized, and eliminated. They are predominantly distributed in the liver, kidneys, and adrenal glands. Major methabolic pathways include acetylation of amino groups, oxidation of methyl groups, and ring hydroxylation. Phenolic metabolites of 2,4-TDA and small amounts of unchanged TDA are excreted in the urine.[15,20]

Rats were given a single oral dose of 60 mg ^{14}C-labeled 2,4-TDI/kg BW. More than 93% of the radioactivity was recovered: in the urine (about 8.0%), feces (81%), cage wash 16 to 39% of the quantitated urinary metabolites existed as *acid-labile conjugates*.[21]

Regulations. ***U.S. FDA*** (1998) approved the use of TDA as an antioxidant in rubber articles intended for repeated use in producing, manufacturing, packing, processing, treating, packaging, transporting, or holding food (total not to exceed 5.0% by weight of the rubber product) in accordance with the conditions prescribed in 21 CFR part 177.2600.

References:

1. Chan, S. C., Birdsell, D. C., and Gradeen, C. Y., Urinary excretion of free toluenediamines in a patient with polyurethane-covered breast implants, *Clin. Chem.*, 37, 2143, 1991.
2. Sepai, O., Henschler, D., Czech, D., Eckert, P., and Sabbioni, G., Exposure to toluenediamines from polyurethane-covered breast prostheses, *Toxicol. Lett.*, 77, 371, 1995.
3. Izmerov, N. F., Sanotsky, I. V., and Sidorov, K. K., *Toxicometric Parameters of Industrial Toxic Chemicals under Single Exposure*, USSR/SCST Commission for UNEP, Center Int. Projects, Moscow, 1982 (in Russian).
4. Weisbrod, D. and Stephan, U., Studies on the toxic, methemoglobin-producing and erythrocyte-damaging effects of diaminotoluene after a single administration, *Z. Gesamte Hyg.*, 29, 395, 1983.
5. Dent, J. G. and Graichen, M. E., Effect of hepatocarcinogens on epoxide hydrolase and other xenobiotic metabolizing enzymes, *Carcinogenesis*, 3, 733, 1982.
6. Burns, L. A., Bradley, S. G., White, K. L., McCay, J. A., Fuchs, B. A., Stern, M., Brown, R. D., Musgrove, D. L., Holsapple, M. P., Luster, M. I., et al., Immunotoxicity of 2,4-diaminotoluene in female B6C3F$_1$ mice, *Drug Chem. Toxicol.*, 17, 401, 1994.
7. Cardy, R. H., Carcinogenicity and chronic toxicity of 2,4-toluenediamine in F344 rats, *J. Natl. Cancer Inst.*, 62, 1107, 1979.
8. Thysen, B., Varma, S., and Bloch, E., Reproductive Toxicity of 2,4-toluenediamine in the rat. 1. Effect on male fertility, *J. Toxicol. Environ. Health*, 16, 753, 1985; 2; Spermatogenic and hormonal effects, *Ibid,* 16, 763, 1985.
9. Varma, S. K., Bloch, E., Gondos, B. et al., Reproductive toxicity of 2,4-toluene diisocyanate in rat., *J. Toxicol. Environ. Health,* 25, 435, 1988.
10. Knickerbocker, M., Re, T. A., Parent, R. A., et al., Teratogenic evaluation of *o*-toluenediamine in Sprague-Dawley rats and Dutch Belted rabbits, *Toxicologist*, 19, 89, 1980.
11. Dybing, E. and Thorgeirsson, S. S., Metabolic activation of 2,4-diaminoanisole, a hair-dye component, *Biochem. Pharmacol.*, 26, 729, 1977.
12. George, E. and Westmoreland, C., Evaluation of the *in vivo* genotoxicity of the structural analogues 2,6-diaminitoluene and 2,6-diaminitoluene using the rat micronucleus test and rat liver USD assay, *Carcinogenesis*, 12, 2233, 1991.
13. Taningher, M., Peluso, M., Parodi, S., Ledda-Columbano, G. M., and Columbano, A., Genotoxic and non-genotoxic activities of 2,4- and 2,6-diaminotoluene, as evaluated in Fischer 344 rat liver, *Toxicology*, 99, 1, 1995.
14. Allavena, A., Martelli, A., Robbiano, L., and Brambillo, G., Evaluation in a battery of *in vivo* assays of four *in vitro* genotoxins proved to be non-carcinogens in rodents, *Teratogen. Carcinogen. Mutagen.*, 12, 31, 1992.
15. *Diaminotoluenes*, Environmental Health Criteria No 74, WHO, Geneva, 1987, 67.
16. *Bioassay of 2,4-Diaminotoluene for Possible Carcinogenicity*, Natl. Cancer Institute Technical Report Series No 162, Bethesda, Maryland, 1979.

17. Cunningham, M. L., Burka, L. T., and Matthews, H. B., Metabolism, disposition, and mutagenicity of 2,6-diaminitoluene, a mutagenic noncarcinogen, *Drug Metab. Dispos.*, 17, 612, 1989.
18. *Bioassay of 2,6-Toluenediamine Dihydrochloride for Possible Carcinogenicity*, Natl. Cancer Institute Technical Report Series No 200, NTP No 80-20, NIH Publ. No 80-1756, Bethesda, Maryland, 1980.
19. Cunningham, M. L., Foley, J., Maronpot, R. R., and Matthews, H. B., Correlation of hepatocellular proliferation with hepatocarcinogenicity induced by the mutagenic noncarcinogen: carcinogen pair - 2,6- and 2,4-diaminotoluene, *Toxicol. Appl. Pharmacol.*, 107, 562, 1991.
20. Waring, R. H. and Pheasant, A. E., Some phenolic metabolites of 2,4-diaminotoluene in the rabbit, rat and guinea pig, *Xenobiotica*, 6, 257, 1976.
21. Timmchalk, C., Smith, F. A., and Bartels, M. J., Route-dependent comparative metabolism of ^{14}C-toluene 2,4-diisocyanate and ^{14}C-toluene 2,4-diamine in Fischer 344 rats, *Toxicol. Appl. Pharmacol.*, 124, 181, 1994.

p-TOLUENESULFONIC ACID

Molecular Formula. $C_7H_8O_3S$
M = 172.20
CAS No 536-57-2
Abbreviation. TSA.

Synonyms. 4-Methylbenzenesulfonic acid; Tosic acid.

Properties. Monoclinic leaflets or prisms. Readily soluble in water and alcohol.

Applications. Used as a catalyst in coatings, paint, polymer, and textile industries. A stabilizer for plastics and pharmaceuticals.

Acute Toxicity. LD_{50} is 400 mg/kg BW in rats.

Chemobiokinetics. Probably, excreted unchanged.

Regulations. ***U.S. FDA*** (1998) approved the use of TSA in resinous and polymeric coatings used as the food-contact surfaces of articles intended for use in producing, manufacturing, packing, processing, preparing, treating, packaging, transporting, or holding food in accordance with the conditions prescribed in 21 CFR part 175.300.

Reference:

Occupational Health Guidelines for Chemical Hazards, F. W. Maskison, R. S. Stricoff, and L. J. Partridge, Eds., DHHS (NIOSH) Publ. No 81-123, Washington, D. C., 1981.

1,3,5-TRIALLYL-*S*-TRIAZINE-2,4,6(*1H,3H,5H*)-TRIONE

Molecular Formula. $C_{12}H_{15}N_3O_3$
M = 249.30
CAS No 1025-15-6
RTECS No XZ1915000
Abbreviation. TTT.

Synonyms. Cyanuric acid, *N,N',N''*-triallyl ester; Triallyl isocyanurate.

Properties. Odorless, colorless, oily liquid. Poorly soluble in water.

Applications. Used as a cross-linking agent in polymerization. A stabilizer for polyvinyl chloride.

Acute Toxicity. LD_{50} is 704 mg/kg BW in rats and 437 mg/kg BW in mice. Animals died in the first hours after administration. In 30 minutes after poisoning with high doses, survivors exhibit twitching of various muscle groups and the extremities. Some animals experience hind limb paralysis and impaired motor coordination. Mice tolerate a 300 mg/kg BW dos; none of the rats died after oral administration of a 400 mg/kg BW dose.[1]

Repeated Exposure revealed moderate cumulative properties. 1/5 LD_{50} caused 5 out of 6 rats to die within two weeks to two months. Daily administration of 1/10 or 1/20 LD_{50} caused no mortality but consistent retardation of BW gain in the treated animals.[1]

Mutagenicity.

In vitro genotoxicity. TTT exhibited mutagenic activity in Chinese hamster lung cells.[2]

Regulations.

U.S. FDA (1998) approved the use of TTT as an accelerator of rubber articles intended for repeated use in producing, manufacturing, packing, processing, treating, packaging, transporting, or holding food (total not to exceed 1.5% by weight of the rubber product) in accordance with the conditions prescribed in 21 CFR part 177.2600.

EU (1990). TTT is available in the *List of monomers and other starting substances which may continue to be used for the manufacture of plastic materials and articles intended to come into contact with foodstuffs pending a decision on inclusion in Section A.*

References:

1. Bidnenko, L. I., Problem of toxicity of cyanuric acid and triallylisocyanurate, in *Hygienic Aspects of the Use of Polymeric Materials and Articles Made of Them*, L. I. Medved', Ed., All-Union Research Institute of Hygiene and Toxicology of Pesticides, Polymers and Plastic Materials, Kiyv, 1969, 299 (in Russian).
2. Sofuni, T., Matsuoka, A., Sawada, M., Ishidate, M., Zeiger, E., and Shelby, M. D., A comparison of carcinogenesis induction by 25 compounds tested by two Chinese hamster cell (CHL and CHO) systems in culture, *Mutat. Res.*, 241, 175, 1990.

TRIETHYLAMINE

Molecular Formula. $C_6H_{15}N$
M = 101.2
CAS No 121-44-8
RTECS No YE0175000
Abbreviation. TEA.

Synonym. *N,N'*-Diethylethanamine.

Properties. Colorless liquid with a penetrating fishy, amine odor. Solubility in water is 19.7 g/l at 65°C or 71 g/l at 25°C.[02] Odor perception threshold is 4.0 mg/l,[1] 0.42 mg/l,[02] or <0.09 mg/l.[010] Taste perception threshold is 3.0 mg/l.

Applications. Used as a catalyst in polyurethane systems.

Exposure. The blood levels of TEA can be increased after ingestion of certain foods, such as fish, and during disease states, such as chronic renal failure. Origin of urinary TEA is via the action of intestinal microflora on precursors within the food (such as choline).

Acute Toxicity. LD_{50} is 460 mg/kg BW in rats and 546 mg/kg BW in mice.[1] Sensitivity of rabbits is likely to be the same. Signs of intoxication include effects on the CNS: excitation, then inhibition, motor coordination disorder, and clonic convulsions.

Repeated Exposure. A number of animals, exposed to 1/10 LD_{50} for 2.5 months, died at the end of the experiment. The treatment led to retardation of BW gain and an increase in liver ascorbic acid contents.

Long-term Toxicity. Rabbits were given 1.0 and 6.0 mg TEA/kg BW. In 3 to 4 months after the treatment onset the higher dose caused liver function impairment.[1]

Reproductive Toxicity. Effect on fertility is reported.

Embryotoxicity. Pregnant CD mice were exposed to 2.5 and 5.0 mmol TEA/kg BW by *i/p* injections on gestation days 1 to 17. The treatment produced a significant decrease in fetal BW; 5 out of 11 mice died.[2]

TEA embryotoxic action was noted in 3-day-old chick embryos.[3]

Teratogenicity. TEA can inhibit fetal development *in vivo* and *in vitro* in mice. It caused neural-tube defects, in culture, it inhibited mouse embryos growth by reducing macromolecular synthesis. These effects may not involve glutathione depletion or generation of free radicals.[4]

Mutagenicity.

In vitro genotoxicity. TEA was negative in four *Salmonella typhimurium* strains.[015]

Carcinogenicity. TEA could be metabolized into *diethylamine*. The exact mechanism for this di-ethylation is not known. Similar to other secondary amines, diethylamine might be nitrosated endogenously to form the carcinogenic compound *N-nitrosodiethylamine*.[5] The latter should be regarded as if it were carcinogenic to man.

Chemobiokinetics was investigated in 4 volunteers. After oral administration, TEA is efficiently absorbed from the GI tract, rapidly distributed, and partially metabolized into *triethylamine-N-oxide*.[6]

Formation of *diethylamine* from TEA also occurs. More than 90% of the dose was recovered in the urine as *triethylamine* and *triethylamine-N-oxide*, and to a lesser extent in feces.[5,6] Exhalation of TEA is minimal.

Regulations. ***U.S. FDA*** (1998) approved the use of TEA in the manufacture of cellophane for packaging food in accordance with the conditions prescribed in 21 CFR part 177.1200.

Standards. ***Russia*** (1995). MAC and PML: 2.0 mg/l.

References:

1. Kagan, G. Z., Comparative evaluation of diethylamine and triethylamine in connection with sanitary protection of reservoirs, *Gig. Sanit.*, 9, 28, 1965 (in Russian).
2. Guest, I. and Varma, D. R., Developmental toxicity of methylamines in mice, *J. Toxicol. Environ. Health*, 32, 319, 1991.
3. Korhonen, A., Hemminki, K., and Vainio, H., Toxicity of rubber chemicals towards 3-day-old chicken embryos, *J. Scand.. Work Environ. Health*, 9, 115, 1983.
4. Guest, I. and Varma, D. R., Teratogenic and macromolecular synthesis inhibitory effects of triethylamine on mouse embryos in culture, *J. Toxicol. Environ. Health*, 36, 27, 1992.
5. Bellander, T., Osterdahl, B. G., Hagmar, L., et al., Excretion of *N*-mononitrosopiperasine in the urine in workers manufacturing piperasine, *Int. Arch. Occup. Environ. Health*, 60, 25, 1988.
6. Akesson, B., Vinge, E., and Skerfving, S., Pharmacokinetics of triethylamine and triethylamine-*N*-oxide in man, *Toxicol. Appl. Pharmacol.*, 100, 529, 1989.

TRIETHYLENETETRAMINE

Molecular Formula. $C_6H_{18}N_4$
M = 146.24
CAS No 112-24-3
RTECS No YE6650000
Abbreviation. TETA.

Synonyms and **Trade Names.** *N,N'*-Bis(2-aminoethyl)-1,2-ethanediamine; 1,8-Diamino-3,6-diazoctane; 1,4,7,10-Tetraazadecane; Trientine; Trien.

Properties. Light-yellow, oily liquid. Readily soluble in water, ethanol, and vegetable oil.

Applications. TETA is used as a curing agent for epoxy resins and as a thermosetting resin.

Exposure. TETA is an orphan therapeutic drug for the treatment of Wilson's disease.

Acute Toxicity. LD_{50} is found to be 4.3 g/kg BW in rats, 1.6 g/kg BW in mice,[1,05] and 5.5 g/kg BW in rabbits. According to Dubinina and Fukalova, LD_{50} for rats and mice is 2.75 and 2.0 g/kg BW, respectively.[2] Acute effect threshold appeared to be 30 mg/kg BW. Poisoning is accompanied by symptoms of CNS and GI tract damage.

Repeated Exposure revealed little evidence of cumulative properties. Rats were given 1/10 and 1/20 LD_{50}. The treatment caused decreased BW gain, reduction in the blood *Hb* level and erythrocyte count, affected liver function, and lowered NS excitability. These changes were found to be reversible.[1]

Short-term Toxicity. B6C3F$_1$ mice and F344 rats received TETA dihydrochloride in the drinking water at concentrations of 120, 600, or 3000 ppm for up to 92 days. Toxicity occurred only in mice in the highest dose group fed a cereal-based diet. Signs of intoxication included increased frequencies of inflammation of the lung and liver periportal fatty infiltration in both sexes and hematopoietic cell proliferation in the spleen of males.[3]

Long-term Toxicity. In a 10-month study, mild changes in the activity of a number of enzymes were noted in rats at a dose level of 0.8 mg/kg BW. Animals recovered subsequently.[1]

Allergenic Effect. TETA is reported to be a weak allergen.

Reproductive Toxicity.

Embryotoxic effect of TETA was observed in 3-day-old chick embryos.[4] Sprague-Dawley rats were fed during pregnancy a purified diet containing 0.17, 0.83, or 1.66% TETA. The treatment caused a dose-dependent increase in the frequency of resorptions and the frequency of abnormal fetuses. A significant decrease in maternal and fetal tissue copper levels was noted.[5]

Teratogenicity.

Observations in man. No malformations were observed among six infants of mothers treated with TETA for Wilson disease.[6]

Animal studies. Teratogenic action was observed on skin application in guinea pigs (Wayton, 1978). Rats received 0.17, 0.83, and 1.66% TETA in their diets throughout pregnancy. The treatment caused an increase in resorption rate at all dosages; fetal abnormalities (hemorrhage and edema) occurred at the two highest levels. TETA teratogenicity may be in part due to induction of copper deficiency, and perhaps through induction of *zinc* toxicity.[5,7]

Pregnant rats received up to 12,000 ppm TETA in their drinking water. Teratogenic effect (exencephaly) was observed in offspring of treated mice.[8,9]

Mice were given 3,000, 6,000, or 12,000 mg TETA *dihydrochloride*/l in drinking water *ad libitum*. Treatment throughout the pregnancy caused a dose-dependent increase in frequency of gross brain abnormalities in live fetus at births including hemorrhages, delayed ossification in cranium, hydrocephaly, exencephaly, and microcephaly.[10]

Mutagenicity.

In vitro genotoxicity. TETA showed evident mutagenicity in *Salmonella typhimurium* bioassay.[11] It caused an increase in SCE in Chinese hamster ovary cells, and in unscheduled DNA synthesis in hepatocytes.

In vivo cytogenicity. TETA did not induce micronuclei in bone marrow cells in Swiss-Webster mice (*in vivo*).[12]

Chemobiokinetics. TETA underwent rapid metabolism after absorption from the GI tract. The main absorption route may be permeation across the plasma membrane of intestinal epithelial cells.

Observations in man. Two healthy adults were given TETA orally. The amount of TETA in the urine was only 1.6 and 1.7% of the dose administered. These results suggested that most of the TETA is metabolized and then excreted in the urine.[13]

Animal studies. Urinary excretion of TETA and its metabolites appeared to be as much as 35% of the administered dose.[14]

Regulations. ***U.S. FDA*** (1998) approved the use of TETA (1) in the manufacture of resinous and polymeric coatings of food-contact surface of articles intended for use in producing, manufacturing, packing, processing, preparing, treating, packaging, transporting, or holding food in accordance with the provisions prescribed in 21 CFR part 175.300; (2) as a modifier for aminoresins for the uncoated or coated food-contact surface of paper and paperboard intended for use in producing, manufacturing, packing, transporting, or holding aqueous, fatty and dry food of the type identified in 21 CFR part 176.170 (c); (3) as a component of the uncoated or coated food-contact surface of paper and paperboard intended for use in producing, manufacturing, packaging, processing, preparing, treating, packing, transporting, or holding dry, aqueous and fatty foods in accordance with the provisions prescribed in 21 CFR part 176.170; and (4) as an accelerator for rubber articles intended for repeated use in contact with food up to 1.5% by weight of the rubber product in accordance with the provisions prescribed in 21 CFR part 177.2600.

Standards. ***Russia***. PML in food: 0.02 mg/l.

References:

1. Stavreva, M. S., *Toxicological and Hygienic Characteristics of Triethylenetetramine and Development of Recommendations for Using It as a Curing Agent for Anti-corrosion Epoxy Coatings for Food Industry*, Author's abstract of thesis, Kiyv, 1977. 23 (in Russian).
2. Dubinina, O. N. and Fukalova, I. A., in *Proc. F. F. Erisman Research Sanitary Institute*, Moscow, 1979, 141 (in Russian).
3. Greenman, D. L., Morrissey, R. L., Blakemore, W., Crowell, J., Siitonen, P., Felton, P., Allen, R., and Cronin, G., Subchronic toxicity of triethylenetetramine dihydrochloride in $B6C3F_1$ mice and F344 rats, *Fundam. Appl. Toxicol.*, 29, 185, 1966.
4. Korhonen, A., Hemminki, K., and Vainio, H., Toxicity of rubber chemicals towards 3-day chicken embryos, *Scand. Work Environ. Health*, 9, 115, 1983.

5. Keen, C. L., Cohen, N. L., Lonnerdal, B., et al., Teratogenesis and low copper status resulting from triethylenetetramine in rats, *Proc. Soc. Exp. Biol. Med.*, 173, 598, 1983.
6. Walsche, J. M., Treatment of Wilson's disease with trientine (triethylenetetramine) dihydrochloride, *Lancet*, 1, 643, 1982.
7. Cohen, N. L., Keen, C. L., Lonnerdal, B., and Hurley, L. S., The effect of copper supplementation on the teratogenic effects of triethylenetetramine in rats, *Drug Nutr. Interact.*, 2, 203, 1983.
8. Tanaka, H., Yamanouchi, M., and Arima, M., The effect of maternal triethylenetetramine dihydrochloride on the mouse fetus, *Teratology*, 44, 98, 1991.
9. Tanaka, H., Yamanouchi, M., Imai, S., and Hayashi, Y., Low copper and brain abnormalities in fetus from triethylenetetramine dihydrochloride-treated pregnant mouse, *J. Nutr. Sci. Vitaminol.* (Tokyo), 38, 545, 1992.
10. Tanaka, H., Inomata, K., and Arima, M., Teratogenic effects of triethylenetetramine dihydrochloride on the mouse brain, *J. Nutr. Sci. Vitaminol.* (Tokyo), 39, 177, 1993.
11. Hulla, J. E., Rogers, S. J., and Warren, J. R., Mutagenicity of a series of polyamines, *Environ. Mutagen.*, 3, 332, 1981.
12. Leung Hon-Wing, Evaluation of the genotoxic potential of alkyleneamines, *Mutat. Res.*, 320, 31, 1994.
13. Kodama, H., Meguro, Y., Tsunakawa, A., Nakazato, Y., Abe, T., and Murakita, H., Fate of orally administered triethylenetetramine dihydrochloride: a therapeutic drug for Wilson's disease, Abstract, *Tohoku J. Exp. Med.*, 169, 59, 1993.
14. Kobayashi, M., Sugawara, M., Saitoh, H., Iseki, K., and Miyazaki, K., Intestinal absorption of urinary excretion of triethylenetetramine for Wilson's disease in rats, Abstract, *Yakugaku Zasshi*, 119, 759, 1990 (in Japanese).

TRIETHYLVANADATE, product of the hydrolysis

Molecular Formula. $C_6H_{15}O_4V$

CAS No 1686-22-2

Synonym. Triethoxyoxovanadium.

Applications. Used as a catalyst in polyethylene production.

Acute Toxicity. LD_{50} is found to be 58 mg/kg BW in rats, and 63 mg/kg BW in mice (calculated as V_2O_5).

Repeated Exposure. Mice were given 12.5 mg/kg BW; rats received the 6.0 mg/kg BW dose. The exposure produced a significant increase in liver weights in mice and female rats. Gross pathology examination failed to reveal lesions in the visceral organs.

Allergenic Effect. Irritant and sensitizing effect was not found in experiments on guinea pigs.

Short-term Toxicity. Doses of 0.6 and 0.08 mg/kg BW do not affect general condition and morphological structure of the organs. A concentration of 0.02 mg/l, in terms of V^{5+}, is considered harmless in man.

Reference:

Mikhailets, I. B., Sukhareva, L. V., and Yevsyukov, V. I., Hygienic evaluation of high-density polyethylene manufactured on the vanadian catalystic system, in *Environment Protection in Plastic Industry and Safety of the Use of Plastics*, Plastpolymer, Leningrad, 1978, 99 (in Russian).

TRIETHYLVANADATE and DIETHYLAMMONIUM CHLORIDE, product of the reaction

Applications. Used as a catalyst in the polyethylene production.

Acute Toxicity. LD_{50} is not attained. It appears to be higher than the dose administered (110 mg/kg BW).

Repeated Exposure. Mice were dosed with 6.0 to 23 mg/kg BW, rats received a 12 mg/kg BW dose. The treatment caused decrease of BW gain and prolongation of hexenal sleep in mice and female rats. Gross pathology examination failed to reveal lesions in the visceral organs.

Long-term Toxicity. Mice were dosed by gavage with 1.2 and 0.12 mg/kg BW. This exposure produced no changes in BW gain, blood formula, and relative weights of the visceral organs. A concentration of 0.02 mg V^{5+}/l is considered to be harmless in man.

Reference:

Mikhailets, I. B., Sukhareva, L. V., and Yevsyukov, V. I., Hygienic evaluation of high-density polyethylene manufactured on the vanadian catalystic system, in *Environment Protection in Plastic Industry and Safety of the Use of Plastics*, Plastpolymer, Leningrad, 1978, 99 (in Russian).

2,4,6-TRIS(DIMETHYLAMINOMETHYL)PHENOL TRIOLEATE

Molecular Formula. $C_{15}H_{30}N_3O.3C_{54}H_{102}O_6$
M = 2811.16
CAS No 67274-16-2
RTECS No SN3550000

Synonym. Trioleate-2,4,6-tris(dimethylaminomethyl)phenol.

Applications. Used as a curing agent in the production of epoxy resins.

Acute Toxicity. In rats, LD_{50} is 11.8 g/kg BW. Labored breathing appeared in 40 minutes after poisoning. In 24 hours there was hindlimb paralysis with subsequent death.

Repeated Exposure. Rats received 1/10 LD_{50} for a month. The treatment caused changes in the hematological indices, a rise in the level of residual nitrogen in the blood, an increase in the spleen relative weight, and an impairment of protein-forming function of the liver (dysproteinemia and positive thymol test).

Reference:

Volodchenko, V. A. and Sadokha, E. R., Toxicologic characteristics of 2,4,6-tris (dimethylamino-methyl)phenol trioleate, *Pharmacologia i Toxicologia*, 37, 363, 1974 (in Russian).

VANADIUM compounds

	CAS No	RTECS No
Vanadium pentoxide, Vanadic anhydride	1314-62-1	YW2450000
Molecular Formula. $O_5.V_2$		
M = 181.88		
Vanadium trioxide	1314-36-7	YW3050000
Molecular Formula. $O_3.V_2$		
M = 149.88		

Properties. The solubility of vanadium pentoxide is 250 mg/l in water and 470 mg/l in blood. At a concentration of 30 mg/l, ammonium metavanadate gives water a just discemible tint; vanadium pentoxide turns it bright orange. The color change threshold is 0.1 mg/l. The taste of water perception threshold is greater than 0.8 mg/l.[1]

Applications. Vanadium compounds are used as catalysts in the production of polymeric materials, and as synthetic rubber additives.

Exposure. Vanadium levels ranged from 7.0 to 90.0 µg/l were determined in red and from 6.6 to 43.9 µg/l in white wines from different regions of France and California. The contribution of wine consumption to daily vanadium dietary intake of the French population was estimated to be 11 µg/day per individual.[2]

Acute Toxicity. LD_{50} of some vanadium compounds in mice is reported to be 23.4 mg vanadium pentoxide/kg BW, 130 mg vanadium pentoxide/kg BW, or 24 mg vanadium chloride/kg BW. Lethal doses caused vomiting and disorders of defecation. Subsequent manifestations of the toxic action included (in a few hours) diarrhea, disturbance of respiratory rate and rhythm, and CNS depression. Hematology analyses revealed leukocytosis, erythrocytosis, an increase in the number of eosinophils, and in globulin and *Hb* concentrations, a reduction in concentrations of albumins and free amino acids. Glutaminealanine transaminase activity was increased two- or threefold. ECG changes were also reported.[3]

However, according to Selyankina, LD_{50} of vanadium pentoxide is likely to be 5.0 mg/kg BW in mice; administration of 10 mg/kg BW causes hindlimb paralysis, increased reflex excitability, and GI tract upsets.[1] Gross pathology examination revealed congestion and capillary stasis in the visceral organs.

Repeated Exposure. Vanadium compounds are known to be used therapeutically for the treatment of various diseases. Repeated administration produces changes indicative of effects on protein metabolism and various changes in enzyme activities in the blood. Vanadium compounds inhibit monoamine oxidase and some effects have been ascribed to the elevated tissue serotonin levels and decreased plasma cholesterol levels.[4]

Rats were dosed with ammonium metavanadate and vanadium pentoxide at doses of 0.5 and 1.0 mg/kg BW (calculated as V^{5+}) for 21 days. The exposure caused a change in oxygen consumption. In rats given 1.0 mg/kg BW, blood concentration of inorganic phosphorus was increased.

Short-term Toxicity. Wistar rats received 25 or 50 ppm vanadium pentoxide in the diet for 35 days after which vanadium pentoxide concentrations in the feed were increased up to 100 and 150 ppm and continued for 68 days. Anemia and a decrease in the amount of cystine in the hair of rats ingesting vanadium were noted at the dose levels of 50 to 150 ppm (2.5 to 7.5 mg/kg BW).[5]

Wistar rats were exposed orally to ammonium metavanadate at a dose of 0.19 mmol vanadium/kg BW, and vanadyl sulfate at the dose of 0.15 mmol vanadium/kg BW in drinking water for 12 weeks. Hematology analyses including hematocrit, *Hb*, erythrocyte, leukocyte, and platelet count, and osmotic fragility of the erythrocytes were carried out by standard methods. No significant hematological toxicity was observed in this study.[6]

In a 3-month study, rats received ammonium metavanadate at a concentration of 200 mg/l (in terms of vanadium) in their drinking water. Animals exhibited retardation of BW gain and anemia. Gross pathology examination revealed parenchymatous dystrophy of the liver and kidneys with the formation of cylinders in the tubules in some animals (Gorski; Zaporovska). $NaVO_3$ was given to Sprague-Dawley rats in their drinking water at concentrations of 5.0, 10, and 50 ppm for 3 months. Dose-dependent accumulation in the kidneys and spleen was noted. Appearance, behavior, food and water consumption and growth of the treated rats were affected. Histology examination revealed only mild though dose-dependent lesions in the kidneys and spleen. The plasma concentrations of *urea* and *uric acid* were increased in the highest exposure group.[7]

Long-term Toxicity. Mortality was observed in rabbits given a dose of 5.0 mg vanadium chloride/kg BW (calculated as V^{5+}) in 3 months after the experiment onset. Gross pathology examination revealed that histological changes in the visceral organs were more pronounced than when vanadium oxide was administered. Nucleic acid contents in the visceral organs were consistently reduced. Other manifestations of the chronic effect of vanadium compounds included increased blood amino acid concentration, elevated blood and urinary inorganic phosphorus levels, and reduced number of blood reticulocytes.[7,8]

Rats were fed 10 or 100 ppm vanadium in their diet (about 17.9 or 179 ppm vanadium pentoxide) for 2.5 years. Growth and survival of animals were not affected. Only significant change reported by Stokinger et al. (1953) was a decrease in the amount of cystine in the hair of rats which ingested ammonium metavanadate. The lower dietary level used in this study was considered to be the NOAEl.[03]

Wistar rats received aqueous solution of ammonium metavanadate. Calculated doses were 1.5 and 5 to 6 mg vanadium/kg BW. The treatment caused a decrease in erythrocyte count, *Hb* level, and hematocrit index. Reticulocytosis was observed at a higher dose level. *L*-ascorbic acid levels was lowered in the plasma erythrocytes. A decrease in δ-aminolevulinic acid dehydratase activity was noted.[9]

Rabbits received 1.0 mg vanadium pentoxide/kg BW (calculated as V^{5+}) for 6 months. The treatment affected the phosphorylation process in hemoglycolysis with the resulting increase of the concentration of inorganic phosphorus in the blood and urine. Histology examination revealed inflammation in the GI tract and congestion in the liver and brain.[1]

Allergenic Effect was observed on skin application in guinea pigs.[6] The threshold concentration for allergenic effect on *i/g* administration was reported to be 0.03 mg/l.[10]

Immunotoxicity. Long-term exposure of mice and rats to low doses of vanadium in drinking water caused depression of phagocytosis, splenotoxicity, enlargement of the spleen, leukocytosis, *T*- and *B*-cell activation.[11]

Reproductive Toxicity.

Embryotoxicity. Vanadium compounds can accumulate in the testes, placenta and fetal body. They are transported across the placenta and excreted in the maternal milk and enter the body of newborn rats and mice.[12]

Pregnant Sprague-Dawley rats were given 5.0, 10, and 20 mg $NaVO_3$/kg BW. There was neither embryolethal nor teratogenic effect in rats exposed orally to a dose of 20 mg/kg BW or lower. Nevertheless, this dose was embryotoxic.[13] However, *i/g* injection of $NaVO_3$ to Swiss mice on days 6 to 15 of gestation revealed the LOAEL for maternal toxicity to be 2.0 mg/kg BW. The NOAEL of 2.0 and 4.0 mg/kg BW was identified for significant embryotoxicity and teratogenicity, respectively.[14]

Similar results were reported by Zhang et al.: *i/g* injections of 5.0 mg vanadium pentoxide/kg BW to NIH mice at different stages of gestation caused no adverse effects on pre-implantation and implantation. There was no teratogenicity and premature birth; however, an embryotoxic effect was noted.

Teratogenicity. Vanadium sulfate pentahydrate given by gavage to pregnant Swiss mice at the dose levels of 37.5, 75, and 150 mg/kg BW on days 6 to 15 of pregnancy caused maternal and embryonic toxicity and teratogenicity. The NOEL for these effects in this study was identified to be 37.5 mg/kg BW.[15]

Gonadotoxicity. Vanadium compounds are shown to alter spermatogenesis and male fertility.[8,16]

Mutagenicity.

In vitro genotoxicity. A number of vanadium compounds produced negative results in reverse mutation assay with *Salmonella typhimurium* (NTP-91) and *E. coli*. Positive results were noted in *Bac. subtilis*.[17]

Vanadium compounds caused SCE and CA in Chinese hamster ovary cells.[18] In contrast to this, according to Galli et al., although V^{4+} and V^{5+} compounds are found to be very toxic in Chinese hamster *V79* cells, and no mutagenic effect was observed in the presence or absence of *S9* fraction, they produced dose-dependent increase in frequency of micronuclei.[19,20]

In vivo cytogenetics. Vanadyl sulfate (SVO_5), sodium ortho-vanadate (Na_3VO_4), and ammonium metavanadate (NH_4VO_3) were found to be positive in micronucleous test, they induced genotoxic effect in bone marrow of mice following intragastric treatment.[21] Sodium *o*-vanadate (Na_3VO_4) and ammonium metavanadate (NH_4VO_3) caused structural CA.[6]

Carcinogenicity. No data indicate that vanadium compounds are carcinogenic in animals or man.[21]

Carcinogenicity classification.

U.S. EPA: D.

Chemobiokinetics. Following ingestion only a small amount of vanadium compounds (0.1 to 2.0%) is found to be absorbed in the GI tract of experimental animals. Distribution occurs predominantly in the liver, kidneys, heart, bone tissue, and muscles. In blood plasma they bind with the globulin fraction.[3]

Accumulation of vanadium occurs in the liver nuclei of rats given low doses of vanadium. Vanadium was incorporated exclusively in the vanadyl form.[22] According to Mravkova et al., vanadium accumulates in heart tissues and influences the mineralization of epiphyseal cartilage. This effect is obviously evident in young animals. There were significant differences in vanadium distribution between young and adult animals.[9]

Vanadium compounds affect lipid biosynthesis and, at high concentrations, slow down serotonin oxidation (Vonk). A possible mechanism by which vanadium compounds exert their effect is by their involvement in the *K-Na* pump mechanism. According to Crans *et al,* vanadium compounds are found to be potent inhibitors of Na^+, K^+-ATPase, ribonucleases, and phosphatases.[23]

Incubation of DNA with vanadyl ion and H_2O_2 led to intense DNA cleavage. Hydroxyl radicals are generated during the reactions of vanadyl ion and H_2O_2. The mechanism for vanadium-dependent toxicity and antineoplastic action is due to DNA cleavage by hydroxyl radicals generating in living systems.[24] Radiolabelled vanadium penetrates into milk and is absorbed by suckling pups.[17]

An excretion occurs via the kidneys and, to a lesser extent, in the feces.

Standards.

Russia (1995). MAC: 0.1 mg/l.

Czechoslovakia (1982). MAC: 0.01 mg/l.

References:

1. Selyankina, K. P., Data on hygienic standardization of vanadium compounds contents in water, *Gig. Sanit.*, 10, 6, 1961 (in Russian).
2. Teissedre, P. L., Krosniak, M., Portet, K., Gasc, F., Waterhouse, A. L., Serrano, J. J., Cabanis, J. C., and Cros, G., Vanadium levels in French and California wines: influence on vanadium dietary intake, *Food Addit. Contam.*, 15, 585, 1998.

3. Ordzhonikidze, E. K., Roshchin, A. V., Kasimov, M. A. et al., Transplacenta penetration of vanadium and its age-dependent distribution in the viscera, *Gig. Truda Prof. Zabol.*, 6, 29, 1977 (in Russian).
4. *Vanadium and Some Vanadium Salts*, Health and Safety Guide, WHO, IPCS, Geneva, 1990, 36.
5. Mountain, J. T., Delker, L. L., and Stokinger, H. E., Studies in vanadium toxicology, *Arch. Ind. Hyg. Occup. Med.*, 8, 406, 1953.
6. Dai, S., Vera, E., and McNeill, J. H., Lack of hematological effect of oral vanadium treatment in rats, *Pharmacol. Toxicol.*, 76, 263, 1995.
7. Domingo, J. L., Llobet, J. M., Tomas, J. M., et al., Short-term toxicity studies of vanadium in rats, *J. Appl. Toxicol.*, 5, 418, 1985.
8. Roshchin, A. V., *Vanadium Compounds*, Moscow, 1984, 26 (in Russian).
9. Mravkova, A., Jirova, D., Janci, H., and Lener, J., Effects of orally administered vanadium on the immune system and bone metabolism in experimental animals, *Sci. Total Environ.*, Suppl. 1, 663, 1993.
10. Rusakov, N. V., Experimental study of the allergic effects produced by chemicals of various classes at their oral administration, *Gig. Sanit.*, 2, 13, 1984 (in Russian).
11. Zaporowska, H., Wasilewski, W., and Slotwinska, M., Effect of chronic vanadium administration in drinking water to rats, *Biometals*, 6, 3, 1993.
12. Edel, J. and Sabbioni, E., Vanadium transport across placenta and milk of rats to the fetus and newborn, *Biol. Trace Elem. Res.*, 22, 265, 1989.
13. Paternain, J. L., Domingo, J. L., Llobet, J. M., et al., Embryotoxic effects of sodium metavanadate administered to rats during organogenesis, *Rev. Esp. Fisiol.*, 43, 223, 1987 (in Spanish).
14. Gomez, M., Sanchez, D. J., and Domingo, J. L., Embryotoxic and teratogenic effects of intraperitoneally administered metavanadate in mice, *J. Toxicol. Environ. Health*, 37, 47, 1992.
15. Paternain, J. L., Domingo, J. L., Gomez, M., Ortega, A, and Corbella, J., Developmental toxicity of vanadium in mice after oral administration, *J. Appl. Toxicol.*, 10, 181, 1990.
16. Arkhangelskaya, L. N. et al., in *All-Union Constit. Toxicol. Conf.*, November 25-27, 1980, Abstracts of Reports, Moscow, 1980, 119 (in Russian).
17. Kanematzu, N., Hara, M., and Kada, T. Rec-assay and mutagenicity studies of metal compounds, *Mutat. Res.*, 77, 109, 1980.
18. Owusu-Yaw, J., Cohen, M. D., Fernando, S. Y., et al., An assessment of the genotoxicity of vanadium, *Toxicol. Lett.*, 50, 327, 1990.
19. Galli, A., Vellosi, R., Fiorio, R., et al., Genotoxicity of vanadium compounds in yeast and cultured mammalian cells, *Teratogen. Carcinogen. Mutagen.*, 11, 175, 1991.
20. Zhong, B.-Z., Gu, Z.-W., Wallace, W. E., Whong, W.-Z., and Ong, T., Genotoxicity of vanadium pentoxide in CH V79 cells, *Mutat. Res.*, 321, 35, 1994.
21. Ciranni, R., Antonetti, M., and Migliore L., Vanadium salts induce cytogenetic effects in *in vivo* treated mice, *Mutat. Res.*, 343, 53, 1995.
22. Leonard, A. and Gerber, G. B., Mutagenicity, carcinogenicity and teratogenicity of vanadium compounds, *Mutat. Res.*, 317, 81, 1994.
23. Crans, D. C., Gottlieb, M. S., Tawara, J., et al., A kinetic method for determination of free vanadium (IV) and (V) at trace levels concentration, *Annal. Biochem.*, 188, 53, 1990.
24. Sakurai, H., Vanadium distribution in rats and DNA cleavage by vanadyl complex: implication for vanadium toxicity and biological effects, *Environ. Health Perspect.*, 102 (Suppl. 3), 35, 1994.

VANADIUM-CYCLOPENTADIENE complexes

Abbreviation. VCPD.

Applications. Used as polymerization and oligomerization catalysts.

Properties and **Acute Toxicity.** Rats were dosed by gavage with 1.0 and 5.0% aqueous solutions of VCPD. In 10 to 15 min after poisoning, the animals became alerted and aggressive, with an unsteady gait, tremor, copious salivation, and exophthalmos. Death was preceded by complete adynamia. Gross pathology examination revealed vascular distention, hemorrhages, and cloudy swelling of the proximal tubular epithelium in the kidneys, etc.

Table. Solubility and toxicity parameters of VCPD

Substances	***Appearance***	***pH at the moment of preparation***	***Water solubility g/100 g***	***LD_{50} mg/kg BW***	***K_{acc}***
Vanadocene dichloride	Light-green crystals	3.55	2.50	87	2.16
Vanadium cyclopentadienyl dilactate	Light-brown powder	3.95	0.95	75	1.3
Vanadocene citrate	Blue-violet crystals	3.42	1.0	80	1.67
Vanadium dichloroacetate	Yellow-green powder	2.0	10.30	120	1.05

Short-term Toxicity failed to reveal evident cumulative properties.

Manifestations of the toxic effect are similar to those in the case of acute poisoning. Gross pathology examination revealed shrinkage and atrophy of separate neurons as well as dystrophic changes in the liver, kidneys, and in the cerebral cortex. In the spleen, the number of lymphoid follicles was reduced and the red pulp depleted of formed elements.

Reference:

Mikhailets, I. B., Sukhareva, L. V., and Yevsyukov, V. I., Hygienic evaluation of high-density polyethylene manufactured on the vanadian catalystic system, in *Environment Protection in Plastic Industry and Safety of the Use of Plastics*, Plastpolymer, Leningrad, 1978, 99 (in Russian).

XYLENEDIAMINE, CYANOETHYLATED

Molecular Formula. $C_{14}H_{22}N_4$

M = 246.4

Properties. Liquid. Soluble in water and alcohol.

Applications. Used as a curing agent in the production of epoxy resins.

Acute Toxicity. LD_{50} is reported to be 3.96 g/kg BW in rats and 1.86 g/kg BW in mice.

Repeated Exposure failed to reveal cumulative properties.

Allergenic Effect was not found to develop either on inhalation or on skin application.

Reference:

Ivanov, V. I., Loit, A. O., and Matyukhin, N. Ya., On toxicology of *m*- and *p*-xylenediamine - the raw material in the production of synthetic fibers, in *Hygiene and Toxicology of High-Molecular-Mass Compounds and of the Chemical Raw Materials Used in Their Synthesis*, Proc. 4th All-Union Conf., S. L. Danishevsky, Ed., Khimiya, Leningrad, 1969, 171 (in Russian).

ZIRCONIUM compounds

Zr

CAS No 7440-67-7

RTECS No ZH7070000

Applications. Zirconium salts are used as curing agents for silicones.

Migration of zirconium from food-contact polystyrene in olive oil was investigated. Zirconium was found to migrate at the level of 650 μg/kg.[1]

Acute Toxicity. LD_{50} of zirconium hydrocarbonate (calculated as zirconium) exceeds 10 g/kg BW.[03] LD_{50} is 1.6 g zirconium acetate/kg BW, and that of zirconium sulfate is 1.25 g/kg BW.[05]

Repeated Exposure. Zirconium compounds are low toxic chemicals. Administration of 20% zirconium hydrocarbonate in the diet of rats for 17 days and 5.0% in the diet of cats was tolerated by animals without signs of intoxication.[05]

Allergenic Effect of zirconium compounds was studied by measuring the level of immunoglobulin-*M*-rosette-forming cells in response to injection of male sheep erythrocytes to the mouse spleen. The adjuvant effect on humoral immunity was more pronounced at doses of 1/50 to 1/100 LD_{50} than at doses of 1/5 to 1/10 LD_{50} (Horiba et al., 1986).

Reproductive Toxicity. Feeding of trace amounts of zirconium by pregnant mice altered behavioral reactions in their offspring.[2]

Mutagenicity.

In vitro genotoxicity. Zirconium chloride and zirconium oxide octahydrate produced a doubling of the SCE frequency, and a 3-point monotonic increase with at least the highest dose tested in this study at the $p < 0.001$ significance level.[3]

Carcinogenicity. Negative results for zirconium sulfate are reported in mice.[09]

Regulations. ***U.S. FDA*** (1998) approved the use of zirconium compounds (1) in resinous and polymeric coatings for food-contact surfaces in accordance with the conditions prescribed in 21 CFR part 175.300; zirconium oxide may be used (2) as a component of paper and paperboard intended for use in producing, manufacturing, packaging, processing, preparing, treating, packing, transporting, or holding aqueous and fatty foods in accordance with the conditions prescribed in 21 CFR part 176.170.

Zirconium citrate, zirconium lactate-citrate and *zirconium lactate* may be used as insolubilizers with protein binders in coatings for paper and paperboard, at a level not to exceed 1.4% by weight of coating solids and in accordance with the conditions prescribed in 21 CFR part 176.170.

References:

1. Fordham, P. J., Gramshaw, J. W., Crews, H. M., and Castle, L., Element residues in food contact plastics and their migration into food simulants, measured by inductively-coupled plasma-mass spectrometry, *Food Addit. Contam.*, 12, 651, 1995.
2. Tsujii, H. and Hoshishima, K., Effect of the administration of trace amounts of metals to pregnant mice on the behavior and learning of their offspring, Abstract, *Shinshu Daigaku Nogakubu Kiyo,* 16, 13, 1979 (in Japanese).
3. Tucker, J. D., Auletta, A., Cimino, M. C., et al., Sister-chromatid exchange: second report of the GENE-TOX Program, *Mutat. Res.*, 297, 101, 1993.

CHAPTER 5. RUBBER INGREDIENTS

ACETOACETIC ACID, ETHYL ESTER

Molecular Formula. $C_6H_{10}O_3$
M = 130.16
CAS No 141-97-9
RTECS No AK5250000
Abbreviation. AAE.

Synonyms. Active acetylacetate; Diacetic ether; Ethyl acetylacetate; Ethyl 3-oxobutirate; 3-Oxobutanoic acid, ethyl ester.

Properties. Colorless liquid with a pleasant fruity, rum odor, and a ripe fruit taste. Soluble in water (1:35).

Applications. Used in the preparation of rubber articles. Flavoring agent in food.

Acute Toxicity. LD_{50} is 3.98 g/kg BW, and exceeded 10 ml/kg BW in rabbits.[03]

Regulations. ***U.S. FDA*** (1993) approved the use of AAE in the manufacture of resinous and polymeric coatings for the food-contact surface of articles intended for use in producing, manufacturing, packing, processing, preparing, treating, packaging, transporting, or holding food in accordance with the conditions prescribed by 21 CFR part 175.300.

ALKANESULFONATE, SODIUM SALT, mixture.

Trade Name. Volgonate.

Properties. White to light-yellow flakes or granules. Readily soluble in water. Odor perception threshold is 6.0 mg/l. Threshold concentration for foam formation is 0.1 mg/l.

Applications. Used as an emulsifier in synthetic rubber manufacture; a base for detergents.

Acute Toxicity. In mice, LD_{50} is 1.7 g/kg BW. A dose of 2.2 g/kg BW caused all animals to die. High doses produced apathy, hypodynamia, and GI tract disturbances.

Repeated Exposure revealed no lethal accumulation. Rabbits received doses of 40 mg/kg BW. The treatment caused bradycardia and retardation of BW gain, acceleration of erythrocyte sedimentation rate, leukocytosis and thrombocytosis, increased catalase and cholinesterase activity, and dysbalance of serum protein fractions.

Long-term Toxicity. Rabbits were given doses up to 0.4 mg/kg BW for 6 months. Toxic manifestations included thrombocytosis and ECG changes at the higher dose level.

Standards. ***Russia.*** Recommended PML in drinking water: 0.1 mg/l (organolept., foam).

Reference:

Taradin, Ya. I., Fetisova, L. N., Kuchmina, N. Ya., et al., Sanitary-toxicological evaluation of alkylsulfonate Volgonate, in *Hygiene and Toxicology of High-Molecular-Mass Compounds and of the Chemical Raw Material Used for Their Synthesis*, Proc. 4th All-Union Conf., S. L. Danishevsky, Ed., Khimiya, Leningrad, 1969, 233 (in Russian).

2-[(2-AMINOETHYL)AMINO]ETHANOL

Molecular Formula. $C_4H_{12}N_2O$
M = 104.18
CAS No 111-41-1
RTECS No KJ6300000
Abbreviation. AEAE.

Synonyms. Ethanolethylenediamine; *N*-Hydroxyethyl-1,2-ethanediamine.

Properties. A colorless liquid with a mild ammoniac odor. Readily soluble in water and alcohol.

Applications. Used in the rubber and resin manufacture.

Acute Toxicity. LD_{50} is 3.0 g/kg BW in rats, 3.55 g/kg BW in mice, 1.5 g/kg BW in guinea pigs, and 2.0 g/kg BW in rabbits.

Regulations. ***U.S. FDA*** (1998) approved the use of AEAE in adhesives as a component of articles intended for use in packaging, transporting, or holding food in accordance with the conditions prescribed by 21 CFR part 175.105.

Reference:

Environ. Space Science (Kosm. Biol. Med.), 2, 289, 1968 (in Russian).

AMINOPHENOLS

Molecular Formula. C_6H_7N0

M = 109.14

	CAS No	RTECS No
***o*-Aminophenol**	95-55-6	SJ4950000
***m*-Aminophenol**	591-27-5	SJ4900000
***p*-Aminophenol**	123-30-8	SJ5075000

Synonym. Hydroxyanilines.

Properties. Water- and alcohol-soluble crystalline substances.

o-A. has no odor or taste. Solubility in water is 1.7% (at 0°C). Colors water at a concentration of 0.01 mg/l.

m-A. White powder with a bitter astringent taste. Solubility in water is 2.6% at 20°C. At a concentration of 0.05 to 0.1 mg/l it turns water yellow, the color remaining for 5 to 6 days. Odor perception threshold (rating 1) is 160 mg/l; taste perception threshold is 1600 mg/l.[1]

p-A. White, crystalline powder, darkening in air. Has no odor or taste. Solubility in water is 1.1% (0°C), higher in hot water. Readily soluble in ethanol. Colors water at a concentration of 0.05 mg/l and reduces its transparency at 10 mg/l.[1]

Applications. Used in the synthetic rubber manufacture.

Acute Toxicity.

o-A. LD_{50} is reported to be 0.5 g/kg BW in rats, 0.6 g/kg BW in mice, 1.5 g/kg BW in rabbits, and 1.9 g/kg BW in guinea pigs.

m-A. In mice, LD_{50} is 420 mg/kg BW. The toxic symptoms include excitation, mucosal irritation, convulsions, and paralysis. Death occurs within days. Gross pathology examination revealed hemodynamic disorders in the visceral organs. Neither Heinz bodies, nor methemoglobinemia were found.

p-A. Excessive toxic effect in animals is observed.

Long-term Toxicity. Aminophenols are capable of causing methemoglobinemia. Ineffective concentration of *o*-A. in water is reported to be 1.0 mg/l.[1]

Reproductive Toxicity.

Teratogenicity. *o*-A. appeared to be teratogenic. Equivocal data are reported for teratogenicity of *m*-A. in rats and hamsters.[2,3]

Rats received oral doses of *p*-A. on days 6 through 15 of gestation. Maternal toxicity and teratogenicity were observed at 250 mg/kg BW, but not at 85 mg/kg and lower doses.[4] Other studies failed to reveal malformations caused by *p*-A.[5]

Embryotoxicity and ***Gonadotoxicity.*** Aminophenols caused sperm head abnormalities in mice. No reduction in reproductive performance and no increase in birth defects were found in a 2-generation rat study in which *o*-A. was introduced in the diet.[6]

Mutagenicity. *o*- and *p*-A. produce SCE in cultured Chinese hamster ovary cells (*V79*) and in micronuclear test on mouse bone marrow. Suggested mechanism is involvement of semi-quinone radicals which can bind to proteins and nucleic acids.[7]

m-A. appeared to be inactive.[7,8] *p*-A. caused a significant increase in the frequency of CA in Chinese hamster ovary and in mouse lymphoma L5178Y cells tests.[9]

Chemobiokinetics. After a single administration to rats, *m*-A. is excreted in the urine, partially in the form of *free* and *bound phenols* and to a lesser extent as *amines*.

Regulations. ***U.S. FDA*** (1993) approved the use of A. in adhesives as a component of articles intended for use in packaging, transporting, or holding food in accordance with the conditions prescribed in 21 CFR part 175.105.

Standards. ***Russia*** (1995). MAC and PML for *p*-A., 0.05 mg/l; for *o*-A., 0.01 mg/l (organolept., color).

References:

1. Trofimovich, E. M., in *Hygiene*, Coll. Papers., Novosibirsk, 1968, 78 (in Russian).
2. Shtannikov, E. V. and Lutsevich, I. N., Delayed effects of aromatic amines on the body, *Gig. Sanit.*, 9, 19, 1983 (in Russian).
3. Rutkowski, J. V. and Ferm, V. H., Comparison of the teratogenic effects of isomeric forms of aminophenol in the Syrian golden hamster, *Toxicol. Appl. Pharmacol.*, 63, 264, 1982.
4. Spengler, J., Osterburg, I., and Korte, R., Teratogenic evaluation of *p*-toluenediamine sulfate, resorcinol and *p*-aminophenol in rats and rabbits, Abstract, *Teratology*, 33, 31A, 1986.
5. Kavlock, R. J. et al., The developmental toxicity of a series of *para*-substituted phenols: A structure study, *Teratology*, 35, 75A, 1987.
6. Re, T. A. et al., Results of teratogenicity testing of *m*-aminophenol in Sprague-Dawley rats, *Fundam. Appl. Toxicol.*, 4, 98, 1984.
7. Wild, D., King, M.-T., Eckhardt, K., et al., Mutagenic activity of aminophenols and diphenols, and relations with chemical structure, *Mutat. Res.*, 85, 456, 1981.
8. Goldblat, M. and Goldblat, J., *Industrial Medicine and Hygiene*, E. R. Meriwether, Ed., London, 1956.
9. Majeska, J. B. and Holden, E., The genotoxic effects of *p*-aminophenol in Chinese hamster ovary and mouse lymphoma cells, Results of a multiple endpoint test, *Environ. Molec. Mutagen.*, 23, 41, 1994.

N′-(3-AMINOPROPYL)-*N,N′*-DIMETHYLPROPANE-1,3-DIAMINE

Molecular Formula. $C_8H_{21}N_3$
M = 159.32
CAS No 10563-29-8
RTECS No TX7000000

Synonym. *N,N′*-Dimethyldipropylenetriamine.

Properties. High-boiling liquid. Soluble in fats, organic solvents, and water.

Applications. Used as an inhibitor of the thermal polymerization of rubber and vulcanization stabilizer. Used in the synthesis of plasticizers, polyvinyl acetate, and corrosion-resistant coatings.

Acute Toxicity. In rats, LD_{50} is 1.4 g/kg BW. Poisoning manifestations comprise excitation with subsequent CNS inhibition and adynamia. Death occurs in 2 to 4 days.

Repeated Exposure. Rats received 1/3 LD_{50} during three consecutive administrations. Disorganization in the intracellular structure of the liver, particularly that associated with detoxifying and energy biotransformation processes, was noted. A. penetrates via the skin to exert pronounced toxic effect.

Reference:

Sidorin, G. I., Lukovnikova, L. V., Stroikov, Yu. N., et al., Toxicity of some aliphatic amines, *Gig. Truda Prof. Zabol.*, 11, 50, 1984 (in Russian).

BENZENESULFONIC ACID, HYDRAZIDE

Molecular Formula. $C_6H_8N_2O_2S$
M = 172.22
CAS No 80-17-1
RTECS No DB6888000
Abbreviation. BSH.

Synonyms and **Trade Names.** Benzenesulfohydrazide; Benzenesulfonyl hydrazide; Porofor BSH; Phenylsulfohydrazide; Porofor ChKhZ 9.

Applications. Used in synthetic rubber manufacture: BSH is a gas-generating agent used in making foam rubber. It has also been used as a rodenticide.

Acute Toxicity. LD_{min} is 50 mg/kg BW in rats.[027]

Reproductive Toxicity.

Teratogenicity. Gavage feeding of BSH to pregnant mice on days 8 to 13 increased resorptions at maternally toxic doses (62 mg/kg BW), but did not increase congenital anomalies in the offspring. Only delayed ossification was found in the fetuses.[1]

Reference:

1. Matschke, G. H. and Fagerstone, K. A., Effects of a new rodenticide benzenesulfonic acid hydrazide on prenatal mice, *J. Toxicol. Environ Health.*, 3, 407, 1977.

2-BENZIMIDAZOLETHIOL

Molecular Formula. $C_7H_{6(7)}N_2S$
M = 150.20
CAS No 583-39-1
RTECS No DE1050000
Abbreviation. 2-MBI.

Synonyms and **Trade Names.** 2-Benzothiazolethiol; 2-Benzothiazolylmercaptan; Captax; 2-Mercaptobenzimidazole.

Properties. Yellow-white powder with a bitter taste. Poorly soluble in water, soluble in alcohol.

Applications. Used as a stabilizer in the production of synthetic rubber, polyethylene, and polypropylene.

Migration Data. 2-MBI leaching from rubber closures into drug preparations was studied by a thin-layer chromatographic method. The lowest and highest concentrations were found in syringe-cartridges in the range of 2.8 to 11.1 μg/l.[1]

A study of extracts from natural rubber Qualitex (0.67 and 1.0 cm^{-1}; 20 and 38°C; 24 hours; distilled water) revealed migration of 2-MBI sodium salt at the level of 3.0 mg/l.[2]

Acute Toxicity. LD_{50} is 270 mg/kg BW in rats, and 1250 mg/kg BW[3] or 105 mg/kg BW in mice.[023] The chemical structure of 2-MBI is partially similar to those of *thiourea* and *ethylenethiourea*, both potent thyrotoxic compounds.[6]

Gross pathology examination revealed morphologic changes in the lungs, cerebral vessels, and liver. The single lethal dose caused thyroid toxicity as measured by a 95% decrease in iodine uptake in rats.[4]

Thyroid enlargement is associated with decreased plasma concentrations of circulating thyroxine and triiodothyronine and increased thyotropin levels in rats receiving a single oral dose.[5]

Repeated Exposure. Wistar rats received doses of 2.0, 10, or 50 mg 2-MBI/kg BW administered by gavage over a period of 28 days. The treatment caused no deaths. A decrease in BW gain and food consumption was observed at the highest dose. Hematology analyses revealed a decrease in leukocyte count and *Hb* level, and increased serum urea nitrogen, cholesterol, phospholipid, β-glutamyl transpeptidase, and the Na^+/K^+ ratio at the highest dose. Histological examination revealed diffuse hyperplasia of the thyroid follicles. The NOEL of less than 2.0 mg/kg BW for the significant decrease in thymus weight was established in this study.[6]

Short-term Toxicity. Rabbits were dosed by gavage with 20 mg/kg BW for 4 months. The treatment caused retardation of BW gain, impairment of the liver functions, hemorrhagic syndrome, and changes in hematology analyses.[023]

Immunotoxicity. 2-MBI is suspected to cause thymic involution.

Reproductive Toxicity.

Embryotoxic effect was observed in rats.[7]

Mutagenic Effect was demonstrated in rat fetal cells;[7] cytogenetic abnormalities and immunodeficiency were described in the progeny of rats.[8]

Carcinogenicity. 2-MBI is suspected of being capable of producing thyroid and liver tumors because it is structurally related to ethylenethiourea.

Chemobiokinetics. 2-MBI is readily absorbed after oral administration. Its half-life in blood is approximately 83 hours. Accumulates in the thyroid.[5] The disposition of ^{14}C-labeled 2-MBI in male F344 rats dosed orally (49 or 0.5 mg/kg BW) was determined. Absorption of the oral dose was evident, since, in 72 hours, most of the radioactivity appeared in the urine. Smaller amounts appeared in the feces. The concentration of total radioactivity was higher in liver and kidney tissue than in blood.

One of the major urinary metabolites was identified as *benzimidazole*, and a minor component was tentatively identified as unchanged MBI. Neither of these could be detected in bile.[9]

Regulations. ***U.S. FDA*** (1998) approved the use of 2-MBI, sodium salt, (1) in adhesives as a component of articles intended for use in packaging, transporting, or holding food in accordance with the conditions prescribed by 21 CFR part 175.105; as (2) a component of a defoaming agent intended for articles which may be used in producing, manufacturing, packing, processing, preparing, treating, packaging, transporting, or holding food, subject to the provisions prescribed in 21 CFR part 176.200; as (3) a slimicide in the manufacture of paper and paperboard that contact food, subject to the provisions prescribed in 21 CFR part 176.300; and (4) in the production of animal glue which may be safely used as a component of articles intended for use in producing, manufacturing, packaging, processing, preparing, treating, packing, transporting, or holding food, in accordance with the conditions prescribed in 21 CFR part 178.3020.

References:

1. Airaudo, C. B., Gayte-Sorbier, A., Momburg, R., and Laurent, P., Leaching of antioxidants and vulcanization accelerators from rubber closures into drug preparations, *J. Biomater. Sci. Polymer*, 1, 231, 1990.
2. Shumskaya, N. I., Provorov, V. N., Tolgskaya, M. S., Yemel'yanova, L. V., and Chernevskaya, N. M., Toxicity of some vulcanizates intended for manufacture of children's toys, in *Hygiene Aspects of the Use of Polymeric Materials and of Articles Made of Them*, L. I. Medved', Ed., All-Union Research Institute of Hygiene and Toxicology of Pesticides, Polymers and Plastic Materials, Kiyv, 1969, 104 (in Russian).
3. Mezentseva, N. V., Mercaptobenzimidasol, in *Toxicology of New Chemical Substances Used in Rubber and Tyre Industry*, Z. I. Israel'son, Ed., Meditsina, Moscow, 180, 1968 (in Russian).
4. Searle, C. E., Lawson, A., and Hemmings, A. W., Antithyroid substances, 1. The mercapto-glyoxolines, *Biochem. J.*, 47, 77, 1950.
5. Janssen, F. W., Young, E. M., Kirkman, S. K., et al., Biotransformation of the immunomodulator 3-(*p*-chlorophenyl)-2,3-dihydro-3-hydroxythiazolo[3,2a] benzimidazole-2-acetic acid and its relationship to thyroid toxicity, *Toxicol. Appl. Pharmacol.*, 59, 355, 1981.
6. Kawasaki, Y., Umemura, T., Saito, M., Momma, J., Matsushima, Y., Sekiguchi, H., Matsumoto, M., Sakemi, K., Isama, K., Inoue, T., Kurokawa, Y., and Tsuda, M., Toxicity study of a rubber antioxidant, 2-mercaptobenzimidazole, by repeated oral administration to rats, *J. Toxicol. Sci.*, 23, 53, 1998.
7. Barilyak, I. R., Embryotoxic and mutagenic effects of 2-mercaptoimidazole, *Physiol. Active Substances*, 6, 85, 1974 (in Russian).
8. Barilyak, I. R. and Mel'nik, E. K., Dynamics of the cytogenetic effect of some chemical preparations, *Reports Ukrainian Acad. Sci.*, 1, 66, 1979 (in Ukrainian).
9. El Dareer, S. M., Kalin, J. R., Tillery, K. F., and Hill, D. L., Disposition of 2-mercaptobenzimidazole in rats dosed orally or intravenously, *J. Toxicol. Environ. Health*, 14, 598, 1984.

2-BENZIMIDAZOLETHIOL, COPPER SALT

Molecular Formula. $C_{14}H_8CuN_2S_4$

M = 396.0

Abbreviation. MBIC.

Synonyms. Captax, copper salt; 2-Mercaptobenzoimidazole, copper salt.

Properties. A yellow, fine-particle, easily dusting substance. Poorly soluble in water.

Applications. Used as a vulcanization accelerator in the rubber industry.

Acute Toxicity. LD_{50} was not attained in mice: administration of 10 g/kg BW caused only 27% animals to die on days 6 to 8 after treatment. High doses produce decreased BW gain and reduction in the relative liver weights.

Long-term Toxicity. Rabbits were dosed by gavage with 50 mg MBIC/kg BW every other day for the first 2 months and daily for further 3 months. The treatment resulted in decrease of BW gain, changes in the hematology analyses and in the balance of serum protein fractions. Gross pathology examination revealed congestion in the visceral organs and changes in the suprarenals and myocardium.

Reference:

Arkhangelskaya, L.N. and Roschina, T. A., 2-Mercaptobenzothiazole, copper salt, in *Toxicology of New Chemical Substances Used in Rubber and Tyre Industry*, Z. I. Israel'son, Ed., Meditsina, Moscow, 1968, 43 (in Russian).

2-BENZIMIDAZOLETHIOL, DIPHENYLGUANIDINE SALT

Molecular Formula. $C_{20}H_{18}N_4S_2$

M = 367.79

Abbreviation. MBIDPG.

Synonym. 2-Mercaptobenzimidazole, diphenylguanidine salt.

Properties. Light-gray, fine-particle powder. Readily soluble in alcohol.

Applications. Used as a vulcanization ultra-accelerator in the rubber industry.

Acute Toxicity. In mice, LD_{50} is 360 mg MBIDPG/kg BW. Poisoned animals look immobile, with unkempt fur. Other manifestations of the toxic action included impaired motor coordination, rigid neck and tail muscles, and labored breathing.

Long-term Toxicity. Rabbits were exposed to 50 mg MBIDPG/kg BW for 5.5 months. The treatment caused retardation of BW gain, leukocytosis, and dysbalance in serum protein composition. Gross pathology examination revealed lymphoid stroma infiltration, connective tissue growth along the interlobular septa, and focal, large-drop, fatty infiltration of the liver cell protoplasm; in the kidneys, there was a marked congestion, glomerular swelling, plasma in the Bowman capsules, and cloudy swelling of the renal tubular epithelium.

Reference:

Arkhangelskaya, L. N. and Roschina, T. A., 2-Mercaptobenzothiazolediphenylguanidine, in *Toxicology of New Chemical Substances Used in Rubber and Tyre Industry*, Z. I. Israel'son, Ed., Meditsina, Moscow, 1968, 51 (in Russian).

2-BENZIMIDAZOLETHIOL, ZINC SALT

Molecular Formula. $C_{14}H_{10}N_4S_2.Zn$

M = 363.38

CAS No 3030-80-6

RTECS No DE1115000

Synonyms and **Trade Name.** Antioxidant ZMB; 2-Mercaptobenzimidiazole, zinc salt; Zinc benzimidazole-2-thiolate.

Properties. White-yellowish powder. Insoluble in water, soluble in ethanol.

Applications. Used as a stabilizer in the synthesis of vulcanizates based on natural and synthetic rubber.

Acute Toxicity. LD_{50} is 540 mg/kg BW in rats, 740 to 860 mg/kg BW in mice.[023] Death occurs as a consequence of severe vascular disorder in the lungs, liver, and brain.

According to other data, LD_{50} is 1000 mg/kg BW in rats.[027]

Repeated Exposure failed to reveal cumulative properties. Rats were administered 1/5 LD_{50} for one month. The treatment caused retardation of BW gain, CNS depression, anemia, and a rise in the blood *metHb* concentration.[023] Histology examination revealed parenchymatous dystrophy of the liver and hyperplasia of the lymphoid follicles in the spleen.

Short-term Toxicity. Rats received 20 mg/kg BW for 4 months. Manifestations of the toxic action included impairment of liver function, expressed as a reduction in the content of amine nitrogen in the urine, a rise in serum lipids content, and a reduction in cholinesterase activity.[1]

Reference:

1. Vorob'yeva, R. S. and Mezentseva, N. V., Mercaptobenzimidasole, zinc salt, in *Toxicology of New Chemical Substances Used in Rubber and Tyre Industry*, Z. I. Israel'son, Ed., Meditsina, Moscow, 184, 1968 (in Russian).

1,2-BENZISOTHIAZOL-3(*2H*)-ONE

Molecular Formula. C_7H_5NOS
M = 151.19
CAS No 2634-33-5
RTECS No DE4620000

Synonym and **Trade Name.** 1,2-Benzisothiazolin-3-one; Proxel PL.

Acute Toxicity. LD_{50} is 1020 mg/kg BW in rats, and 1150 mg/kg BW in mice.

Regulations. ***U.S. FDA*** (1998) approved the use of B. as a biocide in uncured liquid rubber latex not to exceed 0.02% by weight of the latex solids in accordance with the conditions prescribed in 21 CFR part 177.2600.

Reference:

Pharmacol. Research Communications, vol 1, Academic Press, Inc., 1969.

p-BENZOQUINONE

Molecular Formula. $C_6H_4O_2$
M = 108.10
CAS No 106-51-4
RTECS No DK2625000
Abbreviation. BQ.

Synonyms and **Trade Name.** 1,4-Cyclohexadiene dioxide; 1,4-Cyclohexadienedione; 1,4-Dioxybenzene; *p*-Quinone.

Properties. Greenish-yellowish solid with a penetrating chlorine-like odor. Slightly soluble in water, solubility in ethanol exceeded 10%.

Applications. Rubber accelerator. BQ is used in the manufacture of unsaturated polyesters as a polymerization inhibitor. Used in adhesives.

Acute Toxicity. Large oral doses induced local lesions in the GI tract, caused convulsions, respiratory disturbances. Death occurred due to paralysis of the medullary centers.[03] The kidneys are reported to be damaged (IARC 15-259).

Mutagenicity.

In vitro genotoxicity. BQ was negative in *Salmonella* mutagenicity assay in presence or absence of metabolic activation systems.[1] However, according to Hakura et al.,[2] it showed potential mutagenic activity in five different *Salmonella* tester strains in the presence or absence of *S9* mix. Administration to pregnant mice produced little evidence of genotoxicity in bone marrow cells of fetuses.[2]

Carcinogenicity. Neoplasms developed in the hematopoietic system of rodents exposed orally to BQ.[3] Mice were painted with 0.25% solution of BQ in benzene. The treatment increased significantly the incidence of skin papillomas and lung adenocarcinomas. Control animals were painted with benzene (Takizawa, 1940, cit. in IARC 15-258).

Chemobiokinetics. Following ingestion, BQ is readily absorbed. Elimination occurs by conjugation with *hexuronic, sulfuric* and other acids. Excretion of unchanged BQ was also noted.[03]

Regulations. ***U.S. FDA*** (1998) approved the use of BQ in the manufacture of cross-linked polyester resins which may be used as articles or components of articles intended for repeated use in contact with food in accordance with the conditions prescribed by 21 CFR part 177.2420.

References:

1. Hakura, A., Mochida, H., Tsutsui, Y., and Yamatsu, K., Mutagenicity of benzoquinone for Ames *Salmonella* tester strains, *Mutat. Res.,* 347, 37, 1995.

2. Ciranni, R., Barale, R., Marrazzini, A., and Loprieno, N., Benzene and the genotoxicity of its metabolites. I. Transplacental activity in mouse fetuses and in their dams, *Mutat. Res*., 208, 61, 1988.
3. Davidson, K. A. and Faust, R. A., Oral carcinogen potency factors for benzene and some of its metabolites, in *Toxicologist*, Abstracts of SOT 1996 Annual Meeting, Abstract No 585, Issue of *Fundam. Appl. Toxicol.*, 30, Part 2, March 1996.

2-BENZOTHIAZOLETHIOL

Molecular Formula. $C_7H_5NS_2$
M = 167.24
CAS No 149-30-4
RTECS No DL6475000
Abbreviation. 2-MBT.

Synonyms and **Trade Name.** 2-Benzothiazolylmercaptan; Captax; 2-Mercaptobenzothiazole.

Properties. Pale monoclonic needles or leaflets, or easily dusting yellow powder with a characteristic odor and bitter taste. Poorly soluble in water, soluble in alcohol.

Applications. Used in the production of rubber, plastics, and lacquers. The commonest vulcanization accelerator for vulcanizate mixes based on natural and synthetic rubbers.

Migration from vulcanizates based on butadiene-nitrile rubbers into water after a 3-day exposure at 20°C was determined at the level of up to 1.0 mg/l.[1]

2-MBT leaching from rubber closures into drug preparations was studied by a thin-layer chromatographic method. The lowest and highest concentrations were found in syringe-cartridges being in range of 8.3 to 13.8 µg/l for 2-MBT and 2.9 to 9.3 µg/l for 2-MBT disulfide.[2]

Migration of 2-MBT from vulcanizates based on polybutadiene rubber in composition with natural rubber and polyisoprene rubber (accelerator system used in the vulcanizates: Vulkazit-*P*-extra-*N* with sulfenamide) was studied into simulant media at 20°C. 2-MBT concentration amounted to 0.1 mg/l.[3]

A study of extracts of vulcanizates based on synthetic polyisoprene rubber (accelerator system used in the vulcanizates contains diphenylguanidine) revealed migration of 2-MBT at level up to 2.4 mg/l.[028]

Acute Toxicity. In mice, LD_{50} is found to be about 1.5 g/kg BW (administered in suspension in 5.0% gum arabic solution) and 3.15 g/kg BW (in olive oil, males). Poisoning led to convulsion development.[4]

According to other data, LD_{50} is reported to be 0.1 g/kg BW in wild rats, 0.5 g/kg BW in household rats, and 2.0 g/kg BW in mice.[027]

All mice died from a 4.0 g/kg BW dose. Manifestations of the toxic action include restlessness, excitation, impaired motor coordination, and convulsions. Death occurs within 24 hours.[5,6]

Repeated Exposure failed to reveal cumulative properties. The treatment of rabbits with 200 mg/kg BW, dogs with 50 to 100 mg/kg BW (for 6 days), and rats with 1.0 g/kg BW (for 10 days) did not affect general condition of animals while a 45-day administration of 2-MBT altered hematology analyses and functional state of the liver and CNS.[7]

Lethargy, prostration, and retardation of BW gain, as well as high incidences of mortality were reported in rats given 1500 mg 2-MBT/kg BW and more for 16 days.[7]

Short-term Toxicity. Rats were dosed with 50 mg/kg BW for 4 months. Only changes in conditioned reflex activity were marked; BW gain indices and blood morphology did not differ from those in the controls. Rabbits received 20 mg 2-MBT/kg BW every other day for 2 months and then daily for 1.5 months; dogs received 50 to 100 mg/kg BW for 3 to 4 months. No manifestations of the toxic effect were noted in the animals.[8]

In a 13-week gavage study, there were no chemical-related deaths in rats, but retardation of BW gain and dose-dependent hepatomegaly was reported, however, without microscopic pathologic changes. In mice, dose-related clinical signs included lethargy and lacrimation, salivation, and clonic seizure.[7]

Long-term Toxicity. Male F344 rats and $B6C3F_1$ were administered 375 or 750 mg 2-MBT/kg BW, female F344 rats received 188 or 375 mg 2-MBT/kg BW in corn oil by gavage for 103 weeks. The treatment decreased survival in dosed male rats, and in high dosed group of female mice. There was no effect on BW gain in dosed rats. Postgavage lethargy and prostration occurred frequently in dosed rats and mice.

No apparent BA-related non-neoplastic responses were observed in mice. The severity of nephropathy was increased in male rats. Ulcers and inflammation of the forestomach were prevalent.[7]

Mice received the diet containing 30 to 1920 ppm 2-MBT for 20 months. In the highest dose level, retardation of BW gain and cell infiltration in the interstitium of kidney were reported. The NOAELs of 14.6 mg/kg BW in males and 13.5 mg/kg BW in females were established in this study.[5]

Reproductive Toxicity.

Embryotoxicity. Severe embryolethality was noted in rats given 4.0 mg 2-MBT/kg BW on gestation day 8 or 10.[9]

Teratogenicity. Following *i/p* administration of 200 mg 2-MBT/kg BW to Sprague-Dawley rats on days 1 through 15 of gestation, teratogenic effect in fetuses was not observed.[10]

Allergenic Effect. 2-MBT showed weak sensitization potential in guinea pigs.[11]

Mutagenicity. Is found to be negative in *Salmonella typhimurium* with or without metabolic activation. It increased the frequency of CA and SCE in Chinese hamster ovary cells, as well as mutations at the TK locus of mouse L5178Y lymphoma cells.[7]

Carcinogenicity. Carcinogenic effect was found in rats gavaged with 750 mg/kg BW (males) and 375 mg/kg BW (females) doses, but not in mice.[12]

In the above-cited study, there was some evidence of carcinogenicity in rats, indicated by increased incidence of mononuclear cell leukemia, pancreatic acinar cell adenomas, adrenal gland pheochromocytomas, and preputial gland adenomas or carcinomas (combined) but no neoplasms of the forestomach were observed.[7]

Carcinogenicity classification.

NTP: SE* - SE - NE - EE (gavage).

Chemobiokinetics. Rats were given 1/10 LD_{50}. In 1 hour after administration, 0.015 mg 2-MBT/ml serum and 0.072 mg 2-MBT/ml urine were found. Accumulates primarily in the liver, brain, kidneys, and fatty tissues. Excretion is mainly via the kidneys.[13]

Rats were given 0.51 mg 2-MBT/kg BW for a 14-day period prior to a single 0.503 mg/kg BW dose of ^{14}C-MBT. In 96 hours, a small portion of radioactivity remained associated with erythrocytes, most of which was bounded to the membranes. Two metabolites were found in the urine, one of which was *thioglucuronide derivative* of 2-MBT, the other was possibly a *sulfonic acid derivative.* Elimination mainly with the urine and to a lesser extent in the feces.[14]

Regulations. ***U.S. FDA*** (1998) approved the use of 2-MBT (1) as a component of adhesives (as preservative only) to be safely used in contact with food in accordance with the conditions prescribed by 21 CFR part 175.105; (2) as an accelerator in the manufacture of rubber articles intended for repeated use in contact with food (up to 1.5% by weight of the rubber product) in accordance with the conditions prescribed by 21 CFR part 177.2600; (3) as an adjuvant substance in the manufacture of foamed plastics intended for use in contact with food, subject to the provisions prescribed in 21 CFR part 178.3010 (2-MBT.*Na*); and (4) in preparation of slimicides in the manufacture of paper and paperboard that may be safely used in contact with food, subject to the provisions prescribed in 21 CFR part 176.300.

Standards. ***Russia*** (1995). PML in food: 0.15 mg/l; MAC and PML: 5.0 mg/l (organolept., odor).

References:

1. Prokof'eva, L. G., in *Hygiene of Use and Toxicology of Pesticides and Polymeric Materials*, Coll. Sci. Proc., A. V. Pavlov, Ed., All-Union Research Institute of Hygiene and Toxicology of Pesticides, Polymers and Plastic Materials, Kiyv, Issue No 17, 1987, 153 (in Russian).
2. Airaudo, C. B., Gayte-Sorbier, A., Momburg, R., and Laurent, P., Leaching of antioxidants and vulcanization accelerators from rubber closures into drug preparations, *J. Biomater. Sci. Polymer.*, 1, 231, 1990.
3. Stankevich, V. V., Ivanova, T. P., and Prokof'yeva, L. G., Hygienic studies of 'food-grade' resins on the base of a new diene rubber, in *Hygiene and Toxicology of High-Molecular-Mass Compounds and*

of the Chemical Raw Material Used for Their Synthesis, Proc. 6th All-Union Conf., B. Yu. Kalinin, Ed., Khimiya, Leningrad, 1979, 96 (in Russian).
4. Ogawa, Y., Kamata, E., Suzuki, S., Kobayashi, K., Naito, K, Kaneko, T., Kurokawa, Y., and Tobe, M., Toxicity of 2-mercaptobenzothiazole in mice, Abstract, *Eisei Shikenjo Hokoku*, 107, 44, 1989 (in Japanese).
5. Vaisman, Ya. I., Zaitseva, N. V., and Mikhailov, A. V., Hygiene standardization of sewage of rubber production in water bodies, *Gig. Sanit.*, 2, 17, 1973 (in Russian).
6. Vorob'yeva, R. S., Kasparov, A. A., and Mezetseva, N. V., 2-Mercaptobenzothiasole, in *Toxicology of New Chemical Substances Used in Rubber and Tyre Industry*, Z. I. Israel'son, Ed., Meditsina, Moscow, 1968, 28 (in Russian).
7. *Toxicology and Carcinogenesis Studies of 2-Mercaptobenzothiazole in F344/N Rats and B6C3F$_1$ Mice (Gavage Studies)*, NTP Technical Report Series No 332, Research Triangle Park, NC, 1988.
8. Litvinchuk, M. D., Studies on toxicity of 2-mercaptobenzothiazole, *Pharmacologia i Toxicologia*, 4, 484, 1963 (in Russian).
9. Vaitekunene, D. I. and Sanatina, K. G., in *Problems of Epidemiology and Hygiene in the Lithuanian SSR*, Vilnius, 1969, 165 (in Russian).
10. Hardin, B. D., Bond, G. P., Sikov, M. R., et al., Testing of selected workplace chemicals for teratogenic potential, *Scand. J. Work Environ. Health*, 7 (Suppl. 4), 66, 1981.
11. Wang, X. S. and Suskind, R. R., Comparative studies of the sensitization potential of morpholine, 2-mercaptobenzothiazole and 2 other derivatives in guinea pigs, *Contact Dermat.*, 19, 11, 1988.
12. Haseman, J. K. and Clark, A. M., Carcinogenicity results for 114 laboratory animal studies used to assess the predictivity of four *in vitro* genetic activity assays for rodent carcinogenicity, *Environ. Mutagen.*, 16 (Suppl. 18), 15, 1990.
13. Datsenko, I. I. and Korneichuk, E. P., Kinetics of 2-mercaptobenzothiasole metabolism in the animals, *Gig. Sanit.*, 1, 51, 1991 (in Russian).
14. El Dareer, S. M., Kalin, J. R., Tillery, K. F., et al., Disposition of 2-mercaptobenzothiasole and 2-mercaptobenzothiazoledisulfide in rats dosed intravenously, orally, and topically and in guinea pigs dosed topically, *J. Toxicol. Environ. Health*, 27, 65, 1989.

2-BENZOTHIAZOLETHIOL, ZINC SALT

Molecular Formula. $C_{14}H_8N_2S_4Zn$
M = 397.85
CAS No 155-04-4
RTECS No DL7000000
Abbreviation. MBTZ.

Synonyms and **Trade Names.** Bis(2-benzothiazolylthio)zinc; 2-Mercaptobenzothiazole, zinc salt; Vulcacit ZM; Zenite.

Applications. Rubber vulcanization accelerator. A fungicide.

Acute Toxicity. LD_{50} exceeded 5.0 g/kg BW in rats.[027]

Reproductive Toxicity.

Gonadotoxicity and ***Embryotoxicity.*** Administration of MBTZ affected estrus cycle in female rats and reduced fertility. An increase in embryolethality was observed in treated offspring.[1]

Teratogenic effect was not found on administration of 200 mg MBTZ/kg BW in rats.[2]

Carcinogenicity. Carcinogenic agent by RTECS criteria.

Regulations. ***U.S. FDA*** (1998) approved the use of MBTZ (1) in adhesives as a component of articles intended for use in packaging, transporting, or holding food in accordance with the conditions prescribed by 21 CFR part 175.105; (2) in the manufacture of rubber articles intended for repeated use in producing, manufacturing, packaging, processing, preparing, treating, packing, transporting, or holding food in accordance with the conditions prescribed by 21 CFR part 177.2600; (3) in the production of animal glue which may be safely used as a component of articles intended for use in producing, manufacturing, packaging, processing, preparing, treating, packing, transporting, or holding food, in accordance with the conditions prescribed by 21 CFR part 178.3020; and (4) as an adjuvant substance in the manufacture of foamed plastics intended for use in contact with food, subject to the provisions prescribed in CFR 178.3010.

References:

1. Alexandrov, S. E., Effect of vulcanization accelerators on embryolethality in rats, *Bull. Exp. Biol. Med.*, 93, 87, 1982 (in Russian).
2. Hardin, B. D. et al., Testing of selected workplace chemicals for teratogenic potential, *Scand. J. Work Environ. Health*, 7 (Suppl. 8), 66, 1981.

N-(2-BENZOTHIAZOLYLTHIO)UREA

Molecular Formula. $C_8H_7N_2OS_2$
M = 225.29

Synonym. *N*-Carbomoyl-2-benzothiasole sulfenamide.

Properties. Beige powder. Poorly soluble in water, soluble in alcohols and vegetable oils.

Applications. Used in the production of natural and synthetic rubbers as a slow-acting vulcanization accelerator.

Acute Toxicity. In mice, LD_{50} is 5.0 g/kg BW, LD_0 appears to be 4.0 g/kg BW. Lethal doses do not affect behavior or BW gain. Gross pathology examination failed to reveal changes in morphology and weights of the visceral organs.[1]

Repeated Exposure. Rats tolerate 10 administrations of 1.0 g/kg BW. The treatment caused liver and kidney functional impairment, exophthalmos, and retardation of BW gain. There were signs of thyroid function depression.[2]

Long-term Toxicity. Rabbits received 20 mg/kg BW for 6 months. Vascular permeability BW impairment and disturbed liver and kidney functions were revealed.[2]

References:

1. Vorob'yeva, R. S., *N,N''*-Diisipropyl-2-benzothiasolyl sulfenamide, in *Toxicology of New Chemical Substances Used in Rubber and Tyre Industry*, Z. I. Israel'son, Ed., Meditsina, Moscow, 1968, 77 (in Russian).
2. Vorob'yeva, R. S., in *Toxicology of the Components of Rubber Mixes and of Rubber and Latex Articles*, Central Research Institute Petrol. Chem., Moscow, 1970, 19 (in Russian).

2-BIPHENYLOL

Molecular Formula. $C_{12}H_{10}O$
M = 170.22
CAS No 90-43-7
RTECS No DV5775000
Abbreviation. BP.

Synonyms and **Trade Name.** Dowicide 1; 2-Phenyl phenol; 2-Hydroxybiphenyl.

Properties. White, or colorless, or pinkish flaky crystals with a mild odor. Solubility in water amounts to 0.7 g/l at 25°C, soluble in alcohol.

Applications. Used in rubber industry, as a plasticizer and a coating agent. A fungicide.

Acute Toxicity. LD_{50} exceeded 1.0 g/kg BW[019] or 2.7 g/kg BW in rats, 3.5 g/kg BW in guinea pigs,[018] 2.0 g/kg BW in mice, and 500 mg/kg BW in cats.[1]

In cats, oral lethal doses induced hemorrhagic gastroenteritis and hemorrhages in the liver, lungs, and myocardium.[019] In rats, BP caused death from NS depression as does phenol.[019]

Short-term Toxicity. In a 90-day study, F344 rats received diets containing 2.0% BP. The treatment caused focal kidney lesions.[2]

Long-term Toxicity. In a 2-year feeding study, rats receiving 0.2 or 2.0 g BP/kg diet exhibited no ill effects as judged by growth, mortality rate, gross appearance, hematology, urinary sugar and protein values, organ weights, tissue content of BP, and histopathological examination of various tissues. The dose of 20 g BP/kg diet caused slight retardation of BW gain and histological kidney changes.[3]

Reproductive Toxicity.

Embryotoxicity. Doses of 600 mg/kg BW caused an increased frequency of fetal resorptions in Wistar rats given BP on days 6 to 15 of gestation.[4]

Teratogenicity. BP produced no malformations in rats and in mice, even at the maternally toxic dose of 700 mg/kg BW administered on gestation days 6 through 15.[4-6]

Allergenic Effect. Mice were administered doses up to 200 mg BP/kg BW for 10 days. The treatment caused no changes in immune function.[1]

Mutagenicity.

In vitro genotoxicity. BP caused microsomal cytochrome *P-450*-dependent redox cycling and *in vitro* genotoxicity. BP or *phenyl hydroquinone*, its hydroxylated metabolite, is able to covalently bind to DNA.[7] It induced SCE in Chinese hamster ovary cells.[1,4] BP was shown to induce significant increase in micronuclei in Chines hamster lung cells only in the presence of arachidonic acid supplementation.[8]

BP was weakly mutagenic in *Salmonella*/microsome assay,[025] but showed mutagenic activity in the mouse lymphoma test.

In vivo cytogenetics. BP induced no DLM in mice dosed daily with 100 or 500 mg BP/kg BW for 5 days. No aberrations were reported from any dose tested in 4-week old Wistar rats after oral daily doses of 4000 mg *o*-PP/kg (Shirasu et al., 1978). It produced negative response in *Dr. melanogaster*.[9]

Carcinogenicity. 2-year NTP dermal study revealed no evidence of carcinogenicity in Swiss CD-1 mice.[1] BP has been implicated in the induction or promotion of kidney and urinary bladder cancer in F344 rats.[7,8] There was no increase in tumor incidence in mice given 0.65, 1.3 or 2.6% BP in a pelleted diet for 52 weeks.[12]

Male F344 rats were gavaged with 15 to 1000 mg OPP/kg BW. The data obtained were consistent with the hypothesis that BP is an indirect acting carcinogen, and that regenerative hyperplasia due to BP-metabolite cytotoxicity and/or binding of BP metabolites to protein targets may play an important role in BP-induced bladder carcinogenesis.[14]

Carcinogenicity classifications. An IARC Working Group concluded that there is inadequate evidence for the carcinogenicity of BP in *experimental animals* and there were no data available to evaluate the carcinogenicity of BP in *humans*.

IARC: 3;

U.S. EPA: C;

NTP: XX - XX - NE - NE.

Chemobiokinetics. It has been proposed that BP metabolic activation occurs via a two-step process involving the cytochrome *P-450*-mediated formation of *phenylhydroquinone* (PGQ) in the liver and prostaglandin *H* synthesis-mediated oxidation of PGQ to *phenylbenzoquinone* in the urinary tract.

Regulations. ***U.S. FDA*** (1998) approved use BP as a constituent of food-contact adhesives which may safely be used as a component of articles intended for use in packaging, transporting, or holding food (as a preservative only) in accordance with the conditions prescribed by 21 CFR part 175.105.

Recommendations. Joint FAO/WHO Meeting on Pesticide Residues (1990). ADI: 0.02 mg/kg BW.[13]

References:

1. *Toxicology and Carcinogenesis Studies of ortho-Phenylphenol alone and with 7,12-dimethylbenz(a)antracene in Swiss CD-1 Mice (Dermal Studies)*, NTP Technical Report Series No 301, Research Triangle Park, NC, NIH Publ. 86-2557, 1986.
2. Reitz, R. H., Fox, R., Quast, J., Hermann, E., and Watanabe, P., Molecular mechanisms involved in the toxicity of orthophenylphenol and its sodium salt, *Chem.-Biol. Interact.*, 43, 99, 1983.
3. Hodge, H. C., Maynard, E. A., Blanchet, J. R., Spencer, H. C., and Rowe, V. K., *J. Pharmacol. Exp. Ther.*, 104, 202, 1952.
4. Kaneda, M., Teramoto, S., Shingu, A., and Shirasu, Y., Teratogenicity and dominant-lethal studies with *o*-phenylphenol, *J. Pesticide Sci.*, 3, 365, 1978.
5. John, J. A., Murray, F. J., Rao, K. S. and Schwetz, B. A., Teratological evaluation of ortho-phenylphenol in rats, *Fundam. Appl. Toxicol.*, 1, 282, 1981.
6. Teramoto, S. et al., Teratologic study in rats with *o*-phenylphenol, *J. Toxicol. Sci.*, 2, 86, 1987.

7. Pathak, D. N., and Roy, D., *In vivo* genotoxicity of sodium *ortho*-phenylphenol: phenylbenzoquinone is one of the DNA-binding metabolite(s) of sodium *ortho*-phenylphenol, *Mutat. Res.*, 286, 309, 1993.
8. Lambert, A. C. and Eastmond, D. A., Genotoxic effects of the *o*-phenylphenol metabolites phenylhyd- roquinone and phenylbenzoquinone in V79 cells, *Mutat. Res.*, 322, 243, 1994.
9. Woodruff, R. C., Mason, J. M., Valencia, R., Zimmerling, S., Chemical mutagenesis testing in *Drosophila*. V. Results of 53 coded compounds tested for National Toxicology Program, *Environ. Mutagen.*, 7, 677, 1985.
10. Fujii, T. et al., Effects of *pH* on the carcinogenicity of *o*-phenylphenol and sodium *o*-phenylphenate in the rat urinary bladder, *Food Chem. Toxicol.*, 25, 359, 1987.
11. Fukushima, S. et al., Pathological analysis of the carcinogenicity of sodium *o*-phenylphenate and *o*-phenylphenol, *Oncology*, 42, 401, 1985.
12. Mikuriya, H., Fujii T., Yoneyama, M., et al., Toxicity of *o*-phenylphenol by dietary administration to mice for 52 weeks, Abstract, *Ann. Rep. Tokyo Metr. Res. Lab. P. H.*, 40, 281, 1989.
13. *Pesticide Manual*, 10th ed., C. Tomlin, Ed., Crop Protect. Publ., The Royal Soc. Chem., 1995, 1341.
14. Kwok, E. S., Buchholz, B. A., Vogel, J. S., Turteltaub, K. W., and Eastmond, D. A., Dose-dependent binding of *ortho*-phenylphenol to protein but not DNA in the urinary bladder of male F344 rats, *Toxicol. Appl. Pharmacol.*, 159, 18, 1999.

1,3-BIS(2-BENZOTHIAZOLYLTHIOMETHYL)UREA

Molecular Formula. $C_{17}H_{14}N_4OS_4$
M = 418.57
CAS No 95-35-2
RTECS No YS2000000

Synonym. *N,N*-Bis-(2-benzothiasolylmercaptomethyl)urea.

Properties. A cream-colored powder, poorly soluble in water.

Applications. Used as a vulcanization accelerator for latex mixtures.

Acute Toxicity. LD_{50} was not attained in mice. The minimum lethal dose was 4.0 g/kg BW. High doses caused no changes in BW gain or in the relative weights of the visceral organs.

Long-term Toxicity. Rabbits received 20 mg/kg BW for 6 months. The treatment had no effect on behavior and general condition of the animals. Retardation of BW gain and early signs of acute hepatitis appeared to the end of the experiment. Histology examination revealed fatty dystrophy of parenchymatous cells in the liver and cloudy swelling of the renal tubular epithelium.

Reference:

Arkhangel'skaya, L. N., *N,N'*-Bis-(2-benzothiasolylmercaptomethyl)urea, in *Toxicology of New Chemical Substances Used in Rubber and Tyre Industry*, Z. I. Israel'son, Ed., Meditsina, Moscow, 1968, 46 (in Russian).

BIS(DIBUTYLDITHIOCARBAMATO)NICKEL

Molecular Formula. $C_{18}H_{36}N_2NiS_4$
M = 467.51
CAS No 13927-77-0
RTECS No QR6140000

Synonym and **Trade Name.** Dibutyldithiocarbamic acid, nickel salt; Vanguard.

Properties. Dark-green flakes.

Applications. Antioxidant for synthetic rubbers.

Acute Toxicity. LD_{50} is 17 g/kg BW in rats.[027]

Carcinogenicity. Induced tumors of respiratory system in mice orally exposed for 1.5 years. Equivocal carcinogenic agent by RTECS criteria.[1]

Standards. ***U.S. EPA*** (1999). MCL: 0.1 mg/l (*Ni*).

Reference:

1. Natl. Technical Information Service PB223-159.

BIS(DIETHYLDITHIOCARBAMATO)ZINC

Molecular Formula. $C_{10}H_{20}N_2S_2.Zn$
M = 361.90
CAS No 14324-55-1
RTECS No ZH0350000
Abbreviation. DEDTZ.

Synonyms and **Trade Names.** Carbamate EZ; Diethyldithiocarbamate, zinc salt; Diethylcarbamodithioic acid, zinc salt; Ethyl zimate; Ethyl ziram; Vulcacure ZE.

Properties. White powder. Poorly soluble in water and alcohol.

Applications. Used as a vulcanization accelerator in the production of rubber and latexes; as a stabilizer for butyl rubber, and for butadiene and urethane rubbers; heat stabilizer for polyethylene.

Migration Data. A study of aqueous extracts and simulant media (20, 40, 100°C; 1 and 5 days; 0.5 cm^{-1}) revealed migration of the accelerators introduced into the vulcanizates based on smoked-sheet natural rubber and products of their transformation (dithiocarbaminates) at levels from traces to 0.43 mg/l. Migration into water after 3-day exposure at 20°C was determined to be 0.6 mg/l. Release into distilled water from natural rubber (67 and 1.0 cm^{-1}; 20 and 38°C; 24 hours) reached 0.4 to 0.6 mg/l.[1,2,028]

Diethyldithiocarbamates were detected by GC determination at levels up to 4.6 mg/g rubber (as dithiocarbamic acid), in chloroform-acetone extracts from isoprene rubber teats for baby bottles. Dialkyldithiocarbamates can form secondary amines by acid hydrolysis, although their levels in the extracts only made a minor contribution to the total level of measured secondary amine precursors.[3]

Acute Toxicity. LD_{50} is found to be 400 mg/kg BW in rabbits,[5] 200 to 500 mg/kg BW in rats,[4] and 700 to 1400 mg/kg BW in mice.[4-6,027]

Repeated exposure. Wistar rats were given 1.0 g DEDTZ daily for 10 days. Ingestion of large doses may produce degenerative changes of neurons in various brain regions. The activity of various cerebral enzymes may be affected as well.[4]

Short-term Toxicity. Rabbits received 450 mg/kg BW for 3.5 months (daily during the first month, 3 times a week thereafter). The treatment affected metabolic processes and liver function, balance of protein fractions in the blood serum, and hematology analyses.[7]

Reproductive Toxicity. No teratogenic effect or postnatal behavioral changes were observed in rats gavaged with up to 250 mg/kg BW on days 7 to 15 of gestation. Even the maternal toxic doses did not cause increased incidence of abnormalities in newborns.[8]

Long-term Toxicity. Treatment with large doses may cause degenerative changes of neurons in various brain regions as well as alteration in the activity of some hydrolytic enzymes in the brain.[4]

Mutagenicity.

In vivo cytogenetics. DEDTZ was not found to be clearly genotoxic.[9]

In vitro genotoxicity. DEDTZ was found to be positive in *Salmonella* mutagenicity assay in the presence of *S9* metabolic activation system.[5]

Carcinogenicity. Negative results are reported in mice.[09]

Regulations. ***U.S. FDA*** (1998) approved the use of DEDTZ (1) as a component of adhesives for articles coming into contact with food in accordance with the conditions prescribed by 21 CFR part 175.105; and (2) as an accelerator in rubber articles intended for repeated use in producing, manufacturing, packing, processing, treating, packaging, transporting, or holding food (total not to exceed 1.5% by weight of the rubber product) in accordance with the conditions prescribed by 21 CFR part 177.2600.

Standards.

Russia. PML in food: 0.03 mg/l.

U.S. EPA (1999). MCL: 5.0 mg/l (*Zn*).

References:

1. Yemel'yanova, L. V. and Chernevskaya, N. M., Toxicity of some rubber materials intended for production of children's toys, *Hygiene Aspects of the Use of Polymeric Materials and of Articles Made of Them*, L. I. Medved', Ed., All-Union Research Institute of Hygiene and Toxicology of Pesticides, Polymers and Plastic Materials, Kiyv, 1969, 104 (in Russian).

2. Grushevskaya, N. Yu., Determination of vulcanization accelerator of vulkacite-P-extra-N and some transformation products in sanitary and chemical investigation of rubber, *Gig. Sanit.*, 4, 59, 1987 (in Russian).
3. Yamazaki, T., Inoue, T., Yamada, T., and Tanimura, A., Analysis of residual vulcanization accelerators in baby bottle rubber teats, *Food Addit. Contam.*,3, 145, 1986.
4. Kozik, M. B., The activity of some hydrolytic enzymes in the brain after administration of "Cynkotox", *Acta Neuropathol.* (Berlin), Suppl. 7, 56, 1981.
5. You, X. et al., Mutagenicity of fourteen rubber accelerators, *Huanjing Kexue*, 3, 39, 1982.
6. Korablev, M. V., Toxicological characteristics of dithiocarbamic acid derivatives used in the national economy and medicine, *Pharmacologia i Toxi cologia*, 32, 356, 1969 (in Russian).
7. Arkhangel'skaya, L. N., Data for toxicological characteristics of ethyl- and methylzimate, in *Toxicology and Hygiene of High-Molecular-Mass Compounds and of Chemical Raw Material Used for Their Synthesis,* Proc. 2nd *All*-Union Conf., A. A. Letavet and S. L. Danishevsky, Eds., Leningrad, 1964, 78 (in Russian).
8. Nakaura, S., Tanaka, S., Kawashima, K., et al., Effects of zinc diethyldithiocarbamate on the prenatal and postnatal development of the rat, *Bull. Natl. Inst. Hyg. Sci.*, 102, 55, 1984.
9. Tinkler, J., Gott, D., and Bootman, J., Risk assessment of dithiocarbamate accelerator residues in latex-based medical devices: genotoxicity considerations, *Food Chem. Toxicol.*, 36, 849, 1998.

BIS(DIBUTYLTHIOCARBAMOYL) DISULFIDE

Molecular Formula. $C_{18}H_{36}N_2S_4$
M = 408.80
CAS No 1634-02-2
RTECS No JO0800000
Abbreviation. BD.

Synonyms. Tetrabutylthiuram disulfide; Thiuram disulfide tetrabutyl.

Applications. Vulcanization accelerator.

Acute Toxicity. LD_{50} *i/p* exceeded 5.0 g/kg BW in mice.[027]

Regulations. ***U.S. FDA*** (1998) approved the use of BD in adhesives as a component of articles intended for use in packaging, transporting, or holding food in accordance with the conditions prescribed in 21 CFR part 175.105.

BIS(DIETHYLTHIOCARBAMOYL) DISULFIDE

Molecular Formula. $C_{10}H_{10}N_2S_4$
M = 296.52
CAS No 97-77-8
RTECS No JO1225000
Abbreviation. TETD.

Synonyms and **Trade Names.** Antabuse; Disulfiram; Ethyl dithiurame; Ethyl thiram; Tetraethyl-thiuramdisulfide; Teturam; Thiuram E; Thiuram disulfide.

Properties. White powder. Solubility in water is 2.0 mg/l (38°C), in ethanol it is 20 g/l. TETD does not affect the taste, odor, or color of water.

Applications. A vulcanization accelerator in the production of natural rubber and synthetic elastomers isobutylene-isoprene, butadiene, styrene-butadiene, isoprene, and nitrile-butadiene rubber. An activator of thiazole accelerators, plasticizer in neoprene. TETD is used to obtain colored vulcanizates.

Exposure. Used in the treatment of patients suffering from alcoholism.

Migration Data. A study of aqueous extracts and simulant media (20, 40, 100°C; 1 and 5 days; 0.5 cm^{-1}) prepared with vulcanizates based on smoked-sheet natural rubber revealed migration of TETD up to 0.15 mg/l.[1]

Acute Toxicity. LD_{50} is reported to be 8.6 or 3.0 to 3.4 g/kg BW in rats, 3.7 or 12 to 14 g/kg BW and above in mice, and 2.05 or 4.7 g/kg BW in rabbits. In guinea pigs, manifestations of toxic action appear at dose levels of 1.0 to 15 g/kg BW.[2,3,014]

TETD poisoning is accompanied by diarrhea and, in dogs, vomiting. Administration led to paralysis and, in rats, reduced thyroid activity. There was leukopenia in rabbits. Reduced oxygen consumption and prolonged chloroform narcosis time were observed in mice. Histology examination revealed inflammatory changes and hemorrhages in the GI tract mucosa, degenerative and necrotic foci in the liver, brain, spleen, suprarenals, and renal tubular epithelium.

Toxicity of TETD is known to be significantly enhanced after administration of alcohol: LD_{50} in rats is reduced by 75%.

Repeated Exposure failed to reveal cumulative properties.[3] Rats tolerated 30 administrations of 250 mg/kg BW, whereas rabbits died after six *i/g* administrations of 300 mg/kg BW and dogs from 250 to 500 mg/kg BW following 3 to 7 administrations. Oral exposure for 1.5 months affected hematology analyses, liver and CNS functions. Pathological changes were found in the kidneys, liver, thyroid, spleen, and myocardium.

Long-term Toxicity. A dose of 0.1 g/kg BW given for a year caused no mortality in dogs and rabbits. F344 rats were administered technical-grade TETD at dose levels of 300 or 600 ppm in feed. Male mice were exposed to levels of 500 or 2000 ppm, female mice received 100 or 500 ppm TETD in the feed for 108 weeks. Consumption of the diet caused retardation of BW gain. The rate of mortality was not significantly affected.[4] In a 80-week feeding study, alopecia, ataxia, and hindlimb paralyses were noted in rats. The NOAEL of 6.1 mg/kg BW was established in this study.[5]

Reproductive Toxicity effect is reported.[6]

Embryotoxicity. Pregnant CD-1 Swiss mice received 4.9 g/kg BW in corn oil by gavage on days 6 through 13 of gestation and were allowed to deliver. The treatment caused no toxic effects in the dams. Litter size, the birth weight, neonatal growth, and survival of pups were not affected.[7]

TETD caused no ***teratogenic effect*** in rats (Alexandrov, 1974).

Mutagenicity.

In vitro genotoxicity. TETD did not cause CA in Chinese hamster ovary cells (NTP-91).

Carcinogenicity. In the above described study, no tumors occurred in all animals tested.[4]

Carcinogenicity classification.

NTP: N - N - N - N (feed).

Chemobiokinetics.

Observations in man. TETD is rapidly and completely absorbed from the GI tract. Because of high solubility in lipids, TETD is accumulated in fat. A greater part is oxidized mainly in the liver. TETD is excreted predominantly in the urine as *sulfate*, partly free, and partly esterized; 13 and 6.0% of ^{14}C-labeled TETD were eliminated with feces and expired air, respectively.[8,03]

Regulations. *U.S. FDA* (1998) approved the use of TETD (1) in adhesives used as components of articles intended for use in packaging, transporting, or holding food in accordance with the conditions prescribed by 21 CFR part 175.105; and (2) as accelerator in the manufacture of rubber articles intended for repeated use in contact with food up to 1.5 % by weight of the rubber product in accordance with the conditions prescribed by 21 CFR part 177.2600.

Standards. *Russia* (1995). MAC: zero (organolept., odor).

References:

1. Grushevskaya, N. Yu., Determination of vulcanization accelerator of vulkacite-*P*-extra-*N* and some transformation products in sanitary and chemical investigation of rubber, *Gig. Sanit.*, 4, 59, 1987 (in Russian).
2. Korablev, M. V., *Pharmacology and Toxicology of Dithiocarbamic acid Derivatives*, Author's abstract of thesis, Kaunas, 1965, 28 (in Russian).
3. Vaisman, Ya. I., Zaitseva, N. V., and Mikhailov, A. V., Hygiene standardization of rubber production sewage in water bodies, *Gig. Sanit.*, 2, 17, 1973 (in Russian).
4. *Bioassay of Tetraethylthiuram Disulfide for Possible Carcinogenicity*, NTP Technical Report Series No 166, Research Triangle Park, NC, 1979.
5. Lee, C. C. and Peters, P. J., Neurotoxicity and behavioral effects of thiuram in rats, *Environ. Health Perspect.*, 17, 35, 1976.

6. Holck, H. G., Lish, P. M., Sjogren, D. W., et al., Effects of disulfiram on growth, longevity, and reproduction of the albino rat, *J. Pharmacol. Sci.*, 59, 1267, 1970.
7. Hardin, B. D., Schuler, R. L., Burg, J. R., Booth, G. M., Hazelden, K. P., MacKenzie, K. M., Piccirillo, V. J., and Smith, K. N., Evaluation of 60 chemicals in a preliminary developmental toxicity test, *Teratogen. Carcinogen. Mutagen.*, 7, 29, 1987.
8. Neiderhiser, D. H., Wych, G., and Fuller, R. K., The metabolic fate of double-labeled disulfiram, *Clin. Exp. Res.*, 7, 199, 1983.

N,N'-BIS(1,4-DIMETHYLPENTYL)-*p*-PHENYLENEDIAMINE

Molecular Formula. $C_{20}H_{36}N_2$
M = 304.58
CAS No 3081-14-9
RTECS No SS8400000

Synonym and Trade Names. Diafen DMA; *N,N'*-Di(1,4-dimethylpentyl)-*p*-phenylenediamine; Eastozone; Santoflex 77.

Properties. Dark-red, oily liquid. Poorly soluble in water, soluble in alcohol.

Applications. Antioxidant in rubber industry.

Acute Toxicity. LD_{50} is 750 mg/kg BW in rats.[1] According to other data, LD_{50} is 1.6 g/kg BW in rats, and 0.8 g/kg BW in mice.[027]

Administration of the high doses caused lethargy.

Repeated Exposure revealed moderate cumulative properties. 1/5 LD_{50} administered over a period of 10 days caused 50% animal mortality.[2]

References:

1. Izmerov, N. F., Sanotsky, I. V., and Sidorov, K. K., *Toxicometric Parameters of Industrial Toxic Chemicals under Single Exposure*, USSR/SCST Commission for UNEP, Center Int. Projects, Moscow, 1982, 59 (in Russian).
2. Stasenkova, K. P. and Kochetkova, T. A., Assessment of santoflex toxiciyt, in *Toxicology of New Industrial Chemical Substances*, N. F. Ismerov and I. V. Sanotsky, Eds., Meditsina, Moscow, Issue No 12, 1971, 132 (in Russian).

BIS(DIMETHYLTHIOCARBAMOYL) DISULFIDE

Molecular Formula. $C_{18}H_{12}N_2S_4$
M = 240.41
CAS No 137-26-8
RTECS No JO1400000
Abbreviation. TMTD.

Synonym and Trade Names. Accelerator thiuram; Bis[(dimethylamino)carbonothioyl] disulfide; α,α'-Dithiobis(dimethylthio)formamide; Methylthiram; Methylthiuram disulfide; Puralin; Tetramethylene-thiuram disulfide; Tetramethylthiocarbamoyl disulfide; Tetramethylthioperoxydicarbonic diamide; Thiosan; Thiram; Thiuram; Thiuram D; Thiuram M; Vulcafor TMTD; Vulkacit MTIC; Vulkacit thiuram.

Properties. White or yellow powder with an unpleasant odor, poorly soluble in water, soluble in hot alcohol. Does not affect the taste, odor, or color of water.

Applications. The major use is as a vulcanization accelerator in the rubber industry. Vulcanizes polyethylene. Used as a pesticide and a fungicide on seeds.

Migration from butadiene-nitrile vulcanizates into water after a 3-day exposure at 20°C was up to 0.6 mg/l.[1]

Exposure. Human exposure may occur through food and water contaminated with pesticide residues.

Acute Toxicity. LD_{50} is 0.4 to 5.4 g/kg BW in rats, 1.2 to 7.1 g/kg BW in mice, and 0.21 g/kg BW in rabbits.[2,3]

Manifestations of the toxic effect include apathy, paralysis of the extremities, cyanosis, decreased temperature, and slowed respiration. Death occurs within 2 days. Gross pathology examination revealed hemodynamic disturbances in the brain, parenchymatous organs, and GI tract, degenerative and dystrophic changes in the liver, kidneys, and heart, and hemorrhages and ulcerations of the gastric mucosa.[3,4]

TMTD increased sensitivity to alcohol and prolonged hexenal sleep.

Repeated Exposure revealed cumulative properties. K_{acc} is 2.85 (by Kagan). Leukopenia is noted after 6 to 20 administrations of a 30 mg/kg BW dose.[2]

Charles River albino mice received TMTD in the diet at the doses of 54, 108, or 201 mg/kg BW in males and 62, 118, or 241 mg/kg BW in females for 4 weeks. The treatment caused decreased BW gain and anemia in males at all dose levels. Significant reduction in food intake was noted in both sexes at all dose levels.[5]

Short-term Toxicity. TMTD is shown to be a polytropic toxicant, causing significant changes in the peripheral blood, liver, NS, and suprarenals. There are pathological changes in the stomach, pancreas, liver, spleen and other organs.[6]

In a 13-week study, CD-1 rats received 58 and 132 mg TMTD/kg BW. The treatment produced mild elevations of blood biochemical parameters indicating renal or hepatic dysfunction. Female rats fed 67 mg TMTD/kg BW developed neurotoxicity including ataxia and paralysis of the hind legs. Histological investigation found the peripheral nerve to be predominantly affected.[7]

Charles River rats received TMTD in the diet at the doses of 2.5, 25, or 50 mg/kg BW over a period of 13 weeks. BW gain and food consumption were significantly reduced, changes in clinical chemistry and hematology parameters were observed in both sexes at the dose levels of 25 mg/kg BW and 50 mg/kg BW. The NOAEL of 2.5 mg/kg BW in rats and 3.0 mg/kg BW in dogs was determined in this study (Kehoe, 1988).

The NOAEL of 2.2 to 2.3 mg/kg BW based on hematological changes noted in both sexes and even 0.84 mg/kg BW based on increased absolute liver weights and altered clinical chemistry were later established by this author.[5]

Rats were dosed by gavage with 25 mg TMTD/kg BW for 2 to 12 weeks. In the course of treatment animals exhibited a decline in pyridoxine and nicotinic acid utilization, disruption of copper metabolism and ceruloplasmin activity, and the occurrence of secondary changes in the metabolism of serotonine, catecholamines, etc.[8]

Long-term Toxicity.

Observations in man. Thyroid disturbances were reported in a group of subjects exposed occupationally to TMTD.[9]

Animal studies. CD-1 female rats were given 60 mg TMTD/kg BW for 80 weeks. There was a neurological syndrome with onset of ataxia in some animals. Some behavioral changes were observed. Gross pathology examination revealed chromatolysis of motor neurons.[7]

In a 2-year study Wistar rats were exposed to the dietary levels of 3.0, 30, and 300 ppm for 104 weeks. The 300-ppm-dosed group of rats had retarded growth and anemia. In another 2-year study, rats were given TMTD in their diet. The NOAEL of 1.2 and 1.4 mg/kg BW in males and females, respectively, based on lower erythrocytes count, *Hb* and hematocrit levels and degenerative changes of the sciatic nerve was found (Maita et al., 1991).[10]

Rats were given 100, 300, 1000, or 25000 ppm TMTD in their diet for 730 days. The LOEL of 15 mg/kg BW, and the NOEL of 5.0 mg/kg BW for neurological effects were established in this study.[1]

In a 2-year study in Wistar rats, the NOEL of 5.0 mg/kg BW was established on the basis of clinical, biochemical and pathomorphological investigations.[11]

Allergenic response to TMTD that could migrate from the rubber is suspected.[12]

Reproductive Toxicity.

Gonadotoxicity. TMTD decreased fertility and induced damage to the sperm morphology in rodents. Exposure to TMTD induced significant increase in the frequency of sperm shape abnormalities in mice in a linear dose-effect manner.[13,14] Oral administration of 132 mg/kg in the feed for 13 weeks decreased fertility in male rats. The dose of 96 mg/kg administered to females for 14 days prolonged the diestrous phase of the estrous cycle and caused a decrease in BW gain.[15]

Embryotoxicity. TMTD caused embryolethality and embryotoxicity in rats and hamsters. It produced adverse effects on reproduction of CD-1 rats when given at a dietary dose of 132 mg/kg for 13 weeks to males and 30 mg/kg BW or more for 2 weeks to females.[15]

When CD-1 rats were treated during organogenesis or in the peri- and postnatal periods, toxic changes were found in the fetuses or offspring only at levels at which dams or adults experienced significant retardation of BW gain and food consumption.[7]

Doses of 136 to 200 mg/kg BW administered by gavage during the organogenesis period (on gestation days 6 to 15) increased the mortality rate of embryos. Weights of surviving embryos were decreased. However, Swiss-Webster mice exhibited no significant developmental effects after exposure to up to 300 mg/kg BW doses administered by gavage on days 6 to 14 of gestation.[15] Subtoxic maternal doses caused embryonic resorptions as well as subcutaneous hematomas and reduction in fetal weight.[16]

Teratogenicity. Pregnant NMRI mice were given daily oral doses of 10 to 30 mg TMTD/animal on days 5 to 15 or 6 to 17 of gestation. The treatment produced an increased number of resorptions during the intermediate and late stages of organogenesis. Fetal malformations were characterized by cleft palate, micrognathia, wavy ribs, and distorted bones.[17]

Two strains of mice were treated with doses over 250 mg/kg BW on the 12th and the 13th days of gestation. The treatment caused the highest incidence of cleft palate, curved long bones, and micrognathia.[18]

Syrian hamsters were dosed with 250 mg/kg BW or more on day 7 or 8 of gestation. An increased rate of resorptions, decreased fetal weights and increased number of terata are reported.[19] With respect to reproduction and postnatal development, the NOAEL was found to be greater than 8.9 and 14 mg/kg BW in male and female rats, respectively.[20]

Mutagenicity. TMTD has been found to be genotoxic; however, despite its established genotoxicity *in vitro*, it is devoid of appreciable clastogenic and/or aneugenic activity *in vivo*.[21]

In vitro genotoxicity. Positive results were reported in bacterial reversion assays,[18,19,22-32] in tests for the induction of SCE and unscheduled DNA synthesis in cultured human lymphocytes,[33,34] and in forward mutation.[29]

TMTD proved to be a strong chromosome breaking agent in the Chinese hamster epithelial cells and Chinese hamster ovary cells in the presence of *S9* metabolic system.[30]

Positive results with both genetic endpoints were obtained in assays with human lymphocytes in the dose ranges 0.5 to 24 and 0.1 to 8 μg/ml for micronucleus and Comet assays (the single cell gel electrophoresis), respectively.[21]

In vivo cytogenetics. Data on the genotoxic potential of TMTD in mammals are controversial. Injection of 500 and 1000 mg TMTD/kg BW but not of 250 mg TMTD/kg BW increased the incidence of micronuclei in polychromatic erythrocytes in mice.[33]

TMTD was administered by repeated oral intubations to groups of male $B6C3F_1$ mice at 100, 300, and 900 mg/kg BW for 4 consecutive days, or at 300 mg/kg BW for 8 and 12 days. No significant increase of micronucleated splenocytes was observed in treated animals.[5]

TMTD was found to be positive in *Dr. melanogaster*.[31]

Carcinogenicity. No increase in tumor incidence at any sites was reported in a mouse study.[5] A dose-dependent increase in the incidence of mononuclear cell leukemia was reported in F344 rats given 0.1 or 0.05% TMTD in the feed for 104 weeks.[35] A high incidence of tumors of the nasal cavity was found in rats of both sexes given the combined but not separate treatment with TMTD (500 ppm) and sodium nitrite (2000 ppm) for 104 weeks. In addition, a 20% incidence of papilloma of the forestomach was noted in rats.[36] This experiment was not designed as a standard bioassay, since only one dose level was used and only a limited number of rats (24 of each sex) were treated. Beagle dogs were treated with 0.4, 4.0, and 40 mg/kg BW for 104 weeks. The dose of 40 mg/kg BW caused severe toxic symptoms (nausea or vomiting, salivation, occasional clonic convulsion, ophthalmologic changes); 4.0 and 40 mg/kg BW doses produced liver failure and kidney damage.[35] No neoplastic changes were found in mice given TMTD in their diet for 97 weeks.[20]

Carcinogenicity classification. An IARC Working Group concluded that there is inadequate evidence for evaluation of carcinogenicity of TMTD in *experimental animals* and in *humans*.

IARC: 3.

Chemobiokinetics. TMTD is readily absorbed in the GI tract and rapidly distributed to the blood and all organs and tissues. It is likely to be reduced by glutathione to *dithiocarbamate,* which is oxidized or converted into the corresponding *metal complex.*[3]

In 2 days following its ingestion, TMTD is found in the liver and spleen together with its metabolites: *amine salt dimethyldithiocarbamic acid* (*DDCA*) and *tetramethylthiourea*, and, in the lungs, *carbon disulfide* and the *amine salt of DDCA*. TMTD is known to form *N-nitrosodimethylamine* by reaction with nitrite in mildly acid solution. TMTD and the amine salt of DDCA are excreted from the body in the urine and feces, and as CS_2 via the lungs.[37]

Regulations. ***U.S. FDA*** (1998) approved the use of TMTD (1) in adhesives as a component of articles intended for use in packaging, transporting, or holding food in accordance with the conditions prescribed by 21 CFR part 175.105; and (2) as an accelerator for rubber articles intended for repeated use in contact with food up to 1.5% by weight of the rubber product in accordance with the conditions prescribed by 21 CFR part 177.2600.

Recommendations. ***FAO/WHO*** (1992). ADI: 0.01 mg/kg BW.

Standards.

EU (1990). Residue limits in food products: 3.0 mg/kg, but this value should be reviewed.

U.S. EPA (1989). Residue limits in some fruits and vegetables: 7.0 mg/kg.

Russia (1988). MAC and PML: 1.0 mg/l; residues in food are not permitted.

References:

1. Prokof'eva, L. G., in *Hygiene of Use and Toxicology of Pesticides and Polymeric Materials*, Coll. Sci. Proc., A. V. Pavlov, Ed., All-Union Research Institute of Hygiene and Toxicology of Pesticides, Polymers and Plastic Materials, Kiyv, Issue No 17, 1987, 153 (in Russian).
2. Korablev, M. V., *Pharmacology and Toxicology of Dithiocarbamic Acid Derivatives*, Author's abstract of thesis, Kaunas, 1965, 28 (in Russian).
3. Vorob'yeva, R. S. and Kasparov, A. A., Tetramethylthiuramdisulfide, in *Toxicology of New Chemical Substances Used in Rubber and Tyre Industry*, Z. I. Israel'son, Ed., Meditsina, Moscow, 1968, 93 (in Russian).
4. Shumskaya, N. I. and Stasenkova, K. P., *Toxicology of the Components of Rubber Mixes and of Rubber and Latex Articles*, Central Research Institute Petrol. Chem., Moscow, 1970, 5 (in Russian).
5. Takahashi, M., Kokulo, T., Furukawa, F., et al., Inhibition of spontaneous leukemia in Fischer 344 rats by tetramethylthiuram disulfide, *Gann*, 74, 810, 1983.
6. Zhilenko, V. N., Chirkova, Ye. M., Domshlak, M. G., et al. Delayed effects of tetramethyl-thiuramdisulfide under different routes of administration in the body of experimental animals, in *Basic Problems of Delayed Consequences of Exposure to Industrial Poisons*, Proc. Research Institute Occup. Diseases, A. K. Plyasunov and G. M. Pashkova, Eds., Moscow, 1976, 83 (in Russian).
7. Lee, C. C. and Peters, P. J., Neurotoxicity and behavioral effects of thiram in rats, *Environ. Health Perspect.*, 17, 35, 1976.
8. Abramova, J. I. and Friedman, S. N., Development of pathogenetic methods for diagnosis of tetramethylthiuramdisulfide intoxications, *Gig. Truda Prof. Zabol.*, 3, 45, 1973 (in Russian).
9. Kaskevich, L. M. and Bezugly, V. P., Clinical aspects of intoxication induced by tetramethylthiuram disulfide, *Vrachebnoye Delo*, 6, 128, 1973 (in Russian).
10. Maita, K., Tsuda, S., and Shirasu, Y., Chronic toxicity studies with thiram in Wistar rats and beagle dogs, *Fundam. Appl. Toxicol.*, 16, 667, 1991.
11. Knapek, R., Kobes, S., Kita, K., and Kita, I., Chronic toxicity of thiram in rats, *Z. Gesamte Hyg.*, 35, 358, 1989.
12. Rudzki, E., Ostaszewski, K., Grzywa, Z., et al., Sensitivity to some rubber additives, *Contact Dermat.*, 2, 24, 1976.
13. Zdzienicka, M., Hryniewicz, M., and Pienkowska, M., Thiram-induced sperm-head abnormalities in mice, *Mutat. Res.*, 102, 261, 1982.

14. Hemavathi, E. and Rahiman, M. A., Toxicological effects of ziram, thiram, and dithane M-45 assessed by sperm shape abnormalities in mice, *J. Toxicol. Environ. Health*, 38, 393, 1993.
15. Short, R. D., Russel, J. Q., Minor, J. I., et al., Developmental toxicity of ferric dimethyldithiocarbamate and bis(dimethylthiocarbamoyl) disulfide in rats and mice, *Toxicol. Appl. Pharmacol.*, 35, 83, 1976.
16. Vasilos, A. F., Anisimova, L. A., Todorova, E. A., and Dmitrienko, V. D., The reproductive functions of rats in acute and chronic intoxication with tetramethylthiuram disulfide, *Gig. Sanit.*, 6, 37, 1978 (in Russian).
17. Matthiaschk, G., Influence of *L*-cysteine on the teratogenicity of thiram in NMRI mice, *Arch. Toxicol.*, 30, 251, 1973.
18. Roll, R., Teratologische Untersuchungen mit Thiram (TMTD) an zwei Mausestammen, *Arch. Toxikol.*, 27, 173, 1971.
19. Robens, J. F., Teratological studies of carbaryl, diazinine, norea, disulfiran and thiram in small laboratory animals, *Toxicol. Appl. Pharmacol.*, 15, 152, 1969.
20. *Pesticide Residues in Food - 1992*, Toxicology Evaluation, Joint Meeting of the FAO/WHO Panel of Experts, Rome, 21-30 September 1992, 391.
21. Villani, P., Andreoli, C., Crebelli, R., Pacchierotti, F., Zijno, A., and Carere, A., Analysis of micronuclei and DNA single-strand breaks in mouse splenocytes and peripheral lymphocytes after oral administration of tetramethylthiuram disulfide, *Food Chem. Toxicol.*, 36, 155, 1998.
22. Paschin, Y. V. and Bakhitova, L. M., Mutagenic effects of thiuram in mammalian somatic cells, *Food Chem. Toxicol.*, 23, 373, 1985.
23. Hedenstedt, A., Rannug, U., Ramel, C., et al., Mutagenicity and metabolism studies on 12 thiuram and dithiocarbamate compounds used as accelerators in the Swedish rubber industry, *Mutat. Res.*, 68, 313, 1979.
24. Franekic, J., Bratulic, N., Pavlica, M., and Papes, D., Genotoxicity of dithiocarbamates and their metabolites, *Mutat. Res.*, 325, 65, 1994.
25. Crebelli, R., Paoletti, A., Falcone, E., Aguilina, G., Fabri, G., and Carere, A., Mutagenic studies in a tire plant: *in vivo* activity of workers' urinary concentrates and raw materials, *Brit. J. Ind. Med.*, 42, 481, 1985.
26. Moriya, M., Ohta, T., Watanabe, K., et al., Further mutagenicity studies on pesticide in bacterial reversion assays systems, *Mutat. Res.*, 116, 185, 1983
27. Rannung, A.. and Rannung, U., Enzyme inhibition as a possible mechanism of the mutagenicity of dithiocarbamic acid derivatives in *Salmonella typhimurium*, *Chem.-Biol. Interact.*, 49, 329, 1984.
28. Zdzienicka, M., Zielenska, M., Tudek, B., and Szymczyk, T., Mutagenic activity of thiram in Ames tester strains of *Salmonella typhimurium, Mutat. Res.*, 68, 9, 1979.
29. Zdzienicka, M., Zielenska, M., Trojanovska, M., Szymczyk, T., Bignami, M., and Carere, A., Microbial short-term assay with thiram *in vitro, Mutat. Res.*, 89, 1, 1981
30. Mosesso, P., Turchi, G., Cinelli, S., et al., Clastogenic effects of the dithiocarbamate fungicides thiram and ziram in Chinese hamster cell lines cultured *in vitro*, *Teratogen. Carcinogen. Mutagen.*, 14, 145, 1994.
31. Donner, M., Husgafvel-Pursiainen, K., Jenssen, D., and Rannung, A., Mutagenicity of rubber additives and curing fumes, *Scand. J. Work Environ. Health*, 9 (Suppl. 2), 27, 1983.
32. Hemavathi, E. and Rahiman, M. A., Effect of ziram, thiram and dithane M-95 on bone marrow cells of mice - assessed by micronucleous test, *Bull. Environ. Contam. Toxicol.*, 56, 190, 1996.
33. Perocco, P., Santucci, M. A., Gasperi Campani, A., and Cantelli Forti, G., Toxic and DNA-damaging activities of the fungicides mancozeb and thiram (TMTD) on human lymphocytes *in vitro, Teratogen. Carcinogen. Mutagen.*, 9, 75, 1989.
34. Pienkowska, M. and Zielenska, M., Genotoxic effects of thiram evaluated by sister-chromatid exchanges in human lymphocytes, *Mutat. Res.*, 245, 119, 1990.

35. Hasegawa, R., Takahashi, M., Furukawa, F., et al., Carcinogenicity study of tetramethylthiuram disulfide (thiram) in Fischer 344 rats, *Toxicology*, 51, 155, 1988.
36. Lijinsky, W., Induction of tumors of the nasal cavity in rats by concurrent feeding of thiram and sodium nitrite, *J. Toxicol. Environ. Health*, 13, 609, 1984.
37. Antonovich, E. A., in *Pesticide Handbook*, L. I. Medved', Ed., Urozhai, Kiyv, 1974, 222 (in Russian).

BIS(DIMETHYLTHIOCARBAMYL) SULFIDE

Molecular Formula. $C_8H_{12}N_2S_3$
M = 208.305
CAS No 97-74-5
RTECS No WQ1750000
Abbreviation. TMTS.

Synonyms and Trade Names. Dimethyldithiocarbamic acid, anhydrosulfide; Tetramethylthiuram sulfide; Thionex; Thiuram MM.

Properties. Yellow, crystalline powder. Poorly soluble in water, soluble in alcohol.

Applications. Used as a vulcanization accelerator for diene type rubber and rubber mixes. A fungicide.

Acute Toxicity. LD_{50} is 413 mg/kg BW in rats, and 820 to 1150 mg/kg BW in mice.[1,2] Lethal doses cause inhibition, diarrhea, paresis, hindlimb paralysis, and convulsions. Hexenal sleep time was prolonged. The liver seems to be predominantly affected. Dogs, cats, and rabbits tolerate the dose of 100 mg/kg BW. Guinea pigs are likely to be more sensitive.

Repeated Exposure revealed cumulative properties. 70% animals died in 10 days from the doses of 10 mg/kg to 40 mg/kg BW.[2,3] Severe leukopenia was observed in rabbits after 4 to 6 administrations of 60 to 80 mg/kg BW doses.

Short-term Toxicity. Exposure to 20 mg/kg BW for 3.5 months caused impaired liver function and changes in the lungs and blood formula. There was 50% rabbit mortality in this study.[1]

Allergenic Effect. Can produce sensitization of exposed workers in rubber industry.[4]

Carcinogenicity. Negative results are reported in mice.[09]

Regulations. ***U.S. FDA*** (1998) regulates the use of TMTS (1) in adhesives as a component of articles intended for use in packaging, transporting, or holding food in accordance with the conditions prescribed by 21 CFR part 175.105, and (2) in the manufacture of rubber articles intended for repeated use in producing, manufacturing, packaging, processing, preparing, treating, packing, transporting, or holding food in accordance with the conditions prescribed by 21 CFR part 177.2600.

References:

1. Vorob'yeva, R. S. and Mezentseva, N. V., Experimental study on comparative toxicity of new agents accelerating vulcanization, *Gig. Truda Prof. Zabol.*, 7, 28. 1962 (in Russian).
2. Korablev, M. V., *Pharmacology and Toxicology of Dithiocarbamic Acid Derivatives*, Author's abstract of thesis, Kaunas, 1965, 28 (in Russian).
3. Korablev, M. V. and Lukienko, P. I., Effect of dithiocarbamine acid derivatives and structurally similar compounds on hypoxia, *Pharmacologia i Toxicologia*, 30, 186, 1967 (in Russian).
4. Casarett and Doul's *Toxicology*, J. Doull, C. D. Klassen, and M. D. Amdur, 2nd ed., Macmillan Publ. Co., New York, 1980, 607.

BIS(*N*-ETHYLDITHIOCARBAMATO)ZINC

Molecular Formula. $C_{18}H_{20}N_2S_4Zn$
M = 457.98
CAS No 14634-93-6
RTECS No ZH0890000
Abbreviation. BEDZ.

Synonym and **Trade Names.** Accelerator *P*-extra-*N*; Carbamate EFZ; Phenylethyldithiocarbamate, zinc salt.

Properties. White, odorless powder. Poorly soluble in water and alcohol.

Applications. Used as a vulcanizing agent (ultra-accelerator) in the production of vulcanizates.

Migration of BEDZ from vulcanizates based on butadiene-nitrile rubber into aqueous extracts (0.5 cm^{-1}; 38 and 20°C; 1 hour) amounted to 0.03 mg/l.[1] Migration from vulcanizates based on synthetic polyisoprene and butadiene-nitrile rubber into water after 3-day exposure at 20°C was determined at the level of 0.08 mg/l.[2]

A study of extracts of vulcanizates based on polybutadiene rubber SKB-35-45 with natural rubber Pale Crepe revealed migration of the accelerator BEDZ at a concentration of 0.4 mg/l. The accelerator migrates from vulcanizates based on nitrile rubber into aqueous and citric acid extracts [0.5 cm^{-1}; 100, 37, and 20°C (infusion); 1 and 24 hours] at a concentration of 0.1 mg/l.[3]

Investigation of aqueous extracts (0.5 cm^{-1}; 100°C, 4 hours) from SKN-26 rubber revealed migration at the level of 0.14 mg/l. On repeated analyses, no migration was noted.[4]

Acute Toxicity. In mice, LD_{50} is reported to be 17.0 g/kg BW. Three administrations of 2.0 g/kg BW altered activity of some enzymes. Sensitivity to alcohol was increased.[5]

Short-term Toxicity. Rats were dosed by gavage with 500 mg/kg BW for 1.5 months. Manifestations of the toxic action comprised severe impairment of hematological indices and of a number of biochemical parameters in rats. Gross pathology examination revealed congestion in the visceral organs, acute enlargement of the spleen, and distension of the stomach and intestine. In a 3-month study, 10 and 100 mg/kg BW doses were administered to rats. The treatment resulted in retardation of BW gain, anemia, and an increase in serum ceruloplasmin activity. The high dose resulted in increased kidney and spleen weights.[5]

Mutagenicity.

In vitro genotoxicity. Was shown to be mutagenic in *Salmonella* mutagenicity bioassay.[6]

Standards. ***Russia.*** PML in food: 1.0 mg/l.

References:

1. Shumskaya, N. I., in *Toxicology of the Components of Rubber Mixes and of Rubber and Latex Articles*, TZNIITENEFTEKHIM, Moscow, 1977, 58 (in Russian).
2. Prokof'yeva, L. G., in *Hygiene of Use and Toxicology of Pesticides and Polymeric Materials*, Coll. Sci. Proc., A. V. Pavlov, Ed., All-Union Research Institute of Hygiene and Toxicology of Pesticides, Polymers and Plastic Materials, Kiyv, Issue No 17, 1987, 153 (in Russian).
3. Znamensky, N. N. and Peskova, A. V., Study of migration of 3,5-benzopyrene from paraffin packaging in dairy products, in *Hygiene Aspects of the Use of Polymeric Materials and of Articles Made of Them*, L. I. Medved', Ed., All-Union Research Institute of Hygiene and Toxicology of Pesticides, Polymers and Plastic Materials, Kiyv, 1969, 62 (in Russian).
4. Khoroshilova, N. V. and Ol'pinskaya, A. Z., Hygienic evaluation of resins, intended for manufacture of milking apparatus, *Gig. Sanit*, 10, 95, 1988 (in Russian).
5. Yalkut, S. I., On the toxicology of vulkacit *P*-extra-*N*, a rubber vulcanization accelerator, *Gig. Sanit.*, 6, 31, 1971 (in Russian).
6. Hedenstedt, A., Genetic health risk in the rubber industry, mutagenicity studies on rubber chemicals and process vapors, in *Int. Symp. Prevent. Occup. Cancer*, Finnish Institute Occup. Health, Helsinki, 1982.

BIS(PIPERIDINOTHIOCARBONYL) DISULFIDE

Molecular Formula. $C_{12}H_{20}N_2S_4$

M = 320.58

CAS No 94-37-1

RTECS No JO1585000

Synonyms. Dipentamethylenethiuram disulfide; 1,1'-(Di thiodicabonothioyl)bispiperidine.

Applications. Used in the manufacture of resinous and polymeric coatings.

Acute Toxicity. LD_{50} *i/p* is 3.2 g/kg BW in mice.[027]

Regulations. ***U.S. FDA*** (1998) approved the use of B. in the manufacture of resinous and polymeric coatings for the food-contact surface of articles intended for use in producing, manufacturing, packing, processing, preparing, treating, packaging, transporting, or holding food in accordance with the conditions prescribed by 21 CFR part 175.300; *dipentamethylenethiuram hexasulfide* has been approved for use in the manufacture of rubber articles intended for repeated use in producing, manufacturing, packaging, processing, preparing, treating, packing, transporting, or holding food in accordance with the conditions prescribed in 21 CFR part 177.2600.

BUTYLAMINE

Molecular Formula. $C_4H_{11}N$
M = 73.14
CAS No 109-73-9
RTECS No EO2975000
Abbreviation. BA.

Synonym. 1-Aminobutane.

Properties. Colorless, oily liquid with a sour, ammoniac odor. Mixes with water at all ratios. Soluble in alcohol. Odor perception threshold is 0.08 mg/l[010] or 6.0 mg/l; taste perception threshold is 3.5 to 4.0 mg/l.[1] According to other data, odor perception threshold is 2.2 mg/l.[2]

Applications. Used in the production of vulcanizates and synthetic fibers.

Acute Toxicity. LD_{50} in rats, mice and guinea pigs is 430 to 450 mg/kg BW.[1] According to other data, LD_{50} of BA is 370 to 380 mg/kg BW and that of isobutylamine is 230 mg/kg BW.[3] Poisoning produces weakness, ataxia, nasal discharges, dyspnea, salivation, and convulsions. ET_{50} appeared to be 34 hours. Gross pathology examination revealed local irritant effect, fatty infiltration and necrotic lesions in the liver, and pulmonary edema.

Repeated Exposure failed to reveal cumulative properties. Rats and guinea pigs tolerated administration of 1/5 LD_{50}; 1/5 to 1/20 LD_{50} reduced blood cholinesterase activity in rats. The 55 mg/kg BW dose caused an increase in the activity of glutaminate oxalate and glutaminate pyruvate transaminase and a reduction in blood prothrombin index. The dose of 9.0 mg/kg BW was ineffective in the short-term study in rats.[1]

Long-term Toxicity. Rats received 1/1000 LD_{50} for 6 months. The treatment produced no effect on conditioned reflex activity, blood cholinesterase and liver diaminooxidase activity or total serum cholesterol level. Rabbits were dosed by gavage with 0.15 and 8.5 mg/kg BW. Neither dose altered BW gain, prothrombin index, or histaminolytic function of the liver. Only serum transaminase activity was significantly increased at the highest dose level.[1]

Standards. ***Russia*** (1995. MAC and PML: 4.0 mg/l (organolept., odor).

References:

1. Trubko, E. I., Studies on hygienic standard-setting for *N*-butylamines in water bodies, *Gig. Sanit.*, 11, 21, 1975 (in Russian).
2. Rudeiko, V. A., Kuklina, M. N., Romashov, P. G., et al., Hygienic rating of tertiary butylamine in water, *Gig. Sanit.*, 12, 70, 1985 (in Russian).
3. Cheever, R. L., Richards, D. E., and Plotnik, H. B., The acute oral toxicity of isomeric butylamines in the adult male and female rat, *Toxicol. Appl. Pharmacol.*, 63, 150, 1982.

N-tert-BUTYL-2-BENZOTHIAZOLE SULFENAMIDE

Molecular Formula. $C_{11}H_{14}N_2S_2$
M = 238.39
CAS No 95-31-8
RTECS No DL6200000
Abbreviation. BBS.

Synonym and **Trade Names.** *N*-(1,1-Dimethylethyl)benzothiazole sulfenamide; Santocure NS; Vulkacid NZ.

Applications. BBS is used as a rubber accelerator.

Properties. Light buff powder or flakes, sometimes colored blue. Soluble in most organic solvents.

Acute Toxicity. LD_{50} is reported to be about 8.0 g/kg BW in rats. Poisoning is accompanied with CNS depression, a decrease in food intake, and muscle weakness.[1] LD_{50} *i/p* is found to be 5.0 to 6.0 g/kg BW in mice.[027]

Mutagenic activity has been evaluated in a battery of *in vivo* assays.[2]

Regulations. ***U.S. FDA*** (1998) approved the use of BBS in the manufacture of rubber articles intended for repeated use in producing, manufacturing, packaging, preparing, treating, packing, transporting, or holding food in accordance with the conditions prescribed by 21 CFR part 177.2600.

Standards. ***Russia*** (1994). MAC for butylcaptax: 0.005 mg/l (organolept., odor).

References:

1. Acute Toxicity Data, *J. Am. Coll. Toxicol.*, Part B, 1, 104, 1990.
2. Hinderer, R. K., Myhr, B., Jagannath, D. R., Galloway, S. M., Mann, S. W., Riddle, J. C., and Brusick, D. J., Mutagenic evaluation of four rubber accelerators in a battery of *in vivo* mutagenic assay, *Environ. Mutagen.*, 5, 193, 1983.

N-BUTYL-*N*-NITROSO-1-BUTANAMINE

Molecular Formula. $C_8H_{18}N_2O$
M = 158.28
CAS No 924-16-3
RTECS No EJ4025000
Abbreviation. NDBA.

Synonyms. Dibutylnitrosamine; *N*-Nitrosodi-*N*-butylamine.

Properties. Yellow oil. Solubility in water is 0.12%, soluble in organic solvents and in vegetable oil.[020]

Exposure occurs during contact with food products, cigarette smoke, and alcoholic beverages. N. are also by-products of endogenous processes.

Migration Data. A survey of 30 samples of various nipples and pacifiers that was carried out by analysis by GLC-thermal energy analyzer, indicated the presence of *N-nitrosamines*, including NDBA (up to 2796 mg/kg). These *N*-nitrosamines were shown to migrate easily from the rubber products to liquid infant formula, orange juice, and simulated human saliva.[1]

All of the 17 samples of rubber nipples and pacifiers tested using GC-thermal energy analysis after extraction with artificial saliva were found to contain at least two of the following: *N*-nitrosodimethylamine, *N*-nitrosodiethylamine, NDBA and *N*-nitrosopiperidine. Total volatile *N*-nitrosamine levels with a mean content of 7.3 mg/kg were found. Nitrosatable compounds, measured as *N*-nitrosamines after nitrosation, were detected in 15 of the 17 samples, the mean level being 5.0 mg/kg.[2]

Volatile *N*-nitrosamines were determined in cut-up pacifier nipples by analysis by GC-thermal energy analysis. *N*-Nitrosamines and precursors in cut-up and intact nipples were determined after a single extraction with artificial saliva. NDBA was the principal nitrosamine found, at levels up to 1.0 ppm, while dibutylamine was the principal precursor found, at levels up to 3.9 ppm. Amounts of NDBA and DBA found after 15 artificial saliva extractions of intact pacifier nipples totaled 0.8 and 15.6 ppm, respectively. *N*-Nitrosamine levels generally showed a gradual decrease in concentration with each extraction.[3]

Significant penetration of nitrosoamines into boneless hams as a result of the use of elastic nettings containing rubber were noted by Pensabene et al.[4] NDBA was detected at the level of 0.3 ppb at a depth of 1.25 in. Penetration of NDBA from surface into inner parts of the ham is slow but significant.

Migration of *N*-nitrosamines and their amine precursors into artificial saliva from 16 types of children's pacifiers and baby-bottle nipples, bought in Israel, has been investigated. NDBA, *N*-nitrosodiethylamine (NDEA), *N*-nitrosodimethylamine (NDMA), *N*-nitrosopiperidine (NPIP), and *N*-nitrosopyrrolidine

(NPYR) were detected by one method involving dichloromethane extraction at individual levels as high as 369 ppb. Using the second method, which consisted of analysis of *N*-nitrosamines and their amine precursors, NDBA, NDEA, NDMA, and *N*-nitrosomorpholine were detected at concentrations up to 41 ppb, in addition to the three nitrosatable amines dibutylamine, diethylamine, and dimethylamine. Of the samples tested, 50% failed to meet both the U.S. and Germany regulations. A larger percentage, 60%, would not conform to the standard suggested in the U.S., and more than 80% failed to comply with the even stricter Dutch standard.[5]

Immunotoxicity. Nitrosamines are environmental chemicals that alkylate macromolecules in various target tissues and produce cellular necrosis and immunosuppression.[6]

Reproductive Toxicity. Embryotoxic effect (stunted fetuses) was reported.[7]

Mutagenicity.

In vivo cytogenetics. No micronucleated polychromatic erythrocyte induction was observed in male BDF_1 mice after a single *i/p* treatment with up to 600 mg/kg BW (a half of LD_{50} dose).[8] NDBA was administered by gavage to F344 rats at concentrations of 1.0 and 2.0 mmol for 30 weeks. 80% of the rats given 2.0 mmol survived until week 83. Approximately 60% of the animals developed liver carcinomas, 50% forestomach carcinomas, and 35% transitional cell carcinomas of the urinary bladder.[9]

NDBA was positive in *Dr. melanogaster.*

In vitro genotoxicity. NDBA was not found to be mutagenic to *Salmonella typhimurium* strain TA100 and *E. coli* in presence of rat liver *S9* mix or hamster liver *S9* mix.[10] It caused CA in cultured mammalian cells. NDBA made dose- and time-related increases in the number of single-strand breaks in rat hepatocytes.[11]

Carcinogenicity. Statistically significant changes of DNA viscometric parameters, which are considered indicative of DNA fragmentation, were produced in rats treated with single oral doses of 0.083 mg NDBA/kg BW.[12]

NDBA caused a significant increase in incidence of several tumor types in rats, mice, and hamsters exposed by various routes. *S/c* treatment of European hamsters with NDBA led to the development of respiratory tract tumors (predominantly, in the nasal cavity and lungs). Histologically, these tumors appeared to be papillary polyps, papillomas, adenomas, adenocarcinomas, squamous cell carcinomas, and mixed carcinomas.[13]

NDBA produced bladder rather than liver tumors following *s/c* injections to rats. BD rats received 10 to 75 mg NDBA/kg BW in their diet for lifetime. No control data were reported in this study. Liver tumors were found in all four of the surviving animals treated with the highest dose. A high incidence of liver, esophageal, and bladder tumor development was observed in other groups of treated animals.[14]

Squamous-cell carcinomas of the bladder, carcinomas and papillomas of the esophagus were observed in C57B16 mice given drinking water with 60 or 240 mg NDBA/l for their lifetime. The treatment solution was replaced by water for 50% of high-dosed animals, as these mice showed hematuria.[15]

Carcinogenicity classifications. An IARC Working Group concluded that there is sufficient evidence for the carcinogenicity of NDBA in *experimental animals* and there were no data available to evaluate the carcinogenicity of NDBA in *humans*.

The Annual Report of Carcinogens issued by the U.S. Department of Health and Human Services (1998) defines NDBA to be a substance which may reasonably be anticipated to be carcinogen, i.e., a substance for which there is a limited evidence of carcinogenicity in *humans* or sufficient evidence of carcinogenicity in *experimental animals*.

IARC: 2B;

U.S. EPA: B2.

Chemobiokinetics. NDBA was extensively metabolized in the rat, no unchanged NDBA being found in the urine. DNA underwent metabolic transformation in at least three ways. The major pathways demonstrated on the basis of urinary metabolites were ω- and (ω-1)-oxidations of one butyl chain to give *N-butyl-N-(3-carboxypropyl)nitrosamine* (BCPN) and *N-butyl-N-(3-hydroxybutyl)nitrosamine*, respectively. The third minor pathway was (ω-2)-oxidation of the butyl chain to afford *N-butyl-N-(2-hydroxybutyl)* *nitrosamine*. Both hydroxylated metabolites were excreted into the urine as such and as their glucuronic acid conjugates. The ω-oxidation of DBN to BCPN is responsible for the induction of bladder tumors in rats, while the products of the (ω-1)- or (ω-2)-oxidation may be involved in tumor induction in the liver.[16]

Regulations. ***U.S. FDA*** (1998) has set a 10-ppb limit on nitrosoamines in rubber nipples for baby bottles.

Standards. *EPA, State of Minnesota*. MCL: 0.06 μg/l.

References:

1. Sen, N. P., Seaman, S., Clarkson, S., Garrod, F., and Lalonde, P., Volatile *N*-nitrosamines in baby bottle rubber nipples and pacifiers. Analysis, occurrence and migration, *Sci. Publ.,* 57, 51, 1984.
2. Osterdahl, B. G., N-nitrosamines and nitrosatable compounds in rubber nipples and pacifiers, *Food Chem. Toxicol.*, 21, 755, 1983.
3. Thompson, H. C., Jr., Billedeau, S. M., Miller, B. J., Hansen, E. B., Jr., Freeman, J. P., and Wind, M. L., Determination of *N*-nitrosamines and *N*-nitrosamine precursors in rubber nipples from baby pacifiers by gas chromatography-thermal energy analysis, *J. Toxicol. Environ. Health,* 13, 615, 1984.
4. Pensabene, J. W., Fiddler, W., and Gates, R. A., Nitrosamine formation and penetration in hams processed in elastic rubber nettings: *N*-nitrosodibutylamine and *N*-nitrosodibenzylamine, *J. Agric. Food. Chem.*, 43, 1919, 1995.
5. Westin, J. B., Castegnaro, M. J., and Friesen, M. D., *N*-nitrosamines and nitrosatable amines, potential precursors of *N*-nitramines, in children's pacifiers and baby-bottle nipples, *Environ. Res.*, 43, 126, 1987.
6. Haggerty, H. G. and Holsapple, M. P., Role of metabolism in dimethylnitrosamine-induced immunosuppression: A review, *Toxicology*, 63, 1, 1990.
7. *Bull. Exp. Biol. Med.*, 78, 1308, 1974 (in Russian)
8. Morita, T., Asano, N., Awogi, T., Sasaki, Yu. F., et al. Evaluation of the rodent micronucleus assay in the screening of IARC carcinogens (Groups 1, 2A and 2B). The summary report of the 6th collaborative study by CSGMT/JEMS-MMS, *Mutat. Res*., 389, 3, 1997.
9. Lijinsky, W. and Reuber, M. D., Carcinogenesis in Fischer rats by nitrosodipropylamine, nitrosodibutylamine and nitrosobis(2-oxopropyl)amine given by gavage, *Cancer. Lett.,* 19, 207, 1983.
10. Araki, A., Muramatsu, M., and Matsushima, T., Comparison of mutagenicities of *N*-nitrosamines on *Salmonella typhimurium TA100* and *Escherichia coli WP2 uvrA/pKM101* using rat and hamster liver *S9*, *Gann*, 75, 8, 1984.
11. Bradley, M. O., Dysart, G., Fitzsimmons, K., Harbach, P., Lewin, J., and Wolf, G., Measurements by filter elution of DNA single- and double-strand breaks in rat hepatocytes: effects of nitrosamines and gamma-irradiation, *Cancer Res.,* 42, 2592, 1982.
12. Brambilla, G., Carlo, P., Finollo, R., and Sciaba L., Dose-response curves for liver DNA fragmentation induced in rats by sixteen *N*-nitroso compounds as measured by viscometric and alkaline elution analyses, *Cancer Res.*, 47, 3485, 1987.
13. Reznik, G., Experimental carcinogenesis in the respiratory tract using the European field hamster (Cricetus cricetus L.) as a model, *Fortschr.-Med.,* 95, 2627, 1977.
14. Druckrey, H., Preussmann, R., Ivankovic, S., and Schmael, D., Organotropism and carcinogenic activities of 65 different *N*-nitrosodi compounds on BD-rats, *Z. Krebsforsch.*, 69, 103, 1967.
15. Bertram, J. S. and Craig, A. W., Induction of bladder tumors in mice with dibutylnitrosamine, *Brit. J. Cancer*, 24, 352, 1970.
16. Suzuki, E. and Okada, M., Metabolic fate of *N,N*-dibutylnitrosamine in the rat, *Gann,* 71, 863, 1980.

*p-(tert-*BUTYL)PHENOL

Molecular Formula. $C_{10}H_{14}O$
M = 150.24
CAS No 98-54-4
RTECS No SJ8925000

Abbreviation. PTBP.

Synonyms. Butylphen; 4-(1,1'-Dimethylethyl)phenol; 1-Hydroxy-4-*tert*-butylbenzene.

Properties. White solid with disinfectant-like aromatic odor. May float or sink in water.

Applications. Used in the synthetic rubber manufacture.

Acute Toxicity. LD_{50} is 2.95 g/kg BW in rats.[1]

Carcinogenicity. PTBP induced tumors in GI tract.[2] Neoplastic by *RTECS* criteria.

Regulations.

EU (1990). PTBP is available in the *List of authorized monomers and other starting substances which may continue to be used for the manufacture of plastic materials and articles intended to come into contact with foodstuffs pending a decision on inclusion in Section A* (*Section B*).

U.S. FDA (1998) listed PTBP for use (1) in the manufacture of resinous and polymeric coatings for the food-contact surface of articles intended for use in producing, manufacturing, packing, processing, preparing, treating, packaging, transporting, or holding food in accordance with the conditions prescribed by 21 CFR part 175.300, and (2) in the manufacture of phenolic resins for food-contact surface of molded articles intended for repeated use in contact with nonacid food (*pH* above 5.0) in accordance with the conditions prescribed by 21 CFR part 177.241.

Great Britain (1998). PTBP is authorized up to the end of 2001 for use in the production of polymeric materials and articles in contact with food or drink or intended for such contact.

Standards. ***Russia*** (1995). MAC: 1.0 mg/l.

References:

1. Smyth, H. F., Carpenter, C. P., Weil, C. S., Pozzani, U. C., Striegel, J. A., and Nycum, J. S., Range-finding toxicity data: List VII, *Am. Ind. Hyg. Assoc. J.,* 30, 470, 1969.
2. Hirose, M., Inoue, T., Asamoto, M., Tagawa, Y., and Ito, N., Comparison of the effects of 13 phenolic compounds in induction of proliferative lesions of the forestomach and increase in the labeling indices of the glandular stomach and urinary bladder epithelium of Syrian golden hamsters, *Carcinogenesis*, 7, 1285, 1986.

p-CHLOROANILINE

Molecular Formula. C_6H_6ClN

M = 127.58

CAS No 106-47-8

RTECS No BXO700000

Abbreviation. PCA.

Synonyms. 1-Amino-4-chlorobenzene; 4-Chlorobenzenamine.

Properties. Orthorhombic crystals (from alcohol or petroleum ether). Soluble in hot water, freely soluble in organic solvents.

Applications. Widely used in the dye and rubber industries.

Acute Toxicity. LD_{50} is reported to be 310 mg/kg BW in rats,[1] 100 mg/kg BW in mice, and 350 mg/kg BW in guinea pigs.[06]

Repeated Exposure led to cyanosis. *MetHb* formation occurs with or without loss of *Hb*.

Short-term Toxicity. Hematopoietic system appeared to be the target of PCA toxicity. Exposure of F344 rats (5.0 to 80 mg/kg BW) and B6C3F$_1$ mice (7.5 to 120 mg/kg BW) to PCA hydrochloride given in water by gavage for 13 weeks failed to reveal treatment-related effect on organ weights (except spleen). Dose-related secondary anemia due to *metHb* formation and increased hematopoiesis in the liver and spleen, as well as in the bone marrow (in rats but not in mice) were noted. Hemosiderin was found in the kidney, spleen, and liver.[2]

Long-term Toxicity. Aromatic amines produce reversible anemia.[3]

Mutagenicity.

In vitro genotoxicity. PCA produced gene mutations, SCE, and CA in Chinese hamster ovary cells *in vitro*.[4] *m*- and *o*-chloroaniline were found to be negative in *Salmonella* mutagenicity assay (NTP-94).

Carcinogenicity. $B6C3F_1$ mice and F344 rats were administered PCA in the diet and by gavage. Mice were given 2500 and 5000 ppm, rats received 250 and 500 ppm in their diet for 78 weeks. Retardation of BW gain was noted in the treated animals. PCA produced hemangiosarcomas in male and female mice.[5]

$B6C3F_1$ mice were given 3 to 30 mg PCA/kg BW, F344 rats were dosed with 2 to 18 mg PCA/kg BW by gavage in aqueous hydrochloric acid for 103 weeks. Hemangiosarcomas of the spleen and liver and hepatocellular adenomas and carcinomas in male mice are reported.[6,7] Sarcomas of the spleen and splenic capsule were found in male rats in both studies.

Carcinogenicity classifications. An IARC Working Group concluded that there is sufficient evidence for the carcinogenicity of PCA in *experimental animals* and there is inadequate evidence for the carcinogenicity of PCA in *humans*.

IARC: 2B;

NTP: E - N - E* - E (PCA, feed);

CE - EE - SE - NE (PCA hydrochloride, gavage).

Chemobiokinetics. *MetHb* formation in erythrocytes results from conversion of hem iron from the ferrous to *ferric state*.[2]

Standards. ***Russia*** (1995). MAC: 0.2 mg/l.

References:

1. Smyth, H. F., Carpenter, C. P., Weil, C. S., et al., Range-finding toxicity data, List VI, *Am. Ind. Hyg. Assoc. J.*, 23, 95, 1962.
2. Chhabra, R. S., Thompson, M., Elwell, M. R., et al., Toxicity of *p*-chloroaniline in rats and mice, *Food Chem. Toxicol.*, 28, 717, 1990.
3. Linch, A. L., Biological monitoring for industrial exposure to cyanogenic aromatic nitro and amino compounds, *Am. Ind. Hyg. Assoc. J.*, 7, 426, 1974.
4. Anderson, B. E., Zeiger, E., Shelby, M. D., et al., Chromosome aberration and sister chromatid exchange test results with 42 chemicals, *Environ. Molec. Mutagen.*, 16 (Suppl. 18), 55, 1990.
5. *Bioassay of p-Chloroaniline for Possible Carcinogenicity*, Natl. Cancer Institute, Technical Report Series No 189; DHEW Publ. No 79-1745, Bethesda, MD, 1989.
6. *Toxicology and Carcinogenesis Studies of p-Chloroaniline Hydrochloride in F344/N Rats and $B6C3F_1$ Mice (Gavage Studies)*, NTP Technical Report Series No 351, NIH Publ. No 89-2806, Research Triangle Park, NC, July 1989.
7. Chhabra, R. S., Huff, J. E., Haseman, J. K., et al., Carcinogenicity of *p*-chloroaniline in rats and mice, *Food Chem. Toxicol.*, 29, 119, 1991.

3-CHLORO-1,1,1-TRIFLUOROPROPANE

Molecular Formula. $C_3H_4F_3Cl$

M = 129.88

CAS No 460-35-5

RTECS No TX6200000

Synonym and **Trade Name.** Freon 253; 1,1,1-Trifluoro-3-chloropropane.

Properties. Colorless, volatile liquid with an aromatic odor. Solubility in water is 1.33 g/l (20°C). Odor perception threshold is more than 50 mg/l. At this concentration there is no color or taste.

Applications. Used in the production of fluorinated rubber.

Acute Toxicity. LD_{50} is 62 mg/kg BW for mice. Lethal doses cause excitation, motor and coordination disorder with subsequent CNS inhibition, lethargy and death.[1]

Long-term Toxicity. In a 7-month study, rats and rabbits were given the doses up to 0.5 mg C./kg BW. Toxic effect manifestations included increased blood pyruvic acid level, changes in glycemic curves and conditioned reflex activity.[1]

Chemobiokinetics. When inhalation of C. is over, it rapidly disappears from the blood and is further found in the liver, kidneys, and bone marrow. The content of fluorine ion in the visceral organs is increased. Prolonged excretion of a fluorine-containing metabolite occurs via the urine.[2]

Standards. ***Russia.*** MAC and PML: 0.1 mg/l.

References:

1. Selyuzhitsky, G. V., in *Industrial Pollution of Water Reservoirs*, S. N. Cherkinsky, Ed., Meditsina, Moscow, Issue No 8, 1967, 112 (in Russian).
2. Mikhaleva, A. L., *Data to Investigation of Some Fuorochloropropanes*, Author's abstract of thesis, Leningrad, 1967, 14 (in Russian).

CROTONONITRILE

Molecular Formula. C_4H_5N
M = 67.10
CAS No 4786-20-3
RTECS No GQ6322000

Synonyms and **Trade Name.** Allylnitrile; 2-Butenenitrile; Crotonic nitrile; 1-Cyanopropene.

Properties. Colorless liquid with a specific odor. Solubility in water is 4.0 to 5.0%. Odor perception threshold is 7.7 mg/l. Taste may be detected at higher concentrations.

Applications. Used in the production of synthetic rubbers.

Acute Toxicity. Toxicity of C. is associated with its hydrolysis to cyanide ions. The LD_{50} values are reported to be 500 mg/kg BW in rats, 396 mg/kg BW in mice, and 272 mg/kg BW in guinea pigs.[1]

According to Farooqui et al., LD_{50} is 115 mg/kg BW in rats.[2] Signs of intoxication included ataxia, convulsions, mild diarrhea, salivation, lacrimation, and bladder urine retention.

Repeated Exposure failed to reveal cumulative properties. 1/5 and 1/10 LD_{50} given to rats for 1.5 months and to guinea pigs for 2.5 months resulted in increased oxygen consumption, changes in total urinary nitrogen, reduced hippuric acid synthesis, shortening of prothrombin time, increased glutamine transaminase activity, and changes in ECG and weights of the visceral organs to the end of the study. Gross pathology examination revealed circulatory disturbances and dystrophic changes in the visceral organs and NS.[1]

Long-term Toxicity. The treatment affected cholinesterase activity and caused dystrophic changes in the visceral organs of rats.[1]

Chemobiokinetics. The similarity of the acute poisoning picture to that observed after inorganic cyanide poisoning is indirect evidence of detachment of *CN*-groups in the body. A very small portion of cyanide ions is excreted with the urine as *rhodanide ions*. On single oral administration of 57.5 mg radiolabelled C./kg BW, C. was rapidly absorbed through the GI tract and distributed in all the tissues in the Sprague-Dawley rats. The major amounts of radioactivity were measured in bone, kidney, blood, and GI tract. C. is predominantly eliminated with the urine and expired air.[2]

Standards. ***Russia*** (1995). MAC and PML: 0.1 mg/l.

References:

1. Loskutov, N. F. and Piten'ko, N. N., Peculiarities of toxic action and safe levels of crotononitrile and isocrotononitrile in water bodies, *Gig. Sanit.*, 4, 10, 1972 (in Russian).
2. Farooqui, M. Y., Ybarra, B., and Piper, J., Toxicokinetics of allylnitrile in rats, *Drug Metab. Dispos.*, 21, 460, 1933.

1,5,9-CYCLODODECATRIENE

Molecular Formula. $C_{12}H_{20}$
M = 162.26
CAS No 4904-61-4
RTECS No GU2310000

Synonym. A trimer of 1,3-butadiene.

Properties. The technical grade product is a clear, highly mobile liquid with a nauseating odor. Solubility in water is 270 mg/l.

Applications. Used in the production of polybutadiene rubber.

Acute Toxicity. In rats, LD_{50} is 4.5 g/kg BW. All animals tolerated a single administration of 2.0 g/kg BW but all died with 7.0 g/kg BW. Poisoning caused clonic convulsions. At 3.0 g/kg BW, leukocytosis and thrombocytosis developed, catalase and cholinesterase activity increased, STI rose and ECG was altered.

Reference:

Taradin, Ya. I., Pulyakhin, G. T., and Fetisova, L. N., Toxicological characteristics of divinyl oligomers - vinylcyclohexene and cyclododecatriene, in *Hygiene and Toxicology of High-Molecular-Mass Compounds and of the Chemical Raw Material Used for Their Synthesis*, Proc. 4th All-Union Conf., S. L. Danishevsky, Ed., Khimiya, Leningrad, 1969, 235 (in Russian).

4-CYCLOHEXENE-1,2-DICARBOXYLIC ANHYDRIDE

Molecular Formula. $C_8H_8O_3$
M = 152.35
CAS No 85-43-8
RTECS No QW5775000
Abbreviation. CDA.

Synonyms. 4-Cyclohexene-1,2-dicarboxylic acid, anhydride; Maleic anhydride, adduct of butadiene; 1,2,3,6-Tetrahydrophthalic anhydride.

Properties. White crystals. Poorly soluble in water.

Applications. Used as an anti-scorching agent in the rubber industry.

Acute Toxicity. In mice, LD_{50} is reported to be 2.7 g/kg BW.

Short-term Toxicity. Rabbits were given 20 mg/kg BW, initially every other day and then daily (for two months in each case). The treatment affected the protein-forming function of the liver, caused hyperholesteremia, a reduction in alkaline phosphatase and aldolase activity, and leukocytosis. Gross pathology examination revealed stimulation of the reticuloendothelial system.

Regulations. ***U.S. FDA*** (1998) approved the use of CDA in the manufacture of 4,4'-isopropylidene diphenol epichlorohydrin thermosetting epoxy resins (finished articles containing the resins shall be thoroughly cleansed prior to their use in contact with food).

Reference:

Mezentseva, N. V., Tetrahydrophthalic anhydride, in *Toxicology of New Chemical Substances Used in Rubber and Tyre Industry*, Z. I. Israel'son, Ed., Meditsina, Moscow, 1968, 164 (in Russian).

N-CYCLOHEXYL-2-BENZOTHIAZOLE SULPHENAMIDE

Molecular Formula. $C_{13}H_{15}N_2S_2$
M = 264.40
CAS No 95-33-0
RTECS No DL6250000
Abbreviation. CBS.

Synonym and **Trade Names.** Accelerator CS; 2-(Cyclohexylaminothio)benzothiazole; Santocure; Sulfenamide TS; Vulkacit CZ.

Properties. Odorless powder with a color ranging from beige to light-green. Poorly soluble in water, slightly soluble in alcohol.

Applications. Used as a vulcanization accelerator in rubber industry (for mixtures based on natural and synthetic rubber).

Migration Data. Investigation of aqueous extracts (0.5 cm^{-1}; 38 and 20°C; 1 hour) revealed the migration of CBS from vulcanizates based on butadiene-nitrile rubber at concentrations of 0.03 to 0.05 mg/l.[1] According to other data, migration into distilled water and 3.0% lactic acid solution (0.5 cm^{-1}; 38 to 40°C, then 28°C; 1 hour) was found to be at the concentration of 0.1 to 0.3 mg/l.[2]

Acute Toxicity. Mice tolerate a dose of 4.0 g/kg BW;[2] 50% of rats but none of mice died after two administrations of 5.0 g/kg BW. Gross pathology examination revealed acute circulatory disturbances, marked congestion, and parenchymal dystrophy of the visceral organs. There was also fine-drop adipose dystrophy of the liver.[3]

LD_{50} is 5.3 g/kg BW in rats. Poisoning caused CNS depression and change in food consumption.[4] According to other data, LD_{50} exceeded 7.5 g/kg BW in rats.[027]

Short-term Toxicity. Rabbits were given 20 mg/kg BW for 3.5 months, initially every other day and then daily. The treatment produced no effect on BW gain, blood morphology, and relative weights of the visceral organs. Histology examination revealed changes in the liver cell cytology and round cell infiltration along the intertrabecular triads.[5]

Allergenic Effect was shown to develop.[5]

Reproductive Toxicity.

Wistar rats were given 0.001 to 0.5% CBS in the diet on days 0 to 20 of pregnancy. There was no evidence of ***teratogenic effect***, death, or clinical signs of toxicity in animals at any dose level.[6] According to other data, CBS produced death and malformations in the developing chick embryo.[7]

Embryotoxicity. In the above-cited study,[6] there was no increase in the incidences of pre- and post-implantation losses and the number and ratio of live fetuses. An embryonic mortality increased when the dose of 2.0 g CBS/kg BW was administered orally to female rats before the beginning of pregnancy on the 1st or 3rd days of estrus or on the 4th and 11th days of pregnancy.[8]

Sitarek et al. (1996) administered 50, 150, or 450 mg CBS/kg BW to female rats by gavage during organogenesis. Maternal toxicity was noted at the highest dose. The treatment induced fetotoxic effect at all doses. The dose-dependent increase in the frequency of fetuses/litters with internal hydrocephalus was observed.[9]

Carcinogenicity. There was no significant increase in tumor incidence in mice receiving CBS at a dietary level of 692 ppm for approximately 18 months.[10] Equivocal tumorigenic agent by RTECS criteria.

Chemobiokinetics. Rats were given a single oral dose of 250 mg CBS/kg BW. About 65 and 24% of the dose were eliminated into urine and feces, respectively, for 3 days after administration. Biliary excretion appeared to be insignificant. No specific organ affinity was observed in distribution study. Metabolites identified were *cyclohexylamine* and *2-mercaptobenzothiazole*.[11]

Regulations. ***U.S. FDA*** (1998) approved the use of CBS in the manufacture of rubber articles intended for repeated use in producing, manufacturing, packaging, processing, preparing, treating, packing, transporting, or holding food in accordance with the conditions prescribed by 21 CFR part 177.2600.

Standards. ***Russia*** (1994). PML in food: 0.15 mg/l.

References:

1. Shumskaya, N. I., in *Toxicology of the Components of Rubber Mixes and of Rubber and Latex Articles*, TSNIITENEFTEKHIM, Moscow, 1977, 58 (in Russian).
2. Khoroshilova, N. V. and Ol'pinskaya, A. Z., Hygienic evaluation of resins, intended for the manufacture of milking apparatus, *Gig. Sanit*, 10, 95, 1988 (in Russian).
3. Zayeva, G. N., Fedorova, V. I., Fedechkina, N. I., Evaluation of toxicity of sulphenamide BT, M & C, in *Toxicology of New Industrial Chemical Substances*, A. A. Letavet and I. V. Sanotsky, Eds., Meditsina, Moscow, Issue No 8, 1966, 70 (in Russian).
4. Acute Toxicity Data, *J. Am. Coll. Toxicol.*, Part B, 1, 105, 1990.
5. Vorob'yeva, R. S. and Mesentseva, N. V., Sulfenamide C and M, in *Toxicology of New Chemical Substances Used in Rubber and Tyre Industry*, Z. I. Israel'son, Ed., Meditsina, Moscow, 1968, 64 (in Russian).
6. Ema, M., Murai,T., Itami, T., Kawasaki, H. and Kanoh, S., Evaluation of the teratogenic potential of the rubber accelerator *N*-cyclohexyl-2-benzothiazyl sulfenamide in rats, *J. Appl. Toxicol.*, 9, 187, 1989.
7. Korhonen, A., Hemminki, K., and Vainio, H., Toxicity of rubber chemicals toward 3-day chicken embryos, *Scand. Work Environ. Health*, 9, 115, 1983.
8. Alexandrov, S. E., Effect of vulcanization accelerators on embryonic mortality in rats, *Bull. Exp. Biol. Med.*, 93, 107, 1982 (in Russian).
9. Sitarek, K., Berlinska, B., and Baranski, B., Effect of oral sulfenamide TS administration on prenatal development in rats, *Teratogen., Carcinogen., Mutagen.*, 16, 1, 1996.
10. Innes, J. R. M., Ulland, B. M., Valerio, M. G., et al., Bioassay of pesticides and industrial chemicals for tumorigenicity in mice: A preliminary note, *J. Natl. Cancer Inst.*, 42, 1101, 1969.

11. Adachi, T., Tanaka, A. and Yamaha, T., Absorption, distribution, metabolism and excretion of *N*-cyclohexyl-2-benzothiazyl sulfenamide (CBS), a vulcanizing accelerator, in rats, Abstract, *Radioisotopes*, 38, 255, 1989 (in Japanese).

DIBENZYLAMINE

Molecular Formula. $C_{14}H_{15}N$
M = 197.27
CAS No 103-49-1
Abbreviation. DBA.

Synonym. *N*-(Phenylmethyl)benzenemethanamine.

Applications. Used in the production of rubber materials.

Properties. Oil with ammonia odor. Practically insoluble in water, soluble in alcohol.[020]

Migration Data. There is significant penetration of nitrosoamines and their precursor amine DBA into boneless hams as a result of the use of elastic nettings containing rubber. Penetration of DBA from surface into inner parts of the ham is slow but significant. A mean value of 0.38 ppm DBA was found on outer surface of the ham and 0.09 to 0.11 ppm DBA were detected at a depth of 1.0 inch.

Regulations. ***U.S. FDA*** (1998) approved the use of DBA in the manufacture of rubber articles intended for repeated use in producing, manufacturing, packaging, processing, preparing, treating, packing, transporting, or holding food in accordance with the conditions prescribed by 21 CFR part 177.2600.

Reference:

Pensabene, J. W. Fiddler, W. and Gates, R. A., Nitrosamine formation and penetration in hams processed in elastic rubber nettings: *N*-nitrosodibutylamine and *N*-nitrosodibenzylamine, *J. Agric. Food. Chem.*, 43, 1919, 1995.

DIBUTYLAMINE

Molecular Formula. $C_8H_{19}N$
M = 129.25
CAS No 111-92-2
RTECS No HR7780000
Abbreviation. DBA.

Synonym. *N*-Butyl-1-butanamine.

Properties. Colorless, oily liquid with a fishy, amine odor. Solubility in water is 0.5%, readily soluble in alcohol. Odor perception threshold is 0.08[010] or 2.0 mg/l, taste perception threshold is 3.5 to 4.0 mg/l.[1]

Applications. Used in the production of rubber and synthetic fibers. It is found to be a degradation product in rubber materials which were made using dialkyldithiocarbamine acid derivatives as vulcanization accelerators.

Migration Data. There is significant penetration of nitrosoamines and their precursor amine DBA into boneless hams as a result of the use of elastic nettings containing rubber. Penetration of DBA from surface into inner parts of the ham is slow but significant. DBA concentrations showed gradual decrease from the exterior surface of the ham to the interior (means 0.57 to 0.23 ppm).[2] Reported migration levels of 0.2 to 0.5 mg/l usually depend on rubber and contact media compositions.[3]

Volatile *N*-nitrosamines were determined in cut-up pacifier nipples by analysis by GC-thermal energy analyzer. *N*-Nitrosamines and precursors in cut-up and intact nipples were determined after a single extraction with artificial saliva. *N*-nitrosodibutylamine (NDBA) was the principal nitrosamine found at levels up to 1.0 ppm, while DBA was the principal precursor found, at levels up to 3.9 ppm. Amounts of NDBA and DBA found after 15 artificial saliva extractions of intact pacifier nipples totalled 0.8 and 15.6 ppm, respectively. *N*-Nitrosamine levels generally showed a gradual decrease in concentration with each extraction.[4]

Acute Toxicity. LD_{50} is reported to be 290 to 300 mg/kg BW in rats and mice, and 230 mg/kg BW in guinea pigs. Poisoning produced the local irritating effect in the GI tract, fatty infiltration and necrotic lesions in the liver.[1]

Repeated Exposure failed to reveal cumulative properties.[1] Rats and guinea pigs tolerate administration of 1/5 LD_{50} for a month; 1/20 LD_{50} increased oxygen consumption and decreased activity of liver deaminoxidase. Increased activity of serum transaminases and depressed liver histaminolytic activity were found in rabbits fed 1/8 LD_{50}.[1]

Long-term Toxicity. Rats were fed dietary levels of 0.3 mg DBA/kg BW for 6 months. There were no treatment-related changes in conditioned reflex activity, blood cholinesterase and liver diaminoxidase activities, and total serum cholesterol. Rabbits tolerated doses of up to 7.6 mg/kg BW without changes in BW gain, prothrombin index, and liver histaminolytic function. However, the increased activity of serum transaminases was noted with the higher dose.[1]

Mutagenicity.

In vitro genotoxicity. DBA was negative in *Salmonella* mutagenicity assay.[013] It also gave negative results in CA test carried out on cultured Chinese hamster cells.[5]

Chemobiokinetics. DBA metabolism occurs by involving microsome non-specific oxidase system. Activation of peroxide oxidation of lipids takes place during the process of detoxication that may be the cause of dysfunction of the enzyme systems which are of fundamental importance in the mechanism of homeostasis maintaining.[6]

Standards. ***Russia*** (1995). MAC and PML: 1.0 mg/l (organolept., odor).

References:

1. Trubko, E. I., Study on hygiene substantiation of the Maximum Allowable Concentration for *N*-butylamines in water bodies, *Gig. Sanit.*, 11, 21, 1975 (in Russian).
2. Pensabene, J. W., Fiddler, W., and Gates, R. A., Nitrosamine formation and penetration in hams processed in elastic rubber nettings: *N*-nitrosodibutylamine and *N*-nitrosodibenzylamine, *J. Agric. Food. Chem.*, 43, 1919, 1995.
3. Duchovnaya, I. S., Grushevskaya, N. Yu., and Kazarinova, N. F., Methods of detection of secondary aliphatic amines in sanitary-chemical investigations of rubber, in *Hygiene of Use, Toxicology of Pesticides and Polymeric Materials*, Coll. Works, A. V. Pavlov, Ed., All-Union Research Institute of Hygiene and Toxicology of Pesticides, Polymers and Plastic Materials, Kiyv, Issue No 14, 1984, 112 (in Russian).
4. Thompson, H. S., Billedeau, S. M., Miller, B. J., et al., Determination of *N*-nitrosoamines and *N*-nitrosoamine-precursors in rubber nipples from baby pacifiers by gas chromatography - thermal energy analysis, *J. Toxicol. Environ. Health*, 13, 615, 1984.
5. Ishidate, M. and Odashima, S., Chromosome tests with 134 compounds in Chinese hamster cells *in vitro* - a screening for chemical carcinogens, *Mutat. Res.*, 48, 337, 1977.
6. Sidorin, G. I., Lukovnikova, L. V., and Stroikov, Yu. N., Toxicity of some aliphatic amines, *Gig. Truda Prof. Zabol.*, 11, 50, 1984 (in Russian).

DIBUTYLDITHIOCARBAMIC ACID, SODIUM SALT

Molecular Formula. $C_9H_{18}NS_2$

M = 227.39

CAS No 136-30-1

RTECS No EZ3880000

Abbreviation. DBDTS.

Synonyms and **Trade Name.** Butyl namate; Sodium DBDT; *N,N'*-Dibutyldithiocarbamate, sodium salt.

Applications. Vulcanization ultra-accelerator for natural and synthetic rubber and latex compounds.

Migration of dithiocarbamates into distilled water from natural rubber (67 and 1.0 cm^{-1}; 20 and 38°C; 24 hours) reached 0.4 to 0.6 mg/l.[1] A study of aqueous extracts and simulant media (20, 40, 100°C; 1 and 5 days; 0.5 cm^{-1}) revealed migration of the accelerators introduced into the vulcanizates based on smoked-sheet natural rubber and products of their transformation (dithiocarbaminates) at levels from traces to 0.43 mg/l.[2]

Acute Toxicity. LD_{50} is *i/p* 300 mg/kg in mice.[3]

Mutagenicity.

In vitro genotoxicity. DBDTS showed a weak activity in a single *in vitro* test (CA).[4]

Regulations. ***U.S. FDA*** (1998) approved the use of DBDTS in accelerators (total not to exceed 1.5% by weight of rubber product) for the preparation of rubber articles intended for repeated use in producing, manufacturing, packaging, processing, preparing, treating, packing, transporting, or holding food in accordance with the conditions prescribed by 21 CFR part 177.2600.

References:

1. Shumskaya, N. I., Provorov, V. N., Tolgskaya, M. S., Yemel'yanova, L. V., and Chernevskaya, N. M., Toxicity of some vulcanizates intended for manufacture of children's toys, in *Hygiene Aspects of the Use of Polymer Materials and of Articles Made of Them*, L. I. Medved', Ed., All-Union Research Institute of Hygiene and Toxicology of Pesticides, Polymers and Plastic Materials, Kiyv, 1969, 104 (in Russian).
2. Grushevskaya, N. Yu., Determination of vulcanization accelerator of vulkacite-*P*-extra-*N* and some transformation products in sanitary and chemical investigation of rubber, *Gig. Sanit.*, 4, 59, 1987 (in Russian).
3. Natl. Technical Information Service AD277-689.
4. Tinkler, J., Gott, D., and Bootman, J., Risk assessment of dithiocarbamate accelerator residues in latex-based medical devices: genotoxicity considerations, *Food Chem. Toxicol.*, 36, 849, 1998.

DIBUTYLDITHIOCARBAMIC ACID, ZINC SALT

Molecular Formula. $C_{18}H_{36}N_2S_4.Zn$
M = 474.12
CAS No 136-23-2
RTECS No ZHO175000
Abbreviation. DBDTZ.

Synonyms and **Trade Names.** Butazate; Butyl zimate; Butyl ziram; Dibutyldithiocarbamate, zinc salt; Vulcacit IDB.

Properties. White or slightly yellowish powder with a pleasant odor. Poorly soluble in alcohol.

Applications. Used as a vulcanization accelerator for natural rubber and latex; as a stabilizer for rubber-based adhesive systems, isobutylene-isoprene copolymers and polypropylene; as an antioxidant in the production of butyl rubber. It is also used as a stabilizer in a number of rubber and food handling.[1]

Migration of dithiocarbamates into distilled water from natural rubber (67 and 1.0 cm^{-1}; 20 and 38°C; 24 hours) reached 0.4 to 0.6 mg/l.[2] A study of aqueous extracts and simulant media (20, 40, 100°C; 1 and 5 days; 0.5 cm^{-1}) revealed migration of the accelerators introduced into the vulcanizates based on smoked-sheet natural rubber and products of their transformation (dithiocarbaminates) at levels from traces to 0.43 mg/l.[3]

Acute Toxicity. LD_{50} *i/p* exceeded 2.5 g/kg BW in mice.[027]

Short-term Toxicity. In a 3-month study, rabbits were given 50 mg/kg BW, rats received 20 mg/kg BW. The treatment did not cause changes in their general condition (Flinn, 1938). Rat were fed diets containing 100, 500, or 2500 ppm DBDTZ for 17 weeks.[1] Hematological indices, clinical chemistry, and urinalyses were not affected. At a 2500 mg/kg BW dose level, an increase in the relative weights of the liver and kidneys was observed, food intake was reduced. There were no histological changes in the tissues examined. The NOAEL of 500 ppm in the diet was identified in this study, providing an intake is between 41 and 47 mg/kg BW.

Long-term Toxicity. Gross pathology examination failed to reveal lesions in the visceral organs of Wistar rats that received 5.0 mg/kg BW for two years.[05]

Mutagenicity.

In vivo cytogenetics. DBDTZ was non-mutagenic in *in vivo* studies, it was therefore regarded as preferable to either the dimethyl or diethyl analogues (both of which gave equivocal results in an unscheduled DNA synthesis assay in rat liver).[4]

Carcinogenicity. In a 18-month study, mice were initially exposed by oral intubation to a dose level of 1000 mg/kg BW for 3 weeks, and thereafter in the diet at a level of 2600 ppm. No evidence of carcinogenicity was observed.[5]

Regulations. ***U.S. FDA*** (1998) regulates DBDTZ for use (1) in adhesives used as components of articles intended for use in packaging, transporting, or holding food in accordance with the conditions prescribed by 21 CFR part 175.105; (2) as a stabilizer in resinous and polymeric coatings in accordance with the conditions prescribed by 21 CFR part 175.300; (3) in coatings for polyolefin films for use only at levels not to exceed 0.2% by weight of isobutylene-isoprene copolymers provided that the finished copolymers contact food only of the types identified in CFR and at levels not to exceed 0.02% by weight of the polypropylene polymers; (4) in closures with sealing gaskets on food containers (not exceeding 0.8% by weight of the closure-sealing gasket composition and for use only in vulcanized natural or synthetic rubber gasket compositions) in accordance with the conditions prescribed by 21 CFR part 177.1210; (5) as an accelerator of rubber articles intended for repeated use in contact with food (total not to exceed 1.5% by weight of the rubber product) in accordance with the conditions prescribed by 21 CFR part 177.2600; and (6) in the manufacture of antioxidants and/or stabilizers for polymers which may be used in the manufacture of articles or components of articles intended for use in producing, manufacturing, packaging, processing, preparing, treating, packing, transporting, or holding food in accordance with the conditions prescribed by 21 CFR part 178.2010.

References:

1. Gray, T. J. B., Butterworth, K. R., Gaunt, I. F., et al., Short-term toxicity study of zinc dibutyldithiocarbamate in rats, *Food Chem. Toxicol.*, 16, 237, 1978.
2. *Hygiene Aspects of the Use of Polymeric Materials and Articles Made of Them*, L. I. Medved', Ed., All-Union Research Institute of Hygiene and Toxicology of Pesticides, Polymers and Plastic Materials, Kiyv, 1969, 104 (in Russian).
3. Grushevskaya, N. Yu., Determination of vulcanization accelerator of vulcacite-*P*-extra-*N* and some transformation products in sanitary and chemical investigation of rubber, *Gig. Sanit.*, 4, 59, 1987 (in Russian).
4. *1994 The Annual Report of the Committees on Toxicity, Mutagenicity, Carcinogenicity of the Chemicals in Food, Consumer Products and the Environment*, DoH, HMSO, London, 1995, 61.
5. Innes, J. R. M., Ulland, B. M., Valerio, W. G., et al., Bioassay of pesticides and industrial chemicals for tumorigenicity in mice: A preliminary note, *J. Natl. Cancer Inst.*, 42, 1101, 1969.

DIBUTYLNAPHTHALENESULFONIC ACID, SODIUM SALT

Molecular Formula. $C_{18}H_{23}NaO_2S$
M = 342.40
CAS No 25417-20-3
RTECS No QK1440000

Trade Name. Nekal.

Properties. Yellowish-gray powder. Soluble in water and alcohol. Gives water a specific odor of soap stock and a bitter, astringent taste. Perception threshold concentrations for odor vary from 0.16 mg/l[1] to 13 mg/l,[2] that for taste, from 0.08 mg/l[1] to 1.3 mg/l.[2] The threshold concentration for foam formation is 0.5 mg/l.[2]

Applications. Used as an emulsifier in the production of synthetic rubber. A surfactant.

Acute Toxicity. LD_{50} is 1.25 g/kg BW in rats, 1.13 g/kg BW in mice, and 1.5 g/kg BW in guinea pigs. ET_{50} for rats is 13 hours. [1,2]

Long-term Toxicity. Rats were exposed to 0.25 to 25 mg/kg BW doses for 6 months. The highest dose produced an increase in serum cholesterol level. Gross pathology examination revealed morphological changes in animals given 2.5 and 25 mg/kg BW doses.[2]

Reproductive Toxicity and **Mutagenic Effects** were not found.

Regulations. ***U.S. FDA*** (1998) approved the use of D. as an accelerator in rubber articles intended for repeated use in producing, manufacturing, packing, processing, treating, packaging, transporting, or holding food (total not to exceed 1.5% by weight of the rubber product) in accordance with the conditions prescribed by 21 CFR part 177.2600.

Standards. ***Russia*** (1995). MAC and PML: 0.5 mg/l (organolept., foam).

References:

1. Ivanov, V. A. and Khal'sov, P. S., Toxicological and hygienic assessment of harmful effluents of synthetic rubber industry, in *Proc. Voronezh Medical Institute*, Voronezh, Issue No 55, 1966, 56 (in Russian).

2. Yegorova, N. A., Prediction of chronic toxicity of nekal by calculation and experimentation with the aim of setting its maximum allowable concentration in water, *Gig. Sanit.*, 4, 39, 1980 (in Russian).

N,N'-DICYCLOHEXYL-2-BENZOTHIAZOLE SULFENAMIDE

Molecular Formula. $C_{19}H_{26}N_2S_2$
M = 346.59
CAS No 4979-32-2
RTECS No DL6300000

Trade Names. Vulcacit DZ; Sulfenamide DC.

Properties. A yellowish, bitter, odorless powder, almost insoluble in water, soluble in alcohol.

Applications. Used as a vulcanization accelerator in rubber industry.

Acute Toxicity. LD_{50} is 6.42 g/kg BW in rats,[029] and 8.5 g/kg BW in male mice.

Long-term Toxicity. Rabbits were fed 20 mg/kg BW every other day for 2 months and then daily for another 2-month period. The treatment changed the content of amino acids and lipids and reduced blood cholinesterase activity. Histological examination revealed mild surface necrosis and desquamation of the GI tract mucosa and moderate edema of the parenchymatous organs. Histochemical investigation found reduced glycogen content in the liver cells.

Reference:

1. Vorob'yeva, R. S., *N.N'*-Dicyclohexyl-2-benzothiasol sulfenamide, in *Toxicology of New Chemical Substances Used in Rubber and Tyre Industry*, Z. I. Israel'son, Ed., Meditsina, Moscow, 1968, 89 (in Russian).

DIETHYLAMINE

Molecular Formula. $C_4H_{11}N$
M = 73.14
CAS No 109-89-7
RTECS No HZ8750000
Abbreviation. DEA.

Synonyms. Diethamine; *N*-Ethylethanamine.

Properties. A colorless, alkaline liquid with a sharp (musty, fishy, amine) odor. Readily miscible with water, soluble in alcohol. Odor and taste perception thresholds are 10.0 and 8.0 mg/l, respectively.[1] According to other data, odor perception threshold is 0.47 mg/l [02] or 0.02mg/l.[010]

Applications. Used in the production of synthetic rubber.

Migration Data. DEA is found to be a degradation product in rubber materials which were made using dialkyldithiocarbamine acides derivatives as vulcanization accelerators. Reported migration levels of 0.05 to 0.1 mg/l usually depend of rubber and contact media compositions.[2]

Acute Toxicity. LD_{50} is 540 mg/kg BW in rats, and 648 mg/kg BW in mice. No difference in species susceptibility was found. Administration led to CNS inhibition.[1]

See ***Dimethylamine.***

Repeated Exposure revealed no cumulative properties.

Short-term Toxicity. Mice were exposed to 1/10 LD_{50} for 2.5 months. Some animals died; survivors developed reduced BW gain and increased contents of ascorbic acid in the liver.

Long-term Toxicity. Rats and rabbits received 1.0 to 10.0 mg/kg BW for 6 months. The dose of 6.0 mg/kg BW affected the liver carbohydrate function.[1]

Reproductive Toxicity. See ***Dimethylamine.***

Mutagenicity.

In vitro genotoxicity. DEA is reported to be negative in *Salmonella* mutagenicity assay.[3]

Carcinogenicity. Similar to other secondary amines DEA might be nitrosated endogenously to form a carcinogenic compound *N*-nitrosodiethylamine.[4] The latter should be regarded as if it were carcinogenic to man.

See ***Dimethylamine.***

Regulations. ***U.S. FDA*** (1998) approved the use of DEA (1) in adhesives used as components of articles intended for use in packaging, transporting, or holding food in accordance with the conditions prescribed by 21 CFR part 175.105; (2) in poly(2,6-dimethyl-1,4-phenylene) oxide resins which may be used in the manufacture of articles intended for use in contact with food (not to exceed 0.16% as a residual catalyst), and (3) as an activator of rubber articles intended for repeated use in producing, manufacturing, packing, processing, treating, packaging, transporting, or holding food (total not to exceed 5.0% by weight of the rubber product) in accordance with the conditions prescribed by 21 CFR part 177.2600.

Standards. ***Russia*** (1995). MAC and PML: 2.0 mg/l.

References:

1. Kagan, G. Z., Comparative evaluation of diethylamine and triethylamine in connection with sanitary protection of reservoirs, *Gig. Sanit.*, 9, 28, 1965 (in Russian).
2. Dukhovnaya, I. S., Grushevskaya, N. Yu., and Kazarinova, N. F., Methods of detection of secondary aliphatic amines in sanitary-chemical investigations of rubber, in *Hygiene of Use, Toxicology of Pesticides and Polymeric Materials*, Coll. Works, A. V. Pavlov, Ed., All-Union Research Institute of Hygiene and Toxicology of Pesticides, Polymers and Plastic Materials, Kiyv, Issue No 14, 1984, 112 (in Russian).
3. Hedenstedt, A., Genetic health risk in the rubber industry: Mutagenicity studies on rubber chemicals and process vapors, in *Prevention of Occup. Cancer*, Int. Symp. Int. Labour Office, Geneva, 1982, 476.
4. Bellander, T., Osterdahl, B. G., Hagmar, L., et al., Excretion of *N*-mononitrosopiperasine in the urine in workers manufacturing piperasine, *Int. Arch. Occup. Environ. Health*, 60, 25, 1988.

2-DIETHYLAMINOMETHYL BENZOTHIAZOLETHIONE

Molecular Formula. $C_{12}H_{16}N_2S_2$

M = 252.3

Properties. D. has an unpleasant odor and an extremely bitter taste. Hydrolyzes in water. Decomposes in acid and alkaline media with the formation of 2-mercaptobenzothiasole (Captax - *q.v.).*

Applications. Used as a vulcanization accelerator in rubber industry.

Acute Toxicity. Mice tolerate administration of 4.0 g/kg BW. However, a single oral dose of 4.5 g/kg BW resulted in the death of 8 out of 10 animals. Gross pathology examination revealed visceral congestion, point hemor rhages in the gastric mucosa and mucosal edema (necrotic gastritis).

Repeated Exposure. There was neither mortality nor retardation of BW gain in mice gavaged with 2.0 g/kg BW in oil solution every other day over 2 weeks. A dose of 4.0 g/kg BW caused a 50% mortality in mice.

Short-term Toxicity. Rats were dosed by gavage with 100 mg/kg BW for 4 months. The treatment produced a reduction in the temperature and BW gain. The NS was unaffected and there were no histology changes in the treated animals.

Standards. ***Russia***. PML: *n/m.*

Reference:

Toxicology of New Industrial Chemical Substances, N. F. Ismerov and I. V. Sanotsky, Eds., Medgiz, Mos cow, Issue No 6, 1964, 26 (in Russian).

3-DIETHYLAMINOMETHYL-2-BENZOTHIAZOLETHIONE

Molecular Formula. $C_{12}H_{16}N_2S_2$

M = 252.3

Properties. White, crystalline powder with a bitter taste. Poorly soluble in water, soluble in alcohol.

Applications. Used as a vulcanization accelerator.

Acute Toxicity. In mice, LD_{50} is 1.08 g/kg BW (technical grade product) and 1.57 g/kg BW (pure substance). High doses caused restlessness, disturbances of motor coordination, tail rigidity, clonic-tonic convulsions. Death occurs in the first 2 to 3 hours.

Short-term Toxicity. Rats were dosed by gavage with 100 mg/kg BW dose for 4 months. There were no significant functional or morphologic changes in the viscera other than local inflammation of the gastric and duodenal mucous membranes.

Rabbits were administered 20 mg/kg BW (every other day for 2 months and daily for 1.5 months). The treatment produced an increase in the bilirubin index and pathomorphological changes in the liver.

Reference:

Vorob'yeva, R. S. and Mezentseva, N. V., 3-Diethylaminomethylbenzothiasolthion-2, in *Toxicology of New Chemical Substances Used in Rubber and Tyre Industry*, Z. I. Izrael'son, Ed., Meditsina, Moscow, 1968, 35 (in Russian).

N,N'-DIETHYL-2-BENZOTHIAZOLE SULFENAMIDE

Molecular Formula. $C_{11}N_{14}N_2S_2$

M = 238.27

Properties. Brown, oily liquid. Poorly soluble in water, soluble in alcohol. Its bitter taste restricts its use as a vulcanizator of rubber intended for contact with food.

Applications. Used as a slow-acting accelerator in the rubber industry.

Acute Toxicity. LD_{50} appeared to be 4.0 g/kg BW in rats, 4.96 g/kg BW in mice, 6.2 g/kg BW in rabbits,[1] and exceeded 2.0 g/kg BW in marmots.[027]

According to other data, rats and mice tolerated 5.0 g/kg BW dose administered in a starch solution. The same dose of the pure substance appeared to be the minimum lethal dose in mice and the LD_{80} in rats.[2]

Treatment-related effects consisted of general inhibition, apathy, decreased food consumption, and hypothermia. Liver functions were affected (tests on hippuric acid and bromsulphalein elimination). Gross pathology examination revealed vascular disorders with pronounced visceral congestion, liver fatty dystrophy, and parenchymatous dystrophy of the myocardium.

Repeated Exposure demonstrated weak cumulative properties. None of the mice died following 3 administrations of 5.0 g/kg BW dose in 3.0% starch solution. The same dosing produced minimum lethal effect in rats. Gross pathology examination revealed fatty and parenchymatous dystrophy of the central lobular liver cells. Reticuloendothelial cell infiltration around the blood vessels and bile ducts was also observed.[2] A 45-day treatment revealed some effect on the peripheral blood, liver, and CNS functions.[1]

Standards. ***Russia*** (1994). MAC and PML: 0.05 mg/l (organolept., taste).

References:

1. Vaisman, Ya,. I., Zaitseva, N. V., and Mikhailov, A. V., Hygienic standardization of kauchuk manufacture sewage in water bodies, *Gig. Sanit.*, 2, 17, 1973 (in Russian).
2. Zayeva, G. N., Fedorova, V. I., Fedechkina, N. I., Evaluation of toxicity of sulphenamide BT, M & C, in *Toxicology of New Industrial Chemical Substances*, N. F. Ismerov and I. V. Sanotsky, Eds., Meditsina, Moscow, Issue No 8, 1966, 70 (in Russian).

N,N'-DIETHYL-*N,N'*-DIPHENYLTHIOPEROXYDICARBONIC DIAMIDE

Molecular Formula. $C_{18}H_{20}N_2S_4$

M = 392.63

CAS No 41365-24-6

RTECS No XM6885000

Synonym and **Trade Names.** *N,N'*-Diphenyl-*N,N'*-diethylthiuramdisulfide; Thiuram EF.

Properties. White powder. Poorly soluble in ethanol, insoluble in water.

Applications. A vulcanization accelerator for rubber intended for use in food-contact surface. In contrast to other dithiocarbamine accelerators, its use in a rubber in the course of vulcanization does not result in development of toxic substances.

Migration Data. Investigation of aqueous extracts (0.5 cm^{-1}; 100°C, 4 hours) revealed migration of thiuram EF (0.02 mg/l) from vulcanizates based on butadiene-nitrile rubber SKN-26. On repeated analyses, no migration was noted.[1]

Acute Toxicity. Rats tolerate the dose of 12 g/kg BW. Acute action threshold is 0.5 g/kg BW.

Repeated Exposure revealed no cumulative properties.

Long-term Toxicity. Rats received a 100 mg/kg BW dose for 6 months. The treatment caused anemia to develop. Normalization was observed in about 10 months. Doses of 50 and 100 mg/kg BW caused liver function deterioration. The NOAEL of 5.0 mg/kg BW was identified in this study.[2]

Reproductive Toxicity.

Gonadotoxic effect was not observed.[1]

Teratogenicity. SNK mice were given 5.0 and 50 mg/kg BW as oily emulsion during pregnancy. Both doses caused the increased rate of hemorrhages to the fetal organs and tissues and generalized edema. No morphofunctional changes in the fetuses were reported.

Embryotoxicity effect is negligible.[3]

Mutagenicity.

In vivo cytogenetics. A weak mutagenic action on somatic and germ cells was observed in Wistar rats given 50 mg/kg BW for 2.5 months. The NOAEL for mutagenic and genotoxic effect appeared to be 5.0 mg/kg BW.[1]

Standards. ***Russia*** (1988). PML in food: 0.5 mg/kg.

References:

1. Khoroshilova, N. V., Sal'nikova, L. S., Domshlak, M. G., et al., Mutagenic and gonadotropic activities of thiuram EF, *Gig. Truda Prof. Zabol.*, 3, 20, 1980 (in Russian).
2. Khoroshilova, N. V., Toxicological characteristics of the new vulcanization accelerator thiuram EF, *Kauchuk i Rezina*, 11, 50, 1979 (in Russian).
3. Sal'nikova, L. S., Study of the embryotropic effect of thiuram, *Gig. Sanit.*, 3, 88, 1989 (in Russian).

DIETHYLDITHIOCARBAMIC ACID, 2-BENZOTHIAZOLYL ESTER

Molecular Formula. $C_{12}H_{14}N_2S_3$
M = 282.46
CAS No 95-30-7
RTECS No EZ4950000

Synonyms and **Trade Name.** 2-Benzothiazolethiol, diethyldithiocarbamate; 2-Benzothiazolyl-*N,N'*-diethylthiocarbamyl sulfide; 2-(*N,N'*-Diethylthiocarbamyl)benzoathiazole; *N,N'*-Diethylthiocarbamyl-2-benzothiazolyl sulfide; Ethylac.

Applications. Vulcanization accelerator.

Migration of dithiocarbamates into distilled water from natural rubber (67 and 1.0 cm^{-1}; 20 and 38°C; 24 hours) reached 0.4 to 0.6 mg/l.[1] A study of aqueous extracts and simulant media (20, 40, 100°C; 1 and 5 days; 0.5 cm^{-1}) revealed migration of the accelerators introduced into the vulcanizates based on smoked-sheet natural rubber and products of their transformation (dithiocarbaminates) at levels from traces to 0.43 mg/l.[2]

Acute Toxicity. LD_{50} is 6.0 g/kg BW in rats, and 2.7 g/kg BW in rabbits.[027]

References:

1. Yemel'yanova, L. V. and Chernevskaya, N. M., Toxicity of some rubber materials intended for production of children's toys, *Hygiene Aspects of the Use of Polymeric Materials and of Articles Made of Them*, L. I. Medved', Ed., All-Union Research Institute of Hygiene and Toxicology of Pesticides, Polymers and Plastic Materials, Kiyv, 1969, 104 (in Russian).
2. Grushevskaya, N. Yu., Determination of vulcanization accelerator of vulkacite-*P*-extra-*N* and some transformation products in sanitary and chemical investigation of rubber, *Gig. Sanit.*, 4, 59, 1987 (in Russian).

DIETHYLDITHIOCARBAMIC ACID, SODIUM SALT

Molecular Formula. $C_5H_{10}NNaS_2$
M = 176.27
CAS No 148-18-5
RTECS No EZ6475000
Abbreviation. DEDTS.

Synonym and **Trade Names.** Carbamate EN; Dithiocarb; Sodium diethyldithiocarbamate trihydrate; Thiocarb.

Properties. White, crystalline powder. Soluble in water and alcohol.

Applications. Used as a vulcanization accelerator for rubber and vulcanizates based on them, for natural and synthetic latex.

Migration of dithiocarbamates into distilled water from natural rubber (67 and 1.0 cm^{-1}; 20 and 38°C; 24 hours) reached 0.4 to 0.6 mg/l.[1] A study of aqueous extracts and simulant media (20, 40, 100°C; 1 and 5 days; 0.5 cm^{-1}) revealed migration of the accelerators introduced into the vulcanizates based on smoked-sheet natural rubber and products of their transformation (dithiocarbaminates) at levels from traces to 0.43 mg/l.[2]

Diethyldithiocarbamates were detected by the GC determination at levels up to 4.6 mg/g rubber (as dithiocarbamic acid), in chloroform-acetone extracts from isoprene rubber teats for baby bottles. Dialkyldithiocarbamates can form secondary amines by acid hydrolysis, although their levels in the extracts only made a minor contribution to the total level of measured secondary amine precursors.[3]

Acute Toxicity. The LD_{50} values are 1.5 to 2.47 g/kg BW for mice, and 1.5 to 3.32 g/kg BW for rats. High doses cause a brief excitation, diarrhea, hind leg paralysis, and convulsions.[4]

Repeated Exposure. Treatment with large doses produced anemia and leukopenia, CNS affection, inhibition of oxidation enzyme activity, particularly in the brain and liver. Rabbits given 800 mg/kg BW developed leukopenia and decreased *Hb* level. 30% of animals died after 5 to 8 administrations.[1]

Ten Dutch male rabbits received 330 mg DEDTS/kg BW by gavage for 4, 6 or 9 weeks. Histological examination revealed exposure-dependent lesions, e.g., Wallerian degeneration and eosinophilic bodies in the medulla and spinal cord.[5]

Short-term Toxicity. Rats were dosed with 30, 100, or 300 mg DEDTS/kg BW for 90 days. The treatment caused retardation of BW gain and signs of erythropenia.[6]

Long-term Toxicity. In an NTP feeding study, F344 rats received 1,250 and 2,500 ppm DEDTS for 104 weeks, $B6C3F_1$ mice were given 500 and 4,000 ppm DEDTS for 108 and 109 weeks. Reduced BW was observed at higher dose. In females, a significant increase in the incidence of cataracts was noted.[7]

Immunotoxicity. DEDTS restores and regulates the numbers and activities of cells of the *T*-cell lineage.[8]

Mutagenicity.

In vitro genotoxicity. DEDTS was found to be negative in *Salmonella* mutagenicity assay.[013] Addition of diethyldithiocarbamate to splenic and thymic rat lymphocytes resulted in a complete inhibition of scheduled DNA synthesis. Under the same conditions, a strictly dose-dependent inhibition of unscheduled (excision repair) DNA synthesis, a decrease of the sedimentation rate of nucleoids, as well as changes of the thymidine pool were observed.[9]

Carcinogenicity. In the above-cited NTP feeding study, the treatment induced no carcinogenic effect.[7]

Carcinogenicity classifications. An IARC Working Group concluded that there are insufficient data to make an evaluation on carcinogenicity of DEDTS in *experimental animals* and no case reports or epidemiological studies were available to the Working Group for evaluation on carcinogenicity of DEDTS in *humans*.

IARC: 3;

NTP: N - N - N - N (feed).

Standards. ***Russia*** (1995). MAC: 0.5 mg/l.

References:

1. Yemel'yanova, L. V. and Chernevskaya, N. M., Toxicity of some rubber materials intended for production of children's toys, *Hygiene Aspects of the Use of Polymeric Materials and of Articles Made of Them*, L. I. Medved', Ed., All-Union Research Institute of Hygiene and Toxicology of Pesticides, Polymers and Plastic Materials, Kiyv, 1969, 104 (in Russian).
2. Grushevskaya, N. Yu., Determination of vulcanization accelerator of vulkacite-*P*-extra-*N* and some transformation products in sanitary and chemical investigation of rubber, *Gig. Sanit.*, 4, 59, 1987 (in Russian).
3. Yamazaki, T., Inoue, T., Yamada, T., and Tanimura, A., Analysis of residual vulcanization accelerators in baby bottle rubber teats, *Food Addit. Contam.*,3, 145, 1986.

4. Korablev, M. V. and Lukienko, B. I., Effect of dithiocarbamine acid derivatives and structurally similar compounds on hypoxia, *Pharmacologia i Toxicologia*, 30, 186, 1967 (in Russian).
5. Rasul, A. R. et al., *Acta Neuropathol.*, 24, 161, 1973 (cit. in U.S. EPA *Health and Environmental Effects Profile for Sodium Diethyldithiocarbamate*, 1983, 15).
6. Sunderman, F. M., Paynter, O. E., and George, R. B., The effects of the protracted administration of the chelating agent sodium diethyldithiocarbamate (dithiocarb), *Am. J. Med. Sci.*, 254, 46, 1967.
7. *Bioassay of Sodium Diethyldithiocarbamate for Possible Carcinogenicity,* NTP Technical Report Series No 172, Research Triangle Park, NC, DHEW Publ., 1979.
8. Renoux, G., Renoux, M., and Guillaumin, J. M., Immunopharmacology and immunotoxicity of zinc diethyldithiocarbamate, *Int. J. Immunopharmacol.*, 10, 489, 1988.
9. Tempel, K., Schmerold, I., and Goette, A., The cytotoxic action of diethyldithiocarbamate in vitro. Different inhibition of scheduled and unscheduled DNA synthesis of rat thymic and splenic cells, *Arzneimittel-Forsch.*, 35, 1052, 1985.

N,N'-DIFURFURYLIDENE-2-FURANMETHANEDIAMINE

Molecular Formula. $C_{15}H_{12}N_2O_3$
M = 268.29
CAS No 494-47-3
RTECS No LU1925000

Synonyms and **Trade Names.** *N,N'*-Difurfurylidene-1-(2-furyl); Furfuramide; Furfurylamide; Hydrofuramide; Methanediamine; Vulkasol A.

Properties. Straw-yellow, crystalline powder. Poorly soluble in water, readily soluble in alcohol.

Applications. Used as a vulcanization accelerator in the rubber production.

Acute Toxicity. LD_{50} is 400 mg/kg BW in rats and 950 mg/kg BW in mice. Lethal doses caused acute adynamia and respiratory disorder. Death occurs within 3 days. Paralysis of the muscles of the pelvic girdle and a change in the relative weights of the liver and lungs are found in some of the surviving mice. Gross pathology examination revealed pulmonary edema and emphysema.

Long-term Toxicity. In the chronic toxicity study, rabbits were exposed to oral doses of 30 and 50 mg/kg BW. The treatment led to a reduction in the erythrocyte count, and changes in blood *Hb* level and in albumen/globulin coefficient. Proliferative changes were observed in the lungs, liver, heart, and kidney.

Reference:

Arkhangelskaya, L. N. and Roschina, T. A., Furfuramide, in *Toxicology of the New Chemical Substances Used in Rubber and Tyre Industry*, Z. I. Israel'son, Ed., Meditsina, Moscow, 1968, 131 (in Russian).

1,5-DI-2-FURYL-1,4-PENTADIEN-3-ONE

Molecular Formula. $C_{13}H_{10}O_3$
M = 214.23
CAS No 886-77-1
RTECS No RZ2476660
Abbreviation. DFP.

Composition. DFP is formed when furfuryl and acetone (2:1) react in the presence of an alkaline catalyst.

Synonym and **Trade Name.** Bifuron P; 1,3-Difurfurylidene acetone.

Properties. Yellow or orange moisture-free powder. Odor perception threshold is 39.3 mg/l (practical threshold 85 mg/l). Water coloration threshold is 12.5 mg/l.

Applications. Used as a vulcanizing agent and also in the production of resins and molding materials.

Acute Toxicity. LD_{50} is 2.73 g/kg BW in rats, 2.66 g/kg BW in mice, and 1.6 g/kg BW in rabbits.[1] Poisoning caused neurotoxic effect because of penetration through hemato-encephalic barrier (cellular membranes) and produced damage of endocrine glands (hypophysis, adrenal glands). A decrease in erythrocyte ATF activity was noted.[2] According to other data, LD_{50} is 6.5 ml/kg BW in rats.[3]

Long-term Toxicity. Calculated NOAEL appeared to be 0.38 mg/kg BW.[1]

Chemobiokinetics. Rats received a single administration of 2/3 LD_{50} of *furfurilidene acetone* and DFP. Metabolism occurs in the liver. Administration led to formation of highly toxic metabolites (*furyl alcohol, furfural,* and *tetrahydrofuran*).[2]

Standards. ***Russia*** (1995). Tentative MAC: 7.5 mg/l.

References:

1. Usmanov, L. A. and Aliev, E., Hygienic norm-setting for furan compounds in water bodies, *Gig. Sanit.*, 8, 87, 1987 (in Russian).
2. Zavgorodnaya, S. V. and Tarasov, V. V., Molecular mechanism of action of furan polymers, *Gig. Sanit.*, 9, 29, 1983 (in Russian).
3. Carpenter, C. P., Weil, C. S., and Smyth, H. F., Jr., Range-finding toxicity data: list VIII, *Toxicol. Appl. Pharmacol.*, 28, 313, 1974.

1,2-DIHYDRO-2,2,4-TRIMETHYLQUINOLINE

Molecular Formula. $C_{12}H_{15}N$
M = 173.28
CAS No 147-47-7
RTECS No VB4900000
Abbreviation. DTQ.

Synonym and **Trade Names.** Acetonanyl; Flectol A; Vulkanox HS/LG.

Properties. Light tan powder.

Applications. An antioxidant and an antidegradation agent for rubber.

Acute Toxicity. LD_{50} is 2.0 g/kg BW in rats and rabbits,[027] 1.45 g/kg BW in mice. Poisoning led to CNS inhibition.[2]

Long-term Toxicity. Manifestations of toxic action included decreased BW gain, CNS inhibition, disruption of liver functions, and anemia.[1]

Mutagenicity. DTQ was found to be negative in *Salmonella* mutagenicity assay.[015]

Reference:

1. Kel'man, G. Ya., Study of toxicity of certain rubber antioxidants and their hygienic assessment, *Gig. Sanit.*, 2, 25, 1966 (in Russian).

DIISOPROPYLBENZENE

Molecular Formula. $C_{12}H_{18}$
M = 162.28
CAS No 25321-09-9
RTECS No CZ6330000
Abbreviation. DIPB.

Synonym. Bis(1-methylethyl)benzene.

Properties. A mixture of isomers. A clear liquid with a sharp specific odor. Solubility in water is 10 mg/l. Odor perception threshold is 0.1 mg/l for *m*-DIPB and 0.05 mg/l for *p*-DIPB.

Applications. Used in the production of synthetic rubber and plastics.

Acute Toxicity. LD_{50} of *m*-DIPB is 7.4 g/kg BW in rats and 2.1 g/kg BW in mice; LD_{50} of *p*-DIPB is 11.54 and 7.98 g/kg BW, respectively. High doses cause convulsions and narcosis.

Repeated Exposure. *p*-Isomer exhibits more pronounced cumulative properties (in comparison to *m*-DIPB). When 1/10 LD_{50} is administered, K_{acc} appeared to be 1.56 (*p*-DIPB) and 3.0 (*m*-DIPB). The treatment caused no changes in the blood *Hb* level, erythrocyte count, or cholinesterase activity in the survivors.

Long-term Toxicity. In a 6-month study in rats, disturbance of CNS activity and changes in protein-forming function of the liver were noted at the higher doses tested.

Reproductive Toxicity. Following oral and inhalation exposure, disorder of the estrous cycle in rats and mice was noted (Yelisuiskaya, 1970).

Standards. ***Russia*** (1995). MAC and PML: 0.05 mg/l.

Reference:

Sologub, A. M. and Bogdanova, T. P., Experimental study on substantiation of maximum allowable concentration for *m*-diisopropylbenzene and *p*-diisopropylbenzene in water bodies, *Gig. Sanit.*, 9, 18, 1971 (in Russian).

DIISOPROPYLDITHIOCARBAMIC ACID, ZINC SALT

Molecular Formula. $C_{14}H_{28}N_2S_4Zn$

M = 418.05

Synonyms. Diisopropyldithiocarbamate, zinc salt; Zinc *N,N′*-propylene-1,2- bisdithiocarbamate.

Properties. White powder. Poorly soluble in water and alcohol.

Applications. Used as a vulcanization accelerator in the rubber production.

Migration of dithiocarbamates into distilled water from natural rubber (67 and 1.0 cm^{-1}; 20 and 38°C; 24 hours) reached 0.4 to 0.6 mg/l.[1] A study of aqueous extracts and simulant media (20, 40, 100°C; 1 and 5 days; 0.5 cm^{-1}) revealed migration of the accelerators introduced into the vulcanizates based on smoked-sheet natural rubber and products of their transformation (dithiocarbaminates) at levels from traces to 0.43 mg/l.[2]

Acute Toxicity. In mice, LD_{50} is 4.25 g/kg BW. Manifestations of the toxic action include apathy, adynamia, and poor grooming.[3]

Short-term Toxicity. There were no deviations in hydrocarbon and fat metabolism or liver function of the rabbits given 20 mg/kg BW for a month. Histology examination failed to reveal dystrophic changes in the liver and kidneys.[3]

References:

1. Shumskaya, N. I., Provorov, V. N., Tolgskaya, M. S., Yemel'yanova, L. V., and Chernevskaya, N. M., Toxicity of some vulcanizates intended for manufacture of children toys, in *Hygiene Aspects of the Use of Polymeric Materials and of Articles Made of Them*, L. I. Medved', Ed., All-Union Research Institute of Hygiene and Toxicology of Pesticides, Polymers and Plastic Materials, Kiyv, 1969, 104 (in Russian).
2. Grushevskaya, N. Yu., Determination of vulcanization accelerator of vulkacite-*P*-extra-*N* and some transformation products in sanitary and chemical investigation of rubber, *Gig. Sanit.*, 4, 59, 1987 (in Russian).
3. Mezentseva, N. V., Zinc salt of *N,N'*-diisopropyl dithiocarbamoic acid, in *Toxicology of New Chemical Substances Used in Rubber and Tyre Industry*, Z. I. Israel'son, Ed., Meditsina, Moscow, 1968, 127 (in Russian).

2,5-DIMETHYL-2,5-DI(*tert*-BUTYLPEROXY)HEXANE

Molecular Formula. $C_{16}H_{34}O_4$

M = 290.50

CAS No 78-63-7

RTECS No MO1835000

Synonym. (1,1,4,4-Tetramethyltetramethylene)bis(*tert*-butylperoxide).

Properties. Occurs as a light-yellowish, oily liquid with a faint characteristic odor. Poorly soluble in water.

Applications. Used in high-temperature vulcanization and in the plastic manufacture.

Acute Toxicity. LD_{50} is found to be 20 g/kg BW in female rats, and 22.5 g/kg BW in male rats; these values in mice are 7.8 and 8.5 g/kg, respectively. Species differences are evident; species sensitivity coefficient appears to be 0.39. Poisoned animals displayed apathy and adynamia. BW gain and renal function are predominantly affected. Death occurs within 3 to 6 days.

Repeated Exposure revealed cumulative properties. K_{acc} is 2.2 (by Lim).

Allergenic Effect was not noted.

Reference:

Matreshin, A. V. and Klimov, V. S., in *Sanitary Evaluation of the Environmental Condition of the Middle Volga Region*, Moscow, 1982, 115 (in Russian).

4,4'-DIMETHYL-1,3-DIOXANE

Molecular Formula. $C_6H_{12}O_2$
M = 116.60
CAS No 766-15-4
RTECS No JH0525000
Abbreviation. DMDO.

Synonym. 4,4'-Dimethyl-*m*-dioxane.

Properties. A colorless liquid with an odor reminiscent of ether. Soluble in water (20%). Does not affect the color and taste of water at a concentration of 1.0 g/l. Odor perception threshold is 2.5 mg/l.[1]

Applications. Used in the production of isoprene rubber.

Acute Toxicity. Rats and mice died following administration of 4.5 to 5.0 g/kg BW doses. Lethal doses cause CNS inhibition, motor disturbances, and respiratory distress. In mice, the LD_{100} is 3.0 g/kg BW;[1,2] in rats it appears to be 3.7 g/kg BW.

Repeated Exposure. Rats and mice received 25 mg/kg BW for 10 days. The treatment produced no effect on general condition and BW gain, but affected catalase activity and oxygen consumption.

Long-term Toxicity. In a 6-month study, the treatment altered oxygen consumption, hematology analyses, catalase activity, and glycogen-forming function of the liver in rabbits.[1]

Reproductive Toxicity.

Gonadotoxicity. Inhalation exposure to DMDO caused changes in hypophyseal gonadotropic activity and ovary morphology in rats.[3]

Embryotoxic effect was observed at a lower concentration of DMDO in air than that producing general toxicity manifestations.[2]

Standards. ***Russia*** (1995). MAC and PML: 0.005 mg/l.

References:

1. Klimkina, N. V., Hygiene standardization in water bodies of harmful substances from the production of synthetic isoprene rubber, *Gig. Sanit.,* 6, 8, 1959 (in Russian).
2. Smirnov, V. T., Dubrovskaya, F. I., and Kiseleva, T. I., Action of dimethyldioxane on the generative function of experimental animals, *Gig. Sanit.*, 9, 16, 1978 (in Russian).
3. Pashkova, G. A., Specific gonadotoxic action of dimethyldioxane, in *Toxicology of New Industrial Chemical Substances*, N. F. Ismerov and I. V. Sanotsky, Eds., Meditsina, Moscow, Issue No 12, 1971, 64 (in Russian).

DIMETHYLDITHIOCARBMIC ACID, SODIUM SALT

Molecular Formula. $C_9H_{18}NaN_3S_6$
M = 143.21
CAS No 128-04-1
RTECS No FD3500000
Abbreviation. DMDCS.

Synonym and **Trade Names.** Carbamate MN; Carbon S; Dimethyldithiocarbamate, sodium salt.

Properties. Odorless, white, crystalline powder with a bitter taste. Readily soluble in water and al cohol.

Applications. Used as a vulcanization ultra-accelerator for mixes based on synthetic and natural rubbers (for example, butadiene rubber) and natural and synthetic latex. A fungicide.

Migration of dithiocarbamates from smoked-sheet natural rubber vulcanizates for children's dummies was determined to be up to 0.6 mg/l.[07] Migration of dithiocarbamates into distilled water from natural rubber (67 and 1.0 cm^{-1}; 20 and 38°C; 24 hours) reached 0.4 to 0.6 mg/l.[1] A study of aqueous extracts and simulant media (20, 40, 100°C; 1 and 5 days; 0.5 cm^{-1}) revealed migration of the accelerators introduced into the vulcanizates based on smoked-sheet natural rubber and products of their transformation (dithiocarbaminates) at levels from traces to 0.43 mg/l.[2]

Dimethyldithiocarbamates were detected by the GC determination at levels up to 3.2 mg/g rubber (as dithiocarbamic acid) in chloroform-acetone extracts from isoprene rubber teats for baby bottles.

Dialkyldithiocarbamates can form secondary amines by acid hydrolysis, although their levels in the extracts only made a minor cortribution to the total level of measured secondary amine precursors.[3]

Acute Toxicity. Rabbits received the single doses of 0.5 to 1.5 g DMDCS/kg BW. The intake was accompanied by excitation with subsequent CNS inhibition, reduced tactile and pain sensitivity and decreased food consumption. Hematology and ECG changes were noted. Animals given 1.0 and 1.5 g/kg BW died within 24 hours. The lowest dose caused no mortality. Gross pathology examination revealed visceral congestion and pulmonary edema, hyperemia and focal hemorrhages in the gastric mucosa and intestinal swelling. The relative weights of the lungs, heart, and stomach were increased. For mice and rats, the LD_{50} values are 1.5 and 1.0 g/kg BW, respectively.[4]

Repeated Exposure failed to reveal cumulative properties. BW gain was increased in rabbits given 100 mg/kg BW (in an aqueous solution) for 15 weeks, as compared to the controls. Adynamia, changes in blood morphology, in the NS, and cardiovascular systems were noted. The fur of animals was ungroomed and had turned yellow.[4]

Regulations. ***U.S. FDA*** (1998) approved the use of DMDCS in the manufacture of rubber articles intended for repeated use in producing, manufacturing, packaging, processing, preparing, treating, packing, transporting, or holding food in accordance with the conditions prescribed by 21 CFR part 177.2600.

References:

1. Shumskaya, N. I., Provorov, V. N., Tolgskaya, M. S., Yemel'yanova, L. V., and Chernevskaya, N. M., Toxicity of some vulcanizates intended for manufacture of children's toys, in *Hygiene Aspects of the Use of Polymer Materials and of Articles Made of Them*, L. I. Medved', Ed., All-Union Research Institute of Hygiene and Toxicology of Pesticides, Polymers and Plastic Materials, Kiyv, 1969, 104 (in Russian).
2. Grushevskaya, N. Yu., Determination of vulcanization accelerator of vulkacite-*P*-extra-*N* and some transformation products in sanitary and chemical investigation of rubber, *Gig. Sanit.*, 4, 59, 1987 (in Russian).
3. Yamazaki, T., Inoue, T., Yamada, T., and Tanimura, A., Analysis of residual vulcanization accelerators in baby bottle rubber teats, *Food Addit. Contam.*,3, 145, 1986.
4. Pulyakhin, G. T., Toxicological characteristics of dimethyldicarbamate sodium salt, in *Hygiene and Toxicology of High-Molecular-Mass Compounds and of the Chemical Raw Material Used for Their Synthesis*, Proc. 4th All-Union Conf., S. L. Danishevsky, Ed., Khimiya, Leningrad, 1969, 201 (in Russian).

DIMETHYLDITHIOCARBAMIC ACID, ZINC SALT

Molecular Formula. $C_6H_{12}N_2S_4Zn$
M = 305.80
CAS No 137-30-4
RTECS No ZH0525000
Abbreviation. DMDCZ.

Synonyms and **Trade Names.** Carbamate Z; Dimethyldithiocarbamate, zinc salt; Methyl cymate; Methyl zineb; Methyl Ziram; Zimate; Ziram.

Properties. White, crystalline powder. Solubility in water is 65 mg/l, soluble in ethanol (<2.0 g/100 ml at 25°C).

Applications. Used as a vulcanization accelerator for rubber; a pesticide; a foliar fumigant, mainly on fruit and nuts.

Migration Data. Investigation of aqueous, hexane, alcohol (20 and 40%) and acid extracts from vulcanizates based on butadiene-nitrile rubber revealed migration of 0.1 mg DMDCZ/l.[1]

Migration of dithiocarbamates into distilled water from natural rubber (67 and 1.0 cm^{-1}; 20 and 38°C; 24 hours) reached 0.4 to 0.6 mg/l.[1]

A study of aqueous extracts and simulant media (20, 40, 100°C; 1 and 5 days; 0.5 cm^{-1}) revealed migration of the accelerators introduced into the vulcanizates based on smoked-sheet natural rubber and products of their transformation (dithiocarbaminates) at levels from traces to 0.43 mg/l.[2,3]

Dimethyl dithiocarbamates were detected by the GC determination at levels up to 3.2 mg/g rubber (as dithiocarbamic acid) in chloroform-acetone extracts from isoprene rubber teats for baby bottles. Dialkyldithiocarbamates can form secondary amines by acid hydrolysis, although their levels in the extracts only made a minor contribution to the total level of measured secondary amine precursors.[4]

Acute Toxicity.

Observations in man. 0.5 liter of DMDCZ solution of unknown concentration was fatal within a few hours of nonspecific pathology.[5]

Animal studies. The LD_{50} values are reported to be 1.2 to 1.4 g/kg BW in rats, 0.34 to 0.8 g/kg BW in mice, and 0.1 to 0.2 mg/kg BW in rabbits and guinea pigs.[6-8] Administration of the lethal doses was accompanied by depression followed by diarrhea, paresis, hindlimb paralysis and convulsions. Animals usually died on days 2 to 5.

Repeated Exposure revealed a marked cumulative effect (with administration of 1/20 LD_{50}). K_{acc} is 1.2 (by Kagan). Doses of 100 mg/kg BW caused some of the rabbits to die after 3 to 7 administrations. Marked leukopenia is reported in rabbits after administration of 70 mg/kg BW for a week.[7]

A dose of 25 mg/kg BW appeared to be harmless in sheep. Administration of 5.0 and 25 mg/kg diet did not affect hematology analyses or the relative weights of the visceral organs in dogs. No morphological changes are found in the visceral organs and thyroid of rats which had received DMDCZ with the feed (0.25% DMDCZ in the diet) for a month.[9]

Short-term Toxicity. Rabbits were given 450 mg/kg BW for 3.5 months. Hematology analyses were affected. The treatment produced disorders of metabolism, blood protein balance, and liver function.[8]

Long-term Toxicity. Rats and rabbits were gavaged with 10 mg/kg BW for 6 months.[8] The treatment caused retardation of BW gain, a decline in cholinesterase activity, reduction in the content of *SH*-groups, anemia, and leukopenia. Gross pathology examination revealed changes in the viscera. In a 2-year feeding study in rats, epiphyseal abnormalities in the long bones of the hind legs were observed at the highest dose tested (2.0 g/kg in the diet).[10]

Antonovich[11] has reported the LOAEL of 1.0 mg DMDCZ/kg BW in a 9-month oral study. According to other data, the safe dose is 12.5 mg DMDCZ/kg BW for rats, and 5.0 mg DMDCZ/kg BW for dogs (*JECFA,* 1967).

Reproductive Toxicity. DMDCZ caused gonadotoxic, embryotoxic and teratogenic effect. It is capable of altering the period of gestation, reducing fertility, and causing fetal destruction and sterility.[12-14] Minimal maternal toxic dose lies between 9.5 and 16.2 mg/kg BW, the maternal NOEL was found to be 9.5 mg/kg BW.[16]

Embryotoxicity. The doses of 50 and 100 mg/kg BW administered by gastric intubation to pregnant CD rats reduced fetal weight. Doses of 25 mg/kg BW and above given on days 6 to 15 of gestation resulted in embryotoxic effects and maternal toxicity signs. Dose of 100 mg/kg BW was lethal to 50% of pregnant rats. The embryotoxic effect was also evident later, when administration of the substance had ceased.[10]

Gonadotoxicity and ***Teratogenicity.*** *I/p* injection of single and cumulative doses induced significant increase in the frequency of sperm shape abnormalities in mice in a linear dose-effect manner.[1]

Increased incidence of sterility, embryonic deaths, and fetuses with skeletal malformations were observed in the offspring of male mice treated with DMDCZ by gavage in daily doses of 0.1 and 1.0 mg% for 3 weeks. These males were mated with normal females.[15] Newborn rats show curvature of the tail and growth retardation.[12]

Mutagenicity. Mutagenic activity depends more on carbamate than on zinc. According to other data, the reason activity of zinc and other dithiocarbamates damages DNA is apparently inhibition of dismutase and catalase.[17,18]

Observations in man. Increased frequency of CA was seen in peripheral blood lymphocytes of workers who handled and packaged DMDCZ.[17]

In vitro genotoxicity. Mutagenicity in the Ames test is increased in conditions favorable to the formation of anionic radicals of oxygen. According to Franekic et al., DMDCZ appeared to be a direct mutagen in *Salmonella typhimurium* strain *TA100.*[19]

Injection of 50 to 100 mg DMDCZ/kg BW caused an increase in incidence of micronuclei in polychromatic erythrocytes in mice.[20] DMDCZ proved to be strong chromosome breaking agent in the Chinese hamster epithelial cells and Chinese hamster ovary cells in the presence of *S9* metabolism.[18]

In vivo cytogenetics. Equivocal data were observed in *Dr. melanogaster*.[21,22]

Carcinogenicity. B6C3F_1 mice were exposed to the dietary levels of 0.6 and 1.2 g DMDCZ/kg BW over a period of 103 weeks. Increased incidence of benign lung tumors was noted in female mice. There was a dose-dependent increase of thyroid carcinomas in male F344/N rats given 0.3 and 0.6 g DMDCZ/kg BW for 103 weeks.[23]

20 administrations of 75 mg/kg BW to mice for 2.5 months produced a weak carcinogenic effect (adenomata). The RNA content in the liver and lungs increased along with continuation of the period of observation.[24]

Carcinogenicity classifications. An IARC Working Group concluded that there is inadequate evidence for the carcinogenicity of DMDCZ in *experimental animals* and there were no data available to evaluate the carcinogenicity of DMDCZ in *humans*.

IARC: 3;

NTP: P - N - N - E (feed).

Chemobiokinetics. DMDCZ is shown to accumulate in the visceral organs. Its metabolism in the body occurs to form much more toxic compounds: *tetramethylthiuram disulfide, dimethylamine salt of dithiocarbamic acid, carbon disulfide, dimethylamine*.[7] Metabolites were found in the blood, kidneys, liver, ovaries, spleen, and thyroid of female rats 24 hours after oral administration of radiolabelled DMDCZ. When pregnant rats were given 55 mg/kg BW, a metabolic product, *dimethylamine salt* of *dimethyldithiocarbamic acid*, was found in the amniotic fluid, placenta and fetal tissues.[12]

Metabolites are predominantly removed in the urine and feces, and *dimethylamine* and *carbon disulfide* in the expired air. Unchanged DMDCZ is excreted in the feces.[25]

Regulations. ***U.S. FDA*** (1998) approved the use of DMDCZ (1) in adhesives as a component of articles intended for use in packaging, transporting, or holding food in accordance with the conditions prescribed by 21 CFR part 175.105; (2) as an accelerator of rubber articles intended for repeated use in producing, manufacturing, packing, processing, treating, packaging, transporting, or holding food (total not to exceed 1.5% by weight of the rubber product) in accordance with the conditions prescribed by 21 CFR part 177.2600; (3) as an adjuvant substance in the manufacture of foamed plastics intended for use in contact with food, subject to the provisions prescribed in 21 CFR part 178.3010; and (4) in the production of animal glue which may be safely used as a component of articles intended for use in producing, manufacturing, packaging, processing, preparing, treating, packing, transporting, or holding food, subject to the provisions prescribed in 21 CFR part 178.3020.

Recommendations. ***FAO/WHO*** (1990). ADI of 0.02 mg/kg BW for humans was confirmed in 1980 (*Codex Committee on Pesticide Residues*).

Standards.

U.S. Residues in food: 7.0 mg/kg in fruits and vegetables (calculated as *zineb*).

Russia. PML in food: 0.03 mg/l. No residues of DMDCZ (pesticide) is permitted in food products.

References:

1. Prokof'eva, L. G., in *Hygiene of Use and Toxicology of Pesticides and Polymeric Materials*, A. V. Pavlov, Ed., Coll. Sci. Proc. All-Union Research Institute of Hygiene and Toxicology of Pesticides, Polymers and Plastic Materials, Kiyv, Issue No 17, 1987, 153 (in Russian).
2. Shumskaya, N. I., Provorov, V. N., Tolgskaya, M. S., Yemel'yanova, L. V., and Chernevskaya, N. M., Toxicity of some vulcanizates intended for manufacture of children's toys, in *Hygiene Aspects of the Use of Polymer Materials and of Articles Made of Them*, L. I. Medved', Ed., All-Union Research Institute of Hygiene and Toxicology of Pesticides, Polymers and Plastic Materials, Kiyv, 1969, 104 (in Russian).
3. Grushevskaya, N. Yu., Determination of vulcanization accelerator of vulkacite-*P*-extra-*N* and some transformation products in sanitary and chemical investigation of rubber, *Gig. Sanit.*, 4, 59, 1987 (in Russian).

4. Yamazaki, T., Inoue, T., Yamada, T., and Tanimura, A., Analysis of residual vulcanization accelerators in baby bottle rubber teats, *Food Addit. Contam.*,3, 145, 1986.
5. Buklan, A. J., Acute ziram poisoning, *Forensic Med. Expertize*, 17, 51, 1974 (in Russian).
6. Hodge, H. C., Mayunard, E. A., Downs, W., et al., Acute and short-term oral toxicity tests of ferric dimethyldithiocarbamate (febram) and zinc dimethyldithiocarbamate (ziram), *J. Am. Pharmacol. Assoc.*, 41, 662, 1952.
7. Korablev, M. V., *Pharmacology and Toxicology of Dithiocarbamic Acid Derivatives*, Author's abstract of thesis, Kaunas, 1965, 26 (in Russian).
8. Arkhangel'skaya, L. I., Data on toxicological characteristics of ethyl- and methylzimate, in *Toxicology and Hygiene of High-Molecular-Mass Compounds and of the Chemical Raw Material Used for Their Synthesis*, Proc. 2nd All-Union Conf., A. A. Letavet and S. L. Danishevsky, Eds., Leningrad, 1964, 78 (in Russian).
9. Chernov, O. V., in *Toxicology and Pharmacology of Pesticides and Other Chemical Compounds*, Zdorov'ya, Kiyv, 1967, 163 (in Russian).
10. Enomoto, A., Harada, T., Maita, K., et al., Epiphyseal lesions of the femur and tibia in rats following oral chronic administration of zinc dimethyldithiocarbamate (ziram), *Toxicology*, 54, 45, 1989.
11. Antonovich, E. A., in *Pesticides Handbook*, L. I. Medved', Ed., 2 ed., Urozhai, Kiyv, 1977, 187 (in Russian).
12. Martson', L. V. and Ryazanova, R. A., in *Pesticides Handbook*, L. I. Medved', Ed., 2nd ed., Urozhai, Kiyv, 1977, 181 (in Russian).
13. Giavini, E., Vismara, C., and Broccia, M. L., Pre- and postimplantation embryotoxic effects of zinc dimethyldithiocarbamate (ziram) in the rat, *Ecotoxicol. Environ. Saf.*, 7, 531, 1983.
14. Hemavathi, E. and Rahiman, M. A., Toxicological effects of ziram, thiram, and dithane M-45 assessed by sperm shape abnormalities in mice, *J. Toxicol. Environ. Health*, 38, 393, 1993.
15. Cilievici, O., Cracium, C., and Ghidus, E., Decreased fertility, increased dominant lethals, skeletal malformations induced in the mouse by ziram fungicide, *Morphol. Embryol.* (Bucur), 29, 159, 1983.
16. Ema, M., Itami, T., Ogawa, Y., and Kawasaki, H., Developmental toxicity evaluation of zinc dimethyldithiocarbamate (ziram) in rats, *Bull. Environ. Contam. Toxicol.*, 53, 930, 1994.
17. Pilinskaya, M. A., Chromosomal aberrations in persons handling ziram under industrial conditions, *Genetica*, 6, 157, 1971 (in Russian).
18. Mosesso, P., Turchi, G., Cinelli, S., et al., Clastogenic effects of the dithiocarbamate fungicides thiram and ziram in Chinese hamster cell lines cultured *in vitro, Teratogen., Carcinogen., Mutagen.*, 14, 145, 1994.
19. Franekic, J., Bratulic, N., Pavlica, M., and Papes, D., Genotoxicity of dithiocarbamates and their metabolites, *Mutat. Res.*, 325, 65, 1994.
20. Hemavathi, E. and Rahiman, M. A., Effect of ziram, thiram and dithane M-95 on bone marrow cells of mice - assessed by micronucleous test, *Bull. Environ. Contam. Toxicol.*, 56, 190, 1996.
21. Foureman, P., Mason, J. M., Valencia, R., and Zimmering, S., Chemical mutagenesis testing in *Drosophila*, X. Results of 70 coded chemicals tested for the National Toxicology Program, *Environ. Molec. Mutagen.*, 23, 208, 1994.
22. Hemavathy, K. C. and Krishnamurthy, N. B., Genotoxicity studies with Cuman L in *Drosophila melanogaster, Environ. Molec. Mutagen.*, 14, 252, 1989.
23. *Carcinogenicity Bioassay of Ziram in F344/N Rats and B6C3F$_1$ Mice (Feed Study)*, NTP Technical Report Series No 238, Research Triangle Park, NC, 1983.
24. Khitsenko, I. I., and Chernov, O. V., in *Pesticides Toxicology and the Clinical Features of Poisoning*, Proc. All-Union Research Institute of Hygiene and Toxicology of Pesticides, Polymers and Plastic Materials, Kiyv, 1968, 770 (in Russian).
25. Ismirova, N., and Marinov, V., Distribution and excretion of ^{35}S-ziram and metabolic products after 24 hours following oral administration of the preparation to female rats, *Exp. Med. Morphol.*, 11, 152, 1972 (in Russian).

4,4'- (2,3-DIMETHYLTETRAMETHYLENE) DIPYROCATECHOL

Molecular Formula. $C_{18}H_{22}O_4$
M = 302.40
CAS No 500-38-9
RTECS No UX1750000

Synonyms. 2,3-Bis(3,4-dihydroxyphenylmethyl)butane; Dihydronorguaiaretic acid; Nordihydroguaiaretic acid.

Applications. Antioxidant.

Acute Toxicity. LD_{50} is 2.0 g/kg BW in rats and mice, and 0.83 g/kg BW in guinea pigs.[1]

Teratogenicity. Rabbits were given 950 mg/kg BW *s/c* on day 12 of pregnancy. The treatment caused no increase in the rate of developmental abnormalities.[2]

Mutagenicity. D. produced genotoxic and cytotoxic effects on the production of SCE, and on the level of mitotic index in cultured human lymphocytes and in mouse bone marrow cells *in vivo.*[3]

Regulations. ***U.S. FDA*** (1998) approved the use of D. in the manufacture of resinous and polymeric coatings for the food-contact surface of articles intended for use in producing, manufacturing, packing, processing, preparing, treating, packaging, transporting, or holding food (for use only as polymerization cross-linking agent in side seam cements for containers intended for use in contact with food (only of the identified types) in accordance with the conditions prescribed by 21 CFR part 175.300.

References:

1. *Advances in Food Research*, 3, 197, 1951.
2. DeSesso, J. M. and Goeringer, G. C., Ethoxyquin and nordihydroguaiaretic acid reduce hydroxyurea developmental toxicity, *Reprod. Toxicol.*, 4, 267, 1990.
3. Madrigal-Bujaidar, E., Diaz Barriga, S., Cassani, M., Molina, D., and Ponce, G., *In vivo* and *in vitro* induction of sister-chromatid exchanges by nordihydroguaiaretic acid, *Mutat. Res.*, 412, 139, 1998.

DINAPHTHYLMETHANE SULFOACIDS and DINAPHTHYLMETHANE SULFOACIDS, SODIUM SALTS, a mixture

Trade Name. Dispersant NF.

Properties. A brown liquid (product is neutralized with NH_4OH) or a gray powder (product is neutralized with *NaOH*), readily soluble in water, insoluble in alcohol.

Applications. Used in rubber industry as a stabilizer of latexes for medical applications.

Migration level varies from 0.1 to 7.0 mg/l.[1]

Acute Toxicity. LD_{50} is 8.7 g/kg BW for rats, 3.8 g/kg BW for male and 3.4 g/kg BW for female mice. Death occurs in 2 to 5 days after administration. Acute poisoning is accompanied with agitation, convulsions, labored breathing, unkempt fur. Gross pathology examination revealed congestion and flaccidity of the liver, kidneys, and spleen.[1,2]

Repeated Exposure revealed moderate cumulative properties on *i/p* administration of 1/5 or 1/20 LD_{50} to mice for 30 days.

Standards. ***Russia.*** Recommended PML of 2.5 mg/l.[1]

References:

1. Shumskaya, N. I., Mel'nikova, V. V., Smeleva, E. V., et al., Substantiation of permissible migration level for dispersant NF migrating from medical latex devices, *Gig. Sanit.*, 11, 70, 1990 (in Russian).
2. Goryainova, A. N. and Shavrikova, L. N., in *Proc. 9th Sci. Conf. of Junior Hygienists and Sanitarians*, Moscow, 1963, 54 (in Russian).

N,p-DINITROSO-*N*-METHYLANILINE

Molecular Formula. $C_7H_7N_3O_2$
M = 165.17
CAS No 99-80-9
RTECS No BX9350000
Abbreviation. DNMA.

Synonyms and **Trade Names.** Elastopar; Nitrozan K; *N*-Methyl-*N*,4-dinitrozoaniline; *N*-Methyl-*N*,4-dinitrosobenzenamine.

Properties. Khaki-colored, amorphous powder. Poorly soluble in water, on heating, dissolves in alcohol.

Applications. DNMA is used as a modifier and activator of vulcanization in the mixtures of natural and synthetic rubber, and also of polyethylene.

Acute Toxicity. LD_{50} is 0.9 g/kg BW in rats, and 1.85 g/kg BW in mice. Poisoning is accompanied by general depression, adynamia and diarrhea. The animals look ungroomed. *MetHb* formation was found when doses of 0.2 to 0.9 g/kg BW were given. Survivors recovered in a week.[1]

Short-term Toxicity. Administration of 18 and 90 mg/kg BW caused death in the majority of the treated animals. Death occurred earlier with the smaller dose. Rabbits were given 200 mg/kg BW dose by gavage for 4 months. The treatment led to a reduction in alkaline phosphatase activity and an increase in aldolase activity. Gross pathology and histology examination revealed moderate congestion in the lungs, liver, kidneys, and myocardium.[2]

Long-term Toxicity. Chronic exposure affected liver and kidney functions in rats to the end of the experiment (doses not indicated). Gross pathology examination revealed circulatory disturbances and parenchymatous dystrophy of the liver and myocardium and an increase in the relative weights of the brain, liver, and spleen at a 90 mg/kg BW dose level. Observed effects appeared to be dose dependent.[1]

Mutagenicity. DNMA was positive in *Salmonella typhimurium* mutagenicity assay (5.0 μg/plate) without metabolic activation system (*S9*).[3]

Carcinogenicity. A number of nitrosocompounds are shown to have carcinogenic potential. DNMA is considered as an equivocal tumorigenic agent by RTECS criteria.

Carcinogenicity classification. An IARC Working Group concluded that there is limited evidence for the carcinogenicity of DNMA in *experimental animals* and there were no adequate data available to evaluate the carcinogenicity of DNMA in *humans*.

IARC: 3.

References:

1. Shurupova, Ye. A., Belova, G. B., and Piotrovskaya, O. G., Toxicologic characteristics of diphenyloxide-4,4'-disulfohydrazide and *N*,4-dinitroso-*N*-methylaniline, in *Hygiene and Toxicology of High- Molecular-Mass Compounds and of the Chemical Raw Material Used for Their Synthesis*, Proc. 4th All-Union Conf., S. L. Danishevsky, Ed., Khimiya, Leningrad, 1969, 217 (in Russian).
2. Mezentseva, N. V., *N*-Nitroso-*N*-methylparanitroso aniline (elastopar), in *Toxicology of New Chemical Substances Used in Rubber and Tyre Industry*, Z. I. Israel'son, Ed., Meditsina, Moscow, 1968, 187 (in Russian).
3. *Progr. Clin. Biol. Res.*, 141, 407, 1984.

3,7-DINITROSO-1,3,5,7-TETRAAZABICYCLO(3.3.1)NONANE

Molecular Formula. $C_5H_{10}N_6O_2$

M = 186.21

CAS No 101-25-7

RTECS No XA5250000

Abbreviation. DTN.

Synonyms and **Trade Name.** Dinitrosopentamethylenetetramine; 1,5-Methylene-3,7-dinitroso-1,3,5,7-tetraazacyclooctane; Porofor Chkhz-18.

Properties. Light cream-colored powder or needles. Solubility in water is about 1.0%, soluble in ethanol.

Applications. Blowing agent in the production of synthetic and natural rubber, polyvinyl chloride plastisols, epoxy, polyester, and silicone resins.

Acute Toxicity. LD_{50} exceeded 2.9 g/kg BW in rats. Poisoning was accompanied by convulsions, tremor, respiratory system damage.[027]

Mutagenicity.

In vitro genotoxicity. Positive in *Salmonella* mutagenicity assay.[1]

Carcinogenicity. No tumors were found in 15 male and female 4-week old Fischer rats given 9 mg/animal by oral gavage 5 days a week for 1 year.

Carcinogenicity classification. An IARC Working Group concluded that there is inadequate evidence for the carcinogenicity of DTN in *experimental animals* and there were no adequate data available to evaluate the carcinogenicity of DTN in *humans*.

IARC: 3.

Reference:

1. *Progress in Mutat. Res.*, 1, 302, 1981.

DI-2-OCTYL-*p*-PHENYLENEDIAMINE

Molecular Formula. $C_{22}H_{40}N_2$

M = 324.57

CAS No 103-96-8

Synonyms and **Trade Names.** Antozite; *N,N'*-Bis(1-methylheptyl)-1,4-benzenediamine; *N,N'*-Bis(1-methylheptyl)-*p*-phenylenediamine; *N,N'*- Bis(2-octyl)-*p*-phenylenediamine; Elastozone 30; Santoflex 217; Tenemene.

Properties. Thick, dark-brown liquid. Poorly soluble in water, soluble in alcohol.

Applications. Used as an antioxidant and anti-ozone agent in the manufacture of natural, isoprene, butadiene, and styrene-butadiene rubbers.

Acute Toxicity. LD_{50} is 2.77 g/kg BW in mice, and 3.83 g/kg BW in rats. Following administration of the lethal doses, mice died in 2 to 4 days, rats in 3 to 6 days. Histology examination revealed hemorrhagic pneumonia, signs of congestion with the areas of fatty destruction, and tiny focal hemorrhages in the liver and signs of congestion in the spleen and brain.[023]

DIOXIME-*p*-BENZOQUINONE

Molecular Formula. $C_6H_6N_2O_2$

M = 138.13

CAS No 105-11-3

RTECS No DK4900000

Abbreviation. DBQ.

Synonyms. Dioxime 2,5-cyclohexadiene-1,4-dione; *p*-Quinone dioxime.

Properties. Brown, highly inflammable, fine crystalline powder. Solubility in water is 200 mg/l at 20°C. Odor perception threshold is 1.0 g/l; taste perception threshold is 0.5 mg/l; color change threshold is 0.1 mg/l.

Applications. Used as a vulcanization agent in synthetic rubber and acrylic resins production.

Acute toxicity. Single oral dose of 0.25 g/kg BW produced no evidence of hepatotoxicity.[1] The LD_{50} values are reported to be 0.46 or 1.6 g/kg BW in rats, and 1.4 to 1.5 g/kg BW in mice. Poisoning is accompanied by apathy with subsequent death within 5 hours.[2]

Short-term Toxicity. Rabbits were dosed by gavage with 20 mg/kg BW (every other day for 2 months and daily for two further months). The treatment decreased blood prothrombin time and increased the activity of serum aldolase and alkaline phosphatase. There were no changes in behavior, BW gain, or hematology analyses. Gross pathology examination revealed visceral congestion.[3]

Mutagenicity.

In vitro genotoxicity. Appears to be a direct-acting mutagen in *Salmonella thyphimurium* strain *TA98*. DBQ gave positive results in the *rec*-assay using spores of *Bac. subtilis* strains *H17* and *M45*. DNA-damaging activity of DBQ was associated with free hydroxyl groups of this compound.[4]

In vivo cytogenetics. Negative results were observed after oral administration to female rats in both the bone marrow micronucleus test and *in vivo* liver unscheduled DNA synthesis test.[3]

Carcinogenicity. Dietary administration of DBQ was carcinogenic to female F344 rats, causing neoplasms of the urinary bladder. Mutagenic activity of DBQ might play a contributory role to the induction of bladder cancer in rats. No carcinogenic effect was observed in $B6C3F_1$ mice of either sex and in male F344 rats.[5]

Carcinogenicity classification.

NTP: N - P - N - N (feed).

Regulations. ***U.S. FDA*** (1998) approved the use of DBQ in the manufacture of rubber articles intended for repeated use in producing, manufacturing, packaging, processing, preparing, treating, packing, transporting, or holding food in accordance with the conditions prescribed by 21 CFR part 177.2600.

Standards. ***Russia*** (1995). MAC and PML: 0.1 mg/l (organolept., color).

References:

1. Westmoreland, C., Gerge, E., York, M., and Gatehouse, D., *In vivo* genotoxicity studies with *p*-benzoquinone dioxime, *Environ. Molec. Mutagen.*, 19, 71, 1992.
2. Vorob'yeva, R. S. and Mezentseva, N. V., Paraquinone dioxime, in *Toxicology of New Chemical Substances Used in Rubber and Tyre Industry*, Z. I. Israel'son, Ed., Meditsina, Moscow, 1968, 152 (in Russian).
3. Grushko, Ya. M., *Harmful Organic Compounds in Industrial Liquid Effluents*, Khimiya, Leningrad, 1976, 189 (in Russian).
4. Ueno, S. and Ishizaki, M., Mutagenicity of organic rubber additives, Abstract, *Sangyo Igaku*, 26, 147, 1984 (in Japanese).
5. *Bioassay of p-Quinone Dioxime for Possible Carcinogenicity*, NTP Technical Report Series No 179, Research Triangle Park, NC, 1979.

N,N'-DI(OXYDIETHYLENE)THIURAM DISULFIDE

Molecular Formula. $C_{10}H_{16}N_2S_4O_2$

Synonym. Dimorpholinothiuram disulfide.

Properties. White-yellow, crystalline powder.

Applications. Used as a vulcanization accelerator in the rubber industry.

Acute Toxicity. In rats, LD_{50} is 2.6 g/kg BW.

Short-term Toxicity. Rabbits were dosed by gavage with a 20 mg/kg BW dose every other day for the first 2 months and then daily for two further months. This treatment impaired protein balance and fat (cholesterol) and electrolyte (*Na* and *K*) metabolism, caused a reduction in alkaline phosphatase activity, and dystrophic changes and circulatory disturbances in the visceral organs.

Carcinogenicity. Nitroso-compounds are known to be powerful carcinogens.

Reference:

Vorob'yeva, R. S., Dimorpholinothiuramdisulfide, in *Toxicology of New Chemical Substances Used in Rubber and Tyre Industry*, Z. I. Israel'son, Ed., Meditsina, Moscow, 1968, 108 (in Russian).

1,3-DIPHENYLGUANIDINE

Molecular Formula. $C_{13}H_{13}N_3$

M = 211.27

CAS No 102-06-7

RTECS No MF0875000

Abbreviation. DPG.

Synonyms and **Trade Names.** *N,N'*-Diphenylguanidine; DPG accelerator; Melaniline; Vulkacid DC.

Properties. Odorless, crystalline powder or white solid, with a color ranging from white to light yellow or lilac. Solubility in water is 0.08%, soluble in alcohol.

Applications. A primary and secondary vulcanization accelerator widely used in processing of acryl-butadiene, styrene-butadiene, chloroprene, siloxane, butyl rubber, etc. Predominantly used in combination with other accelerators.

Migration from rubber into liquid media varies within the limits 0.02 to 1.3 mg/l. According to the data available, migration from butadiene-nitrile vulcanizates into water after 3-day exposure at 20°C was determined at the level of 2.9 mg/l.[1] Migration of DPG from vulcanizates based on polybutadiene rubber in composition with natural rubber and polyisoprene rubber into simulant media at 20°C amounted to 0.5 mg/l.[2] A study of extracts (water and simulant media; 0.1 and 0.5 cm^{-1}; 1 and 24 hours) from vulcanizates based on synthetic polyisoprene rubbers revealed migration of DPG at the level of 0.1 mg/l^3 or 1.2 mg/l.[028] According to other data, study of extracts (water and model media; 1 hour to 5 days; 20 and 100°C; 0.1 to 2.0 cm^{-1}) revealed migration of 0.05 to 1.0 mg DPG/l.[4]

Acute Toxicity. The LD_{50} values are 190 to 850 mg/kg BW in rats, 258 to 625 mg/kg BW in mice, 507 mg/kg BW in guinea pigs, and 246 mg/kg BW in rabbits.[5-7] The acute action threshold appeared to be 25 mg/kg BW identified for the minimal signs of anemia and an increase in the blood peroxidase activity. The NOEL seems to be 10 mg/kg BW. Sensitivity to DPG decreased with age of animals. Poisoning is manifested by unsteady gait, convulsions, and increased sensitivity to pain. There were also liver damage and an increase in its relative weight. Gross pathology examination revealed marked hyperemia and considerable morphological changes. Death occurs in 1 to 2 days.[7]

Repeated Exposure failed to reveal cumulative properties in rats. Moderate capacity for accumulation was found in mice. CD-1 mice were administered 0.06 to 16 mg DPG/kg BW for 8 weeks. There were no dose-related changes in BW gain, gross pathology and histology findings or organ weights.[8] In a two-week study, F344 rats and $B6C3F_1$ mice received feed containing 250, 500, 750, 1500, or 3000 ppm DPG. All animals survived to the end of the study. Feed consumption and BW gain were lowered in rats that received 750 ppm DPG and more. No compound-related gross lesions were found.[9]

Short-term Toxicity. Oral administration of 32 mg/kg BW to rats for 1 to 3 months produced signs of anemia, reduction in blood catalase and peroxydaze activity, liver dysfunction, and pathomorphological changes. A stable reduction in the content of cholic acids in the bile when 5.0 g/kg BW was given daily to dogs is reported.[10] In a 13-week study, F344 rats and $B6C3F_1$ mice received DPG in their diet. BW was found to be lower in the animals that received the diet containing 1500 and 3000 ppm DPG. The treatment caused changes in hematology parameters. Alkaline phosphatase activity and bile acid concentrations were increased and considered to be an indication of cholestasis in rats.[9]

Long-term Toxicity. Rabbits received 50 mg/kg BW by gavage every other day for 1.5 months and daily for the next 4 months. Toxic action was manifested by exhaustion, a decrease in erythrocyte count and *Hb* level, dysbalance of blood proteins. Gross pathology examination revealed changes in the liver and kidney.[9] Administration of 1/100 LD_{50} to rats for 10 months as solutions in milk and in water starch suspension caused a number of signs of intoxication which were very marked in young animals: retardation of BW gain, hematology changes, increase in leukocyte and phagocyte activity, reduction in the relative liver weight.[10,11]

Reproductive Toxicity.

Gonadotoxicity. Reported data appear to be controversial. A testicular toxicity and fertility study was carried out in CD-1 mice.[11] Animals were treated by daily gavage during 8 weeks of premating period. DPG did not exert any significant adverse effect on fertility at dose levels up to 16 mg/kg BW. Following chronic exposure to DPG in a 0.025% solution of acetic acid *ad libitum*, mice and hamsters exhibited changes in sperm morphology, a decline in the number of spermatozoa and in the weights of the testes. The shape of the seminal ducts was altered, fertility index reduced.[12] In the rats that received the diet containing 3000 ppm DPG, secretary depletion of the seminal vesicles and prostate gland, epididymal hypospermia, spermatogenic arrest, and significant reduction in the absolute weights of the prostate gland, seminal vesicles, and testis were observed. A mean length of the estrous cycle was greater in rats (750 and 1500 ppm) and mice (3000 ppm).[9] Male Swiss rats were administered by gavage 0.06, 0.25, 1,4, or 16 mg/kg BW for 8 weeks prior to mating. The treatment produced no microscopic changes in the testes, in fertility, and reproductive performance.[8]

Embryotoxicity. In the above described study, the treatment caused reduced number of implants per pregnancy and fetal mortality in rats.[12]

Teratogenic effect was not observed in mice.[13]

Mutagenicity. DPG is a direct acting mutagen, it showed negative response in *Salmonella typhimurium* strains *TA98, TA100, TA1535,* and *TA1537* with and without *S9* metabolic activation. Micronucleus test in $B6C3F_1$ mice appeared to be negative in males and equivocal in females.[9,14]

Carcinogenicity. Rats were orally exposed to a 40 mg/kg BW dose once a week for 6 weeks; 6 months of follow-up revealed no tumors.

Chemobiokinetics. DPG is rapidly absorbed and distributed throughout the body tissues: 30 min after administration of 100 mg DPG/kg BW, the substance was found in the blood; in an hour it was discovered in all the visceral organs; after 24 hours, it was found in the urine. DPG excretion with the urine had ceased on the day 6.

Regulations. ***U.S. FDA*** (1998) regulates the use of DPG as an accelerator in the manufacture of rubber articles intended for repeated use in producing, manufacturing, packaging, processing, preparing, treating, packing, transporting, or holding food (up to 1.5% by weight of the rubber product) in accordance with the conditions prescribed by 21 CFR part 177.2600.

Standards. ***Russia*** (1994). MAC: 1.0 mg/l; PML in food: 0.15 mg/l.

References:

1. Prokof'eva, L. G., in *Hygiene of Use and Toxicology of Pesticides and Polymeric Materials*, A. V. Pavlov, Ed., Coll. Sci. Proc. All-Union Research Institute of Hygiene and Toxicology of Pesticides, Polymers and Plastic Materials, Kiyv, Issue No 17, 1987, 153 (in Russian).
2. Stankevich, V. V., Ivanova, T. P., and Prokof'yeva, L. G., Hygienic studies of 'food-grade' resins on the base of a new diene rubber, in *Hygiene and Toxicology of High-Molecular-Mass Compounds and of the Chemical Raw Material Used for Their Synthesis*, Proc. 6th All-Union Conf., B. Yu. Kalinin, Ed., Khimiya, Leningrad, 1979, 96 (in Russian).
3. Vlasyuk, M. G., Toxicological and hygienic characteristics of 'food-grade' resins made of SKI-3P rubber, *Ibid*, p. 64.
4. Kazarinova, N. F. and Ledovskikh, N. G., *Kauchuk i Rezina*, 1, 26, 1978 (in Russian).
5. Bourne, H. G., Yee, H. T., and Seferian, S., The toxicity of rubber additives. Finding from a survey of 140 plants in Ohio, *Arch. Environ. Health* (Chicago), 16, 700, 1968.
6. Vlasyuk, M. G. and Prokof'eva, L. G., Toxicological characteristics of diphenylguanidine, in *Hygienic Aspects of the Use of Polymeric Materials*, Proc. 2nd All-Union Meeting on Health and Safety Monitoring of the Use of Polymeric Materials in Construction, Kiyv, 1976, 60 (in Russian).
7. Arkhangel'skaya, L. N. and Roschina, T. A., Diphenylguanidine, in *Toxicology of New Chemical Substances Used in Rubber and Tyre Industry*, Z. I. Israel'son, Ed., Meditsina, Moscow, 1968, 117 (in Russian).
8. Koeter, H. B. W., Regnier, J. F., and Marwijk, M. W., Effect of oral administration of 1,3-diphenylguanidine on sperm morphology and male fertility in mice, *Toxicology*, 71, 173, 1992.
9. *Toxicity Studies of 1,3-Diphenylguanidine Administered in Feed to F344 Rats and B6C3F$_1$ Mice*, (R. Irwin), NTP Technical Report Series No 42, Research Triangle Park, NC, NIH Publ. 95-3933, September 1995.
10. Vlasyuk, M. G., Data on substantiation of permissible migration level for diphenylguanidine from rubber, *Gig. Sanit.*, 7, 35, 1978 (in Russian).
11. Vlasyuk, M. G., Comparative investigation of sensitivity to diphenylguanidine in the young and adult animals, in *Hygiene of Applications, Toxicology of Pesticides and Clinical Features of Poisoning*, L. I. Medved', Ed., Proc. All-Union Research Institute of Hygiene and Toxicology of Pesticides, Polymers and Plastic Materials, Kiyv, Issue No 9, 1971, 363 (in Russian).
12. Bempong, M. A. and Hall, E. V., Reproductive Toxicity of 1,3-diphenylguanidine: analysis of induced sperm abnormalities in mice and hamsters and reproductive consequence in mice, *J. Toxicol. Environ. Health*, 11, 869, 1983.
13. Yasuda, Y. and Tanimura, T., Effect of diphenylguanidine on development of mouse fetuses, *J. Environ. Pathol. Toxicol.*, 4, 451, 1980.
14. Bempong, M. A. and Mantley, R., Body fluid analysis of 1,3-diphenylguanidine for mutagenicity as detected by *Salmonella* strains, *J. Environ. Pathol. Toxicol. Oncol.*, 6, 293, 1985.

2,2′-DITHIOBISBENZOTHIAZOLE

Molecular Formula. $C_{14}H_8N_2S_4$
M = 332.48
CAS No 120-78-5
RTECS No DL4550000
Abbreviation. DTBT.

Synonyms and **Trade Name.** Altax; Benzothiazole disulfide; Di-2-benzothiazolyl disulfide; Dibenzoyl disul fide; 2,2'-Dithiobis(benzothiazole); 2-Mercaptobenzothiazole disulfide.

Properties. A grayish-yellow free-flowing powder with a faint odor. Poorly soluble in water, soluble in alcohol (2.0 g/l). Perception thresholds for odor and taste are 45 and 50 mg/l, respectively.

Applications. Widely used as a vulcanization accelerator for mixes based on diene-type natural and synthetic rubber, for butyl and isoprene rubber.

Migration from butadiene-nitrile vulcanizates into water after a 3-day exposure at 20°C was determined to be 0.42 mg/l.[1]

Acute Toxicity. LD_{50} is 4.3 g/kg BW in rats, 4.6 g/kg BW in mice, and 6.2 g/kg BW in rabbits.[2] However, according to Radeva,[3] Wistar rats did not die even after ingestion of a 8.0 g/kg BW dose.

Repeated Exposure revealed little evidence of cumulative properties. Rats received 400 and 800 mg/kg BW for 10 days. The treatment increased the amount of total and reduced glutathione in the blood and decreased serum alkaline phosphatase activity.[3]

Reproductive Toxicity.

Gonadotoxicity and ***Embryotoxicity.*** DTBT altered estrous cycle in rats and decreased the rate of conception. Pre- and postimplantation embryolethality were increased.[4] However, according to Ema et al.,[5] there were no pre- or postimplantation loss or adverse effects on pup survival in rats fed up to 1.0% DTBT in their diet (600 mg/kg BW). DTBT was found to be more embryotoxic than 5 other accelerators administered to rats on days 4 and 11 of pregnancy. Nevertheless, according to Alexandrov, it produced no teratogenic effect in rats.[6] In the next study, rats received 5.0, 10, or 20 mg/kg BW on days 1 through 21 of gestation. The treatment with the highest dose produced hydrocephalus, intracerebral hematomas, and disturbances in cranial ossification. Microphthalmia and hydrocephalus occurred in 5.0% of fetuses; 10 mg/kg BW dose had no such effect.[7]

Mutagenicity. DTBT was shown to be positive in mammalian mutagenicity assays.[8]

Carcinogenicity. Negative results are noted in mice.[09]

Regulations. ***U.S. FDA*** (1998) approved the use of DTBT (1) in adhesives as a component of articles intended for use in packaging, transporting, or holding food in accordance with the conditions prescribed by 21 CFR part 175.105; and (2) in the manufacture of rubber articles intended for repeated use in producing, manufacturing, packaging, processing, preparing, treating, packing, transporting, or holding food in accordance with the conditions prescribed by 21 CFR part 177.2600.

Standards. ***Russia*** (1994). MAC and PML: *zero* (organolept., odor), PML in food: 0.15 mg/l.

References:

1. Prokof'eva, L. G., in *Hygiene of Use and Toxicology of Pesticides and Polymeric Materials*, A. V. Pavlov, Ed., Coll. Sci. Proc. All-Union Research Institute of Hygiene and Toxicology of Pesticides, Polymers and Plastic Materials, Kiyv, Issue No 17, 1987, 153 (in Russian).
2. Vaisman, L. I., Zaitseva, N. V., and Mikhailov, A. V., Hygiene standardization of rubber manufacture sewage in water bodies, *Gig. Sanit.*, 2, 17, 1973 (in Russian).
3. Radeva, M., Study upon acute oral toxicity of altax, *Khigiena i Zdraveopazvane*, 4, 326, 1980 (in Bulgarian).
4. Alexandrov, S. E., Effect of vulcanizing accelerants on embryolethality in rats, *Bull. Exp. Biol. Med.*, 93, 87, 1982 (in Russian).
5. Ema, M., Sakamoto, J., Mural, T., and Kawasaki, H., Evaluation of the teratogenic potential of the rubber accelerator dibenzthiazyl disulfide in rats, *J. Appl. Toxicol.*, 9, 413, 1989.
6. Alexandrov, S. E., Embryotoxic effects of vulcanization accelerators, in *Abstracts 13th Sci. Conf. Chemistry and Technology of the Sulfurorganic Compounds of Sulphurous Oils*, 1974, 98 (in Russian).
7. Mirkova, E., Study of the hazard of embryotoxic and teratogenic action of the vulcanization accelerator Altax, *Problems of Hygiene*, Issue No 5, 1980, 83 (in Russian).
8. Hinderer, R. K., Myhr, B., Jagannath, D. R., Galloway, S. M., Mann, S. M., Riddle, J. C., and Brusick, D. J., Mutagenic evaluation of four rubber accelerators in a battery of *in vitro* mutagenic assays, *Environ. Mutagen.*, 5, 193, 1983.

DITHIOCARBAMILIC ACID, ETHYL ESTER

Molecular Formula. $C_9H_{11}NS_2$
M = 197.33
CAS No 13037-20-2
RTECS No FD8892150
Abbreviation. DTAE.

Synonyms. Dithiocarbanilic acid, ethyl ester; Ethyl phenyldithiocarbamate; *N*-Phenyl dithiocarbamate; Phenylcarbamodithioic acid, ethyl ester.

Applications. Vulcanization accelerator.

Migration of dithiocarbamates into distilled water from natural rubber (67 and 1.0 cm^{-1}; 20 and 38°C; 24 hours) reached 0.4 to 0.6 mg/l.[1] A study of aqueous extracts and simulant media (20, 40, 100°C; 1 and 5 days; 0.5 cm^{-1}) revealed migration of the accelerators introduced into the vulcanizates based on smoked-sheet natural rubber and products of their transformation (dithiocarbaminates) at levels from traces to 0.43 mg/l.[2]

Acute Toxicity. LD_0 *i/p* is 300 mg/kg BW in rats.[3] LD_{50} for DTAE zinc salt *i/p* is 600 mg/kg BW in mice.[027]

Reproductive Toxicity. A decrease in viability index was noted in rats exposed orally to the dose of 125 mg/kg BW on days 7 to 15 of gestation.[4]

References:

1. Shumskaya, N. I., Provorov, V. N., Tolgskaya, M. S., Yemel'yanova, L. V., and Chernevskaya, N. M., Toxicity of some vulcanizates intended for manufacture of children's toys, in *Hygiene Aspects of the Use of Polymeric Materials and of Articles Made of Them*, L. I. Medved', Ed., All-Union Research Institute of Hygiene and Toxicology of Pesticides, Polymers and Plastic Materials, Kiyv, 1969, 104 (in Russian).
2. Grushevskaya, N. Yu., Determination of vulcanization accelerator of vulkacite-*P*-extra-*N* and some transformation products in sanitary and chemical investigation of rubber, *Gig. Sanit.*, 4, 59, 1987 (in Russian).
3. Zsolnai, T., The microbial activity of potential isothiocyanate producers, *Arzneimittel-Forsch.*, 16, 1092, 1966.
4. *J. Toxicol. Sci.* (Japanese Soc. of Toxicol. Sciences), Abstract, 8, 337, 1983.

4,4′-DITHIODIMORPHOLINE

Molecular Formula. $C_8H_{16}N_2O_2S_2$
M = 236.38
CAS No 103-34-4
RTECS No QE3325000
Abbreviation. DTDM.

Synonyms and **Trade Name.** Dimorpholino disulfide; *N,N′*-Disulfide morpholine; Morfoline disulfide; Sulfazane.

Properties. White, crystalline powder. Poorly soluble in water, soluble in alcohol. Undergoes decomposition during vulcanization and is converted into *N,N′*-tetrathiodimorpholine (about 60%), which itself is a vulcanizing agent.[1] Exchange processes may occur between DTDM and Altax and Captax with the formation of polysulfide compounds with a variable number of sulphur atoms.[2]

Applications. Used as a vulcanization accelerator in the production of natural and synthetic rubber.

Migration Data. A study of extracts (0.5 cm^{-1}; 20 and 40°C; 1 and 24 hours; distilled water, acid-salt solutions, solutions containing alcohol) revealed migration of DTDM from the vulcanizates intended for contact with food products at the level of 0.1 to 1.0 mg/l.[3] Migration from butadiene-nitrile vulcanizates into water after a 3-day exposure at 20°C appeared to be up to 1.0 mg/l.[4] A special study was carried out with the use of distilled water and model media simulating milk, lactic acid products, and wine (38 to 100°C; 1, 24, 120 hours; 0.5, 0.1, and 0.01 cm^{-1}). Release of DTDM was proportional to its content in the vulcanizate based on α-methylstyrene rubber, the contact time, and temperature, and reached 0.75 mg/l.[5]

Acute Toxicity. In mice, LD_{50} is 2.75 g/kg BW. In rats, reported LD_{50} values are in the range of 7.0 g/kg BW[3] to 4.3 g/kg BW.[6] The acute action threshold (for STI) appeared to be 0.5 g/kg BW. LD_{50} of tetra-dithiodimorpholine is 16.4 g/kg BW in rats.

Repeated Exposure. DTDM exhibits no cumulative properties.

Short-term Toxicity. Administration of 1/10 and 1/20 LD_{50} did not cause mortality in the animals tested. The CNS functions and parenchymatous organs were predominantly affected.[6]

Long-term Toxicity. Mice were dosed by gavage with 5.0 mg DTDM/kg BW and 25 mg DTDM/kg BW for 10 months. The treatment caused no mortality or marked manifestations of intoxication. The higher dose produced changes in the STI and hepatic excretory function. To the end of the experiment there was inhibition of blood pyroracemic acid and cholinesterase activity and an increase in aminotransferase activity. The NOAEL of 5.0 mg/kg BW was established in this study.[6] Administration of 7.0 mg/kg BW for 5 months caused a considerable reduction in the liver weights. Blood formula and catalase and peroxidase activity were not affected. However, hemodynamic disorders in the visceral organs and desquamation of the renal tubular epithelium were found.[3] Rabbits were given 20 mg/kg BW for a 4-month period. The treatment caused disturbance of protein metabolism and increased blood sugar and cholesterol levels. Alkaline phosphatase activity was reduced.[7] Gross pathology examination revealed catarrhal inflammation of the stomach and small intestine and stimulation of the macrophage system.

Mutagenicity.

In vitro genotoxicity. DTDM was negative in *Salmonella* mutagenicity assay at concentrations of 0.001 to 10 mg/ml.[6] It gave positive results in the *rec*-assay using spores of *Bacillus subtilis* strains *H17* and *M45*.[8]

Allergenic Effect. DTDM showed evident sensitization potential in guinea pigs.[9]

Regulations. ***U.S. FDA*** (1998) regulates the use of DTDM (1) in adhesives as a component of articles intended for use in packaging, transporting, or holding food in accordance with the conditions prescribed by 21 CFR part 175.105; and (2) as an accelerator in the manufacture of rubber articles intended for repeated use in producing, manufacturing, packaging, processing, preparing, treating, packing, transporting, or holding food (up to 1.5% by weight of the rubber product) in accordance with the conditions prescribed by 21 CFR part 177.2600.

Standards. ***Russia.*** PML: 0.3 mg/l.

References:

1. Blokh, G. A., *Organic Vulcanization Accelerators and Vulcanizing Agents for Elastomers*, Khimiya, Leningrad, 1978, 155 (in Russian).
2. Petrova, L. F. et al., Toxicological characteristics of dithiomorpholine, *Kauchuk i Rezina*, 12, 29, 1983 (in Russian).
3. Stankevich, V. V. and Shurupova, E. A., Hygienic evaluation of some rubber articles used in the food industry, *Gig. Sanit.*, 9, 24, 1976 (in Russian).
4. Prokof'eva, L. G., in *Hygiene of Use and Toxicology of Pesticides and Polymeric Materials*, Coll. Sci. Proc., A. V. Pavlov, Ed., All-Union Research Institute of Hygiene and Toxicology of Pesticides, Polymers and Plastic Materials, Kiyv, Issue No 17, 1987, 153 (in Russian).
5. Sokol'nikov, Ye. A., *Hygiene Assessment of Vulcanizates Based on SKMS 30 ARKM 15 Synthetic Rubber and Used for Contact with Foodstuffs*, Author's abstract of thesis, Kiyv, 1987, 23 (in Russian).
6. Sokol'nikov, Ye. A., The data to substantiate the permissible migration level of the vulcanization accelerator *N,N'*-dithiodimorpholine from rubber products, *Gig. Sanit.*, 12, 67, 1986 (in Russian).
7. Vorob'yeva, R. S., Toxicology of vulcanization agent *N,N'*-dithiodimorpholine, in *Toxicology and Hygiene of High-Molecular-Mass Compounds and of the Chemical Raw Material Used for Their Synthesis*, Proc. 3rd All-Union Conf., S. L. Danishevsky, Ed., Khimiya, Moscow-Leningrad, 1966, 88 (in Russian).
8. Ueno, S. and Ishizaki, M., Mutagenicity of organic rubber additives, Abstract, *Sangyo Igaku*, 26, 147, 1984 (in Japanese).
9. Wang, X. S. and Suskind, R. R., Comparative studies of the sensitization potential of morpholine, 2-mercaptobenzothiazole and 2 other derivatives in guinea pigs, *Contact Dermat.*, 19, 11, 1988.

4,4′-DITHIODI(*N*-PHENYLMALEIMIDE)

Molecular Formula. $C_{20}H_{12}N_2O_2S_2$
M = 408.44

Synonym. 4,4'-Dithiodiphenyldimaleimide.

Properties. Powder, poorly soluble in water.

Applications. Used as a vulcanizing agent for synthetic rubber.

Acute Toxicity. LD_{50} is 4.3 g/kg BW in mice. In rats, it was not attained.

Repeated Exposure revealed cumulative properties. In mice, K_{acc} is 4.5. A reduction in the content of free *SH*-groups in the blood serum was found in rats that received D. for 2 months.

Reference:

Kel'man, G. Ya., et al., Hygienic standardization of *N,N'-m-* phenylenedimaleimide and 4,4'-dithiodiphenylmaleimidethe - new vulcanization agents for synthetic rubber, in *Hygiene and Toxicology of High-Molecular-Mass Compounds and of the Chemical Raw Material Used for Their Synthesis,* Proc. 6th All-Union Conf., S. L. Danishevsky, Ed., Khimiya, Leningrad, 1979, 261 (in Russian).

1,3-DI-*o*-TOLYLGUANIDINE

Molecular Formula. $C_{15}H_{17}N_3$
M = 239.35
CAS No 97-39-2
RTECS No MF1400000
Abbreviation. DOTG.

Synonyms and **Trade Names.** *N,N'*-Bis(2-methylphenyl)guanidine; 1,3-Bis(*o*-tolyl)guanidine; DOTG accelerator; Vulkacit DOTG/C.

Properties. White powder. Slightly soluble in hot water and alcohol.

Applications. Basic rubber accelerator.

Acute Toxicity. LD_0 is 80 mg/kg BW in rabbits, LD_{50} is 500 mg/kg BW in rats.[027]

Regulations. ***U.S. FDA*** (1998) approved the use of DOTG in the manufacture of rubber articles intended for repeated use in producing, manufacturing, packaging, processing, preparing, treating, packing, transporting, or holding food in accordance with the conditions prescribed by 21 CFR part 177.2600.

6-DODECYL-2,2,4-TRIMETHYL-1,2-DIHYDROQUINOLINE

Molecular Formula. $C_{24}H_{39}N$
M = 341.64

Trade Name. Santoflex DD.

Applications. Antioxidant in the rubber industry.

Acute Toxicity. LD_0 is 40 mg/kg BW in rats.[027]

ELEMI OIL

CAS No 8023-89-0
RTECS No JX8730000
Abbreviation. EO.

Composition. The resin from Manila elemi.

Synonym. Elemi.

Properties. Amorphous, brown mass. Slightly soluble in water, soluble in alcohol.[020]

Acute Toxicity. LD_{50} is 3.37 g/kg BW in rats.

Regulations. ***U.S. FDA*** (1998) approved the use of EO in resinous and polymeric coatings used as the food-contact surfaces of articles intended for use in producing, manufacturing, packing, processing, preparing, treating, packaging, transporting, or holding food in accordance with the conditions prescribed by 21 CFR part 175.300.

Reference:

Elemi oil, *Food Cosmet. Toxicol.,* 14, 755, 1976.

EMULSIFIER STEK

Abbreviation. ES.

Trade Names. Twitchell reagent; Petrov's contact.

Composition. A mixture of sodium salts of petroleum sulfoacids consisting of methane, naphthene, and aromatic hydrocarbons in Baku petroleum. When a 25% aqueous solution of STEK is evaporated, a soapy mass with an odor of kerosene remains.

Properties. Organoleptic threshold concentration is found to be 1.0 mg/l.[1]

Applications. Used as an anionogenic surfactant in synthetic rubber production.

Acute Toxicity. The LD_{50} of unpurified STEK in mice is 1.6 g/kg BW, and that of its active portion is 1.9 g/kg BW. Rabbits given 1.0 g ES/kg BW exhibited leukocytosis, an increase of thrombocyte counts, and decrease in the cholinesterase activity (in 1 to 3 days). 0.5 g/kg BW dose caused leukocytosis and reticulocytosis. Two hours after administration, there was complete loss of conditioned reflexes.[2]

References:

1. Taradin, Ya. I., Fetisova, L. N., et al., in *Toxicology and Hygiene of the Products of Petroleum Chemistry and Petroleum Industry*, Yaroslavl', 1968, 106 (in Russian).
2. Taradin, Ya. I., Shavrikova, L. N., Fetisova, L. N., et al., Toxicological characteristics of emulsifier STEK, in *Toxicology and Hygiene of High-Molecular-Mass Compounds and of the Chemical Raw Material Used for Their Synthesis*, Proc. 3rd All-Union Conf., S. L. Danishevsky, Ed., Khimiya, Moscow-Leningrad, 1966, 94 (in Russian).

ETHANAMINE

Molecular Formula. C_2H_7N

M = 45.09

CAS No 75-04-7

RTECS No KX2100000

Abbreviation. EA.

Synonyms. Aminoethane; Ethylamine; Monoethylamine.

Properties. Colorless liquid or gas with the odor of ammonia. Miscible with water and alcohol at all ratios. Odor perception threshold is 3.9,[1] 4.3,[02] or 0.27 mg/l (70 to 72% in water).[010] Taste perception threshold is 0.5 mg/l.

Applications. Used in the production of synthetic rubber. A stabilizer for rubber latexes.

Acute Toxicity. In rats, LD_{50} is reported to be 400 mg/kg BW[1] or 530 to 580 mg/kg BW (for rats and mice).[2] Rabbits appear to be more susceptible. High doses affected the NS. Gross pathology examination revealed visceral congestion.

Repeated Exposure revealed marked cumulative properties. K_{acc} is 0.54. Oral administration of 90 mg/kg BW for two weeks caused death of 8 rats out of 10.

Long-term Toxicity. In a 6-month study, rats were gavaged with 2.5 mg/kg BW dose. The treatment affected conditioned reflex activity and ascorbic acid content in the spleen and liver.[2]

Mutagenicity.

In vivo cytogenetics. EA is shown to be positive in *Dr. melanogaster* (Dubinin, 1970).

In vitro genotoxicity. EA was negative in *Salmonella* mutagenicity assay.[013]

Standards. ***Russia.*** Recommended MAC and PML: 0.5 mg/l (organolept., taste).

References:

1. Smyth, H. F., Carpenter, C. P., Weil, C. S., et al., Range-finding toxicity data, List V, *Arch. Ind. Hyg. Occup. Med.*, 10, 61, 1954.
2. Gabrilevskaya, L. N. and Laskina, V. P., Experimental substantiation of the MAC for ethylamine, in *Protection of Water Reservoirs against Pollution by Industrial Liquid Effluents*, S. N. Cherkinsky, Ed., Medgiz, Moscow, Issue No 7, 1965, 99 (in Russian).

6-ETHOXY-1,2-DIHYDRO-2,2,4-TRIMETHYLQUINOLINE

Molecular Formula. $C_{14}H_{19}NO$

M = 217.30
CAS No 91-53-2
RTECS No VB8225000
Abbreviation. EDTQ.

Synonyms and **Trade Names.** Antioxidant EC; Antox; Ethoxyquin; 1,2-Dihydro-5-ethoxy-2,2,4-trimethylquinoline; Santoquin; 2,2,4-Trimethyl-6-ethoxy-1,2-dihydroquinoline.

Applications. An antioxidant and an antidegradation agent for rubber. A herbicide.

Acute Toxicity. LD_{50} is 800 mg/kg BW in rats,[1] and 1730 mg/kg BW in mice.[2]

Short-term Toxicity. F344 rats received 0.5% EDTQ-containing diet for 4 weeks. Histological investigation revealed signs of renal papillary necrosis in the male rats, commencing as interstitial degeneration.[1]

Long-term Toxicity. F344 rats received 0.5% EDTQ-containing diet. Histological investigation after 24 weeks of the treatment revealed renal papillary necrosis, active pyelonephritis, and urothelial hyperplasia in the renal pelvis. Spontaneous chronic progressive nephropathy was more evident in females.[3] Administration of 0.2% EDTQ in the diet to rats caused transient retardation of BW gain. Gross pathology examination revealed lesions in the kidneys, liver, and thyroid gland in many males but in none of females.[4]

Reproductive Toxicity.

Teratogenicity. Rats were treated orally on days 6 to 15 of gestation with single daily doses of 125 to 500 mg EDTQ/kg BW. No adverse effect was related to the treatment.[5]

Chemobiokinetics. Sheep were given diets containing 0.5% EDTQ or EDTQ hydrochloride; rats received 0.08 g EDTQ/day/animal. Identified metabolites in the urine were *EDTQ,*, *hydroxylated EDTQ,* and *dihydroxylated EDTQ*.[6]

Rats were given powdered feed containing 0.125 and 0.5% EDTQ hydrochloride. EDTQ was found in different internal organs, predominantly in the liver (84 to 458 μg EDTQ/100 g liver), and in the brain (11 to 92 μg EDTQ/100 g brain).[7]

According to other data, dietary administration of 0.5% EDTQ markedly enhanced rat hepatic UDP-glucuronosyltransferase activities.[8] Major metabolic reaction was deethylation of EDTQ which produced *6-hydroxy-2,2,4-trimethyl-1,2-dihydroquinoline* and *2,2,4-trimethyl-6-quinoline*. Other reactions were hydroxylation to 4 different hydroxylated metabolites and one dihydroxylated metabolite.[9]

Regulation. ***U.S. FDA*** (1998) approved the use of EDTQ as a food additive permitted for direct addition to food for human consumption subject to the provisions prescribed in 21 CFR part 172.140, as long as (1) the quantity added to food does not exceed the amount reasonably required to accomplish its intended physical, nutritive, or other technical effect in food, and (2) when intended for use in or on food it is of appropriate food grade and is prepared and handled as a food ingredient.

References:

1. *Rubber Chem. Technol.*, 45, 627, 1972.
2. *Chem. Abstr.*, 90, 4601c, 1979.
3. Hard, G. C. and Neal, G. E., Sequential study of the chronic nephrotoxicity induced by dietary admini stration of ethoxyquine in Fischer 344 rats, *Fundam. Appl. Toxicol.*, 18, 278, 1992.
4. Garner's *Veterinary Toxicology*, 3 ed., Rev. by E.G. C. Clarke and M. L. Clarke, Williams & Wilkins, Baltimore, MD, 1967, 286.
5. Khera, K. S., Whalen, C., Trivett, G., and Angers, G., Teratologic assessment of maleic hydrazide and daminozide, and formulations of ethoxyquin, thiabendazole and naled in rats, *J. Environ. Sci. Health [B],* 14, 563, 1979.
6. Kim, H. L., Ray, A. C., and Calhoun, M. C., Ovine urinary metabolites of ethoxyquine, *J. Toxicol. Environ. Health*, 37, 341, 1992.
7. Kim, H. L., Accumulation of ethoxyquine in the tissue, *J. Toxicol. Environ. Health*, 33, 229, 1991.
8. Bock, K. W., Kahl, R., and Lilienblum. W., Induction of rat hepatic UDP-glucuronosyltransferases by dietary ethoxyquin, *Naunyn Schmiedebergs Arch Pharmacol.*, 310, 249, 1980.
9. Skaare, J. U. and Solheim, E., Studies on the metabolism of the antioxidant ethoxyquin, 6-ethoxy-2,2,4-trimethyl-1,2-dihydroquinoline in the rat, *Xenobiotica*, 9, 649, 1979.

2-ETHYLANILINE

Molecular Formula. $C_8H_{11}N$
M = 121.19
CAS No 578-54-1
RTECS No BX9800000
Abbreviation. EA.

Synonyms. *N*-Ethylbenzenamine; *N*-Ethylphenylamine, Monoethylaniline.

Properties. Very refracted oily liquid. Poorly soluble in water, miscible with alcohol at all ratios.

Applications. Intermediate in organic synthesis.

Exposure. Rubber destruction product. EA could be detected at concentrations of 3.0 to 4.0 mg/l in the extracts of rubbers that are made using zinc phenylethyldithiocarbamate as a vulcanization accelerator.

Migration of EA from vulcanizates based on smoked-sheet natural rubber into aqueous and citric acid extracts (20, 40, 100°C; 24 hours; 30 days) was observed in a quantity of 2.0 to 3.0 mg/l. A study of aqueous extracts from such vulcanizates (24 hours; 20°C) revealed migration of 0.3 mg EA/l.[1]

A study of extracts from vulcanizates based on synthetic polyisoprene rubber revealed migration of EA up to 2.0 mg/l.[028]

Migration of EA from vulcanizates based on polybutadiene rubber in composition with natural rubber and polyisoprene rubber was studied in model media at 20°C. Concentration of EA in the extracts amounted to 0.05 to 0.1 mg/l.[2] Migration from vulcanizates based on polybutadiene rubber SKB-35-45 with natural rubber Pale Crepe amounted to 2.0 to 5.0 mg EA/l.[3]

EA migrates from vulcanizates based on nitrile rubber into aqueous and citric acid extracts (0.5 cm^{-1}; 100, 37, and 20°C) at a concentration of 0.02 mg/l.[2]

A study of extracts from vulcanizates based on butadiene-nitrile rubber SKN-26 (0.5 cm^{-1}; 20 and 40°C; 1 and 24 hours; distilled water, acid-salt solutions containing alcohol) revealed migration of EA at the concentration of 0.2 to 1.0 mg/l.[4]

Acute Toxicity. LD_{50} is 1.26 g/kg BW in rats, and 0.5 g/kg BW in mice. Poisoning is immediately followed by excitation with subsequent inhibition. Dyspnea, cyanosis, and tonic convulsions are noted to develop. Death occurs within 2 to 4 days. Gross pathology examination revealed congestion in the viscera and distension of the stomach and small intestine. Acute normochromic anemia developed in rats following a single oral administration of 145 mg/kg BW.[5]

Repeated Exposure revealed moderate cumulative properties. K_{acc} is 4.6 (by Kagan).

Short-term Toxicity. Rats were dosed by gavage with 1.5, 6.0, and 15 mg/kg BW for 4 months. By the end of the study, signs of intoxication appeared only at the high dose level. Hypodynamia, loss of appetite, cyanosis, dyspnea, retardation of BW, gain and an increase of the relative weights of the liver and spleen were observed.[5]

Standards. ***Russia.*** MAC: 0.5 mg/l; Recommended PML: 10 mg/l.

References:

1. Grushevskaya, N. Yu., Assessment of vulcanization accelerator of vulcacite-*P*-extra-*N* and some transformation product in sanitary and chemical investigation of rubber, *Gig. Sanit.*, 4, 59, 1987 (in Russian).
2. Stankevich, V. V., Ivanova, T. P., and Prokof'yeva, L. G., Hygienic studies of 'food-grade' resins on the base of a new diene kauchuk, in *Hygiene and Toxicology of High-Molecular-Mass Compounds and of the Chemical Raw Material Used for Their Synthesis*, Proc. 6th All-Union Conf., B. Yu. Kalinin, Ed., Khimiya, Leningrad, 1979, 96 (in Russian).
3. Znamensky, N. N. and Peskova, A. V., Study of migration of 3,5-benzopyren from paraffin packaging in dairy products, in *Hygiene Aspects of the Use of Polymeric Materials and of Articles Made of Them*, L. I. Medved', Ed., All-Union Research Institute of Hygiene and Toxicology of Pesticides, Polymers and Plastic Materials, Kiyv, 1969, 62 (in Russian).
4. Stankevich, V. V. and Shurupova, Ye. A., Hygienic characteristics of rubber containing *N,N*-dithiodimorpholine and intended for contact with food products, *Gig. Sanit.*, 9, 24, 1976

5. Yalkut, S. I., Experimental data on toxicity of vulkacite-*P*-extra-*N*, in *Hygienic Aspects of the Use of Polymeric Materials and Articles Made of Them*, L. I. Medved', Ed., All-Union Research Institute of Hygiene and Toxicology of Pesticides, Polymers and Plastic Materials, Kiyv, 1969, 328 (in Russian).

ETHYLENEBIS(DITHIOCARBAMIC ACID), ZINC SALT

Molecular Formula. $C_4H_6N_2S_4.Zn$
M = 275.75
CAS No 12122-67-7

Synonyms and **Trade Name.** [1,2-Ethanediylbis(carbamodithioato)]zinc; Ethene-*N,N'*-bisdithiocarbamate, zinc salt; Zineb.

Properties. White or pale-yellow, fine crystalline powder with unpleasant odor. Solubility in water at room temperature amounts to 10 mg/l, practically insoluble in common organic solvents.

Applications. Used as an ingredient in rubber manufacture and as a fungicide

Acute Toxicity. LD_{50} is 1.85 g/kg BW in rats. In 30 min. after ingestion of high doses, animals become agitated. Bloody nasal discharge was noted. Death occurred in 3 to 5 days after poisoning. Gross pathology examination revealed swelling of the stomach and intestines, thinning of the mucosa, focal hemorrhages in the stomach.[1]

Repeated Exposure revealed weak cumulative properties.[1]

Long-term Toxicity. Rats received 500 to 10,000 ppm in the diet for 2 years. Thyroid hyperplasia was observed in all rats treated in addition to renal congestion, nephritis and nephrosis, and increased mortality at higher dosages. The LOEL of 25 mg/kg BW was established in this study.[2, 3]

Rats received 0.05 to 0.1% zineb in their diet for two years. No clinical signs of intoxication were observed. Histological investigation revealed hyperplasia of the thyroid gland.[4]

Allergenic Effect. Zineb showed positive patch test reactions in patients.[5]

Mutagenicity. Zineb was ineffective in *Salmonella typhimurium* strain *TA 100*.[6]

Carcinogenicity. Prolonged administration of 3.5 g zineb/kg BW once a week caused adenomas in the lungs of rats to develop. Dose of 0.1 g/kg BW appeared to be ineffective.[017] In the above-cited 2-year study, no malignant neoplasms were noted in treated rats.[4]

Chemobiokinetics. In the body, zineb undergoes degradation to CS_2 which is eliminated with exhaled air.

Regulations. ***U.S. FDA*** (1998) approved use of zineb as a constituent of food-contact adhesives which may safely be used as a component of articles intended for use in packaging, transporting, or holding food in accordance with the conditions prescribed by 21 CFR part 175.105.

Recommendations. ***FAO/WHO*** (1993). ADI: 0.03 mg/kg BW.

References:

1. Kurbat, N. M., *Experimental Characteristics of Ethylenebisdithiocarbamates and Ethers of Dithiocarbamine Acid*, Author's abstract of thesis, Smolensk, 1968 (in Russian).
2. Blackwell-Smith, R., Finnegan, J. K., Larson, P. S., Sahyoun, P. F., Dreyfuss, M. L., and Haag, H. B., Toxicologic studies on zineb and disodium ethylene bis(dithiocarbamates), *J. Pharmacol. Acc. Exp. Ther.*, 109, 159, 1953.
3. *Health and Environmental Effects Profile for Zineb*, Prep. by the Office of Health and Environ. Ass., Environ. Criteria and Assessment Office, Cincinnati, OH for U.S. EPA, the Office of Solid Waste, Washington, D. C., 1984.
4. Smith, I. P., *J. Pharmacol.*, 109, 109, 1966.
5. Kaniwa, M. A., Isama, K., Nakamura, A., Kantoh, H., Itoh, M., Ichikawa, M., Hayakawa, R., Identification of causative chemicals of allergic contact dermatitis using a combination of patch testing in patients and chemical analysis. Application to cases from industrial rubber products, *Contact Dermat.*, 30, 20, 1994.
6. Franekic, J., Bratulic, N., Pavlica, M., and Papes, D., Genotoxicity of dithiocarbamates and their metabolites, *Mutat. Res.*, 325, 65, 1994.

FORMALDEHYDE RESIN, *p-(tert*-BUTYL)PHENOLDISULFIDE

Trade Name. Fenofor BS-2.

Composition. An oligomer is presented by a polycondensation product of 2,2'-dithiobis(*p-tert*-butylphenol) and formaldehyde.

Applications. Used as a vulcanizing agent in rubber industry.

Acute Toxicity. LD_{50} was not attained in mice. Administration of 8.0 g/kg BW caused only one or two animals to die.

Long-term Toxicity. Rabbits were dosed by gavage with 20 mg/kg BW every other day for the first 2 months and daily for 2 months thereafter. The treatment had no effect on protein and fat metabolism. Gross pathology examination revealed mild congestion, signs of edema, and mild destruction and fatty dystrophy in the liver.

Reference:

Vorob'yeva, R. S., in *Toxicology of New Chemical Substances Used in Rubber and Tyre Industry*, Z. I. Israel'son, Ed., Meditsina, Moscow, 1968, 148 (in Russian).

GUAIAC RESIN

CAS No 9000-29-7
RTECS No ME6260000
Abbreviation. GR.

Synonym and **Trade Names.** Guaiac; Guaiac gum; Gum guaiac; Gum guaiacum.

Properties. Brown or greenish-brown, irregular lamps. Insoluble in water, soluble in alcohol.

Acute Toxicity. LD_{50} exceeds 5.0 g/kg BW in rats. Poisoning produced a decrease in BW gain.

Regulations. ***U.S. FDA*** (1998) approved the use of GR (1) in resinous and polymeric coatings used as the food-contact surfaces of articles intended for use in producing, manufacturing, packing, processing, preparing, treating, packaging, transporting, or holding food in accordance with the conditions prescribed by 21 CFR part 175.300; and (2) as a preservative, 0.1% in edible fats and oils. Considered to be *GRAS*.

Reference:

Opdyke, D. L., Monographs on fragrance raw materials, Guaiac gum, *Food Cosmet. Toxicol.,* 12, 919, 1974.

HYDROQUINONE

Molecular Formula. $C_6H_6O_2$
M = 110.12
CAS No 123-31-9
RTECS No MX3500000
Abbreviation. HQ.

Synonyms and **Trade Names.** 1,4-Benzenediol; Benzohydroquinone; Benzoquinol; 1,4-Dihydroxybenzene; 1,4-Dioxybenzene; Hydroquinol; Quinol.

Properties. Light-gray or light-brown crystals (colorless when pure). Solubility in water is 6.0 to 7.0% (20°C). Readily soluble in hot water, ether, and ethanol. Aqueous solutions with up to 3.0 g/l concentrations are odorless. A sweetish taste appears at a concentration of a few g/l. The threshold for effect on the color of water is 0.2 mgl.[1]

Applications. Chemical intermediate in rubber, chemical and cosmetics industries. Antioxidant and stabilizer for materials which polymerize in presence of an oxidizing agent. Polymerization inhibitor of vinyl monomers: styrene, methyl methacrylate, etc.

Exposure. HQ is a known component of cigarette smoke. Low concentrations have been detected in the urine and plasma of humans with no occupational or other known exposure to HQ. Food sources of HQ are wheat products (1.0 to 10 ppm), pears (4.0 to 15 ppm), and coffee and tea (0.1 ppm).[1]

Acute Toxicity. The LD_{50} values were reported to be 340 mg/kg BW in mice, and 720 mg/kg BW[2] or 370 to 390 mg/kg BW in rats. According to other data,[3] LD_{50} for rats is about 1000 mg/kg BW. Mice given three LD_{50} during 2.5 hours died in a few hours. They developed convulsions.[2]

The toxicity of HQ increases by a factor of 2 to 3 when it is given to fasting animals. After oral administration, the animals immediately develop clonic convulsions and motor coordination disorder. Methemoglobinemia occurred. Death within a few minutes.

Repeated Exposure to HQ revealed slight cumulative properties. The amount of microsomal albumen and cytochrome B_5 and *P-450*, the activity of cytochrome-*c*-reductase and liver weights were unchanged after administration of 200 mg/kg BW to rats for 4 days.[4] However, the same dose given to female rats for 14 days is reported to be toxic causing tremors and clonic seizures and death of some animals due to respiratory paralysis after 4 to 5 administrations. 50 and 100 mg/kg BW doses produced no toxic effect.[5]

A 14-day gavage study revealed high mortality at high doses in F344 rats and $B6C3F_1$ mice. Tremors followed by convulsions were observed after dosing in rats (500 and 1000 mg/kg BW) and in mice (250 and 500 mg/kg BW).[5]

Experimental animals were given HQ in corn oil at doses of 63 to 1000 mg/kg BW (F344 rats) and 31 to 500 mg/kg BW ($B6C3F_1$ mice) by gavage for 14 days. CNS, forestomach, and liver were found to be the targets of HQ toxicity. Renal damage was also observed in this study.[6]

Short-term Toxicity. In a 13-week gavage study, the treatment with 100 to 200 mg/kg BW caused hyperplasia of the forestomach and toxic nephropathy in rats.[5]

Rats and mice were exposed orally to 25 to 400 mg HQ/kg BW over a period of 13 weeks. CNS, forestomach, and liver were found to be the target organs of HQ toxicity in both species.[6]

Long-term Toxicity.

Observations in man. 19 volunteers who received 300 to 500 mg HQ daily for 3 to 5 months for experimental purposes exhibited no signs of intoxication.

Animal studies. Rats were dosed by gavage with HQ over a period of 6 months. Treatment with 50 mg/kg caused anemia and leukocytosis. Gross pathology examination revealed dystrophic changes in the small intestine, liver, kidneys, and myocardium of rats given 50 to 100 mg HQ/kg BW.[7]

In a 2-year study in rats, there was no effect on BW gain (diet contains up to 1.0% HQ), no hematology or other pathological changes reported. A 5.0% dose level in the diet caused ~50% loss of BW within 9 weeks; the animals developed aplastic anemia, depletion of the bone marrow, liver cord-cell atrophy, superficial ulceration and hemorrhages of the gastric mucosa.[3]

Reproductive Toxicity.

Embryotoxicity. Addition of 0.003 and 0.3% HQ to the diet of female rats for 10 days prior to mating did not affect reproduction.[9]

HQ is unlikely to be a developmental toxicant. Pregnant rats were dosed by gavage with 30 to 300 mg/kg BW on days 6 to 15 of gestation. The NOEL for maternal and developmental toxicity was identified to be 100 mg/kg BW.[10]

In a two-generation study in CD Sprague-Dawley rats, HQ was administered in an aqueous solution given by gavage at doses of 15, 50, and 150 mg/kg BW. At all dose levels tested, there were no adverse effects on feed consumption, survival, or reproductive parameters for the F_0 or F_1 parental animals. No effect on pup weight, sex distribution, or survival were noted for pups of either generation. No gross lesions or histology changes were found upon postmorten examination.[11]

HQ was administered by gavage to pregnant New Zealand white rabbits on days 6 to 18 of gestation in aqueous solution at dose levels of 25, 75, and 150 mg/kg BW. Doses of 75 and 150 mg/kg affected food consumption and BW gain of dams during the treatment period. The NOAEL was found to be 25 mg/kg BW for maternal toxicity and 75 mg/kg BW for developmental effects.[12] Administration of 115 mg/kg in rat diet throughout gestation period caused an increase of fetal resorptions.[13]

Gonadotoxicity. There was no adverse effect of HQ on the fertility of rats treated through 2 generations with 15 to 150 mg/kg BW.[14]

The dose of 50 to 200 mg HQ/kg BW administered by gavage to female rats for 14 days affected the estrous cycle.[5]

Allergenic Effect. A strong contact sensitizing potential is found in maximized guinea-pig assays.[15] Human allergic reactions observed were rare intermittent asthma or reversible bronchospasm.[16] These findings were considered to be occasional systemic hypersensitivity reactions (risk 1/100)

Mutagenicity.

In vivo cytogenetics. HQ induced micronuclei in bone marrow cells of pregnant mice and, transplacentally, in fetal liver cells.[17] Treatment of mice with various doses of HQ resulted in a dose-dependent increase in the frequency of micronuclei in the bone marrow at all the post-treatment time periods. The frequency of micronucleated polychromatic erythrocytes was significantly higher after administration of 3.125 mg HQ/kg BW at 24-hour post-treatment. A significant increase in the frequency of micronucleated normochromatic erythrocytes was observed after 12.5 mg HQ/kg BW treatment.[18]

However, the more recent study demonstrated that large doses of HQ (0.8% in the diet) may be given orally to mice without induction of micronuclei or bone marrow depression.[26]

In vitro genotoxicity. HQ has demonstrated mutagenic activity in modified micronuclei assay: it induced micronuclei and toxicity signs in cultured human lymphocytes.[19]

HQ exhibited colchicine-like action, caused build-up of metaphases in the small intestine of mice and SCE in human lymphocyte culture.[20] It was negative in *Salmonella* mutagenicity assay.[025] HQ had appreciable mutagenic potential in the mouse lymphoma L5178Y cell assay for mutagens without supplementary metabolic activation system.[8]

Carcinogenicity. Rodents exposed to HQ developed neoplasms in the kidney, hematopoietic system, and liver.[21] In a 2-year study, tubular cell adenomas of the kidney were observed in male rats. Renal tumors appeared to arise from areas of spontaneous progressive nephropathy; the nephropathy itself has been found to be enhanced by HQ. Increases in mononuclear cell leukemia were found in females.[5,22]

In a 2-year oral study, HQ was administered at dose levels of 25 or 50 mg HQ/kg BW (rats) and 50 or 100 mg HQ/kg BW (mice) in deionized water by gavage. The rate of follicular cell hyperplasia of the thyroid gland was increased in dosed mice. Mononuclear cell leukemia in female rats occurred with increased incidences. The rates of hepatocellular neoplasms, primarily adenomas, were increased in dosed female mice.[6]

F344 rats and $B6C3F_1$ mice received 0.8% HQ in the diet for 2 years. The treatment caused renal tubular hyperplasia as well as adenomas, predominantly in males of both species, and was associated with chronic nephropathy in rats. In mice, the incidence of squamous cell hyperplasia of the forestomach epithelium was significantly higher than in the controls. The study indicates potential renal carcinogenicity of HQ in male rats and hepatocar cinogenicity in male mice.[23]

Carcinogenicity classifications. An IARC Working Group concluded that there is inadequate evidence for the carcinogenicity of HQ in *experimental animals* and there were no data available to evaluate the carcinogenicity of HQ in *humans*.

IARC: 3;

NTP: SE - SE - NE - SE* (gavage).

Chemobiokinetics. HQ is readily absorbed from the GI tract and rapidly distributed throughout the body tissues, but maximum accumulation is found in the liver and kidneys. HQ metabolism occurs via conjugation with *glucuronic* and *sulphuric acids*. The major metabolites are reported to be *HQ monosulfate* and *HQ monoglucuronide* which are found in the urine.[4]

HQ was not DNA-reactive in the male F344 rat. It produced enhanced proliferation of the renal tubular epithelium, presumably through toxicity involving glutathione conjugate formation. In the kidney, bone marrow, and other tissues, HQ may induce toxicity by redox cycling and lipid peroxidation.[24] HQ is rapidly excreted in the urine. Excretion level does not depend on the period of exposure. It was suggested that by the oral route there was a potential for a threshold of biological activity as HQ is rapidly conjugated and detoxified through the glutathione pathway.[18,25]

Regulations.

U.S. FDA (1998) approved the use of HQ (1) in adhesives used as components of articles intended for use in packaging, transporting, or holding food in accordance with the conditions prescribed by 21 CFR part 175.105; (2) as an inhibitor in cross-linked polyester resins to be safely used as articles or components of articles intended for repeated use in contact with food in accordance with the conditions prescribed by 21 CFR part 177.2420; (3) in resinous and polymeric coatings for polyolefin films to be safely used as a food-contact surface of articles intended for use in producing, manufacturing, packing, transporting, or holding food in accordance with the conditions prescribed by 21 CFR part 175.320; (4) in the manufacture of polyurethane resins which may be used as food-contact surface of articles intended for use in contact with bulk quantities of dry food of an appropriate type in accordance with the conditions prescribed by 21 CFR part 177.1680; and (5) as an inhibitor for monomer in the uncoated or coated food-contact surface of paper and paperboard intended for use in producing, manufacturing, packing, transporting, or holding aqueous and fatty food in accordance with the conditions prescribed by 21 CFR part 176.170.

EU (1990). HQ is available in the *List of authorized monomers and other starting substances which shall be used for the manufacture of plastic materials and articles intended to come into contact with foodstuffs (Section A)*.

Great Britain (1998). HQ is authorized without time limit for use in the production of polymeric materials and articles in contact with food or drink or intended for such contact. The specific migration of HQ shall not exceed 0.6 mg/kg.

Standards.

Russia (1995). MAC and PML: 0.2 mg/l (organolept., chromaticity).

EU (1990). SML: 0.6 mg/l.

References:

1. Deisinger, P. J., Hill, T. S., and English, J. C., Human exposure to naturally occurring hydroquinone, *J. Toxicol. Environ. Health*, 47, 31, 1966.
2. Mozhayev, Ye. A., in *Protection of Water Reservoirs Against Pollution by Industrial Liquid Effluents*, S. N. Cherkinsky, Ed., Medgiz, Moscow, Issue No 7, 1965, 194 (in Russian).
3. Carlson, A. J. and Brewer, N. R., Toxicity studies on hydroquinone, *Proc. Soc. Exp. Biol.*, New York, Issue No 84, 1953, 684.
4. Divincenzo, G. D., Hamilton, M. L., Reynolds, R. C., et al., Metabolic fate and disposition of ^{14}C-hydroquinone given orally to Sprague-Dawley rats, *Toxicology*, 33, 9, 1984.
5. *Toxicology and Carcinogenesis Studies of Hydroquinone in F344/N Rats and B6C3F$_1$ Mice (Gavage Studies)*, NTP Technical Report Series No 366, Research Triangle Park, NC, 1989.
6. Kari, F. W., Bucher, J., Eustis, S. L., Haseman, J. K., and Huff, J. E., Toxicity and carcinogenicity of hydroquinone in F344/N rats and B6C3F$_1$ mice, *Food Chem. Toxicol.*, 30, 737, 1992.
7. Mozhayev, E. A. et al., *Pharmacologia i Toxicologia*, 29, 238, 1984 (in Russian).
8. McGregor, D. B., Riach, C. G., Brown, A., Edwards, I., Reynolds, D., West, K., and Willington, S., Reactivity of catecholamines and related substances in the mouse lymphoma L5178Y cell assay for mutagens, *Environ. Molec. Mutagen.*, 11, 523, 1988.
9. Ames, S. R., Ludwig, M. I., Swanson, W. J., et al., Effect of DPPD, methylene blue, BHT, and hydroquinone on reproductive process in the rat, *Proc. Exp. Biol. Med.*, 93, 39, 1956.
10. Krasavage, W. J., Blacker, A. M., English, J. C., et al., Hydroquinone: A developmental toxicity study in rats, *Fundam. Appl. Toxicol.*, 18, 370, 1992.
11. Blacker, A. M., Schroeder, R. E., English, J. C., Murphy, S. J., Krasavage, W. J., and Simon, G. C., A two-generation reproduction study with hydroquinone in rats, *Fundam. Appl. Toxicol.*, 21, 420, 1993.
12. Murphy, S. J., Schroeder, R. E., Blacker, A. M., et al., A study of developmental toxicity of hydroquinone in the rabbit, *Fundam. Appl. Toxicol.*, 19, 214, 1992.
13. Telford, I. R., Woodruff, C. S., and Linford, R. H., Fetal resorptions in the rats as influenced by certain antioxidants, *Am. J. Anat.*, 110, 29, 1962.
14. Blacker, A. M., Schroeder, R. E., English, J. C., et al., A two-generation reproduction study with hydroquinone in rats, *Teratology*, 43, 429 (Abstract P39), 1991.
15. Basketter, D. A. and Scholes, E. W., Comparison of the local lymph node assay with the guinea-pig maximization test for the detection of a range of contact allergens, *Food Chem. Toxicol.*, 30, 65, 1992.
16. Choudat, D., Neukirch, F., Brochard, P., Barrat, G., Marsac, J., Conso, F., and Philbert, M., Allergy and occupational exposure to hydroquinone and to methionine, *Brit. J. Ind. Med.*, 45, 376, 1988.
17. Cirrani, R., Barale, R., Marrazzini, A., and Loprieno, N., Benzene and the genotoxicity of its metabolites. I. Transplacental activity in mouse fetuses and in their dams, *Mutat. Res.*, 208, 61, 1988.

18. Jagetia, G. C. and Aruna, R., Hydroquinone increases the frequency of micronuclei in a dose-dependent manner in mouse bone marrow, *Toxicol. Lett.*, 93, 205, 1997.
19. Robertson, M. L., Eastmond, D. A., and Smith, M. T., Two benzene metabolites, catechol and hydroquinone, produce a synergistic induction of micronuclei and toxicity in cultured human lymphocytes, *Mutat. Res.*, 249, 201, 1991.
20. Morimoto, K., Wolf, S., and Koizumi, A., Induction of sister chromatide exchanges in human lymphocytes by microsomal activation of benzene metabolites, *Mutat. Res.*, 119, 355, 1983.
21. Davidson, K. A. and Faust, R. A., Oral carcinogen potency factors for benzene and some of its metabolites, in *Toxicologist*, Abstracts of SOT 1996 Annual Meeting, Abstract No 585, Issue of *Fundam. Appl. Toxicol.*,30, Part 2, March 1996.
22. Whisner, J., Verna, L., English, J. C., and Williams, G. M., Analysis of studies related to tumorogenicity induced by hydroquinone, *Regul. Toxicol. Pharmacol.*, 21, 158, 1995.
23. Shibata, M. A., Hirose, M., Tanaka, H., Asakawa, E., Shirai, E., and Ito, N., Induction of renal cell tumors in rats and mice, and enhancement of hepatocellular tumor development in mice after long-term hydroquinone treatment, *Jap. J. Cancer Res.*, 82, 1211, 1991.
24. *Data on Hygienic Assessment of Pesticides and Polymers*, F. F. Erisman Research Sanitary Hygiene Institute, Moscow, 1977, 174 (in Russian).
25. *1994 The Annual Report of the Committees on Toxicity, Mutagenicity, Carcinogenicity of the Chemi cals in Food, Consumer Products and the Environment*, DoH, HMSO, London, 1995, 61.
26. O'Donoghue, J., Barber, E. D., Hill, T., Aebi, J., and Fiorica, L., Hydroquinone: genotoxicity and prevention of genotoxicity following ingestion, *Food Chem. Toxicol.*, 37, 931, 1999.

5-HYDROXY-1,4-NAPHTHALENEDIONE

Molecular Formula. $C_{10}H_6O_3$
M = 174.16
CAS No 481-39-0
RTECS No QJ5775000
Abbreviation. HND.

Synonym and **Trade Names.** Juglone; 5-Hydroxy-1,4-naphthaquinone; Oil Red BC; Walnut extract; Walnut oil.

Acute Toxicity. LD_{50} is 112 mg/kg BW in rats.[1]

Mutagenicity. *Juglone* was found to be mutagenic to *Salmonella typhimurium* strain *TA2637* with metabolic activation.[2]

Carcinogenicity. When applied on mice skin, HND was found to be neoplastic by RTCES criteria.[3]

Regulations. *U.S. FDA* (1998) approved the use of HND in resinous and polymeric coatings used as the food-contact surfaces of articles intended for use in producing, manufacturing, packing, processing, preparing, treating, packaging, transporting, or holding food in accordance with the conditions prescribed by 21 CFR part 175.300.

References:

1. Priputina, L. S., Ingre, V. G., and Boiko, N. L., Hygienic characteristics of naphthoquinone preservative, *Voprosy Pitania*, 5, 67, 1984 (in Russian).
2. Tikkanen, L., Matsushima, T., Natori, S., and Yoshihira, K., Mutagenicity of natural naphthoquinones and benzoquinones in the *Salmonella*/microsome test, *Mutat. Res.*, 124, 25, 1983.
3. Van Duuren, B. L., Segal, A., Tseng, S. S., Rusch, G. M., Loewengart, G., Mate, U., Roth, D., Smith, A., Melchionne, S., and Seidman, I., Structure and tumor-promoting activity of analogues of anthralin (1,8-dihydroxy-9-anthrone), *J. Med. Chem.*, 21, 26, 1978.

2-IMIDAZOLIDINETHIONE

Molecular Formula. $C_3H_6N_2S$
M = 102.11
CAS No 96-45-7

RTECS No NI9625000

Abbreviation. ETU.

Synonyms. Ethylenethiourea; 2-Imidazoline-2-thiol; 2-Mercaptoimidazoline.

Properties. White crystalline solid. Solubility in water is 20 g/l at 30°C and 90 g/l at 60°C. Half-life in water is 7 to 13 days.[1] Moderately soluble in methanol, ethanol, ethylene glycol.

Applications. Used as an accelerator in synthetic (neoprene) rubber production. It is used also as a part of curing system for polyacrylate rubber and as an intermediate for antioxidants.

Exposure. ETU is a degradation product of ethylenebisdithiocarbamate fungicides such as maneb and zineb. It was detected in food, namely, in vegetables and fruits.

Acute Toxicity. LD_{50} is 780 mg/kg BW in rats, and 3000 mg/kg BW in mice. Poisoning results in decreased serum *SH*-group contents,[2] reduced hexenal-induced sleep and aminopyridinemethylase activity in the liver microsomes.[3]

Repeated Exposure. In a 28-day study, rabbits were exposed to ETU in their drinking water at concentrations of 100 to 300 mg/l. Exposure to ETU decreased BW gain but did not significantly affect urinary sodium, potassium, glucose, or protein excretion and urinary osmolality. High doses resulted in ultrastructural alterations in the epithelial cells or renal proximal tubuli.[4]

Short-term Toxicity. Following oral administration of 80 and 40 mg/kg BW to rats for 3 months, pronounced cumulative effect was noted.[2] Like all urea derivatives, ETU appears to exhibit strumogenic effect.[5]

Administration of 50 to 750 mg/kg diet of rats for 30 to 120 days resulted in decreased BW, thyroid/BW ratio and uptake of I^{131}. Thyroid hyperplasia at higher doses was also noted.[6]

B6C3F$_1$ mice were fed ETU-containing diets for 13 weeks. Dose-related diffuse follicular cell hyperplasia of the thyroid and hepatocellular cytomegaly occurred at a dose level of 75 mg ETU/kg BW in both sexes. No effects were observed at lower doses. The NOAEL in this study was 38 mg/kg BW (NTP-92). CD-1 mice were administered ETU in the diet at doses of 0.16 to 230 mg/kg BW. Follicular cell hyperplasia of the thyroid was observed at a dose of about 20 mg/kg BW. The NOAEL in this study was shown to be 1.7 mg/kg BW in males and 2.4 mg/kg BW in females (O'Hara and DiDonato, 1985).[7]

Beagle dogs received dietary concentrations of ETU for 13 weeks. The NOAEL based on decreased *Hb* and erythrocyte count, packed cell volume, and increased cholesterol level was determined at a dose of approximately 6.0 mg/kg BW (Briffaux, 1991).[7]

Long-term Toxicity. In a 52-week feeding study in dogs, 1.8 mg/kg BW dose caused a reduction in BW gain, hypertrophy of the thyroid with colloid retention, a slight increase in thyroid weight and pigment accumulation in the liver. The NOAEL in this study was 0.18 mg ETU/kg BW (Briffaux, 1992).[7]

Oral lifetime administration of up to 200 mg/kg BW caused upper cholesterolemia in rats and hamsters. The most affected organs are the liver (in hamsters) and thyroid (in rats).[8] In a 2-year oral study in rats, the NOAEL of 0.25 mg ETU/kg BW (lack of thyroid tumors) was reported.[9]

F344 rats received 83 or 250 ppm ETU in the diet for 36 and 105 weeks. The treatment caused a dose-related increase in the incidence and severity of thyroid follicular cell hyperplasia, an increase in glutathione levels, and a decrease in serum thyroxin levels. An 83 ppm level was considered the LOAEL. Perinatal exposure to the highest tested dietary level of 90 ppm resulted in an increased incidence of thyroid follicular cell hyperplasia. B6C3F$_1$ mice received 330 or 1000 ppm ETU in the diet for 105 weeks. The treatment caused a dose-related increase in the incidence of thyroid follicular cell hyperplasia, an increase in glutathione levels, and hepatic hypertrophy. The 1000 ppm level was considered to be the LOAEL.[10]

Reproductive Toxicity.

Teratogenicity. ETU itself, but not its metabolites, appears to be teratogenic. The initial target is the primitive neuroblast that undergoes necrosis with further development of hydrocephalus. High doses caused teratogenic effects in hamsters, mice, guinea pigs, and rabbits.

ETU is a specific *neuroteratogen* that induces communicating hydrocephalus *ex vacuo* at oral doses far lower than those causing any observable toxic symptom or 50% death (LD_{50}) in the rat dam.[11]

Oral administration of 15 to 35 or 100 mg/kg BW, or 200 mg/kg BW during organogenesis period in rats caused specific developmental abnormalities of the CNS and musculo-skeletal, craniofacial systems, etc.[12-14]

Cats given 0.6 g/kg BW on days 16 to 35 of pregnancy exhibited maternal toxicity and a great number of developmental abnormalities (30% animals).[11]

Embryotoxicity. Sprague-Dawley rats received 20 to 60 mg ETU/kg BW on day 11 of gestation. An exposure to the highest dose caused a number of neonatal deaths. Hydronephrosis was observed in this study.[15] A 5-day oral administration of 150 mg ETU/kg BW did not decrease pregnancy percentage and the number of implants in mice.[16] A presumably safe exposure level for ETU of 0.1 mg/kg BW was calculated in rat experiments.[17] In an NTP study, F344 rats and B6C3F$_1$ mice received ETU in the diet prior to breeding through gestation, lactation, and up to 9 weeks post-weaning. No gross fetal abnormalities were noted either in rats or in mice. The NOAELs for developmental toxicity in these studies were calculated as 6.2 mg/kg BW (rats) and 49.5 mg/kg BW (mice).

Mutagenicity. ETU is not found to be a potent genotoxic agent.

In vitro genotoxicity. ETU was mutagenic in *Salmonella* (weak response without activation), it caused CA in cultured mammalian cells, it was positive in *rec*- and several yeast assays.[18] However, according to Franekic et al.[19], ETU was found to be ineffective in *Salmonella* strains *TA100* and *TA98*. DNA synthesis was observed in human fibroblasts.[20]

In vivo cytogenetics. ETU induces micronuclei or SCE in mice, and CA in rats treated *in vivo*. ETU appeared to be positive in DLM test.[18]

Carcinogenicity. The primary effect of ETU in rats is on the thyroid gland. ETU induces mouse liver tumors by a non-genotoxic mechanisms.[18] In rats, but not in hamsters, fed with 60 mg/kg BW (males) and 200 mg/kg BW (females), a carcinogenic effect was shown.[8]

Oral administration of 77 mg/kg BW to mice for 82 weeks caused thyroid and lung cancer.[21] After 80 weeks of feeding rats with a dose of 646 mg/kg BW, thyroid and lung tumors were observed in 100% of cases.[22,23]

In NTP carcinogenicity studies, F344 rats received ETU in the diet for 105 weeks. The treatment caused a dose-related increase in the incidence of thyroid follicular cell neoplasms. In addition to these neoplasms in mice, hepatocellular neoplasms and adenomas of the *pars distalis* of the pituitary gland were also observed.[10]

Carcinogenicity classifications. An IARC Working Group concluded that there is sufficient evidence for the carcinogenicity of ETU in *experimental animals* and there was inadequate evidence for the carcinogenicity of ETU in *humans*.

U.S. The Annual Report of Carcinogens issued by the U.S. Department of Health and Human Services (1998) defines ETU to be a substance which may reasonably be anticipated to be carcinogen, i.e., a substance for which there is a limited evidence of carcinogenicity in *humans* or sufficient evidence of carcinogenicity in *experimental animals*.

IARC: 2B;

U.S. EPA: B2;

NTP: CE - CE - CE - CE (feed).

Chemobiokinetics. After oral administration of 100 mg/kg BW to rats, ETU was rapidly absorbed in the guts and penetrated all the visceral tissues and embryos; it accumulates only in the thyroid and lung alveols.[24] Not detected in the blood probably due to the intracellular transformation. After oral administration of 4.0 mg ^{14}*C*-ETU/kg BW, *imidasol* and *imidasoline* were found to be the main metabolites. Following *i/v* injection, *S-methylethyl thiourea* was detected as a metabolite in cats.[22,25]

ETU is rapidly excreted with the urine.

Regulations. ***U.S. FDA*** prohibited use of ETU as a food additive.

Recommendations.

FAO/WHO (1993). ADI: 0.004 mg/kg BW.

U.S. EPA (1995). Health Advisory for a longer-term exposure is 0.4 mg/l.

References:

1. Voitenko, G. A., Thiourea derivatives, in *Hazardous Substances in Industry*, Organic compounds, Handbook, New data of 1974-1984, A. L. Levina and E. D. Gadaskina, Eds., Khimiya, Leningrad, 1985, 209 (in Russian).

2. Antonovich, E. A., *Toxicology of Dithiocarbamates and Hygienic Aspects of Their Use in Food Production*, Author's abstract of thesis, L'vov, 1975, 32 (in Russian).
3. Lewerenz, H. J. and Plas, R., Effect of ethylene thiourea on hepatic microsomal enzymes in the rat, *Arch. Toxicol.,* Suppl. 1, 189, 1978.
4. Kurttio, P., Savolainen, K., Naukkarinen, A., et al., Urinary excretion of ETU and kidney morphology in rats after continuous oral exposure to nabam or ethylenthiourea, *Arch. Toxicol.,* 65, 381, 1991.
5. Newsome, W. N., Residues of mancozeb, 2-imidazoline and ethylene thiourea in tomato and potato crops after field treatment with mancozeb, *Bull. Environ. Contam. Toxicol.*, 20, 678, 1978.
6. Graham, S. L., Hansen, W. H., Davis, K. J., and Perry, C. H., Effect of 1-year administration of ETU upon the thyroid of the rat, *J. Agric. Food Chem.*, 21, 324, 1973.
7. *Pesticide Residues in Food - 1993.* Toxicity evaluations, Joint FAO/WHO Expert Committee on Food Additives FAO/WHO, Geneva, 20-29 Sept., 1993, 167.
8. Gak, J. C., Graillot, C., and Turhaut, R., Difference in the sensitivity of the hamster and the rat to the effects of long-term administration of ethylene thiourea, *Eur. J. Toxicol.* Appl. *Pharmacol.*, 9, 303, 1976.
9. Graham, S. L., Davis, K. J., Hansen, W. H., and Graham, C. H., Effects of prolonged ethylenethiourea ingestion on the thyroid of the rat, *Food Cosmet. Toxicol.,* 13, 493, 1975.
10. *Toxicology and Carcinogenesis Studies of Ethylene Thiourea in F344 Rats and $B6C3F_1$ Mice (Feed Studies),* NTP Technical Report Series No 388, Research Triangle Park, NC, NIH Publ. No 90-28-43, 1992.
11. Khera, K. S., Ethylenethiourea: A review of teratogenicity and distribution studies and of assessment of reproduction risk, *CRC Crit. Rev. Toxicol.,* 18, 129, 1987.
12. Saillenfait, A. M., Sabate, J. P., Langonne, I., and de Ceaurriz, J., Difference in the developmental toxicity of ethylenethiourea and three-*N,N'*-substituted thiourea derivatives in rats, *Fundam. Appl. Toxicol.,* 17, 399, 1991.
13. Ruddick, J. A., Newsome, W. H., and Nash, L., Correlation of teratogenicity and molecular structure: Ethylenethiourea and relative compounds, *Teratology*, 13, 263, 1976.
14. Chernoff, N., Kavlock, R. J., Rogers, E. N., et al., Perinatal toxicity of maneb, ethylenethiourea and ethyl enebisisothiocyanate sulfide in rodents, *J. Toxicol. Environ. Health*, 5, 821, 1979.
15. Daston, G. P., Rehnberg, B. F., Corver, B., et al., Functional teratogens of the rat kidney, II. Nitrofen ethylenethiourea, *Fundam. Appl. Toxicol.*, 11, 401, 1988.
16. Teramoto, S., Shingu, A., and Shirasu, Y., Induction of dominant-lethal mutation after administration of ethylenethiourea in combination with nitrite of the *N*-nitroso-ethylenethiourea in mice, *Mutat. Res.*, 56, 335, 1978.
17. Gaylor, D. W., Quantitative risk analysis for quantal reproductive and developmental effects, *Environ. Health Perspect.,* 79, 243, 1989.
18. Dearfield, K. L., Ethylenethiourea (ETU). A review of the genetic toxicity studies, *Mutat. Res.,* 317, 111, 1994.
19. Franekic, J., Bratulic, N., Pavlica, M., and Papes, D., Genotoxicity of dithiocarbamates and their metabolites, *Mutat. Res.,* 325, 65, 1994.
20. U.S. EPA GENOTOX Program, 1986.
21. Innes, J. R. M., Ulland, B. M., Valerio, W. G., et al., Bioassay of pesticides and industrial chemicals for tumorigenicity in mice: A preliminary note, *J. Natl. Cancer Inst.*, 42, 1101, 1969.
22. Ulland, B. M., Weisburger, J. H., Weisburger, E. K., et al., Thyroid cancer in rats from ethylenethiourea intake, *J. Natl Cancer Inst.,* 49, 583, 1972.
23. Weisburger, E. K., Ulland, B. M., Nam, J., Gart, J. J., and Weisburger, J. H., Carcinogenicity tests of certain environmental and industrial chemicals, *J. Natl. Cancer Inst.*, 67, 75, 1981.
24. Kato, Y., Odanaka, Y., Teramoto, S., et al., Metabolic fate of ethylenethiourea in pregnant rats, *Bull. Environ. Contam. Toxicol.*, 16, 546, 1976.
25. Iverson, F., Khera, K. S., Hierlihy, S. L., et al., *In vivo* and *in vitro* metabolism of ethylenethiourea in the rat and the cat, *Toxicol. Appl. Pharmacol.*, 52, 16, 1980.

2,2'-IMINODIETHANOL

Molecular Formula. $C_4H_{11}NO_2$
M = 105.14
CAS No 111-42-2
RTECS No KL2975000
Abbreviation. IDE.

Synonyms. 2,2'-Aminodiethanol; Diethanolamine; Diethylolamine; 2,2'-Dihydroxydiethylamine; Di(β-hydroxyethyl)amine.

Properties. The technical product is a glycerol-like, clear, colorless liquid with a faint odor of ammonia. The pure substance occurs in the form of colorless deliquescent prisms. Readily mixes with water and alcohol. Odor perception threshold is reported to be 0.8 mg/l.[1] However, according to other data, it is 22 g/l.[02]

Applications. Used as a secondary vulcanization activator in the rubber industry. IDE has also other applications.

Acute Toxicity. LD_{50} values are reported to be 0.78[2] or 3.5 g/kg BW in rats, 3.3 g/kg BW in mice, and 2.2 g/kg BW in rabbits and guinea pigs. Poisoning caused irritation of the oral and GI tract mucosa and motor excitation. Death occurs within 24 hours from respiratory failure. Gross pathology examination revealed hemorrhages and congestion in the visceral organs,[1] dilatation and degranulation of the endoplasmic reticulum, and mitochondrial swelling in the liver of mice given a lethal dose.[3]

Sprague-Dawley rats exposed to 0.1 to 3.2 g neutralized IDE/kg BW displayed an increase in the relative liver and kidney weights. A single 0.4 g/kg BW dose caused renal tubular epithelium necrosis.[4]

Repeated Exposure. Rabbits and rats were exposed to 1/10 LD_{50} for 2 months. The treatment produced changes in prothrombin-forming and detoxifying functions of the liver and in blood serum enzyme activity. BW gain and gas exchange were reduced. Histology examination revealed dystrophic, necrobiotic, and atrophic changes.[1]

In the 14-day NTP study, no renal toxicity was reported in mice; however, renal degeneration including tubular epithelium necrosis was noted in rats. The NOAEL for this effect appeared to be 160 mg/kg BW.[5] Microscopic examination of bone marrow and red blood cells failed to reveal important morphological changes; the LOAEL for this effect is close to 160 ppm IDE in drinking water.[5]

Regardless of the route of administration, IDE produced normochromic anemia in rats. Normocytic anemia without bone marrow depression or reticulocytosis was observed in male rats exposed to 4000 ppm IDE in drinking water for 7 weeks.[6]

Short-term Toxicity. The mechanism of IDE toxicity is unknown but may be related to its high tissue accumulation and subsequent alteration of membrane phospholipid composition.[5]

$B6C3F_1$ mice were exposed orally to 630 to 10,000 ppm in drinking water for 13 weeks. The treatment produced dose-dependent alterations and necrosis in the liver, nephropathy and tubular epithelial necrosis in males kidneys, cardiac myocyte degeneration in heart, etc. The NOAEL was not achieved for hepatocellular cytological alterations and for acanthosis in the skin.[7]

F344 rats were exposed orally to 160 to 5000 ppm in drinking water for 13 weeks. The treatment caused dose-dependent changes in hematology, kidneys, brain and spinal cord, testis (degeneration in seminiferous tubules), and skin. The NOAEL was not achieved for hematology changes, nephropath,y or hyperkeratosis of the skin.[8]

13-week NTP study in F344 rats and $B6C3F_1$ mice revealed renal tubular epithelium necrosis being less severe than in the 14-day study. The LOAEL for this effect appeared to be 124 mg/kg BW in the more susceptible group of animals tested (female rats). The liver was predominantly affected in mice. Demyelination in the brain is reported to develop in rats but not in mice. Histology examination also revealed cardiac myocyte degeneration and microscopic changes in the salivary glands in mice.[5]

Long-term Toxicity. Rats and rabbits were dosed by gavage with 1.0 to 100 mg/kg BW for 6 months. The highest dose caused retardation of BW gain, reduced oxygen consumption, prolonged prothrombin time, changes in the glycogen-forming and antitoxic functions of the liver and in the blood serum enzyme activity, increased content of free *SH*-groups in the blood homogenate and liver tissue, dystrophic and necrobiotic lesions to the visceral organs. The dose of 2.0 mg/kg BW was found to be ineffective for these changes.[1,6]

Reproductive Toxicity.

Embryotoxicity. Neonatal SD rats were orally dosed with 1.0, 2.0, or 3.0 mM IDE/kg BW on days 5 to 15 postpartum. The treatment at the highest dose level caused an increase in organ/BW ratio and changes in some enzyme activity in liver and kidney.[9]

Gonadotoxicity. In a 13-week NTP study in F344 rats, testicular degeneration appeared to be a direct toxic action of IDE. It was accompanied by a decrease in the testis and epididymis weights; in animals given 2500 ppm IDE in drinking water, there were reduced sperm motility and count.[5]

Mutagenicity.

In vitro genotoxicity. IDE exhibited no mutagenic activity in *Salmonella* mutagenicity assay.[5,10,11] It did not induce SCE and CA,[5,12] and was shown to be negative in mouse lymphoma cells with and without *S9* activation.[5] IDE appeared to be positive in *in vitro* hepatocyte (isolated from rats, hamsters, or pigs) single strand-break as say.[13]

In vivo cytogenetics. IDE appeared to be a mutagen for *Dr. melanogaster*. It did not induce micronuclei in peripheral blood erythrocytes in mice exposed by topical application for 13 weeks.[5]

Carcinogenicity classification.

NTP: NE - NE - NE - NE

Chemobiokinetics. IDE is rapidly and nearly completely absorbed from the GI tract and distributed throughout the body. The highest concentrations of radiolabel are found in the liver and kidney, the target organs for the toxicity of IDE.[5] Incorporation of IDE into microsomal phospholipids and changes in liver enzyme activity were observed in adult rats.[9] The major route of elimination is likely to be urinary excretion.

Regulations. ***U.S. FDA*** (1998) approved the use of IDE (1) as an adjuvant to control pulp content in the manufacture of paper and paperboard for contact with dry food prior to the sheet-forming operation in accordance with the conditions prescribed by 21 CFR part 176.170; (2) in adhesives used as components of articles intended for use in packaging, transporting, or holding food in accordance with the conditions prescribed by 21 CFR part 175.105; (3) as a defoaming agent that may be safely used in the manufacture of paper and paperboard intended for use in producing, manufacturing, packing, transporting, or holding food in accordance with the conditions prescribed by 21 CFR part 176.210; (4) in the manufacture of rubber articles intended for repeated use in producing, manufacturing, packaging, processing, preparing, treating, packing, transporting, or holding food in accordance with the conditions prescribed by 21 CFR part 177.2600, and as (5) a component of the uncoated or coated food-contact surface of paper and paperboard intended for use in producing, manufacturing, packaging, processing, preparing, treating, packing, transporting, or holding dry, aqueous and fatty foods in accordance with the conditions prescribed by 21 CFR parts 176.170 and 176.180.

Standards. ***Russia*** (1994). MAC and PML: 0.8 mg/l (organolept., taste).

References:

1. Diachkov, V. I., Comparative hygienic and toxicological studies on di- and triethanolamine norm-setting in water reservoirs, in *Industrial Pollution of Water Reservoirs*, S. N. Cherkinsky, Ed., Meditsina, Moscow, Issue No 9, 1969, 105 (in Russian).
2. Smyth, H. F., Weil, C. S., West, J., et al., An exploration of joint toxic action, II. Equitoxic versus equivolume mixtures, *Toxicol. Appl. Pharmacol.*, 17, 498, 1970.
3. Blum, K., Huizenga, C. G., Ryback, R. S., et al., Toxicity of diethanolamine in mice, *Toxicol. Appl. Phamacol.*, 22, 175, 1972.
4. Korsrud, G. O., Grice, H. G., Goodman, T. K., et al., Sensitivity of several serum enzymes for the detection of thioacetamide-, dimethylnitrosamine-, and diethanolamine-induced liver damage in rats, *Toxicol. Appl. Pharmacol.*, 26, 299, 1973.
5. *Toxicity Studies of Diethanolamine Administered Topically and in Drinking Water to F344 Rats and B6C3F$_1$ Mice*, NTP Technical Report Series No. 20, Research Triangle Park, NC, NIH Publ. No 92-3342, U.S. Dept. Health and Human Services, October 1992, 68.
6. Vorobyeva, R. S., *N,N'*-Dicyclohexyl-2-benzothiasolsulfen amide, in *Toxicology of New Chemical Substances Used in Rubber and Tyre Industry*, Z. I. Israel'son, Ed., Meditsina, Moscow, 1968, 89 (in Russian).

7. Melnick, R. L., Mahler, J., Bucher, J. R., Hejtmancik, M., Singer, A., and Persing, R., Toxicity of diethanolamine. 2. Drinking water and topical application exposure in B6C3F_1 mice, *J. Appl. Toxicol.,* 14, 11, 1994.
8. Melnick, R. L., Mahler, J., Bucher, J. R., Thompson, M., Hejtmancik, M., Ryan, M. J., and Mezza, L. E., Drinking water and topical application exposure in F344 rats, *J. Appl. Toxicol.*, 14, 1, 1994.
9. Burdock, G. A. and Masten, L. M., Diethanolamine induced changes in the neonatal rat, Abstract, *Toxicol. Appl. Pharmacol.*, 48, A60, 1979.
10. Hartung, R., Rigas, L. K., and Cornish, H. H., Acute and chronic toxicity of diethanolamine, *Toxicol. Appl. Pharmacol.*, 17, 308, 1970.
11. Hedenstedt, A., Genetic health risk in the rubber industry: Mutagenicity studies on rubber chemicals and process vapors, in *Prevention of Occupational Cancer*, Int. Symp., Int. Labour Office, Geneva, 1982, 476.
12. Dean, B. J., Brooks, T. M., Hodson-Walker, G., et al., Genetic toxicology testing of 41 industrial chemi cals, *Mutat. Res.,* 153, 57, 1985.
13. Pool, B. L., Brenndler, S. Y., Liegible, U. M., et al., Employment of adult mammalian primary cells in toxicology: *in vivo* and *in vitro* genotoxic effects of environmentally significant *N*-nitrosodialkyl-amines in cells of the liver, lung, and kidney, *Environ. Mutagen.,* 15, 24, 1990.

ISOCYANIC ACID, METHYLENEDI-*p*-PHENYLENE ESTER

Polymeric 4,4'-diphenylmethane diisocyanate

CAS No 26447-40-5

Abbreviation. PMPDI.

Trade Name. Crude MDDIP.

4,4'-Diphenylmethane diisocyanate

Molecular Formula. $C_{15}H_{10}N_2O_2$

M = 250.27

CAS No 101-68-8

RTECS No NQ9350000

Abbreviation. MPDI.

Synonyms and **Trade Name.** Bis(1,4-isocyanatophenyl)methane; 4,4'-Diisocyanatodiphenylmethane; 4,4'-Ethylenedi(phenylisocyanate); Isonate; Methylenebis(4-isocyanatebenzene); 4,4'-Methylenediphenyl diisocyanate.

Properties. Colorless or pale-yellow crystals or fused solid with a faint odor of oranges.

Applications. Used in the production of polyurethane coatings and elastomers. Vulcanizing agent for rubber.

Acute Toxicity. In rats and mice, LD_{50} is 2.2 g/kg BW. Following administration of the lethal dose animals died within 3 to 5 days. Gross pathology examination revealed changes in the blood vascular and circulatory systems, tiny hemorrhages and areas of dystrophy in the myocardium.[1] At the same time, Wollrich and Roy (1969) considered a single administration of MPDI to be of low toxicity.

Repeated Exposure failed to reveal cumulative properties. K_{acc} is 12.8, when animals were exposed to a mixture of MPDI and polycyclic isocyanates. MPDI was given to rats in corn oil for 5 days (total dose of 4.3 to 5.0 g/kg BW). The treatment caused slight spleen enlargement in two of five rats.[014]

Allergenic Effect was found to develop on *i/g* administration.

Observations in man. Allergic reactions were observed in workers with occupational exposure to MPDI.[4]

Animal studies. MPDI caused intensive skin sensitivity in guinea pigs similar to contact allergy.[2] Phenyl isocyanate has been found to be a potent inducer of both contact sensitization and the humoral immune response (*IgG* antibody production) following dermal application to mice. Karol and Kramarik concluded that phenyl isocyanate may contribute significantly to the sensitization potential of commercial MPDI products in which it is a contaminant.[3]

Mutagenicity.

In vitro genotoxicity. MPDI was found to be mutagenic in bacterial tests.[5]

Carcinogenicity. The risk of cancer associated with occupational exposure to isocyanates has been examined in three industrial cohort studies and in a population-based case-control study of several types of cancer (IARC). No strong association or consistent pattern has emerged.

PMPDI containing 44.8 to 50.2% monomeric MPDI was tested for carcinogenicity by inhalation in rats. An increased incidence of lung tumors was observed.

Carcinogenicity classification. There is inadequate evidence for the carcinogenicity of MPDI or PMPDI in *humans.*There is limited evidence in *experimental animals* for the carcinogenicity of a mixture containing monomeric and PMPDI.

IARC: 3

Chemobiokinetics.

Observations in man. Hydrolyzable *Hb* adducts of *methylenedianiline* (MDA) were found in 10/27 workers exposed to MPDI. MPDI and *N-acetyl*-MPDI were found in urine samples of 23/27 workers after base treatment.[6] The major urinary metabolites of MPDI are MDA and *N-acetyl-4,4-methylenedianiline*, both of which also form haemoglobin adducts in exposed workers and rats.

Regulations.

EU (1990). MPDI is available in the *List of authorized monomers and other starting substances which shall be used for the manufacture of plastic materials and articles intended to come into contact with foodstuffs (Section A)*.

U.S. FDA (1998) approved the use of MPDI in the manufacture of polyurethane resins which may be used as food-contact surfaces of articles intended for use in contact with bulk quantities of dry food of an appropriate type in accordance with the conditions prescribed by 21 CFR part 177.1680.

Great Britain (1998). MPDI is authorized without time limit for use in the production of polymeric materials and articles in contact with food or drink or intended for such contact. The quantity of MPDI in the finished polymeric material or article shall not exceed 1.0 mg/kg (expressed as isocyanate moiety).

Standards. ***EU*** (1990). OM (T): 1.0 mg/kg in finished product (expressed as NCO).

References:

1. Osipova, T. V., in *Problems of Industrial Hygiene and Industrial Pathology in Chemical Industry Workers*, Moscow, 1977, 63; in *Current Problems of Hygiene and Occupational Pathology in Some Sectors of Chemical Industry*, Moscow, 1978, 50 (in Russian).
2. Duprat, P., Gradiski, D., and Marignac, B., Pouvoir irritant et allergisant de deux isocyanates: toluene diisocyanate, diphenylmethane diisocyanate, *Eur. J. Toxicol. Environ. Hyg.*, 9, 43, 1976.
3. Karol, M. N. and Kramarik, J. A., Phenylisocyanate is a potent chemical sensitizer, *Toxicol. Lett.*, 89, 139, 1996.
4. Tse, K. S. et al., A study of serum antibody activity in workers with occupational exposure to diphenylmethane diisocyanate, *Allergy*, 40, 314, 1985.
5. Shimuzu, H. et al., The results of microbial mutations test for forty-three industrial chemicals, Abstract, *Sangyo Igaku*, 27, 400, 1985 (in Japanese).
6. Schutze, D., Sepai, O., Lewalter, J., Miksche, L., Henschler, D., and Sabbioni, G., Biomonitoring of workers exposed to 4,4'-methylenedianiline or 4,4'-methylenediphenyl diisocyanate, *Carcinogenesis*, 16, 573, 1995.

4-ISOPROPOXYDIPHENYLAMINE

Molecular Formula. $C_{15}H_{17}NO$

M = 227.33

CAS No 101-73-5

RTECS No JJ9500000

Synonyms. *p*-Hydroxydipropylamine isopropyl ester; *N*-(4-Isopropoxyphenyl)aniline.

Applications. Antioxidant in the rubber manufacture.

Acute Toxicity. LD_0 is 10 g/kg BW in rats.[027]

Mutagenicity. I. was found to be weakly positive in Chinese hamster lung cells without *S9* fraction and negative in Chinese hamster ovary cells.[1]

Regulations. ***U.S. FDA*** (1998) approved the use of I. in the manufacture of rubber articles intended for repeated use in producing, manufacturing, packaging, processing, preparing, treating, packing, transporting, or holding food in accordance with the conditions prescribed in CFR part 177.2600.

Reference:

1. Sofuni, T., Matsuoka, A., Sawada, M., Ishidate, M., Zeiger, E., and Shelby, M. D., A comparison of chromosome aberration induction by 25 compounds tested by two Chinese hamster cell (CHL and CHO) systems in culture, *Mutat. Res.* 241,175, 1990.

MERCAPTANS

Synonym and **Trade Name.** Thioalcohols; Thiols.

Properties. The lower mercaptans (C_3 - C_4) are highly volatile, colorless liquids with a sharp specific odor. Poorly soluble in water, readily soluble in alcohol. Taste and odor perception thresholds are <0.02 mg/l. Odor threshold concentrations are 0.02 mg/l for amyl mercaptane, dodecyl mercaptane and isopropyl mercaptane, 0.19 µg/l for benzyl mercaptane and ethyl mercaptane, 0.05 µg/l for allyl mercaptane, and 0.1 µg/l for tolyl mercaptane.

Applications. Occurs in rubber and cellophane, various coatings.

N-Butyl mercaptan

Molecular Formula. $C_4H_{10}S$
M = 90.20
CAS No 109-79-5
RTECS No EK6300000
Abbreviation. BM.

Synonyms. *n*-Butanethiol; *n*-Butyl thioalcohol; Thiobutyl alcohol.

Properties. Liquid. Solubility in water is 600 mg/l at 20°C. Odor perception threshold is reported to be 6.0 µg/l[1] or 0.012 µg/l.[02]

Acute Toxicity. Some mice died from a 3.0 g/kg BW dose, some rats from 0.4 g/kg BW.[2] According to other data, LD_{50} for rats is 1.5 g/kg BW.

Following a single administration of 500 mg/kg BW dose, hens exhibited weakness. Hematological changes were noted.[3] Oral administration to hens caused no neuropathological lesions but slightly increased brain acetyl cholinesterase and plasma butyryl cholinesterase activities.[4]

Reproductive Toxicity. Following inhalation exposure to 152 ppm BM for 6 hours daily during organogenesis, BM showed no teratogenic action in rats. At 68 and 152 ppm BM in mice, there was maternal toxicity including deaths. At 68 ppm BM, malformations of cleft palate and hydrocephalus were increased in mice.[5]

References:

1. Baker, K. A., Threshold odors of organic chemicals, *J. Am. Water Work Assoc.*, 55, 913, 1963.
2. Blinova, E. A., Toxicity characteristics of ethylmercaptan on the basis of chronic studies, *Gig. Sanit.*, 1, 18, 1965 (in Russian).
3. Abdo, K. M., Timmons, P. R., Graham, D. G., and Abou-Donia, M. B., Heinz body production and hematological changes in the hen after administration of a single oral dose of *n*-butyl mercaptan and *n*-butyl disulfide, *Fundam. Appl. Toxicol.*, 3, 69, 1983.
4. Abou-Donia, M. B., Late acute effect of *S,S',S''*-tributylphosphorotrithioate (DEF) in hens, *Toxicol. Lett.*, 4, 231, 1979.
5. Thomas, W. C., Seckar, J. A., Johnson, J. T., Ulrich, C. E., Klonne, D. R., Scherdein, J. L., and Kirwin, C. J., Inhalation teratology studies of *n*-butyl mercaptan in rats and mice, *Fundam. Appl. Toxicol.*, 8, 170, 1987.

Methyl mercaptan

Molecular Formula. CH_4S
M = 45.11

CAS No 74-93-1
RTECS No PB4375000
Abbreviation. MM.

Synonyms and **Trade Name.** Mercaptomethane; Methanethiol; Thiomethanol.

Properties. Gas with a sharp odor. Solubility in water is ~14 g/l at 25°C. Odor perception threshold is 0.25 or 1.1 µg/l.[010]

Chemobiokinetics. Metabolism of MM occurs through rapid methylation to form *dimethyl sulfide.* About 40% MM is excreted as CO_2 and 30% MM in the urine as *sulfates.* ^{14}C is detected in the methyl group of *methionine, choline* and *creatine.*[017]

Standards. ***Russia*** (1994). MAC and PML: 0.0002 mg/l (organolept., odor).

METHACRYLIC ACID, 2,3-EPOXYPROPYL ESTER

Molecular Formula. $C_7H_{10}O_3$
M = 142.15
CAS No 106-91-2
RTECS No OZ4375000
Abbreviation. GMA.

Synonyms. 2,3-Epoxy-1-propanol; Glycidyl α-methylacrylate.

Properties. Purified GMA is a clear, colorless liquid. Maximum solubility in water is 3.5 ml/l. It gives water a sharp odor of rotten fruit. Odor perception threshold is 1.4 mg/l. Chlorination of aqueous solution of GMA does not cause a foreign odor to appear.

Applications. Used in printing and in the production of rubber and polymeric materials.

Acute Toxicity. LD_{50} is 600 mg/kg BW in rats, 390 mg/kg BW in mice, 700 mg/kg BW in guinea pigs, and 470 mg/kg BW in rabbits.[1]

Species differences in susceptibility are not found. In rats, lethal doses cause hepatocyte vacuolization and destruction of the plasma membranes with escape of the cell contents into the interstices.

Repeated Exposure failed to reveal cumulative properties. Rats received 1/10 LD_{50} for 30 days. The treatment caused signs of intoxication to appear only on day 17. Decreased food consumption, ungroomed fur, reduced tissue turgor were observed. There were also breakdown of oxidation-reduction processes, and severe hepatic and renal impairment. Gross pathology examination revealed changes in the viscera.[1]

Long-term Toxicity. Rats and guinea pigs were dosed by gavage with 2.8 mg GMA/kg BW for 6 months. Leukopenia and changes in some blood enzyme activity were noted. Processes of cortical inhibition developed (loss of reflexes). Gross pathology examination revealed changes in the liver and myocardium.[1]

Allergenic Effect is reported only on intradermal sensitization.

Reproductive Toxicity.

Gonadotoxic effect is not found: the treatment does not alter the functional state of spermatozoa or testicular morphology.[1]

Mutagenicity.

In vitro genotoxicity. GMA was found to be positive in *Salmonella* mutagenicity assay.[2] Premutagenic lesion of plasmid induced by GMA can be converted into point mutation *in vivo.*[3]

Carcinogenicity. Glycidyl lactate and oleate are reported to induce sarcomas in mice.[4]

Chemobiokinetics. Metabolism of GMA occurs through saponification into the *glycidyl alcohol* and *methacrylic acid.* The acid may form an *acetyl coenzyme derivative*, which then enters the normal lipid metabolism.[03]

Regulations. ***U.S. FDA*** (1998) approved the use of GMA (1) in adhesives as a component of articles (a monomer) intended for use in packaging, transporting, or holding food in accordance with the conditions prescribed by 21 CFR part 175.105; and as (2) a component (a monomer) of the uncoated or coated food-contact surface of paper and paperboard intended for use in producing, manufacturing, packaging, processing, preparing, treating, packing, transporting, or holding dry food in accordance with the conditions prescribed by 21 CFR part 176.180.

Standards. ***Russia*** (1995). MAC and PML: 0.09 mg/l.

Reference:

1. Zdravko, B. I., Manenko, A. K., and Onishchuk, I. N., Data on hygienic regulation of glycidyl methacrylate in water bodies, *Gig. Sanit.*, 2, 67, 1985 (in Russian).
2. Canter, D. A., Zeiger, E., Haworth, S., Lawlor, T., Mortelmans K., and Speck, W., Comparative mutagenicity of aliphatic epoxides in *Salmonella, Mutat. Res.*, 172, 105, 1986.
3. Gao, H., Zuo, J., Xie, D., and Fang, F., Molecular mutagenesis induced by glycidyl methacrylate, *Clin. Med. Sci. J.*, 9, 1, 1994.
4. Wieder, R., McDonough, M., Meranze, D. R., and Shimkin, H.B., Investigation of fatty acids derivatives for carcinogenic activity, *Cancer Res.*, 30, 1037, 1970.

5-METHYL-1,3-BIS(PIPERIDINOMETHYL)HEXAHYDRO-1,3,5-TRIAZINE-2-THIONE

Molecular Formula. $C_{16}H_{31}N_5S$

M = 325.53

Synonym. Bistriazine.

Properties. Poorly soluble in water, readily soluble in alcohol on heating.

Applications. Used as a vulcanization accelerator.

Acute Toxicity. LD_{50} is reported to be 482 mg/kg BW in rats, 984 mg/kg BW in mice, 710 mg/kg BW in rabbits, and 543 mg/kg BW in guinea pigs. Poisoning was accompanied by excitation with subsequent CNS inhibition and convulsions. Death occurred from respiratory arrest.

Short-term Toxicity. Cumulative properties are observed in 4-month rat studies.

Long-term Toxicity. Rats and rabbits were exposed by gavage to oral doses up to 9.6 mg/kg BW for 10 months. The treatment caused anemia and retardation of BW gain, a decline in total serum albumen, and a reduction in the ascorbic acid content in the suprarenals. Changes in the STI value were found at all doses tested. Gross pathology examination revealed necrotic foci in the spleen, parenchymatous dystrophy and necrotic areas in the small intestine mucosa.

Reference:

Verbilov, A. A., Toxicological characteristics of new rubber vulcanization accelerators triazinethione and bistriazine, *Vrachebnoye Delo*, 6, 132, 1974 (in Russian).

5-METHYLHEXAHYDRO-1,3,5-TRIAZINE-2-THIONE

Molecular Formula. $C_4H_9N_3S$

M = 131.2

Synonym. Triazinethione.

Properties. Soluble in water, on heating, soluble in alcohol.

Applications. Used as a vulcanization accelerator.

Acute Toxicity. Substance of very low, if any, acute toxicity. Administration of 15 g/kg BW to mice, rats, and guinea pigs three times at 3-hour intervals caused no mortality or symptoms of intoxication. In 2 days, leukopenia and decrease in the concentration of *SH*-group in the blood serum were noted.[1]

Short-term Toxicity. No accumulation was found in a 4-month study when rats were given 0.5 g/kg BW dose.

Long-term Toxicity. Rats and rabbits were dosed by gavage for 10 months. A 50 mg/kg BW dose affected hippuric acid synthesis in the liver, caused prolongation of narcotic sleep and changes in the sugar curves as a result of galactose loading. Changes in oxidation-reduction enzyme activities and in *SH*-group content in the blood serum are reported. Gross pathology examination revealed hemodynamic disturbances in the liver and areas of superficial necrosis in the stomach and small intestine, with desquamation of the upper mucosal layers.[1,2]

Standards. ***Russia.*** Recommended PML: 10 mg/l.

References:

1. Verbilov, A. A., Toxicological characteristics of triazinethione - a new vulcanization accelerator for rubber, *Gig. Sanit.*, 7, 97, 1974 (in Russian).

2. Verbilov, A. A., Toxicological characteristics of new rubber vulcanization accelerators triazinethione and bistriazine, *Vrachebnoye Delo*, 6, 132, 1974 (in Russian).

MORPHOLINE

Molecular Formula. C_4H_9NO
M = 87.12
CAS No 110-91-8
RTECS No QD6475000

Synonyms. Diethyleneimid oxide; Diethylene oximide; 1-Oxa-4-azacyclohexane; Tetrahydro-*p*-isoxazine.

Properties. Colorless, mobile liquid with a penetrating, characteristic amine odor. Miscible with water, ethanol, castor and pine oil.

Applications. M. is a cheap solvent and intermediate in rubber production. It is used in the manufacture of antioxidants and plasticizers.

Exposure. M. has been detected in samples of foodstuffs and beverages.

Acute Toxicity. LD_{50} is 525 mg/kg BW in mice. Poisoning is accompanied with CNS inhibition.[1]

LD_{50} was also reported to be 1.05 g/kg BW in rats, 0.72 g/kg BW in mice,[2] and 0.09 g/kg BW in guinea pigs.[03] According to other data, LD_{50} is 1.45 g/kg BW in rats.[3,4] LD_{50} of aqueous dilution of M. appeared to be 1.6 g/kg BW in rats and 0.9 g/kg BW in guinea pigs. Administration of undiluted, unneutralized M. caused hemorrhages in the stomach and small intestine in guinea pigs and rats.[5]

Repeated Exposure. Guinea pigs received 0.5 g M./kg BW over a period of 30 days. The treatment caused necrosis of the liver and renal tubules.[5]

Long-term Toxicity. Rats were exposed orally at a level of 1.0 g/kg in the diet. Histological examination revealed fatty degeneration of the liver observed after 270 days of the treatment.[6]

Mutagenicity.

In vitro genotoxicity. M. was negative in *Salmonella* mutagenicity assay.[7] It did not induce micronuclei, CA, or mutations in hamsters and cultural animal cells.[8] It was weakly active in the L5178Y mouse lymphoma assay; it did not produce an increase in the unscheduled DNA synthesis in cultured primary rat hepatocytes or in SCE in Chinese hamster ovary cells.[9]

Carcinogenicity. Swiss mice were fed a diet containing 6.33 g M./kg for 28 weeks (that is a relatively short period for carcinogenicity studies) and observed for a further 12 weeks. Lung adenomas were observed in the treated and untreated animals. There was a significant increase in the incidence of extrapulmonary tumors in treated mice.[10]

$B6C3F_1$ mice were given 0.25 or 1.0% M. oleic acid salt in the drinking water for 96 weeks. Survival of the animals was not affected. Retardation of BW gain was observed at the high dose level. No increase in tumor incidence was reported.[11]

S/c treatment of European hamsters led to the development of respiratory tract tumors (predominantly in the nasal cavity and lungs). Histologically, these tumors appeared to be papillary polyps, papillomas, adenomas, adenocarcinomas, squamous cell carcinomas, and mixed carcinomas.[12]

Carcinogenicity classification. An IARC Working Group concluded that there is inadequate evidence for the carcinogenicity of M. in *experimental animals*; M. is not classifiable as to its carcinogenicity to *humans*.

IARC: 3.

Chemobiokinetics. Following ingestion by rats, M. was distributed over all the body and rapidly excreted.[13] It was not bound to serum proteins and 90% was eliminated unchanged. In the rabbit, kidneys appeared to be the primary route of excretion of M. Its rate of elimination was enhanced by acidification of the urine.[14]

Regulations. ***U.S. FDA*** (1998) approved the use of M. (1) in adhesives as a component of articles intended for use in packaging, transporting, or holding food in accordance with the conditions prescribed by 21 CFR part 175.105; (2) as a component of a defoaming agent intended for articles which may be used in producing, manufacturing, packing, processing, preparing, treating, packaging, transporting, or holding food in accordance with the conditions prescribed by 21 CFR part 176.200; and (3) in the production of animal glue which may be safely used as a component of articles intended for use in producing, manufacturing, packaging, processing, preparing, treating, packing, transporting, or holding food, in accordance with the conditions prescribed by 21 CFR part 178.3020. M. salts of the fatty acids meeting certain CFR requirements nay be used (4) as components of protective coatings applied to fresh fruits and vegetables in compliance with GMP requirements and in accordance with the conditions prescribed in 21 CFR part 172.235.

References:

1. Patel, V. K., Venkatakrishna-Bhatt, H., Patel, N. B., and Jindal, M. N., Pharmacology of new glutamide compounds, *Biomed. Biochim. Acta*, 44, 795, 1985.
2. Ethyl Browning's *Toxicity and Metabolism of Industrial Solvents*, R. Snyder, Ed., 2 ed., vol 2: Nitrogen and Phosphorus Solvents, Amsterdam-NY-Oxford, Elsevier, 1990, 247
3. Smyth, H. F., Carpenter, C. P., Weil, C. S., and Pozzani, U. C., Range-finding toxicity data. List V, *Arch. Ind. Occup. Med.,* 10, 61, 1954.
4. Zayeva, G. I., Timofievskaya, L. A., Bazarova, L. A., and Migukina, N. V., Comparative toxicity of a group of cyclic imidocompounds, in *Toxicology of New Industrial Chemical Substances*, A. A. Letavet, Ed., Meditsina, Moscow, Issue No 10, 1968, 25 (in Russian)..
5. Shea, T. E., The acute and sub-acute toxicity of morpholine, *J. Ind. Hyg. Toxicol.*, 2, 236, 1939.
6. Sander, J. and Burke, G., Induction of malignant tumors in rats by simultaneous feeding of nitrite and secondary amines, *Z. Krebsforsch.,* 73, 54, 1969.
7. Haworth, S., Lawlor, T., Mortelmans, K., Speck, W., and Zeiger, A., *Salmonella* mutagenicity test results for 250 chemicals, *Environ. Mutagen.*, Suppl. 1, 3, 1983.
8. Inui, N., Nishi, Y., Taketomi, M., Mori, M., Yamamoto, M., Yamada, T., and Tanimura, A., Transplacental mutagenesis of products formed in the stomach of hamsters given sodium nitrite and morpholine, *Int. J. Cancer,* 24, 365, 1979.
9. *American Conference of Governmental Industrial Hygienists, Documents of the Threshold Limit Values and Biological Exposure Indices*, 6 ed., vol 1, 2, 3, Cincinnati, OH, 1991, 1059.
10. Greenblatt, M., Mirvish, S., and So, B. T., Nitrosamine studies: induction of lung adenomas and concurrent administration of sodium nitrite and secondary amines in Swiss mice, *J. Natl. Cancer Inst.*, 46, 1029, 1971.
11. Shibata, M.-A., Kurata, Y., Tamano, S. Ogiso, T., Fukushima, S., and Ito, N., 13-week subchronic toxicity study with morpholine oleic salt administered to $B6C3F_1$ mice, *J. Toxicol. Environ. Health,* 22, 187, 1987.
12. Reznik, G., Experimental carcinogenesis in the respiratory tract using the European field hamster (*Cricetus cricetus L.*) as a model, *Fortschr.-Med.,* 95, 2627, 1977.
13. Tanaka, A., Tokieda, T., Nambaru, S., Osawa, M., and Yamaha, T., Excretion and distribution of morpholine in rats, *J. Food Hyg. Soc.*, 19, 329, 1978
14. van Stee, E. W., Wynns, P. C., and Moorman, M. P., Distribution and disposition of morpholine in the rabbit, *Toxicology*, 20, 53, 1981.

2-(MORPHOLINOTHIO)BENZOTHIAZOLE

Molecular Formula. $C_{11}H_{12}N_2OS_2$
M = 252.37
CAS No 102-77-2
RTECS No DL5950000
Abbreviation. MTB.

Synonyms and **Trade Names.** 2-Benzothiazolyl-*N*-morpholino sulfide; *N*-(Oxydiethylene)benzothiazole-2-sulfenamide; Santocure mor; Sulfenamide M.

Properties. Light-yellow powder. Poorly soluble in water, soluble in alcohol.

Applications. Used as a vulcanization accelerator for rubber and vulcanizates. M. is used either alone or in mixtures with other accelerators.

Acute Toxicity. LD_{50} for mice is reported to be 1.87 g/kg BW.[1] According to other data,[2] rats and mice tolerate 5.0 g/kg BW. After administration, there was apathy and adynamia. Death occurs within 8 days. Two doses of 0.5 g/kg BW led to reduced incorporation of radioactive iodine and to histology changes in the thyroid.

Short-term Toxicity. Rabbits were dosed by gavage with 20 mg/kg BW for 3.5 months. The treatment caused some liver impairment and bilirubin index increase indicating cholagogic effect.[3] Histology examination revealed signs of fatty dystrophy of the renal tubular epithelium. Changes in the thyroid and liver dystrophy were found to develop.[2]

Reproductive toxicity.

Teratogenicity. When administered on days 1 through 20 of pregnancy, MTB is shown to cause developmental abnormalities in the musculoskeletal system of rats.[4]

Allergenic Effect. MTB showed evident sensitization potential in guinea pigs.[5]

Mutagenic Effect has been evaluated in a battery of *in vivo* mutagenic assay.[6]

Carcinogenicity. Nitrosocompounds of morpholine are powerful carcinogens.

Standards. ***Russia*** (1995). MAC: 0.5 mg/l.

References:

1. Vorob'yeva, R. S. and Mesentseva, N. V., Sulfenamide C and M, in *Toxicology of New Chemical Substances Used in Rubber and Tyre Industry*, Z. I. Israel'son, Ed., Meditsina, Moscow, 1968, 54 (in Russian).
2. Zayeva, G. N., Fedorova, V. I., Fedechkina, N. I., Evaluation of toxicity of sulphenamide BT, M & C, in *Toxicology of New Industrial Chemical Substances*, A. A. Letavet and I. V. Sanotsky, Eds., Meditsina, Moscow, Issue No 8, 1966, 70 (in Russian).
3. Vorob'yeva, R. S. and Mesentseva, N. V., Experimental study on comparative toxicity of a new agent accelerating vulcanization, *Gig. Truda Prof. Zabol.*, 7, 28, 1962 (in Russian).
4. *Ann. Report Osaka City Inst. Public Health Environ. Sci.*, 43, 90, 1981.
5. Wang, X. S. and Suskind, R. R., Comparative studies of the sensitization potential of morpholine, 2-mercaptobenzothiazole and 2 other derivatives in guinea pigs, *Contact Dermat.*, 19, 11, 1988.
6. Hinderer, R. K., Myhr, B., Jagannath, D. R., Galloway, S. M., Mann, S. W., Riddle, J. C., and Brusick, D. J., Mutagenic evaluation of four rubber accelerators in a battery of *in vivo* mutagenic assay, *Environ. Mutagen.*, 5, 193, 1983.

4-[(MORPHOLINOTHIOCARBONYL)THIO]MORPHOLINE

Molecular Formula. $C_9H_{16}N_2O_2S_2$

M = 248.39

CAS No 13752-51-7

57018-26-5

RTECS No QE7189000

Synonyms and **Trade Names.** Accelerator OTOS; *N*-Oxydiethylene thiocarbamyl-*N*-oxydiethylene sulfenamide.

Applications. Used as a vulcanization accelerator.

Repeated Exposure. Sprague-Dawley rats were exposed to 6.25 to 25 mg/kg BW for 8 weeks. The highest dose decreased BW gain and liver weight.

Long-term Toxicity. Male and female Sprague-Dawley rats were fed dietary levels of 20 to 60 mg/kg BW for 112 weeks. Retardation of BW gain was observed only with the highest dose. Gross pathology examination revealed hydronephrosis, capillary necrosis, hemorrhages in the kidneys and urinary bladder, and epithelial hyperplasia in the ureter. Similar results were reported with 25 to 500 mg/kg BW doses of *N*-oxydiethylene-2-benzothiazole sulfenamide (a vulcanization accelerator).[1]

Reproductive Toxicity. Did not affect the fertility of males.

Mutagenicity. Data available are insufficient for evaluating mutagenic hazard.

In vivo cytogenetics. M. is likely to be negative in DLM assay.[1]

In vitro genotoxicity. M. was positive in mouse lymphoma L5178Y TK+/- forward mutation assay.[2]

Carcinogenicity. A significant increase in the incidence of benign and malignant tumors of the kidneys, urethra and urinary bladder, pineal gland, suprarenals, and mammary gland was observed in the above described chronic experiment.

References:

1. Hinderer, R. K., Lancas J. R., Knezevich A. L., and Auletta, C. S., The effect of long-term dietary administration of the rubber accelerator, *N*-oxydiethylene thiocarbamide-*N*-oxydiethylene sulfenamide, to rats, *Toxicol. Appl. Pharmacol.*, 82, 521, 1968.
2. Hinderer, R. K., Myhr, B., Jagannath, D. R., Galloway, S. M, Mann, S. W., Riddle, J. C., and Brusick, D. J., Mutagenic evaluations of four rubber accelerators in a battery of *in vitro* mutagenic assays, *Environ. Mutagen.*, 5, 193, 1983

1- and 2-NAPHTHOLS

Molecular Formula. $C_{10}H_8O$

M = 144.18

	CAS No	RTECS No
α-Naphthol	90-15-3	QL2800000
β-Naphthol	135-19-3	QL2975000

Synonyms. 1- and 2-Hydroxynaphthalenes.

Properties. Colorless crystals with a characteristic odor. β-N. is poorly soluble in water. Odor perception threshold is 1.29 mg/l [010] or 0.1 to 0.125 for α-N., and 0.75 mg/l[1] or 0.4 to 1.3 mg/l [2] for β-N.

Applications. Used in the production of synthetic rubber, dyes, and antioxidants.

Acute Toxicity. LD_{50} is reported to be 2.6 g α-N./kg and 2.5 g β-N./kg BW in rats, 275 and 98 mg/kg BW in mice, 9.0 and 5.4 g/kg BW in rabbits, and 2.0 and 1.3 g/kg BW in guinea pigs, respectively. Administration of lethal doses resulted in a short excitation with subsequent adynamia, dyspnea, and side position.[2]

Charles River CD-1 mice were orally exposed to 0.5, 1.0, or 2.0 g α-N./kg BW. Acute dosing caused histopathological lesions in the kidney and stomach of mice from all treatment groups. In the kidney, degeneration of the distal tubular epithelium, papillary necrosis, and tubular dilatation were noted. In the stomach, there were vascular congestion and acute inflammatory cell infiltration.[3]

Repeated Exposure revealed cumulative properties. Charles River CD-1 mice received 50, 100, or 200 mg/kg BW for 30 days. Three male mice in the 200 mg/kg BW group showed gastric histopathological effects. In the study, none of the mice showed any kidney lesions or changes in hematology or clinical chemistry parameters.[3]

Long-term Toxicity. The same signs of intoxication were observed. The threshold dose was established for the effect on conditioned reflex activity.[2]

Reproductive Toxicity.

Gonadotoxicity. α-N. is found to inhibit testosterone production by isolated Leydig cells.[4]

Embryotoxicity. Rats received 5.0% α-N. in their diet. The treatment caused mild maternal toxicity but produced no effects on the offspring. A two-generation study in rats revealed no adverse effects on reproduction of a hair-coloring formulation that contained α-N.[5]

Teratogenicity. α-N. was not teratogenic in mice[6] and rats.[7]

Mutagenicity.

In vitro genotoxicity. An effect was found in *Vicia fava* cells; α-N. induced CA in cultured mammalian cells; β-N. was noted to have colchicine-like effect, being evident in a cluster of metaphase cells.[8] α-N. was negative in *Salmonella typhimurium*, BASC and micronucleus tests.[9]

Chemobiokinetics. Metabolism of α-N. and β-N. is likely to occur via partial oxidation to dioxynaphthalenes. α-N. and β-N. are removed predominantly as conjugates with glucuronic acid and partially as conjugates with sulfonic acid (Harkness, 1968).

Oral administration of 45 mg α-N./kg BW resulted in 95% of the administered dose being eliminated within 72 hours after treatment in male mice.[10]

Regulations. ***U.S. FDA*** (1998) approved the use of β-N. as (1) a component of defoaming agent intended for articles which may be used in producing, manufacturing, packing, processing, preparing, treating, packaging, transporting, or holding food in accordance with the conditions prescribed by 21 CFR part 176.200; and as (2) a defoaming agent in the manufacture of paper and paperboard intended for use in producing, manufacturing, packing, transporting, or holding food in accordance with the conditions prescribed by 21 CFR part 176.210.

Standards. ***Russia*** (1994). MAC and PML in drinking water: 0.1 mg α -N./l (organolept., odor) and 0.4 mg β-N./l.

References:

1. Baker, K. A., Threshold odors of organic chemicals, *J. Am. Water Work Assoc.*, 55, 913, 1963.
2. Rakhmanin, Yu. A., Comparative hygienic and toxicological studies of α- and β-naphtol isomers for their norm-setting in water reservoirs, in *Industrial Pollution of Water Reservoirs*, S. N. Cherkinsky, Ed., Meditsina, Moscow, Issue No 9, 1969, 67 (in Russian).
3. Poole, A. and Buckley, P., 1-Naphthol - single and repeated dose (30-day) oral toxicity studies in the mouse, *Food Chem. Toxicol.*, 27, 233, 1989.
4. Wright, K. and Hecht, L. B., 1-Naphthol and 9-phenanthrol inhibit testosterone secretion by isolated leydig cells, *Fed. Proc.*, 46, 525, 1987.
5. Burnett, C. M. and Goldenthal, E. I., Multigeneration reproduction and carcinogenicity studies in Sprague-Dawley rats exposed topically to oxidative hair-coloring formulations containing *p*-phenylene-diamine and other aromatic amines, *Food Cosmet. Toxicol.*, 26, 467, 1988.
6. Courtney, K. D., Gaylor, D. W., Hogan, M. D., and Falk, H. L., Teratogenic evaluation of pesticides: A large-scale screening study, *Teratology*, 3, 199, 1970.
7. Noda, T., Morita, S., Ohgaki, S., Shimizu, M., and Yamada, A., Safety evaluation of chemicals for use in household products (VI). Teratological studies on naphthol AS in rats, *Annu. Rep. Osaka City Inst. Public Health Environ. Sci.*, 47, 80, 1985.
8. Amer, S. M. and Ali, E. M., Cytological effects of pesticides. II. Meiotic effects on some phenols, *Cytolgia* (Tokyo), 33, 21, 1968.
9. Gocke, E., King, M.-T., Eckhardt, K., and Wild, D., Mutagenicity of cosmetic ingredients licensed by the EEC, *Mutat Res.*, 90, 91, 1981.
10. Chern, W. H. and Dauterman, W. C., Studies on the metabolism and excretion of 1-naphthol, 1-naphthyl-β-*D*-glucuronide, and 1-naphthyl-β-*D*-glucoside in the mouse, *Toxicol. Appl. Pharmacol.*, 67, 303, 1983.

1-NAPHTHYLAMINE

Molecular Formula. $C_{10}H_9N$
M = 143.19
CAS No 134-32-7
RTECS No QM1400000
Abbreviation. 1-NA.

Synonyms and **Trade Names.** 1-Aminonaphthalene; 1-Naphthalene amine; Naphthalidam; Naphthalidine; α-Naphthylamine.

Properties. Grey needles, becoming red on exposure to air, or a reddish crystalline mass with an unpleasant odor. Solubility in water is 0.17%, readily soluble in alcohol.

Applications. Used as a vulcanization accelerator for rubber and in the production of dyestuffs.

General Toxicity. Produced *metHb* in dogs.

Reproductive Toxicity.

Gonadotoxicity. Exposure caused no adverse effect in mouse head sperm morphology.[1]

Mutagenicity.

Observations in man. No data are available on the genetic and related effects in humans (IARC, Suppl. 6-406).

In vivo cytogenetics. 1-NA did not induce micronuclei in bone-marrow cells of mice treated *in vivo*, it induced DNA strand breaks in mice, but not in rats.

In vitro genotoxicity. 1-NA increased the incidence of CA in cultured rodent cells, but the results for SCE, mutation, and DNA damages were inconclusive; no cell transformation was induced in Syrian hamster embryo cells (IARC 4-87). However, according to Ishidate and Odashima,[2] it gave negative results in CA test carried out on cultured Chinese hamster cells.

Carcinogenicity. No carcinogenic effect was found following oral administration of 1-NA to hamsters or dogs.[3,4]

In beagle dogs, the effect of highly purified 1-NA has been investigated. 1-NA was given orally, dissolved in corn oil, in gelatin capsules at a dose level of 15 mg/kg BW (this study has limitations due to the small number of animals tested).[1]

1-NA is at least 200 times less potent as a carcinogen than 2-NA.[4]

Carcinogenicity classification. An IARC Working Group concluded that there is inadequate evidence for the carcinogenicity of 1-NA in *experimental animals* and in *humans*.

IARC: 3.

Chemobiokinetics. Metabolism of 1-NA varies in different animals. 1-NA is partly oxidized to *1-amino-2-naphthol*, which is eliminated in the urine but mainly as a *glucuronide*.[5] 0.2% from 5.0 mg dose was shown to be excreted in the urine. A part of 1-NA is excreted unchanged.

Regulations. ***U.S. FDA*** (1998) listed 1-NA for use in adhesives as a component of articles intended for use in packaging, transporting, or holding food in accordance with the conditions prescribed by 21 CFR part 175.105.

References:

1. Wyrobek, A., Gordon, L., and Watchmaker, G., Effect of 17 chemical agents including 6 carcinogen/noncarcinogen pairs on sperm shape abnormalities in mice, *Prog. Mutat. Res.*, 1, 712, 1981.
2. Ishidate, M. and Odashima, S., Chromosome tests with 134 compounds in Chinese hamster cells *in vitro* - a screening for chemical carcinogens, *Mutat. Res.*, 48, 337, 1977.
3. Radomski, J. L., Deichmann, W. B., Altman, N. H., et al., Failure of pure 1-naphtylamine to induce bladder tumors in dogs, *Cancer Res.*, 40, 3537, 1980.
4. Purchase, I. F. H., Kalinowski, A. E., Ishmael, J., et al., Lifetime carcinogenicity study of 1-naphtylamine and 2-naphtylamine in dogs, *Brit. J. Cancer*, 44, 892, 1981.
5. Kleison, D. B. and Ashton, M. I., in *Proc. 3rd Int. Cancer Congress*, Moscow- Leningrad, vol 2, 314, 1983 (in Russian).

2-NAPHTHYLAMINE

Molecular Formula. $C_{10}H_9N$

M = 143.19

CAS No 91-59-8

RTECS No QM2100000

Abbreviation. 2-NA.

Synonyms. 2-Aminonaphthalene; 2-Naphthaleneamine.

Properties. Shiny, colorless, odorless flakes or white to reddish crystals, volatile with stream. Solubility in cold water is 1.17%, readily soluble in hot water and alcohol. A concentration of 0.05 mg/l changes the organoleptic properties of water.

Applications. Is now used for research purposes only. Previously was used as an antioxidant in the rubber industry and dyestuffs. Since epidemiology studies have convincingly shown 2-NA to be a bladder carcinogen in humans, its use in dyestuffs and other industries has long been abandoned.

General Toxicity. The toxic effect is similar to that of aniline. Causes cyanosis and dysuria.

Reproductive Toxicity.

Gonadotoxicity. See ***1-Naphthylamine.***

Mutagenicity.

Observations in man. No data are available on the genetic and related effects in humans.

In vitro genotoxicity. 2-NA induced unscheduled DNA synthesis in human cells *in vitro*. It induced CA, SCE, DNA strand breaks in cultured rodent cells. This chemical induced gene mutations in mouse lymphoma cells *in vitro* and caused morphological transformations of Syrian hamster embryo cells.[1]

In vivo cytogenetics. Mice and rabbits treated with 2-NA had increased incidence of SCE; micronuclei were not induced in bone-marrow cells of mice treated *in vivo*. However, Mirkova reported positive micronucleus responses in $C_{57}BL_6$ mice with single or triple gavage. The discordant results might be due to the different mouse strains and treatment routes.[2]

2-NA gave negative responses in the peripheral blood of ICR mice after a single *i/p* treatment with 100, 200, and 300 mg/kg BW and in *ddY* mice after a single *i/p* treatment with 50, 100, 200, and 300 mg/kg BW.[1]

Mutagenic effect was demonstrated in the mouse spot test. 2-NA induced DNA strand breaks in hepatocytes of treated rats and DNA adducts in bladder and liver cells of dogs *in vivo* (IARC 4-197; IARC, Suppl. 6-410). 2-NA was marginally positive in a *Dr. melanogaster* assay.[3]

Carcinogenicity. The relationship between exposure to 2-NA, as well as to *benzidine*, and bladder cancer risk was established.[4]

When administered orally by gavage or in the diet, 2-NA induced urinary bladder carcinomas in hamsters and non-human primates (IARC 4-97), dogs, and liver tumors in mice.[5,6]

Beagle dogs received daily doses of 400 mg of various mixtures of NA. The treatment caused bladder neoplasms. Administration of pure 2-NA to dogs caused transitional-cell carcinomas of bladder. 1-NA is at least 200 times less potent as a carcinogen than 2-NA.[7]

Carcinogenicity classifications. In 1974, 2-NA was declared (OSHA) to be a suspected human carcinogen. An IARC Working Group concluded that there is sufficient evidence for the carcinogenicity of 2-NA in *experi mental animals* and in *humans*.

The Annual Report of Carcinogens issued by the U.S. Department of Health and Human Services (1998) defines 2-NA to be a known carcinogen, i.e., a substance for which the evidence from human studies indicates that there is a casual relationship between exposure to the substance and human cancer.

IARC: 1;

EU: 1.

Chemobiokinetics. 2-NA metabolism occurs through oxidation to form *2-amino-1-naphthol*. At least 10 metabolites of 2-NA have been identified in the urine of rats and rabbits. In these animals, 70 to 80% of metabolites are removed with the urine, 20 to 30% are excreted in the feces.

References:

1. Morita, T., Asano, N., Awogi, T., Sasaki, Yu. F., et al. Evaluation of the rodent micronucleus assay in the screening of IARC carcinogens (Groups 1, 2A and 2B). The summary report of the 6th collaborative study by CSGMT/JEMS-MMS, *Mutat. Res.*, 389, 3, 1997.
2. Mirkova, E., Activity of the human carcinogens benzidine and 2-naphthylamine in triple- and single-dose mouse bone marrow micronucleus assays: results for a combined test protocol, *Mutat. Res.*, 234, 161, 1990.
3. Vogel, E. W. and Nivard, N. J., Performance of 181 chemicals in a Drosophila assay predominantly monitoring interchromosomal mitotic recombination, *Mutagenesis*, 8, 57, 1993.
4. Budnick, L. D., Sokal, D. C., Falk, H., et al., Cancer and birth defects near the Drake Superfund site, Pennsylvania, *Arch. Environ. Health*, 39, 409, 1984.
5. Romanenko, A. M. and Martynenko, A. G., Morphological peculiarities of vesical tumors induced by 2-naphthylamine in dogs, *Problemy Oncologii*, 18, 70, 1972 (in Russian).
6. Hicks, R. M., Wright, R., and Wakefield, J. S., The induction of rat bladder cancer by 2-naphthylamine, *Brit. J. Cancer*, 46, 646, 1982.
7. Purchase, I. F. H., Kalinowski, A. E., Ishmael, J., et al., Lifetime carcinogenicity study of 1-naphthylamine and 2-naphthylamine in dogs, *Brit. J. Cancer*, 44, 892, 1981.

2,2',2''-NITRILOTRIETHANOL

Molecular Formula. $C_6H_{15}NO_3$

M = 149.2

CAS No 102-71-6
RTECS No KL9275000
Abbreviation. NTE.

Synonyms. Triethanolamine; 2,2',2"-Trihydroxytriethylamine; Tri(2-hydroxyethyl) amine.

Properties. Viscous, glycerine-like, colorless, clear liquid with a faint odor of ammonia. Readily soluble in water and alcohol. Perception threshold for the effect on the organoleptic properties of water is 5.0 mg/l. According to D'yachkov, odor perception threshold is 36 mg/l, taste perception threshold is 1.4 mg/l. Concentration of 1.0 g/l does not change the color of water. According to other data, NTE organoleptic threshold in water is 160 mg/l.[01]

Applications. Vulcanization activator, particularly in the mixes based on styrene-butadiene rubber. NTE is also used in the manufacture of surface-active agents, solvents, and commercial cosmetics. It is the most widely used emulsifier.

Migration of cellulose acetate constituents into contact fluids was found to be negligible (traces). At 40°C, NTE was discovered in the extract in concentration of 1.4 mg/l and was therefore subsequently eliminated from cellulose acetate formulation.[1]

Acute Toxicity. LD_{50} is reported to be 5.2 or 8.4 g/kg BW[3] in rats, 5.4 g/kg BW[2] or 7.8 g/kg BW in mice, and 5.3 g/kg BW in guinea pigs.[3]

Administration of high doses is accompanied by painful and tactile hyperesthesia and diarrhea on the second day. Mice look agitated and rats inhibited. Recovery in survivors occurs in 2 to 3 weeks. Gross pathology examination revealed congestion in the viscera and in the liver.

Repeated Exposure revealed cumulative properties. Retardation of BW gain, excitation, aggression, and changes in the blood morphology were noted in rats given 520 mg NTE/kg BW for 2 months. Mortality reached 75% in this study.[3]

Short-term Toxicity. Rats received 104 and 260 mg/kg BW for 4 months. From the beginning of the experiment animals were agitated and aggressive. The treatment caused anemia, an increase in the level of residual blood nitrogen and other changes which, however, had disappeared by the end of the study.[3]

Rats were given NTE in their feed for 3 months. The dose of 80 mg/kg BW was found to be ineffective. Higher doses (up to 730 mg/kg BW) caused histological changes, liver damage and death.[011]

Long-term Toxicity. NTE dissolved in distilled water at levels of 1.0 and 2.0% was given to male and female $B6C3F_1$ mice *ad libitum* in drinking water over a period of 82 weeks. There were no adverse effects in regard to survival or organ weights.[4]

Renal damage including nodular hyperplasia, pyelonephritis, and papillary necrosis were observed in rats and mice given 1.0 or 2.0% NTE in distilled drinking water for 104 weeks.[5]

Allergenic Effect. Contact sensitizing potential was not found in maximized guinea pig assays.[6]

Mutagenicity.

In vitro genotoxicity. NTE was not shown to be mutagenic in *Bac. subtilis* and in an unscheduled DNA synthesis test.[7,8]

Carcinogenicity. Results of carcinogenicity studies are controversial. NTE is reported to produce an elevated incidence of malignant lymphoid tumors in female ICR-JCL mice fed diets containing 0.3 and 0.03% NTE for their lifespan over that of male mice on the same diet or control mice.[9]

F344/DUCRJ rats were given *ad libitum* NTE dissolved in distilled water (1.0 and 2.0%) for 82 weeks. The dose levels in females were reduced by half from week 69 because of associated nephrotoxicity.[4,5] This study provided no evidence of carcinogenic potential of NTE in rats. Neoplasms developed in all groups of rats including the control group. No dose-related increase of the incidence of any tumor was observed.

Regulations.

U.S. FDA (1998) approved the use of NTE (1) in adhesives as a component of articles intended for use in packaging, transporting, or holding food in accordance with the conditions prescribed by 21 CFR part 175.105; (2) in the manufacture of polyurethane resins to be safely used in the food-contact articles intended for use in contact with dry food in accordance with the conditions prescribed by 21 CFR part 177.1680; (3) as an activator for rubber articles intended for repeated use in contact with food up to 5.0% by weight of the rubber product in accordance with the conditions prescribed by 21 CFR part 177.2600; (4) to adjust *pH* during the manufacture of aminoresins permitted for use as components of paper and paperboard intended for contact with dry food; (5) in resinous and polymeric coatings in a food-contact surface in accordance with the conditions prescribed by 21 CFR part 175.300; (6) as a defoaming agent intended for articles which may be used in producing, manufacturing, packing, processing, preparing, treating, packaging, transporting, or holding food subject to the provisions prescribed in 21 CFR part 176.200; as (7) a component of the uncoated or coated food-contact surface of paper and paperboard intended for use in producing, manufacturing, packaging, processing, preparing, treating, packing, transporting, or holding dry, aqueous and fatty foods in accordance with the conditions prescribed by 21 CFR part 176.210; (8) in surface lubricants employed in the manufacture of metallic articles that contact food in accordance with the conditions prescribed by 21 CFR part 178.3910; and (9) as a substance employed in the production of or added to textiles and textile fibers intended for use in contact with food in accordance with the conditions prescribed by 21 CFR part 177.2800.

EU (1990). NTE is available in the *List of monomers and other starting substances which may continue to be used for the manufacture of plastic materials and articles intended to come into contact with foodstuffs pending a decision on inclusion in Section A* (*Section B*).

References:

1. Selivanov, S. B., *Hygiene Investigation and Assessment of Reverse Osmosis Method for Desalination of Potable Drinking Water in "Filter-press"-type Installations*, Author's abstract of thesis, 2 Moscow Medical Institute, Moscow, 1977, 18 (in Russian).
2. Shurupova, Ye. A., Toxicology of triethyleneamine, in *Hygienic Aspects of the Use of Polymeric Materials and Articles Made of Them*, L. I. Medved', Ed., All-Union Research Institute of Hygiene and Toxicology of Pesticides, Polymers and Plastic Materials, Kiyv, 1969, 323 (in Russian).
3. Diachkov, V. I., Comparative hygienic and toxicologic studies on di- and triethanolamine norm-setting in water reservoirs, in *Industrial Pollution of Water Reservoirs*, Moscow, Issue No 9, 1969, 105 (in Russian).
4. Konishi, Y., Denda, A., Ushida, K., Emi, Y., Ura, H., Yokose, Y., Shiraiwa, K., and Tsutsumi, M., Chronic toxicity and carcinogenicity studies of triethanolamine in B6C3F$_1$ mice, *Fundam. Appl. Toxicol.*, 18, 25, 1992.
5. Mackawa, A., Onodera, H., Tanigawa, H., et al., Lack of carcinogenicity of triethanolamine in Fischer 344 rats, *J. Toxicol. Environ. Health*, 19, 345, 1986.
6. Wahlberg, J. E. and Fregert, S., Guinea-pig maximization test, in *Models in Dermatology*, N. H. I. Maibach and N. Lowe, Eds., Karger, Basel, 2, 225, 1985.
7. *The Fifth Report of the Cosmetic Ingredients*, Review Expert Panel, 1983.
8. Inoue, K., Sunakawa, T., Okamoto, K., et al., Mutagenicity tests and *in vitro* transformation assays on trietanolamine, *Mutat. Res.*, 10, 305, 1982.
9. Hoshino, H. and Tanooka, H., Carcinogenicity of triethanolamine in mice and its mutagenicity after reaction with sodium nitrile in bacteria, *Cancer Res.*, 38 3918, 1978.

p-NITROANILINE

Molecular Formula. $C_6H_6O_2N_2$
M = 138.13
CAS No 100-01-6
RTECS No BY7000000
Abbreviation. NA.

Synonyms. *p*-Aminonitrobenzene; 4-Nitrobenzenamine; *p*-Nitrophenylamine.

Properties. Fine crystalline, yellow powder. Solubility in water is 0.8 g/l (18.5°C) and 2.2 g/l (100°C). Readily soluble in alcohol and oils. Threshold concentration for color (yellow) is 0.05 mg/l. 0.02 to 0.04 mg/l concentrations do not affect the odor or taste of water.

Applications. Used in the production of synthetic resins.

Acute Toxicity. In rats, LD_{50} is 750 mg/kg BW; ET_{50} appeared to be 43 hours. The LD values for guinea pigs and NS mice are 450 and 810 mg/kg BW, respectively.[1]

Repeated Exposure revealed moderate cumulative properties. Retardation of BW gain, reduction in *Hb* content and consistent increase of methemoglobin and sulfohemoglobin levels in the blood were found on days 1 to 5 of the experiment in rats given 1.0 to 15 mg NA/kg BW; 4.0 and 15 mg/kg BW doses caused signs of anemia from day 10 of the experiment.[1]

Short-term Toxicity. Mice were given 1.0, 3.0, 10, 30, or 100 mg NA/kg BW by gavage for 90 days. The LOEL for hepatic effects was 10 mg/kg BW. Munro et al. (1996) suggested the calculated NOEL of 3.0 mg/kg BW for this study.[2]

Long-term Toxicity. Rats were dosed by gavage with up to 1.0 mg/kg BW for 6 months. The treatment affected conditioned reflex and motor activity of animals.

Reproductive Toxicity.

Gonadotoxic effect is found at 1.0 to 15 mg/kg BW dose levels.[1]

Embryotoxicity. Pregnant CD-1 mice received 1200 mg NA/kg BW in corn oil by gavage on days 6 to 13 of gestation. The treatment caused maternal toxicity, decreased the number of viable litters, and reduced pup viability and survival.[3]

Standards. ***Russia*** (1994). MAC and PML: 0.05 mg/l.

References:

1. Dergacheva, T. S. and Mikhailovsky, N. Ya., Substantiating the norm for 4-nitroaniline in water bodies, *Gig. Sanit.*, 4, 82, 1985 (in Russian).
2. *Toxicology and Carcinogenesis Studies of p-Nitroaniline (CAS No. 100-01-6) in B6C3F$_1$ Mice (Gavage Studies),* NTP Technical Report Series No 418, Research Triangle Park, NC, 1993.
3. Hardin, B. D., Schuler, R. L., Burg, J. R., Booth, G. M., Hazelden, K. P., MacKenzie, K. M., Piccirillo, V. J., and Smith, K. N., Evaluation of 60 chemicals in a preliminary developmental toxicity test, *Teratogen. Carcinogen. Mutagen.*, 7, 29, 1987.

3-NITROBIPHENYL

Molecular Formula. $C_{12}H_9NO_2$

M = 199.22

CAS No 2113-58-8

RTECS No DV5570000

Abbreviation. NBP.

Synonyms. *m*-Nitrobiphenyl; 3-Nitro-1,1'-biphenyl.

Applications. Used in rubber industry.

Mutagenicity. 2-NBP was weakly positive in *Salmonella typhimurium* mutagenicity assay.

Carcinogenicity classification. An IARC Working Group concluded that there is inadequate evidence of carcinogenicity of NBP in *experimental animals* and there are no data available to evaluate the carcinogenicity NBP in *humans*.

IARC: 3.

Regulations. ***U.S. FDA*** (1998) approved the use of 3-NBP in resinous and polymeric coatings used as the food-contact surfaces of articles intended for use in producing, manufacturing, packing, processing, preparing, treating, packaging, transporting, or holding food in accordance with the conditions prescribed by 21 CFR part 175.300.

Reference:

Tokiwa, H., Nakagawa, R., and Ohnishi, Y., Mutagenic assay of aromatic nitrocompounds with *Salmonella typhimurium*, *Mutat. Res.*, 91, 321, 1981.

N-NITROSODIBENZYLAMINE

Molecular Formula. $C_{14}H_{14}N_2O$

M = 226.30

CAS No 5336-53-8

RTECS No HQ7100000

Abbreviation. NDBA.

Synonyms. Dibenzylnitrosamine; *N*-Nitroso-*N*-(phenylmethyl)benzenemethanamide.

Exposure occurs during contact with food products, cigarette smoke, and alcoholic beverages. N. are also by-products of endogenous processes.

Migration Data. Significant penetration of nitrosoamines into boneless hams as a result of the use of elastic nettings containing rubber were noted by Pensabene et al.[1] Penetration of NDBA from the surface into inner parts of the ham is slow but significant. NDBA was detected at the level of 61.2 ppb on the outer surface of the ham and 15.6 ppb at a depth of 0.25 inch. Extraction of baby bottle nipples and pacifiers with dichloromethane showed varying levels of nitrosamines, in particular, mean 0.04 mg NDBA/kg.[2]

Acute Toxicity. LD_{50} is 900 mg/kg BW in rats.[3]

Mutagenicity.

In vivo cytogenetics. NDBA is mutagenic to *Salmonella typhimurium* and induces DNA strand breaks in isolated rat hepatocytes. NDBA is inactive in both the rat and mouse bone marrow micronucleus assays and in a rat liver autoradiographic assay for unscheduled DNA synthesis. It is, however, clearly active as a micronucleus-inducing agent and mitogen in the rat liver and is capable of inducing single-strand breaks in the DNA of rat liver.

It is concluded that NDBA is genotoxic to the rat liver *in vivo*.[4]

Carcinogenicity. NDBA is reported to be non-carcinogenic to the rat.[5]

Chemobiokinetics. Nitrosamines are environmental chemicals that alkylate macromolecules in various target tissues and produce cellular necrosis, malignant transformation, and immunosuppression.[5]

References:

1. Pensabene, J. W. Fiddler, W. and Gates, R. A., Nitrosamine formation and penetration in hams processed in elastic rubber nettings: *N*-nitrosodibutylamine and *N*-nitrosodibenzylamine, *J. Agric. Food. Chem.*, 43, 1919, 1995.
2. Sen, N. P., Seaman, S. W., and Kushwaha, S. C., Determination of non-volatile *N*-nitrosamines in baby bottle rubber nipples and pacifiers by high-performance liquid chromatography-thermal energy analysis, *J. Chromatogr.*, 463, 419, 1989.
3. *Z. Krebsforsch.*, 69, 103, 1967.
4. Schmezer, P., Pool, B. L., Lefevre, P. A., Callander, R. D., Ratpan, F., Tinwell, H., and Ashby, J., Assay-specific genotoxicity of *N*-nitrosodibenzylamine to the rat liver *in vivo*, *Environ. Mol. Mutagen.*, 15, 190, 1990
5. Haggerty, H. G. and Holsapple, M. P., Role of metabolism in dimethylnitrosamine-induced immunosuppression: A Review, *Toxicology*, 63, 1, 1990.

NITROSODIETHYLAMINE

Molecular Formula. $C_4H_{10}N_2O$

M = 102.14

CAS No 55-18-5

RTECS No IA3500000

Abbreviation. NDEA.

Synonyms. *N,N'*-Diethylnitrosamine; *N*-Ethyl-*N'*-nitrosoethanamine.

Properties. Slightly yellow liquid. Soluble in water and alcohol.

Exposure occurs during contact with food products, cigarette smoke, and alcoholic beverages. Nitrosamines are also by-products of endogenous processes. NDEA has been detected in trace amounts in tobacco smoke and in processed foods.

Applications. Used as an antioxidant and stabilizer.

Migration Data. Traces of nitrosamines in baby bottle rubber nipples and pacifiers could easily migrate to simulated saliva and milk; 7 out of 42 samples of these products contained greater than 0.03 ppm total volatile nitrosamines.[1]

A survey of 30 samples of various nipples and pacifiers that was carried out by analysis by GLC-thermal energy analyzer indicated the presence of *N-nitrosamines,* including NDEA (up to 88 mg/kg). These

N-nitrosamines were shown to migrate easily from the rubber products to liquid infant formula, orange juice, and simulated human saliva.[2]

All of the 17 samples of rubber nipples and pacifiers tested using GC-thermal energy analysis after extraction with artificial saliva were found to contain at least two of the following: *N*-nitrosodimethylamine, NDEA, *N*-nitrosodibutylamine, and *N*-nitrosopiperidine. Total volatile *N*-nitrosamine levels with a mean content of 7.3 mg/kg were found. Nitrosatable compounds, measured as *N*-nitrosamines after nitrosation, were detected in 15 of the 17 samples, the mean level being 5.0 mg/kg.[3]

Migration of *N*-nitrosamines and their amine precursors into artificial saliva from 16 types of children's pacifiers and baby-bottle nipples bought in shops in Israel has been investigated. *N*-Nitrosodibutylamine, NDEA, *N*-nitrosodimethylamine, *N*-nitrosopiperidine, and *N*-nitrosopyrrolidine were detected by one method involving dichloromethane extraction at individual levels as high as 369 ppb. Using the second method, which consisted of analysis of *N*-nitrosamines and their amine precursors, *N*-nitrosodibutylamine, *N*-nitrosodiethylamine, *N*-nitrosodimethylamine, and *N*-nitrosomorpholine were detected at concentrations up to 41 ppb, in addition to the three nitrosatable amines dibutylamine, diethylamine, and dimethylamine. Of the samples tested, 50% failed to meet both U.S. and German regulations. A larger percentage, 60%, would not conform to the standard suggested in the U.S., and more than 80% failed to comply with the even stricter Dutch standard.[4]

Acute Toxicity. LD_{50} is 210 to 220 mg/kg BW in rats. Poisoning caused hypermotility, diarrhea, and a decrease in BW gain. There was fatty liver degeneration at histological examination.[5,6]

In mice, LD_{50} is 200 mg/kg BW.[7]

Immunotoxicity. Nitrosamines produce cellular necrosis and immunosuppression.[8]

Mutagenicity. Nitrosamines alkylate macromolecules in various target tissues.

In vitro genotoxicity. NDEA made dose- and time-related increases in the number of single-strand breaks in rat hepatocytes.[9]

NDEA was mutagenic when tested on *Salmonella typhimurium* strain *TA100* and *E. coli* in the presence of rat liver *S9* or hamster liver *S9* by preincubation and pour-plate assays.[10]

NDEA was positive in *Neurospora crassa* mutagenicity assay. It induced recessive lethal mutations in cultured mammalian cells,[11] SCE in Chinese hamster cells in presence of a metabolic system (Ishidate *et al*, 1988), as well as gene mutations in mouse lymphoma cells *in vitro* (Amacher and Paillet, 1983).

In vivo cytogenetics. Induced recessive lethal mutations in *Dr. melanogaster* but dominant lethal mutations were not observed in mice. There was no micronuclei induction in bone marrow cells or polychromatic erythrocyte suppression after treatment of E mu-PIM-l transgenic mice with 1.0 and 3.0 mg NDEA/kg BW.[12,13] It induced recessive lethal mutations in *Dr. melanogaster.*[11]

Carcinogenicity. NDEA was carcinogenic in all animal species tested. It induced benign and malignant tumors after administration by various routes, including digestion. It is a potent genotoxic procarcinogen but not mouse lymphomagen. Statistically significant changes of DNA viscometric parameters, which are considered indicative of DNA fragmentation, were produced in rats treated with single oral doses of 0.067 mg NDEA/kg BW.[14]

Pim transgenic mice were exposed to 1.3 mg NDEA/kg BW by gavage for 38 weeks. Increased incidence of malignant lymphoma was seen in high- and low-dose females treated with NDEA. NDEA also produced a high incidence (>70%) of hepatic hemangiosarcomas at the low- and high-dose levels.[15]

Coloworth rats received 0.033 to about 17 ppm NDEA in their drinking water. Positive trends for tumors of the nasopharynx, lower jaw stomach, kidney, ovaries, seminal vesicles, liver, and esophagus were reported. A dose-dependent increase in GI tract and liver tumors has been observed in C57BD mice; tracheal and hepatocellular tumors were noted in Syrian hamsters.[16]

S/c treatment of European hamsters led to the development of respiratory tract tumors (predominantly, in the nasal cavity and lungs). Histologically, these tumors appeared to be papillary polyps, papillomas, adenomas, adenocarcinomas, squamous cell carcinomas, and mixed carcinomas.[17]

Carcinogenicity classifications. An IARC Working Group concluded that there is sufficient evidence for the carcinogenicity of NDEA in *experimental animals* and there were no data available to evaluate the carcinogenicity of NDEA in *humans*.

The Annual Report of Carcinogens issued by the U.S. Department of Health and Human Services (1998) defines NDEA to be a substance which may reasonably be anticipated to be carcinogen, i.e., a substance for which there is a limited evidence of carcinogenicity in *humans* or sufficient evidence of carcinogenicity in *experimental animals*.

IARC: 2A;

U.S. EPA: B2.

Regulations. ***U.S. FDA*** (1998) has set a 10-ppb limit on nitrosoamines in rubber nipples for baby bottles.

References:

1. Sen, N. P., Kushwaha, S. C., Seaman, S. W., et al., Nitrosamines in baby bottle nipples and pacifiers: Occurrence, migration, and effect of infant formulas and fruit juices on *in vitro* formation of nitrosamines under simulated gastric conditions, *J. Agric. Food Chem.*, 33, 428, 1985.
2. Sen, N. P., Seaman, S., Clarkson, S., Garrod, F., and Lalonde, P., Volatile *N*-nitrosamines in baby bottle rubber nipples and pacifiers. Analysis, occurrence and migration, *Sci. Publ.*, 57, 51, 1984.
3. Osterdahl, B. G., *N*-nitrosamines and nitrosatable compounds in rubber nipples and pacifiers, *Food Chem. Toxicol.*, 21, 755, 1983.
4. Westin, J. B., Castegnaro, M. J., and Friesen, M. D., *N*-nitrosamines and nitrosatable amines, potential precursors of *N*-nitramines, in children's pacifiers and baby-bottle nipples, *Environ. Res.*, 43, 126, 1987.
5. *Arzneimittel-Forsch.*, 14, 1167, 1964.
6. Korolev, A. A., Mikhailovskii, N. Ya., and Ilnitskii, A. P., Hygienic evaluation of *N*-nitrosamines and their transformation products in an aqueous environment, *Gig Sanit.*, 8, 11, 1980 (in Russian)
7. Natl. Technical Information Service OTS0555387.
8. Haggerty, H. G. and Holsapple, M. P., Role of metabolism in dimethylnitrosamine-induced immunosuppression: A review, *Toxicology*, 63, 1, 1990.
9. Bradley, M. O., Dysart, G., Fitzsimmons, K., Harbach, P., Lewin, J., and Wolf, G., Measurements by filter elution of DNA single- and double-strand breaks in rat hepatocytes: effects of nitrosamines and gamma-irradiation, *Cancer Res.*, 42, 2592, 1982.
10. Araki, A., Muramatsu, M., and Matsushima, T., Comparison of mutagenicities of *N*-nitrosamines on *Salmonella typhimurium TA100* and *Escherichia coli WP2 uvrA/pKM101* using rat and hamster liver *S9*, *Gann*, 75, 8, 1984.
11. Montesano, R. and Bartsch, H., Mutagenic and carcinogenic N-nitroso compounds: Possible environmental hazards, *Mutat. Res.*, 32, 179, 1976.
12. Armstrong, M. J. and Galloway, S. M., Micronuclei induced in peripheral blood of E μ-PIM-1 transgenic mice by chronic oral treatment with 2-acetylaminofluorene or benzene but not with diethylnitrosoamine or 1,2-dichloroethane, *Mutat. Res.*, 302, 61, 1993.
13. Proudlock, R. J. and Allen, J. A., Micronuclei and other nuclear anomalies induced in various organs by diethynitrosamine and 7,12-dimethylbenz[a]anthracene, *Mutat. Res.*, 174, 141, 1986.
14. Brambilla, G., Carlo, P., Finollo, R., and Sciaba L., Dose-response curves for liver DNA fragmentation induced in rats by sixteen *N*-nitroso compounds as measured by viscometric and alkaline elution analyses, *Cancer Res.*, 47, 3485, 1987.
15. Storer, R. D., Cartwright, M. E., Cook, W. O., Soper, K. A., and Nichols, W. W., Short-term carcinogenesis bioassay of genotoxic procarcinogens in PIM transgenic mice, *Carcinogenesis*, 16, 285, 1995.
16. Peto, R. R., Gray, P., Brantom, P., and Grasso, P., Nitrosamine carcinogenesis in 5120 rodents: Chronic administration of sixteen different concentrations of NDEA, NDMA, NPYR, and NPIP in the water of 4440 inbred rats, with parallel studies on NDEA alone of the effect of age starting (3, 6 or 20 weeks) and of species (rats, mice, hamsters), *IARC Sci. Publ.*, Lyon, 57, 627, 1984.
17. Reznik, G., Experimental carcinogenesis in the respiratory tract using the European field hamster (*Cricetus cricetus L.*) as a model, *Fortschr.-Med.*, 95, 2627, 1977.

N-NITROSODIMETHYLAMINE

Molecular Formula. $C_2H_6N_2O$
M = 74.08
CAS No 62-75-9
RTECS No IQ0525000
Abbreviation. NDMA.

Synonyms. *N,N'*-Dimethylnitrosamine; *N*-Methyl-*N'*-nitrosomethanamine.

Properties. Volatile, yellow, oily liquid of low viscosity. Soluble in water, lipids, alcohols, and other organic solvents.

Applications. NDMA is used primarily as a research chemical, it occurs as an impurity; plasticizer for rubber and acrylonitrile polymers; an antioxidant.

Exposure occurs during contact with food products, cigarette smoke, and alcoholic beverages. Nitrosamines are also by-products of endogenous processes. NDMA intake from a pack of cigarettes is about 0.001 mg. It is also the most commonly detected *N*-nitrosamine in food systems.

Migration Data. Traces of nitrosamines in baby bottle rubber nipples and pacifiers could easily migrate to simulated saliva and milk; 7 out of 42 samples of these products contained greater than 0.03 ppm total volatile nitrosamines (mainly, NDMA and *N*-nitrosodi-*N'*-butylamine).[1]

A survey of 30 samples of various nipples and pacifiers that was carried out by analysis by GLC-thermal energy analyzer indicated the presence of *N-nitrosamines,* including NDMA (up to 70 mg/kg). These *N*-nitrosamines were shown to migrate easily from the rubber products to liquid infant formula, orange juice, and simulated human saliva.[2] All of the 17 samples of rubber nipples and pacifiers tested using GC-thermal energy analysis after extraction with artificial saliva were found to contain at least two of the following: NDMA, *N*-nitrosodiethylamine, *N*-nitrosodibutylamine, and *N*-nitrosopiperidine. Total volatile *N*-nitrosamine levels with a mean content of 7.3 mg/kg were found. Nitrosatable compounds, measured as *N*-nitrosamines after nitrosation, were detected in 15 of the 17 samples, the mean level being 5.0 mg/kg.[3]

Migration of *N*-nitrosamines and their amine precursors into artificial saliva from 16 types of children's pacifiers and baby-bottle nipples bought in shops in Israel has been investigated. *N*-Nitrosodibutylamine, *N*-nitrosodiethylamine, NDMA, *N*-nitrosopiperidine, and *N*-nitrosopyrrolidine were detected by one method involving dichloromethane extraction at individual levels as high as 369 ppb. Using the second method, which consisted of analysis of *N*-nitrosamines and their amine precursors, *N*-Nitrosodibutylamine, *N*-nitroso-diethylamine, NDMA, and *N*-nitrosomorpholine were detected at concentrations up to 41 ppb, in addition to the three nitrosatable amines dibutylamine, diethylamine, and dimethylamine. Of the samples tested, 50% failed to meet both U.S. and German regulations. A larger percentage, 60%, would not conform to the standard suggested in the U.S., and more than 80% failed to comply with the even stricter Dutch standard.[4]

Acute Toxicity. High doses of NDMA produce severe effects, including centrilobular liver necrosis in rodents and fatal liver cirrhosis in humans.[5] Nevertheless, the toxicity of nitrosocompounds is not of great interest because there is no relationship between the toxic effect and carcinogenic activity.[6]

LD_{50} is 40 to 50 mg/kg BW in rats,[5,7] 30 mg/kg BW in hamsters, and 20 mg/kg BW in dogs. Administration of lethal doses causes very pronounced liver lesions in dogs, rabbits, rats, and guinea pigs. Gross pathology examination revealed liver centrilobular necrosis with subsequent hemorrhage. In rats, acute toxic effects occur mainly as cellular necrosis in the liver[8] and kidney.[9]

Mink, and also sheep and cattle, are particularly sensitive to NDMA. A single dosage of 50 mg/kg, which is necrotizing for rat's liver, produces more acute liver damage in cats than in guinea pigs, rats, or monkeys.[10]

Repeated Exposure. Administration of 5.0 mg/kg BW daily produced the same liver damage in guinea pigs and rats as a single 50 mg/kg dose.[10]

Mink received 2.5 to 5.0 mg NDMA/kg BW for 7 to 11 days. Administration led to extensive liver damage and hepatocyte necrosis. Cell proliferation was noted in the bile ducts. Ascites and hemorrhages in the GI tract were found to develop (Carter et al., 1969).

Twelve administrations of 0.5 mg/kg doses to sheep caused death or the development of pronounced anoxia. Masticatory function was inhibited. Cats and lizards received 1.0 mg NDMA/kg BW for 30 days.

The treatment revealed centrilobular necrosis and hemorrhages to be the main alteration detected in the liver of cats. Lizards appear to be unaffected.[7]

Short-term Toxicity. Significant hepatotoxicity was observed in cattle given 0.1 mg/kg BW dose over a period of 1 to 6 months.

Long-term Toxicity. Dose-related mortality and hepatotoxicity were observed following oral exposure of CD-1 mice to 1.0 to 20 ppm NDMA.[11]

Reproductive Toxicity.

Embryotoxicity. A 30 mg/kg BW dose caused embryolethal effect, particularly when given on critical days of gestation.[1]

Immunotoxicity. Chronic exposure to NDMA resulted in a marked and persistent immunodepression of cellular and humoral response in CD-1 mice. Immunodepression of immunoglobulin *M* antibody response to sheep erythrocytes was found to be time related and dose related; cellular immune response was markedly suppressed by 10 and 20 ppm doses; 1.0 ppm NDMA caused no signs of immunotoxicity.[11]

Mutagenicity. NDMA requires metabolic activation to exert genotoxic and carcinogenic effects.[12,13] NDMA is a classical alkylating agent.

In vivo cytogenetics. NDMA was negative in the DLM test but it gave positive results in *Dr. melanogaster* assay.

In vitro genotoxicity. NDMA made dose- and time-related increases in the number of single-strand breaks in rat hepatocytes.[14] It was shown to have positive results in *Salmonella typhimurium* and *B. coli* mutagenicity assays in the presence of rat liver *S9* or hamster liver *S9* by 'preincubation' and 'pour-plate' assays.[15,16]

NDMA gave negative results in CA test carried out on cultured Chinese hamster cells.[17]

Carcinogenicity. Although no direct NDMA-related cancer was reported in humans, nitrosoamines are considered to be potentially epigenetic compounds due to chronic immunodepression.[10]

Statistically significant changes of DNA viscometric parameters, which are considered indicative of DNA fragmentation, were produced in rats treated with single oral doses of 0.022 mg NDMA/kg BW.[18] Carcinogenic potential was revealed in all animals tested: mice, rats, hamsters, guinea pigs, rabbits, ducks, etc.[19-22]

NDMA produced malignant neoplasms predominantly in the liver, kidneys, and respiratory tract. An increase in lung adenoma response in strain *A/J* mice was observed after *i/p* and oral administration.[23] Liver tumors were observed in BD rats which received NDMA in drinking water,[6] and in Porton rats exposed to it in their diet.[22]

Coloworth rats received 0.033 to about 17 ppm NDMA in their drinking water. Positive trends for tumor development of lung, skin, seminal vesicles, lymphatic/hematopoietic system, and liver were reported.[24]

NDMA was shown to be a transplacental carcinogen when it was administered to pregnant rats, mice, and Syrian golden hamsters by several routes.[25] Doses of 2.0 to 50 mg/kg BW were shown to cause tumors, predominantly in the rat liver.

Tomatis and Cefis[26] reported the formation of liver tumors in Syrian golden hamsters after administration of a total dose of 3.6 mg in 5 weeks and a single dose of 1.6 mg. The lowest dose tested in male RF mice (administered via drinking water and corresponding to a dose of 0.4 mg/kg BW) produced lung adenomas in 13 out of 17 animals and hemangiocellular tumors in 2 out of 10 animals.[19]

Carcinogenicity classifications. An IARC Working Group concluded that there is sufficient evidence for the carcinogenicity of NDMA in *experimental animals* and there were no adequate data available to evaluate the carcinogenicity of NDMA in *humans*.

The Annual Report of Carcinogens issued by the U.S. Department of Health and Human Services (1998) defines NDMA to be a substance which may reasonably be anticipated to be carcinogen, i.e., a substance for which there is a limited evidence of carcinogenicity in *humans* or sufficient evidence of carcinogenicity in *experimental animals*.

IARC: 2A;
U.S. EPA: B2;
EU: 2.

Chemobiokinetics. Is slowly absorbed from the stomach but rapidly from the upper part of the small intestine.[27] Oral doses below 40 mg/kg BW were completely metabolized by the liver and did not enter general circulation.[28] The liver and nasal mucosa are likely to be the principal sites of NDMA metabolism and their injury effects may be related to NDMA metabolites. Oxidative *N*-dimethylation to form *formaldehyde* has been shown with liver microsomes of rats, mice, and hamsters.[29]

Metabolic activation of NDMA is mediated by cytochrome *P-450* and culminates in the formation of electrophilic intermediates which may form covalent binds with DNA or other macromolecules. This seems to be a mechanism of mutagenicity and carcinogenicity initiation.

NDMA crosses the placenta in mice and rats.[30]

Regulations. ***U.S. FDA*** (1998) has set a 10-ppb limit on nitrosoamines in rubber nipples for baby bottles. FDA also established action levels of 5.0 ppb *N*-NDMA in malt beverages and 10 ppb in barley malt.

References:

1. Sen, N. P., Kushwaha, S. C., Seaman, S. W., et al., Nitrosamines in baby bottle nipples and pacifiers: Occurrence, migration, and effect of infant formulas and fruit juices on *in vitro* formation of nitrosamines under simulated gastric conditions, *J. Agric. Food Chem.*, 33, 428, 1985.
2. Sen, N. P., Seaman, S., Clarkson, S., Garrod, F., and Lalonde, P., Volatile *N*-nitrosamines in baby bottle rubber nipples and pacifiers. Analysis, occurrence and migration, *Sci. Publ.*, 57, 51, 1984.
3. Osterdahl, B. G., *N*-nitrosamines and nitrosatable compounds in rubber nipples and pacifiers, *Food Chem. Toxicol., 21, 755, 1983.*
4. Westin, J. B., Castegnaro, M. J., and Friesen, M. D., *N*-nitrosamines and nitrosatable amines, potential precursors of *N*-nitramines, in children's pacifiers and baby-bottle nipples, *Environ. Res.*, 43, 126, 1987.
5. Magee, P. N. and Swann, P. F., Nitrosocompounds, *Brit. Med. Bull.*, 25, 240, 1969.
6. Druckrey, H., Preussmann, R., Ivankovic, S., and Schmael, D., Organotropism and carcinogenic activities of 65 different *N*-nitrosodi compounds on BD-rats, *Z. Krebsforsch.*, 69, 103, 1967.
7. Maduagewu, E. N. and Anosa, V. O., Hepatotoxicity of dimethylnitrosamine in cats and lizards, *Toxicol. Lett.*, 9, 41, 1981.
8. Shank, R. C., Toxicology of *N*-nitrosocompounds, *Toxicol. Appl. Pharmacol.*, 31, 361, 1975.
9. Hard, G. C., MacKay, R. L., and Kockhar, O. S., Electron microscopic determination of the sequence of acute tubular and vascular injury induced in the rat kidney by carcinogenic dose of dimethylnitrosamine, *Lab. Investig.*, 50, 659, 1984.
10. Maduagewu, E. N. and Bassir, O., A comparative assessment of toxic effect of dimethylnitrosamine in six different species, *Toxicol. Appl. Pharmacol.*, 53, 211, 1980.
11. Desjardins, R., Fournier, M., Denizeau, F., et al., Immunodepression by chronic exposure to *N*-nitrosodimethylamine in mice, *J. Toxicol. Environ. Health*, 37, 351, 1992.
12. O'Neill, J. P., et al., Cytotoxicity and mutagenicity of dimethylnitrosamine in mammalian cells (CHO/ HGPRT system); enhancement by calcium phosphate, *Environ. Mutagen.*, 4, 7, 1982.
13. Bolognesi, C. et al., A new method to reveal the genotoxic effects of *N*-nitrosodimethylamine in pregnant mice, *Mutat. Res.*, 207, 57, 1988.
14. Bradley, M. O., Dysart, G., Fitzsimmons, K., Harbach, P., Lewin, J., and Wolf, G., Measurements by filter elution of DNA single- and double-strand breaks in rat hepatocytes: effects of nitrosamines and gamma-irradiation, *Cancer Res.*, 42, 2592, 1982.
15. Araki, A., Muramatsu, M., and Matsushima, T., Comparison of mutagenicities of *N*-nitrosamines on *Salmonella typhimurium TA100* and *Escherichia coli WP2 uvrA/pKM101* using rat and hamster liver S9, *Gann*, 75, 8, 1984.
16. Montesano, R. and Bartsch, H., Mutagenic and carcinogenic *N*-nitroso compounds: Possible environmental hazards, *Mutat. Res.*, 32, 179, 1976.
17. Ishidate, M. and Odashima, S., Chromosome tests with 134 compounds in Chinese hamster cells *in vitro* - a screening for chemical carcinogens, *Mutat. Res.*, 48, 337, 1977.
18. Brambilla, G., Carlo, P., Finollo, R., and Sciaba L., Dose-response curves for liver DNA fragmentation induced in rats by sixteen *N*-nitroso compounds as measured by viscometric and alkaline elution analyses, *Cancer Res.*, 47, 3485, 1987.

19. Magee, P. N. and Barness, J. M., Carcinogenic nitroso compounds, *Adv. Cancer. Res.*, 10, 163, 1967.
20. Ishinishi, N., Tanaka, A., Hisanaga, A., et al., Comparative study on the carcinogenicity of *N*-nitrosodiethylamine, *N*-nitrosodimethylamino, *N*-nitroso-*N'*-propylamine to the lung of Syrian golden hamster following intermittent instillation to the trachea, *Carcinogenesis*, 9, 947, 1988.
21. Clapp, N. K. and Toya, R. E., Effect of cumulative dose and dose rate on dimethylnitrosamine oncogenesis in RF mice, *J. Natl. Cancer Inst.*, 45, 495, 1970.
22. Terracini, B., Magee, P. N., and Barness, J. M., Hepatic pathology in rats of low dietary levels of dimethylnitrosamine, *Brit. J. Cancer.*, 21, 539, 1967.
23. Stoner, G. D., Greisiger, E. A., Schut, H. A. J., et al., A comparison of the lung adenoma response in strain *A/J* mice after intraperitoneal and oral administration of carcinogens, *Toxicol. Appl. Pharmacol.*, 72, 313, 1984.
24. Peto, R. R., Gray, P., Brantom, P., and Grasso, P., Nitrosamine carcinogenesis in 5120 rodents: Chronic administration of sixteen different concentrations of NDEA, NDMA, NPYR and NPIP in the water of 4440 inbred rats, with parallel studies on NDEA alone of the effect of age starting (3, 6 or 20 weeks) and of species (rats, mice, hamsters), *IARC Sci. Publ.*, Lyon, 57, 627, 1984.
25. Tomatis, L., Transplacental carcinogenesis, in *Modern Trends in Oncology*, Part I, R. W. Raven, Ed., Butterworths, London, 1973.
26. Tomatis, L. and Cefis, F., The effect of multiple and single administration of dimethylnitrosamine to hamsters, *Tumori*, 53, 447, 1967.
27. Phillips, J. C., Heading, C. E., Lake, B. G., et al., Further studies on the effects of inhibitors on the metabo lism and toxicity of dimethylnitrosamine, *Biochem. Soc. Trans.*, 3, 179, 1975.
28. Diaz Gomes, M. I., Swann, P. E., and Magee, P. N., The absorption and metabolism in rats of small oral doses of dimethyl nitrosamine, *Biochem. J.*, 164, 497, 1977.
29. Argus, M. F., Acros, J. C, Pastor, K. M., et al., Dimethylnitrosamine-demethylase: absence of increased enzyme catabolism and multiplicity of effector sites in repression. Homoprotein involvement, *Chem.-Biol. Interact.*, 13, 127, 1976.
30. Shendrikova, I. A. et al., The mechanism of the transplacental penetration of rat and mouse embryos by *N*-nitrosodimethylamine, *Pharmacol. Toxicol.*, 46, 53, 1983.

N-NITROSODIPHENYLAMINE

Molecular Formula. $C_{12}H_{11}N_2O$
M = 198.22
CAS No 86-30-6
RTECS No JJ9800000
Abbreviation. NDPA.

Synonyms and **Trade Names.** Diphenam N; Diphenylnitrosamine; 4-Nitroso-*N*-phenylbenzenamine; Retarder 2N; Vulcalent A.

Properties. Green or yellow plates or prisms with a bluish luster, or light-yellow, crystalline powder. Slightly soluble in water, soluble in acetone and ethanol. Sensitive to light and undergoes photolytic degradation. Gives water a specific, unpleasant odor (threshold concentration is 0.1 mg/l).

Applications. Used in mixes of natural and synthetic rubber and latex. Slows down the vulcanization and cross-linking of rubber mixes. Used in the production of organic dyestuffs. A stabilizer for polymers, an anti-scorching agent.

Exposure occurs during contact with food products, cigarette smoke, and alcoholic beverages. NDPA. are also by-products of endogenous processes.

Migration Data. *N*-Nitrosamines were shown to migrate easily from the rubber products to liquid infant for mula, orange juice, and simulated human saliva.[1]

Acute Toxicity. All mice died in 24 hours who had been given doses of 4.0 to 5.0 g/kg BW. LD_{50} appeared to be 3.85 g/kg BW.[2]

In rats, LD_{50} is 3.0 g/kg BW.[3]

Toxicity is increased when NDPA is given to animals in sunflower oil. In this case, LD_{50} is reported to be 1.68 g/kg BW in mice, and 1.87 g/kg BW in rats. Signs of cyanosis are noted.[4]

Repeated Exposure revealed insignificant cumulative properties. A 40-day exposure to 70 or 200 mg/kg BW affected the peripheral blood, CNS, and the enzyme activity in the blood and liver.

F344 rats and $B6C3F_1$ mice were fed diets containing up to 46 g/kg for 7 or 11 weeks. Females did not survive doses greater than 16 g/kg diet. Inflammatory lesions were found in the urinary bladder of mice (Shell Chem. Co.).

Short-term Toxicity. Rabbits were dosed by gavage with 20 mg/kg BW for 4 months. The treatment caused retardation of BW gain and increase in serum aldolase activity. Gross pathology examination revealed parenchymatous dystrophy of the renal tubular epithelium, foci of peribronchial pneumonia, and sometimes, emphysema.[2]

Reproductive Toxicity.

Teratogenic effect is found in 2-day-old chick embryos.[5]

Embryolethality is observed in chicken and rats at a 100 mg/kg BW dose level.

Mutagenicity. Nitrosamines alkylate macromolecules in various target tissues.

In vitro genotoxicity. Causes genetic damage *in vitro* (in bacteria, yeast, and cultured mammalian cells).[6] However, according to Ishidate and Odashima, NDPA gave negative results in CA test carried out on cultured Chinese hamster cells.[7]

Mutagenic activity was shown in *Salmonella typhimurium* strain *TA98* in the presence of *S9* microsomal fraction and a co-mutagen.[8] However, according to other communication, NDPA was not found to be mutagenic when tested on *Salmonella typhimurium* strain *TA100* and *E. coli WP2* in the presence of rat liver *S9* or hamster liver *S9* by 'preincubation' and 'pour-plate' assays.[9]

NDPA did not induce unscheduled DNA synthesis (Lake et al., 1978). NDPA was positive in mouse and hamster hepatocytes DNA repair test.[10]

Carcinogenicity. No carcinogenic effect was observed in $B6C3F_1$ mice given diet containing 10 and 20 g NDPA/kg for 101 weeks. However, carcinogenic action was found in F344 rats treated with 1.0 or 4.0 g NDPA/kg diet for 100 weeks. The treatment produced transitional-cell carcinomas of the urinary bladder in animals of the high dose group.[11,12] Carcinogenic potency is at least 100-fold lower than the potency of aliphatic nitrosamines.

Carcinogenicity classifications. An IARC Working Group concluded that there is inadequate evidence for the carcinogenicity of NDPA in *experimental animals* and there were no data available to evaluate the carcinogenicity of NDPA in *humans*.

IARC: 3;
U.S. EPA: B2;
NTP: P - P - N - N (N-NDPA, feed);
P* - N - P* - N (*p*-NDPA, feed).

Chemobiokinetics. NDPA is a powerful nitrosating agent readily donating *NO* to *secondary amines*. After a single dose of NDPA administered orally to rats, the major metabolite was nitrate. Nitrite and diphenylamine were found in minor amounts. A monohydroxylated diphenylamine was detected in a somewhat higher concentration. NDPA is denitrosated to nitric oxide and diphenylamine in the body and nitric oxide is then converted into nitrite and nitrate.[13]

Regulations. ***U.S. FDA*** has set a 10 ppb limit on nitrosoamines in rubber nipples for baby bottles. FDA also established action levels of 5.0 ppb NDMA in malt beverages and 10 ppb in barley malt, which is expected to reduce or eliminate exposure from these sources.

Standards. ***Russia*** (1995). MAC and PML: 0.01 mg/l.

References:

1. Sen, N. P. and Seaman, S., Volatile *N*-nitrosamines in baby bottle rubber nipples and pacifiers, Analysis, occurrence and migration, in *N*-nitroso compounds: Occurrence, biological effects and relevance to human cancer, *IARC Sci. Publ.* No 57, I. K. O'Neill, R. C. von Borstel, C. T. Miller, et al., Eds., IARC, Lyon, 1984, 51.

2. Zhilova, N. A. and Kasparov, A. A., Phthalic anhydride and *N*-nitrosodiphenylamine (vulcalent A), in *Toxicology of New Chemical Substances Used in Rubber and Tyre Industry*, Z. I. Israel'son, Ed., Medtsina, Moscow, 1968, 157 (in Russian).
3. Druckrey, H., Preussmann, R., Ivankovic, S., and Schmael, D., Organotropism and carcinogenic activities of 65 different *N*-nitrosodi compounds on BD-rats, *Z. Krebsforsch.*, 69, 103, 1967.
4. Korolev, A. A., Shlepnina, T. G., Mikhailovsky, N. Ya., et al., Maximum allowable concentration of diphenylnitrosamine and nitroguanidine in water bodies, *Gig. Sanit.*, 1, 18, 1980 (in Russian).
5. Korhonen, A., Hemminki, K., and Vainio, H., Embryotoxicity of 16 industrial amines to the chicken embryo, *J. Appl. Toxicol.*, 3, 112, 1983.
6. McGregor, D., The genetic toxicology of *N*-Nitroso diphenylamine, *Mutat. Res.*, 317, 195, 1994.
7. Ishidate, M. and Odashima, S., Chromosome tests with 134 compounds in Chinese hamster cells *in vitro* - a screening for chemical carcinogens, *Mutat. Res.*, 48, 337, 1977.
8. Wakabayashi, K., Nagao, M., Kawashi, T., et al., in *IARC Sci. Publ.*, 41, 695, 1982.
9. Araki, A., Muramatsu, M., and Matsushima, T., Comparison of mutagenicities of *N*-nitrosamines on *Salmonella typhimurium TA100* and *Escherichia coli WP2 uvrA/pKM101* using rat and hamster liver *S9*, *Gann*, 75, 8, 1984.
10. McQueen, C. A. and Williams G. M., The hepatocyte primary culture/DNA repair test using hepatocytes from several species, *Cell Biol. Toxicol.*, 3, 209, 1987.
11. Cardy, R. H., Lijinsky, W., and Hildebrandt, P. K., Neoplastic and non-neoplastic urinary bladder lesions induced in F344 rats and $B6C3F_1$ hybrid mice by *N*-nitrosodiphenylamine, *Ecotoxicol. Environ. Saf.*, 3, 29, 1979.
12. *Bioassay of N-Nitrosodiphenylamine for Possible Carcinogenicity*, Natl. Cancer Institute Technical Report Series No 164, DHEW Publ. No 79-1720, U.S. Govern. Print. Office, Washington, D. C., 1979.
13. Appel, K. E., Ruhl, C. S., Spiegelhalder, B., and Hildebrandt, A. G., Denitrosation of diphenylnitrosamine *in vivo*, *Toxicol. Lett.*, 23, 353, 1984.

NITROTOLUENE

Molecular Formula. $C_7H_7NO_2$
M = 137.15
CAS No 99-99-0
RTECS No XT2975000
Abbreviations. NT, ONT, MNT, and PNT.

Synonyms. Methylnitrobenzene; Nitrotoluol.

Properties. ONT and MNT are yellow, oily liquids at room temperature. PNT occurs as crystals. Commercial product contains 55 to 60% ONT, 3.0 to 4.0% MNT, and 35 to 40% PNT. Solubility in water at 30°C is 652 mg ONT/l, 498 mg MNT/l, and 442 mg PNT/l. Miscible with ethanol at any ratio. Odor perception threshold is 0.01 mg MNT and PNT/l and 0.05 mg ONT/l.[1]

Applications. Used in the production of agricultural and rubber chemicals, in the synthesis of various dyes, and in cosmetics.

Acute Toxicity. LD_{50} values for laboratory animals are presented in the Table.[2,3,017] Poisoned animals displayed excitability with subsequent inhibition, ataxia, and convulsions.

Table. Mean lethal doses of nitrotoluenes.

Animals	*ONT*	*MNT*	*PNT*	*Mixture of isomers*
	mg/kg BW	mg/kg BW	mg/kg BW	mg/kg BW
Mice	970–2463	330–800	1231–6800	1460
Rats	891–1610	1072–2400	1960–7100	1680
Rabbits	1750	1750–2400	1750	
Guinea pigs		3600		

Repeated Exposure. In a 14-day study, F344 rats and $B6C3F_1$ mice received NT isomers in the diet at concentrations of 388 to 20,000 ppm. The treatment did not affect survival and produced no clinical signs of toxicity. A decrease in BW gain was noted only in high dose groups.[2]

Mice received the doses of 200, 400, or 600 mg NT/kg BW by gavage over a period of 2 weeks. The treatment caused slight to moderate swelling of the hepatocytes in the liver of rats given MNT. No evidence of necrosis was noted. Hematology and serum chemistry analyses were unaffected. Slight decrease in the percentage of *B* lymphocytes in the spleen was observed. In rats given PNT, there was modest dose-dependent increase in liver and spleen weights.[4] Anemia, methemoglobinemia, sulfhemoglobinemia, and Heinz bodies were observed in rats given 1/5 LD_{50} for 1 and 3 months. The treated animals displayed functional changes in the CNS, liver, and kidneys.[017]

Short-term Toxicity. In a 13-week study, F344 rats and $B6C3F_1$ mice received NT in the feed at concentrations of 625 to 10,000 ppm (this corresponds to the doses of 40 to 900 mg/kg BW in rats, and 100 to 2000 mg/kg BW in mice). Neither effects on survival nor clinical signs of toxicity were noted. Retardation of BW gain was noticed only in high dose groups.[5] Histology examination revealed presence of hyaline droplets in the renal tubular epithelial cells, attributed to the increased level of g-2*u*-globulin. The structural changes in the spleen, as well as an increase in the relative liver weights and elevated serum levels of bile acids and liver enzymes were observed. Cytoplasmic vacuolization and oval cell hyperplasia were noted in the liver only in male rats fed ONT. In mice given ONT in the diet, there was an increase in liver weights and degeneration and metaplasia in the olfactory epithelium.[2,5]

Reproductive Toxicity.

Gonadotoxicity. In a 13-week study, changes in the testes comprised degeneration and reduction in sperm concentration, sperm motility, and spermatid number. The treatment increased the length of the estrous cycle in rats.[2]

Mutagenicity. There are known differences in the patterns of metabolism of MNT with the *o*-isomer undergoing a unique series of host and gut microflora-mediated reactions leading to an intermediate with high capacity to bind to hepatic DNA and induce unscheduled DNA synthesis.

In vitro genotoxicity. NT was negative in *Salmonella* mutagenicity assay. All NT isomers caused SCE in Chinese hamster ovary cells.[2,6]

PNT caused CA in Chinese hamster ovary cells and exhibited mutagenic activity in mouse lymphoma test, both with metabolic activation.

In vivo cytogenetics. No increase in micronuclei or CA was observed in bone marrow erythrocytes of mice given PNT *i/p*.[7]

ONT was found to increase unscheduled DNA synthesis *in vivo*.[2]

Carcinogenicity. In a 13-week study, mesotheliomas of the *tunica vaginalis* were observed in 3 out of 10 female rats fed the diet containing 5000 ppm ONT. Mesothelial cell hyperplasia was found to develop in 2 out of 10 male rats given 10,000 ppm ONT in the diet.[2]

Microflora metabolism was found not to be necessary for the mesothelioma response.

Carcinogenicity classification. An IARC Working Group concluded that there is sufficient evidence for evaluation of NT carcinogenicity in *experimental animals* and there is inadequate evidence for evaluation of carcinogenicity of NT in *humans*.

IARC: 3.

Chemobiokinetics. NT-isomers are metabolized being converted to the corresponding *benzyl alcohol* and to *benzoic acid* in the liver. Formation of the *nitrobenzyl alcohol glucuronide* appeared to be the major metabolic pathway only for ONT. *o-Nitrobenzyl glucuronide* is excreted via the bile into the intestine. Its metabolite *o-aminobenzylsulfate* seems to be responsible for binding covalently to DNA.[8] MNT and PNT undergo conjugation with glycine to form the *hippuric acid*, or nitro reduction and acylation. 73 to 86% radiolabelled isomers are excreted in the urine within 72 hours, and 5.0 to 13% with feces.[9]

Standards. ***Russia*** (1995). MAC: 0.01 mg MNT and PNT/l, and 0.05 mg ONT/l (organolept.).

References:

1. Kosachevskaya, P. I., in *Hygiene and Toxicology of Pesticides and Treatment of Poisonings,* L. I. Medved', Ed., Kiyv, Issue No 5, 1967, 92 (in Russian).

2. *Toxicity Studies of o-, m-, and p-Nitrotoluenes Administered in Dosed Feed to F344/N Rats and B6C3F$_1$ Mice*, (J. K. Dunnick), NTP Technical Report, NIH Publ. No 93-3346, Research Triangle Park, NC, 1992.
3. Ciss, M., Dutertre, H., Huyen, N., et al., Etude toxicologique des nitrotoluenes: Toxicite aigue et toxicite subaigue, *Dakar Medical*, 25, 303, 1980.
4. Burns, L. A., White, K. L., McCay, J. A., Fuchs, B. A., Stern, M., Brown, R. D., Musgrove, D. L., et al., Immunotoxicity of mononitrotoluenes in female B6C3F$_1$ mice: II. *Meta*-nitrotoluene, *Drug Chem. Toxi col.*, 17, 359, 1994.
5. Dunnick, J. K., Elwell, M. R., and Bucher, J. R., Comparative toxicities of *o*-, *m*-, and *p*-nitrotoluene in 13-week feed studies in F344 rats and B6C3F$_1$ mice, *Fundam. Appl. Toxicol.*, 22, 411, 1994.
6. Galloway, S., Armstrong, M., Reuben, C., et al., Chromosome aberration and sister-chromatid exchanges in Chinese hamster ovary cells: Evaluation of 108 chemicals, *Environ. Mutagen.*, 10 (Suppl. 10), 1, 1987.
7. Furukawa, A., Ohuchida, A., and Wierzba, K., *In vitro* mutagenicity tests on polyploid inducers, *Envi ron. Mutagen.*, 14 (Suppl. 15), 63, 1989.
8. Chism, J. P. and Rickert, D. E., *In vitro* activation of 2-aminobenzyl alcohol and 2-amino-6-nitrobenzyl alcohol, metabolites of 2-nitrotoluene and 2,6-dinitrotoluene, *Chem. Res. Toxicol.*, 2, 150, 1989.
9. Chism, J. P., Turner, M. J., and Rickert, D. E., The metabolism and excretion of mononitrotoluenes by Fischer 344 rats, *Drug Metab. Dispos.*, 12, 596, 1984.

2-NORBORNENE

Molecular Formula. C_7H_{10}
M = 94.15
CAS No 498-66-8
RTECS No RB7900000

Synonyms. Bicyclo[2.2.1]heptadiene; Norcamphene.

Properties. Crystalline substance. At temperature of 46 to 54°C it is a liquid with a persistent, penetrating odor. Odor threshold concentration is 0.004 mg/l.

Applications. Used in the production of synthetic rubber.

Acute Toxicity. See ***Norbornadiene***.

Repeated Exposure. The non-effective dose was identified to be 0.4 g/kg BW.

Reproductive Toxicity. Not found with the doses used in the subacute experiment.

Regulations. ***EU*** (1990). N. is available in the *List of monomers and other starting substances which may continue to be used for the manufacture of plastic materials and articles intended to come into contact with foodstuffs pending a decision on inclusion in Section A* (*Section B*).

Standards. ***Russia*** (1994). MAC: 0.004 mg/l (organolept., odor).

Reference:

Zholdakova, Z. I., Sil'vestrov, A. E., and Mikhailovsky, N. I., Substantiation of the maximum allowable concentration for norbornene, norbornadiene and dicyclopentadiene in water bodies, *Gig. Sanit.*, 2, 77, 1986 (in Russian).

NORBORNADIENE

Molecular Formula. C_7H_8
M = 92.13
CAS No 121-46-0
RTECS No RB6535000

Synonym. 1,2-Dichloroisobutane.

Properties. Colorless liquid with a penetrating, unpleasant odor. Threshold concentration for affecting the odor of water is 0.4 mg/l.

Applications. Used in the rubber production.

Acute Toxicity. LD_{50} is 0.89 g/kg BW in rats, and 3.85 g/kg BW in mice. Lethal doses cause a short period of excitation with subsequent inhibition and clonic convulsions. Death within 24 hours.

Repeated Exposure failed to reveal cumulative properties. None of rats died from administration of 1/10 to 1/250 LD_{min} for 30 days. Adynamia and CNS depression were noted. The treatment caused inhibition in the activity of LDH, NADP-dependent glucose-6-phosphate dehydrogenase, liver aminotransferase and alkaline phosphatase. An anticholinesterase effect was found. The NOEL of 0.5 g/kg BW was identified in this study.

Reproductive Toxicity not observed.

Standards. ***Russia*** (1995). MAC: 0.4 mg/l (organolept., odor).

Reference:

Zholdakova, Z. I., Sil'vestrov, A. E., and Mikhailovsky, N. I., Substantiation of the maximum allowable concentration for norbornene, norbornadiene and dicyclopentadiene in water bodies, *Gig. Sanit.*, 2, 77, 1986 (in Russian).

N,N'-OCTAALKYLTETRAAMIDOTHIOPYROPHOSPHATES

Synonyms. *N,N'*-Octaethyltetraamidothiopyrophosphate and *N,N'*-Octabutyltetraamidothiopyrophosphate.

Properties. Oils. Soluble in alcohol.

Applications. Used as efficient sulphur vulcanization accelerators with plasticizing ability. They increase the low temperature and heat resistance of vulcanizates.

Acute Toxicity.

Observations in man. The minimal lethal doses of these chemicals in man are 31.8 and 7.0 g/kg BW, respectively.

Animal studies. When administered to mice as an aqueous suspension, LD_{50} of *N,N'*-octaethyltetraamidothiopyrophosphate stabilized with twin-80 (thiopyrophosphate) is 4.75 g/kg BW, and that of *N,N'*-octabtyltetraamidothiopyrophosphate is 1.85 g/kg BW.

Reference:

Davydov, B. Yu., Ratnikova, T. B., Ginak, A. I., et al., Toxicological evaluation of thiopyrophosphates as polyfunctional accelerators, *Kauchuk i Rezina,* 9, 53, 1980 (in Russian).

OCTAFLUOROHEXANEDIOIC ACID, DIETHYL ESTER

Molecular Formula. $C_{10}H_{10}F_8O_4$
M = 346.20
CAS No 376-50-1
RTECS No MO1953000

Synonyms. Diethyl octafluoroadipate; Diethyl octafluorohexane dioate; Diethylperfluoroadipate; Perfluroadipic acid, diethyl ester.

Properties. Liquid. Almost insoluble in water, soluble in alcohol.

Applications. Used in the production of synthetic rubber and plastics.

Acute Toxicity. There was no mortality or visible clinical signs of intoxication in rats given 5.0 to 8.4 g/kg BW doses.

Reference:

Toxicology of New Industrial Chemical Substances, N. F. Ismerov and I. V. Sanotsky, Eds., Meditsina, Moscow, Issue No 12, 1971, 142 (in Russian).

4,4'-OXYBISBENZENESULFONIC ACID, DIHYDRAZIDE

Molecular Formula. $C_{12}H_{14}N_4O_5S_2$
M = 358.42
CAS No 80-51-3
RTECS No DB7321000

Synonyms and **Trade Name.** Celogen OT; Hydrazide SDO; *p,p'*-Oxybis (benzenesulfonyl hydrazide); 4,4'-Oxydibenzene sulfonohydrazide.

Properties. Fine-crystalline, white powder. Does not color or give odor to the articles. Insoluble in water, moderately soluble in ethanol.

Applications. Blowing agent for sponge rubber and expanded plastics.

Acute Toxicity. LD_{50} appeared to be 1.3 g/kg BW.

Reference:

Anonymous, Toxicity profile of 4,4'-oxydibenzene sulfonohydrazide, *Kauchuk i Rezina,* 1, 39, 1987 (in Russian).

PERFLUOROADIPODINITRILE

Molecular Formula. $C_6F_8N_2$
M = 252.08
CAS No 376-53-4
RTECS No AV2800000
Abbreviation. PFAN.

Synonyms. Octafluorohexanedinitrile; Perfluoroadipic acid, dinitrile; Perfluorohexane dinitrile.

PERFLUOROGLUTARONITRILE

Molecular Formula. $C_5F_6N_2$
M = 202.06
CAS No 376-89-6
RTECS No MA5600000
Abbreviation. PRGN.

Synonyms. Hexafluoroglutaronitrile; Hexafluoropentanedinitrile; Perfluoroglutaric acid, dinitrile.

Properties. Colorless, highly mobile liquids. Poorly soluble in water and alcohol.

Applications. Used in the production of high-strength and temperature-stable rubber and plastics.

Acute Toxicity. LD_{50} for PFAN is reported to be 1.95 g/kg BW in mice and 2.92 g/kg BW in rats; LD_{50} for PFGN is 0.98 g/kg BW in mice, and 2.6 g/kg BW in rats. A single administration of the lethal or average lethal dose did not cause formation of cyanides in the blood of animals. This means that splitting of the *CN*-group in the body is absent or low enough to be immediately converted into less toxic rhodanide compounds.

Repeated Exposure revealed no accumulation. Administration of 3 to 4 LD_{50} caused the death of only a few animals.

Reference:

Korshunov, Yu. N., Characteristics of toxic effects of fluorinated dinitriles in laboratory animals, in *Toxicology of New Industrial Chemical Substances*, N. F. Ismerov and I. V. Sanotsky, Eds., Meditsina, Moscow, Issue No 11, 1969, 79 (in Russian).

PEROXYACETIC ACID, *tert*-BUTYL ESTER

Molecular Formula. $C_6H_{12}O_3$
M = 132.18
CAS No 107-71-1
RTECS No SD8925000
Abbreviation. BPA.

Synonyms and **Trade Name.** *tert*-Butyl acetate; *tert*-Butylperacetate; Lupersol 70.

Applications. Used in synthetic rubber manufacture.

Acute Toxicity. LD_{50} is 675 mg/kg BW in rats,[1] and 632 mg/kg BW in mice.[2]

Reproductive Toxicity.

Gonadotoxicity. Rat males were exposed to 1.0 mg BPA/m^3 for 4 hours/day over a period of 17 weeks prior to mating. The treatment damaged spermatogenesis.[3]

Regulations. ***U.S. FDA*** (1998) approved the use of BPA in the manufacture of rubber articles intended for repeated use in producing, manufacturing, packaging, processing, preparing, treating, packing, transporting, or holding food in accordance with the conditions prescribed by 21 CFR part 177.2600.

References:

1. Izmerov, N. F., Sanotsky, I. V., and Sidorov, K. K., *Toxicometric Parameters of Industrial Toxic Chemicals under Single Exposure*, USSR/SCST Commission for UNEP, Center Int. Projects, Moscow, 1982, 30 (in Russian).
2. Babanov, G. P., The toxicity of *para*-tertiary butylphenol, *Gig. Truda Prof. Zabol.*, 7, 31, 1969 (in Russian).
3. Sanotsky, I. V., Ivanov, N. G., Avkhimenko, M. M. et al., Evaluation of toxic and gonadotoxic effects of *tert*-butylperacetate following chronic inhalation exposure, in *Toxicology of New Industrial Chemical Substances*, I. V. Sanotsky and N. F. Izmerov, Eds., Meditsina, Moscow, Issue No 10, 1968, 44 (in Russian).

N,N'-(*m*-PHENYLENE)DIMALEIMIDE

Molecular Formula. $C_{14}H_8N_2O_2$
M = 268.24
CAS No 3006-93-7
RTECS No ON6125000

Synonyms. 1,3-Bismaleimidobenzene; 1,1'-(Phenylenebis)-1*H*-pyrole-2,5-dione.

Properties. An odorless, tasteless, fine crystalline, green powder. Poorly soluble in water, soluble in alcohol.

Applications. Used as a vulcanizing and modifying agent in the rubber industry.

Acute Toxicity. LD_{50} is reported to range from 250 to 1740 mg/kg BW in mice, and from 1370 to 5110 mg/kg BW in rats.[1,2] Manifestations of the toxic effect include apathy and adynamia, bloody nasal discharge, and diarrhea with mucus.

Repeated Exposure revealed marked cumulative properties: K_{acc} is 1.0 (by Lim).[2]

25% mortality rate as well as reticulocytosis and disturbance of liver protein synthesis (hypoproteinemia and hypoalbuminemia combined with increased liver weights) were found in rats given 0.1 g/kg BW for a month. Rats exposed to 137 mg/kg BW for 2 months exhibited decreased BW gain, a brief decline in rheobasis and albumen-globulin coefficient, and increased arterial pressure.[1]

Reproductive Toxicity.

Gonadotoxic effect was not found.

Mutagenicity.

In vivo cytogenetics. Inhalation exposure revealed no cytogenic effect.

Carcinogenicity. 10% animals developed tumors following *i/g* administration. Feeding and skin applications did not produce neoplasms. It is likely to be a weak carcinogen.

Chemobiokinetics. Is partly excreted unchanged via the urine.

References:

1. Kel'man, G. Ya., Rotenberg, Yu. S., Mashbits, F. D., et al., Hygienic standardization of *N,N'-m*-phenylenedimaleimide and 4,4'-dithiodiphenylmaleimidethe - new vulcanization agents for synthetic rubber, in *Hygiene and Toxicology of High-Molecular-Mass Compounds and of the Chemical Raw Material Used for Their Synthesis*, Proc. 6th All-Union Conf., B. Yu. Kalinin, Ed., Leningrad, 1979, 261 (in Russian).
2. Reznichenko, A. K., Vasilenko, N. M., Muzhikovsky, G. L., et al., Toxicity of *N,N'*-meta-phenylenedimaleimide, *Gig. Truda Prof. Zabol.*, 10, 56, 1983 (in Russian).

[PHTHALOCYANINATO(2$^-$)]COPPER

Molecular Formula. $C_{32}H_{16}CuN_8$
M = 576.10
CAS No 147-14-8
RTECS No GL8510000
Abbreviation. PCC.

Synonym and **Trade Names.** Copper phthalocyanine; Cuprolinic blue; Phthalocyanine blue.

Properties. Pigment containing PCC is insoluble in water, oils and majority of organic solvents.

Applications. Used predominantly in the manufacture of rubber and plastics.

Acute Toxicity. Substance of low acute toxicity. The toxicity appears to be due to the presence of soluble copper compounds. Oral administration of 6.0 g/kg BW (in starch) or 15 g/kg BW (in dimethylsulfoxide) caused no mortality in female rats. *I/p* injections damaged kidney function (an increase of proteins level in the urea, a decrease in diuresis).[1]

Repeated Exposure revealed cumulative properties (on low doses given). Female rats received 2.0 g/kg BW (20% solution in starch) for 30 days. No clinical manifestations of toxic action were reported (BW gain, alkaline phosphatase activity in blood serum, relative organ weights were fixed during the treatment). An increase in ceruloplasmin level in blood serum was noted.[1]

Reproductive Toxicity.

Teratogenicity. Multiple malformations were observed in chicks after injection of eggs with a sulfonated phthalocyanine. Injection with cupric chloride in the same study suggested that copper was the active constituent of the phthalocyanine.[2]

See ***Copper compounds***.

Chemobiokinetics. PCC does not penetrate into the hematology system, but there was an increase in copper concentration in blood serum.

Regulations. ***U.S. FDA*** (1998) approved the use of PCC as a component of the uncoated or coated food-contact surface of paper and paperboard intended for use in producing, manufacturing, packaging, processing, preparing, treating, packing, transporting, or holding aqueous and fatty foods in accordance with the conditions prescribed by 21 CFR part 176.170. All batches of this color additive when used for coloring foods shall meet the specifications, uses and restrictions, and labeling regulations contained in 21 CFR part 74 and be certified in accordance with the regulations in 21 CFR part 74.706.

References:

1. Kurlyandsky, B. A., Braude, E. V., Klyachkina, A. M., Torshina, N. L., Khok, S. B., and Zasorina, E. V., Study on copper phthalocyanine-induced toxicity, *Gig. Sanit.*, 92, 1, 1985 (in Russian).
2. Sandor, S. et al., Sulphonated phthalocyanine induced caudal malformative syndrome in the chick embryo, *Morphol. Embryol.* (Bucur), 31, 173, 1985.

1-PIPERIDINECARBODITHIOIC ACID, compound with PIPERIDINE

Molecular Formula. $C_6H_{11}NS_2.C_5H_{11}N$

M = 246.47

CAS No 98-77-1

RTECS No TM5850000

Synonym and **Trade Names.** Accelerator 552; Piperidinium pentamethylenedithiocarbamate; Pip-pip.

Applications. Used in rubber and plastics industries.

Acute Toxicity. LD_{50} *i/p* is 250 mg/kg BW. Poisoning caused degenerative changes in the brain; convulsions were observed in the treated animals.

Regulations. ***U.S. FDA*** (1998) approved the use of P. in adhesives as a component of articles intended for use in packaging, transporting, or holding food in accordance with the conditions prescribed in 21 CFR part 175.105.

Reference:

Mallette, F. S., Studies on toxicity and skin effects of compounds used in rubber and plastics industries: accelerators, activators, and antioxidants, *Arch. Ind. Hyg. Occup. Med.*, 5, 311, 1952.

[(2-PROPENYLOXY)METHYL]OXIRANE

Molecular Formula. $C_6H_{10}O_2$

M = 114.16

CAS No 106-92-3

RTECS No RR0875000

Abbreviation. AGE.

Synonyms. Allyl 2,3-epoxypropyl ether; Allyl glycidyl ether; 1-Allyloxy-2,3-epoxypropane.

Properties. Colorless liquid with a faint specific odor. Water solubility is 172 g/l.

Applications. Used as a stabilizer in the production of vinyl resins and synthetic rubber.

Acute Toxicity. LD_{50} is 433 mg/kg BW for mice, and 700 mg/kg BW for rats. Death occurs within 3 days.[1] According to other data, LD_{50} is 390 mg/kg BW for mice and 1600 mg/kg BW for rats (ACGIH Document, 1986).

Repeated Exposure failed to reveal cumulative properties. K_{acc} is 6.2 (by Lim). The treatment led to alterations in the NS, liver, and kidneys.[1]

Osborne-Mendel rats were fed up to 500 ppm and B6C3F$_1$ mice were fed up to 100 ppm AGE in the diet. All rats exposed to 500 ppm AGE died. No mortality was observed at the next lower (200 ppm) exposure concentration. There was significant mortality in high-dosed mice. Compound-related lesions included respiratory system damage.[2]

Reproductive Toxicity.

Gonadotoxicity. Inhalation exposure revealed reduction in the mating performance of exposed Osborne- Mendel rats. However, sperm motility and number were not affected.[3]

Mutagenicity.

In vitro genotoxicity. AGE is shown to be positive in *Salmonella* mutagenicity assay, creating reversion of the bacteria to histidine independence.[3] AGE induced SCE and CA in Chinese hamster ovary cells both in the presence and absence of metabolic activation.[3]

In vivo cytogenetics. Significant increase in sex-linked recessive lethal mutations was observed in the germ cells of male *Dr. melanogaster*, but no increase in reciprocal translocations occurred in these cells.[2]

Carcinogenicity. Equivocal results were obtained on inhalation NTP studies in Osborne-Mendel rats and B6C3F$_1$ mice.[2]

Carcinogenicity classification.

NTP: EE - NE - SE - EE.

Regulations.

EU (1990). AGE is available in the *List of monomers and other starting substances which may continue to be used for the manufacture of plastic materials and articles intended to come into contact with foodstuffs pending a decision on inclusion in Section A* (*Section B*).

U.S. FDA (1998) approved AGE for use in the manufacture of 4,4'-isopropylidenediphenol epichlorohydrin thermosetting epoxy resins (finished articles containing the resins shall be thoroughly cleansed prior to their use in contact with food) in accordance with the provisions prescribed in 21 CFR part 177.1440.

References:

1. Shugaev, B. B. and Buckhalovsky, A. A., Toxicity of allyl glycidyl ether, *Gig. Sanit.*, 3, 92, 1983 (in Russian).
2. *Toxicology and Carcinogenesis Studies of Allyl Glycidyl Ether in Osborne-Mendel Rats and B6C3F$_1$ Mice (Inhalation Studies)*, NTP Technical Report Series No 376, Research Triangle Park, NC, NIH Publication No 90-2831, 1990.
3. Wade, M. J., Mayer, J. W., and Hine, C. H., Mutagenic action of a series of epoxides in *Salmonella*, *Mu tat. Res.*, 66, 367, 1979.

PYROCATECHOL

Molecular Formula. $C_6H_60_2$

M = 110.11

CAS No 120-80-9

RTECS No UX1050000

Synonyms and **Trade Names.** 1,2-Benzenediol; Catechin; Cathehol; 1,2-Dihydroxybenzene; *o*-Diphenol; *o*-Hydroquinone; 2-Hydroxy phenol; Oxyphenic acid; *o*-Phenylenediol.

Properties. Colorless crystals with faint characteristic phenol odor and sweet and bitter taste. Soluble in 2.3 parts of water.[020]

Applications. Rubber antioxidant.

Exposure. P. is naturally occurring chemical.

Acute Toxicity. In mice, rats, and guinea pigs, LD_{50} is 100 to 260 mg/kg BW. Administration of the lethal doses affected motor activity, caused dyspnea, muscle contraction, and spasticity.[1,2,020]

Reproductive Toxicity.

Embryotoxicity. Oral administration of 1.0 g/kg BW on day 11 of pregnancy affected litter size.[3]

Mutagenicity.

In vitro genotoxicity. P. was not found to be mutagenic in *Salmonella* assay (Howorth et al., 1983), but had appreciable mutagenic potential in the mouse lymphoma L5178Y cell assay for mutagens without supplementary metabolic activation system.[4]

In vivo cytogenetics. P. showed weak capability to induce micronuclei in bone marrow cells of pregnant mice and, transplacentally, in fetal liver cells.[5]

Carcinogenicity. Male F344 rats were administered P. by gastric intubation at doses of 10 to 90 mg/kg BW. The treatment induced up to a 19-fold increase in ornithine decarboxylase activity, and up to a 8-fold increase in replicative DNA synthesis in the pyloric mucosa of the stomach. P. showed tumor promoting activity after *N*-methyl-*N*'-nitro-*N*-nitrosoguanidine initiation in the pyloric mucosa of rat stomach.[6-8]

F344 rats and B6C3F_1 mice received diets containing 0.8% P. for 104 weeks (rats) or 96 weeks (mice). Histological examination revealed glandular stomach adenocarcinomas in 54% ($p < 0.001$) of male and 43% of female rats.[8,9,11]

Glandular stomach carcinomas were noted in about 70% of Wistar, Lewis and CD rats, but in only 10% of WKY rats that were given the same diet for 2 years.[8,10]

Carcinogenicity classification. An IARC Working Group concluded that there is inadequate evidence for the carcinogenicity of P. in *experimental animals* and there were no data available to evaluate the carcinogenicity of P. in *humans*.

IARC: 3.

Chemobiokinetics. P. is readily absorbed from the GI tract. A part of ingested P. is oxidized with polyphenol oxidase to *o-benzoquinone*. Another fraction conjugates in the body with *hexuronic, sulfuric,* and *other acids*.[03] Main metabolites were identified to be *guaiacol* in rats, and *o-hydroxyphenyl-β-D-glucuronide*, *o-hydroxyphenyl sulfate*, and *hydroxyquinol* in rabbits.[12]

Regulations.

EU (1990). P. is available in the *List of authorized monomers and other starting substances which shall be used for the manufacture of plastic materials and articles intended to come into contact with foodstuffs (Section A)*.

Great Britain (1998). P. is authorized without time limit for use in the production of polymeric materials and articles in contact with food or drink or intended for such contact. The specific migration of this substance shall not exceed 6.0 mg/kg.

Standards. *EU* (1990). SML: 6.0 mg/kg.

References:

1. Zholdakova, Z. I., in *Hygiene of Populated Areas*, Reports Sci. Conf., D. N. Kaluzhny, Ed., Zdorov'ya, Kiyv, 1967, 52 (in Russian).
2. Karpushina, Ye. A., Determination of acute threshold effect of benzene and its metabolites (hydroquinone and pyrocatechol), in *Toxicology of New Industrial Chemical Substances*, I. V. Sanotsky and N. F. Ismerov, Eds., Meditsina, Moscow, Issue No 15, 1979, 136 (in Russian).
3. Kavlock, R. J., Structure-activity relationships in the developmental toxicology of substituted phenols, *Teratology*, 41, 43, 1990.
4. McGregor, D. B., Riach, C. G., Brown, A., Edwards, I., Reynolds, D., West, K., and Willington, S., Reactivity of catecholamines and related substances in the mouse lymphoma L5178Y cell assay for mutagens, *Environ. Molec. Mutagen.*, 11, 523, 1988.
5. Cirrani, R., Barale, R., Marrazzini, A., and Loprieno, N., Benzene and the genotoxicity of its metabolites. I. Transplacental activity in mouse fetuses and in their dams, *Mutat. Res.*, 208, 61, 1988.

6. Furihata, C., Hatta, A., and Matsushima, T., Induction of ornithine decarboxylase and replicative DNA synthesis but not DNA single strand scission or unscheduled DNA synthesis in the pyloric mucosa of rat stomach by catechol, *Japan. J. Cancer Res.*, 80, 1052, 1989.
7. Shibata, M. A., Hirose, M., Yamada, M., Tatematsu, M., Uwagawa, S., and Ito, N., Epithelial cell proliferation in rat forestomach and glandular stomach mucosa induced by catechol and analogous, *Carcinogenesis*, 11, 997, 1990.
8. Hirose, M., Yamaguchi, S., Fukushima, S., Hasegawa, R., Takahashi, S., and Ito, N., Promotion by dihydroxybenzene derivatives of *N*-methyl-*N'*-nitro-*N*-nitrosoguanidine-induced F344 rat forestomach and glandular carcinogenesis, *Cancer Res.*, 49, 5143, 1989.
9. Hirose, M., Fukushima, S., Tanaka, H., Asakawa, E., Takahashi, S., and Ito, N., Carcinogenicity of catechol in F344 rats and $B6C3F_1$ mice, *Carcinogenesis*, 14, 525, 1993.
10. Tanaka, H., Hirose, M., Hagiwara, A., Imaida, K., Shirai, T., and Ito, N., Rat strain differences in catechol carcinogenicity to the stomach, *Food Chem. Toxicol.*, 33, 93, 1995.
11. Davidson, K. A. and Faust, R. A., Oral carcinogen potency factors for benzene and some of its metabolites, in *Toxicologist*, Abstracts of SOT 1996 Annual Meeting, Abstract No 585, Issue of *Fundam. Appl. Toxicol.*, 30, Part 2, March 1996.
12. Goodwin, B. L., *Handbook of Intermediary Metabolism of Aromatic Compounds*, Wiley and Sons, Inc., New York, 1976, 11.

RESORCINOL

Molecular Formula. $C_6H_6O_2$
M = 110.12
CAS No 108-46-3
RTECS No VG9625000

Synonyms and **Trade Names.** 1,3-Benzenediol; *m*-Dihydroxybenzene; *m*-Dioxybenzene; 3-Hydroquinone; 3-Hydroxyphenol; Resorcin.

Properties. Colorless crystals with a specific odor. Readily soluble in water (63.7%), acetone, and ethanol. R. produces a very slight effect on the organoleptic properties of water: odor perception threshold is 40 mg/l.

Applications. R. is used in the production of plasticizers and stabilizers, and in the manufacture of rubber products. R. (alone or in combination with phenol) is used to make resins or resin intermediates by reaction with formaldehyde.

Acute Toxicity. LD_{50} is 239 mg/kg BW in mice, and 300 mg/kg BW in rats. Lethal oral doses in rabbits and guinea pigs are reported to be 750 and 370 mg/kg BW, respectively.[1, 018]

Repeated Exposure failed to reveal cumulative properties. No F344 rats died being administered up to 450 mg R./kg BW by gavage for 17 days. High mortality was observed in mice that received 600 mg R./kg BW. No gross or microscopic lesions attributable to R. administration were noted.[2]

Short-term Toxicity. In 13-week gavage studies, very high mortality was observed in rats receiving 520 mg R./kg BW and mice receiving 420 mg R./kg BW. No gross or microscopic lesions attributable to R. administration were observed.[2]

Long-term Toxicity. A chronic exposure to the high doses produced a change in the phagocytic activity of leukocytes, in cholinesterase activity in the brain and liver, in blood eosinophil count, and in the protein fractions balance in the blood serum (Nesmeyanova, 1953). The dose of 5.0 mg/kg BW reduced the number of *SH*-groups in the blood serum. Clinical signs, including ataxia, recumbency, and tremors, were observed in rats and mice in the 2-year gavage studies.[2]

Reproductive Toxicity.

Gonadotoxicity. Causes no sperm-head abnormalities in mice.[3]

Teratogenicity effects were not found in rats, mice, and rabbits.[4-6]

Mutagenicity.

In vitro genotoxicity. R. is shown to produce gene mutations in bacteria and CA in plant cells.[7] It causes no SCE in cultured Chinese hamster ovary cells *V79* and is negative in micronuclei test on mouse bone marrow.[3,8]

According to more recent data, R. induced SCE in Chinese hamster ovary cells but was negative in *Salmonella* mutagenicity bioassay with or without metabolic activation.[2] R. showed a weak activity in the mouse lymphoma L5178Y cell assay for mutagens without supplementary metabolic activation system.[9]

Carcinogenicity. Rats and mice were administered 112 and 225 mg R./kg BW for 2 years (female rats - 50, 100, and 150 mg/kg BW). There was no incidence of neoplasms or non-neoplastic lesions.[2]

Carcinogenicity classifications. An IARC Working Group concluded that there is inadequate evidence for the carcinogenicity of R. in *experimental animals* and there were no data available to evaluate the carcinogenicity of R. in *humans*.

IARC: 3.

NTP: NE - NE - NE - NE (gavage).

Chemobiokinetics. R. is readily absorbed from the GI tract, rapidly metabolized, and excreted by rats. After ingestion of 112 mg/kg BW, the main part of the dose was excreted within 24 hours in the urine, and up to 3.0% in the feces. Accumulation was not observed.[10]

Regulations.

U.S. FDA (1993) approved the use of R. in the manufacture of closures with sealing gaskets in containers intended for use in producing, manufacturing, packing, processing, preparing, treating, packaging, transporting or holding food (0.24%), for use only as reactive adjuvant substance employed in the production of gelatin-bonded cord compositions, for use in lining crown closures in accordance with the conditions prescribed by 21 CFR part 177.1210. The gelatin so used shall be technical grade or better.

EU (1990). R. is available in the *List of authorized monomers and other starting substances which shall be used for the manufacture of plastic materials and articles intended to come into contact with foodstuffs (Section A).*

Great Britain (1998). R. is authorized without time limit for use in the production of polymeric materials and articles in contact with food or drink or intended for such contact. The specific migration of R. shall not exceed 2.4 mg/kg.

Standards. *Russia.* PML: 5.0 mg/l.

References:

1. Zholdakova, Z. I., in *Hygiene of Populated Areas*, Abstracts Sci. Conf., Zdorov'ya, Kiyv, 1967, 52 (in Russian).
2. *Toxicology and Carcinogenesis Studies of Resorcinol in F344 Rats and B6C3F$_1$ Mice (Gavage Studies)*, NTP Technical Report Series No 403, Research Triangle Park, NC, 1992.
3. Wild, D., King, M.-T., Eckhardt, K., et al., Mutagenic activity of aminophenols and diphenols, and relations with chemical structure, *Mutat. Res.*, 85, 456, 1981.
4. Spengler, J., Osterburg, I., and Korte, R., Teratogenic evaluation of *p*-toluenediamine sulfate, resorcinol and *p*-aminophenol in rats and rabbits, *Teratol ogy*, 33, 31A, 1986.
5. Dinardo, J. C., Picciano, J. C., Schnetzinger, R. W., Morris, W. E., and Wolf, B. A., Teratological assessment of five oxidative hair dyes in the rat, *Toxicol. Appl. Phamacol.*, 78, 163, 1985.
6. Kavlock, R. J., Structure-activity relationships in the developmental toxicity of substituted phenols: *in vivo* effects, *Teratology*, 41, 43, 1990.
7. McCann, J., Choi, E., Yamasaki, E., et al., Detection of carcinogens as mutagens in the *Salmonella*/microsome test: Assay of 300 chemicals, in *Proc. Natl. Acad. Sci. USA*, 72, 5135, 1975.
8. Darroudi, F. and Natarajan, A. T., Cytogenetic analysis of human peripheral blood lymphocytes (*in vitro*) treated with resorcinol, *Mutat. Res.*, 124, 179, 1983.
9. McGregor, D. B., Riach, C. G., Brown, A., Edwards, I., Reynolds, D., West, K., and Willington, S., Reactivity of cateholamines and related substances in the mouse lymphoma L5178Y cell assay for mutagens, *Environ. Molec. Mutagen.*, 11, 523, 1988.
10. Kim, Y. C. and Matthews, H. B., Comparative metabolism and excretion of resorcinol in male and female Fischer 344 rats, *Fundam. Appl. Toxicol.*, 9, 409, 1987.

RESOTROPIN

Molecular Formula. $C_{12}H_{18}N_4O_2$

M = 250.31

Composition. A molecular combination of resorcinol and urotropin.

Properties. Fine crystalline powder of a light-pink to light-gray color. Soluble in water and partially soluble in alcohol.

Applications. Used as a modifier of vulcanizate mixtures based on natural, styrene-butadiene, and isoprene rubbers.

Acute toxicity. In mice, the LD_{50} is 1.3 g/kg BW. A 0.5 g/kg BW dose caused clonic convulsions and paralysis of the hind legs in 30 to 40 min. after administration. Death within 24 hours.[1] LD_{50} of a modifier RU-1 (a mechanical mixture of 98.5% R., 1.0% boric acid, and 0.5% alkamone) in mice is 570 mg/kg BW (Zhilova, 1970).

Short-term Toxicity. Rabbits exposed to 20 mg/kg BW for 4 months developed no changes in BW gain, relative weights of the visceral organs and contents of blood albumen, amino acids and prothrombin. The treatment caused a reduction in alkaline phosphatase activity.[1]

In rabbits given 5.0 mg modifier RU-1/kg BW for 3.5 months, manifestations of the toxic effect included a decrease in blood vitamin C contents and increase in cholinesterase activity and blood cholesterol level. Histology examination revealed inflammatory foci around the small bronchi in the lungs, lymphocyte and leukocyte infiltration of the large hepatic vessels, inflammatory changes of the mucosa, and superficial epithelial desquamation in the stomach and intestine,[1] as well as swelling of the renal tubular epithelium (Zhilova, 1970).

Standards. ***Russia*** (1995). PML: 1.0 mg/l (organolept., taste).

Reference:

1. Mezentseva, N. V., Resotropin, in *Toxicology of New Chemical Substances Used in Rubber and Tyre Industry*, Z. I. Israel'son, Ed., Meditsina, Moscow, 1968, 190 (in Russian).

STEARIC ACID

Molecular Formula. $C_{18}H_{36}O_2$

M = 284.54

CAS No 57-11-4

RTECS No WI2800000

Abbreviation. SA.

Synonyms. Heptodecanecarboxilic acid; Octadecanoic acid.

Properties. Amorphous, white or yellowish powder, plate-like or flaky. Poorly soluble in water, soluble in ethanol.

Applications. Used as a vulcanization activator, lubricant, and filler dispersant.

Acute Toxicity. Rats and mice tolerate administration of 5.0 g/kg BW without any signs of intoxication.[1]

Repeated Exposure. The increased liver weights were the only finding in mice given 250 and 1000 mg/kg BW for 1.5 months.[1]

Long-term Toxicity. In a 9-month gavage study, the doses of 20 and 100 mg/kg BW appeared to be harmless for exposed rats.[1]

Allergenic Effect was not found in the guinea pigs studies.

Mutagenicity. SA showed no mutagenic effect.[2]

In vitro genotoxicity. SA appeared to be negative in *Salmonella* mutagenicity assay (NTP-91).

Carcinogenicity. Repeated *s/c* injections in rats and skin application in mice caused sarcomas in sites of application.[3]

Regulations.

EU (1990). SA is available in the *List of authorized monomers and other starting substances which shall be used for the manufacture of plastic materials and articles intended to come into contact with foodstuffs (Section A)*.

U.S. FDA (1998) regulates the use of SA (1) in adhesives used as components of articles intended for use in packaging, transporting, or holding food in accordance with the conditions prescribed by 21 CFR part

175.105; and (2) in the manufacture of resinous and polymeric coatings in a food-contact surface of articles intended for use in producing, manufacturing, packing, transporting, or holding food in accordance with the conditions prescribed by 21 CFR part 175.300.

Affirmed as *GRAS.*

Great Britain (1998). SA is authorized without time limit for use in the production of polymeric materials and articles in contact with food or drink or intended for such contact.

Standards. ***Russia*** (1995). PML: 0.25 mg/l (organolept., turbidity).

References:

1. Komarova, E. N., Toxic properties of some additives for plastics, *Plast. Massy*, 12, 30, 1976 (in Russian).
2. Nakamura, H. and Yamamoto, T., The active part of the [6]-gingerol molecule in mutagenesis, *Mutat. Res.*, 122, 87, 1983.
3. Swern, D., Wieder, R., McDonough, M., Meranze, D. R., and Shimkin, M. B., Investigation of fatty acids and derivatives for carcinogenic activity, *Cancer Res.*, 30, 1037, 1970.

SULPHUR

A = 32.06
CAS No 7704-34-9
RTECS No WS4250000

Trade Names. Black Bird Brand Sulphur; Manox.

Properties. Yellow, gray-yellow or greenish powder. Insoluble in water, poorly soluble in ethanol.

Applications. Used as a vulcanizing agent in rubber industry in the production of vulcanizates based on isoprene, butadiene and other rubbers. Usually it is not a component of chloroprene rubber.

Acute Toxicity.

Observations in man. S. does not cause acute poisoning but some pharmacological preparations used in medicine could exhibit evident toxicity. The lethal oral dose of sedimentated S. (Sulfur praecipitatum, Lac sulfuri cus - Sulfuric milk) is about 12 g.

Repeated Exposure. S. accumulation was noted in the liver, kidneys, and spleen.

Short-term Toxicity. Hypochromic anemia and leukopenia were found in rats given 9.3 mg elemental S./kg BW for 4 months.[1]

Long-term Toxicity. Guinea pigs received daily doses of 40 mg S./animal for 5 months. The treatment induced inflammatory microscopic infiltrations in the parotid glands, dilatation of blood vessels, and leaching of fat bodies from parotid cells.[2]

Chemobiokinetics. Long-term administration of elemental S. increased its amount as well as the quantity of mineral sulfates, common oxidized and neutral S. in the urine, liver, kidney, and spleen.

S. is transformed into sulfides and H_2S which could be reabsorbed from the intestines. Sulfides are predominantly converted into sulfates and excreted via the urine.

Regulations. ***U.S. FDA*** (1998) approved the use of S. (1) in adhesives used as components of articles intended for use in packaging, transporting, or holding food in accordance with the conditions prescribed by 21 CFR part 175.105; (2) in the manufacture of closures with sealing gaskets that may be safely used on containers intended for use in producing, manufacturing, packing, processing, treating, packaging, transporting, or holding food (for use only as a vulcanizing agent in vulcanized natural or synthetic rubber gasket compositions and up to 4.0% by weight of the elastomer contents of the rubber gasket composition) in accordance with the conditions prescribed by 21 CFR part 177.1210; (3) as a defoaming agent in the manufacture of paper and paperboard intended for use in producing, manufacturing, packaging, processing, preparing, treating, packing, transporting, or holding foods in accordance with the conditions prescribed by 21 CFR part 176.210; and (4) as a vulcanizing agent in the manufacture of rubber articles intended for repeated use in contact with foodstuffs (total not to exceed 1.5% by weight of the rubber product) in accordance with the conditions prescribed by 21 CFR part 177.2600.

S. dioxide is considered to be *GRAS* for use with limitations indicated in CFR.

References:

1. Kuziev, R. S., in *Environmental Factors and Their Significance for the Health of Population*, D. N. Kaluzhny, Ed., Zdorov'ya, Kiyv, Issue 1, 1976, 75 (in Russian).
2. Jarzynka, W. and Mietkiewska, B., Microscopic image of guinea pig parotid glands after exposure to powdered sulfur, *Czas. Stomatol.*, 32, 637, 1979 (in Polish).

2,3,5,6-TETRACHLORO-*p*-BENZOQUINONE

Molecular Formula. $C_6Cl_4O_2$
M = 245.89
CAS No 118-75-2
RTECS No DK6825000
Abbreviation. TCB.

Synonyms and **Trade Name.** Chloranil; Quinone tetrachloride; 1,3,4,5-Tetrachloro-1,4-benzoquinone; 2,3,5,6-Tetrachloro-2,5-cyclohexadiene-1,4-dione; 2,3,5,6-Tetrachloroquinone.

Properties. Yellow leaflets or prisms. Almost insoluble in cold alcohol.[020]

Applications. A vulcanization agent. A fungicide.

Acute Toxicity. LD_{50} of technical TCB free of tetrachlorotribenzo-*p*-dioxins is 6.95 g/kg BW in female rats.[1] LD_{50} is 500 mg/kg BW in mice.[2]

Mutagenicity.

In vitro genotoxicity. TCB displayed mutagenicity in five different *Salmonella* mutagenicity tester strains with and/or without *S9* mix.[3]

Carcinogenicity. In 78-week oral study, TCB was questionably carcinogenic in mice (liver tumors were noted at doses of 80 to 215 mg/kg BW).[4]

Regulations. ***U.S. FDA*** (1998) approved the use of TCB in the manufacture of rubber articles intended for repeated use in producing, manufacturing, packaging, processing, preparing, treating, packing, transporting, or holding food in accordance with the conditions prescribed by 21 CFR part 177.2600.

References:

1. Hess, P. et al., Possible formation of tetrachlorotribenzo-*p*-dioxins in production of chloranil, *Ecotoxicol. Environ. Saf.*, 6, 336, 1982
2. Ahlborg, U. G. and Larsson, K., Metabolism of tetrachlorophenols in the rat, *Arch. Toxicol.*, 40, 63, 1978.
3. Hayes, W. J., *Toxicology of Pesticides*, Williams & Wilkins, Baltimore, 1975, 193.
4. Hakura, A., Mochida, H., Tsutsui, Y., and Yamatsu, K., Mutagenicity of benzoquinones for Ames *Salmonella* tester strains, *Mutat. Res.*, 347, 37, 1995.

1,1'-(TETRATHIODICARBONOTHIOYL)BISPIPERIDINE

Molecular Formula. $C_{12}H_{20}N_2S_6$
M = 384.70
CAS No 120-54-7
RTECS No TN4221000

Synonyms and **Trade Names.** Bis(pentamethylenethiuram)tetrasulfide; Bis(piperidinothiocarbonyl) tetrasulfide; 1,1'-(Tetrathiodicarbonothioyl)bispiperidine; Tetron A; Thiuram MT.

Applications. Used in the rubber manufacture.

Acute Toxicity. LD_{50} *i/p* is 200 mg/kg BW in mice.

Regulations. ***U.S. FDA*** (1998) approved the use of T. (1) in adhesives as a component of articles intended for use in packaging, transporting, or holding food in accordance with the conditions prescribed by 21 CFR part 175.105; *dipentamethylenethiuram hexasulfide* has been approved for use (2) in the manufacture of rubber articles intended for repeated use in producing, manufacturing, packaging, processing, preparing, treating, packing, transporting, or holding food in accordance with the conditions prescribed in 21 CFR part 177.2600.

Reference:

Natl. Technical Information Service AD277-689.

2-THIAZOLIDINETHIONE

Molecular Formula. $C_3H_5NS_2$
M = 119.2
CAS No 96-53-7
RTECS No XJ6122000
Abbreviation. TT.

Synonyms. 2-Mercaptothiazoline; Thiazoline-2-thiol.

Properties. A colorless, odorless powder. Poorly soluble in cold water and alcohol, readily soluble in hot water.

Applications. Used as a vulcanization accelerator for mixtures of natural and styrene-butadiene rubber.

Acute Toxicity. LD_{50} is found to be 254 mg/kg BW in male mice.[1]

Short-term Toxicity. Rabbits were exposed to a 20 mg/kg BW dose every other day for 2 months and then daily for the same period of time. The treatment resulted in a decrease of BW gain, reduced cholinesterase and alkaline phosphatase activity, and an increase in serum cholesterol and lipid levels. Gross pathology examination revealed dystrophic changes in the liver. Edema, congestion, and lymphoid elements were found in the myocardium. In the lungs, thickening of the alveolar septa with proliferation of cellular elements and focal desquamation of the epithelium were found. Kidney weights were increased.[1]

Reproductive Toxicity. TT produced no teratogenic effect in rats.[2]

Regulations. ***U.S. FDA*** (1993) approved the use of TT as an accelerator in the manufacture of rubber articles intended for repeated use in producing, manufacturing, packing, processing, treating, packaging, transporting, or holding food (total not to exceed 1.5% by weight of the rubber product) in accordance with the conditions prescribed by 21 CFR part 177.2600.

References:

1. Vorob'yeva, R. S., 2-Mercaptothiasoline, in *Toxicology of New Chemical Substances Used in Rubber and Tyre Industry*, Z. I. Israel'son, Ed., Meditsina, Moscow, 1968, 57 (in Russian).
2. Ruddick, J. A., Newsome, W. H., and Nash, L., Correlation of teratogenicity and molecular structure: Ethylenethiourea and related compounds, *Teratology*, 13, 263, 1976.

THIOBENZOIC ACID, ANHYDROSULFIDE

Molecular Formula. $C_{14}H_{10}O_2S$
M = 242.30
CAS No 1850-15-3
RTECS No DH6843000

Synonyms. Benzoic thioanhydride; Benzoyl sulfide; Dibenzoyl sulfide; Thiobenzoic anhydride.

Properties. White, crystalline powder with a pink tint and an unpleasant odor. Poorly soluble in water, readily soluble in alcohol.

Application. Used as a plasticization accelerator of mixes based on natural and synthetic rubber.

Acute Toxicity. Exposure of rats and mice to 8.0 g/kg BW (in starch solution) caused 100% mortality. The animals exhibit decreased food consumption and progressive weakness; death occurs on days 1 to 17. Gross pathology examination revealed circulatory disturbances and slight fatty degeneration of the liver cells and renal tubular epithelium. The maximum tolerated dose was 4.0 g/kg BW.[1,029]

Reference:

1. Kremneva, S. N., Kochetkova, T. A., and Yakubov, A. T., in *Toxicology of New Industrial Chemical Substances*, A. A. Letavet, Ed., Meditsina, Moscow, Issue No 6, 1964, 55 (in Russian).

3,3′-THIOBIS(5-ISOPROPYLANISOL)

Trade Name. AN-6.

Properties. White, crystalline powder. Poorly soluble in water and alcohol.

Applications. Used as an anti-ozone and anti-scorching action agent in the rubber industry.

Acute Toxicity. Rats and mice tolerate administration of 6.0 g/kg BW. Poisoning is accompanied with signs of adynamia evident on day 3.[023]

4,4′-THIODIANILINE

Molecular Formula. $C_{12}H_{12}N_2S$
M = 216.3
CAS No 139-65-1
RTECS No BY9625000
Abbreviation. TDA.

Synonyms. Bis(4-aminophenyl) sulfide; 4,4'-Diaminodiphenyl sulfide.

Properties. Needle-shaped crystals. Poorly soluble in water.

Applications. Used as a curing accelerator in the rubber industry.

Acute Toxicity. LD_{50} is found to be 9.0 g/kg BW in rats, and 6.2 g/kg BW in mice. Signs of intoxication include reduced motor activity, body temperature, and reflex responses.[1]

Repeated Exposure revealed moderate cumulative properties. K_{acc} is 2.0 (by Lim).

Short-term and **Long-term Toxicity.** Male F344 rats received TDA at dietary levels of 0.5 and 1.0% for 13 or 85 weeks. Changes observed consisted of liver necrosis and atrophy and hyperplasia of the renal tubular epithelium. The phospholipid composition in the liver, kidneys, and spleen was also changed.[2]

Reproductive Toxicity.

Embryotoxicity. Mice were given 100 and 150 mg TDA/kg BW on days 1 to 5 of pregnancy. The treatment completely prevented fetal implantation. The dose of 50 mg/kg BW slightly reduced number of implantation sites (IARC 16-346).

Mutagenicity.

In vitro genotoxicity. TDA was shown to be negative in *Salmonella* mutagenicity assay with and without metabolic activation. It was not mutagenic for liver microsomes of rats which had previously received *Aroclor* 1254.[2] However, according to Endo et al.,[3] TDA has shown mutagenic potential.

Carcinogenicity. TDA induced tumors in the liver, thyroid, colon, and ear canal of male F344 rats, tumors in the thyroid, uterus, and ear canal of female F344 rats, tumors in the liver and thyroid in $B6C3F_1$ mice.[4]

Carcinogenicity classifications. An IARC Working Group concluded that there is sufficient evidence for the carcinogenicity of TDA in *experimental animals* and there were no adequate data available to evaluate the carcinogenicity of TDA in *humans*.

IARC: 2B.

NTP: P* - P* - P*- P* (feed).

References:

1. Gvozdenko, S. I., in *Proc. Rostov-na-Donu Medical Institute*, Issue No 17, 1974, 54 (in Russian).
2. Benjamin, T., Evarts, R. P., Reddy, T. V., et al., Effect of 2,2'-diaminodiphenylsulfide, a resin hardener, on rats, *J. Toxicol. Environ. Health*, 7, 69, 1981.
3. Endo, T. et al., Mutagenicity and chemical structure of diaminoaromatic compounds, *Mutat. Res.*, 130, 361, 1984.
4. *Bioassay of 4,4′-Thiodianiline for Possible Carcinogenicity*, NTP Technical Report Series No 47, Research Triangle Park, NC, DHEW Publ. No 78-847, 1978.

TITANIUM OXIDE

Molecular Formula. O_2Ti
M = 79.90
CAS No 13463-67-7
RTECS No XR2275000
Abbreviation. TO.

Synonyms and **Trade Name.** Flamenco; Titanium dioxide; Titanium peroxide; Titanox.

Properties. White, crystalline powder. Poorly soluble in water. Chemically inactive.

Applications. White pigment; used as a filler in the production of plastics and in the rubber industry.

Short-term Toxicity. Feeding rodents with 0.6 to 0.9 g TO/day caused no signs of intoxication.[1]

Long-term Toxicity. Male and female F344 rats were fed diets containing 1.0 to 5.0% TO coated mica for up to 130 weeks. There were no changes in survival, BW gain, hematology analyses, clinical chemistry parameters, or histopathology.[2]

$B6C3F_1$ mice and F344 rats were fed diets containing 2.5 or 5.0% TO. No difference in BW gain between experimental and control animals was observed in this study.[1]

Mutagenicity.

In vitro genotoxicity. TO did not induce morphological transformation in Syrian hamster embryo cells or mutations in bacteria.[3,4] It was shown to be negative in mouse lymphoma assay.[016]

Carcinogenicity. TO did not produce a significant increase in the frequency of any type of tumor in any species tested in the above described 103-week studies.[2,5]

Carcinogenicity classifications. An IARC Working Group concluded that there is limited evidence for the carcinogenicity of TO in *experimental animals* and there were inadequate data available to evaluate the carcinogenicity of TO in *humans*.

IARC: 3;

NTP: N - N - N - N (feed).

Chemobiokinetics. Absorption from the GI tract after administration by gavage was found to be negligible. [6] According to other data,[03] soluble *titanium* compounds are readily absorbed. Accumulation takes place in the blood and parenchymatous organs. Excretion occurs predominantly with the urine.

Regulations.

U.S. FDA (1998) regulates the use of TO (1) in adhesives as a component of articles intended for use in packaging, transporting, or holding food in accordance with the conditions prescribed by 21 CFR part 175.105; (2) in cellophane for packaging food in accordance with the conditions prescribed by 21 CFR part 177.1200; (3) as a filler for rubber articles intended for repeated use in accordance with the conditions prescribed by 21 CFR part 177.2600; (4) in polysulfide polymer-polyepoxy resins in articles used as a surface intended for contact with dry food subject to the specified conditions prescribed by 21 CFR part 177.1650; (5) in acrylate ester copolymer coating to be safely used as a food-contact surface of articles intended for packaging and holding food, including heating of prepared food in accordance with the conditions prescribed by 21 CFR part 175.210; and (6) as a substance employed in the production of, or added to, textiles and textile fibers intended for use in contact with food in accordance with the conditions prescribed by 21 CFR part 177.2800. *Titanium dioxide*, *titanium dioxide-barium sulfate*, and *titanium dioxide-magnesium silicate* may be used (7) as colorants of the uncoated or coated food-contact surface of paper and paperboard intended for use in producing, manufacturing, packaging, processing, preparing, treating, packing, transporting, or holding aqueous and fatty foods in accordance with the conditions prescribed by 21 CFR part 178.3297.

EU Commission (1982) approved the use of TO as a food color additive with the following specifications limiting impurities: antimony compounds, <100 mg/kg; zinc compounds, < 50 mg/kg; soluble barium compounds, <5.0 mg/kg; and hydrochloric acid-soluble compounds, 3.4 g/kg.

References:

1. Berlin, M. and Nordman, O., Titanium, *Handbook on Toxicology of Metals*, Amsterdam, 1980, 627.
2. Bernard, B. K., Osheroff, M. R., Hofmann, A., and Mennear, J. H., Toxicology and carcinogenesis studies of dietary titanium dioxide-coated mica in male and female Fischer 344 rats, *J. Toxicol. Environ. Health*, 29, 417, 1990.
3. Di Paolo, J. A. and Castro, B. C., Quantitative studies of *in vitro* morphological transformations of Syrian hamster cells by inorganic metal salts, *Cancer Res.*, 39, 1008, 1979.
4. Castro, B. C., Meyers, J., and Di Paolo, J. A., Enhancement of viral transformation for evaluation of the carcinogenic or mutagenic potential of metal salts, *Cancer Res.*, 39, 193, 1979.
5. *Bioassay of Titanium Dioxide for Possible Carcinogenicity*, Natl. Cancer Institute Technical Report Series No 97, Bethesda, MD, 1979.
6. Thomas, R. G. and Archuleta, R. F., Titanium retention in mice, *Toxicol. Lett.*, 6, 115, 1980.

1-*o*-TOLYLBIGUANIDE

Molecular Formula. $C_9H_{13}N_5$
M = 191.27
CAS No 93-69-6
RTECS No DU2800000
Abbreviation. TBG.

Synonyms and **Trade Name.** *N*-(2-Methylphenyl)imidodicarbonimidic diamide; *o*-Toludiguanide; Vulkacit 1000.

Applications. A vulcanization accelerator.

Acute Toxicity. LD_{50} is 800 mg/kg BW in rats; minimal LD is 120 mg/kg BW in marmots, and 80 mg/kg BW in rabbits.[027]

Regulations. ***U.S. FDA*** (1998) approved the use of TBG in adhesives as a component (monomer) of articles intended for use in packaging, transporting, or holding food in accordance with the conditions prescribed in CFR part 175.105.

2,4,6-TRIAMINO- *S*-TRIASINE, compound with *S*-TRIASINETRIOL

Molecular Formula. $C_3H_6N_6.C_3H_3N_3O_3$
M = 255.24
CAS No 37640-57-6
RTECS No XZ1225000
Abbreviation. MC.

Synonyms. Melamine cyanurate; Melamine isocyanurate; 1,3,5-Triazine-2,4,6-(1*H*,3*H*,5*H*)trione, compound with 1,3,5-triazine-2,4,6-triamine.

Properties. White, crystalline substance. Poorly soluble in water and organic solvents.

Applications. Used as a thermostable filler for vulcanizate mixes.

Acute Toxicity. LD_{50} is 2.5 to 5.52 g/kg BW for rats, and 3.46 g/kg BW for mice. Lethal doses cause adynamia and motor coordination disorder. Hematuria and epistaxis, pareses and paralyses were found to develop. Death occurs with signs of respiratory distress. Gross pathology examination revealed severe circulatory disturbances, particularly in the brain.

Repeated Exposure revealed cumulative properties. K_{acc} is 1.23 (by Kagan), when rats are given 1/10 LD_{50}.

Mutagenicity and **Reproductive Toxicity** effects were not observed after inhalation exposure to 1.15 mg MC/m^3.

Reference:

Alexandryan, A. V., Toxicity and sanitary standardization of melamine cyanurate, *Gig. Truda Prof. Zabol.*, 1, 44, 1986 (in Russian).

***S*-TRIAZINE-2,4,6-TRITHIOL**

Molecular Formula. $C_3H_3N_3S_3$
M = 177.27
CAS No 638-16-4
RTECS No XZ2830000
Abbreviation. TTT.

Synonyms. 1,3,5-Triazine-2,4,6-trimercaptan; 2,4,6-Triazinetrithiol; Trithiocyanuric acid; 1,3,5-Trimer captotriazine.

Applications. TTT is used as a rubber curative.

Acute Toxicity. A substance of a low oral and dermal toxicity, and low potential for eye and skin irritation. LD_{50} is 9.5 g/kg BW in rats.

Repeated Exposure. Sprague-Dawley rats were exposed to 625, 2500, and 5000 mg/kg diet for 2 to 30 days. Some effect was noted on BW gain and survival at the higher levels of intake. Main effects comprised unusual lesion of the pinna and the distal portions of the tail. There were purplish discolorations of the ear margin and tip of the tail in some animals (2500 and 5000 mg/kg diet). These changes were apparently site specific and have not been justified by histology examination. The NOAEL with regard to gross pathology lesions in a 30-day feeding study appeared to be 625 mg/kg diet.

Reference:

Koschier, F. J., Brown, D. R., and Friedman, M. A., Effect of dietary administration of the rubber curative trithiocyanuric acid to the rat, *Food Chem. Toxicol.*, 21, 495, 1983.

N,N′,N″-TRICHLORO-2,4,6-TRIAMINO- *N,N′,N″*-1,3,5-TRIAZINE

Molecular Formula. $C_3HCl_3N_6$
M = 229.36
CAS No 7673-09-8
RTECS No XZ1575000
Abbreviation. TCM.

Synonyms and **Trade Name.** Chloromelamine; Retarder TCM; Trichloromelamine; 2,4,6-Tris(chloramino)-1,3,5-triazine.

Properties. Light-yellow powder with the odor of chlorine. Poorly soluble in water.

Applications. Used as a vulcanization accelerator and retarder in the rubber industry.

Acute Toxicity. LD_{50} for mice is reported to be 490 or 3250 mg/kg BW. Gross pathology examination revealed point hemorrhages in the exterior wall of the stomach. It was partially liquefied from the effect of high doses, and easily torn. This effect is evidently associated with the spitting off chlorine and hydrochloride.[1]

Reproductive Toxicity. Embryotoxicity was shown in 2-day-old chick embryos.[2]

Allergenic Effect was not observed (Kostrodymova, 1976).

Mutagenicity.

In vivo cytogenetics. TCM is likely to be mutagenic in *Dr. melanogaster.*[3]

In vitro genotoxicity. TCM does not interact with DNA and does not induce changes in its structure and function (Onoue et al., 1982).

Carcinogenicity. Addition of 0.03% and 0.3% TCM to the diet caused lymphoid tissue tumors in mice.[4]

Standards. ***Russia*** (1995). MAC and PML: 1.4 mg/l (organolept., taste).

References:

1. Arkhangelskaya, L. N. and Roschina, T. A., Trichloromelamine, in *Toxicology of New Chemical Substances Used in Rubber and Tyre Industry*, Z. I. Israel'son, Ed., Meditsina, Moscow, 1968, 167 (in Russian).
2. Korhonen, A., Hemminki, K., and Vainio, H., Embryotoxicity of 16 industrial amines to the chicken embryo, *J. Appl. Toxicol.*, 3, 112, 1983.
3. Dubinin, V. I., *General Genetics*, Nauka, Moscow, 1970 (in Russian).
4. Hoshino, H. and Tanooka, H., Carcinogenicity of triethanolamine in mice and its mutagenicity after reaction with sodium nitrite in bacteria, *Cancer Res.*, Part I, 38, 112, 1978.

1,3,5-TRICHLORO-*S*-TRIAZINE-2,4,6(*1H,3H,5H*)-TRIONE

Molecular Formula. $C_3C_{l3}N_3O_3$
M = 232.41
CAS No 87-90-1
RTECS No XZ1925000

Synonyms. Trichlorocyan; Trichlorocyanuric acid; Trichloroisocyanic acid.

Properties. White powder with a strong odor of chlorine (91.5% active chlorine).

Applications. Used as an anti-scorching agent in the rubber industry. Swimming pool sanitizer.

Acute Toxicity.

Observations in man. A dose of 3.57 g/kg BW appeared to be fatal. Poisoning damaged the GI tract and caused ulceration or bleeding in the stomach.[1]

Animal studies. In mice, LD_{50} is reported to be 2.2 g/kg BW (in sunflower oil) and 0.75 g/kg BW (in vaseline). Gross pathology examination revealed changes in the GI tract.[2]

LD_{50} is 406 mg/kg BW in rats.[3]

References:

1. Deichmann, W. B., *Toxicology of Drugs and Chemicals,* Academic Press, Inc., New York, 1969, 167.
2. Arkhangel'skaya, L. N. and Roschina, T. A., Trichlorocyanuric acid, in *Toxicology of New Chemical Substances Used in Rubber and Tyre Industry*, Z. I. Israel'son, Ed., Meditsina, Moscow, 1968, 171 (in Russian).
3. Vernot, E. H., MacEwen, J. D., Haun, C. C., and Kinkead, E. R., Acute toxicity and skin corrosion data for some organic and inorganic compounds and aqueous solutions, *Toxicol. Appl. Pharmacol.*, 42, 417, 1977.

(3,3,5-TRIMETHYLCYCLOHEXYLIDENE)BIS[(1,1-DIMETHYLETHYL)PEROXIDE]

Molecular Formula. $C_{17}H_{34}O_4$

M = 302.51

CAS No 6731-86-8

RTECS No SD8600000

Synonym and **Trade Name.** 1,1-Bis-(*tert*-butylperoxy)-3,3,5-trimethylcyclohexane; Trigonox.

Applications. T. is used, in particular, in manufacture of rubber as well as in the hardening of unsaturated polyester resins and in the polymerization of styrene.

Acute Toxicity. LD_{50} is 12.0 g/kg BW in rats.[1]

Short-term Toxicity. In a 13-week study, T. was added to MF powdered basal diet and fed *ad libitum* to $B6C3F_1$ mice at dietary concentrations of 0.5 to 4.0%. The treated animals showed a tendency to anemia at 1.0% T. or more in the diet. Significant increase in relative liver weight and decrease in spleen weight were observed in a dose-dependent manner. Swelling of hepatocytes was histologically evident in animals fed 1.0% T. or more in the diet. Atrophy of the red and white pulp was noted in the spleen at dietary levels of 2.0 or 4.0% T. Final BW was in excess of 90% of the control group values only at 0.5% dose level (MTD).[2]

Mutagenicity.

In vitro genotoxicity. T. did not cause mutagenic effect in *Salmonella* bioassay (Machigaki, 1987).

Carcinogenicity. $B6C3F_1$ mice received T. at dietary levels of 0.25 and 0.5% for 78 weeks. No differences were noted in mortality of treated and untreated animals. Spontaneous neoplasms were found in all groups, including the control group at the same rate. Thus, T. exerts no carcinogenic activity in $B6C3F_1$ mice.

References:

1. Mitsui, M., Furukawa, F., Suzuki, J., Enami, T., Nishikawa, A., and Takahashi, M., 13-week subchronic toxicity study of 1,1-bis(*tert*-butylperoxy) 3,3,5-trimethylcyclohexane in mice, Abstract, *Eisei Shikenjo Hokoku,* 110, 42, 1992 (in Japanese).
2. Mitsui, M., Furukawa, F., Sato, M., et al., Carcinogenicity study of 1,1-bis(*tert*-butylperoxy)-3,3,5-trimethylcyclohexane in $B6C3F_1$ mice, *Food Chem. Toxicol.*, 31, 929, 1993.

TRIMETHYLDIHYDROQUINOLINE POLYMER

Molecular Formula. $[\sim C_{12}H_{15}N\sim]_n$

CAS No 26780-96-1

RTECS No TQ2625000

Synonym and **Trade Names.** Algerite MA; Antioxidant HS; 1,2-Dihydro-2,2,4-trimethylquinoline, homopolymer.

Applications. An antioxidant and an antidegradation agent for rubber.

Acute Toxicity. LD_{50} is 2.0 g/kg BW in rabbits.[027]

1,2,3-TRIPHENYLGUANIDINE

Molecular Formula. $C_{19}H_{17}N_3$
M = 287.39
CAS No 101-01-9
RTECS No MF6825000
Abbreviation. TPG.

Synonym. *N,N',N''*-Triphenylguanidine.

Applications. Used in the preparation of synthetic rubber.

Acute Toxicity. LD_0 appeared to be 0.25 g/kg BW. Administration caused narcotic action.

Regulations. ***U.S. FDA*** (1998) approved the use of TPG in the manufacture of rubber articles intended for repeated use in producing, manufacturing, packaging, processing, preparing, treating, packing, transporting, or holding food in accordance with the conditions prescribed by 21 CFR part 177.2600.

Reference:

Review, Natl. Acad. Sci., Natl. Res. Council, Chem.-Biol. Coordinat. Center, 5, 15, 1953.

2,4,6-TRIS(1-AZIRIDINYL)-*S*-TRIAZINE

Molecular Formula. $C_9H_{12}N_6$
M = 204.23
CAS No 51-18-3
RTECS No XZ2100000
Abbreviation. TAT.

Synonyms and **Trade Name.** Persistol; Triethylenemelamine; 2,4,6-Tris(ethyleneimino)- *S*-triazine.

Applications. TAT is used in manufacturing resinous products, and as a modifier for certain adhesives. Chemotherapeutic agent.

Properties. White crystalline powder. Solubility is 40% in water, and 7.7% in ethanol at 26°C.[020]

Toxic Action. TAT produced damage to the lymphoid tissue and bone marrow.[011]

Reproductive Toxicity.

Gonadotoxicity. Caused a prolonged decrease in testes weight, sperm number, and an increased rate of sperm abnormalities.[1]

Teratogenicity. Treatment of mice during pregnancy resulted in an increase in limb, skeletal, and CNS defects.[2]

Mutagenicity. TAT is an alkylating agent; it was found to be positive in the number of mammalian systems.[3,4]

In vivo cytogenetics. CD-1 mice received TAT by *i/p* injection on gestation days 14 to 15 at doses of 0.125, 0.25, and 0.5 mg/kg BW. A significant dose-related increase in both micronuclei and SCE was observed in maternal bone marrow and fetal liver. Weak cytotoxicity was noted in fetal liver cells in micronucleus assay. TAT is transplacental genotoxicant in mice.[5]

Carcinogenicity classification. An IARC Working Group concluded that there is limited evidence for the carcinogenicity of TAT in *experimental animals* and there were no adequate data available to evaluate the carcinogenicity of TAT in *humans*.

IARC: 3.

Chemobiokinetics. No persistent selective uptake by any tissue was reported.[011]

References:

1. Evenson, D. P., Baer, R. K., and Jost, L. K., Long-term effects of testis cells and sperm chromatin structure assayed by flow cytometry, *Environ. Molec. Mutagen.*, 14, 79, 1989.
2. Jurand, A., Action of triethanolmelamine (TEM) on early and late stages of mouse embryos, *J. Embryol. Exp. Morphol.*, 7, 526, 1959.
3. Nath, J. et al., Sister-chromatid exchanges induced by triethanolmelamine: *in vivo* and *in vivo/in vitro* studies in mouse and Chinese hamster bone marrow and spleen cells, *Mutat. Res.*, 206,73, 1988.

4. Hanley, T. R. et al., Triethylenemelamine (TEM): dominant lethal effects in Fischer 344 rats, *Drug Chem. Toxicol.*, 4, 63, 1981.
5. Xing, S. G., Shi, X., Wu, Z. L., Chen, J. K., Wallace, W., Wong, W. Z., and Ong, T., Transplacental genotoxicity of triethylenemelamine, benzene, and vinblastine in mice, *Teratogen. Carcinogen. Mutagen.*, 12, 223, 1992.

CHAPTER 6. SOLVENTS

ACETIC ACID

Molecular Formula. $C_2H_4O_2$
M = 60.06
CAS No 64-19-7
RTECS No AF1225000
Abbreviation. AA.

Synonyms and **Trade Names.** Ethanoic acid; Ethylic acid; Glacial acetic acid; Methanecarboxilic acid; Vinegar acid.

Properties. Rhombic crystals or colorless liquid with vinegar-like pungent odor. Miscible with water and alcohol. Odor perception threshold is 24.3 mg/l (Cooper, 1954).

Applications. Solvent and chemical intermediate for cellulose acetate, vinyl acetate monomer, and acetic esters.

Acute Toxicity. LD_{50} is 3.53 g/kg BW in rats.[020]

Short-term Toxicity. Rats were given orally 0.01% to 0.5% AA in their diet for 105 days.[1] Munro et al. (1996) suggested the calculated NOEL of 726 mg/kg BW for this study.[2]

Reproductive Toxicity. Suckling rats received AA from parturition until the age of 18 days. Male offspring had elevated preweaning BW and were much less active than normals by day 44.[3]

Mutagenicity. AA was negative in *Salmonella* mutagenicity assay[2] and in DNA-cell-binding assay.[4]

Chemobiokinetics. Readily absorbed from the GI tract. In the body, AA is partially converted into *formic acid.*

Regulations.

U.S. FDA (1998) affirmed AA as *GRAS.*

EU (1990). AA is available in the *List of authorized monomers and other starting substances which shall be used for the manufacture of plastic materials and articles intended to come into contact with foodstuffs (Section A).*

Great Britain (1998). AA is authorized without time limit for use in the production of polymeric materials and articles in contact with food or drink or intended for such contact.

Recommendations.

EU (1995). AA is a food additive generally permitted for use in foodstuffs.

Joint FAO/WHO Expert Committee on Food Additives (1999). ADI: No safety concern.

Standards. *Russia* (1995). MAC: 1.0 mg/l.

References:

1. Sollmann, J., *J. Pharmacol. Exp. Ther.,* 16, 463, 1921.
2. Munro, I. C., Ford, R. A., Kennepohl, E., and Sprenger, J. G., Correlation of structural class with No-Observed-Effect Levels: A proposal for establishing a threshold of concern, *Food Chem. Toxicol.,* 34, 829, 1996.
3. Barret, J. and Livesey, P. J., The acetic acid component of lead acetate: its effect on rat weight and activity, *Neurobehav. Toxicol. Teratol.,* 4, 105, 1982.
4. Zeiger, E., Anderson, B., Haworth, S., Lawlor, T., and Mortelmans, K., *Salmonella* mutagenicity tests: V. Results from the testing of 311 chemicals, *Environ. Molec. Mutagen.,* 19 (Suppl. 21), 2, 1992.

ACETIC ACID, BENZYL ESTER

Molecular Formula. $C_3H_{10}O_2$

M = 150.17
CAS No 140-11-4
RTECS No AF5075000
Abbreviation. AABE.

Synonyms. Acetic acid, phenylmethyl ester; α-Acetoxytoluene; Benzyl acetate; Benzyl ethanoate.

Properties. Clear, colorless liquid with a powerful, specific odor (jasmine-like or pear-like). Solubility in water is 250 mg/l; readily soluble in alcohols.

Applications. Is primarily a flavoring component and to a lesser degree, a solvent of cellulose acetate, nitrate cellulose, and natural and synthetic resins. May be safely used as a synthetic flavoring substance for food.

Exposure. AABE has been identified in several fruits and mushrooms. It also occurs naturally in flowers.[020]

Acute Toxicity. LD_{50} is reported to be 2.5 to 2.8 g/kg BW in rats, 0.83 g/kg BW in mice, and 0.22 g/kg BW in guinea pigs and rabbits.[1] However, according to other data, LD_{50} is 3.7 g/kg BW in rats, and 2.64 g/kg BW in rabbits.[2]

Administration of the lethal doses results in CNS depression. Death occurs on days 1 to 5. Acute action threshold was identified to be 50 mg/kg BW for STI.[1] After oral administration of 4.0 g/kg BW, 7 out of 10 treated rats died within 2 hours.[3]

Repeated Exposure failed to reveal evident cumulative properties. K_{acc} is 7.2 (by Lim). Rats were dosed with 50 to 200 mg/kg BW. The treatment caused transient changes in the NS function and blood formula. The NOAEL of 30 mg/kg BW was identified.[1]

F344 rats were exposed by gavage to 0.06 to 1.0 g/kg BW in corn oil and to 0.25 to 4.0 g/kg BW during 14 consecutive days: 3 out of 10 rats receiving 1.0 g/kg BW died. Clinical signs included decreased BW, tremor, ataxia, and sluggishness. No histopathology data are available.[4]

Short-term Toxicity. In a 13-week study, $B6C3F_1$ mice received the doses of 0.045 to 7.2 g AABE/kg BW in their diet. The treatment caused retardation in BW gain; feed consumption was lower in females at the dose of 3.6 g/kg BW and above. No effects on hematology, clinical chemistry, or pancreatic enzyme parameters were noted.[5]

F344 rats received 0.21 to 3.36 g AABE/kg BW in their diet for 13 weeks. Tremor, ataxia and urine stains were observed in animals exposed to 3.36 g/kg BW dose. Decreased serum cholesterol was noted in females dosed with 0.84 g AABE/kg BW and more.[5]

Long-term Toxicity. $B6C3F_1$ mice received 0.33 to 3.0 g AABE/kg feed over the period of 103 weeks. F344 rats were given 3.0 to 12 g AABE/kg feed for the same period of time. The treatment caused dose-related degeneration and atrophy of the olfactory epithelium, cystic hyperplasia of the nasal submucosal glands, and pigmentation of the nasal mucosae epithelium. The NOEL of 0.55 g/kg was identified in this study.[6]

Toxicity of AABE was higher when it was administered by gavage than when it was given in the diet.[7]

Allergenic Effect was not found on dermal applications.[1]

Mutagenicity.

In vitro genotoxicity. AABE was not mutagenic in *Salmonella typhimurium* strains *TA98, TA100, TA1535,* or *TA1537* with or without exogenous metabolic activation (*S9*). However, a positive response was observed in the mouse lymphoma assay for induction of trifluorothymidine resistance in L5178Y cells. It did not induce SCE or CA in Chinese hamster ovary cells in the presence or absence of metabolic activation.[5,6]

In vivo cytogenetics. AABE caused no increase in frequency of either SCE or CA, or in induction of micronucleated erythrocytes in bone marrow cells of male mice treated by *i/p* injection.

It was negative in *Dr. melanogaster* treated with AABE via feed or by injection.[5]

Carcinogenicity. $B6C3F_1$ mice were dosed by oral intubation with 500 and 1000 mg/kg BW in corn oil for 103 weeks; there was an increased incidence of liver adenoma (not statistically significant) and of combined liver adenomas and carcinomas, forestomach tumors. AABE was shown to be a weak promoter of the growth of carcinogen-induced and spontaneous pre-neoplastic azaserine-induced foci in the pancreas of F344 rats fed AABE for 6 months; the gavage vehicle (corn oil) may have been a contributing factor.[4,5,8]

Carcinogenicity classifications. An IARC Working Group concluded that there is limited evidence for the carcinogenicity of AABE in *experimental animals* and there were no adequate data available to evaluate the carcinogenicity of AABE in *humans*.

IARC: 3;

NTP: EE - NE - SE* - SE (gavage);

NE - NE - NE - NE (feed).

Chemobiokinetics. In the body, AABE is hydrolyzed to *benzyl alcohol*. The benzyl radical is oxidized to *benzoic acid*. Gavage administration in corn oil at the dose of 500 mg/kg BW (rats) and 1000 mg/kg BW (mice) induced high benzoic acid plasma concentration. AABE is excreted primarily in the urine as *hippuric* and *benzylmercapturic acids*.[3,9]

Regulations. ***U.S. FDA*** (1998). AABE is a food additive permitted for direct addition to food for human consumption, as long as (1) the quantity added to food does not exceed the amount reasonably required to accomplish its intended physical, nutritive, or other technical effect in food, and (2) when intended for use in or on food, it is of appropriate food grade and is prepared and handled as a food ingredient. (3) Synthetic flavoring substances and adjuvants including BA may be safely used in foods according to the provisions prescribed in 21 CFR part 172.515.

Recommendations. Joint FAO/WHO Expert Committee on Food Additives (1997). ADI of 5.0 mg/kg BW is considered as total *benzoic acid* from all food additive sources.

References:

1. Fursova, T. N., Toxicological and hygienic characteristics of benzyl acetate, *Gig. Sanit.*, 7, 17, 1985 (in Russian).
2. Graham, B. E. and Kuizenga, M. H., Toxicity studies of benzylacetate and related benzyl compounds, *J. Pharmacol. Exp. Ther.*, 84, 358, 1945.
3. von Ottingen, W. F., The aliphatic acids and their esters: toxicity and potential danger, *Arch. Ind. Health*, 21, 28, 1960.
4. Abdo, V. M., Huff, J. E., Haseman, J. K., et al., Benzylacetate carcinogenicity, metabolism, and disposition in Fischer 344 rats and B6C3F$_1$ mice, *Toxicology*, 37, 159, 1985.
5. *Toxicology and Carcinogenesis Studies of Benzyl Acetate in F344 Rats and B6C3F$_1$ Mice (Feed Studies)*, NTP Technical Report Series No 431, Board Draft, NIH, Research Triangle Park, NC, 1992.
6. Mortelmans, K., Haworth, S., Lawlor, T., et al., *Salmonella* mutagenicity tests: II. Results from testing of 270 chemicals, *Environ. Mutagen.*, 8 (Suppl. 7), 1, 1986.
7. *Evaluation of Certain Food Additives and Contaminants*, WHO Tech. Report Series 868, 46th Report of Joint FAO/WHO Expert Committee on Food Additives, Geneva, 1997, 69.
8. Longnecker, D. S., Roebuck, B. D., Curphey, T. J., and MacMillan, D. L., Evaluation of promotion of pancreatic carcinogenesis in rats by benzyl acetate, *Food Chem. Toxicol.*, 28, 665, 1990.
9. Yuan, J. H., Goehl, K., Abdo, K., Clark, J., Espinosa, O., Bugge, C., and Garcia, D., Effect of gavage versus dosed feed administration on the toxicokinetics of benzyl acetate in rats and mice, *Food Chem. Toxi col.*, 33, 151, 1995.

ACETIC ACID, BUTYL ESTER

Molecular Formula. $C_6H_{12}O_2$

M = 116.16

CAS No 123-86-4

RTECS No AF7350000

Abbreviation. AABE.

Synonyms. *n*-Butyl acetate; Butyl ethanoate.

Properties. A liquid with an odor of ether. Water solubility is 1.0% or 6.8 g/l at 25°C.[02] Mixes with alcohol and vegetable oils at any ratio. Odor perception threshold is 1.0 or 0.17 mg/l,[02] or even 0.006 mg/l,[010] taste perception threshold is 0.3 mg/l. Does not affect color and transparency of water.

Applications. Used as a solvent in the manufacture of plastics and lacquers.

Acute Toxicity. LD_{50} is 14.1 g/kg BW in rats, and 4.7 g/kg BW in mice (Sporn et al.). According to other data, LD_{50} is 13.1 and 7.7 g/kg BW, respectively.[1] Gross pathology examination revealed congestion in the viscera and soft brain meninges. Histology examination showed kidney lesions.

Repeated Exposure failed to reveal cumulative properties. Rats were dosed with 0.8 and 1.6 g/kg BW for a month. The treatment caused adynamia, lymphocytosis, depression of leukocyte phagocytic activity and increased vitamin C content in the suprarenals. Gross pathology examination revealed parenchymatous dys trophy in the cerebellum cells, especially evident at a higher dose.[1]

Long-term Toxicity. In a 6-month study, no changes were found in liver function and enzymatic activity of rats exposed to 0.5 mg AABE/kg BW.[1]

Reproductive Toxicity. AABE produced no teratogenic effect in rats, mice, and rabbits.[2,3]

Allergenic Effect is not found in the mouse ear-swelling test.[4]

Mutagenicity.

In vivo cytogenetics. AABE is not shown to be mutagenic in *Dr. melanogaster* (NTP-82).

In vitro genotoxicity. AABE was negative in *Salmonella* mutagenicity assay.[5]

Regulations. ***U.S. FDA*** (1998) regulates AABE (1) as a component of adhesives intended for use in contact with food in accordance with the conditions prescribed by 21 CFR part 175.105; (2) in the manufacture of cellophane to be safely used for packaging food (up to 0.1% by weight) in accordance with the conditions prescribed by 21 CFR part 177.1200; and (3) as an ingredient of resinous and polymeric coatings for food-contact surface of articles intended for use in producing, manufacturing, packing, transporting, or holding food in accordance with the conditions prescribed by 21 CFR part 175.300.

Recommendations. Joint FAO/WHO Expert Committee on Food Additives (1999). ADI: No safety concern.

Standards. ***Russia*** (1988). PML: 0.3 mg/l.

References:

1. Bul'bin, L. A., Substantiation of maximum allowable concentration for butyl acetate in water bodies, *Gig. Sanit.*, 4, 22, 1968 (in Russian).
2. Scheufler, H., Experimental testing of chemical agents for embryotoxicity, teratogenicity and mutagenicity. Ontogenic reactions of the laboratory mouse to these injections and their evaluation - a critical analysis method, *Biol. Rundsch.*, 14, 227, 1976.
3. Hackett, P. L., Brown, M. G., Buschbom, R. L., et al., Teratogenic activity of ethylene and propylene oxide and *n*-butyl acetate, *Gov. Rep. Announce. Ind.*, Issue 26, NTIS/PB 83-258038, 1983.
4. Gad, S. C., Dunn, B. J., Dobbs, D. W., Reilly, C., and Walsh, R. D., Development and validation of an alternative dermal sensitization test: the mouse ear swelling test (MEST), *Toxicol. Appl. Pharmacol.*, 84, 93, 1986.
5. Zeiger, E., Anderson, B., Haworth, S., Lawlor, T., and Mortelmans, K., *Salmonella* mutagenicity tests: V. Results from the testing of 311 chemicals, *Environ. Molec. Mutagen.*, 19 (Suppl. 21), 2, 1992.

ACETIC ACID, ETHYL ESTER

Molecular Formula. $C_4H_8O_2$
M = 88.10
CAS No 141-78-6
RTECS No AH5425000
Abbreviation. AAEE.

Synonyms. Acetic ether; Acetoethyl ether; Ethyl acetate.

Properties. AAE occurs as a clear, colorless liquid with a characteristic odor. Solubility in water is 8.0%, miscible with alcohol at all ratios. Unstable in water. Odor perception threshold is 6.3 mg/l[010] or 20 mg/l, taste perception threshold is 10 mg/l.[1] However, according to Saratikov,[2] odor perception threshold is 2.0 mg/l (20°C); accordng to other data, it is found to be 2.6 mg/l.[02]

Applications. Used as a solvent in the production of materials based on cellulose ethers, alkyd and polyvinyl acetate resins.

Acute Toxicity. LD_{50} is found to be 5.6 g/kg BW,[03] or 6.1 g/kg BW,[2] or 11.3 ml/kg BW[020] in rats, 4.1 g/kg BW in mice,[2] 5.5 g/kg BW in guinea pigs,[2] and 4.9 mg/kg BW[03,019] or 7.65 g/kg BW in rabbits.[2]

Manifestations of the toxic effect comprised a reduction in motor activity and excitability and a narcotic state subsequently supervening. Gross pathology examination revealed visceral congestion and tiny point hemorrhages.

Repeated Exposure failed to reveal cumulative properties. Rats tolerated administration of 1.0 g/kg BW dose for a month without any sign of intoxication or retardation of BW gain.

Short-term Toxicity. Rats were given by gavage 0.3, 0.9, or 3.6 g/kg BW for 90 days. The LOEL of 3.6 g/kg BW and the NOEL of 0.9 g/kg BW were established in this study.[3]

Reproductive Toxicity.

Gonadotoxic effect was observed when AAEE has been inhaled by male rats twice a day for seven days. The exposure led to reduction of testicular and prostate weights and sperm production.[4]

Chemobiokinetics. Following ingestion AAEE is very rapidly hydrolyzed to form *ethyl alcohol.*[1]

Regulations. ***U.S. FDA*** (1998) approved the use of AAEE (1) in the manufacture of cellophane for packaging food in accordance with the conditions prescribed in 21 CFR part 177.1200; (2) in the manufacture of resinous and polymeric coatings for polyolefin films to be safely used as a food-contact surface of articles intended for use in producing, manufacturing, packing, transporting, or holding food in accordance with the conditions prescribed in 21 CFR part 175.320; and (3) in food in accordance with the conditions prescribed in 21 CFR part 173.228.

Affirmed as *GRAS* when used as synthetic flavoring agent.

Recommendations. Joint FAO/WHO Expert Committee on Food Additives (1997): no safety concern at current levels of intake.

References:

1. Gallacher, E. J. and Loomis, T. A., Metabolism of ethylacetate in the rat: Hydrolysis to ethyl alcohol *in vitro* and *in vivo*, *Toxicol. Appl. Pharmacol.*, 34, 309, 1975.
2. Saratikov, A. S., Trofimovich, E. M., Burova, A. V., et al., Substantiation of the maximum allowable concentration for ethylacetate in water, *Gig. Sanit.*, 4, 66, 1983 (in Russian).
3. *Evaluation of Certain Food Additives and Contaminants*, WHO Technical Report Series No 868, 46th Report of Joint FAO/WHO Expert Committee on Food Additives, Geneva, 1997, 69.
4. Yamada, K., Influence of lacquer thinner and some organic solvents on reproductive and accessory reproductive organs in the male rat, *Biol. Pharmacol. Bull.*, 16, 425, 1993.

ACETIC ACID, ISOBUTYL ESTER

Molecular Formula. $C_6H_{12}O_2$
M = 116.18
CAS No 110-19-1
RTECS No AI4025000
Abbreviation. AAIE.

Synonyms. Acetic acid, 2-methylpropyl ester; Isobutyl acetate.

Properties. Clear, colorless liquid. Solubility in water is 5.9 or 6.7 g/l at 25°C; mixes with alcohol at any ratio. Odor perception threshold is 0.15 or 1.0 mg/l,[02] taste perception threshold is 0.5 mg/l. According to other data, odor threshold concentration is 0.35 mg/l.[010]

Applications. Used in the manufacture of plastics and lacquers.

Acute Toxicity. LD_{50} is 4.8 g/kg BW[019] or 15 g/kg BW in rats, 6.68 g/kg BW in mice, 6.66 g/kg BW in guinea pigs, and 3.7 g/kg BW in rabbits.[1]

Long-term Toxicity. The NOAEL appeared to be 5.0 mg/kg BW.

Chemobiokinetics. Following absorption in the GI tract, AAIE undergoes conversion by hydrolysis to *acetic acid* and *isobutanol.*[012]

Regulations. ***U.S. FDA*** (1998) approved the use of AAIE as a food additive for direct addition to food for human consumption as a synthetic flavoring substance and adjuvant in accordance with the following conditions: (1) the quantity added to food does not exceed the amount reasonably required to accomplish its intended physical, nutritive, or other technical effect in food, and (2) when intended for use in or on food, it is of appropriate food grade and is prepared and handled as a food ingredient (21 CFR part 172.515).

Standards. ***Russia.*** PML: 0.5 mg/l (organolept., taste).

Reference:

Meleshchenko, K. F., *Prevention of the Pollution of Water Bodies by Liquid Effluents of Chemical Plants*, Zdorov'ya, Kiyv, 1971, 56 (in Russian).

ACETIC ACID, ISOPROPYL ESTER

Molecular Formula. $C_5H_{10}O_2$
M = 102.13
CAS No 108-21-4
RTECS No AI4930000
Abbreviation. AAIE.

Synonyms. Acetic acid, 1-methylethyl ester; Acetoxypropane; Isopropyl acetate; Paracetat; 2-Propyl acetate.

Properties. Colorless liquid with an intense fruity odor and a sweet apple-like flavor (on dilution). Soluble in 23 parts of water at 27°C. Miscible with alcohol.[020]

Applications. Solvent for cellulose derivatives, plastics, and synthetic coatings. A food additive.

Acute Toxicity. LD_{50} is 3.0 to 6.75 g/kg BW, and 6.95 mg/kg BW in rabbits.[03,020]

Regulations. ***U.S. FDA*** (1998) approved the use of AAIE (1) in adhesives as a component of articles intended for use in packaging, transporting, or holding food in accordance with the conditions prescribed by 21 CFR part 175.105; and (2) in the manufacture of cellophane for packaging food in accordance with the conditions prescribed by 21 CFR part 177.1200.

Reference:

Smyth, H. F., Carpenter, C. P., Weil, C. S., and Pozzani, U. C., Range-finding toxicity data: List V, *Arch. Ind. Hyg. Occup. Med.,* 10, 61, 1954.

ACETIC ACID, METHYL ETHER

Molecular Formula. $C_3H_6O_2$
M = 74.08
CAS No 72-20-9
RTECS No AI9100000
Abbreviation. AAME.

Synonyms and **Trade Name.** Methyl acetate; Methyl acetic ester; Tereton.

Properties. Clear, colorless liquid with a unique odor. Solubility in water is 319 g/l at 20°C[1] or 220 g/l at 25°C;[02] mixes with alcohol at any ratio. Odor and taste perception threshold is 5.0 to 10.0 mg/l. According to other data, odor perception threshold is 3.0 mg/l.[02]

Applications. Used in the manufacture of lacquers and films; a diluent and a solvent for nitrocellulose, acetyl cellulose, and many resins.

Acute Toxicity. LD_{50} is reported to be 2.9 g/kg BW in rats, 2.4 g/kg BW in mice and rabbits, and 3.6 g/kg BW in guinea pigs.[2]

Repeated Exposure revealed cumulative effect. K_{acc} was found to be 3.65 (by Cherkinsky) when 1/5 LD_{50} was administered to animals. A dose of 250 mg/kg BW produced an increase in blood cholinesterase activity and in the number of segmented nucleus neutrophils.[1]

Chemobiokinetics. Inhaled AAME is partly excreted in exhaled air and urine, and partly metabolized.[04]

Regulations. ***U.S. FDA*** (1998) approved the use of AAME (1) as a component of the uncoated or coated food-contact surface of paper and paperboard in accordance with the conditions prescribed in 21 CFR part 176.170; and (2) in adhesives used as components of articles intended for use in packaging, transporting, or holding food in accordance with the conditions prescribed in 21 CFR part 175.105.

Recommendations. Joint FAO/WHO Expert Committee on Food Additives (1999). ADI: No safety concern.

Standards.
Russia (1995). MAC and PML: 0.1 mg/l.

References:

1. Meleshchenko, K. F. et al., in *Hygiene of Populated Areas*, Proc. Sci. Conf., D. N. Kaluzhny, Ed., Zdorov'ya, Kiyv, 1967, 59 (in Russian).
2. Grushko, Ya. M., *Harmful Organic Compounds in the Liquid Industrial Effluents*, Khimiya, Leningrad, 1976, 38 (in Russian).

ACETIC ACID, PENTYL ESTER

Molecular Formula. $C_7H_{14}O_2$
M = 130.19
CAS No 628-63-7
RTECS No AJ1925000
Abbreviation. AAPE.

Synonyms and **Trade Names.** Acetic acid, amyl ester; *n*-Amyl acetate; Chlordantoin; 1-Pentanol acetate; *n*-Pentyl acetate.

Properties. Colorless liquid with persistent banana-like odor. Solubility in water is 1.7 g/l; very soluble in alcohol.[012]

Applications. A solvent. Used in the production of nitrocellulose, lacquers, celluloid, waterproof varnishes, etc.

Acute Toxicity. LD_{50} is 7.4 g/kg BW in rabbits.[029]

Repeated Exposure. Inhalation exposure to vapors of AAPE affected liver serum enzyme findings and histology: degenerative changes in the liver cells with periportal sclerogenic activity (Querci and Mascia, 1970).

Allergenic Effect. AAPE was found to be a possible marginal skin sensitizer in guinea pig maximization test.[1]

Regulations. ***U.S. FDA*** (1998) approved the use of AAPE in adhesives as a component of articles intended for use in packaging, transporting, or holding food in accordance with the conditions prescribed by 21 CFR part 175.105.

Reference:

1. Ballantyne, B., Tyler, T. R., and Auletta, C. S., The sensitizing potential of primary amyl acetate in the guinea pig, *Vet. Hum. Toxicol.*, 28, 213, 1986.

ACETONE

Molecular Formula. C_3H_6O
M = 58.08
CAS No 67-64-1
RTECS No AL3150000

Synonyms. Dimethylformaldehyde; Dimethyl ketone; Ketone propane; Methyl ketone; 2-Propanone; Pyroacetic acid.

Properties. Clear, colorless liquid with a characteristic odor. Mixes with water in all proportions. Odor threshold is 20 mg/l [02] or 40 to 70 mg/l; at these concentrations it does not affect the taste, color, or clarity of water. According to other data, taste threshold is 12 mg/l.[1]

Applications. Used as a solvent in the production of nitro- and acetyl-cellulose, epoxy and vinyl resins, and vulcanizates.

Acute Toxicity.

Observations in man. Ingestion of a high dose resulted in a comatose condition with subsequent convulsions. Poisoning was accompanied by cyanosis of the skin, labored breathing, and increased blood pressure. Signs of cardio-pulmonary insufficiency preceded death. At autopsy, there were pulmonary and cerebral edema and hemorrhages in the renal tubular epithelium.[2]

Animal studies. Dogs tolerated administration of 1.0 g/kg BW without any harmful effect. A dose of 4.0 g/kg BW caused stupor, and 8.0 g/kg BW caused death. In mice, LD_{50} was found to be up to 9.75 g/kg BW. Rabbits tolerated a dose of 1.0 g/kg BW, and LD_{50} varied from 3.8 to 5.3 g/kg BW.[3]

LD_{50}s are reported to be 5.6 mg/kg, 9.1 mg/kg, and 8.5 mg/kg BW in 14-day old rats, young adults rats, and older adults, respectively.[4]

Repeated Exposure revealed slight cumulative ability. K_{acc} is 8.6 (by Cherkinsky). A. exhibits a low level of toxicity. F344/N rats and $B6C3F_1$ mice were given A. in drinking water for 14 days. At concentration of 100,000 ppm, a decreased BW was noted but no histology changes were observed except centrilobular hepatocellular hypertrophy in mice (concentrations 20,000 to 50,000 ppm).[5]

Short-term Toxicity. In a 13-week study, no hematological changes were found in mice given drinking water containing 20,000 to 50,000 ppm A. There was mild hepatocellular hypertrophy in only 2 out of 10 female mice. An increased incidence and severity of nephropathy were the most prominent treatment-related findings.[5]

In a 13-week study in Sprague-Dawley rats, no compound-related effects on BW, ophthalmic, and urinalysis results or mortality were reported.[6]

Long-term Toxicity. Rats were gavaged with 7.0 to 70 mg/kg BW for 6 months. The treatment did not affect general condition, behavior, or BW gain. The highest dose caused the greatest changes in the biochemical indices. Histology examination revealed hemo- and lymphodynamic disorders, dystrophic changes in the myocardium, and parenchymatous dystrophy of the liver and kidneys.[7]

Reproductive Toxicity.

Embryotoxicity.

Observations in man. Female factory workers had undergone a long-term exposure to A. at a concentration below 200 mg/m^3. An increased incidence of complications of pregnancy was reported.[8]

Animal studies. A. can readily cross the placenta and accumulate in exposed fetuses.[9]

Gonadotoxicity.

Observations in man. Slight alterations of semen quality were observed in 25 workers of a reinforced plastic production plant exposed by inhalation to a mixture of A. and styrene compared with 46 age-matched controls from infertility clinic.[10]

Wives of men potentially exposed to A. showed no significant increased risk for spontaneous abortion.[11]

Animal studies. No effect on male Wistar rats fertility and histology of the testes was observed after administration of A. in the drinking water for 6 weeks (1.1 g A./kg BW).[12] However, in male Sprague-Dawley rats, a dose of 3400 mg/kg BW for 13 weeks induced decreased sperm motility and increased proportion of abnormal sperm. Histological examination failed to reveal any testicular changes. No effect of A. on sperm morphology was noted in mice exposed to up to 4.9 g A./kg BW.[9]

Teratogenicity. According to Mast et al.,[13] A. produced no teratogenic effect in rats and mice. No abnormalities were found to develop in rodent embryos in culture after exposure to up to 2.5% A.[14]

Allergenic Effect. Contact sensitizing potential was not found in maximized guinea-pig assays.[15]

Mutagenicity.

In vitro genotoxicity. A. did not show mutagenic potential in *Salmonella* test and did not induce SCE or CA in Chinese hamster ovary cells.[16] A. gave negative response in DNA binding[17] and point mutation assays in mouse lymphoma cells.[18]

Carcinogenicity. There is no evidence of A. carcinogenicity.

Chemobiokinetics. After absorption, A. is rapidly distributed throughout the tissues according to their water content, and is excreted via the kidneys and lungs, and also in the sweat, unchanged or metabolized. Although A. has been considered to be a non-metabolizable compound, it was shown to metabolize by 3 separate gluconeogenic pathways.[19] *Carbon dioxide* formed on oxidation of A. is excreted in the expired air.

Regulations. ***U.S. FDA*** (1998) approved the use of A. (1) as a component of adhesives for food-contact surface in accordance with the conditions prescribed by 21 CFR part 175.105 ; (2) as an ingredient in the manufacture of resinous and polymeric coatings for polyolefin films coming in contact with food in accordance with the conditions prescribed by 21 CFR part 175.320; and (3) as an adjuvant in the preparation of slimicides in the manufacture of paper and paperboard that contact food in accordance with the conditions prescribed in 21 CFR part 176.300.

Recommendations. Joint FAO/WHO Expert Committee on Food Additives: ADI is not specified.

Standards. ***Russia*** (1992). PML in food: 0.1 mg/l; PML in drinking water: 2.0 mg/l.

References:

1. Grushko, Ya. M., *Harmful Organic Compounds in the Liquid Industrial Effluents*, *Khimiya*, Leningrad, 1976, 40 (in Russian).
2. Alexandrova, V. V., Poisoning by liquid acetone, *Sudmedexpertiza*, 3, 57, 1980 (in Russian).
3. Nazarenko, I. V., in *Protection of Water Reservoirs against Pollution by Industrial Liquid Effluents*, S. N. Cherkinsky, Ed., Medgiz, Moscow, Issue No 3, 1959, 221 (in Russian).
4. Kimura, E. T., Ebert, D. M., and Dodge, P.W., Acute toxicity and limits of solvents residue for sixteen or ganic solvents, *Toxicol. Appl. Pharmacol.*, 19, 699, 1971.
5. *Toxicity Studies of Aceton in F344/N Rats and B6C3F$_1$ Mice (Drinking Water Studies),* NTP Technical Report Series No 3, NIH Publ. No 91-3122, 1991.
6. Sonawane, B., de Rosa, C., Rubinstein, R., et al., Estimation of reference dose for oral exposure to acetone, *7th Ann. Meeting Am. College Toxicol.*, November 16-19, 1986, 21.
7. Omel'ynets, N. I., Mironets, N. V., and Martyzchenko, N. V., *Kosm. Biol. Aviokosm. Med.*, 12, 67, 1978 (in Russian).
8. Nizyaeva, I. V., On hygienic assessment of acetone, *Gig. Truda Prof. Zabol.*, 6, 24, 1982 (in Russian).
9. Dietz, D. D., Leininger, J. R., Rauckman, E. J., Thompson, M. B., Chapin, R. E., Morrissey, R. L., and Levine, B.S., Toxicity studies of acetone administered in the drinking water of rodents, *Fundam. Appl. Toxicol.*, 17, 347, 1991.
10. Jelnes, J. E., Semen quality in workers producing reinforced plastic, *Reprod. Toxicol.*, 2, 209, 1988.
11. Taskinen, H., Anttila, A., Lindbohm, M.-L., and Sallmen, M., Spontaneous abortions and congenital malformations among the wives of men occupationally exposed to organic solvents, *Scand. J. Work Envi ron. Health*, 15, 345, 1989.
12. Larsen, J. J., Lykkegaard, M., and Ladefoged, O., Infertility in rats induced by 2,5-hexaedrione in combination with acetone, *Pharmacol. Toxicol.*, 68, 1, 1991.
13. Mast, T. G., Rommereim, R. L., Weigel, R. G., et al., Developmental toxicity study of acetone in mice and rats, *Teratology*, 39, 468, 1989.
14. Kitchin, K. T. and Ebron, M. T., Further development of rodent whole embryo culture: solvent toxicity and water insoluble compound delivery system, *Toxicology*, 30, 45, 1984.
15. Vial, T. and Descotes, J., Contact sensitization assays in guinea pigs: are they predictive of the potential for systemic allergic reactins? *Toxicology*, 93, 63, 1994.
16. Norppa, H. K., The *in vitro* induction of sister chromatid exchanges and chromosome aberrations in human lymphocytes by styrene derivatives, *Carcinogenesis*, 2, 237, 1981.
17. Kubinski, H., Gutzke, G. E., Kubinski, Z. O., DNA-cell-binding (DCB) assay for suspected carcinogens and mutagens, *Mutat. Res.*, 89, 95, 1981.
18. Amacher, D. E., Paillet, S. C., Turner, G. N., Ray, V. A., and Salsburg, D. S., Point mutations at the thymidine kinase locus in L5178Y mouse lymphoma cells. 2. Test validation and interpretation, *Mutat. Res.*, 72, 447, 1980.
19. Morris, J. B. and Covanagh, D. G., Metabolism and deposition of propanol and acetone vapors in the upper respiratory tract of the hamster, *Fundam. Appl. Toxicol.*, 9, 34, 1987.

ALLYL ALCOHOL

Molecular Formula. C_3H_6O

M = 58.08

CAS No 107-18-6

RTECS No BA5075000

Abbreviation. AA.

Synonyms. 2-Propen-1-ol; 3-Hydroxypropene.

Properties. Colorless, transparent liquid with a pungent mustard odor. Readily miscible with water and alcohol. Taste perception threshold is 0.1 mg/l, odor perception threshold is 0.33 to 0.66 mg/l (when heated to 60°C, 0.07 mg/l and 0.75 mg/l, respectively).[1] However, odor perception threshold of 14 mg/l is reported.[02] AA does not alter the transparency or color of water.

Applications. Used in the manufacture of rubber and polymeric materials.

Acute Toxicity. Administration of 1.5 mmol AA/kg BW to starved mice led to the development of hemolysis in nearly 50% of the animals. *Malonic dialdehyde* appears in plasma of the animals showing hemolysis.[2]

Oral LD_{50} values have been reported to be 140 mg/kg BW in rats, 75.5 mg/kg BW in mice, and 90 mg/kg BW in rabbits.[1] Other median lethal doses indicated are: 64, 96, and 71 mg/kg BW, respectively.[06]

Poisoned animals displayed signs of CNS impairment (ataxia) and pulmonary edema. Histology examination showed necrotic and dystrophic changes in the liver, myocardium, and kidneys and edema of the connective tissue.

Repeated Exposure to AA produced general toxic effect with specific liver cell damage. Reduced serum catalase activity was noted. Hematological analysis revealed leukocytosis; erythrocyte counts and *Hb* levels were not altered.

Rats were dosed by gavage with 14 and 28 mg/kg BW for 10 days. Gross pathology examination revealed congestion in the liver, kidneys, and spleen and dystrophic and necrobiotic changes in the myocardium and liver. Severe swelling of the renal tubular epithelium was observed.[1]

Male F344 rats were administered with AA at doses of up to 0.5 ml/kg BW in corn oil by gavage for 14 days. This exposure resulted in significant increase in absolute and relative liver weights, liver glutathione, and periportal hepatocellular vacuolar degeneration.[3]

Short-term Toxicity. Rats received AA in their drinking water for 90 days. Concentrations of 250 ppm and above caused a significant increase in liver and kidney weights.[4]

In a 90-day study, Wistar rats of both sexes received AA with their drinking water at the dose levels of 4.8 mg/kg to 58.4 mg/kg BW. The treatment caused no effect on the hematological parameters or on the results of serum examination. An increase in kidney and liver weight was reported at 7.0 mg AA/kg BW and above dose levels. There were no histopathological changes in the visceral organs of the treated animals. The NOEL of 4.8 to 6.2 mg/kg BW (50 ppm in drinking water) was established in this study.[5]

Long-term Toxicity. Rabbits were dosed with 2.5 mg AA/kg BW for 8 months. The treatment produced protein fractions dysbalance in the blood serum. Microscopic findings included necrotic foci in the liver and moderate congestion and parenchymatous dystrophy of the renal tubular epithelium.[1,6]

Reproductive Toxicity.

Embryotoxicity and ***Teratogenicity.*** AA produced no teratogenic effect in mice.[7] Chronic treatment of male rats caused no efffect on reproductive performance and no abnormalities in their offspring.[8] However, Slott and Hales found increased embryolethality at a dose of 0.1 mg AA and limb defects and other fetal malformations at a dose of 1.0 mg AA.[9] Both doses were injected directly into the amniotic cavity of day 13 rat embryos.

Mutagenicity.

In vitro genotoxicity. AA is found to be positive in *Salmonella* mutagenicity assay and in mutagenicity assay when measured in *V79* cells as resistance to 6-thioguanidine.[10] It was negative in the test for micronucleus induction in rat peripheral erythrocytes (NTP-94).

Carcinogenicity. Negative results are reported in rats.[09]

Chemobiokinetics. AA undergoes transformation into *acrolein* under the action of alcohol dehydrogenase with further conjugation to cellular glutathione to form an *aldehydeglutathione adduct*, which is then gradually transformed into the corresponding acid. A reduction in the intracellular content of *SH*-groups is of considerable importance in the pathogenesis of AA toxic effects.[11] The toxicity of AA (or its metabolite *acrolein*) is dependent on the concentration of glutathione.[12]

Regulations. ***EU*** (1990). AA is available in the *List of authorized monomers and other starting substances which may continue to be used for the manufacture of plastic materials and articles intended to* *come into contact with foodstuffs pending a decision on inclusion in Section A (Section B).*

Recommendations. Joint FAO/WHO Expert Committee on Food Additives (1997). ADI: 0.05 mg/kg BW.

Standards. ***Russia*** (1995). MAC and PML: 0.1 mg/l (organolept., taste).

References:

1. Karmazin, V. E., in *Problems of Environmental Hygiene*, Coll. Works, D. N. Kaluzhny, Ed., Kiyv, Issue No 6, 1966, 108 (in Russian).
2. Ferrali, M., Ciccoli, L., Signorini, C., and Comporti, M., Iron release and erythrocyte damage in allyl alcohol intoxication in mice, *Biochem. Pharmacol.*, 40, 1485, 1990.
3. Berman, E., House, D. E., Allis, J. W., and Simmons, J. E., Hepatotoxic interactions of ethanol with allyl al cohol and carbon tetrachloride in rats, *J. Toxicol. Environ. Health*, 37, 161, 1992.
4. Dunlap, M. K., Kodama, J. K., Wellington, J. S., Anderson, H. H., and Hine, C. H., The toxicity of allyl alcohol. I. Acute and chronic toxicity, *AMA Arch. Ind. Health*, 18, 303, 1958.
5. Carpanini, F. M. B., Gaunt, I. F., Hardy, J., Gangolli, S. D., Butterworth, K. R., and Lloyd, H. G., Short-term toxicity of allyl alcohol in rats, *Toxicology*, 9, 29, 1978.
6. Almeev, Kh. S. and Karmasin, V. E., in *Environmental Factors and Their Significance for the Health of Population*, Zdorov'ya, Kiyv, Issue No 1, 1969, 31 (in Russian).
7. Roschlau, G. and Rodenkirchen, H., Histological examination of the diaplacental action of carbon tetrachloride and allyl alcohol in mice embryos, *Exp. Pathol.*, 3, 255, 1969.
8. Jenkinson, P. C. and Anderson, D., Malformed fetuses and karyotype abnormalities in the offspring of cyclophosphamide and allyl alcohol-treated male rats, *Mutat. Res.*, 229, 173, 1990.
9. Slott, V. L. and Hales, B. F., Teratogenicity and embryolethality of acrolein and structurally related compounds in rats, *Teratology*, 32, 65, 1985.
10. Smith, R. A., Cohen, S. M., and Lawson, T. A., Acrolein mutagenicity in the V79 assay, *Carcinogenesis*, 11, 497, 1990.
11. Ohno, Y., Ormstad, K., Ross, D., et al., Mechanism of allyl alcohol toxicity and protective effects of low-molecular-weight thiols studied with isolated rat hepatocytes, *Toxicol. Appl. Pharmacol.*, 78, 169, 1985.
12. Atzori, L., Dore, M., and Congiu, L., Aspects of allyl alcohol toxicity, *Drug Metabol. Drug Interact.*, 7, 295, 1989.

(ALLYLOXY)PROPANEDIOL

Molecular Formula. $C_6H_{12}O_3$
M = 132.16
CAS No 25136-53-2
RTECS No TY2680000

Synonyms. Glycerol allyl ether; (2-Propenyloxy)propanediol.

Properties. Liquid.

Applications. Used as a plasticizer and a solvent

Acute Toxicity. In mice, LD_{50} is 1.75 g/kg BW. All animals died within 24 hours.

Repeated Exposure failed to reveal cumulative properties.

Reference:

Mel'nikova, L. V., Toxicity and hazard of several chemicals, *Gig. Sanit.*, 1, 94, 1981 (in Russian).

2-AMINOETHANOL

Molecular Formula. C_2H_7NO
M = 61.09
CAS No 141-43-5
RTECS No KJ5775000
Abbreviation. AE.

Synonyms and **Trade Names.** Colamine; Ethanolamine; β-Hydroxyethylamine; MEA inhibitor; Monoethanolamine.

Properties. AE occurs as a colorless, oily liquid having a faint odor of ammonia. Miscible with water and alcohol at any ratio. Odor perception threshold is 625 mg/l[1] or even 20 g/l;[02] taste perception threshold is 700 mg/l. A 15 mg/l concentration of AE in water causes a burning sensation in the mouth.[1]

Applications. Used as a solvent and in the production of dyes.

Exposure. AE is naturally formed in mammals from serine and is a normal constituent of mammalian urine.[03]

Acute Toxicity. LD_{50} is 2.05 g/kg BW in rats, 1.47 g/kg BW in mice, 1.0 g/kg BW in rabbits, and 0.82 g/kg BW in guinea pigs.[2] Poisoned animals displayed labored breathing, motor excitation, and convulsions.

Repeated Exposure. Cumulative properties are not very pronounced. K_{acc} is 5.5. Rats and mice tolerated a total dose of 4.0 g/kg BW administered for 2 months.[3]

Short-term Toxicity. In a 90-day feeding study, dose of 0.64 g/kg BW altered liver or kidney weight. Administration of 1.28 g/kg BW dose caused histological changes with subsequent death of treated animals. The NOEL of 0.32 g/kg BW was identified in this study.[03]

When inhaled, AE caused CNS excitation probably due to acetylcholine inhibition.

Reproductive Toxicity.

Teratogenicity. Malformations and intrauterine growth retardation were more frequent in male than in female offspring at all dose levels.[4]

Embryotoxicity. Pregnant Long Evans rats were exposed to 50, 300, and 500 mg/kg BW oral doses of AE during the critical period of organogenesis. This treatment caused a dose-dependent increase in the intrauterine death.[4]

Pregnant Wistar rats were administered AE as an aqueous solution by gavage at dose levels of 40, 120, and 450 mg/kg BW on days 6 through 15 of gestation. Despite the maternal effects observed at 450 mg/kg BW, no significant fetal effects were observed at this or any dose level tested, nor were there any indications of a treat ment-related effect on postnatal growth or on the viability of offspring.[5]

Mutagenicity.

In vitro genotoxicity. AE demonstrated direct cytotoxicity without cell transformation in Chinese hamster embryonic cells *in vitro.*[6] It was negative in *Salmonella* mutagenicity assay (Hedenstedt, 1973).

Chemobiokinetics. AE metabolism occurs via partial deamination in the body to form *ethylene glycol*, which is then partially oxidized to *oxalic acid.* AE is also methylated to *choline* and converted to *serine* and *glycine. Monomethylaminoethanol* and *dimethylaminoethanol* are intermediates in this conversion to *choline.* Elimination of AE, ingested by rabbits, occurs with the urine at an amount of up to 40%.[03]

Regulations. ***U.S. FDA*** (1998) approved the use of AE (1) in adhesives as a component of articles intended for use in packaging, transporting, or holding food in accordance with the conditions prescribed by 21 CFR part 175.105; (2) as an adjuvant in the preparation of slimicides in the manufacture of paper and paperboard that may be safely used in contact with food in accordance with the conditions prescribed in 21 CFR part 176.300; (3) as a defoaming agent that may be safely used in the manufacture of paper and paperboard intended for use in producing, manufacturing, packing, transporting, or holding food in accordance with the conditions prescribed in 21 CFR part 176.210, (4) as an adjuvant substance in the manufacture of foamed plastics intended for use in contact with food, subject to the provisions prescribed in 21 CFR 178.3010; and (5) in the production of animal glue which may be safely used as a component of articles intended for use in producing, manufacturing, packaging, processing, preparing, treating, packing, transporting, or holding food in accordance with the conditions prescribed in 21 CFR part 178.3120.

Standards. ***Russia*** (1995). MAC and PML: 0.5 mg/l.

References:

1. Rodionova, L. F., Experimental data on substantiation of maximum allowable concentration for ethanolamine in water bodies, *Gig. Sanit.*, 2, 9, 1964 (in Russian).
2. Sidorov, K. K. and Timofievskaya, L. A., Data on substantiation of MAC for monoethanolamine in the workplace air, *Gig. Truda Prof. Zabol.*, 9, 55, 1979 (in Russian).

3. Gurfein, L. N. et al., in *Proc. Final Sci. Conf. Leningrad Sanitary Hygiene Medical Institute*, Leningrad, 1956, 21 (in Russian).
4. Mankes, R. F., Studies on the embryotoxic effect of etanolamine embryopathy in pups contiguous with male siblings *in utero, Teratogen. Carginogen. Mutagen.*, 6, 403, 1986.
5. Hellwig, J. and Liberacki, A. B., Evaluation of the pre-, peri-, and postnatal toxicity of monoethanolamine in rats following repeated oral administration during organogenesis *Fundam. Appl. Toxi col.*, 40, 158, 1997.
6. Inone, K., Sunakawa, T., Okoto, K., et al., Mutagenicity tests and *in vitro* transformation assays on trietanolamine, *Mutat. Res.*, 101, 305, 1982.

BEF SOLVENT

Composition. Mixture of butyl acetate, butanol, butyl butyrate, butyl propionate and isoamyl acetate.

Properties. Liquid with a specific odor. Solubility in water is low.

Applications. Used as a solvent in the production of nitrocellulose, polyester varnishes and enamels, and various resins.

Acute Toxicity. LD_{50} is found to be 5.5 g/kg BW in rats and 4.8 g/kg BW in mice. Lethal doses led to the short period of excitation with signs of severe mucosa irritation, profuse nasal discharge and labored breathing, and subsequent CNS inhibition and lethargy. Death within 1 to 2 days.

Repeated Exposure. Main changes were observed in oxidation-reduction processes. The treatment affected the CNS, liver, and kidney functions.

Allergenic Effect. Not found.

Mutagenic was observed at generally toxic dose levels.

Reference:

Bashkirtsev, A. S., Semenova, V. V., and Karelin, A. O., Toxicity of AKR and BEF solvents, *Gig. Sanit.*, 12, 86, 1987 (in Russian).

BENZENE

Molecular Formula. C_6H_6
M = 78.12
CAS No 71-43-2
RTECS No CY1400000

Synonyms and **Trade Names.** Benzol; Pyrobenzol; Cyclohexatriene; Coal naphtha, phenyl hydrate.

Properties. Clear, colorless liquid or rhombic prisms with a characteristic aromatic odor. Solubility in water is 1.8 g/l at 25°C and is miscible with alcohol and oils. Taste threshold in water is 0.5 to 4.5 mg/l (*EPA*, 1975). Odor threshold is reported to be 10 mg/l.[018]

Applications. Used as raw material in phenolic resins, nylon, polyester resins, polystyrene plastics, and synthetic rubber.

Exposure. B. is a chemical of large production volume and widespread human exposure. Along with other light, high-octane aromatic hydrocarbons, it is a component of motor gasoline. Benzene levels in commercial polymeric products examined were from non-detected to 426 ppb. In three commercial paraffin waxes examined, concentration of 16 to 73 ppb were determined.[1] B. is also present in cigarette smoke. Nevertheless, its use as a solvent has been greatly reduced in recent years. A daily intake from food was estimated to be 0.18 mg. Concentration of B. in the blood in the general population was found to be 262 ng/l (Brugnone et al., 1994).

Migration Data. 29 and 64 mg B./kg were found in two samples of thermoset polyester compounded for the manufacture of plastic cookware. It was established that B. originated from the use of *tert*-butylperbenzoate used as an initiator in the manufacture of the polymer. Migration levels of B. were found to be 1.9 mg and 5.6 mg B./kg in olive oil after extraction for 1 hour at 175°C. Concentration of B. in thermoset polyester cookware purchased from retail outlets were 0.3 to 84.7 mg/kg. Low amounts of B. (less than 0.01 to 0.09 mg/kg) were detected in foods when the articles were used for cooking in microwave or conventional ovens. Migration of B. was found below 0.1 mg/kg, with the highest amounts, migrating from polystyrene to polyvinyl chloride, being 0.2 to 1.7 mg/kg, predominantly in articles of expanded polystyrene.[2]

Migration levels into olive oil at 175°C for samples produced with non-aromatic initiator were less than 0.1 mg/kg. Migration of 5.0 to 50 μg B./dm^2 has been detected from 7/26 samples of retail nonstick cookware covered with polytetrafluoroethylene coatings. B. in a number of these samples was attributed to the use of phenylmethyl silicone ingredient containing 360 mg B./kg.[3]

Migration of B. from PET sheets into food into 8.0% ethanol/water and *n*-heptane was measured. At very high residual concentration of B. (218 mg/kg) in sheets made from unwashed PET, higher amounts of the contaminant migrated into the food simulants (more than 10 μg/kg were found). The results suggest that unwashed recycled PET may not comply with FDA requirements.[4]

Acute Toxicity. In mice and rats, LD_{50} is 1.0 to 10 g/kg BW.[5]

According to Kimura et al., LD_{50}s are reported to be 3.4, 3.8, and 5.6 mg/kg BW in 14-day old rats, young adults rats, and older adults, respectively.[6]

Repeated Exposure damaged blood and blood-forming tissues.[5]

Short-term Toxicity. F344 rats B6C3F_1 mice were gavaged with the doses up to 600 mg/kg in corn oil for 17 weeks. The treatment caused no mortality, but a decrease in final BW in rats and tremor in higher-dosed mice was noted. Clinical pathology examination revealed lymphoid depletion in rats and leukopenia in mice.[7]

Long-term Toxicity.

Observations in man. B. has been linked to nonneoplastic and neoplastic hematopoietic effects, particularly the myelogenous cell types, in humans exposed to B. by inhalation. Hematotoxic effects are well documented and include aplastic anemia and pancytopenia. Davidson et al. suggest that oral exposure to B. is associated with a low hazard to humans because the average intake of B. from drinking water is 0.2 μg/day.[8]

Animal studies. B. was administered to rats and mice by gavage in corn oil for 103 weeks. F344 rats received 50, 100, or 200 mg/kg BW; B6C3F_1 mice were given 25, 50, or 100 mg/kg BW. The treatment caused hematological effects including lymphoid depletion of the splenic follicles (in rats of both sexes) and thymus (in male rats), and bone marrow hematopoietic hyperplasia (in mice). Lymphocytopenia and leukocytopenia in rats and mice were noted even at the lowest dose.[7,9]

Reproductive Toxicity.

Gonadotoxicity.

Observations in man. Paternal exposure to solvents used in petroleum refineries (i.e., B. and gasoline) was reported to be associated with an increased risk for spontaneous abortions.[10] Imputed paternal exposure to B. was also associated with a moderate, but significant, increased risk of delivery of a small-for-gestational age infant, but not of still-birth or preterm delivery.[11]

Animal studies. Testicular atrophy, decreased spermatozoa number, and increased abnormal forms were found in mice but not in rats exposed to the concentration of 300 ppm for 13 weeks.[12]

Embryotoxicity. 1.6 mmol of B. was added to whole rat embryo cultures without any adverse subsequent effect.[13] When mice were gavaged with B., the dose as little as 0.3 ml/kg BW caused embryolethality.

No ***teratogenic effect*** was seen at 1.0 mg/kg BW dose level.[14] B. induced embryo- and fetotoxicity at maternal toxic dose level as low as 30 mg/kg BW.[9,15]

Mutagenicity.

In vitro genotoxicity. B. does not induce mutations or DNA damage in several standard bacterial systems and in the mouse lymphoma cell forward mutation assay. B. has been shown to be capable of interfering with chromosome segregation and inducing chromosomal breakage. This may play a role in carcinogenic effects of B.[16]

In vivo cytogenetics. B. caused CA in a variety of species *in vivo.*[5] Micronucleus induction in peripheral blood was examined during carcinogenicity assays. Micronuclei were increased in polychromatic and normochromatic erythrocytes after 14 weeks of oral treatment with 50 and 100 mg B./kg BW.[17]

Cirrani et al. indicated that B. exhibited capability to induce micronuclei in bone marrow cells of pregnant mice and, transplacentally, in fetal liver cells.[18]

B. does not induce mutations or DNA damage in *Dr. melanogaster.*

Carcinogenicity. B. is a genotoxic procarcinogen.

Observations in man. Epidemiological studies and several case studies proved that exposure to B. at high concentrations may eventually result in leukemia.

Animal studies. Rodents exposed to B. developed neoplasms at numerous sites including Zymbal's gland, Harderian gland, preputial gland, oral cavity, lungs, hematopoietic system, skin, mammary gland, and ovary.

Sprague-Dawley rats were administered 50, 250, and 500 mg/kg BW by gavage for life-span. The treatment produced dose-dependent increase in the incidence of mammary tumors in females and of Zymbal gland carci nomas, oral cavity carcinomas, and leukemias/lymphomas.[19,20]

Schlosser et al.[21] noted that mice are more sensitive than rats to B. toxicity, though neither species has been shown to respond consistently with B.-induced leukemia. In the above described study,[7] B. was found to increase the incidences of Zymbal gland carcinomas, squamous cell papillomas and squamous cell carcinomas of the oral cavity, malignant lymphomas, carcinomas and carcinosarcomas of the mammary gland. Dose-related lymphocytopenia was also observed in the animals of all species and sexes.[7]

Carcinogenicity classifications. An IARC Working Group concluded that there is sufficient evidence for the carcinogenicity of B. in *experimental animals* and in *humans*.

The Annual Report of Carcinogens issued by the U.S. Department of Health and Human Services (1998) defines B. to be a known carcinogen, i.e., a substance for which the evidence from human studies indicates that there is a casual relationship between exposure to the substance and human cancer.

IARC: 1;

U.S. EPA: A;

EU: 1;

NTP: CE* - CE* - CE* - CE* (gavage).

Chemobiokinetics. B. is completely absorbed in the GI tract and widely distributed throughout the body. Following absorption, B. is metabolized to a large number of intermediates. It undergoes metabolic transformations in rats to *phenol*, *hydroquinone, catechol, hydroxyhydroquinone*, and *phenyl mercapturic acid* which are responsible for B. hematotoxicity. Phenolic metabolites were excreted in conjugated form.[22] Enzymes implicated in the metabolic activation of B. and its metabolites include cytochrome *P-450* monooxygenases and myeloper oxidase.

Distribution of B. and its metabolites in the body tissues and bone marrow depends on the balance of activation processes, such as enzymatic oxidation and deactivation processes, like conjugation and excretion.[23]

Regulations. ***U.S. FDA*** (1998) approved B. as a constituent of food-contact adhesives which may be safely used as a component of articles intended for use in packaging, transporting, or holding food in accordance with the conditions prescribed by 21 CFR part 175.105.

Recommendations. ***WHO*** (1996). Guideline value: 0.01 mg/l.

Standards.

U.S. EPA (1999). MCL: 0.005 mg/l.

Canada (1989). MAC: 0.005 mg/l.

Russia (1995). MAC: 0.5 mg/l.

References:

1. Varner, S. L., Hollifield, H. C., and Andrzejewski, D., Determination of benzene in polypropylene food-packaging materials and food-contact paraffin waxes, *J. Assoc., Off. Anal. Chem.*, 74, 367, 1991.
2. Jickells, S. M., Crews, C., Castle, L., and Gilbert, J., Headspace analysis of benzene in food contact materials and its migration into foods from plastics cookware, *Food Addit. Contam.*, 7, 197, 1990.
3. Jickells, S. M., Philo, M. R., Gilbert, J., and Castle, L., Gas chromatographic/mass spectrometric determination of benzene in nonstick cookware and microwave susceptors and its migration into foods on cooking, *J. AOAC Int.*, 76, 760, 1993.
4. Komolprasert, V., Lawson, A. R., and Begley, T. H., Migration of residual contaminants from secondary recycled poly(ethylene terephthalate) into food-simulating solvents, aqueous ethanol and heptane, *Food. Addit. Contam.*, 14, 491, 1997.

5. *Benzene*, Integrated criteria document. W. Sloff, Ed., Natl. Institute Publ. Health Environ. Protect, Bilthoven, The Netherlands, 1988.
6. Kimura, E. T., Ebert, D. M., and Dodge, P.W., Acute toxicity and limits of solvents residue for sixteen organic solvents, *Toxicol. Appl. Pharmacol.*, 19, 699, 1971.
7. *Toxicology and Carcinogenesis Studies of Benzene in F344/N Rats and B6C3F$_1$ Mice (Gavage Studies)*, NTP Technical Report Series No 289, Research Triangle Park, NC, DHHS, 1986.
8. Davidson, K. A. and Faust, R. A., Oral carcinogen potency factors for benzene and some of its metabolites, *Toxicologist*, Abstracts of SOT 1996 Annual Meeting, Abstract No 585, Issue of *Fundam. Appl. Toxicol.*, 30, Part 2, March 1996.
9. Huff, J. E., Haseman, J. K., DeMarini, D. M., Eustis, S., Maronpot, R. R., Peters, A. C., Persing, R. L., Chrisp, C. E., and Jacobs, A. C., Multiple-site carcinogenicity of benzene in Fischer 344 rats and B6C3F$_1$ mice, *Environ. Health Perspect.*, 82, 125, 1989.
10. Lindbohm, M.-L., Hemminki, K., Bonhomme, M. G., Anttila, A., Rantala, K., Heikkila, P., and Rosenberg, M. J., Effect of paternal occupational exposure on spontaneous abortions, *Am. J. Ind. Med.*, 81, 1029, 1991.
11. Savitz, D. A., Whelau, S. A., and Klecker, R. C., Effect of parents' occupational exposure on risk of stillbirth, pattern of delivery and small-for-gestational age infant, *Am. J. Epidemiol.*, 129, 1201, 1989.
12. Ward, C. O., Kuna, R. A., Synder, N. K., Alsaker, R. D., Coate, W. B., and Craig, P. H., Subchronic inhalation toxicity of benzene in rats and mice, *Am. J. Ind. Med.*, 7, 457, 1985.
13. Chapman, D. E., Namkung, M. J., and Juchau, M. R., Benzene and benzene metabolites as embryotoxic agents: Effects on cultured rat embryos, *Toxicol. Appl. Pharmacol.*, 128, 129, 1994.
14. Nawrot, P. S. and Staples, R. E., Embryo fetal toxicity and teratogenicity of benzene and toluene in the mouse, Abstract, *Teratology*, 19, 41A, 1979.
15. *Toxicological Profile for Benzene*, U.S. DHHS, Agency for Toxic Substances and Disease Registry, 1989.
16. Eastmond, D. A., Induction of micronuclei and aneuploidy by the quinone-forming agents benzene and *o*-phenylphenol, *Toxicol. Lett.*, 67, 105, 1993.
17. Armstrong, M. J. and Galloway, S. M., Micronuclei induced in peripheral blood of E mu-PIM-l transgenic mice by chronic oral treatment with 2-acetylaminofluorene or benzene but not with diethylnitrosoamine or 1,2-dichloroethane, *Mutat. Res.*, 302, 61,1993.
18. Cirrani, R., Barale, R., Marrazzini, A., and Loprieno, N., Benzene and the genotoxicity of its metabolites. I. Transplacental activity in mouse fetuses and in their dams, *Mutat. Res.*, 208, 61, 1988.
19. Maltoni, C. and Scarnato, C., First experimental demonstration of the carcinogenic effects of benzene. Long-term bioassays on Sprague-Dawley rats by oral administration, *Med. Lavoro*, 70, 352, 1979 (in Italian).
20. Maltoni, C., Conti, B., and Cotti, G., Benzene: A multipotential carcinogen. Results of long-term bioassays performed at the Bologna Institute of Oncology, *Am. J. Ind. Med.*, 4, 589, 1983.
21. Schlosser, P. M., Bond, J. A., and Medinsky, M. A., Benzene and phenol metabolism by mouse and rat liver microsome, *Carcinogenesis*, 14, 2477, 1993.
22. *Drinking Water & Health*, vol 1, Natl. Research Council, Washington, D. C., Natl. Academy Press, 1977, 688.
23. Medinsky, M. A., Schlosser, P. M., and Bond, J. A., Critical issue in benzene toxicity and metabolism: the effect of interactions with other organic chemicals on risk assessment, *Environ. Health Perspect.,* 102 (Suppl. 9), 119, 1994.

BENZINE

CAS No 8030-30-6
RTECS No SE7555000
DE3030000

Synonyms and **Trade Names.** Coal-tar naphtha; Light ligroin; Naphtha; Petroleum benzine; Petroleum naphtha; Petroleum spirit.

Composition. B. is a complex mixture of distilled hydrocarbons, generally in C_6 to C_{10} range.[03] B. derived from coal contains aromatic hydrocarbons including benzene. N. derived from petroleum source contains olefins and paraffins, and *n*-hexane.

Properties. Clear, colorless, highly flammable liquid with a gasoline odor. Insoluble in water, miscible with alcohol.

Applications. Used as a paint solvent and a degreasing agent.

Repeated Exposure. Rats and mice received 1.0% B. in their feed and 0.1% B. in their drinking water for 1 month. Some specific reactions to the treatment were observed including an increase in intensity of metabolism and energy exchange, and changes in hemopoietic function.[1]

In gavage studies, male rats were exposed to 20 doses of B. hydrocarbon compounds (typically found in unleaded gasoline) administered once daily for a 4-week period. It was suggested that alkane (paraffin) components caused nephrotoxic effect that correlated well with the proportion of branched alkanes contained in each.[2]

Reproductive Toxicity.

Embryotoxicity. Presumed-pregnant Sprague-Dawley rats were exposed to 2150 or 7660 mg light catalytically cracked B. vapors/m^3 for 6 hours/day on days 0 to 19 of gestation. The number of resorptions was increased by ~140% in the group receiving 7660 mg/m^3. There were no other definitive treatment-related changes, indicating minimal toxic effect.[3]

Gonadotoxicity. In a 13-week study in mice and rats, a marginal decrease was noted in the number of sperm per gram of epididymis.[3]

Mutagenicity.

In vitro genotoxicity. Low-boiling B. was not mutagenic in *Salmonella* mutagenicity assay (Calkins et al., 1983).

Carcinogenicity. Coal-derived but not petroleum-derived B. was shown to be carcinogenic on skin application in mice (Witschi et al., 1986).

Regulations. ***U.S. FDA*** (1998) approved the use of B. as (1) a component of a defoaming agent intended for articles which may be used in producing, manufacturing, packing, processing, preparing, treating, packaging, transporting, or holding food in accordance with the conditions prescribed in 21 CFR part 176.200; (2) as a defoaming agent in the manufacture of paper and paperboard intended for use in packaging, transporting, or holding food in accordance with the conditions prescribed by 21 CFR part 176.210. Petroleum B. (refined) may be safely used (32) in food in accordance with the conditions prescribed in 21 CFR part 172.250.

Standards. ***Russia*** (1995). MAC: 0.3 mg/l (organolept., film).

References:

1. Gashev, S. N., Elifanov, A. V., Solov'yov, V. S., and Gasheva, N. A., Impact of crude naphtha on morphological and functional parameters of mice and rats, *Bull. Moscow Soc. Natur. Invest.*, Biol. Dept, Issue No 99, 1994, 23 (in Russian).
2. Halder, C. A., Holdsworth, C. E., Cockrell, B. Y., and Piccirillo, V. J., Hydrocarbon nephropathy in male rats: identification of the nephrotoxic components of unleaded gasoline, *J. Toxicol. Ind. Health*, 1, 67, 1985.
3. Dalbey, W. E., Feuston, M. H., Yang, J. J., Kommineni, C. V., and Roy, T. A., Light catalytically cracked napntha: subchronic toxicity of vapors in rats and mice and developmental toxicity screen in rats, *J. Toxicol. Environ. Health*, 47, 77, 1996.

BENZYL ALCOHOL

Molecular Formula. C_7H_8O

M = 108.13

CAS No 100-51-5

RTECS No DN3150000

Abbreviation. BA.

Synonyms. Benzenemethanol; Benzenecarbinol; α-Hydroxytoluene; Phenylcarbinol; Phenylmethanol; Phenylmethyl alcohol.

Properties. Colorless, transparent liquid with a faint aromatic odor and a sharp burning taste. Water solubility is 40 g/l, readily soluble in organic solvents.

Applications. Used as a solvent in the production of cellulose acetate.

Acute Toxicity. LD_{50} is reported to be 2.0 g/kg BW in male rats, 1.66 g/kg BW in female rats, 1.36 g/kg BW in mice, 2.5 g/kg BW in guinea pigs, and 0.1 g/kg BW in rabbits. Administration of high doses caused NS inhibition. After oral administration of LD_{50}, 50% of rats died in 1 to 2 days, mice in 2 to 3 days.[1]

ET_{50} is 90 min. Acute action threshold is 0.1 mg/kg BW. BA was administered *i/p* to adult (23 to 28 g) and neonatal (2 to 7 g) CD-1 male mice. Doses less than 800 mg/kg BW produced minimal toxic effects within an initial 4-hour observation period. At the end of this time, the LD_{50} was determined to be 1.0 g/kg BW for both age groups. LD_{50} on day 7 was 650 mg/kg BW. Acute toxicity manifestations of BA alcohol include sedation, dyspnea, and loss of motor function and is due to the alcohol itself and not to its metabolite.[2]

Repeated Exposure revealed pronounced cumulative properties. K_{acc} is 2.0 to 3.0. Rats were administered BA at dose levels of 20 to 200 mg/kg BW. Manifestations of the toxic action included retardation of BW gain, an increase in STI, and behavioral changes. Some animals died following repeated intake of 100 mg/kg BW and 200 mg BA/kg BW. Histology examination revealed hyperplasia in the spleen and peribronchial lymphatic follicles. Dose-dependence was noted. Calculated approximate NOEL appeared to be 100 mg/kg BW.[1,3]

Short-term Toxicity. Rats and mice received, by gavage, doses of 50 to 800 mg BA/kg BW for 13 weeks. The treatment at high doses caused mortality, signs of neurotoxicity and hemorrhages, reduction in relative BW gain and histological lesions in the brain, thymus, skeletal muscle, and kidney.[4]

Long-term Exposure. Animals were administered 200 or 400 mg BA/kg BW (F344 rats) and 100 or 200 mg BA/kg BW ($B6C3F_1$ mice) in corn oil by gavage for 103 weeks. The treatment did not affect survival in male rats, and no effect on BW gain was noted. There were no apparent BA-related non-neoplastic responses. Adenomas of the adrenal cortex occurred at an increased incidence in high-dose male mice, but this slight increase was not considered to be related to chemical exposure.[4]

Reproductive Toxicity.

Embryotoxicity. BA could cause neurological abnormalities and death in neonates because of its use in intensive care of premature babies.[5]

Teratogenicity. After injection of 0.01 ml or 0.02 ml BA into the yolk sac of the chick from before incubation up to the 7th day, meningoceles and skeletal defects were observed.[6]

Mutagenicity.

In vitro genotoxicity. BA was not mutagenic in *Salmonella* strains *TA98, TA100, TA1535*, or *TA1537* in the presence or absence of metabolic activation,[4] but it caused SCE in cultured mammalian cells.[016] It showed a weakly positive response in DNA degradation test in primary rat hepatocytes in dose-dependent manner.[7]

Carcinogenicity. BA caused no carcinogenic response in rats and mice.[8]

Carcinogenicity classification.

NTP: NE - NE - NE - NE (gavage).

Chemobiokinetics. Rapid absorption and conversion of BA to its primary metabolite, *benzaldehyde,* has been reported.[2] BA undergoes *in vivo* oxidation to *benzoic acid* in man and rabbits.[8]

Regulations.

U.S. FDA (1998) approved the use of BA (1) in adhesives used as components of articles intended for use in packaging, transporting, or holding food in accordance with the conditions prescribed by 21 CFR part 175.105; (2) in resinous and polymeric coatings used as the food-contact surfaces of articles intended for use in producing, manufacturing, packing, processing, preparing, treating, packaging, transporting, or holding food for use only in coatings at a level not to exceed 4% by weight of the resin when such coatings are intended for repeated use in contact with food only of the types identified in 175.300; and (3) in the manufacture of closures with sealing gaskets on containers intended for use in producing, manufacturing, packing, processing, preparing, treating, packaging, transporting, or holding food at levels not to exceed 1.0% by weight of the closure-sealing gasket composition in accordance with the conditions prescribed by 21 CFR part 177.1210.

EU (1990). BA is available in the *List of authorized monomers and other starting substances which shall be used for the manufacture of plastic materials and articles intended to come into contact with foodstuffs (Section A).*

Great Britain (1998). BA is authorized without time limit for use in the production of polymeric materials and articles in contact with food or drink or intended for such contact.

Recommendations. Joint FAO/WHO Expert Committee on Food Additives (1997). ADI: 5.0 mg/kg BW.[9]

References:

1. Rumyantsev, G. I., Novikov, S. M., Kozeeva, E. E., et al., Experimental studies of biological effects of benzyl alcohol, *Gig. Sanit.*, 7, 81, 1985 (in Russian).
2. McCloskey, S. E., Gershanik, J. J., Lertora, J. J., White, L., and George, J., Toxicity of benzyl alcohol in adult and neonatal mice, *J. Pharmacol. Sci.*, 75, 702, 1986.
3. Rumyantsev, G. I., Novikov, S. M., Fursova, T. N., et al., Experimental study of toxicity of phenylethyl alcohol and phenylethyl acetate, *Gig. Sanit.*, 10, 83, 1987 (in Russian).
4. *Toxicology and Carcinogenesis Studies of Benzyl Alcohol in F344/N Rats and B6C3F$_1$ mice (Gavage Studies)*, NTP Technical Report Series No 343, Research Triangle Park, NC, June 1989.
5. Menon, P. A. et al., Benzyl alcohol toxicity in a neonatal intensive care unit. Incidence, symptomatology, and mortality, *Am. J. Perinatol.*, 1, 288, 1984.
6. Duraiswami, P. K., Experimental teratogenesis with benzyl alcohol, *John Hopkins Hospital Bull.*, 95, 57, 1954.
7. Elia, M. C., Storer, R. D., McKelvey, T. W., et al., Rapid DNA degradation in primary rat hepatocytes treated with diverse cytotoxic chemicals: Analysis by pulsed field gel electrophoresis and implications for al kaline elution assays, *Environ. Molec. Mutagen.*, 24, 181, 1994.
8. The 23 Report of the JECFA, Technical Report Series No 648, WHO, Geneva, 1980, 15.
9. *Evaluation of Certain Food Additives and Contaminants*, WHO Techn. Report Series No 868, 46th Report of Joint FAO/WHO Expert Committee on Food Additives, Geneva, 1997, 69.

BIPHENYL

Molecular Formula. $C_{12}H_{10}$
M = 154.22
CAS No 92-52-4
RTECS No DU8050000
Abbreviation. BP.

Synonyms. Bibenzene; Diphenyl; Phenylbenzene.

Properties. BP occurs as colorless to white crystals, crystalline lumps or white crystalline powder with a characteristic odor.

Applications. Used as a solvent in the synthesis of perfumes and plasticizers. A fungicide.

Migration of BP was noted in a study with *in vitro* specimens made of orthodontic resin.[1]

Acute Toxicity. LD_{50} is 3.28 g/kg BW in rats, 2.4 g/kg BW in rabbits,[014] and exceeded 2.6 g/kg BW in cats.[2] Ingestion of large doses increased rate of respiration, induced lacrimation, anorexia and BW loss, muscular weakness, ataxia with death in coma during a period up to 18 days. Gross pathology examination revealed visceral congestion, myocarditis, hepatitis, nephritis, pneumonia.[019]

Repeated exposure. Rats were given 0.5% BP in the diet for 2 months. The treatment caused polyuria and reversible kidney lesions (focal tubular dilatation).[019]

Long-term Toxicity. Weanling albino rats received 0.001 to 1.0% BP in the diet. Treatment with the dietary concentrations of 0.5 and 1.0% (0.5 to 500 mg/kg BW) reduced *Hb* levels, decreased food consumption and longevity. Histological examination revealed kidney damage, including irregular scarring, lymphocytic infiltration, tubular atrophy, and patchy tubular dilation.[3]

Reproductive Toxicity. Rats were dosed with 125 to 500 mg/kg BW by esophageal intubation on days 6 to 15 of gestation. There were neither ***teratogenic effects*** nor maternal toxicity signs at these dose levels. The dose of 1000 mg/kg BW produced fetal and maternal toxicity.[4]

Developmental defects and mitotic abnormalities were observed in sea urchins.[5]

Embryotoxicity. No adverse effects were observed in weanling rats given up to 0.5% B. in the diet from 60 days before mating through weaning of their offspring.[3]

Mutagenicity.

In vitro genotoxicity. BP showed no mutagenic potential detected in *Salmonella typhimurium* strains *TA 98* and *TA 100* and in Chinese hamster *V79* cells (latter in the absence but not in the presence of metabolic activation).[6] It showed genetic effect in yeast with and without activating system[4] but did not produce an increase in unscheduled DNA synthesis in rat hepatocytes.[7]

Carcinogenicity. Negative results are reported in mice.[09]

Chemobiokinetics. BP is metabolized by activated liver microsomes to its hydroxy derivatives.[8] Following ingestion by rabbits, the main metabolites are *2-hydropxy-, 4-hydroxy-, 3,4-hydroxy,* and *4,4'-hydroxy biphenyl.* Three other phenolic metabolites were present but not identified (Menzie, 1974).

Recommendations. Joint FAO/WHO Expert Committee on Food Additives. ADI: 0.05 mg/kg BW.

Standards. *EPA, State of Minnesota.* MCL: 0.3 mg/l.

References:

1. Lygre, H., Klepp, K. N., Solheim, E., and Gjerdet, N. R., Leaching of additives and degradable products from cold-cured orthodontic resins, *Acta Odontol. Scand.*, 52, 150, 1994 (in Norwegian).
2. *The Pesticide Manual*, 10 ed., C. Tomlin, Ed., Crop Protect. Publ., The Royal Soc. Chem., 1995, 1341.
3. Ambrose, A. M., Booth, A. N., deEds, F., and Cox, A. J., A toxicological study of biphenyl, a citrus fungistat, *Food Res.*, 25, 328, 1960.
4. Khera, K. S., Whalen, C., Angers, G., and Trivett, G., Assessment of the teratogenic potential of piperonyl butoxide, biphenyl, and phosalone in the rat, *Toxicol. Appl. Pharmacol.*, 47, 353, 1979.
5. Pogano, G., Esposito, A., Giacomo, G., et al., Genotoxicity and teratogenicity of diphenyl and diphenyl ether: A study of sea urchins, yeast and *Salmonella typhimurium*, *Teratogen. Carcinogen. Mutagen.*, 3, 377, 1983.
6. Glatt, H., Anklam, E., and Robertson, L. M., Biphenyl and fluorinated derivatives: liver enzyme-mediated mutagenicity detected in *Salmonella typhimurium* and Chinese hamster V79 cells, *Mutat. Res.*, 281, 151, 1992.
7. Abe, S. and Sasaki, M., Chromosomal aberrations and sister chromatid exchanges in Chinese hamster cells exposed to various chemicals, *J. Natl. Cancer Inst.*, 58, 1635, 1977.
8. Burke, M. D. and Prough, R. A., in S. Fleischer and L. Packer, *Methods of Enzymology,* Acad. Press, New York, 52, 319, 1978.

4,4'-BIPHENYLDIOL

Molecular Formula. $C_{12}H_{10}O_2$

M = 186.22

CAS No 92-88-6

RTECS No DV4725000

Synonyms. *p,p*'-Biphenol; 4,4'-Dihydroxydiphenyl.

Acute Toxicity. LD_{50} is 9.85 g/kg BW in rats.[1]

Mutagenicity.

In vivo cytogenetics. B. showed weak capability to induce micronuclei in bone marrow cells of pregnant mice and, transplacentally, in fetal liver cells.[2]

Regulations. ***EU*** (1990). B. is available in the *List of authorized monomers and other starting substances which shall be used for the manufacture of plastic materials and articles intended to come into contact with foodstuffs (Section A).*

References:

1. Carpenter, C. P., Weil, C. S., and Smyth, H. F., Range-finding toxicity data: List VIII, *Toxicol. Appl. Pharmacol.*, 28, 313, 1974.

2. Cirrani, R., Barale, R., Marrazzini, A., and Loprieno, N., Benzene and the genotoxicity of its metabolites. I. Transplacental activity in mouse fetuses and in their dams, *Mutat. Res.*, 208, 61, 1988.

BIS(2-CHLOROETHOXY)METHANE

Molecular Formula. $C_5H_{10}Cl_2O_2$
M = 173.05
CAS No 111-91-1
RTECS No PA3675000
Abbreviation. BCEM.

Synonyms. Bis(β-chloroethyl)formal; Di-2-chloroethyl formal; Formaldehyde bis(β-chloroethyl)acetal; 1,1'-[Methylenebis(oxy)]bis-(2-chloroethane).

Properties. Colorless liquid, soluble in water (7.8 g/l); miscible with alcohol.

Applications. A solvent. Used in the production of polysulfide rubber.

Acute Toxicity. LD_{50} is 65 mg/kg BW in rats[1] and 170 mg/kg BW in guinea pigs.[03]

Carcinogenicity classification.
U.S. EPA: D

Regulations. ***U.S. FDA*** (1998) approved the use of BCEM in the manufacture of polysulfide polymer-polyepoxy resins which may be used as food-contact surface of articles intended for packaging, transporting, holding, or otherwise contacting dry food in accordance with the conditions prescribed in 21 CFR part 177.1650.

Reference:

1. Smyth, H. F. and Carpenter, C. P., Further experience with the range finding test in the industrial toxicological laboratories, *J. Ind. Hyg. Toxicol.*, 30, 63, 1948.

BIS(2-CHLOROETHYL) ETHER

Molecular Formula. $C_4H_8Cl_2O$
M = 143.02
CAS No 111-44-4
RTECS No KN0875000
Abbreviation. BCE.

Synonyms and **Trade Name.** Chlorex; 2-Chloroethyl ether; 2,2'-Dichloroethyl ether; 1,1'-Oxybis(2-chloro)ethane.

Properties. Colorless, clear liquid with a pungent, sweet, chloroform-like odor. Solubility in water is 1.1 wt% at 20°C; soluble in alcohol.

Applications. A solvent.

Acute Toxicity. LD_{50} is 750 mg/kg BW in rats,[03] 136 mg/kg BW in mice, and 126 mg/kg BW in rabbits.[018]

Repeated Exposure. Rats received 13 mg/kg BW for 2 months. The treatment increased retention of bromsulfalein in the blood, increased the blood level of eosinophils and globulins, and decreased that of leukocytes and albumins.[1]

Mutagenicity.

In vitro genotoxicity. BCE appeared to be a direct-acting mutagen producing base pair exchange mutations in *E. coli, Salmonella typhimurium*, and *B. subtilis*.[2]

In vivo cytogenetics. BCE induced the sex-linked recessive lethal mutations in *Dr. melanogaster*.[3]

Carcinogenicity. BCE was not carcinogenic in male and female Charles River CD rats, given by oral administration.[4]

Norpoth et al. did not find convincing increase in tumor incidence associated with BCE treat ment.[5]

Two hybrid mouse strains were administered 100 mg/kg BW from 7 days of age. At 4 weeks, after weaning, the treatment continued through the diet (300 ppm) for up to 18 months. Increased evidence of hepatomas was reported.[6]

Carcinogenicity classification. An IARC Working Group concluded that there is limited evidence for the carcinogenicity of BCE in *experimental animals* and there are no data to evaluate the carcinogenicity of BCE in *humans*.

IARC: 3.

Standards. *EPA, State of Minnesota.* MCL: 0.0003 mg/l.

References:

1. Bakhtizina, G. Z. and Osipova, L. O., in *Proc. Research Institute Labor Hygiene Occup. Diseases*, Moscow, Issue No 5, 1969, 84 (in Russian).
2. Shirasu, Y., Moriya, M., Kato, K., and Kada, T., Mutagenicity screening of pesticides in microbial systems, *Mutat. Res.*, 31, 268, 1975.
3. Foureman, P., Mason, J. M., Valencia, R., and Zimmering, S., Chemical mutagenesis testing in Drosophila. IX. Results of 50 coded compounds tested for the National Toxicology Program, *Environ. Molec. Mutagen.*, 23, 51, 1994.
4. Weisburger, E. K., Ulland, B. M., Nam, J., Gart, J. J., and Weisburger, J. H., Carcinogenicity tests of certain environmental and industrial chemicals, *J. Natl. Cancer Inst.*, 67, 75, 1981.
5. Norpoth, K. et al., Investigations on metabolism, genotoxic effects and carcinogenicity of 2,2'-dichlorodi ethyl ether, *J. Cancer. Res. Clin. Oncol.*, 112, 125, 1986.
6. Innes, J. R. M., Ulland, B. M., Valerio, M. G., et al., Bioassay of pesticides and industrial chemicals for tumorogenicity in mice: A preliminary note, *J. Natl. Cancer Inst.*, 42, 1101, 1969.

BIS(2-METHOXYETHYL) ETHER

Molecular Formula. $C_6H_{14}O_3$

M = 134.17

CAS No 111-96-6

RTECS No KN3339000

Abbreviation. BME.

Synonyms and **Trade Name.** Diethylene glycol, dimethyl ether; Diglyme; 1,1'-Oxybis (2-methoxyethane).

Properties. Colorless liquid with mild odor miscible with water or alcohol.

Applications. Used in the production of plastics.

Acute Toxicity. LD_{50} is 5.4 g/kg BW in rats and 6.0 m/kg BW in mice. Poisoning is associated with general CNS inhibition. Ataxia and respiratory failure were noted.[1]

Reproductive Toxicity.

Embryotoxicity. Timed pregnant New Zealand white rabbits were gavaged with 25 to 175 mg BME/kg BW in distilled water during major organogenesis on gestation days 6 to 19. The dose of 175 mg/kg BW caused maternal toxicity effects and 15% mortality (4.0% in controls). The doses of 100 mg/kg BW and 175 mg/kg BW increased prenatal mortality. At dose level of 25 mg/kg BW, there were no significant signs of maternal toxicity and developmental lesions. The NOAEL of 50 mg/kg BW was identified for developmental toxicity in this study.[2]

Pregnant female CD-1 mice received eight maximum tolerated doses (2.0 g/kg BW, lethal to 10% of the mice). The treatment significantly affected reproductive indices. The treated mice did not deliver any viable pups. Maternal lethality was significant.[3]

Teratogenicity. In the above-cited study, the doses of 100 and 175 mg/kg BW increased incidence of malformed live fetuses (fusion of ribs and hydronephrosis).[2] Time-mated CD-1 mice were orally dosed on gestation day 11 with distilled water (control) or BME at a dose of 537 mg/kg BW. There were no signs of treatment-related maternal toxicity, and intrauterine survival was unaffected by the treatments. There was no treatment-related pattern of gross external malformations other than paw defects. Paw defects were present in BME-treated litters (39.7% of fetuses). Hindpaw defects predominated over forepaw, and syndactyly was the most common malformation. The incidences of oligodactyly and short digits were also significantly increased. The similarity of malformations produced by methyl-substituted glycol ethers is proposed to be attributable to *in vivo* conversion to a common teratogen, *methoxyacetic acid.*[4]

Timed-pregnant CD-1 mice were given doses of 62.5, 125, 250, or 500 mg BME/kg BW by gavage in distilled water during major organogenesis (gestational days 6 to 15). No maternal deaths, morbidity, or treatment-related clinical signs were observed. Oral administration of BME produced selective and profound adverse effects upon fetal growth, viability, and morphological development at greater than or equal to 125 mg/kg BW dose.[5]

Gonadotoxicity. Administration of up to 20 daily oral doses of 684 mg BME/kg BW caused primary and secondary spermatocyte degeneration and spermatidic growth cells after six to eight treatments. In addition, the relative testes weights were significantly reduced. Testicular lactate dehydrogenase-X activity, a pachytene spermatocyte marker enzyme, was significantly decreased.[6] The lack of toxicity of two BME metabolites, namely, *2-(2-methoxyethoxy) ethanol* and *(2-methoxyethoxy)acetic acid*, suggests that the testicular toxicity of BME may be due to *methoxyacetic acid,* a minor metabolite (see below).[7]

Mutagenicity.

In vitro genotoxicity. BME was negative in five *Salmonella typhimurium* strains *TA1535, TA1537, TA97, TA98,* and *TA100* in the presence and absence of rat and hamster liver *S9,* at doses of up to 10 mg/plate.[013]

Chemobiokinetics. Male Sprague-Dawley rats were given a single oral dose of 5.1 mmol ^{14}C-BME/kg BW. The principal metabolite in the urine was *(2-methoxyethoxy)acetic acid.* Other metabolites appeared to be *2-(2-methoxyethoxy)ethanol, methoxyacetic acid, 2-methoxyethanol*, and *diglycolic acid.*[8] Approximately 86 to 90% of the doses was excreted in the urine. Less than 5.0% of the doses was excreted in the feces. Only trace amounts of radiolabel were found in the expired air as volatile organic compounds. The principal urinary metabolites were *(2-methoxyethoxy)acetic acid* and *methoxyacetic acid.* Smaller amounts of *N-(methoxy- acetyl)glycine, diglycolic acid, 2-methoxyethanol, and 2-(2-methoxyethoxy) ethanol* were found. BME metabolism proceeds primarily through an *O*-demethylation pathway, followed by oxidation to *(2-methoxyethoxy)acetic acid.*[7,9]

References:

1. *Toksikologichesky Vestnik,* 1, 27, 1996 (in Russian).
2. Schwetz, B. A., Price, C. J., George, J. D., et al., The developmental toxicity of diethyleneglycol and triethylene glycol dimethyl ethers in rabbits, *Fundam. Appl. Toxicol.,* 19, 238, 1992.
3. Plasterer, M. R., Bradshaw, W. S., Booth, G. M., Carter, M. W., Schuler, R. L., and Hardin, B. D., Developmental toxicity of nine selected compounds following prenatal exposure in the mouse: naphthalene, *p*-nitrophenol, sodium selenite, dimethyl phthalate, ethylenethiourea, and four glycol ether derivatives, *J. Toxicol. Environ. Health,* 15, 25, 1985.
4. Hardin, B. D. and Eisenmann, C. J., Relative potency of four ethylene glycol ethers for induction of paw malformations in the CD-1 mouse, *Teratology,* 35, 321, 1987.
5. Price, C, J., Kimmelm, C, A., George, J. D., and Marr, M. C., The developmental toxicity of diethylene glycol dimethyl ether in mice, *Fundam. Appl. Toxicol.,* 8, 115, 1987.
6. Cheever, K. L., Weigel, W. W., Richards, D. E., Lal, J. B., and Plotnick, H. B., Testicular effects of bis(2-methoxyethyl) ether in the adult male rat, *Toxicol. Ind. Health.,* 5, 1099, 1989.
7. Cheever, K. L., Richards, D. E., Weigel, W. W., Lal, J. B., Dinsmore, A. M., and Daniel, F. B., Metabolism of bis(2-methoxyethyl) ether in the adult male rat: evaluation of the principal metabolite as a testicular toxicant, *Toxicol. Appl. Pharmacol.,* 94, 150, 1988.
8. Richards, D. E., Begley, K. B., DeBord, D. G., Cheever, K. L., Weigel, W. W., Tirmenstein, M. A., and Savage, R. E., Comparative metabolism of bis(2-methoxyethyl)ether in isolated rat hepatocytes and in the intact rat: effects of ethanol on *in vitro* metabolism, *Arch. Toxicol.,* 67, 531, 1993.
9. Daniel, F. B., Cheever, K. L., Begley, K. B., Richards, D. E., Weigel, W. W., and Eisenmann, C. J., Bis(2-methoxyethyl) ether: metabolism and embryonic disposition of a developmental toxicant in the pregnant CD-1 mouse, *Fundam. Appl. Toxicol.,* 16, 567, 1991.

BIS(2-VINYLOXYETHYL) ETHER

Molecular Formula. $C_8H_{14}O_3$

M = 158.22

CAS No 764-99-8
RTECS No KN3850000
Abbreviation. BVE.

Synonym. Diethylene glycol, divinyl ether.

Properties. Colorless, transparent liquid with a specific odor. Readily soluble in water. Odor perception threshold is 1.99 mg/l; taste perception threshold is slightly higher. It does not alter the transparency or color of water and forms no foam.

Applications. Used in the synthesis of polymeric materials.

Acute Toxicity. LD_{50} is 3.73 g/kg BW[1] or 6.39 g/kg BW in rats, and 2.57 g/kg BW in mice.[2] On autopsy, there was congestion in the visceral organs and dystrophic changes in the renal tubular epithelium, liver, and brain.

Repeated Exposure revealed slight cumulative properties. The liver, kidneys, and NS are the target organs for BVE toxicity. Severity and nature of the toxic effect are similar to that of *2-methyl-1,3-dioxolane* (*q. v.*). Morphological changes are not observed.[2]

Chemobiokinetics. BVE metabolism occurs via hydrolysis to form *diethylene glycol* which is then excreted with the urine.

Standards. ***Russia*** (1995). MAC and PML: 1.0 mg/l (organolept., odor).

References:

1. Smyth, H. F., Carppenter, C. P., Weil, C. S., Pozzani, U. C., and Striegel, J. A., Range-finding toxicity data, list VI, *Am. Ind. Hyg. Assoc. J.*, 25, 95, 1962.
2. Buzina, L. Z. and Rudi, F. A., Substantiation of MAC for glycol vinyl ethers in water bodies, *Gig. Sanit.*, 3, 12, 1977 (in Russian).

2-BUTANONE

Molecular Formula. C_4H_8O
M = 72.12
CAS No 78-93-3
RTECS No EL6475000

Synonyms. 3-Butanone; Methyl acetone; Methyl ethyl ketone.

Properties. Colorless liquid with an acetone odor. Water solubility is 26.8%; miscible with alcohol at all ratios. Odor perception threshold is reported to be 1.0 to 2.0 mg/l[1,010] or 8.4 mg/l.[02]

Applications. Used as a solvent in the manufacture of plastics, resins, and lacquers.

Acute Toxicity. Following single oral administration, B. inhibited NS functions and had a mild toxic effect on the liver.[2]

According to Brown and Hewitt,[3] a single administration of 1.08 g B./kg BW to F344 rats in corn oil caused no death or histology alterations in the liver but tubular necrosis in kidneys. LD_{50} is found to be 2.74 to 5.5 g/kg BW in rats, 4.05 g/kg BW in mice, and 13.0 g/kg BW in rabbits.[4-6]

Reproductive Toxicity.

Teratogenicity.

Observations in man. An increase in the incidence of children born with CNS defects following maternal exposure was reported.[7]

Embryotoxicity. Pregnant Swiss mice were relatively insensitive to the toxic effects of B. at the inhaled concentrations up to 3000 ppm. However, the offspring of mice exhibited significant signs of developmental toxicity at the 3000 ppm exposure level. Neither maternal nor developmental toxicity was observed at 1000 ppm of B. or below.[8]

Allergenic Effect. Contact sensitizing potential is not found in maximized guinea-pig assays.[9]

Mutagenicity.

In vitro genotoxicity. B. is unlikely to be mutagenic in mammalian or bacterial cell systems, but caused aneuploidy in yeast.[10] It is found to be negative in *Salmonella* mutagenicity assay with and without rat hepatic homogenate.[11]

Carcinogenicity. There are no data available on carcinogenic effect of B. in experimental animals. Epidemiological studies revealed no carcinogenic effect.

Carcinogenicity classification.

U.S. EPA: D.

Chemobiokinetics. Induction of cytochrome *P-450* systems by B. and/or its metabolites has been reported.[12] B. undergoes reduction to *secondary alcohol.* It is passed with the urine in the form of *glucuronide*. 30-33% of the administered dose is released unaltered through the lungs.[2]

Regulations. ***U.S. FDA*** (1998) regulates B. (1) as a component of adhesives to be safely used in food-contact surface in accordance with the conditions prescribed by 21 CFR part 175.105; (2) as a cellophane ingredient for packaging food (residue limit 0.1%) in accordance with the conditions prescribed by 21 CFR part 177.1200; and (3) as an ingredient in the manufacture of resinous and polymeric coatings for polyolefin films to be safely used as a food-contact surface of articles intended for use in producing, manufacturing, packing, transporting, or holding food in accordance with the conditions prescribed by 21 CFR part 175.320.

Standards. ***Russia*** (1995). MAC and PML: 1.0 mg/l (organolept., odor).

References:

1. Vertebnaya, P. I. and Mozhayev, Ye. A., in *Protection of Water Reservoirs against Pollution by Industrial Liquid Effluents*, S. N. Cherkinsky, Ed., Medgiz, Moscow, Issue No 4, 1960, 76 (in Russian).
2. *Bulletin of the International Register of Potentially Toxic Chemicals*, 6, 18, 1983.
3. Brown, E. M. and Hewitt, W. R., Dose-response relationships in ketone induced potentiation on chloroform hepato- and nephrotoxicity, *Toxicol. Appl. Pharmacol.*, 76, 437, 1984.
4. Smyth, H. F., Carpenter, C. P., Weil, C. S., Pozzani, U. C., and Striegel, J. A., Range-finding toxicity data, list VI, *Am. Ind. Hyg. Assoc. J.*, March-April, 95, 1962.
5. Kimura, E. T., Ebert, D. M., and Dodge, P. W., Acute toxicity and limits of solvent residue for sixteen or ganic solvents, *Toxicol. Appl. Pharmacol.*, 19, 699, 1971.
6. Tanii, H., Tsuji, H., and Hashimoto, K., Structure-activity relationship of monoketones, *Toxicol. Lett.*, 30, 13, 1986.
7. Holmberg, P. C. and Nurminen, M., Congenital defects of the central nervous system and occupational factors during pregnancy: A case-referent study, *Am. J. Ind. Med.*, 1, 167, 1980.
8. Schwetz, B. A., Mast, T. J., Weigel, R. J., Dill, J. A., and Morrissey, R. E., Developmental toxicity of inhaled methyl ethyl keton in Swiss mice, *Fundam. Appl. Toxicol.*, 16, 742, 1991.
9. Gad, S. C., Dunn, B. J., Dobbs, D. W., Reilly, C., and Walsh, R. D., Development and validation of an alternative dermal sensitization test: the mouse ear swelling test (MEST), *Toxicol. Appl. Pharmacol.*, 84, 93, 1986.
10. Yang, R. S. H., The toxicology of methyl ethyl ketone. in *Residue Reviews*, F. A. Gunther & J. D. Gunther, Eds., Springer Verlag, New York, 97, 212, 1986.
11. Florin, I., Rutberg, M., Curvall, M., and Enzell, C. R., Screening of tobacco smoke constituents for mutagenicity using Ames' test, *Toxicology*, 18, 219, 1980.
12. Abdel-Rahman, M. S., Hetland, L. B., and Couri, D., Toxicity and metabolism of methyl *n*-butyl ketone, *Am. Ind. Hyg. Assoc. J.*, 37, 95, 1976.

1-BUTOXY-2,3-EPOXYPROPANE

Molecular Formula. $C_7H_{14}O_2$

M = 130.21

CAS No 2426-08-6

RTECS No TX4200000

Abbreviation. BEP.

Synonyms. Butyl glycidyl ether; 2,3-Epoxypropylbutyl ether.

Properties. A clear, colorless liquid with a sharp, irritant odor. Contains 26% epoxy groups. Poorly soluble in water, readily soluble in oil and alcohol.

Applications. Active diluent for epoxy resins.

Acute Toxicity. In rats and mice, LD_{50} is 1.5 to 2.0 g BEP/kg BW; in guinea pigs, it is 3.0 g BEP/kg BW. Poisoning is accompanied with CNS depression, adynamia, labored breathing and death within 1 to 7 days.[1]

Repeated Exposure failed to reveal evident cumulative properties. K_{acc} is identified to be 8.2 (by Cherkinsky) and 7.7 (by Kagan).

Allergenic Effect. BEP is reported to be a known skin sensitizer.[2]

Mutagenicity.

In vivo cytogenicity. BEP was negative in dominant lethal test in male mice.[3]

In vitro genotoxicity. Was found to be positive in series of mutagenicity assays, including induction of repair-DNA synthesis in human lymphocytes and *Salmonella* mutagenicity test.[4-6]

References:

1. Krechkovsky, E. A., Anisimova, I. G., Zaprivoda, L. P., et al., Toxicological evaluation of butylglycidyl ether, *Gig. Sanit.*, 8, 91, 1986 (in Russian).
2. Thorgeirsson, A., Fregert, S., Magnusson, B., and Berufs, N., Allergenicity of epoxy-reactive diluents in the guinea pig, *Berufs. Dermatosen.*, 23, 178, 1975 (in Norwegian).
3. Whorton, E. B. Jr., Pullin., T. G., Frost, A. F., Onofre, A., Legator, M. S., and Folse, D. S., Dominant lethal effects of *n*-butyl glycidyl ether in mice, *Mutat. Res.*, 124, 226, 1983.
4. Frost, A. F. and Legator, M. S., Unscheduled DNA synthesis induced in human lymphocytes by butyl glycidyl ethers, *Mutat. Res.*, 102, 193, 1982.
5. Canter, D. A., Zeiger, E., Haworth, S., Lawlor, T., Mortelmans, K., and Speck, W., Comparative mutagenicity of aliphatic epoxides in *Salmonella, Mutat. Res.*, 172, 105, 1986.
6. Connor, T. H., Ward, J. B., Jr., Meyne, J., Pullin, T. G., and Legator, M. S., The evaluation of the epoxide diluent, *n*-butyl-glycidyl ether in a series of mutagenicity assays, *Environ. Mutagen.*, 2, 521, 1980.

2-BUTOXYETHANOL

Molecular Formula. $C_6H_{14}O_2$.
M = 118.20
CAS No 111-76-2
RTECS No KJ8575000
Abbreviation. BE.

Synonyms. Butyl cellosolve; Butyl glycol; Ethylene glycol, monobutyl ether.

Properties. Colorless, transparent liquid having a faint, specific odor. Miscible with water - 1:1, or, according to other data, at any ratio at 25°C;[02] soluble in oil and ethanol. Does not affect transparency and color of water and does not form foam and film on the water surface. Odor perception threshold is 9.3 mg/l.[1]

Applications. BE has wide industrial and consumer applications. It is used as a solvent for cellulose ethers, resins, and lacquers. Used in the manufacture of plastics, surface coatings, and cleaners.

Acute Toxicity. LD_{50} is 0.5 ml/kg BW in humans, 775 mg/kg BW in rats, and 320 mg/kg BW in rabbits.[2,3] Grant et al.[4] found less severe signs of hematotoxicity in 4-week-old F344 rats given 500 and 1000 mg/kg BW by gavage than is reported in more recent studies of Ghanayem et al.[5] where 9- to 13-week-old animals have been tested. Histological changes in the liver and kidney were also observed. Two hours after administration of BE, the relative weight of the spleen was more than doubled. A greater sensitivity of older animals to the lethal doses of BE seems to be a relevant explanation of this variation.[5]

The cause of death in rats and rabbits after acute exposure to BE appeared to be secondary to acute intravascular hemolysis, an effect for which guinea pigs and humans are much less sensitive than rats, mice and rabbits. Recent acute toxicity studies in the guinea pig resulted in an acute oral LD_{50} of 1.4 g/kg BW.[6]

Repeated Exposure. Administration of 1/10 LD_{50} to rats caused retardation of BW gain, decrease in glucose and cholesterol concentration in blood serum. The dose dependence was noted.[2] Male Sprague-Dawley rats were exposed to 2000 to 6000 ppm and female rats to 1600 to 4800 ppm of BE in drinking water for 21 days. BW of females was affected at either dose and it was also decreased in males in the high-dose group.[7]

Gross pathology examination of rats exposed to 0.5 and 1.0 g BE/kg BW revealed thymus atrophy, hyperplasia of the spleen and bone marrow, lymphocytopenia, and reticulocytosis.[4]

In 2-week studies, F344 rats and B6C3F_1 mice received BE in their drinking water. Consumption values ranged from 70 to 300 mg/kg BW (rats) and 90 to 1400 mg/kg BW (mice). There were no effects on survival.[8]

Adult male rats were given BE by gavage in doses of 222, 443, or 885 mg/kg BW over a 6-week period. A dose-dependent decrease in BW gain was noted. Feed consumption was decreased at the highest dose. The most significant toxic effects produced were on the red blood cells including a significant dose-dependent decrease in *Hb* concentration, red blood cell counts, and mean corpuscular *Hb* concentration. Gross pathology examination revealed increased spleen weights, splenic congestion, and hemosiderin accumulation in the liver and kidneys.[9]

Short-term Toxicity. In 13-week studies, BE toxicity was limited to the liver and hematopoietic system. The NOAEL for liver degeneration was 1500 ppm in rats.[8]

Immunotoxicity. In the above described studies, thymus weights were reduced in all the treated animals.[7] A decrease in specific antibody production was observed at the low dose.

Reproductive Toxicity.

Embryotoxicity. In a continuous breeding reproduction study Swiss CD-1 mice were given doses of 0.7 to 2.1 g/kg BW in drinking water for 7 days prior and during a 98-day cohabitation period. Effects on reproduction were evident only in the females and occurred at doses which elicited general toxicity.[10] BE did not affect male or female reproduction when applied dermally to male and female rats for 13 weeks prior to mating, then to pregnant females through day 20 of gestation. No differences were seen in the number of live pups per litter, or in the growth or survival statistics of the offspring.[11]

Teratogenicity. Cardiovascular developmental effect was not observed in F344 rats exposed by gavage to 200 mg/kg and 300 mg/kg BW during 3-day period of organogenesis.[12] BE has been inactive for fetoxicity or birth defects when tested in rats, rabbits,[13] and mice.[14,15]

Gonadotoxicity. It was mentioned as possibly causing damage to the testes in rats.[03]

Mutagenicity. Due to its chemical structure, BE is not alerting for likely genotoxic activity.[16] BE has been described as being not genotoxic (Hardin and Lyon, 1984).

In vivo cytogenetics. BE did not induce micronuclei in mouse bone marrow (up to the maximum tolerated dose of 3300 mg/kg BW).[17]

In vitro genotoxicity. BE and its metabolite butoxyacetaldehyde were not mutagenic in a subline of Chinese hamster ovary cells.[18] BE displayed mutagenic potency in *Salmonella typhymurium* strain *TA 97a* with or without *S9* mix at high concentrations.[19] However, according to other data, it was negative in *Salmonella typhimurium* with and without liver *S9* fraction. At high concentrations, it induced SCE in Chinese hamster ovary cells with and without *S9* fraction.[8]

Carcinogenicity. BE is unlikely to be a genotoxic carcinogen to rodents.[16]

Chemobiokinetics. In two days following oral administration, BE is discovered in the stomach, liver, spleen, kidneys, and other visceral organs (independent of the dose). BE metabolism occurs through oxidation to form *butoxyacetic acid* and conjugation with *glucuronic acid* and *sulfates.* Reproductive, teratogenic, and hematotoxic effects of BE are due to the formation of butoxyacetic acid. *Ethylene glycol*, a metabolite of BE, is a result of dealkylation of the ether occurring prior to oxidation to *alkoxyacetic acid. Ethylene glycol* was excreted in urine, representing approximately 10% of the dose.[20]

F344 rats had access for 24 hours to ^{14}C-BE in drinking water. Excretion was found to occur via urine or exhalation as CO_2.[1] Less than 5.0% was exhaled unchanged.[20]

Regulations. ***U.S. FDA*** (1998) approved the use of BE (1) in adhesives as a component of articles intended for use in packaging, transporting, or holding food in accordance with the conditions prescribed by 21 CFR part 175.105; and (2) as a solvent in the manufacture of polysulfide polymer-polyepoxy resins in articles intended for packaging, transporting, or otherwise containing dry food in accordance with specified conditions in accordance with the conditions prescribed in 21 CFR part 177.1650.

Standards. ***Russia.*** Recommended MAC: 0.12 mg/l.

References:

1. *Data on Hygienic Assessment of Pesticides and Polymers*, A. P. Shitzkova, Ed., F. F. Erisman Research Sanitary Hygiene Institute, Moscow, 1977, 236 (in Russian).
2. Yatsina, O. V., Plaksienko, N. F., Pys'ko, G. T., et al., Hygiene regulation of cellosolves in water bodies, *Gig. Sanit.*, 10, 78, 1988 (in Russian).
3. Lomonova, G. V. and Klimova, E. I., Toxicology of acrylic and methacrylic esters of ethylene glycol, *Gig. Truda Prof. Zabol.*, 2, 38, 1974 (in Russian).
4. Grant, D., Slush, S., Jones, H. B., et al., Acute toxicity and recovery in the hemopoietic system of rats after treatment with ethylene glycol monobutyl ether, *Toxicol. Appl. Pharmacol.*, 77, 187, 1985.
5. Ghanayem, B. I., Burka, L. T., and Matthews, H. B., Metabolic basis of ethylene glycol monobutyl ether induced toxicity: Role of alcohol and aldehyde dehydrogenases, *J. Pharmacol. Exp. Ther.*, 242, 222, 1987.
6. Gingell, R., Boatman, R. J., and Lewis, S., Acute toxicity of ethylene glycol mono-*n*-butyl ether in the guinea pig, *Food Chem. Toxicol.*, 36, 825, 1998.
7. Exon, J. H., Hather, G. G., Bussiere, J. L., Olson, D. P., and Talcott, P. A., Effect of subchronic exposure of rats to 2-methoxyethanol or 2-butoxyethanol: thymic atrophy and immunotoxicity, *Fundam. Appl. Toxicol.*, 16, 830, 1991.
8. *Toxicity Studies of Ethylene Glycol Ethers: 2-Methoxyethanol, 2-Ethoxyethanol, 2-Butoxyethanol Administered in Drinking Water to F344/N Rats and B6C3F$_1$ Mice,* NTP Technical Report Series No 26, Research Triangle Park, NC, July 1993.
9. Krasavage, W. J., Subchronic oral toxicity of ethylene glycol monobutyl ether in male rats, *Fundam. Appl. Toxicol.*, 6, 349, 1986.
10. Heindel, J. J., Gulati, D. K., Russell, V. S., Reel, J. R., Lawton, A. D., and Lamb, J. C., IV, Assessment of ethyleneglycol monobutyl ether and ethyleneglycol monophenyl ether reproductive toxicity using continuous breeding protocol in Swiss CD-1 mice, *Fundam. Appl. Toxicol.*, 15, 683, 1990.
11. Auletta, C. S., Schroeder, R. E., Krasavage, W. J., et al., Toxicology of diethylene glycol butyl ether. 4. Dermal subchronic/reproduction study in rats, *J. Am. Coll. Toxicol.*, 12, 161, 1993.
12. Sleet, R. B., Price, C. J., Marr, M. C., et al., Cardiovascular development in F-344 rats following phase-specific exposure to 2-butoxyethanol, Abstract, *Teratology*, 43, 466, 1991.
13. Nolen, G. A., Gibson, W. B., Benedict, J. H., Briggs, D. W., and Schardein, J. L., Fertility and teratogenic studies of diethylene glycol monobutyl ether in rats and rabbits, *Fundam. Appl. Toxicol.*, 5, 1137, 1985.
14. Piccirillo, V. J. et al., NIOSH PB83-257-600, Contract No 210-81-6010, 1983, 108.
15. Schuler, R. L., Hardin, B. D., Niemeier, R. W., Booth, G., Hazelden, K., Piccirillo, V., and Smith, K., Results of testing fifteen glycol ethers in a short-term *in vivo* reproductive toxicity assay, *Environ. Health Perspect.*, 57, 141, 1984.
16. Elliot, B. M. and Ashby, J., Review of the genotoxicity of 2-butoxyethanol, *Mutat. Res.*, 387, 89, 1997.
17. Gollapudi, B. B., Linscombe, V. A., Mcclintock, M. L., et al., Toxicology of diethylene glycol butyl ether. 3. Genotoxicity evaluation in an *in vitro* gene mutation assay and an *in vivo* cytogenetic test, *J. Am. Coll. Toxicol.*, 1993; 12, 155, 1993.
18. Chiewchanwit, T. and Au, W. W., Mutagenicity and cytotoxicity of 2-buthoxyethanol and its metabolite, 2-butoxyacetaldehyde, in Chinese hamster ovary (CHO-AS52) cells, *Mutat. Res.*, 334, 341, 1995.
19. Hoflack, J. C., Lambolez, L., Elias, Z., and Vasseur, P., Mutagenicity of ethylene glycol ethers and of their metabolites in *Salmonella typhimurium, Mutat. Res.*, 341, 281, 1995.
20. Medinsky, M. A., Singh, G., Bechtold, W. E., Bond, J. A., Sabourin, P. J., Birnbaum, L. S., and Henderson, R. F., Disposition of three glycol ethers administered in drinking water to male F344/N rats, *Toxicol. Appl. Pharmacol.*, 102, 443, 1990.

2-(2-BUTOXYETHOXY)ETHANOL

Molecular Formula. $C_8H_{18}O_3$
M = 314.36
CAS No 112-34-5
RTECS No KJ9100000
Abbreviation. BEE.

Synonyms and **Trade Names.** Butoxydiethylene glycol; Butoxydiglycol; Butyl carbitol; Butyl dioxitol; Diethylene glycol, *n*-butyl ether; Diglycol monobutyl ether; Dowanol DB; Poly-Solv DB.

Properties. Liquid with a distinct odor of ether. Soluble in water, oils, and organic solvents. Threshold concentration for the change in organoleptic properties of water is 0.8 mg/l.[01]

Applications. Used as a cellulose nitrate solvent, in the synthesis of resins and plasticizers, and in the production of lacquers.

Acute Toxicity. LD_{50} is 4.5 g/kg BW in rats, and 6.0 g/kg BW in mice.

Repeated Exposure revealed moderate cumulative properties.[1] K_{acc} is 4.4 (by Lim).

Reproductive Toxicity. The doses of 250 to 1000 mg BEE /kg BW were given over a 60-day period prior to mating to male rats, and from the 14th day prior to mating until day 13 or the weaning of the offspring to females. No adverse effect on fertility, embryos, fetuses, or neonates was noted.[2]

Mutagenicity.

In vitro and ***in vivo genotoxicity.*** Shows no mutagenic activity in a set of *in vitro* tests and in experiments on *Dr. melanogaster.*[3]

Chemobiokinetics. BEE forms *diethylene glycol, ethylene glycol,* and *butoxyacetic acid.*

Regulations. ***U.S. FDA*** (1998) approved the use of BEE (1) as a component of adhesives intended for use in contact with food in accordance with the conditions prescribed by 21 CFR part 175.105; (2) in surface lubricants employed in the manufacture of metallic articles that contact food in accordance with the conditions prescribed in 21 CFR part 178.3910; and (3) as a component of the uncoated or coated food-contact surface of paper and paperboard that may be safely used in producing, manufacturing, packing, transporting, or holding dry food in accordance with the conditions prescribed by 21 CFR part 176.180.

References:

1. Krotov, Yu. A., Lykova, A. S., Skachkov, M. A., et al., Sanitary and hygienic characteristics of diethylene glycol ethers (Carbitols), with special reference to air pollution control, *Gig. Sanit.*, 2, 14, 1981 (in Russian).
2. Nolen, G. A., Gibson, W. B., Benedict, J. H., et al., Fertility and teratogenic studies of diethylene glycol monobutyl ether in rats and rabbits, *Fundam. Appl. Toxicol.*, 5, 1137, 1985.
3. Thompson, E. D., Coppinger, W. J., Valencia, R., et al., Mutagenicity testing of diethylene glycol monobutyl ether, *Environ. Health Perspect.*, 57, 105, 1984.

2-(2-BUTOXYETHOXY)ETHANOL, ACETATE

Molecular Formula. $C_{10}H_{20}O_4$
M = 204.30
CAS No 124-17-4
RTECS No KJ9275000
Abbreviation. BEEA.

Synonyms and **Trade Name.** Butoxyethoxyethyl acetate; Butyl carbitol acetate; Diethylene glycol, monobutyl ether, acetate; Diglycol monobutyl ether acetate.

Properties. Clear liquid with mild, not unpleasant odor and bitter taste.[011] Soluble in water, readily soluble in alcohol.

Applications. Solvent for cellulose acetate and polyvinyl acetate; used in the production of coatings and lacquers.

Acute Toxicity. In rats and guinea pigs, ingestion of sublethal doses resulted in marked narcosis.[011]

Chemobiokinetics. BEEA is metabolized by saponification into *diethylene glycol monobutyl ether* and *acetic acid.*

Regulations. ***U.S. FDA*** (1998) approved the use of BEEA in adhesives as a component of articles intended for use in packaging, transporting, or holding food in accordance with the conditions prescribed in 21 CFR part 175.105.

BUTYL ALCOHOL

Molecular Formula. $C_4H_{10}O$
M = 74.12
CAS No 76-36-3
RTECS No EO1400000
Abbreviation. NBA.

Synonyms. 1-Butanol; *n*-Butanol; Butyric alcohol.

Properties. Clear, colorless liquid with a heavy unpleasant odor. Solubility in water is 9.0% at 15°C or 73,000 mg/l at 25°C;[02] mixes with ethyl alcohol, ether, and other organic solvents at all ratios. In small concentrations, it gives an aromatic, but not unpleasant, odor to water. Odor perception threshold is 0.3 mg/l,[010] 1.0 mg/l,[1] or 7.1 mg/l.[02] According to other data, organoleptic threshold is 0.27 mg/l.[01]

Applications. Used as a solvent for coatings, nitrocellulose lacquers, and natural resins; used in rubber vulcanization and in production of foam plastics.

Exposure of general population is principally through NBA natural occurrence in foods and beverages, and its use as a flavoring agent.

Acute Toxicity.

Observations in man. Ingestion of a dose more than 250 ml seems to be fatal although there may be individual variations in sensitivity. NBA does not induce vomiting.[2]

Animal studies. LD_{50} is reported to be 2.68 g/kg BW in mice,[3] 0.7 to 2.1 g/kg BW in rats, 4.2 to 5.3 g/kg BW in rabbits, and 1.2 g/kg BW in hamsters.[2] Its potency for intoxication is approximately 6 times that of ethanol.

Repeated Exposure revealed moderate cumulative properties. K_{acc} is 3.4. NBA is capable of both material and functional accumulation.[4]

Chickens were given NBA starting with 15 mg and rising to 600 mg, for two months. Exposure caused slight anemia of the comb only; BW gain and development were unaffected.[1]

Short-term Toxicity. Rats received by gavage 30, 125, or 500 mg/kg BW for 13 weeks. The treatment caused ataxia and adynamia (at the highest dose) and slight changes in hematology analyses (at 2 higher doses). The NOAEL of 125 mg/kg BW for CNS effects in rats was established in this study.[5]

Reproductive Toxicity.

Embryotoxicity. Mice received the diet containing 1.0% NBA during gestation. The treatment decreased by 50% BW in the offspring in which impaired performance in behavioral tests was also observed.[6]

Teratogenicity. Nelson et al. reported teratogenic effect of NBA administered by inhalation to rats.[3] Female rats were given aqueous solutions of NBA containing 0.24, 0.8, or 4.0% (0.3, 1.0, or 5.0 g/kg BW) for 8 weeks before and during gestation. Control animals received water. The treatment produced developmental anomalies in skeleton and CNS of the fetuses.[7]

Gonadotoxicity. Rats received 300, 1000, and 5000 mg/kg BW with their drinking water for 8 weeks before mating, and for entire period of pregnancy. No effect on estrous cycle duration was observed. In the offspring, a decrease in length of the body, skeletal abnormalities, hydrocephaly, and malformations in brain and kidneys were reported.[7]

Mutagenicity.

In vitro genotoxicity. NBA was shown to be negative in *Salmonella* mutagenicity assay.[8] It did not produce CA in the cultured human lymphocytes.[9]

Chemobiokinetics. Following ingestion NBA is readily absorbed in the GI tract of experimental animals. NBA metabolism is found to occur through rapid and complete oxidation in the body, apparently via the formation of *butyric acid* and via the *aldehyde* to *carbon dioxide*, which is the major metabolite. A small amount (only 2.0 to 4.0%) of the *alcoholic glucuronides* appears in the urine.

Rats were administered *i/p* with 1.0 ml NBA/kg BW. The treatment resulted in the formation of *butyl esters of palmitate, stearate,* and *oleate* in the liver but not in measurable quantities in lung or pancreas.[10]

Regulations.

U.S. FDA (1998) approved the use of NBA (1) as a component of adhesives used in food-contact surface of articles in accordance with the conditions prescribed by 21 CFR part 175.105; (2) in the manufacture of cellophane for food packaging in accordance with the conditions prescribed by 21 CFR part 177.1200; (3) as a component of the uncoated or coated food-contact surface of paper and paperboard intended for use in producing, manufacturing, packaging, processing, preparing, treating, packing, transporting, or holding dry food in accordance with the conditions prescribed by 21 CFR part 176.180; (4) in surface lubricants employed in the manufacture of metallic articles that contact food in accordance with the conditions prescribed in 21 CFR part 178.3910; (5) as a solvent in polysulfide polymer-polyepoxy resins used as the dry food-contact surface in accordance with specified conditions in accordance with the conditions prescribed by 21 CFR part 177.1650; (6) as an ingredient in the manufacture of resinous and polymeric coatings for polyolefin films for food-contact surface of articles intended for use in producing, manufacturing, packing, processing, preparing, treating, packaging, transporting, or holding food in accordance with the conditions prescribed by 21 CFR part 175.320; (7) as a defoaming agent that may be safely used as a component of articles intended for contact with food in accordance with the conditions prescribed in 21 CFR part 176.200; and (8) as a substance employed in the production of or added to textiles and textile fibers intended for use in contact with food in accordance with the conditions prescribed by 21 CFR part 177.2800.

EU (1990). NBA is available in the *List of authorized monomers and other starting substances which shall be used for the manufacture of plastic materials and articles intended to come into contact with foodstuffs (Section A).*

Great Britain (1998). NBA is authorized without time limit for use in the production of polymeric materials and articles in contact with food or drink or intended for such contact.

Recommendations. Joint FAO/WHO Expert Committee on Food Additives (1999). ADI: no safety concern.

Standards. ***Russia*** (1995). MAC: 0.1 mg/l; PML in food: 0.5 mg/l.

References:

1. Nazarenko, I. V., in *Sanitary Protection of Water Bodies against Industrial Sewage Pollution*, S. N. Cherkinsky, Ed., Medgiz, Moscow, Issue No 4, 1969, 65 (in Russian).
2. *Butanols - Four Isomers: 1-Butanol, 2-Butanol, tert-Butanol, Isobutanol*, IPCS, Environmental Health Criteria No 65, WHO, Geneva, 1987, 141.
3. Nelson, B. K., Brightwell, W. S., Khan, A., et al., Lack of selective developmental toxicity of three butanol isomers administered by inhalation to rats, *Fundam. Appl. Toxicol.*, 12, 469, 1989.
4. Rumyantsev, A. P., Lobanova, I. Ya., Tiunova, L. V., et al., Toxicology of Butyl Alcohol, Chemical Industry: Series *Toxicology, Sanitary Chemistry of Plastics*, Issue No 2, 1979, 24 (in Russian).
5. *Butanol: Rat Oral Subchronic Toxicity Study*, U.S. EPA, Office of Solid Waste, Washington, D.C., 1986.
6. Daniel, M. A. and Evans, M. A., Quantitative comparison of maternal ethanol and maternal butanol diet on postnatal development, *J. Pharmacol. Exp. Ther.*, 222, 294, 1982.
7. Sitarek, K., Berlinska, B., and Baranski, B., Assessment of the effect of *n*-butanol given to female rats in drinking water on fertility and prenatal development of their offspring, *Int. J. Occup. Med. Environ. Health*, 7, 365, 1994.
8. McCann, J., Choi, E., Yamasaki, E., and Ames, B. N., Detection of carcinogens as mutagens in the *Salmonella*/microsome test assay of 300 chemicals, *Proc. Natl. Acad. Sci. U.S.*, 72, 5735, 1975.
9. Obe, G., Ristow, M. J., and Herma, J., Chromosomal damage by alcohol *in vitro* and *in vivo*, *Adv. Exp. Med. Biol.*, 85a, 47, 1977.
10. Carlson, G. P., Formation of esterified fatty acids in rats administered 1-butanol and 1-pentanol, *Res. Commun. Mol. Pathol. Pharmacol.*, 86, 111, 1994.

2-BUTYL ALCOHOL

Molecular Formula. $C_4H_{10}O$
M = 74.12
CAS No 78-92-2
RTECS No EO1750000
Abbreviation. SBA.

Synonyms. *sec*-Butanol; Butylene hydrate; 2-Hydroxybutane; Methylethyl carbinol.

Properties. Colorless liquid with a characteristic sweet odor. Solubility in water is 125 g/l at 20°C or, according to other data, 200 g/l at 25°C.[02] Miscible with ethyl alcohol and ether. Odor perception threshold is 0.12 mg/l[010] or 19 mg/l.[02]

Applications. SBA is used primarily as a solvent.

Exposure. It occurs naturally in foods and beverages.

Acute Toxicity.

Observations in man. Consumption is followed by abdominal pain, vomiting, and diarrhea.

Animal studies. In rats, LD_{50} is 6.5 g/kg BW. Administration caused ataxia and narcosis. The potency of SBA for intoxication is approximately 4 times that of ethanol.[1]

In rabbits, LD_{50} is 4.9 g/kg BW.[2]

Short-term Toxicity. Rats were given 0.3, 1.0, or 3.0 % SBA in the diet. The LOEL of 5.0 g/kg BW, and the NOEL of 1.64 g/kg BW were established in this study.[3]

Long-term Toxicity. In a 2-generation rat reproduction study, SBA was administered via drinking water at concentrations of 0.3, 1.0, and 2.0 % (through the first and second generations). The treatment with 2.0% concentration resulted in several changes which represent mild toxicity and are reminiscent of stress lesions.[4]

Reproductive Toxicity.

Embryotoxicity. Addition of 2.0 to 3.0% SBA in drinking water of rats produced significant embryotoxic effect (decreased pup survivability and fetal weight in the F_1 and F_2 offspring). The NOAEL of 1.77 g/kg BW (1.0% solution) was established in this study.[5]

Teratogenicity. In the above described studies, 0.3 and 1.0% SBA concentrations in drinking water produced no effect on growth and reproduction of rats.[4] 2.0% concentration caused significant depression of growth of weaning rats with evidence of retarded skeletal maturation.

Mutagenicity.

In vitro genotoxicity. SBA was negative in the rat liver chromosome assay, in the bacterial mutation assays, and the yeast mitotic gene conversion assay.[6]

Chemobiokinetics. Approximately 97% of the dose is converted by alcohol dehydrogenase into *methyl ethyl ketone*, which is either excreted in the breath and urine or conjugated with a formation of *glucuronide*.[1] However, Traiger and Bruckner believe that only approximately 2.0% SBA is oxidized to *methyl ethyl ketone in vivo*.[7]

According to Dietz *et al.*,[8] the following metabolites are formed in the body: *2-butanone, 3-hydroxy-2-butanone*, and *2,3-butanediol.*

Regulations. ***U.S. FDA*** (1998) approved the use of SBA as a component of the uncoated or coated food-contact surface of paper and paperboard that may be safely used in producing, manufacturing, packing, transporting, or holding dry food in accordance with the conditions prescribed by 21 CFR part 176.180.

Recommendations. ***UK MAFF*** (1978). Recommended residues of butan-2-ol in food should not exceed 30 mg/kg.

Standards. ***Russia*** (1995). MAC: 0.2 mg/l.

References:

1. *Butanols - Four Isomers: 1-Butanol, 2-Butanol, tert-Butanol, Isobutanol*, WHO, IPCS, Environmental Health Criteria No 65, Geneva, 1987, 141.
2. Munch, J. C., Aliphatic alcohols and alkyl esters: narcotic and lethal potencies to tadpoles and to rabbits, *Ind. Med.*, 41, 31, 1972.

3. Cox et al., 1975; cit. in Munro, I. C., Ford, R. A., Kennepohl, E., and Sprenger, J. G., Correlation of structural class with No-Observed-Effect Levels: A proposal for establishing a threshold of concern, *Food Chem. Toxicol.*, 34, 829, 1996.
4. Gallo, M. A., Oser, B. L., Cox, G. E., et al., Studies on the long-term toxicity of 2-butanol, in *16 Ann. Meeting of the Society of Toxicology*, Abstracts of Papers, March 27-30, 1977, Toronto, 1977, 9.
5. *Alpha 2u-globulin: Association with Chemically Induced Renal Toxicity and Neoplasia in the Male Rat*, Risk Assessment Forum, EPA 625/3-91/019F, Washington, D. C., 1991.
6. Brooks, T. M. et al., The genetic toxicology of some hydrocarbon and oxygenated solvents, *Mutagenesis*, 3, 227, 1988.
7. Traiger, G. L. and Bruckner, J. V., The participation of 2-butanone in 2-butanol-induced potentiation of carbon tetrachloride hepatotoxicity, *J. Pharmacol. Exp. Ther.*, 196, 493, 1976.
8. Dietz, F. K., Rodriguez-Jiaxola, M., Traiger, G. J., et al., Pharmacokinetics of 2-butanol and its metabolites in the rat, *J. Pharmacokinet. Biopharmacol.*, 9, 553, 1981.

tert-BUTYL ALCOHOL

Molecular Formula. $C_4H_{10}O$
M = 74.12
CAS No 75-65-0
RTECS No EO1925000
Abbreviation. TBA.

Synonyms. *tert*-Butanol; *tert*-Butyl hydroxide; 1,1-Dimethylethanol; 2-Methyl-2-propanol; Trimethylcarbinol; Trimethylethanol.

Properties. Colorless liquid or white crystals with a camphor-like odor. Mixes with water and alcohol at any ratio. Odor perception threshold is 290 mg/l.[02]

Applications. Powerful solvent. Used in the manufacture of polyolefins, copolymers of methacrylonitrile and methacrylic acid, and also in the production of drugs, perfumes, and lacquers.

Acute Toxicity.

Observations in man. Ingestion is followed by abdominal pain, vomiting, and diarrhea.

Animal studies. LD_{50} is 3.5 g/kg BW in rats, and 3.6 g/kg BW in rabbits. Poisoning exerts a narcotic effect. The primary acute effects in animals are those of alcoholic intoxication. TBA potency for intoxication is approximately 1.5 times that of ethanol.[1,2]

Repeated Exposure. Rats were fed with 20 ml TBA/l liquid diet. The treatment entailed slight ataxia.[1]

Short-term Toxicity. In a 13-week study, F344 rats were given 2.5 mg to 40 mg TBA/ml drinking water. All males and six females given 40 mg/ml died during the study. Urinary tract appeared to be the target organ. Transitional epithelium hyperplasia and inflammation of the urinary bladder were observed in 20 mg and 40 mg TBA/ml drinking water exposed males and 40 mg TBA/ml drinking water exposed females, both in rats and mice. The severity of nephropathy was significantly greater in treated animals than that of controls. Clinical signs of intoxication included ataxia in rats and mice. In females, adynamia was found to develop. Gross pathology examination revealed urinary tract calculi, renal pelvic and ureteral dilatation, and thickening of the urinary bladder mucosa. Histology examination revealed hyperplasia of the transitional epithelium. $B6C3F_1$ mice showed no biologically significant differences in hematology parameters when giving drinking water with up to 40 mg TBA/ml. The NOEL for the urinary tract lesions appeared to be about 800 mg/kg BW (male rats), 1570 mg/kg BW (male mice), or 1450 mg/kg BW (females of both species).[3-5]

Long-term Toxicity. In a 2-year study, F344 rats received up to 10 mg TBA/ml drinking water. Increased incidence of mineralization in the kidney, and of hyperplasia and adenoma was reported. The incidence of follicular cell hyperplasia of the thyroid gland was observed in mice as well as chronic inflammation and hyperplasia in the urinary bladder.[5]

Reproductive Toxicity. Postnatal effects of TBA were found in the offspring exposed *in utero*, though Daniel and Evans[6] found no teratogenicity in rats and mice fed TBA up to 1.0% in the diet.

Mutagenicity. TBA seems to be a non-genotoxic compound.

In vitro genotoxicity. In the NTP study, it was shown to be negative in *Salmonella* mutagenicity assay, mouse lymphoma cells, and *in vitro* cytogenetic assays.[1]

Carcinogenicity. In the NTP study, experimental animals received drinking water containing 0.125 to 1.0% TBA (rats) and 0.5 to 2.0% TBA (mice). In rats, kidneys nephropathy and tubular cell adenomas or carcinomas were observed. Follicular cell hyperplasia and adenomas were noted in thyroid of rats and mice.[5]

Carcinogenicity classification.

NTP: SE - NE - EE - SE.

Chemobiokinetics. TBA is not a substrate for alcohol dehydrogenase and is slowly metabolized by mammals. Possible is conjugation of *hydroxyl group* with *glucuronic acid.*[7] TBA was found to enhance nephropathy in male F344 rats, and it increased renal accumulation of hyaline protein material consistent with α-2 mu globulin disposition.[4]

TBA is excreted in the urine as *glucuronide*, but also in the breath and urine as *acetone* or *carbon dioxide*.[1]

Regulations. ***U.S. FDA*** (1998) approved the use of TBA (1) as a component of a defoaming agent intended for articles which may be used in producing, manufacturing, packing, processing, preparing, treating, packaging, transporting, or holding food in accordance with the conditions prescribed in 21 CFR part 176.200; (2) in surface lubricants employed in the manufacture of metallic articles that contact food in accordance with the conditions prescribed in 21 CFR part 178.3910; and (3) as an indirect food additive when used in preparation and application of coatings for paper and paperboard used in food containers (21 CFR part 176.200) and surface lubricant 178.3910.

Standards. ***Russia*** (1995). MAC: 1.0 mg/l.

References:

1. *Butanols - Four Isomers: 1-Butanol, 2-Butanol, tert-Butanol, Isobutanol*, IPCS, Environmental Health Criteria No 65, WHO, Geneva, 1987, 141.
2. Munch, J. C., Aliphatic alcohols and alkyl esters: narcotic and lethal potencies to tadpoles and to rabbits, *Ind. Med.*, 41, 31, 1972.
3. Lindamood, C. III, Farnel, D. R., Giles, H. D., Prejean, J. D., Collins, J. J., Takahashi, K., and Maronpot, R. R., Subchronic toxicity studies of *tert*-butyl alcohol in rats and mice, *Fundam. Appl. Toxicol.*, 19, 91, 1992.
4. Takahashi, K., Lindamood, C., III, and Maronpot, R. R., Retrospective study of possible *alpha-2 mu* globulin nephropathy and associated cell proliferation in male Fischer 344 rats dosed with *t*-butyl alcohol, *Environ. Health Perspect.*, 101 (Suppl. 5), 281, 1993.
5. *Toxicology and Carcinogenesis Studies of t-Butyl Alcohol in F344/N Rats and B6C3F$_1$ Mice (Drinking Water Studies),* NTP Technical Report Series No 436 (J. Cirvello), Research Triangle Park, NC, May, 1995.
6. Daniel, M. A. and Evans, M. A., Quantitative comparison of maternal ethanol and maternal tertiary butanol diet on postnatal development, *J. Pharmacol. Exp. Ther.*, 222, 294, 1982.
7. Merritt, D. A. and Thomkins, G. M., Reversible oxidation of cyclic secondary alcohols by liver alcohol dehydrogenase, *J. Biol. Chem.*, 234, 2778, 1959.

α-BUTYL-ω-HYDROXYPOLY[OXY(METHYL-1,2-ETHANEDIYL)]

Molecular Formula. $[\sim C_3H_6O\sim]_n$. $C_4H_{10}O$

CAS No 9003-13-8

59029-72-0

RTECS No TR4680000

Synonyms and **Trade Names.** Butoxypolypropylene glycol; Butoxypropanediol polymer; Newpol LB 3000; Polyoxypropylene, monobutyl ether; Stabilene.

Properties. A repellent and a lubricating agent.

Applications. Colorless liquid. Soluble in alcohol.

Acute Toxicity. LD_{50} is 9.1 g/kg BW in rats, and 23.9 g/kg BW in rabbits. According to other data, LD_{50} is 34.5 ml/kg BW in rats.

Regulations. ***U.S. FDA*** (1998) approved the use of B. in adhesives as a component of articles intended for use in packaging, transporting, or holding food in accordance with the conditions prescribed by 21 CFR part 175.105.

Reference:

Pang, S. N. J., *J. Am. Coll. Toxicol.,* 12, 257, 1993.

CARBON TETRACHLORIDE

Molecular Formula. CCl_4
M = 153.84
CAS No 56-23-5
RTECS No FG4900000
Abbreviation. CTC.

Synonyms. Tetrachloromethane; Methane tetrachloride; Perchloromethane.

Properties. Colorless liquid. Water solubility is 800 mg/l at 20°C. Odor perception threshold is 0.52 mg/l.[02]

Applications. Used as a solvent in the manufacture of plastics; a cleaning agent, an intermediate in the production of chlorofluorocarbons.

Exposure. It has been detected in a variety of foodstuffs at levels from 0.1 g/kg to 20 g/kg and less frequently in drinking water. Sources of human exposure from environment include CTC-contaminated air and water.

Acute Toxicity.

Observations in man. No effects were reported after single oral doses of 2.5 to 15 ml CTC (57 to 343 mg/kg BW), although changes may occur in the liver and kidney. Some individual adults suffer adverse effects (including death) from ingestion of as little as 1.5 mg/l (34 mg/kg BW), and doses of 0.18 ml and 0.92 ml may be fatal in children.[1]

Animal studies. LD_{50} values for laboratory animals vary from 1.0 to 12.8 g/kg BW.[1] According to other data, LD_{50} is 6.2 g/kg BW in rats, 12 to 14 g/kg BW in mice, and 5.7 g/kg BW in guinea pigs and rabbits.[2]

Single doses of about 4.0 g/kg BW result in lesions of the renal proximal tubules in rats and pulmonary Clara cells and endothelial cells in rats and mice. Adverse effects in the liver were shown after oral administration of 80 mg CTC/kg BW but not after 40 mg CTC/kg BW.[1,3]

Corn oil delayed and prolonged the GI absorption of CTC and reduced its acute hepatotoxicity (Kim et al., 1990).

Repeated Exposure. After oral administration of large doses, the most severe damage is observed in the liver (including hepatocellular necrosis), kidney, and lung. Lower doses of CTC cause reversible changes.[1]

Hepatotoxic effects were observed in CD-1 mice given 625 to 2500 mg CTC/kg BW for 14 days[4] and in rats given 20 mg CTC/kg BW and higher doses (in corn oil) for 9 days.[3]

Increased liver lipid and triglyceride levels were reported in a 6-week study using doses of 40 and 76 mg/kg BW but not 22 mg/kg BW.[4]

When young adult (8 to 9 weeks old) male F344 rats were given 40 mg CTC/kg BW by gavage for 10 consecutive days, a significant increase in the relative liver weights was noted. Histology examination revealed mild to moderate vacuolar degeneration and minimal to mild hepatocellular necrosis in these livers. When rats were dosed with 20 to 40 mg/kg BW, the serum levels of ALT and AST were elevated; however, no renal effects were observed.[5] Newborn rats seem to be less sensitive to liver damage by CTC than 7-day-old rats.[6]

Short-term Toxicity. CTC hepatotoxicity is a result of the parent compound metabolism to a highly reactive radical intermediate by the cytochrome *P-450* mixed function system. CTC increases lipid peroxidation, fatty infiltration, destruction of the cytochrome *P-450*, and liver necrosis.[7]

In rats given 1.0 mg CTC/kg BW for 12 weeks, no adverse effects were found. Doses of 10 mg/kg and 33 mg/kg BW resulted in enzyme release, centrilobular vacuolization and necrosis in the liver.[3]

Male Sprague-Dawley rats received 1.0, 10, or 33 mg CTC/kg BW by corn oil gavage, 5 days/week for 12 weeks. Dose-dependent liver lesions including mild centrilobular vacuolization and statistically significant increases in serum sorbitol dehydrogenase activity were noted at 10 and 33 mg CTC/kg BW doses. The NOAEL of 1.0 mg/kg BW was established in this study (converted to 0.71 mg/kg BW).[8] The NOAEL in mice gavaged with CTC in corn oil appeared to be at a similar level.[9] CD-1 mice were given the doses of 12 to 1200 mg CTC/kg BW in corn oil by gavage over a period of 3 months. The treatment resulted in increased serum enzyme levels, increased organ weights, and pathological changes.[10]

Long-term Toxicity. The same effects, namely, fatty infiltration, release of liver enzymes, inhibition of cellular enzyme activities, and inflammation and cellular necrosis, were observed following a long-term exposure. No adverse effects were observed in rats of both sexes fed dietary levels of 80 and 200 ppm CTC until final sacrifice in two years. However, survival was below 50% at 21 months and tissues were not microscopically examined in this study.[4]

Immunotoxicity. CTC was administered *i/p* to female B6C3F$_1$ rats for 7 days. Doses of 0.5, 1.0, and 1.5 g/kg BW were found to produce a marked suppression of both humoral and cell-mediated immune functions.[11]

No consistent alterations in the immune parameters examined in F344 rat study were observed.[5] There was no difference in antibody response to sheep red blood cells in another group of rats dosed with 40 to 160 mg/kg BW.

CTC is not immunotoxic in the rat at the dosages that produce overt hepatotoxicity.

Reproductive Toxicity.

Embryotoxicity. There were no reproductive effects in rats fed the diet containing CTC at concentrations of 80 and 200 ppm for two years.[4] However, Narotsky and Kavlock reported CTC to cause high incidence of full-litter resorptions.[12]

CTC is not ***teratogenic.***[1]

Mutagenicity. No data are reported on genetic and related effects of CTC in humans (IARC). It is not genotoxic in the majority of mutagenicity bioassays.

In vitro genotoxicity. Mutagenic effects were not observed in a number of bacterial test systems or in cultured liver cells.[1] It caused cell transformation in Syrian hamster embryo cells.[13]

In vivo cytogenetics. CTC did not induce CA, DNA strand breaks in the cells of rodents treated *in vivo* (NTP-88; IARC 1-53 and 8-371) and unscheduled DNA synthesis in male F344 rats.[14]

Carcinogenicity. CTC demonstrated carcinogenic potential through oral exposure, producing several types of tumors but mainly hepatic neoplasms. Doses of about 30 mg/kg BW or higher administered for 6 months or longer have been found to produce an increased frequency of hepatocellular tumors in mice, rats, and hamsters.[1]

CTC was found to be carcinogenic in the B6C3F$_1$ mice exposed to time-weighted average doses of 1250 and 2500 mg/kg BW for 78 weeks. The incidence of hepatocellular carcinoma was reported to be almost 100% in both sexes. The rate of carcinoma development was substantially lower (about 5.0%) in Osborne-Mandel rats exposed to 47, 94, 80, or 159 mg/kg BW. Because there are doubts regarding the mechanism of tumorogenesis in the liver of this strain of mouse with agents that are known hepatotoxins (such as CTC), the appropriateness of a non-threshold model for extrapolation is questionable.[15]

Syrian golden hamsters were exposed to approximately 10 to 20 mg CTC/day for 43 weeks. Half of the animals died in the course of treatment. Survivors developed liver cell carcinomas.[16] CTC was not shown to be carcinogenic through inhalation exposure in animals. WHO considered it to be non-genotoxic carcinogen.[17]

Carcinogenicity classifications. An IARC Working Group concluded that there is sufficient evidence for the carcinogenicity of CTC in *experimental animals* and there were no adequate data available to evaluate the carcinogenicity of CTC in *humans.*

The The Annual Report of Carcinogens issued by the U.S. Department of Health and Human Services (1998) defines CTC to be a substance which may reasonably be anticipated to be carcinogen, i.e., a substance for which there is a limited evidence of carcinogenicity in *humans* or sufficient evidence of carcinogenicity in *experimental animals.*

IARC: 2B;

U.S. EPA: B2;

EU: 3.

Chemobiokinetics. CTC is absorbed readily from the GI tract and seems to be distributed in all major organs and tissues following absorption. It is converted into *trichloromethyl free radical* which is the main metabolite and undergoes a variety of reactions, including hydrogen abstraction to form *chloroform,* dimerization to form *hexachloroethane* and addition to cellular molecules. Further metabolism of the heme-bound trichloromethyl radical is postulated to result in the eventual formation of *carbonyl chloride (phosgene).*[18] CTC and its volatile metabolites are primarily excreted in exhaled air and also in the urine and feces. The major part of the oral dose is excreted in 1 to 2 days.[1]

Regulations. ***U.S. FDA*** (1998) approved the use of CTC (1) as a component of adhesives for articles intended for use in packaging, transporting, or holding food in accordance with the conditions prescribed by 21 CFR part 175.105; (2) as a component of the uncoated or coated food-contact surface of paper and paperboard that may be safely used in producing, manufacturing, packing, transporting, or holding dry food in accordance with the conditions prescribed by 21 CFR part 176.180; (3) as an ingredient in resinous and polymeric coatings of food-contact surfaces in accordance with the conditions prescribed by 21 CFR part 175.300; and (4) to prevent the transfer of inks employed in printing and decorating paper and paperboard used for food packaging.

Recommendations.

WHO (1996). Guideline value for drinking water: 0.002 mg/l. WHO recommends that no detectable residues (detection limit: 0.01 ppm) be allowed in food or feed, but permits 50 mg/l in cookie cereals.

U.S. EPA (1999): Health Advisory for longer-term exposure is 0.3 mg/l.

Standards.

U.S. EPA (1999). MCL: 0.005 mg/l, MCLG: zero.

Canada (1989). MAC: 0.005 mg/l.

Russia (1995). MAC: 0.002 mg/l.

References:

1. *Criteria Document for Carbon Tetrachloride*, Final Draft, Technical Report No 540-131A, U.S. EPA Office of Drinking Water, Washington, D. C., 1988.
2. Chirkova, V. M., *Carbon Tetrachloride*, Soviet Toxicology Center, Moscow, Issue No 27, 1983, 20 (in Russian).
3. Bruckner, J. V., Kim, H. J., Dallas, C. E., et al., Effect of dosing vehicles on the pharmacokinetics of orally administered carbon tetrachloride, Abstract, *Soc. Toxicol. Ann. Meeting,* 1987.
4. Alumot, E., Nachtomi, E., Mandel, E., et al., Tolerance and acceptable daily intake of chlorinated fumigants in the rat diet, *Food Cosmet. Toxicol.*, 14, 105, 1976.
5. Smelowicz, R. J., Simmons, J. E., Luebke, R. W., et al., Immunotoxicologic assessment of subacute exposure of rats to CCl_4 with comparison to hepatotoxicity and nephrotoxicity, *Fundam. Appl. Toxicol.,* 17, 186, 1991.
6. Dawkins, M. J. R., Carbon tetrachloride poisoning in the liver of newborn rat, *J. Pathol. Bacteriol.,* 85, 189, 1963.
7. Recknagel, R. O., A new direction in the study of carbon tetrachloride hepatotoxicity, *Life Science,* 33, 401, 1983.
8. Bruckner, J. V., MacKenzie, S., Muralidhara, S., Luthra, R., Kyle, G. M., and Acosta, D., Oral toxicity of carbon tetrachloride: Acute, subacute and subchronic studies in rats, *Fundam. Appl. Toxicol.,* 6, 16, 1986.
9. Condie, L. W., Laurie, R. D., Mills, T., Robinson, M., and Bercz, J. P., Effect of gavage vehicle on hepatotoxicity of carbon tetrachloride in CD-1 mice: Corn oil versus Tween-60 aqueous emulsion, *Fundam. Appl. Toxicol.,* 7, 199, 1986.
10. Hayes, J. R., Condie, L. W., and Borcelleca, J. F. Acute, 14-day repeated dosing, and 90-day subchronic toxicity studies of carbon tetrachloride in CD-1 mice, *Fundam. Appl. Toxicol.,* 7, 454, 1986.
11. Kaminski, N. E., Barnes, D. W., Jordan, S. D., et al., The role of metabolism in CCl_4-mediated immunosuppression, *in vitro* studies, *Toxicol. Appl. Pharmacol.*, 102, 9, 1990.

12. Narotsky, M. G. and Kavlock, K. J., A multidisciplinary approach to toxicological screening: II. Developmental Toxicity, *J. Toxicol. Environ. Health.*, 45, 145, 1995.
13. Amacher, E. D. and Zelljadt, I., The morphological transformations of Syrian hamster embryo cells by chemicals reportedly non-mutagenic to *Salmonella typhimurium*, *Carcinogenesis,* 4, 291, 1983.
14. Mirsalis, J. C., Tysen, C. K., and Butterworth, B. E., Detection of genetic carcinogens in the *in vivo-in vitro* hepatocyte DNA repair assay, *Environ. Mutagen.*, 4, 553, 1982.
15. *Carcinogenesis Bioassay of Chloroform*, Carcinogenesis Program, Natl. Cancer Institute Techn. Report, Division of Cancer Cause and Prevention, Bethesda, MD, 1976.
16. Della Porta, G., Terracini, B., and Shubik, P., Induction with carbon tetrachloride of liver cell carcinomas in hamsters, *J. Natl. Cancer Inst.*, 26, 855, 1961.
17. *Revision of the WHO Guidelines for Drinking Water Quality*, Report on the Consolidation Meeting on Organics and Pesticides, Medmenham, UK, January 30-31, 1992, 11.
18. Shah, H., Hartman, S., and Weinhouse, S., Formation of carbonyl chloride in carbon tetrachloride metabolism by rat liver *in vitro*, *Cancer Res.*, 39, 3942, 1979.

CHLOROBENZENE

Molecular Formula. C_6HCl
M = 112.56
CAS No 108-90-7
RTECS No CZ0175000
Abbreviation. CB.

Synonyms and **Trade Name.** Benzene chloride; Monohlorobenzene; Phenyl chloride; Tetrosin SP.

Properties. A colorless, very refractive liquid with a faint, not unpleasant odor. Water solubility is 500 mg/l at 20°C or 110 mg/l at 25°C.[02] Freely soluble in alcohol, ether, benzene. Taste and odor perception threshold is 0.05 mg/l,[02] or 0.01 to 0.12 mg/l,[1,01] or 0.001 to 0.003 mg/l (WHO, 1984).

Applications. CB is used in diisocyanate manufacture, in silicone resin production, and also in the production of perchlorovinyl resins and to obtain lacquers, enamels and glues. A solvent of nitrocellulose.

Exposure. Concentrations of 1.0 to 5.0 µg CB/l are found in water.

Acute Toxicity.

Observations in man. A two-year-old male who swallowed 5.0 ml to 10.0 ml of stain remover which consisted almost entirely of CB became unconscious, did not respond to skin stimuli, showed muscle spasms and became cyanotic. The child made a full recovery.[2]

Animal studies. Rats tolerated an oral dose of 1.0 g/kg BW, but 4.0 g/kg BW was lethal to all. The LD_{50} for rats was reported to range from 2.4 to 3.3 g/kg BW. Guinea pigs tolerated an oral dose of 1.6 g/kg BW, but 2.8 g/kg BW was fatal. LD_{50} is 5.06 g/kg BW. Rats and mice are more sensitive than guinea pigs. LD_{50} is 1.45 to 2.3 g/kg BW in mice, and 2.25 to 2.8 g/kg BW in rabbits. Following administration of lethal doses, ataxia, labored breathing, prostration, or lethargy were observed. Death occurs in the first three days from paralysis of the respiratory centers.[3-5]

The principal morphological effects are hepatic and renal necrosis.

Repeated Exposure. Administration of 1/5 LD_{50} revealed marked cumulative properties. K_{acc} is 1.25 (by Cherkinsky). The treated animals displayed asthenia, adynamia, and anorexia.[1] Neutrophilosis and a reduction in NS excitability was noted in rats that received 1/10 LD_{50}.[4]

In a 14-day study, the dose of 1.0 g/kg BW appeared to be lethal to rats; 0.5 g/kg BW did not affect survival and produced no clinical signs of intoxication.[5]

Short-term Toxicity. Male and female $B6C3F_1$ mice and F344 rats were given CB at the doses of 60 to 750 mg/kg BW by gavage for 13 weeks. Clinical signs of toxicity were not observed. A reduction in BW gain occurred at 250 mg/kg BW and higher doses. A marked increase in the liver weights occurred in a dose-related manner. Histology examinations revealed toxic lesions in the liver, kidney, spleen, bone marrow, and thymus of CB-exposed rats and mice. Porphyrinuria was detected at the higher doses. CB is reported to cause renal necrosis at 250 mg/kg BW.[5,6] TDI was calculated to be 0.086 mg/kg BW from the NOAEL of 60 mg/kg BW.

Beagle dogs were exposed orally to CB by capsule at doses of 27.25, 54.5, or 272.5 mg/kg BW for 13 weeks. Treatment with the highest dose caused BW loss and death. There were also changes in hematology, clinical chemistry, and urinalysis. Histological examination revealed pathologic changes in the liver, kidney, GI tract mucosa, and hematopoietic tissue. The LOAEL appeared to be 54.5 mg/kg BW for slight bile duct proliferation, cytological alterations, and leukocyte infiltration of the stroma (all in the liver). The NOAEL of 27.2 mg/kg BW was established in this study.[7]

Observations in man. Inhalation of CB in the workplace for up to 2 years resulted in CNS disturbances.

Animal studies. Long-term oral exposure to high doses affected mainly the liver, kidney and hematopoietic system. Rabbits and guinea pigs received oral doses of 0.1 and 1.0 mg/kg BW for 11 to 14.5 months. The treatment caused behavioral changes, retardation of BW gain, and altered hemogram. Histology examination revealed visceral congestion, a focal atrophic gastritis with mucosal fibrosis, parenchymatous hepatitis and dystrophy of the renal tubular epithelium (Obukhov, 1955).

Rats were dosed by gavage with 0.1 mg/kg BW for 9 months. Manifestations of the toxic action included CNS and hemopoiesis inhibition, anemia, reticulocytosis, thrombocytosis, and eosinophilia.[1]

There were no CB-related toxic effects in a 2-year study when rats and mice were given the doses of 0.03 to 0.12 mg/kg BW in corn oil for 103 weeks.[6]

Reproductive Toxicity.

Gonadotoxicity. In a two-generation reproduction study, levels of 50 to 450 ppm had no adverse effects on reproductive performance or fertility of male and female rats.[8]

Immunotoxicity. $B6C3F_1$ mice exposed to the mixture of 25 common groundwater contaminants, containing 0.2 mg CB/l, for 14 or 90 days, showed some immune function changes which could be related to rapidly proliferating cells, including suppression of hematopoietic system cells and antigen-induced antibody forming cells. There was no effect on *T*- and *B*-cell numbers in any group. Altered resistance to challenge with an infectious agent also occurred in mice given the highest concentration.[9]

There was an increase in leukocyte-phagocytic activity and in serum γ-globulin at a dose of 0.1 mg/kg BW in rats.[1]

Mutagenicity. The weight of evidence indicates that CB is not mutagenic, although it does bind to DNA *in vitro* and *in vivo.*[10]

In vivo cytogenetics. The *in vivo* level of binding, however, is low. CB did not show mutagenic potential in the sex-linked recessive lethal mutation assay with *Dr. melanogaster.*

In vitro genotoxicity. CB was not mutagenic in several bacterial and yeast systems. It has been found to be non-mutagenic in the Ames test;[025] it is unable to induce unscheduled DNA synthesis in primary cultures of hepatocytes or in mouse lymphoma cell forward mutation assay, but transformed adult rat liver epithelial cells *in vitro.*

Carcinogenicity. CB is related to benzene, because of similarities in structure, metabolism, and hematological effects in rodents; toxic human response could be predicted based upon a rodent model (NTP-85).

CB caused a slight increase in the frequency of neoplastic liver nodules at the highest tested dose of 120 mg/kg BW,[5,6] providing some but not clear evidence of carcinogenicity in male rats. Carcinogenic effect was not observed in female F344/N rats or male and female $B6C3F_1$ mice.

The biological significance of the increased incidence of neoplastic liver nodules (Doull and Abrahamson, 1986) has been questioned by the Environmental Health Committee of the Science Advisory Board (1986). Liver nodules are currently not considered necessarily to be progressive and, consequently, lethal to the host.

Carcinogenicity classifications.

U.S. EPA: D;

NTP: E - N - N - N (gavage).

Chemobiokinetics. Major metabolites of CB are *p-chlorophenyl mercapturic acid, 4-chlorocatechol* and *p-chlorophenol.*[11,12] The main route of excretion is in the urine. The toxicity of chlorinated benzenes could be explained by formation of *mercapturic acid*, for the synthesis of which sulfur-containing amino acids are used. Subsequently, the greater the number of halogen atoms in the benzene molecule, the lesser mercapturic acid is formed in the body and the lower the toxicity.[13]

Regulations. ***U.S. FDA*** (1998) listed CB (1) as a component of adhesives intended for use in packaging, transporting, or holding food in accordance with the conditions prescribed by 21 CFR part 175.105, and (2) in the manufacture of cellophane for packaging food in accordance with the conditions prescribed by 21 CFR part 177.1200.

Recommendations.

WHO (1996). Guideline value for drinking water is identified at the level of 0.3 mg/kg. However, this value far exceeds the lowest reported taste and odor perception threshold in water.[14] The levels of 0.01 to 0.12 mg/l are likely to give rise to consumer complaints of foreign odor and taste.

U.S. EPA (1999). Health Advisory for a longer-term exposure is 7.0 mg/l.

Standards.

U.S. EPA (1999). MCL and MCLG: 0.1 mg/l.

Russia (1995). MAC and PML: 0.02 mg/l.

References:

1. Varshavskaya, S. P., Comparative toxicological features of chlorobenzene and dichlorobenzene (*ortho-* and *para*-isomers) in sanitary protection of water bodies, *Gig. Sanit.*, 10, 15, 1968 (in Russian).
2. Reich, H., Puran (monochlorobenzene) poisoning in 2-year-old child, *Samml. von Vergiftungsfallen*, 5, 193, 1934.
3. Sanotsky, I. V. and Ulanova, I. P., *Criteria of Safety in Assessing the Danger of Chemical Compounds*, Meditsina, Moscow, 1975, 328 (in Russian).
4. Shamilov, T. A., in Proc. Sci. Conf. *Problems Occup. Hyg.*, Azerb., Efendy-zade Institute Ind. Hygiene Occup. Diseases, Sumgait, 1970, 43 (in Russian).
5. Kluwe, W. M., Dill, G., Persing, R., et al., Toxic responses to acute, subchronic and chronic administrations of monochlorobenzene to rodents, *J. Toxicol. Environ. Health.*, 15, 745, 1985.
6. *Toxicology and Carcinogenicity Studies of Chlorobenzene in Fischer 344/N Rats and B6C3F$_1$ Mice*, NTP Technical Report Series No 261, Research Triangle Park, NC, NIH Publ. No 86-2517, 1985.
7. Knapp, W. K., Busey, W. M., and Kundzins, W., Subacute oral toxicity of monochlorobenzene in dogs and rats, Abstract, *Toxicol. Appl. Pharmacol.*, 19, 393, 1971.
8. Nair, R. S., Barter, J. A., and Schroeder, R. E., A two generation reproduction study with monochlorobenzene vapor in rats, *Fundam. Appl. Toxicol.*, 9, 678, 1987.
9. Germolec, D. R., Young, R. S. H., Ackermann, M. P., et al., Toxicology study of a chemical mixture of 25 ground water contaminants. II. Immunosuppression in B6C3F$_1$ mice, *Fundam. Appl. Toxicol.*, 13, 377, 1989.
10. Grilli, S., Arfrllini, G., Colacci, A., et al., *In vivo* and *in vitro* covalent binding of chlorobenzene to nucleic acids, *Japan. J. Cancer Res.*, 76, 745, 1985.
11. Lindsay-Smith, J. R., Shaw, B. A. J., and Foulkes, D. M., Mechanisms of mammalian hydroxylation: Some novel metabolites of chlorobenzene, *Xenobiotica*, 2, 215, 1972.
12. Parke, D. V. and Williams, R. T., The metabolism of halogenobenzenes, (a) *Meta*-dichlorobenzene, (b) Further observations on the metabolism of chlorobenzene, *Biochem. J.*, 59, 415, 1955.
13. Williams, R. T., *Detoxication Mechanisms*, London, 1947.
14. *Revision of the WHO Guidelines for Drinking Water Quality*, Report on the Consolidation Meeting on Organics and Pesticides, Medmenham, UK, January 30-31, 1992.

CHLOROETHANE

Molecular Formula. C_2H_5Cl

M = 64.52

CAS No 75-00-3

RTECS No KH7525000

Abbreviation. CE.

Synonyms and **Trade Names.** Aethylis chloride; Chlorethyl; Chloryl; Ether chloratus; Ether hydrochloric; Ether muriatic; Ethyl chloride; Hydrochloric ether; Narcotile.

Properties. Readily volatile liquid, or gas with an etherial odor and burning taste (at room temperature and pressure), with a characteristic etherial odor and burning taste. Solubility in water is 5.74 g/l at 20°C, soluble in ethanol.[020]

Applications. CE is used in the production of ethyl cellulose (which is used in paper coatings), adhesives and molded plastics, ethylhydroxyethyl cellulose, in the manufacture of polystyrene. It is also used as a refrigerant and a solvent. CE appeared to be a local anaesthetic agent because of its rapid cooling action as it vaporizes.

Acute Toxicity. CE is a potent promoter of cardiac arrhythmia. It potentiated the toxicity of ethyl alcohol in rats.[1]

Allergenic Effect. Allergic contact dermatitis due to exposure to CE has been reported.[2]

Reproductive Toxicity.

Observations in man. In a group of 378 women exposed to CE and other substances, 34% had pathological changes in the genital organs related to exposure duration.[3]

Mutagenicity.

In vitro genotoxicity. CE showed mutagenic activity in HPRT test with Chinese hamster ovary cells in the presence and in the absence of *S9* mix.[4] CE was found to be mutagenic to *Salmonella*,[5] but it was negative in an assay to measure transformation of mouse cells to cancer cells in culture.[5]

In vivo cytogenetics. Both unscheduled DNA synthesis and micronucleus assay failed to detect any indication of genotoxicity in mice exposed to 25,000 ppm CE.[6]

Carcinogenicity. F344 rats and $B6C3F_1$ mice were exposed to CE at concentrations of 15,000 ppm. The treatment caused a high incidence of endometrial uterine carcinomas (female mice but not rats).[4]

Carcinogenicity classification. An IARC Working Group concluded that there is limited evidence for the carcinogenicity of CE *in experimental animals*; CE is not classifiable as to its carcinogenicity to *humans*.

IARC: 3.

Chemobiokinetics. Following inhalation exposure at about 5.0 mg ^{38}Cl-CE to human volunteers, about 30% of radioactivity was eliminated in the breath within 1 hour. CE has been quickly eliminated from the body. Urinary excretion appeared to be insignificant.[7]

Regulations.

EU (1990). CE is available in the *List of authorized monomers and other starting substances which may continue to be used for the manufacture of plastic materials and articles intended to come into contact with foodstuffs pending a decision on inclusion in Section A* (*Section B*).

U.S. FDA (1998) approved the use of CE in the manufacture of resinous and polymeric coatings for the food-contact surface of articles intended for use in producing, manufacturing, packing, processing, preparing, treating, packaging, transporting, or holding food in accordance with the conditions prescribed in 21 CFR part 175.300.

References:

1. Landry, T. D. et al., Ethyl chloride: a two-week inhalation toxicity study and effects on liver non-protein sulfhydryl concentrations, *Fundam. Appl. Toxicol.*, 2, 230, 1982.
2. van Ketel, W. G., Allergic contact dermatitis from propellants in deodorant sprays in combination with allergy to ethyl chloride, *Contact Dermat.*, 2, 115, 1976.
3. Shirokov, O. M., Gynecological morbidity in workers occupied in the manufacture of ethylenediamine and other chlorinated hydrocarbons, *Gig Sanit.*, 7, 107. 1976 (in Russian).
4. Ebert, R., Feddke, N., Certa, H., Wigand, H. J., Regnier, J. F., Marshall, R., and Dean, S. W., Genotoxicity studies with chloroethane, *Mutat. Res.*, 322, 33, 1994.
5. Milman, H. A., Story, D. L., Riccio, E. S., Sivak, A., Tu, A. S., Williams, G. M., Tong, C., and Tylson, C. A., Rat liver foci and *in vitro* assays initiating and promoting effects of chlorinated ethanes and ethylenes, *Ann. NY Acad. Sci.*, 534, 521, 1988.
6. Tu, A. S., Murray, T. A., Hatch, K. M., Sivak, A., and Milman, H. A., *In vitro* transformation of BALB/c-3T3 cells by chlorinated ethanes and ethylenes, *Cancer Lett.*, 28, 85, 1985.
7. Morgan, A., Black, A., and Belcher, D. R., Studies on the absorption of halogenated hydrocarbons and their excretion in breath using ^{38}Cl tracer techniques, *Ann. Occup. Hyg.*, 15, 273, 1970.

2-CHLOROETHANOL

Molecular Formula. C_2H_5ClO
M = 80.52
CAS No 107-07-3
RTECS No KK0875000
Abbreviation. CE.

Synonyms. β-Chloroethyl alcohol; Ethylene chlorohydrin; 2-Ethylene glycol, chlorohydrine.

Properties. Colorless, volatile liquid with an odor of ethyl alcohol. Miscible with water at all ratios. Odor perception threshold is 50 mg/l.

Applications. Used in the epoxy resin manufacture.

Acute Toxicity. LD_{50} is 71 mg/kg BW in rats, and 91 mg/kg BW in mice. CNS functions are altered by high doses.[1]

Repeated Exposure failed to reveal cumulative properties. Rats tolerated administration of 1/5 LD_{50} for 20 days.

Long-term Toxicity. Rats consumed the diet containing 0.01% to 0.08% CE for at least 220 days. The treatment did not affected BW gain. Higher doses, namely 0.12% to 0.24% CE, retarded growth of rats. Histological examination revealed no changes in the visceral organs.[03]

In a 6-month study, rats administered 5.0 mg/kg BW dose displayed hyperholesterolemia, a disorder of some liver functions, and changes in ascorbic acid contents in the liver and blood. The treatment caused nephrotoxic effect and impairment of pancreatic excretion. Gross pathology examination revealed necrotic areas in the GI tract mucosa, parenchymatous hydropic dystrophy of the liver, and renal dilatation.[1]

Reproductive Toxicity.

Embryotoxicity. Pregnant mice were exposed to CE during organogenesis. Doses up to 227 mg/kg BW in the drinking water or 100 mg/kg BW given by gavage produced no malformations. With the gavage dosing, both maternal and fetal weights were reduced.[2]

Teratogenicity. CE was not found to produce teratogenic effect in mice and rabbits.[3]

Mutagenicity.

In vitro genotoxicity. CE was found to be positive in *Salmonella* mutagenicity bioassay.[4]

In vivo cytogenetics. Rats were given by gavage a single or two successive doses equal to 0.5 LD_{50}. CE did not induce genotoxic effect *in vivo*.[5]

CE did not induce sex-linked recessive lethal mutations in *Dr. melanogaster*.[4]

Carcinogenicity. Rats received single doses of 10 or 2.5 mg/kg BW. Carcinogenic effect was not observed.[6]

Chemobiokinetics. Following ingestion of CE by rats, liver glutathione was rapidly depleted and *S-carboxymethylglutathione* was formed. Toxicity of CE is likely due to its conversion to *chloroacetaldehyde in vivo*.[7] CE is found to be oxidized to *chloroacetic acid*.[8]

Rats were administered single oral doses of 5.0 and 50 mg ^{14}C-labeled CE/kg BW. Radioactivity was rapidly eliminated mainly in the urine.[9]

Standards. ***Russia*** (1995). MAC and PML: 0.1 mg/l.

References:

1. Semenova, V. N., Kazanina, S. S., Fedyanina, V. N., et al., Data on substantiation of Maximum Allowable Concentration for ethylene chlorohydrin in water bodies, *Gig. Sanit.*, 8, 13, 1978 (in Russian).
2. Courtney, K. D., Andrews, J. E., and Grady, M., Teratogenic evaluation of ethylene chlorohydrin, *J. Environ. Sci. Health*, B17, 381, 1982.
3. LaBorde, J. B., Kimmel, C. A., Jones-Price, C., et al., Teratogenic evaluation of ethylene chlorohydrin in mice and rabbits, *Toxicologist*, 2, 71, 1982.
4. *Toxicology and Carcinogenesis Studies of 2-Chloroethanol (Ethylene Chlorohydrin) in F344/N Rats and Swiss CD-1 Mice (Dermal Studies),* NTP Technical Report Series No 275, Research Triangle Park, NC, 1985.

5. Allavena, A., Martelli, A., Robbiano, L., and Brambillo, G., Evaluation in a battery of *in vivo* assays of four *in vitro* genotoxins proved to be non-carcinogens in rodents, *Teratogen. Carcinogen. Mutagen.*, 12, 31, 1992.
6. Dunkelberg, H., Carcinogenic activity of ethylene oxide and its reaction products 2-chloroethanol, 2-bromoethanol, ethylene glycol and diethylene glycol. II. Testing of 2-chloroethanol and 2-bromoethanol for carcinogenic activity, *Zentralbl. Bakteriol. Mikrobiol. Hyg. B.*, 177, 269, 1983.
7. Johnson, M. K., Metabolism of chloroethanol in the rat, *Biochem. Pharmacol.*, 16, 185, 1967.
8. Browning, E., *Toxicity and Metabolism of Industrial Solvents*, American Elsevier, New York, 1965, 398.
9. Grunov, W. and Altman, H. J., Toxicokinetics of chloroethanol in the rat after single oral administration, *Arch. Toxicol.*, 49, 275, 1982.

CHLOROFORM

Molecular Formula. $CHCl_3$
M = 119.39
CAS No 67-66-3
RTECS No FS9100000

Synonyms. Formyl trichloride; Methyl trichloride; Trichloromethane.

Properties. Colorless, very volatile, sweet-tasting liquid with an unpleasant odor at 25°C. Solubility in water is 8.0 g/l, readily soluble in organic solvents. Odor perception threshold is 2.4 mg/l.[02]

Applications. Used in the manufacture of fluoropolymers, rubbers, and resins. An important extraction solvent for resins, gums, etc. A cosmetic ingredient.

Exposure. A by-product of drinking water chlorination.

Acute Toxicity.

Observations in man. C. is a CNS depressant; it affects the liver and kidney functions in humans and animals. In humans, LD_0 is approximately 44 g.[022] A fatal dose may be as small as 211 mg/kg BW, with death due to respiratory or cardiac arrest.[1]

Animal studies. The LD_{50} values are 1.25 g/kg BW in rats, 0.1 g/kg BW in mice, 0.82 g/kg BW in guinea pigs, 9.83 g/kg BW in BW in rabbits, and 2.25 g/kg BW in dogs.[2]

According to Kimura et al., LD_{50}s are reported to be 0.3, 0.9, and 0.8 mg/kg BW in 14-day old rats, young adults rats, and older adults, respectively.[3]

In ninety-day-old male F344 rats, C. induced dose-dependent hepatotoxicity. Serum ALT, AST, and sorbitol dehydrogenase were elevated significantly over control at 1.5, 1.0, and 0.5 mmol/kg BW. NOAEL and LOAEL for liver toxicity are 0.25 and 0.5 mmol/kg BW.[4]

Repeated Exposure. C. appears to be both hepatotoxic and neurotoxic in most animal species. According to Larson et al., the target organs of the treatment with C. were liver, kidneys, and nasal passages. Female F344 rats were exposed orally to C. in corn oil at doses of 34, 100, 200, or 400 mg/kg BW for 4 consecutive days or for 3 weeks (5 days/week). After administration of 100 mg/kg BW and more, mild degenerative centrilobular changes and dose-dependent increase in the hepatocyte labeling index were observed. The treatment also caused degeneration of the olfactory epithelium and superficial Bouman's glands.[5]

Administration of 1/30 LD_{50} for 1 month produced adipose dystrophy, cirrhotic, and necrotic lesions in the liver (Miklashevsky et al., 1966).[6]Prolonged exposure to doses more than 15 mg/kg BW can affect the kidney, liver, and thyroid. In a 14-day study, the NOAEL of 125 mg/kg BW was identified in mice (based on elevated serum enzyme levels).[6]

Long-term Toxicity.

Observations in man. A number of epidemiology studies tend to support the finding of an increased risk of bladder, colon, and rectal cancer from exposure to chlorinated water. Positive correlation between C. levels in drinking water and mortality from stomach, large intestine, rectum, and bladder cancer was demonstrated by regression statistical analysis.[7]

Although C. appears to be the single largest constituent in chlorinated water, these studies do not directly prove that C. is a human carcinogen since chlorinated water contains many other chlorination by-products.[8]

Animal studies. Administration of 1/50 LD_{50} for 5 months produced adipose dystrophy, cirrhotic and necrotic lesions of the liver.[6] The LOAEL of 15 mg/kg BW was identified for liver fatty cysts in dogs exposed to C.[9] The calculated ADI and DWEL are 0.01 mg/kg BW and 0.5 mg/l, respectively. Sprague-Dawley rats were dosed by gavage with C. in toothpaste-based vehicle at up to 60 mg/kg BW for 80 weeks. The treatment resulted in retardation of BW gain, a decrease in plasma cholinesterase activity, and a significant decrease in relative liver weights in female rats. The LOAEL of 60 mg/kg BW based on decreases in BW and plasma cholinesterase activity was identified in this study.[10]

$B6C3F_1$ mice were exposed to 600 or 800 mg C./l (86 or 258 mg C./kg BW) via drinking water for 24 or 52 weeks.[11] Manifestations of the toxic action included decreased BW, focal areas of cellular necrosis in the kidneys and liver, and focal areas of hepatic lipid accumulation in the high-dosed group.

Reproductive Toxicity.

Embryotoxicity. Rats received 100 mg/kg BW by gavage on days 6 through 15 of gestation. The treatment caused fetal toxicity associated with evident maternal toxicity.[12]

No significant teratogenicity was found after oral administration of up to 126 mg/kg BW to pregnant rats or up to 50 mg/kg BW to pregnant rabbits. However, pronounced maternal toxicity was observed in animals given 50 mg/kg BW (both rats and rabbits).[13] The maternal NOAEL was 20 mg/kg BW in rats, and 35 mg/kg BW in rabbts, the fetal NOAEL was 50 mg/kg BW in rats.

Mutagenicity. Current review of the genotoxicity data on C. demonstrates a pattern of negative results.[14] Neither C. nor its metabolites appear to interact directly with DNA or possess genotoxic activity (IPCS WHO, 1994).

In vivo cytogenetics. Testing in mammalian systems was largely negative. There is no convincing evidence that C. is genotoxic, as two *in vivo* studies claiming positive results were of poor quality and were outweighed by seven better-conducted studies. C. was shown to cause genotoxic effect in a host-mediated assay in male mice and in sperm head abnormality assay in mice.[15,16]

In vitro genotoxicity. C. caused SCE in cultured human lymphocytes and in mouse bone marrow cells exposed *in vivo.*[17]

C. caused mutagenic effect in *Dr. melanogaster*. A host-mediated assay using mice indicated that C. was metabolized *in vivo* to a form mutagenic to *Salmonella typhimurium* strain *TA1537*. Likewise urine extracts from C.-treated mice were mutagenic.[15]

According to Gocke et al., C. was negative in the *Basc test* on *Dr. melanogaster.*[18]

Carcinogenicity. $B6C3F_1$ mice were exposed to C. in drinking water (300 or 1800 ppm and 120, 240, or 480 ppm) over a period of 1 month. The treatment reduced both hepatotoxicity and the enhanced cell proliferation elicited in response to C. administration by gavage in corn oil to female $B6C3F_1$ mice. Hence, C. administered in drinking water prevents liver cancer induced by C. administered by gavage in corn oil.[19]

C. induced hepatocellular carcinomas in mice when administered by gavage in oil-based vehicles (but not in drinking water), and renal tubular adenomas and adenocarcinomas in male rats regardless of the carrier vehicle. Variability of the obtained result in carcinogenicity testing relates to the species, strain, and sex of the animals tested and also to the vehicle in which C. has been orally administered in each study.

Good correlation was reported between certain cytotoxic effects of C. and the occurrence of liver tumor in female $B6C3F_1$ mice (via drinking water and corn oil gavage).[20]

C. may induce tumor through a non-genotoxic mechanism. In the NCI study (1976), doses of 90 and 180 mg/kg BW were administered in corn oil by gavage to Osborne-Mendel rats and $B6C3F_1$ mice for 78 weeks. The treatment resulted in kidney epithelial tumors in male rats (8.0% in the low dose, 24% in the high dose group, 0% in the control). In another NCI study, male mice received 150 and 300 mg/kg BW and female mice received doses of 250 and 500 mg/kg BW for 78 weeks. Statistically significant incidence of hepatocellular carcinomas in all the treated groups was found.[21]

Roe et al. observed malignant and benign kidney tumors in male ICI mice given not less than 60 mg/kg BW dose by gavage in toothpaste base or in arachis oil for 80 weeks.[22]

Jorgenson et al. found a dose-dependent increased incidence of renal tubular adenomas and adenocarcinomas in male rats (concentration of up to 160 mg C./l in drinking water).[23]

Meanwhile, according to the data reported by EPA Health Effects Laboratory (1985), 1800 ppm C. in drinking water caused no carcinogenicity promotion in Swiss mice. C. should be regarded as a nongenotoxic animal carcinogen.

Carcinogenicity classifications. An IARC Working Group concluded that there is sufficient evidence for the carcinogenicity of C. in *experimental animals* and there were inadequate data available to evaluate the carcinogenicity of C. in *humans*.

The Annual Report of Carcinogens issued by the U.S. Department of Health and Human Services (1998) defines C. to be a substance which may reasonably be anticipated to be carcinogen, i.e., a substance for which there is a limited evidence of carcinogenicity in *humans* or sufficient evidence of carcinogenicity in *experimental animals*.

IARC: 2B;

U.S. EPA: B2;

EU: 3;

NTP: P - N - P - P (gavage).

Chemobiokinetics.

Observations in man. In humans, C. is shown to cross the placenta.[24] *Carbon dioxide* seems to be the major metabolite. Almost the whole oral dose (0.1 g to 1.0 g) administered to humans was excreted in the form of CO_2 or unchanged via the lungs.[25]

Animal studies. C. is readily and rapidly absorbed from the GI tract. More than 90% of ^{14}C-C. administered orally in olive oil to mice, rats, and monkeys (60 mg/kg BW) was absorbed in 48 hours. Accumulation occurs in the fatty tissues and liver. Lesser amounts were found in the blood, brain, kidney, etc. Animals demonstrated a marked species difference in C. metabolism.[26]

Hepatotoxicity of C. is due, at least in part, to its metabolite to *phosgene*, which is known to cross-link DNA.[27] Cytochrome *P-450* in rat liver microsomes metabolizes C. to $COCl_2$ by a rate-determining oxidation of the carbon-hydrogen bond of C. to form presumably *trichloromethanol* ($HOCCl_3$). Formation of CO_2 may be a sequence of C. degradation to *methylene chloride* and then to *formaldehyde*, *formic acid* and CO_2.[28] C. is shown to produce other severe reactive metabolic intermediates in the GI tract.

Regulations. *U.S. FDA* (1998) listed C. as (1) a component of adhesives used as components of articles intended for use in packaging, transporting, or holding food in accordance with the conditions prescribed by 21 CFR part 175.105, and (2) in the manufacture of cellophane for packaging food in accordance with the conditions prescribed by 21 CFR part 177.1200.

Recommendations. *WHO* (1994). TDI: 0.01 mg/kg BW. Guideline value for drinking water: 0.2 mg/l.

Standards.

U.S. EPA (1999). MCL: 0.1 mg/l, MCLG: zero.

Russia (1995). MAC: 0.2 mg/l.

References:

1. *Health Assessment Document for Chloroform*, Final Report, U.S. EPA Office of Research & Development, EPA-600/8-84-004F, Washington, D.C., 1985.
2. *Cahier de Notes Documentaries*, Inst. Natl. Rech. Secur., 26 (Suppl.), 87, 1987.
3. Kimura, E. T., Ebert, D. M., and Dodge, P.W., Acute toxicity and limits of solvents residue for sixteen organic solvents, *Toxicol. Appl. Pharmacol.*, 19, 699, 1971.
4. Keegan, T. E., Simmons, J. E., and Pegram, R. A., NOAEL and LOAEL determinations of acute hepatotoxicity for chloroform and bromodichloromethane delivered in an aqueous vehicle to F344 rats, *J. Toxicol. Environ. Health,* 55, 65, 1998.
5. Larson, J. L., Wolf, D. C., Mery, S., Morgan, K. T., and Butterworth, B. E., Toxicity and cell proliferation in the liver, kidneys and nasal passages of female F344 rats, induced by chloroform administered by gavage, *Food Chem. Toxicol.*, 33, 443, 1995.
6. Munson, A. E., Sain, L. E., Sanders, V. M., et al., Toxicology of organic drinking water contaminants: trichloromethane, bromodichloromethane, dibromochloromethane and tribromomethane, *Environ. Health Perspect.*, 46, 117, 1982.

7. Hogan, M. D., Chi, P., Hoel, D. G., et al., Association between chloroform levels in finished drinking water supplies and various site-specific cancer mortality rates, *J. Environ. Pathol. Toxicol.*, 2, 873, 1979.
8. Wilkins, J. R., Reiches, N. A., and Kruse, C. W., Organic chemical contaminants in drinking water and cancer, *Am. J. Epidemiol.*, 110, 420, 1979.
9. Heywood, R., Sortwell, R. J., Noel, P. R. B., et al., Safety evaluation of toothpaste containing chloroform, III. Long-term study in beagle dogs, *J. Environ. Pathol. Toxicol.*, 2, 835, 1979.
10. Palmer, A. K., Street, A. E., Roe, F. J. C., Worden, A.M., and van Abbe, N. J., Safety evaluation of toothpaste containing chloroform, II. Long-term studies in rats, J. *Environ. Pathol. Toxicol.*, 2, 821, 1979.
11. Klaunig, J. E., Ruch, R. J., and Pereira, M. A., Carcinogenicity of chlorinated methane and ethane compounds administered in drinking water to mice, *Environ. Health Perspect.*, 69, 1986.
12. Ruddick, J. A., Villeneuve, D. C., Chuand, I., and Valli, V. E., Teratogenicity assessment of four trihalomethanes, *Teratology*, 21, 66A, 1980.
13. Thompson, D. J., Warner, S. D., and Robinson, V. B., Teratology studies on orally administered chloroform in the rat and rabbit, *Toxicol. Appl. Pharmacol.*, 29, 348, 1974.
14. Golden, R. J., Holm, S. E., Robinson, D. E., Julkunen, P. H., and Reese, E. A., Chloroform mode of action: Implication for cancer risk assessment, *Regul. Toxicol. Pharmacol.*, 26, 142, 1997.
15. Agustin, J. S. and Lim-Syliano, L., Mutagenic and clastogenic effects of cloroform, *Bull. Phil. Biochem. Soc.*, 1, 17, 1978.
16. Land, P. C., Owen, E. L., and Linde, H. W., Morphologic changes in mouse spermatogen after exposure to inhalation anesthetics during early spermatogenesis, *Anesthesiology*, 54, 53, 1981.
17. Morimoto, K. and Koizumi, A., Trihalomethanes induce sister chromatid exchanges in human lymphocytes *in vitro* and mouse bone marrow cells *in vivo*, *Environ. Mutagen.*, 32, 72, 1983.
18. Gocke, E., King, M.-T., Eckhardt, K., and Wild, D., Mutagenicity of cosmetic ingredients licensed by the European Communities, *Mutat. Res.*, 90, 91, 1981.
19. Pereira, M. A. and Drothatis, M., Chloroform in drinking water prevents development of the liver cancer induced by chloroform administered by gavage in corn oil to mice, *Fundam. Appl. Toxicol.*, 37, 82, 1997.
20. Chiu, N., Orme-Zavaleta, J., Chiu, A., Chen, C., and Blancato, J., Cancer risk assessment for chloroform, *Toxicologist*, Abstracts of SOT 1996 Ann. Meeting, Abstract No. 587, Issue of *Fundam. Appl. Toxicol.*, 30, Part 2, March 1996.
21. *Carcinogenesis Bioassay of Chloroform*, Natl. Cancer Institute Report, NTIS PB-264018, Springfield, 1976.
22. Roe, F. J. C., Palmer, A. K., Worden, A. N., et al., Safety evaluation of toothpaste containing chloroform, I. Long-term studies in mice, *J. Environ. Pathol. Toxicol.*, 2, 799, 1979.
23. Jorgenson, T. A., Meierhenry, E. F., Rushbrook, C. J., Bull, R. J., and Robinson, M., Carcinogenicity of chloroform in drinking water to male Osborne-Mendel rats and female $B6C3F_1$ mice, *Fundam. Appl. Toxicol.*, 5, 760, 1985.
24. Dowty, B. J., Laseter, J. L., and Storer, J., The transplacental migration and accumulation in blood of volatile organic constituents, *Pediatr. Res.*, 10, 696, 1976.
25. Fry, B. J., Taylor, T., and Hathway, D. E., Pulmonary elimination of chloroform and its metabolite in man, *Arch. Int. Pharmacodyn.*, 196, 98, 1972.
26. Brown, D. M., Langley, P. F., Smith, D., et al., Metabolism of chloroform, I. The metabolism of ^{14}C-chloroform by different species, *Xenobiotica*, 4, 151, 1974.
27. Pohe, L. K., George, J. W., Martin, J. L., et al., Deuterium isotope effect in *in vivo* bioactivation of chloroform to phosgene, *Biochem. Pharmacol.*, 28, 561, 1979.
28. Rubinstein, D. and Kanics, L., The conversion of carbon tetrachloride and chloroform to carbon dioxide by rat liver homogenates, *Can. J. Biochem.*, 42, 1577, 1964, Reviewed in U.S. EPA, 1985.

CYCLOHEXANE

Molecular Formula. C_6H_{12}
M = 84.16
CAS No 110-82-7
RTECS No GU6300000
Abbreviation. CH.

Synonyms. Hexahydrobenzene; Hexamethylene.

Properties. Colorless liquid with a penetrating odor. Solubility in water is 55 mg/l;[02] miscible with alcohol. Odor perception threshold is 1.0 g/l,[1] 0.3 mg/l,[2] or 0.011 mg/l.[02]

Applications. A solvent. Used in the production of polyamides and synthetic fibers.

Acute Toxicity. In mice, LD_{50} is 4.7 g/kg BW. Poisoned animals displayed CNS inhibition with subsequent excitation and clonic convulsions. Death within 3 to 5 minutes, possibly as a result of respiratory failure.[3] Rats are less susceptible though they show age-dependent susceptibility to CH: at BW up to 50 g, LD_{50} is 8.0 ml/kg BW; at 80 to 160 g, it is 39 ml/kg BW, and at 300 to 470 g, it is 16.5 ml/kg BW.[4]

Short-term Toxicity. Rats received 200 and 400 mg CH/kg BW. The treatment did not affect general condition, behavior, blood morphology, or liver function. A 400 mg/kg BW dose lowered catalase and cholinesterase activity.

Mutagenicity.

In vitro genotoxicity. CH is found to be positive in *Salmonella* mutagenicity assay.[5]

In vivo cytogenetics. CH caused CA in rats exposed by inhalation.

Chemobiokinetics. CH is metabolized to form metabolites *cyclohexanol* (38%) and *(q)-trans-1,2-cyclohexanediol* (7.0%) which are excreted in rabbits as *glucuronides* in the urine. 30% of administered CH are excreted unchanged via the lungs and 9.0% as CO_2.[021]

Regulations. ***U.S. FDA*** (1998) regulates CH (1) as a component of adhesives used in articles intended for use in packaging, transporting, or holding food in accordance with the conditions prescribed by 21 CFR part 175.105; (2) as a component of the uncoated or coated food-contact surface of paper and paperboard that may be safely used in producing, manufacturing, packing, transporting, or holding dry food in accordance with the conditions prescribed by 21 CFR part 176.180; and (3) as a defoaming agent that may be safely used as a component of food-contact articles in accordance with the conditions prescribed in 21 CFR part 176.200.

Standards. ***Russia*** (1995). MAC and PML: 0.1 mg/l.

References:

1. Savelova, V. A. and Klimkina, N. V., in *Protection of Water Reservoirs against Pollution by Industrial Liquid Effluents*, S. N. Cherkinsky, Ed., Meditsina, Moscow, Issue No 6, 1964, 30 (in Russian).
2. Ogata, M. and Meyake, J., *J. Water Res.*, 7, 1493, 1973.
3. Meleshchenko, K. F., *Prevention of the Pollution of Water Bodies by Liquid Effluents of Chemical Plants*, Zdorov'ya, Kiyv, 1971, 61 (in Russian).
4. Kimura, E. T., Ebert, D. M., Dodge, P. W., et al., Acute toxicity and limits of solvent residue for sixteen organic solvents, *Toxicol. Appl. Pharmacol.*, 19, 699, 1971.
5. McCann, J., Choi, E., Yamasaki, E., et al., Detection of carcinogens and mutagens in the *Salmonella*/microsome test: assay of 300 chemicals, *Proc. Natl. Acad. Sci. USA*, 72, 5135, 1975.

1,2- and 1,4-DICHLOROBENZENES

Molecular Formula. $C_6H_4O_2$
M = 147.01

	CAS No	RTECS No
1,2-Dichlorobenzene	95-50-1	CZ4500000
1,4- Dichlorobenzene	106-46-7	CZ4550000
mixed Dichlorobenzenes	25321-22-6	

Abbreviations. 1,2-DCB, 1,4-DCB and DCBs.

Synonym and **Trade Names.** Di-chloricide; *o*- and *p*-Dichlorobenzene; Paramoth.

Properties. DCB are moderately soluble in water: 140 mg/l *o*-DCB and 79 mg/l *p*-DCB at 25°C.[02] Odor perception thresholds of 0.001 to 0.01 mg/l or 0.024 mg/l were identified for 1,2-DCB, and 0.0003 to 0.03 mg/l or 0.011 mg/l [02] for 1,4-DCB. Taste thresholds of 0.001 and 0.002 mg/l have been reported for 1,2-DCB and 1,4-DCB, respectively.[1,2]

Applications and **Exposure.** Used as plasticizers and solvents in the production of high molecular weight compounds; insecticides.

Acute Toxicity.

1,2-DCB and 1,4-DCB are of low acute toxicity. The acute toxic effects of 1,2-DCB and 1,4-DCB have shown a similar profile in all the species tested and depend on the route of adminstration.[3]

Oral LD_{50} of about 500 to 3860 mg/kg BW has been reported in rats, rabbits and guinea pigs; however, most figures indicate that 1,4-DCB is less acutely toxic than 1,2-DCB. After oral administration, poisoning symptoms included increased lacrimation, salivation, and excitation, followed by ataxia, dyspnea, and death from respiratory paralysis, usually within three days. On autopsy, the animals were found to have enlarged livers with necrotic areas, hemorrhages of the stomach, and necrotic changes to the kidneys and brain. $B6C3F_1$ mice received a single dose of 300, 600, and 1800 mg/kg BW, respectively. The treatment with *o*- and *m*-DCB caused significant elevation of liver weight and ALT activity as well as extensive liver cell necrosis. *p*-DCB induced hepatocyte cell prolifera tion but caused only slight hepatocyte injury.[4]

Repeated Exposure. The essential signs of 1,4-DCB toxicity include such effects as liver and kidney damage, porphyria, and splenic weight changes.

Observations in man. In humans, there were anorexia, liver effects, and blood dyscrasias.

Animal studies. Sprague-Dawley rats were administered 1,2-DCB in corn oil by oral gavage at doses from 37.5 to 300 mg/kg BW for 10 days. Administration of 300 mg/kg BW caused significant decrease in final BW, organ weights (heart, kidney, spleen, tested, and thymus). There was an increase in absolute and relative liver weight, and a significant increase in the incidence of hepatocellular necrosis in males.[5]

Female Wistar rats fed 250 mg 1,2-DCB/kg BW for 3 days showed an increase in the liver weights and microsomal protein content, and enhanced aminopyrine demethylase and aniline hydroxylase activities in the liver.[6]

Short-term Toxicity. DCBs toxicity is dependent on halogen position within the parent compound. 1,2-DCB is the basic hepatic toxicant while 1,4-DCB is the least.

1,2-DCB.

Sprague-Dawley rats were administered 1,2-DCB in corn oil by oral gavage at doses of 25, 100, or 400 mg/kg BW for 90 days. Treatment of males with 400 mg/kg BW caused a significant decrease in BW gain and organ weight (spleen). The only clinical chemistry parameter increased appeared to be a level of ALT activity (100 and 400 mg/kg BW). In the highest dose-treated group, an increase in total bilirubin was also noted. Histological examination of these animals revealed hepatocellular lesions including centrilobular degeneration and hypertrophy, and single cell necrosis. The NOAEL of 25 mg/kg BW was established in this study.[5]

In a 13-week oral study in rats and mice, the doses of 30 to 500 mg 1,2-DCB/kg BW were tested. Animals exhibited histopathology indicative of hepatic centrilobular necrosis and degeneration, as well as lymphoid depletion of the spleen and thymus. The dose of 125 mg/kg BW appeared to be the NOEL.[7]

1,4-DCB

1,4-DCB in similar studies at higher doses produced similar pathology in the spleen, thymus, and liver (necrosis, degenerations, and porphyria). Both species showed hematopoietic hyperplasia of the bone marrow in survivors at the dose of 1500 mg/kg BW. The NOAEL was 150 mg/kg BW in rats and 337.5 mg/kg BW in mice.[7]

Long-term Toxicity. The liver was more sensitive than the kidney for DCB toxicity.[8]

1,2-DCB.

In a 2-year gavage study, exposure to the dose of 120 mg/kg BW affected mainly the liver and kidney of F344 rats. The NOAEL of 60 mg/kg BW was identified; TDI was calculated to be 0.43 mg/kg BW.[7]

Rats were gavaged with 18.8, 188, or 376 mg 1,2-DCB/kg BW for 192 days. Liver and kidney weights were increased at 188 mg/kg dose level; liver pathology and increased spleen weights are reported at 376 mg/kg BW.[9]

1,4-DCB.

In a similar 2-year rat study the LOAEL appeared to be 150 mg/kg BW, and TDI was calculated to be 0.0107 mg/kg BW (UF = 10000, because a LOAEL is used and the toxic endpoint is carcinogenicity).[7] Rabbits received 0.06 mg 1,4-DCB/l in their drinking water for 7 months. This intake produced changes neither in hematological indices nor in the ratio of inorganic to total sulfates in the urine.[10]

Reproductive Toxicity. DCBs showed no evidence of teratogenicity and they appear to be of very low embryotoxic potential.

Embryo- and ***gonadotoxicity.*** 1,4-DCB has been administered orally in olive oil to two generations of rats at the doses of 30, 90, or 270 mg/kg BW. Treatment with the 90 mg/kg BW dose caused massive damage in pups. Exposure to 30 mg 1,4-DCB/kg BW caused no effects on fertility.[11]

Teratogenicity. In an oral study in Sprague-Dawley rats, DCBs were reported to be non-teratogenic at doses of 50 to 200 mg/kg BW given on days 6 to 15 of gestation.[12,13] There was no evidence of teratogenic effects in rabbits exposed by inhalation.[14,15]

Mutagenicity. 1,2-DCB and 1,4-DCB are not considered to be genotoxic.

Observations in man. An increased incidence of CA was found in peripheral blood cells of the workers occupationally exposed to 1,2-DCB vapors (at concentrations about 100 ppm).[16]

In vitro genotoxicity. 1,2-DCB and 1,4-DCB were negative in *Salmonella* mutagenicity assay.[025]

1,4-DCB is unlikely to be genotoxic in a variety of short-term genotoxicity bioassays (*E. coli* WP2 system, etc.). It was not found to induce forward mutations in mouse lymphoma cells, SCE in Chinese hamster ovary cells, or unscheduled DNA synthesis in human lymphocytes. It induced micronuclei in mouse bone marrow (Mavournin et al., 1990), and did not induce CA in mammalian cells *in vitro* (Ishidate et al., 1988).

In vivo cytogenetics. 1,4-DCB did not induce micronuclei in CD-1 mice treated by *i/p* injection or oral gavage at up to 1600 or 2000 mg/kg BW, respectively. A weak response was obtained in NMRI mice after *i/p* injection at dose level of 177.5 mg/kg BW (Mohtashamipur et al., 1987). This compound was negative in cytogenicity studies with rat bone marrow cells and in a DLM study in CD-1 mice. Therefore, it is less likely that it could be carcinogenic by a genotoxic mechanism. 1,4-DCB does not induce unscheduled DNA synthesis in either tissue of $B6C3F_1$ mice following corn oil gavage doses of 300, 600, and 1,000 mg 1,4-DCB/kg BW. It induces cell replication in the male but not the female rat kidney.

1,4-DCB is not genotoxic in the mouse liver or rat kidney at single oral doses comparable to the daily doses given in the National Toxicology Program bioassay (NTP, 1987). The increases in replicative DNA synthesis support the hypotheses that mouse liver tumor formation occurs via stimulation of hepatocyte proliferation and male rat kidney carcinogenesis via increased renal cell proliferation.[17]

Carcinogenicity. 1,2-DCB was administered at doses of 60 and 120 mg/kg BW to F344 rats and $B6C3F_1$ mice in corn oil by gavage for 103 weeks. Under these test conditions, it was not shown to be carcinogenic at doses administered.[14]

1,4-DCB was administered by gavage to F344 rats and $B6C3F_1$ mice in a chronic bioassay. A statistically significant increase in tumor incidence was found in both species of animals. Carcinogenicity of 1,4-DCB was manifested in male rats by an increased rate of renal tubular cell adenocarcinomas; in mice of both sexes carcinogenic effect was manifested by an increased number of hepatocellular carcinomas and hepatocellular adenomas. No evidence of carcinogenicity was found in female rats.[15]

Carcinogenicity classifications. An IARC Working Group concluded that there is inadequate evidence for the carcinogenicity of 1,2-DCB in *experimental animals* and in *humans.*

IARC: 3;

U.S. EPA: D;

NTP: N - N - N - N (gavage).

An IARC Working Group concluded that there is sufficient evidence for the carcinogenicity of 1,4-DCB in *experimental animals* and there were inadequate data available to evaluate the carcinogenicity of 1,4-DCB in *humans.*

The Annual Report of Carcinogens issued by the U.S. Department of Health and Human Services (1998) defines 1,4-DCB to be a substance which may reasonably be anticipated to be carcinogen, i.e., a substance for which there is a limited evidence of carcinogenicity in *humans* or sufficient evidence of carcinogenicity in *experimental animals*.

IARC: 2B;

U.S. EPA: C;

NTP: CE - NE - CE - CE (gavage).

Chemobiokinetics.

Observations in man. Placental transfer has been demonstrated in man.[18] The conjugated *dichlorophenols* appear to be the principal metabolites in humans.

Animal studies. Absorption occurs quite rapidly through the lungs and the GI tract (100% of oral dose) following acute and chronic exposure.[19,20]

DCB is distributed primarily to the fat, or adipose tissues, and blood. The DCBs are primarily metabolized by oxidation in the liver to *dichlorophenols* and their *glucuronide* and *sulfate conjugates*, which are excreted in the urine within 5 to 6 days post exposure.[21] In F344 rats oral study, no covalent binding of ^{14}C-DCB radioactivity could be detected in samples of liver, kidney, lung and spleen. Urine appeared to be the main route of excretion. Main metabolites were *sulfate* and *glucuronide of 2,5-dichlorophenol*. A novel biotransformation pathway was proposed to occur: urinary excretion of a *mercapturic acid* of *chlorophenol*.[22]

1,2-DCB:

Standards.

U.S. EPA (1999). MCL and MCLG: 0.6 mg/l.

Canada (1989). MAC: 0.2 mg/l.

Recommendations.

WHO (1996). Guideline value for drinking water: 1.0 mg/l. However, this value far exceeds the lowest reported taste threshold. The levels of 0.001 to 0.01 mg/l are likely to give rise to consumer complaints of foreign odor and taste.

U.S. EPA (1999). Health Advisory for longer-term exposure: 30 mg 1,2-DCB/l.

Great Britain (1998). 1,4-DCB is authorized without time limit for use in the production of polymeric materials and articles in contact with food or drink or intended for such contact. The specific migration of this substance shall not exceed 12 mg/kg.

1,4-DCB:

Standards.

U.S. EPA (1999). MCL and MCLG: 0.075 mg/l.

Canada (1989). MAC: 0.005 mg/l.

Recommendations.

WHO (1996). Guideline value for drinking water: 0.3 mg/l, but the lowest reported odor perception threshold is 0.0003 mg/l.[23] The levels of 0.0003 to 0.03 mg/l are likely to give rise to consumer complaints of foreign odor and taste.

References:

1. Varshavskaya, S. P., Comparative toxicological features of chlorobenzene and dichlorobenzene in sanitary protection of water bodies, *Gig. Sanit.*, 10, 15, 1968 (in Russian).
2. Zoeteman, B. C. J., Harmsen, K., Linders, J. B. H. J., et al., Persistent Organic Pollutants in River Water and Ground Water of the Netherlands, *Chemosphere,* 9, 231, 249, 1980.
3. Peirano, W. B., *Health Assessment Document for Chlorinated Benzenes*, U.S. EPA, Report No EPA/600/8-84/015F, 1985, 1.
4. Umemura, T., Saito, M., Takagi, A., and Kurokawa, Y., Isomer-specific acute toxicity and cell proliferation in liver of $B6C3F_1$ mice exposed to dichlorobenzene, *Toxicol. Appl. Pharmacol.*, 137, 268, 1996.
5. Robinson, M., Bercz, J. P., Ringhand, H. P., Condie, L. W., and Parnell, M. J., Ten- and ninety-day toxicity studies of 1,2-dichlorobenzene administered by oral gavage to Sprague-Dawley rats, *Drug Chem. Toxicol.*, 14, 83, 1991.

6. Ariyoshi, T., Tadeguchi, K., Iwasaki, K., et al., Relationship between chemical structure and activity, II. Influences of isomers in dichlorobenzene, trichlorobenzene and tetrachlorobenzene on the activities of drug-metabolizing enzymes, *Chem. Pharmacol. Bull.,* Tokyo, 23, 824, 1984.
7. *Toxicology and Carcinogenesis Studies of 1,2-Dichlorobenzene in F344/N Rats and B6C3F$_1$ Mice (Gavage Studies*), Publ. No 86-2511, NTP Technical Report Series No 255, NIH, Bethesda, 1985, 195.
8. Valentovic, M. A., Ball, J. G., Anestis, D., and Madan, E., Acute hepatic and renal toxicity of dichlorobenzene isomers in Fischer 344 rats, *J. Appl. Toxicol.,* 13, 1, 1993.
9. Hollingsworth, R. L., Rowe, V. K., Oyen, F., Torkelson, T. R., and Adams, E. M., Toxicity of *o*-dichlorobenzene. Studies on animals and industrial experience, *AMA Arch. Ind. Health,* 17, 180, 1958.
10. Gurfein, L. N. and Pavlova, Z. K., in *Protection of Water Reservoirs against Pollution by Industrial Liquid Effluents*, S. N. Cherkinsky, Ed., Medgiz, Moscow, Issue No 4, 1960, 117 (in Russian).
11. Bornatowicz, N., Antes, A., Winker, N., and Hofer, H., A two-generation fertility study with 1,4-dichlorobenzene in rats, *Wien Klin. Wochenschr.*, 106, 345, 1994 (in German).
12. Ruddick, J. A., Black, W. D., Villeneuve, D. C., et al., A teratological evaluation following oral administration of trichloro- and dichlorobenzene isomers to the rat, *Teratology*, 27, 73A, 1983.
13. Giavani, E., Broccia, M. L., Prati, M., et al., Teratogenic evaluation of *p*-dichlorobenzene in the rat, *Bull. Environ. Contam. Toxicol.*, 37, 164, 1986.
14. *Toxicology and Carcinogenesis Studies of 1,4-Dichlorobenzene in F344/N Rats and B6C3F$_1$ Mice (Gavage Studies*), NTP Technical Report Series No 319, NIH Publ. No 86-2575, Bethesda, 1987.
15. Hayes, W. C., Hanley, T. R., Gushkov, T. S., Johnson, K.A., and John, J. A., Teratogenic potential of inhaled dichlorbenzenes in rats and rabbits, *Fundam. Appl. Toxicol.,* 5, 190, 202 1985.
16. Zapata-Gayon, C., Zapata-Gayon, N., and Gonzalez-Angulo, A., Clactogenic chromosomal aberrations in 26 individuals accidentally exposed to *ortho*-dichlorobenzene vapors in the National Medical Center in Mexico City, *Arch. Environ. Health*, 37, 231, 1982.
17. Sherman, J. H., Nair, R. S., Steinmetz, K. L., Mirsalis, J. C., Nestmann, E. R., and Barter, J. A., Evaluation of unscheduled DNA synthesis (UDS) and replicative DNA synthesis (RDS) following treatment of rats and mice with *p*-dichlorobenzene, *Teratogen. Carcinogen. Mutagen.,* 18, 309, 1998.
18. Dautz, B. T. and Laseter, J. C., The transplacental migration and accumulation in the blood of volatile organic constituents, *Pediatric Res*., 10, 696, 1976.
19. *Ambient Water Quality Criteria for Dichlorobenzenes*, U.S. EPA 440/5-80-039, PB81-117509, 1980.
20. Ware, S. A. and West, W. L., *Investigation of Selected Potential Environmental Contaminants: Halogenated Benzenes*, EPA 560/2-77-004, 1977.
21. Kimura, R., Hayashi, R., Sato, M., et al., Identification of sulfur-containing metabolites of *p*-dichlorobenzene and their disposition in rats, *J. Pharmacobiodynamics*, 2, 237, 1979.
22. Klos, C., and Dekant, W., Comparative metabolism of the renal carcinogen 1,4-dichlorobenzene on rat: identification and quantitation of novel metabolites, *Xenobiotica*, 24, 965, 1994.
23. *Revision of the WHO Guidelines for Drinking Water Quality*, Report on the Consolidation Meeting on Organics and Pesticides, Medmenham, UK, January 30-31, 1992, 11.

1,2-DICHLOROETHANE

Molecular Formula. $C_2H_4C_{l2}$
M = 98.96
CAS No 107-06-2
RTECS No KI0525000
Abbreviation.. 1,2-DCE.

Synonyms and **Trade Names.** Brocide; *sym*-Dichloroethane; Dutch liquid; Dutch oil; Ethane dichloride; Glycol dichloride.

Properties. Colorless liquid with an odor reminiscent of chloroform. Water solubility is 8820 mg/l (20°C). Odor perception threshold is 2 to 3 mg/l, practical threshold is 5.0 mg/l.[1]

Applications. Used mainly in the production of vinylidene chloride and other plastic intermediates, for clearing, in paints, coatings, and adhesives. It also may be used as a solvent in polyvinyl chloride production.

Exposure. Concentrations found in drinking water in the U.S. reached 0.006 mg/l.

Acute Toxicity.

Observations in man. Clinical symptoms of acute poisoning by ingestion include general weakness, nausea, dizziness, headache, vomiting of blood and bile, etc.,[2] and also unconsciousness, mental disorders, and cerebral and extrapyramidal disorders.[3] Death is most often attributed to circulatory and respiratory failure.[4]

Animal studies. LD_{50} was 680 to 770 mg/kg BW in rats, and 860 mg/kg BW in rabbits.[1,06] After administration, liver damage, myocardial edema, and damage to the coronary vessels were reported. Poisoned animals displayed CNS impairment and multiple hemorrhages. Gross pathology examination revealed dystrophic changes, mainly in the liver, but also in the kidney and other organs. LD_{50} is 5.7 g/kg BW in dogs. Manifestations of poisoning were lacrimation, ataxia, and general anesthesia.[5]

Repeated Exposure. In a 10-day study, Sprague-Dawley rats received 1,2-DCE in corn oil by gavage at dose levels of 10, 30, 100, and 300 mg/kg BW. All female animals died in the high dose group and 2 of the 10 males rats survived. Histological examination revealed multifocal to diffuse inflammation of the mucosal and submucosal layers of the forestomach in the 100 mg/kg BW-dose group. This change was minimal in both males and females.[6]

Rats died after repeated administration of the dose of 300 mg/kg BW given orally for 2 weeks. This dose produced necrosis and fatty changes in the liver.[8]

Short-term Toxicity. In a 13-week study, $B6C3F_1$ mice and F344/N rats were given drinking water containing 1,2-DCE. There was an increase in the liver weights in mice, although histological lesions were not observed. The NOELs were found to be 120 mg/kg BW (male rats) or 150 mg/kg BW (female rats); 780 mg/kg BW (male mice) or 2500 mg/kg BW (female mice), this based on mortality. 9 out of 10 female mice exposed to 1,2-DCE concentration of 8,000 ppm in drinking water died before the end of the study.[7]

In a 90-day study, Sprague-Dawley rats received 1,2-DCE in corn oil by gavage at dose levels of 37.5, 75, and 150 mg/kg BW. No clinical signs of toxic action were observed. There was a decrease in BW gain and total food consumption in high-dose males as well as slight but significant differences in hematological parameters in 75 and 150 mg/kg BW-dosed groups. In females, there was an increase in absolute and/or relative kidney and liver weights at 150 mg/kg BW (liver), and 75 and 150 mg/kg BW (kidney). No differences in mortality, growth, or histopathology were reported.[6]

No effects were observed in rats when the chemical was given orally at a 10 mg/kg BW dose for 90 days.[8]

Rats were given orally 500 to 8000 ppm 1,2-DCE in drinking water for 90 days. The LOEL for organ weight changes was found to be 59 mg/kg BW.[22] Munro et al. (1996) suggested the calculated NOEL of 34 mg/kg BW for this study.[9]

Reproductive Toxicity. In a multigeneration reproduction study, male and female ICR Swiss mice were given the doses of 5.0 to 50 mg/kg BW via their drinking water. The treatment did not result in reproductive effects, as measured by fertility, gestation, viability or lactation indices, pup survival, and BW gain. No statistically significant dose-related developmental effects were observed as indicated by the incidence of fetal visceral or skeletal abnormalities.[10]

Rats received 250 or 500 ppm 1,2-DCE in feed mash for a 2-year period. The treatment caused no effect on fertility, litter size, or fetal weight.[11]

No effect on the adult generations was reported after 25 weeks of dosing as measured by BW gain, fluid intake, or gross pathology examination. 1,2-DCE is capable of crossing the placental barrier in pregnant rats.

Mutagenicity.

In vitro genotoxicity. 1,2-DCE did not induce CA in mammalian cells *in vitro* but induced mutations in human lymphocytes. It was mutagenic to bacteria (Ashby and Tennant, 1991). It is weakly if any mutagenic in *Salmonella* microsome assay system and in DNA polymerase-deficient *E. coli.*[12,13]

In vivo cytogenetics. A weak mutagenic effect was noted in a spot test in mice. Micronucleus induction in peripheral blood was examined during carcinogenicity assays. There was no micronuclei induction or polychromatic erythrocyte suppression after treatment with 100 to 300 mg 1,2-DCE/kg BW.[14]

Negative results were obtained in one DLM assay and two micronucleus assays in mice.[8]

1,2-DCE does not induce sex-linked recessive lethals in *Dr. melanogaster.*[15]

Carcinogenicity. Epidemiology data have not established carcinogenicity of 1,2-DCE, but in mice and rats, it was shown to increase the incidence of several types of tumors. It was not shown to be carcinogenic through inhalation exposure in animals. *Pim* transgenic mice were exposed to 100 to 300 mg/kg BW by gavage for 40 weeks. Increased incidence of malignant lymphoma was seen in high-dose females treated with 1,2-DCE.[16] 1,2-DCE induced circulatory system hemangiosarcomas in male Osborne-Mendel rats which were given oral doses of 97 or 195 mg/kg BW (males), and 149 or 299 mg/kg BW (females) over a period of 78 weeks.[17]

Carcinogenicity classifications. An IARC Working Group concluded that there is sufficient evidence for the carcinogenicity of 1,2-DCE in *experimental animals* and there were no adequate data available to evaluate the carcinogenicity of 1,2-DCE in *humans*.

The Annual Report of Carcinogens issued by the U.S. Department of Health and Human Services (1998) defines 1,2-DCE to be a substance which may reasonably be anticipated to be carcinogen, i.e., a substance for which there is a limited evidence of carcinogenicity in *humans* or sufficient evidence of carcinogenicity in *experimental animals*.

IARC: 2B;

U.S. EPA: B2;

EU: 2;

NTP: P* - P - P - P* (gavage).

Chemobiokinetics. 1,2-DCE is readily absorbed from the GI tract. After oral administration, the adipose tissues, liver and kidney seem to have the highest concentrations of 1,2-DCE.[18] Dechlorination of 1,2-DCE takes place in presence of liver oxidase. It is metabolized into *2-chloroethanol.*[19]

According to Payan *et al*,. 1,2-DCE is extensively metabolized and partially excreted in urine as *thioether compounds*, which include *thiodiglycolic acid.* Male Sprague-Dawley rats were given a single oral dose of labeled ^{14}*C*-DCE (0.125 mmol/kg to 8.08 mmol/kg BW). Amount of radioactivity excreted in urine ranged between 63 and 7.4%.[20]

According to other data, 96% of the radioactivity of a single oral dose of 150 mg/kg BW was eliminated within 48 hours after dosing.[11] The major part of the absorbed chemical is rapidly excreted via the urine, mainly as glutathione conjugates, and via the lungs, as *carbon dioxide* or the unchanged compound.

Regulations.

EU (1990). Banned to certain uses owing to its effects on health and the environment.[21]

U.S. FDA (1998) approved the use of 1,2-DCE (1) in adhesives as a component of articles intended for use in packaging, transporting, or holding food in accordance with the conditions prescribed by 21 CFR part 175.105; and (2) in the manufacture of cellophane for packaging food in accordance with the conditions prescribed by 21 CFR part 177.1200.

Recommendations.

WHO (1996). Guideline values for drinking water: 0.03 mg/l.

U.S. EPA (1999). Health Advisory for longer-time exposure: 2.6 mg/l.

Standards.

U.S. EPA (1999). MCL: 0.005 mg/l; MCLG: zero.

Russia (1995). MAC: 0.02 mg/l.

References:

1. Cherkinsky, S. N., Friedland, S. A., and Trakhtman, N. N., in *Sanitary Protection of Reservoirs against Pollution by Industrial Liquid Effluents*, S. N. Cherkinsky, Ed., Medgiz, Moscow, Issue No 2, 1954, 156 (in Russian).
2. McNally, W. D. and Fostvedt, G., Ethylene dichloride poisoning, *J. Ind. Med.*, 10, 373, 1941.

3. Akimov, G. A. et al., Neurological disorders in acute dichloroethane poisoning, *J. Neuropathol. Psichiatr.*, 78, 687, 1978 (in Russian).
4. Chesnokov, N. Y., Acute dichloroethane poisoning, *Vrachebnoye Delo*, 6, 127, 1976 (in Russian).
5. *FAO Nutrition Meetings* Report Series 48A, Geneva, 1970, 91.
6. Daniel, F. B., Robinson, M., Olson, G. R., York, R. G., and Condie, L. W., Ten and ninety day toxicity studies of 1,2-dichloroethane in Sprague-Dawley rats, *Drug Chem. Toxicol.*, 17, 463, 1994.
7. *Toxicity Studies of 1,2-DCE in F344/N Rats, Sprague-Dawley Rats, Osborne Mendel Rats and B6C3F$_1$ Mice (Drinking Water and Gavage Studies)*, NTP Technical Report Series No 4, Research Triangle Park, NC, NIH Publ. No 91-3123, 1991, 54.
8. *1,2-Dichloroethane*, Health and Safety Guide, WHO, IPCS, Geneva, 1991, 33.
9. Munro, I. C., Ford, R. A., Kennepohl, E., and Sprenger, J. G., Correlation of structural class with No-Observed-Effect Levels: A proposal for establishing a threshold of concern, *Food Chem. Toxicol.*, 34, 829, 1996.
10. Lane, R. W., Riddle, B. L., and Borcelleca, J. F., Effect of 1,2-dichloroethane and 1,1,1-trichloroethane in drinking water on reproduction and development in mice, *Toxicol. Appl. Pharmacol.*, 63, 409, 1982.
11. Alumot, E., Nachtomi, E., Mandel, E., and Holstein, P., Tolerance and acceptable daily intake of chlorinated fumigants in the rat diet, *Food Cosmet. Toxicol.*, 14, 105, 1976.
12. Brem, H., Stein, A., and Rozenkranz, C., The mutagenicity and DNA-modifying effect of haloalkanes, *Cancer Res.*, 34, 2576, 1974.
13. McCann, J., Simmon, V., Streitweisser, D., et al., Mutagenicity of chloroacetaldehyde, *Proc. Natl. Acad. Sci. USA*, 72, 3190, 1975.
14. Armstrong, M. J. and Galloway, S. M., Micronuclei induced in peripheral blood of E mu-PIM-l transgenic mice by chronic oral treatment with 2-acetylaminofluorene or benzene but not with diethylnitrosoamine or 1,2-dichloroethane, *Mutat. Res.*, 302, 61, 1993.
15. Rapoport, I. A., The reactions of gene protein with 1,2-DCE, *Acad. Sci. USSR, Series Biol. Sciences*, Issue No 134, 1960, 745 (in Russian).
16. Storer, R. D., Cartwright, M. E., Cook, W. O., Soper, K. A., and Nichols, W. W., Short-term carcinogenesis bioassay of genotoxic procarcinogens in PIM transgenic mice, *Carcinogenesis*, 16, 285, 1995.
17. *Bioassay of 1,2-DCE for Possible Carcinogenicity*, Natl Cancer Institute Technical Report Series No 55, DHEW Publ. No 78-1361, Washington, D.C., 1978.
18. Reitz, R. H., Fox, T. R., Domoradzki, J. Y., et al., Pharmacokinetics and macromolecular interactions of ethylene dichloride, in *Ethylene dichloride: A potential health risk*? Banbury report No 5, Cold Spring Harbor Laboratory, 1980, 135.
19. Kokarovtzeva, M. G. and Kiseleva, N. I., Chloroethanol (ethylene chlorohydrin) - toxic metabolite of 1,2-dichloroethane, *Pharmacologia i Toxicologia*, 1, 118, 1978 (in Russian).
20. Payan, J. P., Beydon, D., Fabry, J. P., Brondeau, M. T., Ban, M., and de Ceaurriz, J., Urinary thiodiglycolic acid and thioether excretion in male rats dosed with 1,2-dichloroethane, *J. Appl. Toxicol.*, 13, 417, 1993.
21. List of chemicals banned or severely restricted to certain uses by Community legislation owing to their effects on health and the environment, *Off. J. Eur. Commun.*, L251/19, 1992, Annex I.
22. Morgan, D. L., Elwell, M. R., Lilia, H. S., and Murthy, A. S., Comparative toxicity of ethylene chloride in F344, Sprague-Dawley and Osborne-Mendel rats, *Food Chem. Toxicol.*, 28, 839, 1990.

1,2-DICHLOROETHYLENE

Molecular Formula. $C_2H_2Cl_2$
M = 96.95
CAS No 540-59-0
RTECS No KV9360000
Abbreviation. 1,2-DCE.

Synonyms. Acetylene dichloride; 1,2-Dichloroethene; Ethylene chloride.

trans-1,2-Dichloroethylene CAS No 156-60-5

cis-1,2- Dichloroethylene CAS No 156-59-2

Properties. Clear, colorless liquid. Water solubility is 6.3 mg/l (25°C, *trans*-1,2-DCE), and 3.5 mg/l (20°C, *cis*-1,2-DCE). Odor perception threshold is 0.26 mg/l (*trans*-1,2-DCE).[02]

Applications. Solvent and chemical intermediate.

Exposure. Occurs as impurity in commercial grade 1,1-dichloroethylene (vinylidene chloride). Due to their volatility and limited use, levels of either *cis*- or *trans*-1,2-DCE in food are expected to be negligible (*U.S. EPA*, 1983).

Acute Toxicity. In rats, LD_{50} of isomer mixture is 770 mg/kg.[06] LD_{50} of *trans*-1,2-DCE is 1300 mg/kg BW.[1] Single doses of 400 and 1500 mg *cis*-1,2-DCE/kg BW but not of *trans*-1,2-DCE administered to rats produced significant elevations of liver alkaline phosphatase.[2]

Repeated Exposure. The doses of 21 and 210 mg *trans*-1,2-DCE/kg BW were administered by gavage to male CD-1 mice for 14 days. No changes in BW gain, contents of some serum enzymes, or blood urea nitrogen were noted. However, at the 210 mg/kg BW dose-level, fibrinogen level and prothrombin times were significantly decreased.[3]

Short-term Toxicity. CD-1 mice received *trans*-1,2-DCE for 90 days in their drinking water at the dose levels of 17, 175, and 387 mg/kg BW (males) or 23, 224, and 452 mg/kg BW (females). There were no changes in fluid consumption, BW gain, or gross pathology among the experimental animals. In male mice, significant increases in serum alkaline phosphatase were noted at the two highest dose levels. In females, the thymus weight was significantly depressed at 224 and 452 mg/kg BW doses. The NOAEL of 17 mg/kg BW was identified based on normal serum chemistry values in male mice.[3]

It was proposed (*U.S. EPA, WHO*) that the value calculated for the *trans*-isomer also be used for the *cis*-isomer.

Reproductive Toxicity.

Embryotoxicity of *trans*-1,2-DCE was not evident because the dose capable of causing this effects was found to be about twofold higher than the maternally toxic dose.[4]

Rats were chronically exposed to 1,2-DCE. The treatment caused a decrease in litter size.[5] However, no adverse effects on fertility, litter size, neonatal weight, or viability were observed in rats receiving 1,2-DCE in their diet.[6]

Gonadotoxicity. In the above-cited study,[5] disruption in estrous cycle was reported in rats.

Teratogenicity. Human epidemiological studies and previous teratogenic studies using chick embryos and fetal rats have shown an increased incidence of congenital cardiac lesions in animals exposed to DCE. Varying doses of DCE were administered to pregnant rats during fetal heart development. Maternal and fetal variables showed no statistically significant differences between treated and untreated groups. Fetuses of rats did not demonstrate a significant increase in cardiac defects compared with controls. DCE was not found to be a specific cardiac teratogen in the fetus when imbibed by the maternal rat.[7,8]

Immunotoxicity. No significant immunological effects were noted in mice exposed by gavage to 22 or 220 mg/kg BW for 14 days.[9] Administration of 175 mg/kg BW and 387 mg/kg but not 17 mg *trans*-1,2-DCE/kg BW caused a significant decrease in antibody-forming cells of the spleen only in male mice.[10]

Mutagenicity.

In vitro genotoxicity. No mutagenic effect was noted in *Salmonella* test, neither the *cis*- nor *trans*-isomer of 1,2-DCE induced CA or SCE in Chinese hamster lung fibroblast cell line.[11] According to other data, mutagenic effects were noted in *Salmonella* mutagenicity assay.[12]

Animal studies. 1,2-DCE was positive in genotoxicity studies in hamsters and mice.[13,14]

Carcinogenicity classification.

U.S. EPA: D.

Chemobiokinetics. Neutral, lipid-soluble substances are expected to be readily absorbed following oral or dermal exposure (*EPA*, 1984). DCE are metabolized into *epoxides* which can yield *dichloroacetaldehyde*, *dichloroethanol*, and *dichloroacetic acid.*[15,16]

Regulations. ***U.S. FDA*** (1998) approved the use of 1,2-DCE in adhesives as a component of articles intended for use in packaging, transporting, or holding food in accordance with the conditions prescribed by 21 CFR part 175.105. As outlined in the U.S.A.vinyl chloride rule (52 Federal Register 25690) water supplies must test for vinylchloride whenever the 1,2-DCE are found. The vinyl chloride MCL will adequately protect the public against any vinyl chloride that may be produced through the biodegradation of the 1,2-DCE.[17]

Recommendations.

WHO (1996). Guideline value for drinking water: 0.05 mg/l.

U.S. EPA (1999). Health Advisories for longer-time exposure: 11 and 6.0 mg/l for *cis*-1,2-DCE and *trans*-1,2-DCE, respectively.

Standards.

U.S. EPA (1999). MCL and MCLG for *cis*-1,2-DCE: 0.07 mg/l; MCL and MCLG for *trans*-1,2-DCE: 0.1 mg/l.

Russia (1995). MAC and PML: 0.0006 mg/l.

References:

1. Freundt, J. J., Liebaldt, G. P., and Lieberwirth, E., Toxicity studies on *trans*-1,2-dichloroethylene, *Toxicology*, 7, 141, 1977.
2. Jenkins, L. J., Trabulus, M. J., and Murphy, S. D., Biochemical effects of 1,1-dichloroethylene in rats: Comparison with carbon tetrachloride and 1,2-dichloroethylene, *Toxicol. Appl. Pharmacol.*, 23, 501, 1972.
3. Barness, D. W., Sanders, V. M., White, K. L., et al., Toxicology of *trans*-1,2-dichloroethylene in the mouse, *Drug Chem. Toxicol.*, 8, 373, 1985.
4. Hurtt, M. E., Valentine, R., and Alvarez, L., Developmental toxicity of inhaled *trans*-1,2-dichloroethylene in the rat, *Fundam. Appl. Toxicol.*, 20, 225, 1993.
5. Vozovaya, M. A., Development of posterity of two generations obtained from females subjected to the action of dichloroethane, *Gig. Sanit.*, 7, 25, 1974 (in Russian).
6. Alumot, E. et al., Tolerance and acceptable daily intake of chlorinated fumigants in the rat diet, *Food. Cosmet. Toxicol.*, 14, 105, 1976.
7. Johnson, P. D., Dawson, B. V., and Goldberg, S. J., Cardiac teratogenicity of trichloroethylene metabolites, *J. Am. Coll. Cardiol.*,32, 540, 1998
8. Johnson, P. D., Dawson, B. V., and Goldberg, S. J., A review: trichloroethylene metabolites: potential cardiac teratogens. *Environ Health Perspect.*, 106 (Suppl 4), 995, 1998
9. Munson, A. E., Saunders, V. M., Douglas, L. E., et al., *In vivo* assessment of immunotoxicity, *Environ. Health Perspect.*, 43, 41, 1982.
10. Shopp, G. M., Sanders, V. M., White, K. L., et al., Humoral and cell-mediated immune status of mice exposed to *trans*-1,2-dichloroethylene, *Drug Chem. Toxicol.*, 8, 393, 1985.
11. Sawada, M., Sofuni, T., and Ishidate, M., Cytogenetic studies on 1,1-dichloroethylene and its two isomers in mammalian cells *in vitro* and *in vivo*, *Mutat. Res.*, 187, 157, 1987.
12. Davidson, I. W. et al., Ethylene dichloride: A review of its metabolism, mutagenic and carcinogenic potential, *Drug Chem.Toxicol.*, 5, 319, 1982.
13. Storer, R. D. and Conolly, R. B., Comparative *in vivo* genotoxicity and acute hepatotoxicity of three 1,2-dihaloethanes, *Carcinogenesis*, 4, 1491, 1983.
14. Gocke, E. et al., Mutagenicity studies with the mouse spot test, *Mutat. Res.*, 117, 201, 1983.
15. Henschler, D., Metabolism and mutagenicity of halogenated olefins: A comparison of structure and activity, *Environ. Health Perspect.*, 21, 61, 1977.
16. Costa, A. K., The chlorinated ethylenes: Their hepatic metabolism and carcinogenicity, *Abstract of Thesis*, Int. B, 44, 1797, 1983.
17. *Federal Register,* vol 54, No. 97, May 22, 1989, 22084.

DICHLOROMETHANE

Molecular Formula. CH_2Cl_2

M = 84.94

CAS No 75-09-2
RTECS No PA8050000
Abbreviation. DCM.

Synonyms and **Trade Names.** Aerothene MM; Methane dichloride; Methylene chloride; Methylene dichloride; Narkotil.

Properties. DCM occurs as a colorless liquid at room temperature. Water solubility is 20 g/l, miscible with alcohols and oils at all ratios. Odor perception threshold is 7.5 mg/l or 9.1 mg/l;[02] taste perception threshold is 15 mg/l. According to other data, organoleptic threshold is 5.6 mg/l.[01]

Applications. Primarily used in degreasing and cleaning fluids (as a substitute for trichloroethylene), in paint removers and hair lacquers, and for decaffenating coffee. Used as a solvent in the production of cellulose esters, resins and rubber. A substituent of formaldehyde in plastics production and is also used in the manufacture of polyurethanes. DCM has been used as an extraction solvent in food processing.

Acute Toxicity. A dose of 7.5 g/kg BW is very close to LD_{100}.[1] LD_{50} in rats is approximately 2.1 to 3.0 g/kg BW.[2] In mice, it is 5.5 g/kg BW. Doses about 2.1 to 2.15 g/kg BW are lethal for rabbits;[3] doses of 3.0 to 5.0 g/kg BW are lethal for dogs.

Other data suggested that LD_{50} is 1.25 g/kg BW in rats, 1.0 g/kg BW in mice, and 2.0 g/kg BW in rabbits.[4] Poisoned animals exhibit prolonged excitation with subsequent ataxia and CNS inhibition. Animals experienced clonic convulsions. Death occurs from respiratory failure. Gross pathology examination revealed no abnormalities other than cerebral hyperemia and liver dystrophy (in some animals).

Repeated Exposure. Mice tolerated 10 administrations of 750 mg/kg BW. Gross pathology examination revealed dystrophic changes in the liver.

Short-term Toxicity. Wistar rats were exposed to 15 mg/kg BW in drinking water (assuming daily drinking water consumption rates of 12 ml/100 g) for 13 weeks. No treatment-related effects were observed.[5]

Guinea pigs and rats were dosed by gavage with 0.4 and 377 mg/kg BW for 5 and 6 months, respectively. At the higher dose level, only ascorbic acid content in the suprarenals was found to be altered in guinea pigs, but no other functional or structural changes are reported.[6]

Long-term Toxicity. Doses 5.0 to 250 mg/kg BW were administered in deionized water to F344 rats for 104 weeks. An additional group received 250 mg/kg BW for 78 weeks followed by a 26-week recovery period. Doses 125 mg/kg BW and 250 mg/kg affected BW, water and food consumption, produced histomorphologic hepatic changes (cellular alterations and fatty change). Under the experimental conditions of this study, the NOAEL appears to be 5.0 mg/kg BW.[7] The dose of 250 mg/kg BW given in drinking water to $B6C3F_1$ mice for 104 weeks produced hepatocellular alterations (increased fat content in the liver). In this study, the NOAEL was identified to be 185 mg/kg BW.[8]

The NOAEL of 87 mg/m^3 was established in an inhalation study (the equivalent oral dose for rats is about 28 mg/kg BW).[9]

Reproductive Toxicity.

Single doses caused *maternal toxicity*.[10]

Gonadotoxicity.

Observations in man. Testicular pain and infertility histories have been reported in eight out of 34 workers (23.5%) exposed to DCM during priming operations who exhibited a blood carboxyhemoglobin concentration (a marker for DCM exposure) ranging from 1.2 to 17.3%. These findings can't however be ascribed with certainty to DCM, because simultaneous exposure to other chemicals and other confounding factors was not controlled.[11]

According to other data no effect on sperm counts was noted among 4 workers exposed to DCM.[12] A twofold increased risk for spontaneous abortions was reported among wives of rubber manufacturing workers exposed to both DCM and 1,1,1-trichloroethane.[13]

Embryotoxicity. DCM exhibited some potential to induce developmental toxicity.[14] On inhalation exposure, no adverse effects were reported.

Teratogenicity. Exposure to DCM did not induce visceral malformations in fetal mice or rats. In rats, postnatal behavioral development was affected by prenatal exposure.[15]

Mutagenicity. No data were available on the genetic and related effects of DCM in humans (IARC).

In vitro genotoxicity. DCM did not induce unscheduled DNA synthesis in human cells *in vitro*. DCM was positive in *Salmonella* mutagenicity assay[16] and has been shown to transform rat embryo cells and to enhance viral transformation of Syrian hamster embryo cells.[17]

CA, but neither mutations nor DNA damage in rodent cells were observed *in vitro* (IARC, Suppl. 7-195).

In vivo cytogenetics. DCM failed to increase the frequency of either SCE or CA in mouse bone marrow cells following *i/p* exposures to 100 to 2000 mg/kg BW.[18] It did not induce CA in bone-marrow cells of rats or micronuclei in mice treated *in vivo*. DNA single-strand breaks were noted in the livers of $B6C3F_1$ mice immediately following exposure to 4000 to 8000 ppm DCM for 6 hours.

Carcinogenicity. Carcinogenicity studies in DCM have yielded inconsistent and contradictory results.

Observations in man. Epidemiological studies failed to show a positive correlation between DCM exposure and increased cancer incidence.[19,20] Good human epidemiology of workers exposed for many years to DCM (inhalation) did not justify the validity for humans of results obtained in $B6C3F_1$ mice.[21]

Animal studies. In a 2-year study, F344 rats were exposed to 50 and 250 mg/kg BW in their drinking water. Hepatological changes detected in the target dose groups (both sexes) included an increased incidence of foci/areas of cellular alteration. Fatty liver changes were noted at 125 and 250 mg/kg BW dose levels after 78 and 104 weeks of treatment.[7]

DCM was administered to F344 rats in their drinking water. The treatment induced a significant increase in combined hepatocellular carcinoma and neoplastic nodules in female F344 rats of 50 and 250 mg/kg BW dose groups. The analogous study in $B6C3F_1$ mice was considered to be suggestive but not conclusive evidence of DCM carcinogenicity.[22]

DCM is unlikely to be able to induce hepatocellular division in mice.[23]

Carcinogenicity classifications. An IARC Working Group concluded that there is sufficient evidence for the carcinogenicity of DCM in *experimental animals* and there were inadequate data available to evaluate the carcinogenicity of DCM in *humans*.

The Annual Report of Carcinogens issued by the U.S. Department of Health and Human Services (1998) defines DCM to be a substance which may reasonably be anticipated to be carcinogen, i.e., a substance for which there is a limited evidence of carcinogenicity in *humans* or sufficient evidence of carcinogenicity in *experimental animals*.

IARC: 2B;

U.S. EPA: B2;

EU: 3;

NTP: SE - CE* - CE* - CE* (inhalation).

Chemobiokinetics. DCM is expected to be completely absorbed when ingested in mice and rats.[24,25] It is primarily distributed to the liver and fat.[26]

The main metabolites are *carbon monooxide*, formed *via* the cytochrome *P-450* system, and CO_2, formed *via* GSH metabolism of either the parent compound or carbon monooxide. ^{14}C-DCM metabolites did not bind to the RNA or DNA of rat hepatocytes although radioactivity was associated with lipid and protein.[27]

Syrian hamsters metabolize DCM more slowly than mice. According to other data, species such as hamsters and humans having much lower rates of DCM metabolism via DCM-protein cross-links may not generate toxicologically significant concentrations of formaldehyde and DNA-protein cross-links.[28]

Excretion occurs primarily via the lungs.

Regulations. ***U.S. FDA*** (1998) approved the use of DCM (1) in food under the conditions prescribed in 21 CFR parts 173.255; (2) in adhesives used as components of articles intended for use in packaging, transporting, or holding food in accordance with the conditions prescribed by 21 CFR part 175.105; and (3) in the manufacture of cellophane for packaging food in accordance with the conditions prescribed by 21 CFR part 177.1200.

Recommendations.

WHO (1996). Guideline value for drinking water: 0.02 mg/l.

FAO/WHO (1983) withdrew the previously allocated temporary ADI of 0 to 0.5 mg/kg BW and recommended that the use of DCM as an extraction solvent be limited in order to ensure that its residues in food are as low as practicable.

Standards.

U.S. EPA (1999). Proposed MCL: 0.005 mg/l, MCLG: zero.

Canada (1989). MAC: 0.05 mg/l.

Russia (1995). MAC and PML: 7.5. mg/l (organolept., odor).

References:

1. Tugarinova, V. N., Miklashevsky, V. Ye., et al., in *Protection of Water Reservoirs against Pollution by Industrial Liquid Effluents*, S. N. Cherkinsky, Ed., Medgiz, Moscow, Issue No 7, 1965, 41 (in Russian).
2. Kimura, E. T., Ebert, D. M., and Dodge, P. W., Acute toxicity and limits of solvent residues for sixteen organic solvents, *Toxicol. Appl. Pharmacol.,* 19, 699, 1971.
3. von Ottingen, W. F., The aliphatic acids and their esters: toxicity and potential danger, *Arch. Ind. Health*, 21, 28, 1960.
4. *Methylene Chloride*, Environ. Health Criteria No 32, WHO, Moscow, 1988, 55 (in Russian).
5. Bornmann, G. und Loeser, A., Zur Frage einer chronish-toxischen Wirkung von Dichloromethan, *Z. Lebensm.-Untersuch. Forsch.,* 136, 14, 1967.
6. *Hygiene Problems in the Production and Use of Polymeric Materials*, F. F. Erisman Research Sanitary Hygiene Institute, Moscow, 1969, 41 (in Russian).
7. Serota, D. G., Thakur, A. K., and Ulland, B. M., A two-year drinking-water study of dichloromethane in rodents, I. Rats, *Food Chem. Toxicol.,* 24, 951, 1986.
8. Serota, D. G., Thakur, A. K., Ulland, B. M., et al., A two-year drinking-water study of dichloromethane, II. Mice, *Food Chem. Toxicol*., 24, 959, 1986.
9. Haun, C. C., Vernot, E. H., Darmer, K. I., and Diamond, S. S., Continuous animal exposure to low levels of dichloromethane, AMRL-TR-T2-13 in *Proc. 3rd Ann. Conf. Environ. Toxicol*., Aerospace Medical Research Lab., Wright Patterson Air Force Base, Ohio, 1972, 199.
10. Schwetz, B. A., Leong, B. J., and Gehring, P. J., The effect of maternally inhaled trichloroethylene, perchloroethylene, methyl chloroform, and methylene chloride on embryonic and fetal development in mice and rats, *Toxicol. Appl. Pharmacol*., 32, 84, 1975.
11. Kelly, M., Case reports of individuals with oligospermia and methylene chloride exposures, *Reprod. Toxicol.,* 2, 13, 1988.
12. Wells, V. E., Schrader, S. M., McCammon, C. S., Ward, E. M., Turner, T. W., Thun, M. J., and Halperin, T. W., Letter to the Editor, *Reprod. Toxicol.,* 3, 281, 1989.
13. Lindbohm, M.-L., Hemminki, K., Bonhomme, M. G., Anttila, A., Rantala, K., Heikkila, P., and Rosenberg, M. J., Effect of paternal occupational exposure on spontaneous abortions, *Am. J. Ind. Med.,* 81, 1029, 1991.
14. Narotsky, M. G. and Kavlock, K. J., A multidisciplinary approach to toxicological screening: II. Developmental Toxicity, *J. Toxicol. Environ. Health*., 45, 145, 1995.
15. Hatch, G. G., Conclin, P. M., Christensen, C. I., et al., Chemical enhancement of viral transformation in Syrian hamster embryo cells by gaseous and volatile chlorinated methanes and ethanes, *Cancer Res.,* 43, 1945, 1983.
16. Green, T., Proven, W. M., Collinge, D. C., et al., Macromolecular interactions of inhaled methylene chloride in rats and mice, *Toxicol. Appl. Pharmacol*., 93, 1, 1988.
17. Price, P. J., Hassett, C. M., and Mansfield, J. I., Transforming activities of trichloroethylene and proposed industrial alternatives, *In Vitro*, 14, 290, 1978.
18. Westbrook-Collins, B., Allen, J. W., Shariet, Y., et al., Further evidence that dichloromethane does not induce chromosome damage, *J. Appl. Toxicol*., 10, 79, 1990.
19. Friedlander, B. R., Hearne, F. T., and Hall, S., Epidemiological investigation of employees chronically exposed to methylene chloride, *J. Occup. Med*., 20, 657, 1978.

20. Ott, M. G., Skory, L. K., Holder, B. B., et al., Health evaluation of employees occupationally exposed to methylene chloride. General study design and environmental considerations, *Scand. J. Health*, 9 (Suppl. 1), 1, 1983.
21. Hearne, F. T., 1992. Cit. in: P. S. Abelson, Health Risk Assessment, *Regul. Toxicol. Pharmacol.*, 17, 219, 1993.
22. *Addendum to the Health Assessment Document for Dichloromethane (methylene chloride)*, Updated Carcinogenicity Assessment Prep. Carcinogen Assessment Group, OHEA, U.S. EPA 600/882/004F, 1983.
23. Lefevre, P. A. and Ashby, J., Evaluation of dichloromethane an inducer of DNA synthesis in B6C3F$_1$ mouse liver, *Carcinogenesis*, 10, 1067, 1989.
24. Angelo, M. J., Pritchard, A. B., Hawkin, D. R., et al., The pharmacokinetics of dichloromethane, II. Disposition in Fischer 344 rats following intravenous and oral administration, *Food Chem. Toxicol.*, 24, 975, 1986.
25. Angelo, M. J., Pritchard, A. B., Hawkin, D. R., et al., The pharmacokinetics of dichloromethane, I. Disposition in B6C3F$_1$ mice, *Food Chem. Toxicol.*, 24, 965, 1986.
26. McKenna, M. J. and Zempel, J. A., The dose-dependent metabolism of [^{14}*C*]dichloromethane chloride following oral administration to rats, *Food Cosmet. Toxicol.*, 19, 73, 1981.
27. Cunningham, M. L., Gandolfi, A. J., Brendel, K. M., et al., Covalent binding of halogenated volatile solvents to subcellular macromolecules in hepatocytes, *Life Sci.*, 29, 1207, 1981.
28. Casanova, M., Deyo, D. F., and Heck, H. d'A., Dichloromethane (methylene chloride): metabolism to formaldehyde and formation of DNA-protein cross-links in B6C3F$_1$ mice and Syrian golden hamsters, *Toxicol. Appl. Pharmacol.*, 114, 162, 1992.

1,2-DICHLOROPROPANE

Molecular Formula. $C_3H_6Cl_2$
M = 112.99
CAS No 78-87-5
RTECS No TX9625000
Abbreviation. DCP.

Synonyms. α,β-Dichloropropane; Propylene dichloride; α,β-Propylene dichloride.

Properties. Colorless liquid with a sweet, chloroform-like odor. Solubility in water is 0.26% at 20°C, soluble in alcohol, miscible with organic solvents.

Applications. A solvent in plastics, resins, intermediate in the rubber manufacture; solvent mixtures for cellulose esters and ethers.

Acute Toxicity. LD_{50} is 1.9 g/kg BW in Wistar rats,[1] and 2.0 to 4.0 g/kg BW in guinea pigs.[2]

Repeated Exposure. Male rats were gavaged with 100 to 1000 mg DCP/kg BW in corn oil for up to 10 consecutive days. DCP ingestion caused BW and CNS depression. Elevated activity of some serum enzymes occurred only at two highest dose levels. Hepatic nonprotein sulfhydryl levels were decreased and renal nonprotein sulfhydryl levels increased at 24 hours. Resistance developed to DCP hepatotoxicity over the 10 consecutive days of exposure, as reflected by progressively lower serum enzyme levels and by decreases in the severity and incidence of toxic hepatitis. There were a number of manifestations of hemolytic anemia, including erythrophagocytosis in the liver, splenic hemosiderosis, and hyperplasia of erythropoietic elements of the red pulp, renal tubular cell hemosiderosis, and hyperbilirubinemia. Urinalyses and histopathology revealed no evidence of nephrotoxicity.[3]

Short-term Toxicity. Male rats were gavaged for up to 13 weeks with 100 to 750 mg DCP/kg BW. 50% mortality was noted in the highest dosed groups. Histopathology investigation of these animals revealed mild hepatitis and splenic hemosiderosis, as well as adrenal medullar vacuolization and cortical lipidosis, testicular degeneration and a reduction in sperm, and increased number of degenerate spermatogonia in the epididymis in some members of the group. No deaths occurred in the 100 or 250 mg/kg BW groups. The treatment produced a dose-dependent decrease in BW gain. DCP exhibited very limited hepatotoxic potential and no apparent nephrotoxic potential in this study.[3]

Reproductive Toxicity.

Embryotoxicity. DCP was administered *via* oral gavage at dose levels of 10, 30, or 125 mg/kg BW on days 6 through 15 of gestation (rats) or 15, 50, or 150 mg/kg BW on gestation days 7 through 19 (rabbits). Maternal toxicity was observed in both rats and rabbits at the high dose levels. Rats given 125 mg/kg BW of DCP showed clinical signs of toxicity and decreased BW gain. Rabbits given 150 mg DCP/kg BW showed changes in hematological parameters and decreased BW gain.[4]

No indication of ***teratogenicity*** was observed in rat or rabbit fetuses at any dose level. No maternal or developmental effects were observed in rats given 10 or 30 mg/kg BW or in rabbits given 15 or 50 mg DCP/kg BW. The maternal and developmental NOELs in rats and rabbits were 30 and 50 mg/kg BW, respectively.[4]

Mutagenicity.

In vivo cytogenetics. *Dr. melanogaster* males were treated with DCP concentrations that result in approximately 30% mortality. The concentrations of DCP tested by injection (4200 ppm) or feeding (7200 ppm) were negative in this assay.[5]

Carcinogenicity classification. An IARC Working Group concluded that there is limited evidence for the carcinogenicity of DCP in *experimental animals* and there were no adequate data available to evaluate the carcinogenicity of DCP in *humans.*

IARC: 3.

Chemobiokinetics. ^{14}C was found to be rapidly absorbed and excreted after oral administration of 1,2-dichloro-^{14}C-propane to rats during 4 days, 53% were excreted in urine, 6.0% in feces, 42% in expired air as ^{14}C-CO_2.[6]

Standards. ***U.S. EPA*** (1999). MCL: 0.005 mg/l.

References:

1. Smyth, H. F., Carpenter, C. P., Weil, C. S., Pozzani, U. C., Striegel, J. A., and Nyaun, J. S., Range-finding toxicity data: List VII, *Am. Ind. Hyg. Assoc. J.,* 30, 470, 1969.
2. *Farmacol. Chemicals Handbook,* Meister, Willoughby, OH, 1991, 253.
3. Bruckner, J. V., MacKenzie, W. F., Ramanathan, R., Muralidhara, S., Kim, H. J., and Dallas, C. E., Oral toxicity of 1,2-dichloropropane: acute, short-term, and long-term studies in rats, *Fundam. Appl. Toxicol.,* 12, 713, 1989.
4. Kirk, H. D., Berdasco, N. M., Breslin, W. J., and Hanley, T. R., Developmental toxicity of 1,2-dichloropropane (PDC) in rats and rabbits following oral gavage, *Fundam. Appl. Toxicol.*, 28, 18, 1995.
5. Woodruff, R. C., Mason, J. M., Valencia, R., and Zimmering, S., Chemical mutagenesis testing in *Drosophila.* V. Results of 53 coded compounds tested for the National Toxicology Program, *Environ. Mutagen.*, 7, 677, 1985.
6. *Foreign Compound Metabolism in Mammals,* vol 2, A Rev. Literature Publ. in 1970 and 1971, The Chemical Society, London, 1972, 142.

1,3-DICHLORO-2-PROPANOL

Molecular Formula. $C_3H_6Cl_{12}O$

M = 128.99

CAS No 96-23-1

RTECS No UB1400000

Abbreviation. DCP.

Synonyms. 1,3-Dichlorohydrin; α-Dichlorohydrin; 1,3-Dichloroisopropanol; Glycerol α,γ-dichlorohydrin.

Properties. Colorless, oily liquid with an alcoholic odor, darkening on storage. Solubility in water is 12 to 14.5%. Mixes with alcohol at all ratios. Taste perception threshold is 1.0 mg/l. Odor perception threshold is 1.16 mg/l. DCP does not affect coloration of water.

Applications. Used as a solvent in epoxy resin manufacture.

Acute Toxicity. LD_{100} is found to be not lower than 400 to 500 mg/kg BW in rats and mice. Median lethal dose in mice appeared to be 93 mg/kg BW. Poisoning is followed by CNS affection.[1]

According to Smyth et al., LD_{50} is 110 mg/kg BW in rats.[2]

Repeated Exposure revealed cumulative properties.

Short-term Toxicity. Rats tolerate administration of 1/5 LD_{50} for 3 months. Treated animals displayed only reduced ascorbic acid content and decreased relative weights of the suprenals.[1]

Long-term Toxicity and **Immunotoxicity.** In a 6-month study, a 5.0 mg/kg BW dose caused a reduction in immune response and mild changes in the blood formula.[1]

Chemobiokinetics. DCP is not accumulated in biological tissues because of its insolubility in fats. It is readily excreted due to its easy solubility in water. In rats, two mercapturic acid metabolites *N,N'-bis(acetyl)- S,S'-[1,3-bis(cysteinyl)]propan-1-ol* and *N-acetyl-S-(2,3-dihydroxypropyl) cysteine* were discovered in urine. One other oxydative metabolite was *α-chlorolactate*.[3]

Standards. ***Russia*** (1995). MAC and PML: 1.0 mg/l (organolept., taste).

References:

1. Krasovsky, G. N. and Friedland, S. A., in *Industrial Pollution of Water Reservoirs*, S. N. Cherkinsky, Ed., Meditsina, Moscow, Issue No 8, 1967, 129 (in Russian).
2. Smyth, H. F., Carpenter, C. P., Weil, C. S., Pozzani, U. C., and Striegel, J. A., Range-finding toxicity data: List VI, *Am. Ind. Hyg. Assoc. J.*, 23, 95, 1962.
3. Jones, A. R. and Fakhouri, G., Epoxides as obligatory intermediate in the metabolism of α-halohydrins, *Xenobiotica*, 9, 595, 1979.

1,2-DIETHOXYETHANE

Molecular Formula. $C_6H_{14}O_2$
M = 118.20
CAS No 629-14-1
RTECS No KI1225000
Abbreviation. DEE.

Synonyms. Diethyl cellosolve; Ethylene glycol, diethyl ether; Ethyl glume.

Properties. Liquid.

Applications. Used in the manufacture of protective coatings.

Acute Toxicity. LD_{50} is found to be 4.4 g/kg BW in rats, and 2.44 g/kg BW in guinea pigs. The principal effect produced by ethylene glycol ethers in animals at acute exposure to high doses is damage to the kidneys. Upon microscopic examination, tubular degeneration along with almost complete necrosis of the cortical tubules is observed. Additional changes are hematuria, narcosis, and GI tract irritation. Animals exhibited inactivity, weakness, dyspnea, marked testicular toxicity: degeneration of germinal epithelium and testicular atrophy.

Reproductive Toxicity. DEE exhibits adverse developmental effects. Pregnant CD-1 outbred Albino Swiss mice and New Zealand white rabbits were dosed by gavage with DEE dissolved in distilled water during major organogenesis. Doses tested were 50 to 1000 mg/kg BW (mice) and 25 to 100 mg/kg BW (rabbits). No maternal mortality was observed in mice; at the high dose level, decrease in fetal BW and malformations (mainly, exencephaly and fused ribs) were found. The NOAEL for developmental toxicity appeared to be 50 mg/l (in mice) or 25 mg/kg BW (in rabbits).

Mutagenicity.

In vitro genotoxicity. DEE is shown to be positive in *Salmonella* mutagenicity assay (NTP-91).

Reference:

George, J. D., Price, C. J., Marr, M. C., et al., The developmental toxicity of ethylene glycol diethyl ether in mice and rabbits, *Fundam. Appl. Toxicol.*, 19, 15, 1992.

DIMETHOXYMETHANE

Molecular Formula. $C_3H_8O_2$
M = 76.11
CAS No 109-87-5
RTECS No PA8750000
Abbreviation. DMM.

Synonyms and **Trade Name.** Dimethylformal; Formal; Formaldehyde dimethylacetal; Methylal; Methylene dimethyl ether.

Properties. Colorless, highly mobile liquid with a characteristic odor. Solubility in water is 32.3%. Mixes freely with organic solvents. Alcohol (not more than 8.0%) is a component of technical grade M.

Applications. Used as a component and a solvent in the production of ion exchange resins.

Acute Toxicity. LD_{50} is reported to be 9.07 g/kg BW in rats, and 6.95 g/kg BW in mice. Lethal doses cause lethargy in the treated animals.

Chemobiokinetics. DMM metabolism occurs in the body through the rupture of one of the ether bonds and formation of the *dimethyl ether* and *methyl alcohol.*

Recommendations. In view of the equimolar formation of methanol in the body of animals, Tomilina et al. suggested regulating the safe levels of DMM using the MAC of methanol.

Reference:

Tomilina, L. A., Rotenberg, Yu. S., Mashbitz, F. D., et al., Methylal: metabolism and hygienic standardization on the workplace, *Gig. Truda Prof. Zabol.*, 6, 27, 1984 (in Russian).

N,N'-DIMETHYLACETAMIDE

Molecular Formula. C_4H_9NO
M = 87.14
CAS No 127-19-5
RTECS No AB7700000
Abbreviation. DMAA.

Synonyms. Acetic acid, *N,N'*-dimethylamide; Dimethylacetone amid; Dimethylamide acetate.

Properties. A colorless liquid with a faint odor. Readily soluble in water, does not affect its organoleptic properties.

Applications. Used as a solvent in the synthesis of plastics and semi-permeable polyamide membranes for reverse-osmotic water desalination.

Acute Toxicity. LD_{50} values range from 4.2 g/kg BW to 4.85 g/kg BW in mice, and from 4.3 g/kg to 5.2 g/kg BW in rats.[1] Manifestations of the acute toxic effect included CNS inhibition, convulsions and paresis in mice.

Repeated Exposure. DMAA is classified as having moderate cumulative properties.[1] Rats were dosed with 2.0 ml/kg BW for 10 days. The treatment caused 75% animal mortality within 6 to 10 days. Gross pathology examination revealed extensive hemorrhages in the stomach and lungs. Rats tolerate 0.5 ml/kg BW dose (1:5 dilution) ingested for a month without any signs of intoxication. The 90 and 450 mg/kg BW doses, however, were found to induce functional changes in the CNS and affected enzyme-forming function of the liver. A reversible BW loss and anemia were noted. Gross pathology examination revealed hepatic and testicular lesions. A 16 mg/kg BW dose appeared to be ineffective.

CNS was found to be affected in rabbits given 2.0 g DMAA/kg BW by different routes of administration for 21 days. In the animals, variable electroencephalographic changes, dysrhythmias, and electrographic seizures were noted.[2]

Long-term Toxicity. The 6-month treatment affected the CNS and liver functions, as well as hematology analyses of rats.[1]

Reproductive Toxicity.

Embryotoxic effect (over 60% resorptions) was observed at a dose level of 1.5 ml/kg BW in pregnant rats treated with DMAA on day 4 or 7 of gestation (Shepard, 1986).[3]

Teratogenic response was noted in rats but not in rabbits.[2] Pregnant Charles River COBS CD rats received 65, 160, or 400 mg/kg BW with their drinking water on gestation days 6 through 19. The treatment with the highest dose caused significant reduction in maternal BW. Considerable increase in postimplantation loss and in the number of malformations (particularly heart and/or vessel anomalies) was observed. There were no teratogenic effects at the lower doses.[4]

Mutagenicity.

In vitro genotoxicity. DMAA is reported to be negative in *Salmonella* mutagenicity assay.[5]

Chemobiokinetics. In male Crl:CD BR rats and Crl:CD-1 ICR mice, DMAA has been metabolized to *N-methylacetamide*. An absence of *N*-methylacetamide in plasma following repeated 300 and 500 ppm DMAA exposure supported the NOEL of 350 ppm for a chronic inhalation study.[6]

Standards. ***Russia*** (1995). MAC and PML: 0.2 mg/l.

References:

1. Bogdanov, M. V., Korolev, A. A., Kinzirsky, A. S., et al., Experimental substantiation of MAC for dimethylacetamide in water bodies, *Gig. Sanit.*, 6, 76, 1980 (in Russian).
2. *Toxicity and Metabolism of Industrial Solvents*, vol 2, Nitrogen and Phosphorus Solvents, 2 ed., R. Snyder, Ed., Elsevier, Amsterdam-New York-Oxford, 1990, 145.
3. Thiersch, J. B., Investigations into the differential effect of compounds on rat litter and mother, in *Malformations Congenitales Des Mammiferes*, H. Tuchmann-Duplessis, Ed., Masson et Cie, Paris, 1971, 95.
4. Johannsen, F. R., Levinskas, G. J., and Schardein, J. L., Teratogenic response of dimethylacetamide in rats, *Fundam. Appl. Toxicol.*, 9, 550, 1987.
5. Hedenstedt, A., Genetic health risk in the rubber industry: Mutagenicity studies on rubber chemicals and process vapors, in *Prevention of Occupational Cancer*, Int. Symp., Int. Labour Office, Geneva, 1982, 476.
6. Hundley, S. G., Lieder, P. H., Valentine, R., McCooey, K. T., and Kennedy, G. L., Dimethylacetamide pharmacokinetics following inhalation exposure to rats and mice, *Toxicol. Lett.*, 73, 213, 1994.

N,N'-DIMETHYLFORMAMIDE

Molecular Formula. C_3H_7NO
M = 73.10
CAS No 68-12-2
RTECS No LQ2100000
Abbreviation. DMF.

Synonyms. *N,N'*-Dimethylmethanamide; *N*-Formyldimethylamine.

Properties. DMF contains a significant percentage of monomethylformamide. Clear, colorless liquid. Mixes with water and alcohol at all ratios. Soluble in acetone and chloroform. Odor perception threshold is 50 mg/l.[1]

Applications. Used as a solvent for acrylic fibers and polyurethane, in the manufacture of polyvinyl chloride, polyacrylonitrile, and polyurethanes and many vinyl-based polymers and copolymers intended for use as surface coatings.

Exposure. Human exposure occurs primarily through inhalation and dermal absorption.

Acute Toxicity.

Observations in man. Ethanol intolerance is one of the earliest manifestations of excessive exposure to DMF, followed by nausea, vomiting, abdominal pain, and the release of liver cytotoxic enzymes in the plasma.[2]

Animal studies. All rats die after administration of 8.0 g/kg BW, and mice after administration of 5.0 g/kg BW. According to other data, a single administration of 2.25 g/kg BW to rats causes death due to liver necrosis; LD_{50} is found to be 3.0 to 7.0 g/kg BW.[3] LD_{50} is reported to be 3.9 to 6.4 g/kg BW in mice, more than 5.0 g/kg BW in rabbits,[4,5] and 3.0 to 4.0 g/kg BW in guinea pigs (IARC 47-178).

Animals developed general depression, anesthesia, loss of appetite and BW, tremors, convulsions, hemorrhage from the nose and mouth, liver injury, and coma immediately preceding death.[6]

DMF is more toxic in younger than in older rats, with oral LD_{50} of less than 1.0 g/kg BW in newborn, 1.4 g/kg BW in 14-day-old, 4.0 g/kg BW in young adult, and 6.8 g/kg BW in adult animals.[7]

Repeated Exposure. DMF causes dose-related liver injuries in most species tested, including humans.[2] Doses of 620 or 1240 mg DMF/kg were given in the diet to mice over a period of 30 days. The treated animals displayed anorexia and loss of BW.[8]

In Mongolian gerbils given DMF in drinking water with concentration of 10 g/l for 30 days, no changes in BW, liver, or kidney were reported.[9]

Rats received DMF in their drinking water at concentrations of 102 or 497 mg/l for 49 days. Animals exhibited no behavioral changes. Dose-related deviations in cerebral and glial cell enzyme activities were noted.[10]

Short-term Toxicity. Consumption of drinking water containing 17 to 34 g/l DMF for 80 days caused increased mortality due to liver necrosis.[6]

Administration of 160 to 1850 mg DMF/kg diet given for 119 days produced dose-related increases in the liver weights; there were no histological or biochemical changes. The NOEL of 246 to 326 mg/kg was identified in this study.[11]

In a 90-day oral study, slight anemia, leukocytosis and hyperholesterolemia were observed at dose levels of 60 and 300 mg/kg BW. The NOEL is reported to be 12 mg/kg BW.[3]

Reproductive Toxicity. Sprague-Dawley rats were given orally 50 to 300 mg DMF/kg BW on gestation days 6 through 20. The maternal and developmental NOAEL was established to be 50 mg/kg BW.[12]

Two metabolites of DMF, *N-methylformamide* and *formamide*, have been shown to be embryolethal and teratogenic in rodents following oral administration.[13]

Gonadotoxicity. DMF is unlikely to be gonadotoxic to rats; it produces no effect on fertility. Doses of 0.2 to 2000 mg/kg BW produce no sperm abnormalities in mice.[14]

Embryotoxicity. Administration of 193 mg/kg BW by gavage to pregnant mice on days 6 to 15 of pregnancy resulted in fetal weight decrease and malformations.[15]

A dose of 580 mg/kg BW caused embryolethality. In rabbits exposed by gavage to the doses of 47 to 68 mg/kg BW on days 6 to 18 of pregnancy, decreased number of implantations and 3 cases of hydrocephalus were reported.[16]

The rabbits appear to be more sensitive than rats and mice in dermal testing.[17]

Teratogenicity. Administration of 1/20 LD_{50} during pregnancy (460 mg/kg BW over the whole period) caused no changes in the bodies of mothers but disrupted embryogenesis, leading to abnormalities or fetal death.[6]

However, according to other data, oral administration of 200 mg/kg BW dose to pregnant animals caused an increased rate of fetal malformations in the absence of overt maternal toxicity.[16] The NOAEL appeared to be 182 mg/kg BW (mice) and 166 mg/kg BW (rats).

According to Thiersh,[18] the treatment with 0.5 to 2.0 ml/kg BW had no teratogenic effect in experimental rats.

Mutagenicity. DMF was generally found to be inactive, both *in vitro* and *in vivo*, in an extensive set of short-term tests for genetic and related effects.[7,19]

In vitro genotoxicity.

Observations in man. An increased frequency of CA was found in peripheral lymphocytes of industrial workers exposed to DMF.[20] It did not cause mutation in mouse lymphoma cells (NTP-86), or CA and SCE in Chinese hamster ovary cells (NTP-85).

Carcinogenicity.

Observations in man. An excess risk for testicular germ-cell tumors was identified among workers exposed to a solvent mixture containing 80% DMF.[19]

Animal studies. No adequate carcinogenicity studies have been reported. In a 107-week study, BD rats were given the doses of 75 and 150 mg/kg BW until total dose of 38 g/kg BW had been given. The treatment did not produce any tumorigenic effects.[21] However, IARC (47-186) considered this study to be inadequate for evaluation.

Carcinogenicity classification. An IARC Working Group concluded that there is inadequate evidence for the carcinogenicity of DMF in *experimental animals* and there were limited data available to evaluate the carcinogenicity of DMF in *humans*.

IARC: 2B.

Chemobiokinetics.

Observations in man. DMF is reported to be metabolized by sequential *n*-demethylation.[22]

Animal studies. Following ingestion, DMF is readily absorbed in the GI tract of experimental animals.[23] Metabolism occurs predominantly in the liver. The main products of biotransformation are *formamide* and *N-hydroxymethyl-N'-methylformamide*, which is then partially converted to *N-methyl-formamide*.[22-25]

According to other data, DMF altered the hepatic microsomal monooxygenase system and glutathione metabolism due to formation of *monomethylformamide*.[26] Monomethylformamide and formamide are found in the urine.

Regulations. ***U.S. FDA*** (1998) approved the use of DMF (1) in adhesives as a component of articles intended for use in packaging, transporting, or holding food in accordance with the conditions prescribed in 21 CFR part 175.105; and (2) as an adjuvant in the preparation of slimicides in the manufacture of paper and paperboard that may be safely used in contact with food in accordance with the conditions prescribed in 21 CFR part 176.300.

Standards. ***Russia*** (1995). MAC and PML: 10 mg/l.

References:

1. Zamyslova, S. D. and Smirnova, R. D., in *Protection of Water Reservoirs against Pollution by Industrial Liquid Effluents*, S. N. Cherkinsky, Ed., Medgiz, Moscow, Issue No 4, 1960, 177 (in Russian).
2. Scailteur, J. P. and Lauwerys, R. R., Dimethylformamide (DMF) hepatotoxicity, *Toxicology*, 43, 231, 1987.
3. Kennedy, G. L. and Sherman, H., Acute and subacute toxicity of dimethylformamide and dimethylacetamide following various routes of administration, *Drug Chem. Toxicol.*, 9, 147, 1986.
4. Stasenkova, K. P., Toxicity of dimethylformamide, in *Toxicology of New Industrial Chemical Substances*, A. A. Letavet and I. V. Sanotsky, Eds., Medgiz, Moscow, Issue No 1, 1961, 58 (in Russian).
5. Sheveleva, G. A., *A Study of the Influence of Formaldehyde and Dimethylformamide on the Maternal Body, Fetal Development and Offspring*, Author's abstract of thesis, Moscow, 1971, 25 (in Russian).
6. Kennedy, G. L., Biological effects of acetamide, formamide, and their monomethyl and dimethyl derivatives, *CRC Crit. Rev. Toxicol.*, 17, 129, 1986.
7. Kimura, E. T., Ebert, D. M., Dodge, P. W., et al., Acute toxicity and limits of solvent residue for sixteen organic solvents, *Toxicol. Appl. Pharmacol.*, 19, 699, 1971.
8. Aucair, M. and Hameau, N., *Compt. Rend. Soc. Biol.*, 158, 245, 1964.
9. Llewellyn, G. C., Hastings, W. S., and Kimbrough, T. D., The effects of dimethylformamide on female Mongolian gerbils *Meriones Unguicklatus*, *Bull. Environ. Contam. Toxicol.*, 11, 467, 1974.
10. Savolainen, H., Dose-dependent effects of oral dimethylformamide administration on rat brain, *Acta Neuropathol.*, 53, 249, 1981.
11. Becci, P. J., Voss, K. A., Johnson, W. D., et al., Subchronic feeding study of *N,N'*-dimethylformamide in rats and mice, *J. Am. Coll. Toxicol.*, 2, 371, 1983.
12. Saillenfait, A. M., Payan, J. P., Beydon, D., Fabry, J. P., Langonne, I., Sabate, J. P., and Gallissot, F., Assessment of the developmental toxicity, metabolism, and placental transfer of *N,N'*-dimethylformamide administered to pregnant rats, *Fundam. Appl. Toxicol.*, 39, 33, 1997.
13. Kelich, S. L., Mercieca, M. D., and Pohland, R., Developmental toxicity of N-methylformamide administered by gavage to CD rats and New Zealand white rabbits, *Fundam. Appl. Toxicol.*, 27, 239, 1995.
14. Antoine, J. L., Arany, J., Leonard, A., et al., Lack of mutagenic activity of dimethylformamide, *Toxicology*, 26, 207, 1983.
15. *Dimethylformamide*, Environmental Health Criteria Series No 114, WHO, IPCS, Geneva, 1991, 70.
16. Merkle, J. and Zeller, H., Studies on acetamides and formamides for embryotoxic and teratogenic activities in the rabbit, *Arzneimittel-Forsch.*, 30, 1557, 1980.
17. Hellwig, J., von Merkle, J., Klimisch, H. J., et al., Studies on the prenatal toxicity of *N,N'*-dimethylformamide in mice, rats and rabbits, *Food Chem. Toxicol.*, 29, 193, 1991.
18. Thiersch, J. B., *Malformations Congenitales des Mammiferes*, Paris, 1971, 95.

19. *Evaluation of Short-term Tests for Carcinogens*, Serres, F. J. and Ashby, J., Eds., Elsevier Sci. Publishers, Amsterdam, Oxford, New York, 1981, 827 (*Progress in Mutation Research*, vol 1).
20. Ducatman, A. M., Conwill, D. E., and Crawl, J., Germ cell tumors of the testicle among aircraft repairmen workers, *J. Urol.*, 136, 834, 1986.
21. Druckrey, H., Preussmann, R., Ivankovich, S., et al., Organotropic carcinogenic effects of 65 different *N*-Nitrosocompounds on BD rats, *Z. Krebsforsch.*, 69, 103, 1967.
22. Baselt, R. C., *Biological Monitoring Methods for Industrial Chemicals*, 2 ed., PSG Publ. Co., Inc., Littleton, MA, 1988, 128.
23. Massmann, W., Toxicological investigation of dimethylformamide, *Brit. J. Ind. Med.*, 13, 51, 1956.
24. Kimmerle, G. and Eben, A., Metabolism studies of *N,N'*-dimethylformamide, I. Studies in rats and dogs, *Int. Arch. Arbeitsmed.*, 34, 109, 1975.
25. Van den Bulcke, M., Rosseel, M. T., Wijnants, P., Buylaert, W., and Belpaire, F. M., Metabolism and hepatotoxicity of *N,N'*-dimethylformamide, *N*-hydroxymethyl-*N'*-methylformamide, and *N*-methylfor mamide in the rat, *Arch. Toxicol.*, 68, 291, 1994.
26. Imazu, K., Fujishiro, K., and Inoue, N., Effects of dimethylformamide on hepatic microsomal monooxygenase system and glutathione metabolism in rats, *Toxicology*, 72, 41, 1992.

2,6-DIMETHYL-4-HEPTANOL

Molecular Formula. $C_9H_{20}O$
M = 144.29
CAS No 108-82-7
RTECS No MJ3325000
Abbreviation. DMH.

Synonyms. Diisobutyl carbinol; 4-Hydroxy-2,6-dimethyl heptane; *sec*-Nonyl alcohol.

Properties. Colorless liquid, insoluble in water, soluble in alcohol.

Applications. A solvent in the production of urea and melamine synthetic coatings, in the manufacture of rubber and plasticizers.

Acute Toxicity. LD_{50} is 3.56 g/kg BW in rats.

Regulations. ***U.S. FDA*** (1998) approved the use of DMH as a defoaming agent in the manufacture of paper and paperboard intended for use in producing, manufacturing, packaging, processing, preparing, treating, packing, transporting, or holding foods in accordance with the conditions prescribed in 21 CFR part 176.210.

Reference:

Smyth, H. F., Carpenter, C. P., and Weil, C. S., Range-finding toxicity data: List III, *J. Ind. Hyg. Toxicol.*, 31, 60, 1949.

2,6-DIMETHYL-4-HEPTANONE

Molecular Formula. $C_9H_{18}O$
M = 142.27
CAS No 108-83-8
RTECS No MJ5775000
Abbreviation. DMH.

Synonyms and **Trade Names.** Diisobutyl ketone; 2,6-Dimethyl-4-heptanone; 2,6-Dimethylheptan-4-on; 2,6-Dimethylheptan-4-one; Isobutyl ketone; Isovalerone; *sec*-Diisopropyl acetone; Valerone.

Properties. Colorless oil with mild sweet odor. Solubility in water 0.05 wt% at 20°C. Soluble in alcohol.

Applications. Solvent for nitrocellulose and synthetic coatings.

Acute Toxicity. LD_{50} is reported to be 5.8 g/kg BW in rats.[014] Ingestion of up to 3.2 g DMH/kg BW did not cause death in rats. Two to three daily doses of 4.0 g/kg BW given by gavage were fatal to all exposed animals. Manifestations of toxic action included CNS depression, hepatotoxicity, and dehydration. Pulmonary congestion and edema and renal injury were also observed.[03]

Short-term Toxicity. Rats received 2.0 g/kg BW by gavage for 90 days. The treatment reduced glucose levels and increased absolute and relative liver, kidney and adrenal gland weight. Absolute brain and heart weights were decreased.[03]

Mutagenicity.

In vitro genotoxicity. DMH was negative in the rat liver chromosome assay, in the bacterial mutation assays, and in the yeast mitotic gene conversion assay.[1]

Chemobiokinetics. DMH undergoes metabolic conversion common to that of most ketones except acetone. It is reduced to the corresponding *secondary alcohol* which is usually eliminated combined with *glucuronic acid*.[2]

Regulations. ***U.S. FDA*** (1998) approved the use of DMH in adhesives as a component of articles intended for use in packaging, transporting, or holding food in accordance with the conditions prescribed in 21 CFR part 175.105.

References:

1. Brooks, T. M. et al., The genetic toxicology of some hydrocarbon and oxygenated solvents, *Mutagenesis*, 3, 227, 1988.
2. Browning, E., *Toxicity and Metabolism of Industrial Solvents*, American Elsevier, New York, 1965, 436.

p-DIOXANE

Molecular Formula. $C_4H_8O_2$

M = 88.10

CAS No 123-91-1

RTECS No JG8225000

Synonyms. Diethylene dioxide; Diethylene ether; Diethylene oxide; 1,4-Dioxacyclohexane; Dioxyethylene ether; Tetrahydro-1,4-dioxin.

Properties. Volatile, colorless liquid with a mild ethereal odor. Miscible with water, alcohol and majority of organic solvents. Odor perception threshold is 0.8 mg/l[010] or 1.24 mg/l. Taste perception threshold is 0.3 mg/l.[1]

Applications. Solvent for cellulose acetate, ethyl cellulose, benzyl cellulose, lacquers, plastics, resins, and polyvinyl polymers.

Acute Toxicity. A substance of low toxicity. The liver is shown to be a target organ.[2] Laboratory animals given single doses of 3.9 to 5.7 g/kg BW exhibited weakness, CNS depression, coordination disorder, coma, and death. Gross pathology examination revealed hemorrhagic areas in the stomach, bladder distended with urine, enlarged kidneys, slight proteinuria but without hematuria.[03]

Repeated Exposure. Dogs died within 9 days after a total consumption of approximately 3.0 g/kg BW. Histological investigation revealed severe liver and kidney lesions.[014]

Reproductive Toxicity.

Embryotoxicity. D. produced no teratogenic effect in rats gavaged with up to 1.0 ml D./kg BW on days 6 to 15 of gestation.[3] Maternal and fetal weights were reduced at the highest dose.

Mutagenicity.

In vitro genotoxicity. D. does not exhibit mutagenic activity in *Salmonella* assay with and without metabolic activation,[025] and does not react with DNA. It induced neither CA in Chinese hamster ovary cells (Anderson et al., 1990) nor gene mutations in mouse lymphoma cells (McGregor et al., 1991).

In vivo cytogenetics. Female Sprague-Dawley rats were given 168, 840, 2550, or 4200 mg/kg BW of D. for 4 hours before sacrifice. The treatment increased hepatic DNA damage and cytochrome *P-450* content at two higher doses. D. has been considered a weak genotoxic carcinogen in addition to being a strong promoter of carcinogenesis.[4] D. was negative in mouse bone marrow micronucleus assays under conditions of the test used.[5]

According to Ashby (1994), D. gave both positive and negative results in *in vivo* micronucleus assays. A small but statistically significant increase in micronucleus frequency was obtained in CD-1 mice at dose level of 2.0 g/kg BW and the overall responses showed no dose-dependency. A weak positive response was obtained with a single *i/p* treatment at 2.0 g/kg BW dose but not at 4.0 g/kg BW. It did not induce sex-linked recessive lethal mutations in *Dr. melanogaster*.[6]

Carcinogenicity. Administration of D. in drinking water at several dose levels produced liver adenomas and carcinomas in male and female rats, hepatomas in guinea pigs, nasal cavity carcinomas in male and female rats, and gall bladder carcinomas in guinea pigs.[6]

Female Sherman rats received 0.01, 0.1, or 1.0% D. in drinking water for 716 days. At 1.0% dose level but not at two lower doses, an increased incidence of hepatocellular carcinomas, cholangiosarcomas, and squamous-cell carcinomas of nasal cavity were noted (Kociba et al., 1974, cit. in IARC 11-250).

Osborne-Mendel rats and $B6C3F_1$ mice received 0.5 and 1.0% D. in drinking water for 110 and 90 weeks, respectively. Hepatocellular adenomas and squamous-cell carcinomas of the nasal turbinates were found in rats; hepatocellular carcinomas were observed in mice.[7]

Carcinogenicity classifications. An IARC Working Group concluded that there is sufficient evidence for the carcinogenicity of D. in *experimental animals* and there were inadequate data to evaluate the carcinogenicity of D. in *humans*.

The Annual Report of Carcinogens issued by the U.S. Department of Health and Human Services (1998) defines D. to be a substance which may reasonably be anticipated to be carcinogenic, i.e., a substance for which there is a limited evidence of carcinogenicity in *humans* or sufficient evidence of carcinogenicity in *experimental animals*.

IARC: 2B;
U.S. EPA: B2;
EU: 3;
NTP: P - P* - P* - P* (water).

Regulations. ***U.S. FDA*** (1998) regulates D. as an indirect food additive. It is listed for use in adhesives as a component of articles intended for use in packaging, transporting, or holding food in accordance with the conditions prescribed in 21 CFR part 175.105.

References:

1. Amirkhanova, G. F. and Latypova, Z. V., Experimental substantiation of the maximum allowable concentration for acetaldehyde in water bodies, in *Industrial Pollutants of Water Reservoirs*, S. N. Cherkinsky, Ed., Meditsina, Moscow, Issue No 9, 1969, 137 (in Russian).
2. *Bulletin of International Register of Potentially Toxic Chemicals*, 6, 17, September 1983.
3. Giavini, E., Vismara, C., and Brocera, M. L., Teratogenesis study of dioxane in rats, *Toxicol. Lett.*, 26, 85, 1985.
4. Kitchin, K. T. and Brown, J. L., Is 1,4-dioxane a genotoxic carcinogen? *Cancer Lett.*, 53, 67, 1990.
5. Tinwell, H. and Ashby, J., Activity of 1,4-dioxane in mouse bone marrow micronucleus assays, *Mutat. Res.*, 322, 148, 1994.
6. Morita, T., Asano, N., Awogi, T., Sasaki, Yu. F., et al., Evaluation of the rodent micronucleus assay in the screening of IARC carcinogens (Groups 1, 2A and 2B). The summary report of the 6 collaborative study by CSGMT/JEMS-MMS, *Mutat. Res.*, 389, 3, 1997.
7. *Bioassay of 1,4-Dioxane for Possible Carcinogenicity*, NTP Technical Report Series No 80, Research Triangle Park, NC, 1978.

DODECANOIC ACID, 2-(2-HYDROXYETHOXY)ETHYL ETHER

Molecular Formula. $C_{16}H_{32}O_4$
M = 288.42
CAS No 141-20-8

Synonyms. Diethylene glycol, monolauryl ether; Diglycol laurate.

Properties. Oily liquid. Practically insoluble in water, soluble in alcohols and cottonseed oil.[020]

Applications. Used as a plasticizer and emulsifier.

Regulations. ***U.S. FDA*** (1998) approved the use of D. in adhesives as a component of articles intended for use in packaging, transporting, or holding food in accordance with the conditions prescribed in 21 CFR part 175.105.

1,2-EPOXY-3-PHENOXYPROPANE

Molecular Formula. $C_9H_{10}O_2$
M = 150.19
CAS No 122-60-1
RTECS No TZ3675000
Abbreviation. EPP.

Synonyms. (2,3-Epoxypropoxy)benzene; (Phenoxymethyl) oxirane; Phenoxypropene oxide; Phenoxypropylene oxide; Phenylglycidyl ether.

Properties. Colorless liquid. Solubility in water is 2.4 g/l.

Applications. Reactive diluent and plasticizer for epoxy resins.

Acute Toxicity. LD_{50} is 3.85 g/kg BW in rats, and 1.4 mg/kg BW in mice. Oral administration of large doses resulted in CNS depression, caused ataxia and change in motor activity.[014]

Reproductive Toxicity. EPP did not produce evident adverse effects on reproduction on inhalation exposure *in vivo.*[1]

Gonadotoxicity. In a two-generation study and during organogenesis, rat males were exposed to the concentration of 11 ppm. The treatment did not affect fertility.[1]

Mutagenicity. Structurally suspected mutagen.

Observations in man. Negative results were obtained in rats exposed by inhalation.[2-5]

In vitro genotoxicity. EPP was found to be positive in bacterial assays.[2]

Carcinogenicity classification.

Considered to be an animal carcinogen (inhalation).[6]

Regulations. ***U.S. FDA*** (1998) approved the use of EPP in articles made of mineral-reinforced nylon resins or as components of these articles intended for repeated use in contact with nonacidic food (*pH* above 5.0) and at a temperature not exceeding 212°F in accordance with the conditions specified in the 21 CFR part 177.2355.

References:

1. Terrill, J. B., Lee, K. P., Culik, R., and Kennedy, G. L., Inhalation toxicity of phenyl glycidiyl ether: Reproduction, mutagenic, teratogenic and cytogenic studies, *Toxicol. Appl. Pharmacol.*, 64, 204, 1982.
2. Seiler, J. R., The mutagenicity of mono- and difunctional aromatic glycidyl compounds, *Mutat. Res.*, 135, 159, 1984.
3. Terrill, J. B. and Haskell, K. P., Inhalation toxicity of phenylglycidyl ether: mutagenic, teratogenic, cytogenetic and 90-day studies, *Toxicol. Appl. Pharmacol.,* 37, 191, 1976.
4. Canter, D. A., Zeiger, E., Haworth, S., Lawlor, T., Mortelmans K., and Speck, W., Comparative mutagenicity of aliphatic epoxides in *Salmonella, Mutat. Res.*, 172, 105, 1986.
5. von der Hude, W., Carstensen, S., and Obe, G., Structure-activity relationship of epoxides: induction of sister-chromatid exchanges in Chinese hamster *V79* cells, *Mutat. Res.*, 249, 55, 1991.
6. American Conference of Governmental Industrial Hygienists, *Threshold Limit Values (TLVs) for Chemical Substances and Physical Agents and Biological Exposure Indices (BEIs) for 1995-1996*, Cincinnati, OH, 1995, 29.

2-(2,3-EPOXYPROPOXY)FURAN

Molecular Formula. $C_7H_{10}O_3$
M = 154
CAS No 26130-15-4
RTECS No LU1420000

Synonym. Furylglycidyl ether.

Properties. Liquid with a slight odor of ether. Poorly soluble in water; readily soluble in alcohol.

Applications. Used as a diluent in the synthesis of epoxy resins.

Acute Toxicity. In rats, LD_{50} is 1.1 g/kg BW.

Repeated Exposure revealed reduction in BW gain and body temperature.

Reference:

Zolotov, P. A. and Kesel'man, M. L., Toxicity of furyl glycidyl ether, *Gig. Sanit.*, 4, 93, 1984 (in Russian).

1,2-EPOXY-3-(TOLYLOXY)PROPANE

Molecular Formula. $C_{10}H_{12}O_2$
M = 180.22
CAS No 26447-14-3
RTECS No TZ3699000
Abbreviation. ETP.

Synonyms. Cresyl glycidyl ether; [(Methylphenoxy)methyl] oxirane; Tolyl gycidyl ether.

Properties. Clear liquid of low volatility, of a light-straw color and with a sharp, irritating odor. Poorly soluble in water; readily soluble in organic solvents.

Applications. Used as an active diluent in epoxy resin manufacture.

Acute Toxicity. LD_{50} is 5.1 g/kg BW in rats, 1.7 g/kg BW in mice, and 1.65 g/kg BW in guinea pigs. Manifestations of the toxic action include a short period of excitation (20 to 30 minutes) with subsequent CNS inhibition, increased respiratory rate, and adynamia. Death occurs within 8 days after administration. Gross pathology examination revealed dystrophic and destructive changes in the parenchyma of the viscera. Liver lesions and necrosis of the renal tubular epithelium were also found.[1]

Repeated Exposure revealed pronounced cumulative properties. K_{acc} is 1.2 (by Cherkinsky). Rats tolerated daily exposure to 1/20 LD_{50}. The treatment caused anemia and a decrease in rectal temperature, and in the heart relative weight.[1]

Reproductive Toxicity. Following inhalation exposure, gonadotoxic and embryotoxic effects are shown to develop at concentrations lower than the general toxicity level (2.55 mg/m^3).[2]

Mutagenicity.

Allergenic Effect. ETP was positive in the guinea pig maximization test. ETP was found to be more allergenic than butyl glycidyl ether, a known skin sensitizer.[3]

Mutagenicity.

In vivo cytogenetics. ETP did not induce DLM in male mice; body fluids from treated mice were not mutagenic in *Salmonella typhimurium*.[4]

In vitro genotoxicity. ETP caused mutations in the microorganisms, *Salmonella typhimurium* and *S. cerevisiae* (Canter *et al.*, 1986)[5].

References:

1. Krechkovsky, E. A., Anisimova, I. G., Shevchuk, T. M., et al., Toxicity of cresyl glycidyl ether, *Gig. Truda Prof. Zabol.*, 3, 124, 1985 (in Russian).
2. Krechkovsky, E. A., Anisimova, I. G., Vislova, S. V., et al., Experimental study on cresyl glycidyl ether-induced late effects on reproductive function, *Gig. Sanit.*, 10, 81, 1985 (in Russian).
3. Thorgeirsson, A., Fregert, S., Magnusson, B., and Berufs, N., Allergenicity of epoxy-reactive diluents in the guinea pig, *Berufs-Dermatosen.*, 23, 178, 1975 (in Norwegian).
4. Pullin, T. G., *Mutagenic Evaluation of Several Industrial Glycidyl Ethers*, Diss Abstr. Int. Sci. 39, 4795, 1979.
5. Canter, D. A., Zeiger, E., Haworth, S., Lawlor, T., Mortelmans, K., and Speck, W., Comparative mutagenicity of aliphatic epoxides in *Salmonella*, *Mutat. Res.*, 172, 105, 1986.

2-ETHOXYETHANOL

Molecular Formula. $C_4H_{10}O_2$
M = 90.14
CAS No 110-80-5
96231-36-6
RTECS No KK8050000
Abbreviations. EE, EAA (ethoxyacetic acid).

Synonyms and **Trade Name.** Ethyl glycol; Ethyl cellosolve; Ethylene glycol, monoethyl ether; Oxitol.

Properties. Colorless, transparent liquid with a faint sweetish odor. Miscible with water and many organic solvents. Does not affect transparency and color of water; does not form foam and film on the water surface. Odor perception threshold is reported to be 25 mg/l, or 9.0 mg/l,[1] or 190 mg/l;[02] taste perception threshold is 10 mg/l.[2]

Applications. EE has wide industrial and consumer applications. It is used as a solvent in the production of nitro- and acetyl cellulose, natural and synthetic resins. Used in the manufacture of plastics, lacquers, and protective coatings.

Acute Toxicity. LD_{50} is reported to be 3.0 to 5.0 g/kg BW in male rats, 2.3 to 5.4 g/kg BW in female rats, 1.28 to 3.1 g/kg BW in rabbits, and 1.4 g/kg BW in guinea pigs.[3,4]

See ***1,2-Diethoxyethane***.

Repeated Exposure revealed hematological, biochemical, and morphological changes.[4] In 2-week studies, F344 rats and $B6C3F_1$ mice received EE in their drinking water. Consumption values ranged from 200 to 1600 mg/kg BW (rats) and 400 to 2800 mg/kg BW (mice). There were no effects on survival. Decrease in BW gain was reported in rats. The treatment caused thymic and testicular atrophy in males of both species.[5]

Short-term Toxicity. A reduction in blood *Hb* level and changes in the hematocrit, liver, kidneys and ovaries were observed in rats and dogs administered doses of 45 and 750 mg/kg BW over a period of 13 weeks.[1]

Medinsky et al. believe hematotoxic effects to be due to the EAA metabolite of EE.[6] In a 13-week study, histology changes were observed in the testes, thymus and hematopoietic tissues. The treatment caused progressive anemia. Effect on the testes, spleen, and adrenal gland was seen in females only. The NOAEL for all histopathological and hematological effects was 5000 ppm in rats.[5]

Long-term Toxicity. In the NTP-88 study, male and female rats were dosed by gavage with 0.5, 1.0, and 2.0 g/kg BW.[7] High mortality of animals was noted at 2.0 g/kg BW dose level; males exhibited testicular lesions.

Reproductive Toxicity.

Gonadotoxicity. EE is reported to produce testicular atrophy, degenerative changes in the germinal epithelium, pathologic changes in the sperm head, and infertility.[8] Testicular effects produced by EE may be caused by its active metabolite EAA.

Observations in man. EE has been linked with reduction of sperm counts in chronically-exposed workers.[9,10]

Animal studies. Testicular atrophy was observed in male mice given oral doses of 0.5 to 4.0 g/kg BW for 5 weeks. The dose of 0.5 g/kg BW appeared to be ineffective.[11]

Male rats were exposed to 0.25 to 1.0 g/kg BW for 11 days. Decreased testes weight, and spermatocyte depletion and degeneration were noted. 0.25 mg/kg BW appeared to be the NOEL for gonadotoxic effect.[12]

Teratogenicity. Adverse maternal and developmental effects following EE administration were found to develop. An increase in the incidence of abnormal skeletal development was noted in the fetuses of rats exposed to 93 to 186 mg/kg BW on days 1 to 21 of gestation period. The NOAEL for these defects appeared to be 46.5 mg/kg BW.[1]

Embryotoxic effects of EE are reported. Administration of 500 to 1000 mg/kg BW caused decreased liver and testicular weights in young rats on day 11 of administration. A 250 mg/kg BW dose caused no changes.[8]

Mutagenicity.

In vitro genotoxicity. EE gave negative results in mutagenicity assays with *Salmonella typhimurium* strains *TA98, TA100,* and *TA102* either with or without *S9* mix.[13] It was negative in the mouse lymphoma cell mutation assay without *S9,* but was weakly positive in the presence of *S9* rat liver fraction. At high concentrations, it induced SCE in Chinese hamster ovary cells with and without *S9* rat liver fraction.[5, 9]

In vivo cytogenetics. EE was not mutagenic in *Dr. melanogaster*.[14]

Carcinogenicity. Final report of NTP-88 EE carcinogenicity studies has not been published.

Chemobiokinetics. EE is partly broken down in the body. Administration by gastric intubation resulted in two major urinary metabolites in rats, *EAA* and *N-ethoxyacetyl glycine*.[15]

Observations in man. EAA has been detected in the urine of workers exposed to EE and 2-ethoxyethanol acetate.

Animal studies. In rats, metabolism proceeded mainly through oxidation *via* alcohol dehydrogenase to EAA with some subsequent conjugation of the acid metabolite with glycine. According to Medinsky et al., *ethylene glycol*, a metabolite of EE, is a result of dealkylation of the ether occurring prior to oxidation to EAA.

Ethylene glycol was excreted in urine, representing approximately 18% of the dose administered.[6] Excretion of unchanged EE occurs via the urine or exhalation as CO_2.[1] Less than 5.0% EE was exhaled unchanged.

Regulations. ***U.S. FDA*** (1998) approved the use of EE in adhesives as a component of articles intended for use in packaging, transporting, or holding food in accordance with the conditions prescribed in 21 CFR part 175.105.

Standards. ***Russia.*** Recommended MAC: 0.63 mg/l.

References:

1. Stenger, E. G., Aeppli, L., Muller, D., u. a., Zur Toxicologie des Athylenglyckol-Monoathylathers, *Arzneimittel-Forsch.*, 21, 880, 1971.
2. Zyabbarova, S. A., in *Proc. Leningrad Sanitary Hygiene Medical Institute*, Leningrad, Issue No 81, 1965, 70 (in Russian).
3. *Occupational Exposure to Ethylene Glycol Monomethyl Ether and Ethylene Glycol Monoethyl Ether and Their Acetates,* Criteria for a Recommended Standard, U.S. Dept. Health & Human Services, NIOSH, 1991, 296.
4. Yatsina, O. V., Plaksienko, N. F., Pys'ko, G. T., et al., Hygienic regulation of cellosolves in water bodies, *Gig. Sanit.*, 10, 78, 1988 (in Russian).
5. *Toxicity Studies of Ethylene Glycol Ethers: 2-Methoxyethanol, 2-Ethoxyethanol, 2-Butoxyethanol Administered in Drinking Water to F344/N Rats and B6C3F$_1$ Mice,* NTP Technical Report Series No 26, Research Triangle Park, NC, July 1993.
6. Medinsky, M. A., Singh, G., Bechtold, W. E., Bond, J. A., Sabourin, P. J., Birnbaum, L. S., and Henderson, R. F., Disposition of three glycol ethers administered in drinking water to male F344/N rats, *Toxicol. Appl. Pharmacol.*, 102, 443, 1990.
7. Melnick, R. L., Toxicity of ethylene glycol and ethylene glycol monoethyl ether in Fischer 344 rats and B6C3F$_1$ mice, *Environ. Health Perspect.*, 57, 147, 1984.
8. Hardin, B. D., Reproductive toxicity of the glycol ethers, *Toxicology,* 27, 259, 1983.
9. Welch, L. S. and Cullen, M. R., Effect of exposure to ethylene glycol ethers on shipyard painters: III. Hematologic effects, *Am. J. Ind. Med.,* 14, 527, 1988.
10. Welch, L. S., Plotkin, E., and Schrader, S., Indirect fertility analysis in painters exposed to ethylene glycol ethers: sensitivity and specificity, *Am. J. Ind. Med.,* 20, 229, 1991.
11. Nagano, K., Nakayama, E., Koyano, M., et al., Mouse testicular atrophy induced by ethylene glycol monoalkyl ethers, *Jap. J. Ind. Health,* 21, 29, 1979.
12. Foster, P. M., Creasy, D. M., Foster, J. R., et al., Testicular toxicity of ethyleneglycol monomethyl and monoethyl ethers in rats, *Toxicol. Appl. Pharmacol.*, 69, 385, 1983.
13. Hoflack, J. C., Lambolez, L., Elias, Z., and Vasseur, P., Mutagenicity of ethylene glycol ethers and of their metabolites in *Salmonella typhimurium, Mutat. Res.*, 341, 281, 1995.
14. McGregor, D. B., Genotoxicity of glycol ethers, *Environ. Health Perspect.*, 57, 97, 1984.
15. Cheever, K. L., Plotnick, H. B., Richards, D. E., et al., Metabolism and excretion of 2-ethoxyethanol in the adult male rat, *Environ. Health Perspect.*, 57, 241, 1984.

2-ETHOXYETHANOL ACETATE

Molecular Formula. $C_6H_{12}O_3$
M = 132.6
CAS No 111-15-9

RTECS No KK8225000

Abbreviation. EEA.

Synonyms and **Trade Names.** Acetatethyl cellosolve; Cellosolve acetate; 2-Ethoxyethyl acetate; Ethylene glycol, acetacethyl ether; Ethylene glycol, monoethyl ether, acetate; Ethylglycol acetate.

Properties. Colorless, transparent liquid with a mild, non-residual sweet, musty odor. Soluble in water and ethanol. Does not affect *pH*, transparency, and color of water; does not form foam and film on the water surface. Odor perception threshold is 6.9 mg/l or, according to other data, 0.056 mg/l.[010]

Applications. A solvent for cellulose ethers and resins. Used in the manufacture of surface coatings (especially those based on epoxy resins).

Acute Toxicity. LD_{50} is 3.9 to 5.0 g/kg BW in male rats, 2.9 g/kg BW in females, and 1.9 g/kg BW in rabbits and guinea pigs. Poisoning produced changes in the kidneys and GI tract. Death occurs in 2 to 4 days.[1-3]

See ***1,2-Diethoxyethane.***

Repeated Exposure revealed moderate cumulative properties.

Reproductive Toxicity. Embryolethality, visceral and skeletal abnormalities, and reduced fetal weights are found in the offspring of rats treated with EEA.[4]

Male mice were given doses of 0.5 to 4.0 g/kg BW for 5 weeks. The NOAEL for testicular atrophy is reported to be 0.5 g/kg BW.[5] Male rats were exposed orally to 726 mg/kg BW for 11 days. The LOAEL for testicular atrophy and spermatocyte depletion appeared to be 726 mg/kg BW.[6]

Chemobiokinetics. *Ethoxyacetic acid* seems to be the major metabolite of EEA. It has been detected in the urine of the workers exposed to EEA.

Regulations. ***U.S. FDA*** (1998) approved the use of EEA in adhesives as a component of articles intended for use in packaging, transporting, or holding food in accordance with the conditions prescribed in 21 CFR part 175.105.

Standards. ***Russia.*** Recommended MAC: 0.14 mg/l.

References:

1. Yatsina, O. V., Plaksienko, N. F., Pys'ko, G. T., et al., Hygienic regulation of cellosolves in water bodies, *Gig. Sanit.*, 10, 78, 1988 (in Russian).
2. Smyth, H. F., Jr., Seaton, J., and Fisher, L., The single dose toxicity of some glycols and derivatives, *J. Ind. Hyg. Toxicol.*, 23, 259, 1941.
3. *Occupational Exposure to Ethylene Glycol Monomethyl Ether and Ethylene Glycol Monoethyl Ether and Their Acetates,* Criteria for a Recommended Standard, U.S. Dept. Health & Human Services, NIOSH, 1991, 296.
4. Hardin, B. D. and Lyon, J. P., Summary and overview: NIOSH symposium on toxic effects of glycol ethers, *Environ. Health Perspect.*, 57, 273, 1984.
5. Nagano, K., Nakayama, E., Koyano, M., et al., Mouse testicular atrophy induced by ethylene glycol monoalkyl ethers, *Jap. J. Ind. Health*, 21, 29, 1979.
6. Foster, P. M., Creasy, D. M., Foster, J. R., et al., Testicular toxicity of ethyleneglycol monomethyl and monoethyl ethers in rats, *Toxicol. Appl. Pharmacol.*, 69, 385, 1983.

2-(2-ETHOXYETHOXY)ETHANOL

Molecular Formula. $C_6H_{14}O_3$

M = 134.18

CAS No 111-90-0

RTECS No KK8750000

Abbreviation. EEE.

Synonyms and **Trade Names.** Carbitol; Carbitol cellosolve; Diethylene glycol, monoethyl ether; Dowanol; Ethylcarbitol.

Properties. EEE occurs as a colorless or slightly yellowish, very hygroscopic liquid having a characteristic ether odor. Possesses the properties of both alcohol and ether. Readily soluble in water and alcohol. Odor perception threshold is <0.21 mg/l[010] or 8.75 mg/l; practical threshold is 24.3 mg/l. Heating increases odor intensity by 4.5 times. Chlorination of aqueous solutions does not lead to odor intensification or to the appearance of additional odors.[1]

Applications. Used as a solvent for nitrocellulose lacquers and resins used in cosmetics.

Exposure. Use of EEE as a carrier solvent for flavors could lead to carry-over levels as high as 1000 mg/kg in foods as consumed. The estimated intake is 0.25 mg/kg BW; ADI could not be allocated (*JECFA*, 1995).[2]

Acute Toxicity. LD_{50} is reported to be 6.3 to 8.69 g/kg BW in rats, 3.9 to 7.2 g/kg BW in mice, and 3.0 g/kg BW in guinea pigs. Poisoned animals displayed CNS functional disorder with paresis and paralysis of the extremities. High doses affected the urinary tract. Mice and guinea pigs died on day 1 to 2, rats on days 1 to 3.[1,3,020]

Repeated Exposure. EEE exhibited pronounced[1] or moderate[2] cumulative properties. Rats were dosed by gavage with 30 to 750 mg/kg BW for a month. The treatment caused marked signs of intoxication: decreased BW gain, damage of hemopoiesis and oxidation-reduction processes, and changes in the visceral organs.

Short-term Toxicity. Rats were given 0.5 or 5.0% EEE in their diet for 90 days.

The LOEL of 2500 mg/kg BW, and the NOEL of 250 mg/kg BW for renal effects were established in this study.[4]

Long-term Toxicity. Mice were administered doses up to 75 mg/kg BW for 6 months. Blood analyses were changed in the 0.75 mg/kg BW and higher dose groups. The high-dosed animals developed reduced glucose concentration in blood. Pathological changes at the mid and high doses included increased urinary levels of chlorides and proteins, and a reduced amount of creatinine. Gross pathology examination revealed morphological changes in the visceral organs and a reduction in their relative weights (at the dose of 75 mg/kg BW only).[1] Consumption of drinking water containing 1.0 or 5.0% EEE over a period of 2 years caused slight if any toxic effect in rats and mice, respectively.[5]

Reproductive Toxicity.

Gonadotoxicity. In a continuous breeding protocol study, CD-1 mice received 0.25 to 2.5% EEE in their drinking water. This treatment caused no effect on fertility and reproduction in the F_0 or F_1 generation mice despite a 34% decrease in cauda epididymal sperm motility and a decrease in relative liver weights in the F_1 males at 2.5% EEE concentration in drinking water.[6]

The dose of 30 mg/kg BW appears to be ineffective in rats.

Embryotoxic effect was not found to develop following administration of 75 and 750 mg/kg BW doses. Pregnant CD-1 mice received 5.5 g/kg BW in water by gavage on days 6 to 13 of gestation and were allowed to deliver. The treatment caused 14% maternal mortality and a reduction in maternal BW gain. There was no effect on the offspring of treated animals (in the F_0 or F_1 generation).[7]

Teratogenic effect was not observed in rats after inhalation and dermal application.[8]

Mutagenicity.

In vitro genotoxicity. EEE was positive in *Salmonella* mutagenicity assay (1.0 ml/plate), but not in micronuclear test.[9]

Chemobiokinetics. In rabbits, excretion occurs in the form of *glucuronides*.

Regulations. ***U.S. FDA*** (1998) approved the use of EEE (1) as a component of adhesives intended for use in contact with food in accordance with the conditions prescribed by 21 CFR part 175.105; and (2) as a component of the uncoated or coated food-contact surface of paper and paperboard that may be safely used in producing, manufacturing, packing, transporting, or holding dry food in accordance with the conditions prescribed by 21 CFR part 176.180.

Recommendations. ***FAO/WHO*** (1996). No ADI allocated.

Standards. ***Russia*** (1995). MAC and PML: 0.3 mg/l.

References:

1. Kondratyuk, V. A., Sergeta, V. N., Pis'ko, G. T., et al., Experimental derivation of a maximum allowable concentration for the monoethyl ether of diethylene glycol in water bodies, *Gig. Sanit.*, 4, 74, 1981 (in Russian).

2. *Evaluation of Certain Food Additives and Contaminants*, 44th Report of the Joint FAO/WHO Expert Committee on Food Additives, Geneva, 1995, 9.
3. Krotov, Yu. A., Lykova, A. S., Skachov, M. A., et al., Toxicological properties of diethylene glycol ethers (carbitols) with regard to the protection of the atmospheric air, *Gig. Sanit.*, 2, 14, 1981 (in Russian).
4. Gaunt et al., *Food Cosmet. Toxicol.*, 6, 689, 1968, cit. in: *Toxicological Evaluation of Certain Food Additives and Contaminants,* 41st Meeting of the Joint FAO/WHO Expert Committee on Food Additives, WHO Food Additives Series No 32, Geneva, 1993.
5. Browning, E., *Toxicity and Metabolism of Industrial Solvents*, American Elsevier, New York, 1965, 631.
6. Williams, J., Reel, J. R., George, J. D., and Lamb, J. C., Reproductive effects of diethyleneglycol and diethyleneglycol monoethyl ether in Swiss CD-1 mice assessed by continuous breeding protocol, *Fundam. Appl. Toxicol.*, 14, 622, 1990.
7. Hardin, B. D., Schuler, R. L., Burg, J. R., Booth, G. M., Hazelden, K. P., MacKenzie, K. M., Piccirillo, V. J., and Smith, K. N., Evaluation of 60 chemicals in a preliminary developmental toxicity test, *Teratogen. Carcinogen. Mutagen.*, 7, 29, 1987.
8. Nelson, B. K., Setzer, J. V., Brightwell, W. S., et al., Comparative inhalation teratogenicity of four industrial glycol ether solvents in rats, *Teratology*, 25, 64A, 1982.
9. Berte, F., Bianchi, A., Gregotti, C., et al., *In vivo* and *in vitro* toxicity of carbitol, *Bull. Chem. Pharmacol.*, 125, 401, 1986.

2-(2-ETHOXYETHOXY)ETHANOL ACETATE

Molecular Formula. $C_8H_{16}O_4$
M = 176.24
CAS No 112-15-2
RTECS No KK8925000
Abbreviation. EEEA.

Synonyms. Carbitol acetate; Diethylene glycol, monoethyl ether, acetate; Ethoxydiglycol acetate; 2-(2-Ethoxyethoxy)ethanol acetate.

Properties. Colorless liquid with a mild odor and bitter taste. Miscible with water, alcohol, and most oils.

Applications. Solvent for lacquers and coatings. Solvent and plasticizer for cellulose esters, gums, and resins.

Acute Toxicity. LD_{50} is 6.5 g/kg BW in rats (Pesticide Index, 1976), 2.26 g/kg BW in rabbits,[03] and 2.34 g/kg BW in guinea pigs. According to other data, LD_{50} is 11 g/kg BW in rats and 3.93 g/kg BW in guinea pigs (Union Carbide Co., 1965).

Chemobiokinetics. Undergoes hydrolysis in the body to *carbitol* and *an organic acid.*[019]

Regulations. ***U.S. FDA*** (1998) approved the use of EEEA in adhesives as a component of articles intended for use in packaging, transporting, or holding food in accordance with the conditions prescribed by 21 CFR part 175.105.

Reference:

Smyth, H. F., Jr., Seaton, J., and Fisher, L., The single dose toxicity of some glycols and derivatives, *J. Ind. Hyg. Toxicol.*, 23, 259, 1941.

ETHYLENE GLYCOL

Molecular Formula. $C_2H_6O_2$
M = 62.07
CAS No 107-21-1
RTECS No KW2975000
Abbreviation. EG.

Synonyms and **Trade Names.** 1,2-Dihydroxyethane; 1,2-Ethanediol; Ethylene alcohol; Ethylene dihydrate; Glycol; Glycol alcohol; 2-Hydroxyethanol.

Properties. Clear, viscous, odorless, sweet liquid. Readily miscible with water and alcohol. Odor perception threshold is 1.3 g/l; taste perception threshold is 0.45 g/l.[1] However, according to other data, organoleptic threshold is 0.13 g/l.[01]

Applications. EG is the most representative of the glycols used as a painting and plastic solvent, antifreeze, and in synthetic fibers. EG is also used in production of resinous products, especially polyester fibers and resins, and polyethylene terephthalate.

Exposure. EG may also appear as a wine pollutant.

Migration Data. Polyethylene terephthalate bottles were filled with 3.0% acetic acid and stored at 32°C for 6 months. Final concentration of EG was about 100 ppb, which is equivalent to about 94 μg/bottle.[2]

Chocolates, boiled sweets, toffees, cakes, and meat pies were wrapped in regenerated cellulose films that contained various mixtures of glycol softeners. It was shown that higher levels of migration occurred for propylene glycol than for triethylene glycol and the presence of coating reduced the migration of both softeners. EG levels in the food samples were below 10 mg/kg.[3]

Acute Toxicity.

Observations in man. Doses of 1.0 to 2.0 g/kg BW (100 ml) are lethal in humans.[4]

Animal studies. A single administration of 1.0 ml/kg BW did not cause intoxication.[5] LD_{50} is reported to be 13 g/kg BW in rats, 8.05 g/kg BW in mice, 5.0 g/kg BW in rabbits, and 11.0 g/kg BW in guinea pigs. Poisoning is accompanied by a short period of stimulation followed by depression, ataxia, refusal of food, labored breathing, and vomiting. Gross pathology examination revealed hemorrhages in the GI tract walls.[1]

EG is known to be a vascular and protoplasmic poison, causing vascular edema and necrosis. In acute exposure, it appears to act as a typical narcotic; it causes erythrocyte hemolysis and disrupts the oxidation-reduction processes.[6,7]

Repeated Exposure failed to reveal cumulative effect in rats given 1/5 LD_{50} for 20 days. K_{acc} is 8.3 (by Lim).[5]

Rats were given 1.0% EG solution instead of drinking water *ad libitum* for 2 weeks. Hematology and biochemistry analyses showed anemia and an increase in hepatic microsomal cytochrome *P-450.*[8]

EG has been found to be CNS and cardiac depressant of low toxicity; it produced arrythmia and fibrillation.[9] The changes in the neurons consisted of neuronal degeneration, decrease in number of *AchE+* cells, and reactive cellular grouping were observed in culture of nerve cells from Wistar rat embryos which were exposed to EG at doses between 10^{-4} and 10^{-8} M.[10]

Short-term Toxicity. The pathological effects of EG are due to its metabolism resulting in the formation of *oxalic* and *glycolic acids* which are eliminated through the kidney causing renal failure. The toxic effects on the NS are not well known. In some circumstances, convulsions may occur.[10]

In a 13-week study, $B6C3F_1$ mice received dietary concentrations of 3,200 to 50,000 ppm EG. There were no changes in final BW (in males and females) and feed consumption (in males only). Chemical-related kidney and liver lesions (nephropathy and centrilobular hepatocellular hyaline degeneration) were seen only in 25,000 and 50,000 ppm groups of male mice.[11]

Long-term Toxicity. $B6C3F_1$ mice received diets containing 6,250, 12,500, or 25,000 ppm EG (males) and 12,500, 25,000, or 50,000 ppm EG (females) for up to 103 weeks. The treatment did not affect survival, BW gain, and food consumption of the treated animals. No compound-related toxicity was reported in this study.[11]

Sprague-Dawley rats were fed 0.1 to 4.0% EG in their diet for 2 years. The treatment caused an increase in mortality rate, retardation of BW gain, an increase in water consumption, proteinuria, and renal calculi in 1.0 and 4.0% EG-treated females and 4.0% EG-treated males. There was also increased incidence of cytoplasmic crystal deposition in renal tubular epithelium at 0.5 and 1.0% concentrations of EG in the feed. The NOEL of 0.2% EG (approximately 100 mg/kg BW) was established in this study.[12]

Rhesus monkeys were fed a diet containing 0.2 or 0.5% EG for 3 years. Histological examination revealed no changes in the kidney or other visceral organs.[13]

In a 6-month study, monkeys were dosed with 17 to 28 mg EG/kg BW. The treatment resulted in protein precipitation and hydrolytic degeneration of the proximal part of the renal tubules. Doses of 33 to 137 mg/kg BW caused oxalate precipitation in the proximal sections of the tubules with epithelial necrosis.

Hyaline was found in the distal sections of the Henle's loops and the convoluted tubules. The structure of the tubules was destroyed.

In a 6-month study, an increase in the urea and indican contents in the blood serum, a reduction in the prothrombin time, and weakening of hepatic secretory function (bromosulphalein test) were found to develop.[1]

F344 rats and CD-1 mice were fed diets yielding approximately EG dosages of 0.04 g/kg or 0.4 g/kg BW (~1.0 g/kg BW). Higher dose caused retardation of BW gain, an increase in the blood urea nitrogen, in water intake, and mortality rate in males. Urinary calcium oxalate crystals were noted in high dosed rats. Histopathology examination revealed tubular cell hyperplasia, tubular dilatation, and parathyroid hyperplasia.[14]

Allergenic Effect was noted on skin application.[15]

Reproductive Toxicity. Effect on reproduction was found at the toxic dose level.

Embryotoxicity. In a long-term study, mice received 0.25 to 1.0% EG in their drinking water. The treatment caused a slight decrease in the number of litters (in the second generation) per fertile pair and live pups per litter, and in live pups weights in the 1.0% dose group.[16]

Schuler et al. noted the number of dead pups per litter at birth to be elevated and postnatal survival to be reduced in a Chernoff-Kavlock assay design in which pregnant Swiss mice were dosed by gavage with 11 mg undiluted EG/kg BW on days 7 to 14 of gestation.[17]

Sprague-Dawley rats and CD-1 mice were exposed by gavage to 750 to 5000 mg/kg BW doses on days 6 to 15 of pregnancy. The treatment reduced BW gain of dams. There was a dose-dependent increase in post- implantation fetal mortality and reduction in fetal BW in rats (doses of 2500 to 5000 mg/kg BW) and mice (750 mg/kg BW and above).[18]

According to other data, doses of 14 to 1400 mg/kg BW did not produce embryotoxic or teratogenic effect, but functions and morphology of the liver were affected.[19]

No indication of developmental toxicity was found in artificially inseminated New Zealand white rabbits administered EG by gavage on gestation days 6 through 19 at doses of 100 to 2000 mg/kg BW.[20] The NOAEL for developmental toxicity was at least 2.0 g/kg BW in this study. Such NOEL for Swiss mice administered EG by gavage was established at 150 mg/kg BW level.[21]

Teratogenic effect was noted at dose levels of 5.0 g/kg BW in rats, and 3.0 g/kg BW in mice.[22]

CD-1 mice received 1.0% aqueous solution of EG. Developmental abnormalities of the bone system comprised facial bone malformations, reduced size of the skull bones, fusion of the ribs, and malformations of the bones of the thorax and vertebral column.[16]

EG produced gonadotoxic effect in rats. Doses of up to 5.0 mg/kg BW caused a consistent 50 to 66% reduction in cytochrome-*C*-oxidase activity. There was a reduction in alkaline phosphatase activity in the epididymis tissues when 5.0 mg/kg BW was given twice.[22]

EG is not teratogenic in rabbits at doses up to 2.0 g/kg BW. Authors presented data in support of the hypothesis that *glycolic acid*, an acidic metabolite of EG, is the ultimate developmental toxicant in rats, but not in rabbits.[24]

Mutagenicity.

In vivo cytogenicity. In three-generation reproduction and DLM study, rats received 40, 200 mg/kg or 1000 mg/kg diet without evident toxic effects.[23]

In vitro genotoxicity. EG was neither mutagenic in *Salmonella* strains *TA98, TA100, TA1535,* or *TA1537*, nor did it induce SCE or CA in Chinese hamster ovary cells in the presence or absence of metabolic activation.[11]

Carcinogenicity. In 2-year feeding studies, there was no evidence of carcinogenic activity in male mice receiving up to 25,000 ppm or in female mice receiving up to 50,000 ppm EG. Gross pathology examination revealed hepatocellular hyaline degeneration.[11]

No evidence of carcinogenic effect was reported in F344 rats and CD-1 mice fed EG in the diet.[14]

Carcinogenicity classification.

NTP: XX - XX - NE - NE (feed).

Chemobiokinetics. EG is metabolized to form *oxalic acid* in an amount that depends on the administered dose and species of experimental animals.[021] In response to EG exposure, a number of *oxalates* are found in different body tissues.

Administration of 15 ml/kg BW to monkeys resulted in precipitation of oxalates in the renal tubules and in necrosis of the renal tubular epithelium.[5]

Doses of 1.0 and 2.0 mg/kg BW given by *i/v* injection affected the process of *glycolate* metabolism, and the latter is the principal factor determining the toxicity of EG. The compensatory increase of glycolate excretion in the urine is, in its turn, the reason for a less pronounced dose dependence of EG clearance from the blood.[25]

The main end-product of EG metabolism in rabbits is expired CO_2 (60% of the dose over 3 days). 10% EG are excreted unchanged in the urine, and 0.1% as *oxalic acid.* In addition to EG, *glycolic* (*hydroxy-acetic) acid* and small amounts of *oxalic acid* are excreted in the urine.[4]

Regulations.

U.S. FDA (1998) approved the use of EG (1) in adhesives as a component of articles intended for use in packaging, transporting, or holding food in accordance with the conditions prescribed in 21 CFR part 175.105; (2) in the manufacture of resinous and polymeric coatings for food-contact surface of articles intended for use in producing, manufacturing, packing, processing, preparing, treating, packaging, transporting, or holding food in accordance with the conditions prescribed in 21 CFR part 175.300; (3) in the manufacture of resinous and polymeric coatings for polyolefin films intended for use in producing, manufacturing, packing, transporting, or holding food in accordance with the conditions prescribed in 21 CFR part 175.320; (4) in cross-linked polyester resins for repeated use as articles or components of articles coming in contact with food in accordance with the conditions prescribed in 21 CFR part 177.2420; (5) in the manufacture of polyurethane resins which may be used as food-contact surface of articles intended for use in contact with bulk quantities of dry food of an appropriate type in accordance with the conditions prescribed in 21 CFR part 177.1680; (6) as a defoaming agent in the manufacture of paper and paperboard intended for use in packaging, transporting, or holding food in accordance with the conditions prescribed in 21 CFR part 176.210; (7) as a solvent removed by water washing in zinc-silicon dioxide matrix coatings for food-contact surface of articles; and (8) as slimicides in the manufacture of paper and paperboard that may be safely used in contact with food in accordance with the conditions prescribed in 21 CFR part 176.300.

EU (1990). EG is available in the *List of authorized monomers and other starting substances which shall be used for the manufacture of plastic materials and articles intended to come into contact with foodstuffs (Section A*).

Great Britain (1998). EG is authorized without time limit for use in the production of polymeric materials and articles in contact with food or drink or intended for such contact. The specific migration of EG alone or together with diethylene glycol shall not exceed 30 mg/kg.

Standards. ***U.S. EPA*** (1999). Health Advisory for a longer-term exposure is 2.0 mg/l.

Russia (1995). MAC and PML: 1.0 mg/l.

References:

1. Plugin, V. P., Ethylene glycol and diethylene glycol as an object in the sanitary protection of reservoirs, *Gig. Sanit.*, 3, 16, 1968 (in Russian).
2. Kashtock, M. and Breder, C. V., Migration of ethylene glycol from polyethylene terephthalate bottles into 3% acetic acid, *J. Assoc. Off. Anal. Chem.*, 63, 168, 1980.
3. Castle, L., Cloke, H. R., Crews, C., and Gilbert, J., The migration of propylene glycol, mono-, di-, and triethylene glycols from regenerated cellulose film into food, *Z. Lebensmit. Untersuch. Forsch.*, 187, 463, 1988.
4. Balazs, T., Jackson, B., and Hite, M., Nephrotoxicity of ethylene glycol, cephalosporins and diuretics, *Monogr. Appl. Toxicol.*, 1, 487, 1982.
5. McChesney, E. W., Golberg, L., Parekh, C. K., and Min, B. H., Reappraisal of the toxicology of ethylene glycol. II Metabolism studies in laboratory animals, *Food Cosmet. Toxicol.*, 9, 21, 1971.
6. Sanina, Yu. P. and Kochetkova, T. A., Toxicity of ethylene glycol, in *Toxicology of New Industrial Chemical Substances*, A. A. Letavet, Ed., Medgiz, Moscow, Issue No. 7, 1965, 102 (in Russian).

7. Filatova, V. S., Smirnova, E. S., Gronsberg, E. Sh., et al., Data on hygienic norm-setting of ethylene glycol in workplace air, *Gig. Truda Prof. Zabol.*, 6, 28, 1982, (in Russian).
8. Imazu, K., Fujishiro, K., Inoue, N., et al., Effects of ethylene glycol on drug metabolizing enzymes in rat liver, Abstract, *Sanagyo Ika Daigaku Zasshi*, 13, 13, 1991, (in Japanese).
9. Shiedeman, F. E. and Procita, L., The pharmacology of the monomethyl ethers of mono-, di-, and tripropylene glycol in the dog with observations on the auricular fibrillation produced by these compounds, *J. Pharmacol. Exp. Ther.*, 102, 79, 1951.
10. Capo, M. A., Sevil, M. B., Lopez, M. E., and Frejo, M. T., Ethylene glycol action on neurons and its holinomimetic effects, *J. Environ. Pathol. Toxicol. Oncol.*, 12, 155, 1993.
11. *Toxicology and Carcinogenesis Studies of Ethylene Glycol in B6C3F$_1$ Mice (Feed Studies)*, NTP Technical Report Series No 413, Research Triangle Park, NC, 1993.
12. Blood, F. R., Chronic toxicity of ethylene glycol in the rat, *Food Cosmet. Toxicol.*, 3, 229, 1965.
13. Blood, F, R., Elliott, G. A., and Wright, M. S., Chronic toxicity of ethylene glycol in monkey, *Toxicol. Appl. Pharmacol.*, 4, 489, 1962.
14. DePass, L. R., Garman, R. H., Woodside, M. D., et al., Chronic toxicity and oncogenicity studies of ethylene glycol in rats and mice, *Fundam. Appl. Toxicol.*, 7, 547, 1986.
15. Ivanova, T. P., Sensitizing action of some substances migrated from polymeric materials, in *Hygienic Aspects of the Use of Polymeric Materials*, Proc. 2nd All-Union Meeting on Health and Safety Monitoring of the Use of Polymer Materials in Construction, K. I. Stankevich, Ed., Kiyv, 1976, 102 (in Russian).
16. Lamb, J. C., Maronpot, R. R., Gulati, D. K., et al., Reproductive and developmental toxicity of ethylene glycol in the mouse, *Toxicol. Appl. Pharmacol.*, 8, 100, 1985.
17. Schuler, R. l., Hardin, B. D., Niemeier, R. W., et al., Results of testing fifteen glycol ethers in a short-term *in vivo* reproductive toxicity assay, *Environ. Health Perspect.*, 57, 141, 1984.
18. Price, C. G., Kimmel, C. A., Rochelle, W., et al., The developmental toxicity of ethylene glycol in rats and mice, *Toxicol. Appl. Pharmacol.*, 81, 113, 1985.
19. Bariliak, I. R. and Kozachuk, S. Yu., Investigations on cytotoxicity of monoatomic alcohols in bone marrow, *Cytologia i Genetica*, 22, 49, 1988 (in Russian).
20. Tyl, R. W., Price, C. J., Marr, M. C., Myers, C. B., Seely, J. S., Heindel, J. J., and Schwetz, B. A., Developmental toxicity evaluation of ethylene glycol by gavage in New Zealand white rabbits, *Fundam. Appl. Toxicol.*, 20, 402, 1993.
21. Tyl, R. W., Fisher, L. C., Kubena, M. F., et al., Determination of a developmental toxicity NOEL for EG by gavage in Swiss mice, *Teratology*, 39, 487, 1989.
22. Byshovets, T. F., Barilyak, I. R., Korkach, V. I., et al., Gonadotoxic effect of glycols, *Gig. Sanit.*, 9, 84, 1987 (in Russian).
23. DePass, L. R., Woodside, M. D., Moronpot, R. R., and Weil, C. S., Three-generation reproduction and dominant lethal mutagenesis studies of ethylene glycol in the rat, *Fundam. Appl. Toxicol.*, 7, 566, 1986.
24. Carney, E. W., Pottenger, L. H., Bartels, M. J., Jackh, R., and Quast, J. F., Comparative pharmacokinetics and metabolism of ethylene glycol in pregnant rats and rabbits, Abstract P4C27, Abstracts VIII Int. Congr. Toxicol., *Toxicol. Lett.*, (Suppl. 1) 95, 208, 1998.
25. Marshall, T. C., Dose-dependent disposition of ethylene glycol in the rat after intravenous administration, *J. Toxicol. Environ. Health*, 10, 397, 1982.

ETHYLENE GLYCOL DIACETATE

Molecular Formula. $C_6H_{10}O_4$
M = 146.16
CAS No 111-55-7
RTECS No KW4025000
Abbreviation. EGDA.

Synonyms and **Trade Names.** Diacetate-1,2-ethanediol; 1,2-Diacetoxyethane; Ethanediol diacetate; Ethylene diacetate; Ethylene glycol, acetate.

Properties. Colorless liquid with a slight acetic, ester-like odor and bitter taste. Solubility in water is 16.4% at 20°C. Soluble in alcohol.

Applications. Solvent for cellulose esters, resins, and lacquers.

Acute Toxicity. LD_{50} is 6.86 g/kg BW in rats and 4.9 g/kg BW in guinea pigs. Manifestations of toxic actions were noted in the GI tract, kidney, ureter, and bladder.[1]

Repeated Exposure. In rats, minimal dose given for 7 days as 5.0% concentration in drinking water (equal to 6.0 g/kg BW daily), was found to produce kidney damage. Some animals died.[2]

Long-term Toxicity. 1.0 to 3.0% EGDA solution, that was given to rats and rabbits for a prolonged period of time, caused large crystalline deposits and death.[03]

Chemobiokinetics. In the body, EGDA is metabolized into *glycol* and *acetic acid.*

Regulations. ***U.S. FDA*** (1998) approved the use of EGDA in adhesives as a component of articles intended for use in packaging, transporting, or holding food in accordance with the conditions prescribed in 21 CFR part 175.105.

References:

1. Smyth, H. F., Jr., Seaton, J., and Fisher, L., The single dose toxicity of some glycols and derivatives, *J. Ind. Hyg. Toxicol.*, 23, 259, 1941.
2. Browning, E., *Toxicity and Metabolism of Industrial Solvents*, American Elsevier, New York, 1965, 486.

ETHYL ETHER

Molecular Formula.. $C_4H_{10}O$
M = 74.12
CAS No 60-29-7
RTECS No KI5775000
Abbreviation. EE.

Synonyms. Diethyl ether; Diethyl oxide; Ether; Ethoxyethane; Ethyl oxide; 1,1'-Oxybisethane; Sulphuric ether.

Properties. EE occurs as a colorless, highly volatile liquid. Solubility in water is 77.7 g/l at 20°C or 56 g/l at 25°C. Readily soluble in alcohol. Odor perception threshold is 0.3 or 0.75 mg/l.[02]

Applications. Used in the synthesis of butadiene rubber and in the manufacture of plastics.

Acute Toxicity. LD_{50} is 3.56 g/kg BW in rats,[03] and 1.76 g/kg BW in mice. Poisoning is accompanied by narcosis (at EE blood concentration of 100 to 140 mg%), adynamia and motor coordination disorder.[1]

According to Kimura et al., LD_{50} in 14-day-old, young adult, and adult rats is 2.2, 2.4, and 1.7 ml/kg BW, respectively.[2]

Repeated Exposure resulted in the rapid development of habituation.

Short-term Toxicity. Male and female rats were orally dosed by gavage with doses of 500, 2000, or 3500 mg EE/kg BW for 13 weeks. The high dose treatment caused marked toxic effects, including mortality, decreased food consumption, retardation in BW gain. Histological evaluation revealed no exposure-related effects. The NOAEL of 500 mg EE/kg BW was established in this study.[3]

Long-term Toxicity. In a 6-month study, rats were given orally 0.2, 5.0, or 50 mg/kg BW; guinea pigs received 0.2 and 5.0 mg/kg BW. At higher doses, effect on the serum protein fractions ratio and conditioned reflex activity was found to develop. Gross pathology examination revealed catarrhal, desquamatous gastritis, and parenchymatous dystrophy in some animals.[1]

Reproductive Toxicity. EE anesthesia produced some ***embryotoxic*** and ***teratogenic*** effects in mice and rats. It caused increased fetal resorptions and skeletal anomalies, but produced no increase in soft tissue defects.[4]

Mutagenicity.

In vitro genotoxicity. EE does not induce SCE in cultured Chinese hamster ovary cells and is not mutagenic to fungi and bacteria (IARC 11-285). It was found to be negative in *Salmonella* mutagenicity assay (NTP-94).

Chemobiokinetics. EE is rapidly absorbed from inhaled air and passes rapidly into the brain. Accumulation may occur in fatty tissues, partially converted to ^{14}C-*CO*.

Standards. ***Russia.*** PML: 0.3 mg/l (organolept., odor).

References:

1. Amirkhanova, G. F. and Latypova, Z. V., Experimental substantiation of the maximum allowable concentration for diethyl ether in water reservoirs, in *Industrial Pollutants of Water Reservoirs*, S. N. Cherkinsky, Ed., Meditsina, Moscow, Issue No. 9, 1969, 148 (in Russian).
2. Kimura, E. T., Ebert, D. M., and Dodge, P.W., Acute toxicity and limits of solvents residue for sixteen organic solvents, *Toxicol. Appl. Pharmacol.*, 19, 699, 1971.
3. Rat oral subchronic study with ethyl ether, Prep. by Am. Biogenic Corp. for the Office of Solid Waste, U.S. EPA, Washington, D.C., 1986.
4. Schwetz, B. A. and Becker, B. A., Embryotoxicity and fetal malformations of rats and mice due to maternally administered ether, *Toxicol. Appl. Pharmacol.*, 17, 275, 1970.

2-ETHYL-1-HEXANOL

Molecular Formula. $C_8H_{18}O$
M = 130.26
CAS No 104-76-7
RTECS No MP0350000
Abbreviation. 2-EH.

Synonyms. Ethylhexanol; 2-Ethylhexyl alcohol.

Properties. Colorless liquid with musty odor. Odor threshold concentration is 0.075 mg/l.[010]

Applications. Used as a solvent in the production of plastics, in the manufacture of plasticizers, di(2-ethylhexyl)phthalate, and di(2-ethylhexyl)adipate. A lubricant and a finishing compound for paper and textiles.

Exposure. Occurs naturally in food.

Acute Toxicity. LD_{50} is reported to be about 2.0 g/kg BW in rats, mice, and guinea pigs, and about 1.2 g/kg BW in rabbits.[06]

Repeated Exposure. B6C3F$_1$ mice were given by gavage 100 to 1500 mg 2-EH/kg BW for 11 days. The treated animals demonstrated ataxia and lethargy. A dose of 100 mg/kg BW appeared to be ineffective in mice.[1]

Wistar rats were given 1.3 g 2-EH/kg BW for 7 days. Treatment caused an increase in relative liver weight and changes in activity of some enzymes.[2]

Mice received 1,8 g 2-EH/kg BW for 2 weeks. The treatment caused an increase in the relative liver weights and peroxisome proliferation. The same experiment in rats revealed similar but more marked effects.[3]

Short-term Toxicity. B6C3F$_1$ mice received 25 to 500 mg 2-EH/kg BW by gavage over a period of 3 months. Toxic effect was observed at 250 and 500 mg/kg BW dose levels.[1]

F344 rats and B6C3F$_1$ mice were administered 2-EH by oral gavage as an aqueous emulsion at dose levels of 25 to 500 mg/kg BW for 13 weeks. In rats, administration of the highest dose (500 mg/kg BW) reduced BW gain, increased liver, kidney, stomach, and testes weights and produced moderate gross and microscopic changes in the liver and forestomach. No behavioral effects, or spleen or thymus lesions were observed. The NOAEL of 125 mg/kg BW in mice and rats was established in this study.[4]

Long-term Toxicity. The doses of 200 mg 2-EH/kg BW in mice and 50 mg 2-EH/kg BW in rats were found to be ineffective.[1] Oral gavage doses of 2-EH in 0.005% aqueous Cremophor EL (polyoxyl-35 castor oil) were given to rats (50, 150, and 500 mg/kg BW) for 24 months and to mice (50, 200, and 750 mg/kg BW) for 18 months. In rats, the dose of 50 mg/kg produced no dose-related changes, while at 150 nd 500 mg/kg BW there was BW gain reduction and an increased incidence of lethargy. Relative liver, stomach, brain, kidney, and testis weights were increased. Female rat mortality was markedly increased at 500 mg/kg BW. Hematological, gross, and microscopic changes, including tumors, were comparable among all rat groups. In mice, two lower doses produced no changes or adverse trends in liver tumor incidence at the

5.0% significance level. The major effect of chronic dosing was mortality in female rats at 500 mg/kg BW and in male and female mice at 750 mg/kg BW.[4]

Reproductive Toxicity.

Teratogenicity. Dermal application on days 6 to 15 of gestation at dose levels of 0.5 to 3.0 ml/kg BW caused no developmental toxicity in F344 rats at and below levels which produced maternal toxicity.[5] The NOAEL was identified to be at least 2520 mg/kg BW.

Pregnant Wistar rats received repeated oral doses of 130 to 1300 mg 2-EH/kg BW in an aqueous emulsion on days 6 to 15 of gestation. The treatment with the highest dose led to marked maternal toxicity. Skeletal abnormalities were noted in the offspring. The NOAEL for these effects appeared to be 130 mg/kg BW in this study.[6]

Embryotoxicity. Pregnant CD-1 mice were given 1.5 g 2-EH/kg BW in corn oil by gavage on gestation days 6 through 13. The treatment appeared to be lethal for 30% of the dams. Number of viable litters was reduced and there was a decrease in litter size, percentage of survival, and birth weights.[7]

No signs of developmental toxicity were observed in timed pregnant Swiss mice after oral administration of 17 to 190 mg/kg BW doses on gestation days 0 through 17 (by microencapsulation and incorporation in the diet).[8]

Mutagenicity. Nongenotoxic hepatic peroxisome proliferator in the rat.

In vitro genotoxicity. 2-EH is shown to be positive in *Salmonella* mutagenicity assay.[9] Meanwhile, Joint FAO/WHO Expert Committee on Food Additives (1993) does not consider 2-EH to be genotoxic.

Carcinogenicity. 2-EH was found to be a peroxisome proliferator in the rat but not in the mouse at dose level of 500 mg/kg BW. It had not shown carcinogenic activity when administered by gavage to rats for 24 months or to mice for 18 months.[1]

While 2-EH may be a contributing factor in the hepatocellular carcinogenesis in female mice associated with chronic administration of di(2-ethylhexyl)adipate and di(2-ethylhexyl)phthalate, it is unlikely to be the entire proximate carcinogen.[4]

Chemobiokinetics. Following oral administration of up to 300 mg/kg BW, 2-EH was readily absorbed in the GI tract of rats and completely eliminated in the urine within 22 hours; *2-ethylhexanoic acid* is the major metabdite.[10]

Female F344 rats received 500 and 50 mg/kg BW oral doses of ^{14}C-2-EH by single and repeated (50 mg/kg BW) administration. Radioactivity was eliminated rapidly, primarily in the urine. Metabolites were predominantly glucuronides of oxidized metabolites of 2-EH, including *glucuronides of 2-ethyladipic acid, 2-ethylhexanoic acid, 5-hydroxy-2-ethylhexanoic acid,* and *6-hydroxy-2-ethyl-hexanoic acid.*[11] 2-EH is the major metabolite of DEHP.

Regulations.

EU (1990). 2-EH is available in the *List of authorized monomers and other starting substances which may continue to be used for the manufacture of plastic materials and articles intended to come into contact with foodstuffs pending a decision on inclusion in Section A* (*Section B*).

Great Britain (1998). 2-EH is authorized without time limit for use in the production of polymeric materials and articles in contact with food or drink or intended for such contact. The specific migration of this substance shall not exceed 30 mg/kg.

Recommendations. Joint FAO/WHO Expert Committee on Food Additives (1993). ADI: 0.5 mg/kg BW.

References:

1. *Toxicological Evaluation of Certain Food Additives and Contaminants*, WHO Food Additives Series No 32, WHO, IPCS, Geneva, 1993, 35.
2. Lake, B. G., Gangolli, S. D., Grasso, P., et al., Studies on the hepatic effects of orally administered di(ethylhexyl)phthalate in the rat, *Toxicol. Appl. Pharmacol.*, 32, 355, 1975.
3. Keith, Y., Cornu, M. C., Canning, P. M., Foster, J., Lhuguenot, J. C., and Elcombe, C. R., Peroxisome proliferation due to di(2-ethylhexyl)adipate, 2-ethylhexanol and 2-ethylhexanoic acid, *Arch. Toxicol.*, 66, 321, 1992.

4. Astil, B. D., Deckardt, K., Gembardt, Chr., Gingel, R., Guest, D., Hodgson, J. R., Mellert, W., Murphy, S. R., and Tyler, T. R., Prechronic toxicity studies on 2-ethylhexanol in F344 rats and $B6C3F_1$ mice, *Fundam. Appl. Toxicol.*, 29, 31, 1996.
5. Tyl, R. W., Fisher, L. C., Kubena, M. F., et al., The developmental toxicity of 2-ethylhexanol applied dermally to pregnant Fischer 344 rats, *Fundam. Appl. Toxicol.*, 19, 176, 1992.
6. *Chem. Prog. Bull.*, U.S. EPA, Office of Toxic Substances, 12, 20, 1990.
7. Hardin, B. D., Schuler, R. L., Burg, J. R., Booth, G. M., Hazelden, K. P., MacKensie, K. M., Piccirilo, V. J., and Smith, K. N., Evaluation of 60 chemicals in a preliminary developmental toxicity test, *Teratogen. Carcinogen. Mutagen.*, 7, 29, 1987.
8. Price, C. J., Tyl, R. W., Marr, M. C., et al., Developmental toxicity evaluation of DEHP metabolites in Swiss mice, *Teratology*, 43, 457, 1991.
9. Seed, J. L., Mutagenic activity of phthalic esters in bacterial liquid suspension assay, *Environ. Health Perspect.*, 45, 111, 1982.
10. Albro, P. W., The metabolism of 2-ethylhexanol, *Xenobiotica*, 5, 625, 1975.
11. Deisinger, P. J., Boatman, R. J., and Guest, D., Metabolism of 2-ethoxyethanol administered orally and dermally to the female Fischer 344 rat, *Xenobiotica*, 24, 429, 1994.

2-ETHYLHEXYL VINYL ETHER

Molecular Formula. $C_{10}H_{20}O$
M = 156.30
CAS No 103-44-6
RTECS No KO0175000
Abbreviation. EHVE.

Synonym. 2-Ethenoxy-2-ethylhexane.

Acute Toxicity. LD_{50} is 1.35 g/kg BW in rats.

Regulations. ***EU*** (1990). EHVE is available in the *List of authorized monomers and other starting substances which may continue to be used for the manufacture of plastic materials and articles intended to come into contact with foodstuffs pending a decision on inclusion in Section A* (*Section B*).

Reference:

Smyth, H. F., Carpenter, C. P., Weil, C. S., and Pozzani, U. C., Range-finding toxicity data: List V, *Arch. Ind. Hyg. Occup. Med.*, 10, 61, 1954.

ETHYL VINYL ETHER

Molecular Formula. C_4H_8O
M = 72.12
CAS No 109-92-2
RTECS No KO0710000
Abbreviation. EVE

Synonyms and **Trade Names.** Agrisynth; Ethoxyethene; Ether, vinyl ethyl; Vinamar.

Properties. Soluble in water.

Acute Toxicity. LD_{50} is 6.15 g/kg BW in rats.[1]

Mutagenicity.

In vitro genotoxicity. No significant mutagenicity of EVE toward *Salmonella typhimurium* strains *TA100* and *TA100* in the presence of the hepatic *S9* mix fraction was found. Mutagenic activities of vinyl ethers in the presence of the *S9* mix were correlated with stability of their epoxides.[2]

Regulations. ***EU*** (1990). EVE is available in the *List of authorized monomers and other starting substances which may continue to be used for the manufacture of plastic materials and articles intended to come into contact with foodstuffs pending a decision on inclusion in Section A* (*Section B*).

References:

1. Smyth, H. F., Carpenter, C. P., Weil, C. S., Pozzani, U. C., Striegel, J. A., and Nycum, J. S., Range-finding toxicity data: List VII, *Am. Ind. Hyg. Assoc. J.*, 30, 470, 1969.

2. Sone, T., Isobe, M., and Takabatake, E., Comparative studies on the metabolism and mutagenicity of vinyl ethers, *J. Pharmacobiodyn,*, 12, 345, 1989.

HYDRACRYLONITRILE

Molecular Formula. C_3H_5NO
M = 71.08
CAS No 109-78-4
RTECS No MU5250000
Abbreviation. HAN.

Synonyms and **Trade Names.** 2-Cyanoethanol; 2-Cyanoethyl alcohol; Ethylene cyanohydrin; Glycol cyanohydrine; 3-Hydroxy propanenitrile; Methanol acetonitrile.

Properties. Miscible with water and alcohol.

Applications. Solvent for some cellulose esters.

Acute Toxicity. LD_{50} is 3.2 g/kg BW in rats,[03] and 1.8 g/kg BW in mice.[1]

Repeated Toxicity. Rats were exposed to HAN via drinking water. No untoward effects were associated with the ingestion of HAN.

Short-term Toxicity. Rats fed HAN for 90 days exhibited changes in brain and heart weight.[2]

Regulations. ***U.S. FDA*** (1998) approved the use of HAN in adhesives as a component of articles intended for use in packaging, transporting, or holding food in accordance with the conditions prescribed in 21 CFR part 175.105.

References:

1. Sunderman, F. W. and Kincaid, J. F., Toxicity studies of acetone cyanohydrin and ethylene cyanohydrin, *Arch. Ind. Hyg. Occup. Med.*, 8, 371, 1953.
2. Sauerhoff, M. W., Braun, W. H., and Ramsey, J. C., Toxicological evaluation and pharmacokinetic profile of beta-hydroxypropionitrile in rats, *J. Toxicol. Environ. Health.*, 2, 31, 1976.

FORMAMIDE

Molecular Formula. CH_3ON
M = 45.04
CAS No 75-12-7
RTECS No LQ0525000

Synonyms. Carbamaldehyde; Formic acid, amide; Formimidic acid; Methanoic acid, amide; Methan-amide.

Properties. Colorless and odorless liquid. Readily soluble in water and lower alcohols. At a concentration of 25 mg/l does not submit any foreign odor, taste or coloration to water. After chlorination or heating water solutions of F. up to 60°C, organoleptic properties of water are not affected.[1]

Applications. Used as a solvent in the plastic manufacture and in the synthesis of dyes.

Acute Toxicity. LD_{50} is 5.7 to 6.1 g/kg BW in rats, and 2.1 to 3.15 g/kg BW in mice (administered as water solutions).[1,2]

The toxic symptoms include CNS affection (motor coordination disorder, decreased excitability and muscle tonus in a part of animals, respiratory disorders). Some animals developed conjunctivitis. In two days after poisoning, animals displayed hindlimb paresis and lowered pain sensitivity. Death occurs in 3 to 4 days. Histology examination revealed vascular disturbances and signs of parenchymatous dystrophy in the liver. Marked disturbances in the blood and lymph circulation occurred. Gross pathology examination revealed congestion and moderate edema, dystrophic changes in the spinal cord, and parenchymatous dystrophy of the renal tubular epithelium.[2]

Repeated Exposure. When mice were dosed by gavage with 0.3 g/kg BW, cumulative properties were pronounced. K_{acc} is 1.4 (by Kagan).

Long-term Toxicity. In a 6-month study, male rats were administered F. by gavage as water solutions. A dose of 5.0 mg/kg BW caused retardation in bromosulfalein excretion, an increase in protein content in the urine, and shortening of STI.[1]

Reproductive Toxicity.

Gonadotoxicity. Necrosis of the spermatogenic epithelium and a reduction in the spermatozoa count were reported following oral administration of 5.0 to 8.0 g/kg BW. Long-term inhalation exposure to 6 mg/m^3 did not affect fertility of rats.[2]

Embryotoxicity. Pregnant rats were exposed by gavage to 2.0 to 3.0 g F./kg BW. Administration on day 10 of pregnancy increased the rate of resorptions, and, subsequently, embryo weights and size.[3,4]

Teratogenic effect was found in rabbits, rats, and mice, predominantly in the limbs and central nervous system.[5,6]

Regulations. ***U.S. FDA*** (1998) approved the use of F. in adhesives as a component of articles intended for use in packaging, transporting, or holding food in accordance with the conditions prescribed by 21 CFR part 175.105.

References:

1. Saratikov, A. S., Trofimovich, E. M., Novozheeva, T. P., et al., Hygienic regulation of formamide in water bodies, *Gig. Sanit.*, 3, 72, 1987 (in Russian).
2. *Toxicology of New Industrial Chemical Substances*, A. A. Letavet and I. V. Sanotsky, Eds., Meditsina, Moscow, Issue No 9, 1967, 86 (in Russian).
3. Vinogradova, Ye. L., in *Proc. Conf. of Junior Scientists*, Abstracts, Moscow, 1966, 60 (in Russian).
4. Silant'yeva, I. V. et al., in *Current Problems of Industrial Hygiene and Occup. Pathology*, Riga, 1968, 68 (in Russian).
5. Merkle, J. and Zeller, H., Studies on acetamides and formamides for embryotoxic and teratogenic activities in the rabbit, *Arzneimittel-Forsch.*, 30, 1557, 1980.
6. Thiersh, J. B., Investigations into the differential effect of compounds on rat litter and mother, in *Malformations Congenitales des Mammiferes*, H. Tuchmann-Duplessis, Ed., Masson, Paris, 1971, 95.

FURFURYL ALCOHOL

Molecular Formula. $C_5H_6O_2$
M = 98.11
CAS No 98-00-0
RTECS No LU9100000
Abbreviation. FA.

Synonyms. 2-Furancarbinol; 2-Furanmethanol; Furfural alcohol; Furfuralcohol; (2-Furyl) methanol; 2-Hydroxymethylfuran.

Properties. FA occurs as a colorless or slightly yellowish liquid having a faint burning odor and bitter taste. Miscible with water; highly soluble in alcohol and ether.

Applications. Used in the production of urea and furan resins in compositions of corrosion-resistant polymers. A solvent.

Exposure. FA is characterized with high production volume (10 to 20 million pounds), wide human exposure, and its presence in food.

Acute Toxicity. LD_{50} is reported to be 650 mg/kg BW in dogs, 50 to 100 mg/kg BW[018] or 275 mg/kg BW in rats, 50[018] or 160 mg/kg BW in mice. Poisoning produced CNS depression.

Reproductive Toxicity. Inhalation exposure caused no specific gonadotoxic or embryotoxic effect.[1]

Allergenic Effect. Weak evidence.[2]

Mutagenicity. No increase in SCE was associated with exposure in human lymphocytes either *in vitro* or *in vivo.*[3]

In vitro genotoxicity. FA is shown to produce SCE in Chinese hamster ovary cells (NTP-85). Cultured Chinese hamster ovary cells were exposed for 3 hours to FA which induced a relatively high frequency of chromatid breaks and chromatid exchanges in the absence of a liver microsomal activation system. The clastogenic activities of FA were increased, when *S9* mixture has been added.[4]

In vivo cytogenicity. FA was negative in the germ cells of *Dr. melanogaster.*[5]

Carcinogenicity. FA has been predicted to be rodent carcinogen by the computer automated structure evaluation (CASE) and multiple computer automated structure evaluation (MULTICASE) systems.[6]

Carcinogenicity classification.

NTP: NE - NE - CE - CE

Chemobiokinetics. FA is distributed primarily to the liver and kidney, brain being a tissue with the lowest FA concentration. FA is oxidized to *furoic acid* that is excreted as the glycine conjugate *furoyl glycine*. This is the major metabolite, delivered in the urine following feeding FA by rats.[7]

After inhalation exposure, *furoyl glycine* and low concentrations of *2-furoic acid* and *2-furan acrylic acid* were found in the urine (NTP-92). *Furoyl glycine* was the metabolite present in the highest concentration with *furan acryluric acid* next in abundance at about 10% of the *furoyl glycine* concentration.[8]

The urine appeared to be the major route of elimination (up to 90%), followed by feces (2.0 to 4.0%) and exhaled CO_2 (7.0%).[9]

Regulations. ***U.S. FDA*** (1998) regulates the use of FA in adhesives used as components of articles intended for use in packaging, transporting, or holding food in accordance with the conditions prescribed by 21 CFR part 175.105.

Standards. ***Russia.*** PML: 0.5 mg/l (organolept., odor).

References:

1. Gadalina, I. D. and Malysheva, M. V., Substantiation of the MAC for furyl alcohol in the industrialized zone air, *Gig. Truda Prof. Zabol.*, 9, 52, 1981 (in Russian).
2. Chernousov, A. D., On allergenic properties of furan compounds, *Gig. Sanit.*, 6, 28, 1974 (in Russian).
3. Gomez-Arroyo, S. and Souza, V., *In vitro* and occupational induction of sister-chromatid exchanges in human lymphocytes with furfuryl alcohol and furfural, *Mutat. Res.*, 156, 233, 1985.
4. Stich, H. F., Rosin, M. P., Wu, C. H., and Powrie, W. D., Clastogenicity of furans found in food, *Cancer Lett.*, 13, 89, 1981.
5. Rodrigues-Arnaiz, R., Morales, P. R., Moctezuma, R. V., and Salas, R. M. B., Evidence for the absence of mutagenic activity of furfuryl alcohol in tests of germ cells in *Drosophila melanogaster, Mutat. Res.*, 223. 309, 1989.
6. Zhang, Y. P., Sussman, N., Macina, O. T., Rosenkranz, H. S., Klopman, G., Prediction of the carcinogenicity of a second group of organic chemicals undergoing carcinogenicity testing, *Environ. Health Perspect.*, 104S, 1045, 1996.
7. Browning, E., *Toxicity and Metabolism of Industrial Solvents*, American Elsevier, New York, 1965, 383.
8. *NTP Fiscal Year 1995 The Annual Report*, Publ. Health Service, Dept. Health & Human Services, December 1995, 136.
9. Nomeir, A. A., Silveira, D. M., McComish, M. F., and Chadwick, M., Comparative metabolism and disposition of furfural and furfuryl alcohol in rats, *Drug Metab. Dispos.*, 20, 198, 1992.

GLYCEROL

Molecular Formula. $C_3H_8O_3$

M = 92.11

CAS No 56-81-5

RTECS No MA8050000

Synonyms and **Trade Names.** Glyceritol; Glycerine; Glycyl alcohol; Glyrol; 1,2,3-Propanetriol; Trihydroxypropane.

Properties. Syrupy rhombic plates or clear syrupy colorless liquid with a mild odor and sweet warm taste. Soluble in water in all proportions and in alcohol.

Applications. Used as a solvent and plasticizer for alkyd resins and ester gums. A softener or a plasticizer in cellophane.

Acute Toxicity. G. given in very large doses exerts systemic toxic effect: hemolysis, hemoglobinuria, and renal failure.[1]

Manifestations of acute poisoning included also mild headache, dizziness, nausea, vomiting, thirst, and diarrhea. A single dose of 1.5 g/kg BW or less produced very slight diuresis in the man. G. induced tissue dehydration and decrease in cerebrospinal fluid pressure.[2]

Long-term Toxicity. Rats were given 5.0, 10, or 20% G. in their diet for 730 days.

The treatment produced organ weight changes. The LOEL of 6.9 g/kg BW was established for this effect. Munro et al. (1996) suggested the calculated NOEL of 3.4 g/kg BW for this study.[3-5]

Reproductive Toxicity.

Gonadotoxicity.

Observations in man. No significant alterations of sperm motility were observed in workers occupied in G. production for more than 5 years.[3]

Animal studies. G. was shown to impair sperm motility *in vitro*.[6]

Embryotoxicity and ***Teratogenicity.*** Wistar rats received G. by gavage on days 1 through 15 of pregnancy at doses levels less or close to LD_{50}. 1/50 LD_{50} appeared to be non-embryotoxic. No teratogenic activity was found at dose levels of 0.14 and 1.4 g/kg BW. There were observed disturbances in the structure and function of embryonic liver and changes in enzyme activity.[7]

Mutagenicity.

In vitro genotoxicity. G. showed negative or equivocal results in *Salmonella* mutagenicity assay.[025]

In vivo cytogenetics. G. exposure altered chromosomes of bone marrow and male germ cells in rats.[8]

Chemobiokinetics. Following oral administration, G. is rapidly absorbed, distributed throughout the blood with the major part in body fat. G. metabolism is initiated by glycerokinase in liver to *carbon dioxide* and *water*. G. could be utilized in glucose or glycogen synthesis. Partially, it combines with free fatty acids to form *triglycerides*. 80% G. metabolites are available in the liver, 10 to 20% in the kidney. 7 to 14% G. was excreted unchanged in the urine.[2]

Regulations.

EU (1990). G. is available in the *List of authorized monomers and other starting substances which shall be used for the manufacture of plastic materials and articles intended to come into contact with foodstuffs (Section A)*.

EU (1995). G. is a food additive generally permitted for use in foodstuffs.

U.S. FDA (1998) considered G. to be *GRAS* when used in accordance with good manufacturing practice. It is approved for use (1) in the manufacture of resinous and polymeric coatings for the food-contact surface of articles intended for use in producing, manufacturing, packing, processing, preparing, treating, packaging, transporting, or holding food in accordance with the conditions prescribed by 21 CFR part 175.300; (2) in the manufacture of resinous and polymeric coatings for polyolefin films for food-contact surface of articles intended for use in producing, manufacturing, packing, processing, preparing, treating, packaging, transporting, or holding food in accordance with the conditions prescribed by 21 CFR part 175.320; (3) as a defoaming agent in the manufacture of paper and paperboard intended for use in packaging, transporting, or holding food in accordance with the conditions prescribed by 21 CFR part 176.210; (4) in the manufacture of polyurethane resins which may be used as food-contact surface of articles intended for use in contact with bulk quantities of dry food of an appropriate type in accordance with the conditions prescribed by 21 CFR part 177.1680; (5) to be employed in the production of textile and textile fibers intended for use in producing, manufacturing, packaging, processing, preparing, treating, packing, transporting, or holding food in accordance with the conditions prescribed by 21 CFR part 177.2800; and (6) as a component of articles intended for use in packaging, transporting, or holding food in accordance with the conditions prescribed in 21 CFR part 178.3500.

Great Britain (1998). G. is authorized without time limit for use in the production of polymeric materials and articles in contact with food or drink or intended for such contact.

References:

1. Goodman and Gilman's *The Pharmacological Basis of Therapeutics*, A. G. Gilman, L. S. Goodman, and A. Gilman, Eds., 6th ed., Macmillan Publishing Co., Inc., New York, 1980, 952.

2. *American Hospital Formulary Service - Drug Information* 93, McEvoy, G. K., Ed., Bethesda, MD, 1993, 1773.
3. Venable, J. R., McClimans, C. D., Flake, R. E., and Dimick, D. B., A fertility study of male employees engaged in the manufacture of glycerin, *J. Occup. Med.*, 22, 87, 1980.
4. *Toxiciological Evaluation of Certain Food Additives*, 20th Meeting of the Joint FAO/WHO Expert Committee on Food Additives, WHO, Food Additives Series No. 10, Geneva, 1976.
5. Munro, I. C., Ford, R. A., Kennepohl, E., and Sprenger, J. G., Correlation of structural class with No-Observed-Effect Levels: A proposal for establishing a threshold of concern, *Food Chem. Toxicol.*, 34, 829, 1996.
6. Tulandi, T. and McInnes, R. A., Vaginal lubricants: effect of glycerin and egg white on sperm motility and progression *in vitro*, *Fertil. Steril.*, 41, 151, 1984.
7. Bariljak, I. R., Korach, V. I., Spitovskaya, L. D., and Kalinovskaya, L. A., Effects of alcohols on the rat embryos liver, *Reports Acad. Sci. USSR*, 6, 67, 1988 (in Russian).
8. Bariljak, I. R. and Kozachuk, S. I., Mutagenic action of different alcohols in an experiment, *Cytologia i Genetica*, 19, 436, 1985 (in Russian).

GLYCIDALDEHYDE

Molecular Formula. $C_3H_4O_2$
M = 72.07
CAS No 765-34-4
RTECS No MB3150000
Abbreviation. GA.

Synonyms and **Trade Names.** Epihydrine aldehyde; 2,3-Epoxy-1-propanal; 2,3-Epoxypropionaldehyde; Formyloxiran; Glycidal; Oxirane carboxaldehyde.

Properties. Colorless liquid. Miscible with water and organic solvents.

Applications. A cross-linking agent for finishing wool and fat-liquoring of leather and surgical sutures.

Acute Toxicity. In rats, LD_{50} is 230 or 850 mg/kg BW, or according to other data, 50 mg/kg BW in rats, and 250 mg/kg BW in rabbits.[03,020]

Mutagenicity.

In vitro genotoxicity. GA was shown to be positive in *Salmonella* mutagenicity assay.

Carcinogenicity. Negative result were observed in female rats.[09]

Carcinogenicity classifications. An IARC Working Group concluded that there is sufficient evidence for the carcinogenicity of GA in *experimental animals* and there were no adequate data available to evaluate the carcinogenicity of GA in *humans*.

IARC: 2B;
U.S. EPA: B2.

Reference:

McCann, J., Choi, E., Yamasaki, E., et al., Detection of carcinogens and mutagens in the *Salmonella*/microsome test. Assay of 300 chemicals, *Proc. Natl. Acad. Sci. U.S.A.*, Washington, D.C., 72, 5135, 1975.

HEPTANE

Molecular Formula. C_7H_{16}
M = 100.23
CAS No 142-82-5
RTECS No MI7700000

Synonym and **Trade Name.** *n*-Heptane; Skellysolve C.

Properties. Solubility in water is 3.0 mg/l [018] at 20°C (in distillated water 2.2 mg/l), soluble in alcohol.

Applications. A solvent.

Short-term Toxicity. Rats were gavaged with daily doses of 4.0 g/kg BW for 13 weeks. The treatment caused changes in bladder weight, weight loss, or decreased weight gain. Hypoglycemia was also noted.[1]

Mutagenicity.

In vitro genotoxicity. H. was shown to be negative in the rat liver chromosome, in the bacterial mutation assay, and the yeast mitotic gene conversion assay.[2]

Regulations. ***U.S. FDA*** (1998) approved the use of H. (1) in adhesives as a component of articles intended for use in packaging, transporting, or holding food in accordance with the conditions prescribed by 21 CFR part 175.105; and (2) in the manufacture of cellophane for packaging food in accordance with the conditions prescribed by 21 CFR part 177.1200.

References:

1. Natl. Technical Information Service. OTS0571116.
2. Brooks, T. M., Meyer, A. L., and Huston, D. H., The genetic toxicology of some hydrocarbon and oxygenated solvents, *Mutagenesis,* 3, 227, 1988.

HEPTYL ALCOHOL

Molecular Formula. $C_7H_{16}O$
M = 116.21
CAS No 111-70-6
RTECS No MK0350000
Abbreviation. HA.

Synonyms. Enanthic alcohol; 1-Heptanol; Hydroxyheptane.

Properties. A colorless liquid with a strong odor. Solubility in water is 0.9 g/l. Mixes with alcohol at any ratio. Odor perception threshold is 2.0 mg/l, taste perception threshold is 0.5 mg/l.[1]

Applications. Used as a solvent in the production of phenol-formaldehyde resins. HA is used in the production of plasticizers and emulsifiers.

Acute Toxicity. LD_{50} is found to be 6.0 g/kg BW[2] or 1.5 g/kg BW in mice,[3] 0.5 g/kg BW in rats, and 0.75 g/kg BW in rabbits. Clinical signs of intoxication comprised CNS effects and respiratory failure.

Repeated Exposure revealed no cumulative properties in rats exposed to 300 mg/kg BW for 1 month.

Long-term Toxicity. In a 6-month study, gross pathology examination revealed edema and dystrophic changes in the liver and kidneys of rats.[1]

Regulations. ***U.S. FDA*** (1998) approved the use of HA as a food additive for direct addition to food for human consumption as a synthetic flavoring substance and adjuvant in accordance with the following conditions: (1) the quantity added to food does not exceed the amount reasonably required to accomplish its intended physical, nutritive, or other technical effect in food, and (2) when intended for use in or on food, it is of appropriate food grade and is prepared and handled as a food ingredient (21 CFR part 172.515).

Recommendations. Joint FAO/WHO Expert Committee on Food Additives (1999). ADI: No safety concern.

Standards. ***Russia*** (1995). MAC and PML: 0.05 mg/l.

References:

1. Kostovetsky, Ya. J. and Zholdakova, Z. I., in *Protection of Water Reservoirs against Pollution by Industrial Liquid Effluents*, S. N. Cherkinsky, Ed., Medgiz, Moscow, Issue No 7, 1965, 152 (in Russian).
2. Zayeva, G. N. Babina, M. D., Fedorova, V. I., and Scirskaya, V. A., Toxicological characteristics of polivinyl alcohol, polyethylene and polypropylene, in *Toxicology of New Industrial Chemicals*, A. A. Letavet, Ed., Medgiz, Moscow, Issue No 5, 1963, 120 (in Russian).
3. Voskoboinikova, V. B., Substantiation of maximum allowable concentration for floatoreagent IM-68 and hexyl, heptyl and octyl alcohols it is composed of, *Gig. Sanit*., 3, 16, 1966 (in Russian).

2-HEPTYL CYCLOPENTANONE

Molecular Formula. $C_{12}H_{22}O$
M = 182.34
CAS No 137-03-1
RTECS No GY4950000

Synonym. α-Heptyl cyclopentanone.
Properties. Colorless liquid with a specific odor. Soluble in water and alcohol.
Applications. Used as a solvent in the production of vinyl resins, nitrocellulose, and synthetic rubber.
Acute Toxicity. LD_{50} is 10.55 g/kg BW in rats, and 6.8 g/kg BW in mice.
See ***2-Octanone***.

HEXANE

Molecular Formula. C_6H_{14}
M = 86.20
CAS No 110-54-3
RTECS No MN9275000

Synonyms and **Trade Name.** *n*-Hexane; Hexyl hydride; Skellysolve B.

Properties. Colorless, easily evaporated liquid with a specific odor. Solubility in water is 1.0 mg/l; soluble in ethanol. Odor perception threshold is 0.0064 mg/l.[02]

Applications. A component of paints, a solvent in the manufacture of polyethylene, polypropylene, etc.

Acute Toxicity. LD_{50} is found to be 24, 49, and 43.5 mg/kg BW in 14-day-old rats, young adults, and older adults, respectively.[1] The most common toxic response is the development of central and peripheral neuropathy.

Repeated Exposure. Testicular lesions and neurotoxicity are the principal effects of repeated exposure to H.

Short-term Toxicity studies revealed a neuropathic action. CD rats were exposed with 99% pure or technical grade H. by gavage for 17 weeks, or until hindlimb paralysis was observed. The treatment with 4,000 mg technical grade H./kg BW caused severe hindlimb weakness or paralysis, tibial nerve lesions, and atrophy of testicular geminal epithelium.[2]

Wistar rats received 0.04 to 5.0 g commercial H./kg BW for 13 weeks. An increase in relative kidney and liver weights was noted at 0.2 to 5.0 g/kg BW level. Changes in plasma enzymes, indicative of liver damage and elevated cholesterol and triglyceride levels were also observed. Histological examination revealed changes in the liver and kidneys, adrenals, peripheral nerves, spleen, testes, and thymus at the highest dose level tested. The NOAEL of 0.04 g technical grade H./kg BW was established in this study.[3]

Reproductive Toxicity.

Gonadotoxicity. Exposure to high concentrations (1000 to 5000 ppm) caused progressive testicular toxicity.[4]

Embryotoxicity. Inhalation exposure of pregnant F344 rats to 1000 ppm on days 8 to 16 of gestation resulted in depression of postnatal growth of pups up to 3 weeks after birth.[5]

Teratogenicity. Pregnant CD-1 mice exposed by gavage to doses up to 9.9 g/kg BW on days 6 to 15 of gestation gave birth to litters that had no signs of an increased teratogenic effect.[6]

Mutagenicity.

In vivo cytogenetics. Doses of 500 to 2000 mg/kg BW administered *i/p* failed to increase the incidence of SCE in an *in vivo* mouse bone marrow cytogenetic assay.[7] A number of CA were slightly increased.

Chemobiokinetics. Accumulation in the tissues depends on lipid content in these tissues. *n*-H. is oxidized in the liver. Excretion occurs via the lungs and kidneys. The major urinary metabolite in rat is *1-hexanol.* Excretion of H. is related to dose injected.[8]

Regulations. ***U.S. FDA*** (1998) approved the use of H. (1) as a component of adhesives to be used safely in contact with food in accordance with the conditions prescribed by 21 CFR part 175.105; (2) as a component of a defoaming agent that may be safely used as a component of food-contact articles in accordance with the conditions prescribed in 21 CFR part 176.200; (3) in the manufacture of resinous and polymeric coatings for polyolefin films to be safely used as a food-contact surface of articles intended for use in producing, manufacturing, packing, transporting, or holding food in accordance with the conditions prescribed by 21 CFR part 175.320; and (4) in food under the conditions prescribed in 21 CFR part 173.270.

Recommendations.
Joint FAO/WHO Expert Committee on Food Additives. ADI is not specified.
U.S. EPA (1998). Health Advisory for longer-term exposure: 10 mg/l.
References:

1. Kimura, E. T., Ebert, D. M., and Dodge, P.W., Acute toxicity and limits of solvents residue for sixteen organic solvents, *Toxicol. Appl. Pharmacol.*, 19, 699, 1971.
2. Krasawage, W. J., O'Donoghue, J. L., Divincenzo, G. D., and Terhaar, C. J., The relative neurotoxicity of methyl *n*-butyl ketone, *n*-hexane and their metabolites, *Toxicol. Appl. Pharmacol.*, 52, 433, 1980.
3. Til, H. P. et al. (1989), cit in *n-Hexane*, Environmental Health Criteria Series No. 122, WHO, IPCS, Geneva, 1991, 57.
4. Nylen, P. et al., Testicular atrophy and loss of nerve growth factor - immunoreactive germ cell line in rats exposed to *n*-hexane and a protective effect of simultaneous exposure to toluene or xylene, *Arch. Toxi col.*, 63, 296, 1989.
5. Bus, J. S., White, E. L., Tyl, R. W., et al., Perinatal toxicity and metabolism of *n*-hexane in Fisher 344 rats after inhalation exposure during gestation, *Toxicol. Appl. Pharmacol.*, 51, 295, 1979.
6. Marks, T. A., Fisher, P. W., and Staples, R. E. Influence of *n*-hexane on embryo and fetal development of mice, *Drug Chem. Toxicol.*, 3, 393, 1980.
7. *Toxicity Studies of n-Hexane in B6C3F$_1$ Mice (Inhalation Studies),* NTP Technical Report, Research Triangle Park, NC, NIH Publ. No 91-3121, 1991, 32.
8. Dolara, P., Franconi, F., and Basosi, D., Urinary excretion of some *N*-hexane metabolites, *Pharmacol. Res. Commun.*, 10, 503, 1978.

1,6-HEXANEDIOL

Molecular Formula. $C_6H_{14}O_2$
M = 118.17
CAS No 629-11-8
RTECS No MO2100000
Abbreviation. HD.

Synonyms. Hexamethylene glycol; 1,6-Dihydroxyhexene; Hexamethylenediol.
Properties. Crystals. Soluble in water and alcohol.
Applications. Solvent, intermediate for high polymers such as nylon, polyesters, polyurethanes.
Acute Toxicity. LD_{50} is 3.73 g/kg BW in rats.
Regulations.

EU (1990). HD is available in the *List of authorized monomers and other starting substances which may continue to be used for the manufacture of plastic materials and articles intended to come into contact with foodstuffs pending a decision on inclusion in Section A* (*Section B*).

U.S. FDA (1998) approved the use of HD in the manufacture of polyurethane resins which may be used as food-contact surface of articles intended for use in contact with bulk quantities of dry food of an appropriate type in accordance with the conditions prescribed by 21 CFR part 177.1680.

Great Britain (1998). HD is authorized up to the end of 2001 for use in the production of polymeric materials and articles in contact with food or drink or intended for such contact.

Reference:

Carpenter, C. P., Weil, C. S., and Smyth, H. F., Range-finding toxicity data: List VIII, *Toxicol. Appl. Pharmacol.*, 28, 313, 1974.

HEXYL ALCOHOL

Molecular Formula. $C_6H_{14}O$
M = 102.18
CAS No 111-27-3
RTECS No MQ4025000

Abbreviation. HA.

Synonyms. Amylcarbinol; Caproyl alcohol; 1-Hexanol; 1-Hydroxyhexane; Pentyl carbinol.

Properties. A liquid. Solubility in water is 5.9 g/l. Concentrations of 21 and 37 mg/l in water give a rating of 1 for odor and taste, respectively. Odor disappears only after 3 days.[1] According to other data, odor threshold concentration is 0.01 mg/l.[010]

Applications. Used as a solvent for synthetic resins.

Acute Toxicity. In mice, LD_{50} is reported to be 1.95 g/kg BW.

Repeated Exposure failed to reveal cumulative properties.

Long-term Toxicity. The NOAEL has to be not lower than that experimentally established for the more toxic heptyl alcohol.[1,2]

Reproductive Toxicity.

Gonadotoxicity. HA did not induce testicular atrophy, hepatomegaly, peroxisome proliferation, or hypolipidaemia in male rats.[3]

Teratogenicity. HA produced no adverse reproductive effects in rats exposed to vapor at concentration of 14,000 mg/m^3. Inhalation exposure to the concentration of 3,500 ppm caused no increase in incidence of malformation in offspring.[4]

Regulations. ***U.S. FDA*** (1998) approved the use of HA as a food additive for direct addition to food for human consumption as a synthetic flavoring substance and adjuvant in accordance with the following conditions: (1) the quantity added to food does not exceed the amount reasonably required to accomplish its intended physical, nutritive, or other technical effect in food, and (2) when intended for use in or on food, it is of appropriate food grade and is prepared and handled as a food ingredient (21 CFR part 172.515).

Recommendations. Joint FAO/WHO Expert Committee on Food Additives (1999). ADI: No safety concern.

Standards. ***Russia*** (1995). MAC and MPL: 0.01 mg/l (for *normal, secondary,* and *tert*-HA).

References:

1. Voskoboinikova, V. B., Substantiation of maximum allowable concentration for floatoreagent IM-68, and hexyl, heptyl and octyl alcohols it is composed of, *Gig. Sanit.*, 3, 16, 1966 (in Russian).
2. Einhardt, *cit. by* I. V. Nazarenko, in *Protection of Water Reservoirs against Pollution by Industrial Liquid Effluents*, S. N. Cherkinsky, Ed., Medgiz, Moscow, Issue No 4, 1960, 65 (in Russian).
3. Rhodes, C., Soames, T., Stonard, M. D., Simpson, M. G., Vernall, A. J., and Elcombe, C. R., The absence of testicular atrophy and *in vivo* and *in vitro* effects on hepatocyte morphology and peroxisomal enzyme activities in male rats following the administration of several alkanols, *Toxicol. Lett.*, 21, 103, 1984.
4. Nelson, B. K., Brightwell, W. S., Khan, A., Hoberman, A. M., and Krieg, E. F., Teratological evaluation of 1-pentanol, 1-hexanol and 2-ethyl-1-hexanol administered by inhalation to rats, *Teratology*, 37, 479, 1988.

HYDROABIETYL ALCOHOL

Molecular Formula. $C_{20}H_{34}O$

M = 290.54

CAS No 26266-77-3

RTECS No MW3800000

Abbreviation. HA.

Synonyms. Abitol; Dihydroabietyl alcohol; Dodecahydro-1,4a-dimethyl-7-(1-methylethyl)-1-phenanthrenemethane.

Properties. A viscous liquid with a faint wood-like odor prepared from hydrogenated rosin acids.

Acute Toxicity. LD_{50} is 70 g/kg BW in rats.

Regulations. ***U.S. FDA*** (1998) approved the use of HA (1) in adhesives as a component of articles intended for use in packaging, transporting, or holding food in accordance with the conditions prescribed by 21 CFR part 175.105, and (2) as a component of the uncoated or coated food-contact surface of paper and paperboard intended for use in producing, manufacturing, packaging, processing, preparing, treating, packing, transporting, or holding dry food in accordance with the conditions prescribed by 21 CFR part 176.180.

Reference:

Anonymous, Hydroabietyl alcohol, *Food Cosmet. Toxicol.*, 12, 919, 1974.

4-HYDROXY-4-METHYL-2-PENTANONE

Molecular Formula. $C_6H_{12}O_2$
M = 116.18
CAS No 123-42-2
RTECS No SA9100000
Abbreviation. HMP.

Synonyms and **Trade Names.** Acetonyl dimethyl carbinol; Diacetone; Diacetone alcohol; 4-Hydroxy-2-keto-4-methylpentane; 2-Methyl-2-pentanone-4-one; Tyranton.

Properties. Colorless liquid with a faint pleasant odor. Miscible with water and alcohol.[020]

Applications. Solvent for cellulose acetate, nitrocellulose, celluloid, waxes, and resins. HMP is used in preparation of different coating compositions.

Acute Toxicity.

Observations in man. Estimated lethal dose is 30 g.[018]

Animal studies. LD_{50} is found to be 4.0 g/kg BW[1] or 2.56 g/kg BW in rats, and 3.0 g/kg BW in mice. In poisoned animals, excitation is followed by inhibition and death from respiratory arrest.[2] The dose of 2.0 ml/kg BW produced transient liver damage in rats. Manifestations of poisoning in rabbits included CNS depression; the lethal dose appeared to be 5.0 ml/kg BW.[03]

Repeated Exposure. Rabbits received daily doses of 2.0 ml for 12 days. The treatment caused narcotic effect, kidney damage and death of 75% of the animals.[3] Rats received 40 mg/kg BW in their drinking water for a month. The treatment caused renal lesions. The dose of 10 mg/kg BW was ineffective.[03]

Mutagenicity.

In vitro genotoxicity. HMP evoked a weak positive response in the rat liver chromosome assay, but was shown to have negative response in the bacterial mutation assay and the yeast mitotic gene conversion assay.[2] It was negative in *Salmonella* mutagenicity assay (NTP-94).

Regulations. ***U.S. FDA*** (1998) approved the use of HMP in adhesives as a component (monomer) of articles intended for use in packaging, transporting, or holding food in accordance with the conditions prescribed in 21 CFR part 175.105.

References:

1. Smyth, H. F. and Carpenter, C. P., Further experience with the range finding test in the industrial toxicological laboratories, *J. Ind. Hyg. Toxicol.*, 30, 63, 1948.
2. Brooks, T. M., Meyer, A. L., and Huston, D. H., The genetic toxicology of some hydrocarbon and oxygenated solvents, *Mutagenesis*, 3, 227, 1988.
3. Ordjohnikidze, A. K., in *Proc XXIV Moscow Sci. Pract. Conf. on Hygiene Problems*, Moscow, 1969, 102 (in Russian).

ISOBUTYL ALCOHOL

Molecular Formula. $C_4H_{10}O$
M = 74.12
CAS No 78-83-1
RTECS No NP9625000
Abbreviation. IBA.

Synonyms. Isobutanol; Isopropyl carbinol; Fermentation butyl alcohol; 1-Hydroxymethylpropane; 2-Methyl-1-propanol; 2-Methyl propyl alcohol.

Properties. Colorless refractive liquid having a sharp, alcoholic odor similar to that of amyl alcohol. Solubility in water is 8.7 to 9.5%; mixes with alcohols and ethers at any ratio. Odor perception threshold is 0.1 to 10 mg/l.[1] According to other data, organoleptic perception threshold is 0.36 mg/l.[01]

Applications. Used as a solvent in the production of nitrocellulose lacquers and resins; IBA is also used in the manufacture of isobutyl esters (plasticizers), and in the vulcanization of rubber.

Exposure. IBA natural occurrence in food could entail exposure of the population.

Acute Toxicity. LD_{50} is 2.46 g/kg BW [020] or 3.1 g/kg BW in rats, and 3.5 g/kg BW in mice.[2] There are no species differences in susceptibility. The threshold narcotic dose for rabbits is 1.4 g/kg BW, and the LD_{50} is 3.0 g/kg BW.[1] Poisoning produced signs of alcohol intoxication.

Repeated Exposure revealed weak cumulative properties.[2] Rats tolerate administration of 0.6 and 0.3 g/kg BW for a month.

Short-term Toxicity. Rats were given 1.0 mol/l solution of IBA as their sole drinking liquid for 4 months. No adverse effects on the liver were recorded. Exposure to a 2.0 mol/l solution for 2 months produced a reduction in fat, glycogen, and RNA contents, and in the overall size of the hepatocytes.[3]

Rats received by gavage 100, 316, or 1000 mg/kg BW for 13 weeks. The treatment caused a minor decrease in BW gain, decreased serum potassium levels, and hypoactivity (at the highest dose), slight changes in hematology analyses (at 2 higher doses). The NOEL of 316 mg/kg BW was established in this study.[4] This study failed to reproduce the GI effects reported by Hillbom et al.[3]

Long-term Toxicity. The threshold concentration was identified to be more than 100 mg/l.[1]

Immunotoxicity. A positive specific microprecipitation reaction (up to 40%) was noted in 1, 3 and 7 days following inhalation exposure.[2]

Reproductive Toxicity.

Gonadotoxicity. Rats received the dose of 0.05 mg/kg BW and higher with their drinking water. The treatment affected the course of pregnancy in rats.[5]

Mutagenicity.

In vitro genotoxicity. IBA was found to be negative in genotoxicity assays using bacteria, yeast, and mammalian cells.[6]

Carcinogenicity. In a lifetime study, rats were dosed orally twice a week with 0.2 ml IBA/kg BW (160 mg/kg BW). A mean total dose was 29 ml. The treatment resulted in toxic liver damage and hyperplasia of blood-forming tissues. Increase in tumor incidence was noted.[7] Methodology inadequacies in the study questioned its significance.

Chemobiokinetics. IBA is metabolized by alcohol dehydrogenase to *isobutyric acid* via *aldehyde* and may be involved into the *tricarboxylic acid* cycle. The urinary metabolites are *acetaldehyde, acetic acid, isobutyraldehyde, isovaleric acid*, and unmetabolized IBA.[8] Cornish stated that IBA is metabolized to its *aldehydes* and further catabolized to *carbon dioxide* and *water*, using pathways similar to *n-butyl alcohol.*[9] Only a negligible amount is eliminated unchanged or as the *glucuronide* in the urine.

Regulations.

U.S. FDA (1998) approved the use of IBA. (1) in adhesives as a component of articles intended for use in packaging, transporting, or holding food in accordance with the conditions prescribed by 21 CFR part 175.105; (2) as a component of the uncoated or coated food-contact surface of paper and paperboard intended for use in producing, manufacturing, packaging, processing, preparing, treating, packing, transporting, or holding dry food in accordance with the conditions prescribed by 21 CFR part 176.180; (3) as a substance employed in the production of, or added to textiles and textile fibers intended for use in contact with foods in accordance with the conditions prescribed by 21 CFR part 177.2800; (4) as a defoaming agent in the manufacture of paper and paperboard intended for use in packaging, transporting, or holding food in accordance with the conditions prescribed by 21 CFR part 176.210; and (5) as a component of a defoaming agent that may be safely used as a component of articles and in the manufacture of paper and paperboard intended for use in contact with food in accordance with the conditions prescribed by 21 CFR part 176.210.

Regulations. ***EU*** (1990). IBA is available in the *List of authorized monomers and other starting substances which may continue to be used for the manufacture of plastic materials and articles intended to come into contact with foodstuffs pending a decision on inclusion in Section A* (*Section B*).

Recommendations. Joint FAO/WHO Expert Committee on Food Additives (1999). ADI: No safety concern.

Standards. ***Russia*** (1995). MAC: 0.15 mg/l, PML in food: 0.5 mg/l.

References:

1. Nazarenko, I. V., in *Protection of Water Reservoirs against Pollution by Industrial Liquid Effluents*, S. N. Cherkinsky, Ed., Medgiz, Moscow, Issue No 4, 1960, 65 (in Russian).
2. Kushneva, V. S., Koloskova, G. A., and Koltunova I. G., Experimental data for hygienic standardization of isobutyl alcohol in the air of the work environment, *Gig. Truda Prof. Zabol.*, 1, 46, 1983 (in Russian).
3. Hillbom, M. E., Franssila, K., and Forsander, O. A., Effects of chronic ingestion of some lower aliphatic alcohols in rats, *Res. Commun. Chem. Pathol. Pharmacol.*, 9, 177, 1974.
4. *Rat Oral Subchronic Toxicity Study with Isobutyl Alcohol*, U.S. EPA, Office of Solid Waste, Washing ton, D. C., 1986.
5. Nadeyenko, V. G. et al., Embryotoxic effect of isobutanol, *Gig. Sanit.*, 2, 6, 1980 (in Russian).
6. Brooks, T. M. et al., The genetic toxicology of some hydrocarbon and oxygenated solvents, *Mutagenesis*, 3, 227, 1988.
7. Gibel, W., Lohs, K. H., and Wildner C. P., Experimental research on the carcinogenic effect of solvents, using propanol-1, 2-methylpropanol-1, and 3-methylbutanol-1 as examples, *Arch. Geschwulstforsch.*, 45, 19, 1975.
8. Saito, M., Studies on the metabolism of lower alcohols, Abstract, *Nichidai Igaka Zasshi,* 34, 569, 1975 (in Japanese).
9. Cornish, H. H., Solvents and vapors, in Casarett and Doul's *Toxicology*, The Basic Science of Poisons, 2 Ed., Doul, J., Klassen, C. D., and Amdur, M. O., Eds., MacMillan Publ. Co., Inc., New York, 480, 1980.

ISOBUTYL VINYL ETHER

Molecular Formula. $C_6H_{12}O$
M = 100.18
CAS No 109-53-5
RTECS No KO1300000
Abbreviation. IBVE.

Synonym. Isobutoxyethene.

Acute Toxicity. LD_{50} is 17 g/kg BW in rats.[1]

Regulations.

EU (1990). IBVE is available in the *List of authorized monomers and other starting substances which may continue to be used for the manufacture of plastic materials and articles intended to come into contact with foodstuffs pending a decision on inclusion in Section A* (*Section B*).

Great Britain (1998). IBVE is authorized up to the end of 2001 for use in the production of polymeric materials and articles in contact with food or drink or intended for such contact.

Reference:

1. Smyth, H. F., Carpenter, C. P., Weil, C. S., Pozzani, U. C., and Striegel, J. A., Range-finding toxicity data: List VI, *Am. Ind. Hyg. Assoc. J.*, 23, 95, 1962.

ISOPROPYL ALCOHOL

Molecular Formula. C_3H_8O
M = 60.09
CAS No 67-63-0
RTECS No NT8050000
Abbreviation. IPA.

Synonyms and **Trade Names.** Dimethyl carbinol; 1-Methylethyl alcohol; Hartosol; 2-Hydroxypropane; Imsol A; Isopropanol; 2-Propanol; *n*-Propan-2-ol; 2-Propyl alcohol; *sec*-Propyl alcohol.

Properties. Colorless liquid. Mixes with water and alcohol at any ratio. Odor perception threshold is 1.13,[1] 3.2 ,[010] or 160 mg/l.[02]

Applications. Used as a solvent in the production of synthetic resins, e.g., coatings such as phenolic varnishes and nitrocellulose lacquers.

Acute Toxicity.

Observations in man. Ingestion caused an alcoholic intoxication and narcosis.

Animal studies. LD_{50} is 5.0 to 5.5 g/kg BW in rats, and 3.6 to 4.5 g/kg BW in male mice. Poisoning is accompanied by a narcotic effect.[1,2] Death within 24 hours from respiratory failure. Gross pathology examination revealed congestion, edema, and hemorrhages into the interstitial tissues of the parenchymatous organs, which displayed inflammatory and dystrophic changes.[3] According to Kimura et al., LD_{50}*s* are reported to be 5.6, 6.0, and 6.8 mg/kg BW in 14-day-old rats, young adults rats, and older adults, respectively.[4]

Repeated Exposure.

Observations in man. 8 volunteers received daily doses of 6.4 mg/kg BW for 6 weeks. This intake did not affect blood and urinalysis, or liver and kidney functions.[5]

Animal studies. Moderate cumulative properties are noted. K_{acc} is found to be 4.9 and 4.0 (by Lim) for mice and rat, respectively. Rats were dosed by gavage with 115 mg/kg BW in the feed for 2 months. Changes in the serum enzyme activity and dystrophic changes in the liver and brain are reported.

Short-term Toxicity. In a 12-week study, male rats received 1.0, 2.0, 3.0, and 5.0% IPA in their drinking water. The treatment produced a dose-dependent increase in relative organ weights of liver, kidneys, and adrenals. No histopathological changes, apart from a dose-dependent increase in formation of hyaline casts and droplets in the proximal tubules of the kidney were observed. IPA does not seem to cause astroglyosis after 12 weeks of the treatment.[6]

Long-term Toxicity. In a 2-generation reproduction study, IPA was administered to rats via drinking water at 2.0% concentration (through the first and second generation). The treatment resulted in several changes which represent mild toxicity and are reminiscent of stress lesions.[7] Rats received drinking water containing IPA for 27 weeks. At the end of exposure period, all exposed females showed growth retardation.[2]

Reproductive Toxicity.

Embryotoxicity. IPA given orally to timed-pregnant rats at doses as high as 1200 mg/kg BW caused no biologically significant changes in the behavioral tests, in organ weights, and no pathological findings in offsprings that could be attributed to IPA exposure.[11] In a 2-generation study in rats, doses of 100, 500, or 1000 mg/kg BW were administered 10 weeks prior mating. Parental females were dosed during mating, gestation, and lactation. There were no treatment-related histology changes in reproductive tissues or biologically meaningful differences in other reproductive parameters in adults of either generation. The NOAEL appeared to be 500 mg/kg BW in this study.[8]

Gonadotoxicity. Rats given 252 and 1008 mg/kg BW for 2 months exhibited extended duration of the estrous cycle.[9]

Teratogenicity. Timed-pregnant Sprague-Dawley rats were dosed orally with aqueous IPA solutions at dose levels of 400, 800, and 1200 mg/kg BW on gestation days 6 through 15 at a dosing volume of 5.0 ml/kg BW. New Zealand white rabbits were dosed orally at dose levels of 120, 240, and 480 mg/kg BW on gestation days 6 through 18 at a dosing volume of 2.0 ml/kg. No evidence was found of increased teratogenicity at any doses tested in rats and rabbits. The NOAELs for both maternal and developmental toxicity were 40 mg/kg BW in rats, and 240 and 480 mg/kg BW, respectively, in rabbits.[10]

Allergenic Effect was not observed on skin application test.[3]

Mutagenicity. IPA is found to be negative in a variety of genotoxicity studies.

In vivo cytogenetics. Inhalation exposure is reported to produce a mutagenic effect in rats (cytogenetic analysis of bone marrow cells).[12] However, according to Kapp et al., IPA gave a negative response in a bone-marrow micronucleus study in mice.[12]

In vitro genotoxicity. IPA was negative in Chinese hamster ovary cell/HGPRT gene mutation assay.[13]

Chemobiokinetics. IPA is rapidly absorbed and distributed throughout the body, partially as *acetone.* It undergoes oxidation in the body with the formation of *acetic acid.* Following ingestion of an unknown amount of rubbing alcohol, IPA as well as its metabolite *acetone* was found in the spinal fluid of 2 persons at levels similar to those in the serum.[14]

Excretion of both substances is limited and does not exceed 4.0% of the dose in rats, rabbits, and dog.[15,16] The major route of excretion is via the lungs. Rats received by gavage single or multiple 300 and 3000 mg/kg BW doses. 80 to 90% of administered dose were exhaled (as *acetone*, CO_2, and unmetabolized IPA).[17]

Regulations.

U.S. FDA (1998) approved the use of IPA (1) in food under the conditions prescribed in 21 CFR part 173.240; (2) in adhesives as a component of food-contact articles in accordance with the conditions prescribed by 21 CFR part 175.105; (3) in cellophane to be safely used for packaging food (residue limit 0.1%) in accordance with the conditions prescribed by 21 CFR part 177.1200; (4) as a defoaming agent that may be safely used as a component of articles and in the manufacture of paper and paperboard intended for use in contact with food in accordance with the conditions prescribed in 21 CFR part 176.200; (5) in surface lubricants employed in the manufacture of metallic articles that contact food in accordance with the conditions prescribed in 21 CFR part 178.3910; (6) as a substance employed in the production of or added to textiles and textile fibers intended for use in contact with food in accordance with the conditions prescribed by 21 CFR part 177.2800; and (7) as a defoaming agent in the processing of food in accordance with the conditions prescribed in 21 CFR part 173.320.

EU (1990). IPA is available in the *List of authorized monomers and other starting substances which shall be used for the manufacture of plastic materials and articles intended to come into contact with foodstuffs (Section A).*

Great Britain (1998). IPA is authorized, without time limit, for use in the production of polymeric materials and articles in contact with food or drink or intended for such contact.

Standards. *Russia* (1995). MAC and PML: 0.25 mg/l (organolept., odor), PML in food: 0.1 mg/l.

References:

1. Galeta, S. G., *Hygienic Substantiation of the MAC for Isopropyl Alcohol in Water Bodies*, Author's abstract of thesis, Donetsk, 1967, 14 (in Russian).
2. Lehman, A. J. and Chase, H. F., The acute and chronic toxicity of isopropyl alcohol, *J. Lab. Clin. Med.*, 29, 561, 1944.
3. Guseinov, V. G., Toxicologic and hygienic characteristics of isopropyl alcohol, *Gig. Truda Prof. Zabol.*, 7, 60, 1985; in *Proc. 4th Congr. Of Hygienists and Sanitarians of Azerbaidjan*, Baku, 1981, 226 (in Russian).
4. Kimura, E. T., Ebert, D. M., and Dodge, P.W., Acute toxicity and limits of solvents residue for sixteen organic solvents, *Toxicol. Appl. Pharmacol.*, 19, 699, 1971.
5. Wills, J. H., Jameson, E. M., and Coulston, F., Effects on man of daily ingestion of small doses of isopropyl alcohol, *Toxicol. Appl. Pharmacol.*, 15,560, 1969.
6. Pilegaard, K. and Ladefoged, O., Toxic effects in rats of twelve weeks dosing of 2-propanol, and neurotoxicity measured by densitometric measurements of glial fibrillary acidic protein in the dorsal hippocampus, *In Vivo*, 7, 325, 1993.
7. Gallo, M. A., Oser, B. L., Cox, G. E., et al., Studies on the long-term toxicity of 2-butanol, in *16th Annual Meeting Soc. Toxicol.*, March 27-30, 1977, Abstracts of Papers, Toronto, 1977, 9.
8. Bevan, C., Tyler, T. R., Gardiner, T. H., Kapp, R. W., Andrews, L., and Beyer, B. K., Two- generation reproduction toxicity study with isopropanol in rats, *J. Appl. Toxicol.*, 15, 117, 1995.
9. Antonova, V. I. and Salmina, Z. A., Substantiation of the maximum allowable concentration for isopropyl alcohol as regards to its effect on gonads and offspring, *Gig. Sanit.*, 1, 8, 1978 (in Russian).
10. Tyl, R. W., Masten, L. W., Marr, M. C., Myers, C.B., Slauter, R. W., Gardiner, D. E., McKee, R. H., and Tyler, T. R., Developmental toxicity evaluation of isopropanol administered by gavage in rats and rabbits, *Fundam. Appl. Toxicol.*, 22, 139, 1994.
11. Bates, H. K., McKee, R. H., Bieler, G. S., Gardiner, T. H., Gill, M. W., Stronger, D. E., and Masten, L. W., Developmental neurotoxicity evaluation of orally administered isopropanol in rats, *Fundam. Appl. Toxi col.*, 22, 152, 1994.
12. Aristov, V. N., Red'kin, Yu. V., Bruskin, Z. Z., et al., Experimental data on the mutagenous effects of toluene, isopropanol, and sulfur dioxide, *Gig. Truda Prof. Zabol.*, 7, 33, 1981 (in Russian).

13. Kapp, R. W., Marino, D. J., Gardiner, T. H., Masten, L. W., McKee, R. H., Tyler, T. R., Ivett, J. L., and Young, R. R., *In vitro* and *in vivo* assays of isopropanol for mutagenicity, *Environ. Molec. Mutagen.*, 22, 93, 1993.
14. Natowicz, M., Donahue, J., Gorman, L., et al., Pharmacokinetic analysis of a case of isopropanol intoxication, *Clin. Med.*, 31, 326, 1985.
15. Rietbrock, N. and Abshagen, U., Pharmacokinetics and metabolism of aliphatic alcohols, *Arzneimittel-Forsch.*, 21, 1309, 1971.
16. *2-Propanol*, Environmental Health Criteria No 103, WHO, IPCS, Geneva, 1990, 132.
17. Slauter, R. W., Coleman, D. P., Gaudette, N. F., McKee, R. H., Masten, L. W., Gardiner, T. H., Strother, D. E., Tyler, T. R., and Jeffcoat, A. R., Disposition and pharmacokinetics of isopropanol in F344 rats and $B6C3F_1$ mice, *Fundam. Appl. Toxicol.*, 23, 407, 1994.

KEROSENE

CAS No 8008-20-6
RTECS No OA5500000

Synonyms. Coal oil; Fuel oil No 1; Range oil.

Applications. A solvent.

Properties. Pale-yellow or water-white, mobile, oily liquid with a characteristic, not altogether disagreeable odor. Insoluble in water. Miscible with petroleum solvents.[020]

Acute Toxicity. A gastric irritant that produces nausea and vomiting when swallowed. *I/g* administration of 0.35 ml/kg BW caused fatal pneumonia in rabbits.[1]

Repeated oral **Exposure** produced narcotic effect. Gross pathology examination revealed degenerative changes in the liver and kidney and congestion in the lungs (Narsimhan and Gangla, 1967).

Chemobiokinetics. K. is poorly absorbed from the GI tract (Mann, 1977). Following oral administration to monkeys, K. is unlikely to be excreted via the lungs: lungs damage appeared to occur due to aspiration following vomiting (Wolfsdorf and Kundig, 1972).

Regulations. ***U.S. FDA*** (1998) approved the use of K. (1) in adhesives as a component of articles intended for use in packaging, transporting, or holding food in accordance with the conditions prescribed by 21 CFR part 175.105; (2) as a component of a defoaming agent intended for articles which may be used in producing, manufacturing, packing, processing, preparing, treating, packaging, transporting, or holding food in accordance with the conditions prescribed in 21 CFR part 176.200; (3) as a defoaming agent in the manufacture of paper and paperboard intended for use in packaging, transporting, or holding food in accordance with the conditions prescribed by 21 CFR part 176.210; and (4) to be employed in the production of textile and textile fibers intended for use in producing, manufacturing, packaging, processing, preparing, treating, packing, transporting, or holding food in accordance with the conditions prescribed by 21 CFR part 177.2800.

Reference:

1. *Poisoning - Toxicology, Symptoms, Treatments*, J. M. Arena and R. H. Drew, Eds., 5 ed., Charles C. Thomas Publisher, Springfield, IL, 1986, 284.

LACTIC ACID, ETHYL ESTER

Molecular Formula. $C_5H_{10}O_3$
M = 118.15
CAS No 97-64-3
RTECS No OD5075000
Abbreviation. EL.

Synonyms and **Trade Name.** Actylol; Ethyl α-hydroxypropionate; Ethyl lactate; Solactol.

Properties. Colorless liquid with a characteristic odor and milk cream taste.[020] Miscible with water and alcohol at room temperature.

Applications. A solvent for nitrocellulose, cellulose acetate, ethyl cellulose, gums, synthetic polymers and other resins. Used as food additives and in the manufacture of pharmaceuticals and cosmetics, lacquers and dyes, and fragrances.

Acute Toxicity. Lactate esters have an oral LD_{50} greater than 2.0 g/kg BW. Administration of this dose produced no mortality in treated animals. Poisoning caused diarrhea and piloerection for up to 24 hours after administering. Gross pathology examination revealed no changes (Bailey, 1976). LD_{50} for racemic *n*-EL was more than 2.5 g/kg BW in mice and 5.0 g/kg BW in rats. Poisoning caused CNS inhibition (NIOSH, 1977).[1]

Short-term Toxicity. 1/8 rats died when fed a diet containing 5.0% EL.[2]

Reproductive Toxicity. No evidence of ***teratogenicity*** or maternal toxicity of *ethyl-l-lactate* was observed in dermal study (Hoberman, 1989).[1]

Allergenic Effect.

Observations in man. 8.0% EL solution in petrolatum produced no sensitization reactions upon application to volunteers.[3]

Animal studies. Lactate esters may be potential eye and skin irritants, but not skin sensitizers.

Mutagenicity. EL was negative in *Salmonella typhimurium* strains *TA 98, TA100, TA1535, TA1537*, and *TA1538* with and without metabolic activation (Curren and Mecchi, 1988).[1]

Chemobiokinetics. EL undergoes enzymatic hydrolysis into *lactic acid* and *ethyl alcohol.*

Regulations. ***U.S. FDA*** (1998) approved the use of EL in adhesives as a component of articles intended for use in packaging, transporting, or holding food in accordance with the conditions prescribed by 21 CFR part 175.105.

Considered as *GRAS.*

Recommendations.

FAO/WHO (1967). ADI: 100 mg/kg BW.

EU (1974): ADI: 1.0 mg/kg BW.

References:

1. Clary, J. J., Feron, V. J., and van Velthuijsen, J. A., Safety assessment of lactate esters. *Regul. Toxicol. Pharmacol.*, 27, 88, 1998.
2. *Monographs on Fragrance Raw Materials*, D. L. J. Opdyke and C. Letizia, Eds., Special Issue VI, New York, Pergamon Press, New York, 1982, 677.
3. Kligman, A. N., and Epstein, W., Updating the maximization test for identifying contact allergen, *Contact Dermat.*, 1, 231, 1975.

p-MENTH-1-EN-8-OL

Molecular Formula. $C_{10}H_{18}O$

M = 154.24

CAS No 2438-12-2

RTECS No OT0175122

Synonyms. 2-(4-Methyl-3-cyclohexenyl)-2-propanol; α-Terpineol; α,α'-4-Trimethyl-3-cyclohexene-1-methanol.

Properties. When pure, M. occurs as white crystalline powder with peach, floral flavor and sweet, lime taste. Soluble in water (about 2.0 g/l at 15 to 20°C) and alcohol.

Applications. A solvent for resins, lacquers, cellulose esters, and ethers. An antioxidant.

Acute toxicity. On ingestion, M. produced hemorrhagic gastritis. Manifestations of general toxic action include weakness and CNS inhibition. Hypothermia and respiratory failure were noted.[019]

Mutagenicity. M. was found to be negative in *Salmonella* mutagenicity assay.[1]

Regulations. ***U.S. FDA*** (1998) approved the use of M. in resinous and polymeric coatings used as the food-contact surfaces of articles intended for use in producing, manufacturing, packing, processing, preparing, treating, packaging, transporting, or holding food in accordance with the conditions prescribed in 21 CFR part 175.300.

Reference:

Florin, I., Rutberg, L., Curvall, M., and Enzell, C. R., Screening of tobacco smoke constituents for mutagenicity using the Ames' test, *Toxicology*, 18, 219, 1980.

2-MERCAPTOETHANOL

Molecular Formula. C_2H_6OS
M = 78.13
CAS No 60-24-2
RTECS No KL5600000
Abbreviations. 2-ME.

Synonyms and **Trade Names.** 1-Ethanol-2-thiol; Ethylene thioglycol; 1-Hydroxy-2-mercaptoethane; β-Hydroxymercaptoethane; Monothioethylene glycol; 2-Thioethanol; Thioglycol.

Applications. A solvent. An intermediate in the manufacture of plasticizers and stabilizers.

Properties. Colorless mobile liquid with a strong disagreeable odor. Readily miscible with water and alcohol. Odor perception threshold is 0.0001 μg/l.[017]

Acute Toxicity. LD_{50} is 190 to 350 mg/kg BW in mice and 224 mg/kg BW in rats.[017, 020]

Mutagenicity.

In vitro genotoxicity. 2-ME increased SCE in Chinese hamster lung *V79* cells.[1]

Chemobiokinetics. Inhalation exposure increased *organic sulphur* elimination via the urine.[017]

Regulations. ***U.S. FDA*** (1998) approved the use of 2-ME in articles made of acrylonitrile/butadiene/styrene copolymers or as components of these articles intended for use with all foods, except those containing alcohol under the conditions specified in the 21 CFR part 177.1020.

Reference:

1. Speit, G., Wolf, M., and Vogel, W., The effect of sulfhydryl compounds on sister chromatid exchanges, *Mutat. Res.*, 78, 267, 1980.

METHANOL

Molecular Formula. CH_4O
M = 32.04
CAS No 67-56-1
RTECS No PC1400000

Synonym and **Trade Names.** Carbinol; Methyl alcohol; Wood alcohol; Wood spirit.

Properties. Clear, colorless, volatile liquid with a faint alcoholic odor when pure; crude product may exhibit a repulsive, pungent odor. Mixes with water and ethyl alcohol at any ratio. Odor perception threshold is 4.26[010] or 30 to 50 mg/l,[1] or, according to other data,[02] 740 mg/l. At these concentrations, no taste is detected in water.

Applications. Used as a solvent and solvent adjuvant in the manufacture of plastic materials.

Acute Toxicity. In rats, LD_{50} is 10.6 g/kg BW (average over a year, 120 animals). Interspecies variations are negligible.[2] Very young and old animals are more sensitive to M. In rats with BW of up to 50 g, LD_{50} is 7.4 ml/kg BW; at BW of 80 to 160 g, 13 ml/kg BW; and at BW of 300 to 470 g, 8.8 ml/kg BW.[3]

Edema of the optic disc and of other areas of the CNS developed on a single administration of 2.0 g/kg BW.[4]

Administration of 3.5 ml 50% solution to male and female rats caused changes in the CNS, GI tract, parenchymatous, and immunocompetent organs. A lesser dose resulted in mild disturbances of blood circulation. A greater dose produced gross pathological changes in the CNS and lungs.[5]

Mild CNS depression, tremor, ataxia, and recumbence were reported in minipigs YU given a single oral dose by gavage.[6] Animals did not develop optic nerve lesions.

Repeated Exposure. Administration of 10 to 500 mg/kg BW to rats caused focal parenchymatous degeneration in the liver, an increase in the hepatocyte size, and a change in the activity of certain microsomal enzymes.[7]

In a month, a dose of 1.5 mg/kg BW had already affected conditioned reflex activity in rats.[8]

Short-term Toxicity. Sprague-Dawley rats were gavaged with 100, 500, and 2500 mg/kg BW for 90 days. No effect on BW gain, food consumption, gross and microscopic evaluations were observed. Mild signs of toxicity including reduction in brain weights were noted at 2500 mg/kg BW dose level. The NOAEL of 500 mg/kg BW was established in this study.[9]

Long-term Toxicity. In a 5.5-month study, rabbits were dosed by gavage with 2.5 ml MA/kg BW. Animals displayed retardation of BW gain and changes in the optic nerve axons and in the cerebral cortex cells.[8] Rats received 3.25 ml M./kg BW for 6 months. The treatment caused a decline in the heart contraction rate and in skin temperature by the end of the experiment. Myocardial hypoxia developed.[10]

Reproductive Toxicity.

Embryotoxicity. Female mice received a 10% MA aqueous solution for two months; rats were given a 1.0% MA aqueous solution for 6 months. The treatment caused an increase in neonatal mortality compared with the control.[8] Wistar rats were orally given 2.5 g MA/kg BW on days 6 to 15 of gestation. The treatment caused a decrease in fetuses BW, and produced skeletal anomalies.[11] Addition of 8.0 mg MA/ml produced embryotoxic effect in *in vitro* culture system of rat embryos.[12]

Teratogenicity. Rats received doses of 1.3, 2.6, or 5.2 ml/kg BW on day 10 of gestation. Malformations in fetuses were observed, including predominantly undescended testes and eye defects.[13] Wistar rats were given 2.5 g MA/kg BW on gestation days 6 to 15. The treatment led to a decrease in fetal BW and an increase in skeletal abnormalities.[11,14] MA caused craniofacial defects when administered to mice during gastrulation or early organogenesis. MA had no effect on somite number or total protein in embryo culture of C57BL/6J mice.[15]

Mutagenicity. MA produced no effect in *Salmonella* and *E. coli*,[16] or in micronucleus test.[17]

Chemobiokinetics. MA metabolism in the body occurs via oxidation in the direction of the *methanol/formaldehyde/formic acid/carbon dioxide.* Its oxidation and elimination from the body occurs slowly.[1]

A single administration of 2.0 g/kg BW led to development of metabolic adiposis and the accumulation of *formic acid* in the blood of male Macaco monkeys.[4] Acute toxicity in primates is attributed to the conversion of MA to *formate* and subsequent acidosis. Rodents neither accumulate formate nor develop acidosis after M. exposure.[18] High doses of ingested MA are retained in the body for up to 7 to 8 days. In rats, MA is removed unchanged and as carbon dioxide in the expired air. 3.0% MA is excreted in the urine, and further 3.0% as salts or esters of formic acid.[021]

Regulations.

U.S. FDA (1998) approved MA for safe use (1) as a solvent in the manufacture of cross-linked polyester resins which may be used as articles or components of articles intended for repeated use in contact with food in accordance with the conditions prescribed by 21 CFR part 177.2420; (2) in the manufacture of resinous and polymeric coatings for the food-contact surface of articles intended for use in producing, manufacturing, packing, processing, preparing, treating, packaging, transporting, or holding food in accordance with the conditions prescribed by 21 CFR part 175.300; (3) as a component of the uncoated or coated food-contact surface of paper and paperboard intended for use in producing, manufacturing, packaging, processing, preparing, treating, packing, transporting, or holding dry food in accordance with the conditions prescribed by 21 CFR part 176.180; (4) in the poly (2,6-dimethyl-1,4-phenylene)oxide resins as a component of an article intended for use in contact with food (not to exceed 0.02% as residual solvent); (5) in polyurethane resins for use in contact with dry food in accordance with the conditions prescribed by 21 CFR part 177.1680; (6) as a component of adhesives for food-contact surface in accordance with the conditions prescribed by 21 CFR part 175.105; (7) as a component of a defoaming agent intended for articles which may be used in producing, manufacturing, packing, processing, preparing, treating, packaging, transporting, or holding food in accordance with the conditions prescribed in 21 CFR part 176.200; (8) as a defoaming agent in the manufacture of paper and paperboard intended for use in packaging, transporting, or holding food in accordance with the conditions prescribed by 21 CFR part 176.210; (9) as a substance employed in the production of or added to textiles and textile fibers intended for use in contact with food in accordance with the conditions prescribed by 21 CFR part 177.2800; and (10) in cross-linked polyester resins to be safely used as articles or components of articles intended for repeated use in contact with food in accordance with the conditions prescribed by 21 CFR part 177.2420. MA may be present in food under the conditions prescribed in 21 CFR part 173.250.

EU (1990). MA is available in the *List of authorized monomers and other starting substances which shall be used for the manufacture of plastic materials and articles intended to come into contact with foodstuffs (Section A).*

Great Britain (1998). MA is authorized without time limit for use in the production of polymeric materials and articles in contact with food or drink or intended for such contact.

Recommendations. Joint FAO/WHO Expert Committee on Food Additives. ADI is not specified.

Standards. ***Russia*** (1995). MAC: 3.0 mg/l, PML in food: 0.2 mg/l.

References:

1. Nazarenko, I. V., in *Protection of Water Reservoirs against Pollution by Industrial Liquid Effluents*, S. N. Cherkinsky, Ed., Medgiz, Moscow, Issue No. 3, 1959, 209 (in Russian).
2. Trifonov, Yu. A., Tutdyev, A. A., Tiunov, L. A., et al., Correlation of seasonal death of methanol-intoxicated rats with energetic activity of liver mitochondria, *Gig. Sanit.*, 8, 82, 1987 (in Russian).
3. Kimura, E. T., Ebert, D. M., and Dodge, P. W., Acute toxicity and limits of solvent residue for sixteen organic solvents, *Toxicol. Appl. Pharmacol.,* 19, 699, 1971.
4. Martin-Amat, J., Tephly, T. R., McMartin, K. E., et al., Methyl alcohol poisoning. II Development of a model for ocular toxicity of methyl alcohol poisoning using the Rhesus monkey, *Arch. Ophthalmol*., 95, 1847, 1977.
5. Filimonov, V. M., Kalinin, L. V., and Zlobin, A. P., in *Proc 2 All-Union Congr. Forensic Medicine*, Irkutsk-Moscow, 1987, 220 (in Russian).
6. Dorman, D. C., Dye, J. A., Nassise, M. P., et al., Acute methanol toxicity in minipigs, *Fundam. Appl. Toxicol.,* 20, 341, 1993.
7. Skirko, B. K., Ivanitsky, A. M., Pilenitsina, R. A., et al., Study on the toxic action of methanol in experiments with complete and protein deficient nutrition, *Voprosy Pitania*, 5, 70, 1976 (in Russian).
8. Guseva, V. A., in *Proc. 12 Sci.-Pract. Conf. Junior Hygienists and Sanitarians*, Moscow, 1969, 33 (in Russian).
9. *Rat Oral Subchronic Toxicity Study with Methanol*, U.S. EPA, Office of Solid Waste, Washington, D.C., 1986.
10. Rudnev, M. I. and Nozdrachev, S. I., Effect of methanol on some body functions, *Vrachebnoye Delo*, 6, 125, 1975 (in Russian).
11. De-Carvalho, R. R., Delgado, I. F., Souza, S. A. M., Chahoud, I., and Paumgartten, F. J. R., Embryotoxicity of methanol in well-nourished and malnourished rats, *Braz. J. Med. Biol. Res*., 27, 2915, 1994.
12. Andrews, J. E., Ebron-McCoy, M., and Rogers, J. M., Embryotoxic effects of methanol in whole embryo culture, Abstract, *Teratology*, 43, 461, 1991.
13. Youssef, A. F., Baggs, R. B., Weiss, B., and Miller, R. K., Methanol teratogenicity in pregnant Long Evans rats, Abstract, *Teratology*, 43, 467, 1991.
14. Yossef, A. F., Baggs, R. B., Weiss, B., and Miller, R. K., Teratogenicity of methanol following a single oral dose in Long-Evans rats, *Reprod. Toxicol*., 11, 503, 1997.
15. Degitz, S. J., Hunter, E. S., and Rogers, J. M., Developmental toxicity of methanol in C57BL/6J mice as assessed by whole embryo culture, *Teratology*, 55, Abstract *P7*, 1997.
16. De Flora, S., Zannacchi, P., Camoriano, A., Bennicelli, C., and Badolati, G. S., Genotoxic activity and potency of 135 compounds in the Ames reversion test and in bacterial DNA-repair test, *Mutat. Res.,* 133, 161, 1984.
17. Gocke, E., King, M.-T., Eckhardt, K., and Wild, D., Mutagenicity of cosmetic ingredients licensed by the EEC, *Mutat Res.,* 90, 91, 1981.
18. Andrews, J. E., Ebron-McCoy, M., Kavlock, R. J., and Rogers, J. M., Developmental toxicity of formate and formic acid in whole embryo culture: A comparative study with mouse and rat embryos, *Teratol ogy*, 51, 243, 1995.

2-METHOXYETHANOL

Molecular Formula. $C_3H_8O_2$

M = 76.11

CAS No 109-86-4

RTECS No KL5775000

Abbreviations. ME, MAA (2-methoxyacetic acid).

Synonyms and **Trade Names.** Ethylene glycol, monomethyl ether; Methyl cellosolve; Methyl glycol; Methyl oxitol; Monomethyl ethylene glycol ether; Monomethyl glycol.

Properties. Colorless, transparent, volatile liquid with a mild non-residual odor. Soluble in water and ethanol. Does not affect transparency and color of water; does not form foam and film on the water surface. Odor perception threshold is 14.3 mg/l or <0.09 mg/l.[010] According to other data, organoleptic threshold is 100 mg/l.[01]

Applications. ME has wide industrial and consumer applications. It is used as a solvent in the production of nitro- and acetyl cellulose, natural and artificial resins. Used in the manufacture of protective epoxy resin coatings for metals. A surfactant.

Acute Toxicity. LD_{50} is found to be 2.46 to 3.25 g/kg BW in male rats, 3.4 g/kg BW in female rats, 0.89 to 1.425 g/kg BW in rabbits, and 0.95 g/kg BW in guinea pigs.[1,2] High doses caused anuria. Death occurred in rabbits in 2 to 4 days after poisoning.

Observations in man. ME affects hematology parameters and CNS functions.

Repeated Exposure. Gross pathology examination in rats exposed to 100 or 150 mg/kg BW for three weeks revealed reduced weights of the visceral organs, normochromic, normocytic anemia, hemorrhagic changes in the bone marrow and ovarian atrophy.[3] Inflammatory changes were observed in the bladder mucosa.[4]

Decrease in liver weights was observed in Wistar rats administered 300 mg ME/kg BW for 20 days. The activity of cytosolic ADH was increased.[5]

In 2-week studies, F344 rats and $B6C3F_1$ mice received ME in their drinking water. Consumption values ranged from 100 to 400 mg/kg BW (rats) and 200 to 1300 mg/kg BW (mice). There were no effects on survival. Decrease in BW gain in rats was reported. The treatment caused thymic and testicular atrophy in males of both species.[6]

Short-term Toxicity. In 13-week studies, histology changes were observed in the testes, thymus, and hematopoietic tissues. There was progressive anemia, effect on the testes, spleen, and adrenal gland (in females only). The NOAEL in rats was not reached, since testicular degeneration in males and decreased thymus weights occurred at the lowest concentration administered. The NOAEL for testicular degeneration and increased hematopoiesis in the spleen was 2000 ppm in mice.[6]

Long-term Toxicity. Chronic exposure causes hemodynamic injuries and dystrophic changes in the brain, liver, kidneys, myocardium, etc. ME produced hematological disorders in both humans and experimental animas.[7] Medinsky et al. believe hematotoxic effects to be due to the MAA metabolite of ME.[8]

Immunotoxicity. ME has been shown to be immunosuppressive in rats but not in mice, with oxidation of ME to MAA being a prerequisite for immunosuppression. Oral and dermal exposure to ME revealed the ability of the immune system to mount an effective humoral immune response.[15]

MAA, the main metabolite of ME, is shown to cause immunodepressive effect when administered by gavage to young adult F344 rats. Thymic involution is reported.[16]

Repeated high dose oral exposure to MAA does not suppress humoral immunity in the mouse.[9]

Rats given doses greater than 100 mg ME/kg BW displayed significant thymic depression.[10] Dose-related increase in natural killer cells cytotoxic activities and decrease in specific antibody production in rats were reported.[11]

Mice, however, appeared to be insensitive to the immunosuppressive effects of ME and MAA, at the doses producing such effects in rats.[12,13] House did not observe changes in the immunological functions or host resistance in $B6C3F_1$ female mice dosed by gavage 10 times over 2 weeks with 25 to 100 mg ME or MAA per kg BW.[14] Nevertheless, thymic involution was revealed at 50 to 100 mg/kg BW doses.

F344 rats were gavaged with 25 to 200 mg ME/kg BW in distilled water for 4 consecutive days. The treatment caused a reduction in thymus weights at the doses of 50 to 200 mg/kg BW, while spleen weights were reduced in rats that were dosed at 200 mg/kg BW. The lymphoproliferative responses to phytohemagglutinin, pokeweed mitogen, and *Salmonella typhimirium* were increased at the 200 mg/kg BW.

Reproductive Toxicity. ME produces a dose-related embryotoxic and teratogenic effect in mice, hamsters and guinea pigs.

Embryotoxicity. ME is metabolized to the active compound MAA, which readily crosses the placenta and impairs fetal development. Pregnant mice were exposed to 100, 150, or 200 mg ME/kg BW from gestation days 10 to 17. The treatment caused a significant thymus atrophy and cellular depletion in fetal mice. ME inhibited thymocyte maturation. ME-induced immunodepression may result from targeting of multiple hematopoietic compartments.[7]

F344 rats were treated on days 6 to 15 of gestation by dosed feed or gavage (doses 12.5 to 100 mg/kg BW). This exposure caused only a small decrease in maternal BW gain during treatment. Litter weights and postnatal survival were decreased (100 mg/kg BW group); percentage of resorption was increased in 50 and 100 mg/kg BW dosed groups. The number of the live pups was decreased in 25 to 100 mg/kg BW dosed groups.[17]

Monkey were dosed with up to 55 mg ME/kg BW on gestation days 20 to 45. All pregnancies at the highest dose ended in death. One of the fetuses at the highest dose had a missing digit on each forelimb. The LOAEL was considered to be 12 mg/kg BW in this study.[18]

Teratogenicity. Time-mated CD-1 mice were orally dosed on gestation day 11 with distilled water (control) or ME at a dose of 304 mg/kg BW. There were no signs of treatment-related maternal toxicity, and intrauterine survival was unaffected by the treatments. There was no treatment-related pattern of gross external malformations other than paw defects. Paw defects were present in ME-treated litters (68.5% of fetuses). Hindpaw defects predominated over forepaw, and syndactyly was the most common malformation. The incidences of oligodactyly and short digits were also significantly increased. The similarity of malformations produced by methyl-substituted glycol ethers is proposed to be attributable to *in vivo* conversion to a common teratogen, *methoxyacetic acid.*[33]

Bifurcated or split cervical vertebrae were found in the offspring of female mice treated on days 7 to 14 of gestation. The LOAEL of 31 mg/kg BW was established.[19] The NOAEL for induction of malformations after a single administration on gestation day 11 was 100 mg/kg BW.[4]

Gonadotoxicity. ME causes testicular atrophy, degenerative changes in the germinal epithelium, pathological changes in the sperm head, and infertility.[20]

Rats and guinea pigs received a single oral dose of 200 mg/kg BW. The treatment induced spermatocyte degeneration in 24 and 96 hours after dosing, respectively.[21]

ME was found to deplete the spermatocytes of rats and mice which were given a single oral dose of 0.5 to 1.5 g/kg BW. This treatment produced morphologic abnormalities in rat spermatozoa that had been exposed as spermatocytes.[22]

In a 5-week study, male mice were exposed to oral doses of 0.5 to 4.0 g/kg BW. Male Dutch rabbits received the doses of 12.5, 25, 37.5, or 50 mg ME/kg BW in the drinking water for 12 weeks. The treatment caused oligospermia in the highest dose groups. No notable histopathology was found. A marked disruption in spermatogenesis increased above 25 mg ME/kg BW. The NOAEL of 12.5 mg/kg BW was established in this study.[23]

The NOAEL of 0.5 mg/kg BW for testicular atrophy was identified.[24] The ineffective dose for lesions and degeneration in primary spermatocytes and spermatids in male rats dosed for 11 days appeared to be 0.05 g/kg BW.[25]

Mutagenicity.

In vivo cytogenetics. $B6C3F_1$ mice received the doses of 35 to 2500 mg ME/kg BW during acute and subchronic oral exposure. The treatment did not cause induction of CA even after ME administration in cytotoxic doses.[26]

ME is shown to increase the rate of DLM in rats. It produced damage in mouse sperm head morphology following inhalation exposure. Equivocal results are observed in *Dr. melanogaster* (NTP-82).

In vitro genotoxicity. ME gave negative results in mutagenicity assays with *Salmonella typhymurium* strains *TA 98, TA 100,* and *TA 102* either with or without *S9* mix.[27,28] At high concentrations, it induced SCE in Chinese hamster ovary cells with and without *S9* fraction.[6] *Methoxyacetaldehyde*, a metabolite of ME, was mutagenic in a subline of Chinese hamster ovary cells.[29] It displayed mutagenic potency in *Salmonella typhimurium* strain *TA 97a* with or without *S9* mix at high concentrations.[28]

Chemobiokinetics.

Observations in man. MAA is found in the urine of 7 male volunteers exposed to 5.0 ppm ME[30] but it was not found in human urine in acute poisoning cases.

Animal studies. ME is metabolized *via* alcohol dehydrogenase to *methoxyacetaldehyde* and via aldehyde dehydrogenase to MAA, which seems to be a major oxidative metabolite. The urine appears to be a major route of excretion.[16,31]

According to Medinsky et al., ethylene glycol, a metabolite of ME, is a result of dealkylation of the ether occurred prior to oxidation to MAA. *Ethylene glycol* was excreted in urine, representing approximately 21% of the dose administered.[8] Biotransformation of ME has also been detected in testes from Wistar rats and one strain of mice, but not in testes from hamsters, guinea pigs, rabbits, dogs, cats, or humans. Testes from all these species readily converted the aldehyde metabolite of ME to MAA.[32]

ME is removed mainly in the urine, it does not accumulate in the testes.[25] Excretion also occurs via exhalation as CO_2. Less than 5.0% ME was exhaled unchanged.

Regulations. ***U.S. FDA*** (1998) approved the use of ME in adhesives as a component of articles intended for use in packaging, transporting, or holding food in accordance with the conditions prescribed by 21 CFR part 175.105.

Standards. ***Russia*** (1990). PML: 0.6 mg/l.

References:

1. Yatsina, O. V., Plaksienko, N. F., Pys'ko, G. T., et al., Hygienic regulation of cellosolves in water bodies, *Gig. Sanit.*, 10, 78, 1988 (in Russian).
2. *Occupational Exposure to Ethylene Glycol Monomethyl Ether and Ethylene Glycol Monoethyl Ether and Their Acetates,* Criteria for a Recommended Standard, U.S. Dept. of Health & Human Services, NIOSH, 1991, 296.
3. Grant, D., Sulsh, S., Jones, H. B., et al., Acute toxicity and recovery in the hemopoietic system of rats after treatment with ethylene glycol monomethyl and monobutyl ethers, *Toxicol. Appl. Pharmacol.*, 77, 187, 1985.
4. Horton, V. L., Sleet, R. B., John-Greene, J. A., et al., Developmental phase-specific effects of ethylene glycol monomethyl ether in CD-1 mice, *Toxicol. Appl. Pharmacol.*, 80, 108, 1985.
5. Kawamoto, T., Matsuno, K., Kayama, F., Hirai, M., Arashidani, K., Yoshikawa, M., and Kodama, Y., Effect of ethylene glycol monomethyl ether and diethylene glycol monomethyl ether on hepatic metabolizing enzymes, *Toxicology*, 62, 265, 1990.
6. *Toxicity Studies of Ethylene Glycol Ethers: 2-Methoxyethanol, 2-Ethoxyethanol, 2-Butoxyethanol Administered in Drinking Water to F344/N Rats and B6C3F$_1$ Mice,* NTP Technical Report Series No 26, Research Triangle Park, NC, July 1993.
7. Holliday, S. D., Comment, C. E., Kwon, J., and Luster, M. I., Fetal hematopoietic alterations after maternal exposure to ethylene glycol monomethyl ether: prolymphoid cell targeting, *Toxicol. Appl. Pharmacol.*, 129, 53, 1994.
8. Medinsky, M. A., Singh, G., Bechtold, W. E., Bond, J. A., Sabourin, P. J., Birnbaum, L. S., and Henderson, R. F., Disposition of three glycol ethers administered in drinking water to male F344/N rats, *Toxicol. Appl. Pharmacol.*, 102, 443, 1990.
9. Riddle, M. M., Williams, W. C., and Smialowicz, R. J., Repeated high dose oral exposure or continuous subcutaneous infusion of 2-methoxyacetic acid does not suppress humoral immunity in the mouse, *Toxicology*, 109, 67, 1996.
10. Henningsen, G. H., Sendelbach, L. E., Braun, A. G., et al., *Soc. Toxicol., 228 Annual Meeting*, Abstracts, Atlanta, GA, 1989.
11. Exon, J. H., Hather, G. G., Bussiere, J.L., Olson, D. P, and Talcott, P. A., Effect of subchronic exposure of rats to 2-methoxyethanol or 2-butoxyethanol: thymic atrophy and immunotoxicity, *Fundam. Appl. Toxicol.*, 16, 830, 1991.
12. Smialowicz, R. J., Williams, W. C., Riddle, M. M., Andrews, D. L., Luebke, R. W., and Copeland, C. B., Differences between rats and mice in the immunosuppressive activity of 2-methoxyethanol and 2-methoxyacetic acid, *Toxicology*, 74, 57, 1992.

13. Smialowicz, R. J., Riddle, M. M., and Williams, W. C., Species and strain comparisons of immunosuppression by 2-methoxyethanol and 2-methoxyacetic acid, *Int. J. Immunopharmacol.*, 16, 695, 1994.
14. House, R. V., Lauer, L. D., Murray, M. J., et al., Immunological studies in B6C3F_1 mice following exposure to ethylene glycol monomethyl ether and its principal metabolite methoxyacetic acid, *Toxicol. Appl. Pharmacol.*, 77, 358, 1985.
15. Williams, W. C., Riddle, M. M., Copeland, C. B., Andrews, D. L., and Smialowicz, R. J., Immunological effects of 2-methoxyethanol administered dermally or orally to Fischer 344 rats, *Toxicology*, 98, 215, 1995.
16. Smialowicz, R. J., Riddle, M. M., Lueb, R. W., et al., Immunotoxicity of 2-methoxyethanol following oral administration in F344 rats, *Toxicol. Appl. Pharma col.*, 109, 494, 1991.
17. Morrissey, R. E., Harris, M. W., and Schwetz, B. A., Developmental toxicity screen: results of rat studies with diethylhexyl phthalate and ethylene glycol monomethyl ether, *Teratogen. Carcinogen. Mutagen.*, 9, 119, 1989.
18. Scott, W. J., Fradkin, R., Wittfoht, W., et al., Teratologic potential of 2-ethoxyethanol and transplacental distribution of its metabolite, 2-methoxyacetic acid, in non-human primates, *Teratology*, 39, 363, 1989.
19. Nagano, K., Nakayama, E., Oobayashi, H., et al., Embryotoxic effects of ethylene glycol monomethyl ether in mice, *Toxicology*, 20, 335, 1981.
20. Hardin, B. D. and Lyon, J. P., Summary and overview: NIOSH symposium on toxic effects of glycol ethers, *Environ. Health Perspect.*, 57, 273, 1984.
21. Ku, W. W., Wine, R. N., Chae, B. Y., Ghanayem, B. I., and Chapin, R. E., Spermatocyte toxicity of 2-methoxyethanol in rats and guinea pigs: evidence for the induction of apoptosis, *Toxicol. Appl. Pharmacol.*, 134, 100, 1995.
22. Anderson, D., Brinkworth, M. H., Jenkinson, P. C., et al., Effect of ethylene glycol monomethyl ether on spermatogenesis, dominant lethality, and F_1 abnormalities in the rat and mouse after treatment of F_0 males, *Teratogen. Carcinogen. Mutagen.*, 7, 141, 1987.
23. Foote, R. H., Farrel, P. B., Schlafer, D. H., McArdle, M. M., Trouern-Trend, V., Simkin, M. E., Brockett, C. C., Giles, J. R., and Li, J., Ethylene glycol monomethyl ether effects on health and reproduction in male rabbits, *Reprod. Toxicol.*, 9, 527, 1995.
24. Nagano, K., Nakayama, E., Koyano, M., et al., Mouse testicular atrophy induced by ethylene glycol monoalkyl ethers, *Jap. J. Ind. Health*, 21, 29, 1979.
25. Foster, P. M., Creasy, D. M., Foster, J. R., et al., Testicular toxicity of ethyleneglycol monomethyl and monoethyl ethers in rat, *Toxicol. Appl. Pharmacol.*, 69, 385, 1983.
26. Au, W. W., Morris, D. L., and Legator, M. S., Evaluation of the clastogenic effects of 2-methoxyethanol in mice, *Mutat. Res.*, 300, 273, 1993.
27. McGregor, D. B., Genotoxicity of glycol ethers, *Environ. Health Perspect.*, 57, 97, 1984.
28. Hoflack, J. C., Lambolez, L., Elias, Z., and Vasseur, P., Mutagenicity of ethylene glycol ethers and of their metabolites in *Salmonella typhimurium, Mutat. Res.*, 341, 281, 1995.
29. Chiewchanwit, T. and Au, W. W., Mutagenicity and cytotoxicity of 2-buthoxyethanol and its metabolite, 2-butoxyacetaldehyde, in Chinese hamster ovary (CHO-AS52) cells, *Mutat. Res.*, 334, 341, 1995.
30. Groeseneken, D., Veulemans, H., Masschelein, R., et al., Experimental human exposure to ethylene glycol monomethyl ether, *Int. Arch. Occup. Environ, Health*, 61, 243, 1989.
31. Miller, R. R., Metabolism and disposition of glycol ethers, *Drug Metab. Rev.*, 18, 1, 1987.
32. Moslen, M. T., Kaphalia, L., Balasubramanian, H., Yin, Y. M., and Au, W. W., Species differences in testicular and hepatic biotransformation of 2-metoxyethanol, *Toxicology*, 96, 217, 1995.
33. Hardin, B. D. and Eisenmann, C. J., Relative potency of four ethylene glycol ethers for induction of paw malformations in the CD-1 mouse, *Teratology*, 35, 321, 1987.

2-METHOXYETHANOL ACETATE

Molecular Formula. $C_5H_{10}O_3$

M = 118.15
CAS No 110-49-6
RTECS No KL5950000
Abbreviation. MEA.

Synonyms. Ethylene glycol, monomethyl ether, acetate; Methylcellosolve acetate.

Properties. MEA occurs as a colorless liquid having a mild, ether-like odor. Miscible with water, oil, and a number of organic solvents. Odor threshold concentration is 0.34 mg/l.[010]

Applications. Used as an intermediate and a solvent in the production of plastics.

Acute Toxicity. LD_{50} is 3.93 g/kg BW in rats, and 1.25 g/kg BW in guinea pigs.[1]

See also ***1,2-Diethoxyethane.***

Reproductive Toxicity.

Gonadotoxicity. Oral administration of MEA to mice produced testicular toxicity: decreased testes weights and testicular seminiferous tubule atrophy.[2]

Embryotoxicity. In a short-term study, MEA affected fetal development.[3]

Chemobiokinetics. MEA undergoes hydrolysis in the body because it gives rise to toxic effects similar to those caused by ethylene glycol methyl ether. Some *acetaldehyde* may also be formed.[03]

References:

1. *Occupational Exposure to Ethylene Glycol Monomethyl Ether and Ethylene Glycol Monoethyl Ether and Their Acetates,* Criteria for a Recommended Standard, U.S. Dept. Health & Human Services, NIOSH, 1991, 39.
2. Nagano, K., Nakayama, E., Koyano, M., et al., Mouse testicular atrophy induced by ethylene glycol monoalkyl ethers, *Jap. J. Ind. Health,* 21, 29, 1979.
3. Shuler, R. L., Hardin, B. N., Niemeier, R. W., et al., Results of testing 15 glycol ethers in a short-term *in vivo* reproductive toxicity assay, *Environ. Health Perspect.,* 52, 141, 1984.

2-(2-METHOXYETHOXY)ETHANOL

Molecular Formula. $C_5H_{12}O_3$
M = 120.15
CAS No 111-77-3
RTECS No KL6125000
Abbreviation. MEE.

Synonyms and **Trade Names.** Diethylene glycol, monomethyl ether; 3,6-Dioxo-1-heptanol; Diglycol monomethyl ether; Dowanol; 2,2'-Oxybisethanol monomethyl ether; Methoxydiglycol; Methyl carbitol; Methyl dioxytol.

Properties. Colorless liquid with a mild pleasant odor and bitter taste. Miscible with water; soluble in alcohol.

Applications. Solvent for dyes, nitrocellulose and resins.

Acute Toxicity. High doses caused mortality through CNS depression or kidney injury. LD_{50} is 9.21 g/kg BW in rats.[020]

Repeated Exposure. Rats received drinking water containing 3.0 to 5.0% MEE for 11 to 64 days. Only one rat died by the end of treatment.[1]

Wistar rats received 0.5, 1.0, and 2.0 g/kg BW for 20 days. The treatment at the highest dose decreased liver weights and increased hepatic microsomal protein contents, induced cytochrome *P-450,* but not cytochrome *b5* or the reduced form of nicotinamide adenine dinucleotide phosphate cytochrome-*c*-reductase. The activity of cytosolic ADH was not affected.[1]

Wistar rats were exposed to MEE at doses up to 4.0 g/kg BW for 11 consecutive days. The treatment decreased relative weights of the thymus and pituitary gland, erythrocyte and leukocyte counts, *Hb* concentration and hematocrit levels.[2]

Long-term Toxicity. Treatment with 1.83 g/kg BW did not cause death in rats.[1] Gross pathology examination revealed hydropic degeneration of the convoluted tubules in the kidneys comparable to those caused by diethylene glycol and carbitol.[3]

Reproductive Toxicity.

Embryotoxicity. Pregnant rats were given doses of more than 3.0 g MEE/kg BW (administered over gestation days 7 to 17). The treatment caused total resorption of all litters. Rats received doses of 0.2, 0.6, or 1.8 g/kg BW from day 7 to 17 of gestation. Treatment with 1.8 g/kg BW caused significant adverse effects on postnatal development and high mortality.[2]

Teratogenicity. Time-mated Sprague-Dawley female rats received MEE on days 7 to 16 of gestation in distilled water. Resorption of litters occurred at doses of 3.3 or 5.2 g/kg BW. Visceral and skeletal malformations were dose-related.[4]

In the above-cited study, at the dose level of 1.8 g MEE/kg BW, visceral malformations of the cardiovascular system were noted in 28% of the fetuses. The dose of 200 mg MEE/kg BW was found to be ineffective in this study.[2]

Gonadotoxicity. Rats received up to 20 daily doses of 5.1 mmol MEE/kg BW. The treatment caused no gross or histopathological testicular lesions.[5]

Regulations. ***U.S. FDA*** (1998) approved the use of MEE in adhesives as a component of articles intended for use in packaging, transporting, or holding food in accordance with the conditions prescribed in 21 CFR part 175.105.

Standards. ***Russia*** (1995). MAC and PML: 1.0 mg/l (organolept., foam).

References:

1. Kawamoto, T., Matsuno, K., Kayama, F., Hirai, M., Arashidani, K., Yoshikawa, M., and Kodama, Y., Effect of ethylene glycol monomethyl ether and diethylene glycol monomethyl ether on hepatic metabolizing enzymes, *Toxicology*, 62, 265, 1990.
2. Yamano, T., Noda, T., Shimizu, M., Morita, S., and Nagahama, M., Effects of diethylene glycol monomethyl ether on pregnancy and postnatal development in rats, *Arch. Environ. Contam. Toxicol.*, 24, 228, 1993.
3. Browning, E., *Toxicity and Metabolism of Industrial Solvents*, American Elsevier, New York, 1965, 633.
4. Hardin, B. D., Goad, P. T., and Burg, J. R., Developmental toxicity of diethylene glycol monomethyl ether (diEGME), *Fundam. Appl. Toxicol.*, 6, 430, 1986.
5. Cheever, K. L., Richards, D. E., Weigel, W. W., Lal, Y. B., Dinsmore, A. M., and Daniel, F. B., Metabolism of bis(2-methoxyethyl) ether in the adult male rat: evaluation of the principal metabolite as testicular toxicant, *Toxicol. Appl. Pharmacol.*, 94, 150, 1988.

2-[2-(2-METHOXYETHOXY)ETHOXY]ETHANOL

Molecular Formula. $C_7H_{16}O_4$

M = 164.23

CAS No 112-35-6

RTECS No KL6390000

Synonyms and **Trade Names.** Dowanol TMAT; Methoxytriglycol; Triethyleneglycol, monomethyl ether; Triglycol monomethyl ether; 3,6,9-Trioxa-1-decanol.

Properties. Colorless liquid. Soluble in water.

Applications. A solvent and plasticizer intermediate.

Acute Toxicity. LD_{50} exceeded 10 g/kg BW in rats.[1]

Repeated Exposure. Rats were given doses of 0.25 to 1.5 g/kg BW by gavage for 13 days. The LOEL of 1.5 g/kg BW was found for lethality, the NOEL of 1.0 g/kg BW was established in this study (Krasawage et al., 1992).

Short-term Toxicity. Liver and testes appeared to be the target organs of M. toxicity (Gill et al., 1998).

Mutagenicity.

In vivo cytogenetics. M. produced developmental toxicity in *Dr. melanogaster.*

Regulations. ***U.S. FDA*** (1993) approved the use of M. in adhesives as a component of articles intended for use in packaging, transporting, or holding food in accordance with the conditions prescribed by 21 CFR part 175.105.

Reference:

1. Lynch, D. W., Developmental toxicity of triethylene glycol monomethyl ether (TEGMME) in intact *Dr. melanogaster, Teratology*, 55, P54, 1997.

(2-METHOXYMETHYLETHOXY)PROPANOL

Molecular Formula. $C_7H_{16}O_3$
M = 148.23
CAS No 34590-94-8
12002-25-4
112388-78-0
RTECS No JM1575000
Abbreviation. MMEP.

Synonyms and **Trade Names.** 1,4-Dimethyl-3,6-dioxa-1-heptanol; Dipropylene glycol, monomethyl ether; Dowanol DPM; 1-(2-Methoxyisopropoxy)-2-propanol.

Properties. Colorless liquid with a bitter taste and a mild ether-like odor (in moderate concentrations). Miscible with water and many organic solvents.

Applications. A solvent for nitrocellulose and other synthetic resins. Used in the production of water-based surface coatings.

Acute Toxicity. LD_{50} is 5.14 g/kg BW in rats.[014] Poisoning led to behavioral changes. LD_{50} is 7.5 g/kg BW in dogs. Signs of intoxication included respiratory system injury.[1]

Repeated Exposure. No ill-effects were noted in rats given 1.0 g MMEP/kg BW dose for 35 days.[2]

Allergenic Effect was not found when tested in humans.[011]

Regulations. ***U.S. FDA*** (1998) approved the use of MMEP (1) in adhesives as a component of articles intended for use in packaging, transporting, or holding food in accordance with the conditions prescribed by 21 CFR part 175.105; and (2) as a component of the uncoated or coated food-contact surface of paper and paperboard intended for use in producing, manufacturing, packaging, processing, preparing, treating, packing, transporting, or holding dry foods in accordance with the conditions prescribed by 21 CFR part 176.180.

References:

1. Shideman, F. E. and Procita, L., The pharmacology of the monomethyl ethers of mono-, di-, and tripropylene glycol in the dog with observations on the auricular fibrillitaions produced by these compounds, *J. Pharmacol. Exp. Ther.*, 102, 79, 1951.
2. Browning, E., *Toxicity and Metabolism of Industrial Solvents*, American Elsevier, New York, 1965, 659.

1-METHOXY-2-PROPANOL

Molecular Formula. $C_4H_{10}O_2$
M = 90.14
CAS No 107-98-2
RTECS No UB770000
Abbreviation. MP.

Synonyms and **Trade Names.** Dowanol PM; 1-Methoxy-2-hydroxypropane; 2-Methoxy-1-methylethanol; Propasol solvent M; Propylene glycol, methoxy ether; Propylene glycol, monomethyl ether.

Properties. Colorless liquid with a mild, ethereal odor and bitter taste. Soluble in water in all proportions;[011] soluble in methanol.

Applications. Solvent for cellulose and acrylic polymers. Intermediate in the manufacture of lacquers and paints; used in solvent-sealing of cellophane.

Acute Toxicity. LD_{50} is 5.0 g/kg BW in dogs, 5.7 g/kg BW in rats and rabbits,[1,2] and 11.7 g/kg BW in mice. Single administration of the lethal doses caused death by generalized CNS inhibition, probably from respiratory arrest.[3] Large doses caused kidney tubular necrosis to develop.[022]

Repeated Exposure. MP is a CNS and cardiac depressant of low toxicity; it produced arrhythmia and fibrillation.[4]

Rats received doses of 3.0 g/kg BW for 35 days. Histological examination revealed only mild changes in liver and kidneys.[3]

Short-term Toxicity. Rats received 0.5 to 4.0 ml MP/kg BW for 13 weeks. The treatment caused retardation of BW gain, CNS depression, liver enlargement and necrosis, and increase in mortality (at 4.0 ml/kg BW dose level).

Dogs were given 0.5 to 3.0 ml MP/kg BW for 14 weeks. Poisoning affected spermatogenesis and produced kidney injury.[03]

Reproductive Toxicity.

Embryotoxicity. Mice, rats, and rabbits received 0.04 to 2.0 ml MP/kg BW during first 18 to 21 days of gestation. Effect was observed only in rat fetuses: there was a delay in ossification of skull in highest dose (0.8 ml/kg BW).[03]

Teratogenicity. MP is reported to have relatively low acute maternal toxicity and to cause no *terata*.[5,6]

Gonadotoxic effect was not observed in male rats and rabbits exposed to MP.[7]

Mutagenicity.

In vitro genotoxicity. MP was not mutagenic in *Salmonella typhimurium* strains *TA1535, TA100, TA98, TA1537*, or *TA1538* either in the presence or absence of a rat liver microsomal *S9* activation system. There were no evidence for clastogenesis or induction of unscheduled DNA synthesis in Chinese hamster ovary cells or in cultured rat primary hepatocytes.[6]

Chemobiokinetics. Orally administered MP exhibits rapid absorption and dose-dependent metabolism and elimination in rats. Male F344 rats were given *i/g* 90 or 780 mg ^{14}C-MP/kg: 9.0% of the administered dose was retained by the body at 48 hours after exposure, 10 to 20% appeared in the urine, and expired CO_2 accounted for 63% of the dose (ACGIH, 1991).

MP is metabolized to innocuous conjugates and not to active testicular degenerative metabolites.[8] After single oral dose, MP is metabolized to *propylene glycol* presumably by cytochrome *P450*-dependent O-demethylation. Propylene glycol is further metabolized to ^{14}C-CO_2.[9,10]

Regulations. ***U.S. FDA*** (1998) approved the use of MP (1) in adhesives as a component of articles intended for use in packaging, transporting, or holding food in accordance with the conditions prescribed by 21 CFR part 175.105; and (2) as a component of the uncoated or coated food-contact surface of paper and paperboard intended for use in producing, manufacturing, packaging, processing, preparing, treating, packing, transporting, or holding dry foods in accordance with the conditions prescribed by 21 CFR part 176.180.

References:

1. Smyth, H. F., Carpenter, C. P., Weil, C. S., Pozzani, U. C., and Striegel, J. A., Range-finding toxicity data, list VI, *Am. Ind. Hyg. Assoc. J.*, 25, 95, March-April 1962.
2. Stenger, E. G, Aeppli, L., Machemer, L., Muller, D., and Trokan, J., Studies on the toxicity of propylene glycol monomethyl ether, *Arzneimittel-Forsch.*, 22, 569, 1972.
3. *Encyclopedia of Occupational Health and Safety*, vols 1 and 2, Int. Labour Office, McGraw-Hill Book Co., New York, 1971, 620.
4. Shiedeman, F. E. and Procita, L., The pharmacology of the monomethyl ethers of mono-, di- and tripropylene glycol in the dog with observations on the auricular fibrillation produced by these compounds, *J. Pharmacol. Exp. Ther.*, 102, 79, 1951.
5. Schuler, R. L., Hardin, B. D., Niemeier, R. W., Booth, G., Hazelden, K., Piccirillo, V., and Smith, K., Results of testing fifteen glycol ethers in short-term *in vivo* reproductive toxicity assay, *Environ. Health Perspect.*, 57, 141, 1984.
6. *American Conference of Governmental Industrial Hygienists, Inc.*, Documentation of the Threshold Limit Values and Biological Exposure Indices, 6th ed., Cincinnati, OH, 1991, 1311.
7. Landry, T. D. et al., Propylene glycol monomethyl ether: a 13-week inhalation toxicity study in rats and rabbits, *Fundam. Appl. Toxicol.*, 3, 627, 1983.
8. Morgott, D. A. and Nolan, R. J., Nonlinear kinetics of inhaled propylene glycol monomethyl ether in Fischer 344 rats following single and repeated exposures, *Toxicol. Appl. Pharmacol.*, 89, 19, 1987.

9. Miller, R. R., Herrman, J. T., Young, L. L., Landry, T. D., and Calhoun, J. T., Ethylene glycol monomethyl ether and propylene glycol monomethyl ether: Metabolism, disposition and subchronic toxicity studies, *Environ. Health Perspect.*, 57, 233, 1984.
10. Cassaret and Doull's *Toxicology,* M. O. Amdur, J. Doull, and C. D. Klaasen, Eds., 4th ed. Pergamon Press, New York, 1991, 706.

3-[3-(3-METHOXYPROPOXY)PROPOXY]PROPANOL

Molecular Formula. $C_{10}H_{22}O_4$
M = 206.32
CAS No 25498-49-1
RTECS No UB8070000
Abbreviation. MPPP.

Synonyms and **Trade Name.** Dowanol TPM; Glycol ether TPM; Poly-Solv TPM; [2-(2-Methoxy-methylethoxy)methylethoxy] propanol; Tripropylene glycol, monomethyl ester.

Acute Toxicity. LD_{50} is 3.2 g/kg BW in rats. The liver appeared to be the target organ.[1] Administration of 5.0 g/kg BW affected respiratory systems of dogs.[2]

Regulations. ***U.S. FDA*** (1998) approved the use of MPPP (1) in adhesives as a component of articles intended for use in packaging, transporting, or holding food in accordance with the conditions prescribed by 21 CFR part 175.105, and (2) as a component of the uncoated or coated food-contact surface of paper and paperboard intended for use in producing, manufacturing, packaging, processing, preparing, treating, packing, transporting, or holding dry food in accordance with the conditions prescribed by 21 CFR part 176.180.

References:

1. *Arch. Ind. Hyg. Occup. Med.*, 9, 509, 1954.
2. Shideman, F. E. and Procita, L., The pharmacology of the mono methyl ethers of mono-, di-, and tripropylene glycol in the dog with observations on the auricular fibrillitaions produced by these compounds, *J. Pharmacol. Exp. Ther.*, 102, 79, 1951.

2-METHYLBUTANE

Molecular Formula. C_5H_{12}
M = 72.15
CAS No 78-78-4
68923-44-4
92046-46-3
RTECS No EK4430000
Abbreviation. MB.

Synonyms. Ethyldimethylmethane; Isoamylhydride; Isopentane.

Properties. A volatile, colorless liquid with a pleasant odor.[012] Insoluble in water; soluble in oils.

Applications. A solvent. A chemical intermediate and a blowing agent.

Exposure. MB is a component of natural gas. It is contained in liquefied petroleum gas and natural gas as trace constituents.

Repeated Exposure. Male F344 rats were gavaged with twenty doses of 1.0 g MB/kg BW for 4 weeks. The treatment caused weight loss or decreased weight gain and an increase in mortality rate.[1]

Mutagenicity. MB was negative in *Salmonella* mutagenicity assay.[2]

Chemobiokinetics. In rats, mice, rabbits, and guinea pigs, MB is metabolized by hydroxylation and converted into *1-methyl-2-butanol* as the major metabolite and *3-methyl-2-butanol*, *2-methyl-1-butanol*, and *3-methyl-1-butanol* as minor metabolites in the blood and liver tissue. *In vivo* metabolism has not been described.[3]

ICR mice were exposed to about 5.0% I. for 1 hour. It was suggested that MB was metabolized predominantly by liver microsomes.

Regulations. ***U.S. FDA*** (1998) approved the use of MB as an adjuvant that may be used in the manufacture of foamed plastics intended for use in contact with food in accordance with the conditions prescribed by 21 CFR part 178.3010.

References:

1. Halder, C. A., Holdsworth, C. E., Cockrell, B. Y., et al., Hydrocarbon nephropathy in male rats: identification of the nephrotoxic components of unleaded gasoline, *Toxicol. Ind. Health.* 1, 67, 1985.
2. Kirwin, C. J. and Thomas, W. C., *In vitro* microbiological mutagenicity studies in hydrocarbon propellants, *J. Soc. Cosmet. Chem.*, 31, 367, 1980.
3. Chiba, S. and Oshida, S., Metabolism and toxicity of *n*-pentane and isopentane, Abstract, *Nippon Hoigaku Zasshi*, 45, 128, 1991 (in Japanese).

3-METHYL-2-BUTANONE

Molecular Formula. $C_5H_{10}O$
M = 86.15
CAS No 563-80-4
RTECS No EL9100000
Abbreviation. MB.

Synonyms. 2-Acetyl propane; Isopropyl methyl ketone; Methyl isopropyl ketone.

Properties. Transparent liquid with a specific odor. Solubility in water is 60 mg/; soluble in organic solvents. Odor perception threshold is 3.1mg/l.[02]

Applications. Used in the production of synthetic rubber.

Acute Toxicity. In rats, LD_{50} is 3.65 g/kg BW (males) and 3.0 g/kg BW (females). In mice, LD_{50} is reported to be 3.0 g/kg BW. However, according to other data, LD_{50} in rats is only 148 mg/kg BW.[06] Following administration of MB, animals developed adynamia. Death occurred in 24 hours. Congestion and hemorrhages in the lungs, liver, and kidneys were found on the gross pathology examination.

Repeated Exposure failed to reveal evident cumulative properties. K_{acc} is 8.9 (by Lim). MB caused damage to the CNS and liver.

Reference:

Buchalovsky, A. A., Toxicity of methyl isopropyl ketone, *Gig. Sanit.*, 12, 94, 1988 (in Russian).

METHYL CYCLOHEXYL KETONE

Molecular Formula. $C_8H_{14}O$
M = 126.19

Synonyms. Acetocyclohexane; Hexahydrobenzophenone.

Properties. Colorless liquid with a specific odor. Readily soluble in water.

Applications. Used as a solvent in the production of vinyl resins, nitrocellulose, and synthetic rubber.

Acute Toxicity. LD_{50} is 2.5 g/kg BW in rats, and 3.6 g/kg BW in mice.

See ***2-Octanone.***

N-METHYLFORMAMIDE

Molecular Formula. C_2H_4NO
M = 59.07
CAS No 123-39-7
RTECS No LQ3000000
Abbreviation. MF.

Synonyms. Formic acid, *N*-methylamide; Monomethylformamide.

Properties. Liquid. Soluble in water and alcohol.

Applications. Used as a solvent in the production of plastics and synthetic fibers.

Acute Toxicity.

Observations in man. Ingestion of 50 ml by man results in acute poisoning. Manifestations of the toxic effect include liver damage, jaundice, and dyspepsia.[1]

Animal studies. LD_{50} is reported to be 1.58 to 2.6 g/kg BW in mice, and 4.0 g/kg BW in rats.

Reproductive Toxicity.

Teratogenic effect is found in rats, mice, and rabbits. Defects observed were presented by cephalocele, gastroschisis, and skeletal abnormalities.[2-5]

Embryotoxicity. The dose of 100 mg/kg BW administered on day 7 of pregnancy caused 90% mortality. Severe maternal toxicity and subsequent embryolethality were noted in pregnant rats and rabbits exposed to dermal application of 1.0 to 1.5 g MF/kg BW during organogenesis.[6]

Carcinogenicity. A 400 mg MF/kg BW dose caused some animals to die and initiated development of malignant neoplasms in surviving progeny.

References:

1. Vasil'eva, V. N. and Sukharevskaya, G. M., Justification of safe discharge of industrial sewage containing urea, *Gig. Truda Prof. Zabol.*, 12, 53, 1966 (in Russian).
2. Thiersch, J. B., *Malformations Congenitales des Mammiferes*, Paris, 1971, 95.
3. Merkle, J. and Zeller, H., Studies on acetamides and formamides for embryotoxic and teratogenic activities in the rabbit, *Arzneimittel-Forsch.*, 30, 1557, 1980.
4. Rickard, L. B. and Driscoll, C. D., Developmental toxicity study of *n*-methylformamide (MMF) inhalation in the rat, *Teratology*, 41, 586, 1990.
5. Liu, S. L. et al., Developmental toxicity studies with *n*-methylformamide administered orally to rats and rabbits, *Teratology*, 39, 466 1989.
6. Stula, E. F. and Krauss, W. C., Embryotoxicity in rats and rabbits from cutaneous application of amide-type solvents and substituted ureas, *Toxicol. Appl. Pharmacol.*, 41, 35, 1977.

1-METHYL-4-(1-METHYLETHENYL) (*R*)CYCLOHEXENE

Molecular Formula. $C_{10}H_{16}$

M = 136.26

CAS No 138-86-4

RTECS No GW6360000

Synonyms and **Trade Names.** Cinene; Cyclohexene-4-isopropenyl-1-methyl; *d*-Limonene; *p*-Mentha-1,8-diene.

Properties. Liquid with a fresh, citrus odor and taste.

Applications. A solvent. M. is efficient in preclinical treatment of breast cancer; it caused more than 80% of carcinomas (breast cancer) to regress with little host toxicity.

Exposure. M. is a naturally occurring food constituent.

Acute Toxicity. LD_{50} is 5.6 to 6.6 g/kg BW in mice.[1]

Administration of 100 mg M./kg BW produced no evident toxicity.

Repeated Exposure. The doses administered to rats and mice for 16 days ranged from 413 to 6600 mg/kg BW. The treatment caused death and reduction in BW at the two highest doses. No clinical signs or histopathology were observed in survivals.[2]

Wistar rats were fed M. for 4 weeks. The treatment damaged epithelial cells of the renal proximal tubules and caused slight growth depression in males at a minimum nephrotoxic effect level. M. slightly increased kidney weights in males and/or caused slight histopathological changes in the kidneys at a minimum nephrotoxic effect level.[3]

According to Webb *et al.*, M. was shown to produce male rat-specific nephrotoxicity that is manifested as exacerbation of hyaline droplet formation.[4]

Short-term Toxicity. In a 13-week study, rats received 150 and 2400 mg/kg BW; mice were given 125 or 2000 mg/kg BW. Death occurred in the high-dose groups of each exposed species and sex. Nephropathy was reported in the kidney of male rats.[2]

Long-term Toxicity. In a 2-year study, male F344 rats received 75 or 150 mg/kg BW, females were given 300 or 600 mg/kg BW in corn oil by gavage. Survival of the high dose rats was significantly reduced. Kidney was found to be primary target organ (in males). Lesions included exacerbation of the age-related nephropathy, linear deposits of mineral in the renal medulla and papilla, and focal hyperplasia of the transitional epithelium overlying the renal papilla.[2]

Dogs received 1.2 to 3.6 ml M./kg BW for 6 months. The treatment caused frequent vomiting and nausea and decrease in BW, blood sugar and cholesterol. Gross pathology examination revealed no changes in the visceral organs with the exception of the kidney.[1]

Ten adult beagle dogs of both sexes were gavaged twice a day for 6 months with 100 or 1000 mg M./kg BW. Food consumption and BW gain were unaffected by the treatment. No histological findings were observed in the kidneys. There was no evidence of hyaline droplet accumulation nor of any other sign of nephropathy typical of those seen in male rats treated with M.[4]

Male Sprague-Dawley rats were dosed with 277, 554, or 1385 mg M./kg BW by gavage in 1.0% Tween 80 for 6 months. The treatment produced nephrotoxicity consisting of granular cast characteristic of *α-2u-globulin*- mediated nephropathy. No such lesions were observed in females.[5]

Reproductive Toxicity.

Embryotoxicity and ***Teratogenicity.*** Mice were orally given 2.4 g/kg BW for 6 days from day 7 to day 12 of gestation. Poisoning led to a decrease of BW gain and to an increase in the incidence of abnormal bone formation in fetuses. Retardation of the BW gain was noted in male offspring.[6]

Pregnant rats received doses of up to 2.87 mg/kg BW during the organogenesis period. Signs of maternal toxicity and fetal growth reduction were observed but no teratogenic effect was found.[1]

Allergenic Effect. M. was shown to be a strong sensitizer to air-exposed animals in both Freund complete adjuvant test and guinea pig maximization test.[7]

Mutagenicity.

In vitro genotoxicity. M. was found to be mutagenic in the presence and absence of exogenous metabolic activation in *Salmonella;* it did not induce CA or SCE in cultured Chinese hamster ovary cells.[2] According to other data, it was negative in *Salmonella* mutagenicity assay.[025]

Carcinogenicity. 2-year NTP study revealed increased incidence of tubular cell hyperplasia, adenomas, and adenocarcinomas of the kidney in male rats in association with the development of hyaline droplet nephropathy. In contrast, neither kidney tumors nor the associated nephropathy have been found in female rats or mice that received 250 or 500 mg/kg BW (males) and 500 or 1000 mg/kg BW (females).[2]

U.S. EPA Risk Assessment Forum concluded that nephropathy in male rats associated with *α-2u-globulin* accumulation in hyaline droplets is not an appropriate endpoint to determine non-cancer effects potentially occurring in humans.[8]

Hard and Whysner emphasize that M. does not pose any carcinogenic or nephrotoxic risk to humans. The mechanism of M. tumor development does not appear to be possible in humans since neither the quantity nor the type of protein that binds M. or M.-1,2-oxide is present.[9]

Carcinogenicity classifications. An IARC Working Group concluded that there are no data available on the carcinogenicity of M. to *humans* and there is limited evidence for the carcinogenicity of M. in *experimental animals.*

IARC: 3;

NTP: CE* - NE - NE - NE (gavage).

Chemobiokinetics.

Observations in man.. Human volunteers were exposed to M. by inhalation. M. is readily metabolized. Accumulation occurs in adipose tissues.[10]

Seven healthy human volunteers ingested 100 mg/kg BW in a custard. *Dihydroperillic acid* and *perillic acid* were defined to be major metabolites; minor metabolites were *methyl esters* of these acids, while *limonene- 1,2-diol* appeared to be the third major metabolite. Unchanged M. was a minor component.[11]

Animal studies. Following oral administration of ^{14}C-labeled M. to animals and man, 75 to 95% and >10% of the radioactivity was excreted in the urine and feces, respectively, within 2 to 3 days. Major metabolite in urine was *perillic acid 8,9-diol* in rats and rabbits and *8-hydroxy-p-menth-1-en-9-yl-β-d-glucopyranosiduronic acid* in guinea pigs and man. Five other metabolites were discovered in dog and rat urine.[1] The major metabolite associated with *α-2u-globulin* was *M.-1,2-oxide.*[12]

Regulations. ***U.S. FDA*** (1998) considered M. to be *GRAS*. It is approved for use in adhesives as a component of articles intended for use in packaging, transporting, or holding food in accordance with the conditions prescribed by 21 CFR part 175.105.

References:

1. Tsuji, M. et al., Effects on development of rat fetuses and offspring, Abstract, *Oyo Yakuri*, 10, 179, 1975 (in Japanese).
2. *Toxicology and Carcinogenesis Studies of d-Limonene (Gavage Studies)*, 1990, NTP Technical Report Series No 347, Research Triangle Park, NC, NIH Publ. No 90-2802, 1990.
3. Jonker, D., Woutersen, R. A., van Bladeren, P. J., Til, H. P., and Feron, V.J., Subacute (4-week) oral toxicity of a combination of four nephrotoxins in rats: comparison with the toxicity of the individual compounds, *Food Chem. Toxicol.*, 31, 125, 1993.
4. Webb, D. R., Kanerva, R. L., Hysell, D. K., Alden, C. L., and Lehman-McKeeman, L. D., Assessment of the subchronic oral toxicity of *d*-limonene in dogs, *Food Chem. Toxicol.*, 28, 669, 1990.
5. Tsuji, M., Fujisaki, Y., Okubo, A., Arikawa, Y., Noda, K., Hiroyuki, I., and Ikeda, T., Studies on *d*-limonene as a gallstone solubilizer. III. Chronic toxicity in rats, Abstract, *Oyo Yakuri*, 9, 403, 1975 (in Japanese).
6. Kodama, R., Yano, T., Furukawa, K., Noda, K., and Ide, H., Isolation and characterization of new metabolites and species differences in metabolism, Abstract, *Oyo Yakuri*, 13, 863, 1977 (in Japanese).
7. Karlberg, A. T., Skao, L. P., Nilson, U., Galvert, E., and Nilsson, J. L., Hydroperoxides in oxidized *d*-limonene identified as potent contact allergens, *Arch. Dermat. Res.*, 286, 97, 1994.
8. *Alpha-2u-globulin: Association with Chemically-Induced Renal Toxicity and Neoplasia in the Male Rat*, U.S. EPA 625/3-91/019F, Washington, D.C., 1991.
9. Hard, G. C. and Whysner, J., Risk assessment of *d*-limonene: an example of male rat-specific renal tumorigens, *CRC Crit. Rev. Toxicol.*, 24, 231, 1994.
10. Falk-Filipsson, A., Lof, A., Hagberg, M., Hjelm, E. W., Wang, Z., D-limonene exposure to humans by inhalation: uptake, distribution, elimination, and effects on the pulmonary function, *J. Toxicol. Environ. Health*, 38, 77, 1993.
11. Crowell, P. L., Elson, C. E., Bailey, H. H., Elegbede, A., Haag, J. D., and Gould, M. N., Human metabolism of the experimental cancer therapeutic agent *d*-limonene, *Cancer Chemother. Pharmacol.*, 35, 31, 1994.
12. Lehman-McKeeman, L. D., Rodriges, P. A., Takigiku, R., Caudill, D., and Fey, M. L., *d*-Limonene-induced male rats-specific nephrotoxicity: Evaluation of the association between *d*-limonene and α-2μ-globulin, *Toxicol. Appl. Pharmacol.*, 99, 250, 1989.

4-METHYL-2-OXETANONE

Molecular Formula: $C_4H_6O_2$
M = 86.10
CAS No 3068-88-0
RTECS No RQ8050000
Abbreviation. MO.

Synonyms. β-Butyrolactone; 3-Hydroxybutanoic acid, β-lactone; 3-Hydroxybutyric acid, β-lactone; Hydroxybutyric acid lactone.

Applications. Solvent for polyacrylonitrile, cellulose acetate, methyl methacrylate polymers, polystyrene.

Acute Toxicity. LD_{50} is 17.2 ml/kg BW in rats.[1]

Mutagenicity.

In vivo cytogenetics. Significant, dose-dependent, and reproducible positive responses were obtained in bone marrow cells after double *i/p* treatments (1500 mg/kg BW per treatment), although marginal or negative results were observed after double *i/p* or *i/v* treatments.[2]

In vitro genotoxicity. MO induced CA in mammalian cells *in vitro* (Loveday et al., 1989; Asanami et al., 1994) and was mutagenic to bacteria (Zeiger et al., 1992; Asanami et al., 1994).

Carcinogenicity. It was carcinogenic in rats treated by oral administration.
Carcinogenicity classification (*γ-Butyrolactone*).
NTP: NE - NE - NE - NE

References:

1. Smyth, H. F., Carpenter, C. P., Weil, C. S., Pozzani, U. C., Striegel, J. A., and Nyaun, J. S., Range-finding toxicity data: List VII, *Am. Ind. Hyg. Assoc. J.*, 30, 470, 1969.
2. Morita, T., Asano, N., Awogi, T., Sasaki, Yu. F., et al. Evaluation of the rodent micronucleus assay in the screening of IARC carcinogens (Groups 1, 2A and 2B). The summary report of the 6 collaborative study by CSGMT/JEMS-MMS, *Mutat. Res.*, 389, 3, 1997.

2-METHYLPENTANE

Molecular Formula. C_6H_{14}
M = 86.18
CAS No 107-83-5
RTECS No SA2995000
Abbreviation. MP.

Synonyms. Dimethylpropylmethane; Isohexane.

Properties. Colorless liquid. Water solubility is 14 mg/l; soluble in alcohol.

Applications. A solvent.

Repeated Exposure. Rats received MP diluted in olive oil. The treatment caused changes in tubules (including acute renal failure and acute tubular necrosis) and an increase in mortality rate. MP was found to be less neurotoxic than *n*-hexane.[1,2]

Short-term Toxicity. Rats were treated orally with MP diluted in olive oil for 8 weeks. MP was found to be less neurotoxic than *n*-hexane.[1]

Chemobiokinetics. One of the MP metabolites in the body appeared to be *2-methyl-2-pentanol*.[3]

References:

1. Ono, Y., Takeuchi, Y., and Hisanaga, N., A comparative study on the toxicity of *n*-hexane and its isomers on the peripheral nerve, *Int. Arch. Occup. Environ. Health*, 48, 289, 1981.
2. Halder, C. A., Holdsworth, C. E., Cockrell, B. Y., and Piccirillo, V. J., Hydrocarbon nephropathy in male rats: identification of the nephrotoxic components of unleaded gasoline, *Toxicol. Ind. Health.* 1, 67, 1985
3. Perbellini, L. et al., Identification of the metabolites of *n*-hexane, cyclohexane, and their isomers in men's urine, *Toxicol. Appl. Pharmacol.*, 53, 220, 1980.

4-METHYL-2-PENTANONE

Molecular Formula. $C_6H_{12}O$
M = 100.18
CAS No 108-10-1
RTECS No SA9275000
Abbreviation. MP.

Synonyms. Hexone; Isopropyl acetone; Methyl isobutyl ketone; 2-Methyl-4-pentanone; 2-Methyl-propyl methyl ketone.

Properties. Clear colorless liquid with a sweet sharp odor. Solubility in water is 17 g/l at 20°C or 18 g/l at 25°C. Odor perception threshold is 1.3 mg/l [02] or 0.1 mg/l.[010]

Applications. MP is used as a component in the synthesis of cellulose and polyurethane lacquers and paint solvents. It is permitted as a flavoring agent, and is used in food-contact materials.

Exposure. MP occurs naturally in food. Levels of detection in certain foods are of mg/kg range.

Acute Toxicity. LD_{50} is found to be 4.56 g/kg BW in rats, 2.85 g/kg BW in mice, and 1.6 to 3.2 g/kg BW in guinea pigs.[1,2]

According to Zakhari et al.,[3] LD_{50} is 1,9 g/kg BW in mice.

Repeated Exposure. When given for 3 or 7 days by gavage, MP produced a dose-related enhancement of the cholestasis induced by manganese-bilirubin combination.[4]

Short-term Toxicity. In a 90-day study in mice, rats, dogs, and monkeys, only male rats developed hyaline droplets in the proximal tubules of the kidney. In a 90-day gavage study in rats, a NOEL of 50 mg/kg BW was identified.[5]

Inhalation exposure affects the CNS and liver. Sprague-Dawley rats received MP by oral gavage at dose levels of 59, 250, or 1000 mg/kg BW for 13 weeks. The treatment caused general nephropathy at 1000 mg/kg BW, and an increase in liver and kidney weights without histopathological lesions present in the liver.[6]

Reproductive Toxicity.

Embryotoxicity. Some maternal toxicity and retardation of ossification were observed in rats and mice exposed to 3000 ppm MP during organogenesis period.[7]

Mutagenicity. MP appeared to be non-mutagenic in different test systems, *in vivo* and *in vitro.*

In vitro genotoxicity. It was negative in the rat liver chromosome assay, in the bacterial mutation assays, and the yeast mitotic gene conversion assay.[8]

Chemobiokinetics. After administration, MP is widely distributed throughout the body, is readily metabolized to water-soluble excretory products, and can induce metabolic activation in the liver. Excretion of metabolites occurs mainly via the urine.[5]

Male Sprague-Dawley rats were dosed with MP in corn oil solution for 3 days. The following metabolites were found: *4-methyl-2-pentanol, 4-hydroxymethyl isobutyl ketone*, and *2-hexanol.*[9] Formation of *4-hydroxy-4-methyl-2-pentanone* is also reported.[10]

Regulations.

In the ***EU countries*** and in the ***U.S.,*** MP is allowed as a component of food-packaging materials. A limit of 5.0 mg/l is suggested in beverages (***EU***).

U.S. FDA (1998) approved the use of MP (1) in adhesives as a component of articles intended for use in packaging, transporting, or holding food in accordance with the conditions prescribed by 21 CFR part 175.105, and (2) a solvent in the manufacture of polysulfide polymer-polyepoxy resins which may be used as food-contact surface of articles intended for packaging, transporting, holding, or otherwise contacting dry food in accordance with the conditions prescribed by 21 CFR part 177.1650.

Standards. *Russia* (1995). PML: 0.2 mg/l.

References:

1. Tepikina, L. A., Shipulina, Z. V., Pavlov, V. N., and Kostyukovich, A. A., Sanitary and chemical analysis and hygienic assessment of methylisobutyl carbinol and methylisobutyl ketone in ambient air, *Gig. Sanit.*, 5, 8, 1994 (in Russian).
2. Smyth, H. F., Carpenter, C. P., and Weil, C. S., Range-finding toxicity data: list IV, *Arch. Ind. Health Occup. Med.*, 4, 119, 1951.
3. Zakhari, S., Levy, P., Liebowitz, M., and Aviado, D. M., Acute oral, intraperitoneal, and inhalation toxicity of methylisobutyl ketone in the mouse, in *Isopropanol and Ketones in the Environment*, L. Goldberg, Ed., CRC Press, Cleveland, Ohio, Part 3, Chapter 10-14, 1977, 93.
4. Vezina, M. and Plaa, G. L., Potentiation by methyl isobutylketone of the cholestasis induced in rats by a manganese-bilirubin combination or manganese alone, *Toxicol. Appl. Pharmacol.*, 91, 419, 1987.
5. *Methyl Isobutyl Ketone*, Health and Safety Guide, WHO, IPCS, Geneva, 1991, 28.
6. *Methyl Isobutyl Ketone*, Environmental Health Criteria Series No 117, IPCS, WHO, Geneva, 1990, 80.
7. Tyl, R. W., France, K. A., Fisher, L. C., Prittz, L. M., Tyler, T. R., Phillips, R. D., and Moran, E. J., Developmental toxicity evaluation of inhaled methyl isobutyl ketone in Fischer 344 rats and CD-1 mice, *Fundam. Appl. Toxicol.*, 8, 310, 1987.
8. Brooks, T. M., Meyer, A. L., and Hutson, D. H., The genetic toxicology of some hydrocarbon and oxygenated solvents, *Mutagenesis*, 3, 227, 1988.
9. Duguay, A. B. and Plaa, G. L., Tissue concentrations of methyl isobutyl ketone, methyl *n*-butyl ketone and their metabolites after oral or inhalation exposure, *Toxicol. Lett.*, 75, 51, 1995.

10. Granvil, C. P., Sharkawi, M., and Plaa, G. L., Metabolic fate of methyl *n*-butyl ketone, methyl isobutyl ketone and their metabolites in mice, *Toxicol. Lett.*, 70, 263, 1994.

METHYL-2-PYRROLIDINONE

Molecular Formula. C_5H_9NO
M = 99.13
CAS No 872-50-4
26138-58-9
RTECS No UY5790000
Abbreviation. MP.

Synonyms. 1-Methyl-2-hydroxytetrahydropyrrone; 1-Methyl-2-pyrrolidone; *m*-Pyrol.

Properties. A colorless liquid with a weak characteristic odor. Readily soluble in water and alcohol. Odor and taste perception threshold is 40 to 60 mg/l. Practical threshold for odor and taste is 68 to 76 mg/l. A concentration of 1.0 g/l does not affect color or transparency of water.[1]

Applications. Used as a solvent in the production of plastics.

Acute Toxicity. LD_{50} is 3.9 to 7.9 g/kg BW in rats, 5.3 to 7.7 g/kg BW in mice, 3.5 g/kg BW in rabbits, and 4.4 g/kg BW in guinea pigs.[2-4]

Lethal doses led to apathy and adynamia. Gross pathology examination revealed visceral congestion, lesions in the GI tract walls, and parenchymatous dystrophy of the renal tubular epithelium.[1]

Repeated Exposure failed to reveal cumulative properties. K_{acc} is 20.[2]

Rats and rabbits were gavaged with 1.06 g/kg BW. Treatment resulted in the increase of liver glycogen content, blood cholesterol, and total serum bilirubin levels.[2] Gross pathology examination revealed changes in the visceral organs of animals given this dose or 1/10 LD_{50} for 1.5 months.

Short-term Toxicity. Beagle dogs received MP at dosage levels of 25, 79, and 250 mg/kg BW over a period of 13 weeks. There were no statistically significant treatment-related effects reflected in BW gain and food consumption, hematological and clinical chemistry parameters, and ophthalmic, gross and histopathological examinations. However, a dose-dependent decrease in BW and an increase in platelet count that correlated with increased megakaryocytes was observed. Serum cholesterol in males decreased with increasing doses.[5]

Long-term Toxicity. Animals given 0.025 and 0.25 mg/kg BW for 6 months exhibited no effect on blood morphology, cholesterol level, and total serum bilirubin.[2] The NOAEL of 2.5 mg/kg BW (for functional or histology changes) was established in this study.

Reproductive Toxicity.

Embryotoxicity. Inhalation exposure of pregnant rats to MP decreased pup weights.[6]

Teratogenicity. Equivocal results are reported in regard to ability of MP to cause developmental abnormalities following inhalation exposure.[7,8]

Pregnant rats were exposed to MP by inhalation on gestation days 7 to 20. The exposure caused no effects on basal functions of CNS in male pups. Only some specific neurobehavioral reactions were altered.[9]

Mutagenicity.

In vitro genotoxicity. MP is shown to be negative in *Salmonella* mutagenicity assay (Wells, 1988). It produced a dose-dependent increase in aneuploidy in a yeast assay for genotoxicity but only under special treatment conditions.[10, 11]

In vivo cytogenetics. A single oral administration of 3.8 g/kg BW (approximately 80% of LD_{50}) did not result in an increase either in micronucleated erythrocytes or in structural or numerical CA in Chinese hamster bone marrow test.[11]

Chemobiokinetics. Major urinary metabolite in the rat after acid hydrolysis is a derivative of *4-(methylamino) butanoic acid.*

Regulations. ***U.S. FDA*** (1998) approved the use of MP as a slimicide in the manufacture of paper and paperboard that contact food in accordance with the conditions prescribed in 21 CFR part 176.300.

Standards. ***Russia*** (1995). MAC and PML: 0.5 mg/l.

References:

1. Amirkhanova, G. F., Latypova, Z. V., and Tupeeva, R. B., in *Toxicology of New Industrial Chemical Substances*, N. F. Ismerov and I. V. Sanotsky, Eds., Medgiz, Moscow, Issue No. 7, 1965, 28 (in Russian).
2. Meleshchenko, K. F., Sanitary-toxicological characteristics of methylpyrrolidone as a pollutant in water bodies, *Gig. Sanit.*, 6, 84, 1970 (in Russian).
3. Ethyl Browning's *Toxicity and Metabolism of Industrial Solvents*, R. Snyder, Ed., 2nd ed., vol 2, Nitrogen and Phosphorus Solvents, Elsevier, Amsterdam-New York-Oxford, 1990, 240.
4. Kirk-Othmer's *Encyclopedia of Chemical Technology*, 3rd ed., vol 19, M. Grayson and D. Eckroth, Eds., John Wiley & Sons, Inc., New York, 1978, 517.
5. Becci, P. J., Gephart, L. A., Koschir, F. J., et al., Subchronic feeding study in beagle dog of *N*-methylpyrrolidone, *J. Appl. Toxicol.*, 3, 83, 1983.
6. Ulla, H., Prenatal toxicity of *N*-methylpyrrolidone in rats: postnatal study, *Teratology*, 42, 31A, 1990.
7. Lee, K. P. et al., Toxicity of *N*-methyl-2-pyrrolidone (NMP): Teratogenic, subchronic, and two-year inhalation studies, *Fundam. Appl. Toxicol.*, 9, 222, 1987.
8. Jakobsen, B. M. and Hass, U., Prenatal toxicity of *N*-methylpyrrolidone inhalation in rats: a teratogenicity study, *Teratology*, 42, 18A, 1990.
9. Hass, U., Lund, S. P., and Elsner, J., Effects of prenatal exposure to *N*-methylpyrrolidone on postnatal development and behavior in rats, *Neurotoxicol. Teratol.*, 16, 241, 1994.
10. Mayer, V. W. et al., Aneuploidy induction in Saccharomyces cerevisiae by two solvent compounds, 1-methyl-2-pyrrolidinone and 2-pyrrolidinone, *Environ. Molec. Mutagen.*, 11, 31, 1988.
11. Engelhardt, G. and Fleig, H., 1-Methyl-2-pyrrolidinone (NMP) does not induce structural and numerical chromosomal aberrations *in vivo*, *Mutat. Res.*, 298, 149, 1993.

METHYL VINYL ETHER

Molecular Formula. C_3H_6O
M = 58.09
CAS No 107-25-5
RTECS No KO2300000
Abbreviation. MVE.

Synonyms and **Trade Name.** Agrisynth MVE; Methoxyethene.

Properties. Colorless, compressed gas, or colorless liquid. Freely soluble in alcohol; slightly soluble in water.

Applications. A monomer for copolymers used in the manufacture of coatings and lacquers; modifier for al kyl, polystyrene, and ionomer resins; a plasticizer for nitrocellulose and adhesives.

Acute Toxicity. LD_{50} is 4.9 g/kg BW in rats.

Regulations. ***EU*** (1990). MVE is available in the *List of authorized monomers and other starting substances which may continue to be used for the manufacture of plastic materials and articles intended to come into contact with foodstuffs pending a decision on inclusion in Section A* (*Section B*).

Reference:

Deichmann, W. B., *Toxicology of Drugs and Chemicals*, Acad. Press, Inc., New York, 1969, 395.

MONOACETIN

Molecular Formula. $C_5H_{10}O_4$
M = 134.13
CAS No 26446-35-5
RTECS No AK3595000

Synonyms. Acetin; Acetoglyceride; Acetyl monoglyceride; Glycerol acetate; Glycerol monoacetate.

Applications. A solvent for basic dyes. Food additive.

Properties. Colorless liquid with characteristic odor. Soluble in water and alcohol.

Acute Toxicity. LD_{50} *p/c* is 6.6 g/kg BW in rats.[020] Parenteral administration resulted in CNS depression and respiratory failure. Gross pathology examination revealed pulmonary congestion.[011,022]

Mutagenicity.

In vivo cytogenetics. M. was found to be positive in sex-linked recessive lethal mutations test at a dose of 50,000 ppm when administered to males of *Dr. melanogaster* by feeding.[1]

Chemobiokinetics. M. has been largely but not completely hydrolyzed in bowel to *glycerin* and *acetic acid.*[022]

Regulations. ***U.S. FDA*** (1998) approved the use of M. in the manufacture of cellophane for packaging food in accordance with the conditions prescribed in 21 CFR part 177.1200.

Reference:

1. Yoon, J. S., Mason, J. M., Valencia, R., Woofruff, R.C., and Zimmering S., Chemical mutagens testing in *Drosophila.* IV. Results of 45 coded compounds tested for the National Toxicology Program, *Environ. Mutagen.*, 7, 349, 1985.

NITROBENZENE

Molecular Formula. $C_6H_5NO_2$
M = 123.12
CAS No 98-95-3
RTECS No DA6475000
Abbreviation. NB.

Synonym and **Trade Names.** Essence of Mirbane; Mirbane oil; Nitrobenzol; Oil of mirbane.

Properties. A colorless to pale-yellow, oily liquid with a sharp odor of bitter almonds and sweetish, astringent taste. Water solubility is 0.2%, soluble in alcohol, ether, oils. Odor perception threshold is 0.2 to 0.6 mg/l or 0.11 mg/l,[02] taste perception threshold is 0.34 mg/l. According to Kazakova,[1] a sweetish taste appears at 50 mg/l.

Applications. Used as a solvent in the manufacture of plastics.

Exposure. NB is found in a number of water samples.

Acute Toxicity.

Observations in man. Acute effects include neurotoxicity and hepatotoxicity.[2] Transient leukocytopenia was also recorded in man. Is known to produce methemoglobinemia in animals and humans.[3]

NB produced a decrease in erythrocyte and platelet count and circulating *Hb*, hemolytic anemia, and bone marrow hyperplasia.[4]

Animal studies. LD_{50} is identified to be 640 mg/kg BW in rats, 590 mg/kg BW in mice, and 700 mg/kg BW in rabbits.[2]

According to Alekseyeva, LD_{50} appeared to be 600 mg/kg BW in rats and rabbits, and 550 mg/kg BW in mice.[5] The dose of 750 to 1000 mg/kg BW is lethal in dogs. Poisoning affected NS functions, parenchymatous organs, disrupted oxidation-reduction processes, caused changes in the rat blood and lengthened blood clotting time.

Repeated Exposure revealed cumulative properties. NB can cause methemoglobinemia and damage to the liver. F344 rats received doses of 5.0, 25, or 125 mg/kg BW administered by gavage over a period of 28 days. The treatment caused severe anemia and changes in blood biochemistry analyses at two higher doses. Gross pathology and histology examination revealed changes in the liver, cerebellum, and spleen.[6]

$B6C3F_1$ mice were given 30, 100, or 300 mg/kg BW in corn oil by gavage for 14 consecutive days. The treatment induced an increase in liver and kidney weights. Gross histopathological examination revealed significant changes in the spleen, consisting of severe congestion of the red pulp areas.[7]

Long-term Toxicity.

Observations in man. Produced myelotoxic effect and effects on erythrocytes in animals and in the man.

Animal studies. Administration of 0.7 ml/kg BW to rabbits for 23 weeks increased the number of megakaryocytes in the bone marrow and spleen. Degenerative changes in the adrenal glands were also noted.[2] Retardation of BW gain, hematology changes, methemoglobinemia, and damage to the liver were noted in rabbits and guinea pigs gavaged by 1.0 mg/kg BW and more. A dose of 50 mg/kg BW caused brain edema.[1]

The dose of 0.1 mg/kg BW is considered to be close to the NOAEL since it caused only slight local changes in the visceral organs of 3 animals.[1]

Chronic inhalation exposure of B6C3F_1 mice and F344 and CD rats to NB caused methemoglobinemia, anemia, and adaptive or degenerative changes in the nose, liver, and testis.[8]

Immunotoxicity. In the above-cited study, the phagocytic activity of macrophages in the liver was increased with a concomitant decrease in the activities in the spleen and lung. NB-induced hemolysis and liver injury occurred due to alterations in bone marrow activity.[7]

In a 6-month study, a dose of 1.0 mg/kg BW depressed antibody production by 30% after repeated immunization of animals with typhoid vaccine. In 2 to 4 months, phagocytosis was enhanced but fell subsequently by 28%. The NOEL of 0.1 mg/kg BW for immune response was established in this study.[9]

Contact-sensitizing potential is not found in maximized guinea-pig assays.[10]

Reproductive Toxicity.

Gonadotoxicity. NB is a known inhibitor of male fertility. Degenerative testicular lesions in F344 rats were reported.[11]

A single oral dose caused spermatogenic cell necrosis, multinucleated giant cells, and decreased spermatozoa in the epididymis.

Male Sprague-Dawley rats which received oral doses of 60 mg NB/kg BW were mated with normal proestrus females. The testicular and epididymal weights, sperm count, sperm motility ,and progressive motility decreased significantly by day 14 but not by day 7 of treatment. Even after 70 days of treatment, the copulation index was not affected.[12]

In the above-cited study, disorder of spermatogenesis was noted. Atrophy of seminiferous tubule was found in the 125 mg/kg BW group.[6]

Teratogenicity. Pregnant rats were given 7 to 13 subcutaneous injections of 125 mg/kg BW on days 4 to 6 or on days 9 to 12 of gestation. Fetuses displayed a wide range of developmental abnormalities including hydrocephaly, softening of the body skeleton, agenesis of the pelvis and rear extremities, as well as an abnormal placenta metabolism.[13]

According to Khipko et al., administration of 70 mg/kg BW during pregnancy did not affect fetal development.[14] Male Sprague-Dawley rats were orally administered 30 or 60 mg NB/kg BW for 3 weeks. The higher dose decreased weight of epididymis, sperm motility, and viability, and increased incidence of morphological abnormalities. Histological examination revealed degeneration and decreases of spermatids and pachytene spermatocytes.[15]

Mutagenicity.

In vitro genotoxicity. Equivocal results have been reported in *Salmonella* mutagenicity assays.[16-18]

In vivo cytogenetics. Increased sex-linked recessive lethal mutations in *Dr. melanogaster* were observed. Four strains failed to induce unscheduled DNA synthesis in hepatocytes. There were no CA in bone marrow cells of mice given NB by gavage.[19-22]

Carcinogenicity classifications. An IARC Working Group concluded that there is sufficient evidence for the carcinogenicity of NB in *experimental animals* and there are inadequate data for evaluation of the carcinogenicity of NB in *humans*.

IARC: 2B;

U.S. EPA: D.

Chemobiokinetics. Oxidation-reduced cycling can form *nitrosobenzene* and *phenylhydroxylamine* with *Hb* or autoreduction of quinone intermediates such as *p*-aminophenol with ultimate production of *hydrogen peroxide*. The *nitroxide radical* can also be formed via the generation of the *superoxide radical* which dismutates also to *hydrogen peroxide*.[2] Five metabolites were found in the rat bile.[23,24]

Following administration of ^{14}C-NB, more than 60% radioactivity is found in the urine in rats, while in mice it amounts to 35%. 12 to 19% NB is removed in the feces. Apparently, nitroaryl derivatives take part in metabolism of *N*-substituted aryl compounds, reacting with *SH*-groups to form *sulfinic acid*.[25]

Standards.

Russia (1995). MAC and PML: 0.2 mg/l.

EPA of Arizona. MCL: 0.0035 mg/l.

References:

1. Kazakova, M. I., Sanitary-hygienic evaluation of nitrobenzene as a water bodies pollutant, *Gig. Sanit.*, 3, 7, 1956 (in Russian).
2. Beauchamp, R. O., Bus, J. S., Popp, N., et al., A critical review of the literature on nitrobenzene toxicity, *CRC Crit. Rev. Toxicol.*, 11, 33, 1983.
3. Harrison, M. R., Toxic methemoglobinemia. A case of acute nitrobenzene and aniline poisoning treated by exchange transfusion, *Anestesia*, 32, 270, 1977.
4. Parkes, W. E. and Neill, D. W., Acute nitrobenzene poisoning with transient aminoaciduria, *Brit. Med. J.*, 1, 653, 1953.
5. Alekseyeva, N. P., in *Hygienic Problems Associated with the Development of Large-scale Chemistry*, Proc. Sci. Conf., 1st Moscow Medical Institute, Moscow, 1964, 41 (in Russian).
6. Shimo, T., Onodera, H., Matsushima, Y., et al., A 28-day repeated dose toxicity study of nitrobenzene in F344 rats, Abstract, *Eisei Shikenjo Hokoku*, 112, 71, 1994 (in Japanese).
7. Burns, L. A., Bradley, S. G., White, K. L., Jr., McCay, J. A., Fuchs, B. A., Stern, M., Brown, R. D., Musgrove, D. L., Holsapple, M. P., Luster,. M. I., et al., Immunotoxicity of nitrobenzene in female B6C3F1 mice, *Drug Chem. Toxicol.*, 17, 271, 1994.
8. Cattley, R. C., Everitt, J. I., Gross, E. A., Moss, O. R., Hamm, T. E., and Popp, J. A., Carcinogenicity and toxicity of inhaled nitrobenzene in $B6C3F_1$ mice and F344 and CD rats, *Fundam. Appl. Toxicol.*, 22, 328, 1994.
9. *Protection of Water Reservoirs against Pollution by Industrial Liquid Effluents*, S. N. Cherkinsky, Ed., Medgiz, Moscow, Issue No 7, 1965, 269 (in Russian).
10. Stevens, M. A., Use of the albino guinea-pig to detect the skin-sensitizing ability of chemicals, *Brit. J. Ind. Med.*, 24, 189, 1967.
11. Bond, J. A., Chism, J. P., Rickert, D. E., et al., Induction of hepatic and testicular lesions in Fisher 344 rats by single oral dose of nitrobenzene, *Fundam. Appl. Toxicol.*, 1, 389, 1980.
12. Kawashima, K., Usami, M., Sakemi, K., and Ohno, Y., Studies on the establishment of appropriate spermatogenic endpoints for male fertility disturbances in rodent induced by drugs and chemicals. I. Nitro benzene, *J. Toxicol. Sci.*, 20, 15, 1995.
13. Kazanina, S. S., The morphology and histochemistry of the hemochlorial placentas of white rats following nitrobenzene poisoning of the mother, *Bull. Exp. Biol. Med.*, 65, 93, 1968 (in Russian).
14. Khipko, S. E., Vasilenko, N. M., Yashina, L. N., et al., Alternative development of rat embryos on various routes of administration of nitrobezene, *Gig. Truda Prof. Zabol.*, 4, 48, 1987 (in Russian).
15. Matsuura, I., Hoshino, N., Wako, Y., Tani, E., Satou, T., Aoyama, R., and Ikeda, Y., Sperm parameter studies on three testicular toxicants in rats, Abstracts of paper presented at the 35th Annual Meeting of the Japanese Teratology Society, Tokyo, 1995, *Teratology*, 52, P33, October 1995.
16. Haworth, S., Lawlor, T., Mortelmans, K., Speck, W., and Zeiger, E., *Salmonella* mutagenicity test results for 250 chemicals, *Environ. Mutagen.*, (Suppl.) 1, 3, 1983.
17. Suzuki, J., Koyami, T., and Suzuki S., Mutagenicities of mono-nitrobenzene derivatives in the presence of norharman, *Mutat.* Res., 120, 105, 1983.
18. Assmann, N., Emmrich, M., Kampf, G., and Kaiser, M., Genotoxic activity of important nitrobenzenes and nitroanilines in the Ames test and their structure-activity relationship, *Mutat. Res.*, 395, 139, 1997.
19. Butterworth, B. E., Earle, L. L., Strom, S., et al., Induction of DNA repair in human and rat hepatocytes, *Proc. Am. Assoc., Cancer Res.*, 23, 274, 1983.
20. Rapopport, I. A., Mutagenicity of Nitrobenzene, Reports, Series *Biology Sciences,* Issue No 160, 1965, 168 (in Russian).
21. Mirsalis, J. C., Tysen, C. K., and Butterworth, B. E., Detection of genotoxic carcinogens in the *in vivo* and *in vitro* hepatocyte DNA repair assay, *Environ. Mutagen.*, 4, 553, 1982.
22. Feldt, E. G., Assessment of mutagenic hazards of benzene and its derivatives, *Gig. Sanit.*, 7, 21, 1985 (in Russian).

23. Rickert, D. E., Schnell, S. R., and Long, R. M., Hepatic macromolecular covalent binding and intestinal disposition of ^{14}C-dinitrotoluenes, *J. Toxicol. Environ. Health*, 11, 555, 1983.
24. Rickert, D. E., Bond, J. A., Long, R. M., and Chism, J. P., Metabolism and excretion of nitrobenzene by rats and mice, *Toxicol. Appl. Pharmacol.*, 67, 206, 1983.
25. Albrecht, W. and Neumann, H. G., Biomonitoring of aniline and nitrobenzene. Hemoglobin binding in rats and analysis of adducts, *J. Arch. Toxicol.*, 57, 1, 1985.

2-NITROPROPANE

Molecular Formula. $C_3H_7NO_2$
M = 89.11
CAS No 79-46-9
RTECS No TZ5250000
Abbreviation. 2-NP.

Synonyms. Dimethylnitromethane ; Isonitropropane; β-Nitropropane.

Properties. Liquid. Solubility in water is 17 ml/l; 0.6 ml water dissolves in 100 ml of 2-NP. Miscible with organic solvents.[020]

Applications. A solvent for cellulose acetate, vinyl resins, lacquers, synthetic rubbers, fats, and oils.[020]

Acute Toxicity. LD_{50} is 725 mg/kg BW in Sprague-Dawley rats.[1] Plasma activities of some hepatic enzymes were significantly elevated in 48, 72, and 96 hours after *i/p* administration of 2-NP to male BALB/c mice. Administration of 6.7 mg/kg BW caused hepatotoxicity in females. Histological investigation revealed damage in periportal region.[2]

Reproductive Toxicity.

Teratogenicity. Rats were exposed to 170 mg 2-NP/kg BW by *i/p* injections on days 1 to 15 of gestation. The treatment caused developmental abnormalities in the cardiovascular system of pups from mothers treated with 2-NP.[3]

Mutagenicity.

In vivo cytogenetics. Rats have been treated with single oral doses of 2-NP. The treatment caused a progressive increase of liver DNA fragmentation at doses ranging from 0.5 to 8.0 mmol/kg BW. In contrast, DNA fragmentation was absent in lung, kidney, bone marrow, and brain of rats given 8.0 mmol/kg BW.[4]

Did not induce micronucleated polychromatic erythrocytes or micronucleated reticulocytes in male CD-1 mice by double *i/p* treatment of up to 500 mg/kg BW per treatment (a half of LD_{50} dose).[5]

In vitro genotoxicity. Inadequate data were reported in the mouse micronucleus assay (Mavournin et al., 1990). 2-NP was positive in *Salmonella* mutagenicity assay and showed DNA-modifying activity.[6] It was positive in the rat hepatocyte DNA-repair test,[7] but did not induce CA in Chinese hamster ovary cells (Galloway et al., 1987).

Carcinogenicity. 2-NP has been found to be a potent hepatocarcinogen. Sprague-Dawley rats were administered 1.0 mmol 2-NP/kg BW three times per week for 16 weeks. Both benign and malignant liver tumors occurred in 100% of the animals.[8]

Regulations. ***U.S. FDA*** (1998) approved the use of 2-NP in adhesives as a component of articles intended for use in packaging, transporting, or holding food in accordance with the conditions prescribed by 21 CFR part 175.105.

References:

1. Deichman, W. B., *Toxicology of Drugs and Chemicals*, Academic Press, Inc., New York, 1969, 430.
2. Dayal, R., Gescher, A., Harpur, E. S., Pratt, J., and Chipman, J. K., Comparison of the hepatotoxicity in mice and the mutagenicity of three nitroalkanes, *Fundam. Appl. Toxicol.*, 13, 341, 1989.
3. Harris, S. J., Bond, G. P., and Niemeir, R. W., The effects of 2-nitropropane, naphthalene, and hexachlorobutadiene on fetal rat development, Abstract, *Toxicol. Appl. Pharmacol.*, 48, A69, 1979.
4. Robbiano, L., Mattioli, F., and Brambilla, G., DNA fragmentation by 2-nitropropane in rat tissues, and effects of the modulation on biotransformation processes, *Cancer Lett.*, 57, 61, 1991.

5. Morita, T., Asano, N., Awogi, T., Sasaki, Yu. F., et al. Evaluation of the rodent micronucleus assay in the screening of IARC carcinogens (Groups 1, 2A and 2B). The summary report of the 6 collaborative study by CSGMT/JEMS-MMS, *Mutat. Res.*, 389, 3, 1997.
6. Speck, W. T., LeRoy, W. M., Zeiger, E., and Rozencranz, H. S., Mutagenicity and DNA-modifying activity of 2-nitropropane, *Mutat. Res.*, 104, 49, 1982.
7. Williams, G. M., Mori, H., and McQueen, C. A., Structure-activity relationship in the rat hepatocyte DNA-repair test for 300 chemicals, *Mutat. Res.*, 221, 263, 1989.
8. Fiala, E. S., Czerniak, R., Castonguay, A., Conaway, C. C., and Rivenson, A., Assay of 1-nitropropane, 2-nitropropane, 1-azoxypropane and 2-azoxypropane for carcinogenicity by gavage in Sprague-Dawley rats, *Carcinogenesis,* 8, 1947, 1987.

NONYL ALCOHOL

Molecular Formula. $C_9H_{20}O$
M = 144.26
CAS No 143-08-8
RTECS No RB1575000
Abbreviation. NA.

Synonyms and **Trade Names.** 1-Nonalol; 1-Nonanol; Octyl carbinol; Pelargonic alcohol.

Properties. Colorless to yellowish-colored liquid with a strong odor reminiscent of toilet soap. Poorly soluble in water (75 mg/l); readily soluble in alcohol. Odor and taste perception threshold is 0.1 mg/l. Does not affect the color and clarity of water.

Applications. Used in the rubber and dyestuff industry, in the production of plasticizers, emulsifiers, etc.

Acute Toxicity. LD_{50} is 12 to 19 g/kg BW in rats, and 20 g/kg BW in mice.[1,2] Poisoning include increased reflex excitability, tail rigidity, convulsive twitching, motor coordination disorder, and respiratory distress.

Long-term Toxicity. The doses of up to 0.05 mg/kg BW were ineffective for behavior, BW gain, hematology analysis, and vitamin C content in the visceral organs of rats and rabbits. Gross pathology examination revealed signs of parenchymatous dystrophy in the viscera.[3]

Reproductive Toxicity. Inhalation exposure to NA produced no teratogenic effect in rats.[4]

Chemobiokinetics. Following ingestion, NA, like other primary alcohols, undergoes two general reactions. The first is oxidation to the *carboxylic acid derivative* and next the direct conjugation with glucuronic acid. NA undergoes direct glucuronic conjugation to the extent of 4.1%.[5]

Regulations.

EU (1990). NA is available in the *List of authorized monomers and other starting substances which shall be used for the manufacture of plastic materials and articles intended to come into contact with foodstuffs (Section A).*

Great Britain (1998). NA is authorized without time limit for use in the production of polymeric materials and articles in contact with food or drink or intended for such contact.

Recommendations. Joint FAO/WHO Expert Committee on Food Additives (1999). ADI: No safety concern.

Standards. ***Russia*** (1995). MAC and PML: 0.01 mg/l.

References:

1. Yegorov, Yu. L. and Andrianov, L. A., in *Proc. F. F. Erisman Research Sanitary Hygiene Institute*, Moscow, Issue No 9, 1961, 47 (in Russian).
2. Kostovetsky, Ya. J. and Zholdakova, Z. I., in *Protection of Water Reservoirs against Pollution by Industrial Liquid Effluents*, S. N. Cherkinsky, Ed., Medgiz, Moscow, Issue No. 7, 1965, 152 (in Russian).
3. Kostovetsky, Ya. J. et al., in *Problems of Environmental Hygiene*, D. N. Kaluzhny, Ed., Kiyv, Issue No 6, 1966, 66 (in Russian).
4. Nelson, B. K., Brightwell, W. S., Khan, A., et al., Developmental toxicology assessment of 1-octanol, 1-nonanol, and 1-decanol administered by inhalation to rats, *Teratology*, 39, 471, 1989.
5. Ethyl Browning's *Toxicity and Metabolism of Industrial Solvents*, R. Snyder, Ed., 2nd ed., vol 3, Alcohols and Esters, Elsevier, New York, NY, 1992, 186.

1-OCTADECANOL

Molecular Formula. $C_{18}H_{38}O$
M = 270.56
CAS No 112-92-5
RTECS No RG2010000
Abbreviation. OD.

Synonyms and **Trade Names.** Adol; Crodacol; Decyloctyl alcohol; *n*-Octadecyl alcohol; Siponol S; Stearol; Stearyl alcohol.

Properties. Unctuous white flakes or granules with a faint odor and bland taste.[020] Insoluble in water; soluble in ethanol.

Acute Toxicity. LD_{50} is 20 g/kg BW.[1]

Carcinogenicity. Implantation at the dose of 1000 mg OD/kg BW caused tumors in kidney, ureter, and bladder.[2] OD is neoplastic agent by RTECS criteria.

Regulations. ***U.S. FDA*** (1998) approved the use of OD (1) in the manufacture of resinous and polymeric coatings for the food-contact surface of articles intended for use in producing, manufacturing, packing, processing, preparing, treating, packaging, transporting, or holding food in accordance with the conditions prescribed by 21 CFR part 175.300; (2) as a component of a defoaming agent intended for articles which may be used in producing, manufacturing, packing, processing, preparing, treating, packaging, transporting, or holding food in accordance with the conditions prescribed in 21 CFR part 176.200; (3) as a defoaming agent in the manufacture of paper and paperboard intended for use in packaging, transporting, or holding food in accordance with the conditions prescribed by 21 CFR part 176.210; and (4) in the manufacture of cellophane for packaging food in accordance with the conditions prescribed by 21 CFR part 177.1200.

Reference:

1. Kirk-Othmer *Encyclopedia of Chemical Technology*, 3rd ed., M. Grayson and D. Eckroth, Eds., John Wiley & Sons, Inc., New York, 1978, 1, 722.
2. Brian, G. T. and Springberg, P. D., Role of the vehicle in the genesis of bladder carcinomas in mice by the pellet implantation technique, *Cancer Res.*, 26, 105, 1966.

2-OCTANONE

Molecular Formula. $C_8H_{16}O$
M = 128.21
CAS No 111-13-7
RTECS No RH1484000

Synonyms. Hexyl methyl ketone; Oxooctane.

Properties. Colorless liquid with a specific pleasant odor and bitter cheese or camphor taste. Slightly soluble in water; soluble in alcohol.

Applications. Used as a solvent in the production of vinyl and epoxy resins, nitrocellulose, and synthetic coatings.

Acute Toxicity. LD_{50} is 3.1 to 3.8 g/kg BW in mice and 9.2 g/kg BW in rats. Poisoned animals displayed lethargy. Death occurs within 2 to 4 days.[1,2]

Repeated Exposure failed to reveal cumulative properties. Rats were given 20 mg/kg BW for 16 days. The treatment caused a significant elevation of serum lipase activity.[03]

Allergenic Effect. O. did not produce skin sensitization in guinea pigs.[03]

Mutagenicity.

In vitro genotoxicity. O. was shown to be negative *Salmonella* mutagenicity assay.[03]

Regulations. ***U.S. FDA*** (1998) listed O. as a food additive permitted for direct addition to food for human consumption as a synthetic flavoring substance and adjuvant in accordance with the conditions prescribed in 21 CFR part 172.515: (1) the quantity added to food does not exceed the amount reasonably required to accomplish its intended physical, nutritive, or other technical effect in food; and (2) when intended for use in or on food, it is of appropriate food grade and is prepared and handled as a food ingredient.

References:

1. Tanii, H., Tsuji, H., and Hashimoto, K., Structure-toxicity relationship of monoketones, *Toxicol. Lett.*, 30, 13, 1986.
2. Abbasov, D. M. et al., Toxicology of some ketones, in *Proc. Azerb. Research Institute Hygiene Labor & Prof. Diseases*, Issue No. 11, 1977, 145 (in Russian).

OCTYL ALCOHOL

Molecular Formula. $C_8H_{28}O$
M = 130.23
CAS No 111-87-5
RTECS No RH6550000
Abbreviation. OA.

Synonyms and **Trade Name.** Caprylic alcohol; Heptyl carbinol; 1-Hydroxyoctane; Lorol 20; 1-Octanol; *n*-Octanol; Primary octyl alcohol.

Properties. Colorless liquid with a penetrating odor. Solubility in water is 568 mg/l; odor perception threshold is 0.05 mg/l (for primary and secondary OA).[1] According to other data, it appeared to be 0.13 to 0.2 mg/l.[2]

Applications. Used in the production of plastics, lacquers and dyes, synthetic detergents.

Acute Toxicity. LD_{50} of *primary* OA is reported to be 20 g/kg BW in rats, and 15 g/kg BW in mice. That of *secondary* OA is 3.7 g/kg BW in rats, 3.1 g/kg BW in mice, 9.5 g/kg BW in guinea pigs, and 9.3 g/kg BW in rabbits. Intoxication is followed by narcosis. The survivors do not differ from the controls in appearance and behaviour.[1]

Repeated Exposure revealed no cumulative properties. Rats tolerate administration of four LD_{50}s given as daily doses of 1/5 LD_{50}.

Long-term Toxicity. Rats were dosed by gavage with 75 mg/kg BW of *sec*-OA. The treatment affected blood catalase and liver cholinesterase activity. On chronic inhalation exposure (0.9 to 540 mg/m^3) of rats, OA caused NS inhibition, impairment of the oxidation-reduction processes and liver functions as well as disorders in the blood acid-alkali balance.[3]

Immunotoxicity. A 75 mg/kg BW dose of *sec*-OA affected the immunobiological indices in rats.

Reproductive Toxicity.

Inhalation during pregnancy caused neither ***embryotoxic*** nor ***teratogenic effect***.[3,4]

Regulations.

U.S. FDA (1998) approved the use of OA (1) in adhesives as a component of articles intended for use in packaging, transporting, or holding food in accordance with the conditions prescribed by 21 CFR part 175.105; (2) in the manufacture of resinous and polymeric coatings in a food-contact surface of articles intended for use in producing, manufacturing, packing, transporting, or holding food in accordance with the conditions prescribed by 21 CFR part 175.300; (3) as a defoaming agent in the manufacture of paper and paperboard intended for use in packaging, transporting, or holding food in accordance with the conditions prescribed by 21 CFR part 176.210; (4) in the manufacture of cellophane for packaging food in accordance with the conditions prescribed by 21 CFR part 177.1200; and (5) as a lubricant (0.1%) in cellophane to be safely used for packaging food (2-ethylhexyl alcohol) in accordance with the conditions prescribed in 21 CFR part 177.1200.

EU (1990). OA is available in the *List of authorized monomers and other starting substances which shall be used for the manufacture of plastic materials and articles intended to come into contact with foodstuffs (Section A).*

Great Britain (1998). OA is authorized without time limit for use in the production of polymeric materials and articles in contact with food or drink or intended for such contact.

Recommendations. Joint FAO/WHO Expert Committee on Food Additives (1999). ADI: No safety concern.

Standards. ***Russia*** (1995). MAC and PML: 0.05 mg/l (organolept., taste).

References:

1. Korolev, A. A., Krasovsky, G. N., and Varshavskaya, S. P., Hygienic evaluation of amyl alcohol, primary and secondary octyl alcohols as regard to their standardization in water bodies, *Gig. Sanit.*, 9, 88, 1970 (in Russian).
2. Grushko, Ya. M., *Harmful Organic Compounds in Industrial Liquid Effluents*, Khimiya, Leningrad, 138 (in Russian).
3. Krashenina, G. I. and Kosiborod, N. R., Substantiation of the maximum allowable concentration for octyl alcohol in the air, *Gig. Sanit.*, 6, 82, 1986 (in Russian).
4. Nelson, B. K., Brightwell, W. S., Khan, A., et al., Developmental toxicology assessment of 1-octanol, 1-nonanol, and 1-decanol administered by inhalation to rats, *Teratology*, 39, 471, 1989.

OCTYL PHENOL condensed with 1 mole ETHYLENE OXIDE

Molecular Formula. $C_{16}H_{26}O_2$
M = 250.42
CAS No 1322-97-0
RTECS No RI0170000
Abbreviation. OPEO.

Synonym. Octylphenoxyethanol.

Composition.

Acute Toxicity. LD_{50} is 7.1 ml/kg BW in rats.

Regulations. ***U.S. FDA*** (1993) approved the use of OPEO in adhesives as a component of articles intended for use in packaging, transporting, or holding food in accordance with the conditions prescribed by 21 CFR part 175.105.

Reference:

Proc. Sci. Section Toilet Goods Assoc., 20, 16, 1953.

cis-9-OCTADECENOIC ACID

Molecular Formula. $C_{18}H_{34}O_2$
M = 282.45
CAS No 112-80-1
RTECS No RG2275000
Abbreviation. OA.

Synonyms and **Trade Names.** Elaic acid; Oleic acid; Palmitic acid; Pamolyn; Red oil.

Properties. Colorless or nearly colorless liquid with a peculiar lard-like taste and odor. Practically insoluble in water; soluble in alcohol.

Applications. A solvent and a lubricant.

Exposure. Is found in animal and plant fats.

Acute Toxicity. LD_{50} is 74 g/kg BW (Union Carbide Data Sheet, 1964).

Reproductive Toxicity. OA may cross human placenta.

Embryotoxicity. Pregnant pigs fed high fat diets rich in oleic acid. The treatment caused no adverse effects in offspring.[1]

Mutagenicity.

In vitro genotoxicity. OA was negative in five *Salmonella typhimurium* strains in the presence and absence of liver *S9* fraction at doses up to 333 mg/plate.[2,013]

In vivo cytogenetics. Administration via rectum with 35 mg OA/kg BW increased unscheduled DNA synthesis in mice bone marrow cells.[3]

Carcinogenicity. Like other ingredients of dietary fat, OA may cause the promotional effect in carcinogenesis development. Concomitant oral administration of calcium salts, as $CaCO_3$, largely reduced the mitogenic effects of fatty acids on colon epithelium, presumably by forming biologically inert calcium soaps.[3]

Regulations.

EU (1990). OA is available in the *List of authorized monomers and other starting substances which shall be used for the manufacture of plastic materials and articles intended to come into contact with foodstuffs (Section A)*.

U.S. FDA (1998) considered OA to be *GRAS* as substances migrating to food from paper and paperboard products.

Great Britain (1998). OA is authorized without time limit for use in the production of polymeric materials and articles in contact with food or drink or intended for such contact.

Standards. *Russia* (1995). MAC and PML: 0.5 mg/l.

References:

1. Farnworth, E. R. and Kramer, J. K., Fat supplementation to the gestation diet of older sows and its effect on maternal and fetal fat metabolism, *Reprod. Nutr. Dev.*, 30, 629, 1990.
2. Nakamura, H. and Yamamoto, T., The active part of the [6]-gingerol molecule in mutagenesis, *Mutat. Res.*, 122, 87, 1983.
3. Wargovich, M. J., Eng, V. W., and Newmark, H. L., Calcium inhibits the damaging and compensatory proliferative effects of fatty acids on mouse colon epithelium, *Cancer Lett.*, 23, 253, 1984.

OXALIC ACID, DIETHYL ESTER

Molecular Formula. $C_6H_{10}O_4$
M = 146.16
CAS No 95-92-1
RTECS No RO2800000
Abbreviation. OADE.

Synonyms. Diethyl oxalate; Diethyl ethanedioate; Ethyl oxalate.

Properties. Clear, colorless, oily liquid with an aromatic odor. Poorly soluble in water; miscible with ethanol and hydrolyzes in water in presence of alkalis.

Applications. Solvent for cellulose ethers.

Acute Toxicity. LD_{50} is 1.5 g/kg BW in rats, and 2.0 g/kg BW in mice. Poisoning caused disturbed respiration and muscle twitching.[03] Species susceptibility is not evident. Massive renal oxalate deposits and dilatation of tubules were observed in the rats after administration of 400 mg/kg BW oral dose.[1]

According to Timofiyevskaya, LD_{50} is 1.5 g/kg BW in rats, and 2.0 g/kg BW in mice. Poisoning led to CNS depression, thereafter a narcotic effect was observed. Death occurs in weeks.[2]

Repeated Exposure revealed pronounced accumulation. K_{acc} appeared to be 1.1 (by Lim).

Chemobiokinetics. Administration of 0.4 g/kg BW resulted in massive renal *oxalate* deposits and dilatation of tubules.[03] OADE is hydrolyzed in the body to form *oxalic acid* which accounts for its toxic action.[3]

Regulations. *U.S. FDA* (1998) approved the use of OADE in adhesives as a component of articles intended for use in packaging, transporting, or holding food in accordance with the conditions prescribed in 21 CFR part 175.105.

References:

1. Patty, F., Ed., *Industrial Hygiene and Toxicology,* vol 2, Toxicology, 2 ed., Interscience Publ., New York, 1963, 1889.
2. Timofiyevskaya, L. A., Toxicity and hazard of some plasticizers, *Gig. Sanit.*, 5, 87, 1981 (in Russian).
3. *Encyclopedia of Occupational Health and Safety*, Int. Labour Office, vols 1 and 2, McGraw-Hill Book Co., New York, 1971, 478.

PENTYL ALCOHOL

Molecular Formula. $C_5H_{12}O$
M = 88.12
CAS No 71-41-0
RTECS No SB9800000
Abbreviation. PA.

Synonyms and **Trade Names.** Amylol; *n*-Butylcarbinol; Pentanol; Pentasol; Amyl alcohol.

Properties. A yellow liquid with an odor of fusel oil. Odor perception threshold is 0.12[010] or 0.7 mg/l; taste threshold concentration is 0.5 mg/l.[1] According to other data, odor threshold concentration is 1.3 mg/l, and the taste threshold slightly higher.[2]

Applications. PA is used in the production of plastics, varnishes, paints, and synthetic detergents.

Exposure. Maximum daily intake in man is calculated to be 33 mg in the UK and 42 mg in the U.S. and in Europe.[3]

Acute Toxicity. In rats, LD_{50} is 3.0 to 4.25 g/kg BW.[2,4,5]

CNS depression was the only observed effect of a single oral dose administration. Appearance and behavior of survivors did not differ from that in the controls.[2]

Repeated Exposure failed to reveal cumulative properties. Mice were given 1/5 LD_{50}. Overall dose of 12.0 g/kg BW caused no animal mortality.[2]

Short-term Toxicity. Rats were administered PA, dissolved in corn oil, by oral intubation at dose levels of 50 to 1000 mg/kg BW for 13 weeks. No reduction in BW gain, food and water consumption, changes in hematological indices or urinalyses, renal function, organ weights or hystopathology were reported. Butterworth et al. considered the NOAEL to be 1000 mg/kg BW, which is about 2000 times the estimated maximum likely intake by man.[3]

Long-term Toxicity. In a 6-month study, rats received the dose of 95 mg/kg BW. There were changes in blood catalase and liver cholinesterase activity as well as in the immune status of animals.[2]

Liver necrosis was noted in rabbits following oral exposure to PA for up to 1 year.[6] Damage to the gastric mucosa was the only pathological finding observed in another study.[7]

Chemobiokinetics. PA is oxidized to *valeric acid* by aldehyde in rats.[8] Less than 0.03% of a single oral dose was excreted in the urine of rats within 8 hours.[9]

Regulations.

U.S. FDA (1998) has permitted PA for use as (1) a synthetic flavoring substance and additive, in accordance with the conditions prescribed by 21 CFR part 172.515; and (2) as a defoaming agent in the manufacture of paper and paperboard intended for use in packaging, transporting, or holding food in accordance with the conditions prescribed by 21 CFR part 176.210.

EU (1990). PA is available in the *List of authorized monomers and other starting substances which shall be used for the manufacture of plastic materials and articles intended to come into contact with foodstuffs (Section* A).

Great Britain (1998). PA is authorized without time limit for use in the production of polymeric materials and articles in contact with food or drink or intended for such contact.

Recommendations. Joint FAO/WHO Expert Committee on Food Additives (1999). ADI: No safety concern.

Standards. ***Russia*** (1995). MAC and PML: 1.5 mg/l (organolept., odor). Recommended MAC: 0.1 mg/l.[7]

References:

1. Ivanov, V. A. and Khal'zov, P. P., Health and safety aspects of noxious wastes from synthetic rubber manufacture, in *Proc. Voronezh Medical Institute*, Voronezh, Issue No 55, 1966, 56 (in Russian).
2. Korolev, A. A., Krasovsky, G. N., and Varshavskaya, S. P., Hygiene evaluation of amyl alcohol, primary and secondary octyl alcohols as regard to their standardization in water bodies, *Gig. Sanit.*, 9, 88, 1970 (in Russian).
3. Butterworth, K. R., Gaunt, I. F., Heading, C. E., et al., Short-term toxicity of *n*-amyl alcohol in rats, *Food Chem. Toxicol.*, 16, 203, 1978.
4. Jenner, P. M., Hagan, E. C., Taylor, J. M., et al., Food flavoring and compounds of related structure, I. Acute oral toxicity, *Food Cosmet. Toxicol.*, 2, 327, 1964.
5. Grushko, Ya. M., *Harmful Organic Compounds in the Liquid Industrial Effluents*, Khimiya, Leningrad, 1976, 31 (in Russian).
6. Straus, I. and Blocq, P., Etude experimentelle sur la cirrhose du foie, *Arch. Physiol.*, 10, 409, 1887.

7. Markova, M. A. and Sorokina, L. N., in *Integrated Problems of Hygiene and Health Protection in Siberian Regions*, Moscow, 1988, 48 (in Russian).
8. Haggard, H. W., Miller, D. P., and Greenberg, L. A., The amyl alcohols and their ketones: Their metabolic fates and comparative toxicities, *J. Ind. Hyg. Toxicol.*, 27, 1, 1945.
9. Gaiollard, D. and Derache, R., Vitesse de la metabolisation de differents alcohols chez la rat, *Cr. Seanc. Soc. Biol.*, 125, 1605, 1964.

PETROL SOLVENTS

Synonym. Light hydrocarbons, a mixture.

Properties. Colorless, volatile, highly flammable liquids, poorly soluble in water. Odor perception threshold is 0.06 to 0.2 mg/l. White spirit is a clear, oily liquid.

Applications. The solvents are used in the production of plastics and paints, and in the dyestuff industry.

Standards. ***Russia*** (1988). MAC and PML: 0.1 mg/l (organolept., odor).

2-PHENOXYETHANOL

Molecular Formula. $C_8H_{10}O_2$
M = 138.18
CAS No 122-99-6
RTECS No KM0350000
Abbreviation. PE.

Synonyms and **Trade Names.** Ethylene glycol, monophenyl ether; β-Hydroxyethyl phenyl ether; 1-Hydroxy-2-phenoxyethane; 2-Phenoxyethyl alcohol; Phenyl cellosolve; Phenoxethol.

Properties. Colorless, oily liquid with a faint aromatic odor and burning taste. Solubility in water is 26.7 g/l; freely soluble in alcohol.

Applications. Solvent; primarily used in latexes, paints, cosmetics, in the production of plasticizers, and in the manufacture of protective coatings. A bacteriocide and a solvent for cellulose acetate.

Acute Toxicity. LD_{50} is 1260 mg/kg BW in rats. Poisoning produced changes in the kidney, ureter, bladder, and in GI tract.[1] LD_{50} is 933 mg/kg BW in mice. Degenerative changes in the brain and coverings were reported.[2]

See ***1,2-Diethoxyethane.***

Repeated Exposure. Female New Zealand white rabbits were dosed by gavage with 100 to 1000 mg/kg BW for up to 10 consecutive days. The treatment resulted in dose-related intravascular hemolytic anemia (decreased erythrocyte count, *Hb* level, packed cell volume, regenerative erythroid response in the bone marrow, etc.) and renal tubule damage.[3]

Reproductive Toxicity. Swiss CD-1 mice received 0.4 to 4.0 g/kg BW administered via feed for 7 days prior and during a 98-day cohabitation period. Effects on reproduction were only evident in the female and occurred at doses which elicited general toxicity.[4]

Pregnant New Zealand white rabbits were exposed dermally to undeluted PE. Doses of 600 and 1000 mg/kg BW caused maternal death. No embryotoxic or teratogenic effects were observed.[5]

Chemobiokinetics. *Phenoxyacetic acid* was identified as a major blood metabolite.[4]

Regulations. ***U.S. FDA*** (1998) approved the use of PE in adhesives as a component of articles intended for use in packaging, transporting, or holding food in accordance with the conditions prescribed by 21 CFR part 175.105.

References:

1. Smyth, H. F., Jr., Seaton, J., and Fisher, L., The single dose toxicity of some glycols and derivatives, *J. Ind. Hyg. Toxicol.*, 23, 259, 1941.
2. *Toksikologichesky Vestnik.*, 3, 37, 1996 (in Russian).
3. Breslin, W. G., Phillips, J. E., Lomax, L. G., Bartels, M. J., Dittenber, D. A., Calhoun, L. L., and Miller, R. R., Hemolytic activity of ethylene glycol phenyl ether in rabbits, *Fundam. Appl. Toxicol.*, 17, 466, 1991.

4. Heindel, J. J., Gulati, D. K., Russell, V. S., Reel, J. R., Lawton, A. D., and Lamb, J. C., IV, Assessment of ethylene glycol monobutyl ether and ethylene glycol monophenyl ether reproductive toxicity using continuous breeding protocol in Swiss CD-1 mice, *Fundam. Appl. Toxicol.*, 15, 683, 1990.
5. Scortichini, B. H., Quast, J. F., and Rao, K. S., Teratologic evaluation of 2-phenoxyethanol in New Zealand white rabbits following dermal exposure, *Fundam. Appl. Toxicol.*, 8, 272,1987.

2-(2-PHENOXYETHOXY)ETHANOL

Molecular Formula. $C_{10}H_{14}O_3$
M = 182.24
CAS No 104-68-7
RTECS No KM0875000
Abbreviation. PEE.

Synonyms. Diethylene glycol, phenyl ether; Phenoxydiglycol; 2-(Phenoxyethoxy)ethanol; Phenyl carbitol.

Acute Toxicity. LD_{50} is 2.14 g/kg BW in rats.[1]

Regulations. ***U.S. FDA*** (1998) approved the use of PEE in adhesives as a component of articles intended for use in packaging, transporting, or holding food in accordance with the conditions prescribed in 21 CFR part 175.105.

Reference:

Smyth, H. F., Carpenter, C. P., Weil, C. S., et al., Range-finding toxicity data, List V, *Arch. Ind. Hyg. Occup. Med.,* 10, 61, 1954.

PHENYL ETHER

Molecular Formula. $C_{12}H_{10}O$
M = 170.22
CAS No 101-84-8
RTECS No KN8970000
Abbreviation. PE.

Synonyms. Biphenyl oxide; Diphenyl ether; Diphenyl oxide; 1,1'-Oxybisbenzene; Phenoxyben zene.

Properties. Colorless, crystalline solid or liquid with a geranium-like odor. Insoluble in water; soluble in ethanol.

Applications. Used as a solvent in the synthesis of perfumes and plasticizers.

Acute Toxicity. In rats, LD_{50} is reported to be 3.37 g/kg BW.[1]

Reproductive Toxicity. Developmental defects and mitotic abnormalities were observed in sea urchins.[2]

Mutagenicity.

In vitro genotoxicity. PE is shown to be ineffective in *Salmonella* mutagenicity assay.[2]

Chemobiokinetics. PE is metabolized to its *4-hydroxy derivative* by rabbit and trout liver microsomes by *PE hydroxylaze.*[3, 4]

References:

1. Anonymous, *Food Cosmet. Toxicol.,* 12, 707, 1974.
2. Pogano, G., Esposito, A., Giacomo, G., et al., Genotoxicity and teratogenicity of diphenyl and diphenyl ether: A study of sea urchins, yeast and *Salmonella typhimurium*, *Teratogen. Carcinogen. Mutagen.,* 3, 377, 1983.
3. Bray, H. G., James, S. P., Thorpe, W. V., et al., The metabolism of ethers in the rabbit, *Biochem. J.*, 54, 547, 1953.
4. Wong, K., Addison, R. E., and Law, F. C. P., Uptake, metabolism and elimination of diphenyl ether by trout and stickleback, *Bull. Environ. Contam. Toxicol.*, 26, 243, 1981.

PROPIONITRILE

Molecular Formula. C_3H_5N
M = 55.08

CAS No 107-12-0
RTECS No UF9625000
Abbreviation. PN.

Synonyms. Cyanoethane; Ethyl cyanide; Hydrocyanic ether; Propanenitrile; Propylnitrile.

Properties. Colorless liquid with a pleasant, ethereal, sweetish odor. Water solubility is 119 g/l at 40°C.[020] Miscible with alcohol.

Applications. A solvent.

Acute Toxicity. LD_{50} is 39 mg/kg BW in rats.[020]

Reproductive Toxicity.

Embryotoxicity. Rats received by gavage 20, 40, or 80 mg/kg BW on 6 through 19 days of gestation. The treatment caused no *terata*, but at high dose levels, maternal and embryotoxicity were noted.[1]

Sprague-Dawley rats were exposed to PN by inhalation during days 6 to 20 of gestation at concentrations of 50 to 200 ppm. Embryolethality was observed at the highest concentration, fetotoxicity was noted after exposure to 200 ppm in the presence of overt signs of maternal toxicity.[2]

Teratogenicity. Administration of PN to the pregnant hamsters produced severe axial skeletal malformations.[3]

Mutagenicity.

In vitro genotoxicity. PN was positive in *Salmonella* and *E.coli* mutagenicity assays when stimulated by metabolic activation systems (Mueller and Norpoth, 1979).

References:

1. Johannsen, F. R., Levinskas, G. J., Berteau, P. E., and Rodwell, D. E., Evaluation of the teratogenic potential of three aliphatic nitriles in the rat, *Fundam. Appl. Toxicol.*, 7, 33, 1986.
2. Saillenfait, A. M., Bonnet, P., Guenier, J. P., and de Ceaurriz, J., Relative developmental toxicities of inhaled aliphatic mononitriles in rats, *Fundam. Appl. Toxicol.*, 20, 365, 1993.
3. Willhite, C. C., Ferm, V. H., and Smith, R. P., Teratogenic evaluation of aliphatic nitriles, *Teratology*, 23, 317, 1981.

PROPYL ALCOHOL

Molecular Formula. C_3H_8O
M = 60.09
CAS No 71-23-8
RTECS No UH8225000
Abbreviation. NPA.

Synonyms. Ethylcarbinol; 1-Hydroxypropane; *n*-Propanol; 1-Propanol.

Properties. Colorless liquid with a characteristic odor of alcohol. Miscible with water and ethyl alcohol at all ratios. Odor perception threshold is in the range of 23 to 40 mg/l, but according to other data, it exceeds 0.033 mg/l.[010] Taste perception threshold is 12 mg/l.[1,2]

Applications. Used as a solvent in the production of synthetic polymers, such as polyvinyl butyral, cellulose esters, lacquers, and polyvinyl chloride adhesives.

Exposure of human beings may occur through the ingestion of food or beverages containing NPA.

Acute Toxicity. NPA exhibits low toxicity, except in very young rats (LD_{50} is found to be 560 to 660 mg/kg BW).[3] LD_{50} is 1.87-6.5 g/kg BW in rats, 6.8 g/kg BW in mice, and 2.82 g/kg BW in rabbits.[4]

The principal toxic effect appeared to be depression of the CNS. NPA is more neurotoxic than ethanol. In rabbits, LD_{50} for narcosis is 1440 mg/kg BW.

Repeated Exposure. Four daily doses of 2.16 g/kg BW given to rats did not produce death or gross pathological changes in their livers.[4]

Rats were exposed by feeding to doses of 45 mg/kg BW for two months.[1] The treatment produced changes in serum enzyme activity and signs of dystrophy in the liver and brain.

Reproductive Toxicity.

Inhalation exposure caused no ***gonadotoxic action*** in male rats.[5]

Teratogenicity. NPA produced teratogenic effect in rats at high inhalation concentrations that were maternally toxic. The congenital abnormalities included tail and skeletal malformations, and cardiovascular and urinary tract defects.[6]

Allergenic Effect. Contact sensitizing potential was not found in maximized guinea pig assays.[7]

Mutagenicity.

In vitro genotoxicity. NPA did not produce SCE or micronuclei in mammalian cells.[8]

Carcinogenicity. Wistar rats were exposed to the oral doses of 240 mg/kg BW or to subcutaneous doses of 48 mg/kg BW throughout their lifespan. A significant increase in the incidence of liver sarcoma was found in the group dosed subcutaneously.[9] According to IARC, however, because of technical limitations, this study seems to be inadequate.

Chemobiokinetics. NPA is rapidly absorbed and distributed following ingestion. It is oxidized to *propionic acid* and then to CO_2. The traces of *glucuronide conjugates* are found in rabbits.

NPA is excreted in the urine or in expired air.

Regulations.

U.S. FDA (1998) approved the use of NPA in (1) in adhesives as a component of articles intended for use in packaging, transporting, or holding food in accordance with the conditions prescribed by 21 CFR part 175.105; (2) as a component of the uncoated or coated food-contact surface of paper and paperboard intended for use in producing, manufacturing, packaging, processing, preparing, treating, packing, transporting, or holding dry food in accordance with the conditions prescribed by 21 CFR part 176.180; (3) in the production of cellophane to be safely used for packaging food (residue limit 0.1%) in accordance with the conditions prescribed by 21 CFR part 177.1200; and (4) as a defoaming agent in the manufacture of paper and paperboard intended for use in packaging, transporting, or holding food in accordance with the conditions prescribed by 21 CFR part 176.210.

EU (1990). NPA is available in the *List of authorized monomers and other starting substances which shall be used for the manufacture of plastic materials and articles intended to come into contact with foodstuffs (Section A)*.

Great Britain (1998). NPA is authorized without time limit for use in the production of polymeric materials and articles in contact with food or drink or intended for such contact.

Recommendations. Joint FAO/WHO Expert Committee on Food Additives (1999). ADI: No safety concern.

Standards. ***Russia*** (1994). MAC: 0.25 mg/l (organolept., odor), PML in food: 0.1 mg/l.

References:

1. Galeta, S. G., in *Hygiene of Populated Areas*, Proc. Sci. Conf., Kiyv, 1969, 40 (in Russian).
2. *Data on Hygiene Assessment of Pesticides and Polymers*, F. F. Erisman Research Sanitary Hygiene Institute, Moscow, 1977, 281 (in Russian).
3. Purchase, I. F. H., Studies in kaffircorn malting and brewing, XXII. The acute toxicity of some fusel oils found in Bantu beer, *S. A. Med. J.*, 43, 795, 1969.
4. Taylor, J. M., Jenner, P. M., and Jones, W. I., A comparison of the toxicity of some allyl, propenyl, and propyl compounds in the rat, *Toxicol. Appl. Pharmacol.*, 6, 378, 1964.
5. Cameron, A. M., Zahlsen, K., Haug, E., Nilsen, O. G., and Eiknes, K. B., Circulation steroids in male rats following inhalation of *n*-alcohols, *Arch. Toxicol.*, Suppl. 8, 422, 1985.
6. Nelson. B. K., Brightwell, W. S., MacKensie-Taylor, D. R., et al., Teratogenicity of *n*-propanol and isopropanol administered at high inhalation concentrations to rats, *Food Chem. Toxicol.*, 26, 247, 1988.
7. Gad, S. C., Dunn, B. J., Dobbs, D. W., Reilly, C., and Walsh, R. D., Development and validation of an alternative dermal sensitization test: the mouse ear swelling test (MEST), *Toxicol. Appl. Pharmacol.* 84, 93, 1986.
8. *2-Propanol*, Environmental Health Criteria No 103, WHO, IPCS, Geneva, 1990, 132.
9. Gibel, W., Lohs, K., und Wildner, G. P., Experimental study on the cancerogenic activity of propanol-1, 2-methylpropanol-1 and 3-methylbutanol-1, *Arch. Geschwulst Forsch.*, 45, 19, 1975.

See also ***Isopropyl alcohol***.

PYRIDINE

Molecular Formula. C_5H_5N
M = 79.10
CAS No 110-86-1
RTECS No UR8400000

Synonym and **Trade Name.** Azabenzene; Azine.

Properties. Colorless volatile liquid with a powerful, fish-like odor. Miscible with water and alcohol.

Applications. P. is used as a denaturant in alcohol and antifreeze mixtures, as a solvent for paints, rubber, and polycarbonate resins, and as an intermediate in the manufacture of insecticides, herbicides, and fungicides. It is used in the production of piperidine, an intermediate in the manufacture of rubber, and as an intermediate and solvent in the preparation of vitamins and drugs, dyes, textile water repellants, and flavoring agents in food. P. is used in the manufacture of cellophane for packaging food; it is also a food additive (flavoring substance) and a solvent.

Acute Toxicity.

Observations in man. Estimated lethal oral dose is in the range of 1 ounce to 1 pint for an average adult. Manifestations of toxic action include anesthesia and irritation.[03]

Animal studies. LD_{50} is reported to be 1.58 g/kg BW in rats [020] and 0.8 to 1.6 g/kg BW in mice.[018] Large doses caused GI disturbances, kidney and liver damage.

Repeated oral **exposure** resulted in liver and kidney injury.

Short-term Toxicity. Rats and mice received P. by gavage for 90 days. Mortality was noted at the doses of 100 and 200 mg/kg BW in rats, and at 400 mg/kg BW in mice. For both species, the NOAEL of 25 mg/kg BW was determined in this study.[1]

Sprague-Dawley rats were given oral doses of 0.25, 1.0, 10, 25, or 50 mg P./kg BW for 3 months. The treatment caused nonneoplastic hepatic lesions at the highest dose. The NOAEL of 1.0 mg/kg BW for hepatic hyper trophy in female rats was established in this study.[2]

F344/N rats were exposed to P. in drinking water at concentrations of 50, 100, 250, 500, or 1,000 ppm (equivalent to average daily doses of 5.0, 10, 25, 55, or 90 mg P./kg BW). There was evidence of hepatocellular injury and/or altered hepatic function demonstrated by increases of serum ALT and sorbitol dehydrogenase activity and bile acid concentrations in 500 and 1,000 ppm rats. The estrous cycle length of 1,000 ppm females was significantly longer than that of the controls. In the liver, the incidences of centrilobular degeneration, hypertrophy, chronic inflammation, and pigmentation were generally increased in 500 and 1,000 ppm males and females relative to controls. The incidences of granular casts in the kidney and renal tubule hyaline degeneration were increased as compared to the controls, in males exposed to 1,000 ppm.[3]

Long-term Toxicity. F344/N rats were exposed to P. in drinking water at concentrations of 100, 200, or 400 ppm (equivalent to average daily doses of 7, 14, or 33 mg/kg BW) for 103 (males) or 104 (females) weeks. The exposure resulted in increased incidences of centrilobular cytomegaly and degeneration, cytoplasmic vacuolization, and pigmentation in the liver of males and females; periportal fibrosis, fibrosis, and centrilobular necrosis in the liver of males; and bile duct hyperplasia in females.

Male Wistar rats were exposed to P. in drinking water at concentrations of 100, 200, or 400 ppm (equivalent to average daily doses of 8.0, 17, or 36 mg/kg BW) for 103 weeks. The exposure resulted in increased incidences of centrilobular degeneration and necrosis, fibrosis, periportal fibrosis, and pigmentation in the liver, and secondary to kidney disease, mineralization in the glandular stomach, and parathyroid gland hyperplasia.

Male $B6C3F_1$ mice were exposed to P. in drinking water at concentrations of 250, 500, or 1,000 ppm (equivalent to average daily doses of 35, 65, or 110 mg/kg BW) for 104 weeks, and female $B6C3F_1$ mice were exposed to P. in drinking water at concentrations of 125, 250, or 500 ppm (equivalent to average daily doses of 15, 35, or 70 mg/kg BW) for 105 weeks.[3]

Reproductive Toxicity.

Gonadotoxicity. $B6C3F_1$ mice were exposed to P. in drinking water at concentrations of 50, 100, 250, 500, or 1,000 ppm (equivalent to average daily doses of 10, 20, 50, 85, or 160 mg/kg BW for males and 10,

20, 60, 100, or 190 mg/kg BW for females). Sperm motility in exposed male mice was significantly decreased as compared to controls.[3]

Teratogenicity. P. caused muscle/skeletal effects in frogs.[1] No teratogenic findings were noted when P. has been injected into developing chickens.[4]

Mutagenicity.

In vitro genotoxicity. P. was not mutagenic in *Salmonella typhimurium* strain *TA98, TA100, TA1535,* or *TA5137* or in L5178Y mouse lymphoma cells, with or without *S9* metabolic activation. It did not cause SCE or CA in Chinese hamster ovary cells, with or without *S9*. P. showed mutagenic effect in *E. coli.*[03,3,5]

In vivo cytogenetics. No induction of CA or micronuclei was noted in bone marrow cells of male mice administered P. via *i/p* injection.[3]

Results of a single reciprocal translocation test in male *Dr. melanogaster* were negative.

Carcinogenicity. Under the conditions of the above-cited 2-year drinking water studies, there was some evidence of carcinogenic activity of P. in male F344/N rats based on increased incidences of renal tubule neoplasms. There was equivocal evidence of carcinogenic activity of P. in female F344/N rats based on increased incidences of mononuclear cell leukemia, and in male Wistar rats based on an increased incidence of interstitial cell adenoma of the testis. There was clear evidence of carcinogenic activity of P. in male and female $B6C3F_1$ mice based on increased incidences of malignant hepatocellular neoplasms.[3]

Carcinogenicity classification.

NTP: SE - EE - CE - CE

Chemobiokinetics. Can be absorbed in toxic amounts by all routes.[03] P. has been shown to be an effective inducer of cytochrome *P-450* in rat microsomes; it activated metabolism of some specific substrates 24 hours after injection of 200 mg/kg BW dose.[6]

Regulations. ***U.S. FDA*** (1998) approved the use of P. (1) in the manufacture of cellophane for packaging food in accordance with the conditions prescribed by 21 CFR part 177.1200; and (2) as a basic component of single and repeated use food contact surfaces in accordance with the conditions prescribed by 21 CFR part 175.105.

References:

1. *Subchronic Study of Pyridine in Fischer Rats and $B6C3F_1$ Mice*, NTP Quality Assessment Report Prep. by Gulf South Research Institute, 1979.
2. *Pyridine. 90-day Subchronic Oral Toxicity in Rats,* Sponsored by Office of Solid Waste, U.S. EPA, Washington, D.C., 1986.
3. *Toxicology and Carcinogenesis Studies of Pyridine in F344/N Rats, Wistar Rats, and $B6C3F_1$ Mice (Drinking Water Studies),* Draft of NTP Technical Report Series No 470, Research Triangle Park, NC, December 9-10, 1997.
4. Davis, K. R., Schultz, T. W., and Damont, J. N., Toxic and teratogenic effects of selected aromatic amines on embryos of the *Amphibian Xenopus laevis*, *Arch. Environ. Contam. Toxicol.*, 10, 371, 1981.
5. Ishidate, M. and Odashima, S., Chromosome tests with 134 compounds in Chinese hamster cells *in vitro* - a screening for chemical carcinogens, *Mutat. Res.*, 48, 337, 1977.
6. Kozlovskaya, V. E., Grishanova, A. Yu., Mishin, V. M., and Liakhovich, V. V., The effect of pyridine and pyridine-*N*-oxide on the monooxygenase system of rat liver microsomes, *Biochemistry*, 10, 1592, 1992 (in Russian).

4-PYRIDINE ETHANOL

Molecular Formula. C_7H_9ON

M = 123.16

CAS No 5344-27-4

RTECS No UT2971000

Synonyms. 4-Ethanolpyridine; 2-(4-Pyridyl)ethanol; 4-(2-Hydroxyl)pyridine.

Properties. Clear liquid, tinted brown and with the odor of pyridine. Readily soluble in water and polar solvent.

Applications. Used as a solvent and intermediate in paint manufacture, etc.

Acute Toxicity. LD_{50} is 3.5 g/kg BW in rats, and 2.2 g/kg BW in mice. Administration of the high doses led to CNS inhibition. Death occurs in 2 to 3 days.

Repeated Exposure revealed no evident cumulative properties. K_{acc} is 5.9 (by Lim). Liver and CNS are shown to be the target organs.

Allergenic Effect. In a 1-month study, allergenic dermatitis was observed upon skin application of guinea pigs.

Reference:

Shugayev, B. B. and Bukhalovsky, A. A., Toxic effects of pyridine ethanol, *Gig. Truda Prof. Zabol.*, 6, 56, 1984 (in Russian).

STEARIC ACID, MONOESTER with 1,2-PROPANEDIOL

Molecular Formula. $C_{21}H_{42}O_3$
M = 342.63
CAS No 1323-39-3
RTECS No WI4550000

Synonyms and **Trade Names.** Monosteol; NOCA; 1,2-Propanediol, monostearate; Propylene glycol, monostearate; Prostearin.

Acute Toxicity. LD_{50} is 26 g/kg BW in mice.

Regulations. ***U.S. FDA*** (1998) approved the use of S. in adhesives as a component of articles intended for use in packaging, transporting, or holding food in accordance with the conditions prescribed by 21 CFR part 175.105.

Reference:

Acute Toxicity Data, *J. Am. Coll. Toxicol.*, 2, 101, 1983.

1,1,2,2-TETRACHLOROETHANE

Molecular Formula. $C_2H_2Cl_4$
M = 167.84
CAS No 79-34-5
RTECS No KI8575000
Abbreviation. TCE.

Synonyms and **Trade Name.** Acetosol; Acetylene tetrachloride; Bonoform; 1,1-Dichloro-2,2-dichloroethane.

Properties. Colorless or pale-yellow liquid with a suffocating chloroform-like odor. Solubility in water is 2.9 mg/l at 20°C[018] or 1.0 g/350 ml at 25°C.[020] Soluble in alcohol and oil.

Applications. Used as a non-flammable solvent for resins, cellulose acetate, rubber, and polyesters.

Acute Toxicity.

Observations in man. Following ingestion of 3.0 ml TCE by mistake, eight humans became comatose and areflexic. Nevertheless, they have recovered without sequelae.[019] TCE exerts high toxicity because of its slow elimination from the body.

Animal studies. LD_{50} is 0.2 ml/kg BW in rats,[020] and 0.7 g/kg BW in dogs.[018]

Repeated Exposure. TCE was administered by gavage in corn oil to groups of five F344/N rats once a day for 21 days. The doses selected were 0.62 and 1.24 mmol/kg BW. 100% mortality was observed in 1.24 mmol/kg BW group of animals. All rats in 1.24 mmol/kg BW group had diarrhea and considerable BW loss; animals became lethargic. Significant renal toxicity was noted. The liver weight and the renal tubule cell labeling index were increased, indicating replicative DNA synthesis.[1]

Reproductive Toxicity.

DBA and AB Jena mice were injected *i/p* with doses of 300 to 700 mg/kg BW during organogenesis. The treatment caused ***embryotoxic effect*** and low incidence of dose-dependent malformations (exencephaly, cleft palate, anophthalmia, fused ribs and vertebrae).[2] The route of administration used in this study is inadequate for hazard assessment.

Gonadotoxicity. Histological changes have been observed in the testes of rats treated with 8.0 mg TCE/kg BW in peanut oil by gavage for 150 days.[3] However, no effects on gonades were observed in the long-term studies in which rats and mice were administered much higher doses for 78 weeks.[4]

Mutagenicity.

In vitro genotoxicity. TCE is shown to be positive in *Salmonella* mutagenicity assay.[5] According to other data, it gave negative results in this test.[025]

Carcinogenicity. B6C3F_1 mice received time-weighted average doses of 142 or 284 mg/kg BW, Osborne-Mendel rats were given time-weighted average doses of 43 or 76 mg/kg BW for females and 62 or 108 mg/kg BW for males. TCE was administered in corn oil by gavage for 78 weeks, followed by observation periods of 32 weeks for the rats and 12 weeks for the mice. Hepatocellular carcinomas were found in mice. No significant increase of tumors was noted in animals of either sex.[6]

Carcinogenicity classifications. An IARC Working Group concluded that there is limited evidence for the carcinogenicity of TCE in *experimental animals* and there were inadequate data available to evaluate the carcinogenicity of TCE in *humans*.

IARC: 3;

U.S. EPA: C;

NTP: E - N - P* - P* (gavage).

Chemobiokinetics. Following oral administration, TCE is readily absorbed in the GI tract and apparently readily excreted by the lungs.[03] It is found to get metabolic transformation to *trichloroethene* and *tetrachloroethane* (96 and 4.0%, respectively) by rat liver microsomal fractions under anaerobic conditions (Nastainczyk et al., 1982).

Mice were given a single dose of 300 or 600 mg/kg BW. A significant decrease in cytochrome *P-450* and *NADPH*-cytochrome (*P-450*) reductase in hepatic microsomes together with alteration of mixed function oxygenases were noted. Paolini et al.[6] concluded that lipid peroxidation may be of the main mechanisms responsible for T. hepatotoxicity.

Standards.

Russia (1995): MAC and PML: 0.2 mg/l (organolept., odor).

EPA of Minnesota. MCL: 0.002 mg/l.

References:

1. *Renal Toxicity Studies of Selected Halogenated Ethanes Administered by gavage to F344/N Rats*, NTP Technical Report, Research Triangle Park, NC, NIH Publ. 96-3935, February 1996, 52.
2. Schmidt, R., The embryotoxic and teratogenic effect of tetrachloroethane - experimental investigations, *Biol. Rundschau.*, 14, 220, 1976.
3. Gohlke, R., Schmidt. P., and Bahmann, H., 1,1,2,2-Tetrachloroethylene and heat stress in animal experiment. Morphological results, *Z.Gesamte Hyg. Grenzgeb*., 20, 278, 1977.
4. *Bioassay of 1,1,2,2-Tetrachloroethane for Possible Carcinogenicity*, NTP Technical Report Series No 27, National Institutes of Health, U.S. Dept. Health, Education & Welfare, Publ. No 78-827, Bethesda, MD, 1978.
5. Brem, H., Stein, A. B., Rozencranz, H. S., The mutagenicity and DNA-modifying effect of haloalkanes, *Cancer Res*., 34, 2576, 1974.
6. Paolini, M., Sapigni, E., Mesirca, R., Pedulli, G. F., Corongiu, F. P., Dessi, M. A., and Cantelli-Forti, G., On the hepatotoxicity of 1,1,2,2-tetrachloroethane, *Toxicology*, 73, 101, 1992.

TETRACHLOROETHYLENE

Molecular Formula. C_2Cl_4

M = 165.82

CAS No 127-18-4

RTECS No KX3850000

Abbreviation. PCE.

Synonyms and **Trade Names.** Ethylene tetrachloride; Perclene; Perchloroethylene; Tetlen; 1,1,2,2-Tetrachloroethene.

Properties. Colorless, highly volatile, nonflammable liquid with a sweetish taste. Odor reminiscent of ether or chloroform. Water solubility is 150 mg/l at 25°C. Odor perception threshold is 0.17 mg/l [02] or 0.3 mg/l (U.S. EPA, 1987). In Japan and in The Netherlands, odor threshold values are regulated at the levels of 1.8 and 1.2 ppm, respectively.[1]

Applications. Used as a solvent for resins, lacquers, and paints. Primarily used in the dry-cleaning industry. Used in the synthesis of fluorocarbons.

Exposure. PCE is a common environmental and workplace contaminant in the U.S. Trace amounts of PCE are found in food and ground and surface water but seem to be minimal due to PCE high volatility. In the U.S., PCE was found in 93 of 231 food samples at a mean concentration of 0.013 mg/kg.[2] It leached from the plastic lining of drinking water distribution pipes.[3] Concentration of PCE in blood of the general population was found to be 149 ng/l (Brugnone et al., 1994).

Acute Toxicity. LD_{50} is reported to be 3.8 g/kg BW in male and 3.0 g/kg BW in female rats.[4] In mice, LD_{50} is 8.4 to 10.3 g/kg BW5 or 6.4 to 8.0 g/kg BW.[5]

The principal manifestations in humans and animals of acute exposure to PCE are CNS depression, ataxia, respiratory and cardiac arrest. Gross pathology examination revealed fatty infiltration of the liver and heart, and changes in the respiratory and circulatory systems in dogs given 0.3 to 0.4 g/kg BW.[6]

Repeated Exposure. Swiss-Cox mice received calculated doses of 14 to 1400 mg/kg BW in corn oil for 6 weeks.[7] The treatment caused histopathology lesions, including impairment of hepatic triglyceride levels, DNA content, and serum enzyme activity. In mice dosed with up to 100 mg PCE/kg BW, there was an increase in liver triglyceride concentration and in liver relative weights. Histological changes were noted in mice given 0.1 g/kg BW for 11 days,[8] but not in rats exposed to a 0.016 mg/kg BW dose.[5]

Male NMRI mice were orally exposed to PCE (5.0 and 320 mg/kg BW) between days 10 and 16 postnatally. 17-day-old mice were unaffected. At 60 days of age, mice exposed to PCE showed changes in all three spontaneous motor activity variables.[9]

F344 mature rats were given 500 mg PCE/kg BW in corn oil for 4 weeks. In male rats, a trend toward progressive albuminuria and transient increase in *α 2-mu* and *N*-acethylglycosoaminidase were observed. Histological examination revealed no glomerular changes, whereas *α 2-mu* accumulation and mild lesions were noted in the *S2* segment of proximal tubules.[10]

Short-term Toxicity effects included damage to the liver and kidney. Sprague-Dawley rats received doses of 14 to 1400 mg PCE/kg BW via their drinking water for 3 months.[4] The exposure revealed retardation of BW gain in high dose level groups of animals. Rats were given 10 mg PCE/kg BW. The treatment caused 18 to 38% increase in the excretion of 17-ketosteroides with subsequent normalization on the 3 to 4 month of the experiment.[11]

Reproductive Toxicity.

Gonadotoxicity.

Observations in man. When comparing sperm parameters of 34 male dry cleaning workers to those of 48 laundry workers, only slight effects on sperm motility and mor phology were found.[12]

No increase in rate of spontaneous abortions or congenital malformations in offspring of men assumed to be exposed (on the basis of questionnaire) to PCE was noted.[13-15]

Embryotoxicity. PCE crosses rat placenta. It may be metabolized by the uterus, placenta, or fetus, and forms *trichloroacetic acid*, which accumulates in the amniotic fluid after inhalation of PCE.[16]

No fetal toxicity or teratogenicity was found in pregnant mice and rats exposed to a concentration of 300 ppm.[17] However, Narotsky and Kavlock reported PCE caused high incidence of full-litter resorptions.[18]

Mutagenicity.

In vivo cytogenetics. PCE was negative in *Dr. melanogaster* assay.[19]

In vitro genotoxicity. PCE has not been shown to exhibit DNA binding in short-term studies. It induced single-strand DNA breaks in the mouse, but did not cause CA in rat bone marrow or human lymphocytes.[2,20]

PCE was negative in *Salmonella* mutagenicity assay.[026] PCE was negative in the test for micronucleus induc tion in mouse peripheral erythrocytes (NTP-94).

Carcinogenicity.

Observations in man. Evidence of human cancer risk is not available. Ramlow[21] pointed out that the largest and most recent occupational studies have not reported any increased risk of leukemia in PCE exposed groups.[21]

Animal studies. Daily administration of high oral doses (by gavage) increased the incidence of hepatocellular carcinomas in both sexes of $B6C3F_1$ mice, but not in Osborne-Mendel rats.[20] Due to high level of mortality among the rats, these data are not considered to be adequate. Inhalation exposure to PCE resulted in an increased incidence of renal tubular cell adenomas and adenocarcinomas,[22] which could be considered a result of formation of a highly reactive metabolite and cell damage produced by renal accumulation of *α-2μ globulin*, both in male rats only.[23]

Carcinogenicity classifications. An IARC Working Group concluded that there is sufficient evidence for the carcinogenicity of PCE in *experimental animals* and there were inadequate data available to evaluate the carcinogenicity of PCE in *humans*.

The Annual Report of Carcinogens issued by the U.S. Department of Health and Human Services (1998) defines PCE to be a substance which may reasonably be anticipated to be carcinogen, i.e., a substance for which there is a limited evidence of carcinogenicity in *humans* or sufficient evidence of carcinogenicity in *experimental animals*.

IARC: 2A;
U.S. EPA: B2/C;
EU: 3;
NTP: IS - IS - P* - P* (gavage).

Chemobiokinetics. PCE is readily absorbed in the GI tract. The main metabolites are *tetrachloroethene oxide* and *trichlorometabolites*. *Trichloroethylene* and *trichloroacetic acid* (25% of *excreta*) are found in the urine. PCE is not shown to exhibit direct interaction with hepatic DNA.

Oxalic acid is considered to be an important metabolite. Disposition of PCE is a saturable, primarily dose-dependent process in rats. PCE is reportedly transferred to human breast milk following inhalation exposure.[26]

According to Birner et al., PCE is metabolized by cytochrome *P-450* and by glutathione conjugation.[24] Cytochrome *P-450*-dependent oxidation results in the formation of *trichloroacetyl chloride*, which may acylate cellular nucleophils; glutathione conjugation results in the formation of *S-(1,2,2-trichloro-vinyl) glutathione*, which is metabolized to the corresponding *cisteine S-conjugate*. PCE is metabolized to *trichloroacetic acid* in mice, F344 rats, and in humans. Only $B6C3F_1$ mice exhibit peroxisome proliferation and carcinogenicity in the liver.[23]

PCE is removed predominantly *via* exhalation. Following oral administration of 1.0 mg tetrachloro-(^{14}C)ethylene/kg BW to Sprague-Dawley rats, about 70% radioactivity is excreted in expired air as PCE, and 26% as CO_2 and non-volatile metabolites in the urine and feces.[25]

Regulations. *U.S. FDA* (1998) approved the use of PCE (1) in adhesives as a component of articles intended for use in packaging, transporting, or holding food in accordance with the conditions prescribed by 21 CFR part 175.105; and (2) as an adjuvant which may be used in the manufacture of foamed plastics intended for use in contact with food, subject to the provisions prescribed in CFR 178.3010.

Recommendations.

WHO (1996). Guideline value for drinking water: 0.04 mg/l.

U.S. EPA (1999): Health Advisory for a longer-term exposure: 5.0 mg/l.

Standards:

U.S. EPA (1999): MCL:0.005 mg/l, MCLG: zero.

Switzerland (1991). The tolerance limit in food is 0.05 mg/kg and 0.2 mg/kg in the fat of meat and milk.

Russia (1995). PML: 0.02 mg/l.

References:

1. Hoshika, Y., Imamura, T., Muto, G., Van Gemert, L. J., Don, J. A., and Walpot, J. I., International comparison of odor threshold values of several odorants in Japan and in The Netherlands, *Environ. Res.*, 61, 78, 1993.

2. Daft, J. L., Rapid determination of fumigant and industrial chemical residues in food, *J. Assoc. Offic. Anal. Chem.*, 71, 748, 1988.
3. Aschengrau, A., Ozonoff, D., Paulu, C., Coogan, P., Vezina, R., Heeren, T., and Zhang, Y., Cancer risk and tetrachloroethylene-contaminated drinking water in Massachusetts, *Arch. Environ. Health*, 48, 284, 1993.
4. Hayes, J. R., Condie, L. W., and Borcelleca, J. F., The subchronic toxicity of tetrachloroethylene (perchloroethylene) administered in the drinking water of rats, *Fundam. Appl. Toxicol.*, 7, 119, 1986.
5. Valitov, R. B. et al., in *Problems of Medical Chemistry*, Ufa, 1980, 22 (in Russian).
6. *Tetrachloroethylene*, Environmental Health Criteria Series No 31, WHO, Geneva, 1984, 48.
7. Buben, J. A. and O'Flaherty, E. J., Delineation of the role of metabolism in the hepatotoxicity of trichlororthylene and perchlororthylene: a dose-effect study, *Toxicol. Appl. Pharmacol.*, 78, 105, 1985.
8. Schumann, A. M., Quast, J. F., and Watanabe, P. G., The pharmacokinetics and macromolecular interaction of perchloroethylene in mice and rats as related to oncogenicity, *Toxicol. Appl. Pharmacol.*, 55, 207, 1980.
9. Fredriksson, A., Danielsson, B. R., and Eriksson, P., Altered behavior in adult mice orally exposed to tri- and tetrachloroethylene as neonates, *Toxicol. Lett.*, 66, 13, 1993.
10. Bergamaschi, E., Mutti, A., Bocchi, M. C., Alinovi, R., Olivetti, G., Ghiggeri, G. M., and Franchini, I., Rat model of perchloroethylene-induced renal dysfunctions, *Environ. Res.*, 59, 427.
11. Kashin, L. M., in *Endocrine System and Toxic Environmental Factors*, Leningrad, 1980, 139 (in Russian).
12. Eskenazi, B., Wyrobek, A. J., Fenster, L., Katz, D. F., Gerson, J., and Rempel, D. M., A study of the effect of perchloroethylene exposure on semen quality in dry-cleaning workers, *Am. J. Ind. Med.*, 20, 575, 1991.
13. Rachootin, P. and Olsen, J., The risk of infertility and delayed conception associated with exposures in the Danish workplace, *J. Occup. Med.*, 25, 394, 1983.
14. Taskinen, H., Anttila, A., Lindbohm, M.-L., and Sallmen, M., Spontaneous abortions and congenital malformations among the wives of men occupationally exposed to organic solvents, *Scand. J. Work Environ. Health*, 15, 345, 1989.
15. Eskenazi, B., Fenster, L., Hudes, M., Wyrobek, A. J., Katz, D. F., Gerson, J., and Rempel, D. M., A study of the effect of perchloroethylene exposure on the reproductive outcomes of wives of dry-cleaning workers, *Am. J. Ind. Med.*, 20, 593, 1991.
16. Ghantous, H., Danielsson, B. R. G., Dencker, L., Gorczak, J., and Vesterberg, O., Trichloroacetic acid accumulates in murine amniotic fluid after tri- and tetrachloroethylene inhalation, *Acta Pharmacol. Toxicol.*, 58, 105, 1986.
17. Schwetz, B. A., Leong, B. J., and Gehring, P. J., The effect of maternally inhaled trichloroethylene, perchloro-ethylene, methyl chloroform, and methylene chloride on embryonic and fetal development in mice and rats, *Toxicol. Appl. Pharmacol.*, 32, 84, 1975.
18. Narotsky, M. G. and Kavlock, K. J., A multidisciplinary approach to toxicological screening: II. Developmental Toxicity, *J. Toxicol. Environ. Health.*, 45, 145, 1995.
19. Vogel, E. W. and Nivard, N. J., Performance of 181 chemicals in a *Drosophila* assay predominantly monitoring interchromosomal mitotic recombination, *Mutagenesis*, 8, 57, 1993.
20. *Bioassay of Tetrachloroethylene for Possible Carcinogenicity*, Natl. Cancer Institute, Dept. Health, Education & Welfare, NIH 77-813, Washington, D.C., 1977.
21. Ramlow, J. M., Apparent increased risk of leukemia in the highest category of exposure to tetrachloroethylene (PCE) in drinking water, *Arch. Environ. Health*, 50, 170, 1995.
22. *Toxicology and Carcinogenesis Studies of Tetrachloroethylene (Perchloroethylene) in F344/N Rats and B6C3F$_1$ Mice (Inhalation Studies)*, NTP Technical Report Series No 311, Research Triangle Park, NC, U.S. Dept. Health & Human Service, 1986.
23. Green, T., Chloroethylenes: a mechanistic approach to human risk evaluation, *Ann. Rev. Pharmacol. Toxicol.*, 30, 73, 1990.

24. Birner, G., Richling, C., Henschler, D., Anders, M. W., and Dekant, W., Metabolism of tetrachloroethene in rats: identification of *N epsilon*-(dichloroacetyl)-*L*-lysine and *N epsilon*-(trichloroacetyl)-*L*-lysine as protein adducts, *Chem. Res. Toxicol.*, 7, 724, 1994.
25. Pegg, D. G., Zempel, J. A., Braun, W. H., and Watanabe, P. G., Disposition of tetrachloro (^{14}C)ethylene following oral and inhalation exposure in rats, *Toxicol. Appl. Pharmacol.*, 51, 465, 1979.
26. Byczkowski, J. Z. and Fischer, J. W., Lactational transfer of tetrachloroethylene in rats, *Risk Anal.*, 14, 339, 1994.

1-TETRADECANOL

Molecular Formula. $C_{14}H_{30}O$
M = 214.44
CAS No 112-72-1
RTECS No XB8655000

Synonyms and **Trade Name.** Loxanol; Myristic alcohol; *N*-Tetradecyl alcohol.

Properties. White solid or crystals. Insoluble in water, very soluble in alcohols.

Applications. Used in the production of plasticizers, antioxidants, and surfactants. An anti-foaming agent.

Acute Toxicity. LD_{50} exceeded 5.0 g/kg BW in rats.[03]

Regulations. ***U.S. FDA*** (1998) approved use of T. (1) in adhesives as a component of articles intended for use in packaging, transporting, or holding food in accordance with the conditions prescribed by 21 CFR part 175.105; (2) in the manufacture of resinous and polymeric coatings for the food-contact surface of articles intended for use in producing, manufacturing, packing, processing, preparing, treating, packaging, transporting, or holding food in accordance with the conditions prescribed by 21 CFR part 175.300; (3) as a component of a defoaming agent intended for articles which may be used in producing, manufacturing, packing, processing, preparing, treating, packaging, transporting, or holding food in accordance with the conditions prescribed in 21 CFR part 176.200; and (4) as a defoaming agent in the manufacture of paper and paperboard intended for use in packaging, transporting, or holding food in accordance with the conditions prescribed by 21 CFR part 176.210.

TETRAETHYLENE GLYCOL

Molecular Formula. $C_8H_{18}O_5$
M = 194.23
CAS No 112-60-7
RTECS No XC2100000
Abbreviation. TEG.

Synonyms. 2,2'-[Oxybis(ethyleneoxy)]diethanol; 3,6,9-Trioxaundecane-1,11-diol.

Properties. Colorless liquid with a mild odor. Highly soluble in water and alcohol.

Applications. Used in the production of coatings; an intermediate for manufacture of plasticizers and resins; a solvent for nitrocellulose.

Acute Toxicity. LD_{50} is 20 or 32.8 g/kg BW in rats.[1]

Repeated Exposure. Ingestion of the diet containing 5.0% TEG as an energy source for 27 days induced no effect on chicks.[03]

Reproductive Toxicity.

Embryotoxicity. Male rats received up to 50,000 ppm TEG in their drinking water for five days. The treatment did not result in overall decrements in fertility. There was some increase in preimplantation mortality.[2]

Gonadotoxicity. Inhalation and *i/g* exposure of male and female rats to TEG resulted in testicular abnormalities and abnormal sperm. Offspring of treated males had an increased incidence of skeletal and CNS defects (exencephaly, hydrocephaly, etc.). Female rats treated with TEG showed adverse effects on ovarian function.[3]

Chemobiokinetics. Chicks did not metabolize TEG.[03]

Regulations.

EU (1990). TEG is available in the *List of authorized monomers and other starting substances which shall be used for the manufacture of plastic materials and articles intended to come into contact with foodstuffs (Section A).*

Great Britain (1998). TEG is authorized without time limit for use in the production of polymeric materials and articles in contact with food or drink or intended for such contact.

References:

1. Verschueren, K., *Handbook of Environmental Data of Organic Chemicals,* 2nd ed., Van Nostrand Reinhold Co., New York, NY, 1983, 1085.
2. Neeper-Bradley, T. L., Fisher, L. C., Fait, D. L., Kubena, M. F., Neptun, D. A., and Ballantyne, B., Dominant lethal assay of tetraethylene glycol (TTEG) in the drinking water of Fischer 344 rats, Abstract, *Toxi cologist*, 13, 96, 1993.
3. Byshovets, T. F., Bariliak, I. R., Korkach, V. I., and Spitkovskaya, L. D., Gonadotoxic activity of glycols, *Gig. Sanit.*, 9, 64, 1987 (in Russian).

TETRAHYDRO-2-FURANMETHANOL

Molecular Formula. $C_5H_{10}O_2$
M = 102.14
CAS No 97-99-4
RTECS No LU2450000
Abbreviation. TF.

Synonym. Tetrahydrofurfuryl alcohol.

Properties. Colorless liquid with a pronounced odor of diethyl ether. Miscible with water and ethyl alcohol at any ratio. Odor perception threshold is 8.6 mg/l. TF does not alter the taste of water at a concentration of 5.0 g/l.[1,2]

Applications. Used as a solvent in the production of plastics and synthetic fibers.

Acute Toxicity. LD_{50} values are 2.5 g/kg BW in rats, 2.3 g/kg BW in mice, and 3.0 g/kg BW in guinea pigs.[1] Acute poisoning resulted in narcotic effect.

Repeated Exposure revealed cumulative properties.[1] Mice were dosed by gavage with 100 mg/kg BW for 1.5 months. The treatment caused retardation of BW gain, transient hindlimb paralysis, leukocytosis and anemia. Some of the animals died. A 40 mg/kg BW dose appeared to be ineffective.

Long-term Toxicity. Similar changes were found in the 6-month study in mice exposed to oral dose of 10 mg/kg BW.[1] The NOAEL was identified to be 5.0 mg/kg BW.[2]

Allergenic Effect is not observed on chronic oral exposure to a dose of 20 mg/kg BW.[3]

Regulations. ***U.S. FDA*** (1998) approved the use of TF (1) as a component of adhesives to be safely used in food-contact surface in accordance with the conditions prescribed by 21 CFR part 175.105; and (2) as a defoaming agent in the manufacture of paper and paperboard intended for use in producing, manufacturing, packing, transporting, or holding food in accordance with the conditions prescribed by 21 CFR part 176.210.

Standards. ***Russia*** (1995). MAC and PML: 0.5 mg/l (organolept., odor).

References:

1. Pozdnyakova, A. G., in *Proc. Leningrad Sanitary Hygiene Medical Institute*, Leningrad, Issue No 81, 1965, 91 (in Russian).
2. Teplyakova, E. V. et al., in *Proc. Leningrad Sanitary Hygiene Medical Institute*, Leningrad, Issue No 106, 1974, 51 (in Russian).
3. Ilichkina, A. G., Method of investigation of allergenic properties of chemicals in the course of their standard-setting in water bodies, *Gig. Sanit.*, 12, 42, 1979 (in Russian).

THIOPHENE

Molecular Formula. C_4H_4S
M = 84.7

CAS No 110-02-1
RTECS XM7350000

Synonyms and **Trade Names.** Divinylene sulfide; Thiofuran; Thiofurfuran; Thiole; Thiotetrole.

Properties. Colorless liquid with an odor reminiscent of benzene. Soluble in water up to 3.2 g/l (20°C). Odor perception threshold is 2.0 mg/l. At concentration of 500 mg/l, T. is found to have a sweetish taste, with a slightly astringent aftertaste.

Applications. T. is used as a solvent in the production of resins from thiophene-phenol mixtures and formaldehyde. A vulcanization accelerator.

Acute Toxicity. Rats died in 3 hours after administration of 4.0 g/kg BW, and in 4 to 6 days after administration of 2.0 g/kg BW. Manifestations of the toxic action included initial excitation with subsequent general depression and death due to respiratory arrest. Gross pathology examination revealed congestion in the brain and viscera.[1]

Short-term Toxicity. Rats were dosed by gavage with up to 500 mg/kg BW over a period of 4 months. The treatment did not affect the prothrombin-forming function of the liver or serum cholinesterase activity.[1]

Chemobiokinetics. Rats were administered 200 to 300 mg T./kg BW by stomach tube. T. was found to be partially excreted unchanged with air (32%), feces (<1.0%), and urine (40%). Two mercapturic acids, namely, *2-thienylmercapturic acid* and *3-hydroxy-2,3-dihydro-2-thienyl mercapturic acid*, were determined in urine of rats and rabbits (administered 150 to 225 mg T./kg BW).[2]

Standards. ***Russia*** (1995). MAC and PML: 2.0 mg/l (organolept., odor).

References:

1. Rubets, V. I., in *Protection of Water Reservoirs against Pollution by Industrial Liquid Effluents*, S. N. Cherkinsky, Ed., Medgiz, Moscow, Issue No 7, 1965, 219 (in Russian).
2. Bray, H. G., Carpanini, F. M. B., and Water, B. D., The metabolism of thiophen in the rabbit and the rat, *Xenobiotica*, 1, 157, 1971.

TOLUENE

Molecular Formula. C_7H_8
M = 92.14
CAS No 108-88-3
RTECS No XS5250000

Synonyms and **Trade Name.** Methylbenzene; Methacide; Phenylmethane; Toluol.

Properties. Clear, colorless liquid with a characteristic odor. Solubility in water is 470 mg/l at 16°C or 540 mg/l at 25°C, miscible with alcohol at all ratios. Odor perception threshold is 0.17 mg/l[010] or 0.042 mg/l.[02] The reported taste perception thresholds vary from 0.4 to 0.12 mg/l. In Japan and in The Netherlands, odor threshold values are regulated at the levels of 0.92 and 0.99 ppm, respectively.[1]

Applications. Used as a solvent in the production of resins, paints, adhesives, *etc.*

Migration Data. T. is found to migrate from synthetic coating materials commonly used to protect drinking water storage tanks.[2] Release of T. from polystyrene food contact wares into *n*-heptane was not observed at the detection limit of 0.1 ppm.[3]

Exposure. Along with other light, high-octane aromatic hydrocarbons, T. is a component of motor gasoline. There is limited information concerning the oral intake of T. and this intake is likely to be much lower compared to the intake via air.

Acute Toxicity. In rats, oral LD_{50} varies from 2.6 to 7.5 g/kg BW.[4-6] Administration of the high doses led to an increase in the concentration of mediators in the brain with subsequent affection of behavioral responses.

Repeated Exposure failed to reveal cumulative properties. Exposure to T. may cause adverse effects on the NS, kidneys and liver. Several daily doses of 2.0 to 4.0 g/kg BW caused mortality in mice. Administration of 216 to 433 mg/kg BW for 25 days produced no signs of intoxication.[7]

Short-term Toxicity. In NTP study, F344 rats were administered T. in corn oil at dose levels up to 5.0 mg/kg BW for 13 weeks. Liver-to-brain ratio was significantly increased in males receiving the dose of 625 mg/kg BW. Histological examination revealed necrosis of the brain and hemorrhage of the urinary bladder.

A NOAEL of 312 mg/kg BW, adjusted to 223 mg/kg BW for exposure of five days per week, was established in this study. High rate of mortality was observed in mice treated with 2,500 and 5,000 mg/kg BW doses.[8]

Long-term Toxicity. Wistar rats were gavaged with 18 to 422 mg/kg BW given to them for half a year. The highest dose tested was considered to be the NOAEL.[5]

Rabbits received 0.25 and 1.0 mg T./kg BW for 9.5 months, and 10 mg T./kg BW for 5 months. In both studies, there were no changes in general condition, blood, and biochemistry analyses, and in the morphological and histological structure of the visceral organs.[7]

Reproductive Toxicity. The reproductive and developmental toxicity of T. was reviewed recently by Donald et al.[9]

Gonadotoxicity.

Observations in man. Occurrence of testicular atrophy and suppressed spermatogenesis was described in a 28-year-old man who was addicted to T. for 10 years and died as a result of excessive sniffing.[10] High paternal exposure to T. was associated with an increased risk of spontaneous abortion and a slight, nonsignificant increased risk of corgenital malformations in the offspring.[11]

Teratogenicity.

Observations in man. Malformations were described in five children born of women who had sniffed T. during their pregnancies. The anomalies included microcephaly, CNS dysfunction, growth deficiency, and craniofacial changes.[12]

Animal studies.

Embryotoxic and ***fetotoxic effect***, but not teratogenic effect, has been observed at high dose levels in mice and rats, but not in rabbits. In one of two oral studies, embryolethality occurred at dose levels of more than 260 mg/kg BW with a teratogenic effect (increased incidence of cleft palate) at the highest dose-level (870 mg/kg BW) only. Another oral study in mice was limited to behavioral parameters with an effect seen at 400 mg/l in drinking water but not at 80 mg/l.[13,14]

Mutagenicity. T. was not found to be genotoxic.

Observations in man. No changes were found in the peripheral blood lymphocytes of workers exposed daily to T.[15]

In vitro genotoxicity. T. does not cause SCE in human lymphocyte culture,[16] and did not produce CA and SCE in Chinese hamster ovary cells in the presence or absence of exogenous metabolic activation. It was negative in *Salmonella* mutagenicity assay.[8]

In vivo cytogenetics. There is no cytogenic effect in the bone marrow cells of mice on 10 administrations of 0.0001 to 0.1 LD_{50}. When given to males for 5 weeks, it has no effect on the incidence of DLM in mice. Positive results were found in micronuclear test in random-bred SNK mice given 8.0 to 1000 mg T./kg BW in sunflower oil by gavage. The safe concentration of T. in water for genotoxic effect is 20 mg/l.[17]

Carcinogenicity. In a 2-year study, T. was administered to Sprague-Dawley rats in olive oil by stomach tube. An increase in the total number of animals with malignant tumors (types unspecified) was found.[18] The incomplete reporting of tumor pathology in this study was indicated. Available studies are considered by IARC inadequate for evaluation.

Carcinogenicity classifications. An IARC Working Group concluded that there is inadequate evidence for the carcinogenicity of T. in *experimental animals* and in *humans*.

IARC: 3;

NTP: NE - NE - NE - NE (inhalation).

Chemobiokinetics. T. appears to be absorbed completely from the GI tract after oral intake. After absorption, it is rapidly distributed in the body and accumulates preferentially in the highly vascularized organs and adipose tissue successively followed by adrenals, kidneys, liver, and brain. T. metabolism occurs through the oxidation of the methyl group to *benzyl alcohol* by the microsomal mixed-function oxidase system in the liver. Another biotransformation route includes subsequent oxidation, sulfonation and conjugation with glutathione. 80% of the dose is excreted as *hippuric acid* in the urine of man and rabbit. In the lung, part of the resorbed amount of T. is excreted unchanged.[021]

T. crosses the placenta, but is not converted to hippuric acid by the fetus or neonate.[19]

Regulations. ***U.S. FDA*** (1998) approved the use of T. (1) in adhesives as a component of articles intended for use in packaging, transporting, or holding food in accordance with the conditions prescribed by 21 CFR part 175.105; (2) in the manufacture of resinous and polymeric coatings for polyolefin films to be safely used as a food-contact surface in accordance with the conditions prescribed by 21 CFR part 175.320; (3) as a component of the uncoated or coated food-contact surface of paper and paperboard that may be safely used in producing, manufacturing, packing, transporting, or holding dry food in accordance with the conditions prescribed by 21 CFR part 176.180; (4) in the manufacture of semirigid and rigid acrylic and modified acrylic plastics to be safely used as articles intended for use in contact with food in accordance with the conditions prescribed by 21 CFR part 177.1010; (5) as a solvent in polysulfide polymer-polyepoxy resins used as the surface contacting dry food in accordance with the conditions prescribed by 21 CFR part 177.1650; (6) as an adjuvant which may be used in the manufacture of foamed plastics intended for use in contact with food, subject to the provisions prescribed in CFR 178.3010; (7) in poly(2,6-dimethyl- 1,4-phenylene) oxide resins which may be used in the manufacture of articles intended for use in contact with food (not to exceed 0.2% as residual solvent); and (8) in cellophane to be safely used for packaging food (residue limit of 0.1%) in accordance with the conditions prescribed by 21 CFR part 177.1200.

Recommendations.

WHO (1996). Guideline value for drinking water: 0.7 mg/l. The levels of 0.024 to 0.17 mg/l are likely to give rise to consumer complaints of foreign taste and odor.

U.S. EPA (1999). Health Advisory for a longer-term exposure: 7.0 mg/l.

Joint FAO/WHO Expert Committee on Food Additives: ADI is not specified.

Standards.

U.S. EPA (1999). MCL and MCLG: 1.0 mg/l.

Canada (1989). MAC: 0.024 mg/l (organolept.)

Russia (1995). MAC and PML: 0.5 mg/l (organolept., odor).

References:

1. Hoshika, Y., Imamura, T., Muto, G., Van Gemert, L. J., Don, J. A., and Walpot, J. I., International comparison of odor threshold values of several odorants in Japan and in The Netherlands, *Environ. Res.*, 61, 78, 1993.
2. Bruchet, A., Shipert, E., and Alban, K., Investigation of organic coating material used in drinking water distribution systems, *J. Francais d'Hydrologie*, 19, 101, 1988.
3. Ito, S., Hosogai, T., Sakurai, H., Tada, Y., Sugita, T., Ishiwata, H., and Takada, M., Determination of volatile substances and leachable components in polystyrene food contact wares, Abstract, *Eisei Shikenjo Hokoku,* 110, 85, 1992 (in Japanese).
4. van der Heijeden, C. A., Mulder, H. C. M., de Vrijer, F., et al., *Integrated Criteria Document: Toluene Effects,* Appendix to Report No 75847310, Natl. Institute Publ. Health Environ. Protect., Bilthoven, The Netherlands,1988.
5. Wolf, M. A., Rowe, V. K., McCollister, D. D., Hollingsworth, R. L., and Oyen, F., Toxicological studies of certain alkylated benzenes and benzene, *Arch. Ind. Health*, 14, 387, 1956.
6. Kimura, E. T., Ebert, D. M., and Dodge, P.W., Acute toxicity and limits of solvents residue for sixteen or ganic solvents, *Toxicol. Appl. Pharmacol.*, 19, 699, 1971.
7. Abramovich, G. A., Belova, R. S., et al., in *Protection of Water Reservoirs against Pollution by Industrial Liquid Effluents*, S. N. Cherkinsky, Ed., Medgiz, Moscow, Issue No 4, 1960, 109 (in Russian).
8. *Toxicology and Carcinogenesis Studies of Toluene in F344 Rats and $B6C3F_1$ Mice (Inhalation Studies*), NTP Technical Report Series No 371, Research Triangle Park, NC, February 1990.
9. Donald, J. M., Hooper, K., and Hopenhayn-Rich, C., Reproductive and developmental toxicity of toluene: a review, *Environ. Health Perspect.*, 94, 237, 1991.
10. Suzuki, T., Kashimura, S., and Umetsu, K., Thinner abuse and aspermia, *Med. Sci. Law*, 23, 199, 1983.

11. Taskinen, H., Anttila, A., Lindbohm, M.-L., and Sallmen, M., Spontaneous abortions and congenital malformations among the wives of men occupationally exposed to organic solvents, *Scand. J. Work Environ. Health*, 15, 345, 1989.
12. Hersh, J. H., et al., Toluene embryopathy, *J. Pediat.*, 106, 922, 1985.
13. *Toluene*, Environmental Health Criteria Series No 52, WHO, IPSC, Geneva, 1985.
14. Kostas, J. and Hotchin, J., Behavioral effects of low-level perinatal exposure to toluene in mice, *Neurobehav. Toxicol. Teratol.*, 3, 467, 1981.
15. Maki-Paakkanen, J., Husgafvel-Pursiainen, K., Kolliomaki, P. L., et al., Toluene-exposed workers and chromosomal aberrations, *J. Toxicol. Environ. Health*, 6, 775, 1980.
16. Gerner-Schmidt, P. and Freidrich, U., The mutagenic effect of benzene, toluene, and xylene studied by the SCE technique, *Mutat. Res.*, 85, 313, 1978.
17. Feldt, E. G., Assessment of mutagenic hazards of benzene and its derivatives, *Gig. Sanit.*, 7, 21, 1985 (in Russian).
18. Maltoni, C., Conti, B., and Cotti, G., Experimental studies on benzene carcinogenicity at the Bologna Institute of Oncology: current results and ongoing research, *Am. J. Ind. Med.*, 7, 415, 1985.
19. Goodwin, T. M., Toluene abuse and renal tubular acidosis in pregnancy, *Obstet. Gynecol.*, 71, 715, 1988.

TRIBUTYLAMINE

Molecular Formula. $C_{12}H_{27}N$
M = 185.36
CAS No 102-82-9
RTECS YA0350000
Abbreviation. TBA.

Synonym. *N,N'*-Dibutyl-1-butanamine.

Properties. Hygroscopic, colorless, oily liquid with a penetrating odor. Solubility in water is 0.1%; readily soluble in alcohols and fats. Odor and taste perception threshold is 0.9 mg/l.[1]

Applications. A principal solvent in vulcanization processes.

Acute Toxicity. LD_{50} is reported to be 115 mg TBA/kg BW in mice, 455 mg TBA/kg BW in rats, 350 mg TBA/kg BW in guinea pigs, and 615 mg TBA/kg BW in rabbits. Poisoned animals displayed a short period of excitation with subsequent inhibition and adynamia. Death occurs within 2 to 4 days. Histology examination revealed dystrophic changes in the liver, necrosis, white pulp hyperplasia in the spleen and myocardial edema.

Repeated Exposure failed to reveal cumulative properties. Mice tolerate administration of 1/5 LD_{50} for 20 days.[1]

A 3-day administration of 1/3 LD_{50} caused enzyme disorganization of the intracellular structures of the liver cells, especially of those associated with detoxifying and energy biotransformation processes.[2]

Long-term Toxicity. Rabbits were dosed by gavage with 0.6 and 6.0 mg TBA/kg BW for 6 months. The treatment did not affect BW gain, but in the higher dose group, changes were noted in the blood-clotting time system. There was an increase in glutaminate oxalate transaminase activity and decline in blood histamine level as well as in liver deaminoxidase activity. Histology examination revealed splenic hyperplasia at the higher dose, liver fatty dystrophy and intestinal mucosa edema.[1]

Chemobiokinetics. See ***Dibutylamine***.

Standards. ***Russia.*** (1994). MAC and PML: 0.9 mg/l (organolept., odor).

References:

1. Le Din Min, *Hygienic Assessment of Liquid Effluents from the Manufacture of Normal Butylamines and Protection of Reservoirs*, Author's abstract of thesis, Leningrad, 1977, 24 (in Russian).
2. Sidorin, G. I., Lukovnikova, L. V., and Stroikov, Yu. N., Toxicity of some aliphatic amines, *Gig. Truda Prof. Zabol.*, 11, 50, 1984 (in Russian).

1,1,1-TRICHLOROETHANE

Molecular Formula. $C_2H_3Cl_3$

M = 133.42

CAS No 71-55-6

RTECS No KJ2975000

Abbreviation. TCA.

Synonyms and **Trade Name.** Chlorothene; Methylchloroform; Trichloromethylmethane.

Properties. Colorless, nonflammable liquid. Water solubility is 44 mg/l at 25°C. Absorbs some water. Soluble in acetone, methanol, and ether.

Applications. Used as a solvent in the production of plastics, adhesives and coatings.

Exposure. TCA is found in small amounts as a contaminant in various foodstuffs.

Acute Toxicity.

Observations in man. Non-lethal acute intoxication occurred after oral ingestion of a liquid ounce of TCA (0.6 g/kg BW).[1] Such an intake is accompanied with nausea, vomiting and diarrhea. Lethal dose in man is likely to be 5.0 ml.

Animal studies. LD_{50} for several species of animals ranges from 5.7 to 14.3 g/kg BW.[2] It is 14.3 g/kg BW in rats, and 8.6 g/kg BW in guinea pigs.[3]

A single oral dose of about 1.4 g/kg BW depressed some hepatic microsomal metabolic indices in rats (including cytochrome *P-450* and epoxide hydratase).[4]

Repeated Exposure. Rats were gavaged with 0.5 g TCA/kg BW for 9 days. There was relatively little evidence of toxicity. Doses of 5.0 and 10 g/kg BW caused a transient hyperexcitability and protracted narcosis.[5]

TCA was administered by gavage in corn oil to groups of five F344/N rats once a day for 21 days. The doses selected were 0.62 and 1.24 mmol/kg BW. The treatment caused no clinical signs of toxicity. Histological examination revealed no microscopic lesions in the liver and kidneys.[6]

Short-term Toxicity. Doses of 0.5 to 5.0 g/kg BW were administered to rats by gavage for up to 12 weeks. Doses of 2.5 to 5.0 g/kg BW reduced BW gain and produced CNS effects. Approximately 35% of these rats died during the first 7 weeks of the experiment. The dose of 0.5 g/kg BW was ineffective.[5]

Long-term Toxicity. Principal non-carcinogenic effects include CNS inhibition, increased liver weights, and cardiovascular changes. Rats were given 0.75 and 1.5 g /kg BW in corn oil by gavage for 78 weeks. Similarly mice received 2.8 and 5.6 g/kg BW for 78 weeks. Diminished BW gain and decreased survival time were reported in both rats and mice. Selected dose levels were very high in this study, and only 3.0% of the animals survived to the end of the experiment.[7]

Reproductive Toxicity.

Gonadotoxicity.

Observations in man. A twofold increased risk for spontaneous abortions was reported among wives of rubber manufacturing workers exposed to TCA.[8]

Embryotoxicity. There were no dose-dependent effects on fertility, gestation, or viability indices in mice ex posed to TCA at dose levels up to 1.0 g/kg BW in drinking water.[3]

Teratogenicity. Neither maternal toxicity nor significant effect on the morphological development of CD rats were found in the study where animals received 3.0, 10, and 30 ppm TCA in their drinking water for 14 days prior to cohabitation and for up to 13 days during the cohabitation period and during pregnancy.[9]

Mutagenicity.

In vitro genotoxicity. TCA appeared to be mutagenic to various strains of *Salmonella* with metabolic activation.[3] However, positive and negative results in *Salmonella* mutagenicity assays provided minimal evidence of DNA binding in test systems.[6]

TCA did not increase the incidence of CA in the bone marrow cells of cats.[10] TCA was positive in a transformation test with rat embryo cells.[11] It was negative in the test for micronucleus induction in female mouse peripheral erythrocytes. In males, equivocal results were noted (NTP-94).

Carcinogenicity. Rats and mice were gavaged with TCA in corn oil at doses of 375 or 750 mg/kg BW (rats) and 1500 or 3000 mg/kg BW (mice) for 103 weeks. No treatment-related tumors were observed in male rats. The study was inadequate for evaluation of female rats because the high dose was toxic and there was a large number of accidental deaths (NTP-83). However, in mice there was an increased incidence of hepatocellular carcinoma.

Carcinogenicity classifications. An IARC Working Group concluded that there is inadequate evidence for the carcinogenicity of TCA in *experimental animals* and there were no data available to evaluate the carcinogenicity of TCA in *humans*.

IARC: 3;

U.S. EPA: D;

NTP: IS - IS - IS -IS (gavage).

Chemobiokinetics. TCA is metabolized to a very limited extent (no more than 6.0% of the dose). The metabolites include *trichloroethanol,* TCA-*glucuronide,* and *trichloroacetic acid* which are excreted primarily in the urine. Approximately 1.0% of TCA, however, is excreted unchanged by the lungs.[12]

According to D'Urk et al., *acetylene* is found to be a new metabolite of TCA in Sprague-Dawley rats following inhalation at the concentration of 2000 ppm.[13]

Regulations. ***U.S. FDA*** (1998) regulates TCA (1) as a component of adhesives to be safely used in food-contact surface in accordance with the conditions prescribed by 21 CFR part 175.105; and (2) as a cross- linking agent in the manufacture of polysulfide polymer-polyepoxy resins which may be used as food-contact surface of articles intended for packaging, transporting, holding, or otherwise contacting dry food in accordance with the conditions prescribed by 21 CFR part 177.2420.

Recommendations.

WHO (1996). Provisional Guideline value for drinking water: 2.0 mg/l.

U.S. EPA (1999): Health Advisory for a longer-term exposure: 100 mg/l.

Standards. ***U.S. EPA*** (1999). MCL and MCLG: 0.2 mg/l.

References:

1. Stewart, R. D. and Andrews, J. T., Acute intoxication with methyl chloroform vapor, *J. Am. Med. Assoc.,* 195, 705, 1966.
2. Torkelson, T. R., Oyen, F., McCollister, D., et al., Toxicity of 1,1,1-trichloroethane as determined on laboratory animals and human subjects, *Am. Ind. Hyg. Assoc. J.,* 19, 353, 1958.
3. Lane, R. W., Riddle, B. L., and Borzelleca, J. F., Effect of 1,2-dichloroethane and 1,1,1-trichloroethane in drinking water on reproduction and development in mice, *Toxicol. Appl. Pharmacol.*, 63, 409, 1982.
4. Vainio, H., Parkki, M. A., and Marniemi, J. A., Effects of aliphatic chlorohydrocarbons on drug-metabolizing enzymes in rat liver *in vivo*, *Xenobiotica*, 6, 599, 1976.
5. Bruckner, J. V., Muralidhara, S., Mackenzie, W. F., et al., Acute and subacute oral toxicity studies of 1,1,1-trichloroethane in rats, *Toxicologist*, 5, 100, 1985.
6. *Renal Toxicity Studies of Selected Halogenated Ethanes Administered by gavage to F344/N Rats*, NTP Technical Report, Research Triangle Park, NC, NIH Publ. 96-3935, February 1996, 52.
7. *Bioassay of 1,1,1-Trichloroethane for Possible Carcinogenicity*, Natl Cancer Inst. Technical Report Series No 3, January, 1977.
8. Lindbohm, M.-L., Hemminki, K., Bonhomme, M. G., Anttila, A., Rantala, K., Heikkila, P., and Rosenberg, M. J., Effect of paternal occupational exposure on spontaneous abortions, *Am. J. Ind. Med.*, 81, 1029, 1991.
9. George, J. D., Price, C. J., Marr, M. C., et al., Developmental toxicity of 1,1,1-trichloroethane in CD rats, *Fundam. Appl. Toxicol.*, 13, 641, 1989.
10. Rampy, L. W., Quast, J. F., Leong, B. K. J., and Gehring, P. J., Results of long-term inhalation toxicity studies on rats of 1,1,1-trichloroethane and perchloroethylene formulations, in *Proc. Int. Congr. Toxicol.*, Toronto, 1977.
11. Price, P. J., Hassett, C. M., and Mansfield, J. I., Transforming activities of trichloroethylene and proposed industrial alternatives, *In Vitro*, 14, 290, 1978.
12. Monster, A. C., Boersma, G., and Steenweg, M., Kinetics of 1,1,1-trichloroethane in volunteers; influence of exposure concentration and work load, *Int. Arch. Occup. Environ. Health*, 42, 293, 1979.

13. D'Urk, H., Poyer, J., Lee, K. C., et al., Acetylene, a mammalian metabolite of 1,1,1-trichloroethane, *Biochem. J.*, 286, 353, 1992.

1,1,2-TRICHLOROETHANE

Molecular Formula. $C_2H_3Cl_3$

M = 133.40

CAS No 79-00-5

RTECS No KJ3150000

Abbreviation. TCA.

Synonyms. Vinyl trichloride; Ethane trichloride.

Properties. Clear, colorless liquid with a sweet odor. Water solubility is 4.50 g/l at 20°C. Soluble in ethanol and diethyl ether.

Applications. An intermediate in the production of vinylidene chloride; a solvent for natural resins, chlorinated rubbers, adhesives, and coatings laid down on films.

Acute Toxicity. In CD-1 mice, LD_{50} is found to be 378 mg/kg BW (males) and 491 mg/kg BW (females). Signs of toxicity included sedation, gastric irritation, lung hemorrhage, and liver and kidney damage.[1]

Repeated Exposure. In a 14-day study, doses of 3.8 and 38 mg/kg BW caused no toxic effects in CD-1 mice.[2]

Short-term Toxicity. In a 90-day study, CD-1 mice were given 0.02 to 2.0 g/l in their drinking water. A decrease in BW gain and liver glutathione level in males was observed. Decreased hematocrit and *Hb* blood levels were found in females. The LOEL was found to be 46 mg/kg BW.[1,2] Munro et al. (1996) suggested the calculated NOEL of 4.0 mg/kg BW for this study.[3]

Long-term Toxicity. In rabbits, an increase in excretion of 17-ketosteroides was noted on months 1 to 3 of the treatment with a dose of 100 mg/kg BW. Normalization was observed up to the 5th month of experiment.[4]

Immunotoxicity. In the above described studies, antibody-forming function and phagocytic activity of the peritoneal macrophages were altered.[1,2]

Reproductive Toxicity.

Embryotoxicity. The oral dose that killed 10% of the pregnant mice caused no developmental toxicity in the offspring of survivors.[5]

Mutagenicity.

In vitro genotoxicity. TCA was negative in *Salmonella* mutagenicity assays.[6]

Carcinogenicity. TCA is structurally related to *1,2-dichloroethane*, a probable human carcinogen.

$B6C3F_1$ mice were given doses of 150 and 300 mg/kg BW by gavage in corn oil for 8 weeks, then doses of 200 and 400 mg/kg BW for 70 weeks; Osborne-Mendel rats received 35 and 70 mg/kg BW for 20 weeks, then 50 and 100 mg/kg BW for 58 weeks. Hepatocellular neoplasms and adrenal phaeochromocytomas were observed in mice. No increase in tumor incidence was found in rats.[7]

Carcinogenicity classifications. An IARC Working Group concluded that there is limited evidence for the carcinogenicity of TCA in *experimental animals* and there were no adequate data available to evaluate the carcinogenicity of TCA in *humans*.

IARC: 3;

U.S. EPA: C;

NTP: N - N - P* - P* (gavage).

Chemobiokinetics. In rats, TCA is metabolized by hepatic cytochrome *P-450* to *chloroacetic acid*[8] and to *inorganic chloride*. According to Ikeda et al., following acute inhalation exposure or *i/p* injections of TCA to rats, urinary metabolites identified were *trichloroacetic acid, trichloroethanol, chloroacetic acid,* and *thiodiacetic acid.*[9]

Regulations. ***U.S. FDA*** (1998) regulates TCA as a component of adhesives to be safely used in food-contact surface in accordance with the conditions prescribed by 21 CFR part 175.105.

Standards. ***U.S. EPA*** (1999). MCL: 0.005 mg/l; MCLG: 0.003 mg/l.

References:

1. White, K. L., Sanders, V. M., Barness, D. W., Shopp, G, M., and Munson, A. E., Toxicology of 1,1,2-trichloroethane in the mouse, *Drug Chem.Toxicol.*, 8, 333, 1985.
2. Sanders, V. M., White, K. L., Shopp, G. M., et al., Humoral and cell-mediated immune status of mice exposed to 1,1,2-trichloroethane, *Drug Chem. Toxicol.*, 8, 357, 1985.
3. Munro, I. C., Ford, R. A., Kennepohl, E., and Sprenger, J. G., Correlation of structural class with No-Observed-Effect Levels: A proposal for establishing a threshold of concern, *Food Chem. Toxicol.*, 34, 829, 1996.
4. Kashin, L. M., in *Endocrine Systems and Toxic Environmental Factors*, Proc. Sci. Conf., Leningrad, 1980, 139 (in Russian).
5. Seidenberg, J. M., Anderson, D. G., and Becker, R. A., Validation of an *in vivo* development toxicity screen in the mouse, *Teratogen. Carcinogen. Mutagen.*, 6, 361, 1986.
6. Simmon, V. P., Kauhanon, K., and Tardiff, R. G., Mutagenic activity of chemicals identified in drinking water, in *Progress in Genetic Toxicology*, D. Scott et al., Eds., Elsevier/North Holland Biomedical Press, Amsterdam, 1977, 249.
7. *Bioassay of 1,1,2-Trichloroethane for Possible Carcinogenicity*, Natl. Cancer Institute Technical Report Series No 74, DHEW Publ. No 78-1324, U.S. Dept. Health, Education & Welfare, Washington, D.C., 1978.
8. Ivanetich, K. M. and van den Honert, L. N., Chloroethanes: their metabolism by hepatic cytochrome *P-450 in vitro*, *Carcinogenesis*, 2, 697, 1981.
9. Ikeda, M. and Ohtsuji, H., A comparative study of the excretion of Fujiwara reaction-positive substances in urine of humans and rodents given trichloro- or tetrachloro-derivatives of ethane and ethylene, *Brit. J. Ind. Med.*, 29, 99, 1972.

TRICHLOROETHYLENE

Molecular Formula. C_2HCl_3
M = 131.38
CAS No 79-01-6
RTECS No KX4550000
Abbreviation. TCE.

Synonyms and **Trade Names.** Acetylene trichloride; Chlorylene; Ethylene trichloride; Ethynyl trichloride; Trichloran; Trichloroethene; Triclene; Trilene.

Properties. Colorless, volatile, non-combustible liquid with an odor of chloroform. Water solubility is 1.0 g/l at 20°C. Miscible with alcohols and dissolves oils. Organoleptic perception threshold is 0.31 mg/l (WHO, 1987). According to other data, odor perception threshold is 0.5 mg/l; practical perception threshold is 1.0 mg/l. Can be tasted at concentrations 50 to 100 times greater.[1]

Applications. A solvent for resins and rubber, and a component of lacquers and adhesives.

Exposure. In the U.S., TCE was found in 5 of 372 fatty and non-fatty food samples at a mean concentration of 0.051 mg/kg.[2]

Acute Toxicity.

Observations in man. TCE seems to be a classical CNS depressant. Fatal hepatic failure has been observed following the use of TCE as an anesthetic. Oral exposure to 15 to 25 ml TCE resulted in vomiting and abdominal pain, followed by transient unconsciousness.[3]

Animal studies. LD_{50} is 4.92 g/kg BW in rats.[06] In mice, it is reported to be 2.4 g/kg BW (females) and 2.44 g/kg BW (males).[4] Predominantly affects CNS activity and liver functions. Single oral administration of 0.5 to 1.5 g/kg BW caused an increase in the liver weight and a decline in DNA concentration in the liver of rats and mice.

Repeated Exposure. In a 14-day study, male CD-1 mice were given 1/10 and 1/100 LD_{50} (24 and 240 mg/kg BW, respectively). An increase in the liver weights was observed at the higher dose level.[4] The NOAEL of 100 mg/kg BW is reported for minor effects on relative liver weights in a 6-week reproduction study in rats.[4]

Female Swiss mice were dosed with 0.5, 1.0, and 2.0 mg/kg BW for 4 weeks. Increase in the liver size and stimulation of proliferation of sinusoid cells, degeneration and necrotization of hepatocytes were found in all dosed animals. Kidney were affected at the 2.0 mg/kg BW dose level.[5]

B6C3F$_1$ mice were given TCE by gavage for 10 days. No histopathological changes were found in the liver. Moderate changes around central veins were observed in mice that received 1.0 g/kg BW. The author suggested that liver cell DNA synthesis and mitosis are stimulated by TCE and that these effects may be, in part, responsible for transformation of liver cells in these mice.[6] Male NMRI mice were orally exposed to TCE (50 and 290 mg/kg BW) between days 10 and 16 postnatally. 17-day-old mice were unaffected. At 60 days of age, TCE-exposed mice were only affected in the rearing.[7]

Short-term Toxicity. CD-1 mice of both sexes received TCE in their drinking water at concentrations of 0.1 to 5.0 mg/l for 4 to 6 months. A decrease in BW gain was noted in the high-dose group. An increase in liver weights was accompanied by increased non-protein sulfhydrile levels in males; an increase in kidney weights was accompanied by increase in protein and ketone levels in the urine.[4]

F344 rats and B6C3F$_1$ mice were administered TCE in corn oil by gavage for 13 weeks. Survival in mice was greatly decreased at 3.0 and 6.0 g TCE/kg BW dose levels. Histology examination revealed changes in the renal tubular epithelium at 1.0 g/kg (rats) and 3.0 g/kg BW (mice) dose levels.[8]

Long-term Toxicity. Oral administration to mice induced hepatic peroxysome proliferation; however, no such effect was observed in rats.[9]

Mice received 0.1 to 5.0 mg TCE/l in their drinking water for 4 to 6 months. The treatment caused an increase in the liver and kidney weights and in the concentration of ketones and albumen in the urine of mice.[3] The NOAEL of 0.5 mg/kg BW is reported for rats and rabbits.[1] However, in more recent studies, the LOAEL of 500 mg/kg BW in rats and 1000 mg/kg BW in mice was determined for signs of toxic nephrosis.[8]

Immunotoxicity. Exposure to 24 and 240 mg/kg BW for 14 days resulted in depression of the cellular, but not of humoral immunity in mice.[10]

In order to clarify the role of TCE in the pathogenesis of autoimmune responses, mice received *i/p* injection of 10 mmol TCE/kg BW, 0.2 mmol *dichlroloacetyl chloride*/kg BW (one of the metabolites of TCE with strong acylating property), or an equal volume (100 microliters) of corn oil alone (controls). Animals were dosed every 4th day for 6 weeks. TCE and its metabolite induced and/or accelerated autoimmune responses in female *MRL* +/+ mice.[11]

Reproductive Toxicity.

Gonadotoxicity.

Observations in man. No difference was found with respect to sperm count and morphology between controls and metal workers moderately exposed to TCE in degreasing process (mean TCE concentration in urine, 3.7 mg/l).[12]

Animal studies. Testicular and epididymal weights were found to be decreased, but no histology changes were noted in F344 rats fed the diet containing 75 to 300 mg TCE/kg.[13] Reduced sperm motility was observed in CD-1 mice given 750 mg/kg BW in a continuous breeding-fertility study.[14]

Decreased testes weights, subtle sperm morphology abnormalities, and disruption of copulatory behavior were reported in rodents exposed to high levels of TCE (up to 2000 ppm).[15,16]

Embryotoxicity. TCE passes through the placenta and penetrates the fetal blood.[17] Inhalation exposure of rats to relatively high concentrations (1800 ppm) caused an embryotoxic effect manifested as retarded development. B6C3F$_1$ mice were gavaged from day 1 to 5, 6 to 10, or 11 to 15 of gestation with TCE in corn oil at dose levels of 1/10 or 1/100 LD_{50}. No maternal or reproductive effects have been found at either dose level. TCE was not found to cause any effect on reproduction at dose level up to 1/10 LD_{50}.[18]

In contrast with this conclusion, Narotsky and Kavlock pointed out that TCE caused high incidence of full-litter resorptions.[19]

Teratogenicity. TCE was not found to be teratogenic when administered in corn oil at doses of 0.5 and 1.0 g/kg BW[20] or when inhaled by pregnant rats and mice. Human epidemiological studies and previous teratogenic studies using chick embryos and fetal rats have shown an increased incidence of congenital cardiac lesions in animals exposed to TCE. Varying doses of TCE were administered to pregnant rats during fetal heart development. Maternal and fetal variables showed no statistically significant differences between treated and untreated groups. Fetuses of rats did not demonstrate a significant increase in cardiac defects compared with controls. TCE was not found to be a specific cardiac teratogen in the fetus when imbibed by the maternal rat.[21,22]

Mutagenicity. TCE appeared to be mainly negative in gene and chromosome mutation tests *in vivo* and *in vitro*, positive in micronucleus tests *in vitro* and *in vivo*, and positive in recombination tests.[23]

Observations in man. TCE produced SCE and unscheduled DNA synthesis in the human lymphocytes. No induction of sperm abnormalities was noted.

In vitro genotoxicity. TCE showed equivocal response in a number of bacterial strains. It induced transformation of mouse and rat cells but not of Syrian hamster cells. No SCE were observed in Chinese hamster cells *in vitro* or unscheduled DNA synthesis in rat hepatocytes. TCE produced no CA, it was weakly positive in mouse lymphoma assay, but induced DNA strand breaks.[23]

In vivo cytogenetics. TCE induced micronuclei and somatic mutations (in the spot test), sperm anomalies and DNA strand breaks in the kidney and liver, but not lung of mice treated *in vivo*. TCE did not produce DLM or micronuclei in spermatids.[23]

Carcinogenicity.

Observations in man. There is no epidemiological evidence of TCE carcinogenicity: epidemiological data are inadequate to refute or demonstrate a human carcinogenic potential (U.S. EPA, 1985. Health Assessment Document on Trichloroethylene).

Animal studies. TCE carcinogenic potential was evident in mice and rats of both sexes exposed by inhalation and orally. Epichlorohydrin-free TCE was reported to be carcinogenic in $B6C3F_1$ mice when administered in corn oil at 1.0 g/kg BW for 103 weeks.[24]

A hepatocellular carcinogenic response was found in $B6C3F_1$ mice (average daily doses 1.2 and 2.34 g/kg BW for males and 0.87 and 1.74 g/kg BW for females) but not in Osborne-Mendel rats exposed to 0.55 g and 1.1 g technical grade TCE/kg BW.[25] Templin et al.[26] believe that the hepatocarcinogenicity of TCE in mice has been attributed to its metabolite, trichloroacetate.

Henschler et al. concluded that TCE containing *ECH* and *epoxybutane* causes tumors in test animals, but purified TCE was not carcinogenic to ICR/HA mice.[24] However, U.S. EPA concluded that Henschler's study used mice which are known to be less responsive to hepatocellular carcinomas than the mice in several other studies.

Carcinogenicity classifications. An IARC Working Group concluded that there is limited evidence for the carcinogenicity of TCE in *experimental animals* and there were inadequate data available to evaluate the carcinogenicity of TCE in *humans*.

IARC: 2A;
U.S. EPA: B2/C;
ARC: 2
EU: 3;
NTP: N - N - P - P (gavage);
IS - N - P* - P (gavage).

Chemobiokinetics. TCE is readily absorbed in the GI tract. The principal products of TCE metabolism measured in the urine are *trichloroacetaldehyde, trichloroetanol, trichloroacetic acid*, and conjugated derivatives (*glucuronides*) of TCE.[27]

According to Templin et al., TCE has been found in the blood, urine, and bile, primarily as the glucuronide conjugate.[26] In mice, metabolism is rapid and gives rise to relatively large amounts of the acid, which could induce liver peroxisome proliferation and cancer. In rats and humans, the rate of oxidation is limited, and, subsequently, cancer risk to humans should be decreased.[28]

TCE and its metabolites are excreted in exhaled air, urine, sweat, feces, and saliva.

Regulations. ***U.S. FDA*** (1998) approved the use of TCE in adhesives used as components of articles intended for use in packaging, transporting, or holding food in accordance with the conditions prescribed by 21 CFR part 175.105.

Recommendations. ***WHO*** (1996): Provisional guideline value for drinking water 0.07 mg/l. However, the lowest reported odor threshold for TCE is 0.3 mg/l.

Standards.

U.S. EPA (1999). MCL: 0.005 mg/l, MCLG: zero.

Canada (1989). MAC: 0.05 mg/l.

Russia (1995). MAC: and PML: 0.04 mg/l.

References:

1. Miklashevsky, V. E., Tugarinova, V. N., et al., in *Protection of Water Reservoirs against Pollution by Industrial Liquid Effluents*, S. N. Cherkinsky, Ed., Medgiz, Moscow, Issue No 5, 1962, 308 (in Russian).
2. Daft, J. L., Rapid determination of fumigant and industrial chemical residues in food, *J. Assoc. Off. Anal. Chem.*, 71, 748, 1988.
3. Stephens, C. A., Poisoning by accidental drinking of trichloroethylene, *Brit. Med. J.*, 2, 218, 1945.
4. Tucker, A. N., Sanders, V. M., Barness, D. W., Bradshaw, T. J., White, K. L., Sain, L. E., Borzelleca, J. F., and Munson, A. E., Toxicology of trichloroethylene in the mouse, *Toxicol. Appl. Pharmacol.*, 62, 351, 1982.
5. Goel, S. K., Rao, G. S., Pandya, K. P., et al., Trichloroethylene toxicity in mice; a biochemical, hematological and pathological assessment, *Indian J. Exp. Biol.*, 30, 402, 1992.
6. Dees, C. and Travis, C., The mitogenic potential of trichloroethylene in $B6C3F_1$ mice, *Toxicol. Lett.*, 69, 129, 1993.
7. Fredriksson, A., Danielsson, B. R., and Eriksson, P., Altered behavior in adult mice orally exposed to tri- and tetrachloroethylene as neonates, *Toxicol. Lett.*, 66, 13, 1993.
8. *Carcinogenesis Studies of Trichloroethylene (without Epichlorohydrin) in F344/N Rats and* $B6C3F_1$ *Mice (Gavage Studies)*, NTP Technical Report Series No 243, Research Triangle Park, NC, 1990.
9. Elcombe, C. R., Rose, M. S., and Pratt, I. S., Biochemical, histological and ultrastructural changes in rat and mouse liver following the administration of trichloroethylene: possible relevance of species differences in hepatocarcinogenicity, *Toxicol. Appl. Pharmacol.*, 79, 365, 1985.
10. Sanders, V. M., Tucker, A. N., White. K. L., et al., Humoral and cell-mediated immune status in mice exposed to trichloroethylene in the drinking water, *Toxicol. Appl. Pharmacol.*, 61, 358, 1982.
11. Khan, M. F., Kaphalia, B. S., Prabhakar, B. S., Kanz, M. F., and Ansari, G. A., Trichloroethene-induced autoimmune response in female MRL +/+ mice, *Toxicol. Appl. Pharmacol.*, 134, 155, 1995.
12. Rasmussen, K., Sabroe, S., Wohlert, M., Ingerslev, H. I., Kappel, B., and Nielsen, J., A genotoxic study of metal workers exposed to trichloroethylene: sperm parameters and chromosome aberrations in lymphocytes, *Int. Arch. Occup. Environ. Health*, 60, 419, 1988.
13. *Trichloroethylene: Reproduction and Fertility Assessment in F344 Rats When Administered in the Feed*, NTP Technical Report, Research Triangle Park, NC, NTP-86-085, 1986.
14. *Trichloroethylene: Reproduction and Fertility Assessment in CD-1 Mice When Administered in the Feed*, NTP Technical Report, Research Triangle Park, NC, NTP-86-068, 1985.
15. Land, P. C., Owen, E. L., and Linde, H. W., Morphologic changes in mouse spermatozoa after exposure to inhalation anesthetics during early spermatogenesis, *Anesthesiology*, 54, 53, 1981.
16. Zenick, H., Blackburn, K., Hope, E., Richdale, N., and Smith, M. K., Effects of trichloroethylene exposure on male reproductive functions in rats, *Toxicology*, 31, 237, 1984.
17. Laham, S. L., Studies on placental transfer, Trichloroethylene, *Ind. Med. Surg.*, 39, 46, 1970.
18. Cosby, N. C. and Dukelow, W. R., Toxicology of maternally ingested trichloroethylene (TCE) on embryonal and fetal development in mice and TCE metabolites on *in vitro* fertilization, *Fundam. Appl. Toxicol.*, 19, 268, 1992.
19. Narotsky, M. G. and Kavlock, K. J., A multidisciplinary approach to toxicological screening: II. Developmental Toxicity, *J. Toxicol. Environ. Health.*, 45, 145, 1995.
20. *Carcinogenesis Bioassay for Trichloroethylene*, NTP Technical Report No 82-1799 (Draft), Research Triangle Park, NC, 1982.

21. Johnson, P. D., Dawson, B. V., and Goldberg, S. J., Cardiac teratogenicity of trichloroethylene metabolites, *J. Am. Coll. Cardiol.*,32, 540, 1995.
22. Johnson, P. D., Dawson, B. V., and Goldberg, S. J., A review: trichloroethylene metabolites: potential cardiac teratogens. *Environ Health Perspect.*, 106 (Suppl 4), 995, 1998.
23. Fahrig, R., Madle, S., and Baumann, H., Genetic toxicology of trichloroethylene (TCE), *Mutat. Res.*, 340, 1, 1995.
24. Henschler, D., Elsasser, W., Romen, W., and Eder, E., Carcinogenicity study of trichloroethylene, with and without epoxide stabilizers, in mice, *J. Cancer. Res. Clin. Oncol.*, 107, 149, 1984.
25. *Carcinogenesis Bioassay of Trichloroethylene*, Natl Cancer Inst., U.S. Dept. Health Education & Welfare, Publ. Health Service, 1976.
26. Templin, M. V., Stevens, D. K., Stenner, R. D., Bonate, P. L., Tuman, D., and Bull, R. J., Factors affecting species differences in the kinetics of metabolites of trichloroethylene, *J. Toxicol. Environ. Health*, 44, 435, 1995.
27. Ikeda, M., Imamura, T., and Ohtsvji, H., Urinary excretion of total trichlorocompounds, trichloroethanol, trichloroacetic acid, as a measure of exposure to trichloroethylene and tetrachloroethylene, *Brit. J. Ind. Med.*, 29, 46, 1970.
28. Green, T., Chloroethylenes: A mechanistic approach to human risk evaluation, *Annu. Rev. Pharmacol. Toxicol.*, 30, 73, 1990.

1,2,3-TRICHLOROPROPANE

Molecular Formula. $C_3H_5Cl_3$
M = 147.43
CAS No 96-18-4
RTECS No TZ9275000
Abbreviation. TCP.

Synonyms. Allyl trichloride; Glycerol trichlorohydrin; Trichlorohydrin.

Properties. Colorless to straw-colored liquid with odor similar to that of trichloroethylene. Water solubility is 1750 mg/l at 25°C. Soluble in alcohol.

Applications. TCP is used as a solvent and degreasing agent, and in the production of polyamide filaments.

Acute Toxicity. LD_{50} is 0.5 g/kg BM in rats, 0.37 g/kg BW in mice, 0.38 g/kg BW in rabbits, and 0.34 mg/kg BW in guinea pigs. Produced effects included somnolence (general depressed activity), ataxia, and changes in blood analysis.[1]

Repeated Exposure. Sprague-Dawley rats were administered TCP by gavage in corn oil at doses of 0.01 to 0.8 mmol/kg BW for 10 days. The treatment caused no lethality. Retardation of BW gain, thymus atrophy, and mild hepatotoxic response was observed at 0.8 mmol (118 mg/kg BW) dose level. Histological examination revealed cardiopathy: myocardial necrosis and degeneration with marked eosinophilia of affected cells.[2]

Short-term Toxicity. F344 rats were gavaged with doses of 8 to 250 mg TCP/kg BW. All animals tolerated doses of up to 63 mg/kg BW. There was a loss of BW at this dose but not at lower doses. Treatment with two highest doses produced BW loss, hunched appearance, depression, and abnormal eye and urine stains. Necrosis and inflammation of the nasal mucosa were found to develop. 32 mg TCP/kg BW dose appeared to be ineffective.[3] Sprague-Dawley rats were administered TCP by gavage in corn oil at the doses of 0.01 to 0.4 mmol/kg BW for 90 days. The treatment caused no lethality. Retardation of BW gain and mild hepatotoxic response was observed at 0.4 mmol (59 mg/kg BW) dose level. Histological examination revealed cardiopathy: myocardial necrosis and degeneration with marked eosinophilia of affected cells.[2]

Reproductive Toxicity. CD-1 mice were administered TCP by gavage at doses of 30, 60, or 120 mg/kg BW during a 7-day precohabitation and 98-day cohabitation period. The treatment caused impairment of the female reproductive system.[4]

Gonadotoxicity. In the above described study, changes in testicular weight, sperm count, and sperm motility were not dose-dependent.[3]

Rats were given 15 daily doses of 37 mg/kg BW by *i/p* injection. No ***embryotoxic*** and ***teratogenic effects*** were observed.[3]

Mutagenicity.

In vitro genotoxicity. TCP was found to be positive in *Salmonella* mutagenicity assay in the presence of activation with *S9*.[4,5]

In vivo cytogenetics. Male Sprague-Dawley rats received 5 doses of 80 mg/kg BW by gavage. The treatment caused no DLM.[4]

Carcinogenicity. TCP is found to be a multispecies, multisite carcinogen.

F344 rats received TCP at doses of 3.0, 10, or 30 mg/kg BW; $B6C3F_1$ mice were administered TCP at doses of 6, 20, or 60 mg/kg BW by gavage for up to 104 weeks. TCP produced tumors of the oral mucosa and of uterus in female mice and increased the incidence of tumors of the forestomach, liver, and Harderian gland in mice of each sex. In mice, it caused the following kinds of tumors to develop: in the preputial gland, kidney and pancreas of males, in the clitoral gland and mammary gland of females, and in the oral cavity and forestomach of both males and females.[3,6]

Carcinogenicity classifications.

IARC: 2A.

U.S. EPA: B2;

ARC: 2

NTP: CE - CE - CE - CE (gavage).

Chemobiokinetics. In F344 rats and $B6C3F_1$ mice, TCP is found to be rapidly absorbed, metabolized, and excreted after oral administration. Distribution occurs predominantly in the liver, kidney, and forestomach.[7]

^{14}C-TCP was administered to male $B6C3F_1$ and F344 rats by gavage. La et al.[8] found formation of one major *DNA adduct* which was distributed widely among organs examined. Authors concluded that factors in addition to adduct formation may be important in TCP-induced carcinogenesis.[8]

Mercapturic acid was found to be the major metabolite in rat urine, and minor component in mouse urine, *2-(S-glutathionyl)malonic acid*, is the major biliary metabolite in rats.[4,7]

Major route of elimination is *via* urine, followed by exhalation as CO_2 and excretion in the feces (each accounted for 20%).

Regulations. ***U.S. FDA*** (1998) approved the use of TCP in the manufacture of polysulfide polymer-polyepoxy resins which may be used as food-contact surface of articles intended for packaging, transporting, holding, or otherwise contacting dry food in accordance with the conditions prescribed by 21 CFR part 177.1650.

Standards. ***Russia*** (1995). MAC and PML: 0.07 mg/l (organolept., odor).

References:

1. *Proc. Kharkov Medical Institute*, Kharkov, Issue No 124, 1976, 27 (in Russian).
2. Merrick, B. A., Robinson, M., and Condie, L. W., Cardiopathic effect of 1,2,3-trichloropropane after subacute and subchronic exposure in rats, *J. Appl. Toxicol.*, 11, 179, 1991.
3. *Toxicology and Carcinogenicity Studies of 1,2,3-Trichloropropane in F344/N Rats and $B6C3F_1$ Mice (Gavage Studies)*, NTP Technical Report Series No 384, Research Triangle Park, NC, NIH Publ. No 94-2829, 1990.
4. Saito-Suzuki, R., Teramoto, S., and Shirasu, Y., Dominant lethal studies in rats with 1,2-dibromo-3-chloropropane and its structurally related compounds, *Mutat. Res.*, 101, 321, 1982.
5. Stolzenberg, S. J. and Hine, C. H, Mutagenicity of 2- and 3-carbon hologenated compounds in the *Salmonella*/mammalian-microsome test, *Environ. Mutagen.*, 2, 59, 1980.
6. Irwin, R. D., Haseman, J. K., and Eustis, S. L., 1,2,3-Trichloropropane: a multisite carcinogen in rats and mice, *Fundam. Appl. Toxicol.*, 25, 241, 1995.
7. Mahmood, N. A., Overstreet, D., and Burka, L. T., Comparative disposition and metabolism of 1,2,3-TCP in rats and mice, *Drug Metab. Disposit.*, 19, 411, 1991.
8. La, D. K., Lilly, P. D., Anderegg, R. J., and Swenberg, J. A., DNA adduct formation in $B6C3F_1$ mice and F344 rats exposed to 1,2,3-trichloropropane, *Carcinogenesis*, 16, 1419, 1995.

TRIDECANOL

Molecular Formula. $C_{13}H_{28}O$

M = 200.41

CAS No 112-70-9

RTECS No YD4200000

Abbreviation. TD.

Synonyms. *n*-Tridecanol; *n*-Tridecyl alcohol.

Properties. Crystals. Insoluble in water; soluble in alcohol.

Applications. Used in the production of detergents, lubricants, defoaming agent.

Acute Toxicity. LD_{50} is 17.2 g/kg BW in rats. Toxic to rats when administered by the inhalation route; caused pulmonary edema and hemorrhage.[1,2]

Regulations. ***U.S. FDA*** (1998) approved the use of TD (1) in adhesives as a component of articles intended for use in packaging, transporting, or holding food in accordance with the conditions prescribed by 21 CFR part 175.105; and (2) as a defoaming agent in the manufacture of paper and paperboard intended for use in producing, manufacturing, packaging, processing, preparing, treating, packing, transporting, or holding foods in accordance with the conditions prescribed by 21 CFR part 176.210.

References:

1. *Raw Material Data Handbook*, vol 1, Organic Solvents, 1, 114, 1974.
2. Gerarde, H. W. and Ahlstrom, D. B., The aspiration hazard and toxicity of a homologous series of alcohols, *Arch. Environ. Health*, 13, 457, 1966.

3,5,5-TRIMETHYL-2-CYCLOHEXEN-1-ONE

Molecular Formula. $C_9H_{14}O$

M = 138.23

CAS No 78-59-1

RTECS No GW7700000

Abbreviation. TMC.

Synonyms and **Trade Names.** Isoacetoforone; Isoforone; 5-Oxo-1,1,3-trimethylcyclohexene-3; 1,1,3-Trimethylcyclohexen-3-one-5.

Properties. Pale-yellow liquid with a specific odor. TMC is not hydrolyzed in water. Odor perception threshold is 0.9 mg/l. Practical perception threshold is 2.92 mg/l.

Applications. TMC is used as a unique solvent in the manufacture of a number of resins and as a chemical intermediate.

Acute Toxicity. LD_{50} is 2.4 g/kg BW (male rats), and 3.7 g/kg BW (females); in mice, LD_{50} is found to be 2.7 g/kg BW.[1]

According to other data, LD_{50} is 1.87 to 3.2 g/kg BW (Coquet, 1977; Exxon, 1982; Smyth et al., 1970)

Repeated Exposure. Does not exhibit cumulative properties.[1]

Rats and mice received up to 2.0 g TMC/kg BW by gavage in corn oil for 16 days. Pathology examination revealed no gross or histopathology effects. The treatment caused high mortality in rats. All mice died during the studies.[2]

Short-term Toxicity. Rats and mice were gavaged with up to 1.0 g TMC/kg BW doses in corn oil. In a 13-week study, there were no findings related to the treatment either at necropsy or at histological examination, but 1/10 high dose female rats and 3/10 high dose female mice died.[2]

Short-term Toxicity. In a 13-week study in rats and mice, the NOAEL of 0.5 g/kg BW was established.[4]

Long-term Toxicity. Chloride levels in the urine and enzymatic activity in the blood serum were affected in rats. Liver functions were found to be altered.[1]

Administration of 250 or 550 mg/kg BW to F344 rats and $B6C3F_1$ mice in corn oil by gavage for 103 weeks caused proliferative lesions of the kidneys including hyperplasia, and epithelial hyperplasia of the renal pelvis. A more severe nephropathy was noted in low dose animals than in controls or high dose animals. In mice, no non-neoplastic lesions in males or females, or neoplastic lesions in females were considered associated with the treatment. In a 103-week study in rats and mice, the NOAEL of 250 mg/kg BW was established.[3]

Mutagenicity.

In vitro genotoxicity. TMC was found to be negative in *Salmonella* mutagenicity assay in the presence or absence of metabolic activation; it induced SCE only in the absence of S9 fraction, and did not induce CA in Chinese hamster ovary cells.[2]

Carcinogenicity. F344/N rats and B6C3F_1 mice were treated with 250 and 500 mg TMC/kg BW by gavage for 103 weeks. In male rats, renal tubular cell adenomas and adenocarcinomas were found in animals given 250 or 500 mg/kg BW; carcinomas of the preputial gland were also observed at increased incidence in male rats given 500 mg/kg BW. No evidence of carcinogenicity was noted in female rats and mice.[2]

Carcinogenicity classifications.

NTP: SE - NE - EE - NE (gavage);

U.S. EPA: C.

Chemobiokinetics. Following oral administration to the rabbit, allylic methyl group of TMC was oxidized to a *carboxilic acid group*. The product was the only metabolite identified in the urine (Truhaut et al., 1970). A single oral administration of 500 mg TMC/kg BW to rats significantly reduced glutathione in the liver, in the testes, and epididymides.[4]

Standards.

U.S. EPA (1999). Health Advisory for longer-term exposure: 15 mg/l.

Russia (1995). PML: 0.03 mg/l.

References:

1. Eliseyeva, O. I. and Mel'nik, L. A., Toxic effects of 1,1,3-trimethylcyclohexen-3-one-5, *Gig. Sanit.*, 1, 52, 1987 (in Russian).
2. *Toxicology and Carcinogenesis Studies of Isoforone in F344/N Rats and B6C3F1 Mice (Gavage Studies)*, NTP Technical Report Series No 291, Research Triangle Park, NC, January 1986.
3. Bucher, J., Huff, J., and Kluwe, W., Toxicology and carcinogenesis studies of isophorone in F344 rat and B6C3F_1 mice, *Toxicology*, 39, 207, 1986.
4. Gandy, J., Millner, G. C., Bates, H. K., Casciano, D. A., and Harbison, R. D., Effects of selected chemicals on the glutathione status in the male reproductive system of rats, *J. Toxicol. Environ. Health*, 29, 45, 1990.

2,4,6-TRIMETHYL-*S*-TRIOXANE

Molecular Formula. $C_6H_{12}O_3$

M = 132.18

CAS No 123-63-7

RTECS No YK0525000

Abbreviation. TMTO.

Synonyms and Trade Names. *p*-Acetaldehyde; Acetaldehyde, trimer; Elaldehyde; Paraacetaldehyde; Paral; Paraldehyde.

Properties. Colorless, transparent liquid with a characteristic aromatic odor and disagreeable taste. Soluble in 8 parts of water at 25°C or 17 parts of boiling water. Miscible with oils and alcohols.[020]

Applications. Solvent for gums, resins, cellulose derivatives. Rubber activator and antioxidant. Substitute for acetaldehyde.

Acute Toxicity. LD_{50} is reported to be 1.65 g/kg BW in the rat (Figot *et al.*, 1952), and 3.0 to 4.0 g/kg BW in the dog.[018]

Mutagenicity and **Carcinogenicity.** There is no evidence of such effects.[1]

Chemobiokinetics. TMTO is likely to be depolymerized to *acetaldehyde* in the liver and than oxidized by aldehyde dehydrogenase to *acetic acid.* The acid is ultimately converted to *carbon dioxide* and *water*. When hypnotic doses of TMTO were administered, 70 to 80% is metabolized in the liver, 11 to 28% were eliminated exhaled. Only negligible amount of TMTO is excreted with urine.[022]

Regulations. ***U.S. FDA*** (1998) approved the use of TMTO in the manufacture of phenolic resins for food-contact surface of molded articles intended for repeated use in contact with nonacid food (*pH* above 5.0) in accordance with the conditions prescribed by 21 CFR part 177.2410.

Reference:

1. Sittig, M., *Handbook of Toxic and Hazardous Chemicals and Carcinogens*, 2nd ed., Noyes Data Corp., Park Ridge, NJ, 1985, 688.

TURPENTINE

CAS No 8006-64-2

RTECS No YO8400000

Trade Names. Gum spirits of turpentine; Gum turpentine; Oil of turpentine; Spirit of turpentine; Turpentine oil; Turpentine oil, rectifier; Wood turpentine.

Composition. A complex mixture, mainly of $C_{10}H_{16}$ terpene hydrocarbons.

Properties. Clear and colorless or green-yellowish sticky masses. T. exhibits a unique, pleasant odor and a sharp, abrasive taste. Almost insoluble in water. Soluble in alcohol and ether. Odor perception threshold is 0.2 mg/l.

Applications. Used as a solvent in the production of resins, synthetic rubbers, and lacquers.

Acute Toxicity.

Observations in man. A dose of 0.5 g/kg BWseems to be the minimal lethal dose for man.

Animal studies. In rats, LD_{50} is reported to be 5.76 g/kg BW. The urinary system is predominantly affected.

Chemobiokinetics. T. is readily absorbed from the GI tract.[022]

Regulations. ***U.S. FDA*** (1998) approved the use of T. in adhesives used as components of articles intended for use in packaging, transporting, or holding food in accordance with the conditions prescribed by 21 CFR part 175.105.

Standards. ***Russia*** (1995). MAC and PML: 0.2 mg/l (organolept., odor).

2-UNDECANONE

Molecular Formula. $C_{11}H_{22}O$

M = 170.33

CAS No 112-12-9

RTECS No YQ2820000

Synonyms. Methyl nonyl ketone; Nonyl methyl ketone; 2-Hendecanone.

Properties. Colorless liquid with a specific odor. Poorly soluble in water; soluble in alcohol.

Applications. Used as a solvent in the production of vinyl resins, nitrocellulose, and synthetic rubber.

Acute Toxicity. LD_{50} is found to be 5 to 5.5 g/kg BW in rats, and 3.2 to 3.9 g/kg BW in mice.[1] For other data, see ***2-Octanone***.

Reproductive Toxicity. *i/p* administration of 1.0 g/kg BW to mice during pregnancy caused effect on fertility.[2]

References:

1. *Pharm. Chemicals Handbook*, 1991, 201.
2. Carlson, G. L., Hall, I. H., and Piantadose, C., Cycloalkanones. 7. Hypocholesterolemic activity of aliphatic compounds related to 2,8-dibenzylcyclooctanone, *J. Med. Chem.*, 18, 1024, 1975.

2-(VINYLOXY)ETHANOL

Molecular Formula. $C_4H_8O_2$

M = 88.12

CAS No 764-48-7

RTECS No KM5495000

Abbreviation. VE.

Synonyms. 2-(Ethenyloxy)ethanol; Ethylene glycol, vinyl ether; 2-Hydroxyethyl vinyl ether; Vinyloxyethanol.

Properties. Clear, colorless liquid with a specific odor. Soluble in water. Odor perception threshold is 1.92 mg/l; taste perception threshold is slightly higher. Does not alter the color and transparency of water, does not form foam.

Applications. Used in the manufacture of plastics and lacquers and also in the production of protective coatings.

Acute Toxicity. LD_{50} values are 3.9 g/kg BW in rats and 2.9 g/kg BW in mice.

See also ***DEGDVE***.

Chemobiokinetics. VE metabolism in the body occurs via hydrolysis to form *acetaldehyde* which is rapidly oxidized and is therefore not important in the pathogenesis of intoxication. The detached *ethylene glycol* selectively attacks the kidneys - either itself or through its separation prodcts.

Standards. *Russia* (1995). MAC and PML: 1.0 mg/l (organolept.).

Reference:

Busina, A. Z. and Rudi, F. A., On substantiation of Maximum Allowable Concentration for glycol vinyl ethers in water bodies, *Gig. Sanit.*, 3, 12, 1977 (in Russian).

2-(2-(VINYLOXY)ETHOXY)ETHANOL

Molecular Formula. $C_6H_{12}O_3$
M = 132.18
CAS No 929-37-3
RTECS No KM5495500
Abbreviation. VOEE.

Synonyms and Trade Name. Diethyleneglycol vinyl ether; 2-[2-(Ethenyloxy)ethoxy] ethanol; Vinylcarbitol.

Properties. Colorless, transparent liquid with a specific odor. Readily soluble in water. Odor perception threshold is 1.83 mg/l, taste perception threshold is slightly higher. VOEE does not alter transparency or color of water and forms no foam.

Applications. Used in the production of polymeric materials and coatings.

Acute Toxicity. LD_{50} is reported to be 4.93 g/kg BW in rats, and 4.45 g/kg BW in mice.

Repeated Exposure.

See ***Bis(2-vinyloxyethyl) ether***.

Standards. *Russia* (1988). MAC and PML: 1.0 mg/l (organolept., odor).

Reference:

Buzina, L. Z. and Rudi, F. A., Substantiation of MAC for glycol vinyl ethers in water bodies, *Gig. Sanit.*, 3, 12, 1977 (in Russian).

XYLENES (mixed)

Molecular Formula. C_8H_{10}
M = 106.16
CAS No 1330-20-7
RTECS No ZE2100000

Synonym. Xylols.

	RTECS No	CAS No
1,2-Dimethylbenzene	ZE2450000	95-47-6
1,3-Dimethylbenzene	ZE2275000	108-38-3
1,4-Dimethylbenzene	ZE2625000	106-42-3

Properties. Colorless, highly combustible liquid. Water solubility of X. isomers is up to 160 to 198 mg/l at 25°C. Odor threshold concentration is reported to be 0.02 to 2.2 mg/l.[010] The data on the taste threshold value are limited to the concentration range of 0.3 to 1.0 mg/l; these concentrations produced a detectable taste and odor.

In Japan and in The Netherlands, odor threshold values for *m*-X. are regulated at the levels of 0.012 and 0.12 ppm, respectively.[1]

Applications. Used as a solvent in the production of synthetic rubbers, adhesives, and polyether fibers. Used as an intermediate in the manufacture of plastics.

Migration Data. Leaching from synthetic coating materials commonly used to protect drinking water storage tanks is reported.[2]

Exposure. Along with other light, high-octane aromatic hydrocarbons, X. are components of motor gasoline. X. are found in the drinking water in the U.S. and Switzerland. Air seems to be the major source of exposure.

Acute Toxicity.

Observations in man. Accidental human poisonings are followed with adverse effects on the kidneys and liver.

Animals studies. In rats, LD_{50} is found to be 3.6 to 5.8 g/kg (3.57 g *o*-X./kg BW; 4.99 g *m*-X./kg BW; 3.91 g *p*-X./kg BW).[3] *I/p* administration to rats at the dose-level of 1.0 *o*-X./kg BW decreased total pulmonary cytochrome *P-450* content and aryl hydrocarbon hydroxylase activity.[4]

Repeated Exposure. In a 14-day study, rats and mice exhibited shallow breathing and prostration following dosing with 2000 mg/kg BW. Tremor was observed in some animals.[5,6] Wistar rats were injected with 2.0 mg X./kg BW for 5 weeks. The treatment caused changes in peripheral blood (Tannhauser et al., 1995).

Short-term Toxicity. In a 13-week gavage study, no death or clinical signs of toxicity were recorded in rats that received up to 1.0 g/kg BW.[5] The LOAEL was identified to be 200 mg/kg BW. Ultrastructural changes were observed in the liver.[7,8]

In a 17-week study, $B6C3F_1$ mice were given up to 0.6% X. in the diet. All mice given 0.3% X. or more died. Gross pathology and histology examination revealed enlargement and irregularity of hepatocytes at dose levels of 0.15% or more X. in the diet.[6]

Long-term Toxicity. In NTP-86 study, rats were administered 250 and 500 mg X./kg BW in corn oil by gavage for 103 weeks.[9]

The dose of 250 mg/kg BW was considered to be the NOAEL since the mean BW of low-dose and vehicle control male rats and those of dosed and vehicle control female rats were comparable.

Rabbits were given the dose of 48 mg X./kg BW in starch emulsion for 5.5 months. Eosinophilia and lymphopenia were observed. Gross pathology examination revealed adipose dystrophy of the liver and kidneys, and pyelitis. There were no clinical signs of intoxication.[10]

Reproductive Toxicity. X. cross the placenta in mice, rats, and small number of humans.[11-13]

Maternal toxicity with concurrent embryotoxicity and teratogenicity was observed in mice. The NOAEL of 255 mg/kg BW was established in this oral study.[14]

o- and *p*-isomers appear to be more hazardous to the offspring than is *m*-isomer.

Embryotoxic effect was found in rats and mice but no ***teratogenic effect*** was observed in rats.[9,13,15,16]

Mutagenicity.

Observations in man. No increase in frequency of SCE and CA was observed in paint industry workers exposed to X.[17,18]

In vitro genotoxicity. Negative results were shown in *Salmonella* mutagenicity assay[5] and in mammalian somatic cells (both *in vitro* and *in vivo*).

In vivo cytogenetics. X. did not produce CA in bone marrow cells in mice.[19]

Carcinogenicity.

Observations in man. Available information on X. has been reviewed by U.S. EPA and was found to be inadequate for determining potential carcinogenicity in humans.[20]

Animal studies. $B6C3F_1$ mice were fed dietary levels of 0.075 or 0.15% X. for 80 weeks. The treatment caused a significant increase in tumor incidence (malignant and benign liver cell tumors). In males, the incidence of Harderian gland tumors was increased.[21]

However, according to other data, no increase in the incidence of tumors was observed in either mice or rats following the administration of a technical-grade X. There was no evidence of carcinogenicity in a 2-year oral study.[5]

Carcinogenicity classifications. An IARC Working Group concluded that there is inadequate evidence for the carcinogenicity of X. in *experimental animals* and in *humans*.

IARC: 3;

U.S. EPA: D.

NTP: NE - NE - NE - NE (gavage).

Chemobiokinetics. Data on the absorption after ingestion are not available. Adipose tissue is likely to be the site of storage. The general pathways of *m*-X. metabolism involve initial side chain and aromatic hydroxylation catalyzed by cytochrome *P-450*.[22] X. produced an effect on rat hepatic and pulmonary mixed-function oxidase.

X. is converted in the body primarily to *methyl benzoic acid*, that is excreted in the urine as *methylhippuric acid*.

Regulations. ***U.S. FDA*** (1998) approved the use of X. (1) as components of the uncoated or coated food-contact surface of paper and paperboard that may be safely used in producing, manufacturing, packing, transporting, or holding dry food in accordance with the conditions prescribed by 21 CFR part 176.180; (2) as plasticizers for rubber articles intended for repeated use in contact with food up to 30% by weight of the rubber product in accordance with the conditions prescribed in 21 CFR part 177.2600; (3) in adhesives used as components of articles intended for use in packaging, transporting, or holding food in accordance with the conditions prescribed by 21 CFR part 175.105; (4) in the manufacture of semirigid and rigid acrylic and modified acrylic plastics for single and repeated use in food-contact articles in accordance with the conditions prescribed by 21 CFR part 177.1010; and (5) as a solvent in polysulfide polymer-polyepoxy resins in articles intended for packaging, transporting, holding, or otherwise containing dry food in accordance with specified conditions in accordance with the conditions prescribed by 21 CFR part 177.1650.

Recommendations. ***WHO*** (1996). Guideline value for drinking water is 0.5 mg/l. The levels of 0.02 to 1.8 mg/l are likely to give rise to consumer complaints of foreign taste and odor.

Standards.

U.S. EPA (1999). MCL and MCLG: 10 mg/l.

Canada (1989). MAC: 0.3 mg/l (organolept.).

Russia (1995). MAC and PML in drinking water: 0.05 mg/l (organolept., odor).

References:

1. Hoshika, Y., Imamura, T., Muto, G., et al., International comparison of odor threshold values of several odorants in Japan and in The Netherlands, *Environ. Res.*, 61, 78, 1993.
2. Bruchet, A., Shipert, E., and Alban, K., Investigation of organic coating material used in drinking water distribution systems, *J. Francais d'Hydrologie*, 19, 101, 1988.
3. Jori, A. et al., Ecotoxicological profile of xylenes, Working party on ecotoxicological profiles of chemicals, *Prog. Clin. Biol. Res.*, 163B, 301, 1985.
4. Park, S. H., AuCoin, T. A., Silverman, D. M., and Schatz, R. A., Time-dependent effects of *o*-xylene on rat lung and liver microsomal membrane structure and function, *J. Toxicol. Environ. Health*, 43, 469, 1994.
5. *Toxicology and Carcinogenesis Studies of Xylenes (mixed) (60% m-Xylene, 14% p-Xylene, 9% o-Xylene, 17% Ethylbenzene) in F344/N Rats and B6C3F$_1$ Mice (Gavage Studies)*, NTP Technical Report Series No 327, Research Triangle Park, NC, 1986.
6. Maekawa A., Matsushima, Y., Onodera, H., et al., Toxicity and carcinogenicity studies of musk xylol in B6C3F$_1$ mouse, Abstract, *Eisei Shikenjo Hokoku*, 108, 89, 1990 (in Japanese).
7. Bowers, D. E., Cannon, M. S., and Jones, D. H., Ultrastructural changes in the liver of young and aging rats exposed to methylated benzenes, *Am. J. Vet. Res.*, 43, 679, 1982.
8. Janssen, P., van der Heijden, C. A., and Knaap, A. G., Short summary and evaluation of toxicological data on xylene, dated 21 March 1989 (Dutch), *Docum. Toxicol. Advis. Centre, Natl. Institute Public. Health Environ. Protection*, The Netherlands, 1989.
9. Rosen, M. B., Crofton, K. M., and Chernoff, N., Postnatal evaluation of prenatal exposure to *p*-xylene in the rat, Postnatal evaluation of prenatal exposure to *p*-xylene in the rat, *Toxicol. Lett.*, 34, 223, 1986.

10. Rubleva, M. N., in *Industrial Pollution of Water Reservoirs*, S. N. Cherkinsky, Ed., Medgiz, Moscow, Issue No 9, 1960, 100 (in Russian).
11. Ghantous, H. and Danielson, B. R., Placental transfer and distribution of toluene, xylene and benzene and their metabolites during gestation in mice, *Biol. Res. Pregn. Perinat.*, 3, 98, 1986.
12. Dowty, B. J., Laseter, J. L., and Storer, M., The transplacental migration and accumulation on blood of volatile organic constituents, *Pediatr. Res.*, 10, 696, 1976.
13. Ungvary, G., Tatrai, D., Hudak, A., Barcza, G., and Lorincz, M., Studies on embryotoxic effects of *o-, m-* and *p*-xylenes, *Toxicology*, 18, 61, 1980.
14. Marks, T. A., Ledoux, T. A., and Moore, J. A., Teratogenicity of a commercial xylene mixture in the mouse, *J. Toxicol. Environ. Health*, 9, 97, 1982.
15. Hood, R. D. and Utley, M. S., Developmental effects associated with exposure to xylenes: A Review, *Drug Chem. Toxicol.,* 8, 281, 1985.
16. Brown-Woodman, P. D. C., Webster, W. S., Picker, L., et al., Embryotoxicity of xylene and toluene: an *in vitro* study, *Ind. Health.,* 29, 139, 1991.
17. Haglund, U., Lundberg, I., and Zech, L., Chromosome aberrations and sister chromatid exchanges in Swedish paint industry workers, *Scand. J. Work Environ. Health*, 6, 291, 1980.
18. Gerner-Smidt, P. and Friedrich, U., The mutagenic effect of benzene, toluene and xylene studied by the SCE technique, *Mutat. Res.*, 58, 313, 1978.
19. Feldt, E. G., Assessment of mutagenic hazards of benzene and its derivatives, *Gig. Sanit.*, 7, 21, 1985 (in Russian).
20. U.S. *Federal Register*, vol 56, No 20, 1992, 3543.
21. Maekawa A., Matsushima, Y., Onodera, H., et al., Long-term toxicity/carcinogenicity of musk xylol in $B6C3F_1$ mice, *Food Chem Toxicol.,* 28, 581, 1990.
22. Sedivec, V. and Flek, J., The adsorption, metabolism and excretion of xylenes in man, *Int. Arch. Occup. Environ. Health*, 37, 205, 1976.

CHAPTER 7. OTHER ADDITIVES AND INGREDIENTS

ACETAMIDE

Molecular Formula. C_2H_5NO
M = 59.07
CAS No 60-35-5
RTECS No AB4025000

Synonyms. Acetic acid, amide; Acetimidic acid; Ethanamide; Methanecarboxamide.

Properties. Colorless, odorless needles. Solubility in water is 97%, solubility in alcohol up to 25% at 20°C.

Applications. Used as an intermediate in the production of plastics.

Exposure. Overoxidized wine can contain A.

Acute Toxicity. Rats and mice tolerate 7.5 and 8.0 g/kg BW, respectively. In rats, LD_{50} is reported to be 30 g/kg BW. Lethargy and respiratory distress in rats, and excitation (convulsion) accompany poisoning with subsequent CNS inhibition in mice (Garrett and Dangherti, 1951).

Repeated Exposure. Rats were given 400 mg/kg BW for 36 days. Decreased growth without other signs of intoxication or pathological lesions was reported.[1]

Reproductive Toxicity.

Embryotoxic effect is observed in rabbits. Mild teratogenic effect is found in rats, mice, and rabbits.[2,3]

Rats received *i/p* doses of 2.0 g A./kg BW once between days 4 and 14 of gestation. A. did not produce embryotoxic effect, but *N*-methyl acetamide and *N,N*-dimethyl acetamide were lethal to the embryo-fetuses.[11]

Mutagenicity.

In vivo cytogenetics. Single oral administration of A. was negative as a micronucleus-inducing agent in mouse bone marrow in four repeated assays using different sexes of two strains of mice.[4]

Chieli et al. (1987) reported a marked micronucleus induction in female C57BL6 mice at 200 mg/kg BW exposed by oral gavage. However, according to other data, no micronucleus induction was noted in bone marrow or peripheral blood erythrocytes of mice that received single, double or quadruple *i/p* injections of up to 2.0 g or 5.0 g A./kg BW.[5]

A. produced significant increase in *Dr. melanogaster* wing spot somatic mutation and recombination assay.[6]

In vitro genotoxicity. A. showed no mutagenic activity in *Salmonella typhimurium*, in tests for DNA damage in rat hepatoma cells or for DNA repair in isolated rat hepatocytes.[7,025] It induced positive morphological transformation in Syrian hamster embryo cells in presence of metabolic activation.[8]

Carcinogenicity. A. appeared to be hepatocarcinogen in rats and caused malignant lymphomas in mice.[9]

Wistar 1-month-old male rats were fed a diet containing 1.25, 2.5, or 5.0% A. for 4 to 12 months. This treatment caused development of liver tumors (trabecular carcinomas and adenocarcinomas with lung metastases).[7]

Increased incidence of tumors (hepatomas) was noted in the study, where Wistar rats were exposed to 2.5% A. in the diet for one year.[10]

A single dose of 100 and 400 mg/kg BW acts as an initiator in rat liver carcinogenicity.[7]

Carcinogenicity classifications. An IARC Working Group concluded that there is sufficient evidence for the carcinogenicity of A. in *experimental animals* and there were no adequate data available to evaluate the carcinogenicity of A. in *humans*.

IARC: 2B;
EU: 3.

Chemobiokinetics. *N-hydroxyacetamide* seems to be a putative metabolite of A. Treatment of rats with 0.1 and 1.0 g/kg BW dose of *N*-hydroxyacetamide or acetic acid is not accompanied with significant excretion in the urine.[7]

References:

1. Caujolle, F., Chanh, P. H., Dat-Xuong, N., u. a., Toxicity studies on acetamide and its *N*-methyl and *N*-ethyl derivatives, *Arzneimittel-Forsch.*, 20, 9042, 1970.
2. Merkle, J. and Zeller, H., Studies on acetamides and formamides for embryotoxic and teratogenic activities in the rabbit, *Arzneimittel-Forsch.*, 30, 1557, 1980.
3. Fleischman, R. W., Naqvi, R. H., Rosenkrantz, H., and Hayden, D. W., The embryotoxic effects of cannabinoids in rats and mice, *J. Environ. Pathol. Toxicol.*, 3, 149, 1980.
4. Mirkova, E. T., Activities of the rodent carcinogens thioacetamide and acetamide in the mouse bone marrow micronucleus assay, *Mutat. Res.*, 352, 23, 1996.
5. Morita, T., Asano, N., Awogi, T., Sasaki, Yu. F., et al. Evaluation of the rodent micronucleus assays in the screening of IARC carcinogens (Groups 1, 2A and 2B). The summary report of the 6th collaborative study by CSGMT/JEMS-MMS, *Mutat. Res.*, 389, 3, 1997.
6. Batiste-Allentorn, M., Xamena, N., Creus, A., and Marcos, A., Genotoxic evaluation of ten carcinogens in the *Drosophila melanogaster* with spot test, *Experientia*, 51, 73, 1995.
7. Dybing, E., Soderlund, E. J., Gordon, W. P., Holme, J. A., Christensen, T., Becher, G., Rivedal, E., and Thorgeirsson, S. S., Studies on the mechanism of acetamide hepato carcinogenicity, *Pharmacol. Toxicol.*, 60, 9, 1987.
8. Amacher, D. E. and Zelljadt, I., The morphological transformation of Syrian hamster embryo cells by chemicals reportedly nonmutagenic to *Salmonella typhimurium*, *Carcinogenesis* (London), 4, 291, 1983.
9. Jackson, B. and Dessau, F., Liver tumor in rats fed acetamide, *Lab. Investig.*, 10, 909, 1961.
10. Weisburger, J. H., Yamamoto, R. S., Glass, R. M., et al., Prevention by arginine glutamate of the carcinogenicity of acetamide in rats, *Toxicol. Appl. Pharmacol.*, 14, 163, 1969.
11. Thiersch, J. B., Effects of acetamides and formamides on the rat litter *in utero*, *J. Reprod. Fertil.*, 4, 219, 1962.

ACETIC ACID, MANGANESE(2^+) SALT (2:1)

Molecular Formula. $C_4H_6O_4.Mn$
M = 173.04
CAS No 638-38-0
RTECS No AI5770000
Abbreviation. AAM.

Synonyms and **Trade Names.** Acetic acid, manganese salt; Diacetyl manganese; Manganese diacetate; Manganese acetate.

Properties. Brown crystals. Soluble in alcohol. Manganese tetrahydrate is pale red, transparent monoclinic crystals. Soluble in water and alcohol.[020]

Applications. Used in food packaging and as a feed additive.

Acute Toxicity. LD_{50} is 1.08 g/kg BW in rats.[1]

Short-term Toxicity. Rats received feed containing 1.75% AAM (of dry weight). The treatment caused retardation of BW gain, loss of calcium in feces due to its poor absorption, negative phosphorus balance, and severe rickets. Anorexia, sexual impotence, and muscular fatigue were also noted. Manifestation of toxic action included erythrocyte agglutination and hemolysis.[2]

Long-term Exposure. NS is the major organ affected following long-term exposure to manganese.

Male mice were chronically treated with AAM administered in the diet for 12 months. The food intake for the control mice and the mice exposed to manganese was similar, but the manganese treatment reduced normal weight gain in the mice. RBC and WBC count was decreased. Accumulation of manganese in the brain correlated with reduced hypothalamic dopamine levels; and, the amount of manganese accumulated correlated with the intensity of suppression of motor activity. Out of all divalent manganese compounds, AAM seemed to have the greatest toxic effect.[3]

Reproductive Toxicity.

Embryotoxicity. Female rats received from 4 to 1004 mg manganese/kg dry weight diet during pregnancy and weaning. The treatment caused an increase in AAM concentration in fetal liver (at dietary doses of 154 to 1004 mg/kg).[4]

Teratogenic effect was not observed. In the above-cited study, consumption of the diet did not cause teratogenic effect.[4]

Chronic exposure to Mn^{3O4} administered in the diet at the concentration of 1050 ppm *Mn* (284 mg/kg BW) retarded the sexual development and lowered reactive locomotor activity levels in male mice. Testis, seminal vesicle, and preputial gland weights were significantly smaller as a result of *Mn* administration.[5]

Mutagenicity.

In vivo cytogenetics. AAM compounds appeared to be positive when tested by *rec-assay*.[6] Mn^{2+} produced a weak but significant increase in the frequency of CA and SCE in rats and mice receiving it in toxic doses.[7]

Carcinogenicity. No evidence of cancer development was noted in F344 rats given 20 to 200 mg AAM/kg BW via feed (NTP-1993).

Chemobiokinetics. AAM is poorly absorbed from the GI tract owing to low solubility of cationic *manganese*. Ingested Mn^{2+} is converted into Mn^{3+} in alkaline duodenal medium.[2]

Regulations. ***U.S. FDA*** (1998) approved the use of AAM in resinous and polymeric coatings used as the food-contact surfaces of articles intended for use in producing, manufacturing, packing, processing, preparing, treating, packaging, transporting, or holding food in accordance with the conditions prescribed in 21 CFR part 175.300.

Standards. WHO (1996). Provisional Guideline (Mn^{2+}): 0.5 mg/l.

References:

1. Singh, P. P. and Junnarkar, A. Y., Behavioral and toxic profile of some essential trace metal salts in mice and rats, *Ind. J. Physiol. Pharmacol.*, 23, 153, 1991.
2. Venugopal, B. and Luckey, T. D., *Metal Toxicity in Mammals*, Plenum Press, New York, 1978, 264.
3. Komura, J. and Sakamoto, M., Effects of manganese forms on biogenic amines in the brain and behavioral alterations in the mouse: long-term oral administration of several manganese compounds, *Environ. Res.*, 57, 34, 1992.
4. Friberg, L., Nordberg, G. F., Kessler, E., Vouk, V. B., *Handbook of the Toxicology of Metals*, 2 ed., vol 2, Elsevier Sci. Publ., Amsterdam, 1986, 377.
5. Gray, Le, Jr. and Laskey, J. W., Multivariate analysis of the effects of manganese on the reproductive physiology and behavior of the male house mouse, *J. Toxicol. Environ. Health*, 6, 861, 1980.
6. Nioshioka, H., Mutagenic activities of metal compounds in bacteria, *Mutat. Res.*, 31, 185, 1975.
7. Nadeyenko, V. G., Goldina, I. R., Dyachenko, O. Z., and Pestova, L. V., Comparative informative value of chromosomal aberrations and sister chromatid exchanges in the evaluation of environmental metals, *Gig. Sanit.*, 10, 3, 1997 (in Russian).

ACETOXYPROPARGYL

Molecular Formula. $C_5H_6O_2$

M = 98.10

Properties. A clear, colorless liquid with the odor of geraniums. Poorly soluble in water, readily soluble in organic solvents.

Applications. Used in the production of polymeric materials.

Acute Toxicity. LD_{50} is reported to be 250 mg/kg BW in rats, and 520 mg/kg BW in mice. Poisoning is accompanied by a short period of excitation followed by inhibition, labored respiration, ataxia, and adynamia. Death occurs in the state of lethargy.

Repeated Exposure revealed moderate cumulative properties. K_{acc} is 5.0 (by Lim). Dermal applications produced changes in the cortical reflex. A. is a narcotic with a mild irritant effect.

Reference:

Balynina, E. S., Toxicity of acetoxypropargyl, *Gig. Truda Prof. Zabol.*, 6, 55, 1984 (in Russian).

ACETYLENE

Molecular Formula. C_2H_2
M = 26.02
CAS No 74-86-2
RTECS No AO9600000

Synonym and **Trade Names.** Ethine; Ethyne; Narcylen.

Properties. Colorless gas with a faint odor of ether. Solubility in water is 100 vol/100 vol, solubility in alcohol is 600 vol/100 vol at 18°C.[011]

Applications. Used in the manufacture of vinyl chloride, vinylidene chloride, vinyl acetate, acrylates, acrylonitrile, acetaldehyde, perchloroethylene, trichloroethylene, 1,4-butandiol and carbon black.

Repeated Exposure to tolerable concentrations of A. in the air failed to provide evidence of toxic effects.[011] Impurities in commercial A. may produce NS affection and respiratory disorder.[012]

Reproductive Toxicity.

Observations in man. A. was reported as one of several substances present in a workplace environment where female employees experienced complications of pregnancy.[1]

Mutagenicity. A. was not found to be mutagenic in *Salmonella*/microsome assay (Frazier, 1984).

Regulations.

EU (1990). A. is available in the *List of authorized monomers and other starting substances which shall be used for the manufacture of plastic materials and articles intended to come into contact with foodstuffs (Section A).*

Great Britain (1998). A. is authorized without time limit for use in the production of materials and articles in contact with food or drink or intended for such contact.

Reference:

1. Talakina, E. I., Rachkovskaia, L. V., and Skorokhod, L. P., Health protection for pregnant women in acetylene-vinyl acetate manufacture, *Gig. Truda Prof. Zabol.*, 3, 46, 1977 (in Russian).

ACRYLIC ACID, CYCLOHEXYL ESTER

Molecular Formula. $C_9H_{14}O_2$
M = 154.23
CAS No 3066-71-5
RTECS No AS7350000
Abbreviation. AACE.

Synonyms and **Trade Name.** Cyclohexyl acrylate; 2-Propenoic acid, cyclohexyl ester; Sartomer SR 220.

Acute Toxicity. LD_{50} is 9.0 g/kg BW in rats.

Regulations. ***EU*** (1990). AACE is available in the *List of authorized monomers and other starting substances which may continue to be used for the manufacture of plastic materials and articles intended to come into contact with foodstuffs pending a decision on inclusion in Section A* (*Section B*).

Reference:

Carpenter, C. P., Weil, C. S., and Smyth, H. F., Range-finding toxicity data: List VIII, *Toxicol. Appl. Pharmacol.*, 28, 313, 1974.

ADIPIC ACID, DIALLYL ESTER

Molecular Formula. $C_{12}H_{18}O_4$
M = 226.30
CAS No 2998-04-1
RTECS No AV0350000

Abbreviation. DAA.

Synonyms. Allyl adipate; Diallyl adipate; Hexanedioic acid, di-2-propenyl ester.

Properties. Colorless liquid with a faint odor of ether. Soluble in alcohol.

Applications. Used in the manufacture of plastics.

Acute Toxicity. LD_{50} is 415 mg/kg BW in rats, and 180 to 500 mg/kg BW in mice. Administration of 1200 to 1500 mg/kg BW caused 100% mortality in mice within a few days. Signs of intoxication include respiratory distress and lacrimation. Gross pathology examination revealed splenic congestion and enlargement.

Repeated Exposure failed to reveal cumulative properties. K_{acc} is 12.8 (by Lim). The treatment altered the liver (reduced cholinesterase activity, increased relative weights and gross pathology changes) and renal functions and disturbed the oxidation-reduction processes. Animals died on days 21 to 28.

Short-term Toxicity. DAA exhibited a marked hepatotoxicity and caused catarrhal gastritis to develop.

Regulations. *EU* (1992). DAA is available in the *List of authorized monomers and other starting substances which may continue to be used for the manufacture of plastic materials and articles intended to come into contact with foodstuffs pending a decision on inclusion in Section A* (*Section B*).

Reference:

Osipova, T. V. and Kasatkin, A. N., in *Topical Problems of Hygiene and Occupational Pathology in the Chemical Industry,* Proc. F. F. Erisman Research Sanitary Hygiene Institute, Moscow, 1985, 48 (in Russian).

ADIPONITRILE

Molecular Formula. $C_{10}H_8N_2$

M = 108.56

CAS No 111-69-3

RTECS No AV2625000

Abbreviation. AN.

Synonyms. Adipic acid, dinitrile; 1,4-Dicyanobutane; Hexanedinitrile.

Properties. Light, odorless, oily liquid. Poorly soluble in water, soluble in alcohol. Odor perception threshold is 31 mg/l, practical threshold is 63 mg/l.[1]

Applications. Used in the production of plastics and synthetic resins.

Acute Toxicity.

Observations in man. Ingestion of several *ml* of AN resulted in vomiting, chest tightness, headache, weakness, cyanosis, increased heart rate, etc. A man recovered within 4 hours following sodium thiosulfate treatment and then suffered a recurrent episode.[2]

Animal studies. LD_{50} is reported to be 500 or 105 mg/kg BW in rats, 48 mg/kg BW in mice, 20 mg/kg BW in rabbits, and 50 mg/kg BW in guinea pigs.[1,3] LD_{50} of technical grade AN, containing 1.0% of ammonia, is 1.0 g/kg BW in rats.[4]

According to Daniel and Kennedy, LD_{50} is 138 mg/kg BW for fasted rats and 301 mg/kg BW for non-fasted rats. Excitation, convulsions, and respiratory arrest accompany poisoning.[5]

Repeated Exposure revealed little evidence of cumulative properties. Mice were given 10 mg A./kg BW. The treatment caused retardation of BW gain. 25% animals died after 12 administrations. Gross pathology examination revealed dystrophic changes in the viscera. A 250 mg/kg BW dose given to rats for 10 days produced temporary retardation of BW gain; albumen appeared in the urine.[4]

Long-term Toxicity. A 2-year feeding study in rats resulted in adrenal degeneration in females at feeding levels of 0.5, 5.0, and 50 ppm. Other parameters were found to be unaltered.[6]

In a 6-month study in rabbits, the main manifestations of the toxic action included impairment of enzyme indices reflecting the state of the oxidation-reduction systems, and changes in the contents of blood *SH*-groups and blood nucleic acids.[1]

Reproductive Toxicity.

Embryotoxicity. Fertility, as monitored by reproductive performance and litter parameters was normal in both males and females similarly exposed to atmospheres containing up to 493 mg AN/m^3 for 4 or 13 weeks.[7]

Sprague-Dawley rats were administered the doses of 20, 40, and 80 mg/kg BW on days 6 to 19 of gestation. *Maternal toxicity* effect was observed at 40 and 60 mg/kg BW dose levels. The highest dose caused a slight fetotoxicity.[8]

Teratogenic effect has not been observed.[8]

Chemobiokinetics. AN is likely to break down in the body to form CN^-. Its toxicity is possibly determined by the action of intact molecules. According to Ghiringhelley, 83.5% of administered dinitrile is broken down in the body to form *hydrogen cyanide*.[1,2] In guinea pigs, AN was found to be eliminated as *thiocyanate* in the urine.[03]

Standards. *Russia* (1995). MAC and PML: 0.1 mg/l.

References:

1. Klimkina, N. V., Boldina, Z. N., and Sergeev, A. N., in *Industrial Pollution of Water Reservoirs*, S. N. Cherkinsky, Ed., Meditsina, Moscow, Issue No. 8, 1967, 85 (in Russian).
2. Ghiringhelley, L., Toxicity of adipic nitrile. Acute poisoning and mechanism of action, *Med. Lavoro*, 46, 221, 1955.
3. Dieke, S. H., Allen, G. S., and Richter, C. P., The acute toxicity of thioureas and related compounds to wild and domestic Norway rats, *J. Pharmacol.*, 90, 260, 1947.
4. Plokhova, E. I. and Rubakina, A. P., Toxicity characteristics of adipic acid dinitrile, *Gig. Truda Prof. Zabol.*, 9, 56, 1965 (in Russian).
5. Daniel, O. L. and Kennedy, G. L., The effect of fasting on the acute oral toxicity of nine chemicals in rats, *J. Appl. Toxicol.*, 4, 320, 1984.
6. Svirbely, J. L. and Floyd, E. P., Toxicologic studies of acrylonitrile, adiponitrile, and oxydipropionitrile. III. Chronic Studies, U.S. Dept. HEW, OHS, R. A. Taft, Eng. Center J-4614, 1964.
7. Short, R. D., Roloff, W. V., Kier, L. D., and Ribelin, W. E., 13 week inhalation toxicity study and fertility assessment in rats exposed to atmospheres containing adiponitrile, *J. Toxicol. Environ. Health.*, 30, 199, 1990.
8. Johannsen, F. R., Levinskas, G. J., Berteau, P. E., et al., Evaluation of the teratogenic potential of three aliphatic nitriles in the rat, *Fundam. Appl. Toxicol.*, 7, 33, 1986.

α-ALKENE SULFONIC ACID

CAS No 72674-05-6

RTECS No WR7100000

Abbreviation. ASA

Synonym. α-Olefin sulfonate.

Applications. A surfactant.

Reproductive Toxicity.

Embryotoxicity. ASA was given by oral gavage, at dose levels of 0.2 to 600 mg/kg BW during days 6 to 15 of pregnancy, to rats and mice, and days 6 to 18 of pregnancy to rabbits. Administration of the 600-mg/kg BW dose produced maternal toxicity signs, and death in rabbits and mice. Litter parameters were unaffected at the dose of up to 2.0 mg/kg BW in rabbits and mice, and at the dose level of 600 mg/kg BW in rats.

Teratogenicity. Higher incidence of minor skeletal anomalies and variants was observed in rabbits at 300 mg/kg BW, and of major malformations (cleft palate) and minor skeletal anomalies in mice at 600 mg/kg BW.

Regulations. *U.S. FDA* (1998) approved the use of ASA in adhesives as a component of articles intended for use in packaging, transporting, or holding food in accordance with the conditions prescribed in 21 CFR part 175.105.

Reference:

Palmer, A. K., Readshaw, M. A., and Neuff, A. M., Assessment of the teratogenic potential of surfactants. Part II - AOS, *Toxicology*, 3, 91, 1975.

ALKOXYMETHYLENEMETHYL DIETHYL AMMONIUM METHYLSULFATE

Molecular Formula.

$[C_nH_{2n+1}OCH_2N^+(CH_3)(CH_2CH_3)_2][CH_3SO_4^-]$, where $n = 10$ to 18

CAS No 11139-76-7
RTECS No AZ7746200

Synonym and **Trade Name.** Alkamon DS; Synthetic fatty alcohols C_{10}-C_{18}, diethylaminomethyl ether.

Properties. Viscous mass with a color ranging from light-yellow to brown. Readily soluble in water and alcohol. Threshold perception concentrations are reported to be 6.1 mg/l (odor), 11.4 mg/l (taste), and 1.0 mg/l (foam formation).

Applications. Used as an emulsifier, antistatic, finishing agent.

Acute Toxicity. LD_{50} is 380 mg/kg BW in rats, and 300 mg/kg BW in mice. High doses affect the CNS and GI tract. Acute action threshold (STI) appeared to be 18.9 mg/kg BW.

Repeated Exposure revealed evident cumulative properties. K_{acc} is 1.93 (by Kagan) on administration of 1/20 and 1/100 LD_{50}.

Long-term Toxicity. The NOAEL of 0.4 mg/kg BW was identified.

Allergenic Effect was found on skin application of 1:100 solution.

Reproductive Toxicity.

Embryotoxic effect was not found.

Mutagenicity.

In vivo cytogenetics. Administration of 0.08 mg/kg BW for 2.5 months produced no cytologically detectable damage to somatic cells or to germ cells of rats.

Standards. ***Russia*** (1995). MAC and PML: 0.15 mg/l (organolept., odor).

Reference:

Trikulenko, V. I., Biological effect of some new surfactants and substantiation of their safe levels in water bodies, *Gig. Sanit.*, 3, 14, 1978 (in Russian).

ALKYL ARYL SULFONATE

RTECS No AZ8400000
Abbreviation. AAS.

Trade Name. Witconate.

Composition. Synthetic anionic detergent, containing 40% sodium alkyl sulfonate, approximately 2.0% moisture, 1.0% unsulfonated oil, and the balance sodium sulfate.

Properties. Threshold concentration is 0.3 to 0.7 mg/l for odor and 0.4 to 0.6 mg/l for taste. Classified as an anionic surfactant.

Applications. Used as an emulsifier.

Acute Toxicity. LD_{50} is 1.4 to 2.3 g/kg BW in rats, 1.5 to 2.0 g/kg BW in mice, and 1.13 g/kg BW in hamsters. Manifestations of toxic action included change in motor activity, hypermotility, diarrhea.

Short-term Toxicity. 6 volunteers received 100 mg AAS/day with their food over a period of 4 months. Ingestion did not result in any toxic effect.

Reference:

Grushko, Ya. M., *Hazardous Organic Compounds in the Liquid Industrial Effluents*, Khimiya, Leningrad, 1976, 96 (in Russian).

ALKYLBENZENE SULFONATE, SODIUM SALT

Molecular Formula. $C_nH_{2n+1}C_6H_4SO_4Na$, where n = 11 to 14.
M = 340 to 350
CAS No 6841-30-3
RTECS No DB4550000
Abbreviation. ABSS.

Properties. At 20°C, a light-yellow, paste-like liquid which forms layers; at 70°C, it is homogenous. Smells of soap. Soluble in water. Odor perception threshold is 23.8 mg/l; taste perception threshold is 17.3 mg/l. Practical thresholds are 54.4 and 35.3 mg/l, respectively. There are no chlorinated phenol odors on chlorination. Foam formation threshold is 0.4 mg/l.[1]

Applications. Used as a foam-forming surfactant, a wetting agent.

Acute Toxicity. LD_{50} is 3.0 g/kg in rats, 2.1 g/kg in mice, and 1.9 g/kg BW in guinea pigs. LD_{50} of alkylbenzene sulfonate is reported to be 437 mg/kg BW in rats.

Repeated Exposure revealed evident cumulative properties.[2] Rats were given 300 mg ABSS/kg BW for a month. The treatment reduced leukocyte count, blood *Hb* level and cholesterol concentrations, cholinesterase, alkaline and acid phosphatase activity, and increased *SH*-group concentration in the blood serum of rats. This dose was considered to be the LOAEL.

Long-term Toxicity.

Observation in man. Volunteers consumed 100 mg daily doses for 6 months without any toxic effect observed.

Animal studies. In a 6-month study, guinea pigs were given drinking water with a concentration of 2.0 g/l; rats received 500 mg/l for 2 years. No harmful effect was noted in both studies.[3]

Reproductive Toxicity.

Gonadotoxic effect was not observed at dose levels of 12 to 300 mg/kg BW administered for 3 months.

Carcinogenicity. Negative results for linear ABSS are reported in rats.[09]

Regulations. ***U.S. FDA*** (1998) approved the use of ABSS in (1) adhesives as a component of articles intended for use in packaging, transporting, or holding food in accordance with the conditions prescribed in 21 CFR part 175.105; and (2) as emulsifiers and/or surface-active agents in the manufacture of non-food articles or components of articles.

Standards. ***Russia*** (1995). MAC: 0.4 mg/l (organolept., foam).

References:

1. Deichmann, W. B., *Toxicology of Drugs and Chemicals*, Academic Press, Inc., New York, 1969, 690.
2. Kondratyuk, V. A., Pastushenko. T. V., Gun'ko, L. M., et al., Substantiation of the hygienic standard for sodium alkylsulfate in water, *Gig. Sanit.*, 6, 81, 1983 (in Russian).
3. Grushko, Ya. M., *Hazardous Organic Compounds in the Liquid Industrial Effluents*, Khimiya, Leningrad, 1976, 26 (in Russian).

(ALKYLDIOXYETHYLENE)METHYLENE METHYL DIETHYL AMMONIUM BENZENE SULFONATE

Molecular Formula.

$[C_nH_{2n+1}O(C_2H_4O)C_2N^+(CH_3)(CH_2H_5)_2][C_6H_5SO_3^-]$, where $n = 16$ to 18

CAS No 11098-05-8

RTECS No AZ7748000

Trade Name. Alkamon OS-2.

Properties. Yellow or yellow-brown, waxy mass. Readily soluble in water, forming emulsion. Taste perception threshold is 14.8 mg/l, odor is detected at higher concentrations. Threshold concentration for foam formation is 0.3 to 0.5 mg/l.[1]

Applications. Used as an antistatic additive, filler, fabric softener.

Acute Toxicity. LD_{50} is 3.1 to 3.75 g/kg BW in rats, and 2.0 to 2.3 g/kg BW in mice. LD_{50} of 40% oil solution in rats is 2.08 g/kg BW, while for an alkali solution, it is 3.8 g/kg BW; in mice, LD_{50} values are 1.2 g/kg BW and 2.1 g/kg BW, respectively.[1]

Repeated Exposure revealed cumulative properties. K_{acc} is 1.45 (mice) and 1.67 (rats). The treatment led to retardation of BW gain and an increase in the relative weights of the visceral organs.

Long-term Toxicity effects in mice include retardation of BW gain, CNS stimulation, and an increase in the relative liver weight.[2] The NOEL of 0.5 mg/kg BW was reported.[1]

Standards. ***Russia*** (1995). MAC and PML: 0.5 mg/l (organolept., foam).

References:

1. Kordysh, E. A., Ratpan, M. M, and Dekanoidze, A. A., The method of investigation of acute toxicity of detergents having pronounced alkaline and acidic properties, *Gig. Sanit.*, 7, 70, 1978 (in Russian).
2. Mikhailets, I. B., Toxic properties of some antistatics, *Plast. Massy*, 12, 27, 1976 (in Russian).

ALKYLSULFATES, SODIUM SALT
(***Primary***)

Molecular Formula. $C_nH_{2n+1}OSO_3Na$, where $n = 10$ to 20

Abbreviation. AS.

Properties. Water-soluble liquid containing 27 to 32% of the pure substance. The threshold concentration of *primary* AS from sperm-whale fat is 0.3 mg/l for odor and 0.4 mg/l for foam formation. The threshold concentration of *primary* AS from *secondary unsaponifiables* is 0.6 mg/l for odor and 0.5 mg/l for foam formation.[1]

Applications. A base for detergents; used also as an ingredient of coatings.

Acute Toxicity. In rats LD_{50} is 1.7 g/kg BW. Manifestations of the toxic action include circulatory disturbances, reduction in the RNA contents in the liver, myocardium, and lung, and DNA contents in the liver. Increased glycogen contents in all organs is noted.[2,3] For *primary* AS from sperm-whale fat, LD_{50} is 3.82 g/kg BW in rats, 3.44 g/kg BW in guinea pigs, and 2.8 g/kg BW in mice. For *primary* AS from *secondary unsaponifiables*, LD_{50} is 7.0 g/kg BW in rats, 3.75 g/kg BW in mice, and 6.85 g/kg BW in guinea pigs.[1]

Long-term Toxicity. In a 6-month study, a 100 mg/kg BW dose caused changes in enzyme activity at various points in the metabolic chain.

Allergenic Effect. The LOEL is identified to be 10 mg/kg BW (on skin application).

Regulations. ***U.S. FDA*** (1998) approved the use of AS (1) as emulsifiers and/or surface-active agents in the manufacture of articles or components of articles at levels not to exceed 2.0% polyvinyl chloride and/or vinyl chloride copolymers in accordance with the conditions prescribed in 21 CFR part 178.3400; (2) as antistatic and antifogging agents in food-packaging materials at levels not to exceed 3.0% by weight of polystyrene or rubber-modified polystyrene; and (3) as emulsifiers for vinylidene chloride polymers and copolymers at levels not to exceed a total of 2.6% by weight of the coating solids.

Standards. ***Russia*** (1995). MAC and PML: 0.5 mg/l (organolept., foam). In the case of *primary AS* from sperm-whale fat, MAC and PML: 0.3 mg/l (organolept., foam).

References:

1. Rusakov, N. V., Allergenic effects of different chemicals by oral administration to the body, *Gig. Sanit.*, 2, 13, 1984 (in Russian).
2. Mozhayev, E. V., *Pollution of Reservoirs by Surfactants, Health and Safety Aspects*, Meditsina, Moscow, 1976, 94 (in Russian).
3. Voloshchenko, O. I. and Medyanik, L. A., *Safety and Toxicology of Household Chemicals*, Zdorov'ya, Kiyv, 1983, 142 (in Russian).

ALKYLSULFATES, SODIUM SALT
(***Secondary***)

Molecular Formula. $C_nH_{2n+1}CH(CH_3)OSO_3Na$, where $n = 6$ to 16

Abbreviation. AS.

Properties. Light-yellow to amber-colored liquid, readily soluble in water. Threshold perception concentration is 0.6 mg/l for odor and 0.5 mg/l for foam formation.[1]

Applications. Base for liquid detergents; used as a foaming agent in the production of foam concrete, as a dispersant and an emulsifier.

Acute Toxicity. In rats, LD_{50} is 1.65 to 5.63 g/kg BW.[1]

Allergenic Effect. When given *i/g*, the threshold concentration for such an effect was 0.1 mg/l.[2]

Regulations. ***U.S. FDA*** (1998) approved the use of AS as antistatic and antifogging agents in food-packaging materials at levels not to exceed 3.0% by weight of polystyrene or rubber-modified polystyrene.

References:

1. Mozhayev, E. V., *Pollution of Reservoirs by Surfactants, Health and Safety Aspects*, Meditsina, Moscow, 1976, 94 (in Russian).

2. Rusakov, N. V., Allergenic effects of different chemicals by oral administration to the body, *Gig. Sanit.*, 2, 13, 1984 (in Russian).

4-ALLYL-2-METHOXYPHENOL

Molecular Formula. $C_{10}H_{12}O_2$
M = 164.20
CAS No 97-53-0
RTECS No SJ4375000
Abbreviation. AMP.

Synonyms and **Trade Names.** Allylguaiacol; 2-Allyl-2-methoxyphenol; Caryophyllic acid; Eugenol; 1-Hydroxy-2-methoxy-4-allylbenzene; 2-Methoxy-4-allylphenol; 2-Methoxy-4-prop-2-enylphenol.

Properties. Colorless or pale-yellow liquid with spicy odor of cloves and a pungent taste. Practically insoluble in water; soluble in oils and in 70% alcohol (1:2).[020]

Applications. Used predominantly as a fragrance and flavoring agent. A chemical intermediate. Used in perfumes and as a topical dental analgesic.

Acute Toxicity. LD_{50} is 1.9 g/kg BW in rats. Poisoning produced desquamation of the epithelium in rats and guinea pigs, and hemorrhages in pyloric and glandular regions of the stomach.[1] After ingestion of a single oral dose of 250 mg/kg BW, dogs exhibited vomiting, weakness, lethargy, and ataxia. A 500 mg/kg BW dose caused coma and death within 24 hours.[03]

Short-term Toxicity. Rats and mice were given dietary concentrations of up to 12,500 ppm for 13 weeks. The only change in the treated animals was a 10% difference from controls in BW in the highest dose-treated male rats. No histological findings were reported.[2] Rats were given AMP in their diet for 84 days. Munro et al. (1996) suggested the calculated NOEL of 80 mg/kg BW for this study.[3]

Long-term Toxicity. F344 rats received 6,000 or 12,500 ppm AMP, B6C3F$_1$ mice were given 3,000 or 6,000 ppm AMP in the feed for 103 weeks. Only change noted was retardation in BW gain in the highest dose-treated animals.[2]

Mutagenicity.

In vitro genotoxicity. AMP produced genotoxic effects in a mouse micronucleus assay,[4] and in a yeast test system.[5] It was negative in *Salmonella* mutagenicity assay.[6]

In vivo cytogenetics. Rats were given by gavage a single or two successive doses equal to 0.5 LD_{50}. The treatment did not induce genotoxic effect *in vivo*.[6]

Carcinogenicity. AMP is shown to be non-carcinogen in rodents. In a 2-year NTP study, no increase in tumor incidence was reported in F344 rats.[2]

In mice, evidence of carcinogenicity was considered equivocal because of increased incidence of both carcinomas and adenomas of the liver in males at the 3,000 ppm dietary level and because AMP was associated with an increase in the combined incidences of hepatocellular carcinomas or adenomas in females.[2]

Carcinogenicity classifications. An IARC Working Group concluded that there is limited evidence for the carcinogenicity of AMP in *experimental animals* and no adequate data available to evaluate the carcinogenicity of AMP in *humans*.

IARC: 3;

NTP: N - N- E -E (feed).

Chemobiokinetics. AMP has been known to exhibit non-enzymatic peroxidation in liver mitochondria. It did not exhibit cytochrome *P-450* reductase activity but it inhibited *P-450*-linked monooxygenase activities such as aminopyrine-*N*-demethylase, *N*-nitrosodimethylamine demethylase, benzo(a)pyrene hydroxylase, and ethoxyresorufin-*O*-deethylase to different extents. The inhibitory effect of AMP on lipid peroxidation is predominantly due to its free radical quenching ability.[7]

Regulations. ***Great Britain*** (1998). AMP is authorized without time limit for use in the production of polymeric materials and articles in contact with food or drink or intended for such contact. The specific migration of this substance shall not exceed 0.01 mg/kg.

Recommendations. Joint FAO/WHO Expert Committee on Food Additives. Conditional ADI: 5.0 mg/kg BW.

References:

1. Haddad, L. M., *Clinical Management of Poisoning and Drug Overdose*, 2nd ed., W. B. Saunders Co., Philadelphia, PA, 1990, 1471.
2. *Toxicology and Carcinogenicity Studies of Eugenol in F344/N Rats and B6C3F$_1$ Mice (Feed Studies),* NTP Technical Report Series No 223, Research Triangle Park, NC, NIH Publ. 84-1779, 1983.
3. Oser, Unpublished report, 1967, in *Toxicological Evaluation of Some Flavouring Substances and Non-Nutritive Sweetiening Agents*, The 11th Meeting of the Joint FAO/WHO Expert Committee on Food Additives, Geneva, 1967.
4. Woolverton, C. J., Fotos, P. G., Mokas, M. J., and Mermigas, M. E., Evaluation of eugenol for mutagenicity by the mouse micronucleus test, *J. Oral. Pathol.*, 15, 450, 1986.
5. Schiestl, R. H., Chan, W. S., Gietz, R D., Mehta, R. D., and Hastings, P. J., Safrole, eugenol and methyleugenol induce intrachromosomal recombination in yeast, *Mutat. Res.*, 224, 427, 1989.
6. Allavena, A., Martelli, A., Robbiano, L., and Brambillo, G., Evaluation in a battery of *in vivo* assays of four *in vitro* genotoxins proved to be non-carcinogens in rodents, *Teratogen. Carcinogen. Mutagen.*, 12, 31, 1992.
7. Nagababu, E. and Lakshmaiah, N., Inhibition of microsomal lipid peroxidation and monooxygenase activities by eugenol, *Free Radic. Res.*, 20, 253, 1994.

p-AMINOBENZOIC ACID

Molecular Formula. $C_7H_7NO_2$
M = 137.13
CAS No 150-13-0
RTECS No DG1400000
Abbreviation. PABA.

Synonyms and **Trade Names.** Amben; 1-Amino-4-carboxybenzene; Anticanitic vitamin; 4-Carboxyaniline; *p*-Carboxyphenylamine; Chromotrichia factor; Paraminol; Vitamin H.

Properties. Yellowish to red odorless crystals or crystalline powder. 1.0 g PABA dissolves in 170 ml water (6.0 g/l) at 25°C, in 90 ml boiling water, in 8.0 ml alcohol.[020]

Applications. PABA is used in the production of various esters.

Acute Toxicity. LD_{50} is reported to be 6.0[020] or 10.3 g/kg BW in rats.[03]

Reproductive Toxicity.

Embryotoxicity. Mice were given 50 mg/kg BW on days 9 through 11. There was no increase in resorption frequency. However, a 23% resorption rate was reported in rats fed about 25 mg PABA/kg BW.[1,2]

Teratogenicity. Rats received 1.0% PABA in their diet before and during pregnancy. The treatment caused no adverse effects or malformations in newborns.[3]

Carcinogenicity. In the only study available, PABA was tested in mice by skin application; no carcinogenic effects were observed.

Carcinogenicity classification. An IARC Working Group concluded that the available data are insufficient for an evaluation of the carcinogenicity of PABA in *experimental animals*, and that there were no adequate data available to evaluate the carcinogenicity of PABA in *humans*.

IARC: 3

Chemobiokinetics.

Observations in man. Following ingestion of 1.0 g, 82% were excreted in the urine within 4 hours. Main metabolites were *p-aminohippuric acid* and *acetyl-p-aminohippuric acid* (IARC 16-256).

Regulations. ***EU*** (1992). PABA is available in the *List of monomers and other starting substances which may continue to be used pending a decision on inclusion in Section A.*

Standards. ***Russia*** (1995). MAC: 0.1 mg/l.

References:

1. Kato, T., Effect of folate metabolism-related factors on the teratogenic action of sulfonamide in mice, *Congen. Anomal.*, 13, 85, 1973.

2. Telford, I. R., Woodruff, C. S., and Linford, R. H., Fetal resorption in the rat as influenced by certain antioxidants, *Am. J. Anat*, 110, 29, 1962.
3. Ershoff, B. H., Effect of massive doses of *p*-aminobenzoic acid and inositol on reproduction in the rat, *Soc. Exp. Biol. Med.*, 63, 479, 1946.

N-(2-AMINOETHYL)-*N'*-{2-[(2-AMINOETHYL)AMINO]ETHYL}-1,2-ETHANEDIAMINE

Molecular Formula. $C_8H_{23}N_5$
M = 189.36
CAS No 112-57-2
RTECS No KH8585000
Abbreviation. TEPA.

Synonyms and Trade Names. 1,4,7,10,13-Pentaazatridecane; Tetraethylenepentamine; Tetren; 3,6,9-Triazaundecane-1,11-diamine.

Properties. Viscous, hygroscopic, yellow liquid with ammoniacal penetrating odor. Soluble in water and in most organic solvents.

Applications. Used in the manufacture of surfactants and emulsifying agents; a stabilizer for rubber latex.

Acute Toxicity. LD_{50} is 2.1 to 4.0 g/kg BW in rats.[1,019]

Mutagenicity. The overall weight of evidence from the *in vitro* and *in vivo* tests suggested that TEPA had a weak mutagenic potential.

TEPA was found to be positive in the *Salmonella*/microsome preincubation assay. It was tested in as many as five *Salmonella typhimurium* strains (*TA1535, TA1537, TA97, TA98,* and *TA100*) in the presence and absence of rat and hamster liver *S9*, at doses of 0.033 to 10.000 mg/plate.[013]

TEPA was considered active in the induction of SCE in CHO cells. It produced significant increases in the amount of unscheduled DNA synthesis activity, and thus was considered positive in inducing primary DNA damage. In a micronucleus study with Swiss-Webster mice, no clastogenic activity was observed.[2]

Regulations.

U.S. FDA (1998) approved the use of TEPA in adhesives as a component of articles intended for use in packaging, transporting, or holding food in accordance with the conditions specified in the 21 CFR part 175.105.

References:

1. Kirk-Othmer *Encyclopedia of Chemical Technology,* 3rd ed., M. Grayson and D. Eckroth, Eds., vol 1-26, John Wiley and Sons, New York, NY, 1978-1984, vol 7, 591.
2. Leung, H. W., Evaluation of the genotoxic potential of alkyleneamines, *Mutat. Res.,*320, 31, 1994.

6-AMINOHEXANOIC ACID

Molecular Formula. $C_6H_{13}NO_2$
M = 131.20
CAS No 60-32-2
93208-38-9
RTECS No MO6300000
Abbreviation. AHA.

Synonym and **Trade Names.** Acepramin; Amicar; 6-Aminocaproic acid; Capramol; Caprolisin; Epsicapron.

Properties. Fine, white, odorless, and tasteless crystalline powder. 1.0 g AHA dissolves in 3.0 ml of water. AHA is slightly soluble in alcohol.

Applications. Chemical intermediate in the production of Nylon-6.

Acute Toxicity.

Observations in man. Administration of 221 mg/kg BW for 8 days resulted in kidney, ureter, and bladder damage. Acute renal failure, acute tubular necrosis, and hematuria were found to develop.[1]

Animal studies. LD_{50} is 14.3 g/kg BW in mice. Manifestations of acute poisoning included GI tract lesions, rash, hypotension, nasal congestion, and conjunctival suffusion.

Repeated Exposure. Administration of 221 mg/kg BW for 8 days resulted in kidney, ureter, and bladder damage. Acute renal failure, acute tubular necrosis, and hematuria were found to develop.[1]

Long-term Toxicity. Oral administration occasionally causes nasal congestion and conjunctival hyperemia.[2]

Reproductive Toxicity.

Teratogenic effect is reported (Howorka et al., 1970).[3]

Observations in man. There were no apparent adverse fetal effects when AHA was used during human pregnancy.[5]

Allergenic Effect. Contact dermatitis from AHA has been reported.[4]

Regulations. ***EU*** (1990). AHA is available in the *List of authorized monomers and other starting substances which may continue to be used for the manufacture of plastic materials and articles intended to come into contact with foodstuffs pending a decision on inclusion in Section A* (*Section B*).

References:

1. Pitts, T. O., Spero, J. A., Bontempo, F. A., and Greenberg, A., Acute renal failure due to high-grade obstruction following therapy with aminohexanoic acid, *Am. J. Kidney Diseases*, 8, 441, 1986.
2. Grant, W. M., *Toxicology of the Eye*, 2nd ed., Charles C Thomas, Springfield, IL, 1974, 114.
3. Tanaka, M., Niizeki, H., and Miyakawa, S., Contact dermatitis from aminocaproic acid, *Contact Dermat.*, 28, 124, 1993.
4. Schardein, J. L., *Chemically Induced Birth Defects*, 2nd ed., Marcel Dekker, New York, 1996, 107.
5. Willoughby, J. S., Sodium nitroprusside, pregnancy and multiple intracranial aneurysms. *Anaesth. Intensive Care*, 12, 358, 1984.

AMMONIA

Molecular Formulas. NH_3 or NH_4^+
CAS No 7664-41-7
RTECS No BO0875000

Properties. The term *ammonia* refers to non-ionized (NH_3) and ionized (NH_4^+) species. Under room temperature and atmospheric pressure, A. occurs as a colorless gas with a pungent, repulsive smell. The gas dissolves readily in water, forming the ammonium cation. In solution, it forms and is in equilibrim as ammonium ions. Water solubility is 421 g/l at 20°C. Ammonium solutions are alkaline and react with acids to form ammonium salts. Odor perception threshold at alkaline *pH* is approximately 1.5 or 0.037 mg/l.[010] According to other data, taste and odor perception threshold is likely to be 35 mg/l.[1]

Applications. In the U.S.A., 10% A. is used in manufacture of fibers, rubber, and plastics.

Exposure. Ammonium compounds are a natural component of foods. Minor amounts of ammonium compounds are added to food as acid regulators, stabilizers, flavoring agents, and fermentation aids. The levels detected in ground and surface water are not higher than 0.2 mg/l.

Human exposure from environmental sources is negligible in comparison with endogenous synthesis. Daily intake via food and drinking water seems to be about 18 mg.[1]

Migration Data. Water contamination can arise from cement pipe linings.

Acute Toxicity. In rats, LD_{50} is found to be 350 mg/kg BW; in cats, it is 750 mg/kg BW.[2] LD_{50} equal to 4070 to 5020 mg/kg BW is reported for rats.[3]

Guinea pigs orally exposed to watery solutions of different ammonium salts (nitrate, acetate, bromide, chloride, sulfate) at dose levels of 200 to 510 mg/kg BW died of acute lung edema. NS dysfunction (changes in respiration rhythms and depth, weakness and difficulties of locomotion, hyperexcitability to tactile, acoustic and pain stimuli) was shown to develop.[1]

Rabbits were treated by instillation with 0.8 and 1.0 g non-acidic A. carbonate/kg (283 and 354 mg NH_4/kg BW, respectively). Hyperemia of the renal cortex was noted.[1]

Repeated Exposure. A. causes acidotic effect. It is believed that signs of intoxication might be noted at exposure level above 0.2 g/kg BW.

Short-term Toxicity. Male Sprague-Dawley rats received A. via drinking water at a dose of 510 mg NH_4Cl/l. Urinalysis and histology examination failed to reveal changes in the kidneys. Consumption of drinking water with concentration of 10 g NH_4Cl/l for 2.5 months caused no signs of renal damage.[1,4]

Weaned rats were administered orally 0.1 to 5.0 g/kg BW for 30, 60, and 90 days. Only young animals reacted to the 5.0 g/kg BW dose by reduced food consumption and elevated intake of water. No retardation of BW gain or histology changes were noted to develop, and only a mild fatty degeneration of the liver in the adult animals was found after a 90-day exposure.[4]

Long-term Toxicity. A. itself is not of direct importance for health at concentrations below 0.5 mg/l, nor at higher concentrations. Ammonium hydroxide at a dose of 100 mg/kg BW was administered to rabbits by gavage initially every second day, then daily for a maximum of 17 months. The treatment resulted in enlargement of the adrenals. After several weeks of treatment, blood pressure was increased.[1,4]

Sprague-Dawley rats received ammonium chloride in their drinking water at a dose of 478 mg/kg BW (males for 330 days, females for 300 days). Significant decrease of the non-lipid bone mass (the upper thigh) and of the calcium contents was observed. The blood *pH* value and the carbonic acid contents in the plasma decreased. Reduced BW gain with lower fat accumulation was also noted.[4]

Reproductive Toxicity. Administration of 100 to 200 mg/kg BW doses of different A. compounds caused enlargement of the ovaries and uterus. Breast hypertrophy with milk secretion, follicle ripening and formation of the *corpus luteum* was observed in unpuberal female rabbits. Administration of a 200 mg/kg BW dose to pregnant rats in their drinking water inhibited the fetal growth but did not produce any teratogenic effect.[5]

Mutagenicity.

In vitro genotoxicity. A. was shown to be negative in *Salmonella* mutagenicity assay. CA were observed in Chinese hamster fibroblasts without metabolic activation.

In vivo cytogenetics. A. did not induce micronuclei in the bone marrow cells after *i/p* administration.[6,7]

Carcinogenicity. *Ammonium hydroxide* was administered to Swiss mice and C3H mice in their drinking water for their lifetime (the calculated doses were 140 to 270 mg/kg BW). No increase in tumor incidence was found.[4] Negative results are reported for *ammonium chloride* in female mice.[09]

Chemobiokinetics. A. is produced by animal and microbial metabolism; it is a product of the bacterial degradation of amino- and nucleic acids. A. is absorbed from the GI tract, but not through the skin, and is rapidly distributed throughout the body. In the liver, A. is incorporated in urea as a part of the urea cycle; furthermore, urea in the liver is absorbed into blood circulation and transferred to the kidneys in order to be excreted in the urine. A. seems to be a key metabolite in mammalian body and is formed through deamination of aminoacids in the liver and GI tract by the enzymatic breakdown of food products and with the help of microorganisms.

Recommendations. Joint FAO/WHO Expert Committee on Food Additives (1982). No ADI allocated in food.

Regulations.

U.S. FDA (1998) approved the use of different ammonium salts (1) in adhesives as components of articles intended for use in packaging, transporting, or holding food in accordance with the conditions prescribed in 21 CFR part 175.105; (2) as a component of the uncoated or coated food-contact surface of paper and paperboard intended for use in producing, manufacturing, packaging, processing, preparing, treating, packing, transporting, or holding aqueous and fatty foods in accordance with the conditions prescribed in 21 CFR part 176.170; (3) as a component of a defoaming agent intended for articles which may be used in producing, manufacturing, packing, processing, preparing, treating, packaging, transporting, or holding food in accordance with the conditions prescribed in 21 CFR part 176.200; (4) in the manufacture of cellophane for packaging food in accordance with the conditions prescribed in 21 CFR part 177.1200; (5) in the manufacture of polyurethane resins which may be used as food-contact surface of articles intended for use in contact with bulk quantities of dry food of the type identified in 21 CFR parts 177.1680 and 176.170 (c); and (6) in articles made of carboxyl-modified polyethylene resins or as components of these articles intended for use in contact with food.

Ammonium benzoate may be used as a preservative only; ammonium bifluoride and ammonium silicofluoride are allowed to be used only as bonding agents for aluminum foil, stabilizer, or preservative (up to 1.0% by weight of the finished adhesive). Ammonium citrate and ammonium potassium phosphate may be used in the manufacture of resinous and polymeric coatings for the food-contact surface of articles intended for use in producing, manufacturing, packing, processing, preparing, treating, packaging, transporting, or holding food.

In the ***EU***, paints, varnishes, printing inks, adhesives, and similar products that contain A. in solution at concentrations greater than 35% are considered toxic and corrosive and, at concentrations of 10 to 35%, as harmful and irritant, and must be packaged and labeled accordingly.[5] The maximum concentration of A. in finished cosmetic products must not exceed 6.0% calculated as NH_3. If the concentration exceeds 2.0%, the label must read: *CONTAINS AMMONIA.*

Standards.

WHO (1996) did not consider it necessary to recommend the Guideline value for drinking water. Concentration of 1.5 mg/l is likely to give rise to consumer complaints of foreign odor and taste. ***EU*** (1990). MAC: 0.5 mg NH_4/l.

EU (1980). MAC: 0.5 mg/l.

Great Britain (1998). A. is authorized without time limit for use in the production of polymeric materials and articles in contact with food or drink or intended for such contact.

Russia (1995). MAC: 2.0 mg *N*/l.

The Netherlands (1986). Limit value: 10.0 mg N/l.

Czechoslovakia (1975). MAC in drinking water: 0.5 mg/l.

References:

1. *Ammonia*, Environmental Health Criteria Series No 54, WHO, Geneva, 1986.
2. *Ammoniae et solutions aqueuses, Fiche toxicologique* 16, Institute Natl. de Recherche et du Securite (INRS), Cahiers de notes documentaires, Issue 128, 1987, 461.
3. Ishidate, M., Sofuni, T., Yoshikawa, K, et al., Primary mutagenicity screening of food additives currently used in Japan, *Food Cosmet. Toxicol.*, 22, 623, 1984.
4. *Health Effect Assessment for Ammonia*, EPA/600/8-88/017, Cincinnati, 1987.
5. *Ammonia*, IPCS Health Safety Guide No 37, WHO, Geneva, 1990, 30.
6. Reichert, J. und Lochtmann, S., Occurrence of nitrite in water distribution system, *JMF Wasser-Abwasser.*, 125, 442, 1984.
7. Hayashi, M., Kishi, M., Sofuni, T. et al., Micronucleus tests in mice on 39 food additives and 8 miscellaneous chemicals, *Food Cosmet. Toxicol.*, 26, 487, 1988.

AMMONIUM HYDROGEN FLUORIDE

Molecular Formula. $F_2H.H_4N$

M = 57.06

CAS No 1341-49-7

RTECS No BQ9200000

Abbreviation. AHF.

Synonyms. Acid ammonium fluoride; Ammonium bifluoride; Ammonium hydrofrluoride.

Properties. Rhombic or tetragonal crystals (or odorless white solid). Solubility is 41.5 wt% in water at 25°C; 1.73 wt% in 9.0% ethanol at 25°C.

Applications. Used in glass and porcelain industries.

Acute Toxicity. Poisoning caused salivation, nausea, abdominal pain, vomiting, and diarrhea. Other symptoms observed were tonic and clonic convulsions, hypocalcemia and hypoglycemia, respiratory arrest and cardiac failure, etc.[1]

Long-term Toxicity. An important effect of fluoride is dental fluorosis. The major manifestations of chronic ingestion of excessive amounts of fluoride are osteosclerosis and mottled enamel.[1] Ingestion of more than 6 mg/day led to fluorosis development including such symptoms as weight loss, anemia, weakness, etc.[2]

Regulations. ***U.S. FDA*** (1998) approved the use of AHF in adhesives as a component of articles intended for use in packaging, transporting, or holding food in accordance with the conditions prescribed in 21 CFR part 175.105.

References:

1. Goodman and Gilman's *The Pharmacological Basis of Therapeutics*, A. G. Gilman, L. S. Goodman, and A. Gilman, Eds., 6th ed., Macmillan Publ. Co., Inc., New York, 1980, 1546.
2. Dreisbach, R. H., *Handbook of Poisoning*, 9th ed., Lange Medical Publ., Los Altos, CA, 1977, 207.

AMMONIUM IRON (2^+) SULFATE HEXAHYDRATE

Molecular Formula. $H_4N_2.Fe_2O_4S.6H_2O$
M = 356.09
CAS No 7783-85-9
RTECS No BR6500000
Abbreviation. AISH.

Synonym and **Trade Name.** Ammonium ferrous sulfate hexahydrate; Mohr's salt.

Acute Toxicity. LD_{50} is 3.25 g/kg BW in rats.

Regulations. ***U.S. FDA*** (1998) approved the use of AISH in the manufacture of cellophane for packaging food in accordance with conditions prescribed in 21 CFR part 177.1200.

Reference:

Smyth, H. F., Carpenter, C. P., Weil, C. S., Pozzani, U. C., Striegel, J. A., and Nyaun, J. S., Range-finding toxicity data: List VII, *Am. Ind. Hyg. Assoc. J.*, 30, 470, 1969.

2-ANTHRAQUINONESULFONIC ACID, SODIUM SALT

Molecular Formula. $C_{14}H_7O_5S.Na$
M = 310.26
CAS No 131-08-8
RTECS No CB1095550

Synonyms and **Trade Name.** 9,1-Anthraquinone-2-sodium sulfonate; Silver salt; Sodium 2-anthraquinone sulfonate; 2-Sulfoanthraquinone, sodium salt.

Acute Toxicity. LD_{50} is found to be 21 g/kg BW in guinea pigs. Poisoning affects liver functions and caused renal functions depression.[1] LD_{50} *i/p* is 0.73 g/kg BW in rats, and 0.63 g/kg BW in mice.[2]

Regulations. ***U.S. FDA*** (1998) approved the use of A. in adhesives as a component of articles intended for use in packaging, transporting, or holding food in accordance with the conditions prescribed in 21 CFR part 175.105.

References:

1. Zaitseva, N. V. and Kulikov, A. L., Data for the experimental establishment of the maximum permissible concentrations of organosulfur compounds in the water reservoirs, *Gig. Sanit.*, 3, 73, 1980 (in Russian).
2. Information from the Soviet Toxicological Center, *Gig. Sanit.*, 4, 93, 1990 (in Russian).

ASBESTOS

	CAS No
Chrysotile	12001-29-5
Amosite	12172-73-5
Anthophyllite	17068-78-9
Crocidolite	12001-28-4

Composition. A. are fibrous silicate minerals of the serpentine and amphibole mineral groups. The different forms of A. consist of silica (40 to 60%) and also of oxides of iron, magnesium, and other metals. The length of A. fibers in drinking water is 0.5 to 2.0 μ; the diameter is 0.03 to 0.1 μ.

Applications. Used as a filler for A.-cement and A.-filled polymeric materials.

Migration Data. A. gets into water by means of dissolution of A.-containing sheets and pipes in the distribution system. Exfoliation of A. fibers from A.-containing pipes is related to the corrosiveness of the water supply, increasing in aggressive waters.

Exposure. A. may enter foodstuffs from the water and impure talc used in their production (chewing gums, rice sticks, etc.).

Long-term Toxicity.

Observations in man. In the studies of populations in Duluth, Canadian cities, Connecticut, Florida, and Utah, all of which were ecological in nature and in which population mobility could not be adequately assessed, there was no consistent evidence of an association between cancer mortality or incidence and ingestion of A.[1-3] concentrations in food and water. All GI tract tumors and tumors of the esophagus and colon in both sexes were observed in an ecological study in the San Francisco Bay area; the reanalysis of the data taking potential confounding factors into consideration, has undermined the significance of these results.[1,2]

The mobility of the population in this area is high and coincidental geographical associations were not apparent when San Francisco and all other Bay area counties were considered separately. In an analytical epidemiological study that was inherently more sensitive than the ecological studies mentioned above, there was no consistent evidence of a cancer risk due to ingestion of A. in drinking water in Puget Sound, where levels ranged up to 200 MFL (million fibers/l).[2,3]

Animal studies. Information on the transmigration of ingested A. through the GI tract to other tissues is also contradictory.[4,5] Although it is not possible to conclude with certainty that ingested fibers do not cross the intact GI tract, available data indicate that penetration, if it occurs at all, is extremely limited.

Reproductive Toxicity.

Embryotoxicity. CD-1 mice received approximately 4.0 to 400 mg chrysotile/kg on days 1 to 15 of pregnancy. The treatment did not affect survival of the progeny. *In vitro* administration interfered with implantation upon transfer of exposed blastocysts to recipient females, but did not result in a decrease in post-implantation survival. A. was not considered as a teratogen in these studies.[6]

Mutagenicity.

In vivo cytogenetics.

Observations in man. Workers exposed to A. showed a marginal increase in SCE frequency.

Animal studies. A single oral administration of chrysotile did not increase the frequency of micronuclei in mice or CA in bone marrow cells of rhesus monkeys treated *in vivo*.[8]

In vitro genotoxicity. Conflicting results were reported for the induction of CA in cultured human cells exposed to chrysotile: chrysotile transformed BALB/c3T3 mouse cells and rat mesothelial cells and negative results for the induction of SCE by chrysotile and crocidolite; amosite, and crocidolite did not induce DNA strand breaks.[7]

Amosite, anthophyllite, chrysotile and crocidolite induced transformations of Syrian hamster embryo cells. In cultured rodent cells, amosite, anthophyllite, chrysotile, and crocidolite induced CA, and amosite, chrysotile, and crocodolite induced SCE. Chrysotile did not induce unscheduled DNA synthesis in rat hepatocytes. Chrysotile and other forms of asbestos were not mutagenic to bacteria.[7]

Carcinogenicity. There has been great public concern about the adverse health effects resulting from the presence of A. fibers in municipal drinking water supplies.[9]

In 1984, U.S. National Research Council concluded that "the association of asbestos with an increased risk of malignancies other than lung cancer and mesothelioma has not been confirmed in animal studies and has not been observed consistently in human studies".

Observations in man. A. is a known human carcinogen by the inhalation route. Although well studied, there has been little convincing evidence of the carcinogenicity of ingested A. in epidemiology studies of populations with drinking water supplies containing high concentrations of A.[10]

On the basis of available epidemiological data then, for the majority of cancer sites and for all ages and sexes, no excess risk is present, even for high levels of A. in the drinking water.

Animal studies. Condie presented a review of 11 experimental studies of orally administered A. in which the carcinogenic potential of A. following its ingestion has been evaluated. The long-term, high-level ingestion of various types of A. fibers in more than one animal species failed to produce any definite, reproducible, organ-specific carcinogenic effect.[11]

In extensive studies in animal species, A. has not induced consistent increases in the incidence of tumors of the GI tract.[4,12]

A. was administered to Syrian golden hamsters as 1.0% amosite, or short-range or intermediate-range chrysotile in the diet over their lifetime, and also to F344 rats as 1.0% tremolite or amosite in the diet over their lifetime. No treatment-related effects in animals were observed; increase in the incidence of benign epithelial neoplasms in the GI tract of male rats (treated with chrysotile) was not statistically significant. Moreover, no increase in tumor incidence was observed in F344 rats ingesting short-range chrysotile (98% shorter than 0.01 mm) which was composed of fiber sizes more similar to those found in drinking water.[12,13]

The effects of ingested A. fibers were studied in Wistar Han rats. Chrysotile and a mixture of chrysotile/crocidolite in palm oil were given to the animals for 24 months at daily doses of 10, 60 and 360 mg. The results indicate that ingestion of A. fibers at high doses had no toxic effects and did not affect animal survival.[14]

F344 rats were gavaged with a suspension of untreated UICC anthophyllite fibers (50 mg/kg BW) and fibers which had been allowed to adsorb benzo[a]pyrene molecules from aqueous solutions. Whereas anthophyllite fibers failed to induce cytogenetic alterations, fibers pretreated with the polycyclic aromatic solutions caused dose-dependent increase in the SCE frequencies. The observed cytogenetic impact can be explained by a local action of carcinogen molecules accumulated and subsequently transported. The results support the hypothesis that epidemiological evidence of carcinogenicity of A. in potable water may be explained by the cogenotoxic action of the A. fibers and biologically active organic micropollutants adsorbed on their surface.[15]

Carcinogenicity classifications (inhalation exposure). An IARC Working Group concluded that there is sufficient evidence for the carcinogenicity of A. in *experimental animals* and in *humans*.

The Annual Report of Carcinogens issued by the U.S. Department of Health and Human Services (1998) defines A. to be a known carcinogen, i.e., a substance for which the evidence from human studies indicates that there is a casual relationship between exposure to the substance and human cancer.

IARC: 1 (inhalation);

U.S. EPA: A;

EU: 1.

Chemobiokinetics. The fate of absorbed A. fibers is imperfectly understood at present. There was evidence that particles of A., entering the GI tract with drinking water or food, penetrate its walls and enter the bloodstream. On being disseminated throughout the parenchymatous organs, they may stimulate DNA biosynthesis and exert a powerful carcinogenic effect. Now there is considerable disagreement concerning whether or not A. fibers ingested can migrate from the GI lumen through the walls of the GI tract in sufficient numbers to cause adverse local or systemic effects.

Regulations.

U.S. FDA has taken no action to date with regard to A. in food because there is no evidence that the ingestion of small amounts of A. found in food poses any human health risk.[016]

U.S. FDA (1998) regulates A. as (1) a component of adhesives intended for use in contact with food in accordance with the conditions prescribed in 21 CFR part 175.105; (2) in cross-linked polyester resins for use in articles or components of articles intended for repeated use in contact with food in accordance with the conditions prescribed in 21 CFR part 177.2420; (3) in the manufacture of phenolic resins for food-contact surface of molded articles intended for repeated use in contact with nonacid food (*pH* above 5.0) in accordance with the conditions prescribed in 21 CFR part 177.2410; and (4) as a filler in the manufacture of rubber articles intended for repeated use in producing, manufacturing, packing, processing, treating, packaging, transporting, or holding food in accordance with the conditions prescribed in 21 CFR part 177.2600.

Standards. *U.S. EPA* (1998). MCL and MCLG: 7 MFL (million fibers/l).

Recommendations. *WHO* (1996). Ingested A. is not hazardous to health and, for this reason, a health-based Guideline Value for A. in drinking-water is not recommended. A.-cement pipes should not be used in areas where the water supply is aggressive.[10]

References:

1. Wigle, D. T., Cancer mortality in relation to asbestos in municipal water supplies, *Arch. Environ. Health*, 32, 185, 1977.
2. Harrington J. M., Craun, G. F., Meigs, J. W., et al., An investigation of the use of asbestos cement pipe for public water supply and the incidence of gastrointestinal cancer in Connecticut 1935-1973, *Am. J. Epidemiol.*, 107, 96, 1978.
3. *Overall Evaluation of Carcinogenicity*, IARC Monographs, Suppl. 7, An Updating of IARC Monographs, vol 1 to 42, Lyon, 1987.
4. Toft, P., Meek, M. E., Wigle, D. T., et al., Asbestos in drinking water, *CRC Crit. Rev. Environ. Control,* 14, 151, 1984.
5. DHHS Working Group, Report on cancer risk associated with ingestion of asbestos, *Environ. Health. Perspect.*, 72, 253, 1987.
6. Schneider, V. and Maurer, R. R., Asbestos and embryonic development, *Teratology*, 15, 273, 1977.
7. Morita, T., Asano, N., Awogi, T., Sasaki, Yu. F., et al. Evaluation of the rodent micronucleus assay in the screening of IARC carcinogens (Groups 1, 2A and 2B). The summary report of the 6th collaborative study by CSGMT/JEMS-MMS, *Mutat. Res.*, 389, 3, 1997.
8. Montizaan, G. K., Knaap, A. G., and van der Heijden, C. A., Asbestos: toxicology and risk assessment for the general population in The Netherlands, *Food Chem. Toxicol.*, 27, 53, 1989.
9. Macrae, K. D., Asbestos in drinking water and cancer, *J. Royal Coll. Physicians*, London, 22, 7, 1988.
10. *Revision of the WHO Guidelines for Drinking-Water Supply*, Report of Group Meeting on Inorganics, The 2nd Review, Brussels, Belgium, October 14-18, 1991.
11. Condie, L. W., Review of published studies of orally administered asbestos, *Environ. Health Perspect.,* 53, 3, 1983.
12. *Toxicology and Carcinogenesis Studies of Chrysotile Asbestos in Fischer-344 Rats*, Public Health Service, NTP Technical Report Series No 295, U.S. Dept. Health & Human Service, Washington, D.C., NIH Publ. No. 86-2551, 1985.
13. McConnell, E. E., Shefner, A. M., Rust, J. H., et al., Chronic effects of dietary exposure to amosite and chrysotile asbestos in Syrian golden hamsters, *Environ. Health Perspect.*, 53, 11, 1983.
14. Truhaut, R. and Chouroulinkov, I., Effect of long-term ingestion of asbestos fibers in rats, *IARC Sci Publ.*, 90, 127, 1989.
15. Varga, C., Horvath, G., and Timbrell, V., *In vivo* studies on genotoxicity and cogenotoxicity of ingested UICC anthophyllite asbestos, *Cancer Lett.,* 105, 181, 1996.

ASPHALT

CAS No 8052-42-4

RTECS No CI9900000

Synonym and **Trade Names.** Asphalt fumes; Petroleum asphalt; Bitumen; Not tar; Pitch; Road asphalt.

Composition. A mixture of saturated and aromatic hydrocarbons and asphaltenes.

Properties. Amorphous, partially oxidized residue from evaporated petroleum. Black or dark brown cement-like mass, solid or semisolid in consistency, or viscous liquid, with pitch-like odor. Insoluble in water and alcohol.

Exposure. Many workers in the highway construction and roofing industries are potentially exposed to asphalt fumes.

Mutagenicity.

In vivo cytogenetics. Condensates of A. fumes are found to be weakly mutagenic to bacteria and are capable of inducing micronucleus formation in cultured mammalian cells. Male CD rats received 3 intratracheal instillations of A. fume condensates in a 24-hour period. The treatment caused DNA adduct formation in rat lung cells but did not induce DNA adducts in WBCs.[1]

In vitro genotoxicity. A. was found to be negative in *Salmonella* mutagenicity assay.[2]

Carcinogenicity.

Observations in man. There is a possible link between A. and scrotal cancer, Hodgkin disease and lung cancer (Wahlberg, 1974). Nevertheless, according to Chiazze et al., no firm conclusions can be drawn about the risk of cancer in human from asphalt.[3]

Animal studies. Male Parker mice were treated topically with solution of bitumen, a product of oil-refining, for up to 5 weeks. A steady accumulation of DNA adduct, formed *in vivo* by a large number of different chemical compounds present in the applied solution, was seen in skin DNA.[4]

Carcinogenicity classification. An IARC Working Group concluded that there is inadequate evidence that bitumens alone are carcinogenic to *humans*; there is sufficient evidence for the carcinogenicity of extracts of steam-refined bitumens, air-refined bitumens and pooled mixtures of steam- and air-refined bitumens in *experimental animals*; there is inadequate evidence for the carcinogenicity of undiluted air-refined bitumens in *experimental animals*.

Regulations. ***U.S. FDA*** (1998) approved the use of A. in adhesives as a component of articles intended for use in packaging, transporting, or holding food in accordance with the conditions prescribed in 21 CFR part 175.105.

References:

1. Qian, H. W., Ong, T., Nath, J., and Whong, W. Z., Induction of DNA adducts *in vivo* in rat lung cells by fume condensates of roofing asphalt, *Teratogen. Carcinogen. Mutagen.,* 18, 131, 1998.
2. Robinson, M., Bull, R. J., Munch, J., and Meier, J., Comparative carcinogenic and mutagenic activity of coal tar and petroleum asphalt paints used in potable water supply systems, *J. Appl. Toxicol.*, 4, 49, 1984.
3. Chiazze, L., Watkins, D. W. and Amsel, J., Asphalt and risk of cancer in man, *Brit. J. Ind. Med.*, 48, 538, 1991.
4. Schoket, B., Hewer, A., Grover, P. L., and Phillips, D. H., Covalent binding of components of coal-tar, creosote and bitumen to the DNA of the skin and lungs of mice following topical application, *Carcinogenesis*, 9, 1253, 1988.

1,1-AZOBISFORMAMIDE

Molecular Formula. $C_2H_4N_4O_2$

M = 116.08

CAS No 123-77-3

RTECS No LQ1040000

Abbreviation. ABFA.

Synonyms and **Trade Names.** Azodicarbonamide; Azodicarbonic acid, diamide; Azoform A; Porofor ChKhZ 21.

Properties. Fine, yellow-orange crystalline powder. Poorly soluble in cold water, more readily soluble in hot water. Insoluble in ethanol. Does not react with plasticizers and other components of plastics.

Applications. High-temperature blowing (pore-formation) agent for polyvinyl chloride, polyolefins, polyamides, polyepoxys, polysiloxanes, polyacrilonitrile-butadiene-styrene copolymers, etc. A vulcanizing agent and vulcanization accelerator.

Acute Toxicity. No animals died following administration of 1.5 to 2.0 g/kg BW doses. BW loss and changes in the blood morphology accompanied poisoning. In Wistar rats, LD_{50} is 6.4 g/kg BW.[1] According to Joiner, however, this dose is shown to be harmless for rats and mice.[05]

Repeated Exposure revealed cumulative properties: 2 out of 5 rats died after 4 administrations of 1.5 g/kg BW. Gross pathology examination revealed atony of the stomach and intestine. The liver and kidneys looked unusually dark, and the brain was hyperemic. However, according to other data, administration of 1.0 g/kg BW to rats for 8 weeks was found to be harmless for animals.[05]

Rats received 5.0 or 10% ABFA in their diet for 10 days or 4 weeks. The treatment lowered thyroidal iodine uptake and serum protein bound iodine. Goitrogenic effect was not observed.[2]

Long-term Toxicity. Because ABFA is readily reduced to *biurea* during the baking process, feeding studies were performed in which biurea was added to the bread diet at levels of 100 to 1000 times the normal use levels (7.5 ppm in bread dough). The studies elicited no adverse response for the 2-year investigation period.[3]

Dogs and rats received 5.0 and 10% biurea in their diet for a year.[3] No adverse effect was reported in rats. In dogs, evidence of the renal pathology was noted after about 4 months. Histology examination revealed a deposition of calculi in the kidney, ureter and bladder of the majority of animals.

Allergenic Effect. ABFA was shown to cause pulmonary sensitization and dermatitis in people.

Mutagenicity.

In vitro genotoxicity. ABFA was found to be positive in *Salmonella* mutagenicity assays.[013]

Chemobiokinetics. 30% of the dose given by gavage to F344 rats was found to be absorbed in 72 hours after administration. Upon inhalation, ABFA is readily converted into *biurea* under physiological conditions, and biurea was the only ^{14}C-labeled compound present in *excreta.* Excretion occurs predominantly via the urine.[4]

Regulations. ***U.S. FDA*** (1998) approved the use of ABFA (1) as an aging and bleaching ingredient in cereal flour in an amount not to exceed 45 ppm of flour; (2) as a chemical blowing agent in the preparation of rubber articles (total not to exceed 5.0% by weight of rubber product); (3) as an adjuvant which may be used in the manufacture of foamed plastics intended for use in contact with food, subject to the provisions prescribed in 21 CFR part 178.3010; and (4) in the manufacture of closures with sealing gaskets on containers intended for use in producing, manufacturing, packing, processing, preparing, treating, packaging, transporting, or holding food not to exceed 1.2% by weight of the closure-sealing gasket composition in accordance with the conditions prescribed in 21 CFR part 177.1210. It may be used (4) in food in accordance with the conditions prescribed in 21 CFR part 172.806.

Standards. ***Russia.*** PML in drinking water: 0.2 mg/l.

References:

1. Ferris, B. G., Peters, J. M., Burgess, W. A., et al., Apparent effect of an azodicarbonamide on the lungs, A preliminary report, *J. Occup. Med.*, 19, 424, 1977.
2. Gafford, E. T., Sharry, P. M, and Pittman, J. A., Effect of azodicarbonamide (11,1'-azobisfomamide) on thyroid function, *J. Clin. Endocrinol. Metab.*, 32, 659, 1971.
3. Oser, B. L., Oser, M., and Morgareige, K., Studies of the safety of azodicarbonamide as a fluormaturing agent, *Toxicol. Appl. Pharmacol.*, 7, 445, 1965.
4. Mewhinney, J. A., Ayres, P. H., Bechtold, W. E., Dutcher, J. S., Cheng, Y. S., Bond, J. A., Medinsky, M. A., Henderson, R. F., and Birnbaum, L. S., The fate of inhaled azodicarbonamide in rats, *Fundam. Appl. Toxicol.*, 8, 372, 1987.

BENTONITE

CAS No 1302-78-9
11004-12-9
RTECS No CT9450000

Composition. Aluminium silicate clay.

Synonyms and **Trade Names.** Bentonite magma; Magbond; Tixoton; Volclay; Wilkinite.

Properties. Odorless and tasteless, cream to pale-brown powder.

Applications. B. is used as a filler in ceramics, in the production of paper coatings and abrasives. A food additive.

Acute Toxicity.

Observations in man. Probable oral lethal dose is above 15 g/kg BW.[022]

Animal studies. B. did not cause intoxication or death after a single oral administration to domestic animals at doses of up to 4.0 g/kg BW.[1]

Repeated Exposure. Activated B. was added to fodder of chickens at the concentrations of 1.0 to 6.0% for a month. The consumption did not cause any changes in behavior, overall state, or in the clinical, biochemical, and electrolytic composition of blood. Growth suppression was observed.[2]

Carcinogenicity. Mice fed a diet containing 50% B. for 28 weeks exhibited liver tumors.[2] Equivocal tumorigenic agent by *RTECS* criteria.

Chemobiokinetics. In *in vitro* studies, B. caused release of fatty acids in human umbilical vein endothelial cells and cell lysis.[3]

Regulations.

U.S. FDA (1998) approved the use of B. (1) in adhesives as a component of articles intended for use in packaging, transporting, or holding food in accordance with the conditions prescribed in 21 CFR part 175.105; (2) in the manufacture of resinous and polymeric coatings for the food-contact surface of articles intended for use in producing, manufacturing, packing, processing, preparing, treating, packaging, transporting, or holding food in accordance with the conditions prescribed in 21 CFR part 175.320; and (3) as a component of the uncoated or coated food-contact surface of paper and paperboard intended for use in producing, manufacturing, packaging, processing, preparing, treating, packing, transporting, or holding aqueous and fatty foods in accordance with the conditions prescribed in 21 CFR part 176.170. Affirmed as *GRAS.*

EU (1995). B. is a food additive generally permitted for use in foodstuffs.

References:

1. Drumev, D., Donev, B., Dimitrov, K., Dilov, P., and Dzhurov, A., Toxicological studies of Bulgarian activated bentonite, *Vet. Med. Sci.*, 17, 84, 1980.
2. Wilson, J. W., Hepatomas produced in mice by feeding bentonite in the diet, *Ann.. NY Acad. Sci.*, 57, 678, 1954.
3. Murphy, E. J., Roberts, E., and Horrock, L. A., Aluminum silicate toxicity in cell cultures, *Neuroscience*, 55, 597, 1993.

BENZALDEHYDE

Molecular Formula. C_7H_6O
M = 106.12
CAS No 100-52-7
RTECS No CU4375000

Synonyms. Benzenecarbonal; Benzenecarboxaldehyde; Benzoic aldehyde; Phenylmethanal.

Properties. Clear, colorless liquid with the odor of bitter almonds. Solubility in water is 3.0 g/l. Mixes with alcohol at all ratios. Odor perception threshold is 0.002 mg/l. Taste perception threshold is 0.003 mg/l.[1]

Applications. Used in the production of phenol-aldehyde and other resins.

Migration Data. B. has been found to be a degradation product of dibenzoyl peroxide and to migrate as such from rubber articles into contact media.

Acute Toxicity. Rats tolerate administration of the maximum oral dose of 0.5 g/kg BW. LD_{50} appeared to be about 1,3 g/kg BW.[2]

Repeated Exposure. All animals given B. in corn oil by gavage at the dose of 1600 mg/kg BW died during 2-week treatment. Gross pathology examination revealed no lesions attributable to B.[3]

The dose of 435 mg B./kg BW given for 14 days caused the death of 1 out of 6 rats. Addition of 1.0% B. in the diet for 14 days resulted in decreased BW and liver weight gains in rats.[4]

Short-term Toxicity. In a 90-day study rats received 50 to 800 mg/kg BW, mice received 75 to 1200 mg/kg BW. The treatment with high doses caused mortality and decrease in BW. Histological investigation revealed necrotic and degenerative changes in cerebellar and hypocampal regions of brain and renal tubular necrosis in the high dose rats. The NOEL of 400 mg/kg BW was established for rats; the NOEL of 300 mg/kg BW was estalished for male mice.[5]

In a 13-week gavage study (NTP-90), there were lesions involving the brain, forestomach, kidney, and liver of male and female F344 rats and the kidney of male $B6C3F_1$ mice.

Mutagenicity.

In vivo cytogenetics. B. was not found to be mutagenic in *Dr. melanogaster.*

In vitro genotoxicity. B. induced DNA strand breaks in PM2 DNA in the presence of copper chloride. Neither aldehyde nor copper chloride alone showed DNA breakage properties.[6] B. did not induce CA, but increased SCE in Chinese hamster ovary cells.[3] It did not show mutagenic activity in *Salmonella typhimurium.*

Carcinogenicity. Male and female F344 rats were exposed to 200 and 400 mg B./kg for 2 years. The treatment produced no evidence of carcinogenic activity in the animals. Increased incidence of squamous cell papillomas and hyperplasia of the forestomach was noted.[3]

Carcinogenicity classification.

NTP: NE - NE - SE - SE* (gavage).

Chemobiokinetics. After absorption, B. is rapidly converted into *hippuric acid* in the blood, brain, and adipose tissue of rats.[4] In rabbits, rats and dogs, B. is removed from the body with the urine as *hippuric (N-benzoylaminoacetic) acid*.

Regulations. ***U.S. FDA*** (1998) affirmed BA as *GRAS* when used as synthetic flavoring agent.

Standards. ***Russia*** (1995). MAC and PML: 0.003 mg/l (organolept., taste).

Recommendations. Joint FAO/WHO Expert Committee on Food Additives (1997): ADI: 0 to 5.0 mg/kg BW.

References:

1. Klein, L. et al., *Aspects of River Pollution*, London, 1957.
2. Herrman, H., Jungstand, W., und Schnabel, R., Cancerogentest mit *p-N*-Methylnitrosamino benzaldehyd, *Arzneimittel-Forsch.*, 9, 1244, 1966.
3. *Toxicology and Carcinogenesis Studies of Benzaldehyde (Gavage Studies)*, NTP Technical Report Series No 378, Research Triangle Park, NC, 1990.
4. *Monographs on Fragrance Raw Materials*, D. L. J. Opdyke, Ed., Pergamon Press, New York, 1979, 115.
5. Kluwe, W. M., Montgomery, C. A., Giles, H. D., and Prejeau, J. D., Encephalopathy in rats and nephropathy in rats and mice after subchronic oral exposure to benzaldehyde, *Food Chem. Toxicol.*, 21, 245, 1983.
6. Becker, T. W., Krieger, G., and Witte, I., DNA single and double strand breaks induced by aliphatic and aromatic aldehydes in combination with copper (II), *Free Radical Res.*, 24, 325, 1996.

BENZALKONIUM CHLORIDE

M = 360 (average)
CAS No 8001-54-5
8011-91-4
8039-63-2
RTECS No BO3150000
Abbreviation. BC.

Synonyms and **Trade Names.** Alkyl dimethylbenzyl ammonium chloride; Bionol; Catamin AB; Germinol; Marinol; Roccal.

Properties. White or yellowish-white, amorphous powder or gelatinous pieces with aromatic odor and very bitter taste. Readily soluble in water and alcohol.[020]

Applications. Surface-active agent, mainly used as disinfectant (in cosmetics, etc.).

Acute Toxicity.

Observations in man. Concentrated solutions caused necrosis of the mucous membranes, erosions, ucerations, and petechial hemorrhages.

Animal studies. LD_{50} is reported to be 400 mg/kg BW in rats.[1] Lethal doses are in the range of 100 to 700 mg/kg BW. Manifestations of toxic action included visceral congestion, cloudy swelling, and mild pulmonary edema, but no recognized pathological lesions were observed. There was a curare-like paralysis of skeletal muscles after parenteral injections.[019]

Repeated Exposure. Administration of daily doses of 25 mg/kg BW for several weeks was lethal.[2] The treatment produced focal hemorrhagical necrosis of the gastric mucosa in rats. Mortality was associated with chronic diarrhea.[019]

Short-term Toxicity. Sprague-Dawley rats received BC solutions via stomach tube at dose level of 50 (1:20 dilution of 100% BC) and 100 mg/kg BW (1:10 dilution) for 12 weeks. Two rats died at the 100 mg/kg BW dose group; retardation of BW gain was noted. There were neither treatment-related changes in hematology analyses nor gross or histology changes in the viscera.[3]

Long-term Toxicity. Beagle dogs were administered BC via stomach tube in milk or water at dose levels of 12.5, 25, or 50 mg/kg BW for 52 weeks. Four dogs died at 25 and 50 mg/kg BW dose groups. There were no changes in blood, urinalyses, and in histology findings in animals treated with up to 25 mg/kg BW (in milk). Administration of 25 and 50 mg/kg BW in water caused moderate to severe irritation and congestion of the stomach and intestines, but no tumors.[3]

Allergenic Effect. BC was tested in more than 2000 patients regarding its ability to induce allergic contact dermatitis. BC is considered a weak allergen.[4]

Reproductive Toxicity.

Gonadotoxicity. BC appeared to be spermicidal; it caused ultrastructural changes in cervical muscus.[5]

Embryotoxicity. Rats received intravaginal administration of 50 mg/kg BW single dose during pregnancy. The treatment affected litter size, causing fetotoxicity. Exposure to 100 mg/kg BW dose increased post-implantation mortality and caused fetal death. A 200 mg/kg BW dose increased pre-implantation mortality in rats.[6]

Teratogenic effect was not noted in this study.[6]

Mutagenicity.

In vitro genotoxicity. BC was shown to induce SCE in Chinese hamster embryo cells. It was not mutagenic in reverse mutation systems in *Salmonella* and *E. coli* assays.[7]

Carcinogenicity. In dermal study in Swiss mice and New Zealand rabbits, 8.5 or 17% BC was applied to the backs of mice and to the left ear of rabbit, respectively, twice per week. The treatment did not result in tumor formation or systemic toxic effects. There was no significant decrease in the survival, but ulceration and inflammation were observed.[8]

Regulations. ***U.S. FDA*** (1998) approved the use of BC (C_{10}-C_{20}) in adhesives as a component of articles intended for use in packaging, transporting, or holding food in accordance with the conditions prescribed in 21 CFR part 175.105.

Standards. ***Russia*** (1995). MAC: 0.1 mg/l.

References:

1. Cummins, L. M. and Kimura, E. T., Safety evaluation of selenium sulfide antidandruff shampoos, *Toxi col. Appl. Pharmacol.*, 20, 89, 1971.
2. Humphreys, D. J., *Veterinary Toxicology*, 3rd ed., Bailliere Tindell, London, 1988, 204.
3. Coulston, F., Drobeck, H. P., Mielens, Z. E., and Garvin, P. J., Toxicology of benzalkonium chloride given orally in milk or water to rats and dogs, *Toxicol. Appl. Pharmacol.*, 3, 584, 1961.
4. Fuchs, T., Meinert, A., Aberer, W., Bahmer, F. A., Peters, K. P., Lischka, G. G., Schulze-Dirks, A., Enders, F., and Frosch, P. J., Benzalkonium chloride - a relevant contact allergen or irritant? Results of a multicenter study of the German Contact Allergy Group, *Hautarzt.*, 44, 699, 1993.
5. Erny, R. and Siborni, C., The effect of benzalkonium chloride on ovulatory cervical mucus, *Acta Eur. Fertil.*, 18, 109, 1987.
6. Buttar, H. S., Embryotoxicity of benzalkonium chloride in vaginally treated rats, *J. Appl. Toxicol.*, 5, 398, 1985.
7. Shirasu, Y., Significance of mutagenicity testing on pesticides, *Environ. Qual. Safety*, 4, 226, 1976.
8. Stenback, F., Local and systemic effects of commonly used cutaneous agents: lifetime studies of 16 compounds in mice and rabbits, *Acta Pharmacol. Toxicol.*, 41, 417, 1977.

BENZENESULFONIC ACID, HYDRAZINE

Molecular Formula. $C_6H_8N_2O_2S$

M = 172.22

CAS No 80-17-1

RTECS No DB6888000

Synonyms and **Trade Names.** Benzenesulfohydrazide; Benzenesulfonyl hydrazide; Hydrazide BSG; Phenylsulfohydrazide; Porofor BSH; Porofor ChKhZ 9.

Applications. Blowing agent.
Acute Toxicity. LD_0 is 50 mg/kg BW in rats.[027]

1,2,4-BENZENETRICARBOXILIC ACID

Molecular Formula. $C_9H_6O_6$
M = 210.44
CAS No 528-44-9
RTECS No DC1980000
Abbreviation. TMAc.

Synonyms. 1,2,4-Tricarboxy benzene; Trimellitic acid.

1,2,4-BENZENETRICARBOXYLIC ACID, 1,2-ANHYDRIDE

Molecular Formula. $C_9H_4O_5$
M = 192.13
CAS No 552-30-7
RTECS No DC2050000
Abbreviation. TMAn.

Synonyms. Anhydrotrimellitic acid; 1,3-Dihydro-1,3-dioxo-5-isobenzofurancarboxylic acid; 1,3-Di-oxo-5-phthalancarboxylic acid; Trimellitic acid, 1,2-anhydride; Trimellitic anhydride.

Properties. TMAn occurs as white crystals in the form of flakes. Readily hydrolyzed in water to TMAc. Poorly soluble in water. React with alcohols. Soluble in organic solvents.

Applications. TMAn is an epoxy resin widely used in the production of polyester resins, polyimide compounds, adhesives, and dyes. Used as a plasticizer in materials intended to store and cover food and in the synthesis of various anticorrosive surface coatings and pharmaceutical products.

Acute Toxicity.

Observations in man. A person exposed to TMAn by inhalation experienced respiratory failure, anemia, and GI bleeding.[1]

Animal studies. LD_{50} of TMAc is 2.5 g/kg BW in mice. LD_{50} of TMAn is 1.9 to 5.6 g/kg BW in rats and 1.25 g/kg BW in mice.[2,3] Poisoned mice experienced irritation of the gastrointestinal mucosa, accompanied with hyperemia and hemor rhage, and sometimes with perforations.

Repeated Exposure revealed moderate cumulative properties of TMAn. *Hb* level was lowered and CNS activity was affected.[2]

Allergenic Effect is observed due to formation of conjugates between protein groups and TMAn. TMAn is a potent respiratory sensitizer. Dermal sensitization reactions are observed.

Animal studies. TA-specific antibody can be transferred from mother to fetus in rats and guinea pigs.[4] A moderate contact sensitizing potential was found in maximized guinea-pig assays.[5]

In guinea pigs, TMAn induced allergic response in the lung.[6]

Observations in man. Human allergic reactions observed were asthma, influenza-like symptoms and isolated cases of hypersensitive pneumonitis.[7] These findings were considered to be rare systemic hypersensitivity reactions (risk 1:1000).

Reproductive Toxicity effect of TMAn was not found in mice.

Teratogenicity. Rats and guinea pigs were exposed by inhalation for an hour six times during organogenesis to the concentration of 0.5 mg TMAn/m^3. This did not result in teratogenic findings but lung hemor rhage was found to develop.[8]

Mutagenicity.

In vitro genotoxicity. TMAn was negative in *Salmonella* mutagenicity assay.[013]

Chemobiokinetics. Following inhalation exposure, the major retention was found to be in the lymph nodes associated with the lung. TMAn reacts with the *free amino groups on proteins* to form *conjugates.*[9]

Regulations.

EU (1990). TA is available in the *List of monomers and other starting substances which may continue to be used for the manufacture of plastic materials and articles intended to come into contact with foodstuffs pending a decision on inclusion in Section A* (*Section B*).

U.S. FDA (1998) approved the use of TMAn (1) in resinous and polymeric coatings for use only as a cross-linking agent at the level not to exceed 15% by weight of the resin in contact with food under all conditions of use, except that resins intended for use with food containing more than 8.0% alcohol must contact such food only under the specified conditions. (2) TMAc may be used in the manufacture of resinous and polymeric coatings for the food-contact surface of articles intended for use in producing, manufacturing, packing, processing, preparing, treating, packaging, transporting, or holding food in accordance with the conditions prescribed in 21 CFR part 175.300; (3) in the manufacture of resinous and polymeric coatings for polyolefin films for food-contact surface of articles intended for use in producing, manufacturing, packing, processing, preparing, treating, packaging, transporting, or holding food in accordance with the conditions prescribed in 21 CFR part 175.320; and (4) in the manufacture of cross-linked polyester resins which may be used as articles or components of articles intended for repeated use in contact with food in accordance with the conditions prescribed in 21 CFR part 177.2420.

Great Britain (1998). TMAc and TMAn are authorized up to the end of 2001 for use in the production of polymeric materials and articles in contact with food or drink or intended for such contact. The quantity of these substances alone or together in the finished polymeric material or article shall not exceed 5 mg/kg.

References:

1. Rivera, M., Nicotra, M. B., Byron, G. E., Patterson, R., Yawn, D. H., Franco, M., Zeiss, C. R., and Greenberg, S. D., Trimellitic anhydride toxicity. A cause of acute multisystem failure, *Arch. Ind. Med.,* 141, 1071, 1981.
2. Batyrova, T. F. and Uzhdavini, E. R., Toxic properties of trimellitic acid and its anhydride, in *Hygiene and Toxicology of High-Molecular-Mass Compounds and of the Chemical Raw Material Used for their Synthesis*, Proc. 6th All-Union Conf., B. Yu. Kalinin, Ed., Leningrad, 1979, 178 (in Russian).
3. *Trimellitic Anhydride*, Health and Safety Guide No 71, IPCS, WHO, Geneva, 1992, 30.
4. Leach, C. L., Hatoum, N. S., Zeiss, C. R., et al., Immunologic tolerance in rats during 13 weeks of inhalation exposure to trimellitic anhydride, *Fundam. Appl. Toxicol.*, 12, 519, 1989.
5. Basketter, D. A. and Scholes, E. W., Comparison of the local lymph node assay with the guinea-pig maximization test for the detection of a range of contact allergens, *Food Chem. Toxicol.*, 30, 65, 1992.
6. Fraser, D. G., Regal, J. F., and Arndt, M. L., Trimellitic anhydride-induced allergic response in the lung: role of the complement system in cellular changes, *J. Pharmacol. Exp. Ther.,* 273, 793, 1995.
7. Venables, K. M., Low-molecular-weight chemicals, hypersensitivity, and direct toxicity: the acid anhydrides, *Brit. J. Ind. Med.*, 46, 222, 1989.
8. Ryan, B. M., Hatoum, N. S., Zeiss, C. R., and Garvin, P. J., Immuno-teratologic investigation of trimellitic anhydride (TMA) in the rat and guinea pig, *Teratology*, 39, P114, 1989.
9. Thrasher, J. D., Madison R., Broughton, A., et al., Building-related illness and antibodies to albumin conjugates of formaldehyde, toluene diisocyanate, and trimellitic anhydride, *Am. J. Ind. Med.,* 15, 187, 1989.

BENZETHONIUM CHLORIDE

Molecular Formula. $C_{27}H_{42}NO_2.Cl$
M = 448.15
CAS No 121-54-0
RTECS No BO7175000
Abbreviation. BC.

Synonyms and **Trade Names.** Antiseptol; Benzethonium; Benzyldimethyl{2-[2-(*p*-(1,1,3,3-tetramethylbutyl)phenoxy)ethoxy]ethyl}ammonium chloride; Phemerol; Solamine.

Properties. Colorless crystals with mild odor and very bitter taste. Very soluble in water giving a foamy, soapy solution. Soluble in alcohol.[020]

Applications. Used in cosmetics for its antimicrobial and surfactant properties. Biocide.

Acute Toxicity.

Observations in man. Ingestion may cause vomiting, collapse, convulsions, coma.[020]

Animal studies. LD_{50} is 368 mg/kg BW in rats, and 338 mg/kg BW in mice.[1]

Reproductive Toxicity.

Teratogenic effect was not observed in rats.[2]

Immunotoxicity. No statistically significant dose-dependent contact hypersensitivity responses chloride were observed in mice by dermal exposure to BC.[3]

Mutagenicity.

In vitro genotoxicity. BC was negative in *Salmonella typhimurium* strains *TA98, TA100, TA1535,* and *TA1537*, in SCE or CA tests in cultured Chinese hamster ovary cells with and without *S9* metabolic activation system.[4,5]

Carcinogenicity. BC was not carcinogenic in F344/N rats and $B6C3F_1$ mice on dermal application.[4]

Regulations. ***U.S. FDA*** (1998) approved the use of BC in adhesives as a component of articles intended for use in packaging, transporting, or holding food in accordance with the conditions prescribed in 21 CFR part 175.105.

References:

1. *Proc. Soc. Exp. Biol. Med.*, 120, 511, 1965.
2. Gilman, M. R. and DeSalva, S. J., Teratology studies of benzethonium chloride, cetyl pyridinum chloride and chlorhexidine in rats, Abstract, *Toxicol. Appl. Pharmacol.*, 48, A35, 1979.
3. NTP Technical Report on the *Immunotoxicity of Benzethonium Chloride* (CAS No. 121-54-0) *in Female $B6C3F_1$ Mice (contact hypersensitivity studies)*, Research Triangle Park, NC, 1995.
4. *Toxicology and Carcinogenesis Studies of Benzethionium Chloride in F344/N Rats and $B6C3F_1$ Mice (Feed Studies),* NTP Technical Report Series No. 438, Research Triangle Park, NC, July 1995.
5. De Flora, S., Zanacchi, P., Camoirano, A., Bennicelli, C., and Badolati, G. S., Genotoxic activity and potency of 135 compounds in the Ames reversion test and in a bacterial DNA-repair test, *Mutat. Res.*, 133, 161, 1984.

1,2-BENZISOTHIAZOLIN-3-ONE

Molecular Formula. C_7H_5NOS

M = 151.19

CAS No 2634-33-5

RTECS No DE4620000

Trade Name. Proxel PL.

Applications. Used in the production of coatings of food-contact surfaces.

Acute Toxicity. LD_{50} is 1.02 mg/kg BW in rats, and 1.15 mg/kg BW in mice.

Repeated Exposure. Rats were administered orally with 200 mg B./kg BW for 15 days. The treatment caused no damage to the tissues of rats.

Regulations. ***U.S. FDA*** (1998) approved the use of B. (1) in adhesives as a component of articles intended for use in packaging, transporting, or holding food in accordance with the conditions prescribed in 21 CFR part 175.105; (2) as a slimicide in the manufacture of paper and paperboard that contact food as prescribed in 21 CFR part 176.300, and (3) as a component of the uncoated or coated food-contact surface of paper and paperboard intended for use in producing, manufacturing, packaging, processing, preparing, treating, packing, transporting, or holding dry foods in accordance with the conditions prescribed in 21 CFR part 176.180.

Reference:

Bertaccini, G., Crupicciatore, M., and Vitali, T., Pharmacological activities of benzisothiazoline and benzisoxasolone, *Pharmacol. Res. Commun.*, 3, 385, 1971.

BENZO[a]PYRENE

Molecular Formula. $C_{20}H_{12}$

M = 252.32

CAS No 50-32-8

RTECS No DJ3675000

Abbreviations. BaP; PAH (polycyclic aromatic hydrocarbons).

Synonyms. Benz[a]pyrene; 3,4-Benzo[a]pyrene; 1,2-Benzopyrene; 3,4-Benzopyrene.

Properties. Pale-yellow needles. Water solubility is 1.2 µg/l at 25°C, soluble in alcohol.

Occurrence and **Exposure.** PAH have no industrial use. BaP is only a member of a class of more than 100 compounds belonging to the family of PAH. BaP may occur in plastics as a contaminant of starting material and additives. BaP is usually found in drinking water in combination with other PAH. The typical level of BaP in U.S. drinking water is estimated to be 0.00055 µg/l, and daily intake of total PAH is reported to be 0.027 µg, 0.001 µg of which is BaP.[1]

Migration Data. Some food-grade vulcanizates containing blacks (DG-100, PM-70, PGM-33, TG-10, PM-15) can release up to 0.06 to 9.2 µg BaP/l into water and model media simulating foodstuffs. The greatest quantity of BaP was released by low-dispersion blacks PM-15, TG-10, and PGM-33. BaP hardly migrates at all from vulcanizates with high-dispersion blacks PM-70 and DG-100. The release of carcinogenic PAH can be reduced by the combined use of low- and high-dispersion blacks.[2]

PAH migrate from stoppers containing furnace blacks into benzene extracts. The PAH found in extracts were the same as those contained in furnace blacks: BaP, pyrene, chrysene, 1,2-benzopyrene, 1,12-benzo-perylene, 1,2-benantracene, etc. Very few PAH are released into cottonseed oil and unskimmed homogenized milk (59°C, 7 days) from vulcanizates containing 20 wt.% furnace blacks with a specific surface of 24 to 61 m^2/g. Migration of PAH from vulcanizates into aqueous solutions of citric and acetic acids, sodium bicarbonate and sodium chloride (59°C, 6 days and 145°C, 30 min) was not established. However, migration of PAH into fat-containing foodstuffs was reported.[028]

Acute Toxicity. The oral LD_{50} for various PAH is reported to range between 490 and 18,000 mg/kg BW.[5] Effect induced in animals following acute exposure to PAH include inflammation, hyperplasia, hyperkeratosis and ulceration of the skin, pneumonia, damage to the hematopoietic and lymphoid systems, immunosuppression, adrenal necrosis, ovotoxicity and anti-spermatogenic effects.[3]

Repeated Exposure. Male MutaMouse mice were orally administered BaP at doses of 75 and 125 mg/kg BW for 5 consecutive days. Squamous cell papilloma and hyperplasia in the forestomach were induced at incidences of 25 and 50%, respectively, and were induced 26 weeks after the final treatment without any significant alterations in the hematological and plasma biochemical parameters in mice of the 125 mg BaP/kg BW-treated satellite group. Administration of 75 and 125 mg BaP/kg BW induced bronchiolar-alveolar hyperplasia in the lung at incidences of 18 and 9%, respectively. Slight increases were also observed in the weight of the liver and in the levels of urea nitrogen, creatinine, and potassium ion in the plasma biochemical examinations, although no significant pathological alterations were found in the liver and kidney.[4]

Reproductive Toxicity. BaP may cross the placenta of mice and guinea pigs.[5,6]

Gonadotoxicity. BaP minimally inhibits DNA synthesis in the seminiferous tubules and inhibits the progression of spermatocytes through meiosis.[7]

Embryotoxicity. Pregnant CD-1 mice were given oral doses of 10 to 160 mg/kg BW on days 7 through 16 of gestation. No toxic effects were shown in the dams. In the offspring, total sterility was noted in 97% of the animals in groups administered 40 and 160 mg/kg BW.[8]

Teratogenicity. In Swiss mice, BaP metabolite produced malformations, which were presented by exencephaly, ventral wall defects, and phocomelia.[9]

Immunotoxicity. According to Kawabata and White, BaP was found to be immunodepressive.[10]

Pregnant mice were exposed to a single dose of 150 mg BaP/kg BW *i/p*. The offspring was severely immunosuppressed. This effect may have led to the subsequent widespread development of tumors in these animals.

Mutagenicity. The diol-epoxide metabolites of BaP are considerably more mutagenic than the parent compound.

In vitro genotoxicity. BaP was positive in *Salmonella* mutagenicity assay.[025] Mutations have been induced in cultured human lymphoblastoma cells.[11] Induction of SCE chromatid exchanges in Chinese hamsters following *i/p* administration of BaP has also been reported.[12]

In vivo cytogenetics. BaP did not increase the incidence of *Dr. melanogaster* mutant clones in wing spot somatic mutation and recombination assay.[13]

Carcinogenicity. PAH health effect of primary concern is carcinogenicity. Adequate data were identified only for BaP that is one of the most potent carcinogens: primary tumors have been produced in mice, rats, hamsters, guinea pigs, rabbits, ducks, and monkeys following intragastric, subcutaneous, dermal, or intratracheal administration, both in the site of administration and in other tissues.

Observations in man. There is no information from epidemiological studies regarding the carcinogenicity of BaP alone to humans.

Animal studies. Sprague-Dawley rats were fed 0.15 mg BaP/kg diet either every 9th day or 5 times/week. The rats were treated until moribund or dead. The treatment induced an increase in combined incidence of tumors of the forestomach, esophagus, and larynx.[14]

BaP was given to CC57 mice in a solution (triethylene glycol) at a dose of 0.001 to 10.0 mg, 10 times, at weekly intervals. As a result, the animals developed malignant neoplasms of the rumen, predominantly squamous cell carcinoma with keratinization, more rarely without keratinization, and benign papillomata.[15]

The percentage of animals with tumors declined as the dose was reduced. The LOAEL was identified as 0.01 mg; when this was given orally, rumen papillomata developed in 7.7% mice. The NOAEL appeared to be 0.001 mg, which caused only proliferative changes similar to those in the control group. This dose showed no effect on descendants.

Administration of BaP in the diet (0.001 to 0.25 mg/kg food) for a period of 98 to 197 days resulted in an increased incidence of forestomach tumors in CFW mice (papilliomas and squamous cell carcinomas).[11]

Gastric tumors were found in more than 70% mice fed 50 to 250 ppm BaP for 4 to 6 months. There were no gastric tumors in 287 control mice while 178 out of 454 mice fed various levels of BaP developed gastric tumors.[16] According to Turusov et al., the treatment of pregnant mice with BaP results in an increase in incidence of lung adenomas in four subsequent generations of animals.[17]

Forty-one weeks after the final treatments of MutaMouse mice, the doses of 75 and 125 mg BaP/kg BW induced squamous cell carcinoma, papilloma, and hyperplasia in the forestomach, and anemia possibly due to continuous hemorrhage from tumors in the forestomach. The dose of 125 mg BaP/kg BW also produced malignant lymphoma accompanied by a marked increase in leukocyte count and decrease in erythrocyte count and by a remarkable decrease in BW.[4]

Carcinogenicity classifications. An IARC Working Group concluded that there is sufficient evidence for the carcinogenicity of BaP in *experimental animals* and there were no adequate data available to evaluate the carcinogenicity of BaP in *humans*.

The Annual Report of Carcinogens issued by the U.S. Department of Health and Human Services (1998) defines BaP to be a substance which may reasonably be anticipated to be carcinogen, i.e., a substance for which there is limited evidence of carcinogenicity in *humans* or sufficient evidence of carcinogenicity in *experimental animals*.

IARC: 2A;

U.S. EPA: B2;

EU: 2.

Chemobiokinetics. BaP is absorbed principally through the GI tract and lungs. Absorbed BaP is rapidly distributed to the organs and tissues, and may be stored in the mammary and adipose tissue.[18]

Metabolism of BaP occurs primarily in the liver. It is initially converted by mixed function oxidazes resulting in the formation of *diol epoxides*. These metabolites are subsequently detoxified in a series of conjugation reactions. BaP is metabolized with the formation of 27 metabolites, which are excreted through the bile and subsequently in the feces.[021]

Recommendations. There are insufficient data available to derive drinking-water guidelines for individual PAH but only for BaP, one of the most potent among PAH compounds tested to date.

WHO (1996). Guideline value for drinking water: 0.0007 mg/l.[19]

Joint FAO/WHO Expert Committee on Food Additives (1983). ADI: 0 to 5.0 mg/kg BW.

Standards.

U.S. EPA (1998). Proposed MAC is 0.0002 mg/l for BaP, antracene, benz[a]anthracene, benz[b]-fluoranthene, benzo[g,h,i]perylene, benzo[k]fluoranthene, chrysene, dibenz[a,h]anthracene, fluorene, indeno [1,2,3,c,d]pyrene, phenanthrene, and pyrene.

Canada (1989). MAC: 0.00001 mg/l.

France (1990). MAC: 0.00001 mg/l.

Great Britain (1996). MAC of 0.00001 mg/l was adopted as an annual average. It is recommended that the WHO-93 guideline of 0.0007 mg/l should not be adopted, and efforts should be made to decrease concentrations wherever possible.[20]

References:

1. Santodonato, J., Howard, P., and Basu, D., Health and ecological assessment of polynuclear aromatic hydrocarbons, *J. Environ. Pathol. Toxicol.*, 5, 1, 1981.
2. Medvedev, V. I., Migration of polycyclic hydrocarbons from resins used in the food industry, *Voprosy Pitania*, 6, 70, 1973 (in Russian).
3. Montizaan, G. K., Kramers, P. G. N., Janus, J. A., et al., *Polynuclear aromatic hydrocarbons (PAH): Effects of 10 selected compounds*, Integrated Criteria Document, Natl. Inst. Publ. Health Environ. Protection, The Netherlands, 1989.
4. Hakura, A., Sonoda, J., Tsutsui, Y., Mikami, T., Imade, T., Shimada, M., Yaguchi, S., Yamanaka, M., Tomimatsu, M., and Tsukidate, K., Toxicity profile of benzo[a]pyrene in the male LacZ transgenic mouse (MutaMouse) following oral administration for 5 consecutive days, *Regul. Toxicol. Pharmacol.*, 27, 273, 1998.
5. Lu, L. J. and Wang, M. Y., Modulation of benzo[a]pyrene-induced covalent DNA modifications in adult and fetal mouse tissues by gestation stage, *Carcinogenesis*, 11, 1367, 1990.
6. Kihlstrom, I., Placental transfer of benzo[a]pyrene and its hydrophilic metabolites in the guinea pig, *Acta Pharmacol. Toxicol.* (Copenhagen), 58, 272, 1986.
7. Georgellis, A., Toppari, J., Veromaa, T., Rydstrom, J., and Parvinen, M., Inhibition of meiotic divisions of rat spermatocytes *in vitro* by polycyclic aromatic hydrocarbons, *Mutat. Res.*, 231, 125, 1990.
8. MacKenzie, K. M. and Angevine, D. M., Infertility in mice exposed in utero to BaP, *Biol. Reprod.*, 24, 183, 1981.
9. Barbieri, O., Ognio, E., Rossi, O., Aetigiano, S., Rossi, L., Embryotoxicity of benzo[a]pyrene and some of its synthetic derivatives in Swiss mice, *Cancer Res.*, 46, 94, 1986.
10. Kawabata, T. T. and Whitw, K. L., Effects of naphthalene and naphthalene metabolites on the *in vitro* humoral immune response, *J. Toxicol. Environ. Health*, 30, 53, 1990.
11. Danheiser, S. L., Liber, H. L., and Thilly, W. G., Long-term, low-dose BaP-induced mutation in human lymphoblasts competent in xenobiotic metabolism, *Mutat. Res.*, 210, 142, 1989.
12. Raszinsky, K., Basler, A., and Rohrborn, G., Mutagenicity of polycyclic hydrocarbons. V. Induction of sister chromatid exchanges *in vivo*, *Mutat. Res.*, 66, 65, 1979.
13. Batiste-Allentorn, M., Xamena, N., Creus, A., and Marcos, A., Genotoxic evaluation of ten carcinogens in the *Drosophila melanogaster* with spot test, *Experientia*, 51, 73, 1995.
14. Brune, H., Deutsch-Wenzel, R. P., Habs, M., Ivankovic, S., and Schmal, D., Investigation of the tumorigenic response to benzo[a]pyrene in aqueous caffeine solution applied orally to Sprague-Dawley rats, *J. Cancer Res. Clin Oncol.*, 102, 153, 1981.
15. Yanysheva, N. Ya., Chernichenko, I. A., Balenko, N. V., et al., *Carcinogenic Substances and Their Environmental Safety Standards*, Zdorov'ya, Kiyv, 1977, 136 (in Russian).
16. Neal, J. and Rigdon, R. H., Gastric tumors in mice fed benzo[a]pyrene: A quantitative study, *Tech. Rep. Biol. Med.*, 25, 553, 1967.
17. Turusov, V. S., Nikonova, T. V., and Parfenov, Yu. D., Increased multiplicity of lung adenomas in five generations of mice treated with benz[a]pyrene when pregnant, *Cancer Lett.*, 55, 227, 1990.
18. Weyand, E. N. and Bevan, D. R., Species differences in disposition of benzo[a]pyrene, *Drug Metabol. Dispos.*, 15, 442, 1987.
19. *Revision of the WHO Guidelines for Drinking Water Quality*, Report on the Consolidation Meeting on Organics and Pesticides, Medmenham, UK, January 30-31, 11, 1992.
20. *1994 The Annual Report of the Committees on Toxicity, Mutagenicity, Carcinogenicity of the Chemicals in Food, Consumer Products and the Environment*, DoH, HMSO, London, 1995, 61.

BENZOIC ACID, AMMONIUM SALT

Molecular Formula. $C_7H_6O_2.H_3N$
M = 139.17
CAS No 1863-63-4
RTECS No DG3378000
Abbreviation. BAA.

Synonym and **Trade Name.** Ammonium benzoate; Vulnoc AB.

Properties. Colorless rhombic crystals or crystalline powder, odorless or with faint benzoic acid odor. Solubility in water is 196 g/l at 14.5°C; solubility in alcohol is 163 g/l at 25°C.

Applications. A preservative.

Acute Toxicity. LD_{50} is 825 mg/kg BW in rats, and 235 mg/kg BW in mice. Poisoning produced CNS depression, weak reaction to sound, and paresis. Paralyses of anterior extremities, tremor, and side position preceded death.

Repeated Exposure did not reveal cumulative properties (by Lim). In rats, 30 administrations of 82.5 mg/kg BW caused a decrease in *Hb* level, erythropenia and leukocytosis, an increase in urea contents in the urine and blood, in chlorides concentration in the urine, in ALT and alkaline phosphatase activity in blood serum.

Regulations. ***U.S. FDA*** (1998) approved the use of BAA as a preservative only in adhesives as a component of articles intended for use in packaging, transporting, or holding food in accordance with the conditions prescribed in 21 CFR part 177.1420.

Reference:

Korolenko, T. K., Toxicity of sodium benzoate, ammonium benzoate and monoethanolamine benzoate, *Gig. Sanit.,* 1, 75, 86 (in Russian).

2,5-BIPHENYLDIOL

Molecular Formula. $C_{12}H_{10}O_2$
M = 186.22
CAS No 1079-21-6
RTECS No DV4550000
Abbreviation. BPD.

Synonyms. 2,5-Dihydroxybiphenyl; Phenylhydroquinone.

Acute Toxicity. LD_{50} *i/v* exceeded 20 mg/kg BW in mice.

Mutagenicity.

In vitro genotoxicity. Induced DNA damage in cultured human leukocytes and in *E. coli.*[1] Caused SCE in Chinese hamster ovary cells.[2]

Regulations. ***EU*** (1990). BPD is available in the *List of monomers and other starting substances, which may continue to be used pending a decision on inclusion in Section A.*

References:

1. Horvath, E., Levay, G., Pongracz, K., and Bodell, W. J., Peroxidative activation of *o*-phenylhydroquinone leads to the formation of DNA adducts in HL-60 cells, *Carcinogenesis*, 13, 1937, 1992.
2. Tayama, S. and Nakagawa, Y., Sulfhydryl compounds inhibit the cyto- and genotoxicity of *o*-phenylphenol metabolites on CHO-K1 cells, *Mutat. Res.*, 259, 1, 1991.

1,4-BISBENZOXAZOLESTILBENE

Molecular Formula. $C_{28}H_{19}N_2O_2$
M = 428
Abbreviation. BBS.

Trade Name. Hostalux KS.

Properties. A yellow-green powder.

Applications. Used as an optical bleacher in the production of polystyrene plastics; used for inner lining and in refrigerators.

Acute Toxicity. Rats and mice tolerate administration of a 10 g BBS/kg BW dose given in oil suspension. This dose produced no effect on BW gain, STI, or motor activity in rats. Gross pathology examination failed to reveal structural changes in the viscera.

Repeated Exposure revealed no evidence of cumulative properties.

Long-term Toxicity. Mice were exposed to 0.2 and 0.5 mg/kg BW for 4 months; rats were fed the same doses for 6 months. The treatment caused no retardation of BW gain and did not alter conditioned reflex activity, hematology analysis, the activity of certain enzymes, etc. Histology investigation showed no changes in the viscera.

Allergenic Effect was not observed on application of oil paste of BBS to guinea pigs.

Standards. ***Russia***. Recommended PML: *n/m*.

Reference:

Toxicology and Sanitary Chemistry of Polymerization Plastics, Coll. Works, B. Yu. Kalinin, Ed., Leningrad, 1984, 40 (in Russian).

1,4-BIS(BENZOXAZOL-2-YL)NAPHTHALINE

Molecular Formula. $C_{24}H_{14}N_2O$

M = 366

Properties. A yellow-green powder. Poorly soluble in water; soluble in ethanol (32 mg/100 ml).

Applications. Used as an optical bleacher. No more than 0.05% is added to plastics.

Acute Toxicity. Rats and mice tolerate administration of a 10 g/kg BW dose given in sunflower oil suspension. Gross pathology examination failed to reveal structural changes in the visceral organs.

Long-term Toxicity. Rats and mice received 0.5 and 1.0 g/kg BW for 5 months. The treatment caused no mortality and no retardation of BW gain. No changes were observed in STI, motor and conditioned reflex activity, hematology analyses, and enzyme activity in experimental animals as compared to the controls. Observed toxic symptoms included the increase in hexenal-induced sleep duration, in thyroid weights (1.0 g/kg BW dose) and liver weights (both doses), as well as in the lipid contents in the liver. Gross pathology examination revealed morphological changes in the thyroid at doses of 1.0 and 0.5 g/kg BW (males only).

Regulations. ***Russia***. No more than 0.5% is permitted to be used as an additive in polystyrene articles and refrigerator linings (which come into direct contact with food products).

Reference:

Toxicology and Sanitary Chemistry of Polymerization Plastics, Coll. works, B. Yu. Kalinin, Ed., Leningrad, 1984, 40 (in Russian).

2,5-BIS[5'-*tert*-BUTYLBENZO(2')]THIOPHENE

Molecular Formula. $C_{26}H_{26}N_2O_2S$

M = 430.57

Trade Name. Uvitex OB.

Properties. Yellow-green solid. Poorly soluble in water.

Applications. Used as a whitener in the production of impact-resistant polystyrene.

Acute Toxicity. Rats and mice tolerated administration of a 10 g/kg BW dose. Treated animals exhibited no retardation of BW gain or changes in behavior and morphology of the visceral organs.

Repeated Exposure failed to reveal cumulative properties: mice and rats tolerated 16 administrations of 50 or 200 mg/kg BW and 40 administrations of 500 or 2500 mg/kg BW without retardation of BW gain. Decrease in the STI value was found in mice. The acute effect threshold for STI appeared to be 0.5 mg/kg BW (mice) or 1.0 to 2.5 mg/kg BW (rats). Gross pathology examination revealed no changes in the viscera. In rats, an increase in the thyroid relative weights was noted.[1]

Long-term Toxicity. There was no mortality or retardation of BW gain, no changes in blood morphology and serum enzyme activity in mice and rats given 50 mg/kg BW doses in oil suspension for 10 months. A 50 mg/kg BW dose is considered to be the NOAEL.[1]

Reproductive Effects are not found in rats.

Standards. ***Bulgaria.*** Recommended PML in food: 2.5 mg/l.[2]

References:

1. Mikhailetz, I. B. and Slusareva, I. P., Toxicology characteristics of whitener Uvitex OB, in *Environment Protection in Plastic Industry and Safety of the Use of Plastics*, T. N. Zelenkova and B. Yu. Kalinin, Eds., Plastpolymer, Leningrad, 1978, 155 (in Russian).
2. Ganeva, M., Radeva, M., and Mileva, M., Studies on Uvitex OB migration with a view to its hygienic characteristics, *Khigiena i Zdraveopasvane*, 23, 445, 1980 (in Bulgarian).

BIS(*p*-CHLORODIPHENYL)SULFONE

Molecular Formula. $C_{12}H_8Cl_2O_2S$
M = 287.16
CAS No 80-07-9
RTECS No WR3450000
Abbreviation. BCDS.

Synonyms. Bis(4-chlorophenyl) sulfone; 4-Chloro-1-(4-chlorophenylsulfonyl)benzene; 4-Chlorophenyl sulfone; 4,4'-Dichlorodiphenyl sulfone.

Properties. White crystal solid. Poorly soluble in water, readily soluble in organic solvents. DS does not affect color, taste, and odor of water and does not induce foam.

Applications. Used as a component of high temperature plastics.

Acute Toxicity. Lethal doses appeared to be above 20 g/kg BW.[1]

Repeated Exposure failed to reveal cumulative properties. Rats received 1.0 g/kg BW over a period of 20 days. There were no significant morphological changes in the viscera.

Short-term Toxicity. BCDS is being studied by NTP. It is a known inducer of cytochrome *P-450s*, and was shown to cause marked hepatomegaly in NTP prechronic studies.

Long-term Toxicity. In 6-month rat oral study, 2.0 mg/kg BW dose affected liver, kidney, and CNS functions. The NOAEL was found to be 0.02 mg/kg BW in this study.

Reproductive Toxicity.

Embryotoxicity. Rats were given 1.0 g BCDS/kg BW for a month. The treatment caused an increase in the pre- and postimplantation embryonal mortality. No adverse effect was observed at 0.5 mg/kg BW.

Gonadotoxicity. Above described treatment caused no damage in male and female rat gonads.

Mutagenicity.

In vitro genotoxicity. BCDS produce no CA in bone marrow cells of rats.

In vivo cytogenetics. BCDS was shown to be negative in DLM test.

Chemobiokinetics. Facile oral absorption and a relatively simple metabolite pattern, as well as self induction of metabolism were noted in the NTP study with repeated administration.

Regulations. ***Great Britain*** (1998). BCDS is authorized up to the end of 2001 for use in the production of polymeric materials and articles in contact with food or drink or intended for such contact.

Standards. ***Russia*** (1995). MAC: 0.4 mg/l.

Reference:

Pis'ko, G. T., Tolstopyatova, G. V., Barilyak, I. R., Guds', O. V., Korcach, V. I., and Samoilov, A. P., Substantiation of MAC for 4,4-dichlorodiphenylsulphone in water bodies, *Gig. Sanit.*, 2, 84, 1982 (in Russian).

4,4'-BIS(DIMETHYLAMINO)BENZOPHENONE

Molecular Formula. $C_{17}H_{20}N_2O$
M = 268.39
CAS No 90-94-8
RTECS No DJ0250000
Abbreviation. MK.

Synonyms and **Trade Name.** Bis[4-(dimethylamino)phenyl]methanone; Michler's ketone; Tetramethyldiaminobenzophenone.

Properties. White to greenish leaflets. Practically insoluble in water; soluble in alcohol.

Applications. Used in the manufacture of dyes and pigments.

Migration Data. MK was identified in solvent extracts obtained from paper and paper board materials intended for food contact. Level was generally below 1.0 mg/kg paper.[1]

Extracts of paper and board food-contact materials and articles with ethanol containing 0.4% triethylamine, were analyzed using HPLC. Presence of MK (0.06 to 3.9 mg/kg paper) and 4,4'-bis-(diethylamino)benzophenone (0.1 to 0.2 mg/kg paper) was confirmed using GC-MS. Authors conclu- ded that concentrations of MK present in the packaging samples analyzed are unlikely to pose a risk.[2]

Mutagenicity.

In vitro genotoxicity. The mutagenicity of MK for *Salmonella typhimurium* strain *TA100* was compared using activated 9000 X g (*S9*) liver supernatants from 2 animal species. MK was not mutagenic when incubated with phenobarbital-induced $B6D2F_1$ mouse or Osborne-Mendel rat-liver *S9*. A higher percentage of MK became irreversibly bound to mouse-liver macromolecules than to rat-liver macromolecules when incubated at 37°C in the presence of reduced nicotinamide adenine dinucleotide phosphate.[3]

MK was shown to produce abnormalities of mitotic cell division in cultured mammalian cells probably by interference with centrosome replication. There was also some evidence of increased levels of CA at concentrations of 1.5 μg/ml.[4] MK was positive in hepatocyte primary culture/DNA repair test.[5]

Carcinogenicity. Mice were orally exposed to technical grade MK at the levels of 250 or 500 ppm (male F344 rats) and 500 and 1000 ppm (female F344 rats) and 1250 and 2500 ppm ($B6C3F_1$ mice of both sexes) in the feed for 78 weeks. The treatment caused an increase in mortality and in incidence of hepatocellular carcinomas and hemangiosarcomas (in male mice only).[6]

Carcinogenicity classifications. An IARC Working Group concluded that there is limited evidence for the carcinogenicity of MK in *experimental animals* and there were no data available to evaluate the carcinogenicity of MK in *humans*.

The Annual Report of Carcinogens issued by the U.S. Department of Health and Human Services (1998) defines MK to be a substance which may reasonably be anticipated to be carcinogen, i.e., a substance for which there is limited evidence of carcinogenicity in *experimental animals* and no data available to evaluate the carcinogenicity in *humans*.

IARC: 3;

NTP: P - P - P - P.

References:

1. Castle, L., Offen, C. P., Baxter, M. J., and Gilbert, J., Migration studies from paper and board food packaging materials.1. Compositional analysis. *Food Addit. Contam.*, 14, 35, 1997.
2. Castle, L., Damant, A. P., Honeybone, C. A., Johns, S. M., Jickells, S. M., Sharman, M., and Gilbert, J., Migration studies from paper and board food packaging materials. Part 2. Survey for residues of dialkylamino benzophenone UV-cure ink photoinitiators, *Food. Addit. Contam.*, 14, 45, 1997.
3. McCarthy, D. J., Suling, W. J., and Hill, D. L., A comparison of the mutagenicity and macromolecular binding of the carcinogens Michler's ketone and reduced Michler's ketone, *Mutat. Res.*, 119, 7, 1983.
4. Lafi, A., Parry, J. M., and Parry, E. M., The effect of Michler's ketone on cell division, chromosome number and structure in cultured Chinese hamster cells, *Mutagenesis,* 1, 17, 1986.
5. Williams, G. M., Laspia, M. F., and Dunkel, V. C., Reliability of the hepatocyte primary culture/DNA repair test in testing of code carcinogens and noncarcinogens, *Mutat. Res.*, 97, 359, 1982.
6. *Bioassay of Michler's Ketone for Possible Carcinogenicity*, Natl. Cancer Institute Technical Report Series No 181, DHEW Publ. No.79-1137.

1,4-BIS(2,3-EPOXYPROPOXY)BUTANE

Molecular Formula. $C_{10}H_{18}O_4$

M = 202.28

CAS No 2425-79-8

RTECS No EJ5100000

Synonyms. 1,4-Bis(glycidyl ether); 1,4-Butane diglycidyl ether; Butanediol diglycidyl ether; 1,4-Diglycidyloxybutane; [Tetramethylenebis(oxymethylene)]dioxirane.

Applications. Reactive diluent for epoxy resins; flexibilizer for aromatic resins.

Acute Toxicity. LD_{50} is 1.1 g/kg BW in rats.

Allergenic Effect. B. is shown to cause sensitization in guinea pig maximization test.[1]

Mutagencity.

In vitro genotoxicity. B. was found to be mutagenic in *Salmonella* and *E. coli* assays, it induced SCE.

In vivo cytogenetics. B. was mutagenic in the sex-linked recessive lethal assay in *Dr. melanogaster;* it also induced reciprocal translocations.[2]

Regulations. ***EU*** (1990). B. is available in the *List of monomers and other starting substances which shall be used for the manufacture of plastic materials and articles intended to come into contact with foodstuffs* (*Section B*).

Great Britain (1998). B. is authorized up to the end of 2001 for use in the production of polymeric materials and articles in contact with food or drink or intended for such contact.The quantity in the finished polymeric materials or articles shall not exceed 5.0 mg/kg.

Standards. ***EU*** (1990): QM(T) + 5.0 mg/kg in FP (expressed as epoxy).

References:

1. Thorgeirsson, A., Sensitization capacity of epoxy reactive diluents in the guinea pig, *Acta Dermatol. Venerol.* (Stockholm), 58, 329, 1978.
2. Foureman, P., Mason, J. M,, Valencia, R., and Zimmering, S., Chemical mutagenesis testing in *Drosophila.* IX. Results of 50 coded compounds tested for the National Toxicology Program, *Environ. Mol. Mutagen.,* 23, 51, 1994.

N,N'-BIS(2-HYDROXYETHYL)DODECANAMIDE

Molecular Formula. $C_{16}H_{33}NO_3$
M = 287.50
CAS No 120-40-1
RTECS No JR1925000

Synonyms and **Trade Names.** Bis(2-hydroxyethyl)lauramide; Diethanol lauramide; Lauric acid, diethanolamide; Lauryl diethanolamide; Rolamid CD; Synotol L-60.

Properties. Wax or amber liquid. Soluble in alcohol; insoluble in water.

Acute Toxicity. LD_{50} is 2.7 g/kg BW in rats.

Short-term Toxicity. Rats received 0.1, 0.5, or 2.0 ppm B. in their diet for 3 months. The treatment caused no significant ill effects in animals. A decrease in *Hb* level, hematocrit and erythrocyte count was noted in females at the end of the treatment. Relative liver weight increased in females at two higher levels. The NOAEL of 50 mg/kg BW was established in this study.[1]

Reproductive Toxicity. There were no effects on reproductive functions in rats and rabbits treated orally or in rats treated dermally (British Industrial Biological Research Association, 1990).

Mutagenicity.

In vitro genotoxicity. Negative in *Salmonella typhimurium* strains *TA100* and *TA98.*[2]

In vivo cytogenetics. B. produced no cell transformations when administered at the doses of 0.1 to 10 µg/ml to cell culture 7708 prepared from pregnant Syrian golden hamster which were killed on days 13 to 14 of gestation.[2]

Regulations. ***U.S. FDA*** (1998) approved the use of B. as a component of the uncoated or coated food-contact surface of paper and paperboard intended for use in producing, manufacturing, packaging, processing, preparing, treating, packing, transporting, or holding dry food in accordance with the conditions prescribed in 21 CFR part 176.180.

References:

1. Gaunt, I. F., et al., Short-term feeding study of lauric diethylamide in rats, *Food Cosmet. Toxicol.,* 5, 497, 1967.

2. Inoue, K. et al., Studies of *in vitro* cell transformation and mutagenicity by surfactants and other compounds, *Food Cosmet. Toxicol.*, 18, 289, 1980.

BIS(TRICHLOROMETHYL) SULFONE

Molecular Formula. $C_2H_{16}O_2S$
M = 300.78
CAS No 3064-70-8
RTECS No WR4920000

Trade Names. Chlorosulfona; Slimicide E.

Acute Toxicity. LD_{50} is about 0.7 g/kg BW in rats.

Regulations. ***U.S. FDA*** (1998) approved the use of B. (1) as a slimicide in the manufacture of paper and paperboard that contact food as prescribed in 21 CFR part 176.300, and (2) as a component of the uncoated or coated food-contact surface of paper and paperboard intended for use in producing, manufacturing, packaging, processing, preparing, treating, packing, transporting, or holding dry foods in accordance with the conditions prescribed in 21 CFR part 176.180.

BORAX

Molecular Formula. $B_4O_7.2Na.10H_2O$
M = 381.38
CAS No 1303-96-4
RTECS No VZ2275000

Synonym and **Trade Names.** Borax decahydrate; Boricin; Neobor; Polybor; Sodium biborate; Sodium borate anhydrous; Sodium borate, decahydrate; Sodium borate, pentahydrate; Sodium borate, tetrahydrate; Sodium orthoborate; Sodium pyroborate; Sodium tetraborate.

Properties. Colorless crystals or solid. Water solubility is 21.2 g/l at 0°C and 220 g/l at 50°C.

Applications. B. is used in the production of rubber and ceramics, soaps and pharmaceuticals, detergents, and in medicine.

Exposure. B. is found in all animal tissues. The levels of B. in potable water sources have been found to be as high as 1.2 mg/l.[1] In the U.S., a dietary intake of B. in humans is 1.5 mg B./day.[2]

Acute Toxicity. Acute poisoning caused CNS depression, nausea and vomiting, nephrosis, hepatitis, cerebral edema, shock, lowered body temperature, red rash, coma, and death.[03,012]

Long-term Toxicity. In a 2-year study, the NOEL for rats amounted to 154 mg/kg BW, and was 78 mg/kg BW for dogs.[3]

Reproductive Toxicity.

Gonadotoxicity.

Observations in man. Drinking water with high borate concentrations resulted in reduced sperm count and decreased libido.[4]

Animal studies. According to Dixon et al., only high B. levels in drinking water induce testicular damage. The concentration of 6.0 mg/l was found to be ineffective in rats,[5] and testicular toxicity may require B. concentration as high as 150 mg/l.[6] NOAELs for reproductive effects are reported to be 125 and 40 mg B./kg BW for rabbits and mice, respectively. The LOAEL for rats appeared to be 78 mg/kg BW for decrease in fetal weights.[2]

Teratogogenicity. Oral administration of B. to pregnant rats caused CNS abnormalities in the offspring.[7]

Chemobiokinetics. 95% of absorbed B. is eliminated via kidneys and its biological halflife is less than one day.[7]

Regulations. ***U.S. FDA*** (1998) approved the use of B.(1) in adhesives as a component of articles intended for use in packaging, transporting, or holding food in accordance with the conditions prescribed in 21 CFR part 175.105; (2) as a constituent of acrylate ester copolymer coating which may be used as a food-contact surface of articles intended for packaging and holding food, including heating of prepared food, subject to the provisions prescribed in 21 CFR part 175.210; and (3) as a component of the uncoated or coated food-contact surface of paper and paperboard intended for use in producing, manufacturing, packaging, processing, preparing, treating, packing, transporting, or holding dry food for use as a preservative only in accordance with the conditions prescribed in 21 CFR part 176.180.

References:

1. Lee, I. P., Effect of environmental metals on male reproduction, in *Reproduction and Developmental Toxicity of Metals,* T. W. Clarkson et al., Eds., Plenum Press, New York, 1983, 253.
2. Mastromatteo, E. and Sullivan, F., Summary: International symposium on the health effects of boron and its compounds, *Environ. Health Perspect.*, 102 (Suppl. 7), 139, November 1994.
3. Dixon, R. L. et al., Methods to assess reproductive effects of environmental chemicals: Studies of cadmium and boron administered orally, *Environ. Health Perspect.*, 13, 59, 1976.
4. Seal, B. S. and Weeth, H. J., Effect of boron in drinking water on the male laboratory rat, *Bull. Environ. Contam. Toxicol.*, 25, 782, 1980.
5. Abashidze, M. T., Embryotropic action of boron, *Gig. Sanit.*, 4, 10, 1973 (in Russian).
6. *The Pesticide Manual*, 10th ed., C. Tomlin, Ed., Crop Protect. Publ., The Royal Soc. Chem., 1995, 1341.
7. Krasovsky, G. N., Varshavskaya, S. P., and Borisov, A. I., Toxic and gonadotropic effects of cadmium and boron relative to standards for these substances in drinking water, *Environ. Health Perspect.*, 13, 69, 1976.

BORIC ACID

Molecular Formula. BH_3O_3
M = 61.84
CAS No 10043-35-3
RTECS No ED4550000
Abbreviation. BA.

Synonyms. Boracic acid; Orthoboric acid.

Properties. Solid. Solubility in cold water is about 550 mg/l.

Applications. BA is used in the production of antiseptics and flame retardants, as an insecticide and an ingredient in nylon production, as a component of glass, ceramics, cosmetics, and pharmaceuticals.

Exposure. BA is a natural component of foods and an essential plant micronutrient that occurs naturally in fruits and vegetables. Use as food preservatives has been prohibited in domestic commercial foods in the USA. The total daily *boron* intake seems to be 1.0 to 5.0 mg (equivalent to about 0.1 to 0.4 mg BA/kg BW).[1]

Acute Toxicity. In mice, rats, and dogs, LD_{50} range from 2.0 to >6.0 g/kg BW.[2,3] Poisoning caused general depression, ataxia, convulsion, and death. Kidney degeneration and testicular atrophy were observed.[4]

Repeated Exposure. Mortality occurred in mice fed 25, 50, and 100 g/kg food. There were hyperplasia and/or dysplasia of the forestomach in these dose groups. Gross pathology and histology examination revealed no changes in mice exposed up to 12 g BA/kg food.[5] BA was introduced in the diet of mice at concentrations of 0.062 to 10% for 14 days. The treatment caused dose- and time-dependent mortality (2.5 to 10% BA); 10% BA in the diet led to retardation of BW gain.[6]

Short-term Toxicity.

Observations in man. There is limited evidence of adverse reproductive effects of BA compounds.

Animal studies. Rats and dogs received BA at doses of 17.5 to 5250 mg boron/kg food for 3 months. The treatment induced decrease of BW and testicular atrophy. NOAEL has not been established in this study.[3]

In a 13-week study, mice were fed diets containing 1.2 to 20 g BA/kg food. All tested doses caused extramedullary hemopoiesis of the spleen. Administration of more than 5.0 g/kg dose increased mortality, and produced testicular lesions.[5] In another 13-week study, mice received the diet containing 0.012 to 0.2% BA. The treatment caused an increase in mortality and a decrease in BW gain at the highest doses.[6]

Long-term Toxicity. Chronic ingestion of *borates* causes anorexia, retardation of BW gain, vomiting, mild diarrhea, skin rash, alopecia, convulsions, and anemia. Kidneys are predominantly affected due to highest concentration reached during excretion. $B6C3F_1$ mice received 2.5 or 5.0 g/kg diet for 103 weeks. Mortality was significantly increased; 5.0 g/kg dietary dose induced testicular atrophy in males.[5]

In a 2-year study, mice received 0.025 or 0.5% BA in their diet. Survival was reduced after week 63, reduction in BW gain was noted.[6]

Reproductive Toxicity.

Embryotoxicity and ***Teratogenicity.*** Rats received BA or *borax* at a dose of 350 mg/kg BW during pregnancy. There was no reduction in live born or any increase in incidence of malformation in newborns.[3] Fail *et al.* failed to note an increase in incidence of congenital effects in Swiss mice (CD-1) using the continuous breeding protocol.[4]

Pregnant rats and mice fed up to 0.33 or 1.0 g/kg BW, respectively, during embryogenesis or during the entire gestation. Lethality in the dams occurred at the highest doses. Dose of 163 mg/kg BW increased incidence of hydrocephalus and skeletal malformations in rats. No terata was found in mice.[7]

In contrast to these data, Price et al. found the developmental toxicity NOAEL in the rat to be 55 mg/kg BW on gestation day 20 and 74 mg/kg BW on postnatal day 21.[8]

Gonadotoxicity.

Observations in man. Decreased sperm counts and reduced motility in 6 out of 38 workers from a boric acid production plant (airborne exposure levels of 22 to 80 mg B./m^3). A recent study involved more than 500 male workers exposed to sodium borates for 18 years on average did not reveal an effect on fertility as measured by the standardized birth ratio.[9]

In several *animal studies*, the adverse effects on the testes including inhibition of spermiation followed by testicular atrophy, were observed in several species treated with high oral doses (>1000 ppm for up to 90 days).[3,10,11]

In a multigeneration study, rats received 1.17 g boron/kg food. Poisoning caused total sterility. Other effects included lack of spermatozoa in atrophied testes, and decreased ovulation in females. The NOAEL of 17.5 mg boron/kg BW was identified in this study.[3] BA reduced pup weight and changed oestrus cycle.[12]

Mice received 1000, 4500, or 9000 ppm BA in their diet. Decrements in sperm motility at all levels, and test atrophy at two higher doses were observed. Concentration of 9000 ppm caused an inhibition of spermiation (release of mature spermatids), germ cells sloughing and death, and, finally, atrophy. No effect level for these effects was 2000 ppm.[13] Testicular degeneration or atrophy of the seminiferous tubules was observed in rats exposed to 0.5 to 2.0% BA in their diet for 13 weeks.[6]

Mutagenicity.

In vitro genotoxicity. BA is not genotoxic, since tests with prokaryotic and eukaryotic cells were uniformly negative.[6] It was negative in *Salmonella* and mouse lymphoma assays; it did not induce SCE or CA in Chinese hamster ovary cells.[5]

Carcinogenicity. A 2-year NTP study revealed no evidence of carcinogenicity of BA at the levels of 2.5 or 5.0 g/kg diet.[5] The decrease in survival of dosed male mice may have reduced the sensitivity of this study.

Carcinogenicity classification.

NTP: XX - XX - NE - NE (feed).

Chemobiokinetics. Following intragastric administration, BA has been rapidly and completely absorbed.[14]

BA and *borates* are not metabolized, nor do they accumulate in the body except for low deposits in bone. No organic *boron* compounds have been reported as metabolites. Excretion occurs primarily in the urine.

Observations in man. More than 90% of ingested dose of 750 mg BA were eliminated within 4 days (U.S. EPA Doc., 1989)

Animal studies. Rats were exposed to a 1.0 ml oral dose of sodium tetraborate solution at different dose levels ranging from 0 to 4 mg/kg BW (as boron). Authors suggest that orally administered boric acid is rapidly and completely absorbed from the GI tract into the blood stream. The main route of boron excretion from the body is via glomerular filtration. It may be inferred that there is partial tubular resorption at low plasma levels.[15]

Regulations. ***U.S. FDA*** (1998) approved use of BA as (1) a component of food-contact adhesives which may safely be used as a component of articles intended for use in packaging, transporting, or holding food as prescribed in 21 CFR part 175.105; (2) as a component of the uncoated or coated food-contact surface of paper and paperboard intended for use in producing, manufacturing, packaging, processing, preparing, treating, packing, transporting, or holding dry food in accordance with the conditions prescribed in 21 CFR part 176.180; and (3) for use only at levels not to exceed 0.16% by weight of ethylene-vinyl or acetate-vinyl alcohol copolymers complying with 21 CFR part 177.1360(a)(3) and (d) of 178.2010.

Standards. ***Canada.*** MAC: 5.0 mg *boron*/l.

Recommendations. ***WHO*** (1996). Guideline value: 0.3 mg/l.

References:

1. Anderson, D. L., Cunningham, W. C. and Lindstrom, T. R., Concentrations and intakes of *H, B, S, K, Na, Cl*, and *NaCl* in foods, *J. Food Comp. Anal.*, 7, 59, 1994.
2. *Revision of the Guidelines for Canadien Drinking Water Quality*, Draft, Dept. Natl. Health & Welfare, Ottawa, Canada, March 19, 1991.
3. Weir, R. J. and Fisher, R. S., Toxicological studies on boron and boric acid, *Toxicol. Appl. Pharmacol.*, 23, 351, 1972.
4. Fail, P. A., George, J. D., Seely, J. C., Grizzle, T. B., and Heindel, J. J., Reproductive toxicity of boric acid in Swiss (CD-1) mice: Assessment using the continuous breeding protocol, *Fundam. Appl. Toxicol.*, 17, 225, 1991.
5. *Toxicology and Carcinogenesis Studies of Boric Acid in B6C3F$_1$ Mice (*Food Studies), NTP Technical Report Series No 324, Research Triangle Park, NC, NIH Publ. 88-2580, October 1987.
6. Dieter, M. P., Toxicity and carcinogenicity studies of boric acid in male and female B6C3F$_1$ mice, *Environ. Health Perspect.*, 102, (Suppl. 7), 93, November 1994.
7. Heindel, J. J., Price, C. J., Field, E. A., Marr, M. C., Myers, C. B., Morrisey, R. C., and Schwetz, B. A., Developmental toxicity of boric acid in mice and rats, *Fundam. Appl. Toxicol.*, 18, 266, 1992.
8. Price, C. J., Strong, P.L., Marr, M. C., Myers, C. B., and Murray, F. J., Developmental toxicity NOAEL and postnatal recovery in rats fed boric acid during gestation, *Toxicol. Appl. Pharmacol.*, 32, 179, 1996.
9. Tarasenko, N. Y., Kasparov, A. A., and Strongina, O. M., Effect of boric acid on the generative function in males, *Gig. Truda Prof. Zabol.*, 11, 13, 1972 (in Russian).
10. Dixon, R. L., Sherins, R. J., and Lee, I. P., Assessment of environmental factors affecting male fertility, *Environ. Health Perspect.*, 13, 59, 1979.
11. Ku, W. W., Chapin, R. E., Wine, E., and Gladen, B. C., Testicular toxicity of boric acid (BA): relationship of dose to lesion development and recovery in the F344 rat, *Reprod. Toxicol.*, 7, 259, 1993.
12. Heindel, J., Fail, P., George, J., and Grizzle, T., Boric acid, *Environ. Health Perspect.,* 105, Suppl. 1, 1997.
13. Chapin, R. E. and Ku, W. W., The reproductive toxicity of boric acid, *Environ. Health Perspect.*, 102, (Suppl. 7), 87, November 1994.
14. Mastromatteo, E. and Sullivan, F., Summary: International symposium on the health effects of boron and its compounds, *Environ. Health Perspect.,* 102 (Suppl. 7), 139, November 1994.
15. Usuda, K., Kono, K., Orita, Y., Dote, T., Iguchi, K., Nishiura, H., Tominaga, M., Tagawa, T., Goto, E., and Shirai, Y., Serum and urinary boron levels in rats after single administration of sodium tetraborate, *Arch. Toxicol.,*72, 468, 1998.

BORIC ACID, DISODIUM SALT

Molecular Formula. $B_4Na_2O_7$

M = 201.27

CAS No 1330-43-4

1332-28-1

RTECS No ED4588000

Abbreviation. STB.

Synonyms. Anhydrous borates, sodium salt; Borax glass; Disodium tetraborate; Fused borax; Sodium biborate; Sodium boron oxide; Sodium pyroborate; Sodium tetraborate.

Properties. Odorless, white, vitreous granules. Water solubility is 2.48% w/w at 20°C.

Applications. Bleaching agent for textile; an ingredient of detergents. An oxidizer in tooth powder and tooth paste.

Long-term Toxicity. Chronic ingestion of boric acid and boron derivatives causes anorexia, retards BW gain, vomiting, mild diarrhea, skin rash, alopecia, convulsions, and anemia. Kidneys are predominantly affected due to the highest concentration reached during excretion.[1]

Reproductive Toxicity.

Gonadotoxicity. Borate compounds were given to rats and dogs. Chronic feeding caused accumulation in the testes, germ cell depletion and testicular atrophy.[019]

Chemobiokinetics. Borates are not metabolized, neither do they accumulate in the body except for low deposit in bone. No organic boron compounds have been reported as metabolites. Excretion occurs primarily in the urine.

Animal studies. Rat were exposed to a 1.0 ml oral dose of sodium tetraborate solution at different dose levels ranging from 0 to 4 mg/kg BW (as Boron). Authors suggest that orally administered boric acid is rapidly and completely absorbed from the GI tract into the blood stream. The main route of *boron* excretion from the body is via glomerular filtration. It may be inferred that there is partial tubular resorption at low plasma levels.[2]

Regulations. ***U.S. FDA*** (1998) approved the use of STB (1) in resinous and polymeric coatings used as the food-contact surfaces of articles intended for use in producing, manufacturing, packing, processing, preparing, treating, packaging, transporting, or holding food in accordance with the conditions prescribed in 21 CFR part 175.300, and (2) as a component of acrylate ester copolymer coating which may be used as a food contact surface of articles intended for packaging and holding food, including heating of prepared food, subject to the provisions prescribed in 21 CFR part 175.210.

References:

1. Dreibach, R. H., *Handbook of Poisoning*, 12 ed., Appleton & Lange, Norwalk, CT, 1987, 360.
2. Usuda, K., Kono, K., Orita, Y., Dote, T., Iguchi, K., Nishiura, H., Tominaga, M., Tagawa, T., Goto, E., and Shirai, Y., Serum and urinary boron levels in rats after single administration of sodium tetraborate, *Arch. Toxicol.*,72, 468, 1998.

BORIC ACID, SODIUM SALT

Molecular Formula. $BHO_2.Na$

M = 66.81

CAS No 7775-19-1

RTECS No ED4640000

Abbreviation. BAS.

Synonyms. Borosoap; Kodalk; Sodium borate; Sodium metaborate.

Properties. Colorless and odorless hexagonal prisms with saline taste. White pieces or powder.[020] Solubility in water is 260 g/l at 20°C. Decomposes with liberation of H_2O_2.[020]

Applications. Used in the manufacture of adhesives and detergents.

Acute Toxicity. LD_{50} is 2.3 g/kg BW in rats.[1]

Long-term Toxicity. Chronic borates feeding resulted in accumulation in the testes, germ cell depletion, and testicular atrophy.[016] Manifestations of toxic action included anorexia, weight loss, vomiting, mild diarrhea.[2] Gross pathology examination revealed inflammation of the GI tract, toxic degenerative changes in the liver and kidneys, and cerebral edema.[2,3]

On low level of ingestion, there was little more than dry skin and mucous membranes, followed by appearance of a red tongue, patchy alopecia, cracked lips, conjunctivitis, and sometimes periorbital edema (Haddad and Winchester, 1983); other manifestations of toxic action included anorexia, retardation of BW gain, vomiting, mild diarrhea, skin rash, alopecia, convulsions, and anemia.[2]

Chemobiokinetics. BAS and borates are not metabolized, neither do they accumulate in the body except for low deposit in bone. No organic boron compounds have been reported as metabolites. Excretion occurs primarily in the urine. The kidneys are more seriously damaged because of highest concentration reached during excretion.[2]

Regulations. ***U.S. FDA*** (1998) approved the use of BAS in adhesives as a component of articles intended for use in packaging, transporting, or holding food in accordance with the conditions prescribed in 21 CFR part 175.105.

References:

1. *Pesticide Manufacturing and Toxic Materials Control Encyclopedia*, M. Sittig, Ed., Noyes Data Corporation, Park Ridge, NJ, 1980, 682.
2. Dreisbach, R. H., *Handbook of Poisoning*, 12th ed., Appleton & Lange, Norwalk, CT, 1987, 361.
3. Clarke, M. L., Harvey, D. G., and Humphreys, N., *Veterinary Toxicology*, 2nd ed., Bailliere Tindall, London, 1981, 36.

BROMIC ACID, POTASSIUM SALT

Molecular Formula. $BrO_3.K$
M = 167.01
CAS No 7758-01-2
RTECS No EF8725000
Abbreviation. PB.

Synonym. Potassium bromate.

Applications. Food additive, used mainly in bread-making process.

Acute Toxicity. LD_{50} is 290 mg/kg BW in mice.[1]

Mutagenicity.

In vitro genotoxicity. PB was positive in both the Ames test and the chromosome test *in vitro* using a Chinese hamster fibroblast cells.[2]

In vivo cytogenetics. The incidence of CA in bone marrow rat cells increased rapidly and decreased within 24 hours on acute exposure to PB. Dose-response relationships were observed for both *i/p* and oral administrations.[3] PB was found to induce micronuclei after a single *i/p* injection; positive by oral administration.[1]

Carcinogenicity. F344 rats received 250 and 500 ppm PB in their drinking water for 110 weeks. The treatment caused high incidence of renal cell tumors and mesotheliomas of the peritoneum (only in males given 500 ppm),[4] and follicular cell tumors of the thyroid. PB possesses both initiating and promoting activities for rat renal tumorigenesis. Carcinogenic potential seems to be weak in mice and hamsters. Adverse effects are not evident in animals fed bread-based diets made from flour treated with PB.[5]

Carcinogenicity classification. An IARC Working Group concluded that there is sufficient evidence for the carcinogenicity of PB in *experimental animals* and there were no adequate data available to evaluate the carcinogenicity of PB in *humans*.

IARC: 2B.

Regulations. ***U.S. FDA*** (1998) approved the limit of 50 ppm (1) in unfinished bromated flour and enriched bromated flour and of 75 ppm in bakery products alone or with potassium iodate, calcium peroxide. PB may be safely used (2) in the malting of barley under the conditions prescribed in 21 CFR part 172.730.

Standards. ***U.S. EPA*** (1998). MCL: 0.01 mg/l (*bromate*).

References:

1. Hayashi, M., Kishi, M., Sofuni, T., and Ishidate, M., Micronucleus tests in mice on 39 food additives and eight miscellaneous chemicals, *Food Chem. Toxicol.*, 26, 487, 1988.
2. Ishidate, M., Sofuni, T., Yoshikawa, K., Hayashi, M., Nohmi, T., Sawada, M., and Matsuoka, A., Primary mutagenicity screening of food additives currently used in Japan, *Food Chem. Toxicol.*, 22, 623, 1984.
3. Fujie, K., Shimazu, H., Matsuda, M., and Sugiyama, T., Acute cytogenetic effects of potassium bromate on rat bone marrow cells *in vivo*, *Mutat. Res.*, 206, 455, 1988.
4. Kurokawa, Y., Hayashi, Y., Maekawa, A., Takahashi, M., Kokubo, T., and Odashima, S., Carcinogenicity of potassium bromate administered orally to F344 rats, *J. Natl. Cancer Inst.*, 71, 965, 1983.

5. Kurokawa, Y., Maekawa, A., Takahashi, M., and Hayashi, Y., Toxicity and carcinogenicity of potassium bromate - a new renal carcinogen, *Environ. Health Perspect.*, 87, 309, 1990.

2-BROMO-2-(BROMOMETHYL)GLUTARONITRILE

Molecular Formula. $C_6H_6Br_2N_2$
M = 265.96
CAS No 35691-65-7
RTECS No MA5599000

Synonyms. 1,2-Dibromo-2,4-dicyanobutane; Methyldibromoglutaronitrile; 2-Bromo-2-(bromomethyl-pentanedinitrile).

Acute Toxicity. LD_{50} is 515 mg/kg BW in rats.

Short-term Toxicity. In a 13-week oral study, changes in thyroid weight were reported in dogs.

Reproductive Toxicity. Rats received 175 mg/kg BW on days 6 to 15 of pregnancy. An increase in post-implantation mortality was noted.[1]

Allergenic Effect. No sensitizing capacity was observed in the guinea pig maximization test.[2]

Regulations. ***U.S. FDA*** (1998) approved the use of B. as a component of the uncoated or coated food-contact surface of paper and paperboard intended for use in producing, manufacturing, packaging, processing, preparing, treating, packing, transporting, or holding aqueous and fatty foods in accordance with the conditions prescribed in 21 CFR part 176.170.

References:

1. *Toxicol. Lett.*, 18 (Suppl. 1), 158, 1983.
2. Bruze, M., Gruvberger, B., and Agrup, G., Sensitization studies in the guinea pig with the active ingredents of Euxyl K 400, *Contact Dermat.*, 18, 37, 1988.

2-BROMO-2-NITROPROPANE-1,3-DIOL

Molecular Formula. $C_3H_6BrNO_4$
M = 200.01
CAS No 52-51-7
RTECS No TY3385000

Synonym and **Trade Names.** 2-Bromo-2-nitro-1,3-propanediol; Bronidiol; Bronocot; Bronopol.

Properties. Colorless to pale yellow-brown solid. Solubility in water is 59 g/l at 20°C; solubility in ethanol is 500 g/l.

Acute Toxicity. LD_{50} is 180 to 400 mg/kg BW in rats, 250 to 500 mg/kg BW in mice, and 250 mg/kg BW in dogs.

Short-term Toxicity. In a 72-day feeding study, rats received 1.0 g/kg diet. The treatment caused no ill effects.

Regulations. ***U.S. FDA*** (1998) approved the use of B. in adhesives as a component of articles intended for use in packaging, transporting, or holding food in accordance with the conditions prescribed in 21 CFR part 175.105.

Reference:

The Pesticide Manual, 10th ed., Brit. Crop Protect. Council, Worcestershire, the UK, 1994, 125.

BUTYL-(5-CHLOROMETHYL-2-FUROATE)

Molecular Formula. $C_{10}H_{13}ClO_3$
M = 216.66
CAS No 21893-86-7
RTECS No LV1805000

Synonyms. 5-Chloromethyl-2-furancarboxylic acid, butyl ester; Butyl (5-chloromethyl-2-furan-carboxylate).

Properties. A yellowish liquid with an unpleasant odor. Readily soluble in alcohol and oils.

Applications. Used in the production of plastics.

Acute Toxicity. In mice, LD_{50} is 2.0 g/kg BW, LD_{100} appeared to be 2.5 g/kg BW. Poisoning is accompanied with narcotic effect including adynamia with subsequent muscle fibrillation and coordination disorder. Gross pathology examination revealed congestion in the visceral organs and brain. Histologically, there were acute distur bances of blood and lymph circulation and hemorrhages.

Reference:

Stasenkova, K. P and Scirskaya, V. A., Toxicity study of butyl ethers of 2-furancarbonic acid and 5-chlormethyl-2-furankarbonic acid, in *Toxicology of New Industrial Chemical Substances*, A. A. Letavet and I. V. Sanotsky, Eds., Meditsina, Moscow, Issue No 9, 1967, 118 (in Russian).

1,2-BUTYLENE OXIDE

Molecular Formula. C_4H_8O
M = 72.12
CAS No 106-88-7
RTECS No EK3675000
Abbreviation. BO.

Synonyms. 1,2-Butene oxide; Epoxybutane; Ethyl ethylene oxide; Ethyl oxirane.

Properties. Water-white liquid with disagreeable, sweetish odor. Solubility in water is 90 g/l at 25°C.[03]

Applications. Used in the production of glycols and glycol ethers.

Acute Toxicity. LD_{50} is 1170 mg/kg BW in rats.[1] According to other data, it was about 500 mg/kg BW when rats were fed mixed BO as a 30% solution in corn oil by intubation. Death occurred in a day or not at all.[03]

Reproductive Toxicity.

In rats, BO was found to be fetotoxic, but not teratogenic following exposure by inhalation concentrations of 250 and 1000 ppm.[2,3]

Mutagenicity.

In vitro genotoxicity. *cis-* and *trans*-2,3-epoxybutane induced a significant increase of revertants in *Salmonella typhimurium* strains *TA1535* and *TA100*.[4] BO induced forward mutations at the TK locus of cultured mouse L5178Y lymphoma cells, and CA and SCE in Chinese hamster ovary cells with and without metabolic activation.

In vivo cytogenetics. BO was found to be positive in *Dr. melanogaster*.[5]

Carcinogenicity. In 2-year inhalation studies, F344 rats and $B6C3F_1$ mice were exposed to 200 and 400 ppm or 50 and 100 ppm, respectively. Carcinogenic response was observed in male rats (neoplasms of respiratory tract), but not in mice.[3]

Carcinogenicity classification.

NTP: CE - EE - NE - NE (inhalation).

Chemobiokinetics. BO was readily absorbed, extensively metabolized, and rapidly excreted following oral exposure.

Regulations. ***U.S. FDA*** (1993) approved the use of BO as a slimicide in the manufacture of paper and paperboard that contact food, subject to the provisions prescribed in 21 CFR part 176.300.

References:

1. Weil, C. S., Condra, N., Haun, C., and Striegel, J. A., Experimental carcinogenicity of representative epoxides, *Am. Ind. Hyg. Assoc. J.*, 24, 305, 1963.
2. Hardin, B. D., Bond, G. P., Sikov, M. R., Andrew, F. D., Beliles, R. P., and Niemeier, R.W., Testing of selected workplace chemicals for teratogenic potential, *Scand. J. Work Environ. Health.*, 7, 66, 1981.
3. Kimmel, C. A., LaBorde, J. B., and Hardin, B. D., Reproductive and developmental toxicology of selected epoxides, in *Toxicology and the Newborn*, S. Kacew and M. J. Reasor, Eds., Elsevier Sc. Publ., Amsterdam, Chapter 13, 270, 1984.
4. Castelain, P., Criado, B., Cornet, M., Laib, R., Rogiers, V., and Kirsch-Volders, M., Comparative mutagenicity of structurally related aliphatic epoxides in a modified *Salmonella*/microsome assay, *Mutagenesis*, 8, 387, 1993.

5. *Toxicology and Carcinogenesis Studies of 1,2-Epoxybutane in F344 Rats and B6C3F$_1$ mice (Inhalation Studies)*, NTP Technical report Series No 329, Research Triangle Park, NC, March 1988.

BUTYL-2-FUROATE

Molecular Formula. $C_9H_{12}O_3$
M = 168.21
CAS No 583-33-5
RTECS No LV1775000

Synonyms. Butyl-2-furancarboxylate; 2-Furancarboxylic acid, butyl ester; 2-Furylpyromucic acid, butyl ester; 2-Furoic acid, *n*-butyl ester.

Properties. A yellowish liquid with an unpleasant odor. Poorly soluble in water; soluble in alcohols and oils.

Applications. Used in the production of plastics.

Acute Toxicity. In mice, LD_{50} is 1.5 g/kg BW, LD_{100} appeared to be 2.0 g/kg BW. Manifestations of the toxic effect include narcosis, muscle fibrillation, and coordination disorder. Gross pathology examination revealed congestion in the brain and visceral organs.

Reference:

Stasenkova, K. P. and Shirskaya, V. A., in *Toxicology of New Industrial Chemical Substances*, A. A. Letavet and I. V. Sanotsky, Eds., Meditsina, Moscow, Issue No. 9, 1967, 118 (in Russian).

BUTYL LACTATE

Molecular Formula. $C_7H_{14}O_3$
M = 146.21
CAS No 138-22-7
RTECS No OD4025000
Abbreviation. BL.

Synonyms. Butyl-2-hydroxypropanoate; Butyl-α-hydroxy propionate; *n*-Butyl lactate; 2-Hydroxypropanoic acid, butyl ester.

Properties. A colorless liquid. Solubility in water is 45 g/l at 20°C.

Applications. Food additive, used in the manufacture of fragrances, pharmaceuticals and cosmetics, as solvent for nitro- and ethyl cellulose, gums, synthetic polymers.

Acute Toxicity. LD_{50} exceeded 5.0 g/kg BW in rats (Moreno, 1970).[1] According to other data, LD_{50} is greater than 2.0 g/kg BW. Poisoning caused diarrhea and piloerection for up to 24 hours after administering. Gross pathology examination revealed no changes. LD_{50} for racemic *n*-BL was more than 5.0 g/kg BW in rats.[2]

Allergenic Effect.

Observations in man. No sensitization reactions were noted in volunteers.

Animal studies. Lactate esters may be potential eye and skin irritants, but not skin sensitizers.[3]

Chemobiokinetics. BL undergoes enzymatical hydrolysis into lactic acid and butyl alcohol.[3]

Regulations. ***U.S. FDA*** (1998) approved the use of BL in adhesives as a component of articles intended for use in packaging, transporting, or holding food in accordance with the conditions prescribed in 21 CFR part 175.105. Considered as *GRAS.*

Recommendations. ***EU*** (1974). ADI is 1.0 mg/kg BW.

References:

1. Butyl Lactate, Fragrance Raw Material Monographs, *Food. Cosmet. Toxicol.*, 17, 727, 1979.
2. *Cosmetic Ingredient Review* (1997), cit. in #3.
3. Clary, J. J., Feron, V. J., and van Velthuijsen, J. A., Safety assessment of lactate esters. *Regul. Toxicol. Pharmacol.*, 27, 88, 1998.

BUTYL VINYL ETHER

Molecular Formula. $C_6H_{12}O$
M = 100.16

CAS No 111-34-2
RTECS No KN5950000
Abbreviation. BVE.

Synonyms. Butoxyethene; 1-(Ethenyloxy)butane; Vinyl butyl ether.

Properties. Liquid. Soluble in water (3.0 g/l) and alcohol.

Applications. Used in the synthesis of copolymers.

Acute Toxicity. LD_{50} is 10.3 g/kg BW in rats.[03] *n*-BVE administered *i/p* in mice showed no significant effects either on the hepatic non-protein sulfhydryl contents or on the serum GP activity.[1]

Mutagenicity.

In vitro genotoxicity. No significant mutagenicity of *n*-BVE toward *Salmonella typhimurium* strains *TA100* and *TA100* in the presence of the hepatic *S9* mix fraction was found. Mutagenic activities of vinyl ethers in the presence of the *S9mix* were correlated with stability of their epoxides.[2]

Regulations. ***EU*** (1990). BVE is available in the *List of authorized monomers and other starting substances which may continue to be used for the manufacture of plastic materials and articles intended to come into contact with foodstuffs pending a decision on inclusion in Section A* (*Section B*).

References:

1. Isobe, M., Sone, T., and Takabatake, E., Depletion of glutathione and hepato-toxicity caused by vinyl ethers in mice, *J. Toxicol. Sci.*, 20, 161, 1995.
2. Sone, T., Isobe, M., and Takabatake, E., Comparative studies on the metabolism and mutagenicity of vinyl ethers, *J. Pharmacobiodyn,*, 12, 345, 1989.

BUTYNE DIACETATE

Molecular Formula. $C_6H_6O_4$
M = 142.1

Synonym. Acetylene dicarboxylic acid, dimethyl ether.

Properties. A white solid. Poorly soluble in water, soluble in organic solvents.

Applications. Used in the manufacture of plastics.

Acute Toxicity. In mice, LD_{50} is 550 mg/kg BW. High doses cause inhibition, adynamia and motor coordination disorder. Death occurs in a comatose state.

Repeated Exposure failed to reveal pronounced cumulative properties. K_{acc} is 8.0 (by Lim).

Reference:

Balynina, E. S., Toxicity of butine diacetate, *Gig. Truda Prof. Zabol.*, 6, 55, 1984 (in Russian).

BUTYRALDEHYDE

Molecular Formula. C_4H_8O
M = 72.12
CAS No 123-72-8
RTECS No ES2275000
Abbreviation. BA.

Synonyms and **Trade Names.** Butal; Butalide; Butanal; Butural.

Properties. Colorless liquid with sharp odor. Miscible with most organic solvents.

Applications. Used in the production of rubber, plasticizers, polyvinyl butyral, and oil-alcohol-dissolving resins.

Acute Toxicity. LD_{50} is 5.89 g/kg BW in rats.[1]

Mutagenicity.

In vitro genotoxicity. BA is found to be negative in *Salmonella* mutagenicity assay.[013] It induced DNA strand breaks in PM2 DNA in the presence of copper chloride. Neither aldehyde nor copper chloride alone showed DNA breakage properties.[2] BA induced unscheduled DNA synthesis in primary cultures of rat and human hepatocytes. In rat hepatocytes 10 to 100 mM BA induced a modest but significant and dose-dependent increase of net nuclear grain counts, while in human hepatocytes this effect was not detected.[3]

Chemobiokinetics. BA is capable of binding nucleosides and proteins.[4]

Regulations.

U.S. FDA (1998) considered BA to be a direct food additive that may be safely used in foods. It is approved for use in adhesives as a component of articles intended for use in packaging, transporting, or holding food in accordance with the conditions prescribed in 21 CFR part 175.105.

EU (1990). BA is available in the *List of authorized monomers and other starting substances which shall be used for the manufacture of plastic materials and articles intended to come into contact with foodstuffs (Section A*).

Great Britain (1998). BA is authorized without time limit for use in the production of polymeric materials and articles in contact with food or drink or intended for such contact.

Recommendations. Joint FAO/WHO Expert Committee on Food Additives (1999). ADI: No safety concern.

References:

1. Smyth, H. F., Carpenter, C. P., and Weil, C. S., Range-finding toxicity data: list IV, *Arch. Ind. Health Occup. Med.*, 4, 119, 1951.
2. Becker, T. W., Krieger, G., and Witte, I., DNA single and double strand breaks induced by aliphatic and aromatic aldehydes in combination with copper (II), *Free Radical Res.*, 24, 325, 1996.
3. Martelli, A., Canonero, R., Cavanna, M., Ceradelli, M., and Marinari, U. M., Cytotoxic and genotoxic effects of five *n*-alkanals in primary cultures of rat and human hepatocytes, *Mutat. Res.*, 323, 121, 1994.
4. Hemminki, K. and Suni, R., Sites of reaction of glutaraldehyde and acetaldehyde with nucleosides, *Arch. Toxicol.*, 55, 186, 1984.

BUTYRIC ACID

Molecular Formula. $C_4H_8O_2$

M = 88.10

CAS No 107-92-5

RTECS No ES5425000

Abbreviation. BAc.

Synonyms. Butanoic acid; Ethylacetic acid; Propenecarboxilic acid; Propylformic acid.

BUTYRIC ANHYDRIDE

Molecular Formula. C_4H_7O

M = 158.20

CAS No 106-31-0

RTECS No ET7090000

Abbreviations. BAn; PET (polyethylene terephthalate).

Synonyms. Butanoic acid anhydride; Butanoic anhydride; Butyryl oxide.

Properties. BAc is a colorless oily liquid with an unpleasant odor (the odor of rancid oil in diluted solution). Mixes with water and alcohol.[020] BAn is a colorless liquid with a suffocating odor. Soluble in water (40 g/l) and alcohol with decomposition. BAc odor threshold concentration is 0.001 mg/l.[010]

Applications. Used in the production of acetobutyrate cellulose, etrol, etc.

Migration of BAc from PET sheets into food, into 8.0% ethanol/water and *n*-heptane was measured. At very high residual concentration of BAc (147 mg/kg) in sheets made from unwashed PET, higher amounts of the contaminant migrated into the food simulants (more than 10 μg/kg) were found. The results suggested that unwashed recycled PET may not comply with FDA requirements.[1]

Acute Toxicity. LD_{50} of BAc is 8.78 g/kg BW in rats.[2] According to other data, LD_{100} of BAc is 2.5 g/kg BW for rats and mice. Rats tolerate 0.5 g/kg BW. Immediately after administration of the lethal doses, animals assume side position, wheezing and convulsions are observed. Gross pathology examination revealed marked visceral congestion and extensive necrosis of the gastric mucosa. With the small doses, dystrophic changes in the kidneys, catarrhal phenomena in the gastric mucosa and other morphology changes were noted.[2]

Rats tolerate 5.0 g BAn/kg BW, mice tolerate 0.5 g BAn/kg BW. In mice, LD_{50} is 3.5 g BAn/kg BW. Poisoning is accompanied by the symptoms of CNS inhibition and adynamia. Death occurred on days 10 to

15. Gross pathology examination failed to reveal visible changes in the viscera. Histology examination showed mild dystrophy of the liver and kidneys.[3]

Short-term Toxicity. In high doses, BA produced hematological and neuropathological effects.[4]

Long-term Toxicity. ADI for BAc appeared to be 1.25 mg/kg BW.

Reproductive Toxicity.

Embryotoxicity. BAc is suggested to exert embryotoxic effect by alteration of intracellular *pH*.[5] Pregnant rats were exposed to large doses of BA. This treatment reduced pups BW and decreased their viability. These effects seemed related to the impaired respiration in the dams produced by BA.[6]

Teratogenicity. In a frog embryo teratogenesis assay, 50% of the offspring were found to be malformed at the concentration of 400 mg BAc/l. The most commonly seen defects appeared to be microencephally, eye malformations, edema and gut defects.[7]

Mouse embryos exposed to butyrate in culture were found to be malformed (neural tube defects).[8]

Regulations.

EU (1990). BAc and BAn are listed in the *List of authorized monomers and other starting substances which shall be used for the manufacture of plastic materials and articles intended to come into contact with foodstuffs (Section A).*

U.S. FDA (1998) regulates BAn for use (1) as a component of adhesives to be safely used in food-contact surface in accordance with the conditions prescribed in 21 CFR part 175.105; and (2) as a component of the uncoated or coated food-contact surface of paper and paperboard that may be safely used in contact with dry food in accordance with the conditions prescribed in 21 CFR part 176.180. Affirmed as *GRAS* when used as a synthetic flavoring agent.

Great Britain (1998). BA is authorized without time limit for use in the production of polymeric materials and articles in contact with food or drink or intended for such contact.

Standards. *Russia* (1995). MAC: 0.7 mg/l.

Recommendations. Joint FAO/WHO Expert Committee on Food Additives (1999). ADI: No safety concern.

References:

1. Komolprasert, V., Lawson, A. R., and Begley, T. H., Migration of residual contaminants from secondary recycled poly(ethylene terephthalate) into food-simulating solvents, aqueous ethanol and heptane, *Food. Addit. Contam.*, 14, 491, 1997.
2. Stasenkova, K. P. and Kochetkova, T. A., Toxicological characteristics of butyric acid, in *Toxicology of New Industrial Chemical Substances*, A. A. Letavet and I. V. Sanotsky, Eds., Medgiz, Moscow, Issue No 4, 1962, 19 (in Russian).
3. *Toxicology of New Industrial Chemical Substances*, A. A. Letavet and I. V. Sanotsky, Eds., Medgiz, Moscow, Issue No 4, 1962, 29 (in Russian).
4. Blau, C. A., Constantoulakis, P., Shaw, C. M., and Stamatoyannopoulos, G., Fetal hemoglobin induction with butyric acid: efficacy and toxicity, *Blood*, 81, 529, 1993.
5. Brown, N. A., Teratogenicity of carboxylic acids: distribution studies in whole embryo culture, in *Pharmakokinetics in Teratogenesis*, H. Nau and W. J. Scott, Eds., vol 2, CRC Press, Inc., Boca Raton, FL, 1987, 153.
6. Narotsky, M. G., Francis, E. Z., and Kavlock, R. J., Developmental toxicity and structure-activity relationships of aliphatic acids, including dose-response assessment of valproic acid in mice and rats, *Fundam. Appl. Toxicol.*, 22. 251, 1994.
7. Dawson, D. A., Additive incidence of developmental malformation for xenopus embryos exposed to a mixture of ten aliphatic carboxylic acids, *Teratology*, 44, 531, 1991.
8. Hunter, E. S. and Sadler, T. W., Potential mechanisms of ketone body teratogenicity, Abstract, *Teratol ogy*, 31, 35A, 1985.

CALCIUM ACETATE

Molecular Formula. $C_4H_6O_4.Ca$

M = 158.18

CAS No 62-54-4
RTECS No AF7525000
Abbreviation. CA.

Synonyms and **Trade Names.** Brown acetate; Calcium diacetate; Gray acetate; Lime acetate; Vinegar salts.

Properties. Very hygroscopic, rod-shaped crystals. Soluble in water; practically insoluble in ethanol.

Applications. Food stabilizer. Corrosion inhibitor.

Acute Toxicity. LD_{50} is 4.28 g/kg BW in rats.[1]

Long-term Toxicity. CA is known to be a potent phosphate binder.

Mutagenicity. Calcium compounds caused unscheduled DNA synthesis.[2]

Regulations. ***U.S. FDA*** (1998) approved the use of CA (1) in adhesives as a component of articles intended for use in packaging, transporting, or holding food in accordance with the conditions prescribed in 21 CFR part 175.105; and (2) in the manufacture of resinous and polymeric coatings for the food-contact surface of articles intended for use in producing, manufacturing, packing, processing, preparing, treating, packaging, transporting, or holding food in accordance with the conditions prescribed in 21 CFR part 175.320. Affirmed as *GRAS.*

References:

1. Smyth, H. F., Carpenter, C. P., Weil, C. S., Pozzani, U. C., and Striegel, J. A., Range-finding toxicity data: List VII, *Am. Ind. Hyg. Assoc. J.*, 30, 470, 1969.
2. Kaspzak, K. S. and Poirier, A., Effect of calcium and magnesium on nickel uptake and stimulation of thymidine incorporation into DNA in the lungs of strain A mice, *Carcinogenesis,* 6, 1819, 1985.

CALCIUM ETHYL ACETOACETATE

Molecular Formula. $C_{12}H_{18}O_6.Ca$
M = 298.38
CAS No 16715-59-6
RTECS No AK5400000

Synonyms. Bis(hydrogen acetoacetato) calcium, diethyl ester; Ethyl acetoacetate, calcium salt; Ethyl calcioacetoacetate.

Acute Toxicity. LD_{50} is 9.93 g/kg BW in rats.

Regulations. ***U.S. FDA*** (1998) approved the use of calcium ethyl acetoacetate in adhesives as a component of articles intended for use in packaging, transporting, or holding food in accordance with the conditions prescribed in 21 CFR part 175.105.

Reference:

Smyth, H. F. and Carpenter, C. P., Further experience with the range finding test in the industrial toxicological laboratories, *J. Ind. Hyg. Toxicol.*, 30, 63, 1948.

CALCIUM NITRATE (1:2)

Molecular Formula. N_2O_6Ca
M = 164.10
CAS No 10124-37-5
RTECS No EW2985000
Abbreviation. CN.

Synonyms and **Trade Names.** Calcium dinitrate; Calcium saltpeter; Nitric acid, calcium salt; Norway saltpeter.

Properties. White mass, colorless granules or cubic crystals. Readily soluble in water, freely soluble in ethanol.

Acute Toxicity. LD_{50} is found to be 302 mg/kg BW in rats.[1]

Long-term Toxicity.

Observations in man. Clinical manifestations of nitrate toxicity include cyanosis among infants who drink well water.

Animal studies. Addition of 40 g CN/l milk consumed by pigs fed large amounts of whey could lead to toxic effects if the whole of the nitrate converted into *nitrites*, as might occur during a warm season after prolonged storage of the whey.[2]

Regulations. ***U.S. FDA*** (1998) approved the use of CN in adhesives as a component of articles intended for use in packaging, transporting, or holding food in accordance with the conditions prescribed in 21 CFR part 175.105.

References:

1. Diskalenko, A. P., Opopol, N. I., Trofimenko, Yu. N., and Dobrianskaya, E. V., Effect of compounds of nitrates and fluorine entering the body jointly, *Gig. Sanit.*, 12, 66, 1981 (in Russian).
2. Clarke, M. L., Harvey, D. G., and Humphreys, D. J., *Veterinary Toxicology*, 2nd ed., Bailliere Tindall, London, 1981, 67.

CALCIUM OXIDE

Molecular Formula. CaO
M = 56.08
CAS No 1305-78-8
RTECS No EW3100000

Synonyms and **Trade Names.** Burnt lime; Calcia; Lime; Lime, burned; Quick lime.

Properties. Colorless cubic crystals, white or grayish white lumps, or granular powder; odorless; soluble in acids; practically insoluble in alcohol.

Applications. Nutrient and/or dietary supplement food additive. Direct human food ingredient

Regulations. ***U.S. FDA*** (1998) approved the use of calcium oxide with no limitations other than current GMP (21 CFR part 184.1210). Affirmed as *GRAS*.

CALCIUM RESINATE

CAS No 9007-13-0
RTECS No EW3970000
Abbreviation. CR.

Synonyms and **Trade Name.** Limed rosin; Resin acid and rosin acid, calcium salt; Uraprint 62-126.

Properties. Yellowish white or dark-brown amorphous powder or lumps with an odor of rosin. Insoluble in water.

Applications. CR is used in manufacture of porcelains, perfumes, enamels, coatings, and paper.

Regulations. ***U.S. FDA*** (1998) approved the use of CR in resinous and polymeric coatings used as the food-contact surfaces of articles intended for use in producing, manufacturing, packing, processing, preparing, treating, packaging, transporting, or holding food in accordance with the conditions prescribed in 21 CFR part 175.320.

CALCIUM SULFATE

Molecular Formula. $Ca.O_4S$
M = 136.14
CAS No 7778-18-9
23296-15-3
RTECS No WS6920000
Abbreviation. CS.

Synonym and **Trade Names.** Anhydrous calcium sulfate; Crysalba; Thiolite.

Properties. Colorless crystals. Gypsum hemihydrate ($CaSO_4.5H_2O$); after mixing with water is converted into dihydrate, which sets into a firm mass.

Applications. A filler in the production of plastics. Used in construction and medicine.

Toxicity. Oral toxicity has not been studied.

Regulations. ***U.S. FDA*** (1992) regulates the use of CS (1) as a colorant in the manufacture of articles intended for use in producing, manufacturing, packaging, processing, preparing, treating, packing, transporting, or holding food in accordance with the conditions prescribed in 21 CFR part 178.3297. It is used (2) in food at levels not to exceed GMP in accordance with the conditions prescribed in 21 CFR part 184.1230.

Affirmed as *GRAS*.

Recommendations. Joint FAO/WHO Expert Committee on Food Additives: ADI is not specified.

CAPROLACTONE

Molecular Formula. $C_6H_{10}O_2$
M = 114.16
CAS No 502-44-3
RTECS No MO8400000
Abbreviation. CL.

Synonyms and **Trade Names.** 6-Hexanolactone; 1,6-Hexanolide; 6-Hydroxyhexanoic acid lactone; 2-Oxepanone; 2-Oxohexamethylene oxide.

Properties. An intermediate in adhesives, urethane coatings and elastomers. A solvent and a diluent for epoxy resins.

Mutagenicity.

In vitro genotoxicity. CL was not found to be mutagenic in the *i/p* host-mediated assay in *Salmonella;* it did not increase mitotic recombination in *Saccharomyces cerevisiae* D3.

Regulations. ***EU*** (1990). CL is available in the *List of authorized monomers and other starting substances which may continue to be used for the manufacture of plastic materials and articles intended to come into contact with foodstuffs pending a decision on inclusion in Section A* (*Section B*).

Reference:

Simmon, V. F., Rosenkranz, H. S., Zeiger, E., and Poirier, L. A., Mutagenic activity of chemical carcinogens and related compounds in the intraperitoneal host-mediated assay, *J. Natl. Cancer Inst.*, 62, 911, 1979.

CAPRYLIC ACID

Molecular Formula. $C_8H_{16}O_2$
M = 144.21
CAS No 124-07-2
RTECS No RH0175000
Abbreviation. CA.

Synonyms and **Trade Names.** 1-Heptanecarboxilic acid; Octanoic acid; *N*-Octoic acid; *N*-Octylic acid.

Properties. Colorless oily liquid with a faint, fruity-acid odor and slightly sour taste. Solubility in water is 680 mg/l at 20°C; freely soluble in alcohol.[020]

Applications. Chemical intermediate in the production of dyes and lacquers.

Acute Toxicity. In rats, oral administration of 7.5 to 25 g/kg BW in 3 to 5 min. resulted in ataxia, clonic convulsions, and paresis of extremities. Death was preceded by a narcotic state. Gross pathology examination revealed congestion in the visceral organs, degenerative changes in the liver, necrotic zones in the kidneys, and intestinal epithelium.[1]

Repeated Exposure. Addition of 1.0 to 5.0% CA to the diet of dogs caused diarrhea.[011] Rats received repeated doses equal to 1/5 LD_{50}. The treatment caused exhaustion and decreased blood cholinesterase activity. Dystrophy was found in the liver, kidneys, and myocardium.[1]

Reproductive Toxicity.

Embryotoxicity. Pregnant NMRI mice received a single *s/c* injection of 600 mg CA/kg BW in 10 ml water on day 8 of gestation. Examination of fetuses failed to reveal any embryotoxic effect.[2]

Teratogenic effect was not observed in rats.[3]

Regulations.

EU (1992). CA is available in the *List of authorized monomers and other starting substances which shall be used for the manufacture of plastic materials and articles intended to come into contact with foodstuffs* (*Section* A).

Great Britain (1998). CA is authorized without time limit for use in the production of polymeric materials and articles in contact with food or drink or intended for such contact.

U.S. FDA (1998) affirmed CA as *GRAS*.

References:

1. Kuz'menko, S. P. et al., in *Hygiene and Toxicology of Polymers Used in Construction.*, Rostov-na-Donu, Issue 1, 1968, 296 (in Russian).
2. Nau, H. and Loscher, W., Pharmacologic evaluation of various metabolites and analogs of valproic acid: teratogenic potencies in mice, *Fundam. Appl. Toxicol.*, 6, 669, 1986.
3. Suzuki, Y. et al., Negative results of teratological studies on caprylo-hydroxamic acid, Abstract, *Nippon Juigaku Zasshi*, 37, 307, 1975 (in Japanese).

CARBON BLACK

CAS No 1333-86-4

RTECS No FF5800000

Abbreviation. CB.

1) Acetylene black

Synonyms and **Trade Names.** Carbon black, acetylene; CI Pigment Black 7; Explosion acetylene black.; Shawinigan acetylene black.

2) Channel black

Synonyms and **Trade Names.** Atlantic; Black Pearls; Carbolac; Carbomet; Carbon black, channel; CI Pigment Black 7; Collocarb; Impingement black; Kosmolak; Kosmos; Micronex; Monarch; Superba; Texas; Triangle; Witco.

3) Furnace black

Synonyms and **Trade Names.** Aro; Arogen; Atlantic; Black Pearls; Carbon black, furnace; CI Pigment Black 7; Collocarb; Gas-furnace black; Kosmos; Metanex; Monarch; Oil-furnace black; Opal; Special Sterling; Texas; Vulcan.

4) Lampblack

Synonyms and **Trade Names.** Carbon Black BV and V; Carbon black, lamp; CI Pigment Black 6; Durex; Flamruss.

5) Thermal Black

Synonyms and **Trade Names.** Atlantic; Cancarb; Carbon black, thermal; Pigment Black 7; Shell Carbon; Therma-atomic black; Thermax.

Composition. A quasi-graphitic form of carbon of small particle size formed during incomplete combustion or thermal decomposition of hydrocarbons contained in natural or industrial gases (Gas black, Furnace black, Channel black) and in oil and coal treatment products (Lamp black or Activated charcoal, Carboraffin, Medicoal, Norit, Ultracarbon). Structurally, it occupies an intermediate position between crystalline graphite and amorphous carbon. In addition to carbon, it contains hydrogen, sulfur, and mineral substances. *Benz[a]pyrene* concentrations in the benzene extracts of 10 CB samples varied from 0.14 to 35 mg/kg (IARC 65-161).

Properties. The varieties of CB differ in particle size, contents of volatile substances, and absorption properties.

Applications. Used as a filler for most vulcanizates, as well as for polyolefins, etc. It improves the mechanical properties of vulcanizates and is a dye and antioxidant when exposed to heat, light, and in particular, to UV-radiation.

Acute Toxicity. In rats, LD_{50} is 5.0 g/kg BW.[1]

Mutagenicity. Most assays for mutagenicity are negative for C.

In vitro genotoxicity. Commercially produced oil furnace CB appeared not to be genotoxic in five standard assays: *Salmonella typhimurium* reverse mutation test, SCE test in Chinese hamster cells, mouse lymphoma test, cell transformation assay in CH3/10T1/2 cells. Limited cellular toxicity was observed.[2] Extracts of various commercial CB were mutagenic to *Salmonella* in the presence of an exogenous metabolic system.[3]

In vivo cytogenetics. CB did not induce a significant increase in DNA adducts in peripheral lung tissue of rats after two years of inhalation exposure. Commercially produced oil furnace CB was negative in assay for genetic effects in *Dr. melanogaster.*[2]

Carcinogenicity.

Animal studies. No adequate study of the carcinogenicity of CB administered by the oral route was available. Different sources of CB are carriers of resin containing polycyclic aromatic hydrocarbons including benzo[a]pyrene, pyrene, antracene, and fluorene. In one study in female mice treated by inhalation exposure, CB did not increase the incidence of respiratory tract tumors.

Female weanling CF_1 mice received a diet for two years that did or did not (controls) include furnace black (ASTM N-375; 2.05 g/kg diet). Survival at two years was similar in treated mice (84%) and in controls (71%). No increase in tumor incidence was observed.[4] Biological activity of the carcinogens contained in CB may be demonstrated by a solvent capable of eluting them from CB. The extractability of carcinogenic hydrocarbons is proportional to the particle size of CB. Benzene extracts of CB PM-15, PGM-33, and PM-50 appear to be highly carcinogenic (more than 95% of tumors). CB PM-70 produced a moderate effect (37.5%), and CB PM-100 had a slight effect (2.5%). 25 mg/kg and 31 mg benzo[a]pyrene/kg are found in CB TGM-33 and TM-15, respectively, while TG-10 contains 32.8 mg/kg.[5]

Carcinogenicity classification. An IARC Working Group concluded that there is inadequate evidence for the carcinogenicity of CB in *experimental animals*, and sufficient evidence for the carcinogenicity of CB extracts in *experimental animals*, and there were inadequate data available to evaluate the carcinogenicity of CB in *humans*.

IARC: 2B (Carbon Black).

Regulations.

U.S. FDA has banned the use of CB (prepared by the impingement or channel process) for direct use in food, drugs and cosmetics in accordance with the provisions prescribed in 21 CFR part 81.10.

U.S. FDA (1998) approved the use of CB (1) as an indirect food additive as a component of adhesives that come in contact with food in accordance with the conditions prescribed in 21 CFR part 175.105); (2) as a colorant in accordance with the conditions prescribed in 21 CFR part 178.3297; (3) in resinous and polymeric coatings that come in contact with food in accordance with the conditions prescribed in 21 CFR part 175.300); and (4) in rubber articles intended for repeated use that come in contact with food in accordance with the conditions prescribed in 21 CFR part 177.2600); CB (*channel process*) is permitted (5) as a component of polysulfide polymer-polyepoxy resins that come in contact with dry food in accordance with the conditions prescribed in 21 CFR part 177.1650); CB (*channel process* or *furnace combustion process*) is permitted (6) as an optional adjuvant substance: in perfluorocarbon-cured elastomers that come in contact with non-acid food (pH above 5.0) at concentrations not to exceed 15 parts per 100 parts of the terpolymer in accordance with the conditions prescribed in 21 CFR part 177.2400); and (7) in phenolic resins in molded articles that come in contact with non-acid food (*pH* above 5.0) in accordance with the conditions prescribed in 21 CFR part 177.2410).

Russia. Ministry of Health allows CB PM-15 to be used temporarily in vulcanizates intended for food-contact purposes; CB PM-70 and DG-100 are recommended for use in materials coming in contact with food and phar macology products.

Recommendations. *FAO/WHO* provisionally accepted the use of CB from hydrocarbon sources in food-contact materials.

References:

1. *Bull. Int. Register Potent. Toxic Chemicals*, 6, 18, 1983.
2. Kirwin, C. J., LeBlanc, J. V., Thomas, W. C., Haworth, S. R., Kirby, P. E., Thilagar, A., Bowman, J. T., and Brusick, D. J., Evaluation of the genetic activity of industrially produced carbon black, *J. Toxicol. Environ. Health*, 7, 973, 1981.
3. Agurell, E. and Lofroth, G., Presence of various types of mutagenic impurities in carbon black detected by the *Salmonella* assay, in *Short-term Bioassays in the Analysis of Complex Environmental Mixtures*, M. D. Waters, Ed., Plenum Press, New York, 1983, 297.

4. Pence, B. C. and Buddingh, F., The effect of carbon black ingestion on 1,2-dimethyl hydrazine-induced colon carcinogenesis in rats and mice, *Toxicol. Lett.*, 25, 273, 1985.
5. Pylev, L. N. and Iankova, G. D., Studies on possible absorption of benzo[a]pyrene and its contents in the several domestic types of carbon black, *Gig. Truda Prof. Zabol.*, 4, 52, 1974 (in Russian).

CARBONIC ACID, ZINC SALT (1:1)

Molecular Formula. CO_3Zn
M = 125.38
CAS No 3486-35-9
RTECS No FG3375000
Abbreviation. CAZ.

Synonym and Trade Names. Calamine; Natural Smithsonite; Zinc carbonate; Zincspar.

Properties. White or colorless, odorless crystalline powder or solid of rhombohedral structure. Solubility in water is 10 mg/l at 15°C.[020] Insoluble in alcohol. At 140°C is dissociated with the formation of *zinc oxide*. CAZ is poorly soluble in water. Levels in drinking water above 3.0 mg/l give an undesirable astringent taste and may result in discoloration. The threshold perception concentration for the effect on the organoleptic properties of water is 5.0 mg/l.

Applications. A fire-proofing filler for temperature-resistant rubber and plastics. A pigment in the manufacture of porcelains, pottery, and rubber.

Acute Toxicity. Rats tolerate administration of 10 g/kg BW; mice do not die after ingestion of 7.0 g/kg BW dose.[1]

Repeated Exposure. Weanling pigs consuming the feed containing 0.1% CAZ showed no signs of intoxication. There were no changes during later stages of development of pigs.[2]

Long-term Toxicity. Consumption of the diet containing 2500 ppm CAZ caused no ill effects in rats.[3]

Reproductive Toxicity. Mice, exposed to excess (2000 ppm) CAZ in the maternal diet during gestation, lactation, and post-weaning development for 8 weeks, had reduced plasma copper levels, lower hematocrit values, and reduced BW, compared to the controls. Loss of hair was noted.[4] Rats received 0.4% CAZ in their diet. The treatment reduced liver cytochrome oxidase and catalase activity. Excess of CAZ decreased growth in weanling rats, and they produced stillborn pups when they mature.[5]

Regulations. ***U.S. FDA*** (1998) approved the use of CAZ (1) as a filler for preparation of rubber articles intended for repeated use in contact with food; as a component of food-contact adhesives in accordance with the conditions prescribed in 21 CFR part 177.2600; (2) as a colorant in the manufacture of melmine-formaldehyde resins in molded articles in accordance with the conditions prescribed in 21 CFR part 178.3297; and (3) in xylene-formaldehyde resins condensed with 4,4'-isopropylidene diphenol-epichlor-hydrin epoxy resins, ethylene-vinyl acetate copolymers.

References:

1. Spiridonova, V. S. and Shabalina, L. P., Experimental study on zinc nitrate and zinc carbonate toxicity, *Gig. Truda Prof. Zabol.*, 3, 50, 1986 (in Russian).
2. Clarke, M. L., Harway, D. G., and Hamphreys, D. J., *Veterinary Toxicology*, 2nd ed., Bailliere Tindall, London, 1981, 76.
3. Browning, E., *Toxicity of Industrial Metals*, 2nd ed., Apleton-Century-Crofts, New York, 1969, 351.
4. Mulhern, S. A., Stroube, W. B., and Jacobs, R. M., Allopecia induced in young mice by exposure to excess dietary zinc, *Experientia*, 42, 551, 1986.
5. Venugopal, B. and Luckey, T. D., *Metal Toxicity in Mammals*, Plenum Press, New York, 1978, 73.

See ***Zinc compounds***.

α-CARBOXY-1-OXO-3-SULFOPROPYL)-ω-(DODECYLOXY)-POLY(OXY-1,2-ETHANEDIYL), DISODIUM SALT

Molecular Formula. $[\sim C_2H_4O\sim]_nC_{16}H_{30}O_7S.2Na$
CAS No 39354-45-5
RTECS No TR1581150

Synonym and **Trade Name.** Aerosol A 102; Sulfonic acid 4-ester with polyethylene glycol dodecyl ether, disodium salt.

Acute Toxicity. LD_{50} is 30.8 ml/kg BW in rats. Poisoning affected respiratory system and the liver.

Regulations. ***U.S. FDA*** (1998) approved the use of S. in adhesives as a component of articles intended for use in packaging, transporting, or holding food in accordance with the conditions prescribed in 21 CFR part 175.105.

Reference:

Acute Toxicity Data, *J. Am. Coll. Toxicol.*, 1, 107, 1990.

CASTOR OIL

CAS No 8001-79-4

RTECS No FI4100000

Abbreviation. CO.

Synonyms and Trade Names. Aromatic castor oil; Cosmetol; Crystal O; Gold Bond; Oil of Palma, Ricinus oil; Tangantangan oil.

Composition. CO is a natural oil derived from the seeds of the castor bean, *Ricinus communis*. It is comprised predominantly of triglycerides with a high ricinolin contents.

Properties. Pale-yellowish or colorless, transparent, viscous liquid with slight characteristic odor and crude oil taste. Insoluble in water; soluble in absolute ethanol.

Applications. Rubber preservative. Used in the manufacture of resins and plastics, and in the production of alkyds and resinous copolymers. A plasticizer for urethane polymers and nitrocellulose. A component of protective coatings in vitamin and mineral tablets; a diluent in inks for marking gum. CO is used as a drying oil for paints, varnishes, plastics, and resins. It is also used in numerous cosmetics.

Exposure. Apart from its laxative effect, CO is likely to be used without harm.

Acute Toxicity.

Observations in man. CO is considered minimally toxic when administered orally to humans; the estimated LD is 1 to 2 pints of undiluted oil.[022] Large doses cause gripping.

Animal studies. In rat studies, ingestion of 2.0 ml CO produced diarrhea, which is associated with gross damage to the duodenal and jejunal mucosa. The injury is accompanied by release of acid phosphatase into the gut lumen, indicating cellular injury.[1]

Repeated Exposure. Rhesus monkeys were administered 1.0 ml CO/kg BW by gavage for 4 days. Mild morphological changes in the small intestine characterized by lipid droplets along the mucosal epithelium and in the underlying lamina propria were observed.[2]

This was considered a possible indication that CO had reduced lipid metabolism in the intestinal epithelium.

Short-term Toxicity. Dietary concentrations as high as 10% did not affect survival or BW gain in F344 rats or $B6C3F_1$ mice in 13-week studies. No hematology changes and histopathology lesions in the liver or other visceral organs were observed in the treated animals.[3]

Allergenic Effect. Several cases of sensitization to CO in cosmetics were reported including an allergic reaction to make-up remover.[4] A lipstick containing CO may cause contact dermatitis.[5] A hypersensitivity reaction has been associated with ingestion of CO.[6]

Reproductive Toxicity.

Teratogenicity.

Observations in man. Mauhoub et al.[7] observed a three month-old infant with ectrodactyly, vertebral defects, and growth retardation. The mother had taken one castor oil seed during each of the first 3 months of pregnancy.

Gonadotoxicity.

Animal studies. There were no significant changes in male reproductive characteristics (sperm count and motility) or in the length of the estrous cycles of rats or mice given diets containing CO for 13 weeks.[022]

Mutagenicity.

In vitro genotoxicity. CO was negative in *Salmonella* mutagenicity studies; in did not induce SCE or CA in Chinese hamster ovary cells.[3]

In vivo cytogenetics. CO did not induce micronuclei in the peripheral blood erythrocytes of mice evaluated at the end of the 13-week studies.[022]

Chemobiokinetics. Low doses of CO are readily absorbed, but as the oral dose increases, absorption decreases and laxation occurs. At doses of 4.0 g/kg BW in adults, absorption seems to be complete. JECFA considered this dose to be the NOAEL. CA is hydrolyzed by intestinal lipases to glycerol and ricinoleic acid. The absorbed acid undergoes conversions like other fatty acids.[8, 022]

Regulations.

U.S. FDA (1998) regulates use of CO as (1) a component of the uncoated or coated food-contact surface of paper and paperboard intended for use in producing, manufacturing, packing, transporting, or holding aqueous and fatty food of the type identified in 21 CFR part 176.170 (c); (2) in adhesives used as components of articles intended for use in packaging, transporting, or holding food in accordance with the conditions prescribed in 21 CFR part 175.105; (3) as a plasticizer in rubber articles intended for repeated use in contact with food in amounts up to 30% by weight of the rubber product; (4) in cellophane for packaging food in accordance with the conditions prescribed in 21 CFR part 177.1200; (5) in the manufacture of resinous and polymeric coatings in a food-contact surface of articles intended for use in producing, manufacturing, packing, processing, preparing, treating, packaging, transporting, or holding food in accordance with the conditions prescribed in 21 CFR part 175.300; (6) in acrylic and modified acrylic plastics; (7) in cross-linked polyester resins to be safely used as articles or components of articles intended for repeated use in contact with food, subject to the provisions prescribed in 21 CFR part 177.2420; (8) in the manufacture of closures with sealing gaskets (up to 2.0% by weight) used on containers intended for use in producing, manufacturing, packing, processing, preparing, treating, packaging, transporting, or holding food, in accordance with conditions prescribed in 21 CFR part 177.1210; (9) in the manufacture of rubber articles intended for repeated use in producing, manufacturing, packaging, processing, preparing, treating, packing, transporting, or holding food in accordance with the conditions prescribed in 21 CFR part 177.2600; (10) as lubricants with incidental food contact intended for used on machinery used for producing, manufacturing, packaging, processing, preparing, treating, packing, transporting, or holding food as prescribed in 21 CFR part 178.3570; and (11) as a defoaming agent that may be safely used in the manufacture of paper and paperboard and as components of articles intended for use in producing, manufacturing, packing, transporting, or holding food, subject to the provisions prescribed in 21 CFR part 176.200; (12) CO may be used in food in accordance with the conditions prescribed in 21 CFR part 172.876.

Maximum tolerance in food: 500 ppm.

EU (1990). CO is available in the *List of authorized monomers and other starting substances which shall be used for the manufacture of plastic materials and articles intended to come into contact with foodstuffs (Section A).*

Great Britain (1998). CO is authorized without time limit for use in the production of polymeric materials and articles in contact with food or drink or intended for such contact.

Standards. ***Russia*** (1995). MAC: 0.2 mg/l.

Recommendations. Joint FAO/WHO Expert Committee on Food Additives. ADI. man: 0 to 0.7 mg/kg BW.

References:

1. Capasso, F., Mascolo, N., Izzo, A. A., and Gaginella, T. S., Dissociation of castor oil-induced diarrhea and intestinal mucosal injury in rat: effect of *NG*-nitro-*L*-arginine methyl ester, *Brit. J. Pharmacol.*, 113, 1127, 1994.
2. Diener, R. M. and Sparano, B. M., Effects of various treatments on the histochemical profile of the gastrointestinal mucosa in rhesus monkeys, *Toxicol. Appl. Pharmacol.*, 13, 412, 1968.
3. *Toxicity Studies of Castor Oil in F344/N Rats and B6C3F$_1$ Mice (Dosed Feed Studies)*, (R. Irwin), NTP Technical Report Series No 12, Research Triangle Park, NC, NIH Publ. No 92-3131, 1992, 30.

4. Brandle, I., Boujnah-Khouadja, A., and Fousserau, J., Allergy to castor oil, *Contact Dermat.*, 9, 424, 1983.
5. Say, S., Lipstick dermatitis caused by castor oil, *Contact Dermat.*, 9, 75, 1983.
6. McGuire, T., Rothenberg, M. B., and Tiler, D. C., Profound shock following intervention for chronic untreated stool retention, *Clin. Pediatr.*, 23, 459, 1983.
7. Mauhoub, M. E., Khalifa, M. M., Jaswal, O. B., et al., Ricin syndrome, a possible new syndrome associated with ingestion of castor-oil seed in early pregnancy: A case report, *Ann. Trop. Pediatr.*, 3, 57, 1983.
8. *The Pharmacological Basis of Therapeutics*, L. S. Goodman and A. Gilman, Eds., 5th ed., Macmillan Publ. Co., Inc., New York, 1975, 982.

CASTOR OIL, HYDROGENATED

Abbreviation. COH.

Composition. HCO-60 is a *polyoxyethylene* castor oil derivative.

Toxicity. Its toxicity is associated with histamine release from the mast cells. No symptoms occurred in monkeys, rabbits, or guinea pigs treated *i/v* with 50 or 100 mg HCO-60/kg BW; there was no change in plasma histamine levels. Toxicity of HCO-60 is species specific to dogs among all the animals tested.[1]

Mutagenicity. HCO-60, a polyoxyethylene hydrogenated castor oil, induced neither reverse mutation in *Salmonella typhimurium* strains *TA100, TA98, TA1535, TA1537,* and in *E. coli*, nor CA in Chinese hamster V79 cells. No increase in micronucleated polychromatic erythrocytes was elicited in the bone marrow of BDF_1 male and female mice. HCO-60 was not genotoxic.[2]

Regulations.

U.S. FDA (1998) approved the use of COH (1) as a lubricant for *vinyl chloride* polymers used in the manufacture of articles or components of articles authorized for food-contact use complying with 21 part 178.3280; (2) in the manufacture of resinous and polymeric coatings for the food-contact surface of articles intended for use in producing, manufacturing, packing, processing, preparing, treating, packaging, transporting, or holding food (for use only as polymerization cross-linking agent in side seam cements for containers intended for use in contact with food (only of the identified types) complying with 21 CFR part 175.300; (3) as a component of cellophane for packaging food in accordance with the conditions prescribed in 21 CFR part 177.1200; (4) in the manufacture of closures with sealing gaskets used on containers intended for use in producing, manufacturing, packing, processing, preparing, treating, packaging, transporting, or holding food, in accordance with the conditions prescribed in 21 CFR part 177.1210; (5) as a component of paper and paperboard in contact with aqueous and fatty food complying with 21 CFR part 176.170; (6) as a component of cross-linked polyester resins subject to the conditions prescribed in 21 CFR part 177.2420; and (7) as a component of olefin polymers complying with 21 CFR part 177.1520.

Great Britain (1998). Hydrogenated CO is authorized without time limit for use in the production of polymeric materials and articles in contact with food or drink or intended for such contact.

References:

1. Hisatomi, A., Kimura, M., Maeda, M., Matsumoto, M., Ohara, K. and Noguchi, H., Toxicity of polyoxyethylene hydrogenated castor oil 60 (HCO-60) in experimental animals, *J. Toxicol. Sci.,* 18 (Suppl. 3), 1, 1993.
2. Hirai, O., Miyamae, Y., Zaizen, K., Miyamoto, A., Takashima, M., Hattori, Y., Ohara, K., and Mine, Y., Mutagenicity tests of polyoxyethylene hydrogenated castor oil 60 (HCO-60), *J. Toxicol. Sci.*, 19, 89, 1994.

CHLORAL

Molecular Formula. C_2HCl_3O

M = 147.38

CAS No 75-87-6

RTECS No FM7870000

Synonyms. Trichloroacetaldehyde; Trichloroethanal.

CHLORAL HYDRATE

Molecular Formula. $C_2H_3Cl_3O_2$
M = 165.40
CAS No 302-17-0
RTECS No FM8750000
Abbreviation. CH.

Synonyms. Trichloroacetaldehyde monohydrate; 1,1,1-Trichloro-2,2-dihydroxyethane.

Properties. Colorless, oily liquid with a sharp penetrating unpleasant odor. Soluble in water; readily soluble in alcohol. Combines readily with water and becomes completely converted to chloral hydrate (colorless crystals with specific odor). CH solubility in water is 470 g/100 ml. Taste and odor perception threshold is 16 mg C./l; practical threshold is 32 mg C./l. Heating and chlorination of such solutions did not affect odor intensity. The color of solutions remained unaffected even by high concentrations.[1]

Applications. C. used in the production of rigid polyurethane foams and other materials intended for contact with food. CH has been widely used as a sedative, an anesthetic, or hypnotic drug in humans at oral doses of up to 14 mg/kg BW; it is still used in pediatric medicine and dentistry.

Exposure. C. has been formed as a by-product of chlorination when chlorine reacts with humic acids. CH has been found in drinking water at concentrations up to 0.02 mg/l.

Acute Toxicity. LD_{50} of CH is 725 mg/kg in rats, 850 mg/kg BW in mice, 1400 mg/kg BW in rabbits, and 940 mg/kg BW in guinea pigs. According to Sanders et al.,[2] in Sprague-Dawley rats, the LD_{50} values are 1442 mg CH/kg BW (males) and 1265 mg CH/kg BW (females). Administration of the lethal doses led to a brief period of excitation with marked activity and subsequent CNS inhibition and motor coordination disorder. Animals assumed side position, became comatose, and died within a few hours.[1] Gross pathology examination revealed hemodynamic disturbances and parenchymatous dystrophy of the liver and kidneys.[3]

Repeated Exposure. CH has no marked cumulative properties. Rats tolerate administration of 1/5 and 1/10 LD_{50} though the total dose was equal to three LD_{50}.[2]

In a 14-day gavage study, male mice were exposed to the doses of 0.1 and 0.01 LD_{50} of CH. Poisoning caused an increase in liver weights and a decrease in spleen weights at the highest dose level.[2]

Short-term Toxicity. In a 90-day study, mice received daily doses approximately 16 or 160 mg CH/kg BW. The treatment caused a dose-related hepatomegaly in male but not in female mice, accompanied with significant changes in serum chemistry and hepatic microsomal parameters (the latter was observed in both sexes). No other significant toxicology effects were found.[2] The LOAEL was identified to be 96 mg/kg BW for liver enlargement. Mild liver toxicity was noted in rats receiving CH in their drinking water over a period of 90 days.[4]

Long-term Toxicity. B6C3F$_1$ mice were exposed via their drinking water to 1.0 g CH/l (166 mg/kg BW) for 104 weeks. The liver was found to be a primary target organ. An increase in the liver weights and hepatocellular necrosis were noted.[5]

In a 7-month study in rats, CH was found to exhibit a polytropic action. Primary target tissues are the liver and cardiovascular system. Conditioned reflex activity was affected. Diminished bromsulfalein retention by the liver and increased activity of serum transaminases were noted. The NOAEL of 0.01 mg/kg BW was identified in this study.[1] U.S. EPA (1990) considered RfD of 0.0016 mg/kg BW.

Reproductive Toxicity.

Embryotoxicity.

Observations in man. No increase of the adverse effect was found in the offspring of 71 women who took CH as a drug in their four lunar months.[6]

Teratogenicity.

Animal studies. Mice were exposed throughout pregnancy to CH in their drinking water. The dose of 204 mg/kg but not 21.3 mg/kg BW caused behavioral changes (impaired learning retention of a passive avoidance task).[7]

Immunotoxicity. In the above-described studies,[2] no alterations were found in either humoral or cell-mediated immunity. However, female mice treated for 90 days demonstrated a significant inhibition of humoral immunity function.[8]

Mutagenicity.

In vitro genotoxicity. CH was shown to be mutagenic in *Salmonella* assay,[025] but not to bind to DNA. It caused SCE and disruption of chromosomal segregation in cell division,[9] produced aneuploidy in human lymphocytes (Vagnarelli et al., 1990), and cultured hamster cells (Degrassi and Tanzorella, 1988).

No significant damage was found after treatment with CH in the cytoplasmin B micronucleus test in human lymphocyte culture (Vian et al., 1995). CH was negative in SOS chromotest using strain PQ37 of *E. coli.*[10]

In vivo cytogenetics. CH was shown to be mutagenic in *Dr. melanogaster* and positive in aneuploidy tests *in vivo.*[11] It produced a significant increase in the frequency of micronucleated erythrocytes following *in vivo* exposure of the amphibian larvae.[10]

Carcinogenicity. The exposure of male B6C3F_1 mice to up to 1.0 g CH/l in their drinking water[6] resulted in an increased incidence of hepatocellular carcinomas (46% of survivors), hepatocellular adenomas (29%), and combined tumors (71%). A single oral dose of 10 mg CH/kg BW administered to 15-day-old male mice resulted in a significant increase in the rate of liver tumors (including trabecular carcinomas after 48 to 92 weeks of observation.[12]

Carcinogenicity classifications. An IARC Working Group concluded that there is inadequate evidence for the carcinogenicity of C. and CH in *humans* and that of C. in *experimental animals*. There is limited evidence for the carcinogenicity of CH in *experimental animals*.

IARC: 3 (Chloral and Chloral hydrate).

U.S. EPA: C (Chloral hydrate).

Chemobiokinetics. After oral administration CH is rapidly absorbed. It is metabolized by mice liver microsomes and increased amounts of lipid peroxidation products (*maloaldehyde* and *formaldehyde*); the reactions could be inhibited by α-tocopherol or menadione.[13] *Trichloroethanol* appears to be the main metabolite of CH.

Male and female B6C3F_1 mice and F344 rats were treated by gavage with 1 or 12 doses of CH. Metabolites found in plasma were *trichloroethanol* and *trichloroacetic acid.*

The halflife of trichloroethanol and its glucuronide was significantly greater in rats as compared to mice. None of the metabolic parameters appears to account for the hepatocarcinogenicity of CH seen in mice but not rats.[14]

Regulations. ***U.S. FDA*** (1998) regulates CH (1) as a component of the uncoated or coated food-contact surface of paper and paperboard intended for use in producing, manufacturing, packaging, processing, preparing, treating, packing, transporting, or holding dry food of the type identified in 21 CFR part 176.170 (c); (2) cross-linking agent in the manufacture of polysulfide polymer-polyepoxy resins which may be used as food-contact surface of articles intended for packaging, transporting, holding, or otherwise contacting dry food in accordance with the conditions prescribed in 21 CFR part 177.1650; and (3) in adhesives used as components of articles intended for use in packaging, transporting, or holding food in accordance with the conditions prescribed in 21 CFR part175.105

Standards.

U.S. EPA (1999). MCLG of CH: 0.06 mg/l.

Russia (1995). MAC and PML: 0.2 mg/l.

Recommendations. *WHO* (1996). Provisional Guideline Value for CH in drinking water: 0.01 mg/l.

References:

1. Kryatov, I. A., Hygienic evaluation of sodium *p*-chlorobenzene sulfonate and chloral as water pollutants, *Gig. Sanit.*, 3, 14, 1970 (in Russian).
2. Sanders, V. M., Kauffmann, B. M., White, K. L., Douglas, K. A., Barnes, D. W., Sain, L. E., Bradshaw, T. J., Borzelleca, J. F., and Munson, A. E., Toxicology of chloral hydrate in the mouse, *Environ. Health Perspect.*, 44, 137, 1982.
3. Rotenberg, Yu. S., Boitsov, A. N., Klochkova, S. I., et al., Toxicological-hygienic evaluation of chloral, in *Hygiene and Toxicology of High-Molecular-Mass Compounds and of the Chemical Raw Material Used for Their Synthesis*, Proc. 4th All-Union Conf., S. L. Danishevsky, Ed., Khimiya, Leningrad, 1969, 175 (in Russian).

4. Daniel, F. B., Robinson, M., Stober, J. A., Page, N. P., and Olson, G. R., Ninety-day toxicity study of chloral hydrate in the Sprague-Dawley rat, *Drug Chem. Toxicol.*, 15, 15, 1992.
5. Daniel, F. B., De Angelo, A. B., Stober, J. A., Page, N. P., and Olson, G. R., Hepatocarcinogenicity of chloral hydrate, 2-chloroacetaldehyde, and dichloracetic acid in the male $B6C3F_1$ mouse, *Fundam. Appl. Toxicol.*, 19, 159, 1992.
6. Heinonen, O. P., Slone, D., and Shapiro, S., *Birth Defects and Drugs in Pregnancy*, Publishing Sciences Group Inc., Littleton, MA, 1977.
7. Kallman, M. J., Kaempf, G. L., and Balster, R. L., Behavioral toxicity of chloral in mice: An approach to reevaluation, *Neurobehav. Toxicol. Teratol.*, 6, 137, 1984.
8. Kauffmann, B. M., White, K. L., Sanders, V. M., et al., Humoral and cell-mediated immune status in mice exposed to chloral hydrate, *Environ. Health Perspect.*, 44, 147, 1982.
9. Gu, W. Z., Sele, B., Jalbert, P., Vincent, M., Marka, C., Charma, D., and Faure, J., Induction of sister chromatid exchanges by trichloroethylene and its metabolites, *Toxicol. Eur. Res.*, 3, 63, 1981.
10. Fahrig, R., Madle, S., and Baumann, H., Genetic toxicology of trichloroethylene (TCE), *Mutat. Res.*, 340, 1, 1995.
11. Giller, S., Le Curieux, F., Gauthier, L., Erb, F., and Marzin, D., Genotoxicity of chloral hydrate and chloropicrine, *Mutat. Res.*, 348, 147, 1995.
12. Rijhainghani, K. S., Abrahams, C., Swerdlow, M. A., Rao, K. V. N., and Ghose, T., Induction of neoplastic lesions in the liver of C57BL x C_3HF_1 mice by chloral hydrate, *Cancer Detect. Prevent.*, 9, 279, 1986.
13. Ni, Y.-C., Wong, T.-Y., Kadlubar, F. F., and Fu, P. P., Hepatic metabolism of chloral hydrate to free radical(s) and induction of lipid peroxidation, *Biochem. Biophys. Res. Comm.*, 204, 937, 1994.
14. Beland, F. A., Schmitt, T. C., Fullerton, N. F., and Young, J. F., Metabolism of chloral hydrate in mice and rats after single and multiple doses, *J. Toxicol. Environ. Health,* 54, 209, 1998.

CHLORIC ACID, SODIUM SALT

Molecular Formula. $ClNaO_3$
M = 106.44
CAS No 7775-09-9
RTECS No FO0525000
Abbreviation. CAS.

Synonyms and **Trade Names.** Atlacide; Chloracil; Chlorax; Soda chlorate; Sodium chlorate, solution.

Properties. Pale-yellow or white cubic or trigonal crystals, granules, or white or colorless powder. Odorless, with salty taste. Soluble in water and alcohol.

Applications. SC is used in the manufacture of dyes.

Acute Toxicity.

Observations in man. A dose of 100 g CAS is likely to be lethal.

Animal studies. LD_{50} is found to be 1.2 g/kg BW in rats,[1] and 7.2 mg/kg BW in rabbits.[2] Poisoning with chlorates caused hematological changes: Heinz bodies are present, many of the erythrocytes clump and are broken up. Tissue anoxia resulted from methemoglobinemia. Later other signs of injury were noted, including emboli and hematuria, cyanosis, and dyspnea. Death occurs within a few hours to a few days due to severe methemoglobinemia or to acute nephritis.[3] Sudden death, without obvious symptoms, is also reported.[4]

In dogs, LD_{50} is 0.5 to 2.0 g/kg BW in 2 to 4 daily doses.[5] Gross pathology examination of dogs revealed evi dence of marked splenic congestion and moderately severe chronic interstitial nephritis.[4]

Repeated Exposure. Cows were given 0.1 to 0.25 g/kg BW for 5 days. The treatment produced dark-brown urine and methemoglobinemia. In dogs, LD_{50} was 0.5 to 2 g/kg BW after 2 to 4 daily doses.[5]

Short-term Toxicity. Sprague-Dawley rats received drinking water containing 3, 12, or 48 mM CAS for 90 days. The treatment caused no mortality but caused significant BW loss and signs of anemia in the high exposure groups. The NOAEL of 0.36 mM CAS/kg BW in male and 0.5 mM CAS/kg BW in females was established in this study.[6]

Long-term Toxicity. Rats received drinking water containing concentrations of 10 or 100 mg CAS/l for 1 year. A decrease in BW gain was noted after 10 and 11 months; signs of anemia became evident after 9 months; after 3 months, the treatment inhibited the incorporation of ^{3}*H-thymidine* into nuclei of rat testes.[7]

Reproductive Toxicity.

Gonadotoxicity. *P* and F_1 generations of rats received CAS in their drinking water at concentrations of 25, 100, 175, or 250 ppm. The treatment produced no dose-related effects in testis weight, prostate/seminal vesicle weights, non-reproductive organ weights, testicular spermatid counts, sperm production per gram of testis per day, sperm production per gram of testis. Histological changes were not observed in testicular tissues. CAS does not adversely affect spermatogenesis or endocrine function in the *P* and F_1 generation male rats.[8]

Mutagenicity. CAS was positive in *Salmonella* and *Basc* test, and negative in micronucleus test.[9]

Chemobiokinetics. Chlorates are readily absorbed from the GI tract and excreted unchanged via urine.[3]

Regulations. ***U.S. FDA*** (1998) approved the use of CAS (1) in adhesives as a component of articles intended for use in packaging, transporting, or holding food in accordance with the conditions prescribed in 21 CFR part 175.105; as (2) an optional adjuvant substance employed in the production of animal glue in accordance with the conditions prescribed in 21 CFR part 178.3120; and (3) as an adjuvant substance in the manufacture of foamed plastics intended for use in contact with food, subject to the provisions prescribed in 21 CFR part 178.3010.

References:

1. Hayes, W. J., *Pesticides Studies in Man*, Williams & Wilkins, Baltimore/London, 1982, 61.
2. Information from the Soviet Toxicology Center, *Gig. Truda Prof. Zabol.*, 1, 49, 1987 (in Russian).
3. Thienes, C. and Haley, T. J., *Clinical Toxicology*, 5th ed., Lea & Fediger, Philadelphia, 1972, 159.
4. Humphreys, D. J., *Veterinary Toxicology*, 3rd ed., Bailliere Tindall, London, 1988, 29.
5. Buck, E. T. et al., *Clinical and Diagnostic Veterinary Toxicology*, 2nd ed., 1976, 167.
6. McCauley, P. T., Robinson, M., Daniel., F. B., and Olson, G. R., The effects of subchronic chlorate exposure in Sprague-Dawley rats, *Drug Chem. Toxicol.*, 18, 185, 1995.
7. Abdel-Rahman, M. S., Couri, D., and Bull, R. J., Toxicity of chlorine dioxide in drinking water, *J. Environ. Pathol. Toxicol. Oncol.*, 6, 105, 1985.
8. Sprando, R. L., Collins, T. F. X., Black, T. N., Rorie, J., Ames, M. J., and O'Donnell, M., Testing the potential of sodium fluoride to affect spermatogenesis in the rat, *Food Chem. Toxicol.*, 35, 881, 1997.
9. Gocke, E., King, M.-T., Eckhardt, K., and Wild, D., Mutagenicity of cosmetic ingredients licensed by the EEC, *Mutat.Res.*, 90, 91, 1981.

CHLOROACETAMIDE

Molecular Formula. C_2H_4ClNO

M = 93.52

CAS No 79-07-2

RTECS No AB5075000

Abbreviation. CAA.

Synonym and **Trade Name.** 2-Chloroethanamide; Microcide.

Properties. Crystals. Soluble in water (1:10) and absolute alcohol (1:10).

Acute Toxicity. LD_{50} is 155 mg/kg BW in mice, 122 mg/kg BW in rabbits, and 31 mg/kg BW in dogs.[1]

Repeated Exposure. In 30-day oral study, spastic paralysis with or without sensory change was observed in rabbits.[1] Exposure to CAA caused hepatotoxicity manifestations including hydropic degeneration and lipid peroxidation.[2]

Reproductive Toxicity.

Embryotoxicity. Rats received 20 mg CAA/kg BW on single days 7, 11, or 12 of pregnancy. The treatment caused no effects on litter size or fetuses.[3]

Allergenic Effect. Contact dermatitis was observed in workers of print industry occupationally exposed to CAA.[4]

Regulations. ***U.S. FDA*** (1998) approved the use of CAA (1) in adhesives as a component of articles intended for use in packaging, transporting, or holding food in accordance with the conditions prescribed in 21 CFR part 175.105; and (2) as a component of the uncoated or coated food-contact surface of paper and paperboard intended for use in producing, manufacturing, packaging, processing, preparing, treating, packing, transporting, or holding aqueous and fatty food in accordance with the conditions prescribed in 21 CFR part 176.170.

References:

1. Abstract, in *Zhonghua Yixue Zazhi* (Chinese Med. J.), 58, 462, 1978.
2. Anundi, I., Rajs, J., and Hogberg, J., Chloroacetamide hepatotoxicity: hydropic degeneration and lipid peroxidation, *Toxicol. Appl. Pharmacol.*, 55, 273, 1980.
3. Thiersch, J. B., Investigations into the differential effect of compounds on rat litter and mother, in *Malformations Congenitales Des Mammiferes*, H. Tuchmann-Duplessis, Ed., Paris, Masson et Cie, 1971, 95.
4. Jones, S. K. and Kennedy, C. T., Chloroacetamide as an allergen in print industry, *Contact Dermat.*, 18, 304, 1988.

CHLOROACETIC ACID

Molecular Formula. $C_2H_3ClO_2$
M = 94.50
CAS No 79-11-8
RTECS No AF8575000
Abbreviation. CAA.

Synonyms. Chloroethanoic acid; Monochloroacetic acid.

Properties. Colorless or white crystals or monoclinic prisms with characteristic, penetrating odor similar to vinegar. Readily soluble in water; soluble in ethanol.

Applications. Used in the manufacture of dyes; intermediate in the production of carboxymethyl cellulose.

Acute Toxicity.

Observations in man. Probable oral lethal dose is in the range of 50 to 500 mg/kg BW for a 70 kg person.[019]

Animal studies. LD_{50} is 580 mg/kg BW in rats.[1] However, according to other data, LD_{50} of monochloroacetate is 108 mg/kg BW in rats (Hayes, 1973), or 76 mg/kg BW in rats, 255 mg/kg BW in mice, and 80 mg/kg BW in guinea pigs.[020]

Reproductive Toxicity.

Teratogenicity. Pregnant rats received by gavage up to 140 mg/kg BW during organogenesis. Maternal toxicity and an increase in cardiovascular malformations, predominantly levocardia, were noted in offspring at the highest dose level.[2]

Mutagenicity.

In vitro genotoxicity. CAA significantly increased mutant count in mouse lymphoma cell mutation assay.[3] It was found to be negative in five *Salmonella typhimurium* strains.[013]

Carcinogenicity. According to Hayes, *i/g* administration at the dose level of 46.4 mg/kg BW over a period of 1.5 years produced no increase in tumor incidence in mice.[4] Equivocal tumorigenic agent by *RTECS* criteria (on *s/c* injections).

Regulations. ***U.S. FDA*** (1998) approved the use of CAA in adhesives as a component of articles intended for use in packaging, transporting, or holding food in accordance with conditions prescribed in 21 CFR part 175.105.

References:

1. Ismerov, N. F., Sanotsky, I. V., and Sidorov, K. K., *Toxicometric Parameters of Industrial Toxic Chemicals under Single Exposure*, USSR/SCST Commission for UNEP, Center Int. Projects, Moscow, 1982, 33 (in Russian).

2. Smith, M. K., Randall, E. J., Stober, R., and Strober, J. A., Developmental effects of chloroacetic acid in the Long-Evans rat, Abstract, *Teratology*, 41, 593, 1990.
3. Amacher, D. E. and Turner, G. N., Mutagenic evaluation of carcinogens and non-carcinogens in the L5178Y/TK assay utilizing postmitochondrial fractions (S9) from normal rat liver, *Mutat. Res.*, 97, 49, 1982.
4. Hayes, W. J., *Toxicology of Pesticides*, Williams & Wilkins, Baltimore, 1975, 160.

CHLOROHYDROQUINONE

Molecular Formula. $C_6H_5ClO_2$
M = 144.56
CAS No 615-67-8
RTECS No MX4800000
Abbreviation. CHQ.

Synonyms. 2-Chloro-1,4-benzenediol; 1,4-Dihydroxy-2-chlorobenzene.

Properties. White to light-tan, fine crystals. Readily soluble in water and alcohol.

Applications. Bactericide.

Acute Toxicity.

Observations in man. Lethal doses are likely to be 5 to 12 g.[022]

Animal studies. Minimal lethal dose is about 200 mg/kg BW in rats (Kodak Company Reports, 1971).

Short-term Toxicity. CHQ is presumably similar in toxicity to hydroquinone. Exposure to oral doses of 300 to 500 mg for 3 months caused no ill effects.[022]

Mutagenicity.

In vitro genotoxicity. CHQ was negative in *Salmonella* mutagenicity assay.[1]

Reference:

1. Rapson, W. et al., Mutagenicity produced by aqueous chlorination of organic compounds, *Bull. Environ. Contam. Toxicol.*, 24, 590, 1980.

5-CHLORO-2-METHYL-4-ISOTHIAZOLIN-3-ONE

Molecular Formula. C_4H_4ClNOS
M = 149.60
CAS No 26172-55-4
26530-03-0
RTECS No NX8156850
Abbreviation. CMI.

Trade Names. Kathon; Kathon CG.

Applications. Used as a preservative and biocide in cooling water treatments, cosmetics, paints, cleaners, animal dips, and other products.

Allergenic Effect. CMI is a skin sensitizer at concentrations in the ppm range. CMI is one of the most common causes of allergic contact dermatitis. The prevalence of sensitized persons in various studies has ranged from 1 to 6.3%.[1,2]

Reproductive Toxicity. CMI was embryotoxic and fetotoxic but not teratogenic in rabbits.

Mutagenicity.

In vitro genotoxicity. CMI was mutagenic in *Salmonella typhimurium*, and cooling water was mutagenic shortly after treatment with CMI.[3,4]

Kathon biocide, an aqueous solution containing a mixture of *5-chloro-2-methyl-4-isothiazolin-3-one* and *2-methyl-4-isothiazolin-3-one* in approximate ratio of 3:1, produced point mutations in the absence but not in presence of a rat-liver metabolizing system in *Salmonella typhimurium* strain *TA100* and in mouse bone marrow cells in culture. It was negative in the unscheduled DNA synthesis assay in cultured rat hepatocytes, and in the *in vitro* cell transformation assay.[5] CMI did not induce micronuclei in mice.[6]

In vivo cytogenetics. It was found to be negative in *Dr. melanogaster*, and in the cytogenetic assay in mice.[5]

Carcinogenicity. CMI is reported not to be carcinogenic in animal testing.

Regulations. ***U.S. FDA*** (1998) approved the use of CMI in adhesives as a component of articles intended for use in packaging, transporting, or holding food in accordance with the conditions prescribed in 21 CFR part 175.105.

References:

1. Hunziker, N., Pasche, F., Bircher, A, et al., Sensitization to the isothiazolinone biocide, Report of the Swiss Contact Dermatitis Research Group 1988-1990, *Dermatology,* 184, 94, 1992.
2. Perrenoud, D., Bircher, A., Hunziker, T., et al., Frequency of sensitization to 13 common preservatives in Switzerland, *Contact Dermat.,* 30, 276, 1994.
3. Woodall, G. M., Pancorbo, O. C., Blevins, R. D., et al., Mutagenic activity associated with an isothiazolinone biocide used in cooling towers, *Environ. Mutagen.,* 8 (Suppl 6), 92, 1986.
4. Woodall, G. M., Pancorbo, O. C., Blevins, R. D., et al., Mutagenic activity associated with cooling tower waters treated with a biocide containing 5-chloro-2-methyl-4-isothiazolin-3-one, *Environ. Sci. Technol.,* 21, 815, 1987.
5. Scribner, H. E., McCarthy, K. L., Moss, J. N., Hayes, A. W., Smith, J. M., Cifone, M. A., Probst, G. S., and Valencia, R., The genetic toxicology of Kathon biocide, a mixture of 5-chloro-2-methyl-4-isothiazolin-3-one and 2-methyl-4-isothiazolin-3-one, *Mutat. Res.*, 118 129, 1983.
6. Richardson, C. R., Styles, J. A., and Burlinson, B., Evaluation of some formaldehyde-release compounds and other biocides in the mouse micronucleus test, *Mutat. Res.,* 124, 241, 1983.

CHLOROPHENOLS

Abbreviation. CP.

	CAS No	RTECS No
2-Chlorophenol	95-57-8	SK2625000

Molecular Formula. ClC_6H_4OH

M = 128.56

Abbreviation. 2-CP.

Synonym. 1-Chloro-2-hydroxybenzene

2,4-Dichlorophenol	120-83-2	SK2450000

Molecular Formula. $Cl_2C_6H_3OH$

M = 163.00

Abbreviation. 2,4-DCP.

2,4,6-Trichlorophenol	88-06-2	SK2800000

Molecular Formula. $Cl_3C_6H_2OH$

M = 197.44

Abbreviation. 2,4,6-TCP.

Trade Names. Dowicide 25; Omal; Phenachlor.

Properties (*Tables 1 and 2*).

Table 1. Solubility and organoleptic properties.

Properties	*2-CP*	*2,4-DCP*	*2,4,6-TCP*
Water solubility (mg/l)	28,000	4,500	900
Taste threshold (mg/l)	0.0001	0.0003	0.002
Odor threshold (mg/l)	0.01	0.04	0.3

Applications. Used in the production of plasticizers. CP can be produced as a result of chlorination of water containing phenol or lower chlorophenols. May be formed in chlorinated water should phenols from phenolic plastic piping, containers, and coatings migrate in. *o*-CP is a component of the vulcanization accelerator. CP, par ticularly *o*-CP, have an unpleasant, intrusive odor.

Table 2. Odor threshold concentrations of 2-CP, mg/l.[1]

Temperature	*o-CP*	*m-CP*	*p-CP*
25°C	0.002		
30°C	0.0033	0.2	0.033
60°C	0.0025	0.1	0.143

Exposure. The concentrations occurring in drinking water are usually quite low.

Acute Toxicity. In rats, LD_{50} is 670 mg *o*-CP and *p*-CP/kg BW, and 570 mg *m*-CP/kg BW.

2,4-DCP. LD_{50} is 480 to 4500 mg/kg BW for rats,[2,3] and 1000 to 1600 mg/kg BW for mice.[3,4] The liver, CNS, and kidneys are affected.

2,4,6-TCP. LD_{50} is reported to be 770 or 820 mg/kg BW for rats.[5,6] Poisoned animals developed increased body temperature and convulsions.

Repeated Exposure. Sprague-Dawley rats received 13, 64, 129, or 257 mg 2-CP/kg BW in corn oil daily by gavage for 10 consecutive days. Hematological and clinical chemistry, food and water consumption, absolute and relative organ weights, and histopathological findings revealed no compound or sex-related effects.[7]

Administration of 96 mg 2,4-DCP/kg BW to rats resulted in a 50% animal mortality from a total dose of 293 mg/kg BW. After administration, labored respiration, decreased motility, tremor, convulsion, coma, and death of animals were observed. Histology investigation revealed pronounced kidney damage, fatty infiltration of the liver, and intestinal hemorrhages.

2,4,6-TCP. Rats were exposed to a dose of 154 mg/kg BW for a month. The treatment caused no mortality or changes in general condition, behavior or BW gain. There was only some reduction in total gas exchange.

Short-term Toxicity.

2-CP. Sprague-Dawley rats received 17, 50, or 150 mg 2-CP/kg BW in corn oil daily by gavage for 90 consecutive days. There were no significant gross or histopathological findings that were treatment-related to either sex.[7]

2,4-DCP. No differences in BW gain or in terminal organ weights were found in CD-1 mice given doses of 40 to 490 mg/kg BW in their drinking water over a period of 3 months. Hematology changes were observed only in males. Clinical chemistry parameters were altered significantly in females (a decrease in creatinine and an increase in alkaline phosphatase activity).[4]

2,4,6-TCP. Feeding with 80 to 720 mg/kg BW administered in corn oil by gavage for 90 days produced no histopathology changes in Sprague-Dawley rats. The NOAEL of 80 mg/kg BW was identified in this study (Bercz *et al.*, 1989). Sprague-Dawley rats were given doses of 0.3 to 30 mg 2,4,6-TCP/kg BW in drinking water from 3 weeks through breeding (at 90 days), gestation, parturition and lactation. Post-weaning treatment for 12 to 15 weeks established the NOAEL of 0.3 mg/kg BW.[8]

Long-term Toxicity.

2,4-DCP. ICR mice were exposed to 45 to 230 mg/kg BW in the diet. No histopathology changes were observed. The NOAEL of 100 mg/kg BW was defined.[3] Pre- and postnatal treatment of rats with 30 mg/kg BW increased liver and spleen weights; histopathology changes were absent. An exclusively low NOAEL of 0.3 mg/kg BW was established in this study.[9]

2,4,6-TCP. Doses of 250 to 500 mg/kg BW administered to F344 rats, and those of 750 to 1500 mg/kg BW given to mice in the feed for 106 weeks resulted in BW decrease in the dosed groups of animals.[5] Other clinical signs and mortality were common to both dosed and control groups. Reduced liver and kidney weights were reported in a 6-month study in guinea pigs dosed by gavage with 0.5 mg 2,4,6-TCP/kg BW.

Immunotoxicity.

2-CP. The NOAEL of 50 mg/kg BW was identified for the absence of immune (humoral immunity, cell-mediated immunity, and macrophage function) and hematological effects.[10,11]

2,4-DCP. The dose of 30 mg/kg BW enhanced humoral immune response and depressed cell-mediated immunity. Macrophage activity was not altered.[9]

2,4,6-TCP. No immunology effects were observed in mice at a dose level of 30 mg/kg BW.[8]

Reproductive Toxicity.

2-CP.

Embryotoxicity. Rats were mated after 10 weeks of treatment with 0.5 to 50 mg/kg BW. The treatment was continued during mating, gestation, and weaning. The increase in the number of stillbirths as well as the decrease in the size of the litters after exposure to 50 mg/kg BW were reported.[10]

2,4-DCP.

Gonadotoxicity. The doses of 50 to 500 mg/kg BW administered via drinking water to male CD-1 mice for 90 days did not affect sperm motility or ability to penetrate ova.[12]

Embryotoxicity. Female rats were exposed to 0.3 to 30 mg/kg BW via drinking water from 3 weeks through parturition (bred at 90 days), and their progeny weaned at 3 weeks of age. No significant effects on conception, litter size and weight, number of stillborn pups, or survival of weanlings were observed.[9]

Teratogenicity. F344 rats were dosed with 200 to 750 mg 2,4-DCP/kg BW in corn oil on days 6 to 15 of gestation. The highest dose produced maternal toxicity and lowered fetal weights. There were no fetal malformations in any treated group of animals.[13]

2,4,6-TCP.

Embryotoxicity. Oral dose of 1000 mg/kg BW produced gross maternal toxicity (increased mortality, decreased BW gain in the dams) but there were no treatment-related differences in litter survival.[14]

The doses of 3.0 to 30 mg/kg BW and even 500 mg/kg BW were ineffective for reproduction in male and female rats. Reproduction does not appear to be a primary target for the effect of TCP.[9,15]

Teratogenicity. Sprague-Dawley rats were exposed to 2,4,5-TCP by oral gavage on days 6 to 15 of gestation. The treatment produced maternal toxicity and developmental effects.[16]

Mutagenicity.

Observations in man. No data were available on mutagenic or related effects in humans.

2,4-DCP.

In vitro genotoxicity. 2,4-DCP did not show genotoxic potential in *Salmonella* mutagenicity assay or in primary hepatocyte cultures.

2,4,6-TCP.

In vitro genotoxicity. 2,4,6-TCP appeared to be negative in *Salmonella* mutagenicity assay with and without metabolic activation.[025] It did not induce mutation or CA but produced a dose-related increase in hyperploidy and micronuclei. TCP is likely to cause chromosomal malsegregation as its major mode of genotoxic action.[14]

Carcinogenicity.

2-CP significantly altered ethylnitroso-urea-induced tumor incidence and latency in rats.[9]

2,4-DCP. In a 2-year study, doses of 210 to 440 mg 2,4-DCP/kg BW and 430 to 1300 mg/kg BW were given to rats and mice, respectively, in their feed. No evidence of carcinogenicity was reported.[17]

2,4,6-TCP. Mice and rats were given 5000 to 10000 ppm 2,4,6-TCP in the feed for 2 years. An increased rate of lymphomas and leukemias in male rats and of hepatic tumors in mice was observed.[6]

Carcinogenicity classifications. An IARC Working Group concluded that there is limited evidence for the carcinogenicity of CPs, including 2,4,6-TCP, in *humans*.

The Annual Report of Carcinogens issued by the U.S. Department of Health and Human Services (1998) defines 2,4,6-TCP to be a substance which may reasonably be anticipated to be carcinogen, i.e., a substance for which there is a limited evidence of carcinogenicity in *humans* or sufficient evidence of carcinogenicity in *experimental animals*.

IARC: 2B (CPs and 2,4,6-TCP);
U.S. EPA: D (CPs) and B2 (2,4,6-TCP);
EU: 3 (2,4,6-TCP);
NTP: NE - NE - NE - NE (2,4-DCP, feed);
P* - N - P* - P (2,4,6-TCP, feed).

Chemobiokinetics. CPs are well absorbed after oral administration. They do not appear to accumulate in the body tissues of rats. The major metabolite identified is a *glucuronide conjugate* of the parent CP.[18]

2,4-DCP is readily metabolized to *hydroxyphenols* and, before excretion, combines with *glucuronic* and *monochloro-mercapturic acids*.

2,4,6-TCP is, apparently, dechlorinated and combined in the form of *glucuronates* and/or *sulfates*, which are excreted from the body. CP may also combine as *glutathione* after nucleophil attack. CPs are rapidly cleared from the body of rats and excreted in the urine and feces.[17]

Recommendations. ***WHO*** (1991). Guideline values for drinking water:

2-CP: 0.0001 mg/l;

2,4-DCP: 0.0003 mg/l;

2,4,6-TCP : 0.2 mg/l (the lowest reported taste thresholds are 0.002 mg/l).

The levels of 0.0001 to 0.001 mg 2-CP/l, 0.0003 to 0.04 mg 2,4-DCP/l, and 0.002 to 0.3 mg 2,4,6-TCP/l are likely to give rise to consumer complaints of foreign odor and taste.

Standards. ***Russia*** (1995). MAC and PML for 2,4-DCP: 0.002 mg/l. MAC and PML for 2,4,6-TCP: 0.004 mg/l.

References:

1. Grushko, Ya. M., *Hazardous Organic Compounds in the Liquid Industrial Effluents*, Khimiya, Leningrad, 1976, 128 (in Russian).
2. Deichmann, W. B. and Mergard, E. G., Comparative evaluation of methods employed to express the degree of toxicity of a compound, *J. Ind. Hyg. Toxicol.,* 30, 373, 1948.
3. Kobayashi, S., Fukuda, T., Kawaguchi, K., et al., Chronic toxicity of 2,4-dichlorophenol in mice: a simple dosing for checking the toxicity of residual metabolites of pesticides, Abstract, Toho Univ, Japan, *J. Med. Soc.*, 19, 356, 1972 (in Japanese).
4. Borzelleca, J., Hayes, L., Condie, L. M., and Egle, J. L., Acute and subchronic toxicity of 2,4-dichlorophenol in CD_1 mice, *Fundam. Appl. Toxicol.*, 5, 478, 1985.
5. Gabrilevskaya, L. N. and Laskina, V. P., Experimental substantiation of the maximum allowable concentrations for di- and trichlorophenols in water bodies, in *Industrial Pollution of Water Reservoirs*, Meditsina, Moscow, Issue No 9, 1969, 128 (in Russian).
6. *Bioassay of 2,4,6-Trichlorophenol for Possible Carcinogenicity*, Natl. Cancer Institute Technical Report No. 155, Dept. Health, Education & Welfare, Washington, D.C., 1979.
7. Daniel, F. B., Robinson, M., Olson, G. R., York, R. G., Condie, L. W., Ten and ninety-day toxicity studies of 2-chlorophenol in Sprague-Dawley rats, *Drug Chem. Toxicol.*, 16, 277, 1993.
8. Exon, J. H. and Koller, L. D., Toxicity of 2-chlorophenol and 2,4,6-trichlorophenol, in *Water Chlorination: Chemistry, Environmental Impact and Health Effects*, R. Jolley, et al., Eds., vol 5, Lewis Publishers, Inc., 1985, 307.
9. Exon, J. H., Henningsen, J. M., Osborne, C. A., et al., Toxicological, pathological and immunotoxic effects of 2,4-dichlorphenol in rats, *J. Toxicol. Environ. Health*, 14, 723, 1984.
10. Exon, J. H. and Koller, L. D., Effects of transplacental exposure to chlorinated phenols, *Environ. Health Perspect.*, 46, 137, 1982.
11. Exon, J. H. and Koller, L. D., Effects of chlorinated phenols on immunity in rats, *Ind. J. Immunol. Pharmacol.,* 5, 131, 1983.
12. Seyler, D. E., East, J. M., Condie, L. M., et al., The use of *in vitro* methods for assessing reproductive toxicity, Dichlorophenols, *Toxicol. Lett.,* 20, 309, 1984.
13. Rodwell, D. E., Wilson, R. D., Nemec, M. D., et al., Teratogenic assessment of 2,4-dichlorophenol in Fischer 344 rats, *Fundam. Appl. Toxicol.*, 13, 635, 1989.
14. Jansson, K. and Jansson, V., Genotoxicity of 2,4,6-trichlorophenol in V79 Chinese hamster cells, *Mutat. Res.*, 80, 175, 1992.
15. Blackburn, K., Zenic, H., Hope, E., et al., Evaluation of the reproductive toxicology of 2,4,6-trichlorophenol in male and female rats, *Fundam. Appl. Toxicol.*, 6, 233, 1985.
16. Chernoff, N., Setzer, R. W., Miller, D. B., Rosen, M. B., and Rogers, J. M., Effects of chemically induced maternal toxicity on prenatal development in the rat, *Teratology*, 42, 651, 1990.
17. *Toxicology and Carcinogenesis Studies of 2,4-Dichlorophenol in F344/N Rats and* $B6C3F_1$ *Mice,* NTP Technical Report Series No 353, Research Triangle Park, NC, 1989.

18. Carpenter, H. M., Bender, R. C., and Buhler, D. R., Absorption, metabolism and excretion of 2- and 4-chlorophenol in rats, Abstract, *Toxicologist*, 5, 109, 1985.

3-CHLORO-1,2-PROPANEDIOL

Molecular Formula. $C_7H_7ClO_2$
M = 110.55
CAS No 96-24-2
RTECS No TY4025000

Synonyms. 3-Chloro-1,2-dihydroxypropane; α-Chlorohydrine; Glycerol α-chlorohydrin.

Properties. An oily liquid with a characteristic odor. Soluble in water and alcohol. Odor perception threshold is 3.1 mg/l; taste perception threshold is 0.7 mg/l.

Applications. Used in the production of epoxy resins, polysulfide rubbers, and as a solvent.

Acute Toxicity. In mice, LD_{50} is reported to be 135 mg/kg BW. Interspecies differences in sensitivity (rat, mouse, rabbit, guinea pig) are not observed. The CNS is predominantly affected. Gross pathology examination revealed congestion in the viscera.[1]

Repeated Exposure revealed slight cumulative properties. Some animals died after daily exposure to 1/10 LD_{50} for two weeks.

Short-term Toxicity. Administration of 1/5 LD_{50} over a period of three months caused no mortality in rats. Several mice, but not all, died when given 1/10 LD_{50} for 12 to 15 weeks.[1]

Long-term Toxicity. The NOAEL should not be lower than that of the more toxic *glycerol dichlorohydrin* (0.5 mg/kg BW).[1]

Reproductive Toxicity.

Gonadotoxicity. Male CD rats were administered by gavage with 1.0, 5.0, or 25 mg/kg BW over of a period of 2 weeks. At the highest dose, testicular and epididymal lesions were noted. Sperm motion was altered; decreased epididymal sperm concentrations and increased breakage were reported.[2]

C. exerted antiandrogenic action in male hedgehogs. It inhibited RNA synthesis, *protein*, and *sialic acid* in testis, epididymis, and seminal vesicle and decreased fructose concentration in latter.[3]

Standards. ***Russia*** (1995). Recommended MAC and PML: 0.7 mg/l (organolept., taste).

References:

1. Krasovsky, G. N. and Friedland, S. A., in *Industrial Pollution of Water Reservoirs*, S. N. Cherkinsky, Ed., Meditsina, Moscow, Issue No. 8, 1967, 129 (in Russian).
2. Hoyt, J. A., Fisher, L. F., Hoffman, W. P., Swisher, D. K., and Seyler, D. E., Utilization of a short-term male reproductive toxicity study design to examine effects of α-chlorohydrin (3-chloro-1,2-propanediol), *Reprod. Toxicol.*, 8, 237, 1994.
3. Dixit, V. P. and Lohiya, N. K., Mechanism of action of alpha-chlorohydrin on the testes and caput epididymis of rat, gerbil (*Meriones Hurrianae*), bat, and mouse, *Acta Anat.* (Basel), 95, 50, 1976.

4-CHLORO-3,5-XYLENOL

Molecular Formula. C_8H_9ClO
M = 156.62
CAS No 88-04-0
RTECS No ZE6850000

Synonyms and **Trade Name.** Benzytol; 4-Chloro-3,5-dimethylphenol; 2-Chloro-5-hydroxy-1,3-dimetylbenzene; 4-Chloro-1-hydroxy-3,5-dimethylbenzene; 2-Chloro-5-hydroxy-*m*-xylene; *p*-Chloro-*m*-xylenol.

Properties. Crystals from benzene with phenolic odor. Solubility is 333 mg/l at 20°C, more soluble in hot water. Soluble in 95% alcohol and in fixed oils.

Applications. An antiseptic and germicide.

Acute Toxicity. LD_{50} is 3.83 g/kg BW in rats. Administration of large doses produced changes in GI tract, impairment of the respiratory system and kidney functions.[1]

Reproductive Toxicity.

Embryotoxicity and ***Teratogenicity.*** Rats were given C. during days 1 to 19 of pregnancy. The treatment caused fetotoxicity. Abnormalities were observed in the musculoskeletal system.[2]

Allergenic Effect. Contact dermatitis was found in workers occupationally exposed to C.[3]

Mutagenicity.

In vitro genotoxicity. Showed mutagenic activity in *E. coli* strains on ochratoxin A.[4]

Chemobiokinetics. Mongrel dogs received *i/v* and oral single doses of 200 and 2000 mg C., respectively. Low range of absorption was noted. Kidneys were not the major route for rapid elimination of unchanged C. Main metabolites found in the urine were *glucuronides* and *sulfates.*[5]

Regulations. ***U.S. FDA*** (1998) approved the use of C. in adhesives as a component of articles intended for use in packaging, transporting, or holding food in accordance with the conditions prescribed in 21 CFR part 175.105.

References:

1. Acute Toxicity Data, *J. Am. Coll. Toxicol.*, 4, 147, 1985.
2. *Ann. Rep. Osaka City Inst. Publ. Health Environ. Sci.*, 45, 100, 1983.
3. Libow, L. F., Ruzkowski, A. M., and De Leo, V. A., Allergic contact dermatitis from *p*-chloro-3,5-dimethylphenol in Lurosep soap, *Contact Dermat.*, 20, 67, 1989.
4. Malaveille, C., Brun, G., and Bartsch, H., Structure-activity studies in *E. coli* strains on ochratoxin A (OTA) and its analogues implicate a genotoxic free radical and cytotoxic thiol derivatives as reactive metabolites, *Mutat. Res.*, 307, 141, 1994.
5. Dorantes, A. and Stavchansky, S., Pharmacokinetics and metabolic disposition of *p*-chloro-*m*-xylenol (PCMX) in dogs, *Pharmcol. Res.*, 9, 677, 1992.

CITRIC ACID

Molecular Formula. $C_6H_8O_7$

M = 192.12

CAS No 77-92-9

RTECS No GE7350000

Abbreviation. CA.

Synonyms and **Trade Name.** Anhydrous citric acid; Citretten; 1,2-Hydroxy-1,2,3-propanetricarboxylic acid; β-Hydroxytricarboxilic acid.

Properties. Colorless, translucent, odorless crystals, or white, granular to fine crystalline powder with strongly acid taste. Solubility in water is 59.2% at 20°C,[020] soluble in alcohol. Occurs in living animal cells.

Applications. Used in the preparation of foods and medicines, in the manufacture of alkyd resins.

Exposure. CA is a normal component of the body and is found naturally in many foods. The average daily intake has been estimated to be greater than 500 gm/kg BW.[1]

Acute Toxicity. CA has generally been nontoxic in experimental animals.

Allergenic Effect. CA has been reported to have allergenic properties; it could cause contact dermatitis.[1]

Mutagenicity.

In vitro genotoxicity. CA was not mutagenic in *Salmonella typhimurium* in the presence or absence of metabolic activation, and did not induce CA in cultured Chinese hamster fibroblast cells.[1]

Reproductive Toxicity.

Embryotoxicity. Rats received 1.2% CA in their diet over 2 generations. The treatment did not affect reproduction. CA did not affect litter size or survival of mice with prenatal exposure to up to 5% CA in the diet.[1]

Teratogenicity. Administration of large doses of citrate during pregnancy caused no teratogenic effect in the offspring of mice, rats, hamsters, or rabbits.

Observations in man. There was no increase in incidence of birth defects when drugs containing potassium citrate were used by women during pregnancy.[2]

Animal studies. The CA sodium salt did not produce developmental abnormalities in rats.[3] CA potassium salt was not associated with an increased incidence of birth defects.[2]

Gonadotoxicity. CA treatment of human sperm *in vitro* renders them immotile; however, this effect is also seen with other organic acids and is probably a nonspecific manifestation of *pH* alteration.[4]

Long-term Toxicity. Two successive generations of rats received 1.2% CA in their diet over a period of 90 weeks. The treatment caused slight increase in dental attrition. 5.0% CA level in the diet did not affect food consumption but produced retardation in BW gain and reduced survival time in mice.[03] Rabbits tolerated up to 7.7% CA in the diet with no effect.[1]

Chemobiokinetics. Ingestion of CA may impair absorption of calcium and iron.[03] Metabolites appeared to be *oxaloacetic* and *acetic acids* (Fenaroli, 1975).

Regulations.

Great Britain (1998). CA is authorized without time limit for use in the production of polymeric materials and articles in contact with food or drink or intended for such contact.

U.S. FDA (1998) affirmed CA as *GRAS.*

Recommendations. ***EU*** (1995). CA is a food additive generally permitted for use in foodstuffs.

References:

1. Patty's *Industrial Hygiene and Toxicology*, G. D. Clayton and F. E. Clayton, Eds., vol 2E, *Toxicology*, 4th ed., John Wiley & Sons, New York, NY, 1994, 3587.
2. Mellin, G. W., Drugs in the first trimester of pregnancy and fetal life of Homo sapiens, *Am. J. Obstet. Gynecol.*, 90, 1169, 1964.
3. Nolen, G. A., Bohne, R. L., and Buehler, E. V., Effects of trisodium nitrilotriacetate, trisodium citrate, and a trisodium nitrilotriacetate-ferric chloride mixture on cadmium and methyl mercury toxicity and teratogenesis in rats, *Toxicol. Appl. Pharmacol.*, 23, 238, 1972.
4. Brown-Woodman, P. D., Post, E. J., Chow, P. Y., and White, I. G., Effects of malonic, maleic, citric, and caffeic acids on the motility of human sperm and penetration of cervical mucus, *Int. J. Fertil.*, 30, 38. 1985.

COPPER compounds

	CAS No	RTECS No
Copper chloride	7447-39-4	GL7000000
Molecular Formula. $CuCl_2$		
M = 134.44		
Copper naphthenate	1338-02-9	OK9100000
Copper nitrate	3251-23-8	QU7400000
Molecular Formula. $N_2O_6.Cu$		
M = 187.56		

Properties. The amounts of copper entering liquid media will be greater, the more acidic the media is and the lower its calcium contents. Concentration of 0.5 mg/l colors water, 1.0 mg/l increases its turbidity, and 1.5 mg/l affects the taste (astringent after-taste). In distilled water the after-taste is detected at a concentration of 2.6 mg/l.[1] Below concentration of 1.0 mg/l the taste is usually acceptable.

Applications. Copper compounds (powdered) are used in the synthesis of a number of plastics.

Migration Data. Copper plumbing may result in its increased concentrations in drinking water.

Exposure. Copper compounds are essential micronutrients. Copper is a potential toxic metal. Daily intake of copper is 2.0 to 5.0 mg, mainly in food.

Acute Toxicity.

Observations in man. I/g administration of 25 g copper led to clinical signs of poisoning in humans; the same result is observed following 10 g administered to babies.[2]

Acute gastric irritation may be observed in some individuals at concentrations above 3.0 mg/l. Doses of 10 to 20 mg/kg BW may cause nausea and vomiting, so poisoning after ingestion of copper is unlikely to occur.[3]

The acute GI effects were observed in healthy, adult women who received copper at the concentrations of 1.0, 3.0, and 5.0 mg copper/l in their drinking water for a 2-week study period, followed by 1 week of standard tap water. Twenty-one subjects (35%) recorded GI disturbances sometime during the study, nine had diarrhea, some with abdominal pain and vomiting, and 12 subjects presented abdominal pain, nausea, or vomiting. There was no association between aggregate copper in drinking water within the range of 0 to 5 mg/l and diarrhea, but a greater than/equal to 3.0 mg copper/l level was associated with nausea, abdominal pain, or vomiting.[4]

Animal studies. Oral toxicity depends on the animal species and copper anion. LD_{50} values (in mA/kg BW) are 1.900 for copper sulfate, 0.820 for copper chloride, and 3.440 for copper nitrate. According to other data, LD_{50} values for these copper salts are 200, 140, and 940 mg/kg BW, respectively.[5]

Ingestion of excessive copper with water or food leads to mucosal irritation and erosion, extensive capillary lesions, to ultimately necrotic lesions of the liver and kidneys, and to excitation, followed by depression, of the CNS.

Repeated Exposure.

Observations in man. Cases of copper poisoning occurring during the treatment of kidney diseases by hemodialysis were reported.

Animal studies. Administration of copper sulfate to Wistar rats at a dose of 100 mg/kg BW (calculated as Cu^{2+}) caused retardation of BW gain and by chemical changes in the liver after 30 administrations (Dinu and Boghianu, 1977). 20 administrations of this dose may produce anemia. Gross pathology examination revealed necrotic changes in the liver and renal edema and tubular necrosis.[6]

Short-term Toxicity. Oral administration of 10 to 30 mg copper/kg BW or several months was harmless to rats, but 60 to 100 mg/kg BW doses caused nausea and vomiting.[5]

Long-term Toxicity.

Observations in man. Humans seem to be as sensitive to copper as the most sensitive of the animal species investigated, that is, sheep. In both man and sheep, the LOEL is 0.2 to 1.0 mg copper/kg BW. Even 5.0 mg/person or 0.07 mg/kg BW has no effect on the body, other than in people predisposed to Wilson's disease (a combination of liver and brain pathology). Poisonings caused by babies' food prepared on drinking water with high copper concentration (15.5 mg copper/l) were reported in Germany.[2]

Animal studies. Rats were given the doses of 20 to 30 mg copper/kg BW. Incidences of liver and kidney necrosis, and changes in blood and testicular degeneration were observed in a 44-week study.[7] Similar effects could be observed in pigs after oral administration of less than 10 mg/kg BW.[8,9]

In sheep, a daily intake of 1.0 to 2.0 mg copper/kg BW led to such serious adverse effects as hepatitis, fatty liver dystrophy, and renal tubular degeneration (resulting in death).[8] In chronic poisoning of sheep with the dose of 15 mg/kg BW, both the glomerular and the tubular functions of the kidney were impaired.[07]

Reproductive Toxicity.

Teratogenicity. More recent data prove that oral administration of 4.0 mg copper gluconate/kg BW to mice and rats caused no embryotoxic and teratogenic effects.[7]

Injection of 10 mg copper naphthenate/kg BW into rats on days 1 through 15 of gestation produced no *terata*, fetal, or maternal toxicity.[10]

Gonadotoxicity and ***Embryotoxicity.*** Rats were given drinking water containing 1.0 mg copper/l and 10 mg copper/l for 5 months and then throughout pregnancy. The treatment resulted in increased embryolethality and sharply reduced fertility.[11]

Allergenic Effect. Contact sensitizing potential is not found in maximized guinea-pig assays with copper sulfate and copper chloride.[12] Isolated cases of immediate type hypersensitivity reactions (eczematous and urticarial eruptions) were considered to be occasional (risk 1:100).

Mutagenicity.

Observations in man. Plasma copper level was found to be significantly higher in forty women using the copper-containing intrauterine contraceptive device than in that of controls. Shubber et al. found a positive correlation between the long term use of this contraceptive device and DNA damage in the host somatic cells.[13]

In vitro genotoxicity. Like all heavy metals, copper is a weak mutagen and its compounds cause CA in animal and plant cells. The genotoxicity of copper compounds has been reported in animal cell cultures.[14,15]

Tinwell and Ashby reported negative effect in the mouse bone marrow micronucleus assay.[16]

In vivo cytogenetics. The toxic doses of copper are reported to increase the frequency of DLM.[11] Its clastogenic effects have been observed in mice *in vivo.*[17]

Carcinogenicity. Rats were given a dose of 1000 mg copper hydroxyquinoline/kg BW (~181 mg copper/kg BW/day) over a period of 3 weeks. Up to their 78th week of life, animals were then given 25 to 50 mg copper/kg BW. In this study, no significant increase in the rate of tumors could be observed.[07,14]

There were no tumors in rabbits following administration of 12.5 mg copper/kg BW every 2 days (observation up to 16 months).[8]

Carcinogenicity classification.

U.S. EPA: D.

Chemobiokinetics. Copper serves as a co-enzyme for a number of enzymes with oxidase activity: ceruloplasmin, monoaminooxidase, superoxide dismutase, etc. Copper participates in erythrocyte formation and the release of tissue iron. In acute copper intoxication, copper ions exhibit prooxidant activity and capacity to produce lipoperoxidation but, at pathophysiological concentrations, do not induce hemolysis by an oxidative mechanism.[18] Copper metabolism is complex and inadequately studied. In adults, the absorption and retention rates depend on the daily intake and, as a consequence, copper overload is unlikely: with increased ingestion of copper, its resorption is reduced. In contrast, copper metabolism in infants is not well developed and the liver of the newly born child contains over 90% of the body burden with much higher levels than in adults. Long-term ingestion can give rise to liver cirrhosis. Copper ions cause *Hb* autoxidation, which results in the formation of *MetHb* and O^{2-}.[15] This mechanism may induce lipid peroxidation because of the occurrence of reactive oxygen species.[16] Despite copper prooxidant activity and its capacity to produce lipoperoxidation, it is unlikely that copper ions in pathophysiological concentrations induce hemolysis by an oxidative mechanism.[17]

Copper is excreted mainly by the kidneys or in the feces.

Regulations. ***U.S. FDA*** (1998) regulates copper naphthenate as an accelerator in the manufacture of cross-linked polyester resins. Copper gluconate meets the specifications of the Food Chemical Codex. It is used in food with no limitations other than current GMP. It is listed as *GRAS* and as a direct human food ingredient. Copper dimethyldithiocarbamate is approved for use as an accelerator for rubber articles intended for repeated use in contact with food (total not to exceed 1.5% by weight of the rubber product) in accordance with the conditions prescribed in 21 CFR part 177.2600. Copper acetate is listed for use as an antioxidant and/or stabilizer in polymers used in the manufacture of food-contact articles in accordance with the conditions prescribed in 21 CFR part 178.2010.

See also ***Copper sulfate, pentahydrate.***

Standards.

U.S. EPA (1998). Proposed MCL and MCLG (action level): 1.3 mg/l, SMCL: 1.0 mg/l.

Canada (1989). MAC: 1.0 mg/l (organolept.)

Russia (1992). State Standard for Drinking Water: 1.0 mg/l; MAC in milk and diary produce: 1.0 mg/kg; MAC in meat and poultry: 5.0 mg/kg; MAC in mineral water 1.0 mg/kg; MAC in children's food 1.0 mg/kg.

Recommendations.

WHO (1996). Provisional guideline value for drinking water: 2.0 mg/l. However, 1.0 mg/l is likely to give rise to consumer complaints of staining of laundry and sanitary ware.

The ***U.S. EPA*** current MCLG is not adequately protective for infants and children under 10 years of age. Infants and children up to 10 years of age have increased susceptibility to copper toxicity. The proposed drinking water guideline (0.3 mg/liter) for copper will adequately protect health of infants, children, and adults.[19]

Joint FAO/WHO Expert Committee on Food Additives (1982) proposed Provisional Maximum TDI (PMTDI) of 0.5 mg/kg BW.

References:

1. Cohen, J. M., Kamphake, L. J., Harris, E. K., et al., Taste threshold concentrations of metals in drinking water, *J. Am. Water Work Assoc.*, 52, 660, 1960.
2. Ohse, R. and Strubelt, O., Drinking water and copper, *Pediat. Prax.*, 44, 571, 1992.
3. *Guidelines for Drinking-Water Quality*, vol 2, Health Criteria and Other Supporting Information, WHO, Geneva, 1996, 219.

4. Pizarro, F., Olivares, M., Uauy, R., Contreras, P., Rebelo, A., and Gidi, V., Acute gastrointestinal effects of graded levels of copper in drinking water, *Environ. Health Perspect.*, 107, 117, 1999.
5. Grushko, Ya. M., *Hazardous Inorganic Compounds in the Liquid Industrial Effluents*, Khimiya, Leningrad, 1976, 75 (in Russian).
6. Rana, S. V. and Kumar, A., Liver and kidney function in molibden and copper poisoning, *Arch. Hig. Rada. Toksikol.*, 34, 9, 1983 (in Polish).
7. Jannus, J. A., Canton, J. H., van Gestel, C. A. M., et al., *Copper Effects,* Integrated Criteria Document, Natl. Institute Publ. Health and Environ. Hyg., Bilthoven, The Netherlands, Appendix to Report No 758474009, June 1989.
8. Copper, in *Toxicological Evaluation of Certain Food Additives*, Joint FAO/WHO Expert Committee on Food Additives, Food Additives Series No 17, WHO Techn. Rep., 1986, 265.
9. Kline, R. D., Hays, V. M., and Cromwell, G. L., Effects of copper, molybdenum and sulfate on performance, hematology and copper store of pigs and lambs, *J. Anim. Sci.,* 33, 771, 1971.
10. Hardin, B. D., Bond, G. P., Sikov, M. R., Andrew, F. D., Beliles, R. P., and Niemeier, R. W., Testing of selected workplace chemicals for teratogenic potential, *Scand. J. Work Environ. Health,* 7, 66, 1981.
11. Nadeyenko, V. G., Borzunova, E. A., Seliakina, K. P., et al., Embryotoxic and gonadotoxic effects of copper compounds, *Gig. Sanit.*, 3, 8, 1980 (in Russian).
12. Karlberg, A. T., Boman, A., and Wahlberg, J. E., Copper - a rare sensitizer, *Contact Dermat.*, 9, 134, 1983.
13. Shubber, E., Amin, N. S., El-Adhami, B. H., Cytogenetic effects of copper-containing intrauterine contraceptive device (IUCD) on blood lymphocytes, *Mutat. Res.*, 417, 57, 1998.
14. Flessel, C. P., Metals as mutagens, in *Inorganic and Nutritional Aspects of Cancer*, G. N. Schrauzer, Ed., Plenum Press, New York, 1978, 117.
15. Hollstein, M., McCann, J., Angelosanto, F. A., and Nichols, W. W., Short term tests for carcinogens and mutagens, *Mutat. Res.*, 65, 141, 1979.
16. Tinwell, H. and Ashby, J., Inactivity of copper sulfate in a mouse bone-marrow micronucleus assay, *Mutat. Res.*, 245, 223, 1990.
17. Bhunya, S, P. and Pati, P. C., Genotoxicity of an inorganic pesticide, copper sulfate in mouse in vivo test system, *Cytologia*, 52, 801, 1987.
18. Pirion, A., Tallineau, C., Chahboun, S., et al., Copper- induced lipid peroxidation and hemolysis in whole blood: evidence for a lack of correlation, *Toxicology*, 47, 351, 1987.
19. Sidhu, K. S., Nash, D. F., and McBride, D. E., Need to revise the national drinking water regulation for copper, *Regul. Toxicol. Pharmacol.*, 22, 95, 1995.

COPPER 8-HYDROXYQUINOLINE

Molecular Formula. $C_{18}H_{12}CuN_2O_2$
M = 351.86
CAS No 10380-28-6
RTECS No VC5250000
Abbreviation. CHQ.

Synonyms and **Trade Names.** Bioquin; Bis(8-oxyquinoline)copper; Celluquin; Copper oxinate.

Applications. Fungicide in plastic production.

Acute Toxicity. LD_{50} is 9.93 g/kg BW in rats, and 3.94 in mice. Poisoning caused CNS depression, dyspnea, hypomotility, and diarrhea.[1]

Repeated Exposure. K_{acc} is 2.3 (by Lim). Rats received 30 *i/g* administrations of 1.6 g CHQ/kg BW. The treatment caused retardation in BW gain and death of 50% of the animals. Clinical observation and analysis revealed changes in blood morphology and liver functions.

Long-term Toxicity. Rats were given 1500 or 3000 ppm HQ in their diet for 721 days. The treatment produced organ weight changes. The LOEL of 143 mg/kg BW was established.[2] Munro et al. (1996) suggested the calculated NOEL of 73 mg/kg BW for this study.[3]

Mutagenicity. CHQ was positive in *Salmonella*/microsome test, but gave negative results in the *BASC test* on *Dr. melanogaster* and in micronucleus test on mouse bone marrow.[4]

Carcinogenicity. On *s/c* injections to mice for 39 weeks, CHQ has been found to be an equivocal agent by RTECS criteria.

Regulations. *U.S. FDA* (1998) approved the use of CHQ in adhesives as a component of articles intended for use in packaging, transporting, or holding food in accordance with the conditions prescribed in 21 CFR part 175.105.

References:

1. Resnichenko, A. K., Vasilenko, N. M., Muzhikovsky, G. L., and Krasnorutskaya, Ye. P., Toxicity of paraoxybenzoic acid, 3,5-dinitrosalicylic acid, *ortho*-nitroanizole, 2-oxy-1-naphthaldehyde, copper 8-quinolate, polyether P2200, and disperse colorants anthraquinone-1-nitro-2-carbonic acid and 1,5-naphthylaminosulfoacid, *Gig. Sanit.,* 1, 85, 1986 (in Russian).
2. *Toxicology and Carcinogenesis Studies of 8-Hydroxyquinoline (CAS No 148-24-3) in F344/N Rtas and B6C3F$_1$ Mice (Feed Studies),* NTP Technical Report Series No 276, Research Triangle Park, NC, 1985.
3. Munro, I. C., Ford, R. A., Kennepohl, E., and Sprenger, J. G., Correlation of structural class with No-Observed-Effect Levels: A proposal for establishing a threshold of concern, *Food Chem. Toxicol.,* 34, 829, 1996.
4. Gocke, E., King, M.-T., Eckhardt, K., and Wild, D., Mutagenicity of cosmetic ingredients licensed by the European Communities, *Mutat. Res.*, 90, 91, 1981.

COPPER SULFATE

Molecular Formula. $SO_4.Cu$
M = 159.60
CAS No 7758-98-7
RTECS No GL8800000

Copper(2$^+$) sulfate, pentahydrate (1:1:5)

Molecular Formula. $SO_4.Cu.5H_2O$
CAS No 7758-99-8
RTECS No GL8900000
Abbreviation. CS.

Synonym and **Trade Names.** Cupric sulfate; Bluestone; Blue vitriol; Roman vitriol; Salzburg vitriol.

Properties. CS is usually used in *pentahydrate* form. In this form it occurs as large, deep-blue or ultramarine, triclinic crystals, as blue granules, or as a light-blue powder.

Applications. Used in industry, agriculture, and veterinary medicine. A feed additive.

Exposure. Copper, in trace amounts, is essential for life, while, in excess, it is toxic.

Acute Toxicity. LD_{50} is 200 mg/kg BW.[1] Ingestion of excessive copper with water or food leads to mucosal irritation and erosion, extensive capillary lesions, to ultimately necrotic lesions of the liver and kidneys, and to excitation, followed by depression, of the CNS.

Repeated Exposure. No cumulative properties are established. In the 2-week drinking water studies, rats and mice received 300 to 30,000 ppm CS for 15 days. Clinical signs of toxicity observed were attributed to dehydration. The only gross or microscopic change was found to be an increase in the size and number of cytoplasmic protein droplets in the epithelium of the renal proximal convoluted tubules in male rats in the 300 and 1000 ppm CS groups.[5]

In the 2-week feed studies, rats and mice were fed diets containing 1000 to 16,000 ppm CS. No increase in mortality was noted. Retardation of BW gain was observed in the two highest dose groups. Histological examination revealed hyperplasia with hyperkeratosis of the squamous epithelium in the stomach. Inflammation of the liver occurred in rats.[4]

In a 1-month study, Wistar rats received 100 mg CS/kg BW (calculated as copper). The treatment caused retardation of BW gain and biochemical changes in the liver.[2]

Short-term Toxicity. A dose as low as 0.1 mg CS/kg BW given to rats alters the bilirubin level and reduces hippuric acid excretion. Erythrocyte hemolysis is noted. A rise in the concentration of albumens in the urine has been considered as a sign of renal dysfunction.[3]

In the 13-week feed studies, rats received diets containing 500 to 8000 ppm CS, and mice were given 1000 to 16,000 ppm CS for 92 days. The treatment caused no increase in mortality and no clinical signs of toxicity in rats and mice. In mice, there was a dose-related decrease in liver weights. There were changes in hematology, clinical chemistry, and urinalysis of rats in 4000 and 8000 ppm groups. Microcytic anemia with a compensatory bone marrow response was noted. Histological examination revealed hepatocellular and renal tubular epithelial damage in males. Rats in the three highest groups had hyperplasia and hyperkeratosis of the forestomach, and inflammation of the liver. The NOEL for evidence of histological injury to the kidney of females was 500 ppm.[4,5]

Long-term Toxicity. Zlateva (1978) observed coronary sclerosis, focal and diffuse myocardial sclerosis, and other effects in rats given 1/100 and 1/20 LD_{50} of CS for 6 months. In a 1-year rat study, males received copper orally at doses of 0.01, 0.1, and 1.0 mg/kg BW. Higher doses disturbed protein and electrolyte exchange in the body (*Na* and *K*), and vitamin *C* level in the blood.

Reproductive Toxicity. In the above-cited study, CS was found to produce no adverse effects on any parameters measured in rats or mice of either sex.[4]

Teratogenicity. A month prior to mating, mice received CS in their feed (at concentrations of 500 to 4000 ppm). There was a positive effect on the litter size and fetal weight up to the concentration of 1000 ppm (~52 mg/kg BW). At higher concentrations, fetal malformations were found.[6,7]

Allergenic Effect. Contact sensitizing potential is not found in maximized guinea pig assays with CS.[8]

Mutagenicity.

A dose-related increase in chromosome damage in the bone marrow cells or peripheral blood cells was noted in chicks given orally the dose of 10 mg/kg BW.[9]

In vitro genotoxicity. CS was negative in *Salmonella typhimurium*,[10] and did not induce gene conversion and reverse mutations in *Saccharamyces cerevisiae*.[10] It increased the frequency of CA in human cells culture.[12]

Carcinogenicity. There is no direct positive correlation between copper exposure and cancer.[1,13]

Chemobiokinetics. Dose-related increases in copper occurred in all male rats tissues examined, accompanied by increases in *zinc* in the liver and kidney. Plasma *calcium* was significantly reduced in 4000 and 8000 ppm GS groups.[4, 13]

Regulations. ***U.S. FDA*** is developing food-grade specification for CS. In the interim, this ingredient must be of a purity suitable for its intended use. In accordance with CFR requirements, CS is used with no limitation other than current GMP. It is listed as *GRAS* and as a direct human food ingredient.

References:

1. Howell, J. S., The effect of copper acetate on *p*-dimethylaminoazobenzene carcinogenesis in rat, *Br. J. Cancer*, 12, 594, 1958.
2. Dinu, I. and, Boghianu, L., *Stud. Sci. Cers. Biochim.*, 20, 143, 1977 (in Italian).
3. Rana, S. V. and Kumar, A., Liver and kidney function in molibdenum and copper poisoning, *Arch. Hig. Rada. Toksikol.*, 34, 9, 1983 (in Polish).
4. *Toxicity Studies of Cupric Sulfate Administered in Drinking Water and Feed to F344/N Rats and B6C3F*$_1$ *Mice,* NTP Technical Report Series No 29, Research Triangle Park, NC, July 1993.
5. Herbert, C. D., Elwell, M. R., Travlos, G. S., Fitz, C. J., and Bucher, J. R., Subchronic toxicity of cupric sulfate administered in drinking water and feed to rats and mice, *Fundam. Appl. Toxicol.*, 21, 461, 1993.
6. Vodichenska, Ts., The biological effect of copper taken into the body via the drinking water, *Probl. Khigiene,* 13, 29, 1988 (in Bulgarian).
7. *Summary Review of the Health Effects Associated with Copper: Health Issue Assessment,* U.S. EPA, PA 600/8-87/001, Cincinnati, 1987.
8. Karlberg, A. T., Boman, A., and Wahlberg, J. E., Copper - a rare sensitizer, *Contact Dermat.*, 9, 134, 1983.

9. Bhunya, S. P. and Jena, G. D., Clastogenic effect of copper sulfate in chick *in vivo* test system, *Mutat. Res.*, 367, 57, 1996.
10. De Flora, S., Zannacchi, P., Camoriano, A., Bennicelli, C., and Badolati, G. S., Genotoxic activity and potency of 135 compounds in the Ames reversion test and in bacterial DNA-repair test, *Mutat Res.*, 133, 161, 1984.
11. Singh, I., Induction of gene conversion and mitotic gene conversion by some metal compounds in *Saccharomyces cerevisiae*, *Mutat. Res.*, 117, 149, 1983.
12. Jokhadze, T. A. and Lezhava, T. A., *In vivo* and *in vitro* studies of chromosome structural mutations induced by heavy metals' salts during aging, *Genetics*, 30, 1630, 1994 (in Russian).
13. Linder, M. C., Changes in the distribution and metabolism of copper in cancer, A review, *J. Nutr. Growth Cancer*, 1, 27, 1983.

CORN OIL

CAS No 8001-30-7
RTECS No GM4800000
Abbreviation. CO.
Trade Names. Maise oil; Mazola oil.

Properties. Pale-yellow or brownish-yellow, rather viscid liquid with slight disagreeable odor. Slightly soluble in alcohol.[020]

Long-term Toxicity. Male rats receiving a CO vehicle exhibited a higher incidence of pancreatic proliferative lesions and a lower incidence of mononuclear cell leukemia than untreated control males.

Doses of 0.5 g *dichloromethane*/kg BW were administered in 2.5, 5.0, or 10 ml CO/kg BW by gavage to male F344/N rats for 2 years. DCM was chosen because it appeared to cause pancreatic proliferative lesions when administered by gavage in CO vehicle. The treatment caused hyperplasia and adenoma of the exocrine pancreas, decreased incidences of mononuclear cell leukemia, and reduced incidences or severity of nephropathy in male F344/N rats. Further, the use of CO as a gavage vehicle may have a confounding effect on the interpretation of chemical-induced proliferative lesions of the exocrine pancreas and mononuclear cell leukemia in male F344/N rats.[1]

Reproductive Toxicity.

Gonadotoxicity. Ingestion of CO did not affect testicular histology[2] or estrous cycling, including measurements of prolactin and steroid hormones.[3]

Embryotoxicity. Rats received 20% CO in the diet on 10 through 22 day of pregnancy and on 23rd day after birth. The treatment caused metabolic and biochemical changes in offspring.[4]

Teratogenicity. Mice received 0.25 or 0.5 ml CO on days 15 to 17 of pregnancy. The treatment caused specific developmental abnormalities in the immune and reticuloendothelial systems.[5]

Mutagenicity.

In vitro genotoxicity. CO was not mutagenic in *Salmonella typhimurium* strains *TA97, TA98, TA100*, or *TA1535* with or without *S9*.[1]

Regulations. ***U.S. FDA*** (1998) approved the use of CO (1) in the manufacture of resinous and polymeric coatings for the food-contact surface of articles intended for use in producing, manufacturing, packing, processing, preparing, treating, packaging, transporting, or holding food in accordance with the conditions prescribed in 21 CFR part 175.300; and (2) as a defoaming agent in the manufacture of paper and paperboard intended for use in packaging, transporting, or holding food in accordance with the conditions prescribed in 21 CFR part 176.210.

References:

1. *Comparative Toxicology Studies of Corn Oil, Sunflower Oil, and Tricaprylin in male F344/N rats as Vehicles for Gavage*, NTP Technical Report Series No 426, Research Triangle Park, NC, April 19, 1994.
2. Alexander, J. C., Valli, V. E., and Chanin, B. E., Biological observations from feeding heated corn oil and heated peanut oil to rats, *J. Toxicol. Environ. Health*, 21, 295, 1987.

3. Wetsel, W. C., Rogers, A. E., Rutledge, A., and Leavitt, W. W., Absence of an effect of dietary corn oil contents on plasma prolactin, progesterone, and 17-beta-estradiol in female Sprague-Dawley rats, *Cancer Res.*, 44, 1420, 1984.
4. Karnik, H. B, Sonawane, B. R., Adkins, J. S., and Mohla, S., High dietary fat feeding during perinatal development of rats after hepatic drug metabolism of progeny, *Develop. Pharmacol. Ther.*, 14, 135, 1989.
5. Shipman, P. M. and Schmidt, R. R., Corn oil modulates immune function: altered postnatal immune function in mice following its prenatal administration, Abstract, *Teratology*, 29, 57A,1984.

COTTONSEED OIL

CAS No 8001-29-4
RTECS No GN2815000
Abbreviation. CSO.

Properties. Pale-yellow, oily liquid,[020] or yellowish-brown to dark ruby-red semidrying oil. Practically odor less; slightly soluble in alcohol.

Acute Toxicity. Non-toxic on single administration.

Reproductive Toxicity.

Embryotoxicity and ***Teratogenicity.*** Rats were dosed with CSO by *i/p* injections on days 5 to 15 of pregnancy. The treatment produced fetotoxic effects. Malformations were noted in muscular skeletal system.[1]

Gonadotoxicity. In Sprague-Dawley rats fed 30% level of CSO, parental generation was not significantly affected; however, F_1 generation had significantly altered sexual maturity, and character and length of estrus cycles were altered with 20% mortality in newborn.[2]

Carcinogenicity. When given orally to mice for 35 weeks, there were changes indicating the presence of cyclopropene fatty acids in tissue lipids. The incidence of spontaneous mammary tumors in *C3H* mice at 52 weeks was higher in mice fed rations with CSO than in mice fed comparable fatty acid compositions.[3] CSO has been considered an equivocal agent by RTECS criteria.

Regulations. ***U.S. FDA*** (1998) approved the use of CSO (1) in resinous and polymeric coatings used as the food-contact surfaces of articles intended for use in producing, manufacturing, packing, processing, preparing, treating, packaging, transporting, or holding food in accordance with the conditions prescribed in 21 CFR part 175.300; and (2) as a component of articles intended for use in packaging, transporting, or holding food in accordance with the conditions prescribed in 21 CFR part 175.105.

References:

1. Singh, A. R., Lawrence, W. H., and Autin, J., Embryonic-fetal toxicity and teratogenic effects of a group of methacrylate esters in rats, *J. Dent. Res.*, 51, 1632, 1972.
2. Sheehan, E. T. et al., in *Fed. Proc. Fed. Am. Soc. Exp. Biol.*, 26, 800, 1967.
3. Tinsley, I. J., Wilson, G., and Lowry, R. R., Tissue fatty acid changes and tumor incidence in C3H mice ingesting cottonseed oil, *Lipids*, 17, 115, 1982.

CREOZOTE

Synonyms. Wood creozote; Beechwood creozote; Creazote.

Composition. A mixture of phenols obtained from wood tar.

Properties. Almost colorless or yellowish oily liquid with a characteristic smoky odor and caustic, burning taste. Soluble in 150 to 200 parts of water, miscible with alcohol.[020]

Acute Toxicity.

Observations in man. Administration of large doses caused GI irritation, cardiovascular collapse, death.[020]

Carcinogenicity. Male Parkes mice were treated topically with solution of bitumen, a product of oil-refining, for up to 5 weeks. A steady accumulation of DNA adduct, formed *in vivo* by a large number of different chemical compounds present in the applied solution, was seen in skin DNA.

Regulations. ***U.S. FDA*** (1998) approved the use of C. in adhesives as a component of articles intended for use in packaging, transporting, or holding food in accordance with the conditions prescribed in 21 CFR part 175.105.

Reference:

Schoket, B., Hewer, A., Grover, P. L., and Phillips, D. H., Covalent binding of components of coal-tar, creosote, and bitumen to the DNA of the skin and lungs of mice following topical application, *Carcinogenesis,* 9, 1253, 1988.

CROTONIC ACID, VINYL ESTER

Molecular Formula. $C_6H_8O_2$
M = 112.14
CAS No 14861-06-4
RTECS No GQ5850000
Abbreviation. CAVE.

Synonyms. 2-Butenoic acid, ethenyl ester; Vinyl butenoate; Vinyl crotonate.

Acute Toxicity. LD_{50} is 6.5 g/kg BW (Union Carbide Data Sheet, 1971).

Mutagenicity. CAVE induced a clear dose-dependent increase in the number of SCEs/cell at concentrations of 0.125 to 0.5 mM, probably due to formation of *acetaldehyde.*[2]

Regulations. ***U.S. FDA*** (1998) approved the use of CAVE in adhesives as a component of articles intended for use in packaging, transporting, or holding food in accordance with the conditions prescribed in 21 CFR part 175.105.

Reference:

Sipi, P., Jarventaus, H., and Norppa, H., Sister-chromatid exchanges induced by vinyl esters and respective carboxylic acids in cultured human lymphocytes, *Mutat. Res.,* 279, 75, 1992.

CYANODITIOIMIDOCARBOMATE, DISODIUM SALT

Molecular Formula. $C_2H_2N_2S_2.2Na$
M = 164.16
CAS No 138-93-2
RTECS No FD3405000

Synonyms. Cyanodithioimidocarbonic acid, disodium salt; Disodium cyanodithioimidocarbonate.

Regulations. ***U.S. FDA*** (1998) approved the use of C. as an adjuvant substance in the manufacture of foamed plastics intended for use in contact with food, subject to the provisions prescribed in CFR 178.3010.

CYCLOHEXANE

Molecular Formula. C_6H_{12}
M = 84.16
CAS No 110-82-7
RTECS No GU6300000
Abbreviation. CH.

Synonyms. Hexahydrobenzene; Hexamethylene.

Properties. Colorless liquid with a penetrating odor. Solubility in water is 55 mg/l;[02] miscible with alcohol. Odor perception threshold is 1.0 g/l,[1] 0.3 mg/l,[2] or 0.011 mg/l.[02]

Applications. A solvent. Used in the production of polyamides and synthetic fibers.

Acute Toxicity. In mice, LD_{50} is 4.7 g/kg BW. Poisoned animals displayed CNS inhibition with subsequent excitation and clonic convulsions. Death within 3 to 5 minutes, possibly as a result of respiratory failure.[3]

Rats are less susceptible though they show age-dependent susceptibility to CH: at BW up to 50 g, LD_{50} is 8.0 ml/kg BW; at 80 to 160 g, it is 39 ml/kg BW, and at 300 to 470 g, it is 16.5 ml/kg BW.[4]

Short-term Toxicity. Rats received 200 and 400 mg CH/kg BW. The treatment did not affect general condition, behavior, blood morphology, or liver function. A 400 mg/kg BW dose lowered catalase and cholinesterase activity.

Mutagenicity.

In vitro genotoxicity. CH is found to be positive in *Salmonella* mutagenicity assay.[5]

In vivo cytogenetics. CH caused CA in rats exposed to it by inhalation.

An increase in chromatid breaks and SCE was observed in cultured lymphocytes from the children of female chemical laboratory workers when compared with controls. The mothers had been exposed to a number of different solvents including, in some cases, CH.[6]

Chemobiokinetics. CH is metabolized to form metabolites *cyclohexanol* (38%) and (q)-*trans*-1,2-*cyclohexanediol* (7.0%) which are excreted in rabbits as *glucuronides* in the urine. 30% of administered CH are excreted unchanged via the lungs and 9.0% as CO_2.[021]

Regulations. ***U.S. FDA*** (1998) regulates CH (1) as a component of adhesives used in articles intended for use in packaging, transporting, or holding food in accordance with the conditions prescribed in 21 CFR part 175.105; (2) as a component of the uncoated or coated food-contact surface of paper and paperboard that may be safely used in producing, manufacturing, packing, transporting, or holding dry food in accordance with the conditions prescribed in 21 CFR part 176.180; and (3) as a defoaming agent that may be safely used as a component of food-contact articles in accordance with the conditions prescribed in 21 CFR part 176.200.

Standards. ***Russia*** (1995). MAC and PML: 0.1 mg/l.

References:

1. Savelova, V. A. and Klimkina, N. V., in *Protection of Water Reservoirs against Pollution by Industrial Liquid Effluents*, S. N. Cherkinsky, Ed., Meditsina, Moscow, Issue No 6, 1964, 30 (in Russian).
2. Ogata, M. and Meyake, *J. Water Res.*, 7, 1493, 1973.
3. Meleshchenko, K. F., *Prevention of the Pollution of Water Bodies by Liquid Effluents of Chemical Plants,* Zdorov'ya, Kiyv, 1971, 61 (in Russian).
4. Kimura, E. T., Ebert, D. M., Dodge, P. W., et al., Acute toxicity and limits of solvent residue for sixteen organic solvents, *Toxicol. Appl. Pharmacol.*, 19, 699, 1971.
5. McCann, J., Choi, E., Yamasaki, E., et al., Detection of carcinogens and mutagens in the *Salmonella*/microsome test: assay of 300 chemicals, *Proc. Natl. Acad. Sci. USA*, Issue No 72, 1975, 5135.
6. Funes-Cravioto, F., Zapata-Gayon, C., Kolmodin-Hedman, B., et al., Chromosome aberrations and sister-chromatid exchange in workers in chemical laboratories and a rotoprinting factory and in children of women laboratory workers, *Lancet*, 2. 322, 1977.

1,4-CYCLOHEXANEDICARBOXYLIC ACID

Molecular Formula. $C_8H_{12}O_4$

M = 172.20

CAS No 1076-97-7

RTECS No GU9060000

Abbreviation. CHDA.

Synonyms. 1,4-Dicarboxycyclohexane; Hexahydroterephthalic acid.

Regulations. ***U.S. FDA*** (1993) approved the use of CHDA in the synthesis of polyester resin for resinous and polymeric coatings used as the food-contact surfaces of articles intended for use in producing, manufacturing, packing, processing, preparing, treating, packaging, transporting, or holding food in accordance with the conditions prescribed in 21 CFR part 175.300.

1,4-CYCLOHEXANE DIMETHANOL

Molecular Formula. $C_8H_{16}O_2$

M = 144.24

CAS No 105-08-8

RTECS No GU9800000

Abbreviation. CHDM.

Synonyms. 1,4-Bis(hydroxymethyl)cyclohexane; Hexahydro-2-oxo-1,4-cyclohexane dimethanol.

Properties. Liquid. Soluble in water and ethyl alcohol.

Applications. Used in the production of polyester films and protective coatings, polyester resins (foams and elastomers).

Acute Toxicity. The lowest lethal dose is reported to be 3.3 g/kg BW in rats, and 1.6 g/kg BW in mice (Kodak Co, 1971).

Chemobiokinetics. Charles River rats were given ^{14}C-CHDM/kg BW by gavage in water doses of 40 and 400 mg. CHDM was readily absorbed from the GI tract. *4-Hydroxymethylcyclohexane carboxylic acid* was discovered in plasma. Excretion occurs predominantly via urine. The major metabolites of ^{14}C-CHDM identified in the urine were *cyclohexanedicarboxylic acid* and *4-hydroxy-methylcyclohexanecarboxylic acid.*

Regulations.

EU (1990). CHDM is available in the *List of authorized monomers and other starting substances which shall be used for the manufacture of plastic materials and articles intended to come into contact with food-stuffs (Section* A).

Great Britain (1998). CHDM is authorized without time limit for use in the production of polymeric materials and articles in contact with food or drink or intended for such contact.

Reference:

Divincenzo, G. D. and Ziegler, D. A., Metabolic fate of carbon-^{14}C-labeled 1,4-cyclohexanedimethanol in rats, *Toxicol. Appl. Pharmacol.,* 52, 10, 1980.

CYCLOHEXANONE OXIME

Molecular Formula. $C_6H_{11}NO$
M = 113.16
CAS No 100-64-1
RTECS No GW1925000
Abbreviation. CHO.

Synonym and **Trade Name.** Antioxidant D; (Hydroxyimino)cyclohexane.

Properties. White, crystalline powder with an unpleasant, specific odor. Solubility in water is 15 g/l at 18°C; soluble in alcohol. Practical odor perception threshold is 7.8 mg/l. At this concentration, no foreign taste is detected in water.

Applications. CHO is a captive intermediate in the synthesis of caprolactam for the production of Nylon-6 fibers and plastics; also used in a variety of industrial applications.

Acute Toxicity. In mice, LD_{50} is 3.1 g/kg BW when CHO is given to mice in the alcoholic solution. However, administration of the maximum dose possible in terms of solubility caused no mortality in these animals. Poisoning is accompanied by motor excitation with subsequent CNS inhibition, muscular weakness, and hindlimb paresis.[1]

Acute exposure caused neurobehavioral changes (aggression and CNS depression) in rats. Conjunctivitis was found to develop.[2]

Repeated Exposure. In a 2-week study, B6C3F$_1$ mice received CHO in their drinking water at concentrations of 100 to 2,500 ppm. No deaths occurred and no retardation in BW gain was noted. Gross pathology examination revealed no visceral lesions. There was a significant increase in relative spleen weights in the 2,500 ppm CHO group and increases in the relative liver weights of male mice exposed to 312 ppm and greater.[3]

In a 14-day gavage study, a dose-related erythroid hyperplasia in the spleen and bone marrow were noted in Sprague-Dawley and F344 rats. Sprague-Dawley rats received 1.0, 10, and 100 mg CHO/kg BW for two weeks. The treatment caused erythropenia, higher platelet counts, lower *Hb* concentrations and hematocrit levels. General splenic enlargement with hematopoietic cell proliferation were noted. The same hematological changes were observed in F344 rats.[4]

Decreased BW gain and cholinesterase activity, as well as slight reduction in blood *Hb* level were observed in rats given the oral dose of 200 mg CHO/kg BW for 40 days.[1]

Short-term Toxicity. B6C3F$_1$ mice received drinking water containing 625 to 10,000 ppm CHO for 13 weeks. Deaths and retardation of BW gain were noted at the highest dose level. A significant increase in relative spleen weights was observed in 5,000 and 10,000 ppm groups of animals. There was also a significant increase in relative liver weights in 10,000 ppm CHO group. Histologically, hematopoietic cell proliferation was found in the spleen at 5,000 and 10,000 ppm groups. Centrilobular cell hypertrophy in the liver was noted at the higher doses. There was olfactory epithelial degeneration in all exposed animals (the NOEL of 625 ppm was established for this effect).[3]

In a 13-week study, the same hematological changes were observed as were noted in the 2-week study[6] with evidence of splenomegaly and erythroid hyperplasia in the spleen and bone marrow.[5]

Long-term Toxicity. Changes in catalase activity were reported.[1]

Mutagenicity.

In vivo cytogenetics. CHO was negative in *Dr.melanogaster*.[6]

In vitro genotoxicity. CHO was mutagenic in *Salmonella typhimurium* strain *TA1535* with *S9* activation only; but it was negative in strains *TA97, TA98,* and *TA100* with and without *S9* activation. CHO shows equivocal results in CA test in cultured Chinese hamster ovary cells without *S9* activation, but it was negative with *S9* activation system.[3,7] Its mutagenic activity was observed in L5178Y mouse lymphoma cells treated in the absence, but not in presence of rat liver *S9* activation.[7]

Chemobiokinetics. CHO is rapidly absorbed and cleared within 24 hours after a single oral administration of 1.0 to 30 mg ^{14}C-CHO/kg BW, eliminated predominantly in the urine. Three urinary meta- bolites identified are *cyclohexyl glucuronide* and the *monoglucuronides* of *cis-* and *trans-cyclohexane-1,2-diol*.[8]

CHO was found to induce increased liver microsomal activity in rats exposed by gavage to 100 mg CHO/kg BW for 14 days.[9]

Standards. ***Russia*** (1995). MAC and PML: 1.0 mg/l.

References:

1. Savelova, V. A. and Klimkina, N. V., in *Protection of Water Reservoirs against Pollution* by *Industrial Liquid Effluents*, S. N. Cherkinsky, Ed., Meditsina, Moscow, Issue No 6, 1964, 64 (in Russian).
2. Rublack, H. und Henkel, W., The combined action of cyclohexanone and oxime, *Z. Gezamte Hyg.*, 21, 538, 1975.
3. *Toxicity Studies of Cyclohexanone Oxime Administered by Drinking Water to $B6C3F_1$ Mice* (L. T. Burka), NTP Technical Report Series No 50, Research Triangle Park, NC, NIH Publ. 96-3934, April 1996, 42.
4. Derelanko, M. J., Gad, S. C., Powers, W. J., Mulder, S., Gavican, F., and Babich, P. C., Toxicity of cyclohexanone oxime. I. Hematotoxicity following subacute exposure in rats, *Fundam. Appl. Toxicol.*, 5, 117, 1985.
5. Gad, S. C., Derelanko, M. J., Powers, W. J., Mulder, S., Gavican, F., and Babich, P. C., Toxicity of cyclohexanone oxime. II. Acute dermal and subchronic oral studies, *Fundam. Appl. Toxicol.*, 5, 128, 1985.
6. Vogel, E. and Chandler, J. L. R., Mutagenicity testing of cyclamate and some pesticides in Dr. melanogaster, *Experentia*, 30, 621, 1974.
7. Rogers-Back, A. M., Lawlor, T. E., Cameron, T. P., and Dunkel, V. C., Genotoxicity of 6 oxime compounds in the *Salmonella*/mammalian-microsome assay and mouse lymphoma TK+/-assay, *Mutat. Res.*, 204, 149, 1988.
8. Parmar, D. and Burka, L. T., Metabolism and disposition of cyclohexanone oxime in male F344 rats, *Drug Metab. Dispos.*, 19, 1101, 1991.
9. Komsta, E., Secours, V. E., Chu, I., Valli, V. E., Morris, R., Harrison, J., Baranowski, E., and Villeneuve, D. C., Short-term toxicity of nine industrial chemicals, *Bull. Environ. Contam. Toxicol.*, 43, 87, 1989.

CYCLOHEXYLAMINE

Molecular Formula. $C_6H_{13}N$

M = 99.20

CAS No 108-91-8

RTECS No GX0700000

Abbreviation. CHA.

Synonyms and **Trade Name.** Aminocyclohexane; Cyclohexanamine; Hexahydroaniline.

Properties. Colorless liquid with a penetrating odor and a bitter taste. Completely soluble in water and alcohol. Odor perception threshold is 25 mg/l.[02]

Applications. Used in the production of silicon elastomers, plasticizers, rubber chemicals and dyestuffs; an emulsifying agent.

Acute Toxicity. In rats, LD_{50} is 600 mg/kg BW (5.0% solution).

Short-term Toxicity. CHA hydrochloride was added to the diet of mice, Wistar and DA rats over a period of 13 weeks. Addition of 2,000 ppm caused decrease of BW gain and food consumption only in rats.[1]

Retardation of BW gain and other changes were noted in rats, guinea pigs, and rabbits given the 100 mg/kg BW dose for 3 months.[05]

Long-term Toxicity. Wistar rats were fed the diet containing 600, 2,000, or 6,000 ppm CHA hydrochloride for 104 weeks. The treatment caused dose-related increases in mortality and retardation of BW gain. At two higher concentrations, elevation of relative thyroid weights in females and testicular degeneration in males were reported.[2]

Reproductive Toxicity.

Embryotoxicity. In a six-generation study, FDRL parental rats received 15 to 150 mg CHA hydrochloride/kg their diet. In the offspring of rats treated with the higher doses, a decrease in food intake and, subsequently, a significant decrease in BW gain were noted in association with the incidence of testicular atrophy. Reduced fer tility and number of live/young litter were also reported.[3]

Teratogeniciy. No fetal changes were recorded in rats gavaged with the dose of 36 mg CHA/kg BW on days 7 through 13 of gestation,[4] and in mice fed 100 mg CHA/kg BW on days 6 through 11 of gestation.[5]

Gonadotoxicity. A limited toxic effect seems to be testicular atrophy (decreased weight and histology) in rats fed dietary levels of 2,000 ppm or more; in mice, there was no evidence of testicular damage even at a dietary concentration of 3,000 ppm.[6]

At the doses of 200 to 250 mg CHA/kg BW given to rats and dogs, a decrease in spermatocyte number was reported.[7]

Carcinogenicity. Negative results for *CHA hydrochloride* and *CHA sulfate* are reported in rats and mice.[09]

Chemobiokinetics. Possible differences in metabolism in mice and rats (see above) are reported. *Cyclohexanol* was discovered to be the urinary metabolite after administration of CHA hydrochloride to rats, rabbits, guinea pigs (doses of 50 to 500 mg/kg BW, orally or *i/p*), and humans (25 or 200 mg orally) (Renwick and Williams, 1972).

Standards. ***Russia*** (1995). MAC: 0.1 mg/l; MAC of CHA *carbonate* and *chromate*: 0.01 mg/l.

References:

1. Gaunt, I. F., Sharratt, M., Grasso, P., Lansdown, A. B. G., and Gangolli, S. D., Short-term toxicity of cyclohexylamine hydrochloride in the rat, *Food Cosmet. Toxicol.*, 12, 609, 1974.
2. Gaunt, I. F., Hardy, J., Grasso, P., Gangolli, S. D., and Butterworth, K. R., Long-term toxicity of cyclohexylamine hydrochloride in the rat, *Food Cosmet. Toxicol.*, 14, 55, 1976.
3. Oser, B. L., Carson, S., Cox, G. E., Vogin, E. E., and Steinberg, S. S., Long-term and multigeneration toxicity studies with cyclohexylamine hydrochloride, *Toxicologist*, 6, 47, 1976.
4. Tanaka, S., Nakaura, S., Kawashima, K., et al., Teratogenicity of food additives. 2. Effect of cyclohexylamine and cyclohexylamine sulfate on fetal development in rats, Abstract, *Shokuhin Eiseigaku Zasshi*, 14, 542, 1973 (in Japanese).
5. Takano, K. and Suzuki, M., Cyclohexylamine, a chromosome-aberration producing substance: No teratogenicity in the mouse, *Congen. Anomal.,* 11, 51, 1971 (in Japanese).
6. Roberts, A., Renwick, A. G., Ford, G., et al., The metabolism and testicular toxicity of cyclohexylamine in rats and mice during chronic dietary administration, *Toxicol. Appl. Pharmacol.*, 98, 216, 1989.
7. James, R. W. et al., Testicular responses of rats and dogs to cyclohexylamine overdosage, *Food Cosmet. Toxicol.*, 19, 291, 1981.

2-(CYCLOHEXYLAMINO)ETHANOL

Molecular Formula. $C_8H_{17}NO$
M = 143.26
CAS No 2842-38-8
RTECS No KK3530000

Synonym and **Trade Name.** Abromeen E-25; *N*-(2-Hydroxyethyl)cyclohexylamine.

Acute Toxicity. LD_{50} is 38.3 g/kg BW.

Regulations. ***EU*** (1990). C. is available in the *List of monomers and other starting substances which may continue to be used pending a decision on inclusion in Section A.*

Reference:

Deichmann, W. B., *Toxicology of Drugs and Chemicals*, Academic Press, Inc., New York, 1969, 61.

N-CYCLOHEXYL-*p*-TOLUENE SULFONAMIDE

Molecular Formula. $C_{13}H_{19}NO_2S$
M = 253.39
CAS No 80-30-8
RTECS No XT5617000

Synonym and **Trade Name.** *N*-Cyclohexyl-4-methylbenzene sulfonamide; Santicizer IH.

Acute Toxicity. Lethal doses appeared to be more than 0.5 g/kg BW.

Regulations. ***U.S. FDA*** (1998) approved the use of C. as a component of the uncoated or coated food-contact surface of paper and paperboard intended for use in producing, manufacturing, packaging, processing, preparing, treating, packing, transporting, or holding dry foods in accordance with the conditions prescribed in 21 CFR part 176.180.

Reference:

Natl. Acad. Sci, Natl. Res. Council, Chem.-Biol. Coordinat. Center, *Review,* 5, 41, 1953.

1,3-CYCLOPENTADIENE

Molecular Formula. C_5H_6
M = 66.11
CAS No 542-92-7
RTECS No GY1000000
Abbreviation. CPD.

Synonym and **Trade Names.** Pentole; Pyropentylene; *R*-Pentine.

Properties. Colorless liquid with terpene odor. Water solubility is 1.8 g/l at 25°C. Miscible with alcohol.

Applications. Used in the manufacture of unsaturated polyester resins.

Acute Toxicity. LD_{50} of CPD-dimer is 0.82 g/kg BW in rats.[1]

Repeated Exposure. Inhalation of CPD affected liver cells and renal tubular epithelium in rats.[014]

Allergenic Effect. May cause contact dermatitis and sensitization.[2]

Regulations. ***EU*** (1990). CPD is available in the *List of authorized monomers and other starting substances which may continue to be used for the manufacture of plastic materials and articles intended to come into contact with foodstuffs pending a decision on inclusion in Section A* (*Section B*).

References:

1. Smyth, H. F., Carpenter, C. P., Weil, C. S., et al., Range-finding toxicity data, List V, *Arch. Ind. Hyg. Occup. Med.*, 10, 61, 1954.
2. Grant, W. M., *Toxicology of the Eye*, 2nd ed., Charles C Thomas, Springfield, Ill., 1974, 345.

CYCLOPENTENE

Molecular Formula. C_5H_8
M = 68.13
CAS No 142-29-0
RTECS No GY5950000

Acute Toxicity. LD_{50} is 1.65 g/kg BW in rats.

Regulations. ***EU*** (1990). C. is available in the *List of authorized monomers and other starting substances which may continue to be used for the manufacture of plastic materials and articles intended to come into contact with foodstuffs pending a decision on inclusion in Section A* (*Section B*).

Reference:

Smyth, H. F., Carpenter, C. P., Weil, C. S., Pozzani, U. C., Striegel, J. A., and Nycum, J. S., Range-finding toxicity data: List VII, *Am. Ind. Hyg. Assoc. J.*, 30, 470, 1969.

DECANOIC ACID

Molecular Formula. $C_{10}H_{20}O_2$
M = 172.30
CAS No 334-48-5
RTECS No HD9100000
Abbreviation. DA.

Synonyms and **Trade Name.** Capric acid; Caprinic acid; Decylic acid; 1-Nonane carboxylic acid.

Properties. White crystals, needles, or crystalline solid with rancid odor. Poorly soluble in water (15 mg/l at 20°C). Soluble in ethanol.

Applications. Intermediate in the production of food-grade additives, plasticizers, and resins. A lubricant, binder, and defoaming agent in foods.

Acute Toxicity. LD_{50} exceeded 10 g/kg BW in rats.[1]

Short-term Toxicity. Rats were fed 10% DA in their diet for 150 days. The treatment caused no gastric lesions.[2]

Chemobiokinetics. In rats, administration of DA elevated urinary excretion of the C_{10}-dicarboxylic acid (*sebacic acid*) and highly elevated excretion of β-oxidation products C_8- and C_6-dicarboxylic acids (*suberic* and *adipic acids*).[3]

Regulations.

U.S. FDA (1998) approved the use of DA (1) as a component of sanitizing solutions used on food-processing equipment and utensils, and on other food-contact articles as specified in 21 CFR part 178.1010; and (2) as an antioxidant and/or stabilizer in polymers used in the manufacture of articles or components of articles intended for use in producing, manufacturing, packing, processing, preparing, treating, packaging, transporting, or holding food, subject to the provisions and definitions set in 21 CFR part 178.2010.

EU (1990). DA is available in the *List of authorized monomers and other starting substances which may continue to be used for the manufacture of plastic materials and articles intended to come into contact with foodstuffs pending a decision on inclusion in Section A* (*Section B*).

Great Britain (1998). DA is authorized without time limit for use in the production of polymeric materials and articles in contact with food or drink or intended for such contact.

References:

1. Briggs, G. B., Doyle, R. L., and Young, J. A., Safety studies on a series of fatty acids, *Am. Ind. Hyg. Assoc. J.*, 37, 251, 1976.
2. Patty's *Industrial Hygiene and Toxicology,* G. D. Clayton and F. E. Clayton, Eds., Toxicology, 4th ed., John Wiley & Sons Inc., New York, NY, 1994, 3560.
3. Mortensen, P. B., C_6-C_{10}-dicarboxylic aciduria in starved, fat-fed and diabetic rats receiving decanoic acid or medium-chain triacylglycerol. An *in vivo* measure of the rate of β-oxidation of fatty acids, *Biochim.-Biophys. Acta*, 664, 349, 1981.

DECYL ALCOHOL

Molecular Formula. $C_{10}H_{22}O$
M = 158.28
CAS No 112-30-1
RTECS No HE4375000
Abbreviation. DA.

Synonyms and **Trade Names.** Alcohol C-10; Decanol; *N*-Decanol; *N*-Decatyl alcohol; Decylic alcohol; *N*-Nonyl carbinol.

Properties. Colorless to light-yellow liquid with characteristic fatty taste and odor resembling orange flowers. Insoluble in water, solubility in 60% alcohol is 1:3.

Applications. Chemical intermediate in the production of plasticizers, solvents, and synthetic lubricants.

Acute Toxicity. LD_{50} is 4.72 g/kg BW in rats.[012]

Repeated inhalation **Exposure** decreased serum cholinesterase activity in rats and rabbits.[03]

Allergenic Effect.

Observations in man. No sensitization was observed in a maximization test that was carried out on volunteers using a 3.0% DA concentration in petrolatum (Opduke, 1979).

Carcinogenicity. Skin application in mice that had previously received an initiating dose of dimethyl benzanthracene increased incidence of tumors including development of squamous cell carcinomas.[03]

Regulations.

EU (1990). DA is available in the *List of authorized monomers and other starting substances which shall be used for the manufacture of plastic materials and articles intended to come into contact with foodstuffs (Section* A).

U.S. FDA (1998) approved the use of DA (1) in the manufacture of resinous and polymeric coatings for the food-contact surface of articles intended for use in producing, manufacturing, packing, processing, preparing, treating, packaging, transporting, or holding food in accordance with the conditions prescribed in 21 CFR part 175.300; and (2) as a component of the uncoated or coated food-contact surface of paper and paperboard intended for use in producing, manufacturing, packaging, processing, preparing, treating, packing, transporting, or holding aqueous and fatty foods in accordance with the conditions prescribed in 21 CFR part 176.170.

Great Britain (1998). DA is authorized without time limit for use in the production of polymeric materials and articles in contact with food or drink or intended for such contact.

3,3'-DIAMINODIPROPYLAMINE

Molecular Formula. $C_6H_{17}N_3$
M = 131.26
CAS No 56-18-8
RTECS No JL9450000

Synonyms. Aminobis(propylamine); Bis(3-aminopropyl)amine; Dipropylenetriamine; Iminobis(propylamine).

Acute Toxicity. LD_{50} is 738 mg/kg BW in rats, 435 mg/kg BW in mice, 210 mg/kg BW in rabbits, and 258 mg/kg BW in guinea pigs.

Regulations. ***U.S. FDA*** (1998) approved the use of D. as a component of the uncoated or coated food-contact surface of paper and paperboard intended for use in producing, manufacturing, packaging, processing, preparing, treating, packing, transporting, or holding dry food in accordance with the conditions prescribed in 21 CFR part 176.180.

Reference:

Z. Gesamte Hyg. Grenzgebiete, 20, 393, 1974.

DIAMMONIUM HEXAFLUOROSILICATE(2^-)

Molecular Formula. $F_6Si.2H_4N$
M = 178.19
CAS No 16919-19-0
RTECS No VV7800000
Abbreviation. DHS.

Synonyms and **Trade Name.** Ammonium fluorosilicate; Ammonium hexafluorosilicate; Ammonium silicone fluoride; Diammonium; Fluosilicate; Diammonium hexafluorosilicate; Diammonium silicon; Hexafluoride.

Properties. Fine, white, odorless crystalline powder. Solubility in water is up to 186 g/l; slightly soluble in alcohol.

Applications. Used in dental preparations in children.

Acute Toxicity. LD_{50} is 70 mg/kg BW in mice. Poisoning caused peripheral nerve paralysis, ataxia, muscle contraction, or spasticity.[1]

DHS appears to be as highly toxic by the oral route of administration as other fluorides. It provokes a salty or soapy taste in the mouth, excessive salivation, nausea, abdominal cramps, vomiting, diarrhea, thirst, muscle weakness, tremors or spasms, CNS inhibition, shock, and death.

Long-term Toxicity. Chronic excess exposure to fluorides may cause structural changes in bones: fluorosis and osteosclerosis, which involve excessive calcification of the bone and ligaments with limitation of motion. DHS is at least as toxic as sodium silicofluoride, if not more so.

Mutagenicity.

In vitro cytogenetics. Fluorides damage chromosomes (Hodge and Macgregor, 1982).

Reproductive Toxicity.

Teratogenicity. Excess exposure to fluoride (12 to 18 ppm in drinking water) during pregnancy and the first eight years of life can caused mottled teeth (Hodge and Macgregor, 1982). No association has been found between levels of fluoride in the drinking water and birth effects.[2]

Regulations. ***U.S. FDA*** (1998) approved the use of DHS in adhesives as a component of articles intended for use in packaging, transporting, or holding food in accordance with the conditions prescribed in 21 CFR part 175.105.

References:

1. Rumyantzev, G. I., Novikov, S. M., Mel'nikova, N. N., Levchenko, N. I., Kozeyeva, Ye. Ye., and Kochetkova, T. A., Experimental study of biologic effect of salts of silicofluoric acid, *Gig. Sanit.*, 11, 80, 1988 (in Russian).
2. Erickson, J. D., Oakley G. P., Flynt, J. W., and Hay, S., Water fluoridation and congenital malformations: No association, *J. Am. Dent. Assoc.*, 93, 981, 1976.

2,3-DIBROMOPROPIONALDEHYDE

Molecular Formula. $C_3H_4Br_2O$
M = 215.89
CAS No 5221-17-0
RTECS No UE0800000
Abbreviation. DBPA.

Synonyms. Acrolein dibromide; 2,3-Dibromopropanal.

Acute Toxicity. LD_{50} *i/p* is 5.0 g/kg BW in mice.[1]

Mutagenicity.

In vitro genotoxicity. DBPA was found to cause mutations in *Salmonella typhimurium* strain *TA100* both in the absence and presence of a metabolic system, and caused extensive DNA single-stranded breaks as evidenced by alkaline elution of DNA. It produced increased unscheduled DNA repair synthesis in isolated rat hepatocytes, in monolayer cultures.[2,3]

Regulations. ***U.S. FDA*** (1998) approved the use of DBPA as a slimicide in the manufacture of paper and paperboard that contact food in accordance with the conditions prescribed in 21 CFR part 176.300.

References:

1. Marsden, P. J. and Casida, J. E., 2-Haloacrylic acids as indicators of mutagenic 2-haloacrolein intermediates in mammalian metabolism of selected promutagens and carcinogens, *J. Agric. Food Chem.*, 30, 627, 1982.
2. Rosen, J. D., Segall, Y., and Casida, J. E., Mutagenic potency of haloacroleins and related compounds, *Mutat. Res.*, 78, 113, 1980.
3. Gordon, W. P., Soderlund, E. J., Holme, J. A., Nelson, S. D., Iyer, L., et al., The genotoxicity of 2-bromoacrolein and 2,3-dibromopropanal, *Carcinogenesis*, 6, 705, 1985.

2,6-DICHLORO-*p*-PHENYLENEDIAMINE

Molecular Formula. $C_6H_6ClN_2$
M = 177.04
CAS No 609-20-1
RTECS No SS9175000

Synonyms. 1,4-Diamino-2,6-dichlorobenzene; 2,6-Dichloro-1,4-benzenediamine.

Applications. Used in the manufacture of polyamide fibers and polyurethane.

Acute Toxicity. In rats, LD_{50} is reported to be 0.7 g/kg BW.

Mutagenicity.

In vitro genotoxicity. D. manifested mutagenic activity in *Salmonella*/microsome preincubation assay. It was found to cause CA in *in vitro* studies (NTP-92).

Carcinogenicity. No potential for carcinogenicity was observed in male rats given D. at dose levels of 1000 or 2000 ppm in their feed. Meanwhile, there was an increase in tumor incidence in females fed 2000 or 6000 ppm doses. In mice (dose levels of 1000 or 3000 ppm), hepatocellular adenomas (in males), hepatocarcinomas, and adenomas were reported.

Carcinogenicity classifications. An IARC Working Group concluded that there is limited evidence for the carcinogenicity of D. in *experimental animals* and there were no adequate data available to evaluate the carcinogenicity of D. in *humans*.

IARC: 3;
NTP: N - N - P - P (feed).

Reference:

Carcinogenic Bioassay of *2,6-Dichloro-p-phenylenediamine in F344 Rats and B6C3F*$_1$ *Mice (Feed Study),* NTP Technical Report Series No 219, Research Triangle Park, NC, 1982.

N,N'-DIETHYLHYDROXYLAMINE

Molecular Formula. $C_4H_{11}NO$
M = 89.16
CAS No 3710-84-7
RTECS No NC3500000

Synonym. *N*-Hydroxydiethylamine.

Properties. Clear liquid. Readily soluble in water and alcohol.

Applications. Used as a stopper in the process of emulsion polymerization.

Acute Toxicity. In rats, LD_{50} is 1.75 g/kg BW. A 2.15 g/kg BW dose killed 80% mice. Poisoning is accompanied by clonic convulsions. Death within 2 days.

Short-term Toxicity. Rats were given doses of 18 and 180 mg/kg BW over a period of 4 months. The NS as well as liver antitoxic function were predominantly affected.

Reference:

Taradin, Ya. I. and Kuchmina, N. Ya., Toxicological characteristics of diethylhydroxylamine, in *Hygiene and Toxicology of High-Molecular-Mass Compounds and of the Chemical Raw Material Used for Their Synthesis,* Proc. 6th All-Union Conf., B. Yu. Kalinin, Ed., Leningrad, 1979, 289 (in Russian).

N,N-DIETHYL-1,3-PROPANEDIAMINE

Molecular Formula. $C_7H_{18}N_2$
M = 130.27
CAS No 104-78-9
RTECS No TX7350000

Synonyms. 1-Amino-3-(diethylamino)propane; Diethylaminotrimethylenamine; *N,N*-Diethyl-1,3-diaminopropane.

Acute Toxicity. LD_{50} is 1410 mg/kg BW in rats (Union Carbide Data Sheet, 1967).

Regulations. ***U.S. FDA*** (1993) approved the use of D. for use in resinous and polymeric coatings used as the food-contact surfaces of articles intended for use in producing, manufacturing, packing, processing, preparing, treating, packaging, transporting, or holding food at a level not to exceed 6% by weight of the resin when such coatings are intended for repeated use in contact with foods only of the types identified in 175.300.

1,1-DIFLUOROETHANE

Molecular Formula. $C_2H_4F_2$
M = 66.05
CAS No 75-37-6
RTECS No KI1410000

Synonyms and **Trade Names.** Ethylene fluoride; Ethylidene difluoride; Ethylidene fluoride; Freon 152A; Genetron 100; Refrigerant 152A.

Properties. Colorless and odorless gas. Insoluble in water.

Mutagenicity.

In vivo cytogenetics. Genetron-152A exposure increased the mutation rate in progeny of *Dr. melanogaster.*

Regulations. ***U.S. FDA*** (1993) approved the use of D. as a blowing agent in the manufacture of polystyrene in accordance with the provisions prescribed in 21 CFR part 178.3010.

Reference:

Foltz, V. C. and Fuerst, R., *Environ. Res.*, 7, 275, 1974.

1-[(DIIODOMETHYL)SULFONYL]-4-METHYLBENZENE SULFONE

Molecular Formula. $C_8H_8I_2O_3S$
M = 422.02
CAS No 20018-09-1
RTECS No CZ6000000

Synonym and **Trade Name.** Amical 48; Diiodomethyl *p*-tolylsulfone.

Acute Toxicity. LD_{50} is 5.0 g/kg BW.

Reproductive Toxicity.

Embryotoxicity. Rats received 0.125 to 1.0% D. of their diet on days 6 to 15 of pregnancy. The treatment caused an increase in pre-implantation mortality (reduction in number of implants per female, total number of implants per *corpora lutea*). Ingestion of the highest dose resulted in maternal effects (retardation of BW gain, lowered food consumption, etc.).

Teratogenicity. D. has no teratogenic effects on rat offspring even at doses causing maternal toxicity.[1]

Regulations. ***U.S. FDA*** (1998) approved the use of D. in adhesives as a component of articles intended for use in packaging, transporting, or holding food in accordance with the conditions prescribed in 21 CFR part 175.105.

Reference:

1. Ema, M., Itami, T., and Kawasaki, H., Teratological assessment of diiodomethyl *p*-tolyl sulfone in rats, *Toxicol. Lett.*, 62, 45, 1992.

2-DIMETHYLAMINOETHANOL

Molecular Formula. $C_4H_{11}NO$
M = 89.16
CAS No 108-01-0
RTECS No KK6125000
Abbreviation. DMAE.

Synonyms and **Trade Name.** Deanol; Dimethylethanolamine; (2-Hydroxyethyl)dimethylamine.

Properties. Colorless liquid with amine odor. Miscible with water and alcohol.

Applications. Catalyst for curing epoxy resins and polyurethanes, a stabilizer for amino resins.

Acute Toxicity. DMAE is shown to inhibit cholinesterase *in vitro*. Administration of the LD_{50} dose decreased rat brain cholinesterase levels (by 12 to 18%). Red blood cells cholinesterase level was reduced by 45%.[1] LD_{50} is 2.0 g/kg BW in rats.

Repeated Exposure. High daily doses produced epileptic seizures in mice and rats.[022]

Long-term Toxicity. C3H mice received 10 mg or 15 mg DMAE/kg BW in their drinking water. The exposure did not affect average survival. No gross or histological changes were observed in organ structure or morphology of treated animals except for an apparent decrease in the amount of lipofuscin in the liver.[2]

Reproductive Toxicity. Rat dams were fed 1.0% DMAE on choline-deficient diet during their pregnancy. The treatment resulted in most newborn rats dying within 36 hours of birth with elevated brain levels of choline and acetylcholine.[3]

Administration of DMAE to pregnant rats after the period of organogenesis and to the neonates prevents some of the adverse effects of hypoxia on the CNS. Whether this has relevance for human exposure to this agent is unknown.[4]

Mutagenicity.

In vitro genotoxicity. Was found to be negative in *Salmonella* mutagenicity assay.[015]

Carcinogenicity. In the above described lifetime study, DMAE did not induce any neoplasms.[2]

Chemobiokinetics. Following oral administration, DMAE penetrated through the blood-brain barrier. In mice brain, it yielded *phosphoryl* DMAE, which was converted into *phosphatidyl DMAE*.[5]

Regulations.

EU (1990). DMAE is available in the *List of authorized monomers and other starting substances which shall be used for the manufacture of plastic materials and articles intended to come into contact with food-stuffs (Section A).*

Great Britain (1998). DMAE is authorized without time limit for use in the production of polymeric materials and articles in contact with food or drink or intended for such contact. The specific migration of DMAE shall not exceed 18 mg/kg.

Standards. ***EU*** (1990). SML: 18.0 mg/kg.

References:

1. Hartung, R. and Cornish, H. H., Cholinesterase inhibition in the acute toxicity of alkyl-substituted 2-aminoethanols, *Toxicol. Appl. Pharmacol.*, 12, 486, 1968.
2. Stenback, F., Weisburger, J. H., and Williams, G. M., Effect of lifetime administration of dimethylaminoethanol on longevity, aging changes, and cryptogenic neoplasms in CHS mice, *Mech. Aging Develop.*, 42, 129, 1988.
3. Zahniser, N. R. et al., Effects of *N*-Methylaminoethanol and *N,N'*-dimethylaminoethanol in the diet of pregnant rats on neonatal rat brain cholinergic and phospholipid profile, *J. Neurochem.*, 30, 1245, 1978.
4. Gramatte, T. et al., Effects of nootropic drugs on some behavioural and biochemical changes after early postnatal hypoxia in the rat, *Biomed. Biochim. Acta*, 45, 1075, 1986.
5. Myyazaki, H. et al., Comparative studies on the metabolism of β-dimethylaminoethanol in the mouse brain and liver following administration of β-dimethylaminoethanol and its *p*-chlorophenoxy acetate, meclofenoxate, *Chem. Pharmacol. Bull.*, 24, 763, 1976.

N,N'-DIMETHYLDODECYLAMINE

Molecular Formula. $C_{14}H_{31}N$
M = 213.46
CAS No 112-18-5
RTECS No JR6600000
Abbreviation. DDA.

Synonyms and **Trade Name.** Antioxidant DDA; *N,N'*-Dimethyllaurylamine; Monolauryldimethyl-amine.

Properties. Liquid.

Applications. Corrosion inhibitor.

Carcinogenicity. DDA administered to rats in drinking water with nitrite induced tumor of urinary bladder in 10% of treated rats.

Regulations. ***U.S. FDA*** (1998) approved the use of DDA in the manufacture of polyvinyl alcohol films intended for use in contact with food in accordance with the conditions prescribed in 21 CFR part 177.1670.

Reference:

Lijinsky, W. and Taylor, H. W., Feeding tests in rats on mixtures of nitrite with secondary and tertiary amines of environmental importance, *Food Cosmet. Toxicol.*, 15, 269, 1977.

N,N'-DIMETHYLMETHANAMINE

Molecular Formula. C_3H_9N
M = 59.1
CAS No 75-50-3
RTECS No PAO350000
Abbreviation. DMMA.

Synonym. Trimethylamine.

Properties. Gas with a pungent, fishy, ammoniac odor. Readily soluble in water (410 g/l at 19°C) and alcohol. Odor perception threshold appears to be 0.04 mg/l; practical threshold is 0.16 mg/l. At 60°C, practical odor threshold is 0.05 mg/l. However, odor perception threshold is also reported to be 0.0002 mg/l.[02]

Applications. Used as a component of ion exchange resins.

Exposure. DMMA blood levels can increase after ingestion of certain foods, such as fish, and during disease states, such as chronic renal failure.

Acute Toxicity.

Observations in man. Intake of 15 mg DMMA hydrochloride/kg BW induces nausea and ichthyohydrosis (Calvert, 1973).

Animal studies. LD_{50} is 535 mg/kg BW in rats, 460 mg/kg BW in mice, 240 mg/kg BW in rabbits, and 315 mg/kg BW in guinea pigs. Gross pathology examination of rats that received lethal dose revealed lesions in the GI tract including mucosal defects, liver fatty dystrophy, as well as tiny necrotic foci and vacuolization in the renal tubular epithelium.[1]

Repeated Exposure revealed no cumulative properties: administration of 1/10 LD_{50} for a month caused no animal mortality.

Short-term Toxicity. Rats were given 0.08 to 0.62% DMMA in their diet for 84 days. The treatment produced organ weight changes. The LOEL of 267 mg/kg BW, and the NOEL of 138 mg/kg BW were established in this study (Amoore et al., 1978).[2]

Long-term Toxicity. Rabbits were dosed by gavage with 2.4 mg/kg BW for 6 months. The treatment produced anemia, reduction in prothrombin time, and decrease in monoaminooxidase activity as well as an increase in the enzyme capacity of serum aminotransferases.[1]

Reproductive Toxicity.

Embryotoxicity. DMMA can inhibit *in vivo* and *in vitro* fetal development in mice. In culture, DMMA causes neural-tube defects; it inhibits mouse embryos growth by reducing macromolecular synthesis. These effects may not involve glutathione depletion or generation of free radicals.[3]

Mice received *i/v* injections of 2.5 or 5.0 mM/kg BW on days 1 to 17 of gestation. At the highest dose, 5 of 11 dams died; both doses decreased fetal BW.

Teratogenicity. Congenital defects were not observed.[4]

Allergenic Effect. A decline in the agglutinin titer is noted in rabbits immunized with typhoid vaccine.

Mutagenicity.

In vitro genotoxicity. DMMA is found to be negative in *Salmonella* mutagenicity assay.[013]

Chemobiokinetics. DMMA is potential substrate for *in vivo* nitrosation to form *N-nitrosodimethylamine*, a potent carcinogen. *N*-oxidation is the major route of metabolism while *N*-demethylation is negligible and only significant at the higher dose levels.[5]

When ^{14}C- DMMA was given *i/p*, recovery of radioactivity in the urine was essentially complete. Thus, exhalation, fecal excretion, and accumulation in tissues of DMMA and its metabolites were negligible.[6]

Standards. ***Russia*** (1995). MAC and PML: 0.05 mg/l (organolept., odor).

References:

1. Trubko, E. I. and Teplyakova, E. V., Hygienic norm-setting for trimethylamine in water bodies, *Gig. Sanit.*, 8, 79, 1981 (in Russian).
2. Munro, I. C., Ford, R. A., Kennepohl, E., and Sprenger, J. G., Correlation of structural class with No-Observed-Effect Levels: A proposal for establishing a threshold of concern, *Food Chem. Toxicol.*, 34, 829, 1996.
3. Guest, I. and Varma, D. R., Teratogenic and macromolecular synthesis inhibitory effects of triethylamine on mouse embryos in culture, *J. Toxicol. Environ. Health*, 36, 27, 1992.
4. Guest, I. and Varma, D. R., Developmental toxicity of methylamines in mice, *J. Toxicol. Environ. Health.*, 32, 319, 1991.
5. Al-Waiz, M., Mitchell, S. C., Idle, J. R., et al., The relative importance of N-oxidation and N-demethylation in the metabolism of trimethylamine in man, *Toxicology*, 43, 117, 1987.
6. Smith, J. L., Wishnok, J. S., and Deen, W. M., Metabolism and excretion of methylamines in rats, *Toxicol. Appl. Pharmacol.*, 125, 296, 1994.

2,2-DIMETHYL-1,3-PROPANEDIOL

Molecular Formula. $C_5H_{12}O_2$
M = 104.17
CAS No 126-30-7
RTECS No TY5775000
Abbreviation. DMP.

Synonyms and **Trade Names.** Dimethylolpropane; Dimethyltrimethylene glycol; Neol; Neopentilene glycol.

Acute Toxicity. 3.2 g/kg BW is reported to be the lethal dose (Kodak Co. Doc., 1971).

Regulations.

U.S. FDA (1998) approved the use of DMP as a component of cross-linked polyester resins for use as articles or components of articles intended for repeated use in contact with food, in accordance with the conditions prescribed in 21 CFR part 177.2420.

EU (1990). DMP is available in the *List of monomers and other starting substances which may continue to be used pending a decision on inclusion in Section A.*

Great Britain (1998). DMP is authorized up to the end of 2001 for use in the production of polymeric materials and articles in contact with food or drink or intended for such contact.

DIPENTAERYTHRITOL

Molecular Formula. $C_{10}H_{22}O_7$
M = 254.26
CAS No 126-58-9
Abbreviation. DPE.

Synonyms. Bis(pentaerythritol); Dipentek.

Properties. White crystalline compound. Solubility in water is 3.0 g/l; solubility in ethanol is 0.7 g/l at 25°C.

Applications. Used in preparation of paints and coatings.

Toxicity. DPE is considered to be nontoxic.

Regulations.

EU (1990). DPE is available in the *List of authorized monomers and other starting substances which shall be used for the manufacture of plastic materials and articles intended to come into contact with foodstuffs (Section A).*

Great Britain (1998). DPE is authorized without time limit for use in the production of polymeric materials and articles in contact with food or drink or intended for such contact.

DIPHENYLDIETHYLUREA

Molecular Formula. $C_{17}H_{20}N_2O$
M = 269.37

Properties. White or yellowish crystals with an unpleasant odor and bitter astringent taste. Solubility in water is 80 mg/l. Odor perception threshold is 0.5 mg/l.

Applications. Used in the production of plastics, lacquers, and paints.

Acute Toxicity. LD_{50} is 2.75 g/kg BW in rats, and 2.5 g/kg BW in mice. Administration of the lethal doses led to CNS affection and development of cyanosis.

Long-term Toxicity and **Allergenic Effect.** See ***Diphenylamine***.

Reference:

Korolev, A. A., Arsen'yev, M. V., Vitvitskaya. B. R., et al., Experimental data on hygienic substantiation of diphenylamine and diphenyldiethylurea in water bodies, *Gig. Sanit.*, 5, 21, 1976 (in Russian).

N,N'-DI(TETRAZOLYL-5)CARBAMIDE

Molecular Formula. $C_4H_4N_9O$
M = 193.13

Trade Name. Porofor DTK-5.

Properties. Yellow powder. Poorly soluble in water and alcohol.

Applications. Used as a blowing (pore-formation) agent.

Acute Toxicity. In mice, LD_{50} is 3.95 g/kg BW. Administration of 8.0 g/kg BW caused convulsions. Death within 1 to 2 days.

Repeated Exposure. Rats were dosed by gavage with 2.0 g/kg BW for a month. The treatment caused an increase in blood *SH*-group level and leukocyte and reticulocyte count.

Reference:

Gnezdilova, A. I. and Volodchenko, V. A., Toxicity of porofor DTK-5, *Gig. Truda Prof. Zabol.*, 12, 59, 1983 (in Russian).

DODECANEDIOIC ACID

Molecular Formula. $C_{12}H_{22}O_4$
M = 230.31
CAS No 693-23-2
Abbreviation. DDA.

Synonyms. 1,10-Decanedicarboxylic acid; 1,10-Decamethylene dicarboxylic acid; Dicarboxydecane.

Properties. Slightly soluble in hot water.

Chemobiokinetics. Wistar rats were administered *i/g* with DDA. Starved and diabetic ketonic rats had larger excretion of *adipic acid* than unstarved rats. *Suberic acid* excretion did not differ significantly, but the excretion of *sebacic acid* was lower in the diabetic group compared with the unstarved and starved rats.

Regulations.

EU (1990). DDA is available in the *List of authorized monomers and other starting substances which may continue to be used for the manufacture of plastic materials and articles intended to come into contact with foodstuffs pending a decision on inclusion in Section A* (*Section B*).

Great Britain (1998). DDA is authorized up to the end of 2001 for use in the production of polymeric materials and articles in contact with food or drink or intended for such contact.

Reference:

Mortensen, P. B. and Gregersen, N., The biological origin of ketonic dicarboxylic aciduria. II. *In vivo* and *in vitro* investigations of the beta-oxidation of C_8-C_{16}-dicarboxylic acids in unstarved, starved and diabetic rats, *Biochim. Biophys. Acta*, 710, 477, 1982.

DODECYL ALCOHOL

Molecular Formula. $C_{12}H_{26}O$
M = 186.38

CAS No 112-53-8
RTECS No JR5775000

Synonyms. Duodecyl alcohol; 1-Hydroxydodecane.

Properties. At room temperature, colorless liquid with characteristic fatty odor and waxy flavor. Insoluble in water, soluble in alcohol.

Applications. Used in the production of detergents.

Acute Toxicity. LD_{50} seems to be more than 13 to 36 g/kg BW in rats and rabbits.[03]

Regulations. ***U.S. FDA*** (1998) approved the use of D. (1) in the manufacture of resinous and polymeric coatings for the food-contact surface of articles intended for use in producing, manufacturing, packing, processing, preparing, treating, packaging, transporting, or holding food in accordance with the conditions prescribed in 21 CFR part 175.300; and (2) in the manufacture of cellophane for packaging food in accordance with the conditions prescribed in 21 CFR part 177.1200.

DODECYLBENZENESULFONIC ACID

Molecular Formula. $C_{18}H_{30}O_3S$
M = 326.54
CAS No 27176-87-0
RTECS No DB6600000
Abbreviation. DBA.

Synonym and **Trade Names.** Laurylbenzenesulfonic acid; Richonic acid B; Sulfamin acid 1298.

Properties. Light yellow to brown substance.

Applications. Anionic detergent.

Acute Toxicity. LD_{50} is 650 mg/kg BW in rats.

Regulations. ***U.S. FDA*** (1993) approved the use of DBA (1) as a curing catalyst for urea-formaldehyde and triazine-formaldehyde resins used as the basic polymers in coating intended for food-contact use; and (2) as a component of sanitizing solutions used on food-processing equipment and utensils, and on other food-contact articles as specified in 21 CFR part 178.1010.

Reference:

Glohuberg, Ch., Toxicological properties of surfactants, *Arch. Toxicol.*, 32, 245, 1974.

DODECYLGUANIDINE ACETATE

Molecular Formula. $C_{13}H_{29}N_3.C_2H_4O_2$
M = 287.44
CAS No 2439-10-3
96923-04-5
RTECS No MF1750000
Abbreviation. DGA.

Synonyms and **Trade Names.** Carpene; Cyprex; Dodecylguanidine monoacetate; Doguadine; Laurylguanidine acetate; Venturol.

Properties. Colorless crystals or slightly waxed solid. Soluble in hot water and alcohol.

Applications. Used in the manufacture of surface-active agents and lubricants. A solvent in the manufacture of coatings made of urea and melamine resins. Used in the production of plasticizers. An agricultural fungicide.

Acute Toxicity. LD_{50} is 1.1 g/kg BW in rats, 0.23 g/kg BW in mice, 0.54 g/kg BW in rabbits, and 0.18 g/kg BW in guinea pigs.[1]

According to other data, LD_{50} is 0.66 to 1.0 g/kg BW in rats. Poisoning caused CNS inhibition; hypermotility and diarrhea were observed in GI tract.[2] Death occurred in several days after treatment. Gross pathology examination revealed mild irritation of the gut with adhesions of the stomach and liver. Dogs vomited large doses.[019]

Repeated Exposure revealed moderate cumulative properties. K_{acc} appeared to be 4.0 and 5.4 in guinea pigs and mice, respectively, upon administration of 1/20 LD_{50}.[1]

Short-term Toxicity. In a 100-day feeding study, rats were given overall the dose of 3220 mg/kg. The treatment caused no increase in the mortality rate, but induced a marked reduction in food intake and BW gain.[3]

Long-term Toxicity. Rats received 800 mg DGA/kg diet for 2-year period. Slight retardation of BW gain was noted.[2]

Doses of 50 and 200 ppm in the diet were tolerated by rats over 2 years; there were no hematology or microscopic changes in the treated animals. Dogs were fed the diets containing 50, 200, or 800 ppm for 1 year. Slight stimulation of the thyroid gland was noted at the dose level of 800 ppm. Histological changes were also reported. The LOEL of 200 ppm was established in this study.[3]

Reproductive Toxicity. No effect on reproduction or lactation was observed in the above noted study.[3]

Mutagenicity. DGA was mutagenic in bacteria in the presence of a mouse liver microsomal preparation for metabolic activation.[4]

Carcinogenicity. Administration of DGA to pregnant mice produced lymphomas as well as lung and liver tumors in the dams and subsequently in the pups. This effect was seen only with coadministration.[5]

Regulations. ***U.S. FDA*** (1998) approved the use of DGA (1) as a component of the uncoated or coated food-contact surface of paper and paperboard intended for use in producing, manufacturing, packaging, processing, preparing, treating, packing, transporting, or holding aqueous and fatty foods in accordance with the conditions prescribed in 21 CFR part 176.170; (2) in the manufacture of resinous and polymeric coatings for polyolefin films for food-contact surface of articles intended for use in producing, manufacturing, packing, processing, preparing, treating, packaging, transporting, or holding food in accordance with the conditions prescribed in 21 CFR part 175.320; and (3) as a slimicide in the manufacture of paper and paperboard that contact food (DGA hydrochloride) as prescribed in 21 CFR part 176.300.

References:

1. Tovstenko, A. I., Hygienic evaluation of residual amounts of the fungicide capren (dodecylguanidineacetate) in apples, *Voprosy Pitania*, 5, 72, 1973 (in Russian).
2. *The Pesticide Manual*, C. R. Worthing and S. B. Walker, Eds., A World Compendium, 8th ed., The British Crop Protection Council, Thornton Heath, UK, 1987, 329.
3. Levinskas, G. J., Vidone, L. B., O'Grady, J. J., and Shaffer, C. B., Acute and chronic toxicity of dodine (*N*-dodecylguanidine acetate), *Toxicol. Appl. Pharmacol.*, 3, 127, 1961.
4. Greim, H., Bimboes. D., Egert, G., Goggelmann, W., and Kramer, M., Mutagenicity and chromosomal aberrations as an analytical tool for *in vitro* detection of mammalian enzyme-mediated formation of reactive metabolites, *Arch. Toxicol.* 39, 159, 1977.
5. Borzsonyi, M., Pinter, A., Surjan, A., and Torok, G., Carcinogenic effect of a guanidine pesticide administered with sodium nitrite on adult mice and the offspring after prenatal exposure, *Cancer Lett.*, 5, 107, 1978.

N-DODECYL SARCOSINE, SODIUM SALT

Molecular Formula. $C_{15}H_{30}NO_2.Na$

M = 279.45

CAS No 7631-98-3

RTECS No VQ3020000

Abbreviation. DSS.

Synonyms and **Trade Names.** Sodium *N*-dodecyl sarcosone; Sodium lauryl sarcosinate.

Acute Toxicity. LD_{50} *i/v* is 180 mg/kg BW in mice.

Mutagenicity.

In vitro genotoxicity. DSS showed positive response in DNA degradation test in primary rat hepatocytes in dose-dependent manner.

Regulations. ***U.S. FDA*** (1998) approved the use of DSS in adhesives as a component of articles intended for use in packaging, transporting, or holding food in accordance with the conditions prescribed in 21 CFR part 175.105.

Reference:

Elia, M. C., Storer, R. D., McKelvey, T. W., et al., Rapid DNA degradation in primary rat hepatocytes treated with diverse cytotoxic chemicals: Analysis by pulsed field gel electrophoresis and implications for alkaline elution assays, *Environ. Molec. Mutagen.*, 24, 181, 1994.

ELEMI OIL

CAS No 8023-89-0
RTECS No JX8730000
Abbreviation. EO.

Synonyms. Elemi.

The resin from *Manila* elemi.

Properties. Amorphous, brown mass. Slightly soluble in water, soluble in alcohol.[020]

Acute Toxicity. LD_{50} is 3.37 g/kg BW in rats.

Regulations. ***U.S. FDA*** (1998) approved the use of EO in resinous and polymeric coatings used as the food-contact surfaces of articles intended for use in producing, manufacturing, packing, processing, preparing, treating, packaging, transporting, or holding food in accordance with the conditions prescribed in 21 CFR part 175.300.

Reference:

Elemi oil, *Food Cosmet. Toxicol.,* 14, 755, 1976.

ELKAN-120

Synonym. Hydrated gum rosin, glycerin ester.

Properties. Clear, solid resin, dark yellow in color. Poorly soluble in water and alcohol.

Applications. Intended for making superconcentrates of dyes used in coloration of polystyrene plastics.

Acute Toxicity. The dose of 15 to 20 g/kg BW caused no mortality in rats and mice of both sexes. Poisoning did not affect BW gain or motor activity of the treated animals. Gross pathology examination failed to show changes in their viscera.

Repeated Exposure revealed no cumulative properties.

Long-term Toxicity. Rats and mice were dosed by gavage with 0.5 mg/kg BW dose (as 20% oil solution) for 6 and 4 months, respectively. Impaired conditioned reflex activity and leukocytosis, as well as decreased thyroid relative weights are reported. There was no retardation of BW gain or histology changes in the visceral organs of the treated animals.

Recommendations. ***Russia.*** E-120 is recommended for use as a component of pigment superconcentrates.

Reference:

Toxicology and Sanitary Chemistry of Polymerization Plastics, Coll. works, B. Yu. Kalinin, Ed., Leningrad, 1984, 44 (in Russian).

ETHENYLBENZENESULFONIC ACID, HOMOPOLYMER, SODIUM SALT

Molecular Formula. $[\sim C_8H_8O_3S\sim]_n Na$
CAS No 9080-79-9
9003-59-2
RTECS No DB6836000

Synonyms and **Trade Name.** Flexan 500; Sodium carbonate stabilized sulfonated polystyrene, sodium salt; Sodium polystyrene sulfonate.

Acute Toxicity. LD_{50} exceeded 8.0 g/kg BW in rats and 10 g/kg BW in mice.

Regulations. ***U.S. FDA*** (1998) approved the use of E. in adhesives as a component of articles intended for use in packaging, transporting, or holding food in accordance with the conditions prescribed in 21 CFR part 175.105.

1-(ETHENYLOXY)OCTADECANE

Molecular Formula. $C_{20}H_{40}O$
M = 296.60
CAS No 930-02-9
RTECS No RG0300000

Synonyms. Octadecyl vinyl ether; Stearyl vinyl ether; Vinyloctadecyl ether; Vinyl stearate; Vinyl stearyl ether.

Acute Toxicity. LD exceeded 14 g/kg BW in rats.

Regulations.

U.S. FDA (1998) approved the use of VS in adhesives as a component (monomer) of articles intended for use in packaging, transporting, or holding food in accordance with the conditions prescribed in 21 CFR part 175.105.

EU (1990). VS is available in the *List of authorized monomers and other starting substances which may continue to be used for the manufacture of plastic materials and articles intended to come into contact with foodstuffs pending a decision on inclusion in Section A* (*Section B*).

Reference:

Information from the Soviet Toxicology Center, *Gig. Truda Prof. Zabol.*, 11, 51, 1985 (in Russian).

ETHYLBENZENE

Molecular Formula. C_8H_{10}
M = 106.17
CAS No 100-41-4
RTECS No DAO700000
Abbreviation. EB.

Synonyms. Ethylbenzol; Phenylethane.

Properties. Colorless liquid at room temperature with an aromatic odor. Solubility in water is 0.017% at 25°C. Readily soluble in alcohol and in most organic solvents. Odor perception threshold is 0.029 mg/l.[02] Taste perception threshold values vary from 0.072 to 0.2 mg/l.[01]

Applications. EB is primarily used in the production of styrene, of cellulose acetate silk, and in the styrene-butadiene rubber manufacture.

Exposure. EB is distributed widely in the environment because it is often used in foils and as a solvent.

Migration Data. EB may migrate from polystyrene packaging materials into food. Concentrations of 0.0025 to 0.021 mg/l in milk beverage and soup have been reported.[1] Release of EB from polystyrene food-contact wares into *n*-heptane was not observed at the detection limit of 0.1 ppm.[2]

Acute Toxicity. LD_{50} for Wistar rats of both sexes is 3.5 g/kg BW.[3] The maximum tolerated dose for rats is 2.0 g/kg BW. The dose of 1.0 g/kg BW affects the NS function.[4]

Poisoned animals displayed primary excitation with subsequent severe CNS depression, impaired motor coordination, and, later, a progressive decline in body temperature, respiratory and cardiac activity, convulsions, and death from respiratory center paralysis. Histology examination revealed signs of venous congestion with point hemorrhages[2] and moderate changes in the kidneys. Administration of 1.0 g/kg BW to rabbits caused leukocytosis and thrombocytopenia.[5]

Repeated Exposure failed to reveal the evidence of accumulation. The treatment caused changes in hematology analyses (neutrophilia and relative lymphopenia) and in the rate of motor conduction along the sciatic nerve. Histology examination revealed alterations in the structure of the visceral organs (Mitran et al., 1986).

Long-term Toxicity. Rats received doses of 13.6 to 680 mg/kg BW for 6 months. The treatment caused no effects on growth, mortality, behavior, and hematology analysis. The NOAEL of 136 mg/kg BW for histopathologic changes in liver and kidney was reported.[4]

Rabbits received 200 mg/kg BW for the same period of time. The treatment caused retardation of BW gain, marked changes in CNS function, and in blood morphology. Gross pathology examination revealed changes in the visceral organs. In rabbits given water containing 2.0 mg EB/l, there were no such changes.[6]

Immunotoxicity. Simultaneous immunization and poisoning of rats caused some inhibition of the plasmocytic reaction and a stable rise in antibody titre. A decline in the blood serum bactericidal capacity was noted.[7]

Reproductive Toxicity. A mild ***teratogenic*** and ***embryotoxic effect*** was observed in rats following inhalation exposure to 2.4 g/m^3 on days 7 to 15 of pregnancy.[8] EB is shown to cross the placenta.

Mutagenicity. EB did not show mutagenic activity in bacteria, yeasts, insects, mammalian cells *in vitro* and in mammals.

In vitro genotoxicity. No induction of SCE or CA were found in Chinese hamster ovary cells, but a weak positive response was reported for SCE induction in human lymphocytes cultured with *S9*.[1,6-9]

Carcinogenicity. Positive results are noted in rats.[09]

Chemobiokinetics. EB, in a liquid form, is easily absorbed by humans via the skin and via the GI tract. Distribution and excretion are rapid processes. In humans, it is stored in the fat. EB is likely to be metabolized almost completely into *mandelic acid* and *fenylglyoxalic acid*, both compounds being excreted in the urine. Metabolism in experimental animals differs from that in humans in the formation of *benzoic acid* as the major metabolite along with *mandelic* acid. Urinary excretion of metabolites is almost completed within 24 hours.[1,7]

Recommendations.

WHO (1996). Guideline value for drinking water: 0.3 mg/l.[10] The levels of 0.0024 to 0.2 mg/l are likely to give rise to consumer complaints of foreign odor and taste.

U.S. EPA (1999). Health Advisory for a longer-term exposure is 3.0 mg/l.

Standards.

U.S. EPA (1999). MCL and MCLG: 0.7 mg/l.

Canada (1989). MAC: 0.0024 (organolept.).

Russia (1995). MAC and PML: 0.01 mg/l (organolept., taste).

References:

1. ECETOC Joint Assessment of Commodity Chemicals, Issue No 7, *Ethylbenzene,* CAS 100-41-4, Report from the European Chemical Industry, Ecology and Toxicology Center, August 1, 1986.
2. Ito, S., Hosogai, T., Sakurai, H., Tada, Y., Sugita, T., Ishiwata, H., and Takada, M., Determination of volatile substances and leachable components in polystyrene food contact wares, Abstract, *Eisei Shikenjo Hokoku,* 110, 85, 1992 (in Japanese).
3. Zubritsky, K. V., in *Protection of Reservoirs against Pollution by Industrial Liquid Effluents*, S. N. Cherkinsky, Ed., Medgis, Moscow, Issue No 5, 1962 62 (in Russian).
4. Wolf, M. A., Rowe, V. K., McCollister, D. D., Hollingworth, R. L. and Oyen, F., Toxicological studies of certain alkylated benzenes and benzene, *Arch. Ind. Health*, 14, 387, 1956.
5. Faustov, A. S. and Volchkova, R. I., in *Hygienic Characteristics of Liquid Effluents from the SR (Synthetic Rubber) Industry and the Problems of Treatment*, Proc. Voronezh Medical Institute, Voronezh, Issue No 35, 1958, 263 (in Russian).
6. Norppa, H. and Vainio, H., Induction of SCE by styrene analogues in cultured human lymphocytes, *Mutat. Res.,* 116, 379, 1983.
7. U.S. EPA, Office of Drinking Water Health Advisors, *Rev. Environ. Contam. Toxicol.,* 106, 123, 1988.
8. Ungvary, G. and Tatrai, E., On the embryotoxic effects of benzene and its alkyl derivatives in mice, rats and rabbits, *Arch. Toxicol.,* Suppl. 8, 425, 1985.
9. Janssen, P. and van der Heiden, C. A., *Summary and Evaluation of Toxicological Data on Ethylbenzene,* March 21, 1987. Doc. from the Toxicol. Advisory Centre - Natl. Institute Publ. Health and Environ. Prαect., The Netherlands, 1987.
10. *Revision of the WHO Guidelines for Drinking Water Quality*, Report on the Consolidation Meeting on Organics and Pesticides, Medmenham, UK, January 30-31,1992.

ETHYL HEPTAFLUOROISOBUTENYL ETHER

Molecular Formula. $C_6H_5F_7O$

M = 226.09

CAS No 360-58-7

RTECS No KN9950000

Synonyms. 1-Ethoxyperfluoro-2-methyl-1-propene; Perfluoroisobutenylethyl ester.

Properties. A colorless liquid with a specific aromatic odor. Solubility in water is above 400 mg/l; readily soluble in alcohol. Odor perception threshold is 2.8 mg/l, practical odor threshold is 2 to 6 mg/l. At this concentration it does not change water color.

Applications. Used in the manufacture of fluoroplastics.

Acute Toxicity. In mice, LD_{50} is 164 mg/kg BW. Lethal doses cause excitation with subsequent apathy, labored breathing, cyanosis of the paws and tail. Death within 2 days.

Repeated Exposure produced impairment of some hematology parameters. The sugar curve was altered under galactose loading and the prothrombin time was shortened. Gross pathology examination revealed a marked congestion in the viscera, fatty degeneration in the liver, and a reduced glycogen contents in the hepatocytes.

Long-term Toxicity. Rats and rabbits were exposed by gavage to the oral doses of 0.015 to 6.0 mg/kg BW for 7 months. No hematology changes were found in the treated animals. The highest dose caused impairment of prothrombin-forming liver function and distortion of the glycemic curves. No morphological changes were found in the parenchymatous organs and CNS at the lowest dose.

Standards. ***Russia***. Recommended PML: 0.3 mg/l.

Reference:

Shvartsman, I. E., Data on hygienic standardization of perfluoroisobutenyl ether in water bodies, *Gig. Sanit.*, 5, 13, 1964 (in Russian).

4-ETHYL-4-HEXADECYL MORPHOLINIUM ETHYLSUFATE

Molecular Formula. $C_{22}H_{46}NO.C_2H_5O_4S$

M = 465.82

CAS No 78-21-7

RTECS No QF3195000

Synonyms. Cetylethyl moroholinium ethosulfate; Sulphuric acid, 4-ethyl-4-hexadecylmorpholinium, monoethyl ester.

Regulations. ***U.S. FDA*** (1998) approved the use of E. in the production of or added to textiles and textile fibers intended for use in producing, manufacturing, packing, processing, preparing, treating, packaging, transport ing, or holding food, subject to the provisions of 21 CFR part 177.2800.

FATTY (C_{10}-C_{16}) ACIDS, DIETHANOLAMIDES, a mixture

Molecular Formula. $RCON(CH_2CH_2OH)_2$

Abbreviation. FAD.

Synonym. Fatty acids, *N,N'*-di-2-hydroxyethylamides.

Properties. A dark yellow, paste-like substance.

Applications. Used as an antistatic additive.

Migration of fatty acid amides from polyvinyl chloride used in food packaging and containing commonly used fatty acid amide slip additives into fat and aqueous food simulant media was less than 0.05 mg/l (10 day, 40°C).[1]

Acute Toxicity. LD_{50} is 7.5 g/kg BW for rats, and 5.0 g/kg BW for mice.[2]

Long-term Toxicity. Chronic exposure revealed some CNS inhibition in mice. The relative weights of the lungs and kidneys are found to be increased in mice and rats.[2]

Regulations. ***U.S. FDA*** (1998) approved the use of FAD (C_{12}-C_{16}) as a component of the uncoated or coated food-contact surface of paper and paperboard intended for use in producing, manufacturing, packaging, processing, preparing, treating, packing, transporting, or holding dry food in accordance with the conditions prescribed in 21 CFR part 176.180.

References:

1. Cooper, I. and Tice, P.A., Migration studies on fatty acid amide slip additives from plastics into food simulants, *Food Addit. Contam.*, 12, 235, 1995.

2. Mikhailets, I. B., Toxic properties of some antistatics, *Plast. Massy*, 12, 27, 1976 (in Russian).

FERRIC CHLORIDE

Molecular Formula. Cl_3Fe
M = 162.20
CAS No 7705-08-0
RTECS No LJ9100000
Abbreviation. FC.

Synonyms and **Trade Name.** Flores martis; Iron chloride; Iron trichloride.

Properties. Black-brown hexagonal crystals, dark leaflets or plates. Soluble in cold and hot water; readily soluble in alcohol.

Exposure. Iron is a nutrient and/or dietary supplement.

Applications. Used as a food additive, a hemostatic, or treatment agent for hypochromic anemia.

Acute Toxicity. LD_{50} is 450 mg/kg BW in rats,[1] and 895 mg/kg BW in mice.[2]

Reproductive Toxicity.

Gonadotoxicity. Intratesticular injection of 13 mg/kg BW one day prior to mating affects spermatogenesis, testis, epididymus, sperm duct.[3]

Mutagenicity.

In vitro genotoxicity. Elemental and salt forms of iron, including compounds currently being used in dietary supplements and for food fortification, did not induce a positive response in *Salmonella* mutagenicity assay (except for the weak response obtained with ferrous fumarate). In L5178Y mouse lymphoma assay, FC caused an increase in mutant frequency only with *S9*. Ferrous sulfate and ferrous fumarate gave positive responses without *S9*.[4]

Carcinogenicity. F344 rats of both sexes received FC dissolved in distilled water at levels of 0.25 or 0.5% as their drinking water for up to 2 years. The mean BW of the treated rats were lower than control group values for both males and females. A variety of tumors developed in all groups, including the control group, but all these neoplasms were histologically similar to those known to occur spontaneously in this strain of rats, and no statistically significant increase in the incidence of any tumor was found in the treated groups of either sex. It was concluded that under the conditions of this study, FC exerts no carcinogenic potential in F344 rats.[6]

Chemobiokinetics. Free circulating iron damages systemic blood vessels (Ellenhorn and Barceloux, 1988). Iron was shown to cross placenta and concentrate in fetus. In cases of overload, iron is excreted predominantly in the urine and feces. A part of it is excreted via the bile.[5]

In cell, Fe^{2+} is converted to Fe^{3+} in ferritin, the latter not being absorbed until cell is physiologically "depleted." In the blood stream, iron could be quickly oxidized by dissolved oxygen to Fe^{3+}, which complexes with specific *Fe*-transport β1-globulin.[03]

Regulations. ***U.S. FDA*** (1998) approved the use of FC in adhesives as a component of articles intended for use in packaging, transporting, or holding food in accordance with the conditions prescribed in 21 CFR part 175.105. Affirmed as *GRAS*.

References:

1. Volodchenko, V. A., Piatnitzkaya, L. V., and Mantokovsky, V. V., Hygienic characteristics of the technological process for obtaining β-naphthol, *Gig. Sanit.*, 5, 16, 1974 (in Russian).
2. *The Annual Report of Tokyo Metropolitan Res. Lab. Publ. Health*, 27, 159 1976.
3. *J. Reprod. Fertility*, 7, 21, 1964.
4. Dunkel, V. C., San, R. H., Seifried, H .E., and Whittaker, P., Genotoxicity of iron compounds in *Salmonella typhimurium* and L5178Y mouse lymphoma cells, *Environ. Mol. Mutagen.*, 33, 28, 1999.
5. Casarett and Doull's *Toxicology*, 2nd ed., Macmillan Publishing Co., New York, 1980, 446.
6. Sato, M., Furukawa, F., Tolyoda, K., Mitsumori, K., Nashikawa, A., and Takahashi, M., Lack of carcinogenicity of ferric chloride in F344 rats, *Food Chem. Toxicol.*, 30, 837, 1992.

FERROCENE

Molecular Formula. $C_{10}H_{10}Fe$
M = 186.03
CAS No 102-54-5
RTECS No LK0700000

Synonyms. Bis(cyclopentadienyl)iron; Dicyclopentadienyl iron.

Properties. Orange crystals with a camphoric odor. A relatively volatile, organometallic compound. Poorly soluble in water, soluble in alcohol.

Applications. Used as a catalyst or photosensi tizer in the manufacture of plastics. A food additive.

Acute Toxicity. LD_{50} is reported to be 1190 to 2260 mg/kg BW in rats, and 380 to 930 mg/kg BW in mice. Gross pathology examination showed enteritis and deposition of F. in the fatty tissue of the liver.[1]

Repeated Exposure. No mortality or clinical signs of intoxication were observed in F344 rats and $B6C3F_1$ mice exposed to 2.5 to 40 mg F./m^3 for 2 weeks.[2]

Short-term Toxicity. Mice were given 1/20, 1/10, or 1/5 LD_{50} for 3.5 months. Retardation of BW gain and changes in the functional condition of the CNS were observed. A decline in *Hb* blood level and impaired liver function were found in rats administered 1/20 LD_{50} for 3 months. Liver weights were increased and F. accumulation was noted in the liver, spleen, lungs, and suprarenals. Gross pathology examination revealed exudative glomerulonephritis and changes in the thyroid.[1]

Allergenic Effect is not observed.

Mutagenicity.

In vitro genotoxicity. F. appeared to be positive in *Salmonella* mutagenicity assay with metabolic activation. It produced a doubling of the SCE frequency, and a 3-point monotonic increaase with at least the highest dose at the $p < 0.001$ significance level.[3]

However, in NTP-94 studies, F. was reported to be negative in *Salmonella* assay.

In vivo cytogenetics. A long-term administration of 1/10 LD_{50} to rats caused mutagenic effect in the bone marrow cells. 1/20 LD_{50} appeared to be ineffective in this study.[4]

F. was positive in the sex-linked recessive lethal test in *Dr. melanogaster.*[5]

Carcinogenicity. F. has structural similarities to other metallocenes that have been shown to be carcinogenic. F344 rats were given 0.25 or 0.5% F. in their drinking water *ad libitum* for up to 2 years. F. exerted no carcinogenic potential.

Chemobiokinetics. Organic portion of F. is readily absorbed and rapidly metabolized. Carbon portion of the molecule is nearly completely eliminated, primarily in the urine. F. labeled with ^{55}Fe is excreted primarily in the urine. Major metabolite seems to be *ferrocenyl glucuronide* (NTP-92).

Standards. ***Russia*** (1995). MAC (ferrocyanide iron): 0.2 mg/l (organolept., suspension)

References:

1. Mishina, A. D., Putilina, L. V., Yevsyukov, V. I., et al., Toxicity of some photosensitizers of plastics, in *Hygiene and Toxicology of High-Molecular-Mass Compounds and of the Chemical Raw Material Used for Their Synthesis*, Proc. 6th All-Union Conf., B. Yu. Kalinin, Ed., Leningrad, 1979, 275 (in Russian).
2. Sun, J. D., Dahl, A. R., Gillett, N. A., et al., Two-week repeated inhalation exposure of F344 rats and $B6C3F_1$ mice to ferrocene, *Fundam. Appl. Toxicol.*, 17, 150, 1991.
3. Tucker, J. D., Auletta, A., Cimino, M. C., et al., Sister-chromatid exchange: second report of the Gene-Tox program, *Mutat. Res.*, 297, 101, 1993.
4. Grigor'yeva, M. N., Kalinin, B. Yu., Novikova, O. I., et al., Investigation of mutagenic action of some plastics and polymers, in *Hygienic Aspects of the Use of Polymeric Materials*, Proc. 3rd All-Union Meeting on New Methods of Hygienic Monitoring the Use of Polymers, K. I. Stankevich, Ed., Kiyv, December 2-4, 1981, 319 (in Russian).
5. Zimmering, S., Mason, J. M., Valencia, R., et al., Chemical mutagenesis testing in Drosophila. II. Result of 20 coded compounds tested for the National Toxicology Program, *Environ. Mutagen.*, 7, 81, 1985.

FERROCENE DERIVATIVE

Trade Name. FEP-2.

Properties. Dark-brown, oily liquid with a penetrating odor. Poorly soluble in water. Contains diethylferrocene as an impurity.

Applications. Organometallic photosensitizer which is effective in plastics intended for agricultural use. The quantity incorporated into plastics does not exceed 0.1%.

Acute Toxicity. In female rats, LD_{50} is 17.1 g/kg BW for technical grade product. Male mice and rats of both sexes tolerate the administration of 20 g/kg BW of the distilled FEP-2 and technical grade product. No changes in behavior or in BW gain were observed. However, gross pathology and histology examination revealed exudative glomerulonephritis in these animals.

Repeated Exposure failed to reveal marked cumulative properties. K_{acc} is 6. The treatment caused a reduction in *Hb* blood level, and an increase in the relative weights of the thyroid, liver and kidney. Histology examination revealed deposition of dark-brown drops in the viscera.

Short-term Toxicity. Mice were dosed by gavage with 200 and 1000 mg/kg BW, and rats with 500 and 1000 mg/kg BW as an oil suspension for 4 months. Animals looked unkempt with tousled fur. A 1000 mg/kg BW dose caused 50% mortality in mice of both sexes; a dose of 200 mg/kg caused 23% mortality.

Long-term Toxicity. In a 13-month study, rats and mice were dosed by gavage with 50 mg FEP-2/kg BW. The treatment produced a pronounced toxic effect. The NOAEL of 5.0 mg/kg BW was identified in this study.

Reproductive Toxicity.

Embryotoxic and ***teratogenic effects*** were not found upon administration of 1.0 g/kg BW to pregnant rats.

Mutagenicity.

In vivo cytogenetics. Metaphase analysis of bone marrow cells revealed a cytogenetic effect in animals treated with 50 mg/kg BW in a long-term study.The NOEL for this effect is 20 mg/kg BW.

Standards. ***Russia.*** PML in food: 0.5 mg/l; recommended PML: *n/m*.

Recommendations. Joint FAO/WHO Expert Committee on Food Additives. ADI: 5.0 mg/kg BW.

Reference:

Toxicology and Sanitary Chemistry of Polymerization Plastics, Coll. Works, B. Yu. Kalinin, Ed., Leningrad, 1984, 34 (in Russian).

FIBROUS GLASS

RTECS No LK3651000

Abbreviation. FG.

Trade Names. Fiberglass; Glass; Glass fibers.

Composition. A borosilicate of low alkalinity, consisting of calcium, aluminum, and silicone compounds (ACGIN, 1986-1991).

Mutagenicity.

In vitro genotoxicity. FG caused no dose-related increase in SCE levels in Chinese hamster ovary cells, human fibroblasts, or lymhoblastoid cells.[1]

Regulations. ***U.S. FDA*** (1998) approved the use of FG as a component of phenolic resins in molded articles intended for repeated use in contact with non-acid food (*pH* above 5.0), in accordance with the conditions prescribed in 21 CFR part 177.2410.

Reference:

Casey, G., Sister chromatid exchange and cell kinetics in CHO-K1 cells, human fibroblasts and lymhoblastoid cells exposed *in vitro* to asbestos and glass fiber, *Mutat. Res.*, 116, 369, 1983.

FISH OIL

CAS No 8016-13-5

RTECS No LK5150000

Repeated Exposure. Feeding rats a diet containing 20% Norway Haddoch oil for 4 weeks did not affect heart weight or amount of heart lipids.[1]

Reproductive Toxicity. Olsen et al. noted an increased length of pregnancy among fish-eating women from the Faroe Islands. The group of women ingesting fish oil averaged 4 days longer gestation and had newborns 102 g heavier than those taking olive oil.[2]

Chemobiokinetics. Mice were fed a diet supplemented with FO for 4 weeks. FO was found to inhibit tryglyceride formation in their livers.[3]

Regulations. ***U.S. FDA*** (1998) approved the use of FO in resinous and polymeric coatings used as the food-contact surfaces of articles intended for use in producing, manufacturing, packing, processing, preparing, treating, packaging, transporting, or holding food in accordance with the conditions prescribed in 21 CFR part 175.300.

References:

1. Lang, K. and Reimold, W. V., Changes of lipids and fatty acid composition in myocardial tissue of rats under the influence of red-perch oil and coconut fat in a long-term feeding experiment *Z. Ernahrungswiss.*, 10, 325, 1971.
2. Olsen, S. F., Sorensen, J. D., Secher, N . J., Hedegaard, M., Henriksen, T. B., Hansen, H. S., and Grant, A., Randomized controlled trial of effect of fish-oil supplementation on pregnancy duration, *Lancet*, 339, 1003, 1992.
3. Kudo, N. and Kawsshima, Y., Fish oil feeding prevents perfluorooctanoic acid-induced fatty liver in mice, *Toxicol. Appl. Pharmacol.,* 145, 285, 1997.

FLUOROSILICIC ACID

Molecular Formula. $F_6Si.2H$
M = 144.11
CAS No 16961-83-4
1309-45-1
RTECS No VV8225000
Abbreviation. FSA.

Synonyms and **Trade Names.** Dihydrogen hexafluorosilicate; Hexafluosilicic acid; Hydrofluosilicic acid; Sand acid; Silicofluric acid.

Properties. Colorless liquid with a sour, pungent odor. Soluble in cold and hot water. Marketed as aqueous solution only. Fairly strong acid.[020]

Applications. Used in the manufacture of ceramics.

Long-term Toxicity. Ingestion of large doses led to osteosclerosis and mottled enamel, increased density, and calcification of bones.

Carcinogenicity classification. The IARC Working Group concluded that FSA is not classifiable as to its carcinogenicity *to humans*.

IARC: 3.

Regulations. ***U.S. FDA*** (1998) approved the use of FSA in adhesives as a component of articles intended for use in packaging, transporting, or holding food in accordance with the conditions prescribed in 21 CFR part 175.105.

Reference:

Goodman and Gilman's *The Pharmacological Basis of Therapeutics*, A. G. Gilman, L. S. Goodman, and A. Gilman, Eds., 6th ed., Macmillan Publishing Co., Inc., New York, 1980, 1546.

FORMIC ACID

Molecular Formula. CH_2O_2
M = 46.03
CAS No 64-18-6
RTECS No LQ4900000
Abbreviation. FA.

Synonyms. Aminic acid; Formylic acid; Hydrogen carboxylic acid; Methanoic acid.

Properties. White, crystalline powder or colorless, highly caustic liquid with a pungent odor. Miscible with water, alcohol, and ether.[1] Odor perception threshold is 1700 mg/l.[02]

Applications. Used as a plasticizer for vinyl resins and a coagulant for latex.

Migration Data. Thermal degradation of polyethylene during manufacturing may result in the release of FA.

Exposure. FA is present in a free acid state in a number of plants. Human exposure occurs due to consumption of food and water containing FA.

Acute Toxicity.

Observations in man. Ingesting by man is accompanied by severe intoxication (sometimes, hematuria, anuria, uremia, circulatory failure, or pneumonia) and death.[2]

Animal studies. LD_{50} is 1.1 to 1.85 g/kg BW in rats, 0.7 to 1.1 g/kg BW in mice, and 4.0 g/kg BW in dogs.[012]

Repeated Exposure. Young rats received FA in the diet or drinking water at levels 0.5 to 1.0% for 6 weeks. The treatment caused a reduction in BW at the higher dose levels. Hypochromic anemia and mild lymphocytosis developed in rats receiving FA in the diet.[03] The concentration of 124 mg/l has been proposed as a water standard.[2]

Reproductive Toxicity.

Gonadotoxicity. In a 13-week inhalation study, there were no effects on sperm motility, density, and testicular or epididymal weights in rats and mice. No changes were found to develop in the estrous cycle.[1]

Embryotoxicity. Embryotoxic effect and some growth reduction in whole embryo cultures of rats and mice were observed due to addition of 2 to 3 mg FA/ml.[3,4]

Survival of the offspring of female rats exposed to 1.0% FA in their drinking water for up to 7 months was reduced to 50 to 67%.[5]

Mutagenicity. FA is not mutagenic by itself but could give positive results when tested in concentrations that produced non-physiological *pH* levels. It induced or inhibited clastogenic effects probably due to alterations in *pH* caused by the addition of FA to the *in vitro* systems being studied.[6]

In vitro genotoxicity. FA is not shown to be positive in *Salmonella* mutagenicity assay with or without *S9*;[5] it was not reported to induce SCE in Chinese hamster *V79* cells treated with a maximum dose of 2.0 mM,[8] but some increase in SCE in cultured human lymphocytes was found at a concentration of 10 mM.[9]

Chemobiokinetics. Following absorption, FA is oxidized to CO_2 and H_2O, predominantly in the liver, partly metabolized in the tissues, and partly excreted unchanged in the urine.[10] FA is an inhibitor of the mitochondrial cytochrome oxidase causing histotoxic hypoxia. The most significant acid load results from the hypoxic metabolism. Urinary acidification is affected by FA.[11]

Regulations. ***U.S. FDA*** (1998) listed FA as a constituent of paper and paperboard used for food packaging at levels not to exceed GMP. It is considered to be *GRAS*.

Recommendations. Joint FAO/WHO Expert Committee on Food Additives (1999). ADI: No safety concern.

References:

1. *Toxicity Studies of Formic Acid Administered by Inhalation to Fischer 344 Rats and B6C3F$_1$ Mice* (M. Thompson), NTP Technical Report, Research Triangle Park, NC, NIH Publ. 92-3342, July 1992, 41.
2. *Handbook of Toxic and Hazardous Chemicals and Carcinogens*, 2nd ed., Sittig, M., Ed., Noyes Publ., Park Ridge, NJ, 1985, 465.
3. Ebron-McCoy, M. E., Nichols, H. P., and Andrews, J. E., Effects of altered pH of culture medium on embryogenesis in WEC, Abstract, *Teratology*, 49, 393, 1994.
4. Andrews, J. E., Ebron-McCoy, M., Kavlock, R. J., and Rogers, J. M., Developmental toxicity of formate and formic acid in whole embryo culture: A comparative study with mouse and rat embryos, *Teratology*, 51, 243, 1995.
5. Tracor Jitco, Inc., *Scientific Literature Reviews on Generally Recognized as Safe Food Ingredients - Formic Acid and Derivatives*, Prep. by Tracor Jitco, Inc. for FDA, NTIS Publ. PB-228 558, U.S. Dept. of Commerce, Washington, D.C., 1974.

6. Morita, T., Takeda, K., and Okumura, K., Evaluation of clastogenicity of formic acid, acetic acid and lactic acid on cultured mammalian cells, *Mutat. Res.*, 240, 195, 1990.
7. Zeiger, E., Anderson, B., Haworth, S., et al., *Salmonella* mutagenicity tests. V. Results from the testing of 311 chemicals, *Environ. Molec. Mutagen.*, 19 (Suppl. 21.), 2, 1992.
8. Basler, A., Hyde, W. D., and Scheutwinkel-Reich, M., Formaldehyde-induced sister-chromatid exchanges *in vitro* and the influence of the exogenous metabolizing systems *S9* mix and primary rat hepatocytes, *Arch. Toxicol.*, 58, 10, 1985.
9. Siri, P., Jarventaus, H., and Norppa, H., Sister chromatid exchanges induced by vinyl esters and respective carboxylic acids in cultured human lymphocytes, *Mutat. Res.*, 279, 75, 1992.
10. Clay, R. L., Morphy, E. C., and Watkins, W. D., Experimental ethanol toxicity in the primate: analysis of metabolic acidosis, *Toxicol. Appl. Pharmacol.*, 34, 49, 1975.
11. Liesivuori, J. and Savolainen, H., Methanol and formic acid toxicity: biochemical mechanisms, *Pharmacol. Toxicol.*, 69, 157, 1991.

FORMIC ACID, SODIUM SALT

Molecular Formula. $CH_2O_2.Na$
M = 68.02
CAS No 141-53-7
RTECS No LR0350000
Abbreviation. FAS.
Synonyms. Salachlor; Sodium formate.

Properties. White granules or colorless monoclinic crystals. One part of FAS dissolves in about 1 part of water at 20°C. Slightly soluble in alcohol.

Applications. Used in dyeing and printing fabrics.[020]

Acute Toxicity.

Observations in man. Chemical of low toxicity: ingestion of 10 g FAS produced no ill effects in man. It inhibits certain enzymes (catalase) and probably affects brain causing convulsions.[022]

Chemobiokinetics.

Observations in man. Formate ion is extensively oxidized to *formic acid* in the liver.[022]

2.0% of the dose was eliminated as *formic acid* in urine in 24 hours after ingestion or 3.0 to 4.5 g FAS.[1]

Regulations. ***U.S. FDA*** (1998) approved the use of FAS in adhesives as a component of articles intended for use in packaging, transporting, or holding food in accordance with the conditions prescribed in 21 CFR part 175.105.

Reference:

1. Malorny, G., Metabolism studies with sodium formate and formic acid in man, *Z. Ernaehrungswiss.*, 9, 340, 1969.

FUMARIC ACID

Molecular Formula. $C_4H_4O_4$
M = 116.08
CAS No 110-17-8
RTECS No LS9625000
Abbreviation. FA.

Synonyms. Allomaleic acid; 2-Butenedioic acid; 1,2-Ethylenedicarboxilic acid.

Properties. Needles monoclinic prisms or leaflets from water. Colorless odorless crystals or white crystal powder with a fruit acid odor. Solubility in water is 630 mg/100 g at 25°C. Soluble in alcohol.

Applications. Fortifier in paper size resins, unsaturated polyester resins and alkyd surface coating resins.

Acute Toxicity. LD_{50} is 10.7 g/kg BW in rats, LD_0 is 5.0 g/kg BW in rabbits.[03]

Repeated Exposure. A dose of 100 mg FA *monoethyl ester*/kg BW have distinct nephrotoxic effects in the rats: morphological lesions of the glomeruli, glomerular filtration rate was diminished by about 40%.[1]

Long-term Toxicity. Rats were given 0.05, 1.0, or 1.5% FA in the diet for 730 days. The LOEL of 1.08 g/kg BW was established.[2] Munro et al. (1996) suggested the calculated NOEL of 0.72 mg/kg BW for this study.

Carcinogenicity. Treatment with FA (1.0% in the diet and 0.025% in drinking water) enhanced DNA synthesis of hepatocytes and, thus, suppressed the carcinogenesis in the liver of male Donryu strain rats fed *3'-methyl-4-(dimethylamino)azobenzene*.[3]

Regulations.

EU (1990). FA is available in the *List of authorized monomers and other starting substances which shall be used for the manufacture of plastic materials and articles intended to come into contact with foodstuffs (Section A)*.

EU (1995). FA is a food additive generally permitted for use in foodstuffs (Maximum Level in food: 1.0 to 4.0 g/l).

U.S. FDA (1998) approved the use of FA (1) in adhesives as a component of articles (monomer) intended for use in packaging, transporting, or holding food in accordance with the conditions prescribed in 21 CFR part 175.105; (2) in the manufacture of resinous and polymeric coatings for the food-contact surface of articles intended for use in producing, manufacturing, packing, processing, preparing, treating, packaging, transporting, or holding food in accordance with the conditions prescribed in 21 CFR part 175.300; (3) in the manufacture of resinous and polymeric coatings for polyolefin films for food-contact surface of articles intended for use in producing, manufacturing, packing, processing, preparing, treating, packaging, transporting, or holding food in accordance with the conditions prescribed in 21 CFR part 175.320; (4) as a component (monomer) of the uncoated or coated food-contact surface of paper and paperboard intended for use in producing, manufacturing, packaging, processing, preparing, treating, packing, transporting, or holding dry food in accordance with the conditions prescribed in 21 CFR part 176.180; (5) in the manufacture of cellophane for packaging food in accordance with the conditions prescribed in 21 CFR part 177.1200; and (6) in the manufacture of polyurethane resins which may be used as food-contact surface of articles intended for use in contact with bulk quantities of dry food of the type identified in 21 CFR part 176.170 (c).

Great Britain (1998). FA is authorized without time limit for use in the production of polymeric materials and articles in contact with food or drink or intended for such contact.

References:

1. Hohenegger, M., Vermess, M., Sadjak, A., Egger, G., Supanz, S., and Erhart, U., Nephrotoxicity of fumaric acid monoethyl ester (FAME), *Adv. Exp. Med. Biol.*, 252, 265, 1989.
2. *Toxicological Evaluation of Some Antimicrobials, Antioxidants, Emulsifiers, Stabilizers, Flour-treatment Agents, Acids and Bases,* The 9th and 10th Meeting of the Joint FAO/WHO Expert Committee on Food Additives, Geneva, 1967.
3. Kuroda, K. and Akao, M., Inhibitory effect of fumaric acid on 3'-methyl-4-(dimethylamino) azobenzene-induced hepatocarcinogenesis in rats, *Chem. Pharmacol. Bull.* (Tokyo), 37, 1345, 1989.

FURFURILIDENE ACETONE

Molecular Formula. $C_8H_8O_2$

M = 136.16

Abbreviations. FA; DFA (difurfurilidene acetone).

Synonym. Monofurfurilidene acetone.

Composition. A reaction product of furfural and acetone (1:1).

Properties. A yellow, crystalline product. Odor perception threshold is 0.46 mg/l (practical threshold 0.85 mg/l).

Applications. Used to modify furan resins in order to give plasticity to polymers based on them.

Acute Toxicity. LD_{50} is 325 mg/kg BW in rats, 216 mg/kg BW in mice, and 158 mg/kg BW in rabbits. CNS inhibition was found to develop. In 6 to 10 hours, animals became immobile and gave a weak response to external stimuli. Death occurred on day 1 to 4 of observation.[1] Poisoning caused neurotoxic effect because of penetration of hemato-encephalic barrier (cellular membranes) and damage of endocrine glands (hypophysis, adrenal glands).[2]

Chemobiokinetics. Rats received a single administration of 2/3 LD_{50} of FA and DFA. Metabolism occurs in the liver. Administration led to formation of highly toxic metabolites (*furyl alcohol, furfural,* and *tetrahydrofuran*).[2]

Standards. ***Russia***. Tentative MAC: 0.5 mg/l (organolept.).

References:

1. Usmanov, I. A. and Aliev, E. I., Hygienic norm-setting for furan compounds in water bodies, *Gig. Sanit.*, 8, 89, 1987 (in Russian).
2. Zavgorodnaya, S. V. and Tarasov, V. V., Molecular mechanism of action of furan polymers, *Gig. Sanit.*, 9, 29, 1983 (in Russian).

GLUTARALDEHYDE

Molecular Formula. $C_6H_8O_2$
M = 100.13
CAS No 111-30-8
RTECS No MA2450000
Abbreviation. GA.

Synonyms and **Trade Names.** 1,3-Diformylpropane; Glutaral; Glutaric dialdehyde; Glutarol; 1,5-Pentanedial; 1,5-Pentanedione.

Properties. Oily liquid. Soluble in water and ethanol at all ratios.

Applications. Used as a replacement for formaldehyde in the manufacture of paper and paperboard intended for food packaging; a bactericide.

Acute Toxicity. LD_{50} of the 50% aquatic GA solution is reported to be 1.3 ml/kg BW in rats, and 1.59 ml/kg BW in rabbits; that of 25% aquatic solution is 1.87 and 8.0 ml/kg, and that of 5.0% aquatic solution is 3.3 and 16 ml/kg BW, respectively. In mice, LD_{50} is found to be 0.1 g/kg BW.[1]

According to Ohno et al., LD_{50} in the young rats (5 to 6 weeks old) is 283 mg/kg BW; LD_{50} in the old animals (57 to 60 weeks old) is 141 mg/kg BW.[2] Signs of intoxication include piloerection, red periocular and perinasal encrustations, sluggish movement, rapid breathing, and diarrhea. At necropsy, no gross pathology changes were noted in survivors; animals that died displayed distention, congestion and hemorrhage of the stomach and small intestine, lesions in the kidneys, liver, and lungs.[1]

Short-term Toxicity. Gross pathology examination failed to reveal changes in the visceral organs of rats given 0.5, 2.5, and 5.0% GA in the diet for 3 months. This study affirmed the *GRAS* status of GA when it is used in the food industry to cross-link edible collagen sausage casings.[3]

Rats received 0.25% GA in their drinking water for 11 weeks. Animals exhibited no evidence of damage in the NS and CNS. There were no data indicating neurotoxic action of GA.[4]

Immunotoxicity. Contact hypersensitivity has been reported in mice and guinea pigs which resulted from dermal application of 0.3 to 3.3% GA for 5 to 14 days.[5]

Reproductive Toxicity. According to Ballantyne, there were no embryotoxic effects in the offspring of mice treated by gavage with up to 30 mg GA/kg BW on days 7 to 12 of gestation.[1]

In the recent Ema et al. studies,[6] rats were dosed by gastric intubation with 25 to 100 mg/kg BW on days 6 to 15 of pregnancy. The highest dose caused maternal toxicity and decreased fetal weights. Nevertheless, no teratogenicity or postimplantation loss was noted even at the dose of 100 mg/kg BW on days 6 to 15, which induced maternal toxicity.[7]

Mutagenicity. Exhibits mutagenic activity which could be explained as a result of oxidative damage to the DNA; GA is markedly cytotoxic.[8]

In vitro genotoxicity. GA is shown to be negative in *E. coli.*[9] NTP results ranged from no activity to weakly positive in *Salmonella* reversion assay.[8] However, Muller et al. reported GA to be mutagenic in *Salmonella typhimurium* strain *TA102* assay in three different laboratories.[10] GA appeared to be a potent mutagen in the mouse lymphoma cell line.[11]

According to other data, GA did not produce a significant genotoxic effect in *Salmonella typhimurium*, Chinese hamster ovary cells/HGPRT gene mutation system, the SCE test with Chinese hamster ovary cells, the frequency of unscheduled DNA synthesis in primary rat-hepatocyte cultures.[12]

In vivo cytogenetics. GA is shown to be negative in DLM test (30 to 60 mg/kg BW oral doses).[9]

Chemobiokinetics. F344 rats and New Zealand rabbits were exposed *i/v* to 1,5-^{14}C-GA. GA metabolism occurs through oxidation by rat liver mitochondria, and in the kidney. Excreted predominantly as CO_2. Urinary excretion of radioactivity was found to be in the range of 8.0 to 12% in rats and 15 to 28% in rabbits.[1]

Regulations.

EU (1990). GA is available in the *List of monomers and other starting substances which may continue to be used for the manufacture of plastic materials and articles intended to come into contact with foodstuffs pending a decision on inclusion in Section A* (*Section B*).

U.S. FDA (1998) allowed the use of GA (1) as an antimicrobial agent in pigment and filler slurries used in the manufacture of paper and paperboard at levels not to exceed 300 ppm by weight of the slurry solids; (2) as a component of the uncoated or coated food-contact surface of paper and paperboard intended for use in producing, manufacturing, packaging, processing, preparing, treating, packing, transporting, or holding aqueous and fatty foods in accordance with the conditions prescribed in 21 CFR part 176.170; (3) as a slimicide in the manufacture of paper and paperboard that contact food as prescribed in 21 CFR part 176.300; and (4) in adhesives used as components of articles intended for use in packaging, transporting, or holding food in accordance with the conditions prescribed in 21 CFR part 175.105.

Standards. *Russia* (1995). MAC and PML: 0.07 mg/l.

References:

1. Ballantyne, B., Review of toxicological studies and human health effects, in *Glutaraldehyde,* Union Carbide Corp., Danbury, CT, 1986.
2. Ohno, K., Yasuhara, K., Kawasaki, Y., et al., Comparative study of glutaraldehyde acute toxicity in the old and young rats, Abstract, *Bull. Natl. Inst. Hyg. Sci.*, 109, 92, 1991 (in Japanese).
3. Devro, Inc., Glutaraldehyde oral toxicity tests indicate no related lesions, *Food Chem. News.*, 26, 42, 1984.
4. Spencer, P. C., Bischoff, M. C., and Schaumburg, H. H., On the specific molecular configuration of neurotoxic aliphatic hexacarbon compounds causing central peripheral distal axonopathy, *Toxicol. Appl. Pharmacol.,* 44, 17, 1978.
5. Stern, M. C., Holsapple, M. P., McClay, J. A., et al., Contact hypersensitivity response to glutaraldehyde in guinea pigs and mice, *Toxicol. Ind. Health*, 5, 31, 1989.
6. Ema, M., Itami, T., and Kawasaki, H., Teratological assessment of glutaraldehyde in rats by gastric intubation, *Toxicol. Lett.*, 63, 147, 1992.
7. Marks, T. A. et al., Influence of formaldehyde and Sonacide R (potentiated acid glutaraldehyde) on embryo and fetal development of mice, *Teratology*, 22, 51, 1980.
8. Tamada, M., Sasaki, S., Kadono, Y., et al., Mutagenicity of glutaraldehyde in mice, *Bobkin Bobai*, 6, 62, 1978.
9. Beauchamp, R. O., St Clair, M. B. G., Fenell, T. R., et al., A critical review of the toxicology of glutaraldehyde, *CRC Crit. Rev. Toxicol.*, 22, 143, 1992.
10. Muller, W., Engelhart, G., Herbold, B., Jackh, R., and Jung, R., Evaluation of mutagenicity testing with *Salmonella typhimurium TA102* in three different laboratories, *Environ. Health Perspect.*, 101 (Suppl. 3), 33, 1993.
11. Hawort, S., Lawlor, T., Mortelmans, K., et al., *Salmonella* mutagenicity tests results for 250 chemicals, *Environ. Mutagen.*, Suppl. 1, 3, 1983.
12. Slesinski, R. S., Hengler, W. C., Guzzie, P. J., and Wagner, K. J., Mutagenicity evaluation of glutaraldehyde in a battery of *in vitro* bacterial and mammalian systems, *Food Chem. Toxicol.*, 21, 621, 1983.

GLUTARIC ACID

Molecular Formula. $C_5H_8O_4$
M = 132.11
CAS No 110-94-1
RTECS No MA3740000

Abbreviation. GA.

Synonyms. 1,3-Propanedicarboxilic acid; 1,5-Pentanedioic acid; *N*-Pyrotartaric acid.

Properties. Colorless crystals. Water solubility is 429 mg/l at 20°C. Freely soluble in absolute alcohol.

Applications. Chemical intermediate in the production of plastics (polyamides, polyesters).

Acute Toxicity. LD_{50} is 6.0 g/kg BW in mice.[1]

Chemobiokinetics. Rat liver mitochondria metabolized glutarate very slowly compared with glutaryl-coenzyme A.[2]

Regulations.

EU (1990). GA and *glutaric anhydride* are available in the *List of authorized monomers and other starting substances which shall be used for the manufacture of plastic materials and articles intended to come into contact with foodstuffs (Section* A).

Great Britain (1998). Glutaric acid and anhydride are authorized without time limit for use in the production of polymeric materials and articles in contact with food or drink or intended for such contact.

References:

1. *Biochem. J.*, 34, 1196, 1940.
2. Besrat, A., Polan, C. E., and Henderson, L. M., Mammalian metabolism of glutaric acid, *J. Biol. Chem.*, 244, 1461, 1969.

GLYOXAL

Molecular Formula. $C_2H_2O_2$

M = 58.04

CAS No 107-22-2

RTECS No MD2625000

Synonyms and **Trade Names.** Biformal; Diformyl; Ethanedial; 1,2-Ethanedione; Glyoxal aldehyde; Oxal.

Properties. Yellow prisms or light yellow liquid with mild odor. Very soluble in water and al cohol.

Applications. Used in paper industry as a paper coating, and in medicine.

Acute Toxicity. LD_{50} is 0.2 to 0.4 g/kg BW.[011] Single oral administration to cats caused violent vomiting.[1]

Repeated Exposure caused necrosis in pancreas.

Mutagenicity.

In vitro genotoxicity. G. is reported to be mutagenic in *Salmonella typhimurium* strain *TA102* assay in three different laboratories.[2]

Chemobiokinetics. In 24 hours after oral administration, *oxalic acid* was discovered in the urine. It completely disappeared during next two weeks.

Regulations. ***U.S. FDA*** (1998) approved G. for use (1) as a constituent of food-contact adhesives which may safely be used as component of articles intended for use in packaging, transporting, or holding food in accordance with the conditions prescribed in 21 CFR part 175.105; (2) in the manufacture of 4,4'-isopropylidenediphenol epichlorohydrin thermosetting epoxy resin (finished articles containing the resins shall be thoroughly cleaned prior to their use in contact with food) in accordance with the conditions prescribed in 21 CFR part 177.1440; and (3) as a component of the uncoated or coated food-contact surface of paper and paperboard intended for use in producing, manufacturing, packaging, processing, preparing, treating, packing, transporting, or holding aqueous, fatty, and dry food of the type identified in 21 CFR parts 176.170 (c) and 176.180.

References:

1. Browning, E., *Toxicity and Metabolism of Industrial Solvents*, American Elsevier, New York, 1965, 486.
2. Muller, W., Engelhart, G., Herbold, B., Jackh, R., and Jung, R., Evaluation of mutagenicity testing with *Salmonella typhimurium TA102* in three different laboratories, *Environ. Health Perspect.*, 101 (Suppl. 3), 33, 1993.

GUANIDINE

Molecular Formula. CH_5N_3

M = 59.09
CAS No 113-00-8
RTECS No ME7750000

Synonyms and **Trade Name.** Aminoformamidine; Carbamamidine; Iminourea.

Properties. Deliquescent, crystalline mass. Very alkaline. Very soluble in water and alcohol.

Exposure. G. is found in a number of foods, including turnips, mushrooms, corn, rice, and mussels.

Acute Toxicity. A dose of 500 mg/kg BW appeared to be the lethal dose in rabbits.[020]

Regulations. ***U.S. FDA*** (1998) approved the use of G. as a component of the uncoated or coated food-contact surface of paper and paperboard intended for use in producing, manufacturing, packaging, processing, preparing, treating, packing, transporting, or holding dry food in accordance with the conditions prescribed in 21 CFR part 176.180.

Reference:

Handbook of Toxicology, vol 1, W. S. Spector, Ed., Philadelphia, 1956, 152.

GUAR GUM

CAS No 9000-30-0
RTECS No MG0185000
Abbreviation. GG.

Synonyms and **Trade Names.** Cyamopsis gum; Decopra; Guar flour.

Properties. High-molecular-mass polysaccharide.

Applications. Food stabilizer, used in drugs and cosmetics.

Acute Toxicity. LD_{50} is 6.77 g/kg BW in rats.[1]

Short-term Toxicity.

Observations in man. Eight adults consumed at least 30 g of GG for at least 16 weeks without any change in hematological, hepatic, or renal function. Serologic screening revealed no change in lipid, protein, or mineral me tabolism and no change in electrolyte balance.[2]

Animal studies. Rats and mice received GG at dietary levels of 10% for 13 weeks. There was a decrease in food consumption in rats but not in mice. No histological changes were noted in both species tested.[3]

Rats were given 5.0% partially hydrolyzed GG in their diet (4.0 g GG/kg BW) for 13 weeks. The treatment produced no toxic effects.[4]

Long-term Toxicity. GG (83.5 to 91.9% purities) has been added at the levels of 25,000 and 50,000 ppm to the diet of male and females F344 rats and $B6C3F_1$ mice for 103 weeks (MTD, 5.0% of diet). The treatment lowered BW gain. No clinical signs or adverse effects on survival were noted.[3]

Reproductive Toxicity.

Embryotoxicity. GG was administered in the diet (at levels of 1.0 to 15%) that was available *ad libitum* to male and female Osborne-Mendel rats for 13 weeks before mating, during mating, and through gestation. During gestation, the females received doses of 0.7 to 11.8 g GG/kg BW. The treatment caused no behavioral effects in any of the treated dams; no animals died. In dams fed 15% GG in the diet, slightly fewer corpora lutea and implantations than the controls, as well as a slight decrease in the number of viable fetuses/litter were observed. Fetal development or sex distribution were found unchanged.[5]

Teratogenicity. GG was not teratogenic in rats and mice exposed to it in the diet.[2] It was not found to cause developmental abnormalities in chick embryos.[6]

Mutagenicity.

In vivo cytogenetics. Was negative in DLM test in rats.[7]

Carcinogenicity. Under conditions of above described bioassay,[3] GG was not carcinogenic for male and females F344 rats and $B6C3F_1$ mice.[3,8]

Carcinogenicity classification.

NTP: N - N - N - N (feed).

Regulations. ***U.S. FDA*** (1998). GG is considered to be *GRAS* at the levels and uses approved by CFR. GG may be used as a stabilizer in accordance with the provisions prescribed in 21 CFR part 121.101.

Recommendations. ***EU*** (1995). GG is a food additive generally permitted for use in foodstuffs.

References:

1. Graham, S. L., Arnold, A., Kasza, L., Ruffin, G. E., Jackson, R. C., Watkins, T. L., and Graham, C. H., Subchronic effects of guar gum in rats, *Food Cosmet. Toxicol.,* 19, 287, 1981.
2. McIvor, M. E., Cummings, C. C., and Mendeloff, A. I., Long-term injection of guar gum is not toxic in patient with noninsulin-dependent diabetes mellitus, *Am. J. Clin. Nutr.*, 41, 891, 1985.
3. *Carcinogenesis Bioassay of Guar Gum in F344 Rats and B6C3F$_1$ Mice (Feed Study)*, NTP Technical Report Series No 229, Research Triangle Park, NC, March 1982.
4. Koujitani, T. et al., *Int. J. Toxicol.*, 16, 611, 1997.
5. Collins, T. F. X., Welsh, J. J., Black, T. H., Graham, S. L., and O'Donnell, M. W., Study of the teratogenic potential of guar gum, *Food Chem. Toxicol.*, 25, 807, 1987.
6. Verrett, M. J., Scott, W. F., Reynaldo, E. F., Alterman, E. K., and Thomas, C. A., Toxicity and teratogenicity of food additive chemicals in the developing chicken embryo, *Toxicol. Appl. Pharmacol.*, 56, 265, 1980.
7. Newell. G. M. et al., U.S. Natl. Technical Information Service PB No 221815/4, 102, 1972.
8. Melnick, R. L., Huff, J., Haseman, J. K., Dieter, M. P., Grieshaber, C. K., Wyand, D. S., Russfield, A. B., Murthy, A. S. K., Fleischman, R. W., and Lilja, H. S., Chronic effects of agar, guar gum, gum arabic, locust-bean gum, or tara gum in F344 Rats and B6C3F$_1$ Mice, *Food Chem. Toxicol.*, 21, 305, 1983.

GUM ARABIC

CAS No 9000-01-5
RTECS No CE5945000
Abbreviation. GA.

Trade Name. Gum acacia.

Composition. Dried gummy exudate from tropical and subtropical *Acacia senegali trees*. GA consists of several high-molecular-mass polysaccharides, primarily indigestible to both humans and animals, and their salts. The product is subject to seasonal and geographical variations.

Properties. Extremely highly soluble in water.

Applications. Food gum. Used in preparation of chewing gum (5.6 %, CFR). There is no limitation to the use of GA as a food additive.[1]

Acute Toxicity.

Observations in man. Probable oral lethal dose is above 15 g/kg BW.[022]

Animal studies. LD_{50} exceeded 10 g/kg BW

Repeated Exposure. Rats fed GA for 28 days at levels of up to 8.0% in the diet exhibited no changes in heart and liver mitochondrial function.[2] However, according to other data, repeated oral administration of GA to rats caused uncoupling of oxidative phosphorylation in liver and heart mitochondria and partial inhibition of mixed function oxidases of liver endoplasmic reticulum, as measured by 2-biphenylhydroxylation and 4-biphenylhydroxylation.[3]

Short-term Toxicity. Rats consumed GA at dietary levels of up to 20% for 90 days. Males showed significant decrease in BW gain and liver weights at the highest dose level. No such effects were seen in females at the same dose level.[4]

Long-term Toxicity. Male and female F344 rats and B6C3F$_1$ mice were given GA (81 to 86% purities) at levels of 25,000 and 50,000 ppm in the diet for 103 weeks. The treatment caused body weights slightly lower only in dosed female rats comparable with those of controls. No other clinical signs or adverse effects on survival were noted.[5]

Reproductive Toxicity.

Embryotoxicity. GA was administered in the diet at 1.0 to 15% that was available *ad libitum* to male and female Osborne-Mendel rats during premating and mating periods and through gestation. There were no dose-related changes in maternal findings, number of fetuses, fetal viability, visceral or skeletal variations.[6]

Teratogenicity. No teratogenic effects were observed in rats.[6]

Mutagenicity.

In vivo cytogenetics. GA was found to be positive in DLM test in male rats. No increase in incidence of dominant lethals or heritable translocations was observed in mice.[7]

Carcinogenicity. In the above described study,[5] statistically significant increasing trends of hepatocellular carcinomas in mice and in total liver tumors were observed. Nevertheless, under conditions of this bioassay, GA was not considered to be carcinogenic for male and females F344 rats and $B6C3F_1$ mice.[5]

Carcinogenicity classification.

NTP: N - N - N - N (feed).

Chemobiokinetics. GA does not degrade in the intestine but is fermented in the colon under the influence of microorganisms.

Regulations. ***U.S. FDA*** (1998). GA is considered to be *GRAS*. Its use in *chewing gum* as flavoring agent and adjuvant, and as a surface-finishing agent is approved at levels up to 5.6%.

GA is **recommended** as suspending medium for the use in pharmacological and toxicological experiments.

References:

1. Anderson, D. M., Evidence for the safety of gum arabic (*Acacia senegal* (L.) Willd.) as a food additive – a brief review, *Food Addit. Contam.*, 3, 225, 1986.
2. Anderson, D. M. W., Bridgeman, M. M. E., Farquhar, J. G. K., and McNab, C. G. A., The chemical characterization of the test article used in toxicological studies of gum arabic and tragacanth, *Toxicol. Lett.*, 21, 83, 1983.
3. Bachmann, E., Weber, E., Post, M., and Zbinden, G., Biochemical effects of gum arabic, gum tragacanth, methylcellulose and carboxymethylcellulose-Na in rat heart and liver, *Pharmacology,* 17, 39, 1978.
4. Anderson, D. M. W., Ashby, P., Busuttil, A., Easrwood, M. A., Hobson, B. M., Ross, A. H. M., and Street, C. A., Subchronic effects of gum arabic (acacia) in the rat, *Toxicol. Lett.*, 14, 221, 1982.
5. *Carcinogenesis Bioassay of Guar Gum in F344 Rats and $B6C3F_1$ Mice (Feed Study),* NTP Technical Report Series No 227, Research Triangle Park, NC, May 1982.
6. Collins, T. F. X., Welsh, J. J., Black, T. H., Graham, S. L., and Brown, L. H., Study of the teratogenic potential of gum arabic, *Food Chem. Toxicol.*, 25, 815, 1987.
7. Sheu, C. W. et al., Tests for mutagenic effects of ammoniated glycyrrhizin, butylated hydroxytoluene, and gum arabic in rodent germ cells, *Environ. Mutagen.*, 8, 357, 1986.

HEXACYANOFERRATE(3^-), TRIPOTASSIUM

Molecular Formula. $C_6FeN_6.3K$

M = 329.27

CAS No 13746-66-2

RTECS No LJ8225000

Abbreviation. HFT.

Synonyms. Potassium ferricyanate; Potassium ferrocyanide.

Acute Toxicity. LD_0 is 1.6 g/kg BW in rats (Kodak Co. Report, 1971). LD_{50}s are established to be 2.97 or 5.0 g/kg BW in mice, and 5.3 or 6.4 g/kg BW in rats (for potassium ferricyanide and potassium ferrocyanide, respectively).[1]

Reproductive Toxicity.

In inhalation rat study, concentrations of 1.0 and 0.5 mg/m^3, respectively, produced ***gonadotoxic*** and ***embryotoxic*** effect.[1]

Potential **allergen** on epicutaneous and inhalation exposure.[1]

Mutagenicity.

In vitro genotoxicity. HFT induced reverse mutation and mitotic gene conversion in *Saccharomyces cerevisiae.*[2]

Regulations. ***U.S. FDA*** (1998) approved the use of HFT in adhesives as a component of articles intended for use in packaging, transporting, or holding food in accordance with the conditions prescribed in 21 CFR part 175.105.

References:

1. Besedina, V. A., Grin', N. V., Yermachenko, A. B., Govorunova, N. N., and Yermachenko, T. P., Hygienic assessment of potassium ferri- and ferrocyanides as atmospheric air pollutants, *Gig. Sanit.*, 4, 23, 1986 (in Russian).
2. Singh, I., Induction of reverse mutation and mitotic gene conversion by some metal compounds in *Saccharomyces cerevisiae*, *Mutat. Res.*, 117, 149, 1983.

HEXAETHYL CYCLOTRISILOXANE

Molecular Formula. $C_{12}H_{30}Si_3O_3$
M = 306.6

Properties. A clear liquid with a faint specific odor. Poorly soluble in water; soluble in alcohol.

Applications. Used in the production of siloxane polymers.

Acute Toxicity. Rats and mice tolerate *i/g* administration of up to 20 g/kg BW and up to 10 g/kg BW administered *i/p* without any sign of toxicity.

Repeated Exposure failed to reveal signs of accumulation in rats on daily administration of 3.0 g/kg BW dose. Functional and morphological treatment-related changes in animals were absent.

Standards. ***Russia.*** PML: *n/m.*

Reference:

Shugayev, B. B. and Bukhalovsky, A. A., Toxicity of hexaethyl cyclotrisiloxane, *Gig. Sanit.*, 8, 93, 1986 (in Russian).

HEXAFLUOROSILICATE(2$^-$), DIHYDROGEN

Molecular Formula. F_6H_2Si
M = 144.08
CAS No 16961-83-4
RTECS No VV8225000

Synonyms. Fluorosilicic acid; Hexafluosilicic acid; Silicofluoric acid.

Properties. Liquid, when anhydride dissociates almost instantly into SiF_4 and HF, with sour, pungent odor.[020]

Acute Toxicity. Administration of lethal doses led to peripheral nerve paralysis, ataxia, muscle contraction, or spasticity. CNS was predominantly affected.[1]

Repeated Exposure. Rats were given 5.0 mg/kg BW for 30 days. The treatment caused retardation of BW gain, an increase in fluor contents in urine, and an increase in fluor accumulation in the bones. An increase in alkaline phosphorilase activity in blood serum was noted by the end of the experiment. All these changes appeared to be statistically insignificant. Histology investigation revealed only weak changes in the liver and myocard; a tendency in hyperplasia was found in the spleen. K_{acc} = 5.1.

Regulations. ***U.S. FDA*** (1998) approved the use of H. in adhesives as a component of articles intended for use in packaging, transporting, or holding food in accordance with the conditions prescribed in 21 CFR part 175.105.

Reference:

1. Rumyantzev, G. I., Novikov, S. M., Mel'nikova, N. N., Levchenko, N. I., Kozeyeva, Ye. Ye., and Kochetkova, T. A., Experimental study of biologic effect of salts of silicofluoric acid, *Gig. Sanit.*, 11, 80, 1988 (in Russian).

HEXAMETHYLENETETRAMINE

Molecular Formula. $C_6H_{12}N_4$
M = 140.22
CAS No 100-97-0
RTECS No MN4725000
Abbreviation. HMTA.

Synonyms and **Trade Names.** Aminoform; Formamine; Hexaform; Hexamine; Methamin; Methenamine; Resotropin; Urotropine.

Properties. Odorless, colorless, hygroscopic crystals with a caustic, sweet, but subsequently bitter taste. Solubility in water is 81% (12°C); solubility in alcohol is 3.2% at 12°C. Decomposes in a slightly acidic solution, forming ammonia and formaldehyde. Aqueous solutions are clear, colorless, and odorless. Taste perception threshold is 60 mg/l.[1]

Applications. A component of stabilizers, plasticizers, and catalysts, used in the production of amine resins and of foam plastics. A vulcanization accelerator for rubber.

Exposure. HMTA is used in the food industry as an antimicrobial additive (in fish products and in cheese-making).

Acute Toxicity. Rats and mice tolerate doses of 2.0 to 15.0 g/kg BW. Administration of high doses caused no pathological changes in the visceral organs or species differences in susceptibility to HMTA.[1]

However, according to other data, LD_{50} appeared to be 512 mg/kg BW in mice.[06]

Repeated Exposure.

Observations in man. No harmful reactions or complications were observed in patients receiving HMTA as an antiseptic at dose levels of 4.0 to 6.0 g/day for weeks. However, doses of 8.0 g/day for 3 to 4 weeks produced bladder irritation, painful and frequent micturition, albuminuria, and hematuria.[2]

Animal studies. Mice tolerated 10-day treatment with 5.0 g/kg BW dose.

Short-term Toxicity. Rats were gavaged with 400 mg/day for 90 or 333 days.[3] The yellow fur discoloration observed in animals was shown to be a consequence of a reaction between formaldehyde in the urine and kynurenine in the rat hair.[4]

Long-term Toxicity. Cats were fed up to 1250 mg/kg BW daily doses for 2 years. No effect was observed on food consumption and BW gain. There were also no histological changes in animal tissues.[5]

No adverse effects and histopathology were reported in rats given the doses of 200 and 400 mg/day for 1 year.[3]

Mice were given orally 0.5, 1.0, or 1.5% HMTA in their diet for 420 days. The treatment produced organ weight changes. The LOEL of 2.36 g/kg BW, and the NOEL of 1.2 g/kg BW were established in this study.[6]

Rats were dosed by gavage with 10 mg/kg BW for 7 months. The treatment caused liver dysfunction and behavioral effects. 0.1 mg/kg BW dose was considered to be ineffective in this study. Bearing in mind that HMTA may be hydrolyzed with the formation of *formaldehyde*, MAC for HMTA was recommended to be set at the same level as that of formaldehyde.[1]

Reproductive Toxicity.

Embryotoxicity.

Observations in man. Use of HMTA during pregnancy in about 200 patients was not associated with any adverse effects.[7]

Animal studies. Pregnant rats were exposed to 5.0 and 50 mg HMTA/kg BW via their drinking water. In a five-generation study lasting 42 months, no alterations attributable to HMTA were demonstrated in animals or fetuses and placentas at each dose level or at every period of examination.[8]

There were no effects on body growth, survival, reproduction, and viability of the offspring in rats fed a diet containing 400 to 1600 mg/kg for 2 years.[5]

Addition of 1.0% HMTA to drinking water did not cause any embryotoxic effect in Wistar rats (after mating, the treatment of females continued during pregnancy and lactation). No adverse effects were also observed in a three-generation study in rats (there were 1.0 to 2.0% concentrations of HMTA in drinking water).[9]

Teratogenicity. No effects on reproduction or litter numbers were reported in dogs that were fed the doses from 125 to 1875 mg HMTA/kg BW for 32 months. However, 2/3 litters had malformations.[5]

In another dog study (0.06 and 0.125% HMTA in the diet from day 4 to 56 after mating), reproductive function of animals was not affected.[10]

Gonadotoxicity. No differences in fertility were detected in 2 rat generations (the exposed rats consumed about 100 mg HMTA/kg BW).[11]

Mutagenicity.

In vivo cytogenetics. HMTA appeared to be a mutagen for *Dr. melanogaster.*[12]

In vitro genotoxicity. HMTA gave positive results in the *rec*-assay using spores of *Bac. subtilis* strains H17 and M45. The action of HMTA against DNA was due to the electrophilic state of this material.[13]

Carcinogenicity. Carcinomas and adenocarcinomas were reported to occur in female rats after 2 years of feeding with 1.0% HMTA in the diet. However, a further study at the same dose levels showed no differences in tumor incidence between the treated and control animals.[4]

Tumor incidence in rats fed a diet containing 0.16% HMTA from weaning to death was not higher than that observed in control rats.[11]

F344/N rats were given HMTA by gavage in water at doses of 25 or 50 mg/kg BW for two years. There was dose-related increase in the incidences of neoplasms in numerous tissues. Survival of rats was markedly reduced because of the early induction of neoplastic disease. In male rats, mesothelioma arising in the tunica vaginalis and metastasizing to the peritoneum were the major cause of early death. Early death in female rats was associated with mammary gland neoplasms.[14]

Chemobiokinetics. Toxicity of ingested HMTA depends on the dose and the rate at which it is hydrolyzed in presence of protein in the acid contents of the stomach. Intensive breakdown of HMTA is shown to also proceed in the kidneys and urinary bladder. The most toxic metabolite is likely to be *formaldehyde*. HMTA is rapidly absorbed in the GI tract in man and appears in the urine in a few minutes. HMTA enters breast milk in small amounts.[15]

Regulations.

U.S. FDA (1998) approved the use of HMTA (1) as a component of adhesives to be safely used in contact with food in accordance with the conditions prescribed in 21 CFR part 175.105; (2) as a component of the uncoated or coated food-contact surface of paper and paperboard intended for use in producing, manufacturing, packaging, processing, preparing, treating, packing, transporting, or holding dry, aqueous, and fatty foods of the type identified in 21 CFR part 176.170 (c); (3) as an accelerator for rubber articles intended for repeated use in contact with food up to 1.5% by weight of the rubber product; (4) as a polymerization cross-linked agent for protein, including casein in the manufacture of paper and paperboard; (5) in urea-formaldehyde resins in molded articles intended for use in contact with food only as a polymerization control agent in accordance with the conditions prescribed in 21 CFR part 177.1900; (6) in the manufacture of closures with sealing gaskets that may be safely used on containers intended for use in producing, manufacturing, packing, processing, treating, packaging, transporting, or holding food (1.0%) in accordance with the conditions prescribed in 21 CFR part 177.1210; and (7) as a curing agent in the manufacture of phenolic resins to be safely used as a food-contact surface of molded articles intended for repeated use in contact with non-acid food (*pH* above 5.0) in accordance with the conditions prescribed in 21 CFR part 177.2410.

EU (1990). HMTA is available in the *List of authorized monomers and other starting substances which shall be used for the manufacture of plastic materials and articles intended to come into contact with foodstuffs (Section* A).

Great Britain (1998). HMTA is authorized without time limit for use in the production of polymeric materials and articles in contact with food or drink or intended for such contact. The specific migration of this substance shall not exceed 15 mg/kg (expressed as formaldehyde).

Standards.

EU (1990). SML (T) 15 mg/kg (expressed as formaldehyde).

Russia (1995). MAC and PML in drinking water: 0.5 mg/l.

Recommendations. Joint FAO/WHO Expert Committee on Food Additives. ADI: 0.15 mg/kg BW.

References:

1. Krasovsky, G. N. and Friedland, S. A., in *Industrial Pollution of Reservoirs*, S. N. Cherkinsky, Ed., Meditsina, Moscow, Issue No 8, 1967, 140 (in Russian).
2. Goodman, L. S. and Gilman, A., *The Pharmacological Basis of Therapeutics*, 2nd ed., MacMillan, New York, 1955, 1087.
3. Brendel, R., Untersuchungen an Ratten zur Ventraglichkeit von Hexamethylenetetramin, *Arzneimittel-Forsch.*, 14, 51, 1964.

4. Kewitz, H. und Welsh, F., Ein gelber farbstoff aus formaldehyd und Kynurenin bei hexaminbehandelten ratten, *Naunyn-Schmiedeberg Arch. Exp. Pathol. Pharmak.*, 254, 101, 1966.
5. Restani, P. and Corrado, L. G., Oral toxicity of formaldehyde and its derivatives, *CRC Crit. Rev. Toxicol.*, 21, 315, 1991.
6. Della Porta *et al., Food Cosmet. Toxicol.*, 6, 707, 1968; in *Toxicological Evaluation of Some Enzymes, Modified Starches and Certain Other Substances,* The 15th Meeting of the Joint FAO/WHO Expert Committee on Food Additives, WHO Food Additives Series No 1, Geneva, 1972.
7. Gordon, S. F., Asymptomatic bacteriurea of pregnancy, *Clin. Med.*, 79, 22, 1972.
8. Malorny, G., Rietbrock, N., und Schassan, H.-H., Uber den stoffwechsel des trimethylaminoxyds, *Naunyn-Schmiedebergs Arch. Exp. Pathol. Pharmakol.*, 246, 62, 1963.
9. Della Porta, G., Cabral, J. R., and Parmiani, G., Studio della tossicita tranceplacentare e di cancerogenesi in ratti trattati con esametilentetramina, *Tumori*, 56, 325, 1970 (in Italian).
10. Hurni, H. and Ohder, H., Reproduction study with formaldehyde and hexamethylenetetramine in beagle dogs, *Food Cosmet. Toxicol.*, 11, 459, 1973.
11. Natvig, H., Andersen, J., and Rasmussen, E. W., A contribution to the toxicological evaluation of hexamethylene tetramine, *Food Cosmet. Toxicol.*, 9, 491, 1971.
12. Rapoport, J. A., Carbonyl compounds and chemical mechanism of mutation, *Reports of Acad. Sci. USSR*, 54, 65, 1946 (in Russian).
13. Ueno, S. and Ishizaki, M., Mutagenicity of organic rubber additives, Abstract, *Sangyo Igaku*, 26, 147, 1984 (in Japanese).
14. Irwin, R. D., Eustus, S. L., Stefanski, S., and Haseman, J. K., Carcinogenicity of glycidol in F344 rats and B6C3F$_1$ mice, *J. Appl. Toxicol.*, 16, 201, 1996.
15. Allgen, L. G., Holmberg, G., Persson, B., and Sarbo, B., Biological fate of methenamine in man, *Acta Obstet. Gynecol. Scand.*, 58, 287, 1979.

HEXAMETHYLENETETRAMINE and BENZYL CHLORIDE, condensation product

Trade Name. Product PKU-3.

Properties. A dark brown liquid. Slightly soluble in water, readily soluble in alcohol.

Applications. Used in the production of polymeric coatings.

Acute Toxicity. In rats, LD_{50} is 1.4 g/kg BW. Lethal doses caused CNS inhibition. Death occurs within 1 to 4 days.

Reference:

Paustovskaya, V. V., Torbin, V. F., Onikienko, F. A., et al., Toxicity of some chemicals used as additives for polymeric coatings, in *Hygiene and Toxicology of Plastics*, Proc. Kiyv Medical Institute, Kiyv, 1979, 82 (in Russian).

HEXAMETHYLENETETRAMINE, BENZYL CHLORIDE, and UREA, condensation product

Trade Name. Product PKU-M.

Properties. Amber- to orange-colored liquid, with an odor of isopropyl alcohol, benzaldehyde and heavy amines. Poorly soluble in water; readily soluble in alcohol.

Applications. Used in the production of polymeric coatings.

Acute Toxicity. LD_{50} is 2.67 g/kg BW in rats, and 1.8 g/kg BW in mice. CNS inhibition is shown to develop. Death occurs within 1 to 4 days.

Short-term Toxicity. Animals tolerated administration of 53 or 267 mg/kg BW for 4 months.

Reference:

Paustovskaya, V. V., Torbin, V. F., Onikienko, F. A., et al., Toxicity of some chemicals used as additives for polymeric coatings, in *Hygiene and Toxicology of Plastics*, Proc. Kiyv Medical Institute, Kiyv, 1979, 82 (in Russian).

HEXAMETHYLENETETRAMINE and DICHLOROETHANE, condensation product

Trade Name. Product PKD.

Properties. A dark brown liquid with an odor of amines. Poorly soluble in water; readily soluble in alcohol.

Applications. Used in the synthetic coatings production.

Acute Toxicity. Rats tolerate a 10 g/kg BW dose. After administration, signs of CNS inhibition are evident. Recovery is observed in 3 to 5 days.

Repeated Exposure failed to reveal cumulative properties.

Short-term Toxicity. Rats were dosed by gavage with 200 or 1000 mg/kg BW for 4 months. Manifestations of the toxic effect at the higher dose level included retardation of BW gain, changes in hematology analyses, and in the activity of a number of liver enzymes. Gross pathology examination revealed hemodynamic and dystrophic lesions in the visceral organs.

Reference:

Paustovskaya, V. V., Torbin, V. F., Onikienko, F. A., et al., Toxicity of some chemicals used as additives for polymeric coatings, in *Hygiene and Toxicology of Plastics*, Proc. Kiyv Medical Institute, Kiyv, 1979, 82 (in Russian).

HEXANITROCOBALTATE, POTASSIUM SALT

Molecular Formula. $CoN_6O_{12}.3K$

M = 452.29

CAS No 13782-01-9

RTECS No GF9475000

Synonyms and **Trade Names.** Aureolin; Cobaltic potassium nitrite; C. I. Pigment Yellow 40; Hexanitrocobaltiate tripotassium.

Properties. Yellow, crystalline powder. Very poorly soluble in water and alcohol. Odor perception threshold is 5.0 mg/l.

Applications. Used in the manufacture of plastics and dyestuffs.

Acute Toxicity. LD_{50} is 25 g/kg BW in rats, and 23.7 g/kg BW in mice. Death occurs in 2 to 5 days.

Repeated Exposure revealed evident cumulative properties. The treatment with 1/625 LD_{50} over a period of 1.5 months produced a pronounced toxic effect in rats.

Long-term Toxicity. There were marked symptoms of intoxication when rats were dosed by gavage with 0.5 and 5.0 mg/kg BW for 6 months. Changes in hematology analyses, CNS functions, and activity of certain enzymes were observed.

Reproductive Toxicity. Functional condition of spermatozoa was not altered in rats exposed orally to 1/125 LD_{50} for 15 days.

Mutagenicity.

In vivo cytogenetics. A dose of 0.5 mg/kg BW was ineffective in DLM test in rats.

Standards. ***Russia*** (1995). MAC and PML: 1.0 mg/l.

Reference:

Korolev, A. A., Substantiation of hygienic standard for potassium salt of hexahydrocobaltiate in water bodies, *Gig. Sanit.*, 10, 72, 1980 (in Russian).

HYDRIODIC ACID

M = 127.91

CAS No 10034-85-2

RTECS No MW3760000

Abbreviation. HIA.

Composition. A solution of hydrogen iodide in water.

Synonyms. Anhydrous hydriodic acid; Hydrogen iodide.

Properties. Colorless when freshly made, but rapidly turn yellowish or brown upon exposure to light and air. Readily soluble in water and alcohol.

Applications. Used in the preparation of pharmaceuticals and disinfectants.

Regulations. ***U.S. FDA*** (1998) approved the use of HIA as a component of sanitizing solutions used on food-processing equipment and utensils, and on other food-contact articles as specified in 21 CFR part 178.1010.

HYDROFLUORIC ACID

Molecular Formula. FH
M = 20.01
CAS No 7664-39-3
RTECS No MW7875000
Abbreviation. HFA.

Synonyms and **Trade Names.** Hydrofluoride; Hydrogen fluoride; Hydrogen fluoride, anhydrous; Rubiqine.

Properties. Colorless liquid; miscible with water.

Acute Toxicity. Ingestion of about 1.5 g caused sudden death without gross pathological damage.[1] LD *i/p* is 25 mg/kg BW in rats.[2]

Repeated Exposure. Small amounts of HFA led to osteosclerosis.[019]

Reproductive Toxicity. Rats were exposed to HFA on 1 to 22 day of pregnancy by inhalation. The treatment resulted in increase in pre- and postimplantation mortality.[3]

Chemobiokinetics. Fluoride concentrated in various tissues, predominantly in the thyroid, bone, and teeth. Most of the fluoride was excreted in the urine, and a considerable amount was also recovered in the feces.[2]

Regulations. ***U.S. FDA*** (1998) approved the use of HFA in adhesives as a component of articles intended for use in packaging, transporting, or holding food in accordance with the conditions prescribed in 21 CFR part 175.105.

References:

1. Aitbaev, T. Kh., Almaniyazova, V. M., and Abylkasymova, A. S., Isolated and combined effect of small concentrations of hydrogen fluoride and sulfur dioxide under the conditions of chronic experiment, *Gig. Sanit.,* 5, 6, 1976 (in Russian).
2. Jacobs, W. B., The absorption, distribution, excretion, and toxicity of trifluoroamine oxide, *Toxicol. Appl. Pharmacol.*, 13, 76, 1968.
3. Danilov, V. B. and Kas'anova, V. B., Experimental data on embryotoxic effect of hydrogen fluoride in white rats, *Gig. Truda Prof. Zabol.*, 3, 57, 1975 (in Russian).

HYDROGEN PEROXIDE

Molecular Formula. H_2O_2
M = 34.02
CAS No 7722-84-1
RTECS No MX0900000
Abbreviation. HPO.

Synonyms and **Trade Names.** Albone; Hydrogen dioxide; Hydrogen dioxide solution; Hydroperoxide; Peroxide; Superoxol.

Properties. Clear and odorless liquid with slightly acid taste. Miscible with water.

Applications. HPO is an oxidizing agent used in the plastic industry, in production of hardeners and initiators, and in water purification. It is also used for cold sterilization of packaging such as films and bottles before filling with hot or sterilized food. HPO is permitted as a sterilant for polyethylene terephthalate, olefins, and ethylene-vinyl acetate copolymers.

Acute Toxicity. LD_{50} is 1.5 g/kg BW in female rats, and 1.6 g/kg BW in male Wistar-JCL rats.[1]
According to other data, LD_{50} is 2.0 g/kg BW in mice,[2] and 4.0 g/kg BW in rats.[3]

Repeated Exposure. Palatability and toxicity of water were evaluated in C57BL/6N mice given 200, 1000, 3000, or 6000 ppm HPO in distilled water for 14 days. The treatment caused no clinical signs of toxicity. No effect on BW gain was observed at 1000 ppm or less. The highest dose depressed BW and food consumption. Histology examination revealed minimal/mild degenerative and regenerative alterations in the mucosa of the stomach and duodenum at two highest doses. The NOAEL was established to be 1000 ppm or 164 mg/kg BW for males and 198 mg/kg BW for females.[4] Mice received HPO in their drinking water for 3 weeks. Concentration of 0.6%, but not of 0.3%, caused significant decrease in BW. Concentration of more than 1.0% decreased BW of animals, and they died within 2 weeks.[5] Male NMRI mice and male Wistar rats received 0.5% HPO solution for 40 and 50 days, respectively. Depression in water consumption was noted after one week. Only an increase in kidney weights was also reported.[6]

Short-term Toxicity. HPO was administered by gavage to Wistar JCL rats at doses of 56.2, 168.7, or 506 mg/kg BW (in 0.5% solution) for 12 weeks. Signs of anemia, and a marked decrease in BW, serum SGOT, SGPT, and alkaline phosphatase activity were noted; kidney liver and heart weights were decreased without histological changes in high dose group. Slight liver damage was observed at 168.7 mg/kg BW dose.[1]

Long-term Toxicity. Toxic effects occur as an enzymatic conversion of the peroxy bond (i.e., *O-O* bond). These effects occur only after long administration of high dose levels, where the parent compound is continually present.

Rats and rabbits were given HPO by gavage for 6 months. The treatment caused a decrease in BW and blood lymphocyte count at the highest dose level, changes in enzyme activity in the visceral organs, and changes in the GI tract mucosa. The NOEL was found to be 0.05 mg/kg BW.[7] Mice received 0.15% HPO in their drinking water. Focal necrosis in the liver and thickening of the stomach wall and other pathological changes in the GI tract and spleen were observed.[5]

Reproductive Toxicity.

Gonadotoxicity. Administration of 1/10 to 1/5 LD_{50} caused modification of the oestrus cycle in female rats and decrease in sperm motility in males.[7]

Embryotoxicity and ***Teratogenicity.*** Addition of 10% HPO in the rat diet reduced maternal and fetal weights. No malformations were reported.[8]

In the above-cited study, the NOAEL for embryotoxic effect appeared to be 0.005 mg/kg BW.[7]

Mutagenicity.

In vitro genotoxicity. HPO was found to be positive in *Salmonella* mutagenicity assay.[9,10] It caused CA and morphological transformation in mammalian cells.[11,12]

Carcinogenicity. HPO acts as a promoter of carcinogenesis at relatively high doses (in excess of 600 mg/kg BW) after chronic administration in drinking water to experimental animals.

In rats, HPO induced only papillomas, no malignant tumors of the forestomach were noted even at nearly lethal concentration (1.0 to 1.5% HPO in drinking water).[13]

Glandular stomach lesions were seen in mice given 0.1 and 0.4% HPO in their drinking water. These inflammatory responses may lead to carcinogenic transformations.[14]

Carcinogenicity classification. An IARC Working Group concluded that there is limited evidence for the carcinogenicity of HPO in *experimental animals* and there were no data available to evaluate the carcinogenicity of HP in *humans*.

IARC: 3.

Chemobiokinetics. HPO is highly reactive and short lived because of the inherent instability of the peroxide bond. HPO does not appear to produce toxic metabolites: residues are not of toxicological concern because HPO is metabolized rapidly into *oxygen* and *water* (by catalase and glutathione peroxidase) coming in contact with mammalian tissues. For this reason, it is impossible to estimate actual quantitative absorption.

Regulations. ***U.S. FDA*** (1998) approved use of HPO (1) as a constituent of food-contact adhesives which may safely be used as component of articles intended for use in packaging, transporting, or holding food in accordance with the conditions prescribed in 21 CFR part 175.105; and (2) as a sanitizing solution for use on food processing equipment and utensils, and on dairy processing equipment in accordance with the conditions prescribed in 21 CFR part 178.1010. Residual HPO is limited (3) to 0.1 ppm after packaging in accordance with the conditions prescribed in 21 CFR part 178.1005. It is approved for use (4) in washing or to assist in the lye peeling of fruits and vegetables in accordance with the conditions prescribed in 21 CFR part 173.315. Affirmed as *GRAS.*

HPO is currently registered by EPA for a wide variety of uses including agricultural premises and equipment, food handling/storage establishments premises and equipment, etc.

Standards. ***Russia*** (1995). MAC: 0.1 mg/l.

References:

1. Ito, R., Kawamura, H., Chang, H. S., Toida, S., Matsuura, S., Hidano, T., Nakai, S., Inayoshi, Y., Matsuura, M., and Akuzawa, S., Oral safety of hydrogen peroxide acute and subacute toxicity, Abstract, *Toho Igakki Zasshi* (J. Medical Soc. Toho), 23, 531, 1976 (in Japanese).
2. Lyarsky, P. P., Gleiberman, S. E., Pankratova, G. P., Yaroslavskaya, L. A., Yurchenko, V. V., Toxicological and hygiene characteristics of decontaminating preparations produced on the basis of hydrogen peroxide and its derivatives, *Gig. Sanit.,* 6, 28, 1983 (in Russian).
3. Kondrashov, V. A., Comparative toxicity of hydrogen peroxide vapors via inhalatory and skin routes of exposure, *Gig. Truda Prof. Zabol.,* 10, 22, 1977 (in Russian).
4. Weiner, M., Sarver, J., De Gerlache, J., Malinverno, G., Mayr, W., Regnier, J. F., and Trochimowicz, H., Stability, palatability and toxicity of hydrogen peroxide administered to mice in drinking water for 14 days, in *The Toxicologist*, Abstracts of SOT 1996 Annual Meeting, Abstract No 531, Issue of *Fundam. Appl. Toxicol.,*30, Part 2, March 1996.
5. Aoki, M. and Tani, Y., Growth and histopathological changes in mice fed with hydrogen peroxide solution instead of water, Abstract, *Icaku to Seibutsugaku* (Medicine and Biology), 84, 159, 1972 (in Japanese).
6. Kihlstrom, M., Salminen, A., and Vihko, L., Chronic hydrogen peroxide intake and peroxide metabolizing enzyme activities in some tissue of mice and rats, *Experientia*, 42, 1018, 1986.
7. Antonova, V. I., Salmina, Z. A, Latkina, L. L., Bukina, A. P., and Mishina, N. E., Hygienic substitution of MAC for hydrogen peroxide in water bodies, *Gig. Sanit*., 10, 20, 1974 (in Russian).
8. Moriyama, I. et al., Effects of food additive hydrogen peroxide on fetal development, Abstract, *Teratology*, 26, 28A, 1982.
9. Abu-Shakra, A. and Zeiger, E., Effects of *Salmonella* genotypes and testing protocol on hydrogen peroxide-induced mutation, *Mutagenesis,* 5, 469, 1990.
10. Rossman, T. G., Molina, M., Meyer, L., Boone, P., Klein, C. B., Wang, Y., Li, F., Lin, W. C., and Kinney, P. L., Performance of 133 compounds in the lambda prophase induction endpoint of the microscreen assay and comparison with *S. typhimurium* mutagenicity and rodent carcinogenicity, *Mutat. Res*., 260, 349, 1991.
11. Stich, H. F. and Dunn, B. P., Relationship between cellular levels of beta-carotene and sensitivity to genotoxic agents, *Int. J. Cancer*, 38, 713, 1986.
12. Cristavao, L., Teixeira, M., Bras, A., and Rueff, J., DNA and chromosomal damage by hydrogen peroxide in human cells, *EEMS Meeting*, Prague, September 1991, 1.
13. Ito, A., Naito, M., and Watanabe, H., Implication of chemical carcinogenesis in the experimental animals, Tumorigenic effect of hydrogen peroxide in mice, *Ann. Rep. of Hiroshima Univ. Res. Inst. Nuclear Medicine and Biology*, 22, 147, 1981 (in Japanese).
14. Ito, A., Watanabe, H., and Naito, M., Induction of duodenal tumors in mice by oral administration of hydrogen peroxide, *Gann*, 72, 174, 1981.

HYDROXYACETIC ACID

Molecular Formula. $C_2H_4O_3$
M = 76.10
CAS No 79-14-1
RTECS No MC5250000
Abbreviation. HAA.

Synonyms. Glycolic acid; Hydroxyethanoic acid.

Properties. Colorless odorless crystals. Soluble in water and alcohol.

Applications. Used in the production of biodegradable polymers.

Acute Toxicity. Administration of 2.0 ml HAA/kg BW produced transient liver damage,[03] and 4.0 to 6.0 g HAA/kg BW was lethal for rats. Oxalate crystals were found in renal tubules.[1] Doses of 2.4 to 4.0 ml HAA/kg BW caused CNS depression in rabbits.[03]

Repeated Exposure. Rats received 40 mg/kg BW in their drinking water for 30 days. The treatment induced renal lesions. No effects were seen in rats ingesting a 10 mg/kg BW dose.

Oral administration of 2.0 ml HAA daily for 12 days caused CNS depression, kidney damage, and death in 3 of 4 rabbits.[03]

Reproductive Toxicity.

Teratogenicity. Rats were dosed orally with up to 600 mg/kg BW on days 7 to 21 of gestation. The treatment at the highest doses produced maternal toxicity skeletal defects in the fetuses.[2]

Chemobiokinetics. HA may undergo conversion to *glycine* through intermediate oxidation to *glyoxylic acid.*[011]

Administration of 0.5 g/kg BW doses to *Rhesus* monkeys resulted in urinary excretion of 37 to 52% ^{14}C-labeled HA within 96 hours.[3]

Regulations. ***U.S. FDA*** (1998) approved use of HAA as a constituent of food-contact adhesives which may safely be used as component of articles intended for use in packaging, transporting, or holding food in accordance with the conditions prescribed in 21 CFR part 175.105.

References:

1. Casarett and Doull's *Toxicology*, Doull, J., Klassen, C. D., and Amdur, M. D., Eds., 2nd ed., MacMillan Publishing Co., New York, 1980, 482.
2. Munley, S. M. and Hurtt, M. E., Developmental toxicity study of glycolic acid in rats, Abstract, *Teratology*, 53, 117, 1996.
3. McChesney, E. W., Golberg, L., and Harris, E. S., Reappraisal of the toxicology of ethylene glycol. IV. The metabolism of labeled glycolic and glyoxylic acids in the *Rhesus* monkey, *Food Cosmet. Toxicol.*, 10, 655, 1972.

p-HYDROXYBENZOIC ACID

Molecular Formula. $C_7H_6O_3$

M = 138.13

CAS No 99-96-7

RTECS No DH1925000

Abbreviation. HBA.

Synonyms. 4-Hydroxybenzoic acid; 4-Carboxyphenol; *p*-Salicylic acid.

Properties. Crystals. Water solubility is 8.0 g/l; freely soluble in alcohol.

Applications. Intermediate for dyes.

Acute Toxicity. LD_{50} exceeded 2.0 g/kg BW in mice.

Reproductive Toxicity.

Teratogenicity. Produced congenital defects in the skeleton and CNS of rats.[1,2]

Regulations.

EU (1990). HBA is available in the *List of authorized monomers and other starting substances which shall be used for the manufacture of plastic materials and articles intended to come into contact with foodstuffs (Section* A).

Great Britain (1998). HBA is authorized without time limit for use in the production of polymeric materials and articles in contact with food or drink or intended for such contact.

References:

1. Tanaka, S. et al., Effects of salicylic acid and acetylsalicylic acid on the fetuses and offspring of rats, *Teratology*, 6, 121, 1972.

2. Bergman, K., et al., Effects of dietary sodium selenite supplementation on salicylate-induced embryo- and fetotoxicity in the rat, *Toxicology*, 17, 135, 1990.

p-HYDROXYBENZOIC ACID, ETHYL ESTER

Molecular Formula. $C_9H_{10}O_3$
M = 166.17
CAS No 120-47-8
RTECS No DH2190000
Abbreviation. HBAE.

Synonyms and **Trade Names.** Aseptoform E; *p*-Carbethoxyphenol; Ethyl-*p*-hydroxybenzoate; Ethyl *p*-oxybenzoate; Ethyl paraben; Ethyl Parasept; Nipagin A.

Properties. Small, colorless crystals or white powder. Freely soluble in water (0.7 g/l) and alcohol.[020]

Applications. Preservative for pharmaceuticals.

Acute Toxicity. LD_{50} is 3.0 g/kg BW in rats, 5.0 g/kg BW in dogs and rabbits, 2.0 g/kg BW in guinea pigs.[1] In mice, poisoning caused rapid onset of ataxia, paralysis, and deep depression resembling anesthesia. Only rarely was there evidence of increased motor activity. Death usually occurred within 1 hour.[2]

Reproductive Toxicity.

Embryotoxicity. Wistar rats received 10, 1.0, or 0.1% HBAE in their feed. The treatment decreased BW in some fetuses of rats who had received 10% HBAE.[3]

Teratogenicity. No developmental abnormalities were observed in fetuses.[3]

Chemobiokinetics. Cats received orally ^{14}C-labeled HBAE. Two major metabolites, *p*-hydroxyhippuric acid and free *p-hydroxybenzoic acid*, were found in urine (Phillips et al., 1978).

Regulations. ***U.S. FDA*** (1998) approved the use of HBAE in adhesives as a component of articles intended for use in packaging, transporting, or holding food in accordance with the conditions prescribed in 21 CFR part 175.105.

References:

1. FAO Nutrition Meetings Report Series, WHO, Geneva, 53A, 81, 1974.
2. *CRC Handbook of Food Additives,* T. E. Furia, Ed., 2nd ed., 1972, 126.
3. Moriyama, I., Hiraoka, K., and Yamaguchi, R., Teratogenic effects of food additive ethyl-*p*-hydroxy benzoate studied in pregnant rats, *Acta Obstet. Gynaecol. Japan.*, 22, 94, 1975.

7-HYDROXY COUMARIN

Molecular Formula. $C_9H_6O_3$
M = 162.15
CAS No 93-35-6
RTECS No GN6820000
Abbreviation. HC.

Synonym and **Trade Names.** Hydrangin; 7-Oxycoumarin; Skimmetine; Umbelliferone.

Acute Toxicity. LD_{50} *i/v* is 450 mg/kg BW in mice.

Regulations. ***U.S. FDA*** (1998) approved the use of HC in adhesives as a component of articles intended for use in packaging, transporting, or holding food in accordance with the conditions prescribed in 21 CFR part 175.105.

Reference:

Nieschulz, O. und Schmersahl, P., Uber choleretische wirkstoffe aus Artemisia abrotanum, *Arzneimittel-Forsch.*, 18, 1330, 1968.

1-HYDROXY-1,1-DIPHOSPHONOETHANE

Molecular Formula. $C_2H_8O_7P_2$
M = 206.04
CAS No 2809-21-4
106908-76-3

RTECS No SZ8562100

Synonyms. Ethane-1-hydroxy-1,1'-diphosphonate; 1,1,1-Ethanetriol diphosphonate; Etidronic acid; Oxyethylidenediphosphonic acid; Phosphonic acid, 1-hydroxy-1,1-ethanediyl ester.

Acute Toxicity. LD_{50} is 1.8 g/kg BW mice.[1]

Reproductive Toxicity.

Embryotoxicity. Mice were fed 200 mg/kg BW on one day during embryogenesis. The treatment caused decreased fetal weight, decreased skeletal growth, and a number of diverse abnormalities, including angular deformities of long bones.[2]

Teratogenicity. H. was not found to be teratogenic in rabbits and rats.[3] Meanwhile, administration of H. into the yolk sac of chicks produced an angulation deformity of the tibia.[4]

Mutagenicity.

In vitro genotoxicity. H. induced a high frequency of CA in the root tips of barley and onion.[5]

Regulations. ***U.S. FDA*** (1998) approved the use of H. as a component of sanitizing solutions used on food-processing equipment and utensils, and on other food-contact articles as specified in 21 CFR part 178.1010.

References:

1. *Angewandte Chemie*, Int. Ed., 14, 94, 1975.
2. Morio, Y., Congenital angular deformity of the tibia in chick embryo induced by etane-1-hydroxy-1,1-diphosphonate, Abstract, *Nippon Seikeigeka Gakkai Zasshi,* 58, 1253, 1984 (in Japanese).
3. Eguchi, M. et al., Fault of ossification and angular deformities of long bones in mouse fetuses caused by high doses of ethane-1-hydroxy-1,1,diphosphonate (EHDP) during pregnancy, Abstract, *Senten Ito* (Congen. Anom.), 22, 47, 1982 (in Japanese).
4. Nolen, G. A., Buehler, E. V., The effect of disodium etidronate on the reproductive functions and embryogeny of albino rats and New Zealand rabbits, *Toxicol. Appl. Pharmacol.*, 23, 238, 1971.
5. Gaivoronskaya, G. G. et al., Mutagenic and antimutagenic effect of diphosphonic acids, *Tsitologia i Genetika*, 15, 41, 1981 (in Russian).

1-HYDROXYETHYL FERROCENE

Molecular Formula. $C_{12}H_{14}FeO$

M = 230.11

CAS No 1277-49-2

RTECS No LKO730000

Synonyms. 1-Ferrocenylethanol; α-Methylferrocenemethanol.

Applications. Used as a light-sensitizer of plastics.

Acute Toxicity. The LD_{50} values are 380 to 930 mg/kg BW for mice, and 1190 to 2660 mg/kg BW for rats. Gross pathology and microscopic examinations revealed enteritis and deposition of administered substances in the fat tissue and liver, gastric mucosa edema, and increased thyroid and suprarenal activity.

Short-term Toxicity. Experimental animals were given suspensions and emulsions of H. in sunflower oil for 3 months: 1/20, 1/10, and 1/5 LD_{50} (mice), and 1/20 LD_{50} (rats). Gross pathology examination revealed increased liver weights, deposition of H. in the liver, spleen, lungs, and suprarenals, exudative glomerulonephritis, and changes in the thyroid.

Allergenic Effect was not noted.

Reference:

Mishina, A. D., Putilina, L. V., Yevsyukov, V. I., et al., Toxicity of some photosensitizers of plastics, in *Hygiene and Toxicology of High-Molecular-Mass Compounds and of the Chemical Raw Material Used for Their Synthesis*, Proc. 6th All-Union Conf., B. Yu. Kalinin, Ed., Leningrad, 1979, 275 (in Russian).

HYDROXYLAMINE

Molecular Formula. H_3NO

M = 33.03

CAS No 7803-49-8
RTECS No NC2975000

Synonyms. Bis(hydroxylamine); Oxammonium.

Properties. Unstable, large, white flakes or needles. Mixes with water at all ratios to form a weak base, hydroxylamine hydrate.

Applications. Used in the production of polycaprolactam.

Acute Toxicity. The lethal doses of H. chloride are 10 to 75 mg/kg BW for rabbits, and 200 to 300 mg/kg BW for dogs. CNS inhibition, convulsions, and paralysis accompany poisoning. Death occurs from respiratory arrest.

Long-term Toxicity. A four- to five-fold enlargement of the spleen was observed in rats administered 330 to 380 mg/kg BW for 5.5 months. The thyroid shrank to half its size. Erythrocyte hemolysis and methemoglobinemia in mice R1R and blood clotting system disturbances were also noted.

Allergenic Effect. H. is found to increase sensitivity to allergens.

Reproductive Toxicity.

Embryotoxic effect was observed only at high doses.

Teratogenic effect was noted in New Zealand white rabbits but not in rats.[3] The congenital malformations produced in rabbits included abnormalities of the craniofacial region and sternum.[4,6]

Mutagenicity.

In vitro genotoxicity. H. and its derivatives are shown to cause CA in the animal and plant cells. Such an effect seems to be the result of a direct action on DNA as a sequence of a radical formation, or it is caused indirectly.[5] H. is reported to increase slightly the rate of SCE and chromatid aberrations without recombination (Peticone et al., 1977).

Standards. ***Russia*** (1995). MAC of H. sulfate: 0.1 mg/l.

References:

1. Arnol'dova, K. A. and Speransky, N. N., Some aspects of hydroxylamine chloride action on animals, *Gig. Truda Prof. Zabol.*, 12, 39, 1963 (in Russian).
2. Gross, P., Biological activity of hydroxylamine: A review, *CRC Crit. Rev. Toxicol.*, 14, 87, 1985.
3. DeSesso, J. M., Demonstration of the embryotoxic effects of hydroxylamine on the New Zealand white rabbit, *Anat. Rec.*, 196, 45a, 1980.
4. Chaube, S. and Murphy, M. L., The effect of hydroxyurea and related compounds on the rat fetus, *Carcer Res.*, 26, 1448, 1966.
5. Auerbach, Ch., Mutation Research, in *Problems of Mutagenesis*, Mir, Moscow, 1978, 315 (in Russian, translation from English).
6. DeSesso, J. M. and Goeringer, G. C., Developmental toxicity of hydroxylamine: an example of a maternally mediated effect, *Toxicol. Ind. Health*, 6, 109, 1990.

HYDROXYLAMINE SULFATE

Molecular Formula. $H_8N_2O_6S$
M = 164.16
CAS No 10039-54-0
RTECS No NC5425000
Abbreviation. HAS.

Synonyms. Bis(hydroxylamine) sulfate; Oxammonium sulfate.

Properties. Crystals. Freely soluble in water.

Applications. Used in textile chemistry. Dehairing agent for hides.

Acute Toxicity. *I/p* administration of about 100 mg/kg BW may cause death in mice.

Mutagenicity. In the dominant lethal assay, mice were exposed *i/p* with 102 to 112 mg/kg BW. The treatment produced no increase in the incidence of early death or preimplantation losses.

Regulations. ***U.S. FDA*** (1998) approved the use of HAS in adhesives as a component of articles intended for use in packaging, transporting, or holding food in accordance with the conditions prescribed in 21 CFR part 175.105.

Reference:

Epstein, S. S., Arnold, E., Andrea, J., Bass, W., and Bishop, Y., Detection of chemical mutagens by the dominant lethal assay in the mouse, *Toxicol. Appl. Pharmacol.*, 23, 288, 1972.

HYDROXYMETHANESULFINIC ACID, MONOSODIUM SALT

Molecular Formula. CH_3NaO_3S
M = 119.10
CAS No 149-44-0
RTECS No PB0380000

Synonyms and **Trade Names.** Aldanil; Formaldehyde hydrosulfite; Formopan; Hydrolit; Rongalite; Sodium formaldehyde sulfoxylate; Sodium hydroxymethanesulfinate; Sodium methanelsulfoxylate.

Properties. Hard white masses. Readily soluble in water; poorly soluble in dehydrated alcohol.

Applications. Used in polymerization of ethylene compounds.[020] A surfactant.

Acute Toxicity. LD_{50} *s/c* is 4.0 g/kg BW in mice (Rosental, 1934).[020]

Regulations. ***U.S. FDA*** (1998) approved the use of H. in resinous and polymeric coatings used as the food-contact surfaces of articles intended for use in producing, manufacturing, packing, processing, preparing, treating, packaging, transporting, or holding food in accordance with the conditions prescribed in 21 CFR part 175.300; (2) in adhesives as a component of articles intended for use in packaging, transporting, or holding food in accordance with the conditions prescribed in 21 CFR part 175.105; and (3) as a component of the uncoated or coated food-contact surface of paper and paperboard intended for use in producing, manufacturing, packaging, processing, preparing, treating, packing, transporting, or holding aqueous and fatty foods in accordance with the conditions prescribed in 21 CFR part 176.170.

12-HYDROXYSTEARIC ACID

Molecular Formula. $C_{18}H_{36}O_3$
M = 300.54
CAS No 106-14-9
RTECS No WI3850000
Abbreviation. HSA.

Synonym and **Trade Names.** Hydrofol acid 200; 12-Hydroxyoctadecanoic acid; Loxiol G 21.

Properties. Crystals (from alcohol). Insoluble in water.

Applications. Defoaming agent in food and cosmetics.

Mutagenicity.

In vitro genotoxicity. HSA was negative in *Salmonella typhymurium* in the presence and absence of liver fraction *S9*.[1,4]

Carcinogenicity. Repeated *s/c* injections in rats caused sarcomas in sites of application.[2]

Chemobiokinetics. There was no *steatorrea* observed in dogs fed HSA (Kim and Spritz, 1968). Metabolites of HSA in rats are *10-hydroxypalmitic acid* and *8-hydroxymyristic acid*.[3]

Regulations. ***U.S. FDA*** (1998) approved the use of HSA as a defoaming agent used in the manufacture of paper and paperboard intended for use in packaging, transporting, or holding food in accordance with the condi tions prescribed in 21 CFR part 176.210.

References:

1. Nakamura, H. and Yamamoto, T., The active part of the [6]-gingerol molecule in mutagenesis, *Mutat. Res.*, 122, 87, 1983.
2. Swern, D., Wieder, R., McDonough, M., Meranze, D. R., and Shimkin, M. B., Investigation of fatty acids and derivatives for carcinogenic activity, *Cancer. Res.*, 30, 1037, 1970.
3. Binder, R. G. et al., Hydroxystearic acid deposition and metabolism in rats fed hydrogenated castor oil, *Lipids,* 5, 832, 1970
4. Scheutwinkel-Reich, M. et al., Microbiological studies investigating mutagenicity of deep frying fat fradions and some of their components, *Lipids*, 15, 849, 1980.

5-HYDROXY-1-(*p*-SULFOFENYL)-4-(*p*-SULFOPHENYL)AZOPYRAZOLE-3-CARBOXYLIC ACID, TRISODIUM SALT

Molecular Formula. $C_{16}H_9N_4O_9S_2$.3Na
M = 534.38
CAS No 1934-21-0
RTECS No UQ6400000

Synonyms and **Trade Names.** Acid Yellow 23 or T; A. F. Yellow No 4; 3-Carboxy-5-hydroxy-1-sulfophenylazopyrazole, trisodium salt; Tartrazine Yellow; Vondacid tartrazine; Yellow No 5.

Acute Toxicity. LD_{50} *i/p* exceeded 2.0 g/kg BW in rats.[1]

Reproductive Toxicity.

Teratogenicity. H. produced developmental abnormalities in rats, mice, and rabbits.[2,3]

Mutagenicity. H. caused CA in some mammalian test systems.[4,5]

Regulations. ***U.S. FDA*** (1998) approved the use of H. as a component of sanitizing solutions used on food-processing equipment and utensils, and on other food-contact articles as specified in 21 CFR part 178.1010.

References:

1. FAO Nutrition Meetings Report Series 38B, WHO, Geneva, 1966, 88.
2. Collins, T. F., Black, T. N., Brown, L. H., and Bulhack, P., Study of the teratogenic potential of FD & C Yellow No. 5 when given by gavage to rats, *Food Chem. Toxicol.*, 28, 821, 1990.
3. Collins, T. F., Black, T. N., O'Donnell, M. W. Jr., and Bulhack, P., Study of the teratogenic potential of FD & C yellow No. 5 when given in drinking-water, *Food Chem. Toxicol.*, 30, 263, 1992.
4. Giri, A. K., Das, S. K., Talukder, G., and Sharma, A., Sister chromatid exchange and chromosome aberrations induced by curcumin and tartrazine on mammalian cells *in vivo, Cytobios*, 62, 111, 1990.
5. Patterson, R. M. and Butler, J. S., Tartrazine-induced chromosome aberrations in mammalian cells, *Food Chem. Toxicol.*, 20, 461, 1982.

HYPOCHLOROUS ACID, LITHIUM SALT

Molecular Formula. ClO.Li
M = 58.39
CAS No 13840-33-0
RTECS No NH3486000
Abbreviation. HAL.

Synonyms. Lithium chloride oxide; Lithium hypochlorite; Lithium oxychloride.

Reproductive Toxicity. Rats were given 0.5 g/kg BW on days 6 to 15 of pregnancy. The treatment caused signs of fetotoxicity, and produced malformations in skeletal system.[1]

Immunotoxicity. Lithium is associated with systemic autoimmune disease (thyroid).[2]

Mutagenicity. HAL was not found to be a genotoxic substance.[3]

Regulations. ***U.S. FDA*** (1998) approved the use of HAL as a component of sanitizing solutions used on food-processing equipment and utensils, and on other food-contact articles as specified in 21 CFR part 178.1010.

References:.

1. Hoberman, A. M., Deprospo, J. R., Lochry, E. A., and Christian, M. S., Developmental toxicity study of orally administered lithium hypochlorite in rats, *J. Am. Coll. Toxicol.*, 9, 367, 1990.
2. Bagazzi, P. E., Autoimmunity caused by xenobiotics, *Toxicology*, 119, 1, 1991.
3. Weiner, M. L., Batt, K. J., Putman, D. L., Curren, R. D., and Yang, L. L., Genotoxicity evaluation of litium hypochlorite, *Toxicology*, 65, 1, 1990.

HYPOCHLOROUS ACID, SODIUM SALT

Molecular Formula. ClO.Na
M = 74.44
CAS No 7681-52-9

8007-59-8
RTECS No NH3486300
Abbreviation. SHC.

Synonyms and **Trade Names.** Antiformin; Chloros; Chlorox; Parozone; Purin B; Sodium chloride oxide; Sodium hypochlorite; Sodium oxychloride.

Properties. Greenish-yellow liquid with a disagreeable, sweetish odor. Soluble in cold water (294 g/l at 0°C).

Applications. Bleaching agent in paper and textile industry. Disinfectant and sanitizer. Used in laundry bleach.

Acute Toxicity. LD_{50} is 8.9 g/kg BW in rats. Toxic action is due to presence of chlorine. Rats were given 0.1 or 0.5% (males) and 0.2 or 1.0% (females) in their diet for 104 weeks. The treatment led to the reduction in BW gain, but there were no changes in biochemistry and hematological analyses. No histopathology lesions were reported.[1]

Reproductive Toxicity.

Gonadotoxicity. Poisoning caused changes in mouse sperm head morphology (at the doses equivalent to 4.0 to 8.0 mg chlorine/kg BW).[2]

Immunotoxicity. Male Sprague-Dawley rats received dietary levels of 5, 15, or 30 ppm SHC. The treatment caused significant reduction of spleen weight. In the high-dose treated group, delayed type hypersensitivity reactions and oxidative metabolism by microphages were noted. Prostaglandin *E2* production was elevated.[3]

Mutagenicity.

In vitro genotoxicity. The compound has been shown to possess mutagenic activity in several *in vitro* test systems, notably in *Salmonella typhimurium* strain *TA100* with *S9* mix, the CA test with Chinese hamster cells, and the SCE test using human embryo fibroblast cells.[2] SHC was positive in CA tests on activation (*S9*).[4]

Carcinogenicity. F344 rats were given SHC in their drinking water at concentrations of 0.1 and 0.05% for males and 0.2 and 0.1% for females. Rats showed a reduction in BW gain, but hematological and biochemical examination of the blood showed no changes due to treatment, and no significant lesions attributable to the treatment were detected in any tissue in the histopathological investigation. Although a variety of tumors developed in all groups, no dose-related change in either the incidence or latent period of tumors was observed for any organ or tissue in either sex. Under the experimental conditions described, SHC was not carcinogenic in F344 rats.[1,5]

Carcinogenicity classification. An IARC Working Group concluded that there is inadequate evidence of carcinogenicity of hypochlorite salts in *experimental animals*; no data were available from studies in *humans* on the carcinogenicity of hypochlorite salts.

IARC: 3.

Chemobiokinetics. *Di-* and *trichloroacetic acid, chloroform* and *dichloroacetonitrile* were found in the gut contents of fasted and nonfasted rats 1 hour after dosing with SHC. It has been demonstrated that *chloroform* and other *halogenated reaction products* are produced in the stomach after ingestion of SHC by rats.[6]

Regulations. ***U.S. FDA*** (1998) approved the use of SHC in adhesives as a component (monomer) of articles intended for use in packaging, transporting, or holding food in accordance with the conditions prescribed in CFR part 175.105.

References:

1. Hasegawa, R., Takahashi, M., Kokubo, T., Furukawa, F., Toyoda, K., Sato, H., Kurokawa, Y., and Hayashi, Y., Carcinogenicity study of sodium hypochlorite, *Food Chem. Toxicol.,* 24, 1295, 1986.
2. Meier, J. R. et al., Evaluation of chemicals used for drinking water disinfection for production of chromosomal damage and sperm-head abnormalities in mice, *Environ. Mutagen.,* 7, 201, 1985.
3. Exon, J. H., Koller, L. D., O'Reilly, C. A., and Bercz, J. P., Immunotoxicologic evaluation of chlorine-based drinking water disinfectants, sodium hypochlorite and monochloroamine, *Toxicology*, 44, 257, 1987.

4. Matsuoka, A., Hayashi, M. and Ishidate, M., Chromosome aberration tests in 29 chemicals combined with *S9* mix *in vitro*, *Mutat. Res.*, 66, 277, 1979.
5. Kawachi, T., Komatsu, T., Kada, T., Ishidate, M., Sasaki, M., Sugiyama, T., and Tajima, Y., Results of recent studies on the relevance of various short-term screening tests in Japan, in *The Predictive Value of Short-term Screening Tests in Carcinogenicity Evaluation*, G. M. Williams, R. Kroes, H. W. Waaijers and K.W. van de Poll, Eds., Elsevier, Amsterdam, 1980, 253.
6. Mink, F. L., Coleman, W. E., Munch, J. W., Kaylor, W. H., and Ringhand, H. P., *In vivo* formation of galogenated reaction products following oral sodium hypochlorite, *Bull. Environ. Contam. Toxicol.*, 30, 394, 1983.

12-HYDROXYSTEARIC ACID, METHYL ESTER

Molecular Formula. $C_{19}H_{38}O_3$
M = 314.57
CAS No 141-23-1
RTECS No WI4200000
Abbreviation. HSAM.

Synonym and **Trade Name.** Cenwax ME; Methyl 12-hydroxystearate.

Properties. White, waxy solid in form of short flat rods. Insoluble in water; reasonably soluble in organic solvents.

Applications. HSAM is a component of adhesives used in contact with food. Cosmetic ingredient; defoaming agent.

Carcinogenicity. Sarcomas at site of applications were found in rats repeatedly exposed to HSAM by *s/c* injections. Equivocal agent by *RTECS* criteria.

Regulations. ***U.S. FDA*** (1998) approved the use of HSAM as a defoaming agent in the manufacture of paper and paperboard intended for use in producing, manufacturing, packaging, processing, preparing, treating, packing, transporting, or holding foods in accordance with the conditions prescribed in 21 CFR part 176.210.

Reference:

Swern, D., Wieder, R., McDonough, M., Meranze, D. R., and Shimkin, M. B., Investigation of fatty acids and derivatives for carcinogenic activity, *Cancer. Res.*, 30, 1037, 1970

2-IMIDAZOLIDINONE

Molecular Formula. $C_3H_6N_2O$
M = 86.11
CAS No 120-93-4
RTECS No NJ0570000
Abbreviation. IA.

Synonyms. Ethyleneurea; 2-Imidazolidone.

Properties. White lumpy powder. Very soluble in water and hot alcohol.

Applications. Used in the production of high polymers, plasticizers, lacquers, adhesives.

Acute Toxicity. LD_{50} *i/p* is 0.5 g/kg BW in mice .[1]

Reproductive Toxicity.

Teratogenicity. Single dose of 480 mg/kg BW was administered to female Wistar rats on day 12 or 13 of gestation. The treatment caused neither maternal toxicity nor teratogenic effect.[2]

Mutagenicity.

In vitro genotoxicity. Relative mutagenic activity was found in *Salmonella* assay.[3]

Carcinogenicity. IA was shown to induce cancer in rats and mice when added to food with sodium nitrite.[4]

Chemobiokinetics. Following oral administration of ^{14}C-labeled IA, it is converted by dehydrogenation mechanism to *imidazole* which is major urinary metabolite.[5]

Regulations. ***U.S. FDA*** (1998) approved the use of IA in adhesives as a component of articles intended for use in packaging, transporting, or holding food in accordance with the conditions prescribed in 21 CFR part 175.105.

References:

1. *Eur. J. Med. Chem.*, Elsevier, Paris, 17, 235, 1982.
2. Ruddick, J. A., Williams, D. T., Hierlihy, W. L., Khera, K. S., Correlation of teratogenicity and molecular structure: ethylenethiourea and related compounds, *Teratology*, 13, 263, 1976.
3. Seiler, J. P., Ethylenethiourea (ETU), a carcinogenic and mutagenic metabolite of ethylenebisdithiocarbamate, *Mutat. Res.*, 26, 189, 1974.
4. Sander, J. et al., Induction of tumors by nitrite and secondary amines or amides, *Top. Chem. Carcinog.*, Proc. 2nd Int. Symp., 1972, 297.
5. Lawrence, L. and Marshall, W. A., Metabolism of ethyleneurea and ethylene thiourea in the rat, *Toxicol. Appl. Pharmacol.*, 48, A11, 1979.

1,1'-IMINODI-2-PROPANOL

Molecular Formula. $C_6H_{15}NO_2$
M = 133.22
CAS No 110-97-4
RTECS No UB6600000
Abbreviation. IDP.

Synonyms. Bis(2-hydroxypropyl)amine; Bis(2-propanol)amine; Diisopropanolamine; 1,1-Iminobis-2-propanol.

Properties. Crystals. Slightly soluble in water and alcohol.

Acute Toxicity. LD_{50} is 7.6 g/kg BW in male rats, and 4.77 g/kg BW in female rats. Poisoning induced severe CNS inhibition, coordination disorder, hypotermia, and diarrhea. Mice appeared to be more sensitive than rats and rabbits.[1]

Repeated Exposure revealed weak cumulative properties.[1]

Long-term Toxicity. The NOAEL of 0.022 mg/kg BW was established in rats and guinea pigs.[2]

Allergenic Effect. Contact dermatitis due IDP was observed in occupationally exposed workers.[3] Weak sensitization was observed.[1]

Reproductive Toxicity. No ***embryo-*** and ***gonadotoxic effects*** were noted on exposure to threshold concentration (4.0 mg/m^3).[1]

Mutagenicity.

In vitro cytogenetics. IDP was negative in five *Salmonella typhimurium* strains in the presence and absence of liver *S9* fraction at doses up to 10 g/plate.[013]

Carcinogenicity. *S/c* treatment of European hamsters with IDP led to the development of respiratory tract tumors (predominantly in the nasal cavity and lungs). Histologically, these tumors appeared to be papillary polyps, papillomas, adenomas, adenocarcinomas, squamous cell carcinomas, and mixed carcinomas.[4]

Regulations. ***EU*** (1990). IDP is available in the *List of authorized monomers and other starting substances which may continue to be used for the manufacture of plastic materials and articles intended to come into contact with foodstuffs pending a decision on inclusion in Section A* (*Section B*).

Standards. ***Russia*** (1995). MAC: 0.5 mg/l.

References:

1. Burkatskaya, Ye. N., Karpenko, V. N., Pokrovskaya, T. N., Ivanova, Z.. V., Medved', I. L., and Verich, G. Ye., Data on norm-setting maximum allowable concentrations of mono-, di-, and triisopropanolamine in the workplace air, *Gig. Truda Prof. Zabol.*, 7, 46, 1986 (in Russian).
2. Toporkov, V. V., Hygienic substantiation of maximum allowable concentrations of mono-, di-, and triisopropanolamine in water bodies, *Gig. Sanit.*, 3, 79, 1980 (in Russian).
3. Fujmoto, K., Hashimoto, S., Kozuka, T., and Yoshikawa, K., Contact dermatitis due diisopropanolamine, *Contact Dermat.*, 21, 56, 1989.

4. Reznik, G., Experimental carcinogenesis in the respiratory tract using the European field hamster (*Cricetus cricetus L.*) as a model, *Fortschr.-Med.*, 95, 2627, 1977.

IODINE

Molecular Formula. I_2
M = 253.81
CAS No 7553-56-2
RTECS No NN1575000

Synonyms and **Trade Names.** Diiodine; Eranol iode; Iodine colloidal; Iodine crystals; Jod; Molecular iodine.

Properties. Violet solid or rhombic, violet-black crystals with metallic luster, sharp characteristic odor, and acrid taste. Taste threshold is 0.147 to 0.204 mg/l. Water solubility is 3.0 g/l at 25°C.

Exposure. Concerns have been raised over the use of iodine for disinfecting drinking water on extended space flights.

Applications. *Iodine* is the form used in drinking-water disinfection. It has been used to disinfect both drinking water supplies and swimming pools with excellent results.

Acute Toxicity.

Observations in man. LD in adult range from a few tenths of a gram to more than 29 g. The mean LD probably lies between 2.0 and 4.0 g of free iodine.[019]

Repeated Exposure.

Observations in man. Risk assessments have treated the various forms of iodine as if they were toxicologically equivalent. A 14-day repeated-dose study utilizing total doses of iodine in the two forms at either 0.3 or 1.0 mg/kg BW was conducted with 33 male volunteers. Thyroid hormones evaluated included T4, T3, and thyroid-stimulating hormone (TSH). TSH was significantly increased by the high dose of both I_2 and I^- compared to the control. Decreases in T4 were observed with dose schedules with I^- and I_2, but none were statistically significant compared to each other, or compared to the control. This human experiment failed to confirm the differential effect of I_2 on maintenance of serum T4 concentrations relative to the effect of I^- that was observed in prior experiments in rats. However, based on the elevations in TSH, there should be some concern over the potential impacts of chronic consumption of iodine in drinking water.[1]

Short-term Toxicity. Rats received 1 to100 mg iodine or iodide (*NaI*) in drinking water for 100 days. The treatment produced no effect on BW gain and testes. An increase in the thyroid weight of males correlated with increasing contents of iodide but not iodine in the water. A decrease in BW of females was noted at the highest dose of iodide. Both substances affect thyroid hormone status in rats.[2]

Long-term Toxicity. Chronic ingestion of modest doses of dietary iodine is an environmental factor with known or suspected adverse effects on human thyroid.[3]

Reproductive Toxicity.

Embryotoxicity. Minks were given 10 to 320 ppm iodine for 1 to 7 months before breeding. Long-term consumption of 10 to 20 ppm iodine was beneficial for both reproduction and lactation. Ingestion in excess of 80 ppm caused embryotoxicity and embryolethality. Oral exposure of more than 20 ppm to dams resulted in hypertrophy of thyroid glands in kits.[4,5]

Immunotoxicity. Iodine is associated with autoimmune thyroid disease, autoimmune disease, and/or autoimmune responses in experimental animals.[6]

Chemobiokinetics. Food present in the digestive tract rapidly inactivates iodine by converting it into comparatively harmless iodide.[019]

Most fears revolve around effects of iodide on thyroid function. The storage of iodine in the thyroid depends upon the functional state of the gland.[03] It is removed from the blood and incorporated into the organic form in the thyroid gland.[7]

Iodine is absorbed from the lung, converted to iodide in the body, and then excreted, mainly in urine.[8]

Orally administered iodine is rapidly excreted in the urine and in smaller quantities in saliva, milk, sweat, bile, and other secretions.

Regulations. ***U.S. FDA*** (1998) approved the use of an aqueous solution of iodine as sanitizing solution on food-processing equipment and utensils, and on other food-contact articles in accordance with the conditions prescribed in 21 CFR part 178.1010.

References:

1. Robison L. M., Sylvester, P. W., Birkenfeld, P., Lang, J. P., and Bull, R. J., Comparison of the effects of iodine and iodide on thyroid function in humans, *J. Toxicol. Environ. Health,* 55, 93, 1988.
2. Sherer, T. T., Thrall, K. D., and Bull, R. J., Comparison of toxicity induced by iodine and iodide in male and female rats, *J. Toxicol. Environ. Health.,* 32, 89, 1991.
3. Barsano C. P., Environmental factors altering thyroid function and their assessment *Environ. Health Perspect.,* 38, 71, 1981.
4. Jones, R. E., Auberich, R. J., and Ringer, R. K., Feeding supplemental iodine to mink: reproductive and histopathologic effect, *J. Toxicol. Environ. Health.,* 10, 459, 1982.
5. Vorhees, C. V., Butcher R. E., and Brunner, R. L., Developmental toxicity and psychotoxicity of potassium iodide in rats: a case for the inclusion of behavior in toxicological assessment, *Food Chem. Toxicol.,* 22, 963, 1984.
6. Bagazzi, P. E., Autoimmunity caused by xenobiotics, *Toxicology,* 119, 1, 1991.
7. Jones, L. M., et al., *Veterinary Pharmacology and Therapeutics,* 4th ed., Iowa State University Press, Ames, 1977. 802.
8. International Labour Office, *Encyclopedia of Occupational Health and Safety,* vols 1 and 2, Geneva, 1983, 1154.

IODOFORM

Molecular Formula. CHI_3
M = 393.72
CAS No 75-47-8
RTECS No PB7000000
Abbreviation. IF.

Synonyms. Carbon triiodide; Triiodomethane.

Properties. Yellow powder, crystals, or solid with characteristic disagreeable odor. Solubility is 1.0 g/60 ml cold alcohol.[020] Solubility in water is 0.01 g/100 ml at 20°C.

Applications. Used mainly as an antiseptic agent.

Acute Toxicity. LD_{50} is 355 mg/kg BW in rats, 810 mg/kg BW in mice, 487 mg/kg BW in guinea pigs. Poisoning led to drowsiness, hypokinesia, and apathy. Clonic-tonic convulsions and pareses developed. A blood nasal exudate and breathing disturbances followed and led to death (Ivanitsky et al., 1983).

Administration of 2600 mM IF/100 g produced liver damage in rats including enzyme functions disorder followed by development of centrolobular necrosis similar to that caused by CCl_4.[1]

Repeated Exposure. Daily oral administration of 35.5 mg/kg BW caused granulation and vacuolization of the cytoplasm in rat hepatocytes, decreased the number of nucleoli, and induced a moderate proliferation of the Kupffer cells and occurrence of micronecrosis. Histological examination revealed swelling, desquamation, and granular dystrophy of the epithelium of renal glomeruli.[1]

Long-term Toxicity. IF was administered in corn oil by gavage for 78 weeks, followed by an observation period of 34 weeks for Osborne-Mendel rats and 13 or 14 weeks for $B6C3F_1$ mice. Rats received 71 or 142 mg/kg BW (males), and 27 or 55 mg/kg BW (females). Mice were given 47 and 93 mg/kg BW. Dose- dependent mortality was reported only in male rats.[2]

Mutagenicity.

In vitro genotoxicity. IF was positive in *Salmonella* mutagenicity assay.[025]

Carcinogenicity. In the above described bioassay,[2] there was no convincing evidence for the carcinogenicity in both species.

Carcinogenicity classification.

NTP: N - N - N - N (gavage).

Chemobiokinetics. IF is metabolized to *carbon monooxide* by hepatic microsomal mixed function oxidases. This reaction is markedly stimulated by sulfhydryl compounds.[3]

Regulations. ***U.S. FDA*** (1998) approved the use of IF (1) as component of adhesives for single and repeated use in food contact surfaces in accordance with the conditions prescribed in 21 CFR part 175.105; (2) as a polymerization control agent; and (3) in preparation of rubber articles intended for repeated use in producing, manufacturing, packaging, processing, preparing, treating, packing, transporting, or holding food (total amount of I. should not exceed 5.0% by weight of rubber product) in accordance with the conditions prescribed in 21 CFR part 177.2600.

References:

1. Sell, D. A. and Reynolds, E. S., Liver parenchymal cell injury. 8. Lesions of membranous cellular components following iodoform, *J. Cell. Biol.*, 41, 736, 1969.
2. *Bioassay of Iodoform for Possible Carcinogenicity*, NTP Technical Report Series No 110, Research Triangle Park, NC, 1978.
3. Stevens, J. L. and Anders, M. W., Metabolism of haloforms to carbon monooxide. III. Studies on the mechanism of the reaction, *Biochem. Pharmacol.*, 28, 3189, 1979.

IRON(3^+) OXIDE

Molecular Formula. Fe_2O_3
M = 159.7
CAS No 1309-37-1
8011-97-0
RTECS No NO7400000
Abbreviation. FO.

Synonyms and **Trade Names.** Anhydrous iron oxide; Black oxide of iron; C. I. Pigment Red 101 or 102; English Red; Ferric oxide; Iron oxide pigments; Iron Red; Light Red; Mars Brown; Mars Red; Natural Red Oxide; Ochre; Prussian Brown; Red ochre; Red oxide; Rubigo; Sienna; Stone Red; Supra Vitriol Red; Yellow ferric oxide.

Applications. Wet lubricant for packing materials. Drying oil. Used in the production of soaps, detergents, protective coatings, and alkyd resins.

Properties. Insoluble in water.

Acute Toxicity.

Observations in man. Human probable oral lethal dose is 0.5 to 5.0 g/kg BW.[1]

Animal studies. LD_0 *s/c* is 30 mg/kg BW in dogs.

Mutagenicity.

Observations in man. Urine of workers exposed to FO was more mutagenic than that of persons exposed only to mineral oil (Rueff et al., 1982).

Animal studies. Iron oxide has been tested in the *B. subtilis* rec-assay.[2]

Carcinogenicity. Sarcomas were observed at the site of *s/c* massive dose injections of iron-dextran complex in rats and mice.[3]

Carcinogenicity classification. An IARC Working Group concluded that there is evidence suggesting strong lack of carcinogenicity of FO in *experimental animals* and there is inadequate evidence to evaluate the carcinogenicity of FO in *humans*.

IARC: 3

Regulations. ***U.S. FDA*** (1998) approved the use of FO as a component of the uncoated or coated food-contact surface of paper and paperboard intended for use in producing, manufacturing, packaging, processing, preparing, treating, packing, transporting, or holding aqueous and fatty foods in accordance with the conditions prescribed in 21 CFR part 176.17

References:

1. Abdernalden's *Handbuch der Biologischen Arbeitsmethoden.*, 4, 1289, 1935.

2. Kanematsu, N., Hara, M., and Kada, T., Rec assay and mutagenicity studies on metal compounds, *Mutat. Res.*, 77, 109, 1980.
3. Haddow, A. and Horning, E. S., On the carcinogenicity of an iron-dextran complex, *J. Natl. Cancer Inst.*, 24, 109, 1960.

ISOBUTYRALDEHYDE

Molecular Formula. C_4H_8O
M = 72.10
CAS No 78-84-2
RTECS No NQ4025000
Abbreviation. IBA.

Synonyms. 2-Methyl-1-propanal; 2-Methylpropionaldehyde; Isobutanal; Isobutyric aldehyde; Isopropyl formaldehyde.

Properties. Transparent, colorless liquid with fruity taste and extremely sharp odor. Solubility in water is 110 g/l at 20°C; miscible with ethanol.[020]

Applications. Used in the production of cellulose esters. Chemical intermediate for isobutyl alcohol, used in synthesis of resins and plasticizers. Rubber antioxidant and accelerator.

Acute Toxicity. LD_{50} is 3.7 g/kg BW in rats.[1] Oral administration increases relative weight of lungs, liver, and kidneys by 68, 36, and 52%, respectively.[2]

Mutagenicity. Was negative in *Salmonella*[013] and in *Dr. melanogaster*.[3]

Carcinogenicity. IBA has been predicted to be rodent carcinogens by the computer automated structure evaluation (CASE) and multiple computer automated structure evaluation (MULTICASE) system.[4]

Regulations.

EU (1992). IBA is available in the *List of monomers and other starting substances, which may continue to be used pending on desision on inslusion in Section A* (*Section B*).

U.S. FDA (1998) approved the use of IBA as a direct food additive.

References:

1. Smyth, H. F., Carpenter, C. P., Weil, C. S., et al., Range-finding toxicity data, List V, *Arch. Ind. Hyg. Occup. Med.*, 10, 61, 1954.
2. Swintukhovsky, O. A., Toxicological characteristics of isobutyric aldehyde, in *Toxicology and Hygiene of Naphtha-chemistry and Naphtha-chemical Industry Products*, 1972, 187 (in Rus sian).
3. Woodruff, R. C., Mason, J. M., Valencia, R., and Zimmerling, S., Chemical mutagenesis testing in *Drosophila*. V. Results of 53 coded compounds tested for National Toxicology Program, *Environ. Mutagen.*, 7, 677, 1985.
4. Zhang, Y. P., Sussman, N., Macina, O. T., Rosenkranz, H. S., and Klopman, G., Prediction of the carcinogenicity of a second group of organic chemicals undergoing carcinogenicity testing, *Environ. Health Perspect.*, 104S, 1045, 1996.

13-ISOPROPYLPODOCARPA-7,13-DIEN-15-OIC ACID

Molecular Formula. $C_{20}H_{30}O_2$
M = 302.50
CAS No 514-10-3
RTECS No TP8580000
Abbreviation. AA.

Synonyms and **Trade Names.** Abietic acid; Kyselina abietova; Sylvic acid.

Properties. Monoclinic plates from alcohol + water. Insoluble in water; soluble in alcohol.[020]

Applications. Used in the manufacture of plastics and esters (methyl, vinyl, ester gum, etc.).

Migration Data. I. was identified in solvent extracts obtained from paper and board materials intended for food contact. Level was generally below 1.0 mg/kg paper.[1]

Acute Toxicity. LD_{50} *i/v* is about 200 mg/kg BW in mice.

Regulations.

U.S. FDA (1998) approved the use of I. in adhesives as a component of articles intended for use in packaging, transporting, or holding food in accordance with the conditions prescribed in 21 CFR part 175.105.

EU (1990). ICAN is available in the *List of authorized monomers and other starting substances which shall be used for the manufacture of plastic materials and articles intended to come into contact with foodstuffs (Section* A).

Great Britain (1998). AA is authorized without time limit for use in the production of polymeric materials and articles in contact with food or drink or intended for such contact.

Reference:

1. Castle, L., Offen, C. P., Baxter, M. J., and Gilbert, J., Migration studies from paper and board food packaging materials. 1. Compositional analysis. *Food Addit. Contam.*, 14, 35, 1997.

ISOCYANIC ACID, 1,5-NAPHTHYLENE ESTER

Molecular Formula. $C_{12}H_6N_2O_2$
M = 210.20
CAS No 3173-72-6
RTECS No NQ9600000
Abbreviation. ICAN.

Synonyms. 1,5-Diisocyanatonaphthalene; 1,5-Naphthalene diisocyanate.

Carcinogenicity classification. IARC Working Group concluded that there are no experimental or epidemiological data available for evaluation of the carcinogenicity of ICAN in *humans* and in *experimental animals.*

IARC: 3.

Regulations.

EU (1990). ICAN is available in the *List of authorized monomers and other starting substances which shall be used for the manufacture of plastic materials and articles intended to come into contact with foodstuffs (Section* A).

Great Britain (1998). ICAN is authorized without time limit for use in the production of polymeric materials and articles in contact with food or drink or intended for such contact. The quantity in the finished polymeric material or article of any substance within, or any combination of substances within shall not exceed 1 mg/kg (expressed as isocyanate moiety).

Standards. ***Russia*** (1995). MAC: 0.1 mg/l.

ISOPROPYL ALCOHOL, TITANIUM(4^+) SALT

Molecular Formula. $C_3H_8O.1/4Ti$
M = 284.26
CAS No 546-68-9
RTECS No NT8060000
Abbreviation. IAT.

Synonyms. Isopropyl orthotitanate; Isopropyl titanate; Tetraisopropoxytitanium; Tetraisopropyl *o*-titanate; Tetrakis(isopropoxy)titanium; Titanium isopropylate; Titanium tetra-*n*-propoxide; Titanium tetraisopropoxide; Titanium(4^+) isopropoxide.

Properties. Light yellow liquid.

Applications. Used in the production of heat-resistant surface coatings, in paints and plastics. A polymerization and condensation catalyst.

Acute Toxicity. LD_{50} is 7.46 ml/kg BW in rats.[1]

Chemobiokinetics. *Titanium* salts are poorly absorbed. However, detectable amounts of *titanium* can be found in the blood, brain, and parenchymatous organs of humans. Most ingested *titanium* is excreted unabsorbed, predominantly in urine. In man, *titanium* was determined in urine at a rate of about 10 μg/l.[2]

Regulations. ***U.S. FDA*** (1998) approved the use of IAT as an indirect food additive for use only as a compoent of adhesives in accordance with the conditions prescribed in 21 CFR part 175.105.

References:

1. Carpenter, C. P., Weil, C. S., and Smyth, H. F., Jr., Range-finding toxicity data: list VIII, *Toxicol. Appl. Pharmacol.*, 28, 313, 1974.
2. *Handbook of the Toxicology of Metals*, L. Friberg, G. F. Nordberg, E. Kessler, and V. B. Vouk, Eds., 2nd ed., vols 1, 2, Elsevier Publ., Amsterdam, vol 2, 1986, 600.

ITACONIC ACID

Molecular Formula.
M = 130.10
CAS No 97-65-4
Abbreviation. IA.

Synonyms. Methylenebutanedioic acid; Methylenesuccinic acid; 2-Propene-1,2-dicarboxylic acid; Propyle nedicarboxylic acid.

Applications. Used in copolymerization, in production of resins and plasticizers.

Properties. White crystals with a characteristic odor. 1.0 g dissolves in 12 ml water and 5.0 ml alcohol.[020]

Long-term Toxicity. Rats received 1.0% IA in the diet. The treatment caused retardation of BW gain.[011]

Chemobiokinetics. A single large oral dose of IA caused an increased output of *succinic acid* in rabbit urine.[011]

Regulations.

EU (1990). IA is available in the *List of monomers and other starting substances, which may continue to be used pending a decision on inclusion in Section A* (*Section B*).

U.S. FDA (1993) approved the use of IA in the manufacture of cellophane for packaging food. ***Great Britain*** (1998). IA is authorized without time limit for use in the production of polymeric materials and articles in contact with food or drink or intended for such contact.

KAOLIN (*Clay*)

Molecular Formula. $Al_2O_3.2SiO_2.2H_2O$
CAS No 1332-58-7
RTECS No GF1670500

Trade Names. Altowhite; Bentone; Bolus alba; China clay; Porcelain clay; White bole.

Properties. A white or whitish powder consisting of tiny hydrophilic particles with a lamellar structure. Finely-dispersed plastic rock.

Composition. A product of the erosion of feldspars, mica, granites, and other strata. Consists mainly of mineral kaolinite, $Al_2O_3.2SiO_2.2H_2O$. K.-modified is produced by treating kaolin with a reaction product of isopropyl titanate and oleic acid in which 1 mole of isopropyl titanate reacts with 1 to 2 moles of oleic acid. The reaction product will not exceed 8.0 % of K.-modified.

Applications. Filler.

Acute Toxicity. K. seems to be non-toxic. Probable lethal oral dose for human is above 15 g/kg.[022]

Long-term Toxicity. Toxicity manifestations were not found in mice given the diet containing 80% K.[022] However, according to Patterson et al., in Sprague-Dawley rats fed K., anemia was found to develop (significant reduction in *Hb*, hematocrit, and erythrocyte count).[1]

Reproductive Toxicity.

Embryotoxicity. Male Sprague-Dawley rats received the total dose of 590 g/kg BW prior to mating and during the pregnancy. The treatment affected growth of newborns.[1]

Chemobiokinetics. K. was found in tumor tissue using extraction replication technique with electron microscope-microanalysis (Henderson et al., 1975).

Regulations.

U.S. FDA (1998) approves the use of K.-modified (1) in olefin polymers as articles or components of articles intended for use in contact with food in accordance with the conditions prescribed in 21 CFR part 177.1520; (2) in the manufacture of cellophane for packaging food in accordance with the conditions prescribed in 21 CFR part 177.1200; and (3) as a pigment, colorant, or pacifier in olefin polymers at levels not to exceed 40% by weight of the olefine polymers. K. is considered to be *GRAS*.

EU (1995). K. is a food additive generally permitted for use in foodstuffs.

Reference:

1. Patterson, E. C. and Staszak, D. J., Effects of geophagia (kaolin ingestion) on the maternal blood and embryonic development in the pregnant rat, *J. of Nutrition*, 107, 2020, 1977.

LANOLIN

CAS No 8006-54-0

RTECS No OE3201000

Synonyms and **Trade Names.** Adeps Lane; Agnin; Agnolin; Amber lanolin; Laniol; Wool fat; Wool grease; Wool wax, refined.

Properties. Yellowish-white unctuous mass with a slight odor. Practically insoluble in water.[020]

Applications. Used predominantly for pharmaceutical purposes.

Reproductive Toxicity.

Teratogenic action was not noted in rabbits fed L. up to maternally toxic doses on days 2 to 16 of gestation.[1]

Mutagenicity.

In vitro genotoxicity. L. was negative in *Salmonella* mutagenicity assay.[2]

Allergenic Effect. L. may produce dermatitis of allergic-contact etiology.[3] This could depend on availability of natural free alcohols in L.[4]

Chemobiokinetics. L. is very poorly absorbed from the GI tract.

Regulations. ***U.S. FDA*** (1998) approved the use of L. (1) in the manufacture of resinous and polymeric coatings for the food-contact surface of articles intended for use in producing, manufacturing, packing, processing, preparing, treating, packaging, transporting, or holding food in accordance with the conditions prescribed in 21 CFR part 175.300; (2) as a component of the uncoated or coated food-contact surface of paper and paperboard intended for use in producing, manufacturing, packaging, processing, preparing, treating, packing, transporting, or holding aqueous and fatty foods in accordance with the conditions prescribed in 21 CFR part 176.170; (3) as a defoaming agent in the manufacture of paper and paperboard intended for use in packaging, transporting, or holding food in accordance with the conditions prescribed in 21 CFR part 176.210; and (4) in the manufacture of cellophane for packaging food in accordance with the conditions prescribed in 21 CFR part 177.1200.

References:

1. Anonymous, Final report on the safety assessment of laneth-10 acetate group, *J. Am. Coll. Toxicol.*, 1, 1, 1982.
2. Gocke, E., King, M.-T., Eckhardt, K., and Wild, D., Mutagenicity of cosmetic ingredients licensed by the EEC, *Mutat Res.*, 90, 91, 1981.
3. O'Donnell, B. F. and Hodson, C., Allergic contact dermatitis due to lanolin in an adhesive plaster, *Contact Dermat.*, 28, 191, 1993.
4. Clark, E. W., Blondeel, A., Cronin, E., Oleffe, J. U. A., and Wilkinson, D. S., Lanolin of reduced sensitizing potential, *Contact Dermat.*, 7, 80, 1981.

LECITHIN

CAS No 8002-43-5

Synonyms and **Trade Names.** Granulestin; Lecithol; Phosphatidylcholine.

Composition. A mixture of the diglycerides of stearic, palmitic, and oleic acids, linked to the choline ester of phosphoric acid.

Properties. Waxy mass when the acid value is about 20. Pourable, thick fluid when the acid value is around 30. Insoluble in water; soluble in about 12 parts of cold absolute alcohol. Soluble in fatty acids; practically insoluble in cold vegetable and animal oils.[020]

Applications. Edible and digestible surfactant and emulsifier of natural origin.

Exposure. L. is normally present in human milk at concentrations of 100 to 200 nmol/ml.[1]

Reproductive Toxicity. Pregnant rats and their offspring received a diet high in supplemented L. The treatment produced abnormalities of function in the developing CNS.[2]

Regulations.

U.S. FDA (1998) approved the use of L. in the manufacture of resinous and polymeric coatings for the food-contact surface of articles intended for use in producing, manufacturing, packing, processing, preparing, treating, packaging, transporting, or holding food in accordance with the conditions prescribed in 21 CFR part 175.300. Affirmed as *GRAS.*

EU (1995). L. is a food additive generally permitted for use in foodstuffs.

Standards. *Russia* (1994). RAL: 22 mg/l.

References:

1. Zeisel, S. H. et al., Choline, phosphatidylcholine and sphingomyelin in human and bovine milk and infant formulas, *J. of Nutrition*, 116, 50, 1986.
2. Bell, J. M. et al., Perinatal dietary exposure to soy lecithin: altered sensitivity to central cholinergic stimulation, *Int. J. Dev. Neurosci.*, 4, 497, 1986.

LIME SULPHUR

CAS No 1344-81-6

Abbreviation. LS.

Synonyms and **Trade Names.** Calcium polysulfide; Calcium sulfide; Orthorix; Sulka.

Properties. Deep orange malodorous liquid. According to other sources, Calcium sulfide occurs as colorless crystals.

Acute Toxicity.

Observations in man. Probable lethal oral dose is in the range 50 to 500 mg/kg BW for a 70 kg person.[022]

Allergenic Effect. A sensiticizer.[022]

Chemobiokinetics. A bathing agent containing polysulfides was administered orally to rats. Distribution of polysulfides resulted in their highest concentration in blood, liver, lung, and kidneys. Maximum concentration of sulfides was determined in blood.[1]

LS may form *hydrogen sulfide* by decomposition before or after ingestion.[022]

Regulations. *U.S. FDA* (1998) approved the use of calcium sulfide as a component (colorant) of the uncoated or coated food-contact surface of paper and paperboard intended for use in producing, manufacturing, packaging, processing, preparing, treating, packing, transporting, or holding aqueous and fatty foods in accordance with the conditions prescribed in 21 CFR part 176.170.

Reference:

1. Nagata, T., Kage, S., Kimura, K., Kudo, K., and Imamura, T., How to diagnose polysulfide poisoning from tissue samples? *Int. J. Legal Med.*, 106, 288, 1994.

LINSEED OIL

CAS No 8001-26-1

RTECS No OI9690000

Abbreviation. LO.

Trade Name. Groco.

Composition. A drying oil obtained by expression of linseed. Constituents: glycerides of linolenic, linoleic, oleic, stearic, palmitic, and myristic acids.[020]

Properties. Yellowish liquid with peculiar odor and bland taste. Poorly soluble in alcohol.[020]

Applications. Used in the production of artificial rubber; waterproofing agent.

Acute Toxicity. LO was heated for 12 hours at 275°C under nitrogen and fractionated. One fraction was especially toxic: all mice died after oral administration.[1]

Repeated Exposure. LO (preheated to 275°C for 12 hours without oxidation) was fed to rats for 2 months. Animals showed hepatic hypertrophy, accompanied by an increase in liver pyruvic dehydrogenase and α-ketoglutaric dehydrogenase activities and an increase in endogenous respiration.[2] LO heated at 240°C for 24 hours gave reduced growth when fed to chickens at 20% of the diet.[3]

Reproductive Toxicity. Rats were fed LO (preheated to 275°C) during lactation and gestation. Most offspring died within 3 days of birth; none survived past day 13 of lactation. In dead offspring, liver weights and hepatic total lipid contents were markedly increased. There was an accumulation of *oleic acid* and short-chain fatty acids (Potteau, 1975).

Regulations. ***U.S. FDA*** (1998) approved the use of LO in resinous and polymeric coatings used as the food-contact surfaces of articles intended for use in producing, manufacturing, packing, processing, preparing, treating, packaging, transporting, or holding food in accordance with the conditions prescribed in 21 CFR part 175.300. Substances classified as drying oils, when migrating from food-packaging material (as components of finished resins) shall include LO in accordance with the provisions prescribed in 21 CFR part 181.26.

References:

1. Saito, M. and Kaneda, T., Abstract, *Yukagaku,* 25, 79, 1976 (in Japanese).
2. Grandgirard, A. et al., *Ann.. Nutr. Aliment*., 26, 161, 1972.
3. Farrow, G. et al., *Poult. Sci.,* 62, 85, 1983.

MAGNESIUM FLUORIDE

Molecular Formula. F_2Mg
M = 62.31
CAS No 7783-40-6
RTECS No OM3325000
Abbreviation. MF.

Trade Names. Afluon; Irtran 1; Sellaite.

Properties. Colorless substance with slight violet fluorescence. Solubility in water is 87 mg/l (18°C).

Migration of magnesium from acrylonitrile/butadiene/styrene copolymer into food simulants, measured by inductively-coupled plasma-mass spectrometry amounted to 0.05 mg/kg. The concentration found was less than proposed limit of migration.

Acute Toxicity. Doses about 1.0 g/kg BW appeared to be lethal.[020]

Regulations. ***U.S. FDA*** (1998) approved the use of MF in resinous and polymeric coatings used as the food-contact surfaces of articles intended for use in producing, manufacturing, packing, processing, preparing, treating, packaging, transporting, or holding food in accordance with the conditions prescribed in 21 CFR part 175.300.

Reference:

Fordham, P. J., Gramshaw, J. W., Crews, H. M., and Castle, L., Element residues in food contact plastics and their migration into food simulants, measured by inductively-coupled plasma-mass spectrometry, *Food. Addit. Contam*., 12, 651, 1997.

MAGNESIUM NITRATE

Molecular Formula. $N_2O_6.Mg$
M = 148.33
CAS No 10377-60-3
RTECS No OM3750000
Abbreviation. MN.

Properties. Hexahydrate, clear, colorless, deliquesce crystals. Soluble in water; freely soluble in alcohol.[020]

Migration of magnesium from acrylonitrile/butadiene/styrene copolymer into food simulants, measured by inductively-coupled plasma-mass spectrometry amounted to 0.05 mg/kg. The concentration found was less than the proposed limit of migration.

Acute Toxicity. MN is a *metHb*-forming agent. Probable oral lethal dose in humans is 0.5 to 5.0 g/kg BW.[019]

Chemobiokinetics. Nitrate salts, if not properly absorbed, may be reduced to nitrites.

Regulations. ***U.S. FDA*** (1998) approved the use of MN in adhesives as a component of articles intended for use in packaging, transporting, or holding food in accordance with the conditions prescribed in 21 CFR part 175.105.

Reference:

Fordham, P. J., Gramshaw, J. W., Crews, H. M., and Castle, L., Element residues in food contact plastics and their migration into food simulants, measured by inductively-coupled plasma-mass spectrometry, *Food. Addit. Contam.*, 12, 651, 1997.

MAGNESIUM OXIDE

Molecular Formula. MgO
M = 40.31
CAS No 1309-48-4
RTECS No OM3850000
Abbreviation. MO.

Synonyms and **Trade Names.** Animag; Calcined brucite; Calcined magnesia; Magcal; Magnesia; Magnesioum oxime fume.

Properties. White, odorless, very fine powder.[020] Insoluble in alcohol, solubility in water is 86 mg/l at 30°C.

Applications. Used as a rubber accelerator, as well as in the production of animal feed and in cosmetics, and as a catalyst. Used also in paper manufacture.

Migration of *magnesium* from acrylonitrile/butadiene/styrene copolymer into food simulants, measured by inductively-coupled plasma-mass spectrometry amounted to 0.05 mg/kg. The concentration found was less than proposed limit of migration.[1]

Acute Toxicity. Non-toxic on ingestion.

Observations in man. Human LD_{min} (inhalation) is 400 mg/m^3.[03]

Repeated Exposure.

Observations in man. 50 normal volunteers received placebo or MO, 400 mg capsules, twice a day for 60 days to study if MO would produce changes in the lipid profile. No significant differences were noted.[2]

Animal studies. Large oral doses caused diarrhea in Holstein bull calves (Gentry et al., 1978).

Carcinogenicity. On intratracheal administration, MO appeared to be an equivocal tumorigenic agent by *RTECS* criteria.[3]

Chemobiokinetics. Insoluble compounds such as MO are retained for longer periods. Magnesium is retained in skeleton, muscle, and soft tissue. 5 to 10% of magnesium can be absorbed, and its ion is rapidly excreted by kidney.[4]

Regulations. ***U.S. FDA*** (1998) approved the use of MO (1) as a component of the uncoated or coated food-contact surface of paper and paperboard intended for use in producing, manufacturing, packaging, processing, preparing, treating, packing, transporting, or holding aqueous and fatty foods in accordance with the conditions prescribed in 21 CFR part 176.170; (2) in the manufacture of resinous and polymeric coatings for the food-contact surface of articles intended for use in producing, manufacturing, packing, processing, preparing, treating, packaging, transporting, or holding food in accordance with the conditions prescribed in 21 CFR part 175.300; and (3) in adhesives as a component (monomer) of articles intended for use in packaging, transporting, or holding food in accordance with the conditions prescribed in 21 CFR part 175.105.

References:

1. Fordham, P. J., Gramshaw, J. W., Crews, H. M., and Castle, L., Element residues in food contact plastics and their migration into food simulants, measured by inductively-coupled plasma-mass spectrometry, *Food. Addit. Contam.*, 12, 651, 1997.
2. Marken, P. A. et al., Effects of magnesium oxide on the lipid profile of healthy volunteers, *Atherosclerosis*, 77, 37, 1989.
3. Stenback, F. G., Ferrero, A., and Shubik, P., Synergistic effects of diethylnitrosamine and different dusts on respiratory carcinogenesis in hamsters, *Cancer Res.*, 33, 2209, 1973.
4. *The Pharmacological Basis of Therapeutics,* L. S. Goodman and A. Gilman, Eds., 5th ed. Macmillan Publishing Co, Inc., New York, 1975.

MALEIC ACID, DIALLYL ESTER

Molecular Formula. $C_{10}H_{12}O_4$.
M = 196.22
CAS No 999-21-3
RTECS No ON0700000
Abbreviation. DAM.

Synonyms. Diallyl maleate; Maleic acid, diallyl ester.

Acute Toxicity. LD_{50} is 300 mg/kg BW in rats, and about 500 mg/kg BW in mice (Shell Chem. Co., 1961).

Regulations.

U.S. FDA (1998) approved the use of DAM in adhesives as a component of articles intended for use in packaging, transporting, or holding food in accordance with the conditions prescribed in 21 CFR part 175.105.

EU (1990). DAM is available in the *List of authorized monomers and other starting substances which may continue to be used for the manufacture of plastic materials and articles intended to come into contact with foodstuffs pending a decision on inclusion in Section A* (*Section B*).

Great Britain (1998). DAM is authorized up to the end of 2001 for use in the production of polymeric materials and articles in contact with food or drink or intended for such contact.

Reference:

Smyth, H. F., Carpenter, C. P., and Weil, C. S., Range-finding toxicity data: List III, *J. Ind. Hyg. J. Ind. Hyg. Toxicol.*, 31, 60, 1949.

MALEIC ACID, DIETHYL ESTER

Molecular Formula. $C_8H_{12}O_4$
M = 172.20
CAS No 141-05-9
RTECS No ON1225000
Abbreviation. DEM.

Synonyms. 2-Butenedioic acid, diethyl ester; Diethyl maleate; Ethyl maleate.

Properties. Liquid. Insoluble in water. Soluble in alcohol.

Acute Toxicity. LD_{50} is 3.2 g/kg BW in rats. LD_{50} *i/p* is 3.07 g/kg BW. Poisoning is accompanied with lacrimation, functional structural changes of salivary glands. In the GI tract, hypermotility and diarrhea were noted.[1]

Reproductive Toxicity.

Teratogenicity. Rats were dosed *i/p* with 150 or 300 mg DEM/kg BW followed by 55 mg phenytoin/kg BW. DEM enhanced teratogenic effect of phenytoin in this study.[2]

Mutagenicity.

In vitro genotoxicity. DEM was positive in mouse lymphocytes.[3]

Chemobiokinetics. A single oral administration of 500 mg DEM/kg BW affected hepatic and epididymal glutathione, but have insignificant effect on testicular levels.[4]

Regulations. ***EU*** (1990). DEM is available in the *List of monomers and other starting substances which may continue to be used pending a decision on inclusion in Section A.*

References:

1. Andersen, M. E., Thomas, O. E., Gargas, M. L., Jones, R. A., and Jenkins, L. J., The significance of multiple detoxication pathways for reactive metabolites in the toxicity of 1,1-dichloroethylene, *Toxicol. Appl. Pharmacol.*, 52, 422, 1980.
2. Wong, M., Helston, L. M. J., and Wells, P. G., Enhancement of murine phenytoin teratogenicity by the gamma-glutamylcysteine synthetase inhibitor l-buthionine-(S,R)-sulfoximine and by the elutathione depletor diethyl maleate, *Teratology*, 40, 127, 1989.
3. Wangenheim, J. and Bolcsfoldi, G., Mouse Lymphoma L5178Y thymidine kinase locus of 50 compounds, *Mutagenesis*, 3, 193, 1988.
4. Gandy, J., Millner, G. C., Bates, H. K., Casciano, D. A., and Harbison, R. D., Effects of selected chemicals on the glutathione status in the male reproductive system of rats, J. *Toxicol. Environ. Health*, 29, 45, 1990.

MERCAPTOACETIC ACID, 2-ETHYLHEXYL ESTER

Molecular Formula. $C_{10}H_{20}O_2S$
M = 204.32
CAS No 7659-86-1
RTECS No AI7255000

Synonyms. 2-Ethylhexyl(mercaptoacetate); 2-Ethylhexyl(thioglycolate).

Applications. Used in the production of plastics.

Acute Toxicity. LD_{50} is reported to be 0.3 to 1.8 g/kg BW in rats and mice, 0.53 g/kg BW in rabbits, and 0.96 g/kg BW in guinea pigs. Rats are the most sensitive. The toxicity of the substance is evidently determined by presence of thioglycolate. Sublethal dose slightly damages the liver and kidneys.

Reference:

Schmidt, P., Fox, G., Hollenback, K., u. a., Zur akuten Toxicitat des Thioglycolsaure-2-ethylhexyl esters in Tierversuch, *Zeitschr. Gesamte Hyg.*,20, 575, 1974.

METHOX 29-PM-75

Composition. A mechanical mixture of copper hydroxyquinolate with methionine and carbon black.

Properties. A dark blue powder. Insoluble in alcohol.

Applications. Used as a vulcanizing agent in the production of fluoroplastics and as a stabilizer of siloxane rubber.

Acute Toxicity. LD_{50} was not attained.

Repeated Exposure. Rats were dosed by gavage with 1.5 g/kg BW for 1.5 months. The treatment caused leukocytosis and reticulocytosis, reduction in the blood coagulation time, and an increase in the STI value. The decreased levels of residual nitrogen in the blood and of creatinine in the urine were observed.

Reference:

Gnezdilova, A. I. and Volodchenko, V. A., Toxicity of Methox 29-PM-75, *Gig. Truda Prof. Zabol.*, 12, 58, 1983 (in Russian).

2-METHYLFURAN

Molecular Formula. C_5H_6O
M = 82.11
CAS No 534-22-5
RTECS No LU2625000
Abbreviation. MF.

Trade Name. Silvan.

Properties. Highly mobile, colorless to slightly yellowish liquid with an odor of ether. Water solubility is 0.2%. Miscible with alcohol. Odor perception thresholds are 0.63 mg/l at 20°C and 0.21 mg/l at 60°C.[1]

Applications. Used in the production of lacquers, dyes, rubber, and plastics.

Acute Toxicity. LD_{50} is found to be 480 mg/kg BW in rats, 455 mg/kg BW in rabbits, 600 mg/kg BW in mice, and 730 mg/kg BW in guinea pigs. Histological examination revealed degenerative changes in the liver and kidney in rabbits given 200 to 300 mg/kg BW.[1]

Repeated Exposure revealed weak cumulative properties. Rats received 1/10 LD_{50} for a month. None of the animals died.[1]

Long-term Toxicity. Rabbits were dosed with 4.55 and 0.455 mg/kg BW. The higher dose caused retardation of BW gain and changes in hematology analysis (signs of anemia). The NOAEL of 0.455 mg/kg BW was estalished in this study.[1]

Rats received the dose of 5.0 mg/kg BW for 6 months. Gross pathology examination revealed proliferative changes in the liver.[2]

Allergenic Effect. MF exhibited negative response in guinea pigs.

Chemobiokinetics. MF has been shown to be metabolized in mice to a highly reactive product that binds irreversibly to tissue macromolecules. The reactive metabolite is most likely to be *dialdehyde, methylbutenedial*, resulting from the oxidation and ring-opening of the furan moiety.[3]

Standards. ***Russia*** (1995). MAC: 0.5 mg/l (organolept., odor).

References:

1. Trubko, E. I. and Teplyakova, E. V., Hygienic norm-setting for 2-methylfuran (silvan) in water bodies, *Gig. Sanit.*, 2, 68, 1981 (in Russian).
2. Teplyakova, E. V. et al., in *Proc. Leningrad Sanitary Hygiene Medical Institute.*, Leningrad, 105, 51, 1974 (in Russian).
3. Ravindranath, V., Burka, T., and Boyd, M. R., Isolation and characterization of reactive metabolites of 2-methylfuran and 3-methylfuran, *Pharmacologist*, 25, 171, 1983.

2-METHYLHEXANE

Molecular Formula. C_7H_{16}
M = 100.23
CAS No 591-76-4
RTECS No MO3871500
Abbreviation. MH.

Acute Toxicity. LD_{50} is 10 g/kg BW in rats.

Repeated Exposure. Inhalation studies showed inducing nephropathy in male rats. F344 rats were orally exposed to 0.5 g MH/kg BW for 4 weeks. Several animals died. The treatment affected the kidney (acute renal failure and acute tubular necrosis), urethra, and bladder.

Regulations. ***U.S. FDA*** (1998) approved the use of MH in adhesives as a component of articles intended for use in packaging, transporting, or holding food in accordance with the conditions prescribed in 21 CFR part 175.105.

Reference:

Halder, C. A., Holdworth, C. E., Cockrell, B. Y., et al., Hydrocarbon nephropathy in male rats: identification of the nephrotoxic components of unleaded gasoline, *J. Toxicol. Ind. Health*, 1, 67, 1985.

METHYLHYDROQUINONE

Molecular Formula. $C_7H_8O_2$
M = 124.15
CAS No 95-71-6
96937-50-7
RTECS No MX6700000
Abbreviation. MHQ.

Synonyms and **Trade Names.** 2-Methyl-1,4-benzenediol; 2,5-Dihydroxytoluene; 2-Methyl-1,4-benzenediol; Methyl-*p*-hydroquinone; 2,5-Toluenediol; Toluquinol; Tolylhydroquinone.

Acute Toxicity. A dose of 200 mg/kg BW appeared to be lethal in rats.

Mutagenicity. Oral administration of MHQ at the dose of 200 mg/kg BW induced DNA single strand scission, but not unscheduled DNA synthesis.

Regulations. ***EU*** (1990). MHQ is available in the *List of authorized monomers and other starting substances which may continue to be used for the manufacture of plastic materials and articles intended to come into contact with foodstuffs pending a decision on inclusion in Section A* (*Section B*).

Reference:

Furihata, C., Oguchi, S., and Matsushima, T., Possible tumor-initiating and -promoting activity of *p*-methylcatechol and methylhydroquinone in the pyloric mucosa of rat stomach, *Japan. J. Cancer Res.*, 84, 223, 1993.

N-METHYL-*N*-OLEOYLTAURINE, SODIUM SALT

Molecular Formula. $C_{21}H_{40}NO_4S.Na$
M = 425.67
CAS No 137-20-2
RTECS No WX0500000
Abbreviation. MOS.

Synonyms and **Trade Name.** Adinol; Oleoylmethyltaurine, sodium salt; Sodium oleylmethyltauri-de.

Properties. Surfactant.

Acute Toxicity. LD_{50} is 6.63 g/kg BW in mice.

Regulations. ***U.S. FDA*** (1998) approved the use of MOS as a component of the uncoated or coated food-contact surface of paper and paperboard intended for use in producing, manufacturing, packaging, processing, preparing, treating, packing, transporting, or holding aqueous and fatty foods in accordance with the conditions prescribed in 21 CFR part 176.170.

Reference:

J. Am. Pharmacol. Assoc., Sci . Ed., 38, 428, 1949.

2-METHYL-2,4-PENTANEDIOL

Molecular Formula. $C_6H_{14}O_2$
M = 118.20
CAS No 107-41-5
RTECS No SA0810000
Abbreviation. MPD.

Synonyms. 2,2-Dihydroxy-2-Methylpentane; Hexylene glycol; 1,1,3-Trimethyltrimethylenediol.

Properties. Colorless liquid with a mild sweetish odor. Soluble in alcohol and water.

Applications. In textile dye vehicles, cement additive, chemical intermediate.

Acute Toxicity. LD_{50} is found to be 4.7 g/kg BW in rats.[1] Lethal doses caused muscular discoordination, followed by a narcosis that lasted for several hours.[019] Following administration of single doses of 2.0 ml/kg BW, mice got prostration, while higher doses caused irritation of the lungs and large intestine. Gross pathology examination revealed no changes in the kidney, brain, and heart.[03]

Long-term Toxicity. Chronic feeding induced slight liver and kidney damage.[019]

Mutagenicity.

In vitro genotoxicity. MPD was negative in the rat liver chromosome assay, in the bacterial mutation assays, and the yeast mitotic gene conversion assay.[2]

Chemobiokinetics. MPD formed glucuronic acid conjugate.[03] Only 2.5% remained in the body 8 days after dosage. Urine contained 7 metabolites including MPD *glucuronide* (46%), and unchanged MPD (2.5%). MPD is found to convert into *diacetone alcohol* by incubation with rat liver slices.[3] Being fed to rabbits, MPD is eliminated in urine (88% of dose in 5 days).[019]

Regulations. ***U.S. FDA*** (1998) approved the use of MPD (1) in adhesives as a component of articles intended for use in packaging, transporting, or holding food in accordance with the conditions prescribed in 21 CFR part 175.105; (2) as a component of the uncoated or coated food-contact surface of paper and paperboard intended for use in producing, manufacturing, packaging, processing, preparing, treating, packing, transporting, or holding dry food in accordance with the conditions prescribed in 21 CFR part in accordance with the conditions prescribed in 21 CFR part 176.180; (3) as a component of a defoaming agent intended for articles which may be used in producing, manufacturing, packing, processing, preparing, treating, packaging, transporting, or holding food in accordance with the conditions prescribed in 21 CFR part 176.200; (4) as a defoaming agent in the manufacture of paper and paperboard intended for use in packaging, transporting, or holding food in accordance with the conditions prescribed in 21 CFR part 176.210; and (5) in the manufacture of closures with sealing gaskets for food containers intended for use in producing, manufacturing, packaging, processing, preparing, treating, packing, transporting, or holding dry food in accordance with the conditions prescribed in 21 CFR part 177.1210.

EU (1990). MPD is available in the *List of authorized monomers and other starting substances which may continue to be used for the manufacture of plastic materials and articles intended to come into contact with foodstuffs pending a decision on inclusion in Section A* (*Section B*).

References:

1. Smyth, H. F. and Carpenter, C. P., Further experience with the range-finding test in the industrial toxicology laboratory, *J. Ind. Hyg. Toxicol.*, 30, 63, 1948.
2. Brooks, T. M. et al., The genetic toxicology of some hydrocarbon and oxygenated solvents, *Mutagenesis*, 3, 227, 1988.
3. Parke, D. V., *The Biochemistry of Foreign Compounds*, Pergamon Press, Oxford, 1968, 175.

2-METHYL-2-PROPENOIC ACID, HOMOPOLYMER, SODIUM SALT

Molecular Formula. $[\sim C_4H_6O_2\sim]_n .xNa$
CAS No 54193-36-1
RTECS No UD3425900

Synonyms and **Trade Name.** Acrynax; 2-Methyl-2-propenoic acid, homopolymer, sodium salt; Sodium polymethacrylic acid.

Acute Toxicity. LD_{50} *i/v* is 250 mg/kg BW in mice.[029]

Regulations. ***U.S. FDA*** (1998) approved the use of M. in adhesives as a component of articles intended for use in packaging, transporting, or holding food in accordance with the conditions prescribed in 21 CFR part 175.105.

5-METHYL RESORCINOL

Molecular Formula. $C_7H_8O_2$
M = 124.15
CAS No 504-15-4
RTECS No VH2100000

Synonym and **Trade Names.** 3,5-Dihydroxytoluene; Orcin; Orcinol.

Properties. Colorless crystals. Readily soluble in water and alcohol. Odor perception threshold is 2.3 to 2.4 g/l, taste perception threshold is 0.9 g/l.

Applications. Resorcinol substitute in the manufacture of plastics.

Acute Toxicity. LD_{50} is 845 mg/kg BW in rats, 770 mg/kg BW in mice, 2400 mg/kg BW in rabbits, and 1700 mg/kg BW in guinea pigs.

Repeated Exposure failed to reveal marked cumulative properties.

Long-term Toxicity. Rats were dosed by gavage with 33.8 mg/kg BW dose. The treatment resulted in decreased BW gain and slightly reduced serum *SH*-group contents. Histology examination revealed parenchymatous and fatty dystrophy of the liver, kidney, and spleen. The NOAEL was identified to be 1.7 mg/kg BW.

Chemobiokinetics. Upon administration of M., concentration of *phenols* in the urine is consistently increased (up to 500 to 600 mg/l).

Standards. ***Russia*** (1994). MAC and PML: 1.0 mg/l (organolept., chromacity).

Reference:

Veldre, I. A., Experimental data on hygiene norm-setting of 5-methyl resorcin in water bodies, *Gig. Sanit.*, 4, 14, 1969 (in Russian).

METHYLSILANOLATE, SODIUM SALT

Trade Name. GKhZh-11.

Properties. Viscous liquid. Readily soluble in water. Odor perception threshold is 1.5 mg/l. Monomer, which forms a polymer film on surfaces.

Applications. An ingredient of coatings; as a water-alcohol solution, it is used in water-repellency treatments.

Acute Toxicity. See ***Vinylsilanolate, sodium salt.***

Long-term Toxicity. Rats and guinea pigs were dosed by gavage with 0.05 to 1.25 mg/kg BW. Transitional decline in cholinesterase activity was observed. Doses up to 2.5 mg/kg BW produced no changes in animals. The highest dose revealed some evidence of parenchymatous dystrophy of the liver and kidney.[1,2]

Standards. ***Russia*** (1995). MAC and PML: 2.0 mg/l (organolept., odor).

References:

1. Kleshchitskaya, A. L., Krasovsky, G. N., and Friedland, S. A., Hygienic assessment of conditions of discharge of industrial effluents containing silicone organic compounds, *Gig. Sanit.*, 1, 28, 1970 (in Russian).
2. Krasovsky, G. N., Friedland, S. A., Rubleva, M. N., et al., Peculiarities of biological action and hygienic significance of siliconorganic compounds in course of their draining in water reservoirs, in *Industrial Pollution of Water Reservoirs*, S. N. Cherkinsky, Ed., Meditsina, Moscow, Issue No 9, 1969, 38 (in Russian).

N-METHYL-*N*'-VINYLACETAMIDE

Molecular Formula. C_5H_9NO

M = 99.15

CAS No 3195-78-6

RTECS No AC6475000

Abbreviation. MVA.

Synonyms. *N*-Ethenyl-*N*'-methylacetamide; *N*-Vinylmethyl acetamide.

Acute Toxicity. LD_{50} is 2.83 g/kg BW in rats.

Regulations.

EU (1990). MVA is available in the *List of authorized monomers and other starting substances which may continue to be used for the manufacture of plastic materials and articles intended to come into contact with foodstuffs pending a decision on inclusion in Section A (Section B).*

Great Britain (1998). MVA is authorized up to the end of 2001 for use in the production of polymeric materials and articles in contact with food or drink or intended for such contact. The quantity of CA in the finished plastic material or article shall not exceed 5 mg/kg.

Reference:

Smyth, H. F., Carpenter, C. P., Weil, C. S., Pozzani, U. C., and Striegel, J. A., Range-finding toxicity data: List VI, *Am. Ind. Hyg. Assoc. J.*, 23, 95, 1962.

MYRISTIC ACID

Molecular Formula. $C_{14}H_{28}O_2$

M = 228.36

CAS No 544-63-8

RTECS No QH4375000

Abbreviation. MA.

Synonyms. 1-Tridecanecarboxilic acid; Tetradecanoic acid.

Properties. White crystalline solid. Soluble in absolute alcohol.

Applications. MA is used in the production of soaps and lubricants.

Exposure. MA is available in the oils of plants of the family *Myristicaceae.*

Mutagenicity. MA showed no mutagenic effect.[1]

Chemobiokinetics. In rats fed coconut oil, MA was found to be one of the principal fatty acids present in hepatic and adipose tissue triglycerides.[2]

In addition to metabolism by β-oxidation, MA undergoes chain elongation to *palmitic* and *stearic acids*, desaturation to *myristoleic acid,* and incorporation into hepatic neutral lipids (and to a lesser extent, phospholipids).[03]

Regulations.

EU (1990). MA is available in the *List of authorized monomers and other starting substances which shall be used for the manufacture of plastic materials and articles intended to come into contact with foodstuffs (Section A).*

Great Britain (1998). MA is authorized without time limit for use in the production of polymeric materials and articles in contact with food or drink or intended for such contact.

References:

1. Nakamura, H. and Yamamoto, T., The active part of the [6]-gingerol molecule in mutagenesis, *Mutat. Res.*, 122, 87, 1983.
2. Mendenhall, C. L. et al., Effect of ethanol on fatty acid composition of hepatic phosphatidylcholine and phosphatidyl ethanolamine and on microsomal fatty acyl CoA lysophosphatide transferase activities in rats fed corn oil or coconut oil, *Biochem. Biophys. Acta*, 187, 510, 1969.

NAPHTHALENE

Molecular Formula. $C_{10}H_8$

M = 128.17

CAS No 91-20-3

RTECS No QJ0525000

Synonyms and **Trade Names.** Naphthaline; Naphthene; Tar camphor; White resin; White tar.

Properties. Colorless, lamellar crystals with a characteristic odor and a sweetish astringent taste. Readily soluble in organic solvents; poorly soluble in water (34.4 mg/l at 25°C). Odor perception threshold is 0.01 or 0.021 mg/l;[02] taste perception threshold is higher. Odor disappears in 3 to 4 days. The half-life in water is 24 to 48 hours.[1]

Applications. Used in the production of plastics and vulcanizates.

Migration Data. N. contamination was observed in sterilized milk drinks contained in low-density polyethylene bottles. The determination of N. in the packaging material by GC method showed levels of N. ranging from 0.7 to 2.0 mg/kg. Analysis of the milk drinks contained in low-density polyethylene bottles by GC method showed levels of N. in the range 0.01 to 0.03 mg/1. The level of contamination in milk increased with storage time at room temperature (25°C) and, depending on the concentration of N. in the packaging material, values of 0.08 to 0.27 mg/1 were found at the expire date. The amount of N. extracted by milk from the polyethylene at 25°C over a period of 60 days was related to the amount that could be extracted using peanut oil at 66°C for 30 hours.[2]

Acute Toxicity. The LD_{50} values are reported to be 1250 mg/kg BW in rats, 580 mg/kg BW in mice, and 1200 mg/kg BW in guinea pigs. Death within 3 days. Poisoning is accompanied by apathy and adynamia. Doses of 400 and 600 mg/kg BW affected the renal tubular epithelium; a 200 mg/kg BW dose caused selective damage to the bronchial epithelium in mice (O'Brien et al., 1985).

Repeated Exposure revealed the capacity for functional accumulation. A total dose of 354 mg/kg BW given to mice for 8 days caused a 50% mortality (Stillwell et al., 1982). Administration of 27 to 267 mg/kg BW doses for 14 days caused 5 to 10% mortality and retardation of BW gain at the highest dose level. There was a decrease in thymus weights in males and in spleen weights in females.[3]

Rats were dosed with 1/5, 1/25, and 1/125 LD_{50} for a month. At the end of the treatment they developed decreased BW gain, anemia, and reticulocytosis, and a reduction in cholinesterase activity in the blood urea and creatinine levels.[1]

Rabbits were exposed orally to resin-free N. There was distinct alteration of individual fatty acids after treatment with 10 administrations.[4]

Short-term Toxicity. Administration of 5.5 to 133 mg/kg BW doses for 90 days did not cause mortality, retardation of BW gain and any effect on serum enzyme or electrolytes level in rats.[2]

Feeding N. daily to Black-Hooded rats for 79 days caused gradual progressive development of cataract.[5]

Long-term Toxicity. Changes of glomerulonephritis type are observed in rats given 0.15 and 1.5 mg/kg BW.

Reproductive Toxicity.

Embryotoxicity. N. has been found to be embryotoxic. It induced such effect in *S9* supplemented media, namely, in preimplantation mouse embryo culture system.[6] Mice received N. by gavage on days 7 to 14 of gestation. Tolerated dose administered during pregnancy caused high maternal mortality. Embryotoxic effect was slight. The threshold dose for this effect is 0.75 mg/kg BW.[7]

Teratogenicity. Rats were exposed to 395 mg N./kg BW by *i/p* injections on days 1 to 15 of gestation. The treatment caused retardation of cranial ossification and developmental abnormalities in the cardiovascular system of pups from mothers treated with N.[8] According to other data, N. is not found to cause teratogenic effect in rats.[9]

Allergenic Effect was not observed after *i/g* exposure.

Immunotoxicity effect was not found following repeated or short-term exposure.[3] 90-day exposure to N. did not alter humoral and cellular-mediated immune responses in mice. This is explained by Kawabata et al. as being due to the inability of splenocytes to metabolize N.[10]

Mutagenicity.

In vitro genotoxicity. N. produces CA in the somatic cells (threshold dose is 0.015 mg/kg BW). It is mutagenic for microorganisms.[11] N. was found to be positive in *Salmonella* mutagenicity test at concentrations of 5.0 and 10 μg/plate, but negative at 50, 100, or 1000 μg/plate. There was no increase in forward mutation frequency in *Salmonella*.[12] It was negative in a DNA damage *Salmonella typhimurium* strain *TA1535/pSK1002* assay at a concentration of up to 83 μg/ml.[13]

Carcinogenicity. Rats of in-house strains BDI and BDIII received doses of approximately 30 or 60 mg N./kg in their diet. The treatment was stopped after administration of total dose of 10 g N./rat. No carcinogenic effect was reported.[14] Rats received 20 mg N./rat by *i/p* injections once a week for 40 weeks. The treatment produced no increase in the incidence of tumor development.[14]

Carcinogenicity classifications. *U.S. EPA.* No data are available to evaluate the carcinogenic potential of N. in exposed human populations.

U.S. EPA: D;

NTP: XX - XX - NE - SE (inhalation).

Chemobiokinetics.

Observations in man. Workers occupationally exposed to N. were found to have *1-naphtol* (0.89 to 4.86 mg/l) in their urine. N. was metabolized to cytotoxic, protein-reactive, and stable, but not genotoxic, metabolite by human liver cytochrome *P-450* enzymes, which is probably a *1,2-epoxide*.[15]

Animal studies. Following ingestion, N. accumulates in the liver, GI tract, heart, and lungs (in pigs and cattle). Its metabolism occurs with the help of microsomal monooxygenase system. N. forms corresponding *epoxides* that are subsequently converted into *dihydrodiols* which themselves are converted into *phenols* and *catechols*. The latter are conjugated with glucuronides or sulfates.[16] A single oral administration of 500 mg N./kg BW affected hepatic and epididymal glutathione but have an insignificant effect on testicular levels.[17]

In mice, a large part of N. is metabolized to a *glutathione adduct* which is then converted by hydrolysis, deamination, and decarboxylation into corresponding *mercaptolactic* and *mercaptoacetic acids*. *Metylthioesters* are also formed (Stillwell et al., 1982).

Regulations. ***U.S. FDA*** (1998) regulates the use of monosulfonated N. in adhesives used as components of articles intended for use in packaging, transporting, or holding food in accordance with the conditions prescribed in 21 CFR part 175.105.

Standards. ***Russia*** (1995). MAC and PML: 0.01 mg/l (organolept., odor).

Recommendations. ***U.S. EPA*** (1998). Health Advisory for longer-term exposure is 1.0 mg/l.

References:

1. Matorova, N. I., Data on the substantiation of maximum allowable concentration for naphthalin and naphthalene in water bodies, *Gig. Sanit.*, 11, 78, 1982 (in Russian).
2. Lau, O. W., Wong, S. K., and Leung, K. S., Naphthalene contamination of sterilized milk drinks contained in low-density polyethylene bottles, Part 1, *Analyst*, 119, 1037, 1994.
3. Shopp, G. M., White, K. L., Holsapple, M. P., et al., Naphthalene toxicity in CD-1 mice: General toxicology and immunotoxicology, *Fundam. Appl. Toxicol.*, 4, 406, 1984.
4. Guliyeva, S. A. and Samedov, S. I., Fatty acid composition of blood plasma of rabbits treated with naphthalene oil, *Probl. Med. Chem.*, 3, 43, 1991 (in Russian).
5. Rathbun, W. B., Holleschau, A. M., Murray, D. L., Buchanan, A., Sawaguchi, S., and Tao, R. V., Glutathione synthesis and glutathione redox pathway in naphthalene cataract of the rat, *Curr. Eye Res.*, 9, 45, 1990.
6. Iyer, P., Martin, J. E., and Irvin, T. R., Role of biotransformation in the *in vitro* preimplantation embrydtoxicity of naphthalene, *Toxicology*, 66, 257, 1991.
7. Plasterer, M. R., Bradshaw, W. S., Booth, G. M., et al., Developmental toxicity of 9 selected compounds, following prenatal exposure in the mouse: naphthalene, *p*-nitrophenol, sodium selenite, dimethylphthalate, ethylenethiourea, and 4 glycol ether derivatives, *J. Toxicol. Environ. Health*, 15, 25, 1985.
8. Harris, S. J., Bond, G. P., and Niemeir, R. W., The effects of 2-nitropropane, naphthalene, and hexachlorobutadiene on fetal rat development, Abstract, *Toxicol. Appl. Pharmacol.*, 48, A69, 1979.
9. Hardin, B. D., Schuler, R. L., Burg, J. R., Booth, G. M., Hazelden, K. P., MacKenzie, K. M., Piccirillo, V. J., and Smith, K. N., Evaluation of 60 chemicals in a preliminary developmental toxicity test, *Teratogen. Carcinogen. Mutagen.*, 7, 29, 1987.
10. Kawabata, T. T. and Whitw, K. L., Effects of naphthalene and naphthalene metabolites on the *in vitro* humoral immune response, *J. Toxicol. Environ. Health*, 30, 53, 1990.
11. McCann, J., Choi, E., Yamasaki, E., et al., Detection of carcinogens and mutagens in the Salmonella/microsome test: assay of 300 chemicals, *Proc. Natl. Acad. Sci. U.S.A.*, 72, 5135, 1975.
12. Seixas, G. M., Andon, B. M., Hollingshead, P. G., and Thilly, W. G., The azaarenes as mutagens for *Salmonella typhimurium*, *Mutat. Res.*, 102, 201, 1982.
13. Nakamura, S., Oda, Y., Shimada, T., Oki, I., and Sugimoto, K., Sos-inducing activity of chemical carcinogens and mutagens in *Salmonella typhimurium TA1535/pSK1002*: Examination with 151 chemicals, *Mutat. Res.*, 192, 239, 1987.
14. Schmahl, D., Examination of the carcinogenic action of naphthalene and anthracene in rats, *Z. Krebsforsch.*, 60, 697, 1955.
15. Tingle, M. D., Piromohamed, M., Templeton, E., Wilson, A. S., Madden, S., Kitteringham, N. R., and Park, B. K., An investigation of the formation of cytotoxic, genotoxic, protein-reactive, and stable metabolites from naphthalene by human liver microsomes, *Biochem. Pharmacol.*, 46, 1529, 1993.
16. Bieniek, G., The presence of 1-naphthol in the urine of industrial workers exposed to naphthalene, *Occup. Environ. Med.*, 51, 357, 1994.
17. Gandy, J., Millner, G. C., Bates, H. K., Casciano, D. A., and Harbison, R. D., Effects of selected chemicals on the glutathione status in the male reproductive system of rats, *J. Toxicol. Environ. Health*, 29, 45, 1990.

2-NAPHTHALENESULFONIC ACID

Molecular Formula. $C_{10}H_8C_3S$
M = 208.23
CAS No 120-18-3
RTECS No QK1225000
Abbreviation. NSA

Synonym. β-Naphthylsulfonic acid.

Properties. White to slightly brownish, crystalline leaflets. Readily soluble in water.

Applications. Used in the production of β-naphthol.

Acute Toxicity. LD_{50} is 4.44 g/kg BW in rats.

Regulations. ***U.S. FDA*** (1998) approved the use of NSA in adhesives as a component of articles intended for use in packaging, transporting, or holding food in accordance with the conditions prescribed in 21 CFR part 175.105.

Reference:

Lysogorova, I. K., Sanitary-toxicological evaluation of iron compounds, *Gig. Sanit.,* 10, 101, 1974 (in Russian).

NAPHTHENIC ACIDS, SYNTHETIC

Molecular Formula. $R(CH_2)_n(COOH)_m$, where *R* - naphthenic cycloalkyl radical
CAS No 13308-24-5
RTECS No QK8750000

Composition. A mixture of naphthenic (cyclo-alkanecarbonic) acids obtained in oxidation treatment of a naphthene concentrate of petroleum oils from Azerbaijan.

Properties. An oily liquid of a light yellow to brown color with a penetrating odor. Readily soluble in water. Threshold for organoleptic effect is 0.3 mg/l.[1]

Applications. SNA are used in the production of stabilizers and plasticizers.

Acute Toxicity. LD_{50} is 1.75 to 6.42 g/kg BW in rats, and 1.77 to 7.17 g/kg BW in mice. Administration of 20 and 50% oil solutions resulted in a mild stimulation with subsequent CNS inhibition and adynamia. Animals assumed side position. Death occurred in two days. Gross pathology examination failed to reveal pathologic changes but relative liver weights were increased.[2,3]

Repeated Exposure revealed cumulative properties: 50% of animals died from the overall dose of 2.16 g/kg BW having been daily exposed to 1/10 LD_{50}.[4] K_{acc} (by Lim) is reported to be 0.33.[2]

However, according to other data, accumulation does not take place. Some of the animals died only when 1/5 LD_{50} was administered over a period of 1.5 months, but not when 1/10 LD_{50} was given.[3]

Repeated exposure did not affect STI and blood morphology. Consistent BW loss was noted. Rats were dosed with 200 mg SNA/kg BW given as a 10% oil solution for 1 month. The treatment caused retardation of BW gain but did not affect biochemical and hematology indices.

Long-term Toxicity. The NOAEL of 5.0 mg/kg BW was identified.[1]

Allergenic Effect was not found on dermal application.

Standards. ***Russia*** (1995). MAC and PML: 1.0 mg/l (organolept., odor).

References:

1. Golubeva, M. T., Ozerova, V. F., and Gutkovskaya, A. I., in *Protection of Water Reservoirs against Pollution by Industrial Liquid Effluents*, S. N. Cherkinsky, Ed., Moscow, Issue No 3, 1959, 160 (in Russian).
2. Rubinskaya, S. E., *Problems of Occupational Hygiene, Industrial Toxicology and Occupational Pathology*, Sumgait, Issue No 9, 1974, 37 (in Russian).
3. Uzhdaviny, E. R. and Glukharev, Yu. A., Toxic properties of synthetic naphthenic and alkylbenzenoic acids, *Gig. Truda Prof. Zabol.*, 9, 48, 1984 (in Russian).
4. *Toxicological Evaluation of Some New Plasticizers, Additives and Oil Coolants*, Azerb. State Publ., Baku, 1979, 33 (in Russian).

NAPHTHENIC ACID, ZINC SALT

CAS No 12001-85-3
RTECS No QK9275000
Abbreviation. NAZ.

Synonyms. Zinc naphthenate; Zinc uversol.

Acute Toxicity. LD_{50} is 4.9 g/kg BW in rats,[1] and 2.8 g/kg BW in mice.[2]

Reproductive Toxicity. Rabbits were dosed with 940 mg/kg BW on 6 to 15 day of pregnancy. The treatment caused post-implantation mortality; signs of fetotoxicity were reported.[2]

Regulations. ***U.S. FDA*** (1998) approved the use of NAZ in adhesives as a component of articles intended for use in packaging, transporting, or holding food in accordance with the conditions prescribed in 21 CFR part 175.105.

References:

1. Smyth, H. F., Carpenter, C. P., Weil, C. S., Pozzani, U. C., Striegel, J. A., and Nyaun, J. S., Range-finding toxicity data: List VII, *Am. Ind. Hyg. Assoc. J.*, 30, 470, 1969.
2. Natl. Technical Information Service AD-A180-019.

p-(2-NAPHTHYLAMINO)PHENOL

Molecular Formula. $C_{13}H_{10}Cl_2O_2$
M = 269.13
CAS No 97-23-4
RTECS No SM1750000

Synonyms and **Trade Names.** Bis(5-chloro-2-hydroxyphenyl)methane; 5,5'-Dichloro-2,2'-dihydroxydiphenylmethane; 4,4'-Dichloro-2,2'-methylenediphenol; Dichlorophen; Difentan; 2,2'-Dihydroxy-5,5'-dichlorodiphenylmethane; 2,2'-Methylenebis(4-chlorophenol); Panacide.

Properties. Crystals, practically insoluble in water, freely soluble in ethanol.

Applications. Fungicide and hermicide.

Acute Toxicity. LD_{50} is 1.5 g/kg BW in rats, 1.0 g/kg BW in mice, 1.25 g/kg BW in guinea pigs, and 2.0 g/kg BW in dogs.[1-3]

Short-term Toxicity. In a 90-day feeding study, the dose of 2.0 g/kg diet caused no toxic effect in rats.[4]

Mutagenicity.

In vitro genotoxicity. N. was found to be negative in Ames test, *BASC* test, and in micronucleus tests.[5]

Chemobiokinetics. Following ingestion, N. is well absorbed. It was conjugated with *sulfate* and *glucuronic acid.* Excretion occurs via urine and feces.[6]

Regulations. ***U.S. FDA*** (1998) approved the use of N. in adhesives as a component of articles intended for use in packaging, transporting, or holding food in accordance with the conditions prescribed in 21 CFR part 175.105.

References:

1. Gaines, T. B. and Linder, R. E., Acute toxicity of pesticides in adult and weanling rats, *Fundam. Appl. Toxicol.*, 7, 299, 1986.
2. Florestano, H. J., Tuberculocidal activity and toxicity of some diphenylmethane derivatives, *J. Pharmacol. Exp. Ther.*, 96, 238, 1949.
3. *Pesticide Index*, E. H. Frear, Ed., Coll. Sci. Publ., State Coll., PA, 1969.
4. *Pesticide Manual*, C. R. Worthing, Ed., 6th ed., Brit. Crop Protect. Council, Worcestershire, UK, 1979, 172.
5. Gocke, E., King, M.-T., Eckhardt, K., and Wild, D., Mutagenicity of cosmetic ingredients licenced by the EEC, *Mutat Res.*, 90, 91, 1981.
6. Dixon, P. A. F. and Caldwell, J., The fate of dichlorophen in the rat, *Eur. J. Drug Metab. Pharmacokinet.*, 3, 95, 1978.

NICKEL compounds

Ni

CAS No 7440-02-0
RTECS No QR5950000

Applications. Nickel compounds are used in the production of catalysts and in electrolytic coatings in chromium-plated taps and fittings used for tap water.

Exposure. Food has been found to be the main source of nickel intake by man. The highest concentrations of nickel were found in the canned vegetables, sugars and preserves, and bread and cereals food groups, suggesting a contribution from food processing equipment and, possibly, food cans. Mean dietary nickel intakes in the UK (1981-84) were between 0.14 and 0.15 mg/day.[1]

Concentrations found in drinking water because of leaching nickel from *Ni-Cr* plated taps and fittings may reach up to levels of 1.0 mg/l in the tap overnight. Nickel may be leached from kitchen utensils by acidic boiling water.[2]

Acute Toxicity. In mice and rats, LD_{50} is found to be 70 to 140 mg/kg BW.[3]

Repeated Exposure. Sprague-Dawley rats received 0.1, 0.5, or 1.0% nickel sulfate via drinking water for 2 weeks. 4/10 rats died at 1.0% dose level at the end of 2 weeks. Decrease in BW gain was noted in 0.5 and 1.0% groups. Hematocrit value was increased in 0.5 and 1.0% groups of rats. Nickel sulfate produced toxicity to certain organs and tissues. Lesser doses given via drinking water for 4 weeks caused no changes in BW gain and in organ weight ratios. Nickel sulfate induced significant stimulation of lymphocytes *T* and *B* at 0.1% dose level.[4]

Long-term Toxicity. Dogs were given 3.0, 29, or 70 mg nickel chloride/kg in their diet for 2 years. The treatment caused retardation of BW gain. Histological changes in the lungs were found to develop at the highest dose. The NOAEL of 29 mg/kg BW was established in this study.[5]

Wistar rats were fed nickel chloride at the doses of 5.0, 50, or 125 mg/kg BW. Altered BW gain and relative organ weights were observed at two highest doses. There were no hematology and histology findings. The NOAEL of 5.0 mg/kg BW was reported.[5]

Wistar rats were given 100 mg nickel/l (as nickel sulfate) in drinking water for 6 months. Kidney weights were significantly increased. No changes were observed in other parameters. The results suggest that low-level oral exposure to soluble nickel either induces changes of glomerular permeability in female and possibly in male rats, or enhances the normal age-related glomerular nephritis lesions of aging rats. The intake was probably not high enough to induce significant tubular changes.[6]

Reproductive Toxicity. Nickel is known to cross the human placenta and produce teratogenesis and embryotoxicity.[1] It produced lipid peroxidative damage to human placenta incubated with 2.5 mM nickel and this metabolic change may be responsible for decreased placental viability, altered permeability, and potential subsequent embryotoxicity.[7]

Embryotoxicity. In a three-generation study, rats received 12.5, 25, or 50 mg nickel/l in their drinking water. An increase in incidence of stillborn pups was observed at all doses in the first generation. Decreased BW of weanlings was noted in all generations at the highest dose.[5]

Teratogenicity. Gestation exposure to a variety of nickel salts caused malformations and growth retardation. In mice, acephalia, exencephaly, cerebral hernia, open eyelid, cleft palate, micromelia, and skeletal anomalies were observed.[7] High concentrations of nickel chloride in drinking water increased runting, but produced no teratogenic effect in rats.[8]

Nickel acetate was found to cause multiple developmental abnormalities in hamsters and sheep,[9] but not in rats.[10]

Gonadotoxicity. Nickel salts induced sperm head abnormalities in mice.[7]

Allergenic Effect. Nickel is the most common contact allergen in Europe. Once acquired, nickel sensitivity is apparently never lost.[12] When consumed with food, nickel can be a source of chronic urticaria (itching).[13]

Mutagenicity.

Observations in man. An increase in frequency of SCE were observed in workers exposed to elemental chromium-containing, cobalt-containing, and nickel-containing dusts.[16]

In vitro genotoxicity. Divalent nickel ions produced DNA cross-links and single strand breaks.[14,15] Nickel increased frequency of CA and SCE in cultured human lymphocytes.[17] Nickel acetate, chloride, and nitrate were reported to be negative in *Salmonella* mutagenicity assay.[18,19,21]

Nickel oxides did not induce CA in cultured human cells. Nickel sulfate increased the frequency of CA in cultured human cells; nickel sulfate and chloride increased the frequency of SCE; these chemicals transformed cultured mammalian cells causing DNA damage.[20]

In vivo cytogenetics. Nickel chloride, nitrate and sulfate are reported to be genotoxic in mammalian systems. Nickel increased an incidence of CA but not SCE in rats and mice receiving it in toxic doses (2.0 mg/kg BW).[17]

$NiCl_2 \cdot 6H_2O$ did not induce micronucleated polychromatic erythrocytes in ddY mouse bone marrow after *i/p* treatment with 3.2, 6.3, 12.5, and 25 mg/kg BW. Neither did nickel sulfate at 5, 10, and 20 mg/kg BW. Negative response was found in CD-1 mouse bone marrow cells after *i/p* treatment with nickel oxide at 18.1, 36.3, 72.5, and 145 mg/kg BW.

Nickel chloride was inactive in the dominant lethal assay but induced micronuclei *in vivo* (Dhir et al., 1991).

In a single study in *Dr. melanogaster*, nickel sulfate induced aneuploidy and gene mutations.

Carcinogenicity. Nickel compounds could contribute to the development of lung cancer by initiation of the heritable changes (genetic or epigenetic) and by promotion of cell proliferation.

Observations in man. Epidemiological data suggest that different classes of nickel compounds appear to be inhalation carcinogens.

Carcinogenicity classifications. An IARC Working Group made the overall evaluation on *nickel* compounds as a group. There is inadequate evidence in *humans* for the carcinogenicity of metallic nickel and nickel alloys; there is limited evidence for the carcinogenicity of nickel alloys in *experimental animals*.

The Annual Report of Carcinogens issued by the U.S. Department of Health and Human Services (1998) defines *nickel* and certain nickel compounds to be substances which may reasonably be anticipated to be carcinogenic, i.e., substances for which there is a limited evidence of carcinogenicity in *humans* or sufficient evidence of carcinogenicity in *experimental animals*.

Oller et al. indicate lack of consensus in the interpretation of carcinogenicity of any given nickel compound. For example, in 1990 IARC classified soluble nickel as "Group 1" (confirmed human carcinogen). Whereas in 1991 the Directorate General XI (DGXI) of the *Commission of European Communities* (CEC) classified soluble nickel sulfate as "Category 3" (possible human carcinogen), and recently the American Conference of Governmental Industrial Hygienists (ACGIH) has proposed a "Category A4" (not classifiable as human carcinogen) for these soluble compounds.[19]

IARC: 1;

U.S. EPA: A (inhalation and injection);

EU: 3.

NTP: SE - SE - NE - EE (Nickel oxide, inhalation).

Chemobiokinetics. Absorption of soluble nickel from drinking water may be 40 times higher than absorption of nickel from food. It is distributed in all organs, predominantly in the liver, kidneys, lungs.[1,23] Excretion occurs mainly with the urine.

Regulations. ***U.S. FDA*** (1998) approved the use of nickel as a component of the uncoated or coated food-contact surface of paper and paperboard intended for use in producing, manufacturing, packaging, processing, preparing, treating, packing, transporting, or holding dry food in accordance with the conditions prescribed in 21 CFR part 176.180. Affirmed as *GRAS*.

Standards.

EEC (1980). MAC: 0.05 mg/l.

Russia (1995). MAC: 0.1 mg/l.

Recommendations. ***U.S. EPA*** (1999). Health Advisory for a longer-term exposure is 1.7 mg/l.

References:

1. Smart, G. A. and Sherlock, J. C., Nickel in foods and the diet, *Food Addit. Contam.*, 4, 61, 1987.
2. Grandjean, P., Nielsen, G. D., and Andersen, O., Human nickel exposure and chemobiokinetics, in *Nickel and the Skin: Immunology and Toxicology*, T. Menne and H. I. Maibach, Eds., CRC Press, Inc., Boca Raton, Fl, 9, 34, 1989.
3. Coogan, T. P. et al., Toxicity and carcinogenicity of nickel compounds, *CRC Crit. Rev. Toxicol.*, 19, 341, 1990.
4. Bai, C., Chakrabarti, S., Obone, E., Malick, A. M., Lamontagne, L., and Subramanian, K. S., Subchronic toxicity study of nickel sulfate ($NiSO_4$) in the rat, in *The Toxicologist*, Abstracts of SOT 1996 Annual Meeting, Abstract No 905, Issue of *Fundam. Appl. Toxicol.*, 30, Part 2, March 1996.
5. Ambrose, A. M. et al., Long-term toxicological assessment of nickel in rats and dogs, *J. Food Sci. Technol.*, 13, 181, 1976.

6. Vyskocil, A., Viau, C., and Cizkova, M., Chronic nephrotoxicity of soluble nickel in rats, *Hum. Exp. Toxicol.*, 13, 689, 1994.
7. Chen, C. Y. and Lin, T. H., Nickel toxicity to human term placenta: *in vitro* study on lipid peroxidation, *J. Toxicol. Environ. Health*, 54, 37, 1998.
8. Lu, C. C. et al., Teratogenic effects of nickel chloride on embryonic mice and its transfer to embryonic mice, *Teratology*, 9, 137, 1979.
9. Kimmel, G. L. et al., The effect of nickel chloride in drinking water on reproductive and developmental parameters, *Teratology,* 3, 90C, 1986.
10. Ferm, V. N., The teratogenic effects of metals on mammalian embryos, *Advan. Teratol.*, 5, 51, 1972.
11. Sunderman, F. W. et al., Eye malformations in rats: induction by prenatal exposure to nickel carbonyl, *Science*, 203, 550, 1979.
12. Elsner, P. and Burg, G., Irritant reactivity is a better risk marker for nickel sensitization than atopy, *Acta Dermatol. Venerol.* (Stockholm), 73, 214, 1993.
13. Abeck, D., Traenckner, I., Steinkraus, V. et al., Chronic urticaria due to nickel intake, *Acta Dermatol. Venerol.* (Stockholm), 73, 438, 1993.
14. Patierno, S. R. and Costa, M., DNA-protein cross-links induced by nickel compounds in intact cultural mammalian cells, *Chem. Biol. Interact.*, 55, 75, 1985.
15. Sunderman, F. W. et al., Nickel absorption and kinetics in human volunteers, *Proc. Soc. Exp. Biol. Med.*, 191, 511, 1989.
16. Gennart, J. P., Baleux, C., Verellendumoulin, C., et al., Increased sister chromatid exchanges and tumor markers in workers exposed to elemental chromium-containing, cobalt-containing and nickel-containing dusts, *Mutat. Res.*, 299. 55, 1993.
17. Nadeyenko, V. G., Goldina, I. R., Dyachenko, O. Z., and Pestova, L. V., Comparative informative value of chromosomal aberrations and sister chromatid exchanges in the evaluation of environmental metals, *Gig. Sanit.*, 10, 3, 1997 (in Russian).
18. De Flora, S., Zannacchi, P., Camoriano, A., Bennicelli, C., and Badolati, G. S., Genotoxic activity and potency of 135 compounds in the Ames reversion test and in bacterial DNA-repair test, *Mutat. Res.,* 133, 161, 1984.
19. Rossman, T. G., Molina, M., Meyer, L., Boone, P., Klein, C. B., Wang, Y., Li, F., Lin, W. C., and Kinney, P. l., Performance of 133 compounds in the lambda prophase induction endpoint of the microscreen assay and comparison with *S. typhimurium* mutagenicity and rodent carcinogenicity, *Mutat. Res.,* 260, 349, 1991.
20. Morita, T., Asano, N., Awogi, T., Sasaki, Yu. F., et al. Evaluation of the rodent micronucleus assay in the screening of IARC carcinogens (Groups 1, 2A and 2B). The summary report of the 6th collaborative study by CSGMT/JEMS-MMS, *Mutat. Res.*, 389, 3, 1997.
21. Biggart, N. W. and Costa, M., Assessment of the uptake and mutagenicity of nickel chlorite in Salmonella tester strains, *Mutat. Res.,* 175, 209, 1986.
22. Oller, A. R., Costa, M., and Oberdorster, G., Carcinogenicity assessment of selected nickel compounds. *Toxicol. Appl. Pharmacol.,* 143, 152, 1997.
23. Sobti, R. C. and Gill, R. K., Incidence of micronuclei and abnormalities in the head of spermatozoa caused by the salts of the heavy metal, nickel, *Cytologia*, 54, 249, 1989.

NITRIC ACID

Molecular Formula. HNO_3
M = 63.02
CAS No 7697-37-2
RTECS No QU5775000
Abbreviation. NA.

Synonym and **Trade Names.** Aqua fortis; Azitic acid; Hydrogen nitrate.

Properties. Transparent colorless or yellow liquid with characteristic choking odor (sweet to acrid). Reacts violently with alcohol. Soluble in water forming a negative azeotrope. Concentrated NA is a water solution containing 70 to 71% NA.[020]

Applications. Oxidizing agent. Used in pharmaceutical preparations. Used as an intermediate of dye and other chemicals.

Acute Toxicity.

Observations in man. Lethal doses are found to be equal or more than 110 mg/kg BW.[1]

Short-term and **Long-term Toxicity.** There is a possibility that exposure to NA may result in methemoglobin formation, although no direct evidence for this was found. The fetus is more susceptible to methemoglobin formation and is more resistant to its oxidation back to normal *Hb* than is the adult, and the resulting lack of oxygen would also have a greater effect on the fetus.[2]

Reproductive Toxicity. Fetotoxic action was observed in rats given 1.9 g NA/kg BW during pregnancy.[3]

Regulations. ***U.S. FDA*** (1998) approved the use of NA in adhesives as a component of articles intended for use in packaging, transporting, or holding food in accordance with the conditions prescribed in 21 CFR part 175.105.

References:

1. Arena, J. M., *Poisoning: Toxicology, Symptoms, Treatments*, 2nd ed., Springfield, IL, 1970, 73.
2. Mansouri, A., Methemoglobinemia, *Am. J. Med. Sci.*, 289, 200, 1985.
3. *Z. Gesamt. Hyg. Grenzgebiete*, 29, 667, 1983.

NITRIC ACID, ZINC SALT

Molecular Formula. $N_2O_6.Zn$
M = 189.39
CAS No 7779-88-6
RTECS No ZH4772000
Abbreviation. NAZ.

Synonym and **Trade Name.** Celloxan; Zinc nitrate.

Properties. Hexahydrate, colorless, odorless crystals, freely soluble in water and alcohol.

Toxicity. Clinical manifestations of nitrate toxicity include cyanosis among infants who drink well water.

Regulations. ***U.S. FDA*** (1998) approved the use of NAZ in adhesives as a component of articles intended for use in packaging, transporting, or holding food in accordance with the conditions prescribed in 21 CFR part 175.105.

1,1',1''-NITRILOTRI-2-PROPANOL

Molecular Formula. $C_9H_{21}NO_3$
M = 191.31
CAS No 122-20-3
RTECS No UB8750000
Abbreviation. NTP.

Synonyms. Triisopropanolamine; Tri-2-propanolamine; Tris(2-hydroxypropyl)amine.

Properties. White crystalline solid or liquid at room temperature. Very soluble in water.

Applications. Emulsifying agent.

Acute Toxicity. LD_{50} is 4.73 to 6.5 g/kg BW in rats, 2.52 g/kg BW in mice, 11 g/kg BW in rabbits, and 1.58 g/kg BW in guinea pigs. Poisoning caused convulsions and spasms, hypermotility in GI tract, and diarrhea. A decrease in body temperature was noted.[1-4]

Mutagenicity.

In vitro genotoxicity. NTP was negative in four *Salmonella typhimurium* strains *TA98, TA100, TA1535,* and *TA1537* in the presence and absence of liver fraction *S9.*[015]

Regulations. ***U.S. FDA*** (1998) approved the use of NTP (1) as a component of a defoaming agent intended for articles which may be used in producing, manufacturing, packing, processing, preparing, treating, packaging, transporting, or holding food in accordance with the conditions prescribed in 21 CFR part 176.200, and (2) in polyvinyl alcohol films intended for use in contact with food in accordance with the conditions prescribed in 21 CFR part 177.1670.

Standards. ***Russia*** (1995). MAC: 0.5 mg/l.

References:

1. Burkatskaya, E. N., Karpenko, V. N., Pokrovskaya, T. N., Ivanova, Z. V., and Medved', I. L., Substantiation of maximum permissible levels of mono-, di-, and triisopropanolamine in workplace air, *Gig. Truda Prof. Zabol.*, 7, 46, 1986 (in Russian).
2. Toropkov, V. V., Hygienic basis for the maximum permissible concentrations of mono-, di- and triisopropanolamines in the water of reservoirs, *Gig. Sanit.*,3, 79, 1980 (in Russian).
3. Smyth, H. F., Seaton, J., and Fischer, L., The single dose toxicity of some glycols and derivatives, *J. Ind. Hyg. Toxicol.*, 23, 259, 1941.
4. Smyth, H. F. and Carpenter, C. P., Further experience with the range finding test in the industrial toxicological laboratories, *J. Ind. Hyg. Toxicol.*, 30, 63, 1948.

NITROCYCLOHEXANE

Molecular Formula. $C_6H_{11}NO_2$
M = 129.16
CAS No 1122-60-7
RTECS No GV6600000

Properties. A colorless liquid with an aromatic odor. Solubility in water is 2.0% at 20°C. Odor perception threshold is 0.15 mg/l; taste perception threshold is slightly higher.[1]

Applications. Used in the production of polyamides.

Acute Toxicity. Rats tolerate 50 to 70 mg/kg BW but all die from 200 mg/kg BW. A 100 mg/kg dose is lethal for rabbits. LD_{50} in mice is reported to be 54[1] or 250 mg/kg BW.[2] Animals administered high doses experience clonic-tonic convulsions. Death occurs within 2 to 6 minutes after poisoning.[1]

Long-term Toxicity. In a 6-month study, rats and rabbits were exposed to 0.15 to 1.5 mg/kg BW doses. The treatment did not affect behavior, blood morphology, and glycogen-forming function of the liver. Changes in the sugar curves and abnormalities in the visceral organ histology were observed.[1]

Standards. ***Russia*** (1995). MAC and PML: 0.1 mg/l.

References:

1. Savelova, V. A. and Brook, E. S., in *Sanitary-Chemical Control of Water Reservoirs*, F. F. Erisman Research Sanitary Hygiene Institute, Moscow, 1964, 195 (in Russian).
2. *Toxicology of New Industrial Chemical Substances*, A. A. Letavet, Ed., Medgiz, Moscow, Issue No 1, 1961, 85 (in Russian).

1-NITROGUANIDINE

Molecular Formula. $CH_4N_4O_2$
M = 104.09
CAS No 556-88-7
RTECS ME4600000
Abbreviation. NG.

Synonym and **Trade Name.** 2-Nitroguanidine; Picrite.

Properties. Yellowish crystals. Solubility in water is 0.27 g/l. Poorly soluble in ethanol. Gives water a bitter astringent taste (threshold concentration 50 mg/l).

Applications. Used in the synthesis of plastics and dyestuffs.

Acute Toxicity. In rats, LD_{50} is 10.2 g/kg BW (sunflower oil solution). Other laboratory animals appeared to be more sensitive: LD_{50} is 3.85 g/kg BW in mice, and 3.12 g/kg BW in guinea pigs. Following administration, signs of cyanosis were found to develop.[1]

Repeated Exposure revealed evident cumulative properties. The treatment affected predominantly the CNS and blood formula as well as activity of the liver and blood enzymes. Doses of 0.5 and 0.05 mg/kg BW cause changes in STI, in enzyme-forming function of the liver, in blood formula and in the number of blood serum *SH*-groups.[1]

Short-term Toxicity. Rats were given 100, 316, or 1000 mg/kg BW in their diet for 90 days. The treatment produced organ weight changes. The LOEL of 1000 mg/kg BW, and the NOEL of 316 mg/kg BW were established in this study (Morgan et al., 1988).

Reproductive Toxicity.

0.5 mg/kg BW dose is reported to be ineffective for ***gonadotoxic effect***.[1]

Mutagenicity.

In vivo cytogenetics. A 0.5 mg/kg BW dose does not increase the number of CA in the bone marrow cells in mice.[1] NG was not mutagenic in *Dr. melanogaster* sex-linked recessive lethal mutation assay.[2]

In vitro genotoxicity. Meanwhile, Ishidate and Odashima reported that NG did produce CA when incubated *in vitro* with Chinese hamster cells. CA included chromatid or chromosomal breaks and translocations.[3]

NG was not mutagenic in *Salmonella typhimurium* and *E. coli* assays.[4]

Chemobiokinetics. NG degradation product is *nitrosoguanidine*.

Standards.

U.S. EPA (1999). MCL: 0.7 mg/l.

Russia (1995). MAC and PML: 0.1 mg/l.

References:

1. Korolev, A. A., Shlepnina, T. G., Mikhailovsky, N. A., et al., Hygienic substantiation of the maximum allowable concentration for diphenylnitrosamine and nitroguanidine in water bodies, *Gig. Sanit.*, 1, 18, 1980 (in Russian).
2. Gupta, R. K., Korte, D. W., and Reddy, G., Mutagenic potential of nitroguanidine in the *Dr. melanogaster* sex-linked recessive lethal assay, *J. Appl. Toxicol.*, 13, 231, 1993.
3. Ishidate, M. and Odashima, S., Chromosome tests with 134 compounds on Chinese Hamster cells *in vitro*: A screening for chemical carcinogens, *Mutat. Res.*, 48, 337, 1977.
4. McGregor, D. B., et al., Genotoxic activity in microorganisms of tetryl, 1,3-dinitrobenzene and 1,3,5-trinitrobenzene, *Environ. Mutagen.*, 2, 531, 1980.

β-NITROSTYRENE

Molecular Formula. $C_8H_7NO_2$

M = 149.16

CAS No 102-96-5

RTECS No WL5450000

Abbreviation. BNS.

Synonyms and **Trade Names.** (2-Nitrovinyl)benzene; (2-Nitroethenyl) benzene; β-Nitrostyrene; γ-Nitrostyrene.

Acute Toxicity. LD_0 is 710 g/kg BW in mice.

Regulations. ***U.S. FDA*** (1998) approved the use of BNS as a component of the uncoated or coated food-contact surface of paper and paperboard intended for use in producing, manufacturing, packaging, processing, preparing, treating, packing, transporting, or holding dry food in accordance with the conditions prescribed in 21 CFR part 176.180.

Reference:

Schafer, E. W. and Bowles, W. A., Acute oral toxicity and repellency of 933 chemicals to house and deer mice, *Arch. Environ. Contam. Toxicol.*, 14, 111, 1985.

NITROUS ACID, SODIUM SALT

Molecular Formula. $NO_2.Na$

M = 69.00

CAS No 7632-00-0

RTECS No RA1225000

Abbreviation. NAS.

Synonyms and **Trade Names.** Diazotizing salts; Erinitrit; Filmerine; Sodium nitrite.

Properties. Colorless-yellow rhombohedral prisms, crystals, granules, pellets, sticks, or powder with a mild saline taste. Soluble in 1.5 parts cold, 0.6 parts of boiling water.

Applications. A food preservative.

Exposure. Nitrites are found in many leaf vegetables and in meats, fish, and other foods. Average daily intake has been estimated to be 1.5 mg.

Acute Toxicity.

Observations in man. LD_0 was found to be 71 mg/kg BW. Poisoning caused coma, nausea, vomiting, methemoglobinemia-carboxyhemoglobinemia development.[1]

Animal studies. LD_{50} values of 85 to 220 mg NAS/kg BW have been reported for rats and mice.[2]

Short-term Toxicity. One-month-old male rats were given drinking water that contained 200 ppm nitrite for 16 weeks. Except for higher incidence of pulmonary lesions, administration of nitrite or nitrate at these levels had no significant effect on any of the biochemical parameters measured in the blood or the mutagenicity of liver extract.[3]

Wistar rats received drinking water containing 12.5 to 3000 mg potassium nitrite/l for 13 weeks. The treatment caused a dose-related hypertrophy of the adrenal zona glomerulosa at all dose levels. An increase in *MetHb* level was noted in the highest dose group only.[4] The NOEL for potassium nitrite is 50 mg/l in the drinking water, equivalent to about 5.0 mg/kg BW.[5]

Long-term Toxicity. Nitrite may convert *Hb* to *metHb* that led to tissue anoxia. About 20% of *Hb* conversion caused clinical manifestations to develop.

Two-month-old male rats received drinking water containing 200 ppm NAS for 14 months. An increase in *MetHb* level was observed. The red cells hemolysis was less than 5.0%. The treatment caused significant decrease in body and liver weights and plasma vitamin *E* levels. An increase in red cells reduced glutathione levels and inci dence of pulmonary lesions were noted.[3]

Weanling male Wistar rats received drinking water containing 0.2% nitrite for 9 months. The treatment caused higher *in vitro* lipoperoxidation, free lysosomal enzyme activities, and cytosole superoxide dismutase activity in the liver. The authors concluded that induced toxicity was developed through some free radical reactions.[6]

Rats received nitrite in their drinking water for 2 years. The most important effect appeared to be an increased *metHb* level, accompanied by histopathological changes of the lung and heart. The NOAEL of 10 mg nitrite/l was established in this study.[4,7]

Reproductive Toxicity. Rats received 100 mg NAS/kg BW in drinking water during their entire lifespan over three generation. No evidence of chronic toxicity, carcinogenicity, or teratogenicity was reported.[7]

Embryotoxicity. Excessive uptake of NAS is consistent with lactational induction of severe iron deficiency in the neonates.[8]

Teratogenicity. Pregnant ICR mice were given drinking water containing 100 or 1000 mg NAS/l on days 7 to 18 of gestation. The treatment did not cause developmental abnormalities.[9] Fetal erythropoiesis was stimulated by nitrites in CD-1 mice.[10]

Gonadotoxicity. Male mice received 600 or 1200 mg/kg BW for 14 days by stomach intubation. The treatment caused significant changes in sex-chromosomal univalence and fertility, as well as sperm-head abnormalities.[11]

Mutagenicity.

In vitro genotoxicity. NAS could be genotoxic in somatic cells of rats and mice.[12]

No genotoxic effect was noted in mouse bone marrow cells.[13] Treatment of pre-implanted mouse embryos with nitrites *in vitro* caused no effect on micronucleus frequency.[14]

NAS was found to be positive in *Salmonella* strains *TA100* and *TA1535* and in *E. coli.*[14,15] However, according to Chow, Ames' *Salmonella*/microsome test showed an absence of mutagens in the feces extract of any of the animal groups treated with nitrates in their drinking water.[3] NAS was negative in DNA-cell-binding assay.[16]

In vivo cytogenetics. Rats, mice, and rabbits given nitrite in their drinking water for 3 months showed a significant increase in CA frequency. NAS administered in drinking water caused an increase in CA frequency in bone marrow of non-pregnant rats, in rats given it on days 5 to 18 of gestation, and in liver cells of transplacentally-exposed embryos.[17]

Mice received 4.0 mg SN/kg BW daily. The treatment increased the incidence of chromosomal breaks, but not SCE in bone marrow cells.[18,19] Ames' *Salmonella*/microsome test showed absence of mutagens in the feces extract of any of the animal groups.[3]

In mice, NAS induced increases in the number of micronucleated polychromatic erythrocytes.[11]

Carcinogenicity. Formation of *nitrite* from ingested *nitrate* can result in several adverse health effects and implies a genotoxic risk as a consequence of endogenous formation of carcinogenic *N*-nitrosocompounds.

Observations in man. Although it has been known that *nitrosocompounds* are responsible for some human stomach cancers, and that they can be generated *in vivo* from nitrites and nitrates, some epidemiological studies have not shown any correlation between the incidence of this type of cancer and either consumed levels of nitrites and nitrates, or the levels being generated in the body.[8,20]

In the recent communication, twenty-five volunteers consumed a fish meal during 7 consecutive days; a diet low in nitrate was consumed during 1 week before and 1 week after the test week. Nitrate intake at the ADI level in combination with a fish meal containing amines nitrosatable precursors increases *N*-nitrosodimethylamine excretion in urine and thus demonstrates increased formation of carcinogenic *N*-nitrosamines.[21]

Animal studies. NAS induced lymphatic cancer in rats.[22] Nevertheless, other authors did not confirm these findings.[23]

Rats were given 0.125 or 0.25% *nitrite* in drinking water for 2 years and 2.5 or 5.0% NAS in the diet. Neither rats nor mice experienced a carcinogenic effect under conditions of this study in which animals showed a high incidence of spontaneous tumors.[24,25]

Rats received 3.0 g nitrite/l in their drinking water (total dose of 63 g/kg BW). An increase in incidence of forestomach squamous papillomas was noted in this study.[26] NAS was found not to be carcinogenic when fed to rats in the reduced-protein diet for up to 115 weeks.[27]

Chemobiokinetics. Nitrites are capable of nitrosating other compounds with the production of more reactive species.

Regulations. ***U.S. FDA*** (1998) approved the use of NAS (1) in adhesives as a component (monomer) of articles intended for use in packaging, transporting, or holding food in accordance with the conditions prescribed in 21 CFR part 175.105; (2) in the manufacture of resinous and polymeric coatings for the food-contact surface of articles intended for use in producing, manufacturing, packing, processing, preparing, treating, packaging, transporting, or holding food (for use only as polymerization cross-linking agent in side seam cements for containers intended for use in contact with food (only of the identified types) in accordance with the conditions prescribed in 21 CFR part 175.300, (3) as a component of the uncoated or coated food-contact surface of paper and paperboard intended for use in producing, manufacturing, packaging, processing, preparing, treating, packing, transporting, or holding dry food of the type identified in 21 CFR part 176.170 (c), (4) in the manufacture of closures with sealing gaskets used on containers intended for use in producing, manufacturing, packing, processing, preparing, treating, packaging, transporting, or holding food, in accordance with the conditions prescribed in 21 CFR part 177.1210; (5) in the manufacture of rubber articles intended for repeated use in producing, manufacturing, packaging, processing, preparing, treating, packing, transporting, or holding food in accordance with the conditions prescribed in 21 CFR part 175.2600; and (6) as a food preservative and color fixative in accordance with the conditions prescribed in 21 CFR part 172.170.

Recommendations. ***WHO*** (1996). MAC: 3.0 mg/l

Standards. ***U.S. EPA*** (1999). MCL: 5.0 mg/l.

References:

1. Deichman, W. B., *Toxicology of Drugs and Chemicals*, Academic Press, Inc., New York, 1969, 543.

2. *Guidelines for Drinking-Water Quality*, 2nd ed., vol 2, Health criteria and other supporting information, WHO, Geneva, 1996, 313.
3. Chow, C. K., Effect of nitrate and nitrite in drinking water on rats, *Toxicol. Lett.*, 6, 199, 1980.
4. Til, H. P., Falke, H. E., Kuper, C. F and Willems, M. I., Evaluation of the oral toxicity of potassium nitrite in a 13-week drinking water study in rats, *Food Chem. Toxicol.*, 26, 851, 1988.
5. Til, H. P., Kuper, C. F., and Falke, H. E., Nitrite-induced adrenal effects in rats and consequences for the no-observed-effect-level, *Food Chem. Toxicol.*, 35, 349, 1997.
6. Darad, R., De, A. K., and Aiyar, A. S., Toxicity of nitrite and dimethylamine in rats, *Toxicol. Lett.*, 17, 125, 1983.
7. *Drinking Water & Health*, vol 1, Natl. Res. Council, Natl. Acad. Press, Washington, D.C., 1977, 420.
8. Roth, A. C., Herkert, G. E., Berez, J. P., and Smith, M. K., Evaluation of the developmental toxicity of sodium nitrite in Long Evans rats, *Fundam. Appl. Toxicol.*, 9, 668, 1987.
9. Shimada, T., Lack of teratogenic and mutagenic effects of nitrite on mouse fetuses, *Arch. Environ. Health.*, 1, 59, 1985.
10. Globus, M. and Samuel, D., Effect of maternally administered sodium nitrite on hepatic erythropoiesis in fetal CD-1 mice, *Teratology*, 18, 367, 1978.
11. Alavantic, D., Sunjevaric, I., Pecevski, J., Bozin, D., and Cerovic, G., *In vivo* genotoxicity of nitrates and nitrites in germ cells of male mice. I. Evidence for gonadal exposure and lack of heritable effects. *Mutat. Res.*, 204, 689, 1988.
12. Luca, D., Raileanu, L., Luca, V., and Duda, R., Chromosomal aberrations and micronuclei induced in rat and mouse bone marrow cells by sodium nitrite, *Mutat. Res.*, 155, 121, 1985.
13. Seiler, J. P., *In vivo* mutagenetic interaction of nitrite and ethylene thiourea, *Experientia*, 31, 214, 1975.
14. De Flora, S., Zannacchi, P., Camoriano, A., Bennicelli, C., and Badolati, G. S., Genotoxic activity and potency of 135 compounds in the Ames reversion test and in bacterial DNA-repair test, *Mutat. Res.*, 133, 161, 1984.
15. Rossman, T. G., Molina, M., Meyer, L., Boone, P., Klein, C. B., Wang, Y., Li, F., Lin, W. C., and Kinney, P. l., Performance of 133 compounds in the lambda prophase induction endpoint of the microscreen assay and comparison with *S. typhimurium* mutagenicity and rodent carcinogenicity, *Mutat. Res.*, 260, 349, 1991.
16. Kubinsky, H., Gutzke G. E., and Kubinsky, Z. O., DNA-cell-binding (DCB) assay for suspected carcinogens and mutagens, *Mutat. Res.*, 89, 95, 1981.
17. El Nahas, S. M., Globus, M., and Venthamany-Globus, S., Chromosomal aberrations induced by sodium nitrite in bone marrow of adult rats and liver cells of transplacentally exposed embryos, *J. Appl. Toxicol.*, 13, 643, 1984.
18. Banerjee, T. S. and Giri, A. K., Effects of sorbic acid and sorbic acid-nitrite *in vivo* on bone marrow chromosomes of mice, *Toxicol. Lett.*, 31, 101, 1986.
19. Krishna, G. et al., *In vivo* induction of sister chromatid exchanges in mice by nitrosated coal dust extract, *Environ. Res.*, 42, 106, 1987.
20. Forman, D. and Doll, R. Nitrates, nitrites and gastric cancer in Great Britain, *Nature*, 313, 620, 1985.
21. Vermeer, I. T., Pachen, D. M., Dallinga, J. W., Kleinjans, J. C., and van Maanen, J. M. S., Volatile *N*-nitrosamine formation after intake of nitrate at the ADI level in combination with an amine-rich diet, *Environ. Health Perspect.*, 106, 459, 1998.
22. Newberne, P. M., Nitrite promotes lymphoma incidence in rats, *Science*, 204, 1079, 1979.
23. Inai, K., Aoki, Y., and Tokuoka, S., Chronic toxicity of sodium nitrite in mice, with reference to its tumorogenicity, *Gann*, 70, 203, 1979.
24. Maekawa, A., Ogiu, T., Onodera, H., Furuta, K., Matsuoka, C., Ohno, Y., and Odashima, S., Carcinogenicity studies of sodium nitrite and sodium nitrate in F-344 rats, *Food Chem. Toxicol.*, 20, 25, 1982.
25. Lijinsky, W., Induction of tumors in rats by feeding nitrosatable amines together with sodium nitrite, *Food Chem. Toxicol.*, 22, 715, 1984.

26. Mirvish, S. S., Bulay, O., Runge, R. G., and Patil, K., Study of the carcinogenicity of large doses of dimethylnitramine, *N*-nitroso-*L*-proline, and sodium nitrite administered in drinking water to rats, *J. Natl. Cancer Inst.*, 64, 1435, 1980.
27. Grant, D. and Butler, W. H., Chronic toxicity of sodium nitrite in the male F344 rat, *Food Chem. Toxicol.*, 27, 565, 1989.

NONYLPHENOL

Molecular Formula. $C_{15}H_{24}O$
M = 220.39
CAS No 25154-52-3
84852-15-3
RTECS No SM5600000
Abbreviation. NP.

Synonym and **Trade Name.** 2,6-Dimethyl-4-heptylphenol; *p*-Nonylphenol, branched.

Composition. Mixture of monoalkyl phenols, predominantly *p*-substituted.

Properties. Thick, light yellow, straw color liquid with slight characteristic phenolic odor. Poorly soluble in water; soluble in alcohols.

Applications. Chemical intermediate for antioxidants for rubber and plastics, phenolic resins, rubber processing chemicals, nonionic and anionic surfactants, and polyvinyl chloride plasticizers. Antioxidant in the manufacture of polystyrene.

Migration Data. NP was released from plastic centrifuge tubes.

Acute Toxicity. LD_{50} is 1.3 g/kg BW. Poisoning decreased weight gain, caused changes in the liver, hemorrhages.

Clastogenic Effects. Has been shown to have estrogenic properties. Estrogen-like activity of NP was observed *in vitro*. NP induced both cell proliferation and progesterone receptor in human estrogen-sensitive MCF7 breast tumor cells. It also triggered mitotic activity in rat endometrium.[1,2]

Short-term Toxicity. Rats were exposed to 200, 650, or 2000 ppm NP in their diet for 3 months. The treatment caused a small decrease in BW and food consumption in the 2000 ppm dose group. No treatment-related clinical or histopathological changes, including effects on endocrine organs, estrous cycling, or sperm measurements were noted up to 2000 ppm exposure. The NOAEL is considered to be 650 ppm NP in the diet (approximately 50 mg/kg BW).[3]

NP was shown to increase uterine weight in the standard uterotrophic assays. The oral NOEL ranged from approximately 50 to 100 mg/kg BW (Moffat, 1996; Cervan, 1997).[3]

Reproductive Toxicity.

Embryo- and ***Gonadotoxicity.*** Sprague-Dawley rats were exposured to NP by treating via diet at 200, 650, and 2000 ppm. BW gain was reduced by 8 to10% in the 650 and 2000 ppm groups. Vaginal opening was accelerated by approximately 2 days (650 ppm) and approximately 6 days (2000 ppm) in F_1, F_2, and F_3 generations. No consistent changes were seen in pup number, weight or viability, litter indices, or other functional reproductive parameters. Sperm indices were unchanged in F_0 and F_1 males. Testis and epididymis weights were unchanged. These data show that NP had limited effects on the reproductive system in the presence of measurable nephrotoxicity.[4]

Allergenic Effect. Caused skin sensitization reaction (Gaworski, 1979).

Chemobiokinetics. Only about 1.0% of the oral dose entered circulation.5[4]

Recommendations. Non-occupational exposure does not pose any risk for humans. Safe level in drinking water is likely to be 0.001 mg/l (ADI ~ 55 μg/kg BW).[5]

Regulations.

EU (1990). NP is available *in* the *List of authorized monomers and other starting substances which may continue to be used for the manufacture of plastic materials and articles intended to come into contact with foodstuffs pending a decision on inclusion in Section A (Section B).*

U.S. FDA (1998) approved the use of NP (1) in adhesives as a component of articles intended for use in packaging, transporting, or holding food in accordance with the conditions prescribed in 21 CFR part

175.105; (2) as a component of a defoaming agent intended for articles which may be used in producing, manufacturing, packing, processing, preparing, treating, packaging, transporting, or holding food in accordance with the conditions prescribed in 21 CFR part 176.200; as (3) a defoaming agent in the manufacture of paper and paperboard intended for use in packaging, transporting, or holding food in accordance with the conditions prescribed in 21 CFR part 176.210; and (4) in the manufacture of resinous and polymeric coatings for the food-contact surface of articles intended for use in producing, manufacturing, packing, processing, preparing, treating, packaging, transporting, or holding food in accordance with the conditions prescribed in 21 CFR part 175.300.

References:

1. Soto, A. M., Justicia, H., Wray, J. W., and Sonnenschein, C., *p*-Nonylphenol: an estrogenic xenobiotic released from "modified" polystyrene, *Environ. Health Perspect.*, 92, 167, 1991.
2. White, R., Jobling, S., Hoare, S. A., Sumpter, J. P., and Parker, M. G., Environmentally persistent alkylphenolic compounds are estrogenic, *Endocrinology,* 135, 175, 1994.
3. Cunny, H. C., Mayes, B. A., Rosica, K. A., Trutter, J. A., and Van Miller, J. P., Subchronic toxicity (90-day) study with *para*-nonylphenol in rats, *Regul. Toxicol. Pharmacol.,* 26, 172, 1997.
4. Chapin, R. E., Delaney, J., Wang, Y., Lanning, L., Davis, B., Collins, B., Mintz, N., and Wolfe, G., The effects of 4-nonylphenol in rats: a multigeneration reproduction study, *Toxicol, Sci.,*52, 80, 1999.
5. Schlatter, C., Human health risk assessment for the xeno-oestrogen nonylphenol, Commun. to seminar *Mechanistic and Regulatory Aspects of Risk Assessment,* Jerusalem, January 11, 1998.

OCTABROMOBIPHENYL

Molecular Formula. $C_{12}H_2Br_8$
M = 786.34
CAS No 27858-07-7
RTECS No DV5700000
Abbreviation. OBB.

Properties. Powder. Brom contents 80 to 82%.

Applications. Used as a flame retardant for polyurethanes, polyolefins, and ABc-plastics.

Acute Toxicity. OBB is likely to be of low toxicity following a single *i/g* administration. Ingestion of 2.0 g/kg BW produced no functional or morphological changes in the rat.[1]

Repeated Exposure revealed little evidence of cumulative properties. Young rats tolerate intake of 10 mg Firemaster P-B-6 (a mixture of hexa- and octabromobiphenyl)/kg BW over a period of 6 weeks similar to adult animals given 8.0 mg/kg BW for 30 days. The threshold dose appeared to be 80 mg/kg BW.[1]

A 30 mg/kg BW dose of mixed bromobiphenyls administered for 30 days caused signs of anemia and porphyria in rats and mice. In the liver, there were hepatocyte swelling and proliferation of the endoplasmic reticulum. Liver pathologic changes persisted for 120 days.[2]

In a 20-day study (Akllen-Rowlands, 1981), rats were dosed with 1.0 to 6.0 mg/kg BW doses. The treatment resulted in increased liver and thyroid weights. The concentration of thyroxine and thyrotropic hormone in the plasma declined and the contents of I^{131} in the thyroid rose. Adynamia developed after administration of OBB. In cattle (cows) that survived after OBB intoxication, a BW loss and reduced lactation as well as increased susceptibility to infection were noted.[3]

Enlargement of the liver due to hepatocellular hypertrophy was observed in rats fed 100 and 1000 mg OBB/kg BW for 2 and 4 weeks, respectively. After 4 weeks of feeding 1000 mg/kg BW, the bromine concentration in the adipose tissue was 600 times greater than that in the controls.[4]

Long-term Toxicity. Oral exposure to 10 mg/kg BW dose of a mixture of hexa- and octa-bromobiphenyl did not affect BW gain and general condition of young rats dosed for 36 weeks.[5]

Reproductive Toxicity.

Teratogenicity. Exposure to 1.0 g/kg BW on days 6 to 15 of pregnancy resulted in congenital abnormalities in the young rats.[1]

Gonadotoxicity. Rats received 0.001 mg OBB/kg feed. The treatment disrupted the oestrus cycle and impaired reproduction in females causing abortions.[6]

Mutagenicity.

In vivo cytogenetics. According to Norris et al. (1975), administration of OBB did not cause cytogenic changes (DNA damage, mutation, or CA) in rat or mouse bone marrow cells.[1] PCBs did not induce micronuclei in mouse bone marrow cells (IARC 41-278).

Carcinogenicity. A single exposure of young rats to high doses of bromophenyls may cause development of hepatocellular tumors. Gavaging rats with 1.0 g caused gastric tumors to develop.[7]

Commercial PCBs preparation, composed primarily of hexabrombiphenyl, tested by oral administration by gavage, produced malignant hepatic tumors in rats and mice.[8]

Chemobiokinetics. Absorbed OBB is predominantly distributed in the suprarenals and fatty tissues where it can be retained for the lifetime because of its persistence in the lipid-rich tissues.[2] It is metabolized by microsomal enzyme system of the liver. OBB is excreted mainly with the feces. OBB is a powerful inductor for liver microsomal enzyme system, especially for the monooxygenase system. A similar effect was observed in the renal, pulmonary, and suprarenal tissues.

Regulations. ***U.S. FDA*** set guidelines of 0.3 mg/kg in the fat of milk, meat and poultry; 0.05 mg/kg in animal feed.

References:

1. Norris, J. M., Kociba, R. J., Schwetz, B. A., et al., Toxicology of octabrombiphenyl and decarbamodiphenyl oxide, *Environ. Health Perspect.*, 11, 153, 1975.
2. Micelli, J. N. and Marks, B. N., Tissue distribution and elimination kinetics of polybrominated biphenyls (PBB) from rat tissue, *Toxicol. Lett.*, 9, 315, 1984.
3. Gupta, B. N., McConnell, E. E., Harris, M. W., et al., Polybrominated biphenyl toxicosis in the rat and mouse, *Toxicol. Appl. Pharmacol.*, 57, 99, 1981.
4. Stross, J., Smolker, I. A., Isbister, J., et al., The human health effects of exposure to polybrominated biphenyls, *Toxicol. Appl. Pharmacol.*, 58, 145, 1981.
5. Lee, K. P., Herbert, R. R., Aftosmis, J. G., et al., Bromine tissue residue and hepatotoxic effects of octobromobiphenyls in rats, *Toxicol. Appl. Pharmacol.*, 34, 115, 1975.
6. Johnston, C. A., Demarest, K. T., McCormack, K. M., et al., Endocrinological, neurochemical, and anabolic effects of polybrominated biphenyls in male and female rats, *Toxicol. Appl. Pharmacol.*, 56, 240, 1980.
7. Kimbrough, R. D., Korver, M. P., Burse, V. W., et al., The effect of different diets or mineral oil on liver pathology and polybrominated biphenyl concentration in the tissues, *Toxicol. Appl. Pharmacol.*, 52, 442, 1980.
8. *Carcinogenesis Studies of Polybrominated Biphenyl Mixture (Fire Master FF-1)(*CAS No 67774-32-7) *in F344/N Rats and B6C3F$_1$ mice* (Gavage Studies), NTP Technical Report Series No 244, Research Tri angle Park, NC, 1983.

OCTADECAMIDE

Molecular Formula. $C_{18}H_{37}NO$

M = 283.50

CAS No 124-26-5

RTECS No RG0182000

Synonyms. Octadecanamide; Octadecylamide; Stearamide; Stearic amide; Stearic acid amide; Stearylamide.

Properties. Colorless leaflets (from alcohol). Soluble in hot ethanol; insoluble in water.

Applications. Release agent in manufacture of plastics and films.

Chemobiokinetics. Lack of toxic effect depends on relatively rapid hydrolysis (probably in the liver) to corresponding acid. In cats and rabbits, O. is usually excreted unchanged with urine or as simple carboxylic amides.[03]

Regulations. ***U.S. FDA*** (1998) approved the use of O. in adhesives as a component of articles intended for use in packaging, transporting, or holding food in accordance with the conditions prescribed in 21 CFR part 175.300.

OCTADECYLAMINE

Molecular Formula. $C_{18}H_{39}N$
M = 269.58
CAS No 124-30-1
RTECS No RG4150000
Abbreviation. ODA.

Synonym. 1-Aminooctadecane.

Properties. Insoluble in water, soluble in alcohol.

Applications. Anticorrosive agent.

Acute Toxicity. In rats, LD_{50} is 1.0 g/kg BW.[011]

Long-term Toxicity. Prolonged exposure resulted in anorexia, BW loss and some histological changes in the mesenteric lymph nodes, GI mucosa, and liver.[019]

Allergenic Effect. ODA is shown to be a primary skin sensitizer.

OCTAETHYL CYCLOTETRASILOXANE

M = 408.92
Molecular Formula. $C_{16}H_{40}Si_4O_4$
CAS No 1451-99-6
RTECS No GZ4396000

Properties. Clear, water-like liquid with a weak, specific odor. Poorly soluble in water, miscible with alcohol.

Applications. Used in the production of siloxane plastics.

Acute Toxicity. Rats and mice tolerate administration of up to 20 mg/kg BW without any sign of intoxication.

Repeated Exposure to oral dose of 3.0 g/kg BW caused no functional and morphological changes in rats.

Allergenic Effect is not observed.

Reference:

Shugayev, B. B. and Bukhalovsky, A. A., Toxicity of hexaethyl cyclotrisiloxane and octaethyl cyclotetrasiloxane, *Gig. Sanit.*, 8, 93, 1986 (in Russian).

OCTYLPHENOL

Molecular Formula. $C_{14}H_{22}O$
M = 206.32
CAS No 1322-69-6
27193-28-8
RTECS No SM5775000
Abbreviation. OP.

Synonym and **Trade Name.** 1-(*p*-Hydroxyphenyl)octane; *p*-Octylphenol.

Applications. OP is a commercial intermediate used primarily for the production of octylphenol polyethoxylate surfactants. Used in the manufacture of plasticizers and antioxidants. Intermediate for resins and adhesives.

Acute Toxicity. A substance of low toxicity (less toxic than phenol itself).[022]

Reproductive Toxicity.

Gonadotoxicity. In 1993, OP has been hypothesized to be responsible for adverse effects on reproductive systems.[1] Pregnant rats were exposed during pregnancy and lactation to drinking water at the concentration of approximately 0.1 mg/l. Sharpe *et al.* observed a reduction in sperm counts and testes weights in offspring.[2] However, in 1998, these authors reported that they could not reproduce their initial results.[3]

In the most recent study, male rats were exposed to OP during gestation or during the first 21 days of postnatal life. OP was administered via the drinking water or *ad libitum* to rats at dietary concentrations of 0.2, 20, 200, or 2,000 ppm. Effects were observed only at 2,000 ppm dose level, including decreased BW in adults and during the latter portion of lactation in offspring and minor BW-related delays in acquisition of vaginal opening and preputial separation. No effects on reproductive parameters, testes, prostate, or ovary weights or morphology, on sperm counts, motility, morphology, production, or on estrous cyclicity were observed. No estrogen-like effects were evident. The NOAELs for systemic and postnatal toxicity were 200 ppm and at or above 2,000 ppm for reproductive toxicity.[4] This study does not support previous preliminary data[2] on low dose effects of OP.

Carcinogenicity. Alkylphenols bind to the estrogen receptor and exert estrogenic actions on mammalian cells and human breast tumor cells.[5,6]

Chemobiokinetics. OP is found to be toxic to cultured rat and mouse splenocytes, this effect being exerted, at least partially, through calcium-dependent apoptosis.[7]

Regulations.

EU (1990). OP is available in the *List of authorized monomers and other starting substances which may continue to be used for the manufacture of plastic materials and articles intended to come into contact with foodstuffs pending a decision on inclusion in Section A (Section B).*

U.S. FDA (1998) approved the use of OP (1) in adhesives as a component of articles intended for use in packaging, transporting, or holding food in accordance with the conditions prescribed in 21 CFR part 175.105; and (2) in resinous and polymeric coatings used as the food-contact surfaces of articles intended for use in producing, manufacturing, packing, processing, preparing, treating, packaging, transporting, or holding food in accordance with the conditions prescribed in 21 CFR part 175.300.

Great Britain (1998). OP is authorized up to the end of 2001 for use in the production of polymeric materials and articles in contact with food or drink or intended for such contact.

References:

1. Sharpe, R. M. and Skakkebaek, N. E., Are estrogens involved in failing sperm counts and disorders of the male reproductive tract? *Lancet,* 341, 1392, 1993.
2. Sharpe, R. M., Fisher, J. S., Millar, M. M., Jobling, S., and Sumpter, J. P., Gestational and lactational exposure of rats to xenoestrogens results in reduced testicular size and sperm production, *Environ. Health Perspect.,* 103, 1136. 1995.
3. Sharpe, R. M., Turner, K. J., and Sumpter, J. P., Endocrine disruptors and testis development, *Environ. Health Perspect.,* 106, A220, 1998.
4. Tyl, R. W., Myers, C. B., Marr, M. C., Brine, D. R., Fail, P. A., Seely, J. C., and Van Miller, J. P., Two-generation reproduction study with para-*tert*-octylphenol in rats, *Regul. Toxicol. Pharmacol.,* 30, 81, 1999.
5. Davis, D. L. and Bradlow, H. L., Can environmental estrogens cause breast cancer? *Sci. Am.,* October 1995, 166.
6. White, R., Jobling, S., Hoare, S., Sumper, J. P., and Parker, M. G., Environmentally persistent alkylphenolic compounds are estrogenic, *Endocrinology,* 135, 175, 1994.
7. Nair-Menon, J. U., Campbell, G. T., and Blake, C. A., Toxic effect of octylphenol on cultured rat murine splenocytes, *Toxicol. Appl. Pharmacol.,* 139, 437, 1996.

α-OLEFINS C_{10}-C_{14}, mixture

Molecular Formula. CH_2=CH-C_nH_{2n+1}, where n = 8 to 12

M = 163 to 178

Synonym. 1-Alkens.

Composition. Contains 1-decene (α-C_3H_{17}), 1-dodecene (α-$C_{10}H_{21}$), 1-tetradecene (α-$C_{12}H_{23}$).

Applications. Used in the production of plasticizers, alcohols, and synthetic detergents.

Acute Toxicity. LD_{50} is 13 g/kg BW in rats and 17.5 g/kg BW in mice.

Repeated Exposure failed to reveal cumulative properties.

Reference:

Abbasov, D. M. et al., See ***Hexanediamine.***

cis-OLEIC ACID, METHYL ESTER

Molecular Formula. $C_{19}H_{36}O_2$
M = 296.55
CAS No 112-62-9
RTECS No RK0895000
Abbreviation. OAM.

Synonyms and **Trade Names.** 9-Octadecenoic acid (*cis*), methyl ester; Methyl oleate; Methyl-9-octadecenoate; Emerest 2801.

Properties. Clear to amber liquid with faint odor. Insoluble in water; miscible with alcohol.

Applications. Intermediate for detergents and stabilizers. Plasticizer and softener for natural and synthetic rubbers.

Mutagenicity.

In vitro genotoxicity. OAM was tested in five *Salmonella typhimurium* strains and was found to be negative.[013]

Carcinogenicity. OAM promoted skin tumor formation in mice.[1]

Regulations. ***U.S. FDA*** (1998) approved the use of OAM in adhesives as a component of articles intended for use in packaging, transporting, or holding food in accordance with the conditions prescribed in 21 CFR part 175.105.

Reference:

1. Arffmann, E. and Glavind, J., Carcinogenicity in mice of some fatty acid methyl esters. I. Skin application, *Acta Pathol. Microbiol. Scand.*, Sect. A 82A, 127, 1974.

OP-7

Molecular Formula. $C_nH_{2n+1}C_6H_4O(C_2H_4O)_mH$, where $n = 8$ to10; $m = 6$ to 7
CAS No 9004-87-9
11100-29-1
RTECS No RL1070000

Synonym. Monoalkylphenols, polyethylene glycol esters, mixture.

Properties. OP-7 is a viscous, oily liquid with a color ranging from light yellow to brown, and with a bitter taste and specific odor. Water solubility is 5.0 g/l at 25°C. Odor perception threshold is 0.45 mg/l; taste perception threshold is more than 0.9 mg/l; foam formation threshold concentration is 1.0 to 2.0 mg/l.[1]

Applications. Used in the production of synthetic rubber and plastic synthetic coatings. A component of sealing pastes for hermetically sealing food cans.

Acute Toxicity. In rats, the LD_{50} is reported to be 3.0 or 7.9 g/kg BW.[2] Death is within 1 to 2 days.[3]

Long-term Toxicity. Rats that received OP-7 in their drinking water at a concentration of 2.0 g/l developed mild fatty dystrophy.[4]

Standards. ***Russia*** (1995). MAC and PML: 0.1 mg/l (organolept., foam).

References:

1. Grushko, Ya. M., *Hazardous Organic Compounds in the Liquid Industrial Effluents*, Khimiya, Leningrad, 1976, 139 (in Russian).
2. Il'in, I. E., Investigation of toxic products of transformation formed in the course of water chlorination, *Gig. Sanit.*, 2, 11, 1980 (in Russian).
3. Mozhayev, E. V., *Pollution of Reservoirs by Surfactants (Health Safety Aspects)*, S. N. Cherkinsky, Ed., Meditsina, Moscow, 1976, 94 (in Russian).
4. Goeva, O. E., in *Proc. Leningrad Sanitary Hygiene Medical Institute*, Leningrad, Issue No 68, 1961, 111 (in Russian).

OP-10

Molecular Formula. $C_nH_{2n+1}C_6H_4O(C_2H_4O)_mH$, where $n = 8$ to10; $m = 10$ to 12

M = 800 to 820

CAS No 9041-29-6

RTECS No RL1070030

Synonym. Polyethylene glycol, alkylphenyl ester.

Composition. A product of condensation of alkylphenol with 10 molecules of ethylene oxide.

Properties. Brown, oily, viscous liquid with a specific odor. Readily soluble in water (5.0 g/l). When shaken with water, forms a stable foam; when shaken with mineral and vegetable oils, forms stable emulsions. A concentration of 1.8 mg/l gives water an odor rating of 1.0 mg/l; a concentration of 3.6 mg/l corresponds to rating of 2. It can be tasted at a concentration of more than 3.6 mg/l. Foam formation occurs at the concentration of 0.09[1] or 1.0 mg/l.

Applications. Used in the production of synthetic rubbers and plastics, synthetic anti-corrosion coatings, and as an emulsifier in production of latex articles for medical and food purposes.[1]

Migration Data. Distilled water was poured into 250 ml cans and autoclaved. After one month water transparency and color had changed and migration of OP-10 (0.01 to 0.15 mg/l, in proportion to the amount contained in the mix formulation of butadiene-styrene latex) was found.[2]

Acute Toxicity. LD_{50} is reported to be 5.0 g/kg BW for mice, and 3.5 to 5.48 g/kg BW for rats.[2,3] The main sign of poisoning seems to be CNS inhibition. Gross pathology examination revealed congestion in the visceral organs.

Repeated Exposure showed moderate cumulative properties. K_{acc} is 5 (by Lim).

Long-term Toxicity. Rats were dosed by gavage with 0.02 mg/kg BW. Adynamia was found to develop in a month after onset of the treatment. To the 3rd month, a decline in the STI value and a reduction in the lung, kidney and spleen weights were noted.[3]

Standards. ***Russia*** (1995). MAC and PML: 0.1 mg/l (organolept., foam).

References:

1. *Toxicology of the Components of Rubber Mixes and of Rubber and Latex Articles*, TSNIITECHIM, Moscow, 1974, 50 (in Russian).
2. Goyeva, O. Z., in *Proc. Leningrad Sanitary Hygiene Medical Institute*, Leningrad, Issue No 73, 1961, 5 (in Russian).
3. Mel'nikova, V. V. and Zhilenko, V. N., On toxicity of alkylphenyl ether of polyethylene glycol, *Kauchuk i Rezina,* 6, 56, 1980 (in Russian).

OXALIC ACID

Molecular Formula. $C_2H_2O_4$

M = 90.04

CAS No 144-62-7

RTECS No RO2450000

Abbreviation. OA.

Synonym and **Trade Name.** Aktisal; Ethanedioic acid.

Properties. Transparent, colorless crystals or white powder. Solubility in water is 83.4 g/l at 20°C. Very soluble in ethanol.

Occurrence. OA was found in saliva from dentally healthy subjects (0.1 to 0.18 mmol/l), in tooth tartar (3.3 mmol/kg tartar), in human teeth (1.0 mmol/kg). The formed Ca oxalate is proposed to be a physiological protective mechanism for teeth.[1]

Acute Toxicity. Symptoms of acute poisoning included salivation and nasal discharge with progressive weakness, rapid shallow breathing and collapse.[2]

Repeated Exposure. Long-Evans rats received the diet containing 2.5 and 5.0% OA. The treatment reduced absolute organ weights but enhanced organ/body weight ratio (Goldman et al., 1977).

Long-term Toxicity. Rats were given 0.1 to 1.2% OA in their diet for 730 days.[3] Munro et al. (1996) suggested the calculated NOEL of 840 mg/kg BW for this study.[4]

Reproductive Toxicity.

Embryotoxicity. OA crossed the placenta in sheep and caused kidney disturbances in fetal rats when given orally to the dams, but did not cause abortions.[5, 6]

Reproductive toxicity of OA was tested in Swiss CD-1 mice which received 0.05%, 0.1%, or 0.2% OA in their drinking water (89, 162, or 275 mg/kg BW) using RACB protocol. In the F_0 mice, OA reduced the number of litters per pair, adjusted pup weight, and prostate weight in the absence of detected somatic organ changes. In F_1 mice, the treatment caused a reduction in the number of live pups per litter.[7]

Gonadotoxicity. Above described treatment disrupted estrous cycles in Long-Evans rats.[13] Increased sperm abnormalities were seen in the second generation of mice administered 0.2% OA in the drinking water.[7, 8]

Mutagenicity.

In vitro genotoxicity. OA was negative in *Salmonella* mutagenicity assay.[9,10] It exhibited mild clastogenic activity in a plant assay.[11]

Carcinogenicity. An increased incidence of bladder tumors was observed in male mice but not in rats fed diets containing 10 or 20% *xylitol.* This increase has been associated with epithelial hyperplasia and urinary bladder calculi consisting mainly of *calcium oxalate*. Wistar Albino rats were given OA via gastric gavage either with water or with 625 mg *xylitol*/kg BW. The author indicated an increase in the absorption and urinary excretion of dietary *oxalate* in xylitol-treated mice.[12]

Chemobiokinetics. OA salts affected kidney subsequently to blocking of tubules by crystals of *calcium oxalate. Oxalates* were found to crystallize in brain tissue and to cause symptoms of paralysis and other disturbances of CNS. They may induce erythrocytes breakdown. Accumulation of *calcium oxalate* crystals in kidney and urinary tract was noted.[2]

According to Parke, OA is excreted in the urine unchanged.[13]

Regulations. ***U.S. FDA*** (1998) approved the use of OA in the manufacture of phenolic resins for food-contact surface of molded articles intended for repeated use in contact with nonácid food (*pH* above 5.0) in accordance with the conditions prescribed in 21 CFR part 177.2410.

Standards. ***Russia*** (1995). MAC: 0.5 mg/l.

References:

1. Wahl, R. and Kallee, E., Oxalic acid in saliva, teeth and tooth tartar, *Eur. J. Clin. Chem. Clin. Biochem.*, 32, 821,1994.
2. Clarke, E. G. and Clarke, M. L., *Veterinary Toxicology*, The Williams and Wilkins Co., Baltimore, MI, 1975, 258.
3. Fitzhugh and Nelson, 1947, cit. in *Toxicological Evaluation of Some Antimicrobials, Antioxidants, Emulsifiers, Stabilizers, Flour-treatment Agents, Acids and Bases,* The 9th and 10th Meeting of the Joint FAO/WHO Expert Committee on Food Additives, Geneva, 1967.
4. Munro, I. C., Ford, R. A., Kennepohl, E., and Sprenger, J. G., Correlation of structural class with No-Observed-Effect Levels: A proposal for establishing a threshold of concern, *Food Chem. Toxicol.*, 34, 829, 1996.
5. Schardein, J. L., *Chemically Induced Birth Defects*, 2nd ed., Marcel Dekker, Inc.; New York, 1993, 107.
6. Schiefer, B., Hewitt, M. P., and Milligan, J. D., Fetal renal oxalosis due to feeding oxalic acid to pregnant ewes, *Z. Veterinarmed.*, [A] 23, 226, 1976.
7. Lamb, J. C., Gulati, D. K., Barnes, L. H., Welch, M., and Russel S., Oxalic acid, *Environ. Health Perspect.*, 105, Suppl. 1, 1997.
8. Patty's *Industrial Hygiene and Toxicology*, vol 2E, *Toxicology*, C. D. Clayton and F. E. Clayton, Eds., 4th ed. John Wiley & Sons, New York, NY, 1994.
9. Rossman, T. G., Molina, M., Meyer, L., Boone, P., Klein, C. B., Wang, Y., Li, F., Lin, W. C., and Kinney, P., Performance of 133 compounds in the lambda prophase induction endpoint of the microscreen assay and comparison with *S. typhimurium* mutagenicity and rodent carcinogenicity, *Mutat. Res.*, 260, 349, 1991.

10. Haworth, S., Lawlor, T., Mortelmans, K., Speck, W., Zeiger, E., Salmonella mutagenicity test results for 250 chemicals, *Environ. Mutagen,* 5 (Suppl. 1), 1, 1983.
11. Shevchenko, V. V. et al., Clastogenic effect of oxalic acid and its specific features, *Genetika* (Moscow), 21, 779, 1985 (in Russian).
12. Salminen, E. and Salminen, S., Urinary excretion of orally administered oxalic acid in xylitol fed to rats and mice, 3rd Int. Congress Toxicol., San Diego, CA, August-September 1983, *Toxicol. Lett.*, 18, 37, 1983.
13. Parke, D. V., *The Biochemistry of Foreign Compounds*, Pergamon Press, Oxford, 1968, 147.

OXAMINE S-2

Trade Name. Ethamine T/12 (an English analogue of O.).

Composition. A product of the reaction of octadecylamine (stearylamine) with ethylene glycol.

Properties. Waxiform material, light yellow in color.

Applications. Used as an antistatic additive.

Acute Toxicity. In mice, LD_{50} is 2.1 g/kg BW for oxamine, and 4.15 g/kg BW for ethamine. Rats exhibit the same sensitivity.

Repeated Exposure. Accumulation is observed in mice after administration of the high doses (420 mg O./kg and 830 mg ethamine/kg BW). Gross pathology examination revealed marked swelling in the stomach and intestine with congestion in their serous membranes and the mesentery as well as in the brain. There were no morphology changes in the viscera of survivors.

Long-term Toxicity. Administration of 40 or 400 mg/kg BW doses caused some mortality in mice. In rats which received 20 mg/kg for 6 months, the subcortical areas of the CNS were found to be affected. Chronic exposure of rats to 42 mg O./kg BW caused no deviations in their normal condition. Administration of 20 and 40 mg O./kg BW caused no mortality in mice. Animals experienced CNS excitation.

Standards. ***Russia.*** PML: *n/m.*

Reference:

Krynskaya, I. L., Bukevich, G. M., Robachevskaya, Ye. G., et al., Data on toxicology of some antistatic additives to plastic materials, in *Hygiene and Toxicology of High-Molecular-Mass Compounds and of the Chemical Raw Material Used for Their Synthesis*, Proc. 6th All-Union Conf., B. Yu. Kalinin, Ed., Lenirgrad, 1979, 215 (in Russian).

OXANOL KD-6

Molecular Formula. $C_nH_{2n+1}O[\sim C_2H_4O\sim]_mH$, where $n = 8$ to 10; $m = 6$

CAS No 71060-57-6

74565-57-4

RTECS No RP3378500

Synonym. Ethoxylated C_8-C_{10} alcohols.

Composition. Polyethylene glycol based on primary fatty alcohols, monoalkylesters.

Properties. Yellowish liquid, readily soluble in water and alcohol.

Applications. Used in the production of surfactants; an emulsifier and an antistatic.

Acute Toxicity. In rats, LD_{50} is reported to be 2.7 g/kg BW.

Repeated Exposure. K_{acc} appeared to be more than 10 (by Lim).

Allergenic Effect is not observed on repeated skin application in guinea pigs.

Reference:

Timofievskaya, L. A. and Khailurina, Kh. Kh., Toxicity of oxanol KD-6, *Gig. Truda Prof. Zabol.,* 12, 58, 1987 (in Russian).

OXANOL 0-18

Molecular Formula. $C_{18}H_{35}[\sim C_2H_4O\sim]_nH$

Properties. Cream or yellowish, paste-like mass. Readily soluble in water and alcohol.

Applications. Antistatic additive. Emulsifier, pigment dispersant.

Acute Toxicity. LD_{50} is 4.2 g/kg BW in rats, and 2.7 g/kg BW in mice.

Repeated Exposure. O-18 exhibits no signs of accumulation.

Allergenic Effect is observed on skin application of a 12% solution of O-18.

Reference:

Mikhailets, I. B., Toxic properties of some antistatics, *Plast. Massy*, 12, 27, 1976 (in Russian).

2-OXETANONE

Molecular Formula. $C_3H_4O_2$

M = 72.06

CAS No 57-57-8

RTECS No RQ7350000

Synonyms. Hydracrylic acid, β-lactone; 3-Hydroxypropanoic acid, lactone; 3-Propanolide; β-Propiolactone; 1,3-Propiolactone.

Properties. Colorless liquid with a pungent, slightly sweetish odor. Solubility in water is 37%, miscible with alcohol.

Applications. Intermediate in the production of acrylic acid and esters.

Acute Toxicity. LD_{50} was found to be about 50 to 100 mg/kg BW. Single oral dose caused death by generalized CNS depression, and, probably, respiratory arrest. Administration was rapidly followed by the symptoms of acute poisoning: twitching, gasping, convulsions, and collapse.[1,011]

Repeated Exposure. Rats received the dose of 3.0 for 35 days. Histology investigation revealed only mild changes in the liver and kidneys.[1]

Mutagenicity.

In vitro genotoxicity. O. was positive in *Salmonella* mutagenicity assay,[2] and in Chinese hamster ovary cells.[3] It did not induce micronuclei in mouse bone marrow but did in spermatids (Cliet et al., 1993). O. was found to be positive in DNA-cell-binding assay.[4]

In vivo cytogenetics. In the DLM assay, mice were exposed *i/p* with 3 to 10 mg/kg BW. The treatment produced no increase in the incidence of early death or pre-implantation losses.[5] There was no micronucleus induction in *ddY* mouse bone marrow or peripheral blood after double *i/p* treatments of up to LD_{50} per treatment. A reduction of polychromatic erythrocyte to total erythrocyte ratio was observed in the high dose group in the bone marrow test.[6]

Carcinogenicity. Female Sprague-Dawley rats received weekly doses of 10 mg O. in 0.5 ml tricaprylin for 487 days. The treatment caused squamous cell carcinomas of forestomach in 3/5 rats. No such tumors were found in five control animals given weekly doses of 0.5 ml tricaprylin.[7]

Carcinogenicity classification. An IARC Working Group concluded that there is sufficient evidence for the carcinogenicity of O. in *experimental animals* and there are no adequate data available to evaluate the carcinogenicity of O. in *humans*.

The Annual Report of Carcinogens issued by the U.S. Department of Health and Human Services (1998) defines O. to be a substance which may reasonably be anticipated to be carcinogen, i.e., a substance for which there is a limited evidence of carcinogenicity in *humans* or sufficient evidence of carcinogenicity in *experimental animals*.

IARC: 2B;

EU: 2.

Chemobiokinetics. O. can react with chloride ion to form *3-chloropropionic acid*, particularly in blood plasma.[8] β-*Hydroxypropionic acid* appeared to be the hydrolysis product of O.

Regulations. ***U.S. FDA*** (1998) approved the use of O. in adhesives as a component of articles intended for use in packaging, transporting, or holding food in accordance with the conditions prescribed in 21 CFR part 175.105.

References:

1. *Encyclopedia of Occupational Health and Safety*, vols 1 and 2, Int. Labour Office, McGraw-Hill Book Co., New York, 1971, 620.

2. Rossman, T. G., Molina, M., Meyer, L., Boone, P., Klein, C. B., Wang, Y., Li, F., Lin, W. C., and Kinney, P. I., Performance of 133 compounds in the lambda prophase induction endpoint of the microscreen assay and comparison with *S. typhimurium* mutagenicity and rodent carcinogenicity, *Mutat. Res.*, 260, 349, 1991.
3. Ishidate, M. and Odashima, S., Chromosome tests with 134 compounds in Chinese hamster cells *in vitro* - a screening for chemical carcinogens, *Mutat. Res.*, 48, 337, 1977.
4. Kubinsky, H., Gutzke G. E., and Kubinsky, Z. O., DNA-cell-binding (DCB) assay for suspected carcinogens and mutagens, *Mutat. Res.*, 89, 95, 1981.
5. Epstein, S. S., Arnold, E., Andrea, J., Bass, W., and Bishop, Y., Detection of chemical mutagens by the dominant lethal assay in the mouse, *Toxicol. Appl. Pharmacol.*, 23, 288, 1972.
6. Morita, T., Asano, N., Awogi, T., Sasaki, Yu. F., et al. Evaluation of the rodent micronucleus assay in the screening of IARC carcinogens (Groups 1, 2A and 2B). The summary report of the 6th collaborative study by CSGMT/JEMS-MMS, *Mutat. Res.*, 389, 3, 1997.
7. Van Duuren, B. L., Langseth, L., Orris, L., Teebor, G., Nelson, N., and Kuschner, M., Carcinogenicity of epoxides, lactones, and peroxy compounds. IV. The carcinogenic actions of some monofunctional ethyleneimine derivatives, *Brit. J. Pharmacol.*, 9, 306, 1966.
8. Searle, C. E., Experiments on the carcinogenicity and reactivity of β-propiolactone, *Brit. J. Pharmacol.*, 15, 804, 1961.

4,4'-OXYDIPHENOL

Molecular Formula. $C_{12}H_{10}O_3$
M = 202.22
CAS No 1965-09-9
RTECS No SM6040000
Abbreviation. ODP.

Acute Toxicity. LD_{50} *i/p* is 150 mg/kg BW.

Regulations. ***EU*** (1990). ODP is available in the *List of authorized monomers and other starting substances which may continue to be used for the manufacture of plastic materials and articles intended to come into contact with foodstuffs pending a decision on inclusion in Section A* (*Section B*).

PALMITIC ACID

Molecular Formula. $C_{16}H_{32}O_2$
M = 256.48
CAS No 57-10-3
RTECS No RT4550000
Abbreviation. PA.

Synonyms and **Trade Name.** Cetylic acid; Emersol 140; Hexadecanoic acid; Hexadecylic acid; 1-Pentadecanecarboxylic acid.

Properties. White crystalline scales. Insoluble in water; sparingly soluble in cold alcohol; freely soluble in hot alcohol.[020] Occurs as the glyceryl ester in many oils and fats.[020]

Applications. Food-grade additive.

Acute Toxicity. LD_{50} exceeded 10 g/kg BW in rats.[1]

Mutagenicity. PA has been identified as the major component responsible for the antimutagenicity seen in *Salmonella* mutagenicity assay against an established mutagen.[2]

Regulations.

EU (1990). PA is available in the *List of authorized monomers and other starting substances, which shall be used for the manufacture of plastic materials and articles intended to come into contact with foodstuffs (Section A)*.

Great Britain (1998). PA is authorized without time limit for use in the production of polymeric materials and articles in contact with food or drink or intended for such contact.

Recommendations. Joint FAO/WHO Expert Committee on Food Additives (1999). ADI: No safety concern.

References:

1. Briggs, G. B., Doyle, R. L., and Young, J. A., Safety studies on a series of fatty acids, *Am. Ind. Hyg. Assoc. J.*, 37, 251, 1976.
2. Nadathur, S. R., Carney, J. R., Gould, S. J., and Bakalinsky, A. T., Palmitic acid is the major fatty acid responsible for significant anti-*N*-methyl-*N'*-nitro-*N*-nitrosoguanidine (MNNG) activity in yogurt, *Mutat. Res.*, 359, 179, 1996.

PARAFFIN

CAS No 8002-74-2

RTECS No RV0350000

Synonyms. Paraffin wax; Paraffin wax fume; Petroleum wax.

Composition. P. and microcrystalline waxes are complex mixtures of solid hydrocarbons, primarily, alkanes, obtained from petroleum distillation. Present-day paraffins are more highly purified and contain very low levels of free aromatic hydrocarbons. Reinforced wax consists of P. to which certain optional substances required for its production or for imparting desired properties have been added.[1] In three commercial Ps. examined, concentration of 16 to 73 ppb benzene was determined.[2]

Properties. White, semi-translucent, odorless solid. Insoluble in water; soluble in organic solvents.

Applications. Used in the preparation of coatings for paper, food containers, and for medicinal preparations. P. could be a component of non-food articles in contact with food. Reinforced wax is used in producing, manufacturing, packing, processing, transporting, or holding food.

Repeated Exposure. F344 rats and Sprague-Dawley rats were fed paraffinic white oil at 0.2 or 2.0% of the diet for 30, 61, or 92 days. No adverse observations or unscheduled deaths were reported. Mesenteric lymph nodes were enlarged. F344 rats appeared to be more sensitive compared to SD rats.[3]

Paraffin wax is generally regarded as being biologically inert.

Short-term Toxicity. P. produced pathological changes, mainly in the liver and lymph nodes. The effects were inversely related to molecular mass, viscosity, and melting point, but appeared to be independent of oil type processing.[5]

Petroleum-derived P. and microcrystalline waxes are considered non-toxic and non-carcinogenic (JECFA, 1995). In a 13-week study, F344 rats received 0.2 to 5.0% liquid paraffin in the powdered diet. The exposure neither produced effect on BW gain and food intake nor caused hematological or histopathological changes. Hematological, serum biochemical and histopathological examination failed to show obvious toxicity of liquid paraffin. Maximum tolerable dose in F344 rats is 5.0% or more in the diet.[4]

3 waxes (paraffinic-derived, hydrogenation- or percolation-refined with a range of molecular mass) were fed in the diet of rats at doses of around 2.0, 18, 180, or 1900 mg/kg BW. No treatment-related biology was seen with the higher molecular-sizes hydro-carbons (microcrystalline waxes and the higher viscosity oils). In a 90-day feeding study, low-melting-point wax (LMPW), intermediate-melting-point wax (IMPW), microcrystalline high-melting-point wax (HMPW), and high-sulfur wax (HSW) were investigated. The dose levels tested were 2.0 to 2000 mg/kg BW, except for IMPW, for which the lowest dose was 20 mg/kg BW. Neither HSW nor HMPW accumulated in any tissues or produced any effects. The typical effects observed included focal hystiocytosis, increase in the weight of the liver, lymph nodes, spleen and kidney, granulomas or microgranulomas in the liver, and hematological changes typical of a mild, chronic, inflammatory reaction. Biochemical changes exhibited mild hepatic damage.[6]

See the *Table*.

Long-term Toxicity. Groups of 50 males and 50 females of F344 rats were given medium-viscosity liquid paraffin at dietary doses of 2.5 or 5.0% for 104 weeks. Slight increases in food consumption and BW gain were observed in both sexes of the 5.0% group. However, no significant differences between the control and treated groups were noted with regard to clinical signs, mortality, and hematology findings. Granulomatous inflammation in the mesenteric lymph nodes, considered to be a reaction to paraffin absorption, was observed with similar incidence and severity in both sexes of the 2.5 and 5.0% groups.[7]

Table. NOELs and ADI for paraffin waxes and microcrystalline waxes tested in 90-day studies in F344 rats.[6]

Substances	*NOEL (mg/kg BW)*	*ADI (mg/kg BW)*
LMPW	<2	ADI withdrawn
IMPW	<2	ADI withdrawn
HSW	2000	0-20
HNPW	2000	0-20

Carcinogenicity. The carcinogenicity of medium-viscosity liquid paraffin was examined in F344 rats. No statistically significant increase in the incidence of any tumor type was found for either sex in the treated groups. It is concluded that under the present experimental conditions, the high dose, about 2,000 to 200,000 times higher than the current temporary ADI, does not have any carcinogenic potential in F344 rats. Granulomatous inflammation observed in mesenteric lymph nodes were not associated with any development of neoplastic lesions.[7]

Mutagenicity.

In vitro genotoxicity. P. was found to be negative in *Salmonella* mutagenicity assay.[4,8]

Chemobiokinetics. Hydrocarbon waxes consumed in the diet are not absorbed or metabolized in significant amounts.

Regulations. ***U.S. FDA*** (1998) listed P. for use (1) in adhesives as a component of articles intended for use in packaging, transporting, or holding food in accordance with the conditions prescribed in 21 CFR part 175.105; (2) in the manufacture of resinous and polymeric coatings for polyolefin films to be safely used as a food-contact surface in accordance with the conditions prescribed in 21 CFR part 175.320; (3) in the manufacture of resinous and polymeric coatings for the food-contact surface of articles intended for use in producing, manufacturing, packing, processing, preparing, treating, packaging, transporting, or holding food in accordance with the conditions prescribed in 21 CFR part 175.300; (4) as a component of the uncoated or coated food-contact surface of paper and paperboard intended for use in contact with aqueous and fatty food in accordance with the conditions prescribed in 21 CFR part 176.170; (5) as a defoaming agent that may be safely used as a component of articles or in the manufacture of paper and paperboard intended for use in contact with food in accordance with the conditions prescribed in 21 CFR part 176.200; (6) in the manufacture of cellophane to be safely used for packaging food in accordance with the conditions prescribed in 21 CFR part 176.1200; (7) in cross-linked polyester resins to be safely used as articles or components of articles intended for repeated use in contact with food in accordance with the conditions prescribed in 21 CFR part 177.2420; (8) as a constituent of acrylate ester copolymer coating which may safely be used as a food-contact surface of articles intended for packaging and holding food, including heating of prepared food (P.) in accordance with the conditions prescribed by 21 CFR part 175.210; and (9) as a plasticizer in rubber articles intended for repeated use in producing, manufacturing, packing, processing, treating, packaging, transporting, or holding food (total not to exceed 30% by weight of the rubber products) in accordance with the conditions prescribed in 21 CFR part 178.3740. P. may contain any oxidant permitted in food, a total of more than 1.0% by weight of residues of the following polymers: homopolymers and/or copolymers derived from one or more of the mixed *n*-alkyl (C_{12} - C_{18}) methacrylate esters where the C_{12} and C_{14} alkyl groups are derived from coconut oil and the C_{16} and C_{18} groups are derived from tallow; 2-hydroxy-4-*n*-octoxybenzophenone at level not to exceed 0.01% by weight of the P.; poly(alkylacrylate) (CAS No 27029-57-8) as a processing aid in the manufacture of P.

Synthetic P. may be safely used in application and under the same conditions where naturally derived PW is permitted as a component of food in accordance with the conditions prescribed in 21 CFR part 172.275.

Reinforced wax (RW) may include (1) *GRAS* substances; (2) substances subjected to CFR sanctions for use in RW; (3) copolymers of isobutylene modified with isoprene, PW (Types I and II), polyethylene, rosins and rosin derivatives, synthetic wax polymer (not to exceed 5.0% by weight of the PW).

References:

1. *Code of Federal Regulations* 21, part 178.3710, 1998.
2. Varner, S. L., Hollifield, H. C., and Andrzejewski, D., Determination of benzene in polypropylene food-packaging materials and food-contact paraffin waxes, *J. Assoc., Off. Anal. Chem.*, 74, 367, 1991.
3. Firriolo, J. M., Morris, C. F., Trimmer, G. W., Twitty, L. D., Smith, J. H., Freeman, J. J., Comparative 90-day feeding study with low-viscosity white mineral oil in Fischer-344 and Sprague-Dawley-derived CRL:CD rats, *Toxicol. Pathol.*, 23, 26, 1995.
4. Toyoda, K., Kawanishi, T., Uneyama, C., and Takahashi, M., Subchronic toxicity study of liquid paraffin in F344 rats, Abstract, *Eisei Shikenjo Hokoku*, 112, 64, 1994 (in Japanese).
5. Smith, J. H., Mallett, A. K., Priston, R. A., Brantom, P. G., Worrell, N. R., Sexsmith, C., and Simpson, B. J., Ninety-day feeding study in Fischer-344 rats of highly refined petroleum-derived food-grade white oils and waxes, *Toxicol. Pathol.*, 24, 214, 1996.
6. *Evaluation of Certain Food Additives and Contaminants,* The 44th Report of the Joint FAO/WHO Expert Committee on Food Additives, Geneva, 1995, 18.
7. Shoda, T., Toyoda, K., Uneyama, C., Takada, K., and Takahashi, M., Lack of carcinogenicity of medium-viscosity liquid paraffin given in the diet to F344 rats, *Food Chem. Toxicol,* 35, 1181, 1997.
8. De Flora, S., De Renzi G.P., Camoirano, A., Astengo, M., Basso, C., Zanacchi, P., and Bennicelli, C., Genotoxicity assay of oil dispersants in bacteria (mutation, differential lethality, SOS DNA-repair) and yeast (mitotic crossing-over), *Mutat. Res.,* 158, 19, 1985.

PARAFORMALDEHYDE

Molecular Formula. $[\sim CH_2O\sim]_n$
M = 30,000-40,000
CAS No 30525-89-4
RTECS No RV0540000
Abbreviation. PFA.

Trade Names. Aldacide; Formagene; Paraform.

Composition. Polyethers of linear structure, the products of polymerization of formaldehyde or its cyclic trimer (trioxane).

Properties. White crystalline or amorphous powder with light pungent odor of formaldehyde. Slowly soluble in cold water, more readily soluble in hot water; insoluble in alcohol.[020]

Applications. PFA is used in manufacture of aminoresins for coatings, in synthesis of phenolic, urea, resorcinol, and melamine resins when high solids contents are required.[1] Ingredient for adhesives.

Migration Data. PFA produced a significant change in the organoleptic properties of contact water. Migration of low-molecular-mass organic compounds was observed (oxidizability of extracts reached 44 mg O_2/l). Migration of the monomer was not carried out.[2]

Acute Toxicity. LD_{50} is 0.5 g/kg BW in mice, and 5.0 g/kg BW in rats.[07]

Mutagenicity. PFA caused a weak mutagenic effect in plant assays but not in bacterial or mammalian tests.[3]

Regulations. ***U.S. FDA*** (1998) approved the use of PFA (1) in adhesives as a component of articles intended for use in packaging, transporting, or holding food in accordance with the conditions prescribed in 21 CFR part 175.105; (2) as a component of the uncoated or coated food-contact surface of paper and paperboard intended for use in producing, manufacturing, packaging, processing, preparing, treating, packing, transporting, or holding aqueous and fatty foods in accordance with the conditions prescribed in 21 CFR part 176.170; and (3) as a basic component of single and repeated use food contact surfaces (up to 1.0% by weight of closure-sealing gasket composition for food containers intended for use in producing, manufacturing, packaging, processing, preparing, treating, packing, transporting, or holding dry food in accordance with the conditions prescribed in 21 CFR part 177.1210.

References:
1. Kirk-Othmer's *Encyclopedia of Chemical Technology*, 3rd ed., vol 1, M. Grayson and D. Eckroth, Eds., John Wiley & Sons, Inc., New York, 1978, 722.
2. Aksyuk, A. F. and Ashmarina, N. P., Hygienic evaluation of polyformaldehyde articles used in water supply, in *Hygiene Problems in the Production and Use of Polymeric Materials*, Moscow, 1969, 151 (in Russian).
3. Alexseyenok, A. I., The genetic activity of an alkaline solution of formaldehyde, *Tsitologia i Genetika,* 22, 25, 1988 (in Russian).

PENTACHLOROBENZENETHIOL

Molecular Formula. C_6HCl_5S
M = 282.38
CAS No 133-49-3
RTECS No DC1925000
Abbreviation. PCBT.

Synonym. Pentachlorothiophenol.

Acute Toxicity. LD_{50} *i/p* is 100 mg/kg BW in mice.[1]

Reproductive Toxicity. PCBT was not found to be ***embryotoxic*** to the chicken embryo.[2]

Chemobiokinetics. PCBT was determined in kidneys, spleen, and heart of rats given orally 770 mg hexachlorobenzene/kg BW for 1 month.[3]

Regulations. ***U.S. FDA*** (1998) approved the use of PCBT in adhesives as a component of articles intended for use in packaging, transporting, or holding food in accordance with the conditions prescribed in 21 CFR part 175.105.

References:
1. Natl. Technical Information Service AD277-689.
2. Korhonen, A. et al., Embryotoxicity of benzothiazoles, benzenesulfohydrazide, and dithiomorpholine to the chicken embryo, *Arch. Environ. Contam. Toxicol.*, 11, 753, 1982.
3. Richter, E. et al., Differences in the biotransformation of hexachlorobenzene (HCB) in male and female rats, *Chemosphere*, 10, 779, 1981.

PENTACHLOROPHENOL

Molecular Formula. C_6HCl_5O
M = 266.35
CAS No 87-86-5
RTECS No SM6300000
Abbreviation. PCP.

Synonym and **Trade Names.** Acutox; Chlorophen; Dowicide 7; Penchlorol; Penta; Pentacon; Pentasol; Permacide; Permasan; Permatox; Santophen.

Applications. PCP has been used as wood preservative, fungicide, and molluscicide. Chemical intermediate.

Exposure. PCP residues have been found worldwide in soil, water, and air samples; in food products; and in human and animal tissues and body fluids.

Properties. White monoclinic, crystalline solid with phenolic odor, very pungent when hot.[020] Taste perception threshold is 0.03 mg/l (U.S. EPA, 1980). Water solubility is 14 mg/l at 20°C and 20 mg/l at 30°C.

Migration Data. A study of the seals and enamel of Mason lids containing PCP revealed PCP in all the canned foods. Levels were as high as 17 µg/l. Recoveries from foods fortified with PCP ranged from 85 to 94%. The presence of PCP was confirmed by GLC/MS determination of the derivative, pentachloroanisole.[1]

Acute Toxicity. LD_{50} is 146 mg/kg BW in male rats, and 175 mg/kg BW in female rats.[020]

Repeated Exposure. $B6C3F_1$ mice were fed diets containing 20 to 12,500 ppm PCP for 30 days. A decrease in BW and deaths of animals occurred at the highest dose. Histological changes were noted in the liver.[2]

F344 male rats were exposed by oral gavage to PCP given in olive oil twice weekly for 28 days at the doses of 2.0 mg/kg BW per treatment. An increase in BW and in liver and kidney relative weights has been noted.[3]

Male and female F344/N rats were given daily doses of approximately 20, 40, 75, 150, and 270 mg PCP/kg in feed for 28 days. The treatment caused retardation of BW gain and an increase in absolute and relative liver weights of animals at higher dose levels (NTP, 1997). The rate of minimal to mild hepatocyte degeneration in males and females exposed to 40 mg PCP/kg BW or greater and the incidences of centrilobular hepatocyte hypertrophy in the 270 mg PCP/kg BW groups was increased.[4]

Short-term Toxicity. In a 160-day study, cattle fed 20 mg technical PCP/kg BW for 42 days, followed by 15 mg/kg BW for the remainder of the study, developed a decrease in BW, progressive anemia, and immune effects. Only minimal adverse effects were observed after exposure to analytical grade PCP.[5]

Long-term Toxicity. Oral exposure of rats to 30 mg PCP/kg BW caused retardation of BW gain, an increase in specific gravity of the urine in females, and pigmentation of the liver and kidneys in males. Administration of 10 mg/kg BW dose result in pigmentation of the liver and kidneys in females. The NOAEL of 3.0 mg/kg BW was established in this study.[6]

B6C3F$_1$ mice were given diets containing 200 to 1800 ppm various grades of PCP for six months. Histological examination by the end of this treatment revealed hepatocellular karyomegaly, cytomegaly, and degeneration in the liver. Urinary bladder tissues were also affected. In a 2-year feeding study, compound-related nonneoplastic lesions occurred in the liver, spleen, and nose of B6C3F$_1$ mice exposed to either technical grade PCP or EC-7, a technical grade PCP formulation.[2]

RfD is considered to be 0.03 mg/kg BW (U.S. EPA).

Reproductive Toxicity.

Embryotoxicity. Small amounts of PCP cross the placenta.[7] Pregnant Charles River CD rats received 60 mg/kg BW doses during organogenesis. Embryotoxic effect was noted.[5]

Rats were gavaged with 5.0 to 50 mg/kg BW on days 6 to 15 of gestation. A dose of 5.0 mg/kg BW produced no effect, 15 mg/kg BW and higher doses caused embryolethality and toxicity effects.[8]

Pregnant hamsters received oral doses of 1.25 to 20 mg/kg BW and exhibited resorptions and embryotoxicity.[9]

Embryo- and maternal toxicity in rats was observed on oral exposure to 30 mg/kg BW dose.[10]

Teratogenicity. Rats were given 4.0, 13, or 43 mg/kg BW in the diet during mating and pregnancy. The treatment resulted in fetal weight reduction and skeletal variations. At 43 mg/kg BW dose level, maternal lethality was observed.[11]

PCP did not increase congenital malformations in mice.[12] No teratogenic effects were reported in the study of Schwetz et al. (1978).[3]

Immunotoxicity. In the above-cited study, PCP was found to affect immune system functions in F344 male rats. It enhanced *T*-lymphocyte blastomogenesis and suppressed the antibody response against sheep erythrocytes by 39%.[3]

Mutagenicity.

Observations in man. PCP has been found in the semen of male workers and has been associated with CA frequency in the lymphocytes of these workers.[13]

In vivo cytogenetics. PCP was shown to be negative in sex-linked lethal mutation test in *Dr. melanogaster.*[14]

No increase in the frequency of micronucleated erythrocytes was noted in bone marrow of male rats or mice administered PCP by *i/p* injection three times at 24 hour intervals. The highest dose administered to rats (75 mg/kg BW) and mice (150 mg/kg BW) was lethal (NTP, 1997).

In vitro genotoxicity. PCP was found to be negative *in Salmonella* mutagenicity assay. PCP produced an increase in CA in cultured mammalian cells only in presence of metabolic activation, and an increase in SCE only in absence of metabolic activation.[2]

Carcinogenicity.

Observations in man. Some epidemiological observations suggest that exposure to CPs in general and PCP solutions in particular may result in an increased risk for certain malignant disorders such as nasal carcinoma and soft tissue sarcoma.[15]

Animal studies. Mice were fed 100 and 200 ppm technical grade PCP or 100, 200, and 600 ppm Dowicide EC-7 for 2 years. The treatment with the both types of PCP caused an increase in the incidence of adrenal medullary and hepatocellular neoplasms in male mice. The same result was observed in females treated with PCP EC-7.[2]

Sprague-Dawley rats were fed diets containing 8 to 231 ppm 90% pure PCP (Dowicide EC-7) for 22 to 24 months (this corresponds with 1.0 to 30 mg PCP/kg BW). The treatment did not affect survival. A slight increase in pheochromocytomas of the adrenal medulla was noted at the lower dose tested. There was no significant increase in tumor incidence.[6]

No carcinogenic effect was noted in mice that received, by stomach tube, PCP as the commercial product Dowicide-7 at the dose of 46.4 mg/kg BW from 7th day up to 4 weeks of age; subsequently, the mice were fed 130 mg/kg diet for 74 weeks.[16]

F344/N rats were given daily doses of approximately 20, 40, or 75 mg PCP/kg BW for 2 years. A stop-exposure group of rats received 1000 ppm of PCP in feed (about 100 mg PCP/kg BW) for 52 weeks and control feed thereafter for the remainder of the 2-year studies. Based on the increased incidences of mesotheliomas and nasal tumors, there was some evidence of carcinogenic activity of PCP in male rats given a diet containing 1000 ppm for 1 year followed by control diet for 1 year. There was no evidence of PCP carcinogenic activity in stop-exposure female rats.[4]

Carcinogenicity classifications. An IARC Working Group concluded that there is sufficient evidence of carcinogenicity of PCP in *experimental animals* and there is inadequate evidence to evaluate the carcinogenicity of PCP in *humans*.

IARC: 2B;

EPA: B2;

NTP: XX - XX - CE* - CE* (*Dowicide EC-7*, feed);
XX - XX - CE* - SE* (*PCP*, technical grade, feed);
NE - NE - XX - XX (200 to 600 ppm in the diet for 2 years);
SE - NE - XX - XX (1000 ppm in the diet for 1 year).

Chemobiokinetics. PCP is rapidly absorbed in rodents, monkeys, and humans following oral exposure. In humans, accumulation occurred predominantly in the liver, kidney, brain, spleen, and fat. In the mouse, the main storage site is the gall bladder; in the rat, it is the kidney.[5] In rats, dechlorination of PCP is mediated by microsomal enzymes. A single oral administration of 25 mg PCP/kg BW affected hepatic and epididymal glutathione but have insignificant effect on testicular levels.[17]

PCP toxic action depends on its interference with oxidative phosphorylation. The occurrence of *tetrachlohydroquinone* as a metabolite of PCP in humans is still questionable.[18] PCP is excreted in the urine as free PCP or tetrachlorohydroquinone, although *glucuronide conjugates* have also been identified.[14]

Regulations. *U.S. FDA* (1998) approved the use of PCP in adhesives as a component of articles intended for use in packaging, transporting, or holding food in accordance with the conditions prescribed in 21 CFR part 175.105.

Recommendations. *U.S. EPA* (1999). Health Advisory for a longer-term exposure is 1.0 mg/l.

Standards.

Canada (1989). MAC: 0.06 mg/l.

Russia (1995). MAC: 0.01 mg/l.

References:

1. Heikes, D. L. and Griffitt, K. R., Gas-liquid chromatographic determination of pentachlorophenol in Mason jar lids and home canned foods, *J. Assoc. Off. Anal. Chem.*, 63, 1125, 1980.
2. *Toxicology and Carcinogenesis Studies of Two Pentachlorophenol Technical-Grade Mixtures in B6C3F$_1$ Mice*, NTP Technical Report Series No 349, Research Triangle Park, NC, NIH Publ. No 89-2804, 1989.
3. Blackley, B. R., Yole, M. J., Broussen, P., Boermans, H., and Fournier, M., Effect of pentachlorophenol on immune system, *Toxicology*, 125, 141, 1998.
4. Chhabra, R. S., Maronpot, R. M., Bucher, J. R., Haseman, J. K., Toft, J. D., and Hejtmancik, M. R., Toxicology and carcinogenesis studies of pentachlorophenol in rats, *Toxicol. Sci.*, 48, 14, 1999.

5. *Drinking Water and Health*, vol 6, National Research Council, Natl. Academy Press, Washington, D.C., 1986, 392.
6. Schwetz, B. A., Quast, J. F., Keelev, P. A., Humiston, C. G. and Kociba, R. J., Results of 2-year toxicity and reproduction studies of pentachlorophenol in rats, in *Pentachlorophenol: Chemistry, Pharmacology and Environmental Toxicology*, K. R. Rao, Ed., Plenum Press, New York, 1978, 301.
7. Larsen, R. V., Born, G. S., Kessler, W. V., Shaw, S. M., and van Sicle, D. C., Placental transfer and teratology of pentachlorophenol in rats, *Environ. Lett.*, 10, 121, 1975.
8. Schwetz, B. A., Keeler, P. A., and Gehring, P. J., The effect of purified and commercial grade pentachlorophenol on rat embryonal and fetal development, *Toxicol. Appl. Pharmacol.*, 28, 151, 1974.
9. Hinkle, D. K., Fetotoxic effects of pentachlorophenol in the golden Syrian hamster, *Toxicol. Appl. Pharmacol.*, 25, 455, 1973.
10. Schwetz, B. A. and Gehring, P. J., The effect of tetrachlorophenol and pentachlorophenol on rat embryonal and fetal development, *Toxicol. Appl. Pharmacol.*, 25, 455, 1973.
11. Welsh, J. J., Collins, T. F. X., Black, T. N., Graham, S. L., and O'Donnell, M. W., Teratogenic potential of purified pentachlorophenol and pentachloroanisole in subchronically exposed Sprague-Dawley rats, *Food Chem. Toxicol.*, 25, 163, 1987.
12. Courtney, K. D. et al., Teratogenic evaluation of pesticides: A large-scale screening study, *Teratology*, 3, 199, 1970.
13. Schrag, S. D. and Dixon, R. L., Occupational exposures associated with male reproductive dysfunction, *Ann. Rev. Pharmacol. Toxicol.*, 25, 567, 1985.
14. Vogel, E. and Chandler, J. L. R., Mutagenicity testing of cyclomate and some pesticides in *Drosophila melanogaster*, *Experientia*, 30, 621, 1974.
15. Jorens, P. G. and Schepens, P. J., Human pentachlorophenol poisoning, *Hum. Exp. Toxicol.*, 12, 479, 1993.
16. Innes, J. R. M., Ulland, B. M., Valerio, M. G., et al., Bioassay of pesticides and industrial chemicals for tumorigenicity in mice. A preliminary note, *J. Natl. Cancer Inst.*, 42, 1101, 1969.
17. Gandy, J., Millner, G. C., Bates, H. K., Casciano, D. A., and Harbison, R. D., Effects of selected chemicals on the glutathione status in the male reproductive system of rats, *J. Toxicol. Environ. Health*, 29, 45, 1990.
18. Ahlborg, U. G. and Thuberg, T., Effects of 2, 3, 7, 8-tetrachlorodibenzo-*p*-dioxin in the *in vivo* and *in vitro* dechlorination of pentachlorophenol, *Arch. Toxicol.*, 40, 259,1978.

PENTACHLOROPHENOL, POTASSIUM SALT

Molecular Formula. $C_6HC_{15}O.K$
M = 305.42
CAS No 7778-73-6
RTECS No SM6445000

Synonyms. Potassium pentachlorophenate; Potassium pentachlorophenoxide.

Regulations. ***U.S. FDA*** (1998) approved the use of P. in adhesives as a component (monomer) of articles intended for use in packaging, transporting, or holding food in accordance with the conditions prescribed in 21 CFR part 175.105.

PENTACHLOROPHENOL, SODIUM SALT

Molecular Formula. $C_6C1_5O.Na$
M = 288.30
CAS No 131-52-2
RTECS No SM6490000
Abbreviation. PCPS.

Synonyms and **Trade Names.** Dowicide G or G-ST; Mystox D; Sodium pentachlorophenol; Sodium pentachlorophenoxide; Whitophen.

Properties. Buff colored flakes or white powder. Soluble in water (33% at 25°C) and ethanol.

Applications. PCPS is used in the production of adhesives-antimicrobial protection of adhesives based on starch, vegetable, and animal proteins, in construction materials, in water treatment for control of algal, fungal, and bacterial-induced slimes in industrial recirculating water.

Acute Toxicity. LD_{50} is 197 mg/kg BW in rats.[1]

Mutagenicity.

In vivo cytogenicity. PCPS was negative in *Dr. melanogaster* assay.[2]

Reproductive Toxicity.

Pregnant Sprague-Dawley rats were given 5.0 mg pure PCP/kg BW and more. The treatment caused ***embryotoxic effect*** and produced delayed ossification of skull. Hamsters received PCP on days 5 to 10 of gestation. Fetal death and/or resorptions were observed in animals exposed to 5.0 mg/kg BW dose.[3]

See also ***Pentachlorophenol.***

Carcinogenicity classification. An IARC Working Group concluded that there is sufficient evidence of carcinogenicity of PCPS in *experimental animals* and there is inadequate evidence to evaluate the carcinogenicity of PCPS in *humans*.

IARC: 2B

Regulations. ***U.S. FDA*** (1998) approved the use of PCPS (1) in adhesives as a component (monomer) of articles intended for use in packaging, transporting, or holding food in accordance with the conditions prescribed in 21 CFR part 175.105; and (2) as an adjuvant substance in the manufacture of foamed plastics intended for use in contact with food, subject to the provisions prescribed in 21 CFR 178.3010.

References:

1. Izmerov, N. F., Sanotsky, I. V., and Sidorov, K. K., *Toxicometric Parameters of Industrial Toxic Chemicals under Single Exposure*, USSR/SCST Commission for UNEP, Center Int. Projects, Moscow, 1982, 106 (in Russian).
2. Shirasu, Y., Moriya, M., Kato, K., Furuhashi, A., and Kada, T., Mutagenicity screening of pesticides in the microbial system, *Mutat. Res.*, 40, 19, 1976.
3. *Drinking Water & Health*, vol 1, Natl. Res. Council, Natl Acad. Press, Washington, D.C., 1977, 753.

PENTAERYTHRITOL, TETRABENZOATE

Molecular Formula. $C_{38}H_{28}O_8$

M = 552.61

CAS No 4196-86-5

RTECS No RZ2605000

Abbreviation. PETB.

Synonyms. Benzoic acid, tetraester with pentaerythritol; 2,2'-Bis[(benzoyloxy)methyl]-1,3-propanediol dibenzoate.

Acute Toxicity. LD_{50} is 1.16 g/kg BW in rats.

Regulations.

U.S. FDA (1998) approved the use of PETB in adhesives as a component of articles intended for use in packaging, transporting, or holding food in accordance with the conditions prescribed in 21 CFR part 175.105.

Great Britain (1998). Pentaerythritol is authorized without time limit for use in the production of polymeric materials and articles in contact with food or drink or intended for such contact.

PENTANE

Molecular Formula. C_5H_{12}

M = 72.15.

CAS No 109-66-0

RTECS No RZ9450000

Synonym. *n*-Pentane.

Properties. Colorless, flammable liquid with a gasoline-like odor. Slightly soluble in water; miscible with alcohol.

Exposure. P. is a hydrocarbon solvent with an estimated production volume of 50,000 metric tons in Europe. It is contained in liquefied petroleum gas and natural gas as trace constituents.

Acute Toxicity. P. was not found to be acutely toxic by oral or inhalation routes.[1]

Repeated Exposure. P. did not exhibit cumulative toxicity at levels up to 20 g/m^3. The absence of any demonstratable toxicity of P. at high treatment levels indicates that the risk of its adverse health effects is minimal.[2]

Allergenic Effect. P. did not induce skin sensitization.

Reproductive Toxicity. P. did not induce developmental toxicity.[2]

Mutagenicity. P. was not found to be positive in *Salmonella* mutagenicity assay.[3]

Chemobiokinetics. ICR mice were exposed to about 5.0% P. for 1 hour. It was suggested that 2-P. was metabolized predominantly by liver microsomes. Metabolites in the blood and liver tissue were *2-pentanol, 3-pentanol*, and *2-pentanone*.[1]

Regulations. ***U.S. FDA*** (1998) approved the use of P. as an adjuvant which may be used in the manufacture of foamed plastics intended for use in contact with food, subject to the provisions prescribed in 21 CFR 178.3010.

References:

1. Chiba, S. and Oshida, S., Metabolism and toxicity of *n*-pentane and isopentane, Abstract, *Nippon Hoigaku Zasshi*, 45, 128, 1991 (in Japanese).
2. McKee, R., Frank, E., Heath, J., Owen, D., Przygoda, R., Trimmer, G., and Whitman, F., Toxicology of *n*-pentane, *J. Appl. Toxicol.*, 18, 431, 1998.
3. Kirwin, C. J. and Thomas, W. C., *In vitro* microbiological mutagenicity studies in hydrocarbon propel lants, *J. Soc. Cosmet. Chem.*, 31, 367, 1980.

2,4-PENTANEDIONE

Molecular Formula. $C_5H_8O_2$

M = 100.13

CAS No 123-54-6

RTECS No SA1925000

Abbreviation. PD.

Synonyms. Acetoacetone; Acethyl acetone; Acetyl-2-propanone; Diacetylmethane.

Properties. Colorless or slightly yellow, flammable liquid with pleasant, acetone-like odor. Soluble in water (1:8); miscible with alcohol.

Acute Toxicity. LD_{50} is 55 mg/kg BW in rats and 950 mg/kg BW in mice.[03] Poisoning caused CNS depression, and rigidity.[1]

Repeated Exposure. Administration by Lim method caused 70% mortality in rats.[1]

Reproductive Toxicity. Timed-pregnant F344 rats were exposed from days 6 to 15 of gestation to vapors of PD. Maternal toxicity and fetotoxicity were observed at the concentration of 398 ppm PD.

Embryotoxicity and ***teratogenicity*** were not observed at any concentration tested. The concentration of 53 ppm was ineffective for both maternal and developmental toxicity.[2,3]

Mutagenicity.

In vivo cytogenetics. F344 rats were exposed by inhalation to the PD vapor. A weak dominant lethal effect was observed in rats.[3]

In vitro genotoxicity. PD showed mutagenic effect in bacterial assay.[4]

Regulations. ***U.S. FDA*** (1998) approved the use of PD in adhesives as a component of articles intended for use in packaging, transporting, or holding food in accordance with the conditions prescribed in 21 CFR part 175.105.

References:

1. Berezina, Z. V., Toxicity of acetyl acetone, *Gig. Sanit.*, 10, 88, 1987 (in Russian).

2. Tyl, R. W., Ballantine, B., Fisher, L. C., Tarasi, D. J., and Dodd, D. E., Dominant lethal assay of 2,4-pentanedione vapor in Fischer 344 rats, *Toxicol. Ind. Health*, 5, 463, 1989.
3. Tyl, R. W., Ballantyne, B., Pritts, I. M., Garman, R. H., Fisher, L. S., France, K. A., and McNael, D. J., An evaluation of the developmental toxicity of 2,4-pentanedione in the Fischer 344 rat by vapor exposure, *Toxicol. Ind. Health*, 6, 461, 1990.
4. Gava, C. et al., Genotoxic potentiality and DNA-binding properties of acetylacetone, maltol, and their aluminum (III) and chromium (III) neutral complexes, *Toxicol. Environ. Chem.*, 2, 149, 1989.

p-(*tert*-PENTYL)PHENOL

Molecular Formula. $C_{11}H_{16}O$
M = 164.27
CAS No 80-46-6
RTECS No SM6825000
Abbreviation. PTPP.

Synonyms and **Trade Names.** *p-tert*-Amylphenol; 4-(1,1-Dimethylpropyl)phenol; Methyl-2-*p*-hydroxy-phenylbutane; Pentaphen.

Properties. White crystals. Practically insoluble in water; soluble in alcohol.

Applications. Used in the manufacture of oil-soluble resins.

Acute Toxicity.

Observations in man. Probable oral lethal dose is in the range of 0.5 to 5.0 g/kg BW.[022]

Animal studies. LD_{50} is 1.83 g/kg BW in rats.

Regulations.

EU (1990). PTPP is available in the *List of monomers and other starting substances that may continue to be used pending a decision on inclusion in Section A* (*Section B*).

U.S. FDA (1998) approved the use of PTPP in the manufacture of phenolic resins for food-contact surface of molded articles intended for repeated use in contact with nonacid food (*pH* above 5.0) in accordance with the conditions prescribed in 21 CFR part 177.2410.

Standards. *Russia* (1995). MAC: 0.05 mg/l (organolept., color).

PERBORIC ACID, SODIUM SALT

Molecular Formula. $BHO_3.Na$
M = 81.80
CAS No 7632-04-4
RTECS No SC7310000
Abbreviation. PAS.

Synonyms and **Trade Name.** Dexol; Sodium perborate; Sodium peroxoborate.

Properties. White and odorless, amorphous powder with a saline taste.

Applications. Used as antiseptic or bacteriostat. Bleaching agent in the textile manufacture and in laundry detergents.

Acute Toxicity. LD_{50} is 3.25 g/kg BW in mice. Poisoning produced changes in motor activity and GI tract.

Long-term Toxicity. Chronic ingestion of borates causes anorexia, retardation of BW gain, vomiting, mild diarrhea, skin rash, alopecia, convulsions, and anemia. Kidneys are predominantly affected due to highest concentration reached during excretion.[1]

Reproductive Toxicity. Rats and dogs received PAS with their feed. Accumulation occurs in the testes; germ cell depletion and testicular atrophy were reported.[019]

Mutagenicity. PAS is regarded as a direct-acting *in vitro* mutagen.[2]

Chemobiokinetics. Borates are not metabolized, neither do they accumulate in the body except for low deposits in bone. No *organic boron compounds* have been reported as metabolites. Excretion occurs primarily in the urine.

Regulations. *U.S. FDA* (1998) approved the use of PAS (1) in resinous and polymeric coatings used as the food-contact surfaces of articles intended for use in producing, manufacturing, packing, processing, preparing, treating, packaging, transporting, or holding food in accordance with the conditions prescribed in 21 CFR part 175.300, and (2) in adhesives as a component of articles intended for use in packaging, transporting, or holding food in accordance with the conditions prescribed in 21 CFR part 175.105.

References:

1. Dreibach, R. H., *Handbook of Poisoning*, 12th ed., Appleton & Lange, Norwalk, CT, 1987, 360.
2. Seiler, J. P., The mutagenic activity of sodium perborate, *Mutat. Res.*, 224, 219, 1989.

PERILLA OIL

CAS No 68132-21-8
RTECS No SD4384000
Abbreviation. PO.

Synonym. Perilla frutescens oil.

Properties. A clear yellowish liquid.

Applications. Used in fragrances.

Acute Toxicity. LD_{50} is 2.77 g/kg BW in mice, and exceeded 5.0 g/kg BW in rats (Moreno, 1980).[1]

Allergenic Effect. No sensitization reactions were noted in volunteers.

Mutagenicity. PO was negative in *Salmonella*/microsome reversion assay with and without *S9* activation system.[2] A dose of 0.04 mg PO/kg BW produced positive effects in a CA test using Chinese hamster fibroblast cell lines.[3]

Regulations. ***U.S. FDA*** (1998) approved the use of PO in resinous and polymeric coatings used as the food-contact surfaces of articles intended for use in producing, manufacturing, packing, processing, preparing, treating, packaging, transporting, or holding food in accordance with the conditions prescribed in 21 CFR part 175.300.

References:

1. Perilla oil, Fragrance Raw Material Monographs, *Food Chem. Toxicol.*, 26, 397, 1988.
2. Morimoto, I., Watanabe, F., Osawa, T., and Okitsu, T., Mutagenicity screening of crude drugs with *Bacillus subtilis rec*-assay and *Salmonella*/microsome reversion assay, *Mutat. Res.*, 97, 81, 1982
3. Ishidate, M., Sofuni, T., Yoshikawa, K., Hayashi, M., Nohmi, T., Sawada, M., and Matsuoka, A., Primary mutagenicity screening of food additives currently used in Japan, *Food Chem. Toxicol.*, 22, 623, 1984.

PERMANGANIC ACID, POTASSIUM SALT

Molecular Formula. $MnO_4.K$
M = 158.04
CAS No 7722-64-7
RTECS No SD6475000
Abbreviation. PPM.

Synonyms and **Trade Names.** Chameleon mineral; Condy's crystals; Permanganate of potash; Potassium permanganate.

Properties. Dark purple or bronze-like, odorless crystals which have a sweet with astringent aftertaste.[020] Water solubility reaches 63.8 g/l at 20°C.

Applications. Used as human drugs (in tablets), as well as a sanitizer, disinfectant, bactericide, topical antiseptic. Bleaching agent in purifying water.

Acute Toxicity. LD_{50} is 1.09 g/kg BW in rats,[1] 0.75 g/kg BW in mice, and 1.15 g/kg BW in guinea pigs. Poisoning caused CNS inhibition.[2]

Mutagenicity. The major mutagenic agent generated by PPM solutions was found to be Mn^{2+}.[3]

In vivo cytogenetics. Mice received PPM orally for 3 weeks. The treatment caused a significant increase in CA frequency and micronuclei formation. Sperm-head abnormalities showed a significant enhancement as well.[4]

In vitro genotoxicity. PPM was less strongly clastogenic than $MnSO_4$. It appeared to be negative in *Salmonella* mutagenicity assay,[5] but positive in DNA-cell-binding assay.[6]

Regulations. ***U.S. FDA*** (1998) approved the use of PPM (1) in adhesives as a component of articles intended for use in packaging, transporting, or holding food, in accordance with the conditions prescribed in 21 CFR part 175.105; and (2) as a component of sanitizing solutions used on food-processing equipment and utensils, and on other food-contact articles as specified in 21 CFR part 178.1010.

References:

1. Smyth, H. F., Carpenter, C. P., Weil, C. S., Pozzani, U. C., Striegel, J. A., and Nyaun, J. S., Range-finding toxicity data: List VII, *Am. Ind. Hyg. Assoc. J.*, 30, 470, 1969.
2. Shigan, S. A. and Vitvitskaya, B. R., Experimental proof of permissible residual concentrations of potassium permanganate in drinking water, *Gig Sanit.*, 9, 15, 1971 (in Russian).
3. De Meo, M., Laget, M., Castegnaro, M., and Dumenil, G., Genotoxic activity of potassium permanganate in acidic solutions, *Mutat. Res.*, 260, 295, 1991
4. Joardar, M. and Sharma, A., Comparison of clastogenicity of inorganic *Mn* administered in cationic and anionic forms *in vivo*, *Mutat. Res.*, 240, 159, 1990.
5. De Flora, S., Zannacchi, P., Camoriano, A., Bennicelli, C., and Badolati, G. S., Genotoxic activity and potency of 135 compounds in the Ames reversion test and in bacterial DNA-repair test, *Mutat. Res.*, 133, 161, 1984.
6. Kubinsky, H., Gutzke G. E., and Kubinsky, Z. O., DNA-cell-binding (DCB) assay for suspected carcino gens and mutagens, *Mutat. Res.*, 89, 95, 1981.

PEROXYDISULPHURIC ACID, SODIUM SALT

Molecular Formula. $O_8S_2.2Na$
M = 238.10
CAS No 7775-27-1
RTECS No SE0525000

Synonyms. Sodium peroxydisulfate; Sodium persulfate.

Acute Toxicity. Minimal lethal dose *i/v* was found to be 178 mg/kg BW in rabbits.[020]

Regulations. ***U.S. FDA*** (1998) approved the use of P. in adhesives as a component of articles intended for use in packaging, transporting, or holding food in accordance with the conditions prescribed in 21 CFR part 175.105.

PETROLATUM

CAS No 8009-03-8
RTECS No SE6780000

Synonyms and **Trade Names.** Cosmoline; Mineral fat; Mineral jelly.

Composition. P. is obtained from paraffin oils by deparafinisation. P. is a mixture of high-molecular-mass heavy hydrocarbons with residual oil and ceresin.

Properties. P. does not alter the chromacity of water even at a concentration of up to 100 mg/l.

Applications. Used for waterproofing.

Migration. Benzo(a)pyrene is not found in the samples of P.

Repeated Exposure failed to reveal cumulative properties. Mice that received 1.0 g/kg BW for 10 days exhibited neither mortality nor behavioral changes.[1]

Long-term Toxicity. Administration of a 10 mg/kg BW dose caused BW gain decrease and anemia. Gross pathology examination revealed a number of visceral changes. Rats given 2-day water extracts of P. did not exhibit signs of intoxication.[1]

Sprague-Dawley rats were fed diets containing 10% ground wax (petrolatum) for 2 years (polycyclic aromatic hydrocarbons contents up to 0.64 mg/kg). There were no effects on survival rates and BW gain. Histology examination revealed no other wax-associated toxic effects.[2]

Carcinogenicity. Some Amber P. contain trace amounts (about 6 ppm) of polycyclic aromatic hydrocarbons. Nevertheless, skin application of a representative sample revealed no evidence of carcinogenic action in mice.[2]

No increase in tumor incidence was noted in a 2-year study in Sprague-Dawley rats.[3]

Regulations. ***U.S. FDA*** (1998) regulates P. (1) in adhesives as a component of articles intended for use in packaging, transporting, or holding food in accordance with the conditions prescribed in 21 CFR part 175.105; (2) in preparation of pressure-sensitive adhesives which may be used as the food-contact surface of labels and/or tapes applied to raw fruit and raw vegetables in accordance with the conditions prescribed in 21 CFR part 175.125; (3) in the manufacture of resinous and polymeric coatings for the food-contact surface of articles intended for use in producing, manufacturing, packing, processing, preparing, treating, packaging, transporting, or holding food in accordance with the conditions prescribed in 21 CFR part 175.300; (4) in the manufacture of rubber articles intended for repeated use in producing, manufacturing, packaging, processing, preparing, treating, packing, transporting, or holding food in accordance with the conditions prescribed in 21 CFR part 177.2600; (5) as a plasticizer for rubber articles intended for repeated use in contact with food up to 30% by weight of the rubber product in accordance with the conditions prescribed in 21 CFR part 178.3740; (6) as lubricants with incidental food contact intended for used on machinery used for producing, manufacturing, packaging, processing, preparing, treating, packing, transporting, or holding food in accordance with the conditions prescribed in 21 CFR part 178.3570; (7) as a component of a defoaming agent intended for articles which may be used in producing, manufacturing, packing, processing, preparing, treating, packaging, transporting, or holding food in accordance with the conditions prescribed in 21 CFR part 176.200; (8) as a component of the uncoated or coated food-contact surface of paper and paperboard intended for use in producing, manufacturing, packaging, processing, preparing, treating, packing, transporting, or holding dry, aqueous, and fatty foods in accordance with the conditions prescribed in 21 CFR part 176.170; (9) in the manufacture of closures with sealing gaskets for food containers in accordance with the conditions prescribed in 21 CFR part 177.1210; and (10) as a substance employed in the production of or added to textiles and textile fibers intended for use in contact with food in accordance with the conditions prescribed in 21 CFR part 177.2800. P. may be safely used (11) as a component of non-food articles in contact with food (specifications in the *U.S. Pharmacopeia* XX for white P.). It is used or intended for use (12) as a protective coating of the surfaces of metal or wood tanks used in fermentation process (GMP requirements) and may contain any antioxidant permitted in food.

References:

1. Kudrin, L. V. et al., in *Health Protection in the Production and Use of Polymeric Materials*, F. F. Erisman Research Sanitary Hygiene Institute, Moscow, 1969, 139 (in Russian).
2. *Encyclopedia of Occupational Health and Safety*, International Labour Office, vols 1 and 2, McGraw-Hill Book Co., New York, 1971, 1488.
3. The 39th Report of Joint FAO/WHO Expert Committee on Food Additives, Technical Report, WHO, Geneva, 1992, 25.

PETROLEUM SULFONIC ACIDS, SODIUM SALTS

CAS No 68608-26-4
RTECS No WR7150000

Synonyms. Mineral oil sulfonic acids, sodium salts; Oil soluble petroleum sulfonate, sodium salt; Petroleum sulfonate, sodium salt.

Regulations. ***U.S. FDA*** (1998) approved the use of P. in the production of or added to textiles and textile fibers intended for use in producing, manufacturing, packing, processing, preparing, treating, packaging, transport ing, or holding food, subject to the provisions of 21 CFR part 177.2800.

PHENANTHRENE

Molecular Formula. $C_{14}H_{10}$
M = 178.24
CAS No 85-01-8
RTECS No SF7175000

Trade Name. Ravatite.

Properties. Colorless crystals. Solubility in water is 1.6 mg/l. Organoleptic threshold concentration is 0.4 mg/l.[1]

Applications. Used in the production of stabilizers and surfactants.

Occurrence. P. is a major coal tar component.

Acute Toxicity. In mice, LD_{50} is 700 mg/kg BW. There is no difference in sensitivity to P. in rats and rabbits.

Short-term Toxicity. Rats and rabbits were dosed by gavage with 70 mg/kg BW for 3 months. The treatment affected carbohydrate-forming liver function and produced changes in serum *SH*-group contents in rabbits.

Mutagenicity.

In vitro genotoxicity. P. was tested in the 3-methylcholantrene-induced microsomes assay.[2] It was inactive in the gene conversion system and showed a weak SCE effect only at high doses.

Carcinogenicity. A carcinogenic effect is reported in mice.[3]

Carcinogenicity classification. An IARC Working Group concluded that there is inadequate evidence for the carcinogenicity of P. in *experimental animals* and there are no data available to evaluate the carcinogenicity of P. in *humans*.

IARC: 3.

Chemobiokinetics. Sprague-Dawley rats and guinea pigs were given 10 mg ^{14}C-P./kg BW administered by gavage in corn oil. P. is metabolized to free *hydroxylated phenanthrenes* and their conjugates. Enzymatic hydrolysis of glucuronides and sulfates resulted in the formation of free 1,2-, 3,4-, and 9,10-*dihydrodiols* of P. and 1-, 2-, 3-, and 4-*hydroxyphenanthrene* in both species.[4]

According to Kadry et al., absorption of P. from the GI tract is relatively rapid with maximum plasma concentration of radioactivity occurring within 1 hour following P. oral administration. The highest tissue concentration took place in the ileum. Urine was shown to be a primary excretion route of ^{14}C-P. activity. *Phenanthrene quinone* and *9,10-phenanthrene dihydrodiol* appeared to be the main urinary metabolites occurring in the 0 to 12-hour urine.[5]

References:

1. Rakhmanina, N. A., Hygienic standardization of phenantrene and pyrene contents in water bodies, *Gig. Sanit.*, 6, 19, 1964 (in Russian).
2. Marquardt, H., *Comm. Eur. Communities,* (Issue, Eur. 6388, Environ. Res. Programme), 248, 1980.
3. Shubic, P. and Hartwell, G. L., *Survey of Compounds Which Have been Tested for Carcinogenic Activity*, Suppl. 2, PHS No 149, Washington, D.C., 1969.
4. Chu, I., Ng, K. M., Benoit, F. M., and Moir, D., Comparative metabolism of phenanthrene in the rat and guinea pig, *J. Environ. Sci. Health*, 27, 729, 1992.
5. Kadry, A. M., Skowronski, G. A., Turkall, R. M., and Abdel-Rahman, M. S., Comparison between oral and dermal bioavailability of soil-adsorbed phenanthrene in female rats, *Toxicol. Lett.*, 78, 153, 1995.

o-PHENYLENEDIAMINE

Molecular Formula. $C_6H_8N_2$

M = 108.14

CAS No 95-54-5

RTECS No SS7875000

Abbreviation. *o*-PDA.

Synonyms. *o*-Aminoaniline; *o*-Benzenediamine; *o*-Diaminobenzene.

Properties. Grayish or light yellow crystals with a sharp unpleasant odor. Solubility in water is 3.0% at 20°C. Readily soluble in alcohol. Gives water an odor of rotten hay. Odor perception threshold is 436 mg/l at 20°C (rating 1) and 364.5 mg/l at 60°C. Chlorination does not affect the perception threshold. Gives water a color ranging from light yellow to dark brown (chlorination decreases the threshold from 0.05 to 0.01 mg/l). Does not cause suspensions, films, or foam to form in water.[1]

Applications. Used in the paint and dyestuffs industry, and in the production of coatings.

Acute Toxicity. In rats, mice, and guinea pigs, LD_{50} is 660, 470, and 360 mg/kg BW, respectively. CNS is predominantly affected. Excitation is followed by depression, motor coordination disorder and impaired respiratory rhythm. Animals experience bloody nasal discharge, paralysis, and convulsions. Death within the first 24 hours.[1,2]

According to Reznichenko et al.[3], LD_{50} is 800 mg/kg BW in male rats, 500 mg/kg BW in female rats, and 380 mg/kg BW in mice. Rats appeared to be less sensitive. Manifestations of the toxic action included convulsions, paresis of posterior extremities, anorexia, yellow-colored urine.[3]

Repeated Exposure revealed moderate cumulative properties. K_{acc} is 1.48 (by Lim). Administration of 1/10 to 1/50 LD_{50} defined the LD_{50}/LOEL ratio to be 250. According to Reznichenko et al., K_{acc} was 7.3 (by Lim). Rats received daily doses of 160 mg *o*-PDA/kg BW for 30 days (as a 10% water suspension). The treatment caused retardation of BW gain, an increase in the liver and kidney relative weights, and changes in behavior. Hematology changes, and changes in protein and mineral metabolism were also noted.[3]

Long-term Toxicity. Administration of *o*-PDA to rats caused erythrocyte depletion and increased activity of alkaline phosphatase, transaminase, and aldolase.[1,2]

Allergenic Effect is not found on *i/g* administration,[2] but is noted on dermal application.[4]

Reproductive Toxicity.

Gonadotoxicity. *o*-PDA produces a consistent reduction in the number and mobility of spermatozoa. Administration of 0.96 and 9.6 mg/kg BW to rats led to an increase in the number of pathologic forms and tubules with epithelial desquamation and the 12th stage of meiosis.[2]

Embryotoxicity. Dyban et al. (1970) revealed a marked distention and engorgement of the large blood vessels in embryos, and of the medium-sized vessels of the liver, thoracic, and abdominal cavities (15% fetuses). The NOEL for gonado- and embryotoxic effect is reported to be 1.8 mg/kg BW.

Teratogenic effect is observed in chickens on introduction of 0.5 ml into the yolk sac of 4-day-old embryos. Observed malformations included ophthalmic defects, cleft palate, and skeletal abnormalities.[5]

Mutagenicity. The mutagenicity of *o*-PDA was remarkably enhanced by oxidation; major mutagenic oxidation product was *2,3-diaminophenazine* (Watanabe et al., 1990).

In vitro genotoxicity. *o*-PDA was positive in *S. typhimurium* mutagenicity assay.[6]

In vivo cytogenetics. Contradictory data are reported. A single administration of 1/5 or 1/50 LD_{50} of *o*-PDA does not increase the number of CA in bone marrow cells.[1,2]

Sbrana and Loprieno found an increase in the number of CA in murine bone marrow cells and in cultured human lymphocytes.[5]

Carcinogenicity. Wild et al. revealed carcinogenic effect of *o*-PDA and consider it to be potentially genetically harmful to humans.[7]

Carcinogenicity classification.

U.S. EPA: B2.

Standards. ***Russia*** (1995). MAC and PML: 0.01 mg/l (organolept., color).

References:

1. Galushka, A. I., Manenko, A. K., Gzhegotsky, M. I., et al., Hygienic substantiation of maximum admissible concentration for *o*-phenylenediamine and methyl cyancarbamate dimer in water bodies, *Gig. Sanit.*, 6, 78, 1985 (in Russian).
2. Galushka, A. I., Kogut, O. N., Dodoleva, I. K., et al., Toxicity of dimer methyl cyanocarbamate, *o*-phenylenediamine, trilane, *Gig. Sanit.*, 1, 73, 1986 (in Russian).
3. Reznichenko, A. K., Muzhikovsky, G. L., Krasnorutzkaya, Ye. P., Solov'yova, L. M., Popova, D. P., Shevchuk, Ye. A., and Portnaya, G. A., Toxicity of 1,2-phenylenediamine, *Gig. Truda Prof. Zabol.*, 9, 50, 1988 (in Russian).
4. Kurliandsky, B. A., Alexeyeva, O. G., Livke, T. N., et al., Complex comparative assessment of sensitivity effect of phenylenediamine isomers for MAC-setting at workplace air, *Gig. Truda Prof. Zabol.*, 8, 46, 1987 (in Russian).
5. Karnofsky, D. A. and Lacon, C. R., Survey of cancer chemotherapy service center compounds for teratogenic effect in the chick embryo, *Cancer Res.*, 22, 84, 1962.
6. Rossman, T. G., Molina, M., Meyer, L., Boone, P., Klein, C. B., Wang, Y., Li, F., Lin, W. C., and Kinney, P. I., Performance of 133 compounds in the lambda prophase induction endpoint of the

microscreen assay and comparison with *S. typhimurium* mutagenicity and rodent carcinogenicity, *Mutat. Res.*,260, 349, 1991.
7. Wild, D., King, M. T. and Eckhart, K., Cytogenetic effect of *o*-phenylenediamine in the mouse, Chinese hamster, and guinea pigs and derivatives, evaluated by the micronuclei test, *Arch. Toxicol.*, 43, 249, 1980.

PHENYLPHOSPHONOUS DICHLORIDE

Molecular Formula. $C_6H_5Cl_2O_2P$
M= 178.98
CAS No 644-97-3
RTECS No TB2478000
Abbreviation. PPC.

Synonyms. Dichlorophenylphosphine; Phenyldichlorophosphine.

Properties. Colorless, volatile liquid. Readily soluble in water. Odor perception threshold, taking into account heating and chlorination, is found to be 30 mg/l. There is no foreign taste at this concentration in water.

Applications. Used in the manufacture of foam plastics.

Acute Toxicity. LD_{50} is found to be 1630 mg/kg BW in rats, 630 mg/kg BW in mice, and 800 mg/kg BW in rabbits and guinea pigs.

Repeated Exposure failed to reveal cumulative properties. A reduced oxygen consumption and changes in cholinesterase activity are reported in rats given 82 and 165 mg/kg BW for 45 days.

Long-term Toxicity. Rats and rabbits were exposed by gavage to the dose of 30 mg/kg BW for 6 months. The treatment caused an impairment of oxidation-reduction processes and liver function. A change in the liver metabolism was also noted. Gross pathology examination revealed hemodynamic disturbances and dystrophic changes and necrotic areas in the myocardium, liver, and kidneys.

Chemobiokinetics. PPC metabolism in the body occurs through hydrolysis to form *phenol metabolites* which conjugate with glucuronic and sulphuric acids in the liver and, in this form, are excreted via the kidneys.

Standards. ***Russia.*** Recommended PML: 0.3 mg/l.

Reference:

Vorob'yeva, L. V., in *Problems of Water Bodies Protection and Sanitary Engineering*, Proc. Perm' Medical Institute, Perm', 1973, 109 (in Russian).

PHOSPHORIC ACID

Molecular Formula. H_3O_4P
M = 98.00
CAS No 7664-38-2
RTECS No TB6300000
Abbreviation. PA.

Synonyms. Hydrogen phosphate; Orthophosphoric acid.

Properties. Colorless, viscous, odorless, clear syrupy liquid or unstable orthorhombic crystals. Very soluble in water and alcohol.

Applications. An acid catalyst in ethylene manufacture; used also in dental cements and in the production of rubber latex.

Regulations.

U.S. FDA (1998) approved the use of PA in the manufacture of resinous and polymeric coatings for the food-contact surface of articles intended for use in producing, manufacturing, packing, processing, preparing, treating, packaging, transporting, or holding food (for use only as polymerization cross-linking agent in side seam cements for containers intended for use in contact with food (only of the identified types) in accordance with the conditions prescribed in 21 CFR part 175.300. Considered *GRAS* when used in accordance with GMP.

Great Britain (1998). PA is authorized without time limit for use in the production of polymeric materials and articles in contact with food or drink or intended for such contact. The specific migration of PA shall not exceed 7.5 mg/kg.

PHOSPHORIC ACID, TRIPOTASSIUM SALT

Molecular Formula. $O_4P.3K$
M = 212.27
CAS No 7778-53-2
RTECS No TC8450000
Abbreviation. TPP.

Synonyms. Potassium orthophosphate; Potassium phosphate; tribasic; Tripotassium phosphate.

Properties. Colorless or white trigonal crystals. Water solubility is about 90 g/l.

Application. Used in the paper manufacture. Sodium phosphate is a dietary supplement.

Acute Toxicity. LD_{50} is 7.4 g/kg BW in rats.[020] Administration of about 5.0 g/kg BW caused changes in CNS and GI tract. LD_{50} on skin application exceeded 0.3 g/kg BW.

Chemobiokinetics. TPP is poorly absorbed; in the body, phosphorus is converted to *phosphates.* Elimination occurs predominantly in the urine as organic and inorganic phosphate.[03]

Regulations. ***U.S. FDA*** (1998) approved the use of TPP (1) in adhesives as a component of articles intended for use in packaging, transporting, or holding food in accordance with the conditions prescribed in 21 CFR part 175.105, and (2) as a component of sanitizing solutions used on food-processing equipment and utensils, and on other food-contact articles as specified in 21 CFR part 178.1010.

Considered *GRAS* when used in accordance with GMP.

Reference:

Natl. Technical Information Service OTSO571153.

PHOSPHORIC ACID, TRISODIUM SALT

Molecular Formula. $O_4P.2Na$
M = 163.94
CAS No 7601-54-9
96337-98-3
RTECS No TC9490000
Abbreviation. TSP.

Synonyms and **Trade Name.** Sodium phosphate; Sodium phosphate, anhydrous; Sodium phosphate, tribasic; Sodium tertiary phosphate; Trisodium orthophosphate; Tromete.

Properties. Colorless, trigonal crystals. Solubility in water is 15 g/l at 0°C; insoluble in alcohol (trisodium phosphate dodecahydrate).

Applications. Water treatment agent; used in the production of elastomers; emulsifier in processed cheese; food additive.

Acute Toxicity. LD_{50} is 7.4 g/kg BW in rats.[020] The toxicity of TSP relates only to its alkalinity since its ions are normal constituents of all living matter. Its alkalinity is close to that of sodium carbonate (trisodium phosphate dodecahydrate).[03]

Severe hyperphosphatemia and hypocalcemia developed following the administration of a single hypertonic sodium phosphate enema in an adult with mild chronic renal insufficiency.[1]

Reproductive Toxicity.

Embryo- and ***gonadotoxicity.*** Pregnant Sprague-Dawley rats were treated on day 16, 18, or 20 of pregnancy with 1 or 3 mu*Ci*/g BW of ^{32}P supplied as sodium phosphate. There was a significant shortening of the lifespan of most groups of irradiated animals. Testes of male animals irradiated at 3.0 mu*Ci*/g maternal BW were approxi mately a quarter of normal size and showed obvious histological changes.[2]

Carcinogenicity. In the above-cited study, no tumor was induced in an increased frequency in fetuses from treated animals. Tumors that normally occur showed a clear tendency to appear earlier in post-natal life in treated animals.[2]

Chemobiokinetics. TSP is poorly absorbed.

Regulations. ***U.S. FDA*** (1998). Sodium phosphate (mono-, di-, and tribasic) used as a multiple purpose food substance in food for human consumption is *GRAS* when used in accordance with GMP (21 CFR parts 182.1778, 182.5778, 182.6778, and 182.8778).

References:

1. Biberstein, M. and Parker, B. A., Enema-induced hyperphosphatemia, *Am. J. Med.*, 79, 645, 1985.
2. Berry, C. L., Amerigo, J., Nickols, C., and Swettenham, K. V., Transplacental carcinogenesis with radioactive phosphorus, *Hum. Toxicol.*, 2, 49. 1983.

PHOSPHORIC ACID, ZINC SALT

Molecular Formula. $H_3O_4P.Zn$
M = 555.59
CAS No 7779-90-0
1337-79-7
57572-56-2
87502-49-6
RTECS No TD0590000
Abbreviation. PAZ.

Synonyms and **Trade Name.** Acid zinc phosphate; Bonderite; Zinc phosphate.

Properties. Tetrahydrate, white odorless powder. Insoluble in water and alcohol.

Applications. Anticorrosive pigment. Used in dental cements.

Acute Toxicity. Low toxicity depends on poor solubility of PAZ. Administration of *zinc phosphate* water suspension at the dose levels of 0.5 to 10 g PAZ/kg BW caused no changes in appearance and behavior of rats. LD_{50} *i/p* is 0.55 to 0.6 g/kg BW in mice.[1]

Regulations. ***U.S. FDA*** (1998) approved the use of PAZ in adhesives as a component of articles intended for use in packaging, transporting, or holding food in accordance with the conditions prescribed in 21 CFR part 175.105.

Reference:

Spiridonova, V. S. and Shabalina, L. P., Experimental study of zinc phosphates toxicity, *Gig. Sanit.*, 8, 23, 1982 (in Russian).

PHOSPOLYOL II

Composition. A mixture of hydroxypropylated esters of pentaerythritol and alkylphosphoric acid. Contains 10 to 15% of alcoholic and hydroxyl groups, 10% of phosphorus, and 2.5% of volatile compounds.

Properties. A sticky substance, light brown in color with a heavy odor of stale apples. May give water a bitter, sweet, salty, astringent, metallic taste depending on concentration. No correlation is found between P. concentration in water and odor intensity; the product apparently undergoes hydrolysis. The color of water is not changed.

Applications. Used in the production of polyurethane foams and glass-reinforced plastics.

Acute Toxicity. A single administration caused no mortality.

Repeated Exposure revealed pronounced cumulative properties.

Short-term Toxicity. Rats were dosed with 1.36 g/kg BW every other day. The treatment caused some animals to die and produced signs of intoxication in survivors. Gross pathology examination revealed changes in the hypophyseal - adrenal system and in the structure of the visceral organs.

Long-term Toxicity. The NOAEL appeared to be 170 mg/kg BW. However, Tsyganok considers it essential to investigate the toxicity of the products of P. hydrolysis.

Reference:

Tsyganok, L. A., in *Hygienic Significance of Low Intensity Factors in the Populated Locations and in Industrial Areas*, Proc. F. F. Erisman Research Sanitary Hygiene Institute, Moscow, Issue No 8, 1974, 100 (in Russian).

PHOSPOLYOL ACYCLIC

Properties. An organophosphorus compound. Viscous liquid of low volatility.

Applications. Used in the production of polyurethane foams and glass-reinforced plastics.

Acute Toxicity. LD_{50} is 8.59 g/kg BW in rats, and 8.45 g/kg BW in mice. Acute action limit is determined to be 0.43 g/kg BW.

Repeated Exposure failed to reveal cumulative properties. No animals died after fractional administration of the overall dose equal to LD_{50}.

Reference:

Klimov, V. S., Chachanidze, E. I., Eskin, A. M., et al., in *Hygienic Significance of Low Intensity Factors in the Populated Locations and in Industrial Areas,* Proc. F. F. Erisman Research Sanitary Hygiene Institute, Moscow, Issue No 12, 1978, 70 (in Russian).

PHOSTEROL

RTECS No TH5900000

Composition. Compound of hydroxyethyltetraalkyl phosphonate pentaerythritol in which n = 1.5 to 2.0.

Properties. Viscous, non-volatile liquid. Soluble in alcohol.

Applications. Used in the production of polyurethane foams and glass-reinforced plastics.

Acute Toxicity. In mice, LD_{50} is 33.3 g/kg BW.[1]

Repeated Exposure failed to show cumulative properties. Administration of 1/20 LD_{50} for 20 days caused no animal mortality. The treatment caused retardation of BW gain, decrease in cholinesterase activity, changes in the STI, and relative kidney weight.[2]

Long-term Toxicity. Manifestations of the toxic effect in male rats include retardation of BW gain, NS impairment, and some changes in cholinesterase activity and in kidney function. Histology examination revealed changes in the kidney (in tubular section of nephron). The NOAEL appeared to be 614 mg/kg BW.[2]

References:

1. Izmerov, N. F., Sanotsky, I. V., and Sidorov, K. K., *Toxicometric Parameters of Industrial Toxic Chemicals under Single Exposure*, USSR/SCST Commission for UNEP, Center Int. Projects, Moscow, 1982, 76 (in Russian).
2. Kazmina, N. P. and Gaber, S. N., in *Hygienic Significance of Low Intensity Factors in the Populated Locations and in Industrial Areas*, Proc. F. F. Erisman Research Sanitary Hygiene Institute, Moscow, Issue No 8, 1974, 103 (in Russian).

[PHTHALOCYANINATO(2$^-$)] COPPER

Molecular Formula. $C_{32}H_{16}CuN_8$

M = 576.10

CAS No 147-14-8

RTECS No GL8510000

Abbreviation. PCC.

Synonym. Copper phthalocyanine.

Properties. Pigment containing PCC is insoluble in water, oils, and the majority of organic solvents.

Applications. Used predominantly in the manufacture of rubber and plastics.

Acute Toxicity. Substance of low acute toxicity. Oral administration of 6.0 g/kg BW (in starch) or 15 g/kg BW (in dimethyl sulfoxide) caused no mortality in female rats. *I/p* injections damaged kidneys function (an increase of proteins level in the urea, a decrease in diuresis).[1]

Repeated Exposure revealed cumulative properties (on low doses given). Female rats received 2.0 g/kg BW (20% solution in starch) for 30 days. No clinical manifestations of toxic action were reported (BW gain, alkaline phosphatase activity in blood serum, and relative organ weights were fixed during the treatment). An increase in ceruloplasmin level in blood serum was noted.[1]

Reproductive Toxicity.

Teratogenicity. Sulfonated phthalocyanine induced caudal malformative syndrome in the chick embryo.[2]

Chemobiokinetics. PCC does not penetrate into hematology system, but there was an increase in Cu^{2+} concentration in blood serum.

Regulations. ***U.S. FDA*** (1998) approved the use of PCC as a component of the uncoated or coated food-contact surface of paper and paperboard intended for use in producing, manufacturing, packaging, processing, preparing, treating, packing, transporting, or holding aqueous and fatty foods in accordance with the condi tions prescribed in 21 CFR part 176.170.

References:

1. Kurlyandsky, B. A., Braude, E. V., Klyachkina, A. M., Torshina, N. L., Khokhlova, S. B., and Zasorina, E. V., Study on copper phthalocyanine-induced toxicity, *Gig. Sanit.*, 92, 1, 1985 (in Russian).
2. Sandor, S., Prelipceanu, O., and Checiu, I., Sulfonated phthalocyanine induced caudal malformative syndrome in the chick embryo, *Morphol. Embryol.* (Bucur), 31, 173, 1985.

2-PINENE

Molecular Formula. $C_{10}H_{16}$
M = 136.26
CAS No 80-56-8
RTECS No DT7000000

Synonyms. α-Pinene; 2,6,6-Trimethylbicyclo(3.1.1)-2-hept-2-ene.

Properties. Obtained from oil of turpentine. Colorless, mobile liquid with a characteristic odor of pine. Practi cally insoluble in water. Soluble in alcohol.

Applications. Used in the manufacture of plasticizers and synthetic resins, a solvent for protective coatings.

Acute Toxicity.

Observations in man. Ingestion of large doses caused GI tract irritation, delirium, ataxia, kidney damage, coma. LD is about 180 g as *turpentine*.[020] P. irritates skin.

Animal studies. LD_{50} is 3.7 g/kg BW in rats.[012]

Chemobiokinetics.

Observations in man. Toxicokinetics were studied in eight healthy males, average age 31 year,s in an inhalation chamber. P. is readily metabolized and elimination of unchanged P. is very low.

Carcinogenicity. Application caused benign skin tumor. [020]

Regulations.

U.S. FDA (1998) approved the use of P. in resinous and polymeric coatings used as the food-contact surfaces of articles intended for use in producing, manufacturing, packing, processing, preparing, treating, packaging, transporting, or holding food in accordance with the conditions prescribed in 21 CFR part 175.300.

EU (1990). ICAN is available in the *List of authorized monomers and other starting substances which shall be used for the manufacture of plastic materials and articles intended to come into contact with foodstuffs (Section* A).

Great Britain (1998). P. is authorized without time limit for use in the production of polymeric materials and articles in contact with food or drink or intended for such contact.

Reference:

Falk, A. A., Hogberg, M. T., Lof, A. E., Wigaen, S., Helm, E. M., and Wang, Z. P., Uptake, distribution and elimination of α-pinene in man after exposure by inhalation, *Scand. J. Work. Environ. Health*, 16, 372, 1990.

PINE OIL

CAS No 8002-09-3
RTECS No TK5100000
Abbreviation. PO

Synonyms and **Trade Names.** Arizole; Oleum abietis; Terpentinoel.

Composition. Consists mainly of isomeric tertiary and secondary, cyclic terpene alcohols.[020]

Properties. Colorless to pale yellow liquid with turpentine-like odor. Insoluble in water; soluble in alcohol.

Applications. A solvent and a disinfectant.

Acute toxicity.

Observations in man. Ingestion of 4.7 g/kg BW led to excitation development, ataxia, and headache.[1]

Animal studies. LD_{50} is 3.2 g/kg BW in rats. PO cleaner ingestion led to acute toxicity manifestations including mucous membrane and GI irritation. Ataxia and CNS depression were also noted.[2]

Horses received PO by *i/v* injection. Marked histopathological changes in the lungs and respiratory tract were observed. Death occurs due to acute respiratory edema.

Chemobiokinetics. Monoterpenes are poorly resorbed in the GI tract, slowly metabolized and then excreted by the kidneys. The main metabolic pathways are hydratation, hydroxylation, rearrangement, and acetylation. Five metabolites were identified.[1]

After lethal *i/v* injection of 0.1 ml PO/kg BW, 150 to 300 ppm PO were found in blood and tissues of horses. No traces occurred in blood and tissues after injection of 0.033 ml/kg BW (Tobin et al., 1976).

Regulations. ***U.S. FDA*** (1998) approved the use of PO (1) in resinous and polymeric coatings used as the food-contact surfaces of articles intended for use in producing, manufacturing, packing, processing, preparing, treating, packaging, transporting, or holding food in accordance with the conditions prescribed in 21 CFR part 175.300, and (2) in adhesives as a component of articles intended for use in packaging, transporting, or holding food in accordance with the conditions prescribed in 21 CFR part 175.105.

References:

1. Koppel, C., Tenczer, J., Tonnesmann, U., Schirop, T., and Ibe, K., Acute poisoning with pine oil - metabolism of monoterpenes, *Arch. Toxicol.*, 49, 73, 1981.
2. Brook, M. P., McCarron, M. M., and Mueller, J. A., Pine oil cleaner ingestion, *Ann. Emerg. Med.*, 18, 391, 1989.

PIPERAZINE

Molecular Formula. $C_4H_{10}N_2$
M = 86.16
CAS No 110-85-0
RTECS No TK7800000

Synonyms and **Trade Names.** Antiren; 1,4-Diazacyclohexane; 1,4-Diethylenediamine; Diethyleneimine; Hexahydro-1,4-diazine; Piperazidine; Pipersol.

Properties. White to slightly off-white lumps or flakes or colorless, transparent, needle-like crystals with salty taste. Readily soluble in water; poorly soluble in alcohol.

Acute Toxicity. Ingestion of lethal doses caused convulsions and respiratory arrest.[022] P. exhibits its major effect on the CNS. Neurotoxicity in cats and dogs usually was manifested by muscle tremors, ataxia, and/or behavioral disturbances within 24 hours after estimated daily dose(s) between 20 and 110 mg/kg BW.[1]

Tympany, diarrhea, and anorexia were observed in calves given a dose of 0.88 g/kg BW.[2]

Allergenic Effect. P. salts failed to cause systemic or skin sensitivity.[04]

Reproductive Toxicity.

Teratogenicity. Exposure to P. during development stage could result in cleft hand and foot deformities in a child.[5]

Mutagenicity.

In vitro genotoxicity. P. was negative in *Salmonella* mutagenicity assay.[025]

Chemobiokinetics. P. is readily absorbed from the GI tract and partly undergoes destruction. In humans, 15% of 2.0 g dose was found eliminated unchanged with the urine.[3,4]

Regulations. ***U.S. FDA*** (1998) approved the use of P. in adhesives as a component of articles intended for use in packaging, transporting, or holding food in accordance with the conditions prescribed in 21 CFR part 175.105.

Standards. ***Russia*** (1995). MAC and PML: 9.0 mg/l (organolept., odor).

References:

1. Lovell, R. A., Ivermectin and piperazine toxicoses in dogs and cats, *Vet. Clin. North Am. Small Anim. Pract.*, 20, 453, 1990.
2. Clarke, E. G. and Clarke, M. L., *Veterinary Toxicology*, The Williams & Wilkins Co., Baltimore, Ml, 1975, 137.
3. *The Pharmacological Basis of Therapeutics*, L. S. Goodman and A. Gilman, Eds., 5th ed., Macmillan Publ. Co., Inc., New York, 1975, 1027.
4. Williams, R. T., *Detoxication Mechanisms. The Metabolism and Detoxication of Drugs, Toxic Substances and Other Organic Compounds*, 2nd ed., Chapman & Hall Ltd., London, 1959.
5. Meyer, H. H. and Brenner, P., Cleft hand and foot deformity as a possible teratogenic side effect of anthelminthic agent piperazine, *Internist*, 29, 217, 1988.

PITCH RESIN VARNISH

Properties. Gives water a specific odor.

Migration Data. Organoleptic properties of water are unaltered at a pitch concentration in water of 10 mg/l. No migration of benzo[a]pyren into water from pitch samples investigated has been noted.[1]

Applications. Used for making water-tight coatings.

Repeated Exposure. Mice were dosed by gavage with 100 mg/kg BW for 10 days. The treatment caused no mortality or retardation of BW gain. There were no differences in behavior or capacity to work between experimental and control animals.

Short-term Toxicity. Rats received 10 mg/kg BW for 3 months and experienced consistent retardation of BW gain. Gross pathology examination revealed a number of changes in the viscera.[1]

Carcinogenic Effect was not found on skin application (1.0 mg/kg BW) in mice and rabbits.[2]

References:

1. Kudrin, L. V. et al., in *Problems of Water Hygiene and Reservoirs Protection*, Moscow, 1968, 127 (in Russian).
2. Konstantinov, V. G., Filatova, A. S., Kus'minykh, A. J., et al., Hygienic evaluation of petroleum pitch resin varnish, *Gig. Sanit.*, 3, 27, 1973 (in Russian).

POTASSIUM BROMIDE

Molecular Formula. BrK
M = 119.01
CAS No 7758-02-3
RTECS No TS7650000
Abbreviation. PB.

Synonyms. Bromide salt of potassium; Tripotassium tribromide (K_3Br_3).

Properties. Colorless and odorless, cubic crystals or white granules, or powder with strong bitter saline pungent taste. 1.0 g dissolves in 1.5 ml water, 1.0 ml boiling water, 250 ml alcohol, 21 ml boiling alcohol.[020]

Exposure. The increasing environmental concentration of bromine is observed.

Acute Toxicity. Large doses of bromine or bromides are toxic to animals and humans, a major target organ being the thyroid gland.

Observations in man. Ingestion of 12 to 50 g/man caused vomiting, diarrhea, gastritis, anuria, acute nephrosis, uremia, degeneration of neurons and proximal tubules of kidney, fatty degeneration of kidney and liver, edema in brain and kidney, and hemolysis. Death occurs in 5 days.[1]

Animal studies. LD_{50} is 3.07 g/kg BW in rats and 3.12 g/kg BW in mice. Poisoning caused CNS inhibition and ataxia.[2] Ingestion of high doses produced salivation and nausea, or laxative effects in cats and dogs.[3]

Repeated Exposure. Rats were exposed to $KBrO_3$, KBr, or $NaBrO_3$ in their drinking water. Increases in cell proliferation were found in the proximal renal tubules of male rats given $KBrO_3$ or $NaBrO_3$ but not KBr for 2, 4, and 8 weeks.[4]

Long-term Toxicity and **Carcinogenicity.** F344 rats received a diet containing 500 ppm PB for up to 2 years. The treated animals did not show any treatment-related changes.[5]

Reproductive Toxicity. Feeding of brominated vegetable oil to rats interfered with reproductive function and caused maternal and fetal toxicity when given during pregnancy.[6]

Embryotoxicity. Prenatal exposure to PB caused significant delays in rat postnatal development and permanent retardation of their BW and brain weight gain.[7]

Teratogenicity. An increase in birth defects was not reported.[6,7]

Chemobiokinetics. Following oral administration, PB was rapidly absorbed from the lower part of the small intestine. Bromide is widely distributed throughout extracellular fluids including secretions and transudates.[8]

Regulations. ***U.S. FDA*** (1998) approved the use of PB (1) as basic component of acrylic and modified acrylic plastics, semirigid and rigid, which may be safely used as articles intended for single and repeated use food contact surfaces in accordance with the conditions prescribed by 21 CFR part 178.1010; and (2) as antioxidants and/or stabilizers in the manufacture of polymers intended for use in producing, manufacturing, packing, processing, preparing, treating, packaging, transporting, or holding food at levels not exceeding 0.18% PB and 0.005% copper as cupric acetate or cupric carbonate by weight of nylon 66 resins complying with 177.1500.

References:

1. Oinuma, T., Abstract, *Nichidai Igaku Zasshi*, 33, 759, 1974 (in Japanese).
2. Information from the Soviet Toxicology Center, *Gig. Truda Prof. Zabol.*, 10, 57, 1989 (in Russian).
3. Rossoff, I. S., *Handbook of Veterinary Drugs*, Springer Publ. Co., New York, 1974, 475.
4. Umemura, T., Sai, K., Takagi, A., Hasegawa, R., and Kurokawa, Y., A possible role for cell proliferation in potassium bromate (*$KBrO_3$*) carcinogenesis, *J. Cancer Res. Clin. Oncol.*, 119, 463, 1993.
5. Mitsumori, K., Maita, K., Kosaka, T., Miyaoka, T., and Shirasu, Y., Two-year oral chronic toxicity and carcinogenicity study in rats of diets fumigated with methyl bromide, *Food Chem. Toxicol.*, 28, 109, 1990.
6. Vorhees, C. V., et al., Behavioral and reproductive effects of chronic developmental exposure to brominated vegetable oil in rats, *Teratology*, 28, 309, 1983.
7. Disse, M., Joo, F., Schulz, H., and Wolff, J. R., Prenatal exposure to sodium bromide affects the postnatal growth and brain development, *J. Hirnforsch.*, 37, 127, 1996.
8. American Hospital Formulary Service, vols 1 and 2, Washington, D.C., *Am. Soc. Hospital Pharmacol.*, 1984, 28.

POTASSIUM IODIDE

Molecular Formula. IK
M = 166.00
CAS No 7681-11-0
RTECS No TT2975000
Abbreviation. PI.

Trade Names. Knollide; Potide.

Properties. Colorless or white, odorless, cubic crystals, white granules, or powder with saline, strong bitter taste.

Applications. Dietary supplement as a source of dietary iodine. Used for emergent drinking water disinfection, and in the photography process.

Acute Toxicity. LD_0 was found to be 1.86 g/kg BW in rats. Poisoning produced CNS inhibition, muscle weakness, and dyspnea.[1]

Short-term Toxicity. Rats received 1.0, 3.0, 10, or 100 mg iodine or iodide (*NaI*) in drinking water for 100 days. The treatment produced no effect on BW gain and testes. An increase in the thyroid weight of males correlated with increasing contents of iodide but not iodine in the water. A decrease in BW of females was noted at the highest dose of iodide. Both substances affect thyroid hormone status in rats.[2]

Reproductive Toxicity.

Teratogenicity. Pregnant rats were given 25 to 600 mg PI/300 g BW by gavage on day 9 of gestation. About 25% of fetuses with the dose of 300 mg PI were found dead or malformed. Increased incidence of malformations were found in the higher dose groups, the main anomalies being ventricular septal defects, right aortic arch, vascular ring, hypoplasia or incompletion of the lungs and omphalocele, and the occurrence of growth retardation.[3]

Embryotoxicity. PI was given to male and female rats before and during breeding, to females only during gestation and lactation, and to their offspring after weaning (day 21 after birth) through to day 90, at levels of 0.025, 0.05, or 0.1% of the diet. PI significantly reduced litter size and increased offspring mortality, brain, and BW at the highest dose. It was concluded that PI produced evidence of developmental toxicity consistent with a picture of impaired thyroid function.[4]

Chemobiokinetics. On PI ingestion, iodide has been found in saliva, tears, sweat, and milk, some of retained *iodide* is stored in the thyroid and incorporated into thyroxine. PI is eliminated predominantly via the kidneys.[5]

Regulations. ***U.S. FDA*** (1998) approved the use of PI as a component of sanitizing solutions used on food-processing equipment and utensils, and on other food-contact articles as specified in 21 CFR part 178.1010. Considered to be *GRAS.*

References:

1. *J. Pharmacol. Exp. Ther.*, 120, 171, 1957.
2. Sherer, T. T., Thrall, K. D., and Bull, R. J., Comparison of toxicity induced by iodine and iodide in male and female rats, *J. Toxicol. Environ. Health.*, 32, 89, 1991.
3. Lee, J. Y. and Satow, Y., Developmental toxicity of potassium iodide in rats, *Teratology*, 40, 676, 1989.
4. Vorhees, C. V., Butcher R. E., and Brunner, R. L., Developmental toxicity and psychotoxicity of potassium iodide in rats: a case for the inclusion of behavior in toxicological assessment, *Food Chem. Toxicol.*, 22, 963, 1984.
5. Rossoff, I. S., *Handbook of Veterinary Drugs*, Springer Publishing Co., New York, 1974, 475.

PROPIONALDEHYDE

Molecular Formula. C_3H_6O
M = 58.09
CAS No 123-38-6
RTECS No UE0350000
Abbreviation. PA.

Synonyms and **Trade Names.** Methylacetaldehyde; Propaldehyde; Propanal; Propional; Propyl aldehyde.

Properties. Colorless liquid with unpleasant, fruity odor. Miscible with alcohol; soluble in 5 volumes of water at 20°C.

Applications. Used in the production of polyvinyl and other plastics and rubber; in preparation of alkyd resin systems.

Acute Toxicity. LD_{50} is in range of 0.8 to 1.6 g/kg BW.[018]

Reproductive Toxicity.

Embryotoxicity. Injection of up to 1.0 g PA directly into rat amnion on day 13 of gestation increased incidence of embryolethality. No malformations were observed in the fetuses.[1]

Mutagenicity.

In vitro genotoxicity. PA was negative in *Salmonella* mutagenicity assay.[2] It induced unscheduled DNA synthesis in primary cultures of rat and human hepatocytes. In rat hepatocytes 10 to 100 *m*M PA induced a modest but significant and dose-dependent increase of net nuclear grain counts, while in human hepatocytes this effect was not detected.[3]

In vivo cytogenetics. PA was weakly mutagenic to *Dr. melanogaster.*[013]

Regulations.

EU (1990). PA is available in the *List of authorized monomers and other starting substances which shall be used for the manufacture of plastic materials and articles intended to come into contact with foodstuffs (Section A).*

U.S. FDA (1998) approved PA for use as a direct food additive.

Great Britain (1998). PA is authorized without time limit for use in the production of polymeric materials and articles in contact with food or drink or intended for such contact.

Recommendations. Joint FAO/WHO Expert Committee on Food Additives (1999). ADI: No safety concern.

References:

1. Slott, V. L. and Hales, B. F., Teratogenicity and embryolethality of acrolein and structurally related compounds in rats, *Teratology*, 32, 65, 1985.
2. Florin, I. et al., Screening of tobacco smoke constituents for mutagenicity using the Ames' test, *Toxicology*, 15, 219, 1980.
3. Martelli, A., Canonero, R., Cavanna, M., Ceradelli, M., and Marinari, U. M., Cytotoxic and genotoxic effects of five *n*-alkanals in primary cultures of rat and human hepatocytes, *Mutat. Res.*, 323, 121, 1994.

PROPIONIC ACID

Molecular Formula. $C_3H_6O_2$
M = 74.09
CAS No 79-09-4
RTECS No UE5950000
Abbreviation. PA.

Synonyms and **Trade Names.** Carboxyethane; Ethanecarboxylic acid; Ethylformic acid; Luprosil; Propanoic acid; Pseudoacetic acid.

Properties. Colorless oily liquid with slightly pungent disagreeable odor. Miscible with water in all proportions.

Applications. Used in the production of cellulose ethers.

Acute Toxicity. LD_{50} is 1.37 g/kg BW in mice, 1.51 g/kg BW (Shepetova, 1970) or 4.29 g/kg BW in rats.[020]

Repeated Exposure PA was fed to rats for 9 days. The treatment induced a variety of proliferative effects in the midregion of the rat forestomach. The action was not apparent until 21 or 37 days of treatment.[1]

Administration of 1/10 and 1/20 LD_{50} to rats for 2 months led to exhaustion and anemization of the animals (Shepetova, 1970).

Short-term Toxicity. Daily *i/g* administration of PA affected predominantly the rat liver.[2] Inhalation exposure affected nervous and cardiovascular systems, hemopoiesis, respiratory, and parenchymatous organs of Wistar rats. Morphological lesions appeared to be dose-dependent.[3]

Long-term Toxicity. Rats received doses of 1.0, 10, or 50 mg PA/kg BW for 6 months without any manifestations of toxic action (Shepetova, 1970).

Mutagenicity.

In vitro genotoxicity. PA was found to be negative in *Salmonella*/microsome mutagenicity assay, the SOS chromotest, SCE *in vitro* but not in *E. coli* DNA repair assays.[4]

In vivo cytogenetics. PA was negative in the micronucleus test *in vivo*.[4]

Carcinogenicity. Rats received 4.0% PA in their diet. The treatment induced pre-neoplastic/precancerous changes including hyperplasia, hyperplastic ulcers, papillomas, and proliferation of basal cells in the mucosa of the forestomach.[5]

Regulations.

EU (1992): PA is available in the *List of authorized monomers and other starting substances which shall be used for the manufacture of plastic materials and articles intended to come into contact with foodstuffs (Section* A).

Great Britain (1998). PA is authorized without time limit for use in the production of polymeric materials and articles in contact with food or drink or intended for such contact.

Recommendations. Joint FAO/WHO Expert Committee on Food Additives (1999). ADI: No safety concern.

References:

1. Rodrigues, C., Lok, E., Nera, E., Iverson, F., Page, D., Karpinski, K., Clayson, D. B., Short-term effects of various phenols and acids on the Fischer 344 male rat forestomach epithelium, *Toxicology*, 38, 103, 1986.
2. Rotenberg, Yu. S., Klochkova, S. I., Mashbits, F. D., et al., in *24th Moscow Sci.-Pract. Conf. Ind.. Hyg. Problems*, 1969, 68 (in Russian).
3. Melnikova, A. P. and Tokanova, Sh. E., The biological effect of propionic aldehyde and propionic acid on laboratory animals, *Gig. Sanit.*, 4, 74, 1983 (in Russian).
4. Basler, A., von der Hude, W., and Scheutwinkel, M., Screening of the food additive propionic acid for genotoxic properties, *Food Chem. Toxicol.*, 25, 287, 1987.
5. Griem, M. N., Tumorogene Wirkung von Propionsaure an der Vormagenschleimhaut von Ratten im Futterungsversuch (vorlaufige Kurzmitteilung), *Bundesgesundheitsblatt,* 28, 322, 1985.

PROPIONIC ACID, VINYL ESTER

Molecular Formula. $C_5H_8O_2$
M = 100.13
CAS No 105-38-4
RTECS No UF8575000
Abbreviation. PAVE.

Synonyms. Propanoic acid, ethenyl ester; Vinyl propionate.

Acute Toxicity. LD_{50} is 4.76 g/kg BW in rats.[1]

Mutagenicity. PAVE is expected to be hydrolyzed in mammalian cells into *propionic acid* and *acetaldehyde* by carboxyl esterases and could thus be genotoxic through acetaldehyde formation. It induced a clear dose-dependent increase in the number of SCE in cultured (72 hours) human lymphocytes (at concentration of 0.125 to 0.5 mM).[2]

Regulations.

U.S. FDA (1998) approved the use of PAVE in adhesives as a component of articles intended for use in packaging, transporting, or holding food in accordance with the conditions prescribed in 21 CFR part 175.105.

EU (1990). PAVE is available in the *List of authorized monomers and other starting substances which may continue to be used for the manufacture of plastic materials and articles intended to come into contact with foodstuffs pending a decision on inclusion in Section A* (*Section B*).

Great Britain (1998). PAVE is authorized up to the end of 2001 for use in the production of polymeric materials and articles in contact with food or drink or intended for such contact.

References:

1. Smyth, H. F., Carpenter, C. P., Weil, C. S., Pozzani, U. C., and Striegel, J. A., Range-finding toxicity data, list VI, *Am. Ind. Hyg. Assoc. J.*, 25, 95, March-April 1962.
2. Siri, P., Jarventaus H., and Norppa, H., Sister-chromatid exchanges induced by vinyl esters and respective carboxylic acids in cultured human lymphocytes, *Mutat. Res.*, 279, 75, 1992.

SAFFLOWER OIL

CAS No 8001-23-8
RTECS No VN2230000
Abbreviation. SO.

Properties. Straw-colored liquid. Insoluble in water; soluble in oil and fat solvents.

Applications. Used in the production of alkyd resins and paints, and as a dietary supplement. Edible drying oil, intermediate between soybean and linseed oil.

Acute Toxicity. LD_{50} *i/p* exceeded 50 g/kg BW in mice.[1]

Short-term Toxicity. Weanling male rats were *i/g* exposed to 1.0 ml autoxidized SO/day over a period of up to 3 months. The treatment caused a decrease in BW and an increase in the liver weight. An initial increase of plasma triglyceride level in the liver was followed by its decrease.[2]

Long-term Toxicity. SO was administered to F344/N rats at levels of 2.5, 5.0, or 10 ml/kg BW. The treatment caused hyperplasia and adenoma of the exocrine pancreas, decreased incidences of mononuclear cell leukemia, and reduced incidences or severity of nephropathy in male F344/N rats.[3]

Reproductive Toxicity.

Embryotoxicity. Female Sprague-Dawley rats were fed diets containing 10% SO starting on day 14 of gestation. No significant effect on maternal BW gain or food intake was noted. Biochemically, fetuses exhibited no signs of essential fatty acids (EFA) deficiency, even though the diet of the dams lacked EFAs.[4]

Immunotoxicity. Pregnant C57BL/6J mice were fed diets containing 2.0, 8.0, 16, or 24% SO. Dietary fat plays a crucial role in the development of the immune system, and dietary manipulation prenatally and postnatally is an important factor in modulation of the immune response.[5]

Mutagenicity.

In vitro genotoxicity. SO was not mutagenic in *Salmonella typhimurium* strains *TA97, TA98, TA100*, or *TA1535* with or without *S9*.[3]

References:

1. Natl. Technical Information Service, AD691-490
2. Nakamura, M., Tanaka, H., Hattori, Y., and Watanabe, M., Biological effects of autoxidized safflower oil, *Lipids*, 8, 566, 1973.
3. *Comparative Toxicology Studies of Corn Oil, Safflower Oil, and Tricaprylin in male F344/N rats as Vehicles for Gavage*, NTP Technical Report Series No 426, Research Triangle Park, NC, April 1994.
4. Samulski, M. A. and Walker, B. L., Maternal dietary fat and polyunsaturated fatty acids in the developing fetal rat brain, *J. Neurochem.*, 39, 1163, 1982.
5. Erickson, K. L. et al., Influence of dietary fat concentration and saturation on immune ontogeny in mice, *J. Nutr.*, 110, 1555, 1980.

SALICYLIC ACID

Molecular Formula. $C_7H_6O_3$
M = 138.12
CAS No 69-72-7
RTECS No VO0525000
Abbreviation. SA.

Synonyms and **Trade Names.** 2-Hydroxybenzoic acid; Retarder W; Salicylic acid collodion; Salnil.

Properties. White crystalline powder or needles with a sweetish, afterward acrid taste. 1.0 g SA dissolves in 460 ml cold water, 15 ml boiling water, and 80 ml fats or oils.[020]

Applications. Antiseptic and antifungal agent. Food preservative; its use for this purpose is forbidden in some countries. Chemical intermediate in the synthesis of rubber retarders.

Long-term Toxicity.

Observations in man. Chronic exposure to salicylate caused CNS disturbances including headache, dizziness, tinnitus, difficulty in hearing, dimness of vision, mental confusion, lassitude, drowsiness, sweating, and occasionally diarrhea.[1]

Reproductive Toxicity.

Embryotoxicity. Increased frequencies of fetal death and growth retardation were observed among the offspring of pregnant rats, mice and ferrets treated with sodium salicylate in doses similar to or greater than those used clinically.[2-7]

Teratogenicity. Developmental abnormalities of the CNS, eye, skeleton, and heart were observed in rats and mice that received doses two or more times those used clinically.[3,4,6,8]

Chemobiokinetics. SA is readily absorbed. 15% of SA dose in man is metabolized to *salicyluric acid*, and with glucuronic acid to give *o-carboxyphenyl glucuronide* and *o-hydroxybenzoyl glucuronide*. 60% of dose have been excreted unchanged.[9]

According to Laznicek and Laznickova, salicylate and its metabolites (mainly *conjugates with glycine and glucuronic acid*) are eliminated mostly into urine.[10] Both kidney and liver were shown to contribute to the SA metabolism. Glycine conjugate of SA was formed predominantly in the kidney whereas both kidney and liver participated in the formation of glucuronides.

Regulations.

EU (1990). SA is available in the *List of authorized monomers and other starting substances which shall be used for the manufacture of plastic materials and articles intended to come into contact with foodstuffs (Section A)*.

U.S. FDA (1998) approved the use of SA (1) in adhesives as a component of articles intended for use in packaging, transporting, or holding food in accordance with the conditions prescribed in 21 CFR part 175.105; and (2) in resinous and polymeric coatings used as the food-contact surfaces of articles intended for use in producing, manufacturing, packing, processing, preparing, treating, packaging, transporting, or holding food for use only in coatings at a level not to exceed 0.35% by weight of the resin when such coatings are intended for repeated use in contact with foods only of the types identified in 175.300.

Great Britain (1998). SA is authorized without time limit for use in the production of polymeric materials and articles in contact with food or drink or intended for such contact.

References:

1. *The Pharmacological Basis of Therapeutics*, A. G. Gilman, T. W. Rall, A. S. Nies, and P. Taylor, Eds., 8th ed., Pergamon Press, New York, NY, 1990, 651.
2. Warkany, J. and Takacs, E., Experimental production of congenital malformations in rats by salicylate poisoning, *Am. J. Pathol.*, 35, 315, 1959.
3. Beck, F. and Gulamhusein, A. P., The effect of sodium salicylate on limb development, in *Teratology of the Limbs*, Neubert, D., Ed., Walter de Gruyter, New York, 1981, 393.
4. Skowronski, G. A., Abdel-Rahman, M. S., Gerges, S. E., and Klein, K. M., Teratologic evaluation of Alcide (R) liquid in rats and mice, *J. Appl. Toxicol.*, 5, 97, 1985.
5. Bergman, K., Cekan, E., Slanina, P., et al., Effects of dietary sodium selenite supplementation on salicylate-induced embryo- and fetotoxicity in the rat, *Toxicology*, 61, 135, 1990.
6. Fritz, H. and Giese, K., Evaluation of the teratogenic potential of chemicals in the rat, *Pharmacology*, 40 (Suppl. 1), 1, 1990.
7. Khera, K. S., Chemically induced alterations in maternal homeostasis and histology of conceptus: Their etiologic significance in rat fetal anomalies, *Teratology*, 44, 259, 1991.
8. Larsson, K. S. and Eriksson, M., Salicylate-induced fetal death and malformations in two mouse strains, Acta Paediatr. Scand., 55, 569, 1966.
9. Parke, D. V., *The Biochemistry of Foreign Compounds*, Oxford, Pergamon Press, 1968, 175.
10. Laznicek, M. and Laznickova, A., Kidney and liver contribution to salicylate metabolism in rats, *Eur. J. Drug. Metab. Pharmacokinet.*, 19, 21, 1994.

SALICYLIC ACID, METHYL ESTER

Molecular Formula. $C_8H_8O_3$
M = 152.10
CAS No 119-36-8
RTECS No VO4725000
Abbreviation. SAM.

Composition. SAM is the principal constituent of Wintergreen oil, obtained from the leaves of the teaberry (also called checkerberry), *Gaultheria procumbens*, and the bark of the sweet birch, *Betula lenta*.

Synonyms and **Trade Names.** 2-Hydroxybenzoic acid, methyl ester; Fetula oil; Methyl-2-hydroxybenzoate; Methyl salicylate; Oil of wintergreen.

Properties. Colorless oily liquid with characteristic odor and taste. Soluble in water, miscible with alcohol.

Applications. Mainly used in perfumery. SAM is also used as a flavoring in candies.

Acute Toxicity.

Observations in man. Ingestion of relatively small amounts may cause severe poisoning and death. LD_{50} is about 10 ml in children, 30 ml in adults. Main manifestations of poisoning included nausea, vomiting, acidosis, pulmonary edema, pneumonia, convulsions, death.[020]

Animal studies. LD_{50} is 890 mg/kg BW in rats, 1,300 or 2,800 mg/kg BW in rabbits, 1,100 mg/kg BW in mice, 700 mg/kg BW in guinea pigs, and 2,100 mg/kg BW in dogs. Poisoning induced somnolence, affected the respiratory and GI system.[1,011,020]

Repeated Exposure resulted in cumulative toxicity. Dogs received, in capsules, doses of 1.2 g SAM/kg BW and died in 3 to 4 days. Doses of 0.5 to 0.8 g MS/kg BW caused death of animals in a month.[2]

Long-term Toxicity. Dogs fed 350 mg SAM/kg BW for 2 years developed only BW loss and some enlargement of the liver.[2]

Reproductive Toxicity.

Gonadotoxicity and ***Embryotoxicity.*** In the feeding study, SAM did not affect fertility and fecundity of dams or growth and viability of pups in a 3-generation study of Osborne-Mendel rats. In mice, high doses of SAM reduced liter size, but did not increase the incidence of developmental abnormalities,[7] lower doses had no observable adverse effects.[3,4]

Teratogenicity. Administration of about 2.0 mg SAM/kg BW to pregnant rats during organogenesis produced malformations in the offspring, particularly involving NS.[5] Other reported abnormalities in the rat offspring included an increase in dilated renal pelvis and alterations in fetal urine formation.[6] Pregnant rats were given SAM on days 10 and 11 of gestation. The treatment increased the incidence of renal abnormalities (hydronephrosis) and retarded renal development in offspring.[5] SAM induced CNS teratogenicity in hamsters when the pregnant dam was treated orally or with topical application.[8]

Allergenic Effect. Hypersensitivity manifests itself in form of skin rashes and anaphylactic phenomena.[9]

Mutagenicity.

In vitro genotoxicity. SAM has been shown to be negative in *Salmonella* mutagenicity assay.[013]

Chemobiokinetics. Absorbed SAM could undergo conjugation or hydrolysis to *salicylic acid*, predominantly in the liver by microsomal system and mitochondria or in the intestinal tract. Excreted partly as *sulfate*, unchanged SAM was eliminated in the urine.[9, 011]

Regulations. ***U.S. FDA*** (1998) approved use of SAM (1) in adhesives as a component of articles (a monomer) intended for use in packaging, transporting, or holding food in accordance with the conditions prescribed in 21 CFR part 175.105; (2) as a monomer in the manufacture of semirigid and rigid acrylic and modified acrylic plastics in the manufacture of articles intended for use in contact with food in accordance with the conditions prescribed in 21 CFR part 177.1010; and (3) as a component of the uncoated or coated food-contact surface of paper and paperboard intended for use in producing, manufacturing, packaging, processing, preparing, treating, packing, transporting, or holding aqueous and fatty foods in accordance with the conditions prescribed in 21 CFR part 176.170.

References:

1. Jenner, P. M., Hagan, E. C., Taylor, J. M., Cook, E. L., and Fitzhugh, O. G., Food flavorings and compounds of related structure, *Food Cosmet. Toxicol.*, 2, 327, 1964.
2. Clarke, E. G. and Clarke, M. L., *Veterinary Toxicology*, Williams & Wilkins Co., Baltimore, Maryland, 1975, 125.
3. Lamb, J., Reproductive toxicology. Methyl salicylate, *Environ. Health Perspect.*, 105, 321, 1997.
4. Lamb, J., Reproductive toxicology. Methyl salicylate, *Environ. Health Perspect.*, 105, 323, 1997.
5. Warkany J. and Takacs, E., Experimental production of congenital malformations in rats by salicylate poisoning, *Am. J. Pathol.*, 35, 315, 1959.
6. Daston, G. P. et al., Functional teratogens of the rat kidney. I. Colchicine, dinoseb, and methyl salicylate, *Fundam. Appl. Toxicol.*, 11, 381, 1988.
7. Gibson, J. E., *Environ. Health Perspect.*, 15, 121, 1976.
8. Overman, D. O. and White, J. A., Comparative teratogenic effects of methyl salicylate applied orally or topically to hamsters, *Teratology*, 28, 421, 1983.

9. *The Pharmacological Basis of Therapeutics*, L. S. Goodman and A. Gilman, Eds., 5th ed., MacMillan Publ. Co., Inc., New York, 1975, 336.

SESAME OIL

CAS No 8008-74-0
RTECS No VU3940000
Abbreviation. SO.

Synonyms and **Trade Names.** Benne oil; Gingilli oil; Sextra; Teel oil.

Composition. Constituents: olein, stearin, palmitin, myristin, linolein, sesamin, sesamolin.[020]

Properties. Pale yellow, almost odorless oil. Insoluble in water; poorly soluble in alcohol.

Applications. Used in cosmetics.

Acute Toxicity. LD_{50} *i/p* exceeded 50 g/kg BW in mice.[1]

Carcinogenicity. Mice received SO by *s/c* injection at the dose of 2.0 g/kg BW for 1 week. SO was found to be equivocal carcinogen by *RTECS* criteria.[2]

Regulations. ***U.S. FDA*** (1998) approved the use of SO in resinous and polymeric coatings used as the food-contact surfaces of articles intended for use in producing, manufacturing, packing, processing, preparing, treating, packaging, transporting, or holding food in accordance with the conditions prescribed in 21 CFR part 175.300.

References:

1. Natl. Technical Information Service, AD691-490
2. *Proc. Fedn. Am. Soc. Exp. Biol.*, 38, 1450, 1979.

SHALE RESINS

Properties. Products of thermal processing of bituminous shales. A dark, viscous liquid with a specific, irritant odor. Odor perception threshold is 0.02 mg/l.

Applications. Used as a raw material in the manufacture of lacquers, impregnation compounds and anti-corrosion coatings.

Acute Toxicity. LD_{50} of generator shale resin is 10.4 g/kg BW for mice and 9.76 g/kg BW for rats,[1] or 10.4 and 8.15 g/kg BW, respectively.[2]

Poisoning with high doses affects CNS functions. Death occurs from respiratory failure. Gross pathology examination revealed changes in the brain, liver, and GI tract.

Repeated Exposure. Rats were dosed by gavage with 1.0 g of shale generator resin per kg BW for 15 days and with 0.1 mg/kg BW for 4 months without manifestations of the toxic effect[3]

Long-term Toxicity. In a 6-month study, only the highest dose level of 25 mg/kg BW produced a decline in cholinesterase and AST activity in rats.[2]

Reproductive Toxicity.

Gonadotoxic effect was found. Exposure to 125 mg/kg BW dose of shale resin disturbed the oestrus cycle in rats.

Embryotoxicity. Treatment with 1.2 g/kg BW dose during pregnancy led to almost total fetal mortality and embryoresorption. A 0.12 g/kg BW dose was ineffective[3]

Teratogenic effect was also noted.[3]

Mutagenicity.

In vivo cytogenetics. Mice received 1.0% solution of water formed during shale distillation. The treatment induced CA in the bone marrow cells.[4]

Standards. ***Russia***. Recommended MAC: 0.02 mg/l (organolept.).

References:

1. Bidnenko, L. I., Data on toxicity of generator shale resins, in *Hygienic Aspects of the Use of Polymeric Materials and Articles Made of Them*, L. I. Medved', Ed., All-Union Research Institute of Hygiene and Toxicology of Pesticides, Polymers and Plastic Materials, Kiyv, 1969, 280 (in Russian).

2. Veldre, N. A., in *Proc. of the 3rd Congress of Epidemiologists, Microbiologists, Infectionists and Hygienists of Estonia*, Tallinn, 1977, 251 (in Russian).
3. Yanes, Kh. Ya. and Annus, Kh. I., in *Proc. of the 2nd Congress of Epidemiologists, Microbiologists, Infectionists and Hygienists of Estonia*, Tallinn, 1972, 357 (in Russian).
4. Meyne, J. and Deaven, L. L., Cytogenetic effects of shale-derived oils in murine bone-marrow, *Environ. Mutagen.*, 4, 639, 1982.

SILICIC ACID

Molecular Formula. approx. H_2O_3Si
M = 78.11
CAS No 7699-41-4
RTECS No VV8850000
Abbreviation. SA.

Synonym and **Trade Name.** Bio-Sil; Metasilicic acid.

Composition. Partially structured colloidal solution of silicon dioxide.

Properties. White amorphous powder. Insoluble in water.

Applications. Basic inorganic high-molecular-mass flocculent; mode of action: anionic effect.

Acute Toxicity. LD_0 *i/v* is 234 mg/kg BW.[1]

Long-term Toxicity. SA leached out from asbestos is considered to be responsible for some of the pathological alteration associated with asbestos toxicity.[2]

Mutagenicity.

In vitro genotoxicity. Double-stranded DNA showed the formation of an increasing number of strand breaks when incubated with increasing concentrations of SA.[3]

Chemobiokinetics. Data from a rat study in which radiolabeled SA was injected *i/v,* indicated that very little protein binding occurred in the blood stream, and most of the *silicon* was distributed to the kidney, liver, and lungs, with relatively little reaching the reproductive organs or the brain.[4]

Regulations. ***U.S. FDA*** (1998) approved the use of SA in the manufacture of cellophane for packaging food in accordance with the conditions prescribed in 21 CFR part 177.1200. Calcium silicate and sodium metasilicate are considered to be *GRAS*.

Standards. ***Russia.*** MPC: 50 mg/l with respect to SiO_2 concentration.

References:

1. *AMA Arch. Ind. Health.*, 17, 204, 1958.
2. Rahman, Q., Viswanathan, P. N., and Tandon, S. K., *In vitro* dissolution of three varieties of asbestos in physiological fluids, *Work Environ. Health*, 11, 39, 1974.
3. Khan, S. G., Rizvi, R. Y., Hadi, S. M., and Rahman, Q., Strand breakage in DNA by silicic acid, *Mutat. Res.*, 208, 27, 1988.
4. Adler, A. J. et al., Uptake, distribution, and excretion of ^{31}Silicon in normal rats, *Am. J. Physiol.*, 251, E670, 1986.

SILICIC ACID, CALCIUM SALT

Molecular Formula.
M = 116.2
CAS No 1344-95-2
59787-14-3
RTECS No VV9150000
Abbreviation. SAC.

Synonyms and **Trade Names.** Calcium hydrosilicate; Calcium monosilicate; Calcium polysilicate; Calcium silicate; Mineral wool; Microcel; Solex; Wollastonite.

Applications. Reinforcing filler in elastomers and plastics. SAC is used in Portland cement, in lime glass, and a number of other applications.

Properties. White or creamy colored free-floating powder. Insoluble in water.

Regulations. ***U.S. FDA*** (1998) approved the use of SAC as colorants in the manufacture of articles or components of articles intended for use in producing, manufacturing, packing, processing, preparing, treating, packaging, transporting, or holding food, subject to the provisions and definitions set in 21 CFR part 178.3297. *GRAS* when used at levels not exceeding 2.0% in table salt and 5.0% in baking powder in accordance with GMP.

SILICON DIOXIDE

Molecular Formula. SiO_2
M = 60.09
CAS No 7631-86-9
RTECS No VV7310000
Abbreviation. SD.

Synonyms and **Trade Names.** Aerosil, powdered silica gel; Porasil; Silanox; Silica; Ultrasil VN; White car bon; Zorbax.

Properties. SD occurs in both crystalline and amorphous form. Poorly soluble in water. Chemically stable.

Applications. A filler. Used in the manufacture of fire- and acid-proof packing materials.

Acute Toxicity. LD_{50} was found to be above 22.5 g/kg BW in rats and above 15 g/kg BW in mice.[1]

Long-term Toxicity. No pathologic change was found in rats and dogs as the result of feeding of crystalline or amorphous SD (IARC 42-88).

Mutagenicity.

In vitro genotoxicity. Two types of SD, namely Min-U-Sil 5 and Min-U-Sil, induced micronucleus formation in Syrian hamster embryo cells but not CA and SCE in Chinese hamster lung (*V79*) and human lung (*Hel 299*) cells.[2,3]

In vivo cytogenetics. Quartz did not induce micronuclei in mice treated *in vivo.*[3]

Carcinogenicity classification (inhalation). An IARC Working Group concluded that there is sufficient evidence for the carcinogenicity of respirable crystalline SD in *experimental animals.*

The Annual Report of Carcinogens issued by the U.S. Department of Health and Human Services (1998) defines silica crystalline (respirable size) to be a substance which may reasonably be anticipated to be carcinogen, i.e., a substance for which there is limited evidence of carcinogenicity in *humans* or sufficient evidence of carcinogenicity in *experimental animals.*

IARC: 2A.

Regulations. ***U.S. FDA*** (1998) listed SD (1) as a component of adhesives intended for use in contact with food in accordance with the conditions prescribed in 21 CFR part 175.105; (2) as a colorant of the uncoated or coated food-contact surface of paper and paperboard intended for use in producing, manufacturing, packaging, processing, preparing, treating, packing, transporting, or holding aqueous and fatty foods in accordance with the conditions prescribed in 21 CFR part 178.3297; (3) as a component of a defoaming agent intended for articles which may be used in producing, manufacturing, packing, processing, preparing, treating, packaging, transporting, or holding food in accordance with the conditions prescribed in 21 CFR part 176.200; (4) in the manufacture of cellophane for packaging food in accordance with the conditions prescribed in 21 CFR part 177.1200; (5) as a filler in the manufacture of rubber articles intended for repeated use in producing, manufacturing, packaging, processing, preparing, treating, packing, transporting, or holding food in accordance with the conditions prescribed in 21 CFR part 177.2600; (6) in vinyl acetate/crotonic acid copolymer for use as a coating or as a component of coating which is the food-contact surface of polyolefin films intended for packaging food in accordance with the conditions prescribed in 21 CFR part 175.320; and (7) in the manufacture of cross-linked polyester resins which may be used as articles or components of articles intended for repeated use in contact with food in accordance with the conditions prescribed in 21 CFR part 177.2420. It may be used (8) in food in accordance with the conditions prescribed in 21 CFR part 172.480 and (9) as a defoaming agent in the processing of food in accordance with the conditions prescribed in 21 CFR part 173.320.

Affirmed as *GRAS.*

References:

1. *The Agrochemicals Handbook*, D. Hartley and H. Kidd, Eds., 2nd ed., The Royal Society of Chemistry, Lechworth Herts, England, 1987.
2. Nagalakshmi, R., Nath, J., Ong, T., and Whong, W.-Z., Silica-induced micronuclei and chromosomal aberrations in Chinese hamster lung (*V79*) and human lung (*Hel 299*) cells, *Mutat. Res.*, 335, 27, 1995.
3. Morita, T., Asano, N., Awogi, T., Sasaki, Yu. F., et al., Evaluation of the rodent micronucleus assay in the screening of IARC carcinogens (Groups 1, 2A and 2B). The summary report of the 6th collaborative study by CSGMT/JEMS-MMS, *Mutat. Res.*, 389, 3, 1997.

SINTAMID-5

Molecular Formula. $C_nH_{2n+1}O[\sim C_2H_4O\sim]_mH$, where $n = 10$ to 16; $m = 5$ or 6
CAS No 12679-83-3
RTECS No WV4300000
Abbreviation. S-5.

Synonyms. Monoethanolamides of synthetic fatty acids, polyethylene glycol esters; Synthetic fatty acids, *N*-mono-(2-polyethylene glycolethyl)amide.

Composition. Hydroxylation product of synthetic fatty acid (C_{10} - C_{16}) amides.

Properties. Yellow, paste-like mass. Readily soluble in water and alcohol. Threshold concentrations appeared to be 6.7 mg/l for taste and 0.3 mg/l for foam formation.[1]

Applications. S-5 is used as emulsifier, dispersant, antistatic.

Acute Toxicity. LD_{50} is 3.5 g/kg BW in rats, and 17.9 g/kg BW in mice.[1]

However, according to other data, these values are 7.5 and 5.0 g/kg BW, respectively.[2] Poisoning is accompanied by alteration in the CNS and GI tract. Acute action threshold is found to be 1.9 g/kg BW.

Repeated Exposure revealed slight cumulative properties. 1/20 and 1/100 LD_{50} caused less than 50% animal mortality.[1]

Long-term Toxicity. Rats received 100 mg/l in their drinking water. The treatment caused thyroid hypofunction. Gross pathology examination revealed venous congestion and focal hemorrhages in the liver and other visceral organs. No other changes were found.[3]

Carcinogenicity. Benzo[a]pyrene is found in *S*-5, but no carcinogenic effect was observed on dermal application.[4]

Allergenic Effect is found on dermal application.[5]

Standards. ***Russia*** (1995). MAC and PML: 0.1 mg/l (organolept., foam).

References:

1. Trikulenko, V. I., Biological effect of some new surfactants and their safe levels in water reservoirs, *Gig. Sanit.*, 3, 14, 1978 (in Russian).
2. Mikhailets, I. B., Toxic properties of some antistatics, *Plast. Massy*, 12, 27, 1976 (in Russian).
3. Mozhaev, E. A., *Pollution of Reservoirs by Surfactants (Health Safety Aspects)*, Meditsina, Moscow, 1976, 124 (in Russian).
4. Yanysheva, N. Ya., Voloshchenko, O. I., Chernichenko, I. A., et al., On carcinogenic effects of certain surfactants, components of synthetic detergents used by population, *Gig. Sanit.*, 7, 9, 1982 (in Russian).
5. Yes'kova-Soskovets, L. B., Sautin, A. I., and Rusakov, N. V., Allergenic properties of some surfactants, *Gig. Sanit.*, 2, 14, 1980 (in Russian).

SINTONOX 14-19

Molecular Formula. $RCOO(CH_2CH_2O)_nH$, where $n = 7$; $R = C_{13}H_{27}$ - $C_{17}H_{25}$.
M = 550.78

Composition. Higher aliphatic acids, polyethylene glycol esters, mixture.

Properties. Thick syrup-like mass, light brown in color. Readily soluble in water and alcohol.

Applications. Used in the production of plastics.

Acute Toxicity. Administration of 15 to 20 g/kg BW caused no mortality in rats and mice. Administration of high doses is accompanied by apathy and lethargy. Recovery in 2 to 3 days.

Repeated Exposure revealed no cumulative properties. Rats were dosed by gavage with 2.0 g/kg BW for 30 days. No changes in the blood morphology or impairment of kidney function were observed. However, an in crease in prothrombin time was noted.

Reference:

Vasilenko, N. M. and Kudrya, M. Y., Toxicological characteristics of sintonox, *Gig. Truda Prof. Zabol.*, 4, 58, 1983 (in Russian).

SODIUM BISULFATE

Molecular Formula. $HO_4S.Na$
M = 120.06
CAS No 7681-38-1
RTECS No VZ1860000
Abbreviation. SBS.

Synonyms and **Trade Name.** Sodium acid sulfate; fused; Sodium hydrogen sulfate, solid; Sodium hydrogen sulfate, solution; Sodium pyrosulfate; Sulphuric acid, monosodium salt.

Applications. Used as a cathartic.

Reproductive Toxicity.

Teratogenicity. SBS caused a small increase in skeletal anomalies in mice after antepartum treatment of the pregnant animals.[1]

Mutagenicity.

In vitro genotoxicity. SBS was negative in *B. Micrococcus aureus.*[2]

Regulations. ***U.S. FDA*** (1998) approved the use of SS in adhesives as a component of articles intended for use in packaging, transporting, or holding food in accordance with the conditions prescribed in 21 CFR part 175.105.

References:

1. Arcuri, P. A. and Gautieri, R. F., Morphine-induced fetal malformations. 3. Possible mechanisms of action, *J. Pharmacol. Sci.*, 62, 1626, 1973.
2. Clark, J., The mutagenic action of various chemicals on *Micrococcus aureus*, in *Proc. Okla. Acad. Sci.*, 34, 114, 1953.

SODIUM BROMIDE

Molecular Formula. NaBr
M = 102.91
CAS No 7647-15-6
RTECS No VZ3150000
Abbreviation. SB.

Synonyms. Bromide salt of sodium; Trisodium tribromide.

Properties. White crystals, granules, or powder. 1.0 g dissolves in 1.1 ml water, or about 16 ml alcohol.[020]

Acute Toxicity. Bromide has a low acute oral toxicity. LD_{50} is 3500 mg/kg BW in rats.[1]

Short-term Toxicity.

Observations in man. Systemic effects of bromide ion are predominantly mental. Other manifestations of toxic action are as follows: drowsiness, irritability, ataxia, vertigo, confusion, mania, hallucinations, and coma.[022]

Animal studies. In a 3-month study, rats exhibited changes in thyroid weight (goiter) and in erythrocyte count. Weight loss or decreased weight gain were also noted (Vanlogten et al., 1974).

Long-term Toxicity.

Observations in man. A total of 4.0 and 9.0 mg SB/kg BW was administered orally to 45 healthy female volunteers. The experiment lasted for six menstrual cycles: only during the first three cycles was bromide administered. Except for nausea in relation to the intake of bromide, no adverse effects were observed. No significant differences were observed in the thyroid hormones. There were no significant differences in EEG in both treated groups. From this experiment and previous experiments, a NOEL in humans for SB of 4.0 mg/kg BW is proposed.[3]

Animal studies. In a 6-month oral study, alteration of classical conditioning reflexes and changes in serum composition (bilirubin, cholesterol, etc.) were noted.[2]

Wistar rats fed diets of 75 to 19,200 ppm SB showed depression in grooming, in coordination of hind legs, growth retardation, low erythrocyte counts, double count in neutrophil granulocytes, increase in thyroid weight, and decrease in prostate weight (Vanlogten et al., 1974).

Immunotoxicity. Male rats were fed a normal or SB-enriched diet for 4 or 12 weeks. SB concentrations were 20 to 19,200 mg/kg diet. Histopathological examination revealed an activation of the thyroid and a decreased spermatogenesis in the testes. A decrease was noted in the amount of thyroxine in the thyroid, while immunoreactivity for thyroid-stimulating hormone and for adrenocorticotropic hormone was increased. The concentration of thyroxine, testosterone, and corticosterone in the serum appeared to be decreased. The release of growth hormone was suppressed. Most of these changes were restricted to rats on the highest treatment level. It is concluded that SB, at least in high doses, directly disturbs the function of the thyroid, testes, and adrenals.[5]

Reproductive Toxicity.

Gonadotoxicity. Feeding SB to rats for 90 days in concentrations of 75 to 19,200 mg/kg led to a complex of changes in the endocrine system, thyroid activation being the most prominent. Furthermore, in the highest dose groups a decrease in spermatogenesis in the testes and decreased secretory activity of the prostate or a reduction in the number of *corpora lutea* in the ovaries were found. A three-generation reproduction study of the same dietary concentrations showed in the two highest dose groups a decrease in fertility which appeared to be reversible upon bromide withdrawal.[5]

Teratogenicity.

Animal studies. In the above-cited study, gross pathology examination revealed no changes in the offspring. From these studies, a NEL for bromide ion of 240 mg/kg diet was determined, corresponding to a tentative ADI of 0.12 mg/kg BW. This is in good agreement with a preliminary ADI of 0.1 mg/kg BW established in an experiment with human volunteers, but is considerably lower than the ADI of 1.0 mg/kg BW estimated by FAO/WHO.[5]

Chemobiokinetics. Following oral administration, bromide salts are rapidly absorbed from the lower part of the small intestine. *Bromide* is widely distributed throughout extracellular fluids including secretions and transudates. The biological half-life of bromide, and consequently the serum levels, are strongly dependent on chloride intake.[5]

Standard. *EPA, State of Maryland.* MCL: 0.25 mg/l (*Br* ion)

References:

1. *J. Pharmacol. Exp. Ther.*, 55, 200, 1935.
2. Elpiner, L. I., Shafirov, Yu. B., Khovakh, I. M., Shub, O. A., and Gurvich, I. A., Health bases for permissible contents of bromine in drinking water, *Gig. Sanit.*, 1, 13, 1972 (in Russian).
3. van Gelderen, C. E., Savelkoul, T. J., Blom, J. L., van Dokkum, W., and Kroes, R., The no-effect level of sodium bromide in healthy volunteers, *Hum. Exp. Toxicol.*, 12, 9, 1993.
4. Loeber, J. G., Franken, M. A., and van Leeuwen, F. X., Effect of sodium bromide on endocrine parameters in the rat as studied by immunocytochemistry and radioimmunoassay. *Food Chem. Toxicol.*, 21, 391, 1983.
5. van Leeuwen, F. X., den Tonkelaar, E. M., and van Logten, M. J., Toxicity of sodium bromide in rats: effects on endocrine system and reproduction, *Food Chem. Toxicol.*, 21, 383, 1983.

SODIUM FLUORIDE

Molecular Formula. NaF

M = 41.99

CAS No 7681-49-4

39287-69-9

RTECS No WB0350000

Abbreviation. SF.

Synonyms and **Trade Names.** Disodium difluoride; Flura; Flursol; Pedident; Sodium hydrofluoride; Trisodium trifluoride.

Properties. White powder or colorless and odorless crystals with a salty taste. Solubility in water is 43 g/l at 25°C.[020]

Applications. Fungicide, rodenticide. Fluoridation agent in drinking water supply; a component of glues and adhesives; a toothpaste ingredient. Used in the glass manufacture. Leather bleach.

Exposure. Fluoride has cariostatic efficacy. A concentration of 1.0 ppm in water is beneficial for caries prevention and does not appear to exacerbate any diseases.

Acute Toxicity.

Observations in man. LD_0 is 1.7 g/kg BW. Poisoning caused respiratory depression, nausea, or vomiting.[1] According to other data, LD_0 is reported to be 71 mg/kg BW. Manifestations of toxic action included tremor, changes in teeth, and in the musculoskeletal system.[020]

Animal studies. LD_{50} is 32 mg/kg BW in rats, 46 mg/kg BW in mice (IARC 27-273).

Reproductive Toxicity. SF affected reproductive function of mice.[2] Placental transfer of fluorides was reported.[3]

Embryotoxicity. A increase in post-implantation mortality was observed in rats given 10 mg SF/kg BW during pregnancy.[3]

Gonadotoxicity. Rats were exposed *in utero* and during lactation at one of four concentrations (25 to 250 ppm). The treatment did not adversely affect testis structure or spermatogenesis in the rat.[4]

Mutagenicity. Published results are contradictory and often very confusing.[5]

In vitro genotoxicity. SF was reported to be negative in the *Salmonella*/microsome assay strains *TA97a, TA98, TA100, TA102*, and *TA1535* both in the presence and absence of Aroclor induced rat liver microsomes.[6]

SF produced a significant increase in the frequency of CA at the chromatid level in Chinese hamster ovary cells by an indirect mechanism involving the inhibition of DNA synthesis/repair,[7] as well as in cultured rat bone marrow cells.[8,9]

SF stimulated SCE, and unscheduled DNA synthesis in cultured human diploid fibroblasts. SF was found to be capable of inducing neoplastic transformation of Syrian hamster embryo cells in culture.[10] It was positive in L5178Y mouse lymphoma cell forward mutation assay.[11]

In vivo cytogenetics. Chronic fluoride consumption of drinking water containing SF at concentrations of 1 to 75 ppm had no effect on the frequency of SCE in Chinese hamster bone-marrow cells.[12]

Carcinogenicity. A modest increase in osteosarcomas was seen in male rats, but the results were called equivocal. No increases in neoplasm development were seen in mice or in female rats.[13]

Carcinogenicity classification. The IARC Working Group concluded that SF is not classifiable as to its carcinogenicity to *humans*.

IARC: 3.

Regulations. ***U.S. FDA*** (1998) approved the use of SF in adhesives as a component of articles intended for use in packaging, transporting, or holding food in accordance with the conditions prescribed in 21 CFR part 175.105.

References:

1. Abukurah, A. R., Moser, A. M., Baird, C. L., Randall, R. E., Setter, J. G., and Blanke, R. V., Acute sodium fluoride poisoning, *JAMA*, 222, 816, 1972.
2. Messer, H. H., Armstrong, W. D., and Singer, L., Influence of fluoride intake on reproduction in mice, *J. of Nutrition*, 103, 1319, 1973.
3. Teuer, R. C., Mahoney, A. W., and Sarett, H. P., Placental transfer of fluoride and tin in rats given various fluoride and tin salts, *J. of Nutrition*, 101, 525, 1971.
4. Sprando, R. L., Collins, T. F., Black, T., Olejnik, N., and Rorie, J., Testing the potential of sodium fluoride to affect spermatogenesis: a morphometric study, *Food Chem. Toxicol.*, 12, 1117, 1998.

5. Li, Y. M., Dunipace, A. J., and Stookey, G. K., Genotoxic effects of fluoride: a controversial issue, *Mutat. Res.,* 195, 127, 1988.
6. Li, Y., Dunipace, A. J., and Stookey, G. K., Absence of mutagenic and antimutagenic activities of fluoride in Ames *Salmonella* assays, *Mutat Res.,* 190, 229, 1987.
7. Tsutsui, T., Suzuki, N., and Ohmori, M., Sodium fluoride-induced morphological and neoplastic transformation, chromosome aberrations, sister chromatid exchanges, and unscheduled DNA synthesis in cultured Syrian hamster embryo cells, *Cancer Res.,* 44, 938, 1984.
8. Aardema, M. J., Gibson, D. P., and LeBoeuf, R. A., Sodium fluoride-induced chromosome aberrations in different stages of the cell cycle: a proposed mechanism, *Mutat. Res.,* 223, 191, 1989.
9. Khalil, A. M., Chromosome aberrations in cultured rat bone marrow cells treated with inorganic fluorides, *Mutat. Res.,* 343, 67, 1995.
10. Tsutsui, T., Suzuki, N., Ohmori, M., and Maizumi, H., Cytotoxicity, chromosome aberrations and unscheduled DNA synthesis in cultured human diploid fibroblasts induced by sodium fluoride, *Mutat. Res.,*139, 193, 1984
11. Caspary, W. J., Myhr, B., Bowers, L., McGregor, D., Riach, C., and Brown, A., Mutagenic activity of fluorides in mouse lymphoma cells, *Mutat. Res.,* 187, 165, 1987
12. Li, Y. M., Zhangm, W., Noblitt, T. W., Dunipace, A. J., and Stookey, G. K., Genotoxic evaluation of chronic fluoride exposure: sister-chromatid exchange study, *Mutat Res.,* 227, 159, 1989.
13. Bucher, J. R., Hejtmancik, M. R., Toft, J. D. II, et al., Results and conclusions of the National Toxicology Program's rodent carcinogenicity studies with sodium fluoride, *Int. J. Cancer*, 48, 733, 1991.

SODIUM IODIDE

Molecular Formula. INa
M = 149.89
CAS No 7681-82-5
RTECS No WB6475000
Abbreviation. SI.

Synonym and **Trade Name.** Ioduril; Sodium monoiodide.

Properties. Colorless, odorless, cubic crystals or white powder with saline, somewhat bitter taste. Solubility in water up to 1.0 g/700 ml; soluble in alcohol.

Acute Toxicity. LD_{50} is 1.0 g/kg BW in mice.[1]

Short-term Toxicity. Rats received 1.0 to 100 mg iodine/l or iodide (*NaI*) in drinking water for 100 days. The treatment produced no effect on BW gain and testes. An increase in the thyroid weight of males correlated with increasing contents of iodide but not iodine in the water. A decrease in BW of females was noted at the highest dose of iodide. Both substances affect thyroid hormone status in rats.[2]

Immunotoxicity. SI is associated with autoimmune thyroid disease or autoimmune responses in experimental animals.[3]

Reproductive Toxicity.

Gonadotoxicity. Radioactive SI exhibited gonadotoxic effect, it caused sperm abnormalities and micronuclei in mice which were probably due to ionizing radiation in the I^{125} isotope used.[4]

Chemobiokinetics. Iodide can be transferred to breast milk.[5] Fetuses of guinea pigs and sheep accumulated 2 to 6 times more iodide than their mothers.[6]

Regulations. ***U.S. FDA*** (1998) approved the use of SI as a component of sanitizing solutions used on food-processing equipment and utensils, and on other food-contact articles as specified in 21 CFR part 178.1010.

References:

1. Izmerov, N. F., Sanotsky, I. V., and Sidorov, K. K., *Toxicometric Parameters of Industrial Toxic Chemicals under Single Exposure*, USSR/SCST Commission for UNEP, Center Int. Projects, Moscow, 1982, 105 (in Russian).

2. Sherer, T. T., Thrall, K. D., and Bull, R. J., Comparison of toxicity induced by iodine and iodide in male and female rats, *J. Toxicol. Environ. Health.*, 32, 89, 1991.
3. Bagazzi, P. E., Autoimmunity caused by xenobiotics, *Toxicology*, 119, 1, 1991.
4. Lavu, S., Reddy, P. P., and Reddi, O. S., Iodine-125 induced micronuclei and sperm head abnormalities in mice, *Int. J. Radiat. Biol. Relat. Stud. Phys. Chem. Med.*, 47, 249, 1985.
5. Hedrick, W. R., Di Simone, R. N., and Keen, R. L., Radiation dosimetry from breast milk excretion of radioiodine and pertechnetate, *J. Nucl. Med.*, 27, 1569, 1986.
6. Book, S. A., Wolf, H. G., Parker, H. R., and Bustad, L. K., *Health Phys.*, 26, 533, 1974.

SODIUM NITRATE

Molecular Formula. $NNaO_3$
M = 85.01
CAS No 7631-99-4
RTECS No WC5600000
Abbreviation. SN.

Synonyms and **Trade Names.** Cubic niter; Nitric acid, sodium salt; Soda niter; Sodium saltpeter.

Properties. SN occurs as colorless, transparent, trigonal, or rhombohedron crystals, white granules or powder with a saline, slightly bitter taste. Soluble in water (approximately, 1:1 at 25°C); poorly soluble in alcohol, dissoci ated at neutral *pH.*

The nitrate ion (NO_3^-) is the stable form of combined nitrogen for oxygenated systems.

Applications. A food preservative.

Exposure. Nitrates have been found in many crops, vegetables, and drinking water.

Acute Toxicity.

Observations in man. Fatal poisoning were noted after single intakes of 4 to 50 g nitrate. Some cases of methemoglobinemia have been reported in adults consuming high doses by accident or as medical treatment.[1]

Animal studies. LD_{50} values are reported to be 6.6 to 9.0 g/kg BW. LD_{50} is 0.45 g/kg BW in cows (Speiiers et al., 1989).[1]

Short-term Toxicity. One-month-old male rats were given drinking water that contained 400 ppm nitrate for 16 weeks. Except for higher incidence of pulmonary lesions, administration of nitrate at this level had no significant effect on any of the biochemical parameters measured in the blood or the mutagenicity of liver extract.[2]

Long-term Toxicity.

Observations in man. The toxicity of nitrate to humans is thought to be solely due to its reduction to nitrite. Methemoglobinemia has been observed at nitrate levels of 50 mg/l and above, and almost exclusively in infants under 3 months of age.[3]

Animal studies. 2-Month-old male rats were given drinking water that contained 4000 ppm SN for 14 months. The *metHb* level in animals receiving nitrate fluctuated from 1.0 to 35% as compared with 0 to 2.0% for the groups receiving only water.[2] In a 2-year study, rats received dietary concentrations of 1.0 or 5.0% SN upward. The higher concentration caused only growth inhibition. The NOAEL of 1.0% concentration in the feed (500 mg SN/kg BW) was established in this study.[1] Shtabsky and Fedorenko proposed ADI of 5.0 mg/kg BW for adults and 2.5 mg/kg BW for children up to 1 year old.[4]

Reproductive Toxicity.

Embryotoxicity. Pregnant rats were given 1.0 or 0.3% *nitrate* in the diet on days 9 and 10 of gestation. The treatment caused no embryotoxic or teratogenic effects.

Teratogenicity.

Observations in man. A threefold increase in CNS and muskuloskeletal system congenital defects was observed in women taking 5 to 15 ppm nitrates. A fourfold increase was noted in those drinking water containing over 15 ppm. Authors suggested nitrates were converted to *nitrites* and then to *amides* including *N-nitrosamines.*[5]

Gonadotoxicity. No gonadotoxic effect was observed in mice. The reduced fertility was attributed to the general toxicity of the nitrate exposure and not to an adverse effect on germ cells.[6,7]

Only very high nitrate concentrations could impair the reproductive behavior of guinea pigs. The NOAEL for this effect appeared to be 10 g/l.[1]

Mutagenicity.

In vitro genotoxicity. SN induced CA and micronuclei in rat and mouse bone marrow cells.[8] SN was found to be positive in CA test *in vitro* on Chinese hamster cells.[9]

In vivo cytogenetics. One-month-old male rats were given drinking water that contained 400 ppm nitrate for 16 weeks. In the *Salmonella*/microsome test SN showed an absence of mutagens in the feces extract of the animals.[2]

Carcinogenicity. Nitrate did not show carcinogenic properties in laboratory animals.

Chemobiokinetics. On oral administration, SN was rapidly absorbed and excreted unchanged.[019] Transformation of nitrate to nitrite usually occurs within the GI tract.

Regulations. ***U.S. FDA*** (1998) approved the use of SN (1) in adhesives as a component (monomer) of articles intended for use in packaging, transporting, or holding food in accordance with the conditions prescribed in 21 CFR part 175.105; and (2) as a food preservative and color fixative in accordance with the conditions prescribed in 21 CFR part 172.170.

References:

1. *Guidelines for Drinking-Water Quality*, 2nd ed., vol 2, Health Criteria and other Supporting Information, WHO, Geneva, 1996, 313.
2. Chow, C. K., Effect of nitrate and nitrite in drinking water on rats, *Toxicol. Lett.*, 6, 199, 1980.
3. Walton, C., Survey of literature relating to infant methemoglobinemia due to nitrate-contaminated water, *Am. J. Publ. Health*, 41, 1986, 1951.
4. Shtabsky, B. M. and Fedorenko, V. I., On toxicology of sodium nitrite and sodium nitrate, *Toxicol. Vest nik*, 5, 22, 1996 (in Russian).
5. Dorsch, M. M., Scragg, R. K. R., McMichael, A. J., Badhurst, P. A., and Dyer, K. F., Congenital malformations and maternal drinking water supply in rural South Australia: A case-control study, *Am. J. Epidemiol.*, 119, 473, 1984.
6. Alavantic, D., Sunjevaric, I., Pecevski, J., et al., *In vivo* genotoxicity of nitrates and nitrites in germ cells of male mice. I. Evidence for gonadal exposure and lack of heritable effects, *Mutat. Res.*, 204, 689, 1988.
7. Alavantic, D., Sunjevaric, I., Cerovic, G., et al., *In vivo* genotoxicity of nitrates and nitrites in germ cells of male mice. II. Unscheduled DNA synthesis and sperm abnormality after treatment of spermatids, *Mutat. Res.*, 204, 697, 1988.
8. Luca, D. et al., Chromosomal aberrations and micronuclei induced in rat and mouse bone marrow cells by sodium nitrate, *Mutat. Res.*, 155, 121, 1985.
9. Ishidate, M. and Odashima, S., Chromosome tests with 134 compounds in Chinese hamster cells *in vitro* - a screening for chemical carcinogens, *Mutat. Res.*, 48, 337, 1977.

SODIUM POLYACRYLATE

Molecular Formula. $[\sim C_3H_4O_2\sim]_n.xNa$

CAS No 9003-04-7

RTECS No WD6826000

Abbreviation. SPA

Trade Names. Polyco; Rhotex GS.

Acute Toxicity. LD_{50} exceeded 40 g/kg BW in rats.

Regulations. ***U.S. FDA*** (1998) approved the use of SPA in adhesives as a component of articles intended for use in packaging, transporting, or holding food in accordance with the conditions prescribed in 21 CFR part 175.105.

Reference:

Proc. Sci. Section Toilet Goods Assoc., 20, 16, 1953.

SORBITAN, MONOOLEATE

Molecular Formula. $C_{24}H_{44}O_6$
M = 428.68
CAS No 1338-43-8
RTECS No WG2932400
Abbreviation. SO.

Synonym and **Trade Names.** Monodehydrosorbitol monooleate; Montan 80; Sorbitan O; Span 80.

Properties. Amber liquid. Dispersible in water, soluble in oils.

Applications. Emulsifying agent.

Toxicity. Obviously non-toxic.

Short-term Toxicity. Rats received 2.5, 5.0, or 10% Span 80 in the diet for 16 weeks. Ingestion caused a significant increase in kidney weights. Liver enlargement was observed on the highest dietary level.

Regulations. ***U.S. FDA*** (1998) approved the use of SO (1) in adhesives as a component of articles intended for use in packaging, transporting, or holding food in accordance with the conditions prescribed in 21 CFR part 175.105; and (2) in resinous and polymeric coatings used as the food-contact surfaces of articles intended for use in producing, manufacturing, packing, processing, preparing, treating, packaging, transporting, or holding food in accordance with the conditions prescribed in 21 CFR part 175.300. It may be safely used (3) in or on food in accordance with the conditions prescribed in 21 CFR part 173.75.

Reference:

Ingram, A. J. et al., Short-term toxicity study of sorbitan monooleate (Span 80) in rats, *Food Cosmet. Toxicol.*, 16, 535, 1978.

SORBITAN, MONOSTEARATE

Abbreviation. SS.

Trade Name. Span 60.

Composition. SS is a mixture of partial stearic and palmitic esters of sorbitol and its mono- and dianhydrides.

Properties. Cream-colored waxy solid with slight odor and bland taste. Insoluble in water.

Applications. Emulsifier and stabilizer in plastics, cosmetics, in food and drugs. Defoaming agent.

Acute toxicity.

Observations in man. Ingestion of 20 g SS produced no signs of intoxication.[019]

Animal studies. LD_{50} is 31 g/kg BW in rats.[1] Administration of up to 15 g/kg BW appeared to be non-lethal (Brandner, 1973).[2]

Repeated Exposure.

Observations in man. Ingestion of 6 g SS for 1 month caused no ill-effects.[019]

Animal studies. Young rats were fed a diet containing 1.0 or 4.0% SS for 6 weeks. The treatment produced no effect on BW gain. Histology investigation revealed no changes in the liver, kidneys, intestine, and bladder (Krantz, 1946).[2]

Short-term Toxicity. Rats received 5.0 or 15% Span 60 in soybean-meal diet for 14 weeks. Ingestion of the diet caused no signs of toxicity but treated animals exhibited lowered feed consumption an retarded BW gain.[3]

Long-term Toxicity. Mice received SS (Span 60) in their diet at dose levels of 0.5, 2.0, or 4.0% for 80 weeks. No retardation of BW gain or any increase in mortality was noted. An increase in kidney weights and higher incidence of nephrosis were observed at the highest dose level.[2] Mice, hamsters, and dogs tolerated ingestion of Span 60 at dietary levels of 5.0 or 10% for up to 1 year without evident manifestation of its toxic action. 15% dietary level caused GI disturbances and reduced food intake; 20% dietary level caused a small but significant depression of growth rate in males and some enlargement of histologically normal livers and kidneys.[4] Rats were given 2.0 to 25% SS in their diet for 730 days. The LOEL of 7.2 g/kg BW was established.[5] Munro et al. (1996) suggested the calculated NOEL of 3.6 g/kg BW for this study.[6]

Reproductive Toxicity. In the above-cited study, no effect was seen on gestation and fertility at any dose level, but survival of the newborn animals and maternal lactation were slightly diminished at the 20% level.[4]

Carcinogenicity. In the above-cited study,[2] administration of 4.0% *Span 60* in the diet produced no tumors in mice.

Chemobiokinetics. Following absorption, SS is completely excreted in urine.[019]

Regulations. ***U.S. FDA*** (1998) approved the use of SS (1) in resinous and polymeric coatings used as the food-contact surfaces of articles intended for use in producing, manufacturing, packing, processing, preparing, treating, packaging, transporting, or holding food in accordance with the conditions prescribed in 21 CFR part 175.300; (2) as a defoaming agent in the processing of food in accordance with the conditions prescribed in 21 CFR part 173.320; and (3) in or on food in accordance with the conditions prescribed in 21 CFR part 172.842.

References:

1. Eagle, E. and Poling, C. E., The oral toxicity and pathology of polyoxyethylene derivatives in rats and hamsters, *Food Res.*, 21, 348, 1956.
2. Hendy, R. J., Butterworth, K. R., Gaunt, I. F., Kiss, I. S., and Grasso, P., Long-term toxicity study of sor bitan monostearate (Span 60) in mice, *Food Cosmet. Toxicol.*, 16, 527, 1978.
3. Chow, B. F., Burnett, J. M., Ling, C. T., and Barrows, L., Effects of basal diets on the response of rats to certain dietary non-ionic surface-active agents, *J. of Nutrition*, 49, 563, 1953.
4. Allison, J. B., Rosenthal, H. L., and Mills, A. H., Effects of non-ionic surface active agents on the growth of animals, *Proc. Fedn. Am. Soc. Exp. Biol.*, 11, 435 1952.
5. Fitzhugh *et al.*, 1959, cit. in *Toxicological Evaluation of Certain Food Additives and Contaminants*, WHO Food Additives Series No 17, The 26th Meeting of the Joint FAO/WHO Expert Committee on Food Additives, Geneva, 1982.
6. Munro, I. C., Ford, R. A., Kennepohl, E., and Sprenger, J. G., Correlation of structural class with No-Observed-Effect Levels: A proposal for establishing a threshold of concern, *Food Chem. Toxicol.*, 34, 829, 1996.

STARCH, 2-HYDROXYETHYL ETHER

CAS No 9005-27-0
RTECS No WI0410000
Abbreviation. SHE.

Synonym and **Trade Names.** Hespander; Hydroethyl starch; Plasmasteril.

Acute Toxicity. LD_{50} exceeded 50 g/kg BW in rats.

Reproductive Toxicity. Rats received doses of 300 mg/kg BW by *i/p* injections on 16 to 21 days of pregnancy. The treatment caused an increase in incidence of abortions.

Hydroxyethyl starch was shown not to be ***teratogenic***. The dose of 50 g/kg BW *i/p* led to abortion in all pregnant rats.

Regulations. ***U.S. FDA*** (1998) approved the use of SHE in adhesives as a component of articles intended for use in packaging, transporting, or holding food in accordance with the conditions prescribed in 21 CFR part 175.300.

Reference:

Ivankovic, S. and Bulow, I., On the lack of teratogenic effect of the plasma expander hydroxyethyl starch in the rat and mouse, *Anaesthesist*, 24, 244, 1975.

STEARIC ACID, MAGNESIUM SALT

Molecular Formula. $C_{36}H_{70}O_4.Mg$
M = 591.37
CAS No 557-04-0
RTECS No WI4390000
Abbreviation. SAM.

Synonyms. Magnesium stearate; Octadecanoic acid, magnesium salt.

Properties. Fine, white, tasteless, and odorless powder or lumps. Soluble in water (40 mg/l at 25°C) and alcohol (200 mg/l at 25°C).

Applications. Used as a lubricant in cosmetics. Anticaking agent in food industry.

Acute Toxicity. LD_{50} is 1.09 g/kg BW in rats. Poisoning produced a decrease in BW gain and an increase in relative liver weights. An increase in mortality in poisoned animals has been reported.

Short-term Toxicity. Male and female rats received 5.0, 10, and 20% SAM in a semi-synthetic diet for 3 months. Retardation of BW gain and urolithiasis were observed in males in the 20%-group. The treatment caused a decrease in relative liver weight and an increase in iron amount in the liver at the highest dosed group. The NOAEL of 2.5 g/kg BW (5% SAM in the diet) was established in this study.

Regulations. ***U.S. FDA*** (1998) approved the use of SAM in the manufacture of resinous and polymeric coatings for the food-contact surface of articles intended for use in producing, manufacturing, packing, processing, preparing, treating, packaging, transporting, or holding food for use only as polymerization cross-linking agent in side seam cements for containers intended for use in contact with food (only of the identified types) in accordance with the conditions prescribed in 21 CFR part 175.300.

Reference:

Sondergaard, D., Meyer, O., and Wurtzen, G., Magnesium stearate given orally to rats. A short-term study, *Toxicology*, 17, 51, 1980.

SUCCINIC ACID

Molecular Formula. $C_4H_6O_4$
M = 118.09
CAS No 119-15-6
RTECS No WM4900000
Abbreviation. SAc.

Synonyms and **Trade Names.** Amber acid; Butanedioic acid; Dihydrofuramic acid; 1,2-Ethanedicarboxilic acid.

SUCCINIC ANHYDRIDE

Molecular Formula. $C_4H_4O_3$
M = 100.08
CAS No 108-30-5
RTECS No WN0875000
Abbreviation. SAn.

Synonyms and **Trade Name.** Butanedioic anhydride; Dihydro-2,5-furandione; Succinyl anhydride; Succinyl oxide; Tetrahydro-2,5-dioxyfuran.

Properties. SAc occurs as white and odorless minute monoclinic prisms with very acid taste. 1.0 g dissolves in 13 ml cold water or 18.5 ml of alcohol.[020] SAn occurs as colorless odorless needles with burning taste. Insoluble in water, soluble in alcohol.

Applications. SAn is used in the production of plastics and dyes. Food additive.

Acute Toxicity. LD_{50} of SAc is 8.4 g/kg BW in rats.[020] Ingestion of large doses caused vomiting and diarrhea.[011]

Repeated Exposure. F344 rats were administered 47 to 750 mg suspensions of SAn/kg BW in corn oil by gavage for 20 days. Compound-related deaths occurred in males at 375 mg/kg BW or higher doses, and in females at 187 mg/kg BW and higher doses. Inflammation of the upper respiratory tract was observed at high doses. In a 16-day study, all B6C3F_1 mice died being treated with 875 mg/kg BW and higher doses. There were no lesions in animals treated with up to 438 mg SAn/kg BW.

Short-term Toxicity. SAc is reported to produce no systemic toxicity.[011] Significant increase in the incidence of mortality was found in rats which received 200 or 400 mg SAn/kg BW and in mice that received 300 or 600 mg SAn/kg BW for 13 weeks. No gross or microscopic lesions were reported in this study.

Long-term Toxicity. In the 2-year NTP study, rats received 50 or 100 mg SAn/kg BW, mice were given 38 or 75 mg SAn/kg BW for 103 weeks. The treatment caused no significant differences in survival, and there was no increase in the incidence of non-neoplastic lesions in all animal tested.

Mutagenicity.

In vitro genotoxicity. SAn was negative in *Salmonella* mutagenicity assay with or without exogenous metabolic activation, it induced no SCE or CA in cultured Chinese hamster ovary cells in the presence or absence of exogenous metabolic activation.

Carcinogenicity. Rats were *s/c* injected twice weekly with 2.0 mg SAn in 0.5 ml arachis oil for 65 weeks (total dose of 260 mg). The treatment caused local sarcomas (IARC 15-265). In the above described 2-year NTP study, there was no increase in the incidence of neoplastic lesions in rats and mice.

Carcinogenicity classifications (SAn). An IARC Working Group concluded that there is limited evidence for the carcinogenicity of SAn in *experimental animals* and there were no data available to evaluate the carcinogenicity of SAn in *humans.*

IARC: 3;

NTP: NE - NE - NE - NE (SAn, gavage).

Regulations.

EU (1990): SAc and SAn are available in the *List of authorized monomers and other starting substances which shall be used for the manufacture of plastic materials and articles intended to come into contact with foodstuffs (Section A).* ***EU*** (1995). SAc is a food additive generally permitted for use in foodstuffs (maximum levels: 3.0 to 6.0 g/kg).

U.S. FDA (1998) considered SAc to be *GRAS.*

Great Britain (1998). SAc and SAn are authorized without time limit for use in the production of polymeric materials and articles in contact with food or drink or intended for such contact.

Reference:

Toxicology and Carcinogenesis Studies of Succinic Anhydride in F344/N Rats and B6C3F$_1$ Mice (Gavage Studies), NTP Technical Report Series No 373, Research Triangle Park, NC, January 1990.

SUCCINONITRILE

Molecular Formula. $C_4H_4N_2$

M = 80.08

CAS No 110-61-2

RTECS No WN3850000

Abbreviation. SN.

Synonyms and **Trade Names.** Butanedinitrile; Deprelin; *s*-Dicyanoethane; Dinile; Ethylene cyanide; Ethylene dicyanide.

Properties. Crystalline solid with a characteristic odor. Readily soluble in water and alcohol. Does not affect the color or transparency of water. Odor perception threshold is 21 mg/l. Taste perception threshold is 200 mg/l.

Applications. SN is used in the production of plastics and nitron fibers.

Acute Toxicity. LD_{50} is 588 mg/kg BW in rats, 100 mg/kg BW in mice, and 120 mg/kg BW in guinea pigs. Administration of high doses is accompanied with an increase in a concentration of *rhodanides* in the urine. Animals exhibit general inhibition, impaired respiration, and exophthalmos with subsequent convulsions and death from respiratory arrest. Gross pathology examination revealed marked congestion in the brain and all visceral organs.[1]

Short-term Toxicity. Guinea pigs were gavaged with 1/5 and 1/20 LD_{50} for 75 days. The treatment resulted in decreased CO_2 expiration, leukopenia, reduced leukocyte phagocytic activity and an increase in the ascorbic acid contents in the liver, kidneys, and suprarenals. Gross pathology examination revealed parenchymatous dystrophy in the visceral organs, circulatory disturbances, and a reduction in the spleen relative weights.[1]

Reproductive Toxicity.

Teratogenicity. Hamsters received up to 6.24 mmol SN/kg BW on day 8 of pregnancy. Neural tube defects and encephaloceles occurred as well as stunting at doses above 0.147 mmol SN/kg BW and the effect was reversed by thiosulfate.[2]

Standards. ***Russia*** (1995). MAC and PML: 0.2 mg/l.

References:

1. Rubinsky, N. D., *Acetonitrile and Succinonitrile, Their Potential Chemical Hazards, and Conditions of Draining Reservoirs*, Author's abstract of thesis, Khar'kov, 1969, 16 (in Russian).
2. Doherty, P. A., Smith, R. P., and Ferm, V. H., Comparison of the teratogenic potential of two aliphatic nitriles in hamsters: Succinonitrile and tetramethylsucconitrile, *Fundam. Appl. Toxicol.*, 3, 41, 1983.

SUCROSE, DIACETATE HEXAISOBUTYRATE

Molecular Formula. $C_{40}H_{62}O_{19}$
M = 847.02
CAS No 126-13-6; 1338-47-2
RTECS No WN6550000
Abbreviation. SAIB.

Composition. A mixture of esters of sucrose.

Synonyms. Saccharose acetate isobutyrate; Sucrose acetate isobutyrate; Sucrose acetoisobutyrate.

Properties. Clear semi-solid.

Applications. SAIB as a direct food additive in human diets. It has been used for over 30 years in many countries as a 'weighting' or 'ensity-adjusting' agent in non-alcoholic carbonated and non-carbonated beverages. Used as a modifier for lacquers and coatings, and for extrudable plastics.

Acute Toxicity. SAIB has been shown to have very low acute toxicities in rats and monkeys. LD is 25.6 g/kg BW in rats and mice. This dose reduced clearance of bromsulphalein.[1,2] Poisoning resulted in muscle weakness, hypermotility, and diarrhea.[3] According to other data, LD_{50} exceeded 5.0 g/kg BW in rats.[4]

Repeated Exposure.

Observations in man. Dose of 20 mg SAIB/kg BW given orally for 14 days did not result in changes in the activities of serum enzymes nor in the clearance of bromsulfalein.[5]

Short-term Toxicity. Daily doses of 0.5 g/kg BW given for 12 weeks caused changes in liver weights and in phosphatase activity.[3]

An addition of SAIB to the beagle dog diet over a period of 12 weeks was associated with an increase in liver size and elevated serum alkaline phosphatase activity with no evidence of pathological change by light microscopy.[6]

Administration of SAIB to Sprague-Dawley rats did not reveal evidence of any effect on hepatobiliary function, and there was no indication of microsomal enzyme induction. BW gain of male rats fed SAIB was decreased, probably as the result of decreased palatability of the diet; SAIB did not affect BW gain in females.[6]

Long-term Toxicity. Sprague-Dawley rats received 0.38 or 9.38% SAIB in their diet (*w/w*) for 104 weeks. There was no difference in the rate of mortality between treated and control animals. No treatment-related lesions were noted. Gross pathology examination revealed massive hemorrhages at multiple sites of visceral organs.[2]

In a 1-year study in rats and monkeys, up to 2.0 and 2.4 g/kg BW caused no effect on liver functions indicative of biliary excretion. A NOEL of 2.0 g/kg BW was established in rats.[5]

$B6C3F_1$ mice were fed dose levels of 1.25, 2.5, and 5.0 g SAIB/kg BW for 2 years; F344 rats were given dose levels of 0.5, 1.0, and 2.0 g SAIB/kg BW for 1 and 2 years. There were no differences in survival, in clinical chemistry, hematology, organ weights, gross necropsy findings, or light microscopy studies in the 1- or 2-year rat studies between SAIB-treated rats or mice and controls. Electron microscopic examinations of liver sections from high dose level rats from the 1-year study also revealed no effects of SAIB treatment. The dose level of 2.4 g/kg BW was found to be the NOAEL in cynomolgus monkeys.[7]

In dogs, SAIB affected the liver at feeding levels of up to 10% in the diet. On prolonged feeding all effects were reversed when SAIB was withdrawn from the diet. The no-effect level in dogs was near 5.0 mg/kg BW, but no effects were not seen in rats fed up to 4.0 g/kg BW, monkeys fed up to 10 g/kg BW, or humans fed up to 20 mg/kg BW.[1]

Reproductive Toxicity. Rats were given single doses of 20, 200, or 2000 mg/kg BW by gavage. The LOEL for reproductive effects was found to be 2000 mg/kg BW.[3]

Teratogenicity. A three-generation reproduction study in F344 rats and teratology studies in F344 rats and New Zealand white rabbits were carried out. Dietary concentrations to provide dose levels of 0.5, 1.0, and 2.0 g SAIB/kg BW were used for the rat studies, and 0.5, 0.85, and 1.2 g SAIB/kg BW doses in corn oil were administered by gavage in the rabbit studies. No morphological abnormalities of soft tissue or skeleton were observed. The highest dose levels administered, 2.0 g SAIB/kg BW in the rat and 1.2 g SAIB/kg BW in the rabbit, were considered to be the NOAELs.[8]

Mutagenicity.

In vitro genotoxicity. SAIB has been found to be negative in the series of bioassays in *Salmonella typhimurium*, in Chinese hamster ovary cells, and in rat hepatocytes.[2] Testing in CHO cells revealed no evidence for genotoxic activity of SAIB as a mutagen, clastogen, or DNA-damaging agent.[9]

Carcinogenicity. In the above-cited experiment, there were no significant increases in benign or malignant tumors in the long-term rat or mouse carcinogenicity studies.[7,10]

$B6C3F_1$ mice received 1.25, 2.5, or 5.0 g SAIB/kg BW in the diet for 104 weeks. No treatment-related decrease in BW gain or changes in hematology parameters were reported. No histological lesions were observed in kidneys. The treatment caused insignificant increase in the incidence of hyperplasia of the perivascular and peribronchial lymphoid tissue of the lung in treated female mice. There was no increase in tumor incidence. F344 rats were given 0.5, 1.0, or 2.0 g SAIB for 104 weeks. No increase in the incidence of tumors was noted.[2]

Chemobiokinetics. SAIB is hydrolyzed prior to absorption, similarly in rats and man. The metabolism and pharmacokinetic studies in rats, dogs, and humans show that SAIB is extensively metabolized in the GI tract, probably to *sucrose* and partially *acylated sucrose*. Humans handle SAIB more like rats than like dogs. Excretion occurs in exhaled air, urine, and feces.[10]

The absorbed SAIB is readily eliminated in the urine and the bile or, after further metabolism, as *carbon dioxide* and *water*. A considerable portion of ingested SAIB and partially deesterified SAIB is eliminated in the feces.[11,12]

According to other data, on a single oral dose or 7 daily doses of 1.0 g, no SAIB was found in urine.[3]

Regulations. ***U.S. FDA*** (1998) approved the use of SAIB in adhesives as a component of articles intended for use in packaging, transporting, or holding food in accordance with the conditions prescribed in 21 CFR part 175.105.

Recommendations. Joint FAO/WHO Expert Committee on Food Additives (1997). ADI: 20 mg/kg BW.[5]

References:

1. Reynolds, R. C. and Chappel, C. I., Sucrose acetate isobutyrate (SAIB): historical aspects of its use in beverages and a review of toxicity studies prior to 1988, *Food Chem. Toxicol.*, 36, 81, 1998.
2. Phillips, J. C., Kingsnorth, J., Rowland, I., Gangolli, S. D., and Lloyd, A. G., Studies on the metabolism of sucrose acetate isobutyrate in the rat and in man, *Food Cosmet. Toxicol.*, 14, 375, 1976.
3. Krasavage, W. J., DiVincenzo, G. D., Astill, B. D., Roudabush, R. L., and Terhaar, C. J., Biological effects of sucrose acetate isobutyrate in rodents and dogs, *J. Agric. Food Chem.*, 21, 473, 1973.
4. *Toxicological Evaluation of Certain Food Additives and Contaminants*, Food Additives Series No 32, WHO, Geneva, 1993.
5. *Evaluation of Certain Food Additives and Contaminants*, The 46th Report of Joint FAO/WHO Expert Committee on Food Additives, Geneva, 1997, 69.
6. Procter, B. G. and Chappel, C. I., Subchronic toxicity studies of sucrose acetate isobutyrate (SAIB) in the rat and dog, *Food Chem. Toxicol.*, 36, 101, 1998.
7. Blair, M. and Chappel, C. I., 4-week range-finding and 1-year oral toxicity studies of sucrose acetate isobutyrate (SAIB) in the cynomolgus monkey, *Food Chem. Toxicol.*, 36, 121, 1998.
8. Mackenzie, K. M., Henwood, S. M., Tisdel, P. J., Boysen, B. G., Palmer, T. E., Schardein, J. L., West, A. J., and Chappel, C. I., Sucrose acetate isobutyrate (SAIB): three-generation reproduction study in the rat and teratology studies in the rat and rabbit, *Food Chem. Toxicol.*, 36, 135, 1998.

9. Myhr, B. C., Cifone, M. A., Ivett, J. L., Lawlor, T. E., and Young, R. R., Lack of genotoxic effects of sucrose acetate isobutyrate (SAIB), *Food Chem. Toxicol.*, 36, 127, 1998.
10. Mackenzie, K. M., Tisdel, P. J., Hall, R. L., Boysen, B. G., Field, W. E., and Chappel, C. I., Oral toxicity and carcinogenicity studies of sucrose acetate isobutyrate (SAIB) in the Fischer 344 rat and B6C3F_1 mouse, *Food Chem. Toxicol.*, 36, 111, 1998.
11. *Foreign Compounds Metabolism in Mammals*, The Chemical Society, vol 4, A review of the literature published 1974 & 1975, London, 1977, 172.
12. Reynolds, R. C., Metabolism and pharmacokinetics of sucrose acetate isobutyrate (SAIB) and sucrose octaisobutyrate (SOIB) in rats, dogs, monkeys or humans: a review, *Food Chem. Toxicol.*, 36, 95, 1998.

SUCROSE, OCTAACETATE

Molecular Formula. $C_{28}H_{38}O_{19}$
M =678.66
CAS No 126-14-7
RTECS No WN6620000
Abbreviation SOA.

Synonyms. Fructofuranosyl tetraacetate; 1,3,4,6-Tetra-*o*-acetyl-β-D-glucopyranoside.

Properties. Hygroscopic, intensely bitter needles from alcohol. Soluble in water and alcohol.[020]

Applications. Used in the production of plastics and adhesives. Flavoring substance. Used in fragrances.

Acute Toxicity. LD_{50} exceeded 5.0 g/kg BW in rats (Moreno, 1978).[1]

Allergenic Effect. No sensitization reactions were noted in volunteers (Epstein, 1978).[1]

Regulations. ***U.S. FDA*** (1998) approved the use of SOA in adhesives as a component of articles intended for use in packaging, transporting, or holding food in accordance with the conditions prescribed in 21 CFR part 175.105. Considered as *GRAS*.

Reference:

1. Sucrose octaacetate, Fragrance Raw Material Monographs, *Food Chem. Toxicol.*, 20, 827, 1982.

SULFANILIC ACID

Molecular Formula. $C_6H_7NO_3S$
M = 173.20
CAS No 121-57-3
RTECS No WP3895500
Abbreviation. SA.

Synonyms. *p*-Aminobenzene sulfonic acid; Aniline-*p*-sulfonic acid.

Properties. Rhombic plates or monoclinic crystals. Solubility in water is about 10 g/l at 20°C. Insoluble in alcohol.

Applications. Used predominantly in the synthesis of *azo* dyes

Acute Toxicity. LD_{50} is 12.3 g/kg BW in rats.[029]

Mutagenicity. Negative in *Salmonella* mutagenicity assay.[1]

Chemobiokinetics. SA yields *p-acetamidobenzene sulfonic acid* in rabbits (Daniel, 1962). 50% orally administered to rats were found in the urine.[2]

Regulations. ***U.S. FDA*** (1998) approved the use of SA as a component of the uncoated or coated food-contact surface of paper and paperboard intended for use in producing, manufacturing, packaging, processing, preparing, treating, packing, transporting, or holding dry food, subject to the provisions of 21 CFR part 176.180.

References:

1. Chung, K. T. et al., Mutagenicity testing of some commonly used dyes, *Appl. Environ. Microbiol.*, 42, 641, 1981.
2. Scheline, R. R. and Longberg, B., The absorption, metabolism and excretion of the sulfonated azo dye, acid yellow, by rats, *Acta Pharmacol. Toxicol.*, 23, 1, 1965.

SULFONATED 9-OCTADECENOIC ACID (*Z*), SODIUM SALT

CAS No 68443-05-0
RTECS No RG4105000

Synonym. Sulfonated oleic acid, sodium salt.

Regulations. ***U.S. FDA*** (1998) approved the use of S. as a component of sanitizing solutions used on food-processing equipment and utensils, and on other food-contact articles as specified in 21 CFR part 178.1010.

4,4'-SULFONYLDIPHENOL

Molecular Formula. $C_{12}H_{10}O_4S$
M = 250.28
CAS No 80-09-1
RTECS No SM8925000
Abbreviation. SDP.

Synonyms and **Trade Name.** 4,4'-Dihydroxydiphenyl sulfone; 4-Hydroxyphenyl sulfone; Bis(4-hydroxyphenyl)sulfone; Bisphenol S.

Acute Toxicity. LD_{50} is 4.56 g/kg BW in rats and 1.6 g/kg BW in mice.

Short-term Toxicity. In a 1-week study, rats developed signs of renal toxicity including acute renal failure and acute tubular necrosis. The treatment caused weight loss or decreased weight gain.

Regulations.

EU (1990). SDP is available in the *List of monomers and other starting substances, which may continue to be used pending a decision on inclusion in Section A.*

Great Britain (1998). SDP is authorized up to the end of 2001 for use in the production of polymeric materials and articles in contact with food or drink or intended for such contact.

Reference:

Natl. Technical Information Service OTS0534330.

SULFOSUCCINIC ACID, DIHEXYL ESTER, SODIUM SALT

Molecular Formula. $C_{16}H_{30}O_7S.Na$
M = 389.51
CAS No 3006-15-3
RTECS No WN0550000
Abbreviation. SDS.

Synonyms. Dihexyl sodium sulfosuccinate; Dihexyl sulfosuccinate, sodium salt; Sulfobutanedioic acid, 1,4-dihexyl ester, sodium salt.

Acute Toxicity. LD_{50} is 1.75 g/kg BW in rats. Poisoning caused a decrease in BW gain and alterations in GI tract.[1] LD_{50} is reported to be 2.3 g/kg BW in mice.[2]

Regulations. ***U.S. FDA*** (1998) approved the use of SDS. in adhesives as a component of articles intended for use in packaging, transporting, or holding food in accordance with the conditions prescribed in 21 CFR part 175.105.

References:

1. Acute Toxicity Data, *J. Am. Coll. Toxicol.*, Part B, 1, 109, 1990.
2. Information from the Soviet Toxicological Center, *Gig. Truda Prof. Zabol.*, 2, 52, 1983 (in Russian).

SULFOSUCCINIC ACID, 1,4-BIS(2-ETHYLHEXYL)ESTER, SODIUM SALT

Molecular Formula. $C_{20}H_{38}O_7S.Na$
M = 445.63
CAS No 577-11-7
53023-94-2
RTECS No WN0525000

Synonyms and **Trade Names.** Dioctyl sodium sulfosuccinate; Diomedicone; Doxol; Nekal Sulfobutanedioic acid; Soliwax.

Properties. A waxy white solid with a characteristic odor. Soluble in water (15 g/l at 25°C). Soluble in alcohol and vegetable oils.

Applications. Dispersant, emulsifier. Used in the production of adhesives, polymer coatings, paperboard, and emulsifiers, in textile, paint, and pharmaceutical industry. Food additive used in sugar industry. A component of detergents.

Exposure. Maximum consumer exposure is 37 mg/person/day and 200 mg for individuals taking medications containing S.[1]

Acute Toxicity. LD_{50} is 1.9 g/kg BW in rats and 2.6 g/kg BW in mice. Poisoning produced anorexia, vomiting, and diarrhea. CNS inhibition was noted.[2,3]

Long-term Toxicity. Rats received S. in their diet at doses up to approximately 0.87 g/kg BW for 6 months. All animals survived. Diarrhea was occasionally observed at the highest doses. There were no changes in hematology analyses.

Three dogs receiving 0.1 g S./kg BW, three dogs receiving 0.25 g S./kg BW, and three monkeys receiving 0.125 g/kg BW for 6 months survived without manifestations of toxicity for exception of irritation to the stomach. The dose of 0.5 g/kg BW given for 6 months caused anorexia and severe diarrhea in rabbits.[4] No toxicity manifestations were reported in 1-year oral study in dogs.[2]

Reproductive Toxicity.

Embryotoxicity. Groups of 30 male and 30 female rats were fed diets containing 0.1, 0.5, or 1.0% S. for 10 and 2 weeks, respectively. S. administered in the diet to three successive generations of rats at levels of 0.5 and 1.0% caused a reduction in BW for parental males in all generations and for F_1 and F_2 females. Pup weights at the 0.5 and 1.0% dose levels were also lower than those of the control in all three generations, but it did not interfere with development of normal reproductive performance. S. at levels up to 1.0% had no effects on the reproductive function of either sex in any generation and produced no treatment-related antemortem or macroscopic observations.[1]

Rats were fed S. at doses of up to 800 mg/kg BW on days 7 to 17 of gestation. Viability of the fetuses was decreased at the highest dose. No other ill effects were reported.[5]

Teratogenicity.

Observations in man. There was no increase in congenital defects among 792 women who took S. during pregnancy.[6,7] No evidence of an association with malformations was found in offspring of 116 pregnancies that included exposure to dioctyl sulfosuccinate.[8]

Chemobiokinetics. Absorbed in GI tract and excreted in bile (Gilman et al., 1990).

Regulations. ***U.S. FDA*** (1998) approved the use of S. (1) in resinous and polymeric coatings used as the food-contact surfaces of articles intended for use in producing, manufacturing, packing, processing, preparing, treating, packaging, transporting, or holding food in accordance with the conditions prescribed in 21 CFR part 175.300; (2) in the production of or added to textiles and textile fibers intended for use in producing, manufacturing, packing, processing, preparing, treating, packaging, transporting, or holding food, subject to the provisions of 21 CFR part 177.2800.

References:

1. McKenzie, K., Henwood, S., Foster, G., Akin, F., Davis, R., DeBaecke, P., Sisson, G., and McKinney, G., Three-generation reproduction study with dioctyl sodium sulfosuccinate in rats, *Fundam. Appl. Toxicol.*, 15, 53, 1990.
2. Case, M. T., Smith, J. K., and Nelson, R. A., Acute mouse and chronic dog toxicity studies of danthron, dioctyl sodium sulfosuccinate, poloxalkol and combinations, *Drug Chem. Toxicol.*, 1, 89, 1977/78.
3. Goodman and Gilman's *The Pharmacological Basis of Therapeutics*, A. G. Gilman, L. S. Goodman, and A. Gilman, Eds., 6th ed., Macmillan Publishing Co., Inc., New York, 1980.
4. Benaglia, A. E., Robinson, E. J., Utley, E., and Cleverdon, M. A., The chronic toxicity of aerosol-OT, *J. Ind. Hyg. Toxicol.*, 25, 175, 1943.

5. Ichikawa, Y. and Yamamoto, Y., Teratology study of solvents in rats, Abstract, *Gendai Iryo,* 12, 819, 1980 (in Japanese).
6. Aselton, P., Jick, H., Milunsky, A., Hunter, J. R., and Stergachis, A., First-trimester drug use and congeni tal disorders, *Obstet Gynecol.*, 65. 4511, 1985.
7. Jick, H., Holmes, L. B., Hunter, J. R., Madsen, S., and Stergachis, A., First-trimester drug use and congenital disorders, *JAMA,* 246, 343, 1981.
8. Heinonen, O. P. et al., *Birth Defects and Drugs in Pregnancy*, Littleton:Publishing Sciences Group, 1977, 442.

SULPHURIC ACID

Molecular Formula. H_2O_4S
M = 98.08
CAS No 7664-93-9
119540-51-1
127529-01-5
RTECS No WS5600000
Abbreviation. SA.

Synonyms and **Trade Names.** Battery acid; Dihydrogen sulfate; Electrolyte acid; Mattling acid; Oil of vitriol; Vitriol Brown.

Properties. Clear, colorless and odorless, oily liquid when pure, and dark brown when impure. Possesses marked acid taste. Soluble in water and alcohol.

Applications. SA is used as a dehydrating agent in manufacture of ethers and esters, refining of oils, in the leather industry, etc.

Acute Toxicity. Hazard depends primarily on SA acidity.

Observations in man. LD_0 is 135 mg/kg BW.[1]

Animal studies. LD_{50} was found to be 2.14 g/kg BW in rats (50% aqueous solution).[2]

Carcinogenicity classification.

IARC: 1 (occupational exposure).

Regulations. ***U.S. FDA*** (1998) approved the use of SA in manufacture of adhesives as a component of articles intended for use in packaging, transporting, or holding food in accordance with the conditions prescribed in 21 CFR part 175.105. Considered as *GRAS.*

References:

1. Arena, J. M., Springfeld, I. L., and Thomas, C. C., *Poisoning, Toxicology, Symptoms, Treatment,* 2nd ed., 1970, 73.
2. Smyth, H. F., Carpenter, C. P., Weil, C. S., Pozzani, U. C., Striegel, J. A., and Nyaun, J. S., Range-finding toxicity data: List VII, *Am. Ind. Hyg. Assoc. J.,* 30, 470, 1969.

SULPHURIC ACID, MONODODECYL ESTER, AMMONIUM SALT

Molecular Formula. $C_{12}H_{26}O_4S.H_3N$
M = 283.48
CAS No 2235-54-3
RTECS No WT0825000

Synonyms and **Trade Name.** Ammonium dodecyl sulfate; Ammonium lauryl sulfate; Sinopon; Sulphuric acid, lauryl ester, ammonium salt.

Properties. Clear liquid.

Applications. Anionic detergent. Used in food industry.

Acute Toxicity.

Observations in man. Probable lethal dose is about 0.5 to 5.0 mg/kg BW.[019]

Regulations. ***U.S. FDA*** (1998) approved the use of S. (1) in resinous and polymeric coatings used as the food-contact surfaces of articles intended for use in producing, manufacturing, packing, processing, preparing, treating, packaging, transporting, or holding food, subject to the conditions prescribed in 21 CFR part 175.300; and (2) as a component of acrylate ester copolymer coating which may be used as a food contact surface of articles intended for packaging and holding food, including heating of prepared food, subject to the provisions prescribed in 21 CFR part 175.210.

SULPHURIC ACID, MONODODECYL ESTER, SODIUM SALT

Molecular Formula. $C_{12}H_{25}O_4S.Na$
M = 288.42
CAS No 151-21-3
51222-39-0
RTECS No WT1050000

Synonyms and **Trade Names.** Dodecyl sulfate, sodium salt; Duponol; Richonol A; Sipon PD; Sodium lauryl sulfate.

Properties. White or cream-colored crystals, flakes, or powder with a faint odor of fatty substances.[020] Solubility is 100 g/l.

Applications. Surface-active agent for emulsion polymerization. Used in food industry, medicines, and cosmetics.

Acute Toxicity.

Observations in man. Probable lethal dose is about 0.5 to 5.0 mg/kg BW.[019]

Animal studies. LD_{50} is 1.3 g/kg BW in rats.[1]

Short-term and **Long-term Toxicity.** Rats tolerate 1.0% S. in the diet, addition of 4.0% S. decreased growth of animals.[03]

In subacute and chronic feeding tests, even fatally poisoned animals show only diarrhea and intestinal bloating, with no gross lesions outside of the GI tract.[022]

Reproductive Toxicity.

Embryotoxicity. Pregnant mice were treated topically with 6.0 or 60 to 90 mg/kg BW. Retardation of BW gain was observed only in the highest doses.[2]

Teratogenicity. Pregnant mice were dosed orally with 200 mg S./kg BW. The treatment caused an increase in frequency of fetal abnormalities and fetal mortality that have occurred in association with maternal toxicity among the offspring (Oba and Takei).[3]

Mutagenicity.

In vitro genotoxicity. S. caused rapid DNA degradation in primary rat hepatocytes.[4] S. was negative in five *Salmonella typhimurium* strains.[013]

Regulations. ***U.S. FDA*** (1998) approved the use of S. (1) in resinous and polymeric coatings used as the food-contact surfaces of articles intended for use in producing, manufacturing, packing, processing, preparing, treating, packaging, transporting, or holding food in accordance with the conditions prescribed in 21 CFR part 175.300, and (2) in the manufacture of textiles and textile fibers used as articles or components of articles intended for use in producing, manufacturing, packing, processing, preparing, treating, packaging, transporting, or holding food, subject to the provisions prescribed in 21 CFR part 177.2800. It may be used (3) in food in accordance with the conditions prescribed in 21 CFR part 172.822

References:

1. Walker, A. I., Brown, V. K., Ferrigan, L. W., Pickering, R. G., and Williams, D. A., Toxicity of sodium lauryl sulfate, sodium lauryl ethoxysulfate and corresponding surfactants derived from synthetic alcohols, *Food Cosmet. Toxicol.*, 5, 763, 1967.
2. Anonymous, Final report on the safety assessment of sodium lauryl sulfate and ammonium lauryl sulfate, *J. Am. Coll. Toxicol.*, 2, 127, 1983.
3. *Anionic Surfactants: Biochemistry, Toxicology, Dermatology,* 2nd ed., C. Gloxhuber and K. Kuenstler, Eds., Marcel Dekker, New York, 1992, 331.
4. Elia, M. C., Storer, R. D., McKelvey, T. W., Kraynak, A. R., Barmem, J. E., Harman, L. S., De Luca, J. G., and Nichols, W. W., Rapid DNA degradation in primary rat hepatocytes treated with diverse cytotoxic chemicals, *Environ. Molec. Mutagen.*, 24, 181, 1994.

SULPHURIC ACID, TIN(2^+) SALT

Molecular Formula. $O_4S.Sn$
M = 214.75
CAS No 7488-55-3
RTECS No WT1255000
Abbreviation. TS.

Synonyms. Stannous sulfate; Tin sulfate.

Properties. Snow-white, orthorhombic crystals. Soluble in water.

Acute Toxicity. LD_{50} is 2.2 g/kg BW in rats, and 2.15 mg/kg BW in mice. Poisoning is accompanied with excitation later followed by CNS inhibition. Normalization occurs in 2 to 3 days. Gross pathology examination revealed hemorrhages in the stomach and intestinal mucosa, congestion of the visceral organs, and stomach swelling.[1]

Reproductive Toxicity.

Embryotoxicity. Rats were exposed to 0.290 μg TS/m^3 for 24 hours a day during pregnancy. The treatment caused fetal death. An increase in pre-implantation mortality was noted.[2]

Regulations. ***U.S. FDA*** (1998) approved the use of TS in the manufacture of resinous and polymeric coatings for the food-contact surface of articles intended for use in producing, manufacturing, packing, processing, preparing, treating, packaging, transporting, or holding food in accordance with the conditions prescribed in 21 CFR part 175.300.

References:

1. Bessmertny, A. N. and Grin', N. V., Acute toxicity studies of inorganic tin compounds for the purpose of hygienic norm-setting, *Gig. Sanit.*, 6, 82, 1986 (in Russian).
2. Grin', N. V., Govorunova, N. N., Pavlovich, L. V., Bessmertny, A. N., and Besedina, Ye. I., Embryotoxic effects of inhalation exposure to stannous sulfate, *Gig. Sanit.*, 7, 81, 1988 (in Russian).

SUNFLOWER OIL, oxidized

RTECS No WT9700000
Abbreviation. SOO.

Properties. Yellow or pale yellow viscous liquid. Insoluble in water; soluble in alcohol.

Applications. Edible drying oil. Used as a dietary supplement.

Repeated Exposure. Growing male Wistar rats were fed a semi-synthetic diet containing either 15% SO that has been repeatedly used for frying with 19.1% polar material or 15% unused oils with 5.1% polar material for 27 days. The treatment caused a decrease in food efficiency ratio, growth retardation and changes in liver fatty acid composition. Data suggest that frying SO (containing dimers and polymers of triglycerides and oxidized triglycerides) is potentially toxic.

Regulations. ***U.S. FDA*** (1998) approved the use of SOO in resinous and polymeric coatings used as the food-contact surfaces of articles intended for use in producing, manufacturing, packing, processing, preparing, treating, packaging, transporting, or holding food, subject to the provisions prescribed in 21 CFR part 175.300.

Reference:

Lopez-Varela, S., Sanches-Muniz, F. J., and Cuesta, C., Decreased food efficiency ratio, growth retardation and changes in liver fatty acid composition in rats consuming thermally oxidized and polymerized sunflower oil used for frying, *Food Chem. Toxicol.*, 33, 181, 1995.

SURFACTANTS

Properties. White or yellowish powders, pastes or liquids with an aromatic odor. Readily soluble in water with foam formation.

Applications. Used in the production of plastics, vulcanizates, and coatings, and as detergents.

Acute Toxicity. LD_{50} is reported to be 1.1 to 7.0 g/kg BW in rats, 0.85 to 5.0 g/kg BW in mice,[1,2] and 0.85 to 6.85 g/kg BW in guinea pigs. According to other data, in the case of anionic surfactants, LD_{50} is 1.0 to 10.3 g/kg BW, and in the case of non-ionogenic surfactants, it is 3.5 to 9.65 g/kg BW.[3]

Table. The effect of some surfactants (perception thresholds) on the organoleptic indices of water (mg/l).[1]

Surfactants	***Odor***		***Taste***	***Foam***
	Rating 1	***Rating 2***		***Formation***
Anion active				
Alkylsulfate based on secondary non-saponifiable alcohols	0.6	1.0	140	0.5
Secondary alkylsulfate	0.6	1.0	72	0.5
DNS	110	160	125	0.6
Sulfonol NP-3	70	150	60	0.4
Schistose sulfonol	150	230	280	0.7
Chlorine sulfonol	75	100	300	0.5
Sulfonol NP-1	-	200	500	0.5
Alkyl sulfonate	-	200	500	0.5
Azolate B	0.07	0.11	0.1	0.5
DS-RAS	80	150	110	0.5
Non-ionogenic				
Sintanol DS-10	3.0	7.8	2000	0.08
Sintanol MTs-10	24	64	170	0.09
Sintanol DT-7	9	20	400	0.1
Proxanol 186	6	10	6400	0.09
Proxamine 385	14	26	5400	0.09
Sintamide 5	180	300	11000	0.17
Schistose alkylphenol	0.8	1.7	100	0.1
OP-7	0.45	0.9	>0.9	0.1
OP-10	1.8	3.6	>3.6	0.09

Mutagenicity. S. exhibited slight mutagenic effect.

Chemobiokinetics. *I/g* administration with water at concentrations of 100 to 1000 mg/l for a month resulted in an enhanced lipid synthesis in the aorta as evident from tests using labeled cholesterol and ^{14}C-acetate.[1,2]

Standards. ***Russia*** (1988). MAC and PML: 0.1 mg/l (organolept.) for schistose amylphenol, oxanol L-7 and KSH-9, OP-7 and OP-10, proxamine 385 and 186, sintamide-5, and sintanols VN-7, VT-15, DS-10, DT-7 and MC-10. MAC and PML 0.5 mg/l (organolept.) for alkylbenzene sulfonates (ABS), alkylsulfonates, alkylsulfates and the disodium salt of monoalkyl sulfosuccinic acid.

References:

1. Mozhayev, E. A., *Pollution of Reservoirs by Surfactants*, Meditsina, Moscow, 1976, 124 (in Russian).
2. Mozhayev, E. A., Yurasova, O. I., Charyev, O. G., et al., The effect of surfactants on the lipid metabolism in white rats, *Gig. Sanit.*, 2, 85, 1986 (in Russian).
3. Grushko, Ya. M., *Harmful Organic Compounds in the Liquid Industrial Effluents*, Khimiya, Leningrad, 1976, 128 (in Russian).

TALC (powder)

Molecular Formula. $3MgO.4SiO_2.H_2O$

CAS No 14807-96-6

RTECS No WW2710000

Synonyms and **Trade Names**. Agalite; Asbestine; Soapstone; Talcum.

Properties. Fine crystalline powder with a color ranging from white/pale-green to brown. The particles are mainly lamellar in shape.

Applications. Talc is a component of many plastics as a stabilizer, reinforcer, and filler used at up to 70% (*w/w*). Synthetic rubber includes ground talc as a filler in compounding formulations.

Acute Toxicity. LD_{50} is not attained.

Short-term Toxicity. Wistar rats were administered 100 mg talc/day for 101 days. No significant depression of mean lifespan was reported.[1]

Reproductive Toxicity.

Embryotoxicity. Contact of talc-coated surgical gloves with embryo culture medium led to impairment in embryo development.[2]

Teratogenicity effect was not observed in rats and mice given 1.6 g/kg BW, in hamsters given 1.2 g/kg BW, and in rabbits given 0.9 g/kg BW (IARC 42-205).

Mutagenicity.

Observations in man. No data are available on the genetic and related effects in humans.

In vivo cytogenetics. Talc did not induce DLM or CA in bone marrow cells of rats treated with doses of 30 to 5000 mg/kg BW *in vivo*.[3]

In vitro genotoxicity. Talc did not cause CA in human WI38 cells treated with talc at concentrations of 2.0 to 200 μg/ml.[4] It is not mutagenic to yeast or to bacteria in host-mediated assay (IARC 42-185).

Carcinogenicity.

Observations in man. Talc particles were found in stomach tumors from Japanese men, possibly due to ingestion of talc-treated rice.[3] The risk of ovarian cancer is likely to be high among women through daily use of talc-containing products throughout their reproductive years.[5]

Animal studies. Wistar rats were given 50 mg/kg diet or standard diet for life (average survival, 649 days). No increase in tumors incidence was found in this study.[6]

100 mg Italian talc/day was introduced in the diet of Wistar-derived rats for 5 months and caused no tumors in the treated animals.[1]

Carcinogenicity classifications. An IARC Working Group concluded that there is inadequate evidence for the carcinogenicity of talc not containing asbestiform fibers in *experimental animals* and in *humans*; there is inadequate evidence for the carcinogenicity of talc containing asbestiform fibers in *experimental animals* (under inhalation exposure but not in the diet) and there is sufficient evidence for the carcinogenicity of talc containing asbestiform fibers in *humans*.

IARC: 1 (*Talc* containing asbestiform fibers), 3 (T. not containing asbestiform fibers).

NTP: SE - CE - NE - NE (inhalation).

Regulations. ***U.S. FDA*** (1998) affirmed T. as *GRAS*.

Recommendations.

EU (1995). Talc (asbestos free) is a food additive generally permitted for use in cheese (maximum level up to 10 g/kg).

Joint FAO/WHO Expert Committee on Food Additives (1989). ADI is not specified.

References:

1. Wagner, J. C., Berry, G., Cooke, T. J., et al., Animal experiments with talc, in *Inhaled Particles,* W. H. Walton and B. Govern, Eds., vol 4, Part 2, Pergamon Press, Oxford, 1977, 647.
2. Donat, H. et al., Testicular biopsy findings following orchiopexy, *Zentralblat Gynekol.*, 108, 546, 1986.
3. *Mutagenic Evaluation of Compound FDA-71-43; Talc* (PB-245-458), Litton Bionetics, Techn. Informa tion Service, Washington, D.C., 1974.
4. Merliss, R. R., Talc-treated rice and Japanese stomach cancer, *Science*, 173, 1141, 1971.
5. Harlow, B. L., Cramer, D. W., Bess, D. A., and Welsh, W. R., Perineal exposure to talc and ovarian carcer, *Obstet. Gynecol.*, 80, 19, 1992.
6. Gibel, W., Lohs, K., Horn, K.-H., et al., Experimental study of the carcinogenic activity of asbestos fibers, *Arch. Geschwulsforsch.*, 46, 437, 1976.

TALL OIL

CAS No 8002-26-4

Abbreviation. TO.

Synonyms and **Trade Names**. Liquid rosin; Talleol; Tallol; The sap of the pine tree.

Composition. TO consists of fatty acids (saturated and unsaturated) and rosin acids (primary, secondary, neutral and hydroxyacids). Contains a small amount of malodorous substance.

Properties. An oily, dark brown liquid with a soapy odor and taste. Odor perception threshold is 0.16 mg/l; taste perception threshold is 1.25 mg/l.

Applications. TO is used in the manufacture of plastics and synthetic coatings, predominantly alkyd resins and rubber.

Acute Toxicity. LD_{50} in mice is 7.32 g/kg BW; 12.5 g/kg BW appears to be LD_{100}.[1] According to Oshchenkova et al., LD_{50} is 0.78 g/kg BW in rats.[2]

Repeated Exposure. Rats were administered 900 mg/kg BW for 10 days. The treatment caused a slight decrease in BW gain and some changes of leukocyte phagocytic activity.

Long-term Toxicity. In a 6-month study, rabbits were dosed by gavage with 0.03 mg/kg BW, and rats were dosed with 1.5 mg/kg BW. There were no toxic manifestations in the treated animals.[1,2]

Mutagenicity effect was not found.

Regulations. ***U.S. FDA*** (1998) considered TO as *GRAS*. It is listed (1) in adhesives used as components of articles intended for use in packaging, transporting, or holding food in accordance with the conditions prescribed in 21 CFR part 175.105; (2) as a constituent of cotton and cotton fabrics used for dry food packaging at levels not to exceed GMP; (3) in the manufacture of rubber articles intended for repeated use in producing, manufacturing, packaging, processing, preparing, treating, packing, transporting, or holding food in accordance with the conditions prescribed in 21 CFR part 177.2600; and (4) as a defoaming agent that may be safely used in the manufacture of paper and paperboard intended for use in producing, manufacturing, packing, transporting, or holding food in accordance with the conditions prescribed in 21 CFR part 176.210.

Standards. ***Russia.*** Recommended MAC and PML: 0.2 mg/l (organolept., odor).

References:

1. Chen Nai Tun, Experimental data on substantiating the maximum allowable concentration of tall oil in water bodies, *Gig. Sanit.*, 5, 9, 1962 (in Russian).
2. Oshchenkova, E. P., Sadiakhmatov, V., Vinogradova, A. V., et al., *The Use of Sulfite Alkalis and Prehydrolysates in the Economy*, Leningrad, 1985, 59 (in Russian).

TALLOW

CAS No 61789-97-7

Composition. The fat from fatty tissue of bovine cattle and sheep. Chief constituents: stearin, palmitin, and olein.

Properties. White or light colored fat. Insoluble in water.

Applications. Intermediate in plastic manufacture. Could be used as an animal feed.

Acute Toxicity.

Observations in man. T. seems to be nontoxic. Lethal doses are above 15 g/kg BW.[019]

Chemobiokinetics. T. enhanced cholesterol synthesis in rats. It caused marked hyperlipemic effect (Renaud, 1969). Hydrolyzed in bowel to *glycerol* and *fatty acids*.[019]

Regulations. ***U.S. FDA*** (1998) approved the use of T. (1) in adhesives as a component of articles intended for use in packaging, transporting, or holding food (T. blown, oxidized) in accordance with the conditions prescribed in 21 CFR part 175.105; (2) as a constituent of acrylate ester copolymer coating which may be safely used as a food-contact surface of articles intended for packaging and holding food, including heating of prepared food, subject to the provisions prescribed in 21 CFR part 175.210; (3) as a component of the uncoated or coated food-contact surface of paper and paperboard intended for use in producing, manufacturing, packaging, processing, preparing, treating, packing, transporting, or holding aqueous and fatty foods in accordance with the conditions prescribed in 21 CFR part 176.170; (4) as a component of a defoaming agent intended for articles which may be used in producing, manufacturing, packing, processing, preparing, treating, packaging, transporting, or holding food in accordance with the conditions prescribed in 21 CFR part 176.200; and (5) as defoaming agent in the processing of food in accordance with the conditions prescribed in 21 CFR part 173.320 (T. hydrogenated, oxidized or sulfated).

TALLOW ALCOHOL

Composition. A name of commercial mixture of *n*-octadecanol and *n*-hexadecanol.

Properties. Fatty crystalline mass.

Applications. Defoaming agent.

Regulations. ***U.S. FDA*** (1998) approved the use of TA (1) in adhesives as a component of articles intended for use in packaging, transporting, or holding food in accordance with the conditions prescribed in 21 CFR part 175.105; (2) as a component of the uncoated or coated food-contact surface of paper and paperboard intended for use in producing, manufacturing, packaging, processing, preparing, treating, packing, transporting, or holding aqueous and fatty foods in accordance with the conditions prescribed in 21 CFR part 176.170; and (3) as defoaming agent in the processing of food in accordance with the conditions prescribed in 21 CFR part 173.320 (TA hydrogenated).

TERPENE RESIN

CAS No 9003-74-1
RTECS No WZ6405000

Acute Toxicity. LD_{50} exceeded 15 g/kg BW in rats. Poisoning led to CNS inhibition.

Applications. Used in the production of coatings for food-contact surfaces.

Regulations. ***U.S. FDA*** (1998) approved the use of TR in resinous and polymeric coatings used as the food-contact surfaces of articles intended for use in producing, manufacturing, packing, processing, preparing, treating, packaging, transporting, or holding food in accordance with the conditions prescribed in 21 CFR part 175.300.

TETRACHLOROPHTHALIC ACID

Molecular Formula. $C_8H_2Cl_4O_4$
M = 303.90
CAS No 632-58-6
RTECS No TI2795000
Abbreviation. TCPAc.

Synonym. 3,4,5,6-Tetrachloro-1,2-benzenedicarboxilic acid.

TETRACHLOROPHTHALIC ANHYDRIDE

Molecular Formula. $C_8Cl_4O_3$
M = 285.88
CAS No 117-08-8
RTECS No TI3450000
Abbreviation. TCPAn.

Synonyms and **Trade Name.** 1,3-Dioxy-4,5,6,7-tetrachloroisobenzofuran; 4,5,6,7-Tetrachloro-1,3-isobenzofuran dione; Tetrathal.

Properties. TCPAn occurs as a white, odorless, free-flowing powder. Slightly soluble in water.

Applications. TCPAn is used as a flame retardant in unsaturated polyester resins, polyurethane foams, and surface coatings; intermediate for plasticizers.

Acute Toxicity. LD_{50} is found to be greater than 15.8 g TCPAn/kg BW in rats given suspension of TCPAn (1:3) in corn oil (U.S. EPA, 1982). LD_{50} of TCPAc is reported to be more than 4.0 g/kg BW in mice.

Short-term Toxicity. In a 13-week study, F344 rats and B6C3F$_1$ mice received 94 to 1500 mg TCPAn/kg BW by gavage in corn oil. The treatment of rats produced higher rate of mortality at the higher doses, and caused retardation of BW gain. Gross pathology examination revealed dose-dependent increase in kidney, heart, and liver weights, renal tubular necrosis, or dilatation. There were no clinical pathology changes. Neither chemical-related effects nor histopathological changes were found in mice given up to 1.5 g TCPAn/kg BW. The NOAEL with a dose as low as 94 mg/kg BW was not achieved in this study for histopathological lesions in the kidneys. No significant adverse effects were noted in mice given doses as high as 1500 mg/kg BW.[1]

Reproductive Toxicity.

Embryotoxicity. Pregnant Sprague-Dawley rats were given 250 to 2000 mg TCPAn/kg BW on days 6 to 19 of gestation. A slightly increased incidence of skeletal malformations was noted in animals given the 2000 mg/kg BW dose. No signs of embryotoxicity, maternal toxicity, fetotoxicity, or teratogenic effects were found to develop (U.S. EPA, 1982).

Gonadotoxicity. The same treatment caused no adverse changes in sperm morphology and vaginal cytology.[1]

Mutagenicity.

In vitro genotoxicity. TCPAn showed no activity in *Salmonella* mutagenicity assay and did not increase incidence of SCE or CA in Chinese hamster ovary cells.[1]

In vivo cytogenetics. There was no induction of CA in bone marrow cells of mice after *i/p* injection, but an increase in SCE was noted.[2,3]

Carcinogenicity. The non-chlorinated analogue of TCPAn, phthalic anhydride, did not affect tumor incidence (NCI, 1979).

Regulations.

EU (1990). TCPAc is available in the *List of monomers and other starting substances which may continue to be used pending a decision on inclusion in Section A.*

U.S. FDA (1998) approved the use of TCPAc and TCPAn in articles made of urea-formaldehyde resins or as components of these articles intended for use in contact with food in accordance with the conditions specified in 21 CFR part 177.1900.

References:

1. *Toxicity Studies of Tetrachlorophthalic Anhydride Administered by Gavage to F344/N Rats and B6C3F$_1$ Mice*, NTP Technical Report Series No 28, Research Triangle Park, NC, NIH Publ. 93-3351, January 1993.
2. Zeiger, E., Haworth, S., Mortelmans, K., et al., Mutagenicity testing of di(2-ethylhexyl) phthalate and related chemicals in *Salmonella*, *Environ. Mutagen.*, 7, 213, 1985.
3. Galloway, S. M., Armstrong, M. J., Reuben, C., et al., Chromosome aberrations and sister chromatid exchanges in CHO cells: Evaluations of 108 chemicals, *Environ. Molec. Mutagen.*, 10 (Suppl. 10), 1, 1987.

1,1,2,3-TETRACHLOROPROPENE

Molecular Formula. $C_3H_2Cl_4$

M = 179.85

CAS No 10436-39-2

RTECS No UD1925000

Properties. A transparent liquid with a penetrating, unpleasant odor. Almost insoluble in water; readily soluble in fats. Odor perception threshold is 0.17 mg/l (rating 1), practical threshold is 0.3 mg/l (rating 2). Taste detection occurs at higher concentrations.[1]

Applications. Used in the production of thermostable polymers.

Acute Toxicity. LD_{50} is reported to be 1.07 g/kg BW or, according to other data, 0.35 g/kg BW in rats, and 0.8 g/kg BW in mice. Poisoning resulted in CNS affection and dystrophic changes in the liver and other visceral organs.

Repeated Exposure revealed moderate cumulative properties. K_{acc} is 2.8. The treatment affected serum ALT and serum oxidase activity, and the contents of K^+ ions in the erythrocytes and STI.[1]

Rats were administered by gavage with 3.0 to 300 mg T./kg BW for 4 weeks. 100 and 300 mg T./kg BW doses caused reduction in BW and food consumption. The highest dose produced treatment-related necrotic or degenerative lesions in the liver.[2]

Long-term Toxicity. Rats were dosed by gavage with T. for 6 months. Administration of 5.0 mg/kg BW caused leukocytosis, an increased ceruloplasmin activity, phase changes in ALT activity, and reduced glucose-6-phosphatase activity in the liver tissue.[1]

Mutagenicity.

In vivo cytogenetics damage of bone marrow cells was observed on chronic exposure to T. (see above). A dose of 0.01 mg/kg BW is likely to be a NOEL for this effect.[1]

In vitro genotoxicity. T. appeared to be a strong mutagen for *Salmonella typhimurium* and a weak mutagen for *E. coli*. It induced CA in the mammalian cells without metabolic activation.[3]

Standards. ***Russia*** (1995). MAC and PML: 0.002 mg/l.

References:

1. Fedyanina, V. N., Pavlenko, M. N., Kurysheva, N. G., et al., Hygienic substantiation of the MAC for 1,1,2,3-tetrachloropropene in water bodies, *Gig. Sanit.*, 4, 15, 1978 (in Russian).
2. Johannsen, F. R., Levinskas, G. J., and Goldenthal, E. I., Hepatic involvement of two chlorinated propenes following repeated oral exposure in the rat, *Toxicol. Lett.*, 57, 347, 1991.
3. Ellenton, J. A., Douglas, G. R., and Nestmann, E., Mutagenic evaluation of 1,1,2,3-tetrachloro-2-propane, a contaminant in pulp mill effluents, using battery of *in vitro* mammalian and microbial tests, *Canad. J. Genet. Cytol.*, 23, 17, 1981.

TETRADECYL SULFATE, SODIUM SALT

Molecular Formula. $C_{14}H_{29}O_4S.Na$
M = 316.48
CAS No 1191-50-0
RTECS No XB8660000
Abbreviation. STDS.

Synonyms. 7-Ethyl-2-methyl-4-hendecanol sulfate, sodium salt; Myristyl sulfate, sodium salt.

Acute Toxicity. LD_{50} *i/p* is 342 g/kg BW in mice. Manifestations of toxic action included CNS inhibition, a decrease in motor activity, and vessel dilatation.

Regulations. ***U.S. FDA*** (1998) approved the use of STDS in adhesives as a component of articles intended for use in packaging, transporting, or holding food in accordance with the conditions prescribed in 21 CFR part 175.105.

Reference:

J. Am. Pharmacol. Assoc., Sci. ed., 42, 283, 1953.

TETRAHYDROBENZALDEHYDE

Molecular Formula. $C_7H_{10}O$
M = 110.16
CAS No 1321-16-0
RTECS No GW2700000

Synonym. 3-Cyclohexene-1-carboxaldehyde.

Properties. A colorless liquid with a penetrating odor. Odor and taste perception threshold is 0.25 mg/l (rating 1); practical threshold (rating 2) is 0.5 mg/l. A concentration of 20 mg/l does not affect water color or transparency.

Applications. T. is used in the production of epoxy resins.

Acute Toxicity. LD_{50} values are reported to be 1.04 g/kg BW in rats, 1.0 g/kg BW in mice, 1.75 g/kg BW in guinea pigs, and 1.6 g/kg BW in rabbits. Administration of the lethal doses led to a short period of excitation with subsequent CNS inhibition and lethargy. Death occurs within the first 24 hours.

Repeated Exposure failed to reveal cumulative properties. Rats given 1/5 LD_{50} for 20 days did not die. Rats were dosed by gavage with 1/5 and 1/10 LD_{50} for 2 months. The treatment resulted in a reduction of cholinesterase activity in the brain and in Vitamin C contents in the suprarenals. No deviations in general condition, BW gain, or hematology analyses were noted. Gross pathology examination revealed changes in the viscera.

Long-term Toxicity. In a 6-month study, rats were exposed to oral doses of up to 0.05 mg/kg BW. There were no changes in hematology analyses, cholinesterase and phosphorylase activity, blood serum *SH*-group contents, Vitamin C contents in the suprarenals and STI, as well as no pathomorphologic changes in the viscera.

Standards. ***Russia.*** PML in drinking water: 0.25 mg/l (organolept., odor, taste).

Reference:

Meleshchenko, K. F. et al., Substantiation of MAC for tetrahydrobenzaldehyde in water bodies, in *Hygiene of Populated Areas*, D. N. Kaluzhny, Ed., Zdorov'ya, Kiyv, Issue No 11, 1972, 133 (in Russian).

TETRAHYDRO-2*H*-3,5-DIMETHYL-1,3-5-THIADIAZINE-2-THIONE

Molecular Formula. $C_5H_{10}N_2S_2$
M = 162.29
CAS No 533-74-4
RTECS No XI2800000

Synonyms and **Trade Names.** Crag; Dazomet; Dimethylformocarbothialdine; Fennosan B 100; Mylone; Tiazon; Tetrahydro-3,5-dimethyl-2*H*-1,3,5-thiadiazine-2-thione.

Properties. White or colorless, nearly odorless crystals (or with weakly pungent odor). Water solubility is 1.2 g/l at 25°C, and 15 g/kg in ethanol.

Applications. A slimicide in pulp and paper manufacture, a preservative in adhesives and glues.

Acute Toxicity. LD_{50} is 320 mg/kg BW in rats, 180 mg/kg BW in mice, 120 mg/kg BW in rabbits, and 160 mg/kg BW in guinea pigs. Poisoning has been followed by convulsions. A decrease in liver and kidney weights and in body temperature was noted.[1]

Repeated Exposure. Rats were dosed with 2000 ppm in the diet for 30 days. The treatment caused changes in liver and kidney weights. Retardation in BW gain was noted. No histopathological findings were reported.[1]

Long-term Toxicity. In a 2-year oral study, rats received 640 ppm T. in the diet. The treatment did not affect mortality but reduced growth and caused liver and kidney lesions. Even 10 ppm resulted in liver and kidney weight changes, considered to be a response to irritation.[1]

The NOAE is reported to be 20 mg/kg BW.

Allergenic Effect.

Observations in man. Human patch test reveals T. to be a skin sensitizer.

Chemobiokinetics. T. slowly hydrolyzes inside or outside of the body to form *carbon disulfide, formaldehyde,* and *methylamine.* After oral administration of *mylone* to dogs, *carbon disulfide* was found in exhaled air (Menzie, 1969).

Regulations. ***U.S. FDA*** (1998) approved the use of T. in adhesives as a component of articles intended for use in packaging, transporting, or holding food in accordance with the conditions prescribed in 21 CFR part 175.105.

Reference:

1. Smyth, H. F., Carpenter, C. P., and Weil, C. S., Toxicological studies on 3,5-dimethyl-tetrahydro-1,3,5,2*H*-thiadiazine-2-thione, a soil fungicide and slimicide, *Toxicol. Appl. Pharmacol.*, 9, 521, 1966.

N,N,N',N'-TETRAMETHYLETHYLENEDIAMINE

Molecular Formula. $C_6H_{16}N_2$
M = 116.2
CAS No 110-19-9
RTECS No KV7175000

Synonyms and **Trade Names.** 1,2-Bis-(dimethylamino)ethane; *N,N,N',N'*-Tetramethyl-1,2-methane-diamine; *N,N,N',N'*-Tetramethyl-1,2-diaminoethane; Propamine D.

Properties. Oily liquid, colorless or with a yellowish tint. Readily soluble in water or in organic solvents.

Applications. Used in the production of plastics.

Acute Toxicity. LD_{50} is 1.02 g/kg BW in rats, and 0.63 g/kg BW in mice.

Repeated Exposure revealed evident cumulative properties: K_{acc} is 1.1 (by Lim). Rats received 30 administrations of 1/10 LD_{50}. Manifestations of toxic action include changes in the liver, kidney, and NS functions and in hematology analyses. Parenchymatous dystrophy and necrotic processes were noted in the liver.

Mutagenicity.

In vitro genotoxicity. T. appeared to be negative in *Salmonella* mutagenicity assay.[015]

Reference:

Zvezdai, V. I., Alferova, L. I., and Ostrovskaya, I. S., Toxicity of tetramethylethylenediamine, *Gig. Truda Prof. Zabol.*, 6, 57, 1984 (in Russian).

TETRAPROPYLENEBENZENESULFONIC ACID

Molecular Formula. $C_{12}H_{26}O$
M = 186.38
CAS No 27342-88-7
11067-81-5
RTECS No DB7700000
Abbreviation. TPBSA.

Synonyms and **Trade Name.** *N*-Dodecanol; Dodecyl alcohol; Lauryl alcohol.

Properties. Colorless liquid.

Applications. Used in production of detergents.

Acute Toxicity. Oral administration of 24 to 36 ml TPBSA/kg BW to rats and rabbits caused no gross and histology changes in the animals.[011]

Regulations. ***U.S. FDA*** (1998) approved the use of TPBSA (1) in adhesives as a component of articles intended for use in packaging, transporting, or holding food in accordance with the conditions prescribed in 21 CFR part 175.105; (2) in the manufacture of resinous and polymeric coatings for the food-contact surface of articles intended for use in producing, manufacturing, packing, processing, preparing, treating, packaging, transporting, or holding food in accordance with the conditions prescribed in 21 CFR part 175.300; and (3) in the manufacture of cellophane for packaging food in accordance with the conditions prescribed in 21 CFR part 177.1200.

THIOCYANIC ACID, AMMONIUM SALT

Molecular Formula. $CHNS.H_3N$
M = 76.12
CAS No 1762-95-4
RTECS No XK7875000
Abbreviation. TAA.

Synonym and **Trade Names.** Ammonium rhodanate; Ammonium rhodanide; Ammonium sulfocyanide; Ammonium thiocyanate; Rhodanide; Trans-aid; Weedazol.

Properties. White or colorless, odorless or with ammonia odor, monoclinic crystals. Freely soluble in water; soluble in alcohol.

Applications. Used in the production of transparent artificial resins.

Acute Toxicity. LD_{50} is found to be 750, 500, and 500 mg/kg BW in rats, mice, and guinea pigs, respectively.[1] Administration of large doses caused vomiting, extreme cerebral excitement, delirium, convulsions and spasticity of extensor muscles, anuria, or persistent albuminuria.[019]

Chemobiokinetics. *The thiocyanate* ion is slowly excreted in the urine; it is not decomposed to *cyanide* in appreciable quantities.[019] The thiocyanate ion readily diffuses into all tissues. It appears early in saliva and urine.[2]

Regulations. ***U.S. FDA*** (1998) approved the use of TAA in adhesives as a component of articles intended for use in packaging, transporting, or holding food in accordance with the conditions prescribed in 21 CFR part 175.105.

References:
1. Talakin, Yu. N., Chernykh, L. V., Nizharadze, M. Z., Savchenko, M. V., and Ivanova, L. A., Data providing a basis for maximum allowable concentration of ammonium thiocyanate in workplace air, *Gig. Truda Prof. Zabol.*, 10, 51, 1986 (in Russian).
2. Thienes, C. and Haley, T. J., *Clinical Toxicology*, 5th ed., Lea & Febiger, Philadelphia, 1972, 213.

THIOCYANIC ACID, METHYLENE ESTER

Molecular Formula. $C_3H_2N_2S_2$
M = 130.19
CAS No 6317-18-6
RTECS No XL1560000
Abbreviation. TCAM.

Synonyms and **Trade Name.** Methylenebis(thiocyanate); Methylendirhodanid; Methylene dithiocyanate; Methylene thiocyanate; Proxel MB.

Properties. Yellow to light-orange-colored solid. More than 100 mg/ml dissolve in DMSO at 20°C; 10 to 50 mg/ml are soluble in 95% ethanol at 20°C.

Applications. Disinfectant and fungicide.

Acute Toxicity. LD_{50} was about 90 mg/kg BW in Sprague-Dawley rats. Manifestations of toxic action included increasing weakness and adynamia. Gross pathology examination revealed no changes in the viscera.

Repeated Exposure. Rats received TCAM for 14 days. The treatment induced necrotic lesions in the GI tract, and changes in thymus weight. In mice, liver and kidney weights were decreased (NTP-94).

Reproductive Toxicity.

Gonadotoxicity. Administration for 13 weeks prior to mating affected spermatogenesis in rats (NTP-94).

Allergenic Effect has been reported.

Mutagenicity. TCAM was negative in *Salmonella* mutagenicity assay.[025]

Regulations. ***U.S. FDA*** (1998) approved the use of TCAM as a slimicide in the manufacture of paper and paperboard that contact food in according with the provisions prescribed in 21 CFR part 176.300.

Reference:

Andersen, K. E. and Hamann, K., Is Cytox 3522 (10% methylene-bis-thiocyanate) a human skin sensitizer? *Contact Dermat.*, 9, 186, 1983.

THIOCYANIC ACID, METHYL ESTER

Molecular Formula. C_2H_3NS
M = 73.12
CAS No 556-64-9
RTECS No XL1575000
Abbreviation. TCAM.

Synonyms. Methyl sulfocyanate; Methyl thiocyanate.

Properties. Colorless liquid with onion odor. Poorly soluble in water; miscible with alcohol.[020]

Acute Toxicity. LD_{50} is 55 mg/kg BW in rats.[1]

Reproductive Toxicity.

Gonadotoxicity. Male rats were gavaged with 260 mg TCAM/kg BW for 13 weeks prior to mating. Spermatogenesis was affected (NTP-94).

Teratogenicity. Cyanide and chemicals which liberate cyanide has been found to be teratogenic and has affected the fertility of laboratory animals.[2,3]

Chemobiokinetics. The thiocyanate ion is slowly excreted in the urine; it is not decomposed to *cyanide* in appreciable quantities.[019] The *thiocyanate ion* readily diffuses into all tissues. It appears early in saliva and urine.[4]

Regulations. ***U.S. FDA*** (1998) approved the use of TCAM in adhesives as a component of articles intended for use in packaging, transporting, or holding food in accordance with the conditions prescribed in 21 CFR part 175.105.

References:

1. *Developments in Ind. Microbiol.*, 12, 404, 1971.
2. Willhite, C. C., Ferm, V. H., and Smith, R. P., Teratogenic effects of aliphatic nitriles, *Med. J. Aust.*, 1, 1169, 1972
3. Willhite, C. C. and Smith, R. P., The role of cyanide liberation in the acute toxicity of aliphatic nitriles, *Toxicol. Appl. Pharmacol.*, 59, 589, 1981.
4. Thienes, C. and Haley, T. J., *Clinical Toxicology*, 5th ed., Lea & Febiger, Philadelphia, 1972, 213.

THIOCYANIC ACID, SODIUM SALT

Molecular Formula. CNS.Na
M = 81.07
CAS No 540-72-7
RTECS No XL2275000
Abbreviation. TCAS.

Synonyms. Sodium isothiocyanate; Sodium rhodanate; Sodium rhodanide; Sodium sulfocyanate; Sodium thiocyanate.

Acute Toxicity. LD_{50} is 362 mg/kg BW.[1]

Repeated Exposure. Rats received 32 mg TCAS/kg BW in their drinking water for 18 times over a period of 67 days. The treatment induced a stimulation of thyroid gland follicular epithelium. Administration of ST together with 500 mg nitrates/l increased the ST excretion level via the urine.[2]

Carcinogenicity. Rats were treated with 0.08 to 0.32% TCAS in their drinking water over 2 years. TCAS did not show carcinogenic activity in rats, alone or combined with sodium nitrite.[3]

Chemobiokinetics. The *thiocyanate ion* is slowly excreted in the urine; it is not decomposed to *cyanide* in appreciable quantities.[019] The *thiocyanate ion* readily diffuses into all tissues. It appears early in saliva and urine.[4]

Regulations. ***U.S. FDA*** (1998) approved the use of TCAS in adhesives as a component of articles intended for use in packaging, transporting, or holding food in accordance with the conditions prescribed in 21 CFR part 175.105.

References:

1. Izmerov, N. F., Sanotsky, I. V., and Sidorov, K. K., *Toxicometric Parameters of Industrial Toxic Chemicals under Single Exposure*, USSR/SCST Commission for UNEP, Center Int. Projects, Moscow, 1982 (in Russian).
2. Kramer, A., Karnitzky, G., Koch, S., und Hampel, R., Eiflus von Thiocyanat, Cyanid und Nitrat auf die Schilddruse bei alimentarer Gabe, in *III Kongr. Ges. Hyg. und Umweltmed*, Dresden, 30-31 Marz, 1995, *Zentralbl. Hyg. Umweltmed*, 197, 336, 1995.
3. Lijinsky, W. and Kovatsh, R. M. Chronic toxicity tests of sodium thyocyanate with sodium nitrite in F344 rats, *Toxicol. Ind. Health.*, 5, 25, 1989.
4. Thienes, C. and Haley, T. J., *Clinical Toxicology*, 5th ed., Lea & Febiger, Philadelphia, 1972, 213.

THIODIPROPIONIC ACID

Molecular Formula. $C_6H_{10}O_4S$
M = 178.20
CAS No 111-17-1
RTECS No UF7990000
Abbreviation. TDPA.

Synonyms. Bis(2-carboxyethyl) sulfide; 3,3'-Thiobispropanoic acid; Thiodihydracrylic acid; 3,3'-Thiodipropionic acid.

Properties. Nacreous leaflets from hot water. Freely soluble in hot water and alcohol. 1.0 g dissolves in 26.9 ml water at 26°C.[020]

Applications. Used in the production of food packaging materials.

Acute Toxicity. In dogs, *i/v* administration of over 1.0 g TDPA/kg BW caused death that was preceded by a drop in blood pressure, respiratory failure, and cardiac collapse.[04]

Chemobiokinetics. Following injection into rabbits, TDPA has been rapidly excreted unchanged in the urine.[011]

Regulations. ***U.S. FDA*** (1998) approved the use of TDPA (1) in the manufacture of resinous and polymeric coatings for the food-contact surface of articles intended for use in producing, manufacturing, packing, processing, preparing, treating, packaging, transporting, or holding food in accordance with the conditions prescribed in 21 CFR part 175.300; and (2) in the manufacture of antioxidants and/or stabilizers for polymers which may be used in the manufacture of articles or components of articles intended for use in producing, manufacturing, packaging, processing, preparing, treating, packing, transporting, or holding food in accordance with the conditions prescribed in 21 CFR part 178.2010. Affirmed as *GRAS.*

THIOPHENE

Molecular Formula. C_4H_4S
M = 84.14
CAS No 110-02-1
RTECS No XM7350000

Synonyms and **Trade Names.** Divinylene sulfide; Tetrole; Thiofuran; Thiofurfuran; Thiole.

Properties. Liquid with a slight aromatic odor resembling that of benzene. Insoluble in water; miscible with most organic solvents.

Applications. Used in the production of resins.

Occurrence. T. is found in coal tar and in coal gas.

Acute Toxicity. LD_{50} is 1.4 g/kg BW in rats.[1]

Chemobiokinetics. Rabbits were administered 150 mg T./kg BW, rats were given 200 to 300 mg/kg BW by stomach tube. 40% of dose was excreted with urine in form of two *mercapturic acids (thienyl 3-mercapturic acid* and *thienyl 2-mercapturic acid*), 32 to 35% of dose was eliminated unchanged with expired air, and 0.5 to 1.0% with feces.[2]

References:

1. Mikailets, I. B., Mikhailets, G. A., Pel'ts, D. G., and Valiakhmetov, A. V., Toxicology of thiophene, *Gig. Truda Prof. Zabol.,* 10, 57, 1966 (in Russian).
2. Bray, H. G., Carpanini, F. M. B., and Waters, B. D., The metabolism of thiophene in the rabbit and the rat, *Xenobiotica*, 1, 157, 1971.

THIOSULPHURIC ACID, DIAMMONIUM SALT

Molecular Formula. $O_3S_2.2H_4N$
M = 148.22
CAS No 7783-18-8
RTECS No XN6465000
Abbreviation. TSAD.

Synonyms. Ammonium hyposulfite; Ammonium thiosulfate; Diammonium thiosulfate.

Properties. White crystalline powder. Readily soluble in water.

Acute Toxicity. LD_{50} is 2.9 g/kg BW in rats and 1.1 g/kg BW in guinea pigs. The dose of 3.0 g/kg BW was not lethal to mice. Administration caused emphysema and changes in kidney tubules (acute renal fail- ure, acute tubular necrosis, and hemorrhages). Single administration of 200 mg/kg BW caused evident toxic effect in rats.

Repeated Exposure did not reveal cumulative properties. K_{acc} is 10.48 (by Lim).

Allergenic Effect is reported.

Chemobiokinetics. Thiosulfates are poorly absorbed from the GI tract and so act as an osmotic cathartic.[019]

Regulations. ***U.S. FDA*** (1998) approved the use of TSAD in adhesives as a component of articles intended for use in packaging, transporting, or holding food subject to the provisions of 21 CFR part 175.105.

Reference:

Talakin, Yu. N., Chernykh, L. V., Nizharadze, M. Z., Vasilenko, I. V., Gusarenko, V. D., Kondratenko, L. A., and Ivanova, L. A., Primary toxicological evaluation of ammonium thiosulfate, *Gig. Truda Prof. Zabol.*, 6, 54, 1982 (in Russian).

THIOSULFURIC ACID, DISODIUM SALT

Molecular Formula. $O_4S_2.2Na$
M = 176.10
CAS No 7775-14-6
RTECS No JP2100000
Abbreviation. TAS.

Synonyms and **Trade Names.** Burmol; Disodium dithionite; Dithionous acid, disodium salt; Hydrolin; Sodium dithionite; Sodium hydrosulfite; Sodium sulfoxylate.

Properties. White or grayish-white crystalline powder with faint sulfurous odor. Slightly soluble in alcohol.

Applications. Oxygen scavenger for synthetic rubbers.

Toxicity. Incubation of normal human erythrocytes with TAS resulted in the formation of *Heinz bodies.* Addition of superoxide dismutase to the incubation medium increased the formation of *Heinz bodies* by TAS.[1]

Reproductive Toxicity. TAS was not found to be ***embryotoxic*** or ***teratogenic*** in mice, rats, hamsters, or rabbits at maternal doses of up to 550, 400, 400, and 580 mg/kg BW, respectively.[2,3]

Mutagenicity. TAS at a concentration of 0.01% had no apparent mutagenic effect on the *Bac. Micrococcus aureus.*[4]

Regulations. ***U.S. FDA*** (1998) approved the use of TAS in the production of textiles and textile fibers intended for use in producing, manufacturing, packing, processing, preparing, treating, packaging, transporting, or holding food, subject to the provisions of 21 CFR part 177.2800

References:

1. Imanishi, H. et al., Induction of Heinz body formation by sodium dithionite, *Hemoglobin,* 5, 453, 1981.
2. *Teratogenic Evaluation of FDA 71-35 (sodium thiosulfate)*, Food and Drug Research Labs, Inc., NTIS Publ. PB-221, 779, 1972.
3. *Teratogenic Evaluation of FDA 71-35 (sodium thiosulfate)*, Food and Drug Research Labs, Inc., NTIS Publ. PB-264, 820, 1974.
4. Clark, J., The mutagenic action of various chemicals on micrococcus aureus, *Proc. Okla. Acad. Sci.,* 34, 114, 1953.

THIOSULPHURIC ACID, ZINC SALT

Molecular Formula. $O_4S_2.Zn$
M = 193.49
CAS No 7779-86-4
RTECS No JP2105000
Abbreviation. TAZ

Synonyms. Dithionous acid, zinc salt; Zinc dithionite; Zinc hydrosulfite.

Properties. White amorphous solid with the odor of sulphur dioxide. Solubility in water: 400 g/l at 20°C.

Applications. Bleaching agent.

Toxicity. See ***Zinc compounds.***

Regulations. ***U.S. FDA*** (1998) approved the use of TAZ in the production of or added to textiles and textile fibers intended for use in producing, manufacturing, packing, processing, preparing, treating, packaging, transporting, or holding food, subject to the provisions of 21 CFR part 177.2800.

THIOUREA

Molecular Formula. CH_4N_2S
M = 76.12
CAS No 62-56-6
RTECS No YU2800000

Synonym and **Trade Name.** Thiocarbamide; Thiocarbonic acid, diamide.

Properties. Brilliant white or yellowish crystals. Soluble in water (14.2%) and in alcohol.

Applications. Used in the production of plastics, thiourea-formaldehyde resins, and artificial silk.

Acute Toxicity. LD_{50} is 8.0 g/kg BW in rats, 8.1 g/kg BW in mice, and 7.0 g/kg BW in rabbits. There were no species differences in susceptibility. T. caused depression of the thyroid function in animals and humans.

Administration of 5.0 g/kg BW produced severe damage, including follicle necrosis in the thyroid of rats and guinea pigs. Serum triiodothyronine contents in rats rose by 58%.[1] Other manifestations of the toxic effect include depression, diarrhea, paresis, hindlimb paralysis, and convulsions. The soporific effect of some narcotics is prolonged.According to Grin', the LD_{50} was not attained when 1.0 to 10 g/kg BW were given.[2] After administration, apathy, adynamia, refusal of food, and death within 24 hours were observed in the treated animals.

Repeated Exposure revealed moderate cumulative properties in rabbits: animals die after 7 to 12 administrations of 3.0 g/kg BW. The treatment caused anemia and leukocytosis to develop, and a slowing down in the activity of succinate dehydrogenase and cytochrome oxidase. Administration of 4.0 g/kg BW to rabbits led to death after the second dose. Susceptibility is shown to be reduced on repeated administration of a dose of 690 mg/kg BW: repeated administration caused the death of only one animal. The treatment resulted in a reduction of hemopoietic function and in the activity of many tissue respiratory enzymes.[3]

Short-term Toxicity. T. suppressed the thyroid function.

Allergenic Effect. T. was shown to provoke allergic contact dermatitis.

Observations in man. Inhibition of specific resistance in the worker exposed to T. is reported.[4]

Animal studies. I/c injections caused only slight sensitization in guinea pigs.

Reproductive Toxicity.

Gonadotoxic effect was observed at dose levels producing maternal toxicity.[1] Dystrophic changes and desquamation of the testicular epithelium were found in guinea pigs following inhalation exposure to the concentration of 25 mg/m^3 or skin application. The number of cell layers is reduced.

Embryotoxicity. ^{35}S-T. readily crosses the rat placenta.[5] A slight increase in resorptions but no teratogenic effects were observed in rats given 2.0 g/kg BW on day 12 or in mice given 1.0 g/kg BW on day 10 of gestation.[6]

Mutagenicity.

In vivo cytogenetics. T. did not increase the incidence of *Dr. melanogaster* mutant clones in wing spot somatic mutation and recombination assay.[7]

Carcinogenicity. In a rat study, oral administration of 50 ppm in the diet produced liver, thyroid and Zymbal gland tumors.[8] Similar data were obtained in rats exposed to 0.2% T. in their drinking water for 26 months.[9]

In contrast to this, oral administration of T. did not produce thyroid tumor in mice.

Carcinogenicity classifications. An IARC Working Group concluded that there is sufficient evidence for the carcinogenicity of T. in *experimental animals* and there were no adequate data available to evaluate the carcinogenicity of T. in *humans*.

The Annual Report of Carcinogens issued by the U.S. Department of Health and Human Services (1998) defines T. to be a substance which may reasonably be anticipated to be carcinogen, i.e., a substance for which there is a limited evidence of carcinogenicity in *humans* or sufficient evidence of carcinogenicity in *experimental animals*.

IARC: 2B;
U.S. EPA: B2;
EU: 3.

Standards. ***Russia*** (1995). MAC and PML: 0.03 mg/l.

References:

1. Talakin, Yu. N. et al., in *The Endocrine System and Harmful Environmental Factors*, Proc. 2nd All-Union Conf., Leningrad, 1983, 192; *Gig. Truda Prof. Zabol.*, 3, 42, 1986 (in Russian).
2. Grin', N. V., Talakin, Yu. N., Savchenko, M. V., et al., Substantiation of relative safety levels for thiocarbamide and thiocyanogen ammonium effect in the air, *Gig. Sanit.*, 8, 75, 1987 (in Russian).
3. Zhislin, L. E. and Ovetskaya, N. M., Thiourea inhalation toxicity, *Gig. Truda Prof. Zabol.*, 6, 51, 1972 (in Russian).
4. Savchenko, M. V., Thyocyanic ammonium and thiourea effect on the immune system, *Gig. Sanit.*, 11, 29, 1987 (in Russian).
5. Sheppard, T. H. II, Metabolism of thiourea S35 by the fetal thyroid of the rat, *Endocrinology*, 72, 223, 1963.
6. Teramoto, S., Kaneda, M., Aoyama, H., and Shirasu, Y., Correlation between the molecular structure of *n*-alkylureas and *n*-alkylthioureas and their teratogenic properties, *Teratology*, 23, 335, 1981.
7. Batiste-Allentorn, M., Xamena, N., Creus, A., and Marcos, A., Genotoxic evaluation of ten carcinogens in the *Drosophila melanogaster* with spot test, *Experientia*, 51, 73, 1995.
8. Deichmann, W. B., Keplinger, M., Sola, F., et al., Synergism among oral carcinogens. IV. The simultaneous feeding of four tumorigens to rats, *Toxicol. Appl. Pharmacol.*, 11, 88, 1967.
9. Rosin, A. and Ungar, H., Malignant tumors in the eyelids and the auricular region of thiourea-treated rats, *Cancer Res.*, 17, 302, 1957.

THYMOL

Molecular Formula. $C_{10}H_{14}O$
M = 150.21
CAS No 89-83-8
RTECS No XP2275000

Synonyms and **Trade Names**. *p*-Cymen-3-ol; 3-Hydroxy-*p*-cymene; 3-Hydroxy-1-methyl-4-isopropylbenzene; Isopropyl cresol; 2-Isopropyl-5-methylphenol; 3-Methyl-6-isopropylphenol; 5-Methyl-2-(1-methylethyl)phenol; Thyme camphor; Thymic acid.

Properties. Colorless crystals or white crystalline powder with aromatic, thyme-like odor and sweet, medicinal, spicy taste. Solubility in water is about 1.0 g/l. 1.0 g T. dissolves in 1.0 ml alcohol or 1.7 ml olive oil at 25°C.[020]

Applications. An antiseptic. Direct food additive (synthetic flavoring substance and adjuvant).

Acute Toxicity. LD_{50} is 980 mg/kg BW in rats, 1800 mg/kg BW in mice, and 880 mg/kg BW in guinea pigs. Poisoning caused CNS depression, ataxia, and coma.[1] Ingestion caused reversible or irreversible changes to exposed tissue, not permanent injury or death. T. can cause considerable discomfort, gastric pain, nausea, and vomiting. Occasionally convulsions, coma, cardiac and respiratory collapse were noted.[012,022]

Reproductive Toxicity.

Teratogenicity. T. used during pregnancy has not been found to increase an incidence of birth defects[2]

Allergenic Effect was reported.[012]

Chemobiokinetics. Metabolism occurs by conjugation with *glucuronic acid* and *sulfate*.[3]

Regulations. ***U.S. FDA*** (1998) approved the use of T. (1) in adhesives as a component of articles intended for use in packaging, transporting, or holding food in accordance with the conditions prescribed in 21 CFR part 175.105, and (2) as a component of the uncoated or coated food-contact surface of paper and paperboard intended for use in producing, manufacturing, packaging, processing, preparing, treating, packing, transporting, or holding dry foods in accordance with the conditions prescribed in 21 CFR part 176.180.

References:

1. *Food Cosmet. Toxicol.*, 2, 327, 1964.
2. Heinonen, O. P. et al., *Birth Defects and Drugs in Pregnancy*, Publ. Sci. Group, Littleton, MA, 1977.
3. Parke, D. V., *The Biochemistry of Foreign Compounds*, Pergamon Press, Oxford, 1968, 147.

TRIALLYL CYANURATE

Molecular Formula. $C_{12}H_{15}N_3O_3$
M = 249.30
CAS No 101-37-1
RTECS No XZ2080000
Abbreviation. TAC.

Synonyms. Tripropargyl cyanurate; 2,4,6-Tris(allyloxy)-*S*-triasine.

Acute Toxicity. LD_{50} is reported to be 590 mg/kg BW. Poisoning is accompanied with lacrimation, somnolence, changes in the structure or function of the salivary glands.[1]

Mutagenicity. *Triallyl isocyanurate* was negative in Chinese hamster ovary cells with and without *S9 mix*, and positive in Chinese hamster lung cells in presence of *S9*.[2]

Regulations. ***Great Britain*** (1998). TAC is authorized up to the end of 2001 for use in the production of polymeric materials and articles in contact with food or drink or intended for such contact.

References:

1. Acute Toxicity Data, *J. Am. Coll. Toxicol.*, Part B, 1, 42, 1990.
2. Sofuni, T., Matsuoka, A., Sawada, M., Ishidate, M., Zeiger, E., and Shelby, M. D., A comparison of chromosome aberration induction by 25 compounds tested by two Chinese hamster cell (CHL and CHO) systems in culture, *Mutat. Res.* 241,175, 1990.

TRIBUTYLTIN ACETATE

Molecular Formula. $C_{14}H_{30}O_2.Sn$
M = 349.13
CAS No 56-36-0
RTECS No WH5775000
Abbreviation. TBTA.

Synonyms. Acetoxytributyl stannate; Tributyl stannium acetate.

Acute Toxicity. LD_{50} is 99 mg/kg BW in rats and 46 mg/kg BW in mice.[1]

Reproductive Toxicity.

Embryotoxicity and ***Teratogenicity.*** Female Wistar rats received 1.0 mg to 16 mg TBTA/kg BW on days 7 to 17 of pregnancy. Thymus atrophy was observed in treated animals. The exposure to the highest dose resulted in an increase in post-implantation mortality and fetal deaths. Fetal BW was found to be decreased. Craniofacial (nose and tongue) developmental abnormalities were noted in newborns (fetuses with cleft palate, cervical rib, and/or rudimentary lumbar rib).[2]

TBTA increased placental weight in rats.[3]

Regulations. ***U.S. FDA*** (1998) approved the use of TBTA in adhesives as a component of articles intended for use in packaging, transporting, or holding food in accordance with the conditions prescribed in 21 CFR part 175.105.

References:

1. Calley, D. J., Guess, W. L., and Autian, J., Hepatotoxicity of a series of organotin esters, *J. Pharm. Sci.*, 56, 240, 1967.
2. Noda, T., Morita, S., Yamano, T., Shimizu, M., Nakamura, T., Saitoh, M., and Yamada, A., Teratogenicity study of tri-*n*-butyltin acetate in rats by oral administration, *Toxicol. Lett.*, 55, 109, 1991.
3. Itami, T., Ema, M., and Kawasaki, H., Increased placental weight induced by tributyltin chloride in rats, Abstract, *Teratology,* 42, 42A, 1990.

N-[(TRIETHOXYSILYL)METHYL]-1,6-HEXANEDIAMINE

Molecular Formula. $C_{13}H_{32}N_2O_3Si$
M = 292.54
CAS No 15129-36-9
RTECS No MO1450000

Synonym and **Trade Name.** AGM-3; 6-Aminohexylaminomethyltriethoxysilane.

Properties. Clear, light-yellow liquid. Soluble in water and alcohol.

Applications. Used as a water-repelling and finishing agent.

Acute Toxicity. After administration of 0.5 to 3.0 g/kg BW, mice became lethargic but rapidly recovered. General inhibition and adynamia persisted for 2 hours in some animals. A certain number of mice died within 7 days. Gross pathology and microscopic examinations revealed pronounced congestion in the visceral organs and fatty dystrophy of the liver cells, myocardium, and renal tubular epithelium.[1,2]

References:

1. Kulagina, N. K. and Kochetkova, T. A., in *Toxicology of New Industrial Chemical Substances*, A. A. Letavet, Ed., Meditsina, Moscow, Issue No 6, 1964, 95 (in Russian).
2. Kulagina, N. K. and Kochetkova, T. A., in *Toxicology of New Industrial Chemical Substances*, A. A. Letavet and I. V. Sanotsky, Eds., Meditsina, Moscow, Issue No 11, 1969, 118 (in Russian).

3-(TRIETHOXYSILYL)PROPYLAMINE

Molecular Formula. $C_9H_{23}NO_3Si$
M = 221.42
CAS No 919-30-2
RTECS No TX2100000

Synonyms and **Trade Names.** AGM 9; (3-Aminopropyl)triethoxysilane; Silane 1100; Silicone A-1100; Triethoxy (3-aminopropyl)silane; 3-(Triethoxysilyl)-1-propanamine.

Properties. Clear, light yellow liquid.

Applications. Used as a water repellent or finishing agent, and in the preparation of dental and orthopedic materials.

Acute Toxicity. Mice and rats received T. in oil solution. LD_{50} was found to be 1.78 g/kg BW in rats and 4.0 g/kg BW in mice. The treatment caused a transient inhibition.

Regulations. ***U.S. FDA*** (1998) approved the use of T. in resinous and polymeric coatings used as the food-contact surfaces of articles intended for use in producing, manufacturing, packing, processing, preparing, treating, packaging, transporting, or holding food in accordance with the conditions prescribed in 21 CFR part 175.300.

Reference:

Kulagina, N. K. and Kochetkova, T. A., in *Toxicology of New Industrial Chemical Substances*, A. A. Letavet and I. V. Sanotsky, Eds., Meditsina, Moscow, Issue No 11, 1969, 118 (in Russian).

1,2,4-TRIMETHYL-5-CHLOROMETHYLBENZENE

Molecular Formula. $C_{10}H_{13}Cl$
M = 171

Synonym. Monochloromethyl pseudocumol.

Properties. Viscous liquid with a faint odor of chlorine. Poorly soluble in water.

Applications. Used in the production of thermostable plastics.

Acute Toxicity. LD_{50} is 1.2 g/kg BW in rats and 0.75 g/kg BW in mice. Poisoning is accompanied by slowly increasing CNS inhibition, animals assume side position. Death occurs in 1 to 2 days.

Repeated Exposure revealed moderate cumulative properties. K_{acc} appeared to be 4.7 (by Lim). The treat ment affects predominantly the liver and NS functions.

Reference:

Shugayev, B. B. and Bukhalovsky, A. A., Toxicity of trimethylchloromethylbenzene, *Gig. Sanit.*, 3, 92, 1983 (in Russian).

2,2,4-TRIMETHYLPENTANEDIOL-1,3-DIISOBUTYRATE

Molecular Formula. $C_{16}H_{30}O_4$
M = 286.46
CAS No 6846-50-0
RTECS No SA1420000

Synonym and **Trade Name.** Isobutyric acid, 1-isopropyl-2,2-dimethyltrimethylene ester.

Regulations. ***U.S. FDA*** (1998) approved the use of T. (1) as a component of cellophane for use in packaging food in accordance with the conditions prescribed in 21 CFR part 177.1200; and (2) in adhesives as a component (monomer) of articles intended for use in packaging, transporting, or holding food in accordance with the conditions prescribed in CFR part 175.105.

s-TRIOXANE

Molecular Formula. $C_3H_6O_3$
M = 90.08
CAS No 110-88-3
RTECS No YK0350000
Abbreviation. TO.

Composition. Stable, cyclic trimer of formaldehyde possessing characteristic chloroform-like odor.

Synonyms and **Trade Name.** Metaformaldehyde; Triformol; 1,3,5-Trioxacyclohexane; 1,3,5-Trioxane; Trioxymethylene.

Properties. White crystalline solid with characteristic chloroform-like odor. Solubility in water: 172 g/l at 18°C, 211 g/l at 25°C; soluble in alcohols.[020]

Applications. A disinfectant.

Acute Toxicity.

Animal studies. LD_{50} is 8.5 g/kg BW in rats.[1]

Observations in man. Probable oral lethal dose is 50 to 500 mg/kg BW for 70 kg person.[022]

Long-term Toxicity. Rats were *i/g* administered 1/10, 1/40, and 1/80 LD_{50} doses in 20% water solutions of TO at for 7 months. The mortality rate did not show accumulated toxic effects.[2]

Reproductive Toxicity.

Embryotoxicity. Rats were gavaged every other day from days 2 to 20 of gestation with an aqueous solution of TO at daily doses equal to 0.19, 0.58, or 1.16 g/kg BW. Most neonates of females treated with 1.16 g/kg BW died a few days after birth. Treatment of dams with the lesser doses did not affect postnatal growth of their offspring. Behavioral performance in progeny was not affected only in 0.19 mg/kg BW dosed group.[3]

Gonadotoxicity. TO was administered at doses of 0.1 or 0.2 LD_{50} to male rats once a day, 5 days a week, throughout 8 weeks. No increase in the number of preimplantation losses, dead implants and alive fetuses per female was noted in any treated group as compared to an appropriate control group. The test compounds did not affect the fertility of males, although, in some rats, treated microscopic examination of testes revealed focal necrosis of seminiferous epithelium and alteration of spermatogenesis. The study did not reveal induced DLM in germ cells of male rats.[4]

Female rats were gavaged with an aqueous solution of TO for 7 weeks at daily doses equal to 0.19, 0.58, or 1.16 g/kg BW. The females treated with 1.16 g/kg BW exhibited an increase in mean duration of the estrus cycle. Three weeks after cessation of treatment, no alterations of estrous cycle were observed in all animals treated. Thus, only overt signs of TO toxicity were noted in this study.[5]

Mutagenicity.

In vitro cytogenetics. TO was not mutagenic in 5 histidine-requiring strains of *Salmonella typhimurium* strains *TA1535, TA1537, TA1538, TA98,* and *TA100* with and without activation by liver microsomes of rats induced with Aroclor 1254.[6]

In vivo genotoxicity. *I/p* administration produced no CA resulting in erythrocyte micronucleus formation in mice, even at highly toxic doses.[7]

Chemobiokinetics. A single oral administration of 40 mg ^{14}C-TO/kg BW (1.6 MBq/kg BW) to pregnant rats led to distribution of radioactivity predominantly in the liver and plasma.[8]

Regulations.

EU (1990). TO is available in the *List of authorized monomers and other starting substances which may continue to be used for the manufacture of plastic materials and articles intended to come into contact with foodstuffs pending a decision on inclusion in Section A* (*Section B*).

Great Britain (1998). TO is authorized up to the end of 2001 for use in the production of polymeric materials and articles in contact with food or drink or intended for such contact.

References:

1. Czajkowska, T., Krysiak, B., and Popinska, E., Experimental studies of toxic effects of 1,3,5-trioxane and 1,3-dioxolane. I. Acute toxic effect, *Med. Pr.*, 38, 184, 1987 (in Polish).
2. Czajkowska, T. and Krysiak, B., Experimental studies of the toxic effects of 1,3,5-trioxane and 1,3-dioxolane. II. Cumulation of toxic effect, *Med. Pr.*, 38, 244, 1987 (in Polish).
3. Sitarek, K. and Baranski, B., Effects of maternal exposure to trioxane on postnatal development in rats, *Pol. J. Occup. Med.*, 3, 285, 1990 (in Polish).
4. Baranski, B., Stetkiewicz, J., Czajkowska, T., Sitarek, K., and Szymczak, W., Mutagenic and gonadotoxic properties of trioxane and dioxolane, *Med. Pr.*, 35, 245, 1984 (in Polish).
5. Sitarek, K. and Baranski, B., The effect of oral exposure to trioxane on the estrus cycle in rats, *Pol. J. Occup. Med.*, 3, 209, 1990 (in Polish).
6. Kowalski, Z., Spiechowicz, E., and Baranski, B., Absence of mutagenicity of trioxane and dioxolane in *Salmonella typhimurium*, *Mutat. Res.*, 136, 169, 1984.
7. Przybojewska, B., Dziubaltowska, E., and Kowalski, Z., Genotoxic effects of dioxolane and trioxane in mice evaluated by the micronucleus test, *Toxicol. Lett.*, 21, 349, 1984.
8. Sitarek, K. and Baranski, B., Distribution and binding of 1,3,5-trioxane in maternal and fetal rats, *Pol. J. Occup. Med.*, 3, 83, 1990 (in Polish).

TRIS(2-METHOXYETHOXY)VINYLSILANE

Molecular Formula. $C_{11}H_{24}O_6Si$
M = 280.44
CAS No 1067-53-4
RTECS No VV6826000

Synonym and **Trade Name.** Silicone A-172; Vinyltris(methoxyethoxy)silane.

Acute Toxicity. LD_{50} is 2960 mg/kg BW in rats (Union Carbide Data Sheet, 1964).

Regulations. ***EU*** (1990). T. is available in the *List of authorized monomers and other starting substances which may continue to be used for the manufacture of plastic materials and articles intended to come into contact with foodstuffs pending a decision on inclusion in Section A* (*Section B*).

9-VINYLCARBAZOLE

Molecular Formula. $C_{14}H_{11}N$
M = 193.26
CAS No 1484-13-5
RTECS No FE6350000
Abbreviation. VC.

Synonyms. 9-Ethenyl-9*H*-carbazole; *N*-Ethenylcarbazole.

Acute Toxicity. LD_{50} is 50 mg/kg BW in mice, and 100 mg/kg BW in guinea pigs.[029]

Regulations. ***EU*** (1990). VC is available in the *List of monomers and other starting substances, which may continue to be used pending a decision on inclusion in Section A.*

VINYLSILANOLATE, SODIUM SALT

Abbreviation. VSS.

Properties. Readily soluble in water. Odor perception threshold is 1.7 mg/l.

Applications. A component of anti-corrosion coatings. Used to treat glass-reinforced plastics based on polyester resins. 30% water-alcohol solution is a liquid used in water-repellency treatments.

Acute Toxicity. Mice given 0.1 to 0.4 ml die with signs of tetanic convulsions. Gross pathology examination revealed stomach wall lesions characteristic of poisoning by alkalis. The gelatinized silicones released from VSS appear to show a slight toxic effect.[1] LD_{50} in such case is 40 to 50 g/kg BW.

Long-term Toxicity. The NOAEL is reported to be 2.5 mg/kg BW.[2]

Standards. ***Russia*** (1995). MAC and MPC: 2.0 mg/l (organolept., odor).

References:

1. Krasovsky, G. N., Friedland, S. A., Rubleva, M. N., et al., Specificity of biological action and hygienic significance of siliconorganic compounds in course of their draining in water reservoirs, in *Industrial Pollution of Water Reservoirs*, S. N. Cherkinsky, Ed., Meditsina, Moscow, Issue No 9, 1969, 38 (in Russian).
2. Kleshchintskaya, A. L., Krasovsky, G. N., and Friedland, S. A., Hygienic assessment of conditions of discharge of industrial effluents containing silicone organic compounds, *Gig. Sanit.*, 1, 28, 1970 (in Russian).

XYLENESULFONATE, SODIUM SALT

Molecular Formula. $C_8H_{10}O_3S.Na$
M = 209.73
CAS No 1300-72-7
RTECS No ZE5100000
Abbreviation. XSS.

Synonyms and **Trade Names.** Dimethylbenzenesulfonic acid, sodium salt; Hydrotrope; Sodium dimethyl benzene sulfonate; Ultrawet 40SX.

Applications. A hydrotrope in liquid household detergents, in paper and paperboard industry.

Toxicity. Relatively low toxic.

Mutagenicity.

In vitro genotoxicity. XSS was negative in *Salmonella typhimurium* strains *TA98, TA100, TA1535,* and *TA1537* with and without *S9* fraction and negative in mutation assay with mouse lymphoma cells without *S9*.[015] An increase in SCE in Chinese hamster ovary cells was seen in the absence, but not in presence of *S9* fraction. No increase in CA frequency was noted.[1]

Carcinogenicity. No carcinogenic effect on dermal application was observed in rats and mice. XSS has been predicted to be rodent carcinogens by the computer automated structure evaluation (CASE) and multiple computer automated structure evaluation (MULTICASE) system.[2]

Chemobiokinetics. After ingestion, XSS is quickly distributed through the body and readily excreted.[03]
Sulfonates are readily absorbed and distributed through the body, and readily excreted.

Regulations. ***U.S. FDA*** (1998) approved the use of XSS (1) in resinous and polymeric coatings used as the food-contact surfaces of articles intended for use in producing, manufacturing, packing, processing, preparing, treating, packaging, transporting, or holding food in accordance with the conditions prescribed in 21 CFR part 175.300; (2) in adhesives as a component of articles intended for use in packaging, transporting, or holding food in accordance with the conditions prescribed in 21 CFR part 175.105; and (3) as a component of the uncoated or coated food-contact surface of paper and paperboard intended for use in producing, manufacturing, packaging, processing, preparing, treating, packing, transporting, or holding dry foods in accordance with the conditions prescribed in 21 CFR part 176.180; and (4) as a component of sanitizing solutions used on food-processing equipment and utensils, and on other food-contact articles as specified in 21 CFR part 178.1010.

References:

1. *Toxicology and Carcinogenesis Studies of Technical Grade Sodium Xylenesulfonate in F344/N Rats and B6C3F$_1$ Mice (Dermal Studies),* NTP Technical Report Series No 464 Research Triangle Park, NC, April 1996.
2. Zhang, Y. P., Sussman, N., Macina, O. T., Rosenkranz, H. S., Klopman, G., Prediction of the carcinogenicity of a second group of organic chemicals undergoing carcinogenicity testing, *Environ. Health Perspect.*, 104S, 1045, 1996.

XYLENOLS

Molecular Formula. $C_8H_{10}O$
M = 122.17

CAS No 1300-71-6
RTECS No ZE5425000
Abbreviation. DMP.

Synonym. Dimethylphenols.

Properties. Colorless liquid. Solubility in water is 100 mg/l at 21°C. Taste and odor perception thresholds are different in various isomers and vary from 0.12 to 0.25 mg/l. Threshold concentration in chlorinated water is much lower: 0.001 to 0.002 mg/l. Heating accentuates the odor in the presence of 2,5-DMP and 2,6-DMP.

Applications. Used in the production of plastics, rubber and detergents.

Exposure. Concentration of 2,5-DMP in the urine of the workers employed in the distillation of the phenolic fraction of tar and those of non-exposed male workers was found to be 36.7 and 69 µg/l, respectively.[1]

Acute Toxicity. LD_{50} is reported to be 300 to 400 mg/kg BW (2,5-DMP and 2,6-DMP), and 610 to 730 mg/kg BW (3,5-DMP and 3,4-DMP) in rats, 380 to 480 mg/kg BW in mice, and 700 to 1300 mg/kg BW in rabbits (Maazik, 1986).

According to Uzhdavini et al., when 2,4-DMP is administered in oil solution, LD_{50} is 3200 mg/kg BW in rats and 810 mg/kg BW in mice. On administration of 3,5-DMP, LD_{50} appeared to be 2250 and 836 mg/kg BW, respectively. Lethal doses result in ataxia and rapid onset of clonic convulsions and death within 24 hours.[2]

Guinea pigs seem to be much less sensitive: administration of a 1200 mg/kg BW dose provoked only mild symptoms of intoxication.

Repeated Exposure revealed weak cumulative properties only in the case of 3,4-DMP (in rats given 1/5 LD_{50}). Sprague-Dawley rats were given 60 to 1200 mg 2,4-DMP/kg BW for 10 consecutive days. All the high dose treated animals died. The treatment with higher doses produced a significant increase in relative liver weight in females and alterations in hematological and clinical analyses. Histological examination revealed changes in forestomach.[3]

Short-term Toxicity. Rats were dosed by gavage with 1/10 LD_{50} (3,4-DMP and 2,6-DMP) for 10 weeks. The treatment caused retardation of BW gain. Gross pathology examination revealed liver damage. Phenol was not detected in the urine.

Sprague-Dawley rats received 60, 180, or 540 mg 2,4-DMP/kg BW for 90 consecutive days. An increase in mortality occurred in the highest dose group. Clinical chemistry findings included reduced creatinine and increased cholesterol, increased triglycerides, and decreased AST in males only. Histological examination revealed changes in forestomach.[3]

Long-term Toxicity. Rats were exposed by gavage to oral doses of 14 mg 3,4-DMP/kg BW for 8 months. Manifestations of the toxic action included a reduction in BW gain, in erythrocyte count, and in Hb level in the blood. Contents of *SH*-groups in the blood serum and blood pressure were decreased at a dose level of 6.0 mg 2,6-DMP/kg. Gross pathology findings proved changes in the liver, kidney, spleen, and heart.[2]

Regulations.

U.S. FDA (1998) regulates DMP for use as a component in the manufacture of resinous and polymeric coatings for the food-contact surface of articles intended for use in producing, manufacturing, packing, processing, preparing, treating, packaging, transporting, or holding food in accordance with the conditions prescribed in 21 CFR part 175.300.

EU (1990). DMP is available in the *List of monomers and other starting substances which may continue to be used for the manufacture of plastic materials and articles intended to come into contact with foodstuffs pending a decision on inclusion in Section A* (*Section B*).

Great Britain (1998). 2,6-DMP is authorized up to the end of 2001 for use in the production of polymeric materials and articles in contact with food or drink or intended for such contact.

Standards. ***Russia*** (1995): PML: 0.25 mg 2,5-DMP/l (organolept.).

References:

1. Bieniek, G., Concentrations of phenol, *o*-cresol, and 2,5-xylenol in the urine of the workers employed in the distillation of the phenolic fraction of tar, *Occup. Environ. Med.*, 51, 354, 1994.

2. Uzhdavini, E. R., Mamayeva, A. A., and Gilev, V. G., Toxic properties of 2,6- and 3,5-dimethylphenols, *Gig. Truda Prof. Zabol.*, 10, 52, 1979 (in Russian).
3. Daniel, F. B., Robinson, M., Olson, G. R., York, R. G., and Condie, L. W., Ten and ninety-day toxicity studies of 2,4-dimethylphenol in Sprague-Dawley rats, *Drug Chem. Toxicol.,* 16, 351, 1993.

2,6-XYLIDINE

Molecular Formula. $C_8H_{11}N$
M = 121.20
CAS No 87-62-7
RTECS No ZE9275000

Synonyms. 1-Amino-2,6-dimethylbenzene; 2-Amino-1,3-xylene; 2,6-Dimethylamine; 2,6-Dimethylbenzeneamine; 2,6-Xylylamine.

Properties. Clear liquid with pungent odor. Solubility in water is 7500 mg/l at 20°C and 8240 mg/l at 25°C. Soluble in ethanol.

Applications. Chemical intermediate in manufacture of synthetic resins and antioxidants.

Exposure. X. may enter the environment through degradation of some pesticides.

Acute Toxicity. LD_{50} values are reported to be as follows: 2.0 mg/kg BW in male Osborne-Mendel rats given X. in water by gavage;[1] 0.84 g/kg BW in Sprague-Dawley rats,[2] 1.05 g/kg BW in F344 rats,[3] 0.71 g/kg BW in CF mice;[4] 1.2 to 1.3 g/kg BW in F344 and Charles River rats; the last studies revealed marginally toxic effects including changes in the liver, kidneys, and hematopoietic system.[5]

Repeated Exposure. Male F344 rats received 1.57 g X./kg BW by gavage for 5 to 20 days.[3] Sprague-Dawley rats were given 400 to 700 mg X./kg BW for 4 weeks. The treatment induced decrease in BW, lowered *Hb* level, and caused liver enlargement. X. can induce microsomal drug-metabolizing enzyme activity.[6]

Short-term Toxicity. F344 rats received X. in corn oil by gavage at the dose of 310 mg/kg BW for 13 weeks. The treatment increased liver weight, decreased BW, and induced anemia.[5]

Male Osborne-Mendel rats were fed up to 10,000 ppm in the diet for 3 to 6 months. Animals exhibited 25% weight loss, anemia, liver enlargement (without microscopic changes), splenic congestion, and kidney lesions.[1]

Long-term Toxicity. Charles River CD rats received 300, 1000, and 3000 ppm by pre- and postnatal administration of X. in the diet. Retardation of BW gain and lesions in the epithelium of the nasal cavity were noted in this study.[5]

Mutagenicity.

In vitro genotoxicity. Equivocal results for gene mutation were found in bacteria.[015] X. induced SCE and CA in cultured mammalian cells.

In vivo cytogenetics. There was no effect on the frequency of micronucleated cells in bone marrow of mice treated with the doses of 75 and 375 mg/kg BW.[7]

Carcinogenicity. The above-described 2-year NTP study revealed adenomas and sarcomas as well as several sarcomas in nasal cavity, subcutaneous fibromas, and fibrosarcomas in both males and females, and an increase in the incidence of neoplastic nodules in the livers of female rats.[5]

Carcinogenicity classification.

NTP: P - P - XX - XX (feed).

Chemobiokinetics. X. may be metabolized by oxidation. It induced *Hb adduct* formation in humans and rats. Eliminated as *conjugate with glucuronic acid.*[5, 8, 9]

References:

1. Lindstrom, H. V., Bowie, W. C., Wallace, W. C., Velson, A. A., and Fitzhugh, O. G., The toxicity and metabolism of mesidine and pseudocumidine in rats, *J. Pharmacol. Exp. Ther.*, 167, 223, 1969.
2. Jacobson, K. H., Acute oral toxicity of mono- and dialkyl-substituted derivatives of aniline, *Toxicol. Appl. Pharmacol.*, 22, 153, 1972.

3. Short, C. R., King, C., Sistrunk, P. W., and Kern, K. M., Subacute toxicity of several ring-substituted dialkylanilines in the rat, *Fundam. Appl. Toxicol.*, 3, 285, 1983.
4. Vernot, E. H., MacEwen, J. D., Hauh, C. C., and Kinkead, E. R., Acute toxicity and skin corrosion data for some organic and inorganic compounds and aqueous solutions, *Toxicol. Appl. Pharmacol.*, 42, 417, 1977.
5. *Toxicology and Carcinogenesis Studies of 2,6-Xylidine (2,6-Dimethylaniline) in Charles River CD Rats (Feed Studies)*, NTP Technical Report Series No 278, Research Triangle Park, NC, January 1990.
6. Magnusson, G., Mageed, S. K., Down, W. H., Sacharin, R. M., and Jorgeson, W., Hepatic effects of xylidine isomers in rats, *Toxicology,* 12, 63, 1979.
7. Parton, J. W., Beyers, J. E., Garriott, M. L., and Tamura, R. N., The evaluation of a multiple dosing protocol for the mouse bone marrow micronucleus assay using benzidine and 2,6-xylidine, *Mutat. Res.*, 234, 165, 1990.
8. Bryant, M. S., Simmons, H. F., Harrel, R. E., and Hinson, J. A., Hemoglobin adducts of 2,6-dimethylamine in cardial patients administered lidocaine (Abstract 878), *Proc. Am. Assoc. Cancer Res.*, 33, 146, 1992.
9. Sabbioni, G., Hemoglobin binding of monocyclic aromatic amines: molecular dosimetry and quantitative structure activity relationship for the *N*-oxidation, *Chem.-Biol. Interact.*, 81, 91, 1992.

ZINC SULFIDE

Molecular Formula. SZn
M = 97.43
CAS No 1314-98-3
RTECS No ZH5400000
Abbreviation. ZS.

Synonym and **Trade Names.** Albalith; C. I. Pigment White 7; Zinc Blende; Zinc monosulfide.

Properties. White to grayish-white or yellowish powder. Insoluble in water.

Applications. A pigment for dental rubber.

Acute Toxicity. LD_{50} exceeded 2.0 g/kg BW in rats.

Chemobiokinetics. In acidic stomach media, ZS undergoes decomposition to *hydrogen sulfide* with subsequent systemic poisoning.[019]

Regulations. ***U.S. FDA*** (1998) approved the use of ZS (1) in adhesives as a component of articles intended for use in packaging, transporting, or holding food in accordance with the conditions prescribed in 21 CFR part 175.105; and (2) in resinous and polymeric coatings used as the food-contact surfaces of articles intended for use in producing, manufacturing, packing, processing, preparing, treating, packaging, transporting, or holding food in accordance with the conditions prescribed in 21 CFR part 175.300.

CHAPTER 8. POLYMERS

AMINOPLASTICS

Structural Formula.

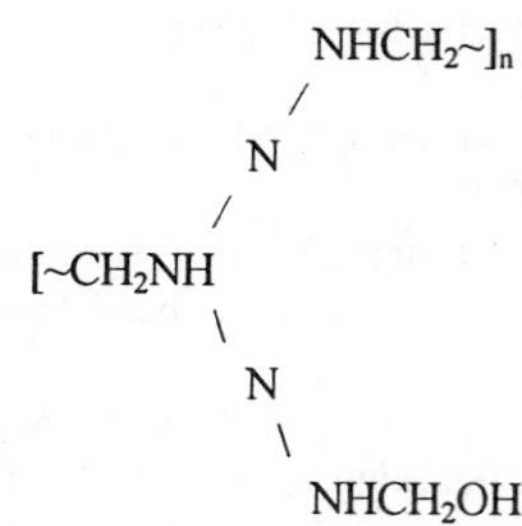

CAS No 9003-08-1
57679-12-6
66829-09-2
RTECS No OS1140000
Abbreviation. MFR.

Synonyms and **Trade Names.** 1,3,5-Triazine-2,4,6-triamine, polymer with formaldehyde; Cymel; Formaldehyde-melamine condensate; Formaldehyde-melamine polymer; Kauramin;. Melaform; Melamine 20 or 366; Melamine, polymer with formaldehyde; Melamine resin; Methylene melamine polycondensate; Mirbane.

Composition. Oligomeric aminoaldehyde resins (aminoplastics) are the products of polycondensation of melamine and formaldehyde.

Properties. White solid substances with good water solubility; insoluble in alcohol.

Applications. Used for production of various aminoplastic molding materials, lacquers, adhesives, and enamels. Used in the manufacture of screw caps for medicine bottles.

Migration Data. A study of aqueous extracts of aminoplastic articles (Karbolit type) revealed migration of 18 to 177 mg formaldehyde/l. Migration of melamine and formaldehyde into food-simulating solvents from cups made of melamine resins were studied under various conditions. Low migration of melamine was observed in any unused cups kept at 60°C for 30 min., 26°C for 1 hour, or cooled at 20°C for several days. The highest migration level of melamine into 4.0% acetic acid was about 40 ppm when the migration test was repeated 7 times at 95°C for 30 min. Migration of melamine from the cups being used at a cafeteria was 0.4 to 0.5 ppm, but that of formaldehyde was undetectable when the cups were kept at 60°C for 30 min with 4.0% acetic acid.[1]

Acute Toxicity. LD_{50} exceeded 10 g/kg BW in rats.[2] The toxicity of MFR esterified through hydroxymethyl (methylol) groups by fractions of isoalcohols, higher alcohols, and by a mixture of butanol with fractions of isoalcohols has been studied. A dose of 5.0 g/kg BW is not lethal to mice and rats. There were signs of a narcotic effect.[3]

Regulations. ***U.S. FDA*** (1998) approved the use of MFR (1) in the manufacture of resinous and polymeric coatings for the food-contact surface of articles intended for use in producing, manufacturing, packing, processing, preparing, treating, packaging, transporting, or holding food, subject to the provisions prescribed in 21 CFR part 175.300, and (2) as the food-contact surface of molded articles intended for use in producing, manufacturing, packing, processing, preparing, treating, packaging, transporting, or holding food, in accordance with conditions prescribed in 21 CFR part 177.1460.

References:

1. Ishiwata, H., Inoue, T., and Tanimura, A., Migration of melamine and formaldehyde from tableware made of melamine resin, *Food Addit. Contam.*, 3, 63, 1986.
2. Acute Toxicity Data, *J. Am. Coll. Toxicol.*, Part B., 1, 162, 1992.
3. Stasenkova, K. P. and Mel'nikova, R. N., Study of toxicity of some isoalcohols, higher alcohols and melamine-formaldehyde resins, in *Toxicology of New Industrial Chemicals*, Moscow, Medgiz, Issue No 3, 1961, 108 (in Russian).

K-79-79 MOLDING MATERIAL based on MELAMINE-FORMALDEHYDE RESIN

Abbreviation. MFR.

Trade Name. Melalite.

Composition. Consists of MFR, sulfite cellulose (filler), zinc stearate, and a dye.

Properties. A finely dispersed molding powder. *Dens.* 1.45 to 1.55. Characterized by high water and heat resistance and increased mechanical strength.

Applications. Used for the manufacture of dishware and food containers.

Migration Data. Articles made of MFR release 1.2 to 47.7 mg formaldehyde/l to water. The optimum temperature for molding MFR dishware is 165°C. The organoleptic properties of foodstuffs stored in MFR salad containers at 3.0°C did not change during storage time specified for these products. There was only a reduction in cucumber brine acidity at 18 to 20°C. Migration of formaldehyde into water and soft drinks from MFR dishware of grades *M* and *MS* within 24 hours at 40°C did not exceed 0.1 mg/l. The release of formaldehyde into a hot 1.0% acetic acid solution from MFR beakers and jugs was higher than the permissible level.[1]

Toxicity. Depends on the level of migration of monomers and additives from the materials.

Regulations. ***U.S. FDA*** (1998) approved the use of MFR (1) in the manufacture of resinous and polymeric coatings for the food-contact surface of articles intended for use in producing, manufacturing, packing, processing, preparing, treating, packaging, transporting, or holding food, subject to the provisions prescribed in 21 CFR part 175.300, and (2) as the food-contact surface of molded articles intended for use in producing, manufacturing, packing, processing, preparing, treating, packaging, transporting, or holding food, in accor dance with conditions prescribed in 21 CFR part 177.1460.

Recommendations. ***Russia***. Aminoplastics B1, B2 and A2 have been recommended for manufacturing pots, funnels, and dishware.[028]

Reference:

1. Kovgan, N. A., *Study of Consumer Properties of Household Melalite Articles*, Author's abstract of thesis, Leningrad, 1972, 22 (in Russian).

UREA-FORMALDEHYDE RESINS

Structural Formula.

```
[~N                              CH2~]n
   \                            /
    N-CONHCH2OCH2NHCO-N
   /                            \
~N                               CH2~
```

CAS No 9011-05-6
39327-95-2
56779-89-6
57608-68-1
57657-45-1
57762-61-5
60267-46-1
60831-80-3

RTECS No YU1610000

Abbreviation. UFR.

Synonyms and **Trade Names.** Anaflex; Carbamol; Formaldehyde-urea condensate; Formaldehyde-urea copolymer; Formaldehyde-urea resin; Formalin-urea copolymer; Karbamol; Paraformaldehyde-urea polymer; Plyamine; Urea-formaldehyde adduct; Urea-formaldehyde copolymer.

Composition. Oligomeric aminoaldehyde resins. The products of polycondensation of urea (carbamide) and formaldehyde.

Properties. White solid substances with good water solubility; insoluble in alcohol. In presence of catalysts they become colorless, light-resistant, readily colored three-dimensional polymers. During processing and in use they release formaldehyde.

Applications. Used for production of different aminoplastic molding materials (carbamide plastics), carbamide adhesives, enamels, and Mipora cellular plastic. Porous Mipora is a hardened foam obtained by mixing aminoplastics with a foaming agent. It is used to insulate refrigerators and to protect sugar-beets from freezing.

Migration Data. Materials of grades MKS-10P and KS-35 are products of the polycondensation of urea, formaldehyde, and polyethylene polyamine (PEPA). In aqueous media they are homogenous products which do not release PEPA. They contain 4.0% free formaldehyde and give water an odor similar to that of sea water. Odor perception threshold is 624 and 434 mg KS/l. With a concentration below 58 and 23 mg KS/l, there was no foam formation.

Migration of formaldehyde from urea-formaldehyde resins can be reduced by treating articles with an aqueous solution of *urea* or melamine (concentration 0.1% at 40 to 90°C for 10 hours) or by washing them with ammonia water (concentration 2.0% at 20°C for 15 to 30 hours or at 50°C for 0.5 to 1 hour).[1]

Acute Toxicity. With single administration of MKS, LD_{50} is 1.82 g/kg BW in mice and 3.43 g/kg BW in rats. With KS, these values are 6.36 and 8.39 g/kg BW, respectively.[2] A dose of 3.43 g MKS/kg BW are not lethal to guinea pigs.

Repeated Exposure. 1/5 to 1/20 of LD_{50} was administered to animals over a period of 1.5 months. There were no changes in the peripheral blood, although some histological changes in the liver were found.[2]

Long-term Toxicity. Doses of 0.34 mg/kg BW or higher caused functional changes in the CNS and deterioration of liver antitoxic function. There was increased permeability of vessel walls and increased lipid contents in the liver cells. A dose of 0.17 mg/kg BW proved to be a NOEL.[2]

Reproductive Toxicity. Doses of 0.34 mg/kg BW or higher reduce the number of pregnant females, increase the number of embryo resorption, reduce defensive reactions.[2]

Regulations. ***U.S. FDA*** (1998) approved the use of UFR (1) in the manufacture of resinous and polymeric coatings for the food-contact surface of articles intended for use in producing, manufacturing, packing, processing, preparing, treating, packaging, transporting, or holding food, subject to the provisions prescribed in 21 CFR part 175.300, and (2) as the food-contact surface of molded articles intended for use in contact with food, in accordance with the conditions prescribed in 21 CFR part 177.1990.

Standards. *Russia.* MPC is 3.0 mg/l (MKS-10) and 0.1 mg/l (KS-35).

References:

1. Petrovsky, K. S. and Braun, D. D., Outlook for improving the health-related properties of plastic arti cles for use in the food industry, *Gig. Sanit.,* 1, 75, 1972 (in Russian).
2. Brodskaya, B. Ya., Rodnikov, A. V., Golovkova, T. V., Kalinina, Z. S., and Nikitina, L. A., Hygienic basis for the maximum permissible concentrations of carbamide resins in the water of reser voirs, *Gig. Sanit.*, 5, 71, 1980 (in Russian).

UREA-FORMALDEHYDE RESIN KSM-O3P

CAS No 9011-05-6

Composition. Contents of solids are 65.6%, that of free formaldehyde are 0.2%.

Synonym and **Trade Names.** Anaflex; Polynoxylin; Methylene-*N,N'*-bis(hydroxymethyl) urea polymer.

Properties. Good water solubility, *pH* = 7.6, *dens.* 1.265.

Applications. It is used as a topical antibacterial, a bulking agent in cosmetics, and to form the outer shell of microcapsules. It is formed by the condensation of urea with formaldehyde and may contain residual formaldehyde at levels between 17 and 30 ppm.

Acute Toxicity. In rats and mice, LD_{50} is 38 and 20 g/kg BW, respectively.

Repeated Exposure. Cumulative properties were not pronounced.

Allergenic Effect. None.

References:

1. Germanova, A. A., *Gig Truda Prof. Zabol.*, 10, 57, 1981 (in Russian).
2. Anonymous, Final report on the safety assessment of Polyoxymethylene Urea, *J. Am. Coll. Toxicol.*, 14, 204, 1995.

UREA-FORMALDEHYDE RESIN UKS-A

Properties. *Dens.* of an aqueous solution of the resin is 1.1 to 1.35; *pH* is 8.0. Water soluble, precipitates after a certain time.

Acute Toxicity. In mice, LD_{50} exceeded 10 g/kg BW.

Repeated Exposure. Cumulative properties were not pronounced.

Reference:

Bruskova, N. I. and Karamzina, N. I., Toxicity of urea-formaldehyde resin UKS-A, *Gig. Sanit.*, 3, 88, 1982 (in Russian).

CELLULOSE

Molecular Formula. $C_6H_7O_2(OH)_3$

CAS No 9004-34-6
9006-02-4
39394-43-9
58968-67-5
61991-21-7
61991-22-8
84503-75-3
99331-82-5

RTECS No FJ5691460

Abbreviation. MCC (microcrystalline cellulose).

Composition. C. is a natural polymer, the main component of the cell walls of the more highly developed plants. It is found in greatest quantity in wood, and in the fibers of cotton, flax, etc.

Application. Used as a filler in the production of some plastics, in fiber form as packing material, and in pharmaceutical practice. C. ethers and esters are widely used synthetic polymers.

Migration Data. Chocolates, boiled sweets, toffees, cakes, and meat pies were wrapped in regenerated C. films that contained various mixtures of glycol softeners. It was shown that higher levels of migration occurred for propylene glycol than for triethylene glycol, and the presence of coating reduced the migration of both softeners. Mono- and diethylene glycol levels in the food samples were below 10 mg/kg.[1]

Short-term Toxicity. Sprague-Dawley rats received 500, 2500, or 5000 mg MCC/kg BW (25% *w/v* in tap water) by oral gavage for 90 days. MCC particles were not observed in any tissue examined. No ill effects were found in any other parameter or organ evaluated. The NOAEL was found to be more than 5000 mg/kg BW, which was the highest dosage tested.[2]

Long-term Toxicity. Chronic ingestion of purified C. over the entire life-spans in rats and mice does not result in any increase in spontaneous disease. Purified C. does not display promotional activity in the mammary gland, the colon, or the bladder of rats and does not significantly alter the absorption or the metabolism of dietary components.[3]

Reproductive Toxicity including deviations in neonate development was not observed in animals exposed orally to purified C.[3]

Carcinogenicity. Ingestion of cellulose was not carcinogenic in lifetime feeding studies in rats and mice.[3]

Chemobiokinetics. Ingestion of purified C. does not significantly alter the absorption or metabolism of the main dietary components.[3]

Regulations.

EU (1990). C. is available in the *List of authorized monomers and other starting substances, which shall be used for the manufacture of plastic materials and articles intended to come into contact with foodstuffs* (*Section A*).

Great Britain (1998). C. is authorized without time limit for use in the production of polymeric materials and articles in contact with food or drink or intended for such contact.

Recommendations. ***EU*** (1995). C. is a food additive generally permitted for use in foodstuffs.

References:

1. Castle, L., Cloke, H. R., Crews, C., and Gilbert, J., The migration of propylene glycol, mono-, di-, and triethylene glycols from regenerated cellulose film into food, *Z. Lebensmitt. Untersuch. Forsch.*, 187, 463, 1988.
2. Kotkoskie, L. A., Butt, M. T., Selinger, E., Freeman, C., and Weiner, M. L., Qualitative investigation of uptake of fine particle size microcrystalline cellulose following oral administration in rats, *J. Anat.*, 189, 531, 1996.
3. Anderson, R. L., Owens, J. M., and Timms, C. W., The toxicity of purified cellulose in studies with laboratory animals, *Cancer Lett.*, 63, 83, 1992.

CELLULOSE ACETATES

CAS No 9004-35-7
9035-69-2 (cellulose diacetate)
9012-09-3 (cellulose triacetate)

Abbreviation. CA.

Composition. CA are partially acetylated cellulose: ester of cellulose and acetic acid ($R = CH_3$). Several CA differ from one another only in the degree of acetylation.

Synonym and **Trade Names.** Acetate cotton; Acetose; Bioden; Cellidor; Cellulose 2,5-acetate.

Properties. White or colorless flakes or powder. Triacetate and tetracetate cellulose are insoluble in water and alcohol. It is easily deformed by heat, acids, bases, and some solvents, but it is resistant to fats and oils. Maximum continuous service temperature for CA is 80°C.

Applications. Used in the manufacture of rubber and celluloid substitutes, and in the production of films (together with acetic acid, acetone, and caprolactam). Intended for use in aqueous and alcoholic media with *pH* 3.0 to 8.0 up to 50°C. CA is used to produce transparent, water resistant, non-combustible film for packing medicines and foodstuffs. CA membranes are used in the dairy industry for utilizing whey and producing high-purity milk proteins for children, in the pharmaceutical industry, for desalination of salty waters, and for concentrating juices.

Migration of CA constituents into contact fluids was found to be negligible (traces). At 40°C, triethanolamine was discovered in the extract at a concentration of 1.4 mg/l and was therefore subsequently eliminated from CA formulation.[1]

Aqueous extracts of film plasticized with triacetin (glycerol triacetate) and triethylene glycol dipropionate are transparent, colorless, and have no sediment. The extracts exhibit a foreign odor and sourish taste, irrespective of temperature or exposure. Triacetin migrates at the level of 1.1 to 1.6 mg/l and triacetin degradation product (acetic acid) migrate at the level of 0.5 to 2.0 mg/l. Triethylene glycol propionate was found in the aqueous extracts at the level of 0.8 to 2.5 mg/l.[2]

No increase in oxidizability level was found in distilled water or in 2.0% citric acid solution after a 10-day contact with MGA type of CA membranes. At 20 and 60°C, the extracts had no foreign odor. Only "MGA-640" CA membrane induced a small change in the taste of extracts. There was no migration of acetone and caprolactam from the membranes.[3]

Migration into quiches of diethyl phthalate at the level of 2.0 to 4.0 mg/kg from CA film purchased from retails and take-away outlets was determined.[4]

Short-term Toxicity. Sprague-Dawley rats were exposed to a dietary admixture of CA at dose levels of 0.5, 2.5, and 5.0 g/kg BW for 3 months. There were no treatment-related changes in hematology, clinical chemistry and urinalysis. Gross pathology and histopathology examination revealed no evidence of an adverse effect related to ingestion of CA.[5]

Long-term Toxicity. Rats received aqueous extracts from MGA CA membrane instead of the drinking water over a period of 8 months. The exposure caused no toxic effect.[1]

Reproductive Toxicity.

Embryotoxicity. There were no effects on reproduction in rats given aqueous extracts of CA membranes as their drinking water.[1] Sound embryotoxic action was observed in rats that consumed aqueous extracts of CA films treated with formamide, over the entire gestation period (Shtannikov et al., 1972).

Regulations. ***U.S. FDA*** (1998) approved use of CA as the basic polymer in the manufacture of resinous and polymeric coatings for the food-contact surface of articles intended for use in producing, manufacturing, packing, processing, preparing, treating, packaging, transporting, or holding food. Affirmed to be *GRAS* as a substance migrating to food from paper and paperboard products.

References:

1. Selivanov, S. B., *Hygiene Investigation and Assessment of Reverse Osmosis Method for Desolination of Drinking Water in "Filter-Press"-Type Installations*, Author's abstract of thesis, 2nd Moscow Medical Institute, Moscow, 1977, 18 (in Russian).
2. Kas'an, V. N., in Proc. IV All-Union Sci.-Techn. Conf. *Synthesis and Applications of Polymeric Materi als Used in Food Industry*, Minsk, 1980, 128 (in Russian).
3. Kodner, M. S. and Mamayeva, Z. A., in Proc. of IV All-Union Conf. *Membrane Methods for Mixture Separation*, April 12-14, 1977, Vladimir, 1977, 414 (in Russian).
4. Castle, L., Mercer, A. J., Startin, J. R., and Gilbert, J., Migration from plasticized films into foods. 3. Migration of phthalate, sebacate, citrate and phosphate esters from films used for retail food packaging, *Food Addit. Contam.*, 5, 9, 1988.
5. Thomas, W. C., McGrath, L. F., Baarson, K. A., Auletta, C. S., Daly, I. W., and McConnell, R. F., Subchronic oral toxicity of cellulose acetate in rats, *Food Chem. Toxicol.*, 29, 453, 1991.

CELLULOSE ACETATE BUTYRATE

CAS No 9004-36-8

Abbreviation. CAB.

Synonym. Cabufocon.

Properties. White powder with a characteristic butyric acid odor. *M. p.* 127 to 240°C. Solubility in water is negligible.

Applications. Used in the manufacture of coatings. Cabufocon is a contact lens material.

Toxicity. Expected to exhibit a low ingestion hazard.

Regulations. ***U.S. FDA*** (1998) approved use of CAB (1) in adhesives as a component of articles intended for use in packaging, transporting, or holding food, subject to the conditions prescribed in 21 CFR part 175.105; (2) in production of hot-melt strippable food coatings which may be applied to food, subject to the conditions prescribed in 21 CFR part 175.230; (3) as the basic polymer in the manufacture of resinous and polymeric coatings for the food-contact surface of articles intended for use in producing, manufacturing, packing, processing, preparing, treating, packaging, transporting, or holding food in accordance with the conditions prescribed in 21 CFR part 175.300; and (4) in the manufacture of cellophane for packaging food in accordance with the conditions prescribed in 21 CFR part 177.1200.

CELLULOSE ACETATE PROPIONATE

CAS No 9004-39-1

Abbreviation. CAP.

Composition. A mixed ester of cellulose and acetic and propionic acids (R = CH_3 and C_2H_5). The techni cal grade product consists of 1.5 to 7.0% acetyl groups and 39 to 45% propionyl groups.

Applications. Used in the production of films.

Migration Data. Aqueous extracts of films plasticized with triacetin (glycerol triacetate) and triethylene glycol dipropionate are transparent, colorless, and have no sediment. The cold and hot extracts exhibited a foreign odor and a sourish taste. Triacetin plasticized films released the plasticizer at the level of 1.1 to 1.6 mg/l.[1]

Reproductive Toxicity. Mice were treated *i/p* with doses of 100 mg CAP/kg BW on days 3 through 7 of pregnancy. The treatment caused an increase in post-implantation mortality.[2]

Regulations. ***U.S. FDA*** (1998) approved use of CAP in production of hot-melt strippable food coatings which may be applied to food, subject to the conditions prescribed in 21 CFR part 175.230.

References:

1. Kas'an, V. N., in Proc. IV All-Union Sci.-Techn. Conf. *Synthesis and Applications of Polymeric Materi als Used in Food Industry*, Minsk, 1980, 128 (in Russian).
2. Guttnerm, J., Klaus, S., und Heinecke, H., Embryotoxicity of intraperitoneally administered hydroxyethylcellulose in mice, *Anat. Anzeiger*, 149, 282, 1981.

HYDROXYETHYL CELLULOSE

Molecular Formula.

$\{\sim C_6H_7O_2\,(OH)_{3-x}[(OCH_2CH_2)_yOH]_x\sim\}_n$

CAS No 9004-62-0

Abbreviation. HEC.

Synonyms and **Trade Names.** Cellosize QP; Hydroxyethyl cellulose ether; Natrosol.

Composition. Cellulose ester, the product of its interaction with ethylene oxide.

Properties. White free flowing powder, without taste and odor. Readily soluble in hot and cold water; insoluble in organic solvents, fat, and oils.

Applications. A thickener in latex paints; protective colloid, binder stabilizer, paper sizing agent.

Acute Toxicity. Administration of a single oral dose of up to 23 g HEC/kg BW (50% solution in corn oil) caused no toxic effect.[03] Ingestion of large quantities induced intestinal obstruction. Toxic action was noted at the doses in excess of 2.0 g/kg (Dreisbach, 1977).

Long-term Toxicity. Rats received 0.2, 1.0, or 5.0% HEC in their diet over a period of 2 years. The treatment produced no adverse effects on growth, food intake, life-span, frequency of extraneous infections, body measurements, kidney and liver weights, hematology analyses, occurrence of neoplasms. Histological examination revealed no changes in the visceral organs.[03]

Allergenic Effect. Skin sensitization seems to be unusual (Dreisbach, 1977).

Chemobiokinetics.

Observations in man. Following ingestion, cellulose ethers are not absorbed but eliminated unchanged with feces.[03]

Regulations. ***U.S. FDA*** (1998) approved use of HEC (1) as a constituent of food-contact adhesives which may safely be used as a component of articles intended for use in packaging, transporting, or holding food in accordance with the conditions prescribed in 21 CFR part 175.105; (2) in the manufacture of resinous and polymeric coatings for the food-contact surface of articles intended for use in producing, manufacturing, packing, processing, preparing, treating, packaging, transporting, or holding food in accordance with the conditions prescribed in 21 CFR part 175.300; and (3) in the manufacture of cellophane for packaging food in accordance with the conditions prescribed in 21 CFR part 177.1200. Water insoluble HEC film may be safely used (4) in packaging food in accordance with the conditions prescribed in 21 CFR part 177.1400.

Standards. ***Russia*** (1995). RAL: 1.0 mg/l.

HYDROXYPROPYL CELLULOSE

CAS No 9004-64-2

RTECS No NF9050000

Abbreviation. HPC.

Synonyms and **Trade Name.** Cellulose hydroxypropyl ether; Klucel; Oxypropylated cellulose; PM 50.

Properties. *M. p.* 100 to 150°C. Easily wetted with water forming gels. Insoluble in water above 45°C.

Applications. HPC is used as a stabilizer and coating agent in a variety of medicines. It is also employed in ophthalmic preparations to treat dry eyes.

Acute Toxicity. LD_{50} is 10.2 g/kg BW in rats.[1]

Reproductive Toxicity.

Teratogenicity. Administration of therapeutic doses of HPC during pregnancy is unlikely to pose a substantial teratogenic risk. Rabbits and rats were fed HPC during pregnancy in doses 8 to 200 times the WHO acceptable daily intake for humans. There was no increase in frequency of malformations among their offspring.[2,3]

Regulations. ***U.S. FDA*** (1998) approved the use of HPC in food in accordance with the conditions prescribed in 21 CFR part 172.870.

References:

1. *FAO Nutrition Meetings Report,* Series No 46A, Geneva, 1969, 131.
2. Kitagawa, H., Satoh, T., Saito, H., et al., Teratological study of hydroxypropylcellulose of low substitution (L-HPC) in rabbits, Abstract, *Oyo Yakuri,* 16, 259, 1978 (in Japanese).
3. Kitagawa, H., Satoh, T., Saito, H., et al., Teratological study of hydroxypropylcellulose of low substitution (L-HPC) in rats, Abstract, *Oyo Yakuri,* 16, 271, 1978 (in Japanese).

HYDROXYPROPYLMETHYL CELLULOSE

CAS No 9004-65-3
8063-82-9
2683-26-5
RTECS No NF9125000
Abbreviation. HPMC.

Synonym and **Trade Names.** Isopto tears; Methocel E, F, K or HG; Propylene glycol ether of methylcellulose; Ultra Tears.

Applications. HPMC is used in contact lens wetting solutions. It has also been used as a sperm immobilizer in vaginal contraceptives.

Acute Toxicity. There was no observed toxicity of hydrophobically modified HPMC after oral administration at a single dose up to 900 mg/kg BW. LD_{50} *i/p* is 5.2 g/kg BW in rats, and 5.0 g/kg BW in mice.[1,2]

Repeated Exposure.

Observations in man. 25 human subjects received three doses of HPMC (*Methocel 65*) in the range of 0.6 to 8.9 g with weekly intervals. Ingested HPMC was eliminated in feces within 96 hours following ingestion.[3]

Animal studies. In a 30-day study, rats developed changes in GI tract (hypermotility and diarrhea).[2]

Short-term Toxicity. Rats were fed Methocel *60 HG* for 4 months. Addition of HPMC up to 3.0% in the diet caused no adverse effects; 10% addition resulted in a slight retardation of BW gain in males; 30% HPMC in the rat diet led to marked growth depression and increase in mortality rate (probably due to poor nutrition) in both sexes.[4]

Male and female SD rats received repeated oral administrations of HPMC of the lowest viscosity grade at doses of 0.5, 1.0, and 2.1 BW over a period of 3 months. Examinations of general conditions, hematology, blood chemistry, ophthalmology, absolute and relative organ weights, autopsy and histopathology revealed only a few, apparently coincidental, statistically significant differences from the control, and no evidence of any dose-dependent changes was found. Males in the 2.1 g/kg BW group showed an insignificant tendency for decreased food consumption and urine volume.[5]

HPMC exhibited extremely low toxicity as has been found for higher viscosity grades.

Reproductive Toxicity.

Gonadotoxicity. HPMC is used as a sperm immobilizer in vaginal contraceptives.

Teratogenicity. HPMC was not found to cause malformations in rat fetuses.[6]

Regulations. ***U.S. FDA*** (1998) approved use of HPMC in food in accordance with the conditions prescribed in 21 CFR part 172.874. in adhesives as a component of articles intended for use in packaging, transporting, or holding food in accordance with the conditions prescribed in 21 CFR part 175.105.

References:

1. Obara, S., Muto, H., Kokubo, H., Ichikawa, N., Kawanabe, M., and Tanaka, O., Studies on single-dose toxicity of hydrophobically modified hydroxypropyl methylcellulose in rats, *Toxicol. Sci.*, 17, 13, 1992.
2. *J. Pharmacol. Exp. Ther.*, 99, 112, 1950.
3. Patty's *Industrial Hygiene and Toxicology*, 4th ed., vol II, Part A, G. D. Clayton and F. E. Clayton, Eds., John Wiley & Sons, Inc., New York, 1993, 511.
4. Hodge, H. C., Maynard, E. A., Wiet, W. G., Blanchet, H. J., and Hyatt, R. E., Chronic oral toxicity of a high gel point methyl cellulose (Methocel HG) in rats and dogs, *J. Pharmacol. Exp. Ther.*, 99, 112, 1950.
5. Obara, S., Muto, H., Shigeno, H., Yoshida, A., Nagaya, J., Hirata, M., Furukawa, M., and Sunaga, M., A three-month repeated oral administration study of a low viscosity grade of hydroxypropyl methylcellulose in rats, *J. Toxicol. Sci.*, 24, 33, 1999.
6. Emerson, J. L., Thompson, D. J., Strebing, R. J., Gerbig, C. G., and Robinson, V. B., Teratogenic studies on 2,4,5-trichlorophenoxyacetic acid in the rat and rabbit, *Food Cosmet. Toxicol.*, 9, 395, 1971.

HYDROXYPROPYLMETHYL CELLULOSE, ACETATE SUCCINATE

Abbreviation. HPMCAS.

Applications. A pharmaceutical excipient.

Acute Toxicity. In the acute toxicity study in rabbits and rats (single oral dose of 2.5 g HPMCAS/kg BW), no deaths or behavioral abnormalities were observed. LD_{50} exceeded 2.5 g/kg BW.[1]

Short-term Toxicity. In the subchronic toxicity study rats were exposed to 0.63, 1.25, or 2.5 g HPMCAS/kg BW daily as a single oral dose for 2 months. No significant behavioral abnormality was observed. There was some decrease in BW gain in rats of both sexes, but the effect was not statistically significant.[1]

Long-term Toxicity. In the 6-month toxicity study, rats were given 1.25 or 2.5 g/kg BW as a single oral dose. No significant behavioral abnormality was observed. There was some decrease in BW gain in male rats, but it was statistically insignificant.[1]

Reproductive Toxicity.

Embryotoxicity. Slc:SD rats were orally administered HPMCAS at dose levels of 625, 1,250 and 2,500 mg/kg BW for a period from day 17 of gestation to day 21 after delivery. No behavioral changes were found in offspring. In the treated group of 2500 mg/kg BW, the liver weight was significantly increased in males and showed a tendency to increase in females compared with the control.[2]

Teratogenicity S1c:SD rats were orally administered HPMCAS at dose levels of 625, 1,250, and 2,500 mg/kg BW for a period of 11 days from day 7 to day 17 of gestation. The incidences of external, internal, and skeletal anomalies were not significantly increased in the fetuses of any treated groups. HPMCAS caused no effects on parturition, lactation, postnatal growth, and reproductive ability of the male and female offspring.[3] HPMCAS was orally administered to New Zealand White rabbits at dose levels of 625, 1,250, and 2,500 mg/kg BW for a period of 13 days from day 6 to day 18 of gestation. The administration of HPMCAS during a period of organogenesis produced no embryotoxic and teratogenic effects as well as no influence on behavior, appearance, and growth of animals.[4]

References:

1. Hoshi, N., Yano, H., Hirashima, K., Kitagawa, H., and Fukuda, Y., Toxicological studies of hydroxypropylmethyl cellulose acetate succinate - acute toxicity in rats and rabbits, and subchronic and chronic toxicities in rats, *J. Toxicol. Sci.*, 10 (Suppl 2), 147, 1985.
2. Hoshi, N., Ueno, K., Igarashi, T., Kitagawa, H., Fujita, T., Ichikawa, N., Kondo, Y., and Isoda, M., Effects on offspring induced by oral administration of hydroxypropyl methylcellulose acetate succinate to the female rats in peri- and postnatal periods, *J. Toxicol. Sci.*, 10, 235, 1985.
3. Hoshi, N., Ueno, K., Igarashi, T., Kitagawa, H., Fujita, T., Ichikawa, N., Kondo, Y., and Isoda, M.,Teratological studies of hydroxypropylmethyl cellulose acetate succinate in rats, *J. Toxicol. Sci.*, 10, 203, 1985.

4. Hoshi, N., Ueno, K., Igarashi, T., Kitagawa, H., Fujita, T., Ichikawa, N., Kondo, Y., and Isoda, M.,Teratological studies of hydroxypropylmethylcellulose acetate succinate in rabbits, *J. Toxicol. Sci.*, 10, 227, 1985.

METHYL CELLULOSE

CAS No 9004-67-5
RTECS No FJ5959000
Abbreviation. MC.

Synonyms and **Trade Names.** Bulkaloid; Cellogran; Cellothyl; Cellulose methylate; Cellulose methyl ether; Cethylose; Cologel; Hydrolose; Mellose; Methalose; Methocel; Methylose; Tylose; Viscosol.

Composition. Cellulose methyl ester.

Properties. Odorless and tasteless, white granules or grayish white powder. Slightly hygroscopic. Soluble in cold water; insoluble in hot water and alcohol. *pH* of aqueous solution exceeded 7. Aqueous solutions are transparent and viscous, and coagulate when heated above 50°C, but, when cooled, again return into solution.

Applications. Food additive. A component of adhesives. Used in vinyl chloride polymerization. Polymer for acetate fibers and rubber. A suspending agent in cosmetics and pharmaceuticals.

Acute Toxicity.

Observations in man. Practically nontoxic. Lethal dose seems to be more than 15 g/kg BW.[019] Tablets of MC caused fecal impaction.

Animal studies. In mice, a single administration of 0.013 mg/kg BW dose in 1.0% solution produced no effect on glutathione level in the liver and ALAT activity in the blood plasma over a period up to 12 hours after administration. Histological investigation revealed no changes in the liver, kidney, and stomach.[1]

Repeated oral administration of MC at a concentration of more than 0.5% to rats caused uncoupling of oxidative phosphorylation in liver and heart mitochondria and partial inhibition of mixed function oxidases of liver endoplasmic reticulum, as measured by 2-biphenylhydroxylation and 4-biphenyl-hydroxylation. MC at a concentration of 0.5% did not alter mitochondrial function and mixed function oxidases.[2]

Long-term Toxicity. Rats that received 440 mg MC/day in their diet and part in their drinking water for 8 months exhibited no manifestations of the toxic effect. The same results were observed in rats fed 6.2 g MC/kg BW for 6 months.[04]

Male and female rats received 1.0, 5.0, or 20% Methocel 65 *HG* (high gel point methyl cellulose) in the diet for 2 years. Retardation of BW gain was the only effect noted in the males fed the 20% diet. Two dogs were given oral doses of 0.1, 0.3, 1.0, or 3.0 g/kg BW for a year without any visible effect. Another dog was fed 50 g/kg BW. This treatment caused some diarrhea, slight weight loss, and a decrease in erythrocyte count. Histology investigation did not reveal any changes in the visceral organs.[3]

Reproductive Toxicity. Rats were given 5.0% MC in their diet through 3 generations. The treatment produced no adverse effects including reproduction.[04]

Teratogenicity. Treatment of pregnant rats may result in thinning of the central tendon of the diaphragm in the newborns.[4]

Chemobiokinetics. MC is not absorbed in the GI tract, it did not undergo hydrolysis into methyl alcohol and cellulose and is excreted unchanged in feces.[011]

Regulations. ***U.S. FDA*** (1998) approved the use of MC (1) as a constituent of acrylate ester copolymer coating which may safely be used as a food-contact surface of articles intended for packaging and holding food, including heating of prepared food; (2) as a constituent of acrylate ester copolymer coating which may be used as a food-contact surface of articles intended for packaging and holding food, including heating of prepared food, subject to the provisions prescribed in 21 CFR part 175.210; (3) in the manufacture of resinous and polymeric coatings for the food-contact surface of articles intended for use in producing, manufacturing, packing, processing, preparing, treating, packaging, transporting, or holding food in accordance with the conditions prescribed in 21 CFR part 175.300; and (4) as a component of a defoaming agent intended for articles which may be used in producing, manufacturing, packing, processing, preparing, treating, packaging, transporting, or holding food in accordance with the conditions prescribed in 21 CFR part 176.200. Affirmed as *GRAS.*

Recommendations. ***EU*** (1995). MC is a food additive generally permitted for use in foodstuffs. It is also recommended as a suspending medium for use in pharmacological and toxicological experi ments.

References:

1. Skoglund, L. A., Ingebrigsten, K., and Natstad, I., Effects of methylcellulose on hepatic glutathione levels and plasma ALAT following single oral administration to male Bom:NMRI mice, *Genet. Pharmacol.*, 18, 497, 1987.
2. Bachmann, E., Weber, E., Post, M., and Zbinden, G., Biochemical effects of gum arabic, gum tragacanth, methylcellulose and carboxymethylcellulose-*Na* in rat heart and liver, *Pharmacology*, 17, 39, 1978.
3. Hodge, H. C., Maynard, E. A., Wiet, W. G., Blanchet, H. J., and Hyatt, R. E., Chronic oral toxicity of a high gel point methylcellulose (Methocel HG) in rats and dogs, *J. Pharmacol. Exp. Ther.*, 99, 112, 1950.
4. Lu, C. C., Mull, R. L., Lochry, E. A., and Christian, M. S., Developmental variations of the diaphragm and liver in Fischer 344 rats, *Teratology*, 37, 571, 1988.

ETHYL CELLULOSE

CAS No 9004-57-3
RTECS No FJ5950500
Abbreviation. EC.

Synonyms and **Trade Names.** Cellulose ethyl ether; Cellulose, triethyl ether; Ethocel; Triethyl cellulose.

Properties. Free-flowing, white to light tan powder or granular, thermoplastic solid.

Applications. Used in the manufacture of coatings and adhesives, lacquers and plastics, and as a binder in the preparation of medicinal tablets. A direct food additive.

Acute Toxicity. LD_{50} exceeded 5.0 g/kg BW in rats and mice. [1]

Short-term Toxicity. Sprague-Dawley rats were administered undiluted Aquacoat ECD ethyl cellulose aqueous dispersion by oral gavage at doses of 903, 2709, or 4515 mg/kg BW for 90 days. The only treatment-related clinical sign observed was pale feces which was noted at the highest treated groups of animals. No statistically significant differences in hematology analyses, BW gain, food consumption, and organ weights were noted among treated animals when compared with controls. Significantly decreased total protein and globulin levels and increases in ALT and AST in male rats receiving 2709 and 4515 mg Aquacoat ECD/kg BW were reported. There were no findings related to the treatment with Aquacoat ECD either at necropsy or at histological examination. The NOAEL for female rats is established in excess of 4.5 g/kg BW; the NOAEL for male rats is 9.0 mg/kg BW.[2]

Long-term Toxicity. Rats received the diet containing 1.2% EC for 8 months (average dose of 182 mg/kg BW). There were no changes in appearance, behavior, and growth of the animals. Gross pathology and histology examination revealed no damage in the visceral organs. [3]

Regulations. ***U.S. FDA*** (1998) approved the use of EC (1) in the manufacture of resinous and polymeric coatings for the food-contact surface of articles intended for use in producing, manufacturing, packing, processing, preparing, treating, packaging, transporting, or holding food in accordance with the conditions prescribed in 21 CFR part 175.300; and (2) in food in accordance with the conditions prescribed in 21 CFR part 172.868. Considered to be *GRAS* as a substance migrating to food from paper and paperboard products.

References:

1. Opdyke, D. L., Monographs on fragrance raw materials, *Food Cosmet. Toxicol.*, 19, 97, 1981.
2. Kotkoskie, L. A. and Freeman, C., Subchronic oral toxicity study of Aquacoat ECD aqueous dispersion in the rat, *Food Chem. Toxicol.*, 36, 705, 1998.
3. Patty's *Industrial Hygiene and Toxicology*, 4th ed., vol 2, Part A, G. D. Clayton and F. E. Clayton, Eds., John Wiley & Sons, Inc., New York, 1993, 513.

ETHYLHYDROXYETHYL CELLULOSE

CAS No 9004-58-4
Abbreviation. EHEC.

Synonym and **Trade Name.** Ethyl 2-hydroxyethyl ether cellulose; Etulos.

Applications. Used in the preparation of surface coatings and rubber materials. A laxative.

Regulations. ***U.S. FDA*** (1998) approved the use of EHEC (1) in adhesives as a component of articles intended for use in packaging, transporting, or holding food in accordance with the conditions prescribed in 21 CFR part 175.105; and (2) in the manufacture of resinous and polymeric coatings for the food-contact surface of articles intended for use in producing, manufacturing, packing, processing, preparing, treating, packaging, transporting, or holding food in accordance with the conditions prescribed in 21 CFR part 175.300.

METHYLETHYL CELLULOSE

CAS No 9004-59-5

Abbreviation. MEC.

Composition. Mixed methyl and ethyl cellulose ester ($R = CH_3$ and C_2H_5).

Synonyms and **Trade Names.** Celacol EM; Cellulose, ethyl methyl ether; Edifas A; Ethyl methyl cellulose.

Properties. Practically odorless, white to pale cream colored fibrous solid or powder. Disperses in cold water to form aqueous solution which undergoes reversible transformation from solution to gel upon heating and cooling.

Applications. An emulsifier, stabilizer, foaming agent.

Long-term Toxicity.

Observations in man. There were no manifestations of toxic action in patients treated with 1.0 to 6.0 g MEC/kg BW (as a laxative).

Animal studies. Rats and mice received 10% of MEC in their diet over periods of 104 and 100 weeks, respectively. The treatment caused no adverse effect on general health or survival of the animals.

Carcinogenic effect was not noted. 30 mg MEC/kg BW is a permissible daily dose.[05]

Animal studies. Rats received 10 or 100 mg MEC/kg feed for 2 years. There were no changes in hematology analyses, survival, and BW gain of animals. Histology investigation revealed no lesions or tumors in the visceral tissues (McElligott and Hurst, 1968).

Regulations. ***U.S. FDA*** (1998) approved the use of MEC in food in accordance with the conditions prescribed in 21 CFR part 172.872.

Reference:

McElligott, T. F. et al., *Food Cosmet. Toxicol.*, 6, 449, 1968.

CARBOXYMETHYL CELLULOSE

CAS No 9004-32-4

RTECS No FJ5700000

Abbreviation. CMC.

Composition. Ester of cellulose and glycol (hydroacetic) acid ($R = CH_2COOH$). CMC is a family of compounds with a polysaccharide backbone and various configurations of carboxymethyl side chains and cross-links.

Applications. CMC is used as a film coating, as an emulsifier in detergents, and as carrying agents (vehicles) for pharmaceuticals and cosmetics and in food industries.

Properties. White solid. Insoluble in water and alcohol.

Acute Toxicity. LD exceeded 5.0 g/kg BW in rats, rabbits, and guinea pigs.[1] CMS is known to be harmless when swallowed. Single doses large enough to case any apparent illness in rats, guinea pigs, and rabbits, were not achieved.[2]

Short-term Toxicity. Wistar rats were fed diets with 2.5, 5.0, and 10% CMC for a 3-month period. Water intake, urine production, and urinary sodium excretion increased with increasing doses of CMC due to their sodium contents of about 7 to 8%. The treatment caused diarrhea and caecal enlargement in the mid- and high-dose groups, a slight increase of plasma alkaline phosphatase, and increased urinary calcium and citrate excretions. These changes were considered to be generic effects that typically are observed in rodent studies with low digestible carbohydrates. The increased occurrence of nephrocalcinosis and hyperplasia of the urothelial epithelium in some of the treated groups was interpreted as an indirect consequence of a more alkaline urine coupled with an increased calcium excretion.[3]

Long-term Toxicity. Rats received 1.0 g CMC/kg diet for 25 months. There were neither toxic effects nor tumors in the exposed animals.[04] Rats fed 5.0% CMC in their diet (as sodium or aluminum salts) for 201 to 250 days exhibited no adverse effects.[2] Rats were fed 1.0 g CMC/kg BW for 25 months. The treatment induced no clinical signs of toxic action or changes in histology, hematology, and urinalysis. Fertility was not affected. Dogs and guinea pigs were exposed orally to 1.0 g CMC/kg BW for 0.5 and 1 year, respectively. No manifestations of the toxic action were observed.[2]

Reproductive Toxicity. 2.0% aqueous solution of CMC (10 ml/kg BW) did not affect reproductive functions in male or female rats.[4]

Teratogenicity.

Observations in man. No epidemiological studies of congenital anomalies among infants born to women who ingested large amounts of CMC during pregnancy have been reported.

Animal studies. No teratogenic effect was observed among the offspring of rats, hamsters, or rabbits treated with CMC during pregnancy in doses of 10 to 50 mg/kg BW.[5-7]

Embryotoxicity. No significant adverse effect was found among the offspring of pregnant rats or rabbits given 30 ml/kg BW *s/c* implants of CMC during pregnancy.[8]

Carcinogenicity. Rats were fed 1.0 g CMC/kg BW for 25 months. No neoplasms were found to develop.[2]

Chemobiokinetics. In rats and humans, ingested CMC has been excreted unchanged. In rats, about 90% of a single oral dose was found in the feces.

Regulations. ***U.S. FDA*** (1998) approved the use of CMC (1) in food in accordance with the conditions prescribed in 21 CFR part 172.870; (2) in adhesives as a component of articles intended for use in packaging, transporting, or holding food in accordance with the conditions prescribed in 21 CFR part 175.105; and (3) in the manufacture of resinous and polymeric coatings for the food-contact surface of articles intended for use in producing, manufacturing, packing, processing, preparing, treating, packaging, transporting, or holding food in accordance with the conditions prescribed in 21 CFR part 175.300. Affirmed as *GRAS* as substance migrating from cotton and cotton fabrics used in dry food packaging.

Recommendations. ***EU*** (1995). CMC is a food additive generally permitted for use in foodstuffs.

References:

1. *Food Research*, Champaign, IL, 9, 175, 1944.
2. Patty's *Industrial Hygiene and Toxicology*, 4th ed., vol II, Part A, G. D. Clayton and F. E. Clayton, Eds., John Wiley & Sons, Inc., New York, 1993, 512.
3. Bar, A., Til, H. P., and Timonen, M., Subchronic oral toxicity study with regular and enzymatically depolymerized sodium carboxymethylcellulose in rats, *Food Chem. Toxicol.*, 33, 909, 1995.
4. Fritz, H. and Becker, H., The suitability of carboxymethylcellulose as a vehicle in reproductive studies, *Arzneimittel-Forsch.*, 31, 813, 1981.
5. Hamed, M. R., Al-Assy, Y. S., and Ezzeldin, E., Influence of protein malnutrition on teratogenicity of acetylsalicylic acid in rats, *Hum. Exp. Toxicol.*, 13, 83, 1994.
6. Kwarta, R. F., Hemm, R. D., Pollock, J. J., Christian, M. S., et al., Levonorgestrel/ethinyl estradiol: Study of developmental toxicity in rabbits, Abstract, *Oyo Yakuri*, 42, 341, 1991 (in Japanese).
7. Robens, J. F., Teratologic studies of carbaryl, diazinon, norea, disulfiram, and thiram in small laboratory animals, *Toxicol. Appl. Pharmacol.*, 15, 152, 1969.
8. Siddiqui, W. H., Schardein, J. L., Cassidy, S. L., and Meeks, R. G., Reproductive and developmental toxicity studies of silicone gel Q7-2159A in rats and rabbits, *Fundam. Appl. Toxicol.*, 23, 370, 1994.

CARBOXYMETHYL CELLULOSE, SODIUM SALT

CAS No 9000-11-7
9004-32-4
RTECS No FJ5950000
Abbreviation. SCMC.

Synonyms and **Trade Names.** Aquaplast; Cellugel; Cellulose glycolic acid, sodium salt; Cellulose sodium glycolate; CM-Cellulose sodium salt; CMC sodium salt; Tylose C.

Composition. Sodium salt of carboxymethyl cellulose.

Properties. Tasteless and odorless, grayish hygroscopic powder. Swells in water. *pH* of 1.0% solution is about 13%.

Applications. Used as a film coating. Emulsifier and stabilizer in the food industry.

Acute Toxicity. LD_{50} of SCMC is 27 g/kg BW in rats and 16 mg/kg BW in guinea pigs.[1]

Repeated oral administration of SCMC to rats caused uncoupling of oxidative phosphorylation in liver and heart mitochondria and partial inhibition of mixed function oxidases of liver endoplasmic reticulum, as measured by 2-biphenylhydroxylation and 4-biphenylhydroxylation.[2]

Short-term Toxicity. Wistar rats were fed diets with 2.5, 5.0, and 10% SCMC for a 3-month period. The treatment-related occurrence of diarrhea and caecal enlargement in the mid- and high-dose groups, a slight increase of plasma alkaline phosphatase, and increased urinary calcium and citrate excretions were considered to be generic effects that typically are observed in rodent studies with low digestible carbohydrates. The increased occurrence of nephrocalcinosis and hyperplasia of the urothelial epithelium in some of the treated groups was interpreted as an indirect consequence of a more alkaline urine coupled with an increased calcium excretion.[3]

Long-term Toxicity. Rats and mice received 1.0 or 10% SCMC in the diet for 104 and 100 weeks, respectively. The treatment produced no signs of absorption; no ***carcinogenic*** effect was observed. A laxative effect was noted.[05]

Reproductive Toxicity. Rats were daily dosed with 10 mg/kg BW (as a 2.0% aqueous solution) for 60 days (males) or 14 days (females) before mating, during entire gestation period, and for 28 days postparturition period. There were no ***embryotoxic*** or ***teratogenic effects*** in the newborns (Fritz and Becker, 1981).

Chemobiokinetics. CMS is known not to be absorbed to a significant extent.

Regulations. ***U.S. FDA*** (1998) approved the use of SCMC in food in accordance with the conditions prescribed in 21 CFR parts 173.310, 182.1745, 150.161, 133.179, 133.134, 182.70, 150.141, and 133.178. Affirmed as *GRAS*.

Recommendations. ***EU*** (1995). SCMC is a food additive generally permitted for use in foodstuffs. It is also recommended as a suspending medium for the use in pharmacological and toxicological experiments.

References:

1. Patty's *Industrial Hygiene and Toxicology*, 4th ed., vol 2, Part A, G. D. Clayton and F. E. Clayton, Eds., John Wiley & Sons, Inc., New York, 1993, 512.
2. Bachmann, E., Weber, E., Post, M., and Zbinden, G., Biochemical effects of gum arabic, gum tragacanth, methylcellulose and carboxymethylcellulose-*Na* in rat heart and liver, *Pharmacology,* 17, 39, 1978.
3. Bar, A., Til, H. P., and Timonen, M., Subchronic oral toxicity study with regular and enzymatically depolymerized sodium carboxymethylcellulose in rats, *Food Chem. Toxicol.*, 33, 909, 1995.

CARBOXYMETHYL HYDROXYETHYL CELLULOSE

CAS No 9004-30-2

Abbreviation. CMHEC.

Trade Name. Tylose CH 50.

Properties. Tasteless and odorless substance.

Applications. Used in the production of coatings and detergents.

Acute Toxicity. A substance of low toxicity even on ingestion of large doses. CMHEC is shown to be non-irritating by ingestion.

Chemobiokinetics. CMHEC is eliminated unchanged with feces.

Regulations. ***U.S. FDA*** (1998) affirmed CMHEC as *GRAS*.

NITROCELLULOSE

Molecular Formula. $[\sim C_6H_7O_2(OH)_{3-x}(ONO_2)_x\sim]_n$

CAS No 9004-70-0

RTECS No QW0970000

Abbreviation. NC.

Trade Names. Collodion; Nitrocel S; Nitrocotton; Parlodion; Pyroxylin.

Composition. Cellulose nitrates. NC containing 10.7 to 12.2% nitrogen forms pyroxylin.

Properties. Colorless, or slightly yellow, or slightly opalescent, syrupy liquid.[020] According to other data, NC occurs as white amorphous powder with the odor of ether. The main shortcoming of NC is its combustibility.

Applications. Used in production of coatings, in preparation of celluloid and nitrocellulose etrol.

Migration of plasticizers from nitrocellulose-coated regenerated cellulose film (RCF) purchased from retails and take-away outlets was studied. Foodstuffs analyzed included cheese, pate, chocolate, and confectionery products, meat pies, cake, quiches, and sandwiches. Levels of plasticizers found to migrate from RCF was as high as: 2.0 to 8.0 mg acetyltributyl citrate/kg in cheese; 0.5 to 53 mg dibutyl phthalate, dicyclohexyl phthalate, butylbenzyl phthalate, and diphenyl(2-ethylhexyl) phosphate/kg in confectionery, meat pies, cakes, and sandwiches.

Acute Toxicity. LD_{50} exceeded 5.0 g/kg BW in rats and mice. 30 to 60 g may be fatal when swallowed, and symptoms are similar to *ethyl alcohol* intoxication except that onset is more rapid and duration is shorter because of its volatility. The stomach becomes promptly distended; this may inhibit breathing.[022]

Regulations.

EU (1990). NC is available in the *List of authorized monomers and other starting substances which shall be used for the manufacture of plastic materials and articles intended to come into contact with foodstuffs* (*Section A*).

U.S. FDA (1998) approved the use of NC (1) in adhesives as a component of articles intended for use in packaging, transporting, or holding food in accordance with the conditions prescribed in 21 CFR part 175.105; (2) in the manufacture of resinous and polymeric coatings for the food-contact surface of articles intended for use in producing, manufacturing, packing, processing, preparing, treating, packaging, transporting, or holding food in accordance with the conditions prescribed in 21 CFR part 175.300; (3) as a component of the uncoated or coated food-contact surface of paper and paperboard intended for use in producing, manufacturing, packaging, processing, preparing, treating, packing, transporting, or holding aqueous and fatty foods in accordance with the conditions prescribed in 21 CFR part 176.170; and (4) in the manufacture of cellophane for packaging food in accordance with the conditions prescribed in 21 CFR part 177.1200.

Great Britain (1998). NC. is authorized without time limit for use in the production of polymeric materials and articles in contact with food or drink or intended for such contact.

Reference:

Castle, L., Mercer, A. J., Startin, J. R., and Gilbert, J., Migration from plasticized films into foods. 3. Migration of phthalate, sebacate, citrate and phosphate esters from films used for retail food packaging, *Food Addit. Contam.*, 5, 9, 1988.

CELLULOID

CAS No 8050-88-2

Trade Names. Pyralin; Zylonite.

Composition. Contains no fillers but contains plasticizers (camphor, castor oil) and dyes (TiO_2).

Properties. Colorless, amorphous mass. Softens in boiling water. The organoleptic properties of aqueous extracts (3 hours, 1:2, distilled water) showed some changes.

Applications. Used in surgery for bandages, in dentistry as substitute for rubber, and in the manufacture of toys and films.

Acute Toxicity. After *i/g* administration to rats (10 to 20 mg extract/kg BW) or to mice (2.0 ml, *i/p*), no deaths were noted.

Repeated Exposure. Rats received 1.0 ml of water extract over for 30 days. The treatment caused only slight changes in enzyme activities. Despite solubility, C. is not absorbed in the GI tract.

Reference:

Kopeikina, N. F., Stankevich, A. I., Zarechenskaya, Z. A., and Popova, Ye. P., On the application of celluloid for production of toys, *Gig. Sanit.*, 1, 29, 1997 (in Russian).

CELLOPHANE

Molecular Formula. $[\sim C_6H_{10}O_5\sim]_n$
CAS No 9005-81-6
9077-41-2
RTECS No FJ4100000

Synonym. Visking cellophane

Composition. Film of regenerated cellulose hydroxide (isolated from cellulose esters). It is formed from alkaline solutions of cellulose xanthate (viscose) or obtained by saponification of acetyl cellulose film.

Properties. C. was introduced in the 1920s as a bread wrap. Highly steam- and moisture-proof, and highly resistant to the action of fats, soda solutions, and diluted sulphuric acid solutions. Glycerine contents amount to 12 to 18%. The odor of cellophane film is due to the presence of volatile sulphur compounds in the material, the source of which may be the residual cellulose xanthate in the film. The cause of the odor is an inadequate degree of washing of the cellophane to remove products accompanying the process of its production: hydrogen sulfide, methyl methacrylate, and substituted sulfides, the presence of which is revealed by an odor with a concentration of the order of tenths of μg/l.

Applications. C. is used as a flexible packaging material. It can be coated with *Saran* for use with oily or greasy products. Lacquered C. and plied-up polyethylene-cellophane films are used to pack food products.

Migration Data have established the presence of these films of the cellulose xanthogenate and products of its decomposition, of carbon bisulfide in particular. With their joint presence the determination of carbon bisulfide and hydrogen sulfide was done colorimetrically, the sensitivity of this method being 0.005 mg/l. Migration of carbon bisulfide from C. into the atmosphere and water was investigated. An interconnection between the amount of the migrated carbon bisulfide and the odor of aqueous C. extracts was disclosed. To improve sanitary and chemical properties of C., a more intensive washing off of sulphur-containing compounds from it is recommended.[1]

Carcinogenicity. Induced malignant tumors at the site of application in rodents following *s/c* imbedding of polymer films.[026] Nevertheless, the results of this study have been later considered inadequate.

Reference:

Tarasova, N. A., Shvagireva, N. A., Beliatskaia, O. N., and Pinchuk, L. M., Hygienic properties of cellophane film intended for food packaging, *Voprosy Pitania,* 6, 76, 1977 (in Russian).

POLYETHYLENE-CELLOPHANE FILM PTS-2

Migration studies showed that formaldehyde is released into water from cellophane: 0.15 mg/l in one day, 0.2 mg/l in three days. It also migrates into media simulating sausage and preserves: 0.06 mg/l in one day, 0.1 mg/l in three days. No formaldehyde is released from polyethylene.

Reference:

Dregval', G. P. and Kuznetsova, V. N., On the problem of sanitary-chemical studies of polyethylene-cellophane film, in *Hygiene and Toxicology of High-Molecular-Mass Compounds and of the Chemical Raw Material Used for Their Synthesis,* Proc. 6th All-Union Conf., B. Yu. Kalinin, Ed., Leningrad, Khimiya, 1979, 71 (in Russian).

ETROL

CAS No 9004-35-7

Composition. Thermoplastic materials containing fillers, plasticizers, and dyes.

Synonyms. A 432-130B; Acetate cotton.

Migration Data. Can give an intense odor to liquid media coming into contact with them.

EPOXY COMPOUNDS

CAS No 61788-97-4

Abbreviation. EC.

Composition. Epoxy resins are made by polymerization of epoxides, and are used in coatings and adhesives.

Migration Data. Food-contact epoxy resins can release different phenolic compounds such as phenol, *m*-cresol, bisphenol F, bisphenol A, 4-*tert*-butylphenol, bisphenol F diglycidyl ether, and bisphenol A diglycidyl ether into foodstuffs. Specific migration of dioctyl phthalate, salicylic acid, and primary aromatic amines from epoxy resin composed of bisphenol A diglycidyl ether, 4,4'-methylenedianiline and additives (plasticizers: dibutyl phthalate, dioctyl phthalate; accelerator: salicylic acid, inorganic fillers) diminished greatly as the curing temperature increased.[1,2] The overall and specific migration of bisphenol A diglycidyl ether monomer and m-xylylenediamine hardener from a BEPOX LAB 889 epoxy system cured at room temperature, into three water-based food simulants was measured. Hydrolysis of bisphenol A diglycidyl ether monomer was observed, giving more polar products.[3]

Acute Toxicity. LD_{50} of several EC were found to be 5.34, 6.64, and 9.18 g/kg BW for resins DE-500, DE-1000, and DE-2000, respectively.[4]

Allergenic Effect. Both allergic and irritant contact dermatitis may be caused by exposure to epoxy resins and their additives. Contact sensitization to epoxy resins is usually caused by the resin itself but hardeners or other additives, such as reactive diluents, plasticizers, fillers, and pigments, can occasionally be responsible. Since completely cured epoxy resins are not sensitizers, epoxy resin sensitization is always due to the presence, in the final polymer, of uncured allergenic low-molecular-mass oligomers.

Contact urticaria, allergic or irritant airborne contact dermatitis caused by volatile compounds released from EC, onychia, and paronychia can occur.[5]

Reproductive Toxicity. Pregnant rats were orally exposed to epoxy resin DE-500 at dose levels of 1/10 or 1/50 LD_{50} and to epoxy resin DE-1000 at dose levels of 1/10 or 1/50, 1/100, 1/250 LD_{50} on 1 to 19 gestation days. The treatment caused ***embryotoxic*** and ***teratogenic*** effects.[6]

Mutagenicity.

In vivo cytogenetics. Administration of epoxy resins DE-500, DE-1000, and DE-2000 at the dose levels of 1/10 or 1/50 LD_{50} caused CA in bone marrow cells.[4]

In vitro genotoxicity. 4 epoxy resin hardeners of unknown carcinogenic activity, i.e., 4-amino-diphenyl ether, 3,4,4'-triaminodiphenyl ether, 3,3'-dichloro-4,4'-diaminodiphenyl ether and 1,3-phenylene di-4-aminophenyl ether were found to be positive in the DNA repair test with rat hepatocytes, suggesting that they may be carcinogens.[7]

Regulations. ***U.S. FDA*** (1998) approved the use of EC (1) in the manufacture of polysulfide polymer-polyepoxy resins which may be used as food-contact surface of articles intended for packaging, transporting, holding, or otherwise contacting dry food in accordance with the conditions prescribed in 21 CFR part 177.1650; and (2) in the manufacture of resinous and polymeric coatings for the food-contact surface of articles intended for use in producing, manufacturing, packing, processing, preparing, treating, packaging, transporting, or holding food, subject to the provisions prescribed in 21 CFR part 175.300.

References:

1. Lambert, C., Larroque, M., Lebrun, J. C., and Gerard, J. F., Food-contact epoxy resin, co-variation between migration and degree of cross-linking, *Food Addit. Contam.*, 14, 199, 1997.
2. Lambert, C. and Larroque, M., Chromatographic analysis of water and wine samples for phenolic compounds released from food-contact epoxy resins, *J. Chromatogr. Sci.*, 35, 57, 1997.
3. Simal, G. J., Lopez, M. P., Paseiro, L. P., Simal, L. J., and Paz, A. S., Overall migration and specific migration of bisphenol A diglycidyl ether monomer and *m*-xylylenediamine hardener from an optimized epoxy-amine formulation into water-based food simulants, *Food Addit. Contam.*, 10, 555, 1993.
4. Yavorovsky, A. P., Bariliak, I. R., and Paustovsky, Yu. A., The cytogenetic activity of some brands of epoxy resins, *Lik. Sprava*, 7-9, 95, 1996 (in Ukrainian).
5. Tosti, A., Guerra., L., Vincenzi, C., and Peluso, A. M., Occupational skin hazards from synthetic plastic, *Toxicol. Ind. Health,* 9, 493, 1993.

6. Yavorovsky, A. P. and Paustovsky, Yu. A., The embryotoxic and teratogenic activity of some brands of epoxy resins, *Lik. Sprava*, 2, 120, 1997 (in Ukrainian).
7. Mori, H., Yoshimi, N., Sugie, S., et al., Genotoxicity of epoxy resin hardeners in the hepatocyte primary/DNA repair test, *Mutat. Res.*, 204, 683, 1988.

EPOXYDIANE RESINS

Structural Formula.

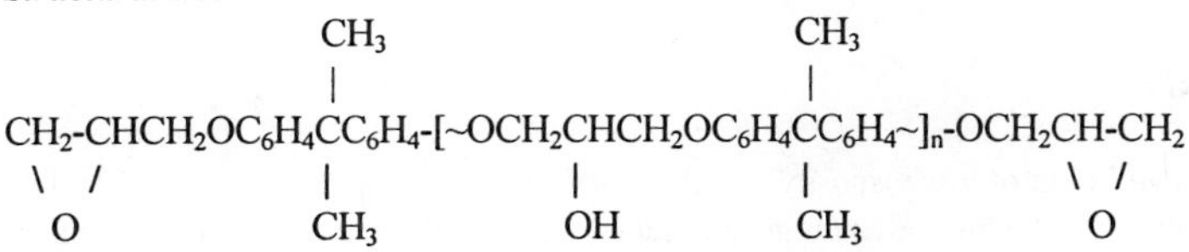

M = 480 to 610

Abbreviation. EDR.

Composition. The most widely used epoxy resins are products of the condensation of diane (4,4'-isopropylidenediphenol, diphenylolpropane, bisphenol A), and epichlorohydrin.

Properties. Soluble and meltable oligomers with glycidol groups at the ends of chains; when cured, they are transformed into three-dimensional polymers. Light-yellow, transparent liquids (less commonly, solids). Primary aliphatic and aromatic polyamines (ethylenediamine, diethylenetriamine, triethylenetetramine, polyethylenepolyamine, *m*-phenylenediamine) are used to cure epoxy resins.

Applications. Used as stabilizers of polyvinyl chloride, as the base of paints and varnishes, and as the binder for high-strength reinforced plastics, polymer concrete, polymer cements, sealants, filling and dipping compounds, and cellular plastics.

Migration Data. The curing agents can migrate from epoxy materials into liquids coming into contact with them.

Regulations. ***U.S. FDA*** (1998) approved the use of EDR (1) in the manufacture of polysulfide polymer-polyepoxy resins which may be used as food-contact surface of articles intended for packaging, transporting, holding, or otherwise contacting dry food in accordance with the conditions prescribed in 21 CFR part 177.1650; and (2) in the manufacture of resinous and polymeric coatings for the food-contact surface of articles intended for use in producing, manufacturing, packing, processing, preparing, treating, packaging, transporting, or holding food, subject to the provisions prescribed in 21 CFR part 175.300.

ED-16 (ED-6) RESIN

Abbreviation. ED-16.

Properties. Transparent viscous product ranging from light yellow to brown in color. Insoluble in water. *Dens.* 1.55.

Applications. Used in the manufacture of adhesives, coatings, and articles used in water piping and the food industry, and also as a thermal stabilizer of polyvinyl chloride and cellulose ethers.

Acute Toxicity. Toxicity decreases with the increase in molecular mass. Administration of an extract and suspension of cured resin and emulsions of uncured ED-16 resin (with additions of anhydrous aluminum oxide, Marshalite, titanium dioxide, dibutyl phthalate, polyethylenepolyamine) had no toxic effect on rats. With *i/g* administration, the minimum lethal dose for mice was 5.0 g/kg BW. Death occurred in 2 to 3 days after diarrhea, emaciation, and increasing weakness. The liver (fatty dystrophy) and less frequently the kidneys and brain were predominantly affected.

Allergenic Effect. The development of an allergy is evidently connected with the number of epoxide groups in the chemical structure of the compound. 40-day oral exposure to ED-16 resin and glass fiber caused an allergenic reaction in the GI tracts of rats. Pronounced sensitizing properties of ED-16 resin have been established in drop tests on patients and in experiments on rabbits.

Regulations. ***U.S. FDA*** (1998) approved the use of ED-16 (1) in the manufacture of polysulfide polymer-polyepoxy resins which may be used as food-contact surface of articles intended for packaging, transporting, holding, or otherwise contacting dry food in accordance with the conditions prescribed in 21 CFR part 177.1650; and (2) in the manufacture of resinous and polymeric coatings for the food-contact surface of articles intended for use in producing, manufacturing, packing, processing, preparing, treating, packaging, transporting, or holding food, subject to the provisions prescribed in 21 CFR part 175.300.

Reference:

Bobrovskikh, L. P., in *Proc. Sci. Works Kazan' Medical Institute,* Kazan', Issue No 20, 1976, 257 (in Russian).

ED-20 (ED-5) RESIN

Abbreviation. ED-20.

Properties. Yellow-brown liquid of low viscosity and fluidity. Insoluble in water. *Dens.* 1.16 to 1.17^{20}.

Applications. Used in the manufacture of coatings and adhesives, as a thermal stabilizer and plasticizer of polyvinyl chloride, etc.

Acute Toxicity. Toxicity decreases with the increase in molecular mass. Administration of the solid high-molecular-mass resins is not lethal to animals even at doses exceeding 30 g/kg BW. Toxic effects are accompanied by slight depression and dyspnea, and in surviving animals by emaciation and diarrhea. No differences were found in the specific sensitivity of laboratory animals. With *i/g* administration, the minimum lethal dose for mice is 5.0 g/kg BW. Death occurred in 2 to 3 days after diarrhea, emaciation, and increasing weakness. The liver (fatty dystrophy) and less frequently the kidneys and brain were predominantly affected.

Allergenic Effect. The development of an allergy is evidently connected with the number of epoxide groups in the chemical structure of the compound. 40-day oral exposure to ED-20 resin and glass fiber caused an allergenic reaction in the GI tracts of rats. Pronounced sensitizing properties of ED-20 resin have been established in drop tests on patients and in experiments on rabbits.

Regulations. ***U.S. FDA*** (1998) approved the use of ED-20 (1) in the manufacture of polysulfide polymer-polyepoxy resins which may be used as food-contact surface of articles intended for packaging, transporting, holding, or otherwise contacting dry food in accordance with the conditions prescribed in 21 CFR part 177.1650; and (2) in the manufacture of resinous and polymeric coatings for the food-contact surface of articles intended for use in producing, manufacturing, packing, processing, preparing, treating, packaging, transporting, or holding food, subject to the provisions prescribed in 21 CFR part 175.300.

Reference:

Bobrovskikh, L. P., in *Proc. Sci. Works, Kazan' Medical Institute*, Kazan', Issue No 20, 1976, 257 (in Russian).

EPOXY COATINGS based on ED-16 and ED-20 RESINS

Migration Data. Epoxy compounds based on ED-20 and ED-16 have no effect on the organoleptic properties of simulant media. Migration of epichlorohydrin and diane varies from virtually zero to several mg/l. Migration of curing agents, especially polyethylene polyamine, is possible. The release of dibutyl phthalate and acetone was not found.[1]

Any migration of curing agent into different simulant media from epoxy coatings cured with *m*-phenylenediamine with 1:1 ratio of epoxide groups and curing agent is unlikely. With a ratio of 1:2, up to 0.4 mg/l of *m*-phenylenediamine is released. With increase in the contents of this curing agent, epoxy coatings give water an odor and taste of rating 3 to 4. With a specific surface of 1:40, migration of epichlorohydrin and diphenylolpropane does not exceed the permissible levels; with a specific surface of 2:1, it reaches 3.0 mg/l. The use of *m*-phenylenediamine as the curing agent instead of polyethylenepolyamine reduces migration of the other components of the composition, for example, epichlorohydrin.[2]

Traces of *m*-phenylenediamine were released from a coating based on epoxy resin into an acetic acid solution and, at room temperature, 0.0125 mg/l of the curing agent was released into water.[3]

Polyethylenepolyamine migrated from steel containers coated with epoxy compounds (ED-16 and ED-20) into liquids simulating wine in a quantity of 0.4 mg/l.[028] After contact at 20°C for 30 to 45 days, diphenylolpropane (below permissible level of 0.01 ppm) and *m*-phenylenediamine (trace quantities) migrated from an epoxy coating based on ED-20 resin into 70° and 96° ethyl alcohol. Migration of *m*-phenylenediamine reached 0.15 mg/l in 96° alcohol extracts (50°C; 10 days). The coating did not alter the organoleptic properties of winemaking products.[028]

Long-term Toxicity. Aqueous extracts of epoxy coatings (with a 1:1 ratio of epoxide groups and curing agent *m*-phenylenediamine) had no toxic effects on rats.[2] Rats drank aqueous extracts (18 to 25°C; 1 month) of epoxy compounds ED-16 and ED-20 plasticized with dibutyl phthalate and MGF-9 polyester (q.v. polyester acrylates) and cured with hexamethylenediamine and polyethylenepolyamine. The treatment caused no changes in the behavior, BW gain, hematology analyses, the contents of ascorbic acid in the organs, the relative organs weights, or the histological structure of visceral organs of the animals.[4]

References:

1. Sheftel', V. O., Hygienic evaluation of epoxy compounds used in water supply, *Gig. Sanit.*, 12, 81, 1974 (in Russian).
2. Myannik, L. Ye., *Toxic and Hygiene Characteristics of m-Phenylenediamine and Epoxy Coatings That Are Cured by It and Used in Food Industry*, Author's abstract of thesis, All-Union Research Institute of Hygiene and Toxicology of Pesticides, Polymers and Plastic Materials, Kiyv, 1981, 21 (in Russian).
3. Kazarinova, N. F., Dukhovnaya, I. S., and Myannik, L. E., Determination of *m*-phenylenediamine in aqueous media, *Gig. Sanit.*, 11, 61, 1974 (in Russian).
4. Dvoskin, Ya. G., Rakhmanina, N. A., and Sandratskaya, S. E., Studies of hazardous effects of polymeric materials intended for use in ship lodgings, in *Hygiene Aspects of the Use of Polymeric Materials in Construction*, Proc. 1st All-Union Meeting on Health and Safety Monitoring of the Use of Polymeric Materials in Construction, Kiyv, 1973, 182 (in Russian).

EPOXY MATERIALS based on ED-20 RESIN

Composition. ED-20 resin and modified curing agents I-6M and PO-300 (condensation products of triethylenetetramine and ethyl acid esters, soybean oil, and dimerised linoleic acid ester).

Migration Data. Migration of epichlorohydrin, diphenylolpropane, and triethylenetetramine into simulant media was observed (Stavreva, 1982). The level of migration of triethylenetetramine depended on its contents in the compound, the contact period of time, and the temperature and composition of the simulant medium. At temperatures of 20 and 38°C, migration of epichlorohydrin and diphenylolpropane did not exceed permissible levels. The components of the coating migrated more intensely into water (0.03 mg/l) than into alcohol and sugar solutions (0.01 mg/l and traces, respectively).

EPOXIDE COATING for hydraulic tunnels

Composition. ED-20 resin and aminoshale curing agent *ASF-10.*

Migration Data. Caused no changes in the organoleptic indices of water quality (0.024 cm^{-1}; 20 and 37°C; 1 and 3 days and 6 months). No migration of epichlorohydrin, polyethylene polyamine, formaldehyde, or alkylresorcinol into water was found. Traces of diane were determined.

Long-term Toxicity. Prolonged consumption of 7-day aqueous extracts had no toxic effect on rats.

Reference:

Petrusha, V. G. and Kessel'man, I. M., Toxicological evaluation of epoxy coating with amino-shale hardener, in *Hygiene and Toxicology of High-Molecular-Mass Compounds and of the Chemical Raw Material Used for Their Synthesis*, Proc. 6th All-Union Conf., B. Yu. Kalinin, Ed., Khimiya, Leningrad, 1979, 153 (in Russian).

EPOXY ADHESIVE COMPOUND for pipes

Composition. Polyamine resin L-20, additives: dibutyl phthalate and polyethylenepolyamine.

Applications. Used in cold and hot water supply.

Migration Data. Investigation of the aqueous extracts (20 to 90°C; 2 days; 0.1 cm^{-1}) revealed migration of diphenylolpropane in a concentration exceeding the permissible level of 0.01 ppm at 90°C. At a lower temperature, there was no migration of chemical substances into water. The taste and odor of extracts were no different from the control.

Reference:

Krat, A. V., Kesel'man, I. M., and Sheftel', V. O., Sanitary-chemical evaluation of polymeric materials used in water supply, *Gig. Sanit.*, 10, 18, 1986 (in Russian).

COMPOUNDS UP 2212 K and UP 2212 KX

Composition. Compound UP 2212 K is obtained by combining UP 2212 epoxy mastic and a curing agent (low-molecular-mass polyamide resin L-20). Compound UP 2212 KX is also obtained by combining UP 2212 epoxy mastic and a curing agent (a mixture of *UP-583* diethylene triaminomethylphenol with UP-0623 dicyanoethylated diethylenetriamine). UP 2212 mastic consists of ED-20 epoxydiane resin (see below), SKTN siloxane rubber, and fillers (lamp black and kaolin).

Properties. Black pasty substances.

Applications. Used as an anticorrosion coating.

Acute Toxicity. The administration of 0.5 to 15 g of both compounds/kg BW to rats in an acetone solution does not kill the animals. The compounds have no toxic effect on the gonads.

Allergenic Effect. With *i/g* administration, the allergenic effect of the compounds was not pronounced.

Reference:

Zhislin, L. E., Chernykh, L. V., Dolzhenko, A. T., Kurennaya, S. S., and Ivanova, L. A., Assessment of the toxicity of new brands of epoxy compounds, *Gig. Truda Prof. Zabol.*, 8, 47, 1979 (in Russian).

POWDER EPOXY COMPOUND UP-2155

Applications. Used as an anticorrosion coating for drinking water tanks.

Migration Data. A study of extracts (20 and 37°C; 1 to 14 days; 0.04 cm^{-1}) revealed an odor of rating 1 to 1.5 and migration of epichlorohydrin (0.025 to 0.05 mg/l), diphenylolpropane (0.004 mg/l), and phenol (0.005 mg/l). Washing of specimens for 2 weeks hardly changed the rate of migration of epichlorohydrin.

Reference:

Krat, A. V., Kesel'man, I. M., and Sheftel', V. O., Sanitary-chemistry evaluation of polymeric materials used in water supply, *Gig. Sanit.,* 10, 18, 1986 (in Russian).

POWDER EPOXY PAINTS P-EP-177, P-EP-534, and P-EP-971

Applications. Used to coat stopcock components.

Migration Data. A study (37°C; 3 days; 0.4 cm^{-1}) revealed no migration of epichlorohydrin, diphenylolpropane, or dicyanodiamide into water. Phenol and formaldehyde migrated into water in trace amounts.

Reference:

Krat, A. V., Kesel'man, I. M., and Sheftel', V. O., Sanitary-chemical evaluation of polymeric materials used in water supply, *Gig. Sanit.,* 10, 18, 1986 (in Russian).

EPOXY LACQUERS 651017 and 913892, EP-547, ENAMEL EP-5147

Applications. Curing agent for cresol-formaldehyde resin. Used to coat cans.

Migration Data. Cans were filled with water, hermetically sealed, sterilized for 1 hour, and then held at 20°C for 1 and 10 days. The lacquers made in Germany do not release cresol or diphenylolpropane. Diphenylolpropane migrates from EP-5147 coatings at the level of 0.02 mg/l.

Reference:

Vas'kovskaya, L. F., Lyaschenko, O. N., Lebedinskaya, N. N., and Pestova, A.G., Determination of trace amounts of cresol and diphenylolpropane used in the food industry, *Voprosy Pitania*, 1, 84, 1985 (in Russian).

ENAMEL EP-2100

Composition. The enamel is a compound of VPFDKE-53 resin (ED-50 epoxy resin, phthalic anhydride, pentaerythrite, butyl cellulose, ethyl cellulose, triethylamine, ionol, rutile titatnium dioxide, green heliohelium, red iron oxide, and strong yellow permanganate pigments, DG-100 black, and upper alcohols C_6-C_8).

Applications. Used to coat articles coming into contact with foodstuffs.

Migration Data. A study of extracts (40 and 80°C; 1 hour) revealed odor of rating 4. Migation of lead reached 1.7 mg/l. At temperatures of 0 and 20°C there was no change in the organoleptic properties of the extracts.[028]

LACQUER EP-527

Composition. Based on epoxy and phenol-formaldehyde resins.

Properties. A dark-yellow, homogenous, transparent liquid.

Applications. Used to coat tin cans for fish and milk products.

Migration Data. The organoleptic indices of the quality of water and a 0.3% lactic acid solution coming into contact with the lacquer were studied (10 days; 20°C). Migration of epichlorohydrin, formaldehyde, chromium, and iron into extracts was not observed. Traces of bisphenol A and phenol were found. The lacquer has no effect on the contents of vitamins *A* and *C* in milk, nor on the general indices of evaporated milk.[028]

EPOXY RESIN UP-650 DIEPOXIDE

Structural Formula.

$(CH_2\text{-}CHCH_2OCH_2)_2\,C_6H_{10}$

  \ /

  O

M = 224.33

Synonym. 1,1'-Bis(glycidyloxymethyl)cyclohexane.

Properties. A light brown, viscous liquid with a specific odor. Insoluble in water and oil, readily soluble in alcohol.

Applications. Used in the production of glass-fiber reinforced plastics.

Acute Toxicity. LD_{50} is 7.72 g UP-650 and 6.98 g UP-650T/kg BW in rats. Poisoning caused stiffness, loss of appetite, bristling of fur, and diarrhea. Later the animals experience convulsion and dyspnea. Deaths occurs in 2 to 3 days.[1,2]

Repeated Exposure revealed pronounced cumulative properties. K_{acc} = 5.3 (by Cherkinsky).[1]

Reproductive Toxicity. Pregnant rats were given *i/g* 770 mg UP-650/kg or 700 mg UP-650T/kg BW on gestation days 1 through 19. The treatment caused ***embryotoxic*** and ***teratogenic*** effects in rats.[2]

References:

1. Yavorovsky, A. P. et al., in *Hygiene and Toxicology of Plastics,* Coll. Sci. Proc. Kiyv Medical Institute, Kiyv, 1979, 70 (in Russian).
2. Yavorovsky, A. P. and Paustovsky, Yu. A., The embryotoxic and teratogenic activity of some brands of epoxy resins, *Lik. Sprava*, 2, 120, 1997 (in Russian).

LACQUER EP-547
(first reserve formulation)

Composition. A light yellow solution of high-molecular-mass E-49 epoxy resin and phenolformaldehyde resin in ethyl cellulose with KO-815 lacquer, orthophosphoric acid, diphenylolpropane, and cyclohexanone added.

Applications. Used to coat fish cans.

Migration Data. There was no change in the organoleptic properties of aqueous extracts of simulant media (120°C, 1 hour, then 20°C, 10 days). Epichlorohydrin and bisphenol A migrated into water in concentrations of 0.36 and 0.06 mg/l, respectively.[028]

LACQUER EP-547
(second reserve formulation)

Composition. Solution of high-molecular-mass ED-05K epoxy resin and phenol-formaldehyde resin in cellosolve with E-30K resin, KO-815 lacquer, orthophosphoric acid, diphenylolpropane, and tetrabutoxytitanium added.

Properties. The color ranges from light to dark yellow.

Applications. Used to coat fish, meat, and milk cans.

Migration Data. There was no change in the organoleptic properties of water or simulant media (110 to 120°C, 1 hour, then 20°C, 10 days). Migration of epichlorohydrin amounted to 0.11 mg/l into water, to 0.21 to 0.39 mg/l into acetic acid solutions, to 0.17 to 0.30 mg/l into citric acid solutions, and to 0.17 mg/l into a lactic acid solution. Bisphenol A release was 0.007 mg/l to water, 0.024 to 0.01 mg/l into acetic acid solution, and 0.006 to 0.1 mg/l into citric acid solution. Release of formaldehyde into water amounted to 0.09 mg/l, into acetic acid solution to 0.12 to 0.24 mg/l, and into citric acid solution to 0.13 to 0.25 mg/l. Migration of phenol was not found.[028]

EPOXY RESIN UP-604

M = 296

Composition. Diglycidyl ester of *cis*-4-methyl-1,2,5,6-tetrahydrophthalic acid. Contains chlorine impurities.

Properties. A light yellow, viscous liquid. *Dens.* 1.2. Soluble in alcohol.

Applications. Used as a binder in adhesives.

Acute Toxicity. For mice, LD_{50} is 2.63 g/kg BW.

Repeated Exposure. K_{acc} is 9.36 (by Lim).

Allergenic Effect. A sensitizing effect was found in guinea pigs (immunoleukolysis reaction, hyperplasia of the lymphoid follicles, etc.).

Reference:

Talakin, Yu. N. et al., Toxicological characteristics of new amine hardeners UP-604/1, *Gig. Truda Prof. Zabol.*, 4, 57, 1986 (in Russian).

ESD-2 EPOXYSHALE COMPOUND

Composition. Based on EIS-1 epoxy resin and *Slamor* shale modifier. Curing agent: polyethylene-polyamine. EIS-1 epoxy resin is the polymeric product of condensation of epichlorohydrin with the alkylresorcinol fraction of shale phenols. Slamor shale modifier is the 280 to 400°C rectification fraction of shale resin cooking distillate. The main components are phenols, neutral oxygen compounds, and hydrocarbons.

Applications. Used to coat the internal surface of hydraulic tunnels.

Migration Data. A study of aqueous extracts (20 and 37°C; 1 and 48 hours) revealed no significant effect of the material on the taste or odor of water. With a specific surface of 1:43 cm^{-1}, no migration of epichlorohydrin, formaldehyde, phenols, or polyethylenepolyamine was found. With a specific surface of 1:4.3 cm^{-1}, the release of diane into water was observed at a level of 0.06 to 0.1 mg/l.

Reference:

Sheftel', V. O., Maximov, Yu. A., Grinberg, I. M., and Bychenkova, L. S., Hygienic evaluation of coatings for the Big Stavropol' channel, in *Hygiene Aspects of the Use of Polymeric Materials*, Proc. 2nd All-Union Meeting on Health and Safety Monitoring of the Use of Polymeric Materials, K. I. Stankevich, Ed., Kiyv, 1976, 165 (in Russian).

EPOXY BINDER UP 2201

RTECS No YQ8578000

Composition. Synthesized on the basis of 3,3-diglycidyldiphenylol glycol epoxy resin and curing agent 3,3-dichloro-4,4-diaminodiphenylmethane. Contents of volatile components in the resin are less than 1.0%; epoxide number is 24.

Properties. A brown, viscous liquid with a good solubility in alcohols. Insoluble in water and mineral oils.

Applications. Used in the synthesis of epoxy glass-fiber reinforced plastics.

Acute Toxicity. LD_{50} is 7.1 g/kg BW in rats, 1.1 g/kg BW in mice, and 1.76 g/kg BW in guinea pigs.

Repeated Exposure. K_{acc} is 4.1 (by Cherkinsky).

Reference:

Yavorovsky, A. P. and Nosenko, G. A., Toxicity of some UP epoxy resins, *Gig. Sanit.*, 8, 88, 1987 (in Russian).

EPOXY BINDER UP-2202A

RTECS No YQ8580000

Composition. Based on diglycidyl phthalate and curing agent triethanolaminotitanate. Contents of volatile components in the resin are less than 1.0%.

Properties. A dark brown, viscous liquid. Soluble in alcohol; insoluble in water and mineral oils.

Applications. Used in the production of glass-reinforced plastics.

Acute Toxicity. LD_{50} is 2.5 g/kg BW in rats, 1.6 g/kg BW in mice, and 1.5 g/kg BW in guinea pigs.

Repeated Exposure. K_{acc} is 13.2 (by Cherkinsky).

Allergenic Effect. A sensitizing effect was observed on skin application.

Reference:

See ***Epoxy binder UP 2201.***

EPOXY BINDER UP-2202B

Composition. Obtained by combining diglycidyl phthalate (100 parts) and *p*-aminobenzaniline (37 parts).

Properties. A dark brown, viscous liquid. Soluble in alcohol; insoluble in water and mineral oils.

Applications. Used in the production of glass-fiber reinforced plastics.

Acute Toxicity. LD_{50} was not reached in mice and rats.

Repeated Exposure. K_{acc} was not established because of the low toxicity of the compound.

Allergenic Effect. A sensitizing effect was observed on skin application.

Reference:

See ***Epoxy binder UP 2201.***

EPOXY BINDER UP-2215

Composition. Based on UP-2215 epoxy resin and curing agent 3,3-dichloro-4,4-diaminodiphenylmethane.

Properties. A light brown, viscous liquid. Soluble in alcohol; insoluble in water and mineral oils.

Applications. Used in the production of glass-fiber reinforced plastics.

Acute Toxicity. LD_{50} was not reached in mice and rats.

Repeated Exposure. K_{acc} was not established because of the low toxicity of the binder.

Allergenic Effect. A sensitizing effect is observed when the compound penetrates the skin.

Reference:

See ***Epoxy binder UP 2201.***

EPOXY BINDER UP-2217

CAS No 86904-48-5

RTECS No YQ8578100

Composition. Obtained by combining modified UP-2222 epoxy resin with aromatic diamine.

Properties. A viscous liquid, soluble in alcohol; insoluble in water and mineral oils.

Applications. Used in the production of glass-fiber reinforced plastics.

Acute Toxicity. LD_{50} is 4040 mg/kg BW in rats.

Repeated Exposure. Administration of 800 mg/kg BW to rats causes adynamia, emaciation, a reduction in rectal temperature and 24-hour diuresis, and proteinuria. Some animals died.

Allergenic Effect. A sensitizing effect was observed on skin application.

Reference:

See ***Epoxy binder UP 2201.***

EPOXY BINDER UP-2218

CAS No 80341-98-6

RTECS No YQ8578110

Composition. Obtained by combining a mixture of UP-650 and UP-643 epoxy resins with UP-605/3R curing agent.

Properties. Reddish-brown, viscous, homogenous liquid. Soluble in alcohol; insoluble in water and mineral oils.

Application. Used in the production of glass-fiber reinforced plastics.

Acute Toxicity. LD_{50} is 4.7 g/kg BW in rats.

Repeated Exposure. Administration of 940 mg/kg BW to rats over a period of 2 weeks is accompanied by adynamia, emaciation, lowered rectal temperature, reduced 24-hour diuresis, and impaired liver function. Some animals died.

Allergenic Effect. A sensitizing effect was observed on skin application.

Reference:

See ***Epoxy binder UP 2201.***

EPOXY COATING PEP-971

Composition. Based on *E-49* epoxy resin and polyvinyl butyral. Modifier: phenol-formaldehyde resin, curing agent: dicyanodiamide.

Properties. Gray substance.

Applications. Used to coat water pipes.

Migration Data. A study of extracts (20 to 60°C; 1 to 10 days; 0.7 cm^{-1}) revealed a specific odor up to rating 2 at 60°C. Migration of phenol at 60°C amounted to more than 0.005 mg/l (1 day) and 0.118 mg/l (10 days). There were no changes in other organoleptic or chemical indices. Washing four times eliminates any adverse effect on water quality. The material does not absorb chlorine dissolved in water.

Long-term Toxicity. Intake of aqueous extracts by rats, rabbits, and guinea pigs over a period of 8 months had no toxic effect.

Allergenic Effect. Extracts had no such effect.

Reference:

Yakovleva, L. Ye. and Pashkina, Ye. N., Hygienic evaluation of polymeric coating PEP-971 used in the water supply, *Gig. Sanit.*, 7, 74, 1984 (in Russian).

EPOXY ADHESIVE EPK-519

Applications. Used for installing water pipes.

Migration Data. A study of aqueous extracts revealed no migration of unsaturated compounds. No significant increase of oxidizability was observed at 60°C. Taste and odor were within rating 1. No migration of phenol, formaldehyde, or epichlorohydrin was observed in athe queous extracts. The *pH* was unaltered.

Recommendations. *Russia.* The adhesive was recommended for installing water pipes.

Reference:

Voitenko, A. M., Tkachuk, N. G., and Petrash, S. A., On hygienic evaluation of synthetic materials contacted with potable water on ships, in *Hygiene and Toxicology of High-Molecular-Mass Compounds and of the Chemical Raw Material Used for Their Synthesis*, Proc. 6th All-Union Conf., B. Yu. Kalinin, Ed., Khimiya, Leningrad, 1979, 144 (in Russian).

EPOXYANILINE RESIN

Structural Formula.

```
CH2--CHCH2NCH2CH--CH2
  \ /      |    \ /
   O      C6H5   O
```

M = 205

Composition. Nitrogen-containing epoxy resin. The product of condensation of epichlorohydrin with aniline in an alkaline medium.

Properties. A viscous liquid with a weak characteristic odor.

Applications. Used in the production of adhesives, impregnation, and coatings, and as the binder for glass-fiber reinforced plastics and compounds.

Acute Toxicity. For mice, LD_{50} is reported to be 2.1 g/kg BW. The animals die within 2 hours of administration. Toxic effects and the results of dissection are the same as with the administration of DEG-1 resin.[1,2]

Allergenic Effect. Pronounced.[3]

References:

1. Shumskaya, N. I. and Tolgskaya, M. S., Toxicological and morphological effects of epoxy resins and their raw materials, in *Toxicology of New Industrial Chemicals*, A. A. Letavet, Ed., Meditsina, Moscow, Issue No 7, 1965, 76 (in Russian).
2. Shumskaya, N. I. and Tolgskaya, M. S., Toxicity of new types of epoxy resins (EA and DEG-1), in *Toxicology of New Industrial Chemicals*, A. A. Letavet and I. V. Sanotsky, Eds., Meditsina, Moscow, Issue No 10, 1968, 110 (in Russian).
3. Bokaneva, S. A., *Epychlorhydrin and Its Toxic and Hygiene Characteristics, and Importance of New Epoxy Resins in Hygiene Regulations*, Author's abstract of thesis, F. F. Erisman Research Sanitary Hygiene Institute, Moscow, 1980, 22 (in Russian).

EPOXYTRIPHENOL RESIN

Composition. The product of condensation of epichlorohydrin and triphenol.

Acute Toxicity. LD_{50} was not reached.

Repeated Exposure. No cumulative properties were noted.

Reference:

Bokaneva, S. A., *Epychlorhydrin and Its Toxic and Hygiene Characteristics, and Importance of New Epoxy Resins in Hygiene Regulations*, Author's abstract of thesis, F. F. Erisman Research Sanitary Hygiene Institute, Moscow, 1980, 22 (in Russian).

EPOXYPHENOL LACQUER EP-5118

Applications. Used to coat cans and lids in the food canning industry.

Migration Data. The organoleptic properties of simulant media were unaltered after contact with lacquered surfaces. Migration of formaldehyde and epichlorohydrin does not occur (with a polymerization temperature of the lacquer of 320°C).

Long-term Toxicity. Young rats received extracts with an epichlorohydrin contents of no more than 0.01 mg/l. The treatment caused no harmful effects in the animals.

Recommendations. ***Russia.*** The lacquer is recommended for coating lids and glass jars (but not metal cans) for children's food.

Reference:

Pestova, A. G., Ivanova, L. N., Petrovskaya, O. G., and Kolomiyets, L. S., Toxicological and hygienic characteristics of lacquer EP-5118 intended for use in canned food industry, *Gig Sanit.,* 10, 84, 1985 (in Russian).

EPOXY-ETHINOL MATERIALS

Trade Name. EP-755 epoxy-ethinol paint.

Major Use. Applied on VL-023 primer. Ethinol (divinylacetylene lacquer) is a solution of linear polymers of divinylacetylene (43%) in xylene and contains an antioxidant (1.5 to 2.5%).

Migration Data. Applied on a primer, EP-755 paint gives water a specific persistent odor of rating 4 to 5. Butanol and polyethylenepolyamine (1.1 to 7.25 mg/l) and epichlorohydrin (2.5 mg/l) migrate into water.[1] During investigation of a similar coating migration of polyethylenepolyamine (up to 5.0 mg/l), dibutyl phthalate (up to 1.0 mg/l), and phenols (up to 0.05 mg/l) into water was found.[1]

Long-term Toxicity. Toxicity tests were conducted over a period of 12 months using aqueous extracts of EP-755 paint. Clinical toxic effects and gross pathology changes in the visceral organs of rats were observed. The authors concluded that morphological and functional changes were variable and transient, and they regarded them as compensatory-adaptive reactions. Daily washing with water over a period of 1 week or treatment with hot air with subsequent double washing eliminates any adverse effect of the coating on water quality.[2]

Reproductive Toxicity, Allergenic Effect. Extracts of a washed coating have no toxic effect on the gonads and cause no sensitization in animals.

References:

1. Yakovleva, L. Ye., Effect of some polymeric anticorrosive materials on water quality, in *Hygiene Aspects of the Use of Polymeric Materials in Construction*, Proc. 1st All-Union Meeting on Health and Safety Monitoring of the Use of Polymeric Materials in Construction, Kiyv, 1973, 165 (in Russian).
2. Rudeiko, V. A. and Pashkina, Y. E., Hygienic evaluation of conditions for using a phosphating primer in the water supply, *Gig. Sanit.*, 6, 16, 1980 (in Russian).

EPOXY-ETHINOL COATING SP-EE-10

Migration Data. The coating may cause an adverse effect on the organoleptic properties and chemical composition of water after prolonged contact. A similar effect was not observed after 12 daily water changes (20°C), and the quality of water was similar to that of the control. No migration of benzo[a]pyrene into water was revealed.

Long-term Toxicity. The toxicity of extracts was investigated in a 6-month study on rats and was found to be insignificant. The authors recommended the coating for use, provided that it is washed beforehand for 6 days.

Reference:

Cherkinsky, S. N., Rubleva, M. N., and Korolev, A. A., Sanitary-toxicological evaluation of new polymeric materials used in water supply systems, in *Hygiene Aspects of the Use of Polymeric Materials in Construction*, Proc. 1st All-Union Meeting on Health and Safety Monitoring of the Use of Polymeric Materials in Construction, Kiyv, 1973, 155 (in Russian).

DIETHYLENE GLYCOL, polymer with 1-CHLORO-2,3-EPOXYPROPANE

Structural Formula.

```
CH2--CHCH2OC2H4OC2H4O-[~CH2CHCH2OC2H4OC2H4O~]n-CH2CH--CH2
  \ /                      |                         \ /
   O                       OH                         O
```

M = 240 to 260

CAS No 25928-94-3

RTECS No ID8970000

Synonyms and **Trade Name.** DEG-1; Epoxydiethylene glycol resin; 2,2'-Oxybisethanol, polymer with (chloromethyl)oxirane.

Composition. Aliphatic epoxy resin. The product of condensation of epichlorohydrin with diethylene glycol.

Properties. Colorless, slightly viscous liquid. Soluble in water and alcohol.

Applications. Used as a binder in the production of impregnations (without solvents), adhesives, and coatings.

Acute Toxicity. For mice, LD_{50} is 2.2 g/kg BW. 10 to 30 min after administration, the animals displayed hollow breathing and convulsions. Death occurred in 1 to 3 days. Gross pathology examination revealed a caustic effect on the gastric mucosa, multiple hemorrhaging in the stomach and small intestine mucosa, plethora and emphysema of the lungs, and parenchymatous dystrophy of the liver and kidneys.

References:

1. Shumskaya, N. I. and Tolgskaya, M. S., Toxicity of new types of epoxy resins (EA and DEG-1), in *Toxicology of New Industrial Chemicals*, A. A. Letavet and I. V. Sanotsky, Eds., Meditsina, Moscow, Issue No 10, 1968, 110 (in Russian).

2. Shumskaya, N. I. and Tolgskaya, M. S., Toxicological and morphological effects of epoxy resins and their raw materials, in *Toxicology of New Industrial Chemicals*, Meditsina, Moscow, Issue No 7, 1965, 76 (in Russian).

EPON RESINS

M = 300 to 6000
CAS No 67076-53-3
RTECS No KC1600000

Trade Names. Epikot; Epoxy resins of the Shell company.

Applications. Used as a stabilizers of PVC.

Acute Toxicity. For rats, mice, and rabbits, LD_{50} is 5.0, 1.87, and 4.01 g/kg BW (EPON-562) and 11.4, 15.6, and 19.8 g/kg BW (EPON-828), respectively. Average lethal doses for rats were 9.0 g EPON-815/kg BW, 13.6 g EPON-920/kg BW, 30 mg both EPON-1001 and EPON-1007/kg BW.[05]

Long-term Toxicity. The addition of 1.0 to 10% cured EPON-1001 and EPON-828 resins to the feed of rats over a period of 6 months led to a significant increase in death rate. There were no histological changes. Cured resins were not lethal. The addition of 0.04 to 5.0% EPON epoxy resins to the feed of rats over a period of 26 weeks caused retardation of BW gain and sometimes to an increase in the kidney weights. There were no histological changes.

Mutagenicity. A mutagenic effect was observed in *Salmonella* mutagenicity bioassay (Anderson et al.)

Reference:

Shumskaya, N. I., Toxicology of epoxy resins and problems of professional hygiene, in *Toxicology of New Industrial Chemicals*, A. A. Letavet, Ed., Medgiz, Moscow, Issue No 2, 1961, 18 (in Russian).

HALOGEN-CONTAINING EPOXY RESIN E-181

CAS No 117277-19-7
RTECS No JW2800000

Properties. Liquid of low viscosity, ranging from amber to brown in color.

Applications. An active diluent of highly viscous epoxy resins used in the production of adhesives and binders for glass-fiber reinforced plastics.

Acute Toxicity. For rats, LD_{50} is 1.65 g/kg BW.

Allergenic Effect. Pronounced.

Reference:

Bokaneva, S. A., *Epychlorhydrin and Its Toxic and Hygiene Characteristics, and Importance of New Epoxy Resins in Hygiene Regulations*, Author's abstract of thesis, F. F. Erisman Research Sanitary Hygiene Institute, Moscow, 1980, 22 (in Russian).

DICYCLOPENTADIENE DIOXIDE

M = 164.2
CAS No 81-21-0
RTECS No PB9625200

Composition. Cycloaliphatic epoxy resin cured with dihydric acid anhydrides in the presence of glyceri-ne to form epoxy resins.

Synonyms and **Trade Names.** Cyclopentadiene dimer diepoxide; Dicyclopentadiene diepoxide; 1,2:5,6-Diepoxyhexahydro-4,7-methanoindan; Epoxide 207; 4,5,6,7,7a-Hexahydro-4,7-methanoindan-1,2:5,6-diepoxy-3a; Octahydro-2,4-Methano-2*H*-indeno (1,2-b:5,6-b')bisoxirene.

Properties. Light yellow, crystalline powder. *M. p.* 184°C. Water solubility is 1.4%.

Applications. Used as the binder for heat-resistant glass-fiber reinforced plastics and as a constituent of heat-resistant adhesives.

Acute Toxicity. For rats, LD_{50} is 210,[1] or 280 mg/kg BW.[2] After administration the animals experience clonic-tonic convulsions and die in 40 to 50 min. Gross pathology and histology examination revealed significant hemodynamic disorder, hemorrhaging in the parenchyma of the liver and kidneys, and extensive emphysema. The epithelial cells of the convoluted renal tubules are swollen, their lumens are filled with albuminous fluid.

Repeated Exposure. The diepoxide has weak cumulative properties. All animals were given a total dose of $7LD_{50}$s. Compared to the controls, accelerated BW gain was observed, and the nervous-muscular excitation threshold was raised.[3]

Allergenic Effect. Did not show any allergenic properties when applied in 10 to 50% solution to skin of guinea pigs for 5 to 15 days.[3]

References:

1. Weil, C. S. et al., *Toxicol. Appl. Pharmacol.*, 4, 1, 1962.
2. Arkhangel'skaya, L. N. and Roschina, T. A., Furfuramide, in *Toxicology of New Compounds Introduced into the Rubber and Tyre Industry*, Meditsina, Moscow, 1968, 131 (in Russian).
3. Shumskaya, N. I. and Mel'nikova, L. V., Toxicology evaluation of dicyclopentadiene dioxide, in *Toxicology of New Industrial Chemicals*, N. F. Izmerov and I. V. Sanotsky, Eds., Meditsina, Moscow, Issue No 14, 1975, 131 (in Russian).

PHENOL-FORMALDEHYDE RESINS

CAS No 9003-35-4

Composition. Resins and plastics based on oligomeric products of polycondensation of phenol and formaldehyde. During processing, the oligomers became hardened with formation of three-dimensional polymers. Fomation of novolak (thermoplastic) and resol (thermosetting) oligomers took place.

Regulations. ***U.S. FDA*** (1998) approved the use of phenol-formaldehyde resins in the manufacture of resinous and polymeric coatings for the food-contact surface of articles intended for use in producing, manufacturing, packing, processing, preparing, treating, packaging, transporting, or holding food, subject to the provisions prescribed in 21 CFR part 175.300.

BAKELITE LACQUER BF-2

Applications. Forms coatings together with ethyl alcohol and fine aluminum powder.

Migration Data. In aqueous extracts ($0.2\ cm^{-1}$; 20 to 60°C) phenolic odor and presence of *phenol* and formaldehyde were found.

Acute Toxicity. 24-hour extracts of BF-2 coating caused increased migration of leukocytes from human oral mucosa (70 to 90% greater than in the control sample) which is apparently due to the presence of phenol (up to 0.5 mg/l).

Repeated Exposure. Rabbits that drank water from a container coated with BF-2 developed decreased BW gain by the end of a 1-month experiment. Gross pathology examination revealed an increase in the relative liver weights. Compared to the control animals, an increased mortality was observed in rats and mice exposed to X-ray irradiation.

Bakelite lacquer toxic effect is likely to be due to the presence of phenol and methyl and ethyl alcohols.

Reference:

Pashkina, Ye. N., Hygienic evaluation of bakelite lacquer BF-2 used as anticorrosive coating for potable water cisterns, in *Hygiene Aspects of the Use of Polymeric Materials in Construction*, Proc. 1st All-Union Meeting on Health and Safety Monitoring of the Use of Polymeric Materials in Construction, Kiyv, 1973, 166 (in Russian).

BAKELITE ADHESIVE BF-4

Applications. Used in the manufacture of gravel filters for artesian wells.

Properties. Water-resistant.

Migration Data. After contact with water for 24 hours, Bakelite gravel filters prewashed with running water for 2 weeks give water an aromatic odor (rating 4 to 5) and a slightly bitter taste (rating 3). Migration of *phenol* amounts to 0.04 to 0.11 mg/l.[028]

BARYTE RESIN VIAM-B

Composition. Water-emulsion resole phenol-formaldehyde resin. Catalysts: caustic soda or barium hydroxide; stabilizer: acetone.

Properties. A brown liquid.

Applications. Used for gluing membranes in desalination units.

Migration of phenol and formaldehyde into water was reported.

Reproductive Toxicity.

Gonadotoxicity. Toxic effect on the gonads and embryos was found. Prolonged action of aqueous extracts on rats caused a slight reduction in the acid resistance and mobility time of spermatozoa. There was an increase in the activity of acid phosphatase and *N*-acetyl-β-*D*-glucosoaminidase of lysosomal origin in the blood serum and testicles of rats.

Mutagenicity. Administration of supersaturated extracts to rats produced an increased number of CA and decreased mitotic activity of cells.

Reference:

Solokhina, T. A. et al., in *Sci. Proc. A. N. Sysin Research General Comm. Hygiene Institute,* Moscow, Issue No 72, 1977, 5 (in Russian).

PHENOL-FORMALDEHYDE RESINS FRV-1A

Composition. Resin FRV-1A is a neutralized solution of the primary products of polycondensation of phenol and formaldehyde with the addition of surfactants and fine aluminum powder.

Properties. A silver-colored, viscous, water soluble liquid with a slight phenol and formaldehyde odor. Limited solubility in ethanol.

Applications. Used to produce foam plastics.

Acute Toxicity. In rats, LD_{50} is 8.0 to 15 g/kg BW.

Mutagenicity.

In vivo cytogenicity. A DLM study established NOEL which was 100 times lower than those causing general toxic effect.

Reference:

Erkis, F. B., in *Hygiene of Populated Locations*, Republ. Coll. Papers, Issue No 17, 1978, 45 (in Russian).

LACQUER FL-559

Composition. A mixture of xylenol- and phenol-formaldehyde and epoxy and alkylepoxyresins in organic solvents.

Properties. Homogenous transparent liquid varying in color from dark-yellow to reddish-brown.

Applications. Used to coat the inner surface of tin cans and the lids of glass containers for foodsuffs.

Migration Data. Organoleptic properties of aqueous extracts and simulant media did not change (120°C, 1 hour; then 20°C, 10 days). Migration of formaldehyde and epichlorohydrin was 0.20 to 0.59 and 0.2 to 0.54 mg/l, respectively. No migration of phenol, diphenylopropane, or lead was found.[028]

Long-term Toxicity. Extracts (120°C, 24 hours) had no toxic effect on adult (10 months) and young (2 months) rats.

LACQUERS FL-560 and FL-56

Composition. Phenol-formaldehyde resin.

Applications. Used in the manufacture of containers for preserved foodstuffs.

Migration of formaldehyde into distilled water is 0.28 to 0.95 mg/l. No release of phenol was found. A reduction in the drying time of coatings to 20 min virtually eliminates formaldehyde migration into water.[028]

ENAMEL FL-5157

Composition. A solution of phenol-formaldehyde resin in butanol and ethyl cellosolve with the addition of mulberry and linseed oil and maleic anhydride. A paste consisting of highly dispersed zinc oxide and dehydrated castor oil is introduced into the enamel.

Applications. Used as a covering layer in fish cans.

Migration Data. No changes were found in the organoleptic properties of aqueous extracts and simulant media (120°C, 1 hour; then 20°C, 10 days). Migration of lead and zinc ions reached 1.0 and 0.29 mg/l, respectively. There was no migration of phenol and formaldehyde.[028]

COATING FD-723

Composition. Phenol-formaldehyde resin.

Migration Data. With a long-term contact and daily replacement of water at 20 to 100°C, the coating gave water foreign odor and taste. A grayish-brown deposit appeared in water. There was migration of phenol and formaldehyde to water.

Recommendations. ***Russia***. The material is not recommended for contact with drinking water.

PHENOL-FORMALDEHYDE FOAM PLASTICS FRP-1A and KFP-20

Composition. Phenol-formaldehyde resin.

Migration Data. A study of 30-day aqueous extracts revealed migration of 0.1 mg formaldehyde/l and 0.045 mg phenol/l from FRP-1A. Migration of these monomers from KFP-20 was 0.16 and 0.012 mg/l, respectively.

Reproductive Toxicity. Rats received 30-day aqueous extracts. The treatment caused changes in the total length and duration of individual phases of the estrus cycle, an increase in the overall pre- and postimplantation loss of embryos, and a definite reduction in BW gain of fetuses. In some embryos, blood effusions in the chest organs were found.

Reference:

Savitsky, I. V., Velikaya, N. V., and Tret'yak, N. P., Embryotropic effect of mixture of substances migrating from phenol-formaldehyde thermo-insulating foam-plastics, in *Hygiene Aspects of the Use of Polymeric Materials*, Proc. 3rd All-Union Meeting on New methods of Hygiene Monitoring of the Use of Polymers in Economy, K. I. Stankevich, Ed., Kiyv, December 2-4, 1981, Kiyv, 1981, 326 (in Russian).

PHENOL-FORMALDEHYDE MOLDING POWDERS

Composition. Phenol-formaldehyde resin.

Migration Data. In extracts of phenolic plastic (molding powder Sp1-342-02 based on SF-342 phenol-aniline-formaldehyde resin), 0.05 mg phenol/l and <0.01 mg formaldehyde/l were found.[1] There was no aniline.Articles based on K-17-23 molding powder can give water a specific odor which increases in the case of a long-term period of contact and in hot water. Migration of low-molecular-mass oxidizing and unsaturated compounds was observed. There was intense release into water of phenol and formaldehyde (respectively 0.15 to 1.3 and 0.3 to 2.1 mg/l at 20°C and 0.25 to 1.5 and 0.9 to 3.0 mg/l at 80°C). Water undergoes a rapid reduction in the contents of residual chlorine with formation of chlorophenol compounds possessing a pungent odor. Phenolic plastics can have a bactericidal effect on water microflora.[2]

Acute Toxicity. Single administration of 0.03 g/kg BW to rabbits and rats reduced blood *Hb* level and sugar contents after 1 week. Subpleural blood effusions, plethora and cell infiltration of the lungs, and dystrophy of the parenchymatous organs were observed. A dose of 0.01 g/kg BW had similar but less pronounced effect.[3] A study of phenolic plastic based on AG-4V molding material revealed odor of rating 0 to 3 at 20°C and 2 to 5 at 80°C. Oxidizability of water and contents of unsaturated compounds in it were increased. Migration of phenol and formaldehyde, was 0.3 to 2.6 and 0.4 to 2.0 mg/l at 20°C, respectively, and 0.6 to 2.8 and 0.7 to 11.1 mg/l at 80°C.

Recommendations. ***Russia***. The materials are not recommended for contact with drinking water.[3, 4]

References:

1. *Hygiene Problems in the Production and Use of Polymeric Materials*, Moscow, 1969, 159 (in Russian).
2. Cherkinsky, S. N. and Rubleva, M. N., Scientific studies for hygienic evaluation of synthetic materials recommended for water supply, in *Hygiene Aspects of the Use of Polymeric Materials*

and of Goods Made of Them, L. I. Medved', Ed., Kiyv, All-Union Research Institute of Hygiene and Toxicology of Pesticides, Polymers and Plastic Materials, 1969, 176 (in Russian).

3. Kakauridze, A. M., in *Sci. Proc. NII Gig. Truda Prof. Zabol. Grusin. SSR*, Tbilisi, Issue No 10, 1966, 213 (in Russian).
4. Sheftel', V. O., Hygienic properties of plastics used in water supply system construction, *Plast. Massy,* 2, 62, 1970 (in Russian).

PIPING OF WOOD CHIPS

Composition. Phenol-formaldehyde resin. RFN-60.

Migration Data. Slightly enhances the color of water coming in contact and reduces its transparency. After contact for 4 hours, water acquires a phenol taste of rating 2. Within 24-hour contact, *pH* of water increased by 0.65. Formaldehyde contents did not exceed 0.1 mg/l. Phenol contents were 0.35 to 1.8 mg/l.

Recommendations. *Russia*. The material is not recommended for contact with drinking water.

Reference:

Bokov, A. N., Nelasova, L. I., Shamshura, I. T. and Shornikov, A. P., Hygienic evaluation of pipes made of wooden particles and synthetic resin by extrusion method, in *Toxicology of High-Molecular-Mass Compounds and of Chemical Raw Material Used for Their Synthesis,* S. L. Danishevsky, Ed., Khimiya, Moscow-Leningrad, 1966, 125 (in Russian).

PLYWOOD PIPING

Composition. Glued with phenol-formaldehyde resin

Migration Data. Gives a specific odor (up to rating 5) and color to water, increases contents of unsaturated compounds. Such piping rapidly absorbs residual chlorine. 0.22 mg phenol/l is released from new piping within 30 min, and 2.85 mg/l within 24 hours. No phenol was released from a prewetted specimen within 30 min, 1.88 mg/l was released within 24 hours.

Recommendation. *Russia*. The material is not permitted to have contact with drinking water.

Reference:

Bezel, L. I. and Klimakov, V. V., Hygienic evaluation of plywood piping intended for water supply, *Gig. Sanit.,* 5, 94, 1960 (in Russian).

p-tert-BUTYLPHENOL-FORMALDEHYDE RESIN

M = 700 to 950

Composition. Phenol-formaldehyde resin.

Trade Names. Fenofor B; Super Bekazite 1001.

Properties. Light-yellow to brown transparent resin. Insoluble in water.

Applications. Phenol-formaldehyde resins, especially the B., are widely used in industry and in numerous materials of everyday use, such as glues and adhesives.[1] Vulcanization agent for natural and synthetic rubber (mainly, for rubber).

Allergenic Effect. Phenol-formaldehyde resins are an important cause of contact dermatitis.[1]

Reproductive Toxicity.

Embryotoxicity. Pregnant rats received 2.5; 5.0, or 10% B. in the diet from days 6 to 15 of gestation. Two higher doses caused a decrease in BW gain and food consumption. The treatment produced no embryotoxic effect, no increase in fetuses or placenta weights, no increase in embryolethality and embryos size, no changes in sex ratio.[2]

Teratogenicity. B. was given orally to pregnant Wistar rats by stomach intubation at the dose levels of 250, 500, and 1000 mg/kg BW during days 7 to 17 of gestation. The treatment caused no changes in general conditions, maternal BW, food consumption, numbers of corpora lutea and implantation ratio. There was no evidence of an increase in fetal death or of malformation attributable to the treatment with B. in any of dose levels examined. PTBPFR had no teratogenic effect in rats.[3]

References:

1. Massone, L., Anonide, A., Borghi, S., and Usiglio, D., Sensitization to *para-tertiary*-butylphenol-formaldehyde resin, *Int. J. Dermatol.*, 35, 177, 1996.
2. Itami, T., Ema, M., and Kawasaki, H., Teratogenic evaluation of *p-tert*-butylphenol formaldehyde resin (novolak type) in rats following oral exposure, *Drug Chem. Toxicol.*, 16, 369, 1993.
3. Tanaka, R., Usami, M., Kawashima, K., and Takanaka, A., Studies on the teratogenic potential of p-*tert*-butylphenolformaldehyde resin in rats, Abstract, *Eisei Shikenjo Hokoku,* 110, 22, 1992 (in Japanese).

FORMALDEHYDE, POLYMER with BENZENAMINE

Molecular Formula. $[\sim C_6H_7N.CH_2O\sim]_n$

CAS No 25214-70-4

RTECS No TQ0600000

Synonym and **Trade Name.** AF 10; Aniline-formaldehyde polymer; Polyamine T.

Properties. A dark brown solid. Poorly soluble in water; soluble in organic solvents.

Applications. Used in the synthesis of polyisocyanates and as a curing agent for epoxy resins.

Acute Toxicity. LD_{50} is 7.46 g/kg BW in rats. According to other data, LD_{50} is 1.05 g/kg BW. Clinical signs of poisoning included general inhibition, coordination disorder, and distention of the abdominal cavity. Death of rats occurs in 15 to 17 days.[1]

Repeated Exposure failed to reveal marked cumulative properties. K_{acc} is 12.2 (by Lim). Rats were given a dose of 200 mg/kg BW for a month. The treatment caused a decline in the level of total and oxidized *Hb*, and of *SH*-groups in the blood serum.

Reference:

Volodchenko, V. A. and Gnezdilova, A. I., Toxicological expertise of polyamine T, *Gig. Sanit.*, 2, 92, 1984 (in Russian).

FORMALDEHYDE, POLYMER with 1,2-ETHANEDIAMINE, and PHENOL

Molecular Formula. $C_9C_{14}N_2O$

CAS No 28985-91-3

RTECS No LP9750000

Composition. Interaction product of phenol, formaldehyde and ethylenediamine.

Trade Name. Agidol AF-2.

Applications. A hardener for epoxy resins.

Acute Toxicity. LD_{50} is 2.3 g/kg BW in mice and 3.7 g/kg BW in rats.

Reproductive Toxicity. A weak effect was observed. Administration of 1/20 LD_{50} caused embryotoxic effect. The treatment affected process of spermatogenesis in rats.

Reference:

Dregval', G. P., Kharchenko, T. P., Kirpichev, V. P., et al., Sanitary- chemical and toxicological studies of the aminophenol hardener agidol AF-2, *Gig. Sanit.*, 11, 23, 1990 (in Russian).

PHENOLIC RESIN S-180

Composition. Following ingredients are used in the manufacture of materials: epoxide compounds, urotropin, polyethylene polyamine (PEPA), and dibutyl phthalate (DBP).

Applications. Used as a binder in the manufacture of fiberglass materials.

Migration Data. A study of aqueous extracts ($1\ cm^{-1}$; 20 to 40°C; 1 to 4 days) did not reveal migration of harmful substances above the permissible levels at 20°C; 0.01 mg phenol/l, 0.01 mg formaldehyde/l, 0.07 mg epichlorohydrin/l, 0.02 mg DBP/l, 0.04 mg PEPA/l, and traces of diphenylpropane were found in water at 40°C. After 24-hour washing, migration of epichlorohydrin and diphenylolpropane amounted to 0.01 mg/l, and neither phenol nor PEPA were released to water.

Regulations. ***U.S. FDA*** (1998) approved the use of phenolic resin as the food-contact surface of molded articles intended for repeated use in contact with non-acid food (*pH* above 5) in accordance with the conditions prescribed in 21 CFR part 177.2410.

Reference:

Kravchenko, T. I., Kharchenko, T. F., Shevchenko, A. M., and Lebed, N. V., Sanitary and chemical studies of fiber glass-reinforced plastic-phenol-lined pipes designed for irrigation systems, *Gig. Sanit.*, 6, 70, 1980 (in Russian)

FLOCCULENTS, POLYMERIC

Applications. High-molecular-mass substances used in the process of water purification to accelerate flocculation and separate out flocs, and to improve water quality.

Toxicity. Polymeric flocculents based on polyacrylamide and methacrylic acid esters are the most toxic.

POLYMETHACRYLIC ACID ESTERS (VA-102 and VA-212)

Acute Toxicity. For mice, rats, and guinea pigs, LD_{50} was 5.0 to 7.0 g/kg BW without differences in specific sensitivity. In clinical terms, the CNS is predominantly affected. Animals die in the first 24 hours after administration.

Repeated Exposure. Cumulative properties were not pronounced. Rats received 0.1 and 1 g VA-102/kg BW. The compound was found to have neurotropic and hepatotropic effects.

Long-term Toxicity. A dose of 1.0 mg/kg BW was a NOEL with respect to the conditioned reflexes of rats. Higher doses affect the liver. The NOAEL of 0.1 mg/kg BW was established.

Standards. *Russia.* MPC: 2.0 mg/l.

Reference:

Trakhtman, M. B., Hygienic assessment of the efficacy of polymethacrylic cationic flocculents and their maximum permissible concentration in drinking water, *Gig. Sanit.*, 9, 15, 1969 (in Russian).

POLY(4-VINYL-*N*-BENZYLTRIMETHYLAMMONIUM)CHLORIDE

(cationic flocculent VA-2)

Structural Formula.

$[\sim CH_2CHC_6H_4CH_2N^+(CH_3)_2\sim]_nCl$, where $n = 100$ to 500

$M = 5 \times 10^4$ to 10^5.

CAS No 26780-21-2

RTECS No DA4525000

Synonyms and **Trade Names.** 4-Ethenyl-*N,N,N*-trimethylbenzenemethanaminium chloride, homopolymer; Trimethyl(*p*-vinylbenzyl) ammonium chloride, polymers; Polystyrene VA; VPK 101.

Composition. Water soluble polyelectrolyte. Obtained by treating polystyrene with monochloro- methyl ether and subsequent amination with trimethylamine. Contains 10 to 14% polymer.

Properties. A viscous liquid ranging in color from light yellow to amber; transparent, homogenous, without foreign inclusions, and water soluble. After VA-2 is added, water acquires a characteristic odor and tart taste. The taste threshold concentration is 5.0 mg/l, the perception threshold is significantly higher.

Applications. Used as a coagulant in drinking water purification and in isolating enzymes and antibiotics.

Acute Toxicity. For mice, rats, and guinea pigs, LD_{50} ranges from 1.5 to 3.2 g/kg BW. Guinea pigs are the most sensitive laboratory animals.

Repeated Exposure. When 0.2 LD_{50} was administered over a period of 20 days, weak cumulative properties were observed.

Long-term Toxicity. A dose of 1.0 mg/kg BW caused a change in hematology analyses (eosinophilia), functional changes in the carbohydrate and protein-forming function of the liver, and active reaction of the reticular tissue of the spleen and liver. A dose of 0.02 mg VA-2/kg BW appeared to be the NOAEL for all tests used in the experiment.

Standards. *Russia.* MPC is 0.5 mg/l.

Reference:

Vitvitskaya, B. R., Permissible levels of residual amounts of new cationic flocculents in drinking water, *Gig. Sanit.*, 5, 13, 1969 (in Russian).

CATIONIC FLOCCULENT VA-3

Composition. Obtained by treating *polystyrene* with pyridine.

Properties. A viscous liquid containing 10 to 14% polymer. After F. VA-3 has been added, water acquires a characteristic aromatic odor and a tart taste (a concentration of 5.0 mg/l corresponds to a rating of 2; the odor perception threshold is significantly higher).

Acute Toxicity. For mice, rats, and guinea pigs, LD_{50} ranges from 1.5 to 3.0 g/kg BW. Guinea pigs are the most sensitive laboratory animals to F. VA-3.

Repeated Exposure. No cumulative properties were found when a dose of 0.2 LD_{50} was administered 20 times.

Recommendations. ***Russia.*** Not recommended for treatment of drinking water.

Reference:

Vitvitskaya, B. R., Permissible levels of residual amounts of new cationic flocculents in drinking water, *Gig. Sanit.*, 5, 13, 1969 (in Russian).

CATIONIC FLOCCULENT VA-2T

CAS No 11111-40-3
RTECS No YV2125285

Composition. Obtained by treating pilyvinyltoluene with monochloromethyl ester and subsequent amination with trimethylamine.

Properties. A viscous liquid containing 10 to 14% polymer. After F. VA-2T has been added, water acquires a characteristic aromatic odor and a tart taste (threshold concentration 3.0 mg/l). The odor perception threshold is at a significantly higher level.

Acute Toxicity. For mice, rats, and guinea pigs, LD_{50} ranges from 1.5 to 3.0 (2.6) g/kg BW. Guinea pigs are the most sensitive laboratory animals to F. VA-2T.

Repeated Exposure. When 0.2 LD_{50} was administered over a period of 20 days, no cumulative properties were found.

Long-term Toxicity. F. VA-2T was administered to rats at the dose of 1.0 mg/kg BW. The treatment caused a pronounced change in the hematology analyses (eosinophilia), functional changes in the carbohydrate and protein-forming function of the liver, and an active reaction of the reticular tissue of the spleen and liver of experimental animals. When F. VA-2T was administered at a dose of 0.1 mg/kg BW, the same changes were found, but they were found to be transient. A dose of 0.02 mg/kg BW appeared to be the NOAEL for all tests used.

Standards. ***Russia.*** MPC: 0.5 mg/l.

Reference:

Vitvitskaya, B. R., Permissible levels of residual amounts of new cationic flocculents in drinking water, *Gig. Sanit.*, 5, 13, 1969 (in Russian).

CATIONIC FLOCCULENT VA-3T

Composition. Obtained by treating polyvinyltoluene with pyridine.

Properties. A viscous liquid containing 10 to 14% polymer. After F. VA-3T has been added, water acquires a characteristic aromatic odor and a tart taste. A concentration of 3.0 mg F. VA-3T/l in the water corresponds to taste rating 2. The odor perception threshold is significantly higher.

Acute Toxicity. For mice, rats, and guinea pigs, LD_{50} ranges from 1.5 to 3.0 g/kg BW. Guinea pigs are the most sensitive laboratory animals.[1]

Repeated Exposure. When 0.2 LD_{50} was administered over a period of 20 days, weak cumulative properties were found.[2]

Recommendations. ***Russia.*** Not recommended for treatment of drinking water.

References:

1. Vitvitskaya, B. R., Permissible levels of residual amounts of new cationic flocculents in drinking water, *Gig. Sanit.*, 5, 13, 1969 (in Russian).
2. Cherkinsky, S. N.,Vitvitskaya, B. R., et al., in *Scientific and Technical Progress and Prophylactic Medicine*, Part 1, Papers 7th Sci. Conf., 1st Moscow Medical Institute, Moscow, 1971, 72 (in Russian).

DIALLYLDIMETHYLAMMONIUM CHLORIDE, POLYMERS

Structural Formula..

```
\__
/  \
     N+ (CH3)Cl-
\__/
/
```

M = 100000 to 1000000.

CAS No 26062-79-3

RTECS No BP6360000

Synonyms and **Trade Names.** Calgon 261; *N,N'*-Dimethyl-*N*-2-propenyl-2-propen-1-ammonium chloride, homopolymer; Poly(dimethyldiallylammonium) chloride; Poly(piperidinum chloride); Quaternium 40; VPK 402.

Composition. Aqueous solution and powder product contain no residual amounts of allyl chloride.

Properties. White, crystalline, odorless substance. Readily soluble in water. Breaks down at 120°C. High stability in water is found at room temperature. VPK-402 gives water an undesirable odor and taste in concentrations of 42 and 25 mg/l, respectively. Concentrations of 1.0 to 2.0 g/l cause frothing and slight opalescence. Trial chlorination at the odor and taste threshold level does not induce chlorophenol odors or the development of taste.

Applications. Used as a flocculent to remove petroleum products and mechanical impurities from water. Used in the preparation of drinking water and to purify sewage, in the production of antibiotics and foodstuffs, and to separate out inactive and albuminous substances.

Acute Toxicity. LD_{50} is 3.0 g/kg BW in rats, 1.12 to 1.72 g/kg BW in mice, and 3.25 g/kg BW in guinea pigs.[1-3]

In clinical terms, effects on the CNS predominate are overall depression, clonic-tonic convulsions, and dyspnea. Gross pathology examination revealed plethora of the visceral organs. Death occurs in 1 to 3 days. For mice and rats, ET_{50} was 19 to 64 hours, and the Shtabsky accumulation index $I_{acc} = 0.03$.[1,2]

Repeated Exposure. The liver and hematopoietic system are mainly affected. Administration of VPK-402 in a dose of 1/10 to 1/250 LD_{50} over a period of 2 months was not lethal to male rats. Erythropenia, leukocytosis, a reduction in the activity of cholinesterase and ALT, a change in conductivity and excitability of the myocardium, increase in the STI with a simultaneous reduction in the cortical reflex index, and a change in the ECG are observed. Gross pathology examination revealed no morphological changes. The LOAEL was equal to 1/50 LD_{50}.[2] No cumulative properties were established. Habituation was noted.[1,2]

Long-term Toxicity. Male rats received doses of 0.15 and 15 mg/kg BW. Greater doses caused leukopenia, an increase in the activity of AST and aldolase in the blood serum, and of ALT and AST in the liver and kidneys, a reduction in the activity of cholinesterase in the liver and brain, and changes in the ECG.[1,2]

Reproductive Toxicity.

Embryotoxicity. A toxic effect was found when a dose of 2.0 mg/kg BW was administered. The treatment increased post-implantation loss. With a dose of 0.2 mg/kg BW, there were no disruptions of this kind.[1,2]

Allergenic Effect. Some sensitization was observed with *i/g* administration. The NOAEL of 0.15 mg/kg BW was established.

Recommendations. ***Russia.*** Recommended MPC: 0.5 and 0.3 mg/l.[2]

References:

1. Vitvitskaya, B. R., Skachkova, I. N., Semenova, A. A., Savonicheva, G. A., and Zakharova, T. A., Hygienic standards for the main components of the sewage from the manufacturing antimicrobial fabrics in the water reservoirs, *Gig. Sanit.*, 10, 69, 1980 (in Russian).
2. Vitvitskaya, B. R., Korolev, A. A., Skachkova, I. N., Savonicheva, G. A., and Sergeyev, S. G., Establishment of the maximum permissible concentration of polydimethyldiallyl ammonium chloride in water reservoirs, *Gig. Sanit.*, 3, 66, 1988 (in Russian).

POLY(1,3-DIMETHYL-5-VINYLPYRIDINE)METHYL SULFATE

Structural Formula.

```
[H3C        CHCH2~ ]n
    \ // \/
     |    |
     \\ /
      N+ -OSO3CH3
      |
      CH3
```

Properties. High-molecular-mass polymer. Odor threshold concentration (20 and 60°C) is 7.0 mg/l. In considerably higher concentrations it gives water a taste and opalescence.

Applications. Used as a flocculent in the preparation of water; mode of action: cationic effect.

Acute Toxicity. LD_{50} is reported to be 2.2 g/kg BW in rats, 1.7 g/kg BW in mice, and 1.97 g/kg BW in guinea pigs. Administration led to adynamia, autonomic disorders, and tremor. Death occurs within 1 to 3 days.

Repeated Exposure. Administration of 1.25 g/kg BW to rats over a period of 20 days was not lethal. There was a reduction in the activity of cholinesterase in the liver and kidney.

Short-term Toxicity. Rats received 80 or 400 mg/kg BW for 3 months. The treatment caused a definite reduction in BW gain, a pronounced reticulocytosis, an increase in the urea level in the blood serum, change in the STI, and a specific reduction in the vitamin C contents in the liver and kidneys.

Long-term Toxicity. Doses of 2.0 and 20 mg/kg BW affected conditioned reflexes. The NOAEL of 0.2 mg/kg BW was established.

Standards. ***Russia.*** MPC: 4.0 mg/l.

References:

1. Vitvitskaya, B. R., Korolev, A. A., Skachkova, I. N., Savonicheva, G. A., and Sergeev, S. G., Establishment of the maximum permissible concentration of polydimethyldiallyl ammonium chloride in water reservoirs, *Gig. Sanit.*, 3, 66, 1988 (in Russian).
2. Vitvitskaya, B. R., Skachkova, I. N., Semenova, A. A., Savonicheva, G. A., and Zakharova, T. A., Hygienic standards for the main components of the sewage from the manuacturing antimicrobial fabrics in the water reservoirs, *Gig. Sanit.*, 10, 69, 1980 (in Russian).

ACTIVE SILICIC ACID

Structural Formula. $xSiO_2 .yH_2O$

CAS No 7699-41-4

RTECS No VV7310000

Abbreviation. ASA.

Composition. Partially structured colloidal solution of silicon dioxide.

Properties. Colorless to gray, odorless powder.

Applications. ASA is a basic inorganic high-molecular-mass flocculent; mode of action: anionic effect.

Acute Toxicity. LD_{50} exceeded 3.0 g/kg BW.

Standards. ***Russia.*** MPC: 50 mg/l with respect to SiO_2 concentration.

FLUOROPLASTICS

FLUOROPLASTIC -1

Structural Formula.. [~CH_2CHF~]$_n$

Synonym and **Trade Name.** Ftorlon-1; Polyvinyl fluoride.

Properties. Solid white product. *Dens.* 1.39.

Applications. Used as films for packaging foodstuffs and as a internal coating for cans.

Migration Data. Organoleptic indices of water (1 cm^{-1}; 100 to 20°C; 3 to 13 months) and oil (1.5 to 4 months) extracts did not differ from the control. Migration of low-molecular-mass compounds was negligible. No fluoromonomer or organic products of breakdown were found in extracts (Petrova et al.).

At the same time, data available indicate that alcohol and freon migrate from F-1 coatings: 0.063 mg/l after 6 days and 0.125 mg/l after 5 months.

Long-term Toxicity. Mice and rats were given aqueous extracts in their drink, and oil extracts were added to their feed (calculated dose 3.0 g/kg BW). A majority of the indices stayed unchanged. The fact that some changes were slight and reversible means that effect of extracts has a threshold value. In view of the aggravated conditions of the experiment, the film was recommended for intended use.

Reproductive Toxicity. Aqueous and oil extracts from F. do not affect the reproductive function of rats or development of the first generation of the offspring.

Reference:

Krynskaya, I. I. and Sukhareva, L. V., Hygienic properties of ftorlon anti-scorching coating, *Plast. Massy*, 12, 38, 1976 (in Russian).

FLUOROPLASTIC-2

Synonym and **Trade Name.** Polyvinylidene fluoride; Ftorlon-2.

Composition. Does not contain any plasticizer or stabilizer.

Properties. White crystalline polymer. *M. p.* 171 to 180°C; *dens.* 1.7625; n^{25} = 1.42.

Applications. Used as coatings and films in the food industry.

Migration Data. Aluminum plates coated on both sides with a suspension of F. mixed with acetone and *N,N'*-dimethylformamide as well as cans were studied. The odor and taste of the extracts (1 cm^{-1}; 1 to 26 months) was found to be insignificant (up to rating 1). The concentration of F^- and dimethyl formamide was 0.01 to 04 and 0.35 mg/l, respectively.

Reference:

Guricheva, Z. G., Kryglova, N. V., and Shaulova, N. S., Sanitary-chemical study of the Fluoroplastic F-2 coating, in *Environmental Protection in Plastics Production and Hygiene Aspects of Their Use*, Leningrad, Plastpolymer, 1978, 130 (in Russian).

FLUOROPLASTIC-2M

Properties. Somewhat less hard than F-2 but more shock-resistant.

Applications. Used in the manufacture of coatings, pipes, etc.

Migration Data. Chromatography revealed 0.35 mg *N,N'*-dimethylformamide/l (solvent used in the production of suspensions based on fluoroplastics) in 2-year extracts. No migration of fluorine monomers was found. Traces of alcohol were found in the extracts.

POLY(TRIFLUOROCHLOROETHYLENE)

Structural Formula. [~$CHClCF_3$~]$_n$

CAS No 9002-83-9

RTECS No KM6555000

Abbreviation. F3.

Composition. PCTFE is primarily oligomers with 3 to 4 monomer units.

Synonyms and **Trade Names.** Daiflon; Polychlorotrifluoroethylene; Fluorolon 3, Fluoroplast 3; Ftorlon 3; Hostaflon; Teflex.

Acute Toxicity. LD_{50} exceeded 9.2 g/kg BW in rats.[1]

Repeated Exposure. Rats were gavaged with F3 for 26 days. Changes were noted in liver weight. Kidney, ureter, bladder, and urine composition were affected. Changes were seen in erythrocyte count.[2]

F3 was given daily for 15 days by oral gavage to four Rhesus monkeys at the doses of 0.725 g/kg BW. No significant indication of peroxisomal proliferation was observed. An increased blood urea nitrogen at 15 days was the only clinical pathological abnormality seen in both monkeys and rats.[1]

References:

1. Jones, C. E., Ballinger, M. B., Mattie, D. R., DelRaso, N. J., Seckel. C., and Vinegar, A., Effect of short-term oral dosing of polychlorotrifluoroethylene (polyCTFE) on the rhesus monkey, *J. Appl. Toxicol.*, 11, 51, 1991.
2. Mattie, D. R., Hoeflich, T. J., Jones, C. E., Horton, M. L., Whitmire, R. E., Godin, C. S., and Andersen, M. E., The comparative toxicity of operational Air Force hydraulic fluids, *Toxicol. Ind. Health*, 9, 995, 1993.

FLUOROPLASTIC-4

Structural Formula. $[\sim CF_2CF_2\sim]_n$
M up to 10,000000
CAS No 9002-84-0
RTECS No KX4025000
Abbreviation. F4.

Synonym and **Trade Names.** Aflon; Fluoroflex; Fluorolon 4; Fluoroplast 4, 4B, 4D, 4M; Ftorlon 4, F-4MB, or 4M; Hostaflon; Polyfen; Polyflon; Polytetrafluoroethylene; Teflon; Tetrafluoroethene homopolymer; Tetran 30.

Properties. White or grayish translucent substance, rigid as a thin film, or soft, waxy, milk-white solid, or white powder. Not wetted or affected by water, does not stick to anything. *Dens.* 2.15 to 2.24; *m. p.* 327 C; $n^{25} = 1.375$.

Applications. Used in the food industry. Used in medicine for vascular plastic surgery and implants, and for a number of other medical applications.[2]

Exposure. In Switzerland, where most pans today are made of stainless steel or teflon-coated aluminum, the average contribution for the use of aluminum utensils to the daily *aluminum* intake of 2 to 5 mg from the diet is estimated to be less than 0.1 mg.[3]

Migration Data. A film of F4 does not affect general indices of water quality. In aqueous extracts (0.25 to 1.0 cm^{-1}; 20 and 100°C; 2 to 240 hours) small changes in organoleptic properties were found. The contents of fluor ions and organofluorine compounds did not exceed 0.3 mg/l in the first extract, and thereafter migration ceased. No formaldehyde, lead, chromium, aluminum, nickel, antimony, or arsenic were found in the extracts.[4]

Migration of 5.0 to 50 μg benzene/dm^2 has been detected from 7/26 samples of retail nonstick cookware covered with polytetrachloroethylene coatings. Benzene in a number of these samples was attributed to the use of phenylmethyl silicone ingredient containing 360 mg benzene/kg.[5]

Long-term Toxicity. A 10-month oral study in rats showed no effect on animals.[6]

Short-term Toxicity. Rats were affected only after consuming aqueous extracts obtained at temperatures higher than 90°C. At lower temperatures, extracts had no harmful effect.[7] When rats ate feed with 25% fine F4 powder over a 90-day period, no toxic effects were observed.[8]

Carcinogenicity. Teflon induced malignant tumors in rodents following *s/c* imbedding of polymer films.[026] Nevertheless, the results of this study were later considered inadequate. *S/c* sarcomas were induced in female mice by implantation of a teflon disc.[9] Teflon is an equivocal tumorigenic agent by RTECS criteria.

Carcinogenicity classification. An IARC Working Group concluded that there is inadequate evidence for the carcinogenicity of AA in *experimental animals* and there were no adequate data available to evaluate the carcinogenicity of AA in *humans*.

IARC: 3

Recommendations. Nonstick coating for cooking utensils F-4D, F-4DU, and F-4MD with or without primers are recommended for coating domestic cooking utensils except in cases where migration of heavy metals is found.[1]

Ftoroplast 4 is recommended as a coating for cooking utensils used at temperatures not exceeding 100°C.

Regulations. ***U.S. FDA*** (1998) approved the use of F4 (1) in adhesives as a component (monomer) of articles intended for use in packaging, transporting, or holding food in accordance with the conditions prescribed in 21 CFR part 175.105; and (2) as articles or components of articles intended to contact food, subject to the provisions of 21 CFR part 177.1550.

References:

1. *Toxicology and Sanitary Chemistry of Polymerization Plastics*, Proc. Sci. Conf., B. Yu. Kalinin, Ed., Leningrad, 1984, 94 (in Russian).
2. Martin, J. N., Brewer, D. W., Ruch, L. V., et al., Successful pregnancy outcome following mid-gestational uterine rupture and repair using Gore-Tex soft tissue patch, *Obstet. Gynecol.* 75, 518, 1990.
3. Muller, J. P., Steinegger, A., und Schlatter, C., Contribution of aluminum from packaging materials and cooking utensils to the daily aluminum intake, *Z. Lebensm. Unters. Forsch.*, 197, 332, 1993.
4. Proklina, T. L. and Shvaiko, I. I., Hygienic evaluation of polymeric materials intended for use in the food industry, *Gig. Sanit.*, 1, 111, 1978 (in Russian).
5. Jickells, S. M., Philo, M. R., Gilbert, J., and Castle, L., Gas chromatographic/mass spectrometric determination of benzene in nonstick cookware and microwave susceptors and its migration into foods on cooking, *J. AOAC Int.*, 76, 760, 1993.
6. Kas'yan, V. N., On the problem of hygienic evaluation of dishes covered with teflon-2, *Gig. Sanit.*, 5, 78, 1981 (in Russian).
7. Khamidullin, R. S., Bronikova, I. A., Petrova, G. A., Moshlakova, L. A., and Taradin, V. V., Hygienic evaluation of fluoroplast-4MD coatings intended for use in food industry, *Gig. Sanit.*, 11, 14, 1985 (in Russian).
8. Loshkareva, V. I., Kucherov, I. S., and Vlasova, L. P., Hygienic evaluation of polymeric film (F-4MB) capacities used for a continuous storage of drinking water supplies, *Gig. Sanit.*, 11, 79, 1983 (in Russian).
9. Menard, S. and Porta, G. D., Incidence, growth and antigenicity of fibrosarcomas induced by Teflon disc in mice, *Tumori*, 62, 565, 1976.

FLUOROPLASTIC-4D

Abbreviation. F-4D.

Trade Name. Ftorlon-4D.

Composition. Dispersed polytetrafluoroethylene. A version of Ftorlon-4. Contains lubricant additives.

Applications. Used in the manufacture of different details, piping, and coatings.

Properties. More chemically stable than any known metal. Not wetted by water and does not swell (zero water absorption).

Migration. A film of F-4D does not affect the general properties of water,[1] nor does a coating of F-4D colored by a mixture of TiO_2 and Fe_2O_3 with the addition of carbon black DT-100.[2]

Long-term Toxicity. Aqueous extracts in the feed of mice and rats over a period of 12 months did not cause changes in their hematological indices, a number of enzyme indices, in immune or NS functions. There were no histological changes. Liquids which were in contact with F-4D at high temperatures were not found to be toxic in a chronic experiment. No signs of sensitization were observed and no tumor development was found.[2]

References:

1. Proklina, T. L. and Shvaiko, I. I., Hygienic evaluation of polymeric materials intended for use in the food industry, *Gig. Sanit.*, 1, 111, 1978 (in Russian).
2. Krynskaya, I. I. and Sukhareva, L. V., Hygienic properties of ftorlon anti-scorching coating, *Plast. Massy*, 12, 38, 1976 (in Russian).

COATING made from a suspension of FLUOROPLASTIC F-4D and DB

Applications. Suspension stabilizers.

Migration Data. Milk and aqueous extracts were investigated (80, then 20°C; 10 days, 100°C; 2 hours in a retort with a reflux condenser). Organoleptic indices were not changed. Release of fluorine ions exceeded 0.03 mg/l. No migration of formaldehyde or salts of heavy metals was found.

Reference:

Toxicology and Sanitary Chemistry of Polymerization Plastics, Proc. Sci. Conf., B. Yu. Kalinin, Ed., Leningrad, 1984, 94 (in Russian).

FLUOROPLASTIC-4MD

Structural Formula.

```
 F   F  CF3 F
 |   |   |   |
[~C - C - C - C ~]n
 |   |   |   |
 F   F   F   F
```

Abbreviation. F-4MD.

Composition. Polytetrafluoroethylene modified with hexafluoropropylene.

Properties. A coating made of F-4MD is smooth, shiny, light-cream in color, and odorless. High heat resistance together with resistance to corrosive media. Thermal oxidation does not occur below 370°C. Insoluble in known solvents. Does not swell in acids or alkalis.

Applications. F. film is used in containers for transporting and storing water. Used as a nonstick coating in the food industry, in medicine, pharmaceutics, and the confectionery industry (the material reduces adhesion of confectionery mass to the machinery).

Migration Data. Extracts were studied over a period of 9 months. Organoleptic indices of water did not change. A decrease in general alkalinity and general hardness was noted at the end of experiment which may be attributed to deposition of hydrocarbonates and calcium and magnesium salts on the walls of the tank. The fluorine contents did not change compared to the control.[1]

There was no change in the organoleptic quality of confectionery in contact with a coating (30 to 40°C; 10 days; specific surface 1.5 cm^{-1}). Harmful substances (organofluoric compounds, fluorine ions, formaldehyde, etc.) did not migrate into water or model media in health hazardous quantity.[2]

Migration of water-soluble organofluorine compounds (up to 0.017 mg/l) was found only the first time the articles were treated with model media. No volatile organofluorine compounds, fluorocarboxyl acids, or propyl alcohol were found in the extracts. There were no changes in the organoleptic properties.[3]

Long-term Toxicity. Rats drank extracts obtained with silvered water (0.1 mg/l) over a period of 6 months. No adverse effect was found, though after a month, a definite decrease in the urea contents of the blood serum was observed. No pathological changes were found.[1]

The toxicity of aqueous extracts was studied in a 15-month experiment with rats. Neither toxic effect nor pathological changes in the visceral organs were found.[3]

Recommendations. In ***Russia***, F-4MD is authorized for use as an nonstick coating in confectionery industry. Tanks made of F-4MD are recommended for prolonged storage of water.

References:

1. Loshkareva, V. I., Kucherov, I. S., and Vlasova, L. P., Hygienic evaluation of polymeric film (F-4MB) capacities used for a continuous storage of drinking water supplies, *Gig. Sanit.*, 11, 79, 1983 (in Russian).
2. *Toxicology and Sanitary Chemistry of Polymerization Plastics*, Proc. Sci. Conf., B. Yu. Kalinin, Ed., Leningrad, 1984, 97 (in Russian).
3. Khamidullin, R. S., Bronikova, I. A., Petrova, G. A., Moshlakova, L. A., and Tararin, V. V., Hygienic evaluation of fluoroplast-4MD coatings intended for use in food industry, *Gig. Sanit.*, 11, 14, 1985 (in Russian).

FLUOROPLASTIC F-10 and F-26B

Abbreviation. F-10 and F26B.

Properties. F-10 is a soft, elastic, organofluorine polymer possessing high thermal and chemical stability. F-26 is a copolymer of vinylidene difluoride with hexafluoropropylene. A highly elastic material.

Applications. Both materials are used in the manufacture of elastic containers for water storage. F-26 is used in the manufacture of lacquers and films with thermal stability up to 200 to 250°C.

Migration Data. In aqueous extracts prepared at 20 and 80°C, an increase in *pH* from 7.4 to 8.36 was noted.

Long-term Toxicity. Aqueous extracts were not found to be toxic in a chronic experiment

Reference:

Mironets, N. V, Panasenko, G. I., Kucherov, I. S., et al., Sanitary-chemical characteristics of fluoroplastic films, in *Hygiene and Toxicology of High-Molecular-Mass Compounds and of the Chemical Raw Material Used for Their Synthesis,* Proc. 6th All-Union Conf., B. Yu. Kalinin, Ed., Khimiya, Leningrad, 1979, 150 (in Russian).

FLUOROPLASTIC-30

Structural Formula.

$[\sim CF_2CFCl\sim]_m\ [\sim CH_2CH_2\sim]_n$

Abbreviation. F-30.

Composition. Copolymer of trifluorochloroethylene with ethylene.

Properties. Possesses high chemical stability.

Applications. Used in the manufacture of containers for long-term water storage.

Migration Data. A study of aqueous extracts (2 cm^{-1}; 20°C; 2 to 24 months) did not show any significant change in their odor or taste. Contents of fluor ions in the extracts did not exceed 0.19 mg/l. No migration of organofluorine compounds was found.

Reference:

Kruglova, N. V., Shaulova, N. S., Guricheva, Z. G. et al., Sanitary-chemical properties of fluoroplastic F-30 intended for prolonged holding of water, in *Hygiene and Toxicology of High-Molecular-Mass Compounds and of the Chemical Raw Material Used for Their Synthesis,* Theses 6th All-Union Conf., B. Yu. Kalinin, Ed., Leningrad, Khimiya, 1979, 147 (in Russian).

FLUOROPLASTIC-32L

Structural Formula..

$[\sim CF_2CFCl\sim]_m\ [\sim CH_2CF_2\sim]_n$

Abbreviation. F-32L.

Composition. Copolymer of trifluoroethylene and vinylidene difluoride

Properties. Soluble polymer with high elasticity. Resistant to corrosive media.

Applications. Used in the manufacture of lacquers and protective coatings based on them.

Migration Data. Lacquer LF-32LN released solvents and other impurities (amyl acetate, isoamyl acetate, cyclohexanone, toluene) into water and saline and acetic acid extracts in quantities ranging from considerable to trace values. Further heating of specimens eliminates migration of solvents. There were no changes in the organoleptic properties.

Reference:

Guricheva, Z. G., Krynskaya, I. L., and Shaulova, N. S., Hygienic properties of ftorlon-1 coating, *Plast. Massy*, 12, 44, 1976 (in Russian).

FLUOROPLASTIC-42

Abbreviation. F-42.

Properties. In compressed form it is a pliable plastic material, yellow to white in color. Transparent in a thin film. One of a few soluble fluorine-containing polymers. Characterized by high strength and chemical stability. *Dens.* 1.91 to 1.93. Insoluble in alcohols.

Applications. F-42L is used in the production of lacquers and coatings.

FTORLON-EPOXIDE LACQUERS LFE-32, LNKh and LFE-42LKh

Abbreviation. FL.

Migration Data. A study of aqueous extracts from ftorlon-1 coating (LFE-42LKh) showed a deterioration of the organoleptic properties. The material released epichlorohydrin and cellosolve in quantities ranging from traces to 0.2 mg/l (both components). No migration of organic substances from LFE-32LNKh was found.

It was **recommended** that cellosolve should be omitted from the formulation; after that hygienic properties of FL were found to be satisfactory.

Reference:

Guricheva, Z. G., Krynskaya, I. L., and Shaulova, N. S., Hygienic properties of ftorlon-1 coating, *Plast. Massy*, 12, 44, 1976 (in Russian).

FLUOROPLASTIC KB, GRAPHITE-REINFORCED

Migration Data. A study of extracts (0.5 cm^{-1}; 20°C; 1 to 30 days) showed insignificant increase in general hardness and fluorine-ion contents of water.

Long-term Toxicity. No significant changes were found in general condition, biological and physiological indices, or histological structure of the visceral organs of animals that drank aqueous extracts of G. for 6 months.

Reference:

Kupyrov, V. N., Kaplina, T. V., Gakal, R. K., Vinarskaya, E. I., and Starchenko, S. N., Hygienic evaluation of films intended for the waterproofing of unit prefabricated swimming pools, *Gig. Sanit.*, 5, 91, 1978 (in Russian).

FURAN RESINS

Composition. Oligomeric products containing a furan ring in their molecules.

Produced with the use of furfural, furyl (furfuryl) alcohol, furyl furfural (furfuryl furfural), and products of condensation of furfural with acetone (furfurylidene) and difurfurylidene acetone (monomer FA).

POLYFURONS

Composition. Furan polymers are the products of polycondensation of difurfurylidene acetone, furfural, and diphenylolpropane in various mass ratios: 3:2:1 in polyfuron-321, 6:2:1 in polyfuron-621.

Properties. Dark brown, highly fluid liquids.

Acute Toxicity. Moderately toxic substances.

Repeated Exposure. Administration of 1/10 and 1/20 LD_{50} was not lethal to rats. There were a weak functional accumulation, changes in hematological indices, STI, and suppressed activity of cholinesterase of the blood serum.[05]

RESINS FA, FAGI, FAFF-31, FAED

Composition. Produced on the basis of monomer FA and formaldehyde or epoxy resin.

Properties. FAFF-31 resin is a dark brown liquid. Density of its solution in acetone is 1.01 to 1.19 at 20°C. FA-15 resin is a dark brown solid.

Migration Data. In contact with water over a period of 2 to 16 days, FAGI resin changed its color and increased oxidizability. As for other resins, migration of furfural and acetone into water was observed.

Recommendations. ***Russia***. Not recommended for use in contact with food and drinking water.

Reference:

Tadzhibayev, N. S. et al., *Hygiene Problems under Conditions of Hot Climate in Uzbekhistan*, Meditsina, Tashkent, Issue No 4, 1970, 57 (in Russian).

FURAN RESIN

CAS No 116958-47-5

RTECS No LV0600000

Trade Name. Furitol-68.

Composition. Phenol, formalin, and furyl alcohol condensation product

Properties. A dark brown, homogenous, viscous liquid with a specific odor. *Dens.* 1.19; $pH = 8.5$.

Migration Data. Formaldehyde, furfural, phenol, and furyl alcohol can be released from the resin into water.

Acute Toxicity. For rats, mice, and rabbits, LD_{50} is 1.7, 0.38, and 2.05 g/kg BW, respectively. Administration caused an increase in motor activity; then there were stiffness and frequent clonic-tonic convulsions.

Repeated Exposure revealed mild cumulative properties.

Reference:

Tarasov, V. V., Appolonova, G. M., and Likho, V. G., Toxicity of furan resin "Furitol-68" and "Furitol-80", *Gig. Sanit.*, 6, 93, 1986 (in Russian).

UREA-FORMALDEHYDE-FURANOL polymer

Molecular Formula. $[\sim C_4H_4O_2.CH_4N_2O.CH_2O\sim]_n$

CAS No 100424-73-5

RTECS No YU1612000

Trade Names. Furitol 80 and 86.

Composition. Product of condensation of urea, formaldehyde, and furyl alcohol (ratio is 18:55:24).

Properties. Homogenous viscous liquid with a dark brown color and specific odor.

Migration Data. Ammonia, formaldehyde, and furyl alcohol are released from the resin.

Acute Toxicity. LD_{50} is 840 mg/kg BW in rats, 2320 mg/kg BW in rabbits, and 1250 mg/kg BW in mice.[1] Poisoning affected CNS (parasympathomimetic effect, depression, ataxia).

Reference:

Tarasov, V. V., Appolonova, G. M., and Likho, V. G., Toxicity of furan resin "Furitol-68" and "Furitol-80", *Gig. Sanit.*, 6, 93, 1986 (in Russian).

FURAN RESIN 2F-S

Composition. Based on FA monomer and phenol alcohol.

Acute Toxicity. LD_{50} is 320 mg/kg BW in mice, 1560 mg/kg BW in rats, and 984 mg/kg BW in rabbits. Acute poisoning led to coordination disorder, adynamia, and paralyses of posterior extremities.

Repeated Exposure revealed moderate functional accumulation: changes in hematology parameters, reduced activity of cholinesterase, increased sugar concentration in the blood, and decreased BW gain.

Reference:

Apollonova, G. M. and Likho, V. G., in *Actual Problems of Hygiene & Professional Pathology under Conditions of Scientific-Technical Progress*, Proc. IV Congr. of Hygienists, Sanitary Inspectors, Epidemiologists, Microbiologists., Infectionists of Uzbekistan, Tashkent, 1980, 171 (in Russian).

PLASTOCONCRETE based on FA MONOMER

Trade Name. Plastobeton.

Composition. Organomineral cement-free P. consists of organic fraction capable of polymerizing in presence of special additives and mineral fillers.

Migration Data. After a brief contact with P. (1.5 kg P./3 liters of water) water acquires a strong odor of acid fermentation and a sour taste. The acidity of the extract amounted to 54 mg-equivalent/l, *pH* exceeded 3.0. Furfural is not found in water. The same material prewetted (55 days) and then washed with running water for a week gave water some odor and taste after exposure for 5 days.[07]

FURAN CONCRETE

Composition. Based on monomer FA, contains benzosulphoacids, sand, and crushed stone.

Migration Data. With mass ratios of the material and water of 1:100 and 1:1000, there were little changes in organoleptic properties of the extract. A certain increase in contents of the solids and oxidizability of water were found.

Reference:

Kochkin, V. P., in *Sci. Proc. Saratov Polytech. Institute*, Saratov, Issue 84, 1975, 28 (in Russian).

5-ETHENYL-2-FURANCARBOXALDEHYDE HOMOPOLYMER

Molecular Formula. $[\sim C_7H_6O_2\sim]_n$
CAS No 32630-50-5
RTECS No LT9930000
Abbreviation. PVF.

Composition. Furfural and polyvinyl alcohol reaction product. Contains 2.0 to 8.0% of furfural impurity and 10 to 12% of furfural groups.

Synonyms. Polyvinyl furfural; 5-Vinyl-2-furaldehyde, polymers.

Properties. Liquid of light to dark yellow color. Soluble in water at all ratios.

Applications. Used as a lubricant in the production of glass-reinforced plastics and glass fibers.

Acute Toxicity. LD_{50} is reported to be 2.08 g PVF/kg BW in rats, and 0.85 g PVF/kg BW in mice. Poisoning with high doses resulted in death from cardiac arrest in four days.

Repeated Exposure revealed evident cumulative properties: K_{acc} appeared to be 2.1 (by Lim). Rats were given 1/5 LD_{50} for 1 month. The treatment led to development of a regenerative-type hypochromic anemia (a decline in *Hb* level and reticulocytosis) and inhibition of carbohydrate-forming function of the liver. The kidney and NS were not affected.

Reference:

Volodchenko, V. A. and Sadokha, E. R., Toxicological evaluation of polyvinylfurfural, *Gig. Truda Prof. Zabol.*, 11, 58, 1985 (in Russian).

ION-EXCHANGE RESINS

Abbreviation. IER.

Properties. Solid, mostly insoluble substances of amorphous or crystalline structure. May give water foreign odor and taste.

Applications. IER are used in contact with drinking water and in the sugar and food industries.

Migration Data. Monomers used for the synthesis of IER (styrene, divinylbenzene, epichlorohydrin, phenols, formaldehyde) and additives (polyethylenepolyamine, other amines, heptane, hydrocarbons, products of monomer and polymer decomposition) as well as inorganic components were found to migrate into water and simulant media. Cationites (KU-1, KU-2, KU-2-8, KU-23, KU-2h) and anionites (AV-17, AV-17-8, AV-17-8p, No. 374, AN-31, EDE-10p, and EDE-10) are the most frequent migrants.[1,2]

After removal of soluble impurities, IER can be used for contact with water and foodstuffs.

Acute Toxicity. Animals survived after a single *i/g* administration of 15 g/kg BW or more.[3]

Repeated Exposure. Administration of 1.0 g/kg BW produced functional and structural changes in the viscera.[4]

Regulations.

U.S. FDA (1998) approved the use of ion-exchange membranes in the processing of food in accordance with the conditions prescribed in 21 CFR part 173.20.

Russia. The following ionites are recommended for use in food industry: cationite KU-2-8 (sugar industry and citric acid production), KU-2-8hS (dairy, starch and syrup, sugar and wine industry), anionite AV-17-8hS dairy, starch and syrup industries (after preliminary purification with solutions of sodium hydroxide and hydrochloric acid and subsequent repeated washing with water), EDE-10p (citric and tartaric acids, etc.).[028]

Polymerization resins of styrene and divinylbenzene and macroporous and purified resins are most suitable for water piping.[2]

The following materials are approved for use in domestic water supply: ionite membranes MA-40, MK-40, MA-41, MAK, MKK, MA-100, MK-100, MKK-1r, MA-41-4, MA-40-OS and MA-40-OS, polyamide membranes Phenylone 2s, acetyl cellulose membranes MGA and MOO, cationites KU-2-12pch, KU-2-16pch, KU-2-8h, Ku-23, and KU-23h, anionites EDE-10 and (in hydroxyl form) AV-17-8h, AV-17-8p, and AV-8pch, mixed-action filters made of Ku-2-8chs and AV-17, KU-2-12pch, and AV-17-8pch, structural materials Miplast, Paronit PON, etc.

References:

1. Kas'yan, V. N., Toxicological and hygienic evaluation of cationites KU-2-8 and KU-2-8chC intended for use in the sugar industry, in *Hygiene and Toxicology of High-Molecular-Mass Compounds and of the Chemical Raw Material Used for Their Synthesis*, Proc. 6th All-Union Conf., B. Yu. Kalinin, Ed., Khimiya, Leningrad, 1979, 73 (in Russian).
2. Omel'yanets, N. I. and Shtannikov, Ye. V., Hygiene of use of ion-exchange resin in water supply, *Ibid*, p.151 (in Russian).
3. Sidorenko, G I., Shtannikov, Ye. V., Rozhnov, G. I., Selivanov, S. B., and Solokhina, T. A., Hygienic evaluation of the polymeric materials used for purposes of desalination, *Gig. Sanit.*, 2, 11, 1978 (in Russian).
4. Shtannikov, Ye. V., Rozhnov, G. I., Sokolov, L. A., and Khonin, B. M., Mutagenic activity of polymers used for water desalination, *Gig. Sanit.*, 2, 17, 1972 (in Russian).

CATIONITE SG-1M

Composition. Copolymer of methacrylate acid and triethylene glycol dimethacrylate.

Applications. Used for water purification.

Migration Data. The study of aqueous extracts (20 to 100°C) revealed a temperature-dependent migration of methacrylic acid and butyl acetate. In practical use, the concentration of methacrylic acid was half the MPC, while butyl acetate was not found. Other indices of water quality conformed to Drinking Water Standards (Russia).

Toxicity. No toxic effect was noted in experimental animals.

Reference:

Akhlustina, L. V. et al., in *Problems of Health Preservation of Population and Environmental Protection against Hazardous Chemical Factors*, Proc. 1st All-Union Toxicol. Conf., Rostov-na- Donu, 1986, 163 (in Russian).

ANIONITE KhAP-4

Composition. Chemically active polymer based on anionite AV-17-10pch in the form of HCO_3 with graft ferric hydroxide.

Applications. Sorption properties for sulfides with simultaneous retention of the ability to enrich water with hydrocarbonates. Used as a hydrogen sulfide absorbent in water recycling systems.

Migration Data. Possible migration of diethylamine into water in a quantity no higher than the MPC (0.05 mg/l). Ammonia nitrogen contents were decreased and stabilized after 2 liters of filtrate have been obtained.

Long-term Toxicity. No toxic effect was observed in rats exposed over a 60-month period to drinking water filtered through KhAP-4.

Recommendations. ***Russia.*** KhAP-4 is recommended for removing *hydrogen sulfide* from water.

Reference:

Omel'yanets, N. I. Naboka, M. V., Gubsky, Yu. V., et al., The prospects of using chemically active polymers for drinking water decontamination from hydrogen sulfide, *Gig. Sanit.*, 5, 64, 1986 (in Russian).

MEMBRANE MA-41

CAS No 9083-37-8

RTECS No OL8500000

Applications. Ionite non-purified membrane.

Acute Toxicity. LD_{50} exceeded 10 g/kg BW in rats.

Reference:

Shtannikov, E. V., Rozhnov, G. I., and Khonin, B. M., Mutagenic activity of polymers used for water desalination, *Gig. Sanit.*, 2, 17, 1972 (in Russian).

NAPHTHALENE and BENZENE COPOLYMER

Molecular Formula. $[\sim C_{10}H_8.C_6H_6\sim]_n$

M = 950 to 1050

CAS No 25748-83-8

RTECS No QK0300200

Synonyms. Benzene-naphthalene copolymer; Copolymer NB; Naphthalene-benzene copolymer.

Acute Toxicity. LD is found to be more than 5.0 g/kg BW in rats and mice.

Repeated Exposure. Rats were given 2.0 g/kg BW over a period of 10 days. The treatment caused pronounced changes in the blood serum composition (e.g., bilirubin, cholesterol) and in dehydrogenase activity.

Reference:

Krapotkina, M. A., Galitskaya, V. A., Abramova, Iu. V., and Gafurova, T. V., Experimental evaluation of the toxic effect of copolymer NB, *Gig. Truda Prof. Zabol.*, 3, 51, 1983 (in Russian).

POLYACRYLATES

Molecular Formula. $[\sim C_3H_4O_2\sim]_n$

M = 168.06

CAS No 9003-01-4

51142-25-7

RTECS No AT4680000

Composition. Polymers of esters of acrylic acid.

Synonyms and **Trade Names.** Acrylic acid polymer; Acrylic resin; Acrysol; Carboset; Carpolen; Haloflex; Poly(acrylic acid); 2-Propenoic acid, homopolymer; Versicol; Zinpol.

Properties. Under normal conditions, P. which have an alkyl radical R from C_1 to C_{12} are, with the exception of polymethyl acrylate, viscous, rubber-like products with low hardness. At normal temperatures they are resistant to water and to diluted solutions of acids and alkalis. Acrylates readily copolymerize with almost all monomers. Copolymers with vinyl chloride, vinylidene chloride, acrylonitrile, styrene, etc., are especially widely used.

Applications. Used in the production of sheets and films, paints, varnishes, and adhesives. P. are widely used in stomatology. They are used as fixing material for bandages, lacquer for lozenges, as a dispersant, and as a basic substance for ion exchange.

Acute Toxicity. LD_{50} is 2.5 g/kg BW in rats,[1] 4.6 g/kg BW in mice, 2.5 g/kg BW in guinea pigs, and exceeded 8.0 g/kg BW in dogs.

Regulations. ***U.S. FDA*** (1998) approved the use of PA (1) in adhesives as a component (monomer) of articles intended for use in packaging, transporting, or holding food in accordance with the conditions prescribed in 21 CFR part 175.105; (2) in the manufacture of resinous and polymeric coatings for the food-contact surface of articles intended for use in producing, manufacturing, packing, processing, preparing, treating, packaging, transporting, or holding food (for use only as polymerization cross-linking agent in side seam cements for containers intended for use in contact with food (only of the identified types), subject to the conditions prescribed in 21 CFR part 175.300; and (3) as a component of the uncoated or coated food-contact surface of paper and paperboard intended for use in producing, manufacturing, packaging, processing, preparing, treating, packing, transporting, or holding dry food of the type identified in 21 CFR part 176.170 (c).

Reference:

1. *Angewandte Chemie*, Int. ed. in English, 14, 94, 1975.

POLYMETHYL METHACRYLATE

Structural Formula.

$[\sim CH_2C(CH_3)\sim]_n$
|
$COOCH_3$

M < 1000000

CAS No 9011-14-7
39404-54-1
53663-63-1
78206-73-2
86438-94-0
87210-32-0

RTECS No TR0400000

Abbreviation. PMMA.

Trade Names. Orgsteklo; Plexiglas.

Properties. Amorphous thermoplastic polymer. *Dens.* 1.19, $n^{20} = 1.492$. Insoluble in water, fats, and alcohol. PMMA obtained by radical bulk polymerization is a colorless, transparent polymer, highly pellucid for rays of visible and UV light. Softens when heated above 120°C.

Applications. Used in the production of various articles, pipes, and containers for water, sheets, packaging (for food products), dishware, kitchen utensils, medical equipment, and instruments.

Migration Data. No changes were found in the organoleptic properties of aqueous extracts of PMMA pipes. There was no migration of methyl alcohol or methyl methacrylate to water. When plasticizer (dibutyl phthalate) contents in PMMA were up to 3.0%, it migrated into water.[07]

Water properties were not adversely affected by specimens of PMMA produced with 0.005% zinc chloride added to the mix formulation.[07]

Long-term Toxicity. Rats received aqueous extracts of PMMA (18 to 20°C; 10 days) for a year. This did not cause changes in their condition, BW gain, hematology analyses, liver, and NS functions. Gross pathology examination revealed no abnormalities in the viscera.[07,028]

Reproductive Toxicity.

Observations in man. An increased incidence of sexual disorders has been reported in women working in the PMMA manufacture.

Carcinogenicity. PMMA induced malignant tumors in rodents following s/c imbedding of polymer films.[026] Nevertheless, the results of this study have been later considered inadequate.

Carcinogenicity classification. An IARC Working Group concluded that there is inadequate evidence for the carcinogenicity of PMMA in *experimental animals* and there were no adequate data available to evaluate the carcinogenicity of PMMA in *humans.*

IARC: 3

Recommendations. *Russia.* Some types of PMMA (SOL, ST-1, etc.) and some of its copolymers are authorized for use in contact with food products (pipes for milk, dishware, etc.).

Regulations. *U.S. FDA* (1998) approved the use of PMMA (1) in adhesives as a component (monomer) of articles intended for use in packaging, transporting, or holding food in accordance with the conditions prescribed in 21 CFR part 175.105; (2) in the manufacture of resinous and polymeric coatings for polyolefin films for food-contact surface of articles intended for use in producing, manufacturing, packing, processing, preparing, treating, packaging, transporting, or holding food, in accordance with the conditions prescribed in 21 CFR part 175.320; and (3) in the manufacture of semirigid and rigid acrylic and modified plastics used as articles intended for use in contact with food in accordance with the conditions prescribed in 21 CFR part 177.1010.

Reference:

Solov'yeva, M. S. et al., in *Problems of Labor Hygiene and Professional Pathology in the Workers Employed in the Chemical Industry*, Moscow, 1977, 29.

2-METHYL-2-PROPENOIC ACID, BUTYL ESTER, HOMOPOLYMER

Structural Formula.
$[\sim CH_2C(CH_3)\sim]_n$
|
$COOC_4H_9$
M = 100,000 to 800,000
CAS No 9003-63-8
RTECS No UD3399000
Abbreviation. PBMA.

Synonyms. Butyl-2-methyl-2-propenoate, homopolymer; Methacrylic acid, butyl ester, polymer; Polybutyl methacrylate.

Properties. White powder with particle size of 20 to 30 μm. The monomer has a specific odor, traces of which are retained by the polymer. *Dens.* 1.06; T_{soft} 16°C; n = 1.483. Poor solubility in water, good in organic solvents.

Applications. Used in the manufacture of lacquers and adhesives.

Acute Toxicity. A single administration of 50 g/kg BW to rabbits does not produce any physiological, biochemical, hematological or morphological changes. No mortality was observed in the treated animals.

Repeated Exposure. Administration of 5.0 g/kg BW to rats for 1.5 months caused no deaths and did not produce toxic effects.

Regulations. ***U.S. FDA*** (1998) approved the use of PBMA in the manufacture of semirigid and rigid acrylic and modified plastics used as articles intended for use in contact with food in accordance with the conditions prescribed in 21 CFR part 177.1010.

Reference:

Zolotov, P. A. and Gorbunov, V. A., Toxicity of polybutyl methacrylate and polycarbonate, *Gig. Sanit.*, 4, 92, 1984 (in Russian).

ACRYLATES and METHACRYLATES COPOLYMERS

ANATERM

Composition. Copolymers of diethylene glycol monoacrylate and ethylene glycol mono- and dimethacrylate and a copolymer of butyl methacrylate and butyl acrylate.

Applications. Used to coat shipping containers for transporting food products.

Migration Data. A study of aqueous and oil extracts revealed no changes in organoleptic properties, nor any migration of ethylene glycol, formaldehyde, or organochlorine compounds.

Long-term Toxicity. Water and oil extracts were not toxic in a 10-month toxicological experiment with rats.

Reference:

Savitsky, I. V., Kryzhanovskaya, Ye. S., Rul', Yu. V., et al., Hygienic evaluation of polymeric materials recommended for ship construction in contact with food, in *Hygiene and Toxicology of Plastics,* Coll. Sci. Proc. Kiyv Medical Institute, Kiyv, 1979, 125 (in Russian).

BUTYL ACRYLATE (22,6%), ACRYLIC ACID (2.5%), and VINYL ACETATE (25%) COPOLYMER

Composition. Contains ED-16 diane epoxy resin.

Properties. A transparent colorless liquid, soluble in ethyl acetate (50%), *B. p.* 77°C, *m. p.* 70°C.

Applications. Used as an adhesive for PVC film.

Acute Toxicity. LD_{50} is 15 g/kg BW in rats, and 12.8 g/kg BW in guinea pigs.

Short-term Toxicity. Administration caused changes in the liver, spleen, kidneys, lungs, and stomach. K_{acc} = 3.3.

Reference:

Mel'nikova, V. V., Toxicity of butyl acrylate, acrylic acid and vinyl acetate copolymer, *Gig. Sanit.*, 9, 92, 1987 (in Russian).

VINYLIDENE CHLORIDE and ACRYLATES COPOLYMERS

Structural Formula.

$[\sim CH_2CCl_2\sim]_m\ [\sim CH_2CH\sim]_n$

|

COOR

Abbreviations. VDCM (vinylidene chloride), MMA (methyl methacrylate).

Trade Name. Diophane 109-D.

Properties. A milky-white aqueous emulsion with a persistent characteristic odor.

Applications. Used in the manufacture of coatings and containers.

Migration Data. A study of aqueous extracts revealed no changes in organoleptic properties and no migration of harmful low-molecular-mass compounds.

Repeated Exposure. Administration of 10% aqueous solution of D. emulsion had no effect on general condition and behavior of rats, nor on BW gain or hematology parameters. Histologically and morphologically there were no changes in the viscera.[1]

Long-term Toxicity. Administration of a 50% solution of sugar in distilled water, of pastry in distilled water, and of vegetable oil, stored in packing made of D. caused no changes in rats over the entire period of observation (6 months). Peripheral blood pattern, antitoxic liver function, and the histological structure of the visceral organs were not different from the controls.[2]

Regulations. ***U.S. FDA*** (1998) approved the use of VDCM + MMA (1) in the manufacture of resinous and polymeric coatings for polyolefin films for food-contact surface of articles intended for use in producing, manufacturing, packing, processing, preparing, treating, packaging, transporting, or holding food, in accordance with the conditions prescribed in 21 CFR part 175.300; and (2) in the manufacture of resinous and polymeric coatings for polyolefin films for food-contact surface of articles intended for use in producing, manufacturing, packing, processing, preparing, treating, packaging, transporting, or holding food, in accordance with the conditions prescribed in 21 CFR part 175.320.

References:

1. Dukhan, D. S., Bezborod'ko, V. D., and Rogovaya, A. B., Sanitary hygienic evaluation of diofane films intended for food packaging, in *Hygiene Aspects of the Use of Polymer Materials and of Goods Made of Them,* L. I. Medved', Ed., Kiev, All-Union Research Institute of Hygiene and Toxicology of Pesticides, Polymers and Plastic Materials, 1969, 53 (in Russian).
2. Rakhmanina, N. A., and Vitvitskaya, B. R., in *Hygienic Problems in Connection with Big Chemistry Development*, Proc. Sci. Conf., Moscow, 1964, 27 (in Russian).

ACRYLONITRILE and METHYL ACRYLATE COPOLYMER

Structural Formula.

$[\sim CH_2CH\sim]_m\ [\sim CH_2C(CH_3)\sim]_n$

| |

CN C_6H_6

Properties. White powder; a 15% solution in dimethyl phthalate should be light yellow.

Applications. Used in the production of nitron fibers.

Acute Toxicity. Low toxic. Administration of 0.5 to 10.0 g/kg BW as a 25% suspension in sunflower oil was not lethal to mice.

Repeated Exposure. When investigated by Lim method, exhibited weak cumulative properties.

Reference:

Shumskaya, N. I. and Mel'nikova, V. V., *Plast. Massy*, 6, 49, 1979 (in Russian).

BUTYL ACRYLATE, STYRENE, METHYL METHACRYLATE, and AMYL ACRYLATE COPOLYMER (40:10:50:0.5)

Trade Name. Inkar-27A.

Properties. White powder insoluble in water and fats, slightly soluble in media with *pH* similar to that in biological media.

Acute Toxicity. LD_{50} was not achieved.

Repeated Exposure. K_{acc} was 0.7 for rats at the threshold level.

Reference:

Shpak, L. I., Study on toxicity of copolymer Inkar 27A, *Gig. Sanit.*, 10, 90, 1982 (in Russian).

α-METHYLSTYRENE, ACRYLONITRILE, and METHYL METHACRYLATE COPOLYMER

Structural Formula.

$$[\sim CH_2CH\sim]_m[\sim CH_2C(CH_3)\sim]_n[\sim CH_2C(CH_3)H\sim]_p$$

with pendant groups CN, C_6H_6, $COOCH_3$ respectively.

Trade Name. Copolymer AMN.

Applications. Used for manufacture of components of water pumps and meters.

Migration Data. Monomers did not migrate into water under modeling conditions corresponding to a water flow rate of 0.3 liter/cm^2.

Reference:

Sheftel', V. O., *Hygiene Aspects of the Use of Polymeric Materials in the Water Supply*, Thesis diss., All-Union Research Institute of Hygiene and Toxicology of Pesticides, Polymers and Plastic Materials, Kiyv, 1977, 76 (in Russian).

ACRYLONITRILE, METHYL ACRYLATE and BUTADIENE COPOLYMER

Structural Formula.

$$[\sim CH_2CH\sim]_m[\sim CH_2CH\sim]_n[\sim CH_2CH{=}CHCH_2\sim]_p$$

with pendant groups CN, $OCOCH_3$ respectively.

CAS No 24968-79-4

Migration Data. Wharton and Levinskas investigated the release of CN^- from a copolymer of acrylonitrile (71%) with methyl acrylate and butadiene into various simulant media. Within 7 days at 75°C, 0.065 mg CN^-/l passed into cola, and 0.04 mg/l into citric acid. Migration of CN^- into 3.0% acetic acid reached 0.075 mg/l (3 days).

Reference:

Wharton, F. D. and Levinskas, G. T., *Chem. Ind.*, 11, 470, 1976.

POLYACRYLAMIDE

Structural Formula.

$$[\sim CH_2CH\sim]_n$$

with pendant group $OC\text{-}NH_2$.

M = 1000000

CAS No 9003-05-8

RTECS No AB3700000

Abbreviation. PAA.

Synonyms. Acrylamide, polymers; 2-Propenamide, homopolymer.

Properties. Solid white product without odor. T_{soft} 180/165°C. Soluble in water; dissociates with formation of a high-molecular ions and is thus a polyelectrolyte. Insoluble in alcohol. Threshold of odor change is 175 mg/l, that of taste change is 10 mg/l.[1]

Applications. Used in the production of adhesives. Food additive. Flocculent for clarification of surface waters used in drinking water supply.

Acute Toxicity. In mice, LD_{50} of different grades of PAA with an anionic effect are: 9.6 g/kg BW for lime PAA, 3.25 g/kg BW for sulfate PAA, 2.5 g/kg BW for ammonium PAA, and 10 g/kg BW for radiation PAA. For aminated PAA with a cationic effect, LD_{50} in mice are as follows: 6.0 g/kg BW for K-4, 7.0 g/kg BW for K-6, and 6.5 g/kg BW for KF-4. LD_{50} of PAA derivatives with an anionic effect are as follows: 8.0 g/kg BW for "Komety" and 9.0 g/kg BW for "Metasa". In case of a product with sodium sulfate impurity, LD_{50} was 4.3 g/kg BW in mice; with AMF grade product, it was 2.5 g/kg BW. Mice survive a dose of 9.0 g/kg BW.[2] Differences in toxicity were due to the presence of impurities and to different monomer contents. According to other data, LD_{50} is 12.95 g/kg BW in mice and 11.25 g/kg BW in rabbits. Poisoning caused convulsions or effects on seizure threshold, and induced ataxia and dispnea.

Repeated Exposure. Administration of 436 mg/kg BW to mice and rats over a period of 1.5 months was not lethal to the animals, but it decreased their BW.[3]

Short-term Toxicity. Rats received a total dose of 250 mg/kg BW over a period of 2.5 months. There were changes in the activity of transferase and cholinesterase in the blood and liver, as well as a change in the ratio of protein fractions of the blood serum. *Hb* level in the blood and STI were reduced.[3]

Long-term Toxicity. Rats and rabbits received 0.01 to 25 mg/kg BW for 10 months. None of the indices was adversely affected. Dose of 3.0 mg PAA/kg BW given during 5 months did not produce any effect; only 100 mg/kg BW dose caused some behavioral changes in rats.[2] Separan-10 and Separan-30 grades of PAA, added to the feed of rats and dogs (1.0% for rats and 5.0 to 6.0% for dogs), produced no toxic effect. The amount of residual monomer in the specimens did not exceed 0.05% (IARC 19-133).

Reproductive Toxicity.

Gonadotoxicity. *I/g* administration of PAA (M = 360,000 to 740,000) affected the structure and functioning of the reproductive organs of males and has an adverse effect on the offspring.

Chemobiokinetics. PAA is retained for a long time in the brain and liver tissues and can be found there even after 2 years.[4]

Regulations. ***U.S. FDA*** (1998) approved the use of PAA (1) in adhesives as a component (monomer) of articles intended for use in packaging, transporting, or holding food in accordance with the conditions prescribed in 21 CFR part 175.105; (2) as a component of the uncoated or coated food-contact surface of paper and paperboard intended for use in producing, manufacturing, packaging, processing, preparing, treating, packing, transporting, or holding dry food of the type identified in 21 CFR part 176.170 (c); (3) as in the manufacture of semirigid and rigid acrylic and modified plastics used as articles intended for use in contact with food in accordance with the conditions prescribed in 21 CFR part 177.1010; and (4) as a film former in the imprinting of soft-shell gelatin capsules in accordance with the conditions prescribed in 21 CFR part 172.255.

Standards. ***Russia***. MPC: 2.0 mg/l.

References:

1. Grushko, Ya. M., *Harmful Organic Compounds in the Industrial Wastes*, Khimiya, Leningrad, 1976, 149 (in Russian).
2. Rakhmanina, N. A. and Vitvitskaya, B. R., in *Hygienic Problems in Connection with Big Chemistry Development*, Proc. Sci. Conf., Moscow, 1964, 27 (in Russian).
3. Perova, N. M., *Toxicol Study of Vinylcaprolactam and Acrylamide Polymers, Proposed for Use in the Medicine, and Their Hygiene Regulations*, Author's abstract of thesis, 1st Moscow Medical Institute, Moscow, 1977, 19 (in Russian).
4. Glubin, P. A., *Lyogkaya promyshlennost'*, 1, 12, 1975 (in Russian).

POLYACRYLONITRILE

Molecular Formula. $[\sim C_3H_3N\sim]_n$
M = 40,000 to 70,000
CAS No 37243-36-0
25014-41-9

RTECS No AT6977900

Abbreviation. PAN.

Synonyms and **Trade Names.** Acrylonitrile-cellulose copolymer; Acrylonitrile, polymers; Cellulose-polyacrylonitrile copolymer; Cellulose, polymer with 2-propenenitrile; Fiber A; Orlon; Polinac; 2-Propenenitrile, homopolymer.

Properties. White linear polymer. Microstructure depends on polymerization. parameters. T_{soft} (with breakdown at 220 to 230°C); *dens.* 1.14 to 1.15. Insoluble in alcohol.

Applications. Used to produce nitron fibers.

Acute Toxicity. A dose of 3.0 g/kg BW had no effect on animals. LD exceeded 3.0 g/kg BW in rats. Poisoring caused changes in liver and adrenal weights.[1]

Repeated Exposure revealed cumulative properties. Causes adiposal dystrophy in the liver and changes in the thyroid gland. Damage to the kidneys is manifested by albuminuria, proliferation of glomerulus and renal epithelial cells, and loss of permeability.[1]

Long-term Toxicity. In a 6-month rat study, oral exposure caused fatty liver degeneration, changes in kidneys, ureter, and bladder (acute renal failure and tubular necrosis), and thyroid hypofunction.[1]

Chemobiokinetics. Hydrolyzed PAN grafted cellulose with ^{14}C-label was administered orally to rats. Examination of respired CO_2, urine, various tissues, and feces indicated that labeled product was not metabolized but was excreted in feces.[2]

Regulations. ***U.S. FDA*** (1998) approved the use of PAN in the manufacture of semirigid and rigid acrylic and modified plastics used as articles intended for use in contact with food in accordance with the conditions prescribed in 21 CFR part 177.1010.

References:

1. Lomonova, G. V., in *Toxicology of High-Molecular-Mass Compounds*, Proc. Sci. Conf., Moscow-Leningrad, 1961, 45 (in Russian).
2. Lai, C. W., Born, G. S., Kessler, W. V., Christian, J. E., Adams, J. W. Jr., et al., Synthesis of ^{14}C-labeled hydrolyzed polyacrylonitrile grafted cellulose and its metabolic fate, *Int. J. Appl. Radiat. Isot.*, 29, 593, 1978.

POLYAMIDES

Abbreviation. PA

Composition. A flexible packaging material made from aminoacids or dimerized vegetable oils.

Properties. White or light yellow, solid, transparent polymers. Insoluble in alcohol; resistant to fats and mineral oils, fungi, and bacteria. Water resistant at 100°C, PA are hydrolyzed at 150°C and under pressure. Alkaline solutions break PA down more actively than water. Polyamides are good oxygen and water barriers, but they have low melting points.

Applications. Used in manufacturing of fibers and films, parts of machinery and equipment in the dairy and meat industry, coatings, filters, food containers, water pipes, medical instruments, etc. P. are approved for use as food contact coating on materials not to exceed room temperatures, as a component of paperboard, sealing gaskets for food containers, and as film coatings. Certain P. are approved for higher temperature applications.

Migration of monomers and other components of PA into liquid media gives them a slightly bitter taste. Migration of low-molecular-mass compounds of caprolactam (LMC) from textile materials with different percentage of polyamide silk (PAS) was studied. Summed up contents of LMC of caprolactam were 5.0 to 7.0 mg/l in the first extracts. Monomer contents in the first extracts reached 1.15 mg/l, decreasing in the following ones to complete disappearance. At the same time dimers, trimers, etc. oligomers contents of caprolactam progressively increased. The presence of triethanolamine was established in the extracts.[1]

Carcinogenicity. Nylon induced malignant tumors in rodents following *s/c* imbedding polymer films.[026] Nevertheless, the results of this study were later considered inadequate.

Regulations. ***U.S. FDA*** (1998) approved the use of PA (1) in adhesives as a component (monomer) of articles intended for use in packaging, transporting, or holding food in accordance with the conditions prescribed in 21 CFR part 175.105; (2) in the manufacture of resinous and polymeric coatings for the food-contact surface of articles intended for use in producing, manufacturing, packing, processing, preparing, treating, packaging, transporting, or holding food for use only as a polymerization cross-linking agent in side seam cements for containers intended for use in contact with food (only of the identified types), in accordance with the conditions prescribed in 21 CFR part 175.300; and (3) in the manufacture of resinous and polymeric coatings for polyolefin films for food-contact surface of articles intended for use in producing, manufacturing, packing, processing, preparing, treating, packaging, transporting, or holding food, in accordance with the conditions prescribed in 21 CFR part 175.320.

Reference:

1. Iordanova, I. and Lolova, D., Migration of chemical substances out of polyamide textile materials, *Problems of Hygiene*, 6, 119, 1981 (in Bulgarian).

POLY-ε-CAPROLACTAM

Molecular Formula. $[\sim C_6H_{11}NO\sim]_n$
M = 10,000 to 35,000
CAS No 25038-54-4
9012-24-2
RTECS No TQ9800000

Synonyms and **Trade Names.** Caproamide polymer; Caprolactam oligomer; ε-Caprolactam polymer; Caprolon B; Capron; Chemlon; Danamid; Durethan; Hexahydro-2*H*-azepin-2-one homopolymer; Nylon-6; Polyamide 6; Polycaproamide; Poly(iminocarbonyl pentamethylene); Tarlon; Ultramid; Widlon.

Composition. Obtained by polymerizing ε-caprolactam in presence of alkaline catalysts with *N-acetyl-ε-caprolactam* added as a co-catalyst or with metallic *Na* and CO_2.

Properties. Polymer of ε-caprolactam. White horny material. T_{soft} 210°C; *m. p.* 225°C; *dens.* 1.13. Water absorption 8.0 to 12%.

Applications. Used in the manufacture of water taps, piping, lids, gauze, capron fibers, and films.

Migration of the monomer into water ranged up to dozens mg/l. 0.3 and 3.0% lactic acid extracts (20°C, 5 days) acquired an unpleasant odor and a slightly bitter taste. Migration of nitrogen-containing substances, including caprolacatm, was established.[028] Caprolon had little effect on water odor and taste. A certain reduction in transparency of aqueous extracts was observed at 37°C. Migration of monomer to water after 3-day contact reached 6.0 to 8.0 mg/l.[1] Volatile compounds were released from Nylon "microwave and roasting bags" (MRB) at cooking temperatures, and non-volatile compounds were extracted with methanol and/or water. Following compounds are defined in MRBs: cyclopentanone - 31.7 mg/bag, 2-cyclopentyl cyclopentanone - 17.4 mg/bag, hexadecane - 2.6 μg/bag, heptadecane - 3.2 μg/bag, octodecane - 3.0 μg/bag, ε-capro- lactam - 5.0 to 35.5 mg/bag. The same non-volatile compounds (monomers and cyclic oligomers) except Nylon-6 heptamer and octamer were found to migrate into olive oil at 175°C for 1 hour. A total of 19.2 mg/bag of non-volatile compounds migrated into olive oil.[2] HPLC was used to quantify seven *Nylon-6 cyclic* monomers and oligomers.

Migration into chicken was 7.48 μg/g or 16% of the total non-volatile compounds contained in the MRB material. GC/MS in the selected ion mode was used for quantification. An average of 14.0 μg/bag migrated, being 0.08% of the total 2-cyclopentyl cyclopentanone present in MRB. Authors concluded that the transference of MRB components into roast chicken can be considered not to present a hazard.[3]

Acute Toxicity. LD_{50} is 3.2 g/kg BW in rats and 1.9 g/kg BW in mice.[4]

Long-term Toxicity. No toxic effect was found in rats.[5]

Regulations. ***U.S. FDA*** (1998) approved the use of Nylon-6 to produce articles intended for use in processing, handling, and packaging food, subject to the provisions of 21 CFR part 177.1500.

References:

1. Sheftel', V. O. and Sova, R. Ye., Use of the method of mathematical design of experiment in sanitary chemical research of epoxy coatings, *Gig Sanit.*, 10, 66, 1974 (in Russian).
2. Soto-Valdez, H., Gramshaw, J. W., and Vandenburg, H. J., Determination of potential migrants present in Nylon 'microwave and roasting bags' and migration into olive oil, *Food Addit. Contam.*, 14, 309, 1997.

3. Gramshaw, J. W. and Soto-Valdez, H., Migration from polyamide 'microwave and roasting bags' into roast chicken, *Food Addit. Contam.*, 15, 329, 1998.
4. Savilov, E. D. and Yanygina, L. F., Sanitary protection of reservoirs in connection with the extraction and enrichment of rare metals, *Gig Sanit.*, 3, 99, 1977 (in Russian).
5. Bornmann, G. und Loeser, A., *Arzneimittlel-Forsch.*, 9, 9, 1959.

ADIPIC ACID, compound with 1,6-HEXANEDIAMINE (1:1)

Molecular Formula. $[\sim NH(CH_2)_6NHCO(CH_2)_4CO\sim]_n$
M = 15,000 to 25,000
CAS No 3323-53-3
RTECS No AV1940000

Synonym and Trade Names. Anid; Polyhexamethylenediamine adipate; Hexanedioic acid, compound with 1,6-hexanediamine (1:1); Nylon-66; Polyamide P-66; Polyhexamethylenedipamide.

Composition. Product of polycondensation of adipic acid and hexamethylenediamine (through AG-salt).

Properties. Nylon-66 has a very high melting point (250 to 260°C) but is difficult to heat-seal. Heat distortion occurs at 205°C; maximum continuous service temperature is 80°C. *dens.* 1.14. Water absorption is 8.0%.

Applications. Used in manufacturing parts of equipment in the food industry.

Migration of low-molecular-mass compounds to water amounts to dozens mg/l.[1]

A weak aromatic odor was noted in 0.3 and 3.0% lactic acid extracts (20°C, 5 days). Migration of a small quantity of nitrogen-containing compounds was established. Migration of hexamethylenediamine was not found.[028]

Migration from Nylon "microwave and roasting bags" (MRB) was investigated: volatile compounds were released at cooking temperatures and non-volatile compounds were extracted with methanol and/or water. The following compounds were defined in MRBs: cyclopentanone, 31.7 mg/bag; 2-cyclopentyl cyclopentanone, 17.4 mg/bag; hexadecane, 2.6 mg/bag; heptadecane, 3.2 μg/bag; octodecane, 3.0 μg/bag; ε-caprolactam, 5.0 to 35.5 mg/bag.[2] Non-volatile compounds (monomers and cyclic oligomers) except Nylon-6 heptamer and octomer were found to migrate into olive oil at 175°C for 1 hour. A total of 19.2 mg of non-volatile compounds/bag migrated into olive oil.[2]

Amount of residual oligomers were determined in *Nylon* food packaging. Migration to a food-simulating liquid (oil) was measured during an oven cooking condition. Total amount of *Nylon-6/66* oligomers that migrated from an oven-baking bag to oil after heating for 30 min at 176°C was 15.5 mg/g (ppm) or 11.9 mg/cm^2, which represented 43% of the total amount of oligomers present in the packaging materials.[3]

Migration of non-volatile and volatile compounds from "microwave and roasting bags" (MRB), made of Nylon-66, into chicken meat, skin, and juices during roasting (200°C; 2 hours) in a conventional oven was determined. HPLC was used to quantify seven *Nylon-66* cyclic monomers and oligomers. Migration into chicken was 7.48 μg/g or 16% of the total non-volatile compounds contained in the MRB material. GC/MS in the selected ion mode was used for quantification. An average of 14.0 μg/bag migrated, being 0.08% of the total 2-cyclopentyl cyclopentanone present in MRB. Authors concluded that the transference of MRB components into roast chicken can be considered not to present a hazard.[4]

Long-term Toxicity. Rats received aqueous extracts of P-66 over a 14-month period. There were no changes in their condition, and health of two generations of ten offspring was unaffected.[5]

Regulations. ***U.S. FDA*** (1998) approved the use of Nylon-66 to produce articles intended for use in processing, handling, and packaging food, subject to the provisions of 21 CFR part 177.1500.

Recommendations. ***Russia***. Approved for use in food industry but not for foods that included alcohol exceeding 20%.[6]

References:

1. Morozova, E. V., *Study of Technical-grade Linear Polymers Interaction with Milk*, Author's abstract of thesis, Moscow, 1972, 23 (in Russian).

2. Soto-Valdez, H., Gramshaw, J. W., and Vandenburg, H. J., Determination of potential migrants present in Nylon 'microwave and roasting bags' and migration into olive oil, *Food Addit. Contam.*, 14, 309, 1997.
3. Begley, T. H., Gay, M. L., and Hollifield, H. C., Determination of migrants in and migration from nylon food packaging, *Food Addit. Contam.*, 12, 671, 1995.
4. Gramshaw, J. W. and Soto-Valdez, H., Migration from polyamide microwave and roasting bags' into roast chicken, *Food Addit. Contam.*, 15, 329, 1998.
5. Babayev, D. A., in *Sci. Proc. Azerb. Research Institute Virusol. Microbiol. Hyg.*, Baku, 1976, 196 (in Russian).
6. Babayev, D. A., Hygienic study of polyamide resins intended for use in the food industry and foodstuffs engineering, *Voprosy Pitania*, 1, 54, 1981 (in Russian).

NYLON-6/66

Molecular Formula. $[\sim C_6H_{16}N_2.C_6H_{11}NO.C_6H_{10}O_4\sim]_n$

CAS No 24993-04-2

Composition. Hexanedioic acid, polymer with hexahydro-2*H*-azepin-2-one and 1,6-hexanediamine.

Applications. Used in the production of food-contact articles.

Regulations. ***U.S. FDA*** (1998) approved the use of Nylon-6/66 in the production of articles intended for use in processing, handling, and packaging food, subject to the provisions of 21 CFR part 177.1500.

POLY-ω-ENANTHAMIDE

Structural Formula. $[\sim NH(CH_2)_6CO\sim]_n$

Composition. Polycondensation product of ω-aminoenanthic acid.

Trade Names. Polyamide-7; Enant; Nylon-7.

Properties. Crystalline powder. *M. p.* 223°C; *dens.* 1.13.

Applications. Used in the manufacture of fibers, etc.

Migration Data. There were neither significant migration of low-molecular-mass compounds nor changes in the organoleptic properties of extracts. Odor appeared only at high temperatures.

Reference:

Morozova, E. V., *Investigation of Technical-grade Linear Polymers Interaction with Milk*, Author's abstract of thesis, Moscow, 1972, 23 (in Russian).

SEBACIC ACID, compound with 1,6-HEXANEDIAMINE (1:1)

Structural Formula. $[\sim NH(CH_2)_6NHCO(CH_2)_8CO\sim]_n$

M = 318.52

CAS No 6422-99-7

RTECS No VS1450000

Synonyms and **Trade Names.** Decanedioic acid, compound with 1,6-hexanediamine (1:1); Hexamethylenediamine sebacate; Nylon-610; Nylon-610 salt; Polyamide P-610; Polyhexamethylene sebacinamide.

Composition. Product of polycondensation of sebacic acid and hexamethylenediamine (through SG-*salt*).

Applications. Used in the production of food-contact articles and in manufacturing various devices to be in contact with water.

Properties. White, odorless, solid, horny, crystalline polymer. *M. p.* 213 to 220°C., *dens.* 1.09 to 1.11.

Migration Data. A study of aqueous extracts (20 to 60°C; 26 days; 500 g/10 liters water) revealed no odor and traces of hexamethylenediamine (up to 0.01 mg/l).[1]

Acute Toxicity. LD_{50} is 10 g/kg BW in rats and 2.85 g/kg BW in mice.[029]

Short-term Toxicity. When rats received aqueous extracts of P. over a 4-month period, there were no changes in their condition, hematological and enzyme indices, or in histology of the viscera.[1]

Regulations. ***U.S. FDA*** (1998) approved the use of Nylon-610 to produce articles intended for use in processing, handling, and packaging food, subject to the provisions of 21 CFR part 177.1500.

Russia. Approved for use in food industry without limitations.[2]

References:

1. Tsapko, V. V. et al., Hygienic evaluation of polyamide materials used in water supply, *Gig. Sanit.*, 1, 101, 1967 (in Russian).
2. Babayev, D. A., Hygienic study of polyamide resins intended for use in the food industry and foodstuffs engineering, *Voprosy Pitania*, 1, 54, 1981 (in Russian).

NYLON-66/610 RESIN

Molecular Formula. $[\sim C_{10}H_{18}O_4.C_6H_{16}N_2.C_6H_{11}NO.C_6H_{10}O_4\sim]_n$

CAS No 25191-90-6

Synonym. Decanedioic acid, polymer with hexahydro-2*H*-azepin-2-one, 1,6-hexanediamine and hexanedioic acid.

Composition. Articles of Nylon- 66/610 are manufactured by the condensation of equal weight mixtures of Nylon 66 salts and Nylon 610 salts.

Applications. Used in the production of food-contact articles.

Regulations. ***U.S. FDA*** (1998) approved the use of Nylon-66/610 in the production of articles intended for use in processing, handling, and packaging food, subject to the provisions of 21 CFR part 177.1500.

POLY-ω-UNDECANAMID

Structural Formula. $[\sim NH(CH_2)_{10}CO\sim]_n$

CAS No 25587-80-8

Trade Names. Polyamide-11; Nylon-11.

Composition. Product of polycondensation of 11-aminoundecanoic acid.

Properties. White, odorless, horny, crystalline material. *Dens.* 1.10; *M. p.* 185°C. Unlike many other aliphatic polyamides, it has low water absorption (up to 1.6%). Insoluble in alcohol.

Applications. Used in the production of food-contact articles, fibers, films, various molded articles, including surgical instruments.

Short-term Toxicity. 5.0 to 10% powder (99.8% P.) added to the feed of Wistar rats and young dogs for 3 months produced no toxic effect (IARC 19-72).

Regulations. ***U.S. FDA*** (1998) approved the use of Nylon-11 to produce articles intended for use in processing, handling, and packaging food, subject to the provisions of 21 CFR part 177.1500.

POLY-ω-DODECANAMIDE

Structural Formula. $[\sim NH(CH_2)CO\sim]_n$

M = 15,000 to 35,000

Trade Names. Polyamide-12; Polyamide P-12; Nylon-12.

Composition. Polymer of ω-aminododecanoic acid lactam.

Properties. Odorless, horny material, transparent in a thin film, color ranging from white to cream. *M. p.* 178 to 180°C; *Dens.* 1.02. Low moisture and water absorption (up to 1.5%). Does not change color and odor of liquids in contact.

Applications. Used in the form of plastics, films, and fibers.

Long-term Toxicity. Rats received aqueous and oil extracts (1.0 cm^{-1}; 20° and 80°C; 3 days) over 12 months with no changes in general condition, functions of the liver, spleen, or gonads. Histological examination revealed no changes.

Reproductive Toxicity. No toxic effect on reproductive function, embryos, or offspring development were reported.

Regulations. ***U.S. FDA*** (1998) approved the use of Nylon-12 to produce articles intended for use in processing, handling, and packaging food, subject to the provisions of 21 CFR part 177.1500.

Recommendations. ***Russia.*** Authorized for packing fatty products, sauces, and pickles.

Reference:

Karplyuk, I. A. and Volkova, N. A., Hygienic-toxicological study of polyamide film P-12 intended for use in the food industry, *Voprosy Pitania*, 1, 63, 1977 (in Russian).

NYLON- 6/12

Molecular Formula. $C_{12}H_{22}O_4.C_6H_{16}N_2$

CAS No 13188-60-8

25191-04-2

Composition. Nylon-6/12 resins are manufactured by the copolymerization of a 1 to 1 ratio by weight of ε-caprolactam and ω-laurolactam.

Applications. Used in the production of food-contact articles.

Regulations. ***U.S. FDA*** (1998). Nylon-6/12 resins with residual ε-caprolactam not to exceed 0.5% by weight and residual ω-laurolactam not to exceed 0.1% by weight is approved for use only as specified in 21 CFR part 177.1500; and (2) for use with nonalcoholic foods at the temperature not to exceed 100°C (212°F) subject to the provisions of 21 CFR part 177.1400.

NYLON-46

Composition. A compound is manufactured by condensation of 1,4-butanediamine and adipic acid.

Applications. Used in the production of food-contact articles.

Regulations. ***U.S. FDA*** (1998) approved the use of Nylon-46 for use only in food-contact mem- brane filters intended for repeated use. The finished membrane filter is intended to contact beverages containing no more than 13% alcohol, under conditions of use E, F, and G listed in Table 2 of 21 CFR part 176.170 (c).

NYLON PA 6-3T

CAS No 26246-77-5

Composition. A compound is manufactured by condensation of 50 mol% 1,4-benzenedicarboxylic acid, dimethyl ester and 50 of an equimolar of 2,2,4-trimethyl-1,6-hexanediamine and 2,4,4-trimethyl-1,6-hexanediamine.

Applications. Used in the production of food-contact articles.

Regulations. ***U.S. FDA*** (1998) approved the use of Nylon PA 6-3T for repeated-use (excluding bottles) in contact with food of type VIA and VIB described in Table 1 of 21 CFR part 176.170 (c) and subject to the provisions of 21 CFR part 177.1500.

POLY-*m*-PHENYLENEISOPHTHALAMIDE

Structural Formula. [~$NHC_6H_4NHOCC_6H_4CO$~]

M = 20,000 to 70,000

Abbreviation. F.

Trade Name. Fenylone.

Composition. Product of polycondensation of *m*-phenylenediamine with isophthalic acid dichloroanhydride (Fenylone P).

Properties. Linear aromatic heat-resistant polyamide. White amorphous substance, crystallized when heated to 340 to 360°C. *M. p.* 430°C. Chemically stable. Fenylone S2 is a copolymer of *m*-phenylene-diamine isophthalic and terephthalic acids.

Applications. Used as membranes for concentrating virus-containing liquids in the production of vaccines. *F. S2* is used, along with other materials, in membranes in desalination units. F. is used in structures exposed up to 200°C.

Migration Data. A study of aqueous F. S2 extracts revealed migration of *N,N'*-dimethylacetamide and phthalic acid. In extracts of membranes (1 cm^{-1}; 40°C; 1 day) there were no changes in *pH* and no migration of formaldehyde, oxidizing substances, hydrogen peroxide, or performic acid. Chromatography analyses revealed no specific chemical components in extracts.[1]

Long-term Toxicity. Rats receiving aqueous F. S2 extracts for 7 to 8 months developed eosinophilia and leukocytosis. Acid resistance and mobility time of spermatozoa were reduced; activity of glcoso-6-phosphate- dehydrogenase in the gonads and blood serum was increased.[2]

Reproductive Toxicity.

Gonadotoxicity. There was a reduction in the spermatogenesis index, and a sharp increase in the number of sperm ducts with a desquamated epithelium. Acid resistance and mobility time of spermatozoa were reduced; activity of glucoso-6-phosphatedehydrogenase in the gonads and blood serum was increased.[2]

Embryotoxicity. A toxic effect on embryos was observed.[2]

Mutagenicity. When rats received aqueous extracts obtained in aggravated conditions, the number of CA in the bone marrow cells increased, and mitotic activity of cells decreased.[2]

References:

1. Guricheva, Z. G., Rotenberg, V. V., Kalinin, B. Yu., Krasheniuk, A. I., and Bogomolov, N. V., Sanitary-chemical evaluation of polysulfonamide membranes, *Gig. Sanit.,* 3, 57, 1983 (in Russian).
2. Solokhina, T. A. et al., in *Sci. Proc. A. N. Sysin Research General Commune Hygiene Institute,* Moscow, Issue No 72, 1977, 5 (in Russian).

POLYIMIDES

Structural Formula.

```
  CO  CO
  / \ / \
N    R    N-R'
  \ / \ /
  CO  CO
```

where R and R' are aromatic or other thermally stable groups.

CAS No P-88-0905
P-91-1257

Properties. Materials based on polyimides possess good chemical, mechanical, and radiation resistance at temperatures ranging from 200 to 300°C.

Applications. Polyimides are used to produce insulating lacquers, fabrics, films, and molding materials and in the manufacture of containers and parts of different designation.

Toxicity. Some of PI reactive monomers may be present in the final product. Oxydianiline, a component used to make polyimide, is carcinogenic in rodents.

PAK-1 LACQUER

Composition and **Migration.** The product of interaction of aromatic tetracarboxylic acid with diamine in *N,N'*-dimethylformamide which migrates from it into water.

Acute Toxicity. LD_{50} is 7.9 g/kg BW in rats, and 7.0 g/kg BW in mice. Gross pathology examination revealed changes in the parenchymatous organs.

Repeated Exposure Cumulative properties are pronounced: $K_{acc} = 2.2$.

Reference:

Kesel'man, M. L., in *Proc. Rostov-na-Donu Medical Institute*, Rostov-na-Donu, Issue No 29, 1977, 26 (in Russian).

POLYIMIDE molding powders

Synonyms. Polyamidoimide powders.

Acute Toxicity. Administration of 15 g/kg BW to rats and 10 g/kg BW to mice was not lethal.

Repeated Exposure Administration of 3.0 g/kg BW to rats was not lethal, but functional changes were observed.

Allergenic Effect was not observed.

Reference:

Kesel'man, M. L., *Hygiene Assessment of Polyimide and Polyamidoimide Molding Powders, Biochemical Indices, and Metabolic Mechanisms of Toxic Effects According to Experimental Data*, Author's abstract of thesis, F. F. Erisman Research Sanitary Hygiene Institute, Moscow, 1980, 21 (in Russian).

POLYETHERS and POLYESTERS

POLYETHERS

POLYALKYLENE MALEINATES

Structural Formula.

$$^{H}[\sim(\sim\underset{\|}{\overset{}{}}OROC\underset{\substack{\|\\O}}{C}H{=}CH\underset{\substack{\|\\O}}{C}\sim)_x\text{-}(\sim OR O\underset{\substack{\|\\O}}{C}R'\underset{\substack{\|\\O}}{C}\sim)_y\sim]_n{}^{OH}$$

where R is a glycol radical, R' is a radical of modifying acid, x = 1 to 5, y = 0 to 5, and n = 1 to 10

Synonym. Polyether maleinates.

Composition. Saturated polyesters of maleic acid and glycols, modified by phthalic acids, cured (cross-linked) by different monomers, for example, styrene. PN-1 is the product of interaction of diethylene glycol with maleic and phthalic anhydrides and styrene, PNM-2 is that of diethylene glycol with the given anhydrides only. In both cases, $R = CH_2\text{-}O\text{-}CH_2$ and $R'{=}O\text{-}C_6H_4$.

Applications. A binder in glass-fiber reinforced plastics and molding materials.

Migration Data. Glass-fiber reinforced plastic pipes manufactured on the basis of PN-1 (isopropylbenzene hydroperoxide as initiator, cobalt naphthenate as accelerator) produce significant deterioration of water organoleptic properties (odor up to rating 3 after 3 hours of contact). After 24-hour contact at room temperature, 0.39 mg/l of styrene was found in the water; after 4-day contact cobalt concentration was 2.0 to 4.0 mg/l.[1]

Kravchenko and Kharchenko have found harmful substances that migrated to water after contact with fiber-glass reinforced plastics (PN-1 and PNM-2 type) reaching concentrations 5 to 10 times higher than permissible level.[2]

References:

1. Sheftel', V. O. and Prusova, G. O., Hygienic evaluation of some polymeric prosthetic appliances, *Vrachebnoye Delo*, 2, 95, 1966 (in Russian).
2. Kravchenko, T. I. and Kharchenko, T. F., in Papers of the 4th All-Union Sci. Techn. Conf. *Developm. Use Polymeric Materials in Food Industry*, Minsk 1980, Moscow, 1980, 77 (in Russian).

POLYMETHYLENE OXIDE

Structural Formula. $[\sim CH_2O\sim]_n$

M = 30,000 to 40,000

Abbreviation. PMO.

Synonyms. Polyformaldehyde; Polyoxymethylene.

Composition. Polyethers of linear structure, the product of polymerization of formaldehyde or its cyclic trimer (trioxane).

Properties. White, crystalline substances with a formaldehyde odor. *Dens.* 1.45^{20}. *M. p.* 173 to 180°C. Chemically stable.

Applications. Used to produce fibers and to replace non-ferrous metals.

Migration of monomer into water was not found. Significant changes were observed in organoleptic properties of water extracts: odor of up to rating 4; migration of low-molecular-mass organic compounds was noted.[1]

Repeated Exposure. Rats received 5.0 g PMO/kg BW in their diet over a period of 14 days. The treatment caused no toxic effect.[2]

Regulations. ***U.S. FDA*** (1998) approved the use of PMO copolymers and homopolymers in food-contact articles in accordance with the conditions prescribed in 21 CFR parts 177.2470 and 177.2480, respectively.

References:

1. *Hygiene Problems in the Production and Use of Polymeric Materials*, Kiyv, 1969, 152 (in Russian).
2. Kokhno, Yu. A. et al., *Polyformaldehyd*, Tekhnika, Kiyv, 1964 (in Russian).

POLYETHYLENE GLYCOLS

Structural Formula. $HO[\sim CH_2CH_2O\sim]_nH$
M = 500,000 to 10,000000
CAS No 25322-68-3
RTECS No TQ3500000
Abbreviation. PEGs.

Synonyms and **Trade Names.** Carbowaxes; 1,2-Ethanediol, homopolymers; Ethylene glycol, homopolymers; Ethylene oxide, polymers; Ethylene polyoxide; Lutrol; Oxyethylene polymer; Polyethylene oxide polymers; Poly(ethyleneoxide)s; Polyhdroxyethylene; Polyoxyethylene; Polyoxyethylenediol; Poly(oxyethylene) glycols; Poly(vinyl oxide).

Properties. Clear, viscous liquids (M = 200 to 600) or waxiform (M = 1000 to 6000) products. Solubility in water is inversely proportional to molecular mass. Liquid PEG are colorless, almost odorless, and miscible with water. Waxiform PEGs (carbowaxes) are soluble in water (50 to 73%). At a concentration of 1.0 g/l, they do not alter the color, odor, or taste of water.

Applications. Used in food and food packaging. Used as plasticizers, solvents, water-soluble lubricants for rubber molds; wetting or softening agents, antistatics in the production of urethane rubber, components of detergents, etc. In medicine, PEGs are used in cosmetics, ointments, suppositories, in ophthalmic solutions and sustained-released oral pharmaceutical applications.

Migration of up to 50 mg PEGs/kg food was observed in chocolates, boiled sweets, toffees, cakes, and meat pies that were wrapped in regenerated cellulose films containing various mixtures of glycol softeners. Analysis of the glycols were performed by capillary GC with flame ionization detection after trimethylsilyl derivatization.[1]

Acute Toxicity. PEGs are generally considered to be inert and possess a low order of toxicity in animals and humans. Administration of 0.5 g high-molecular-mass PEG/kg BW in the form of an aqueous solution caused no visible signs of intoxication. No mortality occurred. Histological examination revealed small areas of round-cell infiltration, expanded vessels in the kidneys, and a plethoric spleen.[2] A dose of 2.5 g/kg BW of PEG with M = 2,000000 and 7,000000 was not lethal to rats or mice. In the last case, the acute effect threshold was 0.5 g/kg BW. In the same doses, PEG synthesized on an organocalcium compound was not lethal. The acute effect threshold was not established.[3]

The mean lethal doses of PEG are presented in the table.[03]

Table. Mean lethal doses of PEGs (g/kg BW).

Polyethylene Glycols	*Mice*	*Rats*	*Guinea pigs*	*Rabbits*
200 (insoluble)	33.9-38.3	28.9	16.9	14.1-19.9
300 (insoluble)	31.0	27.5-31.1	19.6-21.1	17.3-21.1
400 (insoluble)	28.9-35.6	12.9-30.2	15.7-21.3	22.3
600 (insoluble)	35.6-47.0	38.1	28.3	18.9
1000 (50% aqu. sol.)	>50	42.0	22.5-41.0	>50
4000 (50% aqu. sol.)	>50	>50	46.4-50.9	>50
6000 (50% aqu. sol.)	>50	>50	>50	>50
9000 (50% aqu. sol.)	>50	>50	>50	>50

Repeated Exposure. Cumulative properties were observed only in PEG obtained on an organo-aluminum catalyst. Administration of 50 mg/kg BW over a period of two months decreased STI. An increase in BW gain was noted. A dose of 250 mg/kg BW caused a reduction in erythrocyte count and in peroxidase activity of the blood.

Cumulative properties of other PEGs were not pronounced: doses of 20 and 50 mg/kg BW did not kill animals. $K_{acc} = 5$.

Short-term Toxicity. F344 rats received up to 30,000 ppm NF-10 grade Polyox (M = about 100,000) in the diet for 13 weeks. The treatment caused slight increases in food consumption, BW, and BW gain. A dose-related increase in liver weight was not associated with any histopathology.[4]

Long-term Toxicity. Rats and mice received 3.1 g PEG/kg BW in aqueous solutions (M ~ 5,000000) for 12 months. There were no manifestations of toxic action.[2]

A 2-year dietary dosage up to 20,000 ppm NF-10 grade Polyox produced no toxic effects in treated rats. There was no neoplastic and non-neoplastic pathology observed in this study.[4]

Allergenic Effect. Sensitizing properties were not pronounced.

Reproductive Toxicity. In a 2-year feeding study, oral and parenteral administration of PEG caused no effect on reproduction.[5]

See also ***Diethylene glycol.***

Gonadotoxicity. In a 90-day study, administration of the dose of 230 mg PEG-75/kg BW induced testicular tubule degeneration and oligospermia.[6]

Rats received a dose of 3.0 mg/kg BW. The treatment caused no effect on the gonads or embryos, nor did it affect reproduction or the development of offspring.[2]

Rats were injected *i/p* with doses equivalent to 0.5 LD_{50} of PEG-400 and PEG-1500 three times during the gestation period. No increase in pre-implantation mortality was observed.

Embryotoxicity. Mild signs of embryotoxicity together with generally retarded development were reported..

No ***teratogenic*** effect was noted.[7]

Mutagenicity.

In vitro genotoxicity. Polyox showed neither genotoxic activity in *Salmonella typhimurium* and *E. coli* assays, nor did it cause CA in Chinese hamster ovary cells.[7]

In vivo cytogenetics. Polyox was found to be negative in mouse bone marrow micronucleus test.[4]

Carcinogenicity. Exposure to PEG resulted in vaginal tumors and a weak tumor initiator effect in mice.[5,8,9]

Chemobiokinetics. Polyox is not absorbed in the GI tract. It showed high recoveries. Essentially all radiolabel was excreted in the feces.[4] After *i/v* administration, PEG are excreted mainly unchanged.

Regulations.

EU (1990). PEGs are available in the List of authorized monomers and other starting substances which shall be used for the manufacture of plastic materials and articles intended to come into contact with foodstuffs (Section A).

U.S. FDA (1998) regulates PEGs for use (1) in adhesives used as components of articles intended for use in packaging, transporting, or holding food (PEG 200-6000) in accordance with the conditions prescribed in 21 CFR part 175.105; (2) in resinous and polymeric coatings used as the food-contact surfaces of articles intended for use in producing, manufacturing, packing, processing, preparing, treating, packaging, transporting, or holding food in accordance with the conditions prescribed in 21 CFR part 175.300; (3) as a component of the uncoated or coated food-contact surface of paper and paperboard intended for use in producing, manufacturing, packing, transporting, or holding dry, aqueous and fatty food in accordance with the conditions prescribed in 21 CFR parts 176.170 and 176.180; (4) as a component of defoaming agents that may be safely used as components of articles intended for use in contact with food in accordance with the conditions prescribed in 21 CFR part 176.200; (5) PEG (M = 200 to 4600) of as defoaming agent used in the manufacture of paper and paperboard intended for use in packaging, transporting, or holding food in accordance with the conditions prescribed in 21 CFR part 176.210; (6) in the manufacture of cross-linked polyester resins for repeated use in articles or components of articles coming in the contact with food

(PEG-6000) in accordance with the conditions prescribed in 21 CFR part 177.2420; (7) as a substance employed in the production of or added to textiles and textile fibers intended for use in contact with food (PEG 400 to 6000) in accordance with the conditions prescribed in 21 CFR part 177.2800. PEG may be safely used (8) if the additive is an addition polymer of ethylene oxide and water with a mean molecular mass of 200 to 9,500 and if PEG contains no more than 0.2% by weight of the ethylene and diethylene glycols and if its molecular mass is 350 or higher and no more than 0.5% by weight of the total of ethylene and diethylene glycols and if its mean molecular mass is below 350.

PEG monolaurate (PEG-400) containing not more than 0.1% by weight of the ethylene and/or ethylene glycol may be used at a level not to exceed 0.3% by weight of the twine as a finish on twine to be used for tying meat provided the twine fibers are produced from nylon resins.

PEG dilaurate (PEG-200) may be used as a component of the uncoated or coated food-contact surface of paper and paperboard intended for use in producing, manufacturing, packaging, processing, preparing, treating, packing, transporting, or holding dry foods in accordance with the conditions prescribed in 21 CFR part 176.180. PEG alginate is listed for use as a component of the uncoated or coated food-contact surface of paper and paperboard intended for use in producing, manufacturing, packing, transporting, or holding aqueous and fatty food in accordance with the conditions prescribed in 21 CFR part 176.170.

Great Britain (1998). PEGs are authorized without time limit for use in the production of polymeric materials and articles in contact with food or drink or intended for such contact.

Recommendations.

Joint FAO/WHO Expert Committee on Food Additives. ADI: 10 mg/kg BW.

Russia. ADI: 100 mg/kg BW.[2]

Standards. *Russia.* PML: n/m. MAC depends on M (organolept., foam):[3]

0.125 mg/l for PEG with M = 2,000000;

0.1 mg/l for PEG with M = 3,000000;

0.2 mg/l for PEG with M = 5,000000.

References:

1. Castle, L., Cloke, H. R., Crews, C., and Gilbert, J., The migration of propylene glycol, mono-, di-, and triethylene glycols from regenerated cellulose film into food, *Z. Lebensmit. Unters. Forsch.*, 187, 463, 1988.
2. Cherkasova, T. E., Larionov, A. G., Chanyshev, R. O., and Cherkanov, S. P., General toxic action of polyoxyethylene, *Gig. Sanit.*, 12, 86, 87 (in Russian).
3. Larionov, A. G., Cherkasova, T. Ye., and Strusevich, Ye. A., Comparative toxicological evaluation of polyoxyethylene made with different catalysts, in *Hygiene and Toxicology of High-Molecular-Mass Compounds and of the Chemical Raw Material Used for Their Synthesis*, Theses 6th All-Union Conf., B. Yu. Kalinin, Ed., Leningrad, Khimiya, 1979, 80 (in Russian).
4. Ballantyne, B., Leung, H.-W., Hermansky, S. J., and Frantz, S. W., Subchronic, chronic, pharmacokinetic and genotoxicity studies with Polyox water soluble resin, Abstract P1A43, in Abstracts VIII Int. Congr. Toxicol., *Toxicol. Lett.*, Suppl. 1/95, 46, 1998.
5. Smyth, H. F., Carpenter, C. P., and Shaffer, C. B., The toxicity of high molecular weight polyethylene glycols: chronic oral and parenteral administration, *J. Am. Pharmacol. Assoc.*, Sci. Ed., 36, 157, 1947.
6. Smyth, H. F., Carpenter, C. P., Shaffer, C. B., Seaton, J., and Fisher, L., Some pharmacological properties of polyethyleneglycols of high molecular weight ("Carbowax") compounds, *J. Ind. Hyg. Toxicol.*, 24, 281, 1942.
7. Kartashov, V. F. and Belous, A. M., in All-Union Institute Sci.-Technical Information, Dep. No 737-84 (in Russian).
8. Boyland, E., Charles, R. T., and Gowing, N. F. C., The induction of tumors in mice by intravaginal application of chemical compounds, *Br. J. Cancer*, 15, 252, 1961.
9. Field, W. E. H. and Roe, F. J. C., Tumor promotion in the forestomach epithelium of mice by oral administration of citrus oils, *J. Natl. Canc. Inst.*, 35, 771, 1965.

POLYETHYLENE GLYCOL-115

Structural Formula. $HO[\sim CH_2CH_2O\sim]_{115}H$

M = 5080

Abbreviation. PEG-115.

Properties. PEG-115 is a clear, odorless liquid in the molten state; in the solidified form, it is a flaky or paraffin-like white mass. Soluble in many organic solvents. 1.0 kg PEG-115 dissolves in 1.0 liter of water in a few days with periodic stirring, i.e., 50 to 73% aqueous solutions are formed. It gives water an aromatic odor and a specific taste (bitter-sweetish). Odor threshold is 2.3 g/l, taste threshold is 2.35 g/l. The practical limit is 5.8 g/l for odor and 6.98 g/l for taste. Chlorination of the aqueous solutions does not give rise to extraneous odors.

Applications. See ***Polyethylene glycols.***

Acute Toxicity. LD_{50} was not attained in rats and mice (three weeks of observations).

Repeated Exposure failed to reveal toxic effects on administration of 1/10 to 1/250 of the theoretically calculated LD_{50} (40 g/kg BW) for 2 months. The highest dose affected hematology analyses but caused no morphology changes. Administration of 4.0 g/kg BW revealed functional accumulation, evident in impairment of the oxidation-reduction processes and of liver function, and in strain on the body defense mechanisms.

Allergenic effect was not observed in the studies on guinea pigs.

Long-term Toxicity. The NOAEL of 80 mg/kg BW was identified in rats.

Regulations. ***EU*** (1990) and ***Great Britain*** (1998). See ***Polyethylene glycols.***

Standards. ***Russia*** (1995). PML: 0.25 mg/l (organolept., turbidity).

Reference:

Manenko, A. K., A technique for quantitative evaluation of the combined effects produced on the body by two or more hazardous chemical factors, *Gig. Sanit.,* 10, 68, 1982 (in Russian).

POLYETHYLENE GLYCOL 200

Structural Formula. $HO[\sim CH_2CH_2O\sim]_nH$

CAS No 25322-68-3

RTECS No TQ3600000

Abbreviation. PEG-200.

Properties. Odorless liquid. Soluble in water (forming transparent solutions) and in many organic solvents.

Applications. See ***Polyethylene glycols.***

Acute Toxicity. LD_{50} is 28.9 g/kg BW in rats, 34 to 38 g/kg BW in mice, 14 to 20 g/kg BW in rabbits, and 17 g/kg BW in guinea pigs.[03]

Short-term Toxicity. There were no signs of toxicity in rats given 4.0% PEG-200 in their diet for 90 days.[1]

Reproductive Toxicity. In a 2-year feeding study, oral and parenteral administration of PEG caused no effect on reproduction.[2]

Teratogenicity. Oral administration of PEG-200 caused teratogenic effect in mice but not in rats. These included malformations of the skull, paws, and thoracic skeleton.[3]

According to more later data, PEG-200 is reported to cause severe facial malformations in mice. On the other hand, no abnormalities were noted in the rat embryos from mothers given toxic doses. PEG-200 induced devel opmental abnormalities in cultured 10-day old rat embryos.[4]

See also ***Diethylene glycol.***

Chemobiokinetics. Administration of PEG-200 to monkeys (2.0 to 4.0 ml/kg BW) and rats (2.5 to 5.0 ml/kg BW) for 13 weeks led to the deposition of a small quantity of oxalates in the lumen of the proximal tubules of the renal cortex (in monkeys only). No other morphological, biochemical, or hematological changes were found.[5]

Regulations.

U.S. FDA (1998) approved the use of PEG-200 (1) in food in accordance with the conditions prescribed in 21 CFR part 172.820; (2) as a defoaming agent used in the manufacture of paper and paperboard intended for use in packaging, transporting, or holding food in accordance with conditions prescribed in 21 CFR part 176.210; and (3) as a component of articles intended for use in packaging, transporting, or holding food in accordance with the conditions prescribed in 21 CFR parts 175.105 and 178.3750.

EU (1992) and ***Great Britain*** (1998). See ***Polyethylene glycols.***

References:

1. Smyth, H. F., Seaton, J., and Fisher, L., Single dose toxicity of some glycols and derivatives, *J. Ind. Hyg. Toxicol.,* 23, 259, 1941; *J. Am. Pharmacol. Assoc.,* Sci. Ed., 36, 335, 1947; *Ibid,* 44, 27, 1955.
2. Smyth, H. F., Carpenter, C. P., and Shaffer, C. B., The toxicity of high molecular weight polyethylene glycols: chronic oral and parenteral administration, *J. Am. Pharmacol. Assoc.,* Sci. Ed., 36, 157, 1947.
3. Vannier, B., Bremaud, R., Benicourt, M., and Julien, P., Teratogenic effects of polyethylene glycol 200 in the mouse but not in the rat, *Teratology,* 40, 32, 1989.
4. Spezia, F., Lozes, P., Fournex, R., and Vannier, B., Quantitative structure-activity relationship and correlation between the *in vivo* and *in vitro* teratogenic activity of phenothizine derivatives, Abstract, *Teratol ogy,* 46, 28A, 1992.
5. Prentice, D. E. and Majeed, S. K., Oral toxicity of polyethylene glycol (PEG 200) in monkeys and rats, *Toxicol. Lett.,* 2, 119, 1978.

POLYETHYLENE GLYCOL 238

Structural Formula. $HO[\sim CH_2CH_2O\sim]_nH$
CAS No 25322-68-3
RTECS No TQ3610000
Abbreviation. PEG-238.

Properties. Odorless liquid. Soluble in water (forming transparent solutions) and in many organic solvents.

Applications. See ***Polyethylene glycols.***

Acute Toxicity. LD_{50} *i/p* is 9.0 g/kg BW in rats. Poisoning damaged kidneys, ureter, and bladder.

Regulations. *U.S. FDA* (1998), ***EU*** (1990), and ***Great Britain*** (1998). See ***Polyethylene glycols.***

Reference:

Arzneimittel-Forsch., 3, 451, 1953.

POLYETHYLENE GLYCOL 282

Structural Formula. $HO[\sim CH_2CH_2O\sim]_nH$
CAS No 25322-68-3
RTECS No TQ3620000
Abbreviation. PEG-282.

Properties. Odorless liquid. Soluble in water (forming transparent solutions) and in many organic solvents.

Applications. See ***Polyethylene glycols.***

Acute Toxicity. LD_{50} is 31.6 g/kg BW in rats. Poisoning damaged kidneys, ureter, and bladder.

Regulations. *U.S. FDA* (1998), ***EU*** (1992), and ***Great Britain*** (1998). See ***Polyethylene glycols.***

Reference:

Arzneimittel-Forsch., 3, 451, 1953.

POLYETHYLENE GLYCOL 300

Structural Formula. $HO[\sim CH_2CH_2O\sim]_nH$
CAS No 25322-68-3
RTECS No TQ3630000
Abbreviation. PEG-300.

Trade Name. Polyglycol E-300.

Properties. Odorless liquid. Soluble in water (forming transparent solutions) and in many organic solvents.

Applications. See ***Polyethylene glycols.***

Acute Toxicity. LD_{50} is 31 g/kg BW in mice,[03] 17.3 g/kg BW in rabbits, 19.6 g/kg BW in guinea pigs, and 27.5 g/kg BW in rats. Poisoning damaged kidneys, ureter, and bladder.[1]

Short-term Toxicity. There were no signs of toxicity in rats given 4.0% PEG-300 in their diet for 90 days.[2]

Regulations. ***U.S. FDA*** (1998), ***EU*** (1990), and ***Great Britain*** (1998). See ***Polyethylene glycols.***

References:

1. *Arzneimittel-Forsch.*, 3, 451, 1953.
2. Smyth, H. F., Seaton, J., and Fisher, L., Single dose toxicity of some glycols and derivatives, *J. Ind. Hyg. Toxicol.*, 23, 259, 1941; *J. Am. Pharmacol. Assoc.*, Sci. Ed., 36, 335, 1947; *Ibid*, 44, 27, 1955.

POLYETHYLENE GLYCOL 350

Structural Formula. $HO[\sim CH_2CH_2O\sim]_nH$
CAS No 25322-68-3
RTECS No TQ3650000
Abbreviation. PEG-350.

Properties. Odorless liquid. Soluble in water (forming transparent solutions) and in many organic solvents.

Applications. See ***Polyethylene glycols.***

Acute Toxicity. LD_{50} is 22 g/kg BW in rats.[1]

Short-term Toxicity. There were no signs of toxicity in rats given 4.0% PEG-350 in their diet for 90 days.[2]

Regulations. ***U.S. FDA*** (1998), ***EU*** (1990), and ***Great Britain*** (1998). See ***Polyethylene glycols.***

References:

1. Deichman, W. B., *Toxicology of Drugs and Chemicals*, Academic Press, Inc., New York, 1969, 747.
2. Smyth, H. F., Seaton, J., and Fisher, L., Single dose toxicity of some glycols and derivatives, *J. Ind. Hyg. Toxicol.*, 23, 259, 1941; *J. Am. Pharmacol. Assoc.*, Sci. Ed., 36, 335, 1947; *Ibid*, 44, 27, 1955.

POLYETHYLENE GLYCOL 400

Structural Formula. $HO[\sim CH_2CH_2O\sim]_nH$
M = 380 to 420
CAS No 25322-68-3
RTECS No TQ3675000
Abbreviation. PEG-400.

Trade Names. Lutrol 9; Plyglycol E-400 NF.

Properties. Viscous liquid with a slight characteristic odor. *Dens.* 1.110-1.140. Soluble in water (forming transparent solutions) and in many organic solvents.

Applications. See ***Polyethylene glycols.***

Acute Toxicity. LD_{50} is 30.2 g/kg BW in rats,[1] 28.9 g/kg BW in mice,[2] 26.8 g/kg BW in rabbits, and 15.7 g/kg BW in guinea pigs. Poisoning damaged kidneys, ureter, and bladder.[3]

Short-term Toxicity. F344 rats were administered 1.0 to 5.6 g PEG-400/kg BW by gavage for 13 weeks. There were no changes in mortality or hematology or clinical chemistry measurements. Histological examination revealed no microscopic changes in the kidneys or urinary bladder. A slight, reversible renal toxicity may have resulted in male rats treated with 2.5 ml/kg BW and rats of both sexes treated with 5.0 ml PEG-400/kg BW. This conclusion was made due to observed increase in concentration of protein and bilirubin, urinary vascular cell findings and *N*-acetyl-β-*D*-glucosaminidase activity.[4]

Chemobiokinetics.

Observations in man. Following oral ingestion of 5 to 10 g dose, 40 to 50% PEG-400 were recovered in the urine.

Regulations.

U.S. FDA (1998) approved the use of PEG-200 (1) in food in accordance with the conditions prescribed in 21 CFR part 172.820; (2) as a defoaming agent used in the manufacture of paper and paperboard intended for use in packaging, transporting, or holding food in accordance with conditions prescribed in 21 CFR part 176.210; (3) in the manufacture of textiles and textile fibers used as articles or components of articles intended for use in producing, manufacturing, packing, processing, preparing, treating, packaging, transporting, or holding food, subject to the provisions prescribed in 21 CFR part 177.2800; and (4) as a component of articles intended for use in packaging, transporting, or holding food in accordance with the conditions prescribed in 21 CFR parts 175.105 and 178.3750.

EU (1990) and ***Great Britain*** (1998). See ***Polyethylene glycols.***

References:

1. *J. Am. Pharmacol. Assoc.*, Sci. Ed., 39, 349, 1950.
2. *Proc. Eur. Soc. Toxicol.*, 17, 351, 1976.
3. *Arzneimittel-Forsch.*, 3, 451, 1953.
4. Hermansky, S. J., Neptun, D. A., Loughran, K. A., and Leung, H. W., Effects of polyethylene glycol 400 (PEG 400) following 13 weeks of gavage treatment in Fischer-344 rats, *Food Chem. Toxicol.*, 33, 139, 1995.

POLYETHYLENE GLYCOL 425

Structural Formula. HO[~CH_2CH_2O~]$_n$H
CAS No 25322-68-3
RTECS No TQ3700000
Abbreviation. PEG-425.

Trade Name. Polyglycol P-425.

Properties. Odorless liquid. Soluble in water (forming transparent solutions) and in many organic solvents.

Applications. See ***Polyethylene glycols.***

Acute Toxicity. LD_{50} is 0.6 g/kg BW in rats (Dow Chemical Co. Reports MSD-94).

Short-term Toxicity. There were no signs of toxicity in rats given 4.0% PEG-425 in their diet for 90 days.[1]

Regulations. ***U.S. FDA*** (1998), ***EU*** (1990), and ***Great Britain*** (1998). See ***Polyethylene glycols 400.***

Reference:

1. Smyth, H. F., Seaton, J., and Fisher, L., Single dose toxicity of some glycols and derivatives, *J. Ind. Hyg. Toxicol.*, 23, 259, 1941; *J. Am. Pharmacol. Assoc.*, Sci. Ed., 36, 335, 1947; *Ibid*, 44, 27, 1955.

POLYETHYLENE GLYCOL 600

Structural Formula. HO[~CH_2CH_2O~]$_n$H
M = 570 to 630
CAS No 25322-68-3
RTECS No TQ3800000
Abbreviation. PEG-600.

Trade Name. Polyglycol P-600.

Properties. Viscous, slightly hygroscopic liquid with a characteristic odor. *Dens.*^{25} 1.126. Soluble in water (forming transparent solutions) and in many organic solvents.

Applications. See ***Polyethylene glycols.***

Acute Toxicity. LD_{50} is 38 g/kg BW in rats, 35-47 g/kg BW in mice, 19 g/kg BW in rabbits, and 28 g/kg BW in guinea pigs.[03]

Short-term Toxicity. There were no signs of toxicity in rats given 4.0% PEG-600 in their diet for 90 days.

Regulations. ***U.S. FDA*** (1998), ***EU*** (1990), and ***Great Britain*** (1998). See ***Polyethylene glycols 400.***

Reference:

Smyth, H. F., Seaton, J., and Fisher, L., Single dose toxicity of some glycols and derivatives, *J. Ind. Hyg. Toxicol.*, 23, 259, 1941; *J. Am. Pharmacol. Assoc.*, Sci. Ed., 36, 335, 1947; *Ibid*, 44, 27, 1955

POLYETHYLENE GLYCOL 810

Structural Formula. HO[~CH_2CH_2O~]$_n$H
CAS No 25322-68-3
RTECS No TQ3850000
Abbreviation. PEG-810.

Trade Name. Polyglycol P-810.

Properties. Soluble in water (forming transparent solutions) and in many organic solvents.

Applications. See ***Polyethylene glycols.***

Acute Toxicity. LD_{50} *i/v* is 13 g/kg BW in rats.[1]

Short-term Toxicity. There were no signs of toxicity in rats given 4.0% PEG-810 in their diet for 90 days.[2]

Regulations. ***U.S. FDA*** (1998), ***EU*** (1990), and ***Great Britain*** (1998). See ***Polyethylene glycols 400.***

References:

1. *Arzneimittel-Forsch.*, 3, 451, 1953.
2. Smyth, H. F., Seaton, J., and Fisher, L., Single dose toxicity of some glycols and derivatives, *J. Ind. Hyg. Toxicol.*, 23, 259, 1941; *J. Am. Pharmacol. Assoc.*, Sci. Ed., 36, 335, 1947; *Ibid*, 44, 27, 1955

POLYETHYLENE GLYCOL 1000

Structural Formula. HO[~CH_2CH_2O~]$_n$H
CAS No 25322-68-3
RTECS No TQ4025000
Abbreviation. PEG-1000.

Trade Names. Carbowax 1000; Macrogol 1000; Polyglycol P-1000.

Properties. Waxiform solid. Soluble in water (forming transparent solutions) and in many organic solvents.

Applications. See ***Polyethylene glycols.***

Acute Toxicity. PEG-1000 exhibited mild toxicity. LD_{50} is 42 g/kg BW in rats[03] and 22.5 g/kg BW in guinea pigs.[1]

Short-term Toxicity. There were no signs of toxicity in rats given 4.0% PEG-1000 in their diet for 90 days.[2]

Chemobiokinetics. After *i/v* administration, PEG-1000 is excreted unchanged. It is poorly absorbed in rat intestine.

Regulations. ***U.S. FDA*** (1998), ***EU*** (1990), and ***Great Britain*** (1998). See ***Polyethylene glycols 400.***

References:

1. *J. Am. Pharmacol. Assoc.*, Sci. Ed., 39, 349, 1950.
2. Smyth, H. F., Seaton, J., and Fisher, L., Single dose toxicity of some glycols and derivatives, *J. Ind. Hyg. Toxicol.*, 23, 259, 1941; *J. Am. Pharmacol. Assoc.*, Sci. Ed., 36, 335, 1947; *Ibid*, 44, 27, 1955

POLYETHYLENE GLYCOL 1200

Structural Formula. HO[~CH_2CH_2O~]$_n$H
CAS No 25322-68-3
RTECS No TQ4026000
Abbreviation. PEG-1200.

Trade Name. Polyglycol P-1200.

Properties. Waxiform solid. Soluble in water (forming transparent solutions) and in many organic solvents.

Applications. See ***Polyethylene glycols.***

Acute Toxicity. LD_{50} is 1.05 g/kg BW in rats (Dow Chemical Co. Reports MSD-95).

Short-term Toxicity. There were no signs of toxicity in rats given 4.0% PEG-1200 in their diet for 90 days.[1]

Regulations. ***U.S. FDA*** (1998), ***EU*** (1990), and ***Great Britain*** (1998). See ***Polyethylene glycols 400.***

Reference:

1. Smyth, H. F., Seaton, J., and Fisher, L., Single dose toxicity of some glycols and derivatives, *J. Ind. Hyg. Toxicol.*, 23, 259, 1941; *J. Am. Pharmacol. Assoc.*, Sci. Ed., 36, 335, 1947; *Ibid*, 44, 27, 1955

POLYETHYLENE GLYCOL 1250

Structural Formula. HO[~CH_2CH_2O~]$_n$H
CAS No 25322-68-3
RTECS No TQ4027000
Abbreviation. PEG-1250.

Trade Name. Polyglycol P-1250.

Properties. Waxiform solid. Soluble in water (forming transparent solutions) and in many organic solvents.

Applications. See ***Polyethylene glycols.***

Acute Toxicity. LD_{50} is 51.3 g/kg BW in rats. Poisoning damaged kidneys, ureter, and bladder.[1]

Short-term Toxicity. There were no signs of toxicity in rats given 4.0% PEG-1250 in their diet for 90 days.[2]

Regulations. ***U.S. FDA*** (1998), ***EU*** (1990), and ***Great Britain*** (1998). See ***Polyethylene glycols 400.***

References:

1. *Arzneimittel-Forsch.*, 3, 451, 1953.
2. Smyth, H. F., Seaton, J., and Fisher, L., Single dose toxicity of some glycols and derivatives, *J. Ind. Hyg. Toxicol.*, 23, 259, 1941; *J. Am. Pharmacol. Assoc.*, Sci. Ed., 36, 335, 1947; *Ibid*, 44, 27, 1955

POLYETHYLENE GLYCOL 4000

Structural Formula. HO[~CH_2CH_2O~]$_n$H
M = 3000 to 3700
CAS No 25322-68-3
RTECS No TQ4028000
Abbreviation. PEG-4000.

Properties. White, free-flowing powder or creamy white flakes. *Dens.*^{25} 1.20 to 1.21. Soluble in water (forming transparent solutions) and in many organic solvents.

Applications. See ***Polyethylene glycols.***

Acute Toxicity. Waxiform PEG-4000 is non-toxic when administered to rats, mice and rabbits (LD_{50} exceeded 50 g/kg BW). For guinea pigs, it is 46 to 51 g/kg BW.[03]

Repeated Exposure.

Observations in man. 16 healthy male subjects received 20 g PEG-400/day for two 7-day treatment periods. Stool frequency increased significantly but no significant change in consistency was observed. Fecal output of protein increased significantly. Author concluded that no effect on colonic transit and on stool hydration was observed.[1]

Short-term Toxicity. There were no signs of toxicity in rats given 4.0% PEG-4000 in their diet for 90 days.[2] There were no signs of toxicity in the hematology analysis or gross pathology examination in dogs given 10 to 90 mg PEG-4000/kg BW for 43 to 178 days.[2,3]

Long-term Toxicity. Addition of 4.0% PEG-4000 for 2 years and of 2.0% PEG-4000 to the feed of dogs for a year appeared to be harmless.[2,3]

Chemobiokinetics. PEG-4000 is not absorbed.[011] Unchanged PEG is excreted in the urine.

Regulations.

U.S. FDA (1998) approved the use of PEG-4000 (1) in food in accordance with the conditions prescribed in 21 CFR part 172.820; (2) as a defoaming agent used in the manufacture of paper and paperboard intended for use in packaging, transporting, or holding food in accordance with conditions prescribed in 21 CFR part 176.210; and (3) in adhesives as a component of articles intended for use in packaging, transporting, or holding food in accordance with the conditions prescribed in 21 CFR part 175.105.

EU (1990) and ***Great Britain*** (1998). See ***Polyethylene glycols.***

References:

1. Hudziak, H., Bronowicki, J. P., Franck, P., Dubos-Berogin, C., and Bigard, M. A., Low-dose polyethylene glycol 4000: digestive effects. Randomized double-blind study in healthy subjects, *Gastroenterol. Clin. Biol.*, 20, 418, 1996.
2. Smyth, H. F., Seaton, J., and Fisher, L., Single dose toxicity of some glycols and derivatives, *J. Ind. Hyg. Toxicol.*, 23, 259, 1941; *J. Am. Pharmacol. Assoc.*, Sci. Ed., 36, 335, 1947; *Ibid*, 44, 27, 1955
3. Smyth, H. F., Carpenter, C. P., and Shaffer, C. B., The toxicity of high molecular weight polyethylene glycols: chronic oral and parenteral administration, *J. Am. Pharmacol. Assoc.*, Sci. Ed., 36, 157, 1947.

POLYETHYLENE GLYCOL 6000

Structural Formula. $HO[\sim CH_2CH_2O\sim]_nH$

M = 7000 to 9000

CAS No 25322-68-3

RTECS No TQ4100000

Abbreviation. PEG-6000.

Trade Names. Carbowax 6000.

Properties. Powder or creamy-white flakes. *Dens.*[25] 1.21. Soluble in water (forming transparent solutions) and in many organic solvents.

Applications. See ***Polyethylene glycols.***

Acute Toxicity. Waxiform PEG-6000 is non-toxic when administered to rats. LD_{50} is 50 g/kg BW in rats and in guinea pigs. Kidney, ureter and bladder were affected.[1]

Short-term Toxicity. There were no signs of toxicity in rats given 16% PEG-6000 in their diet for 90 days.[2]

Chemobiokinetics. PEG-6000 is not absorbed.[011]

Regulations.

U.S. FDA (1998) approved the use of PEG-4000 (1) in food in accordance with the conditions prescribed in 21 CFR part 172.820; and (2) in adhesives as a component of articles intended for use in packaging, transporting, or holding food in accordance with the conditions prescribed in 21 CFR part 175.105.

EU (1990) and ***Great Britain*** (1998). See ***Polyethylene glycols 4000.***

References:

1. Deichmann, W. B., *Toxicology of Drugs and Chemicals*, Academic Press, Inc., New York, 1969, 747.
2. Smyth, H. F., Seaton, J., and Fisher, L., Single dose toxicity of some glycols and derivatives, *J. Ind. Hyg. Toxicol.*, 23, 259, 1941; *J. Am. Pharmacol. Assoc.*, Sci. Ed, 36, 335, 1947; *Ibid*, 44, 27, 1955

POLYPROPYLENE GLYCOLS

Structural Formula. $[\sim CH_2CH_2CH_2O\sim]_n.H_2O$

CAS No 25322-69-4

Abbreviation. PPGs.

Synonyms and **Trade Names.** Actocol 51-530; α-Hydro-ω-hydroxypoly(oxypropylene); Bloat Guard; Emkapyl; Lineartop E; Methyloxirane, homopolymer; Niax PPG; Poly(propylene oxide); Propylene oxide, homopolymer.

Properties. Clear, lightly colored, slightly oily, viscous liquids.[011] Low-molecular-mass PPGs are soluble in water.

Applications. Solvent, plasticizer, antifoaming agent; used in the production of resins, adhesives, and coatings as well as in polyurethane elastomers and sealants. Rubber lubricant.

Regulations.

U.S. FDA (1998) approved the use of PPGs in adhesives as a component of articles intended for use in packaging, transporting, or holding food in accordance with the conditions prescribed in 21 CFR part 175.105.

EU (1990). PPGs are available in the *List of monomers and other starting substances, which may continue to be used pending a decision on including in Section A.*

Great Britain (1998). PPGs with molecular mass greater than 400 are authorized without time limit for use in the production of polymeric materials and articles in contact with food or drink or intended for such contact.

POLYPROPYLENE GLYCOL (40) BUTYL ETHER

Structural Formula. $[\sim CH_2CH_2CH_2O\sim]_n.C_4H_{10}O$
CAS No 9003-13-8
RTECS No TR6240000
Abbreviation. PPG-40.

Synonym and **Trade Name.** Polyoxypropylene (40) butyl ether; Ucon LB-1715.

Properties. Soluble in water (forming transparent solutions) and in many organic solvents.

Applications. See ***Polypropylene glycols.***

Acute Toxicity. LD_{50} is 34 ml/kg BW in rats.

Regulations. ***U.S. FDA*** (1998). See ***Polypropylene glycols.***

Reference:

Acute Toxicity Data, *J. Am. Coll. Toxicol.*, 12, 257, 1993.

POLYPROPYLENE GLYCOL 150

Structural Formula. $[\sim CH_2CH_2CH_2O\sim]_n.H_2O$
CAS No 25322-69-4
RTECS No TR5300000
Abbreviation. PPG-150.

Trade Name. Polyglycol P-150.

Applications. See ***Polypropylene glycols.***

Acute Toxicity. LD_{50} is 14.8 g/kg BW in rats Poisoning caused excitation and convulsions within minutes after administration. Gross pathology examination 1 to 8 days after exposure failed to reveal any changes in the viscera.

Chemobiokinetics. PPG-150 is rapidly absorbed from the GI tract.

Regulations. ***U.S. FDA*** (1998) and ***EU*** (1990). See ***Polypropylene glycols.***

Reference:

Deichman, W. B., *Toxicology of Drugs and Chemicals*, Academic Press, Inc., New York, 1969, 731.

POLYPROPYLENE GLYCOL 400

Structural Formula. $[\sim CH_2CH_2CH_2O\sim]_n$- H_2O
CAS No 25322-69-4
Abbreviation. PPG-400.

Trade Name. Polyglycol P-400.

Applications. See ***Polypropylene glycols.***

Acute Toxicity. LD_{50} *i/p* is 700 mg/kg BW in mice. Poisoning led to convulsions, excitation, muscle contrac tion or spasticity.

Chemobiokinetics. PPG-400 is rapidly absorbed from the GI tract.

Regulations. ***U.S. FDA*** (1998) and ***EU*** (1990). See ***Polypropylene glycols.***

Reference:

J. Pharmacol. Exp. Ther., 103, 293, 1951.

POLYPROPYLENE GLYCOL 425

Structural Formula. $[\sim CH_2CH_2CH_2O\sim]_n.H_2O$
CAS No 25322-69-4
RTECS No TR5600000
Abbreviation. PPG-425.

Trade Name. Polyglycol P-425.

Applications. See ***Polypropylene glycols.***

Acute Toxicity. LD_{50} is 2.41 g/kg BW in rats (Union Carbide Data Sheet, 1973). Poisoning caused excitation and convulsions within minutes after administration. Gross pathology examination 1 to 8 days after exposure failed to reveal any changes in viscera.[03]

Chemobiokinetics. PPG-425 is rapidly absorbed from the GI tract.

Regulations. See ***Polypropylene glycols.***

POLYPROPYLENE GLYCOL 750

Structural Formula. $[\sim CH_2CH_2CH_2O\sim]_n.H_2O$
CAS No 25322-69-4
RTECS No TR5775000
Abbreviation. PPG-750.

Trade Name. Polyglycol P-750.

Applications. See ***Polypropylene glycols.***

Acute Toxicity. LD_{50} *i/p* is 195 mg/kg BW. Poisoning produced convulsions or effect on seizure threshold, excitation, muscle contraction, or spasticity.

Short-term Toxicity. Rats received the diet containing 0.1% PPG-750 over a period of 100 days. A slight increase in the liver and kidney weights was reported. The treatment did not affect mortality rate, growth, or weights of other organs. Gross pathology and microscopic examination failed to reveal any changes in the viscera.[03]

Chemobiokinetics. PPG-750 is rapidly absorbed from the GI tract.

Regulations. See ***Polypropylene glycols.***

Reference:

J. Pharmacol. Exp. Ther., 103, 293, 1951.

POLYPROPYLENE GLYCOL 1000

Structural Formula. $[\sim CH_2CH_2CH_2O\sim]_n.H_2O$
CAS No 25322-69-4
RTECS No TR5800000
Abbreviation. PPG-1000.

Applications. See ***Polypropylene glycols.***

Acute Toxicity. LD_{50} is 4.2 g/kg BW in rats. Poisoning caused excitation and convulsions within minutes after administration. Gross pathology examination 1 to 8 days after exposure failed to reveal any changes in the viscera.[03,029]

Chemobiokinetics. PPG-1000 is rapidly absorbed from the GI tract.

Regulations.

U.S. FDA (1998) approved the use of PPG-1000 (1) in adhesives as a component of articles intended for use in packaging, transporting, or holding food in accordance with the conditions prescribed in 21 CFR part 175.105, and (2) as a component of the uncoated or coated food-contact surface of paper and paperboard intended for use in producing, manufacturing, packaging, processing, preparing, treating, packing, transporting, or holding aqueous and fatty foods, subject to the provisions prescribed in 21 CFR part 176.170.

EU (1992) and ***Great Britain*** (1998). See ***Polypropylene glycols.***

POLYPROPYLENE GLYCOL 1025

Molecular Formula. $[\sim CH_2CH_2CH_2O\sim]_n.H_2O$

CAS No 25322-69-4
RTECS No TR5950000
Abbreviation. PPG-1025.

Trade Name. Niax PPG.

Applications. See ***Polypropylene glycols.***

Acute Toxicity. LD_{50} is 2150 mg/kg BW in rats (Union Carbide Data Sheet, 1970). Poisoning caused excitation and convulsions within minutes after administration. Gross pathology examination 1 to 8 days after exposure failed to reveal any changes in the viscera.[03]

Chemobiokinetics. PPG-1025 is rapidly absorbed from the GI tract.

Regulations. ***U.S. FDA*** (1998), ***EU*** (1990), and ***Great Britain*** (1998). See ***Polypropylene glycol 1000.***

POLYPROPYLENE GLYCOL 1200

Structural Formula. $[\sim CH_2CH_2CH_2O\sim]_n.H_2O$
CAS No 25322-69-4
RTECS No TR6125000
Abbreviation. PPG-1200.

Applications. See ***Polypropylene glycols.***

Acute Toxicity. LD_{50} *i/p* is 113 mg/kg BW in dogs. Poisoning produced convulsions or effect on seizure threshold, excitation, muscle contraction or spasticity.

Chemobiokinetics. PPG-1200 is rapidly absorbed from the GI tract.

Regulations.

U.S. FDA (1998) approved the use of PPG-1200 (1) in adhesives as a component of articles intended for use in packaging, transporting, or holding food in accordance with the conditions prescribed in 21 CFR part 175.105; (2) as a component of the uncoated or coated food-contact surface of paper and paperboard intended for use in producing, manufacturing, packaging, processing, preparing, treating, packing, transporting, or holding aqueous and fatty foods, subject to the provisions prescribed in 21 CFR part 176.170; and (3) as a defoaming agent in the processing of food in accordance with the conditions prescribed in 21 CFR part 173.320.

EU (1990) and ***Great Britain*** (1998). See ***Polypropylene glycols.***

Reference:

J. Pharmacol. Exp. Ther., 103, 293, 1951.

POLYPROPYLENE GLYCOL 1800

Structural Formula. $[\sim CH_2CH_2CH_2O\sim]_n.H_2O$
CAS No 25322-69-4
RTECS No TR6129000
Abbreviation. PPG-1800.

Applications. See ***Polypropylene glycols.***

Acute Toxicity. LD_{50} is 7.25 g/kg BW in rats. Poisoning caused excitation and convulsions within minutes after administration. Gross pathology examination 1 to 8 days after exposure failed to reveal any changes in the viscera.[03,029]

Regulations. ***U.S. FDA*** (1998), ***EU*** (1990), and ***Great Britain*** (1998). See ***Polypropylene glycol 1200.***

POLYPROPYLENE GLYCOL 2000

Structural Formula. $[\sim CH_2CH_2CH_2O\sim]_n.H_2O$
CAS No 25322-69-4
RTECS No TR6130000
Abbreviation. PPG-2000.

Trade Names. Polyglycol PPG-2000; Polyglycol P-2000; Voranol P-2000.

Applications. See ***Polypropylene glycols.***

Acute Toxicity. LD_{50} is 10.3 g/kg BW in rats (Dow Chemical Company, Reports MSD-395).

Short-term Toxicity. Male rats received 0.1 to 1.0% PPG-2000 in the diet for 100 days. The treatment caused no ill effects such as an increase in mortality, BW gain, or hematology analyses. Pathology and histology examinations revealed no changes in the viscera. At 3.0% dietary level, animals showed a slight decrease in BW.[03]

Allergenic Effect.

Observations in man. PPG-2000 is not a skin sensitizer.[03]

Regulations. ***U.S. FDA*** (1998), ***EU*** (1990), and ***Great Britain*** (1998). See ***Polypropylene glycol 1200.***

POLYPROPYLENE GLYCOL 2025

Structural Formula. [~$CH_2CH_2CH_2O$~]$_n$.H_2O

CAS No 25322-69-4

RTECS No TR6200000

Abbreviation. PPG-2025.

Trade Name. Polyglycol P-2025.

Applications. See ***Polypropylene glycols.***

Acute Toxicity. LD_{50} is 9.76 g/kg BW in rats (Union Carbide Data Sheet, 1960).

Regulations. ***U.S. FDA*** (1998), ***EU*** (1992), and ***Great Britain*** (1998). See ***Polypropylene glycol 1200.***

POLYPROPYLENE GLYCOL 3025

Structural Formula. [~$CH_2CH_2CH_2O$~]$_n$.H_2O

CAS No 25322-69-4

RTECS No TR6210000

Abbreviation. PPG-3025.

Trade Name. Niax PPG 3025.

Applications. See ***Polypropylene glycols.***

Acute Toxicity. LD_{50} is 35.6 ml/kg BW in rats (Union Carbide Data Sheet, 1969).

Regulations.

U.S. FDA (1998) approved the use of PPG-3025 (1) in adhesives as a component of articles intended for use in packaging, transporting, or holding food in accordance with the conditions prescribed in 21 CFR part 175.105, and (2) as a component of the uncoated or coated food-contact surface of paper and paperboard intended for use in producing, manufacturing, packaging, processing, preparing, treating, packing, transporting, or holding aqueous and fatty foods, subject to the provisions prescribed in 21 CFR part 176.170.

EU (1992) and ***Great Britain*** (1998). See ***Polypropylene glycols.***

POLYPROPYLENE GLYCOL 4000

Structural Formula. [~$CH_2CH_2CH_2O$~]$_n$.H_2O

CAS No 25322-69-4

RTECS No TR6215000

Abbreviation. PPG-4000.

Trade Names. Polyglycol P-4000; Voranol P-4000.

Applications. See ***Polypropylene glycols.***

Acute Toxicity. LD_{50} exceeded 15 g/kg BW in rats (Dow Chemical Company, Reports MSD-508).

Regulations. ***U.S. FDA*** (1998), ***EU*** (1990), and ***Great Britain*** (1998). See ***Polypropylene glycol 3025.***

POLYPROPYLENE GLYCOL 4025

Structural Formula. [~$CH_2CH_2CH_2O$~]$_n$.H_2O

Abbreviation. PPG-4025.

Applications. See ***Polypropylene glycols.***

Acute Toxicity. LD_{50} is 56.6 g/kg BW in rats (Union Carbide Data Sheet, 1970).

Regulations. ***U.S. FDA*** (1998) and ***Great Britain*** (1998). See ***Polypropylene glycol 3025.***

POLY(TETRAMETHYLENE GLYCOL)

Structural Formula. $HO[\sim(CH_2)_4O\sim]_nH$
M = 1900 to 2000
CAS No 25190-06-1
RTECS No MD0916000

Synonym and **Trade Name.** α-Hydro-ω-hydroxypoly (oxy-1,4-butanediyl); Polybutylene glycol; Poly(butylene oxide); Polyfurite; Poly(oxy-1,4-butylene) glycol; Poly(oxytetramethylene); Poly(oxytetramethylene)diol; Poly(tetramethylene ether); Poly(tetramethylene ether)diol; Poly(tetramethylene ether)glycol; Poly(tetramethylene oxide).

Properties. Odorless solid. Poorly soluble in water at 20°C and in alcohol.

Applications. P. is used as a basis in the production of polyurethane foams, urethane elastomers and elasticated plastics.

Acute Toxicity. Rats and mice tolerate 10 g/kg BW without visible manifestations of the toxic effect.

Repeated Exposure failed to reveal evident cumulative properties.

Short-term Toxicity. Rats were dosed by gavage with 0.5 g/kg BW and 2.0 g/kg BW for 3 months (total doses of 27.5 g/kg BW and 110 g/kg BW, respectively). The treatment resulted in increased liver weights and in a rise of neuromuscular stimulation threshold. A reduction in diuresis and in excretion of chlorides in the urine was observed only at 2.0 g/kg BW dose level.[1]

Long-term Toxicity. The NOAEL appeared to be 200 mg/kg BW.[2]

References:

1. Stasenkova, K. P., Investigating toxicity of polyfurite, in *Toxicology of the New Industrial Chemical Substances*, N. F. Ismerov and I. V. Sanotsky, Eds., Meditsina, Moscow, Issue No 14, 1975, 146 (in Russian).
2. Shumskaya, N. I. and Stasenkova, K. P., Hygienic investigation of rubber intended for use in contact with food, *Gig. Sanit.*, 8, 28, 1973 (in Russian).

POLY-2,6-DIMETHYL-*n*-PHENYLENE OXIDE

Structural Formula

```
          CH3
          /
      __/
    /    \
[~        ---O~]n.
    \__/
       \
          CH3
```

M = 30,000 to 700,000
Abbreviation. PDMPO

Composition. Aromatic polyether. Product of dehydropolycondensation of 2,6-dimethylphenol.

Trade Name. Arylox.

Properties. White, solid-extinguishing substance. *M. p.* = 250°C. *Dens.* 1.06. Resistant to boiling water, superheated steam, and radioactive radiation. Insoluble in water.

Applications. Used in sanitary engineering and surgery as a dipping material.

Migration of the monomer, solvents (methyl alcohol), and residues of catalysts (copper ions) into water was observed. Migration of methanol from molded PDMPO reaches 12.5 mg/l, and migration of copper ions is 0.15 mg/l. No migration of diphenoquinone, pyridine, or formaldehyde into water was found from PDMPO produced by extrusion with additional vacuum treatment.[1]

Acute Toxicity. No deaths were observed in rats given 25 g/kg BW in 20% solution of glycerine with Twin-80. Manifestations of toxic action included sluggishness, adynamia, and drowsiness. Threshold toxic effect was noted when given 12 g Arylox-100/kg BW or 9.85 g Arylox-200/kg BW.[2]

Repeated Exposure. K_{acc} for Aryloxes appeared to range from 8 to 13.

Allergenic Effect was not found.

Regulations. ***U.S. FDA*** (1998) approved the use of PDMPO in contact with food in accordance with the conditions prescribed in 21 CFR part 177.2460.

References:

1. Glushkov, Yu. T., cit. in *Potential Chemical Hazards of Plastics*, Khimiya, Leningrad, 1977, 131 (in Russian).
2. Larionov, A. G., Marchenko, N. I., and Mezhel'skaya, O. G., Hygienic evaluation of polyphenylene oxide and of aryloxex based on it, *Gig. Sanit.*, 2, 83, 1988 (in Russian).

LAPROLS

Laprol 1601-2-50M

Molecular Formula. $[\sim C_3H_6O\sim]_n \cdot C_3H_8O_3$
CAS No 61036-35-9
RTECS No OE5784512

Synonyms and **Trade Name.** Glycerol-propylene oxide polymer; Polypropylene glycol, 1,2,3-propanetriyls; Voranols.

Composition. Products of the alcoholate copolymerization of ethylene and propylene oxides and various glycols.

Properties. Viscous oily liquids without odor or taste. Poorly soluble in water and oils.

Table 1. Frothing threshold of laprols.

Frothing threshold	*mg/l*
L-402	0.3
L-1502	0.1
L-2507	0.1

Applications. Used in the production of polyurethanes and other plastics.

Acute Toxicity.

Table 2. LD_{50} values of different laprols in mice.[1]

Laprols	*LD_{50}, g/kg BW*
L-402-2-160	above 10
L-5003-2B-10	10
L-3203-4-80	26
L-56	15
L-1601-2-50	10
L-1601-2-50M	5

Table 3. LD_{50} values of different laprols in rats.[2]

Laprols	*LD_{50}, g/kg BW*
L-402	23.4
L-1502	46.7
L-2507	56.7
L-503	21

Treated animals did not develop immediate symptoms of intoxication. In 10 to 14 days, the animals became apathetic and some of them died. Gross pathology examination revealed parenchymatous dystrophy of the liver and kidneys, more evident in the case of L-1601-2-50 and L-1601-2-50 M. Other organs appeared to be unaltered. Rats displayed similar susceptibility to L.[3]

Repeated Exposure to oral doses which in total significantly exceeded the LD_{50} did not have the same effect as a single administration of the whole dose. Histology examination revealed moderate changes in the parenchymatous organs.[3]

Short-term Toxicity. In an oral rat study, there was a reduction in the activity of certain enzymes, and in contents of catecholamines and ascorbic acid in the blood serum. Gross pathology examination revealed no changes in the suprarenals, testes, and pancreas, but the relative weights of the visceral organs were increased. There were changes in the endocrine system. The NOEL appeared to be 46 mg/kg BW for L-402, 23 mg/kg BW for L-503, 46 mg/kg BW for L-1502, and 8.0 mg/kg BW for L-703.[4]

Reproductive Toxicity.

Gonadotoxicity. A dose-dependent decrease in the spermatogenesis index was observed in rats given 1/10 LD_{50} of L-702.[5] Other animals were exposed to 1/100 to 1/10 LD_{50} of L-402, L-503, L-1502, and L-2502 for 48 days. The treatment adversely affected the function and morphology of male rat gonads. L-503 produced the most pronounced effect.[2]

Mutagenicity.

In vivo cytogenicity. A single administration of 1/2 LD_{50} of L-402 and L-1502 had no mutagenic effect. L-2507 caused a slight effect.[2]

Standards. *Russia* (1995). MAC and PML (organolept., foam), for L-202, L-503, L-564, and L-402-2-100: 0.3 mg/l for L-6003-2B-18, I-6003-2B-7, L-4202-2B-30, L-2402, L-2501-2-50, L-2502-2B-40, L-3003-2B-60, L-4003-2-20, L-1502-2-70, and L-2505-2-70: 0.1 mg/l; for L-702: 0.2 mg/l; for L-3003: 10 mg/l.

References:

1. Bruskova, N. I. and Shafranskaya, N. T., Toxicity of polyethyleneglycol (laprol 402-2-100), *Gig. Truda Prof. Zabol.*, 6, 55, 1985 (in Russian).
2. Zhukova, S. V. *et al.*, in *Proc. 4th Conf. Ukrainian Geneticist & Breeders,* Naukova Dumka, Kiyv, vol 5, 1981, 131 (in Russian).
3. Arkhipov, A. S., Dmitrieva, N. V., Kochetkova, T. A., et al., Toxicological investigation of polyesters, *Gig. Truda Prof. Zabol.*, 9, 21, 1974 (in Russian).
4. Zhukov, V. I. et al., in *Endocrine System and Harmful Environmental Factors*, Proc. 2nd All-Union Conf., September 21-23, 1983, Leningrad, 1983, 32 (in Russian).
5. Pereima, V. N., *Ibid.*, p. 156.

POLYESTERS

POLYCARBONATE

Structural Formula

```
          _   CH3  _
        /   \  |  /   \
H--[--O--     --C--     --O-C--]n--Cl
        \__/   |  \_ /
              CH3
```

M = 25,000 to 50,000

CAS No 25766-59-0

Abbreviation. PC.

Composition. Polyester of carbonic acid and bisphenol A.

Properties. A solid, rigid, and transparent material. It is highly break-resistant. *Dens.* 1,7 to 1.22^{20}. n^{20} = 1.56 to 1.65. Insoluble in media with a *pH* similar to biological media. Resistant to water, saline solution, upper alcohols, fats, and oils, but has poor resistance to bases and solvents. PC can be sterilized, it is also cold-resistant. Its steam and gas permeability is higher than that of polystyrene. PC will not tolerate continuous boiling in water. *M. p.* 220 to 270°C.

Maximum continuous service temperature and heat distortion occur at 135°C.

Applications. PC is used in the manufacture of rigid containers, dishware, and other articles. It is used in food and medical industry (syringes, vessels, eyepiece, and denture manufacture), and in the production of baby bottles. Recent legislation in New York allows refill (up to 100 times) of PC plastic bottles for school milk programs and half-gallon milk containers for retail use.

Migration Data. PC did not affect quality of water or other simulant media (1 cm^{-1}; 20 to 30°C; 1 to 15 days). Traces of phenol were detected in extracts at 80°C. Migration of bisphenol *A* to food from molded discs prepared from a composite of bisphenol A-derived PC resins was determined using food-simulating solvents and time and temperature conditions recommended by FDA. The study demonstrates that no detectable bisphenol A was found in the extracts obtained under FDA's most severe default testing condition. The potential dietary exposure to bisphenol A from use of PC resins was determined to be less than 0.25 ppb.[1]

Baby feeding bottles made of PC were investigated for possible release of bisphenol A following domestic practice of sterilization by alkaline hypochlorite, steam, or washing in an automatic dishwasher at 65°C with detergent. Migration of bisphenol A was not detectable in infant feed using a very sensitive LC method with fluorescence detection with a 0.03 mg/kg detection limit.[2]

Acute Toxicity. A dose of 10 g/kg BW was not lethal to rats, mice, or guinea pigs.[3]

Repeated Exposure revealed no cumulative properties.

Long-term Toxicity. Rats and mice received PC extracts in their diet over a period of 12 months. There were no manifestations of toxic action.[4]

Absence of toxic effects both during long-term feeding of PC itself and during administration of aqueous extracts of P. have been reported by other authors.[5-7]

Regulations. ***U.S. FDA*** (1998) approved the use of PC as articles or components of articles intended for use in producing, manufacturing, packaging, processing, preparing, treating, packaging, transporting, or holding food in accordance with the conditions prescribed in 21 CFR part 177.1580.

References:

1. Howe, S. R. and Borodinsky, L., Potential exposure to bisphenol A from food-contact use of polycarbonate resins, *Food Addit. Contam.*, 15, 370, 1988.
2. Mountfort, K. A, Kelly, J., Jickells, S. M., and Castle, L., Investigations into the potential degradation of polycarbonate baby bottles during sterilization with consequent release of bisphenol A., *Food. Addit. Contam.*, 14, 737, 1997.
3. Zolotov, P. A. and Gorbunov, V. A., Toxicity of polybutylmethacrylate and polycarbonate, *Gig. Sanit.*, 4, 92, 1984 (in Russian).
4. Proklina, T. L. and Shvaiko, I. I., Hygienic evaluation of polymeric materials intended for use in the food industry, *Gig. Sanit.,* 1, 111, 1978 (in Russian).
5. Gnoevaya, V. L. et al., in *Toxicology and Hygiene of Polymeric Materials Used in Food Industry*, Meditsina, Moscow, 1980, 61 (in Russian).
6. Shnell, G., *Chemistry and Physics of Polycarbonates*, Mir, Moscow, 1967 (in Russian, translation from English).
7. Bornmann, G. und Loeser, C., *Arzneimittel-Forsch.,* 5, 9, 1959.

PENTAPLAST

Structural Formula. $[\sim CH_2C(CH_2Cl)_2CH_2O\sim]_n$

M = 70,000 to 200,000

Synonym. Poly-3,3'-bis(chloromethyl)oxacyclobutane.

Properties. Colorless, odorless material. *M. p.* 185°C. *Dens.* 1.39 to 141^{20}. Contains 45% chlorine. Noted for chemical and heat resistance.

Applications. Used in the production of coatings, articles, and film of different designation, including those used in contact with water and foodstuffs.

Migration of monomer into aqueous extract was not found. There was a slight increase in oxidizability of aqueous extracts at 80°C.

Long-term Toxicity. There were no changes in general condition, BW gain, hematology analyses or liver functions of rats given aqueous extracts of P. over a period of 12 months. The treatment did not affect

STI, the contents of ascorbic acid in the adrenal glands, or autoflora of the skin. Histological examination revealed no lesions in the viscera.

Reference:

Sheftel', V. O., *Hygienic Aspects of Use of Polymeric Materials and Pesticides*, A Review, All-Union Sci. Institute of Medical Information, Moscow, 1972, 70 (in Russian).

POLYETHYLENE TEREPHTHALATE

Structural Formula

```
         __
        /  \
[~O-C--      --COO-CH2-CH2-O~]n
        \__/
```

M = 20,000 to 40,000

CAS No 29154-49-2

Abbreviation. PET.

Synonyms and **Trade Names.** Amilar; Daiya foil; Dowlex; Ethylene terephthalate polymer; Fiber V; Hostadur; Hostaphan; Lavsan; Lawsonite; Melinex; Mersilene; Nitron lavsan; Poly(oxy-1,2-ethanediyloxy-carbonyl-1,4-phenylenecarbonyl); Polyethylene glycol terephthalate; Terephtahlic acid-ethylene glycol polyester.

Composition. A polyester of terephthalic acid and ethylene glycol can be obtained by the poly= condensation of dimethyl terephthalate (q.v.) with ethylene glycol, and also terephthalic acid with ethylene glycol or ethylene oxide.

Properties. A white or light-cream material. Noted for high heat resistance and chemical stability. When melt-blown, it provides a good barrier for both flavors and hydrocarbons (fat). It is not transparent. PET is resistant to acids, bases, some solvents, and oils and fats. It is difficult to mold. The melting point of unmodified PET is below boiling. Monolayer films will hold a crease, are heat sealable, and transparent. *Dens.* 1.332^{20}. *M. p.* 255 to 265°C. $n^{25} = 1.574$. Insoluble in water.

Applications. PET is used for high-impact resistant containers. It is used for packaging of soda, mouthwash, pourable dressings, edible oils, and peanut butter. It is used for cereal box liners, soda bottles, boil-in-the-bag pouches, and microwave food trays. Modified PETs can be heated in a microwave or in a conventional oven at 180°C for 30 minutes. There has been a moderate amount of concern that additives from these trays may migrate into foods, particularly if the trays are reused in a microwave oven. PET is also used in the production of different bottles, fibers, films for food packaging, and different articles. *Lavsan* fabric is used in the dairy industry for filtering. Used in medicine for plastic vessels and for implantation.

Migration Data. A total of 19 migrants from commercial amber PET bottle wall has been identified by GC/MS analysis: the majority of compounds appeared to be intermediate reaction products or residual monomers of their dehydration and transesterification products. Fatty acids and commonly used plasticizers were also identified.[1]

Quantities of PET cyclic oligomers found in the microwaveable French fries, popcorn, fish sticks, waffles, and pizza ranged from less than 0.012 to approximately 7.0 g/kg.[2]

PET contains detectable amounts of acetaldehyde, which is able to migrate from the polymer into liquid media. With the help of a static headspace GC method, acetaldehyde was found in carbonate mineral water and lemonade. Acetaldehyde concentration ranged between 11 and 7.5 mg/l, while the contents of acetaldehyde in the PET packages ranged from 1.1 to 3.8 μg/g.[3]

Migration of acetaldehyde from PET at 40°C reached a constant level after 4 days which was about 10% of the residual value of acetaldehyde (6.3 mg/kg). At 60°C this level was raised up to 50%.

PET caused no changes in the taste of soft drinks containing carbonic acid when exposed at lower temperature and over a relatively short period of time.[4] *Lavsan* material had no effect on the taste and odor of aqueous extracts, nor on their oxidizability. Migration of antimony ion (catalyst) into water was not observed.[07] Piekacz discovered migration of calcium and magnesium ions from pellets and film. Diethylene glycol and dimethyl terephthalate were not found in extracts.[5]

PET samples including laminates, bottles, and roasting bags, were heated at 120, 150, and 230°C for 50 min, according to sample type. Volatiles released from the material were identified by GC-MS and assessed against a 10 mg/kg migration threshold limit. The main substances identified were not related to PET, but probably came from printing inks and adhesives. Authors concluded that the migration potential of PET in high temperature applications is very low and that the formation of volatiles during use is unlikely to cause any special problems in polymer recovery in recycling schemes.[6]

Migration of ethylene glycol (EG) from PET bottles stored at 32°C for 6 months into the food simulant 3.0% acetic acid was studied by gas-liquid chromatographic procedure and observed at the level of about 94 μg EG/bottle.[7]

Migration of residual contaminants remaining in the extruded PET (benzene, butyric acid, dodecane, octadecane, tetracosane, diazinon, lindane, and cooper ethyl hexonate) into food-simulating solvents, aqueous ethanol, and heptane, resulted in concentrations lower than 0.01 mg/kg. Authors concluded that unwashed recycled PET may not comply with FDA requirements.[8]

Migration of antimony from PET into food simulants, measured by inductively-coupled plasma-mass spectrometry amounted to 4.0 μg/kg. The concentration found was less than proposed limit of migration.[9]

Migration from colored PET bottles for carbonated beverages was studied. PET bottles filled with naturally carbonated mineral water up to 6-month storage released total organic carbon within the EEC and FDA limits. The following migrating substances were identified by GC-MS analysis: acetaldehyde, dimethyl terephthalate, and terephthalic acid.[10]

Acute Toxicity. Neither administration of PET powder nor a single administration of chloroform extracts of PET at a dose of 10 g/kg BW had a toxic effect on rats.[11]

Repeated Exposure. In a 1-month study, rats received wine extracts obtained after several months contact with PET. The treatment produced no harmful effect on animals.[12]

Short-term Toxicity. Rats were given 5.0 to 400 mg technical grade PET/kg BW and 5.0 to 100 mg pure PET/kg BW over a 3-month period. There were no changes in their behavior, BW gain, biochemical indices of blood serum, urine, or hematology analyses, or in relative weights of internal organs.[11]

Long-term Toxicity. No manifestations of toxicity were observed in rats, given aqueous extracts of PVC film reinforced with *Lavsan*.[13]

Mutagenicity.

In vitro genotoxicity. De Fusco et al. studied the mutagenicity in unconcentrated mineral water stored in PET bottles and growing *Salmonella* strains directly in the plastic bottles. Leaching of mutagens after 1 month of water storage in daylight and in the dark in PET bottles used for beverage packaging was noted. This activity was higher after storage in daylight.[14] The mutagenicity test on non-volatile migrant compounds identified in the above-sited study gave negative results.[10]

Carcinogenicity. *Dacron* induced malignant tumors at the site of application in rodents following *s/c* imbedding of polymer films. Nevertheless, the results of this study have been later considered inadequate.[026] *S/c* implantation of pieces of PET graft to mice and Syrian golden hamsters showed no statistical evidence of tumor induction. Observation period was 73 and 82 weeks, respectively.[15]

Regulations. ***U.S. FDA*** (1998) approved the use of PET as components of polyethylene phthalate polymers intended for use in contact with food in accordance with the conditions prescribed in 21 CFR part 177.1630.

References:

1. Kim, H., Gilbert, S. G., and Johnson, J. B., Determination of potential migrants from commercial amber polyethylene terephthalate bottle wall, *Pharmacol. Res.*, 7, 176, 1990.
2. Begley, I. H, Dennison, J. L, and Hollifield, H. C., Migration into food of polyethylene terephthalate (PET) cyclic oligomers from PET microwave packaging, *Food Addit. Contam.*, 7, 797, 1990.
3. Linssen, G., Reitsma, H., und Cozynsen, G., Static headspace gas chromatography of acetaldehyde in aqueous foods and polythene terephthalate, *Z. Lebensm. Untersuch. Forsch.*, 201, 253, 1995.

4. Eberhartlinger, S., Steiner, J., Washuttl, J., and Kroyer, G., The migration of acetaldehyde from polythene terephthalate bottles for fresh beverages containing carbonic acid, *Z. Lebensm. Untersuch. Forsch.*, 191, 286, 1990.
5. Piekacz, H., in *Cz. II Ricz. Panst. Zakl. Hig.*, 22, 295, 1971 (in Polish).
6. Freire, M. T., Castle, L., Reyes, F. G., and Damant, A. P., Thermal stability of polyethylene terephthalate food contact materials: formation of volatiles from retain samples and implications for recycling, *Food Addit. Contam.*, 15, 473, 1998.
7. Kashtock, M. and Breder, C. V., Migration of ethylene glycol from polyethylene terephthalate bottles into 3% acetic acid, *J. Assoc. Off. Anal. Chem.*, 63, 168, 1980.
8. Komolprasert, V., Lawson, A. R., and Begley, T. H., Migration of residual contaminants from secondary recycled poly(ethylene terephthalate) into food-simulating solvents, aqueous ethanol and heptane, *Food Addit. Contam.*, 14, 491, 1997.
9. Fordham, P. J., Gramshaw, J. W., Crews, H. M., and Castle, L., Element residues in food contact plastics and their migration into food simulants, measured by inductively-coupled plasma-mass spectrometry, *Food. Addit. Contam.*, 12, 651, 1997.
10. Monarca, S., De Fusco, R., Biscardi, D., De Feo, V., Pasquini, R., Fatigoni, C., Moretti, M., and Zanardini, A., Studies of migration of potentially genotoxic compounds into water stored in pet bottles, *Food Chem. Toxicol.*, 32, 783, 1994.
11. Otaka et al., cit in *Excerpta Medica*, Sec. 17, 1, 1980, Abstract 284.
12. Bazanova, A. I., Effect of chemical substances extracted from plastics on mammals and micro-organisms, in *Toxicology and Hygiene of High-Molecular-Mass Compounds and of the Chemical Raw Material Used for Their Synthesis*, Proc.3rd All-Union Conf., S. L. Danishevsky, Ed., Khimiya, Moscow-Leningrad, 1966, 113 (in Russian).
13. Kupyrov, V. N., Kaplina, T. V., Gakal, R. K., Vinarskaya, E. I., and Starchenko, S. N., Hygienic evaluation of films intended for the waterproofing of unit prefabricated swimming pools, *Gig. Sanit.*, 5, 91, 1978 (in Russian).
14. de Fusca, R., Monarca, S., Biscardi, D., Pasquini, R., and Fatigoni, C., Leaching of mutagens into mineral water from polyethylene terephthalate bottles, *Sci. Total Environ.*, 90, 241, 1990.
15. Blagoeva, P., Stoichev, I., Balanski, R., Purvanova, L., Mircheva, T. S., and Smilov, A., The testing for carcinogenicity of a polyethylene terephthalate vascular prosthesis, *Khirurgia*, 43, 98, 1990 (in Bulgarian).

OLIGOESTER ACRYLATES

Structural Formula. $CH_2=C(CH_3)COOR_3[\sim COC_6H_4COOR_3\sim]_nCOC(CH_3)=CH_2$

$R = \sim CH_2CH_2O\sim$

M = 540 to 566

Trade Name. MGF-9.

Composition. Oligomeric unsaturated esters and ethers with regularly alternating or end acrylic (methacrylic, chloracrylic) groups.

Properties. Colorless, transparent, non-volatile liquids of different viscosity or solid, colorless, crystalline products. MGF-9 is that of condensation of triethylene glycol, phthalic anhydride, and methacrylic acid.

Applications. A binder in fiberglass reinforced plastics and molding materials.

Migration Data. MGF-9 (containing up to 1.5% benzoyl peroxide) gives water a taste rating 3 after 48 hours at 37°C. Intense migration of styrene into water was observed. Color and transparency of water were not affected.

Acute Toxicity. For mice, LD_{50} is 2.3 g/kg BW. Gross pathology examination revealed plethora and, rarely, necrosis of the gastric and intestinal mucosa. Some animals developed hemorrhagical pneumonia. Plethora of the visceral organs was also found.

Repeated Exposure. Cumulative properties were not pronounced.

Regulations. ***U.S. FDA*** (1998) approved the use of MGF-9 as a component of the uncoated or coated food-contact surface of paper and paperboard intended for use in producing, manufacturing, packaging, processing, preparing, treating, packing, transporting, or holding dry food of the type identified in 21 CFR part 176.170 (c).

Reference:

Dikshtein, E. A. et al., *Gig. Truda Prof. Zabol.*, 1, 63, 1981 (in Russian).

POLYETHYLENE GLYCOL, MONOALKYL ESTERS

Molecular Formula. $C_nH_{2n+1}O(C_2H_4O)_mH$, where $n = 10$ to 18; $m = 8$ to 10.

Abbreviation. PEGA.

Synonym. Sintanol DS-10.

Composition. Hydroxyethylation product of the higher fatty alcohols.

Properties. White or yellowish paste. Readily soluble in water.

Applications. Used as an antistatic additive. Emulsifier. Component of emulsion rubbers and epoxy latexes.

Acute Toxicity. LD_{50} is 1.4 g/kg BW in mice and 1.75 to 5.3 g/kg BW in rats. Manifestations of the toxic effect include apathy, disturbances of the GI tract function and epistaxis. Death within 1-2 days.[1]

Repeated Exposure revealed no cumulative properties.[1] However, according to other data, moderate cumulative properties were found: K_{acc} was identified to be 2.5 to 2.8.[2] Rats received 1/5 LD_{50} for 5 days. The treatment caused an abrupt deterioration in the general condition, apathy, loss of tonus, and decrease in water consumption. In 10 days, the signs of intoxication including vomiting and loss of BW were more pronounced.

Long-term Toxicity. Manifestations of the toxic effect in a 6-month study included retardation of BW gain, CNS stimulation, and increased liver weights[1] as well as changes in serum cholesterol concentration (bromosulphalein test), ascorbic acid contents in the organs, and in arterial pressure. Gross pathology examination revealed changes in the viscera.[2] The NOAEL is unlikely to be less than 40 mg/l.

Carcinogenicity. Yanysheva et al. found benzo[a]pyrene in PEGA, but no carcinogenic effect was observed on dermal application.[3]

Allergenic Effect is noted on dermal application.[4]

Standards. ***Russia*** (1995). MAC and PML: 0.1 mg/l (organolept., foam).

References:

1. Mikhailets, I. B., Toxic properties of some antistatics, *Plast. Massy*, 12, 27, 1976 (in Russian).
2. Mozhayev, E. V., *Pollution of Reservoirs by Surfactants (Health Safety Aspects),* Meditsina, Moscow, 1976, 94 (in Russian).
3. Yanysheva, N. Ya., Voloshchenko, O. I., Chernichenko, I. A., et al., On carcinogenic effects of certain surfactants, components of synthetic detergents used by population, *Gig. Sanit.*, 7, 9, 1982 (in Russian).
4. Yeskova-Soskovets, L. B., Sautin, A. I., and Rusakov, N. V., Allergenic properties of some surfactants, *Gig. Sanit.*, 2, 14, 1980 (in Russian).

POLYETHYLENE GLYCOL, MONODODECYL ETHER

Molecular Formula. $[\sim C_2H_4O\sim]_n.C_{12}H_{26}O$

M unknown

CAS No 9002-92-0

RTECS No MD0875000

Abbreviation. PEGDE.

Synonyms and **Trade Names.** Dodecanol ethylene oxide; Dodecyl alcohol, ethoxylated; Dodecyl poly-(oxyethylene) ether; Emulgen 100; Ethoxylated lauryl alcohol; Hydroxypolyethoxydodecane; Lubrol PX; Polyoxyethylene lauryl alcohol; Rokanol L.

Composition. Product of the reaction of cetyl and stearyl alcohols with ethylene oxide.

Properties. Colorless to yellow liquid with a pleasant odor. Soluble in water.

Applications. A solvent and an emulsifier. Used as an antistatic additive and as a medicine.

Acute Toxicity.

Observations in man. Probable oral lethal dose is 0.5 to 5.0 g PEGDE/kg BW for 70 kg person.[022]

Animal studies. LD_{50} is 1.0 g/kg BW in rats. Administration of high doses produced ulceration or bleeding from stomach and fatty liver degeneration.[1] LD_{50} is 1.2 g/kg BW in mice and 0.4 g/kg BW in guinea pigs. Poisoning induced flaccid paralysis without anesthesia (affected peripheral nerve and sensation). Change in motor activity and in respiratory system was noted.[2]

Repeated Exposure revealed no tendency to accumulation in rats and mice. Mice were dosed by gavage with 8 to 170 mg Lubrol/kg or 13 to 26 mg Oxanol/kg BW. Polytropic action with impaired CNS function were noted.[3]

Rats were dosed with PEGDE for 5 days (total dose of 3.9 g/kg BW). The treatment caused ataxia and dyspnea. Rats were given PEGDE for 22 days (total dose of 14 g/kg BW). Muscle weakness and dyspnea were observed in treated animals before their death.[4]

Mutagenicity.

In vitro genotoxicity. PEGDE was tested over a wide range of doses in four *Salmonella typhimurium* strains and was found to be negative.[015] Treatment with PEGDE produced DNA damage in *E.coli* (dose of 50 mg/l).[5]

Regulations. ***U.S. FDA*** (1998) approved the use of PEGDE as a component of articles intended for use in contact with food in accordance with the conditions prescribed in CFR part 178.3750.

References:

1. McElligott, T. F., Acute toxicity and 5-week feeding studies of lubrol PX in rodents, *Food Cosmet. Toxicol.*, 8, 125, 1970.
2. Zipf, H. F., Wetzels, E., Ludwig, H., and Friedrich, M., General and local toxic effects of dodecyl polyethyleneoxide ethers, *Arzneimittel-Forsch.*, 7, 162, 1957.
3. Krynskaya, I. L., Bukevich, G. M., Robachevskaya, Ye. G., et al., Data on toxicology of some antistatic additives to plastic materials, in *Hygiene and Toxicology of High-Molecular-Mass Compounds and of the Chemical Raw Material Used for Their Synthesis*, Proc. 6th All-Union Conf., B. Yu. Kalinin, Ed., Leningrad, 1979, 215 (in Russian).
4. Berberian, D. A., Gorman, W. G., Drobeck, H. P., Coulston, F., and Slighter, R. G., The toxicology and biological properties of Laureth 9 (a polyoxyethylenelauryl ether), a new Spermicidal agent, *Toxicol. Appl. Pharmacol.*, 7, 206, 1965.
5. Kubinski, H., Gutzke, G. E., and Kubinski, Z. O., DNA-cell-binding (DCB) assay for suspected carcinogens and mutagens, *Mutat. Res.*, 89, 95, 1981.

POLYETHYLENE GLYCOL, MONOPHENYL ETHER

Molecular Formula. $[\sim C_2H_4O\sim]_n.C_6H_6O$

CAS No 9004-78-8

RTECS No MD0906700

Abbreviation. PEGPE

Synonyms. Phenolethylene oxide adduct; α-Phenyl-ω-hydroxypoly(oxy-1,2-ethanediyl); Polyethylene glycol, phenyl ether; Polyoxyethylene, phenyl ether.

Acute Toxicity. LD_{50} *i/p* is 144 mg/kg BW in mice.

Regulations. ***U.S. FDA*** (1998) approved the use of PEGPE in adhesives as a component of articles intended for use in packaging, transporting, or holding food in accordance with the conditions prescribed by 21 CFR part 175.105.

POLYALKYLENE MALEINATES

Structural Formula.

$$^{H}[\sim(OROC\overset{\|}{C}H{=}CH\overset{}{C})_x(OROCR'C)_y\sim]_n{}^{OH}$$

```
H[~(OROCCH=CHC)x(OROCR'C)y~]n OH
        ||      ||      || ||
        O       O       O  O
```

where R is a glycol radical, R' is a radical of modifying acid, $x = 1$ to 5, $y = 0$ to 5, and $n = 1$ to 10

Synonym. Polyether maleinates.

Composition. Saturated polyesters of maleic acid and glycols, modified by phthalic acid, cured (cross-linked) by different monomers, for example, styrene. PN-1 is the product of interaction of diethylene glycol with maleic and phthalic anhydrides and styrene, PNM-2 is that of diethylene glycol with the given anhydrides only. In both cases, $R = CH_2OCH_2$ and $R' = OC_6H_4$.

Applications. A binder in glass-fiber reinforced plastics and molding materials.

Migration Data. During a study of livestock, drinking glass-fiber reinforced plastic pipes manufactured on the basis of PN-1 (isopropylbenzene hydroperoxide as initiator, cobalt naphthenate as accelerator) produced significant deterioration of water organoleprtic properties (odor up to rating 3 after 3 hours of contact). After 24-hour contact at room temperature, 0.39 mg/l of styrene was found in the water. After 4-day contact, cobalt concentration was 2 to 4 mg/l.[1]

Kravchenko and Kharchenko have found harmful substances that migrated to water after contact with glass-fiber reinforced plastics (PN-1 and PNM-2 type) reaching concentrations 5 to 10 times higher than permissible levels.[2]

References:

1. Sheftel', V. O. and Prusova, G. O., Hygienic evaluation of some synthetic prostheses, *Vrachebnoye Delo*, 2, 95, 1966 (in Russian).
2. Kravchenko, T. I. and Kharchenko, T. F., in Proc. 4th All-Union Sci. Techn. Conf. on *Development and Use of Polymeric Materials in Food Industry*, Minsk 1980, Moscow, 1980, 77 (in Russian).

POLYESTER P-2200

M ~ 2200

Composition. A polycondensation product formed of adipic acid, diethylene glycol, and trimethylolpropane in presence of tetrabutoxytitanium, ammonium molybdate, and powdered iron catalysts.

Properties. A viscous homogeneous liquid of light-yellow to light-brown color. Soluble in an alcohol-benzene mixture.

Applications. The basic ingredient for foam-polyurethane intended for use in drinking water filtration.

Acute Toxicity. LD_{50} for rats and mice is not attained. The animals tolerate 10 g/kg BW dose.

Repeated Exposure. Rats received 2.0 g/kg BW for 1.5 months. The treatment produced changes in hematology analysis, disturbances in water-salt exchange, an increase in the relative weights of the suprenals, and a reduction in their ascorbic acid contents.

Reference:

Reznichenko, A. K., Vasilenko, N. M., Muzhikovsky, G. L., et al., Toxicity of paraoxybenzoic acid, polyether P-2200 and other chemicals, *Gig. Sanit.*, 1, 85, 1986 (in Russian).

POLYETHYLENEIMINE

Molecular Formula. $[\sim C_2H_8N_2\sim]_n$
CAS No 26913-06-4
RTECS No TQ7700000
Abbreviation. PEI.

Synonym. Corcat.

Properties. The product of polymerization of ethyleneimine (aziridine). A polymer of linear or branched structure. Slightly colored resin with indefinite odor. *Dens.* 1.07^{20} (linear), 1.05^{20} (branched). Linear P. is virtually insoluble in cold water, but is readily soluble in hot water; branched P. is readily soluble in water. It gives water an unpleasant odor and a bitter-tart taste. The odor perception threshold is 6.14 mg/l (rating 1), and taste perception threshold is 3.8 mg/l.

Applications. Used in physical-chemical purification of waste waters, in paper and pulp industry, and in the production of ion-exchange resins, membranes, and films.

Acute Toxicity. LD_{50} is reported to be 3.3 g/kg BW in rats.[029] No differences were found in the specific or sex sensitivity of small laboratory animals. In the case of PEI with M = 10,000, LD_{50} was 1.15 g/kg BW in mice, 1.35 g/kg BW in rats, and 0.94 g/kg BW in guinea pigs. With M = 40,000, LD_{50} was 1.6 g PEI/kg BW in mice, 2.2 g/kg BW in rats, and 1.4 g/kg BW in guinea pigs.[1] Poisoning predominantly affected CNS.

The LD_{50} of PEI with M = 30,000 was 0.5 to 1.5 g/kg BW.[05] Contamination of PEI with toxic oligomers is possible.

Repeated Exposure. 1/10 and 1/50 LD_{50} of PEI with M = 20,000 to 40,000 was administered to rats over a period of 1.5 months. The animals did not die. Changes were observed in the STI, in erythrocyte and reticulocytes count in the peripheral blood, and in the peroxidase activity in the blood serum.

Long-term Toxicity. PEI with M = 20,000 was administered to male rats in doses of 5.0, 0.5, 0.05, and 0.005 mg/kg BW. A clear dose-effect relationship was observed. With 5.0 mg PEI/kg BW, there were changes in the conditioned reflexes, the STI, and the number of reticulocytes in the peripheral blood, and an increase in the catalase, peroxidase, and aldolase activity in the blood serum. The LOAEL for the effect on conditioned reflexes was 0.05 mg/kg BW. Pathology examination revealed fatty dystrophy of the liver in rats with a dose of 5.0 mg/kg BW.[1]

Reproductive Toxicity.

Gonadotoxicity. Administration of 1/50 and 1/10 LD_{50} over a period of 1.5 months and of 5.0 mg/kg BW over a period of 6 months had no effect on the gonads of males.

Embryotoxicity. PEI had a toxic effect on embryos: administration of 300 and 700 mg/kg BW to rats on the second day of gestation caused significant loss of embryos. A dose of 200 mg/kg BW was the NOEL.[2]

Mutagenicity.

In vivo cytogenetics. In the above-sited study, there was no increase in the number of CA in the bone marrow cells.[2]

Allergenic Effect. Not found.

Standards. ***Russia.*** MPC: 0.1 mg/l.

Regulations.

U.S. FDA (1998) approved the use of PEI (1) in adhesives as a component (monomer) of articles intended for use in packaging, transporting, or holding food in accordance with the conditions prescribed in 21 CFR part 175.105; (2) in the manufacture of resinous and polymeric coatings for polyolefin films for food-contact surface of articles intended for use in producing, manufacturing, packing, processing, preparing, treating, packaging, transporting, or holding food, in accordance with the conditions prescribed in 21 CFR part 175.320; (3) as a component of the uncoated or coated food-contact surface of paper and paperboard intended for use in producing, manufacturing, packaging, processing, preparing, treating, packing, transporting, or holding aqueous and fatty foods, subject to the provisions prescribed in 21 CFR part 176.170; and (4) in the manufacture of cellophane for packaging food in accordance with the conditions prescribed in 21 CFR part 177.1200.

In ***the U.S.,*** up to 5.0% PEI is permitted in food cartons, in ***Germany*** up to 0.5%.[05]

References:

1. Kinzirsky, A. S., Hygienic characteristics of a new polyethyleneimine flocculent and its standardization in the water reservoirs, *Gig. Sanit.*, 7, 19, 1976 (in Russian).
2. Zaugol'nikov, S. D., Haslavskaya, S. L. and Sukhov, Yu. Z., Embryotoxic effect of polyethyleneimine, in *Hygiene and Toxicology of High-Molecular-Mass Compounds and of the Chemical Raw Material Used for Their Synthesis*, Proc. 4th All-Union Conf., S. L. Danishevsky, Ed., Khimiya, Leningrad, 1969, 14 (in Russian).

POLYOLEFINS

POLYETHYLENE

Structural Formula. $[\sim CH_2CH_2\sim]_n$

CAS No 9002-88-4
86089-97-6

RTECS No TQ3325000
KX3270000

Abbreviations. PE; HPPE (high-pressure PE); LDPE (low-density PE); MPPE (medium-pressure PE); LPPE (low-pressure PE). LPPE and MPPE are also referred to as high-density PE (HDPE).

Synonyms and **Trade Name.** Alkathene; Ambythene; Bulen A; Daplen; Epolene C; Ethene polymer; Hostalen; Lupolen; Marlex; Mirason; Mirathen; Neopolen; Petrothene; Politen; Polyethylene AS; Polyethylene resins; Polythene; Polywax 1000; Suprathen; Telcothene; Tenaplas; Valeron; Vestolen; Yukalon.

Properties. PE was introduced in the food industry in the 1950s. It provides mechanical properties (strength, rigidity, abrasion resistance) at low cost. White solid.

Molecular mass, structure, and properties of PE depend on the production method. Pressure at polymerization accounts for production of high-pressure (HPPE), or low-density (LDPE), medium-pressure (MPPE) and low-pressure PE (LPPE). LPPE and MPPE are also referred to as high-density PE (HDPE). Chemical stability increases with density increase.

Applications. PE is used for manufacturing food containers and packaging, various household articles, water pipes, in irrigation and draining systems, in medicine, etc. PE builds up static and tends to cling to itself. Its heat-sealing range is from 120 to 180°C. In general, PE is cleared for use for packaging but not cooking. Certain PE polymers have restricted uses, especially with respect to temperature.

Carcinogenicity classification. An IARC Working Group concluded that there is inadequate evidence for the carcinogenicity of PE in *experimental animals* and there were no adequate data available to evaluate the carcinogenicity of PE in *humans*.

IARC: 3

Regulations. ***U.S. FDA*** (1998) approved the use of PE (1) in adhesives as a component (monomer) of articles intended for use in packaging, transporting, or holding food in accordance with the conditions prescribed in 21 CFR part 175.105; (2) as a component of the uncoated or coated food-contact surface of paper and paperboard intended for use in producing, manufacturing, packaging, processing, preparing, treating, packing, transporting, or holding dry food of the type identified in 21 CFR part 176.170 (c); (3) as a defoaming agent used in articles intended for use in producing, manufacturing, packing, processing, preparing, treating, packaging, transporting, or holding food, subject to the provisions prescribed in 21 CFR part 176.200; (4) in the manufacture of cellophane for packaging food in accordance with the conditions prescribed in 21 CFR part 177.1200; (5) as articles or components of articles intended for use in contact with food, subject to the provisions of 21 CFR part 177.1520; and (6) as a protective coating or component of protective coatings for fresh fruits and vegetables in accordance with GMP and with the conditions prescribed in 21 CFR part 172.260 (*oxidized* PE).

LOW-DENSITY (high-pressure) POLYETHYLENE

Structural Formula. $[\sim CH_2CH_2\sim]_n$

M = 30,000 to 400,000

Abbreviation. LDPE

Composition. Food-wrap grades contain antioxidants to minimize degradation during processing and, in the final films, they are normally present at levels of several hundred ppm. Various kinds of carbon black are used to stabilize LDPE. Its concentration in PE should not exceed 2.5%.

Properties. White, translucent, elastic material, greasy to the touch, without taste or odor, or with a specific odor. Insoluble in water and alcohol. LDPE is strong and clear. It is a good moisture barrier but a poor oxygen barrier, and it will heat seal to itself. Maximum continuous service temperature is 60°C; it heat distorts at 57°C. LDPE is resistant to moisture vapor, acids, bases, and fats and oils. It has poor resistance to solvent and builds up static.

Applications. LDPE is the most widely used food wrapping material. It is used primarily for packaging films and for bread wrapping bags.

Migration Data. During the synthesis, PE may be contaminated by other materials (solvents, catalysts, washing agents, etc.). Small amounts of low-molecular-mass compounds can be released from LDPE into contacting liquid media, generally in safe concentrations but giving foreign taste and odor. There is a direct relationship between the contents of low-molecular mass fraction in polyolefins and the odor. Removal of low-molecular mass fraction results in complete odor disappearance. LDPE articles can also cause formation of rapidly disappearing froth during agitation of aqueous extracts.[1]

LDPE (NP 108-168 grade) intended for prolonged contact with drinking water produces no significant effect on water quality over a long period of time. The smell and taste of extracts is of rating 1.2; oxidizability is slightly increased after 20 to 30-day contact with PE: 0.5 to 0.6 mg O_2/l.[2]

In extracts from LDPE with titanium dioxide and ultramarine added, no significant change in organoleptic properties or an increase in oxidizing ability was observed. There was no migration of formaldehyde, lead, or copper ions into water (formaldehyde was released only into acid medium in a quantity less than 0.5 mg/l). No benzo[a]pyrene or 1,12-benzoperylene were found after a 9-month period of contact.[1]

Some LDPE films do not change the organoleptic properties of model media at 90°C.[3]

Migration of methanol and isopropanol (q.v.), benzene, and polycyclic aromatic hydrocarbons from LDPE has been noted. Catalysts do not generally migrate from PE, but their presence reduces the effectiveness of stabilizers, accelerates oxidative breakdown of PE, and leads to change of its color during service. There are also indications of the possible migration of formaldehyde from PE, but the level of migration is very low.

Migration rates of radiolabeled antioxidants, Irganox-1010 and Irganox-1076, from LDPE were measured at temperatures up to 135°C. Water, 8.0 and 95% aqueous solution of ethanol and corn oil were employed as food simulating liquids. The losses to foods were usually larger than those to water but below those to corn oil.[4]

Butylated hydroxytoluene and Irganox 1010 were radiolabeled. Butylated hydroxytoluene, a much smaller and more volatile molecule than Irganox 1010, migrated more rapidly from polymer into foods. Dry foods can be surpisingly effective sinks for antioxidants under typical storage conditions.[5]

Naphthalene contamination was observed in sterilized milk drinks contained in LDPE bottles. Gas chromatography of the packaging material showed levels of naphthalene ranging from 0.7 to 2.0 mg/kg. Analysis of the milk drinks contained in LDPE bottles by GC showed levels of naphthalene in the range 0.01 to 0.03 mg/1. The level of contamination in milk increased with storage time at room temperature (25°C) and, depending on the concentration of naphthalene in the packaging material, values of 0.08 to 0.27 mg/1 were found at the expiry date. The amount of naphthalene extracted by milk from the LDPE at 25°C over a period of 60 days was related to the amount that could be extracted using peanut oil at 66°C for 30 hours.[6]

Benzophenone-based ultraviolet absorbents (up to 0.5% of a composition) for commodities of LDPE showed a very slight tendency to migrate into aqueous, acid, or dilute alcoholic foods. Very high migration levels were noted for sunflower oil, fat-simulants HB 307 or 50% ethanol. Migration of 2-hydroxy-4-octoxybenzophenone into fat-free foods was observed at the toxicologically insignificant level, nevertheless, the above ultraviolet sorbents should not be used for the ultraviolet stabilization of plastics designed for packaging fat-containing foods.[7] Migration of fatty acid amides in fat simulant medium from LDPE used in food packaging and containing commonly used fatty acid amide slip additives was determined at the level of 1.8 to 3.1 mg/l (10 day, 40°C); migration into aqueous food simulant medium was less than 0.05 mg/l.[8]

Acute Toxicity. Administration of 1.16 g emulsified powder of unstabilized LDPE (emulsifier OP-7) per kg BW did not prove fatal to rats.[3] A dose of 2.5 g LDPE/kg BW given in sunflower oil produced neither effect on general condition and behavior of rats over 14 days of observation nor pathomorphological changes in the intestinal organs.[9]

Short-term Toxicity. The toxicity of apple juice packed in LDPE (P-20-20T grade) packets with secondary cellophane and foil packing was investigated. The material was heated before packing. Juice was stored in the packets for 3 and 6 months and then given to animals for 3 months. BW gain and other indices were similar to those in the control. There was no impairment of liver function. No pathological changes in the viscera were noted.[10]

Long-term Toxicity. In a study with aqueous extracts from PE films from which small amounts of oxidizing substances had been released, in the entire experimental period there were no changes in BW gain, hematology analyses, liver function, behavior, or mass coefficients of the visceral organs.[3]

Rats received aqueous and oil extracts of modified LDPE simultaneously over a period of 15 months. The extracts were obtained at 100°C for 2 hours and 20°C for 24 hours (water) and 6 to 7 days (oil). No toxic effect was observed.[11]

Over a period of one year male rats received, instead of water, aqueous extracts (20°C, 15 days) from LDPE stabilized with gas channel black. No changes were observed in general condition, BW gain, hematology analyses, phagocytic activity of leukocytes, liver functions, etc. Histology of the visceral organs was not significantly different from that of the controls.[9]

In a 12-month toxicological study, aqueous extracts of unstabilized LDPE dyed with phthalocyanine blue pigment (distilled water, 0.6 to 20 cm^{-1}; 20 and 60°C; 10 days; maximum pigment concentration in extracts: 0.57 mg/l at 20°C and 1.18 mg/l at 60°C) were given daily to mice and rats.[12] There were no changes in the studied indices or in histological structure of the visceral organs.

Reproductive Toxicity.

Observations in man. Epidemiological studies of adverse pregnancy outcome revealed no association between contact with PE at work and the incidence of spontaneous abortion.[13]

Gonadotoxicity. An aqueous PE extracts did not affect sperm mobility.[14]

Mutagenicity. Extracts from LDPE (water, 0.9% solution of *NaCl*, 15 and 50% alcohol, 3.0% acetic acid, and peanut oil at 50°C for 72 hours and 212°C for 30 min) were not mutagenic for several *Salmonella typhimurium* strains (Fervolden and Moller, 1978).

Carcinogenicity. PE induced malignant tumors in rodents at the site of application following *s/c* imbedding of polymer films.[026] Nevertheless, the results of this study have been later considered inadequate.

According to Carter and Roe, solid and fragmented PE implants induced development of lymphomas including Hodgkin's disease in rats.[15] Equivocal tumorigenic agent by RTECS criteria.

Regulations. *U.S. FDA* (1998) approved the use of LDPE as articles or components of articles intended for use in contact with food, subject to the provisions of 21 CFR part 177.1520.

Recommendations. *Russia.* LDPE articles are not recommended for packaging fats and fat-containing foodstuffs. Rigid articles of LDPE are used only at room temperature.

References:

1. Vlasyuk, M. G. and Medvedev, V. I., Evaluation of articles made of high-pressure polyethylene intended for use in contact with foods, in *Hygiene and Toxicology of High-Molecular-Mass Compounds and of the Chemical Raw Material Used for Their Synthesis,* Proc. 6th All-Union Conf., B. Yu. Kalinin, Ed., Khimiya, Leningrad, 1979, 63 (in Russian).
2. Sukhareva, L. V. and Kalinin, B. Yu., Hygienic characteristics of water holding cisterns made of polyethylene composition, in *Hygiene Aspects of the Use of Polymeric Materials*, Proc. 3rd All-Union Meeting on New Methods of Hygiene Monitoring of the Use of Polymers in the Economy, K. I. Stankevich, Ed., Kiyv, December 2-4, 1981, Kiyv, 1981, 197 (in Russian).
3. Yurin, V. V, in *Toxicology and Hygiene of High-Molecular-Mass Compounds and of the Chemical Raw Material Used for Their Synthesis*, Proc. 2nd All-Union Conf., A. A. Letavet and S. L. Danishevsky, Eds., Khimiya, Leningrad, 1964, 95 (in Russian).
4. Goydan, R., Schwope, A. D., Reid, R. C., and Cramer, G., High-temperature migration of antioxidants from polyolefins, *Food Addit. Contam.*, 7, 323, 1990.
5. Schwope, A. D., Till, D. E., Ehntholt, D. J., Sidman, K. R., Whelan, R. H., Schwartz, P. S., and Reid, R. C., Migration of BHT and Irganox 1010 from low-density polyethylene (LDPE) to foods and food-simulating liquids, *Food Chem. Toxicol.*, 25, 317, 1987.
6. Lau, O. W., Wong, S. K., and Leung, K. S., Naphthalene contamination of sterilized milk drinks contained in low-density polyethylene bottles. Part 1, *Analyst*, 119, 1037, 1994.
7. Uhde, W. J. and Woggon, H., New results of migration behavior of benzophenone-based UVG absorbents from polyolefins in foods, *Nahrung*, 2, 185, 1976.
8. Cooper, I. and Tice, P. A., Migration studies on fatty acid amide slip additives from plastics into food simulants, *Food Addit. Contam.*, 12, 235, 1995.
9. Sheftel', V. O., Study of toxicity of water extracts from some types of plastic pipes for water supply, in *Toxicology and Hygiene of High-Molecular-Mass Compounds and of the Chemical Raw Material Used for Their Synthesis,* Proc.3rd All-Union Conf., S. L. Danishevsky, Ed., Khimiya, Moscow-Leningrad, 1966, 123 (in Russian).

10. Starikova, T. S. and Fishilevich, S. M., in *Hygiene and Toxicology*, Proc. Sci. Conf. Young Specialists & Hygienists, Zdorov'ya, Kiyv, 1967, 56 (in Russian).
11. *Materials on the Health and Safety Assessment of Pesticides and Polymers,* Coll. Sci. Proc. F. F. Erisman Research Sanitary Hygiene Institute, Moscow, 1977, 87 (in Russian).
12. Kalinin, B. Yu., Krynskaya, I. L., Zimnitskaya, L. P., et al., On toxicity of colored high-pressure polyethylene, in *Hygiene and Toxicology of High-Molecular-Mass Compounds and of the Chemical Raw Material Used for Their Synthesis*, Proc. 4th All-Union Conf., S. L. Danishevsky, Ed., Khimiya, Leningrad, 1969, 38 (in Russian).
13. Lindbohm, R. et al., Spontaneous abortion among women employed in the plastics industry, *Am. J. Ind. Med.,* 8, 579, 1985.
14. Es'kov, A. P., Kaiumov, R. I., Luzhetskii, A. S., Gurilev, O. M., and Riazanov, Yu. N., Method of toxicological evaluation of polymers, *Gig. Sanit.*, 1, 62, 1985 (in Russian).
15. Carter, R. L. and Roe, F. J., Induction of sarcomas in rats by solid and fragmented polyethylene: experimental observations and clinical implications, *Brit. J. Cancer.*, 23, 401, 1969.

POLYETHYLENE foam

Structural Formula. $[\sim CH_2CH_2\sim]_n$

Abbreviations. PEF; LDPE (low-density PE).

Trade Name. Penoplen.

Composition and **Properties.** Contains foaming agents, activators, etc. (azodicarbonamide, zinc oxide, zinc stearate) introduced into LDPE. PEF articles are translucent, dull, and elastic.

Applications. Used for manufacture of drink stopper seals and for packing of products.

Migration Data. Has no effect on the organoleptic properties of extracts at temperatures between -15°C and +110°C. No migration of low-molecular-mass compounds has been found.

Long-term Toxicity. Rats received aqueous and oil extracts from PEF film over a period of 15 months. No harmful effect was found.

Regulations. ***U.S. FDA*** (1998) approved the use of PEF as articles or components of articles intended for use in contact with food, subject to the provisions of 21 CFR part 177.1520.

Reference:

1. Braun, D. D., Moshlakova, L. A., and Zenina, G. V., Hygienic evaluation of new polyolefin class polymeric materials for the food industry, *Gig. Sanit.*, 6, 29, 1980 (in Russian).

LOW-DENSITY POLYETHYLENE, compounds with POLYISOBUTYLENE

Abbreviation. POV.

Composition. These compounds contain different amounts of *polyethylene* (POV-30, POV-50, and POV-90).

Properties. Chemically stable and used for manufacture of seals.

Toxicity. Aqueous extracts from these compounds produced no toxic or cytotoxic effect. Toxic properties have been found in the oil extracts of POV-30.

Reference:

Alyushin, M. G., Artem, A. I., and Trakhman, Yu. G., in *Synthetic Polymers in the Soviet Pharmaceutical Practice,* Meditsina, Moscow, 1974, 142 (in Russian).

POLYETHYLENE films, photodestructable

Composition. *Preparation I:* A ferrocene derivative is used as the photodestruction sensitizer. *Preparation II:* α-Hydroxyethyl ferrocene (HOEF) is used as the photodestruction sensitizer. 80-100 mm thick.

Applications. The film is used in agriculture. It takes 1.5 to 3 months to disintegrate.

Migration Data. *Preparation I*: A study of extracts (water, 10 cm^{-1}, 80°C, 7 hours) revealed migration of the sensitizer which indicates its decomposition within 100 hours of irradiation. There was no migration of formaldehyde into potatoes.[1] *Preparation II*: Extracts (water solutions of citric, lactic, or acetic acids; 2.0 and 10 cm^{-1}) were studied. With HOEF containing in the film about 0.05 to 0.1%, up to 10 mg/l migrated into water; in acid media the organoleptic indices deteriorate sharply (in a number of cases the odor reached rating 5).[2]

Recommendations. ***Russia.*** The film (II) was not recommended for use in contact with liquid food products.

References:

1. Rotenberg, V. V., Guricheva, Z. G., Kondrashkina, N. I., et al., Sanitary-chemistry study of photodestructable polyethylene films, in *Hygiene and Toxicology of High-Molecular-Mass Compounds and of the Chemical Raw Material Used for Their Synthesis,* Proc. 6th All-Union Conf., B. Yu. Kalinin, Ed., Khimiya, Leningrad, 1979, 90 (in Russian).
2. Tarasova, N. A., Shvagiryova, N. A., Belyatskaya, O. N., and Borodulina, M. Z., in *Environmental Protection in Plastics Production and Hygiene Aspects of Their Use,* Plastpolymer, Leningrad, 1978, 94 (in Russian).

LOW-DENSITY POLYETHYLENE

produced by activation polymerization filling

Structural Formula. $[\sim CH_2CH_2\sim]_n$

Abbreviation. LDPE.

Trade Name. Norplast.

Composition. LDPE, fillers: hydrophobised chalk, calcite, kaolin, TiO_2, Celotex.

Applications. Rigid containers.

Migration Data. Articles do not give water or model media any foreign taste or odor at 40 through 80°C. With a filter contents in the material of up to 30%, there is no significant change in the oxidizability of extracts (0.8 to 5.4 mg/l O_2). A higher filler contents (40%) as well as the use of other fillers (asbestos, tuff, perlite, gas channel black, etc.) produce an odor in extracts coming in contact with the articles, and increase migration of low-molecular-mass organic and unsaturated compounds into water. The hygienic properties of articles deteriorate if the drying and processing regimen is not adhered to: the odor of extracts becomes stronger, and formaldehyde (up to 0.62 mg/l) is found to migrate. There was no change in the organic properties of water during experimental use of articles.

Long-term Toxicity. In a 12-month study, no changes were observed in the functional or morphological indices in rats.

Reference:

Braun, D. D., Voronel', T. G., Moshlakova, L. A., Demina, S. Ye., and Chernitsyna, M. A., Hygienic substantiation of the possibility of using norplast based on polyolefins in food outlets of an agribusiness complex, *Gig. Sanit.*, 5, 20, 1988 (in Russian).

LOW-DENSITY POLYETHYLENE, irradiated

Structural Formula. $[\sim CH_2CH_2\sim]_n$

Abbreviation. LDPE

Trade Name. Termoplen.

Produced by γ-irradiation of LDPE. This increases the average molecular mass and heat resistance of LDPE. With a small dose of γ-irradiation, the energy of the accelerated electrons is well below the threshold at which artificial radioactive isotopes are formed by nuclear reactions.

Migration Data. A study of extracts of LDPE at temperatures ranging from -15°C to +125°C did not reveal any change in their organoleptic properties, or any increase in oxidizability or in the contents of formaldehyde.[1]

Long-term Toxicity. 24-hour aqueous extracts were given to rats over a period of 15 months. Gross pathology examination revealed no changes in the viscera.[2]

References:

1. Braun, D. D., Zenina, G. V., and Moshlakova, L. A., Hygienic evaluation of new polyolefin group polymeric materials intended for use in the food industry, *Gig. Sanit.*, 2, 24, 1979 (in Russian).

2. Braun, D. D., Moshlakova, L. A., and Zenina, G. V., Hygienic evaluation of new polyolefin class polymeric materials for the food industry, *Gig. Sanit.*, 6, 29, 1980 (in Russian).

HIGH-DENSITY (low-pressure) POLYETHYLENE

Structural Formula. $[\sim CH_2CH_2\sim]_n$
M = 70,000 to 80,000
Abbreviation. HDPE; LDPE.

Synonym. *Ziegler* HDPE.

Properties. Similar to LDPE but possess greater mechanical strength and elasticity. Color ranges from white to cream; its chemical stability is higher than that of LDPE.

Applications. Differences in the application of HDPE as compared to LDPE are determined by its greater rigidity and heat resistance. HDPE has a maximum continuous service temperature of 70°C but distorts at 60°C. It is resistant to moisture, gases, acids, bases, solvents, and fats/oils. It will build up static. HDPE is used for milk jugs, cleaning supply bottles, and trash bags.

Migration Data. The same substances may migrate from HDPE into water as from LDPE as well as traces of complex organometallic catalysts and solvents. *Ziegler* HDPE to be used in the food industry should be produced only with the use of isopropyl alcohol as a washing agent, migration of which from finished articles can reach 5.5 mg/l. No isopropyl alcohol is released after 16 to 19 washes.[1]

The contents of catalyst (ash contents) can be reduced to a 0.002 to 0.003% minimum by washing. During prolonged (2 to 8 years) service of *Ziegler* HDPE water pipes, there is no increase in migration of chemical substances from them.[2] The oxidizability of aqueous extracts of this HDPE stabilized with P-24 phosphite, 2-hydroxy-4-octyloxy benzophenone, and calcium stearate has proved to be higher than that of extracts of this HDPE with additions of *N,N'*-di-β-naphthyl-*p*-phenylenediamine, gas channel black, and calcium stearate.[3]

In supersaturated extracts from this HDPE (stabilizers - amine compounds, gas black, calcium stearate, phenol, and benzophenone derivatives) a small amount of reducing agents and traces of amine compounds have been found. In the case of long-term (220 day) study, traces of stabilizer were found at the end of the experiment. In supersaturated aqueous extracts (10 cm^{-1}, 20 and 60°C) of unstabilized *Ziegler* HDPE, a very small amount of aluminum and traces of titanium and chlorides (0.3 to 0.7 mg/l) were found. Storage of *Ziegler* HDPE film for 8 months at room temperature had almost no effect on oxidizability or the concentration of chlorides and stabilizer in the extracts.[4]

Vanadium compounds do not migrate into water from *Ziegler* HDPE obtained on homogenic vanadium catalysts (1 cm^{-1}; 80 and 20°C; 1, 3, and 10 days). *Ziegler* HDPE grades 22008-0.40 (injection-molding) and 21708-007 (extrusion) are recommended for use in contact with drinking water.[5] No phenols or amine compounds were found in extracts (0.4 cm^{-1}; 20°C; 9 days) from "Hostalen GM-5010" (Norwegian company Haplast) which in addition to carbon black contains aromatic amines and phenol derivatives as stabilizers.[6] Films, tubing, and injection-molded articles of gas-phase *Ziegler* HDPE can give water and simulant media a faint odor and taste up to rating of 1.5. Migration of Irganox 1010 and 1076, formaldehyde, and chromium was not found.[7]

Benzophenone-based ultraviolet absorbents (up to 0.5% of a composition) for commodities of HDPE showed a very slight tendency to migrate into aqueous, acid, or dilute alcoholic foods. Very high migration levels were noted for sunflower oil, fat-simulants HB 307 or 50% ethanol. Migration of 2-hydroxy-4-octoxy- benzophenone into fat-free foods was observed at the toxicologically insignificant level; nevertheless, the above ultraviolet sorbents should not be used for the ultraviolet stabilization of plastics designed for packaging fat-containing foods.[8]

Migration rates of radio-labeled antioxidants, Irganox-1010 and Irganox-1076, from HDPE were measured at temperature up to 135°C. Water, 8.0 and 95% aqueous solution of ethanol and corn oil were employed as food simulating liquids. The losses to foods were usually larger than those to water but below those to corn oil.[9]

Acute Toxicity. Mice fed 2.5 g of unstabilized *Ziegler* HDPE powder/kg BW and powder stabilized either with gas black, amine compounds and calcium stearate, or with benzophenone derivatives and calcium stearate in their diet, developed no changes in their general condition or BW gain. Histology of the visceral organs was insignificantly different from that in the controls.[3]

Short-term Toxicity. 1.25 to 100% Marlex-50 *Ziegler* HDPE was added to the diet of male and female rats. No manifestations of toxic action were observed.[05]

Long-term Toxicity. Administration of extracts (10 cm^{-1}; 20 and 60°C; 10 days) from several *Ziegler* HDPE specimens with small quantities of chlorides, aluminum, and titanium ions to mice and rats over a period of 16 to 19 months caused slight transient changes in BW gain, STI, behavior, and relative weights of visceral organs. Histological examination revealed no changes attributed to the action of extracts.[3]

Regulations. ***U.S. FDA*** (1998) approved the use of HDPE as articles or components of articles intended for use in contact with food, subject to the provisions of 21 CFR part 177.1520.

References:

1. Braun, D. D., Hygienic evaluation of products made of medium-pressure polyethylene intended for use in the food industry, *Gig. Sanit.*, 3, 36, 1973 (in Russian).
2. Sheftel', V. O. and Sinitsky, V. G., Attempt at the sanitary evaluation of the use of polyethylene pipes in rural water supply lines, *Gig. Sanit.*, 3, 111, 1973 (in Russian).
3. Danishevsky, S. L. and Broitman, A. Ya., Study of mutagenic activity of chemicals on mammals, in *Toxicology of High-Molecular-Mass Compounds and of Chemical Raw Material Used for Their Synthesis*, S. L. Danishevsky, Khimiya, Moscow-Leningrad, 1966, 21 (in Russian).
4. Kalinin, B. Yu., Zimnitskaya, L. P., and Zalesskaya, V. M., Toxic properties of stabilized low-pressure polyethylene, in *Toxicology and Hygiene of High-Molecular-Mass Compounds and of the Chemical Raw Material Used for Their Synthesis*, Proc. 2nd All-Union Conf., A. A. Letavet and S. L. Danishevsky, Eds., Khimiya, Leningrad, 1964, 110 (in Russian).
5. Michailets, I. B., Sukhareva, L. V., and Yevsyukov, V. I., Hygienic characteristics of high-density polyethylene made with help of vanadium catalysts, in *Environmental Protection in Plastics Production and Hygiene Aspects of Their Use,* Plastpolymer, Leningrad, 1978, 99 (in Russian).
6. Sheftel', V. O., *Hygiene Aspects of the Use of Polymeric Materials in the Water Supply*, Thesis Diss., All-Union Research Institute of Hygiene and Toxicology of Pesticides, Polymers and Plastic Materials, Kiyv, 1977, 61 (in Russian).
7. Kalinin, B. Yu. and Sukhareva, L. V., New polymeric materials for use in the potable water supply, in *Hygiene Aspects of the Use of Polymeric Materials*, Proc. 3rd All-Union Meeting on New Methods of Hygiene Monitoring of the Use of Polymers in the Economy, K. I. Stankevich, Ed., Kiyv, December 2-4, 1981, Kiyv, 1981, 181 (in Russian).
8. Uhde, W. J. and Woggon, H., New results of migration behavior of benzophenone-based UVG absorbents from polyolefins in foods, *Nahrung*, 2, 185, 1976.
9. Goydan, R., Schwope, A. D., Reid, R. C., and Cramer, G., High-temperature migration of antioxidants from polyolefins, *Food Addit. Contam.*, 7, 323, 1990.

HIGH-DENSITY HIGH-MOLECULAR-MASS POLYETHYLENE

produced by gas-phase polymerization

Structural Formula. $[\sim CH_2CH_2\sim]_n$

Abbreviation. HDPE.

Properties. Milky-white odorless material with high heat resistance. Can be used at temperatures from -60°C to +100°C.

Applications. In food industry, farming, etc.

Migration Data. No changes were found in organoleptic properties and oxidizability of extracts (obtained at 20 to 100°C). No migration of catalyst residues and formaldehyde was noted.

Long-term Toxicity. Aqueous (2 to 3 days) and oil (14 to 30 days, 100°C) extracts were administered to rats for 12 to 14 months. No significant changes were revealed in general and specific indices studied in the test animals as compared with the controls.

Regulations. ***U.S. FDA*** (1998) approved the use of high-molecular-mass HDPE in the manufacture of articles or components of articles intended for use in contact with food, subject to the provisions of 21 CFR part 177.1520.

References:

Braun, D. D., Voronel', T. G., Moshlakova, L. A., and Demina, S. Ye., Approval of conditions of use in food industry for high density polyethylene, polymerized by gas-phase method, *Gig. Sanit.*, 11, 12, 1985 (in Russian).

HIGH-MOLECULAR-MASS POLYETHYLENE
produced by gas-phase polymerization

Structural Formula. $[\sim CH_2CH_2\sim]_n$

Abbreviation. LDPE (low-density PE).

Trade Name. Norplast.

Composition and **Applications.** q. v. ***LDPE*** (Norplast)

Migration Data. q. v. LDPE (Norplast). Does not impart any taste or odor to water or simulant medium at temperatures up to 100°C. Specimens prepared with disturbance of the temperature processing schedule release 0.07 to 0.26 mg formaldehyde/l.[1]

Toxicity. See ***LDPE*** (Norplast).

Reference:

Braun, D. D., Voronel', T. G., Moshlakova, L. A., and Demina, S. Ye., Approval of conditions of use in food industry for high density polyethylene, polymerized by gas-phase method, *Gig. Sanit.*, 11, 12, 1985 (in Russian).

HIGH-MOLECULAR-MASS POLYETHYLENE
filled by activation polymerization

Structural Formula. $[\sim CH_2CH_2\sim]_n$

Trade Name. Norplast.

Composition and **Applications.** See ***LDPE*** (Norplast)

Migration of up to 0.14 mg isopropyl alcohol/l was determined in the water extracts and food simulants. Migration of formaldehyde was noted at the levels of 0.07 to 0.47 mg/l.

Toxicity. See ***LDPE*** (Norplast).

Reference:

Braun, D. D., Voronel', T. G., Moshlakova, L. A., and Demina, S. Ye., Approval of conditions of use in food industry for high density polyethylene, polymerized by gas-phase method, *Gig. Sanit.*, 11, 12, 1985 (in Russian).

HIGH-DENSITY POLYETHYLENE, irradiated

Structural Formula. $[\sim CH_2CH_2\sim]_n$

Abbreviation. HDPE.

Migration Data. Samples of water pipes made from HDPE subjected to irradiation with doses accelerated electrons of 0.25 and 0.5 J/kg did no alter the taste or odor of water in contact with them for 4 days at 20 and 80°C. There was no significant increase in oxidizability.

Reference:

Sheftel', V. O., *Hygiene Aspects of the Use of Polymeric Materials in the Water Supply*, Thesis Diss., All-Union Research Institute of Hygiene and Toxicology of Pesticides, Polymers and Plastic Materials, Kiyv, 1977, 61 (in Russian).

MEDIUM-DENSITY POLYETHYLENE

Structural Formula. $[\sim CH_2CH_2\sim]_n$

Abbreviation. MDPE.

Properties. Sometimes has a specific odor different from that of other PE.

Applications. Being equally transparent and having a number of properties exceeding those of *Ziegler* HDPE, it possesses higher economical characteristics. It is used in water piping and in the food industry for the manufacture of containers and films. Articles made of rigid MDPE are used in the food industry at up to 60°C, while MDPE films are used at up to 80°C.

Migration Data. Gas chromatography revealed migration of benzine (solvent) from MDPE into water only at 60°C and with exposure for at least 3 days. MDPE, which does not release determinable benzine residues, is odorless in this case. Benzine can be removed by heating for 3 hours (residual pressure 10 to 20 mm *Hg*).[1]

In supersaturated extracts (5 cm^{-1}; 20 and 60°C; 10 days) from unstabilized MDPE (ash contents 0.6 to 0.9 %), 0.005 to 0.007 mg chromium/l (at 60°C) and 0.002 to 0.008 mg aluminum/l (at 20 and 60°C) have been found. Taste, odor, and pH of extracts were not changed.[2] Chromium is not released into simulant media from a polymer with ash contents of up to 0.3%; its migration into 0.1 *N* solution of *HCl* was observed. Oxidized low-molecular fractions of MDPE are released into heptane. With the increase of chromium contents, white color of MDPE changes from white to grayish or brown, especially when it is stabilized with phenol antioxidants, the effectiveness of which is reduced in this case.[3]

Polycyclic aromatic hydrocarbons can migrate from MDPE into fat extracts. However, benzo[a]pyrene contents of MDPE are half that of LDPE or PP.[4]

Acute Toxicity. A dose of 2.5 g MDPE powder/kg BW (ash contents 0.03 and 0.7%) as a suspension in sunflower oil was administered to mice producing neither toxic effect nor changes in the visceral organs histology.[2]

Long-term Toxicity. Toxicity of aqueous extracts (5.0 cm^{-1}; 60°C; 10 days) from unstabilized MDPE film was investigated in mice and rats. There were a slight increase in oxidizability of extracts as well as traces of chromium and aluminum. Administration to rats for 1.5 years caused no changes as compared to the control animals. In mice, there was a slight decrease in BW gain and some behavioral changes. No histological changes were found. In a 14-month study, aqueous and fat extracts of MDPE (ash contents up to 0.04%) were shown to be harmless.[5]

Regulations. ***U.S. FDA*** (1998) approved the use of MDPE as articles or components of articles intended for use in contact with food, subject to the provisions of 21 CFR part 177.1520.

References:

1. Tarasova, N. A., *Study of Medium-Molecular-Mass Polyethylene Contacting Foods*, Author's abstract of thesis, Moscow, 1971, 20 (in Russian).
2. Shumskaya, N. I., Tolgskaya, M. S., Vikherskaya, T. P., et al., On toxicity of new marks of epoxy resins (EA and DEG-1), in *Toxicology of High-Molecular-Mass Compounds and of Chemical Raw Material Used for Their Synthesis,* S. L. Danishevsky, Ed., Khimiya, Moscow-Leningrad, 1966, 42 (in Russian).
3. Petrovsky, K. S. and Braun, D. D., Outlook for improving the health-related properties of plastic articles for use in the food industry, *Gig. Sanit.,* 1, 75, 1972 (in Russian).
4. Golubev, A. A., On peculiarities of the toxic effect of some organosilicone monomers, in *Hygiene and Toxicology of High-Molecular-Mass Compounds and of the Chemical Raw Material Used for Their Synthesis*, Proc. 4th All-Union Conf., S. L. Danishevsky, Ed., Khimiya, Leningrad, 1969, 47 (in Rus sian).
5. Braun, D. D., Hygienic evaluation of products made of medium-pressure polyethylene intended for use in the food industry, *Gig. Sanit*., 3, 36, 1973 (in Russian).

LOW-MOLECULAR-MASS POLYETHYLENE

Structural Formula. $[\sim CH_2CH_2\sim]_n$

M < 10000

Abbreviation. LMPE.

Properties. Waxy resin. Viscous, milky white, colorless substance. Insoluble in water; poorly soluble in alcohol; mixes well with vegetable oil.

Applications. Permitted and used in rubber employed in the food industry.

Acute Toxicity. Administration of 5.0 and 10 g LMPE/kg BW in sunflower oil had no effect on rats: gross pathology examination revealed no changes attributed to the exposure.

Long-term Toxicity. Rats received 0.2 and 0.5 g LMPE/kg BW over a period of 6.5 months. No toxic effect was observed in treated animals. However, 0.2 g/kg BW dose has been considered as the NOEL in chronic toxicity studies.

Regulations. ***U.S. FDA*** (1998) approved the use of LMPE as articles or components of articles intended for use in contact with food, subject to the provisions of 21 CFR part 177.1520.

Reference:

Stasenkova, K. P., Shumskaya, N. I., and Sergeyeva, L. G., Toxicity of low-molecular-mass polyethylene, in *Toxicology of New Industrial Chemicals*, Meditsina, Moscow, Issue 14, 1975, 138 (in Russian).

POLYETHYLENE WAX

CAS No 9002-88-4

Abbreviation. PW.

Composition and **Properties.** Fine, white granules; *M. p.* 110 to 112°C; ash contents 0.02%; no antioxidants. Melts (1:10 and 1:5 parts) of PW with *P*-grade edible paraffin have been investigated.

Applications. Paper with PW coating is used for packaging foodstuffs.

Migration Data. Organoleptic properties of extracts remain unaltered after 3-day (water) and 10-day (milk, oil) contact. Containers made of paper coated with PW-paraffin melt swell in water.

Acute Toxicity. Single administration of 10 g PW/kg BW to mice and rats is not lethal and has no toxic effect over a period of 3 weeks.

Regulations. ***Russia.*** PW-edible paraffin melts are permitted for coating paper and cardboard containers for food products.

Reference:

Maximova, N. S. and Mishina, A. D., Possibility of use of polyethylene wax in the production of food coatings, in *Environmental Protection in Plastics Production and Hygiene Aspects of Their Use,* Plastpolymer, Leningrad, 1978, 111 (in Russian).

POLYETHYLENE, chlorinated

CAS No 64754-90-1

Abbreviation. PEC.

Production. The product of action of gaseous chlorine on a solution or suspension of polyethylene.

Applications. Used as a binder for paints and adhesives and in manufacture of coatings and films.

Migration Data. In extracts (1 cm^{-1}; 20, 40, and 100°C) of SP-12 PEC, with both long-term contact and daily change, water acquired a specific odor and taste. Oxidizability of water was increased.

Long-term Toxicity. General toxic action of aqueous extracts was established in a 6-month study in rats.

Regulations. ***U.S. FDA*** (1998) approved the use of PEC as articles or components of articles that contact food, except for articles used for packing or holding food during cooking, subject to the provisions of 21 CFR part 177.1610.

Reference:

Cherkinsky, S. N., Rubleva, M. N., and Korolev, A. A., Sanitary-toxicological evaluation of the new polymeric materials used in water supply, in *Hygiene Aspects of the Use of Polymeric Materials in Construction*, Proc. 1st All-Union Meeting on Health and Safety Monitoring of the Use of Polymeric Materials in Construction, Kiyv, 1973, 154 (in Russian).

ETYLENE COPOLYMERS

Applications. Ethylene copolymers have been approved for use with aqueous acid and non-acid foods containing free oil that are heat sterilized at temperatures over 100°C if the material is less than 0.02 inch thick or hot-filled, pasteurized at 65°C if the material is between 0.004 and 0.02 inches thick.

ETHYLENE and 1-BUTENE (α-butylene) COPOLYMER

Structural Formula. $[\sim CH_2CH_2\sim]_m[\sim CH_2CH\sim]_nCH_2CH_3$

M = 30,000 to 800,000

Abbreviation. EBC.

Composition. CAO-6 and antioxidant 2246 (CAO-5) are used as stabilizers. Normally it contains 0.2 to 0.3% butylene.

Applications. Used for manufacture of rigid containers, dishware, piping, various articles.

Migration Data. With a certain melt index and minimum ash contents, and provided optimum technology is used for processing, EBC does not give water any foreign odor or taste even at 50 to 60°C. Variable migration of formaldehyde (q.v.) and traces of benzine (if the latter is used as a polymerization agent) are found in the extracts, but migration of hexane is not observed. Residues of the catalyst (chromium compounds) and of stabilizers (phenol and I-butanol) do not migrate into water. Oxidizability of an extract of unstabilized articles is perhaps dependent on migration of solvent while that of an extract of stabilized articles is also dependent on a certain amount of antioxidants.

Long-term Toxicity. Rats received 7-day aqueous extracts of the copolymer over a period of 15 months. No toxic effect was observed.

Recommended. May be used up to 50°C in rigid articles and up to 60°C in films.

Regulations. ***U.S. FDA*** (1998) approved the use of EBC as articles or components of articles intended for use in contact with food, subject to the provisions of 21 CFR part 177.1520.

Reference:

Braun, D. D. and Zenina, G. V., Hygienic evaluation of an ethylene copolymer with butylene intended for use in the food industry, *Gig. Sanit.*, 9, 46, 1975 (in Russian).

ETHYLENE and VINYL ACETATE COPOLYMER

Structural Formula.

$$[\sim CH_2CH_2\sim]_m[\sim CH_2\underset{\displaystyle OCOCH_3}{\underset{|}{C}H}\sim]_n$$

CAS No 24937-78-8

Abbreviations. EVA copolymer; VA (vinyl acetate); LDPE (low-density PE).

Composition. EVA copolymer contains up to 70% vinyl acetate and 30% ethylene. Residual VA contents 0.56%; solid contents 51.5 to 53%. A dye is added to the composition.

Trade Names. Cevilene; Elvax; Sevilen.

Properties. EVA copolymer is flexible packaging material formed from low-density polyethylene and VA. It is more flexible than polyethylene, but it is more permeable to water and gases. EVA copolymers are unstable at high temperatures, but stable at low temperatures. EVA copolymer is cleared for use with fatty foods and for treatment with irradiation as a sterilant (up to 8.0 megarads total) so heat sterilization is not necessary. It stretches very easily so it can be used as a shrink wrap.

Applications. EVA copolymer used to produce various engineering components, adhesives, coatings for paper and cardboard, films, packaging for food products, children's toys, and to manufacture plastic bags for holding solutions in medical delivery devices. Used as an additive to wax coatings and as material for transparent shock-resistant film. Different types contain different proportions of VA.

Migration Data. When EVA copolymer contains 1.5 to 7.15% VA, the organoleptic properties of extracts (1 cm^{-1}; 80 and 20°C; 1, 3, and 7 days) of a 50-μm thick film were satisfactory; no migration of VA was found.[1]

EVA copolymer with 7.0 to 10% VA contents has hygienic properties similar to those of LDPE. Studies revealed no migration of VA into water, simulant media, or food products (20 to 80°C; 1 day to 12 months). The thicker the article, the stronger the odor.[2]

EVA copolymer with VA contents up to 14% possesses satisfactory hygienic properties: migration of VA was 0.1 mg/l. When VA contents are increased to 35%, taste and odor in the aqueous extracts become stronger, and oxidizability increases. However, in most cases, no migration of VA is found and, even in aggravated conditions, it does not exceed the permissible level of 0.2 mg/l. Migration of other additives (plasticizers, dyes, fillers) can occur.[3,4]

Toxicity. A dispersion of EVA copolymer proved to be non-toxic in acute and chronic experiments with mice and rats.[5]

Reproductive Toxicity. EVA copolymer dispersion had no effect on the reproductive function or on offspring development.[5]

Regulations. ***U.S. FDA*** (1998) approved the use of EVA copolymers as articles or components of articles intended for use in producing, manufacturing, packing, processing, preparing, treating, packaging, transporting, or holding food in accordance with the provisions described in 21 CFR part 177.1350.

Recommendations. ***Russia.*** D-13 and D-23 *Miraviten* copolymer is recommended for use in contact with food products at room temperature (open-type containers and films). The copolymer is permitted as a coating for parchment and cardboard used to pack dry and free-flowing products.[3]

References:

1. Boikova, Z. K. and Petrova, L. I., Effect of some technological parameters on hygienic properties of ethylene-vinyl acetate copolymers, in *Environmental Protection in Plastics Production and Hygiene Aspects of Their Use,* Plastpolymer, Leningrad, 1978, 114 (in Russian).
2. Boikova, Z. K. and Petrova, L. I., Sanitary-chemical characteristics of vinyl acetate copolymers and of compositions on their base, in *Hygiene and Toxicology of High-Molecular-Mass Compounds and of the Chemical Raw Material Used for Their Synthesis,* Proc. 6th All-Union Conf., B. Yu. Kalinin, Ed., Khimiya, Leningrad, 1979, 59 (in Russian).
3. Boikova, Z. K., Petrova, L. I., and Stroyeva, I. N., Sanitary-chemical evaluation of thermoglue TK-2P made of ethylene-vinyl acetate copolymer, *Gig. Sanit.*, 5, 71, 1983 (in Russian).
4. Boikova, Z. K., Petrova, L. I., and Slusareva, I. P., Possible use of sevilene in contact with foodstuffs and in toy production, *Gig. Sanit.*, 7, 87, 1983 (in Russian).
5. Maksimova, N. S. and Mikhailets, I. S., *Plast. Massy*, 12, 37, 1976 (in Russian).

ETHYLENE and PROPYLENE COPOLYMER

Structural Formula.

$[\sim CH_2CH_2\sim]_m\ [\sim CH_2CH\sim]_n$ (with CH_3 attached to CH)

M = 80,000 to 500,000

CAS No 9010-79-1

30966-34-8

Abbreviations. CEP; LDPE (low-density PE).

Properties. Insoluble in water and alcohol. CEP containing 20% propylene possesses greater mechanical strength and heat resistance (120 to 125°C) than LDPE, and greater cold resistance than polypropylene. Its chemical stability is similar to that of polyethylene.

Applications. Used in production of plastics, films, and rubber.

Migration Data. In aqueous extracts (20 cm^{-1}; 20 and 60°C; 10 days), the greatest concentrations of formaldehyde and methyl alcohol were found in the first extracts taken and were 0.018 and 0.2 mg/l, respectively. Titanium and aluminum ions were not found in water.[1] The organoleptic index was the limiting index in hygienic assessment of extracts.

Long-term Toxicity. When mice and rats received aqueous extracts of CEP containing 4.0 to 7.0% propylene over a period of 250 days, no functional or organic changes occurred.[2]

Regulations. ***U.S. FDA*** (1998) approved the use of CEP as articles or components of articles intended for use in contact with food, subject to the provisions of 21 CFR part 177.1520

References:

1. Rogovskaya, A. P. and Barabanova, L. N., Sanitary surveillance on the use of polymeric materials and articles made of them in the Ukraine, in *Hygiene and Toxicology of High-Molecular-Mass*

Compounds and of the Chemical Raw Material Used for Their Synthesis, Proc. 6th All-Union Conf., B. Yu. Kalinin, Ed., Khimiya, Leningrad, 1979, 45 (in Russian).
2. *Hygiene and Toxicology of Plastics,* Coll. Sci. Proc. Kiyv Medical Institute, Kiyv, 1979, 45 (in Russian).

ETHYLENE and PROPYLENE COPOLYMER containing 3.0% ethylene

Abbreviation. CEP.

Properties. Japanese film 2061 (thickness 40-50.5 mm)

Migration Data. Causes no significant changes in the organoleptic properties of water in contact with it (up to rating 1). Methyl ether and formaldehyde were not found in extracts (2.0 cm^{-1}; 20 to 80°C; 10 days).

Regulations. ***U.S. FDA*** (1998) approved the use of CEP as articles or components of articles intended for use in contact with food, subject to the provisions of 21 CFR part 177.1520

Reference:

Toxicology and Medical Chemistry of Plastics, Coll. Sci. Tech. Abstracts, NIITEKHIM, Moscow, Issue No 1, 1979, 12 (in Russian).

PROPYLENE and ETHYLENE BLOCK COPOLYMER

Composition. Obtained by successive structural block-copolymerization (monomers: 90 to 93% propylene to 7.0 to 10% ethylene; antioxidants: 0.2 to 0.5% Irganox 1010 and 0.2% calcium stearate).

Properties. Articles are transparent with a grayish tint, smooth-surfaced, and elastic. They have no odor and give no smell or taste to the simulant media.

Applications. Used to manufacture dishware, films, and packaging in the food industry.

Migration Data. It has no significant effect on the composition and properties of extracts.

Long-term Toxicity. In a 14-month study, aqueous and oil extracts were shown to be harmless.

Reference:

Braun, D. D., Voronel', T. G., and Moshlakova, L. A., Hygienic substantiation of the possibility of using new polyolefins (propylene-ethylene block copolymer), *Gig. Sanit.*, 9, 34, 1984 (in Russian).

POLYPROPYLENE

Structural Formula.

[~CH_2CH~]$_n$
        |
      CH_3

M = 75,000 to 200,000

CAS No 9003-07-0
52622-64-7

RTECS No UD1842000

Abbreviations. PP; PE (polyethylene); LDPE (low-density PE).

Composition. Product of polymerization of propylene in presence of complex organometallic catalysts (AlR_3, $TiCl_4$, etc.) in hydrocarbon solvents.

Synonyms and **Trade Names.** Amoco 1010; Bicolene P; Daplen AD; Elpon; Hercules 6523; Hostalen PPH or PPN; Morlen AD; Novolen; Polypropene; Poprolin; Propathene; Trespaphan.

Properties. PP was introduced in the 1950s. It was used successfully in the mid-1960s as a copolymer with PE, but it was eventually replaced by plain PE for bags. Colorless polymer, without odor or taste, resembles LDPE. *Dens.* 0.92 to 0.93. It provides mechanical properties (strength, rigidity, abrasion resistance) at low cost. It is very transparent, but it cracks and breaks at low temperatures. PP is more rigid, stronger, and lighter than PE. It is resistant to water vapor, grease, acids, bases, and solvents, and some PP are resistant to high temperature. Heat distortion occurs at 45°C; maximum continuous service temperature is 40°C. Its chemical stability is similar to that of PE but it is more prone to oxidation. PP exhibits low water absorption and moisture permeability. Shock resistance is intermediate between that of high-impact polystyrene and LDPE.

PP is cleared for irradiation sterilization up to 1.0 megarad total.

Applications. Used to manufacture the same articles as PE. PP-I finely dispersed powder with melt flow index of 10 to 30 g/10 min is used to produce coatings. PP is used for yogurt containers, margarine tubs, and some bottles for pourable foods like syrup. PP is used to make containers for medicines and solutions for injections, syringes, and eye glasses.

PP is used for hot water piping, and also in water pipes operating at a pressure of 10 *MPa.* Its cold resistance is significantly inferior to that of polyethylene piping.[1]

Migration Data. The intensity of migration of organic substances from PP is in inverse relation to the melt flow index and the contents of atactic fraction in the polymer. At 20°C, migration of organic substances amounted to 1.8 mg O_2/l (iodate oxidizability), and migration of isopropyl alcohol (IPA) to 0.5 mg/l. At 60°C, release of IPA reaches 4.5 mg/l, that of methyl alcohol 0.21 mg/l, and that of formaldehyde 0.013 mg/l.[1,2] PP has no significant effect on water taste or odor. Migration of a small quantity of oxidizing substances, chlorides, formaldehyde (0.052 to 0.425 mg/l at 60°C), and methyl alcohol (0.01 to 0.11 mg/l) into water was noted.[3]

Water in contact with PP-3 (PP-4) stabilized with carbon black and antioxidant 2246 (CAO-5) acquired an odor and taste which persisted for 13 hours although their intensity weakened. Migration of low-molecular organic compounds and isopropanol reached was noted (20°C). Methyl alcohol, formaldehyde, and Cl^- ions were found only in extracts at 60°C in concentrations of 0.21, 0.013, and 0.6 mg/l, respectively.[3]

Despite the fact that extracts (3 cm^{-1}; 20 and 60°C; 5 days) of PP sample studied exhibited no toxic properties during prolonged administration to mice and rats, the material was not recommended for application in water piping because of persistent specific odor.[4]

A study was made of the influence on water quality of Propatene-grade PP used to manufacture filters for artesian wells (stabilizers: 3.0% CAO-6, 0.3% dilauryl-3,3'-thiodipropionate, and 0.5% tinuvin). Aqueous extracts (1.0 cm^{-1}) had no odor or taste but they did contain a certain quantity of organic substances. After contact for 10 days, the oxidizability of the extracts increased by 0.16 to 0.24 mg O_2/l. No unsaturated low-molecular-mass compounds were found.[5]

Considering the service conditions of filters, investigations were conducted on 2-day extracts of PP that had been in water for 6 months. There were no aldehydes in the aqueous extracts. Parts of immersion pumps of PP grades 05PO90-V, 04PO90-V, and 04PO90-UP stabilized with dilauryl-3,3'-thiodipropionate (DLTDP), Topanol CA, benzone OA, and calcium stearate were investigated. The organoleptic properties of water remained unchanged. A small increase in oxidizability was noted especially when water temperature was raised and exposure was lengthened compared to standard conditions. Topanol CA and benzone OA migrate into water in concentrations of <0.05 and 0.2 mg/l, respectively, while DLTDP hardly migrates into water.[6]

In extracts (2 cm^{-1}; 20 and 80°C; 1 to 10 days) of Japanese films of P. grades F2062 and F9750 there was no significant change in organoleptic properties and no migration of methyl alcohol or for maldehyde.[7]

Braun and Ishchenko investigated base PP grades 21007-12, -15, -20, and -30, thermally stabilized with Irganox 1010 and calcium stearate. No catalyst residue (*Ti* ions), stabilizers (Irganox 1010), or formaldehyde were found in aqueous extracts or model media.[8]

Migration rates of radiolabeled antioxidants, Irganox-1010 and Irganox-1076 from PP were measured at temperature up to 135°C. Water, 8.0 and 95% aqueous solution of ethanol, and corn oil were employed as food simulating liquids. The losses to foods were usually larger than those to water but below those to corn oil.[9] In certain batches of the polymer, migration of traces of butanol and hexane into extracts was observed. Benzene levels in commercial PP products examined were from non-detected to 426 ppb.[10]

A very low elution of ultra-violet absorbers and light stabilizers of PP into water and aqueous acid solutions simulated by a solution of acetic acid was found. Migration of 3,5-di-*tert*-butyl-4-hydroxy-[2,4-di-*tert*-butylphenyl]benzoate from PP into 50% alcohol attains values in the order of mg/dm^2 of the surface of the plastic and is, moreover, 3 to 20 times higher in case of contact with fats or fat-simulating liquids.[11]

Migration of plasticizers from printing ink (of PP packaging film) into foods has been studied. It was demonstrated that there can be transfer of components from the ink on the outer surface of the film on to the inner food-contact surface. For dicyclohexyl phthalate, this transfer amounted to 6.0% of the total amount of plasticizer available in the printing ink system. The migration of plasticizer increased with storage time of the wrapped product: for dibutyl phthalate, levels increased from 0.2 to 6.7 mg/kg over the period from 0 to 180 days. Migration of dibutyl phthalate from snack products and biscuits wrapped in printed PP film reached 0.02 to 14.1 mg/kg.[12]

Migration of antioxidants (Irgafos 168, Irganox 1076, and Hostanox SE2) from PP films into food simulant media (water, 3.0 % acetic acid, 95% ethanol, olive oil, and heptane) has been studied. PP films (50, 100, and 200 microns thick) were exposed to the simulant media at temperature-time conditions simulating migration under long-term storage. Global migration into aqueous simulant media was independent of film thickness and conditions of exposure. Specific migration into heptane was independent of the polymer mass, though dependent on the thickness. Migration into ethanol was dependent on both mass and thickness. Global migration to fatty food simulant media was dependent on simulant, conditions of exposure, and, in some cases, film thickness. Migration of Irgafos 168 [tris (2,4-di-*tert*-butyl-phenyl)phosphite] into aqueous simulant media was below the detection limit (0.01 mg/dm^2).[13]

Acute Toxicity. After feeding rats with ^{14}C-PP, no traces of the isotope were found in the bodies of animals.[04] Administration of 5.0 g PP/kg BW to mice was not lethal but in mice 2 weeks after administration pathological changes were found in the liver, kidneys, and myocardium.[14]

However, according to other data, administration of 8.0 g unstabilized PP/kg BW was not lethal to mice, and has no toxic effect on them. There were no findings related to the treatment with PP either at necropsy or at histological examination.[15] Similar results were noted by Stasenkova after administration of 10 to 15 g PP/kg BW to mice.

Repeated Exposure. Five administrations of unstabilized PP powder (ash contents 0.03 and 0.1%) in peach oil caused neither significant effect on BW gain and other indices, nor histological changes in mice.[1,2]

Long-term Toxicity. When mice and rats received aqueous extracts (20 and 60°C; 10 days) of unstabilized PP over a period of 15 months, very small changes in BW gain of mice and relative liver weights in rats occurred. Antibody-forming ability in rats was reduced. Aqueous extracts of Propatene-grade PP given for 6 months had no adverse effect on animals.[5]

Reproductive Toxicity. Epidemiological studies revealed no association between occupation in PE production and the incidence of spontaneous abortion.[16]

Carcinogenicity. Aqueous extracts of unstabilized PP or PP stabilized with CAO-6, given to animals for their lifetime, did not cause tumors.[1,2]

Carcinogenicity classification. An IARC Working Group concluded that there is inadequate evidence for the carcinogenicity of PE in *experimental animals* and there were no adequate data available to evaluate the carcinogenicity of PE in *humans*.

IARC: 3

Chemobiokinetics. Rats received ^{14}C-labeled samples of PP (fractions dissolved in vegetable oil). Authors concluded that insoluble fractions of PP are excreted unchanged through the body. No radioactivity has been found in the body.[17]

Regulations. ***U.S. FDA*** (1998) approved the use of PP (1) in the manufacture of resinous and polymeric coatings for the food-contact surface of articles intended for use in producing, manufacturing, packing, processing, preparing, treating, packaging, transporting, or holding food (for use only as polymerization cross-linking agent in side seam cements for containers intended for use in contact with food (only of the identified types) in accordance with the conditions prescribed in 21 CFR part 175.300; (2) in the manufacture of cellophane for packaging food in accordance with the conditions prescribed in 21 CFR part 177.1200; (3) as articles or components of articles intended for use in contact with food, subject to the provisions of 21 CFR part 177.1520; and (4) in adhesives as a component (monomer) of articles intended for use in packaging, transporting, or holding food in accordance with the conditions prescribed in 21 CFR part 175.105.

References:

1. Braun, D. D., Zenina, G. V., and Moshlakova, L. A., Hygienic evaluation of new polyolefin group polymeric materials intended for use in the food industry, *Gig. Sanit.*, 2, 24, 1979 (in Russian).
2. Braun, D. D., Moshlakova, L. A., and Zenina, G. V., Hygienic evaluation of new polyolefin class polymeric materials for the food industry, *Gig. Sanit.*, 6, 29, 1980 (in Russian).
3. Kalinin, B. Yu. and Sukhareva, L. V., Sanitary-chemical and toxicological evaluation of water extracts from unstabilized polypropylene, in *Toxicology and Hygiene High-Molecular-Mass Compounds and of the Chemical Raw Material Used for Their Synthesis,* Proc.3rd All-Union Conf., S. L. Danishevsky, Ed., Khimiya, Moscow-Leningrad, 1966, 121 (in Russian).
4. Komarova, L. V., Sukhareva, L. V., Guricheva, Z. G., and Robachevskaya, Ye. G., Hygienic evaluation of polypropylene intended for production of pipes for potable water supply, in *Hygiene and Toxicology of High-Molecular-Mass Compounds and of the Chemical Raw Material Used for Their Synthesis*, Proc. 4th All-Union Conf., S. L. Danishevsky, Ed., Khimiya, Leningrad, 1969, 44 (in Russian).
5. Trubitskaya, G. P., Hygienic significance of different regimens of thermo-treatment of glass-reinforced plastics made on the base of unsaturated polyether resins, *Ibid*, 1969, 142
6. Sheftel', V. O., *Hygiene Aspects of the Use of Polymeric Materials in the Water Supply*, Thesis Diss., All-Union Research Institute of Hygiene and Toxicology of Pesticides, Polymers and Plastic Materials, Kiyv, 1977, 62 (in Russian).
7. *Toxicology and Medical Chemistry of Plastics,* Coll. Sci. & Techn. Abstracts, NIITEKHIM, Moscow, Issue No 1, 1979, 12 (in Russian).
8. Braun, D. D. and Ishchenko, Ye. O., Data to hygienic evaluation of new sorts of polypropylene intended for use in food industry, *Gig. Sanit.*, 10, 22, 1981 (in Russian).
9. Goydan, R., Schwope, A. D., Reid, R. C., and Cramer, G., High-temperature migration of antioxidants from polyolefins, *Food Addit. Contam.*, 7, 323, 1990.
10. Varner, S. L., Hollifield, H. C., and Andrzejewski, D., Determination of benzene in polypropylene food-packaging materials and food-contact paraffin waxes, *J. Ass., Off. Anal. Chem.*, 74, 367, 1991.
11. Horacek, J. and Uhde, W. J., Plastics from the aspect of hygiene. A contribution to the estimation of ultra-violet absorbers and light stabilizers from the aspect of hygiene, *J. Hyg. Epidemiol. Immunol.*, 24, 133, 1980.
12. Catsle, L., Mayo, A., and Gilbert, J., Migration of plasticizers from printing ink into foods, *Food Addit. Contam.*, 6, 437, 1989.
13. Garde, J. A., Catala, R., and Gavara, R., Global and specific migration of antioxidants from polypropylene films into food simulants, *J. Food Prot.*, 61, 1000, 1998.
14. *Toxicology of New Industrial Chemicals*, Medgiz, Moscow, Issue No 5, 1963, 136 (in Russian).
15. Stankevich, V. V. and Tverskaya, M. Ya., Some problems of toxicodynamics of chlorinated cyclic ethers in the experiment, in *Toxicology of High-Molecular-Mass Compounds and of Chemical Raw Material Used for Their Synthesis,* S. L. Danishevsky, Ed., Khimiya, Moscow-Leningrad, 1966, 55 (in Russian).
16. Lindbohm, R. et al., Spontaneous abortion among women employed in the plastics industry, *Am. J. Ind. Med.*, 8, 579, 1985.
17. Grant, W. M., *Toxicology of the Eye*, 2nd ed., Charles C Thomas, 1974, 50.

POLYPROPYLENE, modified

Structural Formula.

$$[\sim CH_2CH\sim]_n$$
$$|$$
$$CH_3$$

Abbreviation. MPP.

Trade Name. Poprolin.

Composition. Obtained using cross-linking agent (alkaline lignin sulfate), plasticizer (dioctyl sebacinate, dibutyl sebacinate, dibutyl phthalate), etc.

Properties. MPP has greater molecular mass, cold and heat resistance, and breakdown resistance compared to polypropylene.

Migration Data. The optimum contents of lignin, plasticizer, and titanium dioxide in rigid and film articles made of MPP were found to be 1.0, 10, and 20%, respectively. Migration of the washing agent (isopropyl alcohol) reaches 2.0 mg/l. Migration of up to 0.5 mg formaldehyde/l was found in some articles (boxes, trays). After 5 to 9 water changes, isopropyl alcohol, aluminum, and titanium ions are not released into water.

Long-term Toxicity. Administration of aqueous and fat extracts to rats (for 13 months) and mice (for their lifetime) did not cause pathology in the animals. No cocarcinogenic effects were found.

Recommendations. Recommended for use as films up to 80°C, and as containers up to 30°C.

Reference:

Braun, D. D., in *Problems of Hygiene of Nutrition*, Moscow, 1972, 29 (in Russian).

POLYPROPYLENE, modified with special additives

Structural Formula.

[~CH_2CH~]$_n$
|
CH_3

Properties. Obtained as a result of introduction into finished PP of the substances with functional groups that affect the supermolecular structure of the material; another way is to create compositions with different polymeric low-molecular compounds. Modifying additives: organosilicon compounds (silicone liquids, polyethylsiloxane lubricants, etc.) and oxides of metals (titanium, aluminum ions, etc.). Its characteristics depend on the grade of initial PP, formulation of the composition. and production procedure. Stabilizers used (Irganox 1010, Topanol CA, dilaurylthiodipropionate, etc.) are prone to migration. Articles are safe for use at up to 80 and even 100°C.

Migration Data. Traces of isopropyl alcohol were found in extracts.

Toxicity. Addition of oil and aqueous extracts to the diet of rats over a period of 15 months caused no pathological changes.

Recommendations. ***Russia***. Rigid containers based on food-grade PP are recommended for use in contact with foodstuffs at up to 80°C. Films may be used at 100°C.

References:

1. Braun, D. D. and Zenina, G. V., Hygienic evaluation of ethylene copolymer with butylene intended for use in the food industry, *Gig. Sanit.*, 9, 46, 1975 (in Russian).
2. Braun, D. D. and Zenina, G. V., in *Problems of Hygiene of Nutrition*, Moscow, 1975, 61 (in Russian).

POLYPROPYLENE filled by activation polymerization

Structural Formula.

[~CH_2CH~]$_n$
|
CH_3

Synonym. Norplast.

Migration Data. See ***LDPE*** (Norplast). Does not impart any taste or odor to water or simulant medium at temperatures up to 100°C. Migration of up to 0.33 mg isopropyl alcohol/l, up to 0.21 mg butyl alcohol/l, and up to 0.08 mg heptane/l was noted. Specimens prepared with disturbance of the temperature processing schedule release 0.07 to 1.13 mg formaldehyde/l.[1]

Toxicity. See ***LDPE*** (Norplast).

Reference:

Braun, D. D., Voronel', T. G., Moshlakova, L. A., Demina, S. Ye., and Chernitsyna, M. A., Hygienic substantiation of the possibility of using norplast based on polyolefins in food outlets of an agribusiness complex, *Gig. Sanit.*, 5, 20, 1988 (in Russian).

PROPYLENE and ETHYLENE BLOCK-COPOLYMER

CAS No 56453-76-0

Composition. The product is block-polymerized in proportion: propylene, 90 to 93%, ethylene, 7.0 to 10%. Antioxidant: Irganox 1010 (q.v.), 0.2 to 0.5%, calcium stearate, 0.2%.

Properties. The material does not impart any taste or odor to water or model media.

Applications. Used in the production of dishware and films for food packaging.

Migration Data. Produced no effect on quality of simulant media.

Long-term Toxicity. In a 14-month study, there were no signs of toxic action in animals given aqueous or oil extracts.

Reference:

Braun, D. D., Voronel', T. G., and Moshlakova, L. A., Hygienic substantiation of the possibility of using new polyolefins (propylene-ethylene block copolymer), *Gig. Sanit.*, 9, 34, 1984 (in Russian).

POLY-1-BUTENE

Structural Formula.

$[\sim CH_2CH\sim]_n$

$\quad\quad |$

$\quad\quad CH_2CH_3$

M = 50,000 to 100,000

CAS No 9003-29-6

RTECS No EM9032000

Abbreviation. PB.

Synonym and **Trade Names.** Amoco 15H; Oronite 6; Petrofin 100; Poly-α-butylene.

Composition. The product of polymerization of 1-butene using complex catalysts $TiCl_3$, $Al(C_2H_5)Cl$, etc., and solvent systems (benzine, isopropyl alcohol). Ash contents are 0.03%, stabilizer (Irganox 1010) contents of polymer are 0.2%.

Properties. Articles of PB are translucent and have a smooth surface.

Applications. May be used in production of containers, films, and packing, in pipes to be used under pressure at temperatures up to 98°C for supply of hot water, aqueous suspensions, and corrosive foodstuffs. Plasticizer in the manufacture of polyolefins and elastomers, synthetic and natural rubber, component of hot-melt adhesives, sealants, leather and paper coatings; used as a food packaging material, and in cosmetics.

Migration Data. No effect on organoleptic indices of water at temperatures between -15°C and +80°C was found. There was no migration of chemical substances into extracts with the exception of isopropyl alcohol (up to 1.25 mg/l).[1]

Long-term Toxicity. Addition of oil and aqueous extracts (95 to 100°C and 20°C) to the diet of rats over a period of 15 months caused no pathological changes.[1]

Carcinogenicity. *In vitro* tests with animal cells showed that PB can act as a tumor promoter.[2]

Recommendation. ***Russia.*** Material was approved for use at up to 80°C (containers) and up to 100°C (films).

Regulations. ***U.S. FDA*** (1998) approved the use of PB (1) in adhesives as a component (monomer) of articles intended for use in packaging, transporting, or holding food in accordance with the conditions prescribed in CFR part 175.105; (2) as articles or components of articles intended for use in contact with food, subject to the provisions of 21 CFR part 177.1520; and (3) as articles or components of articles intended for use in contact with food, subject to the provisions of 21 CFR part 177.1570.

References:

1. Braun, D. D., Moshlakova, L. A., and Zenina, G. V., Hygienic evaluation of new polyolefin class polymeric materials for the food industry, *Gig. Sanit.*, 6, 29, 1980 (in Russian).
2. Aarsaether, N., Lillehaug, J. R., Rivedal, E., and Sanner, T., Cell transformation and promoter activity of insulation oils in the Syrian hamster embryo cells and in the C3H/10t1/2 mouse embryo fibroblast test systems, *J. Toxicol. Environ. Health*, 20, 173, 1987.

POLYISOBUTYLENE

Structural Formula.

$$[\sim CH_2\overset{\overset{\displaystyle CH_3}{|}}{\underset{\underset{\displaystyle CH_3}{|}}{C}}\sim]_n$$

M = 70,000 to 225,000

CAS No 9003-27-4

RTECS No UD1010000

Abbreviation. PIB.

Synonym and **Trade Name.** 2-Methylpropene, polymer; Oppanol.

Properties. Rubber-like amorphous product with good chemical stability. T_{soft} 90 to 100°C. High elasticity limits the application of PIB but the elasticity decreases when graphite or carbon black is used as a filler.

Applications. Used for coatings in containers and large-diameter pipes, for sealing, in manufacture of adhesives, and in composition with other polymers.

Migration Data. Water quality was not affected in aqueous extracts (1 cm^{-1}; 20 and 37°C; 10 days). Does not significantly increase the oxidizability of water or interact with chlorine dissolved in water, or alter the ability of water to absorb chlorine, but does release unsaturated compounds into water.[1] No migration of catalyst (BF_3) was found.

Migration of PIB, used as wrapping of foods and in reheating in a microwave oven, from polyethylene/polyisobutylene film into foods was investigated by Castle et al. Levels of migration determined were in the range of 8 to 10 mg/kg (into cheese), 1.0 to 5.0 mg/kg (into cake), and from 1.0 to 4.0 mg/kg (into sandwiches).[2]

Long-term Toxicity. When rats received aqueous extracts of PIB for 12 months, their overall condition, hematological, biochemical, physiological, and morphological indices were unaffected.[3]

Regulations.

U.S. FDA (1998) approved the use of PIB (1) as components of articles intended for use in producing, manufacturing, packing, processing, preparing, treating, packaging, transporting, or holding food in accordance with the conditions prescribed in 21 CFR part 177.1420; (2) in the manufacture of resinous and polymeric coatings for the food-contact surface of articles intended for use in producing, manufacturing, packing, processing, preparing, treating, packaging, transporting, or holding food (for use only as polymerization cross-linking agent in side seam cements for containers intended for use in contact with food (only of the identified types), subject to the conditions prescribed in 21 CFR part 175.300; (3) in the manufacture of closures with sealing gaskets used on containers intended for use in producing, manufacturing, packing, processing, preparing, treating, packaging, transporting, or holding food, in accordance with the conditions prescribed in 21 CFR part 177.1210; (4) as a component of the uncoated or coated food-contact surface of paper and paperboard intended for use in producing, manufacturing, packaging, processing, preparing, treating, packing, transporting, or holding dry food of the type identified in 21 CFR part 176.170 (c); (5) in the manufacture of cellophane for packaging food in accordance with the conditions prescribed in 21 CFR part 177.1200; and (6) in adhesives as a component (monomer) of articles intended for use in packaging, transporting, or holding food in accordance with the conditions prescribed in 21 CFR part 175.105.

Russia. High-molecular-mass PIB of grade P-200 with M = 175,000 to 225,000 is permitted for use in the food industry.

References:

1. Sheftel', V. O., Study of toxicity of water extracts from some plastic pipes for water supply, in *Toxicology and Hygiene of High-Molecular-Mass Compounds and of the Chemical Raw Material Used for Their Synthesis,* Proc. 3rd All-Union Conf., S. L. Danishevsky, Ed., Khimiya, Moscow-Lenirgrad, 1966, 123 (in Russian).
2. Castle, L., Nichol, J., and Gilbert, J., Migration of polyisobutylene from polyethylene/polyisobutylene films into foods during domestic and microwave oven use, *Food Addit. Contam.*, 9, 315, 1992.
3. Sheftel', V. O., in *Hygiene of Populated Locations*, Zdorov'ya, Kiyv, 1969, 68 (in Russian).

COATING 4P

Composition. Contains paraffin, polyisobutylene, polypropylene, and polyethylene.

Applications. Intended for coating domestic drinking-water storage tanks.

Migration Data. With exposure for up to 10 days, odor or taste of water in contact with C. did not changed; by 30 days the intensity of paraffin odor was rating 3 with taste rating 2; this level was preserved until the end of the observation period (4 months). C. had no effect on color, transparency, pH, oxidizability, or total hardness of water. Odor can be prevented by additional surface treatment with bentonite clay or water filtration.

Short-term Toxicity. In a 4-month study, there were no changes in animals that had drunk water held in trial tanks.

References:

Maslenko, A. A., Possible use of polymeric anticorrosive coating for ship cisterns, in *Hygiene and Toxicology of High-Molecular-Mass Compounds and of the Chemical Raw Material Used for Their Synthesis*, Proc. 6th All-Union Conf., B. Yu. Kalinin, Ed., Khimiya, Leningrad, 1979, 148 (in Russian).

POLY-4-METHYL-1-PENTENE

Structural Formula.

[~CH_2CH-
|
$CH_2CH(CH_3)$~]$_n$

Abbreviation. PMP.

Properties. Thermoplastic polymer, one of the most stable polyolefins. *M. p.* 230-240°C; *dens.* 0.8320; transparency 90%.

Migration Data. Extracts of PMP stabilized with Irganox-1010 and calcium stearate possessed no odor or taste. PMP stabilized with Irganox 1010 (0.3%), Uvitex (0.1%), and calcium stearate (0.5%) produced no adverse effect on water quality at temperatures from -15 to +121°C with exposure for 1 to 10 days. There was a small increase of water oxidizability and migration of up to 0.16 mg isopropyl alcohol/l.

Long-term Toxicity. 24-hour extracts were studied in a 15-month study in rats. No clinical signs or gross pathology changes were noted. In another study, aqueous extracts (2 cm^{-1}; 20 and 80°C; 3 and 10 days) had no effect on BW gain, peripheral blood composition, or liver function of animals.

Recommendations. ***Russia.*** Templen-grade PMP is permitted in the food industry at a temperature from -15°C to 100°C.

Regulations. ***U.S. FDA*** (1998) approved the use of PMP as articles or components of articles intended for use in contact with food, subject to the provisions of 21 CFR part 177.1520.

References:

1. Komarova, Ye. N. and Baikova, Z. K., *Plast. Massy,* 12, 37, 1976 (in Russian).
2. Braun, D. D., Zenina, G. V., and Moshlakova, L. A., Hygienic evaluation of new polyolefin group polymeric materials intended for use in the food industry, *Gig. Sanit.*, 2, 24, 1979 (in Russian).

4-METHYL-1-PENTENE and 1-HEXENE COPOLYMER

Migration Data. Molded articles containing 5.0 to 12% 1-hexene units and two stabilizer systems: (1) Irganox1010 and calcium stearate; (2) Irganox 1010, Stafor 10, and calcium stearate were investigated. A sample containing less than 0.3 to 2.5% low-molecular-mass fraction did not affect significantly neither taste or odor of extracts nor their chemical composition.

Reference:

Rotenberg, V. V., Kruglova, N. V., and Guricheva, Z. G., Hygienic properties of 4-methyl-pentene-1 and hexene copolymer, in *Environmental Protection in Plastics Production and Hygiene Aspects of Their Use,* Plastpolymer, Leningrad, 1978, 118 (in Russian).

POLY(2-PROPYL-*m*-DIOXANE-4,6-DIYLENE)

Structural Formula.

$[\sim CHCH_2CHCH_2\sim]_n$

|　　|

O -CH-O

|

$CH_2CH_2CH_3$

CAS No 63148-65-2

RTECS No TR4955000

Abbreviation. PVB.

Synonyms and **Trade Name.** Butvar; Polyvinyl butyral resin.

Composition. VL-023 phosphating primer has been prepared on the base of PVB and iditol resin. Pigment: zinc chromate.

Properties. Amorphous white polymer. *Dens.* 1.120; $n^{20} = 1.485$. Soluble in alcohol.

Applications. Used in the preparation of anticorrosion coatings.

Migration Data. Water in contact with VL-023 phosphating primer was colored and had a specific odor. Phenol (up to 2.9 mg/l) and chromium (up to 2.6 mg/l) were found to migrate into water. Other substances found in the contact water were butanol, formaldehyde, and zinc.

Regulations. ***U.S. FDA*** (1998) approved the use of PVB (1) in adhesives as a component (monomer) of articles intended for use in packaging, transporting, or holding food in accordance with the conditions prescribed in 21 CFR part 175.105; (2) as a component of the uncoated or coated food-contact surface of paper and paperboard intended for use in producing, manufacturing, packaging, processing, preparing, treating, packing, transporting, or holding aqueous and fatty foods in accordance with the conditions prescribed in 21 CFR part 176.170; and (3) in the manufacture of resinous and polymeric coatings for the food-contact surface of articles intended for use in producing, manufacturing, packing, processing, preparing, treating, packaging, transporting, or holding food for use only as polymerization cross-linking agent in side seam cements for containers intended for use in contact with food (only of the identified types), subject to the conditions prescribed in 21 CFR part 175.300.

EU (1990). PVB is available in the *List of authorized monomers and other starting substances which may continue to be used for the manufacture of plastic materials and articles intended to come into contact with foodstuffs pending a decision on inclusion in Section A* (*Section B*).

Reference:

Rudeiko, V. A., Pashkina, E. H., Romashov, P. G., and Yakovleva, L. E., Hygienic evaluation of conditions for using a phosphating primer in the water supply, *Gig. Sanit.*, 6, 16, 1980 (in Russian).

POLYURETHANE

Structural Formula. $[\sim CONHRNHCOOR'O\sim]_n$

CAS No 9009-54-5

Abbreviation. PU.

Composition. Polyester of carbamic acids are the products of interaction of isocyanates with polyols (glycols, glycerine, etc.) with low-molecular-mass polyesters, and with diamines. Contain urethane groups *-NH-CO-O-*.

Properties. PU have greater elasticity than polyamides but lower heat resistance.

Applications. PU are manufactured as fibers, foams, lacquers, solid and rubber-like materials. Used at temperatures from -60°C to +110°C. Pipes, hoses, moldings, prosthetic devices, elastic seals, and mattresses are manufactured from PU. PU has a number of biologic applications used to manufacture portions of implantable medical items such as pacemakers, in the vaginal contraceptive sponge (which is no longer on the market in the U.S.).

Migration of low-molecular-mass fractions (LMMF) into physiological solution (10 days, 40°C) from polyetherurethanic endoprostheses produced on the basis of tetrahydrofuran, polyoxypropylene, and toluene diisocyanate has been studied. LMMF values of tetrahydrofuran, polyoxypropylene, and toluene diisocyanate detected in the extract, were about 23, 9.0, and 1.67 mass%, respectively.[1]

Migration of polyethylenepolyamine was found. In an investigation of Styk (polyurethane adhesive) at 80°C, toluene diisocyanate (0.55 mg/l) was noted in contact water.[2]

Long-term Toxicity. 6-month administration of extracts of films based on polyurethaneurea had no significant effect on rats and caused no *mutagenic* effect.[2]

Carcinogenicity. PU and silicone films were implanted *s/c* into rats, and 1- and 2-year adverse tissue responses were studied. *PU* gave higher incidence of the adverse responses including tumor formation in comparison to silicone.[3]

Regulations. ***U.S. FDA*** (1998) approved the use of PU (2) in adhesives as a component (monomer) of articles intended for use in packaging, transporting, or holding food in accordance with the conditions prescribed in 21 CFR part 175.105; and (2) as the food-contact surface of articles intended for use in contact with bulk quantities of dry food of the type identified in 21 CFR parts 176.170 and 177.1680.

References:

1. Budnikov, V. J., et al., Sanitary and chemical assessment of polyetherurethanic endoprotheses, *Gig. Sanit.,* 2, 48, 1997 (in Russian).
2. Kupyrov, V. N., Kaplina, T. V., Gakal, R. K., Vinarskaya, E. I., and Starchenko, S. N., Hygienic evaluation of films intended for the waterproofing of unit prefabricated swimming pools, *Gig. Sanit*., 5, 91, 1978 (in Russian).
3. Nakamura, A., Kawasaki, Y., Takada, K., Aida, Y., Kurokama, Y., Kojima, S., Shintani, H., Matsui, M., Nohmi, T., Matsuoka, A., et al., Difference in tumor incidence and other tissue responses to polyurethanes and polydimethylsiloxane in long-term subcutaneus implantation into rats, *J. Biomed. Mater. Res.,* 26, 631, 1992.

POLYURETHANE FOAM

CAS No 9009-54-5
RTECS No TR7875000
Abbreviation. PUF.

Synonyms and **Trade Names.** Andur; Etheron sponge; Isourethane; Polyfoam plastic sponge; Polyfoam sponge; Polyurethane A; Polyurethane ester foam; Urethane polymers.

Composition. Polyurethane foam based on laprol ester. Material contains laprol poyester, tolylene diisocyanate, diazocyclooctane, tin octoate, foam regulator.

Migration Data. Aqueous extracts (1 to 30 days; 37°C, 0.125 cm^{-1}) had an odor (rating 2 to 3, 20°C) and taste. Migration of tin was found. Changes in water quality were also found in tests on a model filtering unit with a filtration rate of 10 m/hour. There was a change in the color of PUF when stored in chlorinated water for 6 months, but no deterioration in water quality was observed.

Long-term Toxicity. Rats received 15-day aqueous extracts of PUF. A certain inactivation of enzymic systems, changes in the contents of *SH*-groups in the blood serum and in homogenates of the liver, and in the mineral composition of the blood and urine were observed. Morphofunctional changes in the stomach and intestine were found.

Carcinogenicity classification. An IARC Working Group concluded that there is inadequate evidence for the carcinogenicity of AA in *experimental animals* and there were no adequate data available to evaluate the carcinogenicity of AA in *humans*.

IARC: 3

Reference:

Klimkina, N. V., Tsyplakova, G. V., Trukhina, G. M., Tiuleneva, I. S., and Kochetkova, T. A., Hygienic evaluation of polyurethane foam produced on laprol base, *Gig. Sanit.*, 9, 14, 1983 (in Russian).

POLYURETHANE FOAM, RIPOR-TYPE

Abbreviation. PUF.

Applications. Used to store meat, dairy, grain, and other products.

Migration Data. A study of media simulating foodstuffs revealed no increase in their oxidizability.

Long-term Toxicity. Rats received meat and milk that had been stored in cooling chambers with PUF insulation. No differences were noted in the BW gain, enzyme system activity, blood chemistry, and the structure of the visceral organs.

Reference:

Rozenberg, I. A., *Symposium on Urgent Problems of Work Safety and Prevention of Pathology in Industry and Agriculture*, Medical Institute, Riga, 1984, 100 (in Russian).

POLYURETHANE FOAM based on polyether P 2200

Abbreviation. PUF.

Composition. Material contains toluene diisocyanate, dimethylbenzylamine, carbamide, OP-10, sulforicinate, vaseline oil, and water.

Application. Used as an iron absorbent in filtration.

Migration in aqueous extracts (1 to 10 days, 20 and 37°C, 0.5 cm^{-1}) was investigated. After 24 hours, odor of rating 3 to 4 was noted.

Short-term Toxicity. Rats received 15-day aqueous extracts of PUF. No signs of toxic action were observed.

Recommendations. ***Russia.*** Material was recommended as filtering plug for ground water.

Reference:

Tyuleneva, I. S, Hygienic assessment of polyether P-2200-based polyurethane filtering plug, *Gig. Sanit.*, 10, 70, 1988 (in Russian).

POLYURETHANE COATING

Composition. A pre-polymer based on liquid SKD-PG rubber and 2,4-toluene diisocyanate-3,5; cocatalyst is triethylenediamine.

Applications. Used to waterproof the inner surface of reinforced concrete drinking water reservoirs.

Migration Data. A study of extracts (20 and 37°C; 1 to 3 days; 4.1 and 0.05 cm^{-1}) revealed an odor which intensity strengthened with increase in temperature, contact time and specific surface to rating 4 to 5. The intensity of odor decreased after washing and airing of specimens. After 10-day airing there was no perceptible odor. Migration of toluene diisocyanate from unwashed specimens into water amounted to 0.01 to 0.02 mg/l. After specimens had been aired, the release of toluene diisocyanate decreased to trace amounts, and, in extracts obtained with a specific surface of 0.05 cm^{-1}, this substance was not found. With a specific surface of 4.0 cm^{-1}, the quantity of triethylenediamine migration into water amounted to 0.5 mg/l which is lower than the PL. With 0.05 cm^{-1}, the migration does not occur.

Reference:

Krat, A. V., Kesel'man, I. M., and Sheftel', V. O., Sanitary-chemical evaluation of polymeric materials used in water supply, *Gig. Sanit.*,10, 18, 1986 (in Russian).

ELASTOMER SKU-7L

Applications. Used to make tap washers.

Migration Data. An adverse effect on water quality was observed, especially at 85°C (odor of rating 2, toluene diisocyanate contents 0.005 to 0.01 mg/l).

Russia. The material is ***not recommended*** for use.

Reference:

Krat, A. V., Kesel'man, I. M., and Sheftel', V. O., Sanitary-chemical evaluation of polymeric materials used in water supply, *Gig. Sanit.*, 10, 18, 1986 (in Russian).

ENAMELS UR-41, UR-41L, and E-75

Composition. Products of interaction of oligoesters with toluene diisocyanate and trimethylolpropane.

Applications. Used for anticorrosive protection of containers intended to store liquid and free-flowing foodstuffs, drinking water, etc.

Migration Data. A study of distilled water after contact with enamels (20 and 80°C; 1/20 and 1/50 cm^{-1}) revealed no foreign taste; the odor did not exceed rating 1. There was no migration of 2,4- and 2,6-toluene diisocyanate (detection limit of the method was 0.01 mg/l).

Russia. The enamels were ***recommended*** for use after preliminary treatment with hot water or a saline solution.

Reference:

Shuba, P. A., in *Sci. Proc. Hygiene of Use and Toxicology of Pesticides and Polymeric Materials*, All-Union Research Institute of Hygiene and Toxicology of Pesticides, Polymers and Plastic Materials, Kiyv, Issue No 17, 1987, 142 (in Russian).

POLYVINYL ACETATE

Stuctural Formula.

$[\sim CH_2CH\sim]_n$

|

$OCOCH_3$

M = 10,000 to 160,000

CAS No 9003-20-7

76057-08-4

RTECS No AK0920000

Abbreviations. PVAc (polymer), VAc (monomer).

Synonyms and **Trade Names.** Bakelite AYAA; Duvilax; Emultex FR; Gelva; Movinyl; Polysol; Protex; Rhodopas; Soviol; Vinac; Vinyl acetate polymer; Vinyl acetate resin.

Properties. Amorphous, transparent, inert polymer without taste or odor. *Dens.* 1.192. Aqueous dispersions are film-making and adhesive. Soluble in ethanol.

Applications. May be used as a lacquer, film, or adhesive for gluing cardboard packaging for foodstuffs. Adhesive for paper, wood, glass, metals, and porcelain; sealant.

Acute Toxicity. A single administration of the dispersion in doses of 15 g/kg BW (rats) and 20 g/kg BW (mice) did not kill the animals.[028] LD_{50} exceeded 25 g/kg BW in rats and mice.

Migration Data. A study of coatings of concrete containers (based on PVAc with or without $FeCl_3$ showed that extracts were alkaline (pH = 1.3 to 1.6 higher than that of distilled water). Migration of chlorides was 200 mg/l. No formaldehyde, vinyl acetate, or acetic acid was found in the extracts. Odor and taste of 5-day extracts reached rating 5. When water was changed every 2 hours, its properties were satisfactory. The authors recommend the coatings, once washed, for use in water pipes.[1] Migration of VAc from films into water varies from 0.4 to 0.7 mg/l.[2] VAc contents in different layers of cheese matured in colored coatings are 0.2 to 2.0 mg/g.[028]

Repeated Exposure. Mice and rats were given PVAcD unplasticized, plasticized with dibutyl phthalate or dibutyl sebacinate, or modified with maleic anhydride. The daily dose was 1/100 LD_{50}. Reversible morphological changes were found in the GI tract, liver, kidneys, spleen, lungs, and heart. When rabbits were given a 3.0% aqueous suspension of the material (2.0 mg/kg BW for 1, 2, and 3 weeks and 6 months), changes in the liver occurred.[3]

Long-term Toxicity. Rats given different specimens of PVAcD (unplasticized or plasticized with up to 15% dibutyl phthalate) in varying doses for 11 months developed no significant changes. When animals were fed with surface layer of PVAc-coated cheese in doses of 20 g/kg BW (rats) and 40 g/kg BW (mice) for a year, no toxic effects were observed.[028]

Reproductive Toxicity. PVAc was not transferred to the fetus in rabbits.[4]

Embryotoxicity. When male rats received 125 mg unplasticized PVAc/kg BW for 11 months, the orientation reaction of offspring was unaffected. PVAc plasticized with 15% dibutyl phthalate adversely affected the orientation reaction in male, but not female, offspring.[5] Rats and mice given 0.01 LD_{50} dose of PVAc for 12 months developed necrotic changes in the digestive system and organs, but survived.[6,7]

Mutagenicity. According to Andersen, cosmetic-grade PVAc was not mutagenic.[4]

Observations in man. An occupational cytogenetics study reported increased frequencies of CA in persons exposed to PVAc, as compared with an unexposed group.[8]

Carcinogenicity. Cosmetic-grade PVAc emulsion was not carcinogenic in animal testing.[4]

Carcinogenicity classification. An IARC Working Group concluded that there is inadequate evidence for the carcinogenicity of PVAc in *experimental animals* and there were no adequate data available to evaluate the carcinogenicity of PVAc in *humans*.

IARC: 3.

Regulations. ***U.S. FDA*** (1998) approved the use of PVA (1) in adhesives as a component (monomer) of articles intended for use in packaging, transporting, or holding food in accordance with the conditions prescribed in 21 CFR part 175.105; (2) in the manufacture of resinous and polymeric coatings for the food-contact surface of articles intended for use in producing, manufacturing, packing, processing, preparing, treating, packaging, transporting, or holding food for use only as polymerization cross-linking agent in side seam cements for containers intended for use in contact with food (only of the identified types), subject to the conditions prescribed in 21 CFR part 175.300; (3) in the manufacture of resinous and polymeric coatings for polyolefin films for food-contact surface of articles intended for use in producing, manufacturing, packing, processing, preparing, treating, packaging, transporting, or holding food, in accordance with the conditions prescribed in 21 CFR part 175.320; (4) in polyvinyl alcohol films intended for use in contact with food in accordance with the conditions prescribed in 21 CFR part 177.1670; (5) as a component of the uncoated or coated food-contact surface of paper and paperboard intended for use in producing, manufacturing, packaging, processing, preparing, treating, packing, transporting, or holding dry food of the type identified in 21 CFR part 176.170 (c); and (6) in the manufacture of cellophane for packaging food in accordance with the conditions prescribed in 21 CFR part 177.1200.

References:

1. Boikova, Z. K., Petrova, L. I., and Kiseleva, M. N., Possible use of polyvinylacetate dispersion as coating for concrete cistern intended for potable water holding, in *Hygienic Aspects of Polymeric Materials Use*, Proc. 3rd All-Union Meeting on New Methods of Hygiene Monitoring of the Use of Polymers in Economy, K. I. Stankevich, Ed., December 2-4, 1981, Kiyv, 1981, 198 (in Russian).
2. Petrova, G. A. and Moshlakova, L. A., Comparative sanitary and chemical assessment of thermally stable organosilicone polymers intended for use in food industry, *Gig. Sanit.,* 9, 22, 1987 (in Russian).
3. *Toxicology and Medical Chemistry of Plastics,* Coll. Sci. Techn. Abstracts, NIITEKHIM, Moscow, Issue No 4, 1981, 16 (in Russian).
4. Sherbak, B. I., Delayed effects of the effect of chronic poisoning of animals with polyvinylacetate extracts, *Gig. Sanit.,* 4, 99, 1977 (in Russian).
5. Patty's *Industrial Hygiene and Toxicology*, Clayton, G. D. and Clayton, F. E., Eds., vol 2E, *Toxicology*, 4th ed., John Wiley & Sons, New York, 1994, 3797.
6. Subbotin, M. Y. et al., in *Sci. Proc. F. F. Erisman Research Sanitary Hygiene Institute*, Moscow, Issue No 22, 1975, 86 (in Russian).
7. Andersen, F. A., Amended final safety assessment of polyvinylacetate, *J. Am. Coll. Toxicol.*, 15, 166, 1996.
8. Shirinian, G. S. and Arutyunyan, R. M., *Biol. J. Armen.*, 33, 748, 1980 (in Russian).

VINYL ACETATE COPOLYMERS

VINYL ACETATE and *N*-VINYL PYRROLIDONE COPOLYMER

Structural Formula.

```
[~CH2CH~]m [~CH2CH~]n
     |           |
  OCOCH3         N
                / \=O
                |__|
```

Abbreviations. VAc (vinyl acetate), VP (vinyl pyrrolidone), BVE (butylvinyl ether).

Composition. The binary copolymer contains 90% VA and 10% VP. Contents of residual VA are 0.5%. BVE may be added to the reaction mixture to improve elasticity. The tertiary copolymer contains 82% VAc, 10% VP, and 8.0% BVE. VAc residual contents are 1.0%.

Applications. Used as an adhesive.

Migration Data. From copolymer, containing 98% VA, the monomer migrates in quantities of 10.2 and 0.1-0.5 mg/l (with residual VAc contents of 0.7 and 0.04 to 0.1%, respectively).

Acute Toxicity. Aqueous extracts (copolymer /water ratio 1:1; 60°C; 6 hours; continuous agitation) were administered to mice (1.0 ml) and rats (5.0 ml). Extracts of nine specimens of the binary and tertiary copolymers did not kill the animals and did not cause changes in their general condition, or increase in BW gain, or morphological changes in the viscera.[1]

Long-term Toxicity. Extracts of binary and tertiary copolymer (copolymer/water ration 1:100; 60°C; agitation 1 hour; infusion 23 hours) were given to mice and rats as drinking water over a period of 12 months. No changes were found in the animals except for a significant increase in the relative organ masses of the adrenal and thyroid glands of the rats. No histological abnormalities were found.[2,3]

Recommendations. ***Russia***. Types VAP-4, VAP-8, and VAP-12 are recommended for producing the polymeric base of chewing gum.

References:

1. Broitman, A. Ya., Putilina, L. V., and Podval'naya, N. I., Toxicology of vinyl acetate + vinyl pyrrolidone copolymers, *Plast. Massy*, 12, 48, 1976 (in Russian).
2. *Toxicology and Medical Chemistry of Plastics*, Coll. Sci. Techn. Abstracts, NIITEKHIM, Moscow, Issue No 1, 1979, 15 (in Russian).
3. Boikova, Z. K. and Petrova, L. I., Sanitary-chemical characteristics of polyvinylacetate copolymers and compositions made on their base, in *Hygiene and Toxicology of High-Molecular-Mass Compounds and of the Chemical Raw Material Used for Their Synthesis,* Proc. 6th All-Union Conf., B. Yu. Kalinin, Ed., Khimiya, Leningrad, 1979, 59 (in Russian).

VINYL ACETATE and DIBUTYL MALEATE COPOLYMER

CAS No 25035-90-9

Abbreviations. VA (vinyl acetate), DBM (dibutyl maleate).

Synonym and **Trade Name.** 2-Butenedioic acid, dibutyl ester, polymer with ethenyl acetate; Novalen.

Composition. Product of emulsion copolymerization of VA and DBM in an aqueous medium in presence of the initiator calcium sorbate (a salt of sorbic acid) and tartrazine dye.

Applications. Aqueous dispersion used for coating cheese while it matures.

Migration Data. Extracts acquire a strong yellow color. Tartrazine penetrates Rossijski cheese to a depth of 0.1 to 0.5 cm. It was noted that the dye migrated into water and into 3.0% solution of lactic acid. Up to 0.05 mg VA/l migrated into water, and up to 1.36 mg DBM/l migrated into water over a period of 10 days.[1]

Long-term Toxicity. Administration of the copolymer dispersion over a long period of time produced no toxic effect in the experimental animals. The rats did not die when fed with a dose of 200 mg/kg BW for 10 months.

Reproductive Toxicity. Feeding the substance to animals does not affect their reproductive function, nor does it have any ***teratogenic*** effect.

Allergenic Effect was not found.[2]

References:

1. Pestova, A. G. et al., Hygienic evaluation of synthetic coatings, *Gig. Sanit.*, 7, 98, 1978 (in Russian).
2. Zimnitskaya, L. P. and Boikova, Z. K., NIITEKHIM, Dep. No 758-76 (in Russian).

VINYL ACETATE and POLYVINYL ALCOHOL COPOLYMER

Structural Formula.

$[\sim CH_2CH\sim]_m[\sim CH_2CH\sim]_n$

$\quad\quad | \quad\quad\quad\quad |$

$OCOCH_3 \quad\quad OH$

CAS No 25213-24-5

Applications. Used in medicine.

Mutagenicity. Was found negative in *Dr. melanogaster*. It is assumed that the copolymer is not mutagenic in humans.

Reference:

Mikheev, V. S. and Anisimova, L. Ye., in *All-Union Institute Sci.-Techn. Information*, Dep. No 3023-8 (in Russian).

VINYL ACETATE and ETHYLENE COPOLYMERS

Structural Formula.

$[\sim CH_2CH\sim]_m[\sim CH_2CH_2\sim]_n$

$\quad\quad |$

$OCOCH_3$

M = 30,000 to 500,000

CAS No 24937-78-8

Abbreviation. S.; VA (vinyl acetate).

Synonym. Sevilen.

Composition. Contains up to 70% VA and 30% ethylene. Residual VA contents 0.56%; solid contents 51.5 to 53%. A dye is added to the composition.

Applications. Used to produce various engineering components, adhesives, coatings for paper and cardboard, films, packaging for food products, and children's toys. Used as an additive to wax coatings and as material for transparent shock-resistant film. Different types contain different proportions of VA.

Migration Data. When S. contains 1.5 to 7.15% VA, the organoleptic properties of extracts (1 cm^{-1}; 80 and 20°C; 1, 3, and 7 days) of a 50-μm thick film were satisfactory; no migration of VA was found.[1]

S. with 7.0 to 10% VA contents have hygienic properties similar to those of LDPE. Studies revealed no migration of VA into water, simulant media, or food products (20 to 80°C; 1 day to 12 months). The thicker the article, the stronger the odor.[2]

S. with VA contents up to 14% possesses satisfactory hygienic properties: migration of VA was 0.1 mg/l. When VA contents were increased to 35%, taste and odor in the aqueous extracts become stronger. However, in most cases no migration of VA is found and, even in aggravated conditions, it does not exceed permissible level of 0.2 mg/l. Migration of other additives (plasticizers, dyes, filters) can occur.[3, 4]

Toxicity. A dispersion of the copolymer proved to be non-toxic in acute and chronic experiments with mice and rats.[5]

Reproductive Toxicity. C. dispersion had no effect on the reproductive function or on offspring development.[5]

Regulations. ***U.S. FDA*** (1998) approved the use of ethylene-vinyl acetate copolymers as articles or components of articles intended for use in producing, manufacturing, packing, processing, preparing, treating, packaging, transporting, or holding food in accordance with the provisions described in 21 CFR part 177.1350.

Recommendations. ***Russia.*** D-13 and D-23 Miraviten is recommended for use in contact with food products at room temperature (open-type containers and films). The copolymer is permitted as a coating for parchment and cardboard used to pack dry and free-flowing products.[3]

References:

1. Boikova, Z. K. and Petrova, L. I., Effect of some technological parameters on hygienic properties of ethylene-vinyl acetate copolymers, in *Environmental Protection in Plastics Production and Hygiene Aspects of Their Use,* Plastpolymer, Leningrad, 1978, 114 (in Russian).
2. Boikova, Z. K. and Petrova, L. I., Sanitary-chemical characteristics of polyvinylacetate copolymers and compositions made on their base, in *Hygiene and Toxicology of High-Molecular-Mass Compounds and of the Chemical Raw Material Used for Their Synthesis,* Proc. 6th All-Union Conf., B. Yu. Kalinin, Ed., Khimiya, Leningrad, 1979, 59 (in Russian).
3. Boikova, Z. K., Petrova, L. I., and Stroyeva, I. N., Sanitary-chemical evaluation of thermoglue TK-2P made of ethylene-vinylacetate copolymer, *Gig. Sanit.*, 5, 71, 1983 (in Russian).
4. Boikova, Z. K., Petrova, L. I., and Slusareva, I. P., Possible use of sevilene in contact with foodstuffs and in toy production, *Gig. Sanit.*, 7, 87, 1983 (in Russian).
5. Maksimova, N. S. and Mikhailets, I. S., Hygienic properties of SEV dispersion intended for hard cheeses covering, *Plast. Massy*, 12, 37, 1976 (in Russian).

SEVILEN D-13 and D-23-S

Abbreviation. VA.

Allergenic Effect. Types 11104-030 (VA contents 6.0 to 7.0%) and 11306-075 (VA contents 10 to 14%) have satisfactory hygienic properties, do not irritate the surface of the skin or the conjunctiva of the eyes of guinea pigs, and are not allergenic.

Recommendations. ***Russia.*** These types of S. are recommended for use in articles in contact with dry, free-flowing, water- and fat-containing food products at temperatures <80°C, and for unlimited use in children toys.

Reference:

Boikova, Z. K., Petrova, L. I., and Slusareva, I. P., Possible use of sevilene in contact with foodstuffs and in toy production, *Gig. Sanit.*, 7, 87, 1983 (in Russian).

POLYVINYL ACETATE and ETHYLENE, emulsion of COPOLYMER SVED-8PM

Composition. Contains 50% base substance and 50% moisture.

Properties. Viscous white liquid.

Acute Toxicity. LD_{50} exceeded 10 g/kg BW in mice.

Repeated Exposure. No cumulative effect was manifested.

Reference:

Bruskova, N. I. and Rodionova, R. P., Toxicity of polyvinyl acetate-ethylene copolymer emulsion, *Gig. Sanit.*, 3, 88, 1982 (in Russian).

WAX MELT PKS-25

Abbreviation. VA.

Composition. Copolymer of ethylene with vinyl acetate, contains paraffin wax and rosin.

Migration Data. Extracts (4°C, 3 to 4 days; 20°C, 10 days; 40°C, 5 hours; 80°C, 1 hour; 0.7 cm^{-1}) possessed a specific odor of rosin (rating 2). Extracts were colorless, but not transparent, as fragments of the melt which had become detached from the coating were present in water. The resinous odor reached rating 1.5. Taste and odor of food products packed in cardboard and parchment coated with the melt did not alter. VA was not found in water, oil, or sausage meat. In one case, 0.1 mg VA/l was found in 40% ethanol. Absence of any migration of VA is apparently due to its hydrolysis to acetic acid and acetaldehyde, but these were not found in the extracts.

Recommendations. ***Russia.*** The alloy is recommended for use in food packaging.

Reference:

Guricheva, Z. G., Boikova, Z. K., and Petrova, L. I., Hygienic properties of a coating made on an ethylene copolymer base with vinyl acetate, *Gig. Sanit.*, 9, 99, 1973 (in Russian).

ADHESIVE TK-2P

Composition. Contains D-17-KhA miraviten (a copolymer of ethylene with 26-30% VA), glycerin rosin ester, high-purity paraffin oil, and thermal stabilizer Thioalkofen MBP.

Migration Data. The *pH* of extracts (40 and 20°C; 1 day and 10 days; 0.7 cm^{-1}) of the adhesive and of the mixture of paraffin and rosin ester did not change. The *pH* of extracts of the coating was 0.4 to 0.6 lower than the control because of migration of substances from the paper. There was no formaldehyde in extracts of the adhesive, but 0.005 mg formaldehyde/l was found in extracts of the mixture of components. No vinyl acetate or acetic acid were found in the extracts.Taste and odor of food products wrapped in paper with a thermal adhesive coating did not change over a period of 3 days, provided food-grade paraffin is used in its composition.

Recommendations. ***Russia.*** TK-2P thermal adhesive is recommended by the authors for gluing and coating packaging for dry, free-flowing, and water-containing (up to 15%) food products.

References:

1. Boikova, Z. K., Petrova, L. I., and. Stroyeva, I. N., Sanitary and chemical evaluation of the thermo-glue TK-2P produced on the base of ethylene-vinylacetate copolymer, *Gig. Sanit.,* 5, 71, 1983 (in Russian).
2. Boikova, Z. K., Petrova, L. I., and Slyusareva, I. P., Possible use of sevilene in contact with foodstuffs and in toy production, *Gig. Sanit.*, 7, 87, 1983 (in Russian).

POLYVINYL ALCOHOL

Structural Formula.. $[\sim CH_2CH\sim]_n$ (with OH attached to CH)

M = 5,000 to 100,000
CAS No 9002-89-5
RTECS No TR8100000
Abbreviation. PVA.

Synonyms and **Trade Names.** Cipoviol W 72; Ethenol, homopolymer; Ivalon; Lemol; Mowiol NM 14; Polyvinol; Polyviol; Rhodoviol; Vinalak; Vinavilol 2-98; Vinol; Vinyl alcohol polymer.

Properties. White powder. *Dens.* 1.20 to 1.3020; n^{20} = 1.49 to 1.53. Soluble in water when heated. PVA film is greaseproof and gas-permeable. Threshold concentration for foaming is 0.5 mg/l (0.1 mg/l according to Zaitsev), and threshold for odor is 63 mg/l.[1, 2]

Applications. Used in the production of polyvinyl acetates and water-soluble films, such as those used for sausages and other meat products. Special types of PVA are used as plasma substitutes, and also in the perfume and cosmetic industry. PVA is used in the manufacture of adhesives, pastes and emulsions. In gelatinous state (addition of Congo red), it is used as a sealant and also for dressing burns. Waterproof PVA treated with formal dehyde is used for implantation.

Acute Toxicity. Data on toxicity of large doses are contradictory. Rats received 1.3 g/kg BW. During 10-30 days, their food consumption and BW gain did not differ from the controls.[3] LD_{50} for rats exceeded 20 g/kg BW; for mice, it was 14.7 g/kg BW.[2]

Repeated Exposure. K_{acc} = 1.0 (by Shtabsky). When mice were given 5.0 g/kg BW, 6 out of 20 animals died after 7 to 10 days. Gross pathology and histology examination revealed vascular breakdown in the viscera and fine-droplet diffuse fatty dystrophy of liver cells and epithelial cells of the renal ducts.[4] However, according to other data, even 7.5 g/kg BW dose produced no effect in mice.[5] Rats that received 250 and 50 mg/kg BW 17 times over a period of 3 weeks showed a more rapid BW gain than the controls. Intake of 500 mg/kg BW for 1.5 months caused a decrease in the contents of *SH*-groups in the blood and an increase of aldolase activity. On day 30, the activity of alkaline phosphatase had increased. A dose of 50 mg/kg BW did not change the studied indices.[2] When rats received 500 mg/kg 30 times, easily reversible functional shifts, slight dystrophic changes in the viscera, and decreased BW gain were observed. Microscopic examination revealed moderate diffuse fatty dystrophy of the liver.[4]

Short-term Toxicity. Rats received 1.0 g PVA/kg BW orally for 3 months. The treatment caused pronounced changes in the functional state of the liver and kidneys.[6] When mice were exposed to 60 single doses of 250 and 1250 mg/kg BW over a period of 2.5 months, only slight changes in CNS functions were found.[5]

Long-term Toxicity. Rabbits consumed 50 and 500 mg/kg BW during 6 months. There were no changes in cholinesterase activity, prothrombin time, or glycogen contents in the liver. Parenchymatous dystrophy and disturbances in liver function were found with the larger dose.[1]

When mice and rats received 1.0 g PVA-Zh/kg BW and 0.5 g PVA-P/kg BW for two years, their condition, and BW gain, and hematological indices were unaffected. Protein was found in the urine (3 to 7 times the level in the controls), and, by the end of the experiment, functional changes of the kidneys were observed. 0.2 to 0.5 mg% PVA was found in the urine and the blood of the test animals as well as PVA deposits in the kidneys.[5]

However, according to Babayev,[3] rats that consumed 0.5 g PVA/kg BW during 16 months, and mice that consumed the same for 6 months, did not develop functional or morphological changes. During 8 months, mice were given a 1.0% solution instead of water which amounted to 228 g/kg BW. There were no significant changes in any index. Under similar conditions rats received 100 g PVA/kg BW which resulted in only a slightly increased BW gain.[5]

The threshold dose of PVA-Zh in a chronic experiment is acknowledged to be 1.0 g/kg BW. The recommended acceptable daily oral dose is 1.0 mg/kg BW (safety factor 1000).

Extracts of PVA films used for sausages (2.0 cm^{-1}; 2 hours at 90°C, 24 hours at 20°C) were studied in experiments with rats and puppies over a period of 8 months. No changes were found.[7]

Reproductive Toxicity.

Embryotoxicity. After exposure to a dose of 500 mg/kg BW for 6 months rats produced normal offspring for three generations. According to Zaitsev, 100 mg/kg BW dose has no toxic effect on embryos or gonads of rats.[2] However, according to Miyasaki, in pregnant rabbits, PVA crosses the placental barrier and is deposited in the parenchymatous organs of the fetuses.[8]

Gonadotoxicity. PVA exposure of fathers was associated with an increased risk of preterm birth in human pregnancies.[9]

Mutagenicity. No increase in CA frequency was found in a culture of human lymphocytes, nor was any such effect found in rats exposed to PVA-Zh and PVA-P over long periods.[6,9,10]

Carcinogenicity. No evidence of carcinogenicity was found in female $B6C3F_1$ mice on intravaginal administration of 25% PVA solution.[11]

Carcinogenicity classification. An IARC Working Group concluded that there is inadequate evidence for the carcinogenicity of PVA in *experimental animals* and there were no adequate data available to evaluate the car cinogenicity of PVA in *humans*.

IARC: 3

Chemobiokinetics. Following oral administration, PVA enters the bloodstream and is eliminated by the kidneys. Depending on the dose, it can be deposited in the glomerula and ducts and cause specific changes in them.[5] Oral administration of radiolabeled, low-molecular-mass formulations of PVA led to little absorption, but PVA was readily absorbed from the vagina in rat studies. Radioactivity from the absorption of intravaginal PVA was concentrated mainly in the liver.[11]

Recommendations. *Russia* A method for obtaining LPVA-M (medicinal low-molecular-mass PVA) by alkali ethanolysis has been devised in the former USSR. LPVA-M is virtually non-toxic when administered over a short period. A solution of LPVA-M with the addition of 0.9% *sodium chloride* is used in medicine as a detoxicant under the name Polidez.

PVA-P is not recommended for use in the food industry. It has been recommended for use in farm animals.[6]

Standards. *Russia* (1995). MPC: 0.1 mg/l (organolept.)

Regulations.

U.S. FDA (1998) approved the use of PVA (1) in adhesives as a component (monomer) of articles intended for use in packaging, transporting, or holding food in accordance with the conditions prescribed in 21 CFR part 175.105; (2) in the manufacture of resinous and polymeric coatings for the food-contact surface of articles intended for use in producing, manufacturing, packing, processing, preparing, treating, packaging, transporting, or holding food for use only as polymerization cross-linking agent in side seam cements for containers intended for use in contact with food (only of the identified types), subject to the conditions prescribed in 21 CFR part 175.300; (3) as a component of the uncoated or coated food-contact surface of paper and paperboard intended for use in producing, manufacturing, packaging, processing, preparing, treating, packing, transporting, or holding dry food of the type identified in 21 CFR part 176.170 (c); (4) in the manufacture of cellophane for packaging food in accordance with conditions prescribed in 21 CFR part 177.1200; and (5) in polyvinyl alcohol films intended for use in contact with food in accordance with the conditions prescribed in 21 CFR part 177.1670.

EU (1990). PVA is available in the *List of authorized monomers and other starting substances which may continue to be used for the manufacture of plastic materials and articles intended to come into contact with foodstuffs pending a decision on inclusion in Section A* (*Section B*).

References:

1. Lutai, G. F., in *Sci. Proc. Irkutsk Medical Institute*, Irkutsk, Issue No 121, 1973, 10 (in Russian).
2. Zaitsev, N. A. and Skachkova, I. I., Substantiation of hygienic water standards for some polymeric compounds using the principle of stage norm-setting, *Gig. Sanit.*, 10, 75, 1986 (in Russian).
3. Babayev, D. A., *Scientific Principles of Hygiene Study of Polymeric Materials Used in Food Industry and Engineering*, Author's abstract of thesis, Kiyv, 1980, 40 (in Russian).
4. Zayeva, G. N., Babina, M. D., Fedorova, V. I., and Scirskaya, V. A.,Toxicological characteristiks of polivinyl alcohol, polyethylene and polypropylene, in *Toxicology of New Industrial Chemicals*, Medgiz, Moscow, Issue No 5, 1963, 120 (in Russia).
5. Taradin, Ya. I., Shavrikova, L. N., Fetisova, L. N., et al., Toxicological characteristics of emulgator STEK, in *Toxicology and Hygiene of High-Molecular-Mass Compounds and of the Chemical Raw Material Used for Their Synthesis,* Proc.3rd All-Union Conf., S. L. Danishevsky, Ed., Khimiya, Moscow-Leningrad, 1966, 94 (in Russian).
6. *Toxicology and Sanitary Chemistry of Polymerization Plastics*, Proc. Sci. Conf., B. Yu. Kalinin, Ed., Leningrad, 1984, 19 (in Russian).
7. Rogovaya, A. B., Hygienic characteristics of artificial sausage casings based on polyvinyl alcohol, *Gig. Sanit.*, 6, 107, 1977 (in Russian).
8. Miyasaki, K., *Virchov Archiv Pathol. Anat. Histol.*, 365, 351, 1975.
9. Savitz, D. A. et al., Effect of parents' occupational exposures on risk of stillbirth, preterm delivery, and small-for-gestational-age infants, *Am. J. Epidemiol.*, 129, 1201, 1989.
10. Grigorieva, M. N., Kalinin, B. Yu., Novikova, O. I., et al., Genetic study of some plastics and polymers, in *Hygienic Aspects of Polymeric Materials Use*, Proc. 3rd All-Union Meeting on New Methods of Hygiene Monitoring of the Use of Polymers in Economy, K. I. Stankevich, Ed., December 2-4, 1981, Kiyv, 1981, 319 (in Russian).
11. Sanders, J. M. and Matthews, H. B., Vaginal absorption of polyvinyl alcohol in Fischer 344 rats, *Hum. Exp. Toxicol.*, 9, 71, 1990.

POLYMERIC SEMIPERMEABLE MEMBRANE based on mixture of SODIUM POLYVINYL SULFONATE and POLYVINYL ALCOHOL

Toxicity. Contact with the blood does not cause changes in CNS, peripheral blood, liver and kidney function, albumin metabolism, etc. There are no minute or large-scale changes in the structure of the internal organs. *In vitro* experiments showed the absence of any damage to the blood clotting system and of any hemolytic effect on the erythrocytes.

Recommendations. ***Russia.*** The membrane is recommended for use in blood dialysers.[1]

Reference:

Bogdanov, M. E., in *Proc. 2nd All-Union Conf. on Membrane Methods for Separating Mixtures*, Vladimir, 1977, 493 (in Russian).

POLYVINYL CHLORIDE

Structural Formula. $[\sim CH_2CHCl\sim]_n$
M = 300,000 to 400,000
CAS No 9002-86-2
8063-94-3
51248-43-2
93050-82-9
RTECS No KV035000

Abbreviations. PVC (polyvinyl chloride), VCM (vinyl chloride monomer), ATBC (acetyl tributyl citrate), DEHP [di(2-ethylhexyl) phthalate], DEHA [di(2-ethylhexyl) adipate], DOA (dioctyl adipate), ESBO (epoxidized soybean oil), TMSN (tetramethyl succinonitrile).

Synonym and **Trade Names.** Armodour; Astralon; Bakelite; Exon; Hostalit; Igelite; Lucoflex; Lucovyl; Marvinol; Norvinyl; Opalon; Ortodur; Polychlorovinyl; Polytherm; Porodur; Trovidur; Viniplast; Viniplen; Vinnol; Vinoflex; Yugovinyl.

Composition. PVC is manufactured by polymerizing VCM. To increase the heat- and light-aging resistance of PVC, stabilizers are introduced. Different plasticizers (phthalic and phosphoric acids, etc.) are added to give it elasticity. Plasticizer contents can vary from 3.0 to 80% which produce a considerable effect on material properties. With time, plasticizers can migrate to PVC article surface carrying other components of the composition (e.g., stabilizers) with them. PVC materials without plasticizers are referred to as rigid-vinyl plastic.

Properties. The structure of this polymer is relatively tight, but some air will pass through it. PVC has a heat distortion temperature of 72°C and a maximum continuous service temperature of 65°C. Because its shrink and melt points are so low, PVC can be used to shrink wrap foods which will tolerate very little heat. It is resistant to water, acid, bases, some solvents, fats, and oils. It is a carbon-chain linear polymer of amorphous structure. Solid white translucent material. *Dens.* 1.35 to 1.43. The properties of PVC highly depend on the method of its production: in the case of block or suspension polymerization, the polymer is lighter, more readily releases different low-molecular-mass impurities, and possesses higher water and heat resistance than emulsion-synthesized PVC. Even a small increase in the proportion of emulsion PVC in the composition leads to increased intensity of migration of its components.

Applications. PVC is approved for use as the film to wrap fresh red meats because it allows enough air to go through the package to make the meat pigments "bloom" bright red. PVC is prior sanctioned for use in general food-contact applications. The heat sealing range is from 90 to 180°C. It is used in the production of food containers, molded articles, and water pipes.

Migration Data. There are some indications of possible migration of VCM, stabilizers, and plasticizers from PVC. The maximum amount of VCM is released from rigid PVC into fat-containing products and alcoholic drinks (VCM contents in PVC exceed 30 mg/kg): 0.21 mg/l in a dry martini, 0.12 mg/l in gin, 0.25 mg/l in Sherry, 0.19 mg/l in whisky.[1] The authors do not consider given concentrations to be health hazards.

VCM contents of PVC articles in contact with food products and of the products themselves are regulated in EU countries at a level of 1.0 and 0.05 mg/kg, respectively. Migration of VCM into water amounts to 0.01 to 0.2 mg/l (exposure from 1 week to 12 months). When VCM contents in the polymer amount to 10 mg/kg, 0.03 mg/l of the monomer was released into water from PVC bottles within 3 months, but with VCM contents of the polymer of 1.0 mg/kg, no monomer was released into water (Daniels and Proctor, 1975). Ando and Sayato point out that with time there is a greater probability of VCM reacting with chlorine dissolved in water and its transformation into chloroacetic anhydride, chloroacetic acid, etc. There is no migration of VCM into water when its contents in PVC are 2.0 μg/kg.[2]

There is no evidence that VCM can migrate from PVC pipes into water in a quantity that can be a health hazard and stay in water for a long enough time to cause undesirable effects. In contrast to industrial conditions where VCM inhaled with the air can accumulate in the blood and form metabolites with a carcinogenic and mutagenic effect, such accumulation is impossible with occasional entry of VCM into human body in trace amounts with water or food.[3]

VCM tetramer migration has been studied as a representative oligomer that has the potential for migration from PVC packaging. Tetramer levels in PVC bottles for retail beverages ranged from 70 to 190 mg/kg. No tetramer migration was observed into the simulants, such as distilled water, 3.0% acetic acid, 15% ethanol, and olive oil (detection limit of 5 to 10 μg/kg).[4]

60% of the PVC films declared for use in contact with fatty foods showed too high overall migration. Migration of phthalates from PVC into water has been the subject of many investigations.[5-10] In most instances, DEHA made up about 80% of the total amount of plastic constituents migrating to isooctane.[11]

PVC films were exposed to the official food simulant, olive oil, or to isooctane. DEHA were determined by combined capillary GC-MS. Migration exceeding the specific migration limit of 4.0 mg/cm^2 was found in 42 films (89% of the samples) and these films were deemed to be illegal according to their present declared field of application as given by their labeling.[12] Diethyl ether extracts of food-contact PVC were analyzed by GC-MS methods: DEHA, dinonyl phthalate, and other phthalates were present in relatively large quantities (10 times higher than the internal standards). However, 68% of the extracts contained no peaks higher than the internal standards.[13] The static migration test of a PVC film (children's sucking and biting) containing approximately 30% DEHP with saliva simulant gave the lowest values of DEHP; simple shaking increased the amounts of DEHP from 25 to 499 μg/g film.[14]

Migration of DOA from plasticized PVC into both olive oil and distilled water during microwave heating was studied. Migration into olive oil reached equilibrium after heating for 10 min at full power (604.6 mg DOA/l). Migration into distilled water was 74.1 mg/l after 8 min at full power.[15]

Migration of DEHP from plasticized PVC tubing used in commercial milking equipment reached 30 to 50 μg/kg. Retail whole milks from the UK contained 35 μg DEHP/kg.[5] No differences in migrated amounts between food-grade PVC irradiated with γ-radiation and non-irradiated samples of food-grade PVC were observed. The amount of ATBC that migrated into olive oil after 97 hours of contact was non-detectable (<1.0 mg/l at 4 to 5°C). Concentrations of ATBC at 20°C, after 29 and 94 hours were 3.3 and 5.1 mg/l, respectively.[6] At 20°C, traces of plasticizer dioctyl phthalate migrate into water from PVC water pipes containing 8 parts of plasticizer. At 37°C level, of migration amounted to 0.1 to 0.16 mg/l.

The introduction of fillers (chalk, barium sulfate, etc.) into plastisols reduces migration of plasticizers, but elevated contents of ESBO in PVC (> 3.0%) increases migration of dioctyl phthalate.[028] The contents of TMSN, the main decomposition product of 2,2'-azobisisobutyronitrile in PVC products used for food packaging, were examined. The TMSN concentration in 17 PVC products ranged from 0.05 up to 523 mg/kg. The release of TMSN from two PVC products into five kinds of food-simulating solvents (60°C; 30 min.) was observed. When pieces of the bottle were stored in olive oil at 40°C for 120 days, 5.0 mg/kg of TMSN was detected in the oil.[16]

ESBO is used as a plasticizer and heat stabilizer in PVC films and gaskets. Levels of ESBO in fresh retail meat samples wrapped in PVC film ranged from less than 1.0 to 4.0 mg/kg, but were higher (up to 22 mg/kg) in retail cooked meat. Migration into sandwiches and rolls from take-away outlets ranged from less than 1.0 to 27 mg/kg depending on factors such as the type of filling and the length of the contact time prior to analysis. When the film was used for microwave cooking in direct contact with food, levels of ESBO from 5.0 to 85 mg/kg were observed, whereas when the film was employed only as a splash cover for re-heating foods, levels ranged from 0.1 to 16 mg/kg.[9]

Stabilizers calcium stearate, ricinooleate, magnesium and sodium stearate are non-toxic, but their stabilizing effect is low. They are used mainly for their synergistic effect. PVC stabilized with lead, cadmium or organotin is considered unacceptable.

Stabilizers used now form metal chlorides that to some degree are soluble in water. Intense flow of water and its temperature result to more intense washing out of lead stabilizers. When over 2.5% lead stabilizer is introduced in PVC, the amount of lead washed out is proportional to its contents in the composition, and it is washed out only from the surface layer of pipes. The bulk of lead is washed out during the first 2 to 3 days; later the process slows down. Considerable effect on washing out lead salts from PVC pipes has contents of CO_2 dissolved in water. But small amounts of CO_2 can even promote stability of lead because of formation of almost insoluble carbonate $PbCO_3$. On the other hand, lead hydrocarbonate which readily passes into solution, is formed with higher CO_2 concentrations.

Barium compounds are normally used in synergistic mixtures with cadmium compounds. However, their use is limited by their high toxicity.

In some European countries dioctytin compounds are permitted in certain quantities as additives. Migration of organotins from PVC articles (clear food container, rigid pipe, and flexible membrane) into tetrahydrofuran, xylene and methylene chloride was investigated. Methylene chloride extracted >97% of the total extractable organotins in two extractions. There were <0.3 to 4.7 mg butyltins/g and <0.8 to 8.8 mg octyltins/g solvent. In industrial application, pipe samples were <13 to 1.5 mg butyltins/g and 0.7 to 3.0 mg octyltins/g PVC.[17]

An investigation of the migration of a sulphur-containing organotin stabilizer (dioctyltin diisooctyl-thioglycolate) showed that the level of migration is lower when the stabilizer contents in the composition are reduced. The maximum release was observed at 60°C after 5 days contact of PVC with water.[18] At 20°C, 0.36 to 0.52 mg/l were released after 24 hours. Matos et al. found out the low levels of migration of mono- and dioctyl derivatives of organotin stabilizers. After contact for 2 years at 30°C, 3.0 µg *tin* were released from 1.0 dm^2 (10 µg permitted).[19] Concentrations of 0.1 to 0.122 mg dibutyltin-*S,S'*-(isooctylthioglycolate)/l were found in the extracts (1 to 2 cm^{-1}; 20 to 60°C; 1 to 3 days).[10]

Migration of fatty acid amides from PVC used in food packaging and containing commonly used fatty acid amide slip additives into fat and aqueous food simulant was less than 0.05 mg/l (10 day, 40°C).[20]

Acute Toxicity. Rats received powdered PVC dissolved in sunflower oil at the dose of 2.5 g/kg BW for 14 days. There were no acute changes reported.[18]

Repeated Exposure. Capsules containing PVC in the from of 6 x 6 mm lumps, shreds, or powder were administered twice a day for 5 days to male dogs (PVC dose 125 mg/kg BW). No changes in the blood and urine parameters were noted. PVC was excreted with the feces unaltered. No morphological changes were reported.[21]

Long-term Toxicity. Male rats received aqueous extracts of PVC (1 cm^{-1}; 20°C; 15 days) instead of water over a period of 12 months. There were no changes in BW gain, peripheral blood, biochemical indices, or liver function. Lead contents in the liver and bones were slightly higher than in the control animals, though histology of the viscera was not affected.[18] Aqueous extracts of Miplast (emulsion PVC K-62) had no toxic effect on rats during 8 months of exposure.[22]

Aqueous and oil extracts of PVC P-73-M films containing 1.5 parts organotin stabilizers OTS-15 and SSM-9/68 given for 12 months did not have any clinical toxic effect on rats. However, to the end of the study with the oil extracts, there was reduced synthesis of hippuric acid, a slower removal of bromosulfalein from the blood, and depressed activity of mixed-function oxidases.[8]

Reproductive Toxicity. Epidemiological studies of adverse pregnancy outcome showed no association between PVC contact and the incidence of spontaneous abortion.[23] No *terata* was reported in mice.[24]

Immunotoxicity. Local hypersensitivity was observed in guinea pigs during inhalation of a PVC mixture containing dibutyl phthalate and orgnochlorine compounds for 3 weeks.[25]

Carcinogenicity. PVC induced malignant tumors in rodents following s/c imbedding of polymer films.[026] Nevertheless, the results of this study were later considered inadequate. Rats received total dose of 200 g/kg BW for 30 weeks. The material is equivocal tumorigenic agent by RTECS criteria.[26]

Carcinogenicity classification. An IARC Working Group concluded that there is inadequate evidence for evaluation of the carcinogenicity of PVC in *experimental animals* and in *humans*.

IARC: 3

Regulations. ***U.S. FDA*** (1998) approved the use of PVC (1) in adhesives as a component (monomer) of articles intended for use in packaging, transporting, or holding food in accordance with the conditions prescribed in 21CFR part 175.105; (2) in the manufacture of resinous and polymeric coatings for polyolefin films for food-contact surface of articles intended for use in producing, manufacturing, packing, processing, preparing, treating, packaging, transporting, or holding food, in accordance with the conditions prescribed in 21CFR part 175.320; (3) in the manufacture of resinous and polymeric coatings for the food-contact surface of articles intended for use in producing, manufacturing, packing, processing, preparing, treating, packaging, transporting, or holding food for use only as polymerization cross-linking agent in side seam cements for containers intended for use in contact with food (only of the identified types), subject to the conditions prescribed in 21CFR part 175.300; (4) as a component of the uncoated or coated food-contact surface of paper and paperboard intended for use in producing, manufacturing, packaging, processing, preparing, treating, packing, transporting, or holding dry food of the type identified in 21CFR part 176.170 (c); (5) in the manufacture of semirigid and rigid acrylic and modified plastics used as articles intended for use in contact with food in accordance with the conditions prescribed in 21CFR part 177.1010; and (6) in the manufacture of cellophane for packaging food in accordance with the conditions prescribed in 21CFR part 177.1200.

References:

1. Davies, I. W. and Perry, R., *Environ. Pollut. Manag.*, 5, 22, 1975.
2. Ando, M. and Sayato, W., *Water Res.*, 18, 315, 1984.
3. Petersen, J. H., Lillemark, L., and Lund, L., Migration from PVC cling films compared with their field of application, *Food Addit. Contam.*, 14, 345, 1997.
4. Castle, L., Price, D., and Dawkins, J. V., Oligomers in plastics packaging. Part 1: Migration tests for vinyl chloride tetramer, *Food Addit. Contam.*, 13, 307, 1996.
5. Castle, L., Gilbert, J., and Eklund, T., Migration of plasticizer from poly(vinyl chloride) milk tubing, *Food Addit. Contam.*, 7, 591, 1990.
6. Goulas, A. E., Kokkinos, A., and Kontominas, M. G., Effect of gamma-radiation on migration behavior of dioctyladipate and acetyltributylcitrate plasticizers from food-grade PVC and PVDC/PVC films into olive oil, *Z. Lebensm. Unters. Forsch.*, 201, 74, 1995.
7. Sheftel', V. O. and Katayeva, S. E., *Migration of Harmful Substances from Polymeric Materials*, Khimiya, Moscow, 1978 (in Russian).
8. Zinchenko, T. M., *Hygienic Evaluation of Phthalate Plasticizers of PVC*, Author's abstract of thesis, Kyiv, 1988, 20 (in Russian).
9. Castle, L., Mayo, A., and Gilbert, J., Migration of epoxidised soya bean oil into foods from retail packaging materials and from plasticized PVC film used in the home, Food Addit. Contam., 7, 29, 1990.
10. Sheftel', V. O. and Grinberg, I. M., Migration of harmful chemical substances out of polyvinyl chloride materials used in water supply, *Gig. Sanit.*, 8, 78, 1979 (in Russian).
11. Sheftel', V. O., On the risk of vinylchloride migration into water and food-stuffs, *Gig. Sanit.*, 2, 63, 1980 (in Russian).
12. Petersen, J. H. and Breindahl, T., Specific migration of di-(2-ethylhexyl)adipate (DEHA) from plasticized PVC film: results from an enforcement campaign, *Food Addit. Contam.*, 15, 600, 1998.
13. van Lierop, J. B., Enforcement of food packaging legislation, *Food. Addit. Contam.*, 14, 555, 1997.
14. Steiner, I., Scharf, L., Fiala, F., and Washuttl, J., Migration of di(2-ethylhexyl) phthalate from PVC child articles into saliva and saliva simulant, *Food Addit.Contam.*, 15, 812, 1998.
15. Badeka, A. B. and Kontominas, M. G., Effect of microwave heating on the migration of dioctyl-adipate and acetyltributylcitrate plasticizers from food-grade PVC and PVDC/PVC films into olive oil and water, *Z. Lebensmit. Unters.-Forsch.*, 202, 313, 1996.
16. Ishiwata, H., Inoue, T., und Yoshihira, K., Tetramethylsuccinonitrile in polyvinyl chloride products for food and its release into food-simulating solvents, *Z. Lebensm. Unters. Forsch.*, 185, 39, 1987.
17. Forsyth, D. S., Dabeka, R., Sun, W. F., and Dalglish, K., Specification of organotins in poly(vinyl chloride) products, *Food Addit. Contam.*, 10, 531, 1993.
18. Sheftel', V. O., *Hygiene Aspects of the Use of Polymeric Materials in the Water Supply*, Thesis Diss., All-Union Research Institute of Hygiene and Toxicology of Pesticides, Polymers and Plastic Materials, Kiyv, 1977, 70 (in Russian).
19. Matos, C. M., Kroll, Y., Hoppe, H., Romminger, K., and Woggon, H., Migration of organic tin stabilizers of hard PVC packing materials in food, *Z. Ges. Hyg.*, 33, 258, 1987.
20. Cooper, I. and Tice, P. A., Migration studies on fatty acid amide slip additives from plastics into food simulants, *Food Addit. Contam.*, 12, 235, 1995.
21. Johnson, W. S. and Schmidt, R. E., Effects of polyvinyl chloride ingestion by dogs, *Am. J. Vet. Res.*, 38, 1891, 1977.

22. Selivanov, S. B., *Hygienic Study and Evaluation of Reverse Osmosa Method of Potable Water Desalination by Filter-press Installation,* Author's abstract of thesis, Moscow, 1977, 18 (in Russian).
23. Lindbohm, R. et al., Spontaneous abortion among women employed in the plastics industry, *Am. J. Ind. Med.,* 8, 579, 1985.
24. Ungvary, G., Studies on the teratogenicity of PVC, *Acta Morphol. Acad. Sci. Hung.*, 28, 159, 1980.
25. Shevchenko, A. M., Borisenko, N. F., and Pushkar', M. P., *Professional Hygiene in the Manufacture of Polymers and Plastic Materials,* Zdorov'ya, Kiyv, 1978, 16 (in Russian).
26. Costa, V. and Frongia, N., Oncogenic activity of polyvinylchloride, *Pathologica,* 73, 59, 1981.

POVIDEN films

Composition. Contain 8.0% plasticizer.

Migration Data. No migration of plasticizers into aqueous extracts was found. A small quantity of dibutyl sebacate migrates into fat and alcohol mixtures.[028]

POLYVINYL CHLORIDE base anti-corrosion coating PVC-716

Composition. Contains dioctyl phthalate, β-aminocrotonic acid glycol ester, and chromium oxide.

Applications. Used in the production of anticorrosive coating of drinking water pipes on ships.

Migration Data. An aromatic odor of aqueous extracts (20, 37, and 60°C; 1 and 10 days; 0.7 cm^{-1}) was established at 37 and 60°C. No migration of chromium or dioctyl phthalate was found. There was no undesirable odor because of chlorination, but the extracts possessed a greater capacity to absorb chlorine.

Long-term Toxicity. Intake of extracts (20°C; 10 days; 0.7 cm^{-1}) by rats, rabbits, and guinea pigs for 8 months had no effect on animals and caused no histological changes. The coating is recommended for use in contact with cold water only.

Reference:

Selivanov, S. B., *Hygienic Study and Evaluation of Reverse Osmosa Method of Potable Water Desalination by Filter-press Installation,* Author's abstract of thesis, Moscow, 1977, 18 (in Russian).

POLYVINYL CHLORIDE plastisols

Composition. Noted for the presence or absence of accelerators of the altax and thiuram groups. The plasticizers are dioctyl phthalate with triethylene glycol dimethacrylate (TGM-3) or dibutyl ether with polypropylene glycol adipinate (PPA-4).

Applications. Used for manufacture of medicine flasks.

Migration Data. Rapid release of dioctyl phthalate into model media (especially into oil) was established. Captax, thiuram, zinc dimethyl dithiocarbamate, and metals were found in small quantities.

Toxicity. Aqueous, water-alcohol, and specially oil extracts were found to have a toxic effect on animals.

Allergenic Effect was not observed.

Reference:

Vlasyuk, M. G. et al., in *Proc.1st All-Union Conference of Toxicologists on Problems of Health and Environment Protection against Harmful Chemical Factors,* Rostov-na-Donu, 1986, 149 (in Russian).

VINYL CHLORIDE COPOLYMERS

VINYL CHLORIDE and VINYL ACETATE COPOLYMER

Structural Formula.

$[\sim CH_2CHCl\sim]_m[\sim CH_2CH\sim]_n$

$\qquad\qquad\qquad\qquad |$

$\qquad\qquad\qquad\quad OCOCH_3$

M = 10,000 to 70,000

CAS No 34149-92-3

RTECS No TR8090000

Abbreviations. VCM (vinyl chloride monomer), VA (vinyl acetate), PVAc (polyvinyl acetate).

Synonyms and **Trade Names**. Acetic acid, vinyl ester, chloroethylene copolymer; Polyvinyl acetate chloride; Polyvinyl chloride acetate; Vinyl chloride acetate copolymer; Vinyl chloride vinyl acetate copolymer.

Properties. *M. p.* ~ 110°C; *dens.* 1.30 to 1.39.

Applications. Used as plastics, films, and lacquers.

Reproductive Toxicity. 95% VCM + 5.0% VA was added to the feed of rats (1.5 and 12%). Their progeny was kept on the same diet for their lifespan. No effect on the offspring and no histological changes were found.[1]

Carcinogenicity. Sarcomas at site of application were induced following *s/c* implantation of unplasticized PVAc films in female and male mice. The strains and sexes showed marked differences in incidence and mean latency of resulting tumors. Carcinogenic by RTECS criteria.[2]

Carcinogenicity classification. An IARC Working Group concluded that there is inadequate evidence for the carcinogenicity of VCM + VA copolymers in *experimental animals* and there were no adequate data available to evaluate the carcinogenicity of VCM + VA copolymers in *humans*.

IARC: 3.

Regulations. ***U.S. FDA*** (1998) approved the use of VCM + VA copolymers (1) in the manufacture of resinous and polymeric coatings for polyolefin films for food-contact surface of articles intended for use in producing, manufacturing, packing, processing, preparing, treating, packaging, transporting, or holding food, in accordance with the conditions prescribed in 21 CFR part 175.320; (2) in the manufacture of resinous and polymeric coatings for the food-contact surface of articles intended for use in producing, manufacturing, packing, processing, preparing, treating, packaging, transporting, or holding food for use only as polymerization cross-linking agent in side seam cements for containers intended for use in contact with food (only of the identified types), subject to the conditions prescribed in 21 CFR part 175.300; and (3) in the manufacture of cellophane for packaging food in accordance with the conditions prescribed in 21 CFR part 177.1200.

References:

1. Smyth, H. F. and Weil, C. S., Chronic oral toxicity to rats of a vinyl chloride-vinyl acetate copolymer, *Toxicol. Appl. Pharmacol.*, 3, 501, 1966.
2. Brand, I., Buoen, L.C., and Brand, K.G., Foreign-body tumors of mice: strain and sex differences in latency and incidence, *J. Natl. Cancer Inst.*, 58, 1443, 1977.

PAINT HS-720

Composition. Vinyl chloride and vinyl acetate copolymer plasticized with tricresyl phosphate (solvent cyclohexanone).

Migration Data. Water from painted containers had a specific odor, which was strengthened with increase in temperature and contact time. Significant concentrations of cyclohexane and phenol compounds were found in extracts. Tricresyl phosphate migrates into water at 60°C (0.35 mg/l after 10 days). After 6 to 7 days of washing, water quality indices were mainly good, but phenol derivatives still migrate into water after 11-day washing.

Toxicity. Confirmed by functional and morphological changes in animals.

Recommendations. Not recommended for direct contact with water.

Reference:

Romashov, P. G. and Yakovleva, L. Ye., Hygienic aspects of the use of vinyl chloride paint in water supply, in *Hygiene Aspects of the Use of Polymeric Materials*, Proc. 3rd All-Union Meeting on New methods of Hygiene Monitoring of the Use of Polymers in the Economy, K. I. Stankevich, Ed., Kiyv, December 2-4, 1981, Kiyv, 1981, 191 (in Russian).

VINYL CHLORIDE and VINYLIDENE CHLORIDE COPOLYMER

Structural Formula.

$[\sim CH_2CHCl_2\sim]_n [\sim CH_2CCl_2\sim]_m$

CAS No 9011-06-7

68648-82-8

RTECS No KV9900000

Abbreviations. VCM (vinyl chloride monomer), VDC (vinylidene chloride monomer), PVC/PVDC copolymer; ATBC (acetyl tributyl citrate), DOA (dioctyl adipate).

Synonyms and **Trade Names.** Breon 202; Chloroethylene-1,1-dichloroethylene polymer; Daran; 1,1-Dichloroethene polymer with chloroethene; Dow 874; Laplen; Saran 683 or 746; Velon; Viniden 60; Vinyl chloride-1,1-dichloroethylene copolymer.

Properties. Cream-colored powder. Glass-transition temperature 55°C, *dens.* 1.6. Good chemical stability when contents of vinylidene chloride are 40%.

Applications. Used as plastics, paints, and varnish materials with vinylidene chloride contents of up to 85%. Used for manufacture of *Saran*-type film possessing good gas impermeability.

Migration of *DOA* and *ATBC* plasticizers from plasticized PVC/PVDC copolymer (*Saran*) films into both olive oil and distilled water during microwave heating was studied. Migration of DOA into olive oil reached equilibrium after heating for 10 min at full power. Migration into distilled water was 74.1 mg/l after 8 min at full power. The amount of ATBC migrating into olive oil after heating for 10 min was 73.9 mg ATBC/l; into distilled water it was 4.1 mg/l after heating for 8 min. *Saran* should be used with caution in microwave heating, avoiding its direct contact with high fat foodstuffs.[1]

No differences in migrated amounts between food-grade PVDC/PVC copolymer irradiated with γ-radiation and non-irradiated samples of food-grade PVDC/PVC copolymer were observed. After 47 hours of contact, DOA migration was noted at the level of 302.8 mg/l. The amount of ATBC that migrated into olive oil at 4 to 5°C after 97 hours of contact was non-detectable (<1.0 mg/l). Concentrations of ATBC at 20°C, after 29 and 94 hours were 3.3 and 5.1 mg/l, respectively.[2]

Water from containers with the coatings KHS (on the base of VCM/VDC copolymer, solvent: acetone mixed with toluene and butyl acetate*)* applied on VL-023 primer (q.v.) had a specific odor (rating 3 to 4), which was strengthened with increase in temperature and contact time. The *acetone* contents were up to 0.8 mg/l; the chloride contents were up to 3.8 mg/l.[3]

Acute Toxicity. *I/g* administration was not lethal. The acute effect threshold was not established.[4]

Long-term Toxicity. Administration of 5.0% VCM/VDCM copolymer in the feed of dogs for 1 year and in the feed of rats for 2 years caused no manifestation of toxicity.[5] Addition of 1.0% *Saran-1* film in the feed of rats over a period of 2 years had no toxic effect.[05]

Rats received aqueous extracts of VCM/VDCM copolymer coatings KHS-76, KHS-710, and KHS-558 over a period of 10 to 12 months. Functional and morphological changes observed were regarded as temporary compensatory-adaptation reactions. After the coatings had been steam-treated and washed 3 times with water, there was a dramatic improvement in their safety which enabled them to be recommended for application in domestic drinking water piping systems.[3]

Carcinogenicity. Induced malignant tumors in rodents following s/c imbedding of polymer films.[026] Nevertheless, the results of this study were later considered inadequate.

Carcinogenicity classification. An IARC Working Group concluded there were no adequate data available to evaluate the carcinogenicity of AA in *experimental animals* and in *humans*.

IARC: 3.

Regulations.

U.S. FDA (1998) approved the use of VCM/VDCM copolymer (1) in the manufacture of resinous and polymeric coatings for polyolefin films for food-contact surface of articles, in accordance with the conditions prescribed in 21 CFR part 175.320; (2) in the manufacture of resinous and polymeric coatings for the food-contact surface of articles, subject to the conditions prescribed in 21 CFR part 175.300; and (3) in the manufacture of cellophane for packaging food in accordance with the conditions prescribed in 21 CFR part 177.1200.

Russia. An aqueous dispersion of VCM/VDCM copolymer was approved for use (1) as coating for cheeses and sausages, (2) for application to paper for packing frozen and other foodstuffs.

Vinilak A-15 KRP (plasticized) was approved (3) for use in foil and paper for packing of dairy products, (4) for application to the internal surface of large wine containers, and also (5) as a coating for paper and foil for packing foodstuffs.

References:

1. Badeka, A. B. and Kontominas, M. G., Effect of microwave heating on the migration of dioctyladipate and acetyltributylcitrate plasticizers from food-grade PVC and PVDC/PVC films into olive oil and water, *Z. Lebensmit. Untersuch. Forsch.*, 202, 313, 1996.
2. Goulas, A. E., Kokkinos, A., and Kontominas, M. G., Effect of gamma-radiation on migration behavior of dioctyladipate and acetyltributylcitrate plasticizers from food-grade PVC and PVDC/PVC films into olive oil, *Z. Lebensm. Unters. Forsch.*, 201, 74, 1995.
3. Rudeiko, V. A. and Pashkina, Y. E., Hygienic evaluation of conditions for using a phosphating primer in the water supply, *Gig. Sanit.*, 6, 16, 1980 (in Russian).
4. Dubinskaya, A. B., Dronov, I. S., Stepanenko, A. F., et al., Toxic properties of copolymer VCVD-40, acrylate copolymers lacrys-20, lacrys-95, lacrys-2153C, AK-624, *Gig. Sanit.*, 1, 74, 1986 (in Russian).
5. Wilson, R. H. and McCormick, W. E., Toxicology of plastics and rubber plastomers and monomers, *Ind. Med. Surg.*, 323, 479, 1954.

COMPOSITION of PVC and ABS

Applications. Used as a material for food packing.

Migration Data. Aqueous extracts had no odor or taste. Oil extract acquired a specific taste after 2-month contact with a specimen stabilized with Last B-94. Styrene and acrylonitrile were not found in oil, aqueous, alcohol, or acetic acid extracts. Content of plasticizers (dioctyl phthalate and butyl stearate) in the extracts did not exceed 1.3 mg/l.

Long-term Toxicity. Studied in an 11-month experiment with mice and rats. The animals drank 4-day aqueous extracts and were given 4-month oil extracts with their feed (5.0 g/kg BW). No toxic effect was observed.

Regulations. ***U.S. FDA*** (1998) approved the use of PVC-ABS composition in the manufacture of cellophane for packaging food in accordance with the conditions prescribed in 21 CFR part 177.1200.

Reference:

1. Komarova, Ye. N. and Maksimova, N. S., Hygienic properties of PVC-ABS composition, *Plast. Massy*, 12, 29, 1976 (in Russian).

POLYVINYL CHLORIDE, chlorinated

M = 40,000 to 80,000

CAS No 68648-82-8

Abbreviation. PCV (perchlorovinyl).

Synonym and **Trade Name.** Chloroethene homopolymer, chlorinated; Perchlorovinyl.

Composition. Product of limited chlorination of PVC. Contains 62.5 to 64.5% combined chloride. Mol. mass depends on mol. mass of initial PVC.

Properties. Insoluble in water and alcohol. *Dens.* 1.47 to 1.50^{20}

Applications. Used for the manufacture of anticorrosion coatings, adhesives, lacquers, and enamels, and pipes for transporting hot and corrosive liquids. Used as a coating for large wine tanks.

Migration Data. Extracts of PCV coating (0.01 cm^{-1}) revealed no effect on water quality.[1] A bitter taste in water in Igelit water pipes has been reported.[2] The transparency of water decreased significantly over a period of 10-day contact.[3]

Repeated Exposure. Rats received 100 to 1000 mg PCV/kg BW for 10 to 20 day. No toxic effect was observed.

References:

1. Zamyslova, S. D. and Kudrin, L. V., Hygienic assessment of perchlorovinyl hydroinsulating coatings used in water supply, in *Hygiene Problems in the Production and Use of Polymeric Materials*, 1969, 182 (in Russian).
2. Ahrens, W. and Siegert, C., *Wasser-Wirsch.Wasser-Techn.*, 7, 348, 1957.
3. Sheftel', V. O., Leaching of lead stabilizers from PVC-pipes, *Gig. Sanit.*, 10, 105, 1964 (in Russian).

POLYVINYLIDENE CHLORIDE

Structural Formula.. [~CH_2CCl_2~]$_m$
CAS No 9002-85-1

Abbreviations. PVDC (polyvinylidene chloride), ATBC (acetyl tributyl citrate), DBS (dibutyl sebacate).

Applications. PVDC is cleared for use as a food contact surface as a base polymer, in food package gaskets, in direct contact with dry foods, and for paperboard coating in contact with fatty and aqueous foods. PVDC films soften at 120°C so they can be used with food that is boiling, but not necessarily with high-sugar or high-fat foods that may get much hotter than 100°C). It is used in the production of adhesives and synthetic fibers. PVDC and PVC can be copolymerized to produce a family of flexible *Sarans*.

Properties. PVDC was invented in 1941 to protect equipment from outdoor elements. After World War 2, it was approved for food packaging, and it was Prior Sanctioned in 1956 (Society of the Plastics Industry).

PVDC provides protection against transfer of gases, flavors, and odors. It is resistant to oxygen, water, acids, bases, and solvents. The copolymerization results in a film with molecules bound so tightly together that very little gas or water can get through. The heat distortion temperature of the polymer is 45°C and the maximum continuous service temperature is 65°C. *Sarans* are clear and have excellent barrier properties. Their heat sealing range is from 140 to 205°C.

Migration Data. Plasticizers found to migrate from the PVDC films are DBS and ATBC. Levels of migration are 76 to 137 mg DBS/kg in processed cheese and cooked meats; 2.0 to 8.0 mg ATBC/kg were detected in cheese.

Regulations. ***U.S. FDA*** (1998) approved the use of PVDC (1) in adhesives as a component (monomer) of articles intended for use in packaging, transporting, or holding food in accordance with the conditions prescribed in 21 CFR part 175.105; (2) in the manufacture of resinous and polymeric coatings for the food-contact surface of articles intended for use in producing, manufacturing, packing, processing, preparing, treating, packaging, transporting, or holding food, subject to the provisions prescribed in 21 CFR part 175.300. The use of VDC copolymers are approved (3) in VDC copolymer coatings applied on nylon film and used as food-contact surfaces subject to the provisions prescribed in 21 CFR part 175.360; (4) in VDC copolymer coatings applied on polycarbonate film and used as food-contact surfaces subject to the provisions prescribed in 21 CFR part 175.365; (5) as a component of the uncoated or coated food-contact surface of paper and paperboard intended for use in producing, manufacturing, packaging, processing, preparing, treating, packing, transporting, or holding dry food of the type identified in 21 CFR part 176.170 (c); (6) in the manufacture of semirigid and rigid acrylic and modified plastics used as articles intended for use in contact with food in accordance with the conditions prescribed in 21 CFR part 177.1010; (7) in VDC-methylacrylate copolymers as an article or as a component of an article intended for use in contact with food subject to the provisions prescribed in 21 CFR part 177.1990; and (8) in the manufacture of cellophane for packaging food in accordance with the conditions prescribed in 21 CFR part 177.1200.

Reference:

Castle, L., Mercer, A. J., Startin, J. R., and Gilbert, J., Migration from plasticized films into foods. 3. Migration of phthalate, sebacate, citrate and phosphate esters from films used for retail food packaging, *Food Addit. Contam.*, 5, 9, 1988.

POLYVINYL PYRROLIDINONE

Molecular Formula. [~C_6H_9NO~]$_n$
CAS No 9003-39-8
RTECS No TR83700000
Abbreviation. PVP.

Synonyms and **Trade Names.** Haemodez; Haemovinyl; Poly-*N*-vinylbutyrolactam; Povidone.

Properties. Faintly yellow solid or white amorphous, odorless powder. Hygroscopic. Soluble in water giving a colloidal solution. Readily soluble in alcohol.

Applications. PVP is a food additive. Used in medical detoxication practice as a prolongator of the medicines action. A 30% water-salt solution of Haemovinyl is used as a substitute for blood plasma. Used in cast films adher ent to glass, metals, and plastics.

Acute Toxicity.

Observations in man. Probable lethal dose seems to be above 15 g/kg BW.[019]

Animal studies. Mice tolerated a dose of 3.5 g/kg BW.[1] LD_{50} of 25% solution is reported to be in the range of 12 to 15 g/kg BW in mice. LD_{50} of more than 40 g/kg BW in different species is reported by Lefaux.[05] *I/p* administration induced pronounced anemia and reticulocytosis. Small doses are unlikely to be toxic.

Long-term Toxicity. Dogs and rats were given up to 10% PVP in their feed for 2 years. This feeding caused no signs of toxic action.[05]

Reproductive Toxicity.

Teratogenicity. After injection into rabbit yolk on day 9 of gestation, no increase in malformations was observed.[3] There were no defects in rabbits injected with 0.5 ml PVP/kg BW during organogenesis. A reduction in fetal weight was noted.[4]

Gonadotoxicity. Povidone-coated colloidal silica particles were successfully used to fractionate semen samples and select motile, morphologically normal sperm for *in vitro* fertilization.[5]

Allergenic Effect. Laboratory animals well tolerate PVP, apart from dogs, in which it causes histamine anaphylactic shock. PVP with M = 1,000,000, but not 40,000, exhibited pronounced allergenic properties.

Carcinogenicity. Malignant tumors can develop in rats upon repeated injections. Nevertheless, prolonged oral consumption by rats induced no carcinogenic findings.[2]

Carcinogenicity classification. An IARC Working Group concluded that there is limited evidence for the carcinogenicity of PVP in *experimental animals* and there were no data available to evaluate the carcinogenicity of PVP in *humans*.

IARC: 3

Chemobiokinetics. High-molecular-mass PVP is not absorbed from the GI tract. It may produce bulk cathar sis, flatulence, and fecal impaction.[019]

In a man given 1 liter of a 3.5 to 4.5% solution of PVP with M = 80,000, it remains in the cells of the reticuloendothelial tissue of the spleen and liver. PVP with M = 50,000 to 60,000 is excreted entirely unchanged with urine. After 3 days, <15% PVP remains in the body. PVP retention in the body depends on its molecular mass, since only low-molecular-mass PVP is eliminated from the organism.

Absorption and accumulation of PVP occur in the intestinal lymph nodes that prevents substantiation of appropriate *i/g* dose for humans. Accumulation of PVP in the lymphatic system of the treated animals was not noted.[2]

Regulations.

EU (1995). PVP is a food additive generally permitted for use in foodstuffs.

U.S. FDA (1998) approved the use of PVP polymers/copolymers (1) in food in accordance with the conditions prescribed in 21 CFR part 172.874; (2) as defoaming agents of the uncoated or coated food-contact surfaces in the manufacture of paper and paperboard intended for use in contact with aqueous and fatty foods only in accordance with the conditions prescribed in 21 CFR part 176.170, and (3) in the manufacture of resinous and polymeric coatings as the food-contact surface of articles intended for use in producing, manufacturing, packing, processing, preparing, treating, packaging, transporting, or holding food in accordance with the conditions prescribed in 21 CFR part 175.300.

References:

1. Chaplygina, N. I., in *Proc Symp. Physiologically and Optically Active Polymeric Substances*, Riga, 1971, 82 (in Russian).
2. Barnett, L., Report No 40 (Series A, B, C), FAO/WHO Committee, Series No 653, 1982.
3. Claussen, U. and Breuer, H.-W., The teratogenic effects in rabbits of doxycycline, dissolved in polyvinylpyrrolidone, injected into the yolk sac, *Teratology,* 12, 297, 1975.
4. Siegemund, V. B. and Weyers, W., Teratologische untersuchungen eines niedermolekularen polyvinylpyrrolidon-iod-komplexes an kaninchen, *Arzneimittel- Forsch.*, 37, 340, 1987.
5. Pickering, S. J., Fleming, T. P., Braude, P. R., Bolton, V. N., and Gresham, G. A., Are human spermatozoa separated on Percoll density gradient safe for therapeutic use? *Fertil. Steril.*, 51, 1024, 1989.

RUBBER and VULCANIZATES

Abbreviation. PAH (polycyclic aromatic hydrocarbons).

Composition. Treating a rubber with another compound, often sulfur-based, to reduce its tackiness is called *vulcanization*. The rubber vulcanization products (rubber mixes) beside rubber (natural or synthetic polymer), include the following components: vulcanizing agents, vulcanization accelerators, vulcanization activators, and, in some cases, scorch retarders, fillers, plasticizers, mainly antioxidants, and also antiozonants, light stabilizers, etc. (the number of components ranges from 5 to 6 to 15 to 20). The choice of the type of rubber and components and their quantitative ratio in the mix depends on the designation of the V. and also on economic, hygienic, and other considerations (for example, medical grade or food-grade V.).

N-Nitrosamines and precursors are present in rubber products in which the accelerators and stabilizers used in the vulcanization process were derived from dialkylamines. All chemicals used for rubber compounding contain nitrosamines if they are derivatives of secondary amines; e.g., tetramethyl thiuram, zinc-diethyldithiocarbamate, or *N*-oxydiethylene benzothiazolyl sulfenamide. All rubber products containing these dialkyl amine derivatives exhibited considerable levels of the corresponding nitrosamines.[1]

Applications. V. are irreplaceable materials in different applications, including the food industry and water supply.

Migration Data. Some *food-grade V.* containing blacks (DG-100, PM-70, PGM-33, TG-10, PM-15) can release up to 0.06 to 9.2 μg benzo[a]pyrene/l into water and model media simulating foodstuffs. The greatest quantity of benzo[a]pyrene was released by low-dispersion blacks PM-15, TG-10, and PGM-33. Benzo[a]pyrene hardly migrates at all from V. with high-dispersion blacks PM-70 and DG-100. The release of carcinogenic PAH can be reduced by the combined use of low- and high-dispersion blacks.[2]

PAH migrate from stoppers containing furnace blacks into benzene extracts. The PAH found in extracts were the same as those contained in furnace blacks: benzo[a]pyrene, pyrene, chrysene, 1,2-benzopyrene, 1,12-benzoperylene, 1,2-benantracene, etc. Very few PAH are released into cottonseed oil and unskimmed homogenised milk (59°C, 7 days) from V. containing 20 wt.% furnace blacks with a specific surface of 24 to 61 m^2/g. Migration of PAH from V. into aqueous solutions of citric and acetic acids, sodium bicarbonate, and sodium chloride (59°C, 6 days and 145°C, 30 min) was not established. However, migration of PAH into fat-containing foodstuffs was reported.[028]

Traces of nitrosamines in baby bottle rubber nipples and pacifiers could easily migrate to simulated saliva and milk. 7 out of 42 samples of these products contained greater than 0.03 ppm total volatile nitrosamines (mainly, *N*-nitrosodimethylamine and *N*-nitrosodi-*N'*-butylamine).[3]

18 out of 24 samples analyzed (baby bottle rubber nipples and pacifiers) were found to contain varying levels (mean, 41 mg/kg; range, 8 to 146 mg/kg) of *N*-nitrosodibenzylamine. The identity of the compound was confirmed by GC-thermal energy analysis as well as by gas chromatography-mass spectrometry analyses.[4]

According to Havery et al., only one of 189 samples of rubber nipples for babies' bottles was found to be in violation of the *U.S. FDA's* action level of 60 ppb; it contained a total of 137 ppb *N*-nitrosamines.[5]

A survey of 30 samples of various nipples and pacifiers, that was carried out by analysis by GLC-thermal energy analyzer, indicated the presence of the following *N*-nitrosamines: *N*-nitrosodimethylamine (up to 70 mg/kg), *N*-nitrosodiethylamine (up to 88 μg/kg), *N*-nitrosodi-*n*-butyl- amine (up to 2796 mg/kg), *N*-nitrosopiperidine (up to 180 mg/kg), and *N*-nitrosomorpholine (up to 86 mg/kg). These *N*-nitrosamines were shown to migrate easily from the rubber products to liquid infant formula, orange juice, and simulated human saliva.[6]

All of the 17 samples of rubber nipples and pacifiers tested using GC-thermal energy analysis after extraction with artificial saliva, were found to contain at least two of the following *N*-nitrosoamines: *N*-nitrosodimethylamine, *N*-nitrosodiethylamine, *N*-nitrosodibutylamine, and *N*-nitrosopiperidine. Total volatile *N*-nitrosamine levels with a mean contents of 7.3 mg/kg were found. Nitrosatable compounds, measured as *N*-nitrosamines after nitrosation, were detected in 15 of the 17 samples, the mean level being 5.0 mg/kg.[7]

In 1983 and 1985, rubber teats and pacifiers from the Dutch market were analyzed for *N*-nitrosamines and nitrosatable compounds by extraction with an artificial saliva test solution (24 hours, 40°C). Nitrosatable compounds were determined as *N*-nitrosamines (GC-thermal energy analysis) after nitrosation. In 1983, the total contents of *N*-nitrosamines and nitrosatable compounds varied from 4 to 40 and 50 to 3700 mg/kg, respectively (18 samples). In 1985, *N*-nitrosamines and nitrosatable compounds varied from 3 to 94 and 26 to 5100 mg/kg, respectively (20 samples).[8]

Carcinogenicity. Application of 2.0 and 15% solutions of dried benzene extracts of V. stoppers containing PAH to the skin of mice (breed *S57*-black) over 20 months caused papillomata and carcinomata of the skin (in 45.7 and 6.7% of animals, respectively).[3]

Regulations.

U.S. FDA (1998) approved the use of synthetic and natural rubber articles for repeated use in accordance with the provisions indicated in 21 CFR part 177.2600.

According to ***Dutch*** legislation, teats may contain no more than 1.0 and 20 mg/kg *N*-nitrosamines and nitrosatable compounds, respectively.

References:

1. Spiegelhalder, B. and Preussmann, R., *Nitrosamines and Rubber*, IARC Sci. Publ., 41, 231, 1982.
2. Medvedev, V. I., Migration of polycyclic hydrocarbons from resins used in the food industry, *Voprosy Pitania*, 6, 70, 1973 (in Russian).
3. Sen, N. P., Kushwaha, S. C., Seaman, S. W., et al., Nitrosamines in baby bottle nipples and pacifiers: Occurrence, migration, and effect of infant formulas and fruit juices on *in vitro* formation of nitrosamines under simulated gastric conditions, *J. Agric. Food Chem.*, 33, 428, 1985.
4. Sen, N. P., Seaman, S. W., and Kushwaha, S. C., Determination of non-volatile *N*-nitrosamines in baby bottle rubber nipples and pacifiers by high-performance liquid chromatography-thermal energy analysis, *J. Chromatogr.*, 463, 419, 1989.
5. Havery, D. C., Perfetti, G. A., Canas, B. J., and Fazio, T., Reduction in levels of volatile *N*-nitrosamines in rubber nipples for babies' bottles, *Food Chem. Toxicol.*, 23, 991, 1985.
6. Sen, N. P., Seaman, S., Clarkson, S., Garrod, F., and Lalonde, P., Volatile *N*-nitrosamines in baby bottle rubber nipples and pacifiers. Analysis, occurrence and migration, *IARC Sci. Publ.*, 57, 51, 1984.
7. Osterdahl, B. G., *N*-nitrosamines and nitrosatable compounds in rubber nipples and pacifiers, *Food Chem. Toxicol.*, 21, 755, 1983.
8. Ellen, G. and Sahertian, E. T., *N*-nitrosamines and nitrosatable compounds in rubber nipples, pacifiers and other rubber and plastic commodities, *IARC Sci. Publ.*, 84, 375, 1987.

VULCANIZATES based on NATURAL RUBBER

CAS No 9006-04-6
RTECS No VL8020000
Abbreviation. NR.

Composition. NR is *cis*-1,4-polyisoprene, an elastomer obtained from the latex of *Hevea brasiliensis* and other trees.

Synonyms and **Trade Names.** Cautchouc; Gum nafkacrystal; India rubber; Nafka; Nafka crystal gum; Natural latex; Natural rubber; Polyisoprene; Rubber; Thiokol NVT.

Properties. Nearly colorless and transparent in thin layers. Odorless and tasteless. Practically insoluble in water, alcohol; soluble in oil of turpentine.

Applications. In the raw state, rubber has limited application other than being dissolved in solvent for use as adhesives and solutions.[020]

Reproductive Toxicity.

Observations in man. An increased incidence of adverse pregnancy outcome (spontaneous abortion and malformation) was reported in women in a tire manufacturing plant.[1] Other reports found no increase in spontaneous abortion or malformation among women employed as rubber workers.[2,3]

Mutagenicity.

Observations in man. An increase in mutagenic activity was found in the urine of some rubber workers.[4,5]

Regulations.

U.S. FDA (1998) approved the use of NR *latex* for use (1) in adhesives as a component of articles intended for use in packaging, transporting, or holding food in accordance with the conditions prescribed in 21 CFR part 175.105; and (2) in the manufacture of rubber articles intended for repeated use in producing, manufacturing, packaging, processing, preparing, treating, packing, transporting, or holding food in accordance with the conditions prescribed in 21 CFR part 177.2600.

EU (1990). NR is available in the *List of monomers and other starting substances, which may continue to be used pending a decision on inclusion in Section A.*

Great Britain (1998). NR is authorized without time limit for use in the production of polymeric materials and articles in contact with food or drink or intended for such contact.

References:

1. Axelson, O., Edling, C., and Andersson, L., Pregnancy outcome among women in a Swedish rubber plant, *Scand. J., Work Environ. Health,* 9, 79, 1983.
2. Lindbohm, M. L., Hemminki, K., Kyyronen, P., Kilpikari, I., and Vainio, H., Spontaneous abortions among rubber workers and congenital malformations in their offspring, *Scand. J. Work Environ. Health,* 9 (Suppl 2), 85, 1983.
3. Hemminki, K., Niemi, M.-L., Kyyronen, P., Kilpikari, I., and Vainio, H., Spontaneous abortions and reproductive selection mechanisms in the rubber and leather industry in Finland, *Brit. J. Ind. Med.,* 40, 81, 1983.
4. Sorsa, M., Falck, K., Maki-Paakkanen, J., and Vainio, H., Genotoxic hazards in the rubber industry, *Scand. J. Work Environ. Health,* 9,103, 1983.
5. Rendon, A., Rojas, A., Fernandez, S. I., and Pineda, I., Increases in chromosome aberrations and in abnormal sperm morphology in rubber factory workers, *Mutat. Res.*, 323, 151, 1994.

VULCANIZATES based on NATURAL (POLYISOPRENE) RUBBER

Structural Formula.

$[\sim H_2CC{=}CHCH_2\sim]_n$

|

CH_3

M = 40,000

CAS No 9006-04-6

9003-31-0

Abbreviation. NR.

Synonyms and **Trade Names.** Caoutchouc; India rubber; Isoprene D; Isoprene oligomer; Isoprene polymer; 2-Methyl-1,3-butadiene, homopolymer; Natural latex; *cis*-1,4-Polyisoprene; *trans*-1,4-Polyiso- prene; Poly(2-methyl-1,3-butadiene; Poly-1-methylbutenylene.

Composition. NR is a stereoregular polyisoprene (poly-2-methyl-1,3-butadiene) containing 98 to 100% *cis*-1,4-isoprene units. The main types (smoked-sheet and crepes) are produced from NR-latices.The non-rubber components of NR are acetone extract (oleic, linoleic and stearic acids, carotene), albuminous substances, amino acids, ash, and moisture. The V. contain diphenylguanidine, zinc oxide, stearic acid, sulphur, filler, and paraffin oil.

Properties. Nearly colorless and transparent in thin layers. Odorless and tasteless. *Dens.* 0.9. Practically insoluble in water, alcohol; soluble in oil of turpentine. V. based on NR is chemically stable and highly elastic.

Applications. General purpose rubber. Used to replace natural rubber and in the manufacture of "synthetic" natural rubber and butyl rubber. A copolymer in the production of synthetic elastomers. Used in the food industry.

Migration Data. V. contained diphenyl guanidine, zinc oxide, stearic acid, sulphur, filler, and paraffin oil. A study of extracts (water and model media; 1 hour to 5 days; 20 and 100°C; 0.1 to 2 cm^{-1}) revealed migration of diphenyl dicarbamide (0.05 to 0.2 mg/l). The highest level of migration was observed from white V. (filler was whiting); migration into media simulating vegetable pickles and preserves is higher than into distilled water. Migration into milk-simulating media was negligible.[1] Dimethyl dithiocarbamate and diethyl dithiocarbamate were detected by the GC determination at levels up to 3.2 and up to 4.6 mg/g rubber (as dithiocarbamic acid), respectively, in chloroform-acetone extracts from isoprene rubber teats for baby bottles. Dialkyl dithiocarbamates can form secondary amines by acid hydrolysis, although their levels in the extracts only made a minor contribution to the total level of measured secondary amine precursors.[2]

Regulations. ***U.S. FDA*** (1998) approved the use of synthetic and natural rubber articles for repeated use in accordance with the provisions indicated in 21 CFR part 177.2600.

References:

1. Kazarinova, N. F. and Ledovskikh, N. G., *Kauchuk i Rezina*, 1, 26, 1978 (in Russian).
2. Yamazaki, T., Inoue, T., Yamada, T., and Tanimura, A., Analysis of residual vulcanization accelerators in baby bottle rubber teats, *Food Addit. Contam.*,3, 145, 1986.

VULCANIZATE based on NATURAL RUBBER

CAS No 9006-04-6

Composition. A speciment was prepared using the standard formulation based on NR with the use of vulcanization accelerators mercaptobenzothiazole and diphenyl guanidine.

Applications. Used in the production of children's toys.

Migration Data. A study extracts revealed increased migration of MBT (q.v.) into water.

Long-term Toxicity. Rats were given aqueous extracts (0.7 cm^{-1}, 38°C, 24 hours). The treatment decreased oxygen consumption by the end of the study. No retardation in BW gain was observed. There were no changes in the liver and kidney functions, or in biochemistry and hematology analyses. Gross pathology examination revealed plethora and effects of perivascular edema of limited areas and shriveling of the nerve cells of the cortex and subcortical ganglions. Areas of hyperplasia of the lymphoid elements were found around vessels in the lungs. Lesions in the liver and kidney as well as slight catarrhal-desquamation changes were recorded in the intestines.

Reference:

Shumskaya, N. I., Provorov, V. N., Tolgskaya, M. S., Yemel'yanova, L. V., and Chernevskaya, N. M., Toxicity of some vulcanizates intended for manufacture of children's toys, in *Hygiene Aspects of the Use of Polymeric Materials and of Articles Made of Them*, L. I. Medved', Ed., All-Union Research Institute of Hygiene and Toxicology of Pesticides, Polymers and Plastic Materials, Kiyv, 1969, 104 (in Russian).

VULCANIZATE based on NATURAL RUBBER (different vulcanization systems)

CAS No 9006-04-6

Composition. Vulcanization systems: peroxymon F-40 [di(*tert*-butylperoxyisopropyl)benzene] or peroxide P-5 [bis(*n-tert*-butylperoxyisopropylcumyl)peroxide]. Peroxides' contents were 1.0 mg/g rubber 5 mass. parts/100 mass. parts vulcanizate.

Migration Data. Simulant media used in these studies were: 2.0% solution of acetic acid + 2.0% solution of *NaCl*, 2.0% solution of citric acid + 6.0% ethanol/water solution (20 and 40°C, 1 and 24 hours, 0.5 and 1 cm^{-1}).

The substances migrating from vulcanizate-1 were mixture of *m*- and *p*-isomers of bis(2-hydroxypropyl)benzene, 2-(4-acetylphenyl)- and 2-(4-*tert*-butylperoxyisopropylphenyl) propanol-2, and also diacetyl benzene. Predominantly, *m*-isomer and bis(2-hydroxypropyl) benzene (at the concentration of 0.82-6.41 mg/l) were found.

The substances migrating from vulcanizate-2 were identified as 1,4-bis (2-hydroxypropyl)benzene, 2-(2-acetylphenyl) propanol-2, 2-(4-*tert*-butylperoxyisopropylphenyl)propanol-2, and 1,4-diacetyl benzene.

Regulations. ***U.S. FDA*** (1998) approved the use of synthetic and NR articles for repeated use in accordance with the provisions indicated in 21 CFR part 177.2600.

Recommendations. ***Russia.*** Samples of these vulcanizates showed no substantial migration and were recommended for use in the food industry.

Reference:

Kazarinova, N. F. and Ledovskikh, N. G., *Kauchuk i Rezina*, 1, 26, 1978 (in Russian).

VULCANIZATE based on NATURAL RUBBER

CAS No 9006-04-6

Composition. V. contains diphenylguanidine and dithiocarbamates.

Long-term Toxicity. Rats exposed to extracts (over a period of 1 year) as their drinking water after 6 to 8 months developed an increased contents of eosinophilic granulocytes and reticulocytes, decreased oxygen consumption, and elevated relative weights of some visceral organs. There was vascular plethora of the brain, liver, and kidneys.[028]

Regulations. ***U.S. FDA*** (1998) approved the use of synthetic and NR articles for repeated use in accordance with the provisions indicated in 21 CFR part 177.2600.

PARONIT

Composition. V. contains asbestos, sulphur, MBT, carbon black, and zinc oxide.

Applications. Used to make seals in reverse-osmosis and electrolytic distillers.

Migration Data. After contact with PON-type P., water acquired an odor; its color was changed. Migration of unsaturated hydrocarbons was observed. Under the action of phosphates which accelerate aging processes, P. dissociates and releases MBT and thiuram.[1]

Long-term Toxicity. 8-month administration of aqueous extracts obtained in aggravated conditions caused a short-term increase in the activity of blood acid phosphatase, a reduction in the STI, and leukopenia.[2]

Recommendations. ***Russia.*** PON-type P. is recommended for use in electrodialysis distilling units as a structural sealant.

References:

1. Aksyuk, A. F., Tarkhova, L. P., Merzlyakova, N. M., and Zvereva, V. A., Hygienic assessemnt of rubberized asbestos fabric recommended for use as a construction material for electrodialysis desalination units, *Gig. Sanit.*, 3, 19, 1985 (in Russian).
2. Selivanov, S. B., *Hygiene Investigation and Assessment of Reverse Osmosis Method for Analyzing Domestic Drinking Water on "Filter-press"-type Units*, Author's abstract of thesis, A. N. Sysin Research General Comm. Hygiene Institute, Moscow, 1977, 18 (in Russian).

VULCANIZATE based on NATURAL RUBBER (Qualitex)

CAS No 9006-04-6

Composition. Vulcanizate with the addition of sodium salt of 2-mercaptobenzimidazole, accelerator Vulcazit-*P*-extra-*N*, Leikanol dispersing agent, sulphur, and zinc oxide.

Applications. Used to produce children's dummies.

Migration Data. A study of extracts (0.67 and 1.0 cm^{-1}; 20 and 38°C; 24 hours; distilled water) revealed the migration of Leikanol (19 mg/l) and sodium salt of mercaptobenzimidazole (3.0 mg/l).

Short-term Toxicity. Aqueous extracts were studied in a 3- to 4-month experiment on immature male rats weighing 60 to 100 g. Consumption of extracts from the dummies caused retardation of BW gain, reduction in 24-hour diuresis, and an increase in relative kidney weights.

Reference:

Shumskaya, N. I., Provorov, V. N., Tolgskaya, M. S., Yemel'yanova, L. V., and Chernevskaya, N. M., Toxicity of some vulcanizates intended for manufacture of children's toys, in *Hygiene Aspects of the Use of Polymeric Materials and of Articles Made of Them*, L. I. Medved', Ed., All-Union Research Institute of Hygiene and Toxicology of Pesticides, Polymers and Plastic Materials, Kiyv, 1969, 104 (in Russian).

VULCANIZATE based on NATURAL RUBBER (Revultex-2R)

CAS No 9006-04-6

Composition. Vulcanizate based on latex with Lutanol introduced.

Applications. Used to produce children's dolls.

Short-term Toxicity. Aqueous extracts were studied in a 3- to 4-month experiment on immature male rats weighing 60 to 100 g. Drinking aqueous extracts from dolls caused retardation of BW gain, and changes in urinalyses of 24-hour diuresis, chloride contents of urine, etc.

Reference:

Shumskaya, N. I., Provorov, V. N., Tolgskaya, M. S., Yemel'yanova, L. V., and Chernevskaya, N. M., Toxicity of some vulcanizates intended for manufacture of children's toys, in *Hygiene Aspects of the Use of Polymeric Materials and of Articles Made of Them*, L. I. Medved', Ed., All-Union Research Institute of Hygiene and Toxicology of Pesticides, Polymers and Plastic Materials, Kiyv, 1969, 104 (in Russian).

VULCANIZATES based on NATURAL RUBBER (formulations IF-171 and IF-172A)

CAS No 9006-04-6

Applications. Used in high-temperature pasteurization units in the dairy industry.

Migration Data. Migration of accelerators (thiuram and Santocure) into a model solution simulating milk ceased entirely after 10 hours. However, migration of transformation products of vulcanizate components could not be eliminated.

Long-term Toxicity. Male rats received milk extracts of vulcanizates (0.1 cm^{-1}; 120°C; 20 min; cold 24 hours). 9-month administration had no effect on the general state, BW gain, hematological indices, the contents of blood total protein, activity of a number of enzymes, ascorbic acid contents in the adrenal gland, the mass coeffi cients of the visceral organs, etc.

Regulations. ***U.S. FDA*** (1998) approved the use of synthetic and natural rubber articles for repeated use in accordance with the provisions indicated in 21 CFR part 177.2600.

Reference:

Rakhmanina, N. L., Hygienic characteristics of some hot-resistant vulcanizates for pasteurization apparatus, in *Hygiene Aspects of the Use of Polymeric Materials and of Articles Made of Them*, L. I. Medved', Ed., All-Union Research Institute of Hygiene and Toxicology of Pesticides, Polymers and Plastic Materials, Kiyv, 1969, 102 (in Russian).

VULCANIZATES based on NATURAL RUBBER (formulations T-199 and T-199, and 1743 without stabilizer)

CAS No 9006-04-6

Composition. Formulation T-199-x contains Bisalkophene BP; T-199 contains ionol.

Migration Data. Diphenylguanidine migrates into aqueous extracts (0.4 and 0.25 cm^{-1}; 20, 38, 70, and 100°C; 1, 8, and 24 hours) from vulcanizates of the given grades and also grades 191-3F, 10334/16, and 10334/17 in a quantity of 0.16 to 1.04 mg/l. Zinc and dithiocarbamates were not found in the extracts.

Long-term Toxicity. Rats and their progeny received aqueous extracts from vulcanizates of the given grades over a period of 12 and 4 months, respectively. The treatment caused significant changes in the hematology analyses and in the some liver functions. Histological examination revealed disruptions in the viscera.

Reference:

Shurupova, Ye. N. and Vlasyuk, M. G., Toxicological and hygienic characteristics of vulcanizates intended for food industry and children's toys production, in *Hygiene Aspects of the Use of Polymeric Materials and of Articles Made of Them*, L. I. Medved', Ed., All-Union Research Institute of Hygiene and Toxicology of Pesticides, Polymers and Plastic Materials, Kiyv, 1969, 79 (in Russian).

VULCANIZATE based on NATURAL RUBBER (Smoked-sheet)

CAS No 9006-04-6

Composition. *Formulation I:* The mix also contains thiuram, zinc stearate, sulphur, and antimony pentasulfide; plasticizer, transformer oil, based on smoked-sheet NR with sodium butadiene rubber; plasticizer, paraffin oil. *Formulation II:* Vulcanization accelerators: thiuram E or ethyl zinc dimethyldithiocarbamate; fillers: carbon black or lithopone, kaolin, fine colloidal silica.

Applications. Used to produce children's dummies.

Migration Data. A study of extracts of *Formulation I* (0.67 and 1.0 cm^{-1}; 20 and 38°C; 24 hours; distilled water) revealed the migration of dithiocarbamates (0.4 to 0.6 mg/l) and antimony (0.04 to 0.8 mg/l).[1]

A study of aqueous extracts of *Formulation II* and model media (20, 40, 100°C; 1 and 5 days; 0.5 cm^{-1}) revealed migration of the accelerators introduced into the vulcanizates and products of their transformation: thiuram E (up to 0.15 mg/l), ethyl zinc dimethyl dithiocarbamate (up to 0.25 mg/l), carbon disulfide (up to 0.07 mg/l), diethyl ammonium (up to 0.06 mg/l), and dithiocarbaminates (from traces to 0.43 mg/l).[2]

Short-term Toxicity. Aqueous extracts of *Formulation I* were studied in a 3- to 4-month experiment on immature male rats weighing 60 to 100 g. Administration of extracts caused rapid increase in BW gain, damages to the liver and kidney functions, and hematology parameters. In rats receiving aqueous extracts of dummies, manifestations of liver dystrophy were noted.[1]

References:

1. Shumskaya, N. I., Provorov, V. N., Tolgskaya, M. S., Yemel'yanova, L. V., and Chernevskaya, N. M., Toxicity of some vulcanizates intended for manufacture of children's toys, in *Hygiene Aspects of the Use of Polymeric Materials and of Articles Made of Them*, L. I. Medved', Ed., All-Union Research Institute of Hygiene and Toxicology of Pesticides, Polymers and Plastic Materials, Kiyv, 1969, 104 (in Russian).
2. Grushevskaya, N. Yu., Determination of vulcanization accelerator of vulkazite-*P*-extra-*N* and some transformation products in sanitary and chemical investigation of rubber, *Gig. Sanit.*, 4, 59, 1987 (in Russian).

VULCANIZATES based on NATURAL RUBBER (Smoked sheet)

CAS No 9006-04-6

Composition. *Formulation IR-34-7:* Accelerator: Vulkazit-*P*-extra-*N*; filler: carbon black. *Formulation 52-507:* Filler: lithopone, kaoloin, fine colloidal silica; vulcanization accelerator: Vulkazit-*P*-extra-*N*.

Migration Data. A study of aqueous extracts of *formulation IR-34-7* (24 hours; 20°C) revealed migration of monoethylaniline (0.3 mg/l); release of ethylphenyl dithiocarbamic acid derivatives was negligible. A study of aqueous and citric acid extracts of *formulation 52-507* (20, 40, and100°C; 24 hours; 30 days) revealed traces of ethylphenyl dithiocarbamic acid derivatives and migration of monoethyl aniline in a quantity of 2.0 to 3.0 mg/l. Vulkazit-*P*-extra-*N*, thiuram EF, and carbon disulfide were not found.

Reference:

Grushevskaya, N. Yu., Determination of vulcanization accelerator of vulkazite-*P*-extra-*N* and some transformation products in sanitary and chemical investigation of rubber, *Gig. Sanit.*, 4, 59, 1987 (in Russian).

VULCANIZATES based on NATURAL RUBBER (formulations IR-34A and IR-34P)

CAS No 9006-04-6

Long-term Toxicity. Administration of extracts to rats caused a change in the urine protein level (within the range of physiological variations) and also a change in the relative kidneys weights. In rats receiving extracts of IR-34P vulcanizates there were no abnormalities in any of the indices studied compared with the control.

Regulations. ***U.S. FDA*** (1998) approved the use of synthetic and natural rubber articles for repeated use in accordance with the provisions indicated in 21 CFR part 177.2600.

Reference:

Shumskaya, N. I., Mel'nikova, V. V., Chikishev, Yu. G., and Taradai, Ye. P., Toxicity of the extracts of rubber made on varying cautchouc base and containing thiuram in the formulation, *Gig. Sanit.*, 2, 82, 1979 (in Russian).

VULCANIZATES based on SYNTHETIC RUBBER

Composition. Synthetic rubbers are elastic polymers of synthetic latexes.

Applications. Used to replace natural rubber and in the manufacture of "synthetic" natural rubber and butyl rubber. A copolymer in the production of synthetic elastomers. Used in the food industry.

POLYISOPRENE RUBBER, SYNTHETIC

Structural Formula.

$[\sim CH_2C{=}CHCH_2\sim]_n$

|

CH_3

CAS No 9003-31-0

Synonyms and **Trade Names.** Isoprene D; Isoprene polymer; 2-Methyl-1,3-butadiene, homopolymer; *trans*-1,4-Polyisoprene; *cis*-1,4-Polyisoprene; Poly(2-methyl-1,3-butadiene; Poly-1- methylbutenylene.

Applications. General purpose rubber. Used to replace natural rubber and in the manufacture of "synthetic" natural rubber and butyl rubber. A copolymer in the production of synthetic elastomers.

VULCANIZATES based on SYNTHETIC POLYISOPRENE RUBBER (SKI and SKI-3)

Structural Formula.

$[\sim CH_2C{=}CHCH_2\sim]_n$

|

CH_3

M = 200,000

CAS No 9003-31-0

Composition. SKI are synthetic isoprene polymers (2-methyl-1,3-butadiene); the two types differ in the degree of stereoregularity of the chain structure: SKI has an average contact of *cis*-1,4-units of isoprene (65 to 94%), and SKI-3 a high contents (92 to 99%).

Their properties are similar to those of vulcanizates based on NR.

Applications. Intended for use in the food industry.

Migration Data. The vulcanizates contain diphenylguanidine, zinc oxide, stearic acid, sulphur, filler (whiting or carbon black), and paraffin oil. A study of extracts (water and simulant media; 1 hour to 5 days; 20 and 100°C; 0.1 to 2.0 cm^{-1}) revealed migration of diphenylguanidine (0.05 to 1.0 mg/l) and products of its transformation: aniline (up to 0.2 mg/l), triphenyl melamine (up to 0.1 mg/l), tetraphenyl melamine (up to 0.15 mg/l), and triphenyl dicarbamide (up to 0.5 mg/l).

Mechanisms of migration: q.v. vulcanizates based on NR.[1]

Repeated Exposure. Rats received aqueous extracts of SKI-3 (antioxidant - ionol, q.v.) over a period of 2 months. Leukocytosis was observed in treated animals, there was an increase in the content of lactic acid in the blood. Histological analysis revealed plethora and dystrophic changes in the liver, kidneys, and myocardium.[2]

References:

1. Kazarinova, N. F. and Ledovskikh, N. G., *Kauchuk i Rezina*, 1, 26, 1978 (in Russian).
2. Pestova, A. G. and Petrovskaya, O. G., Effect of chemical substances isolated from isoprene rubber SKI-3, *Vrachebnoye Delo*, 4, 135, 1973 (in Russian).

VULCANIZATES based on SYNTHETIC POLYISOPRENE RUBBER [SKI-3P, formulations 17F-54 and 17F-54-(1-7)]

CAS No 9003-31-0

Composition. Thiuram and diphenylguanidine are used as vulcanization accelerators.

Applications. Intended for direct contact with liquid foodstuffs.

Migration Data. A study of extracts (water and simulant media; 0.1 and 0.5 cm^{-1}; 1 and 24 hours) revealed a taste and odor of rating 1 to 2. Thiuram and zinc dimethyldithiocarbamate were not found in extracts, and diphenylguanidine and ionol migrated in a quantity of 0.1 mg/l.

Long-term Toxicity. In mice receiving aqueous extracts of 17F-54-2 and 17F-54-6 vulcanizates over a period of 12 months, there was a single increase in the number of reticulocytes and a reduction in the *Hb* contents in the blood. No other pathological abnormalities were observed.

Recommendations. ***Russia.*** Vulcanizates based on SKI-3P rubber were recommended for the manufacture of rubber hoses feeding liquid foodstuffs.

Reference:

Vlasyuk, M. G., Toxicological and hygienic characteristics of 'food-grade' resins made of SKI-3P rubber, in *Hygiene and Toxicology of High-Molecular-Mass Compounds and of the Chemical Raw Material Used for Their Synthesis,* Proc. 6th All-Union Conf., B. Yu. Kalinin, Ed., Khimiya, Leningrad, 1979, 64 (in Russian).

VULCANIZATES based on SYNTHETIC POLYISOPRENE RUBBER (SKI-3P and SKN-26M, formulations P4-2, P4-9, P4-10)

CAS No 9003-31-0

Migration. A study of extracts revealed migration of Vulkazit (up to 0.08 mg/l), monoethyl aniline (up to 2.0 mg/l), 2-mercaptobenzothiazole (up to 2.4 mg/l), 2-mercaptobenzothiazole, sodium salt (up to 2.8 mg/l), diphenylguanidine (up to 1.2 mg/l), tetraphenyl melamine (up to 0.5 mg/l), aniline (up to 0.2 mg/l), and carbonylamide (up to 0.15 mg/l).[028]

VULCANIZATES based on POLYBUTADIENE RUBBER (SKB and SKD)

POLYBUTADIENE

Structural Formula.

```
[~H2C      CH2]n [~H2C  CH  ]n [~CH2CH~]n
    /\     / \         /\ //\ /        |
    CH=CH              CH CH2          CH
                                       ||
                                       CH2
```

M = 20,000 to 250,000

CAS No 9003-17-2

Abbreviation. PB.

Synonym and **Trade Names.** Alfine; 1,3-Butadiene, polymer; Diene 35 NF; Polyoil 110 or 130.

Composition. SKB (sodium-catalyzed butadiene rubber) is a synthetic non-stereoregular poly-1,3-butadiene. SKD is a synthetic stereoregular poly-1,3-butadiene containing mainly *cis*-1,4-units of butadiene.

Properties. *Dens.* 0.89 to 0.92. Vulcanizates based on SKB possess comparatively low strength and are not noted for good heat resistance. Vulcanizates based on SKD possess high wear resistance, elasticity, and strength.

Regulations. ***U.S. FDA*** (1998) approved the use of PB (1) in adhesives as a component (monomer) of articles intended for use in packaging, transporting, or holding food in accordance with the conditions prescribed in 21 CFR part 175.105; (2) in the manufacture of closures with sealing gaskets used on containers intended for use in producing, manufacturing, packing, processing, preparing, treating, packaging, transporting, or holding food, in accordance with the conditions prescribed in 21 CFR part 177.1210; (3) as a component of the uncoated or coated food-contact surface of paper and paperboard intended for use in producing, manufacturing, packaging, processing, preparing, treating, packing, transporting, or holding dry food of the type identified in 21 CFR part 176.170 (c); and (4) in the manufacture of semirigid and rigid acrylic and modified plastics used as articles intended for use in contact with food in accordance with the conditions prescribed in 21 CFR part 177.1010.

POLYBUTADIENE RUBBER (SKB-35-45), composition with NATURAL RUBBER (Pale Crepe, formulation 52-267)

Applications. Used to make lids for domestic preserves.

Migration Data. A study of extracts revealed migration from the composition of zinc dimethyl dithiocarbamate (2.0 to 3.0 mg/l), *N*-ethyl aniline (2.0 to 5.0 mg/l), and the accelerator Vulkazit-*P*-extra-*N* (0.4 mg/l).[1]

Long-term Toxicity. Rats received extracts as their drinking water over a period of 10 months. There was no animal mortality. No toxic effects were observed.[2]

References:

1. Kazarinova, N. F. and Grushevskaya, N. Yu., Methods of determination of vulcanization accelerators, derivatives of dithiocarbonic acid, under conditions of sanitary-chemical investigation of rubber articles, in *Hygiene Aspects of the Use of Polymeric Materials and of Articles Made of Them*, L. I. Medved', Ed., All-Union Research Institute of Hygiene and Toxicology of Pesticides, Polymers and Plastic Materials, Kiyv, 1969, 533 (in Russian).
2. *Toxicology of the Components of Rubber Mixes and of Rubber and Latex Articles*, TZNIITENEFTEKHIM, Moscow, 1974, 46 (in Russian).

VULCANIZATES based on POLYBUTADIENE RUBBER (SKD-LR)

CAS No 9003-17-2

Composition. A lithium catalyst is used to produce these vulcanizates.

Migration Data. Aqueous extracts were found to have an odor of rating 1.5 to 2, but even under the most rigorous conditions migration of the solvent (hexane-heptene fraction mixed with cyclohexane) is 0.03 mg/l.

Long-term Toxicity. Rats received aqueous extracts (0.1 cm^{-1}; 24 hours) as their drinking water over a period of 10 months. No toxic effects were observed.

Allergenic Effect was not established.

Recommendations. *Russia.* SKD-LR rubber is recommended as a substitute for SKB-rubber in the produc tion of vulcanizates for use in the food industry.

Reference:

Stankevich, V. V., Ivanova, T. P., and Fetisova, L. N., Hygienic studies of SKD-LP rubber intended for production of vulcanizates used in food industry, in *Hygiene and Toxicology of High-Molecular-Mass Compounds and of the Chemical Raw Material Used for Their Synthesis*, Proc. 6th All-Union Conf., B. Yu. Kalinin, Ed., Khimiya, Leningrad, 1979, 97 (in Russian).

VULCANIZATES based on POLYBUTADIENE RUBBER (SKD-LPR), composition with NATURAL RUBBER and POLYISOPRENE RUBBER (SKI-36)

Composition. Two accelerator systems are used in the vulcanizates: Vulkazit-*P*-extra-*N* with sulfenamide (a) and with diphenylguanidine (b).

Migration Data. Migration of *N*-ethyl aniline into model media at 20°C amounted to 0.05 to 0.1 mg/l, and that of mercaptobenzothiazole and diphenylguanidine, was found to be within 0.1 to 0.5 mg/l, respectively.

Long-term Toxicity of aqueous extracts of RS-24 and RS-14a vulcanizates was confirmed in a 10-month experiment.[1]

Allergenic Effect. Aqueous extracts of RS-24 and RS-14a based on SKD-LPR and NR or on SKD-LPR and SKI-3s cause sensitization in the test animals.[2]

References:

1. Stankevich, V. V., Ivanova, T. P., and Prokof'yeva, L. G., Hygienic studies of 'food-grade' resins on the base of a new diene rubber, in *Hygiene and Toxicology High-Molecular-Mass Compounds and of the Chemical Raw Material Used for Their Synthesis*, Proc. 6th All-Union Conf., B. Yu. Kalinin, Ed., Khimiya, Leningrad, 1979, 96 (in Russian).
2. Stankevich, V. V., Ivanova, T. P., Vlasyuk, M. G., and Prokofyeva, L. G., Toxicology hygienic studies of new rubber samples made of synthetic rubber intended for use in contact with food stuff, *Gig. Sanit.*, 8, 66, 1981 (in Russian).

VULCANIZATES based on BUTADIENE-STYRENE RUBBER (SKS and SKMS)

M = 10,000 to 100,000

CAS No 61789-96-6

Composition. SKS and SKMS are synthetic rubbers, copolymers of 1,3-butadiene with styrene or with α-methylstyrene, respectively. They contain fillers.

Acute Toxicity. LD_{50} is about 70 ml/kg BW in rats.

Carcinogenicity classification. An IARC Working Group concluded that there were no adequate data available to evaluate the carcinogenicity of copolymer of butadiene with styrene in *experimental animals* and in *humans*.

IARC: 3.

VULCANIZATES based on BUTADIENE-STYRENE RUBBER (SKS-50 latex)

M = 10,000 to 100,000

CAS No 61789-96-6

Composition. The mix also contains barium sulfate, 2246 (CAO-5) antioxidant, sodium benzoate, OP-10 emulsifier, methyl cellulose, and Leikanol.

Applications. The base for a sealing paste for cans.

Migration Data. Distilled water was poured into 250 ml cans and autoclaved. After one month water transparency and color had changed and migration of OP-10 (0.01 to 0.15 mg/l in proportion to the amount contained in the mix formulation) was found.

Short-term Toxicity. Rats were given water from the autoclaved cans. The toxic effects were observed in the first few weeks of a 3-month experiment. Manifestations of toxicity included: a decrease in BW gain, NS changes, an increase in diuresis and other disorders.

Long-term Toxicity of aqueous extracts of film produced from sealing paste based on *SKS-50p latex* was also studied in a 10-month experiment. Rats received the extracts as their drinking water every day. The treatment caused deterioration in their hematological indices and kidney function. The relative weight of the spleen increased. Histological examination revealed mild dystrophy of the epithelium of the kidney ducts and individual hyaline cylinders in the lumen of the ducts.

Reference:

Toxicology of the Components of Rubber Mixes and of Rubber and Latex Articles, TSNIITEKHIM, Moscow, 1974, 50 (in Russian).

VULCANIZATES based on BUTADIENE-STYRENE RUBBER (SKS-65GP latex)

M = 10,000 to 100,000

CAS No 61789-96-6

Applications. Used for coating cast-iron water pipes.

Migration Data. After the water had been replaced five times, its quality conformed to the drinking water standards. During stagnation for 30 days, satisfactory indices were retained for only 4 days. Migration of styrene and butadiene was not observed.

Reference:

Lastochkina, K. O., Sanitary and hygienic evaluation of polymer-concrete coating on the base of divinyl-styrene latex for water supply pipes, in *Hygiene Problems in the Production and Use of Polymeric Materials*, 1969, 163 (in Russian).

VULCANIZATES based on α-METHYLSTYRENE RUBBER (SKMS 30 ARKM, formulations 155S-3831 and 5S-3019)

Composition. Different vulcanizing systems are employed.

Applications. Used to manufacture articles coming into contact with milk (washers, valves, seals, components of milking apparatus).

Migration Data. A study of extracts (38 to 100°C; 1, 24, or 120 hours; 0.5, 0.1, or 0.01 cm^{-1}) was carried out with the use of distilled water and model media simulating milk, lactic acid products, and wine. The change in the organoleptic properties of water did not exceed rating 2. The model medium simulating milk extracted zinc compounds more intensely. Migration of α-methylstyrene reached 0.03 mg/l. Release of dithiodimorpholine was proportional to its contents in the vulcanizate formulation, the contact time, and temperature, and reached 0.75 mg/l.

Long-term Toxicity. Aqueous extracts (0.2 cm^{-1}; 100°C, then 24 hours at 20°C) were investigated in a 10-month experiment in rats. There were no statistically consistent changes in the condition of test animals, nor were there any pronounced histological changes in their visceral organs compared with the control.

Recommendations. ***Russia.*** The vulcanizates are recommended for the manufacture of articles coming into contact with foodstuffs.

Reference:

Sokol'nikov, Ye. A., *Hygienic Assessment of Vulcanizates Based on SKMS 30 ARKM 15 Synthetic Rubber and Used for Contact with Foodstuffs'*, Author's abstract of thesis, Kiyv, 1987, 23 (in Russian).

VULCANIZATES based on BUTYL-STYRENE RUBBER (LPU-2P sealing paste)

Composition. Vulcanizate contains zinc oxide, sodium, benzoate, ethyl zinc dimethyldithiocarbamate, OP-10, and other components.

Applications. Used in the food industry for sealing cans.

Migration Data. A study of extracts revealed no migration of zinc ions or OP-10 and no change in *pH* above the permissible norms.

Short-term Toxicity. Male and female Wistar rats received aqueous extracts as their drinking water. No toxic effects were observed.

Reference:

Shmeleva, Ye. V., Shumskaya, N. I., Riskina, M. A., and Yefremova, V. M., Toxicity of packing paste LPU-2P, *Gig. Sanit.,* 11, 95, 1987 (in Russian).

VULCANIZATES based on BUTYL RUBBER

CAS No 9010-85-9

Abbreviation. BR.

Composition. BR is synthetic rubber, a copolymer of isobutylene with a small quantity of isoprene (1.0 to 5.0%). BR is 2-methyl-1,3-butadiene polymerized with 2-methyl-1-propene.

Properties. BR possesses higher chemical resistance than NR. The vulcanizates are characterized by low elasticity.

Applications. The resistance of BR to swelling in milk and food fats enables it to be used to manufacture components of milking apparatus and also other articles (e.g. seals) coming into contact with foodstuffs. Material for homemade lids.

Migration Data. A study of extracts of BR formulation 52-268 (1 cm^{-1}; initial temperature 100°C; 1-72 hours) revealed no migration of harmful substances.[1]

Long-term Toxicity. After drinking of aqueous extracts of BR formulation 52-268 over a period of 10 months rats developed impairment of the kidney function (protein and chlorides in the urine). In histological terms there were dystrophic changes in the epithelium of the convoluted renal tubules.[1]

Reproductive Toxicity.

Embryotoxicity. Is found to produce embryotoxic effect when injected into chick eggs.[2]

No effect on fertility of male rats was reported.[1]

Regulations. ***U.S. FDA*** (1998) approved the use of BR (1) in the manufacture of resinous and polymeric coatings for polyolefin films for food-contact surface of articles intended for use in producing, manufacturing, packing, processing, preparing, treating, packaging, transporting, or holding food, in accordance with the conditions prescribed in 21 CFR part 175.320; and (2) in the manufacture of semirigid and rigid acrylic and modified plastics used as articles intended for use in contact with food in accordance with the conditions pre scribed in 21 CFR part 177.1010.

References:

1. *Toxicology of the Components of Rubber Mixes and of Rubber and Latex Articles*, TSNIITEKHIM, Moscow, 1974, 46 (in Russian).
2. Korhonen, A. et al., Toxicity of rubber chemicals towards three-day chick embryos, *Scand. J. Work Envi ron. Health*, 9, 115, 1983.

VULCANIZATES based on BUTADIENE-NITRILE RUBBER (SKN)

M = 2,000000 to 3,000000

Composition. A synthetic rubber, 1,3-butadiene and acrylonitrile polymer. Vulcanization systems - See ***Vul canizates based on natural (polyisoprene) rubber.***

Properties. Exhibit high oil-resistance.

Applications. Have different applications, including food industry and water supply. Plasticizers.

Migration Data.

Table. Migration of rubber ingredients.

Substances	*Content in vulcanizate, parts,%*	*Levels of migration, mg/l*
thiuram	0.2-1.8	up to 0-0.6
zinc dimethyldithiocarbamate		up to 0-0.1
Sulphenamide C	0.25-2.0	up to 0.6
captax		up to 1.0
altax		up to 0.42
dithiodimorpholine	0.5-3.0	up to 1.0
vulcazit P extra N	0.2-1.0	up to 0.08
N-ethylaniline		up to 0.8
diphenylguanidine	0.2-0.5	up to 2.9

Reference:

Prokof'yeva, L. G., in *Hygiene of Use and Toxicology of Pesticides and Polymeric Materials*, Sci. Proc. All-Union Research Institute of Hygiene and Toxicology of Pesticides, Polymers and Plastic Materials, Kiyv, Issue No 17, 1987, 153 (in Russian).

VULCANIZATES based on BUTADIENE-NITRILE RUBBER (SKN)

Migration. Study of aqueous extracts from SKN rubber formulations PB-11, PB -14, PB -19, and PB-27a revealed migration of dithiocarbamates (3.5 mg/l), monoethylaniline (up to 1.0 mg/l), and dioctylphthalate (up to 0.18 mg/l) into the extracts.[028]

Short-term Toxicity. Rats drank aqueous extracts over a period of 3 to 4 months. The treatment affected the kidneys and liver. Changes were noted in relative weights of the visceral organs. In a month after the start of the experiment there was a steady increase in the protein contents and a reduction in the chloride concentration in the urine, decrease in the relative kidney weights, and changes in some liver functions.

Reference:

Shumskaya, N. I., Mel'nikova, V. V., Chikishev, Yu. G., and Taradai, Ye. P., Toxicity of the extracts of rubber made on varying caoutchouc base and containing thiuram in the formulation, *Gig. Sanit.*, 2, 82, 1979 (in Russian).

VULCANIZATES based on BUTADIENE-NITRILE RUBBER

Composition. The vulcanizate contains thiuram, santocure, zinc oxide, furnace black, and dibutyl phthalate.

Applications. Used to make components of milking apparatus.

Migration Data. Thiuram, Santocure, and zinc oxide were not found in the aqueous extracts (0.5 cm^{-1}; 38°C, then 20°C; 2 hours), but after 2-days' exposure they were found in large quantities.[1]

Long-term Toxicity. Over a period of 1 year, rats received 1-hour extracts boiled in a 0.5% saline to drink. There were neither visible signs of toxicity, nor retardation of BW gain. The treatment caused a definite increase in oxygen consumption and a reduction in the hyppuric acid contents in the urine. Changes in the relative liver weight were also noted.[1] Another group of rats received aqueous extracts of vulcanizate of the same grade but based on SKN-26 rubber with phenol-type P-23 antioxidant. This treatment caused a double increase in diuresis and a reduction in the chloride and protein contents of their urine. There were also threshold changes in NS function. Gross pathology examination revealed no changes in the viscera.[2]

References:

1. Stasenkova, K. P., Shumskaya, N. I., Provorov, V. N. et al., Toxicological evaluation of vulcanizate intended for the production of diary apparatus, in *Hygiene Aspects of the Use of Polymeric Materials and of Articles Made of Them*, L. I. Medved', Ed., All-Union Research Institute of Hygiene and Toxicology of Pesticides, Polymers and Plastic Materials, Kiyv, 1969, 95 (in Russian).
2. *Toxicology of the Components of Rubber Mixes and of Rubber and Latex Articles*, TZNIITENEFTEKHIM, Moscow, 1974, 56 (in Russian).

VULCANIZATES based on BUTADIENE-NITRILE RUBBER

Composition. Vulcanization accelerators are thiuram and diphenylguanidine.

Applications. Used to make components of milking apparatus

Migration Data. Thiuram migrates into aqueous extracts (0.5 cm^{-1}; 38 and 20°C; 1 hour) in a quantity of 0.07 mg/l.

Short-term Toxicity. Rats were given aqueous extracts to drink. Changes included a low contents of cholic acid in bile, an increase in the relative liver weight and in 24-hour diuresis, and in the protein contents of the urine. Gross pathology examination revealed moderate plethora and albuminous dystrophy of the liver cells, as well as considerable desquamation of the epithelium in the convoluted renal ducts.

Reproductive Toxicity.

Gonadotoxicity. In an 8-month study, there was an increased post-implantation loss in females paired with test males.

Mutagenicity.

In vivo cytogenetics. Consumption of aqueous extracts over a period of 4 months caused no increase in frequency of DLM in treated animals.

Reference:

Znamensky, N. N. and Peskova, A. V., Study of migration of 3,5-benzopyrene from paraffin packaging in diary products, in *Hygiene Aspects of the Use of Polymeric Materials and of Articles Made of Them*, L. I. Medved', Ed., All-Union Research Institute of Hygiene and Toxicology of Pesticides, Polymers and Plastic Materials, Kiyv, 1969, 62 (in Russian).

VULCANIZATES based on BUTADIENE-NITRILE RUBBER (SKN-26)

Composition. The vulcanization accelerators are vulkazit-*P*-extra-*N* and santocure.

Applications. Used to make components of milking apparatus.

Migration Data. Investigation of aqueous extracts (0.5 cm^{-1}; 38 and 20°C; 1 hour) revealed the migration of santocure (0.03 to 0.05 mg/l), vulkazit-*P*-extra-*N* (0.03 mg/l), and dioctyl phthalate (0.32 mg/l).

Long-term Toxicity. In a 1-year study, consumption of aqueous extracts caused no toxic effect in the experimental animals.

Reference:

Shumskaya, N. I., in *Toxicology of the Components of Rubber Mixes and of Rubber and Latex Articles*, TZNIITENEFTEKHIM, Moscow, 1977, 58 (in Russian).

HEAT-RESISTANT VULCANIZATE based on BUTADIENE-NITRILE RUBBER (SKN-26)

Composition. The vulcanizate contains accelerator vulkazit-*P*-extra-*N*, sulphur, PMG-33, zinc black oxide, and dioctyl phthalate.

Migration Data. *N*-Ethyl aniline (0.02 mg/l) and accelerator vulkazit-*P*-extra-*N* (0.1 mg/l) migrate into aqueous and citric acid extracts [0.5 cm^{-1}; 100, 37, and 20°C (infusion); 1 and 24 hours]. Under the most rigorous simulating conditions, migration of zinc (0.3 to 0.4 mg/l) was observed.

Reference:

Znamensky, N. N. and Peskova, A. V., Study of migration of 3,5-benzopyrene from paraffin packaging in diary products, in *Hygiene Aspects of the Use of Polymeric Materials and of Articles Made of Them*, L. I. Medved', Ed., All-Union Research Institute of Hygiene and Toxicology of Pesticides, Polymers and Plastic Materials, Kiyv, 1969, 62 (in Russian).

VULCANIZATE based on BUTADIENE-NITRILE RUBBER (SKN-26 and SKN-26M, formulations 52-687)

Composition. Vulcanization accelerator: vulkazit-*P*-extra-*N*; fillers: black, lithopone, kaolin, fine collodial silica.

Migration Data. Investigation of aqueous extracts and a 2.0% citric acid solution (20, 40, and 100°C; 1 month) revealed the migration of monoethyl aniline (0.7 to 3.0 mg/l) and traces of ethylphenyl dithiocarbamic acid derivatives. Vulkazit-*P*-extra-*N* and thiuram EF were not found in the extracts.

Reference:

Grushevskaya, N. Yu., Determination of thiuram E, ethylzimate and some of their transformation products for sanitary-chemical investigation of vulcanizates, *Kauchuk i Rezina*, 1, 35, 1987 (in Russian).

VULCANIZATES based on BUTADIENE-NITRILE RUBBER (SKN-26, formulations 52-687 - 52-690)

Composition. Vulcanization accelerators are thiuram and vulcacit-*P*-extra-*N* (*q. v.*). Transforma- tion products appeared to be zimat and *N*-ethyl aniline.

Migration Data. Investigation of aqueous, hexane, alcohol (20 and 40%) and acid extracts revealed migration of zimat (0.1 mg/l), *N*-ethyl aniline (0.13 mg/l), acrylonitrile (1.4 mg/l), dioctyl phthalate (0.75 mg/l), neozone D (0.5 mg/l), benzo[a]pyrene (0.26 μg/l), as well as pyrene, 1,12-benzoperylene and fluorene.

Recommendations. ***Russia.*** The vulcanizates were recommended for use in short-term contact with food.

Reference:

Medvedev, V. I., Hygienic properties of vulcanizates made on the base of butadiene-nitrile rubber with the help of sulfurorganic vulcanization accelerators, in *Hygiene and Toxicology of High-Molecular-Mass Compounds and of the Chemical Raw Material Used for Their Synthesis*, Proc. 6th All-Union Conf., B. Yu. Kalinin, Ed., Khimiya, Leningrad, 1979, 81 (in Russian).

VULCANIZATES based on BUTADIENE-NITRILE RUBBER (SKN-26)

Composition. Anti-aging agent of phenol type P-23 antioxidant (q.v.). Vulcanization accelerator: cyanoacetyl methylurea.

Long-term Toxicity. Rats were given aqueous extracts to drink over a period of 10 months. Only at the very start of exposure were there retardation of BW gain, a decrease in the nervous-muscular excitability threshold, a reduction in 24-hour diuresis, and a change in the permeability of skin capillaries.

Recommendations. ***Russia.*** The vulcanizate is recommended for use in contact with foodstuffs.

Reference:

Shumskaya, N. I., in *Toxicology of the Components of Rubber Mixes and of Rubber and Latex Articles*, TZNIITENEFTEKHIM, Moscow, 1977, 58 (in Russian).

VULCANIZATES based on BUTADIENE-NITRILE RUBBER (SKN-26)

Migration Data. 17 different compounds were found in the extract, their structure differing considerably from the initial components.[1]

Long-term Toxicity. Rats were given aqueous extracts and fat simulant media over a period of 10 months. Signs of toxic action were noted in the treated animals. Extracts of rubber with a greater degree of unsaturation are more toxic.[2]

References:

1. Shumskaya, N. I., *Kauchuk i Rezina*, 9, 24, 1978 (in Russian).
2. Shumskaya, N. I., Mel'nikova, V. V., Chikishev, Yu. G., and Taradai Ye. P., Toxicity of the extracts of rubber made on varying caoutchouc base and containing thiuram in the formulation, *Gig. Sanit.*, 2, 82, 1979 (in Russian).

VULCANIZATES based on BUTADIENE-NITRILE RUBBER (SKN-26)

Composition. The vulcanizate contains accelerator vulkazit-*P*-extra-*N* (0.5 parts) and *N,N'*-dithiodimorpholine (1.5 to 3.0 parts).

Migration Data. A study of extracts (0.5 cm^{-1}; 20 and 40°C; 1 and 24 hours; distilled water, acid-salt solutions, solutions containing alcohol) revealed migration of the following substances: *N,N'*-dithiodimorpholine from all the vulcanizates amounted to 0.1 to 1.0 mg/l, migration of monoethyl aniline (from PS-31 and PS-32 vulcanizates) amounted to 0.2 to 1.0 mg/l.

Short-term Toxicity. Rats received aqueous extracts of PS-31 vulcanizate to drink for 4 months. The treatment caused changes in the hematology indices. Some reduction in relative liver, lungs, spleen, and adrenal glands weights were noted. Gross pathology and histology examination revealed hemodynamic disorders in the viscera, signs of parenchymatous dystrophy of the liver, changes in the lungs, kidneys and spleen.

Recommendations. ***Russia.*** The vulcanizate PS-31, PS-32 and PS-33 are not recommended for use in contact with foodstuffs.

Reference:

Stankevich, V. V. and Shurupova, Ye. A., Hygienic characteristics of rubber containing *N,N'*-dithiodimorpholine and intended for contact with food products, *Gig. Sanit.*, 9, 24, 1976.

VULCANIZATES based on BUTADIENE-NITRILE RUBBER (SKN-26)

Composition. The vulcanizate contains phenol-type P-23 antioxidant and vulcanization accelerator neozone D (BP-1a), thiuram (BP -1), santocure, and other additives.

Migration Data. 17 different compounds were found in the extract, their structure differing considerably from the initial components.[1]

Short-term Toxicity. Rats received 5 ml milk extracts (80°C, then 5°C) 0.1 cm^{-1}; 24 hours) every day over a period of 3 months. The treatment affected kidney functions: a reduction in 24-hour diuresis, and in the protein and chloride contents of the urine. An increase in relative kidney weight was noted. Histological examination revealed dystrophic changes in the kidneys.[2]

Long-term Toxicity. Rats were given aqueous extracts and fat simulant media over a period of 10 months. Extracts of rubbers with a greater degree of unsaturation are more toxic.[1]

References:

1. Shumskaya, N. I., *Kauchuk i Rezina*, 9, 24, 1978 (in Russian).
2. Shakina, L. S., Rogovaya, A. B., Tikhomirova, L. D., and Khazanovich, F. G., Effect of resin on SKN-26 butadiene-nitrile rubber base on the functional state of the kidney, *Vrachebnoye Delo*, 10, 138, 1975 (in Russian).

VULCANIZATES based on BUTADIENE-NITRILE RUBBER (SKN-26)

Composition. Accelerator: carbamate BC.

Migration Data. No migration of dibutyl dithiocarbamic acid derivatives or carbon disulfide was found in aqueous extracts and simulant media (1 and 5 days; 20 and 40°C; 0.5 cm^{-1}). Dibutylamine was found in aqueous extracts (0.25 to 0.3 mg/l and 0.35 to 0.45 mg/l, respectively for two vulcanizates). Migration of carbon disulfide, carbamate BC, and tetrabutylthiuramdisulfide was observed in small quantities only with a specific surface of 2.0 cm^{-1}.

Reference:

Grushevskaya, N. Yu. and Kazarinova, N. F., Determination of the vulcanization accelerator, carbamate-BZ, and its conversion products in sanitary chemical research on rubber, *Gig. Sanit.*, 9, 79, 1978 (in Russian).

VULCANIZATES based on BUTADIENE-NITRILE RUBBER (SKN-26M)

Composition. *Formulation I:* The vulcanizate contains a plasticizer dioctyl sebacinate. *Formulation II:* Vulcanization accelerators are thiuram EF and vulkazit-*P*-extra-*N*, phenol-type P-23 antioxidant, plasticizer: dioctyl phthalate.

Applications. Used in the food industry (*Formulation I)* and to make apparatus for the food industry (*Formulation II)*.

Migration Data. Investigation of aqueous extracts of *Formulation I* and simulant media (0.1 to 2 cm^{-1}; 20 and 100°C; 1 hour and 5 days) revealed migration of diphenylguanidine (0.1 to 0.45 mg/l) and its degradation products: aniline (0.03 to 0.25 mg/l), triphenyl melamine (up to 0.03 mg/l), tetraphenyl melamine (traces), triphenyl dicarbimide (up to 0.2 mg/l).[1] Investigation of aqueous extracts of *Formulation II* (0.5 cm^{-1}; 100°C, 4 hours) revealed mild taste and odor, as well as migration of zinc ions (2.5 mg/l), vulkazit-*P*-extra-*N* (0.14 mg/l), and thiuram EF (0.02 mg/l). On repeated analyses, no migration was noted.[2]

Long-term Toxicity. Rats were given the aqueous extracts to drink for 6 months. There were no significant manifestations of toxic action in the treated animals.[2]

Recommendations. ***Russia.*** Vulcanizates were recommended for use in the food industry.

References:

1. Kazarinova, N. F. and Ledovskikh, N. G., *Kauchuk i Rezina*, 1, 26, 1978 (in Russian).
2. Khoroshilova, N. V. and Ol'pinskaya, A. Z., Hygienic evaluation of resins, intended for manufacture of milking apparatus, *Gig. Sanit*, 10, 95, 1988 (in Russian).

VULCANIZATES based on BUTADIENE-NITRILE RUBBER (SKN-26 and SKN26M)

Composition. Vulcanization accelerators are thiuram E or ethylzimat, fillers: lithopone, kaolin, technical carbon black.

Applications. Used to make apparatus for food industry.

Migration Data. Investigation of aqueous extracts and simulant media (0.5 cm^{-1}; 20°C, 1 day; irrespective the type of accelerator introduced into vulcanizate) revealed migration of the main components: ethylzimat (0.15 to 20 mg/l), thiuram E (0.1 to 0.15 mg/l). Migration of the transformation products was also found: sulfocarbon (0.02 to 0.12 mg/l), diethyl aniline (0.06 to 0.08 mg/l), dithiocarbamates (0.24 to 0.44 mg/l).

Reference:

Grushevskaya, N. Yu. and Kazarinova, N. F., Determination of the vulcanization accelerator, carbamate-BZ, and its conversion products in sanitary chemical research on rubber, *Gig. Sanit.*, 9, 79, 1978 (in Russian).

VULCANIZATE based on BUTADIENE-NITRILE RUBBER (SKN-26MP)

Composition. The vulcanizate contains anti-aging agent P-23, sulphur, sulfenamide C, zinc oxide, stearine, NG-2246, technical carbon black PM-75, dioctyl phthalate, and frygit. One formulation contains thiuram D, another one contains thiuram EF.

Applications. Used in the manufacture of milking apparatus.

Migration Data. Investigation of extracts (distilled water and 3.0% lactic acid solution; 0.5 cm^{-1}; 38-40°C, then 28°C; 1 hour) revealed migration of zinc ions (0.53 to 5.2 mg/l), sulfenamide C (0.1 to 0.3 mg/l). Mild changes of taste and odor of extracts were also observed.

Long-term Toxicity. Male rats received aqueous extracts of the vulcanizates containing thiuram D to drink over a period of 9 months. The treatment caused kidney pathology to develop. No manifestations of toxic action were seen in the similar study with vulcanizate containing thiuram EF.

Recommendations. ***Russia.*** The formulation contained thiuram EF has been recommended for use in the food industry.

Reference:

Khoroshilova, N. V. and Ol'pinskaya, A. Z., Hygienic evaluation of resins, intended for manufacture of milking apparatus, *Gig. Sanit*, 10, 95, 1988 (in Russian).

CHLOROPRENE RUBBER

CAS No 9006-03-5

RTECS No FW7401000

Abbreviation. CR.

Synonym. Chlorinated rubber.

Acute Toxicity. LD_0 was 670 mg/kg BW in rats.

Carcinogenicity classification. An IARC Working Group concluded that there were no adequate data available to evaluate the carcinogenicity of chloroprene rubber in *experimental animals* and in *humans*.

IARC: 3.

Regulations. ***EU*** (1990). CR is available in the *List of authorized monomers and other starting substances which may continue to be used for the manufacture of plastic materials and articles intended to come into contact with foodstuffs pending a decision on inclusion in Section A* (*Section B*).

Reference:

Eisei Kagaku (*Hygienic Chemistry)*, Abstract, 24, 115, 1978 (in Japanese).

POLYSULFIDE RUBBERS

Structural Formula.

$HS[\sim RS_x \sim]_nSH$ - liquid

$[\sim RS_x \sim]_n$ - solid

x -2 or 4

CAS No 63148-67-4
68611-50-7

Abbreviation. PSR.

Composition. PSR are polymers made from dihaloalkanes and sodium polysulfide. PSRs contain varying amounts of lead.

Synonym and **Trade Names.** Thiokol is one of the most common proprietary names for these rubbers. 1,2,3-Trichloropropane, polymer with 1,1'-[methylenebis(oxy)]bis[2-chloroethane] and sodium sulfide; Thiorubber.

Properties. Noted for resistance to sunlight and moisture and gas permeability (because of the absence of double bonds in the macromolecules and a high sulfur contents). The unpleasant odor of some thiokols is due to the presence of thioxane and thiophane.

Applications. Used in rubber and resin manufacture, in the production of protective coatings, sealings of joints, etc. PSRs are used in dental impression materials

Chemobiokinetics. A bathing agent containing polysulfides was administered orally to rats. Distribution of polysulfides resulted in their highest concentration in blood, liver, lungs, and kidneys. Maximum concentration of sulfides was determined in blood.

Regulations. ***U.S. FDA*** (1993) approved the use of PSR in polysulfide polymer-polyepoxy resins in accordance with the conditions prescribed in 21 CFR part 177.1650

Reference:

Nagata, T., Kage, S., Kimura, K., Kudo, K., and Imamura, T., How to diagnose polysulfide poisoning from tissue samples, *Int. J. Legal Med.*, 106, 288, 1994.

POLYSULFIDE MASTICS

Migration Data. After contact with water for 3 to 48 hours (0.024 cm^{-1}; 20 and 37°C), polysulfide mastics gave the water an odor rating 3 to 5. Korolev and Rubleva (1978) also found polysulfide mastics gave water a strong unpleasant odor after several hours contact.

Reference:

Krat, A. V., Kesel'man, I. M., and Sheftel', V. O., Sanitary-chemistry evaluation of polymeric materials used in water supply, *Gig. Sanit.*, 10, 18, 1986 (in Russian).

POLYSULFIDE SEALANTS (formulations U-30, MES-5 and KB)

Migration Data. Investigation of aqueous extracts and simulant media (0.05 and 0.005 cm^{-1}; 18 to 20°C; 10 to 15 days) revealed a specific polysulfide odor of rating 4 after 15 days contact (0.05 cm^{-1}). Migration of dibutyl phthalate (q.v.) and diphenylguanidine (q.v.) into the water was below safe levels, and no migration of epichlorohydrin (q.v.) was found (Korolev and Rubleva, 1978).

Long-term Toxicity. Rats received aqueous extracts to drink over a period of 6 months. There was no retardation of BW gain, no changes in hematology analyses, relative organ weights, in the NS functions, and in cholinesterase, catalase, and ceruloplasmin activity, etc. Histology examination revealed no lesions.

Recommendations. ***Russia.*** All three formulations have been recommended for use in water supply systems (with a specific surface of not less than 0.005 cm^{-1} and water storage time of up to 15 days).

Reference:

Korolev, A. A. and Rubleva, M. N., Hygienic and toxicological evaluation of some thiokol sealants, in *Hygiene Aspects of the Use of Polymeric Materials in Construction*, Proc. 1st All-Union Meeting on Health and Safety Monitoring of the Use of Polymeric Materials in Construction, Kiyv, 1973,168 (in Russian).

RUBBER HYDROCHLORIDE POLYMER

Structural Formula. $[\sim CH_2CH_2CHCl\sim]_n$

CAS No 9006-00-2

RTECS No TP3710000

Abbreviation. RHC.

Synonym and **Trade Names.** Permaseal; Pliofilm; Rubber hydrochloride.

Applications. Used in rubber and plastics industries.

Carcinogenicity. Induced malignant tumors at the site of application in rodents following *s/c* imbedding of polymer films.[026] Nevertheless, the results of this study were later considered inadequate.

Regulations. ***U.S. FDA*** (1998) approved the use of RHC (1) in adhesives as a component of articles intended for use in packaging, transporting, or holding food in accordance with the conditions prescribed in 21 CFR part 175.105; (2) in preparation of pressure-sensitive adhesives for use as the food-contact surface of labels and/or tapes applied to food, in accordance with the conditions prescribed in 21 CFR part 175.125; and (3) in the manufacture of resinous and polymeric coatings for the food-contact surface of articles intended for use in producing, manufacturing, packing, processing, preparing, treating, packaging, transporting, or holding food (for use only of RHC as polymerization cross-linking agent in side seam cements for containers intended for use in contact with food (only of the identified types) in accordance with the conditions prescribed in 21 CFR part 175.300.

SILOXANES

Structural Formula. $H_3Si\text{-}[\sim OSiH_2\sim]_n\text{-}OSiH_3$

Composition. Silicone is a generic term for a group of compounds with the structure R_2SiO. Siloxanes are compounds featuring alternating silicon and oxygen atoms.

Trade Names. Silicones; Polysiloxanes.

Applications. S. are used in food industry, as water repellants and lubricants, and in the production of medical prostheses.

Properties. Silicone polymers may be linear, branched, or cross-linked.[020] Silicones can be gels, resins, gums, or elastomers, depending upon the molecular mass and extent of cross-linking.

Carcinogenicity. *U.S. FDA* banned most uses of S. breast implants in 1992. A newer case-control study failed to find an association between S. breast implants and scleroderma.[1] The *British Medical Devices Agency* found no evidence for increased risk of connective tissue diseases with S. breast implants.[2]

Chemobiokinetics. *In vitro* experiments have demonstrated that *cyclosiloxanes* can migrate out of breast implants, and in mouse experiments *cyclosiloxanes* have been shown to be widely distributed in many organs after a single subcutaneous injection and to persist for at least a year.[3]

Regulations. ***U.S. FDA*** (1998) approved the use of S. (1) in adhesives as a component (monomer) of articles intended for use in packaging, transporting, or holding food in accordance with the conditions prescribed in 21 CFR part 175.105; (2) in the manufacture of resinous and polymeric coatings for polyolefin films for food-contact surface of articles intended for use in producing, manufacturing, packing, processing, preparing, treating, packaging, transporting, or holding food, in accordance with the conditions prescribed in 21 CFR part 175.320; (3) in olefin polymers which may be safely used as articles or components of articles intended for use in contact with food at a level not to exceed 3.0% by weight, subject to the provisions of 21 CFR part 177.1520; and (4) for use as a component of paper and paperboard intended for use in producing, manufacturing, packing, processing, preparing, treating, packaging, transporting, or holding aqueous and fatty foods of the type identified in 21 CFR part 176.170.

References:

1. Burns, C. J., Laing, T. J., Gillespie, B. W. et al., The epidemiology of scleroderma among women.Assessment of risk from exposure to silicone and silica, *J. Rheumatol.*, 23, 1904, 1996.
2. Perkins, L. L., Clark, B. D., Klein, P. J. et al., A meta-analysis of breast implants and connective tissue disease, *Ann. Plast. Surg.*, 35, 561, 1995.
3. Lieberman, M. W., Lykissa, E. D., Barrios, R., Ou, C.N., Kala, G., and Kala, S.V., Cyclosiloxanes produce fatal liver and lung damage in mice, *Environ. Health Perspect.*, 107, 161, 1999.

POLYDIMETHYLSILOXANE

Structural Formula. $[\sim C_2H_6OSi\sim]_n$ or: $[\sim(CH_3)_3\ SiOSi(CH_3)_2O\sim]_n\text{-}Si(CH_3)_3$

CAS No 9016-00-6
63148-62-9

RTECS No TQ2690000
TQ2700000

Abbreviation. PDMS.

Synonym and **Trade Names.** Baysilon; Dow corning 346; GEON; Good-rite; Poly[oxy(dimethyl-silylene)]; Silicone; Siloxanes.

Applications. PDMS is used in the manufacture of coatings, rubber, and plastics, etc. *Silicones* are widely present in consumer articles such as cosmetics and toiletries, processed foods, and household products. It is extensively used in medical practice as a lubricant in tubing and syringes, and in implantable devices (predominantly, PDMS, in the production of medical prostheses, silicone breast and penile implants).[1]

Exposure. Silicone formula bottle nipples and infant pacifiers may serve as sources of S. exposure.[2]

Long-term Toxicity. Silicone fluid and silicone gel were injected *s/c* into female $B6C3F_1$ mice (1.0 ml/mouse) and 6 mm disks of silicone elastomer were implanted *s/c*. There were no treatment-related deaths or overt signs of toxicity during the 180-day exposure. No effects on BW and organ weights, hematology and blood chemistry analyses were noted. The only consistent effect of 180-day exposure to silicone materials was a moderate depression of natural killer activity.[1] However, according to other data, silicone implants and/or their multiple chemical contaminants elicit foreign body reactions associated with granulomatous inflammation and fibrosis. Silicone and its contaminants have the potential for significant toxicity in the implant recipient.[3]

Reproductive Toxicity.

Gonadotoxicity and ***Embryotoxicity.*** Breast implants of silicone gel and envelope produced no effects on parents and offspring of rats and rabbits.[4-6]

Teratogenicity.

Observations in man. In a small study, impaired esophageal motility (low levels of peristalsis in the distal two thirds of the esophagus and decreased lower sphincter pressure) was observed in 6 children from 4 families who were breastfed by mothers with *silicone* breast implants. Such effects were not noted in 3 bottle-fed children of mothers with breast implants.[7]

Animal studies. S/c injection of 5.0 g/kg BW into rats at up to doses of 20 mg/kg BW on 6 to 15 days of pregnancy caused an increase in post-implantation mortality, but no developmental abnormalities were reported.[8]

Mutagenicity.

In vitro genotoxicity. A series of organosilicone compounds representing potential intermediates in the synthesis and degradation of PDMS were evaluated in the Ames bacterial reverse mutation test in *Salmonella*, mitotic gene conversion in *Saccharromyces cerevisiae* D4, and DNA repair in *E. coli* with and without activation system. No evidence of gene mutation was observed.[9,10]

In vivo cytogenetics. None of 6 organosilicone compounds which had been found to have clastogenic activity in the above-sited *in vitro* study, produced significant increase in CA in rat bone marrow clastogenicity assays.[9]

Carcinogenicity. Polyurethane and silicone films were implanted *s/c* into rats, and 1- and 2-year adverse tissue responses were studied. Polyurethane gave higher incidence of the adverse responses including tumor formation in comparison to silicone.[11]

Regulations. ***U.S. FDA*** (1998) approved the use of silicone (1) in the manufacture of resinous and polymeric coatings for polyolefin films for food-contact surface of articles intended for use in producing, manufacturing, packing, processing, preparing, treating, packaging, transporting, or holding food, in accordance with the conditions prescribed in 21 CFR part 175.320; (2) as a defoaming agent in the manufacture of paper and paperboard intended for use in producing, manufacturing, packaging, processing, preparing, treating, packing, transporting, or holding foods in accordance with the conditions prescribed in 21 CFR part 176.210; (3) in adhesives as a component (monomer) of articles intended for use in packaging, transporting, or holding food in accordance with the conditions prescribed in 21 CFR part 175.105; (4) in olefin polymers which may be safely used as articles or components of articles intended for use in contact with food at a level not to exceed 3.0% by weight, subject to the provisions of 21 CFR part 177.1520; and (5) for use as a component of paper and paperboard intended for use in producing, manufacturing, packing, processing, preparing, treating, packaging, transporting, or holding aqueous and fatty foods of the type identified in 21 CFR part 176.170. ***EU*** (1995). PDMS is a food additive generally permitted for use in foodstuffs (up to 10 mg/kg).

References:

1. Bradley, S. G., White, K. L., McCay, J. A., Brown, R. D., Musgrove, D. L., Wilson, S., Stern, M., Luster, M. I., and Munson, A. E., Immunotoxicity of 180 day exposure to polydimethylsiloxane (silicone) fluid, gel and elastomer and polyurethane disks in female $B6C3F_1$ mice, *Drug Chem. Toxicol.*, 17, 221, 1994.
2. Kennedy, G. L., Keplinger, M. L., and Calandra, J. C., Reproductive, teratologic, and mutagenic studies with some polydimethylsiloxanes, *J. Toxicol. Environ. Health*, 1, 909, 1976.
3. Bush, H., Silicone toxicology, *Semin. Arthritis Rheum*, 24 (Suppl. 1), 11, 1994.
4. Siddiqui, W. H. and Schardein, J. L., One generation reproductive study of silicone gel and Silastic II mammary envelope implants in rats, Abstract, *Toxicologist*, 13, 75, 1993.
5. Siddiqui, W. H., Schardein, J. L., Cassidy, S. L., and Meeks, R. G., Reproductive and developmental toxicity studies of silicone gel Q7-2159A in rats and rabbits. *Fundam. Appl. Toxicol.*, 23, 370, 1994.
6. Siddiqui, W. H., Schardein, J. L., Cassidy, S. L., and Meeks, R. G., Reproductive and developmental toxicity studies of silicone elastomer Q7-2423/Q7-2551 in rats and rabbits. *Fundam. Appl. Toxicol.*, 23, 377, 1994.
7. Levine, J. J. and Ilowite, N. T., Scleroderma-like esophageal disease in children breast-fed by mothers with silicone breast implants, *JAMA*, 271, 2136, 1994.
8. Bates, H. K., Filler, R., and Kimmel, C. A., Developmental toxicity study of polydimethyl-siloxane injection on the rat, Abstract, *Teratology*, 31, 50A, 1985.
9. Isquith, A., Matheson, D., and Slesinski, R., Genotoxicity studies on selected organosilicon compounds: *In vitro* assays, *Food. Chem. Toxicol.*, 26, 255, 1988.
10. Marshall, T. C. et al., Toxicological assessment of heat transfer fluids proposed for use in solar energy applications, *Toxicol. Appl. Pharmacol.*, 58, 31, 1981.
11. Nakamura, A., Kawasaki, Y., Takada, K., Aida, Y., Kurokama, Y., Kojima, S., Shintani, H., Matsui, M., Nohmi, T., Matsuoka, A., et al., Difference in tumor incidence and other tissue responses to polyurethanes and polydimethylsiloxane in long-term subcutaneous implantation into rats, *J. Biomed. Mater. Res.*, 26, 631, 1992.

POLYMETHYLHYDROSILOXANE

Structural Formula.

CH_3

|

$[\sim SiHO\sim]_n$, where $n = 9$ to14

$M = (60.13)_n$

Trade Name. GKGh 94M.

Properties. Colorless or light-yellowish liquid with a characteristic odor. Poorly soluble in water, forming a soap-water type emulsion. Odor perception threshold is 2.0 mg/l.

Applications. Used as a water repellent. A component of coatings which are resistant to water and many or ganic solvents. It is applied as a 100% material or as a 50% aqueous emulsion.

Acute Toxicity. Laboratory animals tolerate administration of 40 g/kg (rats, split doses), 80 g/kg (mice), and 5.0 g/kg BW (guinea pigs).

Repeated Exposure. Animals tolerate the maximum possible doses administered for 10 days.

Standards. ***Russia*** (1994). MAC: 2.0 mg/l (organolept., film).

Reference:

Krasovsky, G. N., Friedland, S. A., Rubleva, M. N., et al., Specificity of biological action and hygienic significance of siliconorganic compounds in course of their draining in water reservoirs, in *Industrial Pollution of Water Reservoirs*, S. N. Cherkinsky, Ed., Meditsina, Moscow, Issue No 9, 1969, 38 (in Russian).

POLYETHYLHYDROSILOXANE

Structural Formula.

[~C_2H_5

|

$SiHO_2$~]$_n$, where $n = 9$ to14

M = $(74.16)_n$

CAS No 63148-57-2

Abbreviation. PEHS.

Synonyms and **Trade Name.** GKGh 94; Siloxane and Silicone, methyl hydrogen.

Properties. A colorless or pale-yellow liquid. A faint opalescence is possible. Poorly miscible with water, forms aqueous emulsions. Insoluble in ethyl alcohol. Odor perception threshold is 8.0 mg/l. PEHS is produced in the form of 100% material or 50% aqueous solution.

Applications. A component of coatings and of impregnation treatment for giving water repellent properties. A water repellent in the food industry.

Acute Toxicity and **Repeated Exposure.** See ***Polymethylhydrosiloxane.***

Note. Long-term storage of foodstuffs in glass containers coated with PEHS produces no changes in the or ganoleptic properties of the products or any migration of harmful impurities into them.

Standards. ***Russia*** (1988). MAC and PML: 10 mg/l (organolept., film).

Reference:

Butnikov, N. D. and Il'in, P. T., Polyethylhydrosiloxane, *Molochnaya Promyshlennost'*, 5, 33, 1973 (in Rus sian).

POLYDIETHYLSILOXANE

Structural Formula. [~$(C_2H_5)_2$ SiO~]$_n$

M = 102.21

Synonym. Lubricant No 3.

Properties. Light-yellow, oily liquid with a slight uncharacteristic odor. *Dens.* 0.95 to 0.97. Poor solubility in water. Taste threshold is 10 mg/l.

Applications. A component of nonstick coatings.

Acute Toxicity. With fractional administration, mice stood 80 g/kg BW, guinea pigs 5.0 g/kg BW, and rats 50 g/kg BW.

Short-term Toxicity. Administration of the maximum possible doses for 10 days did not kill the animals.

Long-term Toxicity. Administration of 1.5 g/kg BW caused no abnormalities in the condition of rats and no pathomorphological changes in their visceral organs.

Reference:

Krasovsky, G. N., Friedland, S. A., Rubleva, M. N., et al., Specificity of biological action and hygienic significance of siliconorganic compounds in course of their draining in water reservoirs, in *Industrial Pollution of Water Reservoirs*, S. N. Cherkinsky, Ed., Meditsina, Moscow, Issue No 9, 1969, 38 (in Russian).

K-55 LACQUER

Composition. Obtained by combined hydrolysis of dimethyldichlorosilane, $(CH_3)_2$ $SiCl_2$, and phenyltrichlorosilane, $C_6H_5SiCl_3$.

Migration Data. No changes in taste, odor, color, or transparency were found in aqueous extracts of the coating (1 cm^{-1}; 100°C; 2.5 hours). No migration of phenol, formaldehyde, or unsaturated compounds was observed.

Toxicity. Administration of aqueous extracts (90°C; 2.5 hours) to rats over a period of 3 months caused no changes in the general condition, an increase in BW gain, etc. in the treated animals. Similar data were obtained in a toxicological study of oil extracts (150°C; 30 min).

Reference:

Materials on the Health and Safety Assessment of Pesticides and Polymers, Coll. Sci. Proc. Moscow F. F. Erisman Research Sanitary Hygiene Institute, Moscow, 1977, 73 (in Russian).

"KREOL" COATING

Composition. Based on polydimethyldiphenylsiloxane ($R = CH_3$ and C_6H_5).

Migration Data. When heated to 250°C it releases organic substances, the nature of which is determined by radicals *R* (formaldehyde, phenol, hydrochloric acid, etc.).

Long-term Toxicity. Rabbit were given bread baked in Silumin tins with a "Kreol" coating over a period of 6 months, calculated at 35 g per rabbit (which is 8 to 10 times greater than the norm for an adult human being). Control animals were also given bread. No changes in the activity of blood serum cholinesterase or in the phosphorous contents of the blood were found in test rabbits. Gross pathology examination revealed no abnormalities, but functional changes in the secretion of the gastric mucosa were observed, together with a tendency for lipids to accumulate in the adrenal glands and changes of a functional nature.

Reference:

Materials on the Health and Safety Assessment of Pesticides and Polymers, Coll. Sci. Proc. Moscow F. F. Erisman Research Sanitary Hygiene Institute, Moscow, 1977, 59 (in Russian).

POLYMETHYLPHENYLDICHLOROSILOXANE

Composition. A mixture of organosilicon polymers.

Properties. Poor solubility in water. Odor perception threshold is 10 mg/l.

Applications. Used in the production of lacquers and anticorrosion coatings.

Short-term Toxicity. Administration of 1.0 g/kg BW to guinea pigs over a period of 3 months caused no changes in the *Hb* contents of their blood, in the contents of *SH*-groups in the blood serum, or in the phagocytic activity of leukocytes. Gross pathology examination revealed no changes.

Standards. ***Russia***. MPC: 10.0 mg/l (organolept.).

Reference:

Krasovsky, G. N., Friedland, S. A., Rubleva, M. N., et al., Specificity of biological action and hygienic significance of organosilicone compounds in course of their draining in water reservoirs, in *Industrial Pollution of Water Reservoirs*, S. N. Cherkinsky, Ed., Meditsina, Moscow, Issue No 9, 1969, 38 (in Russian).

SILICONE (*organosilicon*) RUBBER (SKT and SKTV)

Structural Formula

```
         R
         |
R"O [ ~Si-O~ ]n R"
         |
         R'
```

where R and R' are alkyl, alkenyl, or aryl; R" is hydrogen, alkyl triorganosilyl

$M = (3 \text{ to } 8) \times 10^5$.

CAS No 63394-02-5

RTECS No TQ2700000

Synonyms and **Trade Names.** Polysilicone; Silastic; Silastic silicone rubber.

Properties. Transparent, colorless, readily flowing mass without odor or taste. *Dens.* 0.96-0.98. Insoluble in alcohol.

Applications. Used in medicine for the manufacture of blood transfusion pipes and artificial heart valves, and for ear and nose or other facial implants.

Acute Toxicity. LD_{50} exceeded 5.0 g/kg BW in rats.

SILOXANE VULCANIZATES
formulations IRP-1338, IRP-1344, and IRP-1401

Composition. Based on SKTV and SKTV-1 siloxane rubber. The rubber mix contains bis(α,α'-dimethylbenzyl)peroxide (q.v.), stabilizer methyldimethoxyphenylsilane, etc.

Applications. Articles made of vulcanizates are intended for use in the food industry in contact with liquid foodstuffs at medium and high temperatures.

Migration Data. A study of extracts (40 and 100°C; 24 and 4 hours) revealed migration of methyl alcohol, peroxide, and acetophenone (from traces to 0.1 mg/l). As a result, boiling water, milk, juices, and sunflower oil acquired a specific vulcanizate odor of rating 2.

Long-term Toxicity. Administration of aqueous and milk extracts obtained in the first 12 hours of boiling to rats caused an increase in the STI, eosinophilia, a reduction in the activity of cholinesterase of the blood, and an increase in the activity of aldolase of the blood serum. However, after 7 months of the experiment these differences were not found. Gross pathology examination revealed an increase in the relative weights of the liver, lungs, heart, and spleen, but without any changes in their histological structure.

Recommendations. Preliminary boiling of vulcanizates of the given grades for 12 hours makes them toxicologically safe.

References:

Kravchenko, Ye. G., *Vrachebnoye Delo*, 3, 148, 1972; *Effect of Siloxane Vulcanizates on Foodstuffs and Animals and Regulations of Their Use in Food Industry*, Author's abstract of thesis, Kiyv, All-Union Research Institute of Hygiene and Toxicology of Pesticides, Polymers and Plastic Materials, 1972 (in Russian).

COATING based on POLYDIMETHYLSILOXANE RUBBER

Composition. Stabilizer - copper bisdiethyldithiophosphate.

Properties. The coating has a thickness of 30 μm, is elastic and colorless, and has an odor of rating 1.

Applications. Intended for use in the food industry at 80 to 165°C, in particular for baking meatloaves and ham tins and for white and rye bread.

Migration Data. A study of simulant medium coming into contact with coating revealed no significant change in organoleptic properties. Migration of methanol, silanes, siloxanes, copper, phosphorus, and benzine during baking of ham, meatloaf, and rye-white bread was not established. Formaldehyde was found in trace amounts.

A study of thermochemical breakdown of the coating showed that formaldehyde can migrate in a quantity of up to 0.1 mg/l when heating for an hour at 150°C, 0.13 mg/l at 200°C, and 0.45 mg/l at 250°C. No formaldehyde was found in subsequent hours of heating at 150 and 200°C. At 250°C, possible migration of siloxanes in trace amounts was noted.

Reference:

Petrova, G. A. and Moshlakova, L. A., Comparative sanitary and chemical assessment of thermally stable organosilicone polymers intended for use in food industry, *Gig. Sanit.*, 9, 22, 1987 (in Russian).

COATINGS based on COPOLYMER BLOKSIL

Composition. Coating based on block-copolymers of scalariform phenylsilsesquioxane and dimethylsiloxane. Bloksil 20-05 and Bloksil 20-10 are termostabilized by copper bisdiethyl dithiophosphate. Bloksil 20-10 could be thermostabilized by polyphenyl ferrosiloxane.

Applications. Used for baking bread at a maximum stove heating temperature of 330°C.

Properties. Thickness is 50 μm. Surface is smooth, lustrous, and colorless. It is odorless.

Migration Data. A study of extracts (1 cm^{-1}; 200 and 350°C) revealed migration of formaldehyde (0.16 to 1.9 mg/l) and phenol (up to 0.05 mg/l) from the coating Bloksil 20-10, thermostabilized by polyphenyl ferrosiloxane. Five-fold treatment of the samples with simulant media decreased the level of migration lower than PML. Migration of phenol or formaldehyde from the coating Bloksil 20-10 thermostabilized by copper bisdiethyldithiophosphate did not exceed PML levels. Concentration of formaldehyde extracted from Bloksil 20-05 sometimes amounted to 0.2 mg/l, and phenol traces were sometimes found.

Long-term Toxicity. In a 12-month study, male rats received aqueous extracts (obtained by 350°C) of Bloksil 20-10 thermostabilized by polyphenyl ferrosiloxane. The treatment caused only mild changes in the NS. No other findings were noted. Gross pathology examination of treated animals revealed no lesions. No ill effects were observed in rats given extracts prepared at 200°C. There were no changes in rats receiving aqueous extracts of Bloksil 20-05 thermostabilized by polyphenyl ferrosiloxane.

Reference:

Petrova, G. A. and Moshlakova, L. A., Comparative sanitary and chemical assessment of thermally stable organosilicone polymers intended for use in food industry, *Gig. Sanit.*, 9, 22, 1987 (in Russian).

ETHYLSILICATE COATING

Molecular Formula. $C_8H_{20}O_4Si$

M = 208.37

CAS No 78-10-4

RTECS No VV9450000

Composition. Based on ethylsilicate-32, blue powder, and ethyl alcohol.

Synonyms. Ethyl orthosilicate; Ethyl silicate; Silicic acid, tetraethyl ester; Tetraethoxysilane; Tetraethyl orthosilicate.

Migration Data. There were no changes in the organoleptic properties of contacting water at low temperature, but with an increase in temperature the odor increased to rating 4 to 5. Zinc contents exceeded 0.9 mg/l. An increase in the alkalinity of extracts was observed.

Long-term Toxicity. The toxicity of aqueous extracts was investigated in a 10 to 12-month oral study. Morphological and functional changes in animals were transient and/or of compensatory-adaptation nature.

Recommendations. ***Russia***. Washing improved hygienic properties of the coating applied on a VL-023 primer which enabled it to be recommended for use in water piping.

Reference:

Rudeiko, V. A. and Pashkina, Y. E., Hygienic evaluation of conditions for using a phosphating primer in the water supply, *Gig. Sanit.*, 6, 16, 1980 (in Russian).

413-63 MODIFIER (1), KO-921 LACQUER (2), and KO-921 SILANOL LACQUER (3)

Properties. Liquids with a color ranging from light-yellow to brown. Virtually insoluble in water. Good solubility in organic solvents. Frothing threshold concentrations are 2.0 mg/l (1), 0.3 mg/l (2), and 0.05 mg/l (3). Odor threshold is 0.3, 0.1, and 0.06 mg/l, respectively. Heating and chlorination do not alter organoleptic indices.

Applications. Used in the paint and varnish industry as heat-resistant and nonstick coatings.

Acute Toxicity. For rats, mice, and guinea pigs, LD_{50} is 1.0 to 1.7 g/kg BW (1) and 15.0 to 30.0 g/kg BW (2,3).

Repeated Exposure. Moderately cumulative substances.

Short-term Toxicity. Rats received 1/10, 1/50, and 1/250 LD_{50} over a period of 5 months. The greatest dose has a polytropic toxic effect. The NOELs of 6.0 mg/kg BW (1), 128.0 mg/kg BW (2), and 120 mg/kg BW (3) were established.

Reproductive Toxicity.

Embryotoxicity. A dose of 1/20 LD_{50} was ineffective.

Mutagenicity. No effect was observed in *Salmonella* and in *Dr. melanogaster* assays.

Reference:

Bezkopytny, I. N. and Andreiko, E. Yu., Hygienic regulation of a number of silicon organic compounds used in varnish-and-paint industry, *Gig. Sanit.*, 3, 81, 1985 (in Russian).

NONSTICK COATING

Trade Names. Lestosil SMF-*Fe*, Lestosil SM-SDKO, MF-100 Zh, and SKTN-A.

Applications. Intended for use in bread industry.

Migration Data. An investigation was conducted on metal bread-making plates with a coating applied. Extracts were obtained under following conditions: simulant media consisting of water, 2.0% solution of acetic acid, 2.0% *NaCl* solution, and 3.0% solution of lactic acid; 1 hour at 100°C, then 2 hours infusion; 10 cm^{-1}.

Extracts of *SKTN-A* acquired odor rating 3. Simulant medium was clear, transparent, odorless and tasteless. With aggravated (by a factor of 10) experimental conditions, the formaldehyde and phenol contents were 0.1 to 0.25 and 0.01 mg/l, respectively. There was no migration of organic substances or of their transformation products into simulant medium from baking molds coated with Lestosils SMF-Fe, SM-Fe, SMF-SDKO, and SM-SDKO.

Recommendations.

Russia. With the exception of SKTN-A, the coatings investigated were recommended for use in the bread industry.

References:

1. Kravchenko, Ye. G., *Vrachebnoye Delo,* 3, 148, 1972 (in Russian).
2. Kravchenko, Ye. G., *Effect of Siloxane Vulcanizates on Foodstuffs and Animals and Regulations of Their Use in Food Industry,* Author abstract of thesis, All-Union Research Institute of Hygiene and Toxicology of Pesticides, Polymers and Plastic Materials, Kiyv, 1972 (in Russian).

STYRENE POLYMERS

STYRENE POLYMER (POLYSTYRENE)

Structural Formula.

[~CH_2CH~]$_n$
|
C_6H_5

M = 50,000 to 200,000

CAS No 9003-53-6

RTECS No WL6475000

Abbreviation. PS.

Synonym and **Trade Name.** Afcolene; Atactic polystyrene.

Properties. Solid, rigid or flexible colorless material with lustrous smooth surface; transparent in thin layer; no odor or taste. It is resistant to water, oxygen (but not carbon dioxide), weak acids, bases, and fats and oils, but it is damaged by solvents and alcohol. *Dens.* 1.05 to 10.7, $n = 1.59.^{20}$ Insoluble in alcohol. PS is brittle and during long storage can become friable. It can only be heated to 75°C. Possesses a good gas and steam permeability. Heat distorts it at 75°C. PS can be formed into a foam that is less brittle.

Applications. Used in the manufacture of details, articles, water and food containers, syringes, glasses, etc. Thermoformed PS foam is used in the manufacture of plates, cups, bowls, egg cartons, meat trays, and hinged carryout containers. PS foam is approved for food use with aqueous acid or non-acid foods and beverages containing no fat or oils. PS foams are used for hot beverage cups, egg cartons, meat trays, and preparation cups for instant foods such as noodles and soups.

Migration Data. Monomer migration was established in aqueous extracts of different grades of PS. Migration intensified with increase in temperature, contact time, and with content of unpolymerized styrene in the material.

The level of styrene migration from PS cups collected from retail markets in Belgium, Germany, and The Netherlands was investigated in different food systems including: water, milk (0.5, 1.55, and 3.6% fat), cold beverages (apple juice, orange juice, carbonated water, cola, beer and chocolate drink), hot beverages (tea, coffee, chocolate and soup - 0.5, 1.0, 2.0, and 3.6% fat), take away foods (yogurt, jelly, pudding and ice- cream), as well as aqueous food simulants (3.0% acetic acid, 15, 50, and 100% ethanol) and olive oil. Styrene migration was found to be strongly dependent upon the fat content and storage temperature. Drinking water gave migration values considerably lower than all of the fatty foods. Ethanol at 15% showed a migration level equivalent to milk or soup containing 3.6% fat. Maximum observed migration for cold or hot beverages and take-away foods was 0.025% of the total styrene in the cup. Food simulants were responsible for higher migration (0.37% in 100% ethanol).[2]

Studies of *styrene* migration from thermoformed PS foam articles were completed using food oil as the simulant media. Migration was proportional to the square root of time of exposure.[3]

Styrene oxide could in principle be present in PS food packs as a contaminant formed by the oxidation of styrene monomer. Concentrations of styrene oxide were measured in base resins and samples of PS articles intended for food contact. Styrene oxide was not detected in the resins (limit of detection 0.5 mg/kg) but was found in 11 out of 16 packaging samples at up to 2.9 mg/kg. Calculated migration levels expected in packaged foods were from 0.002 to 0.15 μg styrene oxide/kg foods. Hydrolysis of the epoxide group gave rise to the diol as the principal product. Ring opening in aqueous ethanol simulant gave the diol and also the glycol monoethyl ether. Instability of styrene oxide led to formation of hydrolysis products that are less toxic than the parent epoxide.[4]

Release of ethylbenzene, isopropylbenzene, *n*-propylbenzene, and *styrene* from PS food contact wares into *n*-heptane was not observed at the detection limit of 0.1 ppm.[5]

Table. Migration of chemical substances from non-vacuum-treated specimens of polystyrene plastics (20°C; 2 months; 4 cm^{-1}).[5]

Polystyne grade	***Odor rating***	***Concentration in water mg/l***			***Conc. of styrene in sunflower oil, mg/l***
		styrene	***acrylonitrile***	***plasticizers***	
PSSP	0	not found	-	0.04	0.33
	0	not found	-	not found	0.13
UPS-080SL	1.0	0.13	-	not found	1.95
	0.7	traces	-	not found	1.20
UPS-0505	1.0	0.20	-	0.09	-
	0.8	not found	-	0.05	-
UPM-0508	4.0	1.34	-	0.25	-
	3.0	2.32	-	0.12	-
SNP-20P	0	not found	not found	traces	0.15
	0	not found	not found	not found	0.08
SNP-2P	0	not found	not found	traces	not found
	0	not found	not found	not found	not found
SNP-K	0	0.06	not found	0.06	0.15
	0	not found	not found	not found	not found
ABS-1106E	1.0	0.05	not found	0.04	0.13
	0	not found	not found	not found	traces
SNP-2	0	not found	not found	0.11	-
	0	not found	not found	0.05	-

Migration of 2-hydroxy-3-*n*-octyloxybenzophenone and its isooctyl derivative from PS into 50% alcohol attains values in the order of mg/dm^2 of the surface of the plastic and is, moreover, 3 to 20 times higher in case of contact with fats or fat-simulating liquids.[6]

29 and 64 mg benzene/kg were found in two samples of thermoset polyester compounded for the manufacture of plastic cookware. It was established that the benzene originated from the use of *tert*-butylperbenzoate used as an initiator in the manufacture of the polymer. Migration levels of benzene were found to be 1.9 and 5.6 mg/kg in olive oil after extraction for 1 hour at 175°C. Migration levels into olive oil at 175°C for samples produced with non-aromatic initiator were less than 0.1 mg/kg. Concentrations of benzene in thermoset polyester cookware purchased from retail outlets were 0.3 to 84.7 mg/kg. Low

amounts of benzene (less than 0.01 to 0.09 mg/kg) were detected in foods when the articles were used for cooking in microwave or conventional ovens. Migration of benzene was found below 0.1 mg/kg, with the highest amounts, migrating from polystyrene, being 0.2 to 1.7 mg/kg, predominantly in articles of expanded PS.[7]

Migration of fatty acid amides from PS used in food packaging and containing commonly used fatty acid amide slip additives into fat and aqueous food simulant was less than 0.05 mg/l (10 day, 40°C).[8]

Benzophenone-based ultraviolet absorbents (up to 0.5% of a composition) for commodities of PS showed a very slight tendency to migrate into aqueous, acid or dilute alcoholic foods. Very high migration levels were noted for sunflower oil, fat-simulant HB 307 or 50% ethanol. Migration of 2-hydroxy-4-octoxy-benzophenone into fat-free foods was observed at the toxicologically insignificant level, nevertheless, the above ultraviolet sorbents should not be used for the ultraviolet stabilization of plastics designed for packaging fat-containing foods.[9]

Zirconium was found to migrate from food-contact PS in olive oil at the level of 0.65 mg/kg.[10]

Reproductive Toxicity.

Observations in man. Epidemiological studies revealed no association between occupation and adverse pregnancy outcome in plastics workers.[11]

Carcinogenicity. Induced malignant tumors in rodents following s/c imbedding of polymer films.[026] Nevertheless, the results of this study has been later considered inadequate.

Carcinogenicity classification. An IARC Working Group concluded that there is inadequate evidence for the carcinogenicity of AA in *experimental animals* and there were no adequate data available to evaluate the carcinogenicity of AA in *humans*.

IARC: 3

Regulations. ***U.S. FDA*** (1998) approved the use of PS (1) in the manufacture of resinous and polymeric coatings for polyolefin films for food-contact surface of articles intended for use in producing, manufacturing, packing, processing, preparing, treating, packaging, transporting, or holding food in accordance with the conditions prescribed in 21 CFR part 175.320; (2) in adhesives as a component (monomer) of articles intended for use in packaging, transporting, or holding food in accordance with the conditions prescribed in 21 CFR part 175.105; (3) as a component of the uncoated or coated food-contact surface of paper and paperboard intended for use in producing, manufacturing, packaging, processing, preparing, treating, packing, transporting, or holding dry food of the type identified in 21 CFR part 176.170 (c); (4) in the manufacture of semirigid and rigid acrylic and modified plastics used as articles intended for use in contact with food in accordance with the conditions prescribed in 21 CFR part 177.1010; (5) in the manufacture of cellophane for packaging food in accordance with the conditions prescribed in CFR part 177.1200; and (6) as components of articles intended for use in contact with food, subject to the provisions of 21 CFR part 177.1640.

References:

1. Philo, M. R., Fordham, P. J., Damant, A. P., and Castle, L., Measurement of styrene oxide in polystyrenes, estimation of migration to foods, and reaction kinetics and products in food simulants, *Food Chem. Toxicol.*, 35, 821, 1977.
2. Tawfik, M. S. and Huyghebaert, A., Polystyrene cups and containers: styrene migration, *Food Addit. Contam.*, 15, 592, 1998.
3. Lickly, T. D., Lehr, K. M., and Welsh, G. C., Migration of styrene from polystyrene foam food-contact articles, *Food Chem. Toxicol.*, 33, 475, 1995.
4. Ito, S., Hosogai, T., Sakurai, H., Tada, Y., Sugita, T., Ishiwata, H., and Takada, M., Determination of volatile substances and leachable components in polystyrene food contact wares, Abstract, *Eisei Shikenjo Hokoku*, 110, 85, 1992 (in Japanese).
5. *Toxicology and Sanitary Chemistry of Polymerization Plastics*, Coll. Sci. Proc., B. Yu. Kalinin, Ed., Leningrad, 1984, 112 (in Russian).
6. Horacek, J. and Uhde, W. J., Plastics from the aspect of hygiene. A contribution to the estimation of ultra-violet absorbers and light stabilizers from the aspect of hygiene, *J. Hyg. Epidemiol. Immunol.*, 24, 133, 1980.

7. Jickells, S. M., Crews, C., Castle, L., and Gilbert, J., Headspace analysis of benzene in food contact materials and its migration into foods from plastics cookware, *Food Addit. Contam.*, 7, 197, 1990.
8. Cooper, I. and Tice, P.A., Migration studies on fatty acid amide slip additives from plastics into food simulants, *Food Addit. Contam.*, 12, 235, 1995.
9. Uhde, W. J. and Woggon, H., New results of migration behavior of benzophenone-based UVG absorbents from polyolefins in foods, *Nahrung*, 2, 185, 1976.
10. Fordham, P. J., Gramshaw, J. W., Crews, H. M., and Castle, L., Element residues in food contact plastics and their migration into food simulants, measured by inductively-coupled plasma-mass spectrometry, *Food Addit. Contam.*, 12, 651, 1995.
11. Lindbohm, M. L. et al., Spontaneous abortions among women employed in the plastics industry, *Am. J. Ind. Med.*, 8, 579, 1985.

GENERAL PURPOSE POLYSTYRENE

Structural Formula.

$[\sim CH_2CH\sim]_n$

$|$

C_6H_5

Abbreviation. GPPS.

Composition. PSS-grade (suspension) polystyrene is a product of polymerization of styrene in aqueous media.

Applications. GPPS is used for the packaging of aqueous-based, fatty and dry foods.

Migration Data. The amount of *styrene* migrating from GPPS into cooking oil at temperatures ranging from 21 to 82°C was proportional to the square root of the time of exposure, and the total amount of styrene migrating was proportional to the residual levels of styrene in the polymers.[1,2] GPPS has no effect on organoleptic water quality indices. Migration of styrene does not exceed 0.01 mg/l, but amounted to12.5 mg/l into 96% alcohol (20°C; 3 months; 1.0 cm^{-1}). Migration of butyl stearate into water (60°C; 7 months) amounts to 0.6 mg/l, into 40% alcohol (20°C; 5 months; 4.0 cm^{-1}) to 1.14 mg/l, and into 96% alcohol under the same conditions to 9.26 mg/l.[3,4]

Regulations. ***U.S. FDA*** (1998) approved the use of PS in the manufacture of resinous and polymeric coatings for polyolefin films for food-contact surface of articles intended for use in producing, manufacturing, packing, processing, preparing, treating, packaging, transporting, or holding food in accordance with the conditions prescribed in 21 CFR part 175.320.

References:

1. Murphy, P. G., MacDonald, D. A., and Lickly, T. D., Styrene migration from general-purpose and high-impact polystyrene into food-simulating solvents, *Food Chem. Toxicol.,* 30, 225, 1992.
2. Lehr, K. M., Welsh, G. C., Bell, C. D., and Lickly, T. D., The vapor-phase migration of styrene from general purpose polystyrene and high impact polystyrene into cooking oil, *Food Chem. Toxicol.,* 31, 793, 1993.
3. *Toxicology and Medical Chemistry of Plastics,* Coll. Sci. & Techn. Abstracts, NIITEKHIM, Moscow, Issue No 1, 1979, 1 (in Russian).
4. Petrova, L. I., *Investigation of Possible Use of Polystyrene Plastics of Different Composition in Contact with Foodstuffs*, Author's abstract of thesis, Leningrad, 1979, 20 (in Russian).

POLYSTYRENE PSM-grade block-polymerized

Migration of styrene into water at 60°C after 6-month exposure reached 5.5 to 10 mg/l.

Regulations. ***U.S. FDA*** (1998) approved the use of the styrene block polymers as articles or as components of articles intended for use in contact with food subject to the provisions prescribed in 21 CFR part 177.1810.

Reference:

Petrova, L. I., *Investigation of Possible Use of Polystyrene Plastics of Different Composition in Contact with Foodstuffs*, Author's abstract of thesis, Leningrad, 1979, 20 (in Russian).

POLYSTYRENE PSV-grade expanded

Abbreviation. EPS.

Applications. Used to manufacture heat-insulation and sound-proofing plates, packaging, and consumer articles.

Migration Data. Extracts (distilled water; initial temperature 20 and 100°C, then 18 to 20°C; 1 day) did not have taste or odor. Migration of styrene reached 0.08 mg/l.

Immunotoxicity. Mice and rats received 24-hour extracts (1.0 cm^{-1}; 18 to 25°C) of PPS- and PSV-grade EPS over a period of 1 year. There were signs of unspecific immune changes.[1]

Reproductive Toxicity.

Embryotoxicity. Rats developed reduced fertility, their progeny had signs of CNS affection. Consumption of EP extracts containing up to 2.0 mg monomer/l decreased survival rate of embryos (second generation of animals). There were changes in the weight and length of fetuses and placenta.[2]

References:

1. Yefremenko, A. A., *Hygiene Assessment of PSV-grade Expanded Polystyrene Intended for Contact with Liquid Foodstuffs*, Author's abstract of thesis, Leningrad, 1974, 23 (in Russian).
2. *Materials on the Health and Safety Assessment of Pesticides and Polymers,* Coll. Sci. Proc. F. F. Erisman Research Sanitary Hygiene Institute, Moscow, 1977, 59 (in Russian).

STYRENE COPOLYMERS

HIGH-IMPACT POLYSTYRENE

Structural Formula.

$[\sim CH_2CH\sim]_n$

|

C_6H_5

CAS No 9003-55-8

Abbreviation. HIPS.

Composition. Product of styrene copolymerization with rubbers. Contains dibutyl phthalate, 2,2-azobisisobutyronitrile, sodium polymethacrylate or a copolymer of sodium methacrylate with methyl metacrylate, fillers, and dyes. SNP-grade HIPS is obtained by adding SKN butadiene-styrene rubber to a prepared copolymer of *styrene* SN-20 or SN-26 and subsequent mechano-chemical treatment. Residual styrene contents of the polymer are up to 0.01%, *acrylonitrile* contents are up to 0.007%.

Applications. HIPS is used in many food-contact applications for the packaging of aqueous-based, fatty and dry foods.

Migration Data. The amount of styrene migrating from HIPS into cooking oil at temperatures ranging from 21 to 82°C was proportional to the square root of the time of exposure, and the total amount of styrene migrating was proportional to the residual levels of styrene in the polymers.[1]

Aqueous extracts (1 cm^{-1}; 37 and 80°C; 1, 3, and 10 days) from SNP-grade PS acquire odor and taste below rating 1. Migration of styrene ranges from traces to 0.01 mg/l.[2]

Carcinogenicity classification. An IARC Working Group concluded that there were no adequate data available to evaluate the carcinogenicity of butadiene-styrene copolymers in *experimental animals* and in *humans*.

IARC: 3.

References:

1. Murphy, P. G., MacDonald, D. A., and Lickly, T. D., Styrene migration from general-purpose and high-impact polystyrene into food-simulating solvents, *Food Chem. Toxicol.*, 30, 225, 1992.
2. Fetisova, L. N., Taradin, Ya. I., Shlygina, G. S., et al., Sanitary-chemistry assessment of high-impact polystyrene UPAN and perspectives of its use, in *Hygiene and Toxicology of High-Molecular-Mass Compounds and of the Chemical Raw Material Used for Their Synthesis,* Proc. 6th All-Union Conf., B. Yu. Kalinin, Ed., Khimiya, Leningrad, 1979, 103 (in Russian).

HIGH-IMPACT POLYSTYRENE SNP-2P-grade

Abbreviation. HIPS.

Migration Data. Within 5 days at 20°C, 0.43 mg styrene/l was released into 40% alcohol; after 2 hours at 80°C migration was 1.2 mg/l (1.0 cm^{-1}) and 3.0 mg/l (4.0 cm^{-1}).[1]

Long-term Toxicity. Aqueous extracts (1 cm^{-1}; 37 and 80°C; 1 day) from SNP-2P-grade HIPS were studied in a 12-month experiment on rats. No significant changes were found in hematological indices, enzyme activity, *SH*-group content in the blood, and histology of the visceral organs. Exposure of females to extracts for 9 months did not cause significant disturbances in the offspring.[2]

In a 1-year experiment, rats receiving extracts with 0.1 mg/l of PS developed no functional changes.[3]

References:

1. Petrova, L. I., *Investigation of Possible Use of Polystyrene Plastics of Different Composition in Contact with Foodstuffs*, Author's abstract of thesis, Leningrad, 1979, 20 (in Russian).
2. Chernova, T. V., Effect on animals of chemicals leached from high-impact polystyrene SNP-2P, in *Hygiene and Toxicology of High-Molecular-Mass Compounds and of the Chemical Raw Material Used for Their Synthesis*, Proc. 4th All-Union Conf., S. L. Danishevsky, Ed., Khimiya, Leningrad, 1969, 29 (in Russian).
3. Krynskaya, I. L., Zimnitskaya, L. P., and Komarova, Ye. N., Toxicology data of some types of high-impact polystyrene, in *Toxicology and Hygiene of High-Molecular-Mass Compounds and of the Chemical Raw Material Used for Their Synthesis,* Proc. 3rd All-Union Conf., S. L. Danishevsky, Ed., Khimiya, Moscow-Leningrad, 1966, 119 (in Russian).

HIGH-IMPACT POLYSTYRENE UPS-1002 grade

Abbreviation. HIPS.

Composition. Obtained by block-suspension polymerization of styrene in presence of 5.0% polybutadiene rubber, similar to UPS-0803L or UPS-804 (old name).

Migration Data. Has no adverse effect on organoleptic properties of water. Styrene migration is <0.01 mg/l.[1] With a specific surface of 1.0 cm^{-1}, 9.0 mg styrene/l migrate from UPS-0803L grade HIPS into 96% alcohol, and with 4.0 cm^{-1} (3 months; 20°C), 21 mg/l.

Within 6 months at 60°C, ~10 mg styrene/l migrates into water from P grades UPS-0508, UPS-0803L, and UPS-0505. Migration of plasticizer (butyl stearate) from UPS-0803L was ~1.0 mg/l into water (60°C; 7 months), 2.04 mg/l into 40% alcohol (20°C; 5 months; 4.0 cm^{-1}), and 44.55 mg/l into 96% alcohol (under the same conditions).[2]

Regulations. ***U.S. FDA*** (1998) approved the use of HIPS in the manufacture of semirigid and rigid acrylic and modified plastics used as articles intended for use in contact with food in accordance with the conditions prescribed in 21 CFR part 177.1010.

References:

1. Fetisova, L. N., Taradin, Ya. I., Shlygina, G. S., et al., Sanitary-chemistry assessment of high-impact polystyrene UPAN and perspectives of its use, in *Hygiene and Toxicology of High-Molecular-Mass Compounds and of the Chemical Raw Material Used for Their Synthesis,* Proc. 6th All-Union Conf., B. Yu. Kalinin, Ed., Khimiya, Leningrad, 1979, 103 (in Russian).
2. Petrova, L. I., *Investigation of Possible Use of Polystyrene Plastics of Different Composition in Contact with Foodstuffs*, Author's abstract of thesis, Leningrad, 1979, 20 (in Russian).

HIGH-IMPACT POLYSTYRENE UPAN-grade

Abbreviation. HIPS.

Composition. Contains 0.005 to 0.03% residual *styrene*.

Migration Data. Above 20°C, gives water odor (rating 2) and a yellowish color. Styrene migration exceeded 0.01 mg/l.[1]

Reproductive Toxicity. Toxic effect of HIPS on embryos and gonads was described.[2]

Carcinogenicity. Addition of 5.0% HIPS to the diet of rats had a carcinogenic effect (Rinzema).

References:

1. Fetisova, L. N., Taradin, Ya. I., Shlygina, G. S., et al., Sanitary-chemical assessment of high-impact polystyrene UPAN and perspectives of its use, in *Hygiene and Toxicology of High-Molecular-Mass Compounds and of the Chemical Raw Material Used for Their Synthesis,* Proc. 6th All-Union Conf., B. Yu. Kalinin, Ed., Khimiya, Leningrad, 1979, 103 (in Russian).
2. Gnoyevaya, V. L. and Ryazanova, R. A., Possibility of delayed consequences of exposure to polymeric materials used in food industry, in *Basic Problems Concerning Delayed Consequences of Exposure to Industrial Poisons*, Sci. Proc., A. K. Plyasunov and G. M. Pashkova, Eds., Moscow, 1976, 38 (in Russian).

HIGH-IMPACT POLYSTYRENE coatings

Abbreviation. HIPS.

Composition. Xylene or toluene as solvents.

Applications. UPM-0.503 and UPM-0.508 HIPS are used for coating water tanks.

Migration Data. Coatings with xylene give water unpleasant odor (rating 4 to 5). Styrene migration to water does not exceed 0.03 to 0.05 mg/l (20 and 37°C; 1 day; 0.4 cm^{-1}). Coatings with toluene produced no unfavorable effect on water quality. Styrene content in the extracts amounted to 0.02 to 0.03 mg/l.

Reference:

Krat, A. V., Kesel'man, I. M., and Sheftel', V. O., Sanitary-chemistry evaluation of polymeric materials used in water supply, *Gig. Sanit.,* 10, 18, 1986 (in Russian).

ACRYLONITRILE-BUTADIENE-STYRENE COPOLYMERS

Structural Formula.

$$[\sim CH_2CH\sim]_m[\sim CH_2CH{=}CHCH_2\sim]_n[\sim CH_2CH\sim]_p$$

(with CN pendant on the first unit and C_6H_5 pendant on the third unit)

CAS No 9003-56-9
73990-12-2

RTECS No AT6970000

Abbreviation. ABS.

Synonym and **Trade Name.** ABS plastics; 2-Propenenitrile-1,3-butadiene-ethenylbenzene polymer.

Composition. Copolymers of styrene, acrylonitrile, and polybutadiene at a ratio 76:24:7 (ABS-M) and 80:20:20 (ABS-1106E). Residual contents in ABS-M specimens: 0.014 to 0.06% styrene, 0 to 0.008% acrylonitrile, 0.018 to 0.11% ethylbenzene.

Properties. Heat distorts ABS at 92°C. Maximum continuous service temperature is 75°C. ABS is resistant to acids, bases, fats and oils. It does not resist solvents and yellows if exposed to sunlight.

Applications. ABS has been cleared for use with all foods except those containing alcohol and those subjected to no thermal processing. It can be used with foods that are room temperature filled and stored, refrigerated stored, or frozen stored.

Migration Data. ABS plastics can give water odor and taste (rating <3 to 4) and increase the oxidizability of extracts. Migration of *styrene* and acrylonitrile from ABS-1106E can reach 0.04 mg/l, and migration of acrylonitrile from ABS-M is 0.64 mg/l.[1] Another investigation of aqueous extracts and model media (20 to 80°C; 1 to 10 days) revealed no significant effect of ABS on organoleptic indices, nor migration of styrene and acrylonitrile above the PAM. The plastic can be recommended for manufacturing articles to be used in contact with water and foodstuffs.[2]

Petrova[1] established migration of styrene from ABS-1106E into 96% ethyl alcohol at a level of 2.7 mg/l (20°C; 3 months; 4.0 cm^{-1}. Under the same conditions, 10 times less styrene migrated into oil. Migration into water was significantly lower. Migration of butyl stearate was 1.15 mg/l to water (60°C; 7 months), 7.84 mg/l to 40% alcohol (20°C; 5 months; 4 cm^{-1}), and 339.0 mg/l to 96% alcohol under the same conditions.[3]

Wharton and Levinskas studied migration of CN^- from copolymers of acrylonitrile with styrene and butadiene. Here the level of migration to water after 5-day contact of a copolymer containing 67.5%

acrylonitrile was 0.013 to 0.025 mg/l, and with acrylonitrile contents of 72% under the same conditions it was 0.021 mg/l. Within 3 days, 0.085 mg/l of CN^- migrates to 3.0% acetic acid.[4]

Study of the migration of acrylonitrile into water from seven ABS-polymers showed that a linear relationship exists between the concentration of AN in the polymer and the amount of acrylonitrile migrating, for a given set of exposure conditions.[5]

Migration of magnesium *ions* from acrylonitrile/butadiene/styrene copolymer into food simulants, measured by inductively-coupled plasma-mass spectrometry amounted to 0.05 mg/kg. The concentration found was less than proposed limit of migration.[6]

Mutagenicity. ABS-M plastic had no mutagenic effect.[7]

Carcinogenicity classification. An IARC Working Group concluded that there were no adequate data available to evaluate the carcinogenicity of AS copolymers in *experimental animals* and in *humans*.

IARC: 3.

Regulations. ***U.S. FDA*** (1998) approved the use of ABS plastics (1) in the manufacture of resinous and polymeric coatings for the food-contact surface of articles intended for use in producing, manufacturing, packing, processing, preparing, treating, packaging, transporting, or holding food (for use only as polymerization cross-linking agent in side seam cements for containers intended for use in contact with food (only of the identified types), subject to the conditions prescribed in 21 CFR part 175.300; (2) as an article or component of articles intended for use with all foods, except those containing alcohol, under the conditions described in 21 CFR part 176.179; and (3) in the manufacture of semirigid and rigid acrylic and modified plastics used as articles intended for use in contact with food in accordance with the conditions prescribed in 21 CFR part 177.1010.

References:

1. Petrova, L. I., Boikova, Z. K., and Kiseleva, M. N., Use of MSP, ABC-M and ABC-1106E polystyrene plastics in potable water supply, in *Hygiene Aspects of the Use of Polymeric Materials*, Proc. 3rd All-Union Meeting on New Methods of Hygiene Monitoring of the Use of Polymers in the Economy, K. I. Stankevich, Ed., Kiyv, December 2-4, 1981, Kiyv, 1981, 205 (in Russian).
2. Petrova, L. I., Boikova, Z. K., Dokukina, L. F., and Deryagina, G. M., Sanitary-chemistry properties of ABC-M plastic and perspectives of its use, in *Hygiene and Toxicology of High-Molecular-Mass Compounds and of the Chemical Raw Material Used for Their Synthesis,* Proc. 6th All-Union Conf., B. Yu. Kalinin, Ed., Khimiya, Leningrad, 1979, 84 (in Russian).
3. Petrova, L. I., *Investigation of Possible Use of Polystyrene Plastics of Different Composition in Contact with Foodstuffs*, Author's abstract of thesis, Leningrad, 1979, 20 (in Russian).
4. Wharton, F. D. and Levinskas, G. T., *Chem. Ind.*, 11, 470, 1976.
5. Lickly, T. D., Markham, D A, and Rainey, M. L., The migration of acrylonitrile from acrylonitrile/ butadiene/styrene polymers into food-simulating liquids, *Food Chem. Toxicol.*, 29, 25, 1991.
6. Fordham, P. J., Gramshaw, J. W., Crews, H. M., and Castle, L., Element residues in food contact plastics and their migration into food simulants, measured by inductively-coupled plasma-mass spectrometry, *Food. Addit. Contam.*, 12, 651, 1997.
7. Grigorieva, M. N., Kalinin, B. Yu., Novikova, O. I., et al., Genetic activity study of several plastics and polymers, in *Hygienic Aspects of Polymeric Materials' Use*, Proc. 3rd All-Union Meeting *New Methods of Hygiene Monitoring of the Use of Polymers in Economy*, K. I. Stankevich, Ed., December 2-4, 1981, Kiyv, 1981, 319 (in Russian).

ACRYLONITRILE and STYRENE COPOLYMERS

Structural Formula. $[\sim CH_2CH\sim]_m\ [\sim CH_2CH\sim]_n$, with C_6H_5 attached to the CH of the first unit and CN attached to the CH of the second unit.

CAS No 9003-54-7

RTECS No AT6978000

Synonym and **Trade Names.** Acrylonitrile-styrene resin; Cevian; Dialux; Estyrene; Kostil; Lustran; Polysan; Tyril.

Migration Data. After 24 hours at 60°C, 0.053 mg CN^-/l was released into water from copolymers of acrylonitrile (52%) with styrene. Within 5 days, 0.063 and 0.064 mg CN^-/l respectively migrated to 3.0% acetic acid and 8.0 % solution of ethyl alcohol. Within 3 days at 50°C, 0.024 mg CN^-/l migrate from a copolymer containing 80% acrylonitrile into 8.0% ethyl alcohol. Within 3 days, migration from a copolymer containing 8.25% acrylonitrile into 3.0 % acetic acid reaches 0.021 mg/l at 50°C. CN^- was not found in an alcohol extract under the same conditions.[1]

Acute Toxicity. LD_{50} is 1.8 g/kg BW in rats, and 1.0 g/kg BW in mice. Following oral administration, changes were observed in the liver.[2]

Carcinogenicity classification. An IARC Working Group concluded that there were no adequate data available to evaluate the carcinogenicity of AS copolymers in *experimental animals* and in *humans*.

IARC: 3.

Regulations. ***U.S. FDA*** (1998) approved the use of AS copolymers (1) in the manufacture of resinous and polymeric coatings for polyolefin films for food-contact surface of articles intended for use in producing, manufacturing, packing, processing, preparing, treating, packaging, transporting, or holding food, in accordance with the conditions prescribed in 21 CFR part 175.320; and (2) as a component of food packaging materials subject to the provisions of 21 CFR part 177.1040.

References:

1. Wharton, F. D. and Levinskas, G. T., *Chem. Ind.*, 11, 470, 1976.
2. *Ceskoslovenska Hygiena*, 25, 22, 1980 (in Czech).

STYRENE-ACRYLONITRILE-METHYL METHACRYLATE COPOLYMERS

Structural Formula.

```
                                   CH3
                                    |
[~CH2CH~]m[~CH2CH~]n[~CH2C~]p
     |              |              |
    C6H5           CN          OCOCH3
```

Applications. Such films, coatings and semi-rigid containers are cleared for use only for single-use food contact surfaces. They have not been tested for use above 650°C and are used primarily in the pharmaceutical industry.

Migration Data. Copolymers of MS and MSN grades do not alter the organoleptic properties of aqueous extracts, nor those of acid, salt, or oil media by more than rating 1, while grade SAN-A gives water odor of *tert*-dodecylmercaptan (rating 2) and releases acrylonitrile (0.06 to 0.37 mg/l) and methyl methacrylate into water. Migration from other copolymers does not exceed the permissible level of migration. Migration of styrene amounted to 0.06 mg/l from SAN-A and to 0.01 mg/l from MS; no migration from MSN was found. Ethylbenzene and other impurities of initial technical-grade styrene were not found to migrate into extracts.[1]

MSP-grade C. changed the taste and odor of extracts (rating 3 to 4), migration of *styrene* and acrylonitrile amounted to 0.04 mg/l, and migration of methyl methacrylate to 0 to 0.233 mg/l. This was improved after washing and holding in water for 2 weeks.[2]

Migration of CN^- into water from copolymer of acrylonitrile (55%) with styrene, butadiene, and methyl metahcrylate within 5 days at 50°C amounted to 0.18 mg/l.[3]

Recommendations. Of the given grades, only SAN-A is not recommended for contact with foodstuffs and drinking water.[1]

References:

1. *Toxicology and Medical Chemistry of Plastics,* Coll. Sci. and Techn. Abstracts, NIITEKHIM, Issue No 1, Moscow, 1979, 1 (in Russian).
2. Petrova, L. I., Boikova, Z. K., and Kiseleva, M. N., Use of polystyrene plastics MSP, ABC-M and ABC-1106E in the water supply systems, in *Hygienic Aspects of Polymeric Materials' Use*, Proc. 3rd All-Union Meeting on New Methods of Hygiene Monitoring of the Use of Polymers in Economy, K. I. Stankevich, Ed., December 2-4, 1981, Kiyv, 1981, 205 (in Russian).

3. Wharton, F. D. and Levinskas, G. T., *Chem. Ind.*, 11, 470, 1976.

STYRENE and MALEIC ANHYDRIDE, LOW-MOLECULAR-MASS POLYMER

Molecular Formula. $[\sim C_8H_8.C_4H_2O_3\sim]_n$
M = 50,000 to 60,000
CAS No 9011-13-6
RTECS No ON 4240000
Abbreviation. SMA.

Synonym and **Trade Names.** Polybutylene Dylark; 2,5-Furandione, polymer with ethylbenzene; SMA resin 1440-H; Styromal.

Composition. Product of copolymerization of styrene and maleic anhydride in xylene with 1.0% benzoyl peroxide.

Properties. White powder. *Dens.* 1.2. Soluble in oils.

Applications. Emulsifier and stabilizer of lattices, dispersing agent, stabilizer of suspension PVC. Used as a film-forming component in the manufacture of paint and varnish materials and as a molding material with high deformation heat resistance.

Acute Toxicity. LD_{50} of resin SMA 1440-N (35% solution of resin SMA-1440 in ammonia) exceeded 22 ml/kg BW in male rats and 20 ml/kg BW in female rats. Poisoning caused hypermotility, diarrhea, and changes in the GI tract and blood system.[1]

Acute Toxicity. Administration of 10 g/kg BW to rats caused only sluggishness.[2] LD_{50} is 21 g/kg BW in rats. Poisoning caused hypermotility, diarrhea, and changes in the GI tract and blood system.[1]

Repeated Exposure. Rats receiving 2.0 and 5.0 g SMA/kg BW for 1 month developed functional changes in the CNS, liver, and kidney pathology.[2]

Short-term Toxicity. When 1440-N resin (1 to 10,000 mg/kg BW) was added to the feed of young rats over a period of 3 months, there were no changes in the blood chemistry or mass coefficients of the visceral organs. BW gain and activity of ACT were unchanged.

Reproductive Toxicity.

Embryotoxicity. Administration of 4.0 mg/kg BW to rats at various terms of pregnancy caused no abnormalities in the offspring.

Mutagenicity. Inhalation produced no mutagenic effect.[3]

Allergenic Effect. Did not cause sensitization in guinea pigs.

Regulations. ***U.S. FDA*** (1998) approved the use of SMA (1) in adhesives as a component of articles intended for use in packaging, transporting, or holding food in accordance with the provisions prescribed in 21 CFR part 175.105; and (2) as articles or components of articles intended for use in contact with food subject to the provisions of 21 CFR part 177.1820.

References:

1. Winek, C. L. and Burgun, J. J., Acute and subacute toxicology and safety evaluation of SMA 1440-H resin, *Clin. Toxicol.*, 10, 255, 1977.
2. Khromenko, Z. F., Toxicological and hygienic evaluation of styromal, in *Hygiene and Toxicology of High-Molecular-Mass Compounds and of the Chemical Raw Material Used for Their Synthesis*, Proc. 6th All-Union Conf., B. Yu. Kalinin, Ed., Khimiya, Leningrad, 1979, 245 (in Russian).
3. Khromenko, Ye. F. and Roslyakov, Ye. S., in *Sci. Proc. Angarsk Occupational Hygiene Research Institute*, Angarsk, Issue No 9, 1977, 72 (in Russian).

SUBJECT INDEX

A

B

C

D

E

F

G

H

I

J

K

L

M

N

O

P

Q

R

S

T

U

V

W

X

Y

Z

CAS Number INDEX

RTECS Number INDEX